李启虎院士论文选集

李启虎　著

科学出版社
北京

内 容 简 介

本书是作者及合作者在从事数字式声呐设计及水下信息处理研究工作期间公开发表的部分论文的汇编，包括信号处理的基本理论与方法，自适应信号处理，水下目标的检测、估计与识别，数字式声呐设计理论与应用，水声通信和水下目标成像，北极声学，综合性评述及科普等。

本书可供相关专业的大学生、研究生及科技工作者参考。

图书在版编目(CIP)数据

李启虎院士论文选集/李启虎著．—北京：科学出版社，2019.7
ISBN 978-7-03-061773-6

Ⅰ．①李…　Ⅱ．①李…　Ⅲ．①物理学-文集　Ⅳ．①O4-53

中国版本图书馆 CIP 数据核字(2019)第 132297 号

责任编辑：周　涵 / 责任校对：杨　然
责任印制：吴兆东 / 封面设计：无极书装

科 学 出 版 社 出版
北京东黄城根北街 16 号
邮政编码：100717
http://www.sciencep.com
北京凌奇印刷有限责任公司 印刷
科学出版社发行　各地新华书店经销
*
2019 年 7 月第　一　版　开本: 787 × 1092　1/16
2019 年 7 月第一次印刷　印张: 64　插页: 4
字数: 1 515 000
POD定价：　499.00元
(如有印装质量问题，我社负责调换)

李启虎院士简介

李启虎 中国科学院院士，中国科学院声学研究所研究员。曾任中国科学院声学研究所所长、国家 863 计划海洋领域海洋监测主题专家组组长、中国科学院信息技术科学部常委会副主任。长期从事信号处理理论和声呐设计、研制工作。著作有《声呐信号处理引论》《数字式声呐设计原理》和 *Digital Sonar Design in Underwater Acoustics : Principles and Applications* 等，发表论文近 200 篇。1984—1986 年曾应邀在美国 Princeton 大学电子工程和计算机科学系做访问学者。多次应邀在国际会议上作特邀报告或担任会议主席。1989 年获国防科工委“献身国防事业勋章”，1992 年获国家科技进步奖一等奖，2010 年获中国科协“全国优秀科技工作者”称号。

1. 20 世纪 70 年代在所里讲课
2. 1979 年全家福照片（李启虎和黄小萍于 1969 年结婚，一直分居两地，1978 年邓小平副总理批给中国科学院 400 个家属进京名额，得以全家团圆）
3. 1985 年在美国 Princeton 大学做访问学者

1. 中美联合考察(从左至右：黎瑛、张春华、李启虎、美方专家)
2. 1995 年 12 月与老所长汪德昭院士交谈
3. 1998 年访美和李政道教授在他的寓所前合影

1. 2002 年 3 月在南海指挥试验

2. 2007 年在希腊克利特岛和 UAM 会议主席 Bjørnø 教授交谈

3. 2013 年和美国 Scripps 研究所 Kuperman 教授在一起

1. 2013 年和美国工程院院士、MIT 教授 Baggeroer 交谈

2. 2015 年主持中国科学院学部科学技术前沿论坛，与空军指挥学院院长马健以及李启虎为部队培养的博士李敏、杨秀庭合影（从左至右：杨秀庭、马健、李启虎、李敏）

3. 2017 年 7 月亲自下潜艇调研

1. 和实验室领导在一起（从左至右：李启虎、刘纪元副主任、黄海宁主任）
2. 深海高技术发展中的水声学问题——技术科学论坛与学生合影（从左至右：李敏、黄海宁、张春华、李启虎、杨秀庭、刘纪元、唐劲松、李淑秋）

序　言

李启虎院士是我国著名的声呐系统设计及信号处理专家，青年时代求学于北京大学数学力学系，毕业后，分配到中国科学院电子学研究所第七研究室，师从我国著名科学家汪德昭院士，从此开始了国防水声学研究，至今已逾半个世纪。先生不仅在科研方面成绩斐然，而且善于发现人才、培养人才，敢于放手让年轻人承担重要项目，在科研和管理实践过程中得到历练，国内许多著名的专家学者均得到过先生的关心与教诲。今年我们在迎来先生八十华诞的时候，看到先生的研究论文结集出版，这既是先生学术思想的总结，也为后人留下了宝贵的学术财富。

先生指出，只有技术创新才能实现跨越式发展。声呐技术是一门发展迅速、需求推动力强大、应用前景广阔的学科，作为其理论支撑的水声信号处理是一门综合性的边缘学科。它在发展进程中，既有自己的特色，又吸收了雷达、医学成像、通信等其他领域的成果。先生结合我国浅海声传播的特点，创造性地应用信息论、数字信号处理、水声工程等理论，解决了一系列水声信号处理中的问题，有力地推动了我国声呐装备的现代化进程。先生第一次把广义互谱法测延时的算法用于水下目标的被动测距，同时在信号处理机中采用一次相关内插、二次相关、互谱法等多种不同的测延时方法和数据过滤方法，为我国声呐技术的进步做出了重要贡献。先生研究了自适应波束成形的稳态特性，给出了用频率域最优传输函数求解波束指向性的表达式，给出了在海洋噪声背景下检测微弱信号的增益计算方法，解决了在时间上非平稳、空间上不均匀噪声场对经典理论应作的修正的问题，提出了指导声呐设计的重要依据——声呐方程的一种新的表达方式。在水下目标的被动检测中提出利用声信号的相位信息估计目标方位的新方法，还提出了用自适应阵处理方法完全分离在空间上不重叠的多个点源信号的新算法。在数字式声呐设计中首次提出动态波束成形、可编程数字滤波、变采样率运算、类卡尔曼滤波和灰度变换技术，并提出用聚类分析方法设计用于水下目标识别的简易专家系统，引领了我国数字式声呐的发展。

李启虎院士数十年如一日地沿着学术创新的道路深耕不辍，不仅参与研制了我国第一代岸基声呐，而且是我国第一代数字式声呐的总设计师；先生主持研制了我国第一代数字式拖曳线列阵声呐，显著地提高了我国声呐技术水平；先生最先提出和积极推动了我国的合成孔径声呐研究，使得该项研究得到了快速发展，达到国际领先水平；进入 21 世纪后，又高瞻远瞩地提出了北极声学等前沿学科，在他的推动下，一批新技术、新装备及一些前瞻性研究迅速发展起来。

李启虎院士曾担任国家 863 计划海洋监测技术主题专题组组长，在“十一五”海洋监测技术展望中指出：海洋监测技术主题是我国有史以来第一次成规模的、系统的发展计划。虽然海洋监测高新技术研究已取得了较大成绩，但总体水平要赶上发达国家海洋监测技术水平还要有很多路要走。先生主导制定了符合我国国情的发展规划，为我国海洋监测技术的快速稳定发展指明了道路。

李启虎院士的研究涉猎广泛，从信号处理的基本理论与方法到数字式声呐设计理论以及北极声学等，本次搜集整理的百余篇论文，汇集了先生和他的学生通过潜心研究得出的真知灼见，有鲜明的创新性特色。先生发表的论文、撰写的专著已经成为声呐系统设计及水下信号处理领域最重要的参考文献，这些研究成果，对雷达、海洋、信号与信息处理等相关领域的研究，也有相当重要的研究参考价值。李启虎院士研究论文结集出版，为国内外学术同行的研究提供了极大的方便，也为后人系统研究先生的学术思想提供了宝贵的文献资料。

有感于此，我们对本书的出版，表示由衷的祝贺。

2019 年 6 月

前　　言

科学研究必须有科学的态度，也即实事求是的态度，我国的水声学研究及声呐[①]技术发展秉承了这一原则，取得了很大的成绩。我国的水声科学研究、水声信号处理研究和声呐设计起步比发达国家晚，所以长期以来总体上落后于发达国家。美国早期主要开展深海中的声波探测技术研究；而我国是世界最大的沿海国之一，除台湾以东和南海海域之外，沿岸水深一般不超过 200 米，大陆架一般宽达几十公里，大都属于浅海海域。浅海和深海的水声特性有的时候有很大差异，应具体情况具体分析，实事求是地进行科学试验和研究。我国国防水声事业的奠基者——中国科学院院士汪德昭先生于 20 世纪 60 年代初根据我国的实际情况提出了发展水声事业的战略方针：由浅入深，由近及远。实践证明，汪德昭院士提出的这个指导思想是行之有效的，在这一方针指导下我国近海、浅海的水声学研究和独立自主的声呐设计取得了辉煌的成绩。随着我国综合国力的不断提升，面对国际深海领域的激烈竞争，走向深海、走向两极已成为我国面对全球化发展的必然选择，开展相关的水声学研究是实现我国参与国际海洋竞争的关键之一。

1963 年，我从北京大学数学力学系信息论控制论专业毕业，分配到当时的中国科学院电子学研究所工作，跟随汪德昭院士等前辈从事声呐技术研究；1964 年成立中国科学院声学研究所，于是我便在声学研究所工作，一直到现在，除中间到山西农村搞了 10 个月的“四清”工作和 1984—1986 年到美国 Princeton 大学做访问学者之外，从未离开。细数起来，我跨入声呐技术领域已经快 60 年了。声呐技术是一门发展迅速、需求推动力强大、应用前景异常广阔的学科。作为其重要理论支撑的水声信号处理是一门综合性的边缘学科。它在发展进程中，既有自己的特色，又吸收了雷达、自动控制、通信、语音信号处理等其他领域的成果。所以，声呐研究是不断吸收、学习、创新的过程。在迈向耄耋之年时，我的同事、学生积极鼓励我把以往的工作做一些总结，在他们的帮助下我对自己过去的工作做了一些整理，从近 200 篇论文中选出一部分集成本书，由科学出版社出版。本书主要收录的是 20 世纪 70 年代末期至今，我和同事及学生一起公开发表的文章，也记录了我的学术研究兴趣、学科发展脉络及我们在声呐技术发展中所做的有益探索。

---

① “声呐”或“声纳”，译自英文 sonar，sonar 是 sound navigation and ranging 的缩写，意指利用声波探测水下目标的各种技术和设备。

从历史上看，sonar 的译名曾出现过“声拿”“声纳”“水声测位仪”“声呐”等。后来随着发展，国军标(GJB)系统采用了“声纳”一词并沿用至今；1988 年由物理学名词审定委员会审定并由全国自然科学名词审定委员会公布的规范名为“声呐”。因此，目前 GJB 中使用“声纳”，国标(GB)中使用“声呐”，两种用法均符合各自的标准。

本书中各篇论文根据不同的背景对两种译名都有所采用，出于尊重原作的想法，未对两种译名进行统一，特此说明。

本书分为七个部分。第一部分是信号处理的基本理论与方法方面的论文，主要涉及波束形成、左右舷分辨、空域矩阵滤波等算法；第二部分是自适应信号处理方面的论文，主要涉及自适应噪声抵消、自适应波束成形、自适应滤波技术等；第三部分是水下目标的检测、估计与识别方面的论文；第四部分是数字式声呐设计理论与应用，包括数字式声呐设计中的检测性能、系统模拟、信号设计、发射接收及数据采集系统设计等；第五部分是水声通信和水下目标成像方面的论文，包括合成孔径声呐成像、三维成像及水声多输入多输出系统等内容；第六部分是北极声学相关的论文；第七部分是综合性评述及科普类文章。这七个部分较为全面地涵盖了数字式声呐及水声信号处理的主要内容。

本书在策划、编写、成书过程中，得到了中国科学院声学研究所和中国科学院先进水下信息技术重点实验室领导和同事的关心和支持，也得到了科学出版社的大力支持及具体指导，为本书出版创造了许多便利条件，在此一并表示衷心的感谢。特别感谢王小民、张春华、黄海宁、刘纪元、尹力、李淑秋、田杰、李宇、薛山花和籍顺心等同志在成书过程所做的大量工作，没有他们的努力，这本文集不可能在这么短的时间内和读者见面。

自新中国成立以来，几代人为祖国的国防水声事业付出了辛劳与智慧，它的价值体现在一代代声呐装备上，体现在日益强大的国防力量上。本论文集实际上是我们团队的全体成员在声呐信号处理理论、声呐设计、海洋声学、北极声学方面所取得的部分成果的展示，如实地反映了各个阶段的时代特点，也反映出发展的轨迹以及在实践中不断修正和不断深化的认识。分布在全国各地和军内外我们团队的很多成员已经将水声信号处理和声呐技术在国防和国家安全及国民经济的诸多方面开拓了新的应用领域，他们的成果也使我学到了许多新的知识。书中难免存在一些疏漏和不足之处，敬请各位专家和读者不吝批评、指正。

李启虎

2019年6月

目　　录

一　信号处理的基本理论与方法

二 自适应信号处理

三 水下目标的检测、估计与识别

四　数字式声呐设计理论与应用

五　水声通信和水下目标成像

六　北极声学

七　综合性评述及科普

一

信号处理的基本理论与方法

多分层相关器输出的一种简单计算法①

李 启 虎

〔**提要**〕 本文提出的相关器模型，适用于从限幅相关到模拟相关的各种分层的相关器。文中给出一种计算多分层相关器输出的简单方法，并指出分层门限存在最佳值。

1. 引言

本文利用Kendall关于二维正态分布的结果[1]，推导出一个可以间接地计算从限幅相关（只分一层）到模拟相关（无限分层）的各类相关器输出的辅助公式，从而解决了输入信号为高斯过程时相关器输出的计算问题。

我们讨论如图1(a)所示的一般相关器的模型。假定输入信号$x(t)$、$y(t)$是均值为零的联合高斯分布的宽平稳随机过程，τ为延迟时间，则输出

$$R(\tau)=\frac{1}{T}\int_0^T x(t-\tau)y(t)dt \tag{1}$$

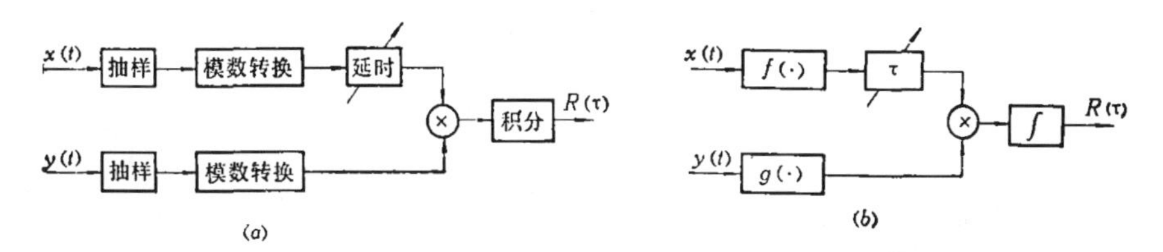

图1 一般相关器模型图

一般而言，模数转换的位数总是有限的，所以，实际上对$x(t)$、$y(t)$施加了某种分层运算。我们将其等效为图1(b)。其中$f(\cdot)$，$g(\cdot)$限定为包括分层运算在内的Borel函数。在不引起混淆时，我们有时候将$f(\cdot)$，$g(\cdot)$统一地记作$B(\cdot)$。例如，限幅运算就是分一层的运算，记作$B_1(x)=\operatorname{sgn}x$。模拟运算就是无限分层的运算，记作$B(x)=x$。

我们知道，只要积分时间T足够长，就有

$$\frac{1}{T}\int_0^T x(t-\tau)y(t)dt\approx E[x(t-\tau)y(t)] \tag{2}$$

等号右方的$E[\cdot]$表示集合平均。本文的目的就是给出一种计算$E[x(t-\tau)y(t)]$的简便方法。

① 电子学报, 1980, (4): 79-81.

2．多分层相关器的输出

根据图1(b)的简化模型，一般要计算

$$R(\tau)=\iint f(x)g(y)p(x,y)dxdy \tag{3}$$

式中，$p(x,y)$为$x(t)$、$y(t)$的联合概率密度函数。根据高斯分布的假设，有

$$p(x,y)=\frac{1}{2\pi\sigma^2\sqrt{1-\rho^2(\tau)}}\exp\left\{\frac{-1}{2(1-\rho^2(\tau))\sigma^2}\times(x^2-2xy\rho(\tau)+y^2)\right\} \tag{4}$$

式中，σ^2为$x(t)$、$y(t)$的方差；$\rho(\tau)$为$x(t)$、$y(t)$的相关系数。所以，实际上$p(x,y)$和$R(\tau)$都是ρ的函数。为简单起见，我们将$R(\tau)$记作$z(\rho)$。

对于二维正态分布，Kendall给出了以下的结果[1]

$$\frac{\partial p(x,y)}{\partial\rho}=\sigma^2\frac{\partial^2 p(x,y)}{\partial x\partial y} \tag{5}$$

利用式（5）及分部积分公式，即可得到

$$\frac{\partial z(\rho)}{\partial\rho}=\sigma^2\iint\frac{\partial f(x)}{\partial x}\quad\frac{\partial g(y)}{\partial y}p(x,y)dxdy \tag{6}$$

这就是计算相关函数的基本公式。因为$\partial f(x)/\partial x$、$\partial g(y)/\partial y$是某些Dirac函数的线性组合，所以式（6）非常便于计算。现举例如下：

〔**例1**〕 模拟相关器

$f(x)=g(x)=B(x)=x$，代入式（6）得

$$z(\rho)=\sigma^2\rho,\qquad 即R(\tau)=\sigma^2\rho(\tau)$$

〔**例2**〕 极性重合相关器

$f(x)=g(x)=B_1(x)=\mathrm{sgn}\,x$，代入式（6）得

$$z(\rho)=\frac{2}{\pi}\arcsin\rho,\qquad 即\ R(\tau)=\frac{2}{\pi}\arcsin\rho(\tau)$$

〔**例3**〕 模拟-分n层相关器

$$f(x)=B(x)=x$$

$$g(y)=B_n(y)=\left\{\mathrm{sgn}\,y+\frac{1}{2}\sum_{k=1}^{n-1}[\mathrm{sgn}(y-kV_0)+\mathrm{sgn}(y+kV_0)]\right\}V_0$$

式中V_0为分层门限，代入式（6）并经过计算可得

$$R(\tau)=\sqrt{\frac{2}{\pi}}\sigma\rho(\tau)V_0\left[1+\sum_{k=1}^{n-1}\exp\left(-\frac{(kV_0)^2}{2\sigma^2}\right)\right] \tag{7}$$

从上面的例子可以得到以下结论：当输入中的一路是模拟量时，不管另一路分多少层，相关器的输出总是和输入的相关系数成正比的。

3．分层的最佳门限

相关器中的分层运算，要用到模数转换。为了合理利用输入信号的动态范围，分层门

限的选取是重要的。曾经有人建议用饱和噪声、抽样噪声来作为选取门限的一个原则[2~3]。我们下面给出的准则是合理选取门限，使分层前后的二阶矩尽量保持一致。

现以分二层的系统为例。设输入信号 $x(t)$ 是正态分布的，其概率密度为

$$p(x)=\frac{1}{\sqrt{2\pi}\,\sigma}\exp\left(-\frac{x^2}{2\sigma^2}\right)$$

分二层的运算是

$$B_2(x)=V_0\left[\frac{1}{2}\operatorname{sgn}(x-V_0)+\operatorname{sgn}x+\frac{1}{2}\operatorname{sgn}(x+V_0)\right]$$

我们的目的是合理选取分层门限 V_0，使得分层前后的二阶矩的绝对误差

$$e=|\sigma^2-\operatorname{Var}[B_2(x)]|$$

成为极小。式中的 σ^2 显然为分层前信号的方差，$\operatorname{Var}[B_2(x)]$ 则是分层后信号的方差。经过计算可知

$$e=\left|4V_0^2-6V_0^2\Phi\left(\frac{V_0}{\sigma}\right)-\sigma^2\right| \tag{8}$$

其中 $\Phi(\cdot)$ 为概率积分。

通过数值计算，可以得知，当 $V_0\approx0.6\sigma$ 时，e 达到极小值。

对于分 n 层的相关器，类似的计算可以给出最佳的分层门限。

4. 讨论

用于计算相关函数的式(6)，不仅适用于分层运算，对其他类型的阶梯函数也合用。例如信号分析中常用到的阈处理：

$$f(x)=\begin{cases}V_0, & |x|<V_0\\ 0, & \text{其他}\end{cases}$$

换句话说，只要出现在式（6）中的 $\partial f(x)/\partial x$、$\partial g(y)/\partial y$ 是Dirac函数形式的线性组合，式(6)均可使计算大为简化。

参 考 文 献

〔1〕 M.G.Kendall, The Advanced Theory of statistics, Charles Griffin Comp. Lit., London, Vol. I, 1958, p.177.

〔2〕 G.A.Gray, G.W.Zeoli, IEEE Trans., Vol.AES-7, No.1, 1971, pp.222-223.

〔3〕 V.G.Hansen, IEEE Trans., Vol.AES-10, No.2, 1974, pp.274-280.

A Simple Method for Calculating the output of Multilayer Correlator

Li Qi-hu

Abstract: The model of correlator considered in this paper is suitable for any kinds of multilayer correlator, including polarity coincidence correlator and analogue correlator.

A simple calculation method of the output of a multilayer correlator is given. In certain sense there exists an optimum threshold for layer operation, the theoretical analysis and practical evolution are presented.

THE SYSTEM GAIN OF MULTICHANNEL WIENER FILTER①

LI Qihu(李启虎)
(Institue of Acoustics, Academia Sinica)

Received March 4, 1987

ABSTRACT

As a pre-processing technique, the Wiener filter can be used to suppress strong interference and noise. The general solution of optimum transfer function has been found for the multichannel system. By using an inverse matrix formula, a simple expression of optimum transfer function can be derived in the case that only one interference is present. The formula for calculating the system gain is presented. The asymptotic performance of system gain as input interference to signal ratio increasing to infinite is evaluated. When the interference is distorted not only in phase but also in amplitude as well, the system gain can be evaluated in some particular cases. The results of numerical calculation are given in this paper.

I. INTRODUCTION

The Wiener filter can be used as a pre-processing technique to obtain processing gain which would enhance a subsequent correlation operation. The aim of the correlation stage is to detect (and obtain a location estimate of) a weak signal embedded in much stronger interference which also overlaps the signal in the frequency domain[1-2].

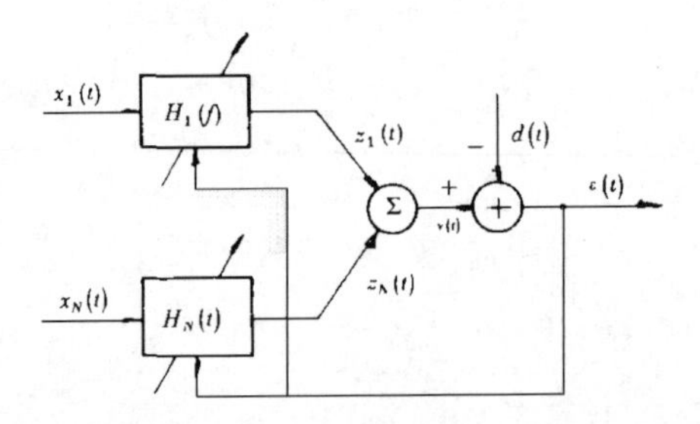

Fig. 1 Block diagram of multichannel nois cancelling system

The block diagram described in this paper is shown in Fig.1. The signal received by the kth element is

$$x_k(t) = s_k(t) + I_k(t) + n_k(t) \qquad k = 1, \cdots, N \tag{1}$$

where $s_k(t)$, $I_k(t)$ and $n_k(t)$ denote signal, interference and noise respectively. N is the number of channels. $s_k(t)$, $I_k(t)$ and $n_k(t)$ are assumed to be stationary random processes with zero mean and independent of each other. $d(t)$ is the desired signal.

Our main purpose is to suppress strong interference and noise. It is necessary to maximize the

① Chinese Journal of Acoustics, 1988, 7(4): 302-311.

output signal to noise (including interference) ratio. This is the MSNR criterion. The solution of this problem is quite complicated except in the case of single frequency[3]. Intuitively, it is easy to understand that, in certain sence, when the noise plus interference is much stronger than the signal, the least mean square (LMS) error criterion would bring the system gain close to the MSNR criterion[4-5].

The solution of optimum transfer function under the LMS criterion is given in Section 2. The formula for calculating the system gain will be derived in Section 3. Generally, the solution of optimum transfer function depends on the input parameters. If there are no other distortions between every two channels except the time delay, then we can get a simple formula for the system gain. The results which illustrate the relationship between the system gain and input parameters, such as the number of channels, input signal to noise ratio and the generalized angle between signal and interference will be given in Section 4.

In Section 5, a special case of two channels will be considered. Besides the difference of time delay, some kinds of amplitude modulation on interference are assumed. In this situation, there will be a different form for the system gain. The theoretical analysis has shown that, if the distortion of interference has sinusoidal modulation form, then the system gain is not sensitive to the time delay of interference.

Section 6 is a summary of results described in this paper. Some problems which we should further study are briefly discussed here.

II. The General Solution of Optimum Transfer Function

The multichannels linear system described in this paper can be considered as a special case of adaptive noise cancelling filter, for which widrow gave the optimum weight coefficients in the time domain[4]. In the frequency domain, the transfer functions $H_1(f), \cdots, H_N(f)$ form a filtering vector. For simplicity of theoretical analysis, we will adopt the vector and matrix notations throughout this paper. Let

$$\mathbf{x}(t) = [x_1(t), \cdots, x_N(t)]^T \tag{2}$$

$$\mathbf{H}(f) = [H_1(f), \cdots, H_N(f)]^T \tag{3}$$

$$\mathbf{z}(t) = [z_1(t), \cdots, z_N(t)]^T \tag{4}$$

where T denotes the transpose operation of vector or matrix. The system output signal is the sum of all $z_k(t)$:

$$y(t) = \sum_{k=1}^{N} z_k(t) \tag{5}$$

The mean square value of output error signal $\epsilon(t)$ is

$$I = E[\epsilon^2(t)] = E[(y(t) - d(t))^2] \tag{6}$$

The concept of adaptive noise cancelling filter is to adjust $\mathbf{H}(f)$ such that I is minimized.

Suppose

$$R_{kl}(\tau) = E[x_k(t)x_l(t-\tau)], \quad 1 \leqslant k, l \leqslant N \tag{7}$$

$$R_k(\tau) = E[x_k(t)d(t-\tau)], \quad 1 \leqslant k \leqslant N \tag{8}$$

are the correlation functions of $x_k(t)$ and $x_l(t)$, $x_k(t)$ and $d(t)$. Then the cross spectrum matrix of input data can be expressed as

$$\Phi_{xx}(f)=\begin{bmatrix}\Phi_{11}(f) & \Phi_{12}(f) & \cdots & \Phi_{1N}(f)\\ \vdots & \vdots & \vdots & \vdots\\ \Phi_{N1}(f) & \Phi_{N2}(f) & \cdots & \Phi_{NN}(f)\end{bmatrix} \tag{9}$$

where

$$\Phi_{kl}(f)=\int R_{kl}(\tau)\exp(-2\pi jf\tau)d\tau,\qquad 1\leqslant k,\ l\leqslant N \tag{10}$$

is the cross spectrum of kth and lth input data.

According to the linear filtering theory of stationary random process, we can express I as:

$$I=E[d^2(t)]-2\int \mathbf{H}^T(f)\Phi_{xd}(f)df+\int \mathbf{H}^T(f)\Phi_{xx}(f)\mathbf{H}^*(f)df \tag{11}$$

where * denotes the complex conjugate. In order to find the optimum transfer function $\boldsymbol{H}_{\text{opt}}(f)$ such that I is minimum, it is necessary to solve the Euler equation for $\boldsymbol{H}(f)$, we have[5]

$$\mathbf{H}_{\text{opt}}(f)=[\Phi_{xx}(f)]^{*-1}\Phi_{xd}(f) \tag{12}$$

where

$$\Phi_{xd}(f)=[\Phi_1(f),\cdots,\Phi_N(f)]^T \tag{13}$$

$$\Phi_k(f)=\int R_k(\tau)\exp(-2\pi jf\tau)d\tau,\qquad 1\leqslant k\leqslant N \tag{14}$$

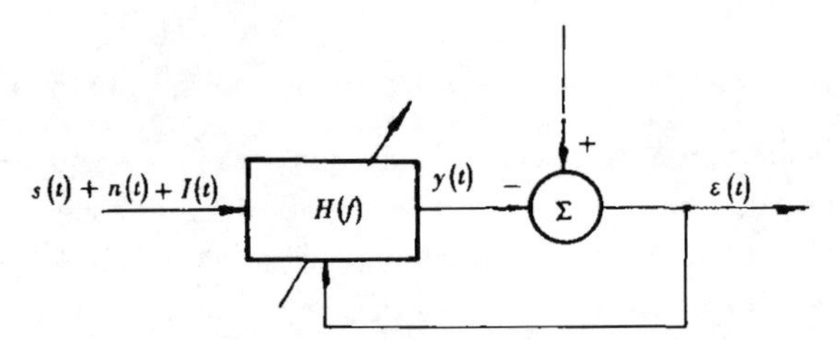

Fig. 2 Single channel Wiener filter

As an example, let us consider the simplest case, i.e. the single channel system (cf. Fig.2). For $N=1$, we have

$$x(t)=s(t)+n(t)+I(t),\quad d(t)=s(t) \tag{15}$$

Obviously,

$$\Phi_{xx}(f)=\Phi_s(f)+\Phi_n(f)+\Phi_I(f) \tag{16}$$

where $\Phi_s(f)$,$\Phi_n(f)$ and $\Phi_I(f)$ are the power spectrum of signal, noise and interference respectively. It follows from equation (12),

$$\mathbf{H}_{\text{opt}}(f)=\frac{\Phi_s(f)}{\Phi_s(f)+\Phi_n(f)+\Phi_I(f)} \tag{17}$$

This is the Wiener solution of single channel.

III. Expression for System Gain

In the vector notation, the input signal $\mathbf{x}(t)$ can be expressed as

$$\mathbf{x}(t)=\mathbf{s}(t)+\mathbf{n}(t)+\mathbf{I}(t) \tag{18}$$

Suppose $\Phi_{ss}(f)$,$\Phi_{nn}(f)$ and $\Phi_{II}(f)$ are the cross spectrum matrix of signal, noise and interference, respectively. Then we have

$$\Phi_{xx}(f) = \Phi_{ss}(f) + \Phi_{nn}(f) + \Phi_{II}(f) \tag{19}$$

According to equation (11), the output power of system is

$$\begin{aligned}[P]_{out} &= E[y^2(t)] = \int \mathbf{H}^T(f)\Phi_{xx}(f)\mathbf{H}^*(f)df \\ &= \int \mathbf{H}^T(f)\Phi_{ss}(f)\mathbf{H}^*(f)df + \int \mathbf{H}^T(f)[\Phi_{nn}(f) + \Phi_{II}(f)]\mathbf{H}^*(f)df\end{aligned} \tag{20}$$

The output signal to noise (including interference) ratio is defined by

$$\left[\frac{\mathrm{S}}{(\mathrm{N}+\mathrm{I})}\right]_{out} = \frac{\int \mathbf{H}^T(f)\Phi_{ss}(f)\mathbf{H}^*(f)df}{\int \mathbf{H}^T(f)[\Phi_{nn}(f) + \Phi_{II}(f)]\mathbf{H}^*(f)df} \tag{21}$$

The input signal to noise ratio is defined by

$$\left[\frac{\mathrm{S}}{(\mathrm{N}+\mathrm{I})}\right]_{in} = \frac{\overline{\sigma_s^2}}{\overline{\sigma_n^2} + \overline{\sigma_I^2}} \tag{22}$$

where

$$\overline{\sigma_s^2} = \frac{1}{N}\sum_{k=1}^{N} \sigma_{sk}^2 \tag{23}$$

and σ_{sk}^2 is the power of the kth signal, the definition of $\overline{\sigma_n^2}$ and $\overline{\sigma_I^2}$ are similar to $\overline{\sigma_s^2}$.

The system gain is then defined by

$$\mathrm{GAIN} = \frac{[\mathrm{S}/(\mathrm{N}+\mathrm{I})]_{out}}{[\mathrm{S}/(\mathrm{N}+\mathrm{I})]_{in}} \tag{24}$$

It is a function of filtering vector $\mathbf{H}(f)$. When $\mathbf{H}(f) = \mathbf{H}_{opt}(f)$ in equation (24), then GAIN $= (\mathrm{GAIN})_{opt}$. Usually, the power of signal, noise and interference from different channels are identical. In this situation $\overline{\sigma_s^2} = \sigma_s^2$, $\overline{\sigma_n^2} = \sigma_n^2$, and $\overline{\sigma_I^2} = \sigma_I^2$, where

$$\sigma_s^2 = \int \Phi_s(f)df, \qquad \sigma_n^2 = \int \Phi_n(f)df, \qquad \sigma_I^2 = \int \Phi_I(f)df \tag{25}$$

It yields

$$\left[\frac{\mathrm{S}}{(\mathrm{N}+\mathrm{I})}\right]_{in} = \frac{\sigma_s^2}{\sigma_n^2 + \sigma_I^2} \tag{26}$$

IV. Point Sources: Nondistorted Channel

From equations (12) and (21) we see that the computation of system gain is quite complicated, because the matrix $[\Phi_{xx}(f)]^{-1}$ is hardly to obtain. In this section we will consider a relativly simple case, for which the signal and interference come from point sources and the channel is distortionless, so that the cross spectrum matrix of input signal and interference $\Phi_{ss}(f)$ and $\Phi_{II}(f)$ can be written as a dyad. The noise field is assumed to be white noise and independet among channels. Let

$$\begin{aligned}x_k(t) &= s_k(t) + n_k(t) + I_k(t) \\ s_k(t) = s(t - \tau_k), &\qquad I_k(t) = I(t - \delta_k), \qquad k = 1, \cdots, N\end{aligned} \tag{27}$$

Define the directional vectors of signal and interference as follows:

$$\mathbf{A} = [e^{-2\pi jf\tau_1}, \cdots, e^{-2\pi jf\tau_N}]^T \tag{28}$$

$$\mathbf{B} = [e^{-2\pi jf\delta_1}, \cdots, e^{-2\pi jf\delta_N}]^T \tag{29}$$

It is easy to show that

$$\Phi_{xx}(f) = \Phi_n(f)\mathbf{I}_0 + \Phi_s^*(f)\mathbf{AA}^{*T} + \Phi_I(f)\mathbf{BB}^{*T} \tag{30}$$

where $\mathbf{I}_0$ denotes the unity matrix.

The key step of finding the system gain is to calculate $[\Phi_{xx}(f)]^{-1}$. Let

$$\mathbf{R} = \mathbf{I}_0 + \left(\frac{\Phi_I(f)}{\Phi_n(f)}\right)\mathbf{BB}^{*T} \tag{31}$$

We have

$$\Phi_{xx}(f) = \Phi_n(f)\left[\mathbf{R} + \left(\frac{\Phi_s(f)}{\Phi_n(f)}\right)\mathbf{AA}^{*T}\right]$$

It is not difficult to prove that

$$\mathbf{R}^{-1} = \mathbf{I}_o - \frac{\mathbf{BB}^{*T}}{N + \Phi_n(f)/\Phi_I(f)} \tag{32}$$

$$[\Phi_{xx}(f)]^{-1} = \frac{1}{\Phi_n(f)}\left[\mathbf{R}^{-1} - \left(\frac{\Phi_n(f)}{\Phi_s(f)}\right)\frac{\mathbf{R}^{-1}\mathbf{AA}^{*T}\mathbf{R}^{-1}}{\mathbf{A}^{*T}\mathbf{R}^{-1}\mathbf{A} + \Phi_n(f)/\Phi_s(f)}\right] \tag{33}$$

Now we have got the explicit expression for $[\Phi_{xx}(f)]^{-1}$. Substituting equation (33) into (21) and (24), the closed-form expression of output signal to noise ratio and system gain can be written as:

$$\begin{aligned}\left[\frac{\mathrm{S}}{(\mathrm{N}+\mathrm{I})}\right]_{\mathrm{opt}} &= \frac{\int \mathbf{H}_{\mathrm{opt}}^T(f)\Phi_{ss}(f)\mathbf{H}^*_{\mathrm{opt}}(f)df}{\int \mathbf{H}_{\mathrm{opt}}^T(f)[\Phi_{nn}(f) + \Phi_{II}(f)]\mathbf{H}^*_{\mathrm{opt}}(f)df} \\ &= \frac{\int \Phi_s(f)|\mathbf{A}^T\mathbf{H}_{\mathrm{opt}}(f)|^2 df}{\int \Phi_n(f)|\mathbf{H}_{\mathrm{opt}}(f)|^2 df + \int \Phi_I(f)|\mathbf{B}^T\mathbf{H}_{\mathrm{opt}}(f)|^2 df}\end{aligned} \tag{34}$$

$$\left[\frac{\mathrm{S}}{\mathrm{N}+\mathrm{I}}\right]_{\mathrm{opt}} = \frac{\int \Phi_s(f)\left[\frac{P(f)}{P(f) + \Phi_n(f)/\Phi_s(f)}\right]^2 df}{\int\left[\frac{\Phi_n(f)U(f)}{[P(f) + \Phi_n(f)/\Phi_s(f)]^2}\right]df + \int\frac{\Phi_I(f)|V(f)|^2}{[P(f) + \Phi_n(f)/\Phi_s(f)]^2}df} \tag{35}$$

where

$$P(f) = \mathbf{A}^{*T}\mathbf{R}^{-1}\mathbf{A} = N - \frac{N^2\cos^2\gamma}{N + \Phi_n(f)/\Phi_I(f)} \tag{36}$$

$$U(f) = \mathbf{A}^{*T}(\mathbf{R}^{-1})^2\mathbf{A} = N - \frac{2N^2\cos^2\gamma}{N + \Phi_n(f)/\Phi_I(f)} + \frac{N^3\cos^2\gamma}{[N + \Phi_n(f)/\Phi_I(f)]^2} \tag{37}$$

$$|V(f)|^2 = |\mathbf{A}^{*T}\mathbf{R}^{-1}\mathbf{B}|^2 = \left[\frac{\Phi_n(f)}{\Phi_I(f)}\right]^2 + \frac{N^2\cos^2\gamma}{[N + \Phi_n(f)/\Phi_I(f)]^2} \tag{38}$$

It is worth to show that

$$\cos^2\gamma = \frac{|\mathbf{A}^T\mathbf{B}|^2}{N^2} \tag{39}$$

is sometimes called generalized angle between signal and interference[6], it is often used in the array signal processing theory.

For some particular environment fields of signal, noise and interference equations (34)—(39) can be simplified. For example the formula derived by Hudson[7] is a special case of equation (35).

Now the system gain can be expressed as

$$(\mathrm{GAIN})_{\mathrm{opt}} = \frac{[S/(N + I)]_{\mathrm{pot}}}{\sigma_s^2/(\sigma_n^2 + \sigma_I^2)} \tag{40}$$

By using equations (35)—(40) we can calculate the system gain of multichannel Wiener filter. It depends on various parameters, such as input noise to signal ratio *NSR*, input interference to signal ratio *ISR*, the number of channels N, and the generalized angle $CR = \cos^2\gamma$etc.

The generalized angle between signal and interference plays an important role in detection problems. The value of $\cos^2\gamma$ depends on the practical geometry of receiving array. For example, if the receiving array consists of N equally spaced elements in a line, then we have

$$\cos^2\gamma = \frac{|\mathbf{A}^T\mathbf{B}|^2}{N^2} = \left[\frac{\sin(N\pi d\sin\theta/\lambda)}{N\sin(\pi d\sin\theta/\lambda)}\right]^2 \tag{41}$$

Where d is the distance of adjacent elements, λ is the wave length of incident signal.

The numerical relationships between the system gain and input parameters are plotted in Fig.3. In these examples, the spectra of signal, noise and interference are all assumed to be flat in the considered frequency band W.

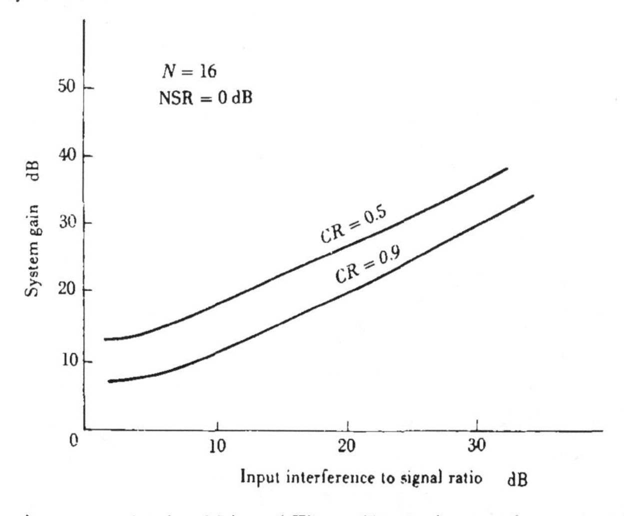

Fig. 3 the system gain of multichannel Wiener filter in the case of strong interference

V. Point Sources: Distorted Channel

Now we consider a more complicated situation. Suppose the interference in different channels appear in some kinds of amplitude distortion and time delay as well. Let

$$x_k(t) = n_k(t) + s(t - \tau_k) + m_k(t) I(t - \delta_k), \qquad k = 1, \cdots, N \tag{42}$$

where $m_k(t)$ is a slowly changed signal as compared with $I(t)$.

Generally, the cross spectrum matrix of interference could not be expressed by a dyad, therefore the inverse matrix formulas (32) and (33) are not suitable in this situation. But in some special cases, the solution of optimum transfer function can be derived in a similar way as described in the previous section.

Let us consider a two-channel system. The input data are

$$x_1(t) = n_1(t) + s(t) + I(t)$$

$$x_2(t) = n_2(t) + s(t - D_1) + m(t) I(t - D_2) \tag{43}$$

It is easy to show

$$\Phi_{nn}(f) = \Phi^{n}(f)\begin{bmatrix} 1 & 0 \\ 0 & 1 \end{bmatrix} \tag{44}$$

$$\Phi_{ss}(f) = \Phi_s(f)\begin{bmatrix} 1 & e^{2\pi j f D_1} \\ e^{-2\pi j f D_1} & 1 \end{bmatrix} = \Phi_s(f)\mathbf{A}\mathbf{A}^{*T} \tag{45}$$

where $\mathbf{A} = [1, e^{-2\pi j f D_1}]^T$.

The cross power spectrum of interference can be derived by using the convolution theorem in the theory of Fourier transformation:

$$\Phi_{II}(f) = \begin{bmatrix} \Phi_I(f) & \bar{m}\Phi_I(f)e^{2\pi j f D_2} \\ \bar{m}\Phi_I(f)e^{-2\pi j f D_2} & \Phi_m(f) * \Phi_I(f) \end{bmatrix} \tag{46}$$

where * denotes the convolution operation, $\bar{m}$ is the mean value of $m(t)$ and $\Phi_m(f)$ is the power spectrum of the modulation function $m(t)$. In the following special cases, the system gain can be calculated directly by using equations (21)—(24). The computation procedure is similar to that described in Section 3. We only show the main results and omit the process of derivation of various expressions.

1. $m(t) = 1$

$$\Phi_{II}(f) = \Phi_I(f)\mathbf{B}\mathbf{B}^{*T} \tag{47}$$

where $\mathbf{B} = [1, e^{-2\pi j f D_2}]^T$. The system gain will depend on $D = D_1 - D_2$ and $\cos^2\gamma = \cos[2\pi f(D_1 - D_2)]$ is a periodic function of $D_1 - D_2$.

2. $m(t) = \alpha$

$$\Phi_{II}(f) = \Phi_I(f)\mathbf{E}\mathbf{E}^{*T} \tag{48}$$

where $\mathbf{E} = [1, e^{-2\pi j f D_2}]^T$. Note that,

$$\left[\frac{\mathrm{S}}{\mathrm{N+I}}\right]_{\mathrm{in}} = \frac{\sigma_s^2}{\sigma_n^2 + \hat{\sigma}_I^2}, \qquad \hat{\sigma}_I^2 = \frac{1}{2}\sigma_I^2(1 + \alpha^2)$$

3. $m(t) = \sin(2\pi f_0 t)$, $f_0 \ll W_I$

$$\Phi_{II}(f) = \begin{bmatrix} \Phi_I(f) & 0 \\ 0 & \frac{\alpha^2}{4}[\Phi_I(f + f_0) + \Phi_I(f - f_0)] \end{bmatrix} \tag{49}$$

$$\approx \Phi_I(f)\begin{bmatrix} 1 & 0 \\ 0 & \alpha^2/2 \end{bmatrix}$$

Now we can obtain the output signal to noise ratio as follows:

$$\left[\frac{S}{N+I}\right]_{opt} = \frac{\int \Phi_s(f)\left[\frac{P(f)}{P(f)+\Phi_n(f)/\Phi_s(f)}\right]^2 df}{\int \frac{\Phi_n(f)U(f)}{[P(f)+\Phi_n(f)/\Phi_s(f)]^2}df + \int \frac{\Phi_I(f)W(f)}{[P(f)+\Phi_n(f)/\Phi_s(f)]^2}df} \tag{50}$$

where

$$P(f) = \frac{1}{a_1} + \frac{1}{a_2} \tag{51}$$

$$U(f) = \frac{1}{a_1^2} + \frac{1}{a_2^2} \tag{52}$$

$$W(f) = \frac{1}{a_1^2} + \frac{\alpha^2}{2a_2^2} \tag{53}$$

$$a_1 = 1 + \frac{\Phi_I(f)}{\Phi_n(f)}, \qquad a_2 = 1 + \frac{\alpha^2{}_I(f)}{\Phi_n(f)} \tag{54}$$

By using equations (50)—(54) we can calculate the $(\mathrm{GAIN})_{opt}$. The numerical results are illustrated in Fig.4.

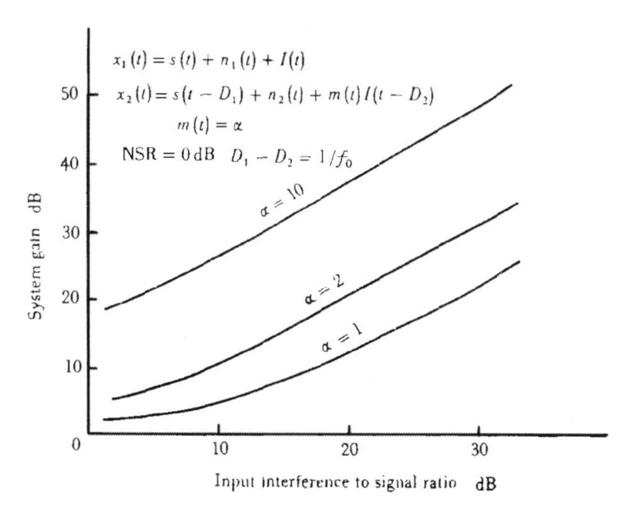

Fig. 4 Modulated strong interference

VI. DISCUSSION

We have derived the general solution of optimum transfer function for multichannel linear system. If the signal and interference come from point sources and the channels are without distortion, the final closed-form expression of the system gain can be obtained by solving the inverse matrix of cross spectrum of input data. Although the results given here are for only single interference, the method canbe easily generalized to multiple interferences by the recursive procedure.

If the distortion of channel has the form of modulation on the interference, then for the two-channel system, the system gain of the Wiener filter can also be obtained. When the modulation function is sinusoidal, the system gain would not be sensitive to the time delay between signal and interference. As the number of channels is larger than 2, it is necessary to find a new method to get a closed-form expression for the system gain.

It is interesting to consider the asymptotic performance as ISR increasing to infinite. For the single frequency signal, from equations (34)—(40), we obtain

$$\left[\frac{\mathrm{S}}{\mathrm{N}+\mathrm{I}}\right]_{\mathrm{out}}=\frac{\sigma_s^2}{\sigma_n^2}\frac{N[1+N(1-\cos^2\gamma)\sigma_I^2/\sigma_n^2]}{1+N\sigma_I^2/\sigma_n^2} \tag{55}$$

$$(\mathrm{GAIN})_{\mathrm{opt}}=\frac{N(1+\sigma_I^2/\sigma_n^2)[1+N(1-\cos^2\gamma)\sigma_I^2/\sigma_n^2]}{1+N\sigma_I^2/\sigma_n^2} \tag{56}$$

when $\cos^2\gamma=1$,

$$\left[\frac{\mathrm{S}}{\mathrm{N}+\mathrm{I}}\right]_{\mathrm{out}}=\frac{\sigma_s^2}{\sigma_n^2}\frac{1+N}{1+N\sigma_I^2/\sigma_n^2}\to 0 \qquad as\ \mathrm{ISR}\to\infty \tag{57}$$

$$(\mathrm{GAIN})_{\mathrm{opt}}=\frac{N(1+\sigma_I^2/\sigma_n^2)}{1+N\sigma_I^2/\sigma_n^2}\to 1 \qquad as\ \mathrm{ISR}\to\infty \tag{58}$$

when $\cos^2\gamma\neq 1$,

$$\left[\frac{\mathrm{S}}{\mathrm{N}+\mathrm{I}}\right]_{\mathrm{out}}\to\frac{\sigma_s^2}{\sigma_n^2}N(1-\cos^2\gamma) \qquad as\ \mathrm{ISR}\to\infty \tag{59}$$

$$(\mathrm{GAIN})_{\mathrm{opt}}\approx 1+N\frac{\sigma_I^2}{\sigma_n^2}(1-\cos^2\gamma)\to\infty, \qquad as\ \mathrm{ISR}\to\infty \tag{60}$$

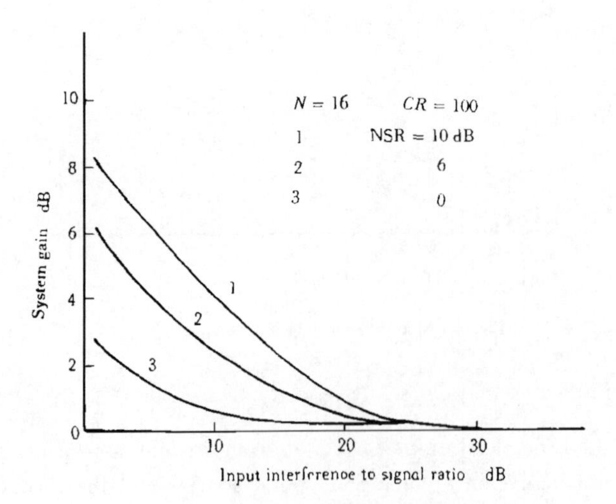

Fig. 5 Asymptotic performance of system gain

Owsley have obtained similar result in a different model[8]. The numerical results are illustrated in Fig.5.

The intuitive explanation is as follows. When the $\cos^2\gamma = 1$, it means that the signal and interference come from same direction, the Wiener filter would not discriminate each other. The system gain will tend to saturate as ISR goes to infinite.

If $\cos^2\gamma \neq 1$, it means that the signal and interference come from different directions and the Wiener filter will form a zero point in the inteference direction such that the output signal to noise ratio is a constant as ISR increasing. Therefore the system gain goes to infinite linearly with ISR.

A part of this work was supported by the ONR under grant N00014–83–k–0275 and AFOSR undergrant AF-AFOSR81-0186. Theauthor would like to express his thanks top Prof. S.C. Schwaertz and Prof Bede Liu fortheir help.

References

[1] Schwartz, S.C., "*Detection and Estimation in Signal-like Noise*", Princeton University, NJ 08544, 1983.

[2] Chen, C.T. and Schwartz, S.C., "*Linear Filtering as a Pre-processing Technique for Correlation Detection and Localization*", Technique Report, Princeton University, NJ 08544, 1983.

[3] Edelblute, D.J. *et al.*, "Criteria for Optimum Signal Detection Theory for Array", *J. Acoust. Soc. Am.* **41**(1967), 199—205.

[4] Widrow, B. *et al.*, "Adaptive Noise Cancelling: Principle and Applications", *Proc. IEEE*, **63**(1975), 1692—1716.

[5] Li, Q.H., "The Equivalent Relationship of Optimum Solution Between Time Domain and Frequency Domain in Adaptive Beamforming System", (in Chinese) *Acta Acoustica of Sinica*, **4**(1979), No.4 296—308.

[6] Cox, H., "Sensitivity Consideration in Adaptive Beamforming", *NATO Advanced Study Institute on Signal Processing*, (1972) Reidal Pub., 619.

[7] Hudson, J.E., "*Adaptive Array Principles*", Ch. 5 Peter Peregrinus Ltd. 1981.

[8] Owsley, N.L., "Overview of Adaptive Array Processing", *NATO Proc. of Advanced Study Institute*, (1984), 31.1—31.17.

用于拖曳式线列阵的一种新的线谱增强系统[①]

李 启 虎

(中国科学院声学研究所)

1987 年 3 月 4 日收到

摘要 拖曳式线列阵是近来声呐信号处理领域中非常活跃的课题. 自适应线谱增强器有可能用于拖曳式线列阵的信号处理系统. 在利用 LMS 迭代算法的自适应线谱增强器中,由于存在迭代噪声,系统增益与输入信噪比有关. 如果不抑制拖曳平台的噪声,线谱增强系统的效用就无法发挥出来,文中提出用自适应噪声抵消系统和自适应线谱增强器级联的概念. 首先把拖曳平台的噪声作为干扰加以抑制,再对输出信号进行线谱增强,从而可以提取微弱信号的线谱分量.

A NEW ARCHITECTURE OF ADAPTIVE LINE ENHANCER FOR TOWED LINE ARRAY

LI QI-HU

(*Institute of Acoustics, Academia Sinica*)

Received March 4, 1987

Abstract Towed line array is recently a very active topic in the field of sonar signal processing. It is possible to use the adaptive line enhancer in the signal processing system of a towed line array. It is shown that the system gain of an adaptive line enhancer in which the LMS iterative algorithm is used, is related to the input signal to noise ratio due to the presence of iterative noise. The line enhancer would be useless, if the platform noise has not been cancelled. A cascade architecture of adaptive noise canceller and adaptive line enhancer is proposed in this paper. The platform noise, as an strong interference is cancelled by the adaptive noise canceller. So that the sinusoidal components in the output signal are enhanced by the followed line enhancer. It is possible to detect the weak line spectrum components from strong noise background. The block diagram of such cascade system and the performance analysis are given. The system simulation shows a good agreement with the theoretical analysis.

一、引 言

拖曳式线列阵系统是近年来声呐信号处理领域中十分活跃的课题[1-4]. 由于利用长线阵接收目标噪声,所以有可能从环境背景噪声中提取微弱信号中的低频线谱分量. 从而实现远距离探测.

自适应线谱增强器 (ALE) 是一种有效的抑制噪声而通过窄带信号分量的系统[4-7],因此它可以应用于拖曳式线列阵的信号处理系统,作为后置处理提取目标噪声中的线谱分量.

一个实用的、由抽头延迟线 (TDL) 构成的 ALE,由于存在迭代噪声,其系统增益依赖于

① 声学学报, 1988, 13(3): 167-173.

输入信噪比．当信噪比很低或信号被强干扰所淹没时，ALE 的性能通常很差． 因此，ALE 不能直接用于波束成形系统．因为环境噪声中所包含的拖曳平台噪声是一个不可忽略的强干扰．

本文指出，对选定的 LMS 算法，实际系统的输入信噪比存在着一个临界值．当信噪比低于这个值时，系统的增益急剧下降．为了使 ALE 正常地工作，对输入信号的预处理就是十分必要的．本文提出用自适应噪声抵消系统（ANC）作为预处理，得到一个 ANC 和 ALE 级联的组合结构．整个系统将工作于系统增益基本一致的状况．

二、ALE 的稳态增益

考虑如图 1 所示的 ALE 系统，系统的解相关时间为 $\Delta = d_0 T_s$，其中 d_0 是某个整数．假定Δ相当大，使得 $z(k) = x(k - d_0)$ 中的噪声分量和 $x(k)$ 独立．可以证明，这种系统的增益大约是 $L/2$[5]．

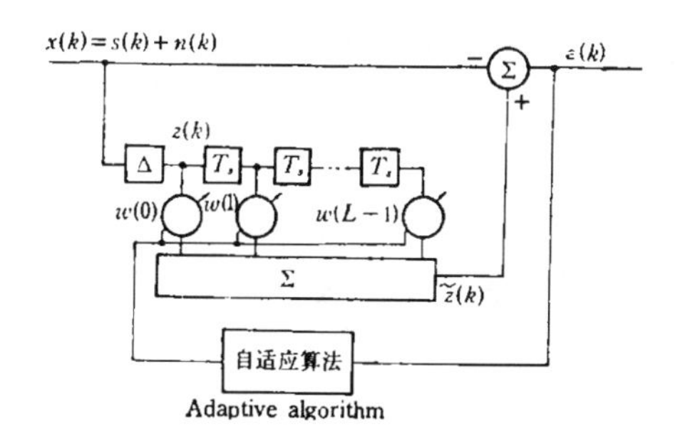

图 1　自适应线谱增强器方框图

Block diagram of adaptive line enhancer

假如我们采用 LMS 算法[1]，权系数的迭代公式是

$$\boldsymbol{W}(j+1) = \boldsymbol{W}(j) - 2\mu[\boldsymbol{Z}^T(j)\boldsymbol{W}(j) - x(j)]\boldsymbol{Z}(j) \tag{1}$$

其中

$$\boldsymbol{Z}(j) = [z(j-d_0), \cdots, z(j-d_0-L+1)]^T$$

是由每个抽头处的信号所构成的信号矢量，而

$$\boldsymbol{W} = [w(0), \cdots, w(L-1)]^T$$

表示加权矢量，我们有

$$\boldsymbol{W}_{\text{opt}} = \boldsymbol{R}_{zz}^{-1}\boldsymbol{R}_{zx} \tag{2}$$

相关矩阵 $\boldsymbol{R}_{zz}$ 的第 (k, l) 个元素是

$$R_z(k-l) = E[z(k)z(l)] = R_s(k-l) + R_n(k-l) \quad k, l = 0, \cdots, L-1 \tag{3}$$

相关矢量 $\boldsymbol{R}_{zx}$ 的第 l 个元素是

$$E[z(k-l)x(k)] = R_s(l+d_0), \quad l = 0, \cdots, L-1 \tag{4}$$

于是 (2) 式可以写成：

$$\sum_{k=0}^{L-1}[R_s(k-l) + R_n(k-l)]w(k) = R_s(d_0+l), \quad l = 0, \cdots, L-1 \tag{5}$$

把(5)式的解记作

$$\boldsymbol{W}_{\text{opt}} = [\mu_0, \cdots, \mu_{L-1}]^T \tag{6}$$

ALE 的输出均方差 (MSE) 是 $I = E[\varepsilon^2(k)]$. 当 $\boldsymbol{W} = \boldsymbol{W}_{\text{opt}}$ 时, I 的极小值为

$$\begin{aligned} I_{\min} &= E[x^2(k)] - \boldsymbol{W}_{\text{opt}}^T \boldsymbol{R}_{zz} \boldsymbol{W}_{\text{opt}} \\ &= E[x^2(k)] - \sum_{k=0}^{L-1} \sum_{l=0}^{L-1} \mu_k \mu_l R_x(k-l) \end{aligned} \tag{7}$$

对于 ALE, (5) 式可以直接求解.

$$\left. \begin{aligned} x(k) &= s(k) + n(k) \\ s(k) &= A\cos(2\pi f k T_s + \theta) \end{aligned} \right\} \tag{8}$$

其中 $n(k)$ 是功率为 σ_n^2 的白噪声, θ 为在 $[0, 2\pi]$ 内均匀分布的随机变量, 于是有

$$\left. \begin{aligned} R_s(k) &= \frac{A^2}{2} \cos(2\pi f k T_s) \\ R_n(k) &= \sigma_n^2 \delta(k) \end{aligned} \right\} \tag{9}$$

方程(5)成为

$$\sigma_n^2 w(l) + \sum_{k=0}^{L-1} w(k) \cdot \frac{A^2}{2} \cos[(k-l)\alpha] = \frac{A^2}{2} \cos[(l+d_0)\alpha], \quad l = 0, \cdots, L-1 \tag{10}$$

其中 $\alpha = 2\pi f T_s$.

仿照文献[7]中的方法,可以从(10)式中得到最佳权系数的解:

$$\mu_k = w_{\text{opt}}(k) \approx Q\cos[(k+d_0)\alpha] \triangleq Q\cos(\zeta + k\alpha), \quad k = 0, \cdots, L-1 \tag{11}$$

其中

$$\zeta = d_0 \alpha,$$

$$Q = \frac{2R_{\text{ALE,in}}}{2 + L R_{\text{ALE,in}}} \tag{12}$$

$R_{\text{ALE,in}}$ 表示 ALE 的输入信噪比,显然

$$R_{\text{ALE,in}} = A^2 / 2\sigma_n^2 \tag{13}$$

TDL 的输出均方值是

$$D = E[\tilde{z}(k)] = \sum_{k=0}^{L-1} \sum_{l=0}^{L-1} \mu_k \mu_l [R_n(k-l) + R_s(k-l)] \triangleq D_n + D_s \tag{14}$$

其中

$$D_n = \sum_{k=0}^{L-1} \sum_{l=0}^{L-1} \mu_k \mu_l R_n(k-l) = \sigma_n^2 \sum_{k=0}^{L-1} \mu_k^2 \tag{15}$$

是 TDL 的输出噪声功率.

$$D_s = \sum_{k=0}^{L-1} \sum_{l=0}^{L-1} \mu_k \mu_l R_s(k-l) = \frac{A^2}{2} \sum_{k=0}^{L-1} \sum_{l=0}^{L-1} \mu_k \mu_l \cos[(k-l)\alpha] \tag{16}$$

是输出信号功率.

$$D_n = \frac{1}{2} \sigma_n^2 Q^2 (L + p) \tag{17}$$

$$D_s = \frac{A^2}{8} Q^2(2Lp + L^2 + q) \tag{18}$$

其中

$$p = \sum_{k=0}^{L-1} \cos(2k\alpha + 2\zeta) = \frac{\cos[2\zeta+(L-1)\alpha]\sin L\alpha}{\sin\alpha} \tag{19}$$

$$q = \sum_{k=0}^{L-1}\sum_{l=0}^{L-1} \cos[2(k-l)\alpha] = \left(\frac{\sin L\alpha}{\sin\alpha}\right)^2 \tag{20}$$

由此得到

$$R_{\mathrm{ALE,out}} = \frac{D_s}{D_n} = \frac{2Lp + L^2 + q}{2(L+p)} R_{\mathrm{ALE,in}} \tag{21}$$

ALE 的系统增益是

$$G_{\mathrm{ALE}} = \frac{R_{\mathrm{ALE,out}}}{R_{\mathrm{ALE,in}}} = \frac{2Lp + L^2 + q}{2(L+p)} \tag{22}$$

容易证明，G_{ALE} 是 p 的递增函数，再由(19)式可知

$$|p| \leqslant \left|\frac{\sin L\alpha}{\sin\alpha}\right| \tag{23}$$

于是

$$\frac{1}{2}\left(L - \left|\frac{\sin L\alpha}{\sin\alpha}\right|\right) \leqslant G_{\mathrm{ALE}} \leqslant \frac{1}{2}\left(L + \left|\frac{\sin L\alpha}{\sin\alpha}\right|\right) \tag{24}$$

这是有关 G_{ALE} 的一个相当好的不等式．因为$|\sin L\alpha/\sin\alpha|$与$L$相比通常很小．例如，当 $f/f_s = 0.2$ 时，$|\sin L\alpha/\sin\alpha| \leqslant |1/\sin\alpha| = 1.05$．对于一个具有 64 个抽头的 ALE 来说，系统增益界于 14.9dB 和 15.1dB 之间．

显然，由 B. Widrow 推导的[5]和 J. T. Rickard 推导的[9]用于计算增益的公式都是(24)式的特例．

对于一个利用 LMS 算法的实际系统，迭代噪声要成为输出噪声的一部分．当利用(1)式计算权系数 $\boldsymbol{W}_{\mathrm{opt}}$ 时，迭代噪声和 TDL 的长度 L，控制收敛因子 μ 及输入噪声有关[8]．它的值为

$$\gamma_n^2 = \mu I_{\min}(1 + R_{\mathrm{ALE,in}})L\sigma_n^2 \tag{25}$$

系统增益是

$$G_{\mathrm{ALE}} = \frac{D_s}{(D_n + \gamma_n^2)R_{\mathrm{ALE,in}}} = \frac{Q^2(2Lp + L^2 + q)}{2Q^2(L + p) + 4\mu I_{\min}(1 + R_{\mathrm{ALE,in}})L} \tag{26}$$

当 $L \gg 1$ 时，(26) 式成为

$$G_{\mathrm{ALE}} \approx \frac{Q^2 L}{2Q^2 + 4\mu I_{\min}(1 + R_{\mathrm{ALE,in}})} \tag{27}$$

这就是文献[8]中的(41)式．

容易证明 $R_{\mathrm{ALE,in}}$ 是 Q 的单调函数．我们感兴趣的情况是 $R_{\mathrm{ALE,in}} \leqslant 1$．在这种情况下，还可以证明 G_{ALE} 是 Q 的单调函数．

当 $L \gg 1$ 时 $Q \approx 2/L$，所以

$$(G_{\mathrm{ALE}})_{\max} = G_{\mathrm{ALE}}\big|_{R_{\mathrm{ALE,in}}=1} = \frac{L}{2(1+L^2\mu I_{\min})} \tag{28}$$

这是系统增益的饱和值. 我们可以求出一个 $R_{\mathrm{ALE,in}}$ 的临界值 $R_{\mathrm{ALE,cr}}$，使得在这一点 G_{ALE} 是饱和值的一半. 也就是

$$G_{\mathrm{ALE}} = \frac{1}{2}(G_{\mathrm{ALE}})_{\max}$$

经过计算，得到

$$R_{\mathrm{ALE,cr}} = \frac{2}{2\sqrt{L^2+2M_0}-L} \tag{29}$$

其中 $M_0 = 1/4\mu\sigma_n^2$ 是以自适应周期为单位的迭代时间常数.

当输入信噪比小于 $R_{\mathrm{ALE,cr}}$ 时，系统增益下降很快. 系统模拟的结果也证实了这一点.

三、ANC+ALE 的级联系统

为了有效利用 ALE，必须使输入信噪比足够高. 为此，我们考虑如图 2 所示的自适应噪声抵消系统 (ANC) 和 ALE 的级联. 这里，原始输入可以是对准某一期望信号的波束输出；参考输入则是来自拖曳平台的噪声. 我们用它来抵消观测波束中的干扰.

如果把 ANC 的增益记作 G_{ANC}，那么整个系统的总增益是

$$G = G_{\mathrm{ANC}} \cdot G_{\mathrm{ALE}} \tag{30}$$

显然，

$$R_{\mathrm{ALE,in}} = R_{\mathrm{ANC,out}} \tag{31}$$

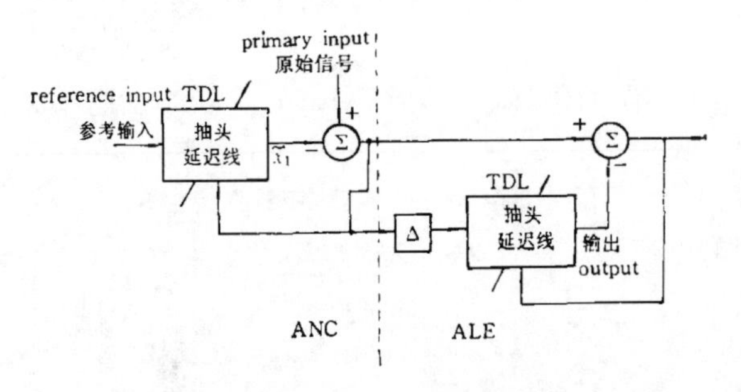

图 2 ANC 和 ALE 的组合结构

Combined architecfure of ANC and ALE

ANC 的频响函数为

$$H(f) = \sum_{l=0}^{L-1} w(l)\exp(-2\pi jlT_s) \tag{32}$$

我们假定 L 足够长，以致 $H(f)$ 和最佳滤波器 $H_{\mathrm{opt}}(f)$ 之间的差异可以忽略. Wiener 滤波器的原始输入是

$$x_2(t) = s(t) + n_2(t) + i_2(t) \tag{33}$$

参考输入是

$$x_1(t) = n_1(t) + i_1(t) \tag{34}$$

其中 $i_1(t)$，$i_2(t)$ 为强干扰. $n_1(t)$，$n_2(t)$ 为互不相关的噪声. ANC 的输出是

$$\begin{aligned} x(t) &= x_2(t) - \tilde{x}_1(t) = s(t) + n_2(t) - \tilde{n}_1(t) + i_2(t) - \tilde{i}_1(t) \\ &\triangleq s(t) + n(t) + i(t) \end{aligned} \tag{35}$$

其中

$$\begin{aligned} n(t) &= n_2(t) - \tilde{n}_1(t) \\ i(t) &= i_2(t) - \tilde{i}_1(t) \end{aligned} \tag{36}$$

输出均方差值是

$$\begin{aligned} J &= E[x^2(t)] = E[x_2(t) - \tilde{x}_1(t)]^2 \\ &= E[s^2(t)] + E[n_2^2(t)] + E(\tilde{n}_1^2(t)] + E[i_2(t) - \tilde{i}_1(t)]^2 \end{aligned}$$

当 $H(f) = H_{opt}(f)$ 时，J 达到极小值 J_{min}[9].

$$\begin{aligned} J_{min} = E[x^2(t)]\big|_{H(f)=H_{opt}(f)} &= E[s^2(t)] + E[n_2(t) - i_2(t)]^2 \\ &- \int_{-\infty}^{\infty} |H_{opt}(f)|^2 \Phi_{x_1}(f)df \end{aligned} \tag{37}$$

其中 $\Phi_{x_1}(f)$ 是 $x_1(t)$ 的自功率谱.

$$(N+I)_{ANC,out} = E[n_2(t) - i_2(t)]^2 - \int_{-\infty}^{\infty} |H_{opt}(f)|^2 \Phi_{x_1}(f)df \tag{38}$$

由此得到 ANC 的输出信噪比

$$R_{ANC,out} = \frac{S_{ANC,out}}{(N+I)_{ANC,out}} = \frac{\sigma_s^2}{\sigma_{n_2}^2 + \sigma_{i_2}^2 - \int_{-\infty}^{\infty} |H_{opt}(f)|^2 \Phi_{x_1}(f)df} \tag{39}$$

ANC 的系统增益为

$$G_{ANC} = \frac{R_{ANC,out}}{R_{ANC,in}} \tag{40}$$

其中

$$R_{ANC,in} = \sigma_s^2 / (\sigma_{n_2}^2 + \sigma_{i_2}^2)$$

(39)式中的最佳传输函数 $H_{opt}(f)$ 由下式给出[9]

$$H_{opt}(f) = \frac{\Phi_{x_1 x_2}^x(f)}{\Phi_{x_1}(f)} \tag{41}$$

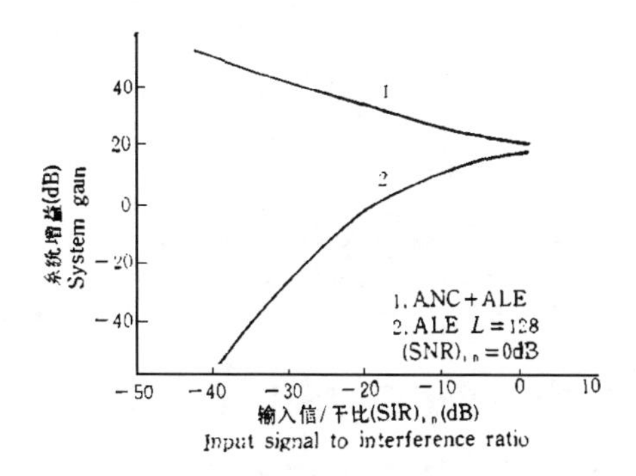

图 3 组合系统和 ALE 性能的比较

Comparison of ANC + ALE and ALE

图 3 给出了 ANC + ALE 和 ALE 的比较. 我们看到, 级联系统大致具有一致的增益. 而 ALE 系统就不同,当输入信/干比低于 15dB 时,系统增益要比 ANC + ALE 系统低 20dB 以上.

四、系统模拟结果

上面的分析为实际系统的设计提供了基础.利用本文提供的模型,我们在 VAX/750 机上进行了系统模拟. 计算结果证实了理论分析. 图 4,5 是部分结果. 图 4 给出了 ALE 系统增益与参数 L, μ 的关系. 图 5 给出了理论计算值和实际模拟结果的比较.

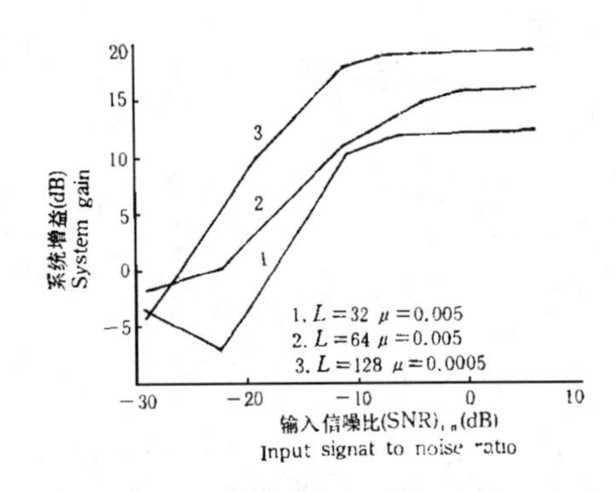

图 4 用 LMS 算法的 ALE 的增益

The system gain of ALE with LMS algorithm

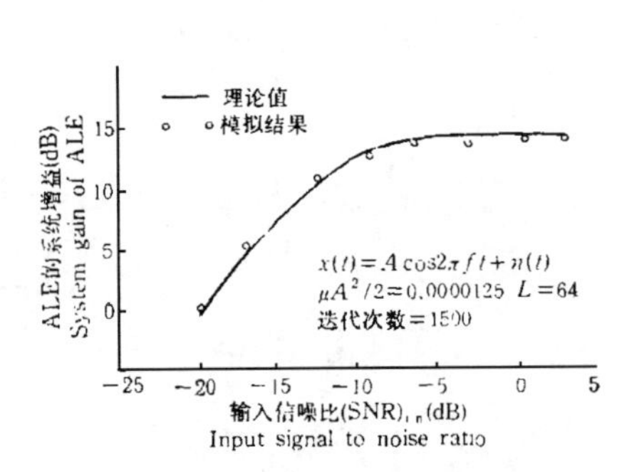

图 5 系统模拟的结果

The results of system simulation

本报告的部分工作是在美国 Princeton 大学作的. 得到美国海军研究办公室 N000 14-83-K-0275 和空军研究办公室 AF-AFOSR81-0186 的资助. 作者对 S. C. Schwartz 教授与 Bede Liu 教授所给予的支持深表感谢.

参 考 文 献

[1] Hinich, M. J. and Rule, W., "Bearing estimation using a large towed array", *J. Acoust. Soc. Amer.*, **58**(1975), 1023—1029.

[2] Bucker, H. P., "Beamforming a towed line array of unknown shape", *J. Acoust. Soc. of Am.*, **68**(1978), 1451—1454.

[3] Kuperman, W. A. *et al.*, "Towed array response to ship noise: a near field propagation problem", in "*Adaptive Methods in Underwater Acoustics*", 71—80 Ed. by H. G. Urban, (D. Reidal Publ. Comp. 1984).

[4] Butler, D., "Beamforming with a distored towed array", *Ibid*. 469—476.

[5] Widrow, B. *et al.*, "Adaptive noise cancelling: principles and applications", *Proc. IEEE* **63**(1975), 1692—1716.

[6] Zeidler, J. R. *et al.*, "Adaptive enhancement of multiple sinusoidal in uncorrelated noise", *IEEE Trans*, **ASSP-26** (1978), 240—254.

[7] Treichler, J. R., "Trnasient and convergent behavior of the adaptive line enhancer", *IEEE Trans.*, **ASSP-27**(1979), 53—62.

[8] Rickard, J. T. and Zeidler, J. R., "Second-order output statistics of the adaptive line enhancer", *IEEE Trans.* **ASSP-27**(1979), 31—39.

[9] 李启虎,"自适应波束成形中稳态特性的研究",声学学报,7(1982), No. 3. 165—173,

实时通用时间压缩式相关器①

李启虎　孙　增　孙长瑜
（中国科学院声学研究所）
1988年8月16日收到

摘要　六十年代以来，时间压缩式相关器在主动声呐的匹配滤波中得到广泛应用，但是由于受数据长度、采样频率等的限制，这种模型不能直接用于被动声呐．否则会引起数据的丢失．本文提出一种交叉存取式时间压缩相关器，利用一个缓冲存贮器可以实时完成任意长数据的相关运算，从而解决了主、被动声呐中所需的相关器设计问题．文中给出了正序、逆序进动型数据链及移位型正序、逆序数据链的流程图和硬件实现的方框图．本文讨论的方法可用于数字式声呐的设计，大大简化系统和提高声呐性能．

Real-time general purpose time-compressed correlator

LI Qihu　SUN Zen　SUN Changyu
(*Institute of Acoustics, Academia Sinica*)
Received August 16, 1988

Abstract The time-compressed correlator has been widely applied in matched filter of active sonar since 1960's. The principles of DELTIC and MACORMATIC can not be directely used in passive sonar due to the limitation of sampling frequency and data length. Otherwise the input data will be missed periodically in the system output. An Alternative Access Time-compressed Correlator (ALATIC) is proposed in this paper. by using a buffer register the correlation operation of arbitrary length can be done in real time. This model is a general purpuse correlator, which can be used in both active and passive sonar. The running positive order and inverse order register and the shift positive order and inverse order model are presented. The hardware implementation and the flow chart graph of ALATIC is also described. The model presented in this paper can be used in digital sonar design.

一、引　　言

相关检测已被证明为从噪声背景中提取信号的有效方法，这种方法自六十年代以来在声呐设计中已得到广泛应用[1-3]．数字式声呐在利用相关技术时，具有特别有利的条件．利用等间隔的时间采样可以获得所需要的数据流，根据不同的延时抽头就能得到不同延时值的相关函数，但是，在设计声呐时，我们立刻就会发现，如果需要在整个时间轴上扫描相关函数的峰值，那么就要设计相当长的延迟线，对应于每一延时间隔，需要有一套相乘器和后置积累器，这

① 声学学报, 1990, 15(2): 146-150.

将使声呐设备非常庞大，甚至是不可取的.

首先寻找到方法解决这一问题的是Anderson 等人[4-7]. 他们提出在主动声呐的匹配滤波器中采用时间压缩的办法，对输入数据进行时间压缩从而使输出数据率提高几个数量级. 它不仅可以在每一采样周期内输出一组相关函数值，同时还可以完成在时间轴上的扫描. 所以可以实时求出相关函数的峰值，实现匹配滤波的时间压缩有好多具体的方法，比如 DELTIC（延迟线时间压缩相关器）MACORMATIC（磁芯存贮时间压缩相关器）等.

但是，当把这种方法用于被动声呐设计时却会出现一系列的问题. 问题之一是主动声呐用匹配滤波器是一种交流相关器，可以将信号外差到接近零频率的地方，从而能使用比较低的采样频率，例如 2kHz，这时采样周期是 500μs. 如果要压缩 128 倍，那么时间压缩器的工作周期为 3μs 左右，容易由硬件实现，对于被动声呐来说，情况就不同，经过升采样运算的系统具有很高的采样频率，比如 100kHz，这时输入采样周期只有 10μs，如果压缩 128 倍，那么工作周期少于 100ns，用于移位寄存的办法就不经济了；问题之二是主动声呐的数据长度有限. 举例来说，如果脉冲宽度是 80ms，以 $T_s = 500\mu s$ 去采样，只有 160 个点，于是积分时间也只需要 $N = 160$. 但被动声呐需要长时间积分，DELTIC，MACORMATIC 等所采用的参考信号反复循环的办法显然不合适.

本文提出一种新的交叉存取式时间压缩式相关器，采用一个外加的缓冲存贮器构成进动、移位两个数据链可以实时得到升采样率下的相关函数值. 这实际上是一种通用的高性能时间压缩式相关器，便于用硬件实现，文中还给出了简化后置积累过程的一种新算法，使得软件编制更容易又简化了硬件设计.

二、交叉存取式时间压缩式相关器的基本原理

相关检测的主要运算是完成两个数据序列的相乘和累加，设数据序列分别是

$$\begin{aligned} &x(k), \quad k = 0,1,2,\cdots \\ &y(k), \quad k = 0,1,2,\cdots \end{aligned} \tag{1}$$

累加次数（或积分时间）为N的相关函数为

$$R(k) = \frac{1}{N}\sum_{i=0}^{N-1} y(i)x(k+i) \quad k = 0,1,\cdots,N-1 \tag{2}$$

如果输入数据的采样周期为 T_s，那么按常规的方法每得到一个相关函数值需要有一个相关器.

主动声呐用的时间压缩相关器只用两个时间压缩环就可同时得到N点相关函数值，见图 1. 其中 $x(k)$ 进入接收时间压缩环，$y(k)$ 进入参考时间压缩环. 这两个环的工作周期是 T_s/N. 如果接收环采用进动式，那么有正序与逆序之分，取决于环的长度是$N-1$ 节还是$N+1$ 节. 在每一采样间隔 T_s 内，相关器可以输出N个相关函数值，出现峰值的时刻就是信号匹配的时间，由此可解算出目标距离.

用这种相关器于被动声呐时会出现两类问题，一个问题是被动声呐通常采用较高的采样频率. 所以必须用非常高速的移位寄存器件以便达到实时运算. 另一个问题是被动声呐需要

较长的积分时间，这时参考环的长度远远小于这个值，我们必须使参考环也变为能更新数据的进动环. 但在更新数据之前又必须完成N点相关函数，这就更加剧了实时运算的困难.

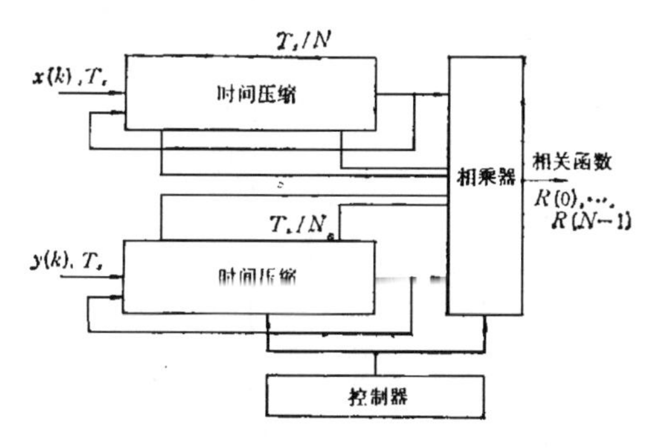

图 1　主动声呐用的时间压缩时间器

我们采用图 2 所示的交叉存取式时间压缩相关器，它成功地解决了上面两个问题，不丢失数据，能在高采样率下实时运算，硬件实现并不困难，软件编程也很容易.

图中X寄存器的长度为 N 节或$N+1$节，根据逆序或正序输出样本来定. $Y1$，$Y2$ 为两个缓冲寄存器. 当一个寄存器处于移位型时间压缩状态时，另一个寄存器开始存贮数据. 举例来说，如果X寄存器存放 $x(0),\cdots,x(N-1)$. $Y1$ 寄存器存放 $y(0),\cdots,y(N-1)$. 这时控制中心开始计算相关函数，每隔 T_s/N 秒，X 寄存器与 $Y1$ 寄存器各输出一个样本，经过 T_s 秒，得到相关函数：

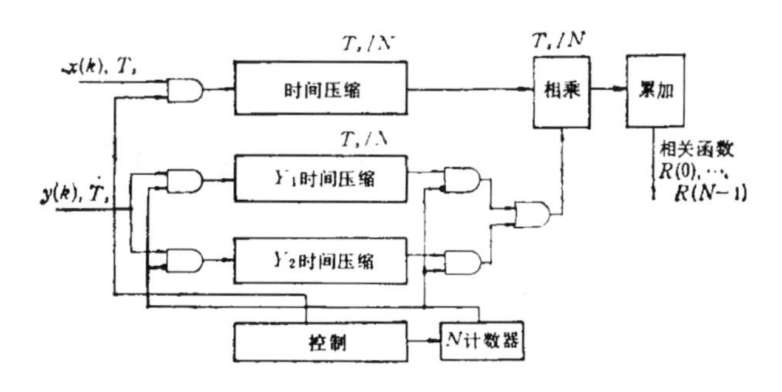

图 2　交叉存取式时间压缩相关器

$$R^{(1)}(0)=\frac{1}{N}\sum_{i=0}^{N-1}x(i)y(i) \tag{3}$$

下一个采样周期得到

$$R^{(1)}(1)=\frac{1}{N}\sum_{i=0}^{N-1}x(i+1)y(i) \tag{4}$$

$$\cdots\cdots$$

每一个采样周期使X寄存器更新一个样本，同时在 $Y2$ 寄存器内存入一个新的样本 $y(N-1+k)$. 在经过N个 T_s 秒之后，得到了N点相关函数

$$R^{(1)}(k)=\frac{1}{N}\sum_{i=0}^{N-1}y(i)x(i+k)\quad k=0,\cdots,N-1 \tag{5}$$

然后 $Y1$ 寄存器变为存贮型，$Y2$ 变为时间压缩型我们得到第二批相关函数：

$$R^{(2)}(k)=\frac{1}{N}\sum_{i=0}^{N-1}y(N-1+i)x(N-1+i+k),\quad k=0,\cdots,N-1 \tag{6}$$

如此类推，把　　$R^{(1)}(k),R^{(2)}(k),\cdots$

累加起来，得到所需的相关函数.

在运算的时候，X寄存器是进动型的，它可以采用正序也可以采样逆序的形式，但以逆序方式较方便，因为对于相关函数来说，正序和逆序无关紧要.

对于 $R^{(I)}(k)$，求和的方式，我们采用一种带反馈的低通滤波方式[8]（见图 3）. 这是一种便于硬件实现的后置积累环. 如果 $u(k)$ 是输入，$v(k)$ 是输出，那么

$$v(k)=\frac{1}{M}u(k)+\left(1-\frac{1}{M}\right)v(k-1) \quad (7)$$

一般可取 $M=2^K$ 的形式，这种后置积累方式具有良好的性能，并且只要改变 M 就可改变积分时间，软件编程非常简单，同时又可避免传统的多次累加引起溢出的问题.

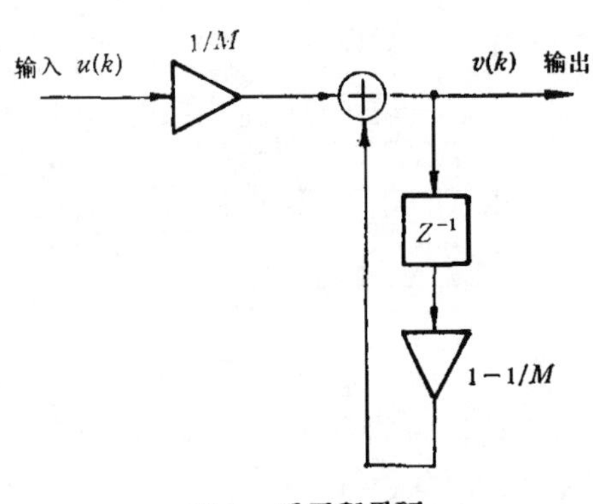

图 3　后置积累环

三、实　例

为了说明本文提出的设计思想，我们以 $N=4$ 为例来给出输入输出关系.（见图 4）.

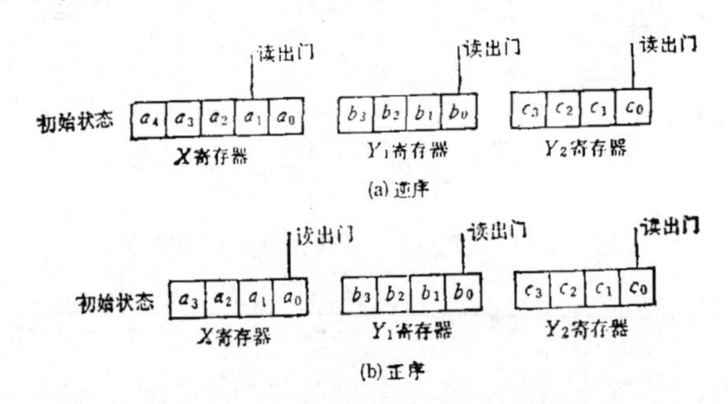

图 4　$N=4$ 的交叉存取式时间压缩相关器

图 4 中的 (a) 是逆序读出的，延时节数为 5，(b) 是正序读出的，延时节数为 4，初始状态随机设定.

以 4 个样本为一组，在逆序情况下，X 寄存器，Y1 寄存器，Y2 寄存器的输出为(取头 48 个样本)：

X 寄存器

$a_1a_2a_3a_4$　$x_0a_1a_2a_3$　$x_1x_0a_1a_2$　$x_2x_1x_0a_1$

$x_3x_2x_1x_0$　$x_4x_3x_2x_1$　$x_5x_4x_3x_2$　$x_6x_5x_4x_3$

$x_7x_6x_5x_4$　$x_8x_7x_6x_5$　$x_9x_8x_7x_6$　$x_{10}x_9x_8x_7$

Y1 寄存器

$b_0b_1b_2b_3$　$b_0b_1b_2b_3$　$b_0b_1b_2b_3$　$b_0b_1b_2b_3$

空　空　空　空

$y_7y_6y_5y_4$　$y_7y_6y_5y_4$　$y_7y_6y_5y_4$　$y_7y_6y_5y_4$

Y2 寄存器

空　空　空　空

$y_3y_2y_1y_0$　$y_3y_2y_1y_0$　$y_3y_2y_1y_0$　$y_3y_2y_1y_0$

空　空　空　空

由此表我们可以看出X寄存器和 Y1，Y2 寄存器交叉相乘求和可以得到逆序的相关函数．类似地我们可以得到正序输出(图 4(b))的结果

参 考 文 献

[1] Knight, W. C. *et al.*, "Digital Signal Processing for *sonar*", *Proc. IEEE,* 69(1981), 1451—1506.

[2] Skitzki, P., "Modern sonar systems", *Electronic Progress,* XVI (1974), No. 3, 20—37.

[3] Glisson, T. H. and Sage, A. P., "On sonar signal analysis", *IEEE Trans.* AES-6 (1970), 37—49.

[4] Stewart, J. L. and Westerfield, E. C., "A theory of active sonar detection", *Proc. IRE,* 47(1959), 872—881.

[5] Allen, W. B. *et al.*, "Diqital compressed-time correlators and matched filters for active sonar", *J. Acoust. Soc. Am.* 36(1964), 121—139.

[6] Anderson, V. C., "DELTIC Correlator", *Harvard Acoust. Lab. Tech. Memo.*, No. 37, 5 Jan. 1956.
Lab. Tech. Memo., No. 37, 5 Jan. 1956.

[7] Stewart, J. L. *et al.*, "Pseudorandom signal-correlation methods of nuderwater acoustic research I: principles", *J. Acoust. Soc. Amer.*, 37(1965) 1079—1090.

[8] 李启虎,"声呐信号处理引论",(海洋出版社,1984年).

论阵形畸变的拖曳式线列阵的工作方式的选取问题①

李 启 虎

（中国科学院声学研究所，北京，100080）

摘要 阵形畸变是拖曳式线列阵声呐性能偏离理想情况的主要原因. 本文讨论在常见阵形畸变方式下，线列阵指向性的计算. 给出在各种不同参数下求解椭圆反函数以确定每个基元的精确坐标的方法. 同时给出一种递推的公式，用于快速求解畸变阵的阵形. 根据对畸变阵指向性与理想指向性的比较，提出拖曳式线列阵的工作方式选取的准则. 从而解决了拖曳式线列阵声呐性能预报这一课题. 计算机模拟结果与理论分析完全一致.

On the problem of choices of operational mode for towed line array with distorted shape

LI Qihu

(Institute of Acoustics, Academia Sinica)

Abstract The distorted shape of array is one of the main reason, which results the performance degeneration from the ideal situation of towed line array. Based on the ordinary array shape distortion, the directivity function of towed line array with distortion is presented in this paper. An algorithm for precisely determining the coordinates of each element by means of the inverse elliptic function is derived. A fast approximation of recursive formula for solving distorted array shape is given. According to the comparison of ideal directivity and directivity of distorted array, a criterion for making decision of operational mode of towed line array is presented. So that the performance prediction problem in towed line array is solved. The results of system simulation in computer shows a good agreement with the theoretical analysis.

一、引 言

由于拖曳式线列阵声呐的水下声系统能远离拖体平台的噪声源，因而可获得较远的作用距离. 目前已在军事上和民用方面得到越来越多的注意. 拖曳式线列阵采用长线阵结构，在拖曳过程中不可避免地会使基阵产生畸变. 最常见又便于分析的畸变形式通常可以用正弦函数给以描述[1-3]. 由于畸变,实际系统的指向性与理想线列阵的指向性会有差异. 当这种差异

① 声学学报, 1991, 16(1): 31-36.

大到一定程度时，系统是无法工作的.

畸变阵的指向性计算是一个相当复杂的问题. 即使局限于二维情况，也很难准确测定实际系统的指向性. 例如，Butler 的简化模型[5]假定基阵接收元只有纵坐标的扰动，而横坐标的间隔不变. 这种假定显然与基阵的不可伸缩性矛盾. 所以也就无法以此来指导实际系统的工作.

有的作者提出，当系统指向性变坏时，应以改变航向使基阵变直从而获得理想的指向性[2] 但是，究竟何时使拖体机动仍旧没有一个切实可行的准则.

本文假定在拖曳过程中只发生二维的畸变. 严格按照基阵的不可伸缩性，求出基阵各基元的坐标. 并以此来计算指向性. 文中首先利用椭圆函数的反函数给出求解畸变阵阵形的公式，然后给出一种简单的递推方法，依次算出每个基元的坐标. 用这两种方法所求得的坐标位置的一致性非常好. 可以直接用于计算畸变阵的指向性.

利用实际计算的结果可以确定一个准则，即拖曳式线列阵首尾接收元时延差的判断准则. 以此来预报拖曳式线列阵声呐系统是否处于正常的工作状态.

本文提出的方法快速、有效，易于在实际系统中实现.

计算机模拟的结果与理论分析的结果一致.

二、畸变阵的指向性函数

拖曳式线列阵一般使用低频段（100—1000 Hz）工作. 对于远场的平面波来说，接收水听器在垂直方向的位移不会产生很大的相移. 所以我们假定阵形畸变只发生在水平面内. 同时假定拖缆中的声学模块是严格不可伸缩的.

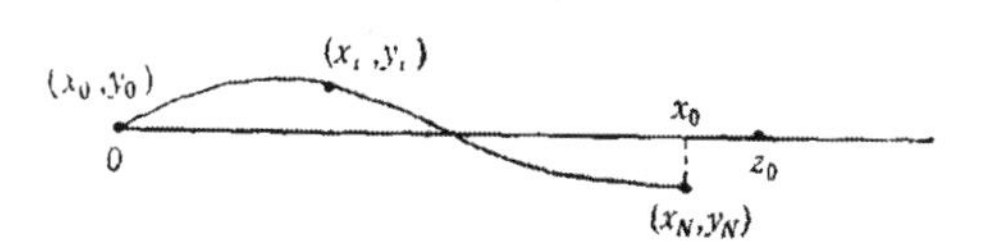

图 1 阵形畸变的拖曳式线列阵

如图 1 所示. 设基阵长为 Z_0. 阵形畸变后为正弦形的一段. 第 i 个基元的坐标是 (x_i, y_i)，$i = 1, \cdots, N$. 若设畸变前的坐标是 (x_{0i}, y_{0i})，那么

$$x_{0i} = (i-1)d,\quad y_{0i} = 0. \tag{1}$$

其中 d 是相邻基元的间隔.

指向角为 θ_0 的波束输出是[5]

$$D(\theta,\theta_0) = \left\{\left(\sum_{i=1}^{N} \sin[\xi_i(\theta) - \tau_i(\theta_0)]\right)^2 + \left(\sum_{i=1}^{N} \cos[\xi_i(\theta) - \tau_i(\theta_0)]\right)^2\right\}^{1/2} \tag{2}$$

其中

$$\tau_i(\theta) = \frac{(i-1)d\sin\theta}{c}$$
$$\xi_i(\theta) = (x_i\sin\theta + y_i\cos\theta)/c \tag{3}$$

c 是声速，θ 为目标入射方向.

x_i 由 t 时刻的阵形来确定. 当阵形是正弦形畸变时，

$$y_i = A\sin px_i \tag{4}$$

其中 A, p 为一对确定畸变形式的参数.

Butler 认为，当扰动 y_i 很小时，不妨认为[5]

$$x_i = (i-1)d$$

这种假定虽然可以使计算简化。但与实际情况并不相附。我们下面将会谈到这一点。

从式(2),(3)可以看出，要计算畸变阵的指向性，关键是确定畸变阵的阵形。同时要弄清楚什么样的畸变是可以容忍的，以什么准则来度量阵形畸变？在得到简单易行的准则之后，我们就有可能对拖曳式线列阵的性能作出预估，确定它应当遵循的工作方式。

根据前面的假定，畸变阵形遵从的函数关系为 $y=f(x)=A\sin px$。为了求出第 i 个阵元的坐标 (x_i, y_i) 必须根据第 i 个阵元所占有的弧长 $(i-1)d$ 来求解 (x_i, y_i)。其办法就是找出 x_i。使它对弧长的积分正好是 $(i-1)d$。也就是

$$(i-1)d = \int_0^{x_i} \sqrt{1+f'^2(x)}dx \tag{5}$$

特别是，当 $i=N$ 时，得到 $(N-1)d = \int_0^{x_N} \sqrt{1+f'^2(x)}dx$，$x_{N-1}$ 为畸变阵端点的横坐标。

由于 $f(x)$ 不是线性函数。所以 $x_i - x_{i-1}$ 就不相同。(5)式实际上是一个椭圆积分。令

$$E(X, A, p) = \int_0^X \sqrt{1+f'^2(x)}dx = \int_0^X \sqrt{1+A^2p^2\cos^2 px}\,dx \tag{6}$$

这个积分可以化为标准的椭圆积分的形式：

$$E(X, A, p) = \int_0^{pX} Q_1 \cdot \frac{1}{p}\sqrt{1-Q_2^2\sin^2 x}\,dx \tag{7}$$

其中

$$\begin{aligned} Q_1 &= \sqrt{1+A^2p^2} \\ Q_2 &= Ap/Q_1 \end{aligned} \tag{8}$$

利用(7),(8)式，依次令

$$E(X, A, p) = (i-1)d,\ i=1,\cdots,N \tag{9}$$

得到

$$\begin{aligned} x_i &= E^{-1}((i-1)d, A, p) \\ y_i &= A\sin px_i \end{aligned} \tag{10}$$

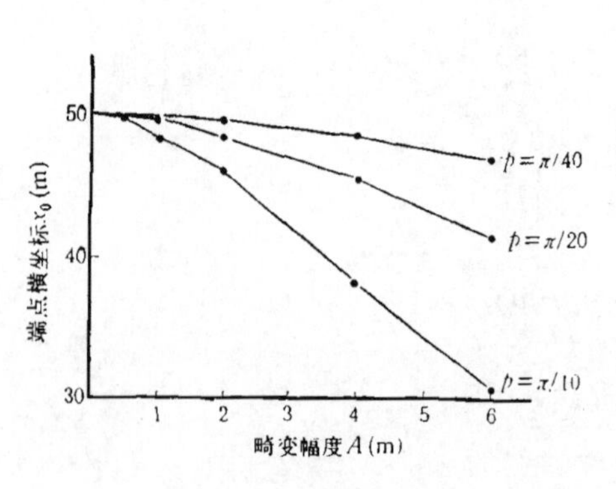

图 2　畸变阵端点横坐标的变化

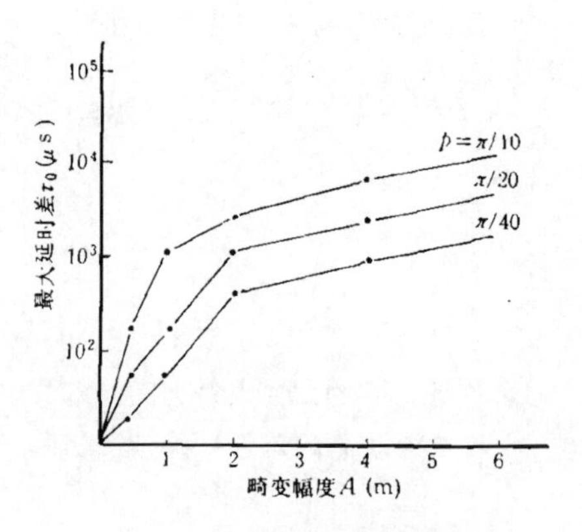

图 3　畸变引起的最大延时差

普通椭圆函数表只在 $0—\pi/2$ 内给值．而按 (10) 式求反函数时，显然有可能使自变量值大大超出 $\pi/2$ 之外．所以在作具体计算时，还必须作适当的变换，我们不在此细述．

以基阵长 $Z_0 = 50$ m 为例，计算基阵的端点坐标 X_0 如图 2 所示．从图中可以看到，当 $A = 2$ m 时，p 值从 $\pi/80$ 至 $\pi/10$ 变化时，端点坐标将从 49.900 m 降到 45.909 m．由此可以大致推想各基元的横坐标的变化．显然，这种变化是无论如何不能被忽略的．

由横坐标变化引起的最大延时差如图 3 所示．在某些情况下，它甚至是毫秒量级的．这些延时误差会对由 (2) 式所给出的指向性产生或多或少的影响．

三、求解畸变阵阵形的递推方法

上一节给出用椭圆函数的反函数求解基元坐标的方法(见 (10) 式)，一般地解决了已知畸变规律情况下的阵形确定问题．这个方法由于要用到椭圆函数，还是比较麻烦的．我们下面给出一种递推的方法，用弧的微分公式逐个把基元的坐标 (x_i, y_i) 解出来．

用 dS 来表示弧的微分．我们有

$$dS^2 = dx^2 + dy^2 = dx^2[1 + f'^2(x)]$$

$$dx = \frac{dS}{\sqrt{1 + A^2p^2\cos^2 px}} \tag{11}$$

对于不可伸缩的基阵，$dS =$ 常数．由此容易将 dx 求出来．显然，它与 x 的横坐标有关．

根据 (11) 式，把微分改写为差分，即 $dS = \Delta S$，$dx = \Delta x$，我们可以得到求解第 i 个基元横坐标的公式：

$$\begin{cases} \Delta x_{i+1} = \dfrac{\Delta S}{\sqrt{1 + A^2p^2\cos^2 px_i}} & i = 0, \cdots, N-1 \\ x_{i+1} = x_i + \Delta x_{i+1}, & i = 0, \cdots, N-1, \quad x_0 = 0 \end{cases} \tag{12}$$

由 (12) 式，可以依次将横坐标 x_i 求出来．然后再由 $y_i = A\sin px_i$ 把纵坐标求出来．

计算表明，当 ΔS 取得足够小时，由 (12) 式求出来的 (x_i, y_i) 和由求解椭圆函数的反函数结果完全一致．

图 4 是按 (12) 求出的基元间隔误差与基元位置的关系．我们看到，由于阵形弯曲的曲率

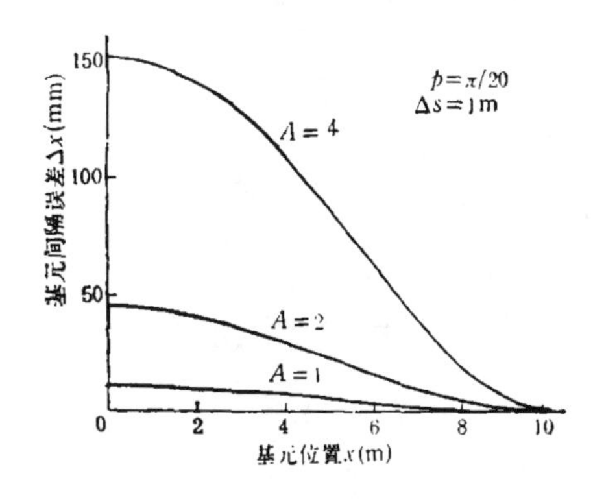

图 4 畸变阵基元间隔的变化

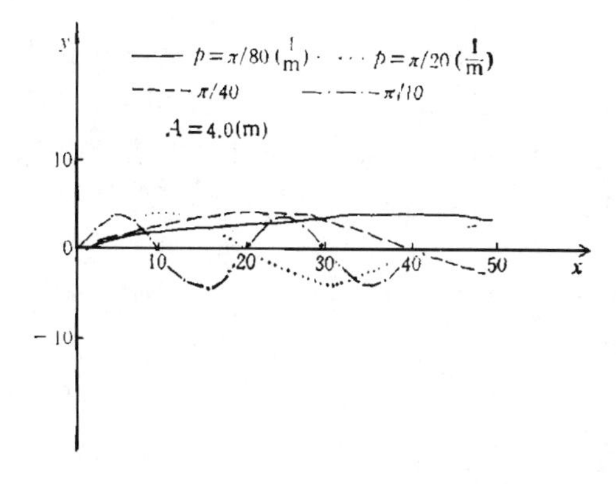

图 5 畸变阵的阵形变化

表 1 拖曳过程中主瓣响应

($N=50$, $d=1$m)

A (m)	4	3	2	1	0.5	0
p (1/m)	$\pi/20$	$\pi/40$	$\pi/80$	$\pi/100$	$\pi/200$	0
主极大值	7.68	14.71	30.03	44.21	49.34	50
端点横坐标(m)	44.58	48.25	48.94	48.98	48.99	49
最大延时差 (μs)	2946	500	40	14	7	0

是沿阵形变化的,所以不同部位基元的横坐标误差也不一样.

由于参数 A, p 的不可预测,以及基元间隔误差的不规则分布,使得拖曳式线列阵的阵形校正几乎是不可能的.

为了观察拖曳式线列阵指向性在拖曳过程中的变化. 我们用计算机模拟了一个拖曳过程. 在拖曳过程中, 阵形畸变 p 值由较大向较小方向变化, 最后变为零, 也就是基阵被拉直了. 图 5 给出了阵形的变化.

在计算机模拟中,用 (12) 式实时求解每个基元的坐标, 再代入 (2) 计算指向性函数. 图 6 是一个具体的例子.

我们要特别指出的是, 除了从图 6 中归一化指向性函数所看到的副瓣方向的变化之外, 还有主瓣响应的变化,见下表.

对照图 6 和表 1 我们可以看出, 当基阵端点基元产生 14 μs 的延时差时, 拖曳式线列阵的性能就已经开始变坏了.

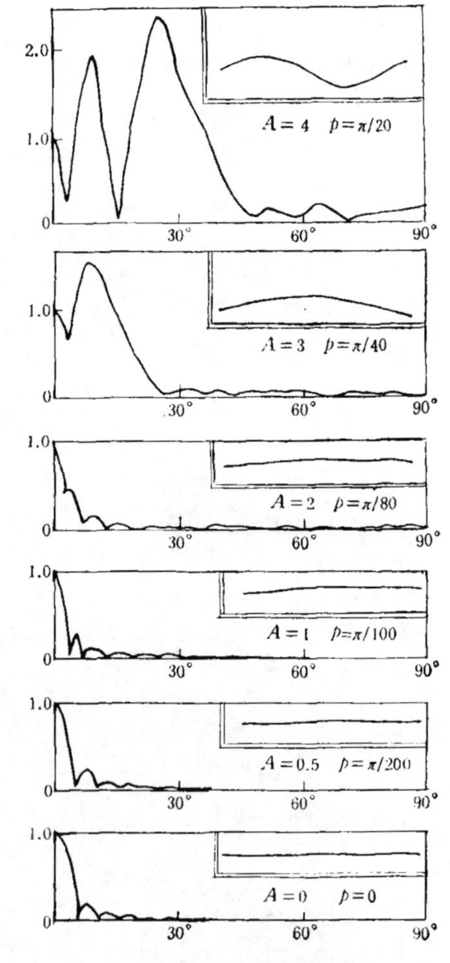

图 6 拖曳过程中指向性的变化

四、畸变阵工作方式的确定

我们前面已经指出, 由于基阵在拖曳过程中的变化受环境、船的机动动作的影响,实际上无法预报畸变阵的阵形. 同时, 从实用观点来讲,也没有必要测定每个基元的位置.

我们实际上只需要测定端点基元的最大延时差的变化就可以预报整个基阵的工作状态. 也就是说, 以此延时差去与一个事先给定的门限值作比较,当超过这个门限值时,我们就预报该拖曳式线列阵处于不适宜状态, 必须调整。 这一原则基于下面这样一个简单的事实(见图 7).

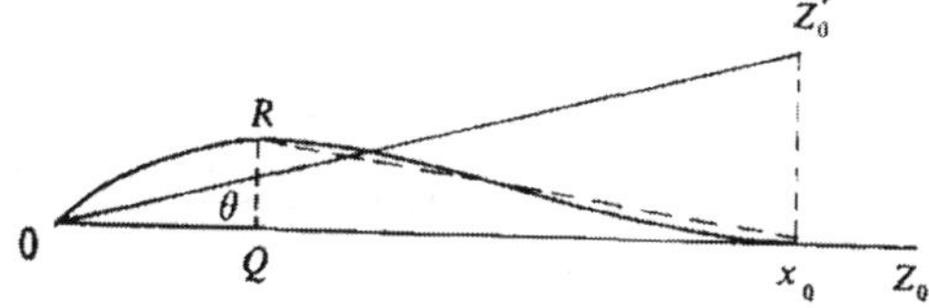

图 7 关于畸变阵最大失真的说明

设某一畸变阵的最大横坐标是 x_0. 理想位置为 ΔZ_0. 设 $\Delta Z_0'$ 的长度正好等于 OZ_0. $Z_0'X_0$ 的长度为 Y. 那么容易证明，任何其它畸变阵的最大纵坐标都不会比 Y 大. 换句话说 OZ_0' 实际上是某种意义上的"最差畸变阵位置"。要证明这一点是很容易的. 我们就不详述了.

$$X_0/Z_0 = \cos\theta,\ 令\ Z_0 - X_0 = \Delta x,$$

我们有

$$\frac{Z_0 - \Delta x}{Z_0} = \cos\theta \approx 1 - \frac{\theta^2}{2}$$

$$Y = Z_0 \sin\theta \approx Z_0 \sqrt{2\frac{\Delta x}{Z_0}} \tag{13}$$

对于任何实际的拖曳式线列阵系统，不难给出 Y/Z_0 的允许范围，从而由 (13) 式可以转化为对 $\Delta x/Z_0$ 或时延的要求.

举例来说，要求 $Y/Z_0 < 0.04$ ($Z_0 = 50\text{m}$, $Y = 2\text{m}$) 则要求 $2\Delta x/Z_0 = 0.0016$ 当 $Z_0 = 50\text{ m}$ 时，允许 $\Delta x = 4\text{ cm}$. 或者最大时延差为 27 μs.

(13) 式也可改写为对延时的门限值即

$$\tau_0 = \frac{\Delta x}{c} = \frac{1}{2c}\left(\frac{Y}{Z_0}\right)^2 \cdot Z_0 \tag{14}$$

其中 c 为声速.

τ_0 可称为警戒门限.

参 考 文 献

[1] Hinich, M. J. and Rule, W., "Bearing estimation using a large towed array", *J. Acoust. Soc. Amer.* **58**(1975), 1023—1029.

[2] Bucker, H., "Beamforming a towed line array of unknown shape" *J. Acoust. Soc. Amer.* **68**(1978), 1451—1454.

[3] Kuperman, W. A. et al., "Towed array response to ship noise: a near field propagation problem", in *Adaptive method in underwater acoustics*", Ed. By H. G. Urban, 1984, 71—80.

[4] Gerken, L., "ASW versus submarine technology battle" USA, 1986.

[5] Butler, D., "Beamforming with distored towed array", in Adaptive *method in underwater acoustics*, Ed. By H. G. Urban, 1984, 469—476.

奇异值分解方法在医学信号处理中的应用——胎儿心电的检测①

The Application of Singular Value Decomposition to the Medical Signal Processing——Detection of Fetal ECG

李淑秋　侯自强　李启虎

(中国科学院声学研究所)

【提要】 奇异值分解已用于从孕妇皮肤测量信号中提取胎儿心电信号。本文着重探索一种在路径并不正交的条件下提取不含母亲心电的胎儿心电的新方法。对工作原理和硬件实现作了简单的介绍。消除了在记录心电图中常遇到的50Hz干扰问题。通过大量实验，确定了可供选择的电极位置。

Abstract: The singular value decomposition (SVD) has been applied to the extraction of the fetal ECG(FECG) out of maternal skin measurement signals. In this pater, we emphatically seek a new method of extracting FECG that is free of maternal ECG under the condition that the path needn't to be orthogonal. The working principle and hardware implementation of this system are introduced briefly. As one of the major problems encountered in recording ECG, the unwanted 50Hz interference is eliminated. Through a lot of experiments, feasible electrode positions are determined roughly.

一、引　言

从50年代起，人们就开始了对胎儿心电图(Fetal ECG)的研究。在临床上，所用的电极有两种，一种是内部电极，常用于破膜之后，在一定程度上受到妊娠期的限制。另一种是腹部皮肤电极，用它记录的胎儿心电信号(FECG)和母亲的心电信号(MECG)作为一种强干扰出现，并且两者的带宽几乎相同，传统的滤波方法对它无能为力。由于胎儿心电信号非常微弱(微伏级)，而母亲心电信号较强(毫伏级)，更增加了处理的难度。

J.Vanderschoot，J. Wandwall和D. Callaerts等人用奇异值分解方法提取胎儿心电信号，做了一些有益的工作。[2,3]在J. Vanderschoot的研究中，要分离出干净的胎儿信号，要求路径必须正交，并试图找到一种保证路径正交的电极位置，这是非常困难的。

本文找到了一种方法，通过对原始的电极信号进行相应的加权，在路径不正交的条件下，也能分离出信噪比较好的FECG信号。

① 电子学报, 1991, 19(1): 91-97.

二、原　　理

用p个电极，在孕妇体表测得了p个电极信号$m_i(t)$，($i=1, 2, \cdots p$)，称为一个测量矢量抽样$m(t)$，它是p维空间的一个矢量，每个电极信号都可以看成是r个源信号(包括母亲各心电信号分量和胎儿心电信号)s_i($i=1, 2, \cdots, r$)的线性组合($p>r$)。

$$\begin{cases} m_1(t)=t_{11}s_1(t)+\cdots+t_{1r}s_r(t)+n_1(t) \\ m_2(t)=t_{21}s_1(t)+\cdots+t_{2r}s_r(t)+n_2(t) \\ \vdots \\ m_p(t)=t_{p1}s_1(t)+\cdots+t_{pr}s_r(t)+n_p(t) \end{cases} \tag{1}$$

写成矩阵形式：

$$\begin{pmatrix} m_1(t) \\ m_2(t) \\ \vdots \\ m_p(t) \end{pmatrix} = \begin{pmatrix} t_{11} & t_{12}\cdots t_{1r} \\ t_{21} & t_{22}\cdots t_{2r} \\ \vdots & \\ t_{p1} & t_{p2}\cdots t_{pr} \end{pmatrix} \begin{pmatrix} s_i(t) \\ s_2(t) \\ \vdots \\ s_r(t) \end{pmatrix} + \begin{pmatrix} n_1(t) \\ n_2(t) \\ \vdots \\ n_p(t) \end{pmatrix} \tag{2}$$

对p个电极($p>r$)的测量信号进行q次采样，得($p\times q$)阵M：

$$M_{p\times q}=T_{p\times r}S_{r\times q}+N_{p\times q} \tag{3}$$

其中传递矩阵T是由信号的传递路径决定的。它与母体的几何形状、电极和源的位置以及身体的导电性等因素有关。传递阵T的第i个列矢量，就称为传递矢量t_i。($i=1,2,\cdots,r$)

对测量阵$M_{p\times q}$，有奇异值分解[1]：

$$M_{p\times q}=U_{p\times p}\Sigma_{p\times p}V^T_{p\times q} \tag{4}$$

其中
$$U^TU=UU^T=V^TV=I_P \tag{5}$$

$$\Sigma=\mathrm{diag}(\sigma_1, \sigma_2, \cdots\sigma_P) \tag{6}$$

且有：
$$\sigma_1\geqslant\sigma_2\geqslant\cdots\geqslant\sigma_p\geqslant0 \tag{7}$$

U(或V)阵称为M阵的左(或右)奇异阵，U(或V)阵的列矢量称为左(或右)奇异基矢量，σ_i($i=1, 2, \cdots, p$)称为奇异值。

当满足下列条件时，

条件1　$NN^T=\sigma_N^2I_p$

条件2　$SS^T=\mathrm{diag}(\sigma_{s_1}^2, \sigma_{s_2}^2, \cdots\sigma_{s_r}^2)=\Sigma_S^2$

条件3　$SN^T=O_{r\times p}$

条件4　传递矢量t_i($i=1\cdots, r$)彼此正交，

可以证明[2]：M阵的奇异值分解的前r个左奇异基矢量u_i($i=1, \cdots, r$)，就是r个传递矢量t_i($i=1, \cdots, r$)的单位矢量，即：

$$u_iu_i^T=t_it_i^T/t_i^Tt_i(i=1, \cdots r) \tag{8}$$

这样，我们只要把M阵在U上投影，就可求出各个源的估计信号，即：

$$\hat{S}=U^TM=\Sigma V \tag{9}$$

由于Σ是对角阵，这意味着V阵的每一列就代表一个源的估计信号。

前三个条件较易满足，条件(4)是要求与信号的传递路径有关的传递矢量彼此正交，我们称它为路径正交。J.Vanderschoot是通过在孕妇的胸部和腹部分别放三个电极，使源与电

极位置空间分离来满足条件(4)的。

但是，按照这种方法常常发现所估计的胎儿信号（FECG）内仍含有母亲的心电信号（MECG）分量。原因是条件(4)没有满足，故在 J. Vanderschoot①的研究中，试图找到一种标准的电极位置来保证路径正交。

我们认为，这样的电极位置很难找到，在胸部和腹部各放三个电极使源与电极位置空间分离，只能迫使母亲信号的传递矢量子空间T_M和胎儿信号的传递矢量子空间T_F达到大致正交，因为腹部的信号内总是含有母亲的MECG信号，所以要达到正交非常困难。如图1。

我们认为从能量的角度来解释奇异值分解的原理，或许能找到解决问题的途径。

$(p\times q)$阵M奇异值分解的左奇异矢量u_i $(i=1,\ \cdots,\ p)$有一递归特性[1,2]：

$\forall x\in R_p,\ x\perp \mathrm{span}(u_1,\ u_2,\ \cdots u_{i-1})$

则有：$\|M^Tx\|/\|x\|\leqslant\|M^Tu_i\|=\sigma_i$ (10)

$\|M^Tx\|^2/x^Tx$ 表示M阵在矢量x方向上能量的平均值。我们知道，u_i也正交于子空间$\mathrm{Span}(u_1,\ u_2\cdots,\ u_{i-1})$。式(10)表明，在所有的与子空间$\mathrm{Span}(u_1,\ u_2,\ \cdots,u_{i-1})$正交的矢量中，只有在$u_i$矢量方向上$M$阵投影的均方根值达到最大，这个最大值就是奇异值$\sigma_i$。

这个递归特性说明，u_i $(i=1,\ \cdots p)$是M阵的左奇异基矢量，在p维空间内，在u_1的方向上，M阵的能量达到最大；在余下的与u_1正交的$(p-1)$维空间内，在u_2的方向上，M阵的能量达到最大；在余下的与u_1，u_2正交的$(p-2)$维空间内，在u_3的方向上，M阵的能量达到最大，依此类推。即u_i代表一组空间滤波器，使输出信号在它的方向上有最大的能量σ_i^2。

可以想象，只有在信号的方向上，能量才会最大，而传递矢量的方向就是信号的方向，这就是u_i矢量方向会逼近于t_i矢量方向的原因。

奇异值分解的性质表明：左奇异基矢量u_i $(i=1,\ \cdots,\ r)$总是彼此正交的(见式5)。显然，当传递矢量t_i $(i=1,\ \cdots,\ r)$彼此正交时，u_i矢量会与t_i矢量的方向重合（见式(8)），这正是我们要求路径正交的原因。

当传递矢量不正交时，u_i逼近于t_i的程度会怎样呢？J. Vanderschoot 对于三维空间二个源信号的例子，作了严格的定量计算。计算表明，u_i逼近t_i的程度和源信号之间的能量比S_i^2/S_{i-1}^2有关(设源信号的次序按能量从大到小排列)。这个能量比越大，u_i就越逼近于t_i，即u_i和t_i的夹角越小。图2表示了这种变化关系(以两维空间为例)。

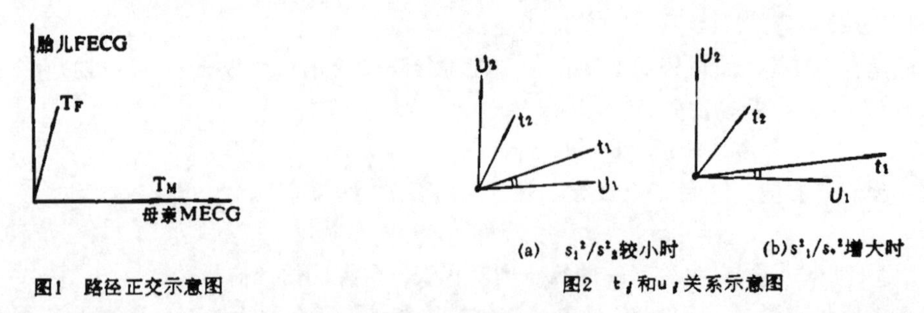

图1 路径正交示意图

图2 t_i和u_i关系示意图

图中矢量t_1的方向表示强信号s_1的方向，长度表示s_1的大小，t_2的方向表示弱信号s_2的方向，长度表示s_2的大小。当$s_1^2/s_2^2=\|\bar{t}_1/\bar{t}_2\|^2$较小时，$t_1$和$u_1$夹角较大；当$s_1^2/s_2^2=\|\bar{t}_1/\bar{t}_2\|^2$

较大时，t_1和u_1的夹角就变小。

t_1，t_2在u_1上的投影就是我们所求的强源s_1的估计信号$\tilde{s}_1$；t_1，t_2在u_2上的投影就是我们所求的弱源s_2的估计信号$\tilde{s}_2$。当t_1和u_1夹角越小时，t_1在u_2上的投影也就越小（u_1，u_2总是正交），即估计信号$\tilde{s}_2$中强源信号s_1的成份也就越小。我们正是利用奇异基矢量的这种性质来解决问题的。我们不再寻找保证路径正交的电极位置，而是在路径不正交的条件下，看能不能通过提高强源和弱源能量比的方法，使左奇异基矢量重新定向，使弱源估计信号含有较少的强源成份，从而使弱源分离出来。

根据上面的分析，我们通过提高母亲信号（MECG）和胎儿信号（FECG）能量比的方法，来提高u_i和t_i的逼近程度，从而提高估计信号 FECG 的信噪比。我们认为胸部信号以母亲信号为主体，增大胸部信号，只是相对增大了母亲的 MECG 信号，同理，减小腹部信号，只是相对减小了胎儿的FECG信号，这样一大一小，就提高了两个信号的能量比。而信号的增大和减小，通过对信号进行加权来实现。

三、实验结果

1．原始电极信号的SVD结果

在得到测量阵M后，经SVD算法，分解出的右奇异阵V^T的各行，就是所求的各个源信号。

图3是构成测量阵M的6路电极信号，1、2、3、通道是腹部信号，4、5、6通道是胸部信号。

图 4 是 SVD 的结果，第 1、2 行是母亲的 MECG 信号分量，5、6行是噪声，3、4 行是母亲信号MECG 和胎儿信号 FECG 的混合信号。右下角是由以上结果推断出的传递矢量与奇异基矢量的关系示意图。测量信号在u_3上的投影就是第 3 行的信号，测量信号在u_4上的投影就是第 4 行的信号，这个结果说明，在路径不正交的条件下，不能很好地分离出胎儿的信号。

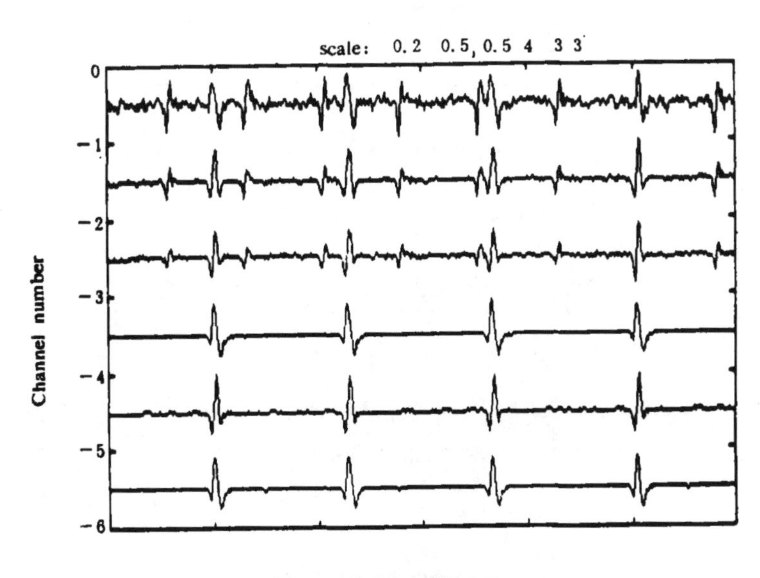

图3　六路原始的电极信号

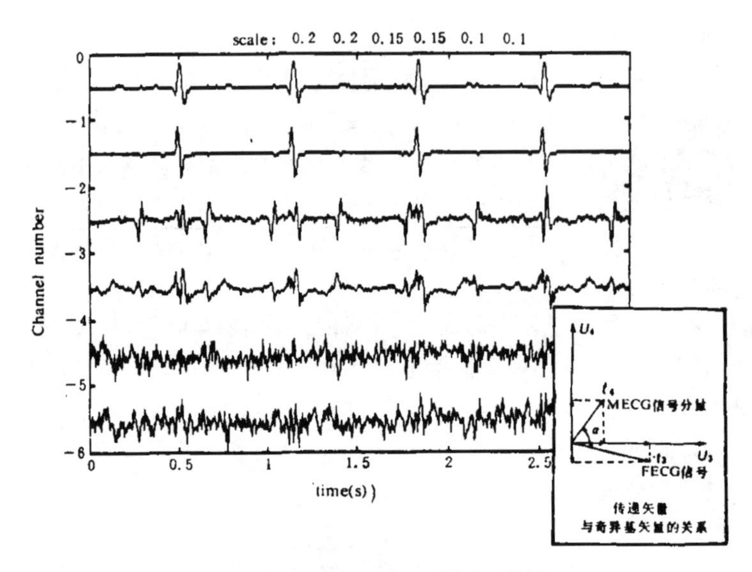

图4 对原始的电极信号SVD的结果

2. 信号加权对SVD的影响

为了增强强源信号MECG和弱源信号FECG的能量比，我们对图3中6路原始电极信号进行加权。为了减小胎儿(FECG)信号，对1、2、3路腹部信号的加权系数采用0.1，0.7，0.7(小于1)，其中1路信号的噪声较大，为了减小噪声，故把加权系数取得较小。为了增大母亲信号(MECG)，对于4、5、6路胸部信号的加权系数取为1.2，1.2，1(不小于1)。对加权后的电极信号，重新进行奇异值分解，结果见图5，第4行是很清楚的胎儿FECG信号，并可由结果推断出各矢量关系。

没有加权以前的各矢量关系，见图5右下角中虚线，加权后的各矢量关系如图中实线。

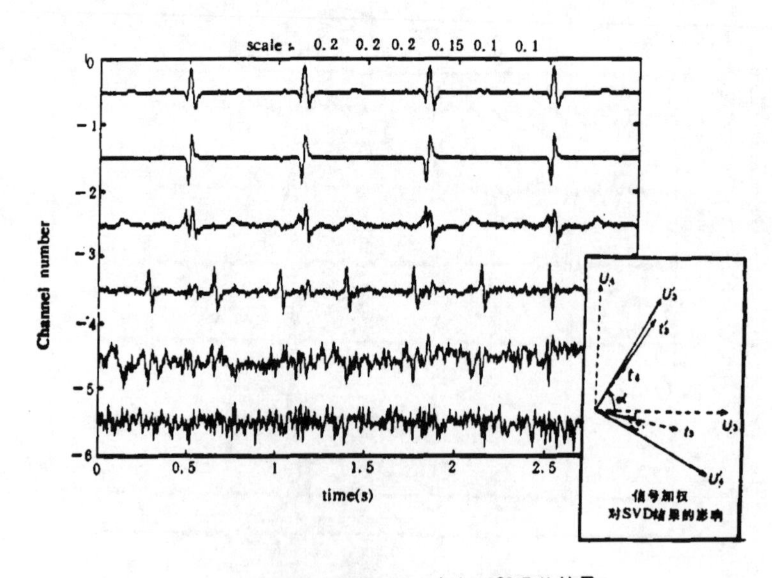

图5 对原始的电极信号进行加权后SVD的结果

可以看到，加权后的t_4伸长为t'_3，t_3缩短为t'_4。加权使得左奇异基矢量重新定向，u'_3非常逼近于t'_3，致使t'_3在u'_4方向上的分量很小，而以FECG信号的成份为主，故第4行上是很清楚的胎儿信号FECG。

需要指出的是，加权系数的选取，遵从增加信号能量比的原则，但每位孕妇的信号存在个体的差异，加权系数也有所不同。不能把胸部信号放得过大，以免MECG信号的第4个分量比FECG信号还大；也不能把腹部信号减得太小，以免FECG信号淹没在噪声里。在6维空间内，让FECG信号处在第4行的位置上比较合适，前3行是MECG信号空间，后2行是噪声空间。

四、实验系统和电极位置

图6是实现胎儿心电检测的系统框图。共有6路前置放大滤波电路，低通滤波器的截止频率是70Hz，4阶Butterworth型，A/D转换12bit采样频率250Hz，并用IBM-PC/XT作系统主机进行工作。

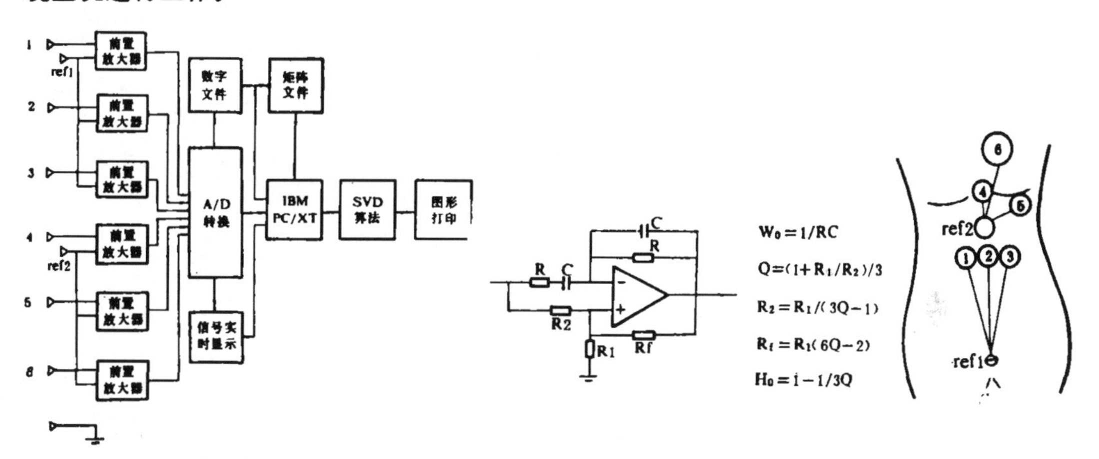

图6　系统框图　　图7　陷波器结构及算式　　图8　电极位置示意图

在接收生物电信号时，一个经常遇到的问题，就是50Hz AC干扰，在滤波电路中，我们采用了50Hz陷波器，这个电路在实验中起到了很好的作用，在接收其它的生物电信号时，也可以采纳。其结构及参数算式如图7所示。

电极位置的选择有两个原则：胸部电极位置，要使母亲的MECG信号非常强，而尽量不含有胎儿信号；腹部电极的位置，要使FECG信号有最大的幅度，这样才能使T_M和T_F准正交。

根据文献[4]，经大量实验证明，大多数孕妇的胎儿心电的最大幅度区域，在孕妇肚脐上方5cm处的范围内，有10%的人在背后。FECG最大幅度的区域和胎盘的位置没有明显关系，而胎儿信号的幅度则和它稍有关系，和产式也有关系，FECG信号的幅度在0～120μV之间，平均幅度值是27.1μV。

我们直接利用这一结论，负电极ref1放在耻骨上方5cm处，三个正电极放在肚脐上方的一个邻近区域内，在我们的实验中，采用这样的电极位置（如图8所示），取得了较好的信号。

参考文献

〔1〕G.H.Golub, C.Reinsch:Singular Value Decomposition and Least squares solution, Numer.Math., Vol.14, pp. 403—420, 1970.

〔2〕J.Vanderschoot, D. Callaerts et al., Two methods for optimal MECG eliminatiou and FECG detection from skin electrode signals, IEEE Trans., Biomed.Engng, Vol.BME—34, No.3, pp.233—243, Mar.1987.

〔3〕D.callaerts, J.Vanderschoot et al.:An on—line adaptive algorithm for signal processing using SVD, in SIGNAL PROCESSING Ⅲ:Theories and Applications, 1986, pp.953—956.

〔4〕J.B.Roche, E.H.Hon:The fetal electrocardiogram Am. J. Obst& Gynec. Vol.92, No.8, pp. 1149—1159, Aug. 1965.

〔5〕李淑秋，侯自强，奇异值分解方法在医学信号处理中的应用，中国科学院声学所硕士论文，1988年9月。

时间压缩式相关器/延迟线 IC 设计①

李云岗　李启虎
（中国科学院声学所）

1990 年 7 月 3 日收到

相关器、延迟线在信号处理领域中有着非常广泛的应用．但到目前为止尚没有看到这类集成电路的报道．依据[1]文提及的新型时间压缩式相关器，本文给出了通用相关器/延迟线的专用集成电路设计，该集成电路（IC）可以完成极性相关和多比特相关，同时还可以作为延迟线应用．可构成 768 级以下任意长延时，在 8kHz 采样频率下单片 IC 最大延时可达 96ms．此 IC 可以多片应用，满足不同场合的应用要求．

一、引　　言

相关检测在信号处理领域，特别是在声呐设计中已得到非常广泛的应用 [2—4]，但也发现一些不足，如可能引起数据丢失，设备庞大等问题．针对被动声呐设计中遇到的一些问题，[1]文提出了一种新的交叉存贮时间压缩相关器，并得到成功的应用．但在电路实现中由于没有专用的相关器集成电路．只好以通用集成电路来实现，虽较以前大有改善，但仍很臃肿，需要一块印制电路板才能实现，且延迟线的长度不能太长，采样频率不能太高．为了进一步缩小体积、增加系统可靠性、提高采样精度，我们设计了一块专用集成电路，它可以实现 256 点相关，既可以完成极性相关，还可以用作多比特相关，同时可以作为延迟线应用，单片可以实现 768 级以下任意长时间延迟，且可多片应用以实现任意长时间延时．

二、交叉存取时间压缩式相关器的基本原理

相关检测的运算是完成两个数据序列的相关和累加，设数据序列分别是

$$x(k), y(k) \quad k = 0, 1、2, \tag{1}$$

则累加次数(或积分时间)为 N 的相关函数为

$$R(k) = \frac{1}{N}\sum_{i=0}^{N-1} y(i)x(k+i)$$

$$k = 0, 1, \cdots N-1. \tag{2}$$

如果输入数据的采样周期为 T_s，那么按常规的方法，每得到一个相关函数需要有一个相关器，然而，主动声呐用的时间压缩相关器只用两个时间压缩环就可同时得到 N 点相关函数值，见图 1，其中 $x(k)$ 进入接收时间压缩环，

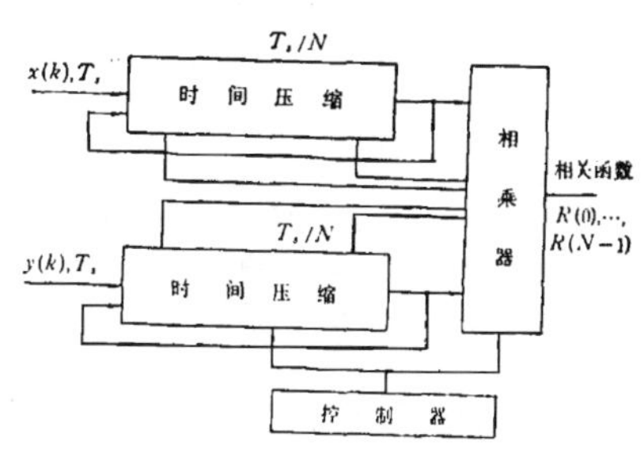

图 1　主动声呐用时间压缩相关器

$y(k)$ 进入参考时间压缩环，这两个环的工作周期都是 T_s/N．如果接收环采用进动式，那么有正序与逆序之分，取决于环的长度是 $N-1$ 节还是 $N+1$ 节，在每一采样间隔 T_s 内，相关器可以输出 N 个相关函数值．出现峰值的时刻就是信号匹配的时间，由此可以解算出目标距离．

用这种相关器于被动声呐时会出现两类问题．一是被动声呐通常采用较高的采样频率，所以必须用非常高速的移位寄存器以便达到实时运算．另一个问题是被动声呐需较长的积分时间，这时参考环的长度远远小于这个值，我们必须使参考环也变为能更新数据的进动环，但在更新数据之前又必须完成 N 点相关函数，这就加剧了实时运算的困难．

我们采用图 2 所示的交叉存取式时间压缩相关器，它成功地解决了上面两个问题，不丢失数据，能在高采样率下实时运算，硬件实现并不困难．软件编程也很容易．

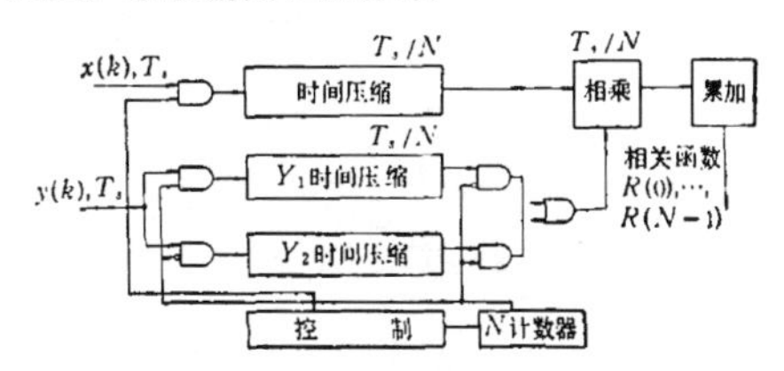

图 2　交叉存取式时间压缩相关器

① 应用声学, 1991, 10(6): 12-15.

图中 x 寄存器的长度为 N 节或 $N+1$ 节，这需根据逆序或正序输出样本来定；Y_1，Y_2 为两个缓冲寄存器，当一个处于移位型时间压缩状态时，另一个寄存器开始存贮数据．举例来说，如果 x 寄存器存放 $x(0),\cdots,x(N-1)$，Y_1 寄存器存放 $y(0)\cdots y(N-1)$，这时控制中心开始计算相关函数，每隔 T_s/N 秒，x 寄存器与 Y_1 寄存器各输出一个样本、经 T_s 秒后，得到相关函数

$$R^{(1)}(0)=\frac{1}{N}\sum_{i=0}^{N-1}x(i)y(i) \tag{3}$$

在下一个采样周期得到

$$R^{(1)}(1)=\frac{1}{N}\sum_{i=0}^{N-1}x(i+1)y(i) \tag{4}$$

每一个采样周期使 x 寄存器更新一个样本，同时在 Y_2 寄存器内存入一个新的样本 $y(N-1+k)$．在经过 N 个 T_s 秒之后得到了 N 点相关函数

$$R^{(1)}(k)=\frac{1}{N}\sum_{i=0}^{N-1}y(i)x(i+k)$$

$$k=0、\cdots N-1. \tag{5}$$

然后 Y_1 寄存器变为存贮器，Y_2 变为时间压缩型，由此得到第二批相关函数

$$R^{(2)}(k)=\frac{1}{N}\sum_{i=0}^{N-1}y(N-1+i)x(N-1+i+k)\quad k=0\cdots N-1 \tag{6}$$

如此类推，把

$$R^{(1)}(k),R^{(2)}(k)\cdots$$

累加起来，得到所需的相关函数

在运算的时候，x 寄存器是进动型的，它可以采用正序也可以采用逆序的形式，但以逆序方式较方便，因为对于相关函数来说，正序与逆序无关紧要．

三、集成电路逻辑设计

我们设计的 IC 既可以完成相关器功能，也可以作为延迟线使用．下面分别予以叙述．先谈相关器．

相关器部分的功能框图如图 3 所示．

为了提高相关器的精度，我们采用了 256 级移位寄存器作为时间压缩环．在 IC 中共有三个时间压缩环，即 X,Y_1 和 Y_2，其中 X 压缩环用来存放样本，Y_1，Y_2 压缩环进行交替工作．当 Y_1 进行相关比较时，Y_2 进行采样；反之，Y_1 采样，Y_2 进行相关比较．相关器的工作状态由

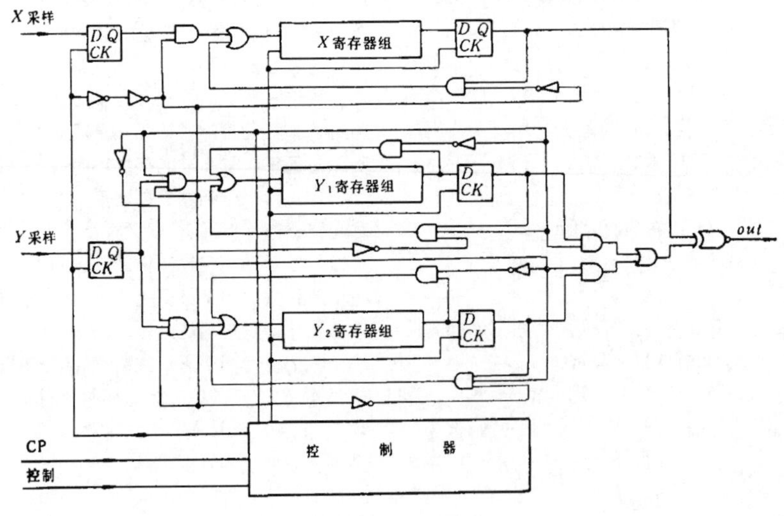

图 3 相关器 IC 原理框图

控制器进行控制。

该IC既可以用作极性相关，也可以进行多比特相关，在作极性相关时，采用一计数器就可完成累加功能形成相关函数，而作多比特相关时需用累加器形成相关函数，为了增加该IC的应用灵活性，这一部分没有集成，便于用户根据自己的需要实现不同功能。

四、延 迟 线

在信号处理及其它领域中延迟线应用十分广泛，但目前专门的延迟线IC尚很少见，我们设计的IC除可实现相关器的功能外还可以完成延迟线的功能，单块IC能实现768级以下任意长度的延时，为实现这一功能，分别设置了1,2,4,8,16,32,64,128,256级等不同长度的延时，用户可以根据实际需要编程，从而完成不同长度的延时，各种组合如表1所示。

由于采样频率的不同可以实现不同的级延时，在表2中给出了不同采样频率下的单级延时，和单片可实现的最大延时，表中给出的值是以系统频率为16MHz和10MHz情况下给出的。

表1 不同长度延时组合表

级 数	组 合	级 数	组 合
1	1	8	8
2	2	9	8+1
3	2+1	⋮	
4	4	208	128+64+4+2
5	4+1	⋮	
6	4+2	768	256+256+128+64+32
7	4+2+1		+16+8+4+2+1+1

表2 不同采样频率下的最大延时

采样频率	单级延时	最大延时	采样频率	单级延时	最大延时
16MHz	62.5ns	48μs	250kHz	4μs	3.072ms
10MHz	100ns	76.8μs	125kHz	8μs	6.144ms
8MHz	125ns	96μs	62.5kHz	16μs	12.288ms
4MHz	250ns	192μs	31.25kHz	32μs	24.576ms
2MHz	500ns	384μs	15.625kHz	64μs	49.152ms
1MHz	1μs	768μs	7.8125kHz	128μs	98.304ms
500kHz	2μs	1.536ms	8kHz	125μs	96ms

若用户需要其它采样频率，只需改变系统频率即可，如用户需4kHz采样频率，只要把系统频率二分频后再送该IC，就可以实现单片IC最大延时为184ms。在表中仅给出了8kHz以上采样频率的状况。

五、结 束 语

本文给出了相关器设计和延迟线设计的基本原理，并给出了IC的原理方块图，该IC可完成极性相关，多比特相关及延时线功能，在工作中曾多次与我所徐俊华、孙增、李士才，孙长瑜等同志多次讨论，使我们获益匪浅，在此表示衷心感谢。

参 考 文 献

[1] 李启虎，孙 增，孙长瑜，声学学报 15(1990)，146—150.

[2] Kinghtet W. C., et al., *Proc. IEEE.*, 69(1981), 1451—1506.

[3] Sbitzb; P., *Electronic Progress*, 16-3 (1974), 20—37.

[4] Glisson T. H. and Sage A. P., *IEEE. Trans.*, *AES*-6 (1970), 37—49.

扩声系统中信源对广播分区的计算机分配[①]

刘秋实 李启虎 李士才 董 立

(中国科学院声学研究所) (中国船舶总公司613厂)

摘要 大型扩声系统要完成多信源对多广播分区的分配。通过本文提出的三级分配网络及相应的自动分配算法，可以使用计算机完成这种分配操作。

从设备角度看，大型扩声系统都具有多个信源，如几个有线、无线传声器，几台遥控放音机、唱机、语音机等；具有多个节目播送地，如体育场馆的看台、竞赛场地、休息室，车站中不同的候车室、大厅、站台等。扬声器根据不同节目播送地的具体条件，合理配置，组成一个个相对独立的广播分区。每个广播分区在同一时刻可以播出不同的节目内容。也就是说，扩声系统要求具有把多个不同的信源同时分配到不同的广播分区的能力。

以往的系统，用机械开关实现这种分配。然而，随着信源与广播分区的增加，开关数量就会非常庞大，操作变得很困难。而对于要求定时自动播音的系统，使用机械开关就根本无法完成。

本文提出一种由微机控制的三级分配网络。分别是：信源-节目通道分配网络，节目通道-扩音机分配网络，扩音机-广播分区分配网络。这三级分配网络将信源、扩大机、广播分区有机地合为一体。通过微机控制各网络通道的开启与关闭，实现信源对广播分区的分配。

一、系统组成

由于不同信源输出功率不一致，因此在进入网络前要加入放大电路，以保证输入信号的一致性。

系统需要的扩音机台数由要求同时播出的节目数、广播分区功率数及扩音机本身额定功率数决定。即要求总的扩音机功率比广播分区额定功率总和大，扩音机台数要求不少于同时播出的节目数。为设计方便，我们选择单台扩音机足以推动任一广播分区。合适的扩音机台数由下面的公式决定：

$$W_A > \{W_L\}_{max}$$
$$nW_A > \Sigma W_L$$
$$n \geqslant m \quad (1)$$

① 微计算机应用, 1991, 12(4): 32-36.

其中W_A为单台扩音机功率，W_L为各广播分区功率，n为扩音机台数，m为要求同时播出的节目数。

例如$m=8$，$\{W_L\}_{max}=270W$，$\Sigma W_L=2700W$，那么可取$W_A=300W$，则有$n>9$，可取n=10。即10台300W的扩音机可以满足系统要求。

图1是计算机控制的扩声系统的框图。

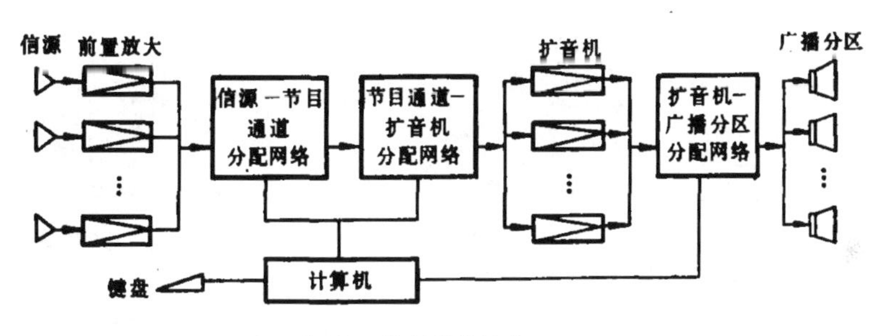

图1　扩声系统组成

为完成由信源到广播分区的自动分配，我们在信源前放输出与扩大机之间引入两级分配网，分别叫信源-节目通道分配网，节目通道-扩大机分配网；在扩大机与广播分区之间引入了一个分配网，叫扩大机-广播分区分配网。为叙述方便三级分配网分别叫s-c-net，c-a-net，a-b-net。上面讲的节目通道可以看成一种虚设备，它决定系统最多可播出的节目数，每一种不同的播出节目要占一个节目通道。

计算机只要向网络控制端送合适的控制码，就能按要求打开或关闭从信源到广播分区的通路，从而实现信源到广播分区的自动分配。

下面具体讨论三级分配网络的设计及有关的软件算法。

二、三级分配网设计

假设整个扩声系统有I个信源，系统要求最多同时播出J套节目，上节公式1中扩音机台数为K台，广播分区为L个。

s-c-net网络是一个有I个输入端，J个输出端的分配网络。根据控制命令的不同，任意的输入端i可以分配到输出端$j_1, j_2, j_3, \cdots, j_l$上去。(其中i为i=1，…，I。$j_1, j_2, j_3, \cdots, j_l$为1，2，…，J中的数，l≤J)。我们可以使用J套输入端并联的I选一电路实现这一网络。每套I选一电路由一个状态选择寄存器控制。计算机通过向该状态选择寄存器写控制码，即可控制该I选一电路的不同工作状态。一个I选一电路有J+1种不同的工作状态，即0为该电路无输出，1，…，J分别选择相应的输入端送输出。因此控制字长度由下式决定：

$$2^{l-1}<I+1\leqslant 2^l \tag{2}$$

例如I=20时，l=5。

值得注意的是这种结构的分配网络要求输入具有最多驱动J个负载的能力，因此前放除了放大之外还要加跟随电路，而每套I选一电路输入端也要加入跟随电路。

I选一电路，由模拟的开关实现，如四选一模拟开关、八选一模拟开关、选一十六模拟开关等等。图 2 给出I＝16时的 I 选一电路实现。请注意为节省硬件我们采用了 5 位编码，第 5 位为 0 时禁止电路输出。

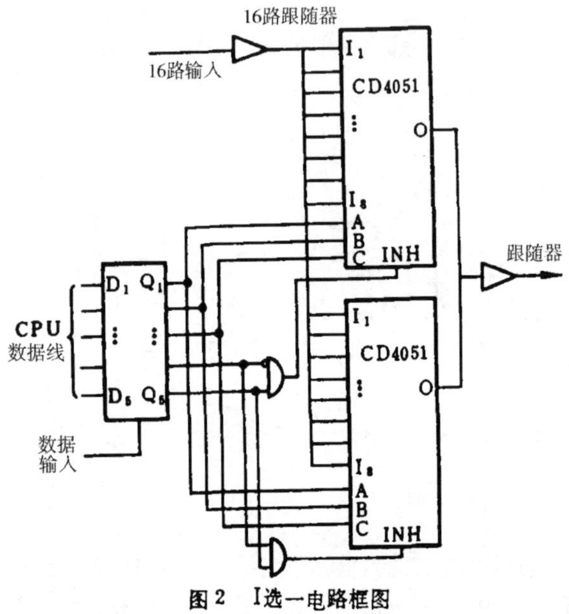

图 2　I选一电路框图

c-a-net是与 s-c-net 实现方法完全一样的 J×K 分配网络。其输入接 s-c-net 的输出，输出接扩音机输入。

a-b-net 网络要完成大功率信号的切换，因而该 K×L 分配网络中K选一电路是使用继电器或可控硅实现的。控制字可用式(2)决定。K选一电路中K个继电器输出端接在一起，选择寄存器输出经过译码用以控制继电器的控制端。L 套K选一电路的输入端并在一起接扩大机输出。整个K×L 网络的 L 个输出端接各广播分区负载。图 3 是K选一电路框图。

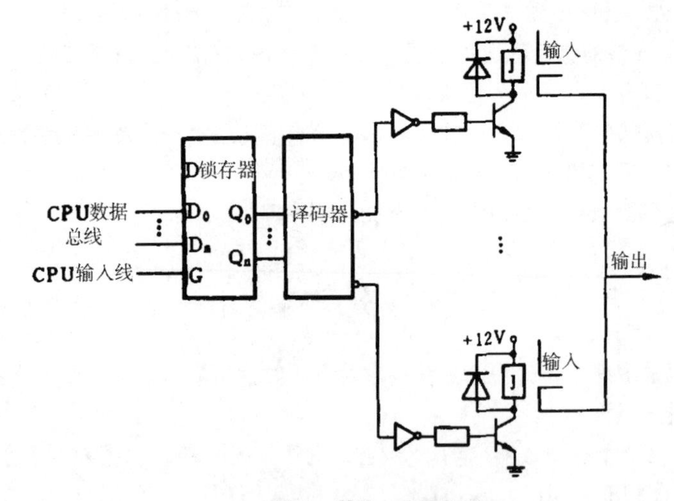

图 3　K选一电路框图

三、计算机自动分配算法

引入上节提出的三种网络，就使系统具备了完成信源到广播分区进行自动切换的硬件能力，然而要完成从接到命令到发送控制字实现自动切换，还必须有一套控制软件。

下面是几个系统功能。

1．信源加入

输入信源号与负载号，系统自动选择空闲的节目通道、选择合适的扩音机，通过三级网络，将信源送到广播分区去广播。

2． 负载删除

给出广播分区号。系统自动删除该广播分区的广播，在广播信源没有负载的情况下删除信源。在可能的情况下释放扩音机及节目通道。

3． 信源删除

给出信源号。系统自动删除该信源所有的广播分区的广播，释放扩音机及节目通道。

在这种系统中扩音机加入与释放由软件自动完成，并按使用次数由小到大优先分配。

整个算法的核心是下面几张表的管理。表1：sct表——s-c-net管理表。sct表的第j行代表第j个节目通道，第j行的内容用于记录该节目通道分配的信源号。记为sct[1··J]，j∈{1,2···,J}。表2：cat表——c-a-net管理表。cat表的第k行代表第k台扩音机，第k行的内容用于记录该扩音机分配到的节目通道号。记为cat[1··K]，k∈{1,2,···,J}。表3：abt表——a-b-net管理表，abt表的第l行代表第l个广播分区，l行的内容用于记录该广播分区分配到的扩音机号。记为abt[1··L]，l∈{1,2···,L}。

上面3张表可以确定信源与广播分区的互联情况。例如sct[2]=1，cat[3]=2，abt[5]=3就表示1号信源通过第2个节目通道与第3号扩音机相联，并送入5号广播分区去广播。而sct[2]=0就表示第2号节目通道空闲。

为实现扩大机按次数自动分配，我们为每台扩大机设置一个使用次数计数器，用act[1··K]表示。每次使用扩大机时将相应计数器加1，而在分配空闲扩大机时，按下面的原则选取扩音机：①扩音机必须空闲；②对在空闲的扩音机中act表的值最小。

为了充分利用扩音机功率，当几个广播分区同时播出一个节目时，如果这几个广播分区功率之和不超过一台扩音机，就使用一台扩音机推动这几个广播分区负载。如果超过再另外分配其余扩音机。为进行功率统计，使用了两个表。扩音机使用功率表awt[1··K]表示当前扩音机推动的负载功率；广播分区额定功率表bwt[1··L]用于表示各广播分区的额定功率。

为减少篇幅，这里仅给出用类PASCAL语言书写的节目通道分配算法和扩音机分配算法，最后给出一个实现信源加入算法。后面的算法调用前面两个算法完成信源加入功能。

算法1：节目通道分配

```
Procedure assign-C(s, c, f)
   s /*INPUT    信源号                          */
   c /*OUTPUT 节目通道号                        */
   f /*OUTPUT 标志,指示分配是否成功              */
   Begin-of-Proc.
1.     For j=1 to J do:
        Begin.
2.       if (sct[j]=s) then
            Begin.
3.            f<-0;
4.            c<-j;
5.            return/*成功返回                   */
             End
          End.
6.    For j=1 to J do:
          Begin.
7.        if (sct[j]=0=0) then
          Begin.
8.            f<-0;
9.            c<-j;
10.           return/*成功返回                   */
          End.
        End.
11.     f<-1;
12.     return
      End-of-Proc.
```

算法2：扩音机分配

```
Procedure assign-a (c, l, a, f)
   c /*INPUT     节目通道                       */
   l /*INPUT     广播分区号                     */
   a /*OUTPUT 扩音机号                          */
   f /*OUTPUT 标志,指定分配是否成功              */
   Begin-of-Proc.
1.     For k=1 to K do:
         Begin.
2.         if (cat[k]=c) then
             Begin.
3.             if (awt[k]+bwt[l]<=wa)
                       /*wa为扩大机额定功率      */
```

```
            Begin.
4.              awt[k]<-awt[k]+bwt[l];
5.              f<-0;
6.              a<k;
7.              return
                End.
          End.
      End.
8.    min<-∞     ;f<-1;
9.    For k=1 to K do:
        Begin
10.       if (cat[k]=0 ) then
          Begin.
11.       if(min>act[k]) then
            Begin.
12.           min<-act[k];
13.           a<-k;
14.           f<-0
            End.
          End.
        End.
15.   if (f=0) then
        Begin.
16.     act[a]<-act[a]+1;
        End.
      return
End-of-Proc.
```

算法3：信源加入

```
Procedure    s-add (s, l, f)
  s /*INPUT   信源号                        */
  l /*INPUT   广播分区号                    */
  f /*OUTPUT 标志,指示分配是否成功          */
  Begin-of-Proc.
1.   if (abt[l]  0)≠then
      Begin.
2.      f<-1; /*  广播分区已占用            */
3.      return
        End.
4.    call assign-c(s, c, f1);
5.    if (f1=0) then
        Begin.
6.      f<-2;/*  无节目通道                 */
7.      return
          End.
8.  else
      Begin.
9.    call assign-a (c, l, a, f2);
10.   if (f2=0) then
        Begin.
11.     f<-3; /*  无扩大机                  */
12.     return
        End.
    End.
13. sct [c]<-s; cat[a]<c; abt [l]<-a;
14. 向 s-c-net中第c个I选一电路写控制字s
15. 向 c-a-net 中第a个J选一电路写控制字c
16. 向 a-b-net 中第l个K选一电路写控制字a
  End-of-Proc.
```

A FAST ALGORITHM FOR BISPECTRA BASED TIME DELAY ESTIMATION①

Wang Jinlin and Li Qihu

Institute of Acoustics, Academia Sinica

Introduction

Time delay estimation is a problem of considerable practical interest in sonar design for a long time. The conventional estimation, GCC processor[4], has poor performance when the signal to noise ratio is small. C. L. Nikias and others show that the performance of estimation can be improved if the higher-order spectra(polyspectra) is used[1]. But not only conventional bispectrum methods, but also parametric bispectrum methods are difficult in implementation, due to complication of calculation, such as: matrix multiplication and inversion. In this paper, bispectra based time delay estimator, which has some genuine advantages in implementation, is developed by using the principle of generalized cross-spectra method[2] proposed by Li. Qihu.

Bispectra based Time Delay Estimation

Let us assume that $x(n), y(n)$ are two sensor measurements satisfying:

$$x(n) = s(n) + w_1(n) \quad (1a)$$

$$y(n) = s(n-D) + w_2(n) \quad (1b)$$

where $\{s(n)\}$ is zero mean non-Gaussian stationary random process with nonzero measure of skewness, $\{w_1(n)\}$ and $\{w_2(n)\}$ are zero-mean possibly correlated stationary random processes, statistically independent of $s(n)$.

The cross correlation function of the noise source

$$r_{12}(\tau) = E\{w_1(n)w_2(n+\tau)\} \quad (2)$$

where $E\{*\}$ denotes ensemble average .The cross-correlation function of $\{x(n)\}$ and $\{y(n)\}$ is given by

$$r_{xy}(\tau) = r_{ss}(\tau - D) + r_{12}(\tau) \quad (3)$$

therefore, all the methods based on cross-correlation will have poor performance without some knowledge of $r_{12}(\tau)$ for the signal model given by (1). We have

$$\begin{aligned} R_{xxx}(\tau,\rho) &= E\{x(n)x(n+\tau)x(n+\rho)\} \\ &= R_{sss}(\tau,\rho) \end{aligned} \quad (4)$$

$$\begin{aligned} R_{xyx}(\tau,\rho) &= E\{(x(n)y(n+\tau)x(n+\rho)\} \\ &= R_{sss}(\tau - D,\rho) \end{aligned} \quad (5)$$

where

$$R_{sss}(\tau,\rho) = E\{s(n)s(n+\tau)s(n+\rho)\}$$

The bispectrum is,by definition, the Fourier transform of the third moment of the data sequence:

$$\begin{aligned} B_{xxx}(\omega_1,\omega_2) &= FT[R_{xxx}(\tau,\rho)] \\ &= FT[R_{sss}(\tau,\rho)] \\ &= B_{sss}(\omega_1,\omega_2) \end{aligned} \quad (6)$$

$$\begin{aligned} B_{xyx}(\omega_1,\omega_2) &= FT[R_{xyx}(\tau,\rho)] \\ &= FT[R_{sss}(\tau - D,\rho)] \\ &= B_{sss}(\omega_1,\omega_2)exp(j\omega_1 D) \end{aligned} \quad (7)$$

where $FT[*]$ denotes the 2-D fourier transform operation.

By using the equation

$$\begin{aligned} C(\omega_1,\omega_2) &= B^*_{xxx}(\omega_1,\omega_2)B_{xyx}(\omega_1,\omega_2) \\ &= |B_{sss}(\omega_1,\omega_2)|^2 exp(j\omega_1 D) \end{aligned} \quad (8)$$

we can extract the imformation D from $C(\omega_1,\omega_2)$

$$\omega_1 D = tg^{-1}\left\{\frac{Im(C(\omega_1,\omega_2))}{Re(C(\omega_1,\omega_2))}\right\} \quad (9)$$

① China 1991 International Conference on Circuits and Systems, 1991, 1: 348-350.

where $Re(*)$ and $Im(*)$ represent the real part and image part of the complex number, respectively.

Fast Algorithm for Estimation

In the digital system, the signal is truncated. The error due to truncation will result in the bias of estimation for time delay. In the following, a brief analysis about the bias and parameters of DFT is carried out.

$$B_{xxx}(\omega_1,\omega_2)=FT[\hat{R}_{sss}(\tau,\rho)] \tag{10}$$
$$=\sum_{m=0}^{N-1}\sum_{n=0}^{N-1}\hat{R}_{sss}(m,n)exp\{-j(\omega_1 m+\omega_2 n)\}$$

where:
$\hat{R}_{sss}(\tau,\rho)$ denotes the estimation of $R_{sss}(\tau,\rho)$

$$B_{xyx}(\omega_1,\omega_2)=FT[\hat{R}_{sss}(\tau+D,\rho)] \tag{11a}$$
$$=\sum_{m=0}^{N-1}\sum_{n=0}^{N-1}R_{sss}(m+D,n)exp\{-j(\omega_1 m+\omega_2 n)\}$$
$$=\sum_{n=0}^{N-1}exp(-j\omega_2 n)\{exp(jn_0\omega_1)$$
$$\sum_{m=0}^{n_0-1}\hat{R}_{sss}(m,n)exp(-j\omega_1 m)[1+\frac{Ae^{j\phi_0}}{|Z(n)|}]\}\}$$

where n_0 denotes the entire part of D, and A is a real number, ϕ_0 is phase factor, they are difined by

$$Ae^{j\phi_0}=\sum_{k=0}^{n_0-1}[\hat{R}_{sss}(N+m,n)-\hat{R}_{sss}(m,n)]$$
$$exp(-j\omega_1 m)exp(j\alpha) \tag{11b}$$

and

$$Z(n)=\sum_{m=0}^{N-1}\hat{R}_{sss}(m,n)exp(-j\omega_1 m) \tag{11c}$$

and α is the phase of $Z(n)$. We can see that, for large $|Z(n)|$, the estimation error is small. So the $|Z(n)|$ can be considered as the signal to noise ratio in measurement of $C(\omega_1,\omega_2)$.For ω_2 space its effection on estimation error can be considered as well-distributed. we then get a new estimator:

$$D=\sum_{l=0}^{n-1}\frac{a_l\psi(\omega_l)}{\omega_l} \tag{12a}$$

where $a_l(l=0,\cdots,N-1)$ are weight coefficients:

$$a_l=\frac{1}{A_0}[\frac{1}{N}\sum_{n=0}^{N-1}|\sum_{m=0}^{N-1}\hat{R}_{sss}(m,n) \tag{12b}$$
$$exp(-j\omega_l m)|^2]$$
$$A_0=\sum_{l=0}^{N-1}[\frac{1}{N}\sum_{n=0}^{N-1}|\hat{R}_{sss}(m,n) \tag{12c}$$
$$exp(-j\omega_l m)|^2]$$
$$\psi(\omega_l)=\frac{1}{N}\sum_{n=0}^{N-1}\phi(\omega_l,\omega_n) \tag{12d}$$

Simulation Results

In this section, we present computer simulation results. We have conducted to investigate the performance of the bispectra based time delay estimation technique. Throughout these computer experiments, we made the data 50 percent overlapped when they are segmented into records, and have chosen the time delay $D=0.5T_s$ or $D=0.2T_s$.Different signal noise ratios have been used for the simulation examples.

The signal to noise ratio is defined as:

$$20log_{10}(\sigma_s/\sigma_n)$$

where σ_s and σ_n are the standard deviation of signal and noise, respectively. Throughout this section we assume that:

(1) The wideband signal $\{s(n)\}$ is one-sided exponentially distributed stochastic process and is generated by the EXPONENTIAL subcommand of the MINITAB software with $E(s^2(n))=1$ and $E(s^3(n))=2$. The normalized band of the signal $\{s(n)\}$ is 0.45.

(2) The noise source $\{w_1(n)\}$ is Gaussian process generated by the NORMAL subcommand of the MINITAB software, whereas $\{w_2(n)\}$ in (1b) is generated from $\{w_1(n)\}$ using the follow FIR system equation

$$w_2(n)=\sum_{i=0}^{10}b_i w_i(n+i)$$

where the parameters $\{b_i\}$ take values:
$\{0.2,0.4, 0.6\ 0.8,1.0,1.0,0.7,0.5,0.3,0.1\}$.

Table 1.($D=0.5T_s$) and Table 2. ($D=0.2T_s$) illustrate the results of these examples with different signal-to-noise ratio,respectively.

Tab.1 The estimation with different SNR corresponding $D = 0.5T_s$.

SNR	10dB	6dB	0dB	-6dB
estimate	0.5063	0.4857	0.5393	0.5679

Tab.2 The estimation with different SNR corresponding $D = 0.2T_s$.

SNR	10dB	6dB	0dB	-6dB
estimate	0.2127	0.2229	0.1874	0.2521

From Tab.1 and discussion aboved, it is apparent that the bispectra based time delay estimation technique has not only a good performance, but also genuine advantages in implementation. Similiar results are illustrated in Tab.2 which corresponds to $D = 0.2T_s$. But when signal to noise ratio is lower than -10dB the same length of data(128*128) cannot get a satisfied results. A longer data need to be processed in order to get good performance. Also the performance can be improved when the windowing technique is used in 2-D fourier transform.

Aknowledgements

This work was supported by State Key Lab. of Acoustics.

References

[1] C. L. Nikias and R. L. Pan "Time Delay Estimation in Unknown Gaussian Spatially Correlated Noise",IEEE Trans. Acoust., Speech, Signal Processing, vol. ASSP-36, pp.706-1714, Nov. 1988.

[2] Li, Q.H "The precise Bearing Method of Digital Split-beam Array System",Acta Acustica, (1984), No. 1, pp.78-91.

[3] C.L. Nikias and M.R. Raghuveer,"Bispectrum estimation: A Digital Signal Processing Framework," Proc.IEEE, vol.75, pp.869-891,July 1987.

[4] C.H. Knapp and G.C. Carter,"The Generalized Correlation Method for Estimation of Time Delay," IEEE Trans. Acoust.,Speech,Signal Processing, vol. ASSP-24, pp.320-327,Aug. 1976.

单片机系统显示控制卡设计与实现①

齐向东 刘秋实 李启虎

(中国科学院声学研究所, 100080)

摘要 本文讨论以8031单片机为控制核心的一种控制板。该板完成了图形、字符、汉字的CRT显示器显示的控制功能，具有16KB的显示缓冲区，显示分辨率为416×200，外部CPU通过控制板提供的10条接口线，控制显示器完成相应内容的显示。

许多工业控制系统及仪器仪表中的显示，不要求很强的人机对话能力，但要求完成现场数据、结果或者状态的显示，即用CRT显示出一定的曲线、图形或者表格等。

因此我们研制了体积小($10\times15\text{cm}^2$)、可靠性高的单片机控制显示系统，采用8031为板上控制核心，加上相应的外围电路，即可接在各种数字控制、测量系统总线上，接收外部CPU命令，完成指定内容的显示。使得显示器易于在工业控制系统及智能仪器仪表中应用，便于仪器等的一体化设计。

一、控制板体系结构及显示原理

1. 控制板主要部件及工作过程

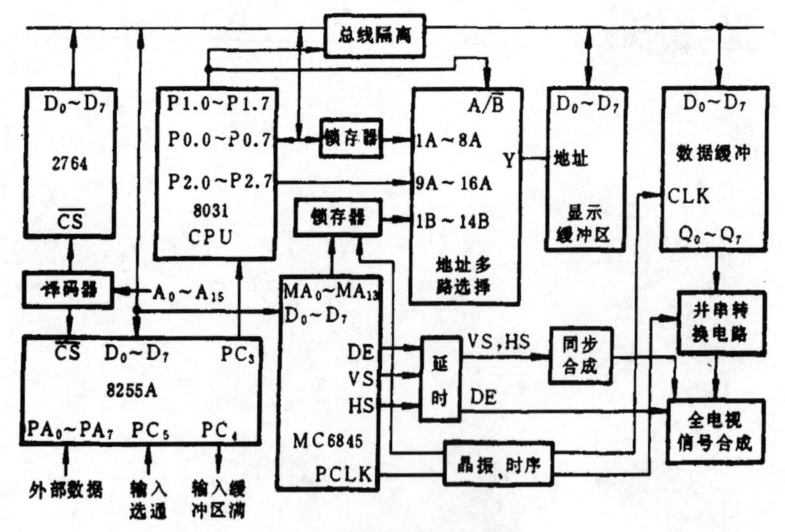

图1 显示控制板结构框图

① 微计算机应用, 1992, 13(2): 44-46.

图 1 是控制板结构框图。板上由8031作为控制核心，通过固化在片外EPROM中的监控程序，8031可以接收外部CPU命令，分析、执行这些命令。显示控制器MC6845产生相应的CRT控制信号。

系统上电或复位后，8031 进入其监控程序，首先初始化其内部工作状态，对MC6845各工作寄存器、8255A的状态寄存器进行初始化，然后等待外部CPU命令。

当外部CPU把命令字写入控制板上8255A的PA口后，通过中断方式8031 读入命令，并产生相应的写显示缓冲区VRAM操作，并等待下一条外部命令。当 8031 不对VRAM进行操作时，由MC6845寻址VRAM，显示当前VRAM中内容。

2. 显示方式

控制板针对工业控制、仪器仪表显示中，字符、汉字显示量较少，大部分为曲线、表格显示的特点，合并图形、字符显示方式为图形显示方式，将常用字符、汉字字模点阵存储在程序存储器EPROM中，供8031执行指令时调用。

二、硬软件设计

1. 总线隔离电路

8031访问外部程序存储器时，如果此时显示控制器MC6845寻址VRAM，则数据总线出现总线竞争，因此控制板采用图 2 所示的总线隔离电路。

2. 时序电路

控制板为完成显示控制功能，要求有严格的时序关系。为此控制板采用图 3 所示的时序发生电路，可以保证数据稳定有序地读写及流程。

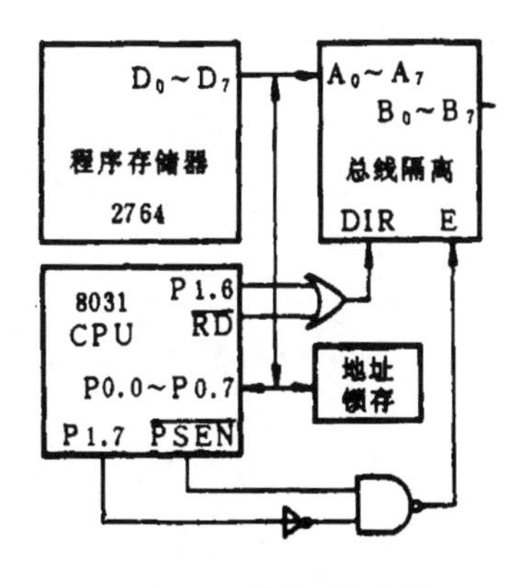

图 2 总线隔离电路

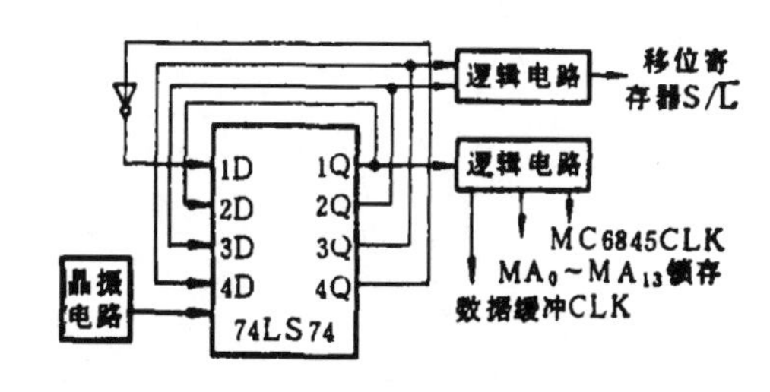

图 3 时序发生电路

3. 全电视信号合成电路

控制板可以直接输出到普通家用电视机的视放端，采用简单的晶体管加法电路，将水平同步、垂直同步、视频信号合成负极性全电视黑白信号，通过调整板上的电位器可获得清晰、稳定的显示。

4. 软件设计

控制板提供最大 64KB 的程序存储空间，因此可以存放字符、汉字点阵字模，以及各种曲线、表格显示子程序，供监控程序调用。

5. 外部接口

控制板采用一片8255A与外部CPU进行并行通讯，接收外部CPU命令。通过两条信号线与外部CPU进行握手应答。

目前本控制板在普通家用黑白电视机改装的基础上可实现准确、清晰的显示。改装的工作是把电视机视放级输入与前级断开，然后将控制板输出接电视机视放级即可。

如果将控制板输出与射频调制器相接，则射频调制器输出可直接接到家用电视机天线，但这样做的代价是成本的提高和清晰度下降。

显示终端也可采用单色监视器。直接将用于全电视信号合成电路输入的水平同步、垂直同步、图象信号送入监视器输入端即可。

从控制板与外部CPU接口线来看，控制板可以方便地与各种总线的控制系统相接，完成显示控制功能。

可编程多路白噪声发生器[①]

李启虎 刘金波 尹 力 赵国英
刘 伟 蒋 宏 陈玉凤
(中国科学院声学研究所，北京 100080)

1991 年 4 月 4 日收到

摘要 伪随机噪声在数字信号处理中具有重要的实用价值，白噪声是产生具有指定概率分布的伪随机噪声的基础．本文给出一种数字式可编程多路白噪声发生器的原理设计，把传统的移位寄存器技术用高速静态 RAM 代替，可以同时产生多路相互独立的白噪声．用 DSP 编程，软件简单，硬件紧凑．系统仿真表明，用本设计所产生的白噪声统计特性好、性能稳定，完全能满足声呐和通信等有关领域的要求，本文给出了原理分析、设计框图，计算机仿真的结果．

Programmable multi-channel white noise generator

LI Qihu LIU Jinbo YIN Li ZHAO Guoying
LIU Wei JIANG Hong and CHEN Yufeng
(*Institute of Acoustics, Academia Sinica*)

Received April 4, 1991

Abstract The pseudo-random data are widely applied in digital signal processing. White noise data is the basis of generating random data,which has the expected probability distribution. The design philosophy of a programmable digital multi-channel white noise generator is proposed in this paper. By using high speed static RAM and digital signal processing chip, In stead of the conventional shift register, the multi-channel independent white noise can be generated. The system simulation on a computer shows that the statistical characteristic of white noise data generated by this method is excellent and stable. It satisfies the requirements in the field of radar, sonar and communication. The analysis of design principle, block diagram and results of system simulation are illustrated.

一、引 言

伪随机噪声在信号处理的理论与实践中都具有重要作用．在信号分析、系统性能评价、保密通信、声呐和雷达的信号模拟方面，都要求有性能符合要求的伪随机噪声．白噪声是产生具

① 声学学报, 1993, 18(3): 204-209.

有指定概率分布密度的伪随机噪声的基础，因而如何产生高质量的白噪声就是一个十分重要的课题.

传统的用模拟方法产生白噪声，由于受到带宽及多路一致性等方面的限制，其应用有局限性.

随着数字计算技术的发展，用数字技术产生白噪声受到越来越多的重视 [1−4]，数字信号处理的理论和实践需要一种能快速产生多至几十路的相互独立的白噪声，并且要求各路噪声的一致性 (包括均值、方差、自相关特性等) 好、周期长，互相关为零. 数字信号处理 (DSP) 芯片的开发为用硬件研制白噪声发生器提供了可靠的基础.

Rader 等曾提出过一种用异或门产生白噪声的快速方法，并用数字硬件实现 [5−6], 但是，用这种方法所产生的白噪声周期较短，不适用于产生多路独立噪声.

Rabiner 等提出过一种用移位寄存器产生白噪声的技术 [1], 但用这种技术所需的运算量相当大，如果需要多路时，设备就十分庞大.

另外，用 M 序列发生器也可以产生性能相当好的伪随机序列 [3−4]. 当反馈函数选取得较好时, 运算量不大, 且周期可以非常大. 但是用这种方法只能得到 0 — 1 序列 (或限幅白噪声). 如果要得到多比特的数字信号，还需进行复杂的硬件设计及软件编程.

本文根据移位寄存器产生白噪声的原理，提出一种用静态 RAM(SRAM) 和 DSP 芯片 TMS320C25 为基础的多通道白噪声发生器. 系统编程容易，硬件简单. 系统仿真表明，所产生的白噪声具有性能稳定、一致性好、相互独立等特点，并且是无周期的、纯随机信号.

我们将在第二节中说明产生多路白噪声的基本原理与设计框图. 在第三节中给出系统仿真的结果，第四节简述用 TMS320C25 进行硬件设计方面的问题.

二、用 SRAM 产生白噪声的原理

图 1 给出了用移位寄存器产生白噪声的原理框图.

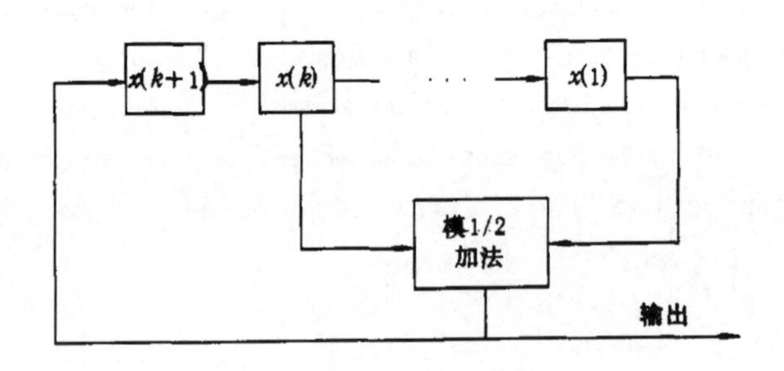

图 1　产生白噪声的方框图

设 $x(1),\cdots\cdots,x(k)$ 是 k 个随机数，在 $(-1/2, 1/2)$ 内均匀分布，相互独立，第 $k+1$ 个随机数由 $X(1)$ 及 $X(k)$ 按模 1/2 相加得到，即

$$x(k+1) = x(k) + x(1) \pmod{1/2} \tag{1}$$

这里模 1/2 相加的意义是

$$x(k+1)=\begin{cases} x(k)+x(1) & \text{如果 -1/2} \le x(k)+x(1) \le 1/2 \\ x(k)+x(1)-1 & \text{如果 } x(k)+x(1) > 1/2 \\ x(k)+x(1)+1 & \text{如果 } x(k)+x(1) < \text{-1/2} \end{cases} \tag{2}$$

显然 $x(k+1)$ 是 $(-1/2,\ 1/2)$ 之间的数，在下一时刻，把 $x(k+1)$ 移到 $x(k)$ 的位置，$x(k)$ 移到 $x(k-1)$ 的位置，$\cdots$, $x(2)$ 移到 $x(1)$ 的位置，然后再由 $x(2)$ 和 $x(k+1)$ 相加 (按模 1/2, 下同) 得到 $x(k+2)$, 等等.

这里我们要特别指出的是，模 1/2 没有特别的意义，只是为了说明问题的方便，实际上我们可以采用其他模. 例如，在定点运算时，可以用“模 1024 ”或“模 512 ”.

利用 (1),(2) 两式可以陆续产生一个白噪声序列. 我们把第一次放在移位寄存器内的 k 个数，称为种子. 由于 $x(1)$ 和 $x(k)$ 相互独立，且在 $(-1/2,\ 1/2)$ 内均匀分布，所以 $x(k)+x(1)$ 就是在 $(-1,\ 1)$ 内有三角形概率分布密度的随机变量. 由于 (2) 式的模 1/2 运算. 使得 $x(k+1)$ 又变成 $(-1/2,\ 1/2)$ 内均匀分布的随机变量 [1] .

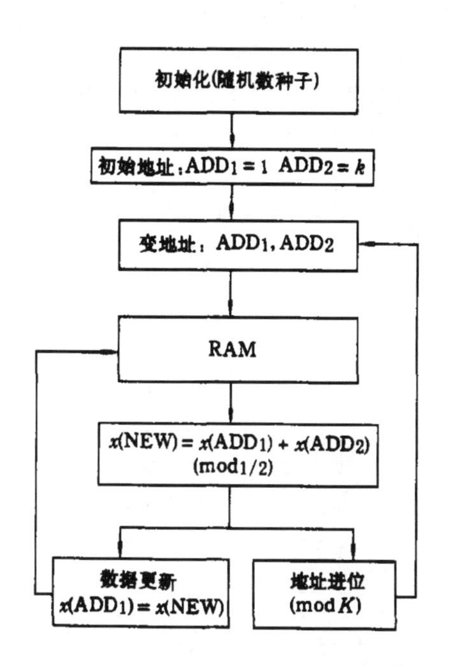

图 2　用 RAM 动态实现白噪声发生器的技术原理

随机数种子的好坏以及 k 的大小，直接决定了所产生的 $x(k)$ 序列的性能. 我们一般可以从随机数表或者用混合同余法来产生原始的种子 [2] .

用这种方法产生白噪声必须在每周期内作 k 次移位，一次加法，运算量是相当大的，并且 k 越大，运算量也越大，不利于产生高质量的白噪声序列.

图 2 给出了一种用 SRAM 产生白噪声序列的软件流程. 它的特点是不管 k 的大小，运算量保持大致的一样. 同时，这种软件编程非常易于用 DSP 芯片完成. 这就为用硬件产生多路相互独立的白噪声打下了基础. 它的基本原理是在容量为 k 的 SRAM 中存入 k 个原始随机种子. 然后在钟频的驱动下，把地址 ADD1 和 ADD2 的数用 1/2 模相加得到新数:

$$x(\text{NEW}) = x(\text{ADD1})+x(\text{ADD2}) \quad (\text{mod } 1/2) \tag{3}$$

地址的更新和数据的更新也采用模 k 的原则，即地址数逢 k 进位.

图 3 给出了用 DSP 芯片实现软件编程的原理框图, 其中 DSP 芯片采用 TMS320C25, 它的存取周期为 100ns, 这为产生多路独立白噪声提供了可能性. 例如，为了产生采样频率为 10kHz(周期 100μs) 的 50 路独立白噪声，可以用一片 TMS320C25 每隔 2μs 产生一个随机数 $x(k)$, 然后用下式得出 50 路独立噪声:

$$y_i(k) = x(50*k+i) \qquad i=1,...,50 \tag{4}$$

$x(k)$ 的间隔为 2μs, 而 $y_i(k)$ 的间隔为 100μs, 由于 TMS320C25 的指令周期为 100ns, 所以在 2μs 内按图 2 的流程产生一个 $x(\text{NEW})$ 是不困难的. 我们将在第四节说明这一点.

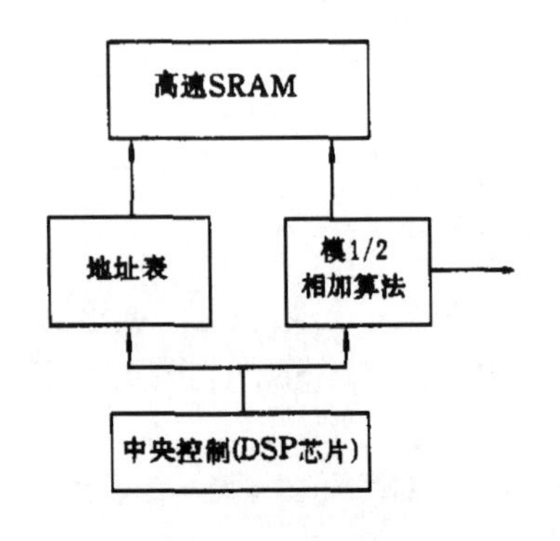

图 3　可编程数字式多路白噪声发生器原理框图

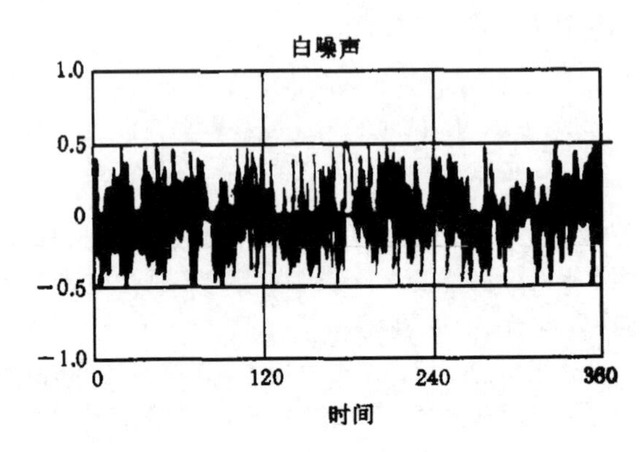

图 4　白噪声样本

三、系统模拟结果

根据图 2 给出的软件流程和图 3 的设计方框图. 我们在计算机上进行了大量的系统模拟. 主要内容包括：随机数种子的产生；白噪声样本的时间特性；白噪声的自相关特性；多路白噪声的互相关特性. 我们下面分开介绍部分结果.

THE SEED ARRAY				
+0.4073	+0.2862	−0.3895	−0.1755	−0.3782
−0.0760	−0.0536	+0.0574	+0.1664	−0.4867
+0.4161	−0.0994	−0.1197	−0.3704	+0.3282
+0.4280	−0.4350	−0.1780	−0.0114	−0.3127
+0.0304	−0.3399	−0.2882	−0.2533	+0.1074
−0.2893	−0.1791	−0.3286	−0.4233	−0.4741
−0.3752	−0.4880	+0.2864	−0.3219	+0.4290
+0.3514	+0.0116	−0.2864	−0.4518	+0.2461
−0.4054	+0.2226	+0.2939	+0.2494	−0.1603
−0.2663	−0.0135	+0.4238	−0.4401	−0.1571

在系统模拟中，取 RAM 容量为 50．随机数种子采用混合同余法产生，见下表，这一组数的均值为 −0.085721, 方差为 0.081168, 具有相当好的统计特性. 我们在整个系统模拟中，一直使用这一数组作为种子.

白噪声的时间样本见图 4, 横坐标为序号，如果以采样间隔 100μs 来理解，这一段样本的长度为 3.6ms．由图可以看出它的幅度分布非常均匀.

图 5 是白噪声自相关函数的实例. 所用的公式为

$$Ra(i) = \frac{1}{M}\sum_{k=1}^{M} x(k) * x(k+i) \tag{5}$$

其中 M 决定积分时间，i 决定延时．图 5(a) 是积分时间 100ms,$i = 0, \cdots, 360$ 的相关函数．图 5(b) 是积分时间为 1 秒的情况．当积分时间增加时，起伏减小，$Ra(\cdot)$ 越来越接近于 Dirac 函数．图 5(c) 是时延超过 1 秒后的相关函数．$Ra(i)$ 只在零值附近起伏．

图 6 给出了 50 路噪声的互相关函数，所用的公式为

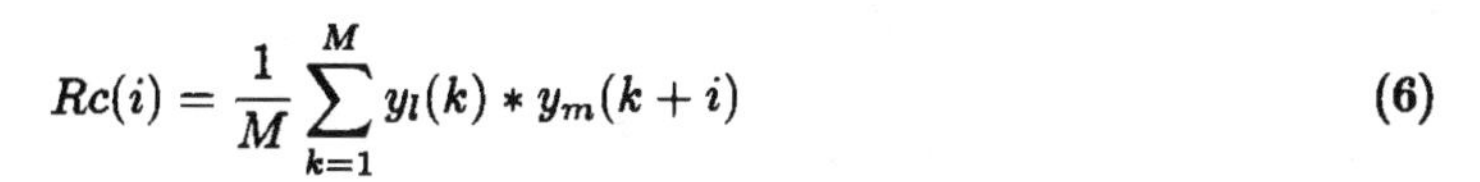

$$Rc(i) = \frac{1}{M} \sum_{k=1}^{M} y_l(k) * y_m(k+i) \tag{6}$$

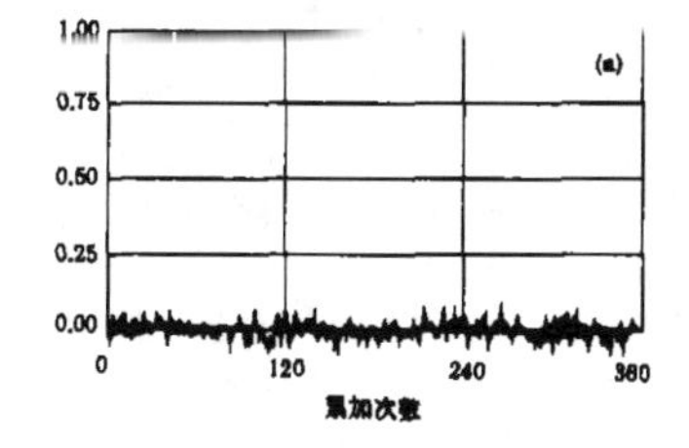

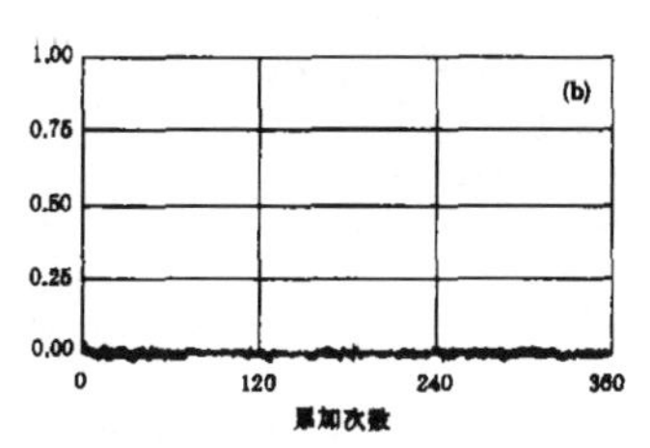

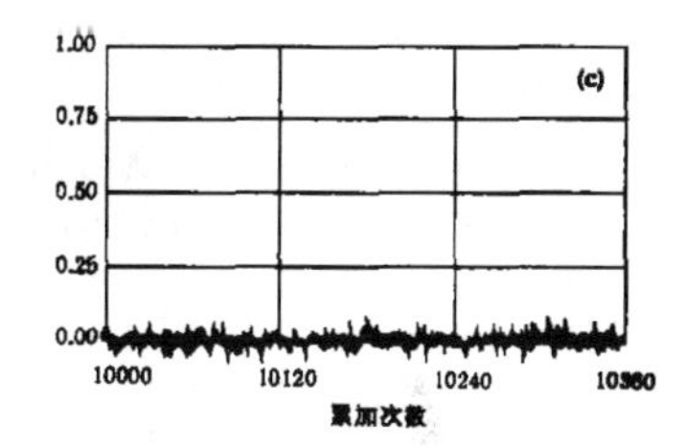

(a) 累加次数 1024(0 — 360 点)；　(b) 累加次数 10240;　(c) 累加次数 1024(10000 — 10360 点)

图 5　自相关函数实例

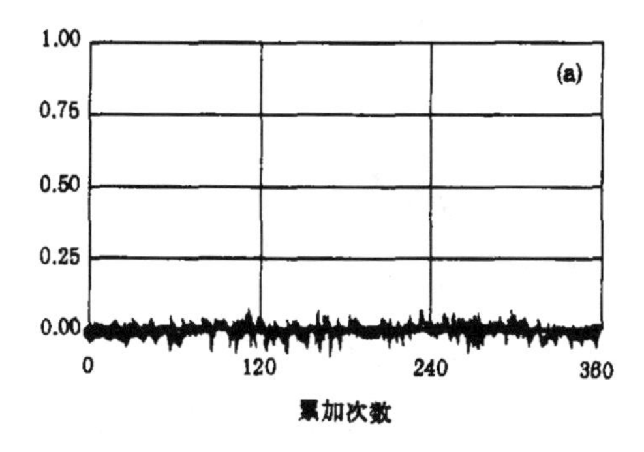

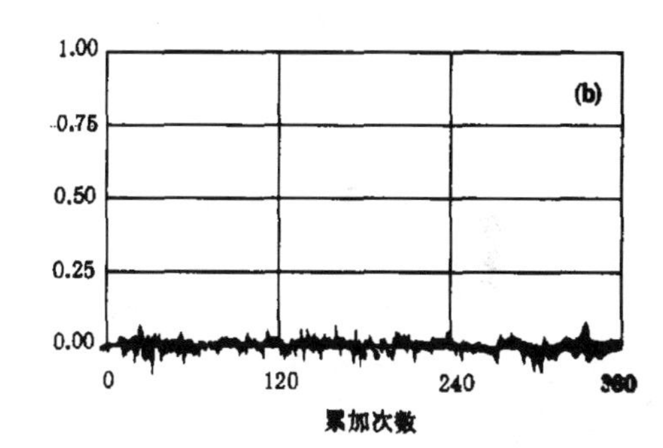

图 6　互相关函数实例

图 6(a) 是 50 路白噪声中第 1 路和第 49 路的互相关函数，图 6(b) 是第 1 路和第 20 路的互相关函数．我们看到，其独立性相当好．

四、用 DSP 芯片硬件实现的问题

本文所提出的方法，特别适用于用高速信号处理芯片去实现．我们选用 Texas Instr 公司生产的第二代数字信号处理 (DSP) 芯片 TMS320C25 进行硬件设计．TMS320C25 是用 CMOS 工艺集成的线宽 2.0μm 的数字信号处理芯片．指令周期是 100ns, 片内有 256 字的程序 RAM 和 288 字的数据 RAM 和 32 比特的算术累加单元，用该芯片完成模 1/2 相加是非常方便的．在加法运算中有两种溢出方式，一种是允许方式；另一种是禁止方式．我们选用禁止溢出加法方式，就可以得到所需的白噪声序列，指令码见下表．

```
BEG3    ADDH    *            *X(N+1)=X(N)+X(1).
        SACH    *+,0,3       *STORE NEW NOISE DATA, (APP)=3.
        SACH    *+,0,4       *OUTPUT RESULT.
        BANZ    BEG3,*-,2    *JUMP TO BEG3 IF (AR3)>0, (ARP)=2.
        B       BEG2
```

参 考 文 献

[1] L.R.Rabiner and Gold, "Theory and application of digital signal processing", Printice-Hall,Inc., 1975.

[2] 李启虎，"声呐信号处理引论"，海洋出版社，1985 年，北京.

[3] 万哲先，"代数与编码"，第二版，科学出版社，1980 年，北京.

[4] 蔡宗蔚，"实用编码技术"，人民邮电出版社，1983 年，北京.

[5] C.M.Rader et al., "A fast method of generating digital random numbers", *Bell System Tech. J.*, **49**(1970), 2303 — 2310.

[6] J.L.Perry et al., "A digital hardware realization of a random number generator", *IEEE Trans.*, **AU-20**(1972), 236 — 240.

声呐信号处理中时域数字多波束成形的新方法①

李启虎 孙长瑜 孙 增 蔡惠智 刘正元 乔 均

(中国科学院声学研究所,北京 100080)

1992年1月27日收到

摘要 本文给出时域上数字多波束成形的一种新方法.它采用并行流水线结构,把数字滤波、升采样以及最优加权结合在一起,利用高速RAM和VLSI DSP芯片动态地构成一种高精度延时补偿的多波束数字信号处理系统.这种系统与传统波束成形比较,具有精度高,运算量少,编程容易,硬件结构紧凑的特点.文中给出设计的基本原理及硬件实现的方法.

A new multi-beamforming method for digital sonar in time domain

LI Qihu SUN Changyu SUN Zeng CAI Huizhi LIU Zhengyuan and QIAO Jun

(*Institute of Acoustics, Academia Sinica*)

Received January 27, 1992

Abstract A new beamforming method in time domain is described. Based on the pipe line architecture, the digital filtering algorithm, rising sampling rate and optimum weighting are combined. The high accuracy time-delay compensation for multi-beam signal processing system is designed by using high speed RAM and VLSI DSP chips. Comparing with the traditional beamforming method, the new method presented in this paper is characterized by high accuracy, less computation complexity, easy programmable and compact hardware architecture. The design philosophy of time-domain RAM-dynamic beamforming method and hardware implementation is given.

一、引 言

波束成形是声呐信号处理系统中的核心环节[1-4],无论是被动声呐还是主动声呐,模拟式声呐或是数字式声呐,都必须有波束成形系统,其目的是为了获得较大的空间增益,同时,也是为了使系统得到较高的精度的目标分辨力.所以,波束成形系统是决定一部声呐性能优劣的核心环节.

六十年代以来,数字多波束技术得到了迅速的发展,为声呐系统提供了360°实时全景跟

① 声学学报, 1993, 18(6): 430-435.

踪的能力，同时，在声呐换能器的布阵方面也有了很大的进展. 基阵的几何形状由直线和圆发展到共形阵、弧形阵、球形阵以及圆柱阵等. 先前，波束成形的技术立足于模拟 LC 延迟线或移位寄存器，它在延时精度补偿和最大延迟时间上，均受到较多的限制，从而，影响了大型基阵的使用[7,8].

近些年来，由于微电子技术和波束成形技术的飞速发展，出现了一些通用或专用的超大规模集成电路(VLSI)芯片，其中有一部分特别适用于声呐信号处理算法的高速或超高速 DSP 芯片，使得由许多水听器组成的大型复杂的基阵，实现多通道多波束成形实时处理系统变成了现实[5-11]. 这种系统可以通过对周围环境噪声场的适应过程，自动调整本系统的有关参数，从而，使系统具有某种“最佳的”抗干扰能力和实时全景检测目标信号的能力，实行多目标全方位的实时跟踪.

为了减少大孔径基阵对波束成形器电子线路要求的复杂性，出现了新的波束成形技术. 如数字内插法波束成形[5]，相移波束成形，频率域波束成形，还有时域和频域混合合成结构[7-9,11]. 对于某些实际应用场合和要求，可根据波束成形器的容许能力来选择上述不同的波束成形技术. 本论文提出一种新的时域波束成形技术，它把数字滤波、升采样、最优幅度加权和最佳后置积累结合在一起，利用 DSP 芯片实现高精度的延时补偿，大大地提高了系统的定向精度和通用性. 由于采用并行流水式体系结构和数字信号处理的新技术，整个系统编程容易，结构紧凑，便于硬件实现.

我们将在第2节中简述波束成形的基本原理，第3节讨论新的时域数字多波束成形的基本原理及其实现方法，第4节是一个简要的结论.

二、波束成形的基本原理

一部声呐系统要通过接收到的水下声波来判定有无目标及其所在的方位，亦就是说，要具有定向定位的能力. 这就需要将几十甚至几百个水听器按一定的几何形状布设成水听器阵列，通过各基元的延时补偿来实现.

图1给出了 N 个基元等间隔排列的线列阵的示意图. 由图1可见，声波除正横(90°)方向入射以外，到达各基元的时间是不同的. 所以，要使在声波入射方向使系统产生最大的响应，就必须人为地将每个基元的信号延时到同一时刻进行求和. 也就是说，系统要完成对所有基元延时量从 A 波阵面到 B 波阵面的补偿. 其各基元延时量由下式给出：

$$T_i=\frac{(i-1)d\sin\theta}{C} \qquad i=1,2,\cdots,N \quad (1)$$

式中：i——基元序号；

d——相邻基元的间隔；

θ——声线入射角；

C——水中声速.

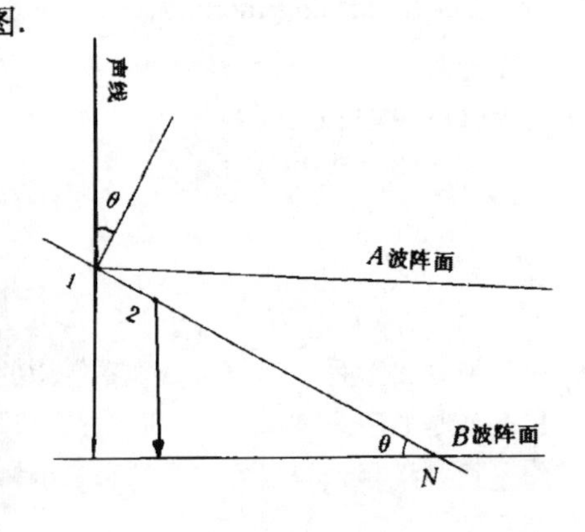

图1 N 个等间隔基元的线列阵

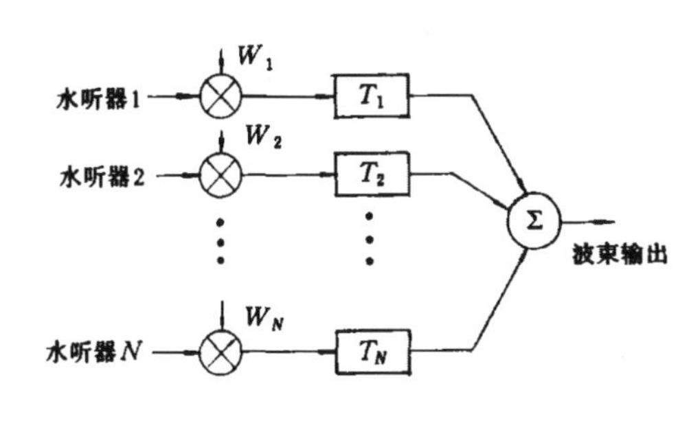

图2 N 个水听器基阵的波束成形原理

由此可见，实现这一处理过程，要进行如下一些运算. 包括对基阵各个基元收到的信号进行加权，延时以及求和. 加权的目的是为了有效地控制波束图的旁瓣结构，以使最大响应方向以外的信号得到充分的抑制. 图2给出了这一过程的实现原理.

图2中的 $W_1, W_2, \cdots, W_N$ 是各通道的加权系数，根据线阵指向性函数的数学表达式，可以借助道尔夫-切比雪夫多项式给出.

在实现波束成形的过程中，精确的延时补偿是最关键的，相比之下，其他运算（如加权、求和、积分）是次要的.

三、时域数字多波束成形的新方法

波束成形可以在时间域也可以在频率域上实现，频域波束成形虽然运算量较节省，但由于数据交换比较复杂，水听器的个数受 FFT 点数的软约束，同时又无法直接获得时间序列的输出，目前已较少使用. 所以，我们主要讨论时间域的波束成形问题. 对于时间延时，可以有多种实现方法，如模拟延迟线，数字移位寄存器，随机存取存储器等. 总而言之，无论哪种方法均要利用各基元间的时延来实现. 因此，对于数字式声呐，首先需将各水听器收到的信号进行模拟——数字（A/D）的转换，变换成数字序列. 在此之前应对每个水听器通道的信号进行前置放大，自动增益控制（AGC）和抗混淆滤波. 模拟到数字的转换处理，包括对每个水听器信号的时间抽样和幅度量化. 时间抽样由每个通道的采样和保持电路完成. 这些电路应对阵中的各水听器信号进行同时采样或几乎同时采样. 如果未能做到同时采样，那么某些水听器之间将产生时滞现象. 这种时滞现象，可在波束成形前对时延进行调整或修正，否则会给波束成形带来误差. 然后 A/D 转换器将这些采样后的数据进行幅度量化，为了尽可能地减少量化误差和硬件的设备量，提高波束成形精度，还需将 A/D 转换后的数字信号序列进行数字内插处理，或称升采样处理. 它包括对等于或略高于奈奎斯特（Nyquist）采样率的输入数字序列信号进行采样，然后对采样到的样本序列间进行补零，再经过一个有限冲激响应（FIR）的数字滤波器，便可得到密采样的时间序列. 其原理如图3所示.

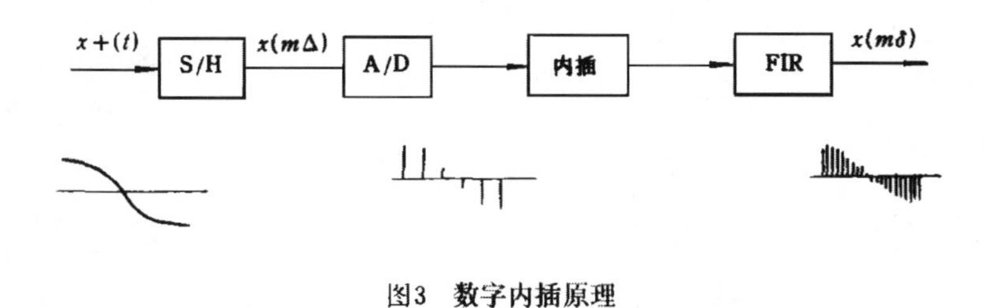

图3 数字内插原理

设输入信号为 $x(t)$，经采样和量化后为 $x(m\Delta)$，其中 Δ 为采样间隔，将 $x(m\Delta)$ 按时间间

隔 $\delta<\Delta$ 进行补零，再通过频率响应为矩形的 FIR 滤波器，便可得到新的时间序列 $x(m\delta)$，δ 即为内插采样后的样本间隔，只要 FIR 滤波器的频率响应特性符合低通滤波的要求，序列 $x(m\delta)$就可逼近 $x(t)$的采样，FIR 的算法为一卷积运算：

$$y(n)=\sum_{i=0}^{N-1}a_i x(m-i) \tag{2}$$

可见它实际上是一个数字窗，其窗口长度由 Δ/δ 来确定，当 Δ/δ 在10的量级时，FIR 滤波器的长度取31就足够了. 如图4所示.

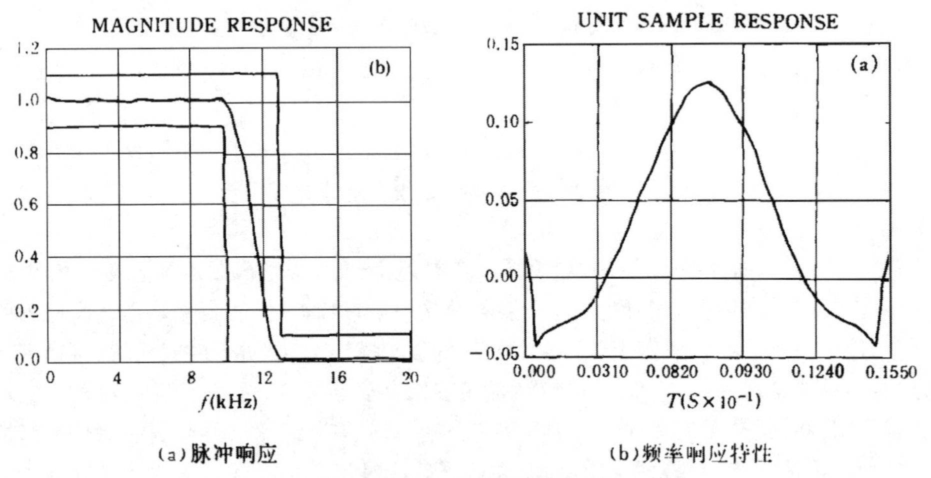

(a)脉冲响应　　(b)频率响应特性

图4　31个系数的 FIR 的脉冲和频率响应特性

这种 FIR 滤波器通常用于输入信号频带 $W=10\text{kHz}$ 的情况，所以取 $f_\Delta=20\text{kHz}$，内插后的采样频率为 $f_\delta=200\text{kHz}$，这样对波束图的主瓣没有多大影响，而旁瓣却得到了一定的抑制. 然后将采样后的样本序列传送至数字多波束成形器、进行波束成形运算. 图5给出了这一整个处理过程的原理.

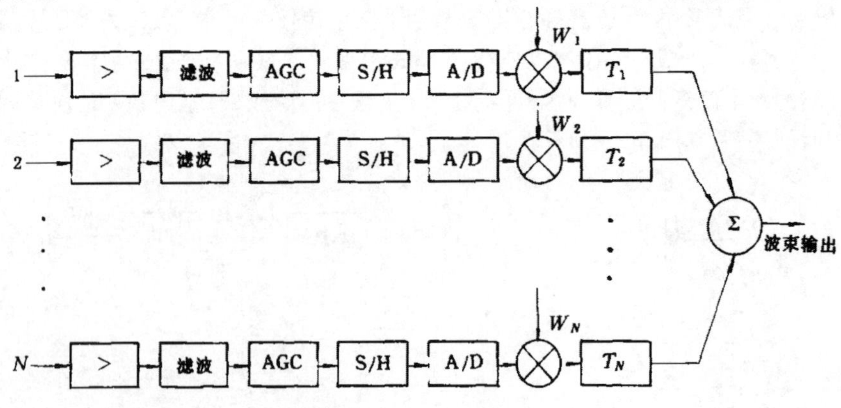

图5　数字多波束成形原理

图5中表示的波束成形器用数学公式可表示成：

$$b(t)=\sum_{n=1}^{N}W_nX_n(t-\tau_n) \tag{3}$$

式中 $b(t)$ 为波束输出，W_n 表示对第 n 个水听器信号的加权因子，X_n 是来自第 n 个水听器的样本，τ_n 是对第 n 个水听器所需调整的延时量，N 为水听器的总数.

水听器信号 X_n 经时间采样，幅度量化和数字内插处理后，式(3)可用离散序列表示成：

$$b(m\Delta)=\sum_{n=1}^{N}W_nX_n(m\Delta-M_n\delta) \tag{4}$$

其中 M_n 为自然数，Δ 为数字内插前的样本间隔，δ 为数字内插后的样本间隔. 图6给出了时域波束成形器的输入与输出样本序列的时间关系. 它的特点是输入是高速的，而输出用于后置积累的序列是低速的，从而便于实现多波束运算.

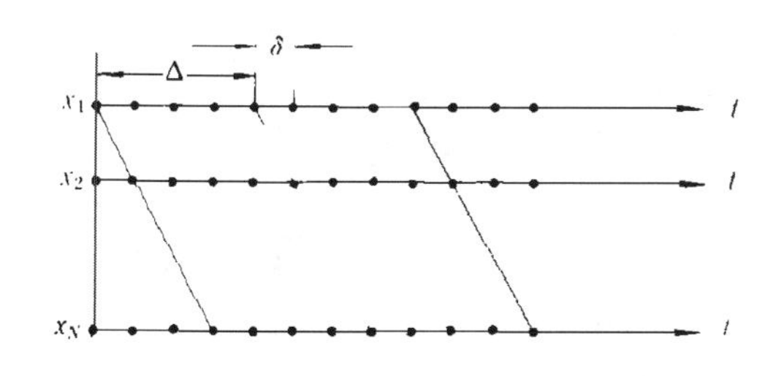

图6 时域波束成形器输入/输出序列的时间关系

根据这一延时关系，我们可用 RAM 作为每个通道信号的延迟线来实现. 也就是说，需将 RAM 与 $x_1,x_2,\cdots,x_N$ 相对应分成 N 块，把输入数据分块高速存入 RAM，再以低速取出进行波束成形. 如果假设 $f_2=\delta^{-1},f_1=\Delta^{-1}$，则存入 RAM 的速率即为 f_2，而取出的速率则为 f_1，这样做的好处在于减少了波束成形器的软硬件开销，而又获得等同于采样频率为 f_2 的波束成形的效果.

假设有一等间隔排列的 $N=64$ 个水听器的线阵，最大开角为±85°，基元间隔为1m. $\delta=50\mu s$，最小声速 $c=1450\text{m/s}$，则最大延时：

$$\tau_{\max}=\frac{(N-1)d\sin 85°}{c}\doteq 43.283\text{ms}$$

折合成系统延时节数为：

$$T_{\max}=43.283\text{ms}/50\mu s\doteq 866\text{节}$$

则 RAM 的最小容量应为：

$$64\times 867=55.5\text{kbt}$$

波束延时 RAM 的存入方法可由数字内插处理器控制，直接按列存入，数据的读出可由 TMS320系列通用的信号处理芯片控制行列计数器和地址表存储器来完成，地址表存储器的内容是依据各水听器间相应的延时，预先计算好后写入的，它的输出与列计数器的内容相加之和作为波束延时 RAM 的读出地址，将数据取至 TMS320处理器的内部 RAM 进行波束成形运算，然后将波束成形结果以数字内插前的样本速率被写入到波束输出缓冲存储器中，待后置处理器使用. 图7给出了这一时域波束成形器的简单原理硬件框图.

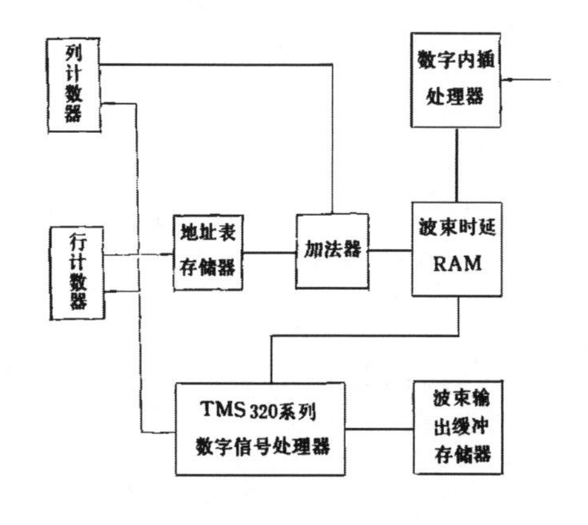

图7 时域数字多波束成形器原理

假如图7中的信号处理器选择(VLSI)TMS320C30来实现，它具有单周期指令执行时间为60ns. 33. 3MFLOPS(million floating-point operations per second). 16. 7MIPS(million instructions per second). 一个4k×32 bit 的单周期双臂取数(dual-access)在片 ROM 块，两个1k×32 bit 的单周期双臂存取在片 RAM 块，64×32 bit 超高速指令缓冲存储器，32 bit 的指令和数据字，24 bit 地址，40/32 bit 浮点/整数乘法器和 ALU，32 bit 桶形移位寄存器，两个串行口支持8/16/32 bit 数据传递，两个32 bit 定时器，两个通用外部标志，四个外部中断，180引脚封装，1μmCMOS 工艺的(VLSI)DSP 芯片. 那么，对于64个基元，在50μs 内可形成13个波束，如果要形成更多的波束，可采取多片 TMS320C30并用的方法即可实现.

四、结　论

本文给出了时域动态 RAM 数字多波束成形的新方法，与过去的模拟延迟线方法，横向滤波器方法，移位寄存器方法，以及频率域波束成形方法相比，大大降低了电路的数量及其复杂性. 而且系统实现了完全数字化可编程控制，所有算法均可由价格低廉的软件编程来完成，数字硬件仅起辅助作用，从而大大提高了系统的灵活性，以及实时处理功能，对于过去不可能做到的任意几何形状的大型基阵的数字多波束成形都可比较方便地实现. 而且通过数字内插法，降低了前级的硬件负担和波束成形器的运算量. 而波束成形的精度得到了有效的提高.

由于这种系统都是由高速或超高速的超大规模的集成电路设计而成. 一般而言，通道数较多，电路设计应尽量紧凑，结构设计要合理可靠. 要充分注意系统的抗干扰能力.

参 考 文 献

[1] 李启虎，"声呐信号处理引论"，海洋出版社，1985，北京.

[2] Knight, W. C., et al., "Digital Signal Procesing for Sonar", Proc. IEEE, **69**(1981), 1451－1506.

[3] Rudnick, P., "Digital beamforming in the frequency domain". *J. Acoust. Soc. Am.*, **46**(1969), 1089－1090.

[4] Horton, C. W., "*Signal processing of underwater acoustics waves*". Unted States Government Printing Office, 1969. 中译本："水声信号处理"，汪元美译，国防工业出版社，1978.

[5] Pridham, R. G. and Mucci, R. A., "Digital interpolation beamforming for low-pass and bandpass signal", Proc. IEEE, **67** (1979), 904－919.

[6] Crochiere, R. E. and Rabiner, L. R., "Interpolation and decimation of digital signal——a tutorial review", Proc. IEEE, **69**(1981), 300－331.

[7] Curtis, T. E. and Ward, R. E., "Digital beamforming for sonar systems", IEEE Proc., **127**(1980), pt. F, 257－265.

[8] Barton, P., "Digital beamforming for radar", IEEE Proc., **127**(1980), pt. F 266－277.

[9] B. Van Veen and K. Buckley., "Beamforming: a versatile approach to spatial filtering", IEEE ASSP Magzine, April 1988. 4－24.

[10] D. A. Gray., "Effect of time delay errors on the beam pattern of a linear array", IEEE Journal of OE vol. OE-10 (1985), 259－279.

[11] J. Allen., "Computer architecture for diyital signal processing", Proc. IEEE, **73**(1985), May, 854－873.

声呐系统的最佳定向精度和最优多目标分辨力研究①

李启虎 尹 力 赵国英

（中国科学院声学研究所，北京）

摘 要 定向精度和多目标分辨力是被动声呐系统的重要技术指标．本文从理论上证明，这两个指标都与声呐系统的指向性函数的主瓣特性有内在联系．定向精度的最佳值与时延估计误差的克拉美-罗（Cramer-Rao）下界有关，在一定的信噪比条件下，定向精度与波束主瓣宽度成反比．多目标的分辨力约等于1.3倍主瓣的宽度．文中以线阵与圆阵为例给出定向精度和多目标分辨力的数学表达式，并以实际计算结果与理论作比较．

关键词 声呐系统 最佳定向精度 最优多目标分辨力 指向性函数 主瓣特性

前 言

目标的定向精度与多目标的分辨力是声呐设计者所关心的最重要的技术指标之一，但是在早期文献[1~5]中却很少讨论这个问题，一般只提出有关指向性的概念．人们深信指向性函数在主瓣附近的性状与定向精度及目标分辨力有关，但是却没有得出它们的内在联系．

随着信号处理技术的发展，我们已明白，定向精度问题是与时延估计问题紧密相连的[6~8]．时延估计的精度可以通过计算克拉美-罗下界求出来，于是定向精度问题便可以根据基阵的几何形状来求解．

本文在理论上证明了定向精度和多目标分辨力与声呐系统指向性的内在联系，求出它们的数学关系，指出在一定信噪比之下，定向精度与指向性函数主瓣宽度成反比．多目标分辨力的问题除与目标的夹角有关外，还与目标之间的强度比有关[9~12]．我们推导了不同强度比之下，声呐系统多目标分辨力与指向性函数主瓣宽度的关系．在两个等强度目标的情况下，多目标分辨力约等于1.3倍指向性函数主瓣宽度．

本文所给的理论结果已经由实际的计算得到证实，所得的结果对声呐系统的设计提供了非常实用的依据，特别是对多波束数字化声呐来说，本文的结果是选择基阵孔径、频率范围和波束数目等参数的出发点．

① 海洋学报, 1996, 18(4): 43-48.

1 声呐系统指向性函数在主瓣附近的性质

直观地说，声呐系统的指向性函数表示它在空间抑制各向同性噪声的能力. 这种空间滤波的作用，在某一方向就是定向精度的基础. 从总体上说，指向性的空间积分就是系统的增益. 我们首先研究声呐系统指向性函数在主瓣附近的性状.

声呐系统的指向性函数与构成该系统的声呐基阵的几何形状有关. 对有 N 个基元的等间隔离散线阵，其指向性表达式为[14]

$$R(\varphi)=\frac{\sin(N\varphi/2)}{N\sin(\varphi/2)}, \tag{1}$$

式中，

$$\varphi=\frac{2\pi d[\sin\theta-\sin\theta_0]}{\lambda}, \tag{2}$$

θ 为目标信号的入射角；θ_0为波束指向角；d 为基元间隔；λ 为信号波长. 当 $\theta\approx\theta_0$时，式（1）具有 $\sin x/x$ 的形状，在 $x=0$处作泰勒展开，得到

$$\frac{\sin x}{x}\approx 1-\frac{x^2}{3}. \tag{3}$$

换言之，直线阵指向性函数在主瓣附近具有抛物线的形状.

对圆阵来说，如果基元数和圆阵直径、信号波长满足一定的关系，其指向性函数可以表达为零阶贝塞尔函数[1]：

$$D(\theta)=\mathrm{J}_0\left(\frac{4\pi r}{\lambda}\sin\frac{\theta-\theta_0}{2}\right), \tag{4}$$

式中，r 为基阵半径；λ 为信号波长. 在 $x=0$附近，

$$\mathrm{J}_0(x)\approx 1-\frac{x^2}{4}, \tag{5}$$

也具有抛物线的形状.

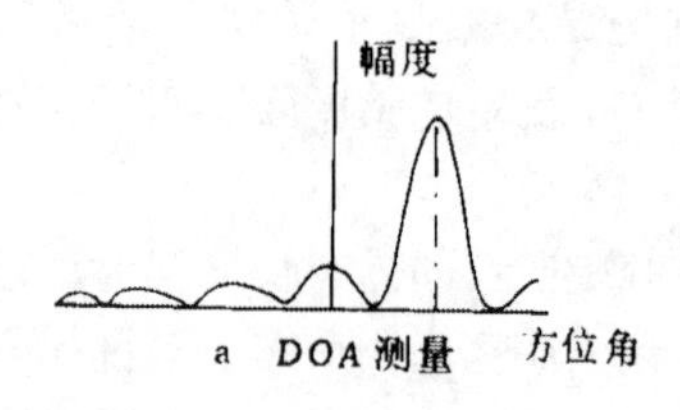

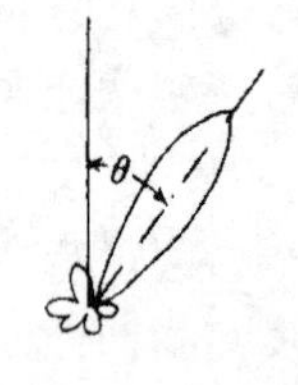

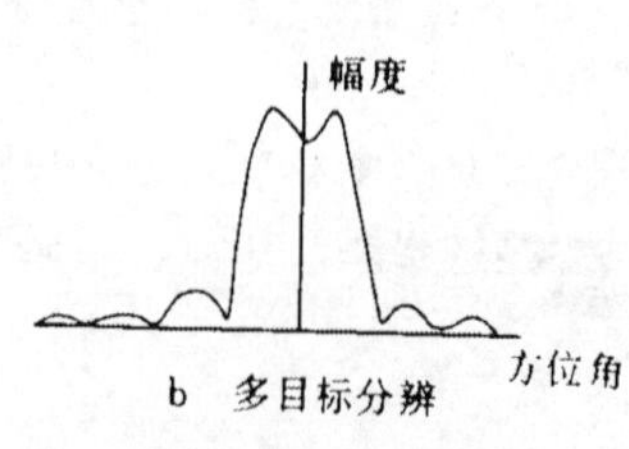

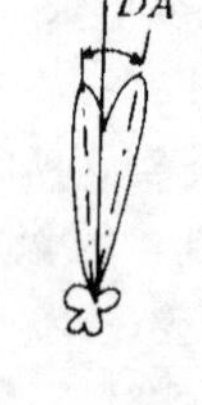

图1 声呐系统中单目标定向和多目标分辨

2 定向精度

见图1a，如果目标的入射方向为 θ，而声呐系统的估计值为 $\hat{\theta}$，则两者的误差 $\hat{\theta}-\theta$ 就是定向精度. 我们通常把 $\hat{\theta}$ 看作是一个统计量，而关心的是 $\hat{\theta}-\theta$ 的均方根值：

$$CA=[E(\hat{\theta}-\theta)^2]^{1/2}, \tag{6}$$

CA 的极小值就是最佳的定向精度，记作 $(CA)_{\mathrm{opt}}$. 我们要证明 $(CA)_{\mathrm{opt}}$是与分裂波束系统的时延估计有关的.

事实上，根据文献 [14] 中的推导，分裂波束系统时延估计的克拉美-罗下界是

$$\Delta\tau_{\min}=\left(\frac{3}{8\pi^2T}\right)^{1/2}\frac{1}{(SNR)_{\text{in}}}\cdot\frac{1}{\sqrt{f_2^3-f_1^3}},\tag{7}$$

式中，T 为积分时间；$(SNR)_{\text{in}}$为输入信噪比，f_1和 f_2为工作频段. 若以 f_{rms}表示信号频率的几何平均值（$f_{\text{rms}}=\sqrt{f_1f_2}$），$W$ 表示带宽，那么式（7）又可以表达为

$$(\Delta\tau)_{\min}=\frac{1}{2\pi}\frac{1}{\sqrt{2TW}}f_{\text{rms}}^{-1}(SNR)_{\text{in}}^{-1}.\tag{8}$$

这就是文献［8］中最近得到的结果.

对线阵或圆阵，$(CA)_{\text{opt}}$和（$\Delta\tau)_{\min}$的关系是不难求出的[14]. 对长为 L 的线阵（或 $Nd=L$）：

$$\begin{aligned}(CA)_{\text{opt}}&=\frac{2(\Delta\tau)_{\min}}{L}C\\&=\frac{C}{L\pi}\frac{1}{\sqrt{2TW}}\frac{1}{f_{\text{rms}}}\frac{1}{(SNR)_{\text{in}}}.\end{aligned}\tag{9}$$

对半径为 r 的圆阵：

$$\begin{aligned}(CA)_{\text{opt}}&=\frac{\pi(\Delta\tau)_{\min}C}{4r}\\&=\frac{C}{8r}\frac{1}{\sqrt{2TW}}\frac{1}{f_{\text{rms}}}\frac{1}{(SNR)_{\text{in}}}.\end{aligned}\tag{10}$$

由式（9）和（10）可看到，定向精度与输入信噪比、基阵孔径、频率等有关.

举例来说，线阵的波束宽度（3dB 点宽度）是 $BW_{\text{li}}=0.88C/L\cdot f_{\text{rms}}$，若 $TW=500$，$(SNR)_{\text{in}}=1$，则由式（9）可知

$$(CA)_{\text{opt}}\approx\frac{1}{80}BW_{\text{li}}.\tag{11}$$

同样条件下，对圆阵而言 $BW_{\text{ci}}=1.12C/\pi rf_{\text{rms}}$，于是由式（10）可得到

$$(CA)_{\text{opt}}\approx\frac{1}{80}BW_{\text{ci}}.\tag{12}$$

有趣的是式（11）和（12）的形式完全一样. 图2给出定向精度与$(SNR)_{\text{in}}$的关系. 由此可以根据指向性函数主瓣的宽度大致确定最佳定向精度，例如，如果波束宽度为10°，那么在信噪比为0dB时，定向精度大约是（1/80）* 10°≈0.125°.

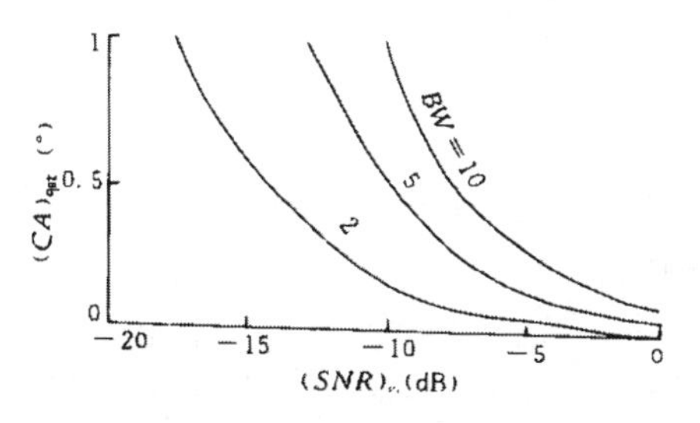

图2 定向精度和（$SNR)_{\text{in}}$的关系

3 多目标分辨力

最简单的多目标情况就是两个目标，图1b 是这种情况的示意图. DA 为两个目标在空间相夹的角. 如果是几个目标同时以很小的 DA 出现，那么情况会非常复杂，但我们下面给出的处理方法可以把多于两个目标的情况逐步化简为双目标的情况. 在本节我们只讨论双目标的情况.

设单目标时的指向性函数在主瓣附近的形状为 $f(x)=1-\alpha x^2$，令 $f(x)=\sqrt{2}/2=0.707$，便可确定波束3dB 点的宽度：

$$BW=\left(\frac{2-\sqrt{2}}{2\alpha}\right)^{1/2}. \tag{13}$$

讨论声呐系统的多目标分辨力，就是要确定在两个信号源强比 $SB:SA$ 在一定值的条件下，最小可分辨的 DA 值究竟是多大?DA 与 BW 有什么关系?为此，先要说明什么是可分辨?我们定义当图1b 中的谷点为-3dB 时，称为两个目标为可分辨．当 $SB:SA\neq1$时，我们限定谷点要与 SA、SB 中较小的那个源强相比．

现在假定在 x_0处出现另一目标，目标源强比是 $SB:SA=\gamma$，在目标 A 方向 $f(x)=1-\alpha x^2$，在 B 方向 $g(x)=f(x-x_0)=1-\alpha(x-x_0)^2$，合成指向性是

$$h(x)=[f^2(x)+r^2g^2(x)]^{1/2}. \tag{14}$$

当 $\gamma=1$时，合成指向性的谷点出现在 $h(x_0/2)$处，这是显然的，但当 $\gamma\neq1$时，谷点的位置必须具体计算．

我们以 BW 为参考量，把 x_0和 x 都以 BW 来归一，计算的结果如图3所示．对两个等强度目标 ($SB:SA=0$dB)，可区分的 DA 大约是$1.3BW$；当 $SB:SA\neq0$dB 时，谷点会向源强较弱的方向偏移（图3b)．图中的横坐标为 $\alpha=x/0.5\cdot BW$，$x_0=DA/0.5\cdot BW$．

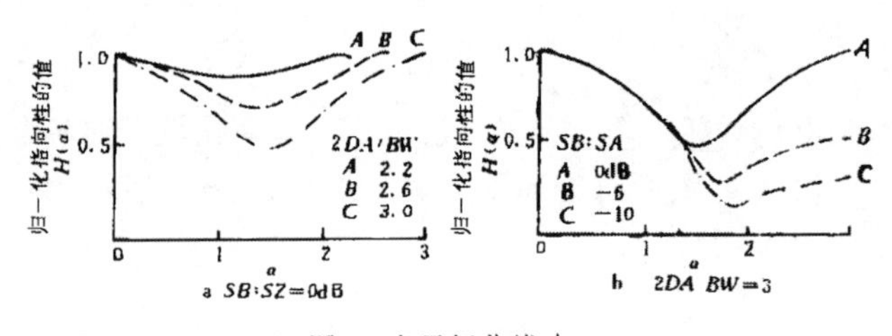

图3　多目标分辨力

4　计算实例

为了验证上几节推导的理论结果，我们进行了计算机模拟实验，同时又以圆阵为例计算了声呐系统的指向性函数．对单目标定向和双目标分辨都进行了大量的测试，效果是令人满意的．

我们选择一个直径2m、基元数为48的均匀离散圆阵，工作基元由2×10的扇面（即360°/48×20=150°）组成．

在频率为4kHz 时，主瓣宽度为 $BW=6°$，在输入信噪化 $(SNR)_{in}=-5$dB 时，用分裂波束进行定向．模拟结果见表1．信号入射角为 $\theta=15°$．10次测量的平均值为14.952 5°，均方误差为$[E(\hat{\theta}-\theta)^2]^{1/2}=0.35°$．

这种情况下的理论值可由图2求出，大约是0.28°．

图4和5给出了一系列双目标时的计算结果，对单频信号，$BW=6°$；对等强度目标最佳分辨力是8°=1.33BW．如果 $SB:SA=-6$dB，最佳分辨力是10°=1.66BW；当 $SB:SA=-20$dB，事实上已难于分清两个目标了（图4b～e)．

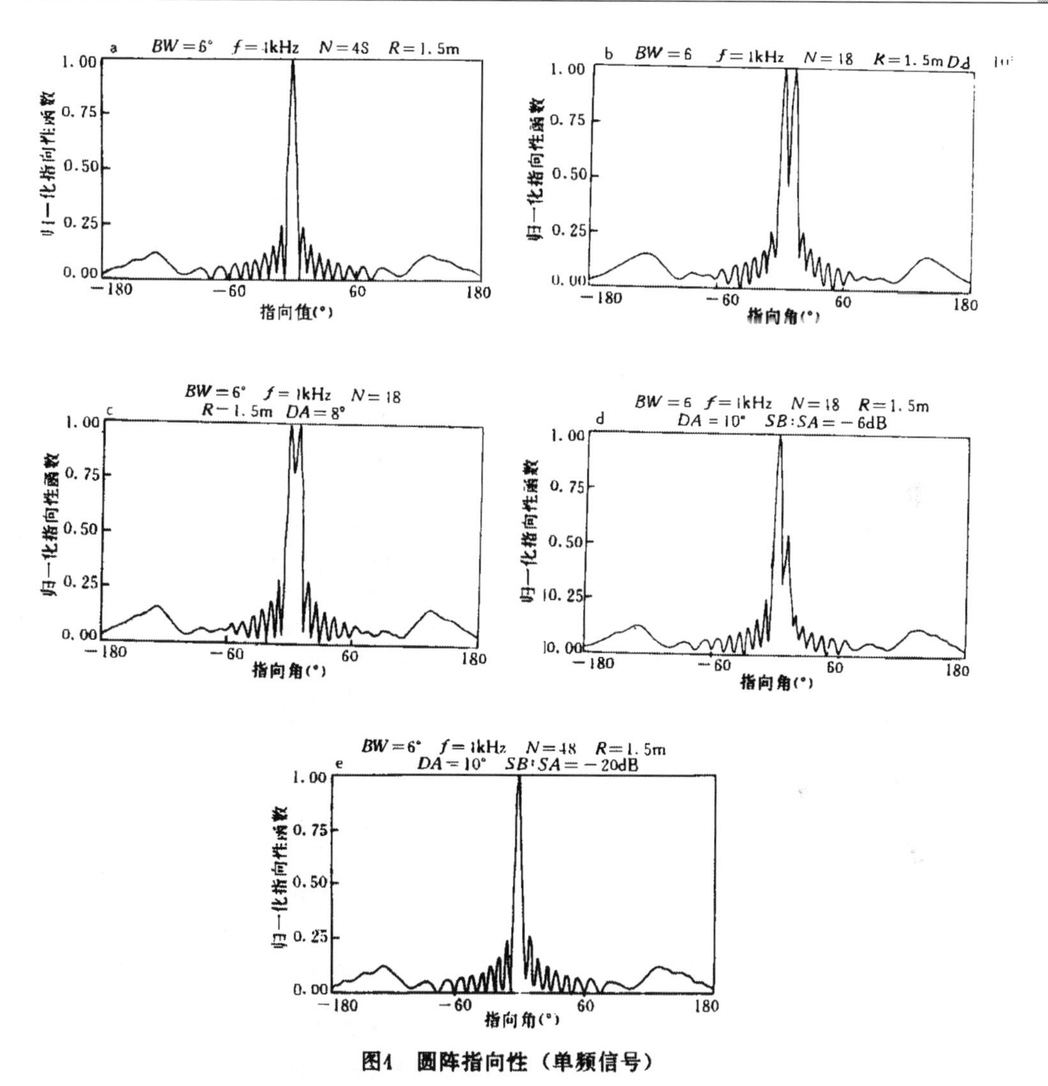

图4 圆阵指向性(单频信号)

表1 精确定向(°)结果

	DOA=15°		(SNR)$_{in}$=-5dB	
15.034 5	15.029 8	14.786 8	14.765 4	14.899 0
15.810 2	14.342 1	15.012 3	14.812 2	15.033 0

对宽带信号，情况完全类似．图5a 是1.25～6.30kHz 的宽带信号的指向性，波束宽度为 $BW=15.5°$．实际可分辨的 $DA=20°=1.29BW$．

实际的计算证实了理论推导，这些结果对声呐设计具有重要的指导意义．

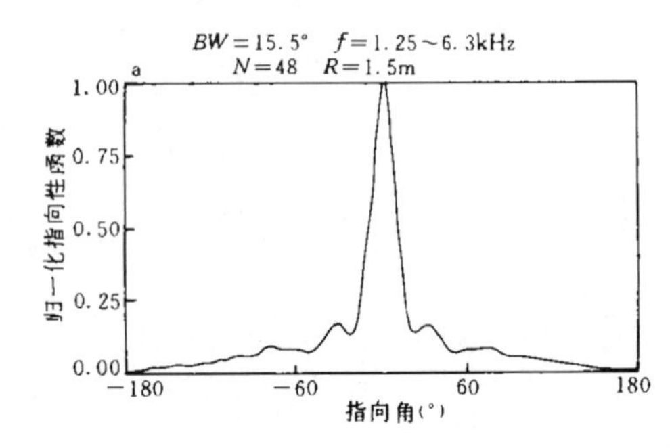

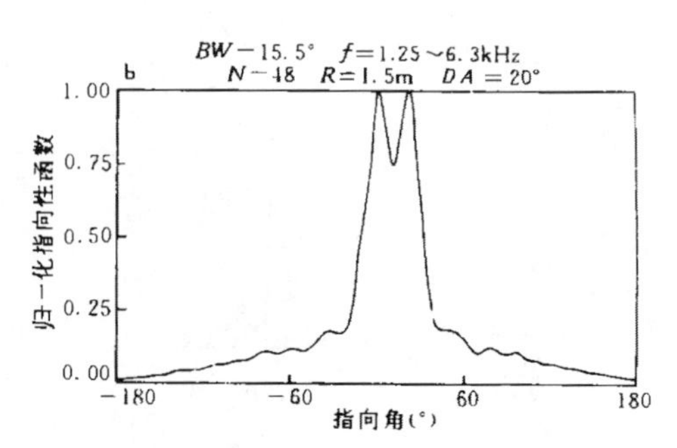

图5　圆阵指向性（宽带信号）

参考文献

1 Horton J W. Fundamentals of Sonar. V. S. Naval Inst., California, USA, 1959

2 Urick R J. Principles of Underwater Sound. 3rd ed. McGraw-Hill, New York, 1983

3 Burdic W S. Underwater Acoustic System Analysis. Prentice-Hall, New York, USA, 1984

4 Albers V M. Underwater Acoustics Handbook — I, II. The Penstate Univ. Press, Pennsylvania, USA, 1965

5 Horton C W. Signal Processing of Underwater Acoustic Waves. Gov. Printing Office, USA, 1969

6 Van Vee B D, K M Buckley. Beam-forming: a versatile approach to spatial filtering. IEEE ASSP Magazine, 1988, **5** (2): 4～24

7 Reddy V U *et al*. Performance analysis of optimum beam-former in the presence of correlated sources and its behavior under spatial smoothing. IEEE Trans. ASSP, 1987, **35**, 927～936

8 Carter C G, R E Robinson. Ocean effects on time delay estimation requiring adaption. IEEE J. Oceanic Engr., 1993, **18**, 367～378

9 Hahn W R. Optimum signal processing for passive sonar range and bearing estimation. J. Acoust. Soc. Am., 1975, **58**, 201～207

10 Kumaresan R, D Tufts. Estimating the angles of arriva of multiple plane waves. IEEE Trans. AES, 1983, **19**(1): 134～139

11 Weinstein E. Optimal source localization and tracking. IEEE Trans. ASSP, 1982, **30**(1): 69～76

12 Raddi S S. Multiple source location—a digital approach. IEEE Trans. AES, 1979, **15**, 95～105

13 Hinich M J, M C Bloom. Statistical approach to passive target tracking. J. Acoust. Soc. Am., 1981, **69**, 738～743

14 李启虎. 声呐信号处理引论. 北京：海洋出版社，1985

基于 ADSP-21060 芯片的多波束形成方法——时域实现与频域模拟[①]

姜 维 孙长瑜 李启虎
(中科院声学所信号处理研究室 北京 100080)
1998 年 9 月 10 日收到

摘要 众所周知，利用阵列处理技术实现多波束形成是大多数声呐和雷达系统的关键步骤. 波束形成技术发展多年，已较成熟. 本文讨论利用被动声呐的阵列输出以检测和定位宽带声源的多波束形成方法. 所用芯片为 AD 公司推出的最新 DSP 芯片 ADSP-21060 SHARC . 对于时域偏重于讨论技术实现，对于频域则讨论实验模拟，并根据芯片特点对两种方法进行了比较.
关键词 多波束形成， ADSP-21060 SHARC, 频域处理，水听器基阵

Multi-beamforming methods in time-domain and frequency-domain based on ADSP-21060 SHARC

Jiang Wei Sun Changyu Li Qihu
(Signal Processing Lab, IAAS, Beijing 100080)

Abstract It's well known that an important step in most Sonar and Radar systems is to implement multi-beamforming with Array-Processing Technology. Beamforming methods have been developed for a long time. This paper presents some multi-beamforming methods with passive Sonar's array output, which can be used to detect and locate wide-band underwater source. In implementation, we use Analog Device Company's newest DSP chip-ADSP-21060 SHARC. We discuss implementation in Time-domain but simulation in Frequency-domain. The comparison of these two methods is presented also in this paper.
Key words Multi-Beamforming, ADSP-21060 SHARC, Frequency-domain processing, Hydrophone array

1 引言

波束形成系统是一种与水听器基阵相连的处理器，可以提供多种形式的空间滤波. 通过波束形成，既可以利用信号的相关特性对其入射方向进行估计，又可取得空间增益，有效地减弱噪声. 迄今为止，可以采用多种方法设计数字多波束形成系统. 采用何种系统显然要依赖于硬件情况以及对该系统的指标要求. 时域的波束形成采用延时相加的办法，很适合于硬

① 应用声学, 1999, 18(6): 6-9.

件电路的实现，因此成为传统方法。由于 DSP 处理器的发展，频域波束形成得以实现。此方法较好地利用了 DSP 处理器的运算能力，因此当前得到广泛采用。

AD 公司的 ADSP-21060 芯片是 ADSP-1000 系列中的一种，有强大的指令集，高精度的浮点运算以及高速运算能力，适于用来设计多波束形成系统。

2 原理与算法

以下我们具体讨论声呐多波束形成的原理和算法，并介绍有关的 ADSP-21060 芯片的特点。首先假定采用 N 个通道的等间距线阵，用 d 表示间距，θ 表示信号入射角。

如果采用时域多波束形成系统，通常要求完成以下操作。N 个通道数据由水听器基阵采集送入处理单元，由于通常只对某些频率段感兴趣，因此第一步往往是滤波。通带的范围可以选择。下一步就是对数据进行延时相加。利用软件编程实现时，通常要求芯片中存有延时表，照该表延时相加。为了保证结果的稳定性，还要进行积累。如果假定入射信号是平面波。以平面上的某一点为参考点，设到达第 i 个基元的信号为 $s[t+\tau_i(\theta_0)]$，这里 θ_0 为信号入射角。如果将这一路信号延时 $\tau_i(\theta_0)$，那么所有 N 路信号都会变成 $s(t)$。将这 N 路信号相加之后便得到 $Ns(t)$，再平方、积分便得到 $N^2\sigma_s^2$(这里 σ_s^2 为信号功率)。如果信号入射方向改变为 θ，那么第 i 路信号经延时 $\tau_i(\theta_0)$ 之后就变成 $s[t+\tau_i(\theta)-\tau_i(\theta_0)]$，系统的输出可表示为 [1]：

$$D(\alpha)=E\left(\sum_{i=1}^{N}s[t+\tau_i(\alpha)-\tau_i(\theta)]\right)^2 \quad (1)$$

时域波束形成系统的框图如下：

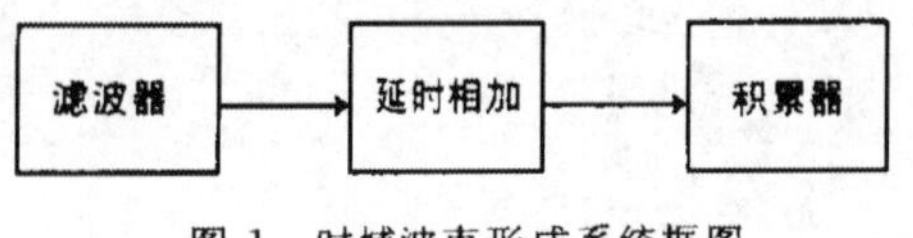

图 1 时域波束形成系统框图

下面介绍频域多波束形成的原理，它涉及宽带波束形成。从概念上说，由于忽略了时域变量，窄带波束的形成较宽带的简单。这一事实使得人们利用窄带分解结构来实现宽带波束形成。频域多波束形成系统的发展与 FFT 算法密切相关。窄带分解通常是对每个水听器通道做 DFT。它的基本原理是将时域上的加权求和转化为频域上的相乘。这种方法每次处理一批样本，称为成批处理。处理过程的第一步是对长度为 M 的各数据段做 DFT 运算，分别得到 N 个通道上的频域数据。第二步将相同频率的 N 个通道的数据向量点乘以波束定向的方向向量，并配以适当加权。第三步综合所得的不同频率的结果并做 IDFT。最后结果是对应于不同波束的宽带时间序列。与时域的波束形成系统相比，频域的波束形成可以省去滤波运算，但仍需要一定数量的积累。如果用 $X_n(k)$ 表示频域向量，不同频带的波束可用下式表示：

$$B(k,\theta)=\sum_{n=0}^{N-1}w_nX_n(k)\exp[-j2\pi nf_k\tau(\theta)] \quad (2)$$

其中为加权系数，为相应的频率。

频域波束形成系统框图如下：

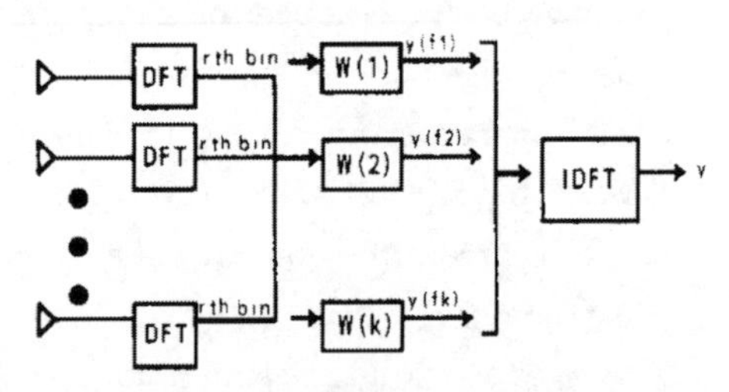

图 2 频域波束形成系统框图

下面介绍 ADSP-21060 的特点：

(1) 计算核心特征。ADSP-21060 的指令执行时间为 25ns，与 TI 公司的 TMS320C40 相同，但它可达 120 MFLOPS，而 C40 只达到 80 MFLOPS。ADSP-21060 具有 32 个数据寄存器，支持 32 个循环堆栈。

(2)I/O 特性。ADSP-21060 具有 10 个 DMA 通道，6 个并行口，2 个串行口，具有

很强的通讯能力.

(3) 存储器. ADSP-21060 具有 128K 32 Bit Words 的片上 RAM .

(4) 该芯片完成 1024 点的复数 FFT 的时间是 0.46 ms.

3 具体实现中的问题

对于多波束形成系统来说, 采用 ADSP-21060 芯片有以下优点:

(1) 拥有 4M 内存, 采样数据可直接存入其中, 提高读写速度.

(2)DMA 控制器独立于中央处理器, 可在传送数据的同时实现波束形成的运算

(3) 拥有强大的并行功能, 可在一个周期内完成取数相加相乘运算.

由于以上优点, 可以在很大程度上提高速度, 减少芯片数目.

在用 ADSP-21060 实现时域和模拟频域波束形成的过程中, 要考虑很多具体问题. 其中很重要的一点是所采用的算法是否能充分利用芯片的特性. 下面分时域和频域两方面讨论.

3.1 时域多波束形成的实现

时域多波束形成系统的滤波部分可以使用 FIR 滤波器, 充分利用芯片的并行能力. 但其关键部位, 即利用延迟表取数据做累加形成波束, 传统算法并不能充分利用芯片的特性.

主要原因如下:

(1) 该芯片对于 DAG 寄存器传输有一限制, 即当 DAG 赋值语句后紧跟一用到同一 DAG 寄存器的指令时, ADSP-21060 要在两个指令中间插入一 NOP 指令. 对于时域波束形成而言, 恰要以延迟节数赋值给 DAG 寄存器, 以此作为偏移量取数. 因此每个取数过程都要加一个 NOP 指令.

(2) 关于寻址方式, ADSP-21060 只提供两种寻址方式: pre-modify 和 post- modify. 相对 TMS320C3x 而言, 灵活性较弱. 具体比较如下:

表 1 ADSP-21060 和 TMS320C3x 寻址方式的比较

	语法	操作	描述
ADSP21060 (以 DM 为例)	ureg=|DM(Mb, Ia)|	addr=|Ia + Mb|	pre-modify I register
	ureg=|DM(Ia, Mb)|	addr=|Ia| Ia=|Ia + Mb|	post-modify I register
TMS320C3x	+ARn(disp)	addr=ARn+disp	with predisplacement add
	++ARn(disp)	addr=ARn+disp ARn=ARn+disp	with predisplacement add and modify
	ARn++(disp)	addr=ARn ARn=ARn+disp	with postdisplacement add and modify
	ARn++(disp)%	addr=ARn ARn=circ(ARn+disp)	with postdisplacement add and circular modify

可以看出 ADSP-21060 缺少 ++ARn(disp) 方式 (with predisplacement add and modify).

- DM 和 PM 并行取数指令不支持 pre-modify 方式, 导致如要利用偏移量取数据, 就不能用并行取数.
- 并行取数指令的数据寄存器必须是 R 寄存器, 不能直接赋给 M 寄存器.

鉴于以上限制, 取延迟节数, 修改 M 寄存器, 以及取数据, 累加不可能在一个指令周期内完成. 以单波束编程, 大约需要 5 个周期完成一次累加.

根据时域波束形成的特点以及芯片特性, 可做如下改进:

- 利用芯片内存大的特性, 可将数据及延迟表存于 PM 和 DM 各一份, 就可避免以上的第四点限制, 实现并行取数指令.
- 利用交叉取数的方法, 避免 DAG 赋值语句后紧跟用同一 DAG 取数的指令, 从而

排除 NOP 的产生.

- 针对延迟表的对称性, 在形成一波束的同时可形成其对称波束, 减少取延迟表数据的次数.
- 对于同一通道, 每次取多个数据, 以减少修改 B 寄存器的次数.

通过这些改进, 可以较充分地利用 ADSP-21060 的大内存和并行取数的特点, 达到实时要求, 效率提高一倍.

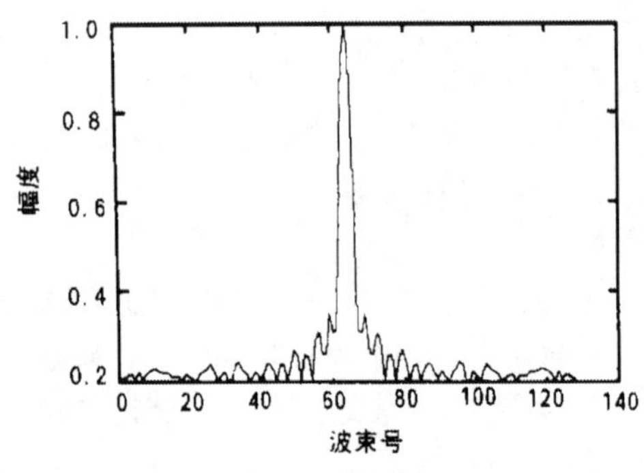

图 3　时域结果

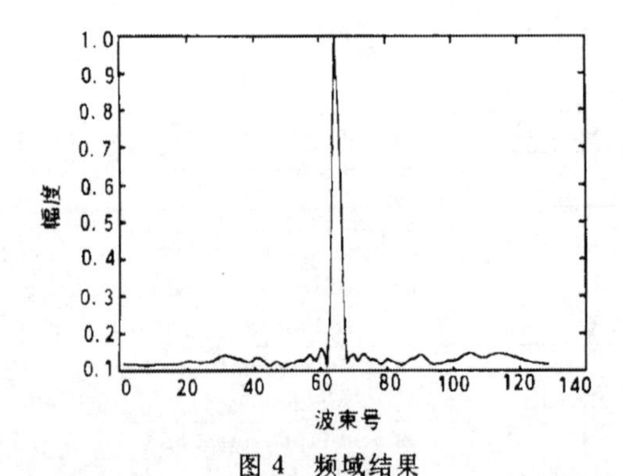

图 4　频域结果

3.2　频域多波束形成的模拟

频域多波束形成大量用到 FFT 和 IFFT, 可以充分发挥 ADSP-21060 的并行特性。前面提到, 该芯片完成 1024 点的复数 FFT 的时间是 0.46ms, 而 TMS320C40-80 需要 0.97ms 才能完成相应处理。因此利用 ADSP-21060 完成频域多波束形成较合适。该算法前面已经介绍, 不再重复。

这里需要强调两点:

(1) 频域算法较之时域灵活, 可根据所要求频带进行处理, 避免了滤波运算。

(2) 频域算法可以避免对输入采样率的严格要求. 时域算法通常要求较高的采样率, 不但对采样设备要求较高, 而且给滤波带来困难。频域算法可用较低的输入采样率。

由于频域波束形成的子算法 (如 FFT) 均有支持程序, 因此不再讨论。这里考虑等式 (2) 的计算方法。有很多算法可以选择, 可分为精确算法与近似算法两种, 如果时间允许, 可采用前者。精确算法中又有直接计算法, Horner 算法, Goertzel 算法等。对于波束等间距的情况, Goertzel 算法可减少一半计算量 [3] .

4　实验模拟结果

通过实验模拟, 分别得到下列图象。图 3 为时域多波束形成模拟结果, 图 4 为频域多波束形成模拟结果。我们生成 129 波束的数据, 图中的横轴就代表波束号。两个实验的模拟数据相同, 均采用单频正弦波, 外加白噪声, 信噪比为 -10db, 信号垂直于基阵入射, 时域算法输入采样率为 20KHz, 频域算法中为 5KHz, 采用 1024 点 FFT.SHARC Simulator 被用来完成模拟实验.

5　结论

由以上讨论可以看出, ADSP-21060 是一种功能强大的数字处理芯片, 用它完成多波束形成可取得较快的速度。虽然对时域多波束形成有些局限, 但经过改进可在规定芯片数目内达到实时要求。对时域多波束形成的加权以及实验误差的影响等问题, 有待进一步讨论.

参　考　文　献

1　李启虎. 声纳信号处理引论. 海洋出版社, 1985

2　Application Note,Considerations for Selecting a DSP Processor, Anolog Device Company

3　Maranda B. *J.Acoust.Soc.Am.*, 1989, **86**(5): 1813-1819.

基于 ADSP-21060 SHARC 高速通用声呐信号处理系统①

孙长瑜　李启虎　李伟昌　潘学宝　秦英达　姜　维

(中国科学院声学所研究所　北京　100080)

1998 年 8 月 13 日收到

摘要　中国科学院声学研究所信号处理研究室长期从事水声信号处理、水声工程和声呐系统的研究和实验工作，研制出了第二代到第四代声呐系统，积累了丰富的经验．本文介绍一种利用 ADSP-21060 SHARC 芯片研发而成的通用信号处理板．它具有体积小、功能完备、运算速度快、并行处理能力强、可靠性高等特点．可任意集成不同的信号处理系统，适用于各种信号处理领域．

PACS 数：　43.60

The research and development of high speed general sonar signal processing system by ADSP-21060 SHARC

SUN Changyu　LI Qihu　LI Weichang　PAN Xuebao　QIN Yingda　JIANG Wei

(*Institute of Acoustics, The Chinese Academy of Sciences* Beijing 100080)

Received Aug. 13, 1998

Abstract　The General Sonar System Lab of IAAS used to underwater acoustics signal processing, underwater acoustics engineering and sonar system's research and experiment for a long term. The second generation to the fourth generation sonar were all borned here. Accordingly, we accumulate abundant experience. This paper presents a general signal processing board which was developed with ADSP-21060 SHARC. It's advantage include small volume, complete function, high operation speed, powerful parallel processing ability and better reliability. It can also be intergrated in any kind of signal processing system and applied to all kinds of signal processing field.

引言

随着现代水声信号处理技术的发展和广泛应用，对信号处理系统的要求也愈来愈高．依靠通用计算机芯片已很难完成或实现高速、高可靠性、大数据量的实时信号处理．随着 VLSI 技术的发展，各种专用 DSP 产品相继问世．并且朝着处理速度更快、并行处理能力更强、处理功能更加完备的方向发展．这就为基于先进的 DSP 芯片开发通用信号处理系统创造了有利条件．

声学所长期从事水声信号处理技术和声呐设备的研究，研制出了第二代到第四代声呐设备，现正在研发第五代声呐设备．由于水声信道的多途，时变 - 空变特性，水声目标背景均相当复杂，尤其在浅海中更加突出．这代声呐其主要特点具有智能功能，能自动与环境相适配．它的运算速度将大大提高，它将是有智能知识库的智能软件系统，因而具有自学习能力，也将具有智能的接口系统来完成人 - 机对话，还具有问题求解与规则推理系统，使整个系统工作在最佳状态下，以适应使用环境．在系统结构上，系统将具有很强的并行处理能力，有非常高的运算速度，以实时实现对多个波束的自适应波

① 声学学报, 2000, 25(2): 150-154.

束形成、匹配滤波、动目标检测、被动定位中的高精度时延估计、谱分析、线谱检测和跟踪等复杂运算。为此，声学所信号处理研究室，为适应现代声呐系统的要求，研发出基于 ADSP-21060 SHARC 芯片的通用信号处理板，依此将非常方便地集成任意功能的声呐系统。

1 声呐技术的发展

水声信号的检测技术是利用声波在水中的传播、反射及其特性，来探测水下固定目标或运动目标。实践证明，在海水中，除声波之外，如光波和电波都衰减非常之快。因此，目前只有声波被广泛用于 (无论是军用或是民用) 水下目标探测。当今水声探测技术已和微电子技术更加紧密的结合起来，推动了水声探测技术的发展。并已成功地用于水下导航、定位、测速、反潜、对抗、水下通信、海洋资源的开发勘探等各个领域。

从声呐设备发展的角度来讲，它与微电子技术的发展息息相关。它经历了从电子管、晶体管和中小规模集成电路，以及现代商业可利用与系统集成技术，声呐设备也随之经历了第一代到第五代的发展过程。第一代声呐系统是由电子管组成的模拟信号处理系统，它一直使用到 50 年代。由于体积庞大、效率低和技术的限制，这代声呐系统未能采用信号处理新技术。第二代声呐是用晶体管设计而成，它主要应用于 60 年代。由于运算速度的限制，这代声呐主要运用了极性重合相关时间压缩技术，声呐系统仍然是体积大、效率低、功能单一和设备繁杂。在 70 年代初，出现了模拟数字混合式声呐，它采用电子计算机作为其主控系统，匹配滤波、波束形成、能量检测以及后置处理均采用了专用的数字硬件来完成。这代声呐系统可用于实现较大孔径的声呐，处理速度可达到几十兆的数据率，有较好性能。第四代声呐源于 70 年代末 80 年代初，实现了全数字化声呐。其特点是提高了硬件的标准化程度，从而大大减少了专用硬件的开销，由中央控制中心来管理各个分系统的动作。各分系统异步工作，提高了硬件的积木式组装性能，即分布式体系结构。

随着水声信道和目标背景等方面研究的深入，以及信号处理技术的进步和微电子技术的更进一步的发展。90 年代初开始研制智能化声呐系统，也就是我们在引言中谈到的第五代声呐。这代声呐的研制成功，将使水声探测技术达到更新更高的水平，无论在军用或民用方面均会得到更广泛的应用。

2 通用声呐信号处理系统的组成

前面讲到的第五代声呐，就是我们这里所要描述的通用声呐系统，一般它的架构如图 1 所示。

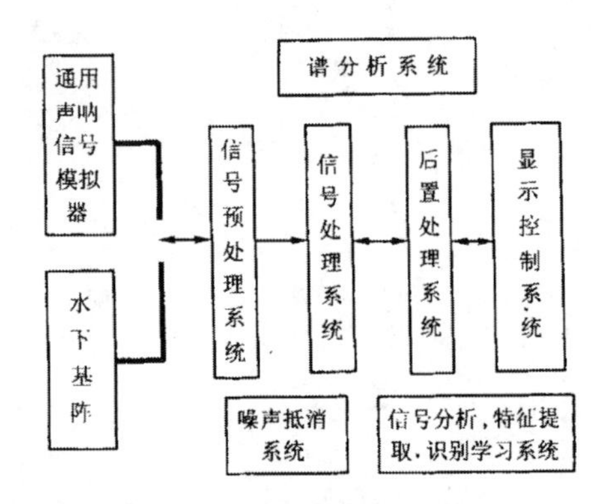

图 1 通用声呐系统的架构

2.1 通用声呐信号模拟器

所谓通用声呐信号模拟器，它能够实现对不同水文条件、海况、信噪比、方位、距离以及各种阵型的信号模拟。也就是由它来相对模拟产生水下基阵的信号，从而在实验室中即可获得相对现场的实验结果，它将对声呐系统设计的功能与验证带来极大的好处，同时可以节省大量的人力、物力和现场实验经费。另外，它还可对新型声呐的方案论证提供依据，为声呐使用部门在基地状态下随时检验声呐系统工作情况。因此，它应是各声呐系统研制单位和使用单位必备的测试设备。

2.2 信号预处理系统

声呐信号预处理系统包括前置放大器、抗混淆滤波、自动增益控制 (AGC) 和模拟数字转换器 (A/D) 。该系统接收来自水下基阵输出或通用信号模拟器的输出信号、放大、滤波和增益控制，可由显控系统控制选择，以避免信号动态过小或波形限幅，然后经对模拟信号的采样产生数字信号 (8bit 、12bit 或 16bit)，量化比特可根据要求设定，以产生合适的量化精度，并提供给后面的信号处理系统。信号的通道数由基阵的水听器个数决定，如果水听器个数为 N，则有 N 路数字信号输出。

2.3 信号处理系统

信号处理系统主要根据任务的要求来完成现代信号处理新技术的一些算法。比如 FIR 滤波、自适应滤波、升样、降样、自适应波束形成、自适应噪声抵消、自适应线谱增强、 FFT 、 ZOOMFFT 、信号分析学习、特征提取、目标识别以及均衡滤波、积累等，最终给出目标的方位、距离、航速、航迹、频谱特性、类型等信息，并实现实时跟踪目标。

系统硬件平台采用基于 ADSP-21060 SHARC 芯片设计而成的通用信号处理板，每秒可完成近两千万条指令运算。由此板可随意组成所需要的高速运算、控制或管理系统，只要嵌入不同功能的软件包即可实现，有关此板的情况将在下节中介绍。

2.4 显控系统

声呐的显控台是声呐的重要组成部分，是声呐系统中的人机接口，显控台的性能好坏将直接影响到声呐的总体性能。总的说来，显控台的显示部分要清晰明了，并显示尽可能多的信息，同时报警部分和重要参数要在容易引起声呐员注意的视觉角度，控制部分的人机接口要简单，便于迅速操作，并具有一定的智能性防误操作。因此，通用显控台的设计原则是：显示画面清晰，显示内容丰富，人机接口友好，控制过程简单，系统的宽容性好。另外，作为现代声呐显控台，系统应是开放的，即必须能与其他设备方便地交换信息，同时要有信息的存储能力。

2.4.1 硬件配置

目前通过计算机技术的飞速发展和不断完善，其应用渗透到各个领域。声呐也不例外，为节省大量的研制经费和大量的人力物力开支，各国都提出商业可利用技术 (COTS)，而且取得了很好的社会效益和经济效益。据此，我们给出通用显控系统的如下配置。

2.4.1.1 主控 CPU 为 Intel 的 80486 以上的芯片。

2.4.1.2 内存应为 32 M 以上。

2.4.1.3 显示卡带有 4 M 以上 Video DRAM，并且有图形加速功能 (2D 图形加速卡) 。

2.4.1.4 建议带有一块 PCI 接口的网卡 (可采用以太网卡，帧类型为 802.3 ，能用同轴电缆构成总线局域网) 。

2.4.1.5 显示器应采用 15 英寸以上超高解析度数控式平面直角显示器。

2.4.1.6 4 G 以上硬盘。

2.4.1.7 跟踪球或鼠标；标准键盘或定制键盘。

2.4.1.8 操作系统采用 Windows(或 NT) 中文版，版本的高低可由设计要求选择。

2.4.1.9 显控系统与信号处理系统可采用标准总线或自定义总线连接。

2.4.2 显示内容

2.4.2.1 目标方位幅度显示和方位历程显示。

2.4.2.2 LOFAR 图显示。

2.4.2.3 DEMON 显示。

2.4.2.4 跟踪历程显示。

2.4.2.5 鱼雷声光报警。

2.4.2.6 侦察波形显示。

2.4.2.7 字符显示等。

以上显示内容可采取功能切换方式显示，根据需要可增减显示内容。

3 基于 ADSP-21060 SHARC 的通用信号处理板的研发

3.1 ADSP-21060 SHARC 芯片的主要特点

四条独立的总线结构，同时可完成双数据存取和指令存取，而不干涉 I/O 操作。

40MIPS 、 25 ns 指令速率、单周期指令。

最大 120MFLOPS ，最低 80MFLOPS 带模和位翻转寻址双数据地址产生器。

零开销程序队列建立单周期循环。

IEEE 1149.1 标准 JIAG 测试访问在片仿真口。

240 引脚散热增强 PQFP 封装。

32 bit 单精度和 40 bit 扩展精度的 IEEE 浮点数据格式或 32 bit 定点数据格式。

并行运算。

双存储器读 / 写和取指，并行单周期乘法和 ALU 运算。

带加减乘法，加速 FFT 蝶形运算。

4 Mbit 双端口片上 SRAM 可由中心处理器和 DMA 独立访问。

片外存储器接口。

可寻址 4Gigawords 。

可编程产生等待状态，支持页方式 DRAM 。

10 个 DMA 通道可在 ADSP-21060 内部存储器、外部存储器、外设、主处理器、串行口和 LINK 口之间传输数据。

6 个 LINK 口点到点连接可形成多处理器阵列。

并行总线和 LINK 口上的数据传输率均为 240 Mbytes/s 。

2 个带有硬件压缩 40 Mbit/s 的同步串行口，具有独立发送和接收功能。

其系统结构框图如图 2 所示。

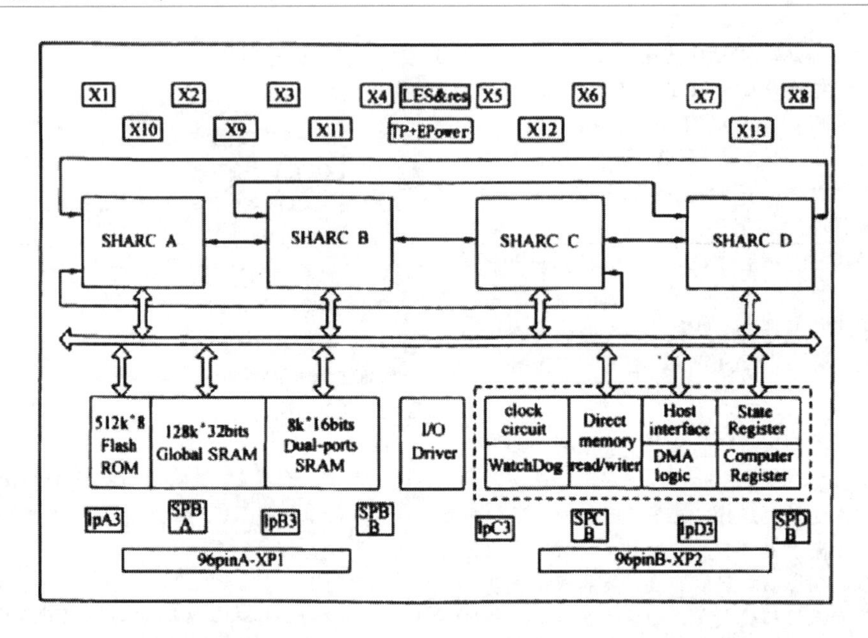

图 2　系统结构框图

3.2 ADSP-21060 通用信号处理板

ADSP-21060 通用信号处理板的主要资源配置如下.

3.2.1 四片 ADSP-21060 SHARC 芯片的并行连接和点到点连接，可以访问板内板外的其他资源.

3.2.2 512 K * 8 bits 的 FlashROM，可以存放各种软件包、运算系数和其他数据表等.

3.2.3 128 K * 32bits 全局静态随机存取存储器 (RAM) 用来存放较大的数据块和全局共享数据.

3.2.4 8 K * 16Bits 双端口静态随机存取存储器 (DP-SRAM) 用来存放较小的输入输出数据块以及运算结果.

3.2.5 超大规模的可编程逻辑阵列 (EPLD), 用来编程产生各种总线驱动逻辑、控制逻辑、译码逻辑、自检逻辑以及接口逻辑等.

3.2.6 每块 SHARC 芯片的 6 个 Link 口 3 个对内 3 个对外由印制板前面板引出，以实现并行处理、阵列处理、流水线处理、分布式处理等各种体系结构.

3.2.7 DMA 总线和 2 个串行口，用来实现数据块传输和串行数据传输的需要.

3.2.8 可与计算机连接的仿真口和仿真器.

3.2.9 印制板采用 6 U(U=44.45 mm) 标准板设计，双 96pin VME 总线标准插头座.

3.2.10 软件资源配有系统开发的所有软件包，可实现 C 语言和汇编语言混合编程，为应用软件的开发带来极大的方便.

通用信号处理板如图 3 所示.

图 3　通用信号处理板

4　结论

中国科学院声学研究所信号处理研究室，ADSP-21060 SHARC 通用信号处理板的研发成功，将使未来声呐系统大大登上一个新的台阶，只需几块板就可组成现代大型声呐多功能信号处理系统，具有很高的系统可靠性 (满足国军标)，适合于各种环境的使用. 它不仅适用于军事领域，更应适用于民用的各个领域，根据各种不同需求可方便地组成大小不同的系统，因而它将有非常广泛的应用前景.

参 考 文 献

1 孙长瑜，李启虎等．“通用声呐信号处理系统”．声学所声呐研究报告，1996

2 潘学宝，孙长瑜，秦英达．通用声呐显控台．声学所声呐研究报告，1997

3 ANALOG DEVICES. ADSP-2106X SHARC User's Manual. 1995

4 ANALOG DEVICES. ADSP-2106X SHARC DSP Microcomputer Family. C2128a-10-11/96

5 ANALOG DEVICES. ADSP-21000 Family C Tools Manual. 1995

6 ANALOG DEVICES. ADSP-21000 Family Assembler Tools & Simulator Manual. 1995

独立观测资料的最佳线性数据融合[①]

李启虎

(中国科学院声学研究所　北京　100080)

2000 年 5 月 8 日收到

摘要　数据融合是声呐信号处理中非常引人注意的一个课题，本文讨论多传感器 (或多基阵) 系统决策级的数据融合问题，给出对同一参数 N 个独立观测资料的最佳线性数据融合算法。证明了，最佳线性数据融合的误差不大于任何一个独立观测的误差，同时给出计算误差的表达式。最后讨论测量时信噪比的影响。

PACS 数：　43.30, 43.60

Optimum linear data fusion for independent observation data

Li Qihu

(*Institute of Acoustics, The Chinese Academy of Sciences* Beijing 100080)

Received May 8, 2000

Abstract　Data fusion is one of the attractive topic in sonar signal processing. Decision level data fusion of multi–sensor (multi–array) system is described in this paper. The optimum linear data fusion algorithm for N independent observations is derived. It is proved that the estimation error of data fusion is not greater than that of individual components. The expression of estimation error and weight coefficients are presented. The results of numerical calculation and some examples are illustrated. The effect of signal to noise ratio for the data fusion is described.

引言

数据融合问题是近年来声呐信号处理领域非常引人注目的问题[1−8]。Hall 给数据融合下的定义为[4]: 对于一个对象的多源信息集成，产生关于该对象的特定的和综合的统一数据。最早讨论与数据融合问题有关概念的是 R.W.Sittler, 他在 1964 年研究雷达多目标跟踪时提出了数据关联的概念[2]。随着雷达、声呐信号处理的发展，数据融合问题引起越来越多的兴趣。一般认为，数据融合问题可以分为 3 个层次，即数据级、特征级及决策级[7]。

目前有关数据融合的研究结果还是较初步的，大多集中于决策级。

本文研究声呐信号处理中一个很典型的情况，即对同一参数有多个独立观测结果的情况，这些观测结果，由于种种原因可能具有不同的误差。对这些观测结果进行数据融合，用最小二乘准则，求出一个综合结果。可以证明，这种最佳的数据融合所具有的观测误差将不大于每一分量的观测误差，这正是我们所期望的结果，我们同时还讨论了观测时信噪比的影响，给出了最优解的数学表达式和数值计算例子。

1　两个独立观察资料的情况

设 x_i 和 x_j 是对某一参数 θ 的两次独立观察结果。

$$\begin{cases} \mathrm{E}[x_i]=\theta, & \mathrm{Var}[x_i]=\sigma_i^2, \\ \mathrm{E}[x_j]=\theta, & \mathrm{Var}[x_j]=\sigma_j^2 \end{cases}$$

即对 θ 的两次测量是无偏的，但是具有不同的测量标准偏差 σ_i 和 σ_j。

数据融合的目的是寻求一个新的估计量:

$$x_{ij}=f(x_i,x_j), \tag{2}$$

使它对 θ 的估计是无偏的，并且具有比 σ_i 和 σ_j 都小的估计偏差。

① 声学声学, 2000, 25(5): 385-388.

最简单的情况是考虑 x_i 和 x_j 的线性加权，即：

$$x_{ij} = ax_i + bx_j, \tag{3}$$

求 a 和 b, 使得

$$\mathrm{Var}[x_{ij}] = \mathrm{E}[x_{ij} - \theta]^2 \qquad \text{极小}, \tag{4}$$

其中 $a + b = 1$ 作为约束条件，因为它是无偏估计的直接要求：

$$\mathrm{E}[x_{ij}] = (a+b)\theta = \theta. \tag{5}$$

根据独立性的假定，求解 (4) 式得到：

$$\mathrm{Var}[x_{ij}] = a^2\sigma_i^2 + b^2\sigma_j^2. \tag{6}$$

用 Lagarange 乘子法，对 (6) 式求条件极值可以得到：

$$a = \frac{\sigma_j^2}{\sigma_i^2 + \sigma_j^2}, \quad b = \frac{\sigma_i^2}{\sigma_i^2 + \sigma_j^2}. \tag{7}$$

由此可知最佳线生加权的数据融合是对两个独立变量的加权，加权参数的大小与分量的标准差成反比。误差越大，加权量越小。这一结果在直观上非常易于理解。

$$x_{ij} = \frac{\sigma_j^2 x_i + \sigma_i^2 x_j^2}{\sigma_i^2 + \sigma_j^2}, \tag{8}$$

直接计算可得到：

$$\mathrm{Var}[x_{ij}] = \left(\frac{1}{\sigma_i^2} + \frac{1}{\sigma_j^2}\right)^{-1}. \tag{9}$$

由此，容易证明：

$$\mathrm{Var}[x_{ij}] \le \left(\frac{1}{\sigma_i^2}\right)^{-1} = \sigma_i^2 = \mathrm{Var}[x_i], \tag{10a}$$

$$\mathrm{Var}[x_{ij}] \le \left(\frac{1}{\sigma_j^2}\right)^{-1} = \sigma_j^2 = \mathrm{Var}[x_j], \tag{10b}$$

(10) 式表明了一个非常重要的结果，以最小均方差为准则的最佳线性数据融合所得到的新的估计量的误差不大于每一分量原来的估计误差。这一特点，使得最佳线性加权的数据融合具有直接的应用前景。

2 推广到 N 个独立观察的情况

N 个独立观察的情况实际上是二维情况的推广。为此，我们把 (8) 式改写一下，

$$x_{ij} = \left(\frac{1}{\sigma_i^2} + \frac{1}{\sigma_j^2}\right)^{-1} \frac{1}{\sigma_i^2} x_i + \left(\frac{1}{\sigma_i^2} + \frac{1}{\sigma_j^2}\right)^{-1} \frac{1}{\sigma_j^2} x_j, \tag{11}$$

这个式子和 (8) 式完全相同，但看起来比 (8) 具有更明确的物理意义，并且在形式上更对称了。

我们立刻想到，如果对 θ 有 N 个独立观察数据 $x_1, \cdots, x_N$，并且都是无偏的，利用它们得到新的最佳线性数据融合 x_{1-N} 应当是：

$$x_{1-N} = \left(\sum_{k=1}^{N} \frac{1}{\mathrm{Var}[x_k]}\right)^{-1} \cdot \sum_{k=1}^{N} \frac{1}{\mathrm{Var}[x_k]} \cdot x_k, \tag{12}$$

新观察数据 x_{1-N} 的误差应当与 (9) 式相对应，即：

$$\mathrm{Var}[x_{1-N}] = \left(\sum_{k=1}^{N} \frac{1}{\mathrm{Var}[x_k]}\right)^{-1}, \tag{13}$$

(12) 和 (13) 式只是一种推测。下面给出正式的证明。

记 $\boldsymbol{x}^{\mathrm{T}} = (x_1, \cdots, x_N)$ 为观测向量。

$\boldsymbol{w}^{\mathrm{T}} = (w_1, \cdots, w_N)$ 为加权向量。

数据融合结果为：

$$y = \boldsymbol{w}^{\mathrm{T}}\boldsymbol{x}. \tag{14}$$

寻求 $\boldsymbol{w}$, 使

$$I = E[\boldsymbol{w}^{\mathrm{T}}(x - Ex)]^2 \tag{15}$$

在条件：

$$\boldsymbol{w}^{\mathrm{T}}\boldsymbol{u} = 1 \tag{16}$$

之下极小，其中 $\boldsymbol{u}$ 表示所有元素为 1 的 N 维向量。

令：

$$R_{xx} = \mathrm{E}[(x - \mathrm{E}x)(x - \mathrm{E}x)^{\mathrm{T}}] \tag{17}$$

为 x 的自相关矩阵。

用 Lagarange 乘子法，求：

$$z = I + \lambda \boldsymbol{w}^{\mathrm{T}}\boldsymbol{u} = \boldsymbol{w}^{\mathrm{T}} R_{xx} w + \lambda \boldsymbol{w}^{\mathrm{T}}\boldsymbol{u} \tag{18}$$

之极小值，可以证明：

$$w_{\mathrm{opt}} = \frac{R_{xx}^{-1}\boldsymbol{u}}{\boldsymbol{u}^{\mathrm{T}} R_{xx}^{-1}\boldsymbol{u}}, \tag{19}$$

$$I_{\min} = w_{\mathrm{opt}}^{T} R_{xx} w_{\mathrm{opt}} = \frac{1}{\boldsymbol{u}^{\mathrm{T}} R_{xx}^{-1}\boldsymbol{u}}. \tag{20}$$

直接计算可以证明，由于 x_i 和 x_j 互不相关 $(i \ne j)$，R_{xx}^{-1} 的第 (i,i) 个元素为 $1/\sigma_i^2$，而其余元素都是 0, 即 R_{xx}^{-1} 是对角矩阵，于是：

$$\boldsymbol{u}^{\mathrm{T}} R_{xx}^{-1}\boldsymbol{u} = \left(\frac{1}{\sigma_1^2} + \cdots + \frac{1}{\sigma_N^2}\right), \tag{21}$$

$$w_{\mathrm{opt}} = \frac{\left(\dfrac{1}{\sigma_1^2}, \cdots, \dfrac{1}{\sigma_N^2}\right)^{\mathrm{T}}}{\left(\dfrac{1}{\sigma_1^2} + \cdots + \dfrac{1}{\sigma_N^2}\right)}, \tag{22}$$

$$I_{\min}=\left(\frac{1}{\sigma_1^2}+\cdots+\frac{1}{\sigma_N^2}\right)^{-1}, \tag{23}$$

这便是 (12) 和 (13) 式。

我们很容易证明：

$$I_{\min}\le\sigma_i^2,\quad i=1,\cdots,N. \tag{24}$$

3 加权系数与信噪比的关系

在前面讨论数据融合时，都是以测量误差作为参数的，有时候，引入信噪比的概念，更方便一些。我们以二维情况为例来说明这一事实，N 维的情况完全是类似的。

设：

$$x_{ij}=w_1x_i+w_2x_j.$$

信号参量为 θ, x_i 对 θ 的测量误差为 σ_i, 当信号功率固定时 σ_i^2 与 $(\mathrm{SNR})_i$ 成反比，同样的结论对 σ_j^2 也成立，于是：

$$x_{ij}=\left(\frac{1}{\sigma_i^2}+\frac{1}{\sigma_j^2}\right)^{-1}\cdot\frac{1}{\sigma_i^2}x_i+\left(\frac{1}{\sigma_i^2}+\frac{1}{\sigma_j^2}\right)^{-1}\cdot\frac{1}{\sigma_j^2}x_j=$$

$$\frac{(\mathrm{SNR})_i}{(\mathrm{SNR})_i+(\mathrm{SNR})_j}\cdot x_i+\frac{(\mathrm{SNR})_j}{(\mathrm{SNR})_i+(\mathrm{SNR})_j}\cdot x_j. \tag{25}$$

这个式子清楚地表明最佳线性数据融合与各分量信噪比的关系。

4 数值例子

前面讨论的结果可以应用于各种实际场合，被测的参数可以是目标的方位，也可以是目标的航速，下面，我们给出一些具体的数值计算结果。

图 1 给出了二维数据融合的计算结果，两个独立变量分别是 X 和 Y, 它们的方差假定为 σ_x^2 和 σ_y^2, 最佳联合估计的方差就是：

$$Z_{xy}=\frac{\sigma_x^2\sigma_y^2}{\sigma_x^2+\sigma_y^2}.$$

从图中可以看出，Z_{xy} 不大于 σ_x^2 和 σ_y^2。

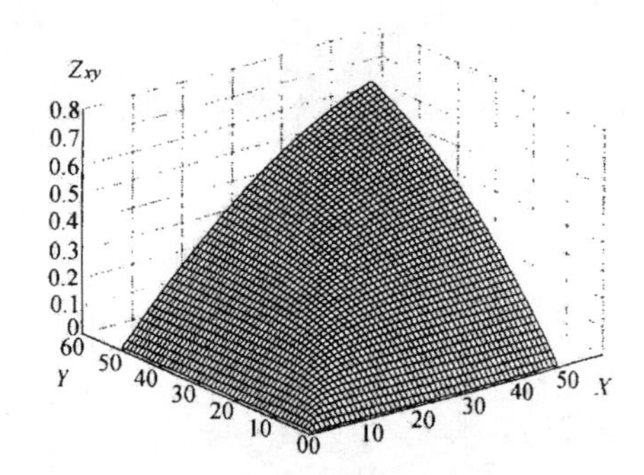

图 1 数据融合：两个独立变量的联合估计误差

$X=50\sigma_x,\ Y=50\sigma_y,\ Z_{xy}=\left(\frac{\sigma_x^2\sigma_y^2}{\sigma_x^2+\sigma_y^2}\right)^{1/2}$

图 2 给出了二维数据融合时的加权系数，我们以 $\sigma_x=0.5$ 和 $\sigma_x=0.9$ 为参数，给出 X 和 Y 的加权系数：

$$w_x=\frac{\sigma_y^2}{\sigma_x^2+\sigma_y^2},\quad w_y=\frac{\sigma_x^2}{\sigma_x^2+\sigma_y^2}.$$

以 σ_y 为变量的情况，可以看到，当 σ_y 由小到大时，W_y 从大到小。

图 3 是同一条件下的联合估计误差，可以看出，在最坏的情况下，最佳数据融合的估计误差也比分量中最小误差还要小一些。

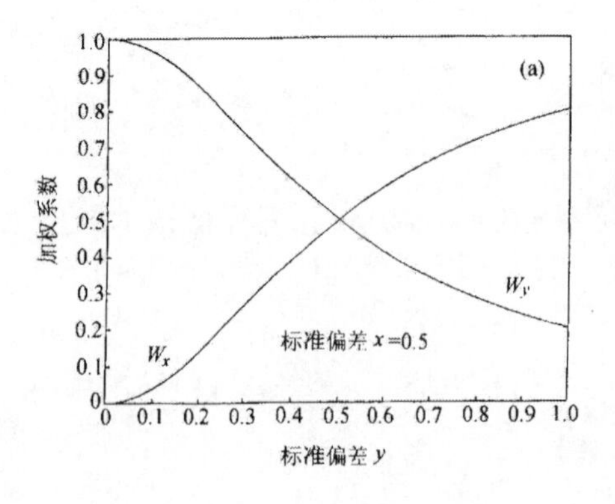

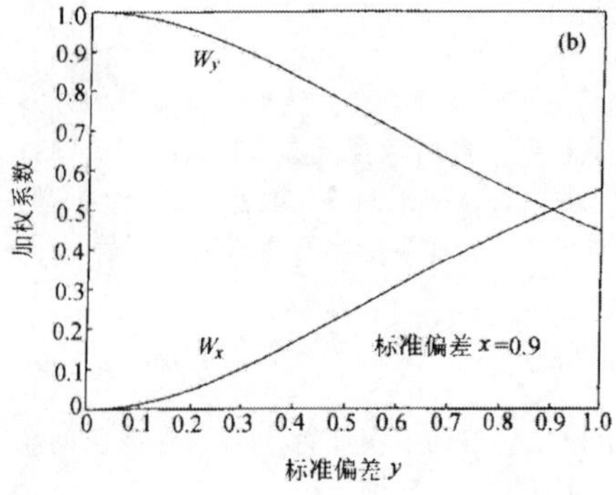

图 2 数据融合：两个独立变量的联合加全系数

$X=\sigma_x,\ Y=\sigma_y,\ w_x=\frac{\sigma_y^2}{\sigma_x^2+\sigma_y^2},\ w_y=\frac{\sigma_x^2}{\sigma_x^2+\sigma_y^2}$

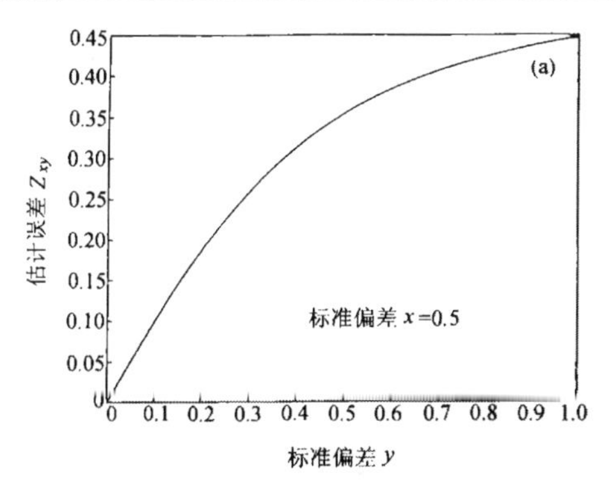

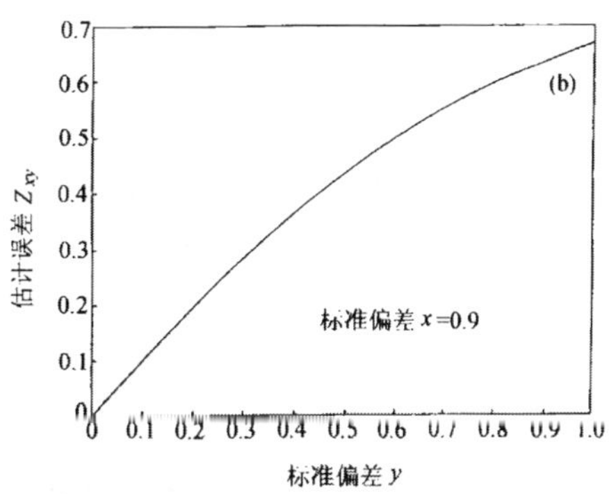

图 3 数据融合：两个独立变量的联合估计误差

$$X=\sigma_x,\ Y=\sigma_y,\ Z_{xy}=\left(\frac{\sigma_x^2\sigma_y^2}{\sigma_x^2+\sigma_y^2}\right)^{1/2}$$

图 4 是一个实际的例子，假定我们用一个线列阵和一个圆柱阵同时对目标进行定向。圆柱阵的定向误差 σ_x 与目标入射角无关，而线列阵的定向误差依赖于目标入射角 (图中 σ_y)，这时最佳联合定向误差显示出很大的优越性，它几乎与目标入射角无关。

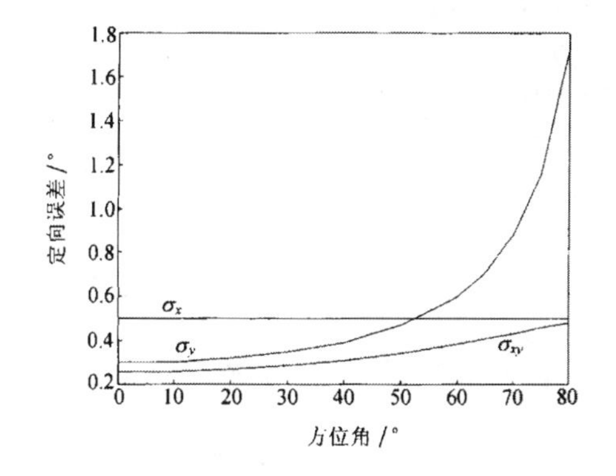

图 4 数据融合：圆柱阵和线列阵联合定向误差

参 考 文 献

1 Chen C T *et al.* The past, present, and future of underwater acoustic signal processing. *IEEE Signal Processing*, 1998; **15**(4): 21—53

2 Sittler R W. An optimal data association problem in surveillence theory. *IEEE Trns. On Military Electronics*, 1964; **8**(2): 125—139

3 Varskney P L. Distributed detection and data fusion. New York Springer, 1996

4 Hall D L, Llinas J. An introduction to multisensor data fusion. Proc. IEEE, 1997; **85**(1): 6—23

5 Dasarathy R V. Sensor fusion potential exploitation-inovarive architecture and illustrative approaches. Proc. IEEE, 1997; **85**(1): 24—38

6 Tenny R R, Samdel N K. Detection with distributed sensors. *IEEE Trans.*, 1981; **AES–17**(4): 501—510

7 Sharma R *et al.* Toward multimodal human-computer interface. *Proc. IEEE*, 1998; **86**(5): 853—869

8 Rick B, Kassan S A. Optimum distributed detection of weak signal independent sensors. *IEEE Trans.*, 1992; **IT–38**(3): 1067—1079

相关观测资料的最佳线性数据融合①

李启虎
(中国科学院声学研究所　北京　100080)

2000 年 7 月 7 日收到

摘要　数据融合是声呐信号处理中非常引人注意的一个课题. 本文讨论多传感器 (或多基阵) 系统决策级的数据融合问题, 给出对同一个参数 N 个互相相关的观测资料的最佳线性数据融合算法. 证明了, 最佳线性数据融合的误差不大于任何一个分量的观测误差, 同时给出计算误差的表达式, 最后给出了数值计算的例子.

PACS 数:　43.30, 43.60

Optimum linear data fusion for dependent observation data

LI Qihu
(*Institute of Acoustics, The Chinese Academy of Sciences* Beijing 100080)

Received Jul. 7, 2000

Abstract　Data fusion is one of the attractive topic in sonar signal processing. Decision level data fusion of Multi-sensor (multi-array) system is described in this paper. Follow the discussion in literature[1], the optimum linear data fusion algorithm for N dependent observations is derived. It is proved that the estimation error of data fusion is not greater than that of individual components. The expression of estimation error and weight coefficients are presented. The results of numerical calculation and some examples are illustrated. The effect of dependence of observation data for the final estimation error is presented.

引言

数据融合问题是近 20 年来声呐信号处理领域非常引人注目的问题, 我们已经在文献 1 中就相互独立的观测资料的最佳线性数据融合问题进行了讨论, 证明了数据融合中的一个基本定理, 即最佳线性融合所得到的观测误差不大于任何一个分量的观测误差.

本文进一步展开数据融合算法的讨论, 把观测数据的独立性的条件去掉, 即允许 N 个观测资料互相相关, 这样可以进一步扩大数据融合的适用范围.

我们要求解在最小二乘法的准则下, 最佳线性数据融合问题. 同时证明, 在 N 个观测结果互相相关的条件下, 最佳数据融合所具有的观测误差将不大于每一个分量的误差. 文中将给出一些数值计算的例子来说明数据的相关性, 实际上是可以充分利用的, 一种好的数据融合算法将远远优于单独观测的结果.

1　两个相关观测资料的情况

假定 x_i 、x_j 是对某一参数的 θ 值的观测结果, 观测是无偏的

$$\mathrm{E}[x_i] = \mathrm{E}[x_j] = \theta, \tag{1}$$

观测的方差分别为:

$$\mathrm{Var}[x_i] = \sigma_i^2, \qquad \mathrm{Var}[x_j] = \sigma_j^2. \tag{2}$$

x_i 、 x_j 的相关函数为:

$$R_{ij} = \mathrm{E}[(x_i - \theta)(x_j - \theta)], \tag{3}$$

我们要求出最佳的线性数据融合:

$$x_{ij} = ax_i + bx_j, \tag{4}$$

① 声学声学, 2001, 26(5): 385-388.

在约束条件 $a+b=1$ 的情况下，使

$$\mathrm{Var}[x_{ij}]=\mathrm{E}[x_{ij}-\theta]^2 \tag{5}$$

极小。

直接计算可知：

$$\mathrm{Var}[x_{ij}]=a^2\sigma_i^2+b^2\sigma_j^2+2abR_{ij}.$$

利用 Lagarange 乘子法，令：

$$u=a^2\sigma_i^2+b^2\sigma_j^2+2abR_{ij}+\lambda(a+b-1).$$

对 u 求偏微商，得到：

$$\frac{\partial u}{\partial a}=2a\sigma_i^2+2bR_{ij}+\lambda=0$$
$$\frac{\partial u}{\partial b}=2b\sigma_j^2+2aR_{ij}+\lambda=0$$
$$\frac{\partial u}{\partial \lambda}=a+b-1=0$$

解决 3 个联立方程，得到：

$$a=\frac{\sigma_j^2-R_{ij}}{\sigma_i^2+\sigma_j^2-2R_{ij}},\quad b=\frac{\sigma_i^2-R_{ij}}{\sigma_j^2+\sigma_j^2-2R_{ij}}, \tag{6}$$

$$x_{ij}=\frac{\sigma_j^2-R_{ij}}{\sigma_i^2+\sigma_j^2-2R_{ij}}x_i+\frac{\sigma_i^2-R_{ij}}{\sigma_i^2+\sigma_j^2-2R_{ij}}x_j. \tag{7}$$

我们还可以把 x_{ij} 写成更为对称的情况：

$$x_{ij}=\left(\frac{1}{\sigma_i^2-R_{ij}}+\frac{1}{\sigma_j^2-R_{ij}}\right)^{-1}\frac{x_i}{\sigma_i^2-R_{ij}}+\left(\frac{1}{\sigma_i^2-R_{ij}}+\frac{1}{\sigma_j^2-R_{ij}}\right)^{-1}\frac{x_j}{\sigma_j^2-R_{ij}}. \tag{8}$$

(8) 式看起来似乎比 (7) 式复杂，但实际上由于 x_{ij} 表达式中，加权系数具有对称性，其物理意义更清楚，也更便于记忆。

$$\mathrm{Var}[x_{ij}]=\frac{\sigma_i^2\sigma_j^2-R_{ij}^2}{(\sigma_i^2-R_{ij})+(\sigma_j^2-2R_{ij})}. \tag{9}$$

我们可以证明：

$$\mathrm{Var}[x_{ij}]\le\mathrm{Var}[x_i],\qquad \mathrm{Var}[x_{ij}]\le\mathrm{Var}[x_j], \tag{10}$$

即最佳线性数据融合的误差不大于每一个分量的误差。

我们以证明：

$$I_{\min}=\mathrm{Var}[x_{ij}]\le\sigma_i^2$$

为例来说明。

为此，只需证明 $\sigma_i^2-I_{\min}\ge 0$ 就行。

事实上

$$\sigma_i^2-I_{\min}=\sigma_i^2-\frac{\sigma_i^2\sigma_j^2-R_{ij}^2}{\sigma_i^2+\sigma_j^2-2R_{ij}}=\frac{(\sigma_i^2-R_{ij})^2}{\sigma_i^2+\sigma_j^2-2R_{ij}} \tag{11}$$

容易证明 $\sigma_i^2+\sigma_j^2-2R_{ij}\ge 0$。

这是因为按 Schwartz 不等式，

$$2R_{ij}=2E[(x_i-\theta)(x_j-\theta)]\le 2\sigma_i\sigma_j\le\sigma_i^2+\sigma_j^2.$$

由此可知 $\sigma_i^2-I_{\min}\ge 0$。

2 N 个相关观测资料的情况

对于相关观测资料，情况要比 [1] 中讨论的独立观测资料复杂一些，因为相关矩阵不再是对角型的，从而矩阵的求逆不具有简单的形式。

仿照 [1] 中的记号，设 N 个相关观测量为 $x_1,\cdots,x_N$。定义观测矢量：

$$\boldsymbol{x}^{\mathrm{T}}=[x_1,\cdots,x_N]$$

加权矢量：

$$\boldsymbol{w}^{\mathrm{T}}=[w_1,\cdots,w_N]$$

观测资料 $x_1,\cdots,x_L$ 的线性组合为：

$$y=\boldsymbol{w}^{\mathrm{T}}x. \tag{12}$$

求：

$$I=E[\boldsymbol{w}^{\mathrm{T}}(x-Ex)]^2 \tag{13}$$

在条件

$$w^{\mathrm{T}}\boldsymbol{u}=1 \tag{14}$$

下的最小值，其中 $\boldsymbol{u}$ 为所有元素为 1 的列向量。

容易证明：

$$I=\boldsymbol{w}^{\mathrm{T}}R_{xx}w, \tag{15}$$

其中：

$$R_{xx}=E[(x-Ex)(x-Ex)^{\mathrm{T}}] \tag{16}$$

是 x 的自相关矩阵，它的第 (i,j) 元素为：

$$R_{ij}=E[(x_i-Ex_i)(x_j-Ex_j)]. \tag{17}$$

令：

$$z=\boldsymbol{w}^{\mathrm{T}}R_{xx}w+\lambda\boldsymbol{w}^{\mathrm{T}}\boldsymbol{u}.$$

对 z 求偏微商得到：

$$\frac{\partial z}{\partial w}=2R_{xx}w+\lambda\boldsymbol{u}=0.$$

得到：

$$w_{\mathrm{opt}}=-\frac{\lambda R_{xx}^{-1}\boldsymbol{u}}{2}$$

代入 $\boldsymbol{w}^{\mathrm{T}}\boldsymbol{u}=1$，得到：

$$-\lambda=\frac{2}{\boldsymbol{u}^{\mathrm{T}}R_{xx}^{-1}\boldsymbol{u}}.$$

注意到 $(R_{xx}^{-1})^{\mathrm{T}} = R_{xx}^{-1}$ 就得到:

$$w_{\text{opt}} = \frac{R_{xx}^{-1}\boldsymbol{u}}{\boldsymbol{u}^{\mathrm{T}}(R_{xx}^{-1})^{\mathrm{T}}\boldsymbol{u}} = \frac{R_{xx}^{-1}\boldsymbol{u}}{\boldsymbol{u}^{\mathrm{T}}R_{xx}^{-1}\boldsymbol{u}}, \tag{18}$$

$$I_{\min} = I[w_{\text{opt}}] = w_{\text{opt}}^{\mathrm{T}} R_{xx} w_{\text{opt}} = \frac{1}{\boldsymbol{u}^{\mathrm{T}}R_{xx}^{-1}\boldsymbol{u}}. \tag{19}$$

(18) 式和 (19) 式给出了 N 个相关观测下，最佳数据融合的完全解，可以证明，当

$$R_{ij} = 0, \qquad i \neq j$$

时， (18) 式和 (19) 式给出了独立观测时的解。

当 $N = 2$ 时，第 1 节的结果是 (18) 式和 (19) 式的特例，事实上，当只有两个相关观测量 x_i 、 x_j 时，

$$R_{xx} = \begin{pmatrix} \sigma_i^2 & R_{ij} \\ Rij & \sigma_j^2 \end{pmatrix},$$

$$R_{xx}^{-1} = \frac{1}{\sigma_i^2\sigma_j^2 - R_{ij}^2}\begin{pmatrix} \sigma_j^2 & -R_{ij} \\ -R_{ij} & \sigma_i^2 \end{pmatrix}, \tag{20}$$

$$R_{xx}^{-1}\boldsymbol{u} = \frac{1}{\sigma_i^2\sigma_j^2 - R_{ij}^2}\begin{pmatrix} \sigma_j^2 - R_{ij} \\ \sigma_i^2 - R_{ij} \end{pmatrix},$$

这就是 (6) ~ (8) 式的结果。

最后，我们来证明，在 N 个相关观测的情况下，最佳数据融合所产生的误差不大于任何一个分量的观测误差。

直接计算 $\boldsymbol{u}^{\mathrm{T}}R_{xx}^{-1}\boldsymbol{u}$ 将有很大的困难，我们可以用另外一种办法来证明这一点，这就是基于这样一来的事实，即最加权向量 w_{opt} 是使所有加权中，误差达到极小的那一个。

为了证明，对任何 i，

$$I_{\min} \leq \mathrm{Var}[x_i] = \sigma_i^2$$

我们取一个新的加权向量 μ，它只有第 i 个分量为 1，而其余皆为零，这显然满足约束条件 (14)，所以有:

$$I_{\min} = I[w_{\text{opt}}] \leq I[\mu] = \mathrm{Var}[x_i] = \sigma_i^2.$$

由此得证。

3 数值计算的例子

以两个相关观测为例，在图 1(a) ~ 图 1(c) 中我们给出了联合估计误差的三维图形，我们发现，当相关系数取负值时，数据融合会带来较小误差的结果。

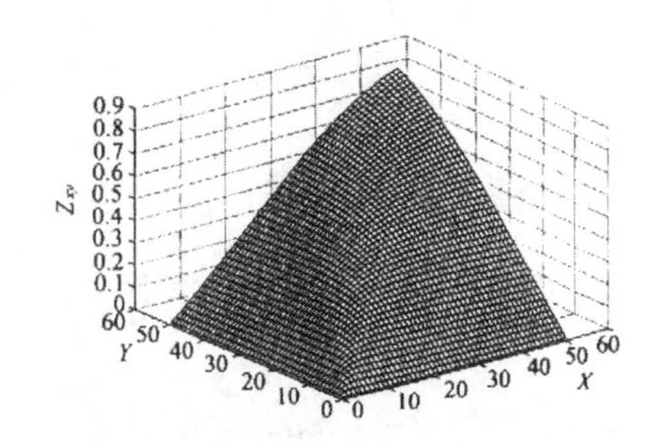

(a) $\rho_{xy} = 0.5$

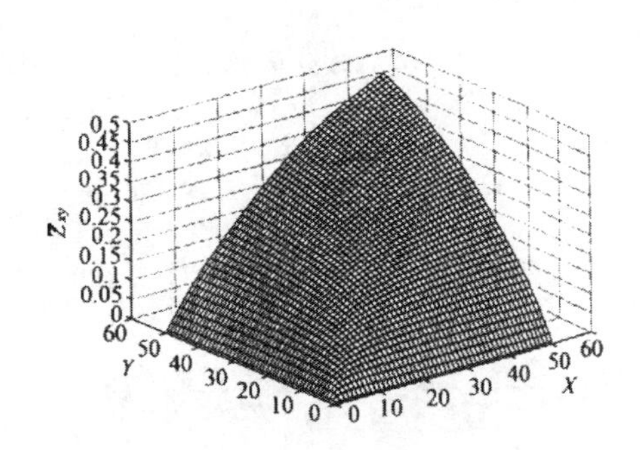

(b) $\rho_{xy} = -0.5$

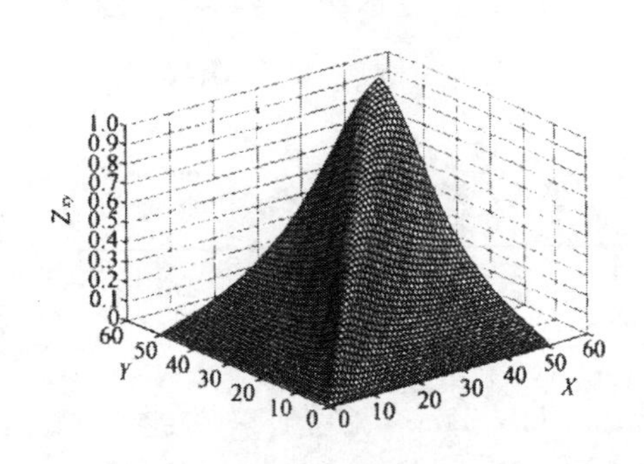

(c) $\rho_{xy} = 0.95$

图 1 数据融合: 两个相关变量的联合估计误差

$$X = 50\sigma_x, Y = 50\sigma_y, Z_{xy} = \left(\frac{\sigma_x^2\sigma_y^2 - (\sigma_x\sigma_y\rho_{xy})^2}{\sigma_x^2 + \sigma_y^2 - 2\sigma_x\sigma_y\rho_{xy}}\right)^{1/2}$$

当相关性很大时，数据融合的效果变差，这也是可以理解的。

图 2(a) ~ 图 2(c) 给出两个相关观测量的情况下，最佳数据融合的加权系数，与独立观测量的情况不同，在这里，加权系数可以取负值，有时还可以取到非常大的值，这是因为权系数的分母可以很小的缘故。

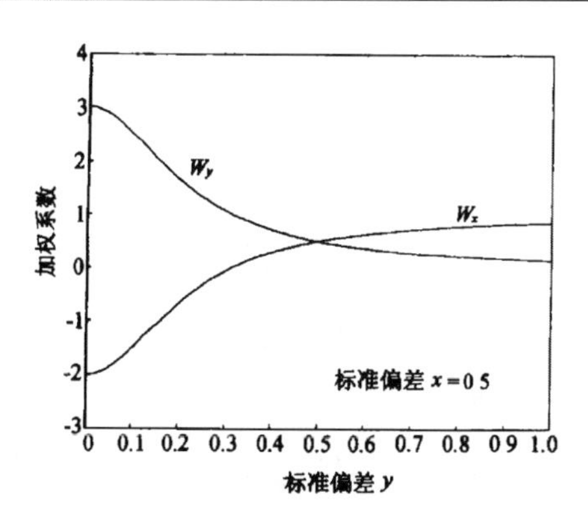

(a) rho = 0.1

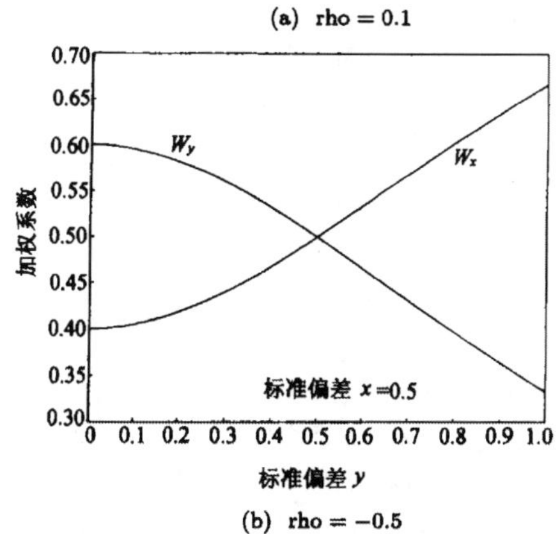

(b) rho = −0.5

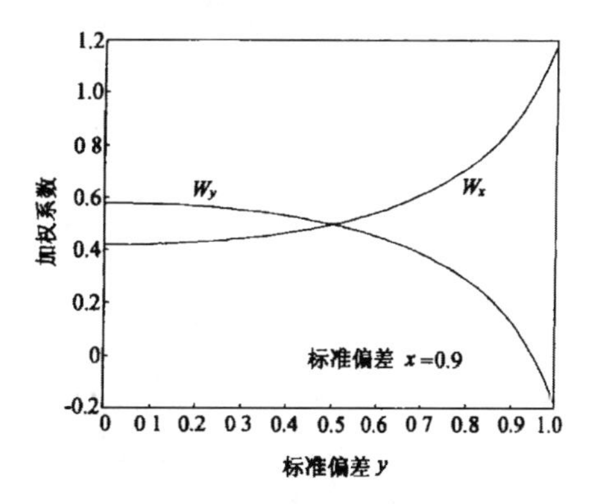

(c) rho = 0.9

图 2 数据融合：两个相关变量的联合加权系数

参 考 文 献

1 李启虎. 独立观测资料的最佳数据融合. 声学学报, 2000; 25(5): 385—388

波束域MVDR高分辨方位估计方法研究①

何心怡[1,2]　黄海宁[2]　叶青华[2]　李启虎[2]　蒋兴舟[1]

(海军工程大学[1]　武汉　430033)　(中国科学院声学所[2]　北京　100080)

摘要：针对MVDR在低信噪比时分辨性能下降问题，提出了波束域MVDR(BMVDR)方法，并提出了在实际声纳系统中用BMVDR代替现有的分裂波束精测定向法，与多波束系统结合成定向设备的新思路.BMVDR不仅能有效解决MVDR在低信噪比下分辨性能下降问题，而且不需要对声纳定向系统做大的改动，BMVDR就能与现有声纳中的多波束系统有效地结合起来，显著地提高多目标分辨性能，并且运算量有所减少.

关键词：MVDR；BMVDR；高分辨方位估计；低信噪比；定向系统

中图法分类号：TB5

0　引　　言

在声纳系统中，传统的定向设备是采用多波束系统(CBF)对目标方位进行粗测，发现目标后，采用分裂波束精测定向法对目标方位进行精测，这种方法，称作多波束定向方法.多波束定向方法算法稳健，但其方位估计精度和分辨能力十分有限，特别是在低信噪比下，存在着多个相互靠近的目标源时，常常无法有效地估计目标方位.

针对多波束定向方法的不足，在20世纪60年代末Capon提出了MVDR高分辨定向方法[1]，即最小方差信号无畸变响应法(minimum variance distortionless response).在理想情况下，MVDR的波束指向函数近似为狄利克δ函数，但在低信噪比时，MVDR分辨能力下降；且其算法是对阵元域的信号直接处理，计算量较大，限制其在声纳系统中的应用.

进入20世纪80年代，高分辨方位估计方法的研究方兴未艾，先后出现了基于信号特征分解法的MUSIC，Mini-Norm，Johnson等方法，基于旋转子空间不变法的ESPRIT，TLS-ESPRIT等方法，并针对这些方法的信噪比门限问题，提出了波束域的BMUSIC[2]，BESPRIT[3]，BRoot-MUSIC[4]等方法.但这些方法存在源个数判别、解相干、阵列校准等问题，短期内难已在实际系统中加以应用.

文中针对MVDR在低信噪比时分辨能力下降问题，提出了一种基于波束域的改进算法：波束域MVDR(BMVDR)方法，可大幅度地提高MVDR方法在低信噪比时的分辨性能，又降低了计算量，且该方法可将声纳多波束系统的输出直接用于方位估计，是一种具有实用价值的高分辨方位估计方法，有望应用于实际声纳系统.

1　MVDR高分辨方法简介

设声纳基阵由M个水听器组成，以$x_i(t)$表示第i个水听器的接收信号，w_i表示对应的加权值，则传统的波束形成器的输出可写为

$$Y_{CBF}(t)=W^HX(t) \tag{1}$$

式中：$W=[w_1,w_2,\cdots,w_M]^T$为权矢量；$X(t)$为基阵输入矢量，此时基阵的输出功率为

$$P_{CBF}(\theta)=E\{|y(t)|\}=W^HRW \tag{2}$$

式中：$R=E\{X(t)X^H(t)\}$为基阵输入协方差矩阵.

在波束形成器的输出功率中，信号源能量不仅在来波方向上有贡献，而且对波束宽度内的其它方向也有不同程度的贡献.而MVDR波束形成法就是在保持来波方向信号源能量不变的前提

① 武汉理工大学学报(交通科学与工程版), 2003, 27(2): 201-204.

下，使信号源能量对波束宽度内的其它方向最小化，实际上是一个约束最佳化问题的解

$$\min_{W} W^H R W \quad 且 \quad W^H \alpha(\theta) = 1 \tag{3}$$

式中：$\alpha(\theta)$是指定方向的方向矢量.

通过式(3)的求解可得出 MVDR 波束形成器的输出功率为

$$P_{\mathrm{MVDR}}(\theta) = \frac{1}{\alpha^H(\theta) R^{-1} \alpha(\theta)} \tag{4}$$

MVDR 波束形成器能够提供最佳的信号保护、干扰消除和噪声降低能力，理想情况下，MVDR 的波束指向性函数近似为狄利克 δ 函数；但它的估计精度受信噪比影响，低的信噪比限制了 MVDR 的高分辨性能. 举一例子说明：按半波长方式布阵的 8 元均匀线列阵(其主瓣宽度约为 13°)，有 3 个等强度的窄带声源，方位角分别为 −20°，10°和 18°，噪声为高斯白噪声，快拍数为 100. 信噪比 SN 为 6 dB 和 0 dB 时，MVDR 和 CBF 波束形成输出对比分别如图 1a)和图 1b)所示. 能看出在 6 dB 时，MVDR 方法能分辨出位于波束宽度内的两个目标；而在 0 dB 时，MVDR 方法性能大幅度下降，分辨不出位于波束宽度内的两个目标.

3 波束域的 MVDR 方法

波束域 MVDR(BMVDR)首先是将阵元(水听器)空间数据转换到波束空间来，如图 2 所示、这一步是波束域方法的关键，不但实现了降维处理，而且可以提高算法的稳健性. 其次，对转换后的数据运用 MVDR 方法进行高分辨定向.

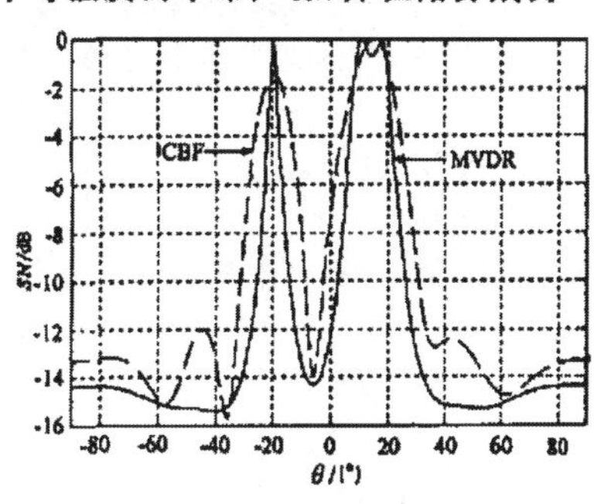

a) SN=6 dB 时

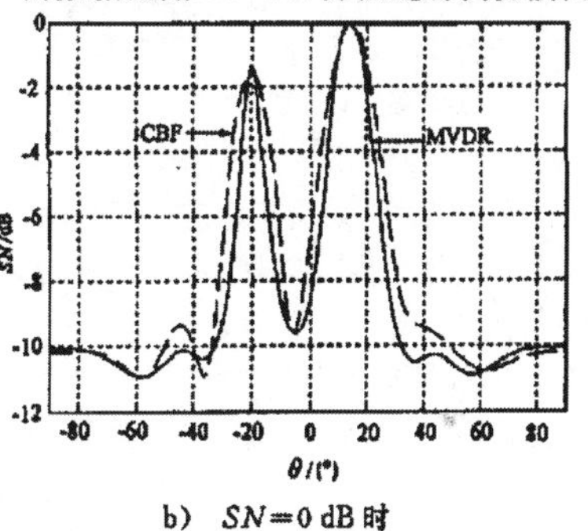

b) SN=0 dB 时

图 1 MVDR 和 CBF 波束输出对比图

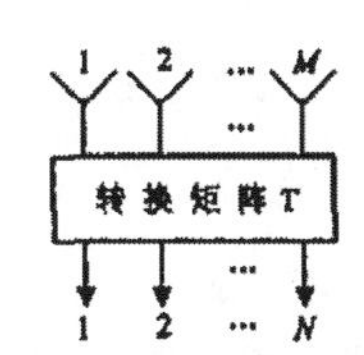

图 2 空间转换示意图

以半波长布阵方式的，由 M 个阵元组成的均匀线列阵的离散空间傅里叶变换 DSFT 可表示为

$$f(u,t) = \sum_{i=0}^{M-1} x_i(t) \mathrm{e}^{-j\pi u} \tag{5}$$

式中：$u=\sin\theta$，θ是与线列阵法线所成的角度.

定义 $M\times 1$ 的 DSFT 波束预形成加权矢量为

$$V = [1, \exp(\mathrm{j}\pi u), \cdots, \exp(\mathrm{j}\pi(M-1)u)]^{\mathrm{T}} \tag{6}$$

并构造 $M\times N$ 的转换矩阵 T 为以下形式

$$T = \frac{1}{\sqrt{M}}\left[v\left(m\frac{2}{M}\right), v\left((m+1)\frac{2}{M}\right), \cdots, v\left((m+N-1)\frac{2}{M}\right)\right] \tag{7}$$

可以看出，T 实际上从 $M\times M$ 的 DSFT 矩阵中连续抽取的一个从第 m 列到第$(m+N-1)$列的 $M\times N$ 子矩阵. 子矩阵中的每一个列向量指向一个不同的方向，N 个向量形成一个扇面. 经 T 矩阵转换后，原来 $M\times 1$ 的声纳基阵输入向量就转换为 $N\times 1$ 的波束空间向量，$N<M$，实现降维处理，减少运算量，并且可以估计方位处在$\left[\arcsin\left(\frac{2m}{M}\right), \arcsin\left((m+N-1)\frac{2}{M}\right)\right]$内的目标源. 可将波束域空间的输入向量表示为

$$Z(t) = T^H X(t) \tag{8}$$

则波束域空间输入向量的协方差矩阵写为

$$R_B = E\{Y(t)Y^H(t)\} = E\{T^H X(t) X^H(t)^{\mathrm{T}}\} \tag{9}$$

因此，BMVDR 波束形成的输出功率是

$$P_{\mathrm{BMVDR}}(\theta) = \frac{1}{\alpha^H(\theta) T R_B^{-1} T^H \alpha(\theta)} = \frac{1}{\alpha^H(\theta) T (T^H R T)^{-1} T^H \alpha(\theta)}$$

4 波束域MVDR与实际声纳系统结合应用的考虑

从以上的推导可知，波束空间转换类似于空间滤波，使得观测扇面外的噪声得到有效降低，转换矩阵 T 相当于传统的多波束形成，即在每个列向量指定的方向上形成一个波束，总共形成 N 个波束. 因此，波束空间的输入向量代表着多波束系统的输出，等于说将 M 个水听器接收信号转变成 N 个波束的输出. M D Zoltowski 经研究认为[4]：当目标源恰好位于某个波束的主极大方向时，可以获得最佳的方位估计效果；但事先不可能知道目标源的方位，为此，他建议各波束重叠50%，这样，波束域方法能够有好的估计性能. 而传统的声纳多波束系统，恰好也是各波束重叠50%.

因此，文中提出了一种新思路：可以考虑将声纳多波束系统与波束域MVDR结合起来，由多波束系统完成对目标方位粗测；由波束域MVDR替代传统的分裂波束精测定向法，完成对目标方位的精测，从而提高整个声纳定向系统的性能. 其实现框图如图3所示. 声纳基阵中的 M 个水听器接收到声信号后，经多波束系统形成 N 个波束，将 N 个波束直接采用BMVDR进行方位精测，给出多目标方位.

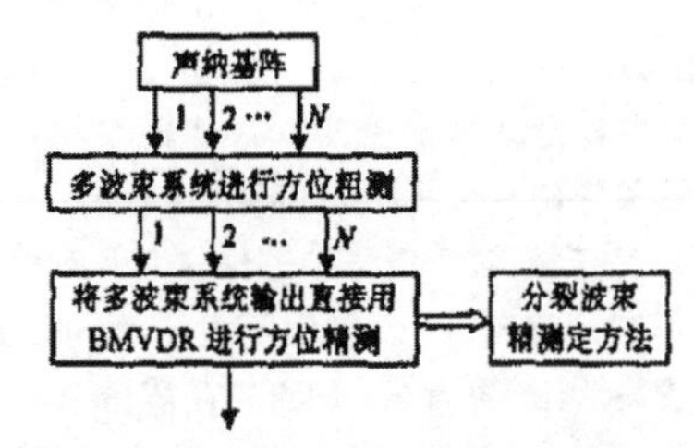

图3 BMVDR与实际声纳系统结合示意图

5 仿真研究

如前例，为覆盖 $-20°\sim20°$ 扇面，将8个阵元全部用于形成波束，相邻两波束重叠50%，共形成6个波束，用其波束形成输出做BMVDR. 图4是信噪比 $SN=0$ dB时，CBF，MVDR和BMVDR的输出结果，可见，BMVDR能将同一个波束宽度内10°和18°两个目标分辨出，在低信噪比时分辨性能比MVDR有明显提高.

另举一个利用部分阵元形成波束的仿真例子：按半波长方式布阵的均匀线列阵共有 $M=24$ 个阵元，利用 $N=16$ 个阵元形成一个波束，相邻两个波束有 $N-1=15$ 个阵元重叠，16个阵元所形成子阵的波束宽度大约为6.36°，相邻两波束重叠50%，那么为覆盖30°扇面需要通过波束转换矩阵形成9个波束（或者说利用多波束系统形成9个波束，这两种方法是等价的）. 在 $\theta_1=-12°$，$\theta_2=5°$ 和 $\theta_3=10°$ 分别存在着三个非相干等强度的窄带声源，快拍数为512. 在信噪比 SN 分别为0 dB和 -12 dB时，多波束系统输出、以16个水听器信号直接做MVDR的波束输出、BMVDR波束输出分别如图5a)和图5b)所示.

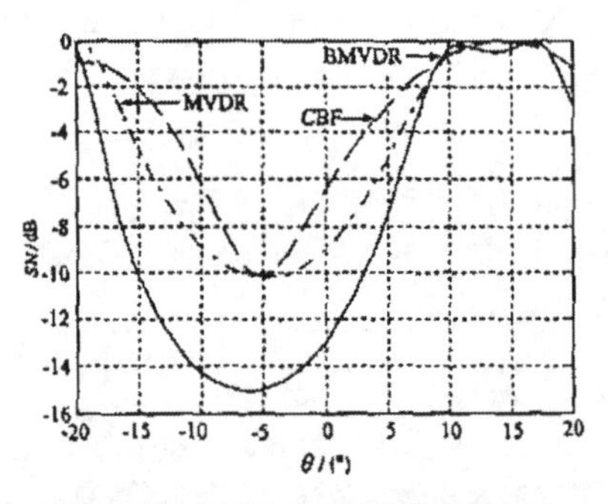

图4 CBF，MVDR和BMVDR的对比图

能够发现：

(1) 信噪比无论是0 dB还是 -12 dB，从分辨性能和谱峰锐度来看，BMVDR都要比MVDR和CBF好很多；

(2) 在 $SN=-12$ dB时，CBF和MVDR都已经不能把在同一个波束宽度内的 $\theta_2=5°$ 和 $\theta_3=10°$ 两个目标区分出来，而BMVDR能够地将这两个目标区分开来.

从运算量上看，在本例中，BMVDR，16个阵元的MVDR和24个阵元的MVDR计算量比较是在PIV1.4G的计算机上，通过MATLAB5.3中的CPUTIME函数测出的，如表1所示. 从表面上看，本例中的BMVDR计算量比16个阵元的MVDR计算量稍大；但实际上BMVDR是利用了24个阵元的信号，而MVDR只利用了16个阵元的信号，而24个阵元的MVDR计算量将近为BMVDR的2倍，也可以说，BMVDR方法比相同阵元数的MVDR计算量要小一半左右，证明了BMVDR能有效地减少计算量.

表1 计算时间比较

	BMVDR	16个阵元的MVDR	24个阵元的MVDR
计算时间/s	0.11	0.09	0.205

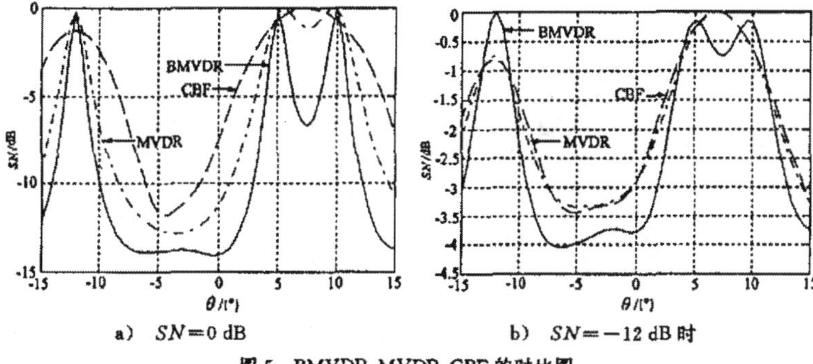

a) $SN=0$ dB b) $SN=-12$ dB 时

图 5 BMVDR,MVDR,CBF 的对比图

参考文献

1 Capon J. High-resolution frequency-wavenumber spectum analysis. Proc. IEEE 57,1969.1 408~1 418
2 Lee H B, Wengrovitz M S. Improved high-resolution direction~finding through use of homogeneous constraints. Proc. IEEE ASSP workshop spectrum estimation and modeling, 1988.152~157
3 Guanghan Xu, Seth D Silverstein, Richard H Roy et al. Beamspace ESPRIT. IEEE Trans. SP, 1994,42(2): 349~355
4 Michael D Zoltowski, Gregory M Kautz, Seth D Silverstein. Beamspace Root~MUSIC. IEEE Trans. SP, 1993,41(1),344~364

High-resolution Direction of Arrival Estimation Method of Beamspace MVDR

He Xinyi[1,2] Huang Haining[2] Ye Qinghua[2] Li Qihu[2] Jiang Xingzhou[1]

(*Navy Engineering University*, *Wuhan*, 430033)[1]

(*Institute of Acoustics*, *Chinese Academy of Sciences*, *Beijing*, 100080)[2]

Abstract

Being aimed at the decrease of MVDR's performance under low SNR, the paper presents beamspace MVDR(BMVDR)method, and puts forward a new thought using BMVDR to replace the accurate orientation method of the split beamers, also brings forward BMVDR as the orientation system instead of muliple beamforming system. BMVDR can not only solve the problem of decrease of MVDR's performance under low SNR, but also does not need to make great change to sonar orientation system. BMVDR can work well with multiple beamforming system in existing sonar, increase performance of the multi-target discrimination, also reduce calculation. After simulation, it is proved that BMVDR is a new practical method for high-resolution direction of arrival estimation.

Key words: MVDR; BMVDR; high-resolution direction of arrival estimation; low SNR; orientation system

基于时延估计的双线阵左右舷分辨技术研究①

何心怡[1,2] 蒋兴舟[1] 黄海宁[2] 李启虎[2] 张春华[2]
（1.海军工程大学，武汉 430033；2.中国科学院声学所，北京 100080）

摘 要：针对双线列阵左右舷模糊问题，文中提出了一套完整的信号处理方案，将波束形成、检测估计、左右舷分辨、镜象源干扰抵消融为一体：两根线列阵分别对应着左/右舷一侧0°～180°范围进行多波束形成；当检测到目标后，对目标方位进行粗测，将目标所处的呈镜象轴对称左右舷两路波束信号取出，以测相关的方法估算出两路信号的时延，根据时延值的符号，判断出目标所处的左右舷；然后根据镜象源所处的位置，采用不同的方法将镜象源干扰抵消掉，从而完成波束形成、左右舷分辨等全过程。经过仿真实验和湖试实验，证明这是一种有效、可行的左右舷分辨技术。

关键词：双线列阵；左右舷模糊；时延估计；相关；镜象源干扰抵消

The Research on the Twin-line Arrays Port/Starboard Distinguishing Technology Based on the Time Delay Estimation

He Xinyi[1,2] Jiang Xingzhou[1] Huang Haining[2] Li Qihu[2] Zhang Chunhua[2]
(1. Navy Engineering University, Wuhan430033 ; 2. Institute of Acoustics, Chinese Academy of Sciences, Beijing100080)

Abstract: To be aimed at twin-line arrays port/starboard blur problem, the paper presents a signal processing solution, bringing the forming of the beams, detection and estimation, the distinguish of the port/starboard and the offset for the mirror source noise as a whole: two line-array corresponds to the 0~180 degree of the left/right shipboard, forming the multi-beams; after detecting the target, estimates the target's rough position, picks out the two beams of the left-right shipboard which are symmetric to the mirror axis, use the correlation estimation to calculate the time delay of the signals. According to the sign of the time delay, judge the left-right shipboard of the target; then uses different methods to counteract the mirror source noise with the position of the mirror source, thus forming the multi-beams, distinguish the left-right shipboard and finishing the whole process. After simulation and lake experiment, it is proved to be an effective and practical technology for port/starboard discrimination.

Key words: twin-line arrays, port/starboard blur, estimation of time delay, correlation, offset for mirror source noise

1 引言

左右舷模糊问题是拖曳线列阵声纳必须要解决的一个难点问题，即使是采用双线列阵的基阵形式也是如此。在这方面，可以借鉴三元水听器组的左右舷分辨技术：几何相移模型或噪声相关模型[1-3]。但是双线列阵在实际使用中，为避免两根线列阵搅在一起，造成拖曳线列阵声纳无法正常工作，往往要求两根线列阵的间距要比较大，这对左右舷分辨提出了更高的要求；而三元水听器组的左右舷分辨技术，无论是几何相移模型还是噪声相关模型，都不适用于两根线阵间距较大的情况。因此，寻找适用于两根线列阵间距较大情况下的左右舷分辨技术尤为重要。

文中提出了一种以时延估计为理论基础的左右舷分辨方法：两根线列阵分别对应着左/右舷一侧0°～180°的范围进行多波束形成；当检测到目标后，先对目标方位进行粗测，将目标所处的呈镜象轴对称左/右舷两路波束信号取出以测相关的方法进行时延估计，根据时延值的正负号，判断出目标所处的左右舷；最后根据镜象源的位置，采用不同的方法将镜象源抵消掉，从而完成波束形成、左右舷分辨等全过程。该方法将波束形成、检测估计、左右舷分辨和镜象源干扰抵消有效地结合起来，经过仿真实验和湖试实验，证明这种方法是有效的和可行的。

① 信号处理, 2003, 19(4): 338-342.

2 时延估计左右舷分辨方法的理论基础

双线列阵示意图如图1所示，左舷的线列阵用线列阵1表示，右舷的线列阵用线列阵2表示；两根线列阵平行拖曳于水面舰艇的尾部，由2×M个分别标识为$(L_1,L_2,\cdots,L_M)$和$(R_1,R_2,\cdots,R_M)$的无指向性水听器组成；两线列阵间距为d，d要足够的大，防止两根线列阵搅在一起。将线列阵1在左舷0°～180°范围内形成N个波束，与左舷N个波束成镜象轴对称的位置上利用线列阵2在右舷0°～180°范围内也形成N个波束。

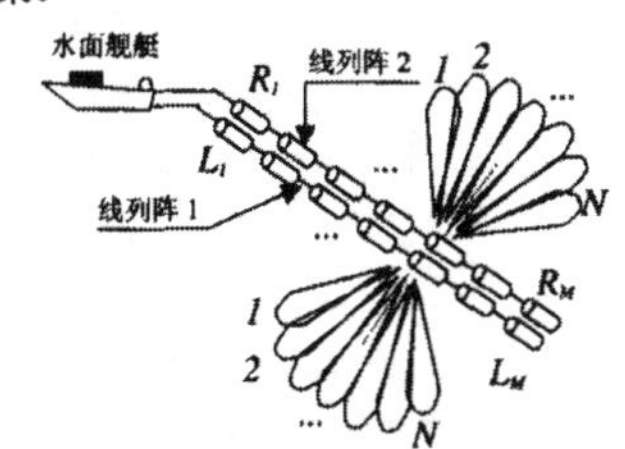

图1 双线列阵示意图

假设在左舷K号波束内有一个目标，其信号入射方向与左舷线列阵轴向夹角为θ，由于构成双线列阵的所有水听器都没有指向性，因此线列阵1和线列阵2都具有圆柱对称指向性，存在着左右舷模糊问题，造成了在右舷K号波束内也将检测到一个目标（即镜象源），信号入射方向与右舷线列阵轴向夹角也为θ。左舷K号波束和右舷K号波束可以用等效声学中心$L_K^{'}$和$R_K^{'}$表示，上述等效声学场景如图2所示。

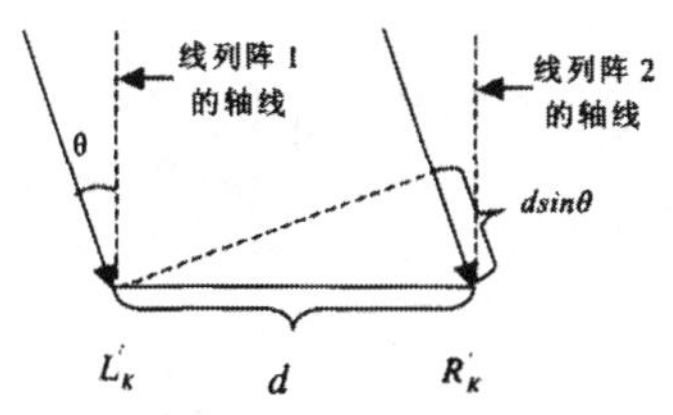

图2 双线阵左右舷分辨时的等效声学场景

假设以f_s的采样频率对左右舷两路波束信号进行A/D转换，那么，可将$L_K^{'}$和$R_K^{'}$的输出表示为：

$$\begin{cases} u_L(n)=s(n)+n_1(n) \\ u_R(n)=As(n-D)+n_2(n) \end{cases} \tag{1}$$

其中：

$s(n)$是目标的信号（被动声纳情况下是辐射噪声，主动声纳情况下是回波）；

$D=Round(d\sin\theta f_s/c)$是$R_K^{'}$接收到目标信号与$L_K^{'}$的整数时延值，符号由目标所处的左右舷决定：当目标位于左舷，符号为正，反之为负；

$n_1(n)$、$n_2(n)$是$L_K^{'}$和$R_K^{'}$信号中的加性噪声，若d大于接收信号中心频率的半波长，可认为$n_1(n)$、$n_2(n)$是相互独立的随机过程；

A是$R_K^{'}$接收到的信号与$L_K^{'}$接收到的信号的幅度比，在一般情况，由于d远小于目标的距离，可把由此带来的传播损失忽略不计，认为A＝1。

计算$u_L(n)$和$u_R(n)$的互相关，得到：

$$\begin{aligned} R_{LR}(\tau)&=E\{u_L(n)u_R(n+\tau)\} \\ &\approx R_{ss}(\tau-D),-\infty<D<\infty \end{aligned} \tag{2}$$

其中，$R_{ss}(\tau)=E\{s(n)s(n+\tau)\}$是信号$s(n)$的自相关序列。

因为$R_{ss}(0)>|R_{ss}(\tau)|,\forall\tau\neq0$，所以$R_{LR}(\tau)$在$\tau=D$处取峰值。在实际中由于采样数据是有限长的，所以$R_{LR}(\tau)$在$D$处不一定取峰值。但对于左右舷分辨技术，$D$估计的精确性并不十分重要，重要的是$D$的符号，符号的正负表明了目标所处的左右舷。当然，还可以采用其它一些方法来提高时延估计的精确性，如利用窗函数来平滑互相关函数，以及采用互累积量或者双谱来估计时延。也可以把时延估计的问题转换到频域，采用求互功率谱法来实现[4, 5]。

此外，如果是主动声纳情况，双线阵的间距d大于单频信号的半波长时，那么发射信号就不能采用单频信号，因为单频信号的周期性会给时延估计带来错误。

3 仿真实验

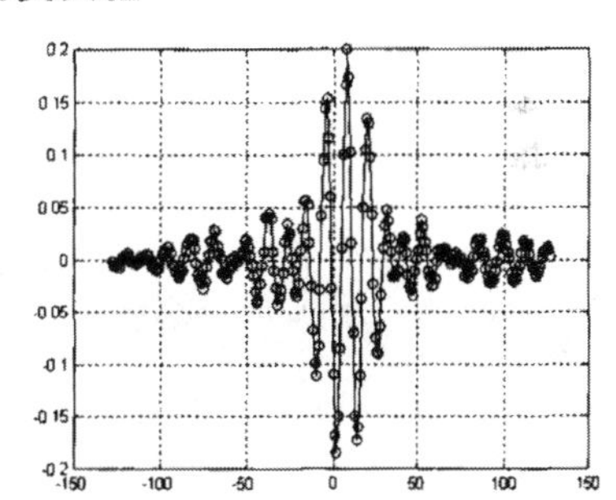

图3 当目标信号为线性调频波、目标位于左舷时，左右波束信号互相关函数

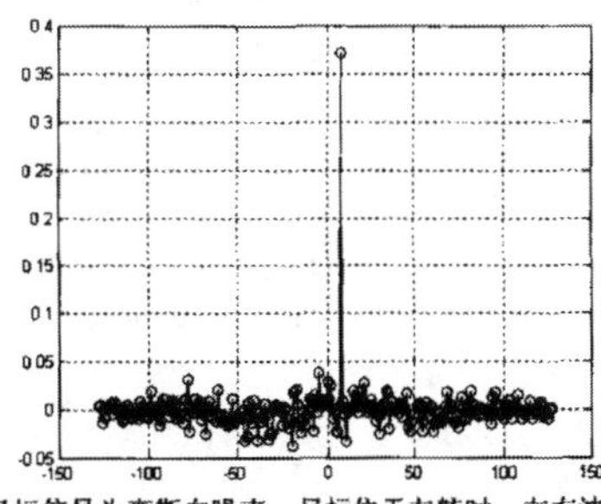

图4 当目标信号为高斯白噪声、目标位于左舷时，左右波束信号互相关函数

假设仿真场景如下：双线列阵间距d=2m，$f_s=12KHz$，

目标位于左舷，目标信号的入射方向与双线列阵的轴向夹角 $\theta = 30^\circ$，波束形成后的输出信噪比为 $S/N = -5dB$，此时对应的时延值为 8。目标信号分别是中心频率为 $1KHz$、带宽为 $500Hz$ 的线性调频波和高斯白噪声，背景噪声均为正态分布的高斯白噪声。图 3、图 4 分别是呈轴向对称的左右两路波束信号的互相关函数图，且估算出的时延值约为 8（四舍五入）。从时延值的符号，可判断目标处于左舷。

若其它条件不变，而目标位于右舷时，图 5 和图 6 分别是两种目标信号的左右波束信号互相关函数，估算出的时延值约为-8（四舍五入）。根据时延值的符号，可判断目标处于右舷。

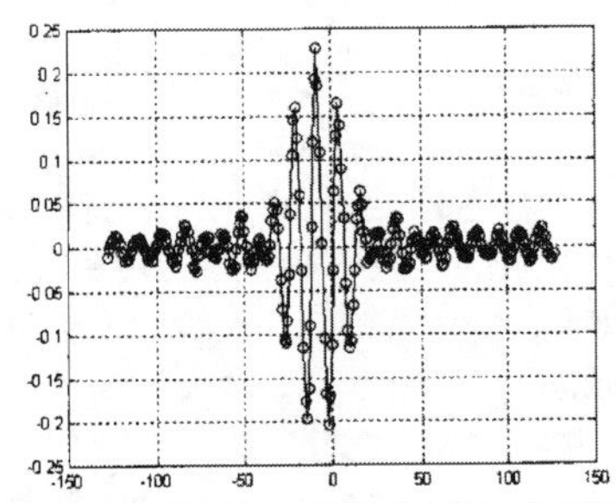

图 5 当目标信号为线性调频波、目标位于右舷时，左右波束信号互相关函数

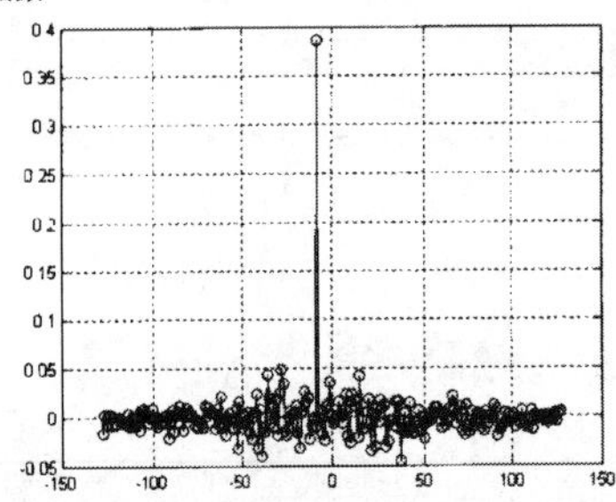

图 6 当目标信号为高斯白噪声、目标位于右舷时，左右波束信号互相关函数

仿真结果与理论分析相符，充分验证了文中提出的基于时延估计的双线列阵左右舷分辨技术的可行性和有效性。

4 湖试实验与分析

2002 年 7 月，在浙江千岛湖鹿鸣岛附近水域进行了双线阵左右舷分辨实验。图 7 是实验方案：将双线阵水平吊放于实验船 1 的水下 8m 处，间距 d=0.5m；在距双线阵左舷 0.29 海里处有艘实验船 2，在实验船 2 水下 5m 处吊放一个发射换能器；在双线列阵的正横方向附近发射周期为 4 秒、频率范围为 900Hz-1100Hz、脉宽为 200ms 的线性调频波；双线列阵附近的信噪比大约为 20dB，采样频率为 12KHz，信号对应的时延点数是 4。实验时，声速梯度呈弱负梯度。现场实验主要是采集数据，然后利用计算机进行后续处理。

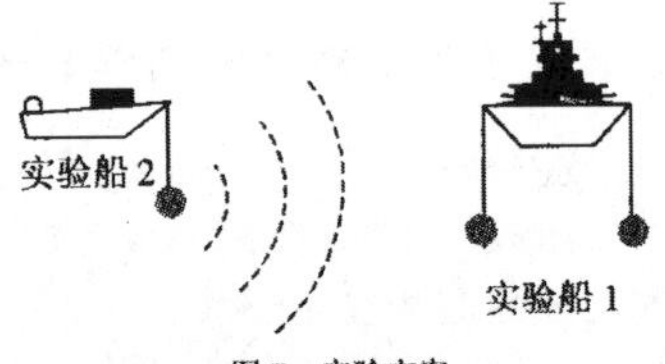

图 7 实验方案

图 8 为采集到的第一组数据中的某路水听器信号的时频分析图。由于实验条件限制，我们不对各水听器的接收信号形成波束，而以两路呈轴对称的水听器接收信号直接做时延估计，这样，与波束形成后的信号做时延估计相比，信噪比低一些，但不影响对时延估计方法的验证。表 1 为时延估计的结果，得到的时延均值是 2.4633，通过时延值的符号可判断出目标位于双线列阵的左舷。图 9 是第一组数据的左右两路水听器信号的互相关函数。时延估计值误差的原因主要在于：混响干扰、多途效应、线列阵存在着阵形畸变、线列阵中各水听器的幅相特性不尽相同以及信号入射方向偏离了双线阵的正横方向。但这些误差，对时延估计左右舷分辨方法的结果影响较小，因为该方法是依据时延值的符号判断目标所处的左右舷，而不是具体的时延值，属于定性的方法，而不是定量的方法，因此方法本身具有较好的稳健性。湖试结果充分验证了基于时延估计的左右舷分辨技术的可行性和有效性。

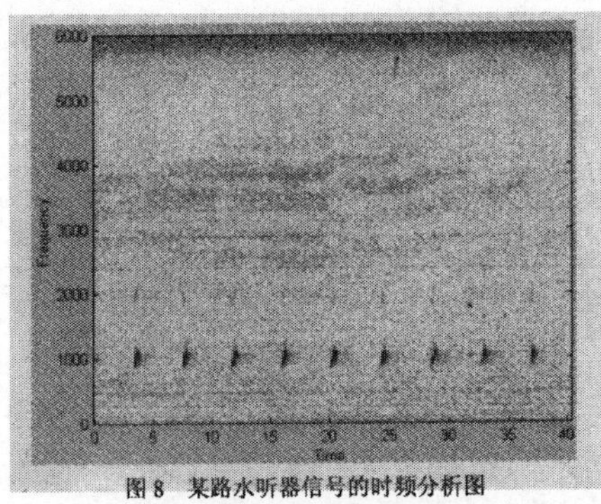

图 8 某路水听器信号的时频分析图

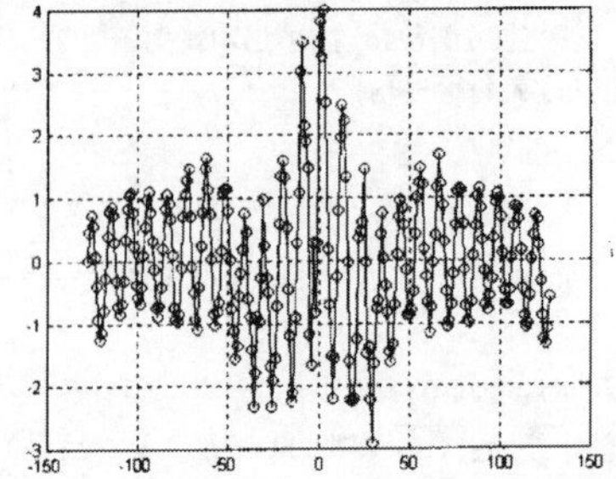

图 9 湖试第一组数据的左右两路水听器信号的互相关函数

表 1 湖试数据的时延估计结果

数据组数	估算出的时延值	理论时延值
1	2.8309	4
2	8.9233	4
3	1.4319	4
4	1.4342	4
5	1.7445	4
6	2.1251	4
7	2.0848	4
8	2.0642	4
9	1.9596	4
10	2.0682	4
11	2.1225	4
12	2.0303	4
13	1.9678	4
14	2.0412	4
15	2.2504	4
16	2.3344	4

5 镜象源干扰问题及可能的解决方法

采用时延估计的方法能够分清目标所处的左右舷，但从原理上并不能消除由于双线列阵水听器无指向性所带来的镜象源干扰问题，因此，这就要求在判断出目标所处的左右舷后，要以一定的信号处理手段去除镜象源的干扰。

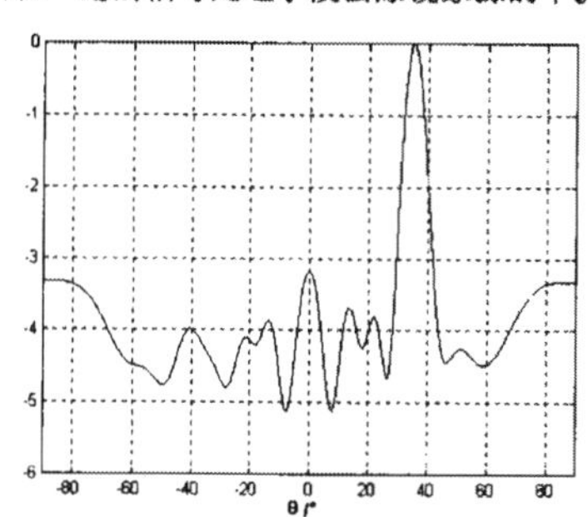

图 10 不存在镜象源干扰时的左舷多波束输出

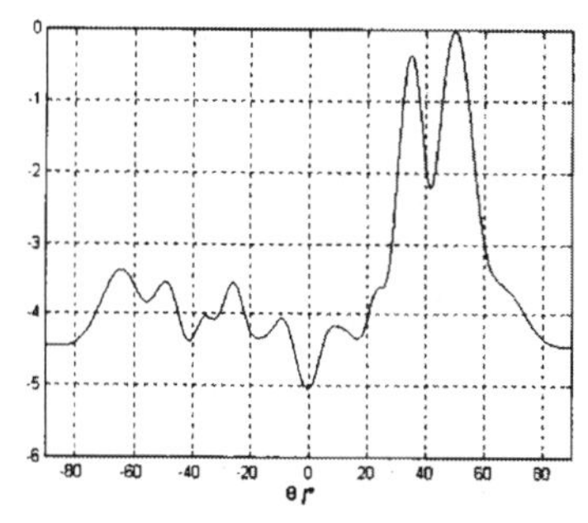

图 11 存在镜象源干扰时的左舷多波束输出（镜象源与目标不在同一个波束内）

假设左舷θ_1存在一个目标，右舷θ_2也存在一个目标。由于镜象源干扰，在左舷θ_2上也会出现一个虚拟的镜象目标，它会干扰、甚至是恶劣影响对左舷θ_1处目标的检测和估计；反之左舷θ_1处目标产生的镜象源对右舷θ_2处目标的干扰也是无法避免的。

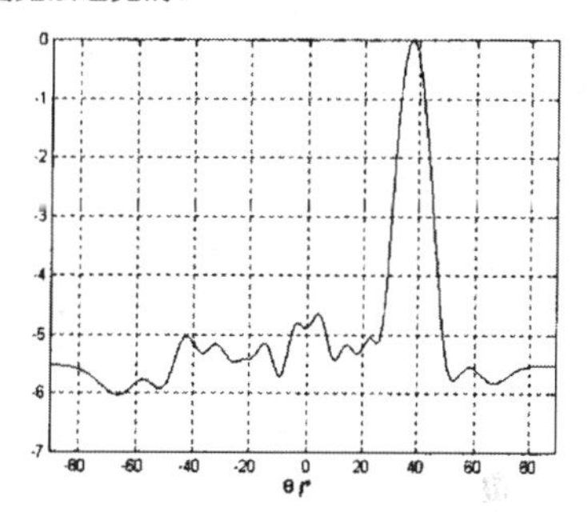

图 12 存在镜象源干扰时的左舷多波束输出（镜象源与目标在同一个波束内）

举一例子说明：假设双线列阵每根线列阵的阵元数都是 14，按半波长方式布阵，经估算，该线列阵主瓣宽度为 7° 左右。若在左舷 35° 存在一个信噪比为-10dB 目标，在右舷 50° 存在一个信噪比为-10dB 目标。如不存在镜象源干扰，左舷线列阵波束形成后的输出应如图 10 所示，仅存在一个目标。而由于存在镜象源干扰，左舷线列阵波束输出变为图 11 所示。甚至当镜象源和目标位于同一波束宽度内（比方说 40°，如图 12 所示），镜象源和目标重叠在一起，对目标方位估计等带来了很大的误差。因此，基于时延估计的左右舷分辨技术，能够在镜象源干扰中区分真实目标所处的左右舷，但无法抵消镜象源的干扰。因此在分辨出目标所处的左右舷后，必须采用一定的技术抑制或去除镜象源干扰。

根据镜象源的位置，可分两种情况去除镜象源干扰：

① 镜象源所处波束内不存在其它真实目标：可采用自适应干扰抵消方法将镜象源信号去除；

② 镜象源所处波束内还有其它真实目标：可先采用高分辨方位估计的方法将真实目标信号和镜象源信号区分出来，如 MVDR 方法等，再利用自适应干扰抵消的方法去除镜象源干扰。

6 结论

文中针对双线列阵间距较大情况下的左右舷模糊问题，提出了一种以时延估计为理论基础的左右舷分辨方法，能够做到在检测到目标的距离上实现对目标的左右舷判别，然后根据镜象源的位置，采用不同的方法去除镜象源干扰。该方法将波束形成、检测估计、左右舷分辨、镜象源干扰抵消合为一体，它的整个信号处理流程如图 13 所示。仿真实验和湖试实验充分证明了这种方法的有效性和可行性。

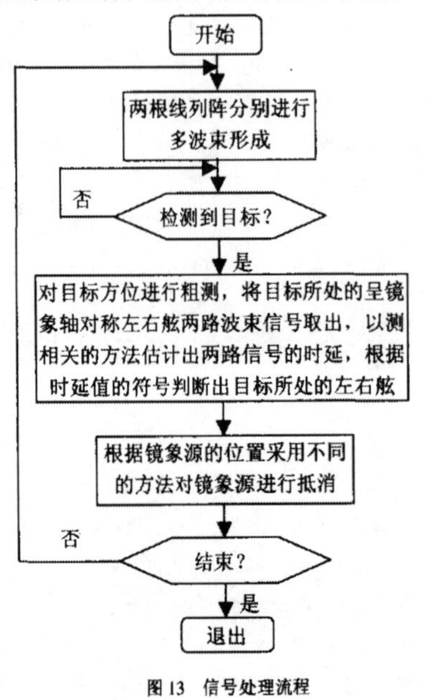

图 13　信号处理流程

参考文献

[1] Jean Bertheas, Gilles Moresco, Philippe Dufourco. Linear hydrophonic antenna and electronic device to remove right/left ambiguity, associated with the antenna. United States Patent, Patent Number：5058082, Oct.15, 1991.

[2] Y.Doisy．Port-starboard discriminatiojn performances on actived towed array systems. UDT95, France：125-129.

[3] G.W.M.Van Mierlo, S.P.Beerens and R.Been ect. Port/starboard discrimination by hydrophone triplets in active and passive towed arrays. UDT97, Germany：176-181.

[4] 李启虎．声呐信号处理引论．北京：海洋出版社，2000：186－187.

[5] Van Trees H L. Detection, estimation, and modulation theory. John Wiley & Sons, 1968.

BROADBAND BEAMSPACE MVDR HIGH RESOLUTION DIRECTION-OF-ARRIVAL ESTIMATION METHOD BASED ON SPATIAL RESAMPLING①

Xinyi He[1], Xingzhou Jiang[1], Qihu Li[2]
[1]Navy University of Engineering, Wuhan, Hubei Province, 430033, China
[2]Institute of Acoustics, Academia Sinica, Beijing, 100080,China

Abstract: Minimum Variance Distortionless Response beamforming is a high resolution Direction-of-Arrival estimation method, but it is not suitable for the broadband case and its computation is complex. This paper raised out the Broadband Beamspace MVDR High Resolution Direction-of-Arrival Estimation Method based on Spatial Resampling implementing on the equispaced linear array. This method focuses the target bearing information of various frequency on some focused frequency by spatial resampling algorithm thus to expand MVDR to the broadband case. Also it reduces the computation complexity and improves the stability of the algorithm by beam transfer at the focused frequency to reduce the computation dimension. After theory analysis and lake experiment, it is proved to be practical and effective.

1. INTRODUCTION

High Resolution Direction-of-Arrival estimation is an attractive research point in underwater acoustics field. Among which MVDR gains much attention because of its simple algorithm and fairly good performance. MVDR is a high-resolution DOA estimation method, which has a bright future [1]. But its algorithm only applies on the narrowband case but not on the broadband case while the receiving signal is usually broadband in underwater acoustics signal processing. Also this method needs to process on the array element space, and needs to calculate the inverse matrix. While the rank of the matrix increases, the computation complexity increases and the possibilities of the error solution increases when inversing the matrix. So, this paper raised out the Broadband Beamspace MVDR High Resolution Direction-of-Arrival Estimation Method based on Spatial Resampling (SR_BMVDR) implemented on the equispaced linear array: Focus the target bearing information of various frequency in the broadband on some referenced frequency (focused frequency), thus to focus all the energy of various frequency on the focused frequency; First get the signals of virtual array elements by spatial resampling on the focused frequency, then transfer the signals to beam space to reduce the computation dimensions, to reduce the computational complexity and improve the stability of the algorithm. After theory analysis and lake experiment, SR_BMVDR is proved to be effective and practical which can be applied in the nowadays sonar systems.

2. BROADBAND MVDR BY SPATIAL RESMAPLING

As to the research on the broadband high-resolution DOA estimation method, there are incoherent methods and coherent methods. The coherent methods have better characteristic than the incoherent methods on the detection and the SNR threshold [2, 3]. The spatial resampling method raised out by Jeferey Krolik and David Swingle is a coherent method by spatial sampling based on the adjustment of the array data. It doesn't need to estimate the target bearing in advance and also it is aimed at the equispaced linear array, which matches the equispaced linear array, the research background of this paper, very well.

Supposed there are D broadband plane-wave sources in the space, the bearing of the i th source relative to broadside of the line array is $\theta_i, i=1,2,\cdots,D$. These signals have the same bandwidth, while the signal and noise are statistically independent. When receiving with an equispaced linear array along the x-axis, which consists of M array elements spaced a uniform distance d meters apart, the equispaced linear array can be treated as the equispaced discrete sampling of the continuous linear array. Then the output $y(m,f)$ of the m th array element at frequency f can be expressed as the discrete sampling of the output $y(x,f)$ of the continuous linear array at $x=md$:

$$y(m,f)=\sum_{i=1}^{D}S_i(f)e^{-j2\pi fmd\sin(\theta_i)/c}+N(md,f), m=0,1,\cdots,M-1 \tag{1}$$

① IEEE Int. Conf. Neural Networks & Signal Processing, 2003, 2: 1326-1329.

$S_i(f)$ is the Fourier transform of the i th signal at frequency f, $N(md,f)$ is the noise component at frequency f, c is the velocity of sound in seawater, can be expressed in vector as:

$$y(f)=A(f)S(f)+N(f) \tag{2}$$

$A(f)=[a(f,\theta_1),a(f,\theta_2),\cdots,a(f,\theta_D)]$ is the direction matrix of the sources, $a(f,\theta_i)=\left[1,e^{-j2\pi fd\sin(\theta_i)/c},\cdots,e^{-j2\pi f(M-1)d\sin(\theta_i)/c}\right]^T$ is the direction vector of the i th source. The output vector of the linear array can be expressed as $Y(f)=[y(0,f),y(1,f),\cdots,y(M-1,f)]^T$, and the covariance matrix of the narrowband $R(f)$ is defined as $E\{Y(f)Y(f)^H\}$. Here $E\{\}$ denotes the expectation operator, and superscripts T and H represent the transpose operator and conjugate transpose operator.

The principle of the spatial resampling is to adjust the spatial sampling distance $d(f)$ virtually, making it the function of the frequency, $d(f)=d\cdot f_0/f$, that means $d(f)\cdot f=d\cdot f_0$, thus to focus all the frequency in the band on the focused frequency f_0 and the output of the virtual array element after the adjustment of the sampling distance is:

$$\begin{aligned}\tilde{y}(m,f)=&\sum_{i=1}^{D}S_i(f)e^{-j2\pi f_0 md\sin(\theta_i)/c}\\&+N(md\,f_0/f,f),m=0,1,\cdots,\tilde{M}-1\end{aligned} \tag{3}$$

Where, $\tilde{M}$ is the number of the array elements which comprise the virtual linear array. The equation above equals to the spatial resmapling of the continuous linear array at $x=md(f)$. To avoid space overlapped, $f/c\le 1/2d(f)$ is required.

If using $Y(e^{j\phi},f)$ and $\tilde{Y}(e^{j\phi},f)$ to represent the space Fourier transforms of $y(m,f)$ and $\tilde{y}(m,f)$:

$$\tilde{Y}(e^{j\phi},f)=\begin{cases}\dfrac{f}{f_0}Y\left(e^{j\tilde{\phi}\frac{f}{f_0}},f\right),|\tilde{\phi}|\le\min\left(\pi,\pi\dfrac{f_0}{f}\right)\\0,\min\left(\pi,\pi\dfrac{f_0}{f}\right)<|\tilde{\phi}|\le\pi\end{cases} \tag{4}$$

This equation means that spatial resampling corresponds to the space frequency mapping actually, it maps the plain-wave signal at $\phi_i=2\pi fd\sin(\theta_i)$ to the $\tilde{\phi}_i$:

$$\tilde{\phi}_i=2\pi fd(f_0/f)\sin(\theta_i)=2\pi f_0 d\sin(\theta_i) \tag{5}$$

This processing result makes the broadband sources have the same space frequency in the overall band thus to make each broadband source have rank-one model.

It can be proved that the pulse response of the space filter to finish this mapping manipulation is [3]:

$$h_f(n,m)=\sin\left[\phi\left(\frac{f_0}{f}n-m\right)\right]\Big/\left[\pi\left(\frac{f_0}{f}n-m\right)\right] \tag{6}$$

$$,n=0,1,\cdots,M-1;m=0,1,\cdots,\tilde{M}-1$$

Thus:

$$\tilde{y}(n,f)=\sum_{m=0}^{\tilde{M}-1}h_f(n,m)y(m,f) \tag{7}$$

It can be inferred that $h_f(n,m)$ is the element at row $n+1$, column $m+1$ of the focused matrix, the focused matrix T_j at frequency f_j is:

$$T_j(n+1,m+1)=\sin\left[\phi\left(\frac{f_0}{f_j}n-m\right)\right]\Big/\left[\pi\left(\frac{f_0}{f_j}n-m\right)\right] \tag{8}$$

The optimal value of $\tilde{M}$ is $\tilde{M}=M\cdot f_l/f_0$, f_l is the lowest frequency in the band. After the spatial resampling we can get the narrowband covariance matrix of the virtual linear array at focused frequency f_0. Using $\hat{R}(f_j)$ to represent the narrowband covariance matrix of the virtual linear array at frequency f_j, the output covariance matrix $\tilde{R}(f_0)$ of the virtual linear array after spatial resampling is:

$$\tilde{R}(f_0)=\sum_{j=l}^{h}T_j(f_j)\hat{R}(f_j)T_j(f_j)^H \tag{9}$$

$f_l,\cdots,f_h$ are the frequency bins in the signal band, and then the output of the space spectrum of broadband MVDR by Spatial Resampling (SR_MVDR) is:

$$\begin{aligned}P_{SR\_MVDR}(\theta)&=\left[a(f_0,\theta)^H\left[\tilde{R}(f_0)\right]^{-1}a(f_0,\theta)\right]^{-1}=\\&\left\{a(f_0,\theta)^H\left[\sum_{j=l}^{h}T_j(f_j)\hat{R}(f_j)T(f_j)^H\right]^{-1}a(f_0,\theta)\right\}^{-1}\end{aligned} \tag{10}$$

$a(f_0,\theta)$ is a $\tilde{M}\times 1$ steering vector at frequency f_0.

3. BROADBAND BEAMSPACE MVDR METHOD BY SAPTIAL RESAMPLING

After the analysis to the equation (10), it can be inferred that the computational complexity of SR_MVDR is determined mostly by the dimensions of $\tilde{R}(f_0)$, whose dimensions equal to the number of array elements which comprise the virtual line array. When the frequency scope of the signal and the focused frequency are set, $\tilde{M}$ is determined by the number of array elements which comprise the physical array. Stergios Stergiopoulos had pointed out: When $\tilde{R}(f_0)$ is larger than 64×64, using various ways to inverse the matrix, such as the eigenvalue method, would get the error result usually and the possibilities of the abnormal results increases [4]. Thus, to reduce the dimension of $\widetilde{R}(f_0)$ is very important to reduce the computational complexity and to improve the stability of the algorithm.

In 1988, H.B.Lee raised out the high-resolution technology in the beamspace, through choosing the pre-process matrix of beam forming delicately to reduce the dimension. Thus it can reduce the computational complexity and improve the stability of the algorithm [5]. Based on the beam space processing principle, this paper raised out the Broadband Beamspace MVDR method based on Spatial Resampling to reduce the computational complexity and to improve the stability of the algorithm.

Figure 1 is the signal process flow of SR_BMVDR. First is to get the output of each virtual linear array element by spatial resampling method introduced in section 2, then the output of each virtual linear array element focused at frequency f_0; then transfer the array element data to beamspace as described in the dashed rectangular in figure 1. This step not only reduces the dimension but also improves the stability of the algorithm. At last, realize the high resolution on the data after beam transfer by MVDR algorithm.

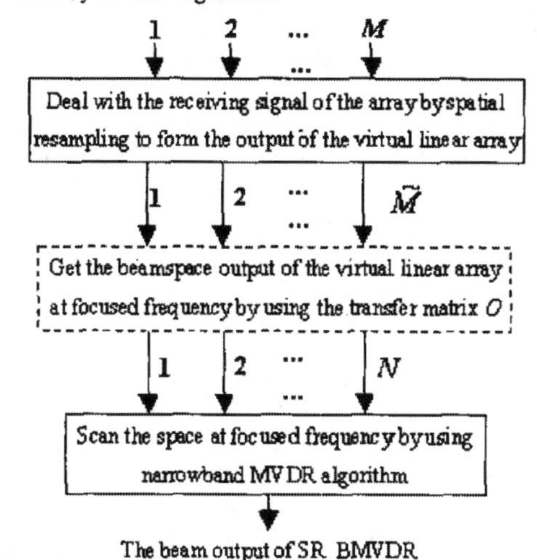

Figure 1 The signal process flow of SR_BMVDR

The discrete space Fourier transform of the equispaced linear array consists of $\widetilde{M}$ array elements:

$$f(u,t)=\sum_{i=0}^{\widetilde{M}-1}x_i(t)e^{-j2\pi iu} \tag{11}$$

$u=\widetilde{d}\sin\theta/\lambda$, $\widetilde{d}$ is the distance between each array element of the virtual linear array, θ is the angel between the normal of the virtual linear array, $\lambda=c/f_0$ is the corresponding wavelength of the focused frequency.

Define the $\widetilde{M}\times 1$ discrete space Fourier transform weight vector as:

$$v=\left[1,\exp(j\pi u),\cdots,\exp\left(j\pi\left(\widetilde{M}-1\right)u\right)\right]^T \tag{12}$$

Construct the transfer matrix O of $\widetilde{M}\times L$ as:

$$O=\frac{1}{\sqrt{\widetilde{M}}}\left[v\left(m\frac{2}{\widetilde{M}}\right),v\left((m+1)\frac{2}{\widetilde{M}}\right),\cdots,v\left((m+L-1)\frac{2}{\widetilde{M}}\right)\right] \tag{13}$$

It can be inferred that O is actually the $\widetilde{M}\times L$ sub matrix from the column m th to the column (m+L-1) of the $\widetilde{M}\times\widetilde{M}$ discrete space Fourier transform matrix. The each column vector of the sub matrix points at different direction, and L vectors form a sector. After the operation with the matrix O, the $\widetilde{M}\times 1$ array input vector is transferred to $L\times 1$ beamspace vector, $L<\widetilde{M}$. It can reduce the dimension and computational complexity, also can estimate the target within the sector $\left[\arcsin\left(\frac{2m}{\widetilde{M}}\right),\arcsin\left((m+N-1)\frac{2}{\widetilde{M}}\right)\right]$, whose center is θ. Then the covariance matrix of the output vector of the virtual line array after beamspace transfer can be:

$$\widetilde{R}_B(f_0)=O^H\widetilde{R}(f_0)O=O^H\left(\sum_{j=l}^{h}T_j(f_j)\hat{R}(f_j)T_j(f_j)^H\right)O \tag{14}$$

So, the output power of the SR_BMVDR beam forming:

$$\begin{aligned}P_{SR\_BMVDR}(\theta)&=\left[a(f_0,\theta)^H O\left[\widetilde{R}_B(f_0)\right]^{-1}O^H a(f_0,\theta)\right]^{-1}\\&=\left\{a(f_0,\theta)^H O\left[O^H\left(\sum_{j=l}^{h}T_j(f_j)\hat{R}(f_j)T(f_j)^H\right)O\right]^{-1}O^H a(f_0,\theta)\right\}^{-1}\end{aligned} \tag{15}$$

4. AUTHENTICATION TEST

The data from the lake experiment at Qian Dao Lake in 2002 is put to the proof: one equispaced line array consists of 16 array elements, the distance between each other is 1m, the data sampling rate $f_s=10\,\text{kHz}$. There is one cooperating target in the experiment: target ship. There are usually some ships passing by in the experiment water area. Whatever the target ship or the ships that passed by, the tonnage is quite small, and their SNR is less than 0dB. The total data length is 440s. The total data points are 12 ×4096 needed to be processed to get the result of each line in the time-bearing plot. The data subsection length is 4096, the overlap ratio between the consecutive subsections is 75%, the processing frequency channel is 500Hz-1000Hz, the corresponding frequency bins are 205, and the snapshots in the frequency domain is 45. The data points which are needed to be processed to get each line in the time-bearing plot are not overlapped, that means the result of each refresh in the time-bearing plot is independent. The focused frequency f_0 equals to 750Hz, the array elements of the virtual linear array are 12, use the beam transfer to turn the covariance matrix of equation (14) to 5 × 5. Conventional Beam Forming(CBF) and SR_BMVDR both form the beam every 1° . Figure 2A and 2B are the time-bearing plot of CBF and SR_BMVDR after process. For convenience, every refresh was normalized, and each result was independent, thus the gray level of each line in the time-bearing plot only shows the beam output each time, not including the history information.

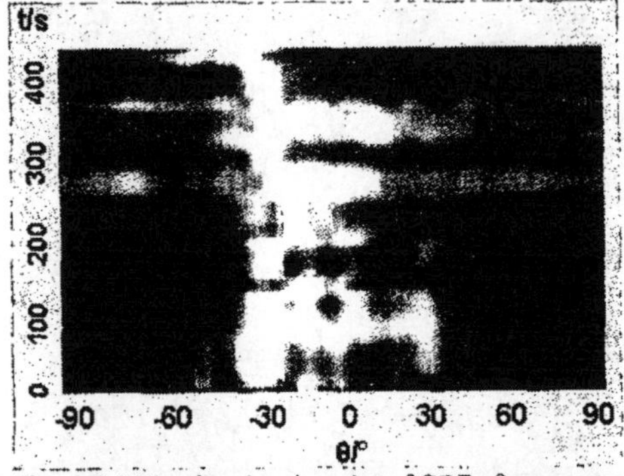

Figure 2A The time-bearing plot of CBF after process

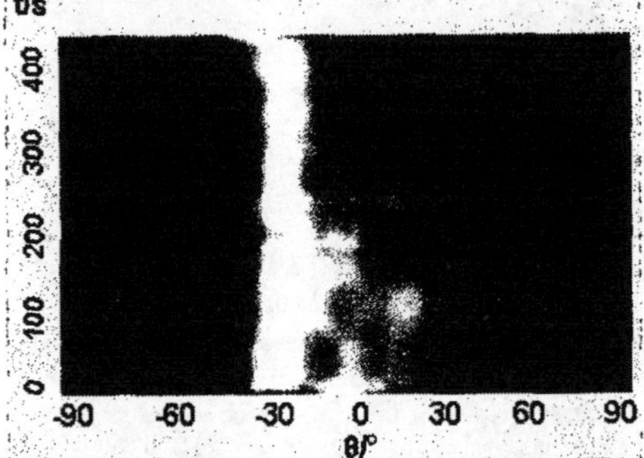

Figure 2B The time-bearing plot of SR_BMVDR after process

Compare figure 2A with figure 2B, CBF's result is faint, and SR_BMVDR's result is clear. The track of target ship in CBF's result is often broken temporarily by the ships that passed by, whereas the track of target ship in SR_BMVDR's result keeps continuous. Figure 3 is the beam forming output result of the SR_BMVDR and CBF after normalization at 240s. The real line is the beam forming output result of SR_BMVDR, and the dash line is the beam forming output result of CBF. It can be inferred that SR_BMVDR and CBF all detect one target on -25° , that is the target ship. SR_BMVDR shows there are two small targets on 6° and 19° , this matches the scene at that time but CBF can not detect them. At the same time, the main lobe of SR_BMVDR is much narrower than that of CBF while the side lobe of the SR_BMVDR is much lower than that of CBF.

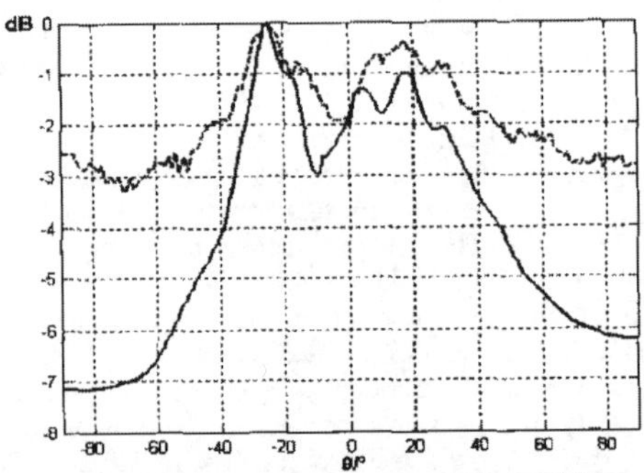

Figure3 Beam forming output result of SR_BMVDR and CBF after normalization at 240 sec.

The computational complexity is compared on the computer whose microprocessor is PIV1.3G, through the CPUTIME function of MATLAB5.3 to measure, when gaining each refresh in the time-bearing plot, CBF takes 13.18 seconds, while SR_BMVDR takes 8.86 seconds, which is only the 67% of CBF. This proves that SR_BMVDR can reduce computational complexity greatly.

5. CONCLUSION

This paper aims at the problem that MVDR doesn't apply on the broadband signal and the computation is complex, raising out SR_BMVDR implemented on the equispaced linear array. SR_BMVDR expands the MVDR algorithm effectively to the broadband and reduces the computational complexity greatly. After theory analysis and lake experiment, this method proves to be practical and effective.

6. REFERENCES

[1] J.Capon, "High-resolution frequency-wavenumber spectum analysis", Proc. IEEE, Vol.57 (8), pp. 1408-1418, 1969

[2] Jeferey Krolik and David Swingle, "Focused Wide-Band Array Processing by Spatial Resampling", IEEE Trans. ASSP, Vol.38 (4), pp. 356-360, 1990

[3] Jeferey Krolik and David Swingle, "The Detection Performance of Coherent Wideband Focusing for a Spatially Resampled Array", IEEE ICASSP, pp. 2827-2830, 1990

[4] Stergios Stergiopoulos, "Extended towed array processing by an overlap correlator", J.Acoust.Soc.Am., 86(1), pp. 158~171, 1989

[5] H.B.Lee and M.S. Wengrovitz, "Improved high-resolution direction-finding through use of homogeneous constraints", Proc. IEEE ASSP workshop spectrum estimation and modeling, pp. 152-157, Aug.3-5 1988

Line spectrum of Ship Noise and High-resolution Direction-of-Arrival Estimation Method of FFTSA Passive Synthetic Aperture Technology①

Xinyi He*, Xingzhou Jiang* and Qihu Li**

*Navy University of Engineering, Wuhan 430033, China
**Institute of Acoustics, Academia Sinica, Beijing 100080,China
Email: hxyisdragon@sina.com

Abstract

Bearing resolution of towed linear array is an important performance index, which is restricted by the physical aperture of towed linear array. To be aimed at this problem, this paper raised out a new method: Using stable line spectrum of ship noise in the low frequency band to expand the effective aperture of towed linear array through FFTSA passive synthetic aperture algorithm, thus to improve bearing resolution greatly. If the integral time required by FFTSA is smaller than the time coherence length of the ocean acoustic channel, it can expand the effective aperture of towed linear array to any size smaller than the space coherence length of the ocean acoustic channel. After simulation, this method's algorithm is simple, easy to be put into practice, and can improve bearing resolution of towed linear array greatly. It is a high-resolution direction-of-arrival estimation method that has a bright future.

1 Introduction

Since the submarines came into the stage of the war, the anti-submarine war became one of the important aspects of naval warfare. In order to be guarded to the underwater circumstances in the long range, towed linear array sonar was invented: a towed neutral buoyancy cable at the tail of ship which has hydrophones that was used to receive the low frequency noise of the submarines thus to achieve the above target. In the report of the Pentagon said: "In the coming 30 years, the most important development in the anti-submarine was the invention of tactics towed linear array sonar."[1]

For towed linear array sonar, high bearing resolution is a very important index. While it uses the Conventional Beam Forming technology (CBF), bearing resolution and signal gain incorporate in the ration between the aperture of towed linear array and signal wavelength, which is restricted by the physical aperture of linear array. In order to expand the detecting range, towed linear array sonar usually works at low frequency band or very low frequency band. That means the higher bearing resolution requires the longer linear array, while too long linear array is unacceptable for towed linear array sonar. Under such conditions, the requirement for high-resolution direction-of-arrival estimation technology is very imminent.

Since the 1980s, many high-resolution direction-of-arrival estimation methods represented by MUSIC, ESPRIT have been developed. But these methods have some problems in the threshold of SNR, judging the number of sources, array calibration, and cannot be put into practice in the near future. In recent years, to be aimed at high-resolution direction-of-arrival estimation problem, Passive Synthetic Array technology (PASA) [2-8] was raised out. Towed linear array is moving; also it is called Moving Towed Array (MTA). Using PASA technology can synthesize virtual linear array, whose aperture is much larger than the physical aperture of towed linear array, thus to break through the restriction of linear array physical aperture to get higher bearing resolution. The algorithm of PASA is simple and stable, and is a high-resolution technology, which can be put into practice.

So, this paper brings forward a method, based on stable line spectrum of ship noise, using FFT Synthetic Aperture algorithm (FFTSA) PASA technology to deal with line spectrum to synthesize vir-

① Proceedings of the 2003 IEEE, International Conference on Robotics, Intelligent Systems and Signal Processing, 2003, 1: 478-483.

tual linear array, whose aperture is larger than the physical aperture of the MTA thus to get higher bearing-resolution. After simulation, this method proves to be effective, practical and have bright future in practice.

2 Analysis of ship noise

Ship noise needed for passive sonar to detect and identify targets, which has the information of targets characteristics. According to spectral characteristic division, ship noise includes stationary continuous spectrum, line spectrum and time-dependent modulated spectrum[9,10]. It can be treated that ship noise approximately obeys Gaussian distribution. Ship noise's statistical property can be depicted through second order moment. Time-dependent modulated spectrum of ship noise is as follows:

$$G(t,f) = G_x(f) + G_L(f) + M(t)M(f)G_x(f) \quad (1)$$

In the equation above, $G_x(f)$is the stationary ergodic Gauss process continuous spectrum, $G_L(f)$ is the discreet distributed line spectrum in frequency domain, $M(t)M(f)G_x(f)$ is the time-dependent power spectrum whose spectrum level is modulated by the time period: $M(t)$ is called modulation function, represents the period time-dependent modulation implemented on the continuous spectrum; $M(f)$ is called modulation depth spectrum, represents the different modulation level of the different frequency composition.

After research[9], in the low frequency band of ship noise, that means, under hundreds of Hz, there are line spectrum existing. These line spectrum are stable, the change in hundreds of seconds is within 1 Hz, and can be 20–30 dB higher than mean value of the spectrum. Using line spectrum to implement direction-of-arrival estimation can achieve favorable effect. The method brought forward by this paper using FFTSA passive synthetic aperture high-resolution direction-of-arrival estimation method is actually based on line spectrum of ship noise.

3 The basic principle of FFTSA algorithm

Passive synthetic aperture (PASA) technology is a new technology developed in recently years to be aimed at direction-of-arrival estimation of towed linear array sonar, the basic principle is as follows[2,3,4]:

- Towed linear array is a moving towed array (MTA) in practice, so PASA is the same as active synthetic aperture radar and active synthetic aperture sonar, depends on the movement of the array to improve bearing resolution performance.

- Discovered in the ocean experiment: In the range of 50 Hz – 1500 Hz, the space coherence length of the ocean acoustic channel reaches 300 times of wavelength, as to the monochromatic signal, the time coherence length is no less than several minutes, sometimes even more than ten minutes.

PASA technology is still now developing. The four mature algorithms are ETAM(Extended-towed-array measurements), ML(Maximum Likelihood algorithm), FFTSA(FFT Synthetic Aperture processing), and one of the passive synthetic aperture methods introduced by Yen and Carey. They all aim at continuous monochromatic signal. ETAM and ML method transact in the array element domain while FFTSA and Yen's method transacts in the beam domain. As to the precision of direction-of-arrival estimation, stability of direction-of-arrival estimation, amount of calculation and the combination with the sonar system in existence, FFTSA algorithm can be treated as the combination of the ETAM and Yen's method. This method has a bright future. Here is the brief introduction to FFTSA algorithm.

Suppose there is a towed linear array of equally spaced hydrophones in the ocean acoustic channel, which consists of N hydrophones, whose space between the consecutive ones is d, there is a far field monochromatic acoustic signals at bearing θ. The sampling period of the signal is Δt, $t_i = i\Delta t$, $i = 1, 2, \ldots, K$, K is the sampling time ordinal number of each hydrophone, and then the received signal is as follows:

$$x_n(t_i) = Aexp[j2\pi f(t_i - \frac{d(n-1)}{c}sin\theta)] + \varepsilon_{n,i} \quad (2)$$

Here $n = 1, 2, \ldots, N$, A is the amplitude. Let $X_n(f) = \sum_{i=1}^{k} x_n(t_i)exp(-j2\pi f t_i)$, which is the Fourier transform of $x_n(t_i)$ at frequency f, $\varepsilon_{n,i}$ is independent, zero mean, Gauss random variable whose variance is σ_ε^2. This random variable represents the ocean background noise. c is the sound velocity in sea water.

Then, the plane-wave response of the MTA at t_0 is:

$$b(f,\theta_s)_{t_0} = \sum_{n=1}^{N} X_n(f)exp[j2\pi f\frac{d(n-1)sin\theta_s}{c}] \tag{3}$$

Where θ_s represents the steering angle.

In equation (2), which describes a monochromatic signal received by the nth hydrophone of the MTA. This frequency f includes the Doppler frequency shift formed by the relative motion between the MTA and the acoustic source. Suppose v is the towing speed of the MTA, compared with the towing speed, the radial speed of the far field acoustic source is usually small, the motion of the acoustic source can be omitted. If the frequency of the stationary field is f_0, the frequency of the received signal can be $f = f_0(1 \pm vsin\theta/c)$, thus equation (2) can approximately be:

$$x_n(t_i) = Aexp[j2\pi f_0(t_i - \frac{vt_i + (n-1)d}{c}sin\theta)] + \varepsilon_{n,i} \tag{4}$$

After τ seconds, the MTA moves $v\tau$. Through choosing parameters v and τ delicately to reach $v\tau = qd$, q represents the number of hydrophone position that the MTA has moved, and then the received signal $x_n(t_i+\tau)$ can be:

$$x_n(t_i) = exp(j2\pi f_0\tau)A\cdot$$

$$exp[j2\pi f_0(t_i - \frac{vt_i + (q+n-1)d}{c}sin\theta)] + \varepsilon_{n,i}^{'} \tag{5}$$

The Fourier transform of the $x_n(t_i+\tau)$ can be:

$$\widetilde{X}_n(f)_{t_i+\tau} = \widetilde{X}_n(f)_{t_i}exp(j2\pi f\tau) \tag{6}$$

$\widetilde{X}_n(f)_{t_i}$ is the DFT of $x_n(t_i)$. If using the phase term $exp(-j2\pi f_0\tau)$ to compensate the signal in equation (5), then the acoustic source information included in $t = t_i$ and $t = t_i + \tau$ two times measurements is equaled to the result of the MTA consists of $q + N$ hydrophones. And the like, the result of M times measurements is equaled to the result of the MTA consists of $(M-1)q + N$ hydrophones.

According to the synthetic aperture technology introduced by Yen and Carey[2], the response of MTA at $t = t_i + \tau$ is

$$b(f_0,\theta_s)_{t_i+\tau} = \sum_{n=1}^{N}\{\sum_{i=1}^{K} x_n(t_i+\tau)exp(-j2\pi f_0 t_i)\}\cdot$$

$$exp[j2\pi f_0\frac{d(n-1)sin\theta_s}{c}] \tag{7}$$

From equation (3) and (5), the equation above can be:

$$b(f_0,\theta_s)_{t_i+\tau} = b(f_0,\theta_s)_{t_i}exp[j2\pi f_0(\tau - \frac{v\tau}{c}sin\theta)] \tag{8}$$

Equation (8) illustrated: Through the phase compensate to the beam output of different times and the accumulation of coherence to expand the effective aperture of the array. Thus we can get:

$$B(f_0,\theta_s)_{M_\tau} = b(f_0,\theta_s)_{t_0}\cdot$$

$$\sum_{m=1}^{M} exp\{[j2\pi f_0(1 - \frac{v\tau}{c}sin\theta)m\tau] - j\phi_m\} \tag{9}$$

Here $\phi_m = 2\pi f_0(1 \pm vsin\theta_s/c)m\tau$ is to create a synthetic aperture, compensate to the phase of the beam output in the mth measure. The time interval between each measure is τ seconds, M is the total number of measurements while expanding the MTA effective aperture, the total integral time $T = M\tau$. Equation (9) can be modified to:

$$B(f_0,\theta_s)_{M_\tau} = \sum_{m=1}^{M} b(f_0,\theta_s)_{t_i+m\tau}exp[-j\phi_m] \tag{10}$$

In the method of Yen and Carey, in order to form a synthetic aperture, a phase compensate factor must be estimated. But the estimation of the phase compensate factor needs to know the relative velocity between the MTA and the acoustic source and the precise frequency f of the acoustic source beforehand, but in the reality, the above two factors could not be known beforehand, this restricts the method's implementation.

Different from Yen's method, ETAM algorithm does not need to know the relative velocity between the MTA and the acoustic source. Its basic principle is that there must be part of hydrophones overlapped in the MTA between two consecutive measurements thus to get the phase compensate factor in the space information of the overlapped hydrophones. Thus to expand the MTA effective aperture in the continuous measurements while it is still deals in the array element domain[3].

Under the basis of Yen's method, Stergios Stergiopoulos brought forward FFTSA method. This method also requires the overlap between two consecutive measurements to get the phase compensate factor without knowing the relative velocity between the MTA and acoustic source beforehand. Thus FFTSA can be put into practice.

The M consecutive measurements from the MTA consists of N hydrophones are Fourier transformed:

$$B(f_s,\theta_s)_{M_\tau}=\sum_{m=1}^{M}b(f_0,\theta_s)_{t_i+m\tau}exp[-j2\pi f_s m\tau] \tag{11}$$

Here $f_s=f_{0s}(1\pm v_s sin\theta_s/c)$, (θ,f_0,v) is the actual signal parameters, (θ_s,f_{0s},v_s) is the variable or steering quantities. According to equation (8), equation (11) can be:

$$B(f_s,\theta_s)_{M_\tau}=b(f_0,\theta_s)_{t_i}\cdot$$

$$\sum_{m=1}^{M}exp[j2\pi f_0(1-\frac{v}{c}sin\theta)m\tau]exp[-j2\pi f_s m\tau] \tag{12}$$

According to equation of (12), the beam power output of (9) is $P(f_s,\theta_s)_{M_\tau}$:

$$P(f,\theta_s)_{M_\tau}=B(f_s,\theta_s)_{M_\tau}B^{+}(f_s,\theta_s)_{M_\tau}$$

$$=\{\frac{sin[\frac{N\pi d}{\lambda}(sin\theta_s-sin\theta)]sin[\frac{M\pi v\tau}{\lambda}(sin\theta_s-sin\theta)]}{sin[\frac{\pi d}{\lambda}(sin\theta_s-sin\theta)]sin[\frac{\pi v\tau}{\lambda}(sin\theta_s-sin\theta)]}\}^2 \tag{13}$$

Where λ is the wavelength corresponds to the source frequency f_0, "+" represents the complex conjugate.

To the M times continuous measurements, FFTSA required that the space sampling frequency is higher than $(Nd/2\lambda)^{-1}$, that is to say, the time interval between the consecutive measurements must satisfy $v\tau\le(Nd)/2$, represents the overlap between the consecutive measurements must be not less than or greater than the half length of the MTA physical array consists of N hydrophones.

Though equation (2)-(13) only discuss the situation of one acoustic source, but to the situation of multi acoustic source, they are all true. FFTSA simplifies Yen's method, using the beam output of several measurements to get the beam output of the MTA after the expanding of the effective aperture to improve bearing resolution and make it easier to be compatible with the sonar system in existence. And also FFTSA doesn't need to know the relative velocity between the MTA and the acoustic source and the received frequency beforehand to estimate the phase compensate factor, it has a bright future.

4 High-resolution Direction-of-Arrival Estimation Method of FFTSA Passive Synthetic Aperture Method based on Line spectrum of ship Noise

As mentioned in the first chapter, there is stable line spectrum in ship noise under hundreds of Hz, line spectrum is the continuous monochromatic signal, which adapts to the PASA technology such as FFTSA. This paper brought forward a new method to improve bearing resolution of towed linear array: Using the stable line spectrum of ship noise and FFTSA to achieve higher bearing resolution, it only requires that the integral time is less than the time coherence length, thus can expand the effective aperture of towed linear array to any size no more than the space coherence length of the ocean acoustic channel. The implementation block diagram is as figure 1:

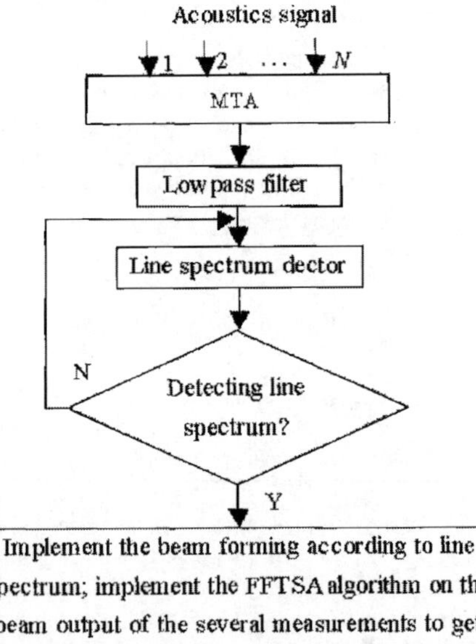

Figure 1: the block diagram of FFTSA high-resolution direction-of-arrival estimation method based on line spectrum of ship noise

First, passing the received signal of the MTA hydrophones through lowpass filter to get the frequency band of line spectrum, then send the signal to line spectrum detector to detect line spectrum; after than, using line spectrum to form the beam; Using FFTSA algorithm to implement on the beam output of the several measurements to get the beam output of the MTA after expansion of the effective aperture to achieve the higher bearing resolution.

5 Simulation Analysis

In order to illustrate the validity and the feasibility of FFTSA high-resolution direction-of-arrival estimation method based on line spectrum of ship noise, the simulation is used to authenticate.

The simulation scene is as follows: a towed linear array consists of 16 hydrophones, the space is $d = 2m$ between two consecutive hydrophones, the central frequency of the corresponding received signal is 375 Hz, the main lobe width of its physical aperture is approximately 6.3°, expanding the MTA to the synthetic aperture of 64 hydrophones, according to FFTSA algorithm, the measure time interval $\tau = 1s$, the total number of measurements $M = 49$, that is, the integral time is $49s$; there are two equal intension ship noise targets: first has line spectrum at 350Hz, the second has line spectrum at 300 Hz and 500 Hz, SNR are also 0dB, the sampling frequency f_s is 2 kHz, using 8 rank Chebyshev lowpass filter, the cutoff frequency is 600Hz.

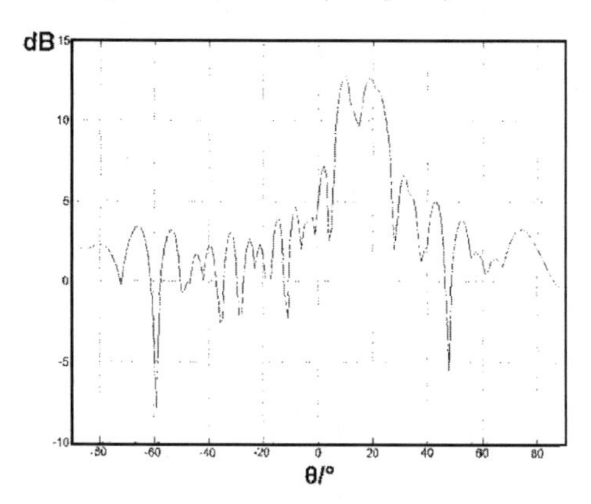

Figure 2: 16 hydrophones physical MTA CBF power output

Figure 2 and figure 3 is when the bearing of the two targets is 10° and 20°, CBF power output of the physical MTA consists of 16 hydrophones and the synthetic MTA consists of 64 hydrophones. At this

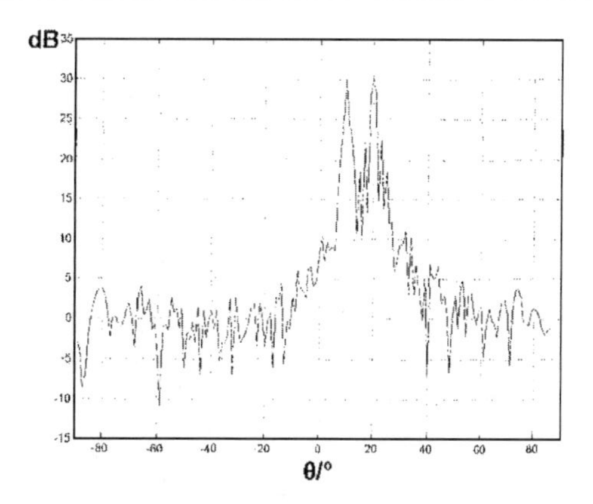

Figure 3: 64 hydrophones synthetic MTA CBF power output

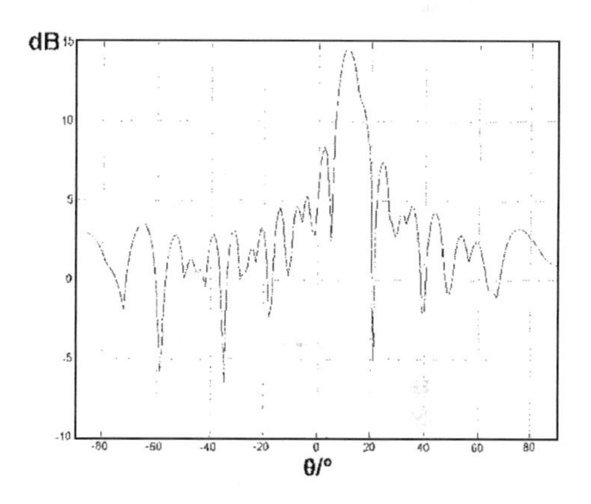

Figure 4: 16 hydrophones physical MTA CBF power output

time the two array both can distinguish two targets, but the acutance of spectrum peak of the synthetic MTA consists of 64 hydrophones is much better than the physical MTA consists of 16 hydrophones; Figure 4 and figure 5 is when the bearing of the two targets is 10° and 13°, CBF power output of the physical MTA consists of 16 hydrophones and the synthetic MTA consists of 64 hydrophones. At this time the physical MTA consists of 16 hydrophones cannot distinguish two targets while the synthetic MTA consists of 64 hydrophones can. Compared with four figures, the power output of the synthetic MTA consists of 64 hydrophones is much larger than the physical MTA consists of 16 hydrophones,

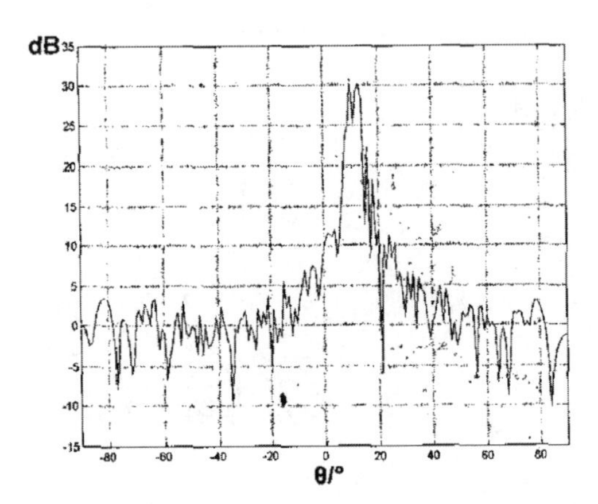

Figure 5: 64 hydrophones synthetic MTA CBF power output

it illustrates that using FFTSA can get the higher signal gain than the physical array.

After all, the simulation proves the validity and feasibility of FFTSA high-resolution direction-of-arrival estimation method based on line spectrum of ship noise.

6 Conclusion

How to improve bearing resolution of towed linear array sonar is an interesting research point, this paper brings forward FFTSA high-resolution direction-of-arrival estimation method based on line spectrum of ship noise, in the continuous measurements, using stable line spectrum composition in the low frequency band of ship noise to form the beam, implementing FFTSA algorithm on the beam output of the several measurements to get the beam output of the MTA after the expansion of the effective aperture thus to improve bearing resolution of towed linear array. Simulation proves this method is effective, practical, simple, and easy to be put into practice and has a bright future.

References

[1] *The research series on towed linear array sonar*, Institute of Acoustics, Academia Sinica, China, 1989.

[2] Nai-Chyuan Yen and William Carey, "Application of synthetic-aperture processing to towed-array data," *J.Acoust.Soc.Am.*, vol. 86, no.2, pp. 754–765, 1989.

[3] Stergios Stergiopoulos, "Extended towed array processing by an overlap correlator," *J.Acoust.Soc.Am.*, vol. 86, no. 1, pp. 158–171, 1989.

[4] Stergios Stergiopoulos, "Optimum bearing resolution for a moving towed array and extension of its physical aperture," *J.Acoust.Soc.Am.*, vol. 87, no. 5, pp. 2128–2140, 1990.

[5] A.H.Nuttall, "The maximum likelihood estimator for acoustic synthetic aperture processing," *IEEE Journal of Oceanic Engineering*, vol. 17, no. 1, pp. 26–29, 1992.

[6] Stergios Stergiopoulos and Heinz Urban, "A new passive synthetic aperture technique for towed arrays," *IEEE Journal of Oceanic Engineering*, vol. 17, no. 1, pp. 16–25, 1992.

[7] Ross Williams and Bernard Harris, "Passive acoustic synthetic aperture processing techniques," *IEEE Journal of Oceanic Engineering*, vol. 17, no. 1, pp. 8–15, 1992.

[8] R.Rajagopal and P.Ramakrishna Rao, "Performance comparison of PASA beamforming algorithms," *IEEE International symposium on signal processing and its applications*, pp. 825–828, 1996.

[9] Urick R, *Principle of underwater sound, 3rd edition*, New York, McGraw-Hill, 1983.

[10] Dezhao Wang and Erchang Shang, *The Underwater Acoustics*, China Science Publish Group, 1981.

矢量水听器线列阵的被动合成孔径技术①

何心怡[1,2]　蒋兴舟[1]　李启虎[2]

（海军工程大学[1]　武汉　430033）（中国科学院声学所[2]　北京　100080）

摘要：针对拖曳线列阵声纳的方位估计问题，提出了由矢量水听器构成线列阵，运用被动合成孔径技术来提高方位估计性能的方法. 该方法利用矢量水听器本身所具有的 $\cos\theta$ 形式的指向性，从而克服了由传统声压水听器构成的线列阵所固有的左右舷模糊问题；利用舰船噪声低频段的线谱成分，采用被动合成孔径技术实现有效孔径的扩展，从而以小物理孔径的拖线阵达到大物理孔径拖线阵才具有的方位分辨力. 通过仿真实验，验证了该方法的可行性和有效性.

关键词：拖曳线列阵声纳；方位估计；矢量水听器；左右舷模糊；线谱；被动合成孔径技术

中图法分类号：TB5；U666.7

0 引　　言

自从潜艇登上战争舞台后，反潜战就成为海战的重要内容之一. 为了能在几十公里甚至更远的距离上对水下环境进行警戒，20 世纪 70 年代出现的拖曳线列阵声纳现已成为水面舰艇和潜艇的最重要的声纳装备之一.

对于拖曳线列阵声纳而言，良好的方位估计性能尤为重要，然而，它的方位估计性能受到左右舷模糊问题和线列阵的物理孔径所限制.

1）传统的拖曳线列阵声纳的基阵是由无指向性的声压水听器构成的直线阵，它具有圆柱对称的指向性，以线阵为轴，在转角为 θ 的圆锥面上对入射信号的响应是完全一致的，存在着方位模糊问题，如图 1 所示. 在一般的情况下，只需要考虑水平面入射信号的方向，这时就要分清信号究竟是来自左舷还是右舷，也就是通常所说的拖线阵左右舷模糊问题.

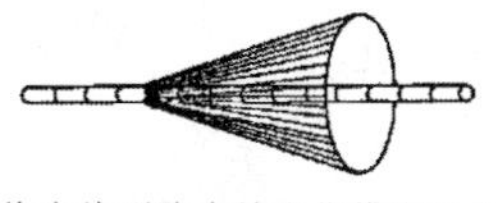

图 1　拖曳线列阵声纳方位模糊问题示意图

2）当拖曳线列阵声纳运用传统的波束形成技术时，其方位分辨率与线列阵的物理孔径和信号波长的比率成正比，被物理孔径所限制. 为达到更远的检测距离，拖曳线列阵声纳通常都工作在低频或甚低频段，因此增加它的方位分辨力意味着要更长的水听器阵列，而更长的水听器阵列意味着更高的成本、增大拖线阵安装的复杂性和给舰艇机动带来更大的负担. 声波是一种兼有标量场和矢量场的物理场，声压是标量，振速是矢量. 然而，长时间以来，声纳系统几乎全是利用声压所携带的信息，声压水听器几乎是水声接收传感器的同义词. 最近几十年来出现的矢量水听器（acoustic vector sensor），将声压水听器和振速传感器组合成一个整体，同时利用了声压信息和振速信息. 单个矢量水听器具有 $\cos\theta$ 形式的指向性，并且该指向性与频率无关，这为其在线列阵中应用并解决左右舷模糊问题提供了可能[1~3]. 最近 20 年来，针对拖曳线列阵的方位估计问题，出现了被动合成孔径技术（passive acoustic synthetic array technology，PASA）[4~8]，它利用拖曳线列阵是运动的拖曳阵列（moving towed array，MTA）这一事实，依靠 MTA 的运动来合成比实际物理孔径大得多的合成孔径，从而显著地提高拖曳线列阵的方位分辨力. 但是，目前国内外有关 PASA 的研究工作都还是着眼于传统的声压水听器，没有将这一工作与矢量水听器联系

① 武汉理工大学学报(交通科学与工程版), 2003, 27(6): 799-803.

起来.

针对拖曳线列阵的方位估计问题,文中提出了矢量水听器线列阵的被动合成孔径技术:由矢量水听器构成线列阵,利用舰船噪声的线谱成分,以被动合成孔径技术中的扩展拖曳线列阵测量方法(extended-towed-array measurements method,ETAM)合成出比实际物理孔径大得多的有效孔径,从而突破阵列孔径的限制,获得比物理孔径高得多的方位分辨力.这样,既克服了传统拖曳线列阵声纳的左右舷模糊问题,又能够大幅度地提高它的方位分辨性能,具有一定的实用前景.

1 矢量水听器信号处理

矢量水听器可以由一个声压水听器和多个相互正交的振速传感器组成,分别测量声场中的声压和相互正交的几个振速分量.一般情况下,仅需要考虑从水平面方向入射的信号,那么仅由一个声压水听器和两个振速传感器构成的矢量水听器就足以完成水平面的定向问题.

$$\begin{cases} p(t) = x(t) \\ v_x(t) = x(t)\cos\theta \\ v_y(t) = x(t)\sin\theta \end{cases} \tag{1}$$

式中:$p(t)$为声压信号;v_x 和 v_y 分别是在水平面上相互正交的两个振速分量;$x(t)$为声压波形;θ 为声波传播的水平方位角.

图 2 中 α 为声线与水平面的夹角,称为声线掠角,仅考虑水平面定向时,$\alpha=0$.振速在 x,y,z 轴上的投影关系见图 2.

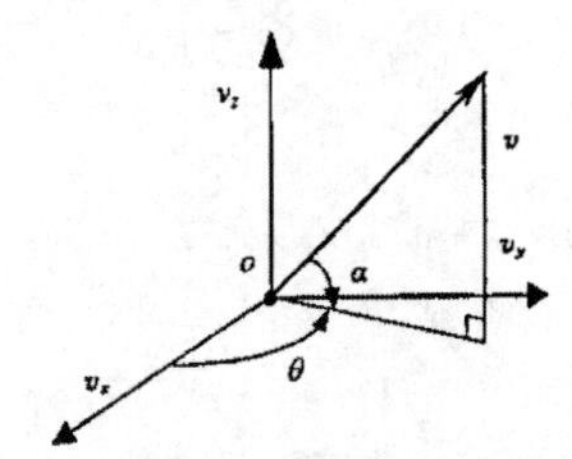

图 2 振速在 x、y、z 轴上的投影示意图

从式(1)可以看出,振速传感器响应声波振速在其轴上的投影分量,所以振速传感器不管它的尺度如何小,均具有 $\cos\theta$ 或 $\sin\theta$ 形式的指向性,并且该指向性与频率无关.

由于矢量水听器有多路输出,可获得比声压水听器更多的额外信息,因此,利用矢量水听器的多路输出进行各种线性组合,可以形成各种双边和单边的心形指向性.那么,根据布里奇(Bridge)乘法定理,即可使由矢量水听器构成的线列阵能够分辨目标所处的左右舷[9].文中提出了一种形成单边心形指向性的线性组合.首先令

$$v_b(t) = v_x(t)\cos\theta_0 + v_y(t)\sin\theta_0 = x(t)\cos(\theta-\theta_0) \tag{2}$$

式中:θ_0 为单个矢量水听器在水平面上的束控方向,只要改变 θ_0 值,$v_b(t)$的指向性就可在水平面内实现电子旋转.

第二步,将 $v_b(t)$与声压信号 $p(t)$进行如下组合

$$v_s(t) = \frac{1}{2}[p(t) - v_b(t)] = x(t)\sin^2\left(\frac{\theta-\theta_0}{2}\right) \tag{3}$$

当 $\theta_0=\pi/2$ 时,$v_b(t)$和 $v_s(t)$的指向性分别如图 3 中的虚线和实线所示.能够看出:$v_b(t)$具有双边的心形指向性,而 $v_s(t)$具有单边的心形指向性,并且都可以通过改变单个矢量水听器的束控方向,实现波束的电子旋转.因此,由矢量水听器构成的线列阵,可从本质上消除左右舷模糊问题.

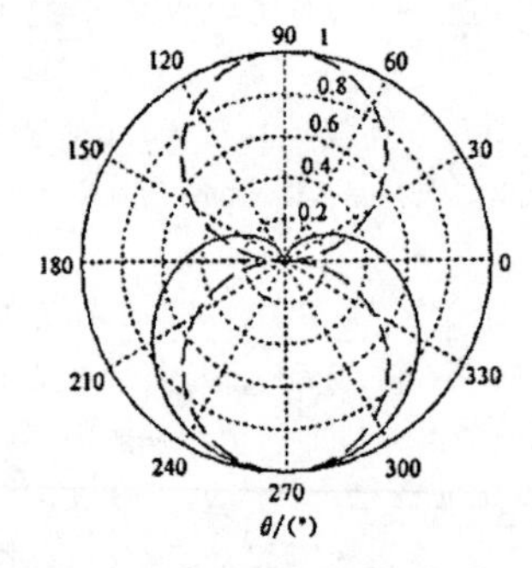

图 3 $v_b(t)$和 $v_s(t)$的指向性图

对于矢量水听器构成的线列阵,可先按式(3)形成单边心形指向性,再按常规的相加阵形成波束,则该线列阵的指向性为

$$P(\theta,\theta_0,\varphi_0) = \sin^2\frac{\theta-\theta_0}{2} \times \frac{\sin\left[\frac{N\pi d(\cos\theta - \cos\varphi_0)}{\lambda}\right]}{N\sin\left[\frac{\pi d(\cos\theta - \cos\varphi_0)}{\lambda}\right]} \tag{4}$$

式中:φ_0 为线列阵在水平面上的束控方向;N 为构成线列阵的矢量水听器的数目;d 为相邻两个矢量水听器的间距;λ 为接收信号的波长.当 $\theta_0=\varphi_0=\pi/2$,$d=\lambda/2$ 时,由 16 个矢量水听器和 16 个声压水听器构成线列阵的指向性分别如图 4 中的实线和虚线所示,能看出,由声压水听器构成的线列阵在 90°和 270°分别有一个最大值,存在着左

右舷模糊问题;而由矢量水听器构成的线列阵仅在270°上有一个最大值,不存在左右舷模糊问题.

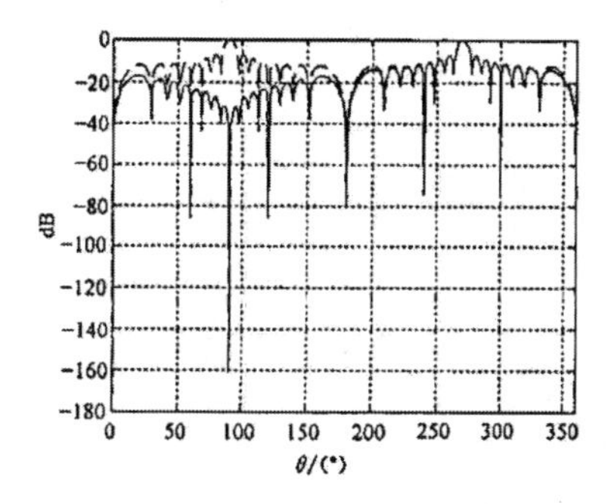

图4 分别由矢量水听器和声压水听器构成的线列阵的指向性图

2 矢量水听器线列阵的ETAM被动合成孔径技术

拖曳线列阵声纳通过接收目标辐射的舰船噪声,实现对目标的检测、估计和识别.根据谱特性划分,舰船噪声包括平稳连续谱、线谱和时变调制谱.研究发现[10],在舰船噪声的低频段,即数百赫兹以下存在的线谱有相当好的稳定性,在数百秒内的变化往往小于1 Hz,线谱通常高出连续谱5~7 dB,有时可高出20~30 dB.在频率范围50~1 500 Hz,海洋声信道的空间相关长度达到了300个波长,对于单频信号(CW),时间相关长度不少于几分钟,甚至可达到十几分钟.因此,稳定强线谱(也就是连续单频信号)的空间相关长度和时间相干长度为合成一个比物理孔径大得多的有效孔径提供了可能.被动合成孔径技术正是利用线谱来大幅度提高拖曳线列阵的方位分辨力.

被动合成孔径技术有3种较成熟的算法,分别是扩展拖曳线列阵测量方法(extended-towed-array measurements method,ETAM)[4]、最大似然法则(maximum likelihood algorithm,ML)[5]和快速傅里叶变换合成孔径处理(FFT synthetic aperture processing,FFTSA)[6],从方位估计精度、稳健性、计算量等方面综合比较,ETAM方法是这3种算法中最好的[7].文中采用ETAM方法扩展矢量水听器线列阵的有效孔径.

假定在远场有一个目标在辐射连续单频信号(也就是线谱,这种情况可扩展到多个声源),由N个等间距的水听器组成的拖曳线列阵,它以速度v沿着直线拖曳航行,相邻两个水听器的间距为d.设采样频率为f_s,对应的采样周期为ΔT_s,则$t_i=i\Delta T_s$,这里$i=1,2,\cdots,M$.M为每个水听器时间序列的数据点数.那么,第n个水听器接收到的目标声压信号可表示为

$$x_n(t_i)=A\exp\left[\mathrm{j}\omega_d\left(t_i-\frac{vt_i+d(n-1)}{c}\sin\beta\right)\right]+\varepsilon_n(t_i)\qquad n=1,2,\cdots,N \tag{5}$$

式中:c为海水中声波的传播速度.通常情况下,目标沿着方位轴的运动速度可以忽略,那么接收到的单频信号的频率可表示为$\omega_d=\omega_0[1\pm(v\sin\beta)/c]$;$\omega_0$为静止时的信号频率;$A$为幅度;$\varepsilon_n(t_i)$为背景噪声,是零均值、方差为$\sigma_N^2$的独立高斯随机变量;$\beta$为目标的方位.

则τ秒后,$x_n(t_i+\tau)$可表示为

$$x_n(t_i+\tau)=\exp(\mathrm{j}\omega_d\tau)\times A\exp\left\{\mathrm{j}\omega_d\left[t_i-\frac{vt_i+v\tau+d(n-1)}{c}\sin\beta\right]\right\}+\varepsilon_n(t_i+\tau)\qquad n=1,2,\cdots,N \tag{6}$$

而拖曳线列阵的第$(n+q)$个水听器在t_i时刻接收到的信号可写为

$$x_{n+q}(t_i)=A\exp\left[\mathrm{j}\omega_d\left(t_i-\frac{vt_i+d(n+q-1)}{c}\sin\beta\right)\right]+\varepsilon_n(t_i) \tag{7}$$

比较式(6)和式(7),能发现:通过合适选择v和τ,使得$v\tau=qd$,那么有

$$x_n(t_i+\tau)=\exp(\mathrm{j}\omega_d\tau)x_{n+q}(t_i) \tag{8}$$

从式(8)可知,若能够估计出相位因子$\omega_d\tau$,就可通过$x_n(t_i+\tau)$合成出第$(n+q)$个水听器的信号$x_{n+q}(t_i)$.对相位因子$\omega_d\tau$的估计,是所有被动合成孔径技术的核心.

ETAM方法是在连续多次测量中,通过合适地选择v和τ,使得$v\tau=qd$,τ代表着相邻两次测量的时间间隔,q代表拖曳线列阵在一次测量中移动的水听器位置数.那么,在两次连续测量中,有$(N-q)$对水听器是重叠的.每对重叠的水听器信号可提供一个相位估计

$$\Psi_n=\arg\{x_{n+q}(t_i)x_n^*(t_i+\tau)\}=\arg\{A^2\exp(-\mathrm{j}\omega_d\tau)\}=-\omega_d\tau\qquad n=1,2,\cdots,N-q \tag{9}$$

则$(N-q)$对水听器信号提供了一个相位修正因子

$$\hat{\Psi}=\frac{1}{N-q}\sum_{n=1}^{N-q}\Psi_n \tag{10}$$

有了相位修正因子 Ψ，就可以合成出虚拟水听器信号，实现拖曳线列阵有效孔径的扩展. 这样，通过 J 次测量，使得合成的拖曳线列阵具有 $N+J\cdot q$ 个物理的和虚拟的水听器，它的有效孔径扩展到 $(N+J\cdot q-1)d$.

在实际应用中，首先要将接收信号通过线谱检测器判断线谱的有无. 当检测到线谱后，将各矢量水听器接收到的信号进行傅里叶变换后，取出线谱频率处的信号成分运用ETAM方法合成孔径. 当然，在矢量水听器线列阵中运用ETAM方法，不能单纯利用声压信号，如果这样做，左右舷模糊问题仍不可避免. 因此，必须要利用声压信号和振速信号的适当组合，使得组合后的信号具有单边的心形指向性，从而使得合成的有效孔径也能分辨目标所处的左右舷. 在文中的仿真实验里，采用的组合信号是式(3)得到的.

3 仿真研究

仿真场景如下. 一个拖曳线列阵由12个矢量水听器构成，相邻两水听器间距1 m，对应接收信号的中心频率为750 Hz，主瓣宽度约为8.46°，以3 m/s的速度拖曳直线航行. 采样频率为3 kHz，线谱检测器中采用的FFT点数为4 096点. 每次测量时间间隔 $\tau=2$ s，每次移动的水听器位置数为6个，测量总次数为6次，也就是在12 s内将拖曳线列阵的有效孔径扩展到47 m. 为了对比，将相同物理孔径的由12个声压水听器构成的线列阵利用其物理孔径进行波束形成. 在线列阵的右舷存在着两个等强度的舰船噪声目标：第一个目标在680 Hz和800 Hz处存在着线谱，第二个目标在770 Hz处存在着线谱，各水听器接收到的背景噪声均为相互独立的500～1 000 Hz的高斯色噪声，信噪比均为−3 dB.

图5是两个目标分别位于0°和10°时，声压水听器线列阵的传统波束形成(conventional beam forming，CBF)输出和矢量水听器线列阵运用ETAM方法得到的合成孔径的CBF对比图；图6是两个目标分别位于0°和6°时，两个线列阵的CBF对比图. 这两个图中，实线均为矢量水听器线列阵运用ETAM方法得到的合成孔径的CBF输出，虚线均为12个声压水听器构成的线列阵的CBF输出. 目标的方位都是指与线列阵右舷法线所成的角度，规定顺时针为正，逆时针为负.

从图5和图6中能够发现：矢量水听器线列阵运用ETAM方法得到的合成孔径的CBF输出不存在左右舷模糊问题；而声压水听器线列阵则存在着左右舷模糊问题，在两个目标的轴向对称方位上存在着等强度的镜像源干扰. 证明了矢量水听器线列阵，运用ETAM方法也能有效地去除左右舷模糊.

图5说明了两目标间隔10°时，两个线列阵都能分辨出两个目标，但矢量水听器线列阵的合成孔径的谱峰锐度远好于声压水听器线列阵的；而在图6的情况时，声压水听器线列阵已不能分辨出两个目标，但矢量水听器线列阵的合成孔径能够分辨.

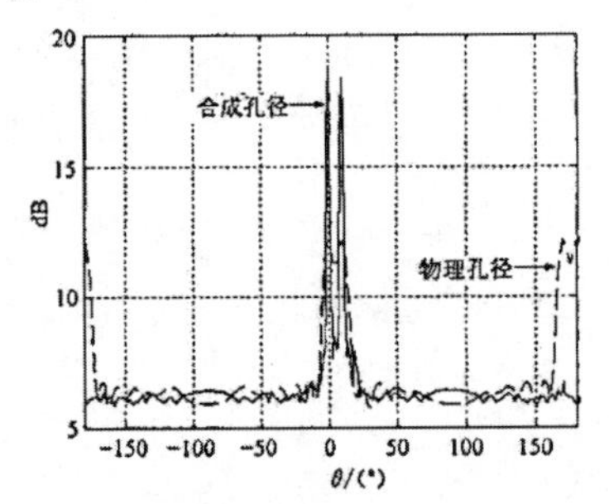

图5 两个目标分别位于0°和10°时，矢量水听器线列阵合成孔径和声压水听器线列阵的CBF输出对比图

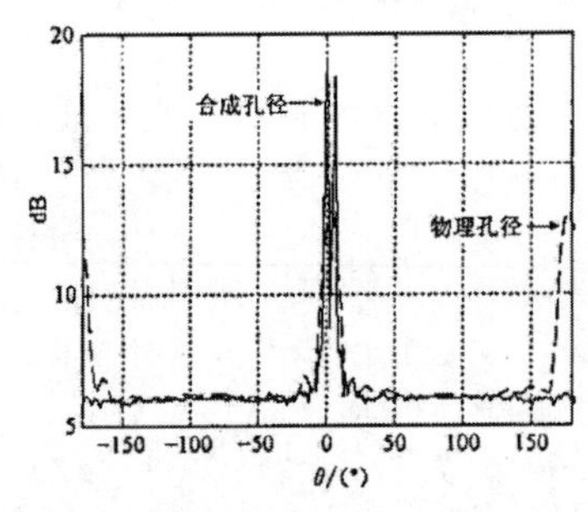

图6 两个目标分别位于0°和6°时，矢量水听器线列阵合成孔径和声压水听器线列阵的CBF输出对比图

总之，仿真实验证明了由矢量水听器构成线列阵，利用舰船噪声的线谱成份，以ETAM方法扩展矢量水听器线列阵的有效孔径，能显著地提高拖曳线列阵的方位分辨力，并且解决了左右舷模糊问题.

4 结 论

文中针对拖曳线列阵声纳的方位估计问题，提出了以矢量水听器构成线列阵，利用舰船噪声低频段的线谱成分，采用被动合成孔径技术中的ETAM方法来提高拖曳线列阵的方位分辨力. 该方法既能够克服传统拖曳线列阵声纳所固有的左右舷模糊问题，又能够以小物理孔径的拖线阵达到大物理孔径拖线阵才具有的方位分辨力，是一种具有实用前景的拖曳线列阵方位估计方法. 通过仿真实验，验证了该方法的可行性和有效性.

参 考 文 献

1 Nickles J C, Edmonds G, Harriss R, et al. A vertical array of directional acoustic sensors. In: Oceans MTS/IEEE, 1992. 340～345

2 D'Spain G L, Hodgkiss W S, Edmonds G L. The simultaneous measurement of infrasonic acoustic particle velocity and acoustic pressure in the ocean by freely drifting Swallow floats. IEEE Journal of Oceanic Engineering, 1991, 16(20): 195～207

3 惠俊英，刘　宏，余华兵等. 声压振速联合信息处理及其物理基础初探. 声学学报，2000，25(4)：303～307

4 Stergios Stergiopoulos. Optimum bearing resolution for a moving towed array and extension of its physical aperture. J. Acoust. Soc. Am., 1990, 87(5): 2128～2140

5 Nuttall A H. The maximum likelihood estimator for acoustic synthetic aperture processing. IEEE Journal of Oceanic Engineering, 1992, 17(1): 26～29

6 Stergios Stergiopoulos, Heinz Urban. A new passive synthetic aperture technique for towed arrays. IEEE Journal of Oceanic Engineering, 1992, 17(1): 16～25

7 Rajagopal R, Ramakrishna Rao P. Performance comparison of PASA beamforming algorithms. In: IEEE International symposium on signal processing and its applications, Gold Coast, Australia, 1996. 825～828

8 Colin M E G D, Groen J. Passive synthetic aperture sonar techniques in combination with tow ship noise canceling: application to a triplet towed array. In: Oceans MTS/IEEE, 2002. 2302～2309

9 李启虎. 声呐信号处理引论. 北京：海洋出版社，2000. 186～187

10 汪德昭，尚尔昌. 水声学. 北京：科学出版社，1981. 517～519

The Passive Synthetic Aperture Technique of Acoustic Vector Sensor Line Array

He Xinyi[1,2]　**Jiang Xingzhou**[1]　**Li Qihu**[2]

(*Navy Engineering University, Wuhan*, 430033)[1]

(*Institute of Acoustics, Chinese Academy of Sciences, Beijing*, 100080)[2]

Abstract

This paper proposes a method to solve the direction-of-arrival estimation of towed line array sonar. By using acoustic vector sensors it composes a line array and passive synthetic aperture technique to improve direction-of-arrival estimation performance. This method overcomes the inherent port/starboard ambiguity which results from line array composed of traditional sound pressure hydrophones with the directivity from acoustic vector sensor. At the same time, this method adopts passive synthetic aperture technique to expend the effective aperture using the line-spectrum composition of the ship noise's low frequency, consequently attains the azimuth resolution of large physical aperture towed line array using small physical aperture towed line array. The simulation shows that this method is feasible and valid.

Key words: towed line array sonar; direction-of-arrival estimation; acoustic vector sensor; port/starboard ambiguity; line-spectrum; passive synthetic aperture technique

拖线阵的阵形畸变与左右舷分辨①

何心怡[1]　蒋兴舟[1]　李启虎[2]　张春华[2]

(1　海军工程大学　武汉　430033)

(2　中国科学院声学研究所　北京　100080)

2002 年 10 月 22 日收到

2003 年 1 月 16 日定稿

摘要　能否利用信号处理方法在单根普通线列阵上实现对目标的左右舷分辨是一个具有实用价值的研究点，文中利用线列阵在拖曳过程中产生的阵形畸变现象来解决单根普通线列阵的左右舷分辨问题。首先建立了拖线阵两种可能的阵形畸变模型：阵列呈圆弧状和阵列呈横向随机误差的类直线阵；然后分析了这两种畸变情况对拖线阵波束形成带来的问题；指出了畸变阵从原理上破坏了直线阵列处理中固有的对称性，按照畸变后的阵形进行波束形成，即可完成对目标的左右舷分辨。在合适的水听器位置标准偏差下，镜像源抑制比能达到 10 dB 以上，并且能显著改善波束形成效果。仿真实验研究充分证明了以上结论。

PACS 数：　43.30, 43.60

The array shape distortion of the towed line array and port/starboard discrimination

HE Xinyi[1]　JIANG Xingzhou[1]　LI Qihu[2]　ZHANG Chunhua[2]

(1　*Navy Engineering University*　Wuhan　430033)

(2　*Institute of Acoustics, The Chinese Academy of Sciences*　Beijing　100080)

Received Oct. 22, 2002

Revised Jan. 16, 2003

Abstract　How to use the method of signal processing to accomplish the port/starboard discrimination on a single line array is a valuable research thesis, the paper use the array shape distortion during the towing of the line array to solve port/starboard discrimination problem on a single line array. First build two models of the possible distortion of the towed line array: the array is traffic circle like and the array is landscape oriented discreet error line-like array; and then analyze the problem brought to the towed line array beam forming by these two conditions; points out that the symmetric features of the line array principle is destroyed by the distortion array. According to the beam forming after distortion, this method can realize the port/starboard distinguishing of the target. With the proper standard error condition of hydrophones, the mirror source restrain ratio can be over 10dB, and can improve the beam forming greatly. Simulation research have proves for that above conclusion.

引言

传统的拖曳线列阵声呐的基阵是由无指向性水听器组成的直线阵，它具有圆柱对称的指向性，以线阵为轴，在转角为 θ 的圆锥面上对入射的信号响应是完全一致的，存在着方位模糊的问题，如图 1 所示。在一般的情况下，只需要考虑水平面入射信号的方向，这时就要分清信号究竟是来自左舷还是右舷，即通常所说的拖线阵左右舷模糊问题。

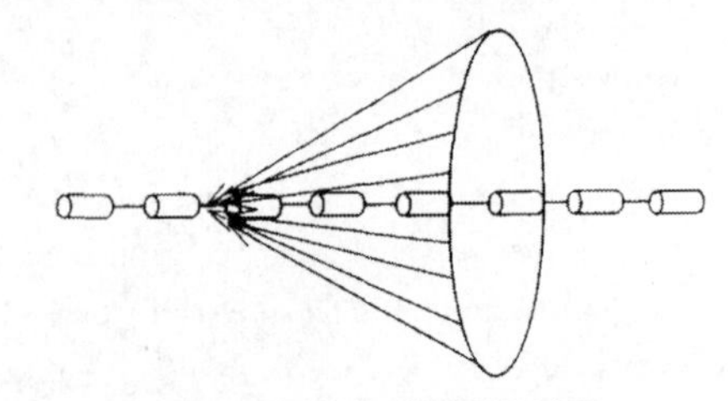

图 1　拖线阵方位模糊问题示意图

① 声学学报, 2004, 29(5): 409-413.

解决左右舷模糊问题的最常见方法是采用本舰机动，在本舰大的机动过程中，根据声呐时间 - 方位历程图中目标方位的变化或根据复杂的跟踪运算法则分辨出目标所处的左右舷[1]。然而，在整个左右舷分辨过程中，基阵的稳定时间比较长，这样既延长了目标捕获时间，又增加了目标丢失的可能性。

最近 10 多年来，拖线阵左右舷分辨技术的进展主要集中在“湿端”，即通过改变阵元结构，通过一定的信号处理方法实现左右舷分辨，如三元水听器组、双线列阵[2−6]。这种方法缺点是：增大了线列阵的物理尺寸、增加了费用。

因此，能否通过信号处理方法在单根普通线列阵上实现对目标的左右舷分辨是一个具有实用价值的研究点，如果能够做到这一点，既可减少拖曳线列阵的物理尺寸，又可减少开发费用。文中的工作正是基于这一思想而进行的。

文中首先建立了两种最常见的阵形畸变模型：阵列呈圆弧状和阵列是具有一定横向随机误差的类直线阵；然后分析了这两种阵形畸变对拖线阵普通波束形成带来的问题：阵列呈圆弧状时，增大了目标方位估计的误差；阵列呈横向随机误差的类直线阵时，对目标方位估计的影响不大，但增大了线列阵的旁瓣高度，对信号检测及参数估计均不利。但是，由于畸变后的阵形从原理上破坏了直线阵列处理中固有的对称性，它不再具有圆柱对称指向性，因此，按照畸变后的阵形进行波束形成，既能够完成对目标的左右舷分辨，又能够改善波束形成效果。通过仿真实验研究，验证了所得结果的正确性。

1 阵列畸变模型

传统的单根线列阵信号处理通常有以下三个假设：(1) 线列阵是由一些无指向性的水听器构成的直线阵，(2) 目标相对于线列阵来说是呈远场关系，并且以平面波传播，(3) 在积分时间内，远场目标相对于线列阵来说是固定的。文中放弃了第一个假设，而认为线列阵是由无指向性水听器构成的非直线阵，即线列阵存在阵形畸变，这也与实际情况相符。

为简化分析，假设拖船朝正北方向航行，以 $h(t)$ 表示航向，此时 $h(t)=0$，$h(t)$ 的反向延长线为 x 轴，线列阵的轴线在 x 轴上，以阵形未畸变时离拖船最近的水听器位置为原点 O，xOy 平面是包含线列阵的水平面。线列阵中水听器数目为 N，依次编号为 $(1,2,\cdots,N-1,N)$，无畸变情况下两相邻水听器间距为 d。那么用 a_i 表示各水听器真实位置，用 $\hat{a}_i=((i-1)d,0)$ 表示无畸变时的水听器位置。

在线列阵的拖曳过程中，阵形畸变是不可避免的。经过分析，将线列阵的阵形畸变归纳为两种模型：

(1) 在拖船的转弯过程中，产生了相对于阵列尺寸的大直径转弯机动，此时阵形畸变成圆弧阵。

(2) 在拖船直线航行过程中，线列阵的水听器产生了横向随机扰动，此时线列阵畸变成具有横向随机误差的类直线阵，也即离散阵。

在直角坐标系中画出这两种阵列形状，分别如图 2(a) 和图 2(b) 所示。

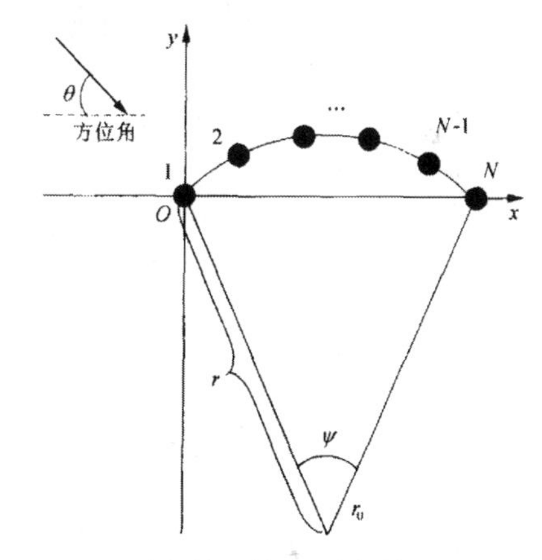

图 2(a) 畸变成圆弧阵示意图

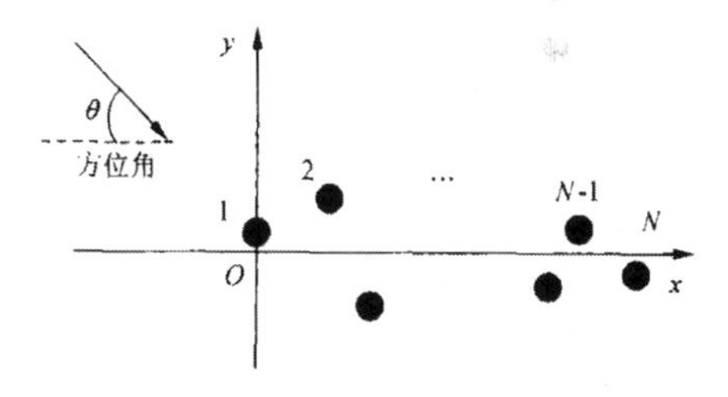

图 2(b) 畸变成类直线阵示意图

图 2(a) 中线列阵的各水听器均匀分布在张角为 ψ，长为 $L=(N-1)d$ 的圆弧上，则圆弧半径为 $r=L/\psi$。图中各水听器位置可表示为：

$$a_i=\begin{cases}2r\sin\left[\dfrac{i-1}{2(N-1)}\psi\right]\cos\left[\dfrac{\psi}{2}-\dfrac{i-1}{2(N-1)}\psi\right],\\[2ex]2r\sin\left[\dfrac{i-1}{2(N-1)}\psi\right]\sin\left[\dfrac{\psi}{2}-\dfrac{i-1}{2(N-1)}\psi\right],\end{cases}\tag{1}$$

图 2(b) 是水听器位置随机扰动，且扰动仅在线列阵的横向方向上发生，则各水听器位置可表示为：

$$a_i=((i-1)d,n_i),\tag{2}$$

其中 n_i 是均值为 0 、正态分布的变量

2 畸变阵按直线阵进行波束形成

设 M 个窄带声源入射信号的方位角分别为 θ_i, $i=1,2,\cdots,M$, 波长为 λ。则线列阵输出的向量为:

$$\boldsymbol{x}(t)=\boldsymbol{A}(\Theta)s(t)+n(t), \tag{3}$$

其中 $\boldsymbol{A}(\Theta)=[a(\theta_1),a(\theta_2),\cdots,a(\theta_N)]$ 为阵列流形, 而 $\boldsymbol{a}(\theta_i)(i=1,2,\cdots,N)$ 称为方向矢量, 将 $\boldsymbol{A}(\Theta)s(t)$ 合称为信号矢量, 表示 t 时刻各声源的输出:

$$\boldsymbol{A}(\Theta)s(t)=\{\exp(\mathrm{j}\langle k_i,a_i\rangle)\}, \tag{4}$$

其中: $k_i=\dfrac{2\pi}{\lambda}[\cos(\theta_i),\sin(\theta_i)]$, λ 是波长。

(3) 式中的 $\boldsymbol{n}(t)=[n_1(t),n_2(t),\cdots,n_N(t)]$ 称为噪声矢量, 表示 t 时刻各水听器接收到的环境噪声。

如果是均匀直线阵, 则:

$$\boldsymbol{a}(\theta_i)=\left[1,e^{-\mathrm{j}2\pi d\cos(\theta_i)/\lambda},\cdots,e^{-\mathrm{j}2\pi(N-1)d\cos(\theta_i)/\lambda}\right]^{\mathrm{T}}. \tag{5}$$

如果是图 2(a) 中的圆弧阵, 若以圆心 r_0 为参考点, 则:

$$\boldsymbol{a}(\theta_i)=\left[e^{-\mathrm{j}\gamma_1},e^{-\mathrm{j}\gamma_2},\cdots,e^{-\mathrm{j}\gamma_N}\right]^{\mathrm{T}}, \tag{6}$$

其中:

$$\gamma_k=-\frac{2\pi r}{\lambda}\cos\left[\theta_i-(k-1)\frac{\psi}{N-1}-\frac{\pi}{2}\right],\quad k=1,2,\cdots,N \tag{7}$$

如果是图2(b)中的离散阵, 以原点O为参考点, 其方向矢量与(6)式相同, 而 $\gamma_k(k=1,2,\cdots,N)$:

$$\gamma_k=-\frac{2\pi}{\lambda}\sqrt{x_k^2+y_k^2}\cos\left[\mathrm{tg}^{-1}\frac{y_k}{x_k}-\left(\frac{\pi}{2}-\theta_i\right)-\frac{\pi}{2}\right],\quad k=1,2,\cdots,N \tag{8}$$

根据最大输出信噪比准则, 则空间波束形成的输出为:

$$y(t)=\boldsymbol{W}^{\mathrm{H}}X(t), \tag{9}$$

其中 $\boldsymbol{W}=[w_1,w_2,\cdots,w_N]^{\mathrm{T}}$ 为权矢量, w_i 代表对应的加权值。阵列的输出功率为:

$$P=\mathrm{E}\{|y(t)|^2\}=\boldsymbol{W}^{\mathrm{H}}\boldsymbol{R}\boldsymbol{W}, \tag{10}$$

其中 $\boldsymbol{R}=\mathrm{E}\{X(t)X^{\mathrm{H}}(t)\}$ 为阵列输出的协方差矩阵。

均匀线列阵的权矢量是:

$$\boldsymbol{W}=\frac{1}{N}\left[1,e^{\mathrm{j}\pi d\cos(\theta)/\lambda},\cdots,e^{\mathrm{j}\pi d(N-1)\cos(\theta)/\lambda}\right]. \tag{11a}$$

而对于圆弧阵和离散阵, 则权矢量 $\boldsymbol{W}=(w_1,w_2,\cdots,w_N)$ 中的各加权值变为:

$$w_i=\frac{1}{N}\left\{\exp\left[\mathrm{j}\frac{2\pi r}{\lambda}\cos\left(\theta_i-(k-1)\frac{\psi}{N-1}-\frac{\pi}{2}\right)\right]\right\}, \tag{11b}$$

$$w_i=\frac{1}{N}\left\{\exp\left[\mathrm{j}\frac{2\pi}{\lambda}\sqrt{x_k^2+y_k^2}\cos\left(\mathrm{tg}^{-1}\left(\frac{y_k}{x_k}\right)-\left(\frac{\pi}{2}-\theta\right)\right)-\frac{\pi}{2}\right]\right\}, \tag{11c}$$

假设一个 64 阵元的直线阵, 按半波长布阵, 目标声源入射方位为 $\theta_0=60°$, 不考虑噪声因素, 若无阵形畸变时, 空间波束形成输出如图 3(a) 所示, 镜像源方位是 $\theta_m=2h(t)-\theta_0=-60°$。

但阵形发生畸变时, 如果还用直线阵的权矢量加权的话, 将产生误差。图 3(b) 和图 3(c) 分别是半径 $r=180$ m 的圆弧畸变阵和随机扰动 n_i 服从于 $(0,(0.1\lambda)^2)$ 分布的类直线阵时的波束形成输出。

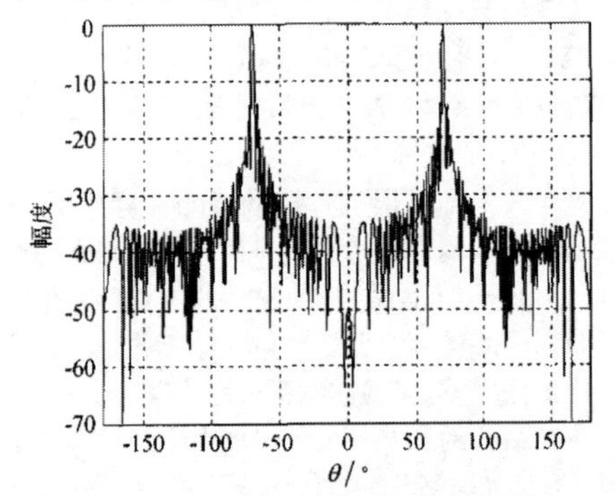

图 3(a) 无阵形畸变时的线列阵波束形成输出

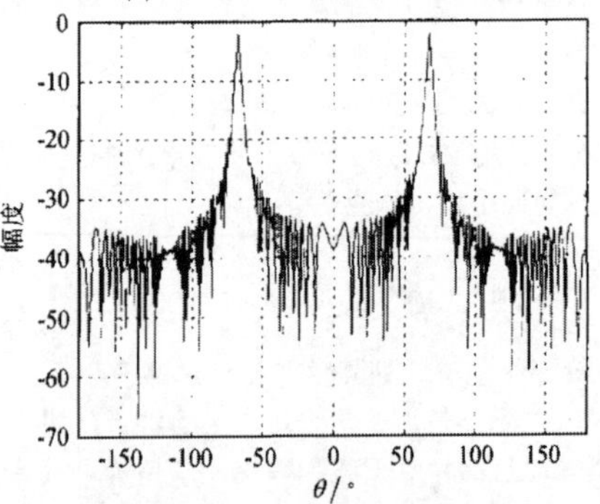

图 3(b) 畸变为圆弧阵时的波束形成输出

图 3(c) 畸变为类直线阵时的波束形成输出

对比图3(a)、图3(b)和图3(c)能够发现，当线列阵畸变时，造成波束形成失真，表现在以下几方面：

(1) 阵列呈圆弧状时，增大了目标方位估计的误差，

(2) 阵列呈类直线阵时，对目标方位估计的影响不大，但增大了线列阵的旁瓣高度，

(3) 波束主瓣变宽，

(4) 基阵增益有所损失，

(5) 有可能会出现高副瓣。

3 利用阵形畸变进行左右舷分辨

如果真实目标方位是 θ_0，则镜像源方位是 $\theta_m = 2h(t)-\theta_0$，用 $\boldsymbol{W}(\hat{a}_i)$ 表示将畸变阵当作直线阵进行波束形成时的权矢量，如式 11(a)，将(9)式重新写为：

$$y(t,\theta,\theta_0,\hat{a}_i) = \boldsymbol{W}(\hat{a}_i)^{\mathrm{H}}X(t) = \frac{1}{N}\sum \exp\{\mathrm{j}(\langle k_0,a_i\rangle\langle k,\hat{a}_i\rangle)\}, \tag{12}$$

其中 $k = \frac{2\pi}{\lambda}(\cos(\theta),\sin(\theta))$，能看出 $|y(t,\theta,\theta_0,\hat{a}_i)| \leq 1$，$|y(t,\theta,\theta_0,\hat{a}_i)| = |y(t,\theta_m,\theta_0,\hat{a}_i)|$；如果 $a_i \neq \hat{a}_i$，那么 $|y(t,\theta_0,\theta_0,\hat{a}_i)| < 1$，如图 3(b) 和图 3(c) 中 $\theta = 60°$ 处所示。

类似地，根据畸变阵的实际阵形进行波束形成时，(9) 式可写成：

$$y(t,\theta,\theta_0,a_i) = W(a_i)^{\mathrm{H}}X(t) = \frac{1}{N}\sum \exp\{\mathrm{j}[\langle (k_0-k),a_i\rangle]\}. \tag{13}$$

由于阵形畸变，无论是圆弧阵和类直线阵均是二维阵列，因而从原理上破坏了直线阵列处理中所固有的对称性，可以看出 (13) 式代表的波束形成器显示了沿线列阵的轴向方向没有对称性。如果没有处理损失时，$|y(t,\theta_0,\theta_0,a_i)| = 1$。

定义镜像源抑制比来描述一个波束形成器对镜像源的抑制能力：

$$\mathrm{SMR} = \frac{|y(t,\theta_0,\theta_0,a_i)|}{|y(t,\theta_0,\theta_0,a_i)|} = \frac{N}{\left|\sum\limits_{i=1}^{N}\exp[\mathrm{j}\langle (k_0-k_m),a_i\rangle]\right|} \tag{14}$$

式中：$k_m = \dfrac{2\pi}{\lambda}[\cos(\theta_m),\sin(\theta_m)]$。

一般地，$\mathrm{SMR} \geq 1$。对于一个使用 (12) 式的直线阵波束形成器时使用等号；或者直线阵使用 (13) 式进行波束形成时 $\mathrm{SMR} = 1$。而当一个非直线阵应用公式 (13) 进行波束形成时 $\mathrm{SMR} > 1$。

可见，SMR 是水听器个数 N 或者说是阵列长度 L、以及水听器位置偏差 $a_i - \hat{a}_i$ 的函数。如果是类直线阵，可以用下式定义的水听器位置的标准偏差 C 来表征水听器偏离直线阵的程度：

$$C = \sqrt{\frac{1}{N}\sum\frac{(a_i-\hat{a}_i)}{\lambda}}. \tag{15}$$

同样，对一个长为 L、半径为 r 的圆弧形阵，对于小的 L/r 时，可以写出 C 和 r 的大致关系[1]：

$$C = \frac{1}{\sqrt{720}}\frac{L}{\lambda}\frac{L}{l}. \tag{16}$$

4 仿真实验研究

为验证理论分析，假设一线列阵由 128 个水听器组成，水听器按半波长布阵，相邻两个水听器间距 0.75 m，接收信号频率为 1 kHz，目标方位角 $\theta = 60°$，线列阵畸变成圆弧阵，水听器位置标准偏差 $C = 0.1$。图 4 为按畸变后的阵形以公式 (13) 进行波束形成的结果，在该情形下 SMR = 7 dB。

图 5 和图 6 分别为 128 个水听器时，不同方位角的 SMR 预测曲线和仿真结果曲线。能看出，仿真研究结果与预测值匹配良好。镜像源抑制比在正横附近最大，而越靠近端射方向 (即线列阵轴向方向)，镜像源抑制比越小，直至为零。这是因为在端射方向，真实目标源和镜像源之间的角度差很小，造成真实目标源波瓣和镜像源波瓣重叠。

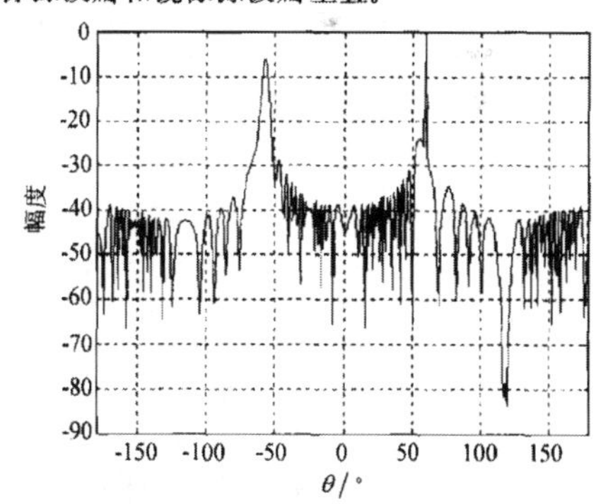

图 4 按畸变后的阵形 (圆弧阵) 进行波束形成的结果

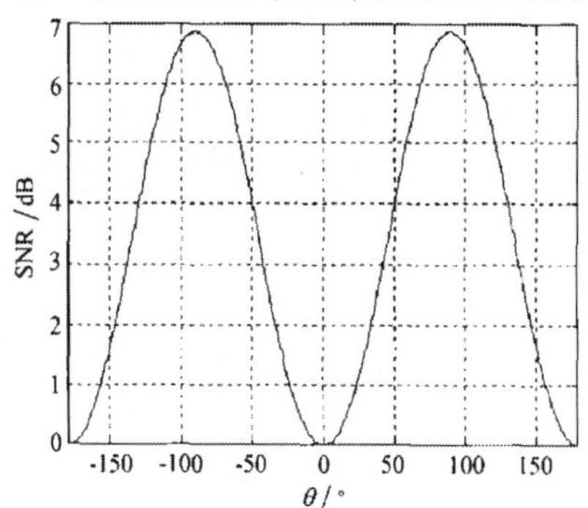

图 5 方位角 - SMR 预测曲线

图 7 是 $\theta = 60°$ 时圆弧阵的 SMR − C 关系曲线，说明了 C 有一个上限，达到这个上限后，就不会有更高的镜像源抑制比。

与图 4 —图 7 仿真条件相同，但线列阵畸变成 $C = 0.1$ 的类直线阵，图 8 为其按畸变后的阵形进行波束形成的结果，此时 SMR > 10 dB 。

分析图4和 图8 能够发现：波束形成均有不同程度的改善，旁瓣也有所降低、主瓣也与理论结果类似；在同样的C下，离散阵比圆弧阵有着更高的镜像源抑制比，这主要因为离散阵比圆弧阵在结构上更复杂，对直线阵列处理中固有的对称性破坏更为彻底。

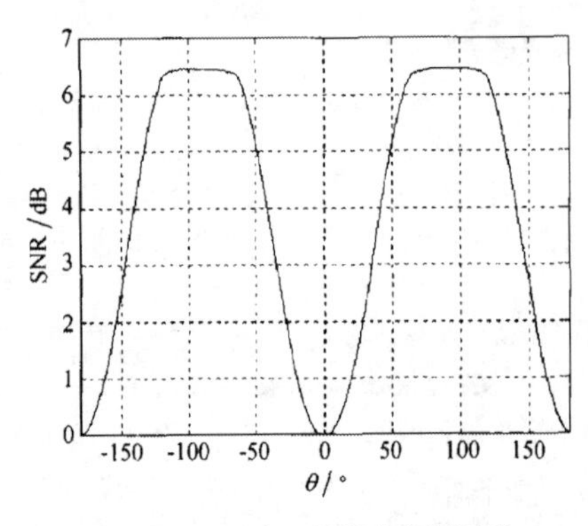

图 6 方位角 - SMR 仿真结果曲线

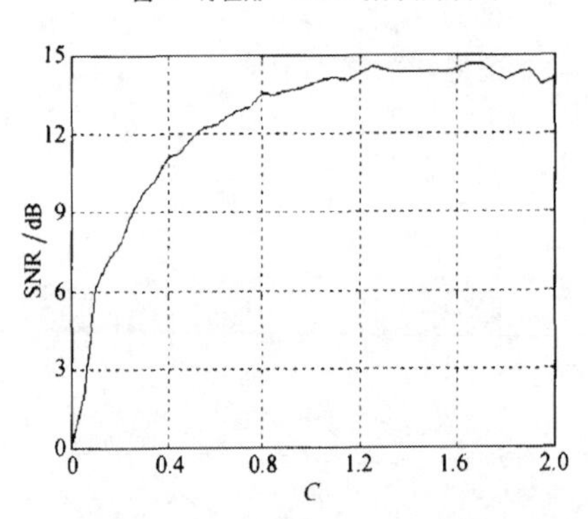

图 7 $\theta = 60°$ 时 SMR − C 曲线

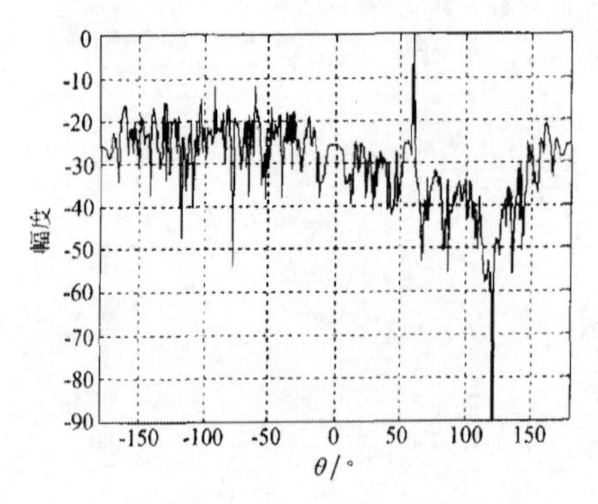

图 8 按畸变后的阵形 (类直线阵) 进行波束形成的结果

整个仿真实验研究充分证明了根据畸变后的阵形进行波束形成，不但能分辨目标所处左右舷，而且对波束形成效果也有所改善。

5 结论

阵形畸变是线列阵实际使用过程中不可避免的现象，文中利用畸变后的阵形从原理上破坏了直线阵列处理中固有的对称性，从而得到显著的镜像源抑制效果，并且改善波束形成效果。该方法可以不改变基阵样式，仅用信号处理方法在单根普通线列阵上实现对目标左右舷分辨。使用该方法的前提是：要估计出畸变后的阵形，这虽然说比较困难，但是有可能实现 [8,9]，也可以在阵形估计精度和左右舷分辨增益大小之间寻找折衷。可通过各种传感器给出的自身平台 (拖船)、拖缆和线列阵的位置信息，采用数据模型来估计单个水听器的位置。

参 考 文 献

1 Grim Steinar Gjonnes, Aagedal. Left/right discrimination in software with conventional, passive thin-line towed submarine arrays . UDT2000, Pacific: 1—6

2 杜选民, 朱代柱, 赵荣荣, 姚 蓝. 拖线阵左右舷分辨技术的理论分析与实验研究. 声学学报, 2000; **25**(5): 395—402

3 杜选民, 朱代柱, 赵荣荣, 姚 蓝. 拖线阵目标左 / 右舷分辨技术研究 (II)— 噪声相关性模型. 舰船科学技术, 1999; **21**(3): 17—20

4 Jean Bertheas, Gilles Moresco, Philippe Dufourco. Linear hydrophonic antenna and electronic device to remove right/left ambiguity, associated with the antenna . United States Patent, Patent Number: 5058082, 1991

5 Doisy Y. Port-starboard discriminatiojn performances on actived towed array systems . UDT95, France: 125—129

6 Van Mierlo G W M, Beerens S P, Been R *et al.* Port/starboard discrimination by hydrophone triplets in active and passive towed arrays . UDT97, Germany: 176—181

7 拖曳式线列阵声纳研究丛书. 北京：中国科学院声学研究所, 1989: 121—124

8 Gray D A, Anderson B D O, Bitmead R R. Towed array shape estimation using Kalman filters-theoretical models. *IEEE Journal of Oceanic Engineering*, 1993; **18**(4): 543—556

9 Ferguson B G. Remedying the effects of array shape distortion on the spatial filtering of acoustic data from a line array of hydrophones. *IEEE Journal of Oceanic Engineering*, 1993; **18**(4): 565—571

一种最佳空间滤波器的实现①

栾经德 李启虎[1] 刘文化 汲长利 陈锦光 彭学礼

(中国海洋开发研究中心 北京 100073)

(1 中国科学院声学研究所 北京 100080)

2003 年 2 月 13 日收到

2003 年 7 月 17 日定稿

摘要 根据最佳检测理论以及绝对最佳空间滤波器概念，给出了一种最佳空间滤波器的完整表达式。利用真实舰船航行辐射噪声作为目标及干扰，正态分布高斯白噪声作为噪声背景，在基元等间隔线阵上实现了最佳空间滤波器。给出了不同情况下常规波束形成及最佳空间滤波器处理基阵输出波束图。海上试验数据的实验室分析证明，此最佳空间滤波器能够有效地抑制弱信号附近空间不同方位的数个强干扰，完成指定弱信号的检测，而且该最佳空间滤波器可以方便、实时地在现役数字化声呐装备上实现。

PACS 数：43.60, 43.30

Realization of one type of optimum spatial filter

LUAN Jingde LI Qihu[1] LIU Wenhua JI Changli CHEN Jinguang PENG Xueli

(*China Marine Development and Research Center* Beijing 100073)

(1 *Institute of Acoustics, The Chinese Academy of Sciences* Beijing 100080)

Received Feb. 13, 2003

Revised Jul. 17, 2003

Abstract Based on the concept of optimum detection theory and optimum spatial filtering theory, the expression of OSF (optimum spatial filter) in frequency domain is derived. The realization of OSF in the case of real target radiated noise in the background of Gaussian white noise for equal spaced line array is illustrated. The beam patterns of CBF (conventional beamforming) system and OSF system are compared. The analysis of at sea experiment data shows that the OSF described in this paper can effectively suppress multiple interferences, which come from in the neighbourhood of target direction and extract the weak signal. This kind of OSF can be easily implemented in digital sonar system.

引言

随着安静型潜艇的出现和消声瓦技术的广泛采用，提高远程弱信号检测能力成为现代声呐的关键技术之一；现代海战中联合作战和多目标威胁的增多，提高声呐方位分辨率成为现代声呐面临的又一关键技术。降低声呐工作频率可以提高声呐作用距离，但必须增大声呐基阵的声孔径，增加通道数量，随之而来的是费用的提高以及受到舰艇安装平台空间的限制。

DICANNE 系统是抑制强方向性背景干扰，检测弱信号的一种方法，但由于干扰波束中剩余信号的存在，要得到纯干扰信号非常困难。而且 DICANNE 系统仅能抑制一个点源干扰信号[1]。现代自适应处理技术是提高弱信号检测能力、提高目标方位分辨率的一种技术[2-4]，但由于水声传输信道的复杂性及不确定性，实时对多水听器、宽带信号进行自适应处理，仍然面临着严重的计算复杂性和实时性问题。M.Simaan, M.T.Hanna, K.M.Buckley 等人利用空间滤波器提高弱信号检测性能，提高目标方位分辨率，开展了绝对最佳空间滤波器方面的理论工作[5-9]，但没有进行工程实践。基于绝对最佳空间滤波器概念，本文给出了抑制强方向性背景干扰、检测弱信号的一种空间滤波器结构，绕开了自适应方法必须通过步进迭代带

① 声学学报, 2004, 29(2): 143-148.

来的严重的计算问题，直接求出问题的最佳解。在基元等间隔线阵上实现了最佳空间滤波器，取得了对弱信号附近空间内数个强干扰源较好的抑制效果。

1 最佳空间滤波器的数学表述

设基阵接收通道数为 N，基本的多通道空间滤波器结构如图 1 所示。

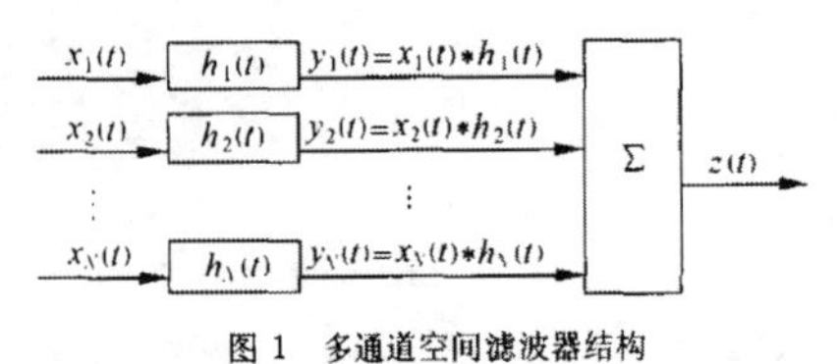

图 1 多通道空间滤波器结构

第 i 个接收通道接收到的信号可以写成下列形式：

$$x_i(t)=p_i(t)+q_i(t), \tag{1}$$

其中 $p_i(t)$ 代表希望检测到的信号分量，$q_i(t)$ 代表所不希望检测到的干扰分量，暂时不考虑背景噪声。

设空间滤波器第 i 个通道的脉冲响应函数为 $h_i(t)$，对应的传递函数为 $H_i(f)$，则空间滤波器的第 i 路输出为：

$$y_i(t)=x_i(t)*h_i(t). \tag{2}$$

基阵滤波器的输出为各路输出之和：

$$Z(t)=\sum_{i=1}^{N}x_i(t)*h_i(t). \tag{3}$$

我们的目的是设计这样一种空间滤波器，它对于所希望的信号分量响应最大，即允许所希望的信号分量完全通过，而对于所不希望的干扰分量响应最小，即拒绝干扰分量通过。

首先讨论一种特殊情况，设空间滤波器的每一通道已经束控到信号方向，即：

$$x_i(t)=p(t)+q(t-\Delta_i). \tag{4}$$

由于已经假设空间滤波器的每一通道已经束控到信号方向，因此信号分量 $p(t)$ 的时间延迟因子已经得到精确补偿， $q(t-\Delta_i)$ 为基阵第 i 通道的干扰分量。将 (4) 式代入 (3) 式，得：

$$Z(t)=\sum_{i=1}^{N}p(t)*h_i(t)+\sum_{i=1}^{N}q(t-\Delta_i)*h_i(t). \tag{5}$$

对基阵滤波器输出 (5) 式进行傅里叶变换，得：

$$Z(f)=\mathrm{FFT}[Z(t)]=P(f)\sum_{i=1}^{N}H_i(f)+Q(f)\sum_{i=1}^{N}\exp(\mathrm{j}2\pi f\Delta_i)H_i(f). \tag{6}$$

记
$$R(f)=\sum_{i=1}^{N}\exp(\mathrm{j}2\pi f\Delta_i)H_i(f), \tag{7}$$

称为干扰响应。

将 (7) 代入 (6)，有：

$$Z(f)=P(f)\sum_{i=1}^{N}H_i(f)+R(f)Q(f). \tag{8}$$

在 (8) 式中，第 1 项为信号响应项，第 2 项为干扰响应项。

定义 若空间滤波器的干扰响应及传递函数满足：

$$\begin{cases}|R(f)|^2\to\min,\\ \displaystyle\sum_{i=1}^{N}H_i(f)=1,\end{cases} \tag{9}$$

则称此空间滤波器为绝对最佳空间滤波器。

由此可以看出，绝对最佳空间滤波器可以最大限度地抑制干扰，使待检测信号最大限度地通过。因此，求解最佳权重矢量 $\boldsymbol{H}_{opt}$，可以转化为在

$$\sum_{i=1}^{N}H_i(f)=1 \tag{10}$$

约束条件下，求 $|R(f)|^2$ 的极值问题。为了使 $|R(f)|^2$ 达到最小，利用拉格朗日乘子法，引入目标函数：

$$u(H(f))=|R(f)|^2+\lambda\left[\sum_{i=1}^{N}H_i(f)-1\right], \tag{11}$$

其中 λ 为待定常数，称为拉格朗日乘子。由极值的必要条件，可以得到下列方程组：

$$\begin{cases}\dfrac{\partial u(H(f))}{\partial H}=0,\\[2mm] \dfrac{\partial u(H(f))}{\partial \lambda}=0,\end{cases} \tag{12}$$

解此方程组 [13]，即可得到绝对最佳空间滤波器的传递函数 $H_{opt}(f)$：

$$H_{i,opt}(f)=\frac{N-\exp(-\mathrm{j}2\pi f\Delta_i)\displaystyle\sum_{n=1}^{N}\exp(\mathrm{j}2\pi f\Delta_n)}{N^2-\left|\displaystyle\sum_{n=1}^{N}\exp(\mathrm{j}2\pi f\Delta_n)\right|^2},\quad i=1,2,\cdots,N \tag{13}$$

在式 (4) 中，空间滤波器的每一通道已经束控到信号方向，每一通道中的信号分量 $p(t)$ 的时间延迟因子已经得到精确补偿。对于一般情况，设第 i 通道信号分量的时间延迟因子为 Δ_i，干扰分量的时间延迟因子为 Δ_i，(4) 式可以写成下述形式：

$$x_i(t)=p(t-\Delta_i)+q(t-\Delta_i),\quad i=1,\cdots,N \tag{14}$$

(6) 式可以改写成:

$$Z(f)=P(f)\sum_{i=1}^{N}\exp(\mathrm{j}2\pi f\varDelta_i)H_i(f)+Q(f)\sum_{i=1}^{N}\exp(\mathrm{j}2\pi f\varDelta_i)H_i(f). \tag{15}$$

记 $\exp(\mathrm{j}2\pi f\varDelta_i)H_i(f)\hat{=}G_i(f).$ (16)

可以得到:

$$Z(f)=P(f)\sum_{i=1}^{N}G_i(f)+Q(f)\sum_{i=1}^{N}\exp[\mathrm{j}2\pi f(\varDelta_i-\varDelta_i)]G_i(f). \tag{17}$$

仿照 (6) 式及 (13) 式, 可以得到:

$$G_{i,opt}(f)=\frac{N-\exp[\mathrm{j}2\pi f(\varDelta_i-\varDelta_i)]\sum\limits_{n=1}^{N}\exp[\mathrm{j}2\pi f(\varDelta_n-\delta_n)]}{N^2-\left|\sum\limits_{n=1}^{N}\exp[\mathrm{j}2\pi f(\varDelta_n-\delta_n)]\right|^2}\quad i=1,\cdots,N. \tag{18}$$

因此, 传递函数的最佳解:

$$H_{i,opt}(f)=\frac{N-\exp[\mathrm{j}2\pi f(\varDelta_i-\varDelta_i)]\sum\limits_{n=1}^{N}\exp[\mathrm{j}2\pi f(\varDelta_n-\delta_n)]}{N^2-\left|\sum\limits_{n=1}^{N}\exp[\mathrm{j}2\pi f(\varDelta_n-\delta_n)]\right|^2}\exp(-\mathrm{j}2\pi f\varDelta_i),\quad i=1,\cdots,N. \tag{19}$$

与 DICANNA 系统不同, 本文采用后置联合检测系统对强方向性干扰实施抑制, 进而完成对指定弱信号的检测。所谓后置联合检测系统, 就是首先对每一通道信号按照常规波束形成方法进行波束形成, 接着在每一波束中, 将波束束控到信号方向, 对形成该波束的所有通道信号进行干扰抑制处理。先常规波束形成, 后干扰抑制处理。

由于干扰抑制系统束控到信号方向, 因此第 i 通道的接收信号可以写成

$$x_i(t)=S[t-\tau_i(\theta_s)+\tau_i(\hat{\theta}_s)]+I[t-\tau_i(\theta_I)+\tau_i(\hat{\theta}_s)], \tag{20}$$

其中 $\hat{\theta}_s$ 是 θ_s 的估计值。仿照 (14) 式及 (19) 式, 可以得到干扰抑制系统最佳空间滤波器传递函数表达式:

$$H_{i,opt}(f)=\frac{N-\exp\{\mathrm{j}2\pi f[\tau_i(\theta_s)-\tau_i(\theta_I)]\}\sum\limits_{n=1}^{N}\exp\{\mathrm{j}2\pi f[\tau_n(\theta_I)-\tau_n(\theta_s)]\}}{N^2-\left|\sum\limits_{n=1}^{N}\exp\{\mathrm{j}2\pi f[\tau_n(\theta_I)-\tau_n(\theta_s)]\}\right|^2}\exp\{\mathrm{j}2\pi f[\tau_i(\hat{\theta}_s)-\tau_i(\theta_s)]\},\quad i=1,\cdots,N \tag{21}$$

在 (21) 式中, 可以设想, 不论 θ_I 定位在哪一个干扰方向上, 最佳空间滤波器只对指定的信号方位 θ_s 响应最大。对于任意一个干扰, 即不管 θ_I 定位在什么方位上, 干扰响应皆为最小。因此, 最优解应该与干扰源的空间分布无关, 最佳空间滤波器应该能够抑制空间内存在的数个干扰源。

干扰抑制系统方框图如图 2 所示。

后置联合检测系统方框图如图 3 所示。

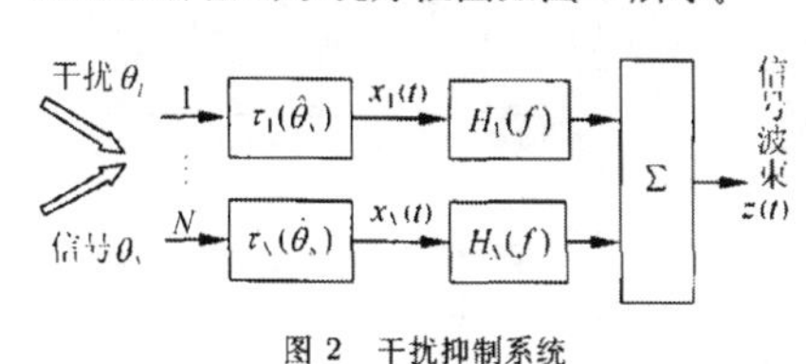

图 2 干扰抑制系统

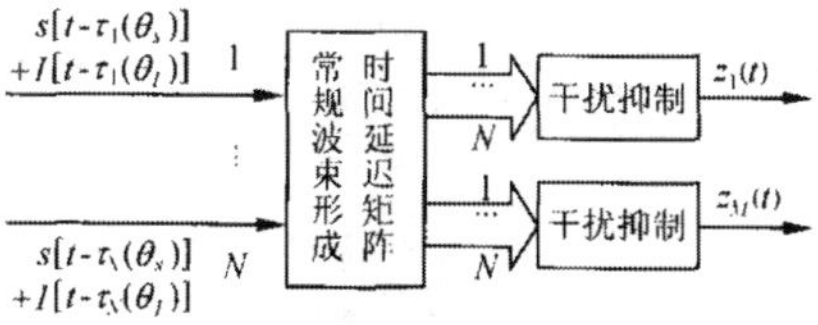

图 3 后置联合检测系统

2 最佳空间滤波器在基元等间隔线阵上的实现

基元等间隔线阵如图 4 所示, 水听器从左到右顺序编号为 $H_1,\cdots,H_N$, 阵元间隔为 d, 为了计算上的方便, 将时间参考点选在 H_1 上, 入射波与基阵法线方向的夹角为 θ 。选取参数如下:

阵元间隔 $d = 1.5$ m,

阵元数 $N = 48$,

$-90° \sim 90°$ 范围内不均匀形成 101 个波束，每一个波束由 48 个基元形成,

最佳空间 FIR 滤波器 H_{opt} 长度取 1024,

信号采样频率 $F_s = 50$ kHz,

信号、干扰采用海上采集的真实目标辐射噪声。对于预检测的目标，将其辐射噪声作为信号，对于不希望检测的目标，其辐射噪声将作为干扰。背景噪声选用正态分布高斯白噪声。

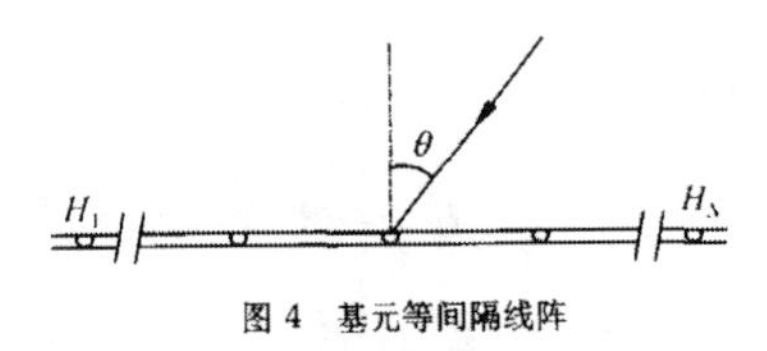

图 4　基元等间隔线阵

2.1　对于空间内存在两个目标情况

目标 1 为水下高速运动目标，方位 30°, 信号较弱，辐射噪声序列幅值为 [1500]；目标 2 为大型水面目标，方位 0°, 信号较强，辐射噪声序列幅值为 [6000]。取正态分布高斯白噪声作为背景噪声，噪声序列幅值取 [15000]。

对目标 1 实施检测，因此目标 1 为信号，目标 2 为干扰，信噪比 $S/N = -20$ dB, 信号干扰比 $S/I = -12$ dB。常规波束形成 $(H = 1)$ 处理基阵输出波束图如图 5(a) 所示，最佳空间滤波器后置联合检测系统基阵输出波束图如图 5(b) 所示。

2.2　对于空间内存在多个干扰情况

目标 1: 水下高速目标, 方位 30°, 信号幅值 [4000]

目标 2: 水面大型目标, 方位 0°, 信号幅值 [6000]

目标 3: 水下大型目标, 方位 −17.46°, 信号幅值 [1500]

目标 4: 水下高速目标, 方位 −30°, 信号幅值 [4000]

目标 5: 水下大型目标, 方位 −53.13°, 信号幅值 [1500]

背景噪声：高斯白噪声，信号幅值 [15000]

将目标 1 作为待检测信号，信噪比 $S/N = -11$ dB。常规波束形成处理基阵输出波束图如图 6(a) 所示，最佳空间滤波器后置联合检测系统基阵输出波束图如图 6(b) 所示。

2.3　对于上述 5 个目标，当目标方位比较接近时

目标 1: 水下高速目标, 方位 30°, 辐射噪声幅值 [4000]

目标 2: 水面大型目标, 方位 28.69°, 辐射噪声幅值 [6000]

目标 3: 水下大型目标, 方位 5.74°, 辐射噪声幅值 [1500]

目标 4: 水下高速目标, 方位 8.05°, 辐射噪声幅值 [4000]

目标 5: 水下大型目标, 方位 0°, 辐射噪声幅值 [1500]

背景噪声：高斯白噪声，信号幅值 [15000]

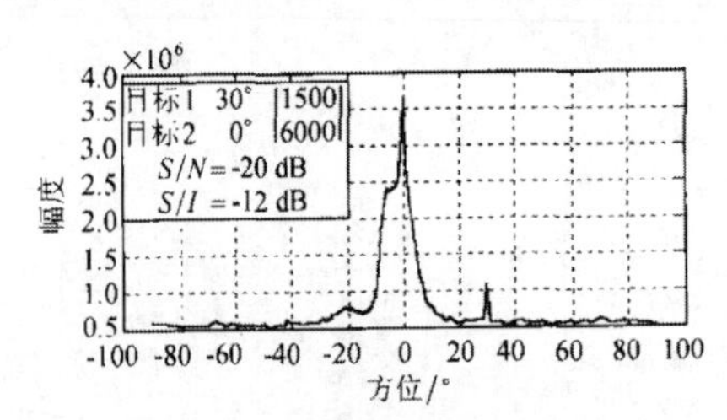

图 5(a)　常规波束形成检测结果 $(H = 1)$

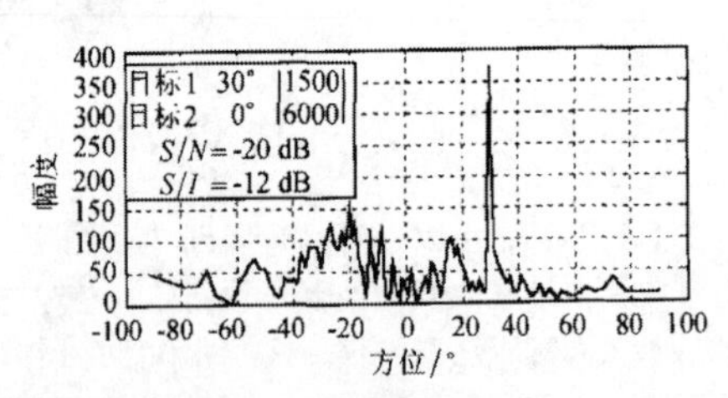

图 5(b)　OSF 后置联合检测结果 $\theta_s = 30°$， $\theta_I = 0°$

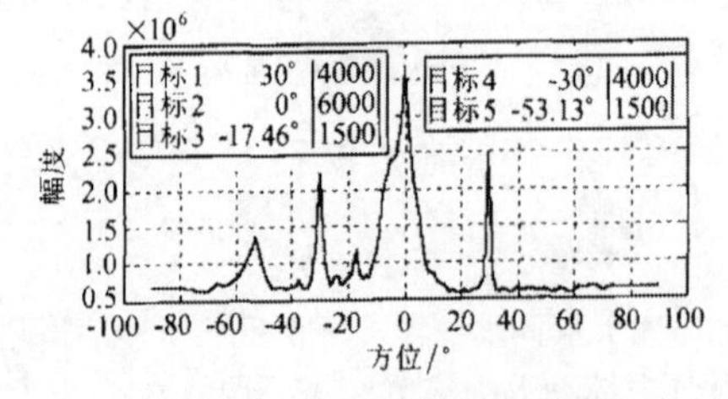

图 6(a)　常规波束形成检测结果 $(H = 1)$

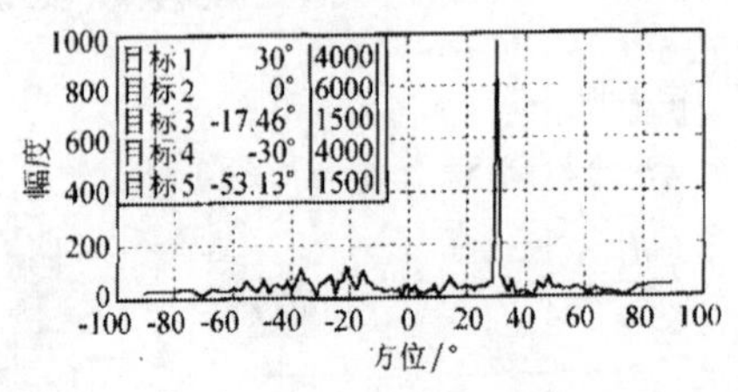

图 6(b)　OSF 后置联合检测结果 $\theta_s = 30°$， $\theta_I = 0°$

最佳空间滤波器后置联合检测系统对目标 1 的检测结果如图 7 中实线 1 所示，对目标 2 的检测结果如图 7 中实线 2 所示。

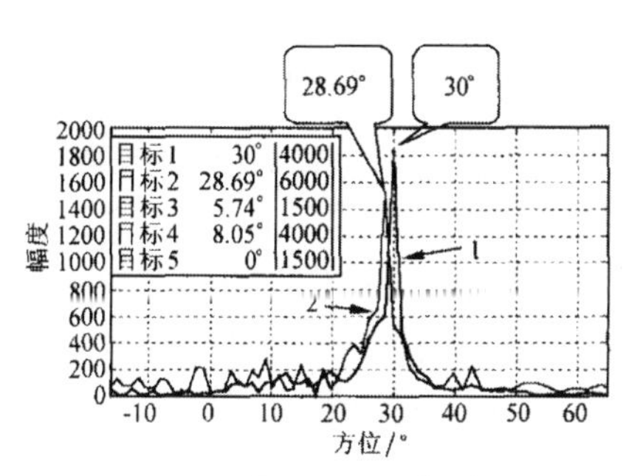

图 7 OSF 后置联合检测结果

实线 1: $\theta_s = 30°$， $\theta_I = 0°$； 实线 2: $\theta_s = 28.69°$， $\theta_I = 0°$

2.4 信噪比大小对系统检测性能的影响

设空间内存在 5 个目标，高斯白噪声作为背景噪声

目标 1: 水下高速目标，方位 30°, 辐射噪声幅值 [4000]

目标 2: 水面大型目标，方位 28.69°, 辐射噪声幅值 [6000]

目标 3: 水下大型目标，方位 5.74°, 辐射噪声幅值 [1500]

目标 4: 水下高速目标，方位 18.66°, 辐射噪声幅值 [4000]

目标 5: 水下大型目标，方位 0°, 辐射噪声幅值 [1500]

背景噪声：高斯白噪声，噪声幅值可变

检测目标 1, 置 $\theta_s = 30°$, $\theta_I = 18.66°$ 。改变信噪比大小。当信噪比 S/N 为 -21 dB, -22 dB 时，最佳空间滤波器后置联合检测系统对目标 1 的检测结果分别示于图 8(a) 和图 8(b) 。由图中可以看出，当信噪比 $S/N = -22$ dB 时，系统已无法完成信号检测。

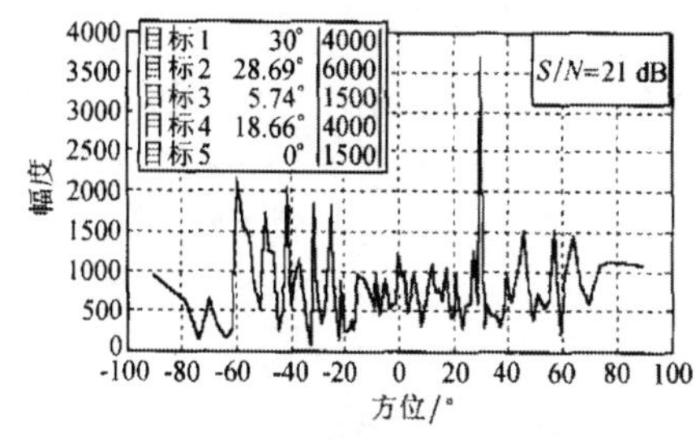

图 8(a) OSF 后置联合检测结果 ($\theta_s = 30°$ ， $\theta_I = 18.66°$)

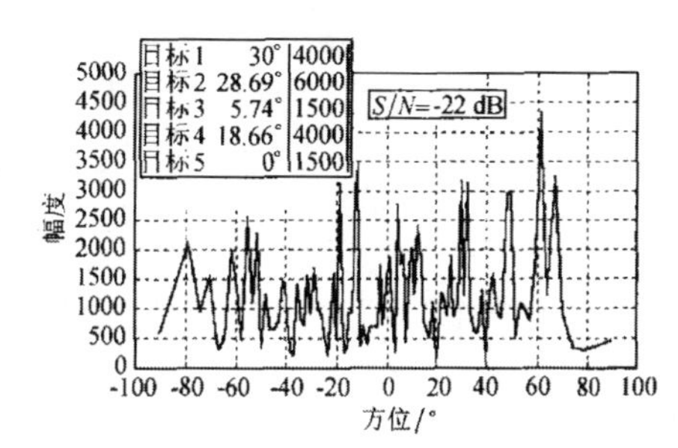

图 8(b) OSF 后置联合检测结果 ($\theta_s = 30°$ ， $\theta_I = 18.66°$)

3 结论

本文所给出的最佳空间滤波器后置联合检测系统能够抑制弱信号周围空间内数个强干扰，完成指定弱信号的检测；当信号与干扰方位非常接近时，系统仍然可以检测出指定目标；系统虚警率较小，检测结果是可信的；系统检测弱信号过程中，干扰方位角 θ_I 的宽容性较好，只要保证 $|\theta_s - \theta_I|$ 不太小，干扰方位角 θ_I 可以任意设置；目标方位角 θ_s 的实际值与估计值存在一定偏差时，系统仍然能够检测出指定目标；当信噪比 $S/N = -21$ dB 时，系统仍然能够抑制多个干扰，完成弱信号检测；当信号干扰比 $S/I = -12$ dB 时，系统仍然能够检测出目标。

参 考 文 献

1 Anderson V C. DICANNE, a realizable adaptive process. *J. Acoust. Soc. Am.*, 1969; **45**(2): 398—405

2 Griffiths L J, Jim C W. An alternative approach to linearly constrained adaptive beamforming. *IEEE Trans. on AP*, 1982; **AP-30**(1): 27—34

3 Shan T, Kailath T. Adaptive beamforming for coherent signals and interference. *IEEE Trans. on ASSP*, 1985; **33**(3): 527—536

4 Griffiths L J. A new approach to partially adaptive arrays. Proc.ICASSP-87, 1987: 1999—2002

5 Simaan M. Optimum array filters for array data signal processing. *IEEE Trans. Acoust., Speech, Signal processing*, 1983; **ASSP-31**(4): 1006—1015

6 Hanna M T, Simaan M. Minimum rejection response array filters in the presence of white noise. *IEEE Int. Conf. Acoust., Speech, Signal processing(Tampa,FL)*, 1985: 1796—1799

7 Hanna M T, Simaan M. Absolutely optimum array filter for sensor arrays. *IEEE Trans*, 1985; **ASSP-33**(6): 1380—1386

8 Hanna M T, Simaan M. Array filters for sidelobe elimination. *IEEE J. of Oceanic Engineering*, 1985; **OE-10**(3): 248—254

9 Buckley K M. Spatial/spectral filtering with linearly-constrained minimum variance beamformers. *IEEE Trans. On ASSP*, 1987; **ASSP-35**(3): 249—266

10 LI Qihu. The performance of absolutely optimum filter of array system. In: Proc. of ICA SSP'87 Dallas, USA, 1987: 2324—2327

11 LI Qihu. A new architecture of spatial filter for suppressing strong directional interference. In: Proc.of IMDEX ASIA, Singapore, 1999: 87—96

12 Jingde Luan. Experimental study of optimum spatial array filter. In: Proc. of UDTeur2001-Conf. PII-12, Hamburg, Germany, 2001

13 栾经德. 声呐信号处理中的最佳空间滤波器研究. 中国科学院声学研究所博士论文, 2002

14 孙进才. 利用三元阵的相干抵消原理及仿真研究. 声学学报, 2001; **26**(6): 537—544

15 朱维庆. 宽带波束形成器的自适应综合. 声学学报, 2003; **28**(3): 283—287

基于离散平面阵波束形成的双线列阵左右舷分辨技术研究①

何心怡[1]，蒋兴舟[2]，李启虎[3]
（1.海军装备研究院，北京 100073； 2.武汉海军工程大学，湖北 武汉 430033；
3.中国科学院声学所，北京 100080）

摘 要： 针对双线列阵左右舷分辨问题，提出了一种基于离散平面阵波束形成的左右舷分辨技术，将双线列阵看作是由无指向性水听器构成的平面阵，依照平面阵进行波束形成，从而在波束形成的同时实时分辨出目标所处的左右舷。运用该方法，最佳的双线阵间距是接收信号中心频率波长的 1/4，得到的结果等同于由具有心形指向性水听器组成的单根线列阵。该方法经过理论分析、仿真实验和湖试实验验证，证明其是简单的、有效的、可行的，有望应用于实际系统。

关键词： 双线列阵；左右舷分辨；离散平面阵波束形成；心形指向性

中图分类号： O427；U666.7 **文献标识码：** A

The research on the twin – line arrays port/starboard distinguishing technology based on the beam forming of the discreet plane array

HE Xin – yi[1]，JIANG Xing – zhou[2]，LI Qi – hu[3]
（1. Naval Academy of Armament, Beijing 100073, China； 2. Navy University of Engineering, Wuhan 430033, China； 3. Institute of Acoustics, Academia Sinica, Beijing 100080, China）

Abstract： To be aimed at twin – line arrays port/starboard distinguishing problem, this paper raised out a technology based on the beam forming of the discreet plane array. It treats the twin – line array as the plane array consists of the non – directional hydrophones, form the beams according to the plane array, thus can distinguish the port/starboard of the target while forming of the beams. Use this technology, the optimum space between two single lines is the one fourth of the wavelength of the center frequency of the receiving signal, the result equals to the single line array consists of the cardioid directed hydrophones. After analysis, simulation and lake authentication, this method proves to be simple, effective and practical, which can be used in the future.

Key words： twin – line array; port/starboard distinguishing; beam forming of the discreet plane array; cardioid directivity

0 引 言

左右舷模糊问题是拖曳线列阵声呐必须要解决的关键性难点问题，为实时完成拖曳线阵声呐的左右舷分辨，通常的解决方法是改变拖曳线列阵声呐的基阵结构，如采用三元水听器组、双线列阵或矢量水听器[1-7]，其中以双线列阵作为拖曳线阵声呐的基阵是一种常见的解决方法。因此，研究适用于双线列阵的左右舷分辨技术是非常重要的。为此，本文提出了一种基于离散平面阵波束形成的双线列阵左右舷分辨技术，经过理论分析、仿真实验和湖试实验，证明该方法是一种简单、有效、可行的左右舷分辨方法，能够在波束形成的同时实时分辨出目标所处的左右舷。

1 利用离散平面阵波束形成法的理论基础

传统的单根普通拖曳线阵是由无指向性阵元构

① 舰船科学技术, 2004, 26(3): 34-38.

成的一维直线阵,以线阵为轴对入射信号的响应是相同的,具有圆柱对称指向性,因此存在左右舷模糊问题。而双线列阵方式是将两根普通线列阵平行拖曳于水面舰艇尾部,如图1所示。

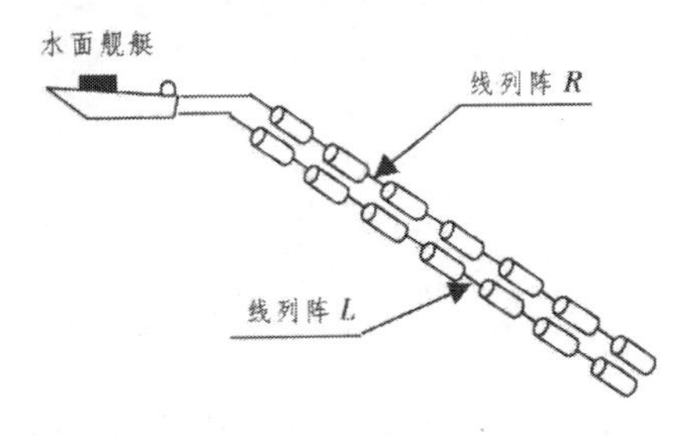

图1 双线列阵结构

注意到:线列阵 L 和线列阵 R 共处于同一平面,可将双线列阵看作是由无指向性水听器构成的平面阵,由于平面阵是二维阵列,从本质上克服了一维直线阵列所固有的圆柱对称指向性,能够分清目标所处的左右舷。以此为基础,文中提出了基于离散平面阵波束形成的双线列阵左右舷分辨技术,其出发点正是利用双线列阵可看作是由无指向性水听器构成的平面阵这一特性。

在三维直角坐标系中建立双线列阵坐标,假设每根线列阵均由 N 个水听器组成,根据离水面舰艇的远近依次编号为 (L_1,L_2,Λ,L_N) 和 (R_1,R_2,Λ,R_N),其中,L 标识着左舷的线列阵 L 的水听器,R 标识着右舷的线列阵 R 的水听器,每根线列阵上相邻两个水听器的间距为 d,两根线列阵的间距为 l。建立坐标系如下:坐标中心 O 点与 L_1 重合;线列阵 L 位于 x 轴上,x 轴的负向指向水面舰艇;y 轴正向指向本舰右舷法线方向,负向指向本舰左舷法线方向;z 轴指向海面。

在图2中,将线列阵 L 和线列阵 R 合在一起看作为由 $2\times N$ 个水听器构成的离散平面阵,各水听器的位置可用矢量 $\vec{r}_i(x_i,y_i,z_i)$ 表示:

$$\vec{r}_i(x_i,y_i,z_i)=\begin{cases}((i-1)d,0,0),\ i=1,2,\cdots,N,\\((i-N-1)d,l,0),\ i=N+1,N+2,\cdots,2N。\end{cases}\tag{1}$$

对于任意方向的入射信号的单位矢量 $\vec{e}(\phi,\theta)$,可以用下式表示:

$$\vec{e}=\vec{i}\sin\theta\cos\phi+\vec{j}\sin\theta\sin\phi+\vec{k}\cos\phi\tag{2}$$

式中:θ 为入射信号与 z 轴的夹角;ϕ 为入射信号在 xoy 平面上的投影与 x 轴的夹角,规定 ϕ 顺时针为正,逆时针为负,见图2所示。如果空间处理的期望信号方向(即波束的主极大方向)用单位方向矢量 $e_0(\phi_0,\theta_0)$ 表示,它可写成:

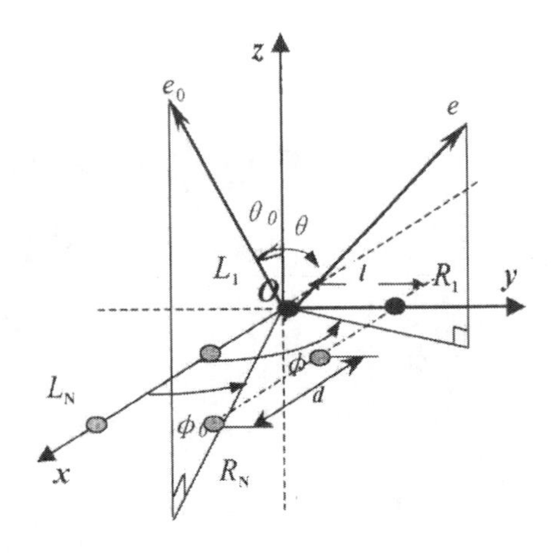

图2 双线列阵坐标

$$\vec{e}_0=\vec{i}\sin\theta_0\cos\phi_0+\vec{j}\sin\theta_0\sin\phi_0+\vec{k}\cos\phi_0。\tag{3}$$

以坐标原点 O 为参考点,则任意方向的入射信号会造成第 i 号水听器相对于参考点 O 的声程差为 ξ_i,那么:

$$\xi_i=\vec{r}_i\cdot\vec{e}=x_i\sin\theta\cos\phi+y_i\sin\theta\sin\phi+z_i\cos\theta,\tag{4}$$

来自主极大方向的信号使第 i 号水听器相对于参考点的声程差为 ξ_{i0},则:

$$\xi_{i0}=\vec{r}_i\cdot\vec{e}_0=x_i\sin\theta_0\cos\phi_0+y_i\sin\theta_0\sin\phi_0+z_i\cos\theta_0。\tag{5}$$

因此,任意方向的入射信号在第 i 号水听器相对于主极大方向的声程差所对应的相位差为

$$\begin{aligned}\Delta\varphi_i=&k(\xi_i-\xi_{i0})=\\&x_i(\sin\theta\cos\phi-\sin\theta_0\cos\phi_0)+\\&y_i(\sin\theta\sin\phi-\sin\theta_0\sin\phi_0)+\\&z_i(\cos\theta-\cos\theta_0),\end{aligned}\tag{6}$$

式中,$k=\dfrac{2\pi}{\lambda}$ 为接收信号的波数,那么由线列阵 L 和线列阵 R 构成的平面阵的归一化指向性函数为

$$b(\phi,\theta,\phi_0,\theta_0)=\left|\frac{1}{2N}\sum_{i=1}^{2N}e^{-j\Delta\varphi_i}\right|。\tag{7}$$

为方便对左右舷分辨问题的描述,仅考虑 $\theta=\dfrac{\pi}{2}$ 时的情况,也就是在 xoy 平面(即水平面)方向的波束指向性,图3是在水平面的双线列阵坐标系,其中 α 为入射声线与双线列阵法线的夹角,规定 α 顺时针为正,逆时针为负,那么有 $\alpha=\phi-\dfrac{\pi}{2}$。

将 $\theta=\dfrac{\pi}{2}$ 和 $\alpha=\phi-\dfrac{\pi}{2}$ 代入式(7),可得到双线列阵基于离散平面阵波束形成的归一化波束指向性图。

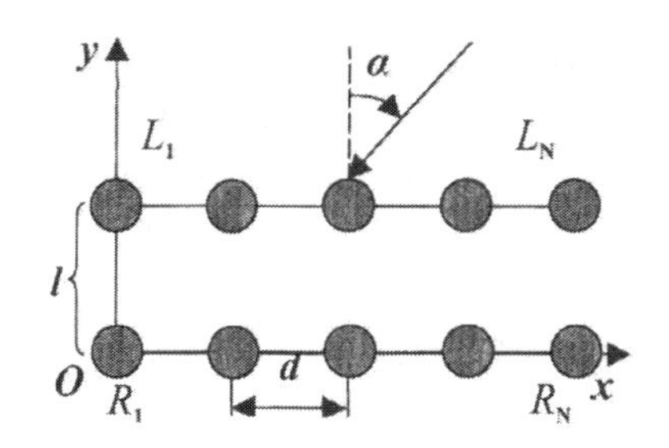

图3 双线列阵在水平面坐标

$$b(f,\alpha,\alpha_0)=\left|\begin{array}{l}\frac{1}{2N}\{1+\exp[-jkl(\cos\alpha_0-\cos\alpha)]\}\cdot\\ \sum_{i=1}^{N}\exp[-jk(i-1)d(\sin\alpha-\sin\alpha_0)]\end{array}\right|。\tag{8}$$

如果入射信号的角度为 α，那么产生镜象源的角度为 $\pi-\alpha$，可如下式定义左右舷分辨增益：

$$G(f,\alpha,\alpha_0)=10\lg\left(\frac{b(f,\alpha,\alpha_0)}{b(f,\pi-\alpha,\alpha_0)}\right)=$$

$$10\lg\left\{\left|\frac{1+\exp[-jkl(\cos\alpha_0-\cos\alpha)]}{1+\exp[-jkl(\cos\alpha_0+\cos\alpha)]}\right|\right\}=$$

$$10\lg\left\{\left|\frac{1+\exp\left[-j2\pi(\cos\alpha_0-\cos\alpha)\frac{l}{\lambda}\right]}{1+\exp\left[-j2\pi(\cos\alpha_0+\cos\alpha)\frac{l}{\lambda}\right]}\right|\right\}。\tag{9}$$

通常情况下，线列阵按半波长方式布阵，即 $d=\frac{\lambda}{2}$，那么在束控方向 α_0 确定时，双线列阵间距 l 的选取直接关系到左右舷分辨增益 $G(f,\alpha,\alpha_0)$ 的大小。更一般的情况，线列阵在360°的空间上形成多个波束，相邻两波束重叠50%，也就是说在360°有 M 个束控方向，α_i，$i=0,1,\cdots,M-1(M>1)$。因此，考查双线列阵基于离散平面阵波束形成方法的左右舷分辨增益时，可近似认为在入射信号的角度上恰好有一个波束的主极大值，即 $\alpha=\alpha_i$，$i\in(0,1,\cdots,M-1)$，那么式(9)可进一步化简为

$$G(f,\alpha,\alpha_i)=G(f,\alpha_i,\alpha_i)=$$

$$10\lg\left[\frac{b(f,\alpha_i,\alpha_i)}{b(f,\pi-\alpha_i,\alpha_i)}\right]=$$

$$10\lg\left\{\left|\frac{2}{1+\exp\left[-j4\pi\cos\alpha_i\frac{l}{\lambda}\right]}\right|\right\}=$$

$$5\lg2-5\lg\left[1+\cos\left(4\pi\cos\alpha_i\frac{l}{\lambda}\right)\right]。\tag{10}$$

从式(10)中能够得出，为使左右舷分辨增益最大，必须使 $1+\cos\left(4\pi\cos\alpha_i\frac{l}{\lambda}\right)$ 最小，也即：

$$\begin{aligned}&4\pi\cos\alpha_i\frac{l}{\lambda}=2n\pi+\pi,\ n=0,1,\cdots\\&\Rightarrow\cos\alpha_i\frac{l}{\lambda}=\frac{n}{2}+\frac{1}{4},\ n=0,1,\cdots\end{aligned}\tag{11}$$

对于拖曳线列阵声呐，一般都希望最好的检测和估计性能在双线列阵法线附近出现，因此，在考虑式(11)时，可以用 $\alpha_i=0$ 的情况来做分析，那么左右舷分辨增益极大值点都处在：

$$l=\left(\frac{n}{2}+\frac{1}{4}\right)\lambda,\ n=0,1,\cdots\tag{12}$$

当 $l>\lambda/2$ 时，对于单频信号将出现相位模糊问题，因此，基于离散平面阵波束形成方法的左右舷分辨技术，最佳的双线阵间距是 $l=\lambda/4$；对于宽带信号，最佳的双线列阵间距是中心频率对应波长的1/4。图4是双线阵间距 $l=\lambda/4$，$N=14$，$\alpha_0=0$ 时的波束指向性图，图5是该条件下的左右舷分辨增益与 α 的关系曲线。从图4和图5中能够看出，在双线列阵的侧射方向附近，左右舷分辨增益相当高，能满足拖曳线列阵声呐在实际使用中的需要。

为进一步分析基于离散平面阵波束形成左右舷分辨方法的机理，将式(8)做以下分解可得：

$$b(f,\alpha,\alpha_0)=\left|\frac{1}{2}\{1+\exp[-jkl(\cos\alpha_0-\cos\alpha)]\}\right|\cdot$$

$$\left|\frac{1}{N}\cdot\sum_{i=1}^{N}\exp[-jk(i-1)d(\sin\alpha-\sin\alpha_0)]\right|=$$

$$b_1(f,\alpha,\alpha_0)\cdot b_2(f,\alpha,\alpha_0);\tag{13.1}$$

$$b_1(f,\alpha,\alpha_0)=$$

$$\left|\frac{1}{2}\{1+\exp[-jkl(\cos\alpha_0-\cos\alpha)]\}\right|;\tag{13.2}$$

$$b_2(f,\alpha,\alpha_0)=$$

$$\left|\frac{1}{N}\cdot\sum_{i=1}^{N}\exp[-jk(i-1)d(\sin\alpha-\sin\alpha_0)]\right|。\tag{13.3}$$

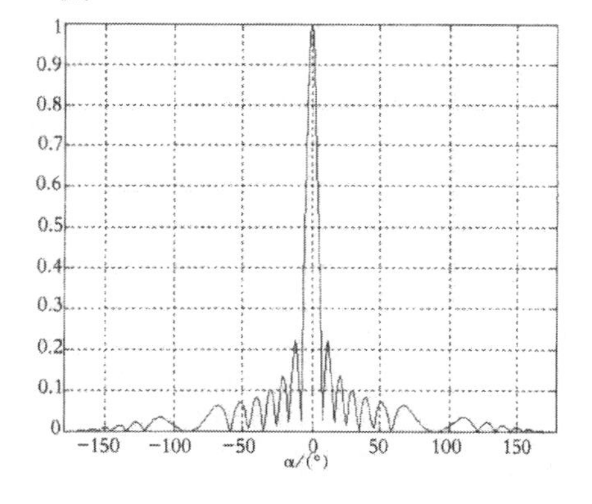

图4 波束指向性

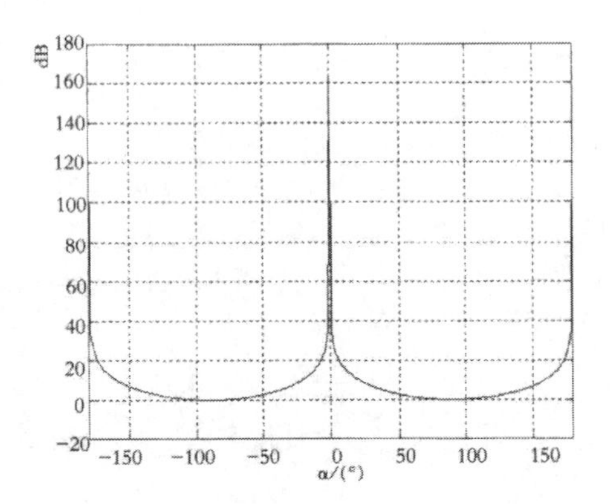

图5 左右舷分辨增益与入射信号角度关系曲线

根据 Bridge 乘法定理可知，基于离散平面阵波束法处理后得到的波束指向性图可认为是由 $b_1(f,\alpha,\alpha_0)$ 和 $b_2(f,\alpha,\alpha_0)$ 的乘积，而 $b_2(f,\alpha,\alpha_0)$ 是 N 元均匀线列阵的波束指向性图；当 $\alpha_0=0, l=\frac{\lambda}{4}$ 时，画出 $b_1(f,\alpha,\alpha_0)$ 的波束指向性函数如图6所示，能够发现它是一个典型的心形指向性图。也就是说，双线列阵运用基于离散平面阵波束形成方法处理后所得到的结果等同于由 N 个具有心形指向性水听器组成的单根线列阵，因而能够实现对目标的左右舷分辨。

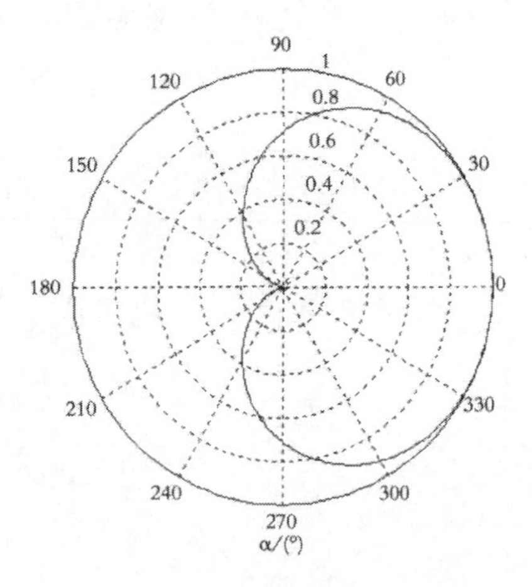

图6 $\alpha_0=0, l=\frac{\lambda}{4}$ 时 $b_1(f,\alpha,\alpha_0)$ 的波束指向性图

2 仿真实验

仿真条件是：双线列阵间距 0.5m，每根线列阵由 14 个水听器组成，相邻两阵元间距 1m。当 750Hz 单频信号从 $\alpha=0°$ 方向入射、信噪比为 0dB 时，经本文方法处理后得到的多波束系统输出如图7所示，得到10dB的左右舷分辨增益；当该信号从 $\alpha=30°$ 方向入射、信噪比为 0dB 时，经本文方法处理后得到的多波束系统输出如图8所示，获得的左右舷分辨增益为7dB。这些结果都与第2节中理论推导的结论相符。

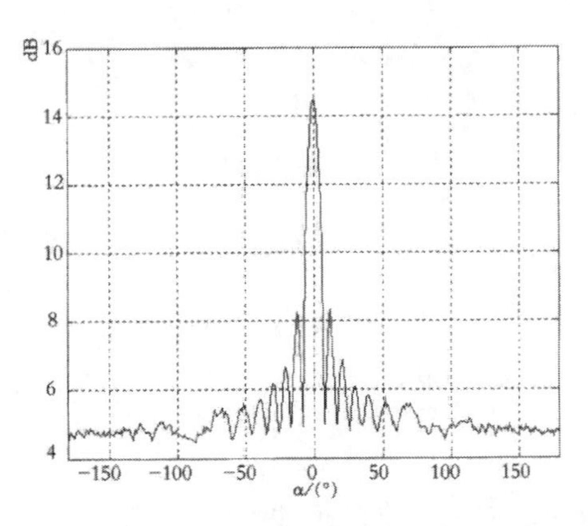

图7 $\alpha=0°$时的多波束系统输出

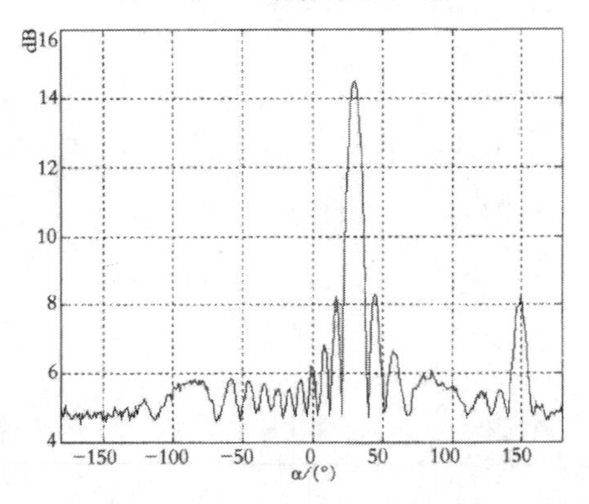

图8 $\alpha=30°$时的多波束系统输出

3 湖试实验

2002年7月，在浙江千岛湖鹿岛附近湖面进行了双线阵左右舷分辨实验。整个实验方案如图9所示：双线列间距 0.5m（对应的中心频率是 750Hz），每根线列阵由 14 个水听器组成，相邻 2 个水听器间距 1m，声速梯度呈弱负梯度。将双线阵水平固定于实验船1的左舷水下 8m 处，而距该实验船左舷 0.29 n mile 处的实验船2水下 5m 处吊放一个发射换能器，发射中心频率为 1 000Hz，带宽为 200Hz，脉宽为 200ms，周期为 4s 的线性调频波，信号的入射方向在双线列阵的正横方向附近。双线列阵附近的信噪比大约为4.675dB，采样频率为12000Hz，现场实验主要

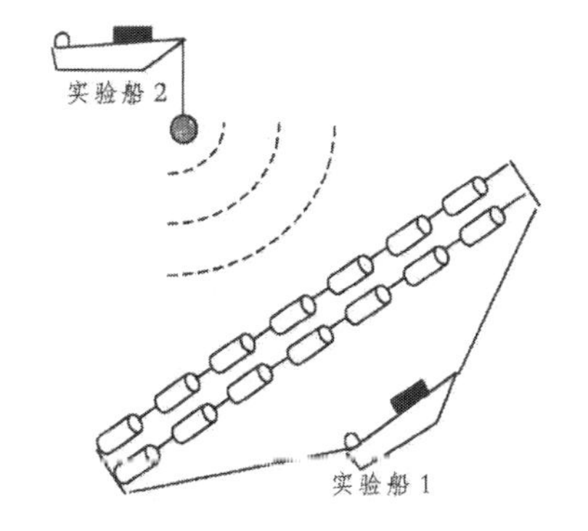

图 9 实验方案

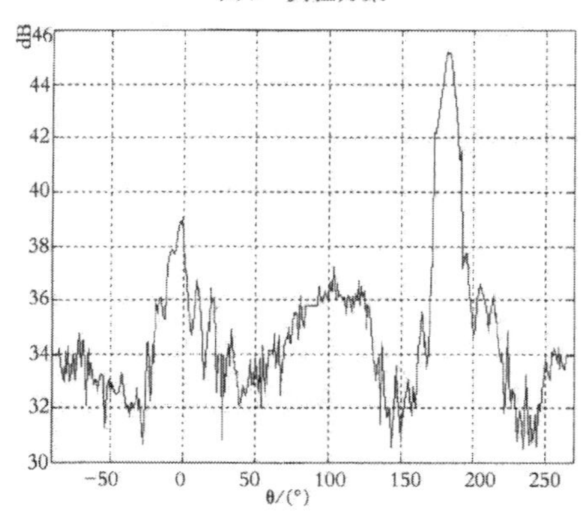

图 10 经本文方法处理后得到的湖试多波束系统输出

是采集数据，然后利用计算机进行后续数据处理。

图 10 是经过离散平面阵波束形成方法处理后的多波束系统输出图，从中可看到：所获得的左右舷分辨增益大约为 6.3dB，考虑到实验中所用的发射信号与 750Hz 有所偏差，湖试实验所取得的左右舷分辨效果是比较理想的。

4 结 语

针对双线列阵左右舷模糊问题，本文所提出的基于离散平面阵波束形成的左右舷分辨技术，经理论分析、仿真实验和湖试实验，证明该方法是一种简单、有效、可行的左右舷分辨技术，有望应用于实际系统。

参考文献

[1] 杜选民，朱代柱，赵荣荣，姚蓝．拖线阵左右舷分辨技术的理论分析与实验研究[J]．北京：声学学报，2000，25(5)：395 – 402.

[2] 杜选民，朱代柱，赵荣荣，姚蓝．拖线阵目标左/右舷分辨技术研究（Ⅱ）– 噪声相关性模型[J]．舰船科学技术，1999，21(3)：17 – 20.

[3] JEAN BERTHEAS, GILLES MORESCO, PHILIPPE DUFOURCO. Linear hyd rophonic antenna and electronic device to remove right/left ambiguity, associated with the antenna. United States Patent, Patent Number: 5058082, 1991, (10).

[4] DOISY Y Port – starboard discrimination performances on actived towed array systems. UDT95, France: 125 – 129.

[5] VAN G W M MIERLO, BEERENS S P, BEEN R, ect. Port/starboard discrimination by hydrophone triplets in active and passive towed arrays. UDT97, Germany: 176 – 181.

[6] GRIM STEINAR GJONNES, AAGEDAL. Left/right discrimination in software with conventional, passive thin – line towed submarine arrays. UDT2000, Pacific: 1 – 6.

[7] SP BEERENS, BEEN R, GROEN J ENOUTARY, DOISY Y. Adaptive port – starboard beamforming of triplet arrays. UDT2000, Pacific: 63 – 68.

一种用于小横截尺寸三元水听器组的左右舷分辨技术①

何心怡[1]，蒋兴舟[2]，李启虎[3]
（1. 海军装备研究院，北京 100073；2. 海军工程大学，湖北 武汉 430033；
3. 中国科学院 声学研究所，北京 100080）

摘　要： 针对小横截尺寸三元水听器组的左右舷分辨问题，文中提出了一种新的移相方法实现左右舷分辨。与原有的三元水听器组左右舷分辨技术相比，该方法不要求三元水听器组的横截尺寸与信号频率相匹配，并且计算量较小，易于实时实现。经过理论分析和仿真实验，证明了文中提出的新的移相方法应用于小横截尺寸三元水听器组，能够有效地解决拖曳线列阵的左右舷分辨问题，具有良好的工程应用前景。

关键词： 左右舷分辨；三元水听器组；小横截尺寸；新的移相方法

中图分类号： TB561.1；TJ830.35；U666　　**文献标识码：** A

The technology of port/ starboard discrimination for the hydrophone triplets of small cross section size

HE Xin—yi[1]，JIANG Xing—zhou[2]，LI Qi—hu[3]
（1. Naval Academy of Armament，Beijing 100073，China；2. Navy University of Engineering，Wuhan 430033，China；3. Institute of Acoustics，Academia Sinica，Beijing 100080，China）

Abstract： To be aimed at the port/starboard discrimination problem of hydrophone triplets of small cross section size，this paper raised out a new phase shifting method to implement port/starboard discrimination. Compared with the original port/ starboard discrimination technology of hydrophone triplets，this method doesn't require the cross section size match the signal frequency，also this method has few computation and easy to be implemented. After theoretical analysis and simulation authentication，it is proved that the new phase shifting method can solve the port/ starboard discrimination problem of the towed linear array effectively when it is implemented on hydrophone triplets of small cross section size. It has a bright future in engineer application.

Key words： port/starboard discrimination；hydrophone triplets；small cross section size；new phase shifting method

0　引　言

为解决拖线阵的左右舷模糊问题，改变基阵结构是常用的方法，如采用三元水听器组[1－5]、双线阵[6]或以矢量水听器构成线列阵等，其中采用三元水听器组的基阵配置样式是一种常见的解决方案。为减小基阵尺寸，便于拖线阵声呐的安装和使用，常常要求三元水听器组的横截尺寸较小（通常是指小于接收信号波长的 1/10，且大于拖线阵流噪声的空间相关距离）。而现有的三元水听器组左右舷分辨技术主要有几何相移模型和噪声相关模型两种。其中，几何相移模型要求三元水听器组的横截尺寸与信号频率相匹配（即要求三元水听器中的两水听器的间距是接收信号波长的 1/3）。频率越低，所需要的基阵尺寸就越大，而声呐所处理的信号都是低频信号，限制了几何相移模型在小横截尺寸三元水听器组中的应用；而噪

① 舰船科学技术，2004，26(5)：34-39.

声相关模型能够满足小横截尺寸三元水听器组接收低频信号时的左右舷分辨需要，但计算量较大。为此，本文提出了一种新的移相方法实现左右舷分辨，其算法简单、计算量小、易于实时实现，能够满足小横截尺寸三元水听器组左右舷分辨的需要。

1 三元水听器组坐标系的建立

在三维直角坐标系中，建立三元水听器组的坐标如图1所示。

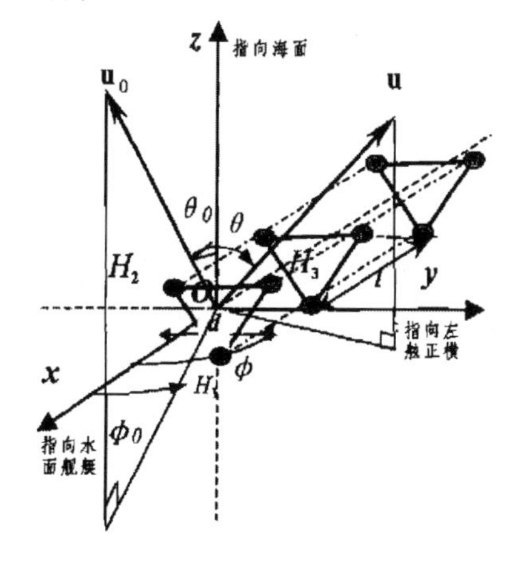

图1 三元水听器组在三维直角坐标系中的示意图

坐标中心 $\boldsymbol{O}$ 点位于拖线阵声学段第一个三元水听器的中心；x 轴指向水面舰艇，M 个三元水听器构成的拖线阵的中轴线恰位于负 x 轴上，相邻两个三元水听器的间距为 l；y 轴正向指向本舰左舷法线方向，负向指向本舰右舷法线方向；z 轴指向海面；接收到的远场单位方向矢量 $\boldsymbol{u}=(\sin\theta\cos\phi,\ \sin\theta\sin\phi,\ \cos\theta)$，方位角 ϕ 和仰角 θ 分别是 x 轴正向到矢量 $\boldsymbol{u}$ 在 xoy 平面上投影的夹角和 z 轴正向到 $\boldsymbol{u}$ 的夹角，并且规定这些角度顺时针为正，逆时针为负；空间处理的期望信号的方向用单位方向矢量 $\boldsymbol{u}_0=(\sin\theta_0\cos\phi_0,\ \sin\theta_0\sin\phi_0,\ \cos\theta_0)$ 表示，相应的方位角为 ϕ_0、仰角为 θ_0。

取出第 i 个三元水听器，重新在三维直角坐标系中画出，如图2所示。可看出，三元水听器是由分布在同一圆周、形成等边三角形的三个无指向性水听器组成，并且三元水听器构成的平面与线阵的轴线相垂直。假设圆周半径为 d，水听器间距为 a，在三维直角坐标系中的圆心坐标为 $(x_i, 0, 0,)$，可写出三个水听器在三维直角坐标系中的坐标为

$$r_{h1,i}=(x_i, 0, -d)=\left(x_i, 0, -\frac{\sqrt{3}}{3}a\right); \quad (1A)$$

$$r_{h2,i}=\left(x_i, -\frac{\sqrt{3}}{2}d, \frac{1}{2}d\right)=\left(x_i, -\frac{1}{2}a, \frac{\sqrt{3}}{6}a\right); \quad (1B)$$

$$r_{h3,i}=\left(x_i, \frac{\sqrt{3}}{2}d, \frac{1}{2}d\right)=\left(x_i, \frac{1}{2}a, \frac{\sqrt{3}}{6}a\right)。 \quad (1C)$$

式中：$x_i=-(m-1)l$ 为第 i 个三元水听器的 x 坐标值。

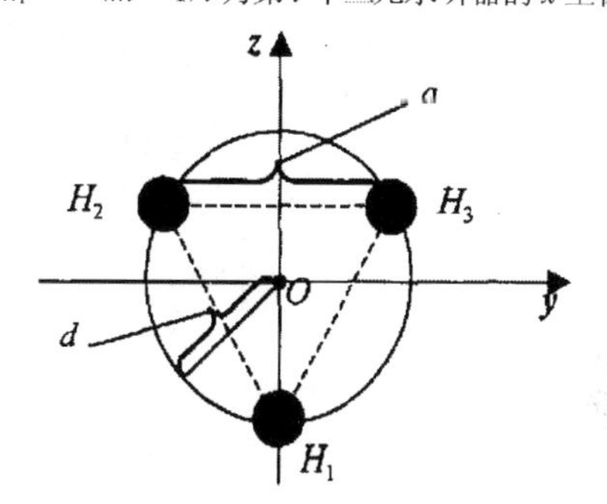

图2 第 i 个三元水听器示意图

要实现左右舷分辨，就是经过一定的处理后，使得每个三元水听器具有心形指向性，那么，根据布里奇(Bridge)乘积定理[7]，即可使三元水听器组具有指向性，能够分辨目标所处的左右舷。

2 原有的三元水听器组左右舷分辨技术概述

2.1 几何相移模型

在一般情况下，只需要考虑从水平面入射信号的方向，这时就要分清信号究竟是来自左舷还是右舷，也就是通常所说的左右舷模糊问题。如图2所示，能够发现：假定束控方向为 ϕ_0，如果信号来自束控方向时，以水听器 H_1 为参考，则 H_2 的输出信号落后(超前) $\frac{1}{2}ka\sin\phi_0$，H_3 的输出超前(落后) $\frac{1}{2}ka\sin\phi$。其中 $k=2\pi f/c$ 为波数，f 为接收信号频率，c 为海水中声速。这样，为保证在束控方向波束输出最大，可先对 H_2、H_3 的输出分别相移 $\frac{1}{2}ka\sin\phi_0$ 和 $-\frac{1}{2}ka\sin\phi$，然后将三路水听器的信号相加。通常是在 $\phi_0=\pm\pi/2$ 方向分别形成左右波束，从而完成左右舷的区分[3]。此时，左右波束的水平指向性函数分别是：

$$B_L(f, \theta, \phi)=B_L\left(f, \frac{\pi}{2}, \phi\right)=\left|1+2\cos[\pi a(\sin\phi+1)/\lambda]\right|; \quad (2A)$$

$$B_R(f, \theta, \phi)=B_R\left(f, \frac{\pi}{2}, \phi\right)=\left|1+2\cos[\pi a(\sin\phi-1)/\lambda]\right|。 \quad (2B)$$

当 $a=\frac{\lambda}{3}$ 时，$\lambda=c/f$，几何相移模型具有最大的左右舷分辨增益和最理想的心形指向性；而当 $a\neq\frac{\lambda}{3}$ 时，则左右舷分辨增益急剧下降，特别是当 $a<<\frac{\lambda}{3}$ 时，已经无法完成左右舷分辨的功能。图3(A)、图3(B)、图3(C)分别是 $a=\frac{\lambda}{3}$，$a=\frac{\lambda}{12}$ 和 $a=\frac{\lambda}{20}$ 时，单个三元水听器的理论水平左右波束指向性图，图中实线均为右波束，虚线均为左波束。

从本节分析可知，采用几何相移模型的三元水听器组左右舷分辨技术，必须满足 $a=\frac{\lambda}{3}$，才能有最大的左右舷分辨增益和理想的心形指向性；当 $a<<\frac{\lambda}{3}$ 时，已无法运用该方法完成左右舷分辨。因此，原有的三元水听器组几何相移模型是无法满足小横截尺寸三元水听器组的左右舷分辨要求。

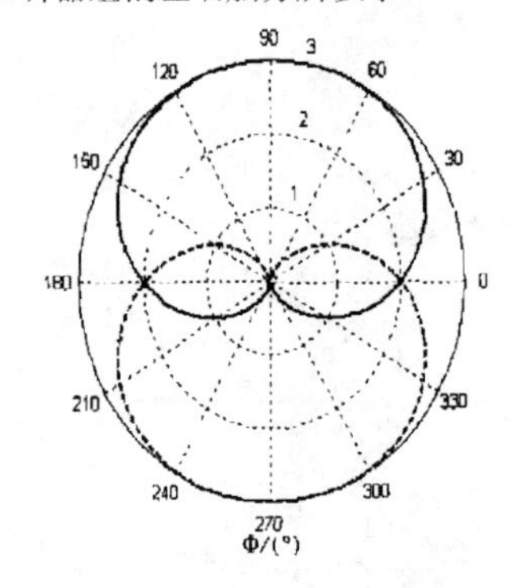

图3(A)　$a=\frac{\lambda}{3}$ 时，单个三元水听器的理论水平左右波束指向性图

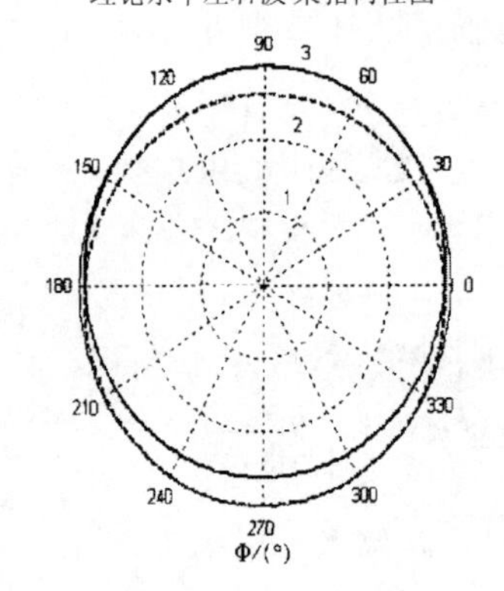

图3(B)　$a=\frac{\lambda}{12}$ 时，单个三元水听器的理论水平左右波束指向性图

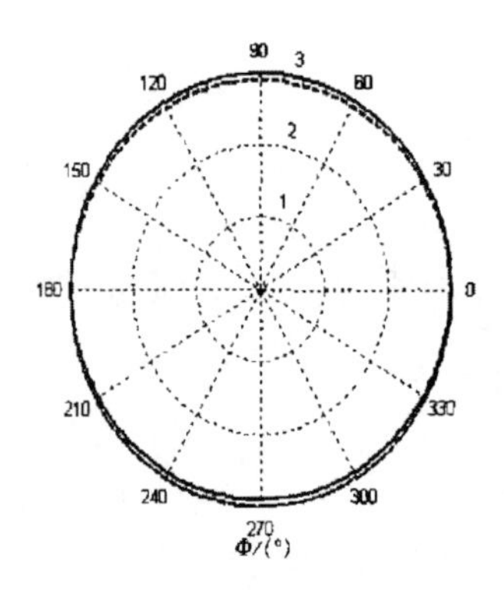

图3(C)　$a=\frac{\lambda}{20}$ 时，单个三元水听器的理论水平左右波束指向性图

2.2 噪声相关模型

三元水听器组的噪声相关模型，即最优波束加权方法，就是将三元水听器接收到的背景噪声分为相关噪声和非相关噪声，构建出三元水听器的噪声相关矩阵，进而推导出最优波束加权系数，使得加权后的左右波束具有心形指向性，因此能分辨目标所处的左右舷[1,2,4,5]。

由于噪声相关模型已应用于三元水听器组的左右舷分辨，因此本节不对整个推导过程作详细阐述，仅把一些中间步骤和结论罗列出来，详细的推导可参见参考文献[1,2,4,5]。

三元水听器组的噪声相关矩阵写为

$$\Gamma_n=\sigma_c^2\Gamma_c+\sigma_n^2I=\sigma_c^2\begin{pmatrix}1+\sigma^2 & \alpha & \alpha\\ \alpha & 1+\sigma^2 & \alpha\\ \alpha & \alpha & 1+\sigma^2\end{pmatrix}。\tag{3}$$

其中，单个三元水听器接收到的背景噪声中的总的相关噪声的功率为 σ_c^2，空间相关函数可用辛克函数表示为 $\alpha=\sin(ka)/(ka)$；而背景噪声中非相关噪声的功率为 σ_u^2；$\sigma^2=\sigma_u^2/\sigma_c^2$ 为非相关噪声功率与相关噪声功率之比。

将图2中三元水听器的三个水听器的位置用仰角表示为：$(\theta_1,\theta_2,\theta_3)=-\pi,-\pi/3,\pi/3)$，将单个三元水听器看作一个小基阵，则三元水听器的基阵响应向量为 $e=(e_1,e_2,e_3)$。

$$e_i=\exp[jkd(\sin\theta_i\sin\theta_0\sin\phi_0+\cos\theta_i\cos\theta_0)]$$
$$i=1,2,3.\tag{4}$$

式中：(θ_0,ϕ_0) 为基阵的束控方向。

当信噪比足够大时，可利用基于最大似然比的最优波束形成器推导方法求出最佳滤波系数为

$$w[\sigma^2 e(f)] = \frac{\Gamma_n^{-1}(f)e(f)}{e^H(f)\Gamma_n^{-1}(f)e(f)}。\tag{5}$$

将左右两个模糊方向所对应的旋转向量表示为 $e_L = e(\theta, -\pi/2)$ 和 $e_L = e(\theta, \pi/2)$，相应的波束形成加权系数分别记作 $w_L(\sigma^2) = w(\sigma^2, e_L)$ 和 $w_R(\sigma^2) = w(\sigma^2, e_R)$。将单个三元水听器接收到的信号经加权后分别形成左右波束，这样，单个三元水听器在 $(\theta, -\pi/2)$ 和 $(\theta, \pi/2)$ 方向的指向性函数可写成：

$$B_L(\sigma^2, f, \theta, \phi) = w_L^H(f, \sigma^2)e(\theta, \phi)；\tag{6A}$$

$$B_R(\sigma^2, f, \theta, \phi) = w_R^H(f, \sigma^2)e(\theta, \phi)。\tag{6B}$$

其中，w_R^H 的右上角 H 表示共轭转置。由式(6)得到的单个三元水听器的指向性函数为心形指向性，能完成左右舷分辨。但噪声相关模型需要计算不同频率的最优波束加权系数 $w(f, \sigma^2)$，并且需要对不同频率的信号分量进行加权，计算量较大。

3 新的移相方法左右舷分辨技术

为克服原有的三元水听器组左右舷分辨技术的不足，文中提出了一种新的移相方法解决小横截尺寸三元水听器组的左右舷模糊问题。

图 2 中三个水听器接收到的远场单位方向信号可分别表示为 $P_i^{(1)}$，$P_i^{(2)}$ 和 $P_i^{(3)}$。

$$P_i^{(1)}(f, \theta, \phi) = \exp\left[jkx_i\sin\theta\cos\phi - j\frac{\sqrt{3}}{3}ka\cos\theta\right]；\tag{7A}$$

$$P_i^{(2)}(f, \theta, \phi) = \exp\left[jkx_i\sin\theta\cos\phi - j\frac{1}{2}ka\sin\theta\sin\phi + j\frac{\sqrt{3}}{6}ka\cos\theta\right]；\tag{7B}$$

$$P_i^{(3)}(f, \theta, \phi) = \exp\left[jkx_i\sin\theta\cos\phi + j\frac{1}{2}ka\sin\theta\sin\phi + j\frac{\sqrt{3}}{6}ka\cos\theta\right]；\tag{7C}$$

考虑从水平面入射信号的方向时，即 $\theta = \frac{\pi}{2}$，假定束控方向为 $\left(\frac{\pi}{2}, \phi_0\right)$，那么束控方向对应的镜象源的方位角为 $\phi_m = -\phi_0$。那么为使从束控方向对应的镜象源方向来的信号经处理后最小，可采用以下的移相方法实现，即：

$$B\left(f, \frac{\pi}{2}, \phi\right) = P_i^{(1)}\left(\frac{\pi}{2}, \phi\right) - P_i^{(2)}\left(\frac{\pi}{2}, \phi\right) \cdot \frac{1}{2}\exp\left[j\frac{1}{2}ka\sin\phi_0\right] - P_i^{(3)}\left(\frac{\pi}{2}, \phi\right) \cdot \frac{1}{2}\exp\left[-j\frac{1}{2}ka\sin\phi_0\right]。\tag{8}$$

注意到，新的移相方法与原有的几何相移模型不同之处在于：式(8)中是三个水听器接收信号经过移相后是相减而不是相加，其目的是使得从束控方向对应的镜象源方向来的信号经处理后输出最小。将式(7)代入式(8)，经过推导可得出：

$$\left|B\left(f, \frac{\pi}{2}, \phi\right)\right| = 1 - \cos\left[\frac{1}{2}ka(\sin\phi + \sin\phi_0)\right] = 1 - \cos[\pi a(\sin\phi + \sin\phi_0)/\lambda]。\tag{9}$$

当束控方向选在三元水听器组的正横方向时，即 $\phi_0 = \pm\frac{\pi}{2}$ 时，束控方向 ϕ_0 和镜象源方向 ϕ_m 的波束输出分别为

$$\left|B\left(f, \frac{\pi}{2}, \phi_0\right)\right| = 1 - \cos\left(\frac{2\pi a}{\lambda}\right)$$
$$\left|B\left(f, \frac{\pi}{2}, \phi_m\right)\right| = \left|B\left(f, \frac{\pi}{2}, -\phi_0\right)\right|。\tag{10}$$

为使束控方向的输出最大，应满足：

$$\frac{2\pi a}{\lambda} = \pi + 2n\pi, n = 0, 1, \Lambda。\tag{11}$$

从式(11)可得，为保证不出现相位模糊，则最佳间距 $a_{opt} = \frac{\lambda}{2}$，此时 $\left|B\left(f, \frac{\pi}{2}, \phi_0\right)\right| = 2$。如果处理的是宽带信号，那么最佳间距应取为中心频率对应波长的 1/2。

注意到：当 $a << \frac{\lambda}{2}$ 时，文中提出的移相方法同样能完成左右舷分辨功能，也就是说该方法适用于小横截尺寸的三元水听器组。如 $a = 0.05\lambda$ 时，绘出单个三元水听器的理论水平左右波束指向性图，如图 4 所示。图中实线为右波束指向性图，虚线为左波束指向性图，仍然是理想的心形指向性，能够满足左右舷分辨的要求，只是基阵的增益比 $a_{opt} = \frac{\lambda}{2}$ 时有所减小。

定义左右舷抑制比如下：

$$r = 10\lg\frac{\left|B\left(f, \frac{\pi}{2}, \phi\right)\right|}{\left|B\left(f, \frac{\pi}{2}, -\phi\right)\right|} = 10\lg\frac{\left|B\left(f, -\frac{\pi}{2}, \phi\right)\right|}{\left|B\left(f, -\frac{\pi}{2}, -\phi\right)\right|} = 0。\tag{12}$$

当 $a = 0.05\lambda$ 时，左右舷抑制比与入射角的曲线如图 5 所示，能够满足拖曳线列阵声呐在实际使用中的需要。

从理论分析可知，新的移相方法算法简单，计算量较小，能够在三元水听器组横截面较小的情况下有效地完成左右舷分辨，付出的代价是基阵的增益有所减小。

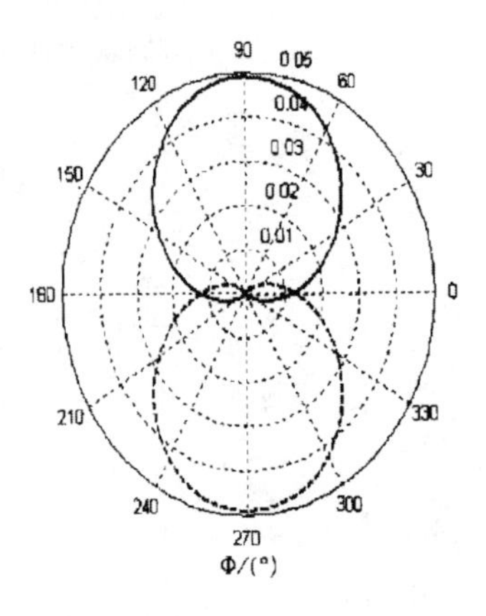

图 4 $a=0.05\lambda$ 时，单个三元水听器的左右波束指向性图

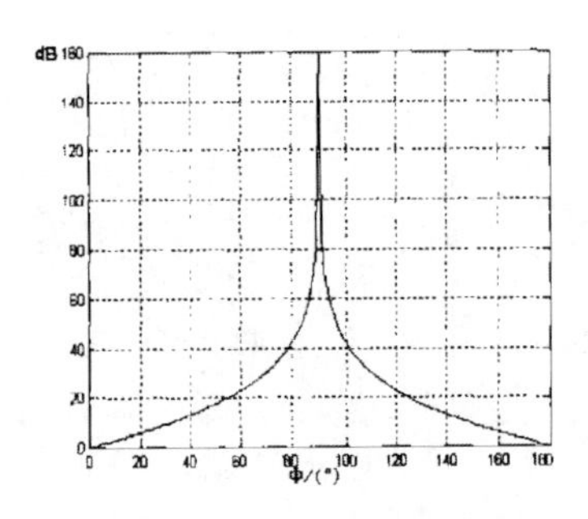

图 5 $a=0.05\lambda$ 时，左右舷抑制比与入射角的关系曲线

4 仿真实验

为验证文中提出的新的移相方法的正确性和有效性，采用仿真实验加以验证。

仿真场景如下：1 个三元水听器组由 16 个三元水听器构成，相邻两个三元水听器间距为 0.75m，对应的接收信号中心频率为 1 000Hz，$a=0.04$m；在右舷0°和右舷30°各有一个等强度的高斯白噪声源，信噪比为 0dB；处理信号频率范围为 750～1 250Hz。图 6～图 9 分别是未经过左右舷分辨处理、经过原有的几何相移模型处理、经过噪声相关模型处理和经过新的移相方法处理后得到的左右波束输出图(图中实线均为右波束，虚线均为左波束，横坐标是与线列阵法线的夹角)。能够看出，三元水听器原有的几何相移模型在小横截尺寸时已无法分辨目标所处的左右舷，如图7所示；而噪声相关模型和新的移相方法都可判

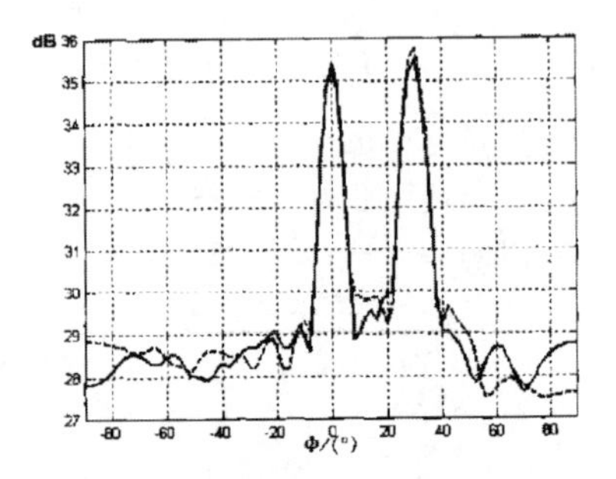

图 6 未做左右舷分辨处理前的左右波束输出

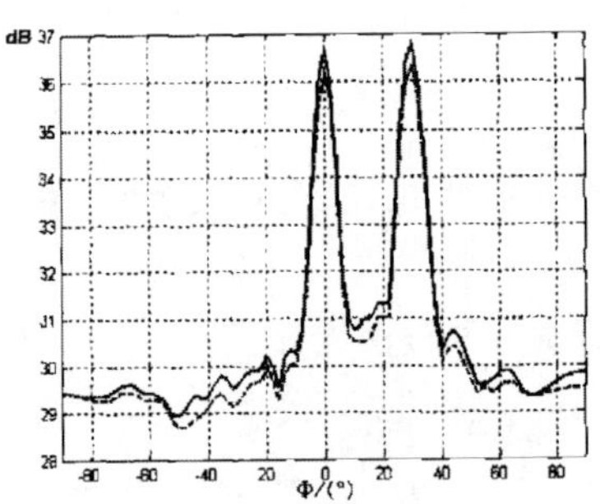

图 7 经过原有的几何相移模型处理后的左右波束输出

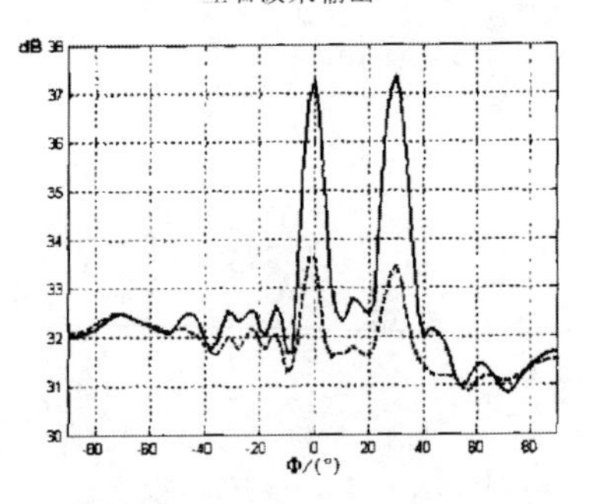

图 8 经过噪声相关模型处理后的左右波束输出

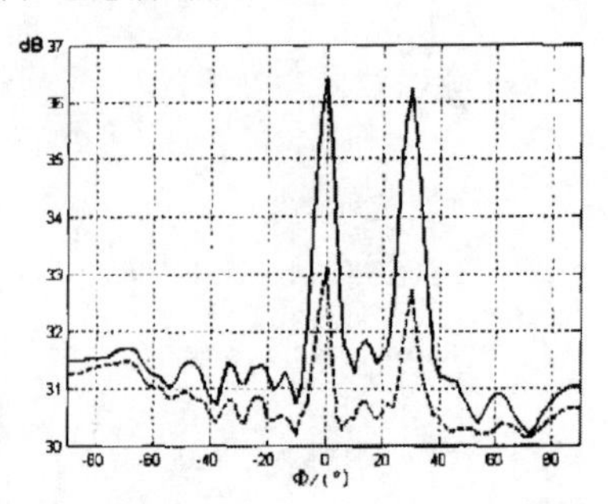

图 9 经过新的移相方法处理后的左右波束输出

定出目标处于右舷，且两个目标的左右舷分辨增益均在3dB以上。

无论是采用什么样的左右舷分辨算法，都可以把左右舷分辨处理看成是在波束形成前的预处理。因此，预处理所需时间的长短，影响着工程实现的实时性。为此，在仿真实验的一次数据分析中（使用512个快拍），把噪声相关模型和新的移相方法所用的计算时间对比，采用在PIV1.3G的计算机上，通过MATLAB5.3中的CPUTIME函数进行测量。得出了以下结果：运用噪声相关模型时，所用的时间为0.05s；而利用新的移相方法，所用的时间仅为0.01s。可看出，新的移相方法比噪声相关模型大大减少了计算量。

5 结 语

针对小横截尺寸三元水听器组的左右舷分辨问题，文中提出新的移相方法，它具有算法简单、计算量小等优点，经过仿真实验验证，证明了它能够有效解决小横截尺寸三元水听器组的左右舷模糊问题，具有良好的实用前景。

参考文献

[1] 杜选民，朱代柱，赵荣荣 等. 拖线阵左右舷分辨技术的理论分析与实验研究[J]. 声学学报，2000，25(5)：395—402.

[2] 杜选民，朱代柱，赵荣荣，姚蓝. 拖线阵目标左/右舷分辨技术研究(Ⅱ)——噪声相关性模型[J]. 舰船科学技术，1999，21(3)：17—20.

[3] JEAN BERTHEAS, GILLES MORESCO, PHILIPPE DUFOURCO. Linear hydrophonic antenna and electronic device to remove right/left ambiguity, associated with the antenna. United States Patent, Patent Number: 5058082, Oct. 15, 1991.

[4] DOISY Y. Port—starboard discrimination performances on actived towed array systems. UDT95, France: 125—129.

[5] VAN MIERIO G W M, BEERENS S P, BEEN R, ect. Port/starboard discrimination by hydrophone triplets in active and passive towed arrays. UDT97, Germany: 176—181.

[6] IMAN W, SCHURMAN. Reverberation rejection with a dual-line towed array. IEEE Journal of Oceanic Engineering, 1996, 21(2): 193—204.

[7] 李启虎. 声呐信号处理引论[M]. 北京：海洋出版社，2000：186—187.

基于子带分解的宽带波束域最小方差无畸变响应高分辨方位估计方法研究①

何心怡[1] 蒋兴舟[1] 李启虎[2]

(1 海军工程大学 武汉 430033)

(2 中国科学院声学研究所 北京 100080)

2003 年 1 月 28 日收到

2003 年 4 月 18 日定稿

摘要 针对最小方差无畸变响应法对宽带不适用、计算量较大和如何与实际声呐系统结合这三个问题，提出了基于子带分解的宽带波束域 MVDR 高分辨方位估计方法：它以子带分解的方法实现 MVDR 在宽带情形下的扩展；利用波束转换实现降维运算，大大减小运算量和提高算法的稳健性；并且提出了在声呐系统使用中的新思路：可将该方法与声呐的多波束系统结合，直接利用声呐多波束系统的输出，可省去从阵元域到波束域的转换，进一步减少运算量；可用于替代传统的分裂波束精测定向系统或单波束系统，提高整个声呐定向系统的性能。经过理论分析、仿真实验和湖试实验，证明了该方法是简单的、有效的、可行的。

PACS 数： 43.30, 43.60

The research on broadband beamspace minimum variance distortionless response high resolution direction-of-arrival estimation method based on sub-band decomposition

HE Xinyi[1] JIANG Xingzhou[1] LI Qihu[2]

(1 *Navy Engineering University* Wuhan 430033)

(2 *Institute of Acoustics, The Chinese Academy of Sciences* Beijing 100080)

Received Jan. 22, 2003

Revised Apr. 18, 2003

Abstract To be aimed at the three problems of the Minimum Variance Distortionless Response (MVDR) method: not suitable for the broadband, complex computation, how to combine with sonar system in existence, this paper brings forward a broadband beamspace high resolution direction-of-arrival estimation method based on sub-band decomposition. It uses the method of sub-band decomposition to realize the expansion of MVDR under broadband condition, uses the beam transforming to realize the dimension amount of calculation reduction as to reduce the amount of calculation and improve the stability of the algorithm, and brings forward a new method to combine this technology with the sonar system in existence, using the direct output of the sonar multi-beam system, which can reduce the computational complexity and leave out the transforming from the array element to beamspace, the method of which can replace the conventional split-beam direction-of-arrival estimation system or the single beam system, which can improve the performance of the whole sonar direction-of-arrival estimation system. After theoretic analysis, simulation and authentication in lake, the method proves to be simple, effective and practical.

① 声学学报, 2004, 29(6): 533-538.

引言

自从Capon提出MVDR(Minimum Variance Distortionless Response) 高分辨方位估计方法以来，MVDR 以其简单的算法、良好的分辨性能得到了广泛关注，是一种具有良好实用前景的高分辨方位估计方法[1]。然而，MVDR 方法也有一些需要改进的地方：(1) 与以 MUSIC, ESPRIT 为代表的高分辨方位估计方法相似，MVDR 也是针对窄带信号情况，当信号为宽带时，该方法失效，而水声信号处理中，信号往往都是宽带。(2) 该方法是在阵元域上进行处理，需要对阵列输出的自相关矩阵求逆，当矩阵的阶数变大时，既增加了计算量，又增大了出现病态解的可能性。(3) 在实际使用时，MVDR 需要对搜索空间进行扫描，扫描步长越小，效果越好，而越精细的扫描需要的计算量越大，因此，需要考虑如何在声呐系统中应用该方法。

针对 MVDR 三个需要改进之处，本文提出了基于子带分解的宽带波束域MVDR高分辨方位估计方法 (Broadband Beamspace MVDR, 简称 BBMVDR)：首先，利用波束变换方法将阵元域信号转换到波束域，实现降维处理，大大减少运算量，提高算法的稳健性。其次，将宽频带进行子带分解，每个子带均采用 MVDR 进行处理，将各窄带的结果进行综合，得到整个工作频带上的总体结果。第三，在实际使用中，可将 BBMVDR 与声呐的多波束系统相结合，利用多波束系统的输出信号直接作为波束变换后的输出，省去从阵元域到波束域的转换，进一步减少运算量；将该方法运用于需要精确扫描的地方，如替代传统的分裂波束精测系统或单波束系统，使它与传统的多波束系统相辅相成，互为补充。经过理论推导、仿真实验和湖试实验，证明了 BBMVDR 高分辨方位估计方法是简单的、有效的、可行的，有望应用于实际系统。

1 MVDR 高分辨方法简介

设声呐基阵由 M 个阵元组成，以 $x_i(t)$ 表示第 i 个阵元的接收信号，w_i 表示对应的加权值，则传统的波束形成器的输出可写为：

$$Y_{\mathrm{CBF}}(t) = \boldsymbol{W}^{\mathrm{H}}\boldsymbol{X}(t), \tag{1}$$

其中 $\boldsymbol{W} = [w_1, w_2, \cdots, w_M]^{\mathrm{T}}$ 为权矢量，$\boldsymbol{X}(t)$ 为基阵输入矢量，此时基阵的输出功率为：

$$P_{\mathrm{CBF}}(\theta) = \mathrm{E}\{|y(t)|^2\} = \boldsymbol{W}^{\mathrm{H}}\boldsymbol{R}\boldsymbol{W}, \tag{2}$$

其中 $\boldsymbol{R} = \mathrm{E}[\boldsymbol{X}(t)\boldsymbol{X}^{\mathrm{H}}(t)]$ 为基阵输入协方差矩阵。

在波束形成器的输出功率中，信号源能量不仅在来波方向上有贡献，而且对波束宽度内的其它方向也有不同程度的贡献。而 MVDR 波束形成法就是在保持来波方向信号源能量不变的前提下，使信号源能量对波束宽度内的其它方向最小化，实际上是一个约束最佳化问题的解：

$$\min_{\boldsymbol{W}} \boldsymbol{W}^{\mathrm{H}}\boldsymbol{R}\boldsymbol{W} \quad 且 \quad \boldsymbol{W}^{\mathrm{H}}\boldsymbol{a}(\theta) = 1, \tag{3}$$

式中 $\boldsymbol{a}(\theta)$ 是指定方向的方向矢量。

通过 (3) 式的求解可得出 MVDR 波束形成器的输出功率为：

$$P_{\mathrm{MVDR}}(\theta) = \frac{1}{\boldsymbol{a}^{\mathrm{H}}(\theta)\boldsymbol{R}^{-1}\boldsymbol{a}(\theta)}. \tag{4}$$

MVDR 波束形成器能够提供最佳的信号保护、干扰消除和噪声降低能力，理想情况下，MVDR 的波束指向性函数近似为狄利克 δ 函数。

比较 (2) 式和 (4) 式能够发现，MVDR 波束形成器与 CBF 波束形成器的差别主要在于：MVDR 多了一个对自相关矩阵求逆的过程，其它的均与 CBF 类似。可以这样讲，MVDR 算法简单、分辨性能良好，但它也存在着如引言中所述的三个不足之处，文中以下的第 2, 3 和 4 节分别就这三方面问题提出了解决方案。

2 基于子带分解的宽带 MVDR 方法

在声呐的应用背景中，接收信号往往是宽带的，对于无源声呐更为如此。而MVDR和后来出现的以MUSIC为代表的基于信号特征分解的高分辨方法、以ESPRIT为代表的基于旋转子空间不变法类似，都是针对窄带信号情况，对于宽带信号不能直接应用。对于宽带高分辨定向方法的研究，可分为相干处理和非相干处理两类。相干处理以H.Wang和M.Kaveh提出的相关信号子空间处理方法为代表，其基本思想就是聚焦(Focusing)[2]。而非相干处理是以子带分解的方法将宽带划分为若干互不重叠的窄带，由所有窄带估计的结果平均得到最后的结果[3]。两类方法相比而言，各有优劣，文中采用的是基于子带分解的非相干方法，它与相干方法相比，方法简单、易于实现。

如果 $x_i(t), 0 \le t \le T$ 为宽带信号，为了对该信号进行子带分解，需要一个带通滤波器组，可以使用 FFT 实现。

假设对基阵输入信号的采样总时间为 T_0 s，现将其分为 U 段，每一段为 ΔT s。对每一段各路水听器

信号在时间上作 N 点 FFT, 其中有效工作频段上有 K 点 (也称频点数, 即 Frequency bin), 也就是有 K 个子带, 便可得到 U 组互不相关的窄带频率分量, 每组有 K 点数据, 即每个频点上有 U 个频域快拍。将 $x_i(t)$ 在 $\omega_i(1 \le i \le K)$ 的频域快拍代入 (4) 式就可得到第 i 个子带的 MVDR 波束输出, 所有子带处理得到的结果经平均就得到整个频带上的总体结果, 从而实现 MVDR 算法在宽带条件下的扩展。基于子带分解的宽带 MVDR 算法的工作过程如图 1 所示。首先, 将接收到的 T_0 s 宽带阵列信号分成 U 段, 每段长 ΔT s; 其次, 运用 FFT 变换在工作频带内形成 K 个子带; 再次, 在每个子带内分别形成频域快拍, 采用 MVDR 算法计算各子带 MVDR 波束输出; 最后, 将 K 个子带的结果平均相加, 从而得到整个工作频带上的 MVDR 波束输出。

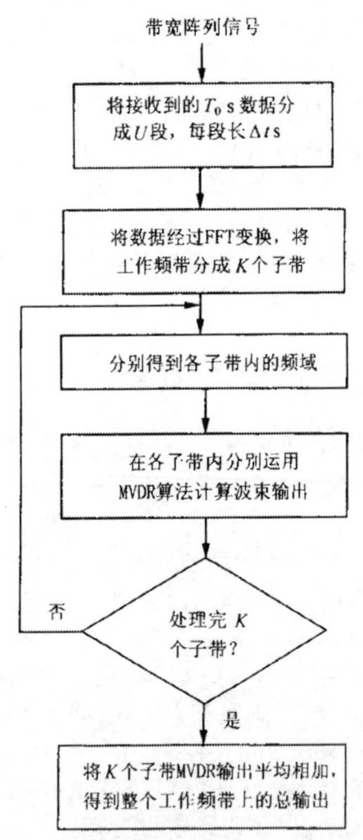

图 1 基于子带分解的宽带 MVDR 算法示意图

3 基于子带分解的宽带波束域 MVDR 方法 (BBMVDR)

Stergios Stergiopoulos 曾指出 [4]: 当自相关矩阵 R 大于 64×64 时, 使用不同的方法求逆, 如特征值法, 常常得不到正确的结果, 并且增大了出现变态解的可能性; 另方面, 大的维数也将增大 MVDR 的运算量。因此, 降低 R 的维数对于 MVDR 算法的稳健性和可实现性是至关重要的, 而 MVDR, 以及上节提出的基于子带分解的宽带 MVDR 算法都是直接对阵元域的信号进行处理, R 的维数与阵元数目等同。

1988年, H.B.Lee 首次提出在波束域实现高分辨定向技术 [5], 其后 XU Guanghan[6], M.D.Zoltowski[7] 先后提出了 BESPRIT 和 Broot-MUSIC, 他们的研究结果表明, 合理地选择波束形成预处理矩阵, 可达到降低运算维数、减少运算量、提高算法的稳健性以及适用于实际系统的目的。因此, 基于波束域处理的思想, 文中提出了基于子带分解的宽带波束域 MVDR 方法 (BBMVDR), 以减少运算量、提高算法的稳健性和可实现性。

BBMVDR 首先是将阵元空间数据转换到波束空间来, 如图 2 所示, 这一步不但实现了降维处理, 而且可以提高算法的稳健性。其次, 对转换后的数据运用上节提出的基于子带分解的宽带 MVDR 方法进行高分辨定向。

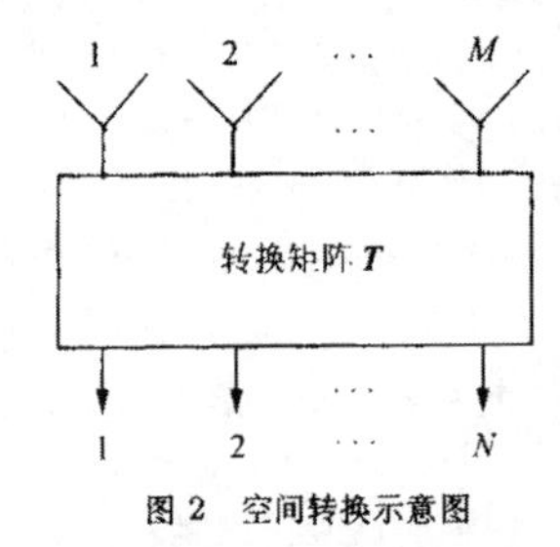

图 2 空间转换示意图

由 M 个阵元组成的均匀线列阵的离散空间傅里叶变换 DSFT 可表示为:

$$f(u,t) = \sum_{i=0}^{M-1} x_i(t) e^{-\mathrm{j}2\pi i u}, \tag{5}$$

其中 $u = d\sin\theta/\lambda$, d 是相邻两阵元的间距, θ 是与线列阵法线所成的角度, $\lambda = c/f$ 为接收信号的波长, c 为声速, f 为接收信号的频率。

定义 $M \times 1$ 的 DSFT 波束预形成加权矢量为:

$$\boldsymbol{v} = [1, \exp(\mathrm{j}\pi u), \cdots, \exp(\mathrm{j}\pi(M-1)u)]^{\mathrm{T}}. \tag{6}$$

并构造 $M \times N$ 的转换矩阵 T 为以下形式:

$$\boldsymbol{T} = \frac{1}{\sqrt{M}}\left\{v\left(\frac{2}{M}\right), v\left[(m+1)\frac{2}{M}\right], \cdots, v\left[(m+N-1)\frac{2}{M}\right]\right\}. \tag{7}$$

可以看出, $\boldsymbol{T}$ 实际上从 $M \times M$ 的 DSFT 矩阵中连续抽取的一个从第 m 列到第 $(m+N-1)$ 列的 $M \times N$ 子矩阵。子矩阵中的每一个列向量指向一个不同的方向，N 个向量形成一个扇面。经 $\boldsymbol{T}$ 矩阵转换后，原来 $M\times 1$ 的声呐阵列输入向量就转换为 $N\times 1$ 的波束空间向量，$N < M$，实现降维处理，减少运算量，并且可以估计方位处在以 θ 为中心、位于 $\{\arcsin(2m/M), \arcsin[(m+N-1)(2/M)]\}$ 扇面内的目标。可将波束域空间的输入向量表示为：

$$\boldsymbol{Z}(t)=\boldsymbol{T}^{\mathrm{H}}\boldsymbol{X}(t). \tag{8}$$

则波束域空间输入向量的协方差矩阵写为：

$$R_D=\mathrm{E}\{\boldsymbol{Y}(t)\boldsymbol{Y}^{\mathrm{H}}(t)\}=\mathrm{E}\{\boldsymbol{T}^{\mathrm{H}}\boldsymbol{X}(t)\boldsymbol{X}^{\mathrm{H}}(t)\boldsymbol{T}\}=\boldsymbol{T}^{\mathrm{H}}\boldsymbol{R}\boldsymbol{T}. \tag{9}$$

因此，波束域 MVDR(Beamspace MVDR，简称 BMVDR) 波束形成的输出功率是：

$$P_{\mathrm{BMVDR}}(\theta)=\frac{1}{\boldsymbol{a}^{\mathrm{H}}(\theta)\boldsymbol{T}\boldsymbol{R}_B^{-1}\boldsymbol{T}^{\mathrm{H}}\boldsymbol{a}(\theta)}=\frac{1}{\boldsymbol{a}^{\mathrm{H}}(\theta)\boldsymbol{T}\left(\boldsymbol{T}^{\mathrm{H}}\boldsymbol{R}\boldsymbol{T}\right)^{-1}\boldsymbol{T}^{\mathrm{H}}\boldsymbol{a}(\theta)}. \tag{10}$$

将 $\boldsymbol{Z}(t)$ 代替 $\boldsymbol{X}(t)$ 作子带分解，进而在各个子带内进行 MVDR 波束形成，最后将各子带的结果平均相加，得到 BBMVDR 最后波束输出。

4 BBMVDR 与实际声呐系统结合应用的考虑

从上节中的推导可知，波束空间转换类似于空间滤波，使得观测扇面外的噪声得到有效降低，转换矩阵 $\boldsymbol{T}$ 相当于传统的多波束形成，即在每个列向量指定的方向上形成一个波束，总共形成 N 个波束。因此，波束空间的输入向量 $\boldsymbol{Z}(t)$ 代表着多波束系统的输出，等于说将 M 个水听器接收信号转变成 N 个波束的输出。M.D.Zoltowski 经研究认为[7]：当目标源恰好位于某个波束的主极大方向时，可以获得最佳的方位估计效果；但事先不可能知道目标源的方位，为此，他建议各波束重叠 50%，这样，波束域方法能够有好的估计性能。而传统的声呐多波束系统，恰好也是各波束重叠 50%。

MVDR 需要对搜索空间进行扫描，扫描步长越小，效果越好，而越精细的扫描需要的计算量越大，因此，如果在声呐系统中应用 MVDR 方法，可以考虑将其应用于需要精细扫描的地方。传统的声呐系统中，采用方位粗测和方位精测相结合的方法对目标方位进行估计：以多波束系统对目标方位进行粗测，在大致判定目标的方位后，采用分裂波束精测定向系统，如采用互功率谱法、正交相关定向法，精确测定目标方位。而如互功率谱法之类的分裂波束定向法，只具备对目标方位精测的能力，没有多目标分辨的能力。此外，因为声呐的多波束系统形成的波束往往较宽，所以声呐中常常以单波束系统对其进行补充，单波束系统还是利用 CBF，但形成的波束更窄，为覆盖搜索空间所需要的波束数目比多波束系统更多。由于声呐的单波束系统和多波束系统常常是相互独立的两套系统，因此大大增加了声呐信号处理机的运算量。

归纳声呐系统中的分裂波束精测系统和单波束系统，可发现它们都需要对方位进行精细扫描。因此，文中提出了一种新思路：可以考虑将声呐多波束系统与 BBMVDR 结合起来，采用 BBMVDR 替代传统的分裂波束精测系统或单波束系统。这样，可以直接利用多波束系统形成的多路波束信号，而不需要再做从阵元域到波束域的转换，进一步减小 BBMVDR 的运算量。而 BBMVDR 的多目标分辨能力和良好的分辨性能使得它能够胜任、甚至是更好地完成分裂波束精测系统或单波束系统的功能，从而提高整个声呐定向系统的性能。实现框图如图 3 所示，其中虚框 1 描述的是 BBMVDR 代替分裂波束精测系统的思路，虚框 2 描述的是 BBMVDR 代替单波束系统的思路。

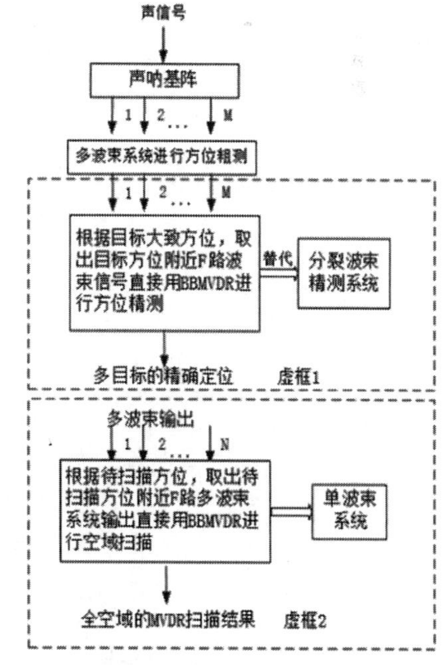

图 3 BBMVDR 替代分裂波束精测系统 / 单波束系统示意图

5 仿真实验

按半波长方式布阵的均匀线列阵共有 $M=24$ 个阵元，相邻两阵元间距 1 m，对应的接收信号中心

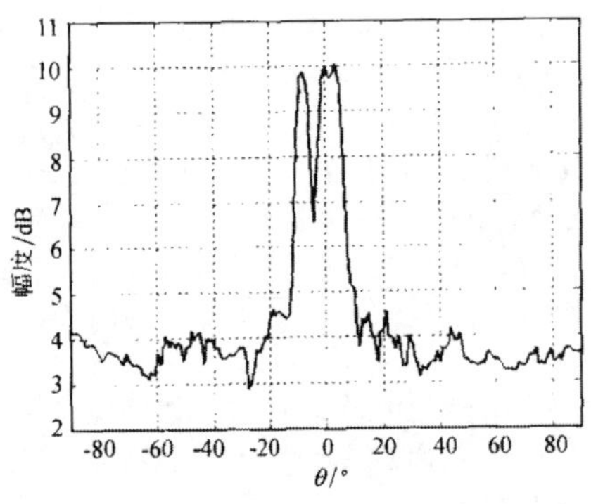

图 4(a) SNR= 0 dB 时, CBF 波束图

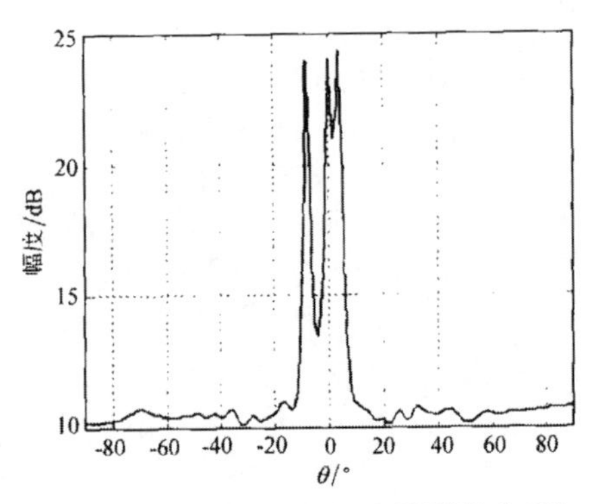

图 4(b) SNR= 0 dB 时, BBMVDR 波束图

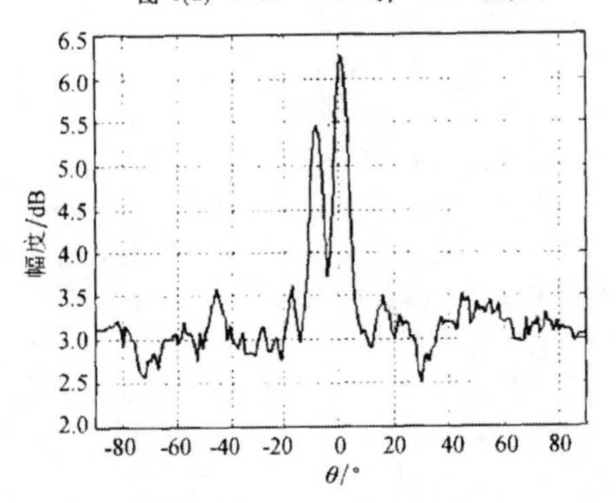

图 5(a) SNR= −10 dB 时, CBF 波束图

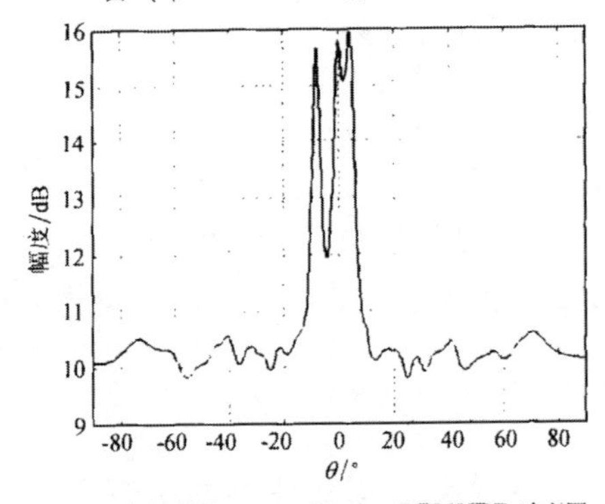

图 5(b) SNR= −10 dB 时, BBMVDR 波束图

频率为 750 Hz 。形成波束时使用全部阵元, 24 个阵元所形成子阵的波束宽度大约为 4.2°, 相邻两波束重叠 50% , 那么为覆盖 180° 空间需要形成 87 个波束。CBF 采用频域波束形成方法, BBMVDR 以待扫描方位附近的 9 路 CBF 输出的波束信号来做处理, 在空间上每隔 1° 形成一个波束, 在 180° 范围内形成 181 个波束。数据采样率 $f_s = 10$ kHz, FFT 长度 $N = 4096$, 频率分辨力 $\Delta f = f_s/N = 2.4414$ Hz, 所用的数据段总点数为 16×4096, 对该数据段进行分段, 每段数据长度为 4096, 相邻两段数据重叠率为 75%, 处理频段为 500~ 1000 Hz, 相应的频点数为 205 个, 频率快拍数为 61 。在 $\theta_1 = -8°$, $\theta_2 = 0°$ 和 $\theta_3 = 4°$ 分别存在着三个高斯白噪声宽带声源, 信噪比 SNR 分别为 0 dB 和 −10 dB 时, CBF 和 BBMVDR 分别如图 4(a) 、图 4(b) 和图 5(a) 、图 5(b) 所示 (图中纵轴单位均为 dB) 。

能够发现: (1) 信噪比无论是 0 dB 还是 −10 dB, 从分辨性能、旁瓣高度和谱峰锐度来看, BBMVDR 都要比 CBF 好很多。 (2) 在 SNR= −10 dB 时, CBF 已经不能把在同一个波束宽度内的 $\theta_2 = 0°$ 和 $\theta_3 = 4°$ 两个目标区分出来, 而 BBMVDR 能够将这两个目标区分开来。

在本例中, BBMVDR 和 CBF 计算量比较是在 PIV1.3G 的计算机上, 通过 MATLAB5.3 中的 CPUTIME 函数测出的, 如表 1 所示, 两种算法都包括了对阵列信号进行 FFT 处理。从表面上看, 本例中 BBMVDR 所需的时间比 CBF 还稍多一点, 但 BBMVDR 在 180° 范围内形成了 181 个波束, 而 CBF 只有 87 个波束, 而在实际使用中, BBMVDR 和 CBF 可共用 FFT 运算, 因此, 在形成相同数目的波束时 BBMVDR 实际运算量比 CBF 少许多, 也就充分地证明了 BBMVDR 能有效地减少计算量。

表 1 计算时间比较

	BBMVDR	CBF
计算时间 /s	16.74	15.28

6 湖试实验

2002 年 12 月, 在千岛湖水域进行了多目标分辨试验。一均匀线列阵由 16 个阵元组成, 相邻两阵元间距 1 m 。实验中有一个合作目标: 目标船, 实验水域附近还常常出现一些过往船只。无论是目标船还是过往船只, 吨位都较小, 信噪比均都低于 0 dB 。

数据采样率 $f_s=10$ kHz, FFT 长度 $N=4096$, 频率分辨力 $\Delta f=f_s/N=2.4414$ Hz, 所用的数据段总点数为 16×4096, 对该数据段进行分段, 每段数据长度为 4096, 相邻两段数据重叠率为 75%, 处理频段为 500~1000 Hz, 相应的频点数为 205 个, 频率快拍数为 61 。为了对比, CBF 采用频域波束形成方法, CBF 和 BBMVDR 都是每隔 1° 形成一个波束, BBMVDR 是利用待扫描方位附近的 9 路 CBF 输出的波束信号来处理。图 6 是 BBMVDR 和 CBF 归一化的波束输出结果, 图中实线为 BBMVDR, 虚线为 CBF 。从图 6 中能看出: 在 −33° BBMVDR 和 CBF 都显示了有一个目标, 这是实验中所用的目标船; 而 BBMVDR 在 15° 和 33° 附近显示了有两个小目标, 这与现场记录的过往船只信息相吻合, 而 CBF 没有分辨出; 此外, BBMVDR 的主瓣宽度比 CBF 窄, 而且它的旁瓣比 CBF 的低多了。细致地比较仿真结果 (如图 4 和图 5) 和湖试结果 (图 6), 能发现湖试结果比仿真结果差一些, 分析原因主要有: (1) 存在着多途效应, (2) 线列阵存在着阵形畸变。但总的说来, BBMVDR 的性能比 CBF 有明显的提高, 与理论分析和仿真实验相符合。

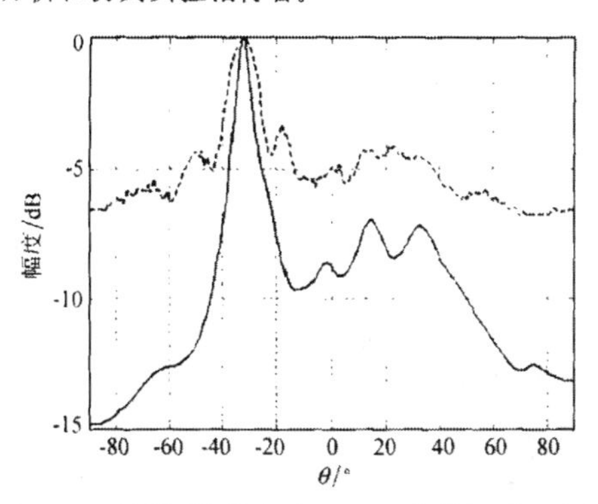

图 6 湖试数据采用 BBMVDR 和 CBF 处理后的对比图

再看计算量对比, 仍采用在 PIV1.3G 的计算机上, 通过 MATLAB5.3 中的 CPUTIME 函数进行测量, 得出了表 2 的结果, 其中带 "*" 上标的 BBMVDR 的计算时间不包括 FFT 运算所需的时间, 因为频域波束形成中就有对各路水听器信号做 FFT, BBMVDR 可以直接利用。表 2 很好地说明了 BBMVDR 能显著地减小运算量。

表 2 计算时间比较

	BBMVDR	BBMVDR*	CBF
计算时间 /s	9.28	6.05	17.48

7 结论

文中针对 MVDR 对宽带信号不适用、计算量较大和与如何与实际声呐系统结合的问题, 提出了基于子带分解的宽带波束域 MVDR 高分辨方位估计方法: 它以子带分解的方法实现 MVDR 在宽带情形下的扩展, 这一步可利用 FFT 完成; 利用波束转换实现降维运算、大大减小运算量和提高算法的稳健性; 并且提出了在实际使用中的新思路: 可将该方法与声呐的多波束系统相结合, 直接利用声呐多波束系统的输出, 可省去从阵元域到波束域的转换, 进一步减少运算量; 它可用于替代传统的分裂波束精测定向系统或单波束系统, 提高整个声呐定向系统的性能。经过理论分析、仿真实验和湖试实验, 证明了该方法是简单的、有效的、可行的, 是一种具有良好实用前景的高分辨方位估计方法。

参 考 文 献

1 Capon J. High-resolution frequency-wavenumber spectum analysis. *Proc. IEEE*, 1969; **57**(8): 1408—1418

2 Wang H, Kaveh M. Coherent signal-subspace processing for the detection and estimation of angles of arrival of multiple wide-band sources. *IEEE Trans. ASSP*, 1985; **33**(4): 823—831

3 Kim, Byung-Chul, Lu, I-Tai. High resolution broadband beamforming based on the MVDR method. IEEE Oceans Conference, 2000: 1025—1028

4 Stergios Stergiopoulos. Extended towed array processing by an overlap correlator. *J. Acoust. Soc. Am.*, 1989; **86**(1): 158—171

5 Lee H B, Wengrovitz M S. Improved high-resolution direction-finding through use of homogeneous constraints. Proc. IEEE ASSP workshop spectrum estimation and modeling, 1988: 152—157

6 XU Guanghan, Silverstein S D, Roy R H, Kailath T. Beamspace ESPRIT. *IEEE Trans. SP*, 1994; **42**(2): 349—355

7 Zoltowski M D, Kautz G M, Silverstein S D. Beamspace Root-MUSIC. *IEEE Trans. SP*, 1993; **41**(1): 344—364

一种确定性盲波束形成的迭代算法①②

杨秦山　李启虎
(中国科学院声学研究所　北京　100080)

2002 年 9 月 13 日收到
2003 年 3 月 11 日定稿

摘要　提出了一种基于信道的信号处理模型的在线迭代盲波束形成算法。该方法利用 Givens 变换，分步迭代地求出期望得到的信息矩阵，忽略无关的运算结果，达到减小运算量，减少内存占有空间的目的。同时，因为没有关于信源个数的先验假设，该方法可以在未知具体信源个数的情况下进行盲波束形成。在线迭代的算法结构也为 DSP 并行实现提供了可能。仿真试验验证了本方法的有效性和正确性。

PACS 数：　43.30, 43.60

An iterative algorithm of deterministic blind beamforming

YANG Qinshan　LI Qihu
(*Institute of Acoustics, The Chinese Academy of Sciences* Beijing 100080)

Received Sept. 13, 2002
Revised Mar. 11, 2003

Abstract　A method based on the model of propagation channel is proposed. In the method, Givens transform is used to calculate the expected information matrix step by step, ignore the irrelative results. In this way, the computational complex and memory needed are reduced. The method can be used without the preinformation of source number. The online structure of the algorithm makes it easy to be realized in parallel multi-DSP. Computer simulation verified the correctness and effectiveness of the method proposed.

引言

在应用需求的推动下，人们对水声通信系统要求的不断提高，需要寻找合适的技术实现通信网络、达到提高系统容量、减少通道间干扰的目的。同时随着用户数目的增加，频带的限制和同信道的干扰将变得非常严重。寻找合适的空分复用技术是建立通信网络、减少干扰提高容量的一种有效方法。这其中盲波束形成又是实现空分复用技术的关键问题。

最初，人们发展了信号源方向估计的算法以及基于方向估计的盲波束形成方法。它们的特点是：先估计每个入射波的方向，根据这个估计结果产生相应的波束形成器恢复来自估计方向上的信号。这就需要先验的信息以及对阵行校正的严格要求。同时，这些方法的应用效果与信道条件的关系很大，通常每个信号源只能有很少的几个传输路径。新提出的一些波束形成方法，它们不是基于常用的物理空间信道模型，而是改而利用信道的信号处理模型 (多输入、多输出模型) 以及信号的性质，估计期望信号的方向矢量。利用信号统计性质的盲处理方法为随机性盲波束形成，利用信号本身确定性质的盲方法为确定性盲波束形成方法 [1]。这些方法可以较少的利用先验信息，但因为需要接收足够的信号才能够进行精确的计算，运算量以及存储量需求较大，同时因为其离线的工作方式，在硬件上实时的实现有一定的难度。

本文提出了一种新的基于信道的信号处理模型的在线迭代盲波束形成算法。新的方法利用 Givens 变换，分步迭代地求出期望得到的信息矩阵，忽略无关的运算结果，达到减小运算量、减少内存占有空间

① 863 青年基金资助项目(2002AA639430).
② 声学学报, 2004, 29(5): 419-424.

的目的。同时，因为没有关于信源个数的先验假设，新方法可以在未知具体信源个数的情况下进行盲波束形成。在线迭代的算法结构也为 DSP 并行实现提供了可能。同时，因为是利用了信道的信号处理模型和信号本身的性质，算法对阵行的畸变具有较好的鲁棒性。相应的结果在本文的计算机仿真中均进行了正确性和有效性的验证。

1 确定性盲波束形成算法

1.1 问题描述

假定 $s_1(t),\cdots,s_d(t)$ 是从不同位置的 d 个信号源发射的信号。则在接收阵上接收到的信号向量是这 d 个信号的线性组合，即：

$$\boldsymbol{x}(t)=\boldsymbol{a}_1 s_1(t)+\cdots+\boldsymbol{a}_d s_d(t), \tag{1}$$

式中 $\boldsymbol{a}_i(i=1,\cdots,d)$ 为各信源发射信号的方向向量，即：

$$\boldsymbol{x}(t)=\boldsymbol{A}\boldsymbol{s}(t), \tag{2}$$

其中矩阵 $\boldsymbol{A}$ 由方向向量 $\boldsymbol{a}_i$ 组成和向量 $\boldsymbol{s}(t)$ 由时间序列 $s_i(t)$ 组成，它们分别为：

$$\boldsymbol{A}=[\boldsymbol{a}_1,\cdots,\boldsymbol{a}_d],\quad \boldsymbol{s}(t)=\begin{bmatrix} s_1(t)\\ \cdots \\ s_d(t)\end{bmatrix}. \tag{3}$$

令矩阵 $\boldsymbol{X}=[x(0),\cdots,x(N-1)]$ 和 $\boldsymbol{S}=[s(0),\cdots,s(N-1)]$，则可以将上式表示成矩阵相乘的形式：

$$\boldsymbol{X}=\boldsymbol{A}\boldsymbol{S}. \tag{4}$$

该式为多输入 - 多输出模型，即 MIMO 模型。该模型是信号分离的一般线性模型，适用于主要路径的时延扩展比信号的逆带宽小的情况，典型的如窄带信号。为了用波束形成达到信号分离的目的，希望得到 $\boldsymbol{A}$ 的左逆矩阵 $\boldsymbol{W}$，使得 $\boldsymbol{W}\boldsymbol{A}=\boldsymbol{I}$，达到信号分离目的 $\boldsymbol{W}\boldsymbol{X}=\boldsymbol{S}$。此时信号源的个数 d 不多于传感器个数 M，即 $d\le M$。

1.2 迭代的确定性波束形成算法

传统的确定性盲波束形成是一种离线 (off-line) 算法。在得到充足的存储数据之前是无法得到相应的结果的。在等待数据的过程中，无法做任何处理。这种情况下，算法将占用大量的数据存储空间、耗费大量的运算能力。

这里，我们将提出一种新的算法，用迭代过程一步步实现整个算法。由于在迭代的每个步骤中只精确的计算出期望得到的矩阵，而忽略无关的运算结果，这就大量地减少了运算量。研究还表明，可以假设信源个数是一个估计量，在迭代的过程中，估计多出的信号源个数的影响会被削弱，最终输出正确的结果。达到减小算法对运算量、存储量要求的目的。而且，迭代的算法很方便在实时性系统中实现。在接收数据的同时，进行迭代过程，得到数据的结果。

1.2.1 迭代算法

经过了迭代处理，可以在不进行预先估计信号源个数的情况下，进行有针对性的奇异值分解，有效降低运算量和实际系统所需的存储量。下降到约 N^2 量级。这里将对整个迭代过程进行详细的阐述。

假设在迭代的第 n 步定义一个 $N\times n$ 维的数据矩阵

$$\boldsymbol{X}_n\equiv\begin{bmatrix} x^{\mathrm{H}}(1) & \cdots & x^{\mathrm{H}}(n)\end{bmatrix}. \tag{5}$$

相应的 SVD 分解为：

$$\boldsymbol{X}_n=\boldsymbol{U}_n\boldsymbol{B}_n\boldsymbol{V}_n^{\mathrm{H}}, \tag{6}$$

式中：$\boldsymbol{U}_n$ 是一个 $N\times d$ 维矩阵，$\boldsymbol{B}_n$ 是一个 $d\times d$ 对角矩阵，$\boldsymbol{V}_n^{\mathrm{H}}$ 是一个 $d\times n$ 维矩阵，$\boldsymbol{U}_n$ 是左奇异向量阵，$\boldsymbol{V}_n^{\mathrm{H}}$ 是右奇异向量阵，根据确定性波束形成算法可以看出，$\boldsymbol{V}_n^{\mathrm{H}}$ 是不需要准确分离出来的向量。可以利用这一点在迭代的过程中显著的降低运算量。

在迭代的第 $n+1$ 步，为了构造出 SVD 分解，给第 n 步得到的信号矩阵加上一列，于是得到：

$$\boldsymbol{X}_{n+1}=[\boldsymbol{X}_n \quad x_{n+1}]. \tag{7}$$

在此基础上，又可以得到[2]：

$$\boldsymbol{X}_{n+1}=\boldsymbol{U}_n[\boldsymbol{B}_n \quad \boldsymbol{q}]\begin{bmatrix}\boldsymbol{V}_n^{\mathrm{H}} & 0\\ 0 & 1\end{bmatrix}, \tag{8}$$

式中 $\boldsymbol{q}=\boldsymbol{U}_n^{\mathrm{H}}\times x_{n+1}$。

这里，使用一系列的 Givens 变换，将列向量 $\boldsymbol{q}$ 化 0 消去。利用 Givens 变换定理[3]，可以通过求 Givens 矩阵的方法，将 (8) 式中的信息保留矩阵无关部分化零消去。

下面用 $d=2$ 情况下的 $\boldsymbol{q}$ 化 0 消去的过程作为例子，对迭代的过程进行详细的说明。

在 $d=2$ 情况下，不失一般性，问题转变为对矩阵 $\boldsymbol{M}=\begin{bmatrix} a & 0 & q_1\\ 0 & b & q_2\end{bmatrix}$ 中 q_1,q_2 的化 0 消除问题。令 $n_1=\frac{q_1}{\sqrt{q_1^2+q_2^2}}$，$n_2=\frac{q_2}{\sqrt{q_1^2+q_2^2}}$ 则 Givens 的左乘矩阵为：

$$\boldsymbol{G}_l=\begin{bmatrix} n_1 & -n_2\\ n_2 & n_1\end{bmatrix}. \tag{9}$$

这样得到形如 $\boldsymbol{N}=\begin{bmatrix}\boldsymbol{A}_1 & k\\ \boldsymbol{A}_2 & \end{bmatrix}$ 的矩阵，其中 $\boldsymbol{A}_1$，$\boldsymbol{A}_2$ 为 1×2 阶的向量。有 $\boldsymbol{M}=\boldsymbol{G}_l*\boldsymbol{N}$。为了达到 q_1，q_2 消 0 的过程，使 k 最小化，构造矩阵 $\boldsymbol{P}$，满足：

$$\left\|\begin{bmatrix}\boldsymbol{A}_1 & k\\ \boldsymbol{A}_2 & 0\end{bmatrix}\boldsymbol{P}\right\|\approx\sigma_{\min}, \tag{10}$$

式中 $\boldsymbol{P}$ 是构造的估计向量，$\sigma_{\min}$ 是设定的极小值。

对 $\boldsymbol{P}$ 进行 Givens 消除，得到：

$$\boldsymbol{W}_1\boldsymbol{P}=\begin{bmatrix}0\\ \vdots\\ 0\\ 1\end{bmatrix}=\boldsymbol{e},$$

式中 $\boldsymbol{W}_1$ 为左乘的 Givens 变换矩阵。于是有：

$$\boldsymbol{N}\boldsymbol{W}_1=\begin{bmatrix}\boldsymbol{A}_1 & k\\ \boldsymbol{A}_2 & 0\end{bmatrix}\boldsymbol{W}_1=\boldsymbol{W}_2^{\mathrm{H}}\boldsymbol{W}_2\begin{bmatrix}\boldsymbol{A}_1 & k\\ \boldsymbol{A}_2 & 0\end{bmatrix}\boldsymbol{W}_1,$$

式中 $\boldsymbol{W}_2$ 为 Givens 矩阵满足：

$$\begin{bmatrix}\boldsymbol{A}_1' & k'\\ \boldsymbol{A}_2' & 0\end{bmatrix}=\boldsymbol{W}_2\begin{bmatrix}\boldsymbol{A}_1 & k\\ \boldsymbol{A}_2 & 0\end{bmatrix}\boldsymbol{W}_1.$$

根据上面的式子，我们得到：

$$k'=\left\|\boldsymbol{X}\boldsymbol{W}_l^{\mathrm{H}}\boldsymbol{e}\right\|=\left\|\begin{bmatrix}\boldsymbol{A}_1 & k\\ \boldsymbol{A}_2 & 0\end{bmatrix}\boldsymbol{P}\right\|\approx\sigma_{\min},$$

得到：

$$\boldsymbol{M}=\boldsymbol{G}_l\boldsymbol{W}_2'\begin{bmatrix}\boldsymbol{A}_1' & k'\\ \boldsymbol{A}_2' & 0\end{bmatrix}\boldsymbol{W}_1', \tag{11}$$

其中有 $k'\approx\sigma_{\min}$。

这样，就达到了对 $\boldsymbol{M}$ 矩阵中，$\boldsymbol{q}$ 的化 0 消去。显然，这个过程同样可以应用在 $d=n$ 的情况。于是针对 (8) 式，消去所得结果里的 0 向量，得到：

$$\boldsymbol{X}_{n+1}=\boldsymbol{U}_n\boldsymbol{G}_L(n)[\boldsymbol{B}_n\ \ q]\boldsymbol{G}_R^{\mathrm{H}}(n)\begin{bmatrix}\boldsymbol{V}_n & 0\\ 0 & 1\end{bmatrix}\equiv\widetilde{\boldsymbol{U}}_n[\widetilde{\boldsymbol{B}}_n]\widetilde{\boldsymbol{V}}_n^{\mathrm{H}}. \tag{12}$$

得到了新的 $\widetilde{\boldsymbol{U}}_n$，迭代的算法描述如下：

(1) 得到的数据向量 $\boldsymbol{x}(n)(n=1,2,\cdots)$，利用迭代的方法得到期望得到的矩阵 $\widetilde{\boldsymbol{U}}_n$；

(2) 令 $\widetilde{\boldsymbol{U}}_1=\boldsymbol{J}_1\widetilde{\boldsymbol{U}}$ 以及 $\widetilde{\boldsymbol{U}}_2=\boldsymbol{J}_2\widetilde{\boldsymbol{U}}$，式中 $\boldsymbol{J}_1=[\boldsymbol{I}_{N-1},0]$，$\boldsymbol{J}_2=[0,\boldsymbol{I}_{N-1}]$；

(3) [1] $\widetilde{\boldsymbol{U}}_1^{+}=(\widetilde{\boldsymbol{U}}_1^{\mathrm{H}}\widetilde{\boldsymbol{U}}_1)^{-1}\widetilde{\boldsymbol{U}}_1^{\mathrm{H}}$，$\widetilde{\boldsymbol{U}}_1^{+}\widetilde{\boldsymbol{U}}_2=\boldsymbol{T}^{-1}\boldsymbol{\Theta}\boldsymbol{T}$，式中，矩阵 $\boldsymbol{\Theta}$ 为 $\widetilde{\boldsymbol{U}}_1^{+}\widetilde{\boldsymbol{U}}_2$ 的特征值组成的对角阵；

(4) $\boldsymbol{W}=\boldsymbol{T}\widetilde{\boldsymbol{U}}^{\mathrm{H}}$；

(5) $\boldsymbol{S}=\boldsymbol{W}\boldsymbol{X}$ 得到信号。

当估计的信号源个数多于实际个数时，整个算法的运算量会相应的增加，但是得到的 $d\times d$ 对角矩阵中将有几个 (个数由多估计的信号源个数决定) 对角线数值的绝对值明显的小。在这种情况下得矩阵 $\widetilde{\boldsymbol{U}}_n$，进而计算得到的分离信号 S 中会有相应的输出结果能量比正常值小很多，同时不会影响真正信号的输出结果，信号源的方位也可以得到。

2 仿真结果及性能分析

这部分将通过仿真研究来讨论新方法的性能。对传统的确定性盲波束形成方法和迭代的盲自适应方法分别进行了仿真试验。根据提出的算法，设置了一个仿真的环境。将窄带信号源作为仿真信号源。用高斯白噪声作为背景噪声。

不失一般性，在仿真过程中采用 32 基元的线阵接收处理信号。

2.1 算法迭代性能描述

本系统给出的系统结构图示于图 1。

为了提高系统迭代的收敛速度，算法采用确定的信源信息作为系统迭代的初值，即在初始化时使用实际接收数据的前几组进行精确的确定性盲波束形成，所得到的信息作为自适应系统输入的初始值。采用这样的初值，调整相应的衰减因子，即可得到较好的系统收敛性能。

一信源处于 47° 方向，发射窄带信号。这里使用迭代的盲波束形成方法对该信源发射信号进行处理。信噪比为 0 dB。图 2(a) 为信号到达角估计和迭

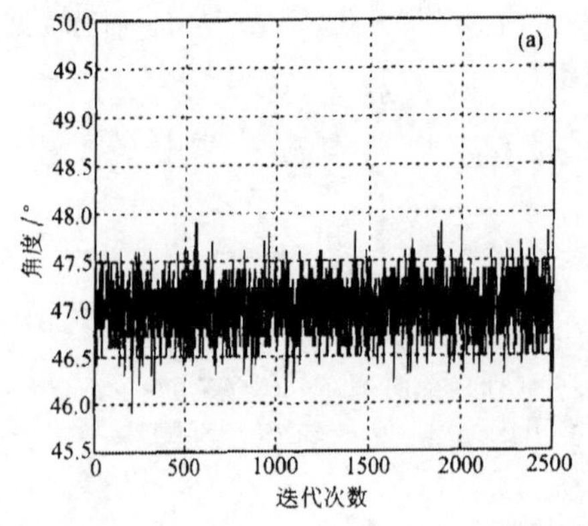

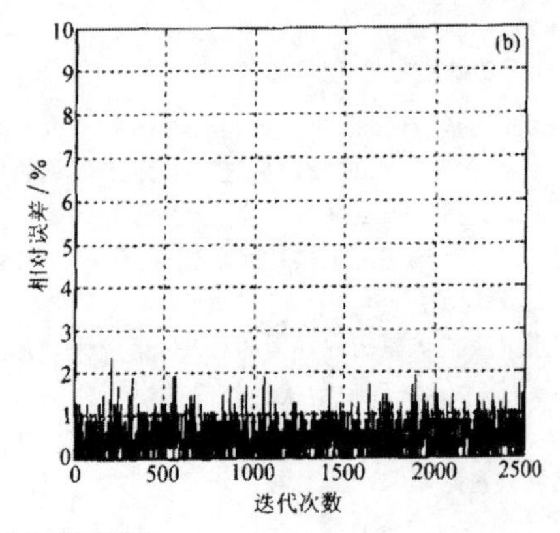

图 2 信号到达角估计结果及误差

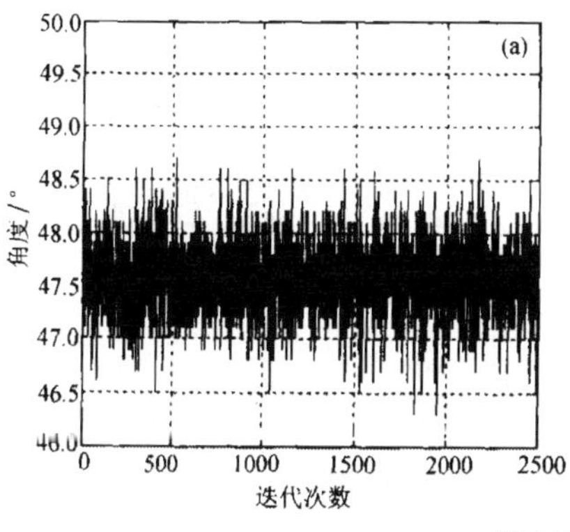

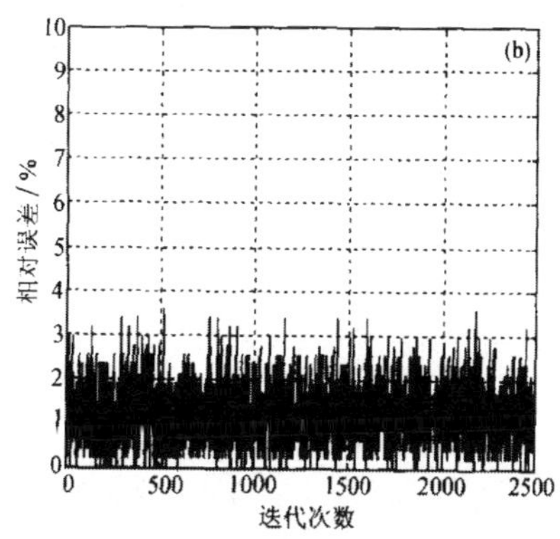

图 3 DOA 估计结果及误差

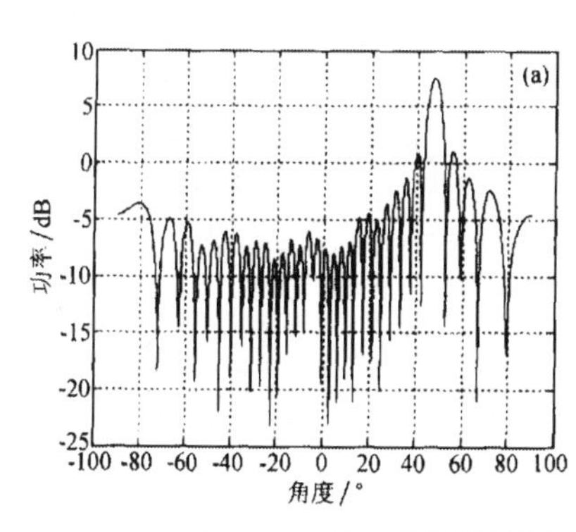

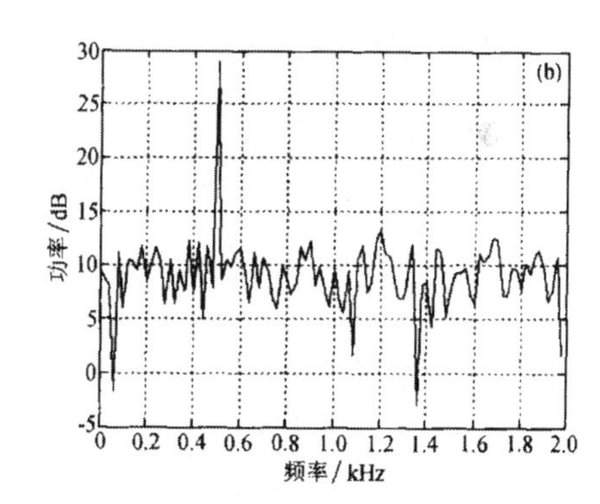

图 4 批处理的盲波束形成方法方位指向性图及信号幅频响应图

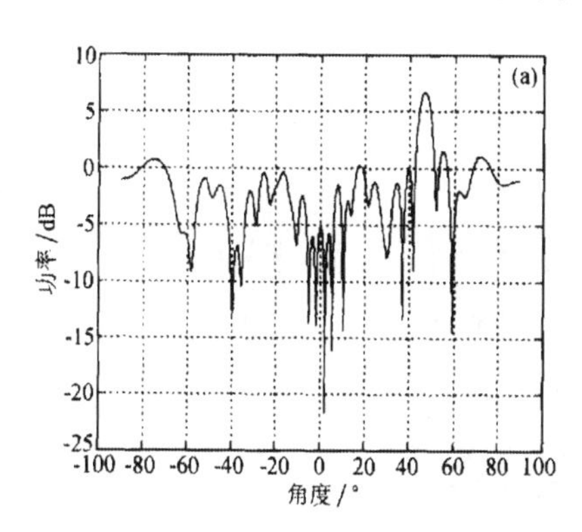

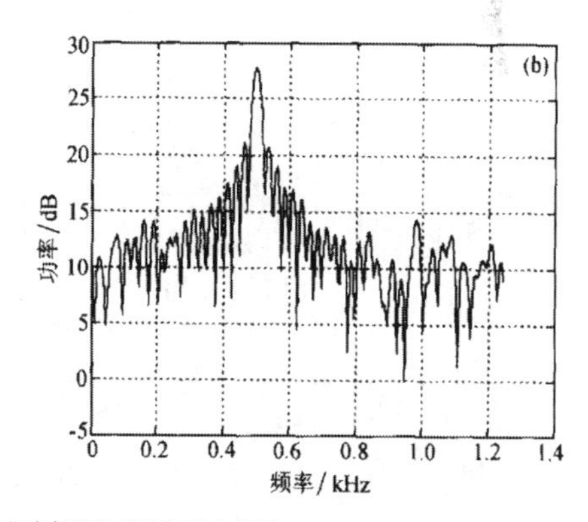

图 5 迭代的盲波束形成方法方位指向性图及信号幅频响应图

代次数的关系图，图 2(b) 是得到的方位估计相对误差与迭代次数的关系图，可以看到，因为采用了准确的初值，迭代的结果在真实结果附近振荡。

2.2 阵形误差性能描述

在同样的仿真条件下，设定每个阵元位置的物理偏差占阵元间隔的 30%，得到的 DOA 估计结果和迭代相对误差结果如图 3。

从图中可以看出，阵元间隔 30% 的误差，使方位估计误差增加了约 1~1.5% 的误差。

2.3 单信源情况下算法性能分析

一信号源处于 47° 方位，发射窄带信号。分别用批处理的盲波束形成方法和迭代的盲波束形成方法对该信号源发射信号进行处理。信噪比为 0 dB。得到的恢复信号频域图和方位图如图 4 和图 5。

可以看到，迭代的盲波束形成方法同样可以达到批处理的盲波束形成方法在方位指向性上的效果。

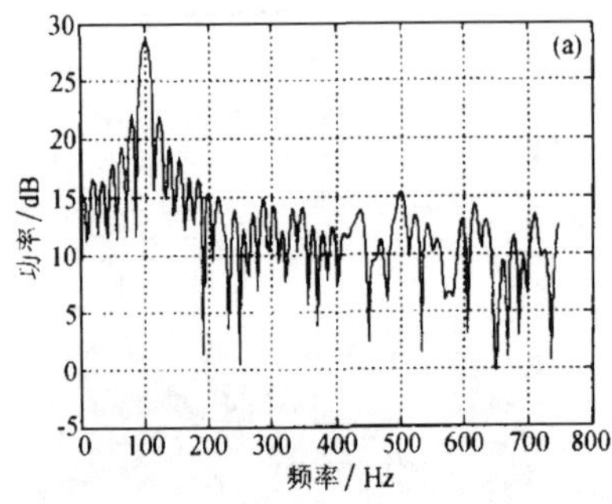

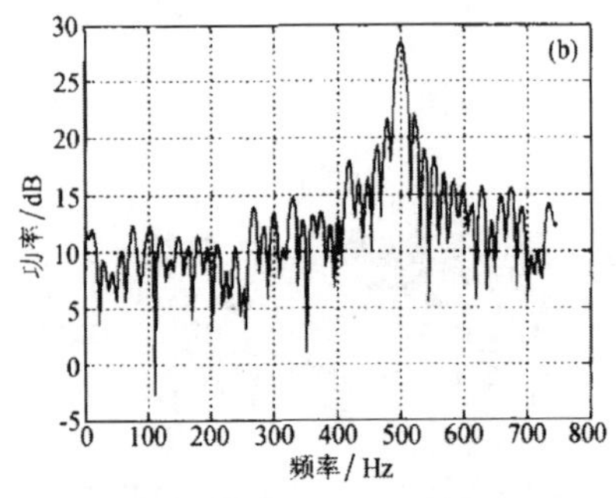

图 6　两信源情况下迭代的盲波束形成方法信号幅频响应图

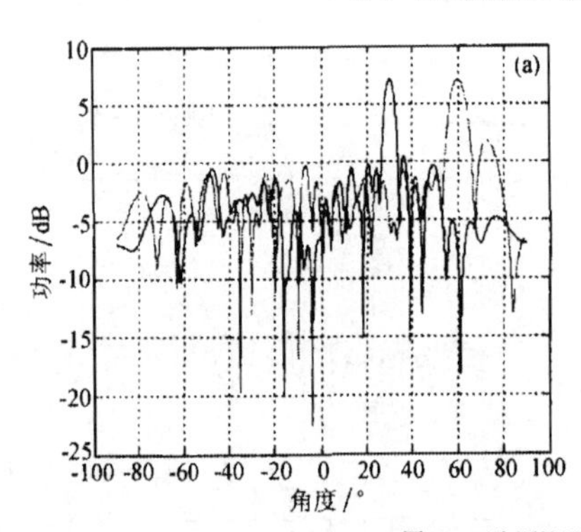

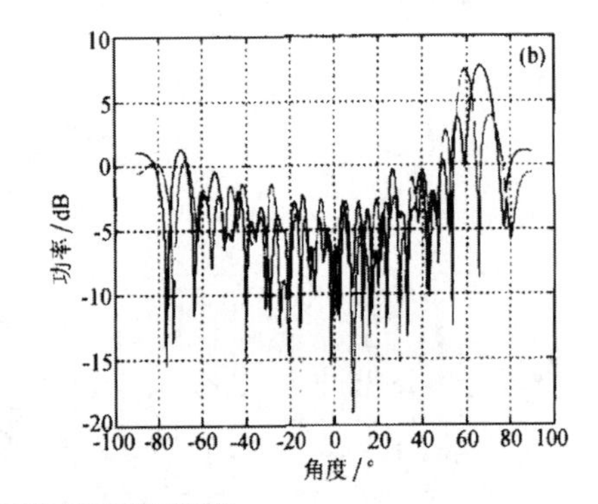

图 7　两信源情况下迭代的盲波束形成方法方位图

((a) 图信源在 30°, 60°, 信噪比 0 dB　(b) 图信源在 65°, 60°, 信噪比 5 dB)

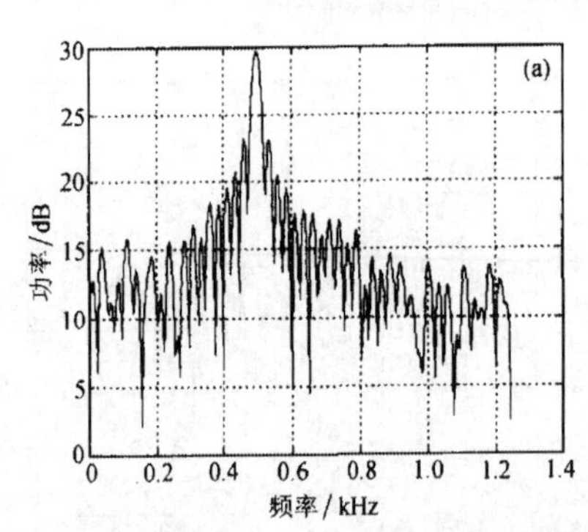

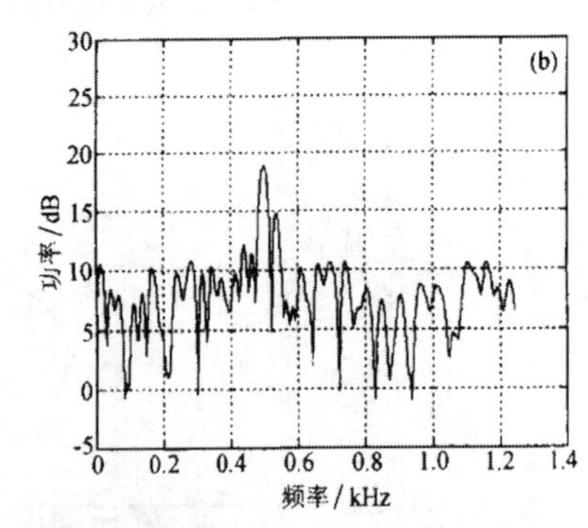

图 8　估计数多于实际信源数情况下迭代后信号幅频响应图

2.4　多信源情况下新方法的信源分辨率性能

两个信号源时，一信号源处于 30° 方位，另一个处于 60° 方位，发射窄带信号。分别用批处理的盲波束形成方法和迭代的盲波束形成方法对该信号源发射信号进行处理。信噪比为 0 dB 。得到的方位图 6 和图 7。

从仿真结果可看到，在多信源情况下，迭代的盲波束形成方法依然可以有效地进行工作。同时，从方位图上可看到，干扰信号方向形成了相应的零点。这也是能够在多信源情况下有效恢复信号的原因。

2.5　估计信源个数不准时新方法的性能

当实际信源为单信源，而估计的信源个数多于实际个数时，对迭代的盲波束形成方法性能进行仿真试验。仿真条件和单信源情况相同，而估计的信源个数为 2, 信源位于 30° 方向。信噪比为 0 dB 。仿真结果示于图 8 和图 9 。

从仿真结果可以看出，当估计信源个数 (2) 大于实际信源个数 (1) 的时候，迭代后得到的信号幅频一路 (信号，图 8(a) 输出能量) 比另一路 (数量估计误差带来的，图8(b)) 高约 12 dB, 可以很轻易地分辨出实际信号的个数。图9中有两条曲线，一条指明信号的实际方位 30°，另一条无明确方向性，且能量很小，可见估计信源数大于实际信源个数时，迭代算法依然有效。

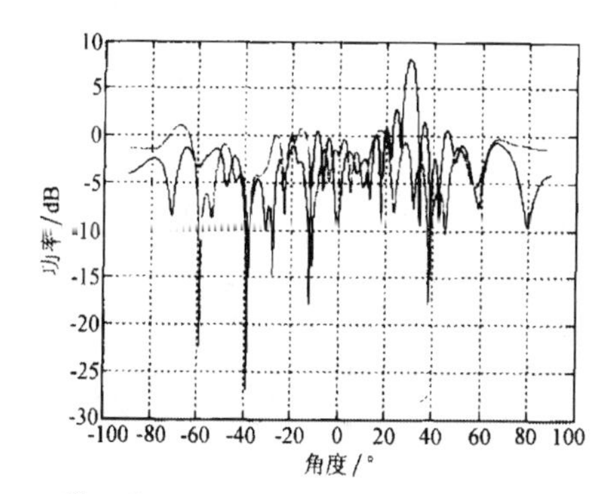

图9　估计数多于实际信源数情况下迭代后方位图结果

3　结论

本文提出的算法是一种迭代的方法。一般地，对于实际使用的复杂情况，盲波束形成需要较大的运算量，即需要较大的硬件负担，而 DSP 并行运算是解决运算量问题的有效方法。不难看出，相对于传统的确定性方法，新的方法可以容易的实现在线实时的运算，这就为 DSP 的实时并行运算提供了可能，在实际使用当中，具有相当的工程实现价值。

盲波束形成方法是实现空分复用技术的关键问题，传统的确定性方法是一种离线方法，所需运算量存储量较大。本文提出了在线迭代的新方法，可以有效减小运算量，缩短运算时间，对阵行误差具有鲁棒性最重要的在为 DSP 硬件的并行实现提供了可能。计算机仿真在单信源、多信源、估计信源个数多于实际信源等多种情况下均得到了较好的效果。

参　考　文　献

1　张贤达，保　铮. 通信信号处理. 北京：国防工业出版社，2000

2　LI Xiaohua, Howard Fan H. Blind channel identification: subspace tracking method without rank estimation. *IEEE Tran, Signal Prosessing*, 2001; **49**(10): 2372—2382

Golud G, Van Load C. Matrix computations, Third ed. Baltimore MD: Johns Hopkins Univ. Press, 1996

4　Stewart G. Updation a tank-revealing ULV decomposition. *SIAM J. Matrix Anal. Appl.*, 1992; **14**: 1535—1541

5　Yang B. Projection approximation subspace tracking. *IEEE trans. Signal Processing*, 1995; **43**: 95—107

6　Blind adaptive multiuser detection in multipath CDMA channels based on subspace tracking. *IEEE Trans. Sifnal Processing*, 1998; **46**: 3030—3044

7　Lutkepohl H. Handbook of matrices. West Sussex, UK, Woley, 1996

8　李启虎. 声呐信号处理引论. 1985

9　杨益新，孙　超. 任意结构阵列宽带恒定束宽波束形成新方法. 声学学报，2001; **26**(1): 55—58

10　单秉彝，郑国伦，邹腾荣. 时频混合分步波束形成试验研究. 声学学报，1999; **24**(6): 582—588

声学学报创刊 40 周年纪念论文

声矢量传感器研究进展①

孙贵青　李启虎
(中国科学院声学研究所　北京　100080)

2004 年 8 月 24 日收到

摘要　声矢量传感器由传统的无指向性声压传感器和偶极子指向性质点振速传感器复合而成，可以同步共点测量声场中一点处的声压和质点振速若干正交分量，由此得到的幅度和相位信息为解决一些水声问题提供了新的思路。因其实际的和潜在的工程应用价值，所以在最近十年间与此相关的声矢量传感器技术备受水声界关注。本文尝试综述声矢量传感器技术近五十年间在物理基础、传感器设计制作、相关工程应用等各方面的发展历史、现状和所取得的一些研究进展。

PACS 数：　43.30, 43.60, 43.85

Progress of study on acoustic vector sensor

SUN Guiqing　LI Qihu
(*Institute of Acoustics, The Chinese Academy of Sciences* Beijing 100080)

Received Aug. 24, 2004

Abstract　Acoustic vector sensor is combined with traditional omni directional pressure sensor and dipole directional particle velocity sensor, which simultaneously and colocately measures orthogonal components of particle velocity as well as pressure at single point in space. It provides new idea for solving underwater acoustic problems using those obtained amplitude and phase information. Because of potential and important value of engineering application, underwater acoustic association pays many attentions to technology corresponding to acoustic vector sensor during last about decade. Based on investigation practice with many years, authors tray to review study developments in physical fundaments, sensor design and manufacture, corresponding engineering applications, and so on during 50 years.

引言

声矢量传感器作为一种新型的水声测量设备，不但可以测量声场中最常见的标量物理量 —— 声压，而且还可以直接、同步测量声场同一点处流体介质质点振速矢量在笛卡儿坐标系下的 x, y, z 轴向投影分量，一般多用三分量和二分量的形式。在结构上它由传统的无指向性的声压传感器和偶极子指向性的质点振速传感器复合而成，质点振速传感器是核心部件，其灵敏度的高低和工作的稳定性等制约声矢量传感器的设计、制作、加工、装配、校准和使用等诸多环节。尽管本文把这种类型的传感器统一称为声矢量传感器，但国内外对此有着不同的称谓，主要的如，俄罗斯将质点振速传感器称为矢量接收器 (vector receiver)，将声矢量传感器称为复合接收器 (combined receiver)；美国等将声矢量传感器还称为声压 - 振速传感器 (presure-velocity sensor or P-V sensor)，还有的称其为声强探头 (sound intensity probe)。声矢量传感器技术的主要应用领域可以覆盖水声警戒声呐、拖曳线列阵声呐、舷侧阵共形阵声呐、水雷声引信、鱼雷探测声呐、多基地声呐、水下潜器的导航定位、分布式传感器网络等。在空气声学中，声矢量传感器可以用于战场警戒探测直升机和隐形飞机，噪声源识别和声强、声功率测量等，此外还有电磁矢量传感器，它的信号处理形式与水声的类似，可以互相借鉴。

声矢量传感器技术是在最近十年间备受水声界关注的研究焦点之一。从上一世纪五十年代中期美国学者发表的有关使用惯性传感器直接测量水中质点振速的经典论文以来 [1]，到在上一世纪七八十年代前苏联的学者利用其研制成功的声矢量传感器 (复合水听器) 开展海洋环境噪声研究 [2]，直至上一世纪

---

① 声学学报, 2004, 29(6): 481-490.

九十年代声矢量传感器技术研究热潮才逐渐兴起。1989 年俄国学者出版了世界上第一部有关声矢量传感器技术的专著“声学矢量 - 相位方法”[3]，较全面地论述了声矢量传感器技术的原理和应用。1991 年的美国声学杂志第 89 卷第 3 期和第 90 卷第 2 期连着刊出美俄两国学者三篇有关声矢量传感器研究方面的论文[4-6]，这种情况以前未曾有过。该技术所蕴含的潜在军事应用前景促使美国海军研究局 (ONR) 于 1995 年资助美国声学学会举行声矢量传感器专题研讨会，并出版了题为“声质点振速传感器：设计、性能和应用”的论文集[7]，它基本反映了当前美国学者在这一领域的研究动态，但迄今为止，它仍是该领域研究的最有价值的参考资料之一，也非常有力地推动了该领域的研究。1997 年俄国学者出版的专著“复合水声接收器”[8] 自成体系，专门论述声矢量传感器的设计、制作和校准等。2001 年美国海军水下战中心 (NUWC) 举办了关于指向性声传感器的研讨会[49]，首次邀请俄国学者参加。2002 年 IEEE 的 OCEANS 设立了“声质点振速传感器”专题[9]，内容涉及低频、高频声矢量传感器的设计、制作和实验，声压和声质点振速的联合信息在匹配场处理中的性能等，这些都反映了最新的研究情况。2003 年出版的“海洋矢量声学”[10] 发展了海洋环境噪声的声压标量场特性的研究，提出了基于声矢量传感器的海上实验、数据处理以及理论分析等一整套方法。

尽管在美国最早出现了基于惯性传感器的现代声矢量传感器设计思想和制作样品的雏形，但在 Rzhevkin 和 Zakharov 的积极倡议和推动下俄国在声矢量传感器技术的基础研究和应用研究两方面要走得更远些，而且还被评为俄罗斯二十世纪十大水声技术之一。

国内的相关工作可追溯到上一世纪九十年代初有关声压梯度水听器和双水听器声强测量等研究工作。但真正较深入开始研究的时间在 1998 年以后，1998 年松花湖实验和 2000 年大连海试是国内最早的两次关于声矢量传感器技术的外场实验，随后的 2002 年密云水库实验和 2003 年东海、南海声矢量传感器线阵实验，作者都有幸参加了这些实验和相关研究工作。

测量声场质点振速的想法很早就有：Rayleigh 于 1882年在其著名的文章[11] 中已经演示了测量声波均方质点振速的可能性，并以此确定声强，这种装置就是空气声学中常说的 Rayleigh 盘。之后的 Olsen 等[13] 人都试图测量声能流密度，但由于质点振速测量的复杂性，这些努力没有得到真正的回报。而现在水声工程中所采用的大多数声矢量传感器工作原理、基本形式和主要的设计理念均基于 Leslie 等人的观点，因此本文重点阐述从 1956 年之后直到现在，即 2004 年的声矢量传感器技术的发展过程以及所取得的一些有价值的研究成果。

鉴于声矢量传感器技术作为一种新兴的水声技术，试图全面综述其过去和现在的技术状态都是比较困难的，尽管如此，但这也从另一个侧面反映了声矢量传感器技术涉及到的许多理论性和工程性问题非常值得深入研讨。正如文献 12 所写的那样，声矢量传感器水声应用才刚刚新兴，还需要做许多工作以评估这类传感器的优势，但显而易见，通过测量完整的声场物理量 (即声压标量和质点振速矢量) 重新深入研究这一课题是非常有益的。作者力求本文立足于能够客观、全面地反映问题，希望能够为正在从事或即将从事该方向研究的科研学者提供些许帮助。

本文首先综述声矢量传感器技术的物理基础，回答为什么需要声矢量传感器，它的应用基础在哪儿等类似的概念性问题；然后是传感器自身的发展概况，从中可以体会到，如何才能拥有高性能的声矢量传感器；最后是基于声矢量传感器及其阵列的信号处理技术，从理论和实验两方面综述声矢量传感器工程应用的这一重要基础。限于篇幅，本文重点综述前两点，至于后者则单独成文在后续文章中给出[58]。

1 物理基础

声矢量传感器作为水声物理量的测量设备，其出现、发展和应用都与水声物理基础息息相关，本节从声能流密度、指向性传感器、信号声压和质点振速之间的相干性、噪声声压和质点振速之间的空间相关性四个方面分别阐述声矢量传感器原理和应用的物理背景。

1.1 声能流密度

由传统的声压水听器测量可以得到声场势能密度，这是最常用的声场能量形式，但是声矢量传感器除此之外还可以得到声场动能密度和声能流，这些概念对于正确理解声矢量传感器测量结果至关重要，而且往往被忽视，因此有必要作简要的概述，而更系统、具体的理论可参见文献 56 。

对于水声学的正问题求解而言，基于速度势的简谐声场理论已经相当完善，原则上可以通过求解含边界条件的亥姆霍兹方程，只要存在速度势函数 Φ 的解析形式，就可以由下式完整地确定声压 p 和

质点振速 $\boldsymbol{v}$ 的解析形式:

$$p=\rho\frac{\partial\Phi}{\partial t},\quad \boldsymbol{v}=-\nabla\Phi,$$

并由此得到如下的声场能量形式:

$$E_p=\frac{p^2}{2\rho c^2},\ \ E_v=\frac{1}{2}\rho v^2,\ \ \boldsymbol{J}=p\boldsymbol{v},$$

它们分别是声压势能密度 E_p, 质点振速动能密度 E_v, 瞬时声能流密度 $\boldsymbol{J}$, 也称瞬时声强，这三者之间的关系由下面的声能守恒方程联系:

$$\frac{\partial E}{\partial t}+\nabla\boldsymbol{J}=0,\ \ E=E_p+E_v\ \text{声波的机械能}.$$

声场的绝大多数研究是集中在与声压有关的声波势能密度上, 而在与质点振速有关的声波动能密度和声能流密度方面的相应研究甚少, 讨论的也仅仅是一些简单的情况, 如平面行波场、驻波场, 球面行波场, 简单波导声场等。从上述声能守恒方程中可以看出，声能流密度更适合于揭示声波能量“流动”的一般性规律[4]。为什么会出现这些现象？归结到一点，那就是缺少相应的质点振速测量设备，在没有声矢量传感器出现之前，基于声压水听器的水声测量技术已经相当完善，这是造成许多水声学正问题求解以声压量和声压势能密度为研究对象的根本原因，因为实验测量是水声学研究的物理基础。由此推想到水声学反问题，出现声压水听器占据统治优势也就不足为奇。是否声学研究的初期就忽视了这些问题？事实显然不是这样的。继 Rayleigh 于 1884 年提出测量声波质点振速以来，之后的 Olsen 等人都试图测量声能流密度，但由于质点振速测量的复杂性使得研究一筹莫展，这种情况一直持续到 Leslie 等人的论文发表。

正因为声能流密度或声强概念在声矢量传感器技术中有着特殊的地位，所以需要进一步认识它的一些应用形式和所对应的物理意义。首先引入复声强，即:

$$\boldsymbol{I}_c(\boldsymbol{r},\omega)=p(\boldsymbol{r},\omega)\boldsymbol{v}^*(\boldsymbol{r},\omega),$$

式中, 符号 ω 表示频率, 上标 $*$ 表示复共轭。$p(\boldsymbol{r},\omega)$ 和 $\boldsymbol{v}(\boldsymbol{r},t)$ 分别是 $p(\boldsymbol{r},t)$ 和 $\boldsymbol{v}(\boldsymbol{r},t)$ 的 Fourier 变换，由公式可见, 复声强定义在频域上。复声强还可以表示为有功声强和无功声强的形式:

$$\boldsymbol{I}_c(\boldsymbol{r},\omega)=\boldsymbol{I}_a(\boldsymbol{r},\omega),+i\boldsymbol{I}_r(\boldsymbol{r},\omega),$$

式中，$\boldsymbol{I}_a(\boldsymbol{r},\omega)$ 称为有功声强，表示向远处传播的声能; $\boldsymbol{I}_r(\boldsymbol{r},\omega)$ 称为无功声强, 表示不传播的声能。对于简谐波, 含频率项的 Dirac 函数不影响问题的讨论, 因此在声压和质点振速的 Fourier 变换 $p(\boldsymbol{r},\omega)$ 和 $\boldsymbol{v}(\boldsymbol{r},\omega)$ 的表达式中忽略该项，得:

$$p(\boldsymbol{r},\omega)=A(\boldsymbol{r})\exp[\mathrm{i}\varphi(\boldsymbol{r})],$$

$$\boldsymbol{v}(\boldsymbol{r},\omega)=\frac{1}{\rho_0(\boldsymbol{r})\omega}\left[\nabla\varphi(\boldsymbol{r})-\mathrm{i}\frac{\nabla A(\boldsymbol{r})}{A(\boldsymbol{r})}\right]p(\boldsymbol{r},\omega).$$

由此可得有功声强和无功声强分别为:

$$\boldsymbol{I}_a(\boldsymbol{r},\omega)=\frac{1}{\rho_0(\boldsymbol{r})\omega}[A^2(\boldsymbol{r})\nabla\varphi(\boldsymbol{r})],$$

$$\boldsymbol{I}_r(\boldsymbol{r},\omega)=\frac{1}{2\rho_0(\boldsymbol{r})\omega}[\nabla A^2(\boldsymbol{r})].$$

由上式可以看出，有功声强与波阵面的传播方向一致，因此，它表示向远处传播的能量；无功声强与声压振幅平方的梯度方向重合。一般，有功声强的方向不一定和无功声强的方向重合，它们之间的关系由两者之间的相位差确定，有功声强和无功声强的相位差定义为:

$$\Delta\phi(\boldsymbol{r},\omega)=\tan^{-1}\frac{\boldsymbol{I}_r(\boldsymbol{r},\omega)}{\boldsymbol{I}_a(\boldsymbol{r},\omega)}.$$

对于简谐声波，有:

$$\Delta\phi(\boldsymbol{r},\omega)=\tan^{-1}\frac{\nabla A(\boldsymbol{r})}{A(\boldsymbol{r})\nabla\varphi(\boldsymbol{r})}.$$

因为有功声强表示向远处传播的声能，所以利用有功声强 $\boldsymbol{I}_a$ 在 x, y, z 上的正交投影 $I_{a,x}$, $I_{a,y}$, $I_{a,z}$ 可以估计声源的水平方位角 ϕ 以及声源 z 轴的夹角 θ:

$$\phi(\boldsymbol{r},\omega)=\tan^{-1}\frac{I_{a,y}(\boldsymbol{r},\omega)}{I_{a,x}(\boldsymbol{r},\omega)},$$

$$\theta(\boldsymbol{r},\omega)=\tan^{-1}\frac{\sqrt{I_{a,x}^2(\boldsymbol{r},\omega)+I_{a,y}^2(\boldsymbol{r},\omega)}}{I_{a,z}(\boldsymbol{r},\omega)}$$

利用上两式可在全空间上对声源进行无模糊定向。

1.2 指向性传感器

传感器的指向性是抑制噪声和干扰的主要指标，常用的声压水听器是无指向性的，即全向接收，不具有噪声和干扰的抑制能力，而声矢量传感器所包含的质点振速传感器则刚好相反，且其尺寸远小于波长。因此除了声能流密度测量需要之外，声矢量传感器还可以利用其良好的低频指向性仅在声场一点处就能够确定声源的方位，而所需的孔径远小于声波波长，这可由下式作出解释:

$$p(\boldsymbol{r})=p(\boldsymbol{r}_0)+(\boldsymbol{r}-\boldsymbol{r}_0)^{\mathrm{T}}\nabla p|_{r=r_0}+\frac{1}{2}(\boldsymbol{r}-\boldsymbol{r}_0)^{\mathrm{T}}\nabla\nabla p|_{r=r_0}(\boldsymbol{r}-\boldsymbol{r}_0)+\cdots$$

上式将声压在声场一点 $\boldsymbol{r}_0$ 处进行 Taylor 级数展开得到的，实质上它定义了一类指向性传感器: 第一项是标量声压传感器，仅测量声压，称为零阶指向性传感器；前二项构成声矢量传感器，同步共点测量声压

和声压梯度 ∇p, 称为一阶指向性传感器；前三项构成声并矢 (dyadic) 传感器[14], 同步共点测量声压、声压梯度 ∇p 和二阶声压梯度 $\nabla\nabla p$, 称为二阶指向性传感器，依此类推可以得到更高阶次的指向性传感器，则相应的指向性也更为尖锐、指向性指数更高。其中：

$$\boldsymbol{r}-\boldsymbol{r}_0=[x-x_0\ \ y-y_0\ \ z-z_0]^{\mathrm{T}},$$

$$\nabla p=\left[\frac{\partial p}{\partial x}\ \ \frac{\partial p}{\partial y}\ \ \frac{\partial p}{\partial z}\right]^{\mathrm{T}},$$

$$\nabla\nabla p=\begin{bmatrix}\dfrac{\partial^2 p}{\partial x^2} & \dfrac{\partial^2 p}{\partial x\partial y} & \dfrac{\partial^2 p}{\partial x\partial z}\\ \dfrac{\partial^2 p}{\partial y\partial x} & \dfrac{\partial^2 p}{\partial y^2} & \dfrac{\partial^2 p}{\partial y\partial z}\\ \dfrac{\partial^2 p}{\partial z\partial x} & \dfrac{\partial^2 p}{\partial z\partial y} & \dfrac{\partial^2 p}{\partial z^2}\end{bmatrix}.$$

上式称为 Hessian 矩阵。考虑到篇幅所限，有关声并矢传感器的内容可参见文献 57 。简谐场中声压梯度 ∇p 与质点振速的关系由下面的 Euler 公式联系：

$$\nabla p=-\rho\frac{\partial \boldsymbol{v}}{\partial t}=\mathrm{j}\omega\rho\,\boldsymbol{v}.$$

将上式代入 Taylor 展开式可得：

$$p(\boldsymbol{r})\approx p(\boldsymbol{r}_0)+\mathrm{j}\omega\rho(\boldsymbol{r}-\boldsymbol{r}_0)\boldsymbol{v}(\boldsymbol{r}_0).$$

这个展开式表明，声压和质点振速的同步、共点测量等价于测量点处半径小于声波波长的球形体元声压，因此单个声矢量传感器可以视为声压传感器的微体积阵，可以证明当其半径约小于声波波长的三分之一时，上式的截断误差是可以接受的。既然是体积阵，则就可以定向，正如上式所指出的那样，真正的声源方位信息蕴含在质点振速中。

实际中，特别是解决水声反问题时，经常通过介质中点接收器记录的声信号以一定精度重构声场。某些情况下，接收器以奈奎斯特间距离散分布，这样做需要相当多的点 (形式上无限) 才有可能精确重构声场。但现实情况下，空间采样总是有限的，而在有限空间区域确定声场参数会导致部分信息丢失，而测量声压的空间导数可以在保证必要重构精度的前提下有效减少接收传感器的数量，因此当构造低频离散阵时，即接收单元间距相当大时，从经济性角度看，这样也比单纯增加测量传感器布放点的数量更好一些。完全重构通常不必要，如，确定声源方位只需测量点的声压和它的梯度三个投影，即上文 Taylor 级数展开式的前两项。

1.3 信号声压和质点振速之间的相干性

为简便起见，若仅考虑在浅海波导中绝热 - 距离无关简正模态展开的远场形式 (忽略模态的连续谱，即 $k_m<\min[K(z)]$, $K(z)=\omega/c(z)$), 则声压 $p(\boldsymbol{r},z)$ 以及声质点振速的水平分量 $v_r(\boldsymbol{r},z)$ 和垂直 $v_z(\boldsymbol{r},z)$ 分量分别为：

$$p(r,z)=\frac{\mathrm{j}}{\rho(z_s)\sqrt{8\pi r}}\exp\left(-\mathrm{j}\frac{\pi}{4}\right)\sum_m\frac{\Psi_m(z_s)\Psi_m(z)}{\sqrt{k_m}}\exp(\mathrm{j}k_m r),$$

$$v_r(r,z)=\frac{\mathrm{j}}{\rho^2(z_s)\omega\sqrt{8\pi r}}\exp\left(-\mathrm{j}\frac{\pi}{4}\right)\sum_m\sqrt{k_m}\Psi_m(z_s)\Psi_m(z)\exp(\mathrm{j}k_m r),$$

$$v_z(r,z)=\frac{1}{\rho^2(z_s)\omega\sqrt{8\pi r}}\exp\left(-\mathrm{j}\frac{\pi}{4}\right)\sum_m\frac{\Psi_m(z_s}{\sqrt{k_m}}\frac{\partial\Psi_m(z)}{\partial z}\exp(\mathrm{j}k_m r).$$

由上式不难看出，当接收位于声源远场时，柱面波波阵面都趋向于平面波波阵面，因此，波阻抗可以近似为平面波阻抗：

$$Z_r=\frac{p}{v_r}\approx\rho c.$$

即声压和质点振速水平分量是高度相干的, 类似工作可参见文献3, 31, 36。在远场平面波近似下，声压与声质点振速水平分量之间的这种强相干性确保声矢量传感器阵列的远程测向能力。声压和声质点振速垂直分量在垂直方向上均为驻波形式。如果不考虑海面噪声源，则在垂直方向上不存在自上而下和自下而上辐射 “流动” 的声能，而是沿垂直分布的 “停驻” 的声波。Gordienko认为[3,51], 在海洋中对于声源远场，声压和水平质点振速在声传播过程中的幅度起伏平均不超过 0.5 dB, 相位起伏约 3° ~ 5°, 这是利用声能流检测远程微弱信号的主要物理基础之一。

但通过模态之间的相干性讨论，Ellisevnin 等基于简单的绝对硬和绝对软等声速模型发现声场一些特殊区域的声能流具有 “涡” 和 “鞍” 奇异结构，在这些声场奇异点上声能流会反向传播， D'Spain 等也认识到了这一点，并提出用声矢量传感器的无功声强测量描述类似结构[50] 。尽管 “涡” 的出现使我们多少感到些吃惊，但可以认识到，基于声压和质点振速的声场相干性研究有助于加深对一些水声物理问题的讨论和理解，会对模基声矢量信号处理有推动作用。

1.4 噪声声压和质点振速之间的空间相关性

声压和质点振速之间的噪声空间相关性研究是声矢量传感器技术的重要基础之一，由此可以定量

评估声矢量传感器 (阵列) 的空间增益，是声呐设计者最关注的问题之一。在一些文章中已经出现相应的理论和实验工作 [15,26,51]，但本节侧重综述各向同性噪声场下的声压和质点振速之间的相关性，这是最常用的，也是最基本的。

考虑经典的三维各向同性噪声模型，即假设噪声点源均匀分布在无限大球面上，相互独立以随机相位辐射单频平面波，则不难证明，声压和声质点振速各正交分量在同一时刻，点 $\boldsymbol{r}=(x,y,z)$ 和 $\boldsymbol{r}'=(x',y',z')$ 之间的空间相关系数为：

$$R_p(t,\boldsymbol{r};t',\boldsymbol{r}')=\mathrm{j}_0(kd)$$

这一经典结论是迄今为止计算阵增益最基本的公式。

$$\begin{cases} R_x(t,\boldsymbol{r};t',\boldsymbol{r}')=\langle v_x(t,\boldsymbol{r})v_x^*(t',\boldsymbol{r}')\rangle=\left[\dfrac{\mathrm{j}_1(kd)}{kd}-k^2(x-x')^2\dfrac{\mathrm{j}_2(kd)}{(kd)^2}\right], \\ R_y(t,\boldsymbol{r};t',\boldsymbol{r}')=\langle v_y(t,\boldsymbol{r})v_x^*(t',\boldsymbol{r}')\rangle=\left[\dfrac{\mathrm{j}_1(kd)}{kd}-k^2(y-y')^2\dfrac{\mathrm{j}_2(kd)}{(kd)^2}\right], \\ R_z(t,\boldsymbol{r};t',\boldsymbol{r}')=\langle v_z(t,\boldsymbol{r})v_x^*(t',\boldsymbol{r}')\rangle=\left[\dfrac{\mathrm{j}_1(kd)}{kd}-k^2(z-z')^2\dfrac{\mathrm{j}_2(kd)}{(kd)^2}\right] \end{cases}$$

上式是质点振速的自相关系数。

$$\begin{cases} R_{xy}(t,\boldsymbol{r};t',\boldsymbol{r}')=\langle v_x(t,\boldsymbol{r})v_y^*(t',\boldsymbol{r}')\rangle=\left[k^2(x-x')(y-y')\dfrac{\mathrm{j}_2(kd)}{(kd)^2}\right], \\ R_{xz}(t,\boldsymbol{r};t',\boldsymbol{r}')=\langle v_x(t,\boldsymbol{r})v_z^*(t',\boldsymbol{r}')\rangle=\left[k^2(x-x')(z-z')\dfrac{\mathrm{j}_2(kd)}{(kd)^2}\right], \\ R_{yz}(t,\boldsymbol{r};t',\boldsymbol{r}')=\langle v_y(t,\boldsymbol{r})v_z^*(t',\boldsymbol{r}')\rangle=\left[k^2(y-y')(z-z')\dfrac{\mathrm{j}_2(kd)}{(kd)^2}\right], \end{cases}$$

上式是质点振速的互相关系数。

$$\begin{cases} R_{px}(t,\boldsymbol{r};t',\boldsymbol{r}')=\langle p(t,\boldsymbol{r})v_x^*(t',\boldsymbol{r}')\rangle=0, \\ R_{py}(t,\boldsymbol{r};t',\boldsymbol{r}')=\langle p(t,\boldsymbol{r})v_y^*(t',\boldsymbol{r}')\rangle=0, \\ R_{pz}(t,\boldsymbol{r};t',\boldsymbol{r}')=\langle p(t,\boldsymbol{r})v_z^*(t',\boldsymbol{r}')\rangle=0. \end{cases}$$

上式是声压和质点振速之间的互相关系数，均为零。其中，$d=|\boldsymbol{r}-\boldsymbol{r}'|$, $k=\omega/c$。$\mathrm{j}_0,\mathrm{j}_1$ 和 j_2 分别为零阶、一阶和二阶球贝塞尔函数。由上面的公式不难得到声矢量传感器均匀直线阵和其它复杂阵形的空间相关系数，以声矢量传感器水平均匀直线阵为例，其声压和质点振速的空间相关系数分别为：

$$\begin{aligned} &R_p(d)=\mathrm{j}_0(kd), \\ &R_y(d)=\mathrm{j}_0(kd)-2\frac{\mathrm{j}_1(kd)}{kd}, \\ &R_x(d)=R_z(d)=\frac{\mathrm{j}_1(kd)}{kd}, \\ &R_{xy}(d)=R_{xz}(d)=R_{yz}(d)=0, \\ &R_{px}(d)=R_{py}(d)=R_{pz}(d)=0. \end{aligned}$$

图 1 给出了相应的曲线。在各向同性噪声场中，声压和质点振速之间、质点振速之间的时空相关性总结如下：在空间同一点处，声压和质点振速之间互不相关，质点振速各投影分量之间互不相关，质点振速的自相关系数是 1/3；在空间不同点处，声压和质点振速之间不相关，质点振速的自相关系数取决于空间两点的间距及其投影的平方，质点振速互相关系数取决于空间间距各投影的乘积。由此也不难得到宽带噪声场中声压和质点振速的空间相关系数。

基于上述结论，通过比较噪声协方差矩阵可以看出，对于声压标量，其噪声协方差矩阵在半波长间距时为主对角线元素相同的对角矩阵，但对于声矢量处理，其噪声协方差矩阵一般不为对角矩阵，即使在半波长间距上，但有时为了理论推导的方便，甚至直接使用对角化的协方差矩阵而不顾及对角化或白化等操作，尽管这有些粗糙，但不影响问题实质的讨论，尤其对于分析宽带检测性能。

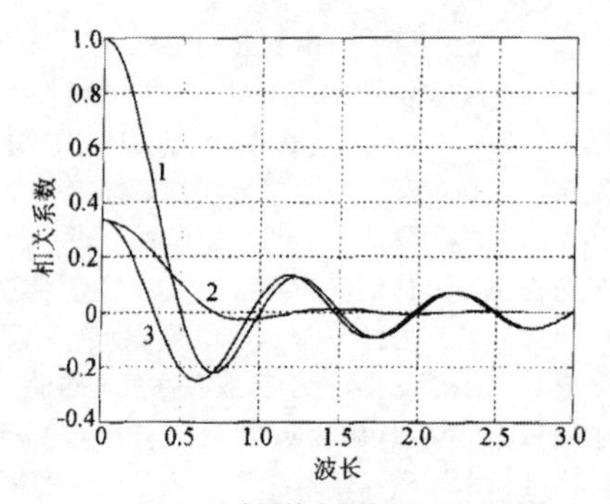

图 1　各向同性噪声场的空间相关系数

在各向同性噪声场中，单个声矢量传感器的归一化噪声协方差矩阵为：

$$\boldsymbol{R}=\begin{bmatrix}1&0&0&0\\0&1/3&0&0\\0&0&1/3&0\\0&0&0&1/3\end{bmatrix}.$$

从上式可以看出，质点振速的噪声功率是声压噪声功率的三分之一，即质点振速的指向性指数为 $3(10\lg 3\approx 4.8\ \text{dB})$，而正如所熟知的那样，声压的指向性指数为 $1(10\lg 1=0\ \text{dB})$，声压和质点振速各投影分量之间互不相关，质点振速各投影分量之间互不相关。

对于海面噪声源，单个声矢量传感器的归一化噪声协方差矩阵为：

$$\boldsymbol{R}=\begin{bmatrix}1&0&0&2/3\\0&1/4&0&0\\0&0&1/4&0\\2/3&0&0&1/2\end{bmatrix}$$

可以发现：声压与质点振速 z 轴投影之间的相关系数为 2/3，即沿 z 轴方向存在自上而下传播的噪声能流，质点振速在水平 x 轴和 y 轴投影的噪声功率都为 1/4 。

有关噪声场中声压和声质点振速之间的空间相关性的一般性分析可参见文献 15 。进一步地，由上述声压和质点振速的空间相关系数还可以得到声矢量传感器阵列的空间增益表达式[31,37,58]。

2 声矢量传感器

本节综述声矢量传感器的一般性分类、工作原理、结构特点、设计原则和性能参数等，对穿插其中的声矢量传感器发展历史也做了简要的概述。

2.1 一般性分类

质点振速传感器是声矢量传感器的核心部件，因此，矢量传感器的分类主要是指质点振速传感器的分类，它原则上分为声压梯度式和惯性式两种类型[16]。惯性式是指将惯性传感器，如加速度计等对振动敏感的传感器安装在刚性的球体、圆柱体或椭球体等几何体中，当有声波作用时，刚性体会随流体介质质点同步振动，其内部的振动传感器拾取相应的声质点运动信息，因此亦称为同振式。声压梯度式多是利用空间两点处声压的有限差分的原理来近似得到声压梯度，这可以通过反相串并联的线路连接在传感器内部实现，而声压梯度与介质质点的加速度之间的关系由 Euler 公式确定，通过计算间接得到介质质点振动信息。惯性式声矢量传感器是对简谐声场中介质质点振动真正意义上的直接测量。由于这两类声矢量传感器的工作机理的差异，则相应的性能参数也明显地不同。一般情况下都习惯将惯性式质点振速传感器统称为质点振速传感器，而无论测量的物理量是声压梯度、质点位移、质点振速，还是质点加速度，但有时根据需要把质点振速传感器细分为声压梯度式、质点位移式、质点振速式和质点加速度式，后两者应用更为普遍。根据所测量的上述物理量投影分量数目，质点振速传感器可以分为：单通道、双通道和三通道，相应地，声矢量传感器有二通道、三通道和四通道。根据换能器的换能原理，质点振速传感器可以分为：压电式、动圈式、电容式、光纤式、磁致伸缩式等。目前总体上看，基于电容、磁致伸缩、光纤换能器的质点振速传感器研究并不普遍，压电式的质点振速传感器因其性能稳定可靠仍占据着当前研究和应用的主导地位。

2.2 声压梯度传感器

声压梯度传感器通常有两种设计理念。最自然的声压梯度传感器是两个小间距分离的无指向性传感器，反相接线使得信号相减，这一理念作为双传声器技术常用于空气声强和阻抗测量，但要注意到，这些只对灵敏度和相位绝对匹配的传感器有效（或通过校准来匹配）且间距远小于波长使由 Euler 方程得到的有限差分近似误差最小。这种思想体现在自上一世纪七十年代起盛极一时的双传声器声强探头[17,18]，在水声中一般称为双水听器探头。因为在继 Rayleigh 提出测量质点振速的想法之后所进行的尝试由于质点振速测量的复杂性和当时技术条件的限制使得研究人员不得不暂时打消直接测量质点振速的念头，继而转向采用这样间接的方法得到质点振速和声能流密度。总体上看，这一时期的声压梯度水听器主要存在两个致命缺陷：一是灵敏度偏低，只能在信噪比较高的条件下使用，如声源的近场声强测量等；二是性能参数不稳定，严重依赖于材料、结构和制作工艺等。很自然地，这大大限制了它的工程应用。尽管后来将分离的一对或多对声压传感器封装在同一个壳体中，如将压电陶瓷柱沿圆周均匀分成四份，或将压电陶瓷球壳均匀分成四份的多模球等，虽然使性能稳定性有了显著的改善，但灵敏度依然太低，指向性也不是很理想，且随频率变化。这类声压梯度水听器较成功的应用实例是航空无线电声呐 DIFAR 浮标 AN/BQQ-53 。另一种声压梯度传感器的设计理念是使隔开的弯曲传感器两侧 (即双迭片) 都受到声波作用，使得纯的电压输出对应于穿过弯曲元件的声压之差。时至今日，还有一些研究人员在此方向继续尝试，随着材料的进步并通过良好的设计和工艺基本上可以保证声压梯度水听器可靠的工作，尽管如此，但随着惯性式声矢量传感器的研制成功[19]，它基本上被排斥在当前声矢量传感器研究的主流之外，因为现在商业化的微型加速度计具有更高的灵敏度、更稳定可靠的性能。

声压梯度传感器实际上是直接测量空间小尺度上的多点声压标量，然后通过线路的反相并联或串联来得到声压梯度的有限差分近似，这与直接测量质点振速和质点加速度的质点振速传感器机理显著不同，因此，有人认为所谓的 “声压梯度” 传感器不应该列入到真正的直接测量声场质点振速的质点振速传感器中，而作为它的过渡角色可能更合适。

2.3 同振式声矢量传感器

在现代水声工程中使用频度较高的一类声矢量传感器是基于惯性传感器的同振式传感器，它的主要优点在于，本身不产生明显的声场畸变，即可以视为点接收器，因此它的指向性比固定式的要好，而且性能参数更稳定，可以用于精确或长时间测量，在不同的应用中，同振式声矢量传感器的平均密度为 $0.9 \sim 1.8\ \mathrm{g/cm^3}$ 。

有关惯性式声矢量传感器的工作最早出现在海军军械实验室 (Naval Ordnance Laboratory)Leslie 等人的工作[1]，他们推导了刚硬、均匀球体在理想水介质声场中运动的数学表达式并证明，这类中性浮力的球体在低频运动时具有与相同位置处水质点相同的振速，即

$$v_1 = v_0 \frac{3\rho}{2\,\bar{\rho} + \rho},$$

其中，v_1 和 v_0 分别为同振球的振动速度与质点振速，ρ 表示水介质的密度，$\bar{\rho}$ 表示水听器的平均密度，当 $\bar{\rho} = \rho$ 时，$v_1 = v_0$，即测量介质质点振速。而且他们还提出在这样的球体中安装拾取振速的传感器以构成对质点振速敏感的水听器，但没有考虑球体密度、流体密度和粘度、柔性悬挂系统等对质点振速传感器的灵敏度指向性和工作频带等性能影响。尽管如此，但这些开创性的工作已经清楚地表明，此类结构的水听器易于制作和校准、性能稳定、具有良好的指向性，更重要的是他们还给出了质点振速传感器设计的基本原则，即中性浮力且质量中心与几何中心重合，以及一些重要的设计思想和实践。时至今日，这些对声矢量传感器研制的关键环节仍然有着借鉴和指导意义。除了球体之外，声矢量传感器还有圆柱体、椭球体、圆盘等多种形式，当 $\bar{\rho} = \rho$ 时不同形状引起的性能差异可以忽略。

现在商业加速度计在 10 Hz~10 kHz 的频带上有平坦的响应，声波对声矢量传感器悬挂系统的影响可能会成为测量声质点振速的水声惯性传感器设计的某些约束。因为实际使用时不得不将惯性传感器悬挂在某些主平台上以确保声矢量传感器悬挂系统和平台不污染测量，这对于声矢量传感器精确测量非常关键，此外还要考虑流体和壳体的密度、流体的粘性所引起的效应。同振式声矢量传感器对流噪声更敏感，在使用时尤其要注意。实验表明[20]，当流足够快以至于在传感器表面形成湍流时质点振速传感器受到的影响甚于声压传感器，但是，如果外裹橡胶层，则可以显著降低流噪声。

2.3.1 质点振速传感器

声波作用下质点振动特性之一是质点振速，记录它的最自然的方法是在测量点放置一个小波长尺寸且平均密度等于水介质密度的物体，此时，物体将象质点一样进行振动，通过测量物体的振动速度，可以记录场的信息。在水声中多采用动圈式结构的质点振速传感器，因为它有良好的低声频和次声频性能。

第一个质点振速水听器由 Kendall 于 1941 年研制出并用于音频范围换能器校准，被命名为 SV-1，工作频带 70 Hz~7 kHz，直径 6.35 mm，灵敏度和阻抗级均比后来研制的 SV-2 更低一些；于 1955 研制的低频质点振速水听器 SV-2 直径 12.7 mm，工作频带 15 Hz~700 Hz，插入 500 Ω 负载时的灵敏度为 −218.4 dB，re 1 V/μPa，它主要用于研究湖底的声学特征[1]。这两个质点振速传感器都采用动圈型换能原理，因其灵敏度偏低所以更适合于实验室校准和大信噪比的水声测量工作。

商业振动速度传感器一般都使用速度灵敏度 ($\mathrm{V/m{\cdot}s^{-1}}$)，但利用速度传感器构成的质点振速传感器经过水声灵敏度校准后通常都采用声压灵敏度来统一表示声压和振速各通道的灵敏度，两者在水介质中的换算关系见下式：

$$M_p = M_v - 243.52\ \mathrm{dB},$$

其中，M_p 是声压灵敏度，0 dB re 1 V/μPa，M_v 是振速灵敏度，0 dB re 1 $\mathrm{V/m{\cdot}s^{-1}}$ 。由上式可知，理论上在工作频带内，质点振速的声压灵敏度与频率无关，即质点振速的声压灵敏度与声压水听器的声压灵敏度类似都是平直的，只是数值有所差异。

2.3.2 质点加速度传感器

实际上，声质点振速测量除了可以使用质点振

速传感器之外还可以使用其它加速度[16,39] 和位移型的惯性传感器，因为在简谐波场中它们之间的微积分关系可以转化为更清晰直观的线性关系。典型的惯性传感器是将压电加速度计埋嵌在小的刚体中，当刚体运动时得到记录的输出电压。这种设计理念很大程度上依赖于无约束硬球在非粘无界流体介质中对声波的响应理论。图 2 的三维质点振动加速度传感器由哈尔滨工程大学研制[34,35]，直径 200 mm，密度工作 1.6 g/cm^3，频带10 Hz～1 kHz，采用中心压缩式加速度计结构，加速度通道声压灵敏度为 −174 dB(1 kHz)，re 1 V/μPa 。

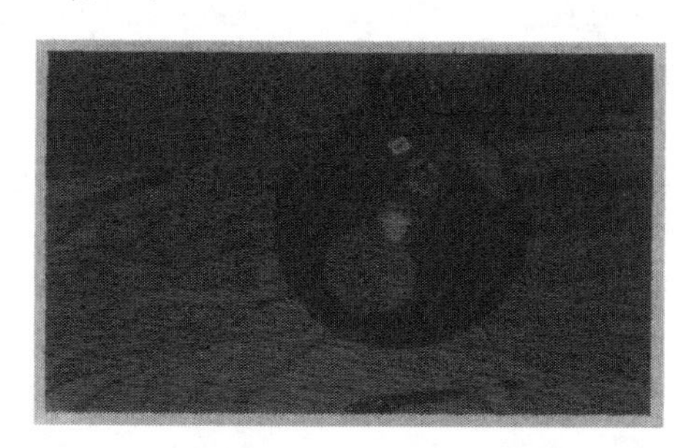

图 2　三维加速度计型质点振速传感器

商业化的振动加速度传感器一般使用加速度灵敏度，但利用加速度传感器构成的质点振速传感器通常采用声压灵敏度量纲来统一表示各通道的灵敏度，两者在水介质中的 1 kHz 处灵敏度换算关系为：

$$M_p = M_a - 187.38 \text{ dB},$$

其中，M_a 是加速度灵敏度，0 dB re 1 V/g 。若对于加速度灵敏度 0 dB re 1 V/(m/s^2)，则上式在 1 kHz 处的声压灵敏度为：

$$M_p = M_a - 167.56 \text{ dB}$$

由此可知，理论上在工作频带内，基于加速度计的质点振速传感器的声压灵敏度与频率成正比例，一般取工作带宽内的一个特征频率 (如 1 kHz) 的声压灵敏度进行标称。例如，若加速度灵敏度为 1200 mV/g，则 1 kHz 的声压灵敏度设计值为 −185.8 dB，100 Hz 的声压灵敏度为 −205.8 dB 。图 3 给出声矢量传感器一个振动加速度通道的水池比较法校准结果，黑点表示校准结果，频率分别为 500 Hz，630 Hz，800 Hz，1000 Hz 和 1250 Hz 。该传感器的照片见图 4 。

现在，商业化的压电加速度计在 10 Hz～10 kHz 的频带上有平坦的响应、灵敏度较高、性能稳定、体积小、重量轻、经济性好等诸多优点，因此越来越多地应用于质点振速传感器中。本节对使用较多的压电加速度计的结构和特点作一简单的介绍。

压电加速度计主要有三种结构模式：剪切 (shear)；弯曲梁(flexural beam)；压缩(compression) 。后者以前常用，现在主要使用前两者。剪切模式可以使传感器尺寸很小，其质量对被测结构体的影响最小，且提供最优的性能。弯曲模式使用敏感的晶体梁，使得传感器尺寸小，重量轻，热稳定性绝佳，价格经济。它对横向灵敏度不敏感也是固有的特点，一般，弯曲模式多用于低频，小加速度的场合，因此在结构测量中经常见到。压缩模式结构简单，硬度高，一般有垂直、倒转和隔离三种基本形式。

压电加速度计根据工作方式分为两类：内置电路的 ICP 加速度计和电荷加速度计。内置电路将灵敏元件产生的高阻抗电荷信号转换成低阻抗的电压信号，可以使用普通的双绞线或同轴电缆将信号传输到测量和记录设备上，这种阻抗转换可以确保在恶劣环境中进行远距离传输。

图 4 的基于商业加速度计的二维同振柱型声矢量传感器照片由哈尔滨工程大学贾志富研究员提供，直径和高度均小于 60 mm，密度 1.6 g/cm^3，内部支架采用环氧和玻璃微珠混合物加工而成 (密度 0.6 g/cm^3)，外部用聚氨酯灌封，工作频带 10 Hz～2 kHz，声压通道灵敏度 −191 dB，振速通道的声压灵敏度 −185.5 dB(1 kHz) 。

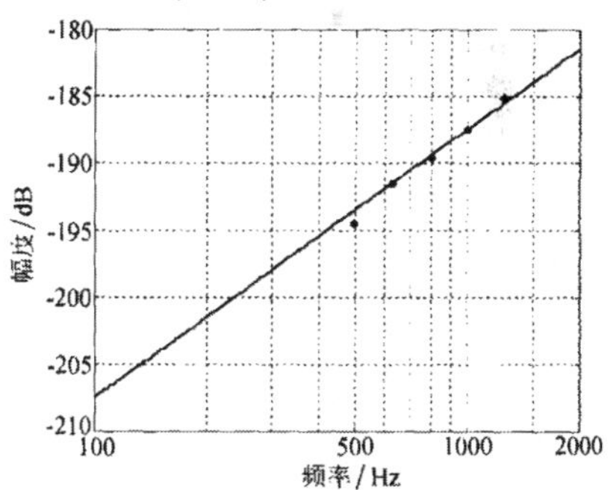

图 3　声矢量传感器振速通道的水池比较法校准结果

图 4　二维同振柱型声矢量传感器 HY-1

图 5 给出美国 Wilcoxon 公司为美国海军研制的可用于拖线阵和岸站的三维微型声矢量传感器[19,22]，使用 PZT-PT 压电单晶体材料，声压传感器灵敏度 −174 dB，振速通道由剪切型加速度计构成，其声压灵敏度 −193 dB(1 kHz)，工作频带 3 Hz~7 kHz，长 71.3 mm，直径 40.7 mm，中性浮力。

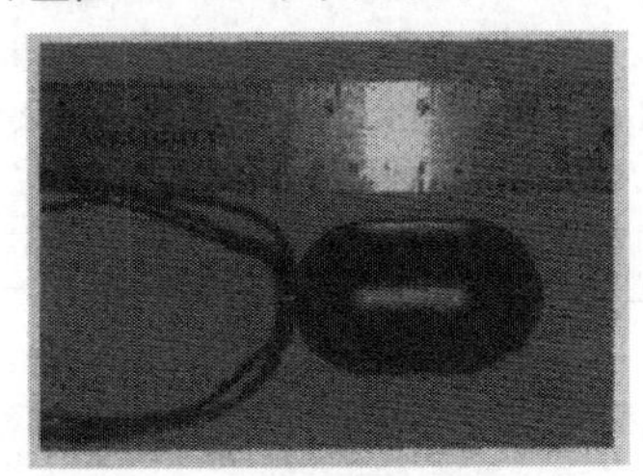

图 5 Wilcoxon 微型声矢量传感器 TV-001

基于 MEMS 技术的低噪声、高灵敏度、宽带的超微型加速度传感器已经研究成功[52−54]。尽管人们更关注低频声矢量传感器的发展，但是高频声矢量传感器[55]也有一定的应用前景，如水声通讯、合成孔径声呐、侦察声呐等。

3 结论

本文尝试综述声矢量传感器的发展历史、现状和所取得的一些研究进展，尽管疏漏在所难免，还是希望一己之言能有助于声矢量传感器技术的深入研究。

参 考 文 献

1 Leslie C B *et al.* Hydrophone for measuring particle velocity. *J. Acoust. Soc. Am.*, 1956; **28**(4): 711—715

2 Zakharov L N *et al.* Phase-gradient measurements in sound fields. *Sov. Phys. Acoust.*, 1974; **20**(3): 241—245

3 Gordienko V *et al.* Vector-phase method in acoustics. Moscow: Nauka, 1989

4 D'Spain G L *et al.* Energetics of the deep ocean's infrasonic sound field. *J. Acoust. Soc.Am.*, 1991; **89**(3): 1134—1158

5 Shchurov V A *et al.* Coherent and diffusive fields of underwater acoustic ambient noise. *J. Acoust. Soc. Am.*, 1991; **90**(2): 991—1001

6 Shchurov V A *et al.* The interaction of energy flows of underwater ambient noise and local source. *J. Acoust. Soc. Am.*, 1991; **90**(2): 1002—1004

7 Berliner M J, Lindberg J F. Acoustic particle velocity sensor: design, performance and applications. Woodbury, NY: AIP Press, 1996, 368 (American Institute of Physics, Woodbury, NY, 1995)

8 Skrebnev G K. Combined underwater acoustic receiver. St.Petersburg: Elmor, 1997

9 IEEE Oceans 2002 Proceedings. Biloxi, MS

10 Shchurov V A. Vector acoustics of the ocean. Vladivostok: Dalhauka, 2003

11 Rayleigh J. Device for measuring the intensity of airborne oscillations. *Phi. Mag.*, 1882; **14**: 186—188

12 Hawkes M *et al.* Acoustic vector-sensor beamforming and Capon direction estimation. *IEEE Trans. on Signal Processing*, 1998; **46**(9): 2291—2304

13 Olsen H F. System responsive to energy flow in sound waves. *U. S. Patent*, 1932; No.1892644

14 Cray B A. Directional point receivers: the sound and the theory. *IEEE Oceans 2002 Proceedings*, 2002(3): 29—31

15 Hawkes M *et al.* Acoustic vector-sensor correlations in ambient noise. *IEEE J. of Oceanic Engineering*, 2001; **26**(3): 337—347

16 McConnell J A. Analysis of a compliantly suspended acoustic velocity sensor. *J. Acoust. Soc. Am.*, 2003; **113**(3): 1395—1405

17 Fahy F J. Measurement of acoustic intensity using the cross-spectral density of two microphone signals. *J. Acoust. Soc. Am.*, 1977; **62**(4): 1057—1059

18 Chung J Y. Cross-spectral method of measuring acoustic intensity without error caused by instrument phase mismatch. *J. Acoust. Soc. Am.*, 1978; **64**(6): 1613—1616

19 Shipps J C *et al.* A miniature vector sensor for line array applications. *IEEE Oceans 2003 Proceedings*, 2003(5): 2367—2370

20 Keller B D. Gradient hydrophone flow noise. *J. Acoust. Soc. Am.*, 1977; **62**(1): 205—208

21 Nehorai A *et al.* Acoustic vector-sensor array processing. *IEEE Trans. on Signal Processing*, 1994; **42**(9): 2481—2491

22 Shipps J C *et al.* The use of vector sensors for underwater port and waterway security. *ISA/IEEE Sensors for Industry Conference Proceedings*, 2004: 41—44

23 Hawkes M *et al.* Acoustic vector-sensor processing in the presence of a reflecting boundary. *IEEE Trans. on Signal Processing*, 2000; **48**(11): 2981—2993

24 Shchurov V A *et al.* Noise immunity of a combined hydroacoustic receiver. *Acoustical Physics*, 2002; **48**(1): 110—119

25 Cray B A *et al.* Directivity factors for linear arrays of velocity sensors. *J. Acoust. Soc. Am.*, 2001; **110**(1): 324—331

26 孙贵青. 矢量水听器检测技术研究. 哈尔滨工程大学博士学位论文，2001

27 孙贵青等. 基于矢量水听器的最大似然比检测和最大似然方位估计. 声学学报，2003; **28**(1): 66—72

28 孙贵青等. 基于矢量水听器的水下目标低频辐射噪声测量方法研究. 声学学报，2002; **27**(5): 429—434

29 冯海泓等. 目标方位的声压振速联合估计. 声学学报，2000; **25**(6): 516—520

30 陈新华等. 声矢量阵指向性. 声学学报，2003; **28**(2): 141—144

31 孙贵青. 声矢量传感器及其在均匀线列阵声呐中的应用. 中国科学院声学研究所博士后研究工作报告，2003

32 Wong K T *et al.* Uni-vector-sensor ESPRIT for multisource azimuth, elevation, and polarization estimation. *IEEE Trans. on Antennas and Propagation*, 1997; **45**(10): 1467—1474

33 Zoltowski M D *et al.* Closed-form eigenstructure-based direction finding using arbitrary but identical subarrays on a sparse uniform Cartesian array grid. *IEEE Trans. on Signal Processing*, 2000; **48**(8): 2205—2210

34 贾志富. 同振球型声压梯度水听器的研究. 应用声学, 1997; **16**(3): 20—25

35 贾志富. 三维同振球型矢量水听器的特性及其结构设计. 应用声学, 2001; **20**(4): 15—20

36 惠俊英等. 声压振速联合信息处理及其物理基础初探. 声学学报, 2000; **25**(4): 303—307

37 孙贵青等. 基于矢量水听器的声压和质点振速的空间相关系数. 声学学报, 2003; **28**(6): 509—513

38 Shchurov V A *et al.* Use of acoustic intensity measurements in underwater acoustics (Modern state and prospects). *Chinese J. of Acoust.*, 1999; **18**(4): 315—326

39 Moffett M B *et al.* A piezoelectric, flexural-disk, neutrally buoyant, underwater accelerometer. *IEEE Trans. Ultrason. Ferroelectr. Fre. Control*, 1998; **45**(5): 1341—1346

40 Silvia M T *et al.* A theoretical and experimental investigation of low-frequency acoustic vector sensors. *IEEE Oceans 2002 Proceedings*, 2002(3): 1886—1897

41 Hawkes M *et al.* Wideband source localization using a distributed acoustic vector-sensor array. *IEEE Trans. on Signal Processing*, 2003; **51**(6): 1479—1491

42 Wong K T *et al.* Beam patterns of an underwater acoustic vector hydrophone located away from any reflecting boundary. *IEEE J. of Oceanic Engineering*, 2002; **27**(3): 628—637

43 Lemon S G. Towed-array history 1917-2003. *IEEE J. of Oceanic Engineering*, 2004; **29**(2): 365—373

44 Lasky M *et al.* Recent progress in towed hydrophone array research. *IEEE Journal of Oceanic Engineering*, 2004; **29**(2): 365—373

45 贾志富. 采用双迭片压电敏感元件的声压梯度水听器. 传感器技术, 1997; **16**(1): 22—24,29

46 陈洪娟等. 压电圆盘弯曲式矢量水听器的设计. 传感器技术, 2002; **21**(8): 23—25

47 陈洪娟等. 基于压电加速度计的同振型矢量传感器的设计. 传感器技术, 2003; **22**(3): 20—21,24

48 陈洪娟等. 采用双迭片压电敏感元件的同振柱型矢量传感器. 应用声学, 2003; **22**(3): 23—26

49 Proceedings of the workshop on directional acoustic sensors. NUWC, Newport, RI. 2001

50 Kuperman W A, D'Spain G A. Ocean acoustic interference phenomena and signal processing. *AIP Conference Proceedings*, 2001, San Francisco

51 Gordienko V A *et al.* Basic rules of vector-phase structure formation of the ocean noise field. *Acoust. Phys.*, 1993; **39**(3): 237—242

52 Liu C H *et al*, Characterization of a high sensitivity micromachined tunneling accelerometer with micro-g resolution. *J. of Microelectromechanical Systems*, 1998; **7**(2): 235—244

53 Bernstein J *et al.* Low noise MEMS vibration sensor for geophysical applications. *J. of Microelectromechanical Systems*, 1999; **8**(4): 433—438

54 Liu C H *et al.* A high precision, wide bandwidth micromachined tunneling accelerometer. *J. of Microelectromechanical Systems*, 2001; **10**(3): 425—433

55 McConnell J A *et al.* Development of a high frequency underwater acoustic intensity probe. *IEEE Oceans 2002 Proceedings*, 2002: 1924—1929

56 Mann J A *et al.* Instantaneous and time averaged energy transfer in acoustic fields. *J. Acoust.Soc.Am.*, 1987; **82**(1): 17—30

57 Cray B A *et al.* Highly directional acoustic receivers. *J. Acoust. Soc. Am.*, 2003; **113**(3): 1526—1532

58 孙贵青, 李启虎. 声矢量传感器信号处理. 声学学报, 2004; **29**(6): 491—498

声学学报创刊 40 周年纪念论文

声矢量传感器信号处理①

孙贵青　李启虎
(中国科学院声学研究所　北京　100080)
2004 年 8 月 24 日收到

摘要　声矢量传感器能够同步、共点、直接测量声场空间一点处声压和质点振速，从而有可能改善传统水声测量或探测设备的性能，为解决一些实际的水声问题拓展了新的思路。尽管声矢量传感器出现在水声领域的历史不算久，但是由于巨大的潜在军事需求牵引，它在最近十数年间的发展势头愈发地强劲，逐渐演变为不断被重视的水声技术之一。在此背景下，本文尝试综述声矢量传感器信号处理研究的当前进展，如信号检测、DOA 估计、波束形成等。
PACS 数：43.60, 43.30

Acoustic vector sensor signal processing

SUN Guiqing　LI Qihu
(*Institute of Acoustics, The Chinese Academy of Sciences* Beijing 100080)
Received Aug. 24, 2004

Abstract　Acoustic vector sensor simultaneously, colocately and directly measures orthogonal components of particle velocity as well as pressure at single point in acoustic field so that is possible to improve performance of traditional underwater acoustic measurement devices or detection systems and extends new ideas for solving practical underwater acoustic engineering problems. Although acoustic vector sensor history of appearing in underwater acoustic area is no long, but with huge and potential military demands, acoustic vector sensor has strong development trend in last decade, it is evolving into a one of important underwater acoustic technology. Under this background, we try to review recent progress in study on acoustic vector sensor signal processing, such as signal detection, DOA estimation, beamforming, and so on.

引言

声矢量传感器由传统的无指向性声压传感器和偶极子指向性质点振速传感器构成，它可以同步、共点、直接测量声场空间一点处的声压和质点振速的若干正交分量，这些信息都有助于改善水声系统的性能，为解决一些水声问题提供了新的思路和方法。此前困扰声矢量传感器的低灵敏度和低可靠性等主要技术瓶颈现在已经被很好地解决，目前的技术工艺使高性能的声矢量传感器不仅成为现实[1−4]，而且已经在实际的工程应用中得到了检验[5,6]，这都极大地促进了声矢量传感器信号处理研究。当然，这些都被强劲的军事应用需求所牵引[1,4,7]，如固定式水声警戒声呐、舷侧阵和共形阵声呐、拖线阵声呐、航空无线电声呐浮标、智能水雷声引信等。

现代声矢量传感器一般由声压传感器和质点振速传感器按照几何中心、质量中心和相位中心的所谓“三心合一”原则组合而成[8]，其中，质点振速传感器是核心，它可以由振动速度、振动加速度、振动位移等惯性换能器构成。由于它们之间的线性关系，所以三者是相互等价的。为了保持信号处理形式的一致，在信号处理中一般不严格区分振动位移、速度和加速度传感器，均统称为质点振速传感器(前提条件是传感器自噪声远小于海洋环境噪声，如零级海况；如果自噪声高于海洋环境噪声，则各传感器得到的等效声压噪声频谱不同，此时必须考虑传感器直接测量的物理量)，测量得到的物理量通过水声校准之后得到相应的声压灵敏度，即声矢量传感器所有测量的数据都统一用声压量纲表示，除非特别指出。文献 54 详细给出了声矢量传感器的物理基础。

自 Nehorai 等[9]首次将声矢量传感器纳入经典的水声信号处理框架以来，声矢量传感器信号处理在

① 声学学报, 2004, 29(6): 491-498.

理论和应用方面不断被丰富, 短短十年间得到的成果就已经印证了他们的预言: 重新审视和研究声矢量传感器及其信号处理是非常有意义的。声矢量传感器信号处理本质上离不开物理问题的讨论[10], 实际上针对某些具体问题可以作些适当简化, 仍能将声矢量传感器信号处理纳入到经典的水声信号处理框架之中, 因此可以有这样的假设: 均匀无限大理想流体介质中, 单源平面波信号入射到 M 阵元声矢量传感器阵列上, 信号和噪声均为独立同分布 (i.i.d) 零均值复高斯过程, 信号和噪声相互独立, 则:

$$\boldsymbol{r} = \boldsymbol{g}s + \boldsymbol{n},$$

其中, $\boldsymbol{r}^{\mathrm{T}} = [\boldsymbol{r}_1^{\mathrm{T}} \quad \boldsymbol{r}_2^{\mathrm{T}} \quad \cdots \quad \boldsymbol{r}_M^{\mathrm{T}}]$, 阵列输出; $\boldsymbol{r}_i^{\mathrm{T}} = [\boldsymbol{p}_i^{\mathrm{T}} \quad \boldsymbol{x}_i^{\mathrm{T}} \quad \boldsymbol{y}_i^{\mathrm{T}} \quad \boldsymbol{z}_i^{\mathrm{T}}]$, 第 i 个阵元输出的声压和声质点振速三个正交分量; $\boldsymbol{g} = \boldsymbol{e} \otimes \boldsymbol{u}$, 符号 $\otimes$ 表示 Kronecker 积,$\boldsymbol{e}^{\mathrm{T}} = [1 \quad e^{-\mathrm{i}\omega\tau_0} \quad \cdots \quad e^{-\mathrm{i}\omega\tau_{M-1}}]$ 为单频平面波方向矢量, $\tau = \tau(\phi, \theta)$, $\phi \in [0, 2\pi)$, 与 x 轴的夹角, $\theta \in [0, \pi]$, 与 z 轴的夹角, $\boldsymbol{u}^{\mathrm{T}} = [1 \quad \cos\phi\sin\theta \quad \sin\phi\sin\theta \quad \cos\theta]$ 为声矢量传感器四个通道的指向性函数。

上述的假设与传统的声压传感器信号处理并无二致, 唯一注意的是, 波束旋转向量 $\boldsymbol{g}$ 不仅含有传统声压传感器阵列的时延信息, 而且还有声矢量传感器自身指向性信息, 它们之间由 Kronecker 积联系在一起, 即 $\boldsymbol{g} = \boldsymbol{e} \otimes \boldsymbol{u}$, 这种表示在声矢量传感器信号处理中被广泛采用。最近, Bihan 等[11] 提出基于四元数 (quaternion) 表示的矢量传感器信号建模和处理的新方法, 并利用该原理构成的奇异值分解 (SVD) 算法 - SVDQ 进行极化地震波分离。为了使上述假设在界面附近成立, Hawkes 等[12] 提出了反射系数模型, 从而保证了模型的形式与上述的假设完全一致。

1 似然比检测

在上述模型的基础上, 依据经典的似然比检测理论可以推导声矢量传感器阵列检测器, 以及检测概率和虚警概率, 这与 Burgess 提出的电磁矢量传感器阵列广义似然比检测类似[13]。

无信号时的似然函数:

$$f_0(\boldsymbol{r}) = \left(\pi^{4M}|\boldsymbol{R}_n|\right)^{-1} \exp\left(-\boldsymbol{r}^{\mathrm{H}}\boldsymbol{R}_n^{-1}\boldsymbol{r}\right),$$

其中, $\boldsymbol{R}_n$ 为噪声协方差矩阵。

有信号时的似然函数:

$$f_1(\boldsymbol{r}) = \left(\pi^{4M}|\boldsymbol{R}|\right)^{-1} \exp\left(-\boldsymbol{r}^{\mathrm{H}}\boldsymbol{R}^{-1}\boldsymbol{r}\right),$$

其中, $\boldsymbol{R} = \boldsymbol{R}_s + \boldsymbol{R}_n = \sigma_s^2\boldsymbol{g}\boldsymbol{g}^{\mathrm{H}} + \boldsymbol{R}_n$ 为信号加噪声的协方差矩阵。

由对数似然比得统计检测量:

$$\lambda = \sigma_s^2 \frac{|\boldsymbol{w}^{\mathrm{H}}\boldsymbol{r}|^2}{1 + \sigma_s^2\boldsymbol{g}^{\mathrm{H}}\boldsymbol{R}_n^{-1}\boldsymbol{g}}.$$

其中, $\boldsymbol{w} = \boldsymbol{r}_n^{-1}\boldsymbol{g}$ 为权向量。若考察上式的数学期望 $\mathrm{E}[\lambda]$, 则有:

$$\mathrm{E}[\lambda] = \frac{\sigma_s^2}{1 + \sigma_s^2\boldsymbol{g}^{\mathrm{H}}\boldsymbol{R}_n^{-1}\boldsymbol{g}}[\boldsymbol{w}^{\mathrm{H}}\boldsymbol{R}\boldsymbol{w}].$$

若忽略式中与最佳预选滤波器有关的系数项 $\sigma_s^2/(1 + \sigma_s^2\boldsymbol{g}^{\mathrm{H}}\boldsymbol{R}_n^{-1}\boldsymbol{g})$, 可得:

$$D_w = \boldsymbol{w}^{\mathrm{H}}\boldsymbol{R}\boldsymbol{w}.$$

令 $\boldsymbol{R}_n^{-1} = \sigma_n^{-2}\boldsymbol{I}_M$, 则上式可以写成如下的形式:

$$D = \boldsymbol{g}^{\mathrm{H}}\boldsymbol{R}\boldsymbol{g}.$$

上式与 Hawkes 等[14] 得到的形式完全相同, 一般将其称为 Cardioid 检测器, 因为单个阵元具有心形指向性。但需注意, $\boldsymbol{R}_n^{-1} = \sigma_n^{-2}\boldsymbol{I}_M$ 中的等号只有在传感器自噪声占优势时才能成立, 而在各向同性海洋环境噪声场中如果传感器自噪声可以忽略不计, 则声压传感器的噪声功率是质点振速传感器的三倍, 尽管此时能够得到更好的检测器, 可能会有一两个分贝的好处, 但这严重依赖于噪声协方差矩阵的具体形式, 实际中反而往往导致性能不稳定。根据传感器类型和所测量的物理量, 可以将上式分解为[15]:

$$D = \frac{1}{4}\boldsymbol{e}^{\mathrm{H}}\boldsymbol{R}_{pp}\boldsymbol{e} + \frac{1}{4}\boldsymbol{g}_v^{\mathrm{H}}\boldsymbol{R}_{vv}\boldsymbol{g}_v + \frac{1}{2}\boldsymbol{e}^{\mathrm{H}}\mathrm{Re}[\boldsymbol{R}_{pv}]\boldsymbol{g}_v,$$

其中, $\boldsymbol{R}_{pp}$, $\boldsymbol{R}_{vv}$ 和 $\mathrm{Re}[\boldsymbol{R}_{pv}]$ 分别为声压、振速和声能流的协方差矩阵; $\boldsymbol{g}_v = \boldsymbol{e} \otimes \boldsymbol{u}$, $\boldsymbol{u}_v^{\mathrm{T}} = [\cos\phi\sin\theta \quad \sin\phi\sin\theta \quad \cos\theta]$ 。仔细观察上式可以发现, 第 1 项是常见的声压水听器阵列常规波束形成器 $D_p = \boldsymbol{e}^{\mathrm{H}}\boldsymbol{R}_{pp}\boldsymbol{e}$, 它表示声压势能密度, 称其为声压检测器; 第 2 项是偶极子指向性传感器阵列的常规波束形成器 $\boldsymbol{D}_V = \boldsymbol{g}_v^{\mathrm{H}}\boldsymbol{R}_{vv}\boldsymbol{g}_v$, 它表示质点动能密度, 因其单个阵元的偶极子指向性, 称其为质点振速偶极子检测器; 第 3 项是声矢量传感器阵列的声能流常规波束形成器 $D_1 = \boldsymbol{e}^{\mathrm{H}}\mathrm{Re}[\boldsymbol{R}_{pv}]\boldsymbol{g}_v$, 它表示声能流密度, 称其为声能流检测器。前两项均属于能量检测器的范畴, 只是后者阵元有偶极子指向性而已, 而声能流检测器可以视为互相关检测器, 但细节上有所区别。由上述分析可知, Cardioid 检测器本质上由这三种基本的检测器线性组合而得; 当然, 由它们还可以演化出其它检测器形式, 如 $D = \frac{1}{2}\boldsymbol{g}_v^{\mathrm{H}}\boldsymbol{R}_{vv}\boldsymbol{g}_v + \frac{1}{2}\boldsymbol{e}^{\mathrm{H}}\mathrm{Re}[\boldsymbol{R}_{pv}]\boldsymbol{g}_v$, 其空间增益、波束图等均由质点振速偶极子和声能流检测器控制。

上述的结果均依赖于理想假设条件: 均匀无限大均匀理想流体 (无粘性, 只有纵波传播)、远场平面波入射、独立同分布高斯噪声。在推导检测器时没有考虑声压和质点振速空间相关系数的信息[17]，这必然会影响最优检测器的形式，因此得到的 Cardioid 等检测器只能是次最优的，但它们的性能更为宽容，对噪声协方差矩阵的微小扰动不敏感，而且简化后所得到的结果很容易给出清晰的物理图景，实验结果验证了这一点[16]。当考虑界面附近的检测问题时，若假设声源来自于远场，则声压和质点振速水平分量之间的平面波关系基本不受影响，因此可以得到水平方位角的估计，但声压和质点振速垂直分量之间有较复杂的行为，Hawkes 等给出简单的反射系数模型以讨论声矢量传感器分布式系统中目标俯仰角测量问题[12]。此外，Burgess 等[13]将广义似然比检测 GLRT(Generalized Likelihood Ratio Test) 引入到电磁矢量传感器阵列检测问题中，给出窄带和宽带检测统计量、参数估计、虚警概率和检测概率的解析式。通过子空间变换压缩检测问题维数到二维，尽管所得结果信噪比有所降低，但参数估计的方差将减小得更大，因此显著改善了检测器的性能，这些都可以借鉴到声矢量传感器信号处理领域中。

2 空间增益

Cray 利用声矢量传感器线阵的指向性函数给出了在各向同性噪声场中它的增益[18]和单个声并矢传感器的增益计算公式[19]，指向性函数利用 Brige 乘积定理很容易得到，因此这种方法在实际中经常采用，但只能限定在各向同性噪声场中，得到的增益常为指向性指数 DI。另外一种方法是利用噪声场中声压和质点振速空间相关系数计算声矢量传感器阵列各种形式的检测器在噪声场中的增益，这种方法更具一般性，得到的增益称为阵增益 AG，指向性指数 DI 和阵增益 AG 都是空间增益。Sun[16,50]给出基于各向同性噪声场中声压和质点振速之间相关系数的声矢量传感器线阵增益。下文的对比分析表明，Cray 和 Sun 的结论基本一致，可以互为证明。以八元水平均匀直线阵为例，声压检测器的空间增益为[20]：

$$G_P(k,\phi)=10\lg\frac{N}{1+2N^{-1}\sum\limits_{i=1}^{N-1}(N-i)R_p(ikd)\cos(ikd\cos\phi)},$$

其中，$R_p(ikd)$ 为声压的空间相关系数[54]。图 1 — 图 3 给出声矢量传感器水平线阵的不同检测器在正横和端射方向上空间增益和频率的关系曲线，横轴表示被半波长对应频率归一化的频率坐标，纵轴表示空间增益 (dB)。这些图中，曲线 1 表示在正横的空间增益；曲线 2 表示在端射的空间增益。图 1 中直线 (数字 3) 表示 $10\lg 8\approx 9$ dB。图 2 和 3 中的直线 (数字 3) 表示 $10\lg 8+10\lg 3\approx 13.8$ dB。

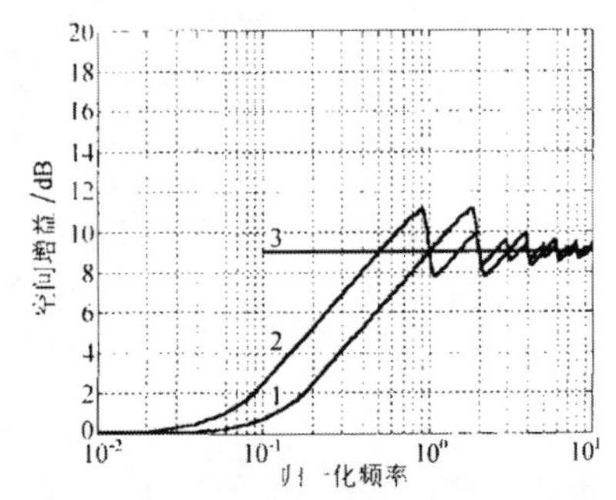

图 1 声压检测器在正横和端射方向空间增益曲线

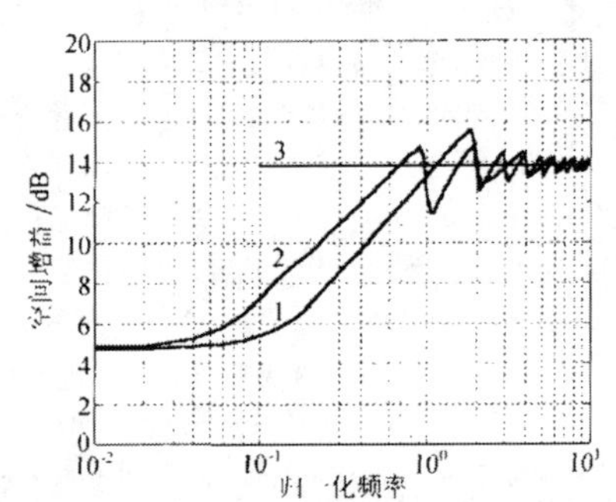

图 2 Cardioid 检测器在正横和端射方向空间增益曲线

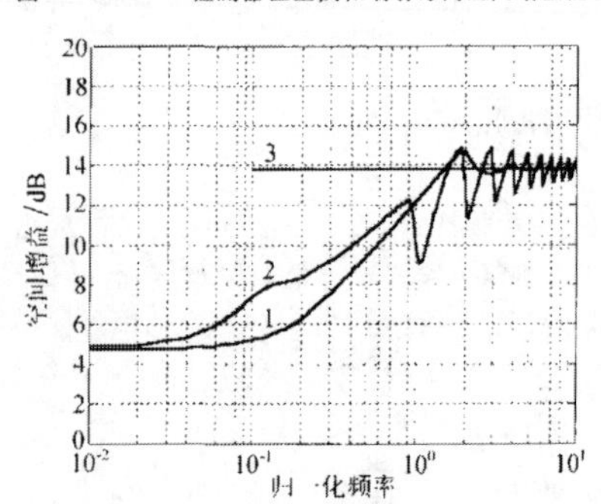

图 3 偶极子检测器在正横和端射方向空间增益曲线

Cardioid 检测器的空间增益为[16]：

$$G_D(k,\phi)=10\lg\frac{3N}{1+1.5N^{-1}\sum\limits_{i=1}^{N-1}(N-i)[R_p(ikd)+R_v(ikd)]\cos(ikd\cos\phi)}.$$

其中，$R_v(ikd)=R_x(ikd)\cos^2\phi+R_y(ikd)\sin^2\phi$，$R_x(ikd)$ 和 $R_y(ikd)$ 具体形式见文献 54。

偶极子检测器的空间增益为[16]:

$$G_V(k,\phi)=10\lg\frac{3N}{1+6N^{-1}\sum_{i=1}^{N-1}(N-i)R_v(ikd)\cos(ikd\cos\phi)},$$

在各向同性噪声场中声能流检测器的空间增益为[16,19]:

$$G_I=10\lg N\sqrt{6WT}.$$

其中 $\sqrt{6WT}$ 是单个声矢量传感器声能流检测器的等效空间增益，对于非各向同性场的使用要慎重，具体数值与噪声空间分布有关，可利用现场测量数据根据 Shchurov 方法进行估计[6,20]。Gordienko 等[21]通过试验证明，声矢量传感器阵列的声能流检测器具有传统意义上的空间增益，并首次给出了声能流检测器的解析形式。

分析对比上述检测器的空间增益可以得出：Cardioid 检测器、偶极子检测器与声压检测器空间增益相比，当频率约大于 1(半波长频率) 的高频部分基本上围绕 $10\lg 3N$ 进行波动，即前两者比声压检测器约高 5 dB，当频率降低时增益趋近于 $10\lg 3\approx 4.8$ dB，这些都是偶极子指向性和心形指向性的必然结果[16,18]；Cardioid 检测器与偶极子检测器的空间增益基本类似，在高频时都围绕 $10\lg 3N$ 振荡，但在频率 1 附近的行为有较大区别；当积分时间达到 30 s 甚至更高时，声能流检测器的空间增益最大，比声压检测器高出约 10 ~ 15 dB[6,20,22]，事实上，仔细观察声能流检测器形式不难看出，互协方差矩阵 $\mathrm{Re}[\boldsymbol{R}_{pv}]$ 仅含有声压和质点振速正交分量之间的互谱实部项，即有功声强，相对于协方差矩阵 $\boldsymbol{R}$ 的主对角线各项 $\boldsymbol{R}_{pp}$ 和 $\boldsymbol{R}_{vv}$，$\mathrm{Re}[\boldsymbol{R}]_{pv}$ 的互谱项有更低的背景噪声，而这需要较长的积分时间才能显现出来，因此声能流检测器有更好的检测性能也就不足为奇。对于单个声矢量传感器声能流检测器简化为：

$$I(\phi,\theta)=I_x\cos\phi\sin\theta+I_y\sin\phi\sin\theta+I_z\cos\theta,$$

其中，$I_x=\mathrm{Re}\{\mathrm{E}[pv_x^*]\}$ 是 x 轴向的声强分量，$I_y=\mathrm{Re}\{\mathrm{E}[pv_y^*]\}$ 是 y 轴向声强分量，$I_z=\mathrm{Re}\{\mathrm{E}[pv_z^*]\}$ 是 z 轴向声强分量，$*$ 表示复共轭。在此之前声能流检测器因其清楚的物理意义已经被广泛使用在水声和空气声的声强测量中。

单个声矢量传感器在孔径小于波长的情况下可以提供目标方位信息，而且具有一定的增益，使用更频繁，因此对于其检测性能的理论分析和实验结果都相对充分。Gordienko 认为[21]，Cardioid 检测器取决于噪声场类型和所使用的振速通道，一般在 3 ~13 dB 之间，这主要归因于振速传感器的余弦指向性；使用声能流处理的矢量特性，可以进一步显著降低检测阈，单频信号在水平各向同性噪声场中由空间滤波得到的附加增益在输出端信噪比 SNR< −20 ~ −25 dB 时才消失，这可以视为声能流检测器与单个声压水听器平方律检测器 (即能量检测器) 相比的抗干扰性能极限；对于噪声背景下的宽带源，声能流检测器的理论极限更高些，−30 ~ −40 dB 完全可能；增益在不同海区有所差别，对于深海约 −30 ~ −40 dB，对靠岸和航运繁忙水域约 −5 ~ −8 dB。Shchurov 等[23]给出了声能流检测器输出信噪比表达式和相对于声压能量检测器的增益表达式，并指出，在准各向同性噪声场中，声能流增益与积分时间的平方根成正比例；200 ~1000 Hz 频带上的深海试验表明，水平方向声能流信噪比与单个声压的相比约高 15 ~ 16 dB，余弦指向性增益在 150 m 深时为 3.8 dB，在 300 m 时为 4.6 dB。D'Spain 等[5]给出了次声频段上的声能流增益，大约 3 ~ 6 dB。Sun 等[22,24]给出了声能流检测器在各向同性噪声场中的增益公式，并给出了数值仿真结果和浅海水域的试验结果，试验表明，靠岸的浅海水域在航运稀少的情况下平均可以得到 10 ~15 dB 的增益。

从上述的试验结果中可以看到，尽管数据的离散性较大，但声能流增益如理论所预计的那样确实存在，只不过在具体海域和具体频带上有所差异。因为不同海区的海洋环境噪声具有一定的区域性，尤其是靠岸的浅海水域，这种区域性更加明显。远处离散目标的低频声和海底地震产生的次声使得噪声场在相当程度上偏离了各向同性假设，可能是导致低声频和次声频的声能流增益减小的主要原因。

3 DOA 的 Cramer-Rao 下界

Nehorai 等首次推导声矢量传感器阵列 DOA 估计的 CRB 一般表达式[9]。Sun[16]基于上文的最大似然检测原理推导出各向同性噪声场中快拍数为 N 时 M 元声矢量传感器水平均匀直线阵的窄带 DOA 估计 Cramer-Rao 下界[31]:

$$\mathrm{CRB}_\phi=\frac{1}{2NM[(M^2-1)k^2\cos^2\phi+9]\sin^2\theta}\left(\frac{1}{4M\,\mathrm{SNR}^2}+\frac{1}{\mathrm{SNR}}\right)$$

其中，$k=2d\pi/\lambda$。为了对比分析，同时给出了 M

元声压水听器均匀直线阵的 Cramer-Rao 下界:

$$\mathrm{CRB}_\phi=\frac{6}{NM(M^2-1)k^2\cos^2\phi}\left(\frac{1}{M\mathrm{SNR}^2}+\frac{1}{\mathrm{SNR}}\right).$$

图 4 给出了信噪比为 0 dB 时声矢量传感器水平线阵 (下面的曲线) 和声压传感器水平线阵 (上面的曲线) 的 DOA 估计 Cramer-Rao 下界。可以看出, 声矢量传感器水平线阵在端射 (图中的 0 度方位) 没有测向模糊, 只是精度低一些。宽带的情况类似于宽带检测, 只要进行频域积分就可得到[16]。Tichavsky 等[25] 推导了不同通道配置的声矢量传感器在空时相关加性噪声背景下窄带信号 DOA 估计的渐近和非渐近 CRB, 基本结论是: 在相同阵元数条件下声矢量传感器阵列具有更低的 CRB; 在 CRB 一定的情况下, 声矢量传感器阵列具有更少的阵元数; 声矢量传感器线阵无左右模糊, 改善了端射的 DOA 估计性能; 水平方位角不影响俯仰角估计的 CRB; 水平方位角的 CRB 随俯仰角增加而减小。

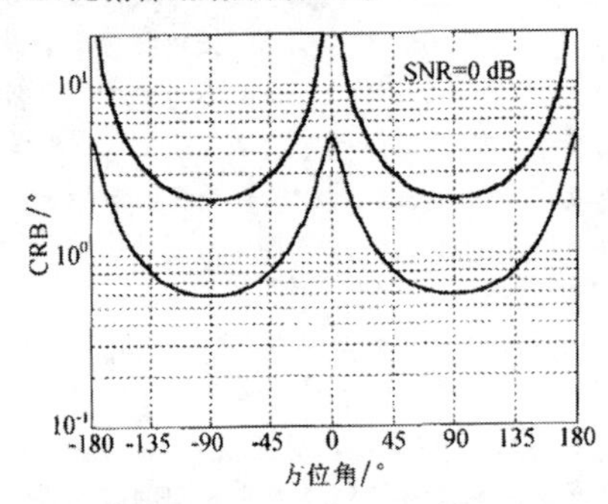

图 4 声矢量传感器水平线阵和声压传感器水平线阵的 CRB 估计 Cramer-Rao 下界对比

4 波束形成器

4.1 常规波束形成

首先仍以水平均匀直线阵为例介绍其指向性函数和波束图, 从而进一步总结声矢量传感器阵列的一些独到之处。

基于 Cardioid 波束形成器得到的指向性函数为:

$$B(\phi;\phi_s)=\left|\frac{\sin\left[N\pi\frac{d}{\lambda}(\sin\phi-\sin\phi_s)\right]}{N\sin\left[\pi\frac{d}{\lambda}(\sin\phi-\sin\phi_s)\right]}\right|^2\frac{[1+\cos(\phi-\phi_s)]^2}{4},$$

其中, 第 1 项为声压传感器均匀直线阵的指向性函数, 第 2 项为单个声矢量传感器的心形指向性函数。根据 Bridge 乘积定理, Cardioid 波束形成器的指向性函数实质上是上述两项的乘积。类似地, 可得到偶极子波束形成器的指向性函数:

$$B_V(\phi;\phi_s)=\left|\frac{\sin\left[N\pi\frac{d}{\lambda}(\sin\phi-\sin\phi_s)\right]}{N\sin\left[\pi\frac{d}{\lambda}(\sin\phi-\sin\phi_s)\right]}\right|^2\cos^2(\phi-\phi_s),$$

其中, 第 2 项为单个声矢量传感器的质点振速偶极子指向性 (余弦平方)。声能流波束形成器的指向性函数为:

$$B_I(\phi;\phi_s)=\left|\frac{\sin\left[N\pi\frac{d}{\lambda}(\sin\phi-\sin\phi_s)\right]}{N\sin\left[\pi\frac{d}{\lambda}(\sin\phi-\sin\phi_s)\right]}\right|^2\cos(\phi-\phi_s),$$

其中, 第 2 项为单个声矢量传感器的声能流余弦指向性, 注意到, 它有正值和负值, 在阵列处理时可以根据相位进行取舍以保留正值。

对比上述三个指向性函数的公式可得[16,47]: 常规的 Cardioid 和声能流波束形成器能够区分左右舷目标, 但端射波束较宽, 对定向精度有较大影响, 这可由 MVDR 波束形成器加以改善; 偶极子波束形成器在正横附近不能很好分辨左右舷目标, 有类似的端射问题; 上述三种常规波束形成器的波束主瓣与声压传感器的基本相当, 只有在阵元数很少时才有明显差别。

4.2 MVDR 波束形成

Hawkes 等除了给出声矢量传感器阵列的常规 Cardioid 波束形成之外, 还给出了 MVDR 波束形成器[14]

$$D_{\mathrm{MVDR}}=\frac{1}{\boldsymbol{g}^{\mathrm{H}}\boldsymbol{R}^{-1}\boldsymbol{g}},$$

其中, $\boldsymbol{R}^{-1}$ 为协方差矩阵的逆矩阵 (若 $\boldsymbol{R}$ 非奇异)。他们利用最优性能界检验了利用声矢量传感器代替传统的声压传感器对 DOA 估计性能的改善。若考虑单源情况下的 CRB, 则可证明, 声矢量传感器阵列具有更小估计误差是下面两个现象的必然结果: 传感器之间更多时延的测量使信噪比显著增加; 声矢量传感器的指向性允许直接测量含在振速场结构中的 DOA 信息。对这两种现象的分析可以确定在阵的尺寸、形状和 SNR 为何条件时, 即在什么情况下利用声矢量传感器阵列最有优势, 并定量研究这些优点。将常规波束形成和 MVDR 波束形成引入声矢量传感器阵列处理中, 消除了所有的测向模糊, 即使是线阵这样简单的结构都可以确定水平方位角和俯仰角, 并且使用空间降采样的等间距阵列用于增加孔径, 从而增加性能, 前提是保证一定的信噪比。他们推导了估计器的均方误差矩阵的大样本近似，由 Monte Carlo 模拟估计它们的性能。在此之前， D'Spain 等[26] 已经开始用 MVDR 波束形成器分析声矢量传感器准垂直阵的数据。

Hawkes 等尽管给出了远离边界情况下声矢量传感器阵列的常规最大信噪比波束图和 MVDR 波束图的解析式，但缺少详细的分析。MVDR 波束形成器是数据有关的波束形成器，它保持期望信号，使用数据中的干扰加噪声信息使波束形成器输出的干扰加噪声方差最小，MVDR 使波束形成器输出的信号和干扰加噪声功率比值 (SINR) 最大，且希望在尽可能宽的空间角度上具有高的 SINR 响应。MVDR 也可以抑制时变干扰源，但性能可能有所退化，尤其阵元较多时，因为它需要较长的积分时间才能得到较满意的干扰抑制效果。Wong 等[33] 详细分析了远离边界的单个声矢量传感器的空间匹配滤波器波束形成器 SMF-Spatial Matched Filter Beamformer(即常规的最大信噪比 SNR 波束形成器) 和 MVDR 波束形成器。以单个指向性干扰为例，使用四种声矢量传感器配置结构 (即 $pxyz$、pxy, xyz、xy)，从理论和数值仿真两方面详细分析了 MVDR 波束形成器的性能：使用四分量 ($pxyz$) 的声矢量传感器可以保障 SMF 单峰，MVDR 宽峰，可以测量水平方位角和俯仰角；若已知质点振速模糊，则无声压传感器的其它两种结构仍然可以保证 SMF 单峰和 MVDR 宽峰；无垂直质点振速传感器的两种结构有助于抑制垂直方向的干扰；使用三分量 (pxy) 的声矢量传感器在 MVDR 波束形成器上有些问题，若指向性干扰强于无指向性的加性噪声，则导致 MVDR 波束图的高度不规则。即使考虑声压传感器的噪声方差是质点振速传感器的三倍，对上述这些特性影响也很小。

4.3 子空间波束形成

子空间波束形成最早由 Wong 和 Zoltowski 引到声矢量传感器信号处理中[39,40]。MVDR 波束形成器的协方差矩阵不但含有噪声子空间，而且含有信号子空间，但根据子空间正交原理，信号子空间可以忽略，由此得到 MUSIC 波束形成器，需注意的是 MUSIC 波束形成器的幅度并不对应于信号的功率，这一点与 MVDR 有明显区别。有关 MUSIC 波束形成的工作可参见文献 41 — 43。

Zoltowski 等[37] 提出 MUSIC 或 MODE 算法改善由于 Nyquist 空间采样定理导致的稀疏均匀矩形网格阵列方向余弦估计的圆模糊问题。传统声压传感器阵列的 ESPRIT 算法需要两个相同的平移子阵以提供空间不变特征。但对于声矢量传感器阵列而言，目标的 DOA 信息含在质点振速的指向性信息中，即方向余弦。因此可以将方向余弦视为 ESPRIT 的不变性，而无须考虑阵元的位置，因此可以用于阵元位置未知的浮标阵 DOA 估计，以及阵元间距大于半波长的稀疏阵 DOA 估计问题，从而扩展阵列的有效孔径。Wong 等[38] 提出使用基于 ESPRIT 二维角度估计方法扩展声矢量传感器稀疏矩形阵 (阵元间距大于半波长) 的有效孔径。声源的 DOA 信息含在质点振速分量中，使用 ESPRIT 得到的低维数本征向量通过分离数据协方差矩阵的信号子空间本征向量截取这些 DOA 信息。当阵元间距超过半波长，这些信息虽然无模糊，但方差较大，可以用于粗略去除 ESPRIT 本征值的周期性相位模糊。仿真表明，随着阵元间距增加到 12 个波长，估计的标准差随之减小了 97%，相对于半波长，估计的标准差明显优于相同孔径半波长间距的声压传感器矩形阵一个数量级；提出的方法可以改善水声通讯性能；可以用于非规则阵形的虚拟阵内插。大孔径阵列增强阵列的方位分辨率和精度。通过增加阵元扩展阵列长度必将增加硬件开销，当然信号处理器的计算量也有明显增加。非均匀阵元间距违反了 ESPRIT(Estimation of Signal Parameter via Rotational Invariance Techniques) 的先决条件：两个相同平移的子阵。阵元间距超过半波长时将导致方向余弦估计的周期性模糊，根据 Nyquist 空间采样定理。使用通常的声压传感器阵列不能解决 ESPRIT 本征值相位的周期性模糊，若没有声源的先验信息。与前述的孔径扩展方法对比，所提出的方法：允许阵元间距超过半波长；解决上文的模糊问题；将 x 轴方向余弦和 y 轴的方向余弦估计自动配对。该方法结合质点振速矢量的 DOA 信息和孔径扩展技术以增加估计精度和分辨率。Wong 等[40] 引入 ESPRIT 源定位算法用于任意间距三维声矢量传感器阵列，且声矢量传感器位置未知的 DOA 估计。该算法的 ESPRIT 本征值与阵形无关，仅依赖于入射信号的方向余弦而非阵列参数，所以阵元位置可以任意，非常适用于声呐浮标等非规则阵列处理。

上述声矢量传感器阵列 DOA 估计算法都是基于窄带假设，大多不能处理相干源问题，在源完全相干时即失效。空间平滑方法虽可用于解决声矢量传感器阵列窄带相干源的问题，但它减小了阵的有效孔径。基于 Wang 和 Kaveh 所提出的声压传感器阵列宽带 DOA 估计相干信号子空间 (Coherent Signal Subspace – CSS) 方法[44]、Chen 等[45] 推导声矢量传感器线阵的宽带最小无失真响应波束形成，得到了声矢量传感器阵列和声压传感器阵列的宽带聚焦矩阵之间关系，给出线阵相应的 MVDR 波束形成仿真结果，研究表明声矢量传感器线阵的 CSS 波束形成有以下特点：多相干源方位分辨；不减小阵列有效孔径；无左右舷模糊；阵元间距可以超过 Nyquist 半波长空间采样上限；有更好的端射性能。

4.4 单个声矢量传感器波束形成

Hawkes 等[9]首次较系统地提出了单个声矢量传感器的声能流法测向算法，并证明声能流 DOA 估计性能基本接近CRB，还给出了声矢量传感器在界面附近的声能流DOA估计算法[12]。类似地工作还可参见文献 22, 24, 46 。此外，他们又将声矢量传感器用于基于方位估计的分布式噪声定位跟踪系统中，给出了相应的性能界[27]。在此之前，利用声能流的矢量特性，Gordienko 等[21]已经在实际中开始使用声能流法测向，并给出了测向精度，分析了$\phi=\tan^{-1}(I_y/I_x)$和$I=I_x\cos\phi+I_y\sin\phi$两种测向算法，并指出，前者在低信噪比下可能导致$40°\sim50°$的估计偏差，后者相对较好。Shchurov 等[28,29]利用声能流研究海洋噪声的动力学特性和噪声能流的传递特性。G'Spain 等[5]利用声能流研究海洋次声的一些性质，给出了声能流检测和 DOA 估计的试验结果。总的说来，使用单个声矢量传感器 DOA 估计精度不高，且多目标分辨能力不足[30]，Hawkes 等证明单个声矢量传感器最多分辨两个目标[31]。Cox 等[32]将声矢量传感器视为最简单的实用超指向性传感器，因为在小于波长的尺寸下能够提供一定增益。理论分析和数值仿真表明，他们提出的自适应Cardioid宽带处理算法可以有效抑制多基地声呐中出现的高指向性干扰。Hui 等[34]提出声矢量传感器抗相干干扰算法以提高声矢量传感器在强干扰下的检测性能。Tichavsky 等[25]提出单个声矢量传感器的 ESPRIT 算法。使用单个声矢量传感器两个时延数据集的方向余弦向量的空间不变性，构成 ESPRIT 算法。Sullivan[35]将匹配场处理 (MFP) 引入到声矢量传感器信号处理中，从而可以无模糊确定目标的方向、距离和深度。他认为，利用声矢量传感器测量得到的更全面、更复杂的声场幅度和相位信息，可能会得到更好的定位结果。常规的 MFP 将声压场视为深度和距离的函数，如果联合质点振速场，则可以发现，它在深度上不仅有幅度变化，而且还有方向变化。基于 Pekeris 模型的数值仿真研究表明：声压和水平质点振速联合使用可以有效降低旁瓣；与声压相比有 3 dB 的信噪比增益；远场时水平质点振速场和声压场差别不大[10,19,36]。Sun 等利用声矢量传感器测量水下目标辐射噪声[48]以及研究深海混响的指向性特征[49]

5 结论

本文尽可能地将一些共性结论和常用的公式进行简要总结，因此有些问题叙述得较多，有些问题则轻描淡写，再者限于篇幅，也不可能对所有的工作逐一介绍，读者可根据自身兴趣对所引参考文献重点解读。

到目前为止，声矢量传感器信号处理还是全新的研究领域[51,52,53]，尽管已经取得了一些进展。这篇综述旨在抛砖引玉，为推动国内相关研究不断深入尽一己之力。

参考文献

1 Berliner M J *et al.* Acoustic particle velocity sensor: design, performance and applications. NY:AIP Press, 1996

2 Skrebnev G K. Combined underwater acoustic receiver. St.Petersburg: Elmor, 1997

3 McConnell J A. Analysis of a compliantly suspended acoustic velocity sensor. *J. Acoust. Soc. Am.*, 2003; **113**(3): 1395—1405

4 Shipps J C *et al.* A miniature vector sensor for line array applications. *IEEE Oceans 2003 Proceedings*, 2003; **5**: 2367—2370

5 D'Spain G L *et al.* Energetics of the deep ocean's infrasonic sound field. *J. Acoust. Soc. Am.*, 1991; **89**(3): 1134—1158

6 Shchurov V A *et al.* Coherent and diffusive fields of underwater acoustic ambient noise. *J. Acoust. Soc. Am.*, 1991; **90**(2): 991—1001

7 Shipps J C *et al.* The use of vector sensors for underwater port and waterway security. *Sensors for Industry Conference 2004 Proceedings*, 2004: 41—44

8 Leslie C B *et al.* Hydrophone for measuring particle velocity. *J. Acoust. Soc. Am.* 1956; **28**(4): 711—715

9 Nehorai A *et al.* Acoustic vector-sensor array processing, *IEEE Trans. on Signal Processing*, 1994; **42**(9): 2481—2491

10 Gulin O E *et al.* On the certain semi-analytical models of low-frequency acoustic fields in terms of scalar-vector description. *Chinese J. of Acoust.*, 2004; **23**(1): 58—70

11 Bihan N Le *et al.* Singular value decomposition of quaternion matrices: a new tool for vector-sensor signal processing. *Signal Processing*, 2004; **84**: 1177—1199

12 Hawkes M *et al.* Acoustic vector-sensor processing in the presence of a reflecting boundary. *IEEE Trans. on Signal Processing*, 2000; 48(11): 2981—2993

13 Burgess K A *et al.* A subspace GLRT for vector-sensor array detection. *ICASSP-94. 1994 IEEE International Conference on Acoustics, Speech, and Signal Processing*, 1994(4): 253—256

14 Hawkes M *et al.* Acoustic vector-sensor beamforming and Capon direction estimation. *IEEE Trans. on Signal Processing*, 1998; **46**(9): 2291—2304

15 Sun G Q *et al.* A conventional beamforming method based on sound intensity for acoustic vector sensor array. *UDT 2004 Proceedings*, 5A.3

16 孙贵青. 声矢量传感器及其在均匀线列阵声纳中的应用. 中国科学院声学研究所博士后研究工作报告, 2003

17 Hawkes M *et al.* Acoustic vector-sensor correlations in ambient noise. *IEEE J. of Oceanic Engineering*, 2001; **26**(3): 337—347

18 Cray B A *et al.* Directivity factors for linear arrays of velocity sensors. *J. Acoust. Soc. Am.*, 2001; **110**(1): 324—331

19 Cray B A *et al.* Highly directional acoustic receiver. *J. Acoust. Soc. Am.*, 2003; **113**(3): 1526—1532

20 Shchurov V A *et al.* Use of acoustic intensity measurements in underwater acoustics (Modern state and prospects). *Chinese J. of Acoust.*, 1999; **18**(4): 315—326

21 Gordienko V *et al.* Vector-phase method in acoustics. Moscow:Nauka, 1989

22 孙贵青等. 基于矢量水听器的最大似然比检测和最大似然方位估计. 声学学报, 2003; **28**(1): 66—72

23 Shchurov V A *et al.* Noise immunity of a combined hydroacostic receiver. *Acoust. Physics*, 2002; **48**(1): 110—119

24 孙贵青. 矢量水听器检测技术研究. 哈尔滨工程大学博士学位论文, 2001

25 Tichavsky P *et al.* Near-field/far-field azimuth and elevation angle estimation using a single vector hydrophone. *IEEE Trans. on Signal Processing*, 2001; **49**(11): 2498—2510

26 D'Spain G L *et al.* Initial analysis of the data from the vertical DIFAR array. *IEEE Oceans 1992 Proceedings*, 1992: 346—351

27 Hawkes M *et al.* Wideband source localization using a distributed acoustic vector-sensor array. *IEEE Trans. on Signal Processing*, 2003; **51**(6): 1479—1491

28 Shchurov V A. Vector acoustics of the ocean. Vladivostok:Dalhauka, 2003

29 Shchurov V A *et al.* The interaction of energy flows of underwater ambient noise and local source. *J. Acoust. Soc. Am.*, 1991; **90**(2): 1002—1004

30 杨士莪. 单矢量传感器多目标分辨的一种方法. 哈尔滨工程大学学报, 2003; **24**(6): 592—595

31 Hochwald B *et al.* Identifiability in array processing models with vector-sensor applications. *IEEE Trans. on Signal Processing*, 1996; **44**(1): 83—95

32 Cox H *et al.* Adaptive cardioid processing. 26th Asilomar conference proceedings, 1992: 1058—1061

33 Wong K T *et al.* Beam patterns of an underwater acoustic vector hydrophone located away from any reflecting boundary *IEEE J. of Oceanic Engineering*, 2002; **27**(3): 628—637

34 惠俊英等. 声压和振速联合信号处理抗相干干扰. 声学学报, 2000; **25**(5): 389—394

35 Sullivan E J. The use of p-v sensor in passive localization. *IEEE Ocean 2002 Proceedings*, 2002: 1898—1902

36 周士弘. 分层介质波导中的声矢量场传播. 哈尔滨工程大学学报, 2004; **25**(1): 38—42

37 Zoltowski M D *et al.* Closed-form eigenstructure-based direction finding using arbitrary but identical subarrays on a sparse uniform Cartesian array grid. *IEEE Trans. on Signal Processing*, 2000; **48**(8): 2205—2210

38 Wong K T *et al.* Extended-aperture underwater acoustic multisource azimuth/elevation direction-finding using uniformly but sparsely spaced vector hydrophones. *IEEE J. of Oceanic Engineering*, 1997; **22**(4): 659—672

39 Wong K T *et al.* Uni-vector-sensor ESPRIT for multisource azimuth, elevation, and polarization estimation. *IEEE Trans. on Antennas and Propagation*, 1997; **45**(10): 1467—1474

40 Wong K T *et al.* Closed-form underwater acoustic direction-finding with arbitrarily spaced vector hydrophones at unknown locations. *IEEE J. of Oceanic Engineering*, 1997; **22**(4): 649—658

41 Wong K T *et al.* Self-initiating MUSIC-based direction finding in underwater acoustic particle velocity-field beamspace. *IEEE J. of Oceanic Engineering*, 2000; **25**(2): 262—273

42 Wong K T *et al.* Root-MUSIC-based azimuth-elevation angle of arrival estimation with uniformly spaced but arbitrarily oriented velocity hydrophones. *IEEE Trans. on Signal Processing*, 1999; **47**(12): 3250—3260

43 张揽月. 基于 MUSIC 算法的矢量水听器阵源方位估计. 哈尔滨工程大学学报, 2004; **25**(1): 30—33

44 Wang H *et al.* Coherent signal-subspace processing for the detection and estimation of angles of arrival of multiple wide-band sources. *IEEE Trans. On Acoust. Speech Signal Process.*, 1985; **33**(4): 823—831

45 Chen H W *et al.* Wideband MVDR beamforming for acoustic vector sensor linear array. *IEE Proceedings Radar, Sonar and Navigation*, 2004; **151**(3): 158—162

46 冯海泓等. 目标方位的声压振速联合估计. 声学学报, 2000; **25**(6): 516—520

47 陈新华等. 声矢量阵指向性. 声学学报, 2003; **28**(2): 141—144

48 孙贵青等. 基于矢量水听器的水下目标低频辐射噪声测量方法研究. 声学学报, 2002; **27**(5): 429—434

49 Sun C Y *et al.* Directional properties of reverberation based on an acoustic vector sensor. *UDT 2004 Proceedings*, 5C.2

50 孙贵青等. 基于矢量水听器的声压和质点振速的空间相关系数. 声学学报, 2003; **28**(6): 509—513

51 李启虎. 水声学研究进展. 声学学报, 2001; **26**(4): 295—301

52 李启虎. 水声信号处理领域若干专题研究进展. 应用声学, **2001**; **20**(1): 1—5

53 李启虎. 进入 21 世纪的声纳技术. 应用声学, 2002; **21**(1): 13—18

54 孙贵青, 李启虎. 声矢量传感器技术的当前进展. 声学学报, 2004; **29**(6): 481—490

一种多目标方位历程实时提取方法①②

郑 援[1,2] 胡成军[2] 李启虎[1] 孙长瑜[1]
(1 中国科学院声学研究所 北京 100080)
(2 海军潜艇学院 青岛 266071)
2003 年 1 月 22 日收到
2003 年 8 月 4 日定稿

摘要 研究多目标方位历程实时提取方法，为目标运动分析和军事辅助决策实时提供多个目标的方位历程数据。利用无源声呐方位历程显示的时空累积效应，提出一种实时自动提取多目标方位历程的方法。通过对无源声呐的动态方位历程显示进行图像降噪、亮点提取、历程扩展和非目标历程剔除四个步骤的数字图像处理，实现多目标方位历程的自动、实时提取。海试和仿真实验数据测试表明，该方法能有效提高信噪比，并能有效解决多目标方位历程交叉时的数据关联问题，计算效率和多目标方位历程提取的正确率较高，可实用。
PACS 数:、 43.30, 43.60

A method to extract multi-target's bearing time tracks on real time

ZHENG Yuan[1,2] HU Chengjun[2] LI Qihu[1] SUN Changyu[1]
(1 *Institute of Acoustics, The Chinese Academy of Sciences* Beijing 100080)
(2 *Navy Submarine Academy* Qingdao 266071)
Received Jan. 22, 2003
Revised Aug. 4, 2003

Abstract To provide multi-target's real-time bearing time history data for target motion analysis and automatic military decision, extraction of multi-target's bearing time tracks on real time is investigated. Taking advantage of the time space accumulation effect of passive sonar's bearing time history display, a method is proposed to extract multiple targets' bearing time tracks on real time. In order to extract multiple tracks from passive sonar's dynamic bearing time history display, the method takes four steps related with digital image processing: noise reduction, distinguished bearing point recognition, bearing time tracks extension and non-target tracks removal. Tests with data from sea trials and simulation experiments show that not only can the method improve the signal-to-noise ratio, but also solve the data association problem effectively when multiple bearing time tracks intersect. The correct extraction ratio of multi-target's bearing time tracks using the method is satisfactory. The method is also efficient, and can be applied in practice.

引言

根据声呐输出信息提取多目标的方位历程，实时提供多个目标的方位历程数据，对于目标运动分析[1-3]或进行军事辅助决策都具有重要意义。多目标方位历程提取的主要难点是克服噪声干扰和方位历程交叉进行正确的数据关联。

目前在声呐信号处理领域，最常用的获取多目标方位历程的方法是对实时波束形成输出进行目标跟踪。用于该类方法的数据关联方法有多种，例如：NN(Nearest Neighbor), JPDA(Joint Probability Data Association)、PDA(Probability Data Association), MHT(Multi Hypothesis Tracking), PMHT (Probabilistic MHT), MCMC(Markov Chain Monte Carlo) 等[4-6]，还可采用多种技术[7-10]有效提高多目标方位的测量精度。但这类方法普遍存在的问题是：在信噪比较低的情况下，难以有效检测目标的存在，从而难以对目标实施有效跟踪；涉及概率计算的方法必须事先掌握目标统计规律，实际应用中难以达到，时空开销比较大，必须限制所能跟踪目标的批次。

而在无源声呐的实际使用过程中存在这样一种现象，就是当无源声呐无法根据实时波束形成输出

① 海装武器装备科研项目和海装青年基金资助项目.

② 声学学报, 2005, 30(1): 83-88.

实施自动跟踪的情况下，声呐专业人员通过仔细查看声呐的方位历程显示，仍可能容易地确定有多少目标存在、目标同我船随时间的相对方位变化等。之所以存在这种现象一方面是由于声呐专业人员进行了智能性的思维活动，另一方面还由于方位历程显示本身比实时波束形成输出在时域和空间上更具有累积效应，可供利用的信息量更大。受此启发，作者考虑从方位历程显示的动态图像中提取多目标的方位历程应能比直接对实时波束形成输出进行跟踪的方法取得更好的效果。

从图像中提取几何曲线已有较为成熟的研究结果，如 Hough 变换及 Radon 变换等[11,12]，通过建立图像空间与参数空间之间的映射，利用统计特性提取几何曲线。但由于目标的方位历程通常是不太规则的几何曲线，且具有不确定的宽度，直接采用上述几何曲线提取方法，时空开销巨大，难以实时处理。

基于上述分析，本文提出并详细介绍一种包括图像降噪、亮点提取、历程扩展和非目标历程剔除四个基本步骤的处理方法，实现动态实时地从无源声呐的方位历程显示中提取多目标的方位历程。

1 原理与方法

本文方法的基本原理是仿照声呐操作人员的思维处理顺序，动态进行多目标方位历程的实时提取：首先利用图像降噪克服噪声干扰，增强图像的信噪比、剔除方位历程显示中由于环境噪声引入的孤立噪声点；然后进行亮点提取，从已经降过噪的方位历程显示中挑选出可能属于目标方位历程的亮点；通过历程扩展进行数据关联，延伸各目标方位历程；最后通过非目标历程剔除去除强目标方位历程引入的旁瓣历程、非孤立的环境噪声历程等非目标方位历程。

1.1 图像降噪

任何一种多目标方位历程提取方法都必须在信噪比高于其检测阈的情况下才能够有效工作。对实时波束形成输出进行目标跟踪的方法，由于其处理对象是某一时刻多个波束的实时输出，是一维的数据，这种情况下，若信噪比低于检测阈，很难找到有效的降噪手段提高信噪比，也就很难实现有效实时跟踪继而提取多目标方位历程。

本文选用的处理对象是方位历程显示，是二维的图像数据，通过图像降噪这一手段，可以较大程度地提高信噪比，从而在实际信噪比较低的情况下，使得降噪后的信噪比高于检测阈，实现多目标的有效检测、跟踪并提取方位历程。

图像降噪是亮点提取的准备工作。图像降噪能够有效去除图像中的噪声像素，但同时也容易造成弱信号丢失、图像边缘模糊、对比度降低，因此本文方法只对信噪比较低的方位历程显示进行图像降噪。

图像降噪分三步进行：图像锐化、灰度值变换、图像卷积去噪。

图像锐化的具体方法是采用非线性增函数对像素点进行处理。令 $\boldsymbol{L}_0$ 为原始像素点灰度值构成的矩阵，$\boldsymbol{L}$ 为锐化后的矩阵，f 为非线性增函数，则有[1]：

$$\boldsymbol{L}=\left\lfloor\frac{f(\boldsymbol{L}_0)-\min[f(\boldsymbol{L}_0)]}{\max[f(\boldsymbol{L}_0)]-\min[f(\boldsymbol{L}_0)]}\right\rfloor\times 255. \tag{1}$$

图像锐化能够扩大图像信噪比，扩大的程度取决于原图像和所选用的非线性增函数，例如经平方函数锐化后新图像的信噪比是原图像信噪比的 2 倍(以 dB 计)。注意，不宜选用锐化性能过强的非线性增函数，否则易发生弱目标丢失。

由于每一时刻波束形成输出值的动态范围不同，导致每帧像素的灰度值分布不同，不能直接采用固定灰度门限降噪，必须经过灰度值变换将像素点的动态灰度值范围限制到两个固定灰度区域。灰度值变换的具体方法如下：令某像素点的灰度值为 x，调整后的灰度值为 y，β 是根据当前灰度值分布和声呐设备参数计算出的动态门限，γ 为降噪使用的固定门限，$\Delta\gamma$ 为两个固定灰度区域之间的距离，则变换方法为：

$$a=0,\quad b=\beta,\quad c=255,\quad a'=0,$$
$$b'=\gamma-\Delta\gamma,\quad b''=\gamma+\Delta\gamma,\quad c'=255.$$

$$y=\begin{cases}(x-a)\dfrac{b'-a'}{b-a}+a', & a\le x\le b,\\ (x-b)\dfrac{c'-b''}{c-b}+b'', & b<x\le c,\end{cases} \tag{2}$$

灰度值变换将各像素点的灰度值变换到两个固定灰度区域，以便进行统一的二维卷积降噪。灰度值变换的效果取决于动态门限的设置方法，动态门限的设置原则是保证动态门限小于有效目标的平均灰度值。

图像卷积去噪是图像降噪的最后一步，用于去除方位历程显示中由背景噪声带来的加性噪声，包括高斯白噪声和“椒盐”状孤立噪声点。图像卷积去噪的具体方法是：使用二维窗对依次经过图像锐化和灰度值变换的方位历程显示图像进行卷积运算，并通过卡门限去除噪声。这也是目前数字图像处理领域较为成熟的图像降噪方法[13]

1 为便于描述，本文以 256 色图像为例，灰度级为 0~255，对于色数不同的情况，可类推。

为了既保证计算效率，又达到降噪的效果，图像卷积去噪总是对高度为 $2w_h+1$ 的最新方位历程显示采用 $(2w_h+1)\times(2w_h+1)$ 点的二维窗进行卷积和卡门限去噪处理， $2w_h+1$ 应根据声呐参数取值为目标方位历程的平均宽度。令 $\omega(x,y)$, x, $y\in[-1,1]$ 为二维窗函数，则用于卷积的二维窗为:

$$\boldsymbol{G}[m,n]=\omega\left(\frac{m-w_h-1}{w_h},\frac{n-w_h-1}{w_h}\right),\quad m=1:2w_h+1,\quad n=1:2w_h+1 \tag{3}$$

令 w_f 为方位历程显示宽度; 令 $\boldsymbol{I}_r$ 为由经过图像锐化和灰度值调整的最新方位历程显示的像素点构成的矩阵, 即 $\boldsymbol{I}_r$ 含有当前行 i、尺寸为 $2w_h+1\times w_f$、行号范围为 $[i-w_h,i+w_h]$; 令 $\boldsymbol{I}_1$ 为卷积后的像素点矩阵; 令 “*” 是二维卷积运算。则卷积操作如下:

$$\boldsymbol{I}_1=\boldsymbol{G}*\boldsymbol{I}_r. \tag{4}$$

令 “·” 为矩阵的点乘运算, 则对卷积后的二维图像卡门限去除噪声点的运算如下:

$$\boldsymbol{I}_2=(\boldsymbol{I}_1>\gamma)\cdot\boldsymbol{I}_1. \tag{5}$$

最后，需要将降噪后的像素点矩阵恢复成降噪前的尺寸:

$$\boldsymbol{I}=\boldsymbol{I}_2[w_h+1:3w_h+1,w_h+1:w_f+w_h]. \tag{6}$$

图像卷积去噪中，二维卷积运算用于提高图像信噪比，卡门限去噪的作用是：为便于后续处理，删除噪声像素点，只留下信噪比较高的有效像素点。二维卷积运算从数字图像处理角度看是一个利用邻域加权平均进行平滑降噪的过程，对图像信噪比提高的程度取决于原始图像和所采用二维窗。例如，在采用 M 点矩形窗的情况下，只具有均值为 0 的高斯白噪声的图像的信噪比将提高 $10\lg(M)$ dB[13] 。但是，由于方位历程显示由多波束能量积分得到，背景噪声均值大于 0 且不一定服从高斯分布，而邻域加权平均的本质又会使得图像卷积去噪在 $2w_h+1$ 大大超出目标方位历程宽度尺寸的情况下，丢失弱目标的可能性增大。因此， w_h 的实际取值受目标方位历程的平均宽度的限制，图像信噪比的提高也通常低于理论值。但毫无疑问，该步骤对于去除加性高斯白噪声和孤立噪声点是行之有效的。

1.2 亮点提取

所谓亮点，是方位历程显示的每一行像素点数据中，可能成为目标方位历程组成部分的像素点组。

经过了图像降噪后，方位历程显示的当前像素行(即：降噪后的最新方位历程显示像素点矩阵的中间像素行)中只保留了信噪比较高的有效像素点，这种情况下，亮点的特征十分明确，即：连续的灰度不为 0 的一组像素点。通过在当前像素行中搜索具有该特征的像素点组，就可以提取出当前像素行中所有的亮点。具体方法不再赘述。

1.3 历程扩展

为动态处理方位历程显示，本文对多目标方位历程的提取采取逐行历程扩展的方式进行，以增强计算的实时性。历程扩展可延伸各目标的方位历程，是本文方法进行数据关联的关键步骤。

令 H_c 数组记录所有当前未结束的目标方位历程，H_e 数组记录所有当前已结束的目标方位历程，每一步的方位历程扩展按照已有方位历程扩展、交叉点标记、方位历程整理、新建方位历程四个步骤进行。

令数组 L_B 存放当前像素行的亮点提取结果，则已有方位历程扩展步骤为 H_c 中的每个方位历程在当前的 L_B 数组中选择可供关联的亮点。数据关联的原则是：邻接、与预估方位最接近。即：只有唯一一个亮点同该方位历程邻接时，采用该亮点进行数据关联；有多个亮点同该方位历程邻接时，预估当前时刻该目标最有可能出现的方位，选择距离此方位最近的亮点进行数据关联；无邻接亮点时，预估当前时刻该目标最有可能出现的方位，在此方位生成一个点，并用此方位点进行数据关联，标记该点为暗点同时记录该方位历程中断一次。

交叉点标记步骤检查当前 L_B 数组，按照“被 2 个以上方位历程使用的亮点是交叉点”这一特征寻找亮点中的交叉点，并在方位历程中对这些点进行标记。

交叉点和暗点的标记以及目标方位的预估是本文方法用于保证多个方位历程交叉时的正确数据关联的关键之处。使用本文方法进行多目标方位历程提取可采用的目标方位预估方法有多种，但不论采用哪种，一个必要的原则是：预估计算时不使用交叉点和暗点。这是因为在多目标方位历程提取时，数据关联的难点在于多个方位历程交叉后如何为每个方位历程寻找到正确的亮点进行关联。交叉点处的亮点由多个方位历程汇集而成，通常发生宽度延展，所确定方位相对每个方位历程的实际方位具有一定的差异，让这样的交叉点参与预估计算，将导致预估结果偏离实际方位，使得各方位历程不能正确扩展。而暗点本身就是通过估计得出的，与实际方位有偏差，因此也不应参与预估计算。这样，通过交叉点和暗点标记，预估计算时不使用这些点，预估计算的准确性得到提高，从而有效解决数据关联的正确性问题。

方位历程整理步骤主要工作是：结束 H_c 中连续

中断时间过长的方位历程，并将符合目标方位历程标准的方位历程放入 H_e 中；为 H_e 中每个最近一段时间内结束的目标方位历程，在 H_c 最近一段时间新出现的方位历程中查找是否有属于同一个目标的方位历程，若有则进行方位历程链接。通过方位历程链接，因为信噪比过低而暂时中断的目标方位历程能够得以有效接续，得到正确提取，而不会被错误地处理为两个目标方位历程。

最后，新建方位历程步骤为 L_R 数组中每个未使用过的亮点开始一个新的目标方位历程。

1.4 非目标历程剔除

即便经过了上述步骤，还是有非目标历程会保留下来，非目标历程主要是强目标方位历程引入的旁瓣历程和非孤立的环境噪声历程，输出提取的目标方位历程时需要对这些非目标历程进行剔除。

本文判定旁瓣历程的标准是：历程的开始时间比某个尚未结束的目标方位历程晚，同该目标方位历程平行，两历程的方位差和强度比均符合声呐波束的旁瓣条件。判定非孤立的环境噪声历程的标准是：方位历程过宽，有效亮点较少，并且灰度值波动较大。

在判定之后，将判别出的非目标历程从输出的目标方位历程中剔除。

2 实验结果

作者对该方法的实用效果进行了测试，测试用的实验数据有两部分：海试数据和仿真数据。海试数据通过在海上实验中对拖曳线阵的多个通道进行实时录音，再经波束形成得到。仿真数据通过对多个具有初始方位、方位变化率的宽带目标进行模拟，产生多通道信号，并经波束形成得到。由于海试数据有限，仿真数据是海试数据的有益补充。

测试用的海试数据是 12 个通道的多通道数据，共有 7 组，每组数据经常规波束形成后，形成 61 个波束，生成 0~180 度的方位历程显示。在测试的 7 组数据中，有 6 组得到了正确的提取结果，1 组因为信噪比过低，只正确提取到部分的目标方位历程。图 1 —图 3 是得到正确结果的某海试数据的一帧原始方位历程显示、本文方法得到的多目标方位历程提取结果[2]以及该方位历程显示帧中某像素行在图像降噪和亮点提取中经历的变换。该海试数据中，共有 8 个目标出现，其平均原始图像信噪比 (以 dB 计) 依次是 1.0657, 1.6793, 0.8945, 0.1316, 0.3690, 0.1376, −0.0485, −0.5439, 图像降噪后各目标的平均图像信噪比依次是 3.2283, 11.0968, 7.9897, 4.0218, 6.7827, 0.5767, 2.5511, 3.2485, 平均提高了 4.4763 dB 。由于信噪比得到有效提高，历程扩展算法又有效提高了弱目标跟踪的连续性，本文方法成功地提取了所有 8 个目标的方位历程。而对该组海试数据，采用直接对波束形成数据进行自动跟踪的方法，即使检测阈低至 1 dB, 也只能实现对 2 个目标的稳定跟踪，得到 2 个目标的方位历程。

由于海试数据所覆盖的信噪比和目标方位历程复杂性的情况有限，作者进行了仿真信号的合成和测试。仿真测试结果表明，在各目标的图像信噪比不低于 0.8 dB 的情况下，本文方法能够进行有效的多目标方位历程提取，并能自动处理多目标方位历程交叉。图 4 和图 5 是某仿真数据的方位历程及相应的提

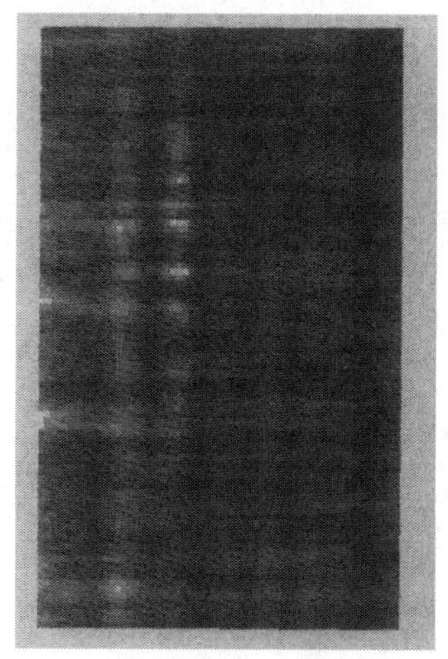

图 1 某海试数据的原始方位历程显示

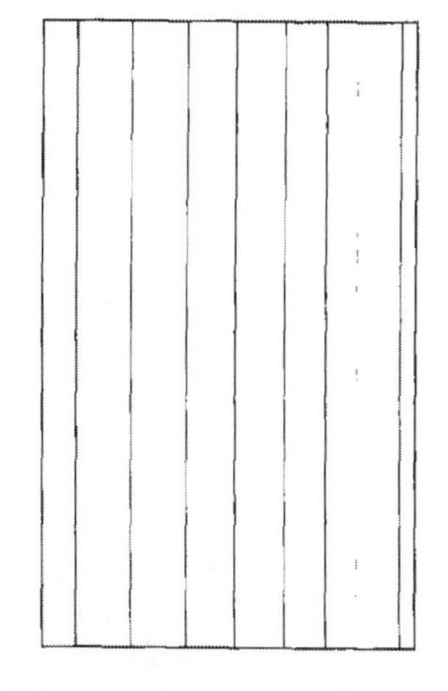

图 2 本文对图 1 海试数据的提取结果

2 为了显示得更清楚，方位历程的提取结果均采用反色显示，即底色为白色，提取出的目标方位历程以黑色显示。

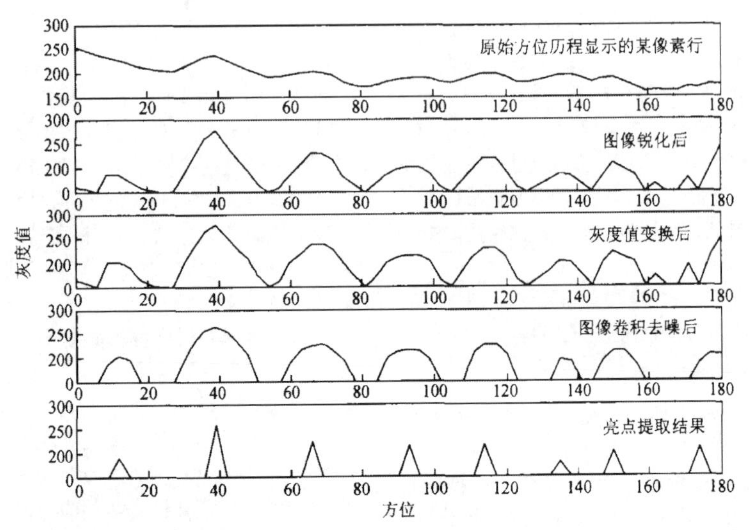

图 3　某像素行在图像降噪和亮点提取过程中的变换

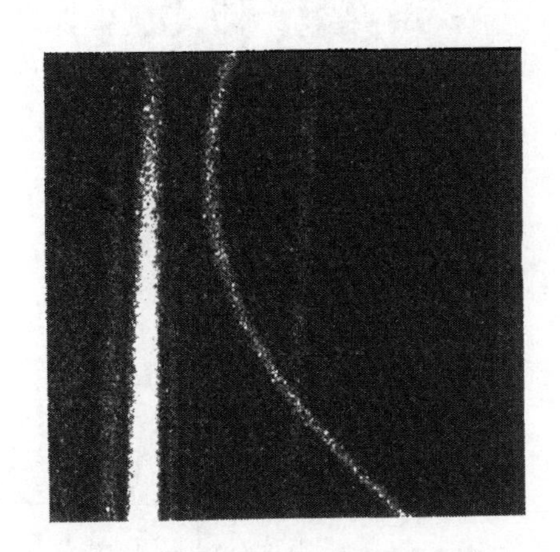

图 4　某仿真数据的原始方位历程显示

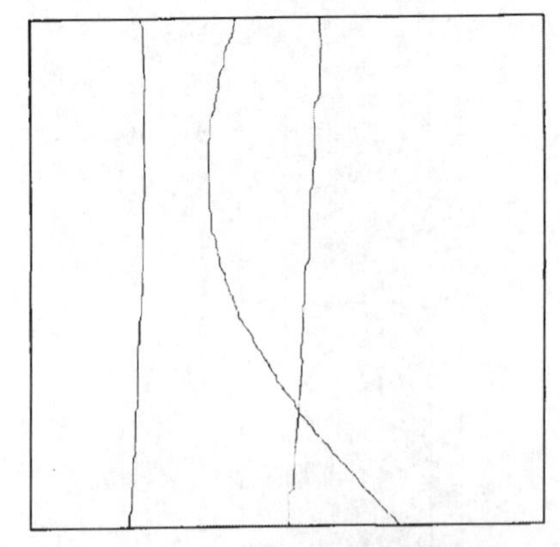

图 5　本文对图 4 仿真数据的提取结果

取结果。在目标图像信噪比低于 0.8 dB 的情况下，对于同样的多目标方位历程和同类型的噪声干扰，提取的正确率随信噪比的降低而下降。

海试数据和仿真数据的测试均在配置为 Intel PIII 800 MHz CPU, 128Mbytes 内存的微机上进行，针对每帧方位历程显示的平均处理时间随数据的不同而不同，但均低于 0.8 s，由于声呐设备方位历程显示的刷新时间一般不短于 1 s，所以本文方法的计算效率达到实时使用要求。

3　结论

本文提出一种多目标方位历程实时提取方法，通过对该方法的研究和检验，可得到如下结论：

(1) 以二维的方位历程显示为处理对象，可利用数字图像处理技术提高信噪比，使得本文方法能工作在信噪比较低的工作环境下，优于直接对实时波束形成输出这样的一维数据进行处理提取多目标方位历程的方法。

(2) 对于多目标方位历程交叉时的自动数据关联问题，本文方法所采用的邻近原则、交叉点和暗点标记、目标方位预估、方位历程链接相结合的策略是行之有效的解决方法。

(3) 对已有海试数据和仿真数据的测试结果表明本文方法的正确率和计算效率较高，无批次限制，可实用。

(4) 当两个目标强度相当、方位历程相接、并在接触点处同时转向使得二者方位历程构成 “X” 状时，本文方法将会误处理成两个完全交叉的方位历程。但由于实际情况中，该现象的发生概率极小，所以不影响本文方法的实用性。

致谢

感谢朱埜研究员、姜维副研究员、王磊博士、黄海宁博士为本文提供的帮助。

参 考 文 献

1 胡友峰, 詹艳梅, 孙进才. 基于状态矢量融合的多基地无源目标运动分析. 声学学报, 2002; **27**(4): 316—320

2 王 燕, 岳剑平, 冯海泓. 双基阵纯方位目标运动分析研究. 声学学报, 2001; **26**(5): 405—409

3 潘志坚, 阎福旺, 刘孟庵, 王广恩. 纯方位水下目标运动分析方法研究. 声学学报, 1997; **22**(1): 87—92

4 Mourad Oussalah, Joris De Schutter. Hybrid fuzzy probabilistic data association filter and joint probabilistic data association filter. *Information Sciences*, 2002; **142**(1-4): 195—226

5 Ding Z, Leung H, Hong L. Decoupling joint probabilistic data association algorithm for multiple target tracking. *IEE Proceedings RADAR, SONAR and Navigation*, 1999; **146**(5): 251—254

6 Bergman N, Doucet A. Markov chain monte carlo data association for target tracking. In: Ali Naci Akansu ed. Proceedings, IEEE International Conference on Acoustics, Speech and Signal Processing (ICASSP 2000), Istanbul, Turkey, 2000, U.S.A: The Printing House, 2000: 705—708

7 冯西安, 黄建国, 张群飞. 依赖于频率变化模型的相干信号子空间聚焦处理方法. 声学学报, 2003; **28**(4): 321—325

8 孙 超, 杨益新. 高分辨目标方位估计算法 —— 递增阶数多参数估计的理论与实验研究. 声学学报, 1999; **24**(2): 210—218

9 杨益新, 孙 超. 波束域加权子空间拟合算法. 声学学报, 2000; **25**(2): 142—145

10 吕仕皖, 林 京, 郭良浩. 浅海声传播相速度对测向精度的影响. 声学学报, 2002; **27**(6): 492—496

11 Enrico Magli, Gabriella Olmo, Letizia Lo Presti. On-board selection of relevant images: an application to linear feature recognition. *IEEE Transactions on Image Processing*, 2001; **10**(4): 543—553

12 XIE Yonghong, JI Qiang. Effective line detection with error propagation. In: Billene Mercer ed. Proceedings, 2001 IEEE International Conference on Image Processing, Thessaloniki, Greece, 2001, U.S.A: The Printing House, 2001: 181—184

13 Castleman K R 著, 朱志刚, 林学訚, 石定机等译. Digital image processing, 数字图像处理. 北京: 电子工业出版社, 1998: 127—137

The Effect of Space-Time joint Correlation on the Underwater Acoustic MIMO Capacity ①

Dazhi Piao, Biao Jiang, Huabing Yu, Changyu Sun, Qihu Li

Institute of Acoustics , Chinese Academy of Sciences

21 West Rd., Beisihuan Zhong Guan Cun, P.O.Box 2712, Beijing, China, 100080

e-mail: piaodazhi@hotmail.com

Abstract **- The joint effect of space and time correlation on the underwater acoustic Multiple-Input-Multiple-Output (MIMO) channel capacity is studied using the multipath space-time correlation model with a moving transmit array and a stationary receive array. The space-time correlation function of the uniform liner array is generated, which can be expressed as the production of the space correlation function and the time correlation function, so the space-time joint correlated complex Gaussian variates can be generated by firstly generating n_T*n_R independent random variates with the desired autocorrelation function, and then multiplied by the square root of their cross-correlation matrix. Using Monte Carlo simulations, the effect of space and time correlation on the underwater acoustic MIMO capacity is studied by the parameters of transmitter velocity, array element spacing, and elevation angle spread. It is shown from the simulation results that in the case of a poor scattering environment and not sufficiently large element spacing , when n transmitter element and n receiver element is employed, the outage capacity will not increase linearly with the increasing of n, and large capacity gains can only be obtained in a rich scattering environment and with sufficiently large element spacing, and the effect of transmitter velocity on the capacity depends on the block length T, when $T \gg T_{coh}$, capacities with different transmitter velocity will be the same.**

I. INTRODUCTION

Besides the urban wireless channel and the indoor wireless local networks, the underwater acoustic channel is another environment of rich scattering, thus suitable for the MIMO technologies, but the MIMO capacity studies based on underwater acoustic channel are still very limited.

It has been shown [1]-[3] that, in a Rayleigh flat-fading environment, a MIMO wireless communication link has a theoretical capacity that increases linearly with the smaller of the number of the transmit and receive array elements, provided that the complex-valued propagation coefficients between all pairs of the transmitter and receiver elements are statistically independent , so the MIMO technology has drawn wide interest in the wireless communication field in the past two decades. However, in some practical environment, there will be some spatial correlations between the propagation coefficients caused by the poor scattering environment or insufficient spacing between the array elements, and the motion of the transmitter or the receiver will effect the temporal correlations of the propagation coefficients.

The capacity of the MIMO system can be studied through field measurements or through ray-tracing to record a large number of typical channel realizations, from which the channel capacity can be computed [4]-[6], and in [7], the effect of the transmit power and signal frequency to the MIMO capacity of the typical shallow water is studied through field measurements. However, results obtained in this manner are only applicable to a particular environment and array configuration, furthermore, it is difficult to study the effect of the environment parameters on MIMO capacity. The MIMO capacity can also be studied through abstract scattering models [8]-[10], from which the essential characteristic of the environment can be illuminated, and its affections on the MIMO capacity can be easily studied.

The "one-ring" or "two-ring" scattering model is generally adopted in the wireless radio communication systems, the "one-ring" model is appropriated when the base station is elevated and unobstructed by local scatters, and the "two-ring" model is appropriated when both of the transmitter and receiver are surrounded by objects. In these models, the scatters are generally assumed to be distributed on the circles around the transmitter or received arrays. However, in the underwater acoustic communication channels, the transmitted and received acoustic rays will be spread both vertically and horizontally, because of the scatters from the sea surface, seabed and the water volume, as well as the unconstant sound speed; furthermore, the moving of the transmit array will influence the temporal correlation characteristic of the channel realizations(in this paper, the temporal characteristic effected by the moving of scatters are not considered), for an example, the communication between UUVs and their mother ships, so the space-time correlation model with acoustic rays spread both vertically and horizontally and the transmitter moving is considered in this article, the space-time joint correlation function is derived first, then the effect of space and time correlations on the underwater MIMO communication systems are studied through Monte Carlo simulations.

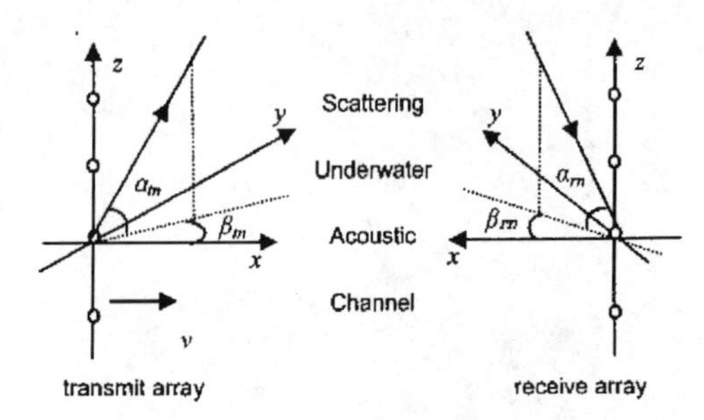

Fig.1. Configuration of the underwater acoustic MIMO communication system

① Oceans-Europe, 2005, 1: 413-418.

II. SPACE-TIME JOINT CORRELATION MODEL

The coherence between the signals received at different array elements with one transmitter element is derived in [11] , in this paper, the results are extended to the MIMO case, the space-time joint correlation function between the coefficients of different transmitter and receiver element pairs is generated

In a MIMO communicaiton system consisting of n_R receiver elements and n_T transmitter elements, when there are multipath between the transmitter and the receiver, and considering a moving transmitter, the received signal of the ith, $i=\{1,...,n_R\}$ receiver element from the jth, $i=\{1,...,n_T\}$ transmitter element can be expressed as

$$h_{i,j}(t)=\sum_{n=1}^{N}A_n\exp(\hat{i}w(t-\tau_{n,i,j}(t))) \tag{2.1}$$

where $\hat{i}$ is the imaginary unit, A_n is the amplitude gain of the nth multipath, N is the number of the multipath, and $\tau_{n,i,j}(t)$ is the time dependent path delay associated with the propagation from the jth transmitter element to the ith receiver element along the nth path.

when the transmitter is moving towards the receiver in a constant speed v along the x axis shown in Fig. 1, Equ. (2.1) can be written as

$$h_{i,j}(t)=\sum_{n=1}^{N}A_n\exp(\hat{i}w(t-\tau_{n,i,j}+(v/c)(t-\tau_{n,i,j})\cos\alpha_m\cos\beta_m)) \tag{2.2}$$

where c is the underwater sound speed, in this paper, it is assumed to be constant as c=1500(m/s), $\tau_{n,i,j}$ is the path delay associated with the propagation from the jth transmitter element to the ith receiver element along the nth path at time instant t=0, α_m , β_m are the elevation angle and azimuthal angle at the transmitter along the nth path, as shown in Fig.1, in this paper, they are treated as random variables with some kind of distribution.

Given the statistical properties of the channel, if N is sufficiently large, the central limit theorem states that $h_{i,j}(t)$ is a zero-mean complex Gaussian-random process, therefore, the envelope of $|h_{i,j}(t)|$ is a Rayleigh-fading process.

The correlation between the two links, T_j-R_j and $T_{i'}-R_{j'}$, for a time delay τ, when normalized by the received signal field variance is defined as

$$\rho_{ij,i'j'}(\tau,t)=E\left[h_{ij}(t)h_{i'j'}{}^{*}(t+\tau)\right]\Big/\sqrt{\Omega_{ij}\Omega_{i'j'}} \tag{2.3}$$

where * is the complex conjugate, Ω_{ij}=E[$|h_{ij}(t)|^2$], substitute Equ.(2.2) into Equ.(2.3) and simplify the result ,

$$\rho_{ij,i'j'}(\tau,t)=\rho_{ij,i'j'}(\tau)=E\left[\sum_{n=1}^{N}\exp(\hat{i}w(1+(v/c)\cos\alpha_m\cos\beta_m)(-\tau+\tau_{n,i',j'}-\tau_{n,i,j}))\right] \tag{2.4}$$

$\tau_{n,i'j'}-\tau_{n,i,j}$ will have different expressions corresponding to different array configurations, when plane waves are assumed , for the vertical transmit array and vertical receive array , we have

$$\tau_{n,i',j'}-\tau_{n,i,j}=(dr\sin\alpha_m(i'-i)+dt\sin\alpha_m(j'-j))/c \tag{2.5}$$

substitute Equ. (2.5) into Equ.(2.4), the correlation function between links of T_i-R_j and $T_{i'}-R_{j'}$ at the time delay τ can be expressed as

$$\rho_{ij,i'j'}(\tau)=[\exp(\hat{i}w(1+\cos\alpha_t\cos\beta_t v/c)\times(-\tau+\frac{i'-i}{c}dr\sin\alpha_r+\frac{j'-j}{c}dt\sin\alpha_t))]_{\alpha_t,\beta_t,\alpha_r} \tag{2.6}$$

where dt and dr are the transmitter and receiver element spacing , α_t , β_t are the elevation angle and azimuthal angle of rays at transmitter, α_r is the elevation angle of rays at the receiver, [•] means statistical averaging. From Equ.(2.6), it is shown that the space-time joint correlation function depends on the velocity of the transmitter, the transmitter and receiver element spacing, the spatial sound distribution at the transmitter and receiver; furthermore, from Equ.(2.6) which can also be seen that the space-time joint correlation function can be expressed as the product of space correlation function and time correlation function,

$$\rho(\tau)=[\exp(\hat{i}w(1+(v/c)\cos\alpha_t\cos\beta_t)(-\tau))]_{\alpha_t,\beta_t} \tag{2.7}$$

$$\rho_{ij,i'j'}=[\exp(\hat{i}w(1+\cos\alpha_t\cos\beta_t v/c)\times((i'-i)dr\sin\alpha_r+(j'-j)dt\sin\alpha_t)/c)]_{\alpha_t,\beta_t,\alpha_r} \tag{2.8}$$

thus the random variables with the space-time joint correlation function of Equ.(2.6) can be generated conveniently.

III. MIMO CAPACITY

Considering a narrow-band single-user communication systems with n_T transmitter and n_R receiver omnidirectional elements, the transmitter is assumed to be moving and the receiver is assumed to be fixed, the relationship between the transmitted signal vector and the received signal vector can be expressed

$$\boldsymbol{y}(t)=\boldsymbol{H}(t)\boldsymbol{x}(t)+\boldsymbol{n}(t) \tag{3.1}$$

where $\boldsymbol{x}(t)$ is the $n_T\times1$ transmitted signal vector, $\boldsymbol{y}(t)$ is the $n_R\times1$ received signal vector, $\boldsymbol{n}(t)$ is the $n_R\times1$ additive white Gaussian noise vector, and $\boldsymbol{H}(t)$ is the $n_R\times n_T$ channel propagation matrix of complex path gains h_{ij}(t) between transmitter element j and receiver element i. The elements of noise vector are assumed to be independent and identically distributed (i.i.d) complex Gaussian-random variables with variance N_0.

When the channel state information is known at the receiver but not known at the receiver the total transmitted power are generally equally allocated to every transmitter element, in this case, the instant MIMO capacity corresponding to each realization of $H(t)$ is expressed as [1][2][3],

$$C_t=\log_2\det(I_{n_R}+(\rho/n_T)H_tH_t^{\dagger})\ \text{bps/Hz} \tag{3.2}$$

where $H^{\dagger}$ means the transpose conjugate of H, det(·) is the matrix determinant, I_{nR} is the $n_R\times n_R$ identity matrix, and ρ is the average signal-to-noise ratio (SNR) at each receiver element.

For a randomly time-varying channel, the definition of the capacity depends on the codeword length T [16], If $T \gg T_{coh}$ (T_{coh} is the channel coherence time) cannot be satisfied, for instance, in the speech-transmission system, for the delay constrains, the transmitted frame spans a fixed number of symbols, the maximum mutual information is not equal to the channel capacity, and the Shannon capacity of the channel may be even zero [17], in such cases, the outage capacity C_{out} is generally used. The outage capacity C_{out} is associated with an outage probability P_{out}, which gives the probability that the channel capacity falls below C_{out}, this can be expressed mathematically as

$$P_{out} = P(C < C_{out}) \tag{3.3}$$

For an example, $C_{0.1}$ means the probability that the channel realization can not support it is 10%. The capacity of a frame comprising N symbols is given by [15]

$$C = \frac{1}{N}\sum_{t=1}^{N} C_t \quad \text{(bps/Hz)} \tag{3.4}$$

IV. SIMULAITON RESULTS

A. Generation of the correlated variables

Exact generation of M Gaussian variables with an arbitrary correlation can be achieved in principle by decomposing the desired $M \times M$ covariance matrix, $R = GG^{\dagger}$, where $G^{\dagger}$ is the conjugate transpose of G, then multiplying M independent Gaussian variates by G, however, this method is computationally intensive and not feasible for long sequences.

Fortunately, the space-time joint correlation function of Equ. (2.6) can be written as the production of the time correlation function Equ.(2.7) and the space correlation function Equ.(2.8), which will greatly facilitate the generation of the correlated variable. Firstly, $n_R \times n_T$ independently Rayleigh fading Gaussian random variables with the autocorrelation function of Equ.(2.7) can be generated, because of Equ.(2.7) is complex valued, the method of generating correlated random process with the correlated quadrature components using AR(Auto Regressive) model [13] is adopted in this paper. Next, these independent Gaussian variables with the desired autocorrelation function are multiplied by the Schur decomposition of their cross-correlation function described in Equ.(2.8), thus the $n_R \times n_T$ complex Gaussian processes with the space-time joint correlation function of Equ.(2.6) can be generated.

B. Simulation results for the 10% outage capacity

Equ.(2.6) indicates that the space-time joint coherence is a function of the velocity of the transmitter v, the transmitter and receiver element spacing dt and dr, the elevation angle spread of the acoustic rays at transmitter α_t and at receiver α_r, and the azimuthal angle spread of the acoustic rays at transmitter β_t. In this article, it is assumed that both of the transmitter and receiver use a vertical array, and all the angles of α_t, α_r and β_t, are assumed to be uniformly distributed around the angle of zero degree, of course, this method of simulation can also be used in the cases of other angle distributions or other array configurations. In all the simulations, the frequency is assumed to be 1000Hz.

Firstly, in order to test the accuracy of the method used to generate the correlated variables, the theoretical and simulated correlation functions are plotted together in Fig.2(a) and Fig.2 (b). Two different values of velocity v=1(m/s) and v=3(m/s) are selected, to demonstrate the effect of transmitter speed on the

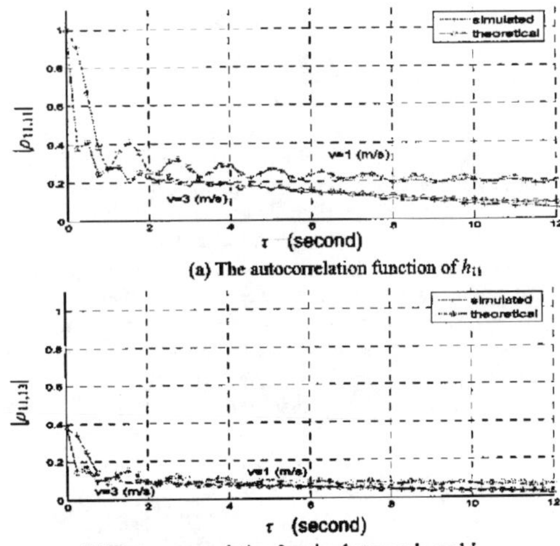

(a) The autocorrelation function of h_{11}

(b) The cross correlation function between h_{11} and h_{13}

Fig.2. Comparisons between the correlation functions using simulated sequence and theoretical value

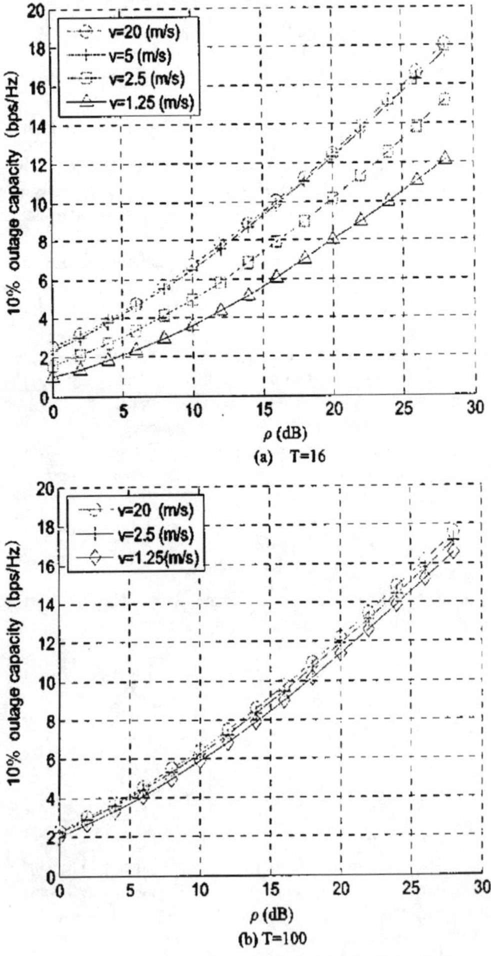

(a) T=16

(b) T=100

Fig. 3. $C_{0.1}$ versus SNR with different transmitter velocity time correlation. Here are the channel parameters of Fig. 2, $\alpha_t \in [-20°,20°]$, $\alpha_r \in [-20°,20°]$, $\beta_t \in [-10°,10°]$, dt=0.5λ, dr=0.5 λ, n_T=n_R=3.

From Fig.3 to Fig.5, the dependence of 10% outage capacity on the average received SNR with different channel parameters are described. In these figures, the number of transmitter and receiver elements is 3, i.e. n_T=n_R=3.

In Fig.3, the dependence of $C_{0.1}$ on SNR with different transmitter velocity is described, using the following parameters that $\alpha_t \in [-20°,20°]$, $\alpha_r \in [-20°,20°]$, $\beta_t \in [-10°,10°]$, dt=0.4λ, dr=0.4λ. It is shown in Fig.3 that with different transmitter velocity, the outage capacity grows with SNR in a similar manner, and with the increase of the velocity the curve is shift upwards, but the distance between curves is getting more smaller when the length of the frame grows larger, and when the frame length $T \gg T_{coh}$, channel capacities with different transmitter velocity will be the same, for in this case the channel realizations are almost temporally independent.

In Fig.4, the dependence of $C_{0.1}$ on SNR with different transmitter and receiver element spacing is described. The simulation parameters are as the follows, $\alpha_t \in [-20°,20°]$, $\alpha_r \in [-20°,20°]$, $\beta_t \in [-10°,10°]$, T=16, v=10(m/s). Increasing the transmitter and receiver element spacing will increase the outage capacity, and different to the result in Fig. 3, with the increasing of dt and dr , the curves in Fig.4 will grow more quickly.

Fig.5 describes the dependence of $C_{0.1}$ on SNR with different elevation angle spread at transmitter and receiver by the parameters of $\beta_t \in [-10°,10°]$, T=100, v=10(m/s). dt=0.4λ, dr=0.4λ. Increasing the elevation angle spread will increase the outage capacity with a similar manner of that in Fig.4, for the reason that increasing either the element spacing or the elevation angle spread will both affect the spatial coherence between the received signals.

The dependence of $C_{0.1}$ on receiver element spacing with different transmitter element spacing is described in Fig.6. with the parameters of $\alpha_t \in [-20°,20°]$, $\alpha_r \in [-20°,20°]$, $\beta_t \in [-10°,10°]$, T=16, v=5(m/s), ρ=10(dB), n_T=n_R=3 . In Fig.7, the dependence of $C_{0.1}$ on elevation angle spread at receiver with different elevation angle spread at transmitter is described. The simulation parameters are, T=16, dt=0.4λ, dr=0.4 λ, v=10(m/s), ρ=10(dB), n_T=n_R=3 . It is shown in Fig.6 and 7 that increasing the array element spacing or the elevation angle spread in either ends

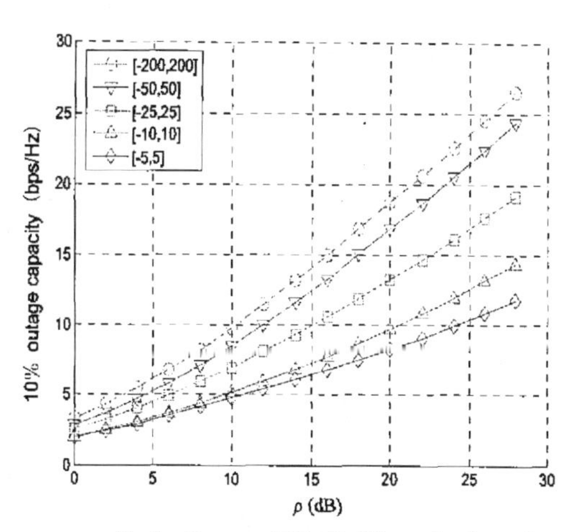

Fig. 5. $C_{0.1}$ versus SNR with different elevation angle spread at transmitter and receiver

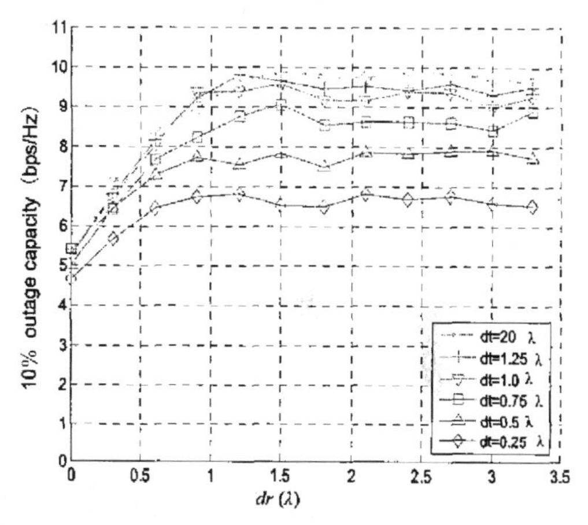

Fig.6. $C_{0.1}$ versus receiver element spacing with different transmitter element spacing

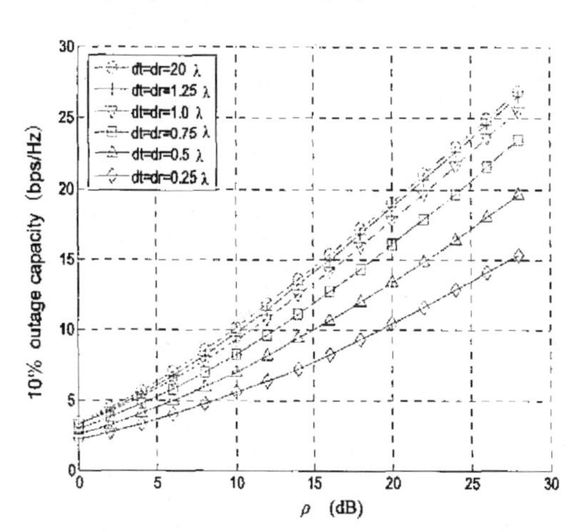

Fig. 4. $C_{0.1}$ versus SNR with different transmit and receive array element spacing

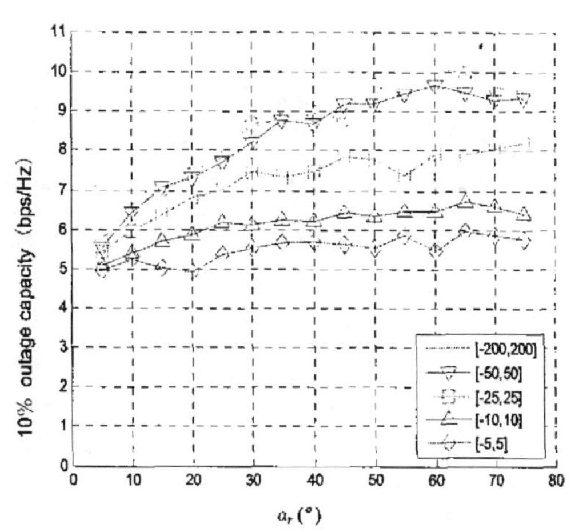

Fig.7. $C_{0.1}$ versus elevation angle spread at receiver with different elevation angle spread at transmitter

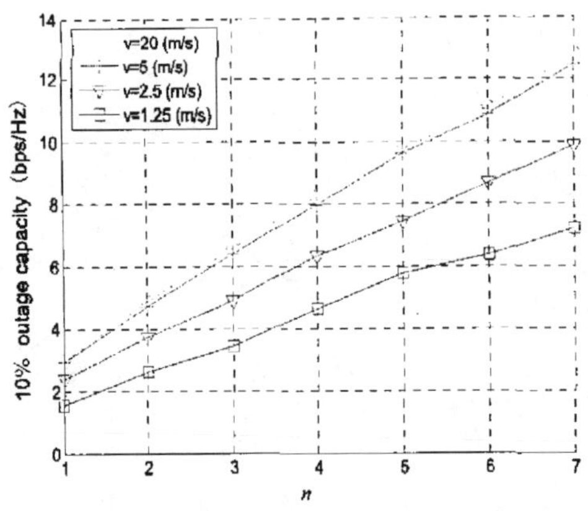

Fig.8. $C_{0.1}$ versus the number of transmitter and receiver element with different transmitter velocity

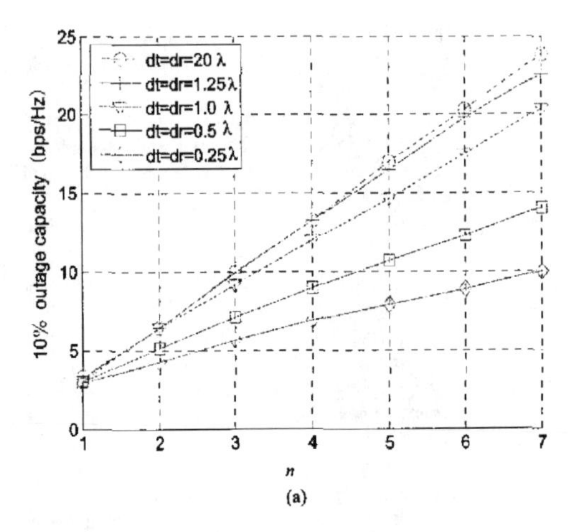

(a)

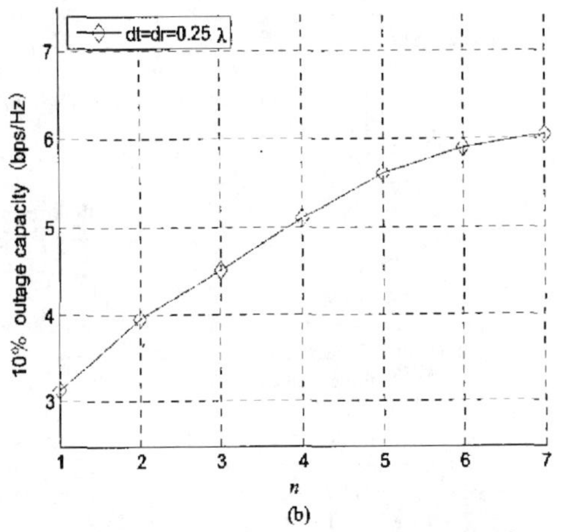

(b)

Fig.9. $C_{0.1}$ versus the number of transmitter and receiver element with different element spacing

of the transmitter or the receiver will increase the outage capacity, and moreover, only when they are increased to sufficiently large in both of the two ends , large capacity gains can be obtained.

In Fig.8, the dependence of $C_{0.1}$ on n with different transmitter velocity is illustrated, here n_T=n_R=n, and the simulation parameters are, $\alpha_t \in [-20^\circ,20^\circ]$, $\alpha_r \in [-20^\circ,20^\circ]$, $\beta_t \in [-10^\circ,10^\circ]$, T=16, ρ=10(dB), dt=0.4λ, dr=0.4λ. Fig.8 shows that with different transmitter velocity, the outage capacity grows almost linearly with the increase of n, and increasing v results in an upwards shift of the curve up to a saturated value, and in the case of Fig.8, the 10% outage capacity of v=5(m/s) has approached almost to the value of v=20(m/s).

In Fig.9, the dependence of $C_{0.1}$ on n with different transmitter and receiver element spacing is described, the parameters of Fig.9(a) are as the follows, $\alpha_t \in [-20^\circ,20^\circ]$, $\alpha_r \in [-20^\circ,20^\circ]$, $\beta_t \in [-10^\circ,10^\circ]$, T=16, ρ=10(dB), v=10 (m/s). It is shown that with different element spacing, the outage capacity will increase with the element number by a different slope, furthermore, when the element spacing are not sufficiently large, the outage capacity will increase not linearly with n, especially for the curve of dt=dr=0.25λ and dt=dr=0.5λ, and this tendency is more obvious in Fig.9(b), where the parameters are the same with that in Fig.9(a), except that $\alpha_t \in [-5^\circ,5^\circ]$, $\alpha_r \in [-5^\circ,5^\circ]$, Fig.9(b) shows that in a not sufficiently rich scattering environment , increasing n when n is large (here, n >6) can only get a small capacity gains. This result is different from the that in [10], where the capacity scales linearly with n in the conditions of different spatial correlation, in that case , with the "two ring" scattering model , the rays are assumed to be distributed in [-π, π) , which is a rich scattering environment . So, we can see that , the scattering environment plays an very important role in the capacity of a MIMO system, in a poor scattering environment, increasing the number of transmitter and receiver element may not get the expected large capacity gain.

V. CONCLUSION

The effect of space-time joint correlation on the underwater acoustic MIMO capacity is studied in this paper. The space-time joint correlation function is derived using a multipath, transmitter moving model, with the angle spread of acoustic rays both vertically and horizontally. The space-time joint correlated variables are generated using vector AR model, by Monte Carlo simulation, the dependence of 10% outage capacity on the SNR with varying parameters of v, dt, dr and elevation angle spread are studied, and the dependence of the outage capacity on the channel parameters at the transmitter and receiver are studied separately, furthermore, the outage capacity dependence on the number of transmitter and receiver element are studied with different space and time correlation conditions.

It is shown form the simulation results that the effect of time correlation on the channel capacity depends on the frame length T, when T is greatly larger than the channel coherence time, the difference between the capacities of different transmitter velocity will disappear, otherwise, capacity will be greatly affected by the transmitter velocity. This is because that channel realizations with large velocity will have smaller temporal correlations, thus bigger outage capacity, and when the frame length T ≫ T_{coh}, the channel realizations will be temporal independent, thus in a average sense, the channel capacities with different transmit velocity will be coincided with each other.

The spatial coherence between the received signals will be affected by the array element spacing and the elevation angle spread, so these parameters will also influence the channel capacity. From the simulation results, it is shown that in a poor scattering environment and not sufficiently large element spacing, large capacity gains can not be obtained, and the MIMO capacity gains depends mainly on the channel scattering conditions, so the rich scattering environment is a precondition where large MIMO capacity can be obtained..

Acknowledgments

The authors are grateful to Geoffrey J. Byers for the useful discussions about the generation of spatially and temporally correlated variables.

REFERENCES

[1] J. H. Winters, J. Salz, and R. D. Gitlin, "The impact of antenna diversity on the capacity of wireless communication system," *IEEE Trans. Commun.*, vol.42, pp.1740-1751, Feb./Mar./Apr. 1994.

[2] I. E .Telatar, "Capacity of Multi-Antenna Gaussian Channels," AT&T Bell Laboratories, Murray Hill, NJ, Tech. Rep.# Bl0 112 170-950-615-07 TM, 1995

[3] G. J .Foschini and M. J. Gans, "On the limits of wireless communications in a fading environment when using multiple antennas," *Wireless Pers. Commun.*, vol.6, no.3, pp. 311-335, 1998.

[4] P. Kyritsi, D. C .Cox, R. A. Valenzuela, and P. W. Wolniansky, "correlation analysis based on MIMO channel measurements in an indoor environment," *IEEE J. Select. Areas Commun.*, vol. 21, pp. 713-720, Jun. 2003.

[5] J. P. Kermoal, L. Schumacher, K. I. Pedersen, P. E. Mogensen, and F. Frederiksen, "A stochastic MIMO radio channel model eith experimental validation," *IEEE J. Select. Areas Commun.*, vol. 20, pp. 1211-1226, Aug. 2002.

[6] C. C. Martin, J. H. Winters, and N. R. Sollenbverger, "MIMO radio channel measurements: Performance comparison of antenna configurations," in *Proc. IEEE VTC '01*, vol. 2, Oct. 2001, pp. 1225-1229.

[7] M. Zatman and B. Tracey, "Underwater acoustic MIMO channel capacity," signals, systems and computers, 2002, conference record of the Thirty-Sixth Asilomar Conference on .volume: 2, 3-6, Nov, 2002, pp. 1364-1368.

[8] D. Shiu, G. J. Foschini, M. J. Gans, and J. M. Kahn, "Fading correlation and its effect on the capacity of multielement antenna systems," *IEEE Trans. Commun.*, vol. 48, pp. 502-513, Mar. 2000.

[9] A. Abdi and M. Kaveh, "A space-time correlation model for multielement antenna systems in mobile fading channels," *IEEE J. Select. Areas Commun.*, vol. 20, pp.550-560,Apr.2002.

[10] G. J. Byers and F. Takawira, "Spatially and temporally correlated MIMO channels: modeling and capacity analysis," *IEEE Trans. Veh. Technol.*, vol. 53, pp. 634-643, May 2004.

[11] W. Jobst and X. Zabalgogeazcoa, "Coherence estimates for signals propagated through acoustic channels with multiple paths," *J. Acoust. Soc. Am.* 65(3), Mar. 1979.

[12] S. M. Kay, *Modern Spectral Estimation*. Englewood Cliffs, N J: PrenticeHall, 1988. pp. 254-256.

[13] K. E. Baddour and N. C. Beaulieu, "Accurate simulation of multiple cross-correlated fading channels," *Proc. IEEE ICC'02*, vol. 1, May 2002, pp. 267-271.

[14] M. A. Khalighi, K. Raoof and G. Jourdain, "Capacity of wireless communication systems employing antenna arrays, a tutorial study," *Wireless Personal Communications* 23: 321-352, 2002.

[15] E. Biglieri, G. Caire, and G. Taricco, "Limiting performance of block fading channels with multiple antennas," *IEEE Trans. Inform. Theory*, vol. 47, pp. 1273-1289, May 2001.

[16] E. Biglieri, J. Proakis and S. Shamai (Shitz), "Fading channels, information-theoretic and communications aspects," invited paper, *IEEE Trans. Inform. Theory*, vol.44, No.6, pp. 2619-2692, 1998.

[17] E .Telatar, "Capacity of Multi-Antenna Gaussian Channels," invited paper, *European Transactions on Telecommunications*, Vol.10, No.6, pp.585-595, 1999.

A Trace Extraction Technique for Fast Moving Underwater Target①

Qihu Li
(Institute of Acoustics, Chinese Academy of Sciences)
P.O.Box 2712, Beijing 100080
People's Republic of China

Abstract

Trace extraction of fast moving underwater target is an interesting topic in passive sonar detection. The application area include harbor protection, torpedo alarming, and so on. The signal processing methods can be used are automatic target tracking (ATT) and target moving analysis (TMA). In this paper, the trace extraction problem is considered as two dimensional image processing problem. The raw data come from the time-bearing history of sonar post processing. The trace of moving targets will be extracted step by step and the random blank pixels and the background noise will be filtered. Some data sieves are established to take away random blank pixels and any accidentally appeared target-like trace. A new algorithm, which is actually a combination of OTA (Order Truncate Average) technique and median filtering, is derived as a data sieve to equalize the background noise. A double check procedure for grazing angle and the change rate of signal to noise ratio is taken to identify the trace of fast moving targets.

1. Introduction

Target moving analysis is one of the interesting topic in passive sonar detection. It has attracted many attention in underwater acoustic field of weak signal detection, target identification and torpedo alarming[1-7].Time/bearing display has been extensively used in digital sonar system and has been proved a very effective technique in target detection. The trace extraction problem can be considered as a specific problem of 2-D image processing. It is different with ordinary image processing, which must concern not only the individual pixel's value, but also the whole picture quality. Any detail information, include color, brightness and contrast have to be taken account in data processing. In TMA, the most important thing is to extract the trace of moving targets. The continuity of the trace is primary concerned. Sometime we have to modify or pad some pixel in the

① Oceans 2005, 1: 7-12.

trace, if necessary. In trace extraction or trace search it focus on the trace of moving target. Any pixels, which is not in a position of trace can be taken away from picture. Some special algorithm has been proposed to deal with trace search, such as subspace algorithm and artificial neural network method. In these algorithm, each trace is considered as a subspace which is unknown exactly before the search procedure finished. There are also some problems in the case of trace crossing, it may be result decision equivocation or lose contact.

In this paper, a top down trace search algorithm is proposed. In any moment the time/bearing display is considered as an image in X-Y plane. Several data sieves are established to filter blank pixel or random noise in whole picture. A background equalization algorithm is derived to eliminate wide range fluctuation, which is typically appear in ocean environment and become serious interference in weak signal detection.

2.The Philosophy of Trace Extraction

As 2-D image processing, the trace extraction in the field of sonar signal processing is quite different with ordinary image processing. The main purpose of trace extraction or trace search focus on the continuous trace of moving target, and then to identify if it is a fast moving target. In this sense, any other pixels appeared in the view window can be taken away regardless its value and position.

There are also some features can be used to modify the trace of moving target. Such as the change of target grazing angle, amplitude variation of beam output in target incidental angle and monotonous increasing or decreasing when target is coming or going. The block diagram of trace extraction of fast moving target is shown in Fig.1. The time/bearing data from multi-beam sonar system is arranged in 2-D image. The time duration of each row is equal to the integration time in post processing, usually is about the magnitude of several seconds. The data processing procedure will be completed in the duration of adjacent sample period T_s. The size of data window is fixed, but the beam data is renewed in duration T_s, and all data row will move one step ahead in this duration.

The real time target trace extraction algorithm consists of

several data sieves. When the multi-beam data pass through these sieves, the target trace will be extracted and the random blank pixels will be removed. And the target trace will be modified according to the trace continuity characteristics. Therefore the features of target trace will be extracted, it include the grazing angle of target and the change rate of input signal to noise ratio (SNR). A double check procedure will be taken to make decision (See Fig.2).

3.Data Sieves and Double Checking Algorithm

Let the time/bearing display data are

$D(kT_s, l\Delta\theta), k=1,\ldots,K; l=1,\ldots,L$

here $\Delta\theta$ is the angle of adjacent beams. The first data sieve is to take away the random pixels, which has no coherence with neighborhood value. In order to simplify the notation, let us write the position function of pixels as $TX(k,l)$ and the amplitude function as $TY(k,l)$, where k represents the order number of rows and l is the order number of columns.

The second data sieve is to choose some " target seed " in the first row, which looks like a contact. Before we doing that, it is necessary to take care the margin point when a local maxima value appears in the end point of this row. As an example, consider the most left point of first row. That is $TX(1,1)$, Put N point in the left of $TX(1,1)$, we have

$TX(1,N), TX(1,N-1),\ldots,TX(1,2), TX(1,1)$

where N is the value of half length of data processing window. Similarly, we put

$TX(1,L-1),\ldots,TX(1,L\_N)$ in the right of $TX(1,L)$. The data length of new row is $L+2N$. The next step of trace search is to track the target. For each target one need to compare the past and future value and make sure if the tracking procedure has to stop. In some cases, a special treatment, called preset tracking subroutine, is necessary. That is two traces of target cross in some point (k_0, l_0), one signal is much stronger than another. In this situation, it is may be lose the weak signal if we continue the old trace search procedure. So we develop a new preset tracking subroutine, which estimate the next target position by least mean square criteria from past data matrix. The third data sieve is to modify the target trace, the position smoothing and amplitude smoothing algorithm are used to get final target trace : $TX'(k,l)$ and $TY'(k,l)$. It is worth to show that most of $TY'(k,l)$ are equal to zero.

4.The Results of Simulation

Based on the algorithm described in previous section, some system simulation have been carried out. The raw data is 128×256 matrix. In the time/bearing display, there are three targets. The scenario is shown in Fig.3. One is fast moving target with fast change of grazing angle, the second target is an ordinary contact, the variation of grazing angle and input SNR are slow, and the third one is a fast approaching target with fast change of input SNR. The background is white Gaussian noise. The result of system simulation is illustrated in Fig.4. The first and third trace are identify as fast moving target.

REFERENCES

1.A.Eriksson et al., " On-line subspace algorithms for tracking moving sources", IEEE Trans. Vol. SP-42 No.9 (1994) 2319-2330

2.C.J.Janffret and C.J.Musso, " New results in target motion analysis", Proc. Of UDT'91, Paris, 239-244

3.D.J.Smith and J.I.Harris, "Line tracking using artificial neural networks and fuzzy inference", Proc. Of UDT'91, Paris 245-252

4.C.D.Bruno, " Torpedo warning receivers", Proc. Of UDT'94, London 44-46

5.Qihu Li et al., "Precise bearing and automatic target tracking in digital sonar", Chinese J. of Acoustics, Vol.15 No.1 (1996) 29-34

6.W.A.Struzinski and E.D.Lowe, " A performance comparison of four noise background normalization scheme proposed for signal detection systems", J.Acoust. Soc.AMer. Vol.76(6) (1984), 1738-1742

7.A.Kumment, " Fuzzy technique implement in sonar systems", IEEE J. of Oceanic Engr. Vol.18 No.4 (1993), 483-490

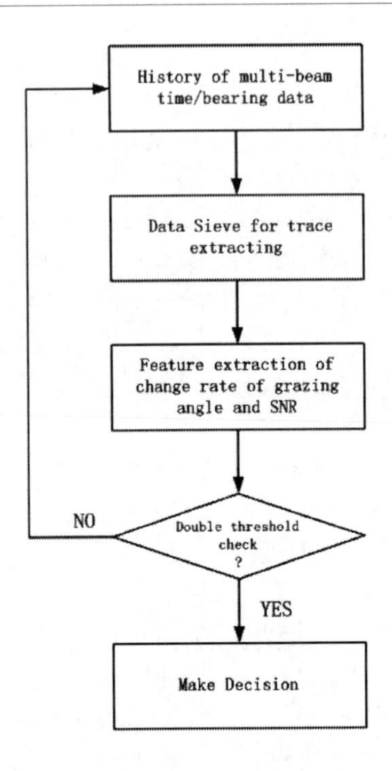

Fig. 1 Block diagram of trace extract of underwater fast moving targets

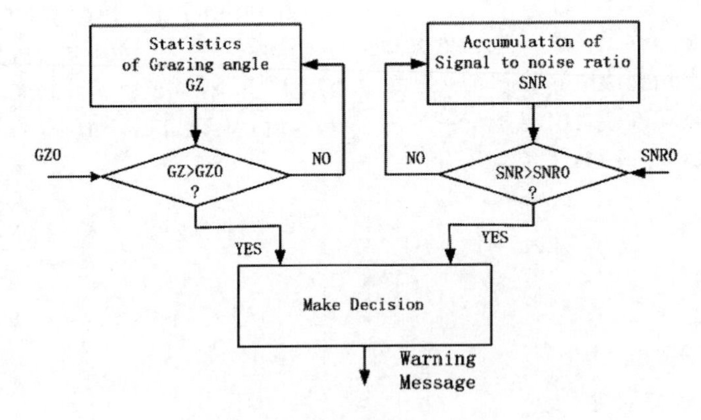

Fig. 2 Double checking algorithm

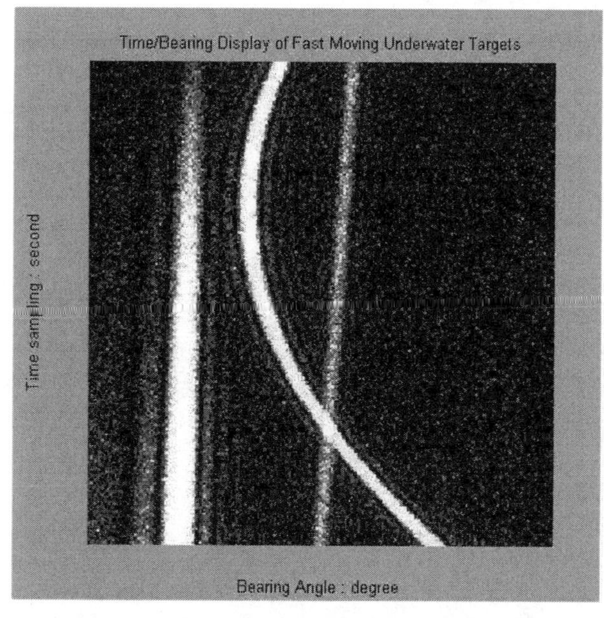

Fig.3 The ting bearing display of three targets

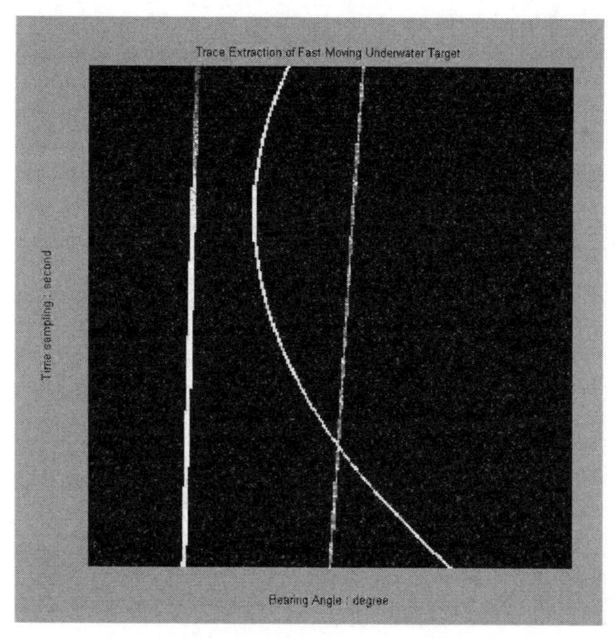

Fig.4 The result of trace extraction

双线列阵左右舷目标分辨性能的初步分析[①]

李启虎
(中国科学院声学研究所　北京　100080)
2006 年 3 月 13 日收到
2006 年 6 月 16 日定稿

摘要　以单水听器阵列构成的单线阵声呐无法区分目标来自左舷或右舷。多线阵为解决左右舷目标的模糊问题提供了一种途径，本文讨论双线列阵左右舷目标分辨的性能，提出把波束指向方向的响应对对称方向的响应的抑制比作为评估左右舷分辨能力的一个参数，并对抑制比的性能作了初步分析，给出了理论表达及实际计算结果。理论分析结果对实际双线列阵设计中的参数选择具有重要意义。
PACS 数：　43.30, 43.60

Preliminary analysis of left-right ambiguity resolution performance for twin-line array

LI Qihu
(*Institute of Acoustics, The Chinese Academy of Sciences* Beijing 100080)
Received Mar. 13, 2006
Revised Jun. 16, 2006

Abstract　It is well known that the single line array sonar can't identify whether the target come from left or right. Multi-line array provides a possible method to solve the left-right ambiguity problem. The performance of left-right resolution for twin-line array is described in this paper. It is shown that the suppression ratio, which is defined as the ratio of the response in the steering direction to opposite symmetrical direction , can be as an index for evaluating the ability of solving left-right ambiguity. The characteristics of suppress ratio is discussed. The theoretical expression for suppression ratio and some numerical results are illustrated. The result of theoretical analysis can be used in design of twin-line sonar.

1　引言

拖线阵早在上世纪初就已引起声呐和海洋开发领域学者的兴趣，特别是近 30 年来，有关的研究报导更加内容丰富 [1-5]。在军事应用领域单线阵难以解决的问题就是所谓的“左右舷模糊”问题，当它探测到目标时，无法判断目标来自声呐平台的左舷或右舷，当然，利用平台的机动可以部分地解决这一问题，但是这不总是可能的。以三基元组构成的单线列阵可以区分目标的左右舷，但一定程度上受到缆的半径和适用频率方面的限制。

多线阵系统为解决左右舷模糊问题提供了一种可能的途径，Zeskind 等已利用多线阵在不同海洋环境中进行多次试验 [2]。但是还没有有关如何评估多线阵区分左右舷目标能力的理论结果。

本文讨论双线阵解决左右舷目标分辨的能力问题，提出利用指向目标方向的系统响应与对称方向的系统响应之比来定量刻划左右舷分辨能力的问题，这个比值我们称之为抑制比。

我们所用的方法虽然是基于双线阵的，但是所遵循的理论分析手段同样适用于多线阵。

2　线阵的指向特性

图 1 给出了等间隔单线列阵的示意图，假定接收基元是 $H_1, H_2, \cdots, H_N$。相邻基元的间隔是 d，把单线阵的基线的法线方向确定为 0° 方向，目标入射信号与法线方向的夹角是 θ。

① 声学学报, 2006, 31(5): 385-388.

这个线列阵的指向性函数是 [6]:

$$D(\theta,\theta_0)=\left|\frac{\sin[N\pi d(\sin\theta-\sin\theta_0)/\lambda]}{\sin[\pi d(\sin\theta-\sin\theta_0)/\lambda]}\right|, \quad (1)$$

其中 λ 为入射信号的波长， θ 为目标方位角 (DOA, Direction of Arrival), θ_0 为声呐波束的指向角。法线右侧，即顺时针方向，θ 为正，法线左侧，即逆时针方向，θ 为负。假定我们认为 H_1 是该线列阵所在平台的船艏方向，那么 $|\theta|\leq\pi/2$ 时，目标处于右舷，当 $|\theta|\geq\pi/2$ 时，目标处于左舷。

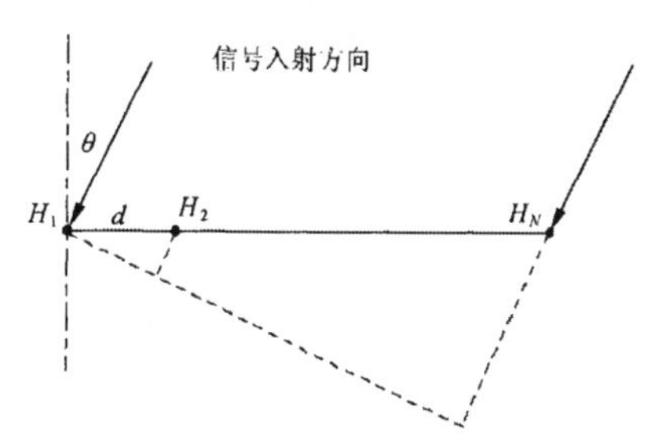

图 1 单线阵指向性计算

从 (1) 式可以看到，无论 θ 为何值，

$$D(\theta,\theta_0)=D(\pi-\theta,\theta_0). \quad (2)$$

也就是说，当目标位于右舷 ($\theta<\pi/2$) 时，与它左舷对称方向 ($\pi-\theta$) 具有相同的指向性响应。这是因为 $\sin\theta=\sin(\pi-\theta)$。这就是所谓左右舷模糊问题，它是单线阵声呐固有的特性。

对于多波束声呐，θ_0 可以在 $-\pi$ 至 π 之间取值。但是由于 $\sin\theta_0=\sin(\pi-\theta_0)$, 我们同样可以发现

$$D(\theta,\theta_0)=D(\theta,\pi-\theta_0). \quad (3)$$

换句话说，我们无须在 $-\pi$ 至此 π 之间形成多波束，而只须在 $-\pi/2<\theta_0<\pi/2$ 之间形成波束即可。这对于软硬件的开销当然是一个可观的节省。但是我们前面已看到，这种节省的代价是左右舷模糊问题的出现。

当然，单线阵的左右舷模糊问题在实际的操作中也是可以解决的。比如拖曳平台的机动就可以为左右舷目标的判断提供依据，假定发现了目标，我们可以让拖曳平台机动，例如往右转。这时，如果目标在左舷，波束最大值响应应向 $|\theta|$ 偏大的方向变化，如果目标在右舷，波束最大值响应应向 $|\theta|$ 变小的方向变化，但是这种方法在使用时受到各种条件的限制。比如，出现多目标时，指向性的多个极大值会妨碍做出正确判断；拖曳平台不允许机动等等。

多线阵声呐为解决左右舷模糊问题提供了一个途径。

我们首先研究双线列阵的情况，见图 2, 这是一个由两条互相平行的线列阵构成的双线列阵，阵与阵之间的距离是 d_0 。两条线列阵的基元分别记做 $H_1,\cdots,H_N$ 和 $Q_1,\cdots,Q_N$, 相邻阵元之间的距离为 d 。可以证明，这样配置的双线列阵的指向性函数为:

$$E(\theta,\theta_0)=\left|\frac{\sin[N\pi d(\sin\theta-\sin\theta_0)/\lambda]}{\sin[\pi d(\sin\theta-\sin\theta_0)/\lambda]}\right| \\ \sqrt{2}\left\{1+\cos\left[\frac{2\pi d_0}{\lambda}(\cos\theta-\cos\theta_0)\right]\right\}^{1/2}\sqrt{2}. \quad (4)$$

(4) 式中的因子

$$F(d_0,\theta,\theta_0)=\left\{1+\cos\left[\frac{2\pi d_0}{\lambda}(\cos\theta-\cos\theta_0)\right]\right\}^{1/2}\sqrt{2}. \quad (5)$$

实际上就是偶极子 H_iQ_i 的指向性。从 (4) 式和 (5) 使可知:

$$E(\theta,\theta_0)=D(\theta,\theta_0)F(d_0,\theta,\theta_0). \quad (6)$$

这实际上就是乘积定理 [6]:

$$D(\theta,\theta_0)=\left|\frac{\sin[N\pi d(\cos\theta-\cos\theta_0)/\lambda]}{\sin[\pi d(\cos\theta-\cos\theta_0)/\lambda]}\right| \\ \{2[1+\cos(2\pi d_0(\sin\theta-\sin\theta_0)/\lambda)]\}^{1/2}$$

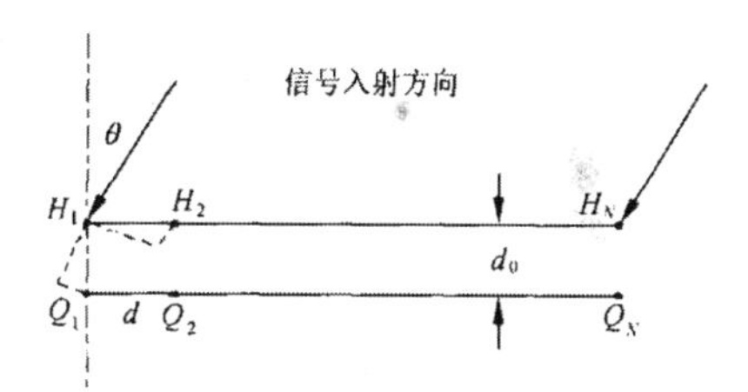

图 2 双线列阵指向性计算

注意 $F(d_0,\theta,\theta_0)$ 之极大值为 2, 也就是说在最好的情况下, $2N$ 个基元的双线列阵声呐的增益是 N 个基元单线列阵声呐的两倍。

一般来说，当 θ 变为 $\pi-\theta$ 时,

$$F(d_0,\pi-\theta,\theta_0)= \\ \sqrt{2}\left\{1+\cos\left[\frac{2\pi d_0}{\lambda}(\cos\theta+\cos\theta_0)\right]\right\}^{1/2}\neq F(d_0,\theta,\theta_0). \quad (7)$$

换句话说，左右舷目标有不同的响应值，从而为解决左右舷模糊问题提供了可能。

3 抑制比的计算

仔细分析 (4) 式和 (5) 式可以发现，双线阵的指向特性是与 d, d_0, λ 等参数密切相关的，尤其是偶极子的指向性函数 $F(d_0,\theta,\theta_0)$ 对参数 d_0 非常敏感。左

右舷目标的分辨力在很大程度上依赖于 d_0。图 3 给出了以 d_0/λ 为参数的偶极子归一化指向性图。我们看到当 $d_0/\lambda = 0.25$ 时线阵区分左右舷目标的能力会相当好。

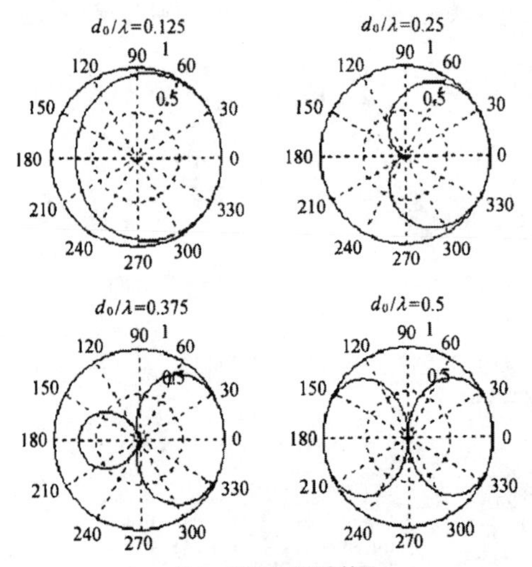

图 3 偶极子的指向性图

如何从数量上描述双线阵对左右舷目标的分辨能力呢？我们自然想到利用 $E(\theta,\theta_0)$ 对 $E(\pi-\theta,\theta_0)$ 的比值来定义这种能力，这个值我们称之为抑制比，即：

$$\mu(\theta) = \frac{E(\theta,\theta_0)}{E(\pi-\theta,\theta_0)} = \frac{F(d_0,\theta,\theta_0)}{F(d_0,\pi-\theta,\theta_0)}. \tag{8}$$

对于单线阵的情况，我们已指出：

$$\frac{D(\theta,\theta_0)}{D(\pi-\theta,\theta_0)} = 1. \tag{9}$$

对于双线列阵，当 $\theta_0 = \theta$(即对准目标时)

$$\mu_0(\theta) = \frac{\sqrt{2}}{\left[1+\cos\left(\dfrac{4\pi d_0}{\lambda}\cos\theta\right)\right]^{1/2}}, \tag{10}$$

这是 d_0 和 θ 的函数。

我们期望在很宽的范围内，$\mu_0(\theta)$ 越大越好，如果 $\cos[(4\pi d_0/\lambda)\cos\theta] = -1$，则 $\mu_0(\theta) = \infty$。

不难证明，如果 $d_0 < \lambda/4$，则不存在抑制比为 ∞ 的解，如果 $d \geq \lambda/4$，则 $(4d_0/\lambda) \geq 1$

当
$$\theta = \cos^{-1}\left(\frac{\lambda}{4d_0}\right) \tag{11}$$

时，抑制比 $\mu_0(\theta) = \infty$

对任何选定的参数 d_0，λ，抑制比也就确定了。我们希望在一个较宽的范围内 $\mu_0(\theta) \gg 1$。

图 4 给出了当 $d_0/\lambda = 0.25$ 时，双线列阵的指向特性，我们分别计算了 $\theta = 0°, 30°, 45°, 60°$ 的情况，从图中可以发现当 θ 大于 60° 时指向特性较差，抑制比较小。

图 5 给出了 $d_0 = \lambda/4$ 时抑制比的情况，图上画出的是 $20\log\mu_0(\theta)$。即抑制比的分贝值。当 $\theta = 1°$ 时，抑制比高达 72 dB。从图中可以看出，当 θ 超过 40° 抑制比已降到 10 dB 之下。

图6给出了 $d_0 = \lambda/8$, $d_0 = \lambda/2$, $d_0 = \lambda$ 的情况。我们看到在这些参数下，双线阵的抑制比并不好，因而从声呐设计角度来说是不可取的。

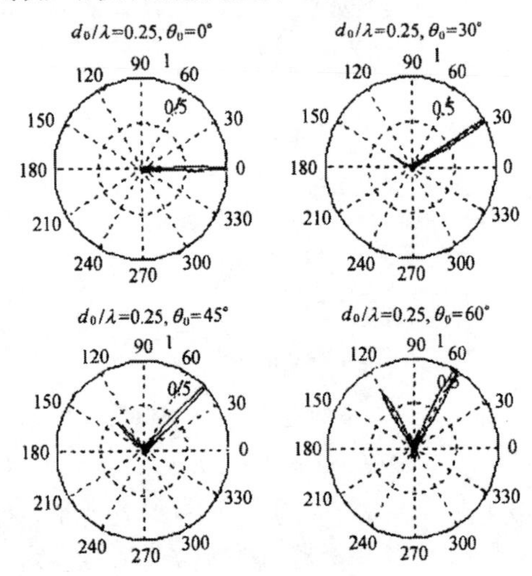

图 4 双线阵的指向性图

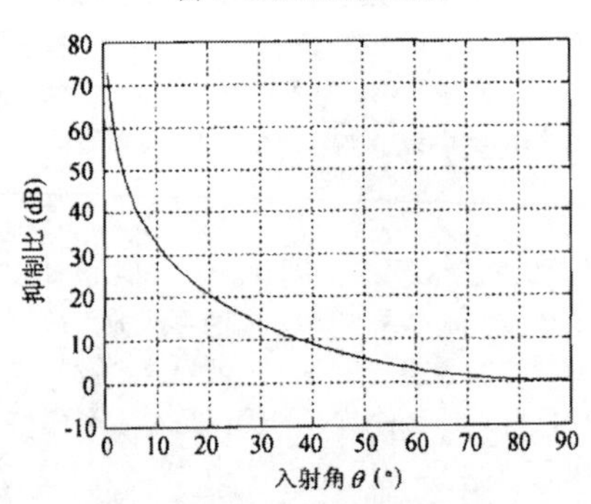

图 5 抑制比随信号入射角的变化 ($d_0 = \lambda/4$)

4 结论

双线阵为解决左右舷模糊问题提供了可能，如果参数选择正确，左右舷目标的抑制比将在一个很宽的范围内满足要求，因而是可行的。

由于多线阵指向性计算也满足乘积定理，所以本文所提供的分析方法完全可以推广至多线阵上去，只需把 (b) 式中的 $F(d_0,\theta,\theta_0)$ 更换成合适的垂直线阵指向性即可。

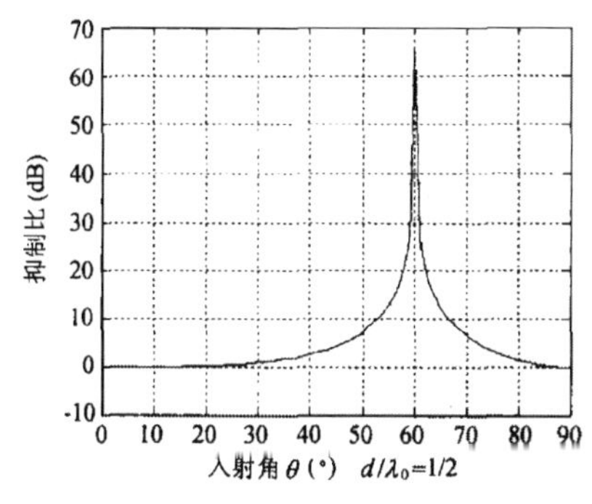

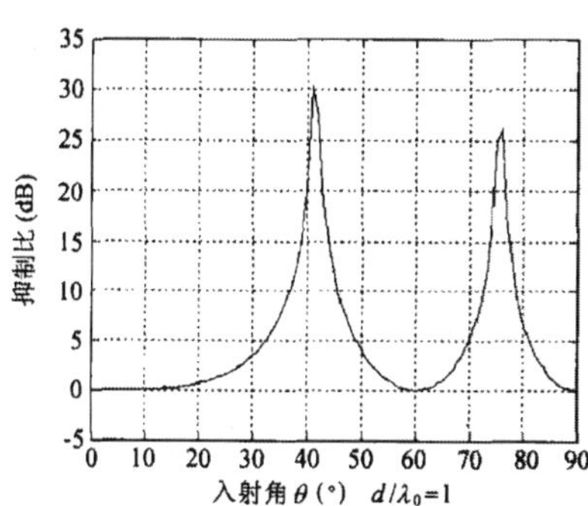

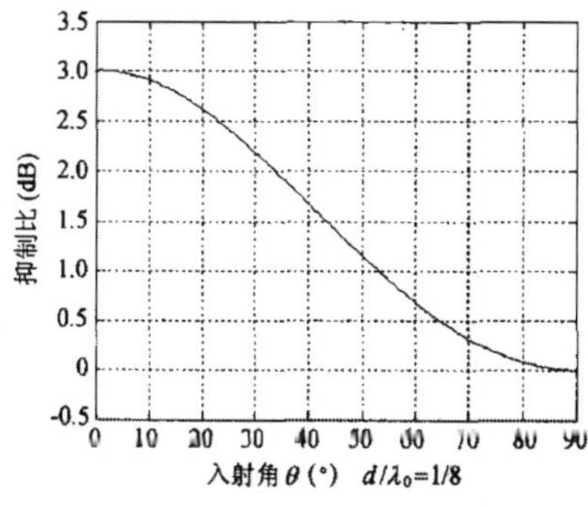

图 6 抑制比和参数 d_0/λ 的关系

参 考 文 献

1 Lemon S G. Towed-array history: 1917-2003. *IEEE J. of Oceanic Engr.*, 2004; **29**(2): 365—373

2 Zeskind R M *et al.* Acoustic performance of a multi-line system towed in several ocean environments. *IEEE J. of Oceanic Engr.*, 1998; **23**(1): 124—128

3 Kaouri K. Left-right ambiguity resolution of a towed array sonar. Univ. of Oxfod, 2000

4 Bucker H. Least-square target detection for a twin-line towed array. *J. Acoust. Soc. Am*, 1994; **95**(3): 1669—1670

5 Lasky M *et al.* Recent progress in towed hydrophone array research. *IEEE J. of Oceanic Engr.*, 2004; **29**(2): 374—387

6 李启虎. 数字式声呐设计原理. 安徽: 安徽教育出版社, 2002

用双线列阵区分左右舷目标的延时估计方法及其实现[①]

李启虎
(中国科学院声学研究所 北京 100080)
2006 年 3 月 20 日收到
2006 年 8 月 30 日定稿

摘要 双线列阵声呐可以利用自身的指向特性区分左右舷目标，但是其性能受线阵间距的约束，且存在分辨盲区。本文提出利用两个单线阵接收信号的时延差区分左右舷目标的方法。该方法对线阵间距具有宽容性，并且盲区明显减小。给出了理论分析结果和实施技术方法，提出了基于内插法的延时精确估计方法，从而可以解决左右舷目标的分辨问题。
PACS 数： 43.30, 43.60

The time-delay estimation method of resolving left-right target ambiguity for twin-line array and its realization

LI Qihu
(*Institute of Acoustics, The Chinese Academy of Sciences* Beijing 100080)
Received Mar. 20, 2006
Revised Aug. 30, 2006

Abstract The directivity function of twin-line array sonar can be used to identify left/right target. But the performance is limited by the distance between two lines and also exists the blind area for target identification. A method, which is based on the time-delay estimation between two line arrays, is proposed to identify left/right target. The method is robust with the variation of distance between two lines, the blind area is considerably decreased. The theoretical analysis results and realization method is described. An algorithm based on interpolation method for precisely estimating time-delay of two line array are derived, therefore the left-right ambiguity problem is solved.

引言

拖曳线列阵声呐的应用引起了军民两个领域研究者的广泛兴趣[1-3,7]，但是单线列阵系统存在左右舷模糊问题，所以有关多线阵声呐的理论工作自然就格外引人注目。我们在文献 4 中已讨论了利用双线阵自身的指向特性解决左右舷模糊的问题，指出左右舷对称方位的抑制比是一个较好的用于描述系统左右舷分辨能力的参数。

但是，由于抑制比本身对双线阵之间距离非常敏感，所以用双线阵指向特性来区分左右舷目标可以说对环境特性是不宽容的。此外，我们还要指出，即使已知双阵之间的距离，左右舷分辨也存在较大的盲区。

本文提出一种新的方法，它是基于对双线阵之间延时差的测量的一种左右舷判决法。这种方法不需要预先知道双线阵之间的距离，因而其性能对环境是较宽容的。

我们给出了理论分析的结果，同时对于实施这种理论技术给出实际的方法，这样就可以由数字式声呐的波束成形系统得出精确的延时估计，从而解决目标左右舷分辨问题。

1 双线列阵区分左右舷目标的盲区

图 1 所示是一个双线列阵，两条线列阵 H_1, H_2, $\cdots$, H_N 和 Q_1, Q_2, $\cdots$, Q_N 相互平行，线列阵之间的间距是 d_0 。等间隔排列的线列阵基元之间的间隔是 d 。我们已经指出[4]，这个线列阵的指向性函数是：

$$E(\theta,\theta_0)=\left|\frac{\sin[N\pi d(\sin\theta-\sin\theta_0)/\lambda]}{\sin[\pi d(\sin\theta-\sin\theta_0)/\lambda]}\right|\times \sqrt{2}\left\{1+\cos\left[\frac{2\pi d_0}{\lambda}(\cos\theta-\cos\theta_0)\right]\right\}^{1/2}. \tag{1}$$

---

① 声学学报, 2006, 31(6): 485-487.

出现目标的那个方向 θ 的系统响应与有可能被误判为目标的那个方向 $\pi-\theta$ 的系统响应之比是:

$$\mu(\theta)=\frac{\mathrm{E}(\theta,\theta_0)}{\mathrm{E}(\pi-\theta,\theta_0)}=\frac{F(d_0,\theta,\theta_0)}{F(d_0,\pi-\theta,\theta_0)},\tag{2}$$

其中:

$$F(d_0,\theta,\theta_0)=\sqrt{2}\left\{1+\cos\left[\frac{2\pi d_0}{\lambda}(\cos\theta-\cos\theta_0)\right]\right\}^{1/2}\tag{3}$$

当 $\theta_0=\theta$, 即声呐系统指向目标时,

$$\mu_0(\theta)=\frac{\sqrt{2}}{\left[1+\cos\left(\frac{4\pi d_0}{\lambda}\cos\theta\right)\right]^{1/2}},\tag{4}$$

当 $d_0=\lambda/4$ 时,

$$\mu_0(\theta)=\frac{\sqrt{2}}{[1+\cos(\pi\cos\theta)]^{1/2}}.\tag{5}$$

在 $[0,\pi/2]$ 范围内, $\mu_0(\theta)$ 是 θ 的单调下降函数, 在 $\theta=0$ 时, $\mu_0(\theta)=\infty$ 。抑制比到达无穷大。把 $\mu_0(\theta)$ 的对数值, 记做 SR:

$$SR=20\log\mu_0(\theta).\tag{6}$$

如果把 $SR\geq 6$ dB 作为可以接受的指标的话, 当目标出现在 [0,50°] 内时, 系统达到指标要求, 但在 [50°, 90°] 范围内, 系统仍无法区分左右舷目标, 从而成为盲区。

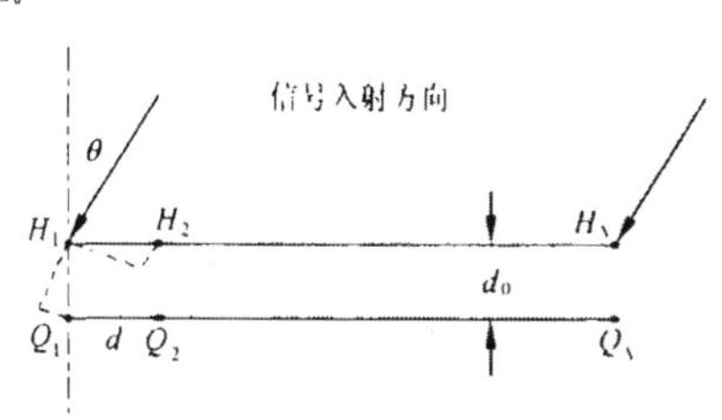

图 1 双线阵指向性计算

盲区的存在, 不仅仅出现在 $d_0=\lambda/4$ 的情况, 当 $d_0<\lambda/4$ 时 $\mu_0(\theta)$ 虽然仍可保证单调下降的特性, 但 $\mu_0(\theta)$ 的值偏小, 抑制比太小而不能满足要求; 当 $d_0>\lambda/4$ 时, $\mu_0(\theta)$ 不再是 θ 的单调函数, 在 $[0,\pi/2]$ 内可能出现一个峰或多个峰值, 这也无法满足要求。

换句话说, 双线列阵如果只利用自身的指向性函数区分左右舷目标, 则由于指向性函数对 d_0 非常敏感并且存在左右舷分辨的盲区, 因此这种方法对环境是不宽容的。

2 利用双线阵延时差区分左右舷目标

双线列阵中的两条线阵在接收目标信号时有一定的延时差, 利用这个延时差就可判断目标来自左舷还是右舷。图 2 是一个示意图, 设目标方位是 θ, 两条线阵的波束成形系统都对准目标, 那么这两个对准波束信号的延时差为:

$$\tau=d_0\cos\theta/c,\tag{7}$$

其中 c 为声速。

假定 H_N, Q_N 为船艏方向, 那么当 $\tau\geq 0$ 时, 目标来自左舷, 当 $\tau\leq 0$ 时目标来自右舷, 从理论上来说, 只要 τ 估计精确, 就能正确区分左右舷。从 (7) 式可以看到, 当 θ 接近 $\pi/2$ 时, τ 的值接近 0 。所以只有当系统能把很小的延时差估计出来时, 左右舷分辨才不存在盲区, 同时延时估计不必事先知道 d_0, 因此这种方法对环境具有很好的宽容性。

图 3 是以 d_0 为参数, 两条线列阵之间的时延值, 我们看到, 当 $d_0=1.5$ m, 入射角 $\theta=80°$ 时, 延时值为 173 μs 。所以延时估计的精度应至少在 10 μs 量级。

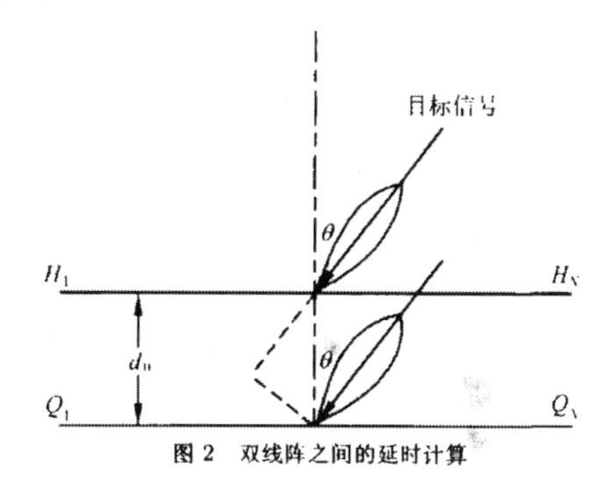

图 2 双线阵之间的延时计算

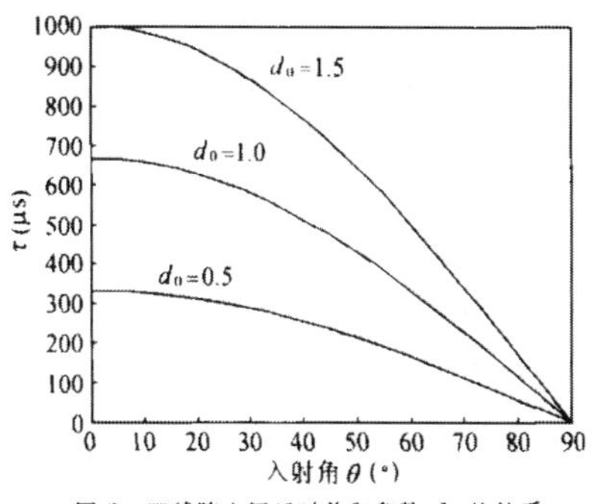

图 3 双线阵之间延时差和参数 d_0 的关系

3 用时延估计区分左右舷目标的系统设计

根据上面的讨论, 我们在图 4 给出了用延时测量区分左右舷目标的方框图。

左基阵和右基阵分别形成自己的波束, 并取同一个方向的波束成形信号。用多路相关器计算不同延时间隔的相关函数值。由于数字式波束成形系统

的采样间隔是量化的，所以为了给出精确的延时值还需作进一步的处理，我们在下面专门来讨论这个问题。

根据所测延时值的正负就可以判别目标来自左舷还是右舷。

作延时估计时延迟线的长度应能覆盖可能出现的最大延时时间。

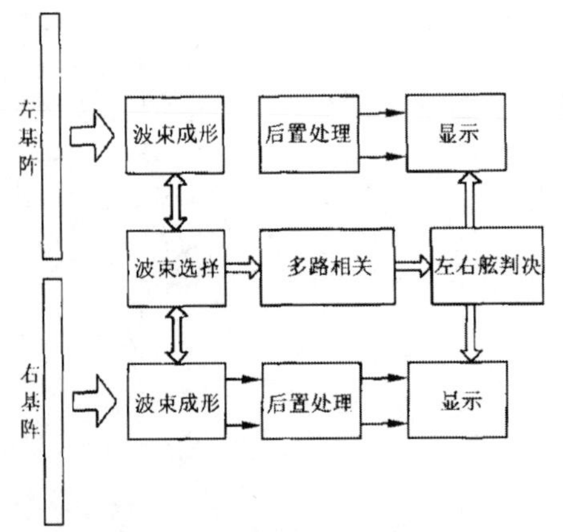

图 4　全波束相关求延时原理方框图

举例来说，如果 $d_0=1.5$ m，则 $\max(\tau)=1000\ \mu s$。实际上我们应把能够测量的延时间隔取得比 1000 μs 稍大一些，因为拖曳线列阵在拖曳过程中不可能严格保持 $d_0=1.5$ m 不变。

另一个重要的现实问题是延时估计的精度问题。

举例来说，如果采样频率为 $f_s=10$ kHz，那么采样周期 $T_s=100\ \mu s$。若直接利用间隔为 100 μs 的相关函数输出值的大小来判别左右舷，显然是不准确的。为此，我们可以采用内插的方法，用它可以得到精确的延时估计值。

假定我们计算了 M 个相关值：

$$R(kT_s),\quad k=0,\cdots,M-1 \tag{8}$$

我们要以这一组值计算出相关函数 $R(\tau)$ 的最大值的位置的估计值[5]。

在 $R(kT_s)$ 中先求得一个最大的值，比如 $R(iT_s)$，实际相关函数的峰值应在

$$\tau_0=iTs+\Delta\tau, \tag{9}$$

取得，其中 $\Delta\tau$ 可以用抛物线内插的方法求出，见图 5。设 $Y_0=R(iTs)$，$Y_1=R[(i+1)Ts]$，$Y_{-1}=R[(i-1)Ts]$，令

$$\Delta\tau=\frac{Y_1-Y_{-1}}{4Y_0-2Y_1-2Y_{-1}}Ts. \tag{10}$$

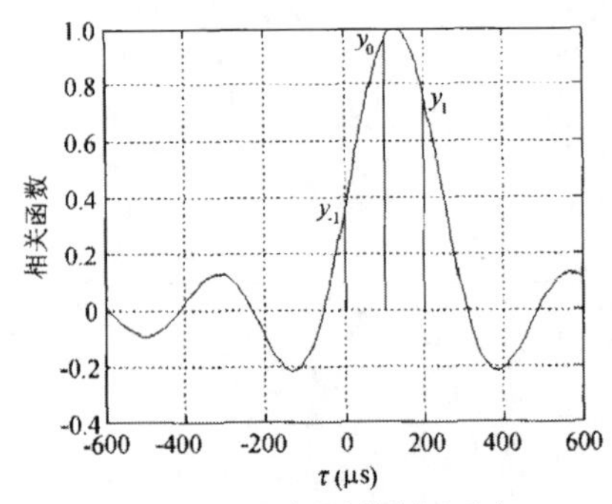

图 5　用内插法精确估计延时

从而

$$\tau_0=iTs+\frac{Y_1-Y_{-1}}{4Y_0-2Y_1-2Y_{-1}}Ts. \tag{11}$$

根据 τ_0 的值就可以判断目标来自左舷还是右舷。

由于相关函数的计算不受入射角 θ 的影响，所以用这种办法判断左右舷目标时盲区会比较小。但是当入射角明显偏向端射方向时，信噪比会下降，这时对相关函数输出的峰值会有影响。

这种影响目前还无法消除，因为当目标出现在偏离法线方向 θ 角的时候，主辨方向的宽度是和 $1/\cos\theta$ 成正比的，θ 越大 $1/\cos\theta$ 也就越大，从而使延时估计的精度下降。

4　结论

用双线列阵指向特性区分左右舷目标的方法对线阵间隔 d_0 较敏感，且存在盲区。

用估计双线列阵之间延时差的办法来判断左右舷目标是可行的。并且对使用环境具有宽容性。

参　考　文　献

1　Lemon S G. Towed-array history. 1917-2003', *IEEE J. of Oceanic Engr.*, 2004; **29**(2): 365—373

2　Kaouri K. Left-right ambiguity resolution of a towed array sonar. Univ. of Oxford, 2000, http://eprints.maths.ox.ac.uk

3　Bucker H. Least-square target detection for a twin-line towed array. *J. Acoust. Soc. Am.*, 1994; **95**(3): 1669—1670

4　李启虎. 双线列阵左右舷目标分辨性能的初步分析. 声学学报, 2006; **31**(5): 385—388

5　李启虎. 数字式声呐设计原理. 安徽: 安徽教育出版社, 2002

6　何心怡等. 基于时延估计的双线阵左右舷分辨技术研究. 信号处理, 2003; **19**(4): 338—342

7　李启虎, 李淑秋, 孙长瑜, 余华兵. 主被动拖线阵声呐中拖曳平台噪声和拖鱼噪声在浅海使用时的干扰特性. 声学学报, 待发表

本舰机动左右舷分辨方法研究①

何心怡[1]　张春华[2]　张　驰[3]　李启虎[2]

(1 海军装备研究院　北京 100073)

(2 中国科学院声学研究所　北京 100080)

(3 海军鱼雷工程办公室　西安 710075)

摘要　针对传统单根线列阵的左右舷分辩问题，经过深入分析，给出了本舰机动左右舷分辨方法中的一些影响因素，并对机动角度和滞后时间得出了定量结论．文中研究的内容对传统拖曳线列阵声纳的左右舷分辨和战术使用具有参考价值．

关键词　本舰机动，单根线列阵，左右舷分辩

Study on the port/starboard discrimination by mother ship maneuverability

HE Xin-Yi[1]　ZHANG Chun-Hua[2]　ZHANG Chi[3]　LI Qi-Hu[2]

(1 Naval Academy of Armament, Beijing 100073)

(2 Institute of Acoustics, Chinese Academy of Sciences, Beijing 100080)

(3 Naval Torpedo Engineering Office, Xi'an 710075)

Abstract　Aimed at the port/starboard discrimination problem of the traditional single-line array, this paper mentioned some relative factors of the port/starboard discrimination method of mother ship maneuverability, and draws quantitative conclusions about the maneuvering angle and the time lag. This paper may be importance to the port/starboard discrimination and its tactics for the traditional towed linear array sonar.

Key words　Mother ship maneuverability, Single-line array, Port/starboard discrimination

① 应用声学, 2006, 25(6): 352-358.

1 引言

拖曳线列阵声纳是现代声纳的重要发展方向之一，其中，相当多的拖曳线列阵声纳是采用传统的单根线列阵作为基阵，以被动方式警戒远距离的水下目标，如广泛装备于外军多型巡洋舰、驱逐舰的 H/SQR-19A 拖曳线列阵声纳。而左右舷模糊问题是传统单根线列阵的主要缺点之一，传统单根线列阵通常以本舰机动的方法解决左右舷模糊问题，但有关本舰机动左右舷分辨方法的定性和定量的分析却鲜见报道。有鉴于此，本文深入分析了本舰机动左右舷分辨方法，给出了它的具体实施方案以及机动角度和滞后时间的定量结论，并且指出该方法是无法有效完成拖曳线列阵声纳的鱼雷报警功能。文中的研究内容对传统拖曳线列阵的左右舷分辨和战术使用具有参考价值。

2 本舰机动左右舷分辨方法的实施方案

如果目标出现在左 (右) 舷 ϕ, 由于传统的单根线列阵的轴向对称性，那么在右 (左) 舷 ϕ 处也将出现一个等强度的镜像源，这就是传统拖曳线列阵声纳的左右舷模糊。经过分析，本舰机动左右舷分辨方法可以这样实现: 首先，确保拖曳线列阵声纳发现并跟踪上目标；其次，本舰机动，朝左 (右) 舷方向转弯 $\Delta\phi$ 角度；最后，根据机动后目标舷角的变化趋势进行判断：如果目标舷角变小，变为 $\phi-\Delta\phi$, 则目标在左 (右) 舷；如果目标舷角变大，变为 $\phi+\Delta\phi$, 则目标在右 (左) 舷。

仿真场景：拖曳线列阵航向 30°, 以 15kn 的速度拖曳航行；远场目标航速 8kn, 航向 210°, 舷角为 40°。发现目标后，拖船以 1°/s 的转弯速率向右舷改变航向，假定拖曳线列阵声纳的每次测量都可获得目标的精确方位，那么目标分别位于左舷和右舷时的目标舷角变化趋势分别如图 1(a) 和图 1(b) 所示。

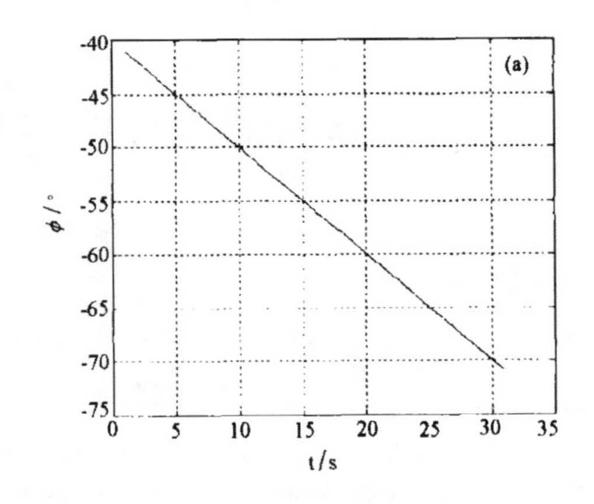

(a) 目标位于左舷时

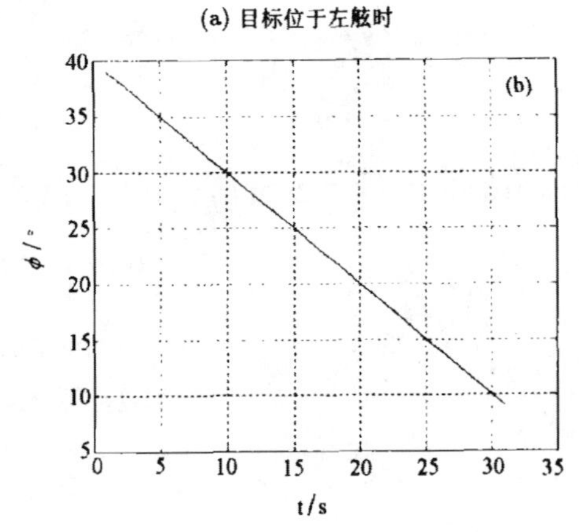

(b) 目标位于右舷时

图 1　目标位于左舷或右舷时目标舷角的变化趋势

3 本舰机动左右舷分辨方法的定量分析

3.1 影响本舰机动左右舷分辨方法的因素

对上述本舰机动左右舷分辨方法，我们认为，有两个方面应深入分析：一是本舰机动应该机动多大的角度；二是本舰机动左右舷分辨方法的实时性。

在本舰机动过程中，拖曳线列阵声纳并不是在每次测量中都可获得目标的精确舷角，目标舷角的测量值 $\hat{\phi}$ 与目标的真实舷角、目标的信噪比和拖曳线列阵的阵形等因素有关，可将测量出的目标舷角 $\hat{\phi}$ 的标准偏差 $\sigma_{\hat{\phi}}$ 表示为:

$$\sigma_{\hat{\phi}} = Function\{\phi, SNR, a_i\} \quad (1)$$

式中：ϕ 是目标的真实舷角，SNR 是目标的信噪比，a_i 表示拖曳线列阵的实际阵形。

(1) 考虑目标真实舷角对目标舷角估计的标准偏差 $\sigma_{\hat{\phi}}$ 的影响。当 $L \gg \lambda$ 时 (L 是线列阵的声学孔径长度，λ 是接收信号的波长)，线列阵的主瓣宽度 BW 可近似表示为：

$$BW \approx \left|\frac{0.88\lambda}{L} \cdot \frac{1}{\cos\alpha}\right| = \left|\frac{0.88\lambda}{L} \cdot \frac{1}{\sin\phi}\right| \quad (2)$$

上式中 α 表示目标与线列阵法线的夹角，可知：在相同信噪比下，目标越靠近线列阵的正横方向，线列阵的主瓣宽度 BW 越小，即线列阵对目标的方位估计精度越高，也就是 $\sigma_{\hat{\phi}}$ 越小。

(2) 其它条件相同时，信噪比 SNR 越大，则 $\sigma_{\hat{\phi}}$ 越小。

(3) 拖曳线列阵的阵形畸变增大了 $\sigma_{\hat{\phi}}$。

(1) 和 (2) 两点，也就是目标的真实舷角和信噪比对方位估计精度的影响，可由拖曳线列阵声纳的测向误差 $\Delta\phi_1$ 表示，可以认为：无论目标的信噪比是多少，只要目标的信噪比大于检测所需的信噪比门限，那么目标的测向误差将不超过 $\Delta\phi_1$，这是由声纳系统本身所决定的。

经过以上分析可知，阵形畸变是影响本舰机动左右舷分辨方法的关键因素。可将本舰机动过程分为两个阶段：一是基阵未稳定阶段，即本舰仍处于机动中或机动后但拖线阵的基阵还未稳定时，此时基阵 (线列阵) 存在阵形畸变；二是基阵稳定阶段，即本舰机动完成，并且基阵也恢复到机动前的未畸变状态。在基阵稳定阶段，不存在因阵形畸变带来的额外的方位估计误差。因此，下一小节将重点分析在基阵未稳定阶段，即存在阵形畸变时，对目标方位估计的影响。

3.2 阵形畸变对本舰机动左右舷分辨方法的影响分析

传统的单根线列阵信号处理通常有以下三个假设：(1) 线列阵是由无指向性的水听器构成的直线阵，并且所有水听器位于同一水平面；(2) 目标相对于线列阵来说是呈远场关系，并且以平面波传播；(3) 在积分时间内，远场目标相对于线列阵来说是固定的。

为简化分析，假设以 $h(t)$ 表示拖船的航向，$h(t)$ 的反向延长线为 x 轴，线列阵的轴线在 x 轴上，以阵形未畸变时离拖船最近的水听器位置为坐标原点 O，xoy 平面是包含线列阵的水平面。线列阵中水听器数目为 M，依次编号为 $(1, 2, \cdots, M-1, M)$，无畸变情况下相邻两个水听器的间距为 d。那么用 a_i 表示各水听器的真实位置，用 $\hat{a}_i = ((i-1)\cdot d, 0)$ 表示无畸变时的水听器位置。

在线列阵的拖曳过程中，由于拖船的横向运动、海流、流体动力等方面影响，阵形畸变是无时不在的，只是程度的差别。经过分析，将本舰机动时线列阵的阵形畸变归纳为以下模型：在拖船的转弯过程中，产生了相对于阵列尺寸的大直径转弯机动，此时阵形畸变成圆弧阵[1]。

在直角坐标系中画出这种阵列形状，如图 2 所示，舷角 ϕ 如图中所示，定义为 x 轴的负向与入射信号方向的夹角，也可称之为水平方

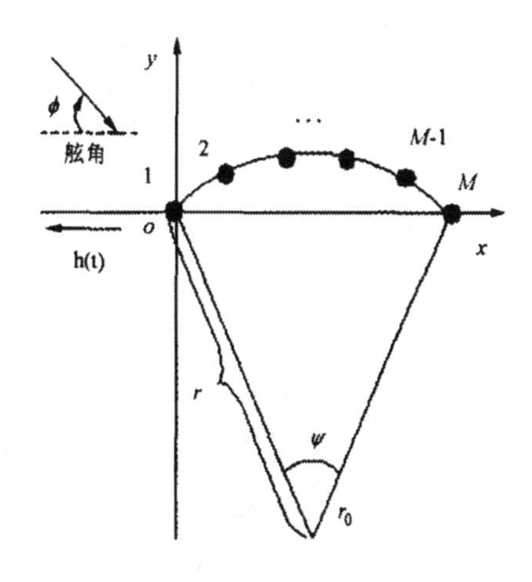

图 2　本舰机动过程中，线列阵畸变成圆弧阵示意图

位角，其中，顺时针为正，代表右舷舷角；逆时针为负，代表左舷舷角。

图 2 中线列阵的各水听器均匀分布在张角为 ψ (单位为弧度)，长为 $L=(M-1)\cdot d$ 的圆弧上，则圆弧半径为 $r=L/\psi$。图中各水听器位置可表示为：

$$a_i = \begin{pmatrix} 2r\sin\left[\frac{i-1}{2(M-1)}\psi\right]\cos\left[\frac{\psi}{2}-\frac{i-1}{2(M-1)}\psi\right] \\ 2r\sin\left[\frac{i-1}{2(M-1)}\psi\right]\sin\left[\frac{\psi}{2}-\frac{i-1}{2(M-1)}\psi\right] \end{pmatrix} \quad (3)$$

Brian G.Ferguson 经过理论分析和海试实验得出结论 [2]：如果拖曳线列阵声纳采用传统波束形成技术，当拖船在直线航行且阵形扰动较小时，则拖曳线列阵是直线阵的假设成立；但当拖船机动时 (指转弯)，阵形畸变是比较大的，则必须要估计阵形，根据阵列形状修正波束形成。因此，一般情况下，拖船在直线航行时，可以假设阵形畸变足够小，即认为拖线阵没有畸变，正如本节开始时提出的假设 1，此时拖曳线列阵声纳是根据无畸变的阵形进行波束形成。

假设只存在一个目标，无阵形畸变时第 i 路水听器信号为 $s(t+\tau_i(\phi))$，其中，$\tau_i(\phi)$ 是第 i 个水听器相对于参考点的时延差。若存在阵形畸变，那么第 i 路水听器的实际接收信号为 $s(t+\tau_i(\phi)+\Delta\tau_i)$，则拖曳线列阵在 ϕ 方向的波束主瓣输出功率为 [3]：

$$P=E\left\{\left[\sum_{i=1}^{M}s(t+\Delta\tau_i)\right]^2\right\} \quad (4)$$

式中 $E\{\cdot\}$ 表示求期望操作。那么无阵形畸变时，即 $\Delta\tau_i=0\ (i=1,2,\cdots,M)$ 时，则：

$$P=M^2\sigma_s^2 \quad (5)$$

式中 σ_s^2 为入射信号的功率。

而存在阵形畸变时，$\Delta\tau_i$ 不全为零，此时波束主瓣输出功率变为：

$$P=\sum_{i=1}^{M}\sum_{j=1}^{M}R_s(\Delta\tau_i-\Delta\tau_j) \quad (6)$$

拖曳线列阵通常工作在较低的频率，所以 $\Delta\tau_i$ 与信号的相关半径相比通常是较小的，可以把 $R_s(\tau)$ 在零点附近展开，即：

$$R_s(\tau)\approx\left[1-\frac{1}{2}(\alpha\tau)^2\right]\sigma_s^2 \quad (7)$$

这里 α 是一个刻画 $R_s(\tau)$ 形状的参数，于是 (6) 式可写成：

$$P\approx M^2\sigma_s^2-\frac{1}{2}\sigma_s^2\sum_{i=1}^{M}\sum_{j=1}^{M}[\alpha(\Delta\tau_i-\Delta\tau_j)]^2 \quad (8)$$

由上式可见，阵形畸变引起了波束主瓣输出功率的下降。

在本舰机动过程中，如果拖曳线列阵声纳还是按照未畸变时的原始阵形进行波束形成，即根据均匀直线阵列计算线列阵的输出功率，那么对方位估计精度是有影响的。为考查阵形畸变对方位估计精度的影响，做如下的仿真实验：假定某型拖曳线列阵声纳的拖缆和隔振段、声阵段、仪表段、零浮力缆等组成的拖线阵湿端部分总长为 1300m，其中声阵段由 160 个水听器组成，相邻两个水听器间距为 1m，对应的接收信号中心频率为 750Hz，那么声阵段总长为 159m 。仅以主动工作方式为例进行说明，其结果可扩展到被动工作方式。假定拖曳线列阵声纳发射 750Hz 单频信号，为判断目标所处的左右舷，进行本舰机动，转弯角度分别为 10°、20° 和 30°；机动后在右舷 80° 有一目标回波，回波的信噪比为 −15dB；假设本舰机动足够快，当本舰机动完成时，而基阵还没有稳定下来，此时阵形呈圆弧状，拖线阵湿端部分所呈的张角等于本舰转弯的角度 (是整个湿端部分对应的张角)，由此可计算出当转弯角度分别为 10°、20° 和 30° 时，畸变阵列 (畸变成圆弧阵) 的波束输出和无畸变阵列的波束输出分别如图 3～图 5 系列所示。

分析对比图 3～图 5，为便于分析对比总结，将它们的条件和结果制成表格，如表 1 所示。

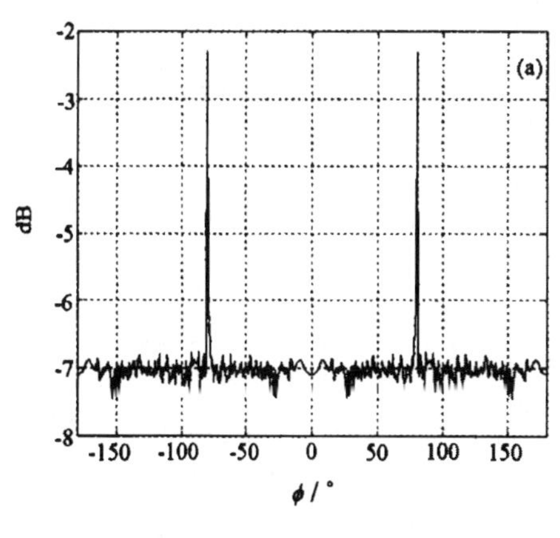

(a) 转弯角度为 10° 时，畸变阵列的波束输出

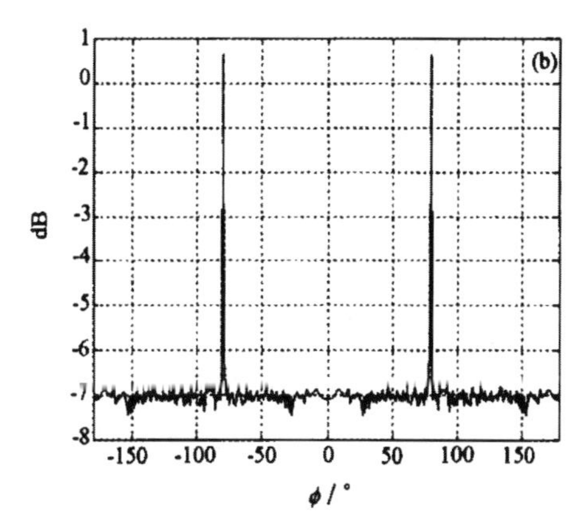

(b) 无畸变阵列的波束输出

图 3

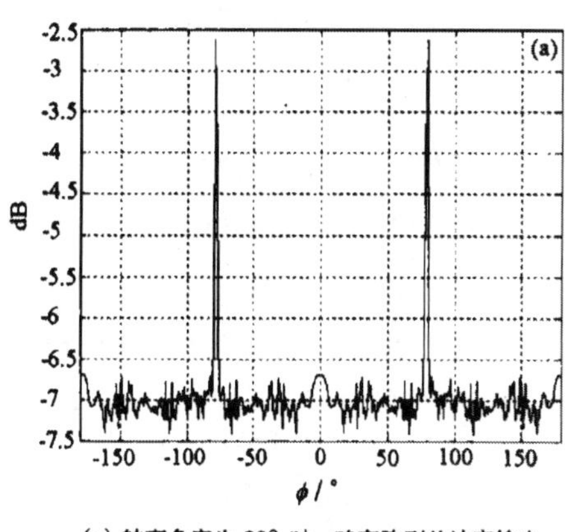

(a) 转弯角度为 20° 时，畸变阵列的波束输出

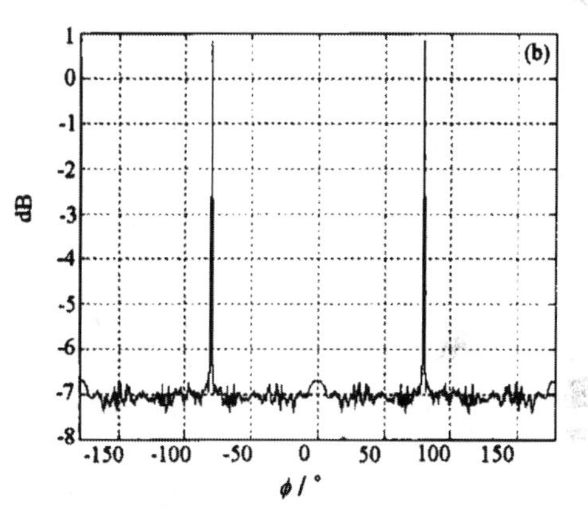

(b) 无畸变阵列的波束输出

图 4

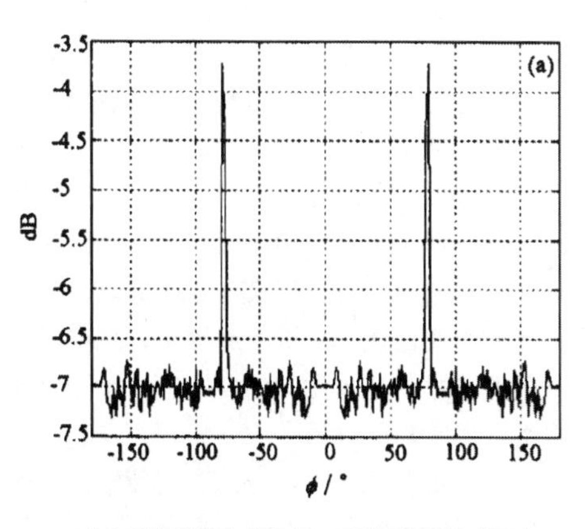

(a) 转弯角度为 30° 时，畸变阵列的波束输出

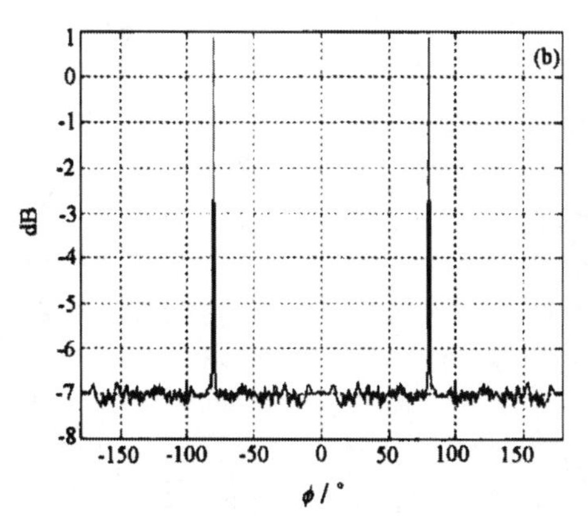

(b) 无畸变阵列的波束输出

图 5

表 1 仿真条件与结果

本舰转弯的角度 (°)	转弯半径 (m)	声阵段对应的张角 (°)	额外的方位估计误差 (°)	阵形畸变时与无畸变时的增益损失 (dB)
10	7448.5	1.2231	0.6	2.9
20	3724.2	2.4462	1.2	3.4
30	2482.8	3.6692	2.5	4.6

能够总结出: 在基阵未稳定阶段, 由于阵形畸变, 将造成波束失真, 表现在以下几方面:

(1) 增大了目标方位估计的误差 (如表 1 所列的额外方位估计误差);

(2) 基阵增益有所损失;

(3) 波束主瓣变宽;

(4) 有可能会出现高副瓣;

3.3 阵形畸变对本舰机动左右舷分辨方法的定量影响

基阵未稳定阶段的波束失真会带来以下几种后果: 一是在本舰机动前 (此时基阵无畸变) 恰好能检测到目标, 而机动后, 在基阵未稳定阶段, 阵形畸变导致基阵增益损失, 造成目标丢失, 更谈不上通过目标的舷角变化趋势判断出目标所处的左右舷; 二是假设在基阵未稳定阶段还能够检测到目标, 但由于目标方位估计误差增大, 造成判断出的目标方位变化趋势不可靠, 从而使得左右舷分辨的结果也不可靠。综上所述, 如果以本舰机动方法进行左右舷分辨, 那么在基阵未稳定阶段是难以可靠地分辨出目标所处的左右舷; 必须要等到本舰机动完成后, 且进入基阵稳定阶段才能有效地判断出目标所处的左右舷, 这样就存在着一个滞后时间。因此, 采用传统的单根线列阵, 以本舰机动的方法进行左右舷分辨, 决定了拖曳线列阵声纳无法在检测到目标的同时判断出目标所处的左右舷; 在发现目标后, 必须要经过一定滞后时间才能判断出目标所处的左右舷, 滞后时间 Δt 的长短取决于本舰机动的时间 t_1 和本舰机动后基阵的稳定时间 t_2, 可能还取决于所需要的声纳的时间-方位历程图的刷新次数所带来的额外时间 t_3 。之所以这样说是因为: 声纳的方位-幅度显示图刷新速率较快, 有时仅依靠声纳的方位-幅度显示图来判断目标舷角的变化趋势, 可靠性会比依靠时间-方位历程图要低 (特别是本舰转弯的角度不太大时); 但声纳的时间-方位历程图刷新速率比较低, 刷新时间-方位历程图中的一行数据, 大约需要 6~7s; 比方说需要刷新五行数据后才能判断出目标所处的左右舷, 那么所需的额外时间 t_3 为 30~35s, 也就是说在本舰机动完成后, 且进入基阵稳定阶段后, 还需要 30~35s 的时间通过声纳的时间-方位历程图来判断目标所处的左右舷, 则总的滞后时间为:

$$\Delta t = t_1 + t_2(+t_3) \tag{9}$$

参考文献 [4] 中曾提到: 本舰机动后, 对于一艘 5000 吨级的驱逐舰, 基阵的稳定时间至少为 5min, 那么总的滞后时间 Δt 至少要大于 5min 。由于滞后时间的存在, 使得采用传统的单根线列阵的拖曳线列阵声纳无法实时完成对目标的左右舷分辨功能。滞后时间对处于远程被动警戒状态的拖曳线列阵声纳也许还可以接受, 但若存在鱼雷攻击时, 这种滞后时间是无法容忍的。因为鱼雷是一种高速的水中兵器, 一般鱼雷航速普遍都在 50kn 左右 (也就是 25m/s), 5min 内鱼雷航行的距离将超过 7700m, 5min 后即使判断出鱼雷所处的左右舷, 也已经是措手不及了; 更何况在中近距离时, 鱼雷的目标舷角变化快, 采用本舰机动方法难以实时有效地分辨出鱼雷所处的左右舷, 无法有效完成鱼雷报警功能。

采用本舰机动方法分辨左右舷时, 本舰航向角的变化角度至少要大于声纳系统的测向误差 $\Delta\phi_1$ 。此外, 如果想在基阵未稳定阶段判断出目标舷角的变化趋势 (假设在基阵未稳定阶段，拖线阵声纳还能够检测到目标)，由于阵形畸变带来的额外的方位估计误差为 $\Delta\phi_2$，为确保左右舷分辨的可靠性，那么本舰航向角的变化角度，也就是转弯角度 $\Delta\phi$ 必须要满足：

$$\Delta\phi \geq \Delta\phi_1 + \Delta\phi_2 \tag{10}$$

4 结论

经过分析，可得出本舰机动左右舷分辨方法的几个要点：

(1) 本舰机动方法是在发现并跟踪上目标后，本舰转弯，通过监视目标舷角的变化趋势判断目标所处的左右舷：如果目标舷角变小，则目标在本舰转向的这一边；如果目标舷角变大，则目标在本舰转向的另一边。

(2) 为可靠地判断出目标所处的左右舷，则机动角度要满足：$\Delta\phi \geq \Delta\phi_1 + \Delta\phi_2$，其中：$\Delta\phi_1$ 是拖曳线列阵声纳本身的测向误差，$\Delta\phi_2$ 是阵形畸变引起的额外的方位估计误差。

(3) 本舰机动方法不能实时完成左右舷分辨功能，在检测到目标后，要经过滞后时间 $\Delta t = t_1 + t_2(+t_3)$ 后才能可靠地判断出目标所处的左右舷，其中：t_1 是本舰机动所需要的时间，t_2 是本舰机动后基阵的稳定时间，t_3 是满足时间-方位历程图的刷新次数的时间。

(4) 由于本舰机动方法不能实时分辨左右舷，因此该方法是无法有效完成拖曳线列阵声纳的鱼雷报警功能。

参 考 文 献

1 何心怡，蒋兴舟，李启虎等. 声学学报，2004，**29**(5): 409~413.

2 Brian G Ferguson. *IEEE Journal Oceanic Engineering*, 1993, **18**(4):565~571.

3 李启虎. 数字式声纳设计原理. 第 1 版. 合肥：安徽教育出版社，2003: 426~442.

4 杜选民. 水面舰鱼雷报警声纳技术研究：[博士学位论文]. 哈尔滨：哈尔滨工程大学，1999.

拖曳线列阵声呐及其左右舷分辨方法概述①

何心怡[1]，张春华[2]，李启虎[2]

（1. 海军装备研究院，北京 100073；2. 中国科学院声学所，北京 100080）

摘　要： 概述了拖曳线列阵声呐的发展概况，重点总结了国内外拖曳线列阵声呐左右舷分辨方法的发展情况，包括传统单根线列阵左右舷分辨方法、三元水听器组左右舷分辨方法、双线阵左右舷分辨方法和矢量水听器线列阵左右舷分辨方法等。本文对于拖曳线列阵声呐的相关研究具有一定的参考价值。

关键词： 拖曳线列阵声呐；左右舷分辨；传统单根线列阵；三元水听器组；双线阵；矢量水听器线列阵

中图分类号： TB565　　　文献标识码： A

Rough introduction of the towed linear array sonar and port/starboard discrimination methods

HE Xin-yi[1]，ZHANG Chun-hua[2]，LI Qi-hu[2]

（1. Naval Academy of Armament，Beijing 100073，China；

2. Institute of Acoustics，Academia Sinica，Beijing 100080，China）

Abstract： This paper introduces the rough development of the towed linear array sonar，summarizes the development of port/starboard discrimination methods of the towed linear array at home and abroad，including traditional single-line array port/starboard discrimination methods，towed linear array made up of hydrophone triplets port/starboard discrimination methods，twin-line array port/starboard discrimination methods and linear array made up of acoustic vector sensors port/starboard discrimination methods. This paper can be referenced for the research on the towed linear array sonar.

Key words： towed linear array sonar；port/starboard discrimination；traditional single-line array；towed linear array made up of hydrophone triplets；twin-line array；linear array made up of acoustic vector sensors

0 引　言

反潜战是现代海战的重要内容之一。在拖曳线列阵声呐出现之前，水面舰艇和潜艇的声呐基阵一般都安装于舰首，基阵的孔径受到安装平台的限制，迫使声呐只能工作在较高频段上，作用距离有限。二战后，随着美苏两个超级大国在全球范围内的竞争，潜艇战和反潜战地位越来越突出，特别是安静型潜艇的出现，迫切需要一种能用于远程警戒的声呐，因此，拖曳线列阵声呐孕育而生。它把接收水听器安装在中性浮力拖缆中，放到远离本舰噪声的地方，用于接收舰船低频噪声，从而实现远距离警戒。当然，拖于舰艇尾部的长达数百米，甚至更长的拖缆对舰艇机动的影响是显而易见的；并且拖曳线列阵的声学段是柔性的，可能存在阵形畸变，对波束形成性能会有所影响。但是，实践证明，拖曳线列阵声呐带来的优点远远超过不便之处，如今，拖曳线列阵声呐已成为水面舰艇和潜艇的最重要的声呐装备之一。美国国防部报告中说到："30 年来，水面舰艇反潜战中最重要的发展是战术拖曳线列阵声呐的问世"[1,2]。

① 舰船科学技术, 2006, 28(5): 9-14.

1 拖曳线列阵声呐的历史与现状

追溯拖曳线列阵声呐的发展历史,可分为2个阶段:第1阶段是从20世纪70年代到冷战结束前;第2阶段是从冷战结束后至今。

最早的军用拖曳线列阵声呐是1975年美国装备于专用警戒船的AN/SQR-14;第1部战术拖曳阵声呐(TACTAS,Tactical Towed Array Sonar)AN/SQR-18由美国EDO公司研制成功。1982年,Gould公司研发的TACTAS—AN/SQR-19安装到DD-980导弹驱逐舰上,如今,该型声呐及其改进型广泛装备于美、日等国的大中型水面舰艇;而Gould公司研制的STASS(Submarine Towed Array Sonar System),型号为AN/BQR-25声呐,也广泛装备于美国海军的核潜艇,如"洛杉矶"级核潜艇[1-4]。

如今美、俄、英、法、意等海军强国都在水面舰艇和核潜艇上装备了此种声呐,典型的有:美国的AN/SQR-19、英国的COMTASS和法国的LAMJPROIE。这一时期,拖曳线列阵声呐基本上以被动探测为主,主要用于远距离监视水下环境。战术型拖曳线列阵声呐一般可探测25~50 n mile远的目标,实现了声呐探测距离质的飞跃。这一阶段,拖曳成列阵声呐的关键技术主要有:阵形估计[5,6],本舰自噪声抵消等。

冷战结束后,海军作战区域开始从深海大洋转向了大陆沿岸浅海区域,对浅海安静型敷瓦潜艇的探测引起了广泛的重视。这时,拖曳线列阵声呐从被动探测转向主/被动联合探测,LFATS(Low Frequency Active Towed Sonar)是现阶段拖曳线列阵声呐的发展重点。Theodore Bick以L-3 LFATS为例系统分析了LFATS若干个关键参数选取:工作频率、信号带宽、声源级、指向性增益、发射信号波形、基阵配置和拖缆尺寸[7]。而Denney重点讨论了LFATS的工作频率选取和大功率发射换能器技术2项内容[8]。为达到隐蔽探测潜艇的目的,采用多基阵的配置方式是LFATS的一个较好的发展方向。Lain Shepherd研究了LFATS采用多基阵配置的可能性,涉及通讯、命令与控制等诸多相关问题[9]。1991年,荷兰皇家海军将研制型号为ALF的LFATS的任务交给了荷兰TNO-FEL公司(Netherlands Organization for Applied Scientific Research,Physics and Electronics Laboratory),并在1995年和1997年成功进行了2次海试[10]。法国和德国为改进F70型护卫舰和F123型护卫舰,联合研制LFTASS(Light Low Frequency Towed Active Sonar System),分别提出了各自的LFATS设计方案,并于1999年先后在德国基尔和法国土伦进行了性能测试,验证了LFTASS的良好性能。虽说后来法、德因研制方案无法统一而分道扬镳,但LFTASS的项目仍在继续,第1艘安装LFTASS的F123型护卫舰于2004年出厂[11,12];与此同时,德国的STN ATLAS公司为中小型水面舰艇开发的ACTA3低频主动拖曳线列阵声呐也已研制成功,1996年1月和1997年2月的2次海试展示了该型LFATS很好的性能[13]。此外,法国海军针对LFATS的发展还专门提出了NATII项目(the second New Array Technology programme)[14-17]。挪威的Kongberg Defence & Aerospace(KDA)为满足挪威海军浅海反潜需要而开发的MSI-2005F综合反潜系统,其核心是一部ATAS(Active Towed Array Sonar,即LFATS)和一部HMS(Hull Mounted Sonar,舷侧阵声呐)[18]。Allied Signal Ocean Systems为小型舰艇开发的LFATS,重量仅为3 750 kg,可探测位于第一会聚区的目标[19]。

这一阶段,拖曳线列阵声呐的关键技术主要有:左右舷分辨、低频宽带大功率发射换能器[2]等。表1列出了4种典型拖曳线列阵声呐的主要性能指标,概括了拖曳线列阵声呐的主要特点。

比较AN/SQR-19、COMTASS和LFTASS,发现后者与前两者有明显的不同:

(1)AN/SQR-19和COMTASS为被动工作方式,而LFTASS为主/被动联合工作方式,这个区别反映了LFATS对探测安静型潜艇的重视。

(2)AN/SQR-19和COMTASS实现左右舷分辨的方法都采用本舰机动方法,但本舰机动方法是无法实时分辨目标所处的左右舷[20,21];而LFTASS都要求能够实时分辨目标所处的左右舷,使用三元水听器组或双线阵的基阵结构,根据一定的信号处理法则都可实现实时分辨目标所处的左右舷[21]。而实时分辨左右舷,能够为水面舰艇反潜战争取更多宝贵的时间,在战术上具有重要意义。能否实时分辨左右舷,是LFATS和传统的拖曳线列阵声呐的主要差别[8]。

(3)LFTASS的最大工作速度比前两者有所提高,更有利于在线列阵拖曳过程中舰艇的战术机动。

表1 4种典型的拖曳线列阵声呐的主要性能指标

声呐型号	AN/SQR-19战术拖曳线列阵声呐	COMTASS战术拖曳线列阵声呐	LFTASS的法国方案	LFTASS的德国方案
生产厂家	美国Gould公司	英国Plessey公司	法国Thomsonsintra公司	德国STN Atlas公司
装备对象	大中型水面舰船	中小型水面舰船或潜艇	F70型护卫舰	F123型护卫舰
基阵	单线阵 阵长:150 m 直径:82.5 mm 基元数:8×48	单线阵 阵长:32 m 直径:63 mm 基元数:32	三元水听器组 阵长:22 m 直径:120 mm 基元数:64	双线阵 阵长:45 m 直径:50 mm 基元数:64
工作方式	被动	被动	主动+被动	主动+被动
工作频段	3~3 000 Hz	5~1 600 Hz	1 400~1 800 Hz	700~1 200 Hz
左右舷分辨功能	本舰机动	本舰机动	以三元水听器组做为基阵实时分辨目标所处的左右舷	以双线阵做为基阵实时分辨目标所处的左右舷

2 拖曳线列阵左右舷分辨方法概述

传统的拖曳线列阵是由无指向性声压水听器构成的直线阵,它具有圆柱对称的指向性,以线阵为轴,在同一转角的圆锥面上对入射信号的响应是完全一致的,存在着方位模糊问题。在一般情况下,只需要考虑水平面入射信号的方向,这时就要分清信号究竟是来自左舷还是右舷,也就是通常所说的拖线阵左右舷模糊问题。左右舷模糊问题是拖曳线列阵声呐的关键性难点问题。

左右舷模糊问题是传统拖曳线列阵的主要缺点之一,以本舰机动的方法解决传统的单根线列阵的左右舷模糊问题似乎广为人知[22-24],也有多种型号的拖曳线列阵声呐正是通过本舰机动方法实现左右舷分辨,如AN/SQR-19、COMTASS和SonacPTA等。Grim Steinar Gjonnes提到:“传统的单根线列阵通过本舰大的自身机动,根据声呐的时间—方位历程图中目标方位的变化或根据复杂的跟踪运算法则,判断出目标所处的左右舷”[22],但本舰机动应如何实施却没有提及。文献[21]给出了本舰机动左右舷分辨方法的实施方案,并对其做了定性和定量的分析。此外,利用线列阵的阵形畸变,采用信号处理方法在传统的单根线列阵上实现对目标的左右舷分辨也是一个具有实用价值的研究点,文献[21]和[25]对该项研究内容做了初步的探索。

然而,采用传统的单根线列阵以本舰机动方法实现左右舷分辨,最大的弊端在于不能实时分辨目标所处的左右舷。因为在机动过程中,线列阵的阵形发生了畸变,造成波束形成性能下降,必须要等到线列阵稳定后才能恢复波束形成性能。这样,从检测到目标到判断出目标所处的左右舷就存在一定的滞后时间,既增大了目标丢失的可能性又延长了捕获目标的时间[20]。因此,快速完成左右舷分辨是拖曳线列阵必须要解决的问题。

为实时完成左右舷分辨,解决的方法主要是在拖曳线列阵声呐的发射端或接收端下功夫。

LFTAS都有发射基阵,如图1是英国宇航公司(BAe)和法国Thomson公司联合生产的主动拖曳线列阵声呐的发射基阵,采用的是弯张型换能器。在发射端进行左右舷分辨的方法是采用左右舷交替发射达到判断目标所处左右舷的目的。TNO-FEL在ALF的研制中就提到了这种方案[10],Denney报道了在DCN SLASM系统,采用一对背靠背的垂直线列阵做发射基阵,通过档板加权的方法使得发射指向性在水平面上形成心形指向性,以左右舷交替发射的方法分辨目标所处的左右舷[8](如图2所示)。但左右舷交替发射方法存在着2个致命的缺点:(1)无法克服LFATS在被动工作方式时的左右舷模糊问题;(2)在主动工作方式时由于发射脉冲的重复周期较长,因此也不能实时分辨目标所处的左右舷。可以说,左右舷交替发射是LFATS在主动工作方式下解决左右舷模

图1 英国宇航公司(BAe)和法国Thomson公司联合生产的主动拖曳线列阵声呐的发射基阵

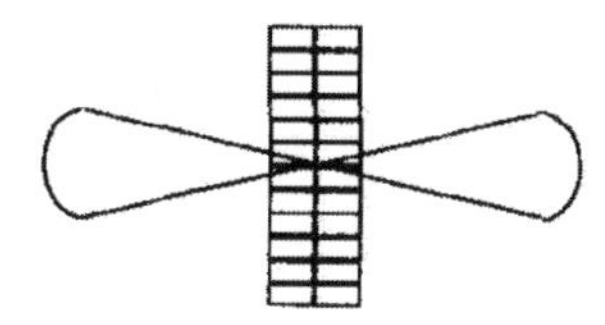

图 2 左右舷交替发射方法采用的发射阵示意图

糊的一种方法，但该方法达不到实时分辨目标所处的左右舷的要求。

在拖曳线列阵的接收端解决左右舷模糊问题，主要是改变拖曳线列阵声呐接收基阵的阵元结构，国外现役装备中采用的方法主要有 3 种：

（1）采用三元水听器组；

（2）采用双线阵方式；

（3）采用矢量水听器。

采用三元水听器组或双线阵的方法，技术上较为成熟，在国外有一定的应用实例报道，如法国“信天翁”专用鱼雷报警声呐[24]、CAPTAS 系统[14-17]、LFTASS 法国方案[11]和挪威的 MSI-2005F 中的 ATAS 拖曳线列阵声呐[18]都是采用三元水听器组的基阵配置来解决左右舷分辨问题的，而荷兰的 ALF 低频主动拖曳声呐[10]、DCN SLASM 系统[8]和 LFTASS 德国方案[12]则是采用双线阵的基阵配置来解决左右舷分辨问题。也有报道说，美国的战略拖曳线列阵声呐 SURTASS 采用了矢量水听器，俄国研制并装备了 BΓA11-9-17/5、BΓA10-4 和 BΓA24-9-6/4 等型号的矢量水听器拖曳线列阵声呐[26]。

1991 年，Jean Bertheas、Villeneuve Loubet 等在其专利中公布了后来被称为三元水听器组几何相移模型左右舷分辨方法[27]。三元水听器组是指线列阵的每个基元是由分布在同一圆周、形成等边三角形的 3 个无指向性水听器组成的，并且单个三元水听器所构成的平面与三元水听器组的轴线相垂直。图 3 是英国宇航公司和法国 Thomson 公司联合生产的拖线阵的基阵，正是三元水听器组。

然而，几何相移模型要求三元水听器中的 2 个水听器的间距是接收信号波长的三分之一，频率越低，所要求的基阵尺寸就越大，这就限制了几何相移模型在三元水听器组中的应用。为减小基阵尺寸，便于拖线阵声呐的安装使用，常常要求三元水听器组的横截尺寸较小（通常是指小于接收信号波长的十分之一，且大于拖线阵流噪声的空间相关距离），为此，Y Doisy 和 G W M Van Mierlo 等人提出了适用于小横截尺寸三元水听器组的噪声相关模型，即最优波束加权方

图 3 英国宇航公司和法国 Thomson 公司联合生产的拖线阵的基阵

法，使得该算法具有更为广泛的适用性[28,29]。S P Beerens 等人发表的三元水听器组流噪声的分析结果和国内某研究所的拖曳线列阵流噪声试验结果，都进一步证实了噪声相关模型[30,31]。杜选民在 Y Doisy 等人的研究基础上详细给出了最优波束加权方法的推导过程[20,23]。

在浅海，混响是 LFATS 的主要干扰场，因此，抗混响是非常重要的，其中发射信号波形设计是一种有效的抗混响技术。1995 年，Cox H 提出了一种新的信号形式：梳状谱信号（comb spectrum waveforms）[32]，引起了广泛的关注[33,34]。在 NATII 项目赞助下，荷兰的 TNO-FEL 和法国的 TMS SAS 公司（Thomson Marconi Sonar SAS）联合进行了梳状谱信号在拖曳线列阵声呐抗混响方面的研究，以荷兰的 ALF 拖曳线列阵声呐和法国的 CAPTAS（Combined Active Passive Towed Array）为物理实体成功进行了海试，取得了明显的抗混响效果[14-17]。其中，CAPTAS 的基阵是三元水听器组，并运用了噪声相关模型左右舷分辨方法。在试验中发现，S P Beerens 提出的基于噪声相关模型的三元水听器组自适应左右舷分辨方法可获得较低的旁瓣，从而显著地改善了混响抑制效果[17,35]。

双线阵是为解决左右舷模糊问题而常用的一种基阵样式，如德国 ATLAS 公司研制的主动拖曳式线列阵声呐就是采用双线阵，如图 4 所示。Joost H de Vlieger 提到了荷兰的 ALF 低频主动拖曳线列阵声呐是采用三维波束形成的方法实现左右舷分辨[10]；而

图4　德国 ATLAS 公司研制的主动拖曳式线列阵声呐的双线阵示意图

Iman W. Schurman 提到了一种解决双线阵左右舷模糊的思路，类似于三元水听器组的几何相移模型[36]。

此外，矢量水听器（Acoustic Vector Sensor）应用于线列阵解决左右舷模糊问题也具有可能性[37-40]。最近几十年出现的矢量水听器，将声压水听器和振速传感器组合成一个整体，不仅利用了声波的声压信息，还利用了声波的振速信息。它的出现给水声工程和水声信号处理领域注入了新的活力，至今仍是美、俄等海军强国重点研究的内容之一。Nehorai 曾提到，矢量传感器与传统的标量传感器相比，其主要优势在于矢量传感器有更多的可用信息，因此在定位精度上优于标量传感器[41,42]。单个矢量水听器具有"8"字形或心形指向性，并且该指向性与频率无关，这为其在线列阵中应用并解决左右舷模糊问题提供了可能[43,44,26]。

文献[21]对三元水听器组、双线阵和矢量水听器的左右舷分辨方法做了较为系统的研究，如应用于大/小横截尺寸三元水听器组的移相方法、应用于双线阵的基于离散平面阵波束形成法、应用于双线阵的时延估计法等等，对于拖曳线列阵左右舷分辨技术的发展具有一定的参考价值。

3　发展展望

可以预见，在今后相当长的一段时期，拖曳线列阵声呐仍将是水声工程的发展重点，LFATS 是其主要发展方向，提高其抗混响能力、抗多途能力和被动测距精度是 3 个重要的努力方向；匹配场技术在拖曳线列阵声呐上的应用是一个富有价值的研究领域，若其能成功应用，将明显提高拖曳线列阵声呐的性能；研究以 LFATS、舰壳声呐、直升机吊放声呐、声呐浮标为基干构建的多基地声呐系统，可有效整合水面舰艇、反潜飞机等平台的反潜探测手段，大幅度提高反潜探测能力。

应用于小横截尺寸三元水听器组、双线阵和矢量水听器线列阵的实时左右舷分辨方法是当前拖曳线列阵声呐左右舷分辨技术的研究重点。

参考文献：

[1]　拖曳式线列阵声呐研究丛书（第 1 版）[M]. 北京：中国科学院声学研究所，1989.

[2]　李启虎. 数字式声呐设计原理（第 1 版）[M]. 合肥：安徽教育出版社，2003.

[3]　姜来根. 21 世纪海军舰船（第 1 版）[M]. 北京：国防工业出版社，1998.

[4]　姜来根. 简明世界舰船手册（第 1 版）[M]. 北京：国防工业出版社，1995.

[5]　YANG T C. A method of range and depth estimation by modal decomposition[J]. J. Acoust. Soc. Amer., 1987, 82(5): 1736-1745.

[6]　HOVER F S, et al. Calculation of dynamic motion and tension in towed underwater cables[J]. IEEE J. Oceanic Engr., 1994, 19(3): 425-437.

[7]　BICK E T. A new approach for towed active ASW sonar[A]. In: UDT96[C]. UK, 1996: 2-8.

[8]　DENNEY M R. Low frequency active sonar-customised design[A]. In: UDT97[C]. Germany, 1997: 1-4.

[9]　SHEPHERD L. Low frequency active sonar a users perspective[A]. In: UDT95[C]. France, 1995: 138-141.

[10]　de VLIEGER J H, van BALLEGOOIJEN E C. ALF, an experimental low frequency active sonar[A]. In: UDT95[C]. France, 1995. 131-136.

[11]　REYNARD F, SCHUMACHER S. German-French Cooperation for a Light Low Frequency Towed Active Sonar System LFTAS[A]. In: UDT2001[C]. Germany, 2001.

[12]　SCHOLZ B. LFTASS-the German solution [A]. In: UDT2001[C]. Germany, 2001.

[13]　LICHT J, BAUER W. Sea trials with a compact ACTAS system[A]. In: UDT97[C]. Germany, 1997: 59-63.

[14]　DOISY Y, DERUAZ L, BEEN R. Sonar waveforms for reverberation rejection, part I: theory[A]. In: UDT2000[C]. Pacific, 2000: 19-24.

[15]　BEEN R, BEERENS S P, et al. Sonar waveforms for reverberation rejection, part II: experimental results[A]. In: UDT2000[C]. Pacific, 2000: 25-29.

[16]　DOISY Y, DERUAZ L, PRUNEL B, et al. Sonar waveforms for reverberation rejection, part III: more experiment results[A]. In: UDT2000[C]. UK, 2000.

[17] van IJSSEIMUIDE S P, et al. Sonar waveforms for reverberation rejection, part IV: adaptive processing[A]. In: UDT2002[C]. Italy,2002:1 - 8.

[18] Integrated ASW System Design[A]. In: UDT2001[C]. Germany,2001.

[19] DUNLAP D R. Long range ASW and compact sensor systems[A]. In: UDT96[C]. UK,1996:278 - 282.

[20] 杜选民. 水面舰鱼雷报警声呐技术研究[D]. 哈尔滨: 哈尔滨工程大学,1999.

[21] 何心怡. 拖曳线列阵波达方位估计方法研究[D]. 武汉: 海军工程大学,2003.

[22] GJONNES G S, AAGEDAL H. Left/right discrimination in software with conventional, passive thin-line towed submarine arrays[A]. In: UDT2000[C]. UK,2000.

[23] 杜选民,朱代柱,赵荣荣,等. 拖线阵左右舷分辨技术的理论分析与实验研究[J]. 声学学报,2000,25(5):395 - 402.

[24] 杜选民,朱代柱,赵荣荣,等. 拖线阵目标左/右舷分辨技术研究(Ⅱ)—噪声相关性模型[J]. 舰船科学技术,1999,21(3):17 - 20.

[25] 何心怡,蒋兴舟,李启虎,等. 拖线阵的阵形畸变与左右舷分辨[J]. 声学学报,2004,29(5):409 - 413.

[26] 孟洪,周利生,惠俊英. 组合矢量水听器及其成阵技术研究[J]. 声学与电子工程,2003,(1):15 - 20.

[27] BERTHEAS J, MORESCO G, DUFOURCO P. Linear hydrophonic antenna and electronic device to remove right/left ambiguity, associated with the antenna[P]. US Patent:5058082,1991 - 10 - 15.

[28] DOISY Y. Port-starboard discrimination performances on actived towed array systems[A]. In: UDT95[C]. France,1995:125 - 129.

[29] van MIERLO G W M, BEERENS S P, BEEN R, et al. Port/starboard discrimination by hydrophone triplets in active and passive towed arrays[A]. In: UDT97[C]. Germany,1997,176 - 181.

[30] BEERENS S P, van IJSSELMUIDE S P, VOLWERK C, et al. Flow noise analysis of towed sonar arrays[A]. In: UDT99[C]. France,1999:200 - 205.

[31] 张国栋,王茂法,顾振福. 线列阵拖曳自噪声测量及分析[J]. 声学与电子工程,2000,(3):8 - 12.

[32] COX H, LAI H. Geometric comb waveforms for reverberation suppression[A]. In: The Twenty-Eighth Asilomar Conference on Signals, Systems and Computers[C]. Los Alamitos CA,1995:1185 - 1189.

[33] COLLINS T, ATKINS P. Doppler-sensitive active sonar pulse designs for reverberation processing[A]. IEE Proceedings - Radar, Sonar and Navigation[C]. London, 1998, 145(6):347 - 353.

[34] DOISY Y, DERUAZ L, BEERENS S P, et al. Target Doppler estimation using wideband frequency modulated signals[J]. IEEE Transactions on Signal Processing, 2000,48(5): 1213 - 1224.

[35] BEERENS S P, BEEN R, GROEN J. Enoutary Adaptive port-starboard beamforming of triplet arrays[A]. In: UDT2000[C]. Pacific,2000:63 - 68.

[36] SCHURMAN I W. Reverberation rejection with a dual - line towed array[J]. IEEE Journal of Oceanic Engineering,1996,21(2):193 - 204.

[37] LESLIE C B, KENDALL J M, JONES J L. Hydrophone for measuring particle velocity[J]. J. Acoust. Soc. Am., 1956,28(4):711 - 715.

[38] Velocity hydrophone[P]. US Patent: 4547870.

[39] SHCHUROV V A. Coherent and diffusive fields of underwater acoustic ambient noise[J]. J. Acoust. Soc. Am, 1991,90(2):991 - 1001.

[40] SHCHUROV V A. The properties of the vertical and horizontal power flows of the underwater ambient[J]. J. Acoust. Soc. Am. ,1991,90(2):1002 - 1004.

[41] NEHORAI A, PALDI E. Acoustic vector sensor array processing[A]. In: The Twenty-Sixth Asilomar Conference on Systems and Computers[C]. Pacific Grove CA,1992: 192 - 198.

[42] NEHORAI A, PALDI E. Acoustic vector sensor array processing[J]. IEEE Transactions on Signal Processing, 1994,42(9):2481 - 2491.

[43] 贾志富. 同振球型声压梯度水听器的研究[J]. 应用声学,1997,16(3):20 - 25.

[44] 贾志富. 三维同振型矢量水听器的特性及其结构设计[J]. 应用声学,2001,20(4):15 - 21.

主被动拖线阵声呐中拖曳平台噪声和拖鱼噪声在浅海使用时的干扰特性①

李启虎　李淑秋　孙长瑜　余华兵

(中国科学院声学研究所　北京　100080)

2006 年 3 月 25 日收到

2006 年 8 月 30 日定稿

摘要　分析了主被动拖线阵声呐中拖曳平台的螺旋桨噪声和主动发射基阵 (拖鱼) 流噪声对被动接收基阵的干扰作用，指出在浅海声学环境下，直达波和海底、海面反射波会成为声呐系统的严重干扰。给出了拖曳平台和拖鱼在各种不同深度和缆长的情况下干扰源的入射角。从理论上计算了不同参数下干扰噪声对系统响应的影响，系统仿真结果可为声呐设计者抑制这种干扰提供途径。

PACS 数：　43.60

The interference characteristics of platform and towed body noise in shallow water for active/passive towed array sonar

LI Qihu　LI Shuqiu　SUN Changyu　YU Huabing

(*Institute of Acoustics, The Chinese Academy of Sciences* Beijing 100080)

Received Mar. 25, 2006

Revised Aug. 30, 2006

Abstract　The interference characteristics of towed platform noise resulted from propeller and towed body noise for active/passive towed array is analyzed. It is shown that, in shallow water environment, the direct wave and bottom/sea surface reflected wave will seriously affect the performance of sonar system. The formula for calculating the direction of arrival (DOA) of interference in various parameters, such as array depth, length of tow cable, is derived. The effect of interference noise for the performance of sonar system is described. The results of system simulation provide the method for reducing the effect of these kind of interferences.

引言

主被动拖曳线列阵声呐的研究工作是近年来声呐领域很引人注目的课题[1-8]。因为安静型潜艇的出现使被动声呐的作用距离下降很多。据国外报导水面舰艇对潜艇威胁的反应时间由小时量级下降到分钟量级。于是发展主动声呐引起了研究者的兴趣。他们认为可以把失去的时间再赢回来[2]。于是水面舰艇使用的主被动声呐就得到了广泛的关注。

但是主被动拖线阵声呐在使用时也受到自身噪声的干扰。产生噪声的来源是多种多样的。如拖曳平台干扰、拖缆流噪声干扰等。

本文讨论的是拖曳平台螺旋桨噪声的干扰和主动发射换能器基阵 (俗称拖鱼) 拖曳噪声的干扰。在浅海环境下，这两种噪声源都会对被动接收功能产生严重影响。除了直达波在舰艏方向的干扰之外，海底反射所产生的回波始终会严重限制被动声呐检测效用的发挥。我们将给出这两种干扰源有可能引起的噪声的入射角 (在理想镜面反射下) 的计算方法，以及这种干扰与有关参数 (接收模块深度、拖鱼深度、海深、缆长以及声学模块长度等) 的关系。当然，海面反射也会带来干扰，本文所用的分析方法同样适用。

系统仿真可以检验这种干扰对实际系统性能的影响，以及提供如何调节某些可以控制的参数最大

① 声学学报, 2007, 32(1): 1-4.

可能地避免这种干扰带来的不利影响。

1 发射拖曳体拖曳噪声分析

图 1 是一个主被动拖线阵声呐在浅海使用时的示意图，拖曳体为 A, 它到被动接收基阵的直线距离为r_0, 被动接收基阵BC的长度为l_0, 一般来说A和BC位于同一深度h_1处。BC至海底的距离假定是h_2。

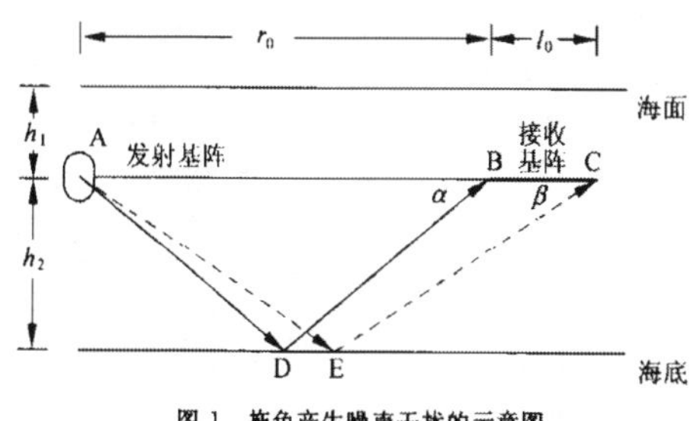

图 1 拖鱼产生噪声干扰的示意图

在主动发射基阵拖曳过程中，导流罩的流噪声会成为系统的噪声干扰源，它对被动接收缆的干扰大致可以分为两个部分，一部分是直达波。如果以 BC 的法线方向为 0° 的话，直达波的入射方向为 −90°，就是端射方向。由于线阵端射方向的波束非常宽，所以由直达波引起的干扰会对系统产生一个检测上的盲区；另一部分干扰则是由拖体流噪声的反射波引起的。严格来说，如果海底是平的，那么由 A 引起的反射波到达接收模块两端 B 和 C 的入射角分别是 α 及 β。但是，实际海底是凹凸不平的，所以反射波的区域会在 $[\alpha,\beta]$ 外延拓。当海底存在可以与拖曳体噪声波长可以比拟的起伏时，反射波也能出现在别的方向，这会使干扰问题变得更加复杂。当然，反射波的强度与底质吸收性能有关，并且传播路径也还和声速剖面有关。

在等声速的假定下，可以证明入射角 α,β 可由下式给出：

$$\alpha=\mathrm{tg}^{-1}(2h_2/r_0),\quad \beta=\mathrm{tg}^{-1}[2h_2/(r_0+l_0)]. \tag{1}$$

对于缆长 $r_0=400$ m, 基阵长 $l_0=60$ m 的线列阵，如果 $h_2=75$ m, 那么 $\alpha\approx 20.5°, \beta\approx 18°$ 。

2 拖曳平台螺旋桨噪声分析

图 2 是拖曳平台螺旋桨产生的干扰噪声的示意图，仍假定基阵长度为 l_0, 螺旋桨 A 至接收基阵 BC 的直线距离为 r_0, 螺旋桨的直达波和基阵所在水平线的夹角为 θ 。螺旋桨噪声干扰的海底反射波的入射角如图所示，分别是 α 和 β 。可以证明，θ,α,β 的计算公式如下：

$$\theta=\mathrm{tg}^{-1}(h_1/r_0), \tag{2}$$

$$\alpha=\mathrm{tg}^{-1}[(2h_2+h_1)/r_0], \tag{3}$$

$$\beta=\mathrm{tg}^{-1}[(2h_2+h_1)/(r_0+l_0)], \tag{4}$$

式 (2) — (4) 与式 (1) 略有差异，主要原因是拖曳平台的螺旋桨噪声与接收阵不在同一水平面上，它们之间的垂直距离 h_1 反映在公式 (2) — (4) 中。

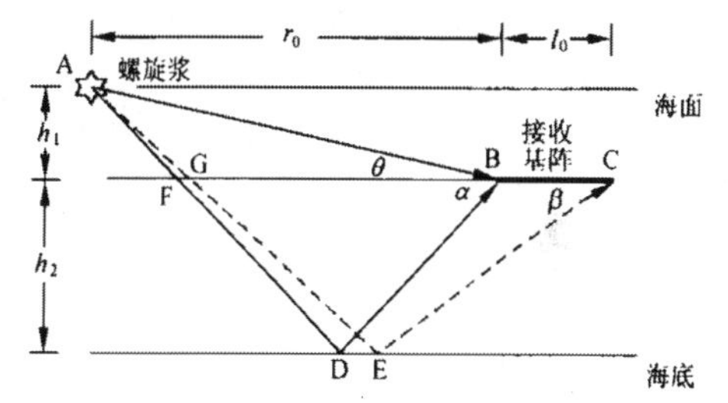

图 2 拖曳平台螺旋桨噪声产生的干扰

举例来说，如果 $h_1=25$ m, $r_0=400$ m, 则 $\theta=3.5°$; 当 $h_2=75$ m, $l_0=60$ m 时， $\alpha=23.6°$, $\beta=20.8°$ 。

我们大致可以看到由螺旋桨产生的干扰源与拖曳平台产生的干扰源比较，从方位上说更加靠近法线方向，所以它对信号检测带来的危害更大一些。当然，干扰的频谱分布与干扰的强度也是应该考虑的因素。

3 系统仿真

为了实际考察拖曳平台噪声和拖曳体噪声的干扰作用，我们进行了一系列的系统仿真，主要的参数如下：

接收基阵 BC 至螺旋桨的垂直距离：$h_1=25$ m;

接收基阵 BC 至海底的距离：$h_2=75$ m;

接收基阵长度：$l_0=100$ m;

拖曳体 A 与接收基阵 BC 处于同一水平面上；

缆长 r_0: 400 ~ 800 m;

目标信号：以 800 Hz 为中心的窄带信号，带宽 $\Delta f=200$ Hz;

螺旋桨干扰噪声：500 ~ 1000 Hz 宽带信号，每一倍频程下降 6 dB;

拖曳体干扰噪声：400 ~ 800 Hz, 宽带信号，每一倍频程下降 6 dB;

信号对螺旋桨直达波干扰的信噪比 $(\mathrm{SNR})_{pd}=-17$ dB;

信号对螺旋桨干扰反射波的信噪比 $(\mathrm{SNR})_{pr}=-9.3\ \mathrm{dB}$;

信号对拖曳体干扰直达波的信噪比 $(\mathrm{SNR})_{td}=-15.3\ \mathrm{dB}$;

信号对拖曳体干扰反射波的信噪比 $(\mathrm{SNR})_{tr}=-7.3\ \mathrm{dB}$;

系统仿真的主要结果归纳如下:

(1) 拖曳体流噪声的存在影响入射角 10 ~ 22° 范围内的目标检测。当缆长变长时,干扰的角度下移并且范围缩小。所以当条件允许时,较长的拖缆会使干扰范围减小 (见图 3)。

(2) 拖曳平台螺旋桨噪声的干扰集中在两个范围,由直达波引起的干扰集中于 0° 附近,影响的角度大约在 ±5° 范围内;由反射波引起的干扰集中于 10 ~ 27° 范围内。当缆长变长时,干扰的角度下移并且范围缩小 (见图 4),这一现象和拖曳体干扰的情况类似。

(3) 拖曳体流噪声干扰和拖曳平台螺旋桨干扰会在系统检测中形成盲区。盲区的范围大致分布在船艏 ±30° 范围内,它会严重影响检测性能。

图5是没有干扰时线列阵的指向特性,目标出现在正横 (0°) 和靠近船艏方向 (−60°),时清晰可辨。

图 6 是在有干扰时的情况,目标出现在 0° 时可以分辨,但当目标出现在 −60° 时,难于分辨。但是,同样情况下,如果缆长加长到 800 m,那么信号又出现了 (见图 7)。

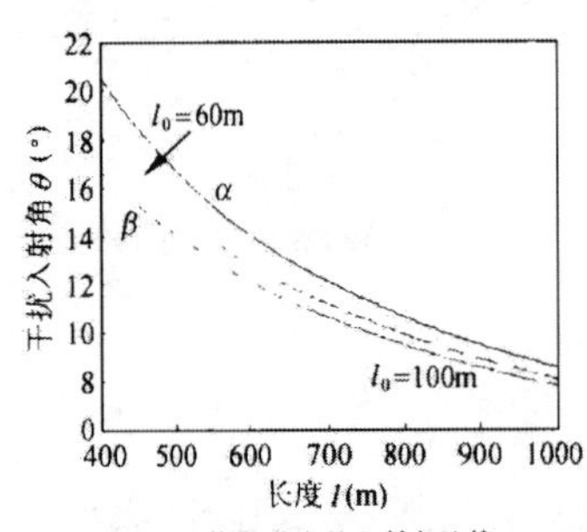

图 3 拖鱼噪声的入射角计算

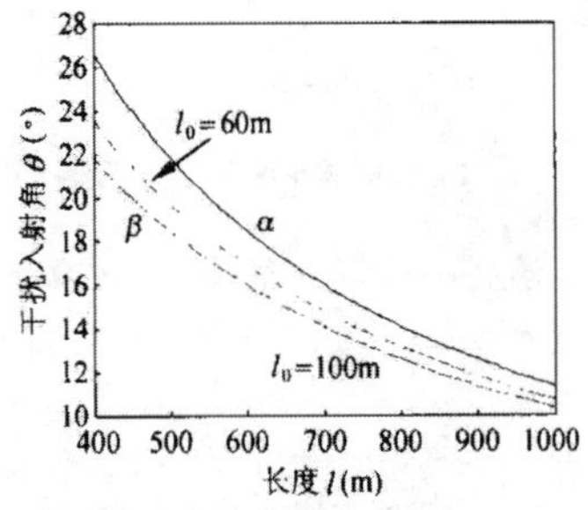

图 4 螺旋桨噪声的入射角计算

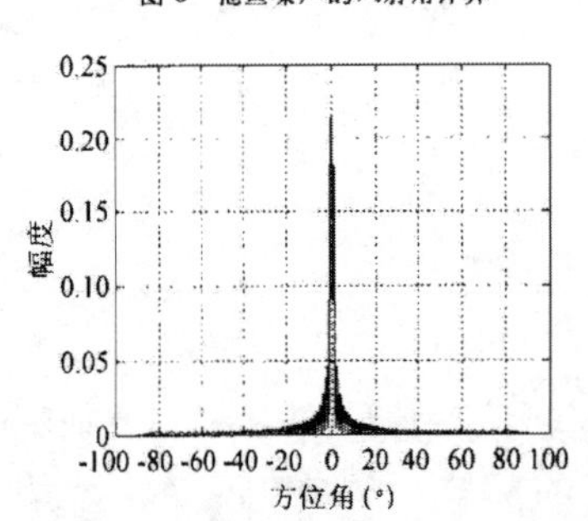

(a) DOA = 0°, $N=100$, $d/\lambda=0.5$, $l_0=100$ m

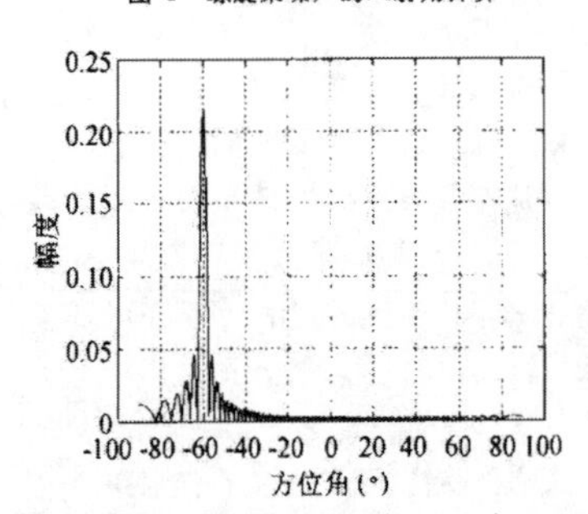

(b) DOA = −60°, $N=100$, $d/\lambda=0.5$, $l_0=100$ m

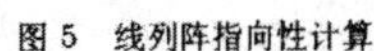

图 5 线列阵指向性计算

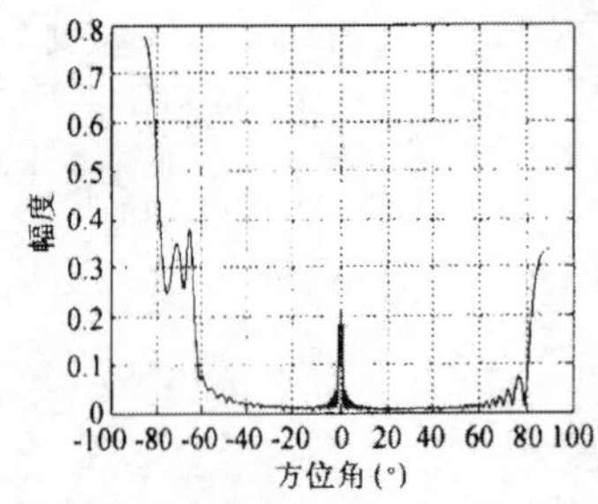

(a) DOA = 0°, $N=100$, $d/\lambda=0.5$, $r_0=400$ m, $l_0=100$ m

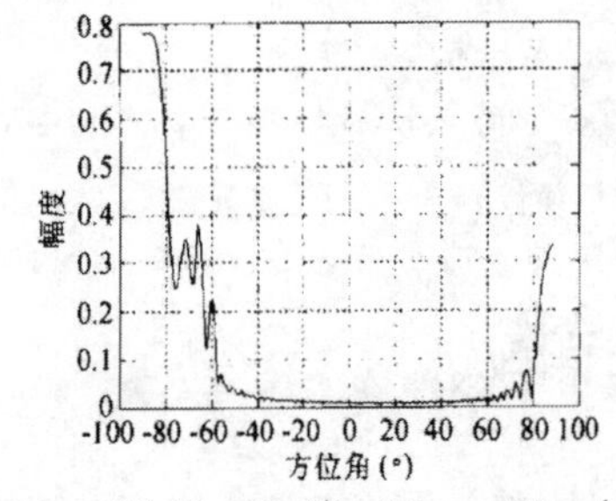

(b) DOA = −60°, $N=100$, $d/\lambda=0.5$, $r_0=400$ m, $l_0=100$ m

图 6 干扰对检测的影响

$(\mathrm{SNR})_{pd}=-17\ \mathrm{dB}$, $(\mathrm{SNR})_{pr}=-9.3\ \mathrm{dB}$, $(\mathrm{SNR})_{td}=-15.3\ \mathrm{dB}$, $(\mathrm{SNR})_{tr}=-7.3\ \mathrm{dB}$.

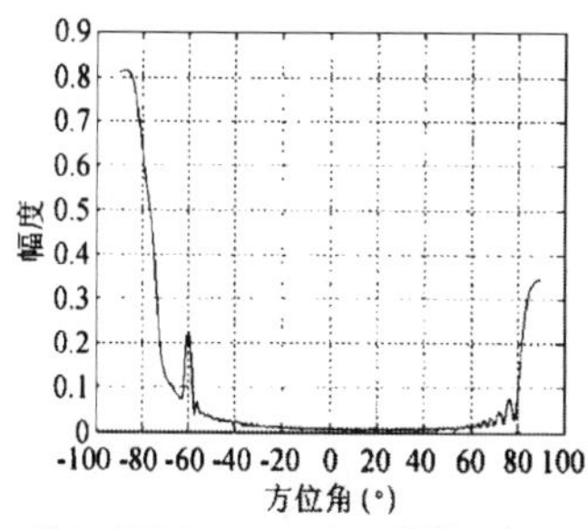

图 7 拖缆加长时对目标的检测情况有改善
$(\mathrm{SNR})_{pd} = -17$ dB, $(\mathrm{SNR})_{pr} = -9.3$ dB,
$(\mathrm{SNR})_{td} = -15.3$ dB, $(\mathrm{SNR})_{tr} = -7.3$ dB.

4 结论

主被动拖曳线列阵声呐在使用过程中会受到来自拖曳平台螺旋桨噪声和拖曳体噪声的干扰。这些干扰会在声呐观测范围内形成盲区，从而使声呐检测性能下降。系统仿真说明了干扰噪声带来的影响以及它与声呐系统有关参数的关系。这为实际系统设计时抑制干扰提供了依据。

参 考 文 献

1 Lemon S G. Towed array history:1917-2003. *IEEE J. of Oceanic Engr.*, 2004; **29**(2): 365—373
2 National Research Council Ed. Technology for the United States Navy and Marine Corps., 2000-2035, becoming a 21st century force, NA Press, USA 1997
3 Jeferry W. DARPA technology transition: 2003, Feb. 2004, DARPA, USA
4 Lasky M *et al.* Recent progress in towed hydrophone array research. *IEEE J. of Oceanic Engr.*, 2004; **29**(2): 374—387
5 Preston J M. Stability of towfish used as sonar platforms. IEEE Oceans'92 Proc.: 888—892
6 Zachow H. System for noise measurements of towed arrays. Proc. Of UDT'1998: 262—267
7 李启虎. 双线列阵左右舷目标分辨性能的初步分析. 声学学报, 2006; **31**(5): 385—388
8 李启虎. 用双线列阵区分左右舷目标的延时估计方法及其实现. 声学学报, 2006; **31**(6): 485—487

失配状态下的双线阵波束形成研究①

王华奎　李启虎　黄海宁　叶青华　李淑秋　张春华
(中国科学院声学研究所　北京　100080)

2006 年 6 月 13 日收到
2006 年 9 月 1 日定稿

摘要　基于双线阵在阵间距失配状态下左右舷分辨性能会急剧下降的事实，利用延时相加和延时相减两种波束形成方法，分别推导了理想状态下双线阵的波束形成和阵间距失配状态下波束形成的方向指向性函数；分析了理想状态下和阵间距失配状态下延时相加和延时相减两种波束形成方法的左右舷分辨性能。理论分析表明延时相减波束形成具有较好的左右舷分辨性能。最后用海试数据验证了理论分析结果的正确性。

PACS 数：　43.30, 43.60

Beamforming methods research on dual-line array with mismatch space between the two line arrays

WANG Huakui　LI Qihu　HUANG Haining　YE Qinghua　LI Shuqiu　ZHANG Chunhua
(*Institute of Acoustics, Chinese Academy of Sciences*　Beijing　100080)

Received Jun. 13, 2006
Revised Sept. 1, 2006

Abstract　The capability of port/starboard discrimination of dual-line array will be seriously degraded when the space between the two lines of the dual-line array is mismatched. The beam pattern of the delay-sum and delay-subtract beamforming is deduced and the port/starboard discrimination of the delay-sum and delay-subtract beamforming is also studied both under the ideal status and the status under which the space between the two lines of the dual-line array is mismatched. Analysis results show the delay-subtract beamforming has better capability of port/starboard discrimination. At last, the sea trial results demonstrate the theoretical results.

引言

在声呐系统中，波束形成是目标方位估计最常用的处理方法。为了检测噪声背景中的安静型目标，被动声呐系统特别是拖曳式线列阵声呐近年来朝大孔径阵列、多基元方向发展以形成窄波束、获得高空间处理增益。但是，在浅海环境中，复杂的环境会使大孔径的单线阵性能急剧下降[1]。为了解决这个问题，国内外发展出了双线阵、多线阵系统[1−7,9−11]。

单线阵由于在水平方向上只有横向孔径，对镜像方向的两个目标源存在左右舷模糊的问题。在线阵一侧的强干扰源，也往往会覆盖另一侧的较弱的信号源。与单线阵不同，双线阵由两根并行的线列阵组成，这种物理结构(既有横向孔径，又有纵向孔径)决定了双线阵有单线阵所不具备的优点。将两根线列阵的接收信号进行适当处理，能将目标的左右空间特性分离开来。拖曳线列阵左右舷分辨问题已有一些理论分析(如文献 2 —文献 7)，这些理论分析都是基于理想情况下(单条线阵在水下不发生畸变，双线阵在水下的相对姿态水平平行、间距已知)进行的。但一般情况下，双线阵在水下真实姿态并非如此[1]。为了简化双线阵在水下的真实姿态，本文所谓的失配是指用于波束形成的双线阵的间距与真实值的不一致。本文讨论了理想状态下两种波束形成方法的方向指向性函数和左右舷分辨性能，给出了双线阵间距失配情况下的两种波束形成方法的方向指向性函数，研究并分析了阵间距失配对双线阵左右舷分辨性能的影响，最后用海试数据验证了分析结果。

1　双线阵波束形成

假定双线阵间距为 D，线阵中各阵元以间距 d 等间隔分布(如图 1)。设双线阵中对应两个阵元接收信

① 声学学报, 2007, 32 (2): 158-164.

号分别为 $s_{A_1}(t), s_{B_1}(t), \cdots, s_{A_N}(t), s_{B_N}(t)$，$S_A(t)=\sum_{k=1}^{N} s_{A_k}(t-(k-1)*\varsigma)$，$S_B(t)=\sum_{k=1}^{N} s_{B_k}[t-(k-1)*\varsigma]$ 分别为扫描方向为 θ 时的两线阵的波束输出，$\varsigma=d\cos(\theta)/c$ 是 θ 方向的声波到达同一线阵相邻两个阵元的时间差，$\tau=D\sin\theta/c$ 是 θ 方向的声波到达两条线阵的时间差，N 为每条线阵的阵元数。对应左右舷两种情况，来波方向为 θ 或 $-\theta$ 时 $S_B(t)=S_A(t-\tau)$ 或 $S_A(t)=S_B(t-\tau)$。

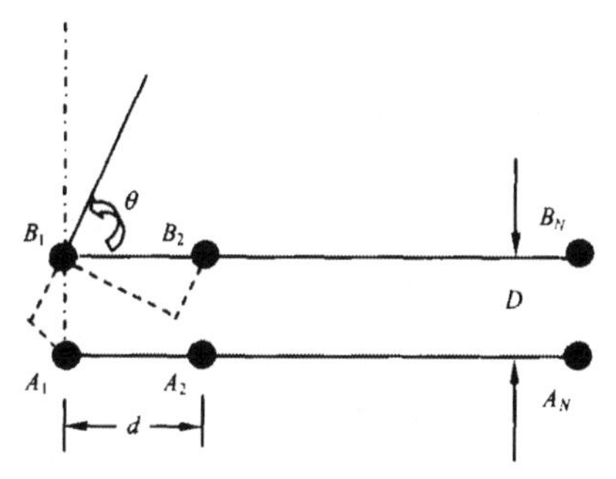

图 1 双线阵结构

1.1 延时相加波束形成

这种方式的输出 $\text{out}(t)=S_A(t-\tau)+S_B(t)$。

根据文献 8 其方向指向性函数为:

$$G_a(\theta,\theta_0)=\left|\frac{\sin\left\{\dfrac{N\pi d}{\lambda}[\cos(\theta)-\cos(\theta_0)]\right\}}{\sin\left\{\dfrac{\pi d}{\lambda}[\cos(\theta)-\cos(\theta_0)]\right\}}\right| \left\{1+\cos\left[\frac{2\pi D}{\lambda}(\sin(\theta)-\sin(\theta_0))\right]\right\}^{1/2}, \tag{1}$$

式中 λ 为信号波长，θ 为波束指向方向，θ_0 为信号入射方向，阵元间距 d 一般为 $\lambda/2$，N 为单阵的阵元数。

令 η 为波束输出中 θ_0 方向和 $-\theta_0$ 方向来波的能量比，则 $\eta=|\frac{G_a(\theta_0,\theta_0)}{G_a(\theta_0,-\theta_0)}|^2$，以 η 来衡量双线阵的左右舷分辨性能，以 $10\lg\eta$ 作为双线阵的左右舷分辨的增益。当单频信号从正横方向 ($\theta_0=90°$) 入射，对于波长 $\lambda=4D$ 的信号，η 最大，有很好的左右舷分辨性能，对于不满足 $\lambda=4D$ 的信号，其左右舷分辨性能会迅速下降。更一般地说，在满足条件 $\lambda=4D\sin\theta_0/m$，$m=1,3,5,\cdots$ 时，左右舷分辨性能最好，偏离此条件时左右舷分辨性能迅速下降。由于利用了两条线阵的信息，在各阵元噪声满足不相关条件时，双线阵的波束输出端有 $10\lg N+10\lg 2$ 的增益。

1.2 延时相减波束形成

这种方式的输出 $\text{out}(t)=S_A(t-\tau)-S_B(t)$。同理，根据文献 8 其方向指向函数为:

$$G_s(\theta,\theta_0)=\left|\frac{\sin\left\{\dfrac{N\pi d}{\lambda}[\cos(\theta)-\cos(\theta_0)]\right\}}{\sin\left\{\dfrac{\pi d}{\lambda}[\cos(\theta)-\cos(\theta_0)]\right\}}\right| \left\{1-\cos\left[\frac{2\pi D}{\lambda}(\sin(\theta)-\sin(\theta_0))\right]\right\}^{1/2}, \tag{2}$$

还用 η 表示波束输出中 $-\theta_0$ 方向和 θ_0 方向来波的能量比，有 $\eta=|\frac{G_s(\theta_0,-\theta_0)}{G_s(\theta_0,\theta_0)}|^2=|\frac{G_s(\theta_0,-\theta_0)}{0}|^2$；还以 $10\lg\eta$ 作为双线阵的左右舷分辨的增益。

从上式可见这种方式在理论上有无穷大的左右来波能量比，因而有极高的左右舷增益。当单频信号从正横方向入射，对于波长 $\lambda=4D$ 的信号，η 表达式与延时相加方式完全一样。当信号不从正横方向入射，或信号频率不满足 $\lambda=4D$ 时，η 平稳下降。因此延时相减方式在信号很宽的频率范围和来波方向范围内进行方位估计均有很高的左右舷抑制能力。同理，$D\sin\theta=(m/4)\lambda$，$m=1,3,5,\cdots$ 时，有最佳的左右舷分辨性能。

1.3 仿真结果分析

仿真得到的图 2 —图 5 为双线阵延时相加波束形成和延时相减波束形成的左右舷分辨性能图 (颜色棒的数值代表 $10\log\eta$，以 dB 为单位，数值越大，也就是颜色越浅代表左右舷分辨性能越强；对应的双线阵的工作频率为 1200 Hz)。从图 2 —图 5 可知：无论双阵间距 D 是 $\lambda/2$ 还是 $\lambda/4$，延时相减几乎在所有的频率和方位上都有很好的左右舷分辨性能 (超过 50 dB)，而延时相加只有在某些特定的频率和方位上有很好的左右舷分辨性能 (超过 50 dB)。双阵间距 D 为 $\lambda/4$ 时，延时相减在所有的频率和方位上都有很好的左右舷分辨性能，而在双阵间距 D 为 $\lambda/2$ 时，延时相减在某些的频率和方位上的左右舷分辨性能则不够理想 (图 3 灰黑色弧线部分)；延时相加在双阵间距 D 为 $\lambda/4$ 时的左右舷分辨性能比双阵间距 D 为 $\lambda/2$ 时的左右舷分辨性能好 (图 4 中白色的区域比图 2 范围更大)。因而，从图 2 —图 5 可得到如下结论：延时相减方式比延时相加方式有更好的左右舷分辨性能，当双阵间距 D 为 $\lambda/4$ 时，延时相减和延时相加的左右舷分辨性能都比双阵间距 D 为 $\lambda/2$ 时的左右舷分辨性能好。

下面分析存在本舰噪声干扰时的情况。在拖曳式双线阵中，本舰噪声相当于一个端射方向目标，即 $S_B(t)=S_A(t)$。扫描方向 θ 接近端射方向时，时间差 $\tau\to 0$，则采用延时相减方式的输出 $\text{out}(t)=S_A(t-\tau)-S_B(t)\to 0$，因此扫描方向接近端射方向时本舰噪声的影响将非常小，即能消除本舰干扰的影

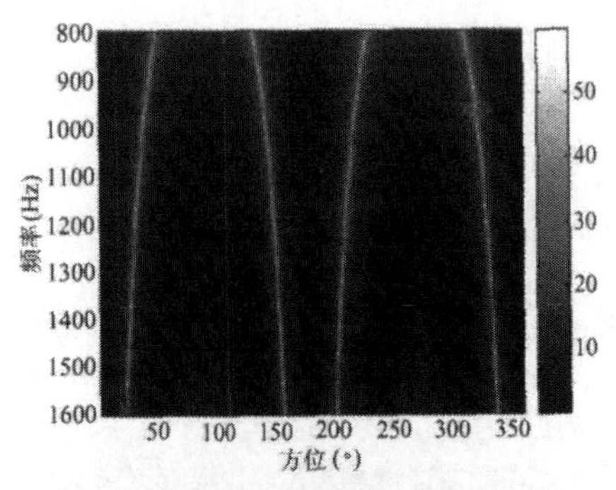

图 2　延时相加左右舷分辨性能示意图 ($D=\lambda/2$)

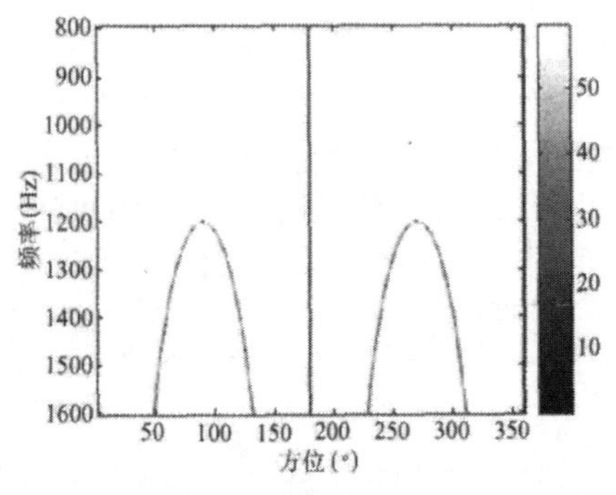

图 3　延时相减左右舷分辨性能示意图 ($D=\lambda/2$)

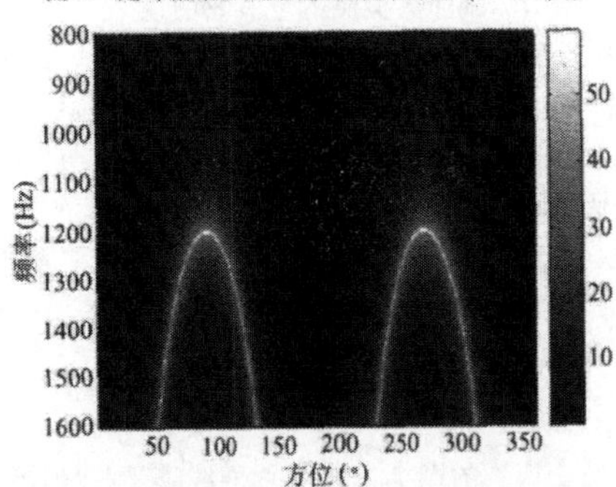

图 4　延时相加左右舷分辨性能示意图 ($D=\lambda/4$)

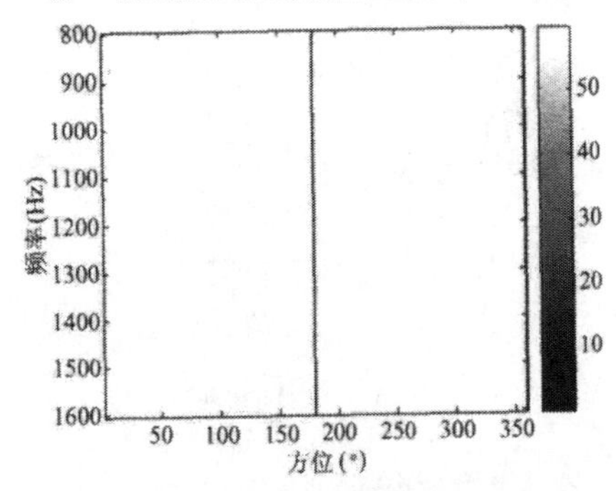

图 5　延时相减左右舷分辨性能示意图 ($D=\lambda/4$)

响. 同时, 端射方向的目标也将丢失。采用延时相加方式不能消除本舰干扰。

2　间距失配的双线阵波束形成

双线阵的实际应用中, 在线阵之间没有安装高频脉冲发生器 (PINGER) 的情况下, 线阵的真正间距是无法获得的。在做波束形成时, 所用的阵间距是不能以当时绞车前端双缆间距为基准的。文献1指出双线阵在水下会有微小夹角, 也就是双线阵的阵间距并不固定。但是, 目前大部分如文献 9 —文献 11 在讨论双线阵的波束形成时, 都把双线阵在水下的姿态简单假设为平行的; 事实上实验表明, 这种假设是值得推敲的 [1]. 本文所指的双线阵间距失配是如文献1 那样认为双线阵在水下有微小夹角, 也就是双线阵的阵间距不固定。为简化模型, 假设单条阵在水下没有畸变, 其模型如图 6 所示, 拖曳方向如图中黑色箭头所示; 阵元 A_1 和 B_1 分别为双线阵拖曳方向的头阵元。

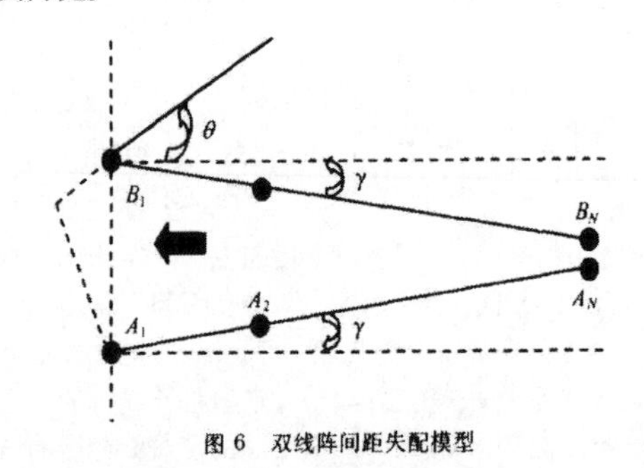

图 6　双线阵间距失配模型

2.1　阵间距失配的延时相加波束形成

根据文献 8 可以推导出其方向指向函数 (以单频信号为例) 为:

$$G_{MA}(\theta,\theta_0)=\left|\frac{\sin^2\frac{N\pi d}{\lambda}[f(\theta,\gamma)-f(\theta_0,\gamma_0)]}{2\sin^2\frac{\pi d}{\lambda}[f(\theta,\gamma)-f(\theta_0,\gamma_0)]}+\frac{\sin^2\frac{N\pi d}{\lambda}[g(\theta,\gamma)-g(\theta_0,\gamma_0)]}{2\sin^2\frac{\pi d}{\lambda}[g(\theta,\gamma)-g(\theta_0,\gamma_0)]}+\right.$$
$$\left.\frac{\sin\frac{N\pi d}{\lambda}[f(\theta,\gamma)-f(\theta_0,\gamma_0)]\sin\frac{N\pi d}{\lambda}[g(\theta,\gamma)-g(\theta_0,\gamma_0)]}{\sin\frac{\pi d}{\lambda}[f(\theta,\gamma)-f(\theta_0,\gamma_0)]\sin\frac{\pi d}{\lambda}[g(\theta,\gamma)-g(\theta_0,\gamma_0)]}\cdot\cos\left\{\frac{2\pi}{\lambda}[(D\sin\theta-D_1\sin\theta_0)-d(N-1)(\sin\gamma\sin\theta-\sin\gamma_0\sin\theta_0)]\right\}\right|^{1/2} \tag{3}$$

其中:

$$f(\theta,\gamma)=\cos(\gamma)\cos(\theta)+\sin(\gamma)\sin(\theta),\quad f(\theta_0,\gamma_0)=\cos(\gamma_0)\cos(\theta_0)+\sin(\gamma_0)\sin(\theta_0),$$
$$g(\theta,\gamma)=\cos(\gamma)\cos(\theta)-\sin(\gamma)\sin(\theta),\quad g(\theta_0,\gamma_0)=\cos(\gamma_0)\cos(\theta_0)-\sin(\gamma_0)\sin(\theta_0).$$

双线阵头阵元之间的真实间距为 D(如图 6 所示阵元 A_1 和 B_1 的间距), 2γ 为双线阵真实向内倾斜的夹角; D_1 为实际修正时的双线阵头阵元之间的间距, $2\gamma_0$ 为实际修正时双线阵向内倾斜的夹角。从上式我们可以得知双线阵失配有三种情况: 双线阵阵头间距失配, 即 D 与 D_1 失配; 双线阵倾斜角失配, 即 2γ 与 $2\gamma_0$ 失配; D 与 D_1 失配, 2γ 与 $2\gamma_0$ 同时失配。

参照 1.1 节的定义, $\eta=\left|\frac{G_{MA}(\theta_0,\theta_0)}{G_{MA}(\theta_0,-\theta_0)}\right|^2$; 并以 $10\lg\eta$ 作为双线阵左右舷分辨的增益。

2.2 阵间距失配的延时相减波束形成

同理根据文献 8 可得其方向指向函数 (以单频信号为例) 为:

$$G_{MS}(\theta,\theta_0)=\left|\frac{\sin^2\frac{N\pi d}{\lambda}[f(\theta,\gamma)-f(\theta_0,\gamma_0)]}{2\sin^2\frac{\pi d}{\lambda}[f(\theta,\gamma)-f(\theta_0,\gamma_0)]}+\frac{\sin^2\frac{N\pi d}{\lambda}[g(\theta,\gamma)-g(\theta_0,\gamma_0)]}{2\sin^2\frac{\pi d}{\lambda}[g(\theta,\gamma)-g(\theta_0,\gamma_0)]}-\frac{\sin\frac{N\pi d}{\lambda}[f(\theta,\gamma)-f(\theta_0,\gamma_0)]\sin\frac{N\pi d}{\lambda}[g(\theta,\gamma)-g(\theta_0,\gamma_0)]}{\sin\frac{\pi d}{\lambda}[f(\theta,\gamma)-f(\theta_0,\gamma_0)]\sin\frac{\pi d}{\lambda}[g(\theta,\gamma)-g(\theta_0,\gamma_0)]}\cos\left\{\frac{2\pi}{\lambda}[(D\sin\theta-D_1\sin\theta_0)-d(N-1)(\sin\gamma\sin\theta-\sin\gamma_0\sin\theta_0)]\right\}\right|^{1/2} \tag{4}$$

参照 1.2 节的定义, $\eta=\left|\frac{G_{MS}(\theta_0,-\theta_0)}{G_{MS}(\theta_0,\theta_0)}\right|^2$; 并以 $10\lg\eta$ 作为双线阵左右舷分辨增益。

2.3 仿真结果分析

图 7 —图 9 是工作频率为 1200 Hz 的双线阵在各种失配状态下对 800~1600 Hz 范围的信号的左右舷分辨性能示意图 (颜色棒的数值代表左右舷分辨性能, 以 dB 为单位, 正值表明能正确判别目标的左右舷, 负值表明不能正确判别目标的左右舷, 图中纵坐标为频率, 横坐标为方位) 。图 7 —图 9 中 (a) 图是失配状况下延时相加得到的左右舷分辨性能示意图, (b) 图是延时相减得到的左右舷分辨性能示意图, (c) 图是两者左右舷分辨性能的比较。对于失配状况下左右舷分辨性能的比较, 我们是这样规定的: 在由频率和角度组成的平面内, 当延时相加和延时相减两种方式都不能正确进行左右舷分辨的区域, 我们用灰色表示; 两者都能正确进行左右舷分辨但是延时相减方式性能不如延时相加方式时, 这样的区域, 我们用黑色表示; 两者都能正确进行左右舷分辨但是延时相加方式性能不如延时相减方式时, 这样的区域, 我们用白色表示。比较图 7 —图 9 延时相加和延时相减在失配状况下左右舷分辨性能图, 发现延时相加方式和延时相减方式在失配状况下, 在由频率和角度组成的平面内, 不能正确进行左右舷判断的区域相同, 也就是说: 在特定的频率和角度上, 延时相加和延时相减的左右舷判决要么同时正确, 要么同时错误。通过对图 7 —图 9 的比较分析, 可以得知: 双线阵阵头间距失配对双线阵左右舷分辨性能影响较大; 双线阵倾斜角的失配对双线阵左右舷分辨性能的影响跟阵头间距失配的影响相比, 相对较小; 双线阵阵头间距和双线阵倾斜角同时失配, 起主要作用的是阵头间距的失配。在各种失配情况下, 延时相加和延时相减的左右舷分辨性能都会

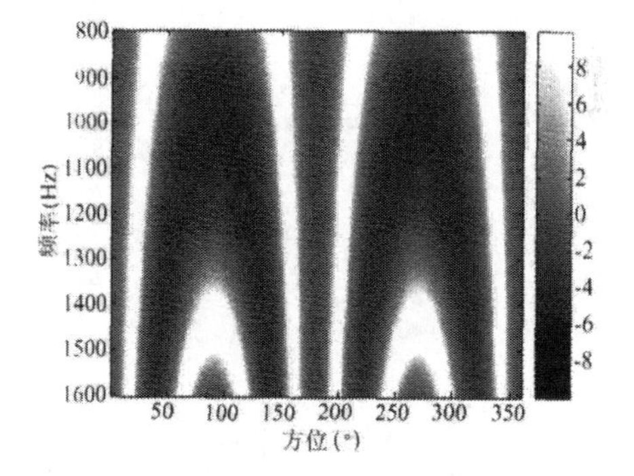

图 7(a) 延时相加左右舷分辨性能图

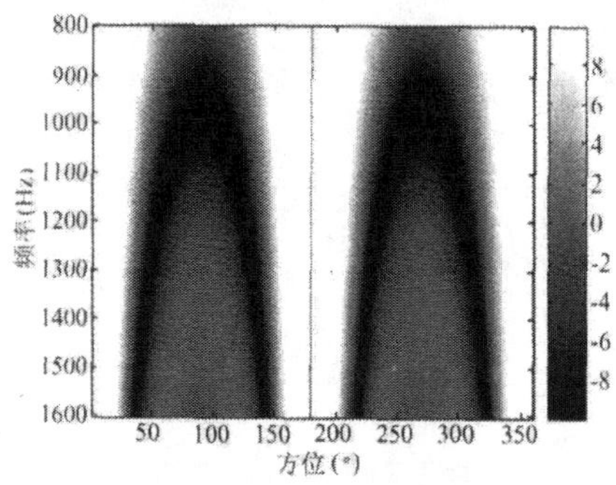

图 7(b) 延时相减左右舷分辨性能图

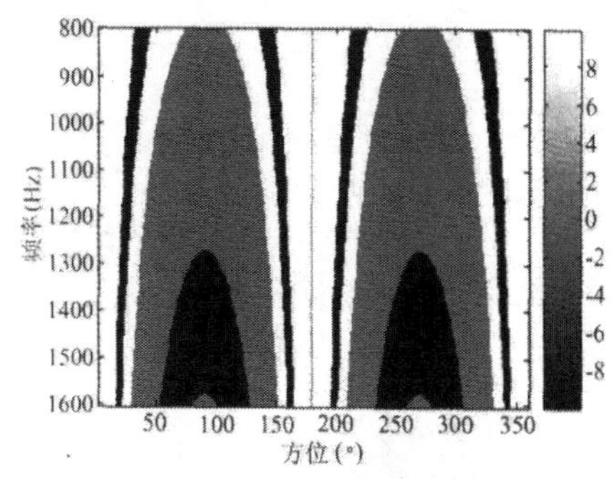

图 7(c) 两者左右舷分辨性能比较图

图 7 阵间距失配 ($D=\lambda/2$, $D_1=3\lambda/4$, $\gamma=\gamma_0=0$)

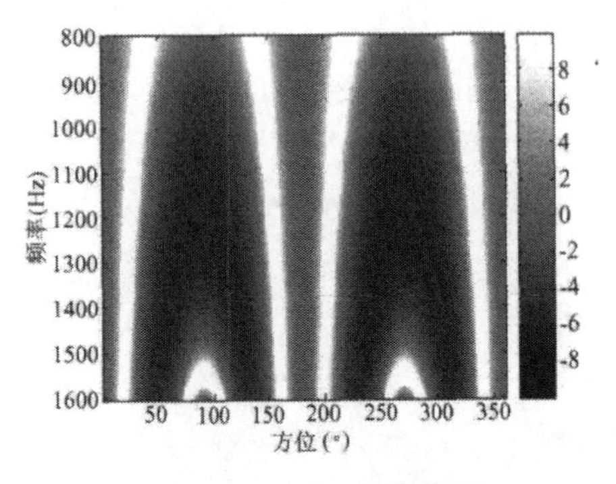

图 8(a) 延时相加左右舷分辨性能图

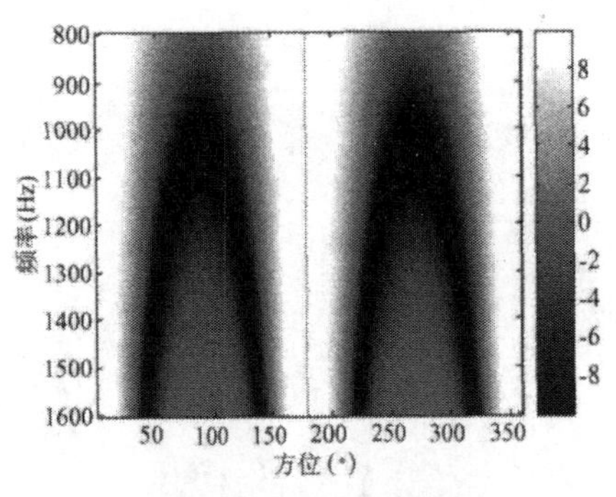

图 8(b) 延时相减左右舷分辨性能图

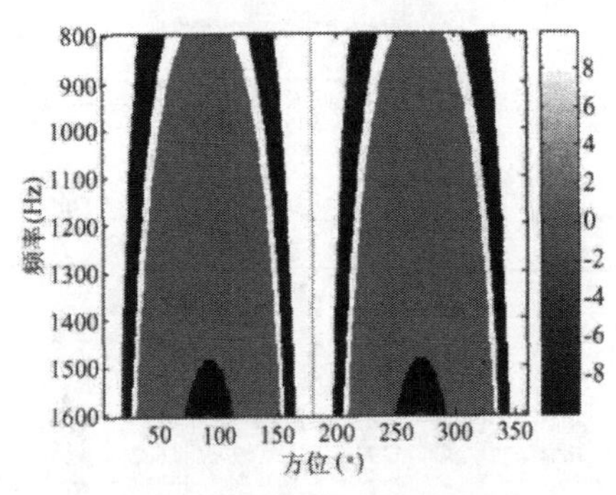

图 8(c) 两者左右舷分辨性能比较图

图 8 阵间距失配 ($D=\lambda/2$, $D_1=3\lambda/4$, $\gamma=0.25°$, $\gamma_0=0$)

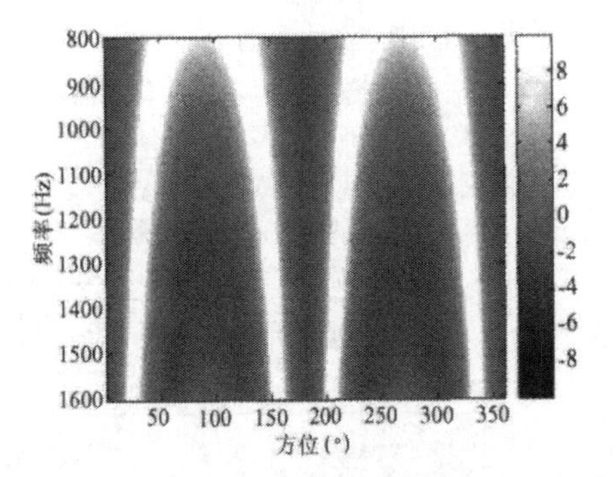

图 9(a) 延时相加左右舷分辨性能图

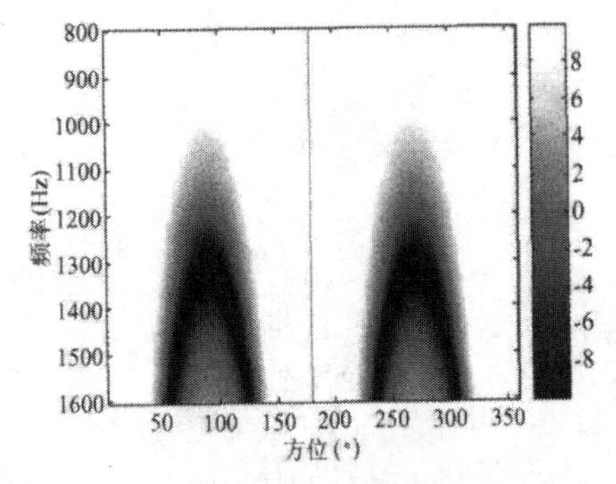

图 9(b) 延时相减左右舷分辨性能图

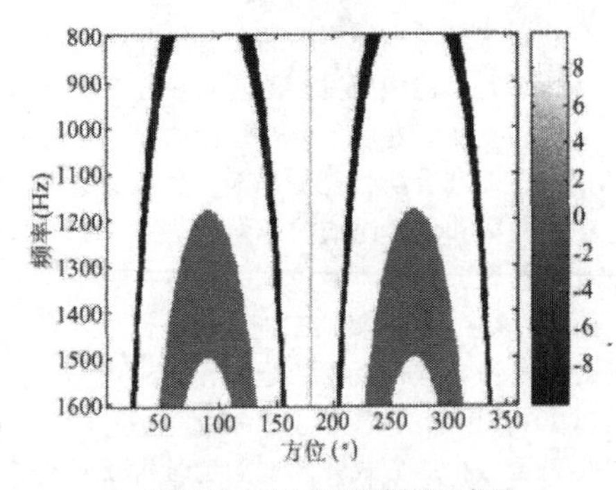

图 9(c) 两者左右舷分辨性能比较图

图 9 阵间距失配 ($D=\lambda/2$, $D_1=\lambda/2$, $\gamma=0.25°$, $\gamma_0=0$

急剧下降，在某些频率和角度上甚至不能正确进行左右舷判决。在阵间距失配情况下，延时相加方式在大部分频率和角度上其左右舷分辨性能都比延时相减的方式差，仅在某些特定的频率和角度上比延时相减方式有好的左右舷分辨性能。此外，在失配情况下，延时相加方式和延时相减方式与各自未失配时相比，延时相减方式对失配更为敏感。

3 海试数据分析

图 10 和图 11 分别是海试中采用延时相减和延时相加方式的原始方位历程图，图中 0° 附近的两根白线是本舰噪声航迹。对应时刻有两个目标①和②(不在同一侧)：实验目标船 (2.1 海里)、远处货船 (10 海里) 分别对应图中斜线 (左右舷) ①目标、 97°(左右舷) 附近②目标。可以看到，延时相加方式 (图 11) 的左右舷分辨性能不如延时相减方式 (图 10)，但也能区分左右。在延时相减方式 (图 10) 中，因为 97° 附近目标是一很远的货船，因此左右舷分辨抑制比极高；而实验目标船距离较近，与远场假设不匹配，左右舷分辨性能稍差，但也能区分出目标的左右舷。由于本舰使用双螺旋桨，分别在拖曳阵两侧，因此从两

图中看不出相减方式在抑制本舰噪声方面的优势。

下面分析双线阵间距 D 的影响。实验时绞车前端双缆间距是 $3\lambda/4$ (λ 是声呐中心频率所对应的波长)，实验时每条线阵各自安装了 3 个深度传感器，根据当时双阵的深度传感器所显示的数据表明，双阵基本上处于同一深度。由于没有安装高频 PINGER，实际水中双阵间距未知。使用当时原始采集数据，在算法中采用延时相加和延时相减方式，并且设置二组间距 $(3\lambda/4, \lambda/2)$ 以及一 0.25° 的补偿角。将双阵间距设置为 $3\lambda/4$ 时，左右舷分辨性能并不是很好 (可以远处货船为判断标准，图 12 —图 15 的上图都是延时相减方式所得到的，而下图都是延时相加方式

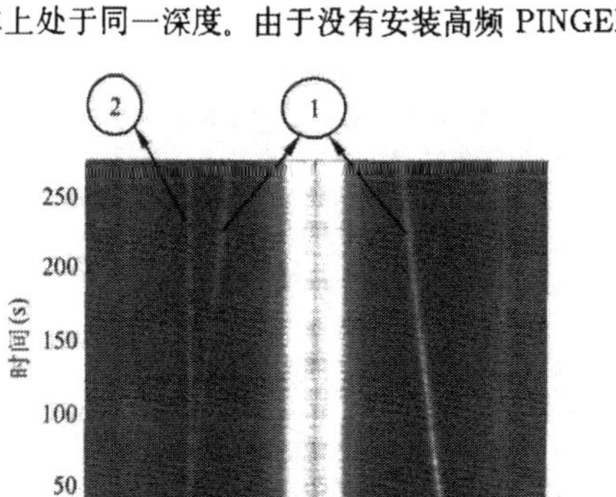

图 10 采用延时相减方式的原始方位历程图

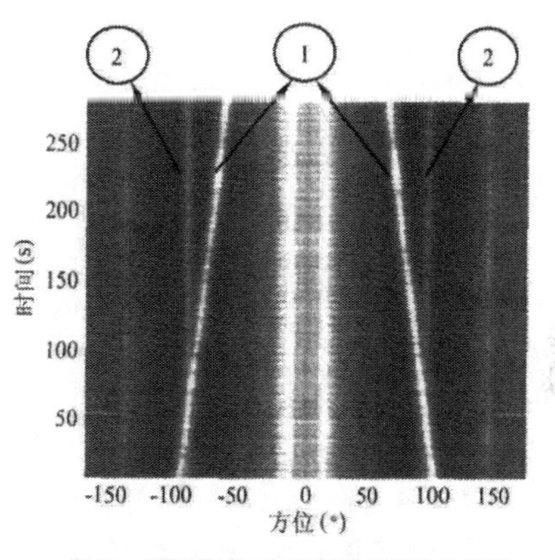

图 11 采用延时相加方式的原始方位历程图

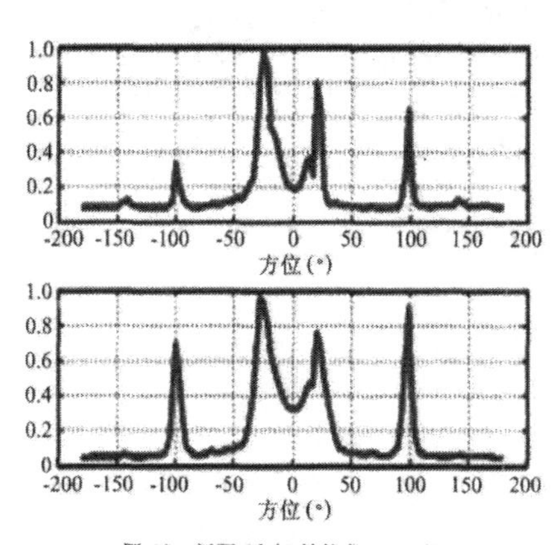

图 12 间距 $3\lambda/4$ 补偿角 $\gamma_0 = 0$

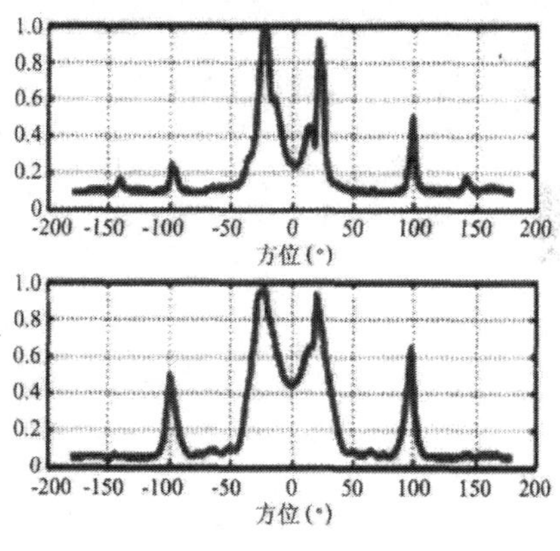

图 13 间距 $3\lambda/4$ 补偿角 $\gamma_0 = 0.25$

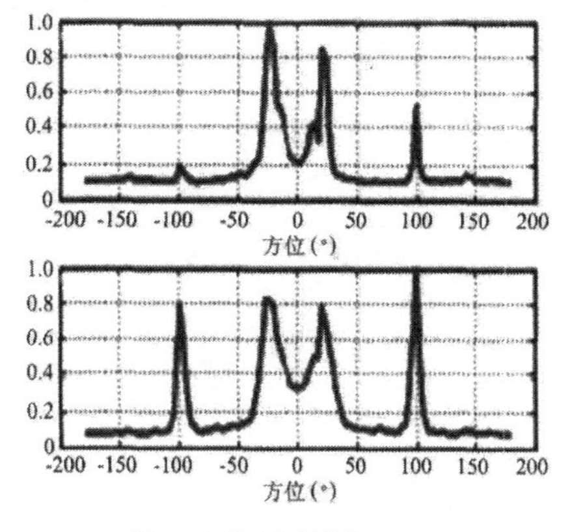

图 14 间距 $\lambda/2$ 补偿角 $\gamma_0 = 0$

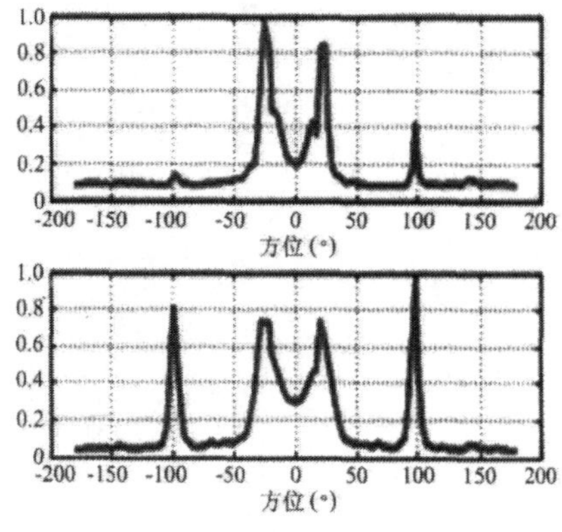

图 15 间距 $\lambda/2$ 补偿角 $\gamma_0 = 0.25$

所得到的) 但也能区分目标的左右舷。而将双阵间距设置为 $\lambda/2$ 时能得到更好的结果，尤其是对延时相减方式更为明显，但是将双阵间距设置为 $\lambda/2$ 再给予一 0.25° 的补偿角时效果则不明显。图 12 一图 15 所显示的结果证实了本文所用双线阵水下模型的有效性，即双线阵在水下会以一微小的角度向内靠拢。

4 结论

在阵间距没有失配的情况下，采用延时相加方式和延时相减方式的双线阵都能准确对目标进行左右舷判决。但相比之下，采用延时相减方式在方位估计中性能得到很大提高 (见第 1 节图 3 和图 5, 对信号频率、入射方向和双线阵间距都不敏感), 左右舷分辨性能比延时相加方式好。

在阵间距失配的情况下，其阵头间距的失配对双线阵左右舷分辨性能影响较大；倾斜角的失配对双线阵左右舷分辨性能影响与阵头间距失配的影响相比，相对较小；而阵头间距和倾斜角同时失配时，起主要作用的是阵头间距的失配。在阵间距失配时，延时相加和延时相减方式的左右舷分辨性能都会急剧下降，在某些方位和频率上甚至不能正确地分辨目标左右舷。虽然延时相减的方式对失配更为敏感，但其左右舷分辨性能依然比延时相加的好。

参 考 文 献

1 Schurman I W. Reverberation rejection with a dual-line towed array. *IEEE, Journal of Oceanic Engineering*, 1996; **21**(2): 193—204

2 杜选民, 朱代柱等. 拖线阵左右舷分辨技术的理论分析与实验研究. 声学学报, 2000; **25**(5): 395—402

3 杜选民, 朱代柱等. 拖线阵目标左右舷分辨技术研究 (II) —— 噪声相关性模型. 舰船科学技术, 1999; **21**(3): 17—20

4 何心怡, 蒋兴舟, 张春华. 基于几何相移模型的双线列阵左右舷分辨技术研究. 应用声学, 2004; **23**(2): 38—44

5 李启虎. 双线阵左右舷目标分辨性能的初步分析. 声学学报, 2006; **31**(5): 385—388

6 李启虎. 用双线阵区分左右舷目标的延时估计方法及其实现. 声学学报, 2006; **31**(6): 481—484

7 李启虎等. 主被动拖线阵声呐中拖曳平台和拖鱼噪声在浅海使用时的干扰特性. 声学学报, 2007; **32**(1): 1—5

8 李启虎. 数字式声呐设计原理. 安徽: 安徽教育出版社, 2002

9 Newhall B K, Allensworth W S, Schurman I W. Unambiguous noise and reverberation measurements from a dual line towed array. *J. Acoust. Soc. Am.*, 1995; **97**(5): 3289

10 Feuillet J P, Allensworth W S, Newhall B K. Nonambiguous beamforming for a high resolution twin-line array. *J. Acoust. Soc. Am.*,1995; **97**(5): 3292

11 Allensworth W S, Kennedy C W, Newhall B K, Schurman I W. Twinline array development and performance in a shallow-water littoral environment. *Johns Hopkins APL Tech. Dig*, 1995; **16**(3): 222—232

被动合成孔径的非对称双线阵相位校正方法研究①

李宇　张扬帆　黄海宁　李淑秋　李启虎　张春华
(中国科学院声学研究所　北京　100080)

2006 年 11 月 12 日收到
2006 年 11 月 24 日定稿

摘要　提出了一种基于被动合成孔径的相位校正方法，有效地解决了因为首阵元未对齐 (非对称) 而引起的双线阵左右舷误判问题。利用被动合成孔径的方法，将其中一个单阵通过相位校正合成为一个虚拟单线阵，新的虚拟单线阵和另一个实际的单线阵构成首阵元对齐的双线阵，恢复了双线阵的对称性，因此可以纠正非对称双线阵导致的左右舷误判。通过数值仿真和海试试验数据的对比分析，说明这种算法可以解决非对称双线阵左右舷误判问题，且方法简单易于实际应用。

PACS 数：　43.30, 43.60

A passive synthetic aperture phase correction algorithm for the asymmetric twin-line array sonar

LI Yu　ZHANG Yangfan　HUANG Haining　LI Shuqiu　LI Qihu　ZHANG Chunhua
(*Institute of Acoustics, The Chinese Academy of Sciences* Beijing 100080)

Received Nov. 12, 2006
Revised Nov. 24, 2006

Abstract　A passive synthetic aperture based on phase correction algorithm for solving the port-starboard discrimination problem in the non-aligned towed twin-line array sonar is described. This method creates a virtual array through applying the estimated phase correction into one array of twin-line array, and because the synthetic virtual array is aligned with the other array in twin-line arrays, the right port-starboard discriminated results can be obtained by array processing based on the new synthetic twin-line array. The effect of proposed method would be shown by simulation and sea-trials results in towed twin-line array sonar. With low extra computational loads, the proposed method is easy to apply into the practice.

引言

近年来，拖曳多线阵声呐特别是双线阵声呐在声呐领域中得到广泛的应用。双线阵声呐的特点众多，除了可以有效地提高探测作用距离和方位分辨能力外，能够判别左右舷也是双线阵声呐所具有的一个重要特点 [1-3]。双线阵可以有效地判断左右舷的一个重要前提是两个线阵必须按严格的矩形配置 [4,5]，即双线阵平行且首阵元必须对齐，这样才能保证双线阵沿中心线是轴对称的，如图 1(a) 所示 (这里主要考虑水平拖曳的情况，图中坐标横轴表示阵的水平运动方向)。但是，在实际运用中由于拖曳和水流的影响，双拖线阵往往出现首阵元没有对齐的非对称现象，即双线阵的配置成为平行四边形，如图 1(b) 所示。这样的现象直接导致左右舷分辨中出现误判，即在一定角度区域内，会出现左右舷判决与实际的左右舷相反的问题 [6]。这样的误判是相当严重的，试验中发现即使是已经知道双线阵由于未对齐而导致的延时 (相位)，并在处理中进行了补偿，误判问题也不能被解决。

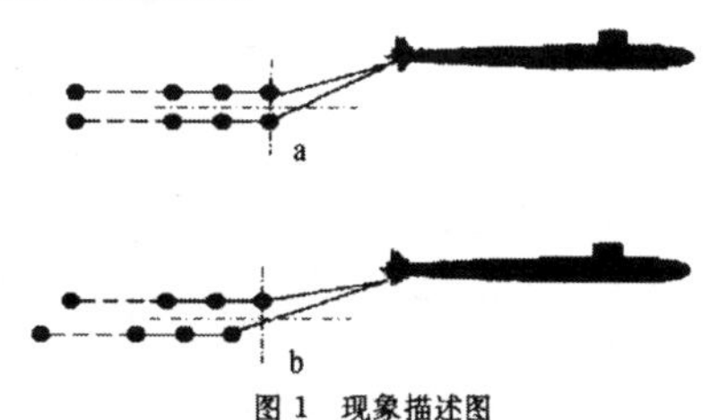

图 1　现象描述图

针对这个问题，本文首先分析了通过简单的相位补偿不能校正左右舷误判的原因，发现是由于拖曳造成的双线阵实际物理坐标系与设定坐标系的不一

① 声学学报, 2007, 32 (3): 264-269.

致性，导致了准确的补偿也无法校正左右舷误判。然后考虑到拖曳阵测量实际上是时空采样，利用这种时空采样特性并基于被动合成孔径的思想[7]，提出了非对称双线阵的相位校正方法。方法首先估计双线阵非对称导致的畸变相位，然后将其中一个单阵通过相位校正合成为一个虚拟单线阵，新的虚拟单线阵和另一个实际的单线阵构成首阵元对齐的双线阵，恢复了双线阵的对称性，因此可以纠正非对称双线阵导致的左右舷误判。最后，利用数据仿真对比了对称理想情况、没有进行非对称相位补偿 (失配情况)、进行非对称相位补偿 (匹配情况) 以及使用描述方法进行相位校正等四种情况下，左右舷分辨增益与目标方位的关系，说明方法可以使非对称双线阵得到与理想对称双线阵一样的左右舷分辨结果，这一结论也通过对海试实测数据的处理得到了验证。另外，还通过仿真分析了双线阵非对称阵畸变程度对左右舷误判严重度的影响。

1 基本原理

1.1 双线阵非对称现象分析

对于阵元数为 N、等间隔 d 布阵且双阵间距为 H 的双线阵 (以靠近船一侧的阵元为首阵元)，如果存在方位为 θ 的远场源目标 (以舷侧方向为参考方向)，t_i 时刻 A 阵通道 n 的输出可以表示为 (暂时忽略噪声的影响)：

$$x_n^A(t_i)=s\left[t_i-\frac{(n-1)d}{c}\cos\theta\right]. \tag{1}$$

若以 A 阵为参考阵，双线阵对称，则同一时刻 B 阵通道 n 的输出可以表示为：

$$x_n^B(t_i)=s\left[t_i-\frac{(n-1)d}{c}\cos\theta-\frac{H}{c}\sin\theta\right]. \tag{2}$$

但是，若双线阵非对称 (阵元未对齐)，如图 2 中实心阵元表示的情况，则同一时刻 B 阵通道 n 的输出为：

$$x_n^B(t_i)_a=s\left[t_i-\frac{(n-1)d}{c}\cos\theta-\frac{H}{c}\sin\theta-\frac{D}{c}\cos\theta\right], \tag{3}$$

对比式 (2) 和式 (3) 可知，非对称情况下，B 阵多了一个时延因子 $(D/c)\cos\theta$，这里称 D 为非对称距离，来描述双线阵非对称的程度，在第 3.2 节中我们会通过仿真试验说明 D 对左右舷误判程度的影响。

如果信号为窄带信号，则单阵的波束形成可以描述为：

$$\boldsymbol{Y}=\boldsymbol{W}\boldsymbol{X}, \tag{4}$$

其中，$\boldsymbol{Y}$ 表示多波束输出向量，$\boldsymbol{Y}=[y_1,y_2,\cdots,y_p]^{\mathrm{T}}$，$p$ 表示输出波束数目，符号 T 表示转置；$\boldsymbol{X}$ 表示数据向量，$\boldsymbol{X}=[x_1,x_2,\cdots,x_N]^{\mathrm{T}}$；$\boldsymbol{W}$ 表示加权矩阵，维数为 $p\times N$，且：

$$\boldsymbol{W}=\begin{bmatrix}1 & \exp[\mathrm{j}2kd\cos\theta_1] & \cdots & \exp[\mathrm{j}(N-1)kd\cos\theta_1]\\ 1 & \exp[\mathrm{j}2kd\cos\theta_2] & \cdots & \exp[\mathrm{j}(N-1)kd\cos\theta_2]\\ \vdots & \vdots & \ddots & \vdots\\ 1 & \exp[\mathrm{j}2kd\cos\theta_p] & \cdots & \exp[\mathrm{j}(N-1)kd\cos\theta_p]\end{bmatrix}$$

其中，k 为波数。

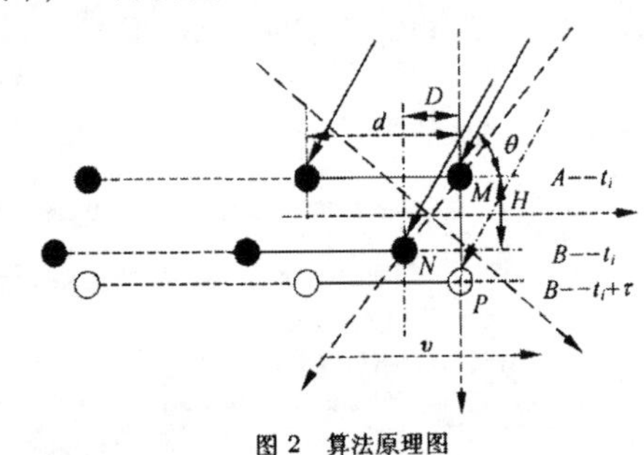

图 2 算法原理图

而双线阵的波束形成过程可以分解为先进行两个单线阵波束形成，然后再进行波束域的双线阵处理，即可以描述为：

$$\boldsymbol{Z}=\mathrm{diag}(\boldsymbol{W}'\boldsymbol{Y}'), \tag{5}$$

其中，$\boldsymbol{Z}$ 表示双线阵多波束输出向量，即360° 方位的波束输出，据此判别左右舷，且 $\boldsymbol{Z}=[z_{-p},\cdots,z_{-2},z_{-1},z_1,z_2,\cdots,z_p]^{\mathrm{T}}$，负的下标表示共轭波束，$\boldsymbol{Y}'$ 表示双线阵波束域数据矩阵，可以表示为：

$$\boldsymbol{Y}'=\begin{bmatrix}Y_A^z & Y_B^z\\ Y_A & Y_B\end{bmatrix}^{\mathrm{T}}. \tag{6}$$

这里，Y_A,Y_B 分别为 A 和 B 阵的单阵波束成形结果，而 Y_A^z,Y_B^z 分别是 Y_A,Y_B 的按角度的倒序排列，即 $Y^z=[y_p,\cdots,y_2,y_1]^{\mathrm{T}}$，$\boldsymbol{Y}'$ 的维数为 $2\times 2p$。

$\boldsymbol{W}'$ 为双线阵加权矩阵，可以表示为：

$$\boldsymbol{W}'=\begin{bmatrix}\boldsymbol{1} & \boldsymbol{W}_d^z\\ \boldsymbol{W}_d & \boldsymbol{1}\end{bmatrix}. \tag{7}$$

这里，$\boldsymbol{1}$ 表示单位向量，维数为$p\times 1$；$\boldsymbol{W}_d$ 和 $\boldsymbol{W}_d^z$ 为双线阵加权向量，且$\boldsymbol{W}_d^z=[\psi_p,\cdots,\psi_i,\cdots,\psi_2,\psi_1]^{\mathrm{T}}$，$\boldsymbol{W}_d=[\psi_1,\psi_2,\cdots,\psi_i,\cdots,\psi_p]^{\mathrm{T}}$，其中 $\psi_i=\exp(\mathrm{j}kH\sin\theta_i)$，$i=1,2,\cdots,p$；$\boldsymbol{W}'$ 的维数为 $2p\times 2$。式 (7) 实际上表示的是双线阵几何相加的情况，而如果是几何相减的情况，双线阵加权矩阵 $\boldsymbol{W}'$ 可以写作：

$$\boldsymbol{W}'=\begin{bmatrix}\boldsymbol{-1} & \boldsymbol{W}_d^z\\ \boldsymbol{W}_d & \boldsymbol{-1}\end{bmatrix}, \tag{8}$$

这里，$\boldsymbol{-1}$ 表示单位向量的负数，维数为 $p\times 1$。

由双线阵加权矩阵的定义可以看出，加权向量共轭对 $\boldsymbol{W}_d$ 和 $\boldsymbol{W}_d^z$ 是旋转对称的，即对方位共轭对的加权相对于船体轴线是旋转对称的，这是因为首阵元相对舷侧法线方向是对齐的，而这样的对称性保证了根据双阵波束形成结果进行左右舷分辨时的对称一致性。如果双线阵出现非对称的情况，即阵元相对舷侧法线方向未对齐，则加权向量共轭对 $\boldsymbol{W}_d$ 和 $\boldsymbol{W}_d^z$ 不是旋转对称的，而是：

$$\boldsymbol{W}_d=[\psi_1^-,\psi_2^-,\cdots,\psi_i^-,\cdots,\psi_p^-]^{\mathrm{T}},$$
$$\boldsymbol{W}_d^z=[\psi_p^+,\cdots,\psi_i^+,\cdots,\psi_2^+,\psi_1^+]^{\mathrm{T}},$$

这里

$$\psi_i^+=\exp[\mathrm{j}k(H\sin\theta_i+D\cos\theta_i)],$$
$$\psi_i^-=\exp[\mathrm{j}k(H\sin\theta_i-D\cos\theta_i)],\ i=1,2,\cdots,p$$

因此在这种情况下，双线阵的左右舷分辨不再具有对称一致性。这种非对称性导致了左右舷分辨中出现误判现象，即在一定方位角度范围内，左右舷分辨结果和实际目标所在位置相反的现象。这样的误判问题即使是按实际的加权向量进行补偿也无法达到理想的分辨结果。

这个现象可以用物理坐标系旋转来解释：在双线阵对称情况，实际坐标系与设定坐标系重合 (这里所谓“设定”是指船体作为参照物的习惯坐标系，即图2中短虚线直角坐标系)，这样加权向量共轭对是旋转对称的，从而保证了左右舷判决时，共轭角度所进行的加权是一致加权，所以也保证了左右舷判决的准确性；非对齐情况下的实际坐标系 (图2中长虚线直角坐标系) 相对于设定坐标系旋转了一定角度，这样实际坐标系中一侧的部分角度就相当于被设定坐标系强制划分到另一侧，破坏了双线阵左右分辨的对称性，而这样的非对称性使得加权向量共轭对不再是旋转对称的，从而导致左右舷判决时共轭角度所进行的加权不一致，这样的非一致加权虽然补偿了时延所产生的相位，但无法校正由于坐标旋转造成共轭角度的失配，因此即使按照实际波达方向进行了相位补偿也会出现左右舷判别的误判现象，而且判别错误区间会根据参考阵列相对于另一个阵列的超前或者滞后的情况，集中出现在相应的一侧。关于这点，我们在第2.1节通过仿真数据进行了量化分析。

1.2 基于被动合成孔径的相位校正方法

通过上节的介绍，可以知道双线阵非对称引起的左右舷误判主要是因为实际坐标系与设定坐标系的不一致，如果能够通过处理使得两坐标系能够重新一致即可解决左右舷误判的问题。

考虑拖曳阵测量实际上是时空采样，即拖曳运动使得每次测量都是不同时间和空间上的采样。因此，如果假设拖曳是水平匀速的话，两条阵在相应的平行空间上可以看成是遍历的。这样如图 2 所示，如果 t_i 时刻 A 阵运动到 M 点，这时 B 阵运动到 N 点，与 A 阵没有对齐，但是总存在一个 $t_i+\tau$ 时刻 B 阵可以运动到与 M 点位置对齐的 P 点，这样不同时刻的双阵在空间上是对齐的，如果能够通过 $t_i+\tau$ 时刻 B 阵虚拟出一个 t_i 时刻就到达 P 点的阵列，那么这个阵列和 A 阵在时间和空间上都是对齐的，这时和双阵对称时一样两坐标系得以恢复一致，可以消除左右舷误判；而根据被动合成孔径理论[7]，如果信号是窄带的话，所需的虚拟阵列和真实阵列之间仅差一个相位校正因子，因此可以通过估计相位校正因子、形成虚拟阵列的方法来解决非对称双线阵左右舷误判的问题。这是基于被动合成孔径的相位校正算法的基本思路，而具体描述如下：

假设双线阵以速度 v 向前运动且在靠近目标，暂时不考虑目标的运动，且信号是窄带信号，且角频率为 ω_0，则式 (1) 至 (3) 可以改写为：

$$x_n^A(t_i)=A\exp\left\{\mathrm{j}\omega_0\left[t_i-\frac{(n-1)d-vt_i}{c}\cos\theta\right]\right\},\tag{9}$$

$$x_n^B(t_i)=A\exp\left\{\mathrm{j}\omega_0\left[t_i-\frac{(n-1)d-vt_i}{c}\cos\theta-\frac{H}{c}\sin\theta\right]\right\},\tag{10}$$

$$x_n^B(t_i)_a=A\exp\left\{\mathrm{j}\omega_0\left[t_i-\frac{(n-1)d-vt_i}{c}\cos\theta-\frac{H}{c}\sin\theta-\frac{D}{c}\cos\theta\right]\right\},\tag{11}$$

而 $t_i+\tau$ 时刻，非对称情况下，B 阵通道 n 的输出为：

$$x_n^B(t_i+\tau)_a=\exp(\mathrm{j}\omega_0\tau)A\exp\left\{\mathrm{j}\omega_0\left[t_i-\frac{(n-1)d-vt_i-v\tau}{c}\cos\theta-\frac{H}{c}\sin\theta-\frac{D}{c}\cos\theta\right]\right\}.\tag{12}$$

对比式 (10) 和式 (12)，可以看出只要满足：

$$D=v\tau,\tag{13}$$

就有：

$$x_n^B(t_i+\tau)_a=\exp(\mathrm{j}\omega_0\tau)*x_n^B(t_i).\tag{14}$$

因此，可以通过对 $t_i+\tau$ 时刻非对称的 $x_n^B(t_i+\tau)_a$ 进行相位校正得到 t_i 时刻对称的 $x_n^B(t_i)$ 。

由于 $\tau = D/v$, 要准确估计 τ, 必须得到准确的 v 和 D 估计。v 的估计可以通过船载速度传感器或者阵上速度传感器信息获得; D 的估计可以根据双线阵的水下深度以及放缆长度通过几何关系求取。另外, D 也可以通过校正双阵相位估计来求取时延。

由式(9)和式(11)可知, 同一时刻非对称的双阵之间除了相差一个非对称相位因子 $\varphi_D=\omega_0(D/c)\cos\theta$, 还相差一个双线阵相位因子 $\varphi_H=\omega_0(H/c)\sin\theta$, 因此可以通过对一个阵进行相位 φ_H 补偿后与另一个阵进行时域相关, 取其相位估计作为 φ_D 的估计, 即:

$$\widehat{\varphi}_D = \arg\{x_{n,t}^A * \mathrm{conj}[x_{n,t}^{B,a} * \exp(\mathrm{j} * \varphi_H)]\}_n, \quad (15)$$

其中: $n = 1,\cdots,N$, $x_{n,t}^A$ 和 $x_{n,t}^{B,a}$ 分别是 $x_n^A(t_i)$ 和 $x_n^B(t_i)_a$ 的简写。

确定 $\widehat{\varphi}_D$ 后即可求取时延 τ 进行相位校正。由式(15) 可以看出对 $\widehat{\varphi}_D$ 的精确估计需要知道目标的方位角, 而在实际情况下目标的方位未知, 但是从式(13) 和式 (14) 可知时延的估计实际上与目标方位无关, 因此可以通过比较强的目标如本舰强干扰来进行时延估计 (如果放缆长度使本舰干扰满足近似远场条件)。这样做还有另一个好处, 就是在考虑环境噪声的影响时, 因为本舰干扰的信噪比比较高, 可以提高所估计相位的准确度。

2 仿真与实验数据分析

2.1 归一化左右舷分辨增益分析

这里通过数据仿真对比对称理想情况、没有进行非对称相位补偿 (失配情况)、进行非对称相位补偿 (匹配情况) 以及使用本文所述算法进行相位校正等四种情况下, 左右舷分辨增益与目标方位的关系。为了更好地分析, 假设方位设定范围 (0° ~ 360°) 是从左舷舰尾顺时针变化到右舷舰尾, 并定义归一化左右舷分辨增益 (PSG) 为目标方位响应与其共轭方位响应的差值, 且如果目标在舰身左舷, PSG 为正; 目标在舰身右舷, PSG 为负。

根据 PSG 的定义, 我们可以得到左右舷分辨增益随方位变化的规律, 如图 3 所示为理想双线阵 (首阵元对齐, 半波长布阵, 未加噪声) 的方位与 PSG 关系图, 图中实线和圆实线分别表示 0.25 倍波长双阵间距时, 双线阵几何相减和几何相加算法的左右舷分辨曲线; 点实线和星实线分别表示 0.75 倍波长双线阵间距时, 双阵几何相减和几何相加算法的左右舷分辨曲线。由图 3 可以看出, 在理想状况下, 两种双线阵算法都可以正确地判别目标的左右舷, 并且有越靠近侧射方向左右舷分辨增益越大, 越靠近端射方向左右舷分辨增益越接近于零的规律, 同时可以得到 0.25 倍波长双阵间距的分辨性能优于 0.75 倍波长双阵间距, 几何相减的左右舷分辨优于几何相加的左右分辨的结论。

而如果出现首阵元非对齐, 即双线阵非对称的情况。无论是使用失配的对称双线阵加权向量还是匹配的非对称双线阵加权向量进行双线阵左右舷分辨, 在一定的角度上都会出现左右舷判别错误的情况。图 4 和图 5 描述了这一现象 (两图中各曲线的定义和参数设定与图 3 相同, 只是增加一个非对称距离 D, 且 D 为 0.25 倍波长距离)。

图 4 为失配的情况, 即实际情况双线阵未对齐但是却按对齐的情况加权进行双阵左右舷判决的情况, 可以看出这种情况下, 双线阵左右舷判别错误相当严重, 以情况最好的 0.25 倍波长双阵间距为例, 出现左右舷判别错误的方位就有两个判别错误区间 (135° ~ 180° 和 315° ~ 360°), 共计 90° 的方位内出现了判别错误。

图5为匹配的情况, 即实际情况双线阵未对齐也

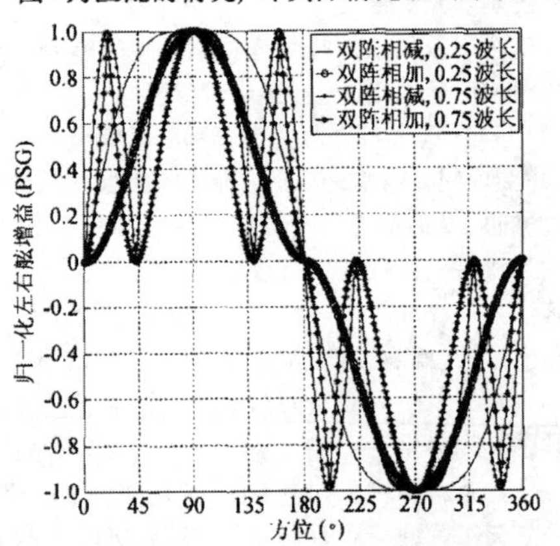

图 3　方位与 PSG 关系图 (标准)

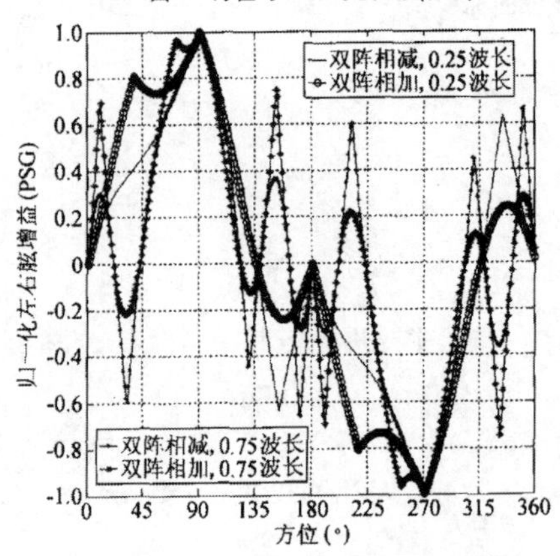

图 4　方位与 PSG 关系图 (权不匹配)

按非对齐的情况加权进行双阵左右舷判决的情况，而正如在第1.1节分析的一样，在这种情况下，双线阵左右舷判别错误也是相当严重，同样0.25倍波长双阵间距为例，仍然有90°的判别错误区间(0°～45°和135°～180°)，与失配情况不同的是匹配时判别错误区间集中在一侧。图5中判别错误区间集中在左舷，是因为数据仿真时按B阵领先于A阵得到的。如果A阵领先于B阵的话，判别错误区间将集中在右舷。

图6是对非对称双线阵进行合成孔径相位校正处理得到左右舷判决的情况(基本参数设定与图4和图5相同)，与图3至图5比较，可以看出经过方法处理后，非对称双线阵左右舷判决的结果与理想对称阵的结果相同，说明经过相位校正解决了左右舷误判问题。

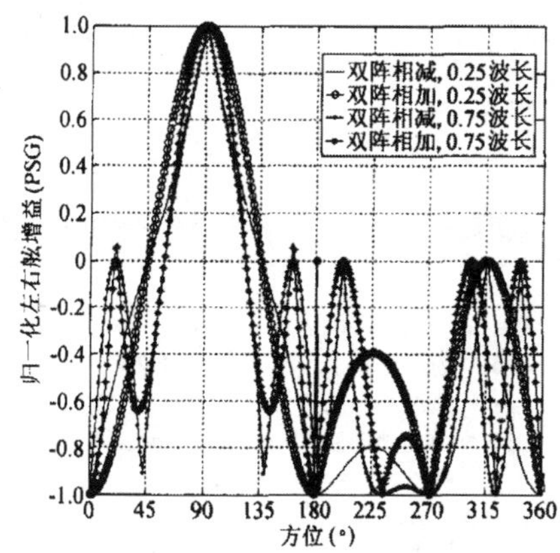

图5　方位与PSG关系图(权匹配)

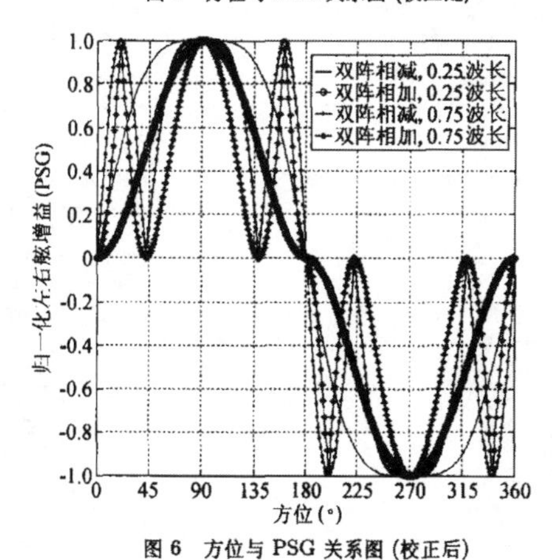

图6　方位与PSG关系图(校正后)

2.2　不同D下的左右舷误判分析

为了进一步分析双线阵非对称对双线阵左右舷分辨的影响，这里通过数据仿真统计了不同非对称距离D下，失配和匹配情况中两种算法左右舷判别错误角度的数目，给出一个非对称影响左右舷分辨严重程度的量化分析，如图7所示，分别比较了0.25倍和0.75倍波长双阵间距H之下不同D时误判角度数的分布。

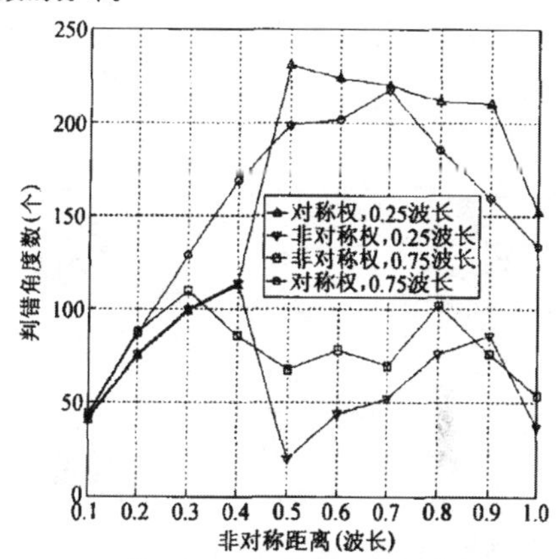

图7　不同D时判错角度数的分布图

由图7可知，即使双线阵所加的权与实际相匹配也存在一定误判角度，而误判角度数目因D的不同而不同；如果未考虑到非对称的情况所加权为失配权，判错角度会显著增加且随D的增加而变化显著。这是几何相加算法的结果，而几何相减算法可以得到一模一样的结果，因此可以得到以下结论：左右舷误判角度数目与采用何种算法无关，仅与加权是否匹配以及H，D有关，加权失配比匹配误判更严重，而H和D取某些值时(如D为半波长，H为0.25倍波长)左右舷误判角度数最少。

2.3　海试数据分析

除了上面的仿真分析，这里通过近期海试数据进一步验证了算法的有效性。使用数据的主要参数为：标准拖曳双线阵，处理频率范围为100~500 Hz。图8和图9是对海试数据的分析结果，为了方便比较左右舷的分辨效果，这里将判断为共轭方位的输出幅值用同帧波束图输出的平均值代替。

图8是没有考虑双线阵不对称现象，而采用普通双线阵几何相加算法得到的方位－历程图，图中可以看到有一个目标从方位69°逐渐运动到144°，在这一过程中总共有两段判别错误的区域，一段是115～144°，另一段是80～87°(通过分析，认为后一段可能是由于此时该目标和另一个一直在85°方位的弱目标相交造成的左右舷模糊)。

图9是同一组数据进行了合成孔径相位校正处理得到的方位－历程图，补偿相位的估计主要是通过方位15°的本舰强干扰得到(利用强干扰的线谱进

行相位估计)，可以看到被分析目标的判错区域获得了校正，即使是后一段由于目标相交所造成的左右舷模糊也可得到校正。

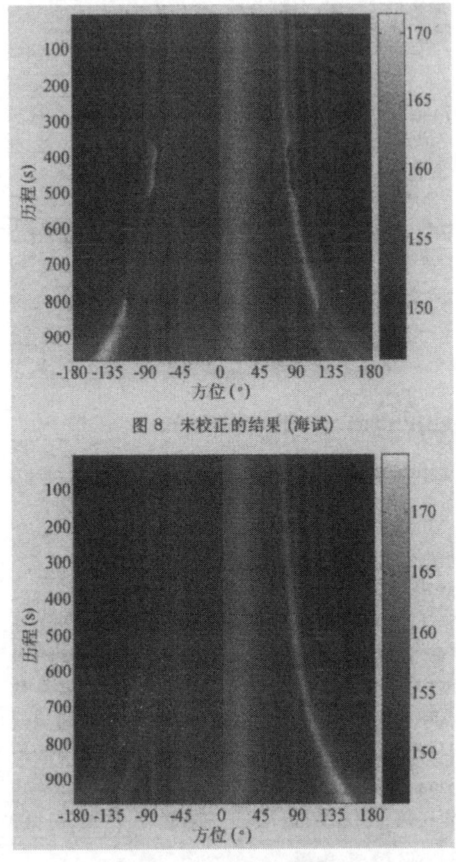

图 8 未校正的结果 (海试)

图 9 校正后的结果 (海试)

3 结论

本文对由于首阵元未对齐而引起的非对称双线阵的左右舷误判问题进行了分析，发现是由于实际与设定物理坐标的不一致性所造成的权向量非对称导致的左右舷误判，而且这种误判不能简单地通过相位补偿的方法得以解决。通过利用拖曳阵时空采样的特点，本文提出了一种基于被动合成孔径的相位校正算法。通过数值仿真和实际数据分析表明，这种方法可以有效地解决非对称双线阵的左右舷误判问题。这种算法简单能够实时实现，适于实际应用。另外，本文还分析了非对称阵畸变程度对左右舷误判严重度的影响，发现如果双阵间距和非对称距离取某些值时 (如 D 为半波长，H 为 0.25 倍波长)，左右舷误判角度数最少。

考虑到方法在估计相位过程中假设了双线阵间距是已知的，而在实际过程中双线阵间距往往是未知的，因此在相位估计过程中存在非对称距离 D 和双线阵间距的联合估计问题，有关这样的问题需要进一步研究。

参 考 文 献

1 Feuillet J P, Allensworth W S. Nonambiguous beamforming for a high resolution twin-line array. *J. Acoust. Soc. Am.*, 1995; **97**(5): 3292

2 Schurman I W. Reverberation rejection with a dual-line array. *IEEE J. Oceanic Eng.*, 1996; **21**(2): 193—204

3 李启虎，李淑秋等. 主被动拖阵声呐中拖曳平台噪声和拖鱼噪声在浅海使用时的干扰特性. 声学学报，2006; **31**(1): 1—4

4 李启虎. 双线列阵左右舷目标分辨性能的初步分析. 声学学报，2006; **31**(5): 385—388

5 李启虎. 用双线列阵区分左右舷目标的延时估计方法及其实现. 声学学报，2006; **31**(6): 485—487

6 LI Qihu, CHEN Xinhua, LI Min *et al.* Effect of shape distortion in solving left-right ambiguity of towed twin-line array. UDT Europe 2006, Hamburg, Germany, 2006: 6A—3

7 Stergiopoulos S, Sullivan E J. Extended towed array processing by an overlap correlator. *J. Acoust. Soc. Amer.*, 1989; **86**(1): 158—171

拖线阵声纳数字式水下数据高速传输的设计①

冯师军[1,2]，李启虎[2]，孙长瑜[2]
(1. 中国科学院研究生院，北京 100039；2. 中国科学院声学研究所，北京 100080)

摘要：通过拓扑结构的分析，提出了一种针对拖线阵声纳的数字式水下数据高速传输的方案；介绍了水下数据传输节点的总体设计，并详细地介绍了 HOT Link 发送接收控制器的设计；然后设计了基于伪随机序列码的误码分析仪，并用其测试了水下数据高速传输节点的误码率。误码测试结果和性能比较结果表明该方案切实可行，并能够方便地实现小型化。

关键词：拖线阵；HOTLink 收发器；误码率；小型化
中图分类号：TN912.35　　文献标识码：A　　文章编号：1000-3630(2007)-03-0362-05

Design of underwater digital high-speed data transmission in towed array sonar

FENG Shi-jun[1,2], LI Qi-hu[1], SUN Chang-yu[1]
(1. Graduate School of Chinese Academy of Science, Beijing 100039, China;
2. Institute of Acoustics, Chinese Academy of Sciences, Beijing 100080, China)

Abstract: In this paper, a scheme of underwater acoustic digital high-speed transmission for towed arrays is proposed. We analyze and build the underwater digital high-speed transmission topology that suits towed arrays, and introduce the system design of transmission node. Additionally, we design control logic of H-OTLink Transmitter/Receiver on FPGA and test the bit error rate (BER) of transmission node. The low BER and comparison with ATM node show that the transmission node is feasible. It is an easy way to realize miniaturization of the towed array system.

Key words: towed array; HOTLink transmitter/receiver; bit error rate; miniaturization

1 引　言

由水下传感器阵列组成的拖线阵声纳一直以来在水声工程中占有重要的地位，各海洋强国都在大力发展拖线阵声纳。随着拖线阵声纳的快速发展，小型水下数据高速传输技术的发展相对缓慢，严重影响了拖线阵声纳性能的提高。早期的拖线阵声纳采用是模拟信号传输，即传感器信号输出经过滤波放大后直接以模拟信号的形式送到基站进行数字变换和信号处理。然而，基于模拟信号传输的拖缆随着水听器数目的增加，其直径粗、重量大、信号衰减畸变严重以及信号间干扰较大等缺点日益突出[1]。

随着无线电电子和通信技术等相关领域技术的发展，数字式水下传输技术开始进入声纳系统，即在线列阵中完成传感器信号的数字变换，然后以一定的编码形式传输到基站，通过解码后进行信号处理。但是由于受水下拖缆内径的限制，数字式水下传输技术难以在拖线阵中得到应用。国内外都在开展数字式水下传输技术的小型化的研究，让水下传输节点适用于空间受限的拖缆。比如，美国 PSI(Planning Systems Incorporated) 公司开发了基于 ATM 技术的水下传输系统 ATM-SONET 节点，该节点被美国海军的最先进的 TB-29A 细线阵采用[2,3]。但是该节点采用的 ATM 网络技术比较复杂，并且昂贵，所以需要寻找一种性价比更高的拖线阵声纳数字式水下数据采集及高速传输的解决方案。

① 声学技术，2007, 26 (3): 362-366.

在功耗、带宽和可行性方面，拓扑结构是一个很重要的因素，所以希望通过研究拓扑结构来寻找一种更适合的解决方案。通过实际需求分析，发现水下数据传输系统的拓扑结构具有以下特点：

- 数据包的传输方向是确定的；
- 传输数据的大小固定；
- 每一时刻的传输负载大致确定。

通过比较，发现环状拓扑结构最适合运用于水下数据传输系统，因为在环形结构中传输线的数量不会随着水听器数目的增加而增加，便于减小了拖缆的直径。但同时也需要速率更高和体积更小的传输节点。本文采用 HOT Link 传输协议设计了一个小型水下数据高速传输节点。

2 传输节点的设计

CYPRESS公司的 HOT Link CY7B933/923 数据收发器是专为点到点串行高速通信而设计的。它适用于光纤、同轴电缆以及双绞线作为通信介质，其传输码率范围为 150Mbit/s~400Mbit/s。HOT Link 收发器功耗较小，驱动能力很强。在不加任何补偿电路情况下，330Mbit/s 码速，对于 50Ω 同轴电缆 (RG-58-A/U) 可以传送 35m；对于 75Ω 同轴电缆 (RG-6-A/U) 可达 304.8m；如果加上光电转换器用光纤传输，可传输数十公里以上或者更远的距离[4-7]。本文采用 HOT Link 收发器 CY7B933/923 实现了水下数据传输节点，如图 1 所示。

图 1 HOT Link 传输节点
Fig.1 HOT Link transmission node

每个传输节点用于接收 A/D 节点和前一个传输节点的数据包，然后将这些数据包传输到下一个传输节点，从而实现环型网络拓扑结构。传输节点包括 RS485 接收电路、FPGA 逻辑电路、HOT Link 接收发送电路和时钟电路，结构框图如图 2 所示。

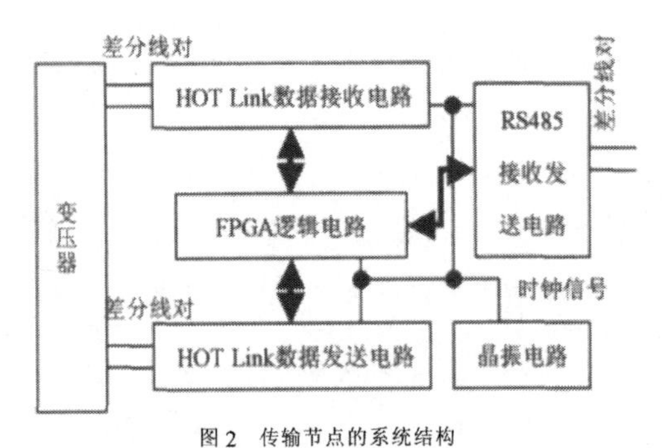

图 2 传输节点的系统结构
Fig.2 The system frame of transmission node

FPGA 逻辑电路包括 HOT Link 接收发送器的控制逻辑、RS485 接收控制逻辑和优先判决逻辑等，如图 3 所示。

RS485 接收控制逻辑用于得到左右 A/D 节点的数据包；HOT Link 接收控制逻辑用于得到上一个传输节点的数据包。这三个数据包通过优先判决逻辑决定发送顺序，然后通过 HOT Link 发送控制逻辑将数据包依序发送到下一个传输节点。

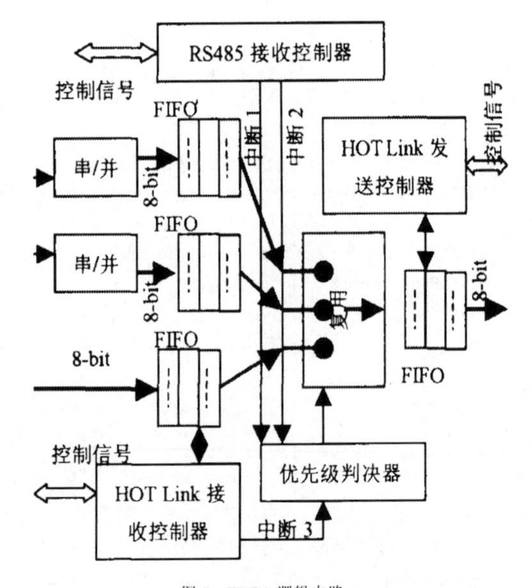

图 3 FPGA 逻辑电路
Fig.3 FPGA logic circuit

2.1 发送器控制器的设计

HOT Link 发送控制器包括三个状态：空闲状态，发送状态和测试状态。HOT Link 发送控制器在复位以后进入空闲状态；当发送允许且 FIFO 没有溢出的时候，进入到发送状态发送数据，不断向 FIFO 中写入数据直到溢出；当发送结束且 FIFO 为空，返回空闲状态；如果测试允许，进入测试状态，测试结束后返回空闲状态。FIFO 的读端口连接到 HOT Link 发送器的并行输出端口。发送控制器的状态机如图 4 所示。

2.2 接收器控制器的设计

HOT Link 接收控制器包括 4 个状态：等待状态，接收状态，帧同步状态和测试状态。HOT Link 接收控制逻辑在复位以后进入帧同步状态，帧同步以后进入等待状态；当 SC/D 为低电平且 FIFO 不为空，进入到接收状态接收数据，不断从 FIFO 中读出数据直到 FIFO 空；当 SC/D 为高电平或 FIFO 为空，返回等待状态；如果测试允许，进入测试状态，测试结束后返回空闲状态。FIFO 的写端口连接到 HOT Link 接收器的并行输出端口。接收控制器的状态机如图 5 所示。

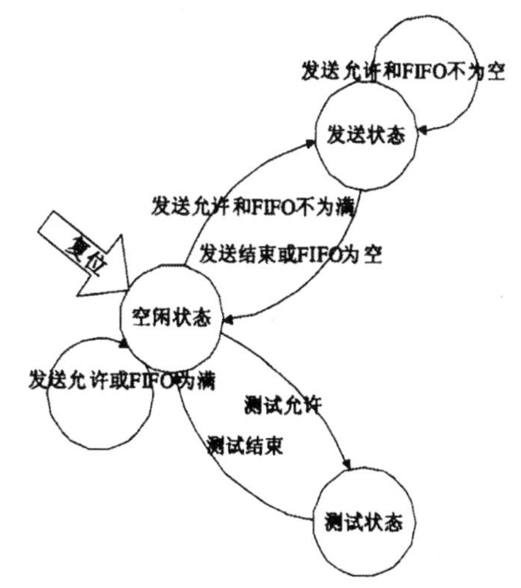

图 4 发送控制器状态机
Fig.4 State transition diagram for transmitter controller

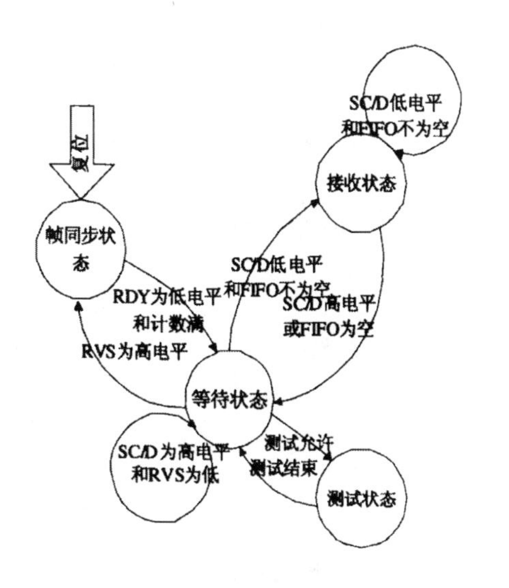

图 5 接收控制器状态机
Fig. 5 State transition diagram for the receiver controller

对于接收器控制器的设计而言，建立同步是必需的。同步即根据接收到的数据流建立同步时序。对于接收器 CY7B933 而言，同步有位同步(Bit Synchronization)和帧同步(Framing)。位同步由芯片内部的接收锁相环(Phase Locked Logic, PLL)自动完成。当检测到数据流时，PLL 根据数据流调整本地的位时钟(Bit-Clock)的相位和频率，以保证在每位数据的中间采集，位同步操作需要知道数据流中的起始特征字符。帧同步则是通过检测接收到的特殊的串行数据流来实现，在 CY7B933 中这种特殊数据称之为间歇符(comma)。如果使用 8B/10B 的编码方式，间歇符是 K28.5，它不会出现在任何 8B/10B 有效编码中。对于 CY7B933 而言，当出现错误而丢失帧同步时，其数据输出引脚 RVS 会变为高电平以指示丢失了数据或收到错误的无效数据，为了控制 CY7B933 重新检测间歇符以建立帧同步，需要将强制帧同步引脚 RF 置为高电平。

接收控制器和发送控制器的后仿真时序图如图 6 所示。其中，RE_RVS 表示接收出错指示；RE_SCD 表示接收端特殊字符或者数据选择；RE_SO 表示接收端的状态输出，即是将 SI 信号的 PECL 电平转换成 TTL 电平输出；RE_RDY 表示接收端的数据输出准备好；RE_BISTEN 表示接收端的内部自检使能；RE_AB 表示接收端的两路差分 PECL 信号(INA± INB ±) 输入选择；RE_RF 表示强制帧同步；RE_MODE 表示接收端工作模式选择；TR_CKW 表示发送端的数据写入时钟；TR_BISTEN 表示发送端的内部自检使能；TR_ENN 表示发送端在 TR_CKW 的下一个上升沿时锁存数据；TR_ENA 表示发送端的并行数据使能；TR_MODE 表示发送端工作模式选择；TR_FOTO 表示发送端光纤模块控制信号。

2.3 误码率测试

误码率即出现错误比特的概率，是表征数据传输系统抗干扰性能的主要技术指标，误码的检测在数据传输中必不可少。在实际测试中，误码率的测试一般采用标准的伪随机二进制码序列 PRBS (Pseudo Random Binary Sequence)。这种码型重复周期长，在较短序列内可以近似地看成随机，因而采用这种码型要比固定码型更可信[8]。

本文在 FPGA 中设计一个误码分析仪来测试 HOTLink 高速串行总线的实际误码率，如图 7 所示。

发送端的伪随机二进制序列(PRBS)发生器产生 m 序列的伪随机序列，通过 HOT Link 串行总线

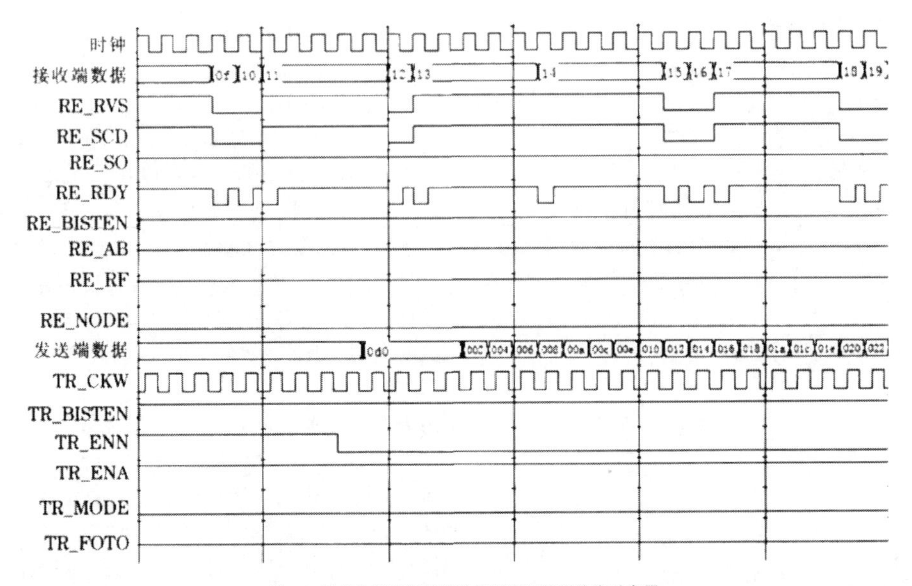

图 6 接收控制逻辑和发送控制逻辑的后仿真时序图

Fig.6 Simulated waveforms for the transmitter & receiver controller

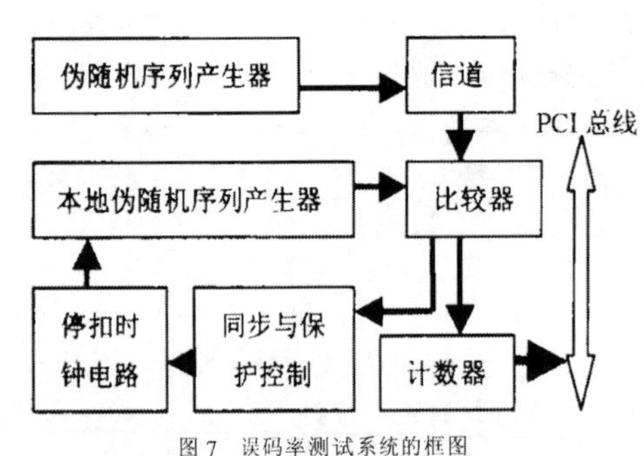

图 7 误码率测试系统的框图

Fig.7 Schematic of error test system

发送到接受端和接受端产生的 m 序列的伪随机序列相比较，然后通过计数器输出结果。其中发送端和接受端的两个伪随机码发生器产生地伪随机序列要求一致。

在接收端，计数比较之前，为了保证本地 m 序列发生器与接收数据缓冲器同步移位输出，需要选择合适的阈值。这里通过 HOT Link 的理论误码率的分析来确定这个阈值。HOT Link 串行总线采用 8B/10B 的编码方式，而在 8B/10B 编码方式中信号波形采用单极性非归零码[5]。理论的误码率采用公式(1)计算[9]:

$$P_b=\int_{A/2\sigma}^{\infty}\frac{1}{\sqrt{2\pi}}e^{-\frac{x^2}{2}}dx=Q\left(\frac{A}{2\sigma}\right) \tag{1}$$

其中 A 信号的幅度值，σ^2 为噪声功率，为均方值，Q函数的表示式见式(2)。

$$Q(x)=\frac{1}{2}\mathrm{erfc}\left(\frac{x}{\sqrt{2}}\right)=\int_{x}^{\infty}\frac{1}{\sqrt{2\pi}}e^{-\frac{y^2}{2}}dy \tag{2}$$

如果 $S=A^2/2$ 是信号平均功率，$E=\sigma^2$ 是噪声平均功率，式(1)能改写为:

$$P_b=Q(\sqrt{S/N})=\frac{1}{2}\mathrm{erfc}\left(\frac{\sqrt{S/N}}{2}\right) \tag{3}$$

当输入 SNR 为 15dB 的时候，由式(3)HOT Link 的理论误码率小于2×10^{-8}，同时随着输入 SNR 的增加，HOT Link 的理论误码率进一步快速减小，故这里选定阈值大小为 2×10^{-8}。

在实验室常温环境中，传输线为 UTP-5 类非屏蔽双绞线，长度为 16m，HOT Link 串行总线的实际误码率小于1.1176×10^{-8}。误码测试结果的统计图如图 8 所示。

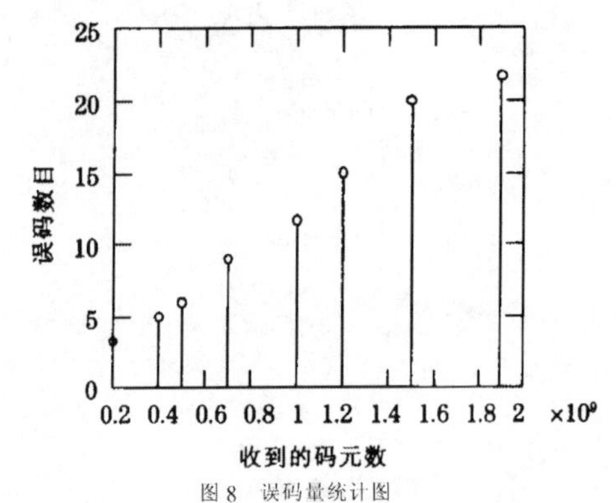

图 8 误码量统计图

Fig.8 Statistical histogram of the error number

3 性能比较

表 1 给出了 PSI 公司的 ATM 节点 ATM-SONET 和 HOT Link 传输节点的性能比较。

表 1 性能比较

Table 1 Performance comparison

节点	HOT Link 节点	ATM-SONET 节点
长度	110mm	111.76mm
直径	20mm	20.32mm
带宽	160- 320Mb/s(可选)	155.52Mb/s
供电	+2.5V +16V	+5.8V +7.3V
功耗	600mA/node	400mA/node

从表 1 可以看出，与复杂的 ATM 网络技术相比较, HOT Link 串行总线是一个不错的替代方案。在满足高带宽的同时, 也能方便实现水下数据传输节点的小型化。

4 总　结

本文在数字式水下数据高速传输系统中采用 HOT Link 串行总线替代 ATM 技术, 在满足系统小型化的同时实现了水下数据的高速传输。采用 UTP-5 类非屏蔽双绞线为传输线在 16m 的传输距离下，实验室测试误码率为 1.1176×10^{-8}。下一步工作是依然采用 HOT Link 串行传输总线，以光纤为媒质实现水听器数据从水下到基站的传输。

参　考　文　献

[1] 李启虎. 数字式声纳设计原理[M]. 安徽：安徽教育出版社，2003.2
LI Qihu. Design Principle of Digital Sonar[M]. Anhui: Anhui Education Publishing Company, 2003. 2.

[2] John Walrod. ATM Telemetry in Towed Arrays[A]. Undersea Defense Technology 1997[C]. Germany, Hamburg 1997 6: 24-26.

[3] Planning Systems Incorporated. ATM-SONET Network Node (-NIC-7)[X]. 2002. 04.

[4] Cypress Semiconductor Corporation. CY7B923/CY7B933 HOTLink Transmitter/Receiver[R]. San Jose, 2003.

[5] Cypress Semiconductor Corporation. HOTLink Design Considerations[R]. San Jose, 2003.

[6] Cypress Semiconductor Corporation. Interfacing the CY7B923 and CY7B933 (HOTLink) to Clocked FIFOs[R]. San Jose, 2003.

[7] Cypress Semiconductor Corporation. HOTLink CY7B933 Pin Description[R]. San Jose, 2003.

[8] 严挺，方志来. 千兆背板总线测试方法[J]. 电子技术应用，2002. 12: 27-30.
YAN Fang, FANG Zhilai. Testing measures for kilomega bus of backboard[J]. Electronic Technology Application, 2002. 12, 27-30.

[9] 曹志刚，钱亚生. 现代通信原理[M]. 北京：清华大学出版社，2000.
CAO Zhigang, QIAN Yasheng.Principle of Modern Digital Communication[M]. Beijing: Tsinghua University Publishing Company, 2000.

微弱信号源的和波束定向方法与分裂波束定向方法的性能比较①②

李启虎　　朴大志

(中国科学院声学研究所　北京 100080)

摘要　本文讨论水下辐射噪声源的精确定向问题，给出被动声纳和波束定向与分裂波束定向方法的性能比较。指出在一定信噪比下，分裂波束精确测向技术比和波束定向技术具有较高的定向精度。但是，随着信噪比的下降，两者趋于一致。推导了估计定向精度的分析表达式，给出在直线阵和圆弧阵情况下，延时估计和声源入射角偏差之间的换算公式和数值模拟结果。同时给出数字式声纳用以计算入射信号左波束和右波束数据的互谱来实现分裂波束定向的方法。

关键词　和波束，分裂波束，精确定向，互谱

Comparison of bearing accuracy of sum beam and split beam in weak signal detection

LI Qi-Hu　　PIAO Da-Zhi

(*Institute of Acoustics, Chinese Academy of Sciences,　Beijing 100080*)

Abstract　Precise bearing of underwater noise source is discussed in this paper. The performance comparison of the sum beam steering and the split beam steering methods in passive sonar is presented. It is shown that, in a certain range of signal to noise ratio, the bearing accuracy of split beam method is much better than the method of sum beam. But with the decrease of signal to noise ratio, the performance of two methods become same. The closed form expression of bearing accuracy of the two methods for estimating DOA (Direction Of Arrival) of the target is derived. In the case of line array or arc array of a sonar system, the relationship of time delay and the bias angle of incidental signal is presented. Some results of system simulation are illustrated. An algorithm is given, which calculate the cross power spectrum of left beam and right beam data for realizing the split beam bearing in digital sonar.

Key words　Sum beam, Split beam, Precise bearing, Cross power spectrum

① 国家自然科学基金资助项目 (60532040).

② 应用声学, 2007, 26 (3): 129-134.

1 引言

水下辐射噪声源的被动精确定向一直是声纳设计中的重要问题。早在上世纪 80 年代, Schultheiss 等就已指出精确定向问题可以归结为统计信号的参数估计问题, 就是信号到达不同基元的时延差的估计。对于时延估计, 存在最优的估计下界 (即 Rao-Cramer 下界), 而普通声纳的"延时－求和－相加"模型就可以实际达到或接近最优下界[1]。

分裂波束是利用普通波束形成中的信号信息实现精确测向的一种有效方法, 已被广泛应用于实际声纳系统[2~4]。但是, 不同定向方法的精度和信噪比之间关系还没有作过系统的分析比较。本文首先把定向精度定义为波束响应对延时差的微分, 然后给出分裂波束定向方法与传统和差定向方法的比较。指出, 一般来说, 分裂波束的定向精度要优于和差定向法。但是随着信噪比的下降, 分裂波束的这种优势会逐渐消失, 两者逐步趋向一致。

对于数字式声纳来说, 由于分裂波束精确定向方法在软硬件上的开销要远大于和差定向法, 所以本文的理论结果可作为实际选用何种定向方法的参考指标, 本文最后给出数字式声纳实现分裂波束定向的具体算法。

2 基本模型

水下辐射噪声源的定向问题实际上是测量不同基元之间所接收信号的延时差。最基本的模型如图 1 所示, 设有两个水听器分别接收到信号 $x(t)$ 和 $x(t+\tau)$, 信号的入射角是 θ。如果两个基元之间的距离是 d, 那么延时 τ 和 θ 的关系由下式表达

$$\tau = d\sin\theta/c \tag{1}$$

其中 c 为声速, 当水听器个数很多时, 可以把波束成形的结果归纳为求解 (1) 式中的延时值, 我们将在第 4 节予以说明。

普通的和路波束定向实际上是利用信息

$$y(t) = x(t) + x(t+\tau) \tag{2}$$

而分裂波束定向则还要利用差路的信息

$$z(t) = x(t) - x(t+\tau) \tag{3}$$

和路信号的均方值

$$u(\tau) = E[y^2(t)] = 2\sigma_x^2[1+\rho_x(\tau)] \tag{4}$$

差路信号的均方值

$$v(\tau) = E[z^2(t)] = 2\sigma_x^2[1-\rho_x(\tau)] \tag{5}$$

从解算延时的角度来说, 我们希望 τ 的微小变化会引起 $u(\tau)$ 或 $v(\tau)$ 的很大变化, 这样就有利于精确定向。从 (4) 我们发现, 如果 $\rho_x(\tau)$ 在 $\tau=0$ 附近是平坦的, 那么 $u(\tau)$ 的变化也很小, 这对定向是不利的。但是从 (5) 式可见, 对 $v(\tau)$ 就不一样, 在 $\tau=0$ 附近 $1/v(\tau)$ 的变化将非常大。所以从直观上来看把差波束的信息引入精确定向是很有必要的。

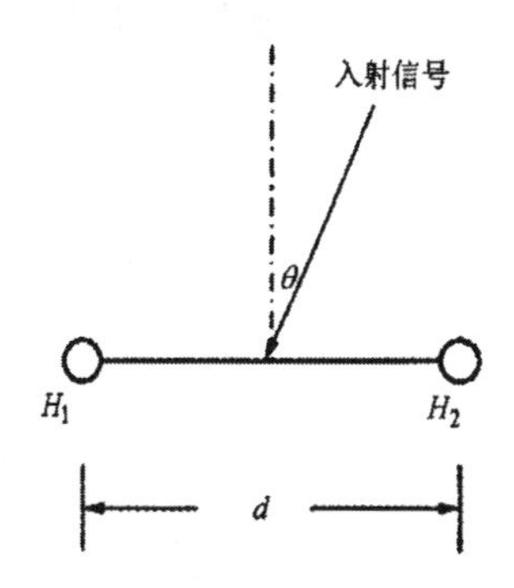

图 1 延时测量的简化模型

定义归一化的和差波束定向函数:

$$p(\tau) = u(\tau)/4\sigma_x^2 = [1+\rho_x(\tau)]/2 \tag{6}$$

$$q(\tau) = 4\sigma_x^2/v(\tau) = 2/[1-\rho_x(\tau)] \tag{7}$$

我们关心的是 $p'(\tau) = dp(\tau)/d\tau$ 和 $q'(\tau) = dq(\tau)/d\tau$, 直接计算得到

$$p'(\tau) = \rho_x'(\tau)/2 \tag{8}$$

$$q'(\tau)=2\rho_x'(\tau)/[1-\rho_x(\tau)]^2 \tag{9}$$

在 $\tau=0$ 附近，$q'(\tau)\gg p'(\tau)$。

3 有噪声的情形

实际接收到的信号总是受到干扰的，即 $x(t)+n_1(t)$ 和 $x(t+\tau)+n_2(t)$, $n_1(t)$ 和 $n_2(t)$ 相互独立。这种情况下，和路信号是

$$y(t)=x(t)+x(t+\tau)+n_1(t)+n_2(t) \tag{10}$$

差路信号是

$$z(t)=x(t)-x(t+\tau)+n_1(t)-n_2(t) \tag{11}$$

由此，

$$u(\tau)=E[y^2(t)]=2\sigma_x^2[1+\rho_x(\tau)]+2\sigma_n^2 \tag{12}$$

$$v(\tau)=E[z^2(t)]=2\sigma_x^2[1-\rho_x(\tau)]+2\sigma_n^2 \tag{13}$$

由于噪声波形无法事先估计，所以就不能把它从 $u(\tau)$ 或 $v(\tau)$ 的表达式中减去。

类似于 (6)、(7) 式重新定义定向函数

$$p(\tau)=\frac{u(\tau)}{4\sigma_x^2+2\sigma_n^2}=\frac{(SNR)[1+\rho_x(\tau)]+1}{2(SNR)+1} \tag{14}$$

$$q(\tau)=\frac{4\sigma_x^2+2\sigma_n^2}{v(\tau)}=\frac{2(SNR)+1}{(SNR)[1-\rho_x(\tau)]+1} \tag{15}$$

其中 $(SNR)=\sigma_x^2/\sigma_n^2$ 是信噪比。决定分辨力大小的是 $p'(\tau)$ 和 $q'(\tau)$，直接计算给出

$$p'(\tau)=\frac{(SNR)\rho_x'(\tau)}{2(SNR)+1} \tag{16}$$

$$q'(\tau)=\frac{2(SNR)+1}{\{(SNR)[1-\rho_x(\tau)]+1\}^2}\cdot(SNR)\bullet\rho_x'(\tau) \tag{17}$$

当 $(SNR)\gg 1$ 时，

$$q'(\tau)\gg p'(\tau) \tag{18}$$

当 $(SNR)\ll 1$ 时，

$$q'(\tau)\approx p'(\tau) \tag{19}$$

当 $SNR=1$ (即 0dB) 时，分裂波束定向精度比和波束定向精度高出一个数量级。

4 系统仿真

我们选取水下目标检测中具有比较典型意义的信号形式，假定它的相关系数是

$$\rho_x(\tau)=\sin\pi f\tau/(\pi f\tau) \tag{20}$$

由 (20) 式所表示的相关系数对应于带宽为 1000Hz 的低通高斯信号。取 $f=1000\text{Hz}$。其相关系数见图 2。

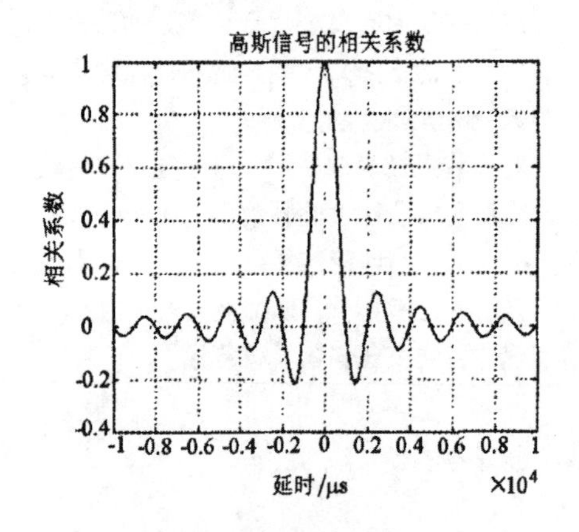

图 2 信号的相关系数

图 3(a) 至图 3(c) 给出输入信噪比为 $SNR=3, 0, -10\text{dB}$ 时和波束定向和分裂波束定向精度的比较。我们可以看到，当输入信噪比由大变小时，分裂波束定向方法的分辨力与和波束定向方法比较，由高变低，最后会趋向于一致，这与我们在第 2 节的分析是完全一致的。

不过我们要指出的是，实际检测目标时，当然不会只用两个基元用于定向。我们可以把这两个基元理解为声纳系统的左、右波束输出。

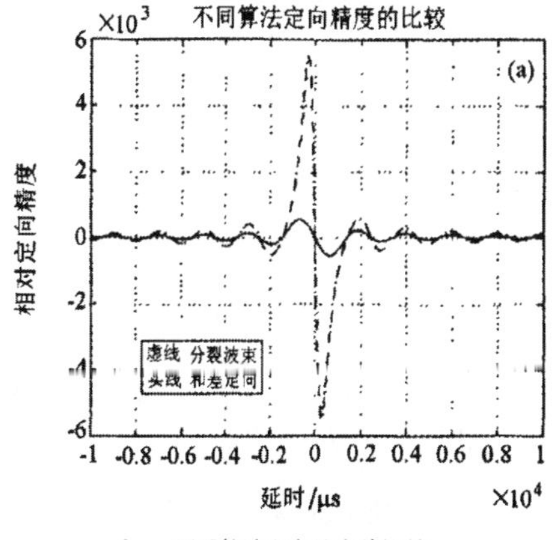

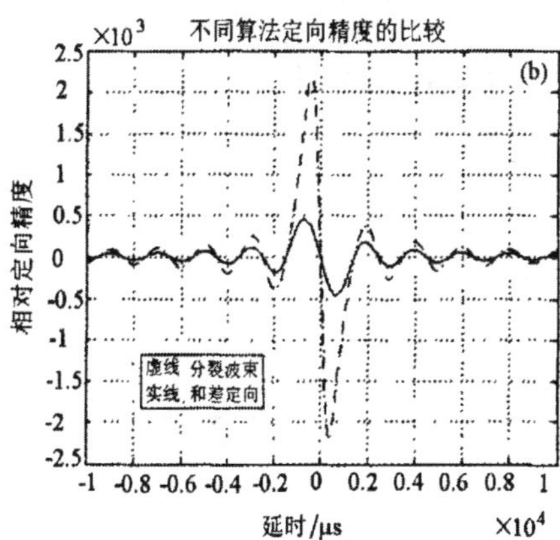

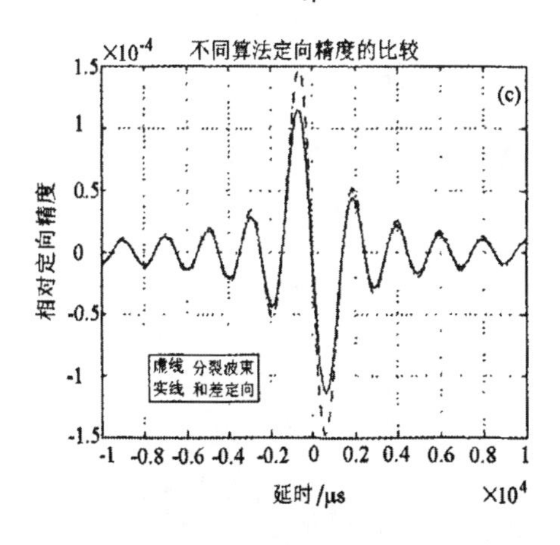

图 3 和波束定向和分裂波束定向精度的比较
(a) $SNR = 3\text{dB}$ (b) $SNR = 0\text{dB}$ (c) $SNR = -10\text{dB}$

这样，参加精确定向的信号的 SNR 一般不会很低，例如，大于 -6dB，在这种情况下，采用分裂波束定向的优点是明显的。

下面我们说明，实际声纳系统定向时，如何把延时转换为角度[5]。

例 1. 直线阵，见图 4，偶数基元，其中第 $1,\cdots,M$ 基元构成左基阵，第 $M+1,\cdots,2M$ 基元构成右基阵。假定信号入射角是 θ，基元间隔是 d。

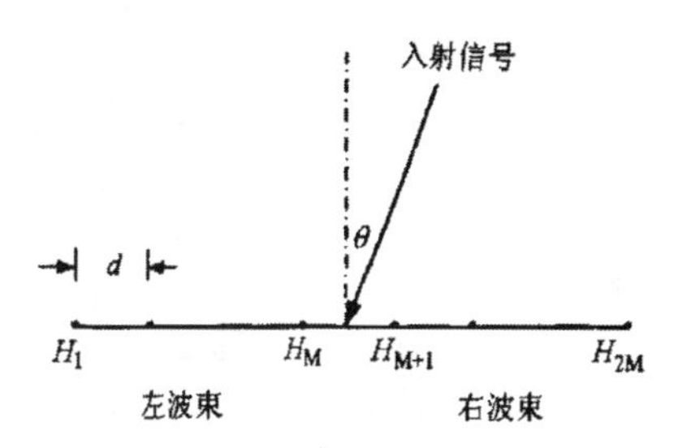

图 4 线列阵分裂波束定向原理

可以证明

$$\tau = Md\sin\theta/c \tag{21}$$

例 2. 圆弧阵，见图 5，偶数基元，其中 $H_1,\cdots,H_M$ 构成左基阵，$Q_1,\cdots,Q_M$ 构成右基阵。总基元数为 N (不妨取 $M/N = 0.25$)，相邻基元之间的夹角是 $\alpha_0 = 2\pi/N$。圆阵半径是 r。

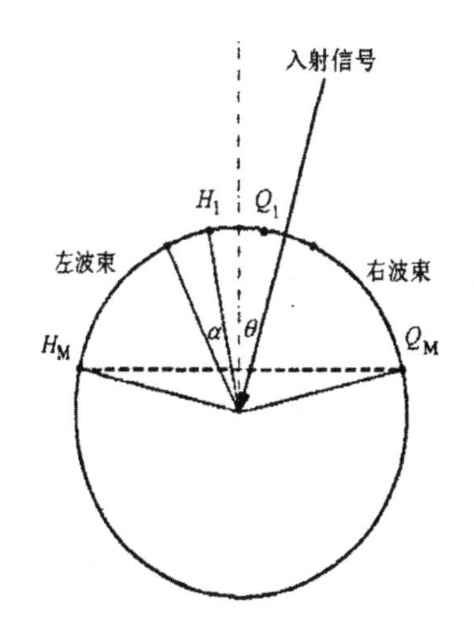

图 5 圆弧阵分裂波束定向原理

可以证明

$$\tau = \frac{2r\theta}{c\pi k}[\sin(k\pi)]^2 \tag{23}$$

其中 $k = M/N$ 。

图 6(a), (b) 给出了两个模拟计算的结果，图 6(a) 是直线阵的情况， $M = 32$, $d = 0.5$m, 我们看到，当左、右波束信号的延时估计精度达到 10μs 时，定向的精度大约是 0.1° 。

图 6(b) 时圆弧阵的情况， $r = 2$m, 如果想达到 0.1° 的定向精度，延时估计精度应当是 $2 \sim 3\mu$s 左右。

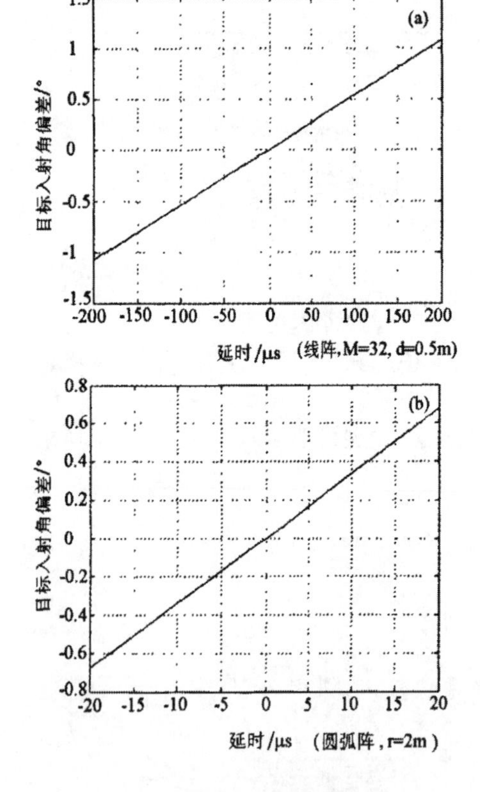

图 6 延时和目标入射角的转换关系
(a) 线列阵 (b) 圆弧阵

5 数字式声纳分裂波束的技术实现

对数字式声纳来说，要实现分裂波束的精确定向可以选取对准目标的那个波束的左、右波束信息，然后计算它的互功率谱，把延时差解出来。

假定左波束为 $x(t)$, 右波束为 $y(t) = x(t+\tau)$, 相应的 Fourier 变换是

$$X(f) = \int x(t)\exp(-2\pi jft)dt \tag{24}$$

$$Y(f) = \int x(t+\tau)\exp(-2\pi jft)dt = X(f)\exp(-2\pi jf\tau) \tag{25}$$

$x(t)$ 和 $y(t)$ 的互动率谱是

$$Z(f) = X^*(f)Y(f) = |X(f)|^2\exp(-2\pi jf\tau) \tag{26}$$

左、右波束之间的延时 τ (以及所对应的相位 $\varphi = 2\pi f\tau$), 可以由下式得出:

$$\varphi = 2\pi f\tau = \mathrm{tg}^{-1}\left[\frac{\mathrm{Im}[Z(f)]}{\mathrm{Re}[Z(f)]}\right] \tag{27}$$

其中 Re 和 Im 分别表示复数的实部和虚部。用:

$$x(0), x(Ts), \cdots, x(NTs), \cdots$$
$$y(0), y(Ts), \cdots, y(NTs), \cdots$$

分别表示 $x(t)$ 和 $y(t)$ 的采样值， Ts 是采样间隔，我们有

$$X_q(l) = \sum_{k=0}^{N=1} x[(q+k)Ts]\exp(-2\pi jkl/N) \tag{28}$$

$$Y_q(l) = \sum_{k=0}^{N=1} y[(q+k)Ts] \cdot \exp(-2\pi jkl/N) \quad l = 0, \cdots, N-1 \tag{29}$$

$$Z_q(l) = X_q^*(l)Y_q(l), \quad \varphi_l = 2\pi\tau l/NTs = \mathrm{tg}^{-1}\left\{\frac{\mathrm{Im}[Z_q(l)]}{\mathrm{Re}[Z_q(l)]}\right\} \tag{30}$$

利用多段 DFT, 可以得到精确、稳定的 φ_l 估计值，设 $Z_{mN}(l)$ 为长度是 N 的第 m 段数据的

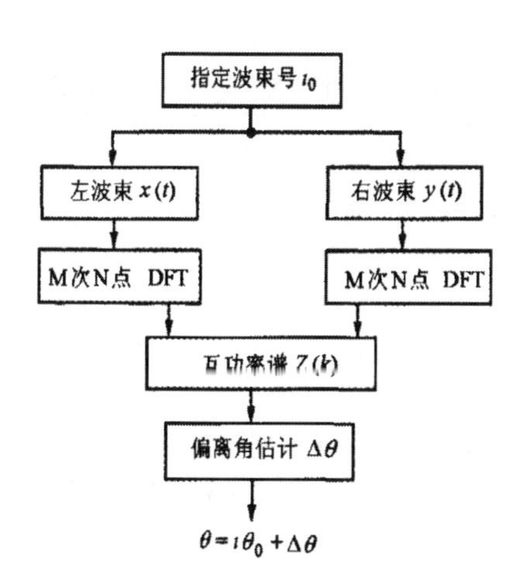

图 7 数字式声纳实现分裂波束定向的算法流程

DFT, 令

$$Z(l) = \frac{1}{M}\sum_{m=1}^{M} Z_{mN}(l) \tag{31}$$

$$\varphi_l = \mathrm{tg}^{-1}\left\{\frac{\mathrm{Im}[Z(l)]}{\mathrm{Re}[Z(l)]}\right\} \tag{32}$$

对每一个 φ_l 进行加权，可以得到 τ 的精确估计，图 7 是这一方法的方框图。

参 考 文 献

1 Schulthess P M, Weinstein E. Detection and estimation: a summary of results. Proc. of NATO ASI on Underwater Acoustics and Signal Processing, La Spizia, Italy, 1980, 379~410.

2 Ed S Stergiopoulos. Advanced Signal Processing Handbook. CRC Press, 2001, USA.

3 Green T J Jr. Robust Passive Sonar. DAPAR Tech., 2000 Sept. Dallas, USA.

4 Jaknbowski W M. AN/BQG-5A Wide Aperture Array. *Sea Technology*, 1996, **37**(11):43~46.

5 李启虎. 声纳信号处理引论. 第 2 版, 海洋出版社, 2000 年, 北京.

阵形畸变对拖曳双线阵左右舷分辨性能的影响①②

张 宾[1,2] 孙贵青[2] 李启虎[2]

(1 青岛科技大学 青岛 266061)

(2 中国科学院声学研究所 北京 100190)

2007 年 11 月 21 日收到

2008 年 3 月 13 日定稿

摘要 分析了阵形畸变对拖曳双线阵左右舷分辨性能的影响。首先由双线阵指向性函数及左右舷抑制比公式，推导并定义了双线阵声呐实现左右舷分辨的两个关键参量 —— 特征方位和特征频率。然后在引入阵形畸变量的基础上，利用阵形畸变时的左右舷抑制比公式，进一步推导得到特征方位及特征频率关于阵形畸变量的解析关系式。理论分析、仿真结果以及海上试验数据都表明，阵形畸变会改变特征方位和特征频率，造成拖曳双线阵左右舷分辨性能下降甚至再次产生左右舷模糊。

PACS 数： 43.60, 43.30

Effects of shape distortion upon left/right discrimination of towed twin-line array

ZHANG Bin[1,2] SUN Guiqing[2] LI Qihu[2]

(1 *Qingdao University of Science and Technology* Qingdao 266061)

(2 *Institute of Acoustics, Chinese Academy of Sciences* Beijing 100190)

Received Nov. 21, 2007

Revised Mar. 13, 2008

Abstract Effects of the distortion upon left/right discrimination of towed twin-line array were analyzed. Two key parameters, the characteristic bearing and characteristic frequency, were derived and defined according to the directivity and the gain of left/right suppression of twin-line array, which determine the performance of left/right discrimination of towed twin-line array. A shape distortion variable was called to formulate the left/right suppression ratio under shape distortion, and the functional relationship between the two key parameters and the shape distortion variable was established. Theoretical analysis and simulation results reveal that the characteristic bearing and characteristic frequency are deviated from original location under shape distortion, resulting in the performance degradation of the sonar left/right discrimination, and even worse bearing ambiguity would appear again. And the sea trial data verified this.

引言

拖曳双线阵声呐是为了解决常规单线列阵声呐左右舷模糊问题应运而生的，它由两条线列阵构成，相较于单线列阵，可以利用更小的孔径实现更大的增益。为了更好地利用双线列阵声呐实现左右舷分辨，国内外一些声呐研究人员给出了多种实现左右舷分辨的方法[1-5]，如基于双线列阵自身指向性的方法以及时延估计方法等等。并且已有海上试验验证了拖曳双线阵可以很好地解决左右舷模糊问题，左右舷抑制比可达到 20 dB 以上[6-8]。

但是在实际当中由于海流的影响或者机械方面的原因，双线列阵的阵形有可能发生畸变。阵形畸变对双线列阵左右舷分辨性能的影响，李启虎首次提出了这一问题，并在理论上分析了双线阵对应阵元错位时左右舷抑制比的变化情况[9-10]。在此基础上

① 国家自然科学基金资助项目 (60532040).

② 声学学报, 2008, 33 (4): 294-299.

本文将通过引入特征方位和特征频率这两个关键参量，通过理论分析、数值仿真和海试数据处理三个方面的研究，系统地阐述阵形畸变对双线列阵左右舷分辨性能的影响，以期能够得到规律性的结论，为拖曳双线阵声呐的实际应用提供有价值的参考和帮助。

1 特征方位和特征频率

由于实际中的目标一般为远场，因此这里只考虑双线列阵的二维波束指向性。图1给出了双线列阵平面坐标 (xOy 平面)，两条线列阵 A 和 B 平行分布，两阵之间的距离为 l，每条线列阵分别由 N 个各向同性的接收水听器等间距分布构成，阵元间距为 d。

1.1 左右舷抑制增益

如图 1 所示的双线阵指向性函数可写为：

$$D_{\rm twin}(\theta,\theta_0)=\left\{\frac{1+\cos[2\pi l(\sin\theta-\sin\theta_0)/\lambda]}{2}\right\}^{1/2}\left|\frac{\sin[N\pi d(\cos\theta-\cos\theta_0)/\lambda]}{N\sin[\pi d(\cos\theta-\cos\theta_0)/\lambda]}\right|. \tag{1}$$

注意到，式 (1) 中第二个乘数因子是单线阵的指向性函数 $D_{\rm single}$，而第一个乘数因子是心形指向性函数 F，即，双线阵指向性函数可根据 Bridge 乘积定理由单线阵指向性函数与一个心形指向性函数相乘而得：

$$D_{\rm twin}(\theta,\theta_0)=F(\theta,\theta_0)D_{\rm single}(\theta,\theta_0), \tag{2}$$

式中：

$$F(\theta,\theta_0)=\left\{\frac{1+\cos\left[2\pi l\left(\sin\theta-\sin\theta_0\right)/\lambda\right]}{2}\right\}^{1/2}. \tag{3}$$

从式 (3) 可以看出，对于一对映像方位 θ 和 $-\theta$，有下面的关系：

$$F(\theta,\theta_0)\neq F\left(-\theta,\theta_0\right). \tag{4}$$

上式表明，正是由于因子 $F(\theta,\theta_0)$ 的存在，双线阵在一对映像方位具有不同的响应值，从而为解决单线阵左右舷模糊问题提供了基础。

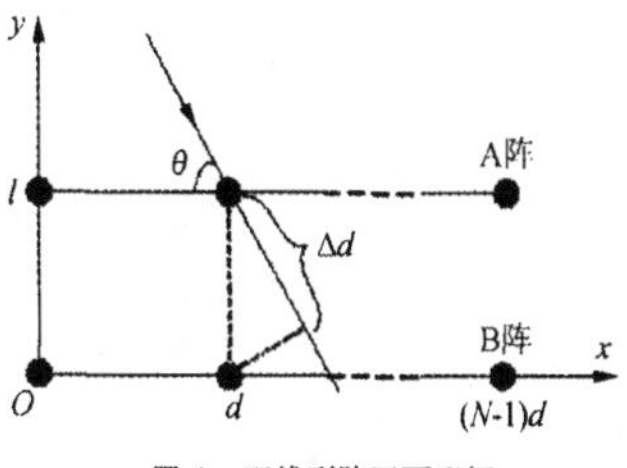

图 1 双线列阵平面坐标

图 2 是单线阵和双线阵在 $\theta_0=90^\circ$ 时的波束图比较，信号频率 500 Hz，声速 1500 m/s，每条线阵各有 20 个半波长等间距布放的阵元，双线阵间距取四分之一波长。从图中两条曲线的对比可以看到，单线阵波束形成器在 θ_0 与其映像方位 (称 θ_0 和 $-\theta_0$ 互为映像方位) 的输出是完全相同的，不能分辨目标是在左侧还是在右侧。而双线阵波束不存在目标左右舷模糊，在 θ_0 的映像方位 $-\theta_0$ 处形成零陷。

左右舷抑制增益是衡量声呐左右舷分辨能力的定量参数，可定义为一对映像方位的响应之比，即：

$$G(\theta,\theta_0)=10\log\left|\frac{D\left(\theta,\theta_0\right)}{D\left(-\theta,\theta_0\right)}\right|^2. \tag{5}$$

根据式 (5) 的定义，可以分别得到如下的单线阵和双线阵的左右舷抑制增益：

$$G_{\rm single}(\theta,\theta_0)=10\log\left|\frac{D_{\rm single}(\theta,\theta_0)}{D_{\rm single}\left(-\theta,\theta_0\right)}\right|^2=0\ {\rm dB}, \tag{6}$$

$$G_{\rm twin}(\theta,\theta_0)=10\log\left|\frac{D_{\rm twin}(\theta,\theta_0)}{D_{\rm twin}\left(-\theta,\theta_0\right)}\right|^2=10\log\left|\frac{F(\theta,\theta_0)}{F\left(-\theta,\theta_0\right)}\right|^2. \tag{7}$$

即单线阵的左右舷抑制增益 $G_{\rm single}$ 在任何方位都为 0 dB，无左右舷分辨能力；而双线阵的左右舷抑制增益 $G_{\rm twin}$ 取决于因子 $F(\theta,\theta_0)$，结合式 (4)，可知双线阵非端射方位的左右舷抑制增益都不为零，具有左右舷分辨能力。为了分析左右舷抑制增益与有关参数的关系，将式 (3) 中 $F(\theta,\theta_0)$ 的表达式带入，得到：

$$G(\theta,\theta_0)=10\log\left\{\frac{1+\cos[2\pi l(\sin\theta-\sin\theta_0)/\lambda]}{1+\cos[2\pi l(\sin\theta+\sin\theta_0)/\lambda]}\right\}. \tag{8}$$

为表示方便，这里将双线阵左右舷抑制增益的下标省略。当声呐扫描波束对准目标方位时 ($\theta=\theta_0$)，

$$G(\theta_0)=10\log\left[\frac{2}{1+\cos(4\pi l\sin\theta_0/\lambda)}\right]. \tag{9}$$

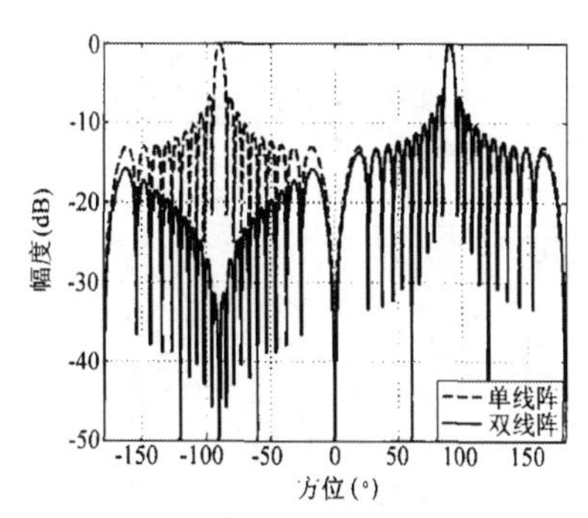

图 2 单线列阵和双线列阵指向性比较

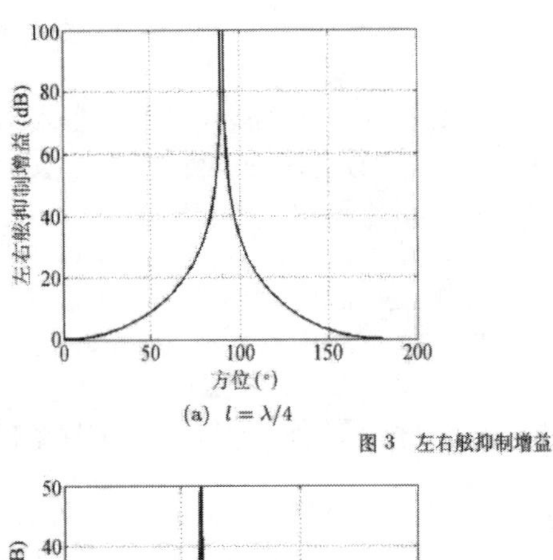

(a) $l=\lambda/4$

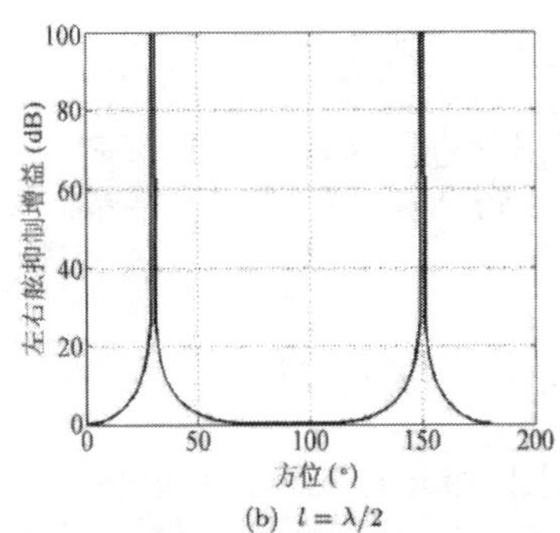

(b) $l=\lambda/2$

图 3　左右舷抑制增益与方位的关系曲线

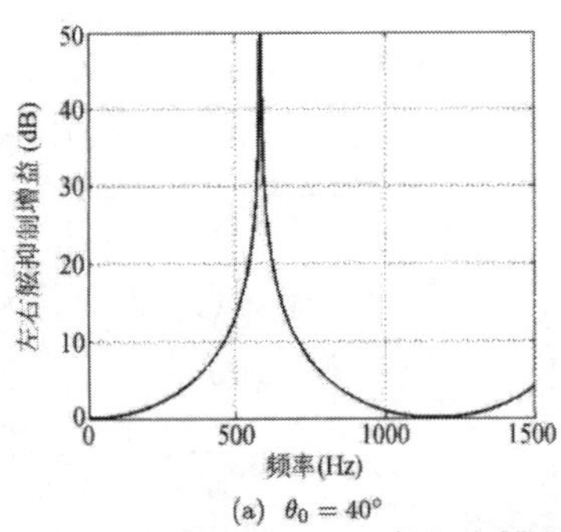

(a) $\theta_0=40°$

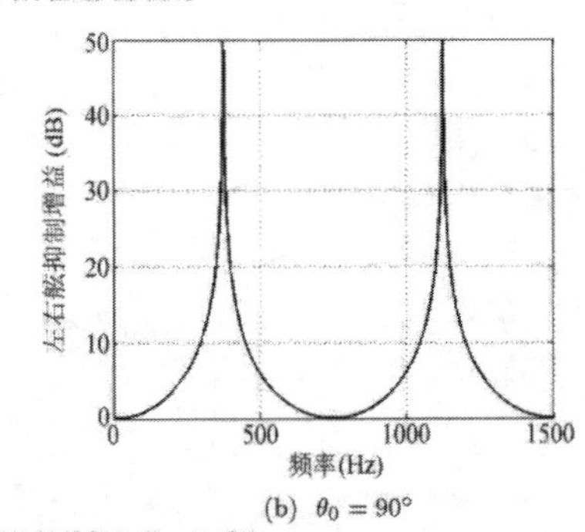

(b) $\theta_0=90°$

图 4　左右舷抑制增益与频率的关系曲线 ($l=\lambda_0/2$)

注意到，左右舷抑制增益与两条线列阵的间距 l、目标方位 θ_0 和波长 λ 都有关。我们期望在尽量大的范围内 $G(\theta_0)$ 达到最大，应有：

$$\cos(4\pi l\sin\theta_0/\lambda)=-1, \tag{10}$$

式 (10) 成立的时候，$G(\theta_0)$ 达到极大值。

1.2　特征方位与特征频率

若使得左右舷抑制增益最大，则有 $\cos(2kl\sin\theta_0)=-1$，进而从以下几个方面来讨论。

(1) 若频率 f 确定，则使得正横方向 ($\theta=90°$) 左右舷抑制比最大的两阵间距应为：

$$l=\left(\frac{n}{2}+\frac{1}{4}\right)\lambda,\quad n=0,1,2,\cdots \tag{11}$$

当 $l>\lambda/2$ 时，对于单频信号会出现相位模糊问题，因此基于双线阵波束形成的左右舷分辨技术，最佳的两阵间距是 $l=\lambda/4$。

(2) 若 f 和 l 确定，则使得左右舷抑制比最大的方位，定义为特征方位：

$$\theta_{\mathrm{ch}}=\arcsin\left[\left(\frac{n}{2}+\frac{1}{4}\right)\frac{\lambda}{l}\right],\quad n=0,1,2,\cdots \tag{12}$$

(3) 若 l 和 θ_0 确定，使得左右舷抑制比最大的频率，定义为 θ_0 方位的特征频率：

$$f_{\mathrm{ch}}=\frac{(2n+1)c}{4l|\sin\theta_0|},\quad n=0,1,2,\cdots \tag{13}$$

图 3 和图 4 分别给出了左右舷抑制增益与方位和频率的关系曲线，图中的峰值位置分别代表了特征方位和特征频率。可见，左右舷抑制增益对目标方位和目标频率都具有选择性，在特征方位处 (目标频率确定) 或特征频率处 (目标方位确定)，左右舷抑制增益达到极大值。图 4 中，两阵间距 $l=\lambda_0/2$，这里 λ_0 为 750 Hz 声呐工作中心频率所对应的波长。

2　阵形畸变时双线阵左右舷抑制比分析

上节的分析是基于两条线阵无畸变的假设条件进行的，即线阵A和B的每个阵元在如图1所示的理想位置处。但是拖曳双线阵声呐在实际工作中，由于机械原因或者海水环境等原因，阵元有可能会偏离理想位置而产生畸变，其中双线阵对应阵元 (图1中A阵与B阵横坐标相同的两个阵元称为一组对应阵元) 发生错位的情况是一种后果较为严重的畸变类型。

双线阵对应阵元错位在实际中是极有可能发生的，主要包括以下两类情况：一是双线阵仍然保持直线平行状态，但是从接收阵始端开始，阵元就发生了错位；二是接收阵始端没有错位，但是两条线阵产生

不同曲线形状 (如正弦曲线) 的畸变，导致对应阵元的错位。本文将主要考虑第一类畸变情况，至于第二类情况在这里暂不做讨论。

2.1 阵形畸变时左右舷抑制增益

假设畸变后各阵元位置如图 5 所示，A 阵向右偏移距离 r。畸变导致信号到达 A 阵各水听器与参考点 O 的声程差改变，

$$\varepsilon_{n0}=x_n\cos\theta_0-y_n\sin\theta_0+r\cos\theta_0, \tag{14}$$

当 A 阵向左偏移时，声程差为：

$$\varepsilon_{n0}=x_n\cos\theta_0-y_n\sin\theta_0-r\cos\theta_0. \tag{15}$$

这里只讨论 A 阵向右偏移的情况，向左偏移的分析与之相似。经推导得到此时的双线阵指向性函数变为：

$$b(\theta,\theta_0)=\left|\frac{1}{2}\{1+\exp[\mathrm{j}kl\sin\theta-\mathrm{j}k(l\sin\theta_0-r\cos\theta_0)]\}\right|\left|\frac{1}{N}\sum_{n=1}^{N}\exp[-\mathrm{j}k(n-1)d(\cos\theta-\cos\theta_0)]\right|. \tag{16}$$

进而得到阵形畸变下的双线阵左右舷抑制比：

$$G=10\log\frac{|1+\exp[\mathrm{j}kr\cos\theta_0]|^2}{|1+\exp[\mathrm{j}kr\cos\theta_0-2\mathrm{j}kl\sin\theta_0]|^2}. \tag{17}$$

若使左右舷抑制比最大，则有：

$$\frac{kr\cos\theta_0-2kl\sin\theta_0}{2}=\frac{\pi}{2}+n\pi,\quad n=0,1,2,\cdots \tag{18}$$

接下来，分别讨论偏移量 r 与特征方位及特征频率的关系。

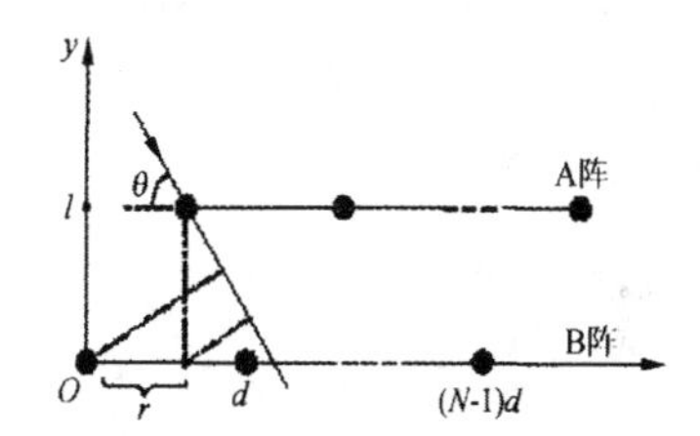

图 5 双线阵对应阵元错位情况下的坐标

2.2 阵形畸变时特征方位与特征频率

解方程 (18)，得到特征方位：

$$\theta_{\mathrm{ch}}=\arccos\left\{\frac{r(2n+1)\pi\pm\sqrt{r^2[(2n+1)\pi]^2-(r^2+4l^2)\{[(2n+1)\pi]^2-4k^2l^2\}}}{k(r^2+4l^2)}\right\}. \tag{19}$$

可见特征方位是关于偏移量 r 的函数，换句话说，偏移量 r 已经影响了特征方位的位置。特别地，令 $l=\lambda/4$，并将波数 $k=2\pi/\lambda$ 带入式 (19)，得到：

$$\theta_{\mathrm{ch}}=\arccos\left\{\frac{r(2n+1)\pi\pm r(2n+1)\pi}{\frac{2\pi}{\lambda}\left[r^2+\frac{\lambda^2}{4}\right]}\right\}. \tag{20}$$

即 $\theta_{\mathrm{ch1}}=\frac{\pi}{2}$ 和 $\theta_{\mathrm{ch2}}=\arccos\left[\frac{(2n+1)r\lambda}{r^2+\lambda^2/4}\right]$，说明 90° 特征方位不受偏移量影响，而另外一个特征方位随偏移量而变化。值得注意的是，在双线列阵无畸变的情况下，当 $l=\lambda/4$ 时，特征方位只有一个，且位于 90°。

由式 (19)，得到阵形畸变情况下的特征频率表达式为：

$$f_{\mathrm{ch}}(r,\theta)=\frac{(2n+1)c}{|4l\sin\theta-2r\cos\theta|},\quad n=0,1,2,\cdots \tag{21}$$

说明特征频率也是偏移量 r 的函数，即偏移量 r 也影响了特征频率的位置。

综合以上分析，对应阵元错位的阵形畸变对拖曳双线阵声呐左右舷分辨有很大的影响，导致特征方位和特征频率的偏离，下面通过实验仿真进一步说明以上的分析结果。

2.3 仿真结果

假设双线阵发生畸变，各阵元位置如图 5 所示，两条线列阵分别由 40 元水听器构成，等间隔半波长分布，设中心频率 750 Hz，并令偏移量 r 从 0 m 增至 1 m，观察偏移量对特征方位的影响。图 6 给出了 0° 至 180° 空间域内特征方位随偏移量的变化。可以看出，90° 方位的左右舷抑制比始终最大，但是偏移量 r 的存在产生了另外一个特征方位，且该方位随着偏移量 r 的增加而向端射方向移动。图中还反映出，偏移量的存在造成近端射部分左右舷模糊，即左右舷抑制比小于零，模糊区域随着偏移量的增加而向正横方向扩展。

接下来固定目标方位，取 90° 和 60° 两个方位，观察随偏移量 r 对特征频率的影响，令两阵间隔为 1 m，结果如图 7 所示。偏移量 r 不影响 90° 方位的特征频率位置，但是 60° 方位的特征频率则随着偏移量 r 的增加而向高频偏移 (这里假设 A 阵向右偏移)。

图 8 给出多个方位上偏移量 r 引起的特征频率具体数值上的相对变化, 结果分别在两幅图中表示, 图 8(a) 中 $\theta \leq 90°$, 图 8(b) 中 $\theta \geq 90°$, 这里考虑 A 阵向右偏移的情况。横坐标为偏移量 r, 从 0 m 增至 0.5 m, 纵坐标为特征频率的相对偏差 (定义为有偏移量时的特征频率与无偏移量时的特征频率之差)。从两图可以看出, 偏移量 r 会导致除正横方位之外的各个方位上的特征频率值发生变化。正横方位的特征频率始终保持不变, 小于 90° 方位的特征频率都向高频移动; 反之, 大于 90° 方位的特征频率都向低频移动。并且随着偏移量 r 的增大, 特征频率的相对偏差也逐渐加大, 越靠近端射方位, 特征频率相对偏差也越大。A 阵向左偏移的情况则正好相反。

2.4 海试数据分析

在某海域进行了拖曳双线阵声呐水下目标探测及左右舷分辨的海上试验[10]。试验过程中海流影响较小, 两线列阵基本上处于水平平行姿态。这里选取被动声呐接收的两组数据进行分析, 其中第1组是阵形无畸变的情况; 第 2 组两条线阵约有 40 cm 的错位。将两组数据得到的左右舷抑制比与理论抑制比曲线进行对比, 结果分别如图 9(a) 和图 9(b) 所示。两图中的实线表示理论结果, “*” 是实际数据得到的结果。图 9(a) 显示在没有畸变的情况下, 实际数据

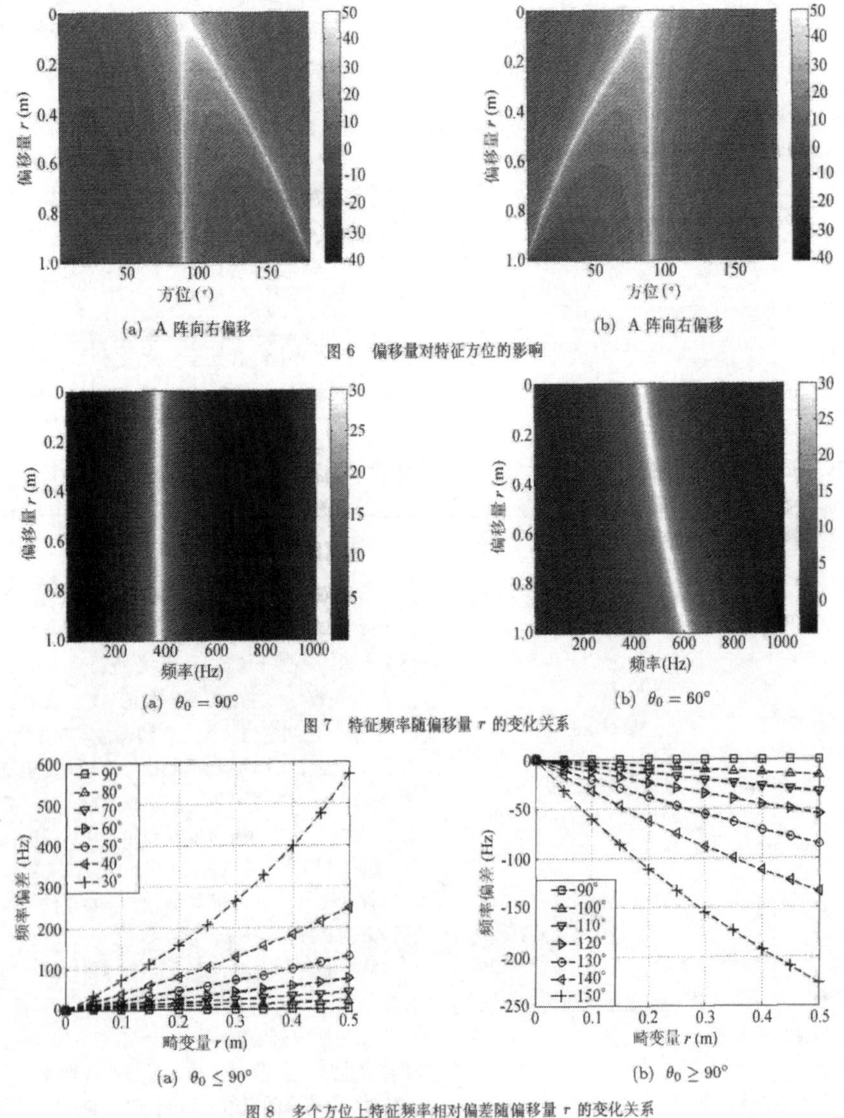

(a) A 阵向右偏移 (b) A 阵向右偏移

图 6 偏移量对特征方位的影响

(a) $\theta_0 = 90°$ (b) $\theta_0 = 60°$

图 7 特征频率随偏移量 r 的变化关系

(a) $\theta_0 \leq 90°$ (b) $\theta_0 \geq 90°$

图 8 多个方位上特征频率相对偏差随偏移量 r 的变化关系

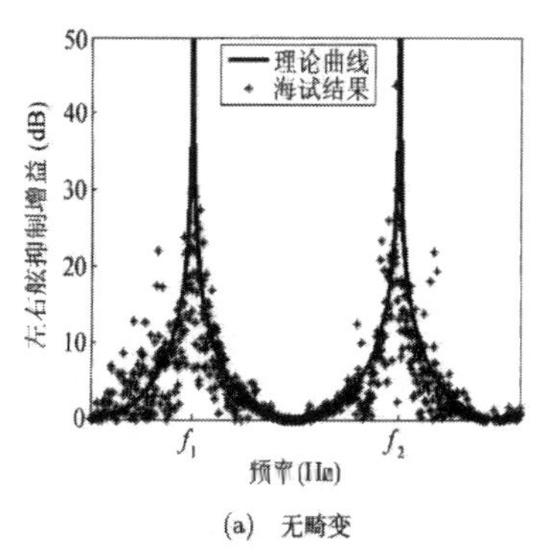

(a) 无畸变

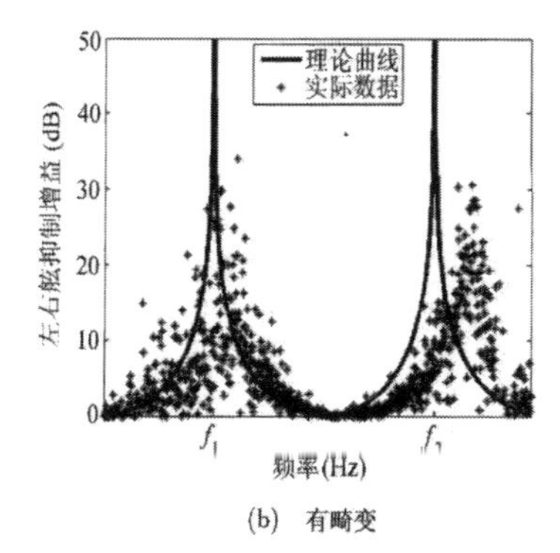

(b) 有畸变

图 9　阵形畸变对特征频率的影响理论曲线与实际结果的对比

结果符合特征频率的规律，在 f_1 和 f_2 处左右舷抑制比达到最大，其分布趋势与理论曲线较为吻合；图 9(b) 中两条线阵发生了错位畸变，造成的结果是实际左右舷抑制比随频率的分布趋势与理论曲线发生偏离，实际特征频率并不在理论值处。

3　结论

阵形畸变对于拖曳双线阵左右舷抑制比有重要的影响。文章推导了双线阵对应阵元错位的畸变情况下的左右舷抑制比公式，并通过大量仿真，得到以下结论：

(1) 对于主动工作模式，双线阵声呐一般取发射信号中心频率对应波长的四分之一作为为双阵间距，此时正横方位的左右舷抑制比始终最大，不受偏移量 r 的影响，但是同时偏移量 r 的存在会产生另外一个特征方位，且该角度随着偏移量 r 的增加而向端射方向移动。

(2) 偏移量造成近端射区域的左右舷模糊，且随着偏移量增加，该区域范围逐渐向正横方位扩展。

(3) 正横方位的特征频率始终保持不变，其它方位的特征频率会受到偏移量的影响而向高频 (小于 90° 方位) 或低频 (大于 90° 方位) 移动，若 B 阵向右偏移，则反之；随着偏移量 r 的增大，特征频率的相对偏差也逐渐加大，并且越靠近端射方位，特征频率相对偏差也越大。

本文的工作将为主被动拖曳双线阵声呐在实际中更好地分辨左右舷目标提供重要的帮助。

参　考　文　献

1　Newhall B K. The equivalence of several types of twin-line beamformers and implications for efficient implementation on the GCBF. Johns Hopkins Appl. Phys. Lab., Laurel, MD, 1992, Tech. Memo. STA-92-476

2　李启虎. 双线列阵左右舷目标分辨性能的初步分析. 声学学报, 2006; **31**(5): 385—388

3　李启虎. 用双线列阵区分左右舷目标的延时估计方法及其实现. 声学学报, 2006; **31**(6): 485—487

4　杨秀庭, 孙贵青, 李　敏, 李启虎. 矢量拖曳线列阵声呐流噪声的空间相关性研究. 声学学报, 2007; **32**(6): 547—552

5　孙贵青, 李启虎. 声矢量传感器信号处理. 声学学报, 2004; **29**(6): 491—498

6　Allensworth W S, Kennedy C W, Newhall B K, Schurman I W. Twinline array development and performance in a shallow-water littoral environment. *Johns Hopkins APL Tech. Dig.*, 1995; **16**(3): 222—232

7　Newhall B K, Allensworth W S, Schurman I W. Unambiguous noise and reverberation measurements from a dual line towed array. *J. Acoust. Soc. Am.*, 1995; **97**(5): 3289

8　Schurman I W. Reverberation rejection with a dual-line towed array. *IEEE J. Ocean. Eng.*, 1996; **21**(2): 193—204

9　LI Qihu, CHEN Xinhua, LI Min, YANG Xiuting, TIAN Tian, HUANG Xiaoping. Effect of Shape Distortion in Solving Left-right Ambiguity of Towed Twin-line Array, UDT Europe 2006, May 2006, Hamburg, Germany

10　李启虎等. 主被动拖线阵声呐中拖曳平台噪声和拖鱼噪声在浅海使用时的干扰特性. 声学学报, 2007; **32**(1): 1—4

拖船噪声抵消与左右舷分辨联合处理方法的研究①

张　宾[1,2]　孙贵青[2]　李启虎[2]

（1 青岛科技大学　青岛　266061）

（2 中国科学院声学研究所　北京　100190）

摘要　利用拖曳双线阵声纳基阵结构的左右舷分辨能力，并结合拖船相对于拖曳线列阵的空间位置关系，本文提出一种拖船噪声抵消与左右舷分辨联合处理的方法，在波束域分别形成左右波束并将映像波束相减，在检测目标并进行左右分辨的同时，可以显著抵消拖船噪声。海试结果验证了该方法的有效性，拖船噪声被抑制掉约 14dB，同时正确分辨出目标的左右舷方位。

关键词　拖曳双线阵，左右舷分辨，噪声抵消

Tow ship noise canceling combined with left/right discrimination

ZHANG Bin[1,2]　SUN Gui-Qing[2]　LI Qi-Hu[2]

(1 *Qingdao University of Science and Technology, Qingdao* 266061)

(2 *Institute of Acoustics, Chinese Academy of Sciences, Beijing* 100190)

Abstract　Based on the left/right discrimination ability of towed twin-line array and the orientation of tow ship relative to the array, a method for tow ship noise canceling combined with left/right discrimination is proposed, which forms left and right beams respectively, and performs subtraction for each pair of image beams. This method achieves target detection, left/right discrimination and tow ship noise canceling at the same time. The results of some sea trial data processing are promising, where the tow ship noise is cancelled by about 14dB, and left/right positions of the targets are discriminated correctly.

Key words　Towed twin-line array, Left-right discrimination, Noise canceling

① 应用声学, 2008, 27(5): 380-385.

1 引言

拖船辐射噪声是拖曳线列阵声纳难以避免的强干扰，如何抑制或抵消拖船干扰一直以来都是拖线阵声纳信号处理领域的一个重点和难点问题。之所以困难主要是由于拖船辐射噪声自身特有的一些性质造成的，如多途传播路径丰富、噪声功率谱起伏大等等，并且相对于几公里甚至几十公里以外的目标而言，拖船噪声是一个近程强干扰，其强度往往比目标强度高出许多，严重影响真实目标的检测，使声纳丧失了端射方向附近约60度的警戒扇面。因此，虽然国内外学者尝试利用多种方法抵消拖船噪声，如自适应噪声抵消、自适应波束形成、后置波束形成干扰抵消，以及匹配场处理等方法，但是实际效果仍不能令人满意[1-3]。

20世纪90年代以来，为了解决常规拖曳单线阵声纳左右舷模糊问题，相继诞生了三元水听器、双线列阵以及矢量水听器阵等基阵结构的拖曳式声纳。特别地，拖曳双线阵声纳具有工作频带低、算法简易、工程可靠性高等优点，自诞生以来受到了越来越多的关注，海上试验也显示了它的潜在优势[4,5]。本文基于双线阵声纳的结构特点，利用双线阵声纳自身具有左右舷分辨的性能，提出一种与常规方法不同的拖船噪声抵消方法，即映像波束相减法。首先将对双线阵左右舷分辨性能进行分析，而后给出映像波束相减法的具体实现，最后利用海试数据对该算法进行验证。

2 双线阵左右舷分辨性能分析

假设由两条互相平行的线列阵构成的双线阵如图1所示分布，分别称两条线列阵为A阵和B阵，二者的间距为l，每条线列阵分别由N个等间隔分布的阵元组成，阵元间距为d，以坐标原点O为参考点。令y轴的原点O端靠近拖曳平台，将y轴正半轴一侧，$\theta\in(0°,180°)$，定义为右舷；将y轴负半轴一侧，$\theta\in(0°,180°)$，定义为左舷。

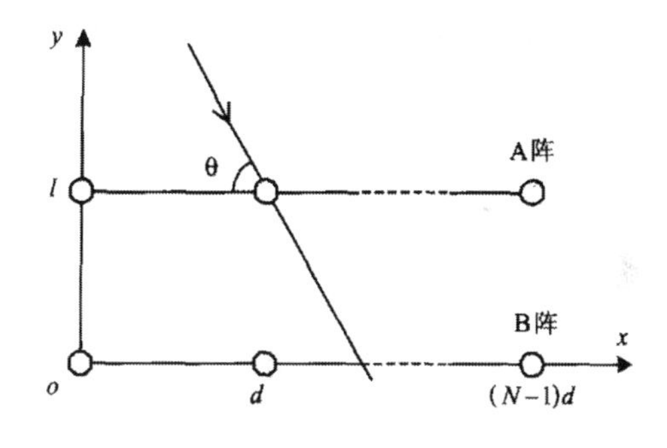

图1 双线阵平面坐标

如图1所示分布的双线阵指向性函数为[6]

$$D_{twin}(\theta,\theta_0)=\left\{\frac{1+\cos[2\pi l(\sin\theta-\sin\theta_0)/\lambda]}{2}\right\}^{1/2}\bullet\left|\frac{\sin[N\pi d(\cos\theta-\cos\theta_0)/\lambda]}{N\sin[\pi d(\cos\theta-\cos\theta_0)/\lambda]}\right| \quad (1)$$

式中θ为声纳波束扫描角，θ_0为目标方位角，λ为与入射信号中心频率对应的波长。注意到式(1)中乘式的第二个因子恰是单线列阵的指向性函数，而第一个因子实际上是心形指向性函数，说明了双线列阵指向性函数是由单线列阵指向性函数与一个心形指向性函数$F(\theta,\theta_0)$相乘得来

$$D_{twin}(\theta,\theta_0)=F(\theta,\theta_0)D_{single}(\theta,\theta_0) \quad (2)$$

这里

$$F(\theta,\theta_0)=\left\{\frac{1+\cos[2\pi l(\sin\theta-\sin\theta_0)/\lambda]}{2}\right\}^{1/2} \quad (3)$$

$$D_{single}(\theta,\theta_0)=\left|\frac{\sin[N\pi d(\cos\theta-\cos\theta_0)/\lambda]}{N\sin[\pi d(\cos\theta-\cos\theta_0)/\lambda]}\right| \quad (4)$$

从式(3)看出，对于一对映像方位θ和$-\theta$有

$$F(\theta,\theta_0) \neq F(-\theta,\theta_0) \tag{5}$$

说明由于因子 $F(\theta,\theta_0)$ 的存在，双线列阵在一对映像方位具有不同的响应值，从而双线列阵的基阵结构具有了左右舷分辨的能力。

图2给出了目标方位 θ_0 分别为90°、60°和30°时的双线阵和单线阵指向性图，左侧为双线阵结果，右侧为单线阵结果。这里取 $d=\lambda/2$，$l=\lambda/4$，$N=40$。从几幅图可以看出单线阵在一对映像方位具有完全相同的响应值，无法辨别目标的真实方位，从而产生左右舷模糊问题；而双线阵则在目标真实方位形成最大响应，在其映像方位无响应（正横方位）或仅有较小的响应（非正横方位），可以实现目标的左右舷分辨，证明了双线阵具有解决左右舷模糊问题的能力。

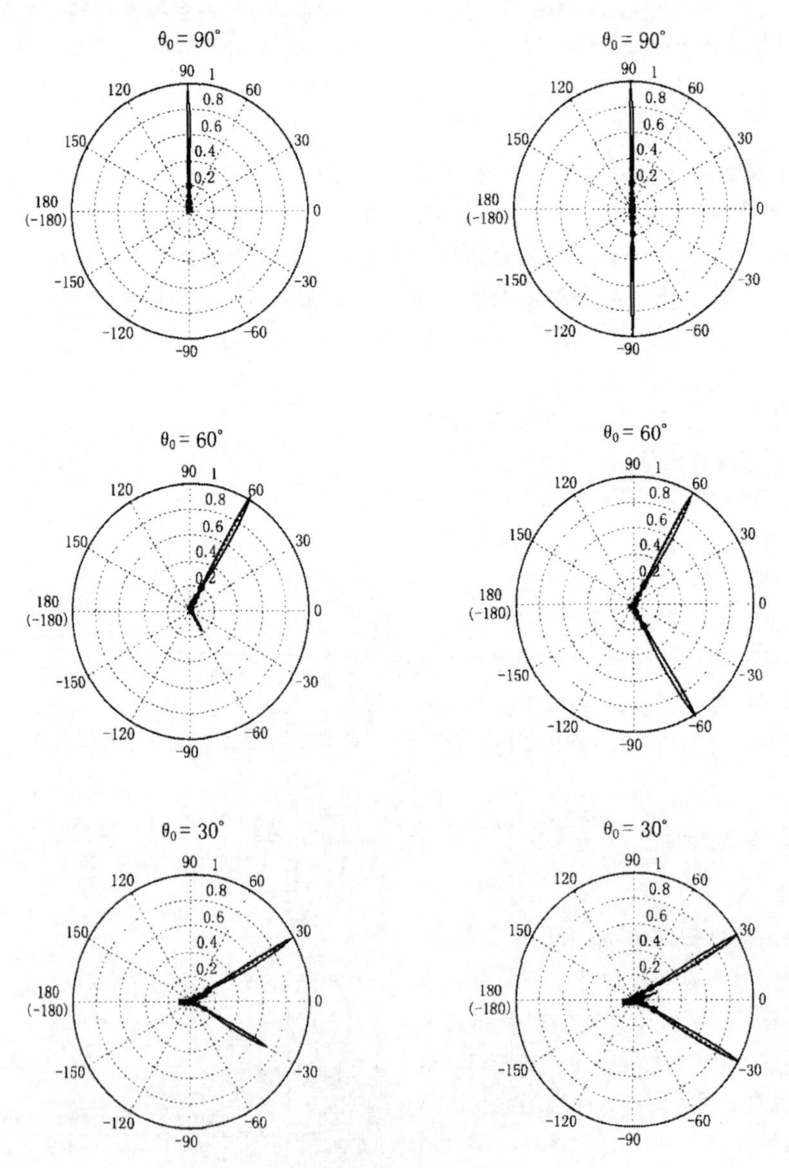

图2　双线列阵与单线列阵的指向性比较

（左侧为双线阵结果；右侧为单线阵结果）

左右舷抑制增益(Gain of Left/Right Suppression)是衡量声纳左右舷分辨能力的定量参数,可定义为一对映像方位的响应之比[7],即

$$G(\theta,\theta_0)=10\lg\left|\frac{D(\theta,\theta_0)}{D(-\theta,\theta_0)}\right|^2 \qquad (6)$$

根据式(4.13)的定义,可以分别得到单线列阵和双线列阵的左右舷抑制增益,

$$G_{single}(\theta,\theta_0)=10\lg\left|\frac{D_{single}(\theta,\theta_0)}{D_{single}(-\theta,\theta_0)}\right|^2=0\text{dB} \qquad (7)$$

$$G_{twin}(\theta,\theta_0)=10\lg\left|\frac{D_{twin}(\theta,\theta_0)}{D_{twin}(-\theta,\theta_0)}\right|^2=10\lg\left|\frac{F(\theta,\theta_0)}{F(-\theta,\theta_0)}\right|^2 \qquad (8)$$

容易看出,单线阵的左右舷抑制增益为0dB,无法判别目标来自于左舷还是右舷,存在左右舷模糊问题。而双线阵的左右舷抑制增益由于因子 $F(\theta,\theta_0)$ 而不等于零分贝,从而使得双线阵声纳具备左右舷分辨的能力。

图3给出了双阵间距 $l=\lambda/4$ 时各扫描方位的左右舷抑制增益,每条线列阵的阵元间距 $d=\lambda/2$。可以看出,对于波长为 λ 的入射信号而言,在正横方位具有最大的左右舷抑制增益,理论上可达到无穷大。虽然随着入射角逐渐向端射靠近,增益会逐渐减小,但是即使在20°(160°),也有近2dB的增益,这充分说明了双线阵结构具有左右舷分辨的能力。

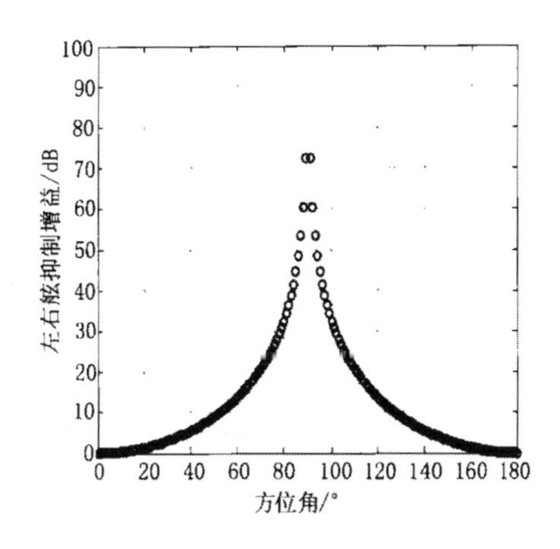

图3 双线阵左右舷抑制增益随角度的变化关系(增益截取至100dB)

3 映像波束相减法

根据上一节的分析,对于非端射方位入射的目标而言,双线阵声纳具有左右舷分辨的能力,并且当目标在52°~128°之间时,左右舷抑制增益可达到10dB以上。而由图4所示,拖船噪声经多途传播从端射方向的垂直面内入射到接收阵,并非在舷侧入射,即无左右舷之分。换言之,对于目标而言,双线阵在左舷和右舷形成的波束会有不同的输出,而拖船辐射噪声的左右舷波束输出应是相同的。基于这样的考虑,本文提出映像波束相减的方法,抵消拖船干扰的同时,检测到目标信号。下面给出该方法的具体实现过程。

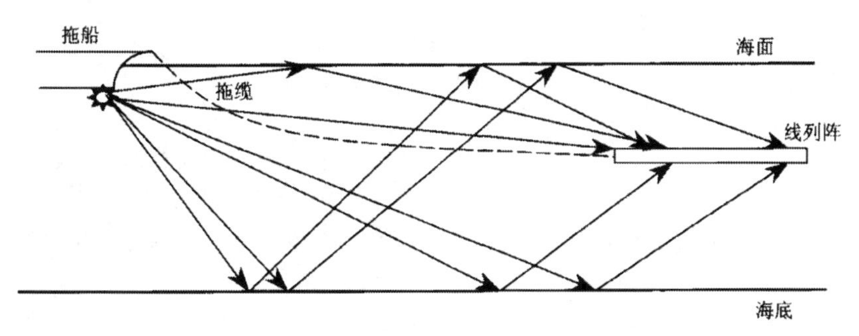

图4 拖船辐射噪声传播示意图

假设一远场平面波入射到如图 1 所示的双线阵，在 $\theta \in (-180°, 180°)$ 空间范围内扫描，对 2N 个阵元进行波束形成，得到

$$y(t,\theta) = W_1^H(\theta)X_1(t) + W_2^H(\theta)X_2(t) \quad (9)$$

式中 $X_1(t)$、$X_2(t)$ 分别为 A 阵和 B 阵的接收信号向量，$W_1(\theta)$、$W_2(\theta)$ 分别为 A 阵和 B 阵的权向量，

$$W_1(\theta) = [\exp(-jkl\sin\theta), \exp(jkd\cos\theta - jkl\sin\theta), \cdots, \exp(jk(N-1)d\cos\theta - jkl\sin\theta)]^T \quad (10)$$

$$W_2(\theta) = [1, \exp(jkd\sin\theta), \exp(jk2d\cos\theta), \cdots, \exp(jk(N-1)d\cos\theta)]^T \quad (11)$$

双线阵波束形成器的功率输出为：

$$P(\theta) = E[y^*(t,\theta) \cdot y(t,\theta)] = |y(t,\theta)|^2 \quad (12)$$

其中，$E[\cdot]$ 表示数学期望，$\{\cdot\}^*$ 表示共轭。当声纳波束扫描到目标所在方位时，即 $\theta = \theta_0$，$P(\theta_0)$ 有极大值。称 θ_0 和 $-\theta_0$ 为一对映像角度，并称 $P(\theta_0)$ 与 $P(-\theta_0)$ 为一组映像波束，对映像波束执行相减操作，理论上可将拖船干扰完全抵消掉，这里视 $\theta_0 \in (0°, 180°)$ 为右舷。

4 海试数据处理结果

本节利用映像波束相减的方法处理拖曳双线阵声纳某次海上试验获取的数据，该段数据中三个目标分别位于 106°、121°和 136°，如图 5 所示。(a) 图给出了全方位的波束输出图，可见双线阵确实具有左右舷分辨性能，能够分辨出三个目标都来自于右舷。而 19°的拖船干扰在两舷的输出则较为接近。利用本节的映像波束相减方法，得到 (b) 图所示结果，端射方位的干扰几乎都被抵消，与理论分析相符，同时三个目标的信息凸现。

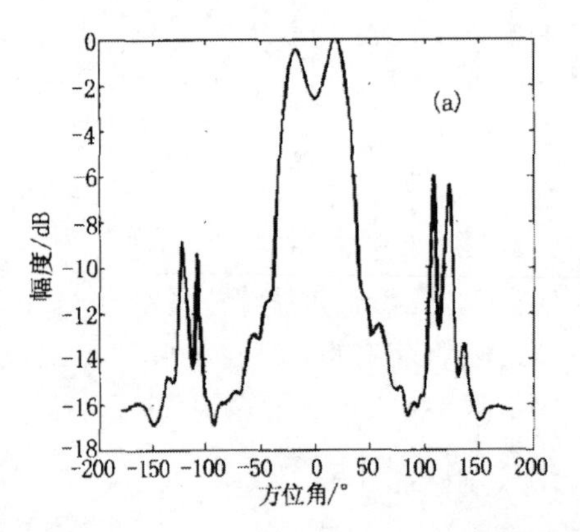

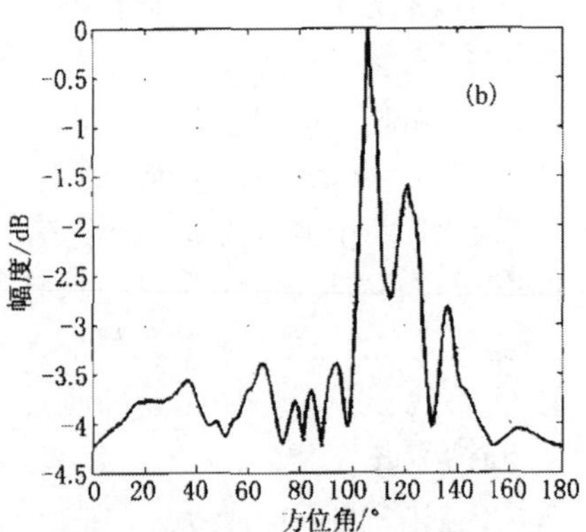

图 5 双线阵映像波束相减法抵消拖船干扰
(a) 双线阵波束输出 (b) 干扰抵消后的输出

5 小结

拖曳双线阵声纳自身的结构特点决定了它能够克服传统拖曳单线阵声纳左右舷模糊的固有缺陷，具备左右舷分辨的能力。基于此，本节给出了映象波束相减的拖船干扰抵消方法，海试数据处理结果显示，该方法不仅将干扰较彻底地抵消掉，背景噪声也得到了均衡处理，使得目标信息更能够凸现出来。映像波束相减的干扰抵消方法，其性能取决于双线阵声纳左右舷分辨的效果，因此寻求更好的双线阵声纳左右舷分辨方法是关键。但是，对于来自端射方向的目标信号仍无法与拖船噪声分离，这是本文方法的一个局限

性。映像波束相减法将双线阵的左右舷分辨与拖船噪声抵消二者结合，是一种值得关注的研究方向。

参 考 文 献

[1] M. K. Robert and S. P. Beerens. Adaptive beamforming algorithms for tow ship noise canceling. Proc. of European Undersea Defense Technology, La Spezia, 2002.

[2] MA Yuanliang, YAN Shefeng, and YANG Kunde. Matched field noise suppression: Principle with application to towed hydrophone line array. Chinese Science Bulletin, 2003, 48(12): 1207 - 1211.

[3] James V. Candy and Edmund J. Sullivan. Canceling tow ship noise using an adaptive model-based approach. Proc. of the IEEE/OES Eighth Working Conference on Current Measurement Technology, 2005: 14 - 18.

[4] T. Warhonowicz, H. Schmidt-Schierhorn, and H. Hostermann, Port/starboard discrimination performance by a twin line array for a LFAS sonar system. Proc. of European Underwater Defense Technology, 1999: 398.

[5] R. M. Zeskind et al. Acoustic performance of a multi-line system towed in several ocean environments. IEEE J. of Ocean Eng., 1998, 23(1): 124 - 128.

[6] 李启虎. 双线列阵左右舷目标分辨性能的初步分析. 声学学报. 2006,31(5):385 - 388.

[7] D. T. Hughes. Aspects of cardioid processing, SACLANTCEN REPORT, SR - 329, 2001.

一种新的二元水听器组左右舷分辨方法①

何心怡[1)] 邱志明[1)] 张春华[2)] 李启虎[2)]

(海军装备研究院[1)] 北京 100073) (中国科学院声学所[2)] 北京 100080)

摘要：为解决拖曳线列阵声呐的左右舷模糊问题，提出了一种新的二元水听器组左右舷分辨方法．该方法以移相处理为基础，将单个二元水听器看作一个小基阵．移相处理的目的是使从镜像源方向来的信号经处理后最小，进而使单个二元水听器具有心形指向性，从而保证二元水听器组能够分辨目标所处的左右舷．经过理论分析和湖试数据验证，证明了这种方法能有效解决大横截尺寸二元水听器组和小横截尺寸二元水听器组的左右舷模糊问题，适用范围广，算法简单，计算量小．

关键词：二元水听器组；左右舷分辨；移相；大横截尺寸；小横截尺寸

中图法分类号：U666.7；TB5

实时分辨左右舷是拖曳线列阵声呐必须要解决的关键性问题之一，采用特殊的基阵样式是一种有效的解决办法，如采用三元水听器组[1-2]、双线阵[3]或矢量水听器线列阵[4]等．笔者曾提出一种基阵结构——二元水听器组来实现左右舷分辨[5]，并将三元水听器组的几何相移模型和噪声相关模型推广到二元水听器组．其中，几何相移模型左右舷分辨方法算法简单，适用于横截尺寸较大的二元水听器组[6]；而噪声相关模型左右舷分辨方法适用于横截尺寸较小的二元水听器组，但算法较复杂、计算量较大[7]．

文中提出了一种新的二元水听器组左右舷分辨方法，它以移相处理为基础，简称为移相左右舷分辨方法，其适用范围广(对横截尺寸较大和横截尺寸较小的二元水听器组均适用)、算法简单、计算量小．经过理论分析和湖试数据验证，证明了该方法是一种有效的左右舷分辨方法．

1 二元水听器组的结构及坐标系的建立

由二元水听器组构成的拖线阵声阵段的结构示意图如图1所示，可认为它是将2根相同的、相互平行的线列阵1和线列阵2制作在同一根缆内．假设每根线列阵均由M个水听器组成，则整个二元水听器组由$2\times M$个水听器组成，根据离拖船的远近依次编号为$(L_1, L_2, \cdots, L_M)$和$(R_1, R_2, \cdots, R_M)$，如图2所示．这样，可以把二元水听器组看作是由$(L_1, R_1), (L_2, R_2), \cdots, (L_M, R_M)$ M个二元水听器构成的，每个二元水听器的2个水听器之间通过物理手段加以固定，单个二元水听器的2个水听器的间距为l，相邻2个二元水听器的间距为d．整个二元水听器组通过中性浮力拖缆，拖曳于水面舰艇的尾部．

在三维直角坐标系中建立二元水听器组的坐标如图2所示：坐标中心O点位于第1个二元水听器(L_1, R_1)的中心；x轴指向水面舰艇，那么二元水听器组的轴线恰位于负x轴上；y轴正向指向本舰左舷法线方向，负向指向本舰右舷法线方向；z轴指向海面；接收到的远场单位方向矢量$\boldsymbol{u}=(\sin\theta\cos\phi, \sin\theta\sin\phi, \cos\theta)$，方位角$\phi$和仰角$\theta$分别是$x$轴正向到矢量$\boldsymbol{u}$在$xoy$平面上投影的夹角和$z$轴正向到$\boldsymbol{u}$的夹角；并且规定这些角度顺时针为正，逆时针为负；空间处理的期望信号的方向用单位方向矢量$\boldsymbol{u}_0=(\sin\theta_0\cos\phi_0\sin\theta_0\sin\phi_0, \cos\theta_0)$表示，相应的方位角为$\phi_0$、仰角为$\theta_0$．

① 武汉理工大学学报(交通科学与工程版), 2008, 32(6): 1098-1101.

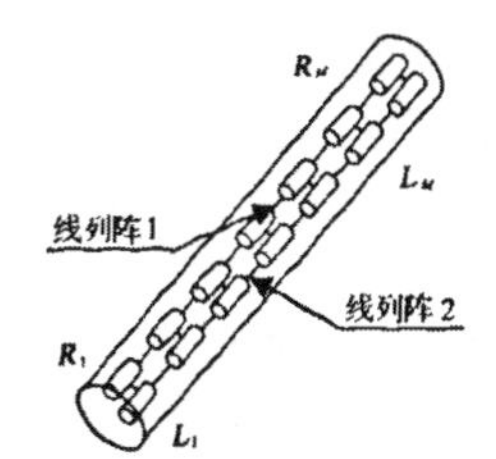

图1　由二元水听器组构成的拖线阵声阵段的结构示意图

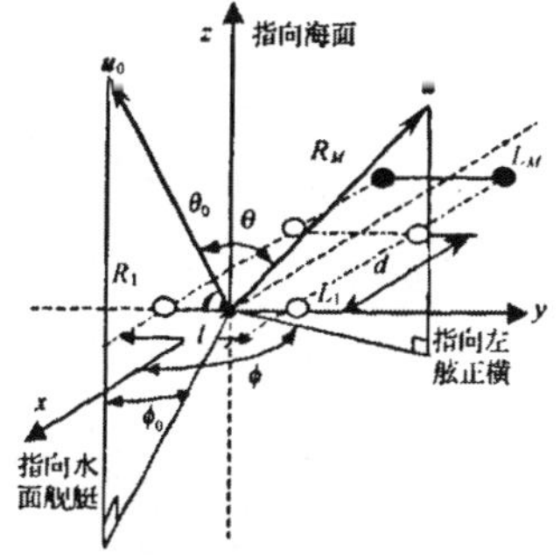

图2　二元水听器组在三维直角坐标系中的示意图

要实现左右舷分辨，就是经过一定的处理后，使得单个二元水听器具有心形指向性. 由布里奇(Bridge)乘积定理可知[8]，此时二元水听器组具有指向性，能够分辨目标所处的左右舷.

2　移相左右舷分辨方法的理论分析

取出单个二元水听器，以1号二元水听器为例，如图3所示.

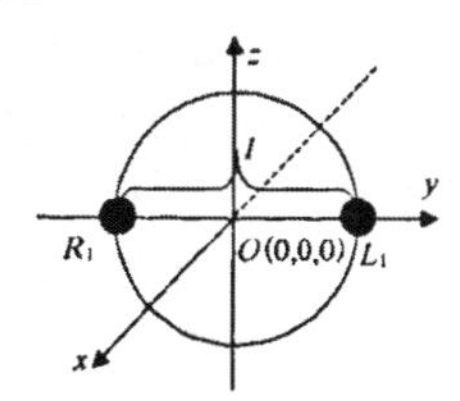

图3　1号二元水听器示意图

图3中的二元水听器的2个水听器坐标分别为$L_1=(0,l/2,0)$和$R_1=(0,-l/2,0)$，2个水听器接收到的远场单位方向信号分别为$P_L^{(1)}$和$P_R^{(1)}$.

$$P_L^{(1)}(f,\theta,\phi)=\exp\left[\mathrm{j}k\sin\theta\sin\phi\cdot(l/2)\right]$$
$$P_R^{(1)}(f,\theta,\phi)=\exp\left[-\mathrm{j}k\sin\theta\sin\phi\cdot(l/2)\right] \tag{1}$$

式中：k为波数，$k=2\pi f/c$. 其中：c为声速；f为频率. 如果将单个二元水听器看作一个小基阵，对二元水听器进行一般波束形成处理时，其指向性函数如下式所示，并不具备左右舷分辨能力.

$$|B(f,\theta,\phi)|=\frac{1}{2}|P_L^{(1)}(f,\theta,\phi)+P_R^{(1)}(f,\theta,\phi)|=\cos\left(k\frac{l}{2}\sin\theta\sin\phi\right) \tag{2}$$

在一般情况下，只需要考虑从水平面入射信号的方向，而这时就要分清信号究竟是来自左舷还是右舷，也就是通常所说的左右舷模糊问题. 此时，$\theta=\frac{\pi}{2}$，假定束控方向为$(\frac{\pi}{2},\phi_0)$，那么束控方向对应的镜像源的方位角为$\phi_m=-\phi_0$. 为使从束控方向对应的镜像源方向来的信号经处理后最小，可通过移相的方法实现，即

$$B'\left(f,\frac{\pi}{2},\phi\right)=\frac{1}{2}P_L^{(1)}\left(\frac{\pi}{2},\phi\right)\times\exp\left[-\mathrm{j}k\sin\phi_0\cdot\frac{l}{2}\right]-\frac{1}{2}P_R^{(1)}\left(\frac{\pi}{2},\phi\right)\exp\left[\mathrm{j}k\sin\phi_0\cdot\frac{l}{2}\right] \tag{3}$$

将式(1)代入式(3)，经过推导可得出

$$\left|B'\left(f,\frac{\pi}{2},\phi\right)\right|^2=1-\cos\left[\frac{2\pi l}{\lambda}(\sin\phi_0+\sin\phi)\right] \tag{4}$$

当束控方向选在二元水听器组的正横方向时，即$\phi_0=\pm\frac{\pi}{2}$时，则束控方向ϕ_0和束控方向对应的镜像源方向$\phi_m=-\phi_0$的波束输出分别为

$$|B'(f,\pi/2,\phi_0)|^2=1-\cos(4\pi l/\lambda)$$
$$|B'(f,\pi/2,\phi_m)|^2=|B(f,\pi/2,-\phi_0)|^2=0 \tag{5}$$

为使束控方向$\phi_0=\pm\pi/2$的输出最大，则应满足

$$\frac{4\pi l}{\lambda}=\pi+2n\pi,\qquad n=0,1,2,\cdots \tag{6}$$

也就是l应满足

$$l=\frac{\lambda}{4}+\frac{\lambda}{2}n,\qquad n=0,1,2,\cdots \tag{7}$$

为保证不出现相位模糊，则最佳间距$l_{opt}=\lambda/4$，此时$|B'(f,\pi/2,\phi_0)|^2=2$. 当然，如果所要处理的信号是宽频带的话，那么最佳间距应取为中心频率对应波长的1/4，从而证明了移相左右舷分辨方法适用于大横截尺寸二元水听器组.

从表面上看，以上分析与小横截尺寸二元水听器的左右舷分辨问题无关，但是，当$1<\lambda/10$时，文中提出的移相方法同样能完成左右舷分辨功能. 如$l=0.0625\lambda$时，单个二元水听器的左右波束指向性图如图4所示，图中实线为右波束指向性图，虚线为左波束指向性图，仍然是理想的心形指向性，能满足左右舷分辨的要求，只是基阵的增益比在$l_{opt}=\lambda/4$时有所减小.

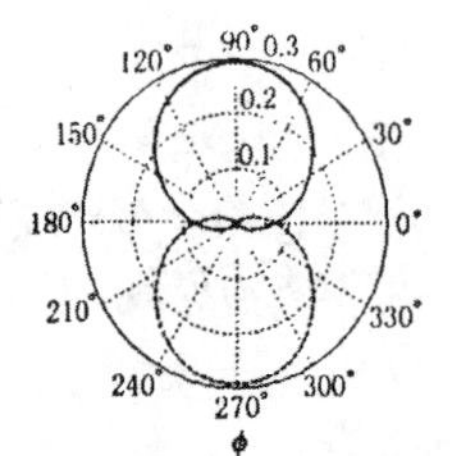

图4 $l=0.0625\lambda$ 时,单个二元水听器的左右波束指向性图

可如下定义左右舷抑制比

$$r = 10\lg \frac{|B'(f,\pi/2,\phi)|}{|B'(f,\pi/2,-\phi)|} = 10\lg \frac{|B'(f,-\pi/2,\phi)|}{|B'(f,-\pi/2,-\phi)|} \tag{8}$$

当 $l=0.0625\lambda$ 时,左右舷抑制比与入射角的曲线如图5所示,能够满足拖曳线列阵声呐在实际使用中的需要.

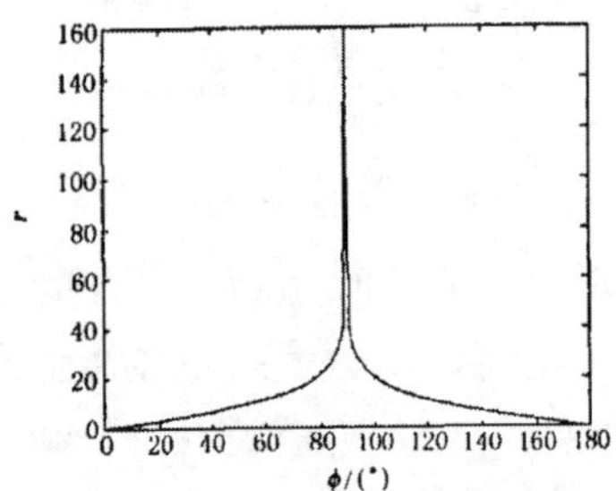

图5 $l=0.0625\lambda$ 时,左右舷抑制比与入射角的关系曲线

3 湖试实验

以2002年12月在某淡水湖进行的双线阵左右舷分辨原理试验中所获取的数据进行验证.从信号处理角度看,这样做与直接用二元水听器组的数据进行验证是等效的.双线阵的2根线阵均由32个均匀分布的水听器构成,相邻2水听器间距0.5 m,双线列阵间距 $l=0.25$ m,平行吊放于实验船水下8 m处静态接收信号.试验中有一个合作目标即目标船,试验水域附近还常常出现一些过往船只,由于目标船和过往船只吨位都很小、航速也较低,大多数情况下接收信噪比均低于0 dB.采样频率为10 kHz.

3.1 湖试试验数据验证一

目的:验证移相左右舷分辨方法对于较大横截尺寸二元水听器组左右舷分辨的有效性.

双线阵间距为0.25 m,是1 500 Hz对应波长的1/4,因此移相左右舷分辨方法设定的频率为1 500 Hz.经过处理,可得到某段噪声数据经过移相左右舷分辨方法处理前后的左右波束水平指向性图如图6和图7所示,图中,实线为右波束,虚线为左波束.对比图6和图7,能够发现:经过移相左右舷分辨方法处理后,能清楚地判断出3个目标分别位于左舷30,84和142°,与现场记录的目标船和过往船只的信息相符.

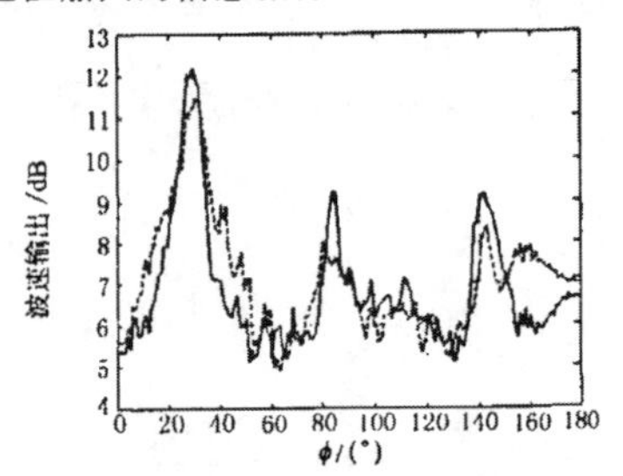

图6 未经左右舷分辨方法处理的左右波束水平指向性图

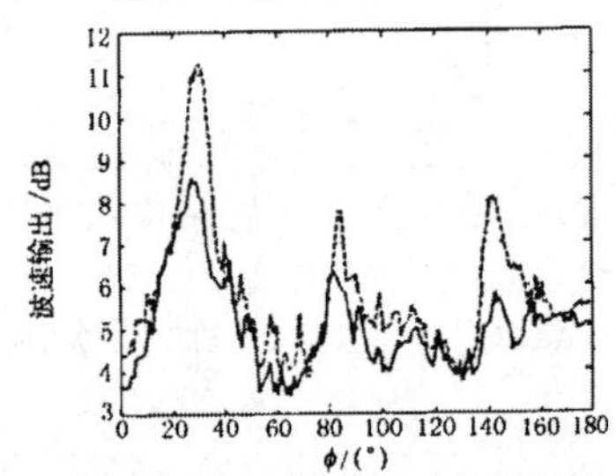

图7 经过移相左右舷分辨方法处理后的左右波束水平指向性图

3.2 湖试试验数据验证二

目的:验证移相左右舷分辨方法对于小横截尺寸二元水听器组左右舷分辨的有效性.

做信号分析时,将双线阵的数据每隔4路取出1路水听器信号,处理的频带范围为250~500 Hz.从信号处理的角度出发,等效为1个由8个二元水听器组成的二元水听器组,其相邻2个二元水听器间距2 m,单个二元水听器的2个阵元间距0.25 m,满足单个二元水听器的2个阵元间距小于接收信号波长的1/10的要求.经过处理后,可得到某段噪声数据未经过左右舷分辨方法处理和经过移相左右舷分辨方法处理后的左右波束水平指向性图如图8和图9所示,同样,实线为右波束,虚线为左波束.

对比图8和图9能够发现在左舷56°附近有一个目标,这是试验中所用的目标船;而在左舷162°附近也有个目标,这与现场记录的过往船只信息相吻合.

4 结束语

文中提出了一种应用于二元水听器组的左右

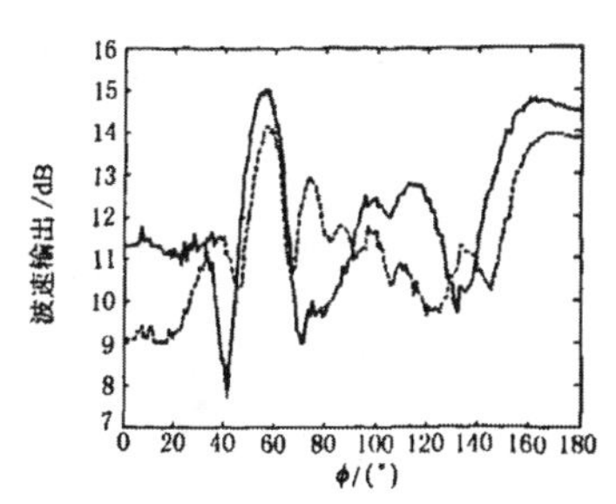

图8 未经左右舷分辨方法处理的左右波束水平指向性图

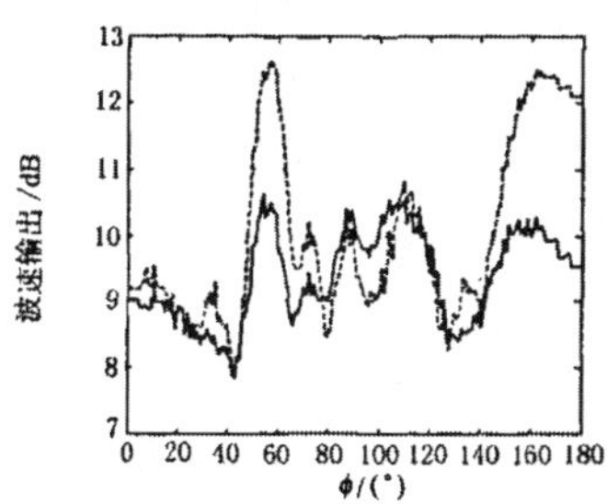

图9 经过移相左右舷分辨方法处理后的左右波束水平指向性图

舷分辨方法——移相左右舷分辨方法，经过理论分析和湖试数据验证，该方法可有效解决横截尺寸较小的二元水听器组和横截尺寸较大的二元水听器组的左右舷分辨问题，算法简单、计算量小．该方法也可扩展应用于双线阵的左右舷分辨．

参考文献

[1] Bertheas J, Moresco G, Dufourco P. Linear hydrophonic antenna and electronic device to remove right/left ambiguity, associated with the antenna: USA, 5058082[P]. 1991-10-15.

[2] 杜选民，朱代柱，赵荣荣，等．拖线阵左右舷分辨技术的理论分析与实验研究[J]．声学学报，2000，25(5)：395-402.

[3] de Vlieger J H, van Ballegooijen E C. ALF: an experimental low frequency active sonar [C] // UDT95. France, 1995: 131-136.

[4] 何心怡，蒋兴舟，李启虎．矢量水听器线列阵的被动合成孔径技术[J]．武汉理工大学学报：交通科学与工程版，2003，27(6)：799-803.

[5] 何心怡，黄海宁，张春华．一种用于拖曳线列阵声呐基阵的水听器组：中国，03105505.2[P]．2003-11-15.

[6] 何心怡，黄海宁，张春华．拖曳线列阵声呐的左右舷分辨方法及装置：中国，03105502.8[P]．2003.

[7] 何心怡，黄海宁，叶青华，等．拖曳线列阵声呐的左右舷分辨方法及装置：中国，03105501.X[P]．2003

[8] 李启虎．声呐信号处理引论[M]．2版．北京：海洋出版社，2000.

A New Towed Linear Array of Hydrophone Two-tuples Port/Starboard Discrimination Method

He Xinyi[1)] **Qiu Zhiming**[1)] **Zhang Chunhua**[2)] **Li Qihu**[2)]

(*Naval Academy of Armament*, *Beijing* 100073)[1)]

(*Institute of Acoustics*, *Chinese Academy of Sciences*, *Beijing* 100080)[2)]

Abstract

In order to solve the port/starboard ambiguity for the towed linear array sonar, a new port/starboard discrimination method of towed linear array of hydrophone two-tuples is presented in this paper. This method is based on the phase shift processing, in which the single hydrophone two-tuples is regarded as a little array, and the aim of phase shift processing is to make that the signal from image source minimized after signal processing, thus make that the single hydrophone two-tuples have cardioid directivity, so that the towed linear array hydrophone two-tuples can discriminate the port/starboard of the target. After theoretic analysis and validation of lake experiment data, it proved that this method can solve the port/starboard ambiguity of the large cross section size towed linear array of hydrophone two-tuples and small cross section size towed linear array made up of hydrophone two-tuples effectively, and it can be applied to a wide range, has simple algorithm and small computation.

Key words: towed linear array of hydrophone two-tuples; port/starboard discrimination; phase shift; large cross section size; small cross section size

Towed line array sonar platform noise suppression based on spatial matrix filtering technology①②

HAN Dong[1,2] LI Jian[3] KANG Chunyu[2] HUANG Haining[1] LI Qihu[1]

(1 *Institute of Acoustics, Chinese Academy of Sciences* Beijing 100190)

(2 *Department of Communication Engineering, Dalian Naval Academy* Dalian 116018)

(3 *College of IOT Engineering, Hohai University* Changzhou 213022)

Received Mar. 29, 2013

Revised Jun. 29, 2013

Abstract The spatial matrix filter was designed and used for solving the problem to detect a weak target who was influenced by the strong nearby platform noise interference of the towed line array sonar. The MFP technology and the DOA estimation technology were combined together by using the sound propagation characteristics of both target and interference. The spatial matrix filter with platform noise zero response constraint was designed by the near-field platform noise normal modes copy vectors and the far-field plane wave bearing vectors together. The optimal solution of the optimization problem for designing the spatial matrix filter was deduced directly, and it was simplified by the generalized singular value decomposition. The total response error to the plane wave bearing vectors and the total response to the platform noise copy vectors were given. The phenomena that strong interferences existed in the bearing course and blind areas existed after filtering were analyzed by the correlation between the platform noise copy vectors and the plane wave bearing vectors. It could be found from simulations that it has less blind area and higher detection ability by using the spatial matrix filtering technology.

PACS numbers: 43.50, 43.60

1 Introduction

The length of towed line array sonar can reach to hundreds of meters, so coherence of the signal field could be utilized sufficiently to achieve higher spatial processing gains. The towed line array sonar is the most important means of detecting the low-frequency, low-noise underwater target. But the distance between the towed platform ship or submarine, and the sonar's sensor array is very limited, and the platform noise forms strong near-field interference. It will generate a wide range of blind detection area in the direction of the sonar's towed platform, and will also affect the detection capabilities of the other directions at the same time[1−2].

① This work was supported by the National Natural Science Foundation of China (60532040, 11374001).

② Chinese Journal of Acoustics, 2013, 32(4): 379-390.

Domestic and foreign scholars attempt to suppress the towed platform noise with a variety of methods, such as the adaptive noise cancellation[3−4], adaptive beamforming[5−6], post beamforming interference cancellation[7−8] and inverse beamforming[9]. But good performance of these interference cancellation techniques are based on the assumption that the received signals are of far-field plane wave propagation, which are unsatisfactory usually, and be worsened in practical environments. By using of twin-line array is an effective method for platform noise suppression until now[10−11], the twin-line array can form left-right symmetrical dipole directivity, it has a deep recess within the vertical plane that contains the middle line of the twin-line array and the towing platform. However, its disadvantage is the existence of blind detection area in the vicinity direction of the towed platform; the sonar loses the detection ability of the signal in the blind area.

The distance of the towed line array and the towed platform are very short, especially in shallow sea. The sound propagation is impacted seriously by the multi-path effects, the platform noise shouldn't be treated as the far-field plane wave signal. For all the methods mentioned above treat the platform noise as far-field plane wave propagation, they are not appropriate for suppressing the platform noise which reflected by the surface and bottom, so they can't improve the direction of arrival (DOA) estimation performance essentially. If we want to reduce the impact of platform noise on the towed line array sonar, we must combine the environment of sound propagation and the multi-path effect between the sound source and the towed array. Matched field processing (MFP) technology is different from the DOA estimation which based on the plane wave propagation property. The plane wave DOA estimation is on the assumption that the signals are propagating on plane wavefront, so it has not the ability of suppressing the underwater sound interference with multi-path effect. While the MFP technology can estimate the underwater source's depth and distance by taking advantage of the underwater environment features and the multi-path effect. The most feasible solution of suppressing the towed line array sonar platform noise is to use for a particular interference suppression method based on the multi-path effect while avoid of its influence on the far-field plane wave signals.

The spatial matrix filtering is an emerging technology which process the array signals in the element space[12−14]. By dividing the spatial space into passband and stopband, the spatial matrix filter has different response on each band. The array signals are processed by the spatial matrix filter before they are used for direction or position of arrival estimation. The spatial matrix filter can pass the signals in the passband without distortion while depress the strong interferences in the stopband at the same time. It can be used in DOA estimation and enhance detection ability of passband signals[15−21]. It can be also used in the MFP and enhance or depress signals in some area of the vertical plane so as to locate the underwater weak target in the presence of the strong interference[22]. In Ref. [23], the technologies of DOA estimation and MFP were combined together; the copy vector of the vertical plane towed ship noise was added into the designs of MVDR beamforming which has zero response to the copy vector. This method utilized the underwater sound propagation characteristics which could suppress the towed ship noise by using the MVDR beamforming. But unfortunately, this method was effective only for MVDR beamforming, and can't be used for robust conventional beamforming or other high-resolution beamforming.

In this paper, the spatial matrix filter technology, the MFP technology and the plane wave DOA estimation technology were combined together. For specific marine environment, the MFP technology was used for the accurate estimation of the relative position of the platform noise and the towed array, and got a copy vector at the same time. The spatial matrix filter with platform noise zero response constraint was designed by the platform noise copy vectors and the plane wave bearing vectors. The optimal solution of the spatial matrix filter design problem was deduced directly. The solution was simplified by the generalized singular value decomposition. The total response error to the plane wave bearing vectors and the total response to the platform noise copy vectors were given. The optimal solution of the spatial matrix filter was verified by the filter response to the platform noise copy vector. The performance of the spatial matrix filter was analyzed by the comparison of the bearing-time picture with and without filtering on the condition of low signal-to-noise ratio (SNR). The phenomena that strong interferences existed in the bearing-time picture and the blind areas exist after using the spatial matrix filtering technique were explained by the relations between the platform noise copy vectors and the plane wave bearing vectors.

2 Design of the platform noise zero response spatial matrix filter

The DOA estimation of towed line array sonar is based on the assumption that the received array signals are of far-field plane wave. But, as we know, the distance between the sonar's platform and the towed sensor array is very limited, the platform noise which incident on the array with multi-path should be treated as near-field signal. Consider an uniform line array with nondirectional sensor elements, the elements amount is N. The received array signals $\boldsymbol{x}(t,\omega)$ of frequency ω are consisted of the far-field plane wave signal and the near-field platform noise.

$$\boldsymbol{x}(t,\omega) = \boldsymbol{A}(\boldsymbol{\theta},\omega)\,\boldsymbol{s}_1(t,\omega) + \boldsymbol{V}(\omega)\boldsymbol{s}_0(t,\omega) + \boldsymbol{n}(t,\omega), \tag{1}$$

where $\boldsymbol{x}(t,\omega) = [x_1(t,\omega),\cdots,x_N(t,\omega)]^{\mathrm{T}}$ are the received array signals, $\boldsymbol{s}_1(t,\omega) = [s_{11}(t,\omega),\cdots,s_{1D}(t,\omega)]^{\mathrm{T}}$ is D-dimensional far-field plane wave signal with frequency ω, $\boldsymbol{s}_0(t,\omega) = [s_{01}(t,\omega),\cdots,s_{0S}(t,\omega)]^{\mathrm{T}}$ is S-dimensional near-field platform noise with frequency ω. $\boldsymbol{A}(\boldsymbol{\theta},\omega) = [\boldsymbol{a}(\theta_1,\omega),\cdots,\boldsymbol{a}(\theta_i,\omega),\cdots,\boldsymbol{a}(\theta_D,\omega)] \in \boldsymbol{C}^{N\times D}$ is the plane wave array manifold matrix with frequency ω, where $\boldsymbol{\theta} = [\theta_1,\cdots,\theta_D]$ are the incident directions of the plane waves, $\boldsymbol{a}(\theta_i,\omega) = [1,\exp(-\mathrm{j}\omega\Delta\sin\theta_i/c),\cdots,\exp(-\mathrm{j}\omega(N-1)\Delta\sin\theta_i/c)]^{\mathrm{T}} \in \boldsymbol{C}^{N\times 1}$ is the direction vector, $(\cdot)^{\mathrm{T}}$ denotes the transpose of an array or vector, Δ denotes the element interval, j denotes the unit imaginary number, $\boldsymbol{V}(\omega) = [\boldsymbol{v}_1(\omega),\cdots,\boldsymbol{v}_S(\omega)] \in \boldsymbol{C}^{N\times S}$ is consisted by the copy vectors of the near-field platform noise interference incident on the array with multi-path propagation, S is the number of platform noise, $\boldsymbol{n}(t,\omega) = [n_1(t,\omega),\cdots,n_N(t,\omega)]^{\mathrm{T}}$ is the N-dimensional environmental noise.

Design a spatial matrix filter $\boldsymbol{H}(\omega) \in \boldsymbol{C}^{N\times N}$, by filtering, we can get:

$$\boldsymbol{y}(t,\omega) = \boldsymbol{H}(\omega)\boldsymbol{A}(\boldsymbol{\theta},\omega)\,\boldsymbol{s}_1(t,\omega) + \boldsymbol{H}(\omega)\boldsymbol{V}(\omega)\boldsymbol{s}_0(t,\omega) + \boldsymbol{H}(\omega)\boldsymbol{n}(t,\omega). \tag{2}$$

The spatial matrix filter can enhance or depress the far-field signal through the role of bearing vector. When $\|\boldsymbol{H}(\omega)\boldsymbol{a}(\theta_i,\omega)\|_F^2$ close to 0, the filter form strong depression to the plane wave signal in direction θ_i with frequency ω, the total response error to the plane wave bearing vectors is $\|\boldsymbol{H}(\omega)\boldsymbol{A}(\boldsymbol{\theta},\omega)-\boldsymbol{A}(\boldsymbol{\theta},\omega)\|_F^2$. Similarly, the total response to the platform copy vectors is $\|\boldsymbol{H}(\omega)\boldsymbol{V}(\omega)\|_F^2$, when $\|\boldsymbol{H}(\omega)\boldsymbol{V}(\omega)\|_F^2$ equal to 0, the platform noise is suppressed completely.

The spatial matrix filter can depress the platform noise by constraining the total response to far-field platform noise copy vectors $\|\boldsymbol{H}(\omega)\boldsymbol{V}(\omega)\|_F^2$, and reduce the impact to the detection of far-field plane wave signal by constraining the total response error to the far-field plane wave bearing vectors $\|\boldsymbol{H}(\omega)\boldsymbol{A}(\boldsymbol{\theta},\omega)-\boldsymbol{A}(\boldsymbol{\theta},\omega)\|_F^2$. The spatial matrix filter is designed by tho optimization problem:

$$\begin{aligned}&\min_{\boldsymbol{H}(\omega)} J[\boldsymbol{H}(\omega)]=\|\boldsymbol{H}(\omega)\boldsymbol{A}(\boldsymbol{\theta},\omega)-\boldsymbol{A}(\boldsymbol{\theta},\omega)\|_F^2\\&\text{Subject to}\quad \boldsymbol{H}(\omega)\boldsymbol{V}(\omega)=\boldsymbol{0}_{N\times S}.\end{aligned}\tag{3}$$

The filter has zero response to the platform noise copy vectors, i.e. the filter response to the platform noise equal to zero.

For convenience, $\boldsymbol{H}(\omega)$, $\boldsymbol{A}(\boldsymbol{\theta},\omega)$ and $\boldsymbol{V}(\omega)$ are abbreviated as $\boldsymbol{H}$, $\boldsymbol{A}$ and $\boldsymbol{V}$ respectively. Now gives the derivation of the optimal solution of the optimization problem. The constraint equation of the optimization problem equivalent to the equations as follows:

$$\begin{cases}\operatorname{Re}(\boldsymbol{H}\boldsymbol{v}_n)=\boldsymbol{0}_{N\times 1}, & 1\leqslant n\leqslant S,\\ \operatorname{Im}(\boldsymbol{H}\boldsymbol{v}_n)=\boldsymbol{0}_{N\times 1}, & 1\leqslant n\leqslant S,\end{cases}\tag{4}$$

where $\operatorname{Re}(\cdot)$ and $\operatorname{Im}(\cdot)$ denote the real and imaginary part of an vector respectively.

Construct a real Lagrange function:

$$\begin{aligned}&L(\boldsymbol{H},\boldsymbol{\lambda}_1,\cdots,\boldsymbol{\lambda}_S,\boldsymbol{\delta}_1,\cdots,\boldsymbol{\delta}_S)=\|\boldsymbol{H}\boldsymbol{A}-\boldsymbol{A}\|_F^2-\sum_{n=1}^{S}\boldsymbol{\lambda}_n^{\mathrm{T}}\operatorname{Re}(\boldsymbol{H}\boldsymbol{v}_n)-\sum_{n=1}^{S}\boldsymbol{\delta}_n^{\mathrm{T}}\operatorname{Im}(\boldsymbol{H}\boldsymbol{v}_n)=\\&\operatorname{tr}(\boldsymbol{H}\boldsymbol{A}\boldsymbol{A}^{\mathrm{H}}\boldsymbol{H}^{\mathrm{H}})-\operatorname{tr}(\boldsymbol{H}\boldsymbol{A}\boldsymbol{A}^{\mathrm{H}})-\operatorname{tr}(\boldsymbol{A}\boldsymbol{A}^{\mathrm{H}}\boldsymbol{H}^{\mathrm{H}})+\operatorname{tr}(\boldsymbol{A}\boldsymbol{A}^{\mathrm{H}})-\\&\sum_{n=1}^{S}\left(\frac{\boldsymbol{\lambda}_n^{\mathrm{T}}}{2}+\frac{\boldsymbol{\delta}_n^{\mathrm{T}}}{2\mathrm{j}}\right)\boldsymbol{H}\boldsymbol{v}_n-\sum_{n=1}^{S}\left(\frac{\boldsymbol{\lambda}_n^{\mathrm{T}}}{2}-\frac{\boldsymbol{\delta}_n^{\mathrm{T}}}{2\mathrm{j}}\right)\boldsymbol{H}^*\boldsymbol{v}_n^*,\end{aligned}\tag{5}$$

where $\operatorname{tr}(\cdot)$ denotes the trace of an matrix, $(\cdot)^*$ denotes the conjugate, $(\cdot)^{\mathrm{H}}$ denotes the conjugate transpose, $\boldsymbol{\lambda}=[\boldsymbol{\lambda}_1,\cdots,\boldsymbol{\lambda}_S]\in R^{N\times S}$ and $\boldsymbol{\delta}=[\boldsymbol{\delta}_1,\cdots,\boldsymbol{\delta}_S]\in R^{N\times S}$ are the Lagrange multipliers.

Seek the partial derivatives of function $L(\boldsymbol{H},\boldsymbol{\lambda}_1,\cdots,\boldsymbol{\lambda}_S,\boldsymbol{\delta}_1,\cdots,\boldsymbol{\delta}_S)$ on $(\boldsymbol{H},\boldsymbol{\lambda}_1,\cdots,\boldsymbol{\lambda}_S,\boldsymbol{\delta}_1,\cdots,\boldsymbol{\delta}_S)$:

$$\begin{aligned}&\frac{\partial L(\boldsymbol{H},\boldsymbol{\lambda}_1,\cdots,\boldsymbol{\lambda}_S,\boldsymbol{\delta}_1,\cdots,\boldsymbol{\delta}_S)}{\partial \boldsymbol{H}}=(\boldsymbol{A}\boldsymbol{A}^{\mathrm{H}}\boldsymbol{H}^{\mathrm{H}})^{\mathrm{T}}-(\boldsymbol{A}\boldsymbol{A}^{\mathrm{H}})^{\mathrm{T}}-\sum_{n=1}^{S}\left(\frac{\boldsymbol{\lambda}_n^{\mathrm{T}}}{2}+\frac{\boldsymbol{\delta}_n^{\mathrm{T}}}{2\mathrm{j}}\right)\boldsymbol{v}_n^{\mathrm{T}}=\\&\boldsymbol{H}^*\boldsymbol{A}^*\boldsymbol{A}^{\mathrm{T}}-\boldsymbol{A}^*\boldsymbol{A}^{\mathrm{T}}-\left(\frac{\boldsymbol{\lambda}}{2}+\frac{\boldsymbol{\delta}}{2\mathrm{j}}\right)\boldsymbol{V}^{\mathrm{T}}=\boldsymbol{H}^*\boldsymbol{A}^*\boldsymbol{A}^{\mathrm{T}}-\boldsymbol{A}^*\boldsymbol{A}^{\mathrm{T}}-\gamma\boldsymbol{V}^{\mathrm{T}},\end{aligned}\tag{6}$$

where $\gamma=\boldsymbol{\lambda}/2+\boldsymbol{\delta}/(2\mathrm{j})\in \boldsymbol{C}^{N\times S}$ is the Lagrange multiplier constructed by $\boldsymbol{\lambda}$ and $\boldsymbol{\delta}$.

$$\frac{\partial L(\boldsymbol{H},\boldsymbol{\lambda}_1,\cdots,\boldsymbol{\lambda}_S,\boldsymbol{\delta}_1,\cdots,\boldsymbol{\delta}_S)}{\partial \boldsymbol{\lambda}_n}=\operatorname{Re}(\boldsymbol{H}\boldsymbol{v}_n),\quad 1\leqslant n\leqslant S\tag{7}$$

$$\frac{\partial L(\boldsymbol{H},\boldsymbol{\lambda}_1,\cdots,\boldsymbol{\lambda}_S,\boldsymbol{\delta}_1,\cdots,\boldsymbol{\delta}_S)}{\partial\boldsymbol{\delta}_n}=\mathrm{Im}(\boldsymbol{H}\boldsymbol{v}_n),\quad 1\leqslant n\leqslant S \tag{8}$$

The stable points $\boldsymbol{H}_{\mathrm{op}},\boldsymbol{\lambda}_{\mathrm{op}},\boldsymbol{\delta}_{\mathrm{op}}$ of the Lagrange function, i.e. the optimal spatial matrix filter and the optimal Lagrange multipliers meet the conditions as follows:

$$\boldsymbol{H}_{\mathrm{op}}^{*}\boldsymbol{A}^{*}\boldsymbol{A}^{\mathrm{T}}-\boldsymbol{A}^{*}\boldsymbol{A}^{\mathrm{T}}-\gamma_{\mathrm{op}}\boldsymbol{V}^{\mathrm{T}}=\boldsymbol{0}_{N\times N}, \tag{9}$$

$$\mathrm{Re}(\boldsymbol{H}_{\mathrm{op}}\boldsymbol{v}_n)=\boldsymbol{0}_{N\times 1},\quad 1\leqslant n\leqslant S, \tag{10}$$

$$\mathrm{Im}(\boldsymbol{H}_{\mathrm{op}}\boldsymbol{v}_n)=\boldsymbol{0}_{N\times 1},\quad 1\leqslant n\leqslant S, \tag{11}$$

where $\gamma_{\mathrm{op}}=\boldsymbol{\lambda}_{\mathrm{op}}/2+\boldsymbol{\delta}_{\mathrm{op}}/(2\mathrm{j})$. Merge formula (10) and (11), we get:

$$\boldsymbol{H}_{\mathrm{op}}\boldsymbol{V}=\boldsymbol{0}_{N\times S}. \tag{12}$$

By formula (9):

$$\boldsymbol{H}_{\mathrm{op}}=\left(\boldsymbol{A}\boldsymbol{A}^{\mathrm{H}}+\gamma_{\mathrm{op}}^{*}\boldsymbol{V}^{\mathrm{H}}\right)\left(\boldsymbol{A}\boldsymbol{A}^{\mathrm{H}}\right)^{-1}. \tag{13}$$

where $(\cdot)^{-1}$ denotes the inverse of full rank square matrix. By its structure, it can be seen that $\boldsymbol{A}$ is a Vandermonde matrix, full row rank, so $\boldsymbol{A}\boldsymbol{A}^{\mathrm{H}}$ is full rank square matrix, $\boldsymbol{A}\boldsymbol{A}^{\mathrm{H}}$ has inverse matrix.

We can get the optimal Lagrange multiplier by bring formula (13) into (12):

$$\gamma_{\mathrm{op}}^{*}=-\boldsymbol{V}\left[\boldsymbol{V}^{\mathrm{H}}\left(\boldsymbol{A}\boldsymbol{A}^{\mathrm{H}}\right)^{-1}\boldsymbol{V}\right]^{-1}. \tag{14}$$

Bring the optimal Lagrange multiplier γ_{op}^{*}, i.e. formula (14) into formula (13), we can get the optimal spatial matrix filter as follows:

$$\boldsymbol{H}_{\mathrm{op}}=\boldsymbol{I}_N-\boldsymbol{V}\left[\boldsymbol{V}^{\mathrm{H}}\left(\boldsymbol{A}\boldsymbol{A}^{\mathrm{H}}\right)^{-1}\boldsymbol{V}\right]^{-1}\boldsymbol{V}^{\mathrm{H}}\left(\boldsymbol{A}\boldsymbol{A}^{\mathrm{H}}\right)^{-1}, \tag{15}$$

where $\boldsymbol{I}_N$ is $N\times N$ dimensional unit matrix. $\boldsymbol{H}_{\mathrm{op}}$ is the stable point of $L(\boldsymbol{H},\boldsymbol{\lambda}_1,\cdots,\boldsymbol{\lambda}_S,\boldsymbol{\delta}_1,\cdots,\boldsymbol{\delta}_S)$, i.e. the global optimal solution of the optimization problem.

3 The generalized singular value decomposition and response error analysis

The optimal spatial matrix filter which obtained from formula (15) can be simplified by the generalized singular value decomposition. The response error to the far-field plane wave bearing vectors and the response to the near-field platform noise copy vectors are analyzed theoretically further.

$\boldsymbol{A}$ is Vandermonde matrix, so it is full row rank. In order to make the spatial spectrum more smoother, the number of the bearing vectors D should more than the sensors number of the array N, so we can get $\mathrm{rank}(\boldsymbol{A})=N$, and the filter response to the plane wave bearing vectors of neighbor beam angle will not produce a substantial hopping. The rank of $\boldsymbol{V}$ is relevant with the platform noise. The towed platform is usually marine ship or underwater submarine, so the platform noise can be approximated by a finite number of point sources. So the number of platform noise copy vectors can be considered less than the number of the sensors of the towed line array, that is $\mathrm{rank}(\boldsymbol{V})=S<N$. By the generalized singular value decomposition

theory, there exist two unitary matrixes $\boldsymbol{U}_A \in \boldsymbol{C}^{D\times D}$, $\boldsymbol{U}_V \in \boldsymbol{C}^{S\times S}$ and a nonsingular matrix $\boldsymbol{Q}_X \in \boldsymbol{C}^{N\times N}$ satisfy the equations as follows:

$$\boldsymbol{U}_A\boldsymbol{A}^{\mathrm{H}}\boldsymbol{Q}_X=\boldsymbol{\Sigma}_A,\ \boldsymbol{\Sigma}_A=\begin{bmatrix} \boldsymbol{I}_{N-S} & \boldsymbol{0}_{(N-S)\times S} \\ \boldsymbol{0}_{S\times(N-S)} & \boldsymbol{Z}_A \\ \boldsymbol{0}_{(D-N)\times(N-S)} & \boldsymbol{0}_{(D-N)\times S} \end{bmatrix}_{D\times N},\ \boldsymbol{Z}_A=\mathrm{diag}\left(\alpha_1,\cdots,\alpha_S\right), \tag{16}$$

$$\boldsymbol{U}_V\boldsymbol{V}^{\mathrm{H}}\boldsymbol{Q}_X=\boldsymbol{\Sigma}_V,\ \boldsymbol{\Sigma}_V=\left[\boldsymbol{0}_{S\times(N-S)},\boldsymbol{Z}_V\right]_{S\times N},\ \boldsymbol{Z}_V=\mathrm{diag}\left(\beta_1,\cdots,\beta_S\right), \tag{17}$$

where $\alpha_k^2+\beta_k^2=1$, $k=1,2,\cdots,S$.

From formulas (16) and (17), we get:

$$\boldsymbol{A}=\boldsymbol{Q}_X^{-\mathrm{H}}\boldsymbol{\Sigma}_A{}^{\mathrm{T}}\boldsymbol{U}_A, \tag{18}$$

$$\boldsymbol{V}=\boldsymbol{Q}_X^{-\mathrm{H}}\boldsymbol{\Sigma}_V{}^{\mathrm{T}}\boldsymbol{U}_V. \tag{19}$$

Bring formulas (18) and (19) into formula (19), we get:

$$\boldsymbol{H}_{\mathrm{op}}=\boldsymbol{Q}_X^{-\mathrm{H}}\begin{bmatrix} \boldsymbol{I}_{N-S} & \\ & \boldsymbol{0}_S \end{bmatrix}\boldsymbol{Q}_X^{\mathrm{H}}. \tag{20}$$

Formula (20) gives the simplified optimal spatial matrix filter by the generalized singular value decomposition. The total response error to the far-field plane wave bearing vectors and the total response to the near-field platform copy vectors are:

$$\left\|\boldsymbol{H}_{\mathrm{op}}\boldsymbol{A}-\boldsymbol{A}\right\|_F^2=\left\|\boldsymbol{Q}_X^{-\mathrm{H}}\begin{bmatrix} \boldsymbol{0}_{N-S} & \\ & \boldsymbol{Z}_A \end{bmatrix}\right\|_F^2, \tag{21}$$

$$\left\|\boldsymbol{H}_{\mathrm{op}}\boldsymbol{V}\right\|_F^2=\boldsymbol{0}_{N-S}. \tag{22}$$

It can be seen from formula (22) that the optimal matrix filter $\boldsymbol{H}_{\mathrm{op}}$ satisfies the constrain Eq. (3), the correctness of the filter is verified.

4 Simulations

The covariance matrix of the output array signals which processed by the spatial matrix filter is as follows:

$$\boldsymbol{R}_y=\mathrm{E}\left[\boldsymbol{y}(t,\omega)\boldsymbol{y}^{\mathrm{H}}(t,\omega)\right]=\boldsymbol{H}\boldsymbol{R}_x\boldsymbol{H}^{\mathrm{H}}+\boldsymbol{H}\boldsymbol{R}_n\boldsymbol{H}^{\mathrm{H}}, \tag{23}$$

where $\boldsymbol{R}_x=E\left[\boldsymbol{x}(t,\omega)\boldsymbol{x}^{\mathrm{H}}(t,\omega)\right]$ is the covariance matrix of the received array signals. One can get the DOA estimation results by using the output covariance matrix $\boldsymbol{R}_y$ without contaminated by the platform noise.

Consider an uniform line array, the sensor number is $N=64$, the interval of neighbor sensors is 4 meters, the detection frequency is 192.5 Hz. The schematic diagram of the towed line array sonar and the sound speed profile (SSP) are given in Fig. 1. The platform noise is propagating on the vertical xz plane. The towed array is parallel with the x axis, and the sensors are all omnidirectional, so it can't determine where the incident signals from between the horizontal plane and the vertical plane. The far-field plane wave signal can be treated as the signal propagating on the horizontal xy plane[2]. The SSP is given by figure 1 too, the seabed speed is 1682 m/s, the seabed medium density is 1.76 g/cm^3, the distance between the first sensor of the towed array and the platform is 400 meters, the towing depth of the array is 25 meters, the depth of the platform noise is 8 meters.

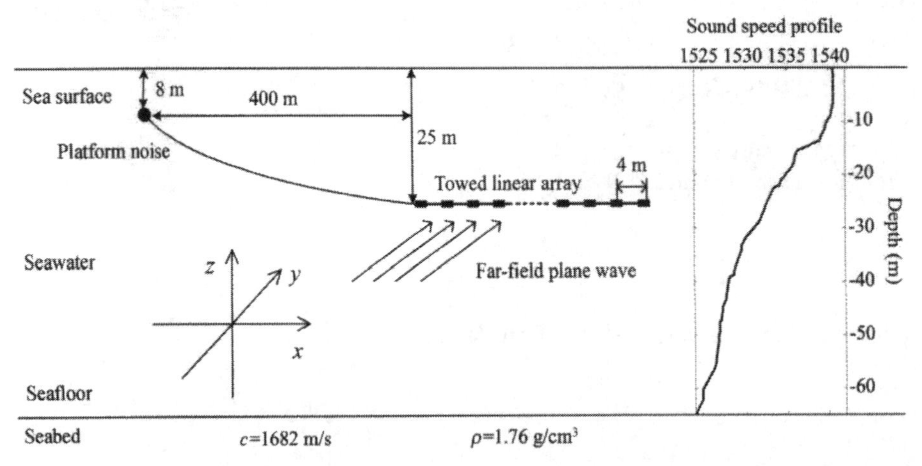

Fig. 1 The schematic diagram of the towed line array sonar and the sound speed profile.

In order to suppress the platform noise by using the spatial matrix filtering technology, we need the plane wave bearing vectors and the platform noise copy vectors. For the given uniform line array, the bearing vectors of specific frequency can be given by the distance of neighbor sensors. The platform copy vectors can be calculated by the marine environment and the sound propagation model. In this paper, we assumed that there was only 1 point sound source in the platform noise interference. For the reason that we didn't know the exact position of the platform noise and the towed array cable was affected by the current, we couldn't take for granted that the comparative distance between the platform noise and the first sensor was exactly the released cable length. By using the MFP technology, we could get the exact estimation of the comparative depth and distance between the platform and the sensor array in given environment. Figure 2 gives the transmission loss of the platform noise. Figure 3 gives the estimation result by using the MFP technology with MV processor, in which the platform position was marked with red circle. The spatial matrix filter was designed by using the far-field plane wave bearing vectors of 192.5 Hz and the platform noise copy vectors calculated in Fig. 3.

The conventional beamforming (CBF) is widely used in the towed line array sonar signal processing for its robustness. In this paper, the platform noise zero response spatial matrix filter was used in CBF to test its performance. Figure 4 gives the towed line sonar CBF result in the case of the received signal was only the platform noise with the interference to noise ratio (INR) varied from 5 dB to 15 dB. The platform noise formed two strong interferences in the vicinity of angle $-76.2°$ and $-67°$ which deviated from the tip direction of the array.

The following simulations were on the assumption that the platform noise was fixed with 10 dB. Suppose that the target moved continuously from $-90°$ to $-60°$ and the relative distance between the platform and the array didn't change. Figures 5–8 give the beamforming results with and without filtering (the subfigures 1(a)-4(a)(1(b)–4(b)) were the orientation courses of

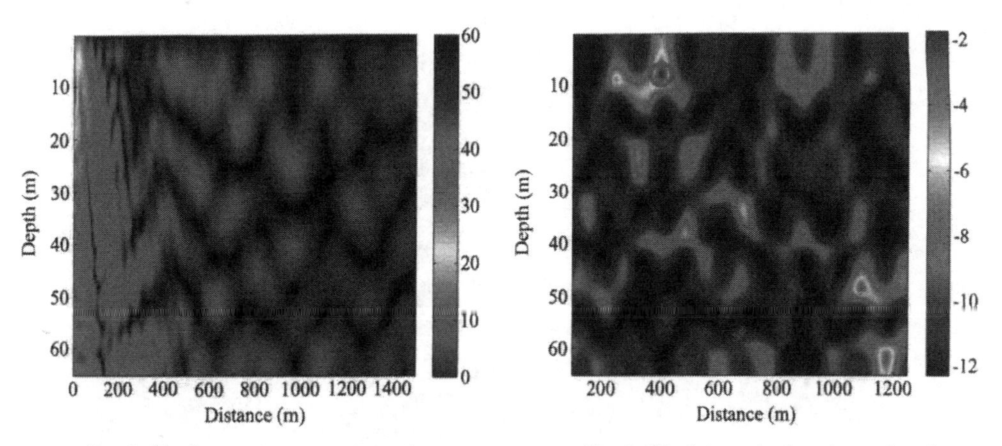

Fig. 2 Platform noise transmission loss.

Fig. 3 Platform noise location estimation result by using MFP.

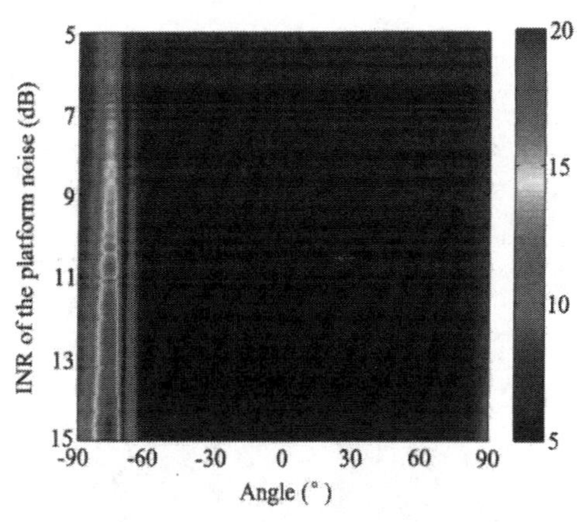

Fig. 4 The influence of the platform noise to CBF.

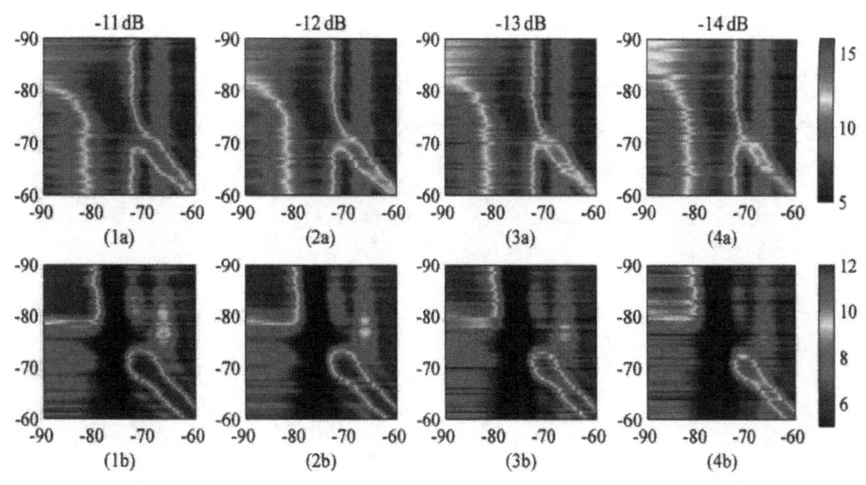

Fig. 5 The bearing-time picture with and without filtering, the SNR varied from −11 dB to −14 dB.

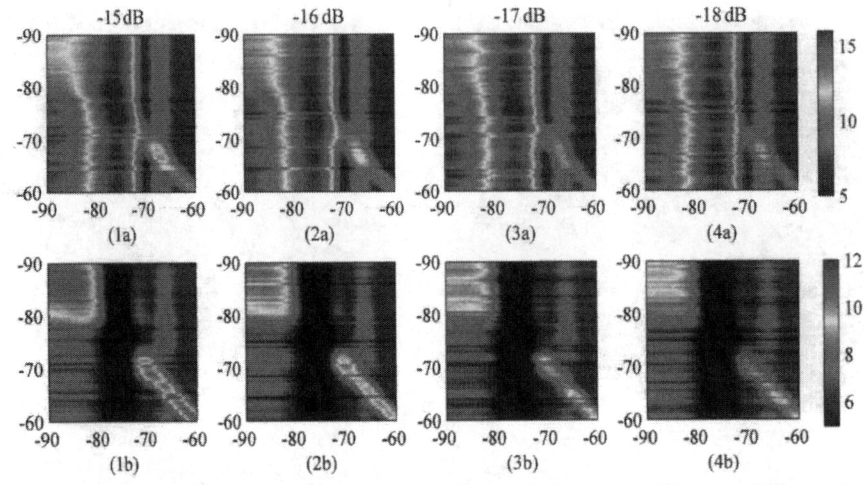

Fig. 6 The bearing-time picture with and without filtering, the SNR varied from −15 dB to −18 dB.

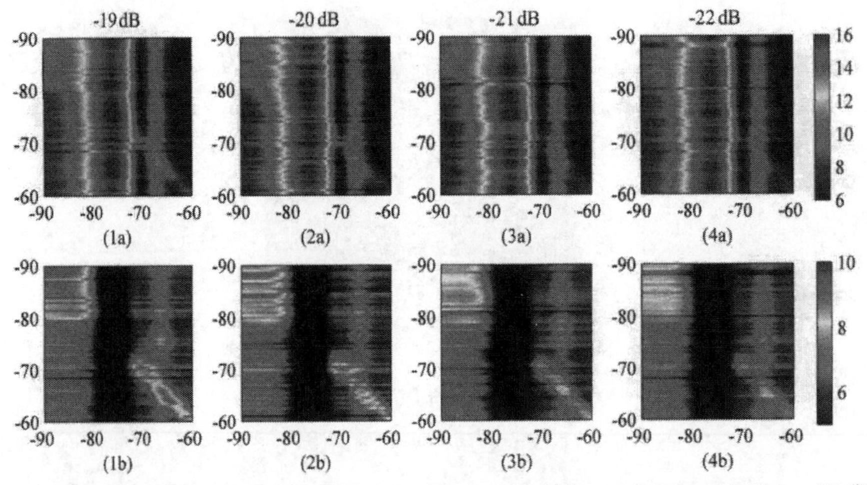

Fig. 7 The orientation course with and without filtering, the SNR varied from −19 dB to −22 dB.

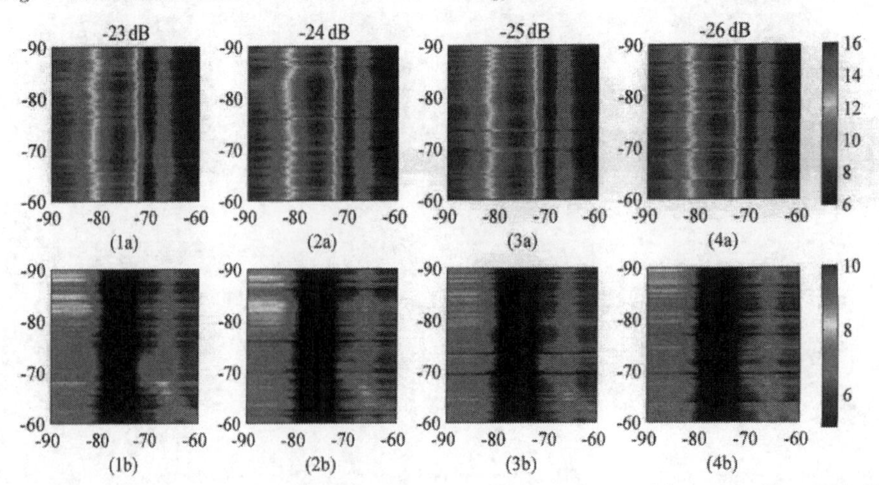

Fig. 8 The bearing-time picturewith and without filtering, the SNR varied from −23 dB to −26 dB.

CBF without (with) filtering, where the abscissa was the beam angle, the ordinate was the orientation angle of the weak target), in which the SNR of the weak target varied from −11 dB to −26 dB. After filtering, a certain width of blind area generated in the vicinity of angle −76.2°. If the target was exactly in the blind area, its orientation course couldn't be detected exactly, whoever it had more precise orientation course. In the case of low SNR (in this simulations, the SNR threshold was about −14 dB), the width of blind area with filtering was less than that of without filtering, and the blind area reduced with the increase of the SNR.

Now analyze the reason why the blind area generated in the vicinity of angle −76.2° after filtering but didn't generate in the vicinity of angle −67°. It can be seen from Fig. 9, there existed great correlation between the platform noise copy vector and the bearing vectors in the vicinity of the angles mentioned above. For this reason, after the platform noise incident on the array with multi-path, the beamforming methods would detect great interference of the platform noise in the vicinity of the two angles. The spatial matrix filter response to the platform noise was set to zero. On the one hand, for the platform noise copy vector and the plane wave bearing vector of angle −76.2° closed to perfectly correlated (the correlation coefficient was $r = 0.9293$ in this angle. Correspondingly $r = 1$ when perfectly correlated); On the other hand, the correlation coefficient between the platform noise copy vector and the bearing vector in the vicinity of −67° was relatively small (the correlation coefficient is $r = 0.3844$), so the influence to the DOA estimation was relatively small after filtering, no blind area occurred. This showed that the position and the magnitude of the platform noise interference was decided by the correlation coefficient between the platform noise copy vector and the bearing vector, and in turn decided the affected degree of spatial matrix filter to the plane wave DOA estimation.

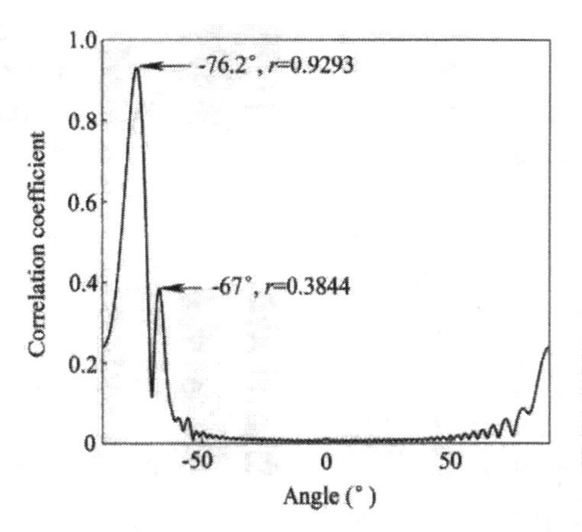

Fig. 9 The correlation coefficient between the platform noise copy vector and the plane wave bearing vectors.

5 Conclusion

In this paper, a spatial matrix filter technology, a MFP technology and a plane wave DOA estimation technology are combined together. The spatial matrix filter which could suppress the platform noise interference was designed. The filter has the minimal response error to the far-field plane wave bearing vectors on condition that it has zero response to the near-field platform noise copy vectors. The filter could remain the ability of detecting the far-field signal on the greatest degree and suppressed the platform noise at the same time. By simulation, the array signals which processed by the spatial matrix filter were used for conventional beamforming, the largest interference of platform noise could be erased perfectly; the detection ability of CBF was enhanced greatly.

The platform noise would form great interference which deviated from the tip direction by the influence of the multi-path propagation characteristics The reason was that the platform noise copy vectors were related with the bearing vectors in the vicinity direction of the platform The degree of the correlation would be depending on the position of the platform noise interference, and influence the strength of interference directly. The filter response to the platform noise was set to zero after processed by the spatial matrix filter, and it resulted in zero response to the plane wave bearing vectors in this area. When a target exists in this area, the filter would results a response closing to zero, and the target couldn't be detected correctly. So we should further research the spatial matrix filter which can suppress the platform noise while pass the far-field plane wave signal distortionless on condition that there are great correlations between the platform noise copy vectors and the plane wave bearing vectors, and obtain the optimal filters while no blind area after filtering.

References

[1] Li Q H, Li S Q, Sun C Y, Yu H B. The interference characteristics of platform and towed body noise in shallow water for active/passive towed array sonar. *Acta Acustica* (In Chinese), 2007; **32**(1): 1–4

[2] Ma Y L, Liu M A, Zhang Z B, Tong L. Response of the towed line array to the noise of the tow ship in shallow water. *Acta Acustica* (In Chinese), 2002; **27**(6): 481–486

[3] Hodgkiss W. Source ship contamination removal in a broadband vertical array experiment. Proc. Oceans'88 on A Partnership of Marine Interests, 1988: 310–314

[4] Candy J V, Sullivan E J. Canceling tow ship noise using an adaptive model-based approach. Proc. OES Eighth Working Conference on Current Measurement Technology, 2005: 14–18

[5] Capon J. High resolution frequency wavenumber spectrum analysis. Proc. IEEE, 1969; **57**(8): 1408–1418

[6] Mio K, Doisy Y, Chocheyras Y. Tow ship noise reduction with active adaptive beamforming on HFM transmissions. UDT, 2002: 320–326

[7] Godara L C. A robust adaptive array processor. *IEEE Trans. Circuits Syst.*, 1987; **34**(7): 721–730

[8] Godara L C. Post beamformer interference canceller with improved performance. *J. Acoust. Soc. Am.*, 1989; **85**(1): 202–213

[9] Colin M, Groen J. Passive synthetic aperture sonar techniques in combination with tow ship noise canceling: application to a triplet towed array. Proc. OES Second Working Conference on Current Measurement Technology, 2002: 2302–2309

[10] Li Q H. The time-delay estimation method of resolving left-right target ambiguity for twin-line array and its realization. *Acta Acustica* (In Chinese), 2006; **31**(6): 485–487

[11] Li Q H. Preliminary analysis of left-right ambiguity resolution performance for twin-line array. *Acta Acustica* (In Chinese), 2006; **31**(5): 385–388

[12] Vaccaro R J, Harrison B F. Optimal matrix-filter design. *IEEE Trans. Signal Processing*, 1996; **44**(3): 705–709

[13] Han Dong, Zhang Xinhua. Optimal matrix filter design with application to filtering short data records. *IEEE Signal Processing Letters*, 2010; **17**(5): 521–524

[14] Han Dong, Yin Junsong, Kang Chunyu, Zhang Xinhua. Optimal matrix filter design with controlled mean-square sidelobe level. *IET Signal Processing*, 2011; **5**(3): 306–312

[15] Ya S F, Hou C H, Ma X C. Matrix spatial prefiltering approach for direction-of-arrival estimation. *Acta Acustica* (In Chinese), 2007; **32**(2): 151–157

[16] Zhu Z W, Wang S, Leung H *et al.* Matrix filter design using semi-infinite programming with application to DOA estimation. *IEEE Trans. Signal Processing*, 2000; **48**(1): 267–271

[17] Wang Shi, Zhu Zhiwen, Henry Leung. Semi-infinite optimization technique for the design of matrix filters. *Statistical Signal and Array Processing*, 1998; **9**(14): 204–207

[18] Macinnes C S. Source localization using subspace estimation and spatial filtering. *IEEE J. Ocean. Eng.*, 2004; **29**(2): 488–497

[19] Hassanien A, Elkader S A, Gershman A B *et al.* Convex optimization based beam-space preprocessing with improved robustness against out-of-sector sources. *IEEE Trans. Signal Processing*, 2006; **54**(5): 1587–1595

[20] Han D, Zhang X H, Kang C Y, Li J. Design of matrix filter with prescribed null. *Acta Acustica* (In Chinese), 2010; **35**(3): 353–358

[21] Han D, Zhang X H. Design of optimal broad band spatial matrix filter. *Acta Acustica* (In Chinese), 2011; **36**(4): 405–411

[22] Vaccaro R J, Chhetri A, Harrison B F. Matrix filter design for passive sonar interference suppression. *J. Acoust. Soc. Am.*, 2004; **115**(6): 3010–3020

[23] Yan S F, Ma Y L. MFP noise suppression: the generalized spatial matrix filtering method. *Chinese Science Bulletin*, 2004; **49**(18): 1909–1912

基于远近场声传播特性的拖线阵声纳平台辐射噪声空域矩阵滤波技术①②

韩　东[1,2]，张海勇[2]，黄海宁[1]，李启虎[1]

（1. 中国科学院声学研究所，北京 100190；2. 海军大连舰艇学院通信系，辽宁大连 116018）

摘　要： 针对拖线阵声纳平台噪声构成近场强干扰影响声纳弱目标探测的问题，利用近场平台噪声的多途传播特性及远场目标信号的平面波传播特性，将匹配场定位技术和平面波目标方位估计技术结合，利用平台噪声到达接收阵的拷贝向量以及平面波方向向量共同设计具有强干扰抑制功能的空域矩阵滤波器，实现了近场平台噪声抑制. 设计两种最优化问题，获得不同的噪声抑制效果，给出它们的最优解，验证了解的正确性，找出解之间的关联. 解释了平台辐射噪声在阵列端首方位附近构成强干扰的原因，同时解释了空域矩阵滤波后存在探测盲区的原因. 由仿真可知，空域矩阵滤波处理可获得更小的探测盲区，同时获得盲区外更高的探测能力.

关键词： 平台辐射噪声；空域矩阵滤波；矩阵滤波器；匹配场处理；波束形成；干扰抑制；噪声抵消；拖线阵声纳

中图分类号： TP911.9　　**文献标识码：** A　　**文章编号：** 0372-2112（2014）03-0432-07

电子学报 URL： http://www.ejournal.org.cn　　**DOI：** 10.3969/j.issn.0372-2112.2014.03.003

Towed Line Array Sonar Platform Radiated Noise Spatial Matrix Filter Based on Far-Field and Near-Field Sound Propagation Characteristics

HAN Dong[1,2], ZHANG Hai-yong[2], HUANG Hai-ning[1], LI Qi-hu[1]

(1. *Institute of Acoustics, Chinese Academy of Sciences, Beijing* 100190, *China*;

2. *Department of Communication Engineering, Dalian Naval Academy, Dalian, Liaoning* 116018, *China*)

Abstract: The detection of far-field weak target is influenced by the strong near-field platform radiated noise interference in the towed line array sonar. The near-field sound is with multiple channel propagation, and the far-field sound is with plane wave propagation. By these propagation characteristics, the matched field processing technology and the direction of arrival estimation technology are combined together. The spatial matrix filters are designed by the near-filed platform noise copy vector and the far-field plane wave bearing vector. Two optimization programs are built to achieve different platform noise suppression effects. The optimal solutions are deduced directly and their relationships are established. In the bearing course, the strong interferences exist without filtering, and the blind areas exist after filtering by the spatial matrix filter. The reason is that there are relativities between the platform copy vector and the plane wave bearing vector. Simulations indicate that it can achieve less blind area and higher detection capability with spatial matrix filter processing.

Key words: platform radiated noise; spatial matrix filter; matrix filter; matched field processing; beamforming; interference suppression; noise cancellation; towed line array sonar

1　引言

多传感器阵列信号处理中，常规以及自适应波束形成技术被广泛使用，实现抑制干扰、增强被检测信号以及目标方位估计的目的[1~4]. 在水下目标探测中，利用大孔径的拖曳线列阵声纳是实现远程、低频、安静型目标探测的最佳途径，由于声纳拖曳平台即舰艇或潜艇距声纳拖线阵的距离有限，平台辐射噪声在阵列端首附近方向构成近程强干扰，导致拖线阵声纳在水听器阵端首方向形成大范围的探测盲区，同时还影响其它方位的目标探测能力[5,6]. 国内外学者尝试利用多种方法抑制平台辐射噪声，如自适应噪声抵消、自适应波束形成、后置

① 基金项目：国家自然科学基金（No.60532040, 11374001）；海军大连航艇学院科研发展基金(No.2009032).

② 电子学报, 2014, 42(3): 432-438.

波束形成干扰抵消、逆波束形成等方法，但这些基于平面波传播理论的干扰抵消技术对拖线阵声纳平台辐射噪声抑制能力十分有限.利用双线阵的联合处理，可在阵列和拖曳平台所在的垂直面内生成很深的凹槽，从而有效抑制平台辐射噪声，但该方法的缺点是会在平台方向附近形成探测盲区，同时双线阵的收放复杂性、阵列相对位置偏差等因素也影响它的性能发挥[7,8].目前被广泛使用的单阵列的拖线阵声纳，急需解决平台辐射噪声抑制问题.

由于平台辐射噪声与拖线阵声纳基阵的距离较近，平台辐射噪声应作为近场强干扰看待.然而现有平台辐射噪声抑制技术其原理都是基于远场平面波信号的降噪或抵消技术，因而并不能对声纳目标方位估计性能产生质的提高.与声场环境、声源到达接收阵的多途传播相结合的干扰抑制技术是解决拖线阵声纳受平台辐射噪声影响的最可行方案.空域矩阵滤波技术是一种新兴的阵元域数据处理技术[9~11]，该技术能获得更高精度的平面波方位估计[12~16]和匹配场定位[17,18]效果，将空域矩阵滤波与平面波方位估计和匹配场定位结合，为平台辐射噪声抑制提供了可行途径.

本文结合远场目标及近场平台辐射噪声的海洋声传播特性，将空域矩阵滤波、匹配场定位和平面波方位估计三种技术结合.利用远场平面波方向向量和近场平台辐射噪声匹配场拷贝向量，建立两个最优化问题，设计了具有近场平台辐射噪声抑制功能的空域矩阵滤波器.推导得出两个最优化问题的最优解，利用广义奇异值分解简化解，分析两个解之间的内在关联.利用仿真，分析本文设计的空域矩阵滤波器性能，及滤波后产生盲区的原因.

2 空域矩阵滤波器设计

2.1 声场模型及空域矩阵滤波最优化问题

拖曳线列阵声纳的目标方位估计，是建立在接收声信号为远场平面波假设条件下的计算结果，然而，由于声纳使用平台与拖线阵列距离有限，平台辐射噪声经海底海面折射反射等效应，应作为近场强干扰对待.考虑由无指向性水听器构成的均匀线列阵，假设阵元数为 N，则频率为 ω 的接收阵列信号 $\boldsymbol{x}(t,\omega)$ 中包含远场平面波及近场平台辐射噪声干扰.

$$\boldsymbol{x}(t,\omega)=\boldsymbol{A}(\theta,\omega)\boldsymbol{s}_1(t,\omega)+\boldsymbol{V}(\omega)\boldsymbol{s}_0(t,\omega)+\boldsymbol{n}(t,\omega) \tag{1}$$

式中，$\boldsymbol{x}(t,\omega)=[x_1(t,\omega),\cdots,x_N(t,\omega)]^{\mathrm{T}}$ 是接收阵列数据，$\boldsymbol{s}_1(t,\omega)=[s_{11}(t,\omega),\cdots,s_{1D}(t,\omega)]^{\mathrm{T}}$ 是 D 维频率为 ω 的远场平面波信号，$\boldsymbol{s}_0(t,\omega)=[s_{01}(t,\omega),\cdots,s_{0S}(t,\omega)]^{\mathrm{T}}$ 是 S 维频率为 ω 的近场平台辐射噪声干扰.$\boldsymbol{A}(\theta,\omega)=[\boldsymbol{a}(\theta_1,\omega),\cdots,\boldsymbol{a}\theta_i,\omega),\cdots,\boldsymbol{a}(\theta_D,\omega)]\in\mathbf{C}^{N\times D}$ 是频率为 ω 的平面波阵列流形矩阵，其中 $\boldsymbol{\theta}=[\theta_1,\theta_2,\cdots,\theta_D]$ 为平面波信号入射方向，假定 $D>N$. $\boldsymbol{a}(\theta_i,\omega)=[1,\exp(-\mathrm{j}\omega\Delta\sin\theta_i/c),\cdots,\exp(-\mathrm{j}\omega(N-1)\Delta\sin\theta_i/c)]^{\mathrm{T}}\in\mathbf{C}^{N\times 1}$，是信号源到接收阵的方向向量，这里 $(\cdot)^{\mathrm{T}}$ 表示转置，Δ 代表阵元间隔，j 为单位虚数，$\boldsymbol{V}(\omega)=[\boldsymbol{v}_1(\omega),\boldsymbol{v}_2(\omega),\cdots,\boldsymbol{v}_S(\omega)]\in\mathbf{C}^{N\times S}$ 是平台辐射噪声干扰经海洋多途到达接收阵的拷贝向量，并假定平台辐射噪声源数目小于阵元数，即 $S<N$. $\boldsymbol{n}(t,\omega)=[n_1(t,\omega),\cdots,n_N(t,\omega)]^{\mathrm{T}}$ 是 N 维环境噪声.设计空域预滤波器 $\boldsymbol{H}(\omega)\in\mathbf{C}^{N\times N}$，利用该滤波器对接收阵列数据滤波：

$$\boldsymbol{y}(t,\omega)=\boldsymbol{H}(\omega)\boldsymbol{A}(\boldsymbol{\theta},\omega)\boldsymbol{s}_1(t,\omega)+\boldsymbol{H}(\omega)\boldsymbol{V}(\omega)\boldsymbol{s}_0(t,\omega)+\boldsymbol{H}(\omega)\boldsymbol{n}(t,\omega) \tag{2}$$

空域矩阵滤波后对平面波信号增强或抑制的效果是通过对阵列流形的作用实现的，当 $\|\boldsymbol{H}(\omega)\boldsymbol{a}(\theta_i,\omega)\|_{\mathrm{F}}^2$ 接近于0时，说明滤波器对 θ_i 方向频率为 ω 的平面波信号有较强的抑制作用，式中 $\|\cdot\|_{\mathrm{F}}$ 表示矩阵 Frobenius 范数.反之，当 $\|\boldsymbol{H}(\omega)\boldsymbol{a}(\theta_i,\omega)-\boldsymbol{a}(\theta_i,\omega)\|_{\mathrm{F}}^2$ 接近于0，说明滤波器对 θ_i 方向频率为 ω 的平面波信号滤波后无影响.经空域矩阵滤波后，远场平面波总体响应及总体响应误差分别为 $\|\boldsymbol{H}(\omega)\boldsymbol{A}(\boldsymbol{\theta},\omega)\|_{\mathrm{F}}^2$ 和 $\|\boldsymbol{H}(\omega)\boldsymbol{A}(\boldsymbol{\theta},\omega)-\boldsymbol{A}(\boldsymbol{\theta},\omega)\|_{\mathrm{F}}^2$.同理，平台辐射噪声总体响应和总体响应误差分别为 $\|\boldsymbol{H}(\omega)\boldsymbol{V}(\omega)\|_{\mathrm{F}}^2$ 和 $\|\boldsymbol{H}(\omega)\boldsymbol{V}(\omega)-\boldsymbol{V}(\omega)\|_{\mathrm{F}}^2$.

为实现平台辐射噪声抑制，可通过约束 $\|\boldsymbol{H}(\omega)\cdot\boldsymbol{V}(\omega)\|_{\mathrm{F}}^2$ 的值接近于0实现，同时约束 $\|\boldsymbol{H}(\omega)\boldsymbol{A}(\boldsymbol{\theta},\omega)-\boldsymbol{A}(\boldsymbol{\theta},\omega)\|_{\mathrm{F}}^2$ 的值接近于0以实现滤波后不影响远场平面波.为方便设计及求解，现将 $\boldsymbol{H}(\omega)$、$\boldsymbol{A}(\boldsymbol{\theta},\omega)$、$\boldsymbol{V}(\omega)$ 分别简记为 $\boldsymbol{H}$、$\boldsymbol{A}$ 和 $\boldsymbol{V}$.通过建立如下最优化问题设计空域矩阵滤波器.

最优化问题1：平台辐射噪声零响应约束空域矩阵滤波器

$$\min_{\boldsymbol{H}_1} J(\boldsymbol{H}_1)=\|(\boldsymbol{H}_1\boldsymbol{A}-\boldsymbol{A}\|_{\mathrm{F}}^2$$
$$\text{Subject to}\quad \boldsymbol{H}_1\boldsymbol{V}=\boldsymbol{0}_{N\times S} \tag{3}$$

这里，空域矩阵滤波器 $\boldsymbol{H}_1$ 对平台辐射噪声的响应设定为零.

最优化问题2：平台辐射噪声响应抑制空域矩阵滤波器

$$\min_{\boldsymbol{H}_2} J(\boldsymbol{H}_2)=\|\boldsymbol{H}_2\boldsymbol{A}-\boldsymbol{A}\|_{\mathrm{F}}^2$$
$$\text{Subject to}\quad \|\boldsymbol{H}_2\boldsymbol{V}\|_{\mathrm{F}}^2\leqslant\varepsilon \tag{4}$$

其中，$\varepsilon>0$ 是空域矩阵滤波器对平台辐射噪声响应的约束.空域矩阵滤波器 $\boldsymbol{H}_2$ 对平台辐射噪声的响应限定为小于或等于 ε，从而实现抑制平台辐射噪声的目的.

2.2 空域矩阵滤波最优化问题求解

现给出该最优化问题1的求解过程及最优解.最优

化问题 1 的约束条件与下式等价：

$$\begin{cases}\mathrm{Re}(\boldsymbol{H}_1\boldsymbol{v}_n)=\boldsymbol{0}_{N\times 1},1\leqslant n\leqslant S\\ \mathrm{Im}(\boldsymbol{H}_1\boldsymbol{v}_n)=\boldsymbol{0}_{N\times 1},1\leqslant n\leqslant S\end{cases}\tag{5}$$

这里 Re(·)及 Im(·)分别表示求向量的实部及虚部.

构造实 Lagrange 函数：

$$\begin{aligned}&L(\boldsymbol{H}_1,\boldsymbol{\lambda}_1,\cdots,\boldsymbol{\lambda}_S,\boldsymbol{\delta}_1,\cdots,\boldsymbol{\delta}_S)\\&=\|\boldsymbol{H}_1\boldsymbol{A}-\boldsymbol{A}\|_{\mathrm{F}}^2-\sum_{n=1}^{S}\boldsymbol{\lambda}_n^{\mathrm{T}}\mathrm{Re}(\boldsymbol{H}_1\boldsymbol{v}_n)-\sum_{n=1}^{S}\boldsymbol{\delta}_n^{\mathrm{T}}\mathrm{Im}(\boldsymbol{H}_1\boldsymbol{v}_n)\\&=\mathrm{Tr}(\boldsymbol{H}_1\boldsymbol{A}\boldsymbol{A}^{\mathrm{H}}\boldsymbol{H}_1^{\mathrm{H}})-\mathrm{Tr}(\boldsymbol{H}_1\boldsymbol{A}\boldsymbol{A}^{\mathrm{H}})-\mathrm{Tr}(\boldsymbol{A}\boldsymbol{A}^{\mathrm{H}}\boldsymbol{H}_1^{\mathrm{H}})+\mathrm{Tr}(\boldsymbol{A}\boldsymbol{A}^{\mathrm{H}})\\&\quad-\sum_{n=1}^{S}\left(\frac{\boldsymbol{\lambda}_n^{\mathrm{T}}}{2}+\frac{\boldsymbol{\delta}_n^{\mathrm{T}}}{2\mathrm{j}}\right)\boldsymbol{H}_1\boldsymbol{v}_n-\sum_{n=1}^{S}\left(\frac{\boldsymbol{\lambda}_n^{\mathrm{T}}}{2}-\frac{\boldsymbol{\delta}_n^{\mathrm{T}}}{2\mathrm{j}}\right)\boldsymbol{H}_1^{*}\boldsymbol{v}_n^{*}\end{aligned}\tag{6}$$

Tr(·)代表矩阵的迹,$(\cdot)^*$ 表示共轭,$(\cdot)^{\mathrm{H}}$ 表示共轭转置,$\boldsymbol{\lambda}=[\boldsymbol{\lambda}_1,\cdots,\boldsymbol{\lambda}_S]\in\mathbf{R}^{N\times S}$和$\boldsymbol{\delta}=[\boldsymbol{\delta}_1,\cdots,\boldsymbol{\delta}_S]\in\mathbf{R}^{N\times S}$是 Lagrange 乘子.

对 $L(\boldsymbol{H}_1,\boldsymbol{\lambda}_1,\cdots,\boldsymbol{\lambda}_S,\boldsymbol{\delta}_1,\cdots,\boldsymbol{\delta}_S)$求关于$(\boldsymbol{H}_1,\boldsymbol{\lambda}_1,\cdots,\boldsymbol{\lambda}_S,\boldsymbol{\delta}_1,\cdots,\boldsymbol{\delta}_S)$的偏导数：

$$\begin{aligned}&\frac{\partial L(\boldsymbol{H}_1,\boldsymbol{\lambda}_1,\cdots,\boldsymbol{\lambda}_S,\boldsymbol{\delta}_1,\cdots,\boldsymbol{\delta}_S)}{\partial \boldsymbol{H}_1}\\&=(\boldsymbol{A}\boldsymbol{A}^{\mathrm{H}}\boldsymbol{H}_1^{\mathrm{H}})^{\mathrm{T}}-(\boldsymbol{A}\boldsymbol{A}^{\mathrm{H}})^{\mathrm{T}}-\sum_{n=1}^{S}\left(\frac{\boldsymbol{\lambda}_n^{\mathrm{T}}}{2}+\frac{\boldsymbol{\delta}_n^{\mathrm{T}}}{2\mathrm{j}}\right)\boldsymbol{v}_n^{\mathrm{T}}\\&=\boldsymbol{H}_1^{*}\boldsymbol{A}^{*}\boldsymbol{A}^{\mathrm{T}}-\boldsymbol{A}^{*}\boldsymbol{A}^{\mathrm{T}}-\left(\frac{\boldsymbol{\lambda}}{2}+\frac{\boldsymbol{\delta}}{2\mathrm{j}}\right)\boldsymbol{V}^{\mathrm{T}}\\&=\boldsymbol{H}_1^{*}\boldsymbol{A}^{*}\boldsymbol{A}^{\mathrm{T}}-\boldsymbol{A}^{*}\boldsymbol{A}^{\mathrm{T}}-\boldsymbol{\gamma}\boldsymbol{V}^{\mathrm{T}}\end{aligned}\tag{7}$$

其中 $\boldsymbol{\gamma}=\frac{\boldsymbol{\lambda}}{2}+\frac{\boldsymbol{\delta}}{2\mathrm{j}}\in\mathbf{C}^{N\times S}$是由 $\boldsymbol{\lambda}$ 和 $\boldsymbol{\delta}$ 所构造的 Lagrange 乘子.

$$\frac{\partial L(\boldsymbol{H}_1,\boldsymbol{\lambda}_1,\cdots,\boldsymbol{\lambda}_S,\boldsymbol{\delta}_1,\cdots,\boldsymbol{\delta}_S)}{\partial \boldsymbol{\lambda}_n}=\mathrm{Re}(\boldsymbol{H}_1\boldsymbol{v}_n),1\leqslant n\leqslant S\tag{8}$$

$$\frac{\partial L(\boldsymbol{H}_1,\boldsymbol{\lambda}_1,\cdots,\boldsymbol{\lambda}_S,\boldsymbol{\delta}_1,\cdots,\boldsymbol{\delta}_S)}{\partial \boldsymbol{\delta}_n}=\mathrm{Im}(\boldsymbol{H}_1\boldsymbol{v}_n),1\leqslant n\leqslant S\tag{9}$$

Lagrange 函数的稳定点$(\boldsymbol{H}_{1\mathrm{op}},\boldsymbol{\lambda}_{\mathrm{op}},\boldsymbol{\delta}_{\mathrm{op}})$,也即最优空域矩阵滤波器及最优 Lagrange 常数满足条件：

$$\boldsymbol{H}_{1\mathrm{op}}^{*}\boldsymbol{H}^{*}\boldsymbol{H}^{\mathrm{T}}-\boldsymbol{H}^{*}\boldsymbol{H}^{\mathrm{T}}-\gamma_{\mathrm{op}}\boldsymbol{V}^{\mathrm{T}}=\boldsymbol{0}_{N\times N}\tag{10}$$

$$\mathrm{Re}(\boldsymbol{H}_{1\mathrm{op}}\boldsymbol{v}_n)=\boldsymbol{0}_{N\times 1},1\leqslant n\leqslant S\tag{11}$$

$$\mathrm{Im}(\boldsymbol{H}_{1\mathrm{op}}\boldsymbol{v}_n)=\boldsymbol{0}_{N\times 1},1\leqslant n\leqslant S\tag{12}$$

其中 $\boldsymbol{\gamma}_{\mathrm{op}}=\frac{\boldsymbol{\lambda}_{\mathrm{op}}}{2}+\frac{\boldsymbol{\delta}_{\mathrm{op}}}{2\mathrm{j}}$.合并式(11)和(12),也即平台辐射噪声零响应的约束条件为：

$$\boldsymbol{H}_{1\mathrm{op}}\boldsymbol{V}_S=\boldsymbol{0}_{N\times S}\tag{13}$$

由式(10)可得：

$$\boldsymbol{H}_{1\mathrm{op}}=(\boldsymbol{A}\boldsymbol{A}^{\mathrm{H}}+\boldsymbol{\gamma}_{\mathrm{op}}^{*}\boldsymbol{V}^{\mathrm{H}})(\boldsymbol{A}\boldsymbol{A}^{\mathrm{H}})^{-1}\tag{14}$$

这里$(\cdot)^{-1}$表示求满秩方阵的逆.由 $\boldsymbol{A}$ 的构造可知,$\boldsymbol{A}$ 是 Vandermonde 矩阵,行满秩,因此 $\boldsymbol{A}\boldsymbol{A}^{\mathrm{H}}$ 是满秩方阵,$\boldsymbol{A}\boldsymbol{A}^{\mathrm{H}}$ 可逆.式(14)带入式(13)可求出所构造的最优 Lagrange 乘子：

$$\boldsymbol{\gamma}_{\mathrm{op}}^{*}=-\boldsymbol{V}[\boldsymbol{V}^{\mathrm{H}}(\boldsymbol{A}\boldsymbol{A}^{\mathrm{H}})^{-1}\boldsymbol{V}]^{-1}\tag{15}$$

将最优 Lagrange 乘子 $\boldsymbol{\gamma}_{\mathrm{op}}^{*}$ 即式(15)带入式(14),获得最优空域矩阵滤波器为：

$$\boldsymbol{H}_{1\mathrm{op}}=\boldsymbol{I}_N-\boldsymbol{V}[\boldsymbol{V}^{\mathrm{H}}(\boldsymbol{A}\boldsymbol{A}^{\mathrm{H}})^{-1}\boldsymbol{V}]^{-1}\boldsymbol{V}^{\mathrm{H}}(\boldsymbol{A}\boldsymbol{A}^{\mathrm{H}})^{-1}\tag{16}$$

式中 $\boldsymbol{I}_N$ 是维数为 $N\times N$ 的单位矩阵. $\boldsymbol{H}_{1\mathrm{op}}$ 是 $L(\boldsymbol{H}_1,\boldsymbol{\lambda}_1,\cdots,\boldsymbol{\lambda}_S,\boldsymbol{\delta}_1,\cdots,\boldsymbol{\delta}_S)$的稳定点,也即是最优化问题 1 的全局最优解.利用与最优化问题 1 相似的求解方法,可得到最优化问题 2 的全局最优解及求解最优 Lagrange乘子 κ_{op}的方程：

$$\boldsymbol{H}_{2\mathrm{op}}=\boldsymbol{A}\boldsymbol{A}^{\mathrm{H}}(\boldsymbol{A}\boldsymbol{A}^{\mathrm{H}}+\kappa_{\mathrm{op}}\boldsymbol{V}\boldsymbol{V}^{\mathrm{H}})^{-1}\tag{17}$$

$$\mathrm{Tr}[\boldsymbol{A}\boldsymbol{A}^{\mathrm{H}}(\boldsymbol{A}\boldsymbol{A}^{\mathrm{H}}+\kappa_{\mathrm{op}}\boldsymbol{V}\boldsymbol{V}^{\mathrm{H}})^{-1}\boldsymbol{V}\boldsymbol{V}^{\mathrm{H}}(\boldsymbol{A}\boldsymbol{A}^{\mathrm{H}}+\kappa_{\mathrm{op}}\boldsymbol{V}\boldsymbol{V}^{\mathrm{H}})^{-1}\boldsymbol{A}\boldsymbol{A}^{\mathrm{H}}]=\varepsilon\tag{18}$$

3 广义奇异值分解误差分析及最优解验证

3.1 利用广义奇异值分解简化最优解表示

由于 $\boldsymbol{A}$ 是 Vandermonde 矩阵,行满秩,由于 $D>N$,故 $\mathrm{rank}(\boldsymbol{A})=N$. $S<N$,故不妨假定平台辐射噪声源数目即为 $\boldsymbol{V}$ 的秩,即 $\mathrm{rank}(\boldsymbol{V})=S$.由广义奇异值分解可知[19],存在酉矩阵 $\boldsymbol{U}_A\in\mathbf{C}^{D\times D}$和 $\boldsymbol{U}_V\in\mathbf{C}^{S\times S}$以及非奇异矩阵 $\boldsymbol{Q}_X\in\mathbf{C}^{N\times N}$,使得

$$\boldsymbol{U}_A\boldsymbol{A}^{\mathrm{H}}\boldsymbol{Q}_X=\boldsymbol{\Sigma}_A,\boldsymbol{\Sigma}_A=\begin{bmatrix}\boldsymbol{I}_{N-S}&\boldsymbol{0}_{(N-S)\times S}\\\boldsymbol{0}_{S\times(N-S)}&\boldsymbol{Z}_A\\\boldsymbol{0}_{(D-N)\times(N-S)}&\boldsymbol{0}_{(D-N)\times S}\end{bmatrix}_{D\times N},$$

$$\boldsymbol{Z}_A=\mathrm{diag}(\alpha_1,\alpha_2,\cdots,\alpha_S)\tag{19}$$

$$\boldsymbol{U}_V\boldsymbol{V}^{\mathrm{H}}\boldsymbol{Q}_X=\boldsymbol{\Sigma}_V,\boldsymbol{\Sigma}_V=[\boldsymbol{0}_{S\times(N-S)},\boldsymbol{Z}_V]_{S\times N},$$

$$\boldsymbol{Z}_V=\mathrm{diag}(\beta_1,\beta_2,\cdots,\beta_S)\tag{20}$$

式中 $\alpha_i^2+\beta_i^2=1,i=1,2,\cdots,S$.由式(19)、(20)可得：

$$\boldsymbol{A}=\boldsymbol{Q}_X^{-\mathrm{H}}\boldsymbol{\Sigma}_A^{\mathrm{T}}\boldsymbol{U}_A\tag{21}$$

$$\boldsymbol{V}=\boldsymbol{A}_X^{-\mathrm{H}}\boldsymbol{\Sigma}_V^{\mathrm{T}}\boldsymbol{U}_V\tag{22}$$

将式(21)、(22)带入式(16)~(18)可得：

$$\boldsymbol{H}_{1\mathrm{op}}=\boldsymbol{Q}_X^{-\mathrm{H}}\begin{bmatrix}\boldsymbol{I}_{N-S}&\boldsymbol{0}_{(N-S)\times S}\\\boldsymbol{0}_{S\times(N-S)}&\boldsymbol{0}_S\end{bmatrix}\boldsymbol{Q}_X^{\mathrm{H}}\tag{23}$$

$$\boldsymbol{H}_{2\mathrm{op}}=\boldsymbol{Q}_X^{-\mathrm{H}}\begin{bmatrix}\boldsymbol{I}_{N-S}&\boldsymbol{0}_{(N-S)\times S}\\\boldsymbol{0}_{S\times(N-S)}&\boldsymbol{Z}_A^2(\boldsymbol{Z}_A^2+\kappa_{\mathrm{op}}\boldsymbol{Z}_V^2)^{-1}\end{bmatrix}\boldsymbol{Q}_X^{\mathrm{H}}\tag{24}$$

$$\mathrm{Tr}\left\{\boldsymbol{Q}_X^{-\mathrm{H}}\begin{bmatrix}\boldsymbol{0}_{N-S}&\boldsymbol{Q}_{(N-S)\times S}\\\boldsymbol{0}_{S\times(N-S)}&\boldsymbol{Z}_A^4\boldsymbol{Z}_V^2(\boldsymbol{Z}_A^2+\kappa_{\mathrm{op}}\boldsymbol{Z}_V^2)^{-2}\end{bmatrix}\boldsymbol{Q}_X^{-1}\right\}=\varepsilon\tag{25}$$

式(25)是由式(18)所得的最优化问题 2 中求解 Lagrange 乘子 κ_{op}的方程.

3.2 最优解验证及滤波器误差分析

利用广义奇异值分解可以从理论上分析最优空域矩阵滤波器对近场平台辐射噪声响应以及对远场平面波响应误差,从而通过检验最优解对近场平台辐射噪声

响应是否等于约束条件验证其正确性.式(23)、(24)给出了广义奇异分解所得的空域矩阵滤波器简化解,利用该简化解和式(21)、(22)可得空域矩阵滤波器对近场平台辐射噪声的响应以及对远场平面波的响应误差:

$$\| \boldsymbol{H}_{1\mathrm{op}}\boldsymbol{A} - \boldsymbol{A} \|_{\mathrm{F}}^{2} = \left\| \boldsymbol{Q}_{X}^{-\mathrm{H}} \begin{bmatrix} \boldsymbol{0}_{N-S} & \boldsymbol{0}_{(N-S)\times S} \\ \boldsymbol{0}_{S\times(N-S)} & \boldsymbol{Z}_{A} \end{bmatrix} \right\|_{\mathrm{F}}^{2} \tag{26}$$

$$\| \boldsymbol{H}_{2\mathrm{op}}\boldsymbol{A} - \boldsymbol{A} \|_{\mathrm{F}}^{2} = \left\| \boldsymbol{Q}_{X}^{-\mathrm{H}} \begin{bmatrix} \boldsymbol{0}_{N-S} & \boldsymbol{0}_{(N-S)\times S} \\ \boldsymbol{0}_{S\times(N-S)} & \kappa_{\mathrm{op}}\boldsymbol{Z}_{V}^{2}\boldsymbol{Z}_{A}(\boldsymbol{Z}_{A}^{2} + \kappa_{\mathrm{op}}\boldsymbol{Z}_{V}^{2})^{-1} \end{bmatrix} \right\|_{\mathrm{F}}^{2} \tag{27}$$

$$\| \boldsymbol{H}_{1\mathrm{op}}\boldsymbol{V} \|_{\mathrm{F}}^{2} = 0_{N-S} \tag{28}$$

$$\begin{aligned} \| \boldsymbol{H}_{2\mathrm{op}}\boldsymbol{V} \|_{\mathrm{F}}^{2} &= \left\| \boldsymbol{Q}_{X}^{-\mathrm{H}} \begin{bmatrix} \boldsymbol{0}_{(N-S)\times S} \\ \boldsymbol{Z}_{A}^{2}\boldsymbol{Z}_{V}(\boldsymbol{Z}_{A}^{2} + \kappa_{\mathrm{op}}\boldsymbol{Z}_{V}^{2})^{-1} \end{bmatrix} \right\|_{\mathrm{F}}^{2} \\ &= \mathrm{Tr}\left\{ \boldsymbol{Q}_{X}^{-\mathrm{H}} \begin{bmatrix} \boldsymbol{0}_{N-S} & \boldsymbol{0}_{(N-S)\times S} \\ \boldsymbol{0}_{S\times(N-S)} & \boldsymbol{Z}_{A}^{4}\boldsymbol{Z}_{V}^{2}(\boldsymbol{Z}_{A}^{2} + \kappa_{\mathrm{op}}\boldsymbol{Z}_{V}^{2})^{-2} \end{bmatrix} \boldsymbol{Q}_{X}^{-1} \right\} \\ &= \varepsilon \end{aligned} \tag{29}$$

由式(28)、(29)可知,空域矩阵滤波器最优解 $\boldsymbol{H}_{1\mathrm{op}}$ 和 $\boldsymbol{H}_{2\mathrm{op}}$分别满足最优化问题1和2的约束条件,从而验证了滤波器最优解的正确性.最优化问题1和最优化问题2是在限定对近场平台辐射噪声不同的响应幅度基础上建立的,最优解之间存在一定的关联.由最优化问题2求解最优Lagrange乘子 κ_{op}的方程式(25)可知,当 $\kappa_{\mathrm{op}}\to\infty$时,最优化问题约束条件 $\varepsilon\to 0$,此时由式(24)可知,最优化问题2与最优化问题1的最优解相等,$\boldsymbol{H}_{1\mathrm{op}}$是 $\boldsymbol{H}_{2\mathrm{op}}$的极限形式.

4 仿真及数据分析

4.1 声场环境及拖线阵结构

考虑一个由阵元数 $N=64$ 组成的水听器均匀线列阵,阵元间距为4m,探测阵列半波长频率即192.5Hz的目标信号.平台辐射噪声与阵列相对位置及海洋声速剖面由图1给出,海底声速为1682m/s,海底介质密度为1.76g/cm³,由点噪声源所构成的平台辐射噪声所在深度为8m,水平拖线阵靠近拖曳平台的第一个水听器位于平台后400m处,拖曳深度为25m,并以该水听器为基准,平台辐射噪声到达该水听器的干噪比(Interference to Noise Ratio,INR)为10dB.

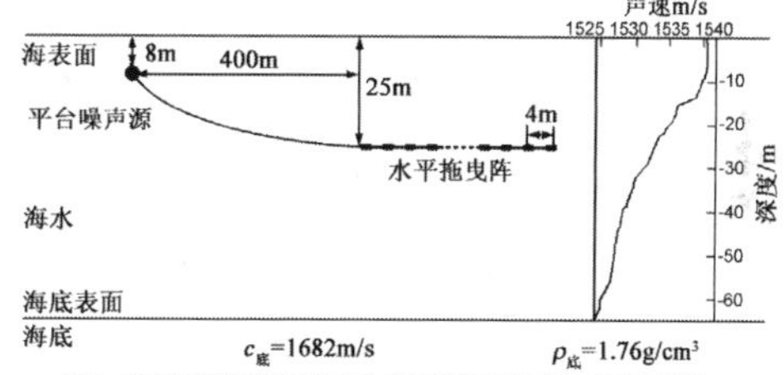

图1 海洋声速剖面及平台辐射噪声相对位置示意图

4.2 拷贝向量与方向向量相关性分析

图2是平台辐射噪声传播损失效果,图3给出利用最小方差MV处理器所得的匹配场定位效果,平台辐射噪声位置用圆圈圈出,通过匹配场定位技术可以获得平台辐射噪声源的正确距离和深度信息.

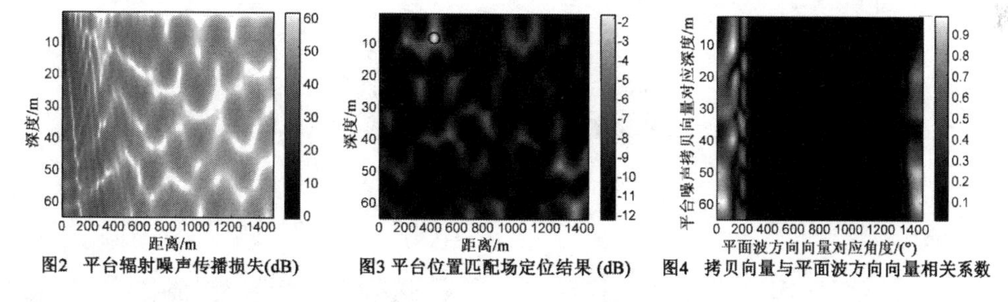

图2 平台辐射噪声传播损失(dB) 图3 平台位置匹配场定位结果 (dB) 图4 拷贝向量与平面波方向向量相关系数

常规波束形成算法(Conventional Beamforming,CBF)由于其目标方位估计的稳健性,在拖线阵声纳信号处理中广泛使用.受海底海面多途传播的影响,平台辐射噪声在 -76.2°及 -67°形成了两个强干扰(参见图7空域矩阵滤波前的方位历程).方位历程上存在强干扰的原因可以通过平台辐射噪声拷贝向量与平面波方向向量的相关性解释.图4给出了阵列首阵元距平台400m条件下,不同深度的平台辐射噪声拷贝向量与平面波方向向量之间的相关系数(取绝对值).在深度为8m时,平台辐射噪声拷贝向量与平面波方向向量在 -76.2°及 -67°附近的相关系数值较大,而与其它方位的相关系数值接近于0,因而形成了平台辐射噪声相当于远场平面波从 -76.2°及 -67°入射到阵列的现象.

4.3 空域矩阵滤波效果

利用匹配场定位所获得的平台辐射噪声位置,利用该点处的拷贝向量和平面波方向向量即可设计空域矩阵滤波器.最优化问题1所对应的空域矩阵滤波器,可直接由式(16)或(23)给出.最优化问题2所对应空域矩阵滤波器,需要在求解最优Lagrange乘子 κ_{op}的基础上给出,确定 κ_{op}的方程为式(18)或(25),图5给出了

κ_{op}与 ε 的函数关系图.

空域矩阵滤波器 $\boldsymbol{H}$ 的效果可以通过滤波器响应 $10\lg(\|\boldsymbol{Ha}(\theta)\|_F^2/N)$ 和响应误差 $10\lg(\|\boldsymbol{Ha}(\theta)-\boldsymbol{a}(\theta)\|_F^2/N)$ 衡量. 一方面,当滤波器响应小于 0dB,说明对角度 θ 上的平面波有一定的抑制效果. 另一方面,当滤波器响应误差远小于 0dB 时,说明对角度 θ 上的平面波失真较小. 从图 6 可知,两种滤波器都在拷贝向量与方向向量相关系数较大值位置产生了一定的抑制效果,对其他方向的平面波方向向量失真较小,零响应约束空域矩阵滤波效果是平台辐射噪声响应约束滤波效果的极限形式.

假设目标从方位 -90°到 -60°连续运动,运动轨迹呈斜对角的形式,平台相对于拖线阵位置保持恒定. 图 7 给出了弱目标信噪比(Signal to Noise Ratio, SNR)为 -5dB、-10dB、-15dB、-20dB 情况下,最优化问题 1 所对应的滤波器滤波前后 CBF 方位历程图(横坐标是波束搜索角,纵坐标是弱目标方位历程角). 通过平台辐射噪声零响应约束空域矩阵滤波处理,-76.2°方位的强干扰被完全滤除,-67°方位的强干扰被大部分抑制滤波后在 -76.2°位置附近产生探测盲区,在盲区外对 CBF 的探测能力有较大的提高作用. 且滤波后的 CBF 盲区宽度要小于未经空域矩阵滤波处理的 CBF 被强干扰覆盖的探测盲区宽度.

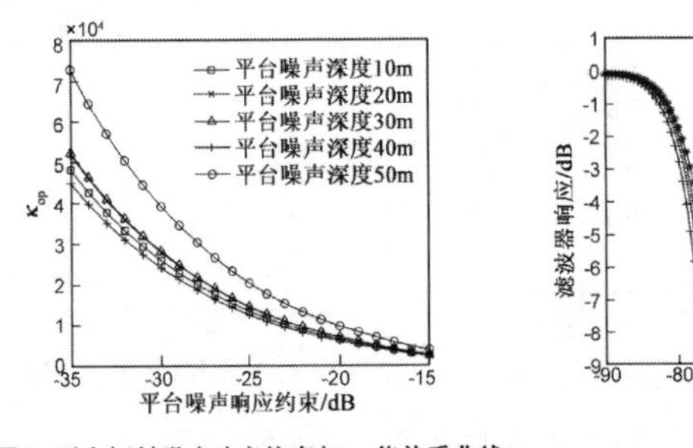

图5 平台辐射噪声响应约束与κ_{op}值关系曲线

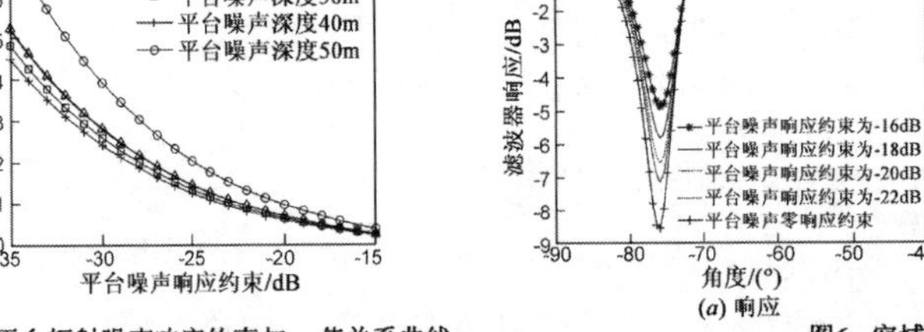

(a) 响应

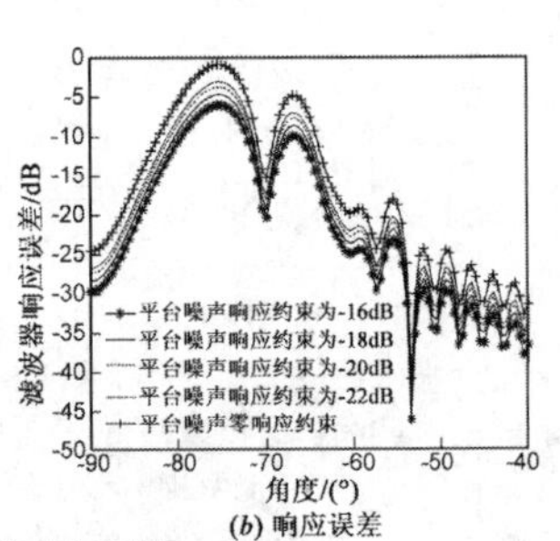

(b) 响应误差

图6 空域矩阵滤波器效果

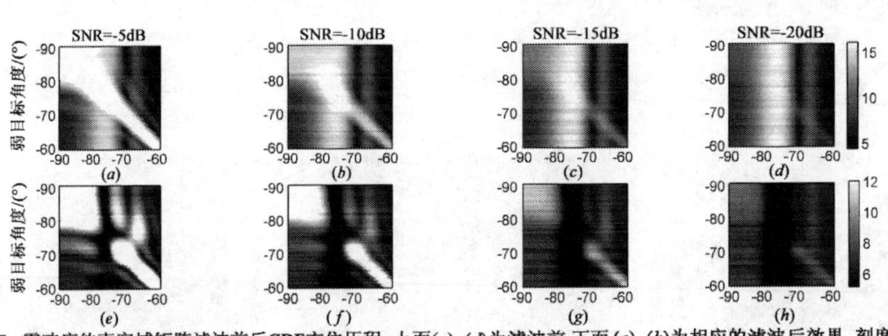

图7 零响应约束空域矩阵滤波前后CBF方位历程. 上面(a)~(d)为滤波前,下面(e)~(h)为相应的滤波后效果. 刻度尺为未归一化的CBF功率值(dB)

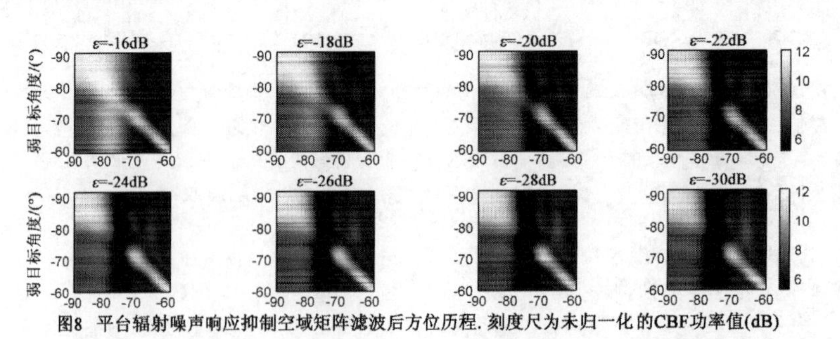

图8 平台辐射噪声响应抑制空域矩阵滤波后方位历程. 刻度尺为未归一化的CBF功率值(dB)

图 8 给出了由最优化问题 2 平台辐射噪声响应抑制约束空域矩阵滤波处理后的 CBF 方位历程图,弱目标信噪比为 - 15dB(未经空域矩阵滤波的 CBF 方位历程见图 7(c)).由图 8 可知,- 67°方位的强干扰被全部抑制,随平台辐射噪声响应的降低,- 76.2°方位的平台辐射噪声强干扰被逐渐抑制直至在 - 76.2°附近产生探测盲区,最终形成图 7(g)的效果.

分析空域矩阵滤波后 - 76.2°附近范围生成探测盲区以及 - 60°不生成盲区的原因:由图 4 可知,平台辐射噪声拷贝向量与 - 76.2°以及 - 60°附近的平面波方向向量存在强相关性,经空域矩阵滤波处理后,滤波器对平台辐射噪声拷贝向量的响应为零或接近于 0,由于拷贝向量与平面波方向向量相关性的原因,导致空域矩阵滤波对该区域的平面波也具有相应抑制作用,抑制能力与相关性成正比.由于在 - 76.2°附近的相关系数值接近于 1,因而在 - 76.2°附近范围产生了探测盲区,在 - 60°附近的相关系数相对较小,因而不产生盲区.

5 结论

本文将空域矩阵滤波、匹配场处理与平面波目标方位估计技术结合,设计两种具有平台辐射噪声抑制功能的空域矩阵滤波器,实现了平台辐射噪声抑制.通过广义奇异值分解,分析了滤波器对平面波方向向量的响应误差,验证了解的正确性,平台辐射噪声零响应约束滤波器是平台辐射噪声响应约束滤波器的极限形式.由平台辐射噪声与平面波方向向量存在相关性,平台辐射噪声经多途传播会在 CBF 方位历程上生成强干扰.同时,空域矩阵滤波处理后会产生一定范围的探测盲区,相关性的强弱是产生盲区的决定性因素,但空域矩阵滤波后的盲区范围要小于未经空域矩阵滤波处理的盲区范围,并且可提高非盲区的探测能力.

参考文献

[1] 郭玉华,常青美,余道杰,岳彩青.一种改进的极化域-空域联合的自适应波束形成算法[J].电子学报,2012,40(6):1279 - 1283.
GUO Yu-hua,CHANG Qing-mei,YU Dao-jie,YUE Cai-qing. An improved polarization-space adaptive beamforming algorithm[J].Acta Electronica Sinica,2012,40(6):1279 - 1283.(in Chinese)

[2] 倪淑燕,程乃平,倪正中.共轭虚拟阵列波束形成方法[J].电子学报,2011,39(9):2120 - 2124.
NI Shu-yan, CHENG Nai-ping, NI Zheng-zhong. Conjugate virtual array beamforming method[J].Acta Electronica Sinica,2011,39(9):2120 - 2124.(in Chinese)

[3] 顾陈,何劲,朱晓华.冲击噪声背景下基于最小均方归一化误差的波束形成算法[J].电子学报,2010,38(6):1430 - 1433.
GU Chen, HE Jin, ZHU Xiao-hua. Minimum mean square "normalized-error" beamforming amid heavy-tailed impulsive noise of unknown statistics[J].Acta Electronica Sinica,2010,38(6):1430 - 1433.(in Chinese)

[4] 李关防,惠俊英.基于经验模态分解的模态域 MVDR 方法研究[J].电子学报,2009,37(5):942 - 946.
LI Guan-fang,HUI Jun-ying.A study of mode domain MVDR algorithm based on empirical mode decomposition[J]. Acta Electronica Sinica,2009,37(5):942 - 946.(in Chinese)

[5] 李启虎,李淑秋,孙长瑜,余华兵.主被动拖线阵声呐中拖曳平台辐射噪声和拖鱼噪声在浅海使用时的干扰特性[J].声学学报,2007,32(1):1 - 4.
LI Qi-hu,LI Shu-qiu,SUN Chang-yu,YU Hua-bing.The interference characteristics of platform and towed body noise in shallow water for active/passive towed array sonar[J].Acta Acustica,2007,32(1):1 - 4.(in Chinese)

[6] 马远良,刘孟庵,张忠兵,童立.浅海声场中拖曳线列阵常规波束形成器对拖船噪声的接收响应[J].声学学报,2002,27(6):481 - 486.
MA Yuan-liang, LIU Meng-an, ZHANG Zhong-bing, TONG Li.Response of the towed line array to the noise of the tow ship in shallow water[J]. Acta Acustica,2002,27(6):481 - 486.(in Chinese)

[7] 李启虎.用双线列阵区分左右舷目标的延时估计方法及其实现[J].声学学报,2006,31(6):485 - 487.
LI Qi-hu. The time-delay estimation method of resolving left-right target ambiguity for twin-line array and its realization[J]. Acta Acustica,2006,31(6):485 - 487.(in Chinese)

[8] 李启虎.双线列阵左右舷目标分辨性能的初步分析[J].声学学报,2006,31(5):385 - 388.
LI Qi-hu.Preliminary analysis of left-right ambiguity resolution performance for twin-line array[J]. Acta Acustica, 2006, 31(5):385 - 388.(in Chinese)

[9] Vaccaro R J, Harrison B F. Optimal matrix-filter design[J]. IEEE Transactions on Signal Processing, 1996,44(3):705 - 709.

[10] Han D,Zhang X H.Optimal matrix filter design with application to filtering short data records[J].IEEE Signal Processing Letters,2010,17(5):521 - 524.

[11] Han D,Yin J S,Kang C Y,Zhang X H.Optimal matrix filter design with controlled mean-square sidelobe level[J]. IET Signal Processing,2011,5(3):306 - 312.

[12] 鄢社锋,侯朝焕,马晓川.矩阵空域预滤波目标方位估计[J].声学学报,2007,32(2):151 - 157.
YAN She-feng, HOU Chao-huan, MA Xiao-chuan. Matrix spatial perfiltering approach for direction-of-arrival estimation[J].Acta Acustica,2007,32(2):151 - 157.(in Chinese)

[13] Zhu Z W ,Wang S,Leung H,et al.Matrix filter design using

semi-infinite programming with application to DOA estimation [J]. IEEE Transactions on Signal Processing, 2000, 48(1): 267-271.

[14] Macinnes C S. Source localization using subspace estimation and spatial filtering[J]. IEEE Journal of Oceanic Engineering, 2004, 29(2): 488-497.

[15] Hassanien A, Elkader S A, Gershman A B, Wong K M. Convex optimization based beam-space preprocessing with improved robustness against out-of-sector sources[J]. IEEE Transactions on Signal Processing, 2006, 54(5): 1587-1595.

[16] 韩东,章新华,孙瑜.宽带最优空域矩阵滤波器设计[J].声学学报,2011,36(4):405-411.
HAN Dong, ZHANG Xin-hua, SUN Yu. Design of optimal broad band spatial matrix filter[J]. Acta Acustica, 2011, 36(4): 405-411. (in Chinese)

[17] Vaccaro R J, Chhetri A, Harrison B F. Matrix filter design for passive sonar interference suppression[J]. Journal of the Acoustic Society of America, 2004, 115(6): 3010-3020.

[18] 鄢社锋,马远良.匹配场噪声抑制:广义空域滤波方法[J].科学通报,2004,49(18):1909-1912.

[19] 张贤达.矩阵分析与应用[M].北京:清华大学出版社,2004.367-380.

THE STUDY OF TIME DELAY ESTIMATION TECHNOLOGY BASED ON THE CROSS-SPECTRUM METHOD①

Zhibo Zhang[a], Changyu Sun[a], Yuan Li[a], Haibo Zheng[a], Bing Li[a], Xizhong Bao[a], Qihu Li[a]

[a]Postal address: No.21 of Beisihuanxi Rd, Beijing, P.R.China, 100190

Contact author: Zhibo Zhang, No.21 of Beisihuanxi Rd, Beijing, P.R.China, 100190, Fax: (8610)82547960, Email: zhangzb@mail.ioa.ac.cn

Abstract: *In Many fields of modern radar and sonar, parameters of the target distance and azimuth often need to be measured accurately, one of its key technologies is the time delay estimation, time delay estimation directly affects the accuracy of acoustic positioning effect. Traditional delay estimation consists mainly of the generalized correlation method , phase-spectrum analysis, parametric model estimation, and adaptive time delay estimation. Cross spectrum method is a common method of time delay estimation, the method in high noise environment, can obtain more accurate estimation of delay; but in low SNR environment, the performance of this method in sharp decline. Cross spectral method is first transformed into the frequency domain, in order to get higher accuracy, often need to segment average, one estimation of the time delay often need thousands of points or even thousands of data points, and the individual outliers will make a serious decline in accuracy. In order to reduce individual outliers brought by the nonstationarity of signal, improve the accuracy of estimation, this paper presents an improved cross spectrum time delay estimation method, theoretical analysis, and gives the specific implementation steps. Computer simulation results show that, the improved method in the low SNR environment, can improve the estimation precision. Cross spectrum time delay estimation method is improved, the optimization in the conventional cross spectral method, less computation complexity, and has a strong practical.*

Keywords: *Passive positioning, time delay estimation, cross-spectrum method*

① UA2014-2nd International Conference and Exhibition on Underwater Acoustics, 2014: 1587-1593.

1 INTRODUCTION

In many fields of modern radar and sonar, parameters of the target distance and azimuth often need to be measured accurately, one of its key technologies is the time delay estimation. For passive sonar system, the way to get the target distance and azimuth is using a sensor array. Through the detection of target sound, it calculates the delay of each sensor signal difference, and gets the distance value according to the geometric positioning principle. Time delay estimation is the foundation of passive acoustic localization, its accuracy is directly related to whether can meet the practical requirements of positioning accuracy. It is also the foundation of extraction of information technology such as multiple paths separation, feature extraction, classification and recognition.

Cross-spectrum method is a common method of time delay estimation, the method in high noise environment, can obtain more accurate estimation of delay; but in low SNR environment, the performance of this method in sharp decline. Aiming at this problem, this paper presents an improved cross-spectrum time delay estimation method, theoretical analysis, and gives the specific implementation steps. Computer simulation results show that, the improved method in the low SNR environment, can improve the estimation precision, less computation complexity, and has a strong practical.

2 INTRODUCTION OF CROSS-SPECTRUM METHOD FOR TIME DELAY ESTIMATION

The basic principle of correlated signal processing is the statistical characteristics of the signal and interference (correlation) difference to improve the output SNR of receiving system. If the source to satisfy the far field condition, in the presence of noise conditions, the mathematical model can be established with two sensors spatially independent detection as follows.

$$x_1(t) = s(t) + n_1(t)$$

$$x_2(t) = s(t-\tau) + n_2(t)$$

$x_1(t)$ and $x_2(t)$ are the received waveforms and two hydrophones, τ is the time difference between two hydrophone signal. $s(t), n_1(t)$ and $n_2(t)$ are the stationary random process. The assumption that the signal $s(t)$ and noise $n_1(t)$ and $n_2(t)$ are uncorrelated, usually the time delay estimation is to be completed on the estimation of τ. To determine the time delay and the common method to determine angles of arrival of signals is calculating the cross-correlation function. The following diagram, gives the basic structure of the cross-correlation time delay estimation:

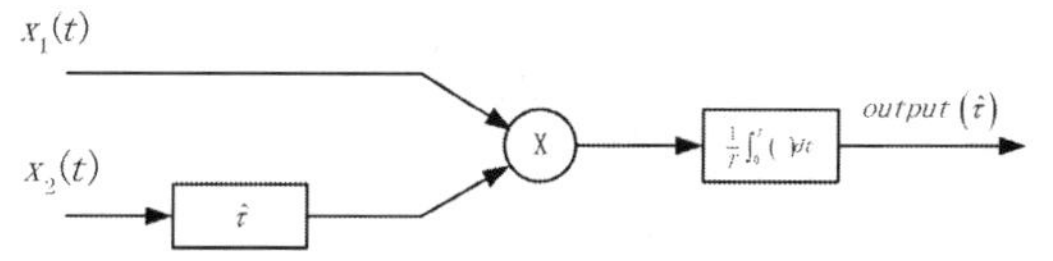

Fig.1: basic cross-correlation time delay estimation structure

The two signals as an example, its correlation is:

$$R_{12}(\tau) = E\left[x_1(t)x_2(t)\right] \triangleq E\left[s(t)s(t-\hat{\tau})\right]$$
$$= R_{ss}\left(\tau - \hat{\tau}\right)$$

The E is expectation. The assumption that the observation time is T, the estimation value for the ergodic process of orthogonal cross-correlation is:

$$R_{12}(\tau) = \frac{1}{T}\int_0^T x_1(t)x_2(t-\tau)dt$$

In the formula, $R_{ss}(\cdot)$ is the autocorrelation function of the source signals. According to the characteristics of autocorrelation function:

$$\left|R_{ss}(\tau-\hat{\tau})\right| \le R_{ss}(0)$$

That is, when $\tau = \hat{\tau}$, $R_{ss}(\cdot)$ or $R_{12}(\cdot)$ to get maximum value, which $x_1(t)$ and $x_2(t)$ have the greatest similarity at this time and, taking $\hat{\tau} = \tau$:

$$R_{12}(\tau_0) = \max[R_{12}(\tau)]$$

As the time delay estimation, this is the basic correlation time delay estimation method.

According to Wiener–Khinchin theorem, stationary random signal power spectrum density and autocorrelation function are the Fourier transform each other. So:

$$X_{12}(f,\tau_{12}) = \int_0^T R_{12}(\tau)e^{-j2\pi f\tau}\,d\tau$$
$$= |x_{12}(f,\tau_{12})|e^{j2\pi f\tau_{12}}$$
$$= |x_{12}(f,\tau_{12})|e^{j\psi(f)}$$
$$= \langle x_1(f)x_2(f,\tau_{12})\rangle$$

In the formula, $X_{12}(f,\tau_{12})$ is the cross-power spectral function of the correlation function of $x_1(t)$ and $x_2(t-\tau_{12})$. $x_1(f)$ and $x_2(f,\tau_{12})$ respectively is the Fourier transform of the corresponding time waveform of $x_1(t)$ and $x_2(t-\tau_{12})$. $\varphi(f)$ is the function of phase frequency characteristic as the cross-power spectrum. The time delay estimation $\hat{\tau}_{12}$ is:

$$\hat{\tau}_{12} = \frac{1}{2\pi}\frac{d\psi(f)}{df}$$

It is time delay estimation $\hat{\tau}_{12}$ is $1/2\pi$ of the slope of the cross-spectral phase frequency function. The solution of the slope is done by the least square method.

The least square estimation of single parameters is discussed below:

If an estimator is a single parameter x, and it has gotten m linear observations of x, the observation equation is:

$$z_i = f_i x + n_i \qquad i = 1,2,.....m$$

f_i is the known factor, and n_i is the observation noise of the i observations.

In this single parameter case, the estimation rule for least squares estimation is to obtain the estimator $\hat{x}$, which can make the square of the error between observation value z_i and the corresponding $h_i\hat{x}$ is minimum. That is:

$$J(\hat{x}) = \sum_{i=1}^{m}(z_i - f_i\hat{x})^2$$

Thus the calculation formula for the least squares estimation $\hat{x}_{ls}$ under single parameter was obtained by the method of finding the minimum value.

$$\hat{x}_{ls} = \frac{\sum_{i=1}^{m} f_i z_i}{\sum_{i=1} f_i}$$

In the formula, f_i is the line corresponding frequency value in the cross spectral function, z_i is the line corresponding argument.

Only when the value τ_{12} is smaller, the estimation accuracy is higher. When the value τ_{12} is greater, the estimation accuracy is reduced. In particular, when the phase measurement is multi-value, large time delay estimation is worried for the blur of the phase estimation multi-value.

In order to solve the blur of the phase estimation multi-value, usually, usually there are two steps. At first, the time domain signal processing gets the coarse delay, after compensation of the coarse delay, phase measurement is performed by cross-spectral method, that is to solve the estimated blur of multi-value, but also can improve the phase measurement accuracy.

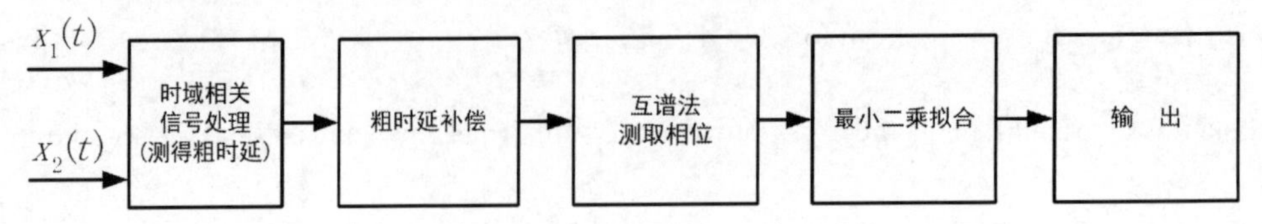

Fig.2: cross-spectrum estimation signal processing flow chart

3 INTRODUCTION OF THE IMPROVED CROSS-SPECTRUM METHOD FOR TIME DELAY ESTIMATION

The improved algorithm is for the least squares method of two-step method in measuring the time delay estimation. Iterative time delay estimation algorithm is proposed in this paper, can improve the delay estimation accuracy at low SNR conditions, needs less computing.

Based on the time domain signal processing to solve the coarse delay, and compensation coarse delay, the tow signals become $y_1(t)$ and $y_0(t)$. Then they are transformed into $Y_1(\omega)$ and $Y_2(\omega)$ in frequency domain . A real time delay of signals $y_1(t)$ and $y_2(t)$ is τ .

Set: $Z(\hat{\tau}) = \sum_{\omega} real\left(Y_1(\omega)(Y_2(\omega)e^{-j\omega\hat{\tau}})^*\right)$When $\hat{\tau}$=τ, $Z(\hat{\tau})$ gets maximum value.

Step 1: After correlation signal processing, the time delay value corresponding to the maximum and the 2nd maximum of two signals $y_1(t)$ and $y_2(t)$ are τ_1^1 (τ_1^1=0) and τ_2^1.

Step 2: For the precision delay correction, real time delay between two signals will set between τ_1^1 and τ_2^1 .

Step 3: frequency domain compensation delay are as follows:

$$\begin{cases} Z_1^1 = \sum\limits_{\omega} real\left(Y_1(\omega)(Y_2(\omega)e^{-j\omega\tau_1^1})^*\right) \\ Z_2^1 = \sum\limits_{\omega} real\left(Y_1(\omega)(Y_2(\omega)e^{-j\omega\tau_2^1})^*\right) \\ Z_3^1 = \sum\limits_{\omega} real\left(Y_1(\omega)(Y_2(\omega)e^{-j\omega\tau_3^1})^*\right) \end{cases}$$

Comparing Z_1^1 , Z_2^1 and Z_3^1 , the delay corresponding to the maximum value and the 2nd maximum value respectively is τ_1^2 and τ_2^2.

Step 4: repeat step 1. Set up after M iterations, the delay corresponding to the maximum value and the 2nd maximum value respectively is τ_1^m and τ_2^m .

4 ALGORITHM SIMULATION

The method has been done in the computer system simulation.

The simulation conditions: signal processing frequency band is 500~2000Hz; signal frequency sampling is 10kHz; time delay difference of signal 1 and signal 2 is 0.013455s; 200 separate statistics under different SNR

After measuring the coarse time delay by time domain cross-correlation, and the time delay compensation for the original signal, the conventional cross-spectral method and the improved cross-spectral method are used for processing The simulation results as shown below:

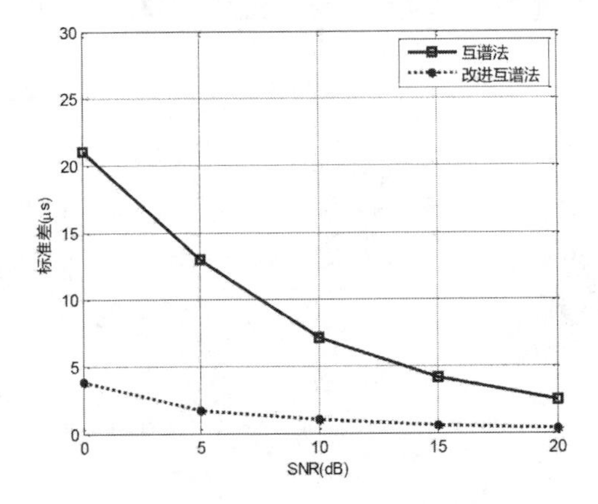

Fig.3: under different signal to noise ratio of cross spectrum method and improved cross spectral method results

As you can see, the time delay estimation accuracy of the improved cross-spectral method has been greatly improved compared with the conventional method. With the SNR reducing, the delay estimation accuracy of the conventional method decreased more seriously, and the improved method decreased not obviously. In low SNR condition, the time delay estimation accuracy has the very big promotion compared with the conventional method.

5 CONCLUSION

Cross-spectrum method is a common method of time delay estimation. In high noise environment, the method can obtain more accurate estimation of delay; but in low SNR environment, the performance of this method in sharp decline. Aiming at this problem, this paper presents an improved cross-spectrum time delay estimation method, makes theoretical analysis, and gives the specific implementation steps. Computer simulation results show that in the low SNR environment the improved method can improve the estimation precision, is less computation complexity, and has a strong practical.

REFERENCES

[1] **Struzinski W A,Lowe E D**, A performance comparison of four noise background

normalization schemes proposed for signal detection systems, *J.Acoust.Soc.Amer.*, 76 (6), 1738-1742, 1984.
[2] **Struzinski W A,Lowe E D**, A performance comparison of four noise background normalization schemes proposed for signal detection systems, *J.Acoust.Soc.Amer.*, 78 (3), 936-941, 1985.
[3] **LI Q H,PAN X B**, A new algorithm of background equalization, *J.Acta Acustica.*, 25 (1), 5-9, 2000.

THE CONTROL METHODS OF ACOUSTIC ARRAY'S POSTURE BASED ON THE OCEANOGRAPHIC BUOY ①

Linyu Wang[a], Shijun Feng[b], Qihu Li

[a] NO.21, Bei-Si-huan-Xi Road, Institute of Acoustics, Chinese Academy of Sciences Beijing, China
[b] NO.21, Bei-Si-huan-Xi Road, Institute of Acoustics, Chinese Academy of Sciences Beijing, China

Linyu Wang, NO.21, Bei-Si-huan-Xi Road, Institute of Acoustics, Chinese Academy of Sciences ,Beijing, China,100190; 010-82547955;wlyhyh@163.com

Abstract: *In order to monitor marine environment and detect the surface targets, an innovative scheme is proposed based on the oceanographic buoy which utilizing an acoustic array with diameter of 9 meters (the acoustic array is a horizontal circle array with 16 hydrophones uniformly distributed on the circle). The posture control is crucial for acoustic array in practical application. A stable structure design and a signal processing azimuth compensation method are adopted. In the stable structure design an elastic suspension system is used to reduce the influence of the waves and the gravity torque of the system make the acoustic array return to the horizontal equilibrium state. While the azimuth compensation is realized in the beam domain by using the angles measured by posture sensor. The performances of the proposed methods were tested in lake trial in May 2013. In the experiment, the acoustic array was tilted 30 degrees. On one hand, effect of the stable structure can be verified; On the other hand, the real targets were detected and the measurement results were compared with the real target to validate the feasibility of the azimuth compensation method. The results indicate that the stable structure design can make the acoustic array return to the horizontal equilibrium state in 20 seconds, and the azimuth compensation method can make the deviation less than 3 degrees between the*

① UACE2015-3nd Underwater Acoustics Conference and Exhibition, 2015: 573-580.

monitoring results and the real target. The performance of the two methods will be further testified in sea trials (below 3 level sea state).

Keywords: *acoustic array, stable structure design, azimuth compensation method*

1 INTRODUCTION

Oceanographic buoy[1] is a modern observing equipment that can automatically collect information of ocean environment at any time and accomplish acquisition data, process data, transmit data. Nowadays the oceanographic buoys are deployed in the sea area of China , the diameters of the buoys are up to 10 meters, and the mast heights are up to15 meters[2].

In order to make full use of the ocean buoys, the relevant departments put forward a kind of idea: the acoustic array is installed on the Ocean buoys to monitor ships in the adjacent sea area. To verify the feasibility of this idea, We design a set of acoustic feature acquisition system. This project is a great challenge for us, there are many difficulties to be solved, of which there are two main difficulties: One is how to design an adaptive acoustic array on the ocean buoy; the other is how to effectively reduce the influence of oceanographic buoy swing on acoustic array.

This paper is organized as follows: In Section II, the system is described; the solutions of two difficulties are discussed in Section III, as well as the results of lake trial and analysis of the trial results are provided in Section IV and the future work is given in Section V.

2 SYSTEM DESCRIPTION

The acoustic feature acquisition system consists of the acoustic array, control and signal processing module and mechanical structure and so on. The block diagram of the acoustic feature acquisition system is show in Fig.1.

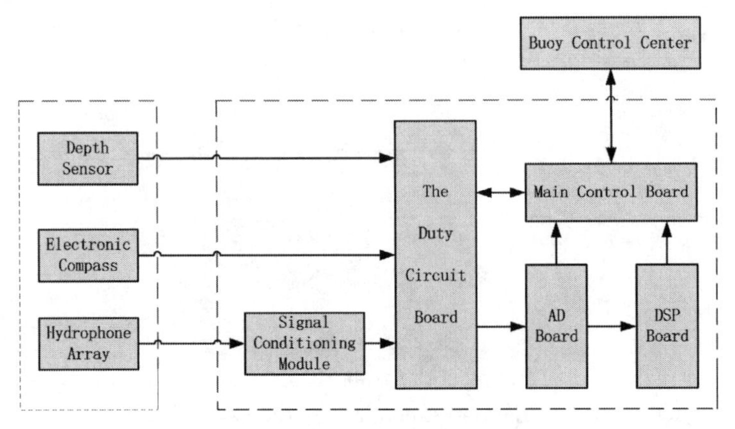

Fig.1The block diagram of the acoustic feature acquisition system

The acoustic array consists of 16 hydrophones, an electronic compass and a depth sensor and so on. The acoustic array is a horizontal circle array with 16 hydrophones uniformly distributed on the circle. The hydrophones are responsible for collecting shipping noise; The electronic compass can provide acoustic array's attitude and direc tion in real time; depth sensor can provide real-time acoustic array's depth.

There are mainly four parts in this control and signal processing module: main control board, A/D board, DSP board and the duty circuit board. The main control board is the core of the control and signal processing module, it is responsible for control work of the whole system; Analog signal is amplified and filtered and converted digital signal though the A/D aboard; The DSP board is responsible for the real-time signal processing; The duty circuit board is responsible for power management of the whole system, and receive real-time attitude sensor's and depth sensor's data.

The above two parts are installed on the dedicated mechanical structure, mechanical structure have three parts: array frame, junction box and electronic cabin.

3 SOLUTION OF DIFFICULTY

A. Structure Design of The Acoustic Array

Due to the acoustic array be installed on the ocean buoy, we design a circular acoustic array[3]. The diameter of the acoustic array is 9 meters, it is a horizontal circle array with 16 hydrophones uniformly distributed on the circle. Array frame design is shown in fig.2. Eight frameworks and eight connectors are sequentially connected to form an array frame which is rigid octagona frame. Two hydrophone units are installed on the every framework. Hydrophone unit have three parts: a hydrophone damping mount and dome. Electronic compass and depth sensor are installed on the connectors. The design of the array frame is convenient for manufacturing and assembly , and can increase the intensity of the acoustic array.

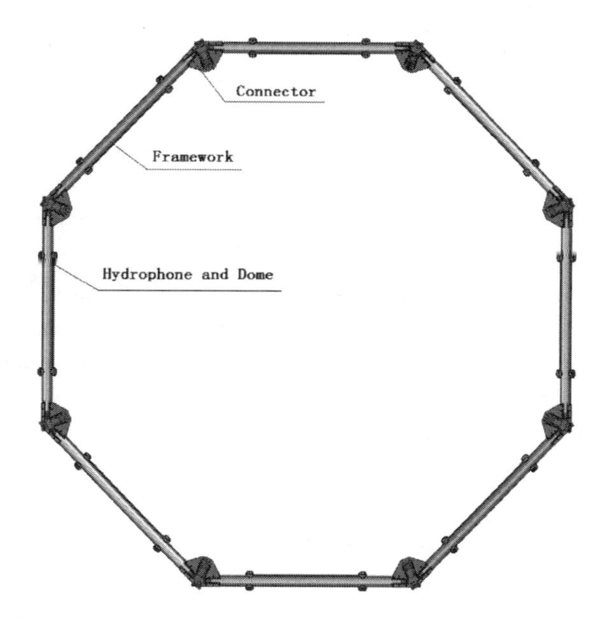

Fig.2 The schematic diagram of the array frame

B. Vibration Control Method Of The Acoustic Array

The bottom of the ocean buoy has four hanging points. They are uniformly distributed on the circle whose diameter is 0.6 meters. Every connector of the array frame has one hanging point. The four guide mechanism are used to connect ropes, the guide mechanism is shown in fig.3.The array frame connect flexibly with the ocean buoy through the rope. The design of the array frame can reduce effectively vibration of the buoy to effect on the acoustic array.

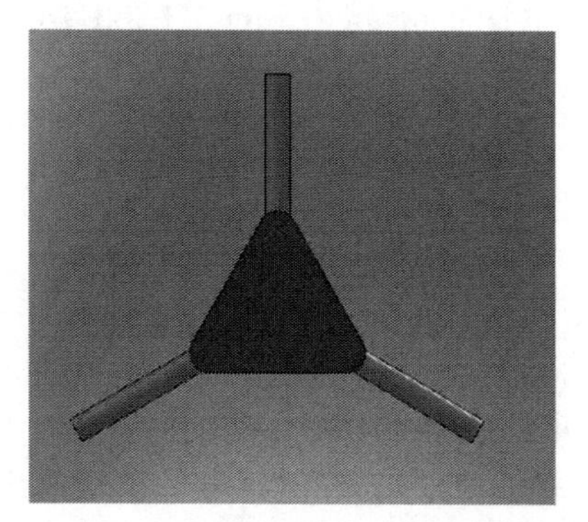

Fig.3 The schematic diagram of the guide mechanism

Taking a guide mechanism for example, we illuminate array frame and ocean buoy of connection mode. A rope connect from the ocean buoy to the guide mechanism, one rope is transformed into two ropes on the guide mechanism. Two ropes connect to the hanging point of the array frame. This connection mode can reduce the winding and increase the stability of array frame.

The hydrophone is installed on the framework though the damping mount, its outside is dome. This design can reduce the vibration of the acoustic array to effect on the hydrophone, and reduce the flow noise.

C. Azimuth Compensation Method Of The Acoustic Array

Under normal circumstances, the acoustic array is horizontal. Due to currents and the buoy's swing, the acoustic array will roll and pitch, produce dip relative to a horizontal plane . The delay time of beamforming is calculated according to the acoustic array in a horizontal plane . If the inclination angle between the horizontal plane and acoustic array increase, target azimuth estimation of beamforming will deviate. For the normal work of the acoustic array, We need to utilize the method of azimuth compensation for the acoustic array[4].

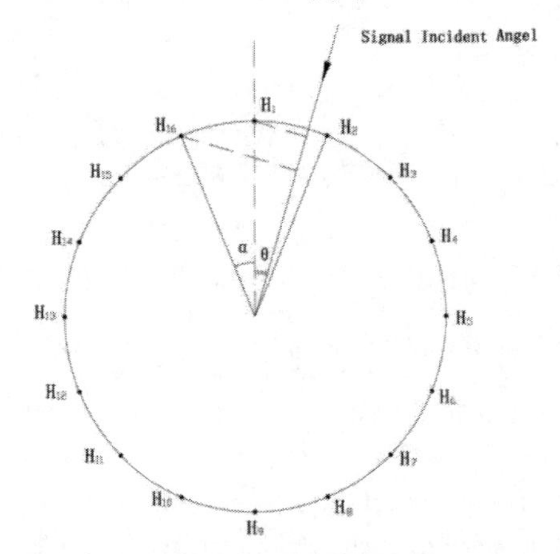

Fig.4 The schematic diagram of circular array in the horizontal plane

Figure 5 shows a discrete and uniformly spaced circular array in the horizontal plane. The array element is numbered from H1 to H16 in a clockwise direction, The direction of center O and the array element H1 is selected as 0 degree direction. The delay time of beamforming follows the standard equation.

$$\tau_i(\theta)=r\cos[\theta-(i-1)\alpha]/c \quad (1)$$

Where $\tau_i(\theta)$ is delay time Hi relative to the point O, α is two adjacent array elements included angle, $\alpha = \pi/8$, r is the acoustic array's radius, θ is signal incident angel, C is the speed of sound in water.

If the acoustic array and the horizontal plane have angle, r in the equation (1) is not a constant, it has different values corresponding to different array element. The purpose of azimuth compensation method is according to the change of acoustic array orientation, determine the value of r, make the delay time $\tau_i(\theta)$ accurate, ensure the target bearing correct.

The acoustic array azimuth compensation method as follow:

1) An electronic compass is installed on the acoustic array, electronic compass can provide real-time change angle between the acoustic array and the horizontal plane;

2) The coordinates of acoustic array's 16 element and the center O should be determined in the horizontal plane;

3) According to the electronic compass providing angle, we make the acoustic array's 16 elements and the center O project to the horizontal plane, and calculate them the coordinate and ri value.

Taking an array element as an example, we introduce the calculating method of ri value. As shown in fig.6, the coordinate of an array's element is (xi, yi), the projection coordinate of the horizontal plane is (xi ', yi'). The center coordinate is (x0, Y0), projection coordinate is (x0 ', Y0'). β is x axis and x' axis include angle , γ is y axis and y'axis include angle .

$$x_i' = x_i \cos\beta, \quad y_i' = y_i \cos\gamma 。 \tag{2}$$

$$x_0' = x_0 \cos\beta, \quad y_0' = y_0 \cos\gamma 。 \tag{3}$$

$$r_i = [(x_i' - x_0')^2 + (y_i' - y_0')^2]^{1/2} 。 \tag{4}$$

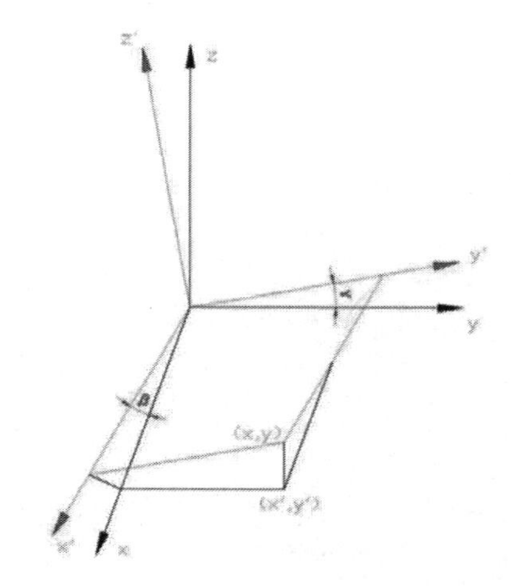

Fig.5 The schematic diagramof coordinate transformation

The ri input equation (1) to calculate τ i((o, then perform beam forming, give the target bearing.

4 LAKE TRIAL RESULTS

In 2014 June, the lake trial was conducted in Qiandao Lake. Because the acoustic array is too big, it is not suitable for lake trial. Thus, we designed the lessened ratio of acoustic array with diameter of 2 meters. The aim of the lake trial certificate the performance of the overall hardware platform, and the signal processing algorithm is efficient and reliable. The most important task would verify the azimuth compensation method of acoustic array is effective.

The experiment method is shown in fig.6. The depth of the acoustic array immersed into was about 15 meters, a transmitting transducer transmits signal (simulation of shipping noise)at the distance of 30 meters , the acoustic array have two ropes that is used to adjust the acoustic array posture.

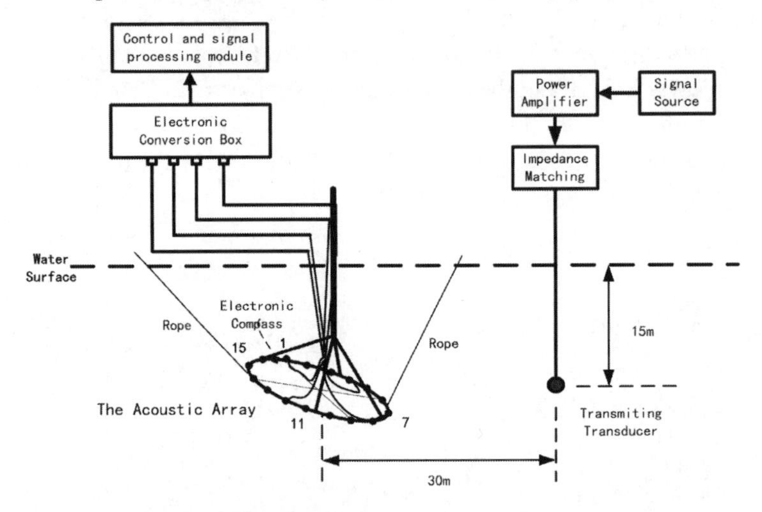

Fig.6The configuration of lake trial

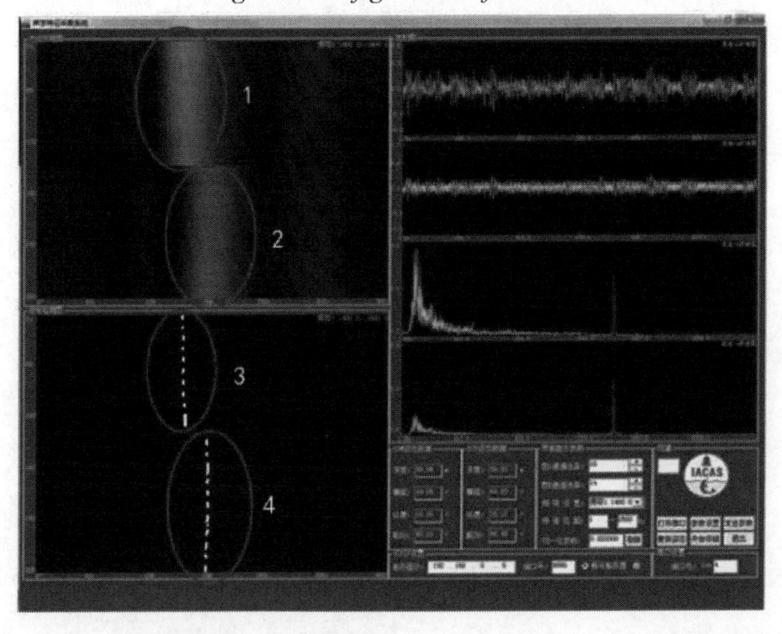

Fig.7The acoustic array roll 44 degrees and pitch 16 degrees, the comparison of results between uncomposated and composated

In fig.7 1 is uncompensated bearing time , 2 is compensated bearing time, 3 is uncompensated bearing discrimination results, 4 is compensated bearing discrimination results. The compensated bearing discrimination results and the true target range is basically the same.

5 CONCLUTION AND FURTURE WORK

Lake trial results show that the azimuth compensation method of the acoustic array is effective, hardware and software of the system work properly, and achieve the desired requirements. Future work would include a sea trial, verify the feasibility of practical application of the system.

REFERENCES

[1] **Honglei, Dai**, Development Status and Trend of Ocean Buoy in China, Meteorological, Hydrological and Marine Instruments, no. 2, pp. 118–125,Jun 2014.
[2] **Chongjiao, Jiang**, Research of Ocean Buoys Monitoring System in China, Ocean Development and Management, no. 11, pp. 13–18, 2013.
[3] **Yusheng, Wang**, Discrete Baffle Thinned Cylinder Array, vol. 20, no. 1, pp. 49–59,Jan 1996.
[4] **Qihu, Li**, Sonar signal Processing, Ocean Publisher, pp. 172-177, Jun 1996.

二
自适应信号处理

自适应波束成形中最优解在时域和频域中的等价关系①

李 启 虎
(中国科学院声学研究所)

在 Wiener 的最小均方差准则下，给出频域上最佳滤波器的解. 证明了，当把 Widrow 的抽头延迟线看作横向滤波器时，可以用任意精确度逼近频域上的最优解. 在某种条件下，时域上最优解的加权系数就是频域上最优传输函数的 Fourier 展开式的系数. 给出这种等价关系的一般表达式和两种不同的证明. 最后给出把这些结果用于实际声呐设计的若干例子.

一、引 言

自从 B. Widrow 于 1967 年提出[1]自适应波束成形的实时运算法以来，很多作者进行了各方面的研究[2~15]. 这些研究大致上可以分为两类: 一类是按 Wiener 最小均方差准则，根据 Widrow 提出的算法，在时间域上求解最优加权系数，并寻求实时的叠代方法，以实际达到最优解[2~4]; 另一类是按最大信噪比准则，在频域上求解最优滤波器的传输函数. Edelblute 等指出[12~14]，对于单频信号来说，最大信噪比准则下的解是与最小均方差准则下的解一样的. 但是对于一般信号，频域上的解与时域上的解的关系仍旧是未知的.

本文根据平稳过程的线性过滤理论，用和 Burg[16] 稍为不同的方法，给出频域上的最优解. 我们证明了，如果把 Widrow 的抽头延迟线模型看作横向滤波器，只要抽头延迟线足够长，它就可以用任意的精确度逼近频域上的最优解. 换句话说，频域上的最优解是时域上最优解的极限形式.

我们还指出，在不太苛刻的条件下，时域上最优解的加权系数就是频域上最优传输函数的 Fourier 展开式的系数. 我们给出时域–频域的等价关系，从而为实现自适应波束成形提供了一种新的途径. 本文给出这种等价关系的直接的和间接的两种证明.

当把频域上的解用于研究自适应波束成形的稳态特性时，将会带来很大好处，在时域上某些难以解决的问题，利用本文建立的关系式可以比较容易地得到解决.

最后我们将给出把本文的结果用于实际设计的若干例子.

二、最优传输函数

考虑如图 1 所示的信号处理模型. 设输入 $x_1(t)$, $\cdots$, $x_K(t)$ 是宽平稳的随机过程，均值为零.

① 声学学报, 1979, (4): 296-308.

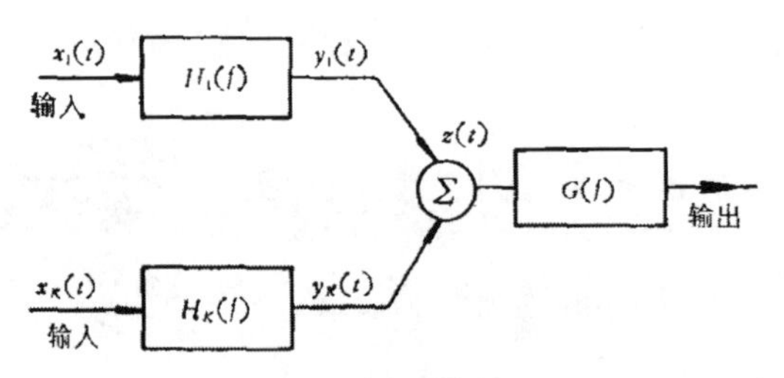

图 1　信号的滤波

我们利用矢量记号

$$\mathbf{X}^T(t) = [x_1(t), \cdots, x_K(t)] \tag{1}$$

$$\mathbf{H}^T(f) = [H_1(f), \cdots, H_K(f)] \tag{2}$$

这里字母 T 表示转置.

Edelblute 等指出[12—14], 在最大信噪比准则下, 最优传输函数 $\mathbf{H}_{\text{opt}}(f)$ 和后置滤波器 $G(f)$ 的解是

$$\begin{aligned} \mathbf{H}_{\text{opt}}(f) &= [\mathbf{K}^*_{xx}(f)]^{-1}\mathbf{A}^*(f) \\ G(f) &= \phi_d^{1/2}(f) \end{aligned} \tag{3}$$

其中 $\mathbf{K}_{xx}(f)$ 表示输入的互谱矩阵, $\mathbf{A}(f)$ 表示输入信号的指向矢量. $\phi_d(f)$ 表示信号的平均功率谱密度. 如果按 Wiener 的最小均方差准则, 我们应当求出最优传输函数矢量 $\mathbf{H}_{\text{opt}}(f)$, 使得

$$I = E\{[z(t) - d(t)]^2\} \tag{4}$$

达到极小. 其中 $d(t)$ 是期望信号, $\mathbf{H}(f)$ 应当按环境噪声场的变化实时调整. 所以我们得到了如图 2 的频域上自适应模型.

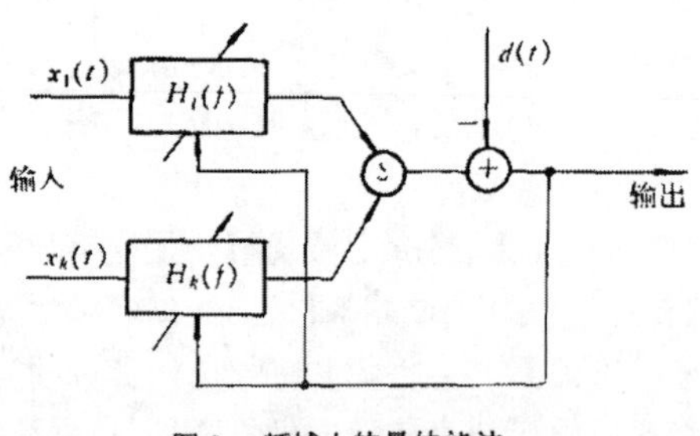

图 2　频域上的最佳滤波

在本节中,我们首先简单地回顾平稳过程的线性过滤理论,然后求出频域上的最优解.

设 $x(t)$ 是一个宽平稳的随机过程, 用 $R_{xx}(\tau)$ 及 $K_{xx}(f)$ 分别表示它的自相关函数及平均功率谱,那么

$$\begin{aligned} K_{xx}(f) &= \int_{-\infty}^{\infty} R_{xx}(\tau)\exp(-2\pi jf\tau)d\tau \\ R_{xx}(\tau) &= \int_{-\infty}^{\infty} K_{xx}(f)\exp(2\pi jf\tau)df \end{aligned} \tag{5}$$

若 $H_1(f)$, $H_2(f)$ 是两个线性系统，输入为 $x(t)$，相应的输出分别为 $y_1(t)$, $y_2(t)$. 则

$$K_{y_1y_2}(f)=H_1(f)K_{xx}(f)H_2^*(f) \tag{6}$$

其中 $K_{y_1y_2}(f)$ 表示 $y_1(t)$ 和 $y_2(t)$ 的互功率谱密度，它是互相关函数

$$R_{y_1y_2}(\tau)=E[y_1(t)y_2(t-\tau)] \tag{7}$$

的 Fourier 变换：

$$K_{y_1y_2}(f)=\int_{-\infty}^{\infty}R_{y_1y_2}(\tau)\exp(-2\pi jf\tau)d\tau \tag{8}$$

作为 (6) 式的一个特例，我们今后经常要用到

$$K_{yy}(f)=|H(f)|^2K_{xx}(f)=H(f)K_{xy}(f)=H^*(f)K_{yx}(f) \tag{9}$$

下面我们来讨论与图 2 中的输入、输出有关的量.

定义输入的互相关矩阵

$$\mathbf{R}_{xx}(\tau)=\begin{bmatrix}R_{11}(\tau)&\cdots&R_{1K}(\tau)\\ \vdots&&\vdots\\ R_{K1}(\tau)&\cdots&R_{KK}(\tau)\end{bmatrix} \tag{10}$$

这个矩阵的第 (k,l) 个元素 $R_{kl}(\tau)$ 是 $x_k(t)$ 和 $x_l(t)$ 的互相关函数：

$$R_{kl}(\tau)=E[x_k(t)x_l(t-\tau)].$$

容易证明

$$\mathbf{R}_{xx}(-\tau)=\mathbf{R}_{xx}^T(\tau) \tag{11}$$

定义 $\mathbf{K}_{xx}(f)\triangleq F[\mathbf{R}_{xx}(\tau)]$ 为 $\mathbf{X}(t)$ 的互谱矩阵，即 $\mathbf{K}_{xx}(f)$ 之第 (k,l) 个元素

$$K_{kl}(f)=F[R_{kl}(\tau)] \tag{12}$$

显然

$$\mathbf{K}_{xx}(-f)=\mathbf{K}_{xx}^*(f)=\mathbf{K}_{xx}^T(f) \tag{13}$$

所以 $\mathbf{K}_{xx}(f)$ 是 Hrmite 矩阵.

现在我们将 (4) 式中的量用频域上的 $\mathbf{H}(f)$, $\mathbf{K}_{xx}(f)$, $\mathbf{K}_{xd}(f)$ 等量表示出来：

$$K_{zz}(f)=F[R_{zz}(\tau)]=\mathbf{H}^T(f)\mathbf{K}_{xx}(f)\mathbf{H}^*(f) \tag{14}$$

又

$$E[z(t)d(t)]=R_{zd}(0)=\int_{-\infty}^{\infty}K_{zd}(f)df=\int\mathbf{H}^T(f)\mathbf{K}_{xd}(f)df \tag{15}$$

由此得到

$$\begin{aligned}I&=E\{[d(t)-z(t)]^2\}=E[d^2(t)]+E[z^2(t)]-2E[d(t)z(t)]\\&=E[d^2(t)]+\int_{-\infty}^{\infty}\mathbf{H}^T(f)\mathbf{K}_{xx}(f)\mathbf{H}^*(f)df-2\int_{-\infty}^{\infty}\mathbf{H}^T(f)\mathbf{K}_{xd}(f)df\end{aligned} \tag{16}$$

上式中，由于 $E[d^2(t)]$ 是一个常数，所以欲求 I 之极小值，仅须求下面一个泛函：

$$J=\int_{-\infty}^{\infty}\mathbf{H}^T(f)\mathbf{K}_{xx}(f)\mathbf{H}^*(f)df-2\int_{-\infty}^{\infty}\mathbf{H}^T(f)\mathbf{K}_{xd}(f)df \tag{17}$$

之极小值即可. 记住这是一个实函数，第一、二项分别都是实的. 引入记号

$$\mathrm{Re}[\mathbf{H}(f)]=\mathbf{U}(f),\qquad \mathrm{Im}[\mathbf{H}(f)]=\mathbf{V}(f)$$
$$\mathrm{Re}[\mathbf{K}_{xd}(f)]=\mathbf{P}(f),\qquad \mathrm{Im}[\mathbf{K}_{xd}(f)]=\mathbf{Q}(f)$$
$$\mathrm{Re}[\mathbf{K}_{xx}(f)]=\mathbf{A}(f),\qquad \mathrm{Im}[\mathbf{K}_{xx}(f)]=\mathbf{B}(f)$$

代入(17)式,展开之,令虚部积分为零,得到

$$J \triangleq \int_{-\infty}^{\infty} F(\mathbf{U}, \mathbf{V}, f)df$$
$$= \int [\mathbf{U}^T\mathbf{A}\mathbf{U} - \mathbf{V}^T\mathbf{B}\mathbf{U} + \mathbf{U}^T\mathbf{B}\mathbf{V} + \mathbf{V}^T\mathbf{A}\mathbf{V} - 2(\mathbf{U}^T\mathbf{P} - \mathbf{V}^T\mathbf{Q})]df$$

为求出泛函 J 的极小值,我们必须解 $F(\mathbf{U}, \mathbf{V}, f)$ 的 Euler 方程: $\nabla_U F = 0$, $\nabla_V F = 0$;即

$$\nabla_U F = \mathbf{A}\mathbf{U} + \mathbf{A}^T\mathbf{U} - \mathbf{B}^T\mathbf{V} + \mathbf{B}\mathbf{V} - 2\mathbf{P} = 0$$
$$\nabla_V F = -\mathbf{B}\mathbf{U} + \mathbf{B}^T\mathbf{U} + \mathbf{A}\mathbf{V} + \mathbf{A}^T\mathbf{V} + 2\mathbf{Q} = 0 \tag{18}$$

根据(13)式,我们知道

$$\mathbf{A}^T = \mathbf{A}, \quad \mathbf{B}^T = -\mathbf{B}$$

代入(18)便得到

$$2\mathbf{A}\mathbf{U} + 2\mathbf{B}\mathbf{V} - 2\mathbf{P} = 0 \tag{19}$$
$$2\mathbf{A}\mathbf{V} - 2\mathbf{B}\mathbf{U} + 2\mathbf{Q} = 0 \tag{20}$$

把(20)乘以 j 和(19)相加,得到

$$2\mathbf{A}\mathbf{H} - 2\mathbf{B}j\mathbf{H} - 2\mathbf{K}_{xd}^* = 0$$

即

$$\mathbf{K}_{xx}^*(f)\mathbf{H}(f) - \mathbf{K}_{xd}^*(f) = 0 \tag{21}$$

故得到

$$\mathbf{H}_{\text{opt}}(f) = [\mathbf{K}_{xx}^*(f)]^{-1}\mathbf{K}_{xd}^*(f) \tag{22}$$

这就是频域上最优滤波器的传输函数.

把(22)代入(16)就得到均方输出之极小值为

$$I_{\min} = E[d^2(t)] - \int \mathbf{H}_{\text{opt}}^T(f)\mathbf{K}_{xd}(f)df \tag{23}$$

三、时域解与频域解的等价关系

Widrow 提出的时域自适应波束成形的模型如图 3 所示. 假定每一条延迟线有 L 节抽头,利用向量记号:

$$\hat{\mathbf{X}}^T = (x_{11}, \cdots, x_{1L}, \cdots, x_{K1}, \cdots, x_{KL})$$
$$\hat{\mathbf{W}}^T = (w_{11}, \cdots, w_{1L}, \cdots w_{K1}, \cdots, w_{KL}) \tag{24}$$

我们已知

$$\hat{I} = E\left\{\left[d(t) - \sum_{k=1}^{K}\sum_{l=1}^{L} w_{kl}x_{kl}\right]^2\right\} \tag{25}$$

之极小值是

$$\hat{I}_{\min} = E[d^2(t)] - \hat{\mathbf{W}}_{\text{opt}}^T\hat{\mathbf{R}}_{xd} \tag{26}$$

其中

$$\hat{\mathbf{W}}_{\text{opt}} = \hat{\mathbf{R}}_{xx}^{-1}\hat{\mathbf{R}}_{xd} \tag{27}$$

我们看到(27)和(23)在形式上有某种类似之处. 当通道数 K 一定时,增加延迟线的节数 L,

$\hat{I}_{\min}$ 作为 KL 的函数显然单调下降. 由于它是一个正数，所以必然有极限. 本节中我们首先要证明 $\hat{I}_{\min}$ 的极限就是 $I_{\min}$. 然后再进一步建立起 $\hat{\mathbf{W}}_{\mathrm{opt}}$ 和 $\mathbf{H}_{\mathrm{opt}}(f)$ 的关系来.

为建立时域和频域的等价关系，我们把抽头延迟线看作是一个横向滤波器. （见图 4）

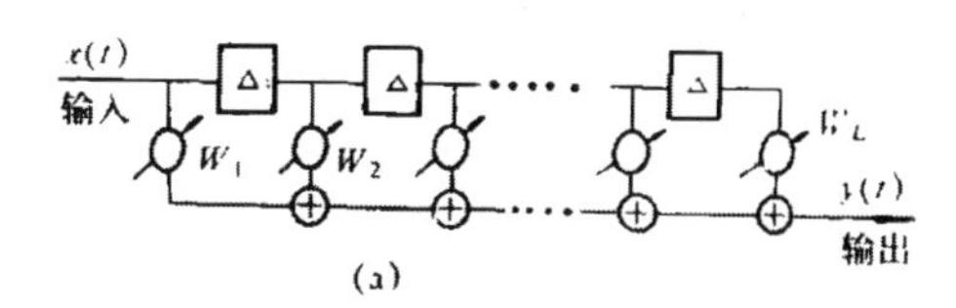

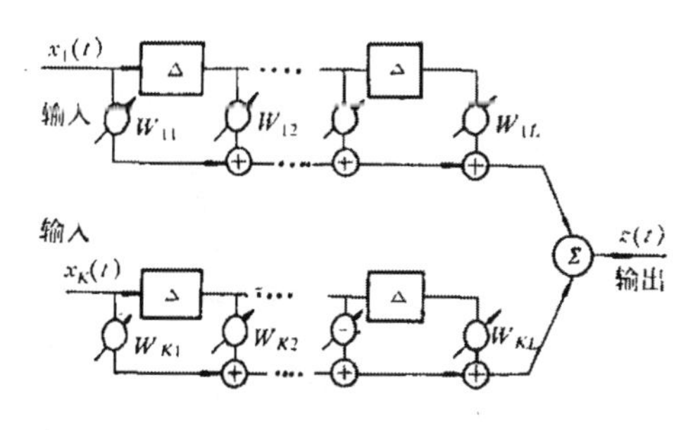

图 3 时域上最佳加权

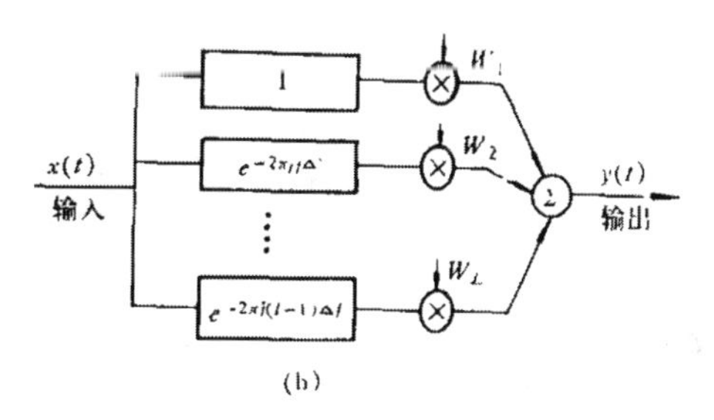

图 4 抽头延时线和横向滤波器的等价关系

在时域上

$$y(t)=\sum_{l=1}^{L} w_l x[t-(l-1)\Delta] \tag{28}$$

在频域上

$$H(f)=\sum_{l=1}^{L} w_l \exp[-2\pi jf(l-1)\Delta] \tag{29}$$

由此可知，抽头延迟线只不过是一种特殊的线性系统. 由于 $I_{\min}$ 是 I 对一切线性系统的极小值，而 $\hat{I}_{\min}$ 是在横向滤波器的约束下的极小值，所以必然有

$$\hat{I}_{\min} \geqslant I_{\min} \tag{30}$$

此式对任何 KL 成立，故

$$\lim_{KL\to\infty} \hat{I}_{\min} \geqslant I_{\min} \tag{31}$$

根据 (29) 的这种表达式，$\hat{\mathbf{H}}(f)$ 的第 k 个分量是

$$\hat{H}_k(f)=\sum_{l=1}^{L} w_{kl} \exp[-2\pi jf(l-1)\Delta] \tag{32}$$

设 $\hat{\mathbf{W}}_{\mathrm{opt}}$ 的分量为 μ_{kl}，则由 (26) 得到

$$\begin{aligned}\hat{I}_{\min} &= E[d^2(t)]-\sum_{k=1}^{K}\sum_{l=1}^{L}\mu_{kl}E[x_k(t-(l-1)\Delta)d(t)]\\ &= E[d^2(t)]-\int_{-\infty}^{\infty}\sum_{k=1}^{K}\sum_{l=1}^{L}\mu_{kl}K_{x_kd}(f)e^{-2\pi jf(l-1)\Delta}df\\ &= E[d^2(t)]-\int_{-\infty}^{\infty}\hat{\mathbf{H}}_{\mathrm{opt}}^{T}(f)\mathbf{K}_{xd}(f)df \end{aligned} \tag{33}$$

对照 (33) 及 (23) 我们看到，在限定线性滤波器为横向滤波器的条件下，均方差的极小值在形

式上和线性滤波器的极小值完全相似. 一般来讲

$$\hat{\mathbf{H}}_{\mathrm{opt}}(f) \neq \mathbf{H}_{\mathrm{opt}}(f)$$

我们发现, Widrow 的解 (27) 建立了权系数 $\hat{\mathbf{W}}_{\mathrm{opt}}$ 和输入互相关矩阵的明确关系, 但是 $\hat{\mathbf{W}}_{\mathrm{opt}}$ 在频域上和输入互谱矩阵的关系则是不明确的.

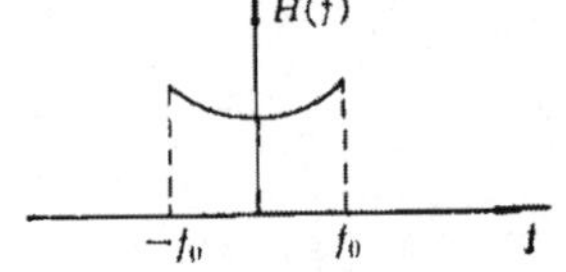

如果 $\mathbf{H}_{\mathrm{opt}}(f)$ 是带宽有限的, 那么我们可以将其展成 Fourier 级数. 设 $\mathbf{H}_{\mathrm{opt}}(f)$ 的某个分量为 $H(f)$, 当 $|f| \geqslant f_0$ 时 $H(f)=0$. 将 $H(f)$ 以 $2f_0$ 为周期延拓为 $G(f)$, (见图 5), 那么

$$G(f)=\sum_{l=-\infty}^{\infty} w_l \exp(-2\pi j f l \Delta), \quad \Delta=\frac{1}{2f_0} \tag{34}$$

$$w_l=\frac{1}{2f_0}\int_{-f_0}^{f_0} G(f)\exp(2\pi j f l \Delta)df \tag{35}$$

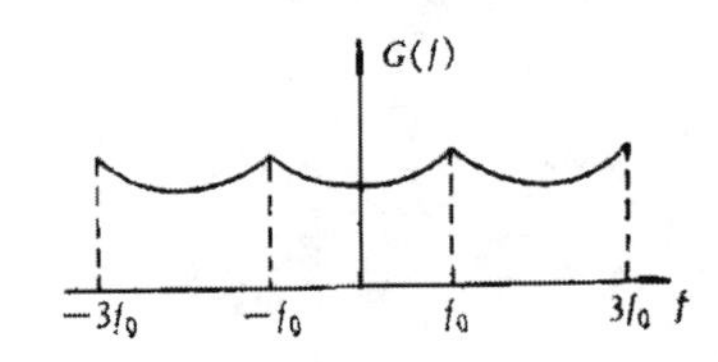

图 5 $H(f)$ 的周期延拓

从三角级数的理论我们知道, $G(f)$ 展开式的前 n 项和:

$$G_n(f)=\sum_{l=-n}^{n} w_l \exp(-2\pi j f l \Delta) \tag{36}$$

均方收敛于 $G(f)$, 并且 $G_n(f)$ 是所有项数为 $2n+1$ 的三角级数中与 $G(f)$ 均方误差最小的一个.

现在, 我们将 Δ 理解为抽头间隔, 那么 $\mathbf{G}_n(f)$ 就对应于某一横向滤波器. 对于它来说, 均方输出

$$I_n=E[d^2(t)]+\int_{-\infty}^{\infty}\mathbf{G}_n^T(f)\mathbf{K}_{xx}(f)\mathbf{G}_n^*(f)df-2\int_{-\infty}^{\infty}\mathbf{G}_n^T(f)\mathbf{K}_{xd}(f)df \tag{37}$$

显然, 当延迟线抽头一样时

$$\hat{I}_{\min} \leqslant I_n \tag{38}$$

如果我们能证明

$$I_n \to I_{\min} \quad (\text{当 } n \to \infty) \tag{39}$$

则也就证明了

$$\hat{I}_{\min} \to I_{\min} \quad (\text{当 } n \to \infty) \tag{40}$$

换句话说 (30) 中的不等号可以改为等号.

为此, 我们要分别在时域和频域中引进两个重要的公式.

在时域上, 利用 (26) 式, 对任何权矢量 $\hat{\mathbf{W}}$

$$\begin{aligned}
\hat{I} &= E[d^2(t)]-2\hat{\mathbf{R}}_{xd}^T\hat{\mathbf{W}}+\hat{\mathbf{W}}^T\hat{\mathbf{R}}_{xx}\hat{\mathbf{W}} \\
&= \hat{I}_{\min}+\hat{\mathbf{R}}_{xd}^T\hat{\mathbf{W}}_{\mathrm{opt}}-2\hat{\mathbf{R}}_{xd}^T\hat{\mathbf{W}}+\hat{\mathbf{W}}^T\hat{\mathbf{R}}_{xx}\hat{\mathbf{W}} \\
&= \hat{I}_{\min}+\hat{\mathbf{R}}_{xd}^T\hat{\mathbf{W}}_{\mathrm{opt}}-2\hat{\mathbf{R}}_{xd}^T\hat{\mathbf{W}}+(\hat{\mathbf{W}}-\hat{\mathbf{W}}_{\mathrm{opt}})^T\hat{\mathbf{R}}_{xx}(\hat{\mathbf{W}}-\hat{\mathbf{W}}_{\mathrm{opt}}) \\
&\quad +\hat{\mathbf{W}}_{\mathrm{opt}}^T\hat{\mathbf{R}}_{xx}\hat{\mathbf{W}}_{\mathrm{opt}}-\hat{\mathbf{W}}_{\mathrm{opt}}\hat{\mathbf{R}}_{xx}\hat{\mathbf{W}}_{\mathrm{opt}} \\
&= I_{\min}+(\hat{\mathbf{W}}-\hat{\mathbf{W}}_{\mathrm{opt}})^T\hat{\mathbf{R}}_{xx}(\hat{\mathbf{W}}-\hat{\mathbf{W}}_{\mathrm{opt}})
\end{aligned} \tag{41}$$

在频域上, 利用 (16) 式, 对任何传输函数 $\mathbf{H}(f)$,

$$
\begin{aligned}
I &= E\left[d^2(t)\right] + \int_{-\infty}^{\infty} \mathbf{H}^T(f)\mathbf{K}_{xx}(f)\mathbf{H}^*(f)df - 2\int_{-\infty}^{\infty} \mathbf{H}^T(f)\mathbf{K}_{xd}(f)df \\
&= I_{\min} + \int_{-\infty}^{\infty} (\mathbf{H}_{\text{opt}} - \mathbf{H})^T \mathbf{K}_{xx}(\mathbf{H}_{\text{opt}} - \mathbf{H})^* df - 2\int_{-\infty}^{\infty} \mathbf{H}^T\mathbf{K}_{xd}df \\
&\quad + \int_{-\infty}^{\infty} \mathbf{H}^T\mathbf{K}_{xx}\mathbf{H}_{\text{opt}}^* df + \int_{-\infty}^{\infty} \mathbf{H}_{\text{opt}}^T\mathbf{K}_{xx}\mathbf{H}^* df \\
&= I_{\min} + \int_{-\infty}^{\infty} (\mathbf{H}_{\text{oPt}} - \mathbf{H})^T\mathbf{K}_{xx}(\mathbf{H}_{\text{oPt}} - \mathbf{H})^* df
\end{aligned} \tag{42}
$$

在证明过程中，我们利用了 $\int_{-\infty}^{\infty} \mathbf{H}^T(f)\mathbf{K}_{xd}(f)df$ 是一实数的事实。

现在，我们利用 (42) 式，令 $\mathbf{H}(f) = \mathbf{G}_n(f)$，则

$$
I_n = I_{\min} + \int_{-\infty}^{\infty} (\mathbf{H}_{\text{opt}} - \mathbf{G}_n)^T\mathbf{K}_{xx}(\mathbf{H}_{\text{opt}} - \mathbf{G}_n)^* df \tag{43}
$$

由此式，我们立即可以证明

$$
\lim_{n\to\infty} I_n = I_{\min}
$$

这是因为 $(\mathbf{H}_{\text{opt}} - \mathbf{G}_n)^T\mathbf{K}_{xx}(\mathbf{H}_{\text{opt}} - \mathbf{G}_n)^*$ 是一个有界的双线性型．因而[17]

$$
0 \leqslant I_n - I_{\min} \leqslant M\|\mathbf{H}_{\text{opt}}(f) - \mathbf{G}_n(f)\|^2 \tag{44}
$$

这里 M 是一个常数，$\|\ \|$ 表示 L_2 中的距离，由于 $\mathbf{G}_n(f)$ 均方收敛于 $\mathbf{H}_{\text{opt}}(f)$，所以 (39) 得证，故 (40) 也就得证了．

总之，我们已经得到如下的结论：

定理 1.

如果频域上的最优解 $\mathbf{H}_{\text{opt}}(f)$ 是带宽有限的，上限记为 f_0，以 $\Delta = \dfrac{1}{2f_0}$ 为间隔作 $\mathbf{K}$ 条抽头延迟线，记时域上的最优解为 $\hat{\mathbf{H}}_{\text{opt}}(f)$，则

$$
\begin{aligned}
\hat{I}_{\min} &= E\left[d^2(t)\right] - \int_{-\infty}^{\infty} \hat{\mathbf{H}}_{\text{opt}}^T(f)\mathbf{K}_{xd}(f)df \\
&\to E\left[d^2(t)\right] - \int_{-\infty}^{\infty} \mathbf{H}_{\text{opt}}^T(f)\mathbf{K}_{xd}(f)df = I_{\min} \quad (\text{当 } L \to \infty)
\end{aligned}
$$

我们要注意，一般来说当延迟线节数一定时 $\hat{\mathbf{H}}_{\text{opt}}(f) \neq \mathbf{G}_n(f)$．但是在加了某种假定之后，我们可以证明：

定理 2.

如果输入信号、干扰都具有矩形的功率谱密度，各通道间噪声相互独立．那么 $\hat{\mathbf{H}}_{\text{opt}}(f) = \mathbf{G}_n(f)$．也就是说时域上最优加权系数就是频域上最优传输函数的 Fourier 展开式的系数．

关于这个定理，我们将给出两种证明．一种是直接的，一种是间接的．为了简单起见，我们仅对一维的情况予以证明．至于多维的情况，还要作一些附加的假设，我们将在附录中给出详细的说明．

证明 1.

设 $x(t) = s(t) + r(t) + n(t)$，$s(t)$，$r(t)$，$n(t)$ 分别代表信号，干扰，噪声．其功率分别用 σ_s^2，σ_r^2，σ_n^2 表示．期望信号 $d(t) = s(t)*h(t)$，其中 $h(t)$ 的谱 $M(f)$ 带宽有限．

显然，$H_{\text{opt}}(f) = M(f)$．$\hat{\mathbf{R}}_{xx}$ 的主对角线元素是 $\sigma_s^2 + \sigma_r^2 + \sigma_n^2$，其余元素皆为零．因为当

$k \neq l$ 时

$$R_{kl} = R_{ss}[(k-l)\Delta] + R_{rr}[(k-l)\Delta]$$
$$= \int_{-\infty}^{\infty} K_{ss}(f) e^{-2\pi jf(k-l)\Delta} df + \int_{-\infty}^{\infty} K_{rr}(f) e^{-2\pi jf(k-l)\Delta} df = 0.$$

所以

$$\hat{\mathbf{R}}_{xx} = \begin{bmatrix} \sigma_s^2 + \sigma_r^2 + \sigma_n^2 & & 0 \\ & \ddots & \\ 0 & & \sigma_s^2 + \sigma_r^2 + \sigma_n^2 \end{bmatrix}$$

又

$$\hat{\mathbf{R}}_{xd}^T = [R_{xd}(0), R_{xd}(\Delta), \cdots, R_{xd}((L-1)\Delta)]$$
$$R_{xd}((l-1)\Delta) = R_{sd}((l-1)\Delta)$$
$$= \int_{-\infty}^{\infty} K_{sd}(f) e^{2\pi jf(l-1)\Delta} df = \int_{-\infty}^{\infty} K_{ss}(f) M^*(f) e^{2\pi jf(l-1)\Delta} df$$
$$= \sigma_s^2 \cdot a_{l-1} \qquad 1 \leqslant l \leqslant L.$$

其中 a_l 表示 $M(f)$ 的 Fourier 展开式系数. 把 $\hat{R}_{xx}$ 和 $\hat{R}_{xd}$ 的表达式代入 (27) 就得到

$$\hat{\mathbf{W}}_{\text{opt}} = \hat{\mathbf{R}}_{xx}^{-1}\hat{\mathbf{R}}_{xd} = \begin{bmatrix} a_0 \\ a_1 \\ \vdots \\ a_{L-1} \end{bmatrix} \tag{45}$$

证明 2.

由 (43) 式,我们有

$$I_n = I_{\min} + \int_{-\infty}^{\infty} |H_{\text{opt}}(f) - G_n(f)|^2 K_{xx}(f) df$$
$$= I_{\min} + \frac{1}{2f_0}(\sigma_s^2 + \sigma_r^2 + \sigma_n^2) \int_{-\infty}^{\infty} |H_{\text{opt}}(f) - G_n(f)|^2 df$$

由于 $G_n(f)$ 是 $H_{\text{opt}}(f)$ 在均方意义下的最佳逼近,所以当把上式中的 $G_n(f)$ 换为 $\hat{H}_{\text{opt}}(f)$ 时, I_n 必然增加, 即

$$I_n \leqslant \hat{I}_{\min} \tag{46}$$

由 (38) 式即得到

$$I_n = \hat{I}_{\min} \tag{47}$$

由三角级数展开的唯一性, 得知 $\hat{H}_{\text{opt}}(f) = G_n(f)$.

这个定理说明了, 如果 $s(t)$, $r(t)$, $n(t)$ 的谱都具有矩形形状, 当抽头延迟线增加时,已经在时域上调好的权系数是不变的,每增加一节延迟线,我们仅须再计算一个权系数就行.

四、在设计中的应用

在上一节证明定理 1,2 时我们隐含了一个假设,即对期望信号的估计是可以实现的.在实际情况中,如果由于某种原因使信号发生畸变,那么我们一般要求抽头延迟线具有正、负延时,这在物理上是不能实现的. 但只要我们进行适当的预处理,这个困难就易于克服. 举例来说,

我们要调节图 6(a) 中的 $H(f)$ 使输出极小，应当有 $H_{opt}(f)=\exp(2\pi jf\Delta)$. 这是无法实现的. 但是如果我们如图 6(b) 那样将 $x(t)$事先延迟 $T_0(T_0\geqslant\Delta)$ 则 $H_{opt}(f)=\exp[-2\pi jf(T_0-\Delta)]$ 就是可以实现的了. 从理论上讲，为实现 $G_n(f)$，我们应将其事先乘以因子 $\exp(-2\pi jfT_0)$，T_0 足够大，那么

$$G_n(f)\exp(-2\pi jfT_0)=\sum_{l=-n}^{n}w_l\exp(-2\pi jf(l\Delta+T_0)) \tag{48}$$

就可以实现.

抽头延迟线的间隔 Δ 应取作带宽两倍的倒数，即 $\Delta=1/2f_0$. 若 $\Delta>1/2f_0$，这是显然不行的. 若 $\Delta<1/2f_0$，在同样抽头个数下也不见得优于 $\Delta=1/2f_0$，它与输入信号有关. 所以为了达到一般的匹配，仍以取 $\Delta=1/2f_0$ 为好.

前面已指出，要用 $\hat{H}_{opt}(f)$ 来实现 $H_{opt}(f)$，延迟线的长度必须足够. 下面我们举两个例子来说明实际情况下 $G_n(f)$ 的收敛是非常快的.

例 1.

$$H_{opt}(f)=\begin{cases}\dfrac{f_0^2}{4}-f^2 & |f|\leqslant\dfrac{f_0}{2}\\ 0 & \text{其他}\end{cases}\quad(\text{见图 7(a)})$$

我们有:

$$H_{opt}(f)=\frac{f_0^2}{6}-\sum_{n=1}^{\infty}\frac{f_0^2}{n^2\pi^2}(-1)^n\cos 2n\pi\frac{f}{f_0}\quad |f|\leqslant\frac{f_0}{2}.$$

按 Parseval 等式

$$\frac{1}{f_0}\int_{-f_0/2}^{f_0/2}|H_{opt}(f)|^2df=\frac{a_0^2}{2}+\frac{1}{2}\sum_{k=1}^{\infty}(a_k^2+b_k^2) \tag{49}$$

在本例中

$$\frac{1}{f_0}\int_{-f_0/2}^{f_0/2}|H_{opt}(f)|^2df\approx 0.03333f_0^4.$$

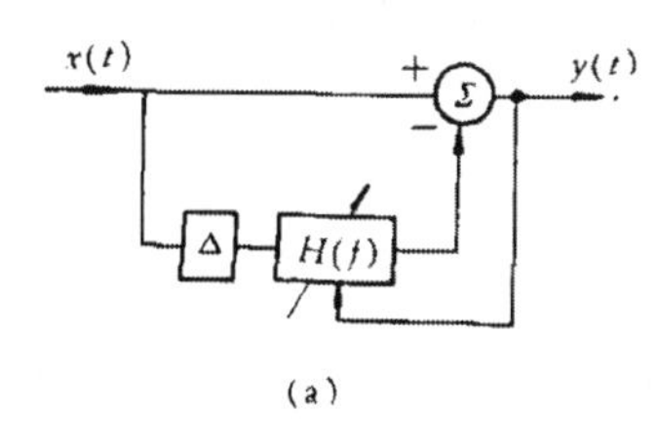

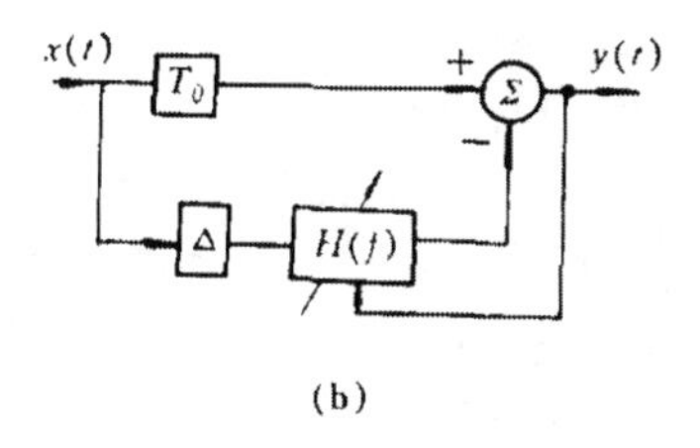

图 6 信号的预处理

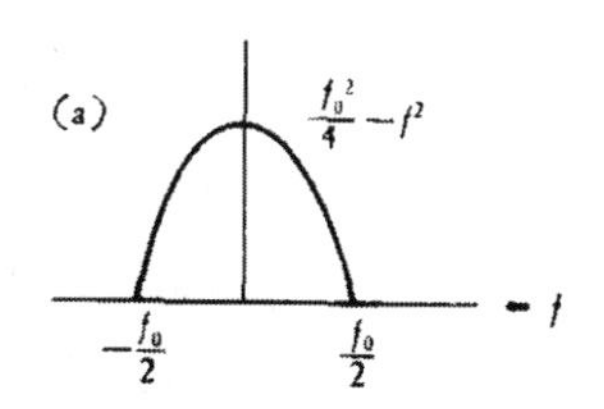

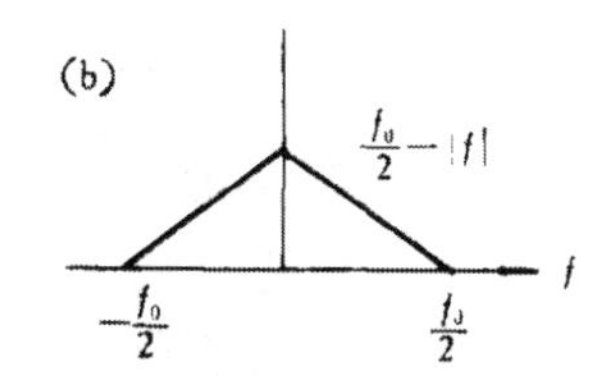

图 7 带宽有限的谱

我们来计算 (49) 右端取一项,二项,三项时的数值,以及占总功率的百分比,得到:

取一项	0.02777	83.3%
取二项	0.03292	98.7%
取三项	0.03324	99.7%

例 2.

$$H_{opt}(f)=\begin{cases}\frac{f_0}{2}-|f| & |f|\leqslant\frac{f_0^2}{2}\\ 0 & \text{其他}\end{cases}\quad(\text{见图 }7(b))$$

我们有

$$H_{opt}(f)=\frac{f_0}{4}+\sum_{n=1}^{\infty}\frac{2f_0}{(2n-1)^2\pi^2}\cos(2n-1)\frac{2\pi f}{f_0}\quad |f|\leqslant\frac{f_0}{2}.$$

$$\frac{1}{f_0}\int_{-f_0/2}^{f_0/2}|H_{opt}(f)|^2df=\frac{1}{12}f_0=0.08333f_0$$

取一项	0.06250	75%
取二项	0.08299	99.0%
取三项	0.08324	99.8%

我们看到，这两种类型的谱展成 Fourier 级数时收敛都非常快，实际上仅须 5 节延迟线就行．所以在实际声呐中,延迟线的长度仅需比空间校正滤波器的等效时间长度稍长就行了．

最后我们举一个例子来说明频域最优解在噪声抵消法方面的应用．考虑如图 8 的系统．

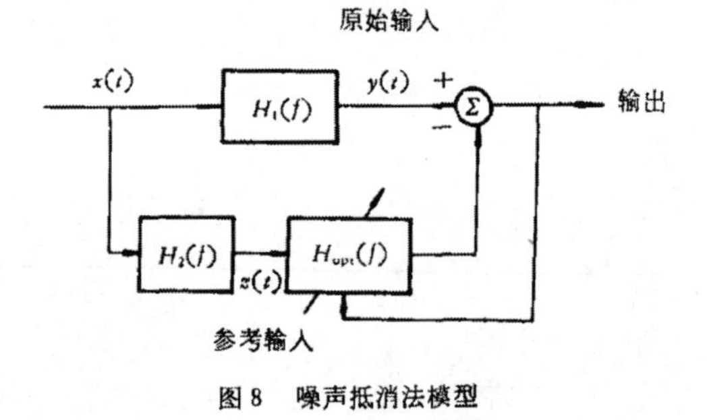

图 8　噪声抵消法模型

根据 (8), (9) 式

$$K_{zz}(f)=|H_2(f)|^2K_{xx}(f)$$
$$K_{zy}(f)=H_2(f)K_{xx}(f)H_1^*(f)$$

代入 (22) 得到

$$H_{opt}(f)=[K_{zz}^*(f)]^{-1}K_{zy}^*(f)=H_1(f)/H_2(f)\tag{50}$$

由此我们可以看到,如果 $H_1(f)$ 是带宽有限的,而 $H_2(f)$ 在它的通带内无零点，那么 $H_{opt}(f)$ 也就是带宽有限的．如前所述，可以由延迟线来实现．我们曾以实际的系统进行了模拟实验．在一个专用的波束成形计算机上进行了模拟,实验结果见图 9．一般情况下，理论与实际的附合情况良好(图 9(a))．但当 $H_2(f)$ 有零点时,延迟线的长度就显得不足,所以不能用少数几节延迟线求出最优解来(图 9(b))．如果 $H_2(f)$ 的零点恰好是 $H_1(f)$ 的零点，那么我们仍可得到 $H_{opt}(f)$ (图 9(c))．

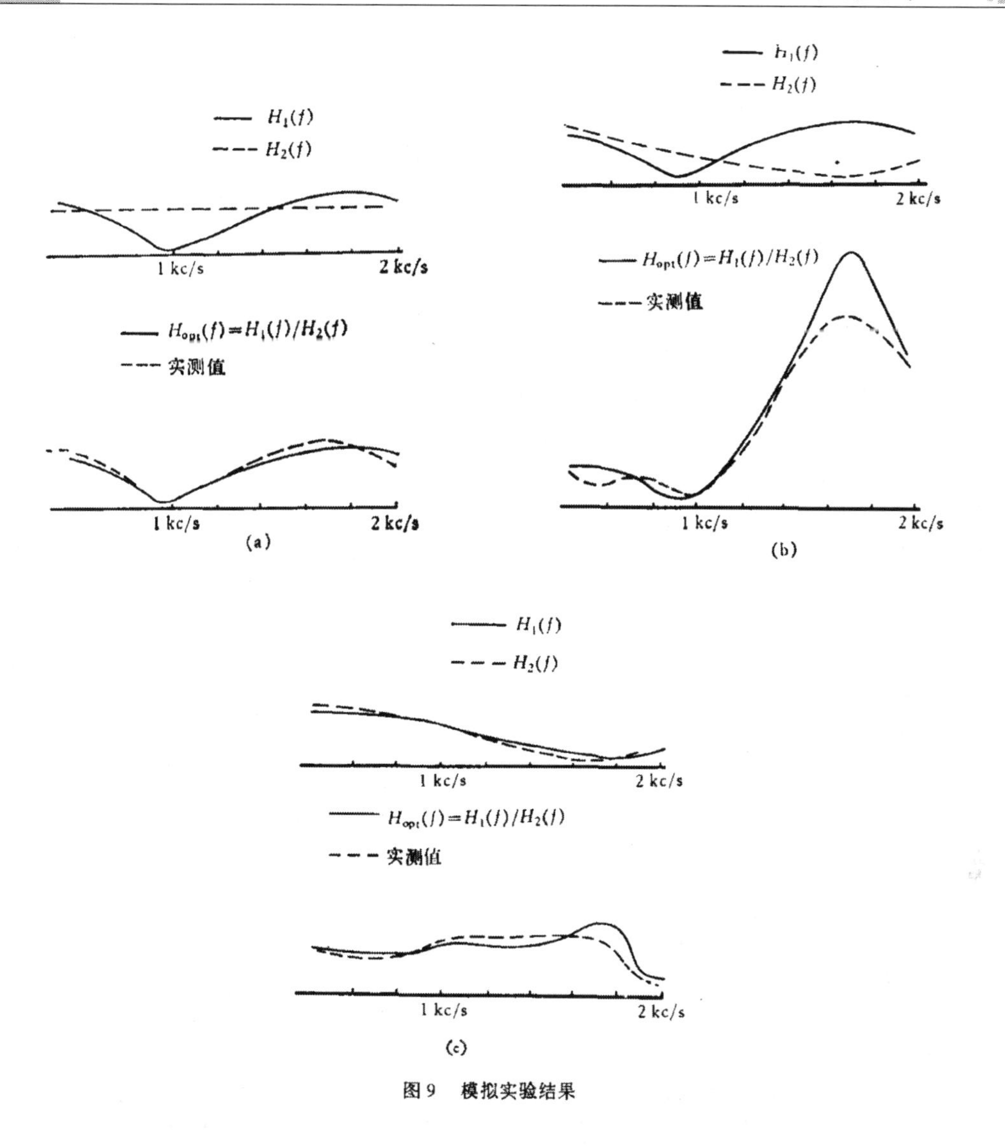

图 9 模拟实验结果

五、结　论

在 Wiener 的最小均方差准则下，时域的最优解均方逼近频域最优解. 并以后者作为下确界. 在某种条件下,时域解的加权系数就是最优滤波器的 Fourier 展开系数.

时域-频域的等价关系为自适应波束成形设计提供了一个工具.

本文在写作过程中得到侯自强同志的指导与帮助，关于横向滤波器延迟时间间隔的设计首先是由他在实验中发现的，文末的噪声抵消法的结果也是由他提供的. 作者和孙允恭同志进行过多次有益的讨论,从他那里得到许多帮助. 此外,在与本组其他一些同志的讨论中也得到不少启发,在此一并表示衷心的感谢.

附　　录

多维情况下，定理 2 的证明.

设 $\mathbf{X}(t)$ 由信号 $s(t)$，一个干扰 $r(t)$ 和各向同性噪声构成，$s(t)$ 之定向矢量

$$\mathbf{A}^T(f) = [e^{2\pi j f \tau_1}, \cdots, e^{2\pi j f \tau_K}],$$

$r(t)$ 之定向矢量为

$$\mathbf{B}^T(f) = [e^{2\pi j f \rho_1}, \cdots, e^{2\pi j f \rho_K}].$$

$K_{ss}(f)$ 和 $K_{rr}(f)$ 的形状都是矩形的，则

$$\begin{aligned}\mathbf{K}_{xx}(f) &= \mathbf{K}_{ss}(f) + \mathbf{K}_{rr}(f) + \mathbf{K}_{nn}(f) \\ &= K_{ss}(f)\mathbf{A}(f)\mathbf{A}^{*T}(f) + K_{rr}(f)\mathbf{B}(f)\mathbf{B}^{*T}(f) + K_{nn}(f)\mathbf{U}\end{aligned}$$

$\mathbf{U}$ 为单位矩阵. 于是:

$$\begin{aligned}I_n &= I_{\min} + \int_{-\infty}^{\infty} (\mathbf{H}_{\text{opt}} - \mathbf{G}_n)^T \mathbf{K}_{xx}(f)(\mathbf{H}_{\text{opt}} - \mathbf{G}_n)^* df \\ &= I_{\min} + \frac{\sigma_s^2}{2f_0}\int_{-\infty}^{\infty} |\mathbf{A}^T(f)(\mathbf{H}_{\text{opt}} - \mathbf{G}_n)|^2 df + \frac{\sigma_r^2}{2f_0}\int_{-\infty}^{\infty} |\mathbf{B}^T(f)(\mathbf{H}_{\text{opt}} - \mathbf{G}_n)|^2 df \\ &\quad + \frac{\sigma_n^2}{2f_0}\|\mathbf{H}_{\text{opt}} - \mathbf{G}_n\|\end{aligned}$$

如果我们假定当 $k \neq l$ 时

$$\int_{-\infty}^{\infty} (H_{\text{opt}}^{(k)} - G_n^{(k)})(H_{\text{opt}}^{(l)} - G_n^{(l)})^* \exp(2\pi j f(\tau_k - \tau_l)) df = 0$$

$$\int_{-\infty}^{\infty} (H_{\text{opt}}^{(k)} - G_n^{(k)})(H_{\text{opt}}^{(l)} - G_n^{(l)})^* \exp(2\pi j f(\rho_k - \rho_l)) df = 0.$$

则上式成为

$$I_n = I_{\min} + \frac{1}{2f_0}(\sigma_s^2 + \sigma_r^2 + \sigma_n^2)\|\mathbf{H}_{\text{opt}} - \mathbf{G}_n\|$$

定理 2 于是得证.

参　考　文　献

[1] B. Widrow, et al., "Adaptive Antenna Systems Proc." *IEEE*, **55** (1967), No. 12, 2143—2159.

[2] L. J. Griffiths., "A Simple Adaptive Algorithm for Real-Time Processing in Antenna Arrays", *Proc. IEEE*, **57** (1969), No. 10, 1696—1704.

[3] Chang, J. H. and Tuteur, F. B., "Optimum Adaptive Array Processor", *Proc. of the Symp. on Computer Processing in Communication, Microwave Res. Inst. Symposis Series* **19** (1969), 695—710.

[4] Widrow, B. and McCool, J. M., "A Comparison of Adaptive Algorithms Based on the Methods of Steepest Descent and Random Search", *IEEE Trans.* AP-24. (1976), No. 5, 615—637.

[5] Cox, H., "Sensitivity Consideration in Adaptive Beamforming", *Proc. NATO Advanced Study Institute on Signal* Processing (1972).

[6] Frost, O. L., III, "An Algorithm for Linearly Constrained Adaptive Array Processing", *Proc. IEEE*, **60** (1972), No. 8, 926—935.

[7] Widrow, B. et al., "Adaptive Noise Cancelling:g: Principle and Applications", *Proc. IEEE*, **63** (1975), No. 12, 1692—1716.

[8] Widrow, B. et al., "Stationary and Nonstationary Learning Characterics of the LMS Adaptive Filter", *Proc. IEEE*, **64** (1976), No. 8, 1151—1161.

[9] Keating, P. N. and T. Sawatari, ''Holographic Adaptive Processing——A Comparision with LMS Adaptive Processing'', *Acoustical Holography*, 7 (1976), (Plenum Press, New York and London), 537—548.

[10] Griffiths, L. J., ''An Adaptive Beamformer which Implements Constrain using auxiliary Array Preprocessor'', *Proc. of the NATO Advanced Study Institute Series Reidel Publishing Comp.*, (1977). 517—524.

[11] Applebaum, S. P., and Chapman, D. J., ''Adaptive Arrays with Main Beam Constrain'', *IEEE. Trans*, AP-24 (1976), No. 5, 650—661.

[12] Edelblute D. J. et al., ''Criteria for Optimum-Signal-Detection Theory for Array'', *J. A. S. A.* **41** (1967), No. 1, 199—205.

[13] Cox, H. ''Resolving Power and Sensitivity to Mismatch of Optimum Array Processor'' *J. A. S. A.* **54** (1973), No. 3, 771—785.

[14] Cox, H., ''Optimum Arrays and the Schwarz Inequality'', *J. A. S. A.* **45** (1969), No. 1, 228—232.

[15] Dentino, M. et al., ''Adaptive Filtering in the Frequeney Domain'', *Proc. IEEE*, **66** (1978), No. 12, 1658—1659.

[16] Burg, J. D. ''Three Dimensional Filtering with An Array of Seismometers'', *Geophysics*, **29** (1964), No. 5, 693.

[17] Hardy, G. H. et al., ''*Inequalities*'', CH 8 (Cambridge University Press 2nd Edition, 1952).

THE EQUIVALENT RELATION OF OPTIMUM SOLUTION BETWEEN TIME-DOMAIN AND FREQUENCY-DOMAIN IN ADAPTIVE BEAMFORMING

Li Qi-hu

(*Institute of Acoustics, Academia Sinica*)

Under the criteria of Wiener's least mean square error, the optimum solution of frequency-domain is given. If we consider the Widrow's tapped daley line as a transversal filter, it can approximate the optimum transfer function with required accurancy.

Under certain conditions, the optimum weight cofficients of time-domain just equal the coefficients of Fourier expansion of optimum transfer function.

two different proofs for those are given.

In the end, some practical examples applying the results to be described in this paper are shown.

多波束自适应噪声抵消法引论[①]

李启虎

（中国科学院声学研究所）

本文给出按 Wiener 最小均方差准则设计的两种噪声抵消系统模型．给出计算输出信噪比的一般表达式．提出了把这种模型用于自适应波束成形时的实际检验准则．

提出一种多波束自适应噪声抵消系统．分析了它的主要性能．

本文的全部讨论都在频域上进行，文中给出利用信号平均功率谱密度和最佳线性滤波器的传输函数计算系统指向性的方法．给出了多波束噪声抵消法在自适应前后的指向性公式．

实际的例子说明这种系统在抑制为数不多的干扰时具有很大的优越性．

一、引　言

目前关于声纳中自适应波束成形的研究大部分限于单波束系统．从理论上讲，Widrow 等提出的[1-3]实时算法似乎可以用于多波束系统．但是在实际上存在着一系列的困难．主要来自两个方面：一个是加权系数的最优解 $\boldsymbol{W}_{\text{opt}}=\boldsymbol{R}_{XX}^{-1}\boldsymbol{P}$ 依赖于引导信号或引导矢量，对于不同的波束指向，$\boldsymbol{W}_{\text{opt}}$ 就不一样．由于 $\boldsymbol{W}_{\text{opt}}$ 数量极大，要实时地求解逆矩阵 $\mathbf{R}_{XX}^{-1}$ 以获得多波束的 $\boldsymbol{W}_{\text{opt}}$ 是非常困难的；另一个是 Widrow 算法是一种不带约束的自适应系统，这种单波束系统的主瓣往往是有偏的．我们的实践已证实了这一点．国外也有类似的报道[4]．所以直接应用这种方法于多波束系统还有不少问题．

关于多波束的自适应系统，有些研究者已提出了一些设想[5-7]，其中包括时域的、频域的和波束域的．它们的共同问题是系统的设备和波束个数成正比，目前难于实时完成．

我们在[8]中曾建立了时域和频域的等价的关系，这使得矩阵求逆问题大为简化，但是当传输函数出现极点时，对 FFT 的要求就非常之高．

本文根据 Widrow[9] 提出的噪声抵消法，将其应用于多波束自适应系统的研究．首先分析这种系统的输出特性，给出计算输出信噪比的一般表达式．

我们提出关于干扰噪声相抵的概念，指出只有在参考输入和原始输入中存在相抵的噪声时，噪声抵消法才是有效的．我们还给出检验相抵特性的一般公式．然后分析多波束自适应噪声抵消法的主要性质．给出自适应前后的指向特性．

文末给出实际的例子．

二、噪声抵消法模型

我们研究图 1 中所给出的噪声抵消法．$n_0(t)$ 为参考输入，$S(t)+n(t)$ 为原始输入，

① 声学学报, 1980, (3): 221-230.

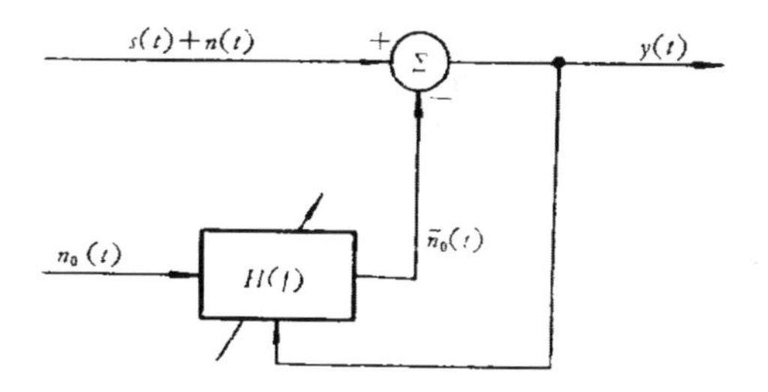

图1 噪声抵消法模型

按 Wiener 准则，调节 $H(f)$ 使输出均方值 $E[y^2(t)]$ 极小.

显然[8]

$$H_{\mathrm{opt}}(f)=\frac{K_{n_0n}^*(f)}{K_{n_0n_0}(f)} \tag{1}$$

其中 $K_{n_0n_0}(f)$ 表示 $n_0(t)$ 的平均功率谱，$K_{n_0n}(f)$ 表示 $n_0(t)$ 与 $n(t)$ 的互功率谱. 我们来计算 $H(f)$ 达到最佳时的输出信噪比.

$$L_{\mathrm{out}}=E[S(t)]^2/E[n(t)-\tilde{n}_0(t)]^2 \tag{2}$$

$G=L_{\mathrm{out}}/L$，这里 L 为输入信噪比.

容易证明

$$L_{\mathrm{out}}=\int K_{ss}(f)df\Big/\left[E[n(t)]^2+\int K_{n_0n_0}(f)|H_{\mathrm{opt}}(f)|^2df-2\int H_{\mathrm{opt}}(f)K_{n_0n}(f)df\right] \tag{3}$$

利用 (1) 代入 (3)，经过必要的简化，可以得到

$$L_{\mathrm{out}}=\int K_{ss}(f)df\Big/\left[\sigma_n^2-\int(|K_{nn_0}(f)|^2/K_{n_0n_0}(f))df\right] \tag{4}$$

为了估计 L_{out}，G，我们引用一个互谱不等式[10]，对任何两个平稳的随机过程 $x(t)$，$y(t)$，有*

$$|K_{xy}(f)|^2\leqslant K_{xx}(f)K_{yy}(f) \tag{5}$$

利用此式，我们先来估计 L_{out} 中的分母.

$$\int\frac{|K_{nn_0}(f)|^2}{K_{n_0n_0}(f)}df\leqslant\int\frac{K_{n_0n_0}(f)K_{nn}(f)}{K_{n_0n_0}(f)}df=\sigma_n^2$$

这一结果从直观上也易于理解，因为

$$0\leqslant E[n(t)-\tilde{n}_0(t)]^2=\sigma_n^2-\int\frac{|K_{nn_0}(f)|^2}{K_{n_0n_0}(f)}df \tag{6}$$

下面假定 $K_{n_0n_0}(f)$ 是矩形的(或单频信号)

$$K_{n_0n_0}(f)=\begin{cases}\sigma_{n_0}^2(1/\Delta f) & |f|\leqslant\Delta f/2\\ 0 & \text{其他}\end{cases} \tag{7}$$

则

$$L_{\mathrm{out}}=\sigma_s^2\Big/\left[\sigma_n^2-\Delta f\cdot\frac{1}{\sigma_{n_0}^2}\int|K_{n_0n}(f)|^2df\right] \tag{8}$$

利用 Parseval 等式:

---

* 原书中对于该式的证明似有错. 但这个式子本身没有错，我们可以用别的办法加以证明.

$$\int |K_{n_0n}(f)|^2 df = \int (R_{n_0n}(\tau))^2 d\tau \triangleq \sigma_{n_0}^2\sigma_n^2\hat{\tau}_{n_0n} \tag{9}$$

这里 $\hat{\tau}_{n_0n}$ 称为 $n_0(t)$ 与 $n(t)$ 的互相关半径. 显然,

$$\hat{\tau}_{n_0n}|_{\max} = 1/\Delta f.$$

我们今后将 $\hat{\tau}_{n_0n}$ 简记作 $\hat{\tau}$, 而将 $1/\Delta f$ 记作 $\tau_{\max}$, 由 (8) 式可得

$$L_{\text{out}} = L/(1-\hat{\tau}/\tau_{\max}), \quad G = 1/(1-\hat{\tau}/\tau_{\max}) \tag{10}$$

由此可知这种形式的噪声抵消系统, 其增益与原始输入的信噪比无关. 图 2 给出了 G 与 $\hat{\tau}/\tau_{\max}$ 的关系. 从图上可以看出 $\hat{\tau}/\tau_{\max} \leqslant 0.5$ 时, 增益非常低.

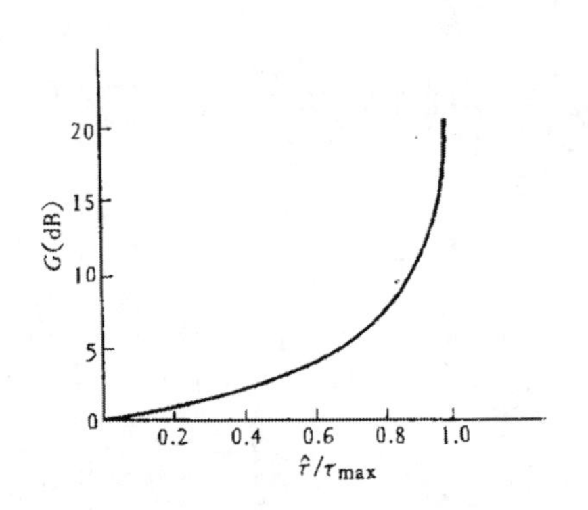

图 2 增益 G 与互相关半径 $\hat{\tau}$ 的关系

除了图 1 形式这种噪声抵消法之外, 还可以考虑另一种噪声抵消法, 那就是以 $n_0(t)$ 为原始输入而以 $S(t)+n(t)$ 为参考输入. 这种类型的噪声抵消法实用价值很低, 仅仅当参考输入的信噪比接近于零, 而 $\hat{\tau}$ 接近于 $\tau_{\max}$ 时才有一定的增益. 在声纳中一般尽量避免这种情况的出现, 我们在此不详述.

从上面分析计算中我们已经发现, 噪声抵消法的运用是否有效, 关键是估计 $\int |K_{n_0n}(f)|^2 df$ 这个量. 根据互谱不等式 (5), $\hat{\tau}_{n_0n}$ 之极大值是

$$\tau_{\max} = \left(\int K_{n_0n_0}(f)K_{nn}(f)df\right)\Big/(\sigma_{n_0}^2\sigma_n^2) \tag{11}$$

我们称 $\hat{\tau}_{n_0n}$ 达到 $\tau_{\max}$ 的两个噪声 $n(t)$ 和 $n_0(t)$ 为相抵的.

对于上述的噪声抵消系统, 只要 $n(t)$ 和 $n_0(t)$ 相抵, 就可以使增益达到极大值.

例如, $n_0(t)$ 和 $n(t) = \sum_{i=1}^{K} n_0(t-\tau_i)$ 就是相抵的. 事实上,

$$K_{nn}(f) = K_{n_0n_0}(f)\left|\sum_{i=1}^{K} e^{-2\pi jf\tau_i}\right|^2$$

$$K_{n_0n}(f) = \sum_{i=1}^{K} K_{n_0n_0}(f)e^{-2\pi jf\tau_i}$$

$$|K_{n_0n}(f)|^2 = K_{n_0n_0}^2(f)\left|\sum_{i=1}^{K} e^{-2\pi jf\tau_i}\right|^2 = K_{n_0n_0}(f)K_{nn}(f)$$

所以 $n_0(t)$ 与 $n(t)$ 相抵.

三、频域上的多波束系统

我们下面的讨论都要在频域上进行, 把最佳滤波器的解和信号平均功率谱结合在一起, 将会简化普通指向性的计算. 而自适应波束成形只不过是将普通波束成形变为必要的最佳线性滤波, 因而它的计算也不是困难的.

图 3 给出时域和频域上普通波束成形的等价关系.

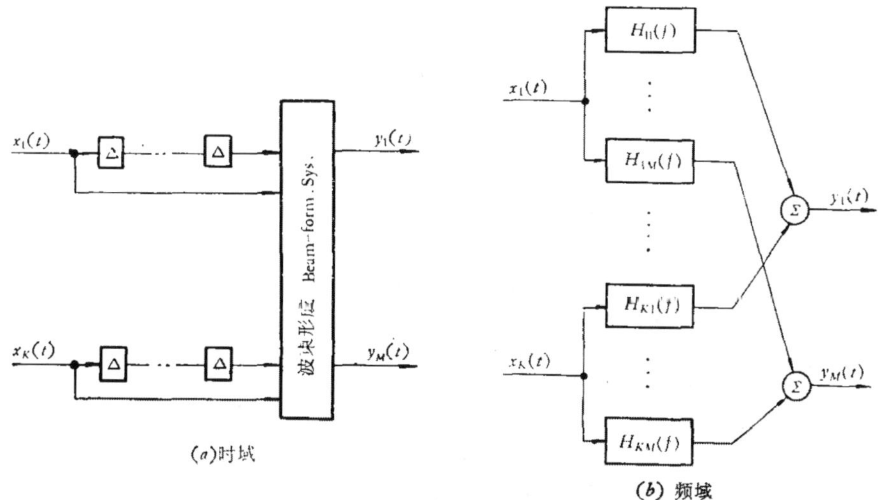

图3 时域和频域上的普通波束成形

图中 $H_{KM}(f)$ 可以理解为横向滤波器. $y_M(t)$ 为第M号波束，$M=1,\cdots,M$.

令 $\boldsymbol{H}_l^T(f)=[H_{1l}(f),\cdots,H_{Kl}(f)]$, $l=1,\cdots,M$. 每一个 $\boldsymbol{H}_l(f)$ 可以形成一个波束.

输入的互谱矩阵为

$$\boldsymbol{K}_{xx}(f)=\begin{bmatrix} K_{x_1x_1}(f),\cdots\cdots,K_{x_1xK}(f) \\ \vdots \qquad\qquad \vdots \\ K_{xKx_1}(f),\cdots\cdots,K_{x_kxK}(f) \end{bmatrix}$$

容易证明

$$K_{y_ly_m}(f)=\boldsymbol{H}_l^T(f)\boldsymbol{K}_{xx}(f)\boldsymbol{H}_m^*(f) \tag{12}$$

多波束的直流输出是一个M维向量

$$\boldsymbol{D}=\begin{bmatrix} E[y_1^2(t)] \\ \vdots \\ E[y_M^2(t)] \end{bmatrix}, \quad E[y_l(t)]^2=\int K_{y_ly_l}(f)df. \tag{13}$$

有的时候,我们为计算方便,不具体指明第几号波束,而是用 $y(t)$, $\boldsymbol{H}(f)$ 来代替 $y_l(t)$, $\boldsymbol{H}_l(f)$.

四、多波束自适应噪声抵消系统

我们首先来看一下单波束的噪声抵消系统，见图4，设输入 $x_k(t)$ 由信号 $s_k(t)$，干扰 $u_k(t)$，噪声 $n_k(t)$ 构成. 并假定 $\sigma_u^2 \gg \sigma_n^2 \gg \sigma_s^2$，图中所画的 $y_0(t)$ 是主波束输出，$y_1(t),\cdots,y_M(t)$ 则分别对准M个干扰. 我们的目的是将 $y_0(t)$ 中干扰由 $y_1(t),\cdots,y_M(t)$ 抵消掉. 由于 $\sigma_u^2 \gg \sigma_s^2$，所以在普通波束成形之后，$y_M(t)$ 中之干扰/信号比更大，可以认为它就是我们在第二节中所提出的模型.

下面,我们只考虑一个干扰的情况,因为多个干扰的计算是完全一样的.

设信号来自 θ_0 方向,干扰来自 θ_1 方向. 则它们的指向矢量是

$$A^T(f,\theta_0)=(e^{2\pi jf\tau_1(\theta_0)},\cdots,e^{2\pi jf\tau_K(\theta_0)})$$
$$A^T(f,\theta_1)=(e^{2\pi jf\tau_1(\theta_1)},\cdots,e^{2\pi jf\tau_K(\theta_1)})$$

因此,它们的普通波束成形矢量分别是

$$\boldsymbol{H}_0(f)=\boldsymbol{A}^*(f,\theta_0),\quad \boldsymbol{H}_1(f)=\boldsymbol{A}^*(f,\theta_1)$$

我们把 $y_0(t)$, $y_1(t)$ 分为信号部分 $s(t)$,干扰部分 $u(t)$,噪声部分 $n(t)$:

$$y_0(t)=s^{(0)}(t)+u^{(0)}(t)+n^{(0)}(t)$$
$$y_1(t)=s^{(1)}(t)+u^{(1)}(t)+n^{(1)}(t)$$

按前面的分析,关键是要考察 $K_{u^{(1)}u^{(0)}}(f)$,我们有

$$\begin{aligned}K_{u^{(1)}u^{(0)}}(f)&=\boldsymbol{H}_1^T(f)\boldsymbol{K}_{uu}(f)\boldsymbol{H}_0^*(f)=\boldsymbol{A}^{*T}(f,\theta_1)K_{uu}(f)\boldsymbol{A}(f,\theta_1)\boldsymbol{A}^{*T}(f,\theta_1)\boldsymbol{A}(f,\theta_0)\\&=K_{uu}(f)K\left(\sum_{i=1}^{K}e^{2\pi jf[\tau_i(\theta_1)-\tau_i(\theta_0)]}\right)\end{aligned}\tag{14}$$

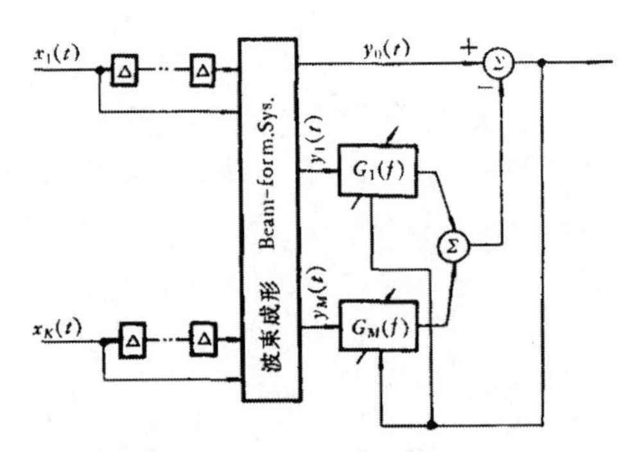

图4 单波束自适应抵消系统

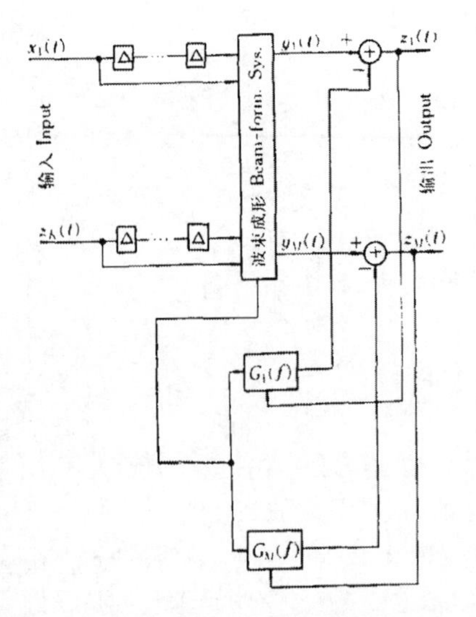

图5 多波束自适应噪声抵消系统

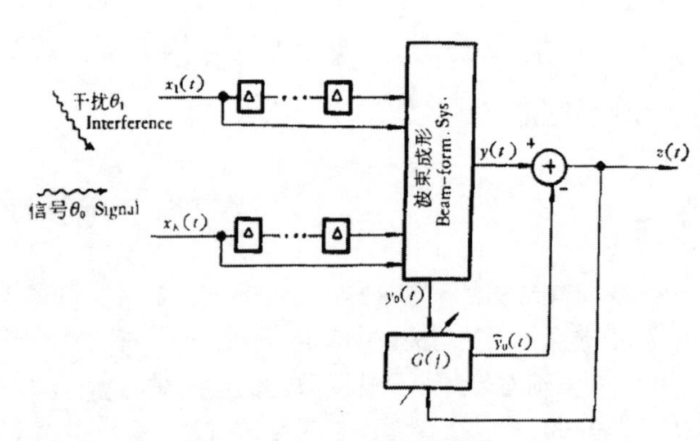

图6 多波束自适应系统的等效图

另一方面:

$$K_{u^{(1)}u^{(1)}}(f)=\boldsymbol{A}^{*T}(f,\theta_1)K_{uu}(f)\boldsymbol{A}(f,\theta_1)\boldsymbol{A}^{*T}(f,\theta_1)\boldsymbol{A}(f,\theta_1)=K^2K_{uu}(f) \tag{15}$$

$$K_{u^{(0)}u^{(0)}}(f)=\boldsymbol{A}^{*T}(f,\theta_0)K_{uu}(f)\boldsymbol{A}(f,\theta_1)\boldsymbol{A}^{*T}(f,\theta_1)A(f,\theta_0)=K_{uu}(f)\left|\sum_{i=1}^{K}e^{2\pi jf[\tau_i(\theta_0)-\tau_i(\theta_1)]}\right|^2 \tag{16}$$

比较(14),(15),(16),立刻可知

$$|K_{u^{(1)}u^{(0)}}(f)|^2=K_{u^{(1)}u^{(1)}}(f)K_{u^{(0)}u^{(0)}}(f),$$

所以 $u^{(1)}(t)$ 和 $u^{(0)}(t)$ 相抵,因而这种系统可取得最大增益.

利用这种单波束的自适应噪声抵消系统,我们自然可以设计同类的多波束系统,见图5,我们这里只画了一个干扰的情况,如果要抵消N个干扰,则要求有 $N\times M$ 个自适应滤波器.

我们下面要计算自适应前后的指向性.

令

$$\boldsymbol{A}^T(f,\theta)=(e^{2\pi jf\tau_1(\theta)},\cdots,e^{2\pi jf\tau_K(\theta)}),$$

我们有

$$\boldsymbol{K}_{xx}(f)=K_{ss}(f)\boldsymbol{A}(f,\theta_0)\boldsymbol{A}^{*T}(f,\theta_0)+K_{uu}(f)\boldsymbol{A}(f,\theta_1)\boldsymbol{A}^{*T}(f,\theta_1)+K_{nn}(f)\boldsymbol{U} \tag{17}$$

这里 $\boldsymbol{U}$ 为单位矩阵.

因为我们要计算的是多波束输出,所以可以任意选定一个波束,设它指向θ方向,这时的输出我们记作 $y(t)$(见图6). 将 $y_0(t)$ 经过相应的最佳滤波器 $\boldsymbol{G}_{\rm opt}(f)$ 的输出记作 $\tilde{y}_0(t)$.

普通波束成形的指向性是输出 $y(t)$ 的均方值,即

$$E[y^2(t)]=\int K_{yy}(f)df=\int\boldsymbol{A}^{*T}(f,\theta)K_{xx}(f)\boldsymbol{A}(f,\theta)df \tag{18}$$

易知

$$K_{yy}(f)=K_{ss}(f)\left|\sum_{i=1}^{K}e^{2\pi jf[\tau_i(\theta)-\tau_i(\theta_0)]}\right|^2+K_{uu}(f)\left|\sum_{i=1}^{K}e^{2\pi jf[\tau_i(\theta)-\tau_i(\theta_1)]}\right|^2+K_{nn}(f)\cdot K \tag{19}$$

当干扰和信号都是单频的情况,(18)就成为

$$E[y^2(t)]=\sigma_s^2K^2D^2(\theta,\theta_0)+K^2\sigma_u^2D^2(\theta,\theta_1)+K\sigma_n^2 \tag{20}$$

指向性函数是

$$E(\theta)=\{\sigma_s^2K^2D^2(\theta,\theta_0)+K^2\sigma_u^2D^2(\theta,\theta_1)+K\sigma_n^2\}^{1/2} \tag{21}$$

其中 $D(\theta,\theta_0)$ 是定向于θ,目标出现在θ_0方向的归一化指向性函数. 当θ_1和θ_0靠得较近时,这种指向性当然无法区分目标.

我们下面来看噪声抵消以后的情况. 前已证明,$y_0(t)$ 中的干扰与 $y_i(t)$ 中的干扰是相抵的. 我们要计算的是 $z(t)=y(t)-\tilde{y}_0(t)$ 的均方值.

$$E[z^2(t)]=E[y^2(t)]+E[\tilde{y}_0^2(t)]-2E[\tilde{y}_0(t)y(t)] \tag{22}$$

为计算 $E[z^2(t)]$,除了(18)式已经求出 $E[y^2(t)]$ 之外还应计算 $E[\tilde{y}_0^2(t)]$ 及 $E[y(t)\tilde{y}_0(t)]$,为此只需求出 $K_{\tilde{y}_0\tilde{y}_0}(f)$ 和 $K_{\tilde{y}_0y}(f)$ 就行.

我们知道[8]

$$G_{\rm opt}(f)=[K^*_{y_0y_0}(f)]^{-1}K^*_{y_0y}(f),\quad 所以$$

$$K_{\tilde{y}_0\tilde{y}_0}(f)=G_{\rm opt}(f)K_{y_0y_0}(f)G^*_{\rm opt}(f)=[K^*_{y_0y_0}(f)]^{-1}|K_{y_0y}(f)|^2 \tag{23}$$

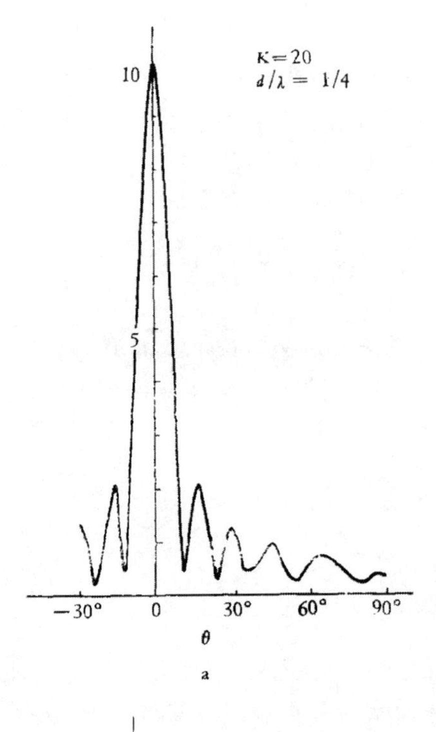

a

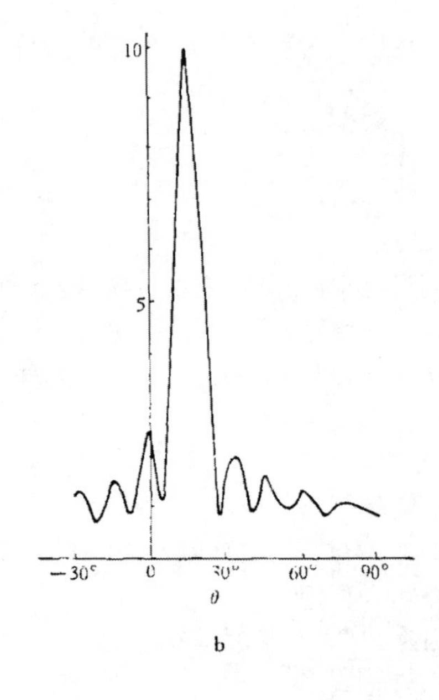

b

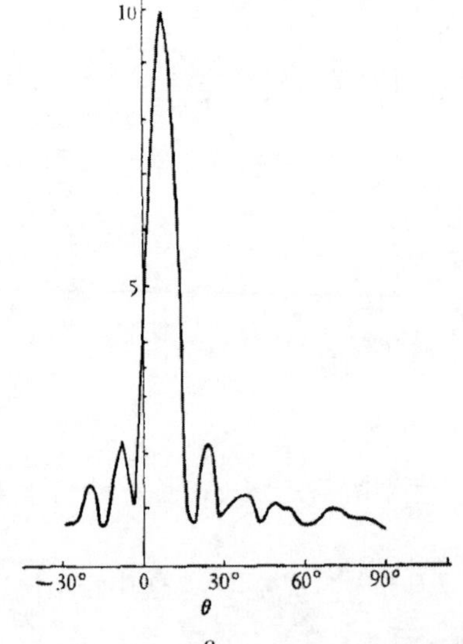

c

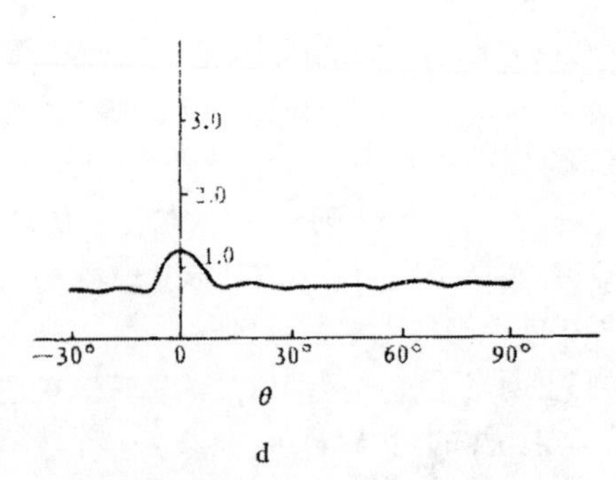

d

图 7　多波束自适应噪声抵消法的指向性

$$K_{\tilde{y}_0 y}(f) = G_{\mathrm{opt}}(f) K_{y_0 y}(f) = [K^*_{y_0 y_0}(f)]^{-1} |K_{y_0 y}(f)|^2 \tag{24}$$

由此得到

$$E[z^2(t)] = \int K_{yy}(f)df - \int [K^*_{y_0 y_0}(f)]^{-1} |K_{y_0 y}(f)|^2 df \tag{25}$$

因为

$$K_{y_0y_0}(f)=\boldsymbol{A}^{*T}(f,\theta_1)\boldsymbol{K}_{xx}(f)\boldsymbol{A}(f,\theta_1)$$
$$=K_{ss}(f)\left|\sum_{i=1}^{K}e^{2\pi jf[\tau_i(\theta_1)-\tau_i(\theta_0)]}\right|^2+K^2K_{uu}(f)+KK_{nn}(f)$$

若我们假定 $\sigma_u^2\gg\sigma_s^2$，$K\sigma_u^2\gg\sigma_n^2$，则

$$[K_{y_0y_0}^{*}(f)]^{-1}\approx 1/(K^2K_{uu}(f))\tag{26}$$

另一方面，$K_{y_0y}(f)=\boldsymbol{A}^{*T}(f,\theta_1)\boldsymbol{K}_{xx}(f)\boldsymbol{A}(f,\theta)$，所以

$$|K_{y_0y}(f)|^2=K_{ss}^2(f)\left|\sum_{i=1}^{K}e^{2\pi jf[\tau_i(\theta_1)-\tau_i(\theta_0)]}\right|^2\left|\sum_{i=1}^{K}e^{2\pi jf[\tau_i(\theta_0)-\tau_i(\theta)]}\right|^2$$
$$+K^2K_{uu}^2(f)\left|\sum_{i=1}^{K}e^{2\pi jf[\tau_i(\theta_1)-\tau_i(\theta)]}\right|^2+K_{nn}^2(f)\left|\sum_{i=1}^{K}e^{2\pi jf[\tau_i(\theta_1)-\tau_i(\theta)]}\right|^2+R\tag{27}$$

这里 R 表示交叉项，它们在对 f 积分时近似于零. 由 (26)，(27) 经过简化得到

$$\int[K_{y_0y_0}(f)]^{-1}|K_{y_0y}(f)|^2df$$
$$\approx K^2\sigma_s^2D^2(\theta,\theta_0)+K^2\sigma_s^2\cdot(\sigma_s^2/\sigma_u^2)D^2(\theta_1,\theta_0)D^2(\theta,\theta_0)-(\sigma_n^4/\sigma_u^2)D^2(\theta,\theta_0)\tag{28}$$

把 (20)，(28) 代入 (25) 得到自适应后的指向性：

$$F_0(\theta)\triangleq E[z^2(t)]/K^2\sigma_s^2=\{D^2(\theta,\theta_0)-(\sigma_s^2/\sigma_u^2)D^2(\theta_1,\theta_0)D^2(\theta,\theta_0)$$
$$-(\sigma_n^2/K^2\sigma_s^2)[K-(\sigma_n^2/\sigma_u^2)D^2(\theta,\theta_0)]\}^{1/2}\tag{29}$$

当 $\sigma_u^2\gg\sigma_s^2$ 时，即使 θ_1 和 θ_0 靠得很近，使 $D(\theta_1,\theta_0)\approx 1$，这时我们仍有 $F_0(\theta)\approx D(\theta,\theta_0)$.

图 7 由 (a) 至 (d) 给出一个实际例子.

我们计算的是一个基元数 $K=20$ 的等间隔线阵，$d/\lambda=1/4$，$\sigma_u^2/\sigma_s^2=100$，$\sigma_u^2/\sigma_n^2=10$，(a) 为无干扰、无噪声时普通波束成形的指向性. (b) 为干扰出现在 $\theta_1=16°$ 时的情况，这时在 $\theta_0=0°$ 的地方勉强可以怀疑有一个目标. (c) 是干扰出现在 $\theta_1=8°$ 时的情况，我们根本无法判断什么地方有信号. (d) 是自适应抵消后的指向性，效果是相当好的. 由于还存在着背景噪声，整个零线电平还比较高.

五、结　　论

前面提出的自适应多波束噪声抵消系统对于主瓣较宽的指向性特别适用，也就是说当一个声纳基阵在非常规的条件下使用于低频状态时，由于主瓣太宽，抗干扰性能就极差，这种情

多波束自适应方案	用于普通波束成形的延迟线数	用于自适应波束成形的滤波器数	备　注〈K 为通道数〉
噪声抵消法	K	5×30	至多抑制 5 个干扰
时域无约束系统	0	$K\times30$	
频域无约束系统	0	K 个 FFT	或用 K 维矩阵求逆
波束域空间处理	K	K 维矩阵求逆	
带约束的求逆矩阵法	0	K 维矩阵求逆迭代算法	仅适用于信/干比较大的情况
辅助基阵自适应法	K_0	$K_0\times30$	$K_0<30$

况下，噪声抵消法的应用可以在为数不多的自适应滤波器的条件下抑制干扰.

下面我们来比较一下，要同时抑制5个干扰时，如果波束数为30，几种多波束自适应系统所需的抽头延迟线(或横向滤波器)的数目(见上页表).

由此可知，就目前的条件来看，多波束噪声抵消法还是一种比较有实用价值的方法.

本报告的研究工作是在侯自强同志的倡议下开展起来的. 文中提到的模拟实验结果由孙允恭、李云言、杨瑞民同志一起提供，作者表示衷心的感谢.

参考文献

[1] Widrow, B. et al., "Adaptive Antenna Systems", *Proc. IEEE*, 55(1967), No. 12, 2143—2159.

[2] Griffiths, L. J., "A Simple Adaptive Algorithm for Real-Time Processing in Antenna Arrays", *Proc. IEEE*, 57(1969), No. 10, 1696—1704.

[3] Frost, O. L. III, "An Algorithm for Linearly Constrained Adaptive Array Processing", *Proc. IEEE*, 60(1972), No. 8, 926—935.

[4] Compton, R. T. Jr. "An Experimental Four-Element Adaptive Array", *IEEE, Trans. on Antennas and Propagation*, AP-24(1976), 697—706.

[5] Vural, A. M., "An Overviews of Adaptive Array Processing for Sonar Applications", EASCON 75, *Rec. IEEE* Electron Aerosp. Syst. Conv. Washington D. C. (1975), 34a—34m.

[6] Owsley, N. L., "A Recent Trend in Adaptive Spatial Processing for Sonar Arrays: Constrained Adaption", *NATO Advanced Study Institute on Signal Processing*, (Longhbrough, 1972).

[7] Dentino, M. et al., "Adaptive Filtering in the Frequency Domain", *Proc. IEEE* 66(1978), No. 12, 1658—1659.

[8] 李启虎"自适应波束成形中最优解在时域和频域中的等价关系", 声学学报, (1979), No.4, 288—300.

[9] Widrow, B. et al., "Adaptive Noise Cancelling: Principle and Applications", *Proc. IEEE*, 63 (1975), No. 12, 1692—1716.

[10] Bendat, J. S. and Piersol, A. G., "Measurement and Analysis of Random Data" (John Wiley & Sons, Inc. New York, 1958), 84.

AN INTRODUCTION OF MULTI-BEAM ADAPTIVE NOISE CANCELLING SYSTEM

Li Qi-hu

(*Institute of Acoustics, Academica Sinica*)

Two noise cancelling models designed with Wiener criteria of least mean square are shown in this paper. The general formula for calculating output signal-to-noise ratio is presented and the practical test rule for adaptive beamforming which employ those models are given.

We give a multi-beam adaptive noise cancelling system. Analyse their important performances, compare with the other available multi-beam adaptive system.

All the evolution are considered in the frequency domain in this paper, we have given a method for calculating the function of directivity by using the mean power spectrum density of signal and optimum transfer function of system.

The directivity formula of multi-beam noise cancelling before and after adaptiving are given.

Some practical examples show the advantage of this model in suppressing interferences when the number of the interference are not many.

Captions of figures

1. The noise cancelling model.
2. The relation between gain G and radius of cross correlation.
3. The conventional beamforming in time-domain and frequency-domain.
(a) Time-domain (b) Frequency-domain.
4. Single-beam adaptive noise cancelling system.
5. Multi-beam adaptive noise cancelling system.
6. The equivalent diagram of multi-beam adaptive system.
7. The directivity of multi-beam adaptive noise cancelling system (a) Conventional beam forming without interference and noise (b) Conventional beamforming for $\theta_0=0°$, $\theta_1=16°$ (c) Conventional beamforming for $\theta_0=0°$, $\theta_1=8°$ (d) After noise cancelling.

自适应噪声抵消滤波器抵消能力的研究①

The Study of Cancelling Ability of an Adaptive Noise Cancelling Filter

李 启 虎
（中国科学院声学研究所）

〔摘要〕 本文根据 Wiener 最小均方差准则，讨论了自适应噪声抵消滤波器的性能，导出了极限抵消能力和系统增益的表达式，给出了应用噪声抵消法必须满足的基本条件，给出了用横向滤波器逼近最佳线性系统的原理与设计方法。同时给出了根据噪声抵消原理设计的自适应波束成形系统抗平面波干扰增益的计算方法。最后提到用计算机进行系统模拟的框图和某些仿真结果。

Abstract: Under the Wiener least mean square (LMS) criteria, the performance of an adaptive noise cancelling filter is considered. Using the solution of optimum transfer function in frequency domain, we prove the limit cancelling ability and system gain are proportional with $1/\hat{\tau}$, where $\hat{\tau}$ is the correlation radius between input noise. The elementary conditions that must be satisfied in application of noise cancelling are given.

In time domain, a transverse filter can approximate the optimum linear system with arbitrary accuracy. The representation of their relation is given. The results given in this paper can be easily applied to adaptive beamforming systems.

A method (which is based on the inverse matrix of cross-power density spectrum) for calculating the ability of adaptive beamforming system to reject planar interference is given. A block diagram of system simulation by computer and some experimental results are shown at the end of the paper.

一、前 言

近十年来自适应滤波技术已在雷达、声纳及其他许多不同的领域中得到应用[1~4]。实际系统的应用方法可分为两类。一类是自适应波束成形，其求和后的输出具有很高的信噪比，可以有效地抑制干扰。另一类是自适应噪声抵消滤波器，系统求和后的输出的信噪比达到最大值。这两种系统虽然在形式上很不一样，但从理论上都可归结为噪声抵消模型。

本文利用频域上自适应滤波器最佳传输函数的解，分析噪声抵消法的主要性能，提出关

① 电子学报, 1981, (4): 16-24.

于噪声完全抵消的概念。指出，只有在参考输入中存在与原始输入完全相抵消的噪声时，噪声抵消法才有可能取得最大增益。同时给出输出信噪比与输入噪声间的相关半径的关系。

文中分析了在时域上用横向滤波器实现最佳自适应滤波的方法。这种滤波器与频域上最佳滤波器之间的差异取决于信号的性质和抽头延迟线的长度。我们将给出度量这种差异的一个简单的表达式，从而为横向滤波器的设计提供了理论依据。

文中还给出自适应波束成形系统抗干扰增益的计算方法和某些结果。

二、噪声抵消系统的增益

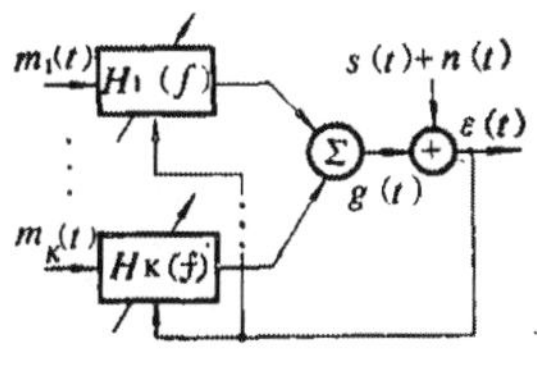

图 1 自适应噪声抵消系统模型

考虑如图 1 所示的噪声抵消系统模型。$s(t)+n(t)$称为原始输入。$m_k(t)$，$k=1, 2, \cdots, K$ 称为参考输入。假定 $S(t)$、$n(t)$、$m_k(t)$ 为平稳随机过程，均值为零。$n(t)$与$m_k(t)$相关，而$S(t)$与$m_k(t)$和 $n(t)$都相互独立。按 Widrow 提出的方法[4]，调节 $H_1(f), \cdots, H_K(f)$，使 $\varepsilon(t)=s(t)+n(t)-g(t)$的均方值 $I=E[\varepsilon^2(t)]$极小。这里 $g(t)$ 表示参考输入经滤波后的合成输出。字母 E 表示集合平均。当 $E[\varepsilon^2(t)]$ 达到极小时，$E[n(t)-g(t)]^2$也达到了极小，从而输出信噪比达到了极大。$m_k(t)$与 $m_l(t)$ 的互相关函数$R_{k,l}(\tau)$及$m_k(t)$与 $n(t)$的互相关函数 $R_k(\tau)$ 分别为：

$$R_{k,l}(\tau)=E[m_k(t)m_l(t-\tau)], \quad 1\leqslant k, \ l\leqslant K \tag{1}$$

$$R_k(\tau)=E[m_k(t)n(t-\tau)], \quad 1\leqslant k\leqslant K \tag{2}$$

我们知道，按最小均方差准则，最佳传输函数[5]

$$H_{\mathrm{opt}}(f)=[K_{mm}^*(f)]^{-1}K_{mn}^*(f) \tag{3}$$

式中 $*$ 表示复共轭，$H_{\mathrm{opt}}(f)$表示由 $H_1(f), \cdots, H_K(f)$构成的滤波矢量。$K_{mm}(f)$ 表示参考输入的互谱矩阵，它的第(k,l)个元素是 $m_k(t)$ 与 $m_l(t)$的互功率谱，即

$$[K_{mm}(f)]_{k,l}=\int_{-\infty}^{\infty}R_{k,l}(\tau)e^{-2\pi jf\tau}\,d\tau, \quad 1\leqslant k, \ l\leqslant K \tag{4}$$

$K_{mn}(f)$是参考输入与原始输入的互谱矢量，它的第 k 个元素是 $m_k(t)$ 与 $n(t)$ 的互功率谱，即

$$[K_{mn}(f)]_k=\int_{-\infty}^{\infty}R_k(\tau)e^{-2\pi jf\tau}\,d\tau, \quad 1\leqslant k\leqslant K \tag{5}$$

对于单通道的噪声抵消系统，由于 $K_{mm}(f)$是实函数，所以由式(3)便知

$$H_{\mathrm{opt}}(f)=K_{mn}^*(f)/K_{mm}(f) \tag{6}$$

在这样的情况下，$H_{\mathrm{opt}}(f)$、K_{mn}^* 和 K_{mm} 均为标量。

下面以单通道的噪声抵消系统为例来分析这种系统的性能。现令噪声抵消之前、后的信噪比分别为 η，Q 及系统的增益为 G，那么

$$\eta=\sigma_s^2/\sigma_n^2 \tag{7}$$

$$Q=E[s^2(t)]/E[n(t)-g(t)]^2 \tag{8}$$

$$G=Q/\eta \tag{9}$$

式中，$\sigma_s^2=\int_{-\infty}^{\infty}K_s(f)df$，$\sigma_n^2=\int_{-\infty}^{\infty}K_n(f)df$，分别表示信号和噪声的功率。为计算增益，关键的问题是估算式(8)中的分母。根据平稳过程的线性过滤理论，易知[6]

$$E[n(t)-g(t)]^2=E[n^2(t)]+E[g^2(t)]-2E[n(t)g(t)]$$
$$=\sigma_n^2+\int_{-\infty}^{\infty}K_{mm}(f)|H(f)|^2df-2\int_{-\infty}^{\infty}H(f)K_{mn}(f)df$$

当 $H(f)$ 达到最佳值 $H_{opt}(f)$ 时，把式(6)代入上式得到

$$E[n(t)-g(t)]^2=\sigma_n^2-\int_{-\infty}^{\infty}H_{opt}(f)K_{mn}(f)df$$
$$=\sigma_n^2-\int_{-\infty}^{\infty}\frac{|K_{mn}(f)|^2}{K_{mn}(f)}df \tag{10}$$

把式(10)代入式(8)并把所得结果代入式(9)，即可得：

$$Q=\sigma_s^2\Big/\left(\sigma_n^2-\int_{-\infty}^{\infty}\frac{|K_{mn}(f)|^2}{K_{mm}(f)}df\right)$$
$$G=1\Big/\left[1-\frac{1}{\sigma_n^2}\int_{-\infty}^{\infty}\frac{|K_{mn}(f)|^2}{K_{mm}(f)}df\right] \tag{11}$$

这就是噪声抵消系统增益(或称抵消能力)的表达式。此外，利用 Parseval 等式[7]，有

$$\int_{-\infty}^{\infty}|K_{mn}(f)|^2df=\int_{-\infty}^{\infty}(R_{mn}(\tau))^2d\tau\triangleq\sigma_m^2\sigma_n^2\int_{-\infty}^{\infty}\rho_{mn}^2(\tau)d\tau$$

式中，$\rho_{mn}(\tau)$ 为 $m(t)$ 及 $n(t)$ 的相关系数，它与 $m(t)$，$n(t)$ 的相关半径 $\hat{\tau}$ 的关系为：

$$\hat{\tau}=\int_{-\infty}^{\infty}\rho_{mn}^2(\tau)d\tau \tag{12}$$

根据互谱不等式[8]

$$|K_{mn}(f)|^2\leqslant K_{mm}(f)K_{nn}(f) \tag{13}$$

由此得到

$$\hat{\tau}=\int_{-\infty}^{\infty}\frac{|K_{mn}(f)|^2df}{\sigma_m^2\sigma_n^2}\leqslant\int_{-\infty}^{\infty}\frac{K_{mm}(f)K_{nn}(f)}{\sigma_m^2\sigma_n^2}df\triangleq\hat{\tau}_{\max} \tag{14}$$

显然，如果式(13)中的等号成立，则增益 G 的分母达到极小值，即G达到极大值。事实上

$$\frac{1}{\sigma_n^2}\int_{-\infty}^{\infty}\frac{|K_{mn}(f)|^2}{K_{mm}(f)}df\leqslant\frac{1}{\sigma_n^2}\int_{-\infty}^{\infty}K_{nn}(f)df=1 \tag{15}$$

我们称使式(13)取等号的 $m(t)$和 $n(t)$为完全相抵消的随机过程。也可以把它理解为$m(t)$、$n(t)$具有潜在的最大相关性。例如，若$n(t)=\sum_{i=1}^{N}m(t-\tau_i)$，则$n(t)$就和$m(t)$完全相抵消。这是因为

$$K_{nn}(f)=K_{mm}(f)\left|\sum_{i=1}^{N}e^{-2\pi jf\tau_i}\right|^2$$
$$K_{mn}(f)=K_{mm}(f)\sum_{i=1}^{N}e^{-2\pi jf\tau_i} \tag{16}$$

不难验证$|K_{mn}(f)|^2=K_{mm}(f)K_{nn}(f)$，即 $m(t)$ 和 $n(t)$完全相抵消。

在一些特殊情况下，增益的计算具有更简单的形式。设$K_{mm}(f)$是矩形谱，即

$$K_{mm}(f)=\begin{cases}\sigma_m^2/2\Delta f, & |f\pm f_0|\leqslant \Delta f/2\\ 0, & \text{其它}\end{cases} \tag{17}$$

代入式(14)便得$\hat{\tau}_{\max}=1/\Delta f$；代入式(11)便可得：

$$G=1/(1-\hat{\tau}/\hat{\tau}_{\max}) \tag{18}$$

此式表示增益G与输入信号相关半径比 $\hat{\tau}/\hat{\tau}_{\max}$ 之间的关系，如图2所示。可以看出，当 $\hat{\tau}/\hat{\tau}_{\max}\leqslant 0.5$ 时，噪声抵消系统的增益非常有限。

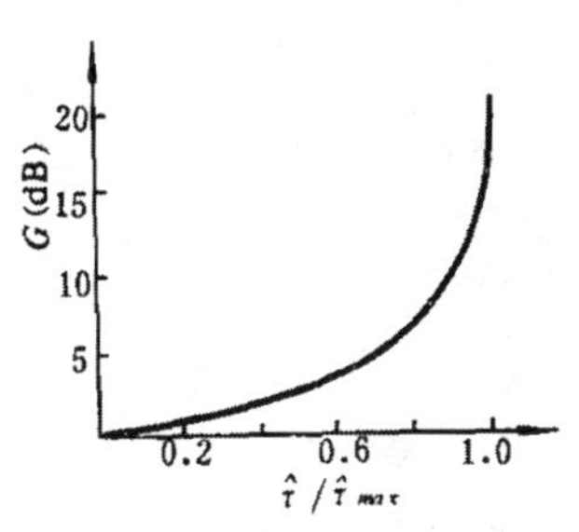

图2　增益与相关半径的关系

三、噪声抵消原理的实现

我们用抽头延迟线来实现式(6) 中的 $H_{opt}(f)$。图3给出了抽头延迟线在时域和频域中的等价关系。时域上的脉冲响应函数 $h(t)$ 及频域中的系统传输函数$H(f)$分别为：

$$h(t)=\sum_{l=1}^{L} w_l \delta[t-(l-1)\Delta] \tag{19}$$

$$H(f)=\sum_{l=1}^{L} w_l e^{-2\pi j f(l-1)\Delta} \tag{20}$$

式中，w_l为待调节的实加权系数，Δ为抽头延迟线的延时间隔，L为延迟线的总节数。一般讲，当Δ 选择适当，只要L足够大，用抽头延迟线构成的 $H(f)$ 总可以逼近 $H_{opt}(f)$，但收敛的性质和 $H_{opt}(f)$的形状有关。

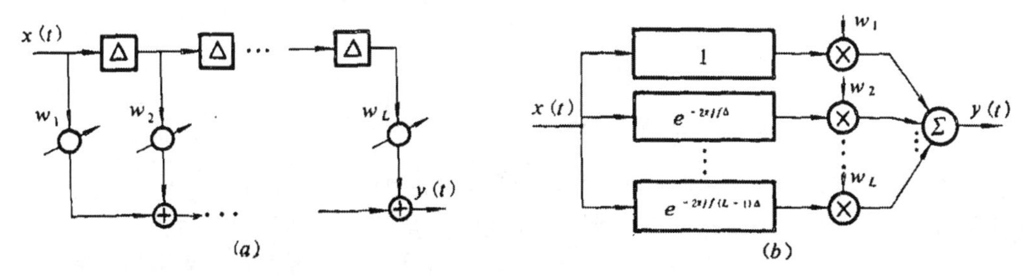

图3　时、频等价关系

设式(20)中$H(f)$达到 $H_{opt}(f)$时，$I=E[n(f)-g(t)]^2$ 达到极小值 $I_{\min}$，在一般情况下有[5]：

$$\begin{aligned} I &= E[n^2(t)]+\int_{-\infty}^{\infty} H(f)K_{mm}(f)H^*(f)\,df-2\int_{-\infty}^{\infty} H(f)K_{mn}(f)df \\ &= I_{\min}+\int_{-\infty}^{\infty} |H_{opt}(f)-H(f)|^2 K_{mm}(f)df \end{aligned} \tag{21}$$

如果把 $H(f)$ 取为 $H_{opt}(f)$ 的 Fourier 展开式的前 N 项的部分和，则当 $N\to\infty$ 时，$\|H_{opt}(f)-H(f)\|^2\to 0$。符号 $\|\cdot\|$ 表示平方可积 Hilbert 空间中的距离，即 $\|H_{opt}(f)-H(f)\|^2=\int_{-\infty}^{\infty}|H_{opt}(f)-H(f)|^2df$。由此可知，用抽头延迟线去模拟最佳传输函数是可行的。若 $m(t)$ 与 $n(t)$ 完全相抵消，在理想情况下，$I_{\min}=0$，这时

$$Q=\sigma_s^2\Big/\int_{-\infty}^{\infty}|H_{opt}(f)-H(f)|^2K_{mm}(f)df \tag{22}$$

$$G=1\Big/\frac{1}{\sigma_n}\int_{-\infty}^{\infty}|H_{opt}(f)-H(f)|^2K_{mm}(f)df \tag{23}$$

式(23)是设计抽头延迟线时必须考虑的重要条件。如果 $K_{mm}(f)$ 具有式(17)的形状，则有：

$$G=1\Big/\frac{\sigma_m^2}{\sigma_n^2\Delta f}\|H_{opt}(f)-H(f)\|^2 \tag{24}$$

式(24)的计算并不困难，因为 $\|H_{opt}(f)-H(f)\|^2$ 就是 $H_{opt}(f)$ 的 Fourier 展开式前 N 项的截断误差，可以通过 Parseval 等式方便地予以计算[7]。还可以进一步将式(24)简化成

$$\|H_{opt}(f)\|^2=\int_{-\infty}^{\infty}|H_{opt}(f)|^2df=\frac{\Delta f}{\sigma_m^2}\sigma_n^2 \tag{25}$$

并代入式(24)，得

$$G=1\Big/\frac{\|H_{opt}(f)-H(f)\|^2}{\|H_{opt}\|^2}\triangleq\frac{1}{\zeta} \tag{26}$$

噪声抵消系统的抵消能力完全由 $H(f)$ 对 $H_{opt}(f)$ 的逼近程度表达出来。我们称 ζ 为时频相对失真，例如，若希望系统具有 40dB 的抵消能力，应要求 $G=10^{-4}$，据此，不难确定出延迟线的长度。图 4 给出了增益 G 与时频相对失真 ζ 的关系。

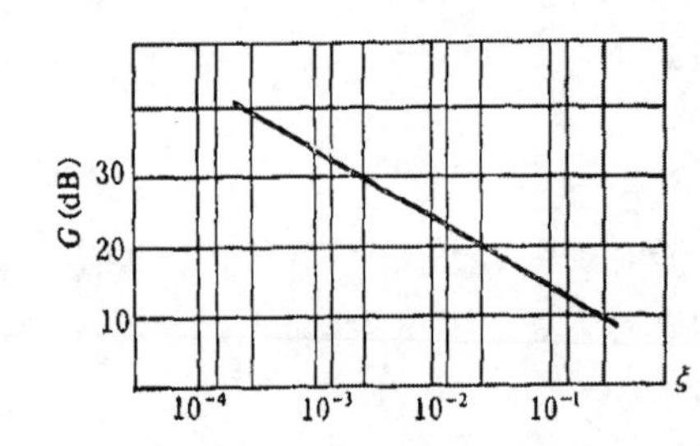

图 4　增益 G 与时频相对失真 ζ 的关系

对于信号处理中常见的频谱形式，在 $H_{opt}(f)$ 展成 Fourier 级数时，收敛往往很快，所需的延迟节数并不多。但为了使噪声抵消系统适应各类信号，必须有足够的余量。

四、自适应波束成形系统的抗干扰增益

声纳中使用的自适应波束成形的概念，在原理上和噪声抵消法一样，所不同的是原始输入是事先选定的期望信号 $d(t)$，在自适应过程中实际抵消的是信号而不是噪声。输出的有用信号不是 $\varepsilon(t)$ 而是 $y(t)$，对这种系统抗干扰能力的研究是十分重要的。这里只考虑平面波干扰的情况。设输入信号 $x_1(t),\cdots,x_K(t)$ 中包含信号、噪声和若干个干扰，于是输入的互谱矩阵可以表示为[9]：

$$K_{xx}(f)=K_n(f)I_0+K_s(f)AA^{*T}+\sum_{p=1}^{M}K_{rp}(f)B_pB_p^{*T} \tag{27}$$

式中，$K_n(f)$，$K_s(f)$，$K_{rp}(f)$ 分别表示噪声、信号和第 p 个干扰的平均功率谱密度。I_0 为单位矩阵，T 表示矩阵的转置，M 为干扰的个数。信号的指向矢量

$$A^T=[e^{2\pi jf\tau_1},\cdots,e^{2\pi jf\tau_K}] \tag{28}$$

式中，τ_k 是信号到达第 k 个接收器的时延，$k=1,2,\cdots K$。第 p 个干扰的指向矢量

$$B_p^T=[e^{2\pi j\delta_{1p}},\cdots,e^{2\pi j\delta_{Kp}}] \tag{29}$$

式中，$1\leqslant p\leqslant M$，δ_{kp} 为第 p 个干扰到达第 k 个接收器的时延。类似式(3)，可以得到最佳传输函数

$$H_{opt}(f)=[K_{xx}^*(f)]^{-1}K_{xd}^*(f) \tag{30}$$

这里，$K_{xd}(f)$为输入信号 $x_1(t),\cdots,x_K(t)$和期望信号 $d(t)$ 的互谱矢量。为计算系统的增益，我们首先应当计算输出信噪比。对于一般的传输函数 $H(f)$，其输出信号和输出噪声的功率分别为：

$$\left.\begin{aligned}\overline{S}^2&=\int_{-\infty}^{\infty}H^T(f)K_{ss}(f)H^*(f)df\\ \overline{N}^2&=\int_{-\infty}^{\infty}H^T(f)K_{nn}(f)H^*(f)df\end{aligned}\right\} \tag{31}$$

式中，$K_{ss}(f)=K_s(f)AA^{*T}$ 为信号的互功率谱矩阵；$K_{nn}(f)=K_n(f)I_0+\sum_{p=1}^{M}K_{rp}(f)B_pB_p^{*T}$，由此得输出信噪功率比 Q及系统增益 G分别为

$$Q=\frac{\overline{S}^2}{\overline{N}^2}=\frac{\int_{-\infty}^{\infty}H^T(f)K_{ss}(f)H^*(f)df}{\int_{-\infty}^{\infty}H^T(f)K_{nn}(f)H^*(f)df} \tag{32}$$

$$G=\frac{Q}{\eta},\quad \eta=\sigma_s^2\Big/(\sigma_n^2+\sum_{p=1}^{M}\sigma_{rp}^2) \tag{33}$$

式中 η 为输入信噪比。如果把式(30)中的$H_{opt}(f)$代替式(32)中的$H(f)$，并利用式(33)，就得到自适应波束成形系统的抗干扰增益 G_{ad}，如果用普通波束成形的滤波矢量 $H_{cv}(f)=A^*$ 代替式(32)中的 $H(f)$，就得到普通波束成形系统的增益 G_{cv}。不难看出，计算 G_{ad} 的关键在于求解 $K_{xx}^{-1}(f)$。下面，我们以M＝1（即只有一个干扰）为例来说明如何计算 $K_{xx}^{-1}(f)$（此方法可立即推广到$M>1$ 的情况）。于是有：

$$K_{xx}(f)=K_n(f)I+K_r(f)BB^{*T}+K_s(f)AA^{*T} \tag{34}$$

式中，A为由式(28)所定义的信号指向矢量，B是由下式给出的干扰指向矢量：

$$B^T=[e^{2\pi jf\delta_1},\cdots,e^{2\pi jf\delta_K}] \tag{35}$$

如令 $R=I_0+[K_r(f)/K_n(f)]BB^{*T}$，则有 $K_{xx}(f)=K_n(f)\{R+[K_s(f)/K_n(f)]AA^{*T}\}$，于是，不难验证

$$R^{-1}=I_0-BB^{*T}/\{K+[K_n(f)/K_r(f)]\} \tag{36}$$

$$K_{xx}^{-1}(f)=\frac{1}{K_n(f)}\left\{R^{-1}+\frac{R^{-1}AA^{*T}R^{-1}}{A^{*T}R^{-1}A+[K_n(f)/K_s(f)]}\right\} \tag{37}$$

把式(36)代入式(37)，就可以得到由已知量表达的 $K_{xx}^{-1}(f)$。对于一般的传输函数 $H(f)$，有：

$$\left.\begin{aligned}&H^T(f)K_{ss}(f)H^*(f)=K_s(f)|H^T(f)A|^2\\ &H^T(f)K_{nn}(f)H^*(f)=\|H(f)\|^2K_n(f)+K_r(f)|H^T(f)B|^2\end{aligned}\right\} \tag{38}$$

将它们代入式(32)，有：

$$Q=\frac{\bar{S}^2}{\bar{N}^2}=\frac{\int_{-\infty}^{\infty}K_s(f)\,|H^T(f)A|^2df}{\int_{-\infty}^{\infty}K_n(f)\,\|H(f)\|^2\,df+\int_{-\infty}^{\infty}K_r(f)\,|H^T(f)B|^2df} \tag{39}$$

对于普通波束成形系统，滤波矢量对准信号，即 $H(f)=H_{cv}(f)=A^*(f)$，故有

$$Q_{cv}=K^2\sigma_s^2/[K\sigma_n^2+\int_{-\infty}^{\infty}K_r(f)\,|A^{*T}B|^2df] \tag{40}$$

$$G_{cv}=Q_{cv}\Big/\left(\frac{\sigma_s^2}{\sigma_n^2+\sigma_r^2}\right) \tag{41}$$

若不存在干扰，则 $K_r(f)=0$，于是得到

$$G_{cv}=K \tag{42}$$

式(38)中的$|A^{*T}B|^2$经归一化处理后，称之为信号矢量和干扰矢量的广义夹角[10]，用γ来表示

$$|A^TB|^2=K^2\cos^2\gamma$$

对于自适应波束成形系统，$H_{opt}(f)$由式(30)给出，注意到$K_{xx}(f)$为Hermite矩阵，即，$K_{xx}^*(f)=K_{xx}^T(f)$，以及 $K_{xd}(f)=K_s(f)A$，于是有：

$$\begin{aligned}H_{opt}^T(f)A&=K_s(f)A^{*T}K_{xx}^{-1}(f)A\\&=A^{*T}R^{-1}A\Big/\left(\frac{K_n(f)}{K_s(f)}+A^{*T}R^{-1}A\right)\end{aligned} \tag{43}$$

$$|H_{opt}^T(f)A|^2=\frac{P(f)}{P(f)+[K_n(f)/K_s(f)]} \tag{44}$$

式中
$$P(f)=A^{*T}R^{-1}A=K-\frac{K^2\cos^2\gamma}{K+[K_n(f)/K_s(f)]} \tag{45}$$

K为基元个数。类似的计算可以得到$\|H_{opt}(f)\|^2$及$|H_{opt}(f)B|^2$，所以代入式(39)后，有：

$$Q_{ad}=\frac{\int_{-\infty}^{\infty}K_s(f)\left\{\frac{P(f)}{P(f)+[K_n(f)/K_s(f)]}\right\}^2df}{\int_{-\infty}^{\infty}\frac{K_n(f)U(f)}{\{P(f)+[K_n(f)/K_s(f)]\}^2}df+\int_{-\infty}^{\infty}\frac{K_r(f)\,|V(f)|^2}{\{P(f)+[K_n(f)/K_s(f)]\}^2}df} \tag{46}$$

式中

$$\begin{aligned}U(f)&=A^{*T}(R^{-1})^2A\\&=K-\frac{2K^2\cos^2\gamma}{K+[K_n(f)/K_r(f)]}-\frac{K^3\cos^2\gamma}{\{K+[K_n(f)/K_r(f)]\}^2}\end{aligned} \tag{47}$$

$$\begin{aligned}|V(f)|^2&=|A^{*T}R^{-1}B|^2\\&=\frac{[K_n(f)/K_r(f)]^2K^2\cos^2\gamma}{\{K+[K_n(f)/K_r(f)]\}^2}\end{aligned} \tag{48}$$

从式(45)至(48)不难看出自适应波束成形系统的增益为

$$G_{ad}=Q_{ad}\Big/\left(\frac{\sigma_s^2}{\sigma_n^2+\sigma_r^2}\right) \tag{49}$$

图6、7分别给出$K=15$时G_{cv}和G_{ad}随$\cos^2\gamma$变化的情况。计算时，假定$K_s(f)$，

$K_n(f)$，$K_r(f)$为矩形谱。比较这两图可以看出，在一般情况下 G_{ad} 比 G_{cv} 高出 10dB左右。σ_n^2/σ_r^2 越小，自适应波束成形系统的优越性越明显。即这种系统抗强点源干扰的能力远优于普通波束成形系统。

对于具体的接收阵，广义夹角 γ 可以用信号的入射角 θ 表达出来。例如，对于线阵，从它的指向性公式，我们有[11]：

$$\cos^2\gamma=\left[\frac{\sin\left(K\dfrac{\pi d}{\lambda}\sin\theta\right)}{K\sin\left(\dfrac{\pi d}{\lambda}\sin\theta\right)}\right]^2 \tag{50}$$

式中，d为相邻基元的间隔，λ 为入射信号的波长。由此可以求出 θ 与 γ 的关系，再由式

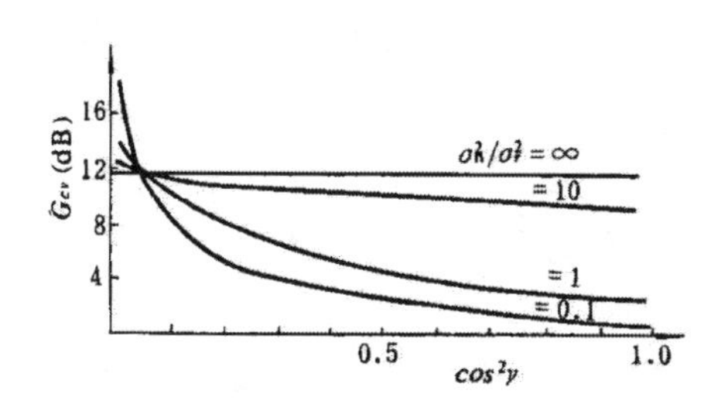

图 5　普通波束成形系统抗干扰增益

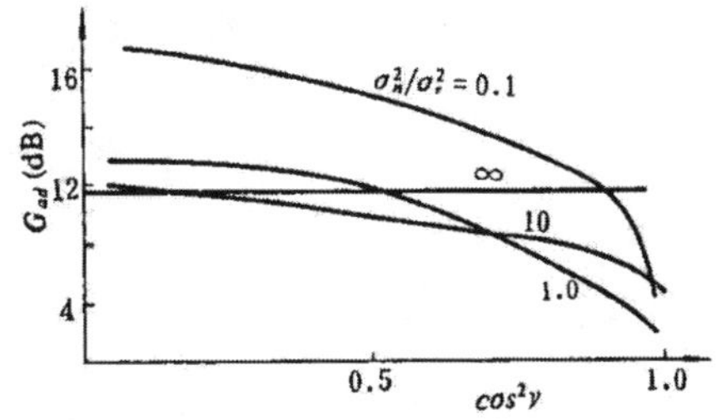

图 6　自适应波束成形系统抗干扰增益

(46)得出相应的增益，具体例子见图 7 。可以看出，对于同样一个基阵，自适应系统的增益比普通波束成形系统的高得多，因而对目标的分辨力也强得多。

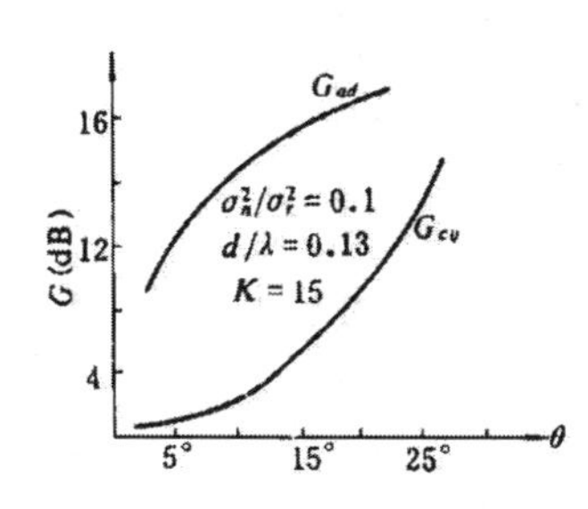

图 7　等间隔线阵的增益

五、噪声抵消系统的计算机模拟

根据前面关于噪声抵消系统的理论分析，用 DJS-6计算机进行模拟，如图 8 所示。利用计算机产生均匀分布的随机数，然后将它按一定规则相加、卷积，以便得到能满足给定要求的 Gauss 随机噪声。抽头延迟线的模拟则利用计算机模拟移位寄存器来实现。权系数的迭代公式是[1]：

$$W(k+1)=W(k)-\mu[X^T(k)W(k)-d(k)]X(k) \tag{51}$$

式中，$W(k)$为第 k 步的 L 个权系数构成的矢量，$X(k)$为第 k 步的 L 个输入样本，$d(k)$ 为第 k 步的期望信号，μ 为控制收敛因子。

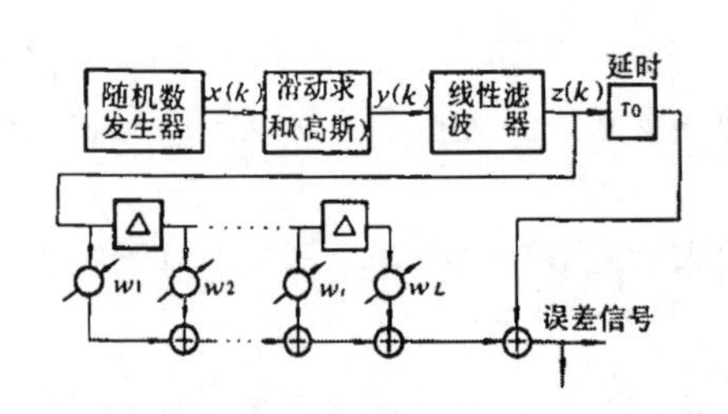

图 8 计算机模拟宽带噪声抵消法框图

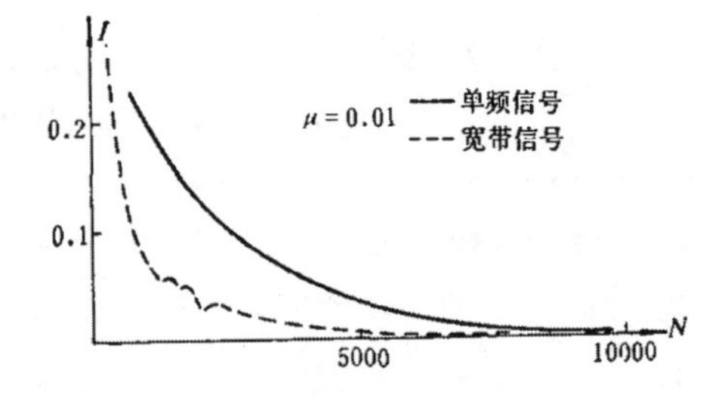

图 9 噪声抵消的学习曲线

随着迭代的进行，均方值 $I=E[\varepsilon^2(k)]=E[X^T(k)W(k)-d(k)]^2$ 逐步下降至 $I_{\min}$，I 随 k 变化的曲线称为系统的学习曲线。图 9 中给出了单频信号和宽带信号的两条学习曲线，前者权系数只有两个，后者用 16 个权系数。两种情况下系统的增益都在 70dB以上。

参 考 文 献

〔1〕B. Widrow, et al, Proc. IEEE Vol. 55, No. 12, pp. 2143-2159, 1967.
〔2〕L. J. Griffiths, Proc. IEEE Vol. 57, No. 10, pp. 1696-1704, 1969.
〔3〕O. L. Frost, Proc. IEEE Vol. 60, No. 8, pp. 926—935, 1972.
〔4〕B. Widrow, et al, Proc. IEEE Vol.63, No. 12, pp. 1692-1716, 1957.
〔5〕李启虎，自适应波束成形中最优解在时域与频域中的等价关系，声学学报，1979年，第 4 期，第296—308页.
〔6〕A. D. Whalen, Detection of signals in noise, Academic Press, New York and London, 1971.（A.D.惠伦，噪声中的信号检测，刘其培、迟惠生译，科学出版社，1977年，北京)。
〔7〕A. Papoulis, The Fourier integral and its application, McGraw-Hill Co., 1965.
〔8〕J. S. Bendat, Principles and applications of random noise theory, John Wiley Sons. Inc., 1958.
〔9〕J. H. Chang, and F. B. Tuteur, Optimum adaptive array Processor, Proc. of Symp. on Computer Processing in Commu., Vol. 19, New York, pp. 695-710, 1969.
〔10〕H. Cox, Sensitivity considerations in adaptive beamforming, NATO Advanced study institute on signal Processing, p. 619, 1972.
〔11〕J. W. Horton, Fundamentals of Sonar, U.S Naval Institute, Annapolis Maryland, 1959.（声纳原理，冯秉铨等译，国防工业出版社，1965.，北京)。

自适应波束成形系统的极限分辨力和最佳加权系数的动态①

李 启 虎
(中国科学院声学研究所)

本文研究自适应波束成形声呐设计中的几个问题．其一是自适应波束成形系统的极限分辨力，我们给出这种系统输出信噪比和增益的一般表达式．提出一种计算自适应波束成形系统抗平面波干扰能力的计算方法．

作为信号和干扰之间广义夹角的函数，给出计算系统增益的公式及相应的数值计算结果．

指出，极限分辨力不仅依赖于广义夹角而且与干扰/噪声比有关．文中给出具体的例子．

本文研究的另一个问题是最优加权系数的动态问题，根据频域上最优传输函数的解，可以给出最优加权系数的一个上界．

本文的结果可以直接应用于实际自适应波束成形的雷达或声呐的设计．

引 言

B. Widrow 指出[1]，对于信号提取来说，最小均方差(LMS)准则是有效的，对于信号检测来说，最大信噪比(MSN)准则较为有利．目前有关 MSN 准则下，最佳基阵处理系统增益的研究已有不少具体结果[2-4]，但是，这种系统由于具体求解比较复杂，难以实时完成．因此自适应波束成形方法仍旧采用 LMS 准则[5-8]．我们自然关心这种系统的抗干扰的性能．这是本文研究的第一个问题．J. H. Chang 和 F. B. Tuteur 曾提出一种定性估计抗一个平面波干扰的方法[6]．

本文利用 LMS 准则在频域上的解[9]，研究自适应波束成形系统的稳态特性．首先指出，计算增益的关键是求解输入资料的互谱矩阵 $\mathbf{K}_{xx}(f)$ 的逆．然后提出逐次求解 $\mathbf{K}_{xx}^{-1}(f)$ 的一种方法．我们在第一节中将给出当存在一个干扰时，$\mathbf{K}_{xx}^{-1}(f)$ 的解．至于多个干扰下的解，将在附录中给出．

在第二节中，我们利用 H. Cox 提出的广义夹角[4]的概念，把自适应系统的增益作为广义夹角的函数表达出来．同时将其与普通波束成形作比较．在大多数情况下，我们发现，自适应增益要比普通波束成形增益高得多．

在第三节中，我们给出一个例子，指出如何在实际应用中将广义夹角换算为干扰和信号的夹角，从而求出自适应系统的极限分辨力．

本文研究的第二个问题是最优加权系数的动态问题．这是声呐设计中所关心的问题之一．从表面上看，它似乎与本文研究的第一个问题无关．但是，在第四节中我们将指出，这实

① 声学学报, 1981, (3): 172-180.

际上是对最优传输函数的模的估计问题. 所以最终也归结为 $\mathbf{K}_{xx}^{-1}(f)$ 的求解. 利用 Parseval 公式,我们将给出 $\mathbf{W}_{opt}$ 的一个上界.

一、自适应波束成形系统的增益

考虑如图 1 所示的自适应波束成形系统. 设输入 $x_1(t), \cdots, x_K(t)$ 为宽平稳随机过程,均值为零. $d(t)$ 为期望信号. $H_1(f), \cdots, H_K(f)$ 为 K 个线性滤波器. 按 LMS 准则,我们要求出

$$I = E[z(t) - d(t)]^2 \tag{1}$$

的极小值.

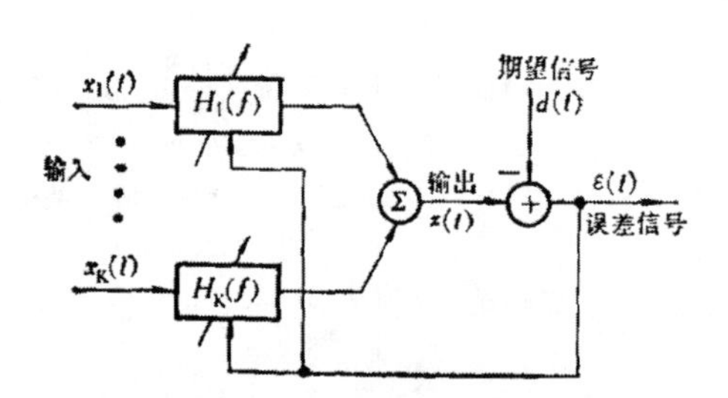

图 1 自适应波束成形方框图

Block-diagram of adaptive beamforming

为书写简单起见,我们将采用向量记号. 用 $\mathbf{H}^T(f) = [\mathbf{H}_1(f), \cdots, \mathbf{H}_K(f)]$ 表示滤波向量,T 表示转置. 用 $\mathbf{K}_{xx}(f)$ 表示输入 $X^T(t) = [x_1(t), \cdots, x_K(t)]$ 的互谱矩阵.

我们知道[9],使 (1) 式极小的最佳传输函数的解是

$$\mathbf{H}_{opt}(f) = [\mathbf{K}_{xx}^*(f)]^{-1}\mathbf{K}_{xd}^*(f) \tag{2}$$

其中 * 表示复共轭, $\mathbf{K}_{xd}(f)$ 表示输入资料和期望信号的互功率谱向量.

我们以后只考虑平面波干扰.

一般情况下,输入 $\mathbf{X}(t)$ 中包含信号、噪声和若干个干扰,所以 $\mathbf{K}_{xx}(f)$ 可以表达为

$$\mathbf{K}_{xx}(f) = K_n(f)\mathbf{I} + K_s(f)\mathbf{A}(f)\mathbf{A}^{*T}(f) + \sum_{k=1}^{N} K_{rk}(f)\mathbf{B}_k(f)\mathbf{B}_k^{*T}(f) \tag{3}$$

其中 $K_n(f), K_r(f), K_s(f)$ 分别表示噪声、干扰和信号的平均功率谱, $\mathbf{I}$ 为单位矩阵.

$$\mathbf{A}^T(f) = (e^{2\pi jf\tau_1}, \cdots, e^{2\pi jf\tau_K}) \tag{4}$$

是信号的指向矢量.

$$\mathbf{B}_k^T(f) = (e^{2\pi jf\delta_{1k}}, \cdots, e^{2\pi jf\delta_{Kk}}) \quad k = 1, \cdots, N \tag{5}$$

是第 k 个干扰的指向矢量.

为计算系统的增益,我们应当首先计算输出信噪比. 它定义为输出信号和输出噪声(包括干扰)功率的比值.

由文献 [9]，输出信号为

$$S_{ad}^2 = \int \mathbf{H}_{opt}^T(f)\mathbf{K}_{ss}(f)\mathbf{H}_{opt}^*(f)df \tag{6}$$

输出噪声为

$$N_{ad}^2 = \int \mathbf{H}_{opt}^T(f)\mathbf{K}_{nn}(f)\mathbf{H}_{opt}^*(f)df \tag{7}$$

输出信噪比定义为

$$\rho_{ad} = \frac{S_{ad}^2}{N_{ad}^2} = \frac{\int \mathbf{H}_{opt}^T(f)\cdot\mathbf{K}_{ss}(f)\cdot\mathbf{H}_{opt}^*(f)df}{\int \mathbf{H}_{opt}^T(f)\cdot\mathbf{K}_{nn}(f)\cdot\mathbf{H}_{opt}^*(f)df} \tag{8}$$

输出信噪比与输入信噪比的比值，就是系统的增益.

$$G_{ad} = \rho_{ad}/L \tag{9}$$

其中

$$L = \sigma_s^2 \Big/ \left(\sigma_n^2 + \sum_{k=1}^{N} \sigma_{rk}^2\right) \tag{10}$$

$$\sigma_s^2 = \int K_s(f)df, \quad \sigma_n^2 = \int K_n(f)df, \quad \sigma_{rk}^2 = \int K_{rk}(f)df \tag{11}$$

显然，如果我们把 (8) 式中的 $\mathbf{H}_{opt}(f)$ 换成普通波束成形的滤波矢量 $\mathbf{H}(f)$，则 (9) 式可以给出普通波束成形系统的增益.

不难看出，G_{ad} 的计算，关键在于求解 $\mathbf{K}_{xx}^{-1}(f)$，下面，我们以只有一个干扰的情况为例，说明如何求出 $\mathbf{K}_{xx}^{-1}(f)$ 的简单表达式，而这里所用的方法可以立刻推广到 $N \geqslant 1$ 的情况.

设

$$\mathbf{K}_{xx}(f) = K_n(f)\mathbf{I} + K_s(f)\mathbf{A}(f)\mathbf{A}^{*T}(f) + K_r(f)\mathbf{B}(f)\mathbf{B}^{*T}(f) \tag{12}$$

令

$$\mathbf{R} = \mathbf{I} + \frac{K_r(f)}{K_n(f)}\mathbf{B}\mathbf{B}^{*T} \tag{13}$$

则

$$\mathbf{K}_{xx}(f) = K_n(f)\left(\mathbf{R} + \frac{K_s(f)}{K_n(f)}\mathbf{A}\mathbf{A}^{*T}\right)$$

显然

$$\mathbf{R}^{-1} = \mathbf{I} - \frac{\mathbf{B}\mathbf{B}^{*T}}{K + [K_n(f)/K_r(f)]} \tag{14}$$

又

$$\begin{aligned}\mathbf{K}_{xx}^{-1}(f) &= \frac{1}{K_n(f)}\cdot\left(\mathbf{R} + \frac{K_s(f)}{K_n(f)}\mathbf{A}\mathbf{A}^{*T}\right)^{-1} \\ &= \frac{1}{K_n(f)}\cdot\left[\mathbf{R}^{-1} - \frac{\mathbf{R}^{-1}\mathbf{A}\mathbf{A}^{*T}\mathbf{R}^{-1}}{\mathbf{A}^{*T}\mathbf{R}^{-1}\mathbf{A} + K_n(f)/K_s(f)}\right]\end{aligned} \tag{15}$$

把 (14) 代到 (15) 中，我们便可以得到由已知量表达出来的，便于计算的 $\mathbf{K}_{xx}^{-1}(f)$ 解.

为了计算 (9)，我们还需作进一步的推导．对于一般的 $\mathbf{H}(f)$，我们有

$$\mathbf{H}^T(f)\mathbf{K}_{ss}(f)\mathbf{H}^*(f) = K_s(f)|\mathbf{H}^T(f)\mathbf{A}(f)|^2 \tag{16}$$

所以

$$S^2 = \int K_s(f)|\mathbf{H}^T(f)\mathbf{A}(f)|^2 df \tag{17}$$

又

$$\mathbf{K}_{nn}(f) = K_n(f)\mathbf{I} + K_r(f)\mathbf{B}(f)\mathbf{B}^{*T}(f)$$

所以

$$\mathbf{H}^T(f)\mathbf{K}_{nn}(f)\mathbf{H}^*(f) = K_n(f)\|\mathbf{H}(f)\|^2 + K_r(f)|\mathbf{H}^T(f)\mathbf{B}(f)|^2 \tag{18}$$

$$N^2 = \int K_n(f)\|\mathbf{H}(f)\|^2 df + \int K_r(f)|\mathbf{H}^T(f)\mathbf{B}(f)|^2 df \tag{19}$$

输出信噪比

$$\rho = \frac{S^2}{N^2} = \frac{\int K_s(f)|\mathbf{H}^T(f)\mathbf{A}(f)|^2 df}{\int K_n(f)\|\mathbf{H}(f)\|^2 df + \int K_r(f)|\mathbf{H}^T(f)\mathbf{B}(f)|^2 df} \tag{20}$$

(20) 式不仅可以用于计算自适应波束成形系统的增益 G_{ad} (令 $\mathbf{H}(f) = \mathbf{H}_{opt}(f)$)，也可以计算普通波束成形系统的增益 G_{CV} (令 $\mathbf{H}(f) = \mathbf{A}^*(f)$).

二、广义夹角

从以上的分析可知，为了计算增益，我们要估计 $|\mathbf{H}^T(f)\mathbf{A}(f)|^2$ 及 $|\mathbf{H}^T(f)\mathbf{B}(f)|^2$．它们分别称为波束指向矢量 $\mathbf{H}(f)$ 与信号指向矢量 $\mathbf{A}(f)$ 及干扰指向矢量 $\mathbf{B}(f)$ 的广义夹角．对于普通波束成形系统，这些量是容易估计的．因为

$$\mathbf{H}(f) = \mathbf{A}^*(f) \tag{21}$$

于是输出信噪比

$$\rho_{CV} = \frac{K^2\sigma_s^2}{K\sigma_n^2 + \int K_r(f)|\mathbf{A}^{*T}(f)\mathbf{B}(f)|^2 df} \tag{22}$$

系统增益

$$G_{CV} = \frac{\rho_{CV}}{\sigma_s^2/(\sigma_n^2 + \sigma_r^2)} \tag{23}$$

特别是，若不存在干扰，则 $K_r(f) = 0$，推知

$$G_{CV} = K \tag{24}$$

(22) 及 (23) 中的 $|\mathbf{A}^T(f)\mathbf{B}(f)|^2$ 经归一化处理之后，就得到信号指向矢量 $\mathbf{A}(f)$ 和干扰指向矢量 $\mathbf{B}(f)$ 的广义夹角．我们用 $\cos^2\gamma$ 来表示：

$$|\mathbf{A}^{*T}(f)\mathbf{B}(f)|^2 = K^2\cos^2\gamma \tag{25}$$

显然 $0 \leqslant \cos^2\gamma \leqslant 1$．它实际上反映干扰和信号在空间上的相互关系，我们将在下节给出具体结果．

对于自适应波束成形，$\mathbf{H}_{opt}(f)$ 由 (2) 式给出，注意到 $\mathbf{K}_{xx}(f)$ 是 Hrmite 矩阵，即 $\mathbf{K}_{xx}^*(f) = \mathbf{K}_{xx}^T(f)$．以及 $\mathbf{K}_{xd}(f) = K_s(f)\mathbf{A}(f)$，我们有

$$\mathbf{H}_{\text{opt}}^T(f)\mathbf{A}(f) = K_s(f)\mathbf{A}^{*T}(f)\boldsymbol{K}_{xx}^{-1}(f)\mathbf{A}(f)$$
$$= \frac{K_s(f)}{K_n(f)} \cdot \mathbf{A}^{*T}\left(\mathbf{R}^{-1} - \frac{\mathbf{R}^{-1}\mathbf{A}\mathbf{A}^{*T}\mathbf{R}^{-1}}{\mathbf{A}^{*T}\mathbf{R}^{-1}\mathbf{A} + K_n/K_s}\right)\mathbf{A}$$

把 $\boldsymbol{R}^{-1}$ 的表达式 (14) 代入上式,并予以简化,得到

$$|\mathbf{H}_{\text{opt}}^T(f)\mathbf{A}(f)|^2 = P^2(f)/[P(f) + K_n(f)/K_s(f)]^2 \tag{26}$$

其中

$$P(f) = \mathbf{A}^{*T}\mathbf{R}^{-1}\mathbf{A} = K - K^2\cos^2\gamma/(K + K_n(f)/K_r(f)) \tag{27}$$

所以输出信号(参看 (17) 式)是

$$S_{ad}^2 = \int K_s(f)[P(f)/(P(f) + K_n(f)/K_s(f))]^2 df \tag{28}$$

类似的计算可以得到输出噪声的表达式(参看 (19) 式):

$$N_{ad}^2 = \int K_n(f)U(f)[1/(P(f) + K_n(f)/K_s(f))]^2 df$$
$$+ \int K_r(f)[|Q(f)|^2/(P(f) + K_n(f)/K_s(f))^2]df \tag{29}$$

其中

$$U(f) = \mathbf{A}^{*T}(\mathbf{R}^{-1})^2\mathbf{A}$$
$$= K - \frac{2K^2\cos^2\gamma}{K + (K_n(f)/K_r(f))} + \frac{K^3\cos^2\gamma}{[K+(K_n(f)/K_r(f))]^2} \tag{30}$$

$$|Q(f)|^2 = |\mathbf{A}^{*T}\mathbf{R}^{-1}\mathbf{B}|^2$$
$$= \frac{(K_n(f)/K_r(f))^2K^2\cos^2\gamma}{(K + K_n(f)/K_r(f))^2} \tag{31}$$

代回 (8) 及 (9) 式,得到自适应系统的输出信噪比及增益:

$$\rho_{ad} = \frac{\int K_s(f)[P(f)/(P(f) + K_n(f)/K_s(f)]^2 df}{\int K_n(f)\dfrac{u(f)\cdot df}{[P(f) + K_n(f)/K_r(f)]^2} + \int K_r(f)\dfrac{|Q(f)|^2\cdot df}{[P(f) + K_n(f)/K_r(f)]^2}} \tag{32}$$

$$G_{ad} = \rho_{ad}/[\sigma_s^2/(\sigma_n^2 + \sigma_r^2)] \tag{33}$$

$P(f)$, $U(f)$, $Q(f)$ 作为广义夹角的函数,可以分别由 (27),(30),(31)式给出,它显然与输入的干扰/噪声比有关.

图 2 及图 3 给出了 $K = 15$ 时 G_{CV} 与 G_{ad} 的计算结果, 其中 $K_n(f)$, $K_r(f)$ 假定为矩形宽带谱,广义夹角 $\cos^2\gamma$ 取其带宽的中心频率从积分号内提出来.

从图中可以看到, G_{CV} 及 G_{ad} 都与 σ_n^2/σ_r^2 有关, σ_n^2/σ_r^2 越小, G_{ad} 就越大, 而 G_{CV} 却正好相反,这一点是易于理解的. 当 σ_n^2/σ_r^2 相当大时, 自适应波束成形系统的优越性就微乎其微了. 但是我们一般设计自适应系统的目的是对付强干扰,所以上述情况并不会出现.

当 σ_n^2/σ_r^2 在 0.1 左右时,只要广义夹角比较小, G_{ad} 一般比 G_{CV} 高出 5—14 分贝.

当参数 K, σ_n^2/σ_r^2, $\cos^2\gamma$ 改变时, 我们仍可以由 (22), (23), (32), (33) 得到 G_{CV} 及 G_{ad} 的精确估计.

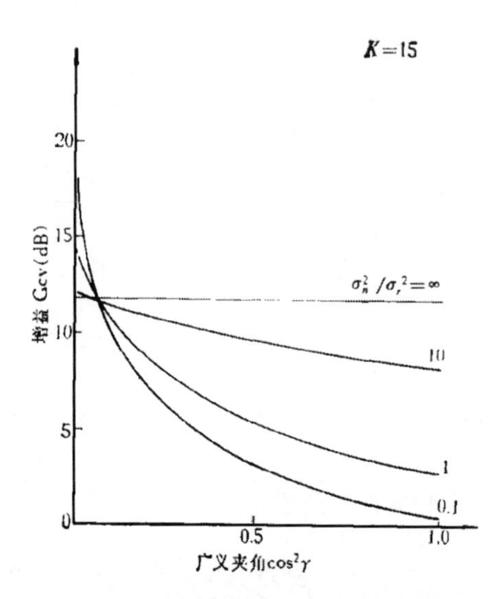

图 2 普通波束成形系统的抗干扰增益

The gain of conventional beamforming system to reject inteference

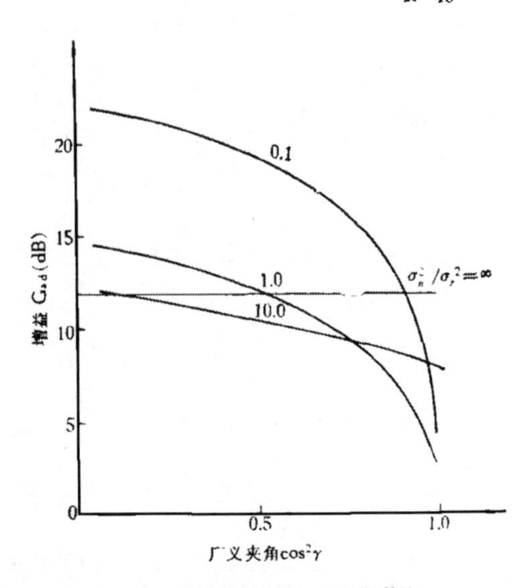

图 3 自适应波束成形系统抗干扰增益

The gain of adaptive beamforming system to reject inteference

三、极 限 分 辨 力

在设计声呐时，我们感兴趣的是，当干扰与信号的夹角θ为多大时，自适应系统仍能有效抑制干扰，这就是极限分辨力问题.

我们知道，普通波束成形的分辨力很差，一般是大于主瓣半宽的. 但是自适应波束成形系统具有高得多的分辨力. 对任何具体的声呐基阵，我们不难计算$\cos^2\gamma$和θ的关系. 然后根

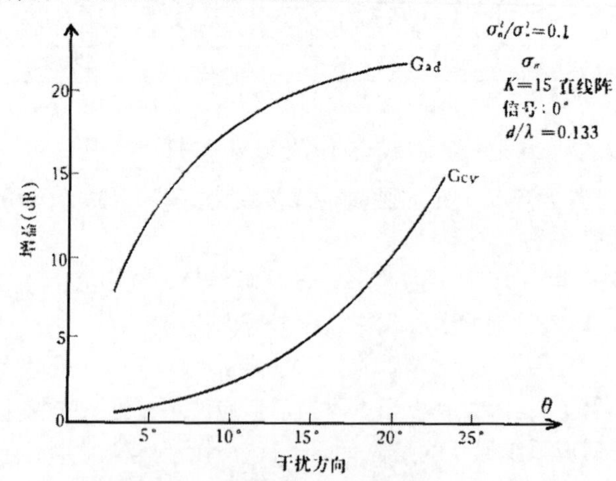

图 4 普通波束成形系统和自适应波束成形系统的极限分辨力

The limit resolution ability of conventional beamforming and adaptive beamforming system

据第二节中的分析就可以得到相应的 G_{CV} 及 G_{ad} 值.

我们以 $K=15$ 元的等间隔直线阵为例予以说明,易知

$$\cos^2\gamma=[\sin(K\pi d\sin\theta/\lambda)/K\sin(\pi d\sin\theta/\lambda)]^2 \tag{34}$$

由此可知 θ 和 $\cos^2\gamma$ 的关系,再由(23),(33)得出 G_{CV} 及 G_{ad} 来,图 4 给出这种基阵在 $d/\lambda=0.133$ 时自适应波束成形下的极限分辨力.

从图中我们看到,当干扰处于 3° 时,已落在普通波束成形的主瓣之内,普通波束成形无法分辨,而自适应系统仍有足够的增益.

四、最佳权系数的动态

我们知道,频域上最佳自适应滤波器的解是[9]

$$\mathbf{H}_{opt}(f)=[\mathbf{K}_{xx}^{*}(f)]^{-1}\mathbf{K}_{xd}(f) \tag{35}$$

时域上最佳加权系数的解是[1]

$$\mathbf{W}_{opt}=\mathbf{R}_{xx}^{-1}\mathbf{R}_{xd} \tag{36}$$

当我们把抽头延迟线看作横向滤波器时,$\mathbf{W}_{opt}$ 和 $\mathbf{H}_{opt}(f)$ 的关系是明确的[9]. 即 $\mathbf{H}_{opt}(f)$ 的第 k 个分量的 Fourier 展开式的系数就是 $\mathbf{W}_{opt}$ 中第 k 组权系数.

自适应系统中权系数的动态是一个重要参数. 动态设计过大,将使机器复杂且无必要;动态过小将引起溢出,而得不到真正的 Wiener 解. 所以,我们一定要针对可能遇到的干扰场的环境,大致估计一下 $\mathbf{W}_{opt}$ 的动态.

从理论上讲,对 $\mathbf{W}_{opt}$ 动态的控制可以在迭代过程中自行调节. 因为,由(36)式我们可以看出,如果在迭代过程中,出现 $\mathbf{W}$ 溢出的情况,则我们可以采用两种办法使迭代过程继续下去. 第一种是增加输入 $X(t)$ 矢量的动态,从而使 $\mathbf{R}_{xx}^{-1}$ 减小;第二种办法是减小 $\mathbf{R}_{xd}$ 的动态,以避免 $\mathbf{W}$ 溢出. 这两种办法在实施过程中都有问题,前者可能使输入通道过于庞大,后者可能使 $\mathbf{R}_{xd}$ 量化的比特数减至 1,从而得不到 Wiener 最佳解.

所以,最好的办法是事先合理估计 $\mathbf{W}_{opt}$ 的动态.

所谓 $\mathbf{W}_{opt}$ 的动态问题,我们是指 $\mathbf{W}_{opt}$ 的每个分量 W_i, $i=1,\cdots,N_0$ 的大小范围,而不是指其任何两个分量的比值,因为后者在实用上并无意义.

对 $\mathbf{H}_{opt}(f)$ 的每一个分量用一下 Parseval 公式,我们有

$$(1/\Delta f)\int\|\mathbf{H}_{opt}(f)\|^2df=\|\mathbf{W}_{opt}\|^2=\sum_{i=1}^{N_0}w_i^2 \tag{37}$$

设 w_{max} 为 $|w_i|$ 中之最大值,则由(37)可知

$$w_{max}\leqslant\left[(1/\Delta f)\int\|\mathbf{H}_{opt}(f)\|^2df\right]^{1/2} \tag{38}$$

关于 $\|\mathbf{H}_{opt}(f)\|^2$ 的计算,我们在第一节中已提及,即

$$\begin{aligned}\|\mathbf{H}_{opt}(f)\|^2&=\mathbf{K}_{xd}^{*T}(f)\mathbf{K}_{xx}^{-1}(f)\mathbf{K}_{xx}^{-1}(f)\mathbf{K}_{xd}(f)\\&=K_d^2(f)\mathbf{A}^{*T}(f)[\mathbf{K}_{xx}^{-1}(f)]^2\mathbf{A}(f)\\&=\frac{K_d^2(f)}{K_s^2(f)}\cdot U(f)\cdot\frac{1}{(p(f)+K_n(f)/K_s(f))^2}\end{aligned} \tag{39}$$

把(27),(30)代入(39)得到

$$\|\mathbf{H}_{opt}(f)\| = \frac{K_d^2(f)}{K_s^2(f)} \frac{K\{(K+K_n/K_r)^2 - 2K\cos^2\gamma(K+K_n/K_r) + K^2\cos^2\gamma\}}{[K^2(1-\cos^2\gamma) + K\cdot K_n/K_r + (K+K_n/K_r)K_n/K_s]}$$

$$\leqslant \left(\frac{K_d}{K_s}\right)^2 \cdot \frac{K\{(K+K_n/K_r)^2 + K^2/4\}}{[K\cdot K_n/K_r + K\cdot K_n/K_s]^2} \leqslant \left(\frac{K_d}{K_s}\right)^2 \cdot \frac{(K+1)^2+K^2/4}{K\cdot(K_n/K_s)^2}$$

$$\leqslant (3K/2)(K_d(f)/K_n(f))^2 \tag{40}$$

如果 $K_n(f)$ 是矩形的话,则

$$w_{\max} \leqslant \sqrt{3K/2}\cdot\sigma_d^2/\sigma_n^2 \tag{41}$$

σ_d^2/σ_n^2 实际上就是估计量的动态和输入动态的比值. (41) 给出一切 w_i 的一个上界,它与通道数 K 的方根成正比.

五、结　　论

按 LMS 准则设计的自适应波束成形系统,具有比普通波束成形系统高得多的增益,它在数值上和干扰与信号的广义夹角有关,同时又与干扰/噪声比有关. 对于强干扰来说,自适应波束成形的极限分辨力远优于普通波束成形系统.

最优加权系数的动态的上界和 $\sqrt{K}$ 成正比,又与期望信号与输入资料的功率比成正比.

附　　录

如果有多个干扰,上面提出的计算方法仍旧有效. 这儿我们给出具体推导方法.

$\mathbf{K}_{xx}(f)$ 由(3)式给出:

$$\mathbf{K}_{xx}(f) = K_n(f)\mathbf{I} + \sum_{k=1}^{N} K_{rk}\mathbf{B}_k(f)\mathbf{B}_k^{*T}(f) + K_s(f)\mathbf{A}(f)\cdot\mathbf{A}^{*T}(f)$$

令

$$\mathbf{R} = \mathbf{I} + \sum_{k=1}^{N}[K_{rk}(f)/K_n(f)]\mathbf{B}_k(f)\mathbf{B}_k^{*T}(f)$$

则

$$\mathbf{K}_{xx}(f) = K_n(f)[\mathbf{R} + [K_s(f)/K_n(f)]\mathbf{A}\mathbf{A}^{*T}]$$

容易证明

$$\mathbf{K}_{xx}^{-1}(f) = \frac{1}{K_n(f)}\left(\mathbf{R}^{-1} - \frac{\mathbf{R}^{-1}\mathbf{A}\mathbf{A}^{*T}\mathbf{R}^{-1}}{\mathbf{A}^{*T}\mathbf{R}^{-1}\mathbf{A} + K_n(f)/K_s(f)}\right)$$

至于 $\mathbf{R}^{-1}$ 可以这样求:

$$\mathbf{R} = \mathbf{I} + \sum_{k=1}^{N-1}[K_{rk}(f)/K_n(f)]\mathbf{B}_k\mathbf{B}_k^{*T} + [K_{rN}(f)/K_n(f)]\mathbf{B}_N\mathbf{B}_N^{*T}$$

再令

$$\mathbf{R}_1 = \mathbf{I} + \sum_{k=1}^{N-1}(K_{rk}/K_n)\mathbf{B}_k\mathbf{B}_k^{*T}$$

则易求出 $\mathbf{R}^{-1}$, 这时 $\mathbf{R}_1$ 已经比 $\mathbf{R}$ 少了一个矩阵,如此继续,最后可以将 $\mathbf{K}_{xx}^{-1}(f)$ 表示为已知矩阵的表达式.

本文部分初稿曾经侯自强同志审阅,作者表示衷心感谢.

参 考 文 献

[1] Widrow, B. et al., "Adaptive Antenna Systems" *Proc. IEEE*, **55**(1967), No. 12, 2143—2159.

[2] Cox, H., "Optimum Arrays and the Schwarz Inequality" *J. A. S. A.* **45**(1969) No. 1, 228—232.

[3] Edelblute, D. J. et al., "Criteria for Optimum Signal-Detection Theory for Array" *J. A. S. A.* **41** (1967), No. 1, 199—205.

[4] Cox, H., "Sensitivity Consideration in Adaptive Beamforming" *Proc. NATO Advanced Study Institute on Signal Processing* (1972).

[5] Griffiths, L. J., "A Simple Adaptive Algorithm for Real-Time Processing in Antenna Arrays" *Proc. IEEE.* **57**(1969), No. 10, 1696—1704.

[6] Chang, J. H and Tuteur, F. B., "Optimum Adaptive Array Processor" *Proc. of the Symp on Computer Processing in Communication*, Micrcwave Res. Inst. Symposis Series **19**(1969), 695—710.

[7] Frost, O. L. III., "An Algorithm for Linearly Constrained Adaptive Array Processing" *Proc. IEEE.* **60**(1972), No. 8, 926—935.

[8] Widrow, B. et al., "Adaptive Noise Canceling: Principle and Application" *Proc. IEEE.* **63**(1975). No. 12, 1692—1716.

[9] 李启虎,"自适应波束成形中最优解在时域和频域中的等价关系",声学学报,1979年第4期,296—308.

THE LIMIT RESOLUTION ABILITY AND DYNAMIC RANGE OF OPTIMUM WEIGH COEFFICIENTS OF ADAPTIVE BEAMFORMING SYSTEM

LI Qi-hu

(*Institute of Acoustics, Academia Sinica*)

Some problems of designing adaptive beamforming sonar are considered.

One is the limit resolution ability of an adaptive beamforing system. The general presentation of output signal to noise ratio and gain of such system are given.

A Method for calculating the ability of adaptive beamforming to reject more than one planan wave interference is shown.

As a function of generalized angle between signal and interference, the formula for evaluating system gain is driven. The corresponding numerical results are presented.

We conclude that the limit resolution ability depends not only on the generalized angle, but also on the interference to noise ratio, some concrete examples are shown.

The other problem considered in this paper is the dynamic range of optimum weight coefficients. In term of the solution of optimum transfer function in frequency domain, a upper bound of optimum weight coefficients is presented.

The results given in this paper can be easily used in the design of a practical adaptive beamforming radar or sonar system.

自适应波束成形中稳态特性的研究①

李 启 虎

（中国科学院声学研究所）

本文给出按 LMS 准则设计的自适应系统达到稳态时，系统指向性的一般表达式．作为信号和干扰之间的广义夹角的函数，提出计算主瓣宽度、干扰方向零点深度和极限分辨力的方法．指出在一般情况下，指向性极大值对信号方向的偏离值．各种情况下都给出数值计算的例子．

一、引　　言

以 B. Widrow 提出的最速下降法为基础的迭代算法，已经在自适应系统中得到广泛应用[1−7]．自适应波束成形理论研究有两个主要的课题．一个是把迭代作为一个过程来研究，这就是 B. Widrow 等所要建立的“自适应过程的统计理论”[6]．另一个是研究当自适应达到稳态时系统所具有的性质．这两个问题的解决对自适应波束成形系统的设计来说都是十分重要的．

由于加权系数的最优解依赖于输入的互相关矩阵，直接求解十分困难．所以在迭代达到稳态之前，要计算系统的指向性等稳态特性是不容易的．

目前有关的文献都是根据具体的环境场的情况，通过计算机进行系统模拟，在迭代的不同阶段将权系数输出，然后计算指向性[1,5,7]．

J. H. Chang 等曾试图在极特殊的假设下分析自适应系统的稳态特性[3]．他指出，在自适应达到稳态时，指向性将在干扰方向形成一个零点．W. E. Rodgers 最近分析了双基元阵的自适应系统抗干扰的能力[9]．

本文利用频域上最佳传输函数的解，给出稳态指向性的一般表达式．对于具体的环境场参数，不用进行迭代运算就可以事先分析指向性的主要特性．首先指出，在一般情况下，指向性的极大值方向对信号所在的方位角有一偏离，偏离角的大小与干扰所在的方位角有关；自适应系统在干扰所在的方向形成一个零点，零点的深度实际上就表示该系统抑制干扰的能力．然后分析主瓣的宽度，我们将指出，自适应系统指向性的主瓣宽度与普通波束成形大体一样．最后我们给出稳态增益的一般表达式，将由此来计算系统的极限分辨力．

在各种情况下，我们都给出数值计算的例子，说明如何按本文所提供的分析方法用于实际声呐的设计．

① 声学学报, 1982, 7(3): 165-173.

二、稳态指向性的表达式

图 1 给出频域上波束成形的方框图. 我们假定输入 $x_1(t), \cdots, x_K(t)$ 为平稳的随机过程. 用矢量表示为

$$\boldsymbol{X}^T(t) = [x_1(t), \cdots, x_K(t)] \tag{1}$$

其中 T 表示转置. 设 $\boldsymbol{X}(t)$ 的每一个分量都由信号及 M 个干扰和各向同性的噪声构成. 信号的方位角为 θ_0, 干扰的方位角分别为 $\theta_1, \cdots, \theta_M$. 信号与干扰的指向矢量分别用 $\boldsymbol{A}(f, \theta_0)$ 与 $\boldsymbol{A}(f, \theta_m)$ $m = 1, \cdots, M$ 来表示.

$$\boldsymbol{A}^T(f, \theta_0) = [e^{2\pi jf\tau_1(\theta_0)}, \cdots, e^{2\pi jf\tau_K(\theta_0)}] \tag{2}$$

$$\boldsymbol{A}^T(f,\theta_m) = [e^{2\pi jf\tau_1(\theta_m)}, \cdots, e^{2\pi jf\tau_K(\theta_m)}] \quad m = 1, \cdots, M \tag{3}$$

输入的互谱矩阵可以表示为

$$\boldsymbol{K}_{xx}(f) = K_n(f)\boldsymbol{I} + K_s(f)\boldsymbol{A}(f,\theta_0)\boldsymbol{A}^{*T}(f,\theta_0) + \sum_{m=1}^{M} K_{rm}(f)\boldsymbol{A}(f,\theta_m)\boldsymbol{A}^{*T}(f,\theta_m) \tag{4}$$

其中 $K_n(f)$, $K_s(f)$, $K_{rm}(f)$ 分别表示噪声、信号、第 m 个干扰的平均功率谱密度, $\boldsymbol{I}$ 表示单位矩阵.

K 个滤波器的传输函数 $H_1(f), \cdots, H_K(f)$ 也用矢量 $\boldsymbol{H}(f)$ 来表示:

$$\boldsymbol{H}^T(f) = [H_1(f), \cdots, H_K(f)] \tag{5}$$

我们用 $\boldsymbol{A}(f, \theta)$ 来表示波束的指向矢量,那么在图 1(a) 的普通波束成形时,

$$\boldsymbol{H}_{cv}(f) = \boldsymbol{A}^*(f, \theta_0) \tag{6}$$

所以,频率为 f 时,单频指向性是[10]:

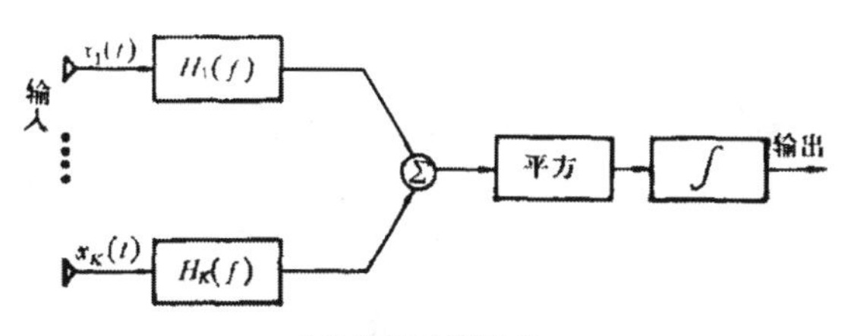

(a) 普通波束形成

Conventional beamforming

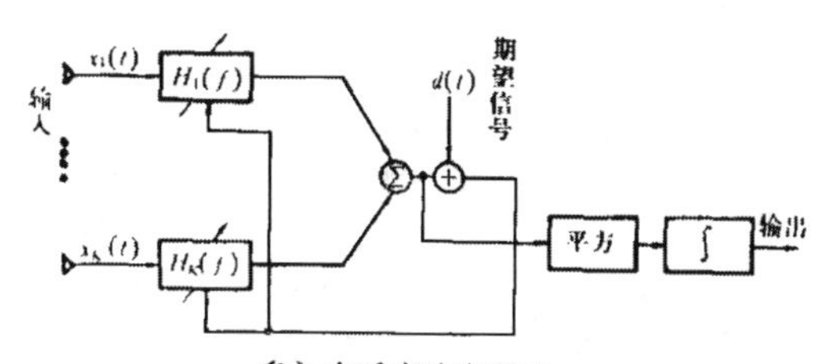

(b) 自适应波束形成

Adaptive beamforming

图 1 频域上波束形成

The beamforming in frequency-domain

$$D_{cv}(\theta)=\boldsymbol{H}_{cv}^{T}(f)\boldsymbol{K}_{ss}(f)\boldsymbol{H}_{cv}^{*}(f) \tag{7}$$

其中 $\boldsymbol{K}_{ss}(f)$ 为信号的互谱矩阵，

$$\boldsymbol{K}_{ss}(f)=K_s(f)\boldsymbol{A}(f,\theta_0)\boldsymbol{A}^{*T}(f,\theta_0)$$

代入(7)便得到

$$D_{cv}(\theta)=K_s(f)\,|\boldsymbol{A}^{*T}(f,\theta_0)\boldsymbol{A}(f,\theta_0)|^2 \tag{8}$$

对于自适应系统(见图 1(b))，当它达到稳态时，

$$\boldsymbol{H}_{opt}(f)=[\boldsymbol{K}_{xx}^{*}(f)]^{-1}\boldsymbol{K}_{xd}^{*}(f) \tag{9}$$

这里 $\boldsymbol{K}_{xd}(f)$ 表示输入 $\boldsymbol{X}(t)$ 与期望信号的互谱矢量，所以它的指向性为

$$D_{ad}(\theta)=\boldsymbol{H}_{opt}^{T}(f)\boldsymbol{K}_{ss}(f)\boldsymbol{H}_{opt}^{*}(f) \tag{10}$$

当我们对期望信号估计无误时，应当有

$$\boldsymbol{K}_{xd}(f)=K_s(f)\boldsymbol{A}(f,\theta_0)$$

将其代入(10)式，即得到：

$$D_{ad}(\theta)=K_s^3(f)\,|\boldsymbol{A}^{*T}(f,\theta_0)\boldsymbol{K}_{xx}^{-1}(f)\boldsymbol{A}(f,\theta)|^2 \tag{11}$$

由此可知，$D_{ad}(\theta)$ 的计算，关键在于 $\boldsymbol{K}_{xx}^{-1}(f)$.

对于(4)式这种类型的环境场，我们知道[8]

$$\boldsymbol{K}_{xx}^{-1}(f)=\frac{1}{K_n(f)}\left(\boldsymbol{I}-\sum_{i,j=0}^{M}B_{ij}\boldsymbol{A}_i\boldsymbol{A}_j^{*T}\right) \tag{12}$$

其中 $\boldsymbol{A}_i$ 表示 $\boldsymbol{A}(f,\theta_i)$，$i=0,1,\cdots,M$. B_{ij} 是下面定义的矩阵 $\boldsymbol{B}$ 的第 (i,j) 个元素：

$$\boldsymbol{B}=(\boldsymbol{I}+\boldsymbol{DG})^{-1}\boldsymbol{D}$$

$$(\boldsymbol{G})_{ij}=\boldsymbol{A}_i^{*T}\boldsymbol{A}_j,\ (\boldsymbol{D})_{ij}=\frac{K_{ri}(f)}{K_n(f)}\delta_{ij} \tag{13}$$

δ_{ij} 表示 Kroneker δ.

把(12)式用于理论分析是方便的，但在具体计算时，我们采用以下方法较为实用. 下面以 $M=1$ 加以说明.

当 $M=1$ 时，

$$\boldsymbol{K}_{xx}(f)=K_n(f)\boldsymbol{I}+K_s(f)\boldsymbol{A}_0\boldsymbol{A}_0^{*T}+K_r(f)\boldsymbol{A}_1\boldsymbol{A}_1^{*T} \tag{14}$$

这种矩阵的逆矩阵可以分两步来求，即

令

$$\boldsymbol{Z}=\boldsymbol{I}+\frac{K_r(f)}{K_n(f)}\boldsymbol{A}_1\boldsymbol{A}_1^{*T},$$

则

$$\boldsymbol{Z}^{-1}=\boldsymbol{I}-\frac{\boldsymbol{A}_1\boldsymbol{A}_1^{*T}}{K+K_n(f)/K_r(f)} \tag{15}$$

$$\boldsymbol{K}_{xx}(f)=K_n(f)\left(\boldsymbol{Z}+\frac{K_s(f)}{K_n(f)}\boldsymbol{A}_0\boldsymbol{A}_0^{*T}\right)$$

容易验证

$$\boldsymbol{K}_{xx}^{-1}(f)=\frac{1}{K_n(f)}\left(\boldsymbol{Z}^{-1}-\frac{\boldsymbol{Z}^{-1}\boldsymbol{A}_0\boldsymbol{A}_0^{*T}\boldsymbol{Z}^{-1}}{\boldsymbol{A}_0^{*T}\boldsymbol{Z}^{-1}\boldsymbol{A}_0+K_n(f)/K_s(f)}\right) \tag{16}$$

将这些表达式代入(11)式，经过必要的化简，可以得到

$$D_{ad}(\theta)=\frac{K_s^3(f)}{K_n^2(f)}\left[1-\frac{p}{p+K_n(f)/K_s(f)}\right]^2|q(\theta)|^2 \tag{17}$$

其中

$$p=\boldsymbol{A}_0^*\boldsymbol{Z}^{-1}\boldsymbol{A}_0=K-\frac{K^2\cos^2\gamma}{K+K_n(f)/K_r(f)} \tag{18}$$

$$q(\theta)=\boldsymbol{A}_0^{*T}\boldsymbol{Z}^{-1}\boldsymbol{A} \tag{19}$$

$\boldsymbol{A}$ 表示 $\boldsymbol{A}(f,\theta)$; $\cos^2\gamma$ 为 $\boldsymbol{A}_0$, $\boldsymbol{A}_1$ 之间的广义夹角，即

$$|\boldsymbol{A}_0^{*T}\boldsymbol{A}_1|^2=K^2\cos^2\gamma \tag{20}$$

在(17)式中，我们看到，对于单频信号而言，当信号，干扰的方位角 θ_0, θ_1 给定，以及它们的信/干比 $K_s(f)/K_r(f)$ 与干/噪比 $K_r(f)/K_n(f)$ 给定之后，$D_{ad}(\theta)$ 仅与 $|q(\theta)|^2$ 有关. 所以只要研究 $|q(\theta)|^2$ 的特性便可知道自适应达到稳态时系统的指向性. 我们首先来求 Max $|q(\theta)|^2$. 为此，

令

$$\left.\begin{aligned}&\boldsymbol{A}_i^{*T}\boldsymbol{A}=K|\cos\gamma_i|e^{2\pi jf\varphi_i}\quad i=0,1\\&\boldsymbol{A}_0^{*T}\boldsymbol{A}_1=K|\cos\gamma|e^{2\pi jf\varphi}\\&K+K_n(f)/K_r(f)=a\end{aligned}\right\} \tag{21}$$

代入(19)式，经过计算，可得

$$|q(\theta)|^2=K^2\cos^2\gamma_0+\frac{1}{a^2}K^4\cos^2\gamma\cos^2\gamma_1$$
$$-\frac{2}{a}K^3|\cos\gamma\cos\gamma_0\cos\gamma_1|\cos 2\pi f(\varphi_0-\varphi-\varphi_1) \tag{22}$$

为了说明问题简单起见，我们分析两种有实用意义的情况:

i) $\theta_1\approx\theta_0$, $K_r(f)\gg K_n(f)$, 这时

$\varphi_1\approx\varphi_0$, $\varphi\approx 0$, $a\approx K$. 故

$$|q(\theta)|^2\approx K^2(|\cos\gamma_0|-|\cos\gamma\cos\gamma_1|)^2 \tag{23}$$

对于任何具体的基阵，$\cos\gamma_0$, $\cos\gamma_1$, $\cos\gamma$ 都不难通过数值计算求出来，由此便可求出 Max $|q(\theta)|^2$ 来.

我们以 $K=15$, $d/\lambda=0.2$ 的直线阵为例，设 $\theta_0=0°$, 则

$$\cos^2\gamma_0=\left[\frac{\sin\left(K\dfrac{\pi d}{\lambda}\sin\theta\right)}{K\sin\left(\dfrac{\pi d}{\lambda}\sin\theta\right)}\right]^2 \tag{24}$$

类似地可以得出 $\cos\gamma$, $\cos\gamma_1$ 的表达式，将其代入(23)，进行计算，便可以得到 $|q(\theta)|^2$ 极大值所在的方位角 $\hat{\theta}$. 我们称 $\theta_b=\hat{\theta}-\theta_0$ 为偏离角. 偏离角 θ_b 与干扰方位角 θ_1 的关系如下表:

干扰方向 θ_1	2°	4°	6°	8°
偏离角 θ_b	−12°	−10°	−8°	−6°

由此可见，自适应波束成形系统不同于普通波束成形系统，它的极大值方向对信号所在的方位角有一偏离(见图 2(a))．这一点已为实验所证实，国外 Compton 也曾作过报道[11]

ii) θ_1 与 θ_0 相隔较远，这时 $|\cos\gamma|\approx 0$，于是

$$|q(\theta)|^2 \approx K^2\cos^2\gamma_0 \tag{25}$$

这实际上就是(7)式，所以不存在主瓣方向的偏离问题．

根据(23)式还可以估计主瓣的半宽度，图 2(b) 给出了具体的计算结果．我们看到，主瓣半宽度 $\Delta\theta$ 基本上与广义夹角无关．稳态时的 $\Delta\theta$ 与普通波束成形的 $\Delta\theta$ 大致一样．

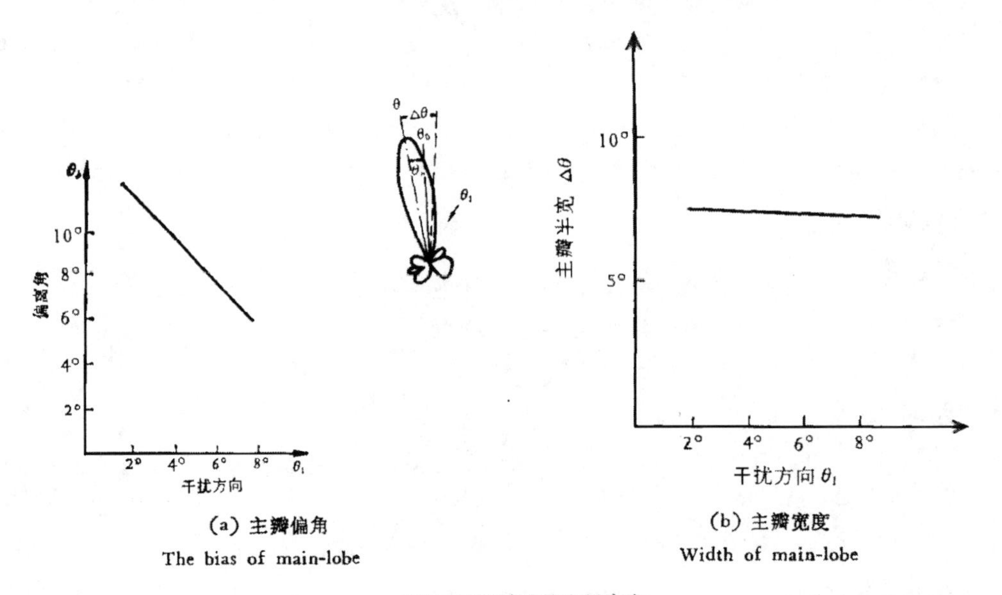

(a) 主瓣偏角

The bias of main-lobe

(b) 主瓣宽度

Width of main-lobe

图 2 主瓣方向和主瓣宽度

The direction of main-lobe and the width of main-lobe

三、干扰方向的零点深度

自适应波束成形系统会在干扰方向形成一个零点，这是人们早已知道的，我们本节中要研究的是这个零点的深度与哪些参量有关．

如果我们将(23)式对 θ 求微商，令第一个因子为零，就有

$$|\cos\gamma_0| - |\cos\gamma\cos\gamma_1| = 0 \tag{26}$$

显然，当 $\theta=\theta_1$ 时 $\cos\gamma_0=\cos\gamma$，$\cos\gamma_1=1$，故(26)成立．可见，θ_1 为 $|q(\theta)|^2$ 的一个零点．我们称 $|q(\theta_1)|^2$ 与 $|q(\theta_0)|^2$ 的比值为零点深度，用 ζ 来表示其平方根：

$$\zeta=\left(\frac{|q(\theta_1)|^2}{|q(\theta_0)|^2}\right)^{1/2} \tag{27}$$

它实际上反映出自适应系统的抑制干扰的能力．由(16)，(19)式不难计算出

$$|q(\theta_0)|^2=\left(K-\frac{K^2\cos^2\gamma}{K+K_n(f)/K_r(f)}\right)^2$$

$$|q(\theta_1)|^2 = K^2\cos^2\gamma\left(1 - \frac{K}{K + K_n(f)/K_r(f)}\right)^2$$

故

$$\zeta = \frac{\cos\gamma\left(1 - \dfrac{K}{K + K_n(f)/K_r(f)}\right)}{1 - \dfrac{K\cos^2\gamma}{K + K_n(f)/K_r(f)}} \tag{28}$$

这是以广义夹角为变量的零点深度表达式，我们有时候用 $20\lg\zeta$ 来表示. 图 3 给出了 $K = 15$ 时的情况.

对于具体的基阵，由于 $\cos^2\gamma$ 与干扰方位角 θ_1 有关，所以我们可以通过类似于(24)式的计算获得全方位的零点分布图. 图 4 给出了一个 $K = 15$, $d/\lambda = 0.2$ 的线阵的零点深度分布. 为了比较起见，我们在图中还给出了同一假定之下普通波束成形的指向性. 从图 4 我们可以确定这种自适应系统抑制干扰的能力.

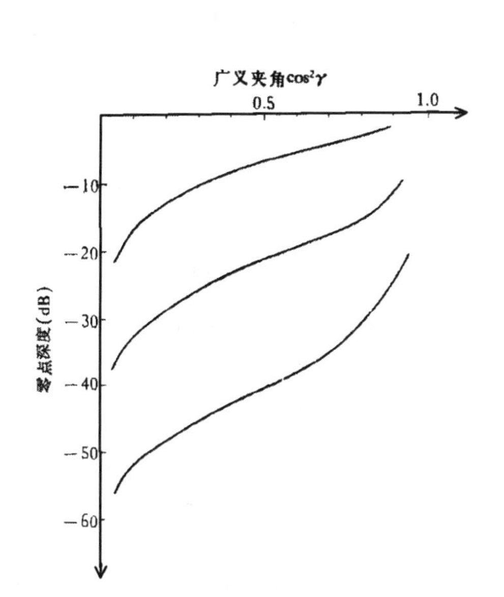

图 3 零点深度与广义夹角的关系

The relation between deepness of zero point and generlized angle

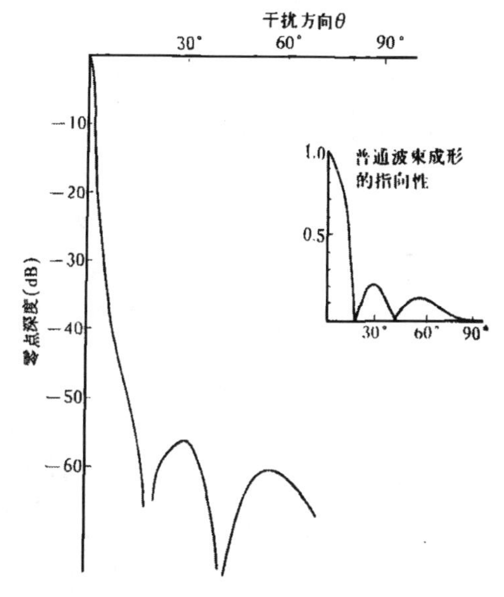

图 4 线阵的零点深度分布 ($K = 15$, $d/\lambda = 0.2$)

The distribution of deepness of zero point for linear array, $K = 15$, $d/\lambda = 0.2$

四、系统的极限分辨力

前已谈到，自适应波束成形系统当干扰与信号在方向上非常接近时，主瓣方向将产生偏离. 这种偏离实际上是自适应系统具有较高分辨力的内在原因. 下面我们来计算自适应系统的增益，进而分析它的极限分辨力. 仍旧以一个干扰为例子.

仅有信号时的输出为

$$S^2=\int \boldsymbol{H}^T(f)\boldsymbol{K}_{ss}(f)\boldsymbol{H}^*(f)df=\int K_s(f)|\boldsymbol{H}^T(f)\boldsymbol{A}_1|^2df \tag{29}$$

仅有噪声(包括干扰,下同)时的输出为

$$\begin{aligned}N^2&=\int \boldsymbol{H}^T(f)\boldsymbol{K}_{nn}(f)\boldsymbol{H}^*(f)df\\&=\int K_n(f)\|\boldsymbol{H}(f)\|^2df+\int K_r(f)|\boldsymbol{H}^T(f)\boldsymbol{A}_1|^2df\end{aligned} \tag{30}$$

其中 $\boldsymbol{K}_{nn}(f)$ 表示噪声加干扰的互谱矩阵.

输出信噪比是

$$\rho=\frac{S^2}{N^2}=\frac{\int K_s(f)|\boldsymbol{H}^T(f)\boldsymbol{A}_0|^2df}{\int K_n(f)\|\boldsymbol{H}(f)\|^2df+\int K_r(f)|\boldsymbol{H}^T(f)\boldsymbol{A}_1|^2df} \tag{31}$$

输出信噪比对输入信噪比 $L=\sigma_s^2/(\sigma_n^2+\sigma_r^2)$ 的比值就定义作系统的增益:

$$G=\rho/L \tag{32}$$

其中

$$\sigma_s^2=\int K_s(f)df,\quad \sigma_n^2=\int K_n(f)df,\quad \sigma_r^2=\int K_r(f)df.$$

在普通波束成形的情况,根据(6)式,有

$$\rho_{cv}=\frac{K^2\sigma_s^2}{K\sigma_n^2+\int K_r(f)K^2\cos^2\gamma df} \tag{33}$$

特别是,若不存在干扰,则 $K_r(f)=0$. 于是

$\rho_{cv}=KL$, 从而 $G_{cv}=K$. 这是我们所熟知的普通波束成形系统抗独立噪声场的增益.

对于自适应波束成形系统,根据(9)式,再仿照第一节中求逆矩阵的方法,代入(31)式,可以得到

$$\rho_{ad}=\frac{\int K_s(f)\left(\frac{p}{p+K_n(f)/K_s(f)}\right)^2df}{\int K_n(f)\frac{u}{(p+K_n(f)/K_s(f))^2}df+\int K_r(f)\frac{|g|^2}{(p+K_n(f)/K_s(f))^2}df} \tag{34}$$

其中 p 由(18)式给出,

$$u=\boldsymbol{A}_0^{*T}(\boldsymbol{Z}^{-1})^2\boldsymbol{A}_0=K-2\frac{K^2\cos^2\gamma}{K+K_n(f)/K_r(f)}+\frac{K^3\cos^2\gamma}{(K+K_n(f)/K_r(f))^2} \tag{35}$$

$$g=\boldsymbol{A}_0^{*T}\boldsymbol{Z}^{-1}\boldsymbol{A}_1=\frac{\frac{K_n(f)}{K_r(f)}\boldsymbol{A}_0^{*T}\boldsymbol{A}_1}{K+K_n(f)/K_r(f)} \tag{36}$$

由此可知,自适应系统的增益 G_{ad} 是广义夹角 $\cos^2\gamma$ 的函数. (34)式不难通过数值计算给出具体结果. 我们在图 5 中给出了 $K=15$ 时的结果.

我们看出,一般情况下, G_{ad} 比 G_{cv} 大 10—15 分贝. 当干扰比信号和噪声都强得多时,普通波束成形的增益非常之低,而自适应系统却正好在这种情况下显示出它的优越性来.

根据广义夹角 $\cos^2\gamma$ 和 θ_1, θ_0 的关系式,我们就能计算出系统的极限分辨力. 图 6 是对

于一个 $K=15$, $d/\lambda=0.13$ 的线阵的计算结果．从图中我们看到，当干扰目标处于普通波束成形的主瓣之内时，普通波束成形系统无法确认信号，因为这时它的增益很低．而自适应系统在这种情况下仍有 8—12 分贝的增益，足以区分出信号来，所以它的极限分辨力高得多．

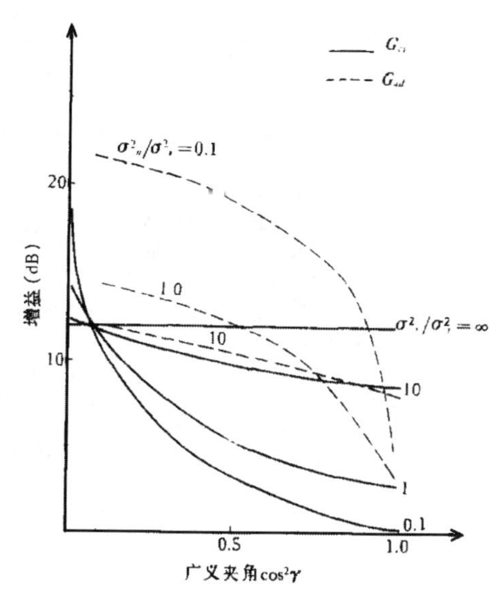

图 5　增益与广义夹角的关系

The relation between gain and generalized angle

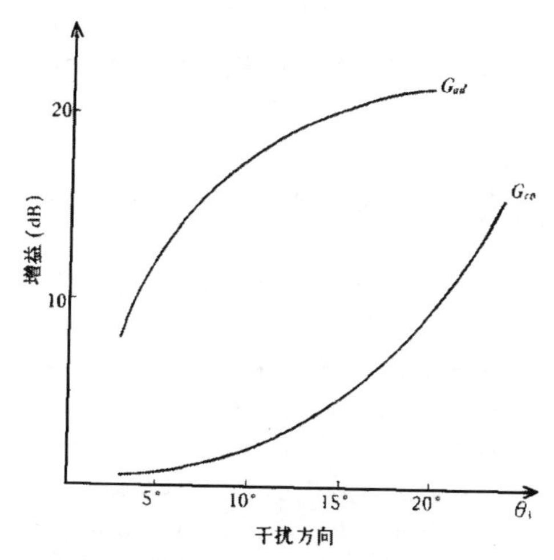

图 6　线阵的极限分辨力 ($K=15, \theta_0=0°, d/\lambda=0.13, \sigma^2_n/\sigma^2_r=0.1$)

The limit resolution ability of linear array

五、结　　论

自适应系统的稳态特性可以通过频域上的最优解加以分析．极大值方向、主瓣半宽度、干扰方向零点的深度以及极限分辨力都是广义夹角的函数．自适应系统在外界出现干扰时，能在干扰方向形成一个零点，作为代价，它的主极大方向由于受到干扰的“排斥”而偏离信号方向，干扰与信号越接近，这种偏离就越大。

参　考　文　献

[1] Widrow, B. et al., "Adaptive Antenna Systems", *Proc. IEEE*, **55**(1967), No. 12, 2143—2159.

[2] Griffiths, L. J., "A Simple Algorithm for Real-Time Processing in Antenna Arrays", *Proc. IEEE*, **57** (1969), No. 10, 1696—1704.

[3] Chang, J. H. and Tuteur, F. B., "Optimum Adaptive Array Processor", *Proc. of the Symp. on Computer Processing in Communication*, Microwave Res. Inst. Symposis Series, **19** (1969), 695—710.

[4] Widrow, B. and McCool, J. M., "A Comparison of Adeptive Algorithms Based on the Methods of Steepest Descent and Random Search", *IEEE. Trans.*, **AP-24** (1976), No. 5, 615—637.

[5] Frost, O. L., III "An Algorithm for Linearly Constrained Adaptive Array Processing", *Proc. IEEE*, **60** (1972), No. 8, 926—935.

[6] Widrow, B. et al., "Stationary and Nonstationary Characterics of the LMS Adaptive Filter", *Proc. IEEE*, **64** (1976), No. 8, 1151—1161.

[7] Broder, A., "Adaptive Sonar Beamforming", *Sperry Technology*, **1** (1973), No. 4, 31—34.

[8] Keating, P. N., "A Rapid Approximation to Optimal Array Processing for the Case of Strong Localized Interference", *J. A. S. A.*, **65** (1979), No. 2, 456—462.

[9] Rodgers, W. E. and Compton, R. T. Jr., "Adaptive Array Bandwidth With Tapped Delay-line Processing", *IEEE Trans.*, **AES-15** (1979), No. 1, 21—28.

[10] 李启虎，“自适应波束成形中最优解在时域和频域中的等价关系”，声学学报，1979年，第4期，296—308.

[11] Compton, R. T. Jr., "An Experimental Four-Element Adaptive Arrays", *IEEE Trans.*, **AP-24** (1976), No. 5, 697—706.

A STUDY OF STEADY-STATE PERFORMANCE IN ADAPTIVE BEAMFORMING

Li Qi-hu

(Institute of Acoustics, Academia Sinica)

The general representation of steady state directivity of adaptive system designed under the LMS criteria is given.

As the function of generalized angle between signal and interference, the methods for calculating the width of main beam, the deepness of zero point of supressing interference, and the limit resolution ability are presented.

In general case, the maximum value of directivity have some deviations from the signal direction. The numerical examples are given in each considered situation.

一种简单的抗强干扰自适应波束成形方法①

李启虎
（中国科学院声学研究所）

1981年9月1日收到

摘　要

本文讨论在有强点源干扰时的自适应波束成形问题。利用频域上最佳传输函数的表达式，证明了在仅有强点源干扰时，最佳传输函数有一个简单的标准形式。它等于干扰和期望信号的指向性矢量的加权和，其加权系数可以很容易地由解一个线性方程组给出。这个线性方程组的阶数等于强干扰的个数加1.

文中给出系数矩阵的计算方法。本文提出的方法不需要迭代运算，而只需实时地解一个低阶的线性方程组。从而使 Widrow 提出的自适应波束成形方法大大地简化了。

实际的数值计算说明，用本文提出的方法进行波束成形效果是好的。文中给出了部分例子及计算结果。

一、引　言

B. Widrow 提出的自适应波束成形系统[1]，对输入数据率及系统的处理速度要求很高。例如一个具有 K 个基元 L 个抽头的延迟线，加权系数有 KL 个。在进行迭代运算时，每一采样周期内要完成 KL 次乘法；为了使由抽头延迟线构成的时域系统的增益尽量接近最佳的系统增益，必须使延时间隔尽量小一些，这将使 KL 进一步增大，所以系统的实时调节更加困难。

对 Widrow 算法可以在两个方面进行工作。一方面是寻求新的算法，使得迭代次数减少或收敛加快；M. Dentino[2] 和 E. R. Ferrara[3] 指出，如果把 Widrow 的时域模型转到频域上用 FFT 来进行，那么当输入样本点足够多时，运算次数就会大为减少，从而减轻了系统实时运算速度的要求。B. L. Lewis 等[4]则提出了一种改进 Widrow 迭代算法的新方法，用这种算法收敛会更快且对控制收敛因子的依赖更少一些。Lunde[5] 曾提出一种直接求最佳权系数 $\boldsymbol{W}_{\text{opt}}$ 的方法，但他的方法依赖于对输入数据互相关矩阵的粗略估计，当估计不合适时，迭代甚至不收敛。

另一方面就是在某种特殊的情况下直接求解 $\boldsymbol{W}_{\text{opt}}$，在仅有强点源干扰的情况下，Keating 曾提出一种在约束条件下求解输入互谱矩阵 $\boldsymbol{K}_{xx}^{-1}(f)$ 的办法。根据这种办法不用实时进行迭代运算，而仅需计算一个与干扰个数同一量级的逆矩阵，从而大大地简化了计算。

本文讨论的就是强点源干扰下的自适应波束成形问题。我们首先证明，最佳传输函数 $\boldsymbol{H}_{\text{opt}}(f)$ 可以写成一种标准形式。这种标准形式仅与干扰方向有关，在所有 KL 个加权系数中，有贡献的权系数的个数等于信号加干扰的个数，它们可以由求解一个复的线性方程组给出来。

① 声学学报, 1983, 8(3): 159-167.

我们提出的方法是利用普通波束成形系统给出强点源干扰的方位，然后通过一个专用机给出复的系数矩阵. 再通过另一个专用机把复的系数矩阵化为求解实的线性方程组并且给出最佳传输函数 $\boldsymbol{H}_{opt}(f)$. 在整个自适应过程中不需要权系数的实时迭代，所以在工程实现上要简单得多.

实际的计算结果是令人满意的，这种方法可以精确地在干扰方位形成零点，抑制干扰的增益一般在 40 至 70 分贝之间.

当把本方法用于实际声呐的设计时，必须首先用普通的多波束定向系统给出强点源干扰的方位，然后再由专用机给出自适应系统的传输函数.

二、记 号、模 型

我们考虑如图 1 所示的自适应波束成形模型，其中 (a) 为 B. Widrow 提出的带抽头延迟线的自适应波束成形系统，(b) 为频域上的等价模型.

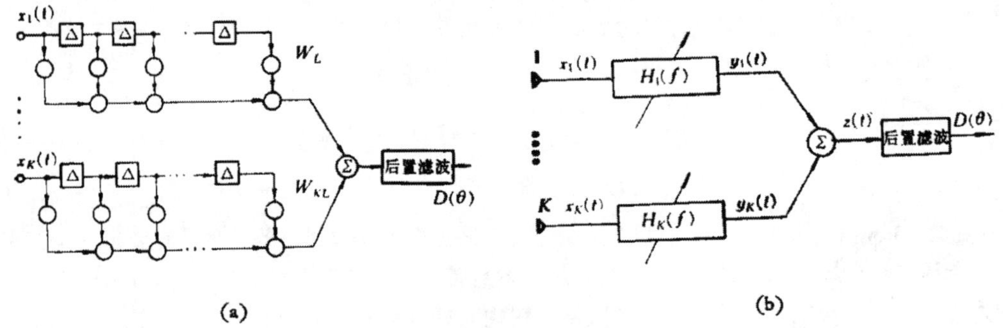

图 1　自适应波束成形模型

Block diagram of traditional adaptive beamforming system

设基阵由 K 个基元构成，第 k 个基元的信号为 $x_k(t)$，用矢量 $\boldsymbol{X}(t)$ 来表示系统的输入信号:

$$\boldsymbol{X}(t)=\begin{bmatrix} x_1(t) \\ \vdots \\ x_K(t) \end{bmatrix} \tag{1}$$

$\boldsymbol{X}(t)$ 的互谱矩阵记作 $\boldsymbol{K}_{xx}(f)$，它的第 (k, l) 个元素是 $x_k(t)$ 和 $x_l(t)$ 的互功率谱，即

$$[\boldsymbol{K}_{xx}(f)]_{(k,l)}=\int_{-\infty}^{\infty} R_{kl}(\tau)e^{-2\pi jf\tau}d\tau \tag{2}$$

其中 $R_{kl}(\tau)$ 是 $x_k(t)$ 和 $x_l(t)$ 的互相关函数. $\boldsymbol{K}_{xx}(f)$ 是一个 Hermite 矩阵，即

$$\boldsymbol{K}_{xx}^{*}(f)=\boldsymbol{K}_{xx}^{\mathrm{T}}(f) \tag{3}$$

设对于从 θ 方向入射的声信号，第 k 个水听器相对于某个参考点的延时为 $\tau_k(\theta)$. 则 θ 方向的波束指向矢量为

$$\boldsymbol{A}(\theta)=\begin{bmatrix} \exp(-2\pi jf\tau_1(\theta)) \\ \vdots \\ \exp(-2\pi jf\tau_K(\theta)) \end{bmatrix} \tag{4}$$

如果信号来自 θ_0 方向，信号的平均功率谱是 $K_s(f)$. 则输入信号的互谱矩阵是

$$\boldsymbol{K}_{ss}(f) = K_s(f)\boldsymbol{A}(\theta_0)\boldsymbol{A}^{*T}(\theta_0) \tag{5}$$

也就是说，它是信号方向的波束指向矢量的一个并矢矩阵.

现在假定在环境噪声场中，除了各向同性相互独立的噪声 $n_1(t), \cdots, n_K(t)$ 之外，还有 N 个干扰：$r_1(t), \cdots, r_N(t)$，它们的自功率谱分别是 $K_{r1}(f), \cdots, K_{rN}(f)$. 如果第 l 个干扰来自 θ_l 方向，那么整个输入信号的互谱矩阵是

$$\boldsymbol{K}_{xx}(f) = K_n(f)\boldsymbol{I} + K_s(f)\boldsymbol{A}(\theta_0)\boldsymbol{A}^{*T}(\theta_0) + \sum_{l=1}^{N} K_{rl}(f)\boldsymbol{A}(\theta_l)\boldsymbol{A}^{*T}(\theta_l) \tag{6}$$

假定图 1 (b) 中的滤波器的传输函数为 $H_1(f), \cdots, H_K(f)$ 则该系统的滤波矢量为

$$\boldsymbol{H}(f) = \begin{bmatrix} H_1(f) \\ \vdots \\ H_K(f) \end{bmatrix} \tag{7}$$

系统的输出为

$$D(\theta) = \{E[z^2(t)]\}^{1/2} = \left\{\int_{-\infty}^{\infty} \boldsymbol{H}^T(f)\boldsymbol{K}_{xx}(f)\boldsymbol{H}^*(f)df\right\}^{1/2} \tag{8}$$

按普通波束成形的办法，如果使 $\boldsymbol{H}(f)$ 依次定向于从0°至 360° 的角，则我们从 (8) 式便可得到指向性函数 $D_{cv}(\theta)$ 来.

令

$$\boldsymbol{H}_{cv}(\theta) = \boldsymbol{A}^*(\theta) \tag{9}$$

则

$$D_{cv}(\theta) = \left\{\int_{-\infty}^{\infty} \boldsymbol{H}_{cv}^T(f)\boldsymbol{K}_{xx}(f)\boldsymbol{H}_{cv}^*(f)df\right\}^{1/2} \tag{10}$$

对于单波束系统，如果用 (5) 式中的 $\boldsymbol{K}_{ss}(f)$ 来代替(10)中的 $\boldsymbol{K}_{xx}(f)$，就得到

$$D_{cv}(\theta) = \left\{\int_{-\infty}^{\infty} K_s(f)|\boldsymbol{A}^{*T}(\theta_0)\boldsymbol{A}(\theta)|^2 df\right\}^{1/2} \tag{11}$$

(11) 式所表示的，就是通常的指向性函数.

我们知道[7]，自适应波束成形系统的最佳传输函数为

$$\boldsymbol{H}_{opt}(f) = [\boldsymbol{K}_{xx}^*(f)]^{-1}\boldsymbol{K}_{xd}^*(f) \tag{12}$$

其中 $\boldsymbol{K}_{xd}(f)$ 为期望信号 $d(t)$ 与 $\boldsymbol{X}(t)$ 的互功率谱. 在正确匹配的情况下(即对信号方向 θ_0 及谱 $K_s(f)$ 都估计正确)，我们有

$$\boldsymbol{K}_{xd}(f) = K_s(f)\boldsymbol{A}(\theta_0) \tag{13}$$

于是，自适应单波束系统的指向性为

$$\begin{aligned} D_{ad}(\theta) &= \left\{\int_{-\infty}^{\infty} \boldsymbol{H}_{opt}^T(f)\boldsymbol{K}_{ss}(f)\boldsymbol{H}_{opt}^*(f)df\right\}^{1/2} \\ &= \left\{\int_{-\infty}^{\infty} K_s(f)|\boldsymbol{H}_{opt}^{*T}(f)\boldsymbol{A}(\theta)|^2 df\right\}^{1/2} \\ &= \left\{\int_{-\infty}^{\infty} K_s(f)|\boldsymbol{A}^{*T}(\theta)\boldsymbol{H}_{opt}(f)|^2 df\right\}^{1/2} \end{aligned} \tag{14}$$

按照 Widrow 的设计[1]，我们要用迭代的方法计算出 KL 个权系数 $w_i, i = 1, \cdots, KL$ 才能最后得到 $D_{ad}(\theta)$.

下面我们要证明,在有强干扰的情况下,$\boldsymbol{H}_{opt}(f)$ 的形式非常简单,它仅仅和信号指向矢量 $\boldsymbol{A}(\theta_0)$ 与干扰指向矢量 $\boldsymbol{A}(\theta_l)$, $l=1,\cdots,N$ 的加权和有关,也就是说,在 KL 个权系数中,仅有 $2(N+1)$ 个系数是有贡献的. 而这些系数仅由 $\theta_0,\cdots,\theta_N$ 便可决定.

三、理论分析

为书写简单起见,令

$$\boldsymbol{A}_l=\boldsymbol{A}(\theta_l),\ l=0,\cdots,N \tag{15}$$

$$K_s(f)=K_{r0}(f) \tag{16}$$

于是 (6) 式成为

$$\boldsymbol{K}_{xx}(f)=K_n(f)\boldsymbol{I}+\sum_{l=0}^{N}K_{rl}(f)\boldsymbol{A}_l\boldsymbol{A}_l^{*T} \tag{17}$$

为求出 $\boldsymbol{H}_{opt}(f)$,必须求出 $\boldsymbol{K}_{xx}^{-1}(f)$,在一般情况下这是十分困难的. 但是,如果所有的干扰都是强点源干扰,即 $K_{rl}(f)\gg K_n(f)$, $l=1,\cdots,N$ 则 $\boldsymbol{H}_{opt}(f)$ 有以下的标准形式:

$$\boldsymbol{H}_{opt}^*(f)=\sum_{l=0}^{N}w_l\boldsymbol{A}_l \tag{18}$$

其中 w_l 是复的加权系数,由下述约束确定: 它使得 $D_{ad}(\theta)$ 在 θ_0 的响应为 1,而使 $D_{ad}(\theta_0)$ 在 θ_l 的响应为零. 根据 (14) 及 (18) 式,我们知道,这等价于下面一个复的线性方程组:

$$\boldsymbol{QW}=\boldsymbol{\Gamma} \tag{19}$$

其中

$$\boldsymbol{W}=\begin{bmatrix}w_0\\w_1\\\vdots\\w_N\end{bmatrix},\quad \boldsymbol{\Gamma}=\begin{bmatrix}1\\0\\\vdots\\0\end{bmatrix} \tag{20}$$

都是 $N+1$ 维矢量. $\boldsymbol{Q}$ 是 $(N+1)\times(N+1)$ 的系数矩阵,其第 (k,l) 个元素是

$$(\boldsymbol{Q})_{kl}=\boldsymbol{A}_k^{*T}\boldsymbol{A}_l\triangleq q_{kl} \tag{21}$$

我们称 $\boldsymbol{Q}$ 为干扰指向角矩阵.

下面我们就来证明以上的论断. 为此,只需证明 (19) 的解确实是在强干扰下 $\boldsymbol{H}_{opt}(f)$ 的近似表达式.

根据 Keating 的结果[6],我们知道 $\boldsymbol{K}_{xx}^{-1}(f)$ 可以表示成

$$\boldsymbol{K}_{xx}^{-1}(f)=\frac{1}{K_n(f)}\left[\boldsymbol{I}-\sum_{k,l=0}^{N}B_{k,l}\boldsymbol{A}_k\boldsymbol{A}_l^{*T}\right] \tag{22}$$

其中 $B_{kl}=(\boldsymbol{B})_{kl}$, $0\leqslant k,l\leqslant N$,

$$\boldsymbol{B}=\boldsymbol{M}^{-1}\boldsymbol{D},\quad \boldsymbol{M}=\boldsymbol{I}+\boldsymbol{DQ} \tag{23}$$

$$(\boldsymbol{D})_{kl}=r_k\cdot\delta_{kl} \tag{24}$$

δ_{kl} 表示 Kroneker δ. 我们把 (22) 式所表示的 $\boldsymbol{K}_{xx}^{-1}(f)$ 代入 (12) 便有:

$$\begin{aligned}\boldsymbol{H}_{opt}^*(f)&=K_s(f)\boldsymbol{K}_{xx}^{-1}(f)\boldsymbol{A}_0\\&=\frac{K_s(f)}{K_n(f)}\left[\left(1-\sum_{l=0}^{N}B_{0l}q_{l0}\right)\boldsymbol{A}_0-\sum_{k=1}^{N}\sum_{l=0}^{N}B_{kl}q_{l0}\boldsymbol{A}_k\right]\end{aligned}$$

$$\approx \frac{K_s(f)}{K_n(f)}\left[(\boldsymbol{Q}^{-1}\boldsymbol{D}^{-1})_{00}\boldsymbol{A}_0 + \sum_{k=1}^{N}(\boldsymbol{Q}^{-1}\boldsymbol{D}^{-1})_{k0}\boldsymbol{A}_k\right]$$

$$= \sum_{k=0}^{N}(\boldsymbol{Q}^{-1})_{k0}\boldsymbol{A}_k \tag{25}$$

显然，(19) 式的解是

$$\boldsymbol{W} = \boldsymbol{Q}^{-1}\boldsymbol{\Gamma} \tag{26}$$

由于 $\boldsymbol{\Gamma}$ 的特殊形式，我们易知

$$w_l - (\boldsymbol{Q}^{-1})_{l0} \tag{27}$$

由此便证明了 (18) 式.

由以上的叙述可以看出，为求得稳态的 w_l 值，仅须解一个 $N+1$ 阶的复的线性方程组就行了. $|w_l|$ 是 $\boldsymbol{A}_l$ 的幅度加权，$\arg(w_l)$ 则是 $\boldsymbol{A}_l$ 的相位加权，也就是在相应的延时抽头 $\tau(\theta_l) - \arg(w_l)$ 处乘以系数 $|w_l|$ 就行.

对于 (19) 式的求解，我们并不采用 $\boldsymbol{Q}$ 直接求逆的方法；虽然 $\boldsymbol{Q}$ 的维数并不高，但是由于它是一个复的矩阵，求逆是不方便的. 我们下面将提出用分块矩阵的方法，将 (19) 式化为实的线性方程组来解.

令 $\boldsymbol{Q}_r$, $\boldsymbol{Q}_i$ 分别表示 $\boldsymbol{Q}$ 的实部和虚部，即 $\boldsymbol{Q} = \boldsymbol{Q}_r + j\boldsymbol{Q}_i$，容易验证

$$\boldsymbol{Q}^{-1} = (\boldsymbol{Q}_r + \boldsymbol{Q}_i\boldsymbol{Q}_r^{-1}\boldsymbol{Q}_i)^{-1} - j(\boldsymbol{Q}_i + \boldsymbol{Q}_r\boldsymbol{Q}_i^{-1}\boldsymbol{Q}_r)^{-1} \tag{28}$$

我们看到，虽然 $(N+1)$ 这个值一般并不大，但是由于 $\boldsymbol{Q}$ 求逆的过程要用到四次实矩阵的求逆运算，四次实矩阵的相乘，所以仍然是比较复杂的.

(19) 式如果用 $\boldsymbol{Q}$ 的实部及虚部和 $\boldsymbol{W}$ 的实部和虚部来表示，就是

$$(\boldsymbol{Q}_r + j\boldsymbol{Q}_i)(\boldsymbol{W}_r + j\boldsymbol{W}_i) = \boldsymbol{\Gamma} \tag{29}$$

把 (29) 式展开，令方程两边的实部与虚部分别相等，就得到以下这样的 $2(N+1)$ 维实的方程组：

$$\begin{pmatrix} \boldsymbol{Q}_r & -\boldsymbol{Q}_i \\ \boldsymbol{Q}_i & \boldsymbol{Q}_r \end{pmatrix}\begin{pmatrix} \boldsymbol{W}_r \\ \boldsymbol{W}_i \end{pmatrix} = \begin{pmatrix} 1 \\ 0 \\ \vdots \\ 0 \end{pmatrix} \tag{30}$$

这里右边是 $2(N+1)$ 维矢量，它除了第一个元素为 1 之外，其余全为零. (30) 的维数虽然比 (19) 增加一倍，但是因为它已是实的线性方程组，我们有很多现成的简单解法，所以要得出 $\boldsymbol{W}_r$, $\boldsymbol{W}_i$ 并不困难.

利用 (25) 式，不难证明 $D_{ad}(\theta_0) = 1$，以及 $D_{ad}(\theta_l) = 0$, $l = 1, \cdots, N$. 事实上，按 (14) 式，

$$D_{ad}(\theta_0) = \left\{\int_{-\infty}^{\infty} K_s(f)\left|\boldsymbol{A}_0^{*T}\left(\sum_{l=0}^{N}(\boldsymbol{Q}^{-1})_{l0}\boldsymbol{A}_l\right)\right|^2 df\right\}^{1/2}$$

$$= \left\{\int_{-\infty}^{\infty} K_s(f)\left|(\boldsymbol{Q}^{-1})_{00}\, q_{00} + (\boldsymbol{Q}\boldsymbol{Q}^{-1})_{00} - (\boldsymbol{Q}^{-1})_{00}\, q_{00}\right|^2 df\right\}^{1/2}$$

$$= \left\{\int_{-\infty}^{\infty} K_s(f) df\right\}^{1/2} = \sigma_s$$

$$
\begin{aligned}
D_{ad}(\theta_l) &= \left\{\int_{-\infty}^{\infty} K_s(f)\left|\boldsymbol{A}_l^{*T}\sum_{l=0}^{N}(\boldsymbol{Q}^{-1})_{l0}\boldsymbol{A}_l\right|^2 df\right\}^{1/2} \\
&= \left\{\int_{-\infty}^{\infty} K_s(f)\left|(\boldsymbol{Q}^{-1})_{00}\, q_{l0} + \sum_{k=1}^{N}(\boldsymbol{Q}^{-1})_{k0}\, q_{lk}\right|^2 df\right\}^{1/2} \\
&= \left\{\int_{-\infty}^{\infty} K_s(f)\left|(\boldsymbol{Q}^{-1})_{00}\left[q_{l0} + \frac{(\boldsymbol{Q}\boldsymbol{Q}^{-1})_{l0}}{(\boldsymbol{Q}^{-1})_{00}} - \frac{q_{l0}(\boldsymbol{Q}^{-1})_{00}}{(\boldsymbol{Q}^{-1})_{00}}\right]\right|^2 df\right\}^{1/2} \\
&= \left\{\int_{-\infty}^{\infty} K_s(f)\,|(\boldsymbol{Q}^{-1})_{00}[q_{l0} - q_{l0}]|^2 df\right\}^{1/2} = 0
\end{aligned}
$$

因此，在单波束指向性系统中，对准信号时，系统响应为 σ_s，对准 N 个干扰时，响应为零．这正是我们所希望的．

四、系 统 设 计

按照上面的理论分析，我们得到如图 2 所示的抗强干扰声呐的方框图．这是普通波束成形与自适应波束成形的结合．先用普通的波束成形方法构成一个多波束系统，由该系统提供 L 个干扰的方向，然后把观测者所欲观测的方向 θ_0 加到一起．总共有 $L+1$ 个方向 $\theta_0, \theta_1, \cdots, \theta_L$．把这 $L+1$ 个方向送给专用机 SP_1，它可以迅速地计算出干扰指向角矩阵 $\boldsymbol{Q}$ 的实部 $\boldsymbol{Q}_r$ 和虚部 $\boldsymbol{Q}_i$．然后把 $\boldsymbol{Q}_r, \boldsymbol{Q}_i$ 输入到专用机 SP_2，这是一个按高斯主元素消去法解线性方程组的专用机．由它解得 $L+1$ 对权系数 w_{lr}，w_{li}，于是得到 $w_l = w_{lr} + jw_{li}$，$l=0, \cdots, L$．以及 $\boldsymbol{H}_{\text{opt}}^*(f) = \sum_{l=0}^{N} w_l \boldsymbol{A}_l$，将其代入 (14) 便可很容易地算出 $D_{ad}(\theta)$．显然，这种系统要比每个抽头延迟线都要调节的系统简单得多．

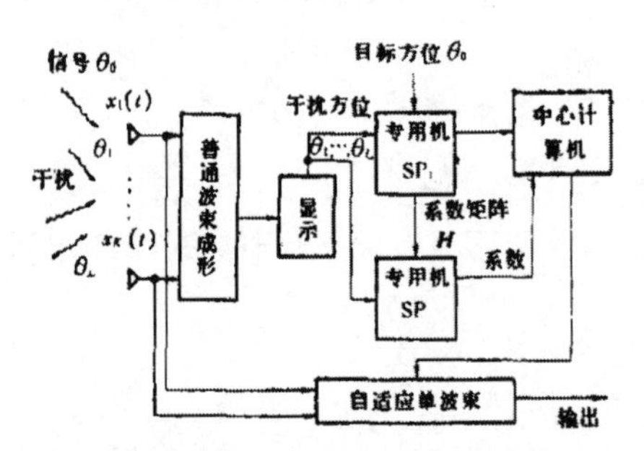

图 2　抗强干扰自适应声呐系统原理方框图
Block diagram of adaptive beamforming system for supressing strong interference
图中系数为权系数 $w_1 \cdots w_L$

当然，在 θ_0 的扫描过程中，不能令 $\theta_0 = \theta_l$，$l=1, \cdots, L$．否则矩阵 $\boldsymbol{Q}$ 就是奇异矩阵，(19) 无解．至于 θ_0 与 θ_l 的接近程度多大才是允许的，我们将在下一节中说明．

五、计 算 实 例

根据图 2 所述的信号处理程序，我们曾作过实际的计算，下面介绍计算的方法及部分结果。

我们将 w_l 的幅度记作 a_l，相位记作 φ_l，则

$$
\begin{aligned}
a_l &= (w_{lr}^2 + w_{li}^2)^{1/2} \\
\varphi_l &= \mathrm{tg}^{-1}(w_{li}/w_{lr})
\end{aligned}
\tag{31}
$$

由于一般计算机中反正切函数只取主值，所以由 w_{li}/w_{lr} 求 φ_l 时应经过一定的变换．首先求出 $\varphi_l^{(0)} = \mathrm{tg}^{-1}|w_{li}/w_{lr}|$，则

$$\varphi_l=\begin{cases}\varphi_l^{(0)} & 当\ w_{li}\geqslant 0,\ w_{lr}\geqslant 0\\ \pi-\varphi_l^{(0)} & 当\ w_{li}\geqslant 0,\ w_{lr}\leqslant 0\\ \pi+\varphi_l^{(0)} & 当\ w_{li}\leqslant 0,\ w_{lr}\leqslant 0\\ 2\pi-\varphi_l^{(0)} & 当\ w_{li}\leqslant 0,\ w_{lr}\geqslant 0\end{cases}$$

由此得出第 l 个指向矢量与权系数的乘积为:

$$w_l\boldsymbol{A}_l=a_l\cdot\begin{bmatrix}\exp(-2\pi jf\tau_1(\theta_l)+\varphi_l)\\ \vdots\\ \exp(-2\pi jf\tau_K(\theta_l)+\varphi_l)\end{bmatrix}\tag{32}$$

根据(32)式及(14)式便可很快地将指向性 $D_{ad}(\theta)$ 计算出来.

图 3 给出了四种指向性曲线. 这是一个基元个数 $K=32$ 的圆阵,计算中取 $f=1.5\text{kHz}$. 其中(a)是无干扰时的指向性. (b)是干扰在 15° 方向的情况. 这时系统在 15° 方向形成一个零点. (c)及(d)是多个干扰的情况,如果干扰落在主瓣之内,则极大值方向要发生偏移,这从图 3(d)看得十分清楚.

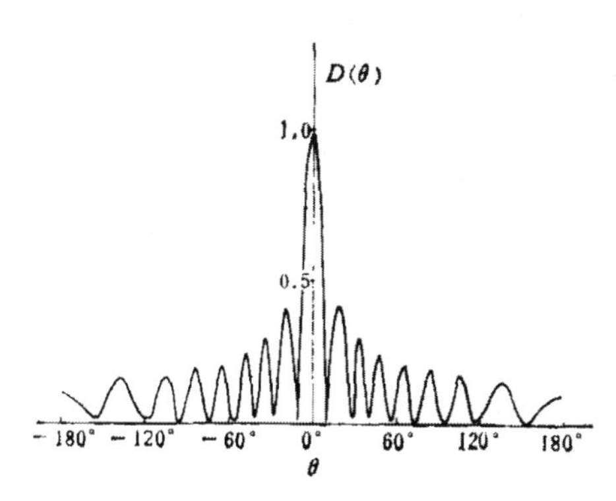

(a) 无干扰时指向性
信号方向 $\theta_0=0°$

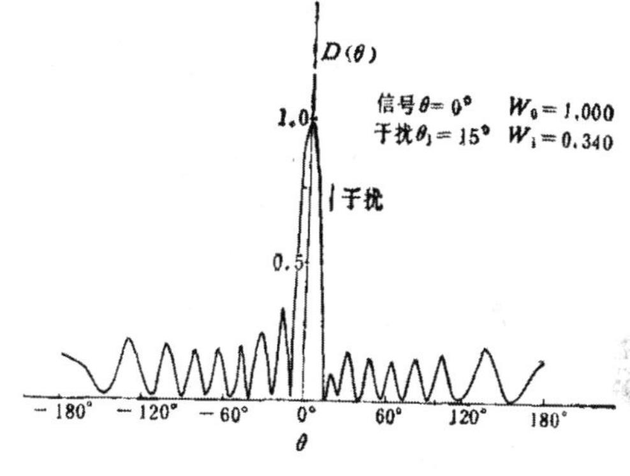

(b) 有一个强干扰时的指向性
信号 $\theta_0=0°$ $w_0=1.000$ 干扰 $\theta_1=15°$ $w_1=0.340$

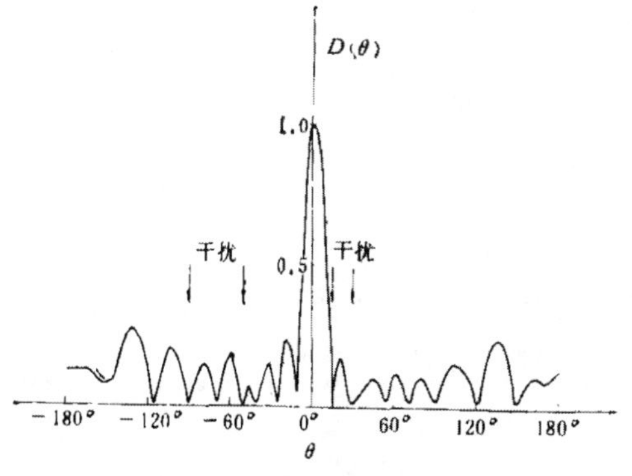

(c) 有四个强干扰时指向性
信号 $\theta_0=0°$ $w_0=1.000$ 干扰 $\theta_1=15°$ $w_1=0.2580$
$\theta_2=30°$ $w_2=-0.1605$ $\theta_3=-50°$
$w_3=0.1500$ $\theta_4=-90°$ $w_4=0.1158$

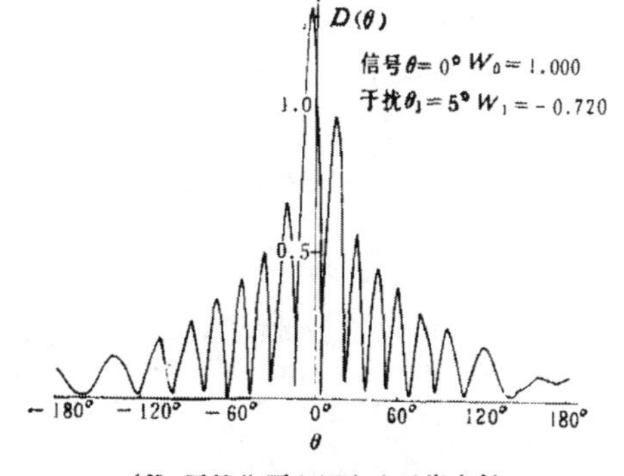

(d) 干扰位于主瓣之内时指向性
信号 $\theta_0=0°$ $w_0=1.000$
干扰 $\theta_1=5°$ $w_1=-0.720$

图 3 抗强干扰自适应波束成形系统的指向性
The directivity of adaptive beamforming system in case of strong localized interference

表 1 给出干扰的方位角及 w_i 的数值，在所有情况下，w_i 之虚部近似于零.

表 1

θ	w				
	w_0	w_1	w_2	w_3	w_4
0°	1	0	0	0	0
15°	1	0.340	0	0	0
30°	1	0.278	−0.167	0	0
−50°	1	0.259	−0.135	0.135	0
90°	1	0.258	−0.160	0.150	0.116
0°	1	0	0	0	0
5°	1	−0.720	0	0	0
20°	1	−0.679	0.1179	0	0
−15°	1	−0.656	0.1050	0.0833	0

从理论上讲，指向性中干扰方向是一个真正的零点. 但是，在实际计算时会有各种误差. 因为任何系统总是字长有限的. 我们在计算时为了模拟实际情况，计算精度为 10^{-4}. 所以带来一定的误差. 表 2 给出干扰方向的零点深度(用分贝表示).

表 2

干扰方向	15°	30°	−50°	−90°	5°	20°	−15°
零点深度 (dB)	−77.1 −44.2 −42.7 −46.8	−58.9 −36.7 −49.9	−39.8 −58.9	−53.0	−50.7 −46.2 −58.3	−53.6 −67.4	−38.2

六、结　　论

本文所提出的方法比实时迭代求权系数要简单一些，在那些对抗强点源干扰感兴趣的场合，这是可行的.

用于解线性方程组的专用机 SP_2，每次所解的线性方程组的阶数是一个变量，它等于当时环境场下干扰个数加 1.

如果整个系统都在频域上进行运算，利用 FFT 后还会更简单一些.

参　考　文　献

[1] Widrow, B. et al., ''Adaptive antenna systems'' *Proc. IEEE* 55(1967), 2143—2159.

[2] Dentino, M. et al., ''Adaptive filtering in the frequency domain'' *Proc. IEEE*, 66(1978), 1688—1689.

[3] Ferrara, E. R., ''Fast implementation of LMS adaptive filter'' *IEEE Trans.* ASSP-28(1980), 474—475.

[4] Kretschmer, Jr. F. F. and Lewis, B. L., ''An im proved algorithm for adaptive processing'', *IEEE Trans.* AES-14(1978), 172—179.

[5] Lunde, E. B., ''The forgotten algorithm in adaptive beamforming'', *Proc. of NATO* Advanced Signal Processing Institute, 1976.

[6] Keating, P. N., "A rapid approximation to optimal array processing for the case of strong localized interference", *JASA*, **65**(1979), 456—462.
[7] 李启虎,"自适应波束成形中权系数在时域与频域中的等价关系",声学学报,**5** (1979),No.4, 296—308.

A SIMPLE METHOD OF ADAPTIVE BEAMFORMING FOR SUPRESSING STRONG INTERFERENCE

Li Qi-hu
(*Institute of Acoustics, Academia Sinica*)
Received September 1, 1981

Abstract

In the traditional adaptive beamforming model proposed by widrow there are K elements and KL weight coefficients. It is necessary that each of KL weight coefficients are iterated in every sample period.

A special situation of adaptive beamforming is considered in this paper, that is the case of strong localized interference.

Using the optimum transfer function in frequency domain we proved that in the case of strong localized interference there is a canonical from for optimum transfer function. It can be expressed in term of directive vector of interference and signal. The weight coefficients of those vectors can easily calculated by a linear equation, the order of this equation is equal the number of interference plus one.

The calculation method of coefficient matrix of linear equation is given. A rapid adaptive beamforming can be obtained without iterative operation.

The adaptive beamforming system described in this paper have considrably redused the calculation of tapped delay line.

Some practical examples are shown.

SIGNAL SEPARATION THEORY BY USING ADAPTIVE ARRAY①

LI QIHU (李启虎)

(*Institute of Acoustics, Academia Sinica*)

Received May 25, 1982

ABSTRACT

Signal separation technique by using adaptive arrays is a new area in array data processing. The model considered in this paper is that of an array consisting of N elements, and the number of signal sources, which are separated in spatial location, is M and $M \leqslant N$. In the case of narrow band signals, the pre-envelope form of incidental signal can be obtained from the original signal and its Hilbert transform. In terms of the conventional beamforming system, the directional bearing angle matrix G is formed. By using a special implementation the inverse matrix of G is derived. At last, a matrix transform is operated on the received signal so that the output signal is separable, i.e. each output channel has only one incidental signal. In the case of wide band signals, the DFT of input signal or heterodyne is required. The block diagram of signal separation technique are given. Some important results of this technique are derived. The results of system simulation experiments in digital computers prove that this theory is effective for separating signals.

I. INTRODUCTION

Adaptive processing of array signals has been applied in various fields[1-9], which include the suppressing of strong interference, noise cancelling and improving of array directivity.

A new possible application of adaptive array is considered in this paper, i.e. signal separation theory. The basic idea of this theory is: the incident signals of array are separated before arrival in each element. The output signal of each element contains the information of time delay among incident signals. By using this information and applying a suitable transform on the received signals, the additive incident signals can be separated again in the output.

H. Mermoz considered the design of propagation model of underwater acoustic signals by using the time delay data[10-11]; C. Giraudon made a similar analysis[12]. In a recent paper[13], C. M. Hackett proposed an algorithm for separating signals by means of calculation of eigenvalue of input data correlation matrix. But it can't obtain the original incident signals.

The method considered here is based on the assumption that the bearing angle matrix of incident signals can be precisely estimated.

First, each received signal of array elements is transformed by Hilbert transform to get an orthogonal signal, and then to form its pre-envelope signal. A real processing inverse matrix of bearing angle matrix is calculated. Finally the demodulation operation is performed on the pre-envelope signal, so that the signal is separable in output channels.

① Chinese Journal of Acoustics, 1983, 2(1): 81-92.

We will review the basic concept of the narrow band signal and its Hilbert transform[14-15] in Section II. In section III we show a method to establish the equivalent relationship between the complex bearing angle matrix and its real generated matrix. The block diagram of signal separation technique for narrow band and wide band signals will be presented in Section IV. In order to separate the wide band signals efficiently by using DFT, a method of line spectrum tracking technique is given. This is an iterative method to compute DFT of time series data. In Section V, some results of computer simulation are presented. Some problems which must be studied further will be given in Section VI.

II. Model and Notation

The model of signal processing system of array data considered in this paper is shown in Fig. 1. Suppose a received array consists of N elements. There are M incident signals. We always assume $M \leqslant N$ throughout this paper. For simplicity, let $M=N$, otherwise we can choose M elements from N elements to form an auxiliary array. Let the incident signals be $c_1(t),\ldots,c_N(t)$. The signal of k-th element is

$$x_k(t)=\sum_{l=1}^{N} c_l(t+\tau_{kl}) \tag{1}$$

where τ_{kl} is the time delay of the l-th component of the k-th element signal.

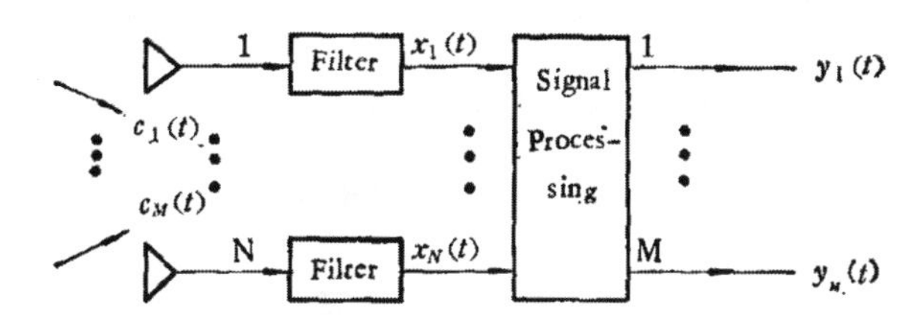

Fig. 1. The block diagram of adaptive array signal processing system.

If $y_k(t)$ is the result of certain processing on $x_1(t), \ldots, x_N(t)$, the final result which we desire is $y_k(t)=c_k(t),\ k=1,\ldots,N$.

Now let us briefly review the Hilbert transform of narrow band signal and its pre-envelope representation. The Hilbert transform of a narrow band signal $s(t)$ is defined as

$$\hat{s}(t) \triangleq H[s(t)]=\frac{1}{\pi}\int_{-\infty}^{\infty}\frac{s(\tau)}{t-\tau}d\tau \tag{2}$$

where the integral means the Cauchy principal value, i.e.

$$\int_{-\infty}^{\infty} = \lim_{\varepsilon\to 0}\left[\int_{-\infty}^{-\varepsilon}+\int_{\varepsilon}^{\infty}\right]$$

It is easy to show that

$$s(t)=H^{-1}[\hat{s}(t)]=-\frac{1}{\pi}\int_{-\infty}^{\infty}\frac{\hat{s}(\tau)}{t-\tau}d\tau \tag{3}$$

A stationary narrow band signal can be expressed as[15]

$$s(t)=u(t)\cos 2\pi f_0 t-v(t)\sin 2\pi f_0 t \tag{4}$$

where $u(t)$ and $v(t)$ are stationary signals with low frequency spectrum. Denote the Fourier transforms of $u(t)$, $v(t)$ by $U(f)$, $V(f)$ respectively, and assume $U(f)=V(f)=0$ for $|f|\geqslant W$.

If
$$f_0 \geqslant W \tag{5}$$
then
$$\hat{s}(t)=H[s(t)]=u(t)\sin 2\pi f_0 t+v(t)\cos 2\pi f_0 t \tag{6}$$
The $\hat{s}(t)$ is called orthogonal signal of $s(t)$. In a general situation, the condition (5) is easy to satisfy.

Now, assume the incident signal of array is a narrow band signal. We have
$$c_l(t)=u_l(t)\cos 2\pi f_l t-v_l(t)\sin 2\pi f_l t,\ l=1,\ \cdots,\ N \tag{7}$$
Using the vector notation, $c(t)=[c_1(t),\ \cdots,\ c_N(t)]^T$, $u(t)=[u_1(t),\ \cdots,\ u_N(t)]^T$, $v(t)=[v_1(t),\ \cdots,\ v_N(t)]^T$, $f=[f_1,\ \cdots,\ f_N]^T$. It is assumed that for each l, the condition (5) is satisfied. For the l-th signal the time delay $\tau_{1l},\ \cdots,\ \tau_{Nl}$ forms a time delay vector as follows: $\tau_l=[\tau_{1l},\ \cdots,\ \tau_{Nl}]^T$.

We express the bearing direction vector of beam as
$$d_l=[\exp(j2\pi f_l\tau_{1l}),\ \cdots,\ \exp(j2\pi f_l\tau_{Nl})]^T.$$
Now we can rewrite the incident signal in the vector form:
$$c(t)=u(t)\cos 2\pi ft-v(t)\sin 2\pi ft \tag{8}$$
Note that $u(t)\cos 2\pi ft$ represents a vector, the l-th component is $u_l(t)\cos 2\pi f_l t$; $v(t)\sin 2\pi ft$ also represents a vector. The Hilbert transform of $c(t)$ is
$$\hat{c}(t)=u(t)\sin 2\pi ft+v(t)\cos 2\pi ft \tag{9}$$
Thus, the equation (1) can be expressed as
$$x(t)=\sum_{l=1}^{N}c_l(t+\tau_l) \tag{10}$$
Our main purpose is to separate $c_1(t),\ \ldots,\ c_N(t)$ from $x(t)$.

III. Main Theoretical Result

For the narrow band signal, the changes of $u(t)$ and, $v(t)$ are slow, as compared with the center frequency. According to the characteristic of array signal processing, we have
$$\begin{aligned}c(t+\tau)&=u(t+\tau)\cos 2\pi f(t+\tau)-v(t+\tau)\sin 2\pi f(t+\tau)\\&\approx u(t)\cos 2\pi f(t+\tau)-v(t)\sin 2\pi f(t+\tau)\end{aligned} \tag{11}$$
It is easy to show that
$$\begin{aligned}x(t)&=\sum_{l=1}^{N}c_l(t+\tau_l)\approx\sum_{l=1}^{N}u_l(t)\cos 2\pi f_l(t+\tau_l)-\sum_{l=1}^{N}v_l(t)\sin 2\pi f_l(t+\tau_l)\\&=\sum_{l=1}^{N}c_l(t)\,\mathrm{Re}[d_l]-\sum_{l=1}^{N}\hat{c}_l(t)\,\mathrm{Im}(d_l)\end{aligned} \tag{12}$$
where the Re and Im denote the real part and image part of a complex number, respectively.

Let
$$G=[d_1,\ \cdots,\ d_N] \tag{13}$$
We will call it the bearing angle matrix of incidental signals. This complex matrix will play an important role in the signal separation theory. We express the G by means of its real part G_r and image part G_i:

$$G=G_r+jG_i \tag{14}$$

That is

$$G_r=[\mathrm{Re}(d_1),\ \cdots,\ \mathrm{Re}(d_N)],\ G_i=[\mathrm{Im}(d_1),\ \cdots,\ \mathrm{Im}(d_N)] \tag{15}$$

and

$$[G_r]_{kl}=\cos 2\pi f_l\tau_{kl},\ [G_i]_{kl}=\sin 2\pi f_l\tau_{kl} \tag{16}$$

We can rewrite the equation (12) as

$$x(t)=G_r\,c(t)-G_i\,\hat{c}(t) \tag{17}$$

As we know, the pre-envelope signal $c'(t)$ of $c(t)$ can be expressed by means of $c(t)$ and its Hilbert transform $\hat{c}(t)$:

$$c'(t)=c(t)+j\,\hat{c}(t) \tag{18}$$

Similary, the pre-envelope signal $x'(t)$ of $x(t)$ can be expressed as:

$$x'(t)=x(t)+j\,\hat{x}(t) \tag{19}$$

where the $\hat{x}(t)$ is the Hilbert transform of $x(t)$. From equation (17) we can immediately obtain

$$\hat{x}(t)=G_r\hat{c}(t)+G_i\,c(t) \tag{20}$$

Based on (17)—(20), it is easy to show

$$x'(t)=G\,c'(t) \tag{21}$$

So far we have performed the transformation from real signal form to complex signal from in array signal processing. This is a key step in the signal separation theory, because from equation (21) we see that in the complex field the reception of incidental signal of an array is essentially a linear transform on the incidental signal. The matrix G of this transform can be estimated by using the conventional beamforming system or other methods.

If we can compute the inverse matrix of G in real time, then we can set

$$y'(t)=G^{-1}x'(t) \tag{22}$$

So we have

$$y'(t)=c'(t) \tag{23}$$

In other words, the signal is separated, and the real part of $y'(t)$ is precisely the incident signal $c(t)$, i.e. in the output system each channel has only one incident signal.

In the following we will show how to get G^{-1}. According to the equation (14) $G=G_r+jG_i$, if G_r^{-1} and G_i^{-1} exist, then the G^{-1} can be calculated by means of G_r^{-1} and G_i^{-1},

$$G^{-1}=(G_r+G_iG_r^{-1}G_i)^{-1}-j(G_i+G_rG_i^{-1}G_r)^{-1} \tag{24}$$

It is quite complicated. For one thing, G_r^{-1} and G_i^{-1} do not necessarily exist simultaneously. Another thing is that the calculation of (24) is trouble some. We would adopt a new algorithm of partition matrix, by which an equivalent relationship between the complex matrix and its generated real matrix is established. This method is quite similar to the one-to-one correspondance between complex number and matrix[16].

Let

$$G_0=\begin{bmatrix} G_r & G_i \\ -G_i & G_r \end{bmatrix} \tag{25}$$

The matrix G_0 is called the generated matrix of G. Obviously, G_0 is a $2N\times 2N$ matrix. It is easy to justify that G_0^{-1} is the generated matrix of G^{-1}. So it is sufficient to compute the inverse matrix of real G_0.

Let

$$F_0=G_0^{-1}=\begin{bmatrix} F_r & F_i \\ -F_i & F_r \end{bmatrix}.$$

The F_0 is the generated matrix of

$$F=F_r+jF_i \tag{26}$$

After performing the transform F on $x'(t)$, we obtain $y'(t)=Fx'(t)$, i.e., $y'(t)=c'(t)$.

$$y(t)=\mathrm{Re}[y'(t)]=F_r x(t)-F_i\hat{x}(t) \tag{27}$$

This is the final result of signal separation. Here $x(t)$ is known, $\hat{x}(t)$ is the Hilbert transform of $x(t)$, and F_r and F_i can be computed from the bearing angle matrix by using a series transform. The flow chart of signal separation system is shown in Fig 2.

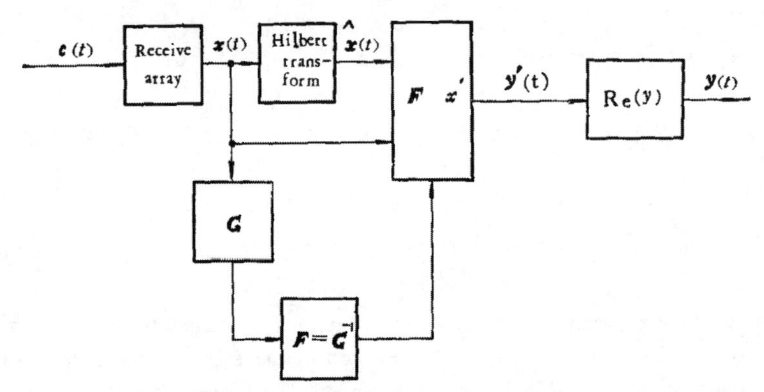

Fig. 2. The flow chart of signal separation system.

IV. The Implementation of Signal Separation Technique

From the above analysis we can see that, for the incident signal of form (8), if the

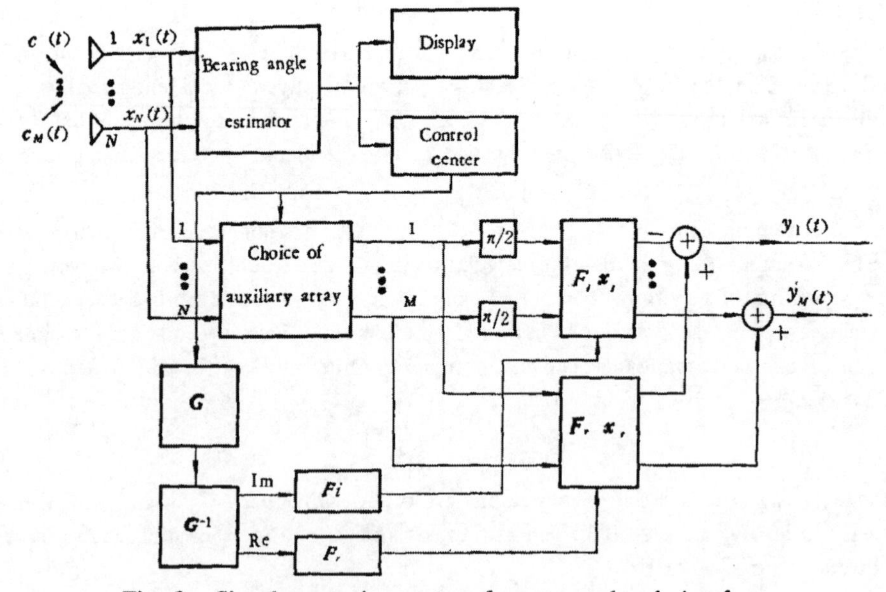

Fig. 3. Signal separation system for narrow band signal.

bearing angle matrix can be precisely estimated then the incident signal will be separated eventually. The block diagram of signal separation system for narrow band signal is shown in Fig. 3.

The number of elements of array is assumed to be N, and the number of incidental signals is M, $M \leqslant N$. The bearing angle estimates of incident signals can be obtained by using the conventional beamforming system. We choose the M channels to form an auxiliary array (the principle of choice for auxiliary array will be described in Section VI). The bearing angle matrix $\boldsymbol{G}$ and its inverse $\boldsymbol{G}^{-1}$ will be calculated by a special-purpose computer. At the same time, the signal $\boldsymbol{x}(t)$, which comes from the auxiliary array, is used to get its Hilbert transform. The resultant will construct the pre-envelope signal $\boldsymbol{x}'(t)$. A linear transform $\boldsymbol{F}$ is performed on $\boldsymbol{x}'(t)$ to get $\boldsymbol{y}'(t)=\boldsymbol{F}\boldsymbol{x}'(t)$. The real part of $\boldsymbol{y}'(t)$ is then the incident signal $\boldsymbol{c}(t)$.

If the incident signal is wide band, then we can not adopt the method presented in Fig. 4. An indirect method is to heterodyne the input signals (Fig. 4 (a)). But there is an obvious

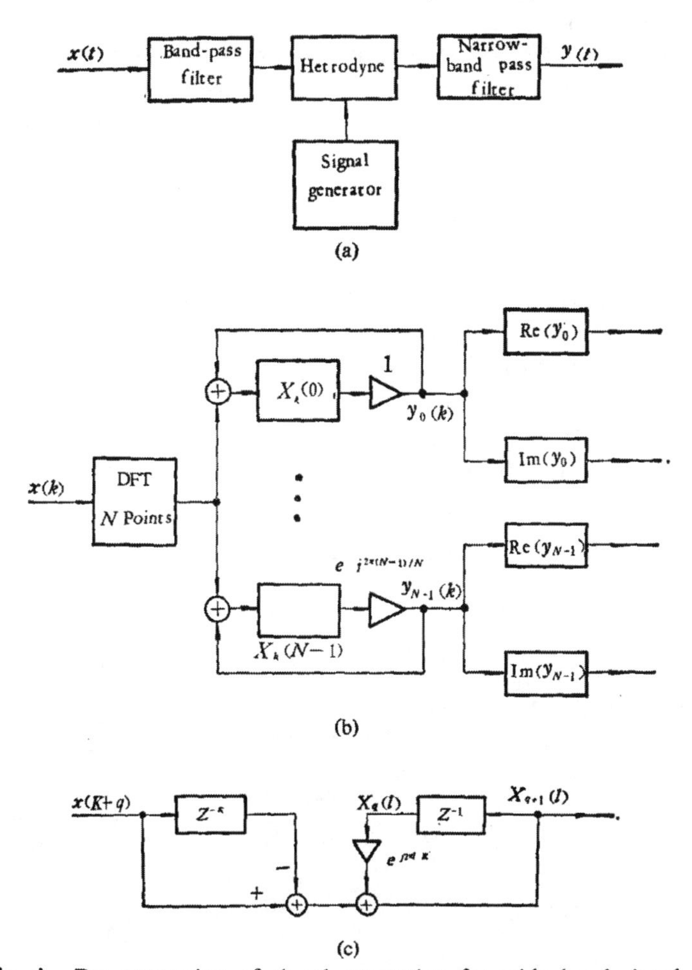

Fig. 4. Pre-processing of signal separation for wide band signal.

limitation in such processing. In general, the signal after heterodyning has the form $x(t)=u(t)\cos 2\pi f_0t-v(t)\sin 2\pi f_0t$; in order to apply the equation (13), the approximate formulas $u(t+\tau)\approx u(t)$ and $v(t+\tau)\approx v(t)$ must be valid. This mean that

$$\tau_{\max}\ll\frac{1}{W} \tag{28}$$

Another method is frequency partition for wide band signal. In the case of digital system, it is necessary to make DFT for input data $x(k)$, $k=0,1,2,\ldots$. For each component of $x(k)$, use a signal separation system of narrow band signal, in the output of system sum up the corresponding components of each incident signal to get wide band separated signal (Fig. 4 (b)).

There are two algorithms for DFT we can adopt. One is batch-input batch-output. In each computation period there are K input points to get K spectrum lines, and then we can obtain by IDFT one output point. Of course, we can use FFT, but the efficiency is very low. If we desire to get sequential output, we have to overlap the samples. And the efficiency is further decreased.

Another algorithm presented here is the method of spectrum tracking, by which not only the sequential output can be obtained, but also the calculation the DFT in each computation period is avoided. Let the input data be $x(k)$, $k=0, 1, 2, \ldots$. In time q, the instataneous DFT of length K is

$$X_q(l)=\sum_{k=0}^{K-1}x(q+k)e^{-j2\pi kl/K} \tag{29}$$

It is easy to prove that the l-th component $y_q(l)$ of sequence $x(k)$ is

$$y_q(l)=X_{q_0}r(l)\cos(2\pi lq/K)-X_{q_0}i(l)\sin(2\pi lq/K) \tag{30}$$

where $q_0\triangleq[q/K]K$, and $[x]$ denote the maximum integer which is not less than x. $X_{q_r}(l)$ and $X_{q_i}(l)$ indicate the real part and image part of $X_q(l)$, respectively.

In order to obtain the output $y_q(l)$, $q=0, 1, 2, \ldots$, we have to calculate the DFT of time series: $x(0),\ldots, x(K-1)$; $x(1),\ldots, x(K)$; …. It is not necessary to perform FFT for each set of $x(k)$. Because in each step, only one sample must be renewed, their DFT have some internal liaison. In fact,

$$X_q(l)=\sum_{k=q}^{K+q-1}x(k)\exp(-j2\pi l(k-q)/K) \tag{31}$$

$$X_{q+1}(l)=\sum_{k=q+1}^{K+q}x(k)\exp(-j2\pi l(k-q-l)/K) \tag{32}$$

multiply the factor $\exp(-j2\pi l/K)$ in both sides of equation (32), it follows

$$X_{q+1}(l)=(X_q(l)+x(k+q)-x(K))e^{-j2\pi l/K} \tag{33}$$

This recursive formula shows how to get $X_{q+1}(l)$ from $X_q(l)$. It considerably simplifies the calculation.

Combining the equations (33) and (30), we obtain the pre-processing system given in Fig. 4 (c).

V. The Results of Computer System Simulation

According to the previous theory and technique of signal separation, we have done a lot of computer simulation experiments in the 108—II computer. A part of these results will be presented below.

We choose a $N=4$ line array, the distance between two elements is d. The ratio of d to the wave length λ of incident signal d/λ is chose to be $0.5\sim1$. Three cases are considered, that is $M=2$, 3, and 4. The average powers and incident angles are as follows:

Signal	1	2	3	4
Bearing angle	0°	30°	60°	90°
Power	1	4	9	16

The calculated result of bearing angle matrix and its inverse are given in Table 1.

As shown previously, if the input signals are sinusoidal signals with independent phase, the signals can be wholly separated, whether the frequency is the same or different. Fig. 5 shows the results of the same frequency. Fig. 6 is the result of different frequencies ($M=2$ and $M=4$, respectively).

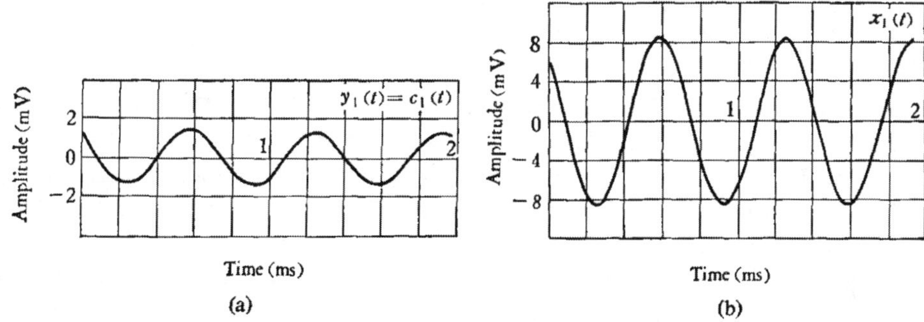

Fig. 5. Signal separation of signals with the same frequency.

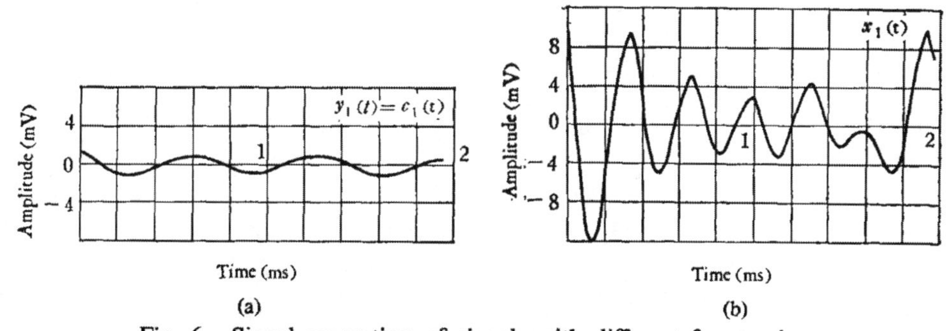

Fig. 6. Signal separation of signals with different frequencies.

From the graph we can see the incident signals are precisely separated. The computation result shows the difference between $y(t)$ and $c(t)$ is less than 10^{-7}.

In Fig. 7 the result of signal separation for amplitude modulation signal is given. The ratio of carrier frequency to envelope frequency is chosen to be 25—100. The maximum time delay $\tau_{max}\approx 1/f_0$, the bandwidth of envelope signal is Δf, $\Delta f/f_0\approx 1.5\times10^{-2}$.

We see that the result of signal separation is good. The graph in Fig. 7 (a) is the poorest case. If we are only interested in the envelope, then there is almost no difference between $y(t)$ and $c(t)$ (Fig. 7 (b)).

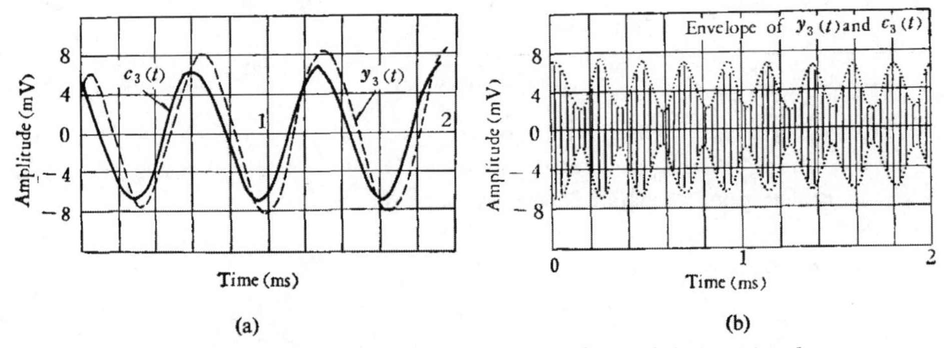

Fig. 7. Signal separation of amplitude modulation signal.

We have also performed the system simulation for narrow band Gaussian signal. The first step is to use the modified mixture congruential method to generate random numbers[17], which are uniformly distributed in the interval [0, 1], and then use these random numbers to get a Gaussian random series by moving average. The ratio of correlation time to the period of center frequency is 25 (Fig. 8).

The result of signal separation is given in Fig. 9. From these graphs we can see that the effect of signal separation is quite good, but when the number of incidental signals is increased, the error will appear due to the differences of incidental signal power.

Besides the above experiment performed in the computer, we have made some justification for wide band signals using equation (33). The theoretical analysis and the practical result are consistent.

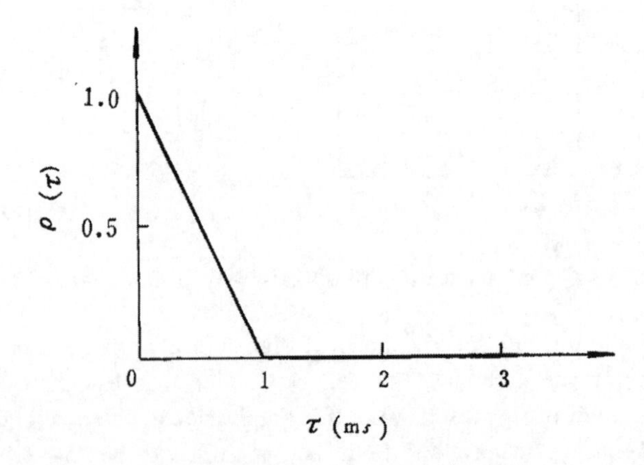

Fig. 8. Correlation coefficient of Gaussian random number.

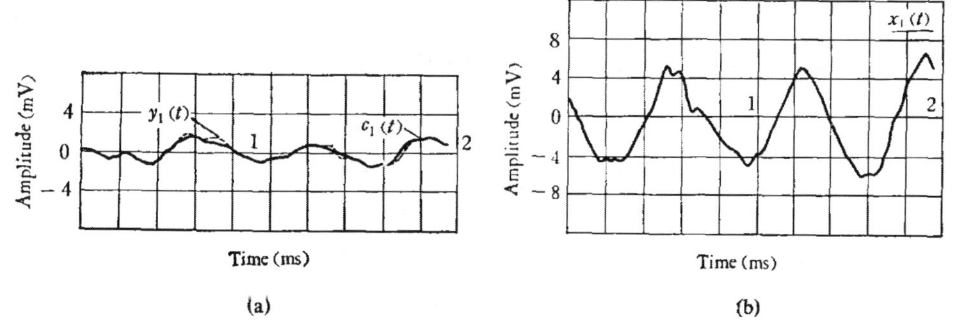

Fig. 9. Signal separation of narrow band Gaussian signal.

VI. DISCUSSION

The signal separation technique is a new topic. A lot of problems must be further studied. Here we describe two problems, which are related to the theory presented above.

1. Does the inverse matrix of bearing angle matrix exist? It is impossible to give a general answer. Let us see a simple example. Consider a two element array with $d/\lambda=1$. The incident angles are assumed to be θ_1 and θ_2. Choose one element as reference point, we have

$$\tau_{kl}=\frac{(k-1)d}{\lambda}\sin\theta_l, \qquad k,\ l=1,\ 2.$$

$$G=\begin{bmatrix} 1 & 1 \\ \exp(j2\pi\frac{d}{\lambda}\sin\theta_1) & \exp(j2\pi\frac{d}{\lambda}\sin\theta_2) \end{bmatrix}.$$

If $\theta_1=0,\ \theta_2=\pi/2$, then $G=\begin{bmatrix} 1 & 1 \\ 1 & 1 \end{bmatrix}$, G^{-1} doesn't exist.

If $\theta_1=0,\ \theta_2=\pi/6$, then $G=\begin{bmatrix} 1 & 1 \\ 1 & -1 \end{bmatrix}$, G^{-1} exists.

In the system simulation, we found that in the almost all situations, G^{-1} does exist. If the number of elements is large, then we can choose a suitable auxiliary array so that the difference among the elements of G^{-1} is as small as possible. Thus the error of calculation can be reduced.

2. The effect of signal separation depends on the estimates of vector d_j, when the noise is present, the error will appear. We can use the scanning method to find the optimum bearing angle in the case of $M\leqslant N$.

The author wishes to express his thanks to Mr. SUN LIANSHENG, Mr. WANG YUJIANG, Mr. ZHU ZHIRONG and Mrs. LIU FENQING for their kind assistance in the computer simulation.

Table 1

SIGNAL SEPARATION

MARCH, 1982

INCIDENTAL ARRIVAL ANGLE

1	2	3	4
0	30	60	90

FREQUENCY F=1.5kHz

MATRIX G

+1.0000	+1.0000	+1.0000	+1.0000	+0.0000	+0.0000	+0.0000	+0.0000
+1.0000	+0.0000	−0.9127	−1.0000	+0.0000	+1.0000	+0.4086	+0.0000
+1.0000	−1.0000	+0.6661	+1.0000	+0.0000	+0.0000	−0.7458	−0.0000
+1.0000	−0.0000	−0.3033	−1.0000	+0.0000	−1.0000	+0.9529	+0.0000
+0.0000	−0.0000	−0.0000	+0.0000	+1.0000	+1.0000	+1.0000	+1.0000
+0.0000	−1.0000	−0.4086	−0.0000	+1.0000	+0.0000	−0.9127	−1.0000
+0.0000	−0.0000	+0.7485	+0.0000	+1.0000	−1.0000	+0.6661	+1.0000
+0.0000	+1.0000	−0.9529	−0.0000	+1.0000	−0.0000	−0.3033	−1.0000

INVERSE MATRIX OF G

+0.0983	+0.4017	+0.4017	+0.0983	−0.1517	−0.1517	+0.1517	+0.1517
+0.2500	+0.3838	−0.2500	−0.3858	−0.3858	−0.2500	+0.3858	+0.2500
+1.1112	+0.1727	−1.1112	−0.1727	−0.1727	+1.1112	+0.1727	−1.1112
−0.4602	−0.9602	+0.9602	+0.4602	+0.7102	−0.7102	−0.7102	+0.7102
+0.1517	+0.1517	−0.1517	−0.1517	+0.0983	+0.4017	+0.4017	+0.0983
+0.3858	+0.2500	−0.3858	−0.2500	+0.2500	+0.3858	−0.2500	−0.3858
+0.1727	−1.1112	−0.1727	+1.1112	+1.1112	+0.1727	−1.1112	−0.1727
−0.7102	+0.7102	+0.7102	−0.7102	−0.4602	−0.9602	+0.9602	+0.4602

References

[1] Anderson, V. C., "Dicanne, a realisable adaptive system", *J. Acoust. Soc. Am.*, **45** (1969), 39—51.

[2] Wang, H. S. C., "Interference reduction by amplitude shading of sonar transducer array", *J. Acoust. Soc. Am.*, **61** (1977), 76—87.

[3] Brennan, L. E. and Reed, I. S., "Theory of adaptive radar", *IEEE Trans.* **AES-9**, (1973), 237—252.

[4] Gabriel, W. F., "Adaptive array—an introduction", *Proc. IEEE*, **64** (1976), 239—272.

[5] Griffiths, L. J., "A simple algorithm for real time processing in antenna arrays", *Proc. IEEE*, **57** (1969), 1696—1704.

[6] Widrow, B. et al., "Adaptive antenna systems", *Proc. IEEE*, **55** (1967), 2143—2159.

[7] Susans, D. E., "An adaptive antenna system for rejection wideband interference", *IEEE Trans.*, **AES-16** (1980), 452—459.

[8] Zahm, C. L., "Application of adaptive arrays to suppress strong jammers", *IEEE Trans.*, **AES-9** (1973), 260—271.

[9] Bienvenue, G. and Vernet, J. L., "Enhancement of antenna performance by adaptive processing", *Proc. of NATO Advanced Inst.* on signal Processing and Underwater Acoustics, Loughborough Academic Press, 1973.

[10] Mermoz, H., "Resources et limitations de la matrice interspectral en traitment spatial", Huitieme Colloque sur le Traitment du Signal et ses Applications, Nice France, **1** (1981), 305—310.
[11] Mermoz, H., "Complementarity of propagation model with array processing", *Proc. of NATO Advanced Study Inst.* on signal Processing and Underwater Acoustics, Italy Reidel, (1977), 463—468.
[12] Giraudon, C., "Optimum antenna processing: a modular approach", Ibid., 401—410.
[13] Hackett, C. M. Jr., "Adaptive arrays can be used to separate communication signal", *IEEE Trans.* **AES-17** (1981), 234—247.
[14] Whalen, A. D., "*Detection of signal in noise*", (Academic Press New York, 1971).
[15] Rice, S. O., "Mathematical analysis of random noise", *BSTJ*, **23** (1944), 282—332, **24** (1945), 46—156.
[16] Barnet, S., "*Matrix methods for engineers and scientists*", Ch. 2, (McGraw-Hill Book Comp., 1979).
[17] Hou, C. H. and Wu, Z. D., "High resolution spectrum estimate of narrow-band signal in wide-band coherence noise" *Acta Acustica*, **6** (1981), 337—347. (in Chinese)

THE PERFORMANCE OF THE OPTIMUM ARRAY FILTER FOR SENSOR ARRAYS①②

QIHU LI

INSTITUTE OF ACOUSTICS, ACADEMIA SINICA
P.O.BOX 2712, BEIJING, PEOPLE'S REPUBLIC OF CHINA

ABSTRACT

A theoretical analysis of the performance of optimum array filter (AOAF) for sensor arrays, which is proposed by Hanna and Simaan is presented. The beam pattern of array processing system not only depends on the rejection response, but also the model mismatch. The filter, which is designed to have the optimum rejection response in the case of model mismatch is derived. The receive response $M(f)$ and the rejection response $P(f)$ of AOAF system can be obtained as a function of directional vector of the signal incidental angle θ.

It is shown that the beam pattern of DICANNE, proposed by V.C.Anderson, is a special case of AOAF when the estimation bias of the incidental angle of interference is equal to zero.

INTRODUCTION

Beamforming is a fast-evolving specialty of underwater acoustics. It is an integral part of sonar system design. It provides not only the means to meet system performance specification but also offers relief in coping with the difficulties that are inherent in the mechanical and electrical design of large array structures.

A new method for eliminating undesired coherent signal while simultaneously providing a all-pass condition for the desired signal is recently proposed by Hanna and Simaan [1-2]. The signal processing system used in this method, according to Hanna and Simaan, is "Absolutely Optimum Array Filters" (AOAF) with zero width main lobe and no side lobes rejection responses. The block diagram of such system is shown in Fig.1.

Hanna's conclusion is based on the assumption that the incidental angles of signal and interference can be obatined. In the practice, the incidental angles, especially for the signal are unknown. It is naturally to scan the bearing angle for the interference instead of the fixed parameters in the filter design. Therefore the beam pattern of array processing system not only depends on the rejection response, but also the model mismatch. The first attempt of entirely cancelling the interference was made by Anderson in DICANNE [3]. And a subsequent analysis was present in [4]. In his system, two mechanic-electrical compensators are used to steering the signal and interference separately.

In any practical application situation, the incidental angle of signal and interference are unknown in advance. We should use the estimate value to design optimum filter. The performance of such optimum filter, the sensitivity of system response with respect to the model mismatch will be substantially concerned in the system design.

THE TRANSFER FUNCTION OF AOAF

The Fourier transform of the output of filter-sum system in Fig.1 is

$$\begin{aligned} Z(f) &= F[z(t)] \\ &= S(f)\sum_{i=1}^{N} H_i(f) + U(f)\sum_{i=1}^{N} e^{2\pi jf\tau_i}H_i(f) \end{aligned} \tag{1}$$

where $S(f)$ and $U(f)$ are the Fourier transform of signal and interference, and $H_i(f)$ is the frequency response of i-th filter. The square value of rejection response $R(f)$ is

$$|R(f)|^2 = |\sum_{i=1}^{N} e^{2\pi jf\tau_i}H_i(f)|^2 \tag{2}$$

Subjecting to the constraint

$$\sum_{i=1}^{N} H_i(f) = 1 \tag{3}$$

the solution of Min $(|R(f)|^2)$ is given by

$$H_i(f) = \frac{N - e^{-2\pi jf\tau_i}\sum_{i=1}^{N} e^{2\pi jf\tau_i}}{N^2 - |\sum_{i=1}^{N} e^{2\pi jf\tau_i}|^2} \tag{4}$$

which is called absolutely optimum filter. It is straightforward to calculate

$$\begin{aligned} R(f) &= \sum_{i=1}^{N} e^{2\pi jf\tau_i}H_i(f) \\ &= \begin{cases} 1 & \text{for } f\tau_n = k_n \quad n = 1, ..., N \\ 0 & \text{otherwise} \end{cases} \end{aligned}$$

and

① A part of this work is supported by the Office of Naval Research under grant N00014-83-K-0075 and by the Airforce Office of Scientific Research of USA under grant AF-AFOSR 81-0186.

② ICASSP87, 1987, 12: 2324-2327.

$$\sum_{i=1}^{N} H_i(f) = 1 \tag{6}$$

where k_n represent an integer.

Now we get

$$Z(f) = S(f) + R(f)U(f) \tag{7}$$

For the most of the array geometry, in no cases $f\tau_n = k_n$, for $n = 1, ..., N$ will be satisfied, except the equi-spaced linear array.

In designing the optimum filter, it is necessary to have a pre-delay line to steer the signal and also we have to use the estimation value for the indencital angle of interference in equation (4). Assume the incidental angle of signal and interference are α and β, respectively. The input signal of $i-th$ element is

$$x_i(t) = s[t + \tau_i(\alpha)] + u[t + \tau_i(\beta)] \tag{8}$$

where $\tau_i(\alpha)$ and $\tau_i(\beta)$ are the time difference between i–th element with a given reference point. Suppose the time delay compasention of i–th element is $\tau_i(\hat{\alpha})$, then

$$\begin{aligned} y_i(t) &= x_i[t - \tau_i(\hat{\alpha})] \\ &= s[t + \tau_i(\alpha) - \tau_i(\hat{\alpha})] + u[t + \tau_i(\beta) - \tau_i(\hat{\alpha})] \end{aligned} \tag{9}$$

where $\hat{\alpha}$ is the estimate value of α. According to equation (4), it is easy to prove that the filter, which is designed to have the optimum rejection response, should be

$$H_i(f) = \frac{N - e^{-2\pi jf[\tau_i(\hat{\beta})-\tau_i(\hat{\alpha})]}\sum_{i=1}^{N} e^{2\pi jf[\tau_i(\hat{\beta})-\tau_i(\hat{\alpha})]}}{N^2 - |\sum_{i=1}^{N} e^{2\pi jf[\tau_i(\hat{\beta})-\tau_i(\hat{\alpha})]}|^2} \tag{10}$$

where $\hat{\beta}$ is the estimate value of β.

Now we have

$$Z_{opt}(f) = S(f)M_{opt}(f) + U(f)P_{opt}(f) \tag{11}$$

where

$$M_{opt}(f) = \sum_{i=1}^{N} e^{2\pi jf[\tau_i(\alpha)-\tau_i(\hat{\alpha})]}H_i(f)$$

$$\tag{12}$$

$$P_{opt}(f) = \sum_{i=1}^{N} e^{2\pi jf[\tau_i(\beta)-\tau_i(\hat{\alpha})]}H_i(f)$$

We will call $M_{opt}(f)$ receive response and $P_{opt}(f)$ rejection response. In the ideal situation, if $\hat{\alpha} = \alpha$ and $\hat{\beta} = \beta$, we have

$$M_{opt}(f) = 1$$

and

$$P_{opt}(f) = \begin{cases} 1 & \text{for } f[\tau_i(\beta) - \tau_i(\alpha)] = k_i \\ 0 & \text{otherwise} \end{cases}$$

That means the receive response is equal to one and the rejection response has zero mainlobe width and no sidelobes.

Let

$$D(\theta) = [e^{2\pi jf\tau_1(\theta)}, ..., e^{2\pi jf\tau_N(\theta)}]^T \tag{13}$$

represent the directional vector of angle θ, where T denotes the transpose operation of vector or matrix. It is starightforward to calculate that

$$M_{opt}(f) = 1 + \frac{[D(\hat{\alpha}) - D(\alpha)]^T[D^*(\hat{\beta}) - NI]D^*(\hat{\alpha})}{N^2 - |D^T(\hat{\beta})D^*(\hat{\alpha})|^2} \tag{14}$$

$$P_{opt}(f) = \frac{[D(\hat{\beta}) - D(\beta)]^T[D^*(\hat{\beta})D^T(\hat{\beta}) - NI]D^*(\hat{\alpha})}{N^2 - |D^T(\hat{\beta})D^*(\hat{\alpha})|^2} \tag{15}$$

where I denote the unit matrix of order N.

The equations (14) and (15) will be useful in evaluating the performance of the directivity and the sensitivity of system with respect to the model mismatch.

COMPARISION WITH CBF AND DICANNE SYSTEM

For the conventional beamforming (CBF) system, the transfer function of the filter is

$$H_i(f) = \frac{1}{N} \tag{16}$$

The Fourier transform of the system output is

$$Z_c(f) = S(f)M_c(f) + U(f)P_c(f) \tag{17}$$

where $M_c(f)$ and $P_c(f)$ are the receive response and rejection response of CBF system,

$$M_c(f) = D^T(\alpha)D^*(\hat{\alpha}) \tag{18}$$

$$P_c(f) = D^T(\beta)D^*(\hat{\alpha}) \tag{19}$$

For the DICANNE system in Fig.2, it is clear that

$$z(t) = \sum_{i=1}^{N} s[t + \tau_i(\alpha) - \tau_i(\hat{\alpha})] -$$

$$-\frac{1}{N}\sum_{i=1}^{N}\sum_{m=1}^{N} s[t + \tau_m(\alpha) - \tau_m(\beta) + \tau_i(\beta) - \tau_i(\hat{\alpha})]$$

We have

$$Z_{DIC}(f) = S(f)[M_{DIC}(f) + M'_{DIC}(f)]$$

Where

$$M_{DIC}(f) = D^T(\alpha)D^*(\hat{\alpha})$$

$$M'_{DIC}(f) = \frac{-1}{N}D^T(\alpha)D^*(\beta)D^T(\beta)D^*(\hat{\alpha})$$

As the estimation value $\hat{\beta}$ is equal to β,

$$Z_{DIC}(f) = \frac{N^2 - |D^T(\beta) - D^*(\hat{\alpha})|^2}{N} Z_{opt}(f) \qquad (20)$$

This means the directivity of DICANNE system is identical to AOAF system except a constant factor.

In order to compare the representation of the directivity of AOAF with that of CBF shown in equations (18) and (19) We rewrite the formula for $M_{opt}(f)$ and $P_{opt}(f)$ in the vector form

$$M_{opt}(f) = C_0 D^T(\alpha)[NI - D^*(\hat{\beta})D^T(\hat{\beta})]D^*(\hat{\alpha}) \qquad (21)$$

$$P_{opt}(f) = C_0 D^T(\beta)[NI - D^*(\hat{\beta})D^T(\hat{\beta})]D^*(\hat{\alpha}) \qquad (22)$$

where

$$C_0 = \frac{1}{N^2 - |D^T(\hat{\beta})D^*(\hat{\alpha})|^2} \qquad (23)$$

NUMERICAL RESULTS

The numerical results sre shown in Fig.3 and Fig.4. The calculation is based on the equations (14)-(20). It is clear that the beam pattern of AOAF is much better than CBF as the interference is present. The target resolution is considerably improved in the case of AOAF. A circle array with radius R is taken in computer simulation. The element number is $N = 32$. The deepness of zero point in the beam pattern vs ISR (interference to signal ratio) is given in Fig.4.

CONCLUSION

The general expression of transfer function for AOAF system is derived. The formula of beam pattern of CBF and DICANNE are compared with AOAF. It is shown that the DICANNE system is a special case of AOAF as the estimation value $\hat{\beta}$ is exact equal to β.

A part of this work was completed in the Department of Electrical Engineering, Princeton University. The author would like to thank Prof.Bede Liu and Prof.S.C.Schwartz in Princeton University for their encouragement and support.

References

1. M.T.Hanna and M.Simaan, "Array filters for sidelobe elimination ", *IEEE Journal of Ocean Engr.* Vol.OE-10 (1985), 248-254

2. M.T.Hanna and M.Simaan, "Absolutely optimum array filters for sensor arrays", *IEEE Trans.* Vol.ASSP-33 (1985), 1380-1386

3. V.C.Anderson, "DICANNE, a realizable adaptive processor", *J.Acoust.Soc.Amer.*, Vol.45 (1969), 406-410

4. V.C.Anderson, "Sidelobe interference suppression with an adaptive null processor ", *J.Acoust.Soc.Amer.* , Vol.69 (1981), 185-190

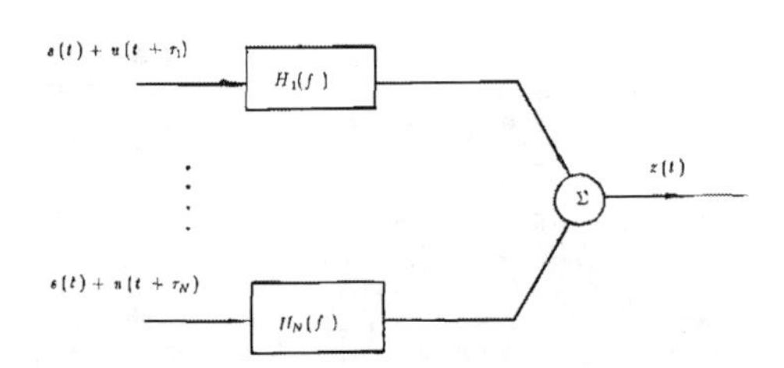

Fig.1 Filter-and -sum array processing

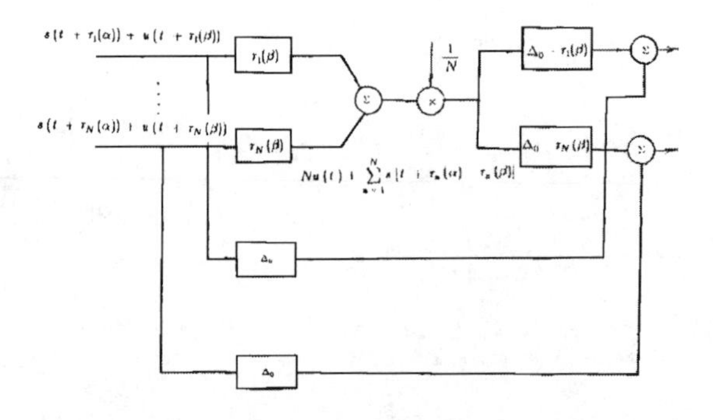

Fig.2 DICANNE system

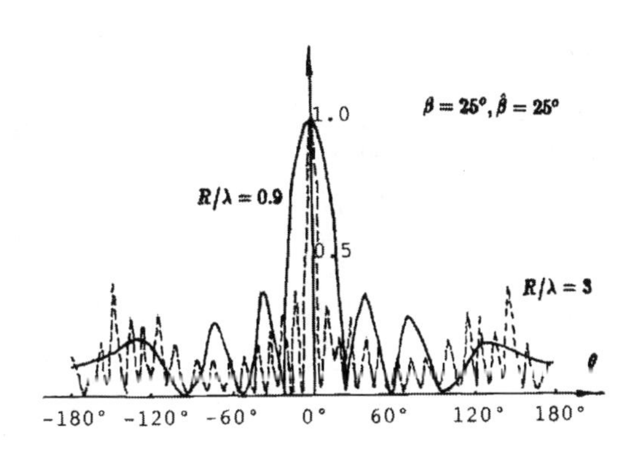

Fig.3(a) Beam pattern of AOAF

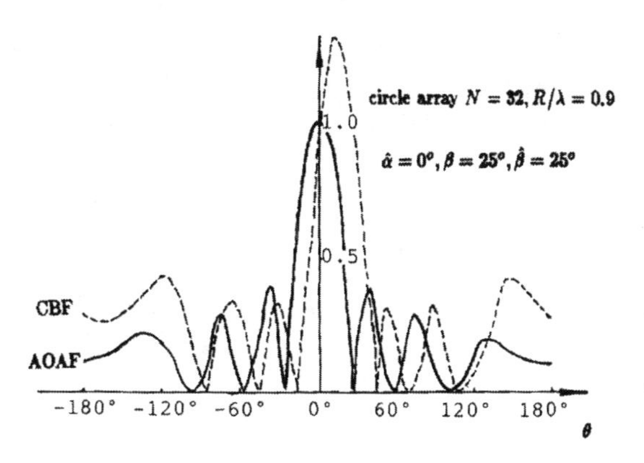

Fig.3(b) Beam pattern of CBF and AOAF

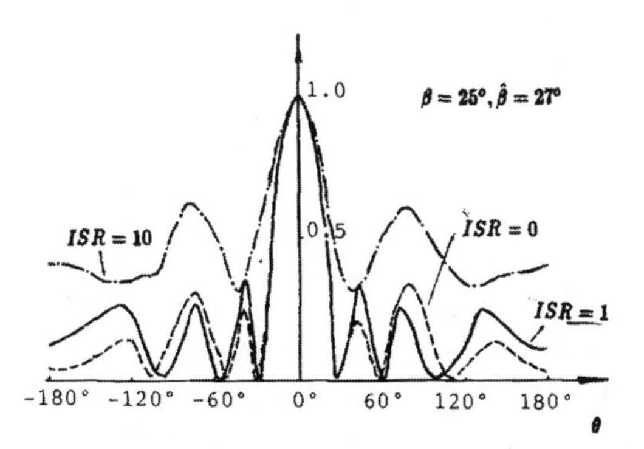

Fig.3(c) Beam pattern of AOAF with estimate bias

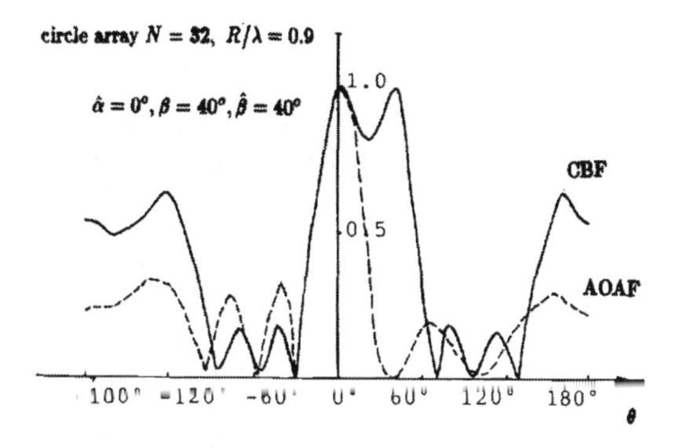

Fig.3(d) Comparision of CBF and AOAF

Fig.3 Comparision of beam pattern of CBF and AOAF

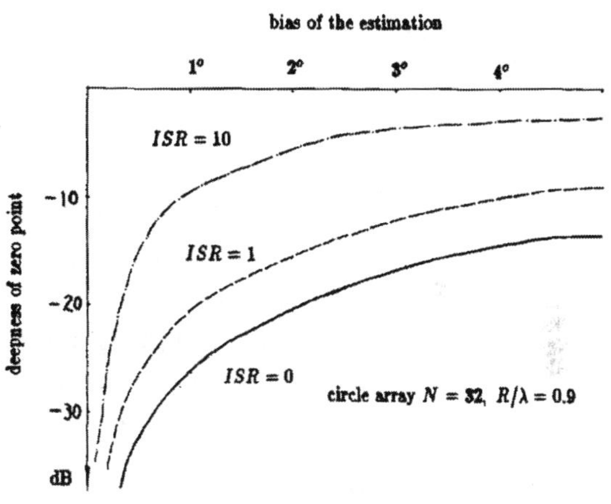

Fig.4 The deepness of zero point vs the estimate bias

自适应滤波器在时延估计中的应用——广义二次内插时延估计法①

王劲林　李启虎
（中国科学院声学研究所北京，100080）

1989 年 3 月 20 日收到

摘要　两个或多个接收信号之间的延时估计问题在许多工程应用中具有重要的意义．本文提出一种在接收信号的统计特性的先验知识了解甚少，或不了解的情况下，仍能准确地估计出接收信号间时延的方法，并推导了这一估计的方差．该方法是基于自适应噪声抵消器，在其收敛时，用权系数进行内插估计，与传统的用采样函数内插不同，本方法对 F. A.Reed 等人提出的二次内插法进行了改进，保证了估计精度，在实现上也比较简单．计算机模拟实验与海上实验结果与理论上的分析是一致的．

Application of adaptive filter in time delay estimation ——generalized quadratic interpolation method

WANG Jinlin　and　LI Qihu
(Institute of Acoustics, Academia Sinica)

Received March 20, 1989

Abstract　The time delay estimation (TDE) of two different received signals from the same source has attracted many interests of researchers in the field of signal processing. A method described for precisely estimating time delay in this paper is based on the assumption that little priori knowledge on statistical characteristics is available for the received signal. The variance of the estimate is derived. The basic architecture of this method is to use the adaptive noise canceller, in the steady state, and to interpolate the weight coefficients by using a generalized quadratic interpolation martrx. The formula of the time delay estimate is presente. The method proposed by F.A. Reed is a special case of this method. The hardware implementation is much easier than that of the conventional time delay estimation method. The results of the system simulation and the experimental results at sea show a good agreement with the theoretical analysis.

一、引　言

在理想情况下，远场点源的信号，在噪声场中被两个传感器接收到，可以记为：

① 声学学报, 1992, 17(3): 208-216.

$$x_1(t) = s(t) + n_1(t) \tag{1a}$$

$$x_2(t) = \alpha s(t + D) + n_2(t) \tag{1b}$$

其中：$s(t)$ 为点源信号，$n_1(t)$，$n_2(t)$ 为噪声，D为两路信号之间的时延，α 为两路信号幅度间的相对衰减．设 $s(t)$，$n_1(t)$，$n_2(t)$．相互独立，在以下具体讨论中，不妨设 $\alpha = 1$．

估计两信号间的相对时延D具有很大的实际应用价值，它已经被广泛用于雷达、声呐等领域．近年来，时延估计理论发展很快，提出了很多适用于各种不同情况的估计方法[1-3]，这些方法大多是通过预滤波器对信号频谱进行不同形式的加权，以获得最佳的处理效果．但在出现近场干扰，或缺少有关信号的统计特性的先验知识，如具有动态谱的信号时，对信号频谱进行预处理的时延估计方法的性能将大大降低．自 1967 年，B. Widrow[4] 提出了自适应滤波技术以来，自适应理论在时延估计领域中得到广泛的应用．1980 年 F. A. Reed 等人[2]提出了用自适应噪声抵消器进行时延估计的方法，由于自适应滤波器能够根据环境噪声场的变化，不断地自动调节自身的参数以适应周围环境，提取有用信息，因此该方法在缺乏信号统计特性的先验知识的情况下，仍能得到较好的效果．

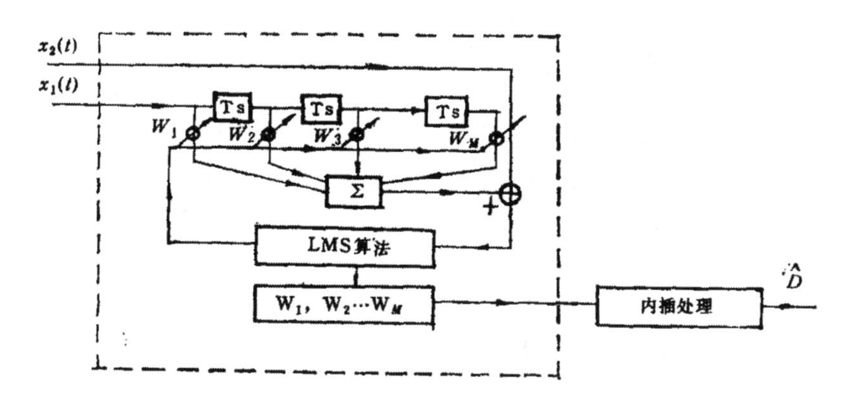

图 1　广义二次内插时延估计器

图 1 的虚线框中为横向自适应滤波器的结构图，ω_1，ω_2，$\cdots$，ω_m 为滤波器系数，M为延时线阶数，Δs 为延时节拍的长度．

将两路接收信号作为自适应滤波器的两路输入信号，考虑到自适应滤波器在时间上的离散性，且长度有限，F. A. Reed 等人[1]提出了用采样函数作为内插函数，对权系数进行内插求解时延估计 $\hat{D}$ 的方法．

设时间 t，自适应算法收敛，权系数矢量 $\boldsymbol{w}$ 用下式进行内插．

$$f(t) = \sum_{m=1}^{M} W_m(m\Delta_s) \frac{\sin[2\pi\beta(t - m\Delta_s)]}{2\pi\beta(t - m\Delta_s)} \tag{2}$$

其中：β——所处理频段的中心频率．

得到连续函数 $f(t)$ 具有下述性质：

$$f(k\Delta_s) = W_k \quad k = 1, 2, \cdots, M \tag{3}$$

以方程

$$\frac{df(t)}{dt}=0 \tag{4}$$

的解作为时延估计 $\hat{D}=t_z$.

这一估计的方差为[1]:

$$V_{ar}^{1/2}(\hat{D})=\frac{1}{\sqrt{2}}\left[\frac{\mu P_n}{P_s^2}\frac{(2P_s+P_n)(P_s+P_n)}{2-2\mu M(P_s+P_n)}\right]^{1/2}K_I \tag{5a}$$

$$K_I=M\sum_{m_1=1}^{M}\sum_{m_2=1}^{M}f'(t_z-m_1\Delta_s)f'(t_z-m_2\Delta_s)\left\{\frac{\sin\left[\frac{\pi k_0}{M}(m_1-m_2)\right]}{\sin\left[\frac{\pi}{M}(m_1-m_2)\right]}\right.$$

$$\left.\cdot\cos\left[\frac{k_0-1}{M}\pi(m_1-m_2)\right]\right\}^{1/2}\Bigg/\sum_{m=1}^{M}f''(t_z-m\Delta_s)\left\{\frac{\sin\left[\frac{\pi k_0}{M}(\Delta-m)\right]}{\sin\left[\frac{\pi}{M}(\Delta-m)\right]}\right.$$

$$\left.\cdot\cos\left[\frac{k_0-1}{M}\pi(\Delta-m)\right]\right\} \tag{5b}$$

其中: $\Delta=t_z/\Delta_s$

P_s, P_n 分别为信号与噪声的功率.

μ 为自适应算法的步长.

t_z 为延时估计值,即(4)的解.

可以看出: 该方法有以下缺陷.

(1) 对于最佳情况,输入过程应是理想的限带随机过程.

(2) 即使存在一些单纯由于干扰引起的权系数,但在内插运算时,所有权系数均被考虑,这将导致误差增大.

(3) 为了提出(4)的零点,须采用重复二分法求根,这使运算速度大大降低.

鉴于这些原因, F. A. Reed 等人[2]在 1981 年提出了二次内插时延估计方法,即选取模最大的权系数 w_{K_0} 及其相邻的两个权系数 w_{k_0-1} w_{K_0+1},用过这三点的二次曲线的顶点对应的时间作为时延的估计,即

$$\hat{D}=K_0\Delta_s+\frac{w_{K_0+1}-w_{K_0-1}}{w_{K_0+1}+w_{K_0-1}-2w_{K_0}}\cdot\frac{\Delta_s}{2} \tag{6}$$

估计方差为[2]

$$V_{ar}(\hat{D})=\frac{1}{2}\frac{\mu P_n(2P_s+P_n)(P_s+P_n)}{P_s^2(2-2\mu M(P_s+P_n))}K_q^2 \tag{7a}$$

$$K_q^2=\frac{2K_0M\Delta_s^2\left(1+12\left(\frac{D}{\Delta_s}-K_0\right)^2\right)}{\left[\sum_{m=-1}^{1}\frac{\sin\left[\frac{\pi k_0}{M}\left(\frac{D}{\Delta_s}-m\right)\right]}{\sin\left[\frac{\pi}{M}\left(\frac{D}{\Delta_s}-m\right)\right]}\cos\left[\frac{K_0-1}{M}\pi\left(\frac{D}{\Delta_s}-m\right)\right]\right]^2} \tag{7b}$$

这种方法利用三个权系数对时延进行估计，当 Δ_s/T_s 这一比值较大时，只考虑三个权系数将丢失很多信息，导致误差增大. T_s 为信号采样周期.

综上所述，我们应考虑用适当数目的权系数来估计时延，使速度与性能比达到最佳.

二、广义二次内插法的提出

鉴于二次内插法的缺陷，且考虑到估计方法在技术上的易实现性，我们仍然采用二次内插法，不同之处在于首先对自适应滤波器收敛时的权系数矢量 $\boldsymbol{W}$ 进行预处理，然后选取 $\widetilde{W}_{k_0-1}$ $\widetilde{W}_{k_0}$，$\widetilde{W}_{k_0+1}$ 作二次内插求解时延估计 $\hat{D}$.

$$\widetilde{\boldsymbol{W}} = A\boldsymbol{W} \tag{8}$$

其中：

$$\widetilde{\boldsymbol{W}} = \begin{bmatrix} \widetilde{W}_1 \\ \widetilde{W}_2 \\ \vdots \\ \widetilde{W}_M \end{bmatrix} \quad A = [a_{ij}]\begin{matrix} i = 1, \cdots M \\ j = 1, \cdots M \end{matrix}$$

当选择线性变换 A 具有下述性质时

$$|\widetilde{W}_{k_0}| > |\widetilde{W}_k| \quad k \neq k_0, \quad 1 \leqslant k; k_0 \leqslant M \tag{9}$$

即保证了经过预处理后模最大的权系数位置不变. 此时时延估计为

$$\hat{D} = k_0\Delta_s + \frac{\widetilde{W}_{k_0+1} - \widetilde{W}_{k_0-1}}{\widetilde{W}_{k_0+1} + \widetilde{W}_{k_0-1} - 2\widetilde{W}_{k_0}} \cdot \frac{\Delta_s}{2} \tag{10}$$

下面我们推导这一估计的方差.

式(10)给出时延估计的方差为：

$$V_{ar}^{1/2}(\hat{D}) = \left.\frac{V_{ar}(df(x)/dx)}{\left[\dfrac{d}{dx} \quad E(df(x)/dx)^2\right]}\right|_{x=D} \tag{11}$$

其中：

$$\begin{aligned} f(x) = {} & \widetilde{W}_{k_0} \\ & + \left[\frac{\widetilde{W}_{k_0+1} - \widetilde{W}_{k_0-1}}{2\Delta_s} - \frac{k_0}{\Delta_s}(\widetilde{W}_{k_0+1} + \widetilde{W}_{k_0-1} - 2\widetilde{W}_{k_0})\right] x \\ & + \left(\frac{\widetilde{W}_{k_0+1} + \widetilde{W}_{k_0-1} - 2\widetilde{W}_{k_0}}{2\Delta_s^2}\right) x^2 \end{aligned} \tag{12a}$$

考虑到有一些权系数是单纯由噪声引起的，及计算上的简洁性，我们只取 W_{k_0} 两边各 m_0 个权系数参与运算，$1 \leqslant k_0 - m_0 < k_0 + m \leqslant M$

$$\widetilde{W}_{k_0-1} = \sum_{k=-m_0}^{m_0} a_{k_0-1,k_0+m} W_{k+m} \tag{12b}$$

$$\widetilde{W}_{k_0} = \sum_{k=-m_0}^{m_0} a_{k_0,-k_0+m} W_{k+m} \tag{12c}$$

$$\widetilde{W}_{k_0+1}=\sum_{k=-m_0}^{m_0}a_{k_0+1,k_0+m}W_{k+m} \tag{12d}$$

假设输入信号及噪声在 $[-B_s,B_s]$ 内谱是平坦的条件下,即当

$$S_s=\begin{cases}P_s & -B_s\leqslant f\leqslant B_s\\ 0 & \text{其他}\end{cases}$$

$$S_n=\begin{cases}P_n & -B_s\leqslant f\leqslant B_s\\ 0 & \text{其他}\end{cases}$$

当我们选取的线性变换矩阵具有下述对称性质时

$$\begin{cases}a_{k_0+1,k_0+m}=a_{k_0-1,k_0-m}\\ a_{k_0,k_0+m}=a_{k_0,k_0-m} & -m_0\leqslant m\leqslant m_0\end{cases} \tag{13}$$

可以得到(10)估计的方差:

$$\mathrm{Var}(\hat{D})=\frac{1}{2}\frac{\mu P_n(2P_s+P_n)(P_s+P_n)}{P_s^2[2-2\mu M(P_s+P_n)]}K^2 \tag{14a}$$

$$K^2=\sum_{m=-m_0}^{m_0}(a_{k_0+1,k_0+m}-a_{k_0-1,k_0+m})^2 2k_0M\Delta_s^2\left[1+\frac{\Delta_s\sum_{m=-m_0}^{m_0}\alpha_m^2\left(\frac{D}{\Delta_s}-k_0\right)^2}{\sum_{m=-m_0}^{m_0}(a_{k_0+1,k_0+m}-a_{k_0-1,k_0+m})^2}\right]\Bigg/$$

$$\sum_{m=-m_0}^{m_0}\alpha_m\frac{\sin\left[\frac{\pi k_0}{M}\left(\frac{D}{\Delta_s}-m-k_0\right)\right]}{\sin\left[\frac{\pi}{M}\left(\frac{D}{\Delta_s}-m-k_0\right)\right]}\cdot\cos\left[\frac{k_0-1}{M}\pi\left(\frac{D}{\Delta_s}-m\right)\right]^2 \tag{14b}$$

其中:

$$\alpha_m=a_{k_0-1,k_0+m}+a_{k_0-1,k_0+m}-2a_{k_0,k_0+m}$$

以下我们介绍广义二次内插时延估计器的几种实现形式

(1) $A=I$ 时,就是 F. A. Reed 提出的二次内插时延估计法.

(2) $a_{k_0\pm1,j}=0\quad j>k_0+2,\ j<k_0-2$

$a_{k_0,j}=0\quad j>k_0+2,\ j<k_0-2$

时,

$$(a_{k_0,k_0-2},\ a_{k_0,k_0-1},\ a_{k_0,k_0},\ a_{k_0,k_0+1},\ a_{k_0,k_0+2})=\frac{1}{35}(-3,12,17,12,-3)$$

$$(a_{k_0-1,k_0-2},\ a_{k_0-1,k_0-1},\ a_{k_0-1,k_0},\ a_{k_0-1,k_0+1},\ a_{k_0-1,k_0+2})=\frac{1}{35}(2,27,12,-8,2)$$

$$(a_{k_0+1,k_0-2},\ a_{k_0+1,k_0-1},\ a_{k_0+1,k_0},\ a_{k_0+1,k_0+1},\ a_{k_0+1,k_0+2})=\frac{1}{35}(2,-8,12,27,2)$$

这组系数满足(13)条件,选用这组系数就是五点三次平滑二次内插时延估计法.

(3) $a_{k_0\pm1,j}=0\quad j\neq k_0\pm1$

$a_{k_0\pm1,j}=1\quad j=k_0\pm1$

时

$$(a_{k_0,k_0-2},\ a_{k_0,k_0-1},\ a_{k_0,k_0},\ a_{k_0,k_0+1},\ a_{k_0,k_0+2})=\frac{1}{35}(-3,\ 12,\ 17,\ 12,\ -3)$$

此时为五点一次平滑二次内插时延估计法．在第三节中，将介绍该方法的实验结果．

以上我们介绍了广义二次内插法的几种实现形式，在实际应用中，应根据具体情况，选取适当的预处理方法，以获取最佳效果．

三、实　　验

1. 计算机模拟实验

图 2 给出了计算机模拟实验的框图．用计算机产生一组宽带信号，加入白噪声，通过调节噪声功率获得不同信噪比的输入信号，再经过自适应算法及五点一次平滑二次内插时延估计法求得时延估计 $\hat{D}$．

图 3、4、5 分别给出五点阵平滑处理后阵元数为 16，信号带宽为 1 kHz，采样频率为 80 kHz 时信噪比为 −5dB，−10 dB，−15 dB 的估计结果，表 1 给出这三个实验结果的数据比较．图 3、4、5 中的直线为真实时延．曲线为瞬时时延估计．

从图 3、4、5 及表 1 可以看出信噪比较低的情况下，经过预处理的二次内插时延估计法仍能得到精度较高的结果．但信噪比小时，收敛速度较慢，起伏也相对大一些．

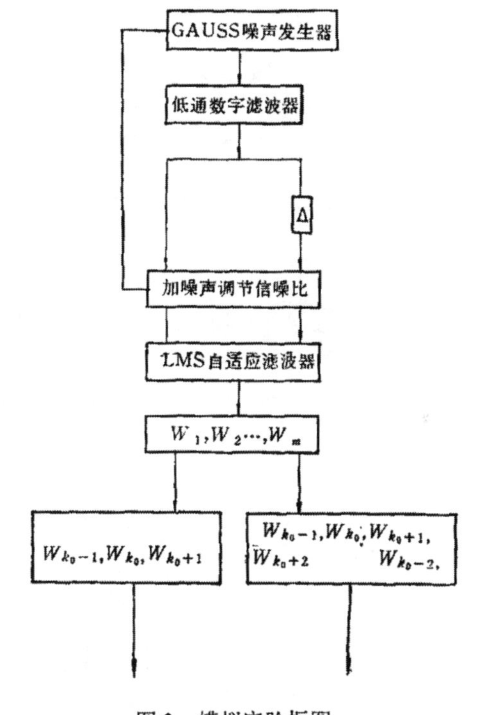

图 2　模拟实验框图

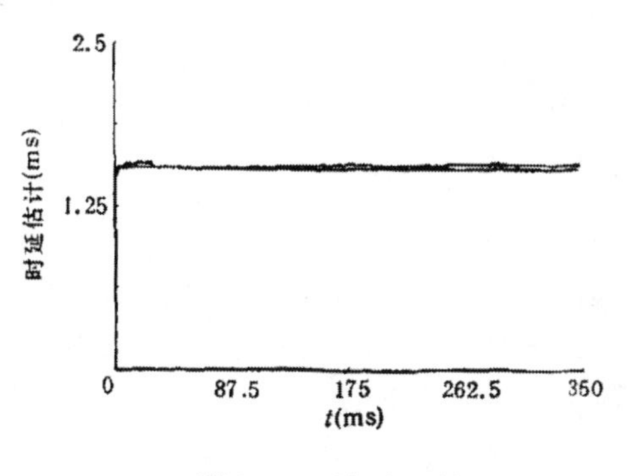

图 3　−5 dB 实验结果

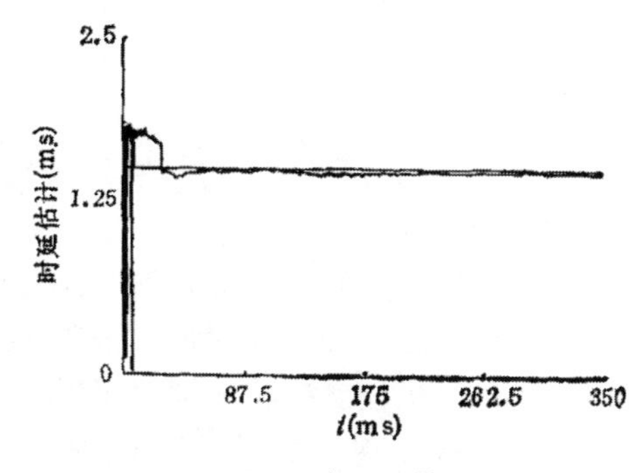

图 4　−10 dB 实验结果

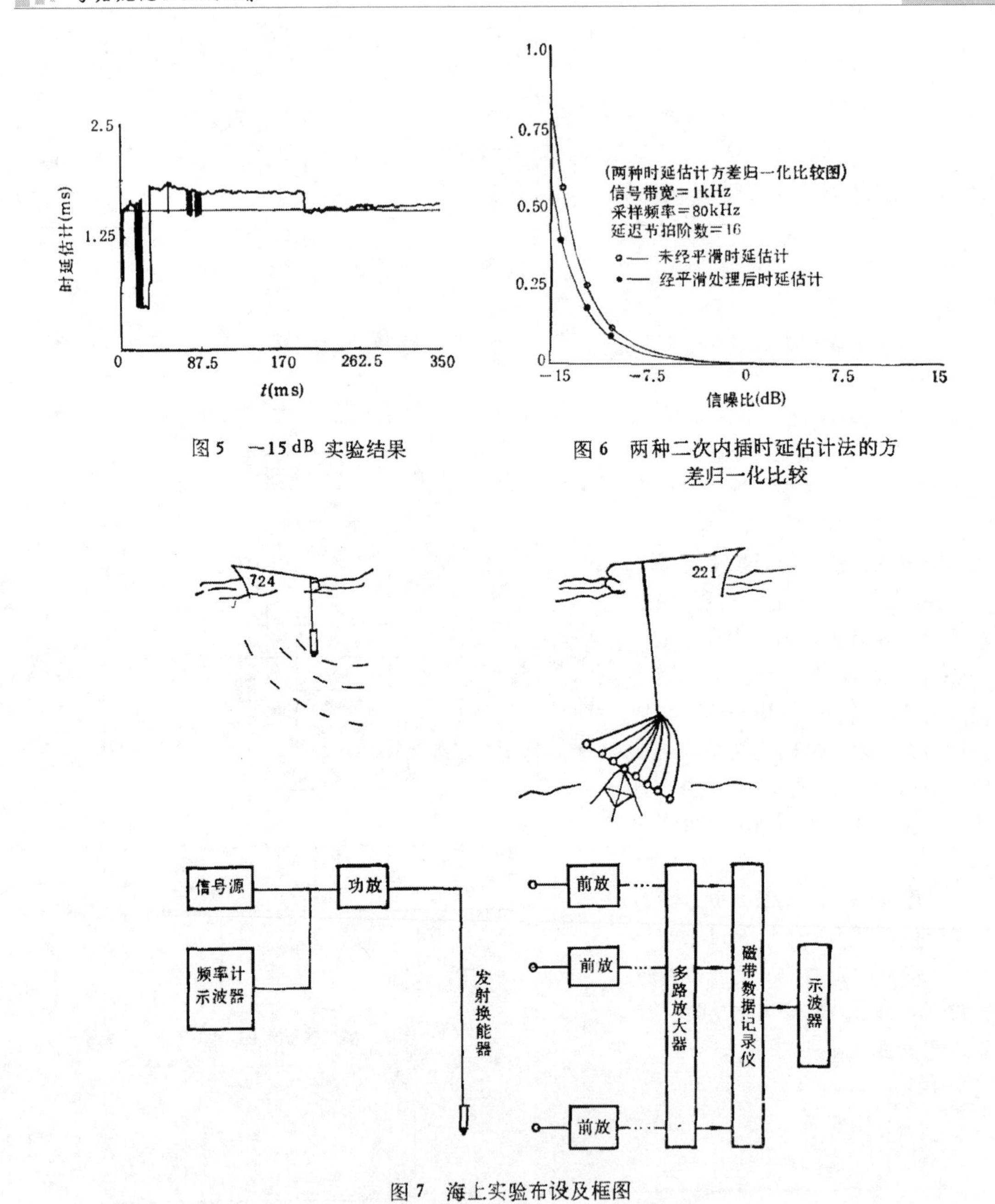

图 5　−15 dB 实验结果

图 6　两种二次内插时延估计法的方差归一化比较

图 7　海上实验布设及框图

2. 方差比较

图 6 给出了经过五点一次平滑的二次内插时延估计与未经过预处理的二次内插方法的估计方差的比较. 图 6 中所对应的时延是落在延迟线某一节拍的中部，因此经过预处理的估计方法的方差较小，但应当指出，一旦真实延时是延时节拍的整数倍，或接近整数倍时，由于这一延时抽头的权系数几乎包含了所有时延的信息，而其它权系数几乎是单纯由噪声引起的，因此考虑过多的权系数也要引起误差增大，在这种情况下，未经预处理的二次内插法的估计方差也

较小.

表 1 不同信噪比自适应时延估计精度

信噪比	真实时延 (ms)	信号长度 (ms)	时延估计 (ms)	相对精度%
−5dB	1.5375	300	1.5537	1.05
−10dB	1.5375	300	1.5168	1.34
−15dB	1.5375	300	1.6067	4.50

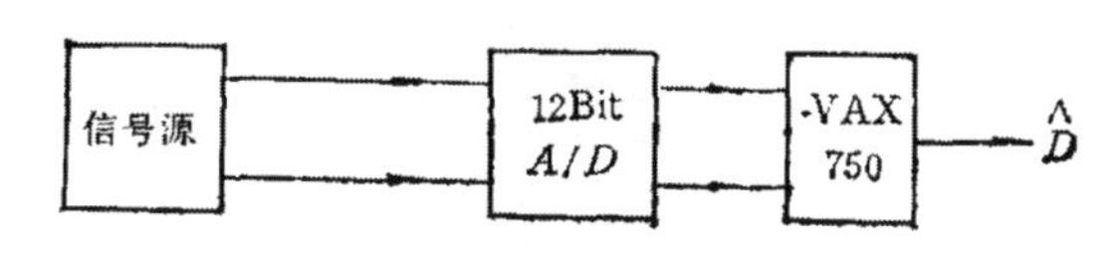

图 8 数据处理框图

3. 海上实验

在南海某海域,以实验船 *A* 作为接收船,其尾部将一个长 6 m, 每间隔 1 m 有一水听器的接收基阵置入约 100 m 深的海底,发射船 B 在距 A 400 m 处开机航行,水听器所收信号通过电缆送至 TEAC-280 磁带记录器中. 具体实验布设见图 7.

由于船 *B* 航速为 6 节,接收基阵与发射船距离相对基阵长度较大,在 500 ms 内信号的最大延时改变不会超过 2 μs, 这对于估计误差是可以接受的.

图 8 为数据处理框图, 所录信号通过 A/D 转换,送至 VAX 750 计算机, 对基阵两端水听器的接收信号的时延估计,结果绘于图 9.

图 9 给出了用五点一次平滑二次内插估计得到的时延, 与用 CF-920 信号相关器所测时延比较, 图中直线为用互相关法测得的时延, 可以看出两种方法的结果是一致的. 因此, 经过预处理的二次内插时延估计法对在实际海洋环境中的信号时延估计的精度是较高的.

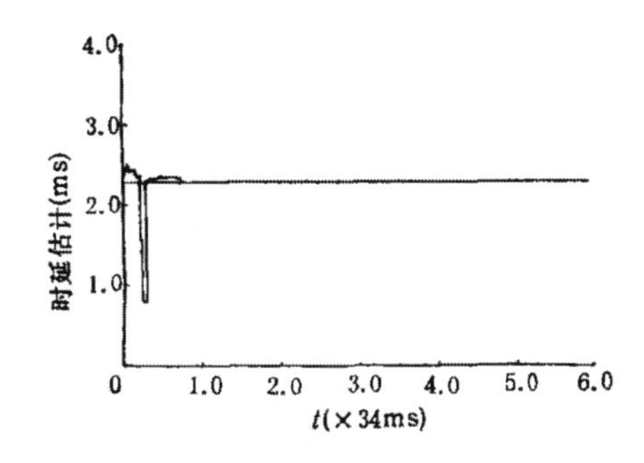

图 9 海上实验结果(直线为用 CF-920 谱分析仪测得的时延值)

四、结 论

本文详细介绍了经过预处理的广义二次内插时延估计方法,推导了估计方差,在理论上和实验上说明了这一方法的优越性, 并通过海上实验进一步验证了本文所提出的理论结果. 但从(14)式可知,是否存在着最佳预处理,仍有待进一步研讨.

海上实验的数据处理得到了陈庚副教授,宋健宁副教授的大力帮助,在此表示衷心感谢.

参 考 文 献

[1] F. A. Reed et al., *IEEE Trans.* ASSP -29, (1981) 561—576.
[2] P. L. Feintuch et al., "Adaptive Tracking System Study-Phase 3", Hudges Aircraft Co., Fullern, CA, Oct., 1980.
[3] C. H. Knapp & G. C. Cater, *IEEE Trans.*, ASSP-24, No. 4, 1976.
[4] B. Widrow et al., *IEEE* 55 Dec. 1967.
[5] 李启虎,"声呐信号处理引论",海洋出版社1985,北京

自适应滤波技术在水声信号处理中的应用①

蔡惠智　李启虎　孙　增　孙长瑜
（中国科学院声学研究所　北京 100080）

1992 年 1 月28日收到

自适应滤波理论是出现于六十年代的信号处理理论，根据这一理论系统可以自行调节本身的参数以适应周围环境，抑制干扰并检测出有用信号．本文根据 Widrow 的自适应算法，提出一种用于声呐系统的自适应滤波技术，并成功地用于被动声呐检测．被动声源的测距问题是声呐信号处理中的难题，利用自适应噪声抵消技术可以精确地测定不同传感器所接收到的信号时延差，利用内插算法准确地给出目标的距离．本文阐述了自适应滤波技术在水声中的一般应用，特别是被动测距中的应用，给出了硬件设计框图及原理，同时给出硬件、软件调试说明和实际测量的结果．本文所提出的系统，结构简单、性能好、编程容易，其性能优于国外某同类系统．

一、前　　言

近几年来，自适应滤波技术是声呐信号处理领域中最引人注目的课题之一[1-3]，这是因为在声呐工作环境中总存在着各种各样的干扰．普通的波束成形系统，当处于各向同性、均匀的噪声场时，可能具有相当好的检测能力．但是一旦出现近场干扰，或者背景噪声有某种不平稳性，声呐的检测能力就会迅速下降，以至完全失去检测能力．因此，抗干扰声呐的研究始终是声呐设计者的一个紧迫任务[4]．

所谓自适应滤波，就是声呐能够根据环境噪声场的变化，不断地自动调节本身的参数以适应周围的环境，抑制干扰并检出有用信号，换句话说，声呐的信号处理系统能够逐点实时地“学习”，降低基阵对噪声(包括干扰)的灵敏度而同时最大限度地提高对信号的灵敏度．

由于自适应滤波算法需要大量的软件编程和复杂的硬件支持，因此目前只有少数几个国家将其用于声呐中[2]．本文叙述我国自行设计的第一个用于被动测距声呐的自适应滤波系统，它采用目前先进的数字信号处理（DSP）VLSI 芯片和先进的算法构成一个实时噪声抵消系统．这个系统硬件公用，软件编程容易，便于实现多功能转换．在实验室测试及海上试验录音的回放试验中表明性能先进，其体积小，从而可靠性大大提高．

本文给出了这种自适应噪声抵消系统的原理、设计和硬件构成．同时给出某些模拟实验和实测结果．

理论分析和实际结果较好地相符．

二、自适应滤波器的原理

一个最简单的自适应滤波器的方框图如图 1.

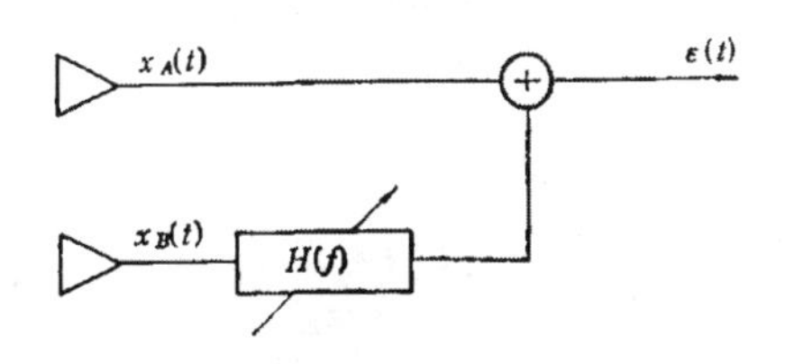

图 1　简单的自适应滤波器的方框图

其中第二路 $\overline{X}_B$ 经过滤波器 $H(f)$ 以后可以写为：

$$Y=\overline{W}^T\overline{X}_B=\overline{X}_B^T\overline{W} \tag{1}$$

$\overline{W}$ 为自适应滤波器的权系数向量．

记 I 为：

① 应用声学, 1993, 12(2): 14-18.

$$I = E[(x_A(t) - \overline{W}^T \overline{X}_B)^2] \quad (2)$$

为了求出 I 的极小值，我们将 I 对 W 求梯度，而令梯度等于零，从而求出最佳权系数矢量 $\overline{W}_{opt}$，经过一系列的运算可以得到

$$I_{\min} = E[x_A^2(t)] - \overline{W}_{opt}\overline{R}_{xd} \quad (3)$$

其中 $\overline{R}_{xd}$ 为表示矢量 $\overline{X}_B$ 与矢量 $\overline{X}_A$ 的相关函数矢量。

(3) 式不能直接计算，根据 Widrow[4] 提出的最速下降法，要求 $\overline{W}_{opt}$，权系数 $\overline{W}$ 的每次迭代都应沿着误差函数的负梯度方向。这样就可以得出 LMS 算法的数学表达式：

$$\overline{W}(j+1) = \overline{W}(j) - \mu[x_A(j) - \overline{X}_B^T(j)\overline{W}(j)]\overline{X}_B(j) \quad (4)$$

(4)式是物理可实现的，也是本文用的最基本的算法。其中 μ 为收敛系数。

如果：

$$x_A(t) = S(t) + n(t) \quad (5)$$

$$x_B(t) = n(t - \Delta) \quad (6)$$

(6) 式中的 Δ 是 B 路换能器接收到的噪声相对于 A 路的延时。

从上面的叙述可以看出：使 I 最小，也就是使与信号叠加在一起的噪声最小，这就是自适应噪声抵消器的原理。可见本文叙述的系统可以很方便地运用在噪声抵消系统中。

三、自适应滤波技术用于被动声呐测距的原理

被动声呐测距站的接收基阵配置见图 2。

普通的三点法测距基于公式：

$$R = \frac{L^2\cos^2\phi}{C(\tau_{12} - \tau_{23})} \quad (7)$$

$$\phi = \sin^{-1}\left[\frac{C(\tau_{12} - \tau_{23})}{2L}\right] \quad (8)$$

其中 L 为两个子阵之间的距离，τ_{12},τ_{23} 分别是第一个与第二个及第二个与第三个子阵之间的延时差。C 为声速。从 (7) 式和 (8) 式可以看出，测距与测向精度关键在于测量延时差 τ_{12}、τ_{23}。自适应滤波器的原理就是经过多次迭代，使其中一路信号的延时及幅度与另一路接近。即：

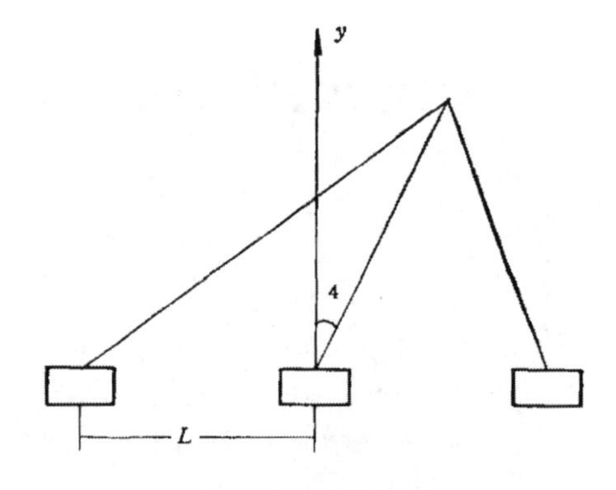

图 2 接收基阵配置图

$$X_A(j) - \overline{X}_B^T(j)\overline{W}(j) \to 0 \quad (9)$$

由此 $\overline{W}(j)$ 应该为：

$$\overline{W}^T(j) = \left(0, \cdots 0, \cdots \frac{X_A(m)}{X_B(m)}, \cdots 0\right) \quad (10)$$

m, m' 分别为第一个与第二个及第二个与第三个子阵间延时差相对于采样周期的倍数。以 τ_{12} 为例：

$$\tau_{12} = m \times \tau \quad (11)$$

其中 τ 为采样频率的倒数。

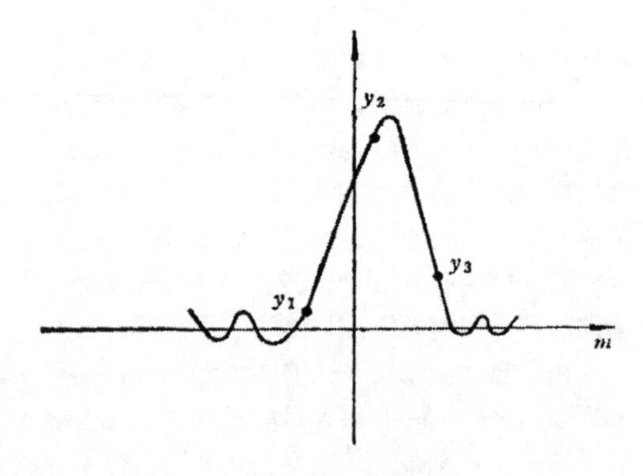

图 3 $\overline{W}$ 的表现形式

从(7)式和(8)式可以看出 τ_{12} 和 τ_{23} 的精度由 m 和 τ 决定，由于技术上的原因，τ 不可能

无穷小，这样m可能不是整数，权系数矢量 $\overline{W}$ 可能不是(10)式中的理想形式，而可能有图 3 所示的形式。

图中 Y_2 为 $\overline{W}$ 矢量中的最大值，Y_1 和 Y_3 为其左右的二个次极大值。

为了提高m的准确度，我们用三点内插法来求出m的修正值．计算公式为：

$$m = \frac{Y_3 - Y_1}{4Y_2 - 2Y_1 - 2Y_3} \tag{12}$$

四、硬 件 实 现

由于在被动声呐测距站中 τ_{12} 和 τ_{23} 可能比较大，若完全靠 $\overline{W}$ 来求得 τ_{12} 和 τ_{23}，硬件实现就比较困难，而且还会导致权的噪声加大，给精确测定m带来困难．我们在自适应滤波器的前级设置一个预延时系统，它根据前面声呐粗测的结果完全由硬件产生一个预延时．自适应滤波器和内插则由二片 TMS320C25 完成．整个系统的框图如图 4。

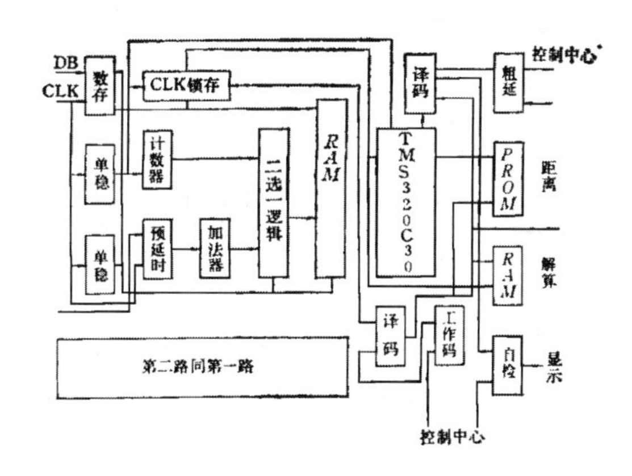

图 4 硬件框图

五、软 件 流 程

软件编制中最关键的问题就是式 (4) 中 μ 的取值问题．我们把 (4) 式中所有(除 μ 以外)的数全扩大 2^{16} 倍就是也把所有的数变成整数．

$$\begin{aligned}\overline{W}(j+1)\cdot 2^{16} = \overline{W}(j)\cdot 2^{16} - \mu[X_A(j) \\ \cdot 2^{16} - \overline{X}_B^T(j)\cdot 2^{16}\cdot \overline{W}(j) \\ \cdot 2^{16}]\overline{X}_B(j)\cdot 2^{16}\end{aligned} \tag{13}$$

为了保证(13)式的右边与左边有相同的数量级，μ 至少取 2^{-32} 量级的数，由于 μ 在正常时(未扩 2^{16}) 取 10^{-2} 左右的数，所以此时 μ 应取 2^{-35} 量级的数。

软件流程如图 5。

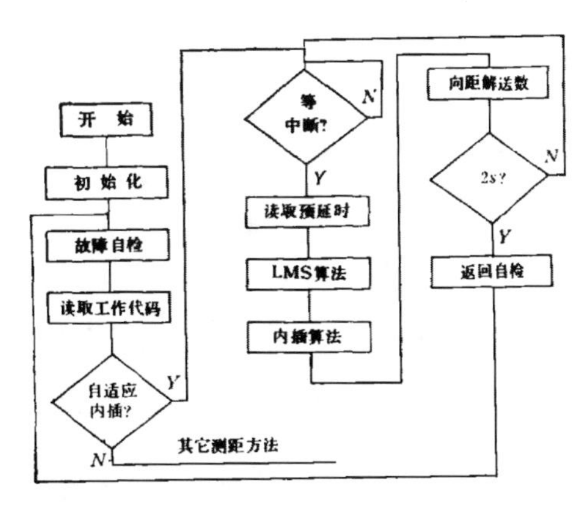

图 5 软件流程图

六、实验结果

实验分成模拟器验证和海上录音回放两种方式．两子阵间的距离 L 为 22m．模拟器由计算机产生三路延时可调、信噪比可调、方向从 210° 连续变化到 330° 的信号．验证的条件是保持距离 15km，海况 6 级，信噪比为−10dB、方向从 210°连续变化到 330°，要求每隔 2° 测量一次距离和方位．从下面给出的部分测量结果可以看出本系统完全符合在大角度下误差不超过 10%，小角度误差不超过 3%的要求．部分测量结果如表 1.

海上录音回放是通过录音机把海上实际录下的三路信号放出，潜艇两子阵间的距离是 22m，由本文提及的声呐系统进行测距和测向验证．由于海上录音回放的情况比模拟器的情况复杂得多，我们对距离解算进行了修正．主要是加上反馈跟踪、累加和剔除飞点．实际测量结果表明运用自适应滤波技术的被动声呐站能很好地完成对实际目标的测距和测向，其精度高于在同样条件下的引进设备（八十年代我国从法国引进的飞尼龙声呐）．实际测量结果如表 2 和图 6 所示．

表 1　模拟器的部分测量结果

方位	210	212	214	216	218	220	222	224	226	228	230	232	234	236
距离	14.3	14.5	14.9	15.0	15.0	15.3	15.5	15.7	15.9	15.9	15.9	16.1	16.5	16.2
方位	238	240	242	244	246	248	250	252	254	256	258	260	262	264
距离	16.0	16.0	15.9	15.8	15.5	15.4	15.4	15.5	15.5	15.4	15.4	15.4	15.5	15.3
方位	266	268	270	272	274	276	278	280	282	284	286	288	290	292
距离	15.2	15.1	15.2	15.1	15.0	15.0	14.9	14.8	14.7	14.6	14.6	14.6	14.5	14.6
方位	294	296	298	300	302	304	306	308	310	312	314	316	318	320
距离	14.7	14.6	14.5	14.7	14.2	14.6	14.7	14.6	14.5	14.5	14.5	14.5	14.6	14.6
方位	322	324	326	328	330									
距离	14.6	14.5	14.4	14.5	14.4									

注：方位的单位是度．　距离的单位是 km 为该系统的测量结果．　标准距离是 15km

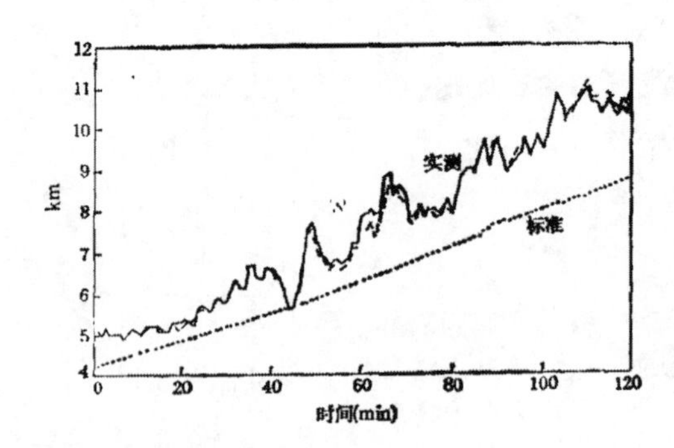

图 6　实测测量结果曲线

表 2　实际测量结果

实测	5.1	5.0	5.0	5.0	4.9	5.0	5.2	5.1	5.0	5.2	5.2	5.1	5.1
标准	4.26			4.37			4.93						
实测	5.3	5.3	5.4	5.4	5.7	5.8	5.9	6.3	6.2	6.7	6.7	6.4	6.6
标准	4.8			4.96			5.23				5.39		
实测	6.5	6.3	5.6	5.7	6.9	7.6	7.4	7.0	6.7	6.8	6.7	6.6	7.1
标准	4.8			5.71						6.09			6.25
实测	7.7	7.4	8.3	8.6	8.6	8.3	8.1	7.8	8.0	8.0	7.9	7.8	8.0
标准	4.8				6.63							7.02	
实测	8.4	9.0	9.0	8.9	9.7	9.2	9.6	9.2	8.9	9.3	9.5	9.3	9.5
标准	7.18			7.35			7.63						8.03
实测	9.5	9.2	10.2	10.6	10.2	10.6	11.0	10.6	10.3	10.4	10.6	10.4	10.6
标准			8.13			8.23		8.37			8.39		8.74

注：实测和标准的单位是 km.

七、总　　结

从上面的实验结果可以认为把自适应滤波器技术应用到被动声呐系统中是十分成功的，这在我国是首创，同时也达到世界的当今水平.在研制的过程中也遇到自适应 LMS 定点算法的截断误差的问题，我们也对此进行过研究，解决此问题的根本办法就是采用浮点算法。我们正在运用 TMS320C30 全浮点 DSP 芯片进行这方面的研究.

参 考 文 献

[1] Hinich M. J. and Bloom M. C., *J. Acoust. Soc. Am.*, **9**(1981), 738—743.

[2] Gerkin L., ASW Versus Submarine Technology, Amer. Sci. Corp. 1986 USA.

[3] Burdic W.S., Underwater Acoustics System Analysis, Beemtice-Hall, 1984, USA.

[4] Widrow B. and Steams S. D., Adaptive Signal Processing, Prentice-Hall, Inc. 1985, USA.

三

水下目标的检测、估计与识别

U1.7

An Alternative Method For Locating a Passive Source By Using The Bearing Estimations ①

QIHU LI

(INSTITUTE OF ACOUSTICS, ACADEMIA SINICA)
P.O.BOX 2712. BEIJING
PEOPLE'S REPUBLIC OF CHINA

ABSTRACT

The traditional method of passive ranging sonar system is to estimate the time-delay from a three point array, which is usually obatined by cross-correlation method. The experimental results at sea show that the time dispersion due to signal propagation fluctuation is an unneglectable factor in passive sonar design. One can improves the ranging accuracy by enlarging the distance of sub-array. But the new problem is the time-fluctuation will increase with enlarged distance. An alternative method for locating a passive source is proposed in this paper. By using three sub-array , each has their own split-beam system, the range of a passive source can be estimated by individuale bearing data provided the distance of three sub-array is large enough. The theoretical analysis shows that the new method is actually robust for the time fluctuation in the process of signal propagation.

INTRODUCTION

The problem for locating a passive source in sonar design has been interested by many authors. The traditional method is to estimate the time delay between the output signal of three-point array and resolve the range from the time difference data [1-3]. The ranging accuracy strongly depends on the time delay estimation, which is usually obtained by cross-correlation method.

For example, if an ± 10 % relative ranging error in low frequency is desired in the distance of 100 cables for a 45 meters long line array, the accuracy of time delay estimation must be equal to 2 μs .

In most cases, this requirement is hardly satisfied. The experimental results at sea show that the time fluctuation, especially in shallow water, due to the dispersion in propagation is serious [4-6]. The typical value is in the order of 10 μs in the low frequency band . So that the ranging accuracy by using time delay estimation is not an effect method in the most cases.

Although one can improves the ranging accuracy by enlarging the distance of sub-array. But the new problem is the time-fluctuation in signal propagation will increase with enlarged distance.

An alternative method for locating a passive source is proposed in this paper. By using three sub-array, each has their own split-beam system, the range of a passive source can be estimated by the individuale bearing data provided the distance of three sub-array is large enough.

The theoretical analysis shows that the new method is actually robust for the time-fluctuation in the signal propagation in certain sense.

The relative ranging accuracy depends on the choice of estimate. A practical method by using split-beam in each sub-array is proposed to get higher bearing estimation [7].

The numerical computation proves the new method for locating a passive source by using bearing estimation is useful in the case that time-fluctuation due to propagation is unneglectable.

THEORETICAL RESULTS

The block diagram of passive ranging system is given in Fig.1. H_1, H_2 and H_3 represent three sub-array. For simplification it is assumed that these array are in same horizontal line. Denote the source by S, the length of sub-array is L and the distance between two adjacent sub-array is K. The parameters of α, θ, ϕ, and R is illustrated in Fig.1.

It is clear that ,

$$\phi_1 = \frac{\pi}{2} - \theta_1, \quad \phi_2 = \frac{\pi}{2} - \theta_2 \tag{1}$$

In the traditional ranging system, $K = L$. The distance of the target R_2 is represented by the time differences τ_{12} and τ_{23} , which is [3]

$$R_2 = \frac{L^2 \cos^2 \theta_2}{c(\tau_{23} - \tau_{12})} \tag{2}$$

where c is the sound velocity

$$\tau_{12} = \tau_1 - \tau_2, \quad \tau_{23} = \tau_2 - \tau_3$$

and τ_i is the arrival time of i-th receive array.

① ICASSP87, 1988, 5: 2634-2637.

For simplification, let $R_2 = R$ and $\theta_2 = \theta$. Equation (2) becomes

$$R = \frac{L^2 \cos^2 \theta}{c\left(\tau_{23} - \tau_{12}\right)}.$$

It is easy to show that

$$dR = \frac{c\, L^4 \cos^4 \theta \left(d\tau_{23} - d\tau_{12}\right)}{c^2 \left(\tau_{12} - \tau_{23}\right)^2 L^2 \cos^2 \theta}$$

$$= c\frac{R^2}{L_e^2}\, d\left(\tau_{23} - \tau_{12}\right) \tag{3}$$

Where $L_e = L\cos\theta$ is the equivalent length of receive array.

The relative ranging error is defined by

$$\xi_1 = \frac{dR}{R} = \sqrt{2}\, c\frac{R^2}{L_e^2}\, d\tau_p / R \tag{4}$$

where $d\tau_p$ represents the time fluctuation due to signal propagation plus the measurement error. Usually, $d\tau_p$ can be considered as the sum of two components :

$$d\tau_p = \tau_0 + d\tau \tag{5}$$

where τ_0 is the stable part in τ_p , it represents the time fluctuation in the process of signal propagation and $d\tau$ is the measurement error.

From equation (4) we can see that the relative ranging error strongly depends on $d\tau_p$ as well as R/L_e. The quantity τ_0 is usually much larger than $d\tau$. For example, if $\xi_1 = 10\ \%$ is desired in 100 cables range for a 45 meters long line array, the accuracy of time delay estimation $d\tau_p$ must be equal to 2 μs . In most cases, this requirement is hard to achieve by traditional method, especially in shallow water. The typical value of τ_0 is about $10\mu s$[5−6].

In order to decrease ξ_1 , one can increase the length L of line array, such that we can decrease the ratio R/L_e . But the term τ_0 in representation of $d\tau_p$ will also increases, even faster than the decreasing of R/L_e.

An alternative method proposed here is to increase the distance of three sub-array and use the bearing information.

From Fig.1, by using the sinusoidal theorem, we have

$$\frac{R_2}{\sin\phi_1} = \frac{K}{\sin(\phi_1 + \phi_2)} \tag{5}$$

or

$$R_2 = K\frac{\sin\phi_1}{\sin(\phi_1 + \phi_2)} \tag{6}$$

According to equation (1), the equation (6) can be rewritten as

$$R_2 = K\frac{\cos\theta_1}{\sin(\theta_1 - \theta_2)} \tag{7}$$

Similarly, we have

$$R_2 = K\frac{\cos\theta_3}{\sin(\theta_2 - \theta_3)} \tag{8}$$

If we can precisely estimate the incidental angle θ_i then the range R_2 can be solved by equations (7) or (8).

The bearing angle θ_i , $i = 1, 2, 3$ can be estimated by cross-power spectrum of split beam system [7], it actually doesn't depend on the distance K . The estimation bias of θ_i depends on L , so that it is not sensitive to the stable part of time fluctuation τ_0 .

Actually,

$$\sin\theta = \frac{c(\tau_{12} + \tau_{23})}{L} \tag{9}$$

It yields

$$d\theta = \frac{c\, d\left(\tau_{12} + \tau_{23}\right)}{L\cos\theta}$$

$$= \frac{\sqrt{2}\, c\, d\tau}{L\cos\theta} \tag{10}$$

From equation (7) we have

$$R=K\frac{\cos\theta_1}{\sin(\theta_1-\theta_2)}\approx K\frac{\cos\theta}{\theta_1-\theta_2}$$

It is easy to show that

$$dR=2\frac{R^2c\ d\tau}{K_e\ L_e} \tag{11}$$

where $K_e=K\cos\theta$.

The relative ranging error is defined by

$$\xi_2=\frac{dR}{R}=\frac{2\,c\,R\,d\tau}{K_e\,L_e} \tag{12}$$

Comparing with equation (4), for $r_0=0$, we have

$$\xi_1=\sqrt{2}\,c\frac{Rd\tau}{L_e^2} \tag{13}$$

$$\xi_2=2\,c\frac{R}{L_e}\frac{1}{L_e}\frac{L_e}{K_e}d\tau \tag{14}$$

When $K_e >> L_e$, it follows that

$$\xi_2 << \xi_1 \tag{15}$$

This means that the new method has less ranging error than the traditional method provided the distance K much large than the length L of the line array.

This fact is useful, for example, in the towed array design, in which the sub-array can be seperated in a large distance.

NUMERICAL RESULTS

According to the theory described in previous section, some numerical results are presented here.

The relationship between relative ranging error and the accuracy of time delay estimation is given in Fig.2. For a line array with $L=45m$, when $\theta=0$ (i.e. $L_e=45m$) in the distance of $R=200$ *cables* , $d\tau=5\mu s$ will result about 20% ranging error. It means that if the time fluctuation is in the order of $10\mu s$ then the traditional ranging method has very poor performance.

Fig.3 is the numerical results of equation (4) by setting $L=45m$ $R=200$ *cables*. As we can see, for the fixed $\Delta\tau$, when the bearing direction θ increases, the relative ranging error also increases because $L_e=L\cos\theta$

Fig.4 demonstrates the performance of a line array with 32 elements and the distance of adjacent elements $d=\lambda/2$. By using the cross power spectrum method the bearing accuracy can be represented as a function of the bearing direction.

Fig.5 is the comparison of equation (12) and (14), for $K_e/L_e=10$, $R=100$ *cables* and $L_e=45m$. It is clear that even for $r_0=0$, the relative ranging error ξ_2 of new method is much less than the ξ_1 of traditional method.

Fig.6 is the relationship between ξ_1 , ξ_2 *vs*.K_e/L_e This figure can be used to choose the parameter in practical sonar design.

CONCLUSION

An alternative method for passive source ranging by using the bearing estimation has been discussed here. In the case that time fluctuation in signal propagation is unneglectable, the traditional ranging method has poor performance in the sense of ranging accuracy. The new method is actually robust for the time-fluctuation. It is useful in the case, for which a long line array is acceptable in sonar design.

REFERENCES

1.G.C.Carter, " The time delay estimation for passive sonar signal processing ", *IEEE Trans.* Vol.ASSP-29 (1981) 463-470

2.P.M.Schultheiss and J.P.Iannielle, " Optimum range and bearing estimation with randomly pertubed array ", *JASA* Vol.68 (1980), 167-173

3.P.M.Schultheiss, "Locating a passive source with array measurements - a summary results " , *Proc. ICASSP'79* (1979), 967-970

4.R.J.Urick, " Multipath propagation and its effects on sonar design and performance in the real ocean ", *NATO ASI Proceedings* Reidal, 1976

5.R.Scholz, " Horizontal spatial coherence measurements with explosive and CW-sources in shallow water ", *Ibid.*

6.R.H.Zhang et al., " Spatial correlation and temporal stability of long-range sound fiel in shallow water ", (*In Chinese*) *Acta of Acoustica Sinica* (1981) No.1, 9-19

7.Q.H.Li, " A precise time delay estimation method for split-beam array system ", *Proc. of ICASSP'86 , Tokyo* 1816-1819

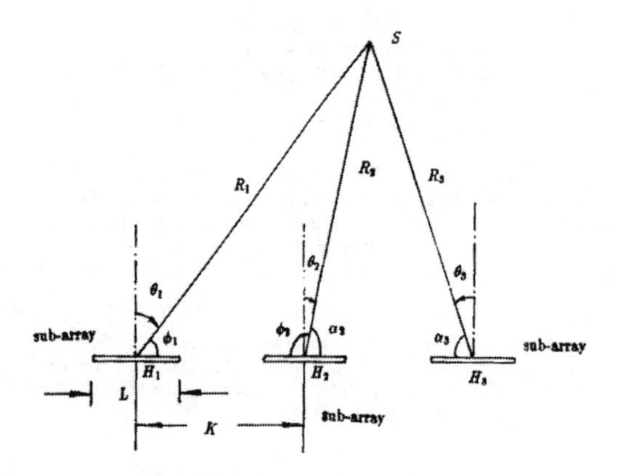

Fig.1. The geometry of a passive ranging system

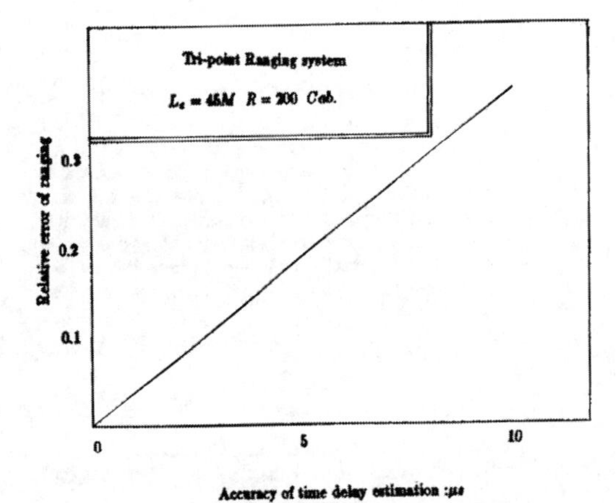

Fig.2. The relationship between ranging accuracy and time delay estimation

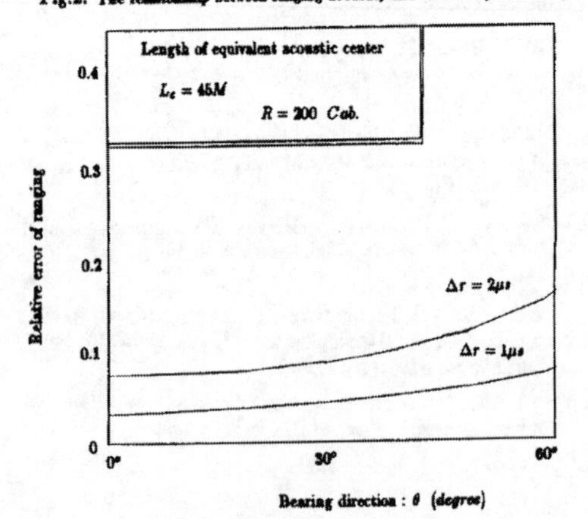

Fig.3 Ranging accuracy vs bearing direction

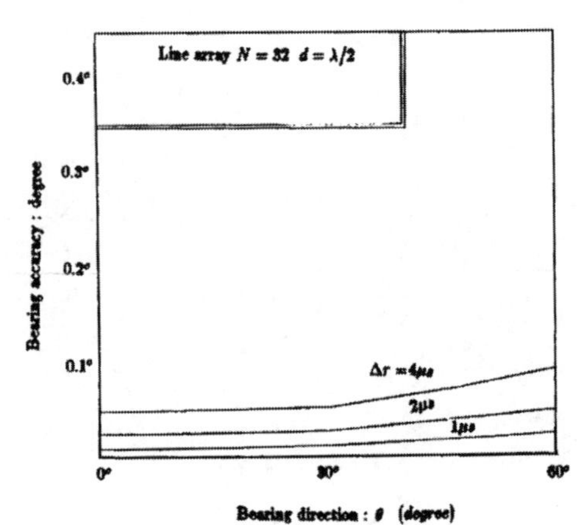

Fig.4 Bearing accuracy vs bearing direction of line array

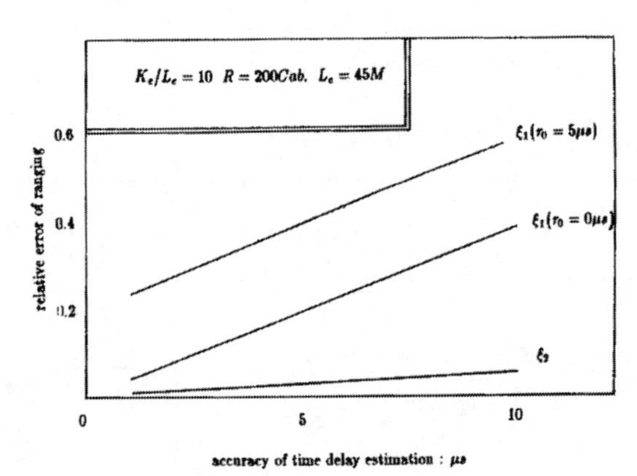

Fig.5 The effect of the time fluctuation in ranging system

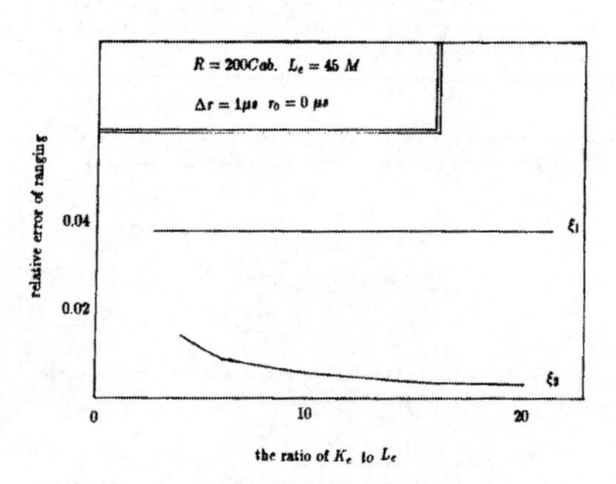

Fig.6 The comparison of relative error of ranging ξ_1 and ξ_2

被动测距声呐中后置处理置信级的设定①

李启虎 李士才 孙增 孙长瑜
(中国科学院声学研究所)

1988 年 8 月 16 日收到

精确测定目标信号到达不同水听器的延时差是被动测距声呐正常工作的基础，由于海水中声传播的多途效应，往往使延时差随机起伏，被动声呐的后置处理系统应当采用合理的算法消除起伏的影响．本文提出一种类似于 Kalman 滤波的数据平滑算法，把测量数据的方差反馈到系统的输入端．同时用筛选的方法把由偶然因素引起的，实际上不可能是由目标信号的延时差过滤掉，从而给出置信度作为指挥决策的依据，这对提高被动测距声呐的测距精度具有重要意义．

一、引言

被动声呐的测距是通过测延时来实现的．传统的测距方法是利用一个三点阵（或三个互相分离开的子阵），根据目标信号到达三点的时间差来解算距离[1-2]

从纯粹的几何角度来看，测距问题似乎十分简单，但是近年来的实际研究结果表明并非如此，这是因为水声信道的空变、时变特性，多途效应，使得信号在传播过程中引起畸变．这种畸变会影响声呐的性能，对测距声呐影响更大[3-5]，尤其是在浅海，根据多年的实际测量，延时起伏的量级不容忽视[4]．

一部工作于低频段的测距声呐，如果要使测距精度达到 10%，那么，要求测延时精度在微秒的量级．但是实际的延时起伏有时候可能在 10 微秒量级，于是精确测距就非常困难[6]．

本文提出一种数据平滑算法．它基于对数据的筛选和平均，然后把方差反馈到输入端，使后置处理变为一种类 Kalman 滤波器[7]，而它的算法和运算量远远少于 Kalman 滤波算法．

置信度的设定是一个非常重要的课题．它是衡量被动测距声呐性能的指标之一．本文提出的方法，很容易编程，并且能够把由各种偶然因素引起的随机起伏量过滤掉，从而提高后置处理系统输出数据的稳定性及可信度．

二、算法的基本依据

被动测距的基本几何配置见图 1，其中 H_1，H_2，H_3 表示三个水听器（或为三个子阵的某一个供测距用的输出波束），设 $H_1H_2 = H_2H_3 = d$．传统测距方法是依据目标 s 的辐射信号到达 H_1，H_2，H_3 的延时差来计算距离 R_2．设 H_1，H_2 之间的延时差为 τ_{12}，H_2，H_3 之间的延时差为 τ_{23}．为方便起见，令 $R_2 = R$，那么

$$R = \frac{d^2\cos^2\varphi}{c(\tau_{12} - \tau_{23})} \tag{1}$$

实际的海上试验已证明，即使在信噪比较大的情况下，$\tau_{12}-\tau_{23}$ 的起伏仍然相当大．因此，用一次测量的结果来估算距离 R 是不准确的，这样作的结果，往往使距离估算值方差很大，以

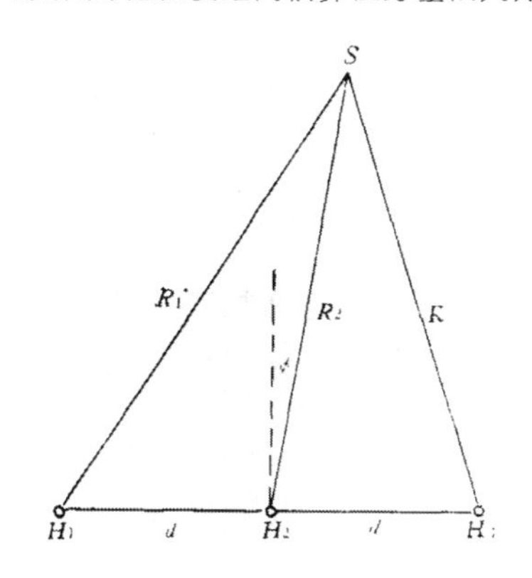

图 1 被动测距站原理示意图

① 应用声学, 1989, 8(3): 4-7.

致于不能相信，举例来说，如果 $d=20\text{m}$，目标距离 $R=10\text{km}$。如果要求测距精度是 10%，那么要求测延时精度大约为 $1.5\mu s$。

由于种种原因，即使系统本身的测延时精度非常高，信号的传播起伏与畸变，会使 $\tau_{12}-\tau_{23}$ 的起伏远远超过 $1.5\mu s$。这就要求后置处理系统能够对数据进行必要的平滑及筛选，把那些实际上不可能是由目标信号引起的起伏去掉，从而提高估算的可信度。

为说明问题起见，记 $\tau=\tau_{12}-\tau_{23}$。假定 τ 的输出速率是 0.5 秒一次，那么 5 秒钟内就有 10 个数据。如果目标航速为 10 节，则在 5 秒钟内目标只能运动 25m。在这样小范围的 ΔR 内，由它引起的 $\Delta\tau$ 的变化显然可以计算出来。比如说是 $0.5\mu s$。如果实际测量的 $\Delta\tau$ 大大超过 $0.5\mu s$。那么这样数据的可信度就非常低。所以测量方差可以作为设置置信度的数据。

下面我们来推导这一结果。

我们知道[6]

$$\{\text{Var}[R-\hat{R}]\}^{1/2}$$

$$=c\left\{\text{Var}(\tau_{12}+\tau_{23})\left(\frac{R}{Le}\right)^4\right\}^{1/2} \tag{2}$$

将 (2) 式两边除以 R。并把基阵的等效长度 Le 换成 $d\cos\varphi$。则有：

$$\frac{\sigma_\tau}{\dfrac{(d\cos\varphi)^2}{CR}}=\frac{\sigma_\tau}{\tau}$$

于是得到

$$\frac{\sigma_R}{R}=\frac{\sigma_\tau}{\tau} \tag{3}$$

其中 σ_R，σ_τ 分别表示测距方差和测延时方差。我们用差分 ΔR，$\Delta\tau$ 分别来代替它们，得到：

$$\Delta\tau=\frac{\Delta R}{R}\tau=\frac{\Delta R}{d^2\cos^2\varphi}\cdot C\cdot\tau^2 \tag{4}$$

这个式子是导出实际算法的基本依据，其中 τ，φ 是实际的测量值，ΔR 是事先按目标运动速度与积分时间确定的一个范围，由此可将 $\Delta\tau$ 解出来。

表 1 给出目标径向航速，积分时间和距离差分值 ΔR 的关系。

表 1　距离差分表

目标航速(节)			6	8	10	12	18
积分时间	30s	ΔR (m)	90	120	150	180	270
	60s		180	240	300	360	540
	120s		360	480	600	720	1080

三、类 Kalman 滤波

设 τ 的测量数据序列为 $\tau^{(1)}, \tau^{(2)}, \cdots, \tau^{(k)}, \cdots$ 积分时间为 T_0，令

$$\bar{\tau}_k=\bar{\tau}_{k-1}\left(1-\frac{1}{T_0}\right)+\frac{1}{T_0}\tau^{(k)} \tag{5}$$

不妨取 $\bar{\tau}_0=0$，注意，其中 T_0 以次数计算单位。根据 ΔR，$\bar{\tau}_k$，φ_k 求出置信电平

$$\Delta\tau_k=\frac{\Delta R}{d^2\cos^2\varphi_k}\cdot C\cdot\bar{\tau}_k^2 \tag{6}$$

设定置信限

$$(\Delta\tau)_c=3\cdot\Delta\tau_k \tag{7}$$

再求出：

$$N_k=\{i:|\tau^{(k-i)}-\bar{\tau}_k|\geqslant(\Delta\tau)_c,$$
$$i=0,1,\cdots,9.\ \text{的个数}\} \tag{8}$$

置信度

$$CL=\left(1-\frac{N_k}{10}\right)100\% \tag{9}$$

软件框图见图 2。

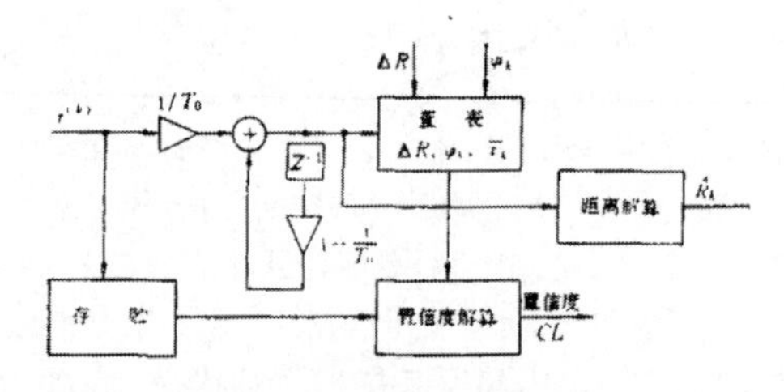

图 2　后置处理软件流程

四、系统模拟与硬件实现

由于本文提出的算法非常简单。因此，硬件实现也是非常容易的。积分时间 T_0 的取法是令 $T_0=2^N$，N 为一个正整数。

硬件的工作只是用TMS32010解算距离和置信度，其他工作都由软件完成。ΔR，$\bar{\tau}_k$，φ_k则是一个以ΔR为参数的表，事先存在ROM内，随时调用。图3及图4给出了R与τ，ΔR与$\Delta\tau$的关系。

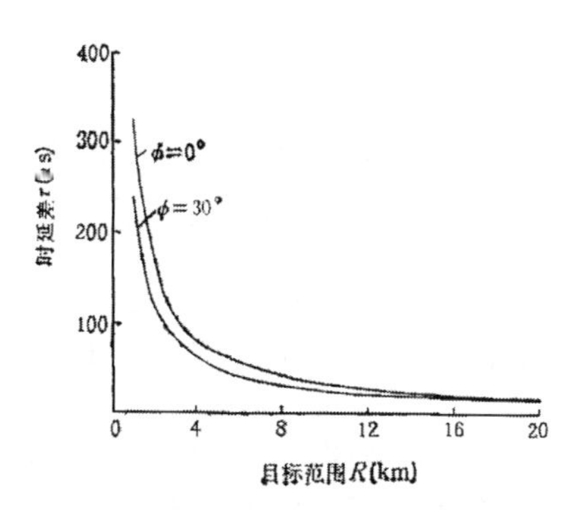

图3 距离R与延时差τ的关系

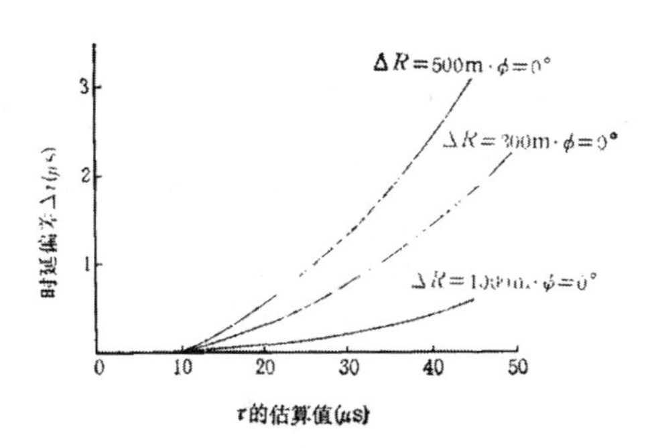

图4 距离差分与延时差分的关系

我们看到，当目标距离太近时，时延差相当大。因而，同样ΔR下所引起的延时差分也大，由此使后置处理的置信电平CL也变大.(见图5)，从τ轴看过去，置信域象是一个喇叭，这是我们意料之中的事。

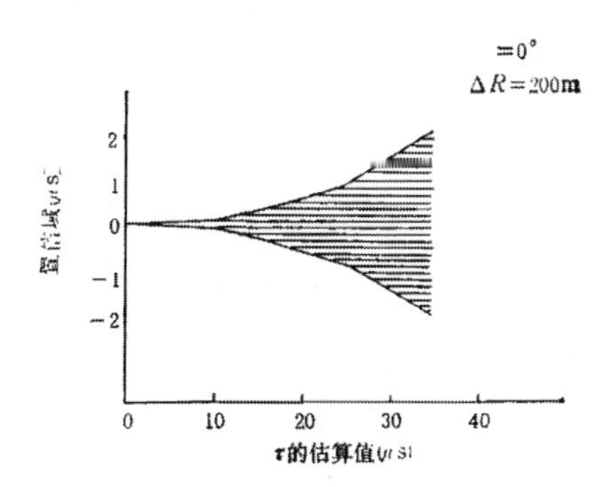

图5 被动测距系统的置信域

参 考 文 献

[1] Quaij. A. H., *IEEE Trans.*, **ASSP-29** (1981), 527—533.

[2] Hahn. W. R., *J. Acoust. Soc. Am.*, 58(1975), 201—207.

[3] Urick, R. J., Multipath Propogation and its effects on Sonar design and performance in the real ocean, Proc. of NATO ASI, 1976.

[4] 张仁和等，声学学报，1(1981)，9—19.

[5] Jobst. W., and Pominijanni, L., *J. Acoust. Soc. Am.*, 65(1979), 62—69.

[6] 李啓虎，声呐信号处理引论，海洋出版社，1985，北京.

[7] Shanmugan. K. S., and Breipohl. A. M., Random signals detection estimation and deta analysis, John Wiley, New York, 1988.

Application of Adaptive Filter Technique in distance measurement of a passive sonar①

LI QIHU, CAI HUIZHI, SUN ZENG, SUN CHANGYU

(Institute of Acoustics, Academia Sinica)
Beijing, P.O.Box:2712, China

Abstract

The theory of adaptive filter technique,that is AFT, as a theory of signal processing appeared in 1960's. In the recent 20 years, AFT is the most concentrated subject in the field of sonar signal processing[1-3]. Because there are always various interferences in the working surroundings of sonar, for the universal system of beam-forming and distance measurement of sonar, its performance would probably quite good, but only when the interference field is in a uniform and all the same in all direction situation. Otherwise in the presence of interferences near to array or some unstable noise in the background, the ability for detecting sonar signal would be drastic reduced even lost. Therefore, for the sonar designer the research about anti-interference in sonar is always very important mission[4]. According to AFT theory system, the parameters could be adjusted themselves in order to adapt to environment around so as to acquire the expected signal and repress the interference. In this paper, a AFT based on Widrow's calculation for sonar system was developed and used successfully in the distance measurement of a passive sonar. The distance measurement is always a difficult problem in sonar signal processing. But by using AFT the signal's delay-times received from several individual sensors can be exactly measured, and the distance of objectives also can be accurately calculated through the exact delay-times, at the same time, in order to enchance the accuracy of measurment, interpolation is imperative.

For the calculation of adaptive filter, quite big amount of software and complexed hardware were needed, therefore this technique was used for sonar by only a few countries[2]. In this paper, the first adaptive filter in a passive sonar distance measurement designed by China is described, in this design, most advanced IC of digital signal processing(DSP) was combined with advanced calculation method. Because in this system all of hardware is communal and software is easy, therefore it is convinent to achieve transformation of functions. In this paper, the diagrames of hardware were shown, the debug method of software and the results of laboratory experiments and practical measurements were also illustrated. the new system discribed in this paper possesses following advantages: the structure is simple, programme and debug are easy and performance is superior to other systems and the results are similar with theoretical analyses.

① Oceans 92 Proceedings, 1992, 1: 375-380.

The general method of three points for distance measurement based on following equations

$$R = \frac{L^2 cos^2\theta}{C(\tau_{12} - \tau_{23})} \tag{7}$$

$$\Psi = Sin^{-1}[\frac{C(\tau_{12} - \tau_{23})}{2d}] \tag{8}$$

in the equation the L=distance between two subarray,τ_{12} and τ_{23} is the delay-times detween subarrays respectively, c is the speed of sound. from equations (7) and (8) it can be seen that the accuracy of measurement of either distance or direction depend on measurement of τ_{12} and τ_{23}. The principle of AFT is that through extenting the delay time of one channel make the delay times of two channels equal.So τ_{12} and τ_{23} can be getten. That is

$$X_A(j) - X_B^T(j)W(j) \in 0 \tag{9}$$

therefore W(j) should be

$$W(j) = (0, ..., 0, ..., X_A(m)/X_B(m), ..., 0) \tag{10}$$

m is multiple of delay-time of signals received from two different channels. Therefore

$$\tau_{12}(\tau_{23}) = m * \tau \tag{11}$$

where τ is the reciprocal of sampling ratio.

From (7) and (8) it can be seen that the accuracys of τ_{12} and τ_{23} are decided by τ and m. Because of technical reason the value of τ can't be unlimited to incline to zero, therefore maybe the value of m is not an integer. The weight of array may not be an ideal form as in equation (10). Maybe as in Fig. 3.

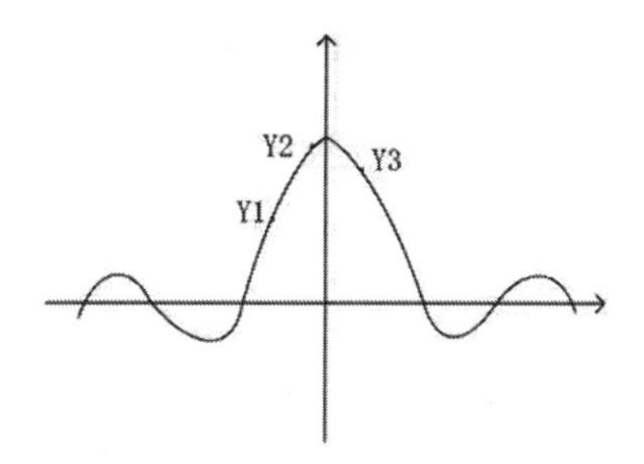

Fig. 3. The form of weight.

Y_2 is maximum in array W, Y_1 and Y_3 is submaximum on the right and left of Y_2 respectively.

In order to rise the accuracy of m,we use interpolation,the calculation of interpolation is list below:

$$m = \frac{Y_3 - Y_1}{4Y_2 - 2Y_1 - 2Y_3} \tag{12}$$

3. IMPLEMENT OF HARDWARES

Because τ_{12} and τ_{23} may be relative large, covering τ_{12} and τ_{23} only with W will be difficult and it will cause increasing of the noise of W and bring a difficulty to determine m accuracy. A pre-delay time system is set in front of adaptive filter and this system is prodused only by hardware. AFT and interpolation are completed with two pieces of TMS320c25.

The scheme of this system as in Fig. 4.

4. SOFTWARE

The key step of software programming is to determine the value of u. If on the two sides of

1. THEORETICAL PRINCIPLE OF AFT

A scheme of AFT is shown as in Fig. 1.

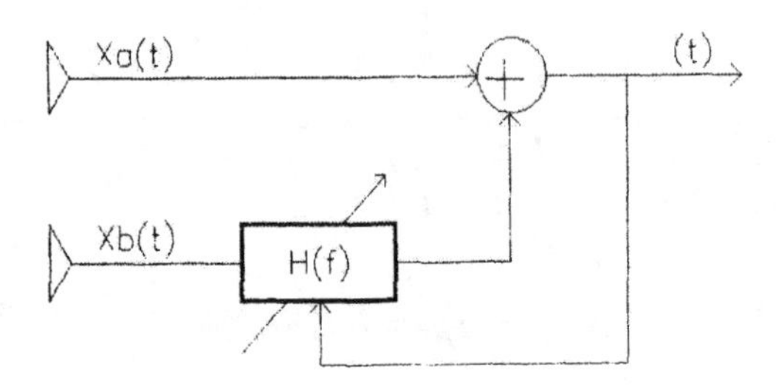

Fig. 1. A scheme of AFT.

$X_B(t)$ is the second channel result after passing filter H(f), it can be written as

(1) $$Y = W^T \vec{X}_B = X_B^T W$$

Recording I as

(2) $$I = E[(X_A - W^T X_B)^2]$$

The principle of Adaptive filter means that after n times alternating a value of W_{opt} can be found and made I minimum. We can obtained by mathematical rearrangement that

(3) $$I_{min} = E[X_A^2(t)] - W_{opt} R_{xd}$$

R_{xd} represent a vector of relative function $X_A(t)$ and $X_B(t)$.

Equation (3) can not be used directly for calculation,but according to LMS method suggested by Widrow,W can be choosen to be a value(usually to be zero) and from it the value of W_{opt} can be found. Each convolution the weight W is extend in the direction of negtive grade of error function E(t). in this case,the mathematical representation of LMS calculation can be found as:

(4)
$$W(j+1) = W(j) - \mu[X_A(j) - X_B^T(j)W(j)]X_B(j)$$

Equation (4) is physically achievable and also used in this work.

if

(5) $$X_A(t) = S(t) + n(t)$$

(6) $$X_B(t) = n(t - \Delta)$$

it can be seen that when I is minimum,noise in signal is also smallest,that's the principle of adaptive noise cancelling system, therefore it's very convinient to change the system in this paper into a adaptive noise cancelling system which is similar to the system adopted in this work.

2. THE PRINCIPLE OF APPLICATION OF AFT in passive sonar measurement of distance

The disposition of based array for reception in the station of distance measurement passive sonar is shown in Fig(2)

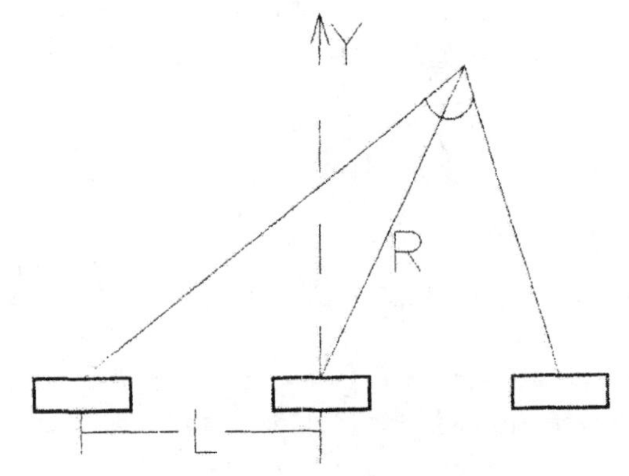

Fig. 2. The disposition of based array.

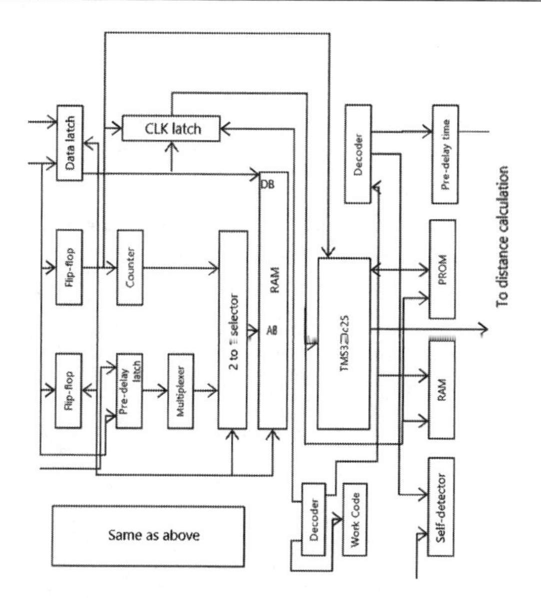

Fig. 4. The hardware scheme of this system.

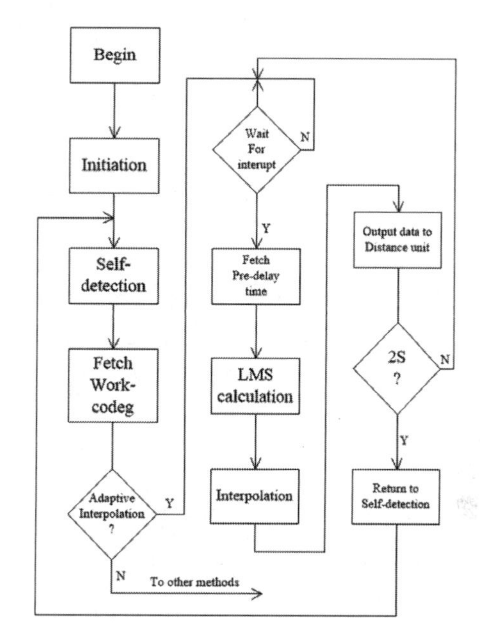

Fig. 5. The programme scheme.

equation (4) times 2^{16}, all items on the two sides of equation (4) are integeral.

$$(13)\quad W(j+1)2^{16}=W(j)2^{16}-\mu 2^{16}[X_A(j)2^{16}-X_B^T(j)2^{16}W(j)2^{16}]X_B(j)2^{16}$$

In order to insure

$$\mu 2^{16}[X_A(j)2^{16}-X_B^T(j)2^{16}W(j)2^{16}]X_B(j)2^{16}$$

only has a 2^{16} dimension and that μ is a number requesting 10^{-2} dimension when all numbers are integer, so *mu* is a number requesting 2^{-35} dimension.

The programme is shown in Fig.(5).

5. EXPERIMENTAL RESULTS

Experiments include two parts,that are the test of simulator and playing the tapes recorded on the sea in submarine. In the experiment of first part signals,whose delay-times between each other are adjustable in three channels,signal to noise ratio,which also can be adjusted,direction varied from 210^0 to 330^0 continuely are generated controled by computer. The test is complited under following conditions:

(. 1) distance is 15 kilometers

(. 2) marine condition is 6 grade

(. 3) signal to noise ratio is -10 dB

(. 4) direction varied from 210^0-330^0 arbitrary, test for each 2 degree

From the partial results (Table (1)) it is testified that this system satisfies the demands namely the error less than 10% for larger angle and 3% for smaller angle.

TABLE 1

The experiment results of simulator

Dir	210	212	214	216	218	220	222	224
Dis	14.5	14.3	14.9	15.0	15.3	15.0	15.5	15.7
Dir	226	228	230	232	234	236	238	240
Dis	15.9	16.1	16.5	16.2	16.5	16.0	16.0	15.9
Dir	242	244	246	248	250	252	254	256
Dis	15.8	15.7	15.6	15.4	15.4	15.5	15.5	15.5
Dir	258	260	262	264	266	268	270	272
Dis	15.4	15.3	15.3	15.2	15.1	15.2	15.1	15.0
Dir	274	276	278	280	282	284	286	288
Dis	15.0	15.0	14.9	14.8	14.8	14.7	14.6	14.6
Dir	290	292	294	296	298	300	302	304
Dis	14.7	14.6	14.5	14.7	14.3	14.6	14.7	14.6
Dir	306	308	310	312	314	316	318	320
Dis	14.5	14.5	14.5	14.6	14.6	14.5	14.7	14.5
Dir	322	324	326	328	330			
Dis	14.5	14.6	14.5	14.4	14.2			

* Dir=Direction
* Dis=Distance

In the experiment of second part, playing the tapes recorded on the sea was completed as follows:

The signals of three channels recorded on the sea were replayed and the measurements of distance and direction were confirmed by the sonar system mentioned above. Because the real situation was much more complicated than simulation, we revise the results according to following regulation.

(. 1) automatic following the track

(. 3) rejecting the results which exceed the maximum

(. 3) accumulation and smooth

The real measurements show that this sonar system applying AFT can accuratly finish the measurements of distance and direction of real target. the results are shown in Fig.(6) and Table(2).

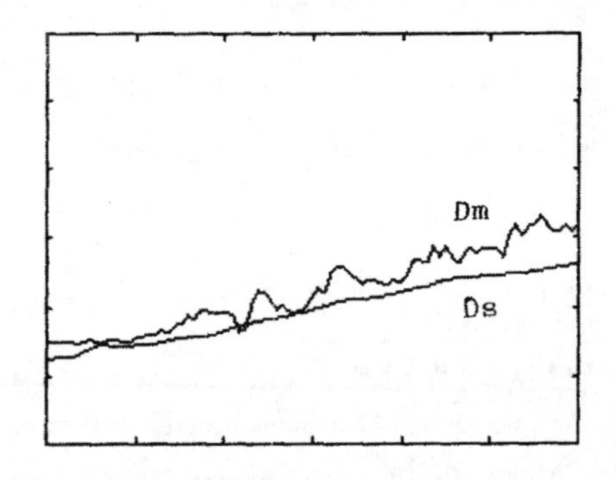

Fig. 6. The picture of real measurements.

7. REFERENCES

(. 1) M.J.Hinich and M.C.Bloom, "Statistical approach to passive target tracking," J.Acoust. Soc.Amer., Vol.09(1981)738-743.

(. 2) L.Gerkin,"ASW versus Submarine Technology Battle," Amer.Scie.Corp.,1986 USA.

(. 3) W.S.Urdic,"Underwater acoustics system analysis," Beemtice-Hall,1984 USA.

(. 4) B.Window and S.D.Steams,"Adaptive signal processing," Promtive-Hall Inc., 1985 USA.

Table 2

The results of real measurements

Dm	5.1	5.0	5.0	5.0	5.0	4.9	5.0	5.2
Ds	4.26			4.37				4.93
Dm	5.1	5.0	5.2	5.2	5.1	5.1	5.3	5.3
Ds							4.8	
Dm	5.4	5.4	5.7	5.6	5.8	5.9	6.3	6.2
Ds		4.96				5.23		
Dm	6.7	6.7	6.4	6.6	6.5	6.3	5.6	5.7
Ds		5.39						5.71
Dm	6.9	7.6	7.4	7.0	6.7	6.8	6.7	6.6
Ds						6.09		
Dm	6.7	7.1	7.7	7.4	8.3	8.6	8.6	8.3
Ds		6.25					6.83	
Dm	8.1	7.8	8.0	8.0	7.9	7.8	8.0	7.9
Ds						7.02		
Dm	8.4	9.0	9.0	8.9	9.7	9.2	9.6	9.2
Ds	7.18			7.35			7.63	
Dm	8.9	9.3	9.5	9.3	9.5	9.5	9.5	9.2
Ds						8.03		
Dm	10.2	10.6	10.2	10.5	10.6	11.0	10.6	10.3
Ds	8.13				8.23		8.37	
Dm	10.4	10.6	10.4	10.6				
Ds		8.59		8.74				

* Dm=Distance of measurement using this Sonar
* Ds=Standard Distance recorded on the sea

Fuzzy Logic For Underwater Target Noise Recognition①

Jinlin Wang, Qihu Li and Wei Wei
Institute of Acoustics, Academia Sinica
P.O.Box 2712, Beijing 100080
People's Republic of China

Abstract—The underwater target noise recognition is one of the most important topic in modern sonar design. In this paper, the target noise are divided into five types, and their characteristics in frequency domain are described. The fuzzy logic algorithm for passive classi- fication is presented. An expert system based on this inference lo- gic for target noise recognition is introduced. Some test data for this system show that this expert system has a good performance and can be considered as an excellent tool of auxiliary decision -maker for sonar operator.

I. Introduction

The underwater target noise classification is very important topic in the modern sonar system design. In fact, it is a branch of artificial intelligent(AI).

By theory, the target recognition is a typical problem of cluster analysis. There are two methods for solving this problem[7],namely:

(1) Statistical classifier
(2) Expert system approach

There are many successful examples by using method (1), such as: Bayes cluster analysis, nonparameter rank test, which has been applied in finger signature identification.

As mentioned by Horst Bendig[7], for underwater target noise recognition, there must be three basis requirements for statistical cluster, but it is difficult to satisfy in case of underwater target noise recognition. This is the reason why the expert system approach is favored.

By ES, we mean a new kind of software that simulates the problem-solving behavior of a human expert. By using inference mechanism, the expert system can simulate the problem-solving strategy of a human expert according to the architecture of knowledge base.

In real world, the expert knowledge contains many uncertain factors. These factors make big difficult in knowledge representation. For instance,the Doppler shift due to ship moving will result in the uncertainty of the frequency of line spectrum. Since 1965, the fuzzy set theory was introduced by Zadeh, L. A.[4], the theory has been used in many fields, such as: industry control, medical diagnosis. Fuzzy set theory provides a natural method for dealing with the linguistic terms by which an expert will describe a domain. An imprecise numeric term can be effectively described by a fuzzy number[1]. Hence, the use of fuzzy set theory in ES has caused an evolution of system design.

An expert system for ship radiated noise recognition named EXPLORE developed recently in Institute of Acoustics is a system based on fuzzy logic. The target noises are divided into five types: Submarine, Merchantship, Fishing vessel, Surface warship, and Oceanic noise, by using the membership functions

① Second IEEE International Conference on Fuzzy Systems, 1993, 1: 588-593.

which are represented by some numerical features of target noise. Using the criteria of maximum dependence principle of membership, EXPLORE will judge that which type mentioned above the input noise is belong to.

In section II, we will discuss the characteristics of warship briefly, then the membership functions will be constructed in section III, according to the analysis in section II. Finally, some test results will be shown in section IV.

II. Characteristics Analysis

On the point of principle, the radiated noise can be analyzed in time domain or in frequency domain. Due to the Fourier transformation, there is one to one correspondence relationship between the two representations. For simplicity of algorithm, most results are discussed in frequency domain.

The EXPLORE discussed here is based on the frequency characteristics of radiated noise. It is necessary to introduce some related results briefly. For target noise recognition, the sole information source is the radiated noise from moving ship.

The source of radiated noise can be divided into three categories: machinery noise, propeller noise, and under water dynamic noise due to target moving. In general, the first two are main radiate noise source[5]. From the point on spectrum domain, the mixture of these sources will evolve strong line spectra and continuous spectrum. In higher frequency band, the spectrum level will decrease by 6-8dB per octave. In lower band, the spectrum level will increase with frequency increasing. This property results in a peak in radiate noise spectrum. Specially, for warship, this peak locates in the band between 100-1000Hz. Figure 1.

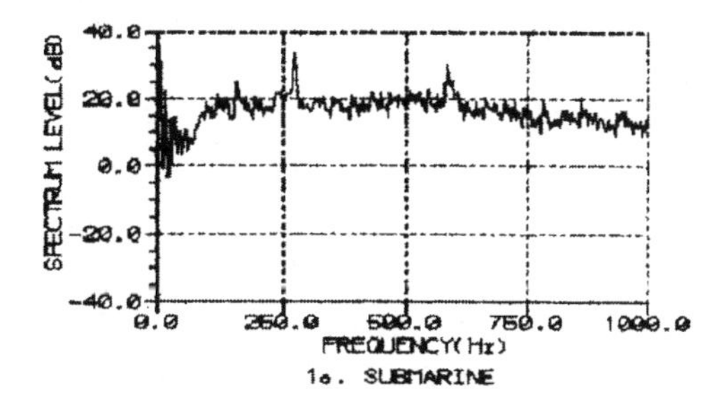

1a. SUBMARINE

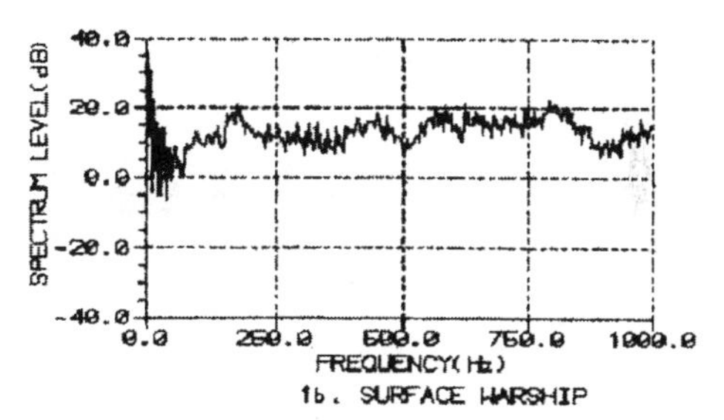

1b. SURFACE WARSHIP

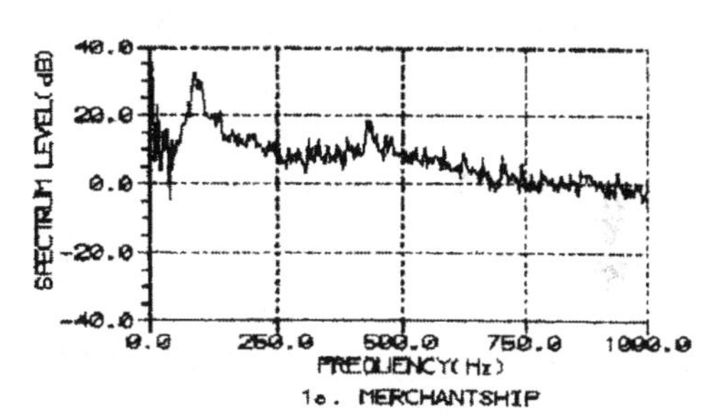

1c. MERCHANTSHIP

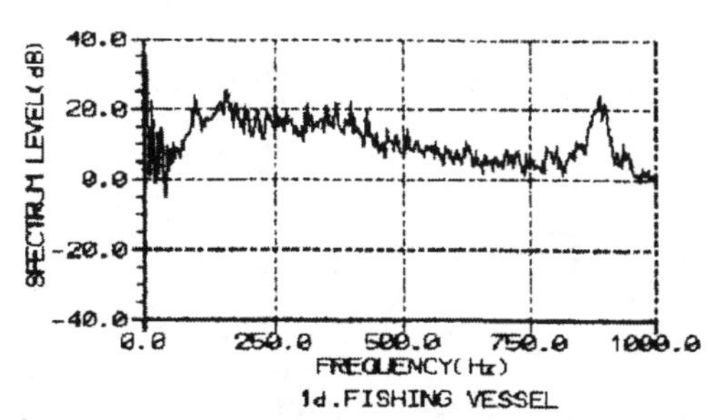

1d. FISHING VESSEL

Fig. 1 The average power spectrum in band 0-1KHz of the various ships.

shows the spectra of the four types in band 0--1000Hz, respectively. From Figure 1a., we can see that there are strong line spectra with level higher than 10dB above average spectrum value in the band 200--700Hz for submarine. It is because that the magnetic field in electrical machinery of submarine generates period vibration between stator and rotor. For other types, because their power device is diesel, the frequency of line spectrum, if existed, is below 200Hz. So the line spectrum is selected as identification characteristics for submarine.

For merchantship and fishing vessel, we choose the position of the strongest frequency band as the characters, because there is a peak with stable position in band 0--1000Hz. This result can be seen in Figure 1c.,and 1d.

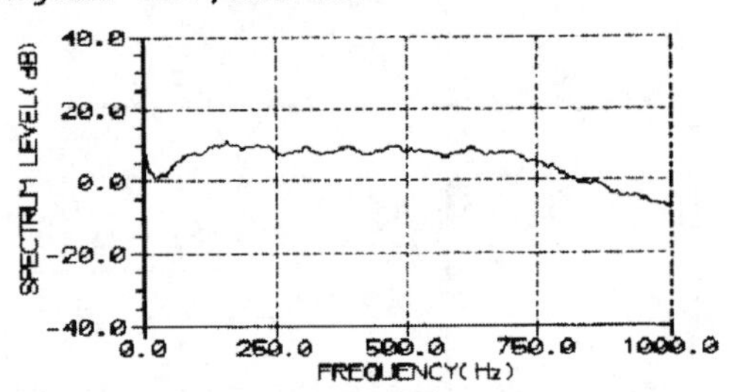

Fig.2 The Average Power Spectrum On 0-1KHz of The Oceanic Noise.

When the target is far away from the receiver, the signal received by sonar array is actually ambient noise due to tide, gush, and so on. In lower frequency band (<500Hz), the oceanic noise can be thought as Gaussian noise, and the curve of the spectrum in decibel scale is a horizontal line, approximately. In higher band(>500Hz),the spectrum level of oceanic noise is decreased by 6-8dB per octave, These feature mentioned above is selected as identification character of ambient noise. Figure 2. show the spectrum of oceanic noise.

III. Decision Function Generation

Although there are many methods based on statistics for pattern recognition theory. In practice, these algorithms have not got success result. There will usually be a much large set of manifestation with vary degrees of equivocation[1]. To find the best explanation requires information search as well as mechanisms for reasoning with uncertain evidence. Uncertainty in EXPLORE arises from noise and distortion uncertainty in subjective inference rules obtained from human experts and the use of qualitative or linguistic terms often used by human experts. A formal way to copy with the uncertainty mentioned above is the theory of fuzzy sets[7].

There are two classic methods for fuzzy pattern recognition, one is based on the maximum dependence principle of membership, the other is based on the principle of the rule of nearest range. In EXPLORE we choose first method.

<u>Definition 3.1:</u>

Suppose the average power spectrum set in band 0-500Hz of a target noise is:

$$ASPL=\{X_0(0), X_0(1), \cdots, X_0(L_0-1)\} \quad (1)$$

where L_0 is the number of line spectrum. And

$$APSH=\{X_1(0), X_1(1), \cdots, X_1(L_1-1)\} \quad (2)$$

is the average power spectrum in band 500Hz-1000Hz.

Denote the $\overline{X_0}$ as the mean value of the set **ASPL**,

$$\overline{X_0} = \frac{1}{L_0}\sum_{i=0}^{L_0-1} X_0(i) \quad (3)$$

then the horizontal factor of the target noise is defined as:

$$HF = \frac{L_x}{L_0} \quad (4)$$

The slope factor of the target noise is defined as:

$$SF = \frac{L_h}{L_1} \quad (5)$$

where L_x is the number of element in **ASPH**, which satisfies :

$$| X_0(i) - \overline{X_0} | \le T_0 \quad (6)$$

T_0 is the threshold of spectrum fluctuation. And L_h is the number of element in ASPH, which satisfies:

$$X_1(i) - X_1(i+1) \ge T_1 \quad (7)$$

T_1 is the threshold of spectrum attenuation.

The symbol **U** denotes a universe of discourse, which is a collection of the spectrum characteristics of the ship radiated noise received by sonar array. That is:

$$U = LS \cup CS \quad (8)$$

LS is the character set of line spectrum which will be described lately. **CS** is the character set of continuous spectrum, which contains the position information of the strongest spectrum band **MAX_SP**, the horizontal factor **HF** and slope factor **SF**.

A fuzzy subset **A** of **U** is a set of order pair:

$$A = \{(u_i, \mu_A(u_i))\} \quad (9)$$

where the $\mu_A(u)$ represent the grades of membership which indicated the degree of membership.

For EXPLORE, there are five fuzzy subsets of **U** defined by five membership functions. The subsets are:

A_1 ={ spectrum characters of known submarine}
A_2 ={ spectrum characters of merchantship}
A_3 ={ spectrum characters of fishing vessel}
A_4 ={ spectrum characters of oceanic noise}
A_5 ={ the others}

The correspondence membership functions can be defined according to the analysis in section II.

u_0 is target noise to be recognized. $u_0 \in U$.

Suppose there are M_s submarines in knowledge base,specially,the character set of line spectrum is :

$$LS_i = \{(d_i(0), b_i(0)), (d_i(1), b_i(1)), \cdots, (d_i(L_i-1), b_i(L_i-1))\} \quad (10)$$

That is there are L_i line spectrum in the spectrum set of i-th submarine, the frequency and level are denoted as : $(d_i(j), b_i(j))$, $0<j<L_i-1$, $0<i<Ms$. The membership function $\mu_{A1}(u)$ is formulated as followed.

<u>Definition 3.2:</u>

The generalized intersection subset of LS_i to LS_j is denoted by S_{ij},

$(x,y) \in LS_i$ if existing $(u,v) \in LS_j$ such as: $|x-u| < PW$

then $(x,y) \in S_{ij} \subset LS$.

where **PW** is the threshold of the width of line spectrum.

<u>Definition 3.3:</u>

The order statistic set of the line spectrum set LS is defined by:

$$RLS = \{R(0), R(1), \cdots, R(L-1)\} \quad (11)$$

where **R(i)** is the order via level of ith line spectrum in overall line spectrum set.

The membership function of the fuzzy subset A_1 is defined as:

$$\mu_{A_1}(u)=\underset{0\le i\le Ms}{Max}(f_{i0}(PW)) \quad (12)$$

where

$$f_{i0}(PW)=\frac{1}{L_{i0}}\sum_{k=0}^{L_{i0}-1}[1-\frac{|RS_{0i}(k)-RS_{i0}(k)|}{L_{i0}} \quad (13)$$

and $RS_{0i}(k)\in RS_{0i}$ which is the order statistic set of generalized intersection subset of Generalization interior set of RLS_0 to RLS_j, L_{i0} is the element number of the set RS_{i0}. And $RS_{i0}(k)\in RS_{i0}$ which is a generalized intersection subset of RLS_j to RLS_0.

Suppose there are L_m merchantships and L_f fishing vessel in knowledge base and their strongest spectrum band position are: **Max_SP_M$_i$** and **Max_SP_F$_i$**, respectively. The membership function of fuzzy subset A_2, A_3 are defined as followed, respectively.

$$\mu_{A_2}(u)=\underset{0\le i\le L_m}{Max}(f_{M_i}(PW)) \quad (14)$$

where

$$f_{m_i}(PW)=[1-\exp(-\alpha|Max\_SP\_M_i-Max\_SP|)] \quad (15)$$

and

$$\mu_{A_3}(u)=Max(f_{F_i}(PW)) \quad (16)$$

where

$$f_{F_i}(PW)=[1-\exp(-\beta|Max\_SP\_F_i-Max\_SP|)] \quad (17)$$

where $0\le\alpha\le1$, $0\le\beta\le1$ are control factor.

For oceanic noise , the $\mu_{A_4}(u)$ is:

$$\mu_{A_4}(u)=WL\times HF+WH\times SF \quad (18)$$

where $WL+WH=1$, $0\le WL,WH\le1$ are weighing factor.

For fuzzy subset A_5, according to its definition, we know that A_5 is a complemental subset of union set of A_1,A_2,A_3,A_4. That is

$$A_5=\neg(\bigcup_{i=1}^{4}A_i) \quad (19)$$

where $\neg$ denotes the complemental operator. So:

$$\mu_{A_5}(u)=1-\underset{1\le i\le 4}{Max}(\mu_{A_i}(u)) \quad (20)$$

So far, we define five subset on universe discourse **U** according to analysis in section II. It is obviously that the membership function correspondent to any subset can express the characteristics of identification of the type specialized.

IV. The Experimental Result and Conclusion

A series test have been carried out for EXPLORE. The input signals originally are collect in some at sea experiment. It include a wide variety of warship and civil vessel class.

By adding the oceanic noise with different amplitude,we can get the input signals with different signal to noise ratio (SNR), and test the performance of the EXPLORE. Table 1.,2.,3. show the test results with different SNR.

It must point out that the SNR indicated in Table 1.,2.,3 is not real SNR, because the signal may contain the oceanic noise unavoidably, which were recorded at the real sea environment. The real SNR is lower than that shown in Table

Table 1. The test results without noise

TYPE	NUMBER OF SAMPLE	RATE OF RECOGNITION
SUBMARINE	151	86%
MERCHANTSHIP	51	88%
FISHING VESSEL	19	86%
SURFACE WARSHIP	130	90%
OCEANIC NOISE	11	73%

Table 2. The test results at 6dB

TYPE	NUMBER OF SAMPLE	RATE OF RECOGNITION
SUBMARINE	71	77%
MERCHANTSHIP	17	100%
FISHING VESSEL	6	100%
SURFACE WARSHIP	69	90%

Table 3. The test results at 3dB

TYPE	NUMBER OF SAMPLE	RATE OF RECOGNI-TION
SUBMARINE	62	52%
MERCHANTSHIP	22	73%
FISHING VESSEL	8	88%
SURFACE WARSHIP	90	74%

1.,2.,3.

From the Table 1.and Table 2., we can find that the performance is better at 6dB than that without noise. It is because that the line spectrum generated by some interference is strong enough to result in the fault judgement of submarine when there is not noise to added, which yield false alarm.

For fuzzy subset A_5, we divide it into two type, that is new submarine and surface warship according to the line spectrum level. In fact, the EXPLORE can judge six type of radiate noise.

Reference:

[1] Abraham Kandel, "Fuzzy expert systems", CRC Press, 1991, LONDON

[2] M.S. Fox, " AI and expert sys tem myths, legends and facts", IEEE expert Vol. 5, No.1(1990)

[3] Ed. by T. Yurban and P.R.Watl ins," Applied expert systems", North Holland, 1988

[4] Zadeh, L.A., "Fuzzy set", Inf. Control, 1965

[5] Wang Dezhao and Sang Erchang, " Underwater Acoustics",Academic Press, 1981,Beijing(In chinese)

[6] Ed. by C.H. Chen,"Pattern Rec ognition and Signal Process ing", SIJTHOFF & NOORDHOFF,1978

[7] Ed. by H.G. Urban,"Adaptive Methods in Underwater Acous tics", NATO ASI Series, 1985

[8] R.K.Blashfield et al.," Cluster analysis software", Ibid, 245 -264

[9] Dieter Nebendahl, "Expert Sys tems", John Wiley & Sons Limit ed, 1987.

An Application of Expert System in Recognition of Radiated Noise of Underwater Target①

Qihu Li, Jinlin Wang and Wei wei

Institute of Acoustics, Academia Sinica
P.O.Box 2712, Beijing 100080, P.R. China

Abstract

An expert system, named "EXPLORE", is recentely developed. It is used to identify radiated noise of underwater targets, including surface ship, submarine, speed gunboat, cargo, ambient noise and fishing vessels. The theory of cluster analysis and fuzzy logic are applied in this system. A multiple feature extraction technique of signal is derived, which can best describe the characteristcs of the signature of underwater noise sources. A special metric is established in Hilbert space of feature vector, by which the decision is deducted. The experimental test for this system, including several hundred samples, show that the EXPLORE system is intelligent in recognition underwater target, even in relative small signal to noise ratio.

1.Introduction

The problem of underwater radiated noise identification is one of the attractive and difficult topic in passive sonar design [1-2]. The development of artificial intelligence, especially the theory and practice of expert system is pushing the recognition technique to a new level [3-5].

Theoretically, the problem of target noise recognition is a typical cluster analysis topic. In statistics, there are several succesful examples in this category, e.g. the Bayes test, non-parameter rank test in finger print recognition and ECG signal identification. But there is a main problem in using cluster analysis in sonar signal processing due to the difficult of features extraction of underwater radiated noise. Although the mechanism of underwater noise has been extensively studied.

In order to find an effective algorithm to identify underwater target noise, a large amount of original data are analyzed in the time domain and frequency domain. It is found that the average power spectrum of radiated signal of underwater targets can be categoried based some rules. The cluster analysis method and fuzzy set theory can be used in feature extraction and analysis.

The characteristics vector of noise signature forms a Hilbert space. A suitable metric, which is based on the distance between characteristc vector, is established to make final decision and confidence level calculation.

① Oceans 95, 1995, 1: 404-408.

A knowledge based expert system, called EXPLORE, has been developed. It can identify a wide variety of underwater noise, including fighting ship, cargo, fishing vessel and submarine from the background of ambient noise. The EXPLORE is a programmble and self-learning system, once a new class of signal is found, it will be added in the database automatically. The induction reasoning procedure is easily realized by hardware.

2. The Characteristcs Analysis of Underwater Noise

The noise sources of underwater targets such as surface ship, submarine and torpedo can be categoried as three classes [7] :

1.mechanical noise of main engine and auxiliary machine

2.Properller noise

3.hydro dynamic noise of moving vechicle

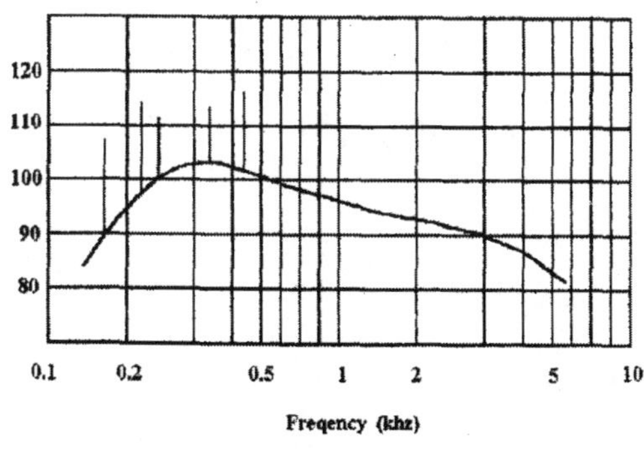

Fig.1 Typical sample of average power spectrum of noise of underwater target

The plot of typical frequency spectrum of underwater target is shown in Fig.1. It consists of a continue broadband spectrum and some line spectrum. The distribution of line spectrum, which is usaully much stronger than the average level in frequency band, strongly depends on the type of vechicle.

The feature extraction of the signature of underwater noise is the key point in target identification. It must be satisfy some basic requirement, that is

A.The value of signal characteristics must be the statistical average results

B.In same condition (same vessel, same speed etc.) the repeateness probability of characteristics must be close to 1.

C.The intersection set of different signal classes should be as small as possible.

D.The algorithm of feature extraction must be simple and easy to implement.

Based on a large amount of data analysis, it is found that the following six parameters can be considered as the signal features, that is

A. The number N_{ls} of line spectrum.

B. The average duration A_{ls} of adjacent line spectrum.

C. The position P_l of line spectrum, which has maximum power level.

D. The average level L of line spectrum over average power spectrum.

E. The position B_m of maximum spectrum band.

F. The slop S_b of spectrum from maximum spectrum band.

There are some new definition we have to show before we describe the feature extraction. For example, what kind of spectrum can be considered as " line spectrum " ? How calculate the exact position of line spectrum ? What is the " maximum spectrum band ", and so on. It is obviously, some concepts are ambiguous. The fuzzy analysis is a right choice to solve these kind of problems. A series of mathematical explicit expression for these concept will be given in the next section.

3. Decision Algorithm

The original data in frequency domain must be treated before the feature is extracted. A 4 – 32 times average of 1024 point FFT is taken for separate disjointed segment time data. The sampling frequency is about 5 times of high cutoff frequency of input signal. based on the fuzzy analysis, A pre-smoothing algorithm is used to remove any incident appearing line-spectrum-like valuc and make the overall power spectrum much smooth. In this way the characteristic vector will be easily extracted as true "representative" for its class.

Dividing the input data in M segment, each with N point :

$$x_m(0), ..., x_m(N-1) \qquad m = 1, ..., M \quad (1)$$

Suppose the DFT of $m-th$ segment is

$$X_m(l) = \sum_{k=0}^{N-1} x_m(k) \exp(-2\pi jkl/N) \quad (2)$$

The average spectrum of input signal is

$$X(l) = \frac{1}{M} \sum_{m=1}^{M} |X_m(l)| \quad (3)$$

Define" overall average " of spectrum

$$A = \frac{2}{N} \sum_{l=0}^{N/2-1} X(l) \quad (4)$$

The line spectrum is the value l , for which $X(l) \geq A+B_1$, where B_1 is a parameter to be adjusted. Let the set of line spectrum is

$$V = \{X(l) : X(l) \geq A + B_1\} \quad (5)$$

Rearrange the value of $X(l)$ in set V according to the order of amplitude from big to small as $v_1, ..., v_p$, where $v_i \in V$ and $v_i = X(l_i)$. We have

$$P1 = l_1, \quad N_{ls} = p, \quad A_{ls} = \frac{l_p - l_1}{p-1} \quad (6)$$

Select parameter B_2 and construct " local average " A_i in the neighbourhood of $i-th$ line spectrum

$$A_i = \frac{1}{2B_2} \sum_{k=1}^{B_2} [X(l_i - k) + X(l_i + k)] \quad (7)$$

The power of $i-th$ line spectrum is defined as $Q_i = X(l_i) - A_i = v_i - A_i$ so that

$$L = \frac{1}{p} \sum_{i=1}^{p} Q_i \quad (8)$$

Let

$$C_l = \frac{1}{B_3} \{X(l) + \sum_{k=1}^{B_3} [X(l-k) + X(l+k)]\} \quad (9)$$

If

$$C_{l_0} = max\{C_l : l = 0, ..., N/2-1\}$$

We have

$$B_m = l_0, \quad S_b = \frac{1}{B_4} [C_{l_0} - C_{l_0+B_4}] \quad (10)$$

where the parameter of B_2, B_3 and B_4 can be adjusted in the algorithm.

The six parameters are derived from the smoothed data. A metric is established to measure the distance between any two vectors. Suppose a new class with a feature vector

$$C = (N_{ls}, A_{ls}, P_1, L, B_m, S_b), \quad (11)$$

the feature vector of some "standard" class is

$$C_0 = (N_{ls0}, A_{ls0}, P_{10}, L_0, B_{m0}, S_{b0}), \quad (12)$$

The distance between C and C_0 is defined by

$$D(C, C_0) = [w_n|N_{ls}-N_{ls0}|^2+w_a|A_{ls}-A_{ls0}|^2 +\ldots + w_s|S_b - S_{b0}|^2]^{1/2} \quad (13)$$

where $w_n, w_a, \ldots$ is a set of non-negative value, which can be adjusted in decision process.

The value of confidence level of decision is inversely propotional to the $D(C, C_0)$. The procedure of decision making is illustrated in Fig.2.

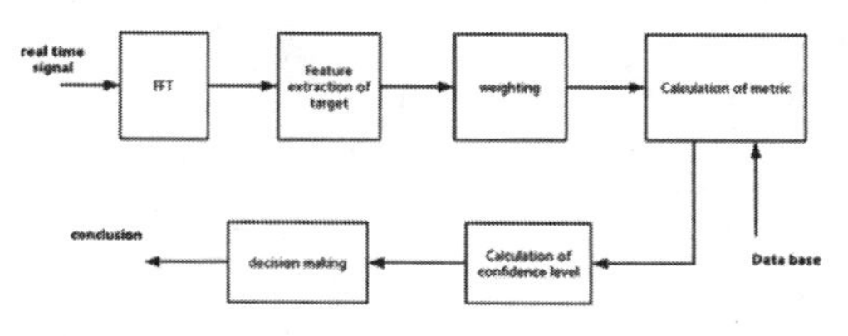

Fig.2 Cluster analysis method for target recognition

The real time data is iuput to carry out pre-processing. Therefore the digital output signal can be processed to get information in frequency domain. The fuzzy analysis algorithm is applied to extract the feature vector of input signal. A mother board match method is used to calculate the distance between this signal and each sample from data base. A pre-designated threshold value is compared with the distance and the minimum value, if exist, is found out. Finally, the decision is concluded.

The blockdiagram of EXPLORE system is shown in Fig.3

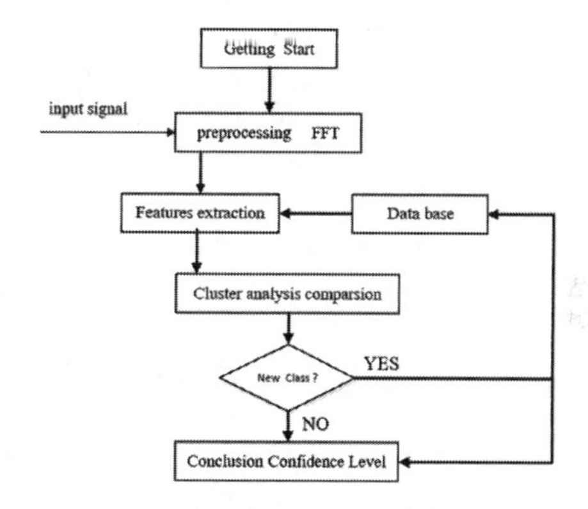

Fig.3 Block diagram of "EXPLORE" expert system

4. System Simulation

The EXPLORE expert system has been developed in our Lab. The platform of this system is an IBM 486 PC with 540 MB harddisk. A DSP board with TMS320C25 is pluged into this computer. The main function of this board is to take data acquisition and real time FFT analysis. The pre-processing and post-processing can be carried out almost in real time. The database consists of several hundred samples from a wide variety of underwater targets and ambient noise and sea animals. The " standard " sample is stored in the basic library. In the learning mode, the EXPLORE can find some new classes, extract the features and add it to this library. In the identifying mode, the EXPLORE will extract the features of input data, compare with the feature vector from basic library and make decision.

The data in time domain and frequency domain can be displayed in the screen and get hardcopy, if necessary.

A series system simulation is tested. In most cases, the rate of correct indentification is better than 75% as $(SNR)_{in}$ higher than 3 *db*. It is worth to show that the value of weight coefficients $w's$ and the parameter B_i strongly affect the performance of the expert system. It is necessary to adjust them very carefully.

REFERENCES

[1] M.Deuser and D.Midlleton, "On the classification of underwater acoustic signal", *J.Aocust.Soc. Amer.*, **Voll.58** (1985)

[2] M.S.Fox, "AI and expert system : mgths, legends and facts", *IEEE, Expert* **Vol.5** (1990), 8-22

[3] M.P.Tarltar, "The use of an expert system toolkit in sonar system design", *NAVAL Forces* **NO.XII** (1989) 62-66

[4] K.Parsays and M.Chignell, "Expert systems for experts", *John Wiley Sons. Inc.* 1988

[5] B.Zerr, "Automatic classification of objects in sonar images", *Proc. of UDT'90, LONDON 1990*, 137-142

[6] J.A.Hartigen, "Clustering algorithms", *Wiley*, 1975

[7] D.Ross, "Mechanics of underwater noise", *Penisula*, 1988

两种数据融合方法在一个目标识别问题上的应用①②

郑　援[1,2]　胡成军[2]　李启虎[1]
(1 中国科学院声学研究所　北京 100080)
(2 海军潜艇学院　青岛 266071)

2002 年 6 月 14 日收到

摘要　本文将贝叶斯模型和 TBM 模型这两种数据融合方法分别应用于一个水下目标识别问题，并对应用效果进行了分析比较。

关键词　数据融合，目标识别，贝叶斯模型，　TBM

Applying two data fusion methods to single target identification problem

ZHENG Yuan[1,2]　HU Chengjun[2]　LI Qihu[1]
(1 Institute of Acoustics, Chinese Academy of Sciences, Beijing 100080)
(2 Navy Submarine Academy, Qingdao 266071)

Abstract　The Bayesian model and the transferable belief model (TBM) are applied separately to solve an underwater target identification problem. Results from the two data fusion methods are analyzed and compared.

Key words　Data fusion, Target identification, Bayesian model, TBM

1　引言

数据融合 [1] 又称信息融合，是利用计算机技术对按时序获得的若干传感器的观测信息在一定准则下加以自动分析、综合以完成所需的决策和估计任务而进行的信息处理过程。数据融合作为一种数据综合和处理技术，实际上是许多传统学科和新技术的集成和应用，涉及模式识别、决策论、不确定性推理、估计理论、最优化技术等，并在上述技术基础上建立其基础理论以及多种算法和模型。数据融合技术已在军事和民用的多个领域得到广泛应用。

水下目标识别过程中，通过数字信号处理，通常可以提取多个特征，为了获得客观的识别结果，有必要对多个特征进行数据融合以便获得较为精确的识别结果。这是一个不确定推理的过程，解决这一问题的数据融合方法有多种，贝叶斯模型和 TBM 模型 (Transferable Belief Model) 是其中最常用的两种。本文将这两种数据融合方法分别应用于 T 类水下目标的

① 海装青年基金项目，海装装备科研项目和总装技术基础项目资助.
② 应用声学, 2003, 22(5): 25-30.

识别问题，并对两种方法的应用结果进行了比较分析。

2 两种数据融合方法

TBM[2−4] 是针对信任度的量化表示而提出的一种模型，是对著名的 Dempster-Shafer 证据理论的解释，已经成为数据融合领域的一个重要流派。TBM 对 Dempster-Shafer 证据理论的贡献，除了 TBM 的基本原理，还包括开放世界概念 (Open World Concept)、GBT 法则 (General Bayesian Theorem)、教条级 (Credal Mental Level) 和对等级 (Pignistic Mental Level) 两种不同的信任度、最小提交原理 (Least Commitment Principle)、组合的析取规则 (Disjunctive Rules of Combinations) 等。TBM 不仅能够以静态方式表示量化的信任度，而且能以动态的方式表示信任度的变化：利用组合规则组合不同传感器收集来的数据；当数据来自不太可靠的数据源时，进行一定程度上的折扣；用 GBT 法则对不同目标集上的信任度进行转换；为了决策，将信任度赋值转换为对等的概率等。其应用于识别的基本工作流程如图 1 所示。

贝叶斯模型 [5,6] 建立在概率论和条件独立的基础上，在由原因和结果构成的概率关系网中进行推理，这种网也称为贝叶斯网。令 $V_1, V_2 \cdots V_k$ 是贝叶斯网中的节点，贝叶斯网规定每个节点 V_i 条件独立于任何由其兄弟或它们的后代节点构成的节点子集，有 D 分离和条

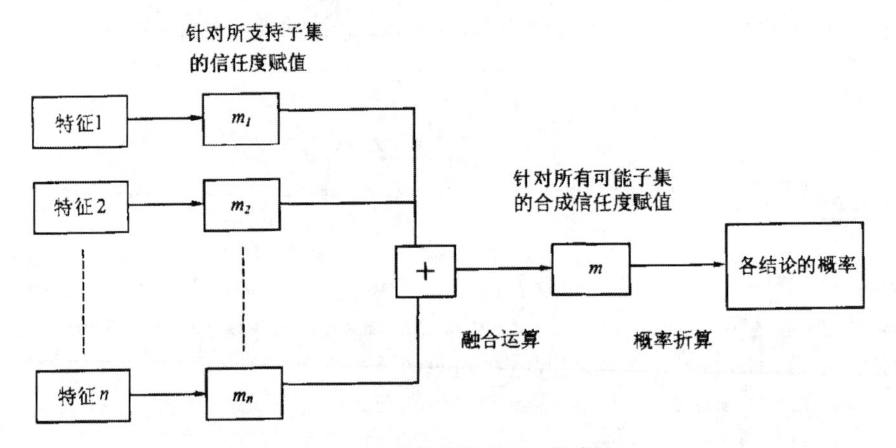

图 1 TBM 模型的基本工作流程

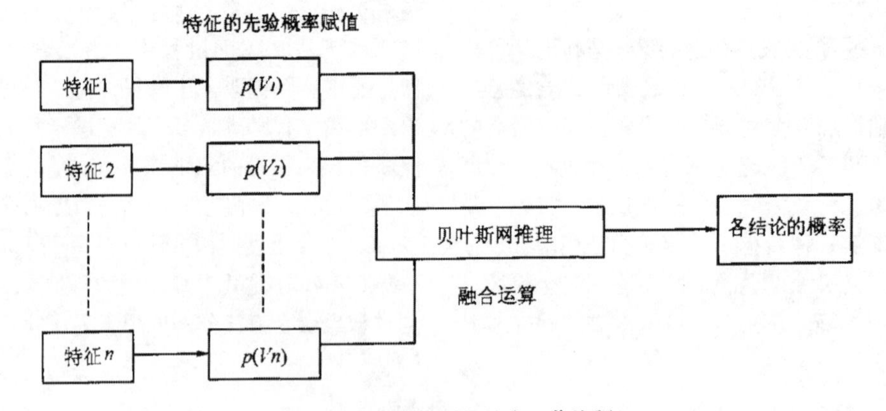

图 2 贝叶斯模型的基本工作流程

件独立存在。贝叶斯网最重要的运算法则就是贝叶斯法则，即：$p(V_i|V_j)=\dfrac{p(V_j|V_i)p(V_i)}{p(V_j)}$[5]。另外，假设 $P(V_i)$ 是 V_i 的直接双亲，给定条件独立性假设，可得出网中所有节点的联合概率为：$p(V_1,V_2,\cdots,V_k)=\prod_{i=1}^{k}p(V_i|P(V_i))$[5]。贝叶斯网有三种推理模式：因果推理、诊断推理、辩解。其应用于识别的基本工作流程如图 2 所示。

由于这两种模型均可用于解决不确定性问题，其效果往往取决于具体问题，很难从理论上给出定量的比较。本文通过这两种数据融合方法在一类目标识别问题中的应用确定其解决问题的有效性，并对这两种方法进行比较。

3　目标识别问题描述

本文的目标识别问题是根据来自声呐的跟踪听测波束信号，判别是否有 T 类目标存在。在对信号进行数字采样、信号分析后，可以得到如下相互独立的特征 [7]:

(1) Hpe: 双谱图中是否有峰值存在？

(2) Hph: 双谱图中最高峰值位置是否较高？

(3) Dh: 信号的信噪比变化是否较快？

(4) Rh: 对应螺旋桨转速的调制谱频率是否较高？

第一个特征主要用于确定是否存在信噪比比较高的目标，对于信噪比比较低的信号，后 3 个特征很难有实际意义。

根据信号处理得到的上述特征，我们需要判定这段信号中是否有 T 类目标存在。容易看出仅通过单一特征难以做出最后的判决，必须对上述多个特征进行数据融合。

下文中，我们把 TBM 模型和贝叶斯模型用于对上述 4 个相互独立的特征进行数据融合，介绍具体计算方法，给出仿真结果及分析结论。

4　采用贝叶斯模型进行数据融合

贝叶斯模型解决这一识别问题的关键一步在于找到特征（证据）同结论之间正确的推理网络关系，以及其中各节点之间的条件概率关系。针对 T 类目标的识别问题，我们给出图 3 所示的贝叶斯网。图中，中间节点 Ta 表示信号中是否存在信噪比比较高的目标；中间节点 Tg 表示信号是否具有 T 类目标辐射噪声的主要特点。

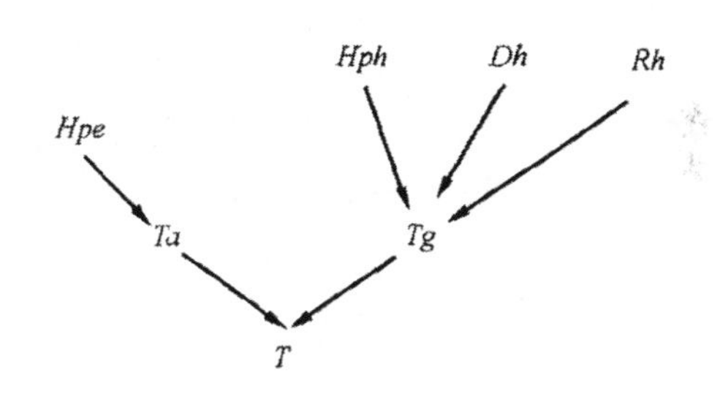

图 3　用于 T 类目标识别的贝叶斯网

根据上述网络结构可知，要在这样的贝叶斯网中进行推理，必须事先已知下述 14 个条件概率。按照已有的统计数据和专家的识别经验，并考虑每个特征本身的可靠程度、在最后判决中所占权重，我们按下面的数值给出这 14 个条件概率：

$$p(Ta|Hpe)=0.98;$$
$$p(Ta|\neg Hpe)=0.0;$$

$$p(Tg|Hph,Dh,Rh)=1.0;$$
$$p(Tg|\neg Hph,Dh,Rh)=0.85;$$
$$p(Tg|Hph,\neg Dh,Rh)=0.92;$$
$$p(Tg|\neg Hph,\neg Dh,Rh)=0.36;$$
$$p(Tg|Hph,Dh,\neg Rh)=0.96;$$
$$p(Tg|\neg Hph,Dh,\neg Rh)=0.45;$$
$$p(Tg|Hph,\neg Dh,\neg Rh)=0.54;$$
$$p(Tg|\neg Hph,\neg Dh,\neg Rh)=0.0;$$

$$p(T|Ta,Tg)=1.0;$$

$$p(T|\neg Ta, Tg) = 0.0;$$

$$p(T|Ta, \neg Tg) = 0.05;$$

$$p(T|\neg Ta, \neg Tg) = 0.0;$$

上式中，“$\neg$”的含义是“非”，则 $p(Ta|\neg Hpe) = 0.0$ 的意义是：在 Hpe 不发生的情况下，Ta 发生的概率是 0.0，即双谱图中无峰值存在的情况下，有信噪比较高的目标存在的概率是 0.0。

根据信号分析的结果，我们能够得到 Hpe、Hph、Dh、Rh 的先验概率，并计算出 $\neg Hpe$、$\neg Hph$、$\neg Dh$、$\neg Rh$ 的概率 $1-p(Hpe)$、$1-p(Hph)$、$1-p(Dh)$、$1-p(Rh)$ 作为输入，通过贝叶斯模型计算“是否存在目标 T”这一判决的概率。基本运算公式如下：

$$p(Ta) = \sum_{Hpe} p(Ta|Hpe) \times p(Hpe) \tag{1}$$

$$p(\neg Ta) = 1 - p(Ta) \tag{2}$$

$$\begin{aligned} p(Tg) &= \sum_{Hph,Dh,Rh} p(Tg|Hph, Dh, Rh) \times p(Hph, Dh, Rh) \\ &= \sum_{Hph,Dh,Rh} p(Tg|Hph, Dh, Rh) \times p(Hph) \times p(Dh) \times p(Rh) \end{aligned} \tag{3}$$

$$p(\neg Tg) = 1 - p(Tg) \tag{4}$$

$$\begin{aligned} p(T) &= \sum_{Ta,Tg} p(T|Ta, Tg) \times p(Ta, Tg) \\ &= \sum_{Ta,Tg} p(T|Ta, Tg) \times p(Ta) \times p(Tg) \end{aligned} \tag{5}$$

公式 (1)、(2) 根据 Hpe 的概率，以及 Hpe 同 Ta 的条件概率关系计算中间节点 Ta 发生、不发生的概率。公式 (3)、(4) 根据 Hph、Dh、Rh 的概率、条件独立性以及同 Tg 的条件概率关系计算中间节点 Tg 发生、不发生的概率。公式 (5) 根据 Ta、Tg 的概率、条件独立性以及同 T 的条件概率关系计算最终结果 T 类目标存在的概率。

部分仿真结果如表 1 所示。

表 1 中 T? 栏是该组数据对应的正确结果，$p(T)$ 栏是采用贝叶斯模型对该组数据进行计算得出的 T 类目标存在的概率。以 0.6 为门限，采用贝叶斯模型对上述 20 组数据的正确识别率为 85%。从表中数据可知，特征 Hpe、Hph、Dh、Rh 对结论的支持程度是不同的，很多情况下甚至是矛盾的，难以直接采用以做最后的判决，但采用贝叶斯模型仍得到了比较令人满意的结果。

表 1 贝叶斯模型部分仿真结果

$p(Hpe)$	$p(Hph)$	$p(Dh)$	$p(Rh)$	$p(T)$	T?
0.0	1.0	0.0	1.0	0.0000	否
0.5	1.0	1.0	1.0	0.4900	否
0.7	1.0	1.0	1.0	0.6860	是
1.0	1.0	1.0	1.0	0.9800	是
1.0	0.0	1.0	1.0	0.8403	是
1.0	1.0	0.3	1.0	0.9279	是
1.0	1.0	1.0	0.0	0.9428	是
1.0	0.5	1.0	0.5	0.8078	是
1.0	1.0	0.0	0.0	0.5517	否
1.0	0.5	0.4	1.0	0.7510	否
1.0	0.1	0.1	0.1	0.1746	否
1.0	0.3	0.3	0.3	0.4190	否
1.0	0.5	0.5	0.5	0.6402	否
1.0	0.7	0.7	0.7	0.8211	是
0.7	0.7	0.7	0.7	0.5748	是
0.5	0.5	0.5	0.5	0.3201	否
0.5	0.8	0.5	0.5	0.3815	否
0.8	0.6	0.5	0.7	0.5863	否
0.2	0.9	0.2	0.2	0.1279	否
0.9	0.9	0.2	0.2	0.5754	否

5 采用 TBM 模型进行数据融合

使用 TBM 模型不需要给出条件概率，但必须给出可能判定结果的全集，对本文的识别问题而言，判决的全集 $H = \{T, O, N\}$，其中 T 代表 T 类目标，O 代表其他类目标，N 代表没有任何目标。

这样，对于每种特征，它所支持的基本信任度赋值只针对 H 的某些子集。根据信号分析结果得到的 Hpe、Hph、Dh、Rh 的先验概率，我们进行如下的基本信任度赋值：

(1) 对于特征 Hpe，其基本信任度赋值为：

$$m1(\{T, O\}) = p(Hpe);$$

$$m1(\{N\}) = 1 - p(Hpe);$$

(2) 对于特征 Hph, 其基本信任度赋值为:

$$m2(\{T,N\}) = p(Hph);$$
$$m2(\{O,N\}) = 1 - p(Hph);$$

(3) 对于特征 Dh, 其基本信任度赋值为:

$$m3(\{T,N\}) = p(Dh);$$
$$m3(\{O,N\}) = 1 - p(Dh);$$

(4) 对于特征 Rh, 其基本信任度赋值为:

$$m4(\{T,N\}) = p(Rh);$$
$$m4(\{O,N\}) = 1 - p(Rh);$$

考虑到每种特征在基本信任度赋值上存在的不同可靠程度 α, 需要按照下面的 TBM 模型的折算信任度公式 [3] 对各基本信任度赋值进行折算:

$$m^*(A) = \alpha \times m(A), \quad \forall A \neq H \tag{6}$$
$$m^*(H) = 1 - \alpha + \alpha \times m(H) \tag{7}$$

在本文的问题中，根据统计数据，得到 $\alpha_1 = 0.98$、$\alpha_2 = 0.9$、$\alpha_3 = 0.75$、$\alpha_4 = 0.6$ 。这样，我们计算出折算后的基本信任度赋值 $m1^*$、$m2^*$、$m3^*$、$m4^*$ 。

下面，利用 TBM 模型的合取公式 [2,3], 对来自不同特征的基本信任度赋值进行数据融合:

$$m_{1(\cap)2}(A) = \sum_{B\cap C=A} m_1(B)m_2(C), \quad \forall A \subseteq H \tag{8}$$

我们得到 4 个特征融合后的基本信任度赋值: $m = m1^*(\cap)m2^*(\cap)m3^*(\cap)m4^*$ 。

最后，为了确定 T 类目标存在的概率，还需利用下面的 TBM 模型的 BetP 公式 [3] 将信任度赋值折算到对等的概率:

$$BetP^H(A) = \sum_{X\subseteq H} \frac{|A\cap X|}{|X|}\frac{m^H(X)}{(1-m^H(\phi))}, \quad \forall A \subseteq H \tag{9}$$

表 2 的仿真条件和实验数据同表 1 相同，但为采用 TBM 模型进行融合后得到的结果。该表内容同表 1 唯一不同的就是 $p(T)$ 栏是采用 TBM 模型对该组数据进行计算得出的 T 类目标存在的概率。以 0.6 为门限，采用 TBM 模型对上述 20 组数据的正确识别率也达到 85% 。

表 2　TBM 模型部分仿真结果

$p(Hpe)$	$p(Hph)$	$p(Dh)$	$p(Rh)$	$p(T)$	T?
0.0	1.0	0.0	1.0	0.0025	否
0.5	1.0	1.0	1.0	0.4975	否
0.7	1.0	1.0	1.0	0.6925	是
1.0	1.0	1.0	1.0	0.9851	是
1.0	0.0	1.0	1.0	0.4562	是
1.0	1.0	0.3	1.0	0.9195	是
1.0	1.0	1.0	0.0	0.9165	是
1.0	0.5	1.0	0.5	0.7808	是
1.0	1.0	0.0	0.0	0.4562	否
1.0	0.5	0.4	1.0	0.6437	否
1.0	0.1	0.1	0.1	0.0285	否
1.0	0.3	0.3	0.3	0.1700	否
1.0	0.5	0.5	0.5	0.4842	否
1.0	0.7	0.7	0.7	0.8045	是
0.7	0.7	0.7	0.7	0.4594	是
0.5	0.5	0.5	0.5	0.1590	否
0.5	0.8	0.5	0.5	0.2387	否
0.8	0.6	0.5	0.7	0.4303	否
0.2	0.9	0.2	0.2	0.0438	否
0.9	0.9	0.2	0.2	0.4095	否

6　结语

针对本文的 T 类目标识别问题，从数值结果上看:

(1) 采用贝叶斯模型和 TBM 模型进行目标识别，能够解决采用单一特征难以判决的问题，并且判决准确性都比较高;

(2) 尽管两种数据融合方法在 T 类目标识别问题上得到的识别率相同，但单次识别结果 $p(T)$ 之间的平均距离为 0.1032, 存在一定差距，所以识别率的相同应视作巧合。

另外，从数据融合过程本身来看:

(1) 贝叶斯模型涉及的条件概率个数同特征、中间节点数目和贝叶斯网络结构有关，当贝叶斯网络规模较大时容易发生组合爆炸；而TBM 模型的基本概率赋值只针对有限子集，能够有效避免这一问题；

(2) 贝叶斯模型中的贝叶斯网络结构、条件概率和 TBM 中的可靠性折扣都需要根据以往经验性结果得出，直接影响融合结论的有效性，如果这些数值选择不当，数据融合的效果甚至不如不采用数据融合的好；

(3) 由于贝叶斯模型可从输入先验概率直接到达结论的概率，更直观，容易控制，在确信不发生组合爆炸的情况下，贝叶斯模型更容易实现；

(4) 贝叶斯模型在缺少已知条件概率的情况下难以奏效；而TBM模型在缺少足够的条件概率的情况下，仍可以给出比较满意的结果。

根据本文的数据融合结果，我们认为采用数据融合方法利用多个特征对目标进行识别是可行的，但需要根据问题的规模、能否得到足够的先验知识选择适当的数据融合方法，不能简单地认为哪种数据融合方法在任何情况下都优于另一种。

参 考 文 献

1 刘同明，夏祖勋，解洪成编著. 数据融合技术及其应用. 北京：国防工业出版社， 1998. 1–16.

2 Philippe Smets. What is Dempster-Shafer's model? In: Advances in the Dempster-Shafer Theory of Evidence, New York: Wiley, 1994. 5–34.

3 Philippe Smets. Data Fusion in the Transferable Belief Model. In: Proc. 3rd International Conference Information Fusion, France: Paris, 2000. 21–33.

4 Philippe Smets, Robert Kennes. *Artificial Intelligence*, 1994, **66**(2): 191–234.

5 Nils J. Nilsson 著，郑扣根，庄越挺译，潘云鹤校. Artificial Intelligence A New Synthesis, 人工智能. 北京：机械工业出版社， 2000. 197–213.

6 张尧庭，陈汉峰编著. 贝叶斯统计推断. 北京：科学出版社， 1991. 1–16.

7 杨日杰. 鱼雷报警声纳合成环境的研究. 西北工业大学博士学位论文. 西安， 1999. 11–43.

基于舰船噪声线谱特征的 ETAM 方法仿真研究①

何心怡[1,2]，蒋兴舟[1]，李启虎[2]
([1]海军工程大学，湖北武汉 430033；[2]中国科学院声学所，北京 100080)

摘 要：拖曳线列阵的方位分辨率是一项重要性能指标，而它被拖曳线列阵的阵列孔径所限制。针对这一问题，文中提出了基于舰船噪声线谱特征的 ETAM 方法，利用舰船噪声在低频段的稳定线谱成份，采用 ETAM 被动合成孔径算法扩展拖曳线列阵的孔径，从而达到显著提高方位分辨率的目的。只要 ETAM 的积分时间小于海洋声信道的时间相关长度，可将拖曳线列阵的等效孔径扩展到不大于海洋声信道的空间相关长度的任意尺寸。经仿真实验验证，该方法是有效的、可行的，且算法简单、易于实时实现，是一种具有良好应用前景的高分辨方位估计方法。

关键词：拖曳线列阵；线谱；被动合成孔径；ETAM；方位估计；高分辨；仿真
文章编号：1004-731X (2004) 01-0066-04 中图分类号：TN911.72；TP391.9 文献标识码：A

The Simulation Research on the ETAM Method Based on the Line-spectrum Features of the Ship Noise

HE Xin-yi[1,2], *JIANG Xing-zhou*[1], *LI Qi-hu*[2]
([1] Navy University of Engineering, Wuhan 430033, [2] Institute of Acoustics, Academia Sinica, Beijing 100080, China)

Abstract: The direction-of-arrival resolution of the towed linear array is an important performance index, which is restricted by the array aperture of the towed linear array. Aming at this problem, this paper has raised out the ETAM method based on the line-spectrum features of the ship noise. This method uses the stable part of the line-spectrum of the ship noise in the low frequency band, and uses the ETAM passive synthetic aperture algorithm to expand the aperture of the towed array. Thus the direction-of-arrival resolution can be improved greatly. If the integral time of the ETAM is smaller than the time correlative length of the ocean acoustic channel, it can expand the equivalent aperture of the towed array to any length, which is smaller than the space correlative length of the ocean acoustic channel. Proven by simulation, this method is useful and practical, and the algorithm is also easy and simple to carry out. It is a high-resolution direction-of-arrival estimation, which has a very good appliance perspective.

Keywords: towed line array; line-spectrum; passive synthetic aperture; ETAM; direction-of-arrival estimation; high-resolution; simulation

引 言

第二次世界大战以来，反潜战成为海战的重要内容之一。为了能在远距离上（几十公里甚至更远）对水下环境进行警戒，拖曳线列阵声纳孕育而生了：它把接收水听器安装在中性浮力电缆中，拖在船后，用于接收潜艇低频噪声，从而实现远距离上对水下目标的警戒。美国国防部报告中说到：“三十年来，水面舰艇反潜战中最重要的发展是战术拖曳线列阵声纳的问世”[1]。

在声纳系统中，高的方位分辨率是非常重要的，而当拖曳线列阵声纳运用传统的波束形成技术时，其方位分辨率与阵列孔径和信号波长的比率成正比，被阵列物理尺寸所限制。为增大检测距离，拖曳线列阵声纳通常都工作在低频或甚低频段，因此为增加它的方位分辨率意味着要更长的水听器阵列，而太长的水听器阵列对于拖曳线列阵声纳是不实际的。在这种情况下，迫切需要高分辨方位估计技术。

进入 20 世纪 80 年代，出现了一大批以 MUSIC、Mini-Norm、ESPRIT 为代表的高分辨方位估计方法，但这些方法存在信噪比门限、源个数判别、解相干、阵列校准等问题，短期内难已在实际系统中加以应用。近年来，针对拖曳线列阵的高分辨方位估计问题，提出了被动合成孔径技术（Passive synthetic array，简称 PASA）[2-8]，因为拖曳线列阵是运动的，也称作运动的拖曳线列阵(Moving towed array，简称 MTA)，所以通过 PASA 处理能得到比实际物理孔径大得多的合成阵列，从而突破阵列孔径的限制，获得更高的方位估计能力。PASA 算法简单、稳健性好、是一类具有实用前景的高分辨算法。但是，在国外的多篇有关 PASA 的文献中都未提及 PASA 技术如何在实际中应用。

为此，本文提出了一种以舰船噪声线谱特征为基础，采用 PASA 中的扩展拖曳线列阵测量算法（Extended-towed-array measurements，简称 ETAM）对线谱成份进行处理，从而合成出比 MTA 物理孔径大得多的阵列，实现方位估计的高分辨。经过仿真实验，证明了这个方法是有效的、可行的，具有良好的实用前景。

① 系统仿真学报, 2004, 16(1): 66-69.

1 舰船噪声特性分析

舰船噪声是被动声纳检测和进行目标识别所需的信号，有关目标及其特性的信息就包含于舰船噪声之中。根据噪声的来源，可把舰船噪声分为机械噪声、螺旋桨噪声和水动力噪声三类。就谱特性划分，舰船噪声包括平稳连续谱、线谱和时变调制谱[9-10]。

根据普遍采用的假设，认为舰船噪声近似服从高斯分布，其统计特性可通过二阶矩来描述，可把舰船噪声时变功率谱表示为：

$$G(t,f)=G_x(f)+G_L(f)+M(t)M(f)G_x(f) \tag{1}$$

在上式中，$G_x(f)$ 为平稳各态历经高斯过程的连续谱，$G_L(f)$ 为在频率上离散分布的线谱，$M(t)M(f)G_x(f)$ 是谱级受到周期调制的时变功率谱：$M(t)$ 称为调制函数，代表连续谱所受到的周期时变调制；$M(f)$ 为调制深度谱，反映不同频率成份所具有的不同的调制程度。

经研究发现[9]，在舰船噪声的低端，即数百赫兹以下，存在着线谱，这些线谱有相当好的稳定性，在数百秒内的变化往往小于 1Hz，线谱有时可高出平均谱 20-30dB。若能利用这些线谱进行方位估计，往往能有好的效果。因此，本文所提出的基于线谱特征的 ETAM 高分辨方位估计方法即以利用舰船噪声中的线谱为出发点。

2 ETAM 算法概述

被动合成孔径（PASA）技术是最近十几年来发展起来的一种新技术，它是利用以下两个基本点[2-4]：

① 在实际使用中，拖曳线列阵是一个运动的拖曳阵列(MTA)，因此，PASA 与主动合成孔径雷达(SAR)和主动合成孔径声纳(SAS)一样，是依靠阵列的运动来改善方位分辨性能；

② 海洋试验中发现：在频率范围 50Hz~1500Hz，海洋声信道的空间相关长度达到了 300 个波长，对于单频信号（CW），时间相关长度不少于几分钟，甚至可达到十几分钟。

PASA 技术仍处于发展之中，它有三种较成熟的算法，分别是 ETAM（Extended-towed-array measurements，即扩展拖曳线列阵测量）、ML（Maximum Likelihood algorithm，即最大似然法则）和 FFTSA（FFT Synthetic Aperture processing，即快速傅里叶变换合成孔径处理），它们都是针对连续单频信号进行处理的，从方位估计精度、稳健性、计算量等方面综合比较，ETAM 算法是这三种算法中最好的[7]。

假定远场有两个声源在连续辐射单频信号（这种情况可扩展到多个声源），拖曳线列阵是由 N 个等间距的水听器构成，d 是相邻两水听器间距，它以速度 V_T 沿着直线航行，在第 n 个水听器接收到的信号可表示为：

$$y_n(\Theta,t_i)=\sum_{k=1}^{2}A_k\cos\left[\omega_k\left(t_i+\frac{d(n-1)}{c}\cos\beta_k\right)+\phi_k\right]+\varepsilon_{n,i}(0,\sigma_N),\ n=1,2,\cdots,N \tag{2}$$

在上式中，$\omega_k=\Omega_k\left[1\pm(V_k-V_T\cos\beta_k)/c\right]$ 定义为接收到的单频信号频率，它包括了多普勒频移；Ω_k 是静止时的单频信号频率；A_k 是幅度；V_k 是沿着 β_k 方向第 k 个目标沿着方位轴的速度；c 是平面波的传播速度。如果采样频率为 f_s，对应的采样周期为Δt，则 $t_i=i\Delta t$，这里$i=1,2,\cdots,M$，M 是每个水听器时间序列的数据点数。在等式 (2) 中，$\varepsilon_{n,i}(0,\sigma_N)$ 是零均值，方差为σ_N^2 的独立的高斯随机变量，表示着背景噪声。矢量 Θ 包括了(2)式中所有未知的参数。

以 MTA 的运动方向为坐标轴，可给出 ETAM 测量过程如图 1 所示。在第 0 次测量时（即$l=0$），也就是$t=0$ 时，N 阵元的 MTA 接收到了声信号；在第 1 次测量时（即$l=1$），对应于$t=\tau$，通过合适选择V_T 和 τ，使得$\tau V_T=qd$，q 代表 MTA 在一次测量中移动的水听器位置数，这时，MTA 移动了距离τV_T。在表达式(2)中，从 t_i 改变到t_{li}，l 代表着第 l 组测量时间。扩展拖曳线列阵处理的积分时间 T 写为 $T=J\tau$，这里 $l=1,2,\cdots,J$。

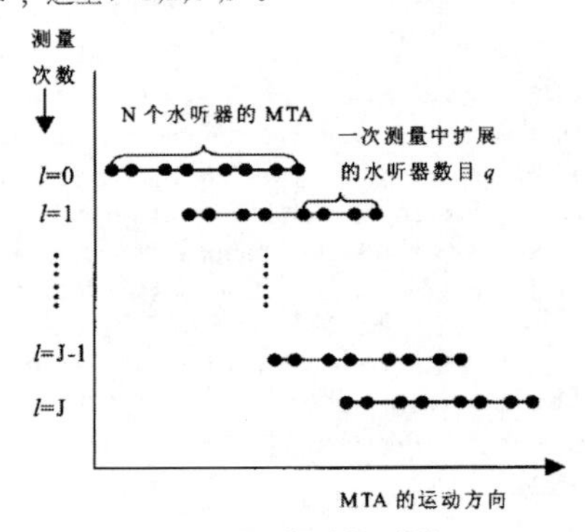

图 1　ETAM 测量过程示意图

如图 1 所示，在两个连续的测量$(l,l+1)$中，有$(N-q)$对空间采样是重叠的，每对代表着空间中相同的位置，也就是说在两个连续的测量中有$(N-q)$个水听器是重叠的。这$(N-q)$对空间采样提供给 ETAM 法则一个相位修正因子$\tilde{\psi}(\omega_m,\tau)=\omega_m\tau+\phi$，通过它将连续空间测量结合在一起，从而扩展 MTA 的物理孔径。相位修正因子 $\tilde{\psi}(\omega_m,\tau)$ 定义为：

$$\tilde{\psi}(\omega_m,\tau)=\arg\left(\frac{1}{N-q}\sum_{y=1}^{N-q}\left[Y_{l,n_0}(\omega_m)\,|\,Y^{H}_{(l+1),n}(\omega_m)\right]_y\right) \tag{3}$$

这里 H 代表着对水听器空间采样的时间序列的傅里叶变换 $Y_{(l+1),n}(\omega_m)$ 在感兴趣的频率 ω_m 处的复共轭；$n_0=q+1,q+2,\cdots,N$，对应着 MTA 的第 l 次测量；它们和 MTA 的第$(l+1)$次测量中的 $n=1,2,\cdots,N-q$ 是处于相同位置的。

那么，在时间 $t_{(l+1)i}$，有以下等式：

$$y_n(\Theta,t_{(l+1)i})=\exp\left[-j\tilde{\psi}(\omega_m,\tau)\right]\ y_n(\Theta,t_i),\quad n=1,2,\cdots,N-q \tag{4}$$

因此，在估算出相位修正因子$\tilde{\psi}(\omega_m,\tau)$后，就可通过它合成出虚拟的阵元，使得 MTA 的孔径得到扩展。

在 ETAM 方法中关键的参数是：$\tau=qd/V_T$ 是两个连续

测量的时间间隔，q 代表了在 τ 秒 MTA 移动过的水听器位置数目，或者说是 MTA 在每个连续测量中扩展的物理孔径所代表的水听器数，J 是要求达到所需要的扩展合成孔径尺寸所要的测量次数。因此，扩展拖曳线列阵处理的最大积分时间是 $T \le J\tau$。换句话说，在 ETAM 法则中的处理时间是 $T=J(M\ddot{A}t)$，M 是在每个测量中的每个水听器时间序列的数据点数（也就是快拍数）。

运用 ETAM 方法得到扩展物理孔径的时间序列后，即可进行后继的波束形成处理，Stergios Stergiopoulos 经过分析认为[2]：波束形成采用传统波束形成法（CBF）能有最好的结果，原因在于 CBF 实质上是个稳健的能量估计器。这样，扩展的拖曳线列阵的 CBF 表达式如下所示：

$$P_{CBF}(\theta)=W^{H}R_{YY}W \tag{5}$$

这里：$W=[w_1,w_2,\Lambda\ ,w_M]^T$ 为权矢量，w_i 表示第 i 路水听器信号对应的加权值，根据不同的加权方式而定；R_{YY} 是扩展后的拖曳线列阵空间采样的自相关矩阵，它可写为：

$$R_{PP}=E\{Y(t)Y^{H}(t)\} \tag{6}$$

其中：$Y(t)=[y_1(t),y_2(t),\Lambda\ ,y_n(t)]^T$，
$n=1,2,\Lambda\ ,(N+J*q)$。

因为在(5)式中所运用的是扩展后的拖曳线列阵，它具有 $N+J*q$ 个阵元，其等效孔径由 $(N-1)*d$ 扩展为 $(N+J*q-1)*d$，因此方位分辨率就会有显著地改善。

归纳本节所述，能够发现 ETAM 算法有以下四个特点：

(1) 在积分时间内，在 MTA 的两个连续测量中，空间采样必须是要有重叠的，这样才能够估算出相位修正因子，进而合成出虚拟阵元；

(2) 在重叠水听器的互相关估计的平均值提供了相位修正因子，这个相位修正因子在合成孔径的过程中，将 MTA 连续测量的结果用于扩展物理阵列的有限孔径；

(3) 因为重叠相关器是直接估计相位，因此它不需要对源的频率进行精确估计；

(4) 因为相位修正因子从空间采样中直接得到估计的，所以在 ETAM 测量期间所发生的相位变化引起的错误可直接用重叠相关器进行直接补偿。

但是，在国外有关 PASA 的文献中（当然也包括有关 ETAM 的文献），均未指出 PASA 的具体应用场合。

3　基于舰船噪声线谱特征的 ETAM 高分辨方位估计方法

如第 1 节所述，舰船噪声谱在数百赫兹下存在着稳定的线谱，线谱就是连续的 CW 信号，而 PASA 技术处理对象都是连续单频信号，线谱正好是 PASA 的适用对象，因此，文中提出了一种提高拖曳线列阵方位分辨率的新思路：利用舰船噪声的稳定线谱成份，采用 ETAM 算法实现方位估计的高分辨，只要 ETAM 的积分时间小于海洋声信道的时间相关长度，可将拖曳线列阵的等效孔径扩展到不大于海洋声信道的空间相关长度的任意尺寸。其实现框图如图 2 所示。

首先，将 MTA 各水听器接收到信号经过低通滤波器，取出线谱所处的频率段，然后将滤波后的信号送往线谱检测器进行线谱检测；由于线谱频率是未知的，线谱检测器可采用 FFT 实现，以 1Hz 左右的分辨率来分析整个低频频段；检测到线谱后，启用 ETAM 算法，计算出在线谱所处频率的相位修正因子，合成虚拟孔径；最后，将扩展后的 MTA 运用 CBF 进行波束形成，获取更高的方位分辨率。

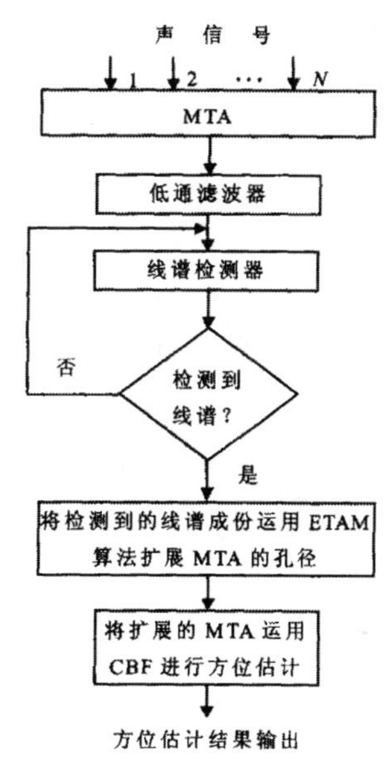

图 2　基于舰船噪声线谱特征的 ETAM 高分辨方位估计方法实现框图

4　仿真研究

为证明本文提出的基于舰船噪声线谱特征的 ETAM 高分辨方位估计方法的有效性和可行性，采用仿真实验来验证。

仿真条件如下：一个拖曳线列阵由 N=16 个阵元构成，相邻两水听器间距 d=1.5m，对应接收信号的中心频率为 500Hz，以 3m/s 的速度拖曳直线航行，其物理孔径的主瓣宽度约为：

$$BW \approx \frac{0.88 \cdot \lambda \cdot 180}{N \cdot d \cdot \pi}=6.3° \tag{7}$$

将 MTA 扩展成 64 阵元的合成孔径，根据 ETAM 算法，每次测量时间 $\tau=2s$，每次移动的阵元数为 q=4，总共需要的测量次数为 J=12，那么在 24s 内可将 MTA 扩展到所要求的孔径；存在着两个等强度的舰船噪声目标：第一个目标分别在 300Hz、400Hz 和 500Hz 处存在着线谱，第二个目标分别在 350Hz 和 550Hz 处存在着线谱，信噪比均为 0 分贝；采样频率 f_s 为 2KHz，采用 6 阶椭圆低通滤波器，截止频率为 600Hz，线谱检测器中所采用 FFT 的点数为 2048 点。

图 3-图 5 中，实线均为合成孔径的 CBF 输出，虚线均

为物理孔径的CBF输出。图3是当两个目标方位角分别是0°和8°时，16阵元MTA和合成的64阵元MTA的CBF对比图，此时两个阵列都能分辨出两个目标，但合成的64阵元MTA的谱峰锐度远好于16阵元MTA；图4是当两个目标方位角分别是0°和3°时，16阵元MTA和合成的64阵元MTA的CBF对比图，此时16阵元MTA已不能分辨出两个目标，但合成的64阵元MTA能够分辨；图5是当两个目标方位角分别是0°和3°时，64阵元MTA和合成的64阵元MTA的CBF对比图，两条曲线基本重合，说明了合成的64阵元MTA的方位分辨性能与实际的64阵元相似。

总之，仿真实验证明了基于舰船噪声线谱特征的ETAM高分辨方位估计方法的有效性和可行性。

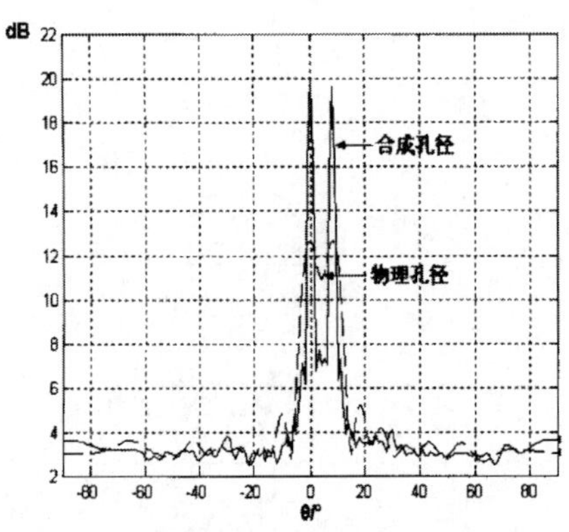

图3　16阵元MTA与合成的64阵元MTA的CBF对比图

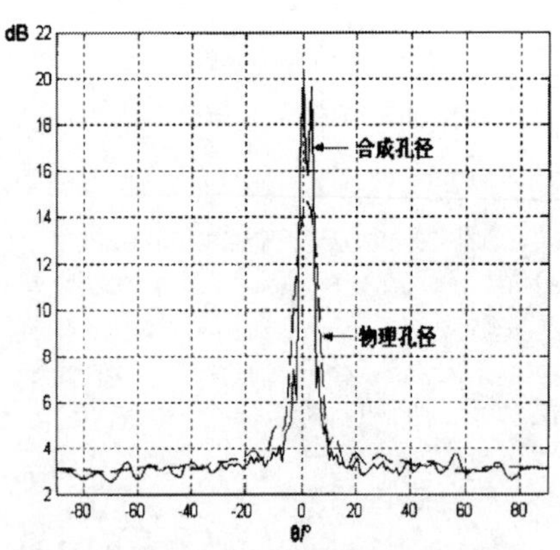

图4　16阵元MTA与合成的64阵元MTA的CBF对比图

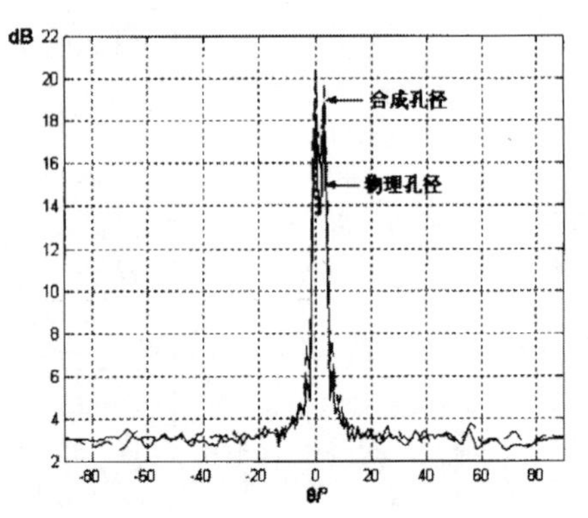

图5　64阵元MTA与合成的64阵元MTA的CBF对比图

5　结论

文中提出了基于舰船噪声线谱特征的ETAM高分辨方位估计方法，利用舰船噪声在低频段的稳定线谱成份，采用ETAM被动合成算法依靠线谱成份扩展线列阵的孔径，然后再利用CBF进行方位估计，从而大幅度提高拖曳线列阵的方位分辨性能。仿真实验证明了该方法是有效的、可行的，且方法简单、计算量较小、易于实现，有良好的应用前景。

参考文献：

[1] 拖曳式线列阵声纳研究丛书[M]. 北京：中国科学院声学研究所，1989.

[2] Nai-Chyuan Yen，William Carey. Application of synthetic-aperture processing to towed-array data[J]. J Acoust Soc Am, 1989, 86(2): 754-765.

[3] Stergios Stergiopoulos. Extended towed array processing by an overlap correlator[J]. J Acoust Soc Am, 1989, 86(1): 158-171.

[4] Stergios Stergiopoulos. Optimum bearing resolution for a moving towed array and extension of its physical aperture[J]. J Acoust Soc Am, 1990, 87(5): 2128-2140.

[5] Nuttall A H. The maximum likelihood estimator for acoustic synthetic aperture processing[J]. IEEE Journal of Oceanic Engineering, 1992, 17(1): 26-29.

[6] Stergios Stergiopoulos，Heinz Urban. A new passive synthetic aperture technique for towed arrays[J]. IEEE Journal of Oceanic Engineering, 1992, 17(1): 16-25.

[7] Ross Williams，Bernard Harris. Passive acoustic synthetic aperture processing techniques[J]. IEEE Journal of Oceanic Engineering 1992, 17(1): 8-15.

[8] Rajagopal R, P Ramakrishna Rao. Performance comparison of PASA beamforming algorithms[C]. IEEE International symposium on signal processing and its applications, Gold Coast, Australia, 1996: 825-828.

[9] 尤立克 R J. 水声原理[M]. 哈尔滨：哈尔滨船舶工业学院出版社，1990.

[10] 汪德昭，尚尔昌. 水声学[M]. 北京：科学出版社，1981.

基于声矢量传感器的分布式定位系统在水下宽带声源定位中的应用①

杨秀庭　李启虎　陈新华

(中国科学院声学研究所　北京 100080)

摘要　提出了基于声矢量传感器的分布式浮标网络定位系统，研究了不同应用背景下单个声矢量传感器的测向算法，推导了目标 DOA 估计的 Cramer-Rao 界，给出了分层海水介质中多个声矢量传感器的几何定位算法。数值仿真结果表明：(1) 系统的定位性能强烈依赖于接收信噪比；(2) 该系统适用于单一强声源的定位。

关键词　声矢量传感器，浮标网络，分布式定位系统

Application of distributed positioning system based on acoustic vector sensors

YANG Xiu-Ting　LI Qi-Hu　CHEN Xin-Hua

(*Institute of Acoustics, Chinese Academy of Sciences,　Beijing* 100080)

Abstract　A new architecture of distributed buoy network based on acoustic vector sensors has been presented in this paper. In order to explore its performance, algorithms of DOA estimation using a single acoustic vector sensor in different scenarios are studied, and the Cramer-Rao lower bound of the DOA estimator in free space is also derived. Finally, with respect to jointly localize the underwater wideband target in horizontally stratified ocean, a new method based on ray theory is given. Numerical simulations show that the performance of this distributed system depends significantly on the signal to noise ratio at the receivers and it is only applicable to localize the single dominant source.

Key words　Acoustic vector sensor, Buoy network, Distributed positioning system

① 应用声学, 2005, 24(6): 340-345.

1 引言

利用信号与噪声的时空差异，从噪声中提取目标信号，实现目标的测向或定位是声纳系统的基本功能 [1]。传统的基于时延估计的单基阵被动式声纳系统，由于其定位精度依赖于基阵孔径，对于远程定位和较大海区的水下环境监视，存在检测概率低，定位精度差的弱点。为此，基于多基地测量的分布式定位系统正日益成为一种发展趋势 [5,7]。

与传统的标量传感器不同，声矢量传感器 (acoustic vector sensor, AVS) 具有空间指向性 [2](余弦指向性)，在一定的信噪比条件下，单个 AVS 即可完成目标的精确测向任务。与多个传感器构成的基阵相比，单个 AVS 的体积要小得多，因此可用来构建浮标网络监测系统，如图 1 所示。由于 AVS 的测向能力，该浮标网络不仅可以检测目标，同时还可以实现目标定位，对于现代反潜战所要求的大范围水下监视，具有重要的现实意义。

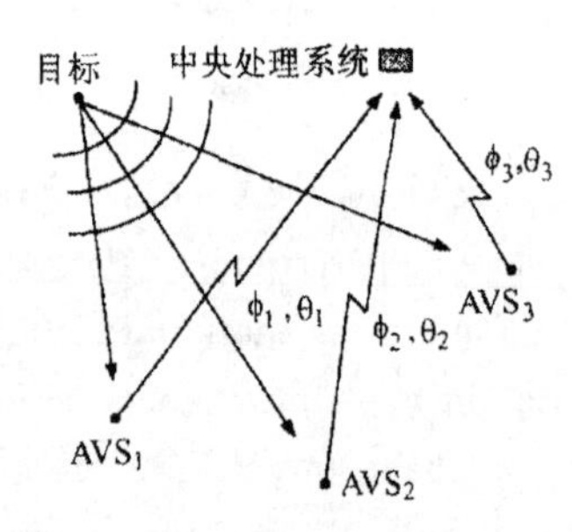

图 1 分布式定位系统的空间拓扑结构

2 测量模型

假设在各向同性、静态的海水介质中仅存在单一宽带声源，其空间位置矢量为 $\boldsymbol{r}(x,y,z)$，各 AVS 的空间位置矢量为 $\boldsymbol{r}_1,\boldsymbol{r}_2,\cdots,\boldsymbol{r}_m$，其中 m 为传感器数。

2.1 自由空间

此时无多途效应，AVS 只接收直达信号，输出为

$$\boldsymbol{y}(t)=\begin{bmatrix}y_p(t)\\ \boldsymbol{y}_v(t)\end{bmatrix}=\boldsymbol{a}(\phi,\theta)p(t)+\boldsymbol{e}(t) \tag{1}$$

式 (1) 中，$y_p(t)$，$\boldsymbol{y}_v(t)$ 为 AVS 声压通道和质点振速通道输出信号的复包络；$\boldsymbol{a}(\phi,\theta)$ 为方位向量 ($\boldsymbol{a}(\phi,\theta)=[1,\boldsymbol{u}^T]^T=[1,\cos\theta\cos\phi,\cos\theta\sin\phi,\sin\theta]^T$，$\phi$，$\theta$ 分别为目标的水平方位角和俯仰角。为表述简洁，以后用 $\boldsymbol{a}$ 表示)；$p(t)$ 为接收的声压信号，$\boldsymbol{e}(t)$ 为背景噪声向量，两者均为零均值的高斯随机过程，并满足

$$E\{p(t)p^*(\tau)\}=\sigma_s^2\delta_{t,\tau}$$
$$E\{\boldsymbol{e}(t)\boldsymbol{e}^H(\tau)\}=\sigma^2\boldsymbol{I}\delta_{t,\tau} \tag{2}$$

(2) 式中，$\delta_{t,\tau}$ 为函数，$\boldsymbol{I}$ 为单位矩阵，“*”代表共轭，“H”代表共轭转置，σ_s^2，σ^2 分别为信号与噪声的方差。

2.2 近边界区

当 AVS 布放于海面附近水层中时，除接收直达信号外，还同时接收来自海面的反射信号，反射波和直达波之间存在一定的传播时延，此时 AVS 的输出为

$$\boldsymbol{y}(t)=\begin{bmatrix}y_p(t)\\ \boldsymbol{y}_v(t)\end{bmatrix}=\boldsymbol{a}p(t)+\mathrm{Re}\cdot\boldsymbol{a}_rp(t-\tau)+\boldsymbol{e}(t) \tag{3}$$

其中，$\boldsymbol{a}_r$ 为海面反射波的方位向量，有 $\boldsymbol{a}_r=[1,\cos\phi_r\cos\theta_r,\sin\phi_r\cos\theta_r,\sin\theta_r]^T$，$\phi_r$，$\theta_r$ 分别为海面反射波的水平方位角和俯仰角。τ 为直达波和反射波之间的传播时延，$\Re$ 为海面声反射系数。

3 单个 AVS 的 DOA 估计算法

3.1 自由空间

此时通过计算声强来估计目标方位

$$\hat{\boldsymbol{I}}=\frac{1}{N}\sum_{t=1}^{N}\mathrm{Re}\{y_p(t)\boldsymbol{y}_v^*(t)\},\quad \hat{\boldsymbol{u}}=\hat{\boldsymbol{I}}/|\hat{\boldsymbol{I}}| \tag{4}$$

其中，$\hat{\boldsymbol{I}}$ 为声强估计值，$\hat{\boldsymbol{u}}$ 为单位方向矢量的估计值。AVS 的方位估计性能可由均方角度误

差 (mean square angular error, MSAE[4]) 来描述

$$\mathrm{MSAE}=\frac{1}{M}\sum_{m=1}^{M}[\cos^{-1}(\hat{\boldsymbol{u}}_m\cdot\boldsymbol{u})]^2 \quad (5)$$

(5) 式中的 $\hat{\boldsymbol{u}}_m$ 为第 m 次测量所得的单位方向矢量, M 为总测量次数。角误差估计的 Cramer-Rao 下界为

$$\mathrm{CRB}_b=(1+2SNR)/(2N\cdot SNR^2) \quad (6)$$

上式中, N 为快拍数; $SNR=\sigma_s^2/\sigma^2$ 。

3.2　近海面区

单个 AVS 通过声强测量进行目标方位估计的算法, 只适用于自由空间中单一目标的测向 [5,6] 。当 AVS 布放于海面附近区域时, 海面反射波的强度可与直达信号相比拟, 声强测向法将失效。由于信源为宽带高斯信号, 此时直达波和海面反射波可看作具有一定相关性的两个独立信号, 理论上可通过波束形成和空间谱估计方法来分辨这种多途结构。本文中比较了基于单个 AVS 的常规波束形成 (CBF) 、最小方差无畸变波束形成 (MVDR) 和多重信号分类 (MUSIC) 三种方法的方位分辨能力。

4　多个 AVS 的几何定位算法

由于海水介质的非均匀性, 水下声传播存在折射效应, 传播路径为一曲线 [8], AVS 测量所得的目标方位, 实则指向一个虚源, 因此根据方位射线交汇法 [3] 得到的目标位置估计, 会存在固定的偏差。为提高目标定位精度, 必须修正声波的折射效应。

每个 AVS 所测的目标方位及其自身的空间位置, 可决定一条本征声线, 目标的空间位置坐标应满足该声线方程。背景噪声的存在, 使得各声线一般不会恰好交汇于空间某一确定点, 因此, 多个 AVS 的联合定位, 实际上可以归结为用最小二乘方法确定目标的位置坐标, 即

$$\hat{\boldsymbol{r}}=\arg\min\Big\{\sum_{t=1}^{m}d_i^2w_i\Big\} \quad (7)$$

(7) 式中, $\hat{\boldsymbol{r}}$ 为目标空间位置矢量的估计值, w_i 为加权系数, d_i 为目标至第 i 条声线的距离, 由下式确定

$$d_i^2=\arg\min|f(\boldsymbol{r}_i,\phi_i,\theta_i)-\boldsymbol{r}|^2 \quad (8)$$

(8) 式中, $\boldsymbol{r}_i$ 为第 i 个 AVS 的空间位置矢量, (ϕ_i,θ_i) 为所测的水平方位角和俯仰角, $f(\boldsymbol{r}_i,\phi_i,\theta_i)$ 为该 AVS 所确定的本征声线方程。

4.1　目标的水平坐标分量估计

如图 2 所示, 由于背景噪声的影响, 各声线在水平面内的投影交汇于某一区域。此时可利用加权最小二乘算法 (weighted least square, WLS) 或二次加权最小二乘算法 (reweighted least square, RWLS)[6], 求得目标的水平位置坐标。

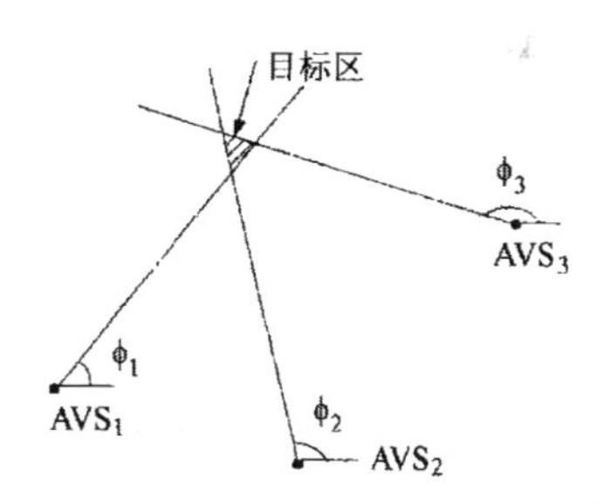

图 2　声线在水平面内的投影交汇

设: AVS 的平面位置坐标为 $\boldsymbol{r}_{h,i}$, 所测目标的水平方位角为 $\hat{\phi}_i, i=1,\cdots,m$, 则 WLS 和 RWLS 解的形式为

$$\begin{bmatrix}\hat{x}\\\hat{y}\end{bmatrix}=\left[\Big(\sum_{i=1}^{m}w_i\Big)I-\hat{U}W\hat{U}^T\right]^{-1}M\boldsymbol{w} \quad (9)$$

其中,

$$\boldsymbol{w}=[w_1,w_2,\cdots,w_m]^T,\quad W=\mathrm{diag}\{\boldsymbol{w}\},$$

$$\hat{U}=[\hat{\boldsymbol{u}}_1,\hat{\boldsymbol{u}}_2,\cdots,\hat{\boldsymbol{u}}_m],\quad \hat{\boldsymbol{u}}_i=[\cos\hat{\phi}_i,\sin\hat{\phi}_i]^T,$$

$$M=[(I-\hat{\boldsymbol{u}}_1\hat{\boldsymbol{u}}_1^T)\boldsymbol{r}_{h,1},\cdots,(I-\hat{\boldsymbol{u}}_m\hat{\boldsymbol{u}}_m^T)\boldsymbol{r}_{h,m}]$$

4.2 目标的垂直坐标分量估计

由式 (9) 求得目标的水平坐标分量 $(\hat{x},\hat{y})$ 后，根据式 (7)，垂直坐标分量应满足下式

$$\hat{z}=\arg\min\left\{\sum_{i=1}^{m}w_i[(\hat{x}-f_1(z_i))^2+(\hat{y}-f_2(z_i))^2+(z-z_i)^2]\right\} \tag{10}$$

式 (10) 中，$f_1(z_i)$、$f_2(z_i)$ 分别为

$$f_1(z_i)=x_{i,0}+\cos\phi_i\cdot\int_{z_{i,0}}^{z_i}\frac{pc(u)du}{\sqrt{1-p^2c^2(u)}},$$
$$f_2(z_i)=y_{i,0}+\sin\phi_i\cdot\int_{z_{i,0}}^{z_i}\frac{pc(u)du}{\sqrt{1-p^2c^2(u)}} \tag{11}$$

式 (11) 中，$(x_{i,0},y_{i,0},z_{i,0})$ 为第 i 个 AVS 的位置坐标；$p=\cos\theta_i/c_{r_i}$ 为声线常数，c_{r_i} 为 AVS 处的声速；z_i 为由 $(x_{i,0},y_{i,0},z_{i,0})$ 和 (ϕ_i,θ_i) 所决定的本征声线上距目标最近点的垂直坐标分量。式 (10) 可通过 Newton-Raphson 迭代法求解。

4.3 定位性能

多个 AVS 的联合几何定位，同样可以用式 (5) 所定义的 MSAE 来描述系统的测向性能。由于斜距和方位角 (r,ϕ,θ) 可以完全确定目标的空间位置，因此可进一步引入均方距离误差 (mean square range error, MSRE)[4] 来描述系统的测距性能

$$\text{MSRE}=\frac{1}{M}\sum_{m=1}^{M}(\hat{r}_m-r)^2 \tag{12}$$

其中，r 为目标的真实斜距，$\hat{r}_m$ 为其第 m 次测量结果。

5 数值仿真

本文针对水下宽带目标的静态定位进行了仿真研究，以此来考察基于 AVS 的分布式定位系统在不同应用背景下的工作性能。仿真的工作频带为 3000~6000Hz，采用 Nyquist 抽样率。

5.1 单个 AVS 的方位估计

图 3 给出了自由空间中方位估计的均方角度误差与信噪比的关系 (目标方位 [45°, 30°])。可以看出，由于观测时间很短，低信噪比所对应的测向误差很大 (0dB 时最优下界近似为 3°)，不具备实用价值。而当信噪比达到 30dB 时，150 个独立抽样 (对应的观测时间仅为 0.025s) 的测向均方根误差低达 0.2°，600 个独立抽样时则可达 0.1°，此时的测向精度可满足一般的定位要求。

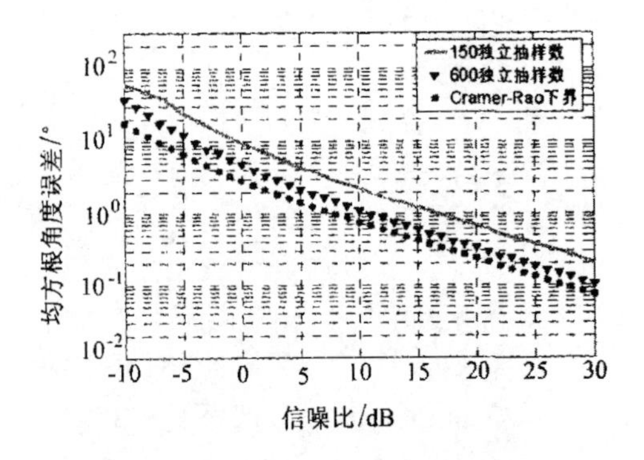

图 3 单个 AVS 的测向性能

图 4 给出了单个 AVS 进行多途结构 (直达和海面反射) 分辨时的空间谱。从中可以看出，各算法分辨力次序为：MUSIC 最优，MVDR 次之，CBF 最差。信噪比为 30dB 时，

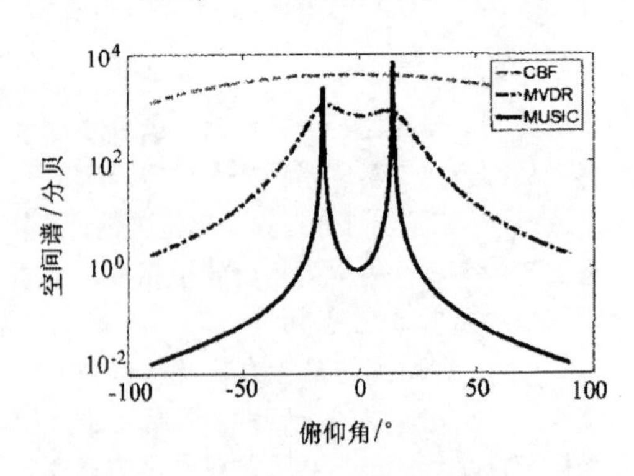

图 4 不同算法对直达和海面反射信号的方位分辨能力

MUSIC 方法得到的空间谱存在两个尖锐的谱峰，MVDR 虽能分辨，但谱峰要宽得多，而 CBF 则仅能识别出直达波和反射波的合成方位。因此，在声强测向法失效的情况下，可通过 MUSIC 方法分辨信号的多途结构，从而正确地提取出目标直达波方位。

5.2多个 AVS 的几何定位

在各 AVS 完成目标测向后，将测量数据和传感器自身的空间位置信息发送至处理中心进行定位融合解算。仿真条件为：

目标位置：(0, 0, 300)，(单位：m)。

传感器位置：(1000, 0, 30)，(0, 1000, 30)，(−1000, 0, 30)，(0, −1000, 30)，(单位：m)。

为考察声速不均匀性对目标定位的影响，仿真中使用 Munk 声速剖面。

图 5~6 给出了自由空间中分布式系统的定位性能。虽然利用单个 AVS 测向面临种种不利条件 (如几何孔径为零，高机动目标定位、跟踪时所允许的积分时间很短)，但由于目标信号具有高强度和大带宽的特点，系统仍可以提供满意的定位精度。图中横坐标为坐标原点处的信噪比，根据仿真条件，由于声波的几何扩展，此时各 AVS 处的信噪比比原点处约低 10.8dB。从图中可以看出，当 AVS 工作于 30dB 的信噪比时，目标定位的角度误差约为 0.53°，距离误差约为 1.3m。WLS 和 RWLS 的误差曲线几乎重合，因而此时采用计算量较小的 WLS 算法即可。

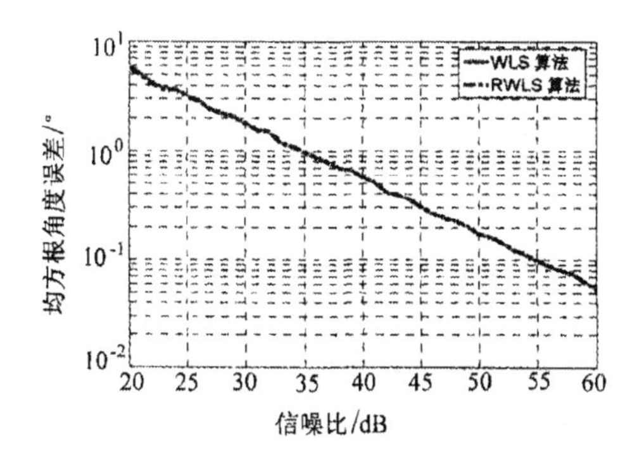

图 5 自由空间中系统的 $\mathrm{MSAE}^{1/2}$

(快拍数：150;Monte Carlo 模拟次数：200)

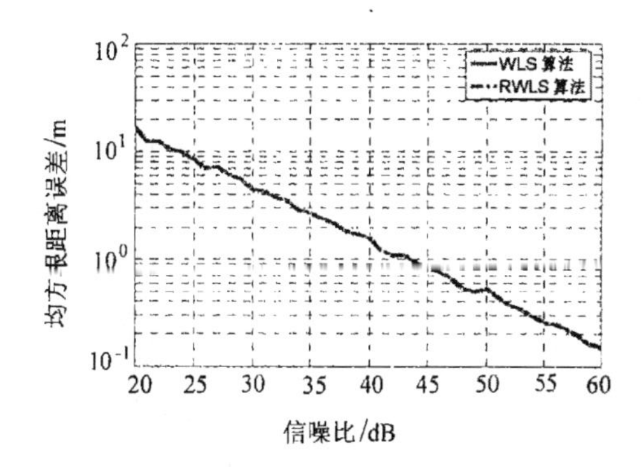

图 6 自由空间中系统的 $\mathrm{MSAE}^{1/2}$

(快拍数：150;Monte Carlo 模拟次数：200)

当 AVS 位于海面附近时，由于海面反射信号的干扰，基于声强测量的目标定位算法已不再适用，此时分辨信号的多途结构是实现目标定位的前提。如图 7~8 所示，目标定位分两步进行：首先通过 MUSIC 方法提取本征声线所对应的目标方位，而后再利用声线交汇方法确定目标的空间位置。该情况下系统的定位性能与信噪比之间的依变关系和自由空间情况类似，WLS 和 RWLS 两种定位算法的结果差异也很小。

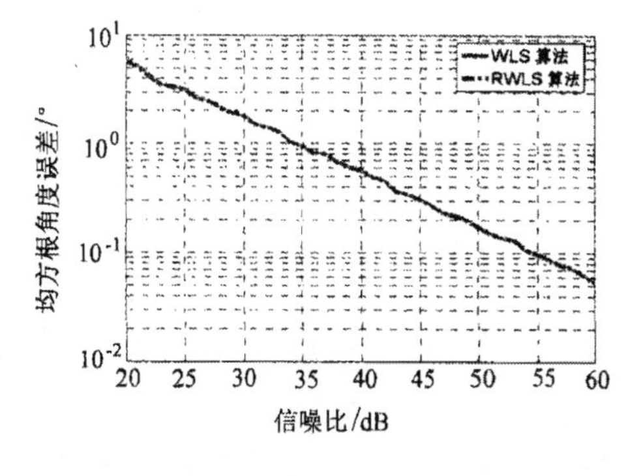

图 7 AVS 位于海面附近时系统的 $\mathrm{MSAE}^{1/2}$

(快拍数：150;Monte Carlo 模拟次数：200)

6 研究结论

研究表明：(1) 基于声强测量的分布式定位算法适用于宽带目标的快速定位，硬件实现简单，高信噪比时具有良好的定位性能；(2) 基于多目标分辨的定位算法在更大的信噪比范围内具有更高的定位精度，AVS 的布放更为灵活，缺点在于计算量大，要求信号多途之间的相干性较弱；(3) 为修正声波的折射效应，本文提出的基于声线跟踪的分步定位算法简单、实用，在单一强声源应用背景下具有理想的定位性能；(4) WLS 和 RWLS 两种算法适用于多个 AVS 的联合定位，在高信噪比条件下，采用前者即可。

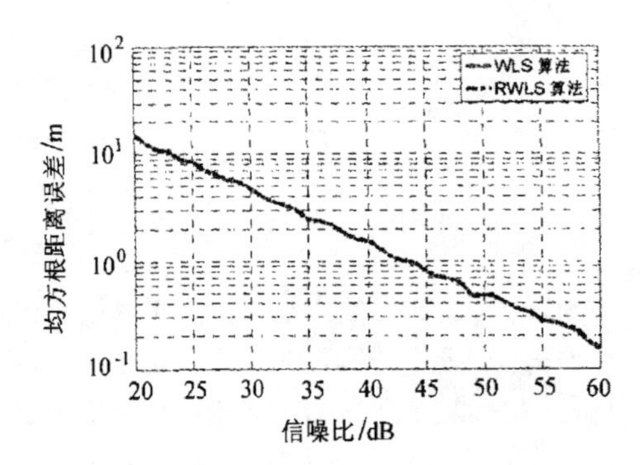

图 8 AVS 位于海面附近时系统的 $\mathrm{MSAE}^{1/2}$
(快拍数：150;Monte Carlo 模拟次数：200)

参 考 文 献

1 李启虎著. 数字式声纳设计原理. 合肥：安徽教育出版社，2002. 1~10.

2 孙贵青著. AVS 及其在均匀线列阵声纳中的应用. 博士后研究报告，北京：中科院声学所，2003. 1~6.

3 孙仲康等著. 单多基地有源无源定位技术. 北京：国防工业出版社，1996. 187~194.

4 Nehorai A, Hawkes M. *IEEE Trans. Signal Processing*, 2000, **48**(6):1737~1749.

5 Hawkes M, Nehorai A. *IEEE Trans. Signal Processing*, 2003, **51**(6):1479~1491.

6 Hochwald B, Nehorai A. *IEEE Trans. Signal Processing*, 1996, **44**(1):83~95.

7 Kozick R J, Sadler B M. *IEEE Trans. Signal Processing*, 2004, **52**(3):601~616.

8 Brekhovskikh L M, Yu P. Lysanov, Fundamentals of Ocean Acoustics. Third Edition, New York: Springer, 2003. 35~56.

一种基于功率谱特征参量的水中目标辐射噪声非母板匹配分类识别方法①

田 甜 李启虎 孙长瑜

(中国科学院声学所 北京 100080)

摘要 本文阐述了一种非母板匹配的水中目标辐射噪声分类识别的思想、技术路线及算法的计算机实现关键技术。该算法基于提取水中目标辐射噪声信号功率谱中的最强谱带位置、线谱位置、线谱根数、线谱平均间隔等特征参量，通过特定处理后，直接给出识别结果，不需要附带庞大数据库。经试验验证，该技术突破了传统的母板匹配目标识别的算法思想，能够适用于复杂海况，做到宽容性匹配。识别效率高，识别率达到试验预期要求。

关键词 非母板匹配，特征参量，宽容性，识别效率，识别率

An algorithm for the non-motherboard-matching identification of the radiated noise of underwater target

TIAN Tian LI Qi-Hu SUN Chang-Yu

(*Institute of Acoustics, Chinese Academy of Sciences, Beijing 100080*)

Abstract In this paper, a new idea about the identifying of the noise of underwater target is described. It's based on the pick-up of the characteristic parameters (such as the position of the frequency band 、 the line spectrum and others) of the noise signal radiated from an underwater target. The algorithm is non-motherboard-matching, which is completely different from the earlier ones in this area. It gives the result directly without the need of searching the motherboard database. As a result, this arithmetic matches most situations of the ocean. The efficiency of the identification is in a high level. And the identifying rate accords with our anticipation in an ocean.

Key words Non-motherboard-matching, Characteristic parameters, Allowance, Efficiency of ientification, Identifying rate

① 应用声学, 2006, 25(5): 265-269.

1 引言

水面/水下被动目标辐射噪声识别一直是水声信号处理领域中一项难度很大的课题，这是因为它不仅包含了船舶类型和大小的识别，而且船舶噪声又随海况和工况的不同是时变的；另外，分类器在实际工程中的可应用性也一直困扰着人们。但另一方面，它又是声纳设计的一个重要和必要的内容，是主被动声纳必须具备的功能。对它的研究一直受到国内外学术界及军事部门极大的关注。由于对这些目标的分类识别直接关系到作战指挥系统的决策能力，所以目标识别技术至关重要。

目前理论方面的文章都是基于人工神经网络理论、混沌理论、聚类分析等方面的，其基本框架是需要一批大信噪比的样本对其加以训练，需要“学习”。而实际情况往往是微弱信号，信噪比很低，那种在大信噪比下学习得到的知识、规则用在微弱信号下，不能发挥有效作用。并且，即使这种依靠大信噪比样本学习的办法可行，如何得到样本呢？特别是敌方水下目标辐射噪声的样本是高度机密的，我们无法得到。

另一方面，目前应用于实际环境的目标识别系统大都是基于母板匹配的最小邻近原则来进行识别的。需要附带一个庞大的母板数据库，在识别时需要将提取到的信号特征参量与数据库中的母板数据进行逐一对比，得到判决结果。这种方法一则大大牺牲了识别效率，二则在匹配的宽容性上大打折扣。海洋的水声信道非常复杂，信号在这样的信道中经过长距离传输后能够被正确识别，则对系统的宽容性要求很高。因此综合来看，目前的目标识别系统仍然处于摸索阶段，尚不能满足充当声纳决策系统的实际应用要求。

因此，我们在非母板匹配的全新目标识别思想上做出了一些尝试，力图找到避免上述不足的方法，提高匹配的宽容性，同时提高识别效率。

我们的基本思路是：通过对相当数量的试验数据进行研究、归纳和总结，寻找到一个适合判决的目标特征参量向量，并总结出这个向量的判决阈值向量，在具体识别过程中，不需要再将提取到的信号特征与母板数据进行比较匹配，而是通过判决阈值筛，直接得到判决结果。这样，在判决效率和匹配的宽容性上均有较大提高。我们在最近某型声纳预研项目的海上试验中对这种算法思想进行了验证，取得了较为理想的效果。

下面具体阐述这种非母板匹配的算法的关键技术思想。

2 算法描述

2.1 时域检测 [1]

时域检测利用声纳系统后置处理的方位历程显示结果。我们用人工干预的方法，人为选定波束号，将该波束的单波束信号作为目标识别的信号源。设采样周期是 Ts，音频输出和后置处理输出的波束数均为 M (两者可以不同，但经过适当内插处理，即可实现两者相同，并且不影响后面的应用)，第 i 路信号是

$$x_i(k) = x_i(kT_s), \quad i = 1, \cdots, N.$$

设波束间隔为 $\theta_0 = 360°/M$，第 m 号波束的延时量为 $\tau_i\ (m\theta_0)$，于是，第 m 号波束的音频信号表达式应为

$$u_m(k) = \sum x_i[kTs - \tau_i(m\theta_0)]\}/N, \quad m = 1, \cdots, M \tag{1}$$

图 1 是时域检测的原理框图。在现有声纳系统中，听测仍然是一个重要的侦察手段。但本文所述的目标识别模块采用的是方位历程显示和目标搜索后得到的单波束信号；输出信噪比解算是为了获得识别结果的置信度，不是本文讨论的主要问题。

2.2 频域特征提取

在时域检测求得 m_0 号波束作为候选波束之后，我们利用 $u_{m0}(k)$ 的信息，提取频域特

征。

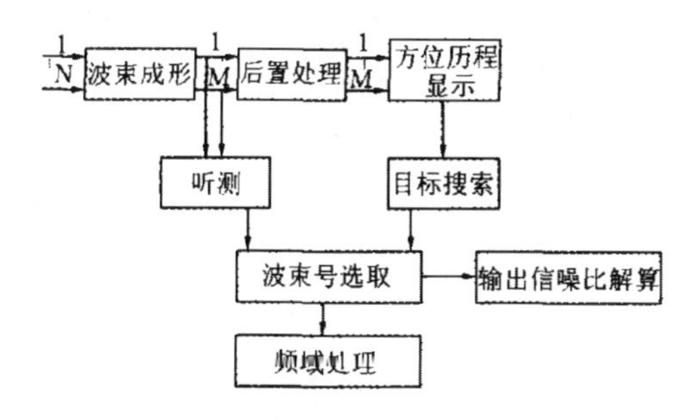

图 1 时域检测原理图

2.2.1 频谱分析

第一步是对信号做谱分析和平滑处理。假设信号采样率为 10kHz, 对信号作 2048 点 FFT, 这样, 分辨率为 10000/2048=5Hz, 足够小, 能够用于线谱提取。然后对频谱序列进行 16 次平滑, 平滑的目的是减少毛刺, 同时增强目标特征量 [2] 。

设输入的时域信号为 $u_{m0}(k)$, 经过 2048 点 FFT 后, 频域序列表示为 $U_{m0}(l)$, 其中 $l=0,1,2,\cdots,2047$ 。 16 次平滑之后, 得到:

$$\overline{U}_{m0}(l)=\left(\sum_{p=0}^{15}|U_{m0}^{(p)}(l)|\right)/16,\quad l=0,\cdots,2047 \tag{2}$$

$\overline{U}_{m0}(l)$ 序列就是我们需要进行分析并从中提取目标特征参量的序列。

2.2.2 均衡处理 [1,3,4]

第二步是 OTA 均衡处理。均衡处理的目的是去掉无关的直流背景分量, 而保留最强谱带和线谱等特征信息。具体算法如下:

由于频段方面的选择, 我们不需要从 $l=0$ 开始, 一直处理到 $l=1023$. 例如, 如果用频段 200Hz~10,000Hz, 则 l 从 11 至 633 。

在极端情况下, 如果需要对 $\overline{U}_{m0}(l)$ 从 $l=0$ 作起, 那么先要对 $\overline{U}_{m0}(l)$ 进行数据扩展, 以避免 OTA 处理的边缘效应。为简单起见, 我们把 $\overline{U}_{m0}(l)$ 中作均衡处理, 并拿去作频域分析、特征提取的数据记作

$$Y(1),\cdots,Y(P)$$

假定均衡处理的窗口长度为 $2K+1$ $(K\ll P)$ 对 $Y(1),\cdots,Y(P)$ 的数据扩展处理是

$$\begin{aligned}&Y(K+1),\cdots,Y(2),Y(1),Y(2),\cdots,\\&\quad Y(P),Y(P-1),\cdots,Y(P-K)\end{aligned}$$

实际作均衡时仍从 1 到 P 进行, 由于数据事先已经进行了扩展, 就不必担心边缘点的问题。这个新的序列我们记作

$$\begin{aligned}&Z(1),Z(2),\cdots,Z(K),Z(K+1),\cdots,\\&\quad Z(K+P),Z(K+P+1),\cdots,Z(2K+P)\end{aligned}$$

显然 $Z(K)=Y(1),\cdots,Z(K+P)=Y(P)$, 等等。

对 K 至 $K+P$ 的每一个 $Z(K)$ 作点到点的均衡处理, 算法如下:

对于 K 取数组

$$Z(k-K),\cdots,Z(k),\cdots,Z(k+K) \text{ 共}2K+1\text{个点},$$

把它们重新从小至大排列, 得到

$$V(1)\leq V(2)\leq\cdots\leq V(2K+1)$$

求出

$$\overline{V}=\sum_{k=K+1}^{2K+1}V(K) \tag{3}$$

如果

$$\begin{aligned}&Z(k)\geq\alpha\overline{V},\ \text{令}\ NEWZ(k)=Z(k)^{-\overline{V}}\\&Z(k)<\alpha\overline{V},\ \text{令}\ NEWZ(k)=0\end{aligned} \tag{4}$$

其中, α 是一个门限值, 我们称之为均衡门限。再重新设置 $Y(1),\cdots,Y(P)$ 的值, 得到

$$\begin{aligned}&NEWY(1)=NEWZ(K)\\&NEWY(P)=NEWZ(K+P)\end{aligned} \tag{5}$$

为方便起见, 我们把新的 $NEWY(p)$ 系列, 仍记作

$$Y(p),\quad p=1,\cdots,P$$

图 2 中图组是均衡前后的一段舰船辐射噪声谱, 可以看出, 直流背景分量被很好地去掉, 留下了线谱等有用的信息。

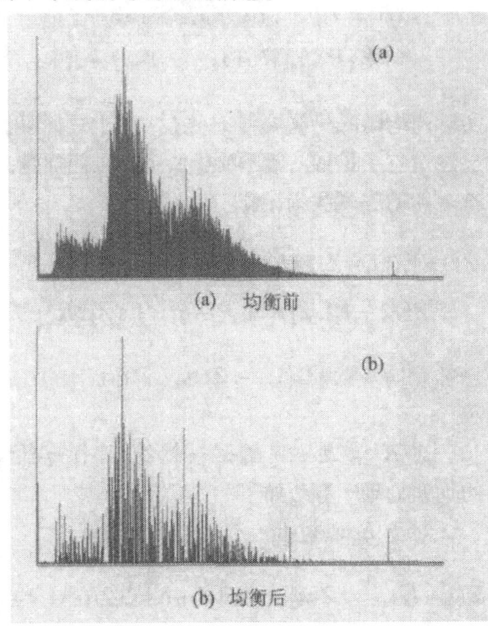

图 2 均衡处理效果图

2.2.3 特征提取

第三步是特征提取, 我们提取的是最强谱带位置、线谱根数等 4 个目标特征参量, 由于篇幅限制, 在此我们只介绍最强谱带位置和线谱的提取算法。

先确定最强谱带位置, 流程如下:

对 $Y(l)$, 求出 $Y(l+k)$ 中第一个使 $Y(l+k)=0$ 的 k, 设为 k_0, 如果 $k_0 \geq M$ (M 为一个常数, 表示窗口长度), 令 $k_0-l=A_1$, 这样, 我们就求出了第一个谱带位置 A_1, 即 $Y(l),\cdots,Y(A_1)$; 如果 $k_0<M$, 则不算作谱带, 继续向下搜索, 再从 $Y(k_0+l)$ 开始, 搜索第一个非零的 $Y(k)$, 设为 $Y(p_1)$, 再求 $Y(p_1+k)$ 中第一个使其为 0 的 k, 设为 k_1, 如果 $k_1 \geq M$, 则令 $k_1-l=A_2$, 我们求得第二个谱带 A_2, 即 $Y(p_1),\cdots,Y(p_1+A_2)$ 。以此类推, 直到 $Y(P-M+1)$ 为止。

假设我们找到了 q 个谱带, $A_1,\cdots,A_q$, 把他们中使 $Y(p)$ 最大的 p 选出来, 设为 p_0, 于是 $Y(p_0)$ 就是最大谱带的谱值, p_0 就成为最强谱带位置。记作 $x_1=p_0$ 。

然后介绍线谱的确定方法: 对任一 $Y(k)$, $k=1,\cdots,P$, 如果

$$\left\{Y(k)-\sum_{i=1}^{N}[Y(k-i)+Y(k+i)]/2N\right\} \div \left\{\sum_{i=1}^{N}[Y(k-i)+Y(k+i)]/2N\right\} \geq \beta \tag{6}$$

我们就认为 $Y(k)$ 是线谱。其中 β 是一个门限值, 我们称之为线谱门限, N 是一个常数, 表示窗口长度。

在所有线谱中强度最大的线谱, 其谱值记为 $Y(v_1)$, 则 v_1 就是最强线谱位置, 记作 $x_2=v_1$ 。

2.2.4 判决 [5~7]

目标特征参量提取完毕之后, 就进入判决过程。通过对大量试验数据的分析归纳和总结, 我们得到了一个判决阈值筛, 对于上面提到的 x_1 和 x_2, 我们分别对水面舰艇、潜艇、商辅船进行研究, 得到这些目标类型中这两个目标特征参数的表现特点, 总结出各自的判决阈值。判决方法可以通过表 1 中的方式表示出来:

表 1 判决规则示意

判决	潜艇	水面舰艇	商辅船	海洋噪声
判据	$x_1<a_1$ $x_2<b_1$ $\cdots$	$x_1 \geq a_2$ $x_2 \leq b_2$ $\cdots$	$x_1<a_3$ $x_2 \leq b_3$ $\cdots$	其它

表中, a_1 、 b_1 等参数就是通过研究, 总结出来的判决阈值筛。例如, 潜艇的最强谱带频率较之水面作战舰艇偏低, 则 a_1 为潜艇谱带位置的上限, 在其他条件确定的情况下, 只要满足最强谱带位置数值小于该上限的, 就可以判定为潜艇。

这种判决方法仅仅是将对信号处理的结果进行几个非常简单的数值比较, 就可以得出

结论，而不需要对母板数据进行繁冗的匹配计算，大大提高了识别效率。

频域分析和判决的总体流程如下图所示：

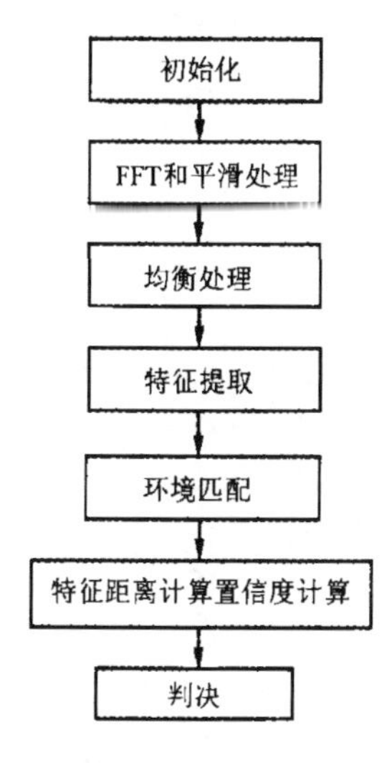

图 3 频域处理和判决流程

图 3 中，环境匹配这个流程是为了有效抑制本舰噪声，增加信号信噪比而专门设置的，有其一套特殊的算法，不是本文讨论的范畴。

3 实验结果

上述算法思想应用到实验室的仿真环境中，做到了较高识别率。其中，水面舰艇识别率 96.5%；潜艇识别率 87.2%；商辅船识别率 98%；

在某型声纳预研项目的海上试验中，我们将该算法制成功能模块，参与了试验，在实际条件有限的情况下，测得对本舰自噪声判定成功率 100% (识别三次，正确三次，每次对一段 8 秒长的数据进行识别，下同)；对目标水面舰艇判定成功率 66.7% (识别三次，正确二次)；对目标潜艇判定成功率 66.7% (识别三次，正确二次)。

4 结束语

目标识别技术是有相当难度的一个研究课题，上述非母板匹配的目标识别思想是我们通过对现有技术的研究分析，找到不足之处之后，做出的一些创新的尝试。通过一系列的试验，我们已经摸索到了一些方法，但远不是完备的。不过我们应该看到，在传统的母板匹配方法效果停滞不前的情况下，非母板匹配这个思想应该值得发展。

参 考 文 献

1 李启虎. 数字式声纳设计原理，合肥：安徽教育出版社，2002. 360~370.

2 沈凤麟，叶中付，钱玉美. 信号统计分析与处理，合肥：中国科学技术大学出版社，2001. 439~446.

3 胡广书. 数字信号处理，北京：清华大学出版社，2003. 118~121.

4 程佩青. 数字信号处理教程，第 2 版. 北京：清华大学出版社，1995. 215~252.

5 章新华. 哈尔滨工程大学博士后研究工作报告，1999.

6 Rajagopal R, Saankaranarayanan B, Ramakrishna Rao P. *IEEE ICASSP-90*, **5**:2911~2914.

7 Foley D H. *IEEE trans IT*, 1992, **18**(2):618~626.

矢量水听器阵列 MVDR 波束形成器的性能研究①

杨秀庭　孙贵青　陈新华　李启虎

(中国科学院声学研究所　北京 100080)

摘要　针对矢量水听器阵列的最佳阵处理，研究了高斯背景噪声条件下 MVDR 波束器的输出性能，给出了系统输出信干噪比与各影响参数之间的解析关系式，并对三种典型情况进行了具体探讨，着重分析了信噪比、干扰强度、相关系数和样本长度对系统输出 SINR 的影响，证明了低信噪比时 MVDR 波束形成器为最佳空域滤波器，分析还表明，MVDR 波束形成器对强信号存在抑制现象、高信噪比时输出的 SINR 为一常数。计算机仿真结果证明了理论分析的正确性。

关键词　矢量水听器，MVDR，波束形成

Performance of MVDR in vector hydrophone array

YANG Xiu-Ting　SUN Gui-Qing　CHEN Xin-Hua　LI Qi-Hu

(*Institute of Acoustics Chinese Academy of Sciences, Beijing 100080*)

Abstract　The output performance of MVDR in the vector hydrophone array applications is studied in this paper. Theoretical analysis gives the relations between SINR and the parameters of consequence, Three typical cases are exploited in which are analyzed the influences of SNR, SIR, the correlation coefficient, and snapshots. It is proved that in the small SNR scenario, MVDR beamformer is the optimal spatial filter. It is also shown that the MVDR beamformer suppresses the signal at high SNR, and the output SINR is a constant. The simulations verify the correctness of the theoretical analysis.

Key words　Vector hydrophone, MVDR, Beamforming

1　引言

三维矢量水听器由三组轴向相互正交的振速水听器和一个无指向性的声压水听器复合而成，能够同步、共点地测量声场的声压和质点振速，较之传统的声压水听器，获取了更多的声场信息。这些丰富的声场信息为矢量阵的信号处理带了更多的选择[3]，近年来已受到越来越多的关注[3~6]。

对于分布于空间不同位置的多个水听器所

① 应用声学, 2007, 26(1): 8-15.

构成的阵列，根据平面波信号到达各个传感器之间存在的传播时延，进行时延补偿后再相加输出的波束形成方法 (CBF) 具有良好的稳健性，并已被证明为单目标情况下均匀噪声场中的最佳空域匹配滤波器 [7]。另外一种流行的波束形成方法是最早由 Capon 提出的 MVDR[1,2] (Minimum Variance Distortionless Response)，该方法旨在保持期望信号不发生幅值畸变的条件下，使整个系统输出的能量最小，借此可以将系统所受干扰和噪声的影响降至最低。到目前为止， MVDR 方法在矢量传感器阵列信号处理中已有所应用 [3~7]。 M. Hawkes[3] 研究了声矢量传感器阵列 MVDR 方法关于目标方位估计的渐近性能；陈华伟 [4] 等研究了声矢量传感器阵列的宽带 MVDR 波束形成，给出了目标左右舷分辨和栅瓣抑制的仿真结果；孙贵青等 [5] 把 MVDR 方法引入矢量水听器阵列的实验数据分析，取得了远优于 CBF 的处理结果；田坦等 [6] 讨论了波束域 MVDR 方法，对比了阵元域和波束域 MVDR 的处理性能和计算量。

鉴于 MVDR 波束形成器在矢量水听器阵列信号处理中具有广泛的应用前景，有必要对其工作性能进行深入分析，给出其理论边界，以便更好地指导在工程实际中的应用。

2　问题描述

假设有 J 个窄带平面波信号在各向同性，均匀，静态的水下环境中传播，信号中心频率为已知的 ω_0。为以后讨论问题方便，假设信号和噪声均为零均值高斯随机过程，且信号和噪声之间互不相关。 M 个矢量水听器的空间位置矢量为 $\boldsymbol{r}_i, (i=1,\cdots,M)$，则矢量水听器阵列的数据接收模型为

$$\begin{aligned}\boldsymbol{x}(t) &= \boldsymbol{a}(\theta_1)s_1(t) + \sum_{k=2}^{J}\boldsymbol{a}(\theta_k)s_k(t)t\boldsymbol{n}(t)\\ &= \boldsymbol{a}(\theta_1)s_1(t) + \boldsymbol{v}(t)\end{aligned} \tag{1}$$

上式中， $s_1(t)$ 为期望信号， $\boldsymbol{v}(t) = \sum_{k=2}^{J}\boldsymbol{a}(\theta_k)s_k(t) + \boldsymbol{n}(t)$ 为干扰加噪声向量； $a(\theta)$ 为矢量阵的阵列流形，为

$$\begin{aligned}\boldsymbol{a}(\theta) = \Big[&\exp\Big(-j\frac{2\pi}{\lambda}\boldsymbol{r}_1^T\boldsymbol{u}),\cdots,\\ &\cdot\exp\Big(-j\frac{2\pi}{\lambda}\boldsymbol{r}_M^T\boldsymbol{u}\Big)\Big]^T\\ &\otimes\boldsymbol{h}(\theta) = a_P(\theta)\otimes\boldsymbol{h}(\theta)\end{aligned} \tag{2}$$

其中，“$\otimes$”为矩阵的 Kronecker 积； $\boldsymbol{h}^T(\theta) = [1,\boldsymbol{u}]$， $\boldsymbol{u} = [\cos\theta\cos\phi,\cos\theta\sin\phi,\sin\theta]^T$ 为信号的方向向量。为表述方便，以下 a 用表示 $a(\theta)$。

系统输出能量可用数据的协方差矩阵表示

$$\begin{aligned}R &= E[\boldsymbol{x}(t)\boldsymbol{x}^H(t)]\\ &= \sigma_s^2\boldsymbol{a}(\theta_1)\boldsymbol{a}^H(\theta_1) + \boldsymbol{a}(\theta_1)\boldsymbol{r}^H + \boldsymbol{r}\boldsymbol{a}^H(\theta_1) + Q\end{aligned} \tag{3}$$

其中， $\sigma_s^2 = E[s(t)s^*(t)]$ 为信号方差， $\boldsymbol{r} = E[s_1^*(t)v(t)]$ 为信号和干扰加噪声之间的相关向量， $Q = E[\boldsymbol{v}(t)\boldsymbol{v}^H(t)]$ 为干扰加噪声的协方差矩阵。

实际接收数据的协方差矩阵，可由最大似然估计 [7] 得到

$$\widehat{R} = \frac{1}{N}\sum_{i=1}^{N}\boldsymbol{x}(t_i)\boldsymbol{x}^H(t_i) \tag{4}$$

因此，实际抽样数据的 MVDR 波束形成器，其最优权向量为

$$\boldsymbol{w}_{\text{opt}} = \frac{\widehat{R}^{-1}a(\theta_1)}{a^H(\theta_1)\widehat{R}^{-1}a(\theta_1)} \tag{5}$$

3　性能分析

实际接收的信号总是限带的，并只有有限的观测时间。波束形成器的实际工作性能和理想情况总是存在一定的差异，为更清晰地看出矢量阵 MVDR 波束形成器在有限时间带宽条

件下的实际工作性能，有必要给出其与各影响参数之间的解析关系。从下面的分析可知，这些常见的参数包括：信噪比，干噪比，信干比，信号与干扰加噪声之间的相关向量，时间带宽积（通常以独立抽样数表示），信号和干扰间的方位相近程度等。

3.1 有限抽样数据的最优权向量

由式 (3)，有限抽样的数据协方差矩阵可表示为

$$\widehat{R}=\hat{\sigma}_s^2\boldsymbol{a}_1\boldsymbol{a}_1^H+\boldsymbol{a}_1\hat{\boldsymbol{r}}^H+\hat{\boldsymbol{r}}\boldsymbol{a}_1^H+\widehat{Q} \tag{6}$$

上式中，$\hat{\sigma}_s^2$、$\hat{\boldsymbol{r}}$、$\widehat{Q}$ 分别为有限抽样条件下信号方差、相关向量和干扰加噪声协方差矩阵的极大似然估计，给出如下

$$\hat{\sigma}_s^2=\frac{1}{N}\sum_{i=1}^N|s_1(t_i)|^2,\quad \hat{\boldsymbol{r}}=\frac{1}{N}\sum_{i=1}^N s_1^*(t_i)\boldsymbol{v}(t_i),$$
$$\widehat{Q}=\frac{1}{N}\sum_{i=1}^N\boldsymbol{v}(t_i)\boldsymbol{v}^H(t_i) \tag{7}$$

为求解抽样数据协方差矩阵的逆阵，把 (6) 式写为

$$\widehat{R}=D+\boldsymbol{b}\boldsymbol{b}^H \tag{8}$$

其中，$D=\widehat{Q}-\hat{\sigma}_s^{-2}\hat{\boldsymbol{r}}\hat{\boldsymbol{r}}^H$，$\boldsymbol{b}=\hat{\sigma}_s\boldsymbol{a}_1+\hat{\sigma}_s^{-1}\hat{\boldsymbol{r}}$。根据矩阵求逆引理，有

$$\widehat{R}^{-1}=D^{-1}-D^{-1}\frac{\boldsymbol{b}\boldsymbol{b}^H}{1+\boldsymbol{b}^HD^{-1}\boldsymbol{b}}D^{-1} \tag{9}$$

再次利用矩阵求逆引理，得到

$$D^{-1}=\widehat{Q}^{-1}+\widehat{Q}^{-1}\frac{\hat{\boldsymbol{r}}\hat{\boldsymbol{r}}^H}{\hat{\sigma}_s^2-\hat{\boldsymbol{r}}^H\widehat{Q}^{-1}\hat{\boldsymbol{r}}}\widehat{Q}^{-1} \tag{10}$$

结合 (9)、(10)，有

$$\begin{aligned}\widehat{R}^{-1}=&\frac{1}{\hat{\sigma}_s^2-\hat{\boldsymbol{r}}^H\widehat{Q}^{-1}\hat{\boldsymbol{r}}}[(\hat{\sigma}_s{}^2-\hat{\boldsymbol{r}}^H\widehat{Q}^{-1}\hat{\boldsymbol{r}})\widehat{Q}^{-1}\\&+\widehat{Q}^{-1}\hat{\boldsymbol{r}}\hat{\boldsymbol{r}}^H\widehat{Q}^{-1}]\cdot\{\boldsymbol{I}-\boldsymbol{b}\boldsymbol{b}^H[(\hat{\sigma}_s^2\\&-\hat{\boldsymbol{r}}^H\widehat{Q}^{-1}\hat{\boldsymbol{r}})\widehat{Q}^{-1}+\widehat{Q}^{-1}\hat{\boldsymbol{r}}\hat{\boldsymbol{r}}^H\widehat{Q}^{-1}]/\\&\{(\hat{\sigma}_s^2-\hat{\boldsymbol{r}}^H\widehat{Q}^{-1}\hat{\boldsymbol{r}})+\boldsymbol{b}^H[(\hat{\sigma}_s^2\\&-\hat{\boldsymbol{r}}^H\widehat{Q}^{-1}\hat{\boldsymbol{r}})\widehat{Q}^{-1}+\widehat{Q}^{-1}\hat{\boldsymbol{r}}\hat{\boldsymbol{r}}^H\widehat{Q}^{-1}]\boldsymbol{b}\}\}\end{aligned} \tag{11}$$

把式 (11) 代入 (5)，可得

$$\boldsymbol{w}_{\text{opt}}=\frac{\widehat{Q}^{-1}\boldsymbol{a}_1}{\boldsymbol{a}_1^H\widehat{Q}^{-1}\boldsymbol{a}_1}-\left[\boldsymbol{I}-\frac{\widehat{Q}^{-1}\boldsymbol{a}_1\boldsymbol{a}_1^H}{\boldsymbol{a}_1^H\widehat{Q}^{-1}\boldsymbol{a}_1}\right]\widehat{Q}^{-1}\hat{\boldsymbol{r}} \tag{12}$$

从上式可以看出，在有限抽样数据条件下，最优权向量由两部分组成，右边第一项为信号无畸变条件下系统输出最小干扰加噪声能量时的最优权向量，右边第二项则为考虑期望信号与干扰加噪声之间的相关性而引入的附加项。记

$$\widehat{P}=\boldsymbol{I}-\frac{\widehat{Q}^{-1}\boldsymbol{a}_1\boldsymbol{a}_1^H}{\boldsymbol{a}_1^H\widehat{Q}^{-1}\boldsymbol{a}_1} \tag{13}$$

上式给出的 $\widehat{P}$ 称为斜投影矩阵，满足幂等性，即 $\widehat{P}=\widehat{P}\widehat{P}$。故式 (12) 可写为

$$\boldsymbol{w}_{\text{opt}}=\frac{\widehat{Q}^{-1}\boldsymbol{a}_1}{\boldsymbol{a}_1^H\widehat{Q}^{-1}\boldsymbol{a}_1}-\widehat{P}\widehat{Q}^{-1}\hat{\boldsymbol{r}} \tag{14}$$

如果信号与干扰加噪声不相干，当满足渐近条件时 ($N\gg1$)，式 (12) 右边第二项渐趋于零。然而，对于有限的抽样数据，即便两者理论上严格互不相关，该项一般也不为零，而是一个随机起伏的量。因此，式 (14) 所给出的有限抽样数据条件下的最优权向量也是一个随机量，此时对系统性能的评价必须建立在对大量随机实验的统计平均之上。

3.2 MVDR 波束形成器输出的 SINR

由于 MVDR 波束形成方法是在保证信号无失真的条件下对干扰加噪声作最大程度的抑制（此时系统输出的信干噪比最大），因此可以将波束形成器输出的信干噪比 (**SINR**, signal to interference plus noise ratio) 作为评价指标，一般可表示为 [2]

$$\text{SINR}=\frac{|E[\hat{s}_1(t)s_1^*(t)]|^2}{E[|\hat{s}_1(t)|^2]E[|s_1(t)|^2]-|E[\hat{s}_1(t)s_1^*(t)]|^2} \tag{15}$$

由于信号、干扰和噪声均为高斯的，通过直接计算，式 (15) 可进一步写为

$$\textbf{SINR}=\frac{\sigma_s^4+2\sigma_s^2\text{Re}\{E[\boldsymbol{w}_{\text{opt}}^H\hat{\boldsymbol{r}}]\}+|E[\boldsymbol{w}_{\text{opt}}^H\hat{\boldsymbol{r}}]|^2}{\sigma_s^2E[\boldsymbol{w}_{\text{opt}}^H\boldsymbol{v}(t)\boldsymbol{v}^H(t)\boldsymbol{w}_{\text{opt}}]-|E[\boldsymbol{w}_{\text{opt}}^H\hat{\boldsymbol{r}}]|^2} \tag{16}$$

上式中，$\sigma_s^2 = E[s_1(t)s_1^*(t)]$ 为期望信号的方差；$\hat{\boldsymbol{r}} = s_1^*(t)\boldsymbol{v}(t)$ 为信号与干扰加噪声的相关向量。上式推导过程中，使用了期望信号无畸变条件，即 $\boldsymbol{w}_{\rm opt}^H\boldsymbol{a}_1 = 1$。

容易验证，有下式成立

$$\widehat{Q}^{-1}\widehat{P}^H = \widehat{P}\widehat{Q}^{-1} \tag{17}$$

对于矢量水听器阵列，有以下关系 [7]

$$\widehat{Q}^{-1} = \frac{N}{N-4M}Q^{-1} + O(N) \tag{18}$$

从而，有

$$E[\boldsymbol{w}_{\rm opt}^H\hat{\boldsymbol{r}}] \approx \frac{\boldsymbol{a}_1^H Q^{-1}\boldsymbol{r}}{\boldsymbol{a}_1^H Q^{-1}\boldsymbol{a}_1} - \frac{N}{N-4M}\boldsymbol{r}^H P Q^{-1}\boldsymbol{r} \tag{19}$$

进而，得到

$$\begin{aligned}|E[\boldsymbol{w}_{\rm opt}^H\hat{\boldsymbol{r}}]|^2 \approx& \frac{|\boldsymbol{a}_1^H Q^{-1}\boldsymbol{r}|^2}{\boldsymbol{a}_1^H Q^{-1}\boldsymbol{a}_1)^2} \\ &- \frac{2N}{N-4M}\frac{\boldsymbol{r}^H P Q^{-1}\boldsymbol{r}}{(\boldsymbol{a}_1^H Q^{-1}\boldsymbol{a}_1)} \\ &\cdot \mathrm{Re}(\boldsymbol{a}_1^H Q^{-1}\boldsymbol{r}) \\ &+ \left(\frac{N}{N-4M}\boldsymbol{r}^H P Q^{-1}\boldsymbol{r}\right)^2\end{aligned} \tag{20}$$

根据斜投影矩阵的定义，不难验证，有 $\boldsymbol{a}_1^H P = 0$, 从而得到

$$\begin{aligned}&E[\boldsymbol{w}_{\rm opt}^H\boldsymbol{v}(t)\boldsymbol{v}^H(t)\boldsymbol{w}_{\rm opt}] \\ &\approx \frac{1}{\boldsymbol{a}_1^H Q^{-1}\boldsymbol{a}_1} + \left(\frac{N}{N-4M}\right)^2 \\ &\quad\cdot \mathrm{tr}\{PQ^{-1}E[\hat{\boldsymbol{r}}\hat{\boldsymbol{r}}^H]\}\end{aligned} \tag{21}$$

其中，

$$E[\hat{\boldsymbol{r}}\hat{\boldsymbol{r}}^H] = \frac{1}{N^2}\sum_{i,j=1}^{N} E[s_1^*(t_i)\boldsymbol{v}(t_i)s_1(t_j)\boldsymbol{v}^H(t_j)] \tag{22}$$

由于信号和干扰加噪声均为高斯随机变量，因此式 (22) 中的数学期望满足以下关系

$$\begin{aligned}&E[s_1^*(t_i)\boldsymbol{v}(t_i)s_1(t_j)\boldsymbol{v}^H(t_j)] \\ &= E[s_1^*(t_i)\boldsymbol{v}(t_i)]E[s_1(t_j)\boldsymbol{v}^H(t_j)] \\ &\quad + E[s_1^*(t_i)s_1(t_j)]E[\boldsymbol{v}(t_i)\boldsymbol{v}^H(t_j)]\end{aligned} \tag{23}$$

把上式代入 (22), 得

$$E[\hat{\boldsymbol{r}}\hat{\boldsymbol{r}}^H] = \boldsymbol{r}\boldsymbol{r}^H + \frac{1}{N}\sigma_s^2 Q \tag{24}$$

由式 (21), 有

$$\begin{aligned}\mathrm{tr}(P) &= \mathrm{tr}\left(\boldsymbol{I} - \frac{Q^{-1}\boldsymbol{a}_1\boldsymbol{a}_1^H}{\boldsymbol{a}_1^H Q^{-1}\boldsymbol{a}_1}\right) \\ &= 4M - \frac{\boldsymbol{a}_1^H Q^{-1}\boldsymbol{a}_1}{\boldsymbol{a}_1^H Q^{-1}\boldsymbol{a}_1} = 4M-1\end{aligned} \tag{25}$$

根据 (24)、(25), 式 (21) 可化为

$$\begin{aligned}&E[\boldsymbol{w}_{\rm opt}^H\boldsymbol{v}(t)\boldsymbol{v}^H(t)\boldsymbol{w}_{\rm opt}] \\ &\approx \frac{1}{\boldsymbol{a}_1^H Q^{-1}\boldsymbol{a}_1} + \left(\frac{N}{N-4M}\right)^2\boldsymbol{r}^H P Q^{-1}\boldsymbol{r} \\ &\quad + \frac{N(4M-1)}{(N-4M)^2}\sigma_s^2\end{aligned} \tag{26}$$

把 (19)、(20)、(26) 一并代入式 (16), 得到

$$\mathrm{SINR} = \frac{S}{I+N} \tag{27}$$

其中，S 代表信号功率，由下式给出

$$\begin{aligned}S \approx& \sigma_s^2 + 2\frac{\mathrm{Re}\{\boldsymbol{a}_1^H Q^{-1}\boldsymbol{r}\}}{\boldsymbol{a}_1^H Q^{-1}\boldsymbol{a}_1} \\ &- \frac{2N}{N-4M}\boldsymbol{r}^H P Q^{-1}\boldsymbol{r} + \frac{|\boldsymbol{a}_1^H Q^{-1}\boldsymbol{r}|^2}{\sigma_s^2(\boldsymbol{a}_1^H Q^{-1}\boldsymbol{a}_1)^2} \\ &- \frac{2N}{N-4M}\sigma_s^2\frac{\boldsymbol{r}^H P Q^{-1}\boldsymbol{r}}{(\boldsymbol{a}_1^H Q^{-1}\boldsymbol{a}_1)}\mathrm{Re}(\boldsymbol{a}_1^H Q^{-1}\boldsymbol{r}) \\ &+ \frac{1}{\sigma_s^2}\left(\frac{N}{N-4M}\boldsymbol{r}^H P Q^{-1}\boldsymbol{r}\right)^2\end{aligned} \tag{28}$$

$I+N$ 为干扰加噪声的功率，满足下式

$$\begin{aligned}I+N \approx& \frac{1}{\boldsymbol{a}_1^H Q^{-1}\boldsymbol{a}_1} + \left(\frac{N}{N-4M}\right)^2\boldsymbol{r}^H P Q^{-1}\boldsymbol{r} \\ &+ \frac{N(4M-1)}{(N-4M)^2}\sigma_s^2 - \frac{|\boldsymbol{a}_1^H Q^{-1}\boldsymbol{r}|^2}{\sigma_s^2(\boldsymbol{a}_1^H Q^{-1}\boldsymbol{a}_1)^2} \\ &+ \frac{2N}{N-4M}\frac{\boldsymbol{r}^H P Q^{-1}\boldsymbol{r}}{\sigma_s^2(\boldsymbol{a}_1^H Q^{-1}\boldsymbol{a}_1)}\mathrm{Re}(\boldsymbol{a}_1^H Q^{-1}\boldsymbol{r})\end{aligned}$$

$$-\frac{1}{\sigma_s^2}\left(\frac{N}{N-4M}\boldsymbol{r}^H P Q^{-1}\boldsymbol{r}\right)^2 \tag{29}$$

以上利用较大抽样数据时满足的一些近似关系，得到了有限抽样数据条件下 MVDR 波束形成器输出的信干噪比的一般关系式。

3.3 几种典型特例

式 (27~29) 给出了系统输出信干噪比的一般形式，为更明确地给出 MVDR 波束形成器的工作性能和相关参数的依变关系，下面就矢量水听器均匀线列阵的几种典型情况进行具体讨论。

3.3.1 接收数据中仅含噪声和期望信号

当接收数据中除噪声外仅含期望信号时，干扰加噪声的协方差矩阵退化为单纯的噪声协方差矩阵，即

$$Q=E[\boldsymbol{v}(t)\boldsymbol{v}^H(t)]=I_M\otimes \boldsymbol{R}_n=\sigma_P^2 I_M\otimes\Lambda_n \tag{30}$$

其中，$\Lambda_n=\begin{bmatrix}1 & \boldsymbol{0}_{1\times 3}\\ \boldsymbol{0}_{3\times 1} & \eta\boldsymbol{I}_3\end{bmatrix}$，$\eta=\sigma_v^2/\sigma_P^2$ 。此时信号与干扰之间的互相关向量为零向量，即 $\boldsymbol{r}=\boldsymbol{0}$、根据式 (27~29)，系统输出的信干噪比为

$$\begin{aligned}\mathbf{SINR}&\approx\frac{\sigma_s^2}{\dfrac{1}{M(1+1/\eta)}\sigma_P^2+\dfrac{N(4M-1)}{(N-4M)^2}\sigma_s^2}\\&=\frac{SNR}{\dfrac{1}{M(1+1/\eta)}+\dfrac{N(4M-1)}{(N-4M)^2}SNR}\end{aligned} \tag{31}$$

讨论：

(1) 当 $1/[M(1+1/\eta)]\gg N(4M-1)SNR/(N-4M)^2$ 时，$\mathbf{SINR}\approx M(1+1/\eta)SNR$; 此时系统输出的信干噪比随信噪比呈线性变化，MVDR 波束形成器为最佳空域滤波器，阵增益达到最大。

(2) 当 $1/[M(1+1/\eta)]\ll N(4M-1)SNR/(N-4M)^2$ 时，$\mathbf{SINR}\approx(N-4M)^2/[N(4M-1)]$, 即在高信噪比条件下，系统输出信干噪比趋于一特定的常数，并没有按预想的随信噪比增加而增加，此时波束形成器对大信号具有抑制效应。

3.3.2 接收数据中除噪声和期望信号外，仅含一个干扰

为了探讨干扰对系统输出性能的影响，不失一般性，可假设接收数据中仅存在单一干扰，信号与干扰之间的相关系数可以是幅值为 $[0,1]$ 之间任何一个复常数，此时有

$$Q=E[\boldsymbol{v}(t)\boldsymbol{v}^H(t)]=\sigma_i^2\boldsymbol{a}_2\boldsymbol{a}_2^H+\sigma_p^2 I_M\otimes\Lambda_n \tag{32}$$

上式中，σ_i^2 为干扰功率，$\boldsymbol{a}_2$ 为干扰的方向响应向量。利用矩阵求逆引理，得

$$\begin{aligned}Q^{-1}&=(\sigma_i^2\boldsymbol{a}_2\boldsymbol{a}_2^H+\sigma_p^2 I_M\otimes\Lambda_n)^{-1}\\&=\frac{1}{\sigma_p^2}\Big(I_M\otimes\Lambda_n^{-1}-INR\\&\quad\cdot\frac{I_M\otimes\Lambda_n^{-1}\boldsymbol{a}_2\boldsymbol{a}_2^H I_M\otimes\Lambda_n^{-1}}{1+M(1+1/\eta)\cdot INR}\Big)\end{aligned} \tag{33}$$

其中，$INR=\sigma_i^2/\sigma_p^2$ 为干噪比。根据斜投影矩阵的定义，可进一步得到

$$\begin{aligned}P=&I_{4M}-\big\{I_M\otimes\Lambda_n^{-1}\boldsymbol{a}_1\boldsymbol{a}_1^H\\&-INR(\boldsymbol{a}_2^H I_M\otimes\Lambda_n^{-1}\boldsymbol{a}_1 I_M\\&\otimes\Lambda_n^{-1}\boldsymbol{a}_2\boldsymbol{a}_1^H)/[1+M(1+1/\eta)INR]\big\}\\&\cdot\big\{M(1+1/\eta)-[INR|\boldsymbol{a}_1^H I_M\otimes\Lambda_n^{-1}\boldsymbol{a}_2|^2]\\&\cdot/[1+M(1+1/\eta)INR]\big\}^{-1}\end{aligned} \tag{34}$$

信号与干扰之间的互相关系数定义如下

$$\rho=\frac{E[s_1^*(t)s_2(t)]}{\sqrt{E[|s_1(t)|^2]E[|s_2(t)|^2]}}=\frac{E[s_1^*(t)s_2(t)]}{\sigma_s\sigma_i} \tag{35}$$

则

$$\boldsymbol{r}=E[s_1^*(t)\boldsymbol{v}(t)]=E[s_1^*(t)s_2(t)\boldsymbol{a}_2]=\rho\sigma_s\sigma_i\boldsymbol{a}_2 \tag{36}$$

把上述 (33~36) 代入 (27~29)，得

$$\mathbf{SINR}\approx\frac{w_1-w_2+w_3-w_4+w_5}{w_6+w_2-w_3+w_4-w_5} \tag{37}$$

其中

$w_1=SNR+2\sqrt{SNR\cdot INR}\cdot\mathrm{Re}[\rho\cdot c_1]/c_5$,

$$w_2 = -2\frac{N}{N-4M}|\rho|^2 SNR\cdot INR\frac{c_3c_5-c_2}{c_4c_5},$$
$$w_3 = |\rho|^2 INR\frac{c_2}{c_5^2},$$
$$w_4 = -2\frac{N}{N-4M}|\rho|^2\sqrt{SNR\cdot INR^3}\frac{c_3c_5-c_2}{c_4c_5},$$
$$w_5 = \left(\frac{N}{N-4M}\right)^2|\rho|^4 SNR\cdot INR^2\left(\frac{c_3c_5-c_2}{c_4c_5}\right)^2,$$
$$w_6 = \frac{c_4}{c_5}+SNR\frac{N(4M-1)}{(N-4M)^2}$$
$$c_1 = \boldsymbol{a}_1^H I_M\otimes\Lambda_n^{-1}\boldsymbol{a}_2 S,$$
$$c_2 = |\boldsymbol{a}_1^H I_M\otimes\Lambda_n^{-1}\boldsymbol{a}_2 S|^2,$$
$$c_3 = M(1+1/\eta),\quad c_4 = 1+M(1+1/\eta)INR$$
$$c_5 = M(1+1/\eta)\cdot[1+M(1+1/\eta)INR] - INR|\boldsymbol{a}_1^H I_M\otimes\Lambda_n^{-1}\boldsymbol{a}_2|^2 \tag{38}$$

通过以上两式，给出了单一干扰条件下系统输出的信干噪比与各参数之间的解析关系。在实际问题中，通常还会遇到信号和干扰互不相关的情况。

3.3.3 仅存在一个干扰，且信号和干扰互不相关

根据相关系数的定义，此时 $\rho=0$，根据式(37)，得

$$\mathbf{SINR}\approx SNR\cdot\left\{\frac{\mathcal{M}_1}{\mathcal{M}\cdot[\mathcal{M}_1]-INR|\boldsymbol{a}_1^H I_M\otimes\Lambda_n^{-1}\boldsymbol{a}_2|^2}+\frac{N(4M-1)}{(N-4M)^2}SNR\right\}^{-1},$$
$$\mathcal{M}=M(1+1/\eta),$$
$$\mathcal{M}_1 = 1+M(1+1/\eta)INR \tag{39}$$

由上式可知，当信号和干扰之间互不相关时，情况要简单得多，此时可作下述讨论：

(1) 当 $\frac{\mathcal{M}_1}{\mathcal{M}\cdot[\mathcal{M}_1]-INR|\boldsymbol{a}_1^H I_M\otimes\Lambda_n^{-1}\boldsymbol{a}_2|^2}\gg\frac{N(4M-1)}{(N-4M)^2}SNR$ 时，有

$$\mathbf{SINR}\approx\frac{1}{\mathcal{M}_1}SNR\cdot\mathcal{M}\cdot[\mathcal{M}_1]-INR|\boldsymbol{a}_1^H I_M\otimes\Lambda_n^{-1}\boldsymbol{a}_2|^2 \tag{40}$$

即，在低信干比时，若保持干噪比不变，系统输出的信干噪比随信噪比呈线性变化。由于干扰的影响可以忽略，此时 MVDR 波束形成器为最佳空域滤波器。

(2) 当 $\frac{\mathcal{M}_1}{\mathcal{M}\cdot[\mathcal{M}_1]-INR|\boldsymbol{a}_1^H I_M\otimes\Lambda_n^{-1}\boldsymbol{a}_2|^2}\ll\frac{N(4M-1)}{(N-4M)^2}SNR$ 时，有

$$\mathbf{SINR}\approx\frac{(N-4M)^2}{N(4M-1)} \tag{41}$$

由上式可以看出，在高信干比条件下，系统输出的信干噪比渐趋于某一特定常数，系统工作性能和无干扰情况相同。

4 仿真结果

本节通过具体算例来研究和验证 MVDR 波束形成器的输出性能和各影响因素之间的关系。基本参量为：矢量水听器阵元数为 $M=8$，阵间距 $d=1/2\lambda$，目标方位为 $\theta_S=(30°,60°)$，干扰方位为 $\theta_I=(45°,50°)$，独立仿真次数为 1000 。

(1) 无干扰情况下 **SINR** 与信噪比和快拍数的关系

两种典型噪声场情况下，**SINR** 与信噪比和快拍数的关系如图 1 所示，独立快拍数分别为 500, 1000 和 2000 。从图中可以看出，理论值和仿真结果符合很好，噪声强度对输出性能的影响主要体现在低信噪比情况下，高信噪比时两者趋于一致；快拍数对低信噪比时系统输出性能的影响并不显著，在高信噪比条件下，更多的快拍数则对应着更高的输出信干噪比。

(2) 单干扰情况下 **SINR** 与信噪比和相关系数的关系。

图 2 给出了单个干扰情况下， **SINR** 与信号 - 干扰间相关系数和快拍数的关系，此时 $\eta=1$，信干比为 −10dB，快拍数分别为 500 和 2000 。从中可见，相关系数越大，高信噪比条件下的滤波增益越小，仿真值和理论值差异也

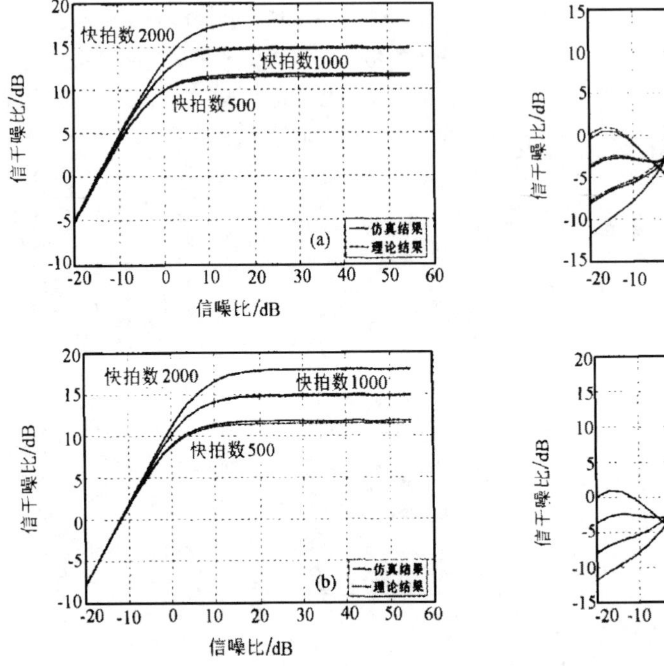

图 1　无干扰情况下 **SINR** 与信噪比和快拍数的关系

(a) $\eta = 1/3$　(b) $\eta = 1$

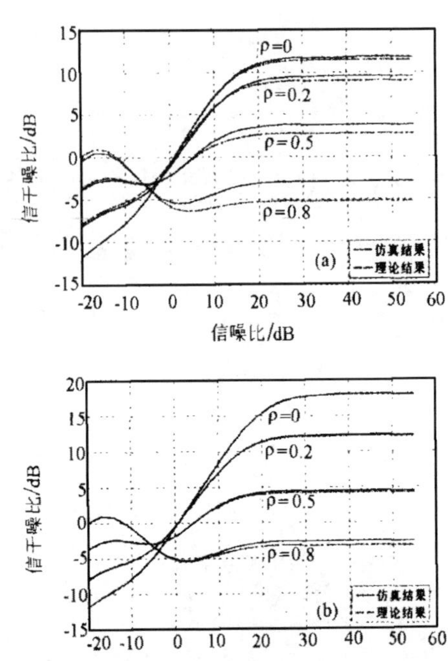

图 2　单干扰情况下 **SINR** 与信噪比和相关系数的关系

(a) $N = 500$　(b) $N = 2000$

越大, 但两者的差异随快拍数的增加而减小。

(3) 单干扰情况下干扰强度对 **SINR** 的影响

信干比 SIR ($\mathrm{SIR} = \sigma_s^2/\sigma_i^2$) 分别为 -10dB 和 10dB, 快拍数为 2000, 系统输出的 SINR 如图 3 所示。从图 3 可以看出, 在低信噪比时, 干扰强度对 **SINR** 影响较为显著, 但在高信噪比时, 干扰的影响被平抑, **SINR** 趋于一常数。

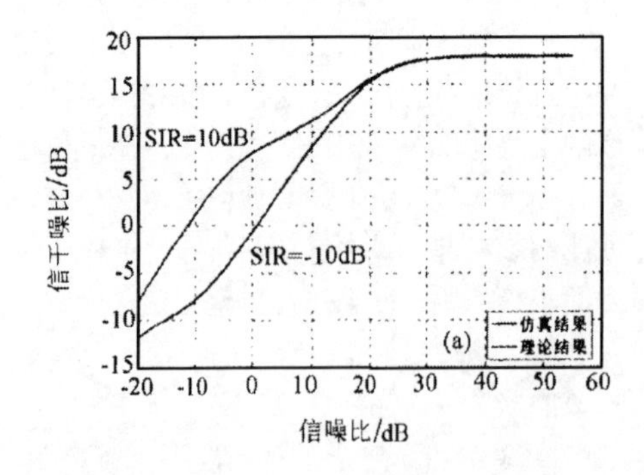

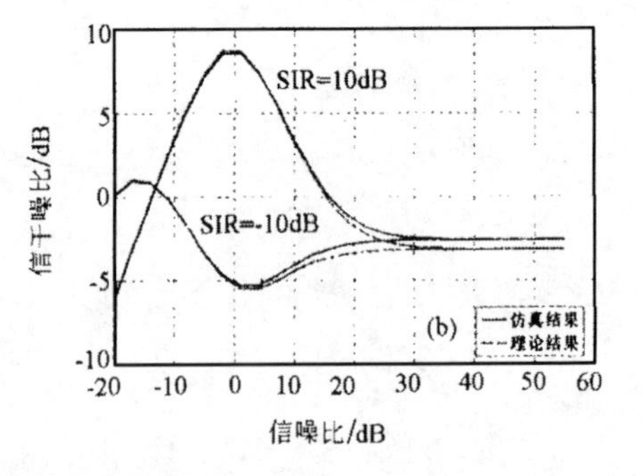

图 3　单干扰情况下干扰强度对 **SINR** 的影响

(a) $\rho = 0$　(b) $\rho = 0.8$

5 结论

本文对矢量水听器阵列 MVDR 波束形成器的滤波性能进行深入研究，给出了有限样本条件下波束形成器输出信干噪比的理论公式，仿真结果验证了理论推导的正确性。通过对三种典型情况的讨论，可以得出以下结论：

(1) 在低信噪比和小干扰时，MVDR 波束形成器是最大输出信噪比准则下的最佳空域滤波器；而在高信噪比时，滤波性能退化，存在对强信号的抑制现象，输出的信干噪比为一取决于阵元数和独立样本数的常数。

(2) 信号和干扰之间的相关性对 MVDR 波束形成器的滤波性能有重大影响，相关性越强，滤波性能越差。

(3) 干扰强度对滤波性能的影响随信噪比的增加而削弱，高信噪比时，干扰被完全抑制。

参 考 文 献

1 Capon J. *Proc. IEEE*, 1969, **57:1408～1418**.

2 Wax M, Anu Y. *IEEE Trans. on Signal Processing*, 1996, **44**(4):928～937.

3 Hawkes M, Nehorai A. *IEEE Trans. on Signal Processing*, 1998, **46**(9):2291～2304.

4 Chen H W, Zhao J W. *IEE Proc.-Radar Sonar Navig*, 2004, **151**(3):158～162.

5 孙贵青，李启虎，蔡惠智．声学技术，2003, 398～401.

6 田坦，齐娜，孙大军．哈尔滨工程大学学报，2004, **25**(3):295～298.

7 庄钊文，肖顺平，王雪松等著．极化敏感阵列信号处理，国防工业出版社，2005, 132～133.

一种改进的 WSF 算法在单矢量水听器多目标方位估计中的应用①

杨秀庭，孙贵青，陈新华，李启虎

（中国科学院声学研究所，北京 100080）

摘要：文章针对单个矢量水听器的多目标方位估计，提出了一种基于加权子空间拟合（WSF）的算法，该算法首先对单个矢量水听器接收数据作一任意的时间延迟，而后仿照 ESPRIT 算法的思路求解阵列响应矩阵，从中抽取各目标的波达方位。该算法在保留 WSF 算法分辨力高、估计方差小的优点的同时，通过结合 ESPRIT 算法的思想，克服了 WSF 算法计算量大，需迭代求解的缺点。由于该算法和声源频率无关，因而可直接应用于宽带声源的测向，并避免了传统 ESPRIT 算法中因估计延时相位而导致的频率-方位模糊问题。文中通过数值仿真和推导 Cramer-Rao 下界，给出了该算法的性能评价，数值仿真和湖试实验结果也充分验证了该算法的有效性。

关键词：矢量水听器；波达方位估计；加权子空间拟合

中图分类号：P731.2　　　文献标识码：A　　　文章编号：1000-3630(2007)-02-0165-04

Multi-source DOA estimation using a modified WSF algorithm in a single vector hydrophone applications

YANG Xiu-ting, SUN Gui-qing, CHENG Xin-hua, LI Qi-hu

(Institute of Acoustics, Chinese Academy of Sciences, Beijing 100080, China)

Abstract: This paper develops a new algorithm based on WSF with respect to estimate the directions of targets in the multi-source scenario using a single vector hydrophone. This method exploits the sources' DOA information embedded in the sources' velocity field by an ESPRIT algorithm and reduces the DOA variances via weighted subspace fitting techniques. It has high resolution as other eigenstructure-based algorithms, moderate computational costs, suffers no frequency-DOA ambiguity, and can be directly applied to wideband source scenarios. The Cramer-Rao lower bound of mean-square-angular-error is deduced to evaluate its performance .Both numerical simulation and lake test results verify the efficacy of this proposed scheme.

Key words: vector hydrophone; DOA estimation; weighted subspace fitting

1 引　言

被动方位估计是水声探测设备的一个传统任务，近年来，利用声场中质点的声振速信息来提高水声探测设备的工作性能，构成了水声研究领域一个新的热点方向[1-8]。矢量水听器由声压标量传感器和三组互相正交的质点振速传感器组成，可以同步、共点地测量声场的声压和质点声振速沿笛卡儿坐标系中三个垂直方向上的分量[6]。在平面波声场中，声压、质点声振速和目标方位之间受欧拉方程约束，因此，通过全面测量声场的声压和声振速信息，单个矢量水听器就可以具备由传统的声压标量水听器所构成的平面阵所具有的二维（水平方位角和俯仰角）测向功能[6,7]。

① 声学技术，2007, 26(2): 165-168.

自美国的DIFAR垂直阵投入使用以来,矢量传感器阵列的信号处理研究迅速发展[8]。D'Spain等[2]使用LCMV波束形成方法对矢量水听器阵列进行了研究。Shchurov等[3]应用矢量水听器阵列测量了环境噪声场。Nehorai和Paldi[4]首次给出了矢量水听器的测量模型,提出了MSAE(均方角误差)的测向性能评估方法和并推导了相应的方差下界。Hochwald和Nehorai[5]研究了单个矢量水听器的多目标辨识能力。Wong和Zoltowski[6]根据矢量水听器所具有的无模糊方位分辨能力,探讨了空间降采样矢量水听器阵列的测向性能。Hawkes和Nehorai[7]应用MVDR方法研究了矢量水听器阵列的方位估计性能,并详细讨论了Cramer-Rao下界。孙贵青[8]深入研究了矢量水听器阵列的常规波束形成和自适应波束形成方法,并给出了试验结果。

本文中提出了一种适用于单个矢量水听器的基于加权子空间拟合(WSF, Weighted Subspace Fitting)[1,9]的二维方位估计算法(ESPRIT-WSF),该算法既避免了单纯使用WSF时的较为繁琐的迭代求解,又可获得比ESPRIT算法更优的估计性能。总体计算量与ESPRIT算法相当,远逊于需二维搜索的MUSIC算法。其方位估计与声源频率无关,避免了基于时延估计的常规子空间算法可能导致的方位模糊问题,该算法可直接应用于宽带声源。此外,该算法还可推广应用于具有任意几何形状的空间降采样矢量水听器阵列[6]。

2 信号模型

设有d个中心频率为f_k, k=1, …, d的窄带平面波,在静态、非色散、各向同性的流体介质中传播。各平面波的方位由水平方位角和俯仰角构成的二维向量确定($\phi\in(0, 2\pi]$, $\vartheta\in[-\pi/2, \pi/2]$),因此,确定目标的方位,实际上可归结为估计DOA参数向量$\theta=[\theta_1^T, \cdots, \theta_d^T]^T$,其中$\theta_k=[\theta_k, \vartheta_k]^T$,为第k个声源的波达方向。假设信号和矢量水听器各通道的噪声均为相互独立的零均值复高斯随机过程,并满足

$$E\{p(t_m)p^H(t_n)\}=\delta_{mn}\mathrm{diag}[\sigma_{S,1}^2, \cdots, \sigma_{S,d}^2] \quad (1)$$

$$E\{e(t_m)e^H(t_n)\}=\delta_{mn}\mathrm{diag}[\sigma_p^2, \cdots, \sigma_v^2 I_3] \quad (2)$$

上式中,$\sigma_{S,k}^2$为矢量水听器所接收的第k个声源的声功率;σ_p^2、σ_v^2分别为矢量水听器声压和振速通道的噪声方差,δ_{mn}为Kronecker delta函数,I_3为3×3的单位阵。

2.1 单矢量水听器的输出

在自由空间中,d个窄带平面波作用在单个矢量水听器上,产生的信号输出为

$$y(t)=h(\theta)p(t)+e(t) \quad (3)$$

其中:

$$h(\theta)=[h(\theta_1), \cdots, h(\theta_d)],\quad h(\theta_k)=[1, u_k^T]^T$$
$$=[1, \cos\phi_k\cos\vartheta_k, \sin\phi_k\cos\vartheta_k, \cos\vartheta_k]^T \quad (4)$$

$$p(t)=[p_1(t), \cdots, p_d(t)]^T,\quad p_k=a_k e^{j(2\pi f_k t+\varphi_k)} \quad (5)$$

上式中,a_k为第k个窄带信号的振幅,ϕ_k为初始相位。

3 DOA估计算法

3.1 输出数据的初步处理

将式(3)的输出作时间为$\Delta\tau$的延迟,即

$$y'(t)=y(t+\Delta\tau)=h(\theta)p(t+\Delta\tau)+e(t+\Delta\tau) \quad (6)$$

根据式(4)、(5)、(6)式可写为

$$y'(t)=h(\theta)\boldsymbol{\Phi}p(t)+e(t+\Delta\tau) \quad (7)$$

其中,$\boldsymbol{\Phi}=\mathrm{diag}[e^{j\omega_1}, \cdots, e^{j\omega_d}]$, $\omega_k=2\pi f_k\Delta\tau$。为方便,将$h(\theta)$记为h。

$$x(t)=\begin{bmatrix} y(t) \\ y'(t) \end{bmatrix}=\begin{bmatrix} h \\ h\boldsymbol{\Phi} \end{bmatrix}p(t)+\begin{bmatrix} e(t) \\ e(t+\Delta\tau) \end{bmatrix}=Ap(t)+n(t) \quad (8)$$

对式(8)的协方差矩阵作特征值分解

$$R=E\{x(t)x^H(t)\}=APA^H+I_2\otimes\begin{bmatrix} \sigma_p^2 & 0 \\ 0 & \sigma_v^2 I_3 \end{bmatrix}$$
$$=U_S\boldsymbol{\Lambda}_S U_S^H+U_N\boldsymbol{\Lambda}_N U_N^H \quad (9)$$

上式中,$\boldsymbol{\Lambda}_S$、$\boldsymbol{\Lambda}_N$、U_S和U_N分别为信号与噪声的特征值和特征向量矩阵,"⊗"为Kronecker积。由式(8)可推得,$U_S=AT$, T为一非奇异矩阵。令$U_S=[U_{S1}^T, U_{S2}^T]^T$, U_{S1}、U_{S2}均为4×d阶矩阵,则两者之间存在以下关系

$$U_{S1}=hT \quad U_{S2}=h\boldsymbol{\Phi}T \quad (10)$$

令$\boldsymbol{\Psi}=T^{-1}\boldsymbol{\Phi}T$,则从式(10)中可得下式

$$U_{S1}\boldsymbol{\Psi}=U_{S2} \quad (11)$$

由于$\boldsymbol{\Psi}$和$\boldsymbol{\Phi}$相似,故二者具有相同的特征值。

3.2 ESPRIT-WSF算法

WSF属于子空间算法的一种,从参数的最大似然估计(ML)衍生而来,具有很高的分辨力,但其计算量很大,需求解某一特定矩阵迹函数的极值来实现目标的波达方向估计,通常采用牛顿类型算法迭代求解。但这类求极值算法对初值较为敏感,且容易收敛至局部极值点。ESPRIT算法也是一种高分辨

力的子空间方法，它是把阵列分解为两个完全相同的子阵，通过求解两子阵间的过渡矩阵，从中获得目标的方位估计。由于式(8)模型中的旋转矩阵并不包含目标的方位信息，因而无法应用常规的ESPRIT算法。本文中所提出的ESPRIT-WSF算法，充分利用了矢量水听器数据测量的特点(目标方位含于声振速信息中)，采用ESPRIT算法的思想，通过求解阵列响应矩阵来估计目标方位，并应用WSF方法改善了算法的估计性能。

3.2.1 $\boldsymbol{\Psi}$ 的求解

式(11)中给出了 $\boldsymbol{\Psi}$ 与信号特征向量矩阵 U_{S1}、U_{S2} 之间的关系，由ESPRIT算法可知，其总体最小二乘解如下

$$\boldsymbol{\Psi}_{TLS}=-V_{12}V^{-1}_{22} \tag{12}$$

其中，V_{12}、V_{22} 由下式给出

$$\begin{bmatrix} W^{1/2}U^{H}_{S1} \\ W^{1/2}U^{H}_{S2} \end{bmatrix}[U_{S1}W^{1/2}\ \ U_{S2}W^{1/2}]=\begin{bmatrix} V_{11} & V_{12} \\ V_{21} & V_{22} \end{bmatrix}L\begin{bmatrix} V^{H}_{11} & V^{H}_{12} \\ V^{H}_{21} & V^{H}_{22} \end{bmatrix} \tag{13}$$

式(12)中，$L=\mathrm{diag}\{\lambda_1,\cdots,\lambda_{2d}\}$，$\lambda_1\geqslant\lambda_2\geqslant\cdots\geqslant\lambda_{2d}$；加权矩阵W为一正定Hermite矩阵，取为[1]

$$W=(\Lambda_S-\overline{\sigma^2}I)^2\Lambda^{-1}_S \tag{14}$$

上式中，$\overline{\sigma^2}$ 为噪声的平均方差，由式(9)，可得

$$\overline{\sigma^2}=\frac{1}{8-d}\sum_{i=d+1}^{8}\Lambda_N(i,i) \tag{15}$$

同理，由式(11)也可直接得到 $\boldsymbol{\Psi}$ 的最小二乘解

$$\boldsymbol{\Psi}_{LS}=(W^{1/2}U^{H}_{S1}U_{S1}W^{1/2})^{-1}W^{1/2}U^{H}_{S1}U_{S2}W^{1/2} \tag{16}$$

3.2.2 非奇异矩阵T的求解

对 $\boldsymbol{\Psi}$ 进行特征值分解，得到

$$\boldsymbol{\Psi}=U\boldsymbol{\Phi}U^{-1}=T^{-1}\boldsymbol{\Phi}T \tag{17}$$

观察上式最右边等号的两侧，容易得到

$$T=aU^{-1} \tag{18}$$

其中a为一无关紧要的常数。

3.2.3 二维波达方向估计

根据式(10)，可得以下关系式

$$h(\theta)=\frac{1}{2a}(U_{S1}U+U_{S2}U\boldsymbol{\Phi}^{-1}) \tag{19}$$

结合式(2)，可得各目标的波达方位(水平方位角和俯仰角)

$$\phi_k=\tan^{-1}(\bar{h}_{k,3}/\bar{h}_{k,2})$$

$$\vartheta_k=\sin^{-1}\bar{h}_{k,4}=\cos^{-1}\left(\sqrt{\bar{h}^2_{k,2}+\bar{h}^2_{k,3}}\right) \tag{20}$$

式(20)中，$\bar{h}_k=h_k/\|h_k\|$，$h_{k,i}$为$\bar{h}(\theta_k)$向量的第i个分量。

4 估计方差的最优下界

以上已经给出了使用单个矢量水听器进行多目标方位估计的算法。由于该算法需要估计阵列响应矩阵的所有元素，因而无法给出其渐近性能的简明公式。为此，本文通过方位估计的均方角误差(MSAE[4])的最优下界与仿真结果之间的对比，给出该算法性能的直观评价。当各信号之间互不相关时，单矢量水听器MSAE的最优下界为

$$MSAE^{CR}_k=CRB(\phi_k)\cos^2\vartheta_k+CRB(\vartheta_k)=\frac{1+2\rho_k}{4N\rho^2_k} \tag{21}$$

上式中，$\rho_k=\sigma^2_{S,K}/\sigma^2_v$ 为第k个声源所对应的功率，N为的独立信号样本数。

5 数值仿真

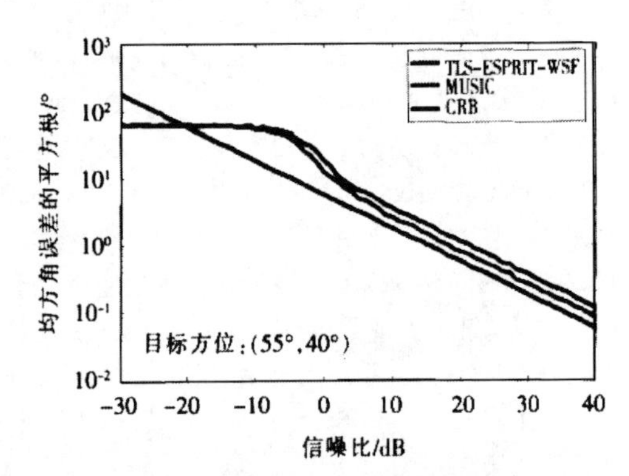

图1 单目标时TLS-ESPRIT-WSF算法的性能

Fig.1 Performance of TLS-ESPRIT-WSF in single source scenario

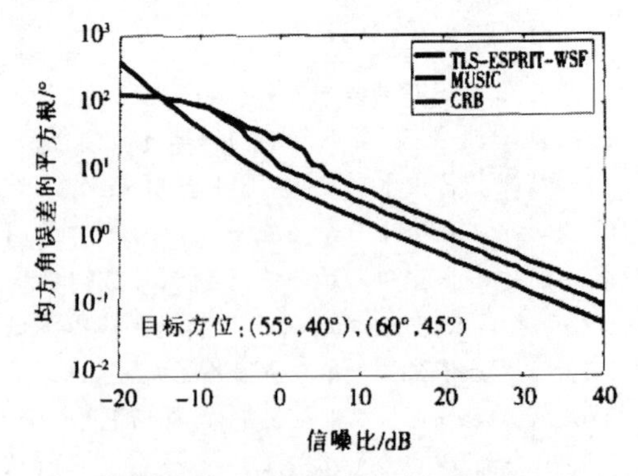

图2 多目标时TLS-ESPRIT-WSF算法的性能

Fig.2 Performance of TLS-ESPRIT-WSF in multi-source scenario

为检验本文算法的性能，仿真中考虑了两个窄带信源，其二维方位分别为：(55°,40°)，(60°,45°)，中

心频率分别为 500Hz 和 750Hz，且信号强度相等；系统采样频率为 4 000Hz, 快拍数为 200, 两组数据之间的时延差 $\Delta\tau$=0.0125s; Monte-Carlo 模拟次数为 1 000。

图 1 为单目标(目标方位：(55°, 40°))时MUSIC、本文算法的估计性能与 Cramer-Rao 界的比较。可以看出，本文算法和 MUSIC 算法均能提供目标方位的渐近无偏估计，但其估计的性能稍逊于 MUSIC。图 2为两目标信源((55°, 40°), (60°, 45°))时 TLS-ESPRIT 和 TLS-ESPRIT-WSF 两种算法的性能对比。从中可以看出，本文算法的性能要优于常规的 TLS-ESPRIT 算法。

6 湖试验证

为进一步检验算法的实用性，本文对单个矢量水听器的湖试试验数据进行了处理。信号为频率为 1 000Hz 的单频信号，信噪比 23.6dB, 测量时间为 48 秒。图 3、4 分别为 MUSIC 和本文算法的测向结果，从两图中可见，这两种算法结果基本一致。

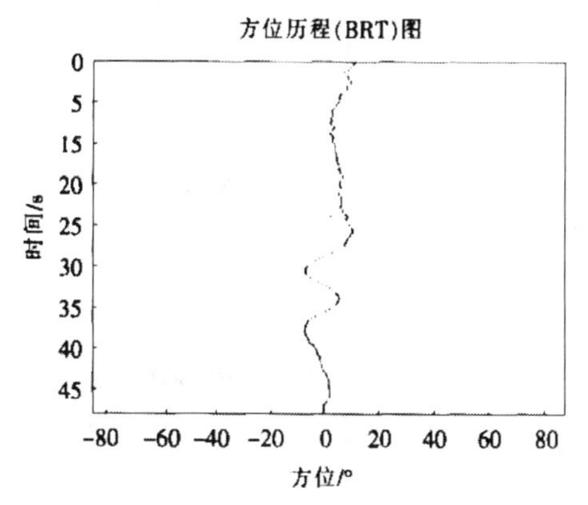

图 3 MUSIC 算法的方位历程图

Fig.3 Waterfall of MUSIC algorithm

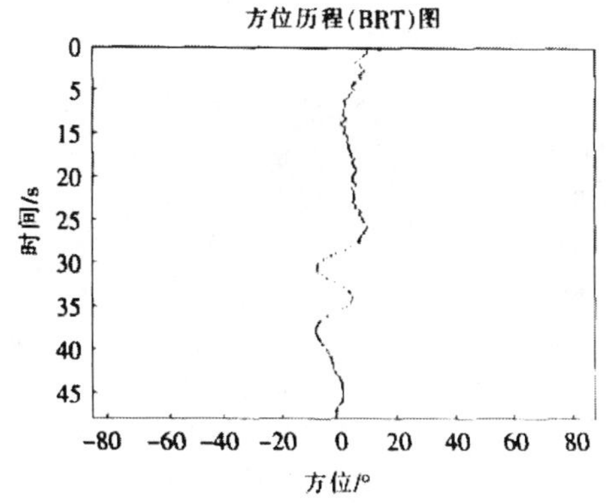

图 4 TLS -ESPRIT-WSF 算法的方位历程图

Fig.4 Waterfall of TLS-ESPRIT-WSF algorithm

7 结 论

本文提出了一种基于单个矢量水听器的多目标二维波达方位估计的新算法，该算法充分利用了矢量水听器的工作特点，通过求解阵列响应矩阵，来间接估计目标的方位。该算法属于信号子空间算法的一种，能提供真实方位的渐近无偏估计，具有很高的分辨力和较小的计算量，并与目标信号的频率无关，同时适用于窄带和宽带声源。数值仿真和湖试实验结果表明，该算法在性能上接近 MUSIC 算法，并具有比文献[6]中算法更小的估计方差。

参 考 文 献

[1] Viberg M, Ottersten B, Sensor srray processing based on subspace fitting[J]. IEEE Transactions on signal processing, 1991, 39(4): 1110-1121.

[2] G L D′ Spain et al. Initial analysis of the data from the vertical DIFAR array[A]. Proc. IEEE Oceans Conf[C]. 1992: 346-351.

[3] A. Shchurov et al. Ambient noise energy motion in the near surface layer in ocean wave-guide[J]. J. phys., pt. 2, 1994, 4(5): 1273-1276.

[4] Nehorai A, Paldi E. Acoustic vector-sensor array processing[J]. IEEE Transactions on signal processing, 1994, 42(9): 2491.

[5] Hochwald B, NehoraiA. Identifiability in array processing models with vector sensor applications[J]. IEEE Transactions on signal processing, 1996, 44(1): 83-95.

[6] WONG K T, Zoltowski M D. Extended-aperture underwater acoustic multisource azimuth/elevation direction-finding using uniformly nut sparsely spaced vector hydrophones[J]. IEEE Journal of Oceanic Engineering, 1997, 22(4): 659-672.

[7] Hawkes M, Nehorai A. Acoustic vector-sensor beamforming and capon direction estimation[J]. IEEE Transactions on signal processing, 1998, 46(9): 2291-2304.

[8] 孙贵青，李启虎. 声矢量传感器研究进展[J]. 声学学报, 2004, 29(6): 491-498.
SUN Guiqing, LI Qihu. Advances in acoustic vector sensor research[J]. ACTA of Acoustics, 2004, 29(6): 491-498.

[9] 王永良. 空间谱估计理论与算法[M]. 清华大学出版社. 2004. 152-154.
WANG Yongliang. Theory and algorithms of Spatial Spectrum Estimation[M]. Tsinghua University Press, 2004. 152-154.

矢量拖曳式线列阵声呐流噪声影响初探①

杨秀庭[1,2]，孙贵青[1]，李 敏[1]，李启虎[1]
(1. 中国科学院声学研究所，北京 100080；2. 海军大连舰艇学院，大连 116018)

摘要：研究矢量拖曳式线列阵声呐护套管内的流噪声场。根据湍流边界层脉动压力谱的 Corcos 模型，采用频率-波数域上的二维谱方法，分析了流噪声对矢量拖曳式线列阵声呐的影响。给出了护套管内流噪声场声压和质点声振速的一般表达式，并由波数积分法得到护套管内点接收器、有限几何尺寸水听器和水听器阵列输出的流噪声各分量的功率谱。数值结果表明，矢量拖曳式线列阵声呐对流噪声的轴向分量十分敏感，在流噪声控制背景下，矢量阵的工作性能将退化为常规阵。

关键词：矢量水听器；流噪声；拖曳式线列阵
中图分类号：O427.5　　文献标识码：A　　文章编号：1000-3630(2007)-05-0775-06

Primary investigations of the flow-induced noise in the application of vector hydrophone towed linear array

YANG Xiu-ting[1,2], SUN Gui-qing[1], LI Min[1], LI Qi-hu[1]
(1. Institute of Acoustics, the Chinese Academy of Sciences, Beijing 100080, China
2. PLA Naval Academy, Dalian 116018, China)

Abstract: Flow noise in the interior fluid of a vector hydrophone towed linear array is analyzed. Following Corcos's model which describes the cross spectrum of the turbulent boundary layer pressure fluctuations, a method in the frequency-wave number field spectrum is applied to analyze the influence of flow noise. General mathematic forms of acoustic pressure and particle velocity in the flow noise field are given with regard to the excitation of the cable jacket. Furthermore, power spectrum of flow noise is calculated for point hydrophone, finite hydrophone and hydrophone array. Numerical results show that performance of the vector hydrophone towed linear array is very sensitive to the axial component of flow noise, and it will degrade to the conventional array in a flow noise controlled background.

Key words: vector hydrophone; flow noise; towed linear array

1 引 言

长期以来，拖曳式线列阵声呐在水下反潜战中一直扮演着重要角色[1]。在拖曳阵工作时，由于拖缆和海水间的相对运动，在拖缆护套的外表面附近存在海水介质微团的湍流脉动，由此激发出的水动力噪声，一般称之为流致噪声，或简称为流噪声。流噪声的强度是影响拖曳式声呐实际工作性能的重要因素。

为探索流噪声的产生机理，以便采取有效的降噪措施，相关的研究工作一直受到水声界的关注。Sung H.Ko 等[2]研究了镶体阵湍流边界层的流噪声影响，给出了 Corcos 湍流模型[3]下湍流脉动引起的流噪声的数值计算结果，并分析了外裹橡胶层的降噪效果；Andrew Knight[4]研究了常规拖线阵的流噪声降噪措施，通过设计波数域上的低通滤波器，以更好地滤除流噪声。Jasong I. Gobat[5]对声呐浮标中矢量水听器的流噪声进行了测量。Gerald C.Lauchle 等[6]研究了球形矢量水听器在流噪声作用下的加速度谱，给出了不同雷诺数下加速度谱的经验公式，并指出小尺寸矢量水听器受流噪声的影响更大。

① 声学技术，2007, 26(5): 775-780.

汤渭霖等[7]研究了常规拖曳式线列阵声呐流噪声的产生机理，指出：护套管内的流噪声主要来自于湍流脉动压力的直接传递和拖缆护套对脉动压力的共振辐射。栾桂冬等[8]通过把拖缆的声学段看成多层同轴圆管系统，分析了护套内部的流噪声场。葛辉良等[9]利用20元密排线阵对流噪声进行了实验研究，并根据实验数据拟合了湍流脉动压力谱的归一化迁移速度。

随着矢量水听器在水声工程应用中的逐步推广，基于矢量水听器的拖曳式线列阵声呐已成为现实的研究方向，但到目前为止，矢量拖曳阵的流噪声研究尚属空白。本文在文献[7]的基础上，采用频率-波数域上的二维谱方法分析了拖缆护套管内的流噪声场，利用波数积分法计算了护套管内流噪声场中点接收器、有限几何尺寸水听器和水听器阵列所接收流噪声的功率谱，并分析了流噪声对矢量拖曳式均匀线列阵声呐目标检测和左/右舷分辨性能的影响。

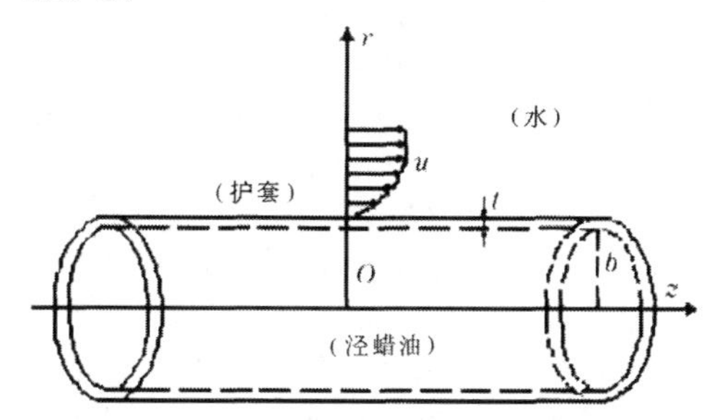

图1 拖曳式线列阵声呐拖缆护套的结构示意图

Fig. 1 The geometry of towed linear array cable jacket

2　流噪声的理论模型

2.1　假　设

在矢量拖曳式线列阵声呐中，矢量水听器安装在由软护套(为聚氨酯材料[10])包裹的拖缆内。为降低不同材料界面处声波的反射损失，以保证阻抗匹配，在护套内一般还需灌注轻蜡油。当拖缆在海水中拖曳时，海水和拖缆间的相对运动会产生流噪声，按其机理，可分为两类[10]：一类是拖缆外表面处湍流边界层中的脉动压力，主要由湍流的微尺度运动及表面的涡脱落(与表面粗糙度有关)产生，共有两种作用模式，一是经护套的作用直接传递到内部流体，二是经护套的耦合激励，产生再辐射；另一类是由拖船尾流、涡流等不稳定性所引起的拖缆和尾缆的振动，这两种振动即使经过隔震模块衰减，仍可引起线阵的振动，并由此可产生三种传播形式的波：一是在液体介质中传播的呼吸波；二是在护套壳内传播的扩展波；三是结构共振模态产生的再辐射。大量的实验表明，在低频带，湍流边界层脉动压力是流噪声的主要来源[7]，在高速拖曳时尤为如此。

如图1所示，当湍流边界层中的脉动压力作用在护套外壁时，所引起的振动将在护套管及内外的流体中传播。为探求流噪声的物理本质，并能获得声场的简明描述，需作以下假设：(1)边界层内的湍流充分发展，为空间均匀，时间平稳的随机过程；(2)拖缆长度为无穷大，两端影响可以忽略，声场关于护套管是柱对称的；(3)护套管为线性系统，满足线性叠加原理。

对于一般的拖曳式线列阵声呐，上述假设可近似满足。

2.2　拖缆护套管内的流噪声场

由于湍流边界层的作用，护套外壁面处会产生相应的脉动压力，利用傅立叶变换，可将其展为频率-波数的二维谱函数[2]，即

$$p(a,z,t)=\left(\frac{1}{2\pi}\right)^2\int\int_{-\infty}^{+\infty}s(k_z,\omega)e^{j(k_z z-\omega t)}d\omega dk_z \quad (1)$$

其中，a为护套外径，$p(a,z,t)$为外壁处的脉动压力，$s(k_z,\omega)$为相应的频率-波数谱，k_z为z轴方向的波数。

考虑到护套和内外流体边界上应力、应变的连续性条件，由文献[11]，可得壁面压力波$s(k_z,\omega)\cdot e^{j(k_z-\omega t)}$在护套管内流体中所激发声场的位移势为

$$\varnothing_1(r,k_z,\omega)=G(k_z,\omega)J_0(k_1\omega)s(k_z,\omega)e^{j(k_z z-\omega t)},\ r\leqslant b \quad (2)$$

其中，$G(k_z,\omega)$为一与护套几何尺寸和护套及内外流体物性有关的函数[11]。

由位移势函数与声压、质点振速之间的关系，利用傅氏积分，得

$$p(r,z,t)=\left(\frac{1}{2\pi}\right)^2\int\int_{-\infty}^{+\infty}\rho_1\omega^2G(k_z,\omega)J_0(k_1\omega)s(k_z,\omega)e^{i(k_z z-\omega t)}dk_z d\omega \quad (3)$$

$$u(r,z,t)=\left(\frac{1}{2\pi}\right)^2\int\int_{-\infty}^{+\infty}\omega G(k_z,\omega)[jk_1J_1(k_1r)e_r+k_zJ_0(k_1r)e_z]s(k_z,\omega)e^{j(k_z z-\omega t)}dk_z d\omega \quad (4)$$

式中，e_r和e_z分别为径向和轴向的单位向量。

2.3　护套外壁处的湍流脉动压力谱

湍流边界层压力起伏产生的噪声，具有一定的随机特征，可根据傅立叶变换，展成频率-波数域上的二维谱函数。Corcos研究了平面湍流边界层的压

力起伏现象，并给出了半经验公式[3]

$$P_w(k_z,\omega)=\frac{1}{\pi}\frac{\alpha k_c}{\alpha^2 k_c^2+(k_z-k_c)^2}P(\omega) \tag{5}$$

其中，$P(\omega)$为脉动压力的自谱；α为模型常数，与壁面的光滑程度[7]有关，光滑表面$\alpha=0.116$，粗糙表面$\alpha=0.32$；$k_c=\omega/u_c$为迁移波波数，迁移速度u_c与拖曳速度U相适应，一般取为$u_c=\beta U$的形式，β为频率、雷诺数的函数。

由式(3)～(5)，可以给出护套内部流体中流噪声的空间分布规律

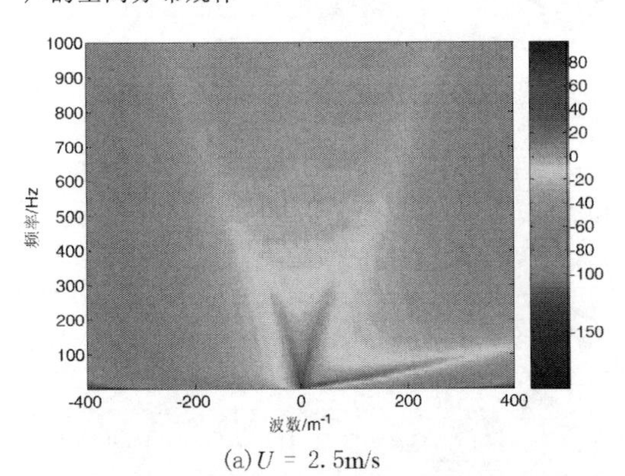

(a) U = 2.5m/s

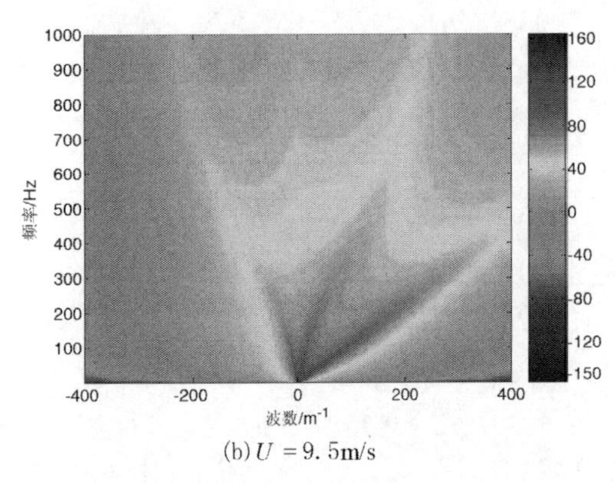

(b) U = 9.5m/s

图2 不同拖曳速度时护套内壁处流噪声声压的二维谱

Fig.2 The two-dimensional spectrum of the acoustic pressure flow noise at the interior wall of the cable jacket at different towed speed

3 数值计算

采用波数积分法计算护套管内流体中的流噪声功率谱，接收点半径为r=20.35mm。护套及内外流体介质的物性和计算参数由表1给出（水听器的结构参数取自美国Wilcoxon公司的TV-001型三维矢量水听器）。

表1 计算参数

Table 1 Baseline data used in the present study

计算参数	点接收器	有限水听器
护套壁厚	3mm	…
护套内径	30mm	…
护套密度	1200kg/m^3	…
护套衰减因子	$\gamma_c=0.03$	…
	$\gamma_c=0.3$	…
护套压缩波波速	2239.3m/s	…
护套切变波波速	104m/s	…
水听器尺寸	d=0	d=40.7mm
	l=0	l=71.3mm
内外流体中的声速	1 500m/s	…
拖曳速度U	2.5, 9.5m/s	…
迁移速度u_c	0.75U	…

图2为护套内壁处由Corcos模型给出的流噪声声压的二维谱。由图可知，经粘弹性护套作用后，流噪声的能量主要集中在低波数区和迁移峰附近。

3.1 点接收器的流噪声功率谱

根据式(3)、(4)，图3给出了不同拖曳速度下流噪声声压和质点振速的功率谱。由于矢量水听器在使用时需进行灵敏度校准(对于平面波，根据欧拉方程对质点振速进行归一化，即$v=p/\rho c$)，此时质点振速的量纲已被校准为与声压一致。从图中可以看出，经声阻抗归一化后，流噪声的振速分量显然被大大放大了，在低于200Hz的频段上，流噪声质点振速的谱级比声压谱级约高30dB，这说明矢量水听器对低频流噪声是十分敏感的。此外，拖曳速度对流噪声的影响是非线性的，图中(a)、(b)两种拖曳速度的噪声级相差约30dB。

3.2 有限几何尺寸矢量水听器的流噪声功率谱

无体积的点接收器属于理想假设，实际的水听器总是具有一定几何尺寸的。因此，矢量水听器各通道输出的噪声为某个接收灵敏面上噪声的平均值。推广文献[12]的结果，得到单个矢量水听器输出的流噪声的声压和质点振速为

$$p(r,z,t)=\left(\frac{1}{2\pi}\right)^2\int\int_{-\infty}^{+\infty}h_{p,a}(r,k_z,\omega)s(k_z,\omega)e^{i(k_z z-\omega t)}dk_z d\omega \tag{6}$$

$$u(r,z,t)=\left(\frac{1}{2\pi}\right)^2\int\int_{-\infty}^{+\infty}h_{r,a}(r,k_z,\omega)s(k_z,\omega)e^{j(k_z z-\omega t)}dk_z d\omega \tag{7}$$

$$u_z(r,z,t)=\left(\frac{1}{2\pi}\right)^2\int\int_{-\infty}^{+\infty}h_{z,a}(r,k_z,\omega)s(k_z,\omega)e^{j(k_z z-\omega t)}dk_z d\omega \tag{8}$$

式中，$h_{p,a}(r,k_z,\omega)$、$h_{r,a}(r,k_z,\omega)$和$h_{z,a}(r,k_z,\omega)$分别

为考虑矢量水听器实际接收表面时流噪声场各分量的等效传递函数，定义为

$$h_{p,a}(r,k_z,\omega)=h_p(r,k_z,\omega)a_p(k_z) \tag{9}$$

$$h_{r,a}(r,k_z,\omega)=h_r(r,k_z,\omega)a_{u_r}(k_z) \tag{10}$$

$$h_{z,a}(r,k_z,\omega)=T_z(k_z,\omega)\frac{2J_1(k_1r)}{k_1rJ_0(k_1b)}a_{u_z}(k_z) \tag{11}$$

其中，$a_p(k_z)$、$a_{u_r}(k_z)$ 和 $a_{u_z}(k_z)$ 是与矢量水听器的接收面有关，称为水听器函数，为

$$a_p(k_z)=a_{u_r}(k_z)=\frac{\sin(k_zl/2)}{k_zl/2} \tag{12}$$

$$a_{u_z}(k_z)=\cos(k_zl/2) \tag{13}$$

$h_p(r,k_z,\omega)$、$h_r(r,k_z,\omega)$ 分别为[11]

$$h_p(r,k_z,\omega)=\rho_1\omega^2G(k_z,\omega)J_0(k_1r);$$

$$h_r(r,k_z,\omega)=j\omega k_1G(k_z,\omega)J_1(k_1r) \tag{14}$$

图 4 为不同拖曳速度下有限矢量水听器接收的流噪声功率谱。在波数域上，水听器函数是一个低通滤波器，可更好地抑制流噪声。与图 3 相比，流噪声各分量的谱级下降 5dB–10dB 不等。

3.3 矢量拖曳式均匀线列阵的流噪声功率谱

由多个矢量水听器组成的基阵，可进一步抑制流噪声的影响。对于 M 元矢量均匀拖线阵，输出流噪声的各分量为

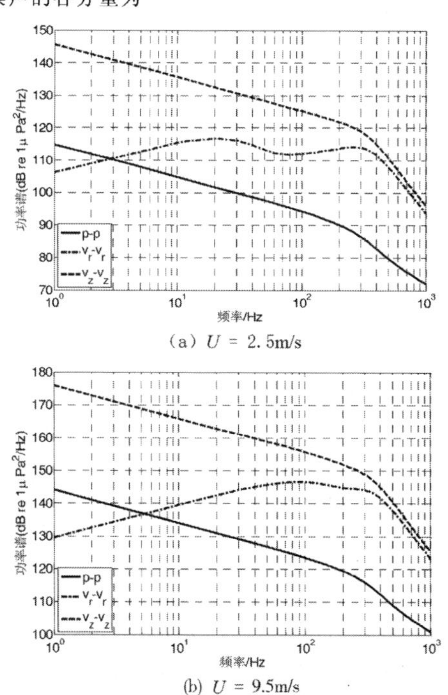

(a) U = 2.5m/s

(b) U = 9.5m/s

图 3 护套内部点接收器的流噪声功率谱

Fig. 3 The power spectrum of the flow noise received by point vector hydrophone in the interior fluid

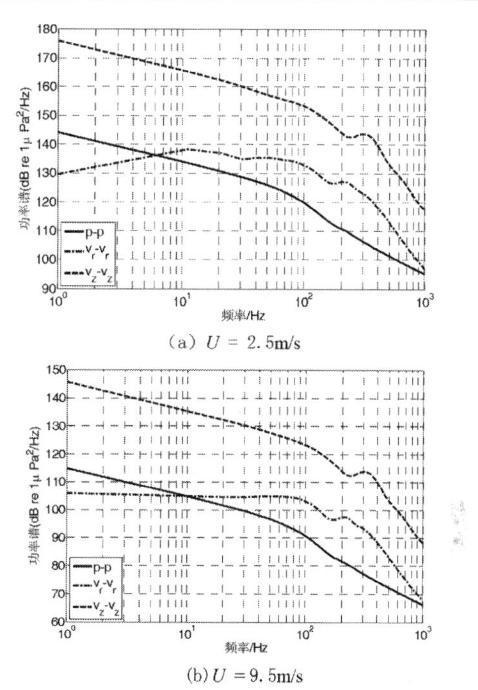

(a) U = 2.5m/s

(b) U = 9.5m/s

图4 护套内部有限几何尺寸矢量水听器的流噪声自功率谱

Fig. 4 The power spectrum of the flow noise received by finite hydrophone in the interior fluid

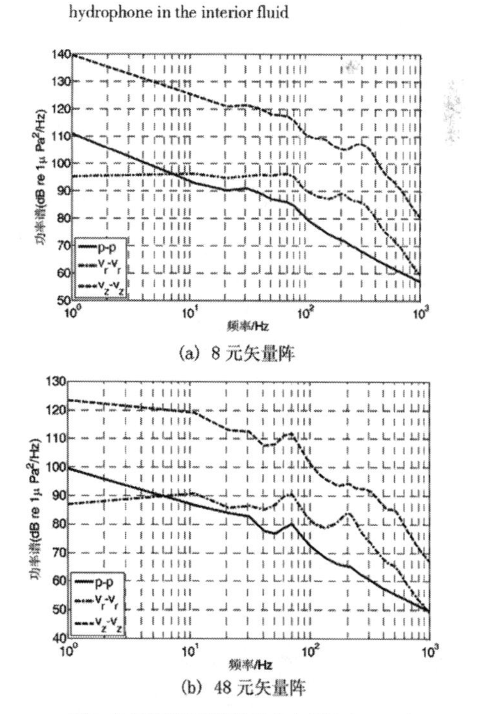

(a) 8 元矢量阵

(b) 48 元矢量阵

图5 矢量拖线阵的流噪声自功率谱（2.5m/s）

Fig. 5 The power spectrum of the flow noise received by the vector towed linear array with different elements

$$p(r,z,t)=\left(\frac{1}{2\pi}\right)^2\int\int_{-\infty}^{+\infty}h_{p,A}(r,k_z,\omega)s(k_z,\omega)e^{j(k_z z-\omega t)}dk_z d\omega \tag{15}$$

$$u_r(r,z,t)=\left(\frac{1}{2\pi}\right)^2\int\int_{-\infty}^{+\infty}h_{r,A}(r,k_z,\omega)s(k_z,\omega)e^{j(k_z z-\omega t)}dk_z d\omega \tag{16}$$

$$u_z(r,z,t)=\left(\frac{1}{2\pi}\right)^2\int\int_{-\infty}^{+\infty}h_{z,A}(r,k_z,\omega)s(k_z,\omega)e^{j(k_z z-\omega t)}dk_z d\omega \tag{17}$$

上式中，各项定义为

$$h_{p,A}(r,k_z,\omega)=h_{p,a}(r,k_z,\omega)A(k_z) \tag{18}$$

$$h_{r,A}(r,k_z,\omega)=h_{r,a}(r,k_z,\omega)A(k_z) \tag{19}$$

$$h_{z,A}(r,k_z,\omega)=h_{z,a}(r,k_z,\omega)A(k_z) \tag{20}$$

$$A(k_z)=\frac{\sin(Mk_z l/2)}{M\sin(k_z l/2)} \tag{21}$$

其中，$A(k_z)$称为基阵函数。与水听器函数类似，基阵函数$A(k_z)$在波数域上也是一个低通滤波器，因此，可进一步抑制流噪声的高波数分量。

图5给出了不同阵元数时流噪声的功率谱。对比图3、4可知，采用阵结构，可带来10dB～20dB的降噪好处。

4 流噪声对矢量拖曳式均匀线列阵声呐工作性能的影响

拖曳阵工作时的背景噪声很复杂，包括环境噪声，拖船噪声(含各种振动)，水听器自噪声和流噪声等，在不同的海区、海况和拖曳速度下，其工作性能会有很大差异。从前面的计算结果可以看出，矢量拖曳式线列阵声呐流噪声的强度与拖曳速度密切相关，不同拖曳速度对应的流噪声强度显著不同，因此，在实际应用中需评估流噪声对矢量拖曳式均匀线列阵声呐工作性能的影响。

J.Groen 等人[1]根据常规三元组拖曳式线列阵声呐的海试结果，给出了中心工作频率为1 500Hz，带宽500Hz情况下流噪声与环境噪声（声压分量）的相对强度，在5节航速下，该值约为-10dB。三元组单线阵在单根拖缆内集成了三条相互平行的常规线阵，各水听器距护套内壁更近，因此所接收的流噪声也更强。单个矢量水听器集成了一个声压传感器和三个振速传感器，在护套内的布放与常规单线阵中的水听器相似，在当前的制作工艺水平下，可获得比三元组强度更小的流噪声(声压分量)。由图5，在400Hz以上的频带上，流噪声声压和质点振速轴向分量的谱级之差小于18dB。为方便计，令

$$\sigma_{f,v_z}^2=\lambda\sigma_{a,p}^2 \tag{22}$$

其中，σ_{f,v_z}^2为流噪声质点振速轴向分量的谱级，$\sigma_{a,p}^2$为海洋环境噪声声压分量的谱级，λ为比例系数。

根据护套管内流噪声场的特点，与质点振速轴向分量的自谱相比，其它自谱和互谱分量可忽略不计。因此，在各向同性的海洋环境噪声场中，单个矢量水听器由$\sigma_{a,p}^2$归一化的噪声互谱密度矩阵为

$$R_n\approx\sigma_{a,p}^2\begin{bmatrix}1&0&0&0\\0&1/3+\lambda&0&0\\0&0&1/3&0\\0&0&0&1/3\end{bmatrix} \tag{23}$$

因此，当目标的二维方位为(ϕ,θ)时(ϕ为水平方位角，θ为俯仰角)，单个矢量水听器的最大空间增益为[1]

$$G_{max}=4-\frac{9\lambda}{3\lambda+1}\cos^2\theta\cos^2\phi \tag{24}$$

此时，单个矢量水听器的检测性能与流噪声强度及目标的方位有关，增益在0dB–6dB之间。

再看矢量拖曳式线列阵声呐的目标左/右舷分辨性能。以Cardioid方法的CBF和ABF为例，这两种波束形成器的左/右舷抑制比为[1]

$$r_{CBF,Cardioid}=\frac{4M\cdot SNR+4/3+\lambda\cos^2\theta\cos^2\phi}{M(1+\cos\theta\cos^2\phi+\sin^2\theta)\cdot SNR+4/3+\lambda\cos^2\theta\cos^2\phi} \tag{25}$$

$$r_{ABF,Cardioid}=1+M\cdot SNR\left(G_{max}-\frac{G^2}{G_{max}}\right) \tag{26}$$

式中，SNR为接收端声压分量的信噪比；G_{max}如式(24)所定义，G为

$$G=1+3\left(\frac{\cos^2\phi}{3\lambda+1}-\sin^2\phi\right)\cos^2\theta+3\sin^2\theta \tag{27}$$

由式(24)~(26)可知，当矢量阵处于高速拖曳状态时，流噪声的强度远大于环境噪声($\lambda>>1$)，目标的质点振速信息完全淹没在流噪声中，此时矢量阵的目标检测和左/右舷分辨性能会严重下降，趋近于常规阵。对于ABF，矢量阵最多可获得4.8dB的左/右舷抑制比(在正横方向)。

5 结 论

本文采用波数积分法计算了矢量拖曳式线列阵声呐的流噪声强度。从数值结果可以看出，矢量拖线阵对流噪声较为敏感，矢量水听器经灵敏度校准后，流噪声振速的谱级要显著高于声压谱级，因此，流噪声的抑制对矢量阵尤为重要，当流噪声的振速

分量较强时，矢量阵的检测性能和左/右舷分辨性能均受到影响，极端情况下矢量阵的性能优势会完全丧失，此时退化为常规阵(受流噪声影响，振速信息不可用)。单个水听器及其成阵后，对流噪声均能进行更大程度的抑制，因为此时的水听器函数和基阵函数都是波数域上的低通滤波器，高波数区的流噪声能量可被进一步滤除。

参　考　文　献

[1] Groen J. Adaptive port/starboard beamforming of triplet sonar arrays[J]. IEEE Journal of Oceanic Engineering, 2005, 30(2): 348-359.

[2] Ko S H, Nuttall A. Analytical evaluation of flush-mounted hydrophone array response to the corcos turbulent wall pressure spectrum[J]. JASA, 1991, 90(1): 579-588.

[3] Corcos G M. The structure of the turbulent pressure field in boundary layer flows[J]. J. Fluid Mech, 1964, 18: 353-378.

[4] Andrew Knight. Flow noise calculations for extended hydrophones in fluid- and solid-filled towed arrays[J]. J A S A, 1996, 100(1): 245-251.

[5] Gobat J I, Grosenbaugh M A. Modelling the mechanical and flow-induced noise on the surface suspended acoustic receiver[J]. OCEANS'97, 2: 748-754.

[6] Lauchle G C. Flow-induced noise on underwater pressure-vector acoustic sensors[J]. OCEANS'02, 3: 1906-1910.

[7] 汤渭霖，吴一. TBL压力起伏激励下粘弹性圆柱壳内的噪声场：Ⅰ噪声产生机理[J]. 声学学报，1997，22(1)：60-69.
TANG Weilin, WU Yi. Interior noise field of a viscoelastic cylindrical shell excited by TBL pressure fluctuations: Ⅰ. Production mechanism of the flow noise[J]. Acta Acoustica, 1997, 22(1): 60-69.

[8] 栾桂冬. 入射声波和边界层压力起伏下同轴多层圆管内场的解. 工程声学[M]. 北京大学出版社，1996: 184-189.
LUAN Guidong. The solutions of the sound field in the interior fluid of the coaxial multilayer cylindrical shells excited by the incident sound wave and the pressure fluctuations of the turbulent boundary layer. Engeering Acoustics[J]. Press of the Peking University, 1996: 184-189.

[9] 葛辉良. 利用线阵估计湍流边界层起伏压力模型参数[J]. 哈尔滨工程大学学报，2003，24(6)：600-603.
GE Huiliang. Estimation of the parameters of the turbulent boundary layer fluctuation pressure model by the use of a linear array[J]. Journal of Marine Science and Application, 2003, 24(6): 600-603.

[10] 张道礼. 拖曳线列阵声呐护套用聚氨酯的声学性能及材料体系[J]. 功能材料，2002，33(2)：145-147.
ZHANG Daoli. Acoustic properties and material systems of polyurethane for use as towed linear sonar jacket[J]. Journal of Functional Materials, 2002, 33(2): 145-147.

[11] 杨秀庭. 矢量水听器及其在声呐系统中的应用研究[D]. 中科院声学研究所，2006，6: 49-73.
YANG Xiuting. The vector hydrophone and its applications to the sonar systems[D]. The Institute of Acoustics, Chinese Academy of Science, Bei-jing, 2006. 49-73.

[12] 吴一，汤渭霖. TBL压力起伏激励下粘弹性圆柱壳内的噪声场：Ⅱ有限水听器和阵[J]. 声学学报，1997，22(1)：70-78.
WU Yi, TANG Weilin. Interior noise field of a viscoelastic cylindrical shell excited by TBL pressure fluctuations: Ⅱ. Finite hydrophone and array[J]. Acta Acoustica, 1997, 22(1): 70-78.

The performance of port/starboard beamforming using vector hydrophone linear arrays①

LI Min[1], YANG Xiu-ting[2], LI Qi-hu[1]

(1. *Institute of Acoustics*, *The Chinese Academy of Sciences*, *Beijing*, 100080, *China*;
2. *PLA Naval Academy*, *Dalian* 116018, *China*)

Abstract: The vector information of the acoustic field which contains the bearing of the source can be measured by the particle velocity sensors, and this makes it possible for the vector hydrophone linear array (VHLA) sonars to have the ability of solving the notorious port/starboard ambiguity suffered by the scalar single-line array. In this paper, two beamformers based on the generalized cardioid processing (GCP) and the sound intensity processing (SIP) are analyzed theoretically to explore the performance of port/starboard discrimination, and the results are compared to the experimental data obtained in sea trials. Analysis shows that the adaptive beamformer outperforms the convectional one. Furthermore, with respect to the ability of signal detection, GCP is a better one for the VHLA signal processing.

Key words: Port/Starboard ambiguity; Beamforming; Vector hydrophone

1 INTRODUCTION

In the past decades, the towed linear array sonar has been verified to be the most effective sonar equipment for the ASW[1]. However, it suffers an inborn deficiency in discriminating the port/starboard bearing of the target for it only measures the acoustic pressure of the sound field. There are two ways to settle the notorious port/starboard (PS) ambiguity. The first way is to make self-ship maneuver, which can discriminate the port/starboard bearing of the target by purposefully changing the track of self-ship. The major drawback of this method is that it usually consumes too much time. The second way is to increase the number of the towed linear arrays, such as twin-line arrays, triplet arrays, and multi-line arrays with even more lines[2-5], which enhances the systematic complexity and operational difficulty of the sonar equipment.

The vector hydrophone techniques provide a new approach to settle the PS ambiguity of the linear array sonar. A vector hydrophone can measure the acoustic pressure and particle velocity simultaneously and co-locatedly. As a result, the vector hydrophone can be deemed as a zero-aperture subarray with frequency-independent cosine directivity. By using the latter, the vector hydrophone linear array (VHLA) has the ability to solve the PS ambiguity, which provides it an attractive future in the ASW.

The signal processing methods jointly using the acoustic pressure and particle velocity can be divided into two categories[6]: One is the generalized cardioid processing (GCP) method, and the other is the sound intensity processing (SIP) method. In this paper, the theoretical and experimental performances of PS discrimination of the VHLA have been analyzed with respect to the GCP and SIP in terms of two beamformers: the conventional beamformer (CBF) and the Capon's beamformer (MVDR, minimum variance distortionless response)[8].

① 声学技术, 2007, 26(6): 1135-1139.

2 PORT/STARBOARD REJECTION

The ability to solve the PS ambiguity is usually quantified by means of the PS rejection. As the output of the beamformer is the signal power, it is convenient to define the PS rejection as follows

$$r=\frac{w^H_{S+}S_x w_{S+}}{w^H_{S-}S_x w_{S-}} \tag{1}$$

Where w_{S+} and w_{S-} are beamformer vectors steering to the two ambiguous directions, S_x is the cross spetrum density matrix of the array.

The physical meaning of **PS** rejection defined in Eq.(1) is the ratio of signal power output by the beamformer steering to the pair of ambiguous directions.

3 THEORETICAL ANALYSIS

The following assumptions should be made for theoretical analysis:

(1)The bearing of the source is θ_s, and the steering vectors corresponding to the two ambiguous directions are denoted by $\boldsymbol{a}_{S+}$ and $\boldsymbol{a}_{S-}$.

(2)The noise received by different channels is homogeneous and mutually independent.

(3)Both the signal and the ambient noise are band limited white Gaussian noise with spectrum level of σ_S^2 and σ_P^2. Noise in the particle velocity channel has a spectrum level of $\sigma^2_V=\eta\sigma^2_P$, and for the isotropic noise field: $\sigma^2_V=1/3\sigma^2_P$[7].

The CBF and MVDR beamformers are very popular in array processing, so the subsequent discussion of **PS** discrimination of the VHLA will be focoused on these two beamformers.

3.1 PS rejection of CBF

The cross spectrum matrix and covariance matrix of the VHLA can be written as

$$S_{pv}=\sigma^2_S\,\boldsymbol{a}_{S+,p}\boldsymbol{a}^H_{S+,v}$$
$$S_x=\sigma^2_S\,\boldsymbol{a}_{S+}\boldsymbol{a}^H_S+\sigma^2_P\Sigma \tag{2}$$

Where Σ is the cross spectrum density matrix of the noise, $\boldsymbol{a}_{S+}$, $\boldsymbol{a}_{S+,p}$ and $\boldsymbol{a}_{S+,v}$ are the response vectors of the VHLA and its subarrays of pressure and particle velocity respectively, i.e.

$$\boldsymbol{a}_{S+,p}=[1,e^{-\frac{2\pi}{\lambda}d\cos\theta_s},\cdots,e^{-\frac{2\pi}{\lambda}(M-1)d\cos\theta_s}]^T$$
$$\boldsymbol{a}_{S+,v}=\boldsymbol{a}_{S+,p}\otimes[\cos\theta_S,\sin\theta_S]^T$$
$$\boldsymbol{a}_S=\boldsymbol{a}_{S+,p}\otimes[1,\cos\theta_S,\sin\theta_S]^T \tag{3}$$

Where d is the element space, λ is the wavelength of the narrowband signal, M is the number of elements, "$\otimes$" denotes the Kronecker product.

Substituting Eq.(2) into Eq.(1), and after some manipulations, we obtain

GCP:

$$r_{CBF,GCP}=\frac{4M\cdot SNR+(1+\eta)}{M(1+\cos2\theta_S)^2\cdot SNR+(1+\eta)} \tag{4}$$

SIP:

$$r_{CBF,SIP}=\frac{1}{\max(\cos2\theta_S,0)} \tag{5}$$

Where $SNR=\sigma_S^2/\sigma_P^2$ is the signal to noise ratio in the pressure channel, $w_{S\pm}$, $w_{S\pm,P}$ and $w_{S\pm,V}$ are the beamforming weight vectors of the VHLA and its pressure and particle velocity subarrays respectively when the beams steering to the source and its ambiguous direction.

Figure 2 demonstrates the relations of the PS rejection versus bearing and the signal to noise ratio (SNR) by the CBF. Fig.2(a) is the PS rejection versus source bearing when the SNR equals 0dB. Fig.2(b) gives the rela-tions between the PS rejection and the SNR when the source bearing equals 25°.

When CBF is used, by analyzing Eqs.(3)-(5) associated with Fig.2, we find that:

For the **GCP** method, the performance of **PS** discrimination maximizes in the abeam direction, and degrades toward the endfire. The optimal area of **PS** discrimination for the **SIP** method is [45°,135°], with $r_{CBF,SIP}\to\infty$, and it also approach 0dB in the endfire.

As shown in Fig.2(b), When $SNR\to0$, the **PS** rejection of the **GCP** method tends to 0dB, but the **PS** rejection of the **SIP** method is a constant determined by the bearing of target, which means that the latter is better in **PS** discrimination in weak signal scenarios. When $SNR\to\infty$, $r_{CBF,GCP}\to4/(1+\cos2\theta_S)^2$ the curves of **PS** rejection of the **GCP** method coincidence with the combined directivity pattern of the single vector hydrophone[6].

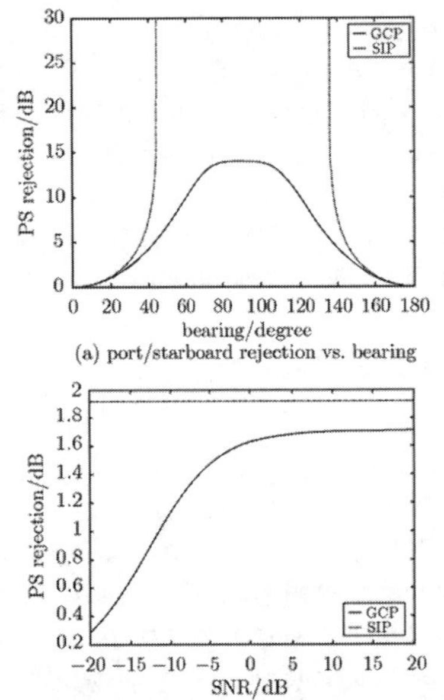

(b)port/starboard rejection vs. SNR with $\theta_S = 25°$

Fig.2 The performance of port/starboard discrimination of a 8-element VHLA by the conventional beamforming.

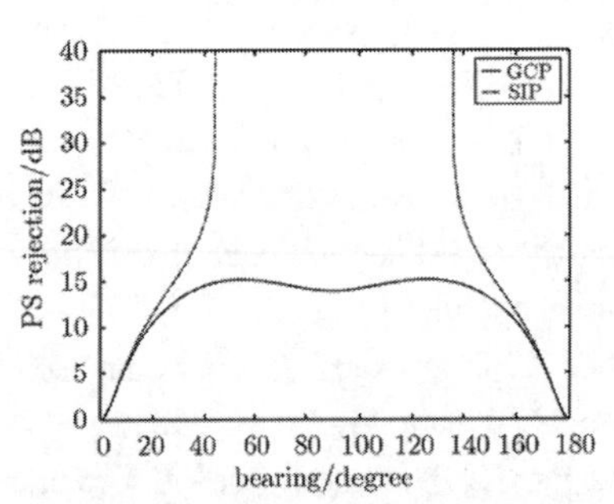

(a) port/starboard rejection vs. bearing with SNR = 0dB

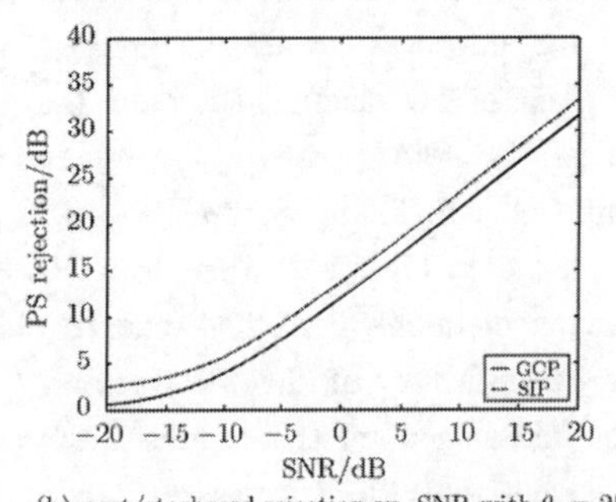

(b) port/starboard rejection vs. SNR with $\theta_S = 25°$

Fig.3 The performance of port/starboard discrimination of a 8-element VHLA by the MVDR beamforming.

2.2 PS rejection of MVDR

The filter vector for the MVDR beamformer is given by[7]

$$w_{MVDR}=\frac{S^{-1}\boldsymbol{a}}{\boldsymbol{a}^H S^{-1}\boldsymbol{a}} \tag{6}$$

where S is the covariance matrix, and $\boldsymbol{a}$ is the array response vector in the steering direction. After some direct manipulations, inversions of the covariance matrixes for the VHLA and its subarrays of pressure and particle velocity are given by

$$S_x^{-1}=\frac{1}{\sigma_P^2}\left[\Sigma^{-1}-\frac{SNR}{1+M(1+1/\eta)\cdot SNR}\Sigma^{-1}\boldsymbol{a}_{S+}\boldsymbol{a}^H_{S+}\Sigma^{-1}\right] \tag{7}$$

$$S_p^{-1}=\frac{1}{\sigma_P^2}\left[I-\frac{SNR}{1+M\cdot SNR}\boldsymbol{a}_{S+,p}\boldsymbol{a}^H_{S+,p}\right] \tag{8}$$

$$S_v^{-1}=\frac{1}{\eta\sigma_P^2}\left[I-\frac{SNR/\eta}{1+M/\eta\cdot SNR}\boldsymbol{a}_{+,v}\boldsymbol{a}^H_{+,v}\right] \tag{9}$$

where S_x is defined in Eq.(2), S_p and S_v are the cross spectrum density matrixes for the subarrays of pressure and particle velocity respectively. Substituting Eqs.(7)-(9) into Eq.(6), and calculating the **PS** rejections by Eq.(1), we obtain

GCP:

$$r_{Capon,GCP}=1+M\cdot SNR\left[(1+1/\eta)-\frac{(1+1/\eta\cos2\theta_S)^2}{(1+1/\eta)}\right] \tag{10}$$

SIP:

$$r_{Capon,SIP}=\frac{1+M\cdot SNR\cdot\sin^2 2\theta_S/\eta}{\max(\cos2\theta_S,0)} \tag{11}$$

When $\cos2\theta_S\to1$(near the endfire), following Eq.(10), we have

$$r_{Capon,GCP}\approx 1+M\cdot SNR(\sin^2 2\theta_S)/\eta$$

Comparing the above with Eq.(11), we find that the performances of these two methods are approximately identical in the endfire direction.

Fig.3 shows the performances of MVDR beamformer akin to Fig.2. We can see:

(1)The performances of the **PS** discrimination for the MVDR beamformer are correlated with the spatial statistics of the noise field. For the isotropic ambient noise, its influence is denoted by the parameter η.

(2)As shown in Fig.3(a), the **PS** rejection of the **SIP** method asymptotically approaches infinity in the angular domain of [45°,135°], and equals 1 in the endfire direction, which is similar to the CBF scenario. For the **GCP** method, the bearing with optimal **PS** discrimination differs from the abeam direction as in the CBF, but approachs the beaing of 55°, and shares the same ambiguity area as the **SIP** method.

(3)As shown in Fig.3(b), the performance of the **PS** discrimination improves with increasing of SNR since the weighted vector of the MVDR beamformer is correlated with the latter.

4 EXPERIMENTAL VERIFICATION

In order to verify the theoretical performance of the **PS** discrimination by the two beamformers demonstrated above, we use the experimental data of a two-dimentional VHLA with 8 elements, which was recorded in the San Ya sea trial in May 2002.

Fig.4 shows the wideband CBF results of the sea trial. From these figures we can see that the performances of **PS** discrimination of the **GCP** and **SIP** methods are almost identical, but the noise level of the **GCP** method is a little higher. In the total observation period of 25 minutes, five targets have been tracked with two in the left side and three in the right. To be consistant with the theoretical analysis, both of the **GCP** and **SIP** methods have optimal performances of **PS** discrimination in the abeam direction, and serve critical ambiguity in the endfire.

We also find that the spatial resolution of the wideband CBF is low, especially near the endfire area, the angular area within ±30° was control by the self noise. Furthermore, the noise level output by the CBF is high so that the **PS** discrimination of the weak signals is not satisfying. In order to suppress the noise level to modify the **PS** discrimination in multi-target scenarios, it is necessary to adopt the adaptive beamformer.

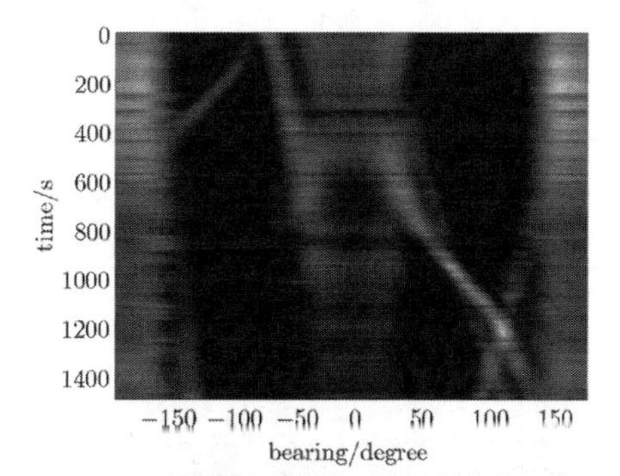

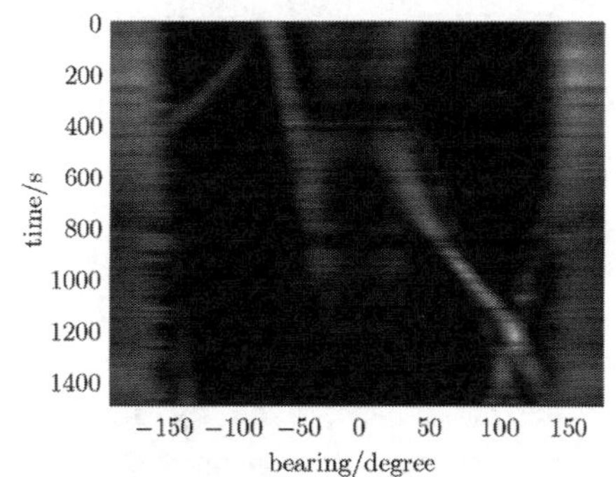

Fig.4 Time-bearing plot of the GCP and SIP methods by CBF using a 8-element VHLA. The result of GCP is shown on the left, and SIP on the right.

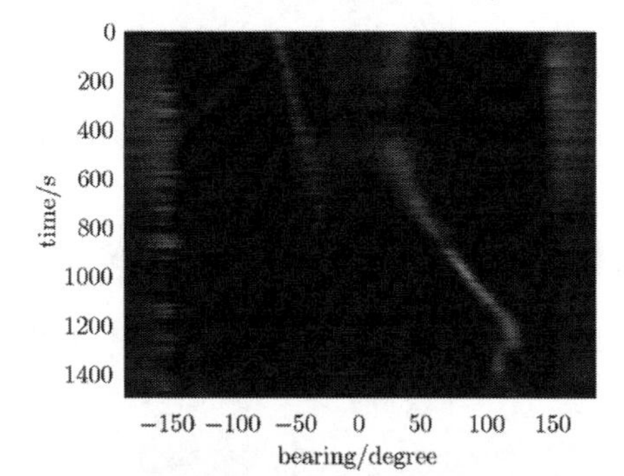

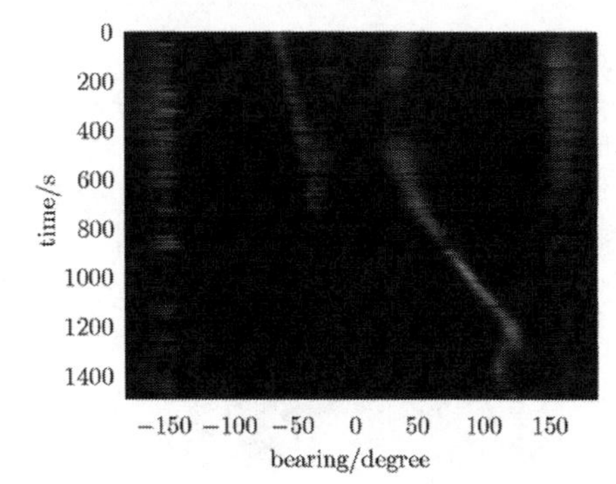

Fig.5 Time-bearing plot of the GCP and SIP methods by the MVDR beamformer using a 8-element VHLA. The result of GCP is shown on the left, and SIP on the right.

Fig.5 shows the spatial spectrum output by the MVDR beamformer. From these figures we can see that the noise levels are much lower than those of the CBF, and the bearing-time tracks are much cleaner either, especially near the endfire area. The spatial resolution of the MVDR beamformer is much higher than that of the CBF, which leads to narrower tracks, and the abilities of the **PS** discrimination of the **GCP** and **SIP** are almost identical.

5 CONCLUSIONS

The vector hydrophone has a frequency-independent cosine directivity, which leads to a outperformance for the VHLA in solving the **PS** ambiguity over the conventional multi-line arrays, the latter depends on the frequency band and the distance between lines.

In this paper, the PS rejection is defined by the output power of the ambiguous directions, the performance of **PS** discrimination of the VHLA under the **GCP** and **SIP** methods has been intensively explored by theoretical analysis and experimental verification, and the results show a good consistency. In the isotropic noise field, the **SIP** method has theoretically better ability in **PS** discrination, but in practical applications the performance of the **SIP** method will degrade due to the fluctuation of the realistic ambient noise and the difference of the spatial directivity, so the GCP is a robust one for VHLA signal processing.

References

[1] LI Qihu. The Principles for Designing Digital Sonars[M]. Anhui Educational Press, 2002. 434-442.

[2] Schurman I W. Reverberation rejection with a dual-line towed array[J]. IEEE Journal of Oceanic Engineering, 1996, **21**(2): 193-204.

[3] DU Xuanmin, ZHU Daizhu. The Theoretical Analysis and Experimental Study of Port/Starboard Discrimination Techniques of the Towed Triplet Arrays[s]. ACTA ACOUSTICA, 2000, **25**(5): 395-402.

[4] He Xinyi. Investigation of Port-Starboard Discrimin-ation Techniques of the Towed Twin Arrays based on Time Delay Estimation[J]. Signal Processing, 2003, **19**(4): 338-342.

[5] Groen. Adaptive port/starboard beamforming of triplet sonar arrays [J]. IEEE Journal of Oceanic Enginee-ring, 2005, **30**(2): 348-359.

[6] SUN Guiqing. The Acoustic Vector Sensor and Its Applications to Uniform Linear Array Sonar[R]. Post Doctor Report, Institute of Acoustics, Chinese Academy of Science, August 2003L28-32.

[7] Van Trees H L. Optimum Array Processing[M]. John Wiley & Sons, New York, 2002. 440-445.

矢量线列阵波束形成算法的左右舷分辨性能

李 敏[1]，杨秀庭[2]，李启虎[1]

(1. 中国科学院声学研究所，北京 100080；2. 海军大连舰艇学院，大连 116018)

摘要：研究矢量均匀线列阵波束形成算法的左右舷分辨性能。水听器可同步共点地测量声场的声压和质点振速，为一具有指向性的空间共点阵，从而能够解决单线阵声呐的目标左右舷模糊问题。文章分析了采用不同波束形成方法时矢量阵广义 Cardiod 处理和声强处理的目标左右舷分辨性能，并利用海试数据进行了验证。研究结果表明：自适应波束形成具有比常规波束形成更佳的左/右舷分辨性能，且对矢量阵处理而言，广义 Cardioid 处理更为稳健和实用。

关键词：矢量阵；波束形成；左右舷分辨

中图分类号：TG 43.20　　**文献标识码**：A　　**文章编号**：1000-3630(2007)-06-1135-05

矢量拖曳线列阵声呐流噪声的空间相关性研究①②

杨秀庭[1,2] 孙贵青[1] 李 敏[1] 李启虎[1]
(1 中科院声学研究所 北京 100080)
(2 海军大连舰艇学院 大连 116018)
2006 年 9 月 19 日收到
2007 年 5 月 18 日定稿

摘要 根据湍流边界层脉动压力谱的 Corcos 模型，采用频率－波数域上的二维谱分析方法对矢量拖曳线列阵声呐拖缆护套管内的流噪声场进行了研究，得到了流噪声声压和振速的一般表达式．在此基础上，计算了管内流噪声场的空间相关性．数值计算结果表明：(1) 流噪声的轴向相关性随阵元间距的增加而迅速衰减，不同矢量阵元所接收的流噪声可近似认为是彼此互不相关的；(2) 在拖缆的径向上，流噪声的低频分量是高度相关的，而其高频分量的相关性则可近似忽略．
PACS 数： 43.20, 43.60

On investigation of the spatial correlation of the flow induced noise in the application of vector hydrophone towed arrays

YANG Xiuting[1,2] SUN Guiqing[1] LI Min[1] LI Qihu[1]
(1 *Institute of Acoustics, Chinese Academy of Sciences* Beijing 100080)
(2 *PLA Naval Academy* Dalian 116018)
Received Sept. 19, 2006
Revised May 18, 2007

Abstract Following the Corcos's model of cross spectrum of the turbulent boundary layer pressure, a method in the frequency – wave number domain is presented to analyze the flow induced noise in the interior hose of the vector hydrophone towed linear array. As a result, the general forms of the acoustic pressure and particle velocity have been obtained, and the spatial correlations of the flow induced noise have also been calculated. The numerical results show that, the spatial correlations of flow induced noise drop rapidly with the increase of the axial separation between the elements, so the flow induced noise received by different vector hydrophones can be considered as independent. Furthermore, the low frequency component of the flow induced noise is highly correlated in the radial direction, but the radial correlations of the high frequency component can be neglected at the same time.

1 引言

拖曳线列阵声呐是当前最为有效的反潜装备，有工作深度可调、本舰自噪声影响小和基阵孔径不受工作母船限制等优点。然而在拖曳线列阵工作时，由于拖缆和海水间的相对运动，在拖缆护套外表面附近存在海水介质微团的湍流脉动，由此激发出的水动力噪声，一般称之为流致噪声，或简称为流噪声．流噪声是影响拖曳式声呐实际工作性能的重要因素．

目前关于拖曳线列阵流噪声的研究成果尚不多见。Corcos[1] 较早地对平板壁面边界层内湍流脉动压力进行了理论和实验探索，给出了脉动压力的频率－波数谱．Shashaty[2] 通过把拖缆软护套外壁处的湍流脉动压力看成是内部流噪声的激励源，计算了有限尺寸水听器的流噪声，但该方法的缺陷在于完全忽略了护套的声学特性，仅考虑了湍流脉动压力的直接透入部分，而没有计及护套在湍流作用下的激励响应成分．Francis 等人 [3] 分析了均匀、多层嵌套式弹性圆柱壳内部的流噪声场，分别给出了内部装填固体和充液时拖缆护套的降噪性能．汤渭霖 [4] 研究了粘弹性柔性护套内部流噪声的产生机理，给出了护套的几何参数和力学参数对内部流噪声强度的影响．上述研究结论都是针对基于声压测量的常规拖曳线列阵的．

① 国家自然科学基金(60532040)、国家安全重大基础研究(613660201, 5132102ZZT22)和所长择优基金(GS08SJJ09)资助项目.

② 声学学报, 2007, 32(6): 547-552.

由于矢量水听器的性能优势，其在水声工程中的应用日趋广泛，矢量拖曳阵正逐渐纳入人们的视野．到目前为止，流噪声对矢量拖曳线列阵性能的影响仍是一个尚待研究的课题，本文在文献 4 的基础上，通过把护套外壁处所受的湍流脉动压力在频率 - 波数域上进行二维谱展开，利用实轴积分法来研究矢量拖曳线列阵声呐拖缆护套内部流噪声场的声压和质点振速的空间相关性．理论计算结果给出了拖缆护套内部流噪声场空间相关性的基本规律，对于矢量拖曳线列阵的实际工程应用具有理论指导意义和参考价值．

1 流噪声模型

拖曳线列阵中的矢量水听器不与海水介质直接接触，而是安装在由软护套 (一般为聚氨酯材料 [5]) 包裹的拖缆内，为降低不同材料界面处声波的反射损失，以保证阻抗匹配，在护套内还需灌注轻蜡油．当拖缆在海水中拖曳时，海水和拖缆间的相对运动会产生流噪声．大量的实验表明：湍流边界层脉动压力是拖曳线列阵流噪声的主要来源 [4]，在高速拖曳时尤为如此．

1.1 物理假设

如图 1 所示，当流体流经拖缆时，由于流体的粘性，会在护套的外壁附近产生边界层，当雷诺数充分大时，边界层会由层流状态发展为湍流，此时流场的压力表现出脉动特征．该脉动压力作用在护套外壁时，所引起的振动将在护套及内外的流体中传播，这属于典型的管壁受迫振动问题．为求解声场，需作以下假设：

(1) 边界层内的湍流充分发展，为空间均匀、时间平稳的随机过程．

(2) 拖缆长度为无穷大，两端影响可以忽略，声场关于护套管是柱对称的．

(3) 护套管为线性系统，满足线性叠加原理．

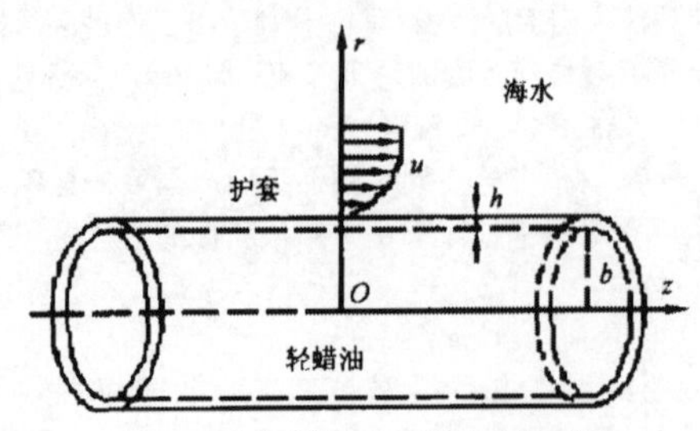

图 1 拖曳线列阵声呐声学模块的基本结构

对于一般的拖曳线列阵，工作时的拖缆长度较大，因此以上 (1)、(2) 两点假设可近似满足，护套的声学特性可保证 (3)．

护套外壁面处所受的湍流脉动压力，利用傅里叶变换，可将其展为频率 - 波数的二维谱函数 [6]，即

$$p(a,z,t)=\left(\frac{1}{2\pi}\right)^2\iint_{-\infty}^{+\infty}s\left(k_z,\omega\right)e^{\mathrm{j}(k_z z-\omega t)}\mathrm{d}\omega\mathrm{d}k_z, \tag{1}$$

其中，a 为护套外径，$p(a,z,t)$ 为外壁处的脉动压力，$s(k_z,\omega)$ 为相应的频率 - 波数谱，k_z 为 z 轴方向的波数．

1.2 声场方程及解的形式

声在护套内外的流体介质中传播时，其位移势方程的基本形式为：

$$\Delta\phi-\frac{1}{c^2}\frac{\partial^2\phi}{\partial t^2}=0, \tag{2}$$

式中，ϕ 为位移势，Δ 为拉普拉斯算子，c 为声速．根据假设 (2)，式 (2) 在柱坐标系下的通解为：

$$\phi=\left[A\mathrm{J}_0\left(kr\right)+B\mathrm{N}_0\left(kr\right)\right]e^{\mathrm{j}(k_z z-\omega t)} \quad 或$$
$$\phi=\left[A\mathrm{H}_0^{(1)}\left(kr\right)+B\mathrm{H}_0^{(2)}\left(kr\right)\right]e^{\mathrm{j}(k_z z-\omega t)}, \tag{3}$$

其中，r 为径向坐标分量，A,B 为 z 和 t 的函数，由初始条件和边界条件决定；$\mathrm{J}_0(\cdot)$ 和 $\mathrm{N}_0(\cdot)$ 分别为零阶第一、二类贝塞尔函数，$\mathrm{H}_0^{(i)}(\cdot)$ 为零阶第 i 类汉克尔函数；k 为波数，满足 $k^2+k_z^2=(\omega/c)^2$．

根据柱对称声场的特点，护套内外流体介质中的声场位移势为 [7]：

$$\phi_1=D\mathrm{J}_0\left(k_1 r\right)e^{\mathrm{j}(k_z z-\omega t)}, \tag{4}$$

$$\phi_2=C\mathrm{H}_0^{(1)}\left(k_2 r\right)e^{\mathrm{j}(k_z z-\omega t)}, \tag{5}$$

式中，$k_1^2+k_z^2=(\omega/c_1)^2$, $k_2^2+k_z^2=(\omega/c_2)^2$, c_1, c_2 分别为内、外流体中的声速．

护套与流体的力学特性不同，一般可将其看作弹性介质或粘弹性介质 (存在声吸收)．声 (或振动) 在护套中的传播具有压缩波和切变波两种形式，其位移势方程的形式与 (2) 一致．因此，其解为：

$$\phi_c=\left[A_1\mathrm{J}_0\left(k_c r\right)+B_1\mathrm{N}_0\left(k_c r\right)\right]e^{\mathrm{j}(k_z z-\omega t)}, \tag{6}$$

$$\phi_s=\left[A_2\mathrm{J}_0\left(k_s r\right)+B_2\mathrm{N}_0\left(k_s r\right)\right]e^{\mathrm{j}(k_z z-\omega t)}, \tag{7}$$

其中，ϕ_c,ϕ_s 分别为压缩波和切变波的位移势函数，k_c，k_s 为对应的波数，满足 $k_c^2+k_z^2=(\omega/c_c)^2$ 和 $k_s^2+k_z^2=(\omega/c_s)^2$；压缩波和切变波的波速由介质材料决定，为：

$$c_c=\left[(\lambda+2\mu)/\rho\right]^{1/2},\quad c_s=(\mu/\rho)^{1/2}, \tag{8}$$

其中，λ 和 μ 为介质的拉梅常数，ρ 为介质密度．

护套内外各区域波动方程的解 (4) — (7) 中的常数，由边界条件决定. 这里所研究的声场，其边界为护套的内、外壁. 在弹性 (或粘弹性) 介质和流体的界面处，由连续性条件，存在三点约束: (1) 界面两侧介质的法向应力保持连续，界面处为受力平衡态; (2) 由于流体无法承受剪应力，故界面处的剪应力为零; (3) 法向位移连续.

对于一般的弹性介质，柱坐标系下由声波作用所引起的法向和切向的位移为[6]:

$$u_r = \frac{\partial \phi_c}{\partial r} + \frac{\partial}{\partial}\left(\frac{\partial \phi_s}{\partial r}\right), \tag{9}$$

$$u_z = \frac{\partial \phi_c}{\partial z} - \frac{\partial}{r\partial r}\left(r\frac{\partial \phi_s}{\partial r}\right). \tag{10}$$

法向应力和切向应力为:

$$\tau_{rr} = (\lambda + 2\mu)\frac{\partial u_r}{\partial r} + \lambda\left(\frac{\partial u_z}{\partial z} + \frac{u_r}{r}\right), \tag{11}$$

$$\tau_{rz} = \mu\left(\frac{\partial u_r}{\partial z} + \frac{\partial u_z}{\partial r}\right). \tag{12}$$

因此，根据式 (9) — (12), 并利用边界条件，经计算，得到壁面压力波 $s(k_z,\omega)e^{j(k_z z-\omega t)}$ 在护套内部所激发声场的位移势为:

$$\begin{aligned}&\phi_1(r,k_z,\omega) = \\ &\quad G(k_z,\omega)\,\mathrm{J}_0(k_1 r)\,s(k_z,\omega)\,e^{j(k_z z-\omega t)},\ r \le b\end{aligned} \tag{10}$$

其中，$G(k_z,\omega)$ 为与护套几何尺寸和护套及内外流体物性有关的函数[8].

根据位移势函数与声压和质点振速之间的关系，利用傅氏积分，得:

$$p(r,z,t) = \left(\frac{1}{2\pi}\right)^2 \iint_{-\infty}^{+\infty} \rho_1\omega^2 G(k_z,\omega)\,\mathrm{J}_0(k_1 r)\,s(k_z,\omega)\,e^{j(k_z z-\omega t)}\mathrm{d}\omega\mathrm{d}k_z, \tag{14}$$

$$\boldsymbol{v}(r,z,t) = \left(\frac{1}{2\pi}\right)^2 \iint_{-\infty}^{+\infty} \omega G(k_z,\omega)\left[jk_1\mathrm{J}_1(k_1 r)\,\boldsymbol{e}_r + k_z\mathrm{J}_0(k_1 r)\,\boldsymbol{e}_z\right]s(k_z,\omega)\,e^{j(k_z z-\omega t)}\mathrm{d}\omega\mathrm{d}k_z, \tag{15}$$

其中，$p(r,z,t)$ 和 $\boldsymbol{v}(r,z,t)$ 分别为拖缆护套内部流噪声场的声压和质点振速分量; $\boldsymbol{e}_r$ 和 $\boldsymbol{e}_z$ 分别为径向和轴向的单位向量.

2 湍流脉动压力谱 - Corcos 模型

湍流边界层压力起伏产生的噪声，具有一定的随机特征. Corcos 研究了平面湍流边界层的压力起伏现象，并给出了频率 - 波数域上的半经验公式[1], 即窄带湍流脉动压力的波数谱为:

$$P_w(k_z,\omega) = P(\omega)\int_{-\infty}^{+\infty} e^{-\alpha k_c|\zeta|}e^{j(k_z-k_c)\zeta}\mathrm{d}\zeta, \tag{16}$$

其中，$P(\omega)$ 为脉动压力的自谱; α 为模型常数，与壁面的光滑程度有关[4], 光滑表面 $\alpha = 0.116$, 粗糙表面 $\alpha = 0.32$; $k_c = \omega/u_c$ 为迁移波波数，迁移速度 u_c 与拖曳速度 U 相适应，一般取为 $u_c = \beta U$ 的形式，β 为频率、雷诺数的函数. 计算式 (16) 右侧的积分项，得到:

$$P_w(k_z,\omega) = \frac{1}{\pi}\frac{\alpha k_c}{\alpha^2 k_c^2 + (k_z - k_c)^2}P(\omega). \tag{17}$$

3 护套内部流噪声场的空 - 时相关函数

3.1 空 - 时相关函数的一般形式

在柱对称情况下，由随机过程互相关函数的定义，护套内流体中流噪声声场的空 - 时互相关矩阵的各元素为:

$$\left\{\begin{aligned}
R_{11} &= E[p(r_1,z,t)\,p^*(r_2,z+L,t-\tau)] = \left(\frac{1}{2\pi}\right)^2\iint_{-\infty}^{+\infty} P_w(k_z,\omega)\,h_p(r_1,k_z,\omega)\,h_p^*(r_2,k_z,\omega)\,e^{j(k_z L-\omega\tau)}\mathrm{d}\omega\mathrm{d}k_z \\
R_{12} &= E[p(r_1,z,t)\,v_r^*(r_2,z+L,t-\tau)] = \left(\frac{1}{2\pi}\right)^2\iint_{-\infty}^{+\infty} P_w(k_z,\omega)\,h_p(r_1,k_z,\omega)\,h_r^*(r_2,k_z,\omega)\,e^{j(k_z L-\omega\tau)}\mathrm{d}\omega\mathrm{d}k_z \\
R_{13} &= E[p(r_1,z,t)\,v_z^*(r_2,z+L,t-\tau)] = \left(\frac{1}{2\pi}\right)^2\iint_{-\infty}^{+\infty} P_w(k_z,\omega)\,h_p(r_1,k_z,\omega)\,h_z^*(r_2,k_z,\omega)\,e^{j(k_z L-\omega\tau)}\mathrm{d}\omega\mathrm{d}k_z \\
R_{22} &= E[v_r(r_1,z,t)\,v_r^*(r_2,z+L,t-\tau)] = \left(\frac{1}{2\pi}\right)^2\iint_{-\infty}^{+\infty} P_w(k_z,\omega)\,h_r(r_1,k_z,\omega)\,h_r^*(r_2,k_z,\omega)\,e^{j(k_z L-\omega\tau)}\mathrm{d}\omega\mathrm{d}k_z \\
R_{23} &= E[v_r(r_1,z,t)\,v_z^*(r_2,z+L,t-\tau)] = \left(\frac{1}{2\pi}\right)^2\iint_{-\infty}^{+\infty} P_w(k_z,\omega)\,h_r(r_1,k_z,\omega)\,h_z^*(r_2,k_z,\omega)\,e^{j(k_z L-\omega\tau)}\mathrm{d}\omega\mathrm{d}k_z \\
R_{33} &= E[v_z(r_1,z,t)\,v_z^*(r_2,z+L,t-\tau)] = \left(\frac{1}{2\pi}\right)^2\iint_{-\infty}^{+\infty} P_w(k_z,\omega)\,h_z(r_1,k_z,\omega)\,h_z^*(r_2,k_z,\omega)\,e^{j(k_z L-\omega\tau)}\mathrm{d}\omega\mathrm{d}k_z
\end{aligned}\right. \tag{18}$$

式中，各项定义为：

$$\begin{cases} h_p(r_1,k_z,\omega)=\rho_1\omega^2 G(k_z,\omega)\,\mathrm{J}_0(k_1 r) \\ h_r(r_1,k_z,\omega)=\mathrm{j}\omega k_1 G(k_z,\omega)\,\mathrm{J}_1(k_1 r) \\ h_z(r_1,k_z,\omega)=\omega k_z G(k_z,\omega)\,\mathrm{J}_0(k_1 r) \end{cases}$$

护套外壁处湍流脉动压力的频率-波数谱 $P_w(k_z,\omega)$ 与其空-时互相关函数及互谱密度函数之间的关系如下：

$$P_w(k_z,\omega)=\int_{-\infty}^{+\infty}R_w(\zeta,\tau)\,e^{-\mathrm{j}(k_z\zeta-\omega\tau)}\mathrm{d}\zeta\mathrm{d}\tau=\int_{-\infty}^{+\infty}P_w(\zeta,\omega)\,e^{-\mathrm{j}k_z\zeta}\mathrm{d}\zeta, \tag{19}$$

其中，$R_w(\zeta,\tau)=E[p(z,t)\,p^*(z-\zeta,t-\tau)]$ 为外壁处湍流脉动压力的空-时互相关函数；$P_w(\zeta,\omega)$ 为对应的互谱密度函数，由具体的湍流模型决定。

3.2 互谱密度矩阵

根据式 (18)，护套内部流噪声场的空-时互相关矩阵 $\boldsymbol{R}(r_1,r_2,L,\tau)$ 为：

$$\boldsymbol{R}(r_1,r_2,L,\tau)=\frac{1}{2\pi}\int_{-\infty}^{+\infty}\boldsymbol{P}(r_1,r_2,L,\omega)\,e^{-\mathrm{j}\omega\tau}\mathrm{d}\omega, \tag{20}$$

其中，空间互谱密度矩阵 $\boldsymbol{P}(r_1,r_2,L,\omega)$ 为：

$$\boldsymbol{P}(r_1,r_2,L,\omega)=\frac{1}{2\pi}\int_{-\infty}^{+\infty}\boldsymbol{P}(r_1,r_2,k_z,\omega)\,e^{\mathrm{j}k_z L}\mathrm{d}k_z. \tag{21}$$

根据复变函数的有关知识，在满足约当引理的前提下，即：

$$\lim_{k_z\to\infty}|\boldsymbol{P}(r_1,r_2,L,\omega)|\to 0.$$

式 (21) 的积分为被积函数在复平面的上半平面和实轴上极点的留数再加上分支点割线的积分。从前面已经给出的护套传递函数的复杂性可以看出，由式 (21) 获得明晰的解析解是十分困难的，在此我们采用实轴积分法 (或离散波数法) 来计算流噪声场的空间相关特性。

3.3 护套内部流噪声场的空间相关性

考虑到轴对称假设，拖缆护套内部流噪声场的空间相关性可分为轴向相关性和径向相关性，下面利用实轴积分法来进行具体计算。护套及内外流体介质的物性和计算参数由表 1 给出。

表 1 护套及内外流体介质参数

拖曳速度 U: 5 m/s	迁移速度 u_c: 0.75U
压缩波波速: 2239.3 m/s	切变波波速: 104 m/s
压缩波衰减因子 γ_c: 0.03	切变波衰减因子 γ_s: 0.3
PU 护套密度: 1200 kg/m³	护套内外流体的声速: 1500 m/s
护套内径: 30 mm	护套壁厚: 3 mm

3.3.1 轴向相关性

图2给出了拖曳速度为5 m/s 时，半径为 20.4 mm 处理想点接收器所接收流噪声的轴向相关性。由于

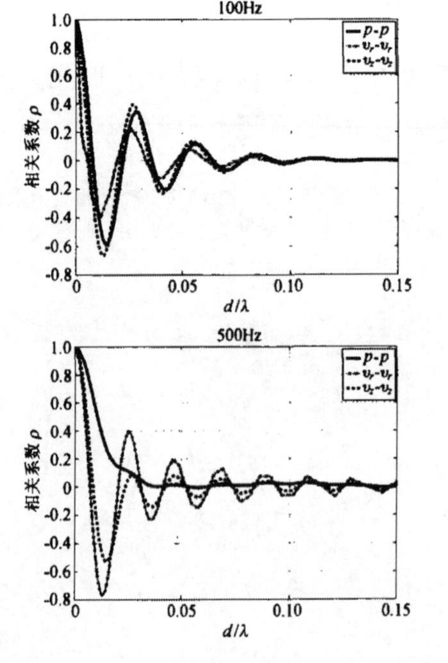

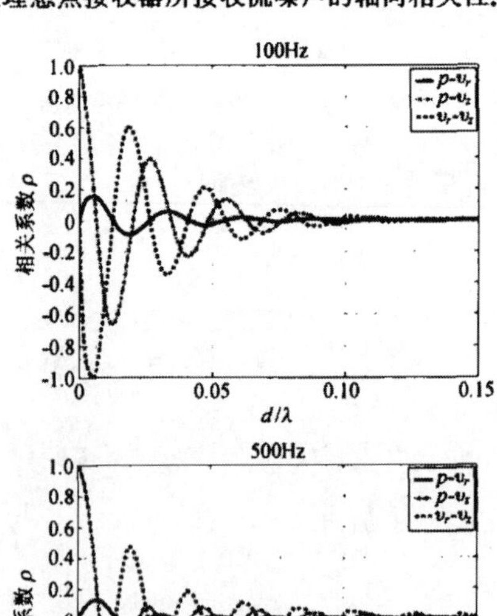

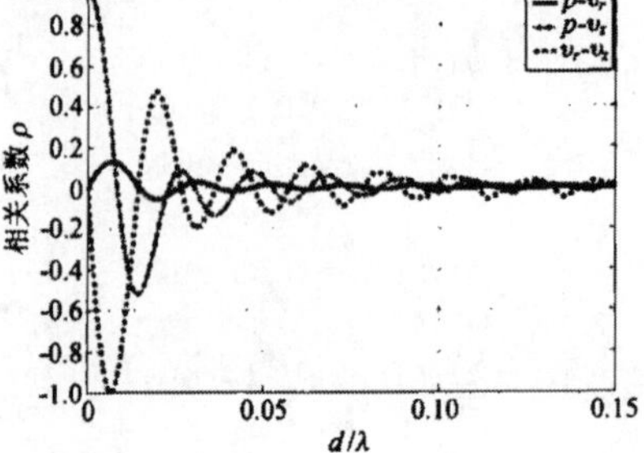

图 2

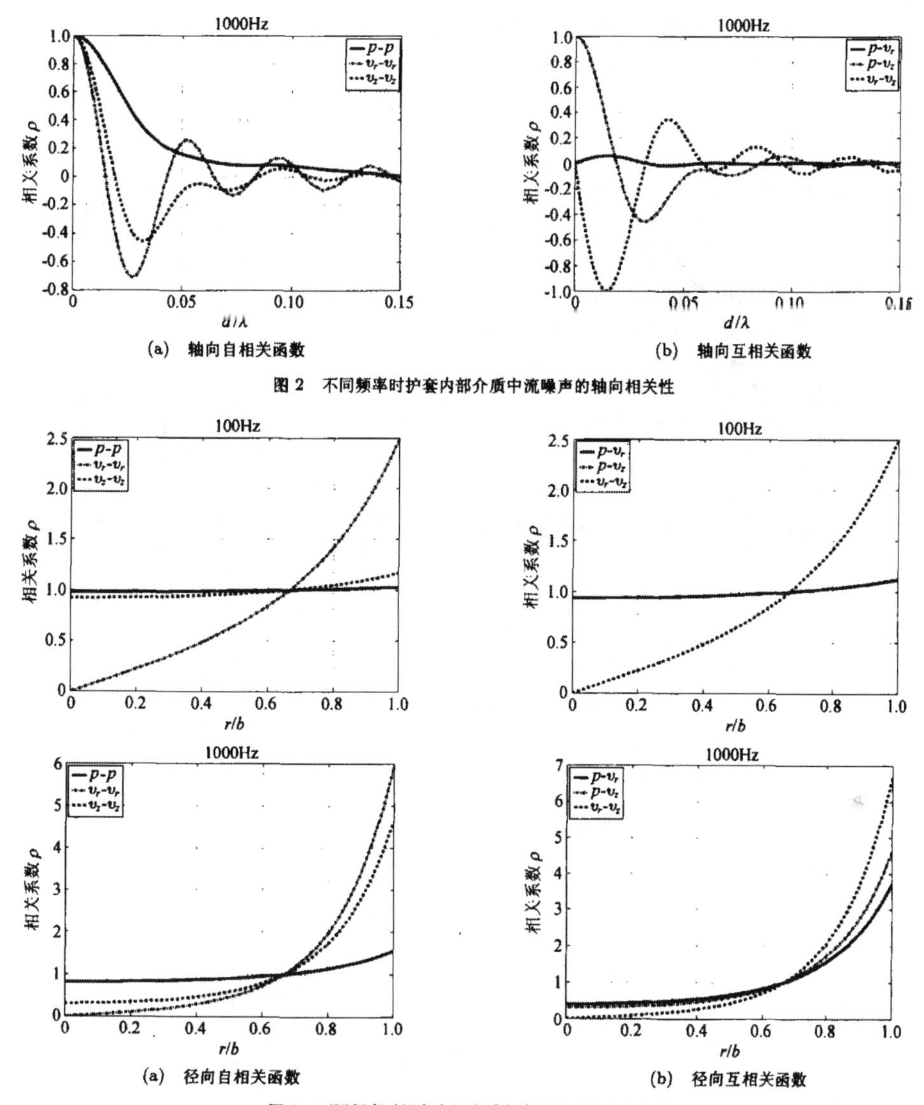

(a) 轴向自相关函数 (b) 轴向互相关函数

图 2 不同频率时护套内部介质中流噪声的轴向相关性

(a) 径向自相关函数 (b) 径向互相关函数

图 3 不同频率时护套内部介质中流噪声的径向相关性

流噪声各分量互谱的谱级差异甚大，因此，需对各互谱分量的计算结果进行归一化处理，通常把归一化参考点选为轴向坐标原点，即 (0,20.4 mm)．分析图 2 可知，对于不同频率的窄带流噪声，若噪声场中任意两点的轴向间距大于 0.1 倍波长时，其相关系数的量值很小，即，此时这两点处流噪声的声压、质点振速彼此间的相关性可以忽略．由此可知，对于通常以半波长进行空间采样的均匀拖曳线列阵声呐，各阵元所接收的流噪声可看成是互不相关的．

3.3.2 径向相关性

图 3 给出了拖曳速度为 5 m/s 时流噪声的径向相关性．同样地，由于流噪声各分量谱级上的差异，需要进行归一化处理．与轴向不同的是，在径向上流噪声并非空间均匀，由于受贝塞尔函数调制，窄带流噪声径向相关性的变化规律与轴向情况差异很大，其相关系数的数值与参考点的具体选取有关，这里我们依然取 (0,20.4 mm) 为参考点．从图中可以看出，护套管内径向上流噪声的强度与其距护套管壁的距离有关，距护套管壁越近，流噪声越强；质点振速的径向分量受轴对称条件约束，在管轴处其量值为零．流噪声径向相关系数的频率依赖性：在低频时，声压和质点振速轴向分量的径向相关性很强，而质点振速径向分量的径向相关系数则随径向位置和间距变化

较大；在高频时，流噪声的径向相关性有所下降，其量值随径向间距的增大较低频时下降更为迅速。

4 结论

本文针对矢量拖曳线列阵声呐流噪声的空间相关性进行了研究。通过把护套看成无限长的粘弹性圆柱壳体，利用实轴积分法，分析了护套内部流噪声场的空间相关性，计算结果表明：

(1) 在拖缆的轴向方向上，流噪声的空间相关性随着空间距离的增大而迅速衰减，当两点的轴向间距大于 0.1 倍波长时，流噪声的相关系数近似为零。因此，对于阵间距为半波长的窄带矢量阵，不同阵元所接收的流噪声可认为是互不相关的。

(2) 由于流噪声受贝塞尔函数调制，因此在径向上具有很强的相关性，其相关系数随频率的升高而缓慢下降。

本文研究从理论上说明了流噪声空间相关性的变化规律，为矢量拖曳线列阵声呐的实际应用提供了理论参考。

参考文献

1 Corcos G M. The structure of the turbulent pressure field in boundary layer flows. *J. Fluid Mech.*, 1964; **18**: 353—378

2 Shashaty A J. The effective lengths for flow noise of hydrophones in a ship-towed linear array. *J. Acoust. Soc. Am.*, 1982; **71**(4): 886—890

3 Francis S H, Slazk M, Berryman J G. Response of elastic cylinders to convective flow noise I. Homogeneous, layered cylinders. *J. Acoust. Soc. Am.*, 1984; **75**(1): 166—172

4 汤渭霖，吴 一. TBL 压力起伏激励下粘弹性圆柱壳内的噪声场：I 噪声产生机理. 声学学报，1997; **22**(1): 60—69

5 张道礼等. 拖曳线列阵声呐护套用聚氨酯的声学性能及材料体系. 功能材料，2002; **33**(2): 145—147

6 张海澜等. 井孔的声场和波. 北京：科学出版社，2004: 226—233

7 何祚镛，赵玉芳. 声学理论基础. 国防工业出版社，1981: 96—102

8 杨秀庭. 矢量水听器及其在声呐系统中的应用研究. 博士学位论文，中国科学院声学研究所，2006: 70—72

水下目标辐射噪声中单频信号分量的检测：理论分析①②

李启虎 李 敏 杨秀庭
(中国科学院声学研究所 北京 100190)
2007 年 10 月 29 日收到
2008 年 2 月 18 日定稿

摘要 在理想情况下高斯噪声背景下高斯信号的最佳检测问题已受到很多关注，它是被动声呐采用能量检测的理论基础之一。但是理论与实验都已证明，舰船辐射噪声中存在着单频信号分量。在小信噪比下，单频信号的检测必须采用有别于宽带能量检测的方法。本文讨论在小信噪比下检测单频信号分量的几种不同的理论方法，包括自相关检测、FFT 分析、自适应线谱增强等，计算了它们的增益。指出分段 FFT 检测具有明显的优势，对可能出现的单频信号分量的频率漂移具有较好的宽容性。
PACS 数： 43.60

The detection of single frequency component of underwater radiated noise of target: theoretical analysis

LI Qihu LI Min YANG Xiuting
(*Institute of Acoustics, Chinese Academy of Sciences* Beijing 100190)
Received Oct. 29, 2007
Revised Feb. 18, 2008

Abstract In the ideal assumption, the optimum detection problem of Gaussian signal in the background of Gaussian noise have attracted many research works, it is the one of theoretical basis of passive sonar detection. The theoretical analysis and experiment proved that there exist single frequency signal component in radiated noise of underwater target. In the situation of low signal to noise ratio, it is necessary to adopt new detection method, which is different with broadband energy detection. Several kinds of theoretical method for detecting single frequency component are described in this paper, including autocorrelation detection, FFT analysis, adaptive line enhancing etc. The system gain are derived. It is showed that the segment FFT detection has obvious advantage, it is more robust for the frequency shifting, which might have in signal propagation.

引言

水下目标辐射噪声的检测、定向和识别一直是被动声呐面临的重要课题。在传统的宽带检测体制下，已经证明了，在高斯噪声背景下高斯信号的最佳检测器是能量检测器[1]。

但是，理论与实验都已证明，水下目标的辐射噪声中具有丰富的单频分量，特别是在低频率段(100 Hz 以下)，由于螺旋桨转动时切割水体会产生低频信号，这些低频分量有的是直接以可加性的形式出现在辐射信号中，有的则是被船体本身的振动调制到较高的频段[2-3]。总而言之，水下目标辐射噪声中单频分量的检测是一个十分重要的课题，它有可能为低噪声、安静型潜艇的检测提供一种新的途径。

低频信号的检测问题当然与低频信号在海洋中传播的问题联系在一起，只有从理论与实践中找出低频信号的传播特性，才能为低频信号的检测提供依据。

S.V.Burenkov 等俄罗斯科学家在上世纪九十年代曾对 228 Hz 低频信号的传播进行过试验，其接收距离大于 9000 km[4]。他们有两个重要发现：一个是信号的幅度衰落现象比较严重，在观察的 5 min 内，起伏值超过 20 dB; 另一个是信号的相位具有“令人难以置信的稳定性”。美国 D.F.Worcester 和 R.C.Spindel

① 国家自然科学基金项目(60532040).
② 声学学报, 2008, 33(3): 193-196.

在一份总结多国参加的 ATOC(Acoustic Thermometry of Ocean Climate) 计划执行情况的报告中[5], 提到, 美国在 1995 — 1999 年之间利用美国海军的 14 个 SOSUS 接收阵, 记录 57 Hz, 75 Hz 信号的传播数据, 最远距离为 3900 km, 单频信号的传输结果令人满意。他们特别指出, 带宽为 1~2 Hz 滤波器所提取的信息可以用于目标识别。

由此可见对甚低频信号 (频率在 100 Hz 以下) 进行检测的前景还是比较乐观的。

本文将探讨高斯宽带噪声背景下, 单频信号分量的检测问题。首先给出待检测信号的模型。然后讨论对这种信号模型的信号检测方法, 包括能量检测、自相关检测、 FFT 分析和自适应线谱增强。

文中还专门讨论了针对长时间积分的两种 FFT 分析方法, 指出, 对于传播过程中有可能产生频率漂移的信号, 一种分段 FFT 技术有利于微弱单频信号的提取。

1 信号模型

设水下目标辐射噪声可以表达为:

$$x(t)=\sum_{k=1}^{N}A_k\cos(2\pi f_kt+\varphi_k)+n(t), \tag{1}$$

其中 A_k 为单频信号分量的幅度, φ_k 为单频信号的随机相位, $n(t)$ 为高斯随机噪声。在式 (1) 中我们假定 N 个分量相互独立, 并且 φ_k 和 $n(t)$ 也相互独立, 同时假定 φ_k 在 $[0,2\pi]$ 内均匀分布。

假定

$$E[n(t)]=0,\quad \mathrm{Var}[n(t)]=\sigma_n^2. \tag{2}$$

由此可知式 (1) 中第 k 个单频信号分量的信噪比是:

$$(\mathrm{SNR})_{\mathrm{in},k}=\frac{A_k^2}{2}\Big/\sigma_n^2. \tag{3}$$

在实际应用中, 我们面临的是微弱信号的检测, 也就是

$$(\mathrm{SNR})_{\mathrm{in},k}\ll 1 \tag{4}$$

的情况。

由于单频信号分量的个数在我们下面的有关检测方法的讨论中没有特殊的作用, 所以我们把信号模型简化为:

$$x(t)=A\cos(2\pi f_0t+\varphi)+n(t), \tag{5}$$

其中 f_0 为我们关心的单频信号分量的频率, φ 是随机相位, 在 $[0,2\pi]$ 内均匀分布, 与 $n(t)$ 相互独立。

我们下面讨论的方法, 虽然是针对式 (5) 模型展开的, 但是不难立刻推广到 $N>1$ 的情形。

2 几种检测模型的比较

2.1 能量检测器

由式 (5) 表达的信号的输入信噪比是:

$$(\mathrm{SNR})_{\mathrm{in}}=\sigma_s^2/\sigma_n^2=A^2/2\sigma_n^2, \tag{6}$$

其中 $\sigma_n^2=E[n^2(t)]$ 是噪声的功率。

假如噪声的功率已知, 那么能量检测不失为一种好的检测方法, 令

$$\begin{aligned}y(t)=&\frac{1}{T}\int_{t-T}^{t}x^2(u)\mathrm{d}u=\\&\frac{1}{T}\int_{t-T}^{t}[A\cos 2\pi f_0u+n(u)]^2\mathrm{d}u.\end{aligned} \tag{7}$$

能量检测器的输出信噪比是:

$$(\mathrm{SNR})_{\mathrm{out}}=\frac{E\left[y|_{\text{有信号}}\right]-E\left[y|_{\text{无信号}}\right]}{\left\{\mathrm{Var}\left[y|_{\text{无信号}}\right]\right\}^{1/2}}. \tag{8}$$

它实际上是信号直流跳变与平均噪声起伏值的比。不难证明[6], 能量检测器的系统增益:

$$G=\frac{(\mathrm{SNR})_{\mathrm{out}}}{(\mathrm{SNR})_{\mathrm{in}}}=\left\{\frac{T}{2\int_0^T\left(1-\frac{\tau}{T}\right)\rho_n^2(\tau)\mathrm{d}\tau}\right\}^{1/2}, \tag{9}$$

其中 $\rho_n(\tau)$ 为噪声相关系数。

式 (9) 也可以写成:

$$G=\sqrt{2TW}. \tag{10}$$

这个式子的物理意义是清楚的, 即能量检测器的系统增益是带宽时间乘积的开根号。在带宽一定的条件下, 积分时间每增加一倍, 增益能提高 $\sqrt{2}$ 倍 (1.5 dB)。

能量检测器检测单频信号的前提是假定噪声功率已知, 或至少可以用某种统计方法加以预测, 这一点在实际使用中不一定总能做到。

2.2 自相关检测器

另外一种可行的办法就是自相关检测法, 图 1 给出了这种检测法的方框图。因为单频信号无论经过多长的延时, 其相关函数总会在某一相位上取得极大值, 而宽带噪声的相关函数, 随着延时值的增加很快衰减, 由此就容易把单频信号提取出来。

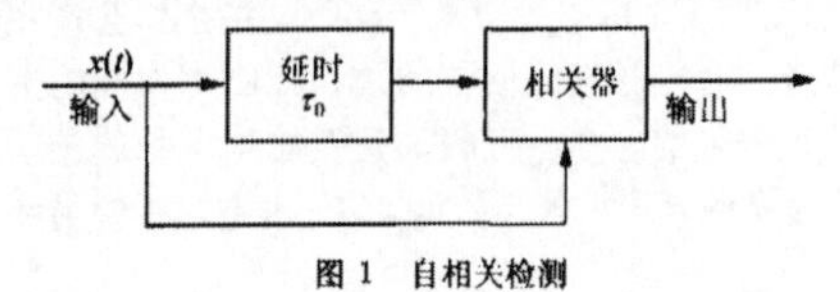

图 1 自相关检测

自相关检测的模型如下:

$$z(t)=\frac{1}{T}\int_{t-T}^{t}x(u)x(u-\tau_0)\mathrm{d}u. \tag{11}$$

我们看到:

$$E[z(t)]=E[x(t)x(t-\tau_0)]=\frac{A^2}{2}\cos 2\pi f_0\tau_0+R_n(\tau_0) \tag{12}$$

当 τ_0 足够大, 而 $f_0\tau_0=$ 整数, 那么:

$$E[z(t)]\approx\frac{A^2}{2}.$$

这就为自相关检测系统提供足够的输出信噪比。可以证明, 这种系统的增益也可以用式 (10) 来表示。

2.3 窄带滤波 (FFT 分析) 器

在宽带噪声背景下检测单频信号的一种较好的方法是窄带滤波。假定信号的频率为 f_0, 如果能把一个窄带滤波器的中心频率对准 f_0, 那么它只让信号分量通过而把大部分噪声滤除。

设 $n(t)$ 在 $[0,W]$ 内的功率谱密度为 $\sigma_n^2/2W$, 即:

$$K_n(f)=\begin{cases}\sigma_n^2/2W, & |f|\le W,\\ 0, & \text{其他}\end{cases} \tag{13}$$

那么噪声功率是 $\sigma_n^2=\int_{-W}^{W}K_n(f)\mathrm{d}f$ 。

设滤波器的带宽为 $\Delta f/2$, 即:

$$[-f_0\pm\Delta f/4],\quad [f_0\pm\Delta f/4].$$

该滤波器输出的信号功率为 $(A^2/2)=\sigma_s^2$, 噪声功率 $\Delta f\sigma_n^2/2W$, 于是:

$$(\mathrm{SNR})_{\mathrm{out}}=\frac{\frac{A^2}{2}}{\Delta f\sigma_n^2/2W}=(\mathrm{SNR})_{\mathrm{in}}\frac{2W}{\Delta f}. \tag{14}$$

系统增益为:

$$G=\frac{2W}{\Delta f}\approx 2WT. \tag{15}$$

由此可知窄带滤波的增益比能量检测或自相关检测要大得多。

图 2 是理论计算的结果。

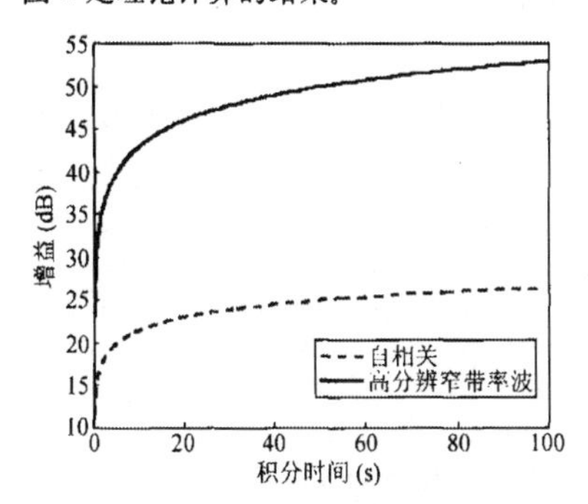

图 2 窄带滤波和自相关检测增益的比较

实现窄带滤波的方法比较多, 用 FFT 计算到达信号的频谱就是一种。

设 f_s 为采样频率, N 为采样点数, 则频率分辨率为:

$$\Delta f=f_s/N=1/T_s, \tag{16}$$

其中 T_s 为采样周期, $T_s=1/f_s$ 。

把式 (16) 代入式 (15), 可以得到:

$$G=2WNT_s \tag{17}$$

作 FFT 分析时, 只要 f_s 足够大, 分析的频带就会覆盖所检测的未知频率, 从而达到自动提取信号单频分量的目的。

2.4 自适应线谱增强器 (ALE)

除了对信号进行 FFT 分析提取信号频率信息的方法之外, 自适应线谱增强器 (ALE, Adaptive Line Enhancer) 也是一种增强单频信号分量的有效方法, 其原理如图 3 所示。

原始信号为 $x(t)=s(t)+n(t)$, 假定 $s(t)$ 中以单频信号分量为主, $n(t)$ 是宽带噪声, 在对 $x(t)$ 做延迟 τ_0 后, 得到信号:

$$y(t)=x(t-\tau_0)=s(t-\tau_0)+n(t-\tau_0). \tag{18}$$

通过自适应滤波器 $H(f)$ 的调节, 使得 $H(f)$ 的输出 $z(t)$ 和 $x(t)$ 之差:

$$\varepsilon(t)=z(t)-x(t) \tag{19}$$

之均方值极小, 由于 $n(t)$ 和 $n(t-\tau_0)$ 无关 (当 τ_0 足够大), 而 $s(t)$ 和 $s(t-\tau_0)$ 始终相关, 由此使 $\varepsilon(t)$ 极小也就等价于把 $n(t)$ 抑制到最低, 从而达到增强单频信号分量的作用 [7] 。

通常线谱增强系统可以在 FFT 之前应用, 这样的处理会使全系统的增益更高。

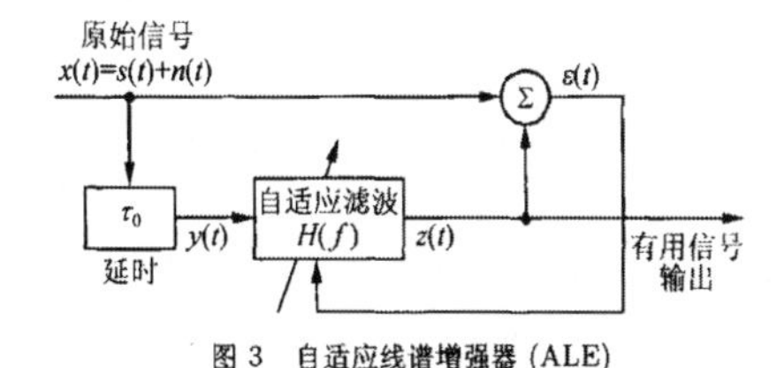

图 3 自适应线谱增强器 (ALE)

3 DFT 分析参数的选取

前面已经指出窄带滤波是提取信号中单频分量的有效方法, 对于数字式声呐来说, 当然需要对信号进行采样及 A/D 转换, 仍旧假定输入信号是:

$$x(t)=A\cos(2\pi f_0t+\varphi)+n(t).$$

设采样间隔为 T_s, 那么信号样本为:

$$x(kT_s) = A\cos(2\pi f_0 kT_s + \varphi) + n(kT_s). \tag{20}$$

为讨论简单起见, 把 (20) 式改写为:

$$x(k) = s(k) + n(k), \quad k = 0, 1, 2, \cdots \tag{21}$$

信号的 N 点 DFT 为:

$$X(l) = \sum_{k=0}^{N-1} x(k)\exp(-2\pi \mathrm{j}kl/N). \tag{22}$$

频率分辨率是:

$$\Delta f = f_s/N = 1/NT_s.$$

从式 (17) 我们已经知道, N 越大, 系统增益就越大。但是, 前提条件是单频信号分量 $A\cos(2\pi f_0 kT_s + \varphi)$ 的频率是稳定的, 且背景噪声是平稳的。这两个条件缺一不可, 因为一旦频率 f_0 有某种漂移, 当积分时间加长时, Δf 变小, DFT 分析无法提取频率 f_0 所在的幅度谱值; 如果背景噪声不平稳, 噪声的起伏值就不会随着时间的加长而下降, 从而增益也不会随着时间的增长而提高。

我们下面给出一种分段 DFT 的算法, 可以较好地解决这一问题。

我们首先把信号的样本分为互不重叠的 M 段, 每一段长为 N_0(比如 $N_0 = 1024$), 于是把信号样本序列标注为:

$$x(mN_0 + k), \quad k = 0, \cdots, N_0 - 1$$
$$M = 0, \cdots, M - 1$$

第 m 段的 N_0 点 DFT 是:

$$Z_m(l) = \sum_{k=0}^{N_0-1} x(mN_0 + k)\exp(-2\pi \mathrm{j}kl/N_0), \quad (l = 0, 1, \cdots, N_0 - 1) \tag{23}$$

每一段的 DFT 的频率分辨率是:

$$\Delta f = 1/N_0 T_s.$$

它显然低于长为 MN_0 点的 DFT 的频率分辨率, 在满足一定条件下对 M 段 $Z_m(l)$ 求和, 得到:

$$Z(l) = \frac{1}{M}\sum_{m=0}^{M-1} Z_m(l).$$

假定:

$$\mathrm{Int}[f_0/\Delta f] = n_0, \tag{24}$$

这里 $\mathrm{Int}[x]$ 表示和 x 最接近的整数。

在这种情况下, 可以证明 $Z(n_0)$ 中的信号成分和噪声成分的比是易于直接计算的。事实上

$$Z(n_0) = \frac{1}{M}\sum_{m=0}^{M-1} Z_m(n_0) = \frac{1}{M}\sum_{m=0}^{M-1}\sum_{k=0}^{N_0-1} x(mN_0+k)\exp(-2\pi \mathrm{j}kn_0/N_0) \tag{25}$$

利用式 (25) 和式 (26) 不难证明, 这种分段 DFT 的输出信噪比为:

$$(\mathrm{SNR})_{\mathrm{out}} = (\mathrm{SNR})_{\mathrm{in}}\, 2N_0 M W T_s. \tag{27}$$

从而增益为:

$$G = 2WN_0 MT_s, \tag{28}$$

这与式 (15) 和式 (17) 完全一致。

当 (24) 式不能得到满足时 (这是大多数的情况) 增益达不到式 (15) 所给出的值, 但是数值仿真的结果证明: 适当选择参数, 仍可以得到较高的增益而对频率漂移较少敏感。

分段 DFT 的优点是在保证检测增益不断随时间的增长而提高的条件下, 又保证分辨单元的一定宽度, 这对于长距离传播的单频信号的检测会有明显的好处。

4 结论

宽带噪声背景下单频信号的检测具有重要的理论与实际意义。本文分析了能量检测、自相关检测和窄带滤波的不同检测方法, 分别给出了检测增益的理论计算公式, 指出分段 DFT 分析技术是一种具有相当应用价值的检测手段。

参 考 文 献

1 McDonough R N, Whalen A D. Detection of signals in noise. 2nd Ed. Academic Press, USA, 1995

2 Urick R J. Principles of underwater sound. 3rd Ed. McGraw Hill, USA, 1983

3 Ross D. Mechanics of underwater noise. Pergmon Press, NY, 1976

4 Burenkov S V *et al.* Heard Island feasibility test: long-range sound transmission from Heard Island to Krylov underwater mountain. *J. Acoust. Soc. Am.*, 1994; **96**(4): 2458—2463

5 Worcester P F, Spindel R C. North Pacific acoustic laboratory. http://www.apl.washington.edu

6 李启虎. 数字式声呐设计原理. 安徽: 安徽教育出版社, 2003

7 李启虎. 用于拖曳式线列阵的一种新的线谱增强系统. 声学学报, 1998; **13**(3): 167—173

8 李启虎. 水下目标辐射噪声中单频信号分量的检测: 数值仿真. 声学学报, 2008; **33**(4)

水下目标辐射噪声中单频信号分量的检测: 数值仿真①②

李启虎 李敏 杨秀庭
(中国科学院声学研究所 北京 100190)

2007 年 10 月 29 日收到
2008 年 3 月 12 日定稿

摘要 继续深入讨论强干扰背景噪声下单频信号分量的检测问题。根据理论分析进行了一系列的数值仿真，验证了理论中关于处理增益与时间成正比的结论，在存在频率漂移时，分段 DFT 分析的宽容特性。同时讨论检测系统的参数选择问题，包括采样间隔、分段 DFT 的长度、频率分辨力、单频信号分量的频率漂移对检测系统性能的影响。

PACS 数: 43.60

The detection of single frequency component of underwater radiated noise of target: digital simulation

LI Qihu LI Min YANG Xiuting
(*Institute of Acoustics, Chinese Academy of Sciences* Beijing 100190)

Received Oct. 29, 2007
Revised Mar. 12, 2008

Abstract The detection problem of weak single frequency component in the background of strong noise is discussed. Based on the theoretical analysis, a series of digital simulation are carried out to verify the theoretical conclusion, including: the system gain, the robust performance of segment DFT analysis in the case of existence of frequency shifting. The selection principle of various parameters is discussed, including sampling duration, length of segment DFT, frequency resolution etc. The effect of frequency shifting on the system performance is also illustrated.

1 引言

在文献 1 中已说明了检测水下目标辐射噪声中单频分量的重要意义以及国外研究者在低频水声信号传播特性方面的实验工作。

在本文中，我们将在文献 1 中所描述的理论框架内进行必要的数值仿真，并根据实际水声信道可能对低频信号传播所产生的畸变进行模拟。对这些模拟结果的分析和归纳将有助于实际声纳系统的设计。

数值仿真的基本内容如下:

(1) 不同信噪比下高斯噪声背景下单频信号的时域波形。

(2) 检测性能和积分时间的关系。

(3) 分段 DFT 和不分段 DFT 性能的比较。

(4) 单频信号有频率漂移时不同检测方法的性能比较。

2 检测性能和积分时间的关系

作为系统模拟的基本数据，我们构造了一个高斯背景下的单频信号的模型:

$$x(k) = s(k) + n(k), \tag{1}$$

$$s(k) = A\cos(2\pi f_0 k T_s + \varphi), \tag{2}$$

其中 A 为信号幅度，用于调节信噪比; $n(k)$ 是均值为 0，方差为 1 的高斯噪声; φ 是一个在 $[0, 2\pi]$ 中均匀分布的随机相位。在整个模拟过程中，取 $f_0 = 300$ Hz, $T_s = 200\ \mu s$。

图 1 给出了原始信号在 $(\text{SNR})_{\text{in}} = A^2/2\sigma_n^2 =$ 0 dB, −10 dB, −20 dB 时的时间波形。

---

① 国家自然科学基金项目(60532040).

② 声学学报, 2008, 33(4): 289-293.

从图 1 可以看出，在 $(\mathrm{SNR})_{\mathrm{in}}=0$ dB 时隐约可以看出有一正弦波出现，但在 $(\mathrm{SNR})_{\mathrm{in}} \ll 0$ dB 时，几乎无法确认有正弦信号的存在。

对 $x(k)$ 的长为 N 点的 DFT 是：

$$X(l)=\sum_{k=0}^{N-1} x(k)\exp(-2\pi \mathrm{j} kl/N),\ l=0,1,\cdots,N-1 \tag{3}$$

因为 $x(k)$ 是实数，所以我们无需计算从 0 至 $N-1$ 的谱值，通常只需计算从 0 至 $(N/2)-1$ 就可以了。

图 2 给出了一系列 DFT 分析的结果，相关的仿真内容见表 1。

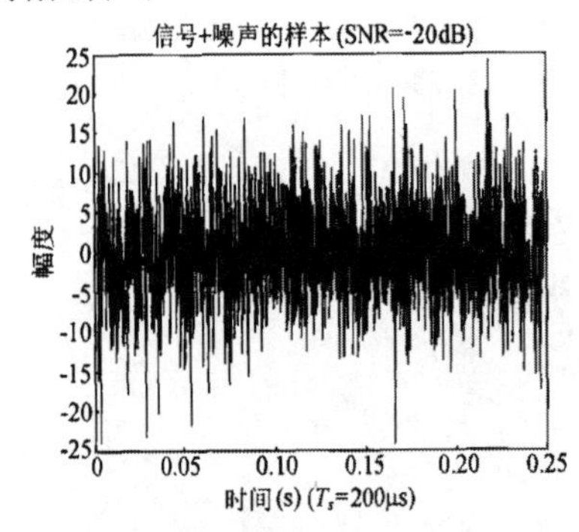

(a) $(\mathrm{SNR})_{\mathrm{in}}=-20$ dB

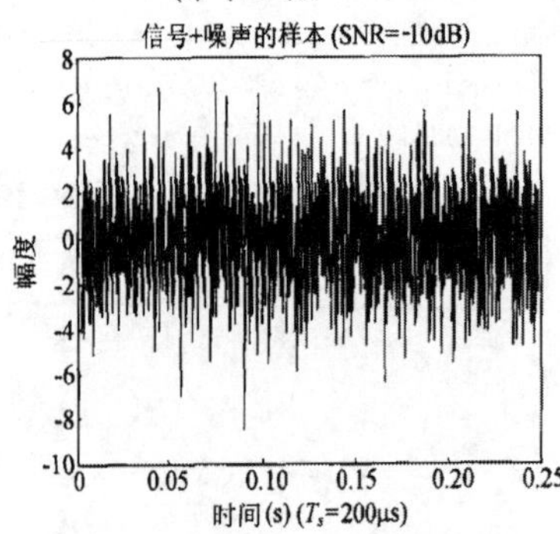

(b) $(\mathrm{SNR})_{\mathrm{in}}=-10$ dB

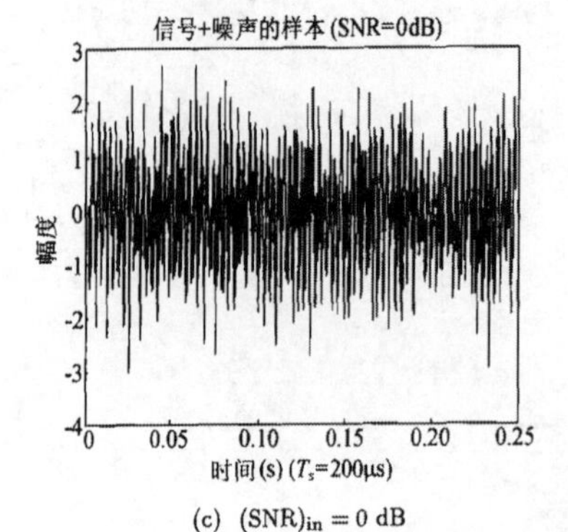

(c) $(\mathrm{SNR})_{\mathrm{in}}=0$ dB

图 1　数值仿真的时域波形

表 1　不分段 DFT 数值模拟内容

图号	$(\mathrm{SNR})_{\mathrm{in}}$	分段数	DFT 长度
2(a)	0 dB	1	1024
2(b)	−10 dB	1	1024
2(c)	−20 dB	1	1024
2(d)	−20 dB	1	16×1024
2(e)	−20 dB	1	64×1024

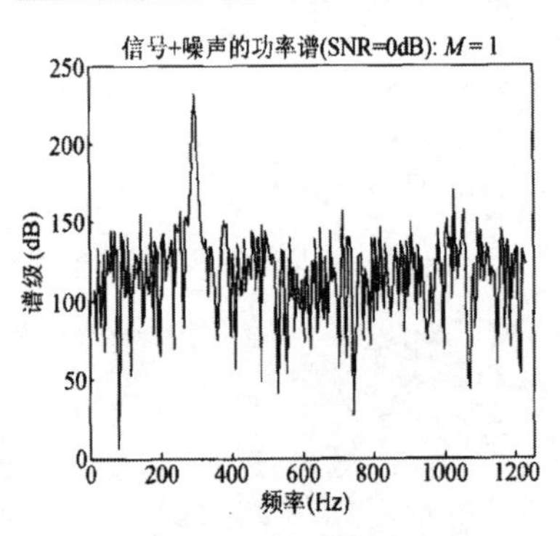

(a) $M=1$, $(\mathrm{SNR})_{\mathrm{in}}=0$ dB

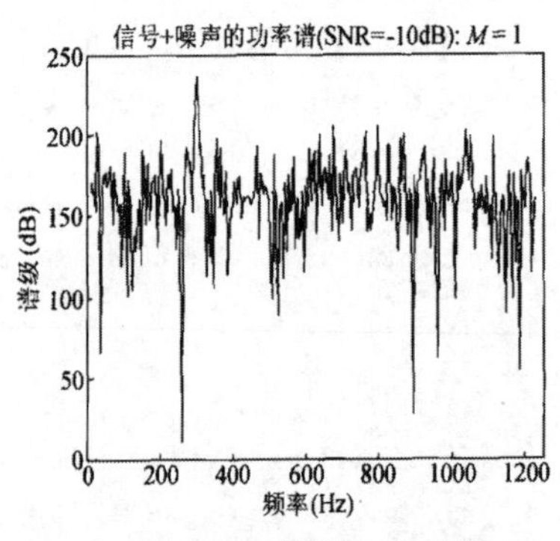

(b) $M=1$, $(\mathrm{SNR})_{\mathrm{in}}=-10$ dB

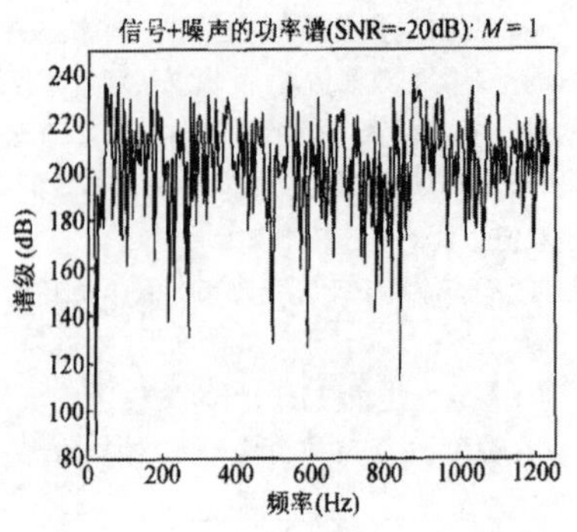

(c) $M=1$, $(\mathrm{SNR})_{\mathrm{in}}=-20$ dB

图 2

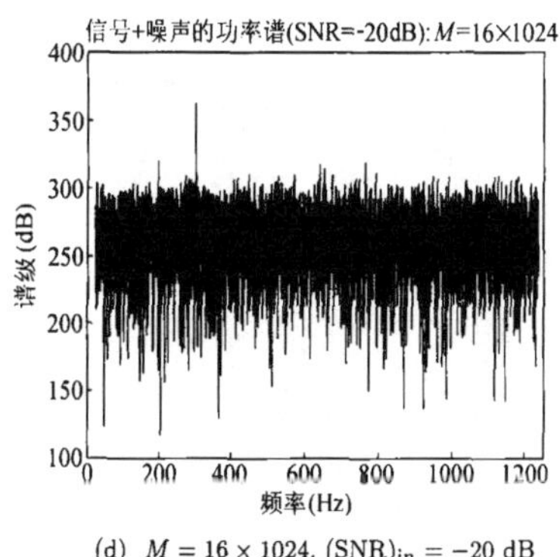

(d) $M = 16 \times 1024$, $(\mathrm{SNR})_{\mathrm{in}} = -20$ dB

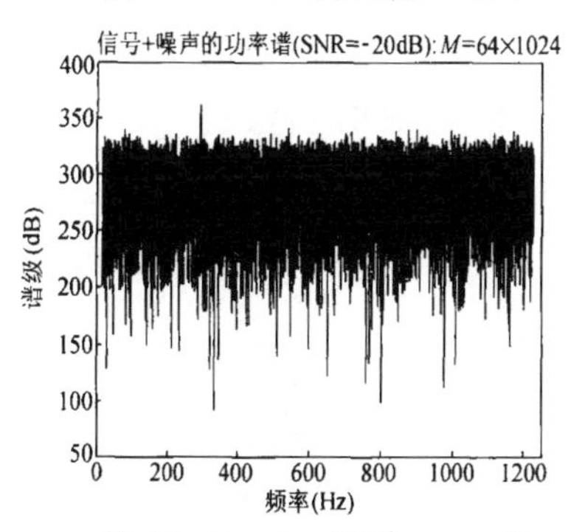

(e) $M = 64 \times 1024$, $(\mathrm{SNR})_{\mathrm{in}} = -20$ dB

图 2 DFT 分析结果

我们首先来估算一下当 $N = 1024$ 点时的系统增益，由于 $T_s = 200\ \mu s$，所以 $f_s = 5$ kHz，不妨认为 $W = 1000$ Hz，这时系统增益是：

$$G = 2\tau W = 2NT_sW \approx 26 \text{ dB}.$$

我们从图 2 中可以看到，对于 −20 dB 的信号检测效果已比较差了。但是当把数据长度由 1024 点 (积分时间 0.2 s) 增加至 16×1024(积分时间 3.2 s) 和 64×1024 点 (积分时间 13.1 s) 时，检测性能迅速改善。比较图 2(d) 和图 2(e) 我们发现，当积分时间由 3.2 s 增加到 13.2 s 时，信号谱的幅度增加得不明显，只是整个频谱的起伏减少了。

图 3(a)—图 3(c) 是分段 DFT 的情形，相关内容见表 2。

表 2　分段 DFT 模拟内容

图号	$(\mathrm{SNR})_{\mathrm{in}}$	分段数	DFT 长度
3(a)	−20 dB	16	16×1024
3(b)	−20 dB	64	64×1024
3(c)	−20 dB	64	64×1024

从图中可以看出，分段 DFT 的效果是好的，当 $M \times N_0(N_0 = 1024$, M 为分段数) 和不分段 DFT 长度 N 相同时，两者的性能也是类似的。分段数的增加对检测性能的改善明显，同时又保持了每一分辨单元的频率宽度。

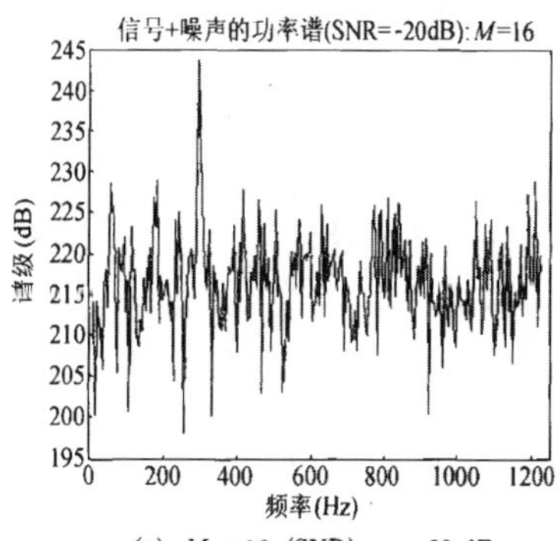

(a) $M = 16$, $(\mathrm{SNR})_{\mathrm{in}} = -20$ dB

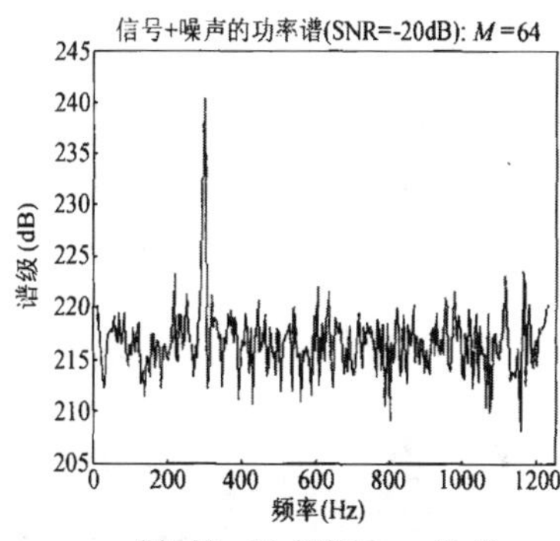

(b) $M = 64$, $(\mathrm{SNR})_{\mathrm{in}} = -20$ dB

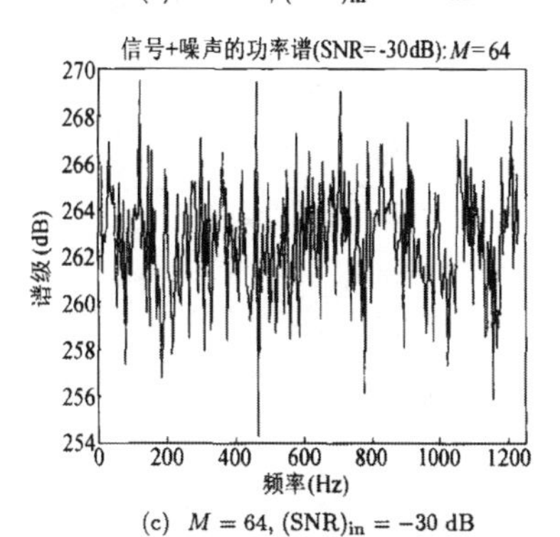

(c) $M = 64$, $(\mathrm{SNR})_{\mathrm{in}} = -30$ dB

图 3　分段 DFT 数值模拟结果

3　频率漂移的影响

文献 1 中曾提到俄罗斯科学家对低频信号进行水声传播实验时，信号频率的稳定性非常好，但考虑到长距离的信道，对单频信号的传播总会引起一定程度相位失真，所以有必要进行一些数值仿真，看一下当存在这类畸变时，对检测系统会有什么要求。

目前还缺乏实际信道中频率漂移的模型，我们假定在传播过程中频率随时间有某种微小的变化：

$$f(t) = f_0 + \Delta f(t), \tag{4}$$

其中 f_0 是频率不变的部分，$\Delta f(t)$ 则是随时间变化的微小起伏。根据国外的研究结果 (见文献 1)，这种变化是非常小的。模拟的情况是，频率漂移为 $0.3\% f_0$，并且在前 100 s 就达到漂移的最大值 (见图 4)。

系统模拟的内容见表 3，模拟的结果见图 5。

表 3　频率漂移模拟内容

图号	$(\text{SNR})_{\text{in}}$	分段数	DFT 长度	备注
5(a)	−20 dB	1	64×1024	有频率漂移
5(b)	−20 dB	64	1024	有频率漂移
5(c)	−24 dB	64	1024	无频率漂移
5(d)	−24 dB	64	1024	有频率漂移
5(e)	−24 dB	1	64×1024	无频率漂移
5(f)	−24 dB	1	64×1024	有频率漂移

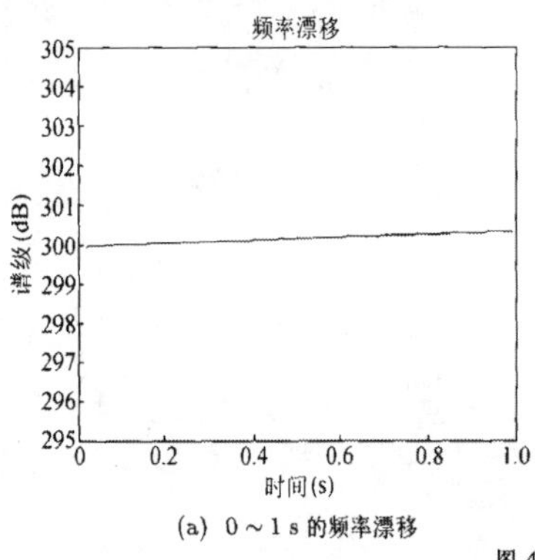

(a) 0 ~ 1 s 的频率漂移

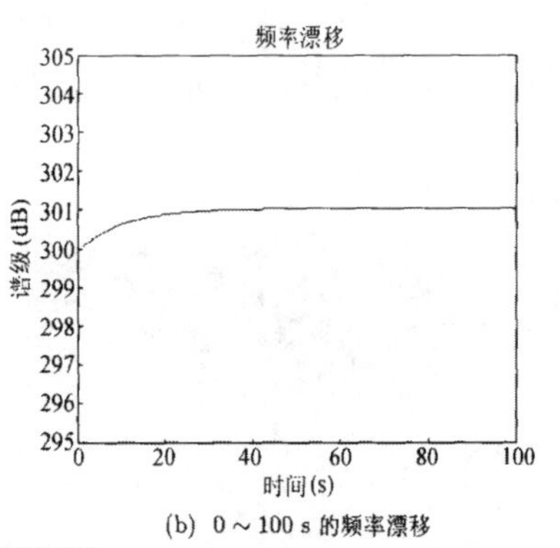

(b) 0 ~ 100 s 的频率漂移

图 4　频率漂移的数值模拟

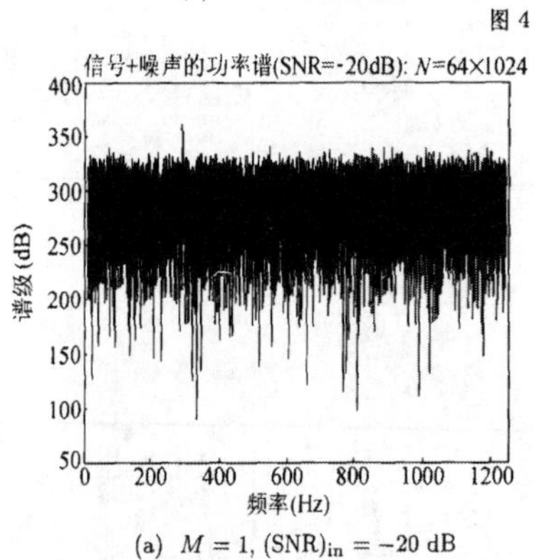

(a) $M = 1$, $(\text{SNR})_{\text{in}} = -20$ dB

64 × 1024 点 DFT 有频率漂移

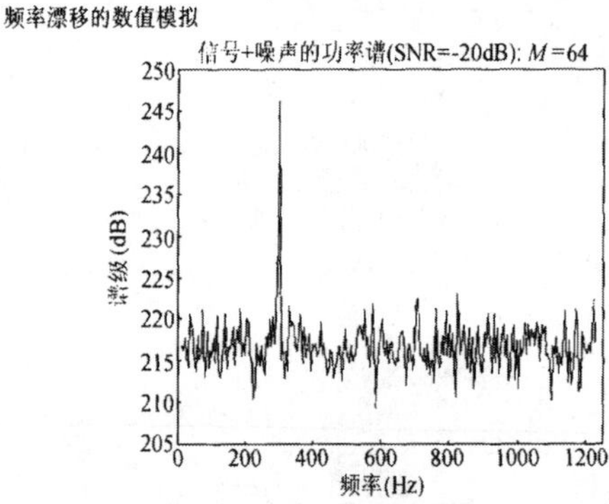

(b) $M = 64$, $(\text{SNR})_{\text{in}} = -20$ dB

64 × 1024 点 DFT 有频率漂移

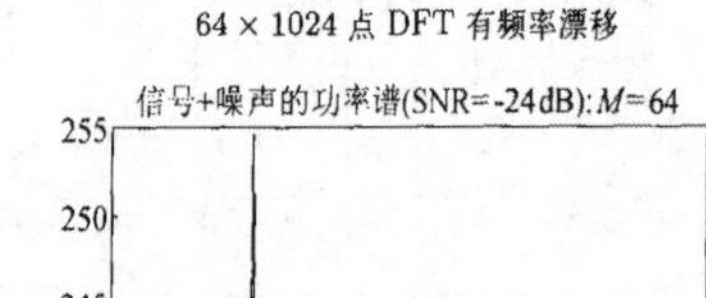

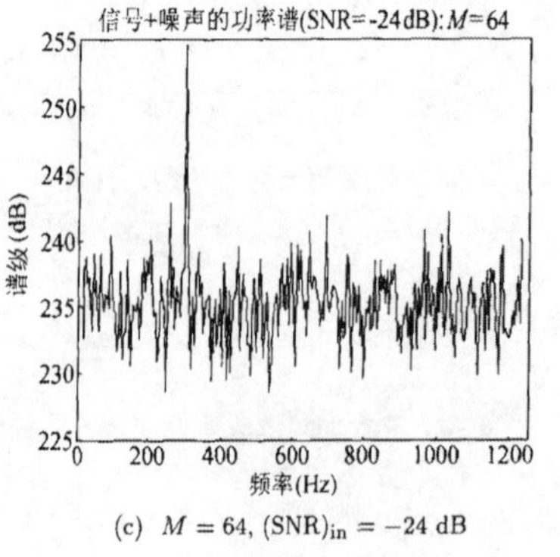

(c) $M = 64$, $(\text{SNR})_{\text{in}} = -24$ dB

64 × 1024 点 DFT 无频率漂移

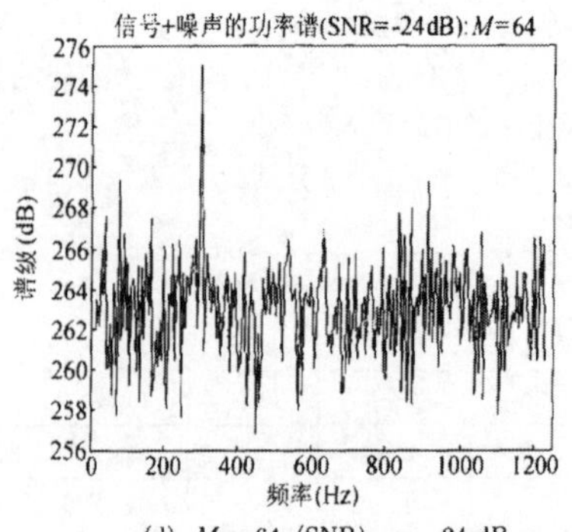

(d) $M = 64$, $(\text{SNR})_{\text{in}} = -24$ dB

64 × 1024 点 DFT 有频率漂移

图 5

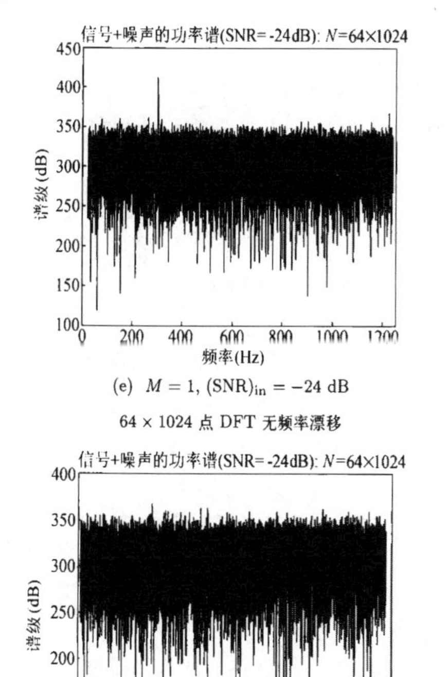

(e) $M = 1$, $(\mathrm{SNR})_{\mathrm{in}} = -24$ dB

64 × 1024 点 DFT 无频率漂移

(f) $M = 1$, $(\mathrm{SNR})_{\mathrm{in}} = -24$ dB

64 × 1024 点 DFT 有频率漂移

图 5　有频率漂移时的情况

前面已指出，对于 64 × 1024 点的 DFT, 相当于积分时间 13.1 s, 而频率分辨力是:

$$\Delta f = 0.0763 \text{ Hz}$$

也就是说每一频率分辨单元的宽度只有 0.0763 Hz, 而对于分段的 DFT, 同样是 13.1 s 的数据，每一段的频率分辨力都是:

$$\Delta f = 4.88 \text{ Hz}$$

这样的频率分辨力对频率漂移是不敏感的。

从图 5 可以看出, 当有频率漂移时, 对 $(\mathrm{SNR})_{\mathrm{in}} = -20$ dB, 64×1024 不分段 DFT 的检测结果还是可以的。但当 $(\mathrm{SNR})_{\mathrm{in}} = -24$ dB 时, 同样的算法, 检测结果已不行了 (图 5(f))。但是，换成分段 DFT 算法，同样的数据，检测性能明显改善 (见图 5(d))。

4　结论

数值仿真的结果较好地验证了理论分析的正确性:

(1) 窄带 DFT 分析可以有效地从噪声背景中提取单频信号分量，检测增益与积分时间成正比。

(2) 当 (系统增益 + 输入信噪比)≥ 10 dB 时，可以从频谱分析图中较好地识别出信号以及它的频率，从输出信噪比的角度来看，自动检测和识别也是可能的。

(3) 分段 DFT 和不分段 DFT 在总的数据长度一样的条件下，具有相同的检测增益，但是分段 DFT 检测方法对信号的频率漂移具有较好的宽容特征。因此是一种比较理想的检测单频信号分量的方法。

参 考 文 献

1　李启虎等. 水下目标辐射噪声中单频信号分量的检测: 理论分析. 声学学报, 2008; **33**: 193—196

一种水下 GPS 系统及其在蛙人定位导航中的应用①

李 敏[1]，李启虎[1]，杨秀庭[2]

(1. 中国科学院声学研究所，北京 100190；2. 海军大连舰艇学院，辽宁大连 116018)

摘要：研究了一种适用于蛙人导航的水下 GPS 系统。针对蛙人执行水下任务所需的高精度导航，提出了一种由主动声纳浮标作为定位基站的 GPS 定位系统，介绍了该系统基于延时测量的定位原理和求解方法，给出了 Kalman 滤波器和扩展 Kalman 滤波器的设计，并通过数值仿真进行了验证，结果表明：为提高定位精度，在定位解算的基础上进行滤波平滑是必要的。

关键词：水下 GPS；导航；Kalman 滤波

中图法分类号：TB56　　**文献标识码**：A　　**文章编号**：1000-3630(2008)-06-0812-04

A type of underwater GPS system and its application to diver positioning and navigation

LI Min[1], LI Qi-hu[1], YANG Xiu-ting[2]

(1. *Institute of Acoustics, The Chinese Academy of Sciences, Beijing* 100190, *China*;
2. *PLA Naval Academy, Dalian* 100190, *Liaoning, China*)

Abstract: In order to supply the navigation service with high accuracy for the diver in underwater actions, a scheme of underwater GPS system which takes the active sonobuoy as a communication beacon, is proposed. The principle of target positioning is based on the measurements of sound propagation delay between the target and beacons, and the algorithm for solving the nonlinear equations of target′ s position is also introduced. Thereafter, the Kalman filter and the extended Kalman filter are designed for improving the accuracy of position estimation, and their performance is demonstrated via numerical simulations. The results show that it is necessary to integrate the smoothing techniques such as the Kalman filtering into the positioning solver to improve the navigation accuracy.

Key words: underwater GPS; navigation; Kalman filtering

1 引 言

近年来，为应对来自水下的恐怖威胁，反蛙人探测技术已成为水声工程领域的一个研究热点。为探测蛙人，保障重要的军事基地、港口和驻泊舰艇的安全，各主要海军国家都研制和装备了相应的反蛙人探测系统。由于采用被动方式探测水下蛙人这类安静型小目标需要足够密集的监视网点，在基阵布放和系统联网上较为复杂，当前的反蛙人探测系统以主动工作方式为主。

为有效地对抗水下蛙人，目前采用的主要作战样式是派遣己方蛙人部队进行拦截、围堵和格斗，以杀死或擒获敌方蛙人。这种作战样式的特点是兵力投放灵活，防御范围大，作战效果好，可在近距离至反蛙人声纳的最大探测距离上解决水下威胁。

为达成作战目的，提高作战效果，必须在港口、锚地等不同复杂程度的海区环境下为己方蛙人提供高可靠、高精度的定位导航系统。为此，本文提出了一种依托于主动式反蛙人探测系统的水下 GPS 定位导航系统，该系统只需在既有的反蛙人探测系统内嵌入水声通信模块，即可为蛙人在港口和近岸等浅海海区内进行水下作业时提供高精度的定位和导航服务，具有较小的系统复杂度和较低的硬件成本。

① 声学技术, 2008, 27(6): 812-815.

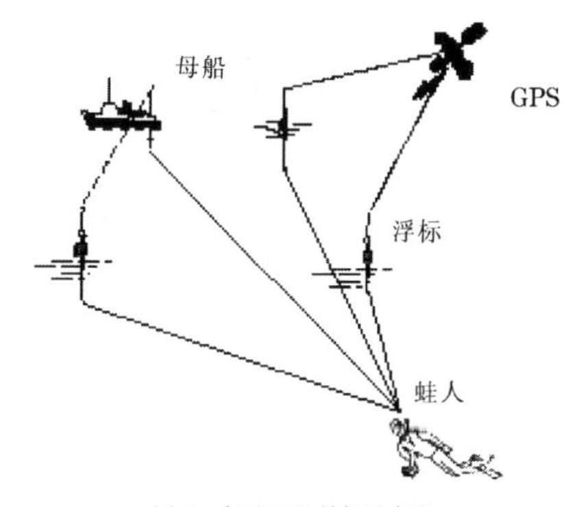

图 1 水下 GPS 结构示意图
Fig.1 The scheme of underwater GPS system

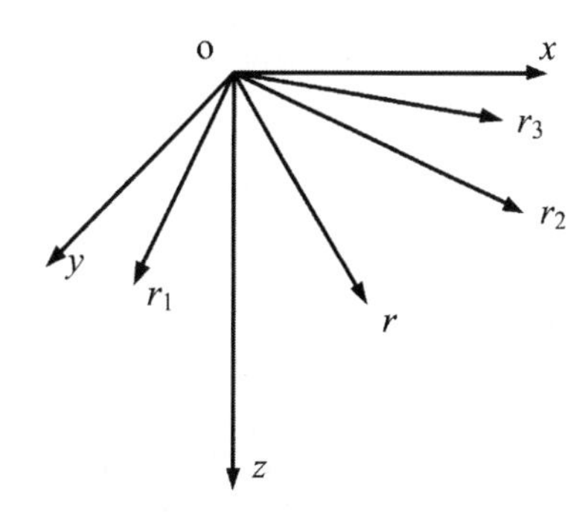

图 2 水下目标及 GPS 信标的位置关系概略图
Fig.2 The schematic positions of target and GPS beacons

2 系统构成

图 1 给出了一种母船(舰船)驻泊时采用的反蛙人探测系统的示意图,以此为例,来探讨用于蛙人定位导航的水下 GPS 系统。在各声纳浮标上加装 GPS 接收机和三维姿态传感器,以记录信标的实时位置信息;母船上安装 DGPS 系统,用于改进各浮标的位置估计精度和同步校准;各浮标上加装用于水下通信的编码调制等通信模块,可在母船上控制,以决定所发射的声脉冲是否携有信标识别码(用于确定所接收声讯息由哪个信标发射)和信标的实时位置信息。

在接收端,由水下蛙人携带被动定位导航系统,内设接收、解调、解码、定位解算、滤波和显示等模块,以一定的时间间隔显示蛙人在水下的位置信息,从而可为蛙人提供实时、高精度的声学导航服务。

3 定位算法

下面根据几何定位关系,给出水下 GPS 系统的定位算法。

3.1 数学描述

为实现水下目标的三维定位,最少需采用 4 个 GPS 信标构成定位导航系统,如图 2 所示。以工作母船上的 GPS 信标作为坐标原点,其余各信标的坐标为 $\boldsymbol{r}_i(x_i,y_i,z_i)$,$i$=1,3,目标位置为 $\boldsymbol{r}(x,y,z)$,在等温层条件下,声传播延时差与目标及信标的空间位置坐标之间满足以下关系

$$\begin{cases}|r-r_1|=ct_{10}\\|r-r_2|=ct_{20}\\|r-r_3|=ct_{30}\end{cases}\tag{1}$$

式中,c 为声速;$t_{i0}(i=1,2,3)$ 分别为 1、2、3 号信标到目标的声传播延时与参考信标到目标的声传播延时之差。通过求解式(1)即可获得目标的水下位置。

3.2 求解定位方程组

式(1)所给出的定位方程组是非线性的,可通过 Newton 迭代算法求得数值解。

先将式(1)写为

$$\begin{cases}|r-r_1|-ct_{10}=0\\|r-r_1|-ct_{20}=0\\|r-r_1|-ct_{30}=0\end{cases}\tag{2}$$

并令

$$F(\boldsymbol{r})=\begin{bmatrix}|r-r_1|-ct_{10}\\|r-r_1|-ct_{20}\\|r-r_1|-ct_{30}\end{bmatrix}\tag{3}$$

迭代时,根据式(4)进行更新[1]

$$\boldsymbol{r}_{new}=\boldsymbol{r}_{old}-F'^{-1}(\boldsymbol{r}_{old})F(\boldsymbol{r}_{old})\tag{4}$$

式中,“′”为求导运算。

需要注意的是,Newton 算法对初值选取是敏感的,选择不当会影响到算法的收敛[1]。

3.3 滤波平滑

对于象蛙人这类低机动的水下目标,可通过 Kalman类的各种滤波方法来进一步提高位置估计精度[2]。

取状态向量 $\boldsymbol{X}=[x,v_x,y,v_y,z,v_z]^T$ 和观测向量 $\boldsymbol{Z}=[x,y,z]^T$,并建立状态方程和观测方程。

状态方程:

$$\boldsymbol{X}(k)=\boldsymbol{\Phi}(k,k-1)\boldsymbol{X}(k-1)+\boldsymbol{G}(k)\boldsymbol{w}(k)\tag{5}$$

观测方程:

$$\boldsymbol{Z}(k)=\boldsymbol{H}(k)\boldsymbol{X}(k)+\boldsymbol{n}(k)\tag{6}$$

在式(5)、(6)中,$\boldsymbol{w}(k)$ 为状态转移噪声,若水下目标的运动加速度不为零,需要附加该项;$\boldsymbol{n}(k)$ 为测量噪声;$\boldsymbol{\Phi}$ 和 $\boldsymbol{H}$ 分别为状态转移矩阵和观

测矩阵。

在水下 GPS 定位导航系统中,声传播延时差是直接观测量,即

$$\boldsymbol{\tau}=\begin{bmatrix}\tau_{10}\\ \tau_{20}\\ \tau_{30}\end{bmatrix}=\frac{1}{c}\cdot\begin{bmatrix}|r-r_1|\\ |r-r_2|\\ |r-r_3|\end{bmatrix} \tag{7}$$

由式(7),通过 Newton 迭代法可以解算出各时刻所对应的目标位置坐标,即观测向量 $\boldsymbol{Z}(k)$,这相当于观测向量已知。

3.3.1 Kalman 滤波

滤波方程:

$$\hat{\boldsymbol{X}}(k)=\boldsymbol{\Phi}(k,k-1)(k-1)+$$
$$\boldsymbol{G}(k)[\boldsymbol{Z}(k)-\boldsymbol{H}(k)\boldsymbol{\Phi}(k,k-1)\hat{\boldsymbol{X}}(k-1)] \tag{8}$$

增益方程:

$$\boldsymbol{KG}(k)=\boldsymbol{N}_\varepsilon(k|k-1)\boldsymbol{H}^T(k)\cdot$$
$$[\boldsymbol{H}(k)\boldsymbol{N}_\varepsilon(k|k-1)\boldsymbol{H}^T(k)\boldsymbol{\Phi}(k,k-1)+\boldsymbol{N}_n(k)]^{-1} \tag{9}$$

方差方程:

$$\boldsymbol{N}_\varepsilon(k)=[\boldsymbol{I}-\boldsymbol{G}(k)\boldsymbol{H}(k)]\boldsymbol{N}_\varepsilon(k|k-1) \tag{10}$$

初始条件:

$$\hat{\boldsymbol{X}}(0)=m_X(0),\ \boldsymbol{N}_\varepsilon(0)=\boldsymbol{N}_X(0) \tag{11}$$

式(9)中,$\boldsymbol{N}_n(k)=\boldsymbol{E}\{\boldsymbol{n}(k)\boldsymbol{n}^T(k)\}$。

通过式(8)~(11),可求得各时刻目标位置向量的估计值。

3.3.2 扩展 Kalman 滤波

以上讨论了线性方程组的 Kalman 滤波,但方程组(1)本质上是非线性的,可使用扩展Kalman 滤波器进行滤波。考虑式(12)的非线性离散方程模型

$$\boldsymbol{X}_{k+1}=f_k(\boldsymbol{X}_k)+g_k(\boldsymbol{X}_k)\boldsymbol{w}_k$$
$$\boldsymbol{Z}_k=h_k(\boldsymbol{X}_k)+\boldsymbol{v}_k \tag{12}$$

式中,$\{\boldsymbol{w}_k\}$和$\{\boldsymbol{v}_k\}$均为零均值的高斯白噪声过程。

比较式(5)、式(6)和式(12),若令

$$\boldsymbol{F}_k=\frac{\partial f_k(\boldsymbol{X})}{\partial \boldsymbol{X}}\Big|_{X=\hat{X}_{k/k}}$$

$$\boldsymbol{H}^T{}_k=\frac{\partial h_k(\boldsymbol{X})}{\partial \boldsymbol{X}}\Big|_{X=\hat{X}_{k/k}}$$

$$\boldsymbol{G}_k=g_k(\hat{\boldsymbol{X}}_{k/k}) \tag{13}$$

将非线性函数 $f_k(\boldsymbol{X}_k)$、$g_k(\boldsymbol{X}_k)$ 在 $\hat{\boldsymbol{X}}_{k/k}$ 附近,$h_k(\boldsymbol{X}_k)$在$\hat{\boldsymbol{X}}_{k/k-1}$ 附近分别展成泰勒级数

$$f_k(\boldsymbol{X}_k)=f_k(\hat{\boldsymbol{X}}_{k/k})+\boldsymbol{F}_k(\boldsymbol{X}_k-\quad)+\cdots$$
$$g_k(\boldsymbol{X}_k)=g_k(\hat{\boldsymbol{X}}_{k/k})+\cdots=\boldsymbol{G}_k+\cdots$$
$$h_k(\boldsymbol{X}_k)=h_k(\hat{\boldsymbol{X}}_{k/k-1})+\boldsymbol{H}^T{}_k(\boldsymbol{X}_k-\hat{\boldsymbol{X}}_{k/k-1})+\cdots \tag{14}$$

式(14)中,略去一阶以上高阶导数项,并假定$\hat{\boldsymbol{X}}_{k/k}$和$\hat{\boldsymbol{X}}_{k/k-1}$已知,则得到以下关系式

$$\boldsymbol{X}_{k+1}=\boldsymbol{F}_k\boldsymbol{X}_k+\boldsymbol{G}_k\boldsymbol{w}_k+\boldsymbol{u}_k$$
$$\boldsymbol{Z}_k=\boldsymbol{H}^T{}_k\boldsymbol{X}_k+\boldsymbol{v}_k+\boldsymbol{y}_k \tag{15}$$

式中,

$$\boldsymbol{u}_k=f_k(\hat{\boldsymbol{X}}_{k/k})-\boldsymbol{F}_k\hat{\boldsymbol{X}}_{k/k}$$

$$\boldsymbol{y}_k=h_k(\hat{\boldsymbol{X}}_{k/k-1})-\boldsymbol{H}^T{}_k\hat{\boldsymbol{X}}_{k/k-1}$$

因此,扩展 Kalman 滤波方程为

$$\hat{\boldsymbol{X}}_{k/k}=\hat{\boldsymbol{X}}_{k/k-1}+\boldsymbol{L}_k[\boldsymbol{Z}_k-h_k(\hat{\boldsymbol{X}}_{k/k-1})]$$

$$\hat{\boldsymbol{X}}_{k+1/k}=f_k(\hat{\boldsymbol{X}}_{k/k})$$
$$\boldsymbol{L}_k=\boldsymbol{\Sigma}_{k/k-1}\boldsymbol{H}_k\boldsymbol{\Omega}_k^{-1}$$
$$\boldsymbol{\Omega}_k=\boldsymbol{H}^T{}_k\boldsymbol{\Sigma}_{k/k-1}\boldsymbol{H}_k+\boldsymbol{R}_k$$
$$\boldsymbol{\Sigma}_{k/k}=\boldsymbol{\Sigma}_{k/k-1}-\boldsymbol{L}_k\boldsymbol{H}^T{}_k\boldsymbol{\Sigma}_{k/k-1}$$
$$\boldsymbol{\Sigma}_{k+1/k}=\boldsymbol{F}_k\boldsymbol{\Sigma}_{k/k}\boldsymbol{F}^T{}_k+\boldsymbol{G}_k\boldsymbol{Q}_k\boldsymbol{G}^T{}_k \tag{16}$$

给定初始条件为

$$\boldsymbol{\Sigma}_{0/-1}=\boldsymbol{P}_0,\ \hat{\boldsymbol{X}}_{0/-1}=\overline{\boldsymbol{X}}_0 \tag{17}$$

4 数值仿真

针对水下蛙人的导航问题,取状态向量为 $\boldsymbol{X}=[x,v_x,y,v_y,z,v_z]^T$,观测向量为 $\boldsymbol{Z}=[\tau_{10},\tau_{20},\tau_{30}]^T$,则有

状态方程:

$$\boldsymbol{X}_{k+1}=\boldsymbol{F}_k\boldsymbol{X}_k=\begin{bmatrix}\boldsymbol{I}_3 & \Delta t\cdot\boldsymbol{I}_3\\ 0_{3\times3} & \boldsymbol{I}_3\end{bmatrix}\boldsymbol{X}_k \tag{18}$$

式中,$\boldsymbol{I}_3$ 为 3×3 的单位阵,Δt 为相邻两次观测的时间间隔。

观测方程:

$$\boldsymbol{Z}_k=\frac{1}{c}\cdot\begin{bmatrix}|r-r_1|\\ |r-r_2|\\ |r-r_3|\end{bmatrix}+\begin{bmatrix}n_{\tau_{10}}\\ n_{\tau_{20}}\\ n_{\tau_{30}}\end{bmatrix} \tag{19}$$

式中,$n_{\tau_{i0}}(i=1,2,3)$为声传播延时的测量噪声,其均值为 0,方差为 σ_τ。

将式(19)在$\hat{\boldsymbol{X}}_{k/k-1}$ 处进行一阶展开,可得线性观测方程如下

$$\boldsymbol{Z}_k=\boldsymbol{Z}_{\hat{X}_{k/k-1}}+\boldsymbol{H}^T{}_k{}^*[\boldsymbol{X}_k-\hat{\boldsymbol{X}}_{k/k-1}]+\boldsymbol{n}_k \tag{20}$$

其中,

$$\boldsymbol{H}^T{}_k=\frac{1}{c}\begin{bmatrix}\frac{\Delta x_1}{r_1}-\frac{x}{r}v_x & \frac{\Delta y_1}{r_1}-\frac{y}{r}v_y & \frac{\Delta z_1}{r_1}-\frac{z}{r}v_z\\ \frac{\Delta x_2}{r_2}-\frac{x}{r}v_x & \frac{\Delta y_2}{r_2}-\frac{y}{r}v_y & \frac{\Delta z_2}{r_2}-\frac{z}{r}v_z\\ \frac{\Delta x_3}{r_3}-\frac{x}{r}v_x & \frac{\Delta y_3}{r_3}-\frac{y}{r}v_y & \frac{\Delta z_3}{r_3}-\frac{z}{r}v_z\end{bmatrix} \tag{21}$$

式中,Δx_i、Δy_i 和 $\Delta z_i(i=1,2,3)$ 分别为目标空间

位置坐标与信标位置坐标在各方向上的坐标分量的差值。

下面利用仿真算例对定位算法的定位性能予以说明。仿真条件：目标运动状态未知，运动加速度服从均值为零、方差为 σ_a 的高斯分布，目标运动的起点为(150,200,20)，初始运动速度为(0.5,0.7,0)，蛙人所携带的被动式导航接收机的解算频率为 0.2Hz，总观测时间为 100s。

图 3 给出了不同测时精度下的定位跟踪结果。从图 3 中可以看出，扩展卡尔曼滤波具有更好的定位跟踪性能，而卡尔曼滤波受噪声影响较大，定位精度较低。因此，先由延时测量值解算出目标空间坐标后，再采用卡尔曼滤波来提高目标定位精度的做法并不理想。要提高扩展卡尔曼滤波器的估计精度，必须增加时域采样频率，以减小非线性方程在线性化时的截断误差，获得更高程度的线性化近似；此外，还要求在观测方程中，满足高斯性质的观测误差也必须足够小，这样才能保证在递推计算过程中具有较低水平的累积误差，否则扩展卡尔曼滤波方法的估计精度将迅速下降。

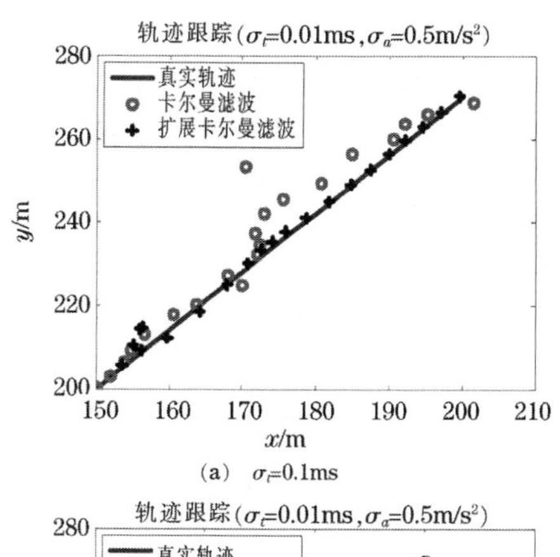

(a)　σ_t=0.1ms

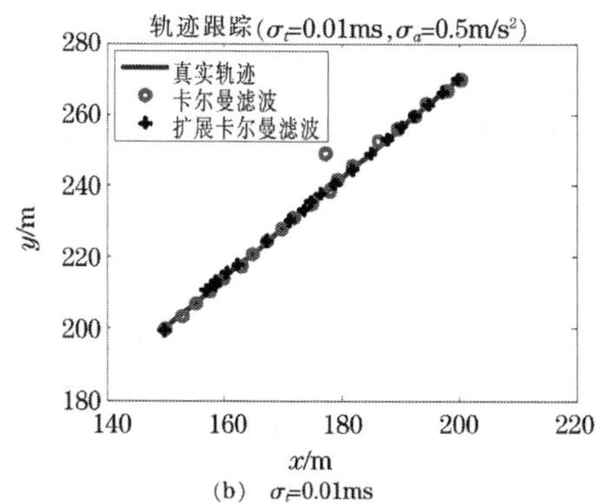

(b)　σ_t=0.01ms

图 3　不同测时精度下的定位跟踪性能

Fig.3　The positioning performance with different time measuring accuracy

5　结　论

本文提出了一种可用于蛙人定位导航的水下 GPS 系统，该系统通过在主动发射端加载水声通信模块，从而可与反蛙人探测声纳共享同一套主动发射系统，降低了反蛙人系统的整体复杂度。蛙人的水下位置由其所携带的水声接收机根据所获取的信标位置坐标和声传播相对延时进行定位解算后获得。为提高基于声延时测量的几何定位算法的性能，分别采用卡尔曼滤波和扩展卡尔曼滤波两种方法对初测结果进行了平滑，并结合仿真算例进行了说明，从中可以看出：

(1)延时测量精度是影响系统定位精度的主要因素，当测量精度较低时可能会产生野值[3]，从而严重降低滤波平滑的性能。

(2)扩展卡尔曼滤波器的性能更佳，其估计精度与时间采样率和延时测量精度有关。

此外，为提高深度方向的定位精度，一般要结合实际水文条件进行修正。

参　考　文　献

[1] 关治，陈景良. 数值计算方法[M]. 北京. 清华大学出版社，1990. 536-541.
Guang Zhi, Chen Jingliang. Methods of Nemerical Computation[M]. Beijing: Tsinghua University Press, 1990. 198-200.

[2] B. D. O. 安德森著，卢伯英，译. 最佳滤波[M]. 国防工业出版社，1983. 12: 178-200.
Anderson B D O. Optimum Filtering[M]. National Defense Industry Press, 1983: 178-200.

[3] 李小民. GPS 在水下定位应用中的几个关键问题研究[R]. 中科院声学研究所博士后研究报告，2002. 34-40.
Li Xiaomin. Investigations on several key problems for the applications of GPS in underwater positioning[R]. Report of the post doctor research, IOA of Chinese Science Academy, 2002. 34-40.

分布式浮标阵水下高速运动声源三维被动定位①②

李 敏 孙贵青 李启虎

(中国科学院声学研究所 北京 100190)

2008 年 10 月 27 日收到

2009 年 3 月 24 日定稿

摘要 给出一种用于垂直爬升弹道三维坐标精确测量的分布式浮标网络超长基线宽带被动定位技术。考虑四个呈正方形布局的浮标节点，利用水平方位角交汇的定位方法测量上浮弹道的水平 x、y 坐标，水平方位角估计由灵巧的声矢量传感器替代常规阵得到，并给出定向误差小于 0.4° 的湖试结果；利用直达声和海面反射声之间的时延差提取弹道的垂直坐标 z，并给出了高速运动目标垂直爬升弹道的海试测量结果，弹道测量误差小于 10 m。

PACS 数: 43.30, 43.60, 43.85

Three dimensional passive localization based on distributed buoy array for underwater moving sound source with high speed

LI Min SUN Guiqing LI Qihu

(*Institute of Acoustics, Chinese Academy of Sciences* Beijing 100190)

Received Oct. 27, 2008

Revised Mar. 24, 2009

Abstract A very long baseline wideband passive localization method based on distributed buoy network for accurately measuring 3D vertical climbing trajectory was presented. Four buoy nodes distributed squarely are considered in this paper. Horizontal position x and y can be estimated by use of the localization methods based on azimuths crossing. Azimuth estimation can be obtained using smart acoustic vector sensor instead of conventional antenna. A result from a lake test shows that azimuth estimation error is lower than 0.4 deg. And vertical position z can be extracted from TDD (Time Delay Difference) between direct and surface reflect paths, and give on-sea test results for climbing mobile target with high speed, which shows that trajectory measurement error is lower than 10 m.

引言

在水声靶场中经常需要评估测试一些具有高速垂直上升弹道特征的装备。空间分布式多传感器网络为声源高精度三维定位提供了必要的技术手段。传统的分布式被动声定位技术多采用空间节点之间的声信号到达时延差估计 (TDD)[1-2]，与此相关的被动时延和 Doppler 估计性能已经被广泛深入研究了几十年。为了达到精确的时延估计，阵列之间的信号相干性是需要的[3]，相干值与声源信号带宽、加性噪声级和观测时间有关。通常情况下，空间分布式测量节点间距从十几米到几十米，一般不超过几百米，否则，水声信号相干性随着传感器之间的间距增加而减弱，这种相干性损失显著影响时延估计精度[4]。因此，该方法的测量精度受限于空间间距，为了扩大作用距离不得不增加空间节点数目，从而可能造成更大的通讯压力和更复杂的融合处理，导致系统宽容性和可靠性降低。

为了增加作用距离，避开时延估计中所遭遇的信号相干性减弱的问题，可以利用方位三角交汇的定位方法。在每个独立的空间节点上测量声源方位，并将方位信息传到融合中心，在融合中心完成方位三角交汇，这种分散式处理方法降低了通讯带宽和系统

① 国家自然科学基金(60532040)和国家重大安全基础研究(613660201)资助项目.

② 声学学报, 2009, 34(4): 289-295.

复杂度. 定位算法和方位估计算法本质上是相互独立的. 一般, 方位估计由无指向性传感器阵列和单个声矢量传感器等得到, 文献 5 利用分布式声矢量传感器测量的方位角发展了通用的加权最小平方算法 (Weighted Least-Square Algorithm: WLS) 以计算目标的三维位置. 使用声矢量传感器替代传统的无指向性传感器阵列的优势在于: (1) 声矢量传感器快速宽带源估计算法优于传统阵列处理; (2) 声矢量传感器大大简化了湿端规模, 降低了系统海上布放回收的难度. 使用声矢量传感器的定位系统结构更紧凑, 工作效率更高, 尤其对于低频宽带声源被动定位.

无论是多用于近距离声源定位的空间节点之间时延差估计 (TDOA) 法, 还是多用于远距离声源定位的方位估计 (DOA) 三角法, 它们在三维定位中的深度坐标估计精度都需要进一步改善. 时延差方法主要是由于大孔径下不能保证高精度的时延差估计. 方位三角法主要在于信号多途和声速剖面对俯仰角估计精度的影响, 文献 6 提出极化平滑算法 (Polarization Smoothing Algorithm: PSA) 解决多径环境下 DOA 估计的相干源问题, 文献 5 建议使用射线跟踪方法修正俯仰角估计偏差. 尽管声源的深度信息本质上体现在接收信号的多径结构中, 但仅依赖多径结构同样难以保证声源的三维定位精度, 尤其在大深度或复杂水文条件下.

1　弹道测量原理

本文发展一种基于分布式浮标网络的超长基线宽带被动定位技术用于该类弹道的三维坐标精确测量. 四个呈正方形布局的声矢量传感器浮标节点分别估计声源水平方位, 通过无线链路将水平方位信息传到融合中心, 利用水平方位三角定位方法确定上浮弹道的水平 x 和 y 坐标; 再利用单个节点接收信号的直达声和海面反射声之间的时延差提取弹道的垂直坐标 z . 系统工作示意图和测量原理框图分别如图 1 和 2 所示.

1.1　水平坐标与 DOA 三角定位

每个浮标提供目标的水平方位估计 ϕ_i 和本地位置坐标信息 (x_i, y_i, z_i), 然后将信息传送到数据融合中心, 进行数据关联和目标位置解算. 假设数据已经关联, 则由加权最小二乘 (WLS) 算法可得声源位置的估计值为:

$$\begin{bmatrix} \widehat{x} \\ \widehat{y} \end{bmatrix} = \boldsymbol{A}^{-1} \cdot \boldsymbol{C}, \tag{1}$$

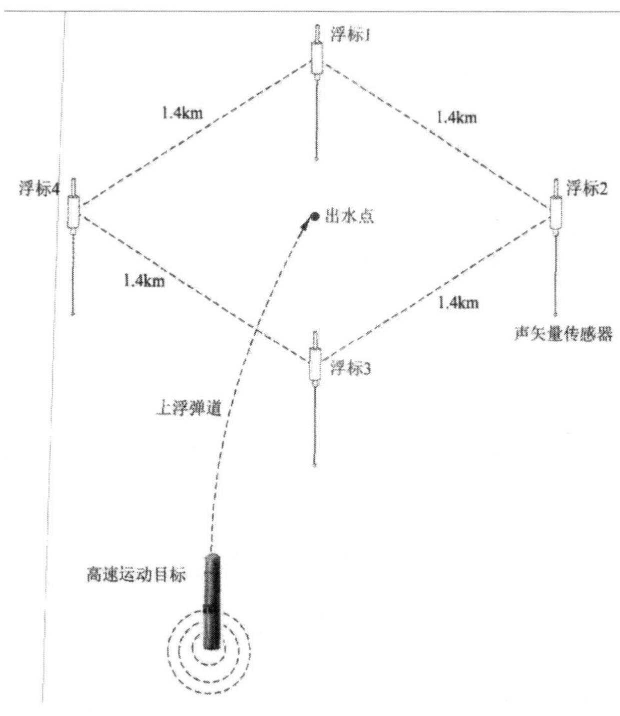

图 1　水下高速目标弹道被动测量示意图

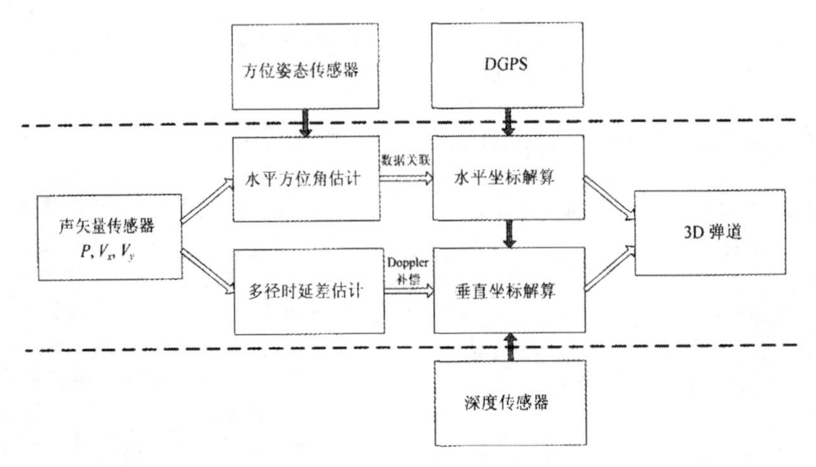

图 2 水下高速目标三维爬升弹道被动测量原理框图

其中,

$$
\boldsymbol{A}=\boldsymbol{A}(w_i,\phi_i)=\begin{bmatrix}\sum\limits_i w_i\sin^2\phi_i & -\sum\limits_i w_i\cos\phi_i\sin\phi_i\\ -\sum\limits_i w_i\cos\phi_i\sin\phi_i & \sum\limits_i w_i\cos^2\phi_i\end{bmatrix}, \tag{2}
$$

$$
\boldsymbol{C}=\boldsymbol{C}(w_i,\phi_i,x_i,y_i,z_i)=\begin{bmatrix}\sum\limits_i w_i(\sin^2\phi_i x_i-\cos\phi_i\sin\phi_i y_i)\\ \sum\limits_i w_i(-\cos\phi_i\sin\phi_i x_i+\cos^2\phi_i y_i)\end{bmatrix}, \tag{3}
$$

第 i 个浮标权值 w_i 的选择和该算法的细节可参见文献 5 。对于大信噪比，不加权和加权的性能基本一致。图 3 给出了水平方位角三角定位法的示意图。从上述定位算法中可以看出，定位本身与水平方位角估计之间是相对独立的两个过程，因此水平方位角估计可由传统的无指向性传感器得到，本文则使用声矢量传感器。为了定量检验上述算法的性能，我们假设算法中的四个随机变量相互独立，即：

$$
\begin{cases}\phi_i=\phi_i^s+\phi_i^n,\\ x_i=x_i^s+x_i^n,\\ y_i=y_i^s+y_i^n, z_i=z_i^s+z_i^n,\end{cases} \tag{4}
$$

其中，ϕ_i^s 为声源真方位，ϕ_i^n 为声源方位估计噪声，ϕ_i^n 服从零均值方差为 σ_i^ϕ 高斯分布；(x_i^s,y_i^s,z_i^s) 为接收器真坐标，(x_i^n,y_i^n,z_i^n) 为接收器坐标测量噪声，(x_i^n,y_i^n,z_i^n)服从零均值方差为$(\sigma_i^x,\sigma_i^y,\sigma_i^z)$高斯分布。

如果考虑四个浮标呈图 1 的正方形配置，目标水平距离各浮标均为 1 km，且目标水平坐标 x 和 y 定位的均方根误差不超过 10 m，则需要水平方位角估计的均方根误差一般控制在 0.6° 以下，基于公式 (1) — (4) 的数值仿真结果如图 4 所示，其中深黑色表示 x 坐标定位标准差，浅灰色表示 y 坐标定位标准差，测向标准差 0.6°，暂不考虑 DGPS 定位精度。图 4 表明，按照图 1 配置，此分布系统水平 360° 全向定位误差均匀，没有明显的定位盲点或盲区。

若综合考虑其它因素，如接收器位置扰动、数据关联等，则水平方位角估计的均方根误差一般要小于 0.5° 。由于被测目标具有很强的宽带辐射噪声，

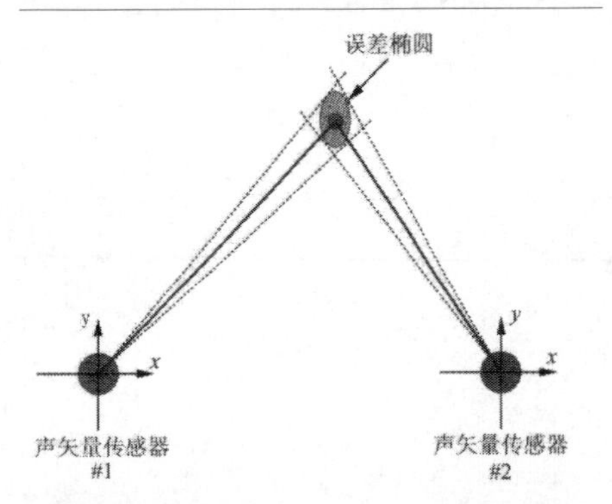

图 3 声矢量传感器的水平方位三角定位法原理示意图

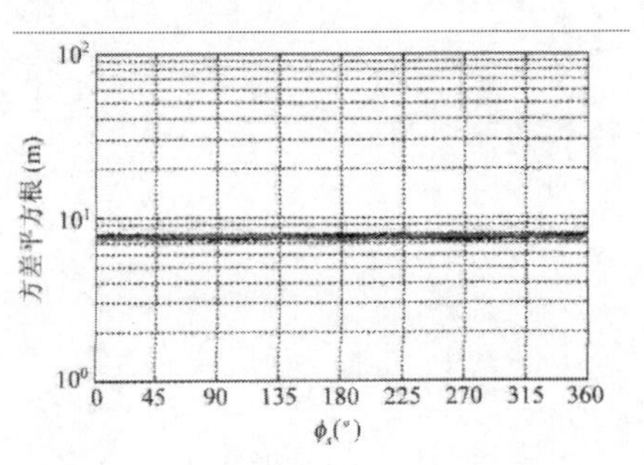

图 4 四元浮标阵水平坐标定位结果仿真

因此基于声矢量传感器的高精度水平方位角估计是完全可以达到的，详见下文。

注意到，在上述分析中并没有考虑数据关联，这主要因为在该系统中数据关联引起的误差对上述算法的影响基本可以忽略，实际上只需要简单的迭代，算法就可以快速收敛到指定的误差范围内。

1.2　声矢量传感器和水平方位角估计

使用声矢量传感器代替传统声压传感器阵列的分布式浮标系统能够完成高精度的三维定位功能。这种传感器测量一点处的标量声压和声质点振速矢量的正交分量，单个声矢量传感器可以快速确定宽带源的方位，使得它在定位问题上特别吸引人。本节首先概要介绍该系统中所使用的声矢量传感器，然后再分析声矢量传感器水平方位角估计性能[7–9]。

图 5 和图 6 是二维同振柱加速度型声矢量传感器的声压灵敏度和指向性图的水池校准结果。该声矢量传感器直径和高度均小于 60 mm，密度 1.6g/cm^3，内部支架采用环氧和玻璃微珠混合物加工而成 (密度 0.6 g/cm^3)，外部用聚氨酯灌封，工作频带 10 Hz~5 kHz。声压通道的声压灵敏度 −191 dB，x 和 y 振速通道的 1 kHz 处声压灵敏度均为 −187.5 dB。

图5给出了质点振速通道的声压灵敏度比较法水池校准结果，实线表示理论值，圆点表示测量值 (测量频率点为 500 Hz, 630 Hz, 800 Hz, 1 kHz, 1.25 kHz)。频率 500 Hz 处校准误差增加的原因在于声源辐射功率降低，信噪比减小。图 6 给出频率在 1 kHz 时声压通道和质点振速通道 x 和 y 的指向性图测量结果，最外围实线圆圈表示声压的全向指向性理论结果，圆圈内部相互正交的两个实线偶极子分别表示质点振速 x、y 分量的理论指向性，对应原点表示 x、y 的测量值。径向每个格表示 10 dB。从上述测量结果可以看出，测量值和理论值吻合得非常好，振速通道偶极子指向性图上偶极子峰所对应的两个最大值之间相差不到 0.5 dB，偶极子凹沟所对应的两个最小值之间相差不到 2 dB，偶极子最小值可达 25~30 dB，其它点与理论值相差不超过 1 dB。

由图 6 可看出，声矢量传感器由无指向性声压传感器和偶极子指向性质点振速传感器复合而成，且其指向性在传感器的工作频带内与频率无关，这些确保声矢量传感器能够无模糊估计空间宽带目标的方位。声矢量传感器最常用的定向方法是声能流方法，其水平方位估计器为：

$$\phi = \tan^{-1}\left(\overline{PV_y}/\overline{PV_x}\right), \tag{5}$$

其中，$\overline{PV_x}$ 和 $\overline{PV_y}$ 分别为声能流 x 分量和 y 分量，$\tan^{-1}(\cdot)$ 表示四象限的反正切运算。这种方法有别于传统阵列处理中的常规波束成形，无需空间方位搜索，因此计算量大大降低，很容易在浮标电子线路上实现，尤其适合于高信噪比下单目标方位估计问题。声能流测向的原理如图 7 所示。

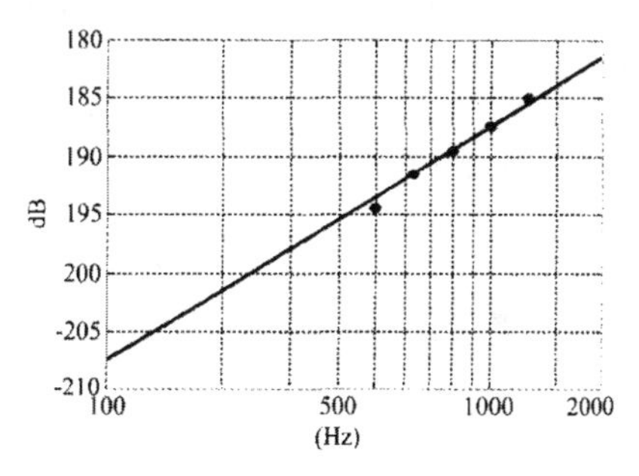

图 5　声矢量传感器振速通道的水池比较法校准结果

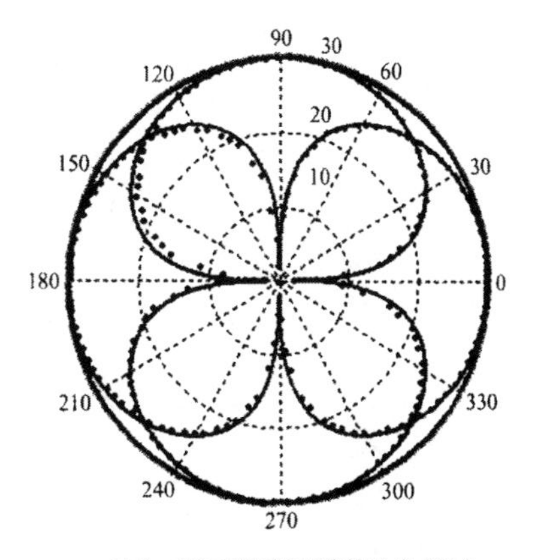

图 6　声矢量传感器指向性图 (1 kHz)

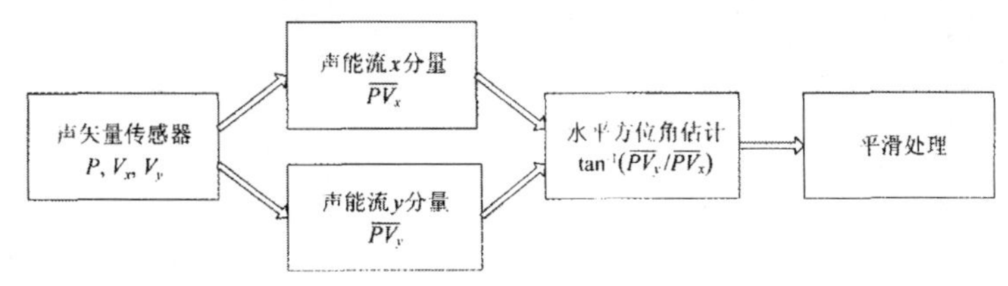

图 7　水平方位角估计的原理框图

上述水平方位角估计器在高信噪比条件下逼近单声矢量传感器水平方位角估计的 CRLB(Cramer-Rao Lower Bound), 即[7]:

$$\mathrm{CRLB}_{\phi}=\frac{1}{6B\cdot T\cdot \mathrm{SNR}\cdot \sin^{2}\theta}, \tag{6}$$

其中, B 为等效带宽, T 为积分时间, SNR 为信噪比。图 8 给出了水平方位角估计的方差下界, 其中俯仰角 $\theta=90°$, 单位带宽信噪比 SNR = 20 dB, 横轴的时间带宽积 BT 从 1 变化到 1000 。当时间 $T=0.1$ s, 带宽 $B\approx 1400$ Hz 时, 水平方位角估计的方差下界约为 0.2°; 带宽 $B\approx 300$ Hz 时, 水平方位角方差下界约为 0.4° 。所用的声矢量传感器上限频率可达 5 kHz, 因此, 带宽 1.4 kHz 是有保障的。上浮目标的实测数据表明, 适当选择工作频带时, 信噪比可达 30 dB, 甚至 40 dB 。

图 9 给出声矢量传感器高精度测向的湖试结果。发射调频脉冲信号 (690~810 Hz), 积分时间 50 ms, 由声能流方位估计器得到的测向方差约 0.4°, 而 $\sqrt{\mathrm{CRLB}}=0.35°(B=120$ Hz, $T=50$ ms, SNR = 28 dB) 。方位估计序列经过平滑处理后, 噪声被显著压低。若使用五点平滑, 则方差可以降到 0.2° 附近。

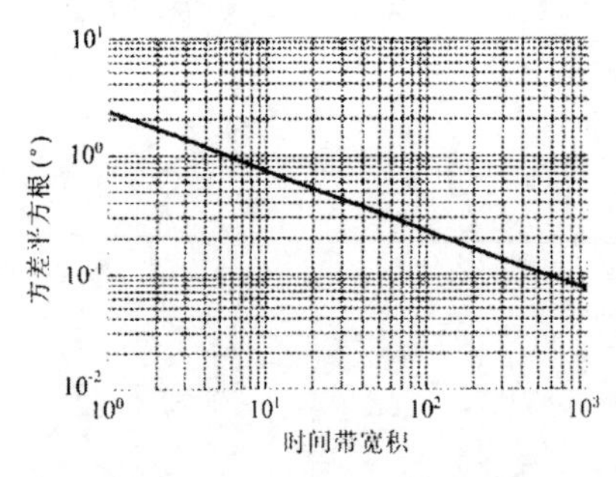

图 8 水平方位角估计方差与时间带宽积的关系

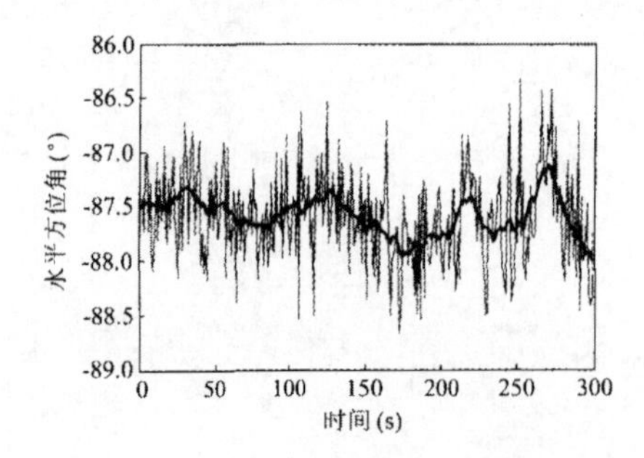

图 9 声矢量传感器精确测向的湖试结果

(细线: 实测结果, 粗线: 平滑结果)

1.3 弹道深度坐标与多径时延差估计

由于多径干扰造成俯仰角分辨率不高, 使得垂直坐标 z 测量精度低。实际上, 深度信息本质上体现在接收信号的多径结构中。因此, 本文采用多径时延差估计达到弹道垂直坐标 z 的精确测量 (如图 10 所示), 并给出了高速运动目标上浮弹道的外场测量结果。

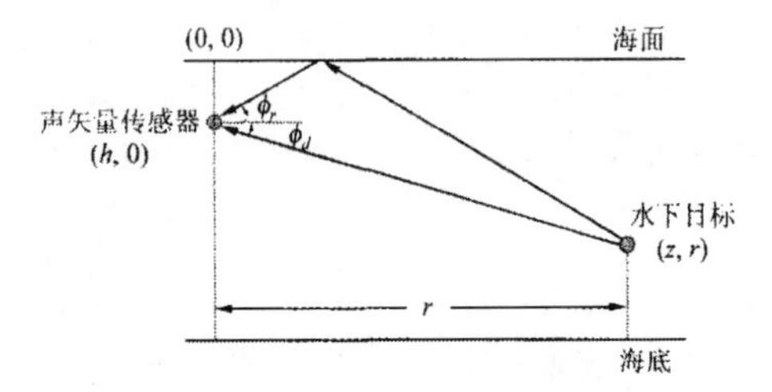

图 10 海面反射和直达声的多径示意图 (等声速剖面)

在海洋声学中, 不同多径之间的到达时延差含有声源的位置信息。若多径到达被单个传感器接收, 一个直接估计时延的方法是对输出信号进行自相关处理。得到的自相关历程图峰值对应不同的时延差。它们的位置可用于估计时延。单个可分多径的时延估计性能在文献 10 中通过对比最大似然估计器的 CRLB 和自相关估计器的方差被详细分析过, 下文仅扼要地引用一些有用的公式 (7) — (10) 。

接收信号的自相关器 $R_r(\tau)$ 为:

$$R_r(\tau)=\frac{1}{T}\int_0^T r(t)r(t-\tau)\mathrm{d}t. \tag{7}$$

自相关器 $R_r(\tau)$ 的时延估计方差 $\mathrm{var}[\hat{\tau}]$, 当 $2BT\gg 1$, $T\gg\tau_0$, 且信号和噪声均为低通白谱 (频带 $[0,B]$) 时有:

$$\mathrm{var}[\hat{\tau}]\cong\frac{3}{8\pi^2a^2B^3T}\left\{\left[(1+a^2)^2+a^2\right]+2(1+a^2)\mathrm{SNR}^{-1}+\mathrm{SNR}^{-2}\right\}, \tag{8}$$

其中, B 为接收信号的有效带宽, T 为自相关函数的积分时间, a 为海面反射系数, 假设与频率无关。

同样条件下, 最大似然估计器 (MLE) 的方差下界, 即 CRLB 在高信噪比条件下为:

$$\mathrm{CRLB}\cong\frac{3}{8\pi^2a^2B^3T}(1-a^2). \tag{9}$$

在低信噪比时, 自相关估计器和最大似然估计器性能相当, 即:

$$\mathrm{var}[\hat{\tau}]=\mathrm{CRLB}\cong\frac{3}{8\pi^2a^2B^3T\cdot\mathrm{SNR}^2}. \tag{10}$$

图 11 给出自相关估计器和最大似然估计器性能对比，当信噪比 SNR = 20 dB，积分时间 $T=0.1$ s，海面反射系数 $a=-0.5$ 时，如若满足 0.1 ms 的精度，则有效带宽约需 700 Hz。图11中实线表示互相关器的时延估计，虚线表示时延估计的克拉美罗下界。

声源的深度信息主要体现在接收信号的多途结构中，这为估计声源深度提供了理论背景。如果收发之间水平距离、接收深度、海面反射和直达声之间的声程差已知，一般可由下式直接估计声源深度[8]：

$$z=(c\tau)\sqrt{\frac{r^2+h^2-0.25(c\tau)^2}{4h^2-(c\tau)^2}}. \tag{11}$$

由上式可知，深度估计误差主要由水平距离、接收深度、时延和声速所决定。限于篇幅仅给出深度估计误差分析的典型结论，如图 12 所示。图中，点线表示时延估计误差 0.2 ms 时的深度误差曲线；虚线表示接收深度 40 m，误差 1 m 时所对应的深度误差曲线；点划线表示水平距离 1 km，误差 10 m 时所对应的深度误差曲线；实线表示声速剖面引起的深度误差。由深度估计误差分析可以归纳如下一些结论：

(1) 接收深度及其误差对大于 140 m 的大深度估计影响最大；接收深度越深，影响越小。

(2) 等声速剖面时，时延估计误差对小于 150 m 的中小深度影响最大；水平距离越小，影响越小；接收深度越大，影响越小。

(3) 非等声速剖面时，在深度 30~140 m 区间上，声速剖面的影响大于时延估计误差的影响。

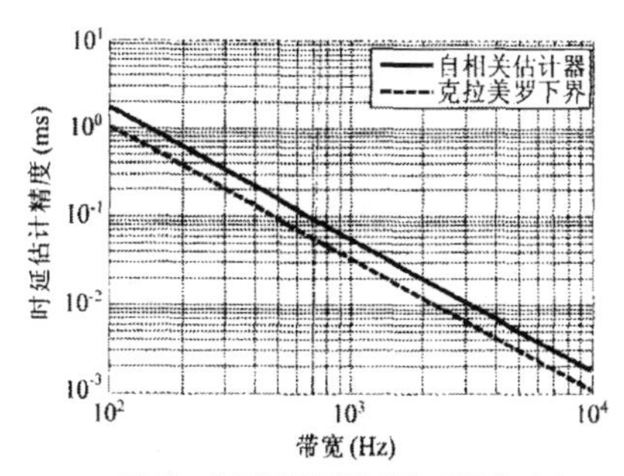

图 11　自相关估计器性能和 CRLB 对比

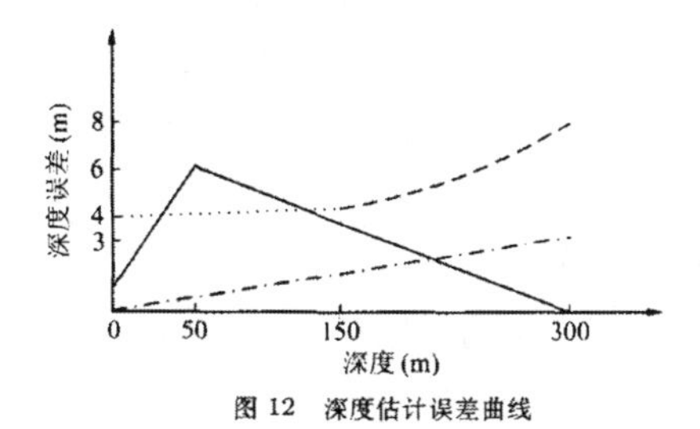

图 12　深度估计误差曲线

(4) 水平距离及其误差主要影响大深度估计误差；水平距离越大，影响越小。

图 13 给出海上实测多径时延差自相关估计器输出的归一化自相关系数历程图，处理频带 1~3 kHz，接收水听器从船舷布放在水下约 5 m，目标和接收水听器之间的水平距离 900 m，目标定深水下 300 m，目标弹道设置为垂直爬升，爬升速度设置 50 节。图中间 0 ms 附近的深色线条表示直达声和海面反射声的多径功率迭加，当目标速度低，Doppler 频移小时，深色线条相对较细些；当目标运动加快时，由于 Doppler 效应该线条随之变宽，即时延估计精度降低。利用所得到的多径时延差和公式 (11) 可以解算出目标爬升弹道的深度坐标，如图 13 所示。由图 14 可得，目标爬升弹道的深度估计方差不超过 10 m(未平滑处理)，目标爬升平均速度约 50 节。

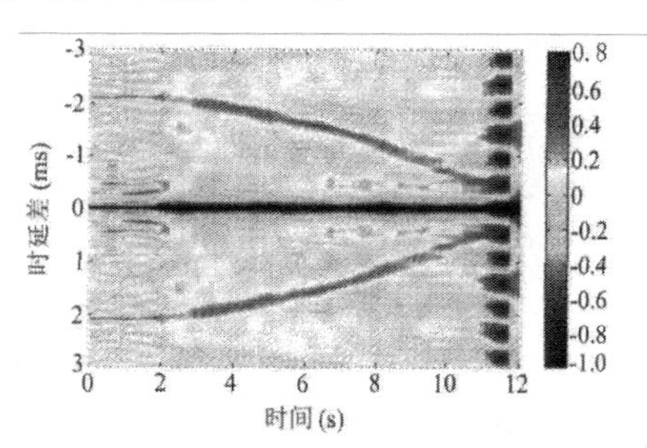

图 13　海试多径时延差的归一化自相关系数历程图

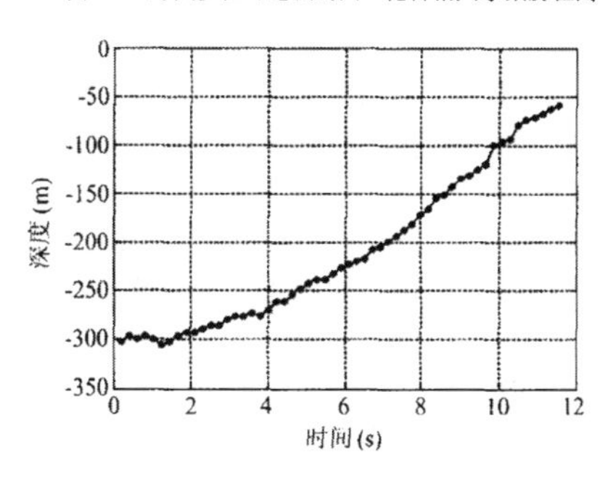

图 14　海试垂直爬升弹道估计历程

2　结论

由上文分析可知，声矢量传感器空间分布式大孔径浮标阵能够精确测量水下高速爬升目标的三维弹道。弹道水平坐标可由水平方位 DOA 估计三角定位法确定，当四个呈正方形浮标相邻之间距离达到 1414 m 时，水平坐标精度不超过 10 m，如果声矢量传感器测向精度不低于 0.5°(有湖试结果支持)。弹道垂直坐标可由直达声和海面反射声之间的多径时延差以及水平坐标得到，由海试结果可知，实测垂直上浮弹道精度不低于 10 m。

尽管本文没有提及 Doppler 效应，但从实测结果看， Doppler 效应对多径时延差估计精度还是有较明显的影响，应该在后置处理中进行适当的补偿。本文仅考虑单个宽带强辐射噪声目标的三维定位，采用声能流测向算法完全有效，但它不能用于多个宽带目标的情况。原则上，声矢量传感器可以利用 MVDR 或 MUSIC 等算法分辨最多两个宽带目标的方位。

利用声矢量传感器构成的空间分布式浮标在满足定位精度前提下，可以显著改善系统的可靠性，提高布放回收操作的安全性和效率，降低系统正常运行和维修保养的成本。

参 考 文 献

1 Chan Y T, Ho K C. A simple and efficient estimator for hyperbolic location. *IEEE Transaction on Signal Processing*, 1994; **42**(8): 1905—1915

2 Xerri B *et al.* Passive tracking in underwater acoustic. *Signal Processing*, 2002; **82**: 1067—1085

3 张 宾，孙贵青，李启虎．阵形畸变对拖曳双线阵左右舷分辨性能的影响．声学学报， 2008; **33**(4): 294—299

4 Kozick R J, Sadler B M. Source localization with distributed sensor arrays and partial spatial coherence. *IEEE Trans. on Signal Processing*, 2004; **52**(3): 601—616

5 Malcolm Hawkes, Arye Nehorai. Wideband source localization using a distributed acoustic vector sensor array. *IEEE Transaction on Signal Processing*, 2003; **51**(5): 1479—1491

6 Dayan Rahamin, Joseph Tabrikian, Reuven Shavit. Source localization using vector sensor array in a multipath environment. *IEEE Trans. on Signal Processing*, 2004; **52**(11): 3096—3103

7 孙贵青．矢量水听器检测技术研究．哈尔滨工程大学博士学位论文， 2001

8 孙贵青，李启虎．声矢量传感器研究进展．声学学报， 2004; **29**(6): 481—490

9 孙贵青，李启虎．声矢量传感器信号处理．声学学报， 2004; **29**(6): 491-498

10 Ianniello J P. High-resolution multipath time delay estimation for broad-band random signals. *IEEE Trans. on ASSP*, 1988; **36**(3): 320—327

声矢量传感器线阵的左右舷分辨①②

孙贵青，张春华，黄海宁，李启虎
（中国科学院 声学研究所，北京 100190）

摘 要：声矢量传感器由声压传感器和惯性传感器复合而成，可同时测量声场一点处的声压标量和质点振速矢量的各正交分量，为解决传统声压水听器线阵固有的左右舷模糊难题提供了新的方法和手段. 利用声矢量传感器线阵指向性函数研究了3种常规波束形成器的左右舷抑制比，然后简要分析了MVDR波束形成器的左右舷抑制比. 理论分析和海试结果均表明：1）左右舷抑制比除了都不可避免地与信噪比有关之外，相比于具有一定左右舷分辨能力的声压水听器多线阵和三元组声压水听器线阵，声矢量传感器线阵的左右舷分辨性能主要取决于与频率无关的矢量传感器阵元指向性；2）声矢量传感器线阵MVDR波束形成器的左右舷分辨性能大大优于常规波束形成器.

关键词：声矢量传感器线阵；左右舷分辨；波束形成
中图分类号：TN911.7 **文献标志码**：A **文章编号**：1006-7043(2010)07-0848-08

Left-right resolution of acoustic vector sensor line arrays

SUN Gui-qing, ZHANG Chun-hua, HUANG Hai-ning, LI Qi-hu
(Institute of Acoustics, Chinese Academy of Sciences, Beijing 100190, China)

Abstract: An acoustic vector sensor (AVS) combines a pressure hydrophone with inertial sensors. It can simultaneously measure pressure scalar and orthogonal components of a particle velocity vector at a given point in a sound field. This provides new methods and means for solving left/right ambiguities intrinsic in pressure hydrophone linear arrays. In this paper, we first studied the left-to-right suppression ratio (LRSR) of the three conventional beamformers using an AVS line array's directivity functions. The LRSR was then briefly analyzed for a minimum variance distortionless response (MVDR) beamformer. Both theoretical analysis and results from sea trials showed that the LRSR significantly relates to the signal-to-noise ratio (SNR), but mostly rests with directivity of the AVS array element. This contrasts with multi-line and triplet arrays, which possess more or less certain left-right resolution. In addition, the LRSR of an MVDR beamformer for an AVS line array significantly excels those of conventional beamformers.

Keywords: acoustic vector sensor line array; left-right resolution; adaptive beamforming

自1917年出现拖线阵雏形以来，历经约一个世纪的发展，其在声呐中的地位已被普遍认可[1-2]. 拖线阵是将水听器和相关配套的放大滤波甚至连同数据采集和传输等电子设备封装在塑料软管中，由拖缆在水中拖曳的声阵. 随着技术的不断更新发展，拖线阵软管直径由早先的粗缆变成现今的细缆（缆径约30 mm）和超细缆（缆径约8 mm）；由早期模拟器件构成的模拟缆发展成更为先进的数字缆和光纤缆；水听器阵元由最初的常规压电陶瓷声压水听器衍生出声矢量传感器和光纤水听器. 拖线阵声呐设计是一个综合权衡的过程，根据声呐技术指标要求需要对声呐参数指标进行优化组合，如噪声或干扰背景下的检测增益、定向/定距/定深三维定位性能

① 基金项目：国家安全重大基础研究基金资助项目（613110010201, 613660201）.
② 哈尔滨工程大学学报，2010, 31(7): 848-855.

以及左右舷分辨能力等. 在传统的被动拖线阵声呐设计中对左右舷分辨性能要求并不高,一般通过本舰机动改变舷角来分辨目标是来自左舷还是右舷,通常需要几分钟到十几分钟的时间. 随着现代潜艇隐身技术的进步,潜艇辐射噪声已经与海洋环境噪声相当甚至更低,主/被动拖线阵声呐应运而生,它的一个基本要求是在单次声脉冲发射之后需要立即知晓目标的左右舷信息以提高反潜效率. 由于存在固有的左右舷模糊,由传统声压水听器构成的拖曳单线阵不能胜任这样的任务. 目前已有的具有实时左右舷分辨能力的3类主要拖线阵声呐分别是美国的以拖曳双线阵(twinline)[3]和拖曳三线阵(triline)等代表的拖曳多线阵 multiline,即在船后平行拖曳多条声压水听器细线阵,利用目标到达每条线阵上的时延差进行左右舷分辨;法国、德国和荷兰发展的三元组水听器拖曳单线阵(triplet array)[4],即由在软管圆周上均匀分布的3个水听器构成的三元组水听器阵元(triplet)替换单个水听器阵元,使得单个阵元具有指向性,可以分别形成左舷波束和右舷波束,缆径较粗,外径一般可达90~100mm,工作频率下移缆径随之增大;声矢量传感器拖曳单线阵[5]不但汇集了上述2类声呐的优点,如单线阵、有指向性的阵元,较细的缆径(外径一般为40~60mm),而且具有更强的检测性能,已被列入未来拖线阵声呐发展的重点之一. 因此将其作为文章的主要对象,从理论分析和海试结果2方面研究声矢量传感器线阵的左右舷分辨能力.

拖线阵声呐信号处理一般可分为本舰噪声和拖曳噪声抵消、阵形估计和补偿、左右舷分辨等. 本舰噪声是拖曳平台本身的辐射噪声,对端射附近扇面的警戒有较大影响;拖曳噪声包含湍流边界层激励的流噪声和缆的抖动噪声,对水听器测量微弱信号有较大影响. 拖线阵作为柔性阵,拖曳过程阵形不可能保持理想的直线状态,如何估计阵形并在后置处理中进行补偿在一些要求连续稳定跟踪的机动过程中显得尤为重要. 然而,上述这些问题均不是本文所要研究的内容,感兴趣的读者可参见相关文献[6-7]. 文中所关注的问题主要涉及到拖线阵左右舷分辨能力. 双线阵和三元组单线阵是最早提出并付诸实施的2种方案[3-4],国内随后开展的一些研究工作进一步深入剖析了这些方案. 文献[8]分析了三元组单线阵的左右舷分辨原理和性能,并给出了相应算法和试验结果. 文献[9-10]研究了双线阵左右舷分辨原理和性能,并给出了多种实现算法. 在此基础上,文献[6]通过双线阵指向性函数和左右舷抑制比研究给出了拖曳双线阵左右舷分辨的2个关键参量——特征频率和特征方位,并分析了双线阵对应阵元存在错位时的阵形畸变对左右舷分辨性能的影响. 文献[11]重点综述了双线阵和三元组单线阵的左右舷分辨,对矢量传感器线阵左右舷分辨的潜力有所期待. 文献[12]通过对 MVDR 波束形成器的研究,指出了矢量传感器线阵左右舷分辨潜力. 文献[13]在试验中验证了矢量传感器线阵各种常规波束形成器和自适应波束形成器对左右舷分辨性能的影响. 文献[14]随后进一步理论推导了矢量传感器线阵波束形成器和左右舷抑制比之间的解析关系,并得到了试验验证. 在此尝试对矢量传感器线阵左右舷分辨原理和性能进行全面的总结和归纳,作为下一阶段研究的基础.

1 理论基础

有关矢量传感器线阵的研究成果散见于诸多文献之中,为便于叙述和保持文章结构的合理性,本节首先总结性地给出声矢量传感器线阵的3种常规波束形成器和 MVDR 波束形成器,然后再根据左右舷抑制比的定义进行左右舷分辨的性能推导和分析,并给出了一些常用公式和图表以方便读者参考引用和进一步的深入研究.

1.1 波束形成器

根据似然比准则不难推导出矢量传感器线阵的 Cardioid 常规波束形成器[13]:

$$\boldsymbol{D} = \boldsymbol{g}^{\mathrm{H}}\boldsymbol{R}\boldsymbol{g}. \tag{1}$$

式中:$\boldsymbol{R}$ 为线阵接收数据的协方差矩阵,方向扫描向量 $\boldsymbol{g}=\boldsymbol{e}\otimes\boldsymbol{u}$,符号⊗表示 Kronecker 积,$\boldsymbol{e}^{\mathrm{T}}=[1\quad \mathrm{e}^{-\mathrm{i}\omega\tau_0}\quad\cdots\quad \mathrm{e}^{-\mathrm{i}\omega\tau_{M-1}}]$ 为单频平面波的方向扫描向量,时延 $\tau=\tau(\phi,\theta)$,与 x 轴的夹角 $\phi\in[0,2\pi)$,与 z 轴的夹角 $\theta\in[0,\pi]$,$\boldsymbol{u}^{\mathrm{T}}=[1\quad \cos\phi\sin\theta\quad \sin\phi\sin\theta\quad \cos\theta]$ 为声矢量传感器阵元4个通道的指向性函数或方向扫描向量. 忽略质点振速垂直分量,可得 Cardioid 常规波束形成器的指向性函数:

$$B_{\mathrm{Cardioid}}(\phi) = \left|\frac{\sin[N\pi\frac{d}{\lambda}(\cos\phi-\cos\phi_0)]}{N\cdot\sin[\pi\frac{d}{\lambda}(\cos\phi-\cos\phi_0)]}\right|\cdot$$

$$\left[\frac{1+\cos(\phi-\phi_0)}{2}\right]^2. \tag{2}$$

式中：N 为阵元数，d 为阵元间距，ϕ 为波束扫描角度，ϕ_0 为信号方位角. 为方便推导左右舷抑制比，一般假设端射为 0°. 因其单阵元具有心形指向性 $\left[\frac{1+\cos(\phi-\phi_0)}{2}\right]^2$，故将式(2)称为 Cardioid 常规波束形成器.

矢量传感器一般由声压水听器和质点振速传感器构成[15-16]，因此根据矢量传感器内部各传感器所测物理量可将式(1)展开为声压、质点振速和声能流 3 项[17]：

$$D = \frac{1}{4}\boldsymbol{e}^{\mathrm{H}}\boldsymbol{R}_{pp}\boldsymbol{e} + \frac{1}{4}\boldsymbol{g}_v^{\mathrm{H}}\boldsymbol{R}_{vv}\boldsymbol{g}_v + \frac{1}{2}\boldsymbol{e}^{\mathrm{H}}\mathrm{Re}[\boldsymbol{R}_{pv}]\boldsymbol{g}_v. \tag{3}$$

式中：$\boldsymbol{R}_{pp}$、$\boldsymbol{R}_{vv}$ 和 $\mathrm{Re}[\boldsymbol{R}_{pv}]$ 分别为声压、振速和声能流的协方差矩阵；$\boldsymbol{g}_v = \boldsymbol{e}\otimes\boldsymbol{u}_v$，$\boldsymbol{u}_v^{\mathrm{T}} = [1 \quad \cos\phi\sin\theta \quad \sin\phi\sin\theta \quad \cos\theta]$. 式(3)表明，声矢量传感器线阵 Cardioid 常规波束形成器既包含了传统的声压常规波束形成器以及振速常规波束形成器，又含有前两者合成所得的声能流波束形成器. Cardioid 常规波束形成器的展开式第 1 项就是常见的声压线阵常规波束形成器，简称声压常规波束形成器：

$$D_P = \boldsymbol{e}^{\mathrm{H}}\boldsymbol{R}_{pp}\boldsymbol{e}; \tag{4}$$

第 2 项是质点振速常规波束形成器：

$$D_V = \boldsymbol{g}_v^{\mathrm{H}}\boldsymbol{R}_{vv}\boldsymbol{g}_v; \tag{5}$$

第 3 项是声能流常规波束形成器：

$$D_I = \boldsymbol{e}^{\mathrm{H}}\mathrm{Re}[\boldsymbol{R}_{pv}]\boldsymbol{g}_v. \tag{6}$$

前 2 项均属于能量检测器的范畴，只是后者阵元有偶极子指向性而已，而声能流检测器可以视为互相关检测器，但细节上有所区别. 由上述分析可知，Cardioid 常规波束形成器本质上由这 3 种基本的常规波束形成器线性组合而得；当然，由它们还可以衍生出其他形式的波束形成器[18-19]，如

$$d = \frac{1}{2}\boldsymbol{g}_v^{\mathrm{H}}\boldsymbol{R}_{vv}\boldsymbol{g}_v + \frac{1}{2}\boldsymbol{e}^{\mathrm{H}}\mathrm{Re}[\boldsymbol{R}_{pv}]\boldsymbol{g}_v. \tag{7}$$

式(7)的空间增益、波束图等性能主要由质点振速常规波束形成器和声能流常规波束形成器所控制. 限于篇幅，文中重点研究 Cardioid 及其 3 种基本常规波束形成器的左右舷分辨能力.

在多目标或多干扰环境下，常规波束形成器的表现难尽人意. 为了最大限度地抑制干扰，依据最大信干比准则可得矢量传感器线阵的 Cardioid 自适应波束形成器[12]：

$$D = \boldsymbol{w}^{\mathrm{H}}\boldsymbol{R}\boldsymbol{w}. \tag{8}$$

式中：$\boldsymbol{w} = \dfrac{\boldsymbol{R}^{-1}\boldsymbol{g}}{\boldsymbol{g}^{\mathrm{H}}\boldsymbol{R}^{-1}\boldsymbol{g}}$ 为自适应波束形成器的权向量. 实际使用较多的自适应波束形成器是 MVDR 波束形成器的一些变种，文中使用对角加载 MVDR 波束形成器进行数据处理以增加矩阵求逆运算的稳定性. 自 Capon[20] 首次提出以来，历经四十几年的发展，随着宽容性和稳健性的显著提高，自适应波束形成器逐渐加入到声呐信号处理主流中，尤其是针对近岸浅海等复杂环境下的一些处理，从传统的平面波抵消到匹配场干扰抑制，无不体现出自适应波束形成器的优势. 尽管还存在一些失配问题，但瑕不掩瑜. 即便如此，适当兼顾稳健性，也会获得优于常规波束形成器的处理效果，况且自适应处理技术本身也处于不断发展完善的过程中；因此，只要根据实际情况适时合理地采用自适应波束形成器，就可以达到事半功倍之效. 文中矢量传感器线阵的海试结果在一定程度上验证了这一点. 此外，三元组水听器拖线阵的海试结果也表明自适应波束形成器确实有效地提高了左右舷分辨的性能[21]. 这在某种程度上体现了一种技术发展的趋势.

1.2　左右舷抑制比

利用声矢量传感器线阵不同波束形成器表达式，根据左右舷抑制比的定义[9]可以得到不同波束形成器的左右舷抑制比表达式. 一般将指向性函数直接表示的左右舷抑制比称为无噪声的左右舷抑制比：

$$\eta(\phi_0) = \left|\frac{B(\phi)}{B(-\phi)}\right|^2_{\phi=\phi_0} = \frac{|\boldsymbol{w}_+^{\mathrm{H}}\boldsymbol{R}_s\boldsymbol{w}_+|_{\phi=\phi_0}}{|\boldsymbol{w}_-^{\mathrm{H}}\boldsymbol{R}_s\boldsymbol{w}_-|_{\phi=-\phi_0}}. \tag{9}$$

式中：$\boldsymbol{R}_s$ 表示信号协方差矩阵，下标“+”表示水平方位角真值，“-”表示水平方位角的镜像值(相对 0°端射对称的水平方位角). 由此定义式可得一些常用的左右舷抑制比表达式，如表 1 所示.

表 1 典型线阵指向性函数和左右舷抑制比汇总表

Table 1 Directional functions and LRSRs of typical line arrays

波束形成器		指向性函数	左右舷抑制比
CBF	声压单线阵	$B(\phi,\phi_0)$	1
	声压双线阵	$B(\phi,\phi_0)\cdot\cos\left[\pi\frac{D}{\lambda}(\sin\phi-\sin\phi_0)\right]$	$\left[\cos(2\pi\frac{D}{\lambda}\sin\phi_0)\right]^{-2}$
	声压三线阵	$B(\phi,\phi_0)\cdot\left\{1-\frac{4}{3}\sin^2\left[\pi\frac{D}{\lambda}(\sin\phi-\sin\phi_0)\right]\right\}$	$\left[1-\frac{4}{3}\sin^2(2\pi\frac{D}{\lambda}\sin\phi_0)\right]^{-1}$
	声压多线阵	$B(\phi,\phi_0)\cdot\frac{\sin\left[M\pi\frac{r}{\lambda}(\sin\phi-\sin\phi_0)\right]}{M\cdot\sin\left[\pi\frac{r}{\lambda}(\sin\phi-\sin\phi_0)\right]}$	$\left[\frac{\sin(2M\pi\frac{D}{\lambda}\sin\phi_0)}{M\cdot\sin(2\pi\frac{D}{\lambda}\sin\phi_0)}\right]^2$
	声矢量线阵振速	$B(\phi,\phi_0)\cdot\cos^2(\phi-\phi_0)$	$[\cos(2\phi_0)]^{-2}$
	声矢量线阵 Cardioid	$B(\phi,\phi_0)\cdot\left[\frac{1+\cos(\phi-\phi_0)}{2}\right]^2$	$\left[\frac{1+\cos(2\phi_0)}{2}\right]^2$
	声矢量线阵声能流	$B(\phi,\phi_0)\cdot\lvert\cos^2(\phi-\phi_0)\rvert$	$\lvert\cos(2\phi_0)\rvert^{-1}$
ABF	声压双线阵	-	$1+2N\cdot\mathrm{SNR}\cdot\left[1-\cos^2(2\pi\frac{D}{\lambda}\sin\phi_0)\right]$
	声矢量线阵 Cardioid	-	$1+2N\cdot\mathrm{SNR}\cdot\left[1-(\frac{1+\cos2\phi_0}{2})^2\right]$

注：$\frac{\sin\left[N\pi\frac{d}{\lambda}(\cos\phi-\cos\phi_0)\right]}{N\cdot\sin\left[\pi\frac{d}{\lambda}(\cos\phi-\cos\phi_0)\right]}$为声压水听器单线阵的指向性函数，$d$ 表示阵元间距，D 表示 2 条水平线阵之间的间距.

1.2.1 左右舷抑制比估计与信噪比之间的关系

实际上，声呐进行左右舷抑制比计算不可避免地会受到噪声的影响，左右舷抑制比估计 $\hat{\eta}$ 一般由式(10)完成：

$$\hat{\eta}(\phi_0)=\frac{\lvert w_+^{\mathrm{H}}Rw_+\rvert_{\phi=\phi_0}}{\lvert w_-^{\mathrm{H}}Rw_-\rvert_{\phi=-\phi_0}}. \tag{10}$$

式中：$R=R_s+R_n$ 为数据协方差矩阵，R_n 为噪声协方差矩阵. 因此，式(10)在估计左右舷抑制比时考虑到信噪比的影响.

令 $w=g$，则可得 Cardioid 常规波束形成器的左右舷抑制比估计 $\hat{\eta}$：

$$\hat{\eta}^{-1}=\frac{N\cdot\mathrm{SNR}}{N\cdot\mathrm{SNR}+1}\cdot\eta^{-1}+\frac{1}{N\cdot\mathrm{SNR}+1}. \tag{11}$$

式中：$\mathrm{SNR}=\frac{\sigma_s^2}{\sigma_n^2}$为信噪比，$\sigma_s^2$ 为信号功率，σ_n^2 为噪声功率，η 为无噪声的左右舷抑制比. 由式(11)可得，在高信噪比($\mathrm{SNR}\gg1$)情况下，$\hat{\eta}\approx\eta$；低信噪比($\mathrm{SNR}\ll1$)时，$\hat{\eta}\approx1$，即无法区分左右舷目标. 此结论对任何常规波束形成器的左右舷抑制比都成立.

1.2.2 自适应波束形成器的左右舷抑制比

为简便计算，重点推导单源情况下 MVDR 波束形成器的左右舷抑制比通用表达式. 单源情况下所推导的公式和图具有一定的典型性. 多源情况和对角加载对自适应波束形成器左右舷抑制比的影响详细推导可参见文献[14]. 单源和多源情况、对角加载与否除了极个别的细节稍有出入之外，其左右舷抑制比基本相同. 单源情况下，N 元声压水听器线阵 MVDR 波束形成器的左右舷抑制比为

$$\hat{\eta}=1+N\cdot\mathrm{SNR}\cdot(1-\eta^{-1}). \tag{12}$$

式中：η 为常规波束形成器的无噪声左右舷抑制比，具体形式可参见表 1. 注意到，式(12)始终为正数且最小值为 1. 对照式(13)、(14)，式(12)的表示形式具有一定的普遍性，细微区别在于一些参量数值的选取上. 举 2 个常用的例子.

例 1 N 元声矢量传感器线阵 Cardioid 的无对角加载 MVDR 波束形成器的左右舷抑制比为

$$\hat{\eta}=1+4N\cdot\mathrm{SNR}\cdot\left[1-(\frac{1+3\cos2\phi_0}{4})^2\right]. \tag{13}$$

式(13)表明,矢量传感器线阵左右舷抑制比与频率无关,这是缘于矢量传感器阵元指向性与频率无关的特性.

例2 声压水听器双线阵(2条N元声压水听器线阵)的无对角加载MVDR波束形成器的组左右舷抑制比为

$$\dot{\eta} = 1 + 2N \cdot \mathrm{SNR} \cdot \left[1 - \cos^2\left(2\pi \frac{D}{\lambda}\sin\phi_0\right)\right]. \tag{14}$$

式(14)表明,声压水听器线阵左右舷抑制比与频率有关. 当$D=\frac{\lambda}{4}$(λ为线阵工作中心频率所对应的波长)时,即最优阵间距情况下,声压水听器双线阵在正横方向上($\phi_0=0°$)具有最佳的左右舷分辨性能(如图1所示);当$D=\frac{\lambda}{2}$时,声压水听器双线阵在$\phi_0=\pm30°$和$\phi_0=\pm150°$处得到最佳的左右舷分辨效果,而在$\phi_0=0°$和$\phi_0=\pm90°$处无法分辨左右舷.

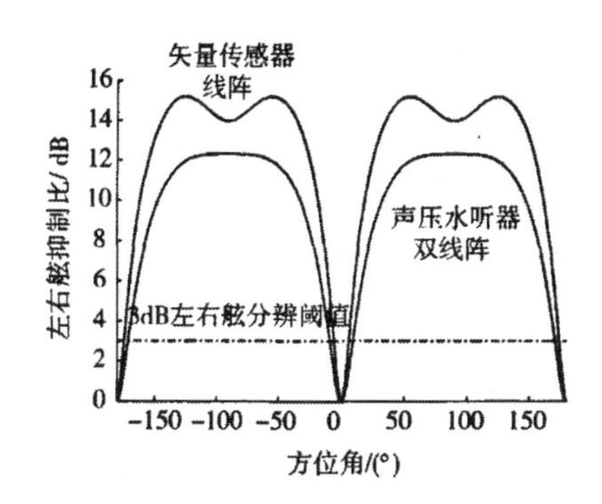

图1 声矢量传感器线阵和声压水听器双线阵MVDR波束形成器左右舷抑制比

Fig. 1 LRSR comparison between MVDR beamformers for AVS line array and hydrophone twin-line array

由式(13)、(14)可分别计算得声矢量传感器线阵和声压水听器双线阵MVDR波束形成器左右舷抑制比. 已知单位带宽内信噪比0 dB,8元声矢量传感器线阵和2条8元线阵构成的声压水听器双线阵,结果如图1所示. 声压水听器双线阵仅计算了两阵间距$D=\frac{\lambda}{4}$的情况. 由图1可知,矢量传感器线阵左右舷分辨效果稍优于声压水听器双线阵至少2 dB,若按照3 dB左右舷分辨阈值,则矢量传感器左右舷可分辨的方位扇面应大于声压水听器双线阵. 但必须注意到,当$D\neq\frac{\lambda}{4}$或频率降低时声压水听器双线阵左右舷分辨能力都或多或少地退化,尽管也存在能使左右舷抑制比最大化的特征频率和特征方位[6],但总体而言,矢量传感器线阵的左右舷分辨性能优于声压水听器双线阵,除了2个端射方位(0°和180°)之外(端射处的左右舷分辨问题转化为前后艏艉分辨,即使是声压水听器单线阵也能够很好地解决此问题). 因此,矢量传感器线阵具有水平360°全景方位无模糊分辨能力,且与频率无关(见公式(13)).

2 海上试验数据分析

2003年3~4月在三亚海域进行了8元声矢量传感器线阵海上试验,阵元间距1 m. 试验使用一艘水面船和一艘水下航行器2个合作目标进行方位交汇,图2给出了整个试验过程矢量传感器线阵自适应波束形成器的方位历程显示. 根据图2和理论结果选取3个典型方位,端射附近28°、45°和正横附近87°的数据重点研究矢量传感器线阵常规波束形成器左右舷分辨效果,即Cardioid、质点振速和声能流3种常规波束形成器.

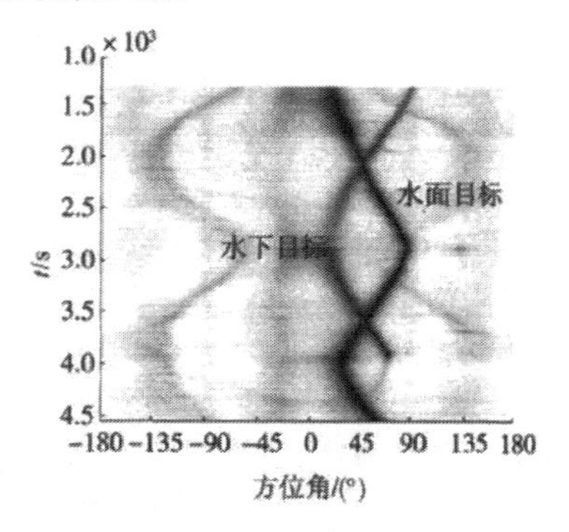

图2 方位交叉分辨的BTR显示

Fig. 2 BTR display for bearing crossing

图3分别给出端射附近28°时Cardioid(a)、质点振速(b)和声能流(c)3种常规波束形成器的左右舷抑制比与频率的关系. 此时,这3种常规波束形成器的左右舷抑制比一般在2~4 dB,质点振速相对好一些,其他2个基本相当. 这主要因为质点振速的波束相对其他2个较窄一些.

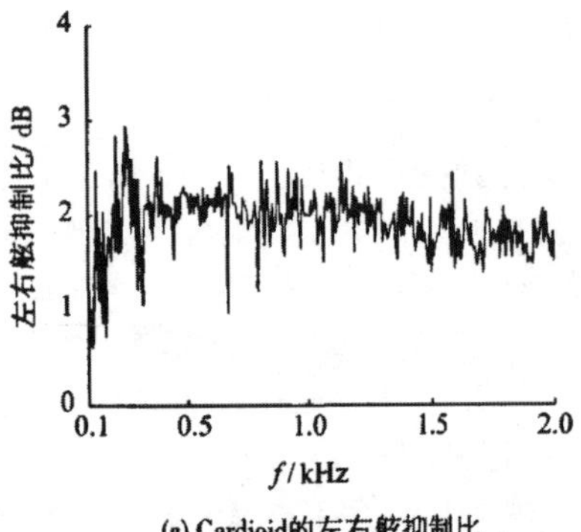

(a) Cardioid的左右舷抑制比

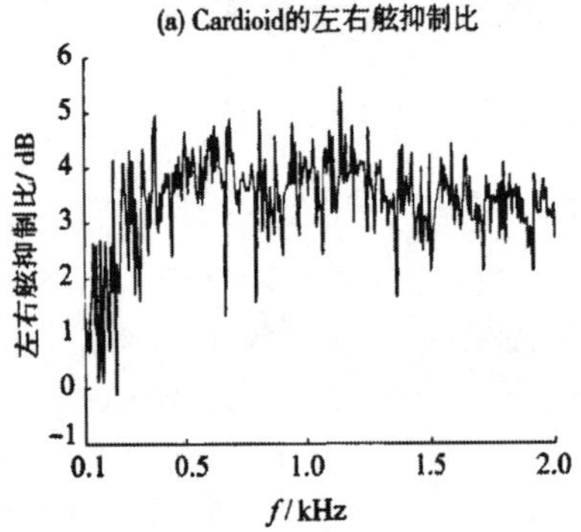

(b) 质点振速的左右舷抑制比

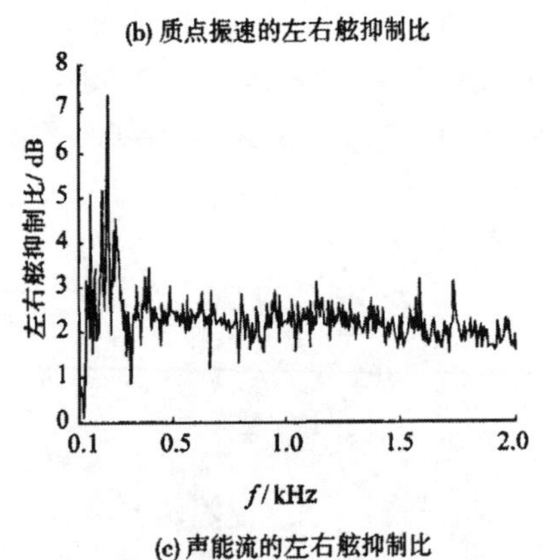

(c) 声能流的左右舷抑制比

图3 水平方位角28°时左右舷抑制比与频率的关系

Fig. 3 LRSR vs frequency when azimuth 28°

图4分别给出45°时Cardioid(a)、质点振速(b)和声能流(c)3种常规波束形成器的左右舷抑制比与频率的关系. 质点振速的左右舷抑制比最大,一般分布在15 dB附近,因为理论上在45°时质点振速常规波束形成器左右舷抑制比最佳. 声能流较Cardioid好一些,约高3 dB,主要原因在于声能流的单边偶极子波束比Cardioid的心形波束要窄. 在45°时理论上声能流左右舷抑制比处于临界点,因此数值起伏较大. 图5分别给出正横附近87°时Cardioid(a)、质点振速(b)和声能流(c)3种常规波束形成器的左右舷抑制比与频率的关系. 此时,声能流的效果最佳,150 dB多的数值主要是计算机误差小量造成的,而非实际数值,因为在理论上,声能流常规波束形成器通过相位分辨可在正横附近90°扇面内达到理想的左右舷分辨效果. Cardioid一般在15 dB附近,Cardioid波束是背向为零点的单边波束,理论上在正横能达到最优的左右舷分辨. 而质点振速基本不能分辨左右舷,因为理论上具有双边偶极子波束的质点振速在正横附近左右舷分辨效果最差.

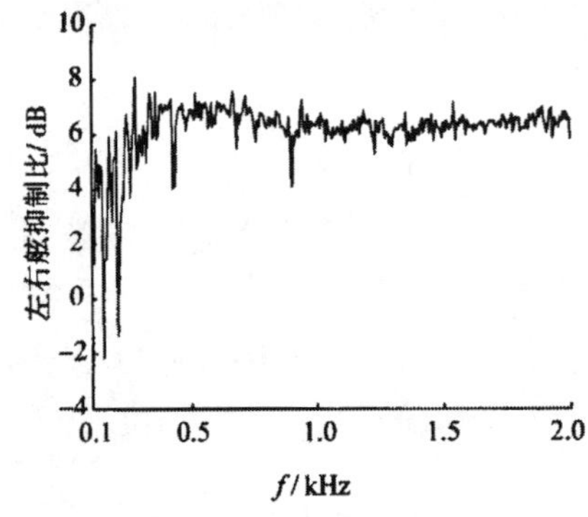

(a) Cardioid的左右舷抑制比

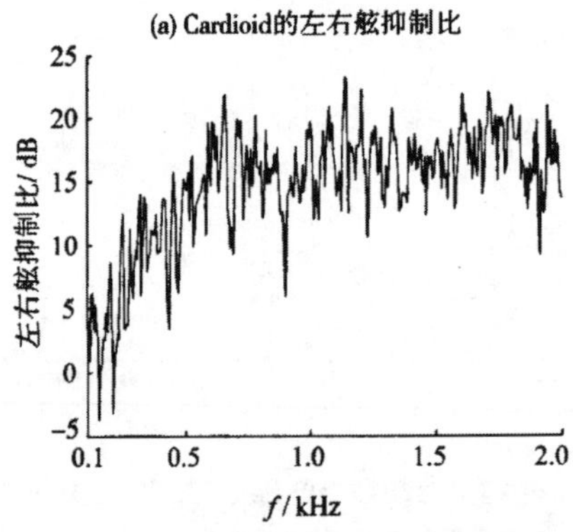

(b) 质点振速的左右舷抑制比

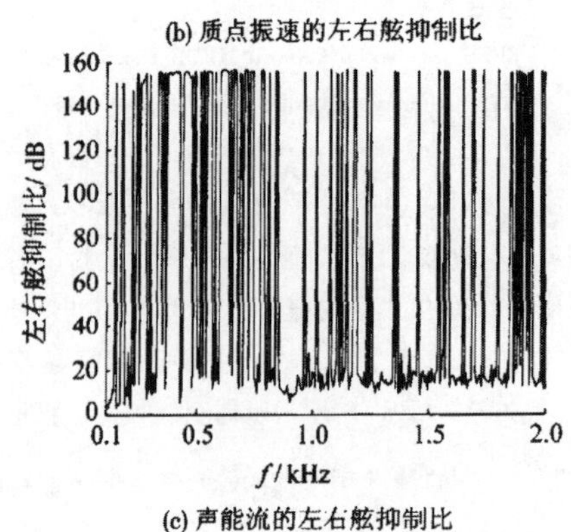

(c) 声能流的左右舷抑制比

图4 水平方位角45°时左右舷抑制比与频率的关系

Fig. 4 LRSR vs frequency when azimuth 45°

由图3～5的分析可知：在相当宽的频带内，左右舷抑制比都接近或达到理论值. 表2中的实验值是从100 Hz～2 kHz频带上的平均结果. 单纯考虑常规波束形成器的左右舷抑制比，显然声能流最优，在正横附近90°扇面内无任何模糊，实际左右舷抑制比可达20 dB以上；Cardioid在正横附近90°扇面内左右舷抑制比均超过6 dB；质点振速则在45°和135°附近有最佳的左右舷分辨能力；在端射附近这3种常规波束形成器的左右舷分辨性能均严重下降，但可用自适应波束形成器加以改善，图2已经充分地验证了这一点，与理论完全相符，式(13)可清楚、彻底地解释任何疑问.

表2　矢量传感器线阵CBF的左右舷抑制比海试结果汇总表

Table 2　LRSR sea-trial results of AVS line array's CBF

参数		左右舷抑制比		
		28°	45°	87°
Cardioid	理论值	2.2	6.0	40.0
	实验值	1.9	5.9	14.5
质点振速	理论值	5.1	40.0	0.1
	实验值	3.5	13.9	0.1
声能流	理论值	2.5	40.0	40.0
	实验值	2.2	14.8	40.0

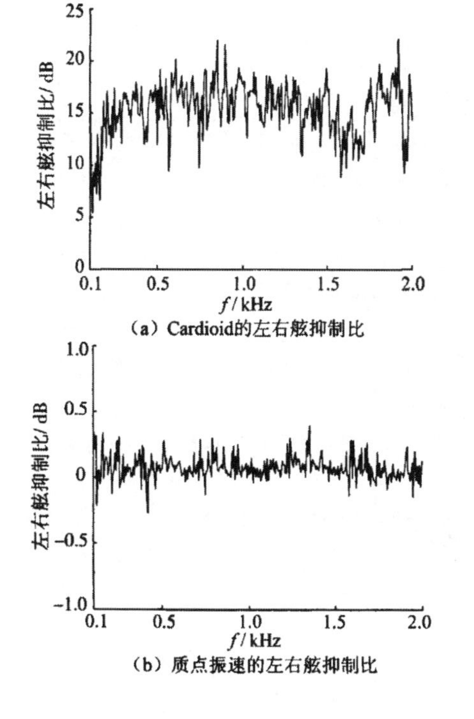

(a) Cardioid的左右舷抑制比

(b) 质点振速的左右舷抑制比

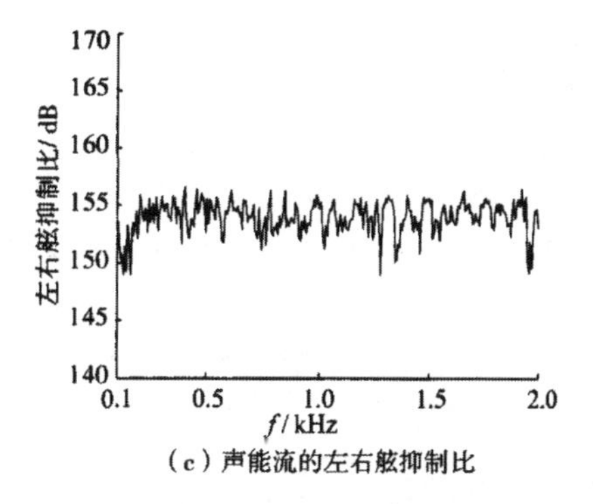

(c) 声能流的左右舷抑制比

图5　水平方位角87°时左右舷抑制比与频率的关系

Fig. 5　LRSR vs frequency when azimuth 87°

3　结　论

通过对矢量传感器线阵3种常规波束形成器和MVDR波束形成器左右舷抑制比的理论研究和海上试验数据分析，可得如下结论：

1)无论是常规波束形成器还是MVDR波束形成器，矢量传感器线阵左右舷抑制比均与频率无关，除了与信噪比有关之外，仅取决于阵元的指向性函数.

2)矢量传感器线阵的MVDR波束形成器左右舷分辨性能远优于3种常规波束形成器的结果.

基本结论1)可拓展成解决左右舷分辨问题的一般性原理，即左右舷分辨能力主要取决于阵元指向性函数. 因为无论是声压水听器双线阵还是三元组水听器单线阵，均可视为阵元具有一定左右舷分辨能力指向性函数的单线阵，指向性函数已清楚表明该假设的合理性.

基本结论2)也具有相当的普适性. MVDR波束形成器的左右舷分辨能力一般都优于常规波束形成器，矢量传感器线阵、声压水听器双线阵和三元组水听器单线阵都已验证了这一点. 而且MVDR为代表的自适应波束形成器在左右舷抑制、检测能力和干扰抑制等综合指标上占据绝对优势，在某种程度上应该是不二之选.

参考文献：

[1] LEMON S G. Towed-array history: 1917-2003[J]. IEEE J of Oceanic Eng, 2004, 29(2): 365-373.

[2] LASKY M. Recent progress in towed hydrophone array re-

search[J]. IEEE J of Oceanic Eng, 2004, 29(2): 374-387.

[3]ALLENSWORTH W S, KENNED W C Y, NEWHALL B K, et al. Twinline array development and performance in a shallow-water littoral environment[J]. Johns Hopkins APL Tech Dig, 1995, 16(3): 222-232.

[4]DOISY Y. Port-starboard discrimination performance on activated towed arrays systems[C]// UDT95. Nice, France, 1995: 125-129.

[5]BERLINER M J, LINDBERG J F. Acoustic particle velocity sensor: design, performance and applications[C] // AIP Conference Proceedings 368 (American Institute of Physics). Woodbury, NY: AIP Press, 1996.

[6]张宾,孙贵青,李启虎. 阵形畸变对拖曳双线阵左右舷分辨性能的影响[J]. 声学学报,2008, 33(4):294-299.
ZHANG Bin, SUN Guiqing, LI Qihu, Effects of shape distortion upon left/right discrimination of towed twin-line array[J]. Acta Acoustics, 2008, 33(4): 294-299.

[7]张宾,孙贵青,李启虎. 拖船噪声抵消与左右舷分辨联合处理方法的研究[J]. 应用声学,2008, 27(5):380-385.
ZHANG Bin, SUN Guiqing, LI Qihu. Towship noise canceling combined with left/right discrimination[J]. Applied Acoustics, 2008, 27(5): 380-385.

[8]杜选民,朱代柱,赵荣荣,等. 拖线阵左右舷分辨技术的理论分析与实验研究[J]. 声学学报,2000, 25(5):395-402.
DU Xuanmin, ZHU Daizhu, ZHAO Rongrong, et al. Theoretical analysis and experimental research on port/ straboard discrimination in towed line array[J]. Acta Acoustics, 2000, 25(5): 395-402.

[9]李启虎. 双线列阵左右舷目标分辨性能的初步分析[J]. 声学学报,2006, 31(5): 385-388.
LI Qihu. Preliminary analysis of left-right ambiguity resolution performance for twin-line array[J]. Acta Acoustics, 2006, 31(5): 385-388.

[10]李启虎. 用双线列阵区分左右舷目标的时延估计方法及其实现[J]. 声学学报, 2006, 31(6): 485-487.
LI Qihu. The time-delay estimation method of resolving left-right target ambiguity for twin-line array and its realization[J]. Acta Acoustics, 2006, 31(6): 485-487.

[11]何心怡,张春华,李启虎. 拖曳线列阵声呐及其左右舷分辨方法概述[J]. 舰船科学技术,2006, 28(5):9-14.
HE Xinyi, ZHANG Chunhua, LI Qihu. Rough introduction of towed linear array sonar and port/starboard discrimination methods[J]. Journal of Ship Science and Technology, 2006, 28(5): 9-14.

[12]HAWKES M, NEHORAI A. Acoustic vector-sensor beamforming and Capon direction estimation[J]. IEEE Trans on Signal Processing, 1998, 46(9): 2291-2304.

[13]孙贵青. 声矢量传感器及其在线列阵声呐中的应用[R]. 北京:中国科学院声学研究所博士后工作报告, 2003:105-111.

[14]杨秀庭. 矢量水听器及其在声呐系统中的应用研究[D]. 北京:中国科学院声学研究所,2006:41-50.
YANG Xiuting. Acoustic vector sensor and its application in sonar system[D]. Beijing: Institute of Acoustics, Chinese Academy of Sciences, 2006:41-50.

[15]SKREBNEV G K. Combined underwater acoustic receiver[M]. St. Petersburg: Elmor Press, 1997:24-36.

[16]孙贵青,李启虎. 声矢量传感器研究进展[J]. 声学学报,2004, 29(6):481-490.
SUN Guiqing, LI Qihu. Progress of study on acoustic vector sensor[J]. Acta Acoustics, 2004, 29(6): 481-490.

[17]孙贵青,李启虎. 声矢量传感器信号处理[J]. 声学学报,2004, 29(6):491-498.
SUN Guiqing, LI Qihu. Acoustic vector sensor signal processing[J]. Acta Acoustics, 2004, 29(6): 491-498.

[18]惠俊英,惠娟. 矢量声信号处理基础[M]. 北京:国防工业出版社,2009:24-25.

[19]杨德森,洪连进. 矢量水听器原理及应用引论[M]. 北京:科学出版社,2009:129-136.

[20]CAPON J. High-resolution frequency-wave number spectrum analysis[C]// Proc IEEE Aug. 1969, 57(8): 1408-1418.

[21]GROEN J, BEERENS S P, BEEN R. Adaptive port-starboard beamforming of triplet sonar arrays[J]. IEEE J of Oceanic Eng, 2005, 30(2): 348-359.

多基地声纳定位误差最小的模板法①

李　嶷，李启虎，孙长瑜
（中国科学院声学研究所，北京 100190）

摘　要：研究了基于定位误差最小化的多基地声纳接收基地配置方法。根据定位误差的几何分布情况，*提出了配置多基地声纳的"模板法"*。"模板法"利用已布设网络提供的目标距离信息和新增接收基地的测量误差信息，对"定位精度几何扩散因子"的几何分布图进行特征提取，得到"模板"。将"模板"与真实的声纳环境匹配，可确定新增接收基地的布放位置。仿真分析说明，该方法可解决攻击模式下目标的更准确定位问题，运算量小，实时性高。

关键词：信息处理技术；多基地声纳；定位误差；配置算法；攻击模式
中图分类号：TP391.9　　**文献标识码**：A　　**文章编号**：1004-731X (2011) 11- 2465-06

Deployment Algorithm of Multistatic Sonar Based on Minimum Localization Error

LI Yi, LI Qi-hu, SUN Chang-yu
(Institute of Acoustics, Chinese Academy of Sciences, Beijing 100190, China)

Abstract: A deployment algorithm of multistatic sonar based on minimum localization error was studied. The method of template for deploying multistatic sonar was brought forward according to the geometrical distributing of localization error. The template method made use of the measurement error of the new receiver and the target's distance information which are provided by the existed network. The template could be obtained by extracting the characteristics of Geometrical Dilution of Precision (GDOP) geometrical figure. At last, the position of the new receiver could be determined by matching the template and the real multistatic condition. Simulation results show that the method can localize the targets more accurately under the attacking mode, and the computational loads are small.

Key words: information processing; multistatic sonar; localization error; deployment algorithm; attacking mode

引　言

多基地声纳是一种基于网络的检测系统，因此传感器的配置是其中一个关键问题。

文献[1]根据声纳的检测覆盖范围，提出用粒子群聚优化算法（Particle Swarm Optimum，PSO）来配置多基地声纳，运算量比贪婪算法明显减少。文献[2]则提出用更少粒子的顺序粒子群聚优化算法（Sequential Particle Swarm Optimum，S-PSO）来配置传感器，其覆盖范围和运算量又比 PSO 方法更优。文献[3]则在网格的基础上，提出用模拟退火算法来配置传感器，它可较有效解决不同范围大小的传感器布放问题。文献[4]将环境视为非各向同性的环境，提出用基因算法来配置浮标，并证明以前提出的一些模型在非各向同性环境下性能会降低。文献[5]则以多基地模糊函数为依据，讨论了在不同的几何布局条件下，能实现对目标的分辨能力。

传感器的配置问题在移动通信网、机械设备的故障诊断、艺术馆内报警器的放置等方面[6-8]都涉及到，但它们的应用场景和需解决的关键问题各不相同。多基地声纳一般应用在宽广的区域，没有障碍物遮挡，出于应战的需要，其配置要求快速、准确。目前，针对多基地声纳配置的研究相对较少，尤其没见到针对定位误差最小化提出的多基地声纳配置方法。

本文以"定位精度几何扩散因子"GDOP（Geometrical Dilution of Precision）作为评判标准，提出攻击模式下多基地声纳的配置算法，该算法在理论上能较大提高目标的定位精度，运算量小，具有一定的工程实用性。

1　多基地声纳定位原理

多基地声纳中每个发射基地与每个接收基地均构成一

① 系统仿真学报, 2011, 23(11): 2465-2470.

组收发分置的双基地声纳，因此多基地声纳可视为由一组组双基地声纳构成。通过测量发射信号和目标反射信号的到达时间、方位，每组双基地声纳均能测量出目标坐标。因此，在多基地声纳系统中，针对同一个目标，可得到 N 个坐标测量值，其中 N 为系统中双基地声纳对数。

假设在多基地声纳系统中，针对某个目标 T，第一组双基地声纳测量得到的目标坐标为 (x_T^1, y_T^1)，第二组双基地声纳测量得到的目标坐标为 (x_T^2, y_T^2)，依此类推，第 N 组双基地声纳测量得到的目标坐标为 (x_T^N, y_T^N)。

另外，假设多基地声纳中每组双基地声纳的观测结果相互独立，且每组双基地声纳对 x_T、y_T 的观测是无偏的，只是测量得到的方差不一样，即有：

$$E(x_T^1) = x_T，E(x_T^2) = x_T，……，E(x_T^N) = x_T \quad (1)$$

$$E(y_T^1) = y_T，E(y_T^2) = y_T，……，E(y_T^N) = y_T \quad (2)$$

$$D(x_T^1) = D(x_T^1)，D(x_T^2) = D(x_T^2)，……，D(x_T^N) = D(x_T^N) \quad (3)$$

$$D(y_T^1) = D(y_T^1)，D(y_T^2) = D(y_T^2)，……，D(y_T^N) = D(y_T^N) \quad (4)$$

那么，选用多基地声纳的目的是为了寻求新的估计量 $x_T = f(x_T^1, x_T^2, \cdots, x_T^N)$ 和 $y_T = f(y_T^1, y_T^2, \cdots, y_T^N)$。新的估计量应保证对目标坐标的估计是无偏的，且定位误差最小。

2 多基地声纳定位误差

2.1 多基地声纳定位误差理论分析

用“定位精度几何扩散因子”GDOP（Geometrical Dilution of Precision）来衡量目标定位误差与几何分布的关系。由于水声环境中的深度值通常远远小于距离值，因此可近似认为对目标的定位是在二维平面内的定位，那么 GDOP 可用二维方式表达，可得到目标在某点处的定位误差：

$$GDOP = [\sigma_x^2 + \sigma_y^2]^{\frac{1}{2}} \quad (5)$$

其中，σ_x^2 和 σ_y^2 分别表示计算得到的目标坐标在 x 轴和 y 轴方向上的方差。从式（5）可看出，当 $GDOP$ 大时，目标定位误差大；当 $GDOP$ 小时，目标定位误差小。

对于多基地声纳，考虑最简单的线性组合情况，得到目标坐标：

$$\begin{cases} x_T = a_1 x_T^1 + a_2 x_T^2 + \cdots + a_N x_T^N \\ y_T = a_1 y_T^1 + a_2 y_T^2 + \cdots + a_N y_T^N \end{cases} \quad (6)$$

其中，a_1、a_2、……、a_N 为待确定的常数。

由于要求估计是无偏的，因此可将 $a_1 + a_2 + \cdots + a_N = 1$ 作为约束条件。有

$$E(x_T) = (a_1 + a_2 + \cdots + a_N)\, x_T = x_T \quad (7)$$

$$E(y_T) = (a_1 + a_2 + \cdots + a_N)\, y_T = y_T \quad (8)$$

此时，多基地声纳的 $GDOP$ 值为：

$$GDOP = \sqrt{\sigma_{x_T}^2 + \sigma_{y_T}^2} = \sqrt{D(x_T) + D(y_T)} \quad (9)$$

进一步推得：

$$GDOP = \sqrt{a_1^2 GDOP_1^2 + a_2^2 GDOP_2^2 + \cdots + a_N^2 GDOP_N^2} \quad (10)$$

其中，$GDOP_k^2$（$k = 1, 2, \cdots, N$）是第 k 组双基地声纳测量得到的目标坐标的 $GDOP$ 值平方。

根据约束条件为等式的条件极值求解方法，引进修正函数 F：

$$F = a_1^2 GDOP_1^2 + a_2^2 GDOP_2^2 + \cdots + a_N^2 GDOP_N^2 + \lambda(a_1 + a_2 + \cdots + a_N - 1) \quad (11)$$

利用拉格朗日乘数法对修正函数求极值，可得：

$$a_1 = \frac{E}{GDOP_1^2}，\quad a_2 = \frac{E}{GDOP_2^2}，\quad ……，\quad a_N = \frac{E}{GDOP_N^2} \quad (12)$$

其中

$$E = \left(\frac{1}{GDOP_1^2} + \frac{1}{GDOP_2^2} + \cdots + \frac{1}{GDOP_N^2}\right)^{-1} \quad (13)$$

从上式容易证明：

$$E < \min(GDOP_1^2, GDOP_2^2, \cdots, GDOP_N^2) \quad (14)$$

因此

$$GDOP = \sqrt{E} < \sqrt{\min(GDOP_1^2, GDOP_2^2, \cdots, GDOP_N^2)} \quad (15)$$

从上面的分析可看出，当多基地声纳中每组双基地声纳的观测结果相互独立，且每组双基地声纳对目标的观测是无偏的，只是测量得到的方差不一样时，多基地声纳的定位性能必然优于其中任何一组双基地声纳。实际工作中的大部分情况可近似认为满足上述条件。

2.2 多基地声纳定位误差仿真分析

双基地声纳的定位方法有三种，分别是：方位交汇、距离交汇和方位/距离交汇。它们的定位性能与发射、接收基地的坐标测量误差，方位角测量误差以及时间测量误差有关。以方位交汇定位系统为例，定义声呐阵所指的0°方向与 x 轴方向一致，定位示意图如图 1 所示。

方位交汇定位系统的发射基地能检测出目标反射信号的到达角度 θ_{TS}，接收基地能检测出目标反射信号的到达角度 θ_{TR}。设发射、接收基地的坐标 (x_S, y_S) 和 (x_R, y_R) 能通过 GPS 定位系统确定，那么可推导出目标坐标 (x_T, y_T) 为：

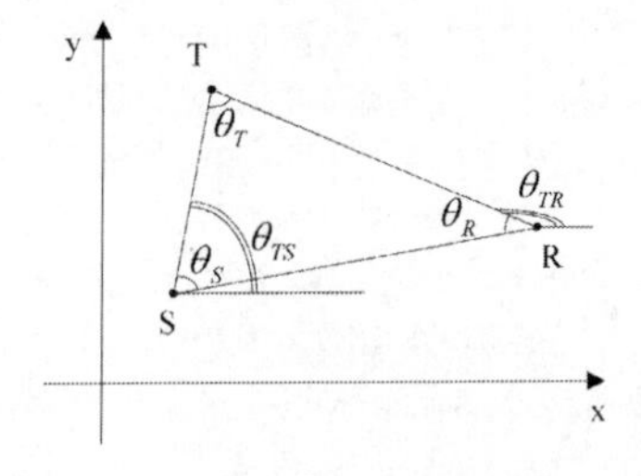

图 1　方位交汇定位系统原理图

$$\begin{cases}\dfrac{y_T - y_S}{x_T - x_S} = tg\theta_{TS} \\ \dfrac{y_T - y_R}{x_T - x_R} = tg\theta_{TR}\end{cases} \tag{16}$$

假设发射、接收基地的位置可通过 GPS 确定，测量误差非常小，基本可忽略，因此只考虑方位角测量误差。另外，假设各测量误差是零均值、彼此不相关的高斯白噪声，那么可推导出方位交汇定位方法的定位误差为：

$$\begin{cases}\sigma_{x_T}^2 = \dfrac{\sec^4\theta_{TS}[tg\theta_{TR}(x_R - x_S) + (y_S - y_R)]^2}{(tg\theta_{TS} - tg\theta_{TR})^4}\sigma_{\theta_{TS}}^2 + \\ \qquad \dfrac{\sec^4\theta_{TR}[tg\theta_{TS}(x_S - x_R) + (y_R - y_S)]^2}{(tg\theta_{TS} - tg\theta_{TR})^4}\sigma_{\theta_{TR}}^2 \\ \sigma_{y_T}^2 = \dfrac{\sec^4\theta_{TS}[tg^2\theta_{TR}(x_R - x_S) + tg\theta_{TR}(y_S - y_R)]^2}{(tg\theta_{TS} - tg\theta_{TR})^4}\sigma_{\theta_{TS}}^2 + \\ \qquad \dfrac{\sec^4\theta_{TR}[tg^2\theta_{TS}(x_S - x_R) + tg\theta_{TS}(y_R - y_S)]^2}{(tg\theta_{TS} - tg\theta_{TR})^4}\sigma_{\theta_{TR}}^2 \\ \sigma_{x_T y_T}^2 = \dfrac{\sec^4\theta_{TS}tg\theta_{TR}[tg\theta_{TR}(x_R - x_S) + (y_S - y_R)]^2}{(tg\theta_{TS} - tg\theta_{TR})^4}\sigma_{\theta_{TS}}^2 + \\ \qquad \dfrac{\sec^4\theta_{TR}tg\theta_{TS}[tg\theta_{TS}(x_S - x_R) + (y_R - y_S)]^2}{(tg\theta_{TS} - tg\theta_{TR})^4}\sigma_{\theta_{TR}}^2\end{cases} \tag{17}$$

其中(x_S, y_S)、(x_R, y_R)、(x_T, y_T)分别表示发射基地、接收基地和目标的坐标。

定位精度：

$$GDOP = \sqrt{\sigma_{x_T}^2 + \sigma_{y_T}^2} \tag{18}$$

从上面公式可看出，$GDOP$ 是一个与声纳几何布局、以及声纳测量误差有关的量。

假设双基地声纳中发射基地坐标为（-1500m，-1000m），接收基地坐标为（1000m，1500m），方位角测量标准差 σ 均为 0.5°。那么，由上面公式可得到如图 1a 所示的双基地声纳方位交汇定位方式下 $GDOP$ 对数值的几何分布图。

另外，还可在图 1a 的基础上增加一个接收基地，由此构成三基地声纳。增加的接收基地坐标为（1500m，-1500m），三基地声纳在方位交汇定位方法下的 $GDOP$ 对数值几何分布图如图 1b 所示。对比图 1a 和图 1b 可看出，当多基地声纳中每组双基地声纳的观测结果相互独立，且每组双基地声纳对目标的观测是无偏的，只是测量得到的方差不一样时，三基地声纳的定位性能明显优于双基地声纳，尤其是基线附近的定位不确定性可完全克服。

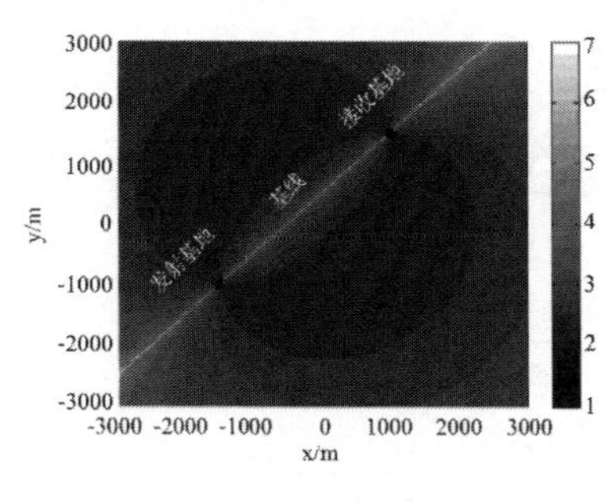

a) 双基地声纳

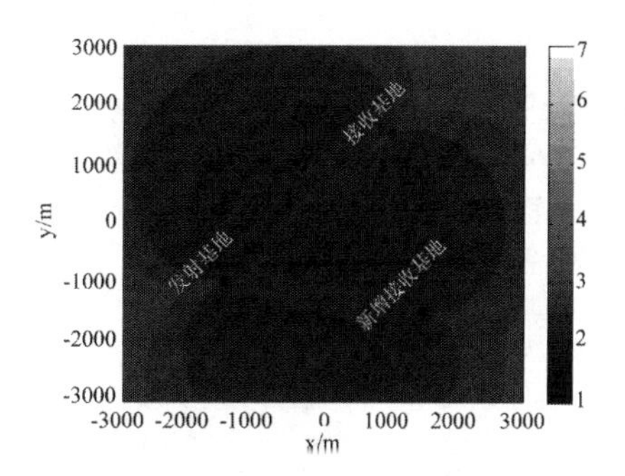

b) 三基地声纳

图 1 定位误差几何分布图

3 多基地声纳配置算法

由于多基地声纳可视为由不同组的双基地声纳构成，因此对于只有一个发射基地、多个接收基地的情况，增加一个接收基地也就等效于增加了一组双基地声纳。下面的讨论均针对这种情况进行。

此处仍假设多基地声纳中每组双基地声纳的观测结果相互独立，且每组双基地声纳对目标的观测是无偏的，只是测量得到的方差不一样。那么，增加一个接收基地后，多基地声纳的 $GDOP$ 值变为：

$$GDOP\_new = \left(\frac{1}{GDOP_1^2} + \frac{1}{GDOP_2^2} + \cdots + \frac{1}{GDOP_N^2} + \frac{1}{GDOP_{N+1}^2}\right)^{-1/2} \tag{19}$$

其中 $GDOP_{N+1}^2$ 表示新增的双基地声纳的定位误差。

将式（15）与式（18）比较可知：

$$GDOP\_new < GDOP \tag{20}$$

上式说明当多基地声纳中每组双基地声纳的观测结果相互独立，且每组双基地声纳对目标的观测是无偏的，只是测量得到的方差不一样时，只要增加一个接收基地，多基地声纳的定位性能就会提高 $(GDOP^{-2} + 1/GDOP_{n+1}^2)^{-1/2} - GDOP$。

但是，由于受到成本的限制，通常接收基地不能布设太多。因此，如何利用有限的接收基地，来得到尽可能小的定位误差成为人们研究的主要方向。

实际工作中，有时先期已布设有声纳阵，能大概测量出目标位置，但还需要通过飞机投放浮标等方式增加新的接收基地，以保证对目标的定位精度进一步提高。本文称这种工作方式为攻击模式下接收基地的配置问题，我们的目标是寻求最优的接收基地配置算法，以保证得到最小的目标定位误差，即保证公式 20 中的 $GDOP_{N+1}^2$ 最小。

3.1 算法原理

由随机信号的参数估计理论可知，当知道的先验知识越多，做出的估计就越准确。本文就在充分利用先验知识的基础上，提出一种新的接收基地配置算法——“模板法”。

从前面的分析可知，当多基地声纳中只有一个发射基地时，增加一个接收基地等效于增加了一对双基地声纳，目标附近的 $GDOP$ 值变为式(19)。在这种情况下，由于先期已经布设有声纳阵，所以多基地声纳中除新增的这个接收基地外，其它接收基地和发射基地的位置及性能已经确定，因此式(19)中的 $GDOP_1^2$ 、$GDOP_2^2$ 、……、$GDOP_N^2$ 为定值。那么，式(19)中剩下的变量就只有 $GDOP_{N+1}^2$ 了。如果要求目标的定位误差 $GDOP\_new$ 最小，就应保证 $GDOP_{N+1}^2$ 最小。

$GDOP_{N+1}^2$ 是一个与新增双基地声纳的几何布局和测量误差有关的变量。实际工作中，由于先期已经布设了一些声纳阵，因此我们对海洋环境、接收信号强度、目标强度等有了一些先验知识。利用这些先验知识，能大概判断出新增接收基地的测量误差大小。另一方面，先期布设的声纳阵能帮助我们粗略确定目标位置，所以，也可将发射基地到目标的距离 l 视为已知条件。这样，$GDOP_{N+1}^2$ 中需确定的变量就只有新增接收基地的位置了。

3.2 算法模型

根据前面的分析，可以将新增接收基地的测量误差作为已知条件，由此能计算出双基地声纳在不同基线长度下的 $GDOP$ 几何分布图；另外，由于发射基地到目标的距离 l 也认为已知，所以还可做出一个以发射基地为圆心，半径为 l 的圆，目标必然落在此圆上。将 $GDOP$ 几何分布图与圆重合，可得到如图 2 所示的图形。

从图 2 容易看出，当双基地声纳的基线长度一定时，寻找新增双基地声纳的最小定位误差点等效于寻找图 2 中圆周上的最小定位误差点。当目标与此点重合时，目标的定位误差最小。

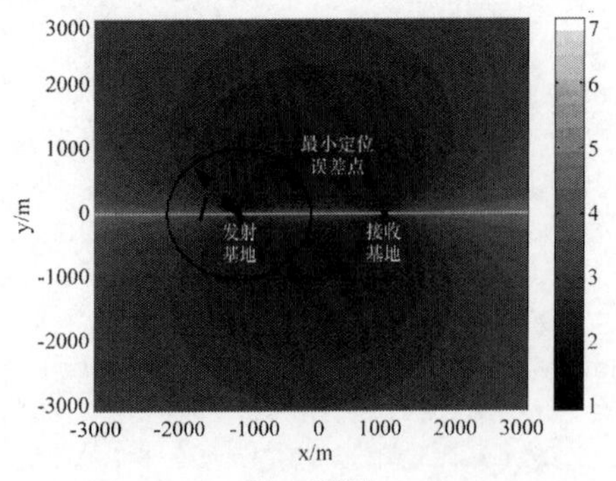

图 2 重叠图

同理，仍将新增接收基地的测量误差和发射基地到目标的距离 l 作为已知条件，可找到不同基线长度下的最小定位误差点和对应的双基地声纳几何布局图。

最后，再寻找不同基线长度下所有最小定位误差点中的极小值，它即是最小 $GDOP_{N+1}^2$ 点，此时对应的双基地声纳几何布局图为新增双基地声纳的最优几何布局。

由此，提出了“模板”的概念。每幅“模板”均对应特定测量误差和特定距离 l，且均由三要素构成，即：发射点、接收点和最小 $GDOP$ 值点，其示意图如图 3 所示。

“模板”的提取过程实际上是一个降维处理的过程，它将声纳的配置问题简化为只有五个分量的向量空间匹配问题，这五个分量分别是：新增接收基地的测量误差、发射点到目标的距离 l 、发射点坐标，接收点坐标和最小 $GDOP$ 值点坐标。

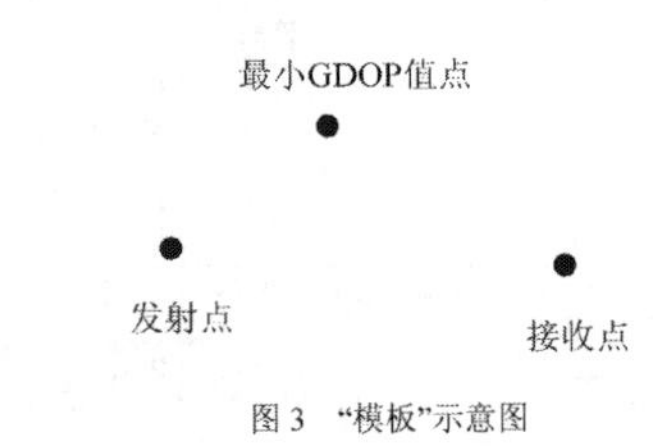

图 3 “模板”示意图

当大概确定目标位置后，就可将数据库中的“模板”与实际的多基地声纳进行匹配了。根据最小邻近准则，匹配的过程其实也是搜索的过程，即根据先期估计的新增接收基地测量误差大小，寻找“模板”中“最小 $GDOP$ 值点”到发射点的距离与多基地声纳中目标到实际发射点的距离相等的“模板”，其原理如图 4 所示。当找到此“模板”后，就能根据“模板”的几何分布图确定新增接收点的最优位置了，即“模板”中的接收点位置对应最优的接收点布放位置。

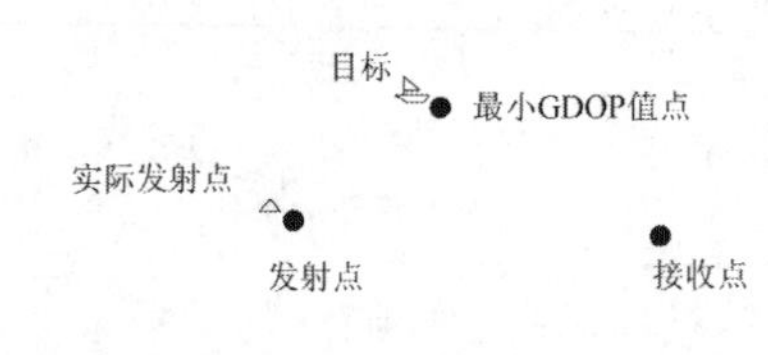

图 4 匹配示意图

由于“模板法”中的“模板”利用了一些先验知识，因此“模板”在布阵早期就能计算出来，并存储在数据库中。另外，也可根据海洋环境的变化，不断地对数据库进行补充和完善。上述计算过程在目标出现前就可完成，不受实时性限制。当目标出现后，只需要简单地进行边长匹配即可，运算量非常小，实时性高。

3.3 算法仿真

以方位交汇定位系统为例。假设前期已布设了三基地声呐阵，声呐阵中发射、接收基地的分布如图 1b 所示，方位角测量标准差 σ 均为 0.5°，目标位于（-1000m，1000m）。此时目标的定位误差 *GDOP* 为 23.7078m。

为了更精确地确定目标坐标，我们需要增加一个接收基地，并确定此接收基地的最优位置。

由于原有三基地声呐阵的方位角测量标准差 σ 均为 0.5°，且新增接收基地位于原三基地声呐阵所在海域，所以可近似认为新增双基地声呐的方位角测量标准差 σ 也为 0.5°。另外，通过前面的叙述可计算出发射基地到目标的距离约为 2061.5m。

下面将数据库中如表 1 所示的 8 个“模板”分别与声呐阵进行匹配，首先，寻找方位角测量标准差 σ 为 0.5° 的“模板”；然后，在此基础上，寻找发射基地到目标的距离最接近 2061.5m 的“模板”。对于本节所述的情况，容易发现 2 号模板最匹配。

由 2 号模板的参数可确定新增的第三个接收基地坐标为（-863.63m，954.04m），且增加该基地后，目标附近的 *GDOP* 值近似变为 14.32m。很明显，相对于两个接收基地的情况，增加第三个接收基地后，目标定位误差明显减小，*GDOP* 值近似减少 9.4m。

从上面的仿真可看出，系统测量误差和目标距离的先验知识越充分，数据库中存放的模板越多，匹配效果会越好，目标定位误差会趋近于最小。

表 1　“模板”参数表（单位：m；发射基地坐标固定为（0，0））

测量误差	发射点到目标的距离	接收点坐标	最小 *GDOP* 点坐标	最小 *GDOP* 值	模板号
$\sigma_{\theta_{TS}}=0.5°$ $\sigma_{\theta_{TR}}=0.5°$	2100	(2110,0)	(2090,204.7)	18.41	1
	2050	(2060,0)	(2040,202.24)	17.98	2
	2000	(2010,0)	(1990,199.75)	17.54	3
	1950	(1960,0)	(1940,197.23)	17.1	4
$\sigma_{\theta_{TS}}=\sqrt{5}°$ $\sigma_{\theta_{TR}}=\sqrt{8}°$	2100	(2110,0)	(2090,204.7)	82.58	5
	2050	(2060,0)	(2040,202.2)	80.63	6
	2000	(2010,0)	(1990,199.75)	78.68	7
	1950	(1960,0)	(1940,197.23)	76.73	8

3.4 误差分析

“模板”法中的误差主要由两方面产生。第一，多基地声纳系统的测量误差导致对真实目标的定位产生偏差，最后引起对距离 l 估计不准确；第二，由于先验知识不充分，导致对“模板”的测量误差估计不准确，最后产生了一个非最优的模板。

要得到“模板”法中误差的解析解非常困难，因此，下面用蒙特卡罗仿真分析方法对误差进行分析。

仍以方位交汇定位系统为例，假设前期已布设了三基地声纳阵，声纳阵中发射、接收基地的分布如图 1b 所示，目标位于（-1000m，1000m）。现在需要增加一个新的接收基地并确定该基地的布放位置。

本文分别对方位角测量标准差 σ 为 0.25° 和 σ 为 2.5° 的情况进行分析。另外，假设“模板”方位角测量误差的标准差理论值也分别为 0.25° 和 2.5°，且方位角的估计值是以理论值为中心，服从标准差为 g 的高斯分布。

按照实际工作中确定新增接收基地位置的算法流程进行仿真，可得到不同误差组合情况下的新增接收基地坐标，其算法流程图如图 5 所示。

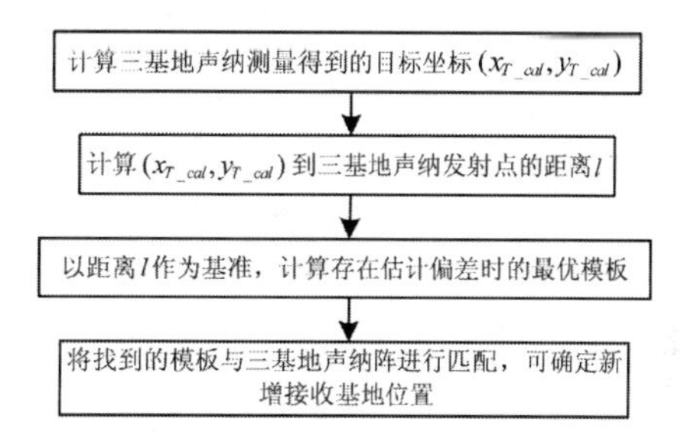

图 5　新增接收基地位置计算流程图

本文假设发射基地到接收基地的距离超过 3000m 后，接收基地就很难检测到目标的反射信号了，即当发射基地与接收基地之间的距离超过 3000m 后就无法实现目标定位了。在此条件下，对每一种误差组合进行 200 次计算，选取平均值作为该误差组合情况下新增接收基地的布放位置，可得到如图 6a 所示的误差仿真结果。从图 6a 可看出，随着原有多基地声纳阵测量方差的增大，计算得到的新增接收基地坐标越偏离最优值；另一方面，从图 6a 还可看出，随着模板测量误差估计方差的增大，计算得到的新增接收基地坐标会更离散。

在找到新增接收基地的坐标后，还需要比较“模板”法与随机布放新增接收基地方法对目标定位性能的差异。其步骤如下：第一，首先利用前面找到的新增接收基地坐标，计算出利用“模板”法增加接收基地后目标的定位误差 *GDOP*。第二，对“随机布放”新增接收基地的方法进行仿真。假设新增接收基地是在观测区域内随机布放的，且布放规律满足平均

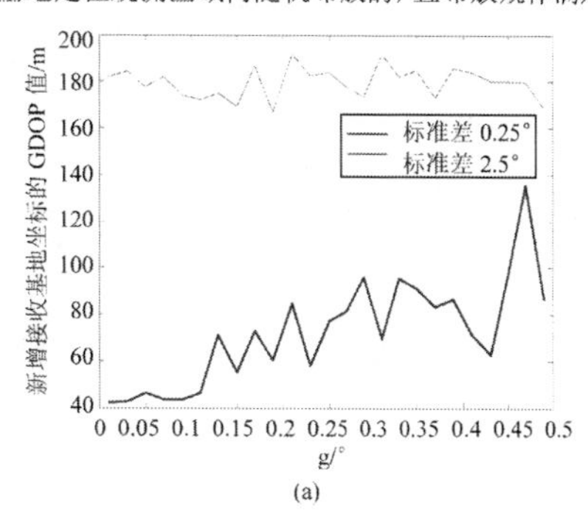

(a)

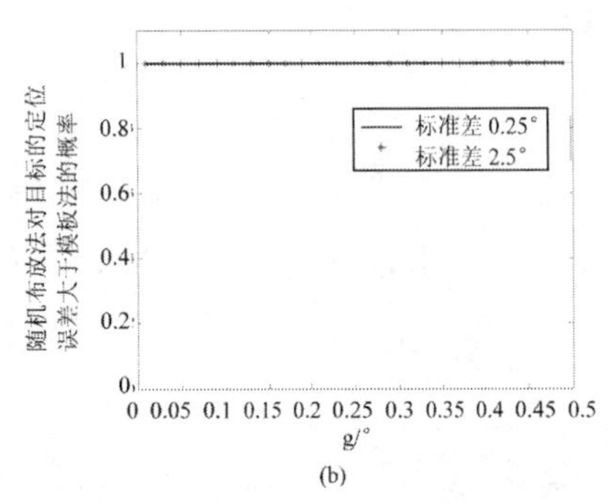

(b)

图 6 “模板”法误差仿真结果

分布，由此可计算出这种情况下的目标定位误差 $GDOP$。本文分别对“随机布放”方法下每种误差组合情况仿真 1000 次。第三，对于不同误差组合情况，分别统计随机布放方法对目标的定位误差超过“模板”法对目标的定位误差的次数 m，并计算 m 占每种误差组合情况下总仿真次数 1000 的比例，仿真结果如图 6b 所示。从图 6b 可看出，“模板”法的优势非常明显，它基本可以保证“模板”法对目标的定位性能百分之百地优于随机布放新增接收基地的方法。

4 结论

多基地声纳可视为由一组组双基地声纳构成。当多基地声纳中每组双基地声纳的观测结果相互独立，且每组双基地声纳对目标的观测是无偏的，只是测量得到的方差不一样时，多基地声纳的定位性能远比任何一组双基地声纳的定位性能好。而且，随着接收基地的不断增加，多基地声纳的定位性能也不断提高。

实际工作中，由于成本限制，多基地声纳的接收基地不能无限制增加，所以我们必须利用仅有的几个接收基地来获得尽量小的目标定位误差。

对于先期已经布设有声纳阵，且目标位置已基本确定的情况，本文提出用“模板”法来进行接收基地优化配置，它能有效解决目标的更准确定位问题。“模板”法通过对 GDOP 几何分布图进行特征提取，达到降维的目的，使声纳配置问题得到充分简化。“模板”的产生基于对海洋环境和多基地声纳系统的了解，当得到的先验知识越多时，“模板”法的效果越好。此方法运算量小，实时性强，对于解决工程实践问题具有一定的指导作用。

参考文献：

[1] Patrick N Ngatchou, Warren L J Fox, Mohamed A El-sharkawi. Multiobjective Multistatic Sonar Sensor Placement [C]// IEEE Congress on Evolutionary Computation, 2006. USA: IEEE, 2006: 2713-2719.

[2] Patrick N Ngatchou, Warren L J Fox, Mohamed A El-sharkawi. Distributed Sensor Placement with Sequential Particle Swarm Optimization [C]// Swarm Intelligence Symposium, Proceedings 2005 IEEE. USA: IEEE, 2005: 385-388.

[3] Frank Y S Lin, P L Chiu. A Near-Optimal Sensor Placement Algorithm to Achieve Complete Coverage/Discrimination in Sensor Networks [J]. IEEE Communications Letters (S1089-7798), 2005, 9(1): 43-45.

[4] Donald R, DelBalzo, Erik R Rike, David N McNeal. Optimizing multistatic sonobuoy placement [J]. JASA (S0001-4966), 2005, 117(4): 2625.

[5] Bradaric I, Capraro G T, Wicks M C. Sensor Placement for Improved Target Resolution in Distributed Radar Systems [C]// Radar Conference, RADAR'08. USA: IEEE, 2008: 1-6.

[6] Mohamed Younis, Kemal Akkaya. Strategies and Techniques for Node Placement in Wireless Sensor Networks: A Survey [J]. Ad Hoc Networks (doi:10.1016/j.adhoc2007.05.003), 2008, 6(4): 621-655.

[7] Josh M Beal, Amit Shukla, Olga A Brezhneva, *et al.* Optimal Sensor Placement for Enhancing Sensitivity to Change in Stiffness for Structural Health Monitoring [J]. Optimization and Engineering (S11081-007-9023-1), 2008, 9(2): 119-142.

[8] T S Michael, Val Pinciu. Art Gallery Theorems for Guarded Guards [J]. Computational Geometry (doi:10.1016/S0925-7721(03)00039-7), 2003, 26(3): 247-258.

浅海波导中水下目标辐射噪声干涉条纹的理论分析和试验结果[①]

李启虎 王 磊 卫翀华 李 嶷 马雪洁 于海春

(中国科学院声学研究所 北京 100190)

2010 年 7 月 1 日收到

2010 年 8 月 20 日定稿

摘要 浅海波导不变量的研究是近 30 年来水声学中引人注意的课题之一。水下目标辐射噪声的直达波和海面、海底反射波之间的干涉现象可以认为是浅海波导的典型例子，直达波和反射波之间的干涉条纹对了解浅海波导的机理是非常有效的。本文给出由水下目标辐射噪声直达波和海面、海底反射波产生的干涉现象的理论分析，指出由海面反射波引起的干涉条纹集中在高频段，而由海底反射引起的干涉条纹既可能在低频段也可能在高频段，它与海底的声学特性密切相关。文中给出了干涉条纹图和目标位置、海洋环境以及接收传感器等参数的关系，指出干涉条纹具有双曲线的形状，而一族双曲线的渐近线为直线。海试给出了若干有趣的结果，部分数据处理的例子证实了理论分析的正确性。本文所述的理论与试验结果表明在一定条件下，干涉条纹图可以用于目标检测与识别。

PACS 数：43.60

Theoretical analysis and experimental results of interference striation pattern of underwater target radiated noise in shallow water waveguide

LI Qihu WANG Lei WEI Chonghua LI Yi MA Xuejie YU Haichun

(*Institute of Acoustics, Chinese Academy of Sciences* Beijing 100190)

Received Jul. 1, 2010

Revised Aug. 20, 2010

Abstract Waveguide invariant in shallow water is an attractive topic in recent three decades. The interference phenomena of direct wave of radiated noise of underwater target and reflection wave from sea surface and sea bottom can be considered as a typical case of shallow water waveguide. The interference striation pattern of direct wave and its reflection is the effective and comprehensive figure for better understanding the essence of shallow water waveguide invariant. The theoretical analysis of interference phenomena generated by direct wave of radiated noise of underwater target and its reflection wave from sea surface and sea bottom is presented in this paper. It is shown that the interference wave resulted by sea surface reflection produces striation pattern centered at high frequency band. But the interference of nulling frequency resulted by sea bottom reflection may be at low frequency or high frequency, it strongly depends on the acoustic behavior of sea bottom. The relationship between main parameters of interference striation pattern and target, receiver, and environment is derived. It is shown that the interference striation has the shape of hyperbolic curve. The equation set of the hyperbolic curve and its asymptotic line are presented. The at-sea experiment carried out in South China sea shows some interesting results. A part of data processing results are illustrated in this paper. The results expressed in this paper show that the interference striation pattern can be used, in some conditions, as a potential means for target recognition.

① 声学学报, 2011, 36(3): 253-257.

引言

浅海中的波导不变量的概念是由 Chuprov 等提出的[1−2]．该理论指出，在一定条件下，浅海中宽带信号在传播过程中由于受海面和海底反射的影响会产生干涉． Kuperman 等推广了波导不变量的概念，把它应用于垂直阵和水平阵的信号处理中[3−5]．Kapolka 指出，如果波导不变量 $\beta = 1$，那么双径的射线模型就等价于等温层下的波导理论，他研究了这种情况下由海面反射引起的干涉现象[6]．

宽带信号传播过程中的干涉现象可以由简正波理论推导出来，对干涉声场有贡献的简正波的数目是频率的函数．二维的 LOFAR(Low Frequency Analysis Record) 图可以用于描绘干涉条纹图像．

在浅海的情况，干涉条纹的情况可以从水下目标辐射噪声的直达波和海面、海底反射波的干涉计算得到．干涉信号的表达式可以分别在海面反射或海底反射的情况下求出来．干涉条纹中的抵消频率与具体环境参数密切相关，包括声源 / 接收器的深度、海深等．

干涉条纹的距离 - 频率相关性的公式可以由双曲线来表达，而它的渐近线也可以从声源 / 接收器的深度信息中计算得到．干涉条纹图可以应用于检测水下目标辐射噪声，而干涉条纹的形状对于目标识别也是有辅助作用的．

海试结果对理论分析进行了验证．波导不变量以及干涉条纹图为水下目标的检测和识别提供了有用的途径．

1 浅海干涉波形理论分析

假定声源的频率为 f，幅度为 $A(f)$，处于深度为 d 的位置．在距离 r 处，深度 h 处有一无方向性的接收器，那么它所接收的声压可以表达为[3]:

$$p(r,d,h,f) = A(f)\sum_{n=1}^{N} B_n(d,h)\exp[-\mathrm{j}(2\pi f t - k_n r + \pi/4)], \tag{1}$$

式中，k_n 是水平波数，$B_n(d,h)$ 是特征函数的函数．接收声压谱的强度是 $p(r,d,h,f)$ 的均方平均值．Kuperman 指出，声压强度由两部分组成：第一部分随距离和频率缓慢变化；而第二部分由于简正波的干涉而振荡．

浅海中水下目标辐射噪声传播过程中的干涉现象，可以用劳埃德镜 (Lloyd's mirror) 加以分析．Kapolka 计算了直达波和海面反射波的干涉条纹图，可以证明，由海面反射所形成的干涉条纹集中于相对较高的频段．实际上，由海底反射波引起的干涉条纹可能会集中在低频段．

由于低频信号的传播损失通常远小于高频信号 (特别是在浅海如此)，所以出现在低频段的干涉条纹在目标检测与识别方面具有更加重要的意义．

图 1 给出了我们所考虑的浅海环境，各种参数的意义如下： d: 声源至海面的距离， h: 接收器至海面的距离， b: 声源至海底的距离， k: 接收器至海底的距离， r: 声源至接收器的水平距离， r_d: 从声源到接收器的直线距离，r_b: 海底反射波至接收器的距离， r_s: 海面反射波至接收器的距离．

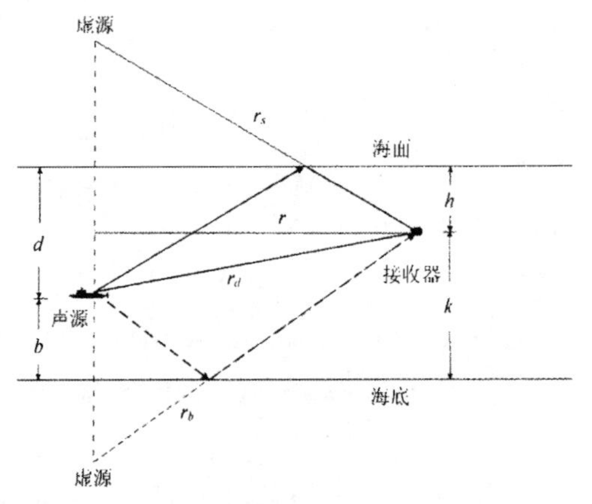

图 1 直达波和海面、海底反射波

在计算波导干涉时，声程差起着重要的作用．对于海面反射波，声程差为：

$$\Delta r_s = r_s - r_d. \tag{2}$$

对于海底反射波，声程差为：

$$\Delta r_b = r_b - r_d. \tag{3}$$

假如声源的直达波为正弦信号：

$$x(t) = A\cos(2\pi f t + \theta), \tag{4}$$

式中， f 为频率， θ 为在 $(0, 2\pi)$ 内均匀分布的随机相位．容易指出，由海面和海底反射的波为：

$$x_s(t) = A_s\cos[2\pi f t + \theta + \pi + 2\pi f \Delta r_s/c], \tag{5}$$

$$x_b(t) = A_b\cos[2\pi f t + \theta + \varphi + 2\pi f \Delta r_b/c], \tag{6}$$

在式 (5) 中，相位 π 表示海面是水声反射的"软边界"．在式 (6) 中， φ 是附加相移，它取决于海底的特性和信号的入射角．

容易证明：

$$\Delta r_s \approx 2hd/r, \quad 当\ r \gg (h+d), \tag{7}$$

$$\Delta r_b \approx 2kb/r, \quad 当\ r \gg (k+b), \tag{8}$$

由式 (4) 和式 (5) 还可以看出，如果 $A_s = A$, 并且

$$\Delta r_s/\lambda = n, \quad n = 1, 2, \cdots \tag{9}$$

那么，直达波将完全被海面反射波所抵消，这里 $\lambda = c/f$ 表示信号波长， n 为整数。式 (9) 还可以表示为：

$$\frac{2hd}{r\lambda} = n, \quad n = 1, 2, \cdots \tag{10}$$

这表示在某种条件下，直达波与海面来的反射波将互相抵消。

同样的结论对海底反射波所引起的干涉现象也是成立的，只要附加相移 φ 满足一定的条件。这个条件是：

$$\Delta r_b/\lambda = n + (\pi - \varphi)/2\pi, \quad n = 0, 1, 2, \cdots \tag{11}$$

或者

$$\frac{2kb}{r\lambda} = n + \frac{\pi - \varphi}{2\pi}, \quad n = 0, 1, 2, \cdots \tag{12}$$

2 波导的干涉条纹图

根据式 (10) 和式 (12) 不难计算各种情况下的干涉条纹图。对于海面反射，当距离固定时，抵消频率可以由下式求出：

$$f_n = n\frac{cr}{2hd}, \quad n = 1, 2, \cdots \tag{13}$$

例如，如果 $h = 8$ m, $d = 45$ m, $r = 1000$ m, 那么

$$f_n = n \times 2083 \text{ Hz}.$$

显然

$$\Delta f_n = f_{n+1} - f_n = \frac{cr}{2hd}, \quad n = 1, 2, \cdots \tag{14}$$

我们看到 Δf_n 与 n 无关。这表明在 (f, r) 二维图中，当 r 固定时，抵消频率 f_n 是等间隔的。

另外，容易得到：

$$\Delta\lambda_n = \lambda_{n+1} - \lambda_n = -\frac{1}{n(n+1)}\frac{2hd}{r}. \tag{15}$$

在海底反射的情况，直达波和反射波之间的关系与海面反射有很大的不同。正如式 (6) 所给出的，海底反射波中有一附加相移。一般来说，如果水中声速 c_1 小于海底介质中的声速 c_2, 那么存在一个临界角 (见图 2):

$$\theta_c = \arccos(c_1/c_2). \tag{16}$$

当信号入射角 θ 小于 θ_c 时，就会发生全反射，附加相移就依赖于 θ 。于是可以认为 $A_b = 1$, 容易证明：

$$\varphi = -2\arctan[\sqrt{\cos^2\theta - (c_1/c_2)^2}/(\rho_2\sin\theta/\rho_1)], \tag{17}$$

式中， ρ_1 是海水密度， ρ_2 是海底底质密度。由于 θ 是距离 r 的函数，所以附加相移也是 r 的函数。

从式 (4) 、式 (6) 、式 (17) 我们得知，如果

$$\varphi + 2\pi\Delta_b/\lambda = \pi + 2n\pi, \quad n = 0, 1, 2, \cdots \tag{18}$$

或者

$$\varphi + 2\pi kb/r\lambda = \pi + 2n\pi, \quad n = 0, 1, 2, \cdots \tag{19}$$

那么直达波有可能被海底反射波抵消。注意，附加相移与距离有关。

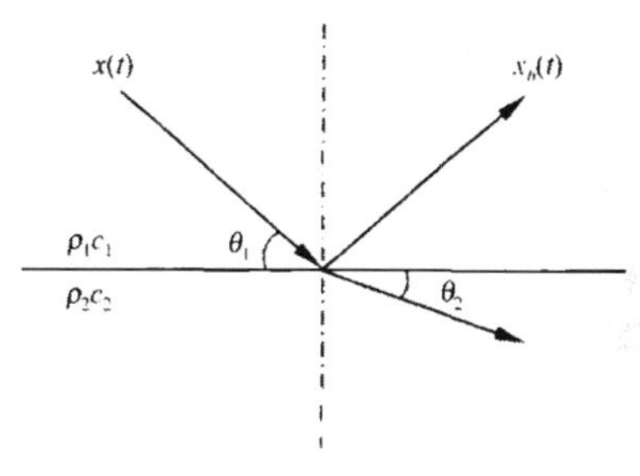

(a) 海底反射波

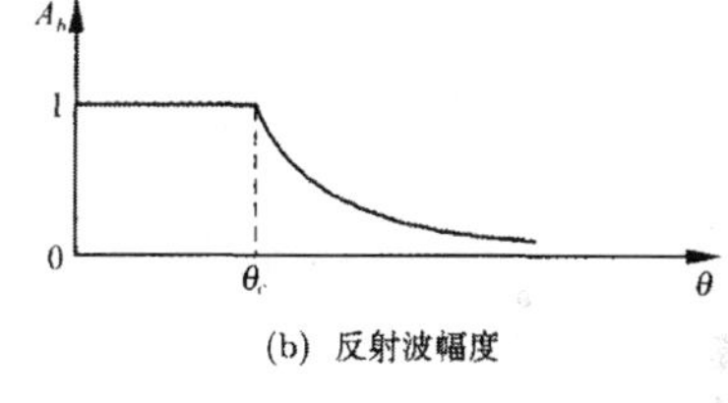

(b) 反射波幅度

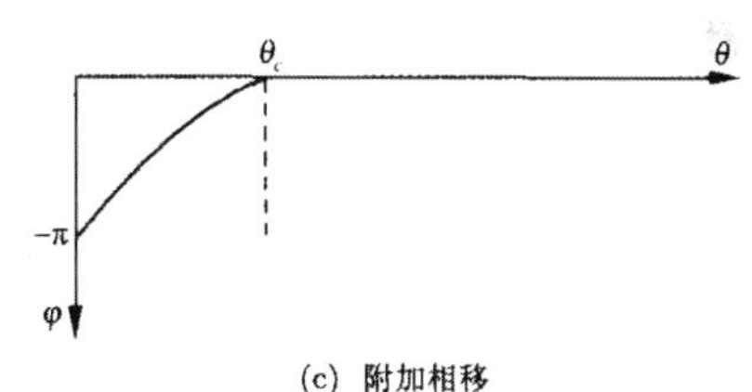

(c) 附加相移

图 2 海底反射波说明

从式 (19) 得到：

$$f_n = \frac{\pi - \varphi + 2n\pi}{2\pi}\frac{cr}{2kb}, \quad n = 0, 1, 2, \cdots \tag{20}$$

$$\lambda_n = \frac{2\pi}{\pi - \varphi + 2n\pi}\frac{2kb}{r}, \quad n = 0, 1, 2, \cdots \tag{21}$$

举例来说，如果 $\varphi = 0.9\pi$, $k = 70$ m, $b = 25$ m, $n = 0$, $r = 1000$ m, 我们有 $f_0 = 21.4$ Hz 。这个结果表明，在海底反射的情况下，干涉现象可能在低频段发生。由于附加相移 φ 依赖于距离 r, 所以与海面反射的情况相比，海底反射的抵消频率与距离的关系更加复杂。

干涉条纹图可以方便地从运动目标的轨迹求解出来[6] 。图 3 是一个运动目标的例子。假定目标以匀速 v 沿直线运动，我们有：

$$r^2 = r_0^2 + (vt)^2. \tag{22}$$

对于海面反射，根据式 (13), 我们有:

$$r = \frac{2hd}{nc} f. \tag{23}$$

第 n 条干涉条纹应当满足:

$$r_0^2 = \left(\frac{2hd}{nc}\right)^2 f^2 - v^2 t^2, \quad n = 1, 2, \cdots \tag{24}$$

这是一族变量 t 和 f 的双曲线方程。当然，我们可以求出它的渐近线，即:

$$t = \frac{1}{v} \frac{2hd}{nc} f, \tag{25}$$

同样的方法可以计算出海底反射的结果，但要稍微复杂一些。

图 4 给出了频率和距离 (时间) 的双曲线族. 值得指出的是，随着距离由近及远，双曲线的形状是变化的。

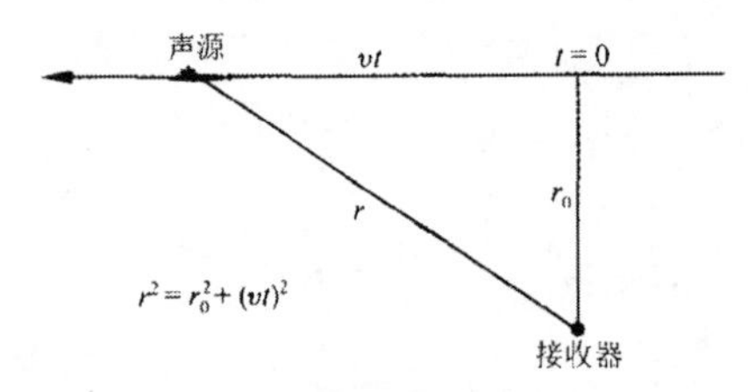

图 3 运动目标轨迹

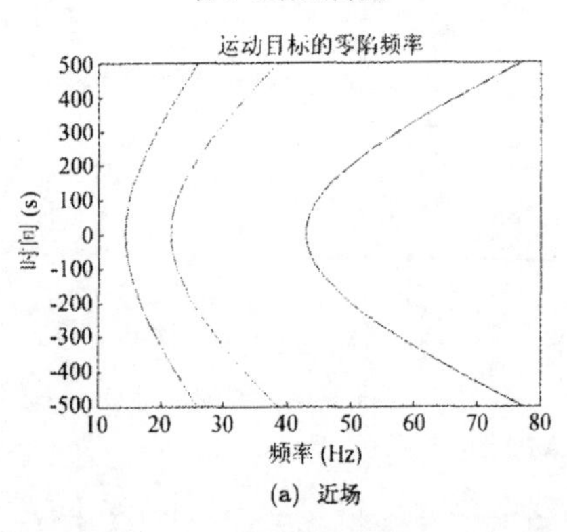

(a) 近场

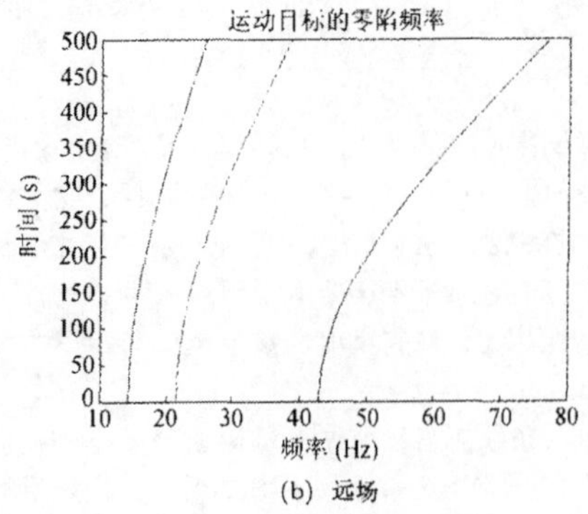

(b) 远场

图 4 干涉条纹图的双曲线族

3 系统模拟和海试结果

我们知道 LOFAR 图是检测水下目标噪声的很有效的工具，特别是用于干涉条纹的演示。如果存在干涉条纹，表明目标存在。用 LOFAR 图检测目标的最少可检测信噪比通常低于常规检测的需求。时间的积累和干涉条纹不仅可以用于目标检测，也可以用于目标分类识别。

图 5 是部分系统模拟的结果。正如我们所看到

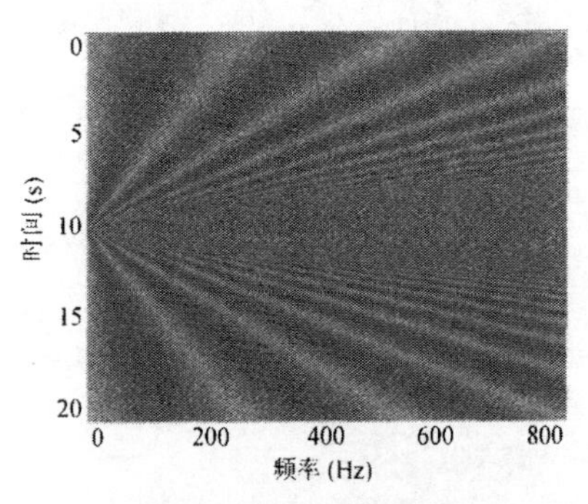

(a) SNR = 0 dB, 近场

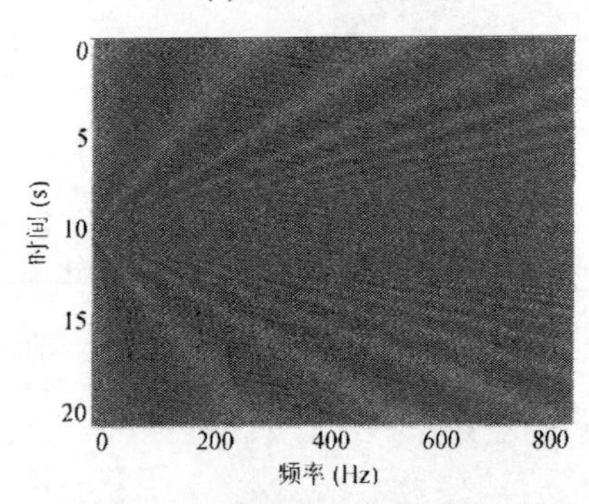

(b) SNR = −6 dB, 近场

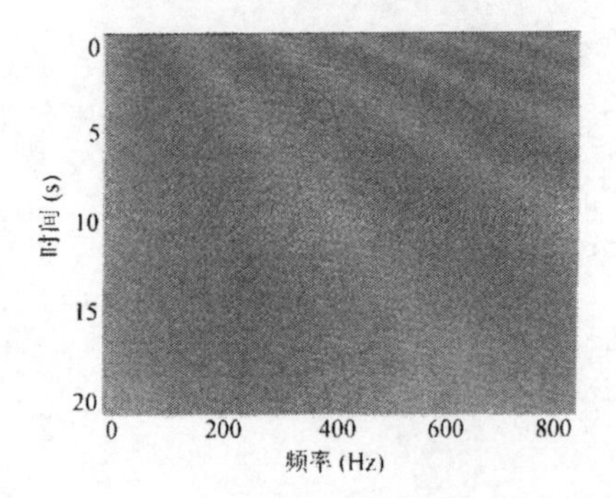

(c) SNR = −6 dB, 远场

图 5 干涉条纹图的系统模拟结果

的，在输入信噪比 $(\mathrm{SNR})_{\mathrm{in}} = 0$ dB 的情况，不管是近场还是远场都是容易提取辐射噪声的。

图 6 给出了一些海上试验结果。这是一些水面船舶辐射噪声的 LOFAR 图，远场情况的距离为 11 km 。

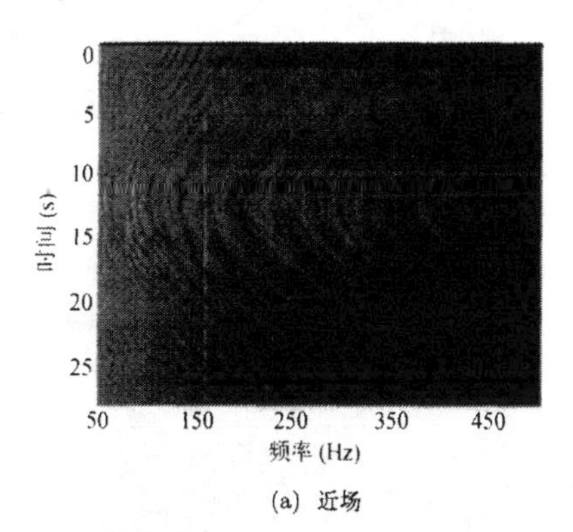

(a) 近场

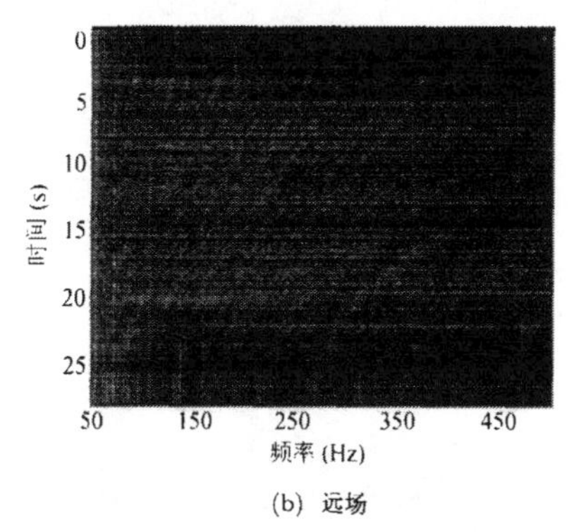

(b) 远场

图 6 干涉条纹图的海试结果 (声源：水面船舶辐射噪声)

4 结论

基于双路径射线模型，对由海底和海面反射产生的传播干涉现象进行了研究。海面和海底干涉的抵消频率是不同的。海底反射波中存在的附加相移在干涉条纹的形成中起着重要作用。已经推导了干涉条纹应满足的方程。系统模拟和海试结果指出了这种模型和理论分析的正确性。

参 考 文 献

1 Chuprov S D. An invariant of the spatial-frequency interference pattern of the acoustic field in a layered ocean. In: Proc. of the Russian Academy of Sciences, 1981; **257**: 475—479

2 Brekhovskikh C M, Lysanov Yu P. Fundamentals of ocean acoustics. 2nd Ed. Springer-Verlag, Berlin, 1991

3 Kuperman W A *et al.* The generalized waveguide invariant concept with application to the vertical array in shallow water. In: Proc. Ocean Acoustic Phenomena and Signal Processing, ONR Workshop, 2002 New York, 33—66

4 Rouseff D, Spindel R C. Modeling the waveguide invariant as a distribution. *Ibid*, 137—148

5 Baggeroer A B. Estimation of the distribution of the interference invariant with seismic streamers. *Ibid*, 151—167

6 Kapolka D. Equivalence of the waveguide invariant and two path ray theory methods for range prediction based Lloyd's mirror patterns. In: Proc. of Acoustics'08, Paris, 3637—3641

7 An L *et al.* Calculating the waveguide invariant by the 2-D Fourier transform ridges of Lofargram image. *Journal of Chinese Electronics & Information Technology*, 2008; **30**(12): 2930—2933

利用等离子体声源测量浅海低频段水声信道特性①②

王 磊[1] 姜晔明[2] 黄逸凡[3] 周士弘[2] 闫克平[3] 李启虎[1]

(1 中国科学院声学研究所 北京 100190)

(2 中船重工集团715所 杭州 310012)

(3 浙江大学工业生态与环境所 杭州 310028)

2010年4月13日收到

2011年6月22日定稿

摘要 浅海水声信道的结构特征及其变化特性是浅海环境水下远程探测和通讯应用的基础. 为了有效测量低频水声信道特性, 我们在实验中采用了拖曳等离子体声源和垂直阵接收系统. 通过实验测量与信道模型仿真输出的比较, 对等离子体声源特性, 特别是脉冲波形、空间指向性和电声转换效率, 以及在信道中的传播特性等进行测量分析. 结果表明: 实验测量与信道模型仿真符合良好, 实验中采用的等离子体声源的发射源级和拖曳姿态稳定, 波形一致性好, 可以满足走航式连续水下信道测量要求.

PACS数: 43.30, 43.60

Measurements of the low-frequency underwater channel features in a littoral environment using a plasma sound source

WANG Lei[1] JIANG Yeming[2] HUANG Yifan[3] ZHOU Shihong[2] YAN Keping[3] LI Qihu[1]

(1 *Institute of Acoustics, Chinese Academy of Sciences* Beijing 100080)

(2 *The 715th Research Institute of China Shipbuilding Industry Corporation* Hangzhou 310012)

(3 *Industrial Ecology and Environment Research Institute, Zhejiang University* Hangzhou 310028)

Received Apr. 13, 2010

Revised Jun. 22, 2011

Abstract In the shallow water environment, the propagation features of the acoustic channel and its variability are important for many littorial applications, such as long-range underwater target detection and communication. In order to measure the low frequency (LF) channel characteristics, a towed plasma sound source (PSS) and a vetical received array were used in our shallow water acoustic experiment, performed in Sept. 7, 2008. Through the comparision of the measured data with the wideband channel simulation model outputs,based on the KRAKEN Program, we analysised the PSS and channel features, including the source's waveforms, directivities, energe transform efficiency and the channel multipath structure. Experiment results show that the numerial analysis is well comformed of the measurements and the used PSS system is suitable for the continuously mobile underwater channel measurements task, due to its good stability of sourse level, towed posture and consistency of pulse waveforms.

引言

在实际海洋环境下, 浅海水声信道的结构特征及其变化特性是当前水声研究的一个重点, 对水下远程探测和水声通讯等工程应用有着基础性意义. 为了对低频水声信道进行有效的实验测量, 对传播多途的时空分离是一个关键[1]. 在声源及信号形式的选择上, 宽带长脉冲和高功率短脉冲各有其具体应用场合, 在信道频散及时变特性突出环境, 脉冲压缩方法测量性能将严重下降, 而爆炸声源激励的脉冲波形具有瞬态声压级高和宽频带的优点, 通常可

① 国家自然科学基金(60971114)和国家安全重大基础研究项目(613660203, 61311002)资助项目.

② 声学学报, 2012, 37(1): 1-9.

以直接得到信道的多途或模态结构的清晰图像。然而，爆炸声源，如 TNT 声弹和高压气枪，同时存在可控性及重复性差，气泡脉冲干扰严重等缺点[2]. 因此，在对激励波形要求较高的信道测量实验中（例如信道时间稳定性测量），一般采用低频宽带换能器[1,3−5]，通过匹配滤波分离多途或传播模态，同时改善信噪比[6]。近年来，随着等离子体声源技术的不断成熟并应用于工程实践，同时在很大程度上改善了爆炸声源可控性和一致性方面的缺陷，在水声信道测量、地层反演[7]及水下强声对抗等方面具有明显的应用前景。

等离子体声源 (Plasma Sound Source, PSS) 是将电能转化为声能的一种装置，通过水下高压脉冲放电产生等离子体，汽化水体生成脉动气泡，产生声辐射。其物理过程被称为液电效应，最早由前苏联 Yutkin 等人进行系统的研究[8]。根据电极间电离通道的不同，水下放电可以分为电弧放电和电晕放电 (corona discharge) 两种方式[9−10]. 前者在放电时两电极间存在着明显的电弧通道，激励电压通常需要达到 10 kV 以上，脉冲声压级较高，能量主要在数十赫兹频带内。第二种放电方式电极间无电弧通道导通，可在稍低的电压等级及电导率较高的电解质液体 (如海水) 中产生较强的声脉冲。电晕放电在电极间注入的能量较小，对电极的烧灼轻微，生成的气泡直径为厘米量级，声能主要在数百赫兹频带内，工程上常采用多电极组阵方式提高源级。由于电晕放电方式在安全性、工程适用性等方面明显优于电弧放电，是当前等离子声源应用的重点，本文讨论除特别说明外均指工作于电晕放电方式的等离子体声源。

对于爆炸声源，气 (汽) 泡脉动在声能转化的过程中起着关键性作用，等离子体声源本质上属于真空内爆声源。随着放电时电极间注入能量增大或放电深度减少，生成的汽泡直径增加，声脉冲的频带下移，反之则向高频方向移动。其声源特性与 R.J.Urick 用真空气瓶进行的水下破裂实验结果一致[11]。通过高速摄像对等离子体汽泡及声脉冲发射过程进行直接观察[12−13]，一般只有前 2、3 次汽泡脉动产生显著的声脉冲，时间上对应于气泡破裂时刻。由于气泡的主要成份是水蒸气，当控制气泡直径在厘米量级时，随着汽泡中心等离子体状态很快地减弱消失，气泡脉冲即停止，从而避免了一般爆炸声源的汽泡脉动干扰问题，在应用中可以得到一致性很好的发射波形。等离子体声源发射时能量的转化和气泡动力学过程较为复杂，包括电离解、黑体辐射、质量传递和气泡动力过程等环节[14−15]。实际应用中等离子体声源的电声转换效率与注入能量、电压和电极配置等因素相关，文献中给出了电弧放电方式下电声效率可达 25% ~ 30%[16]，接近水面处发射效率约为 5% ~ 10%[17]。

国内一些单位，如中科院电工所、浙江大学和西北工业大学[18−19]等均在等离子体声源领域开展了系统性研究，为了进一步考察等离子体声源在水声工程中的应用潜力和工作性能，在北海某海区进行的联合海试中，采用等离子体声源对浅海信道多途结构进行了测量。通过传播实验数据分析，对等离子体声源特性和浅海信道结构进行了较为深入的分析，并通过简正波模型[20]理论分析对实验结果进行分析解释。

1 实验概况

作为浅海远程传播原理性实验的一部分，在 2008 年 9 月 7 日 8:00 — 13:00 进行了拖曳等离子体声源信道测量实验。实验在北海某海区进行，配置如图 1 所示：接收船锚定就位，舷侧布设一 32 元垂直阵作为接收端，阵元间距 1.25 m，第一阵元距水面约 3.8 m，距垂直阵水平距离 30 m 处布放一标准水听器，灵敏度 −201 dB，深度 18 m。接收船于舷侧布设一条温度链，测得的声速分布如图 2 所示，属于典型的夏季

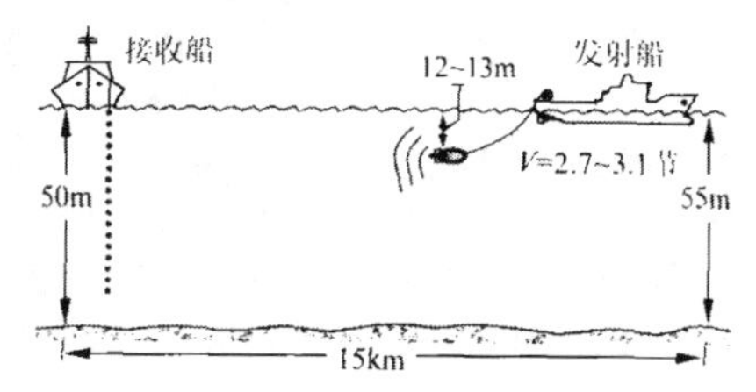

图 1　实验信道及测量系统

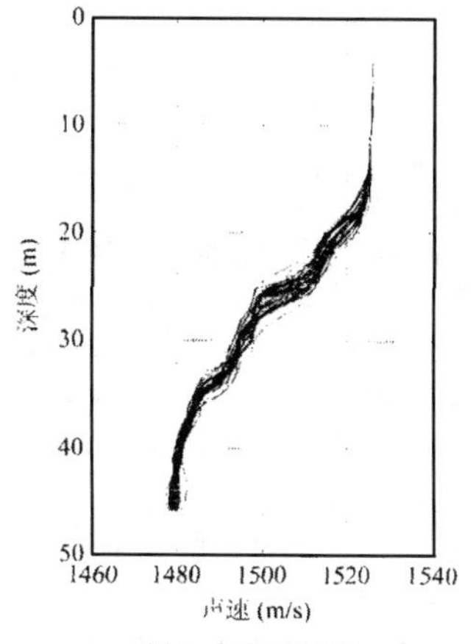

图 2　实验声速剖面

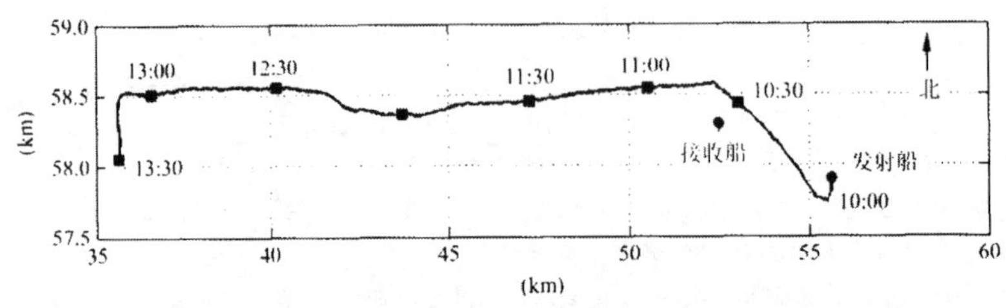

图 3 实验航迹图

(a) 等离子体声源发射机

(b) 等离子体声源发射阵 (1.45 m × 0.8 m)

图 4 实验采用的等离子体声源系统

负跃层条件，实验期间受到一定程度内波的影响．实验期间，发射船沿相应轨迹作低速拉距航行，运动轨迹如图 3 所示．发射声源为浙江大学研制的大功率等离子体声源，其发射机和水下发射阵分别如图 4(a) 和图 4(b) 所示，通过发射电极组成面阵，长宽分别为 1.45 m 和 0.8 m; 实验中采用浮体和配重方式保证发射面的水平姿态，单次电脉冲激励能量为 6 kJ. 间隔为 3 s: 在航速稳定期间 (约 3 节). 拖曳等离子体声源的深度约在 12 ~ 13 m 之间．实验海况 2 级，海深在 48 ~ 55 m 间，海底底质取样主要为较硬的沙砾底质，历史上已作过海底参数反演，海底表层声速范围 1625 ~ 1655 m/s. 密度为 1.8×10^3 kg/m^3. 吸收损失 0.6 ~ 0.8 dB/λ .

2 等离子体脉冲波形特征的测量分析

等离子声脉冲的波形及幅度由较多因素决定，激励能量、电极形式和发射深度都会直接对气泡过程产生影响，因此需要在实验中进行测量分析．为了尽可能降低传播的影响，选取了发射船经过接收阵时的近程测量数据进行分析．图 5(a) 为单水听器记录接收脉冲的波形堆叠图形 (根据脉冲最大幅度进行规一化). 接收深度为 20 m ．图 6 给出了近程测量过程等离子体声源相对接收阵的位置和姿态，同时标出了海深的测量点位置 (图中 ⋆ 处). 可以看到近程海底有一定的倾斜．在发射船移动轨迹上以符号 ◇ 表示时刻，相邻 ◇ 间的时间是 1 分钟，可以看出发射船在这段时间内处于加速状态，其速度变化由图 5(a) 右侧的曲线给出．显然，近程接收脉冲主要由 6 个多途控制，通过与图 5(b) 的多途到达时间仿真结果比较，可以确定主要的 6 个多途分别是 D - “直达”、

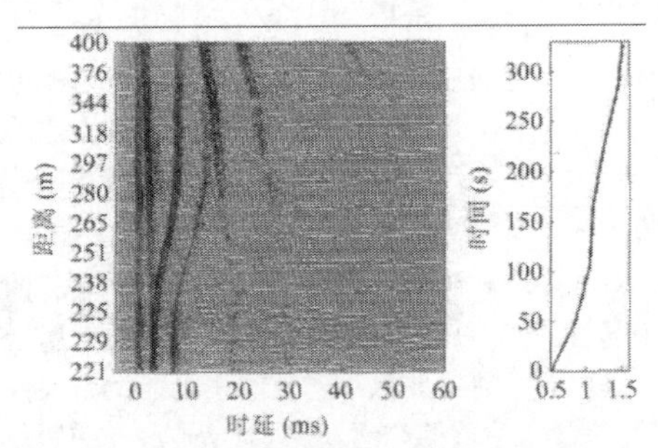

(a) 单水听器接收的等离子体脉冲波形

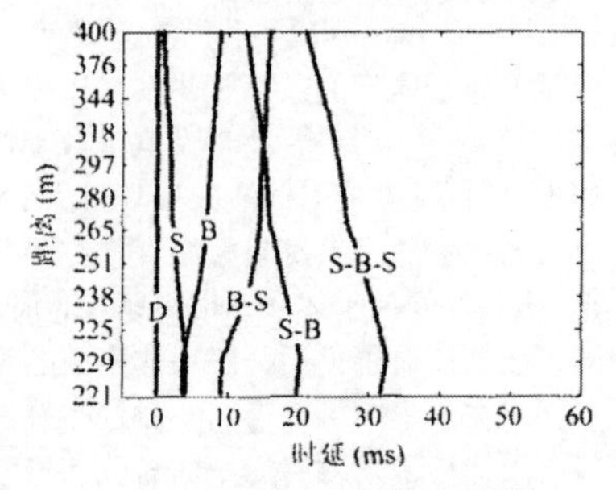

(b) 近场多途达到时间仿真结果

图 5 近场多途演变结构及仿真结果

S - "海面反射"、B - "海底反射"、BS "海底海面反射"、SB "海面海底反射" 以及 SBS - "海面海底海面反射"。需要指出的是，多途到达时刻的仿真需要在三维倾斜海底条件下进行计算，限于实验条件，还需对加速过程中声源深度变化进行假定，并对距离和接收深度等几何参数进行微调，经多次反复得到多途时间演变结构的仿真。计算的主要步骤参见附录 A，其中声速剖面简化为等声速结构。

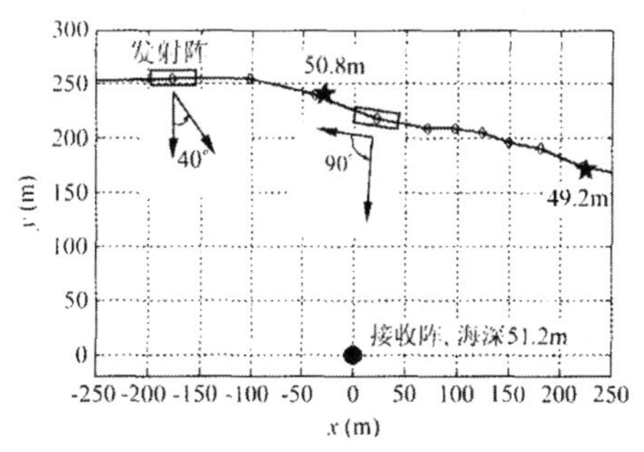

图 6　近场传播的几何图示

实验等离子体声源由数千个微小电极组合而成，实际上构成了一个面阵声源。由于主要能量交换过程，即等离子体气泡破裂过程在微秒量级完成，在电极间距大于气泡直径条件下，多电极声源发射脉冲可以由单电极波形线性叠加得到。实验采用的等离子体声源电极间距约 1 cm，略小于气泡直径，因此声阵发射脉冲只能近似由波束形成方法描述。实验室得到的单电极发射波形如图 7(a) 所示。利用单电极发射波形对面阵进行波束形成：

$$x(t)=\iint s(t)\,\delta(t-\tau_s)\mathrm{d}S, \tag{1}$$

其中 $s(t)$ 是单电极激励波形，$\mathrm{d}S$ 为发射阵单位面元，$\delta(t)$ 为狄拉克冲激函数，面元 $\mathrm{d}S$ 的相对时延因子为 $\tau_s=\boldsymbol{r}\cdot\boldsymbol{n}$；$\boldsymbol{r}$ 为面元的空间位置矢量，$\boldsymbol{n}$ 为发射的空间指向向量；图 7(b) 和图 7(c) 分别是实验中不同发射阵姿态角时接收直达波与仿真波形的比较。当接收端位于发射阵长轴正横方向时 90° 直达波与仿真波形符合良好，在与发射阵长轴成 50° 夹角方向时，仿真波形除中间波宽度基本一致外，两侧波则与接收波形有较大差异，可能的原因是发射阵处于加速状态，两次测量时声阵的深度不同，因此单电极发射的波形出现了一些差异。此外，从图 7(c) 的接收波形可以看出，当直达波与海面反射波时延相差只有 2 ms 时，就可以在波形上清晰分离。

图 8 分别给出了 16 个直达脉冲的波形及频谱，接收端位于发射阵长轴正横方向附近。可以看到左图波形具有良好的一致性；右图给出直达波形的谱，为了减少发射角度变化的影响，去掉了第 1, 2, 15 和 16 号脉冲的谱，可以看到脉冲 −3 dB 带宽约为 550 ~ 1150 Hz，中心频率约为 850 Hz。

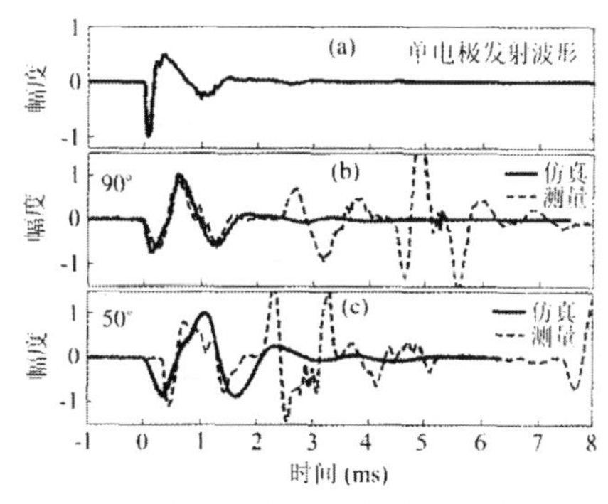

图 7　等离子体脉冲的波形仿真

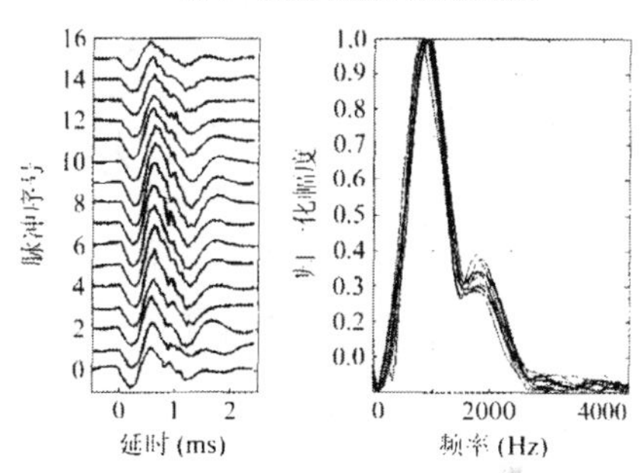

图 8　直达脉冲的波形及频谱

3　等离子体声源效率估算

对等离子体声源辐射声能的测量需要进行空间全向的声压测量，限于实验条件，这里近似给出声源源级和电声转换效率的估算。根据图 7 中波形的比较，声阵发射脉冲可以近似由波束形成方法描述，假定面阵的几何位置如图 9(a) 所示，数值计算得到声源的归一化空间指向性见图 9(b)，幅度为发射脉冲能量。图 9(c) 和图 9(d) 分别为声源在 xOy 和 yOz 平面上指向性，以分贝表示。实验测得面阵水平正横方向 ($\theta=90°$, $\varphi=90°$ 的直达脉冲声压级约为 174 dB，220 m 传播距离对应的传播损失为 46 dB，因此在该方向上声源脉冲声压级为 220 dB。根据图 9(b) 的声源指向性函数对空间角进行积分，可以给出一次发射的总辐射声能：

$$E_p=\int_0^{2\pi}\mathrm{d}\theta\int_0^{\pi}H(\theta,\varphi)\sin\varphi\,\mathrm{d}\varphi.$$

其中，$H(\theta,\varphi)$ 是发射脉冲的空间能量分布函数，θ 和 φ 分别为图 9(a) 球坐标系中方向角。对归一化指向性图积分可得 $E_p/4\pi=0.5296$，因此，空间平均辐

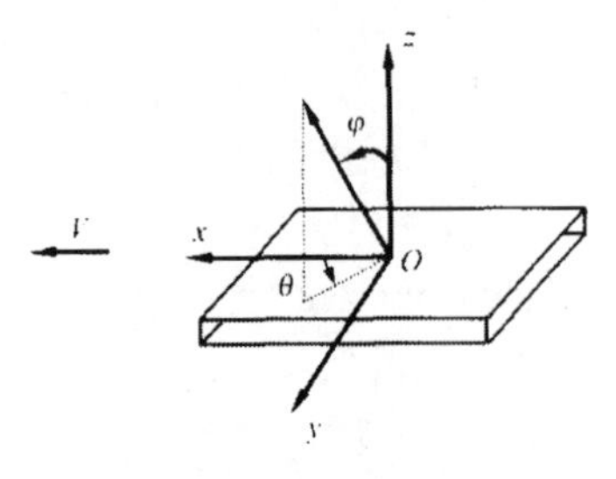

(a) 等离子体声源坐标图示

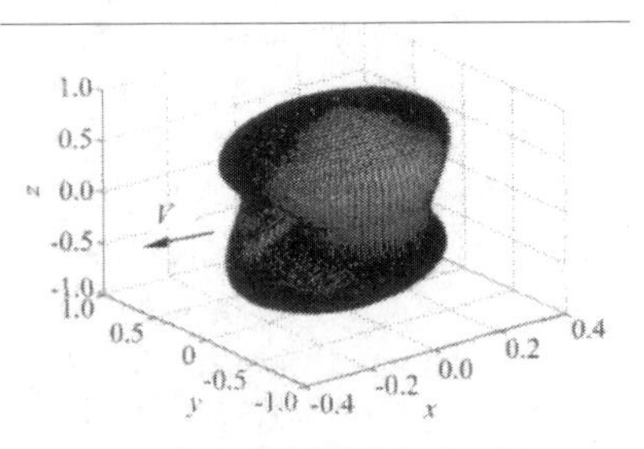
(b) 声源三维空间指向性 (归一化)

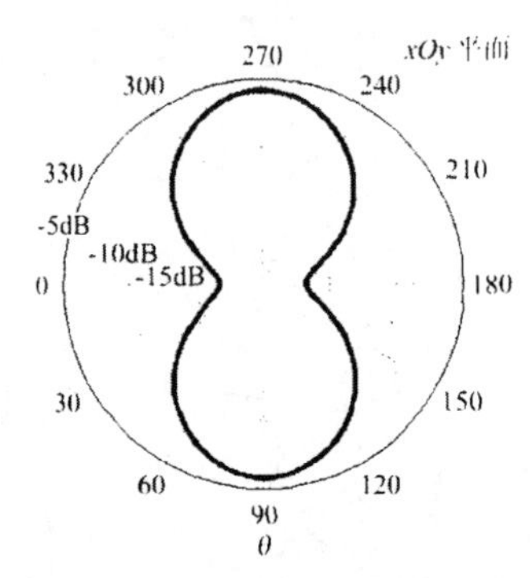

(c) 声源在水平面 xOy 上的指向性

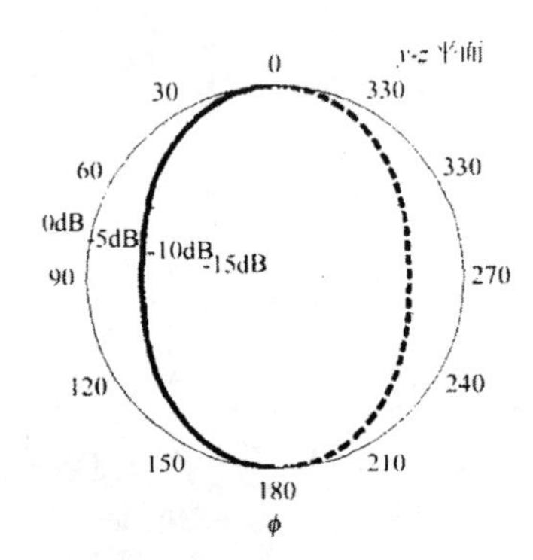

(d) 声源在 yOz 平面的指向性

图 9 矩形面阵声源的空间指向性估算

射声能约为最大脉冲能量的一半. 由图 9(d) 声源指向性曲线, 水平面阵垂直方向 $\phi = 0°$ 的辐射声压级比水平正横方向高约 6 dB . 在 z 轴方向, 各电极发射脉冲可近似认为相干叠加, 辐射脉冲能量最大, 根据计算结果, 空间平均辐射声能约降低 3 dB. 因此可以得到电声转换效率: $\eta = 10^{(PL-170.7)/10}\tau/6000 = 4.6\%$, 其中平均声压级 $PL = 223.3$ dB. τ 为平均等效脉宽, 取值为 1.5 ms .

4 信道传播损失的测量与估计

通过不同距离的接收脉冲能量可以得到信道的宽带传播损失曲线. 由于海底较硬且相对平整, 接收脉冲的扩展随距离增加而增加, 前向混响的影响相对较小. 根据波形观察, 在能量计算中将时间宽度取为 150 ms; 利用脉冲到达前 1 s 数据估计环境噪声级, 同时忽略前向散射的影响. 得到的宽带传播损失曲线如图 10 所示, 实线为 20 m 深度处标准水听器实测传播衰减, 虚线为模型仿真结果, 仿真模型采用 KRAKEN 计算程序 [21]. 声速取测量时间段的平均剖面, 海底表面声速取 1650 m/s, 密度为 1.8×10^3 kg/m^3, 吸收损失取 0.8 dB/λ. 对 300 ~ 1.2 kHz 内各模态求和, 计算公式为:

$$TL(r, z) \cong -20\log\frac{1}{\rho_w}\sqrt{\frac{2\pi}{r}}$$
$$\sqrt{\sum_{i=1}^{I} w_i\left(\sum_{m=1}^{\infty}\left|\Psi_m(z_s, \omega_i)\Psi_m(z, \omega_i)\frac{\exp(\mathrm{j}k_{rm,i}r)}{\sqrt{k_{rm,i}}}\right|^2\right)}. \tag{2}$$

其中 Ψ_m 为模态函数, m 与 i 分别指简正波模态和频率的标号. $k_{rm,i}$ 为第 i 个频率的第 m 号简正波的水平波数, ρ_w 为水体密度. 为了简单, 假定脉冲频谱在频带内是均匀分布的, 即频率加权因子 $w_i \equiv 1/I$, I 为频率划分数. 参见图 10. 可以看到模型输出与实测的传播衰减的规律是一致的, 这也反映了匀速拖曳时等离子体声源的源级及声阵姿态的稳定性.

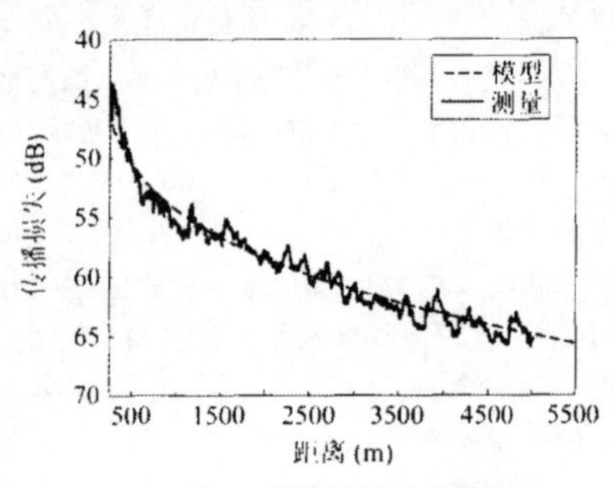

图 10 信道传播损失的测量

5 信道传播结构特性及与模型的比较

5.1 宽带传播波形计算

由于等离子体声源的信号频带较低，可以根据宽带简正波模型[20] 对信道测量结果进行仿真，信道传递函数表示为简正波形式：

$$H(r,\omega)\approx\frac{\mathrm{i}}{\rho(z_s)\sqrt{8\pi r}}e^{-\mathrm{j}\pi/4}\sum_{m=1}^{M}\Psi_m(z_s)\Psi_m(z_d)\frac{e^{\mathrm{j}k_m(\omega)r}}{\sqrt{k_m(\omega)}}, \tag{3}$$

根据发射脉冲谱 $S(\omega)$，通过逆傅里叶变换就可以计算信道输出响应：

$$x(r,t)=\int_{-\infty}^{\infty}S(\omega)H^*(r,\omega)\exp(\mathrm{j}\omega t)\mathrm{d}\omega, \tag{4}$$

其中，信道模态函数 Ψ_m 由 KRAKEN 模型计算，频带取 0.3 ~ 1.2 kHz 。由于海底在接收端附近深度为 50 m，至 15 km 时深度增加到约 55 m，可以近似采用水平分层介质模型。假定海底为半无限空间，平均海深设为 51 m，海底声速为 1650 m/s，密度为 1.8×10^3 kg/m^3，吸收损失取 0.8 dB/λ 。考虑到声源速度较低，计算中忽略运动多普勒影响。

5.2 信道传播波形结构的比较

为了全面的展现信道传播的多途结构，这里采用两种方式给出声源运动过程中多途的变化特性。第一种方法将一个阵元记录的多次脉冲波形，补偿传播延时后以堆叠的方式绘出，可以清晰地表现随声源运动，信道多途结构的演变过程。图 11(a) 给出了第 15 号阵元 (深度 21.3 m) 记录的脉冲多途变化结构。第二种方法直接给出一定距离处，声阵各阵元接收脉冲波形。如图 12(a) 和图 13(a)，分别为距离 580 m 和 1500 m 时的多途垂直结构。

通过宽带简正波模型和近场实测的脉冲波形可以计算脉冲传播波形，图 11 至图 13 的 (b) 图给出了相应的模型仿真结果。由于脉冲频带大于 300 Hz，同时海底较硬，因此海底地层分层结构对传播的影响较小，传播衰减的敏感参数是海底表层声速，实际计算也表明增加沉积层的作用不明显。另一方面，对于信道多途结构而言，敏感因素则主要是平均海深和声源及接收的深度。对比图 11 的 (a) 和 (b)，可以看到 4 个多途群的相对延时已基本一致，而后面的前多途

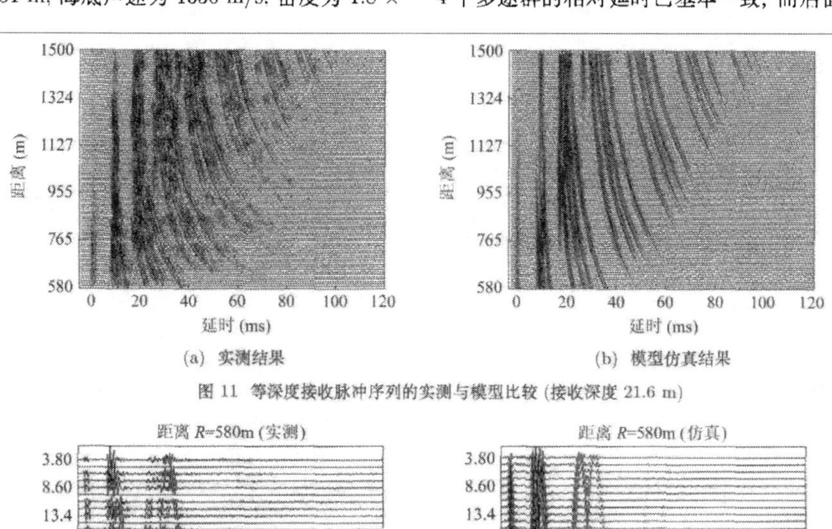

(a) 实测结果 (b) 模型仿真结果

图 11 等深度接收脉冲序列的实测与模型比较 (接收深度 21.6 m)

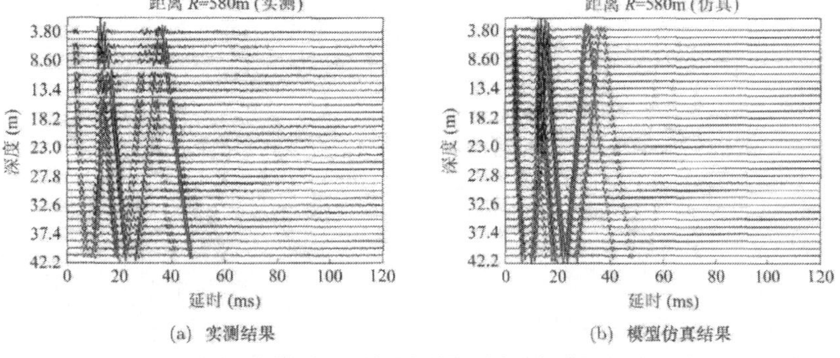

(a) 实测结果 (b) 模型仿真结果

图 12 垂直阵接收脉冲序列的实测与模型比较 (接收距离 580 m)

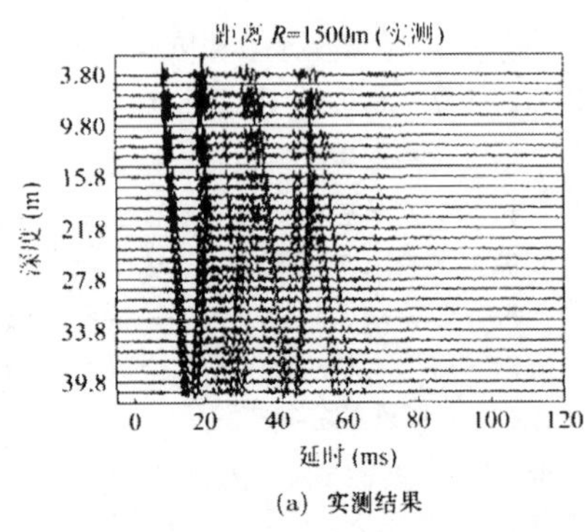

(a) 实测结果

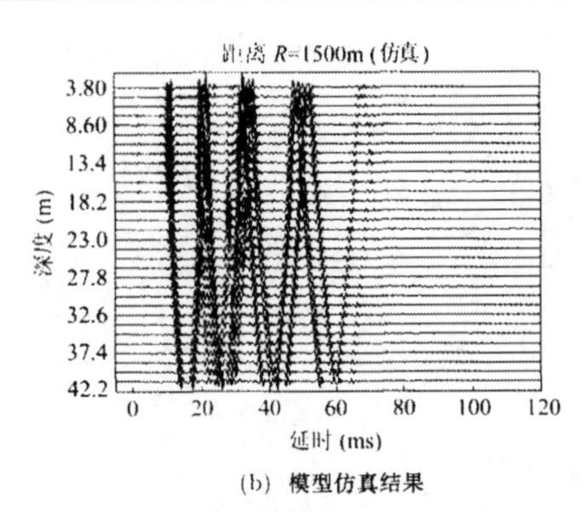

(b) 模型仿真结果

图 13 垂直阵接收脉冲序列的实测与模型比较 (接收距离 1500 m)

群分界不明显，且与仿真结果有一定的差异。考虑到这些多途群一方面受到信道随机扰动的影响较大，另一方面对信道几何结构参数变化十分敏感，包括受海底倾角的影响，因此在水平分层近似下没有对高阶模态或多途作进一步的仿真逼近。

图 11(b) 和 12(b) 分别给出了 580 m 和 1500 m 距离处，垂直阵接收信号与模型估计的比较。实测信号的 2, 6, 10 通道阵元失效，可以看到模拟信号与接收信号相符良好。主要的区别有两点，其一在于仿真的多途波形更"清晰"，而实际脉冲多途由于在海底海面反射时受到界面起伏调制，发生波形畸变，因此实际波形会出现展宽和模糊现象。其二，对比图 12(a) 和图 12(b)，在仿真中各阶模态的损失一般随模态阶数的升高而单调增加，因此后到达多途的传播衰减要大于先前到达的多途，而实际接收多途在相对强度分布会与这种情况会有一定的差异。通常，这是由声源的空间指向性和较为复杂的海底结构导致的。

总之，通过模型仿真可以对等离子体声源在不同距离上实测信道多途结构进行比较准确的描述，包括不同距离的多途变化和垂直剖面的多途结构。这也说明了基于简化平均信道参数的模型仿真能够较好的模拟实验信道的演变过程。

6 结论

通过数据分析，实验采用的等离子体声源具有发射强度稳定，无气泡脉冲干扰和波形一致性好的优点，可以对浅海信道的多途结构进行良好的测量。由于等离子体声源常采用电极组阵方式提高源级，因此由电极阵形状产生了明显的发射指向性，实际发射波形可近似由波束形成方法得到。

考虑到等离子体声源在能量转化过程中，大部分能量以光、热以及流体动能等形式消耗，因此电声转换效率不高，实验估算等离子体声源的电声转换效率约为 5%。

利用等离子体声源波形对信道多途结构的区分能力，实验测得低频段近程信道主要有 6 个声线途径，通过射线追迹可以确定各声线多途的传播路径。同时，利用声源的低频宽带特性，可以对较远距离传播的多途群结构进行直接观测。通过与宽带简正波模型仿真结果比较，观测到前几个多途群更多地依赖平均信道参数，与信道仿真结果符合良好；而后到达的多途群则对信道的微观结构，包括随机因素更为敏感，表现为多途群位置偏移和扩展，与信道模拟结果的差异比较明显，信道仿真需要更加精确的环境信息支撑。

实验等离子体声源的最大辐射能量方向朝向海底，这对海底地层反演是有利的。下一步工作可考虑改变声阵形式或拖曳方式，改善水平传播特性，这对于反演海底地声参数 (包括水平折射斜坡海底条件) 是有利的。

致谢

2008 年北海联合实验的组织单位 - 中船重工集团 715 所提供了本文工作有关的实验数据。在 973 课题组内与孙超教授和赵航芳研究员进行了有益的讨论，在信道仿真结果解释分析方面得到了陈耀明研究员的指导，在此一并致谢。

参 考 文 献

1 Roux R. The structure of raylike arrivals in a shallow-water waveguide. *J. Acoust. Soc. Am.*, 2008; **124**(6): 3430—3439

2 张岩，李风华，李整林，张仁和. 爆炸信号中气泡脉动去除方法及其应用. 声学学报，2009; **34**(2): 124—130

3 Jee Woong Choi, Peter H. Dahl, Measurement and simulation of the channel intensity impulse response for a site in the East China Sea. *J. Acoust. Soc. Am.*, 2006; **119**(5): 2677—2685

4 Simons D G *et al.* Analysis of shallow-water experimental acoustic data including a comparison with a broad-band normal-mode-propagation model. *IEEE Journal of Oceanic Engineering*, 2001; **26**(3): 308—323

5 Andrews M, Chen T, Ratilal P. Empirical dependence of acoustic transmission scintillation statistics on bandwidth, frequency, and range in New Jersey continental shelf. *J. Acoust. Soc. Am.*, 2009; **125**(1): 112—124

6 王宁，张海青，王好忠，高大治．内波、潮导致的声简正波幅度起伏及其深度分布．声学学报，2010; **35**(1): 38—44

7 Bierbaum S. Greenhalgh S. A high-frequency downhole sparker sound source for crosswell seismic surveying. *Exploration Geophysics*, 1998; **29**: 280—283

8 尤特金．液电效应．于家珊译．北京：科学出版社，1962

9 孙鹤鸿，左公宁．传输式大功率电火花震源在浅水中及井中的压力波形及频谱分析．应用声学，2000; **19**(6): 40—46

10 刘强，孙鹤鸿．水中脉冲电晕放电等离子体特性及气泡运动．高电压技术，2006; **32**(2): 54—56

11 Urick R J. Implosions as sources of underwater sound. *J. Acoust. Soc. Am.*, 1963; **35**(6): 2026—2027

12 Buogo S, Cannelli G. Implosion of an underwater spark-generated bubble and acoustic energy evaluation using the Rayleigh model. *J. Acoust. Soc. Am.*, 2002; **111**(6): 2594—2600

13 HUANG Yifan *et al.* Plasma sparkers for marine service application. Conference Proceedings, UDT Europe 2009, CANNES, FRANCE, 2009; **8C.3**: 1—6

14 Roberts R *et al.* The energy partition of underwater sparks. *J. Acoust. Soc. Am.*, 1996; **99**(6): 3465—3475

15 Olson A H, Sutton S P. The physical mechanisms leading to electrical breakdown in underwater arc sound sources. *J. Acoust. Soc. Am.*, 1993; **94**(4): 2226—2231

16 Cook J A, Gleeson A M, Roberts R M, Rogers R L. A spark- generated bubble model with semi-empirical mass transport. *J. Acoust. Soc. Am.*, 1997; **101**(4): 1908—1920

17 Cannelli G B, D'Ottavi E, Prosperetti A. Bubble activity induced by high-power marine sources. Oceans '90, Washington, D.C., 1990: 533—537

18 方明，黄建国，雷开卓，李宁．水下等离子体声源聚焦声场分布特性研究．电声技术，2009; **33**(12): 39—42

19 MA Tian, HUANG Jianguo, LEI Kaizhuo *et al.* Simulating underwater plasma sound sources to evaluate focusing performance and analyze errors. *J. Marine. Sci. Appl.*, 2010(9): 75—80

20 Jensen F, Kuperman W, Porter M, Schmidt H. Computational ocean acoustics. New York: Springer-Verlag, 1997

21 Porter M B. The KRAKEN Normal Mode Program. NRL/MR/5210-92-692, 1992

附录 1 近距离倾斜海底多途到达时刻的计算

为了分析声源特性，需要对近场声线途经进行区分．由于实验过程中对三维海底地形、等离子体声源深度的记录欠缺，需要在分析中对相关参数进行模拟推断，即先对实验有关参数进行预测，然后根据实验结果与模拟结果的对比确认真实的传播参数和过程．

附图 1 给出了海底反射声线的路径，求取声源相对于海底面的镜像点坐标，是倾斜海底条件声线路径计算的关键步骤．对于近场传播，忽略声速剖面的影响，并将附近海底近似为一平面．这样，相对于海底平面的虚源坐标可以通过坐标旋转方式计算．如附图 2 所示，选择大地坐标系 xOy，其中 x 轴和 y 轴分别沿正南和正东方向（参照文中图 6），坐标原点为接收水听器在海底的投影；倾斜海底平面可以通过对原坐标系 xOy 旋转得到：首先选择一过原点的旋转轴，球坐标系下该轴的方向坐标 (φ,θ)，然后旋转一定的角度 α，使得海底平面与坐标系旋转后的 $x'Oy'$ 平面重合．显然，在新坐标系下，虚源的坐标可以直接给出．下面列出基本步骤：

(1) 通过坐标旋转变换得到声源在旋转坐标系下的坐标；

(2) 直接得到新坐标系下声源虚像的坐标；

(3) 将旋转坐标系旋转回原坐标系，得到虚源像在原坐标系下的位置坐标．

下面给出求解虚源的公式：

首先给定旋转轴的方向矢量：

$$n=[n_x,n_y,n_z]^{\mathrm{T}}=[\cos\varphi\cos\theta,\cos\varphi\sin\theta,\sin\varphi]^{\mathrm{T}},$$

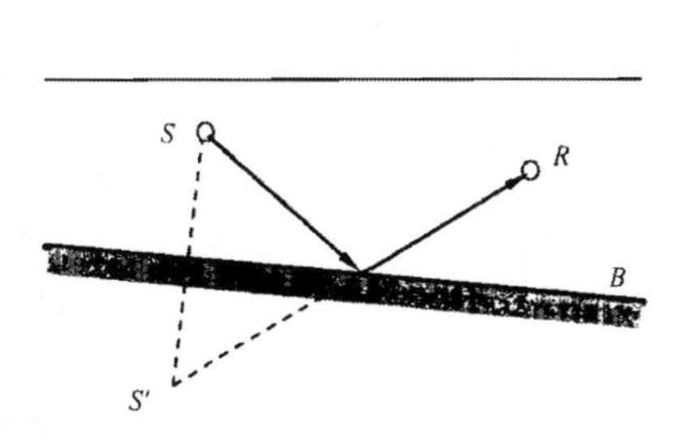

附图 1 声线几何结构

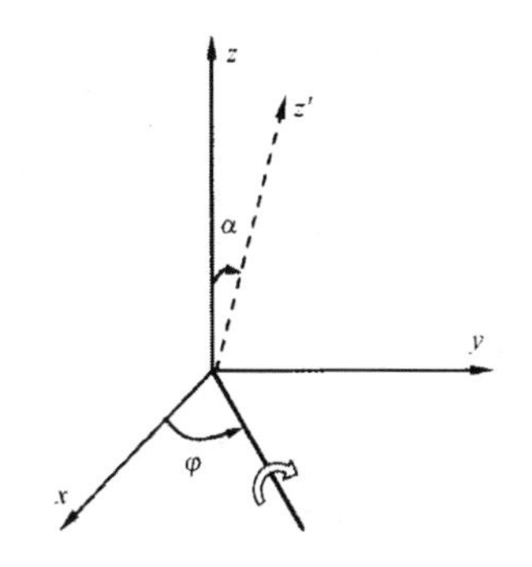

附图 2 坐标旋转的几何图示

坐标系旋转后的坐标变换矩阵：

$$
\boldsymbol{M}=\begin{bmatrix}
\cos\alpha+n_x^2(1-\cos\alpha) & n_xn_y(1-\cos\alpha)-n_z\sin\alpha & n_xn_z(1-\cos\alpha)+n_y\sin\alpha \\
n_xn_y(1-\cos\alpha)+n_z\sin\alpha & \cos\alpha+n_y^2(1-\cos\alpha) & n_yn_z(1-\cos\alpha)-n_x\sin\alpha \\
n_xn_z(1-\cos\alpha)-n_y\sin\alpha & n_yn_z(1-\cos\alpha)-n_x\sin\alpha & \cos\alpha+n_z^2(1-\cos\alpha)
\end{bmatrix}.
$$

则坐标变换后，声源坐标为：

$$
\boldsymbol{s}'=\boldsymbol{M}\,\boldsymbol{s},
$$

其中，s 为原大地坐标系下声源坐标矢量，s' 为旋转坐标系下声源位置坐标矢量。

在多途模拟中，根据环境参数测量，设海深为 50 m，接收深度 17 m，倾斜海底由原大地坐标旋转给出，旋转轴 OM 的球坐标系下方向为 $(\pi/6,0)$，旋转角 α 为 -3(右手系)。参见文中图 5，可以看到模拟结果与实测基本一致，因此可以推断，这样的环境参数假定是基本合理的，误差主要来自于海底倾面参数和平面近似假定，以及声源深度的粗略估计。

水下声信号未知频率的目标检测方法研究①

陈新华，鲍习中，李启虎，孙长瑜
(中国科学院 声学研究所，北京 100190)

摘要：提出了频率未知情况下基于阵列信号处理的一种目标检测方法。针对接收信号进行FFT分析，对处理频带的每一个频率单元进行波束形成，利用噪声对应频率单元波束输出的最大值随机。基于目标对应频率单元波束输出最大值基本一致的特点，统计各频率单元的方位(DOA)估计结果，从而实现对目标的检测。仿真结果表明，该方法的检测性能与频率已知时的检测性能一致，同时实验结果证明了该方法的有效性。为弱线谱目标检测提供了一个新的思路。

关键词：信息处理技术；FFT分析；线谱检测；方位估计
中图分类号：TB566 **文献标志码**：A **文章编号**：1000-1093(2012)04-0471-05

Research on Detection of Underwater Acoustic Signal with Unknown Frequency

CHEN Xin-hua, BAO Xi-zhong, LI Qi-hu, SUN Chang-yu
(Institute of Acoustics, Chinese Academy of Sciences, Beijing 100190, China)

Abstract: A target detection method was presented based on the processing of array signal with unknown frequency. First, the received signals were processed using FFT. Then the beam-forming of every frequency unit was processed. The target was detected by calculating statistically the direction-of-arrival (DOA) results of every frequency unit, as DOA of noise frequency unit is random, and DOAs of target are nearly consistent. The simulation results shows that the detection performance of the method is consistent with the performance of processing the signal with known frequency. And the trial results prove that the algorithm is valid. A new idea is provided for the detection of weak line spectrum target.

Key words: information processing; FFT analyzing; frequency line detecting; direction-of-arrival estimation

0 引言

被动声纳因不对外辐射信号，故隐蔽性较强。一直是对水下目标进行探测的重要手段之一。但随着隐身技术的不断提高，对被动声纳的性能需求也越来越严格。常规被动声纳检测技术基于宽带能量积分的信号检测方法，已远远不能满足远程探测的需要。理论和实验证明，水下目标辐射噪声中具有丰富的单频分量，特别是在低频段，由于螺旋桨转动时切割水体会产生低频信号，低频分量一部分会直接以加性形式出现在辐射信号中，另有部分则被船体本身的振动调制到较高的频带[1-2]，通常线谱谱级比连续谱谱级要高出10～25 dB。这为实现水下目标远程探测提供了一种可能。

本文将探讨高斯宽带噪声背景下，如何利用目标辐射噪声中的线谱实现目标检测的问题。文中分别给出了基于宽带能量检测和基于线谱检测的理论介绍和性能分析，最后给出了在阵列信号处理条件下，利用线谱方位稳定性的目标检测方法，并进行性能分析。

1 能量检测器

在传统的宽带检测体制下，从高斯噪声背景中检测高斯信号的最佳检测器是能量检测器[3]。

① 兵工学报，2012，33(4)：471-475.

过程如下:

对输入信号在时间上进行平方积分。参见(1)式。然后与设定门限比较,从而确定是否检测到信号

$$W = \sum_{n=1}^{N} s^2(n) , \quad (1)$$

式中: $s(n)$ 为声纳所接收到信号; n 为观测的样本序列; N 为能量积分点数。

能量检测器的处理增益为

$$G = 10\lg \sqrt{2BT}, \quad (2)$$

式中: B 为积分带宽; T 为积分时间。

由处理增益公式可以看出,能量检测器与时间带宽积有关,时间带宽积增大一倍,处理增益增大 1.5 dB.

2 FFT 分析

在高斯噪声背景中检测单频信号,一种有效的方法是对输入信号进行 FFT 分析[3],然后对 FFT 输出的谱线强度与设定门限比较,从而实现单频信号的检测。

FFT 分析处理增益为

$$G = 10\lg 2BT. \quad (3)$$

其处理增益大于能量检测器处理增益,所以在检测具有强线谱的目标时常使用基于 FFT 分析的检测方法。

3 基于线谱方位稳定性的目标检测

目标方位(DOA)估计有多种方法,本文利用线列阵阵列信号处理方法对目标方位进行估计[4]。阵列信号处理,即对输入信号进行空间处理,获得空间增益,提高后续信号处理的输入信噪比。

对水下目标进行检测通常有两种情况:一种情况是宽带信号,一般采用能量检测器;另一种情况是线谱信号,一般使用 FFT 分析。但通常情况下目标辐射信号中的线谱位置是未知的,即目标辐射信号的线谱频率大小未知,因此为了利用最大输入信噪比检测信号,需要对分析频带内的每个频率单元进行处理,每一个频率单元对应一个波束输出,需要对每一个频率单元的结果进行循环分析判断才能得到对目标的正确检测,在实际工程应用中是不可取的。

如何使检测结果输出简便,而且利用较高的输入信噪比和获得足够大的处理增益,这是本文需要探讨的主要问题。

下面分析是基于以下两个假设:第一是线谱信噪比足够大使得该线谱能够稳定检测目标;第二是在统计时间内,目标方位变化较慢。

设频率单元共 M 个,记为 $f_i, i=1,\cdots,M$, 波束预成方位 L 个,记为 θ_j, $j=1,\cdots,L$.

首先对接收的基元信号进行 FFT 分析和相位补偿,对每一个频率单元进行波束形成,各频率单元的波束输出记为 $R(f_i,\theta_j)$,$i=1,\cdots,M$, $j=1,\cdots,L$, 为$M\times L$ 维矩阵,对其每行求最大值,则最大值所在位置对应着该频率单元的 DOA 估计,记为 $\theta(f_i)$, $i=1,\cdots,M$. 对于噪声对应的频率单元,$\theta(f_i)$ 是随机的,对于目标线谱对应的频率单元 $\theta(f_i)$ 应当是目标 DOA,是稳定的。

上述信号处理过程重复 N 次,即连续处理 N 帧信号后再进行下一步的处理,则每个频率单元对应 N 个最大值所在方位,记为 $\theta_k(f_i)$, $i=1,\cdots,M$, $k=1,\cdots,N$,分别计算每个频率单元的 DOA 方差,记为 $\delta_\theta(f_i)$, $i=1,\cdots,M$. 由于噪声是随机的,则噪声对应频率单元的 DOA 是随机的,方差较大,而对于目标线谱对应的频率单元的波束输出最大值对应的 DOA 应当是基本不变的,方差很小。

最后对每一个方位的输出值进行统计计算,作为最后的波束输出。计算过程如下,首先将最后的波束输出置 0,即 $R_{out}(\theta_j)=0$, $j=1,\cdots,L$, 所有频率单元的所有 DOA 测量值均参与计算,当某一个频率单元的某一帧 DOA 估计为 $\theta_k(f_i)$ 时,则在 $\theta_k(f_i)$ DOA 对应值上累加该频率单元对应的 DOA 方差的倒数,即

$$R_{out}(\theta_k(f_i))_n = R_{out}(\theta_k(f_i))_{n-1} + 1/\delta_\theta(f_i);$$
$$i=1,\cdots,M; k=1,\cdots,N; n=1,\cdots. \quad (4)$$

以此计算直到每一个频率单元每一帧的方位估计结果均参加运算,最后得到每一个方位的方差倒数累计值,作为最终的波束输出。信号检测流程图如图 1 所示。

实现上述算法分为以下 5 步:

1) 对长线阵的各阵元接收信号作 FFT,然后对每一个频率单元进行频域波束形成,从而得到每一个频率单元的波束输出,记为 $R(f_i,\theta_j)$;

2) 对每一个频率单元的波束输出求最大值,即每一个频率单元的方位估计结果;

3) 更新接收信号,重复进行第 1) 步、第 2) 步,直到重复次数达到预先设定值 N,则每一个频率单元均得到 N 个方位估计结果,记为 $\theta_k(f_i)$;

4) 分别对每一个频率单元的方位估计结果进行方差计算,对应结果记为 $\delta_\theta(f_i)$;

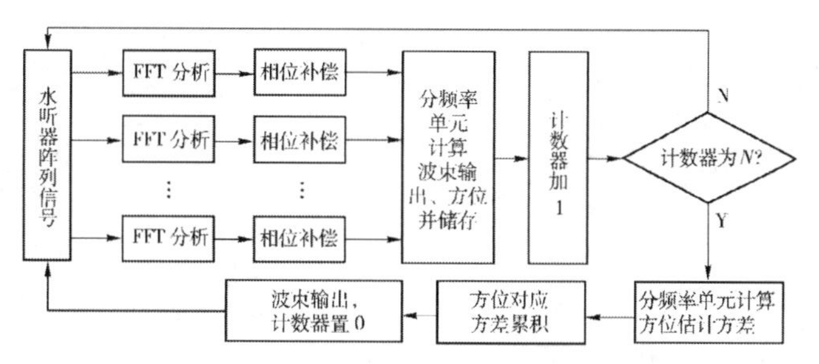

图1 基于方位稳定性的目标检测流程图

Fig. 1 Flow chart of target detection based on the orientation stability

5) 对所有方位估计结果对应的方差进行累计计算，作为最终的波束输出，例如当某一频率单元某一时刻的方位估计结果为 $\theta_k(f_i)$ 时，则其波束输出值 $R_{\text{out}}(\theta_k(f_i))$ 累加该频率单元对应的 DOA 估计方差的倒数，如(4)式所示。

下面给出波束输出中，目标方位的输出值与其它方位输出值的大小关系。设最小和最大频率单元为 f_1、f_M，目标线谱占其中一个频率单元，最小和最大预成方位为 θ_1、θ_L，对目标方位的实际估计值的最小值和最大值分别为 $\hat{\theta}_{\min}$、$\hat{\theta}_{\max}$，共进行 N 帧信号统计。假设每一个频率单元的方位估计结果均服从均匀分布，噪声和信号方位的方差分别为 δ_{n}、δ_{s}，有

$$\begin{cases}\delta_{\text{n}} = \dfrac{\theta_L - \theta_1}{\sqrt{12}},\\ \delta_{\text{s}} = \dfrac{\hat{\theta}_{\max} - \hat{\theta}_{\min}}{\sqrt{12}}.\end{cases} \tag{5}$$

首先对噪声频率单元进行统计，即对$(M-1)$个频率单元进行统计，

$$R_{\text{out}}(\theta_k) = (M-1)\cdot\frac{N}{\theta_L - \theta_1}\cdot\frac{1}{\delta_{\text{n}}}, k = 1, \cdots, L. \tag{6}$$

即对于噪声的波束输出，每个预成方位输出值是相等的，然后进一步将线谱估计方位的结果累加到(6)式表示结果中，有

$$R_{\text{out}}(\hat{\theta}_k) = (M-1)\cdot\frac{N}{\theta_L - \theta_1}\cdot\frac{1}{\delta_{\text{n}}} + \frac{N}{\hat{\theta}_{\max} - \hat{\theta}_{\min}}\cdot\frac{1}{\delta_{\text{s}}}. \tag{7}$$

进一步化简有

$$R_{\text{out}}(\hat{\theta}_k) = (M-1)\cdot\frac{N}{\sqrt{12\delta_{\text{n}}^2}} + \frac{N}{\sqrt{12\delta_{\text{s}}^2}}. \tag{8}$$

因此当线谱估计方位方差很小时，即每帧方位估计结果均接近于目标方位真值，即 δ_{s} 值较小，比较(6)式和(8)式可以看出目标真实方位附近的波束输出值将远远大于其它方位波束输出值。

4 仿真分析

假设目标相对于水平长线阵的方位为90°，辐射信号包括高斯带限白噪声和线谱成份，白噪声带宽为60～300 Hz，线谱频率为100 Hz，线谱谱级与白噪声平均谱级比为13 dB. 目标辐射信号对应的频谱如图2所示。干扰为带限白噪声，目标辐射噪声谱级与干扰噪声谱级比为－20 dB，则长线阵各阵元接收信号的线谱谱级与干扰噪声平均谱级比为－7 dB.

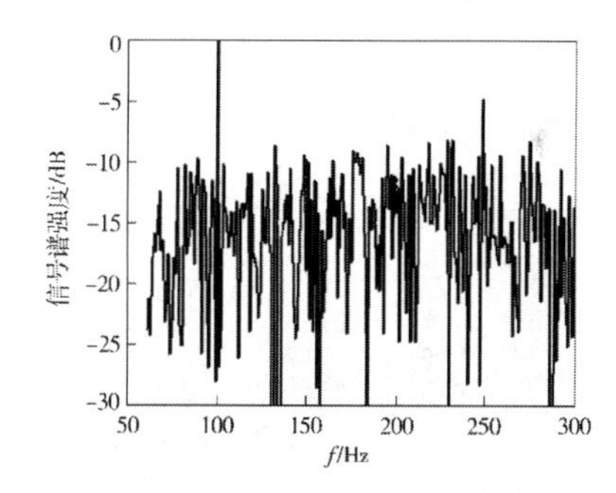

图2 目标辐射信号谱分析输出

Fig. 2 Spectral analysis output of target radiation signal

线列阵阵元数为32，阵元间距为8 m. 分别对3种方法进行了仿真分析：第1种方法基于单线谱检测，即已知线谱位置，只对该线谱进行处理，利用该线谱进行波束形成，从而对目标进行检测；第2种方法是基于宽带能量检测，即整个频带采用波束形成对目标进行检测；第3种方法是基于方位稳定性，即对每一个频率单元处理，估计每一个频率单元对应的目标方位，进行方位统计。仿真结果如图3～图5所示。从图4可以看出，基于能量检测的目标方位历程在仿真条件下目标不明显，而从图3、图5方位历程输出可以有效地实现目标的检测。

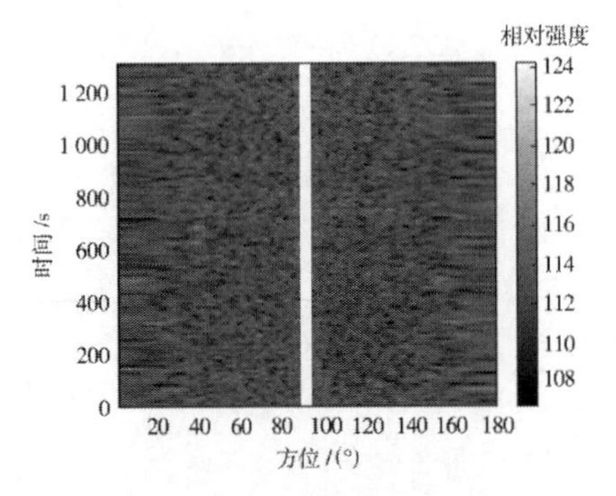

图 3　基于线谱检测的方位历程图

Fig. 3　Time-bearing tracks based on line spectrum detection

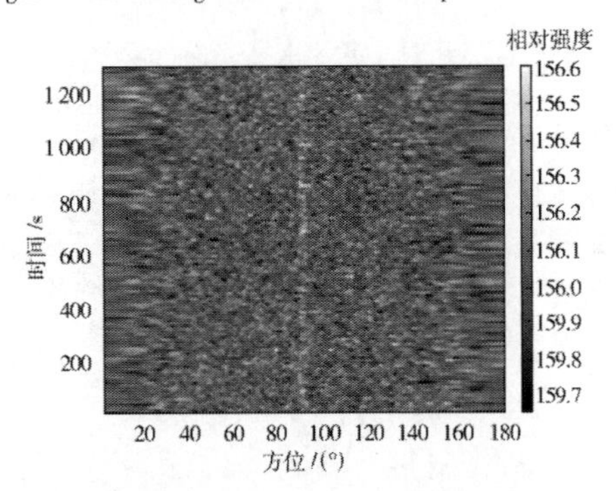

图 4　基于能量检测的方位历程图

Fig. 4　Time-bearing tracks based on energy detection

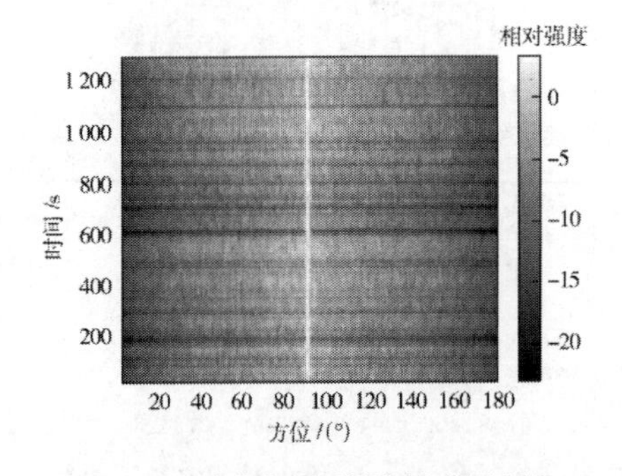

图 5　基于方位稳定性检测的方位历程图

Fig. 5　Time-bearing tracks based on the orientation stability

图 6、图 7 为目标辐射噪声谱级与干扰噪声谱级比降低为 -33 dB 的基于线谱检测和基于方位稳定性的方位历程估计结果。从结果可以看出，该信噪比条件下基于线谱检测和基于方位稳定性的目标检测方法均不能很有效地实现对目标的估计。

5　实验结果

下面给出利用宽带检测和线谱检测的方位历程

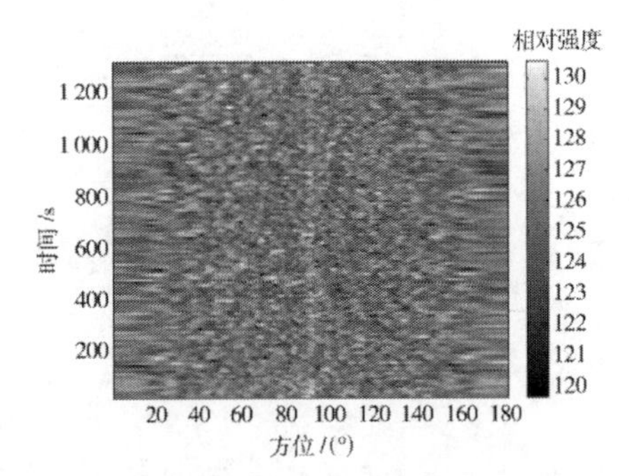

图 6　基于线谱检测的方位历程图

Fig. 6　Time-bearing tracks based on line spectrum detection

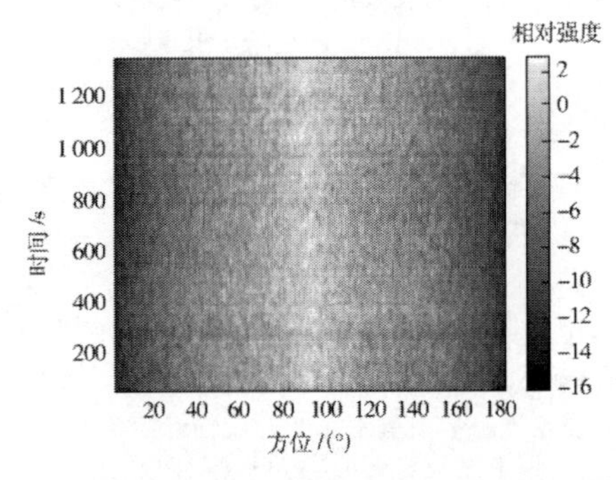

图 7　基于方位稳定性检测的方位历程图

Fig. 7　Time-bearing tracks based on the orientation stability

估计实验结果。简单说明实验用水下目标辐射信号特性，通过对水下目标近距离辐射信号的谱分析，结果表明，辐射信号中含有低频线谱成份，因保密要求，这里省略目标辐射信号的谱分析结果。

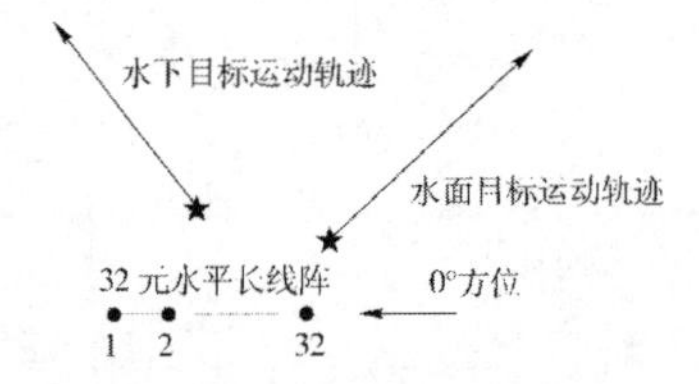

图 8　实验长线阵及目标运动态势图

Fig. 8　Schematic diagram of the experimental long-term array and target motion

实验长线阵及目标运动态势图如图 8 所示。实验采用 32 元水平长线阵接收信号，阵端向方位设为 0°，实验时水下目标运动方位相对于长线阵大约在 100°左右，并且在 20° ~ 40°方位有大型水面目标通过，分别采用基于能量检测和线谱检测的方法进行目标检测，方位历程估计结果分别如图 9、图 10 所示。从图9可以看出,基于能量检测的方位历程图在100°方位未明显检测到目标,而图10所示的基于线谱检测的方位历程图在100°能够明显发现目标,而且干扰背景相对于基于能量检测的干扰背景被压低。

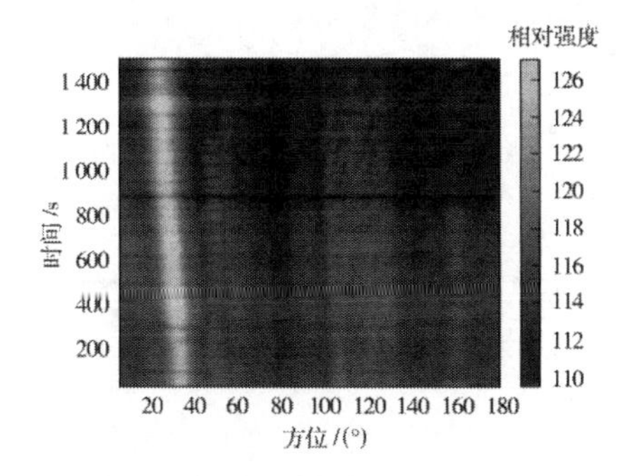

图9　基于能量检测的方位历程图

Fig. 9　Time-bearing tracks based on energy detection

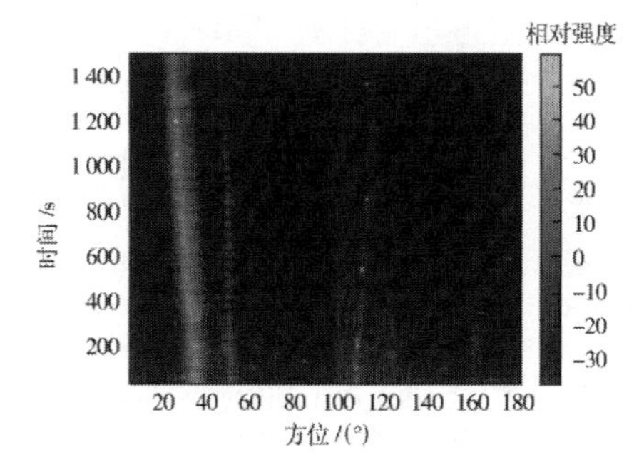

图10　基于方位稳定性检测的方位历程图

Fig. 10　Time-bearing tracks based on the orientation stability

6　结论

本文介绍了一种针对目标辐射信号中线谱不确知的目标检测方法。通过对接收信号FFT分析后,再对每一个频率单元均进行方位估计,利用噪声频率单元方位估计结果随机,而目标线谱方位估计结果一致的特点,对每一个方位估计结果进行方位统计,实现对目标的检测。通过仿真和实验数据验证了该方法的检测性能远高于常规能量检测性能,与已知线谱处理性能基本一致。

参考文献(References)

[1]　Urick R J. Principles of underwater sound [M]. New York: McGraw-Hill Book Company, 1983.

[2]　Ross D. Mechanics of underwater noise [M]. New York: Pergmin Press, 1976.

[3]　李启虎, 李敏, 杨秀庭. 水下目标辐射噪声中单频信号分量的检测: 理论分析[J]. 声学学报, 2008, (3): 193 - 196.
LI Qi-hu, LI Min, YANG Xiu-ting. The detection of single frequency component of underwater radiated noise of target: theoretical analysis [J]. Acta Acoustica, 2008, (3): 193 - 196. (in Chinese)

[4]　李启虎. 数字式声纳设计原理[M]. 安徽: 安徽教育出版社, 2003.
LI Qi-hu. Design principle of digital sonar [M]. Anhui: Anhui Education Press, 2003. (in Chinese)

Kraken 声场建模下目标辐射噪声模拟技术研究①②

解　恺，丁雪洁，李　宇，黄海宁，李启虎

(中国科学院声学研究所，北京 100190)

摘　要：水面目标辐射噪声模拟技术是测试被动声呐系统、目标识别定向及水声软对抗的关键技术。通过对辐射噪声信号特性的描述及分析，建立了水面目标频域线谱及连续谱分量的数理仿真模型。针对水声信道影响因素，将 Kraken 简正波声场模型与信号本身特性有机结合。以 1/3 倍频程滤波带宽算法为基础，完成了对宽带 FIR 滤波器组的设计。采用现在应用广泛的 GPU(Graphic Processing Unit)并行加速平台，实时实现了目标噪声模拟器整体架构，大大降低了系统硬件资源的消耗。通过对模拟器输出信号进行数理分析，证明了该模拟器能较好地实现水面目标的辐射噪声特性，具备良好的科研应用价值。

关键词：辐射噪声；Kraken 简正波；GPU；1/3 倍频程

中图分类号：TN911.7　　**文献标志码**：A　　**文章编号**：1003-2029(2012)03-0024-05

在实际的水声信号处理领域中，水面目标在海水中的辐射噪声信号一直是人们研究的热点，它是被动式声探测设备的唯一信息来源，这一信息可用于对水面目标进行声探测、识别以及对目标的定向、定位和跟踪等。目标辐射噪声的模拟主要是研究和分析水面目标的声学特征，并能够高逼真地将这些特征模拟出来，建立相应的数理模型，利用计算机技术进行可编程设计，根据不同的需要进行参数设置[1]。

对于目标辐射噪声的模拟及研究在许多文章中都有讨论，可绝大部分都是对目标信号本身特点的模拟，没有引入周围传播环境的影响作用。而关于宽带噪声的滤波处理，大多采用特定频率响应的自适应滤波器。这种方式受到收敛速度及运算复杂度的制约，会给模拟系统的实时实现带来困难。针对以上问题，本文在线谱仿真部分增加了 Kraken 简正波声场模型对目标所在传播信道的仿真，将环境影响因素与信号本身特性进行结合。在连续谱仿真部分，设计出三分之一倍频程滤波器组，完成滤波处理，简化了系统硬件实现的难度。另外，本文采用了目前引起广泛关注的 GPU 并行加速平台代替传统的 CPU 硬件平台，大幅提高了运算速度，为模拟系统的实时实现提供了有力保障。

1　Kraken 简正波建模

构成所有的传播数学模型的理论基础是波动方程。它给出的是同一类物理现象的共性即泛定方程。当介质质点瓣振动位移比波长小很多，同时声压幅度值也远小于介质的静压力时，可以近似推导出均匀声传播的波动方程：

$$\nabla^2 P-\frac{1}{c^2}\frac{\partial^2 P}{\partial t^2}=0 \tag{1}$$

式中：$\nabla^2=\partial^2/\partial x^2+\partial^2/\partial y^2+\partial^2/\partial z^2$，是拉普拉斯算子；$P$ 是势函数；c 是声速。

简正波理论认为声波在介质中的场可以分解为许多不同阶次的相互独立传播、互不影响的简正波，在接收点所收到的声波是这些不同阶次的简正波经过信道传播到达接收点后相互叠加的结果。为得到一般条件下波动方程的解，只能利用数值算法。简正波解是波动方程的一种精确的积分解，它常假定海洋信道为柱面对称的分层介质，若设距离为 r，深度为 z，则声场的全解可写为[2]：

$$p(r,z,t)=p(r,z)\exp(j\omega t) \tag{2}$$

假设分层介质是圆柱对称的，那么声场函数可表示成深度函数 $F(z)$和 $S(r)$的乘积。

$$p(r,z)=F(z)*S(r) \tag{3}$$

用 ε^2 作为分离变量常数对变量进行分离，代入波动方程可得简正波深度方程：

$$\frac{\mathrm{d}^2F}{\mathrm{d}z^2}+(k^2-\varepsilon^2)F=0 \tag{4}$$

设 $\rho(z)$为海水密度，$c(z)$为声速，$k_m{}^2$ 为分离常数，k_m 对简正波来说是水平波数，$u_m(z)$相应于 k_m 的简正波深度函数，则(4)式可变为：

$$\rho(z)\frac{\mathrm{d}}{\mathrm{d}z}\left[\frac{1}{\rho(z)}\frac{\mathrm{d}u_m(z)}{\mathrm{d}z}\right]+\left[\frac{\omega^2}{c^2(z)}-k_m^2\right]u_m(z)=0 \tag{5}$$

Kraken[3]方法是采用有限差分方法求解具有边界条件的简正波方程，在此我们假定海底为粗糙硬海底、即在深度

① 国家自然科学基金资助项目(10904160)；中国科学院知识创新工程重大方向项目资助.

② 海洋技术, 2012, 31(3): 24-28.

$z=D$ 的海底处，质点振动速度为 0。即意味着岩石层足够深，在达到底质层之前，声波已被有效地衰减掉。

在此边界条件下，简正波的解可写为：

$$p(r,z)=\frac{i}{\rho(z_s)\sqrt{8\pi r}}\exp(-\frac{i\pi}{4})\sum_{m=1}^{\infty}\frac{u_m(z_s)u_m(z)}{\sqrt{k_m}}\exp(ik_m r) \tag{6}$$

Kraken 方法将整个深度区间 $0\leqslant z\leqslant D$ 等分成 N 个间隔，构成等间隔点 $z_j=jh(j=0,1,\cdots,N)$ 网格，其中 $h=D/N$ 为网格宽度。数目 N 应选择得足够大，保证对模式有充分的抽样。于是利用有限差分近似，可将式(5) 连续问题化成标准的特征值问题：

$$Au_m=h^2k_m^2u_m \tag{7}$$

式中：$h^2k_m^2$ 是矩阵 A 的第 m 个特征值；μ_m 是 N 个相应于特征值的特征向量，近似等于模型函数，A 为 $N\times N$ 非对称三对角矩阵。

$$A=\begin{bmatrix} a_1 & 1 & 0 & & & & & \\ 1 & a_2 & 1 & 0 & & & & \\ & \cdot & \cdot & \cdot & & & & \\ & 0 & 1 & a_n & 1 & 0 & & \\ & & & \cdot & \cdot & \cdot & \cdot & \\ & & & & 0 & 1 & a_{N-1} & 1 \\ & & & & & & 2 & a_N \end{bmatrix} \tag{8}$$

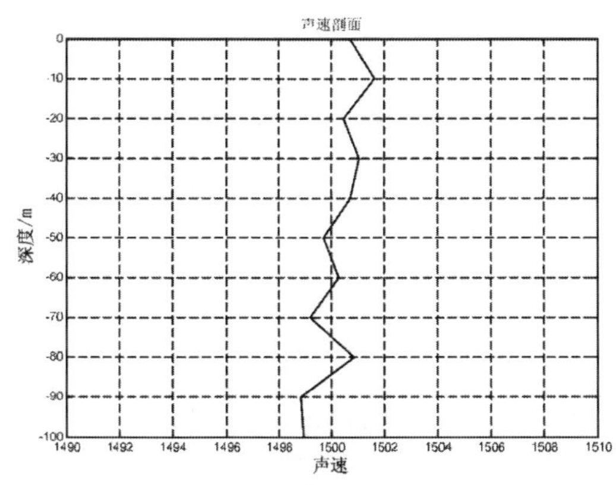

图 1　扰动声速剖面

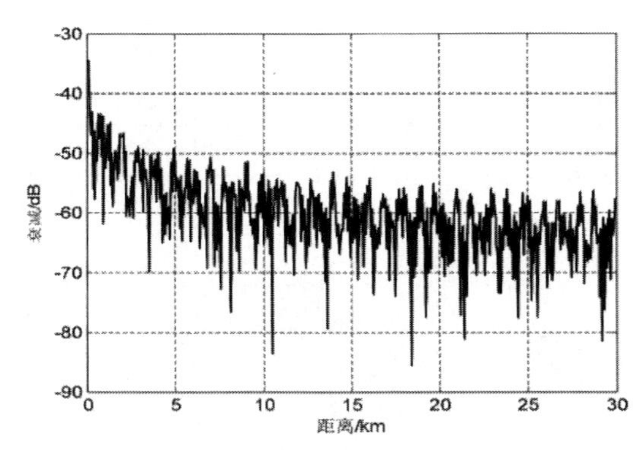

图 2　Kraken 声压传播损失

$a_i=-2+\frac{h^2\omega^2}{c^2(z_i)}$，$i=1,\cdots,N$，$z_i$ 是列向量，这些元素是网格点上特征函数的近似值。求解出上述(7)式的代数特征值问题，便可得到特征值和特征函数，代入声压表达式(6)，便可求出所要计算的海洋声场声压值。

图 1 为 Kraken 仿真建模所需的声速剖面输入条件。表明了 0–100 m 水深范围内，传播声速的扰动情况。图 2 表示了 0–30 km 传播距离下，应用该声场模型各网点的声压传播损失情况。

2　目标辐射噪声建模及实现

水面目标辐射噪声的功率特征比较复杂，因其由众多噪声源的综合作用而产生。而且在水面目标的不同航行状态下，对辐射噪声都有不同的影响。噪声谱主要有两种不同的类型[4]：一种是具有非连续谱的单频噪声，这种噪声由出现在离散频率上的线谱组成。另一种是具有连续谱的宽带噪声，其噪声级是频率的连续函数。

在此仿真模型中，引入 Kraken 简正波声场模型对目标线谱分量水声传播环境的模拟。采用有限差分方法作为理论依据。对于宽带连续谱的设计，提出了一种新的以 1/3 倍频程 FIR 滤波器组为基础的实现方案。

2.1　模拟线谱分量

水面目标辐射噪声线谱是一些幅值明显高出相邻连续谱，并有稳定的频率成分的谱线，其功率谱可以高出连续谱 10–25 dB。产生线谱的声源通常是：不平衡的旋转部件、往复部件、螺旋桨叶片共振以及一些结构部件或空腔被激励谐振等[5]。

螺旋桨叶片切割所有进入螺旋桨及其附近的不规则流动，使螺旋桨噪声含有离散的、分布在叶片速率的倍数上的“叶片速率”谱，其在 1–100 Hz 的频段内是目标的主要噪声源。其频率为：

$$F_m=(R_{sp}+\Delta R_{sp})\cdot N_{vane}\cdot m_{har} \tag{9}$$

式中：F_m 代表叶片速率线谱的第 m 次谐波频率(Hz)；N_{vane} 是螺旋桨的叶片数；R_{sp} 为转速(r/s)；ΔR_{sp} 为螺旋桨转速变化量。

对于线谱模拟的工程实现，可直接把一些谐波信号的集合叠加于连续谱噪声上，这些谐波均以一组特殊的参数设置来选定。

$$S(t)=\sum_{L=1}^{M}a_L\cos\left\{\left[2\pi\frac{f_L}{T}t+\theta_L(t)\right]+\varphi_{in}\right\} \tag{10}$$

式中：f_L 表示任意线谱信号的中心频率；$\theta_L(t)$ 为任意信号的初始相位；φ_{in} 是$[0,2\pi]$上的均匀随机量，表示该目标线谱分量声场作用下的相位起伏情况；a_L 表示线谱的幅度，它是一个与时间、航速、航深、水温、海况等有关的多元函数。图 3 表示辐射噪声信号线谱分量的频域形式，图 4 为距离 20 km 处，水声信道作用下该线谱分量的相位漂移情况。

2.2　模拟连续谱分量

连续谱噪声主要是水动力噪声、机械噪声和螺旋桨噪声组成的。水面目标辐射噪声为不确定性信号，其主体是宽带随机信号，有时也包含谐波成分。为了用统计参数来描述，通常假定目标辐射噪声主体为平稳的、各态历经的随机过程。连续谱的主要频段从几赫兹到数千赫兹，低端为 6–12 dB/oct 的正斜率，高端为 6 dB/oct 左右的负斜率，在几十至几百赫兹之间出现平直谱或峰值，声源级随距离的衰减量与水深和底质有关，但远场一般近似为 6 dB/oct[6]。

本文将从目标辐射噪声的频率特性出发，采用特定频率响应滤波器进行目标辐射噪声宽带连续谱分量的模拟。可将高斯白噪声序列通过一低通滤波器，再将其输出通过一个满足噪声频谱特性要求的特定频率响应的滤波器来产生。

满足上述条件的特定频响滤波器的脉冲响应函数应具有以下形式：

$$H(t)=\sqrt{2L}\cdot u(t)\exp(-Lt) \tag{11}$$

$u(t)$是在 $t=0$ 和 $t=L$ 的单位阶跃响应函数，该函数的频率响应大约具有 6–8 dB /oct 的衰减速率。

滤波器的通频带宽度为f_2-f_1，即在f_2-f_1的频率范围内的信号全部通过。中心频率为$f_c=(f_2\times f_1)^{1/2}$。

对于噪声来讲，由于 1/3 倍频程谱能够很好地体现噪声带宽的能量分布情况。因此引入 1/3 倍频程谱的设计思想尤为重要。由上所述，任一组 1/3 倍频程滤波器计算方法如下：

$$f_u/f_l=2^{1/3}=1.2599 \tag{12}$$

$$f_m=\sqrt{f_u\cdot f_l} \tag{13}$$

$$B=f_u-f_l \tag{14}$$

$$B=Kf_m \tag{15}$$

式中：f_u 为上边频；f_l 为下边频；f_m 为中心频率；B 为 1/3 倍频程带宽；K 为 0.231(常数)。

1/3 倍频程 FIR 数字滤波器的设计任务就是确定在满足滤波器频率特性要求下的各子区间的中心频率点 f_m 及滤波衰减系数。确定满足要求的各子通带区间中心频率点，简单的做法就是把它们所属总区间分为若干段，对所有组合按上式算法要求利用带通滤波器一般设计方法进行计算。最后计算滤波器的输出信号的和，验证符合要求说明设计结束，否则再进行其他组合试探，直到所有输出信号的和满足要求为止，图 5 为滤波器幅频响应图。从滤波器幅频特性曲线来看，完全符合设计要求，验证了工程可实现性。

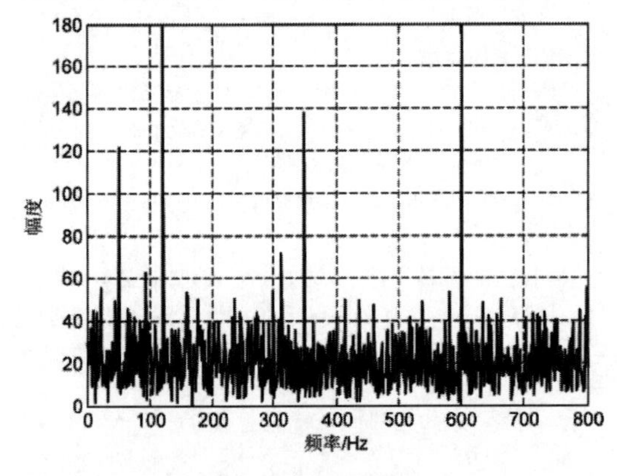

图 3 频域线谱形式

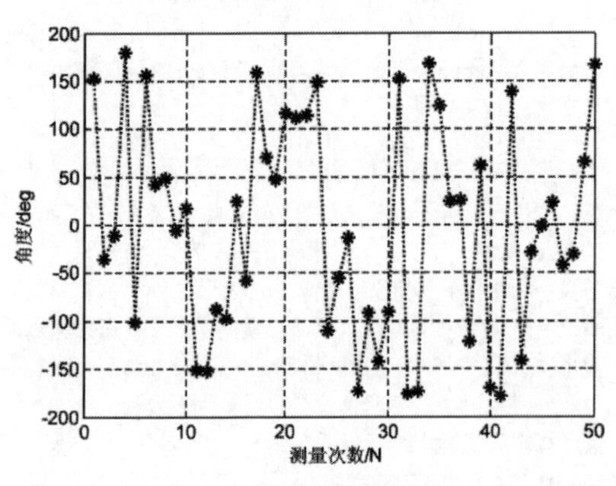

图 4 线谱分量相位漂移情况

3 GPU 加速的目标噪声模拟器设计

3.1 目标辐射噪声表达形式

单通道、随机单阵元的模拟水面目标辐射噪声数学模型为：

$$Y_{redi}(nT_S)=A_{sig}(nT_S)+N_{con}(nT_S)+[1+\mu\times A_{sig}(nT_S)]\times m(nT_S) \tag{16}$$

$$A_{sig}(nT_S)=\sum_{L=1}^{M}a_L\cos\left[2\pi\frac{f_L}{T}t+\theta_L(t)\right] \tag{17}$$

式中：$Y_{redi}(nT_s)$为模拟产生的辐射噪声输出信号；$A_{sig}(nT_s)$是一系列以周期信号为模型的线谱分量的叠加；a_t 表示目标输入线谱的幅度能级；M 为模拟目标的线谱数量，其中，部分参数前已叙述，在此不再赘述；$m(nT_s)$为目标辐射噪声的调制分量，其中心频率在 1 kHz 左右；μ 为线谱精细结构调节参数，范围为 0–1之间；$N_{con}(nT_s)$为连续谱分量对应的时域波形，用宽带平稳随机过程进行模拟。

图 6 为根据蒙特卡罗仿真方式平均产生的 kraken 声场模拟条件下，目标辐射噪声频谱特性。

3.2 基于矩阵并行的 GPU 处理平台

目前的图形处理器(GPU)，提供了良好的可编程特性，能

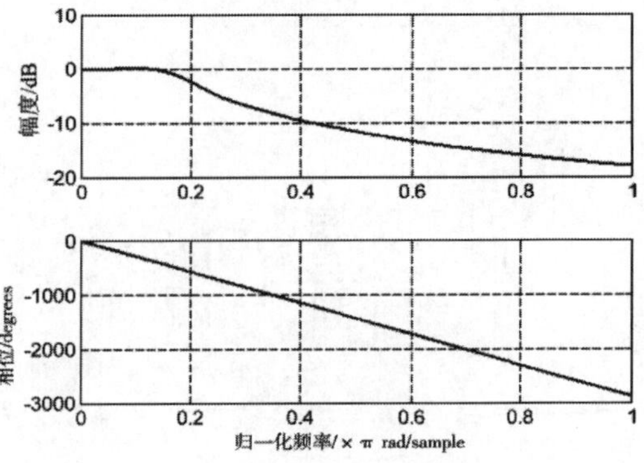

图 5 滤波器组幅频响应

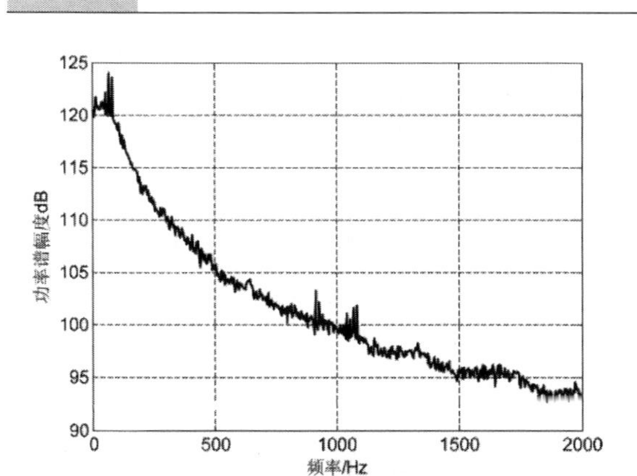

图 6 水面目标辐射噪声

获得极具潜力和性价比的并行加速性能，具有高精度的数据描述、高效的逻辑处理和计算性能。在具体的实现上，使用了NVIDIA公司开发的CUDA并行编程平台。CUDA并行模型是基于单指令多数据结构 (Single Instruction Multiple Data, SIMD)，在向量、矩阵运算方面性能优势明显。

CUDA的编程模型将一个程序的不同部分放在主机端(Host)或设备端(Device)执行，设备端作为主机端的协处理器。

在GPU上实现矩阵并行处理算法是利用了子矩阵的概念，如图7所示，设执行的运算为 $A*B=C$，其中 A 的列数为A.width，B 的行数为B.height，设子矩阵的Block尺寸为BLOCK_SIZE × BLOCK_SIZE。整个运算过程中读取global memory中 A 和 B 矩阵的次数各自为：B.width/BLOCK_SIZE和A.height/BLOCK_SIZE。如果不利用子矩阵的概念，整个运

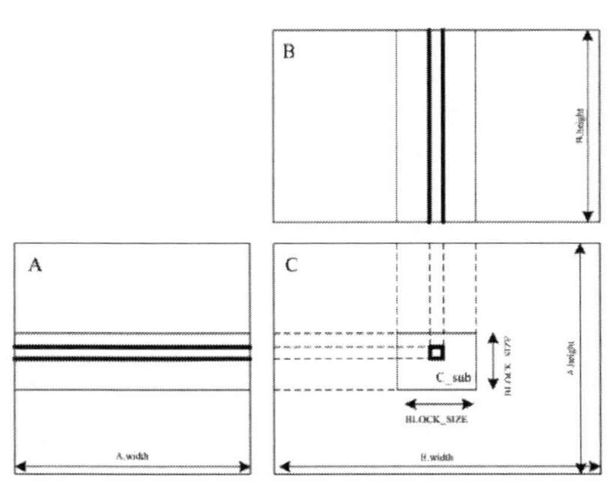

图 7 GPU 子矩阵并行

算过程中读取 A 和 B 的次数分别为B.width和A.height。

估算GPU和CPU的运算速度：以Matrix Size=2048*2048，Block Size=16*16为例。运算量：2048^3=8G次乘法、8G次加法，定义1次乘法和1次加法为1次运算，则GPU运行时间766 ms，速度约为10 G/s；CPU运行时间167 347 ms，速度约为50 M/s。GPU运算速度为CPU的200倍。当BLOCK_SIZE较小，或者Matrix Size较小时，GPU的运算速度达不到CPU的200倍，因为这时数据在不同存储器上的传输占据了运算时间的大部分。

辐射噪声GPU模拟平台的程序流程如图8所示。首先将各种所需模拟的目标参数进行初始化输入，待连续谱滤波器组系数更新完毕，产生宽带噪声随机副本，进入多通道模拟目标噪声循环模块，最终完成数据传输显控。

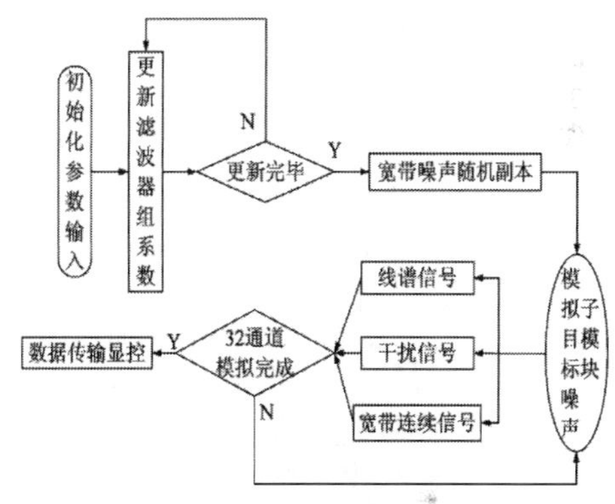

图 8 GPU 模拟器程序框图

4 结论

水面目标辐射噪声的模拟具有重要的用途，如舰艇指挥系统的方案论证、仿真实验、模拟训练以及对声纳系统在实验室阶段的测试等。目标噪声模拟器的设计研究有着重要的现实意义及经济意义。本文在目标辐射噪声数理建模的基础上，加入了Kraken简正波声场模型对目标线谱分量传播环境的仿真，使模拟器的设计更加具有全面性及真实性。

笔者提出了基于1/3倍频程的宽带FIR滤波器组设计方法，使模拟器的设计资源得到了简化，提高了效率。同时通过GPU并行加速处理，更有力保障了整个目标噪声模拟器的实时实现，大大节约了系统的运算耗时，具有非常广泛的应用前景。

参考文献:

[1] 杜选民,姜亚浩. 舰船辐射噪声模拟技术研究[J]. 声学技术, 1999,18(l):10-14.
[2] Green M D, Rice J A. Channel-tolerant FH-MFSK acoustic signaling for undersea communication and networks [J]. IEEE J Oceanic Eng, 2000, 25(1): 28-39.
[3] Jensen F B, Kuperman W A, Porter M B. Computational Ocean Acoustics[M]. New York: American Institute of Physics, 1994.
[4] Urick R J. Principles of Underwater Sound[M]. New York: McGraw-Hili,1975.
[5] LI Qin, YUAN Bing-cheng, MING Xing. Simulation technique of radiated noise from underwater target and its implement of simulator [C]//PEITS, 2nd International Conference, 2009:357-360.
[6] 李启虎.数字式声纳设计原理[M].安徽:安徽教育出版社,2002.

The Simulation Technique of Target Radiated Noise under Kraken Acoustic Field Modeling

XIE Kai, DING Xue-jie, LI Yu, HUANG Hai-ning, LI Qi-hu
(*Institute of Acoustics, Chinese Academy of Sciences, Beijing* 100190, *China*)

Abstract: The simulator design of surface target radiated noise is the key technology to test passive sonar systems, target oriented identification and underwater acoustic soft countermine. The mathematical simulation model of the line and continuous spectrum in frequency domain from surface target was established based on the description and analysis of radiated noise signal characteristic. According to the influence factors of the underwater acoustic channel, Kraken normal acoustic field model combined with the signal characteristics was built. The design of broadband FIR filter banks was completed based on one-third octave filter bandwidth algorithm. The overall structure of the target noise simulator was real-timely implemented using the widely used GPU (Graphic Processing Unit) accelerated parallel platform, which could greatly reduce the consumption of the system hardware resources. It is proved that the simulator can better achieve the characteristics of surface target radiated noise after the mathematical analysis of simulator output. It has a good scientific application value.

Key words: radiated noise; Kraken normal model; GPU; one-third octave

The Simulator Design of GPU accelerated Radiated Noise from Surface Target①

Kai Xie, Qihu Li, Xuejie Ding, Haining Huang
Institute of Acoustics, Chinese Academy of Sciences, Beijing, China
xk_0308@163.com

***Abstract*—The simulator design of surface target radiated noise is the key technology to test passive sonar systems, target oriented recognition and underwater soft against. The mathematical simulation model of the line and continuous spectrum in frequency domain from surface target has been established based on the description and analysis of radiated noise signal characteristic. Delay acoustic orientated method and one-third octave filter bandwidth algorithm for the simulation of moving target position and information are proposed. The design of broadband FIR filters has been completed. The overall structure of the target noise simulator is real-time implemented by using the widely used GPU (Graphic Processing Unit) accelerated parallel platform. It greatly reduces the consumption of the system hardware resources. It is proved that the simulator can better achieve the characteristics of surface target radiated noise after the mathematical analysis of simulator output. It has a good scientific application value.**

***Index Terms*—radiated noise; GPU; simulator; one-third octave; delay acoustic orientated method**

I. INTRODUCTION

Radiated noise signal of surface target in the sea is always the research hotspot in actual acoustics signal processing [1]. It is the only information source of passive sound detection. It can be used to detect, recognize surface targets and orientate, position or track them. The simulation of target radiated noise is mainly to research and analyze acoustic characteristics of surface target. These characters can be simulated vividly and established relevant mathematics model. The simulator is based on computer technology for programmable design and changed parameter settings for different requirements.

The research of target radiated noise has been discussed in many papers, but the majority of them are the simulations for static target signal. Most of the methods for broad band noise filter disposal are using adaptive filter of specific frequency response. This approach will bring difficulties to achieve real-time simulation system by the constraints of convergence speed and computational complexity. To solve the above problem, the simulation of dynamic target has been achieved by a delay of sound pressure oriented method in the part of line spectrum simulation. In the part of simulation continuous spectrum, a one-third octave filter has been designed to complete filtering. It simplifies the difficulty of the system hardware. Otherwise, the GPU parallel accelerated platform which is led to widespread concern currently replaces the traditional CPU hardware platform in this paper. It increases the operation speed substantially and provides a strong guarantee to achieve real-time simulation system.

II. THE CHARACTERISTICS OF TARGET NOISE

The power characteristic of surface target radiated noise is more complex arising from the combined effects large number of noise sources [2]. The radiated noise of surface targets has different effects in different state of navigation. There are two main different types of noise spectrum: One is a single-frequency noise with non-continuous spectrum. This noise is in components of line spectrums which are on discrete frequencies. Another is a broadband noise with continuous spectrum. The noise level is a continuous function of frequency. A large number of measurements and analysis show that the line spectrums of radiated noise spectrums are distributed in the low frequency components. The frequencies and amplitude of the line spectrums from different target noise are not the same. These line spectrums are the main features of the target type identification. The continuous spectrum of target radiated noise has a peak value which frequency is varies of type, speed and depth, usually appears in around 200Hz. When the frequency is lower than the peak frequency, the spectrum level of noise increases with frequencies; when it is more than this peak, the level has an attenuated trend, approximately attenuated 6dB per octave.

From the above analysis, surface target radiated noise is uncertainty signal. It is mostly broadband random signal, and sometimes also contains harmonic components. Radiated noise is usually assumed as a main stationary ergodic random process.

III. MODELING AND IMPLEMENTATION OF RADIATED NOISE

The oriented information of a moving target is simulated by equal spaced linear array technology. The delay acoustic orientation has been used as a theoretical basis. A new implementation what is based on one-third octave FIR filter is proposed for the design of broadband continuous spectrum.

A. The Form of Radiated Noise Signal

The mathematic model of surface target radiated noise in single channel is shown as follows:

$$Y_{redi}(nT_S)=A_{sig}(nT_S)+N_{con}(nT_S)+[1+\mu\times A_{sig}(nT_S)]\times m(nT_S) \qquad (1)$$

$Y_{redi}(nT_S)$ is simulated output signal of radiated noise.

① 2012 2nd International Conference on Consumer Electronics, Communications and Networks (CECNet), 2012: 720-723.

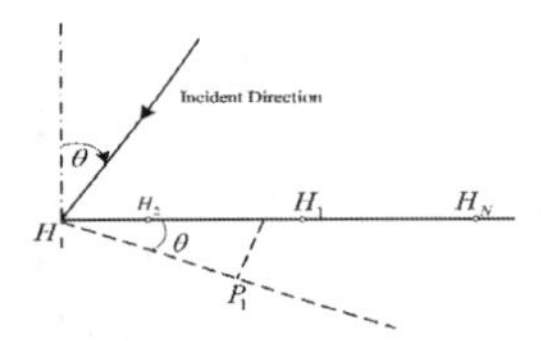

Fig.1 delay acoustic orientation

$A_{sig}(nT_S)$ is $S(t)=\sum_{L=1}^{M} a_L \cos\left[2\pi \frac{f_L}{T} t+\theta_L(t)\right]$. It is a series of adding of line spectrum which are formed by periodic signals. a_L shows the amplitude level of input line spectrum, M is line spectrum number of simulated targets, f_L shows the frequency of arbitrary line spectrum signal, $\theta_L(t)$ is oriented information of arbitrary moving target.

$m\,(nT_S)$ is modulated component of the target radiated noise, its center frequency is around at 1kHz. μ is adjusting parameter of line spectrum fine structure, range between 0 and 1. $N_{con}(nT_S)$ is time-domain waveform corresponding to the continuous spectrum component. It is simulated by broadband Stationary random process.

B. Modeling and Implementation of Line Spectrum Component

The line spectrum of Surface target radiated noise is the spectral component which amplitude is significantly higher than some of the adjacent continuous spectrum and has a stable frequency. Its power spectrum can be 10~25dB higher than the continuous spectrum. The sound sources which generated line spectrum are usually: unbalanced rotating parts, reciprocating parts, and resonant propeller blades.

Propeller blades cut all the irregular flow which comes into or near the propeller. The propeller noise contains discrete "blade speed" spectrum which is distributed in multiples of blade speed. The frequency band of 1 to 100 Hz is the main source of target noise [4]. Its expression of frequency is:

$$F_m=(R_{sp}+\Delta R_{sp})\cdot N_{vane}\cdot m_{har} \tag{2}$$

F_m represents m second harmonic frequency of the blade rate spectrum(unit : Hz), N_{vane} is the number of propeller blades, R_{sp} is the rotational speed(r/s), ΔR_{sp} is the propeller speed variation.

Some harmonic signal can be directly added on the noise of continuous spectrum, for the achievement of line spectrum engineering simulation. These harmonics are selected by a special set of parameters.

$$S(t)=\sum_{L=1}^{M} a_L \cos\left\{\left[2\pi \frac{f_L}{T} t+\theta_L(t)\right]+\varphi_{in}\right\} \tag{3}$$

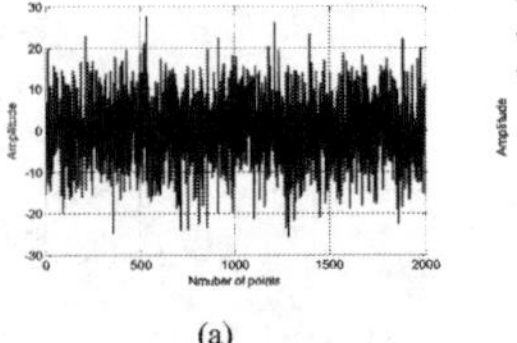

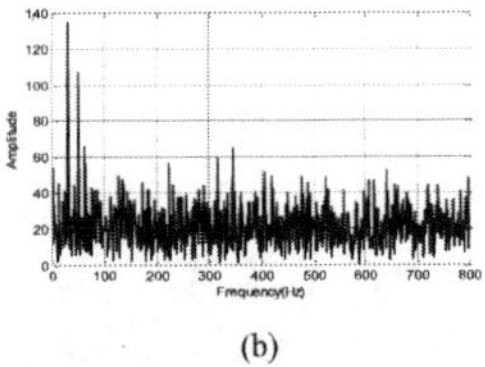

(a) (b)

Fig.2 Time and Frequency domain form of line spectrum

Part of the parameters has been described before, so it will not be explained. φ_{in} is uniform random on the amount of [0,2π]. It indicates the phase information of mobile orientation of target line spectral component. a_L is the magnitude of line spectrum. It is a multi-function of time, speed, depth, and water temperature sea conditions. Its change is random in a certain range. A uniform distributed random amount can be used to approximate the amplitude of line spectrum.

As mentioned earlier, $\varphi_i(t)$ is the information expression for the target mobile location. The program uses delay sound pressure directional method to simulate the reception of spaced linear array for a moving target. The following is the theoretical basis.

Figure 1 shows a spaced line array. Hydrophones are numbered from left to right as $H_1, H_2, \ldots, H_N$. The interval of primitives is assumed as d. The reference point of time is clicked on H_1 for computational convenience. Incident wave is assumed as single-frequency signal $A\cos 2\pi f t$. Its angle between the direction of array normal is θ, so the received signal of the i-th primitive H_i is ahead of H_1. It is caused by the sound path difference H_iP_i.

$$H_iP_i=(i-1)d\sin\theta \tag{4}$$

If line array is oriented at the direction of θ, so the signal of i-th primitive should be delayed,

$$\tau_i(\theta_0)=2\pi(i-1)\frac{d}{\lambda}\sin\theta \tag{5}$$

Received signal of i-th primitive is:

$$s_i(t)=A\cos\left\{2\pi f\left[t+\frac{(i-1)d\sin\theta}{c}\right]\right\} \tag{6}$$

C is sound speed. Because of the signal has single frequency, the phase difference of output signal between H_i and H_1 is :

$$\varphi_i=2\pi(i-1)\frac{d}{\lambda}\sin\theta=(i-1)\psi \tag{7}$$

$\lambda=c/f$ is signal wavelength. After $s_i(t)$ is summed, we can get that:

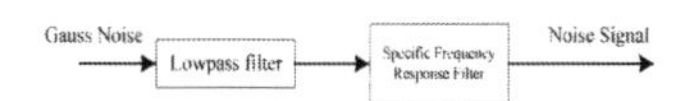

Fig.3 Diagram of broadband continuous spectrum

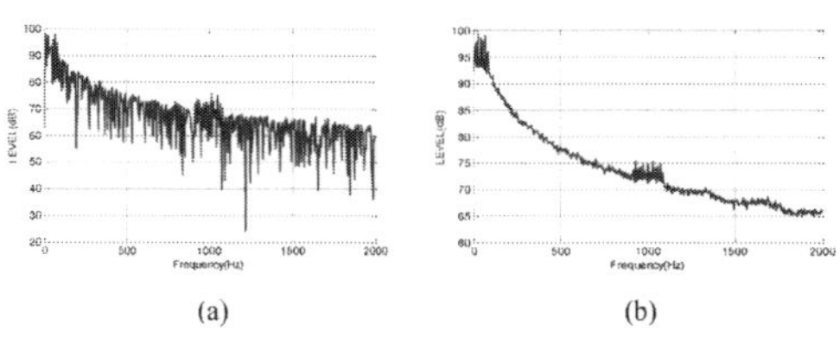

Fig.4 simulation of continuous spectrum component

$$s(t)=\sum_{i=1}^{N} s_i(t)=\sum_{i=1}^{N} A\cos\left[2\pi ft+(i-1)\psi\right] \qquad (8)$$

Figure2 (a) shows the time domain form of noise signal line spectrum. Figure2 (b) is the frequency domain form of this line spectrum component.

C. *Modeling and Realization of Continuous Spectrum*

The noise of continuous spectrum is mainly consisted of hydrodynamic noise, mechanical noise and propeller noise. The radiated noise of surface target is uncertain signal. It is mainly a broadband random signal, sometimes also includes the harmonic content. The uncertainty of waveform in time domain is most obvious and the amplitude is also greatly fluctuant. It is usually assumed that the target radiated noise is mainly steady, ergodic random process in order to be described by the statistical parameters.

The main frequency band of continuous spectrum is from a few Hz to thousands of Hz. It has a positive slope about 6dB/oct to 12 dB/oct at the low part and a negative slope about -6dB/oct at the high part. The flat or peak spectrum appears between tens to hundreds of hertz. The attenuation of Source level with distance is related to water depth and bottom nature, but the approximation for the far-field is generally about 6dB/oct.

Power spectrum that is one of the statistical parameters in frequency-domain can best express the frequency structure of a random signal. It is more suitable for the simulation of continuous spectrum component from the radiated target noise. Therefore, the paper started from the frequency characteristics of target radiated noise. The specific frequency response filter can be used for the simulation of broadband continuous component. The Gaussian white noise sequence is used to go through a low pass filter, and then the output is generated by the filter of specific frequency response which meets the requirements of noise spectrum characteristics. The simulation block diagram is shown in Figure 3。

The impulse response function of low pass filter which fulfills the condition above should have the form below:

$$H(t)=\sqrt{2L}\cdot u(t)\exp(-Lt) \qquad (9)$$

$u_1(t)$ is the unit step response function which is at t =0 and t =L. The frequency response of the function has a 6~8dB /oct decay rate. The simulation of the continuous spectrum components is shown in Figure4 (a) through the above design. Figure4 (b) shows the average of the spectrum characteristics of target noise based on Monte Carlo simulated method.

IV. THE BANDWIDTH ALGORITHM OF ONE-THIRD OCTAVE FILTER

A. *Theoretical Basis*

The frequency range of filter pass band width is $f_2' - f_1'$. It indicated that the signal which frequency range is $f_2 \sim f_1$ all can pass though. The Center frequency is $f_c=(f_2\times f_1)^{1/2}$。 The band segmentation and center frequency values of the filter are also provided. The general provisions are "n" octave band filter. N is defined by the following formula: $f_2 / f_1 = 2^n$。

When n=1, $f_2 / f_1 = 2$. This frequency range which is determined by the frequency ratio is called as 1/1 octave. When n=1/3,

$f_2 / f_1 = 2^{1/3}$, so called 1/3 octave. The spectrum in this band which is ($f_2 - f_1$), called 1/3 octave spectrum.

For the noise, because of 1/3 octave spectra can well reflect the energy distribution of the noise bandwidth, the analysis of 1/3 octave spectrum is particularly important. Commonly, the energy distribution of the noise from each frequency band provide a reference for the noise control according to analysis of 1/3 octave band spectrum. The reasonable measures should be taken for the purpose of noise reduction.

The computing method for each group of 1/3 octave filter is shown below:

$$F_u / f_l = 2^{1/3} = 1.2599 \qquad (10)$$

$$f_m=\sqrt{f_u\cdot f_l} \qquad (11)$$

$$B=f_u-f_l \qquad (12)$$

$$B=Kf_m \qquad (13)$$

f_u is up edge of frequency, f_l is down edge of frequency, f_m is center frequency, B is bandwidth of 1/3 octave，K is 0.231(constant).

B. *The Design of 1/3 Octave FIR Digital Filter*

The design task of 1/3 octave FIR digital filter is to determine center frequency f_m and filter attenuation coefficient in each sub-interval which meets the requirements of the filter frequency characteristic. The results for the total output signal

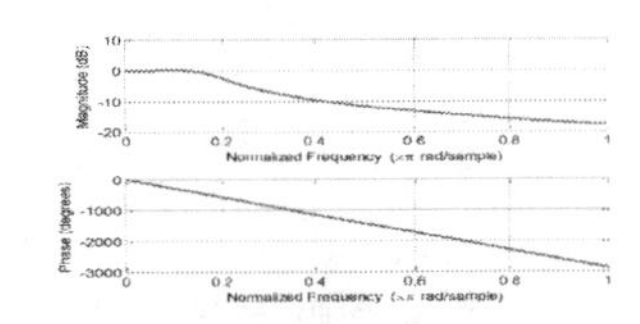

Fig.5 The character of continuous spectrum filter

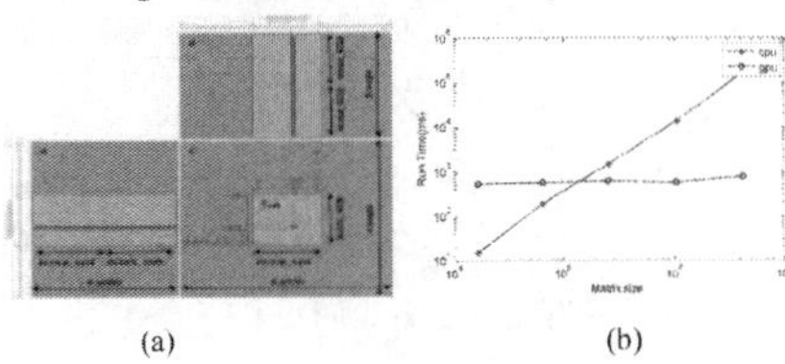

(a) (b)

Fig.6 Sub matrix method of GPU

can be verified after the achievement of 1/3 octave FIR digital filter design.

The simple approach which meets the requirements of the pass band center frequency of each sub-range is that their respective total range is divided into several sections. All the combinations are calculated according to the requirement of algorithms which use general design methods of band-pass filter. The frequency interval of the pass-band and stop-band has been divided finer; number of combinations which meet the requirements is more. Finally, output signal of the filter has been summed. If the validation meets the requirements, the design is over. Otherwise the test is done for other combinations until all of the output signals are satisfactory. Figure 5 shows the filter frequency response graph. The design absolutely fulfills the demands from the Amplitude-frequency characteristic curve of the filter. It is verified the realization of the project.

V. GPU Accelarated Platform Based On Parallel Matrix

The current graphics processors (GPU) provide good programmable features. It can obtain great potential and cost-effective parallel acceleration. It has the data description with high accuracy, efficient logic processing and computing performance. The NVIDIA CUDA parallel programming platform has been used in specific realization. Single Instruction Multiple Data, SIMD) CUDA parallel model has significant performance advantages in the vector or matrix operations based on the single instruction multiple data structures (SIMD).

CUDA programming model puts different parts of a program on the host (Host) or device-side (Device) to carry out. The device-side is as the coprocessor of the host. For example, GPU executes the core (kernel) function, CPU performs other part.

The achievement of parallel matrix processing algorithm on the GPU is using the concept of the sub-matrix, which is shown in Figure6 (a). It is assumed that operation is A*B=C, where the number of columns for A is A.width, the number of lines for B is B.height. According to the definition of matrix multiplication, these two values are equal. The block size of the sub-matrix is supposed BLOCK_SIZE × BLOCK_SIZE. It shows that the number of threads in a block is BLOCK_SIZE ×BLOCK_SIZE actually. Matrix A and B has been read for $\frac{B.width}{BLOCK\_SIZE}$ and $\frac{A.height}{BLOCK\_SIZE}$ times during the whole operation process in the global memory. If the concept of sub-matrix is not used, matrix A and B has been read for B.width and A.height times during the whole operation process. The result of testing time is shown in figure 6 (b).

The speed estimate of GPU and CPU operation: For example, Matrix Size=2048*2048, Block Size=16*16. The computation is 2048^3=8G multiplications, 8G additions. It is defined that one multiplication and one addition are one computing. The running time of GPU is 766ms and the speed is 10G/s around; the running time of CPU is 167347ms and the speed is 50M/s around. The computing speed of GPU is 200 times more than the CPU.

When BLOCK_SIZE or matrix Size is smaller, the computing speed of GPU is less 200 times than the CPU. That is because most of the time has been consumed when data is transmitting on a different memory operations. But it can be seen from the figure that GPU computing time is very stable in the process of increasing the size of the matrix operation. If the computing data is long enough, the advantages will become more obvious.

VI. Conclusion

Surface target radiated noise simulation has important uses, such as ship command system program demonstration, simulated experiment, training simulation and the test of sonar system in the laboratory. The design and research of target noise simulator has important practical and economic significance. In this paper, the location simulation of the moving radiated target has been added on the basis of noise mathematical modeling. It makes the simulator design more comprehensive and authentic. The design method based one-third octave broadband FIR filters is made. It makes the design resources of the simulator be streamlined and improved efficiency. The real-time achievement of the entire target noise simulator is protected more effectively by the accelerated GPU parallel processing. The simulator of radiated noise has a significant reduction of time-consuming in computing system and also has a very wide range of applications.

References

[1] Du Xuan-min, Jiang Ya-hao, "Research on simulation of radiating noise from vessels," Technical acoustics. 1999, vol. 18, No.1, pp:10-14

[2] Urick R J. Principles of Underwater Sound , New York: McGraw-Hili,1975.

[3] Li Qi-hu, The design principles of digital sonar, Anhui: Anhui Education Press, 2002.

[4] Li Qin, Yuan Bing-cheng, Ming Xing, "Simulation technique of radiated noise from underwater target and its implement of simulator", PEITS, 2009 2nd International Conference, pp: 357－360

[5] R. W. Fischer , and N. A. Brown, "Factors Affecting the Underwater Noise of Commercial Vessels Operating in Environmentally Sensitive Areas," IEEE Trans . Oceans. 2005, pp. 18-23

基于目标辐射噪声的信号起伏检测算法研究①②

解 恺* 丁雪洁 孙贵青 黄海宁 李启虎
(中国科学院声学研究所 北京 100190)

摘 要：该文以起伏声场背景下的目标辐射噪声模型分析为前提，针对目标方位估计和窄带线谱的检测问题，提出一种基于信号起伏相位差分对齐的宽窄带综合相干检测算法。利用周期线谱信号与宽带噪声间的时间相关半径与起伏相位均匀性差异，抑制背景噪声的能量干扰；将聚焦波束域输出信号的起伏相位对齐增益结合阵元域处理的空间相干增益，提高对不同来波方向目标的辨识能力。通过仿真分析与海试实验结果验证了该文所提算法可明显增强目标线谱分量的检测信噪比增益和相关目标所在波束方位的相对能量谱级，在抗复杂信道检测、识别领域具有良好的实际应用前景。

关键词：水声信号检测；辐射噪声；起伏相位对齐；线谱检测；方位估计

中图分类号：TB566 **文献标识码**：A **文章编号**：1009-5896(2013)04-0844-08

DOI: 10.3724/SP.J.1146.2012.01008

The Signal Fluctuating Detection Algorithm Based on the Target Radiated Noise

Xie Kai Ding Xue-jie Sun Gui-qing Huang Hai-ning Li Qi-hu
(*Institute of Acoustics, Chinese Academy of Sciences, Beijing* 100190, *China*)

Abstract: The analysis of the target radiated noise model is made as a precondition under the background of fluctuating sound field firstly. A broad and narrow band integrated coherent detection algorithm based on phase fluctuation and differential alignment is proposed for the issue of target azimuth estimation and narrow-band line-spectrum detection. The background noise energy disturbances can be restrained by using the differences of temporal correlated radius and phase fluctuation uniformity for the periodic spectrum signals and broadband noise; The ability to identify different DOA targets can also be improved by combining phase fluctuation aligned gain of the output signal in focused beam domain and spatial coherent gain of the signal processing in array domain. It is proved that the proposed algorithm can significantly enhance the detection SNR gain of the target line spectral component and relative energy spectrum level of the coherent target's beam azimuth by analyzing the simulation and sea experiment results. It has better application prospects in the detection and identification areas of anti-complex channel.

Key words: Underwater acoustic signal detection; Radiated noise; Fluctuating phase alignment; Frequency line detecting; Direction-of-arrival estimation

1 引言

水面目标在海水中的辐射噪声信号一直是实际水声信号处理领域研究的热点，被动式声探测设备可利用它对水面目标进行声检测、识别以及对不同目标的定向和跟踪等。被动声呐系统接收到的目标辐射噪声信号可以描述为目标舰船内部的机械运转产生的窄带数据结构嵌入在由不确定信道噪声、螺旋桨噪声及水动力噪声等众多声源综合作用形成的宽带噪声形式中[1]。

水声信号检测处理中最常采用的是常规能量检测方法，因为其在非相关噪声场中是单一目标的优良检测器。被动声呐系统一般通过宽带波束形成得到目标方位-时间历程结果(BTR)，又结合窄带LOFAR分析共同实现目标检测，因此联合宽带与窄带检测手段是很必要的。于是众多国内外学者进行了有效、有针对性的研究[2-5]。文献[2]以实测海杂波数据为基础，采用目标所处频段的局部Hilbert谱脊线计算平均带宽，将其作为检测统计量实现对微弱目标的检测。文献[5]则提出一种稳健的Capon自适应波束形成算法，通过求解波达角及测量校准误差，避免了对自适应波束形成器所需相应参数的

① 国家自然科学基金(10904160)和中国科学院知识创新工程重大方向项目资助课题.
② 电子与信息学报, 2013, 35(4): 844-851.

选择调整，抵制了目标消减作用。然而由于声源本身特性、多径干扰、波浪及海水温度微结构等原因，导致传播的声信号在幅度和相位上总是充满起伏的。对于这种目标辐射噪声和背景的起伏效应给目标检测及方位估计带来的影响，目前大部分检测方法则很少提及。

为了克服这种起伏干扰带来的目标估计统计特性的偏差和检测性能的下降，针对目标辐射噪声中窄带线谱的频率检测和辐射信号的方位估计问题，本文提出基于目标漂移相位旋转差分处理的时频联合起伏相干算法和波束域对齐相干处理算法。首先将目标辐射噪声信号仿真模型与 Kraken 简正波声场模型进行有机结合，使模拟信号融入不确定环境因素影响，可以更全面、真实地接近实际情况。然后利用目标信号与宽带噪声在起伏环境下的相位均匀性差异，联合二者在时间域、空间域的不同相干特性，结合仿真和海试数据分析结果，验证了本文所提算法相比传统能量检测方式，可显著提高对目标信号特征的检测增益及波达方位的估计能力。同时也证明了算法对起伏海洋信道的鲁棒性和干扰噪声的抑制作用。

2 辐射背景建模

2.1 不确定水声环境声场模型

众所周知，海洋是一个复杂多变的环境，海水的温度、盐度、密度、压力、海底底质、潮汐、海浪等物理量的变化将导致声信号传播的不确定性。针对这一特殊的水声信道影响因素，本文采用 Kraken 简正波模型来模拟目标辐射背景下的声场环境。

简正波理论认为声波在介质中的场可以分解为许多不同阶次的相互独立传播、互不影响的简正波。Kraken[6]方法是采用有限差分方法求解具有边界条件的简正波方程，在此我们假定海底为粗糙硬海底，即在深度 $z=D$ 的海底处，质点振动速度为 0。即意味着岩石层足够深，在达到底质层之前，声波已被有效地衰减掉。

在此边界条件下，简正波的解可写为

$$\boldsymbol{p}(r,z)=\frac{i}{\rho(z_s)\sqrt{8\pi r}}\exp\left(-\frac{i\pi}{4}\right)\cdot\sum_{m=1}^{\infty}\frac{\boldsymbol{u}_m(z_s)\boldsymbol{u}_m(z)}{\sqrt{\boldsymbol{k}_m}}\exp(i\boldsymbol{k}_m r) \tag{1}$$

Kraken 方法将整个深度区间 $0\le z\le D$ 等分成 N 个间隔，构成等间隔点 $z_j=jh(j=0,1,\cdots,N)$ 网格，其中 $h=D/N$ 为网格宽度。数目 N 应选择得足够大，保证对模式有充分的抽样。于是利用有限差分近似，可将式(1)连续问题化成标准的特征值问题：

$$\boldsymbol{A}\boldsymbol{u}_m=h^2\boldsymbol{k}_m^2\boldsymbol{u}_m \tag{2}$$

其中 $h^2\boldsymbol{k}_m^2$ 是矩阵 $\boldsymbol{A}$ 的第 m 个特征值，$\boldsymbol{u}_m$ 是 N 个相应于特征值的特征向量，近似等于模型函数，$\boldsymbol{A}$ 为 $N\times N$ 非对称三对角矩阵。

$$\boldsymbol{A}=\begin{bmatrix}\boldsymbol{a}_1 & 1 & 0 & & & & \\ 1 & \boldsymbol{a}_2 & 1 & 0 & & & \\ & & \cdots & & & & \\ & 0 & 1 & \boldsymbol{a}_n & 1 & 0 & \\ & & & \cdots & & & \\ & & & & 0 & 1 & \boldsymbol{a}_{N-1} & 1 \\ & & & & & & 2 & \boldsymbol{a}_N\end{bmatrix} \tag{3}$$

是三对角矩阵，$\boldsymbol{a}_i=-2+\dfrac{h^2\omega^2}{c^2(\boldsymbol{z}_i)}$，$i=1,\cdots,N$，$\boldsymbol{z}_i$ 是列向量，这些元素是网格点上特征函数的近似值。求解出式(2)的代数特征值，便可得到特征值和特征函数，代入声压表达式(1)，便可求出所要计算的海洋声场声压值。

2.2 目标辐射噪声信号模型

水面目标辐射噪声由众多噪声源综合作用而产生。单通道、随机单阵元的模拟水面目标辐射噪声数学模型为

$$\begin{aligned}Y_{\text{redi}}(nT_S)&=A_{\text{sig}}(nT_S)+N_{\text{con}}(nT_S)\\&\quad+[1+\mu\times A_{\text{sig}}(nT_S)]\times m(nT_S)\end{aligned} \tag{4}$$

其中 $Y_{\text{redi}}(nT_S)$ 为模拟产生的辐射噪声输出信号。$A_{\text{sig}}(nT_S)$ 是一系列以周期信号为模型的线谱分量的叠加。$m(nT_S)$ 为目标辐射噪声的调制分量，其中心频率在 1 kHz 左右。μ 为线谱精细结构调节参数，范围为 0~1 之间。$N_{\text{con}}(nT_S)$ 为连续谱分量对应的时域波形，用宽带平稳随机过程进行模拟。基于以上分析，分别建立线谱、连续谱分量的数学模型。

水面目标辐射噪声线谱是一些幅值明显高出相邻连续谱，并有稳定的频率成分的谱线，产生线谱的声源通常是：不平衡的旋转往复部件、螺旋桨叶片共振以及一些结构部件或空腔被激励谐振等[7]。式(5)为线谱分量的数学模型，f_L 表示任意线谱信号的中心频率，$\theta_L(t)$ 为任意信号的初始相位。φ_{in} 是 $[0,2\pi]$ 上的均匀随机量，表示该目标线谱分量声场作用下的相位起伏情况。a_L 表示线谱的幅度。

$$S(t)=\sum_{L=1}^{M}a_L\cos\left\{\left[2\pi\frac{f_L}{T}t+\theta_L(t)\right]+\varphi_{\text{in}}\right\} \tag{5}$$

连续谱噪声主要是水动力噪声、机械噪声和螺旋桨噪声组成的。连续谱的主要频段从几赫兹到数千赫兹，在几十至几百赫兹之间出现平直谱或峰值，

声源级随距离的衰减量与水深和底质有关，但远场一般近似为 6 dB/oct。

因为 1/3 倍频程谱能够很好地体现噪声带宽的能量分布情况。因此本文将从目标辐射噪声的频率特性出发，采用 1/3 倍频程带宽滤波器组进行目标辐射噪声宽带连续谱分量的模拟。该滤波器组的最终输出应具有与期望的舰船辐射噪声频谱形状相同的频率响应。满足上述条件的滤波器组的总输出脉冲响应函数应具有以下形式[8]：

$$H(t)=\sqrt{2L}\;u(t)\exp(-Lt) \tag{6}$$

$u(t)$ 是在 t=0 和 $t=L$ 的单位阶跃响应函数，该函数的频率响应大约具有 6~8 dB/oct 的衰减速率。1/3 倍频程带宽滤波思想体现在 $f_u/f_l=2^{1/3}$，每一子滤波器的通频带宽度为 $B=K\sqrt{f_u\cdot f_l}$，其中 f_u 为上边频，f_l 为下边频，K 为 0.231(常数)。

3　时频联合起伏相干算法

根据本文 2.1 节所述的不确定水声环境背景，声传播的波导效应和多径效应实际上不但在空间分布上不均匀，而且是随机时变的。因此，声信号在传输过程中也将是随机起伏的，针对 2.2 节中目标辐射噪声信号的线谱分量的检测，结合不确定水声信道，本文提出一种基于时间-频率检测域的联合起伏相干算法，此种新型信号处理技术利用线谱信号与宽带噪声之间的时间相关性差异与相位偏移性差异，可以显著改善线谱信号的信噪比增益。整个算法处理系统原理框图如图 1。

基于起伏相干的时频联合检测算法描述如下：

假设原始信号由宽带噪声和线谱信号组成，其表达式为

$$x(t)=A\cos(2\pi f_0t+\theta_{\text{in}})+n(t) \tag{7}$$

把原始信号延迟 Δ，与自适应滤波器的输出相减，调节滤波器，使误差信号 $\varepsilon(t)$ 极小。其中假设时域滤波输出信号为 $y(t)=B\cos(2\pi f_0t+\theta_{\text{out}})+n'(t)$，

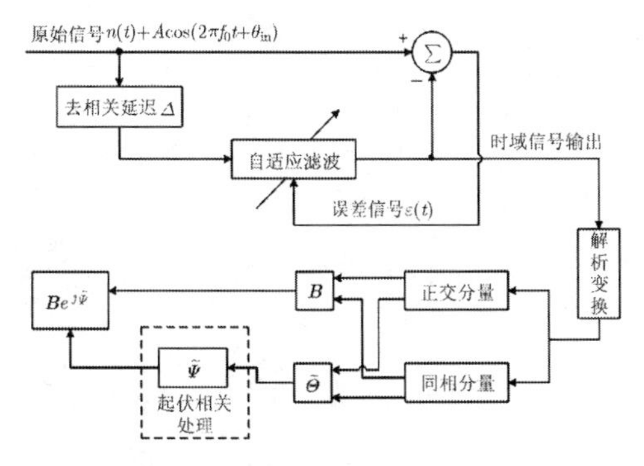

图 1 时频联合起伏相干算法原理框图

通过调节自适应滤波器，可以使时域信号输出 $y(t)$ 中 $n'(t)$ 均方差尽量小，而实际线谱分量 $B\cos(2\pi f_0t+\theta_{\text{out}})$ 则尽量接近原信号 $A\cos(2\pi f_0t+\theta_{\text{in}})$，于是 $y(t)$ 中的 SNR 信噪比达到极大。

进入复平面频域处理部分，将时域输出信号 $y(t)$ 进行正交解析变换，此时引入复平面信号 $\boldsymbol{z}(t)$，称为实信号 $y(t)$ 的解析表示。则有

$$\boldsymbol{z}(t)=y(t)+j\widetilde{H}[y(t)] \tag{8}$$

同时把 $\boldsymbol{z}(t)$ 的实部叫做 $y(t)$ 的同相分量，而把 $\boldsymbol{z}(t)$ 的虚部叫做 $y(t)$ 的正交分量。于是，以 $\boldsymbol{B}(t)$ 表示 $\boldsymbol{z}(t)$ 的瞬时包络，由式(9)给出：

$$\boldsymbol{B}(t)=\sqrt{\mathrm{Re}^2[\boldsymbol{z}(t)]+\mathrm{Im}^2[\boldsymbol{z}(t)]}=\sqrt{y^2(t)+H^2[y(t)]} \tag{9}$$

以 $\widetilde{\boldsymbol{\Theta}}(t)$ 表示 $\boldsymbol{z}(t)$ 的瞬时相位，由式(10)给出：

$$\widetilde{\boldsymbol{\Theta}}(t)=\arctan\left\{\frac{\mathrm{Im}[\boldsymbol{z}(t)]}{\mathrm{Re}[\boldsymbol{z}(t)]}\right\}=\arctan\left\{\frac{H[x(t)]}{x(t)}\right\} \tag{10}$$

综上得到复解析信号 $\boldsymbol{z}(t)$ 的极坐标表示形式：

$$\boldsymbol{z}(t)=\boldsymbol{B}(t)\cdot e^{j\widetilde{\boldsymbol{\Theta}}(t)}=\boldsymbol{B}(t)\cos\widetilde{\boldsymbol{\Theta}}(t)+j\boldsymbol{B}(t)\sin\widetilde{\boldsymbol{\Theta}}(t) \tag{11}$$

图 2，图 3 为根据我国南海海试实验数据得到的线谱与噪声信号的相位差分对齐结果。由图 2(b)可以看出，随着采样时刻的变化，复信号矢量 $\boldsymbol{z}(t)$ 的瞬时相位 $\widetilde{\boldsymbol{\Theta}}(t)$ 具有一种固定变化率的相位偏离现象，且该起伏现象可类比于在时间尺度上含有“固有斜率”的线性分布情况。根据这种特殊状况，我们引出一种基于复数坐标轴旋转变换的相位差分对齐处理方法。

设定偏移相位旋转角速率为 $\widetilde{\omega}(t)$，即表示瞬时相位 $\widetilde{\boldsymbol{\Theta}}(t)$ 在时间轴上的“线性斜率”。于是得到瞬时偏移相位角速度为[9]

$$\widetilde{\omega}(t_i)=\frac{\widetilde{\boldsymbol{\Theta}}(t_i)-\widetilde{\boldsymbol{\Theta}}(t_{i-1})}{t_i-t_{i-1}} \tag{12}$$

因为该“斜率”趋于稳定，于是得到

$$\widetilde{\omega}(t_i)=\widetilde{\omega}(t_{i-1})\approx\widetilde{\omega}(t)=\frac{\mathrm{d}\widetilde{\boldsymbol{\Theta}}(t)}{\mathrm{d}t} \tag{13}$$

定义采样时刻 t_i，瞬时偏移相位参量 $\boldsymbol{\Phi}(t_i)=\widetilde{\boldsymbol{\Theta}}(t_i)-\widetilde{\boldsymbol{\Theta}}(t_{i-1})$，如图 4 所示[10]。又因为采样时间间隔 $t_i-t_{i-1}=t_{i-1}-t_{i-2}=\Delta t=$ 常数，联立式(12)，式(13)可得偏移相位量 $\boldsymbol{\Phi}(t_i)$ 与 $\boldsymbol{\Phi}(t_{i-1})$ 相等。假设复信号矢量 $\boldsymbol{z}(t_i)$ 被不确定水声环境带来的起伏干扰量为 $\widetilde{\boldsymbol{\Psi}}(t_i)$，那么可得到

$$\begin{aligned}\boldsymbol{\Phi}(t_i)&=\left[\widetilde{\boldsymbol{\Theta}}(t_i)-\widetilde{\boldsymbol{\Psi}}(t_i)\right]-\widetilde{\boldsymbol{\Theta}}(t_{i-1})\\&=\widetilde{\boldsymbol{\Theta}}(t_{i-1})-\widetilde{\boldsymbol{\Theta}}(t_{i-2})=\boldsymbol{\Phi}(t_{i-1})\end{aligned} \tag{14}$$

所以起伏干扰因子 $\widetilde{\boldsymbol{\Psi}}(t_i)$ 可表示为

$$\widetilde{\boldsymbol{\Psi}}(t_i)=\widetilde{\boldsymbol{\Theta}}(t_i)-2\widetilde{\boldsymbol{\Theta}}(t_{i-1})+\widetilde{\boldsymbol{\Theta}}(t_{i-2}) \tag{15}$$

然后取 $\widetilde{\boldsymbol{\Psi}}(t_i)$ 替代原信号矢量 $\boldsymbol{z}(t_i)$ 的特征参数

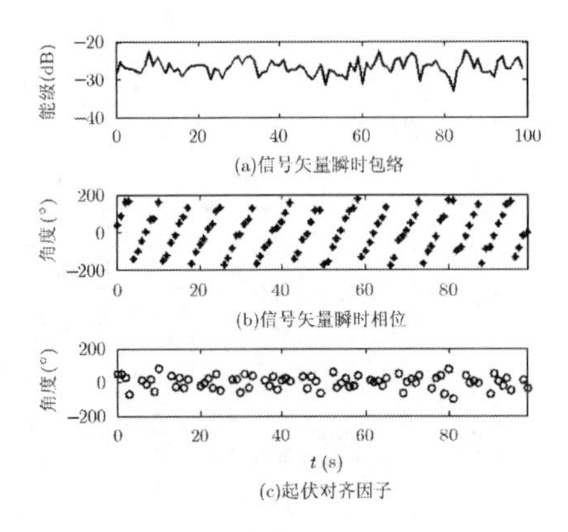

图 2 线谱信号相位起伏对齐处理

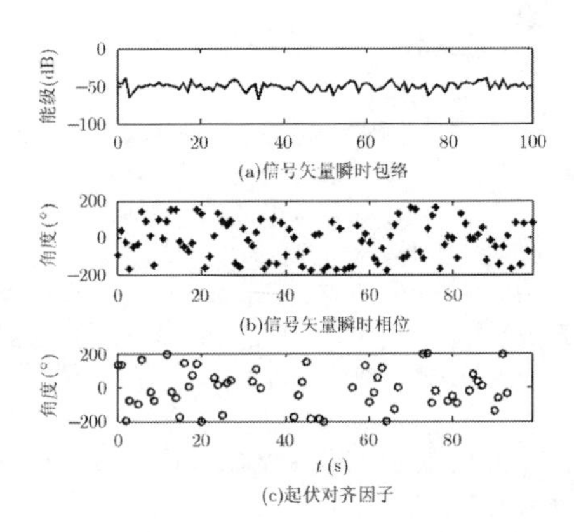

图 3 辐射噪声相位起伏对齐处理

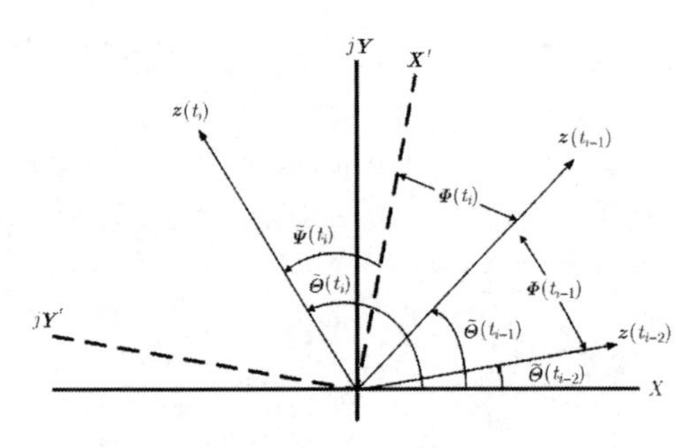

图 4 复平面极坐标轴旋转差分处理

$\widetilde{\boldsymbol{\Theta}}(t_i)$，得到新的信号矢量 $\boldsymbol{z}'(t_i)$，即相当于将基准复坐标轴 $\boldsymbol{X}-j\boldsymbol{Y}$ 旋转变换到 $\boldsymbol{X}'-j\boldsymbol{Y}'$ 轴。图 2(c)表明，由于线谱分量信号具有完好的相位相干特性，所以受到起伏干扰后的相位变化率基本固定，那么 $\widetilde{\boldsymbol{\Psi}}(t_i)$ 应趋于 0，于是可定义该参量为对齐补偿因子。而对于辐射噪声信号来说，补偿因子 $\widetilde{\boldsymbol{\Psi}}(t_i)$ 是随机分布的，如图 3(c)所示。通过累计相关处理，能量则相互抵消，而不是线谱信号能量在相关求和过程中的同相叠加，于是实现了系统增益。

旋转后的声压矢量集合 $\{\boldsymbol{z}'(t_i)\}$ 在 $\boldsymbol{X}'-j\boldsymbol{Y}'$ 的正半平面上得到对齐处理。该矢量 $\boldsymbol{z}'(t_i)$ 如下所示：

$$\boldsymbol{z}'(t_i)=\boldsymbol{B}(t_i)\cos[\widetilde{\boldsymbol{\Psi}}(t_i)]+j\boldsymbol{B}(t_i)\sin[\widetilde{\boldsymbol{\Psi}}(t_i)]=\boldsymbol{B}(t_i)e^{j\widetilde{\boldsymbol{\Psi}}(t_i)} \tag{16}$$

以功率对比为例，因为幅度集 $\{\boldsymbol{B}(t_i)\}$ 对于 $\{\boldsymbol{z}(t_i)\}$ 和 $\{\boldsymbol{z}'(t_i)\}$ 来说是相等的，故矢量集 $\{\boldsymbol{z}(t_i)\}$ 的能量检测 AVG{ $z'(t)$ }可表示为

$$\mathrm{AVG}\{\boldsymbol{z}'(t)\}=\sum_i^N \boldsymbol{B}^2(t_i)\Big/N \tag{17}$$

其中 N 表示整个采样期间所经历的时间片段的总和，i 表示其中的一段采样片段。而基于相位旋转的差分对齐(Rotational Differential Average, RDA) $\mathrm{RDA}\{\boldsymbol{z}'(\boldsymbol{B},\widetilde{\boldsymbol{\Psi}})\}$ 的平均功率由信号矢量 $\boldsymbol{z}'(t_i)$ 得来。

$$\begin{aligned}\mathrm{RDA}\{\boldsymbol{z}'(\boldsymbol{B},\widetilde{\boldsymbol{\Psi}})\}&=\left\{\mathrm{Re}\left[\frac{\sum_i^N \boldsymbol{z}'(t_i)}{N}\right]\right\}^2+\left\{\mathrm{Im}\left[\frac{\sum_i^N \boldsymbol{z}'(t_i)}{N}\right]\right\}^2\\&=\left\{\frac{\sum_i^N \boldsymbol{B}(t_i)\cos[\widetilde{\boldsymbol{\Psi}}(t_i)]}{N}\right\}^2\\&\quad+\left\{\frac{\sum_i^N \boldsymbol{B}(t_i)\sin[\widetilde{\boldsymbol{\Psi}}(t_i)]}{N}\right\}^2\end{aligned} \tag{18}$$

“Re”和“Im”是旋转后矢量 $\boldsymbol{z}'(t_i)$ 的实部和虚部。$\widetilde{\boldsymbol{\Psi}}(t_i)$ 由式(12)~式(15)得出的对齐补偿因子，$\boldsymbol{z}'(t_i)$ 由式(16)表示。

4 波束域对齐相干处理算法

对未知目标方位分布的探知一直是水声信号处理的重要组成部分。根据 2.2 节分析所得，对包含窄带线谱分量和宽带连续谱分量的目标辐射噪声信号方位的检测，结合不确定环境背景下的相位漂移现象，本文提出一种基于起伏相位对齐相干的宽带频域波束形成方法。该检测技术利用接收信号波束域的相位均匀性差异和阵元域不同基元间的空间相干性差异，可以增加空域阵列的输出增益，有效提高不同入射方向目标的分辨能力和波达方位的估计能力。整个算法流程框图如图 5 所示：

基于起伏相位旋转差分的波束域对齐相干算法描述如下：

假设一个基阵由 M 个基元构成，入射信号是平

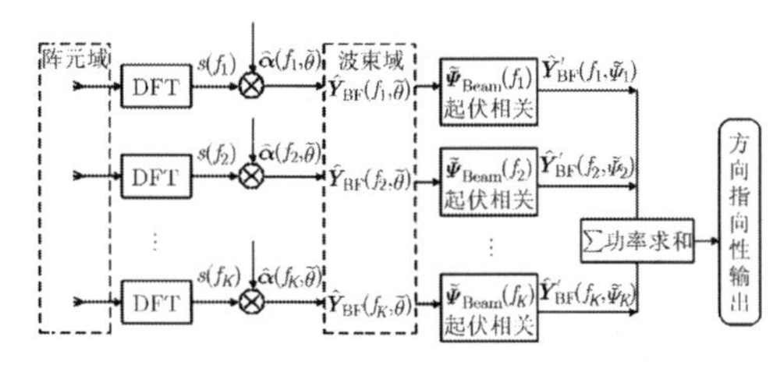

图 5　波束域对齐相干算法流程图

面波。以平面上的某一点为参考点，设到达信号为 $s(t)$，θ 为该信号入射角。针对水面目标辐射噪声信号的宽带频谱特性，其可以被看成若干相邻频率窄带信号之和。假定宽带信号 $s(t)$ 是由 K 个窄带信号叠加而成，即

$$s(t)=\sum_{i=1}^{K}s_{f_i}(t) \tag{19}$$

其中 $s_{f_i}(t)$ 是中心频率为 f_i 的窄带信号。于是对于单个宽带信号入射时的阵列输出模型为

$$\boldsymbol{y}_{\mathrm{BF}}(t)=\sum_{i=1}^{K}\left[\widehat{\boldsymbol{\alpha}}(f_i,\theta)s_{f_i}(t)+n_{f_i}(t)\right] \tag{20}$$

$\widehat{\boldsymbol{\alpha}}(f_i,\theta)$ 称为阵列对来波信号(指定频点 f_i)的方向导引矢量。由于信号到达每个阵元的时刻有差异，同一平面波在各阵元输出端的响应有不同的延迟时间。那么由 M 个无方向性阵元组成的接收阵列，每个阵元上得到的远场信号可表示为 $r_1(t),r_2(t),\cdots,r_M(t)$，则单一频点，该阵列的波束输出信号为

$$\boldsymbol{y}_{\mathrm{BF}}(f_0,t)=\sum_{m=1}^{M}\left[e^{-j2\pi f_0\tau_m(\theta)}r_m(f_0,t)+n_m(f_0,t)\right] \tag{21}$$

式中 $n_m(f_0,t)$ 是阵元 m 上的加性噪声，$\tau_m(\theta)$ 是来波信号 $r_m(f_0,t)$ 到达阵元 m 的相对时延。定义波束扫描导引相角范围 $\theta_1,\theta_2,\cdots,\theta_D$，将式(21)变换到频域形式，得到单一频点、指定扫描相角的阵列频域波束响应为

$$\begin{aligned}\widehat{\boldsymbol{Y}}(f_0,\theta_i)&=\sum_{m=1}^{M}e^{-j2\pi f_0\tau_m(\theta_i)}R_m(f_0)+N_m(f_0)\\&=\sum_{m=1}^{M}\widehat{\boldsymbol{\alpha}}_m(f_0,\theta_i)R_m(f_0)+N_m(f_0),\\&\quad i=1,2,\cdots,D\end{aligned} \tag{22}$$

将指定扫描相角 θ_i 扩展到全部导引矢量，则单一频点的波束域输出信号向量为

$$\widehat{\boldsymbol{Y}}(f_j,\tilde{\theta})=\left\{\widehat{\boldsymbol{Y}}(f_j,\theta_1),\widehat{\boldsymbol{Y}}(f_j,\theta_2),\cdots,\widehat{\boldsymbol{Y}}(f_j,\theta_D)\right\},\quad j=1,2,\cdots,K \tag{23}$$

式(23)表示将原 M 维阵元域信号映射变换到 D 维波束域信号，完成了对接收阵列数据 $s(f,t)$ 从阵元域到波束域的重构。根据软件无线电思想，可得某采样时刻 t_i，单频点波束域复信号矢量 $\widehat{\boldsymbol{Y}}_{\mathrm{BF}}(f_0,t_i)$ 的瞬时幅度包络为

$$\boldsymbol{A}_{\mathrm{Beam}}(f_0,t_i)=\sqrt{\mathrm{Re}^2[\widehat{\boldsymbol{Y}}_{\mathrm{BF}}(f_0,t_i)]+\mathrm{Im}^2[\widehat{\boldsymbol{Y}}_{\mathrm{BF}}(f_0,t_i)]} \tag{24}$$

同理，$\widehat{\boldsymbol{Y}}_{\mathrm{BF}}(f_0,t_i)$ 的瞬时波束漂移相位可表示为

$$\widetilde{\boldsymbol{\Theta}}_{\mathrm{Beam}}(f_0,t_i)=\arctan\left\{\frac{\mathrm{Im}[\widehat{\boldsymbol{Y}}_{\mathrm{BF}}(f_0,t_i)]}{\mathrm{Re}[\widehat{\boldsymbol{Y}}_{\mathrm{BF}}(f_0,t_i)]}\right\} \tag{25}$$

由此，单频点波束输出信号的极坐标表示形式为

$$\begin{aligned}\widehat{\boldsymbol{Y}}_{\mathrm{BF}}(f_0,t_i)&=\boldsymbol{A}_{\mathrm{Beam}}(f_0,t_i)\cdot e^{j\widetilde{\boldsymbol{\Theta}}_{\mathrm{Beam}}(f_0,t_i)}\\&=\boldsymbol{A}_{\mathrm{Beam}}(f_0,t_i)\cos\widetilde{\boldsymbol{\Theta}}(f_0,t_i)\\&\quad+j\boldsymbol{A}_{\mathrm{Beam}}(f_0,t_i)\sin\widetilde{\boldsymbol{\Theta}}(f_0,t_i)\end{aligned} \tag{26}$$

于是在复数极坐标平面，应用坐标旋转变换的差分对齐处理思想(在此不再赘述)，将阵列输出信号矢量的瞬时起伏差异进行对齐补偿，同样利用信号与噪声在波束域的相位相干性差异来提高空域阵列的处理增益。此时，根据第 3 节中式(12)~式(15)所示，由已求得的波束漂移相位 $\widetilde{\boldsymbol{\Theta}}_{\mathrm{Beam}}(f_0,t_i)$ 即可得到波束域对齐补偿因子为

$$\begin{aligned}\widetilde{\boldsymbol{\Psi}}_{\mathrm{Beam}}(f_0,t_i)=&\widetilde{\boldsymbol{\Theta}}_{\mathrm{Beam}}(f_0,t_i)-2\widetilde{\boldsymbol{\Theta}}_{\mathrm{Beam}}(f_0,t_{i-1})\\&+\widetilde{\boldsymbol{\Theta}}_{\mathrm{Beam}}(f_0,t_{i-2})\end{aligned} \tag{27}$$

则对齐补偿后的波束信号 $\widehat{\boldsymbol{Y}}'_{\mathrm{BF}}(f_0,t_i)$ 可表示为

$$\begin{aligned}\widehat{\boldsymbol{Y}}'_{\mathrm{BF}}(f_0,t_i)&=\boldsymbol{A}_{\mathrm{Beam}}(f_0,t_i)\cos\left[\widetilde{\boldsymbol{\Psi}}_{\mathrm{Beam}}(f_0,t_i)\right]\\&\quad+j\boldsymbol{A}_{\mathrm{Beam}}(f_0,t_i)\sin\left[\widetilde{\boldsymbol{\Psi}}_{\mathrm{Beam}}(f_0,t_i)\right]\\&=\boldsymbol{A}_{\mathrm{Beam}}(f_0,t_i)\cdot e^{j\widetilde{\boldsymbol{\Psi}}_{\mathrm{Beam}}(f_0,t_i)}\end{aligned} \tag{28}$$

综上所述，仍然以功率对比为例，M 路阵元信号经空域滤波后得到的方向指向性能量检测输出功率为

$$\mathrm{AVG\_BF}\{\boldsymbol{y}_{\mathrm{AP}}(f,t)\}=\sum_{j}^{K}\left[\frac{\sum_{i}^{N}\boldsymbol{A}_{\mathrm{Beam}}^2(f_j,t_i)}{N}\right] \tag{29}$$

其中 N 表示整个采样期间所经历的时间片段的总和，K 表示按频点划分所叠加的窄带信号的个数，如式(19)。而波束域对齐相干处理算法的输出平均功率为

$$\begin{aligned}\mathrm{RDA\_BF}\left\{\boldsymbol{Y}'_{\mathrm{BF}}(f,t)\right\}=&\sum_{j}^{K}\left\{\mathrm{Re}\left[\frac{\sum_{i}^{N}\boldsymbol{Y}'_{\mathrm{BF}}(f_j,t_i)}{N}\right]\right\}^2\\&+\sum_{j}^{K}\left\{\mathrm{Im}\left[\frac{\sum_{i}^{N}\boldsymbol{Y}'_{\mathrm{BF}}(f_j,t_i)}{N}\right]\right\}^2\end{aligned} \tag{30}$$

5 仿真及试验结果分析

5.1 仿真实验讨论

依据 2.1 节所述的不确定水声环境，采用 Kraken 简正波模型建立仿真系统。模拟水深 100 m，信号传播距离范围 0 到 30 km，信源与接收基阵布放深度均为 20 m。设定扰动声速剖面由不同深度处的瞬时声速传播数值构成，其大体符合均值为 1500 m/s，方差 1~2 m/s 的随机高斯分布。图 6 表示应用该声场模型，模拟目标辐射噪声在接收距离 20 km 处，不同采样时刻下的相位起伏情况。图 7 所示是以 2.2 节描述的理论模型为基础，结合不确定声场环境产生的目标辐射噪声功率频谱图，基本反映了目标辐射信号在水声起伏信道中的传播特性。

第 3 节所述时频联合起伏相干算法的仿真性能结果，如图 8。其中模拟目标线谱分量的中心频率为 100 Hz，初始相位100°，接收信噪比条件为-10 dB，时域自适应滤波器阶数 128 阶，步长设定 0.001。本文所提 RDA(时频起伏相干算法)比起传统的 AVG(频谱能量检测法)提高了 15 dB 左右的检测信噪比增益，对宽带噪声的功率抑制性较为明显。图 9 表示根据复数坐标轴下相位旋转差分思想得到的对齐补偿比较结果。对齐处理后的相位补偿因子如图 9(b)，“替代”相位被“聚敛”到 0 相位轴附近且起伏程度也得到了缓解。而噪声干扰的对齐补偿因子往往在整个相位域内随机分布，于是线谱信号的能量分布在频域得到了同相叠加，提高了系统处理增益。

图 10 和图 11 对比说明了第 4 节所述波束域对齐相干处理算法的方位估计性能。模拟的宽带连续谱分量的频率范围 0 到 750 Hz，波束扫描角度 0°到 180°，仿真的目标信号入射方向为60°。经过波束输出能量归一化处理后可看出两种目标方位估计方法的噪声背景能量分布有很大差距，图 11 中的背景颜色较深，二者大约相差 13 dB 左右。而进一步选取频段 300~500 Hz 范围的宽带信号进行累计相关性能对比，如图 12。通过定量分析证明了波束域 RDA 方位估计算法可获得更高的波束检测信噪比增益。

5.2 海试试验结果及分析

为验证本文所提算法在真实水声信道中的检测性能，于是本节结合我国三亚南海某海域的海试数据处理结果进一步加以说明。此次试验海域水深 120 m 左右，试验母船沿水平方向以 6 kn 航速拖曳 80 元长线阵接收信号，阵端向方位设为 0°，接收阵入水深度 50 m。试验过程中有多艘过往船只在该区域经过，其中目标实验船相对于长线阵大约在 80°至 90°方向范围运动，并且在 20°到 40°方位有大型水面船只通过。

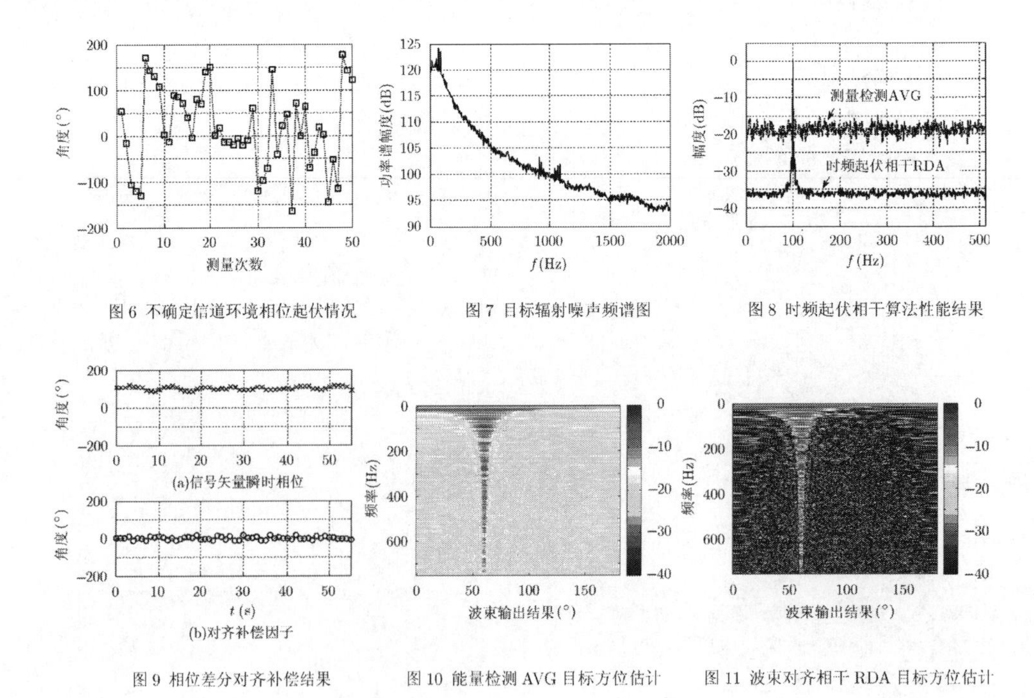

图 6 不确定信道环境相位起伏情况

图 7 目标辐射噪声频谱图

图 8 时频起伏相干算法性能结果

图 9 相位差分对齐补偿结果

图 10 能量检测 AVG 目标方位估计

图 11 波束对齐相干 RDA 目标方位估计

图 13 为在连续采样时间历程下，应用本文第 3 节所提时频联合起伏相干算法处理所得的海试数据频谱平均分析结果。从中可明显看出，时频 RDA 检测方法对于多个目标线谱分量频率特性的提取，在频谱信噪比增益上均优于传统能量检测 AVG 方法，其中目标实验船线谱信号(归一化中心频率 0.4)输出谱级提高了大约 9 dB。该实验目标的相角对齐结果如图 14 所示，旋转差分所需的起伏处理周期为 20 s，实验结果与理论分析效果基本吻合，在此不再赘述。图 15 和图 16 表示在全部累积时间历程下，应用两种检测方法的海试数据 LOFAR 对比结果。此种对比方式是图 13 所示性能结果的时域扩展，能更好的判别目标线谱分量的时间累积效应。在统一最大能量归一化前提下，本文所提时频联合起伏相干算法的干扰背景相对于能量检测方法被较大幅度压低，即证明了该算法具有提高信号检测性能上的优势。

图 17 和图 18 所示为应用本文第 4 节所提波束域对齐相干处理算法所得的目标海试数据方位历程对比结果。在图 17 中，由于受到实验过程中过往船只的影响，应用常规能量检测 AVG 估计所得的目标所在方位运动轨迹(80°至 90°)相对于其他船只并不明显，其波束输出能级比起近场大功率干扰船只甚至更低，对目标方位的判断很不明确。而从图 18 应用波束域对齐 RDA 估计法得到的时间累积方位历程图可看出，目标所在方位波束输出能级较远场其他干扰船只的检测信噪比明显增加，尤其降低了近场干扰的相对能量谱级。被压低的背景噪声相对强度也可验证该算法对宽带干扰噪声的抑制性能。

为更直观说明这种方位估计差异，本文特选取时间历程 10 s 处的二者方位输出能级结果进行比较，如图 19 所示。比起能量检测 AVG 方位估计法，波束域对齐相干 RDA 算法在目标方位 83°的波束输出信噪比最大且增益提高了 8 dB 左右，而其他干扰船只的方位输出相对强度均得到有效抑制，足以说明此算法可以明显提高目标方位检测的有效性和准确性。图 20 表示归一化中心频率 0.4 的被测实验目标，“对准”来波方向 90°的波束域信号对齐补偿结果。正是这种对波束输出信号频域“线性”相位的对齐处理，实现了目标来波方向能量的同相叠加，再结合空域滤波带来的阵列相干增益，很大程度提高了对所测目标方位的辨识能力。

6 结束语

本文以 Kraken 简正波不确定声场背景下的目标辐射噪声数理模型为仿真条件，提出一种基于信号起伏相位差分对齐的宽窄带综合相干检测算法。其中时频联合起伏相干算法利用周期性线谱信号与

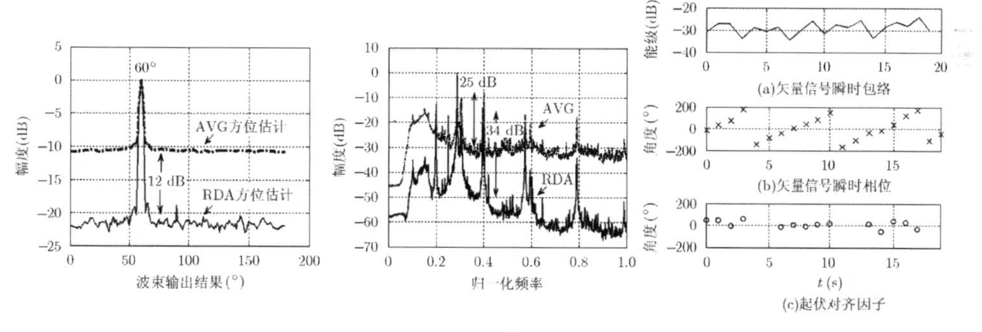

图 12 来波信号方向波束对齐性能结果

图 13 时频起伏相干算法海试结果分析

图 14 目标实验船线谱分量对齐补偿结果

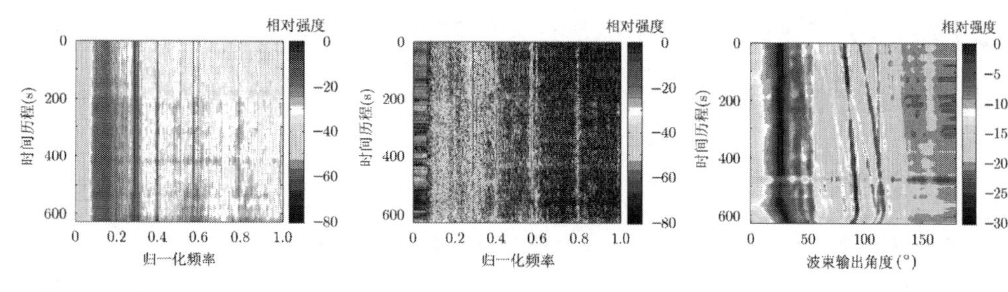

图 15 目标信号 AVG 检测 LOFAR 结果

图 16 目标信号 RDA 检测 LOFAR 结果

图 17 目标 AVG 方位估计历程图

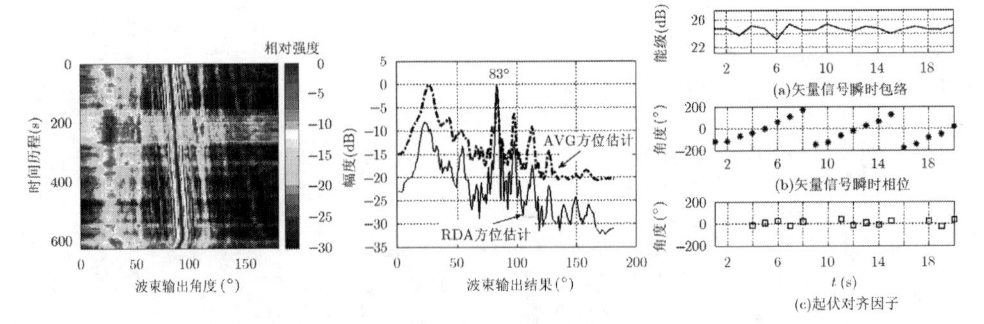

图 18 目标 RDA 方位估计历程图　　图 19 波束输出能量对比结果　　图 20 聚焦目标来波方向的对齐补偿结果

非周期性噪声之间的时间相关性差异与起伏相位偏移性差异，抑制背景噪声的能量干扰，提高了检测线谱输出的信噪比增益；而波束域对齐相干处理算法将阵元域接收到的宽带信号变换到波束域的方位输出形式，将不同阵元间的空间相干增益与波束信号起伏相位的对齐处理增益有机结合，增强了对不同来波方向目标的辨识能力。通过仿真分析及我国南海海试数据的处理验证，该算法对于目标线谱信号的检测增益比起标准的能量检测方法可提高 9 dB 左右的性能，且干扰背景的相对能量也被大幅降低。另外该算法可明显增强相关目标所在波束方位的相对能量谱级，而抵消非相关干扰的波束输出能量，改善了目标来波方位估计的精准性。由此说明本文所提算法在声呐系统检测与识别等领域具有积极的现实意义。

参 考 文 献

[1] Filho W S, de Seixas J M, and de Moura N N. Preprocessing passive sonar signals for neural classification[J]. *IET Radar, Sonar & Navigation*, 2011, 5(6): 605-612.

[2] 张建, 关键, 董云龙, 等. 基于局部 Hilbert 谱平均带宽的微弱目标检测算法[J]. 电子与信息学报, 2012, 34(1): 121-127.
Zhang Jian, Guan Jian, Dong Yun-long, *et al*.. Weak target detection based on the average bandwidth of the partial hilbert spectrum[J]. *Journal of Electronics & Information Technology*, 2012, 34(1): 121-127.

[3] Tucker J D and Azimi-Sadjadi M R. Coherence-based underwater target detection from multiple disparate sonar platforms[J]. *IEEE Journal of Oceanic Engineering*, 2011, 36(1): 37-51.

[4] 王静, 黄建国. 水下小孔径阵列自适应匹配滤波检测方法[J]. 电子与信息学报, 2011, 33(6): 1385-1389.
Wang Jing and Huang Jian-guo. Adaptive matched filter detection method on underwater small aperture array[J]. *Journal of Electronics & Information Technology*, 2011, 33(6): 1385-1389.

[5] Somasundaram S D and Parsons N H. Evaluation of robust capon beamforming for passive sonar[J]. *IEEE Journal of Oceanic Engineering*, 2011, 36(4): 686-695.

[6] Jensen F B, Kuperman W A, and Porter M B. Computational Ocean Acoustics[M]. New York: American Institute of Physics, 1994, Chapter 5.

[7] 李启虎. 数字式声呐设计原理[M]. 合肥: 安徽教育出版社, 2002: 141-147.
Li Qi-hu. The Design Principles of Digital Sonar[M]. Hefei: Anhui Education Press, 2002: 141-147.

[8] Xie Kai, Li Qi-hu, Ding Xue-jie, *et al*.. The simulator design of GPU accelerated radiated noise from surface target[C]. 2nd IEEE International Conference on Consumer Electronics, Communications and Networks, Yichang, China, 2012: 720-723.

[9] Venugopal S, Wagstaff R A, and Sharma J P. Exploiting phase fluctuations to improve machine performance monitoring[J]. *IEEE Transactions on Automation Science and Engineering*, 2007, 4(2): 153-166.

[10] Wagstaff R A. Exploiting phase fluctuations to improve temporal coherence[J]. *IEEE Journal of Oceanic Engineering*, 2004, 29(2): 498-510.

THE STUDY OF PASSIVE RANGING TECHNOLOGY BASED ON THREE ELEMENTS VECTOR ARRAY①

Haibo Zheng[a], Yuan Li[a],Bing Li[a], Zhibo Zhang[a],Xizhong Bao[a],Qihu Li[a]

[a]Institute of Acoustics, Chinese Academy of Sciences,21 North 4th Ring Road, Beijing, China, 100190

Haibo Zheng, Institute of Acoustics, Chinese Academy of Sciences,21 North 4th Ring Road, Beijing, China, 100190,86-010-82547960, zhenghaibo@mail.ioa.ac.cn

ABSTRAC: *Traditional passive ranging technology based on three elements array is well known. Changes in the curvature of a spherical wave front leads to relative time delay of each primitive. By measuring the relative time delay of each base element, the target range and azimuth are estimated. Delay estimation accuracy, target distance, orientation, aperture of array, array installation accuracy and other factors have an impact on the ranging accuracy. The most critical factor is the delay estimation accuracy. Published research literature on ternary array of passive positioning, mainly focused on the sound pressure information. It is well known that sound wave has both scalar quantity and vector field, while traditional acoustic pressure sensor system merely makes use of its acoustic pressure information. Vector hydrophone, also called combined sensor, is combined by traditional and omni-directional pressure hydrophone and natural dipole independent on frequeney, which can co-locating and simultaneously measures pressure(scalar field)and particle velocity(vector field)of acoustic field. This paper presents the passive ranging technology based on three elements vector array, representing Traditional passive ranging technology based on three elements array, more fully applied sound field information to improve the accuracy of delay estimation, and ultimately improve the passive ranging accuracy.*

Key Words: *Three elements array, Passive Location, Vector sensor*

---

① UA2014-2nd International Conference and Exhibition on Underwater Acoustics, 2014: 1595-1601.

1. INTRODUCTION

There are three main types of currently developed passive ranging sonar,as follows:passive ranging technology based on three elements array,target motion analysis (TMA) and matched field processing (MFP).Passive ranging technology based on three elements array is the object of this thesis, which mainly use the change of wavefront curvature based on spherical wave or cylindrical wave to complete the measurement.By measuring the relative delay of each element, the target distance and direction are estimated.

Delay estimation accuracy, target distance, orientation, aperture arrays, arrays installation accuracy and other factors make an appreciable difference to Ranging accuracy.Delay measurement accuracy is the most critical factor.With decreasing distance, changing the curvature of wave front is more and more big, the measurement accuracy of time delay，that is the ranging precision, is also more and more high.This is easy to achieve the high-precision tracking of the rapid short-range goal. It has important implications for improving melee combat capability.

It is well known that sound wave has both scalar quantity and vector field,while traditional acoustic pressure sensor system merely makes use of its acoustic pressure information. Vector hydrophone, also called combined sensor, is combined by traditional and omni-directional pressure hydrophone and natural dipole independent on frequency, which can co-locating and simultaneously measures pressure(scalar field) and particle velocity(vector field) of acoustic field. No doubt, the more the information, the better the signal processing effect will be.

Passive ranging technology based on three elements vector array, compared to passive ranging technology based on three elements scalar array, more fully applied sound field information to improve the accuracy of delay estimation, and ultimately improve the passive ranging accuracy.

2. THE PRINCIPLE OF PASSIVE RANGING TECHNOLOGY BASED ON THREE ELEMENTS SCALAR ARRAY[1]

Assume that the target is a point source, waves rippled outward by spherical wave propagation.The model of passive ranging technology based on three elements array as shown Fig.1.

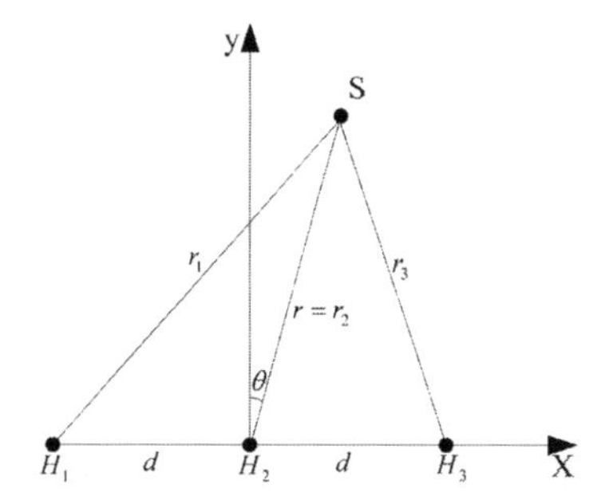

Fig.1:The model of passive ranging technology based on three elements array

In Figure 1, S is defined as the sound source. H_1, H_2 and H_3 denote the 3 elements of passive sonar array.Inter-element spacing: $\overline{H_1H_2} = \overline{H_2H_3} = d$.The angle between the target and the y axis is θ.The distance of target to each array element: $\overline{SH_1} = r_1$, $\overline{SH_2} = r_2 = r$, $\overline{SH_3} = r_3$.Among them, r_2 is to measure the target distance(r).In general, r is greater than d.Therefore, only a slight difference between r_1, r_2, and r_3 by the actual calculation results.The tiny difference is that we measure objects.

Just calculate the delay difference(τ_{12}) of t_1(the time required for signal transmission to H_1) and t_2(the time required for signal transmission to H_2),and the delay difference(τ_{23}) of t_2(the time required for signal transmission to H_2) and t_3(the time required for signal transmission to H_3), $r - r_1$ and $r_3 - r$ can be calculated.From a purely geometrical point of view, that of d, $r - r_1$ and $r_3 - r$,r can be solved out.

Take H_2 as the origin of coordinates to establish the coordinate system is shown in Figure 1. Distance from the target to the array elements:

$$\begin{aligned} r_1 &= \sqrt{r^2 + d^2 + 2rd\sin\theta} \\ r_2 &= r \\ r_3 &= \sqrt{r^2 + d^2 - 2rd\sin\theta} \end{aligned} \tag{1}$$

Assuming the speed of sound is c, delay difference is respectively expressed ,as follows:

$$\tau_{12}=\tau_1-\tau_2=\frac{1}{c}(r_1-r_2)$$
$$\tau_{23}=\tau_2-\tau_3=\frac{1}{c}(r_2-r_3) \tag{2}$$
$$\tau_{13}=\tau_{12}+\tau_{23}$$

Where τ_{12} denotes the delay difference between the array element 1 and the array element 2, τ_{23} denotes the delay difference between the array element 2 and the array element 3.

$$c\tau_{12}=\sqrt{r^2+d^2+2rd\sin\theta}-r \tag{3}$$

$$c\tau_{23}=r-\sqrt{r^2+d^2-2rd\sin\theta} \tag{4}$$

Equation (3) and (4) respectively respectivelyweremoved, and then squaring on both sides of the equation.

$$2r[c\tau_{12}-d\sin\theta]=d^2-c^2\tau_{12}^2 \tag{5}$$

$$2r[d\sin\theta-c\tau_{23}]=d^2-c^2\tau_{23}^2 \tag{6}$$

Dividing formula (5) and (6),the target bearing can be estimated accurately.

$$\theta=\sin^{-1}\left[\frac{cd^2\tau_{13}-c^3\tau_{12}\tau_{23}\tau_{13}}{2d^3-c^2d\left(\tau_{23}^2+\tau_{12}^2\right)}\right] \tag{7}$$

Adding formula (5) and (6),the target distance can be estimated accurately.

$$r=\frac{d^2}{c(\tau_{12}-\tau_{23})}-\frac{c\left(\tau_{12}^2+\tau_{23}^2\right)}{2(\tau_{12}-\tau_{23})} \tag{8}$$

3. THE PRINCIPLE OF PASSIVE RANGING TECHNOLOGY BASED ON THREE ELEMENTS VECTOR ARRAY

In this paper, the program uses a two-dimensional vector sensors,which can co-locating and simultaneously measures pressure(p) and particle velocity(v_x 、 v_y) of acoustic field. Particle velocity has dipole directivity[2-3].

$$\begin{cases} p(t)=x(t) \\ v_x(t)=\cos\theta_s x(t) \\ v_y(t)=\sin\theta_s x(t) \end{cases} \tag{9}$$

In addition, the velocity component weighted combination can make the directional rotation in two-dimensional space.

$$v_c(t)=v_x(t)\cos\psi+v_y(t)\sin\psi=x(t)\cos(\theta_s-\psi) \tag{10}$$

If $\psi=\theta_s, v_c$ has a maximum value.By changing the values ψ, can scan the full range of directivity, which is realized in the horizontal plane directivity of electron spin.

In this paper, the sound pressure and particle velocity appropriate combination to improve the signal to noise ratio of array element domain, the combination of the following form:

$$0.5p(t)+v_c(t)=x(t)\left[0.5+\cos(\theta_s-\psi)\right] \tag{11}$$

Passive Ranging Technology Based on Three elements Vector Array implementation steps:

Step 1: measure delay difference from each other by sound pressure signal of three elements scalar array;

Step 2: using sound pressure signal delay difference from step 1, according to equation (7), measure target position ψ_0;

Step 3: from step 2, the calculated position ψ_0, according to the formula (10) and (11), acoustic pressure, velocity signals are combined for each element of three elements vector array, obtain combined signals;

Step 4: from step 3,processing these combined signals,measure delay difference from each other;

Step 5: using delay difference from step 4,according to the formula (7) and (8), locate the target.

Signal processing flow chart is as follows:

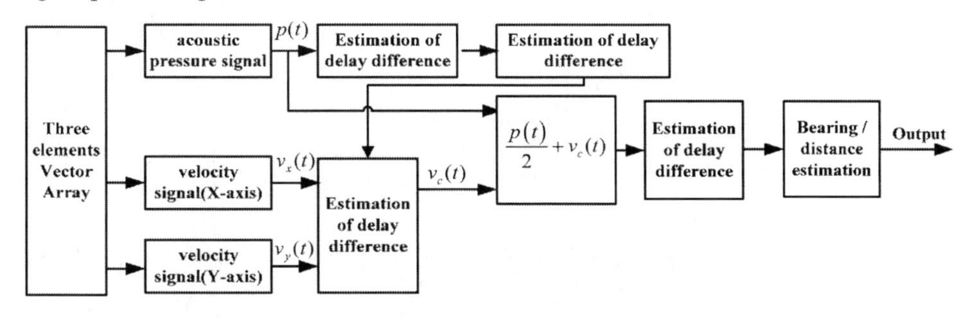

Fig. 2:Signal processing flow chart of passive ranging technology based on three elements vector array

4. SIMULATION

The method has been simulated by the computer system.Simulation conditions:

Target distance:1000m, target azimuth: 30° , the array spacing:20m,the frequency range of signal processing:100~3000Hz,signal processing integration time:1s.Under different SNR conditions, 200 independent statistics.

Simulation results are shown below:

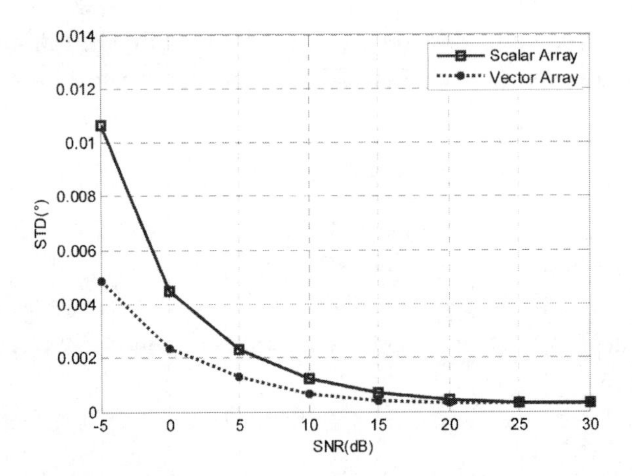

Fig.3:The comparison between azimuth estimation of vector Array and azimuth estimation of scalar array under different SNR

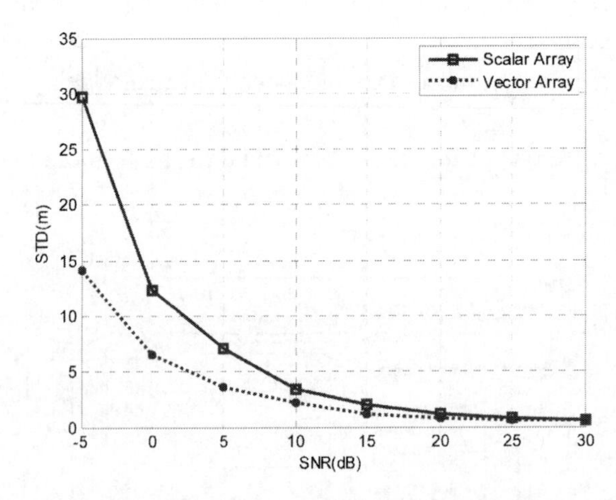

Fig.4:The comparison between the ranging of vector Array and the ranging of scalar array under different SNR

The simulation results can be seen: The passive ranging technology based on three elements vector array, representing traditional passive ranging technology based on three elements array, has been some improvement in positioning accuracy.The results demonstrate the effectiveness of the algorithm.

5. CONCLUSIONS

This paper presents the passive ranging technology based on three elements vector array, representing traditional passive ranging technology based on three elements array, more fully applied sound field information to improve the accuracy of delay estimation, and ultimately improve the passive ranging accuracy. This paper propose the passive ranging technology based on three elements vector array, completes the theory analysis, and gives the specific implementation steps. Simulation show that in the low SNR conditions, weak target signal path is enhanced by the proposed method.The improved method can improve the positioning accuracy, less computation required, this method has a strong practical prospects.

REFERENCES

[1] **Prabhakar S.Naidu**, Broadband source localization in shallow water Signal Processing 72,1999:107-116.

[2] **Hochwald and A. Nehorai,** Identifiability in axray processing models with vector-sensor applications *IEEE Transactions on Signal Processing*, vo1.44, pp.83-95, January l996.

[3] **A.Banner.** Simple velocity hydrophone for bioacoustic application,*J. Acoust. Soc. Amcr,* vol. 53. ,no.4, pp. 1134-1135,1972.

UNDERWATER ACOUSTIC COMMUNICATION SYSTEM SIMULATION BASED ON GAUSSIAN BEAM METHOD①

Bing Li[a] , Yuan Li[b], HaiboZheng,Zhibo Zhang, Qihu Li

[a] NO.21, Bei-Si-huan-Xi Road, Institute of Acoustics, Chinese Academy of Sciences Beijing, China
[b] NO.21, Bei-Si-huan-Xi Road, Institute of Acoustics, Chinese Academy of Sciences Beijing, China

Bing Li, NO.21, Bei-Si-huan-Xi Road, Institute of Acoustics, Chinese Academy of Sciences ,Beijing, China,100190; 010-82547960;libingbj@163.com

Abstract:*Due to dynamic multipath propagation structure and Doppler spreading, the underwater acoustic channel make the performance of underwater communication systems seriously degraded . Comparing with costly sea trials,underwater acoustic channel models provide a reliable and cheaper tool for predicting the performance of communication systems.The accuracy and efficiency of the channel model is essential for prediction. The ray theory based on Gaussian beam method has merits of high efficiency, clarity in physical meaning, and being easy to be parallel processing . In this paper, we proposed a novel simulator based on Gaussian beam method to estimate impulse response function of underwater mobile communication systems. Simulator operates periodically .At the input port of the simulator ,the evaluated communication systems will transmit a frame of communication signal every fixed intervals.According to the relative position relation between transducer and hydrophone in the acoustic field, simulator will estimate the current impulse response function . At the output port, a frame of communication signal will be acquired and processed with a mosaic method.When transmission is over,the BER will be calculated. Parameters such as depth ,sound profile,SNR and SL of transducer can be set up in the simulator.Simulations and trial are performed to validate this method.*

① UA2014-2nd International Conference and Exhibition on Underwater Acoustics, 2014: 1579-1586.

The results indicate that it help to assess the performance of underwater communication systems different in modulation method.

Keywords: *underwater acoustic communication; multipath propagation; Doppler spreading ;ray model; simulation of underwater acoustic communication.*

INTRODUCTION

As electromagnetic waves cannot propagate over long distances in seawater, acoustics is the favorite technology for this specific scenario.

Yet,underwater is a challenging environment for communication.The main reason lies:1.acoustic wave propagation is strongly inhomogeneous because of irregular sea boundaries, as well as variations in sound speeds over a section of ocean. 2. □Doppler spread due to fluctuations in the environment or relative motion between the transmitter and the receiver, sea surface, etc.As a result, the acoustic signals are affected by time varying multipath, which may create severe inter symbol interference (ISI) and large Doppler shifts and spreads. These highly space, time and frequency dependent features pose numerous obstacles for any attempts to establish reliable and long-range underwater acoustic communications.

Recently, the necessity of underwater acoustic communication and demand for transmitting and receiving various data such as voice or high resolution data are increasing as well. UNDERWATER ACOUSTIC (UWA) communication systems have to be designed to operate in a variety of conditions that differ from the nominal ones due to the changes in system geometry and environmental conditions.

As ocean deployment can be expensive and quite difficult, it may not often be practical to perform extensive field testing prior to full deployment. To allocate the appropriate resources (power, bandwidth) before system deployment, as well as to design appropriate signals and processing algorithms, it is necessary to have a relatively accurate channel model.

The contribution of this work is the development of an improved communication model Based On Gaussian Beam Method. This model allows for generation of parameters to be based on a particular target environment,rather than a general set of characteristics.Furthermore, we utilize data from a lake trial to verify the proposed model.

This paper is organized as follows: Section II describes related work and provides motivation for an improved underwater simulation model; Section III describes the proposed model and provides the methodology used in its development; Section IV describes the verification of the model; Section V provides results; and Section VI provides conclusions and indicates areas for future consideration.

I. II. BACKGROUND

A. Related Work

The proposed model is built upon the BELLHOP beam tracing program developed by Porter [1], as implemented in the Acoustic Toolbox [2]. The BELLHOP program computes the acoustic field by tracing the paths of beams as they leave a source. The central ray of a beam follows the standard equation,

$$\frac{\mathbf{d}}{\mathbf{ds}}\left(\frac{\mathbf{1}}{\mathbf{c(r,z)}}\frac{\mathbf{dr}}{\mathbf{ds}}\right)=-\frac{\mathbf{1}}{\mathbf{c^2(r,z)}}\nabla\mathbf{c(r,z)} \tag{1}$$

where r is the range and z is the depth using cylindrical coordinates, s is the arc-length and c(r, z) is the speed of sound at a particular point .

The BELLHOP program provides path computations that connect a given source-receiver pair. These paths are dependent on a description of the environment provided by the user. These paths represent direct, reflected, and refracted routes that acoustic waves follow [3] [4].

B. The Need for an Improved Model

Some of the existing simulation models are based on overly simplified conditions that treat the ocean as a whole,the others normally do not consider any unique attributes of a specific area in the ocean. Since underwater channel characteristics are highly space-time variable, it is not easy to build a physical model in which all phenomena are considered.

Beam tracing tools, such as Bellhop, use ray theory to provide an accurate deterministic picture of a UWA channel for a given geometry and signal frequency, but they do not take into account random channel variation.

In order to emulate mobile communication systems, in our model time-indexed channel impulse response (CIR) will be called for to process the communication data packet according to the relative position relation between AUV-receiver pairs.

II. III. MODEL DEVELOPMENT

A. Introduction to the Proposed Model

As the ocean environment varies greatly with geography [4], it is desirable to tailor the simulation model to the area of target deployment. The environmental parameters factored into the initialization of the model are the Sound Speed Profile (SSP), the contour and roughness of the bottom, the sound propagation characteristics of the sea-floor and surface half space, and the range of positions of any possible source and receiver nodes.

Once parameters have been collected, the model will provide physical layer modeling for a simulation package. The model will provide functions that will return path loss, propagation delay, and bandwidth for a given transmitter-receiver pair, as well as the calculation of the effective distance of a transmitted signal.

B. Development

The proposed model derives its parameters from an analytical model that can provide an accurate prediction of channel response. In our research, the BELLHOP beam tracing program was used [3]. This model is an efficient ray tracing tool for performing two-dimensional analysis of an ocean environment. The BELLHOP program was selected because of its efficiency and accuracy in generating arrival data consisting of amplitude, phase, and delay for each path between a given transmitter-receiver pair. The proposed model then processes these arrivals to generate the required delay, attenuation, and signal fading characteristics. The BELLHOP program is used as a preprocessing tool to provide data to the next step.

1) Environmental Description: To best utilize the ray tracing capabilities of the BELLHOP program, a precise description of the physical characteristics of the target environment is important: These are the depth of the section of the ocean,information concerning the bottom contour and roughness, and information about the surface.

2) Path Collection: With all the environmental parameters set, the BELLHOP program is then used to collect the multipath information for each possible link. In order to gather data for the entire target area and to determine how the acoustic signals will be received for any transmitter-receiver pair, a series of arrival statistics are collected. The input to the model includes the operational frequencies and the possible locations of all transmitter-receiver pairs. The transmitters are defined by their depth, while the receivers are described with their depth and relative horizontal displacement from the transmitters.The selection of frequency range and positions is completely dependent on the communication system's specifications.The number of paths and delays will remain constant for a certain SSP and a given transmitter-receiver pair, but the attenuation is frequency dependent.

C.the procedure of Simulation:

First,construct a 2-D coordinates.According to the velocity of AUV and it's track, compute the relative position relation between AUV-receiver pairs via time and save these data in a chart.

Second, repeatedly call for BELLHOP program to compute the channel impulse response (CIR) between AUV-receiver pairs in different location and save these time-indexed function in the chart.

Third, setup up parameters of underwater mobile communication systems.The parameters include modulation mode , carrier frequency, data rate , sampling frequency and so on.

Four, run the simulator after the data file for transmission is choosed. According to modulation mode, data file will be decomposed to segments,be transformed to a wave form and then put into data packets. Synchronization code and training code are also added to every packets.Utilizing the look-up table , simulator will compute the convolution of the input stream and the channel impulse response (CIR).

Five, according to SNR,noise is added to every packets.

Last, with a mosaic method all packets will be integrated into a whole data file for processing by receiver and the BER will be calculated.

Although BELLHOP's preprocessing is time-consuming, higher resolution can be obtained.

IV. MODEL VERIFICATION

We executed the simulation for estimate the performance of the QPSK technique in underwater mobile communication. Data from a lake trial is utilized to verify the model.The lake trial has been done in QIANDAO LAKE.The right picture in Figure 1 shows the satellite picture of the trial site from google. It can be deduced that underwater circumstance in QIANDAO LAKE is complicated.Thus,it's difficult to use parameters to describe the bottom's feature.

The experiment of QPSK technique was performed in underwater channel between an AUV and a 10 element Vertical Receiver Array (VRA). The bottom picture of Figure 2 shows the configuration of lake trial. The velocity of AUV is about 0.5m/s. Carrier frequency is fc = 10 kHz, data rate is 5 kbps and sampling frequency is 100 kHz.Transmitter signal is designed as the top picture of Figure 2. The chirp signal is used for synchronization.We transmit the training symbol before the data symbol to modify the coefficient of line equalizer .Fig.3 shows the diagram of channel multipath structure at different positions using the method of match filter and the corresponding CIRs in position 1 and position 8. In our model , SSP acquired in the lake trial is used for Bellhop program. The top boundary is described as a vacuum .In order to simplify the model , the bottom boundary is treated as perfectly rigid. The SSP of this target location is shown in the left picture in Fig. 1. Fig.4 shows the Rays calculated by Bellhop in the trial site. Fig.5 shows the CIRs in different positions calculated by simulator.

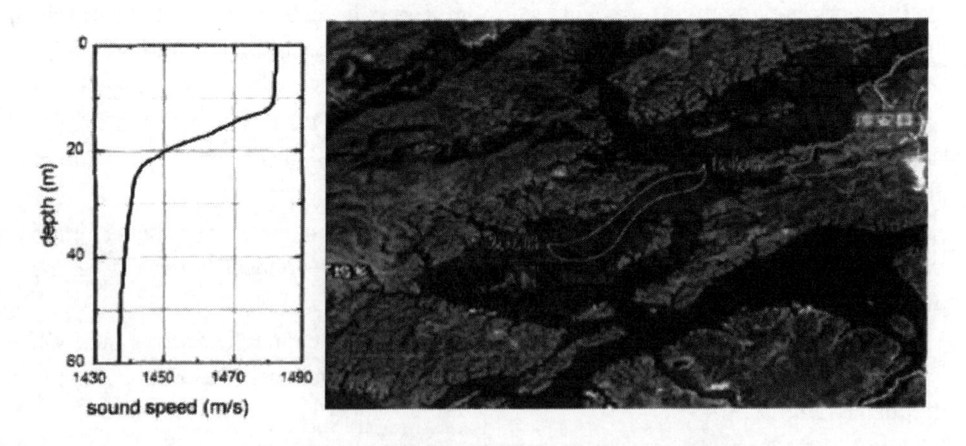

Figure 1 the satellite picture of the trial site and the SSP

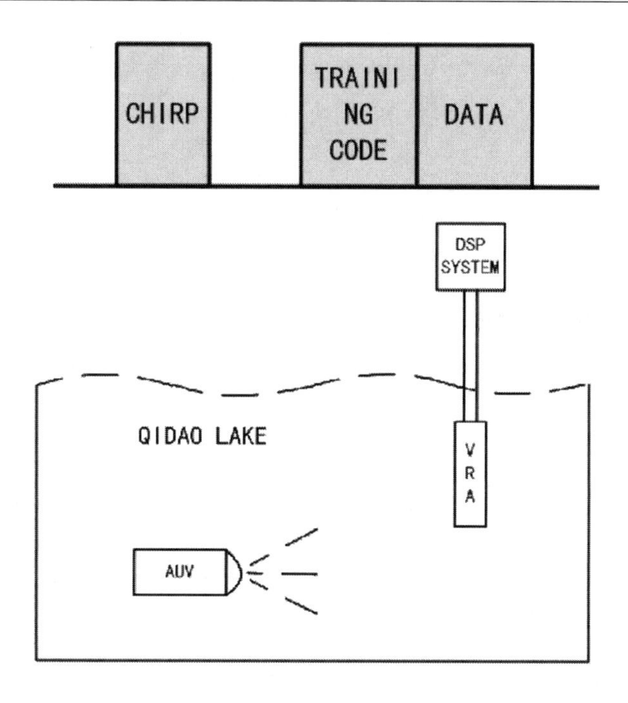

Fig.2 The configuration of lake trial and The formation of data packet

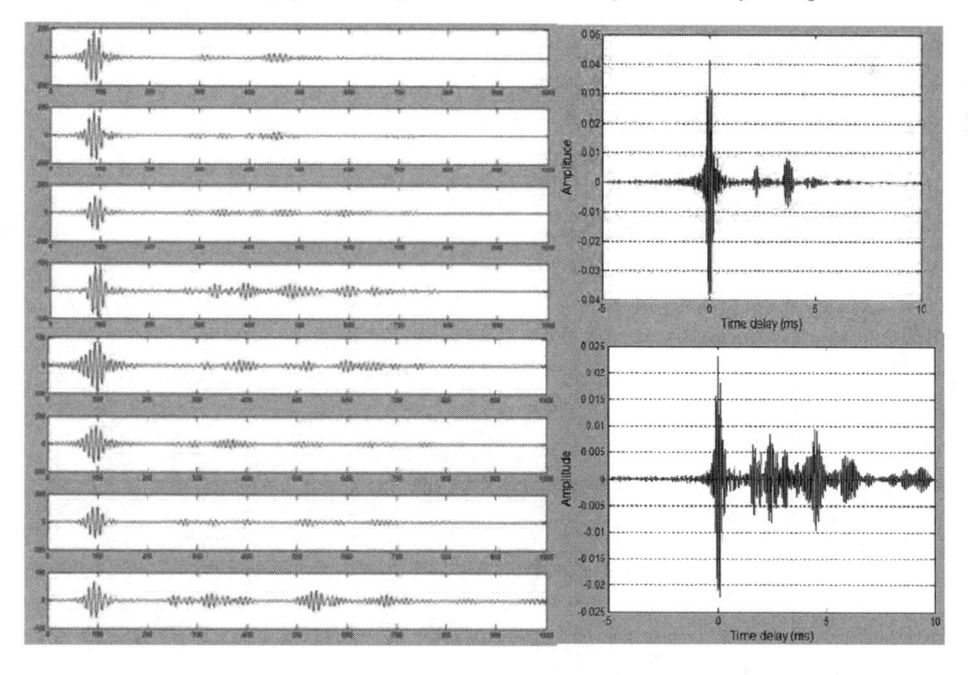

Fig.3 the diagram of channel multipath propagation structure at at different positions and the corresponding CIRs in position 1 and position 8

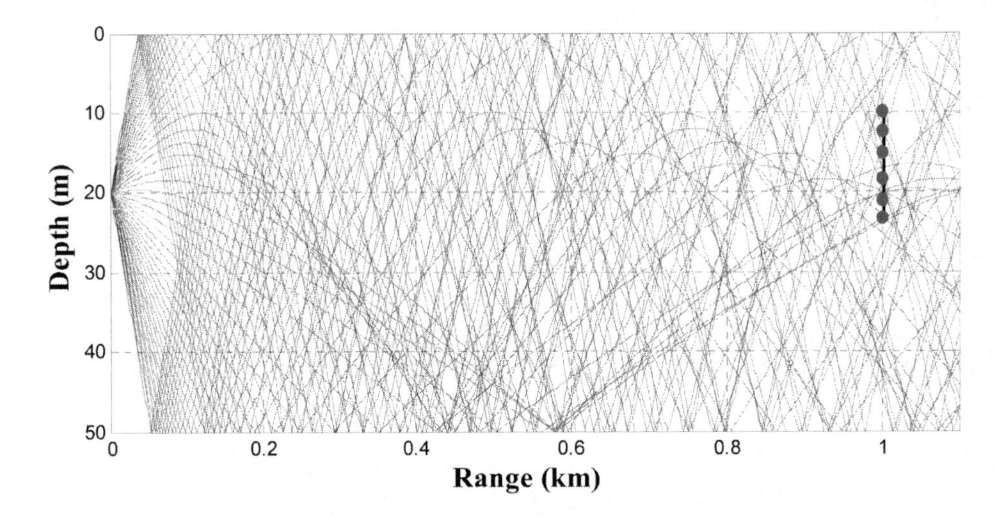

Fig4 the Rays calculated by Bellhop in the trial site

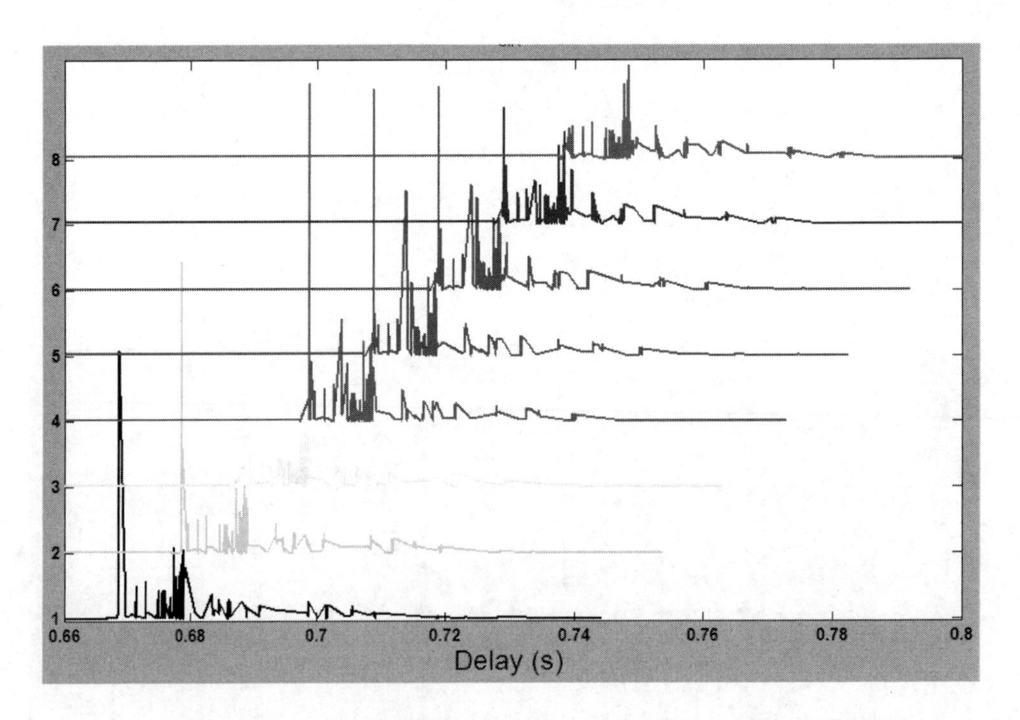

Fig.5 the CIRs in different positions calculated by simulator.

III. V. RESULTS

To gauge the value of the proposed model, comparisons were performed.Fig.3 shows the CIRs calculated by the data from the lake trial and Fig.5 show the CIRs calculated by simulator .It's inferred that the difference in CIRs is caused by bottom's feature .The variations of CIRs in different positions are obvious both in Fig.3 and Fig.5 which manifest the highly space and time dependent features of underwater acoustic channel.

IV.　VI. CONCLUSIONS AND FUTURE WORK

Comparisons lake trial results with simulator results indicate that the proposed model can do help for testing a mobile communication system deployed in a specific area in the ocean. The proposed model provides a great tool for understanding the nature of communication links in an ocean environment. Future work will include extending the model with a more detailed representation of the time-variability of a particular ocean location.

REFERENCES

[1] **H. B. M. Porter**, Gaussian beam tracing for computing ocean acoustic fields,Acoustical Society of America, vol. 82, no. 4, pp. 1349–1359,October 1987.

[2] **M. Porter**,Ocean acoustics library, Online, December 2007.

[3] **R. J. Urick**, Principles of Underwater Sound, 3rd ed., J. D. D. Heiberg, Ed. McGraw-Hill, 1983.

[4] **Y. L. Leonid Brekhovskikh**, Fundamentals of Ocean Acoustics, L. F.G. Ecker, W. Engl, Ed. Springer-Verlag, 1982.

纪念马大猷先生诞辰 100 周年

水下目标被动测距的一种新方法：利用波导不变量提取目标距离信息[①②]

李启虎
(中国科学院声学研究所　北京　100190)

2014 年 7 月 21 日收到
2014 年 12 月 31 日定稿

摘要　水声学中波导不变量的研究是近 30 年来引人注目的课题之一．水下目标辐射噪声的直达波和海面、海底反射波之间的干涉现象中隐含有下水目标的距离信息．提取这种距离信息就为水下目标的被动测距提供了一种新的途径．理论分析和实际海试都证明，甚至单水听器的 LOFAR (Low Frequency Analysis Record) 图都隐含着目标的距离和运动信息．本文给出利用波导不变量提取目标距离信息的理论推导，证明了在形成干涉条纹的外界条件具备时，利用多个水听器构成的基阵也能以较大增益提供目标距离信息．虽然组成基阵的每一水听器出现干涉条纹的条件是有差异的．这种差异在波束成形时可以加以利用和补偿．本文提出的理论和部分仿真、海试结果为水下目标被动测距和目标识别提供了一种新的途径．

PACS 数：43.30, 43.60
DOI:10.15949/j.cnki.0371-0025.2015.02.004

A new method of passive ranging for underwater target: distance information extraction based on wave guide invariant

LI Qihu
(*Institute of Acoustics, Chinese Academy of Sciences* Beijing 100190)

Received Jul. 21, 2014
Revised Dec. 31, 2014

Abstract　The study of wave guide invariant in underwater acoustics is one of attracted topics in recent 30 years. The interferences of direct wave and reflect wave from sea surface and sea bottom of underwater target radiated noise inherent the information of target distance. Extraction of these distance information will provide a possible new way in passive ranging for underwater target. The theoretical analysis and the results of at sea experiments show that the LOFAR (Low Frequency Analysis Record) figure inherently contains the range and moving information of passive acoustic sources, even in the situation that the receiver is only one single hydrophone. The theoretical analysis of extraction of target distance information by using wave guide invariant is presented in this paper. It is shown that, based on the interference striation pattern of target, the hydrophone array system is possible to extract the distance information with quite high array gain. Although the mathematical constrain conditions in forming interference striation pattern are different for individual array element, but it is proved that the differences of time delays between array elements can be used in compensation of beamforming. The theoretical analysis, system simulation and some results of at sea experiment show a new way in passive ranging and target recognition.

引言

单水听器接收的水下目标辐射噪声的直达波中并不包含目标的距离信息．所以传统上，目标的被动测向、测距必须依靠多水听器所收到的信号，利用信号延时差的信息，实现对目标的测向和测距．但是，波导不变量的理论研究和实验研究结果有可能给水下目标辐射噪声的测距提供一种新的途径，并且有可能会使对目标的探测距离有较大幅度的提高．这是

① 2015 年 3 月 1 日是我国著名物理学家、声学事业的开拓者马大猷先生 100 周年诞辰，作为后学之辈，特发表此文以表达对马先生的崇敬和怀念之情．

② 声学学报, 2015, 40(2): 138-143.

因为能显示波导不变量干涉条纹的 LOFAR 图, 只需要有较低的信噪比就可以实现对运动目标的辨识.

水声学中波导不变量的研究是近 30 年来引人注目的课题之一. 无论是在浅海还是深海, 水下目标的直达波和海面、海底反射波之间的干涉现象为水下目标的检测和识别提供了非常有利的条件. 浅海中的波导不变量的概念是由 Chuprov 等提出的[1−2]. 并在以后的研究中得到广泛、深入的分析[3−8].

本文将详细说明水下目标辐射噪声所提供的距离信息, 以及如何用信号处理的方法提取这种信息. 同时指出, 干涉条纹的出现, 不仅限于浅海的条件, 在深海的情况, 当条件具备时也会出现由海底反射而形成的干涉条纹. 这种现象类似于深海会聚区. 但是不需要满足海深大于会聚区 "临界深度" 的要求, 当出现这种现象时, 声呐的有效作用距离会大幅增加.

1 波导干涉现象所提供的距离信息

当水听器接收到从同一声源发出的直达波和可能的反射波信号时, 这两个信号的时延就包含了声源和接收水听器之间的距离信息, 见图 1, 假定声速剖面是等温层.

这是一个声源在水体中除直达波之外, 还存在海底反射和海面反射波的情况. 声源 S 至接收水听器 R 的距离为 r_d. 海面反射波的传播距离为 r_s, 海底反射波的传播距离为 r_b. 它们可以分别看作来自虚源 S_1 和 S_2 的声波.

我们先来分析具有相对延时差 τ_0 的两个不同信号叠加时其延时信息如何可以提取出来.

假定 $x(t)$ 是接收点所接收到的声源的原始信号. 它的频谱记作:

$$F[x(t)] = X(f), \tag{1}$$

其中 F 表示 Fourier 变换. 假定某一反射信号 (来自海底或海面) 为 $x(t+\tau_0)$, 那么:

$$F[x(t+\tau_0)] = \exp(-2\pi \mathrm{j} f\tau_0)X(f), \tag{2}$$

$x(t)$ 和 $x(t+\tau_0)$ 的互功率谱:

$$K_x(f) = |X(f)|^2 \exp(-2\pi \mathrm{j} f\tau_0), \tag{3}$$

τ_0 的信息可以从互功率谱 $K_x(f)$ 的实部和虚部之比求出来, 事实上:

$$2\pi f\tau_0 = \mathrm{tg}^{-1}\left[\frac{\mathrm{Im}(K_x(f))}{\mathrm{Re}(K_x(f))}\right], \tag{4}$$

其中, Im( ) 为复数的虚部, Re( ) 为复数的实部.

由此可见, 在某种情况下, 延时信息 τ_0 可以求出来. 而延时信息通常与距离信息相关 (当然, 不是一一对应的).

我们现在来看, 接收点 R 所收到的直达波加反射波的情况, 为简化起见, 我们只选择一种反射波 - 海面的或海底的进行讨论. 分析的方法可以类推到两种反射波同时存在甚至出现多途效应的情况, 但是计算要复杂得多. 假定直达波是 $x(t)$, 而反射波为 $x(t+\tau_0)$. 这时接收到的信号是:

$$y(t) = x(t) + Ax(t+\tau_0), \tag{5}$$

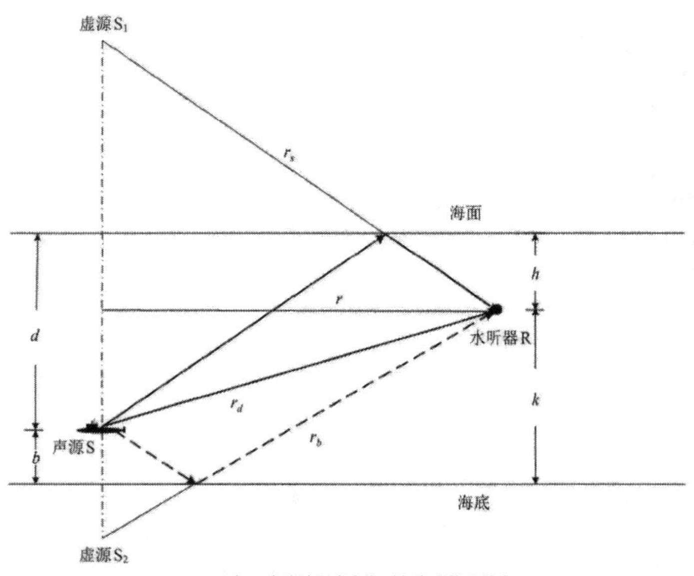

图 1 直达波和来自海面与海底的反射波

其中A表示某个幅度衰减值，$y(t)$的 Fourier 变换是：

$$Y(f)=F[y(t)]=X(f)+A\exp(-2\pi \mathrm{j} f\tau_0)X(f), \quad (6)$$

容易计算出 $y(t)$ 的功率谱是：

$$\begin{aligned}&|Y(f)|^2=Y(f)Y^*(f)=\\&|X(f)|^2+A\exp(-2\pi \mathrm{j} f\tau_0)|X(f)|^2+\\&A\exp(2\pi \mathrm{j} f\tau_0)|X(f)|^2+A^2|X(f)|^2=\\&|X(f)|^2[A^2+2A\cos(2\pi f\tau_0)+1],\end{aligned} \quad (7)$$

可以看出，τ_0 的信息存在于 $y(t)$ 的功率谱 $|Y(f)|^2$ 之中．从理论上讲是可以解出来的．

从式 (7) 我们可以看到，如果对接收信号 $y(t)$ 做 LOFAR 图，横轴为频率，纵轴为观察时间，那么 LOFAR 图上将出现干涉条纹，它的表现形式为：

$$f\tau_0=C, \ (C为常数) \quad (8)$$

随着常数 C 的变化，$\cos(2\pi f\tau_0)$ 的值也会变化，从而在 $|Y(f)|^2$ 的图上产生明暗相间的图像．

实际上这正是以波导不变量 β 为参数的渐近线[8]．因为，按照定义[1-2]，

$$\beta=\frac{df/dr}{f/r}, \quad (9)$$

而 LOFAR 图上干涉条纹是以

$$Af^2+Bt^2=C \quad (10)$$

的形式出现的．其中 A, B, C 为常数．而时间 t 和 r 是线性函数．对式 (10) 求微分得到：

$$\frac{df/dt}{f/t}=常数 \quad (11)$$

式 (11) 的左端就是波导不变量 β．

当然，实际条纹的出现还与很多其它因素有关，比如延时和距离的关系，信噪比的影响等．我们下面予以专门讨论．

先来分析理想海底反射时，反射波信号延时和距离的关系．见图 2．

假定发射信号源位于 A，接收器位于 B．接收器和发射器之间的垂直距离是 h_1，水平距离为 r_0，接收器和海底的距离为 h_2．假定 $r_0\gg h_1$，经过简单的数学推导，可以得到：

$$\theta=\mathrm{tg}^{-1}\left(\frac{h_1}{r_0}\right)\approx\frac{h_1}{r_0}, \quad (12)$$

$$\alpha=\mathrm{tg}^{-1}\left(\frac{2h_2+h_1}{r_0}\right), \quad (13)$$

容易证明：

$$r_1=AB\approx r_0,$$
$$r_2=AD+DB=\frac{r_0}{\cos\alpha}\approx r_0\left[1+\frac{(2h_2+h_1)^2}{2r_0^2}\right], \quad (14)$$

由于，

$$\tau_0=\frac{r_2-r_1}{c}, \ (c 为声速)$$

所以，

$$\tau_0=\frac{(2h_2+h_1)^2}{2cr_0}, \quad (15)$$

这是 τ_0 和 r_0, h_1, h_2 的对应关系．举例来说，如果接收器和发射换能器在同一深度 $h_1=0$，而 $r_0=5000$ m，$h_2=50$ m，那么 $\tau_0=0.66$ ms．由此可知，对声呐接收端来说，如果能估计出声源的深度，那么就可以以一定精度估计声源的距离．

当存在噪声的情况，式 (5) 的 $y(t)$ 除了包含信号项 $x(t)$ 之外，还要附加一项与 $x(t)$ 相互独立的噪声 $n(t)$，即

$$y_1(t)=y(t)+n(t)=x(t)+Ax(t+\tau_0)+n(t), \quad (16)$$

$$|Y_1(f)|^2=|Y(f)|^2+|N(f)|^2, \quad (17)$$

其中 $|Y(f)|^2$ 由式 (7) 给出，$|N(f)|^2$ 是 $n(t)$ 的功率谱密度．

我们从式 (7) 已看到，$|Y(f)|^2$ 的 LOFAR 图是频率 f 和延时 τ_0 的函数，而 τ_0 实际上可以从式 (12) 转换为距离 r_0，所以 r_0 实际上也和目标运动有关，

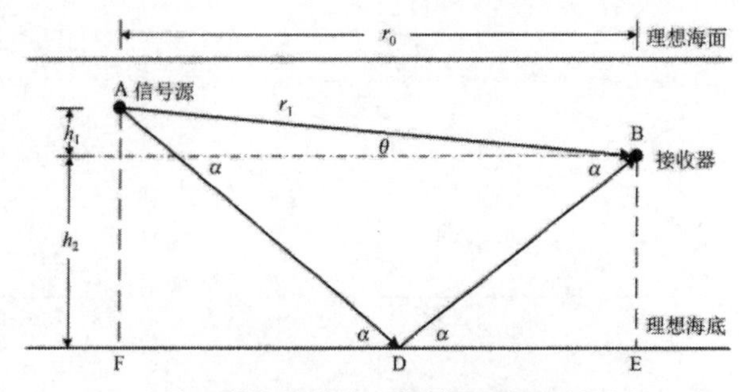

图 2 理想海底反射时，反射波信号延时和距离的关系

它是时间的函数 (当然还和速度有关)．在从 LOFAR 图提取干涉条纹后，就不难求出距离 r_0．图 3 给出了以 h_1 为参数的 τ_0-r_0 关系图.

从理论上讲，干涉条纹的出现不仅限于浅海条件下的水声传播，在深海反射会聚区的情况，同样会出现干涉条纹. 当然延时量会远比浅海的大. 图 4 是计算结果.

我们看到在 3000 m 水深时，干涉条纹出现的条件是延时量在 50～500 ms 之间. 由于延时量级明显大于浅海的情况，因而产生干涉现象的频率将向低端移动. 这正好是传播衰减较小的频率，有利于远距离检测. 当然，反射波会聚区不同于直达波声线在深海中发生弯曲而形成的会聚区. 因为后者要求海深大于临界深度 (一般在 4000 m 以上)，并有足够的 "深度余量"[9].

2 延时 (距离) 估计的误差分析

根据式 (12)，直达波和反射波的延时 τ_0 和距离 r_0 的关系是：

$$\tau_0=\frac{(2h_2+h_1)^2}{2cr_0},$$

令：

$$\alpha=\frac{(2h_2+h_1)^2}{2c},\tag{18}$$

则：

$$\tau_0=\frac{\alpha}{r_0},\tag{19}$$

由此不难推导出延时估计相对误差的表达式：

$$\frac{d\tau_0}{\tau_0}=-\alpha\frac{dr_0}{r_0^2}\Big/\left(\alpha\frac{1}{r_0}\right)=-\frac{dr_0}{r_0},\tag{20}$$

由此可知，对 τ_0 的估计的相对误差，就是对 r_0 估计的相对误差.

3 干涉波导现象的基阵信号处理

前面讨论的是单水听器接收直达波信号和反射波信号形成干涉的情况. 在实际声呐的应用中，当把基阵用于目标检测目的时，我们更关心各水听器之间的延时差会不会因为干涉条件不同而产生差异. 因为声呐检测的核心运算是 "延时－求和－平方" 运算，也就是波束成形[10−11]．这种波束成形系统是基于预延时补偿的信号处理系统. 如果不同水听器形成干涉的条件无法充分利用预延时，则基阵的增益也就无法获得. 从而大大限制了利用干涉条纹实际探测目标的能力.

图 5 以线阵为例来分析基阵不同水听器收到的直达波和反射波的延时差.

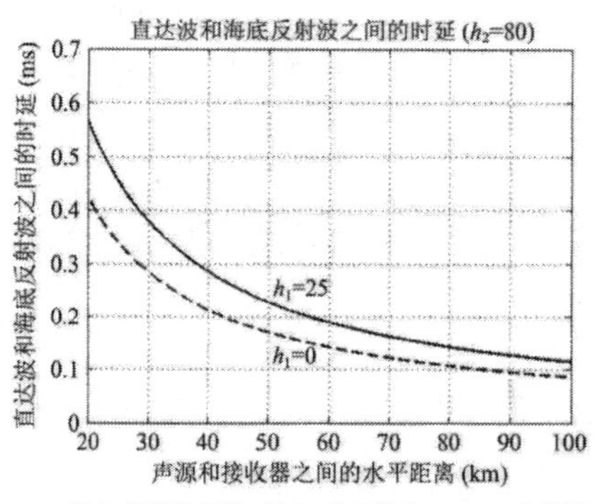

图 3 浅海条件下，以 h_1 为参数的 τ_0-r_0 关系图

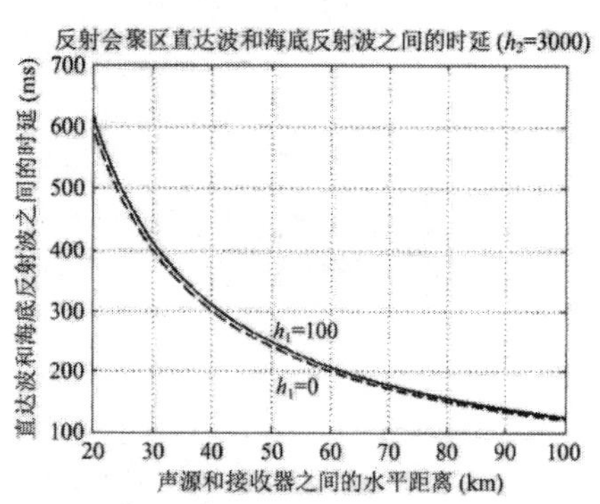

图 4 反射会聚区，以 h_1 为参数的 τ_0-r_0 关系图

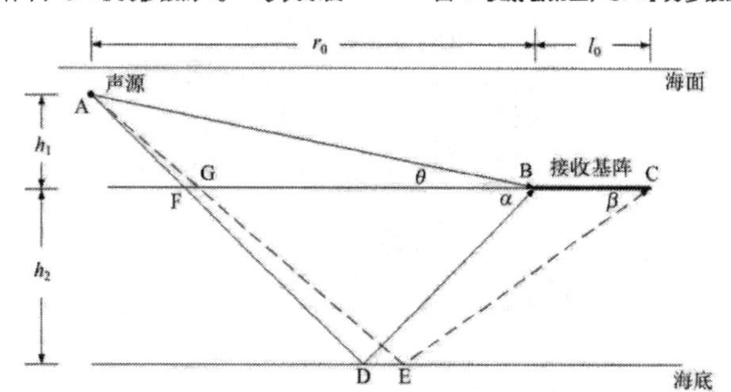

图 5 线阵条件下，不同水听器收到的直达波和反射波示意图

图 5 中，BC 代表一个线阵，长为 l_0 。我们着重考察基阵两端 B 、C 处直达波和反射波的延时。

按图中所给的参数，不难得出：

$$\theta=\mathrm{tg}^{-1}\left(\frac{h_1}{r_0}\right), \tag{21}$$

$$\alpha=\mathrm{tg}^{-1}\left(\frac{2h_2+h_1}{r_0}\right), \tag{22}$$

$$\beta=\mathrm{tg}^{-1}\left(\frac{2h_2+h_1}{r_0+l_0}\right), \tag{23}$$

当 $r_0 \gg l_0$ 时，$\alpha \approx \beta$ 。

我们从式 (12) 已知，对 B 点的水听器：

$$\tau_b=\frac{(2h_2+h_1)^2}{2cr_0}, \tag{24}$$

对 C 点的水听器：

$$\tau_c=\frac{(2h_2+h_1)^2}{2c(r_0+l_0)}, \tag{25}$$

若 $r_0 \gg l_0$, 则 $\tau_c \approx \tau_b$ 。

换句话，整个线列阵中各水听器接收到的信号都会满足同一干涉条件，波导现象将会同时出现。

4 系统仿真和海试结果分析

LOFAR 图是检测水下目标噪声的有效方法之一，特别是当出现信号干涉条纹时，可以确认目标的存在。它所需要的信噪比，通常低于常规检测的要求，这是因为 LOFAR 图的纵向包含有时间积累所带来的增益。干涉条纹的出现不仅可以用于目标的检测，也可以用于目标分类识别。

图 6(a)、图 6(b)、图 6(c) 分别给出了近场目标(距离在 10 km 以内)和远场目标(距离为 30 km)时，干涉条纹的情况。信噪比的大小，对目标干涉条纹的对比度起到了至关重要的作用。在某种情况下，整体调整一幅图的灰度，会对 LOFAR 图的外观起到有益的作用，但对对比度不会有改善。

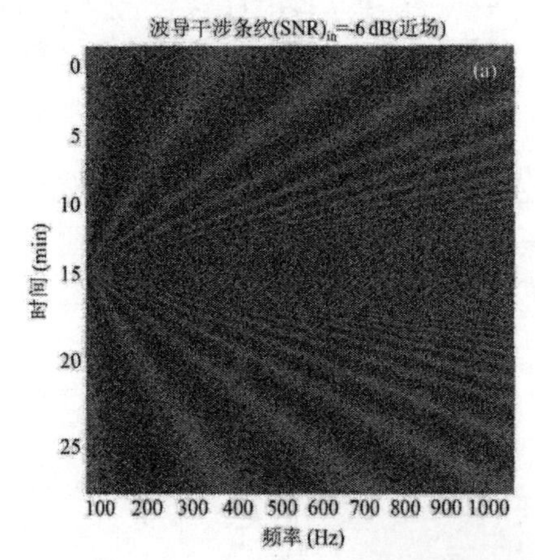

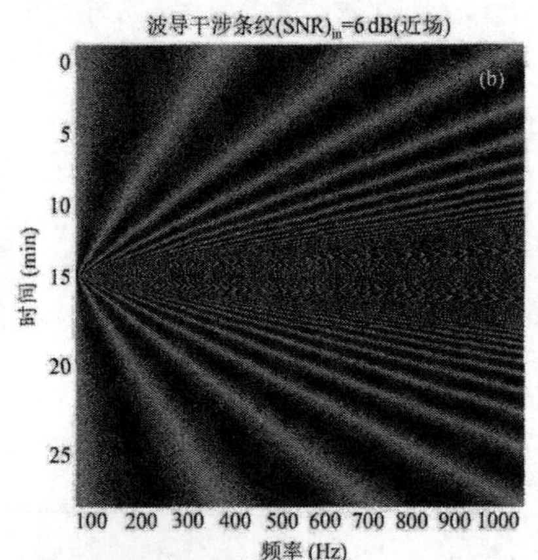

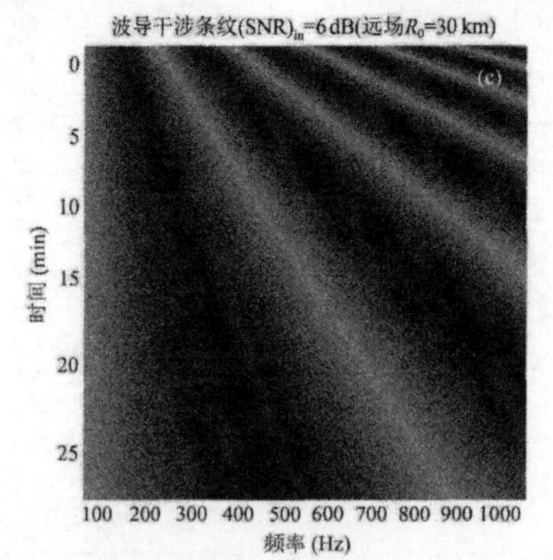

图 6 信噪比一定的条件下，不同近场 / 远场目标的波导干涉条纹

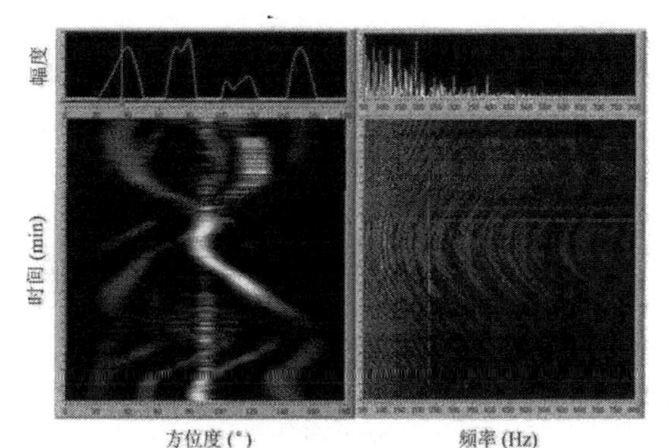

图 7 某大型舰辐射噪声的海试波导干涉条纹

图 7 是一次海试的结果，是一艘水面舰艇辐射噪声的 LOFAR 图. 左边是波束成形的结果，左上方是幅度显示，左下方是方位历程显示. 右侧是对准目标波束的 LOFAR 图分析的结果.

从图中可以看出，目标从接近基阵到远离基阵非常清晰地显示在 LOFAR 图上.

5 结论

理论分析、计算机仿真以及实际海试都已证明，水下目标辐射噪声的干涉现象可以提供目标的距离信息. 无论是浅海还是深海，只要具备产生波导干涉的条件，利用 LOFAR 图进行目标距离估计就有可能.

从信号检测角度来看，利用大孔径基阵进行被动测距也是可能的. 因为形成干涉条纹的延时对构成这个基阵的各接收元来说，几乎是一样的，从而可以统一补偿.

本文的研究结果可以为被动测距声呐的设计提供一种完全新的技术.

参 考 文 献

1 Chuprov S D. An invariant of the spatial-frequency interference pattern of the acoustic field in a layered ocean. In: Proc. of the Russian Academy of Sciences, 1981; **257**: 475—479

2 Brekhovskikh C M, Lysanov Yu P. Fundamentals of ocean acoustics. 2nd Ed. Springer-Verlag, Berlin, 1991

3 Kuperman W A *et al.* The generalized waveguide invariant concept with application to the vertical array in shallow water. In: Ocean Acoustic Phenomena And Signal Processing, ONR Workshop, New York, 2002: 33—66

4 Rouseff D, Spindel R C. Modeling the waveguide invariant as a distribution. Ibid, 137—148

5 Baggeroer A B. Estimation of the distribution of the interference invariant with seismic streamers. Ibid, 151—167

6 Kapolka D. Equivalence of the waveguide invariant and two path ray theory methods for range prediction based Lloyd's mirror patterns. In: Proc. of Acoustics'08, Paris, 3637—3641

7 An L *et al.* Calculating the waveguide invariant by the 2-D Fourier transform ridges of Lofargram image. *Journal of Electronics & Information Technology*, 2008; **30**(12): 2930—2933

8 LI Qihu *et al.* Theoretical analysis and experimental results of interference striation pattern of underwater target radiated noise in shallow water waveguide. *Chinese Journal of Acoustics*, 2011; **30**(1): 73—80

9 Urick R J. Sound propagation in the sea. peninsula, Los Altos, CA 1982

10 Li Q H. Digital sonar design in underwater acoustics, principles and applications. Springer Verlag, Berlin, 2011

11 Van Veen B D, Buckley, K M. Beamforming: a versatile approach to spatial filtering. *IEEE ASSP Magazine*, 1988; **5**(2): 4—24

纪念魏荣爵先生诞辰100周年

无源声呐多目标检测中反波束成形递推算法及其应用①②

李启虎　薛山花　卫翀华
(中国科学院声学研究所　北京　100190)
2016 年 3 月 14 日收到
2016 年 7 月 18 日定稿

摘要　随着声呐检测能力的提高，多目标干扰下微弱信号的检测问题日益突出。当声呐方位历程显示上出现多个干扰轨迹时，弱目标的检测显得十分困难。自适应噪声抵消 (Adaptive Noise Canceling, ANC) 技术为抑制多个干扰提供了理论基础，但是求解稳态最佳滤波矩阵存在着技术实现上的困难。

本文提出用一种反波束成形 (Inverse Beamforming, IBF) 递推算法，在阵元域逐一抵消多个强干扰，从而增强并提取出微弱目标信号。文中给出了递推求解由逆矩阵所表达的最佳滤波矢量的理论推导和相应的公式。利用 IBF 算法处理海试数据得到了较好的结果，显著改善了强干扰下对微弱信号的检测，甚至在普通波束成形 (CBF) 中未能显示出来的信号都可以被检测出来。

PACS 数：43.60, 02.10
DOI：10.15949/j.cnki.0371-0025.2016.05.026

Iterative inverse beamforming algorithm and its application in multiple targets detection of passive sonar

LI Qihu　XUE Shanhua　WEI Chonghua
(*Institute of Acoustics, Chinese Academy of Sciences* Beijing 100190)
Received Mar. 14, 2016
Revised Jul. 18, 2016

Abstract　With the increasing of detection ability of passive sonar, the weak signal detection problem in multiple interferences becomes more and more important. In the time/bearing record (TBR) display of sonar detection, when there exist multiple interferences traces, the identification of weak signal is difficult or impossible. The adaptive noise cancellation technique provides the theoretical basis of suppressing strong interferences. But the solution for finding the steady-state optimum filter matrix is quite difficult due to the real time calculation of inverse matrix of input data correlation matrix.

The iterative inverse beamforming (IBF) algorithm for solving the optimum filter matrix is derived in this paper, by which, the optimum filter can be finally expressed as a series of sum of simple matrix of input data.

Based on the algorithm proposed in this paper, some examples of at sea experiment are provided. The strong interferences are canceled and the weak signal is emerged, even it is not appeared in conventional beamforming (CBF) processing.

引言

随着声呐检测性能的持续改善，在同一海洋声学环境下，出现在无源声呐显示屏上的目标数也随之增加，多目标分辨问题日益成为无源声呐检测中的一个突出问题[1-3]。特别是在微弱目标信号受到强干扰影响，无法在声呐方位历程显示中提取检测参数时，寻求一种新的有效的强干扰抵消方法就显得十分必要。特别是拖曳线列阵声呐的使用，拖曳平

① 国家自然科学基金项目(11304343)资助.
② 声学学报, 2016, 41(5): 744-749.

台噪声是一种必须面对的近场强干扰，它往往使声呐对其它目标的检测受到极大影响[4-5]．

以灰度表征目标强度的方位历程显示是目前无源声呐终端显示的主要方式[6]．当出现多个目标时，由于背景噪声的存在，会使得对弱目标的检测变得十分困难．为了改善对微弱信号的检测，可以采用两种不同的途径．一种是阵元域的处理方法，如自适应噪声抵消，分频段过滤不同目标噪声源等[7-13]，另一种途径是后置处理，把方位历程显示图作为一种特殊的图像，快速提取其不同目标的轨迹，以达到检测和识别的目的[14]．

本文讨论的是阵元域的最佳滤波器，就是用求解最佳自适应噪声抵消矩阵的方法，对每一个被认为是干扰的目标，用普通波束成形 (Conventional Beamforming, CBF) 构成干扰波束，然后将该波束按不同延时分配给各阵元，构成反波束形成 (IBF) 信号，在阵元域把所有干扰抵消之后，再构成新的声呐处理结果．使得最后出现在方位历程显示中的图像只剩下没有被抵消的待检测目标．

关于用 IBF 方法抵消强干扰已研究多年[12-14]，当把这种方法具体应用于拖曳线列阵声呐时会取得良好的结果[15-16]．

1 多目标检测的理论模型

本文讨论的无源声呐在多目标出现的情况下，检测模型如图 1 所示．

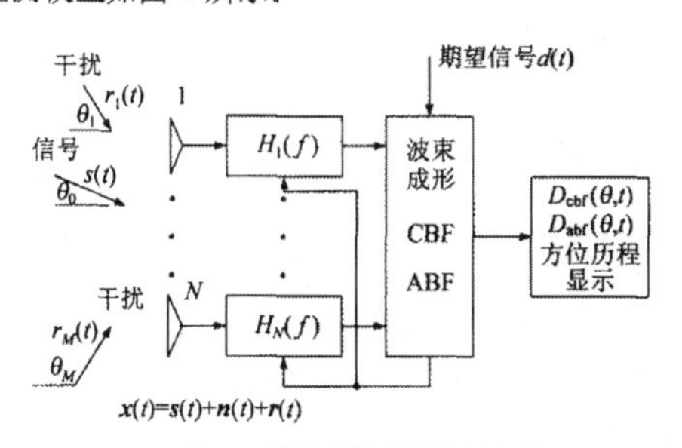

图 1 无源声呐多目标检测方框图

接收基元为 N 个，待检测信号假定为 $s(t)$，同时可能有多达 M 个的干扰 $r_1(t),\cdots,r_M(t)$．背景噪声 $n(t)$ 假定为相互独立、相同分布的高斯噪声 $n_1(t),\cdots,n_N(t)$．

用矢量来表示环境场，用到以下记号．基阵的接收信号：

$$\boldsymbol{x}(t)=\boldsymbol{n}(t)+\boldsymbol{s}(t)+\sum_{m=1}^{M}\boldsymbol{r}_m(t), \tag{1}$$

其中：

$$\boldsymbol{s}^{\mathrm{T}}(t)=[s_1(t),\cdots,s_N(t)], \tag{2}$$

$$\boldsymbol{r}_m^{\mathrm{T}}(t)=[r_{m1}(t),\cdots,r_{mN}(t)], \tag{3}$$

$$\boldsymbol{n}^{\mathrm{T}}(t)=[n_1(t),\cdots,n_N(t)], \tag{4}$$

$$\boldsymbol{x}^{\mathrm{T}}(t)=[x_1(t),\cdots,x_N(t)], \tag{5}$$

T 表示矢量或矩阵的转置．

信号或期望信号 $\boldsymbol{s}(t)$ 的指向矢量是：

$$\boldsymbol{A}_0^{\mathrm{T}}=[\exp(2\pi \mathrm{j} f\tau_1(\theta_0)),\cdots,\exp(2\pi \mathrm{j} f\tau_N(\theta_0))]. \tag{6}$$

第 m 个干扰信号 $\boldsymbol{r}_m(t)$ 的指向矢量：

$$\boldsymbol{A}_m^{\mathrm{T}}=[\exp(2\pi \mathrm{j} f\tau_1(\theta_m)),\cdots,\exp(2\pi \mathrm{j} f\tau_N(\theta_m))]. \tag{7}$$

$\boldsymbol{x}(t)$ 的互功率谱矩阵是：

$$\begin{aligned}\boldsymbol{K}_{xx}(f)=&K_n(f)\boldsymbol{I}+K_s(f)\boldsymbol{A}(f,\theta_0)\boldsymbol{A}^{*\mathrm{T}}(f,\theta_0)+\\&\sum_{m=1}^{M}K_{rm}(f)\boldsymbol{A}(f,\theta_m)\boldsymbol{A}^{*\mathrm{T}}(f,\theta_m),\end{aligned} \tag{8}$$

其中 $K_n(f)$, $K_s(f)$, $K_{rm}(f)$ 分别是噪声、信号和第 m 个干扰的功率谱．$\boldsymbol{I}$ 表示单位矩阵．图 1 中 N 个滤波器的传输函数构成矢量：

$$\boldsymbol{H}^{\mathrm{T}}(f)=[H_1(f),\cdots,H_N(f)]. \tag{9}$$

在普通波束成形 (CBF) 的情况下：

$$\boldsymbol{H}_{\mathrm{cbf}}(f)=\boldsymbol{A}^*(f,\theta), \tag{10}$$

它的指向性函数是：

$$D_{\mathrm{cbf}}(\theta)=\boldsymbol{H}_{\mathrm{cbf}}^{\mathrm{T}}(f)\boldsymbol{K}_{xx}(f)\boldsymbol{H}_{\mathrm{cbf}}^*(f), \tag{11}$$

$\boldsymbol{K}_{xx}(f)$ 的表达式由式 (8) 确定．

当出现多个干扰 (即 M 较大) 或者干信比较大 (即 $K_{rm}(f)$ 远大于 $K_s(f)$ 时)，基于式 (11) 的方位历程显示，会出现多个条纹交叉，它实际上是一系列 $|\boldsymbol{A}^{\mathrm{T}}(f,\alpha)\boldsymbol{A}^*(f,\beta)|^2$ 型的函数的叠加结果．大量干涉条纹及其副瓣以 $\sin x/x$ 形式出现，使得对微弱目标的检测变得非常困难．

图 2 是 $M=7$，再加 1 个信号的情况．由于声呐显示系统采用的是 256 级灰度显示[5-6]，当干扰过强时，信号有可能无法出现在屏幕上．

对干扰的抑制，较有效的办法是自适应波束成形，借助于滤波矩阵的调节，抵消干扰，提取信号．

最佳滤波矢量为[5]：

$$\boldsymbol{H}_{\mathrm{opt}}(f)=[\boldsymbol{K}_{xx}^*(f)]^{-1}\boldsymbol{K}_{xd}^*(f), \tag{12}$$

其中 $\boldsymbol{K}_{xd}(f)$ 表示输入 $\boldsymbol{x}(t)$ 和期望信号 $\boldsymbol{d}(t)$ 的互谱矩阵．

类似于式 (11), 可以得到最佳指向性函数是:

$$D_{\mathrm{opt}}(\theta)=D_{\mathrm{abf}}(\theta)=K_s(f)|\boldsymbol{A}^{*\mathrm{T}}(f,\theta_0)\boldsymbol{K}_{xx}^{-1}(f)\boldsymbol{A}(f,\theta)|^2, \tag{13}$$

这是干扰、噪声受到抑制, 信号得到加强之后的指向特性。

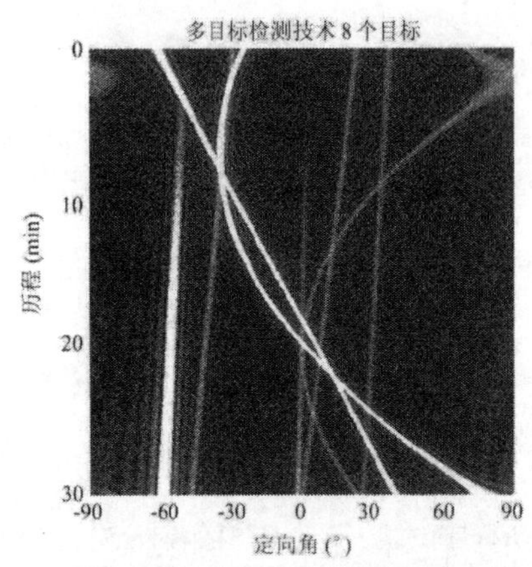

图 2 检测过程中典型的方位历程显示

2 求解最佳滤波器矩阵的递推算法

从式 (13) 可看出, 从理论角度来说, 求解 $D_{\mathrm{opt}}(\theta)$ 的关键是计算 $\boldsymbol{K}_{xx}^{-1}(f)$. 对于式 (5) 所给出的环境场, $\boldsymbol{K}_{xx}^{-1}(f)$ 的直接计算是困难的。可以证明 [5,12]:

$$\boldsymbol{K}_{xx}^{-1}(f)=\frac{1}{K_n(f)}\left(\boldsymbol{I}-\sum_{i,j=0}^{M}B_{ij}\boldsymbol{A}_i\boldsymbol{A}_j^{*\mathrm{T}}\right), \tag{14}$$

其中 $\boldsymbol{A}_i=\boldsymbol{A}(f,\theta_i)$, $i=1,\cdots,M$, B_{ij} 是矩阵 $\boldsymbol{B}$ 的第 i,j 个元素:

$$\boldsymbol{B}=(\boldsymbol{I}-\boldsymbol{D}\boldsymbol{G})^{-1}\boldsymbol{D}, \tag{15}$$

$$(\boldsymbol{G})_{ij}=\boldsymbol{A}_i^{*\mathrm{T}}\boldsymbol{A}_j, \tag{16}$$

$$(\boldsymbol{D})_{ij}=\frac{K_{rm}(f)}{K_n(f)}\delta_{ij}, \tag{17}$$

$\delta_{i,j}$ 为 Kroneck δ, 即:

$$\delta_{ij}=\begin{cases}1, & i=j,\\ 0, & i\neq j.\end{cases}$$

式 (14) 不难验证, 只需证明

$$\boldsymbol{K}_{xx}(f)\boldsymbol{K}_{xx}^{-1}(f)=\boldsymbol{I} \tag{18}$$

就行。

直接把式 (14) 用于计算是有困难的. 本文给出一种逐次递推的方法. 先从 $M=1$ 入手, 然后逐步推广到 $M=2$, $M=3\cdots\cdots$ 的情况。每进行一步都用到上一步的结果。

先假定 $M=1$, 即只有一个强干扰的情况。这时, $\boldsymbol{K}_{xx}^{-1}(f)$ 比较容易计算。因为,

$$\boldsymbol{K}_{xx}(f)=K_n(f)\boldsymbol{I}+K_s(f)\boldsymbol{A}_0\boldsymbol{A}_0^{*\mathrm{T}}+K_{r1}(f)\boldsymbol{A}_1\boldsymbol{A}_1^{*\mathrm{T}}. \tag{19}$$

令:

$$\boldsymbol{Z}=\boldsymbol{I}+\frac{K_{r1}(f)}{K_n(f)}\boldsymbol{A}_1\boldsymbol{A}_1^{*\mathrm{T}}, \tag{20}$$

容易证明:

$$\boldsymbol{Z}^{-1}=\boldsymbol{I}-\frac{\boldsymbol{A}_1\boldsymbol{A}_1^{*\mathrm{T}}}{N+K_n(f)/K_{r1}(f)}, \tag{21}$$

$$\boldsymbol{K}_{xx}^{-1}(f)=\frac{1}{K_n(f)}\left[\boldsymbol{Z}^{-1}-\frac{\boldsymbol{Z}^{-1}\boldsymbol{A}_0\boldsymbol{A}_0^{*\mathrm{T}}\boldsymbol{Z}^{-1}}{\boldsymbol{A}_0^{*\mathrm{T}}\boldsymbol{Z}^{-1}\boldsymbol{A}_0+K_n(f)/K_s(f)}\right]. \tag{22}$$

把式 (22) 代入式 (13), 就可以得到只有一个强干扰时, 最佳指向性函数:

$$D_{\mathrm{opt}}(\theta)=\alpha|\boldsymbol{A}_0^{*\mathrm{T}}\boldsymbol{Z}^{-1}\boldsymbol{A}_0|^2=\alpha q(\theta), \tag{23}$$

其中常数 α 与 θ 无关。

容易验证 $q(\theta)$ 在 $\theta=\theta_1$ 时, 取值为零。即在干扰方向形成一个零点. 为了说明 $M=2$ 时如何利用 $M=1$ 的已有成果, 我们把式 (19) 改写成:

$$\begin{aligned}\boldsymbol{K}_{xx}(f)&=\\ K_n(f)&\left[\boldsymbol{I}+\frac{K_s(f)}{K_n(f)}\boldsymbol{A}_0\boldsymbol{A}_0^{*\mathrm{T}}+\frac{K_{r1}(f)}{K_n(f)}\boldsymbol{A}_1\boldsymbol{A}_1^{*\mathrm{T}}\right]=\\ K_n(f)&\left[\frac{K_s(f)}{K_n(f)}\boldsymbol{A}_0\boldsymbol{A}_0^{*\mathrm{T}}+\boldsymbol{Z}\right].\end{aligned} \tag{24}$$

在 $M=2$ 时, 我们要把 $\boldsymbol{K}_{xx}(f)$ 的形式也改造成式 (24) 的形状, 只不过用另一个已知矩阵来替代 $\boldsymbol{Z}$, 接下来的解算步骤就完全一样了.

假定 $M=2$,

$$\begin{aligned}\boldsymbol{K}_{xx}(f)&=K_n(f)\left[\boldsymbol{I}+\frac{K_s(f)}{K_n(f)}\boldsymbol{A}_0\boldsymbol{A}_0^{*\mathrm{T}}+\right.\\ &\left.\frac{K_{r1}(f)}{K_n(f)}\boldsymbol{A}_1\boldsymbol{A}_1^{*\mathrm{T}}+\frac{K_{r2}(f)}{K_n(f)}\boldsymbol{A}_2\boldsymbol{A}_2^{*\mathrm{T}}\right]=\\ &K_n(f)\left[\frac{K_{r2}(f)}{K_n(f)}\boldsymbol{A}_2\boldsymbol{A}_2^{*\mathrm{T}}+\boldsymbol{P}\right].\end{aligned} \tag{25}$$

式 (25) 和式 (24) 的形式完全一样, 只不过用 $\boldsymbol{P}$ 代替了 $\boldsymbol{Z}$, 因而式 (25) 的求逆也是没有问题的, 而

$$\boldsymbol{P}=\boldsymbol{I}+\frac{K_s(f)}{K_n(f)}\boldsymbol{A}_0\boldsymbol{A}_0^{*\mathrm{T}}+\frac{K_{r1}(f)}{K_n(f)}\boldsymbol{A}_1\boldsymbol{A}_1^{*\mathrm{T}}. \quad (26)$$

由此，$M=2$ 时的情况完满解决。依次类推，就可以实际解决存在多个干扰时最佳指向性的计算问题。

3 IBF 算法的技术实现和部分海试结果

在多目标的检测过程中，要实现实时抵消多个干扰，可以采取 IBF 的方法，它可以在干扰方向形成所希望的零点，从而把信号提取出来。这种方法的级联框图见图 3 。

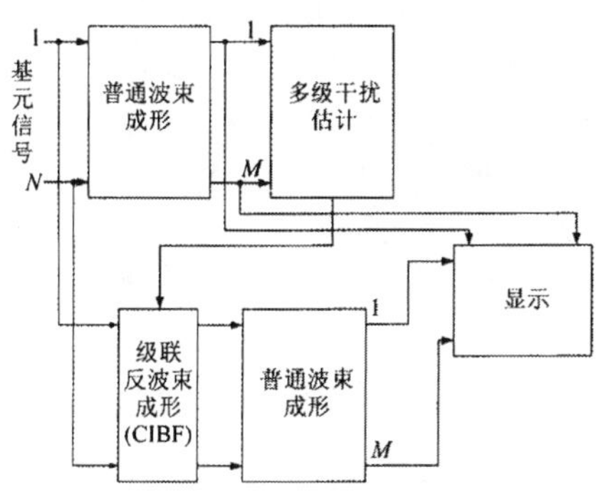

图 3 级联 IBF 方框图

当我们指定 M 个干扰方向时，在每一个方向都用普通波束成形 (CBF) 的方法，求得一个干扰波束的信号。然后依次把阵元域的信号减去这个干扰信号。由此得到"纯净"的信号[12-13,15-16]。M 个干扰的抵消可以采用并行的方法，也可以采取级联的方法。前者需要较大的硬件开销，而后者相对来说，对硬件的要求较低。

为了说明问题起见，我们以拖曳线列阵为例，把拖曳平台噪声作为干扰予以抵消。

我们仍假定待检测的信号出现在 θ_0 方向，而拖曳平台噪声来自 θ_r 方向。

第 i 路的时域信号为 $x_i(t)$。

我们在图 4 中只画出 $i=1,\cdots,N$ 中的一路信号，为了在物理上可实现的延迟，当我们需要把来自任意方向 θ_r 的干扰抵消掉时，必须给 $x_i(t)$ 足够大的预延时 τ_0。τ_0 一般应等于基阵孔径与声速的比，$x_i(t)$ 变为 $x_i(t-\tau_0)$。

用CBF形成拖曳平台噪声的波束，记作 $x_{\mathrm{tow}}(t)$。干扰抵消后的信号为：

$$x_{i,\mathrm{new}}(t)=x_i(t-\tau_0)-x_{\mathrm{tow}}(t-\tau_i(\theta_r)), \quad (27)$$

从频域上看，就是

$$X_{i,\mathrm{new}}(f)=X_i(f)\exp(-2\pi\mathrm{j}\tau_0 f)-X_{\mathrm{tow}}(f)\exp(-2\pi\mathrm{j}\tau_i(\theta_r)),$$

对 $x_{i,\mathrm{new}}(t)$ 做 CBF，就得到了抵消干扰后的新的处理结果。应当指出的是，在相减过程中假定各基元信号的幅度是一致的，波束信号也衰减到相对应的基元信号的幅度。

在有多个干扰需要抵消的情况下，采用图 5 所示的逐级抵消算法。用 CBF 方法求出具有最大幅度(强度) 响应的干扰所在的方位 $\theta_{\max}$，然后对准 $\theta_{\max}$ 方向形成干扰波束。

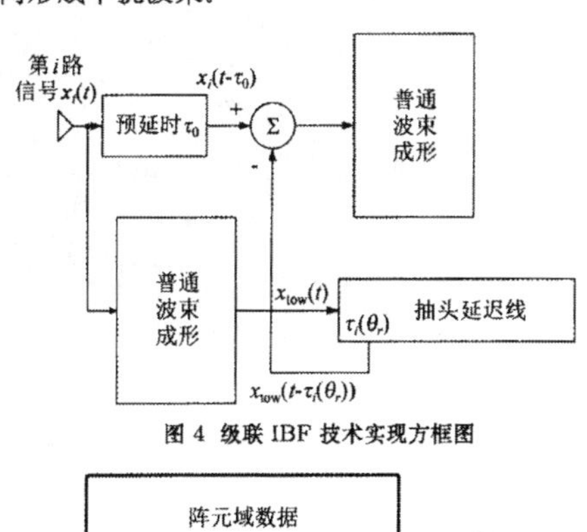

图 4 级联 IBF 技术实现方框图

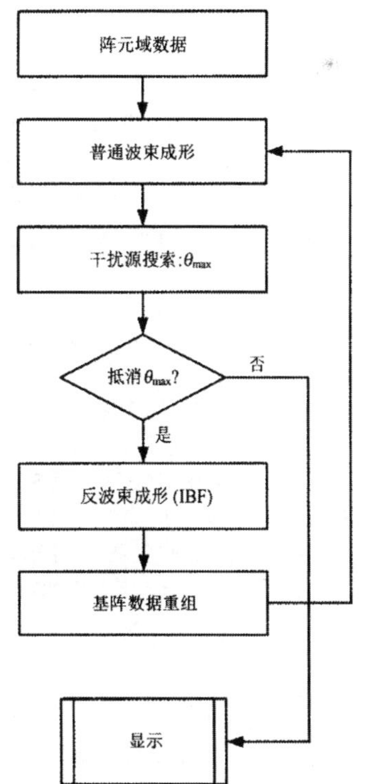

图 5 级联 IBF 运算流程

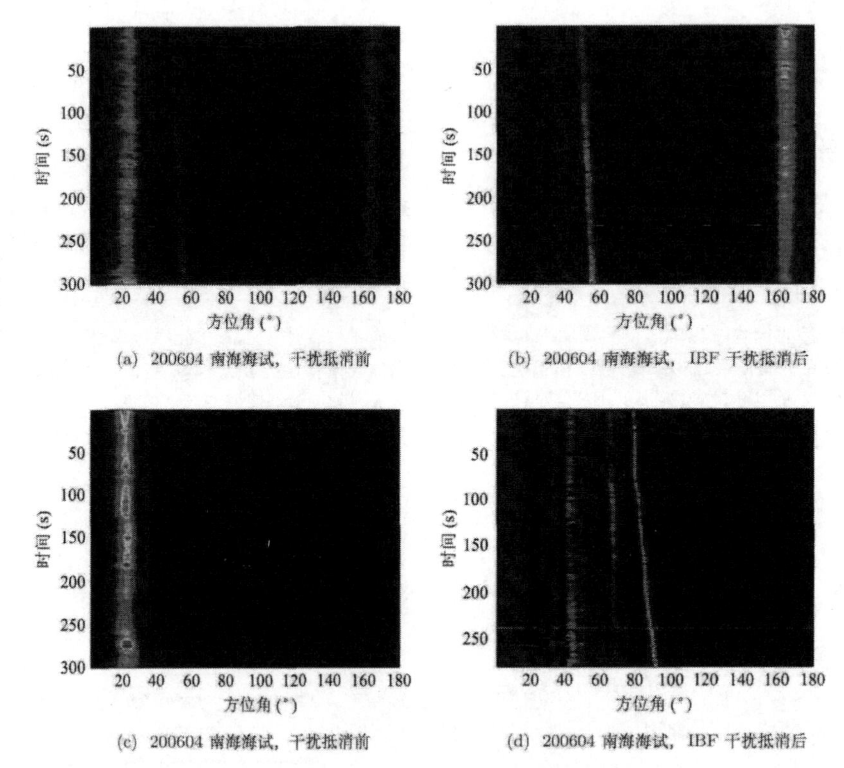

(a) 200604 南海海试，干扰抵消前　(b) 200604 南海海试，IBF 干扰抵消后

(c) 200604 南海海试，干扰抵消前　(d) 200604 南海海试，IBF 干扰抵消后

图 6 实际海上试验数据 IBF 处理

从阵元域把干扰减去. 然后再求出第 2, 第 3, ⋯, 个需抵消的干扰, 最后把所抵消的干扰全部从阵元域抵消掉.

图 6 是一次海试信号对比处理结果，我们可以看出有效利用 IBF 技术会显著改善多目标的检测结果.

在上面的图中，左侧拖船噪声干扰约在 20°, 目标出现在 58°, 163° 方向, 不是很明显. 当使用了 IBF 方法后，58°, 163° 的信号有所增强，并且在 132° 方向可以隐约发现另一个目标的存在.

图 6 的下面, 左右分别是 IBF 处理前后的情况. 在 IBF 处理前, 在 95° 方向隐约可以判断有一目标. 但是在 IBF 之后，在 72°, 95°, 115° 都出现了目标，并且具有相当高的信噪比. 也就是说，原来没有在方位历程图中出现的目标，由于 IBF 技术的应用，使目标显现出来.

4 结论

无源声呐多目标检测是水声领域的一个重要课题. 利用 ANC 和 IBF 技术可以有效抵消干扰信号，使声呐方位历程显示中更有效地突显微弱信号. 同时有可能把本来甚至并没有发现的目标检测出来. 如果把信号检测和后置的信号识别工作结合起来，会产生更好的结果.

参 考 文 献

1 Estrada R F, Starr E A. 50 years of acoustical signal processing for detection: coping with the digital revolution. *IEEE Annals of the History of Computing*, 2005; **27**(2): 65—78
2 National Research Council. Technology for the United States Navy and Marine Corps, 2000-2035, Becoming a 21th century force. NA Press, USA, 1997
3 Chen T. The past, present, and future of underwater acoustic signal processing. *IEEE Signal Processing*, 1998; **15**(4): 21—53
4 Kennedy F D Jr. Experimentation: the key to transformation. *Undersea Warfare*, 2002; **5**(1): 3—10
5 Li Q H. Digital sonar design in underwater acoustics: principles and applications. Springer-Verleg, Berlin, Germany, 2012
6 Rasmussen R A. Studies related to the design and use of time/bearing sonar display. AD690540, 1968
7 Ainslie M A. Principles of Sonar Performance Modeling. Springer Praxis, UK, 2010

8 Li Q H, Schwartz S C. A Kind of pre-processing of robust wiener filter. Technical Report, Dept. of EECS, Princeton Univ. USA, 1985

9 Cadzow J A. Multiple source location-the signal subspace approach. IEEE Transactions on Acoustics Speech & Signal Processing, 1990; **38**(7): 1110—1125

10 Satish A, Kaskyap R L. Multiple target tracking using maximum likelihood principle. IEEE Transactions on Signal Processing, 1995; **43**(7): 1677—1695

11 Messer H. The use of spectral information in optimal detection of a source in the presence of a directional interference. *IEEE J. of Oceanic Engr.*, 1994; **19**(3): 416—424

12 Keating P N. A rapid approximation to optimal array processing for the case of strong localized interference. *J. Acoust. Soc. Am.*, 1979; **65**(2): 456—462

13 李启虎. 自适应波束成形中稳态特性的研究. 声学学报, 1982; **7**(3): 165—173

14 Li Q H. A trace extract technique for fast moving target in underwater acoustics. In: Proc. of UDT'1999, Nice, France, 1999: 177—180

15 Robert M K, Beerens S P. Adaptive beamforming algorithms for tow ship noise canceling. In: Proc. of UDT'2002, La Spezia

16 张宾等. 拖船噪声抵消与左右舷分辨联合处理方法研究. 应用声学, 2008, **27**(6). 000 085

四

数字式声呐设计理论与应用

数字式分裂波束阵系统的精确定向方法①

李 启 虎
(中国科学院声学研究所)

1982年8月23日收到

摘 要

本文讨论数字式声呐分裂波束精确定向的问题. 利用 Piersol 提出的互谱法估计时延的原理,给出线阵与圆阵分裂波束系统的信号入射角与等效声中心的时延解算公式. 利用这些公式可以根据输入数据,直接给出目标的准确入射角的数值. 文中给出估计精度的表达式. 它与累加次数、DFT 的长度及采样周期有关. 同时,还分析了由于对随机信号的截断所带来的误差. 提出了对估计量加权、对输入数据加窗和合理选择 DFT 参数,以提高估计精度的方法. 文中还给出系统的方框图及部分计算机模拟的结果. 本文所提供的方法可以在数字式声呐中用于实时跟踪多个目标.

一、引 言

精确定向问题一直是被动声呐设计中的一个重要课题[1-5]. 传统的极大值法或和差定向法,由于系统的指向性在极大值附近通常是很平滑的,所以这种定向法的精度不能达到最佳的 Cramèr-Rao 下界. Schulthesis 等指出[1,6],如果采用分裂波束系统,定向精度可以提高. 对于直线阵来说,定向误差实际上接近 Cramèr-Rao 下界.

由于数字式声呐的发展,对于精确定向提出了新的更高的要求[7]. 因为数字多波束系统的波束指向通常不能连续转动,所以就要求系统能根据输入数据实时计算信号的入射角,并且具有实时跟踪多个目标的能力. Bendat、Piersol 提出一种用计算两个信号的互谱来估计它们的时间差的方法[8,9],我们将这一方法用到阵信号处理中. 首先分析基于 DFT 的时延估计的精度与数字处理系统的主要参数(采样频率、DFT 长度、平均累加次数等)的关系,然后给出线阵与圆弧阵的分裂波束系统等效声中心和信号入射角之间的关系. 我们给出圆弧阵的一个简易解算公式. 计算表明,在信号偏离角较小的情况下,这个公式具有很高的精度;从而建立起声呐系统的输出与信号入射角之间的直接关系.

本文最后讨论提高估计精度的几种方法. 即按信号平均功率谱对估计量加权; 对输入数据加窗及不增加 DFT 计算量而对时间序列补零的方法. 文中给出系统的方框图及部分计算机模拟的结果. 理论分析与实际的系统模拟表明,基于 DFT 的数字式分裂波束系统可以用于实时跟踪多个目标.

二、基于 DFT 的时延估计法

图 1 给出两路信号时延估计的基本情况. 设两个接收元的信号来自同一信源(不一定是

① 声学学报, 1984, 9(4): 225-238.

平面波),分别用 $x(t)$ 及 $y(t)=x(t+\tau)$ 来表示,其中 τ 表示时延. 在阵信号处理理论中,准确地估计 τ 值是一个十分重要的课题. τ 的信息存在于 $x(t)$ 和 $y(t)$ 的互谱之中.

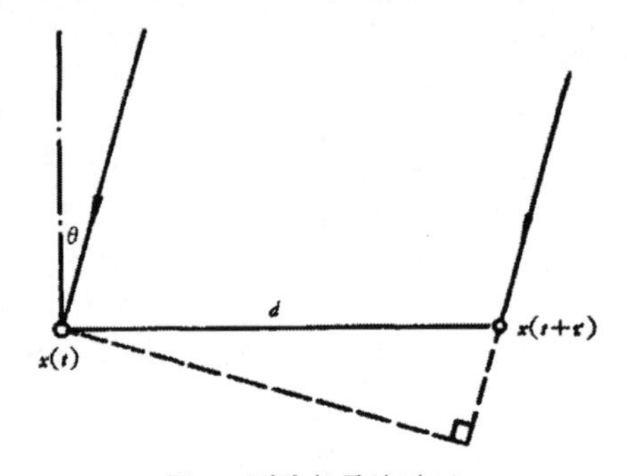

图 1 两路信号的时延

The diagram of time delay problem of two receivers

如果 $x(t)$ 是有规信号,用 $X(f)$ 表示其 Fourier 变换,即

$$X(f)=\int_{-\infty}^{\infty} x(t)\exp(-2\pi jft)dt \tag{1}$$

那么 $y(t)$ 的 Fourier 变换 $Y(f)$ 为

$$Y(f)=\int_{-\infty}^{\infty} x(t+\tau)\exp(-2\pi jft)dt=\exp(2\pi jf\tau)X(f) \tag{2}$$

利用 $Z(f)=X^*(f)Y(f)=\exp(2\pi jf\tau)|X(f)|^2$ 就可以将 τ 的信息提取出来:

$$2\pi f\tau=\mathrm{tg}^{-1}\left[\frac{\mathrm{Im}(Z(f))}{\mathrm{Re}(Z(f))}\right] \tag{3}$$

这里 Im(·), Re(·) 分别表示复数的虚部及实部.

如果 $x(t)$ 是均值为零的平稳随机信号,那么 $x(t)$ 和 $y(t)$ 的互功率谱是

$$K_{xy}(f)=\lim_{T\to\infty}\frac{1}{T}E[X_T^*(f)Y_T(f)] \tag{4}$$

其中 $X_T(f)$, $Y_T(f)$ 分别表示 $x(t)$, $y(t)$ 以区间$[0, T]$截断之后的 Fourier 变换,即

$$\left.\begin{aligned} X_T(f)&=\int_0^T x(t)\exp(-2\pi jft)dt \\ Y_T(f)&=\int_0^T y(t)\exp(-2\pi jft)dt \end{aligned}\right\} \tag{5}$$

类似地可以得到

$$2\pi f\tau=\mathrm{tg}^{-1}\left[\frac{\mathrm{Im}(K_{xy}(f))}{\mathrm{Re}(K_{xy}(f))}\right] \tag{6}$$

此式表明, 我们可以用计算 $x(t)$, $y(t)$ 的有限区间的互谱来求出相位 $\varphi=2\pi f\tau$, 然后将 φ 除以 $2\pi f$ 就得到时延值 τ.

在数字式系统中,输入数据是离散的样本序列,我们可以通过 FFT 计算出互谱来. 因而能迅速给出时延值. 但是,由于对信号进行截断,形为 (3) 式的公式不能严格成立. 所以由截断引起的误差就导致时延估计的偏离. 下面我们具体分析误差与 DFT 诸参数的关系.

设采样间隔为 T_s, 那么样本序列是

$$x(0),\ x(T_s),\ \cdots,\ x(kT_s),\ \cdots$$
$$y(0),\ y(T_s),\ \cdots,\ y(kT_s),\ \cdots$$
$$y(kT_s) = x(\tau + kT_s) \tag{7}$$

起点在 q，长度为N的 DFT 是

$$\left.\begin{aligned} X_q(l) &= \sum_{k=0}^{N-1} x[(q+k)T_s]\exp(-2\pi jkl/N) \\ Y_q(l) &= \sum_{k=0}^{N-1} x[(q+k)T_s+\tau]\exp(-2\pi jkl/N) \quad l=0,\ \cdots,\ N-1 \end{aligned}\right\} \tag{8}$$

容易看出，一般来说 $Y_q(l) \neq \exp(2\pi jf_l\tau)X_q(l)$. 其中 f_l 表示第 l 条谱线，$f_l = l/NT_s$. 这是因为 $x(kT_s)$, $y(kT_s)$ 都不是周期序列.

当 T_s 比 $x(t)$ 的相关半径小得多时，我们有

$$\begin{aligned} Y_q(l) &= \sum_{k=0}^{N-1} x[\tau + (q+k)T_s]\exp(-2\pi jkl/N) \\ &\approx \sum_{k=0}^{N-1} x[Int(\tau/T_s)T_s + (q+k)T_s]\exp(-2\pi jkl/N) \end{aligned}$$

其中 $Int(\tau/T_s)$ 表示 τ/T_s 的整数部分，把它记为 n_0，于是

$$Y_q(l) \approx \exp(2\pi jn_0l/N)\sum_{k=n_0}^{N-1+n_0} x[(k+q)T_s]\exp(-2\pi jkl/N)$$

$$= \exp(2\pi jn_0l/N)\cdot\left\{X_q(l) + \sum_{k=0}^{n_0-1}\{x[(k+N+q)T_s] - x[(k+q)T_s]\}\exp(-2\pi jkl/N)\right\} \tag{9}$$

当 $\tau \ll NT_s$ 时，显然有

$$Y_q(l) \approx \exp(2\pi j\tau l/NT_s)X_q(l) \tag{10}$$

令

$$Z_q(l) = X_q^*(l)\cdot Y_q(l) \tag{11}$$

就有

$$Z_q(l) \approx \exp(2\pi jl\tau/NT_s)\cdot|X_q(l)|^2 \tag{12}$$

$$\varphi_l = 2\pi f_l\tau = \mathrm{tg}^{-1}\left[\frac{\mathrm{Im}(Z_q(l))}{\mathrm{Re}(Z_q(l))}\right] \tag{13}$$

为了得到足够准确的 φ_l 估计值，我们应按(11)式进行多次平均．设平均次数为 K，那么有

$$Z(l) = \frac{1}{K}\sum_{m=1}^{K} Z_{mN}(l) \tag{14}$$

只要 K, N 足够大，把 (14) 代替 (13) 中的 $Z_q(l)$ 我们就可以得到 φ_l 的精确估计．对于不同的 f_l，可以得出一组 φ_l，由此可以得到不同的 τ 的估计值 τ_l. Piersol 建议用最小二乘法对 (f_l, φ_l), $l = 1,\ \cdots,\ N-1$ 进行拟合，以求出对 τ 的最优估计 $\hat{\tau}_{ms}$:

$$\hat{\tau}_{ms} = \frac{1}{2\pi}\frac{\sum_{l=1}^{N-1} f_l\varphi_l}{\sum_{l=1}^{N-1} f_l^2} \tag{15}$$

这种估计的误差为[8-10]

$$\mathrm{Var}(\hat{\tau}_{ms}) = \frac{1}{2\pi}\left\{\sum_{l=1}^{N-1}\frac{f_l^2}{\mathrm{Var}(\varphi_l)}\right\}^{-1/2} \tag{16}$$

在假定 $\mathrm{Var}(\varphi_l)$ 与 l 无关的条件下，有

$$\mathrm{Var}(\hat{\tau}_{ms}) = \frac{1}{2\pi}\left\{\sum_{l=1}^{N-1} f_l^2\right\}^{-1/2}\sqrt{\mathrm{Var}(\varphi_l)}$$

$$\approx \frac{T_s}{2\pi}\sqrt{\frac{3}{NK}}\sqrt{\mathrm{Var}(\hat{\varphi})} \tag{17}$$

其中 $\hat{\varphi}$ 表示由N点 DFT 按(13)式得到的一次估计量．由(17)式可以看出 $\hat{\tau}_{ms}$ 之估计精度和 $\sqrt{NK}$ 成反比而与 T_s 成正比．

三、阵信号的分裂波束精确定向

前面讨论的时延估计方法仅适用于两个接收基元的情况．众所周知，对阵信号处理，分裂波束能提供近于最优的定向精度．对于数字式声呐，我们总是先用多波束系统进行大致的定向，然后再用分裂波束的方法进行精测．

图2给出了线阵与圆弧阵的分裂波束系统产生左、右两路信号的原理．其中 $l(t)$，$r(t)$ 分别表示左、右波束．在一般情况下，不存在 $r(t)=l(t+\tau)$ 这种关系式．所以我们还不能立刻将上一节阐述的方法应用于分裂波束系统．我们应当先计算出左、右波束的等效声中心，假想在这两个声中心放置接收基元，然后利用上一节的结果．

由于任何宽带信号都可以由 DFT 将其分离为若干个窄带信号之和，所以我们下面的讨论仅局限于某一号窄带信号．设入射信号为

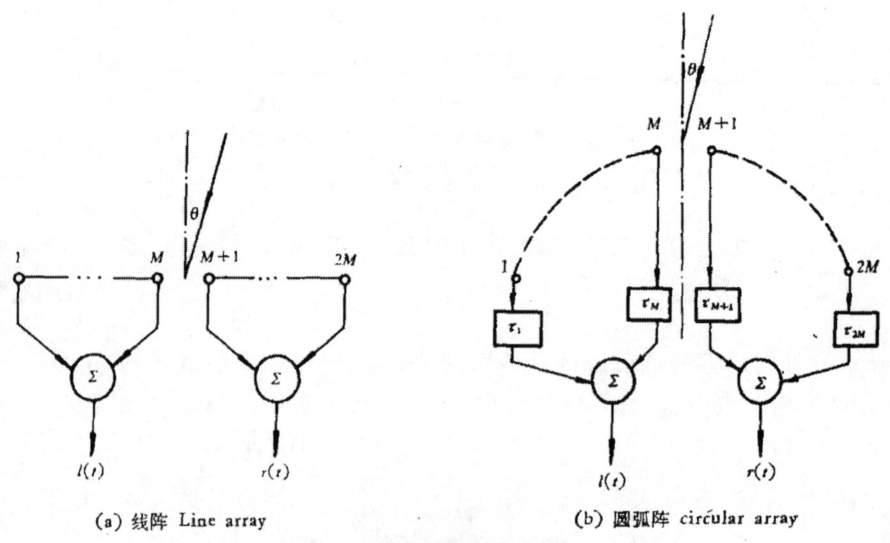

图2 分裂波束阵系统方框图

Block diagram of split-beam array system

$$S(t)=U(t)\cos(2\pi ft)-V(t)\sin(2\pi ft),$$

其中 $U(t)$, $V(t)$ 相对于 f 而言是慢变化的信号. 第 i 个基元信号的相对时延为 $\tau_i(\theta)$, 其中 θ 为信号入射角. 第 i 路信号是

$$\begin{aligned} S_i(t) &= S[t+\tau_i(\theta)] \\ &\approx U(t)\cos\{2\pi f[t+\tau_i(\theta)]\}-V(t)\sin\{2\pi f[t+\tau_i(\theta)]\} \\ &\triangleq U_i(t)\cos(2\pi ft)-V_i(t)\sin(2\pi ft) \end{aligned} \tag{18}$$

其中

$$\left.\begin{aligned} U_i(t) &= U(t)\cos[2\pi f\tau_i(\theta)]-V(t)\sin[2\pi f\tau_i(\theta)] \\ V_i(t) &= U(t)\sin[2\pi f\tau_i(\theta)]+V(t)\cos[2\pi f\tau_i(\theta)] \end{aligned}\right\} \tag{19}$$

由(18),(19)式易知,对于分裂波束阵系统,如果左路信号由 1 至 M 号基元构成,记作 $l(t)$;右路信号由 $M+1$ 至 $2M$ 号基元构成,记作 $r(t)$, 那么它们分别为:

$$\left.\begin{aligned} l(t) &= \left[\sum_{i=1}^{M}U_i(t)\right]\cos(2\pi ft)-\left[\sum_{i=1}^{M}V_i(t)\right]\sin(2\pi ft) \\ r(t) &= \left[\sum_{i=M+1}^{2M}U_i(t)\right]\cos(2\pi ft)-\left[\sum_{i=M+1}^{2M}V_i(t)\right]\sin(2\pi ft) \end{aligned}\right\} \tag{20}$$

于是

$$\left.\begin{aligned} l(t) &= A\cos(2\pi ft+\alpha) \\ r(t) &= B\cos(2\pi ft+\beta) \end{aligned}\right\} \tag{21}$$

其中

$$\left.\begin{aligned} A &= \left\{\left[\sum_{i=1}^{M}U_i(t)\right]^2+\left[\sum_{i=1}^{M}V_i(t)\right]^2\right\}^{1/2} \\ B &= \left\{\left[\sum_{i=M+1}^{2M}U_i(t)\right]^2+\left[\sum_{i=M+1}^{2M}V_i(t)\right]^2\right\}^{1/2} \\ \alpha &= \mathrm{tg}^{-1}\left[\sum_{i=1}^{M}V_i(t)\Big/\sum_{i=1}^{M}U_i(t)\right] \\ \beta &= \mathrm{tg}^{-1}\left[\sum_{i=M+1}^{2M}V_i(t)\Big/\sum_{i=M+1}^{2M}U_i(t)\right] \end{aligned}\right\} \tag{22}$$

由此得出 $l(t)$, $r(t)$ 的相位差是

$$\varphi=\beta-\alpha \tag{23}$$

φ 与 θ 的关系要根据基阵的几何形状定出来. $\tau=\varphi/2\pi f$ 就是 $l(t)$, $r(t)$ 的等效声中心的时延. 如果能精确地估计出 τ 或 φ 来,也就能精确地估计出 θ 来. 对于实际的声呐系统,我们希望给出信号入射角 θ 与时延 τ 的简单关系式, 以便能实时地按 τ 值解算出 θ 来. 下面分别就线阵及圆弧阵给出这种关系(见图 3). 在作理论推导时,我们以单频信号为例. 即把 $U(t)$, $V(t)$ 取作常数. 因为实际系统要用 DFT 来求出各窄带分量, 每一组 DFT 得到的幅度谱在该组时间长度内都是常数. 只有当转移到下一组 DFT 时,幅度谱才变成另一组数.

1. 直线阵

设信号入射角为 θ, 基元间隔为 d, 则

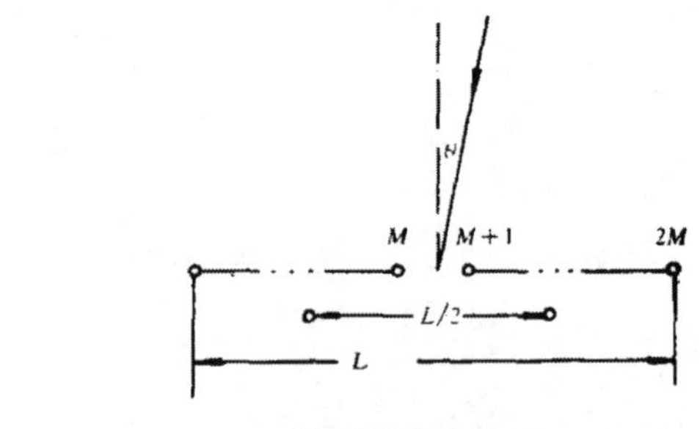

(a) 线阵 Line array

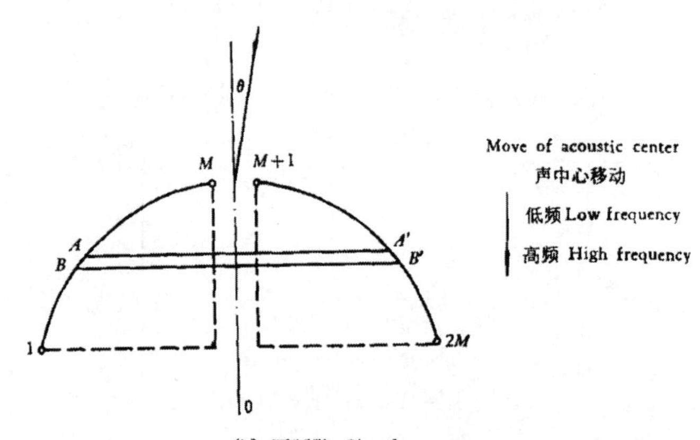

(b) 圆弧阵 Circular array

图 3 分裂波束系统等效声中心的位置

The position of equivalent acoustic center of split-beam system

$$\tau_i(\theta) = \frac{(i-1)d\sin\theta}{c} \tag{24}$$

其中 c 为声速。记 $\alpha' = (2\pi d/\lambda)\sin\theta$，则 $2\pi f\tau_i(\theta) = (i-1)\alpha'$

$$\left.\begin{aligned}
&\sum_{i=1}^{M} U_i(t) = \sum_{i=1}^{M}\cos[(i-1)\alpha'],\\
&\sum_{i=1}^{M} V_i(t) = \sum_{i=1}^{M}\sin[(i-1)\alpha']\\
&\alpha = \mathrm{tg}^{-1}\left[\sum_{i=1}^{M}\sin[(i-1)\alpha'] \Big/ \sum_{i=1}^{M}\cos[(i-1)\alpha']\right] = M\pi d\sin\theta/\lambda
\end{aligned}\right\} \tag{25}$$

类似地可以得到

$$\beta = 3M\pi(d/\lambda)\sin\theta \tag{26}$$

所以有

$$\varphi = 2\pi Md\sin\theta/\lambda \tag{27}$$

$$\tau = \varphi/2\pi f = Md\sin\theta/\lambda \tag{28}$$

由此可将 θ 和 τ 的关系解算出来，

$$\theta = \sin^{-1}[(\tau/Md)c] \tag{29}$$

当基元数M较大时，$Md \approx L/2$，L 为基阵总长度．于是

$$\theta \approx \sin^{-1}\left(\frac{2\tau}{L} c\right) \tag{30}$$

图 4 给出了 τ，θ 的关系．当 θ 很小时，它们是近于线性的．由图中可以看出，对长为 3m 的线阵，如果要求定向精度为 0.5°，则对时延的估计精度应为 8.7μs.

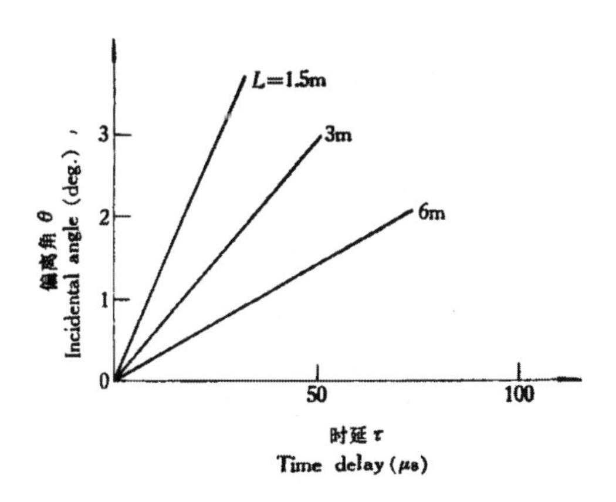

图 4 线阵的信号入射角 θ 与时延 τ 的关系

The relationship between incident angle θ and time delay τ for line array

2. 圆弧阵

对于圆弧阵来说，等效声中心的计算要复杂一些．并且我们可以预见，当频率不同时，等效声中心是移动的(图 3(b))．这将给系统设计带来一定的困难．

考虑一个均匀分布的圆阵，相邻基元的夹角为 $(2\pi/N_0) = \alpha_0$．圆阵半径为 r_0．以阵中心为参考点，分裂波束左半部第 i 路信号的时延为

$$\tau_i(\theta) = \frac{r_0}{c}\cos\left[\theta + \frac{\alpha_0}{2} + (i-1)\alpha_0\right] \qquad i = 1, \cdots, M \tag{31}$$

右半部第 i 路信号的时延为

$$\tau_i(\theta) = \frac{r_0}{c}\cos\left[\theta - \frac{\alpha_0}{2} + (i+1)\alpha_0\right] \qquad i = -1, \cdots, -M \tag{32}$$

于是

$$\begin{aligned}\sum_{i=1}^{M} U_i(t) &= \sum_{i=1}^{M} \cos\{2\pi f[\tau_i(\theta) - \tau_i(0)]\} \\ \sum_{i=1}^{M} V_i(t) &= \sum_{i=1}^{M} \sin\{2\pi f[\tau_i(\theta) - \tau_i(0)]\}\end{aligned} \tag{33}$$

代入 (22) 式，得到

$$\alpha = \mathrm{tg}^{-1}\left[\sum_{i=1}^{M} V_i(t) \Big/ \sum_{i=1}^{M} U_i(t)\right] \tag{34}$$

类似地可以得到右波束信号的相位

$$\beta = \mathrm{tg}^{-1}\left[\sum_{i=-1}^{-M} V_i(t) \Big/ \sum_{i=-1}^{-M} U_i(t)\right] \tag{35}$$

由 (33)—(35) 式去计算 $\varphi = \beta - \alpha$ 是很麻烦的,由于 φ 并不是 f 的线性函数,所以也就无法用 $\tau = \varphi/2\pi f$ 来估计时延. 但是在精确定向问题中, θ 一般很小,在此情况下,可以求出一个 φ 与 θ 的近似表达式. 使圆弧阵也象直线阵一样有一个简单的 τ、θ 解算公式.

设 $\theta \approx 0$, 于是由 (32) 及 (33) 式得出

$$\sum_{i=1}^{M} U_i(t) = \sum_{i=1}^{M} \cos 2\pi f\left\{\frac{r_0}{c}\left[\cos(\theta + (i-0.5)\alpha_0) - \cos((i-0.5)\alpha_0)\right]\right\}$$
$$\approx M \tag{36}$$

上边的近似式中,用了三角函数的和差化积公式,并假定

$$\frac{r_0}{c} 2\pi f\theta \ll 1 \tag{37}$$

另外

$$\sum_{i=1}^{M} V_i(t) = \sum_{i=1}^{M} \sin 2\pi f\left\{\frac{r_0}{c}\left[\cos(\theta + (i-0.5)\alpha_0) - \cos((i-0.5)\alpha_0)\right]\right\}$$
$$\approx -2\pi f\frac{r_0}{c}\theta \sum_{i=1}^{M} \sin\left[\frac{1}{2}\theta + (i-0.5)\alpha_0\right]$$
$$= -2\frac{r_0}{c} f\theta N_0\left[\sin\left(\frac{M}{2}\alpha_0\right)\right]^2 \tag{38}$$

于是

$$\alpha = \mathrm{tg}^{-1}\left[\sum_{i=1}^{M} V_i(t) \Big/ \sum_{i=1}^{M} U_i(t)\right] \approx \frac{-(r_0/c) f\theta N_0}{M}\left[\sin\left(\frac{M}{\alpha}\alpha_0\right)\right]^2$$

类似地

$$\beta \approx \frac{2(r_0/c) f\theta N_0}{M}\left[\sin\left(\frac{M}{2}\right)\alpha_0\right]^2$$

于是

$$\varphi = \frac{4(r_0/c) f\theta N_0}{M}\left[\sin\left(\frac{M}{2}\alpha_0\right)\right]^2 \tag{39}$$

$$\tau = \frac{2r_0\theta N_0}{\pi c M}\left[\sin\left(\frac{M}{2}\alpha_0\right)\right]^2 \tag{40}$$

这就是 θ 与 τ 的近似关系式,由(40)式可以看出,当 θ 较小时,圆弧阵等效声中心也是不动的. 它的几何意义见图 5. 图中直线 BB' 的长度等于 $\overset{\frown}{AA'}$, 声中心既不在 AA' 也不在 BB',它的长度为 x, 正好使 AA' 成为 $\overset{\frown}{AA'}$ 和 x 的比例中项,即 $(\overline{AA'})^2 = \overset{\frown}{AA'} \cdot x$, 所以

$$x = \frac{2N_0 r_0[\sin(M\alpha_0/2)]^2}{M\pi}, \qquad \tau = \frac{x}{c}\theta$$

式(40)是一个近似程度相当好的式子. 图 6 给出按 (40) 式计算的值与按 (33)—(35) 式计算值的比较. 我们可以看出, 当 $\theta \leqslant 3°$ 时, 近似程度非常好, 误差完全可以忽略. 由此得出与 (29) 类似的 τ、θ 解算式子:

$$\theta = \frac{\pi M \tau c}{2 r_0 N_0 [\sin(M \alpha_0 / 2)]^2} \tag{41}$$

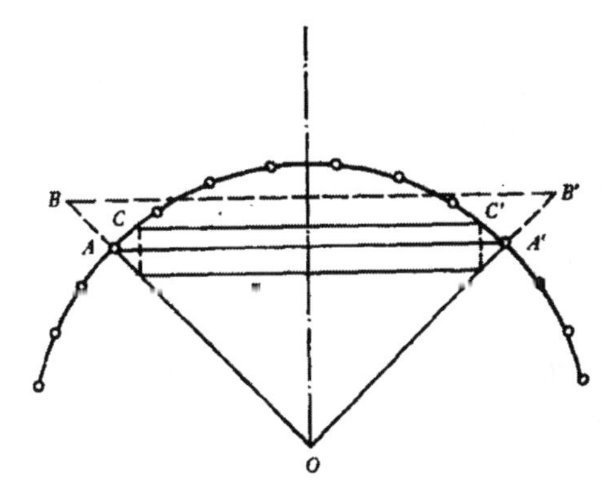

图 5　圆弧阵等效声中心的几何意义

The geometric explanation of equivalent acoustic center of circular array

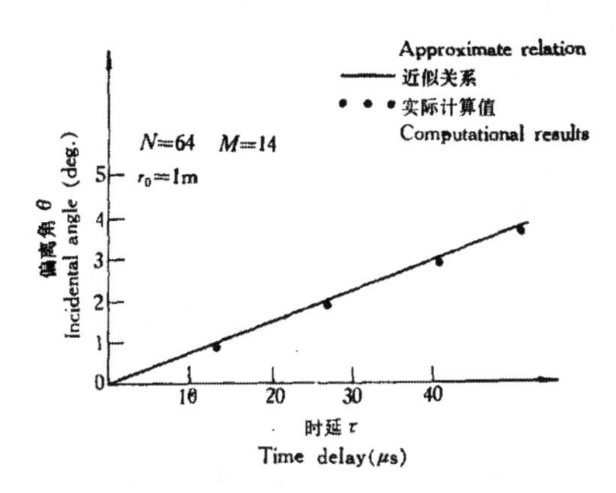

图 6　圆阵信号入射角 θ 与时延 τ 的关系

The relationship between incident angle θ and time delay τ for circular array

——近似关系 Approximate relation　· · · 实际计算值 computational results

四、改善定向精度的若干方法

根据前面的分析可以看出，对于线阵与圆阵，为了精确定向，应当由 $l(t)$，$r(t)$ 分别进行 K 次 DFT，然后在各频率 f_l 上计算出 φ_l 值．由一组 φ_l 值得到 τ，再解算 τ 得到 θ．从 $\{\varphi_l\}$ 得到 τ 可以采取不同的方法．式(15)给出的最小二乘估计 $\hat{\tau}_{ms}$ 仅仅是方法之一，它实际上是对 φ_l 用频率 f_l^2 来加权平均；另一种方法是等权平均：

$$\hat{\tau}_{ew} = \frac{1}{N-1} \sum_{l=1}^{N-1} \frac{\varphi_l}{2\pi f_l} \tag{42}$$

当 φ_l 的估计误差相互独立时，$\hat{\tau}_{ew}$ 是一个相当好的无偏估计量．

如果我们考虑到输入信号的谱的分布，那么显然要采用别的加权方法，即

$$\hat{\tau}=\sum_{l=1}^{N-1} w_l \frac{\varphi_l}{2\pi f_l}.$$

其中 w_l 是加权系数，且 $\Sigma w_l=1$. 加权系数的选择应考虑到测量 φ_l 时引入的误差的大小. 按(9)式，

$$\begin{aligned}Y_q(l)=&\exp(2\pi j\tau l/NT_s)\left\{X_q(l)+\sum_{k=0}^{n_0-1}[x((k+N+q)T_s)\right.\\&\left.-x((k+q)T_s)]\exp(-2\pi jkl/N)\right\}\\&\triangleq X_q(l)\exp(2\pi j\tau l/NT_s)\left(1+\frac{Ae^{j\varphi_0}}{|X_q(l)|}\right)\end{aligned} \tag{43}$$

其中 A 为某一实数，φ_0 为相位，它们由下式确定:

$$Ae^{j\varphi_0}=\sum_{k=0}^{n_0}[x((k+N+q)T_s)-x((k+q)T_s)]\exp(-2\pi jkl/N)\cdot\exp j\alpha$$

其中 α 为 $X_q(l)$ 的相位.

由此可以看出，当 $|X_q(l)|$ 较大时，我们令 $Y_q(l)=X_q(l)\exp(2\pi j\tau l/NT_s)$ 的误差就小，否则误差就大. 所以 $|X_q(l)|$ 可以认为是实际测量 φ_l 时的信噪比，于是最优权的权系数应为(见附录):

$$w_l=\frac{1}{W_0}|X_q(l)|^2,\qquad W_0=\sum_{l=1}^{N-1}|X_q(l)|^2 \tag{44}$$

我们得到一个新的估计量

$$\hat{\tau}_{pw}=\frac{1}{W_0}\sum_{l=1}^{N-1} w_l\frac{\varphi_l}{2\pi f_l} \tag{45}$$

因此，我们有三种关于 τ 的估计量 $\hat{\tau}_{ms}$，$\hat{\tau}_{ew}$ 和 $\hat{\tau}_{pw}$. 下一节将给出例子说明它们的精度.

由(9)式我们还发现，误差的另一个来源是 DFT 在截断时边缘样本的影响. 所以对数据加窗可以提高估计的精度. DFT 中常用的减少泄漏的窗口我们这里都可以应用.

五、系统模拟的结果

图 7 给出了分裂波束精确定向系统的方框图. 我们曾在计算机上进行系统模拟实验.

从理论上讲，如果输入信号 $x(t)$ 的带宽为 W，那么采样频率应为 $f_s=2W$，采样周期 $T_s=1/f_s$. 这时 N 点 DFT 的频率分辨力是 $1/NT_s$. 由于输入是实信号，所以在 N 条谱线中只有 $N/2$ 条是独立的. 参加拟合的可以有 $N/2$ 个分量. 但是我们前面已指出，在频带的边缘部分由于 $|X(l)|^2$ 较小，估计误差相对较大. 从另一方面来说，随机信号经过截断之后，得到长为 NT_s 的序列. 由于 DFT 计算的是循环序列的谱，故有效带宽仅为 $1/T_s$，可能仅有几条谱线有较大的值. 为了克服这一缺点，我们在不增加 DFT 容量和计算量的前提下，采用补零的方法. 把序列补上 N 个零点，得到长为 $2NT_s$ 的 DFT. 于是谱线间隔变为 $1/2NT_s$.

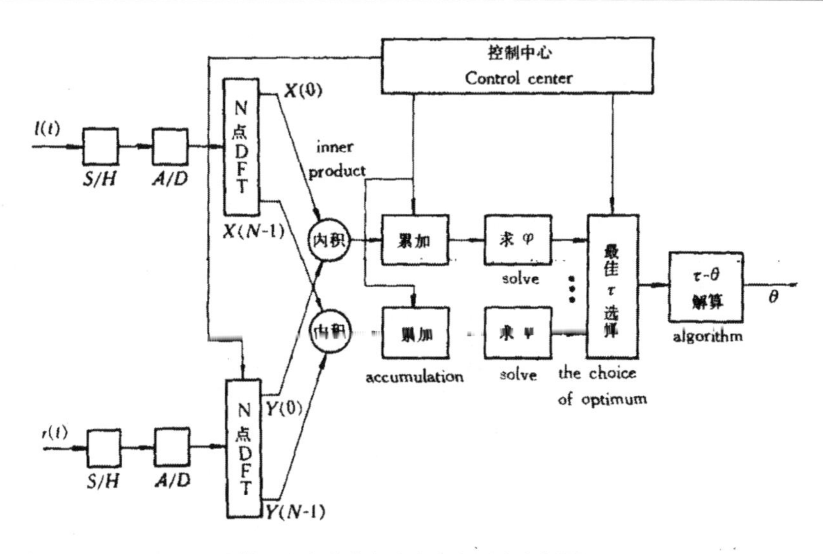

图 7 分裂波束精确定向系统方框图

The block diagram of precise bearing of split-beam array system

这时有资格参加拟合的谱线为N条，但是我们仅取 $N/2$ 条，它们相对来说有较大的谱值．也就是说，我们计算的是

$$\left.\begin{aligned} X(l) &= \sum_{k=0}^{N-1} x(kT_s)\exp(-2\pi jkl/2N) \\ Y(l) &= \sum_{k=0}^{N-1} y(kT_s)\exp(-2\pi jkl/2N) \quad l=0,\cdots,\frac{N}{2}-1 \end{aligned}\right\} \tag{46}$$

这种方法可使估计精度进一步提高．

下面以具有随机相位的周期方波为例来说明系统模拟的结果．

信号周期为 $64T_s$，延时 $\tau \leqslant 0.5T_s$，方波宽度 $8T_s$． 相位在 $[0, T_s]$ 内均匀分布． 当 $\tau = 0.5T_s$ 时，三种估计量的计算结果见表 1．

表 1 τ 的三种估计量

Three estimates of time delay τ

累加次数 K Number of accumlation	8	12	16
$\hat{\tau}_{ew}$	0.598	0.481	0.787
$\hat{\tau}_{ms}$	0.471	0.358	0.589
$\hat{\tau}_{pw}$	0.578	0.503	0.670

当 $\tau = 0.3T_s$ 时，互功率谱的一组典型值见表 2． 由此表我们可以求得 $\hat{\tau}_{ew} = 0.2582$，$\hat{\tau}_{ms} = 0.2242$，$\hat{\tau}_{pw} = 0.2732$．

前面已指出，对数据加窗可以改善估计精度，表 3 给出我们采用余弦窗口的模拟计算结果．

表 2 互功率谱的典型值

Typical Values of cross-power spectrum of input data

l	1	2	3	4	5	6	7
$\|X(l)\|^2$	830	794	729	629	502	364	236
$\hat{\varphi}$	0.03157	0.05889	0.08291	0.10037	0.11357	0.12492	0.13763
$\hat{\tau}$	0.32158	0.30500	0.28152	0.25560	0.23137	0.21208	0.20026

表 3 加窗对估计精度的改善

Improvement of accurancy by setting window

τ 的真值 True value of τ		0.1	0.2	0.3	0.5
$\hat{\tau}_{ms}$	未加窗 without window	0.1373	0.1377	0.2242	0.5889
	加 窗 with window	0.0680	0.1655	0.3283	0.4724

除了以上模拟实验之外，我们还在不同信噪比下作了实验．假定左、右波束的噪声相互独立，且与信号也独立．当 SNR 减小时，应使累加次数增加，当信噪比超过 10dB 时，$K=8$，$N=64$ 便可以得到较好的估计；当信噪比为 -10 dB 时，K 应大于 64，否则误差就较大．Piersol 的研究已指出这一点[8]．

最后要说明的是，由于采样带来了时延补偿的量化误差，使我们得不到真正的同步波束，它对估计精度会有一定的影响．但实际计算表明，在信号偏离角 $\theta \leqslant 3°$ 时，这种误差可以忽略(图 8)．

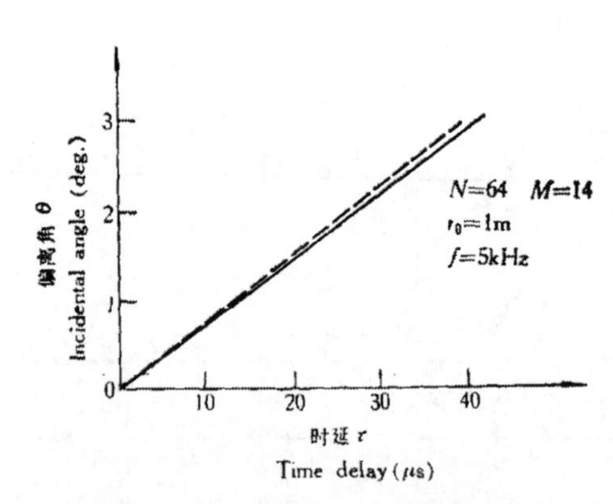

图 8 圆阵分裂波束量化误差的影响

The effect of quantitative error of circular array

——精确延时 Precise time-delay ----$T_s=32\mu s$

六、结 论

按本文提出的线阵、圆弧阵的 θ 与 τ 的解算公式，配合数字多波束系统，用 DFT 可以实时跟踪多个目标．例如，若信号频带为10kHz，采样间隔取为30μs，则能够估计的最大时延为15μs．相对于孔径为2m左右的基阵，最大信号偏离角为1.5°．如果 DFT 为64点，则要求在 $64 \times 30\mu s \approx 2ms$ 内完成 DFT，若实际硬件完成64点 FFT 的时间为0.5ms，那么用它就可跟踪2个目标。

本文的工作是在侯自强同志的倡议下进行的．同时曾与王忠斌、赵真、项定长、赵国英、陈玉凤等同志进行过讨论，作者对他们表示衷心的感谢。

附　　录

设 y 是待估计的量．对 y 进行 N 次独立观测，每次观测时的信噪比为 L_i，$i=1,\cdots,N$，观测值为 y_i．取 y_i 的加权和作为 y 的估计值：

$$\hat{y}=\sum_{i=1}^{N} w_i y_i,$$

在

$$\sum_{i=1}^{N} y_i=1$$

的条件下，使得 $\mathrm{Var}[\hat{y}]$ 极小．我们应取 $w_i=L_i/L$，其中

$$L=\sum_{i=1}^{N} L_i.$$

证明：这是一个条件极值问题，用 Lagrange 乘子法，令

$$P=\mathrm{Var}[\hat{y}]-\lambda\left(\sum_{i=1}^{N} w_i\right)$$

对 P 求微商，得到

$$2w_i/L_i-\lambda=0,\quad i=1,\cdots,N$$

再利用

$$\sum_{i=1}^{N} w_i=1,$$

得到

$$w_i=L_i/L,\quad i=1,2,3,\cdots,N.$$

参　考　文　献

[1] McDonarld, V. H. and Schulthesis, P. H., "Optimum passive bearing estimation in spatially incoherent noise enviroment" *J. Acoust. Soc. Am.*, **46** (1969), 37—43.

[2] Hahn, W. R., "Optimum signal processing for passive sonar range and bearing estimation", *J. Acoust. Soc. Am.*, **58** (1975), 201—207.

[3] Carter, G. C., "Time delay estimation for passive sonar signal processing", *IEEE Trans.*, **ASSP-29** (1981), 463—470.

[4] Pasupthy, S. and Alford, W. J., "Range and bearing estimation in passive sonar", *IEEE Trans.*, **AES-16** (1980), 244—250.

[5] Knapp, C. H. and Carter, G. C., "The generalized correlation method for estimation of time delay", *IEEE Trans.*, **ASSP-24**, (1976), 320—327.
[6] Usher, T Jr., "Random bearing errors in split-beam system", *J. Acoust. Soc. Am.*, **37** (1965), 912—920.
[7] Knight, W. C. *et al.*, "Digital signal processing for sonar", *Proc. IEEE*, **69** (1981), 1451—1506.
[8] Piersol, A. G., "Time delay estimation using phase data", *IEEE Trans.*, **ASSP-29**, (1981), 471—477.
[9] Bendat, J. S. and Piersol, A. G., "*Engineering applications of correlation and spectral analysis*", (John Wiley and Sons, 1980).
[10] Bendat, J. S., "Statistical errors in measurement of coherence function and input/output quantities", *J. Sound and Vibration*, **59** (1978) 405—421.

PRECISE BEARING METHOD OF DIGITAL SPLIT-BEAM ARRAY SYSTEM

LI QI-HU
(Institute of Acoustics, Academia Sinica)

Received August 23, 1982

ABSTRACT

A precise bearing method of digital split-beam array system is considered in this paper. This method is based on the principle of time delay estimation in terms of cross-spectrum of input data, which is proposed by A. G. Piersol. In the case of array processing there are more than two elements. We have to find the equivalent acoustic center. For a line array, the equivalent acoustic center is mdependent of frequency. But for a circle array the center is dependent on frequency. We derived a simple formula for circle array in the case of small incident angle. The precise value of angle of incident signal can be calculated directly from input data.

The estimation bias depends on the length of DFT, sampling frequency and the number of accumulation. Some measures for improving the accuracy of angle bearing are described in this paper. One method is to weight the estimation value in each frequency bin of DFT in terms of their power spectrum values. In some cases it has been proved that choosing a suitable data window is an efficient method.

The block diagram of digital split-beam system is illustrated. Some results simulated in computer are presented. For a digital sonar it is possible to track more than one targets simultaneously by using the method described in this paper.

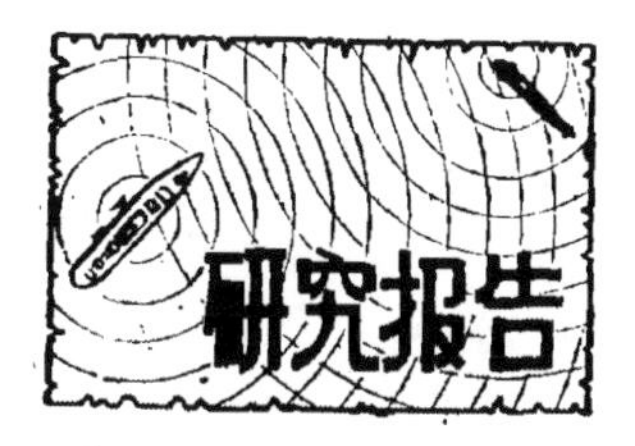

数字多波束系统检测性能的研究①

李　启　虎
（中国科学院　声学研究所）

1983年5月5日收到

本文讨论时域数字多分层、多波束系统的检测性能．首先指出，在有强干扰存在时，一比特量化的限幅系统检测性能将严重下降．进而分析了分层比特数对单波束检测性能的影响．文中提出估计数字式多波束系统检测性能的系统总增益的概念．指出系统均匀性在检测弱信号时的作用．从理论上说明数字多波束系统的模/数转换、后置积累时间以及输出动态的选取准则．

一、引　　言

近年来，数字信号处理技术的迅速发展，推动了声呐系统从模拟处理向数字处理过渡[1-3]．由于数字信号处理技术具有很多优点，比如容易实现多波束全景观察、便于整机控制、能够实现故障自检以及有利于采用 FFT、自动跟踪、自动判决等新的信号处理方法．所以，数字化是声呐现代化的重要标志．

但是，信号的数字处理也带来一些新的问题．例如，对输入信号的分层，要求充分利用信号的动态，所以就出现了分层级差如何与自动增益控制（AGC）配合的问题；对信号的采样、时延补偿、后置积累等都会有新的要求．

早在六十年代初，V. C. Anderson 就提出了一种简单的数字式的多波束系统[4]．用移位寄存器实现了全景观察．但是由于这种系统仅实现了一比特量化，即使不存在干扰，系统的检测性能仍比模拟系统低 1.9dB[5-6]．如果存在强的干扰，则这种系统的性能要严重下降，甚至完全丧失检测微弱信号的能力．这就是所谓强干扰抑制弱目标的问题．因此，这种一比特量化的系统的实际应用受到极大的局限．

近年来，已有不少数字式波束成形方法被提出来[7-10]． R. G. Pridham 等提出的数字内插波束成形方法是属于时域的；P. Rudnick 等所提出的方法则是频域的．无论哪一种方法，都需要对输入信号进行时间量化与幅度量化．A. M. Vural 曾计算过分层运算对检测性能的影响．他发现，即使在不存在干扰的情况下，计算起来也是相当麻烦的[11]．本文首先给出在强干扰下为保证系统的检测性能所需要的分层数目．然后从理论上探讨数字多波束、多分层系统的检测性能．指出，对于微弱信号的检测，系统的均匀性 η 是一个十分重要的量．对于实际系统，由于我们不能做到输出波束完全一致．因此，并不象传统的理论所预计的那样，增长积分时间可以使增益变得越来越大．我们将给出后置积累时间、输出动态与系统均匀性之间的明确关系．从而为数字式多波束、多分层系统的设计提供依据．

二、多分层、多波束系统

本文讨论图1所示的多分层、多波束系统．设输入有 N 路信号．每一路信号都经过 AGC 之后再用频率 f_s 去采样．采样之后的信号经多路混合器依次选通（选通频率是 Nf_s），经 A/D 转换变为数字信号，形成多波束的时间序

① 应用声学, 1984, 3(4): 12-16.

列．每一波束分别经后置积累系统得到多波束的直流输出．

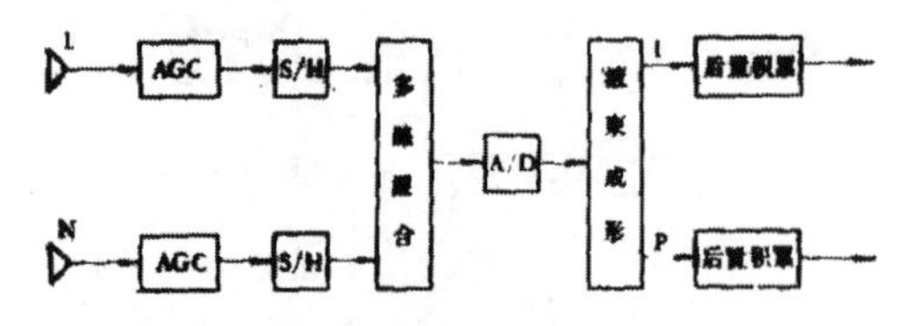

图1 数字多分层、多波束系统方框图

ACC 的作用是使送入 A/D 的信号具有足够的动态，以取得最佳的量化信杂比．一般来讲，如果输入信号为高斯分布的随机信号，均值为零，方差为 σ^2，那么 AGC 的输出动态应保持在 $\pm 4\sigma$ 左右[1]．

我们关心的是，从检测观点来说，分层的比特数应当如何选取才能使数字系统的性能接近模拟系统．图2给出了分层的两种方式．第一种具有零状态，第二种则零状态是过渡的．这两种分层的输入-输出关系分别是：

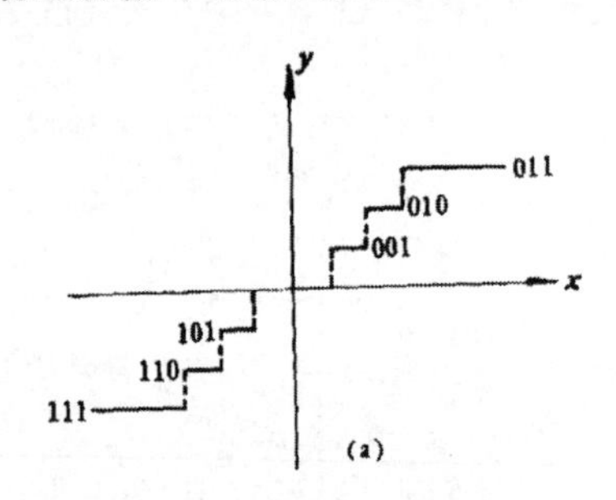

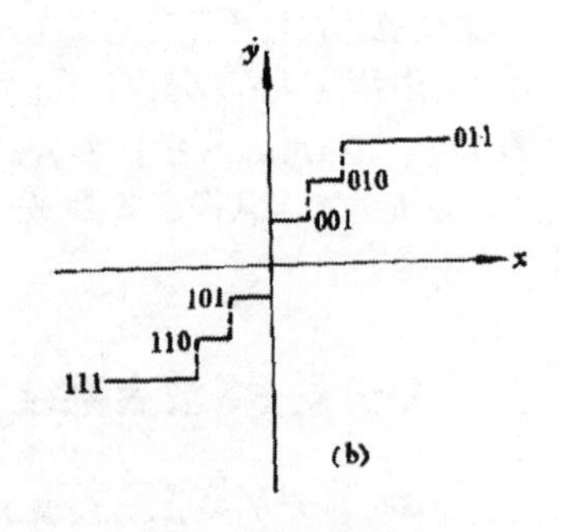

图2 分层的两种方式

$$y=\frac{\Delta}{2}\left\{\sum_{k=1}^{M-1}[\operatorname{sgn}(x-k\Delta)+\operatorname{sgn}(x+k\Delta)]\right\} \tag{1}$$

$$y=\frac{\Delta}{2}\left\{2\operatorname{sgn}x+\sum_{k=1}^{M-1}[\operatorname{sgn}(x-k\Delta)+\operatorname{sgn}(x+k\Delta)]\right\} \tag{2}$$

其中 Δ 为分层级差，M 为分层数．如果 V_0 是信号动态，m 为量化比特数，那么 Δ, M, V_0, m 之间的关系如下：

$$\Delta=V_0/(2^{m-1}-1),\quad M=2^{m-1} \tag{3}$$

由(1)或(2)所表示的这两种分层方式，在计算上差异很小，我们今后分析时，一律以(1)式为准．

我们考虑高斯噪声背景下高斯信号的检测．众所周知，这种情况下的最佳检测器是平方检测器[12]．因此，在我们比较不同分层下系统的性能时，可以只比较相关检测器的性能．

设 $x(t)$, $y(t)$ 为相关检测器的两个输入信号，遵从高斯分布，均值为零．该检测器的输出可以通过下式简单地求出来[13]．

$$R=\sigma^2\int_{-\infty}^{\infty}\int_{-\infty}^{\infty}\frac{\partial f}{\partial x}\frac{\partial g}{\partial y}p(x,y)dxdy \tag{4}$$

其中 $f(x)$, $g(y)$ 表示某种分层运算，由(1)确定．

$$p(x,y)=\frac{1}{2\pi\sigma^2\sqrt{1-\rho^2}}\exp\times\left[-\frac{1}{2\sigma^2(1-\rho^2)}(x^2+y^2-2xy\rho)\right] \tag{5}$$

ρ 为 $x(t)$, $y(t)$ 的相关系数，σ^2 为它们的方差．

如果 $x(t)$, $y(t)$ 中除了相互独立的噪声 $n_1(t)$, $n_2(t)$ 之外，不存在干扰，那么

$$x(t)=s(t)+n_1(t),\quad y(t)=s(t)+n_2(t).$$

在模拟相关的情况下，$f(x)=g(x)=x$．这时输出直流为 $R_{s+n}=\sigma_s^2+\sigma_n^2$．无信号时输出直流为 $R_n=\sigma_n^2$．所以模拟相关器之输出直流跳变是

$$\Delta R=R_{s+n}-R_n=\sigma_s^2 \tag{6}$$

输出起伏是

$$D=V_{ar}\left[\frac{1}{T}\int_0^T n_1(t)n_2(t)dt\right]=\sigma_n^4\frac{2}{T}\int_0^T\left(1-\frac{\tau}{T}\right)\rho_n^2(\tau)d\tau \tag{7}$$

其中 $\rho_n(\tau)$ 为噪声的归一化自相关函数．由此求得模拟相关器之检测增益为

$$G=\Delta R/\sqrt{D}=1\Big/\sqrt{\frac{2}{T}\int_0^T\left(1-\frac{\tau}{T}\right)\rho_n^2(\tau)d\tau}$$

$$\triangleq\sqrt{\frac{T}{\hat{\tau}_n}} \tag{8}$$

其中 $\hat{\tau}_n=2\int_0^T\left(1-\frac{\tau}{T}\right)\rho_n^2(\tau)d\tau$ 为输入噪声的等效相关半径．

多分层相关检测系统的输出，即使在无干扰的情况，计算起来也是相当麻烦的．我们现在利用(3)式，不仅可以方便地计算无干扰时的直流输出 $R_{s+n}^{(M)}$，而且也可以计算有干扰时的输出 $R_{s+n+r}^{(M)}$．

设 $f(x)$，$g(x)$ 由(1)式给出：

$$f(x)=g(x)=\frac{\Delta}{2}\sum_{k=1}^{M-1}[\mathrm{sgn}(x-k\Delta)+sgn(x+k\Delta)].$$

易知

$$\frac{\partial f}{\partial x}=\frac{\partial g}{\partial x}=\Delta\sum_{k=1}^{M-1}[\delta(x-k\Delta)+\delta(x+k\Delta)] \tag{9}$$

把(9)式代入(4)，再予以简化，得到

$$R^{(M)}=\Delta^2\int_0^\rho\frac{2}{\pi\sqrt{1-x^2}}\sum_{k=1}^{M-1}\sum_{l=1}^{M-1}\exp\left[-\frac{\Delta^2}{2(1-x^2)\sigma^2}(k^2+l^2-2klx)\right]dx \tag{10}$$

这里 ρ 为输入 $x(t)$ 及 $y(t)$ 的相关系数．当无信号时

$$\rho|_{r+n}=\frac{L_{rn}}{1+L_{rn}},$$

当有信号时

$$\rho|_{s+r+n}=\frac{L_{rn}+L_{sn}}{1+L_{rn}+L_{sn}}$$

其中 $L_{sn}=\sigma_s^2/\sigma_n^2$，$L_{rn}=\sigma_r^2/\sigma_n^2$ 分别表示输入的信噪比及干噪比．

由(10)式，我们可以得出有干扰情况下，分 M 层的多分层相关器的输出直流跳变

$$\Delta R^{(M)}=R_{s+r+n}^{(M)}-R_{r+n}^{(M)} \tag{11}$$

把这个值与模拟情况下的值[见(6)式]作比较，便可知道分层带来的损失．

以限幅相关为例，

$$\Delta R^{(1)}=\frac{2}{\pi}\sin^{-1}\left(\frac{L_{rn}+L_{sn}}{1+L_{rn}+L_{sn}}\right)-\frac{2}{\pi}\sin^{-1}\left(\frac{L_{rn}}{1+L_{rn}}\right) \tag{12}$$

图3给出了无干扰时，限幅相关检测器与模拟相关检测器性能的比较．我们看到，在最好的情况下，前者的性能仍有1.9dB的损失．输入信噪比越大，增益的损失也越大．幸好我们通常只对小信号的检测感兴趣，所以限幅检测的这个缺陷还是可以容忍的．当有强干扰时，情况就不同了．图4是干噪比为0，0.1，1，10时的计算结果．我们从图中可以看到，当有强干扰时，限幅检测器对微弱信号的检测性能很差．

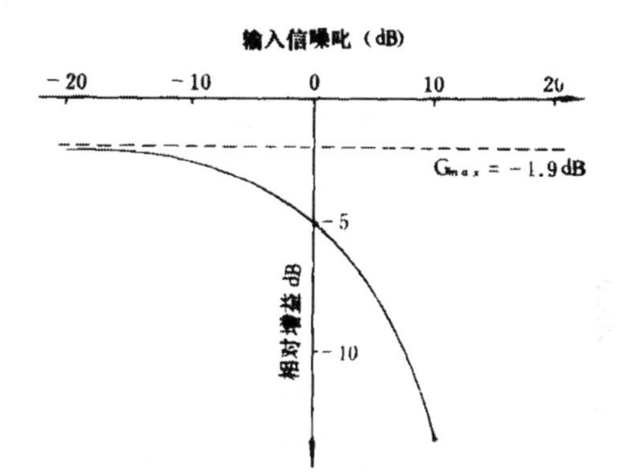

图3　限幅相关的相对增益

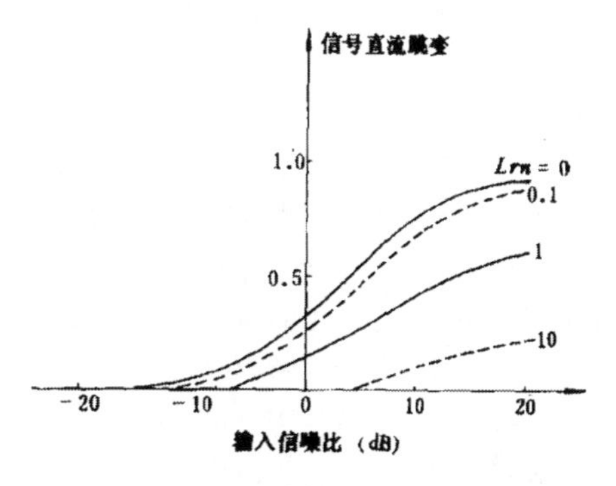

图4　有干扰时限幅系统的检测增益

对于声呐设计者来说，我们总希望它具有抗近场干扰的能力．因此，必须采用多分层．

但是，分层太多会使设备复杂化．所以应当选取合适的分层数，一方面使检测性能接近模拟系统，同时又不会使设备太复杂．

图5是根据(10)式计算的结果．我们看到在 $L_{rn}=1$ (0dB) 时，5比特量化（包括符号位）是足够的．

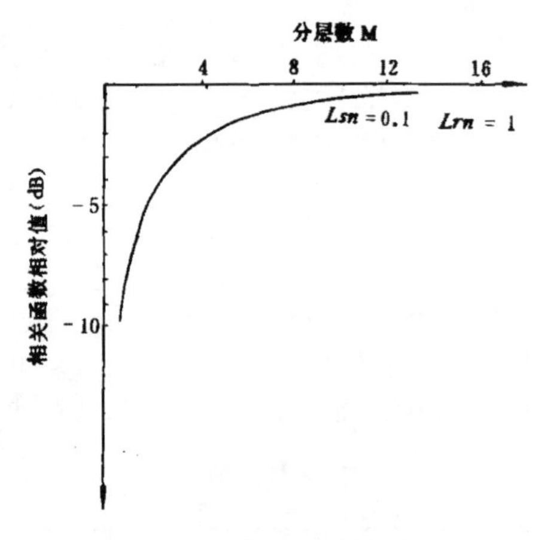

图5　分层数对检测性能的影响

三、多波束系统的检测增益

我们先讨论单波束的检测增益．定义作波束指向对准目标时，该波束输出的直流跳变与输出起伏的比值．我们在前一节已经提到，平方检测器是一种最佳检测器．但是对于数字系统来说，平方检测器要求在每一采样间隔作一次乘法．于是 P 个波束就有 P 次乘法．如果我们改用绝对值检波，那么可以省去乘法运算．对于高斯信号与噪声来说，平方检测和绝对值检测的性能差异是可以忽略的[14]．所以，我们下面的计算就假定后置积累是绝对值检波．

设输入 N 路信号是 $x_1(t),\cdots,x_N(t)$．$x_i(t)$ 是均值为零、方差为 σ_x^2 的高斯信号．当有信号时 $x_i(t)=s(t)+n_i(t)$；当无信号时 $x_i(t)=n_i(t), i=1,\cdots,N$．$n_i(t)$ 表示相互独立的噪声．

令 $y(t)=x_1(t)+\cdots+x_N(t)$，则绝对值检波完成以下运算：

$$R=E[|y(t)|] \tag{13}$$

由于 $x_i(t)$ 是高斯信号，所以 $y(t)$ 仍是高斯信号．我们可以计算它的均值和方差为：

$$E[y(t)]=0,\ \sigma_y^2 \triangleq V_{ar}[y(t)]=N^2\sigma_s^2+N\sigma_n^2 \tag{14}$$

根据高斯信号的特征，容易证明

$$R=E[|y(t)|]=\sqrt{\frac{2}{\pi}}\sigma_y \tag{15}$$

从(13)，(15)式，我们可以计算出全噪声时的输出直流量

$$R_n=\sqrt{\frac{2}{\pi}}N\sigma_n,\quad \sigma_n=\frac{1}{4}V_0 \tag{16}$$

其中 V_0 假定为 AGC 输出、A/D 输入的上限．当有信号时输出直流量是

$$R_{s+n}=\sqrt{\frac{2}{\pi}(N^2\sigma_s'^2+\sigma_n'^2)},$$
$$\sqrt{\sigma_s'^2+\sigma_n'^2}=\frac{1}{4}V_0 \tag{17}$$

其中 $\sigma_s'^2$ 与 $\sigma_n'^2$ 表示原来功率为 σ_s^2 与 σ_n^2 的信号与噪声经自动增益控制后，所得到的新的信号功率与噪声功率．由于自动增益控制器不会改变信号与噪声的相对幅值，所以输入信噪比 L_{sn} 将保持不变．

R_{s+n} 与 R_n 的比值就给出了系统输出的动态范围：

$$U_{s+n}=R_{s+n}/R_n=\sqrt{N^2L_{sn}+1}/\sqrt{L_{sn}+1} \tag{18}$$

这里 L_{sn} 就是上一节定义的输入信噪比．U_{s+n} 与 L_{sn} 的关系见图6，图中给出 $N=20$ 的情况．这条曲线我们称之为检测曲线．在已知单波束输出起伏 σ_0 的条件下，只要输出直流跳变大于 $k\sigma_0$，则我们就可断定有目标．其中 k 是一个规定检测门限的量，通常取 $k=3$．由此便可确定出最小可检测信号．

关于单波束的输出起伏 σ_0 与增益 G 可参照(7)，(8)式进行计算．若我们把单波束的结果直接用到多波束系统中去，那么好像随着积分时间的增长，系统的增益似乎可以无限增加．这无论在理论上还是在实际上都是不可能的．原因之一是，多波束系统对信号的检测并不只是对某一单波束输出观测的结果；而是在各波束

输出直流之间进行比较的结果. 如果没有信号时 P 个波束的输出之间一致性很差，那么产生虚警的可能性就很大. 这时即使每一波束本身输出起伏很小，但也丝毫不能对检测弱信号带来好处. 因为单波束的输出起伏已被淹没在波束的不均匀中了.

为了刻划系统的检测性能，我们定义系统的总的输出起伏为

$$\sigma_{总}^2 = \sigma_0^2 + \sigma^2 \tag{19}$$

其中 $\sigma_0^2 = N\sigma_n^2\sqrt{\frac{\hat{\tau}_n}{T}}$ 表示每一波束的输出量在其直流量附近的起伏，它与积分时间有关.

$$\sigma^2 = \frac{1}{P}\sum_{i=1}^{P}\left(\mu_i - \sum_{i=1}^{P}\mu_i\right)^2 \tag{20}$$

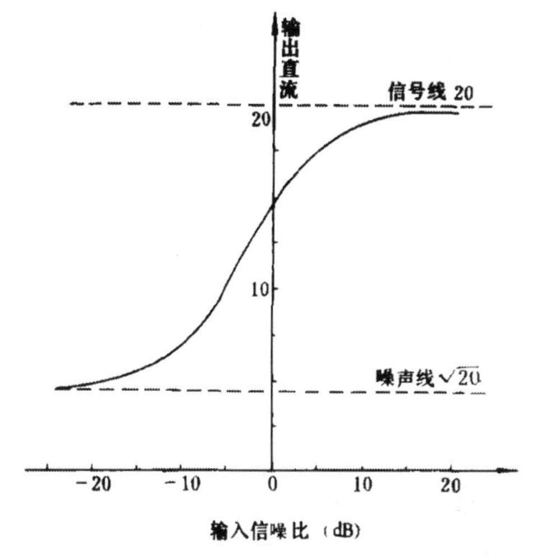

图 6 后置积累为绝对值检波时的检测曲线

σ^2 表示系统各波束之间的不均匀性，μ_i 为第 i 个波束输出的直流量. 按照传统的增益的定义，我们有

$$G = \frac{N^2\sigma_s^2/\sigma_0^2}{\sigma_s^2/\sigma_n^2} = N\sqrt{\frac{T}{\hat{\tau}_n}} \tag{21}$$

经过修正之后的系统增益为

$$G' = \frac{N^2\sigma_s^2/\sigma_{总}^2}{\sigma_s^2/\sigma_n^2} = \frac{N}{\sqrt{\frac{\hat{\tau}_n}{T}} + \frac{\sigma^2}{N\sigma_n^2}} \tag{22}$$

容易看出，如果系统输出各波束之间非常一致，那么仅有噪声时，$\sigma^2 \approx 0$. 这时增加积分时间确实会使系统的增益提高. 但是，当 σ^2 可以与 σ_0^2 比拟时，在 T 增加到一定程度之后，再加长 T 就不会使 G' 有明显的改善了.

我们称 $\sigma^2/N\sigma_n^2 = \eta$ 为系统的均匀性. 图 7 给出了系统增益与 η 的关系. 我们假定输入噪声的等效相关半径是 $\hat{\tau}_n = 200\ \mu s$. 从图上我们看到，当均匀性 η 为 1% 时，积分时间 $T=$ 3s，系统的增益已接近饱和. 再增加积分时间，不会使系统的性能有所改善. 如果均匀性提高到 0.1%，那么积分时间增加到 6s 仍是有价值的.

对于数字化系统，η 选定之后，也就决定了后置积累器的输出动态. 例如，当 $\eta = 0.1\%$ 时，则输出至少应为 10 比特，因为 $2^{10} = 1024$. 能够反映出千分之一的变化.

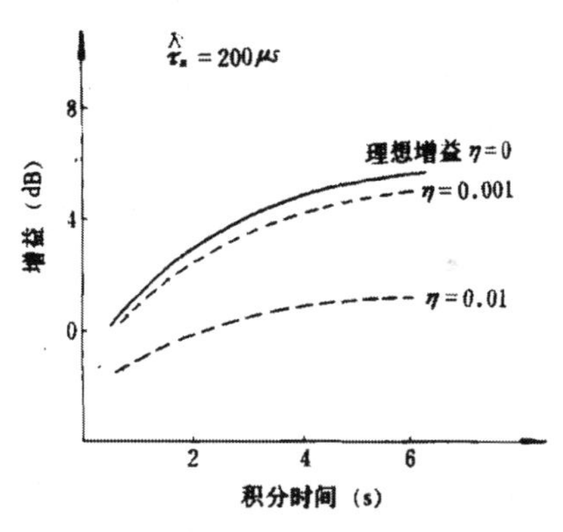

图 7 波束均匀性对系统总增益的影响

四、结　论

为了保证数字式多分层、多波束系统在有强干扰的情况下仍具有接近于模拟系统的性能，A/D 系统的量化比特数应在 5 比特以上. 系统的总增益和波束之间的均匀性有关. 尤其是对微弱信号的检测，只有当系统的均匀性达到一定指标后，增加后置积累时间才是有效的.

影响各波束输出均匀性的因素很多，如海洋噪声的各向异性，系统本身各通道之间的不

一致性等,这是值得进一步探讨的问题.

参 考 文 献

[1] W. C. Knight et al., *Proc. IEEE*, **69**(1981), 1451—1506.

[2] P. Skitzki, Electronic Progress, XVI(1974), 20—37.

[3] T. E. Curtis and R. J. Ward, *IEE Proc.*, **127** (1980), Pt. F 257—265.

[4] V. C. Anderson, *JASA*, **32**(1960), 860—870.

[5] P. Rudnick, *JASA*, **32**(1960), 871—877.

[6] J. Wolloff, *IRE Trans.*, IT-**8**(1962), 5—10.

[7] R. G. Pridham and R. A. Mucci, *Proc. IEEE*. **67** (1979), 904—919.

[8] P. Rudnick, *JASA*, **46**(1969), 1089—1090.

[9] H. J. Whitehouse and J. M. Speiser, Proc. of the NATO Advanced Study Inst. on Signal Processing, 1977, D. Reidel Pub. Comp., Boston, 669—702.

[10] J. F. Dix et al., *IEE Proc.*, **127**(1980), Pt. F 125—131.

[11] A. M. Vural, *JASA*, **46**(1969), 293—313.

[12] *W. W. Peterson, IRE Trans.*, PGIT-4 (1954), Sept. 171—212.

[13] 李启虎,电子学报, 4(1980).

[14] L. Camp, Underwater Acoustics, ch. 10, John-Wiley and Sons, 1970, New York.

相位谱的快速近似计算法①

李启虎
（中国科学院声学研究所）

1982年12月23日收到

时间序列的相位谱，等于它的 DFT 的虚部及实部的比值的反正切. 本文给出一种简单的近似方法，可以快速地由它的 DFT 给出相位谱. 将这一方法与 Robertson 提出的计算功率谱的方法结合起来，会使很多实际应用场合的计算相当简化.

一、引　言

在很多谱分析的应用场合，我们希望计算时间序列的相位谱和功率谱. 如果用

$$Z = A + jB$$

来表示实的时间序列的傅里叶变换，那么功率谱与相位谱分别由以下两式给出

$$P = |Z| = (A^2 + B^2)^{1/2} \tag{1}$$

$$\Phi = \mathrm{tg}^{-1}(B/A). \tag{2}$$

G. H. Robertson 曾提出一种由 A、B 快速计算 P 的近似方法[1]. 他指出，当 $|B| \leqslant |A|$ 时，令

$$P' = |A| + \frac{1}{2}|B| \tag{3}$$

将得到 $P = (A^2 + B^2)^{1/2}$ 的相当好的近似值. 由于(3)式易于用硬件实现，因而在数字式声纳中可以获得广泛应用. 最近，W. C. Knight 等指出[2]，如果用

$$P' = |A| + \frac{3}{8}|B| \tag{4}$$

则近似程度会更好一些. 只是硬件实现比(3)式稍微复杂些. 我们看到，(3)、(4)式都避开了平方与开方的运算，显然比(1)式简单得多.

类似的问题也出现在由(2)式计算相位的过程中. 我们知道，很多实际问题，例如时延估计，需要系统实时给出相位谱[3]. 如果用硬件实现，求反正切则相当复杂. 本文提出一种快速的由 A、B 计算 Φ 的近似方法. 它的精度高，硬件实现也比较容易. 将它与[1]中提出的计算功率谱的方法结合起来，会使 FFT 的后置处理大为简化.

二、反正切的近似计算

先考虑 $|B| \leqslant |A|$ 的情况. 于是我们要寻求当 $|x| \leqslant 1$ 时 $y = \mathrm{tg}^{-1}x$ 的近似式. 如果按泰勒展开式，便有

$$y = x - \frac{1}{3}x^3 + \frac{1}{5}x^5 \cdots\cdots$$

显然我们不能只取前几项作为 y 的近似式子. 因为 x 的取值范围不一定局限于零点附近. 另一方面，我们也不希望取的项数太多，否则乘法的次数太多，影响系统的实时性能.

在 $1 \geqslant x \geqslant 0$ 时，令

$$y_1 = x - \frac{1}{5}x^2 \tag{5}$$

容易验证，$y \approx y_1$(见图1). 计算表明，用 y_1 来近似 y，绝对误差 $(y - y_1)$ 不超过2%，相对误差 $([(y - y_1)/y])$ 不超过4%.

在(2)式中，如果令 $A = \cos\theta$, $B = \sin\theta$, $0 \leqslant \theta \leqslant \pi/4$. 那么，当 θ 在 $(0, \pi/4)$ 内均匀分布时，大约有50%的场合相对误差小于2% ($\approx$ 0.2dB). 由此可知(5)式的近似程度相当好. 类似可以给出 $0 \geqslant x \geqslant -1$ 时 $y = \mathrm{tg}^{-1}x$ 的近似式为

① 应用声学, 1984, 3(2): 42, 43.

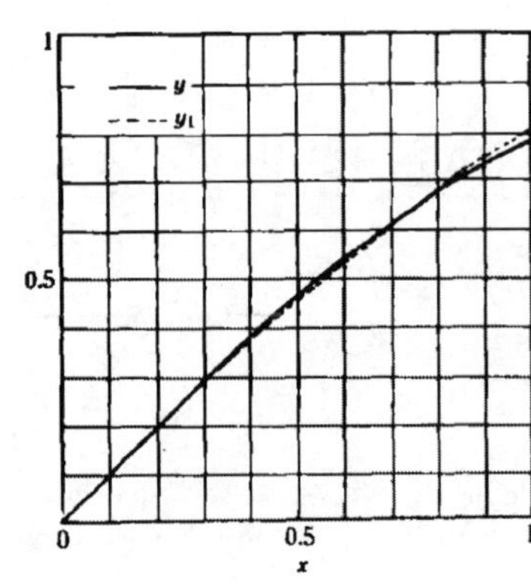

图 1　用 $y = x - \frac{1}{5}x^2$ 近似 $y = \mathrm{tg}^{-1}x$

$$y_1 = x + \frac{1}{5}x^2 \tag{6}$$

(5)、(6)两式可以合并为一个式子

$$y_1 = x - \frac{1}{5}x^2 \,\mathrm{Sgn}\, x \tag{7}$$

式中 $\mathrm{Sgn}\, x$ 为符号函数。

当 $|x| \geqslant 1$ 时，我们不能直接利用(7)式。但是，这时 $1/|x| \leqslant 1$，于是利用正切与余切函数的关系，将(7)式稍作变换就行了。以 $x \geqslant 1$ 为例，由于 $y = \mathrm{tg}^{-1}x$，所以 $x = \mathrm{tg}\, y$，于是 $\frac{1}{x} = ctg y = \mathrm{tg}\left(\frac{\pi}{2} - y\right)$。由此得到

$$\frac{\pi}{2} - y = \mathrm{tg}^{-1}(1/x) \qquad x > 1 \tag{8}$$

在(8)式的右端，$1/x$ 的值小于1，所以便可应用(5)式，于是得到

$$y \approx y_1 = \frac{\pi}{2} - \left[\frac{1}{x} - \frac{1}{5}\left(\frac{1}{x^2}\right)\right], x \geqslant 1 \tag{9}$$

类似地可以得到

$$y \approx y_1 = -\frac{\pi}{2} - \left[\frac{1}{x} + \frac{1}{5}\left(\frac{1}{x^2}\right)\right]$$

$$x \leqslant -1 \tag{10}$$

三、结　论

根据[1]中提出的简化公式与本文提出的方法，我们在 FFT 后面加上一个简单的后置处理，便可同时获得近似程度相当好的功率谱及相位谱(图 2)。

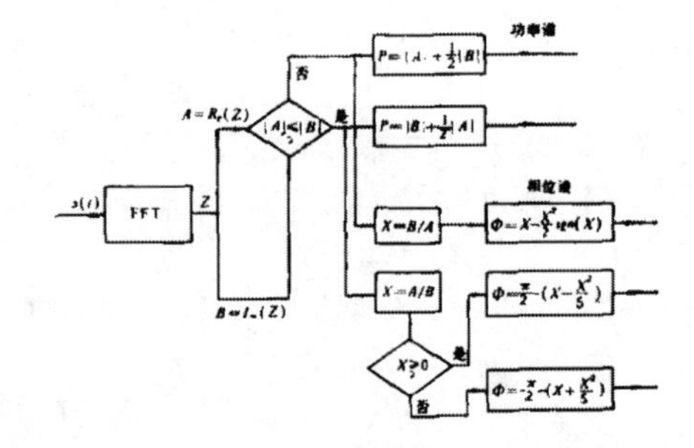

图 2　快速计算功率谱与相位谱的方框图

参考文献

[1] G. H. Robertson, *Bell System Tech. J.*, 50-8 (1971), 2849—2852.

[2] W. C. Knight et al., *Proc. IEEE* 69-11 (1981), 1451—1506.

[3] A. G. Piersol, *IEEE Trans.*, ASSP-29 (1981), 471—477.

声呐设计中的系统模拟技术(I)[①]

李 启 虎
(中国科学院声学研究所)

1984年1月23日收到

摘要 数字信号处理理论与计算机技术的发展已使系统模拟技术成为声呐设计的新的有力的手段.本文叙述在声呐设计中利用系统模拟技术的必要性以及系统模拟技术的主要功能. 文中首先讨论在计算机上模拟声呐环境场的原理及方法. 给出一种快速产生带限高斯随机序列的算法和用变采样率产生高精度基阵信号的原理.同时给出数字式多波束、多分层的波束成形系统的计算机模拟的公式.文中给出在计算机上进行系统模拟的某些结果并与理论值作了比较.

THE SYSTEM SIMULATION TECHNIQUE
IN SONAR DESIGN (I)

LI QI-HU
(Institute of Acoustics, Academia Sinica)

Received January 23, 1984

Abstract The system simulatien technique has become one of powerful approach in sonar design due to the development of the digital signal processing theory. The necessacity of using system simulation technique in sonar design and the main functions of this technique are described in this paper. The principle and method of simulation of sonar environment field in digital computer is firstly shown here. A fast algorithm for generating band-limited gaussian random series and a method for generating high precies array signal by using the method of changing sample rate are proposed.

The formula of computer simulation for modern digital multibeam, multi-layered beamforming system is presented. The sim ulation method and some results carried out in digital computer are given.

一、引 言

当我们按照给定的指标设计一部声呐时，往往面临一种非常复杂的局面：有很多互相关联的参数需要我们作出选择；有很多可供应用的技术需要我们决定取舍. 我们要估价，为了提高某一技术指标而增加设备的复杂性是否值得. 我们又要在各种代价中权衡利弊，作出折衷. 但是，我们不可能在方案设计阶段就研制各种硬件. 这样作，人力、物力的浪费是显而易见的，

① 声学学报, 1986, 11(4): 214-222.

并且研制周期也比较长.

信号处理理论与计算机技术的发展使我们有可能在一台通用的数字计算机上模拟声呐系统.

用计算机去模仿声呐系统中某一部分的信号处理方法或某一硬件是人们早已采用的办法[1-3]. 尤其是对于数字式声呐的设计,由于它可以采用许多新的信号处理手段,整个系统往往很复杂[4-5]. 如果不事先在计算机上进行系统模拟,那么必然不能得到完善的设计. 所以,在今后声呐技术的发展中,系统模拟技术将会起到越来越重要的作用[6].

所谓系统模拟就是由数字计算机模仿我们所设计的声呐系统,按设计框图进行数据处理,把各种必要的中间结果和最后结果输出来,根据这些结果来估价我们设计方案的优劣.

图 1 给出了声呐系统模拟的基本框架. 其中声呐环境场(包括信号、噪声、混响及干扰)的模拟是整个系统模拟的基础. 我们要提出一种快速产生基阵信号的算法,从而为在计算机上模拟声呐系统提供可靠的基础.

为了模拟数字多波束系统,我们还要推导出采样、分层和指向性的计算公式,说明在计算机上模拟的具体方法及模拟结果.

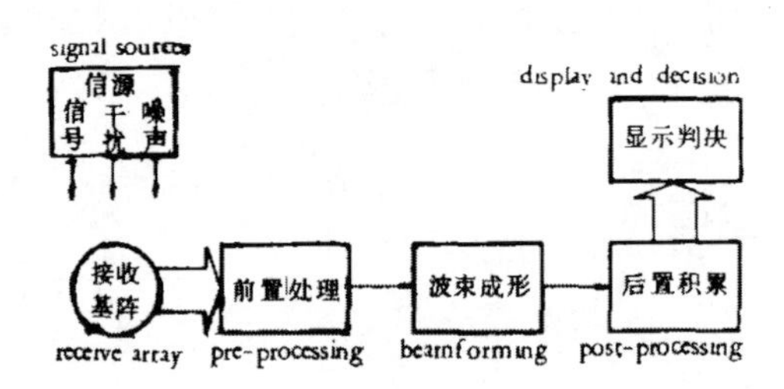

图 1　声呐系统模拟的组成

The framework of sonar system simulation

二、声呐环境场的模拟

声呐环境场指的是声呐基阵所接收的目标信号、干扰与海洋噪声. 我们假定这些量是线性可加的.

设一个基阵由N个基元构成,目标信号源为 $s(t)$,入射角为 θ_0;第m个干扰源是 $r_m(t)$,入射角为 θ_m, $m=1,2,\cdots,M$. 于是第 i 个基元所接收到的信号为

$$x_i(t)=s[t-\tau_i(\theta_0)]+n_i(t)+\sum_{m=1}^{M} r_m[t-\tau_i(\theta_m)] \tag{1}$$

其中 $\tau_i(\theta_0)$ 表示目标信号到达第 i 个基元时的相对延时,$\tau_i(\theta_m)$ 表示第m个干扰到达第 i 个基元时的相对延时,$n_i(t)$ 表示第 i 个基元所接收到的背景噪声.

如果是混响背景,那么还应当在 $x_i(t)$ 中加上混响,这时 $s(t)$ 就代表回波信号.

从系统模拟角度来说,干扰是一种特殊形式的信号,所以它们的模拟可用相同方法来进行. 因此,声呐环境场的模拟由三部分构成,即信号、噪声与混响. 下面我们分别予以讨论.

我们先规定一下所采用的记号．用 T_s 代表采样间隔．第 kT_s 时刻的信号用 $x(kT_s)$ 或 $x(k)$ 来表示．在系统模拟中，T_s 仅仅有相对的意义，我们可以认为它的单位是 ms，也可以认为它的单位是 μs．它只有在与别的量作比较时才有意义．

信号 在主动声呐信号的模拟中，如果信号是单频的，那么模拟是十分容易的．因为在这种情况下，延时 $\tau_i(\theta_0)$ 可以换算为相位 φ_i．令

$$s(k) = A\cos(2\pi fkT_s + \varphi) \quad k = 1, 2, \cdots \tag{2}$$

就得到单频信号序列．其中 f 是信号频率，φ 是我们所要求的相位，幅值 A 可以用于调节基元信号的输入信噪比．在(2)式中，如果把 T_s 用采样频率 $f_s = 1/T_s$ 来表示，则

$$s(k) = A\cos(2\pi kf/f_s + \varphi) \quad k = 1, 2, \cdots. \tag{3}$$

如果信号是窄带的(主动声呐情况)或宽带的(被动声呐情况)，那么我们要模拟的就是具有给定频谱特性的带限高斯随机序列．其标准方法如下所述．

首先要产生在 [0, 1] 中均匀分布的随机变量序列 $u(k)$．在一般的计算机中，都有产生 $u(k)$ 的子程序，它们通常都是用混合同余法产生的[7-9]．

$$u(k+1) = \lambda u(k) + c \quad (\mathrm{Mod}2^L) \tag{4}$$

其中 λ，c 都是常数，L 是正整数；$\mathrm{Mod}2^L$ 表示与 2^L 同余．

由于水声信号处理中往往需要非常长的随机序列，如果标准子程序所产生的 $u(k)$ 不满足我们的要求，那么我们应当自行产生性能较好的 $u(k)$．使得 $u(k)$ 具有尽量长的周期与独立性．下面的参数将给出性能相当好的 $u(k)$ 序列[10]：

$$\lambda = 257, \quad c = 12345678911, \quad u(0) = 12137516145.$$

从 $u(k)$ 产生带限高斯随机序列的传统方法是数字滤波(见图 2)．即由 $u(k)$ 分段求和得到一个具有高斯分布的白噪声序列 $w(k)$：

$$w(k) = \left\{\sum_{i=1}^{P} u[(k-1)P + i] - P/2\right\} \Big/ \sqrt{P/12} \tag{5}$$

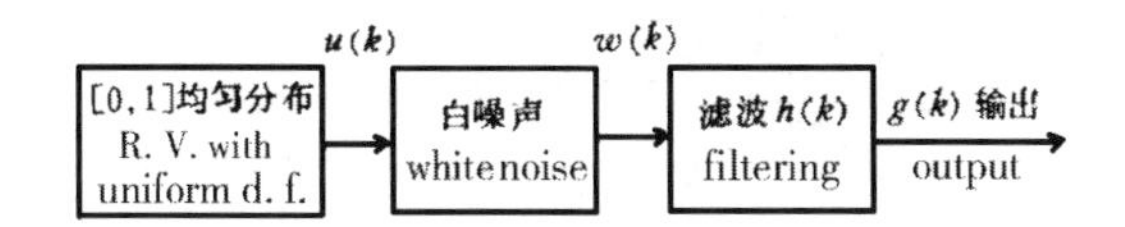

图 2 产生高斯随机序列的标准方法

The canonical method of generating gaussian random series

这里在 $u(k)$ 序列中，每隔 P 个数就产生一个 $w(k)$，(5)式中所表示的是经过归一化的白噪声高斯序列．它的均值为零，方差为 1．在实用中，取 $P \geq 30$，就可以得到性能相当好的白噪声高斯序列．

把 $w(k)$ 通过一个脉冲响应为 $h(k)$ 的数字滤波器，我们就得到一个具有指定频谱的高斯随机序列 $g(k)$：

$$g(k) = h(k) * w(k) = \sum_{i=0}^{M-1} h(k) w(k-i) \tag{6}$$

这里 M 为卷积窗口的长度．$h(k)$ 通常是由所指定的频谱经 Fourier 反变换而得到的．从(6)式

可以看到，为了得出一点 $g(k)$，需要作 P 次加法和 M 次乘法．当 M 较大时，运算量是非常大的．

下面我们提出一种快速产生 $g(k)$ 的方法．根据声呐信号的特点，$g(k)$ 为带通信号．对于带通随机信号，只要它的带宽 Δf 小于中心频率 f_0，就可以写成[11]：

$$g(k)=p(k)\cos(2\pi f_0 kT_s)-q(k)\sin(2\pi f_0 kT_s) \tag{7}$$

这里 $p(k)$，$q(k)$ 是两个相互独立的随机序列，带宽为 Δf．如果我们对 $g(k)$ 谱的形状没有特殊要求，那么从 $u(k)$ 可以采用滑动求和的方法．滑动窗长度的倒数就决定了带宽 Δf．图 3 给出了这种方法的方框图及实际产生的 1024 点的 $g(k)$．中心频率 $f_0=5\text{kHz}$，带宽 $\Delta f=4$ kHz．用这种方法每产生一个 $g(k)$ 只需作两次乘法，运算速度大为加快．

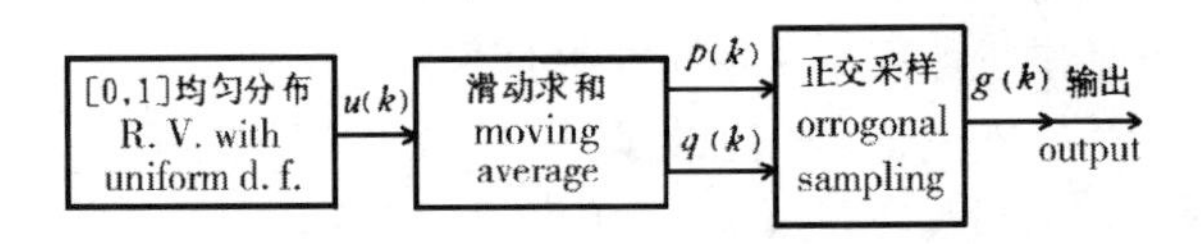

(a) 方框图 block diagram

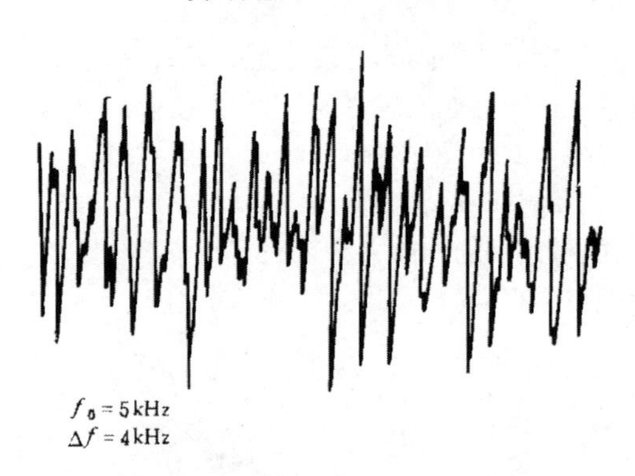

(b) 信号的波形 signal waveform

图 3 产生高斯随机序列的简单方法

A simple method of generating gaussian random series

在得到序列 $g(k)$ 之后，我们就可以按实际需要模拟基阵信号，它的一般形式为 $s(t+\tau)$．为了精确模拟相对延时差 τ 以及很好地与声呐预处理中的采样相衔接，我们必须采取相应的措施．一种比较好的方法就是降采样率技术[12]．

举例来说，如果我们要产生的两个基元信号的相对时延是 $3\mu s$，而声呐信号处理系统的采样间隔为 $8\mu s$．那么我们不能一开始就按 $T_s=8\mu s$ 产生随机序列 $g(k)$．因为这时 $g(k)$ 和 $g(k+1)$ 的相对时延为 $8\mu s$．我们可以用 $T_s'=1\mu s$ 的采样间隔产生一个随机序列 $g'(kT_s')$，这时 $g'(kT_s')$ 与 $g'(kT_s'+3T_s')$ 的相对时延为 $3\mu s$．再对它们降采样率，令 $T_s=8T_s'$，得到 $g(kT_s)=g'(8kT_s')$ 以及 $\tilde{g}(kT_s)=g'(8kT_s'+3T_s')$．这样我们便知道 $g(k)$ 和 $\tilde{g}(k)$ 的相对时延为 $3\mu s$，而采样间隔又是 $8\mu s$．这种方法的原理方框图见图 4．

噪声 我们假定各基元所接收到的信号相互独立，那么用分段滑动平均的方法很容易给出 N 路相互独立的噪声来．由于噪声的独立性；我们无须用降采样率的方法形成人为的时延．

例如，若 $g(k)$ 是我们按(7)式所产生的高斯随机序列，相关半径为 N_0T_s。令

$$n_i(k) = g(iN_0 + k), \quad i = 1, 2, \cdots, N \tag{8}$$

这代表了 N 路相互独立的随机噪声。

混响 混响是一个非平稳随机过程，它的平均振幅按一定规律随时间衰减[13]。根据 Middleton 的分析[14]，在每一时刻，混响的振幅都遵从 Rayleigh 分布。所以，我们从高斯随机序列 $g(k)$ 出发，先产生两个相互独立的高斯随机序列 $v_1(k)$，$v_2(k)$，再得出遵从 Rayleigh 分布的随机序列

$$r(k) = [v_1^2(k) + v_2^2(k)]^{1/2} \tag{9}$$

对 $r(k)$ 进行振幅调制，得

$$r_0(k) = A(k)r(k) \tag{10}$$

其中

$$A(k) = \frac{\alpha \exp(-\beta k T_s)}{kT_s} \tag{11}$$

α，β，T_s 为待定常数。$r_0(k)$ 就是我们所需的混响信号。

图 5 给出了动态为 80dB 的混响的模拟结果。我们同时还给出了一次海上实测的记录。从图上可以看出模拟的情况是良好的。

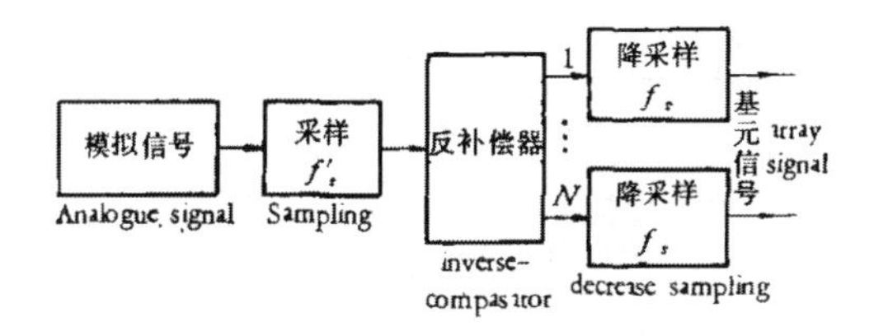

图 4 产生基阵信号的方框图

The block diagram of generating array signals

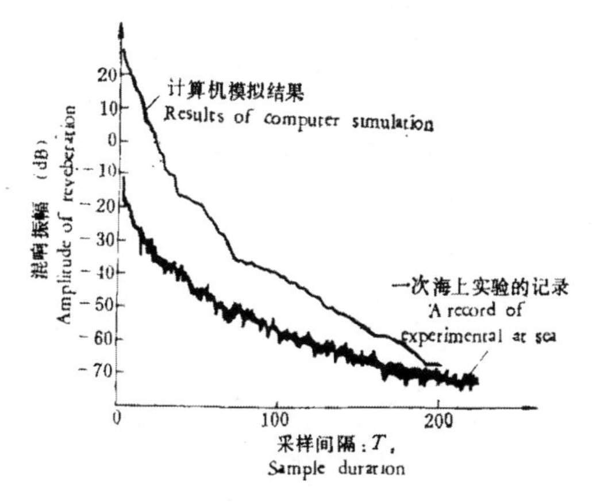

图 5 混响振幅的计算机模拟

Computer simulation of the amplitude of reverberation

三、多分层、多波束系统的模拟

多分层、多波束系统是数字式声呐的核心部件[6,15-16]．目前已有不少数字式波束成形的方法．但是，任何一种数字式波束成形系统都不可能得到时延被精确补偿的所谓同步波束[15]．另一方面，实际系统的后置积累时间总是有限的，因而系统的指向性和理想的积分时间为无穷的稳态指向性总有一定的差异．

本节研究如何在计算机上来模拟实际的系统，以便对有关声呐参数的选择提供依据．我们以圆阵指向性的计算为例来说明模拟的全过程．

参看图 6，假定基阵由 N 个等间隔的基元构成．相邻基元之间的夹角为 $\alpha = 2\pi/N$．如果基阵是加了后档的，那么每一个基元在阵上就有一定的指向性．所以在形成某一波束时，不必让所有 N 个基元都参加工作． 假定参加定向的基元数为 $2M$，这样一个扇面实际上构成一个圆弧阵．设声波的入射角为 θ． 以圆心为参考点，左边第 i 个基元的相对时延为

$$\tau_{il}(\theta) = \frac{r}{c}\cos[\theta + (i - 0.5)\alpha] \quad i = 1, \cdots, M \tag{12}$$

所以，左边第 i 个基元的信号为

$$s_{il}(t) = \cos[2\pi f(t + \tau_{il}(\theta))] \quad i = 1, \cdots, M \tag{13}$$

类似地可以得到右边第 i 个基元的相对时延

$$\tau_{ir}(\theta) = \frac{r}{c}\cos[\theta - (i - 0.5)\alpha] \quad i = 1, \cdots, M \tag{14}$$

右边第 i 个基元的信号为

$$s_{ir}(t) = \cos[2\pi f(t + \tau_{ir}(\theta))] \quad i = 1, \cdots, M \tag{15}$$

(12) 与 (14) 式中 r 为圆阵半径，c 为水中声速．

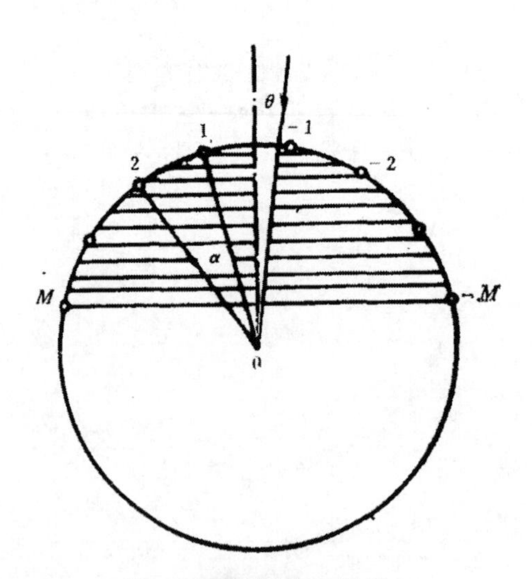

图 6　圆阵指向性的计算

The computation of directivity of circular array

于是，理想的稳态指向性为

$$D_0(\theta)=\left\{E\left(\sum_{i=1}^{M} w_{il}(\theta)\cos[2\pi f(t+\tau_{il}(\theta)-\tau_i)]\right.\right.$$
$$\left.\left.+\sum_{i=1}^{M} w_{ir}(\theta)\cos[2\pi f(t+\tau_{ir}(\theta)-\tau_i)]\right)^2\right\}^{1/2} \tag{16}$$

其中

$$\tau_i=r\cos[(i-0.5)\alpha]/c \tag{17}$$

是为使系统定向于 0° 所必须的精确延时补偿，$w_{il}(\theta)$ 和 $w_{ir}(\theta)$ 分别表示由于基阵加档而引起的左右两边基元的指向性函数。把 (15) 式展开，求总体平均得到，

$$D_0(\theta)=\left\{\left[\sum_{i=1}^{M} w_{il}(\theta)\cos 2\pi f\Delta_{il}(\theta)\right]^2+\left[\sum_{i=1}^{M} w_{ir}(\theta)\sin 2\pi f\Delta_{ir}(\theta)\right]^2\right\}^{1/2} \tag{18}$$

其中

$$\Delta_{il}(\theta)=\tau_{il}(\theta)-\tau_i,\quad \Delta_{ir}(\theta)=\tau_{ir}(\theta)-\tau_i \tag{19}$$

实际系统与这个公式所表达的稳态指向性的差异在于:

1）积分时间有限。

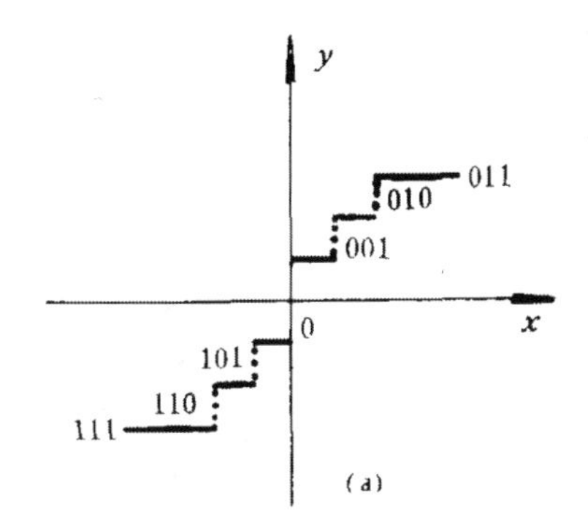

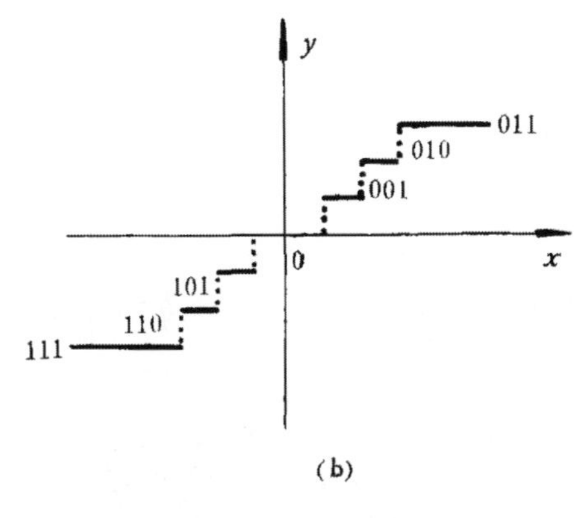

(b)

图 7　分层的两种方式

Two modes of quantization

2）由于采样，使时延不能得到精确补偿.

3）由于 A/D 转换，使系统存在幅度量化误差.

为了推导实际系统的指向性计算公式，关键是研究 A/D 转换器的影响. 参看图 7，A/D 转换有两种方式，其中 (a) 的零状态是过渡的，(b) 具有零状态.

这两种分层的输入、输出关系如下：

$$y=\frac{\Delta}{2}\left\{2\operatorname{sgn}x+\sum_{k=1}^{Q-1}\left[\operatorname{sgn}(x-k\Delta)+\operatorname{sgn}(x+k\Delta)\right]\right\} \tag{20a}$$

$$y=\frac{\Delta}{2}\left\{\sum_{k=1}^{Q-1}\left[\operatorname{sgn}(x-k\Delta)+\operatorname{sgn}(x+k\Delta)\right]\right\} \tag{20b}$$

这里 Δ 为分层级差，Q 为分层数. 当分层数 Q 比较大时，(20a) 与 (20b) 式的差异是不大的，我们以后的分析以 (20a) 式为例. (20a) 式虽然在理论上很好地刻划了 A/D 转换器的输入、输出关系，但是当 Q 比较大时，由 x 到 y 的计算量非常大. 所以，在作实际的系统模拟时采用以下的简化式子：

$$y=\Delta\{\mathrm{INT}(x/\Delta)+(\operatorname{sgn}x+1)/2\} \tag{21}$$

其中 INT(.) 表示取整运算，在一般计算机中已有标准子程序. 利用这一公式，计算可大为简化. 实际系统的指向性公式为

$$D(\theta)=\left\{\frac{1}{K}\sum_{k=1}^{K}\left[\sum_{i=1}^{M}w_{il}(\theta)U_i(k)+\sum_{i=1}^{M}w_{ir}(\theta)V_i(k)\right]^2\right\}^{1/2} \tag{22}$$

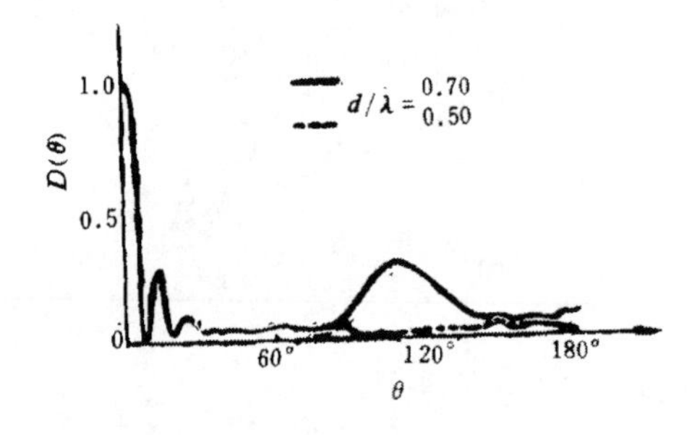

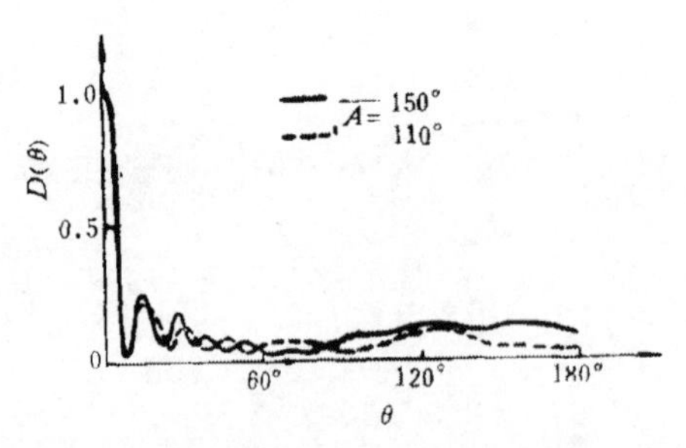

(a) 基元间隔/波长的影响 The effect of d/λ

(b) 工作扇面的影响 The effect of operational fan area A

图 8 基元间隔与工作扇面对指向性的影响

The effects of displacement of array sensors and operational fan area for directivity of system

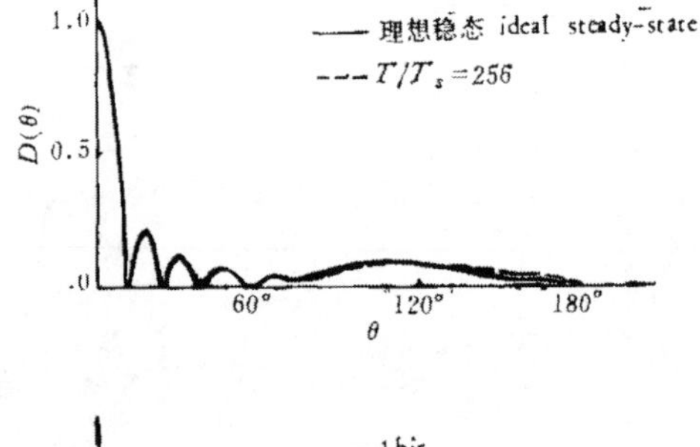

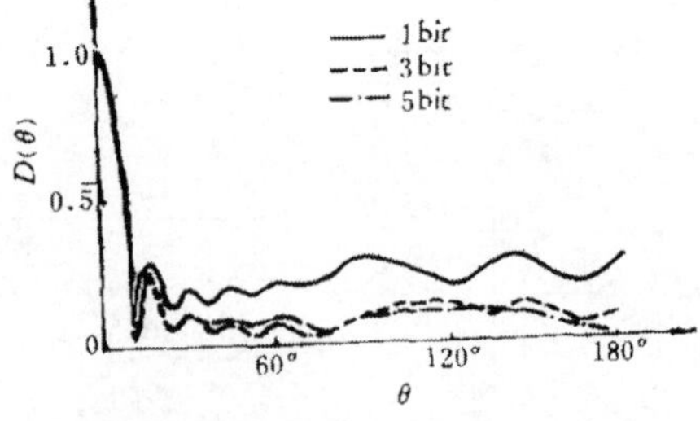

(a) 积分时间的影响 The effect of integral time

(b) 量化的影响 The effect of quantization

图 9 积分时间与幅度量化对指向性的影响

The effects of integral time and amplitude quantization for directivity of system

式中K为后置积累的次数，

$$U_i(k) = \Delta\{\text{INT}[x_{il}(k)/\Delta] + (\text{sgn}[x_{il}(k) + 1]/2)\} \tag{23a}$$

$$V_i(k) = \Delta\{\text{INT}[x_{ir}(k)/\Delta] + (\text{sgn}[x_{ir}(k) + 1]/2)\} \tag{23b}$$

$$x_{il}(k) = \cos(2\pi f[kT_s + \Delta_{il}(\theta)]) \tag{24a}$$

$$x_{ir}(k) = \cos(2\pi f[kT_s + \Delta_{ir}(\theta)]) \quad i = 1, \cdots, M \tag{24b}$$

其中KT_s为积分时间．如果我们还考虑采样引起的量化误差，就要在(24)式中再加上相应的量．

利用(22)—(24)式，我们可以作一系列系统模拟实验．

取$N = 36$．图8说明基元间隔与工作扇面对指向性的影响．在图8(a)中，当$d/\lambda = 0.70$时，除第一副瓣之外，在120°左右还出现了一个高度达33%的相当宽的次极大．当$d/\lambda = 0.5$时，这个次极大值即消失．在图8(b)中，我们看到工作扇面A对指向性也有控制作用，当$A = 110°(M = 12)$时，明显比$A = 150°(M = 16)$好一些．

图9说明积分时间和幅度量化对指向性的影响．在9(a)中，当$T/T_s = 256$时，实际指向性已接近稳态指向性；在9(b)中，我们看到幅度量化对指向性有很大影响，当量化比特数超过5 bit时，实际指向性已接近稳态值．

延时补偿的量化误差对指向性的影响不大，主要是影响定向精度，我们将在下一篇文章中进行讨论．

参　考　文　献

[1] Anarld, C. R., "Digital simulation of a conformal DIMUS sonar system", AD 277557, April, 1962.

[2] James, P. W., "Computer-aided design of sonar arrays for minimum side-lobe level", AD 716488, July, 1970.

[3] Hudson, J. E., "Monte-carlo simulation of an active sonar", *Radio Electr. Engr.*, **40**(1970), 265.

[4] Bartram, J. F. et al., "Fifth generation digital sonar signal processing", *IEEE Trans.*, **OE-2**(1977), 337—343.

[5] Seynaeve, R., "High speed programmable digital signal processing systems for underwater research", *Proc. of NATO ASI on Underwater Acoustics and Signal Processing*, 1980, 643.

[6] Knight, W. C. et al., "Digital signal processing for sonar", *Proc. IEEE.*, **69**(1981), 1451—1506.

[7] Naylor, T. H. et al., "Computer simulation techniques", (John Wiley and Sons. Inc., New York, 1966).

[8] Palston, A., Wilf, H. S., "Mathematical methods for digital computers", (John Wiley and Sons. Inc., 1968).

[9] Basse, S., "Computer algorithms: introduction to design and analysis", (Addison-Wesley Pub. 1978).

[10] 侯朝焕、武振东，"宽带相干噪声中窄带信号的高分辨力谱估计"，声学学报，**6**(1981)，337—347.

[11] Walen, A. D., "Detection of signal in noise", (Academic Press, New York, 1971).

[12] Crochiere, R. E., Rabiner, L. R., "Interpolation and decimation of digital signals—a tutorial review", *Proc. IEEE.*, 69(1981), 300—331.

[13] 汪德昭、尚尔昌，"水声学"，第6章，(科学出版社，1981年).

[14] Middleton, D., "A statistical theory of reverberation and similar first-order scattered field", *IEEE Trans.*, **IT-13**(1967), 372—392.

[15] Pridham, R. G. and Mucci, R. A., "A noval approach to digital beamforming" *J. Acoust. Soc. Am.*, **63**(1978), 425—434.

[16] Curtis, T. E. and Ward, R. J., "Digital beam forming for sonar system", *IEE Proc.*, **127**(1980), Part F, 249—256.

声呐设计中的系统模拟技术（II）[①]

李　启　虎
（中国科学院声学研究所）
1984年1月23日

摘要　本文继续前文讨论声呐设计中的系统模拟技术．数字多波束声呐对目标的精确定向问题提出了新的要求．本文给出利用多波束输出实时计算信号入射角的几种方法，即内插法、微分相关法及互谱法．系统模拟的结果指出这些方法可以用于实时跟踪多个目标．

后置积累是声呐信号处理的重要组成部分．文中给出两种数字式积累器的系统响应，比较了它们的性能．指出指数型平均积累器是一种便于实现而性能又接近理想平均器的后置积累系统．我们还给出在计算机上进行模拟的结果．最后讨论声呐设计中系统模拟所面临的新问题．

THE SYSTEM SIMULATION TECHNIQUE IN SONAR DESIGN (II)

LI QI-HU
(*Institute of Acoustics, Academia Sinica*)
Received January 23, 1984

Abstract　The system simulation technique described in previous paper is studied further in this paper. The new requirement of preeise bearing target has been arised due to the multi-beam sonar. Three methods for real-time estimating the incidental signal angle by using the output data of beam are given in this paper. Those are the methods of interpolation-difference correlation and cross-spectrum. The results of system simulation show that these methods can be used to track multiple target in real-time.

The post-accumulator is one of important part of signal processing of sonar. The frequency response of two kinds of digital accumulator and the relationship between input and output are given here. The various performance are compared. It is shown that the exponential average accumulator is a device which can be easily implemented and have the behavior similar to ideal moving average accumulator. The results of system simulation carried out in digital computer are given.

Some new thesis of system simulation what we must deal with in sonar design are given in the end of this paper.

一、引　言

本文继续讨论前文提出的声呐设计中的系统模拟问题．

一个是对目标的精确定向问题．这个问题的实质就是近年来在信号处理领域中十分活跃

① 声学学报, 1986, 11(5): 297-305.

的时延估计问题．我们给出三种不同的精确定向方法；另一个是数字式声呐的后置积累问题．根据数字运算的特点，我们提出采用指数式平均的方法．这种方法实际上是对模拟系统RC积分器的数字模拟，它的性能接近理想的平均器．

在声呐设计中全面地利用系统模拟技术是一个新的课题，还有不少问题需要解决．我们在最后将予以简单的说明．

二、数字式声呐的精确定向问题

对目标的方位角作出精确的估计是现代声呐必须具备的一个重要功能[1-2]．模拟式声呐依靠旋转单波束系统使其对准目标而实现定向．但是数字式声呐是一种预形成波束系统，它的波束是不能转动的．所以精确定向问题必须采用别的方法．

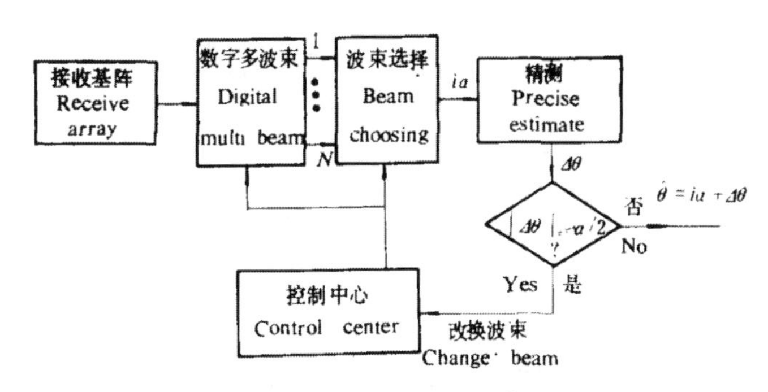

图 1 数字多波束系统对目标的自动跟踪

Automatic tracking of multi-beam system for target

图 1 说明了数字多波束系统对目标进行自动跟踪的原理．我们假定在 360° 范围内波束是均匀分布的．先由普通的多波束系统确定目标的大致方位，例如在 $i\alpha$ 附近．然后再由这个波束的输出数据作精确测量，求出修正角 $\Delta\theta$．那么，目标的精确方位 θ 的估值由下式确定：

$$\hat{\theta} = i\alpha + \Delta\theta \tag{1}$$

当目标运动时，不断根据输出的值来修正波束号码．当 $\Delta\theta > \alpha/2$，波束号码从 i 变为 $i+1$；当 $\Delta\theta < -\alpha/2$ 时，波束号码从 i 变为 $i-1$．用这种办法便可实现对目标的自动跟踪．只要硬件的运算速度允许，我们甚至可以实时跟踪多个目标．

对目标方位角的精测和对目标的被动测距一样，实际上是测量两个分裂波束之间的时延[3-6]．模拟式声呐所用的办法当然应当借鉴．但是，利用数字式声呐的特点我们可以找出更有效的方法来．

下面我们给出三种数字式分裂波束定向法，它们的系统模拟方法见图 2．其中内插法与差分相关法来源于模拟式声呐[3-5,7]，而互谱法则基于 FFT，用计算分裂波束的互谱实现精确定向的一种新方法[6,8-10]．

设多波束系统一共有 N 个波束，相邻波束之间的夹角为 $\alpha = 2\pi/N$．我们不妨假定目标信号出现在 0 号波束与 1 号波束之间，于是信号入射角 θ 满足 $0 \leqslant \theta < \alpha/2$．不然的话，我们总

可以选取合适的波束号码,使相对于该波束的信号入射角局限于 $0\leqslant\theta\leqslant\alpha/2$ 之间.

内插法 设 0 号波束与 1 号波束的左、右两个分裂波束为 $l_0(k)$, $r_0(k)$, $l_1(k)$, $r_1(k)$. 令

$$Z_0=\{E[(l_0(k)-r_0(k))^2]\}^{1/2},\quad Z_1=\{E[(l_1(k)-r_1(k))^2]\}^{1/2} \tag{2}$$

由于分裂波束的输出在零点附近的响应是线性的[12],所以只要根据该直线上的两个点 $(0, Z_0)$ 与 (α, Z_1) 就可以将信号入射角 θ 估计出来,我们把它记作 $\hat{\theta}_1$:

$$\hat{\theta}_1=\frac{Z_0}{Z_0+Z_1}\alpha \tag{3}$$

差分相关法 众所周知,分裂波束的正交相关法的输出在零点附近的响应也是线性的.因而也可以用内插法求出信号入射角. 对模拟系统来说,正交相关可以由求微分 (即移相 $\pi/2$) 来实现. 但是对数字系统来说, 我们就只能用差分来代替微分. 差分相关的输出由下式确定:

$$\begin{aligned}v_0(k)&=r_0(k)[l_0(k+1)-l_0(k-1)]\\ v_1(k)&=r_1(k)[l_1(k+1)-l_1(k-1)]\end{aligned} \tag{4}$$

对 $v_0(k)$, $v_1(k)$ 进行累加,得到

$$V_0=E[v_0(k)],\quad V_1=E[v_1(k)]$$

对 V_0, V_1 进行内插便得到方位角的估计值

$$\hat{\theta}_2=\frac{V_0}{V_0-V_1}\alpha \tag{5}$$

注意这个内插公式与 (3) 式不同的是分母中 V_0, V_1 是相减的, 而在 (3) 式中 Z_0 与 Z_1 是相加的. 这是因为在得到 Z_0、Z_1 时我们用的是平方检波.

研究 (4) 式, 便会知道,当采样间隔 T_s 较大时,由差分代替微分的误差就比较大. 这是造成差分相关法估计误差的主要来源之一.

互谱法 若入射信号对 0 波束的偏离角 $\theta=0°$, 那么该波束左、右分裂波束的输出就一样. 所以它们的互功率谱只有实部, 虚部为零. 当偏离角 $\theta\neq0$ 时, 互功率谱中的虚部不是零. 根据虚部及实部的比值就可以将 θ 估计出来[6]. FFT 为这种估计提供了快速而有效的实

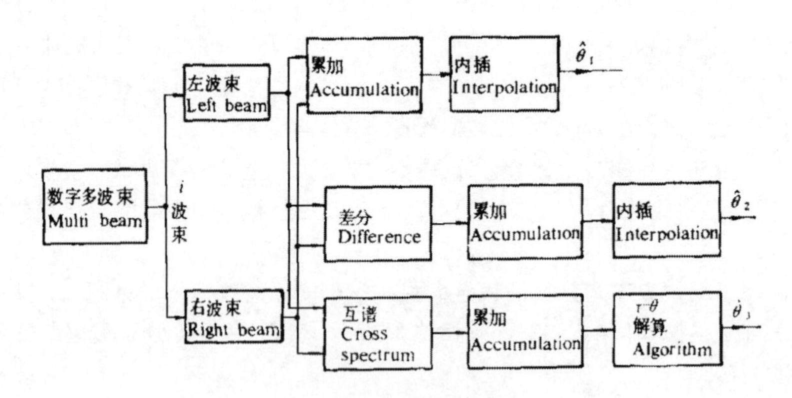

图 2 几种数字式精确定向方法的方框图

Block diagram of several methods for precise bearing

时计算法．我们有：

$$\hat{\theta}_3 = \frac{\pi M \hat{\tau} c}{2Nr\left[\sin\left(\frac{M}{2}\alpha\right)\right]^2} \tag{6}$$

其中 c 为声速，$\hat{\tau}$ 是用互谱法而获得的时延估计值．

在作系统模拟时，我们首先要按前文第 2 节所阐述的方法产生 $2M$ 路信号．然后用变采样率的方法模拟声呐信号处理．

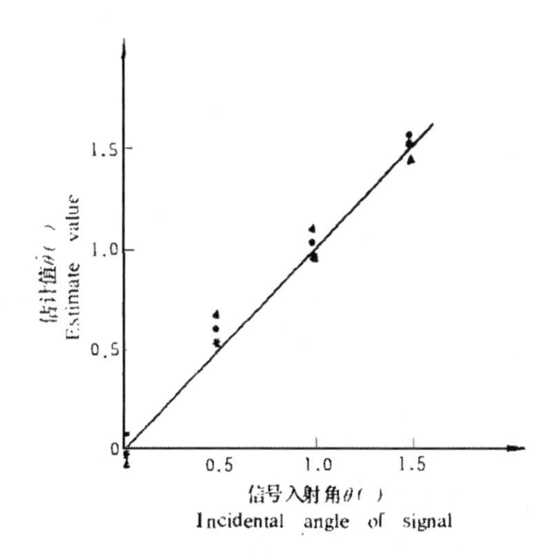

图 3　精确测向的系统模拟结果

Results of system simulation of precise bearing

··· 内插法 (interpalation)；▲▲▲ 差分相关法 (difference correlation)

*** 互谱法 (cross-spectrum)；—— 理想结果 (ideal results)

表 1　定向精度与积累时间的关系

The relationship between accuracy of estimates and average time

积累时间 (T_s) (average time)	估计法 (Estimates)		
	内插法 $\hat{\theta}_1$	差分相关法 $\hat{\theta}_2$	互谱法 $\hat{\theta}_3$
	均方根误差(°) (bias)		
256	0.109	0.153	0.070
512	0.120	0.124	0.142
1024	0.098	0.132	0.050
2048	0.087	0.111	0.010

图 3 给出了 $T_s/T_s' = 8$ 时的系统模拟结果．总的积累时间为 $2048T_s$，输入信噪比在分裂波束处测量为 0dB，约相当于基元处 SNR $= -10$dB． 用互谱法估计时，DFT 的点数为 64．从模拟结果来看，以 $\hat{\theta}_3$ 的互谱法为最好．随着积累时间的增加，定向精度明显提高．下表给出在 SNR $= -10$dB 时，定向精度和积累时间的关系．

三、后置积累的系统模拟

普通声呐中所用的后置积累方式一般是能量检测器．因为它是从高斯背景噪声中检测高斯信号的最佳检测器．但是在数字式声呐中，我们宁愿采用绝对值检波．这是因为这种运算不用乘法,可以大大简化硬件．容易证明,对于高斯信号来说，绝对值检波与平方检波的性能仅有很小的差异[12].

下面我们来讨论两种实用的数字式后置积累系统．

滑动平均 这是对滑动积分器的模拟． 用 $x(k)$ 及 $y(k)$ 分别表示系统的输入与输出，那么

$$y(k)=\frac{1}{L}\sum_{l=0}^{L-1}x(k-l) \tag{7}$$

其中 L 为积累次数．

这种处理方式,实际上是每一次更新一个样本．它的脉冲响应函数是

$$h(k)=\frac{1}{L}\sum_{l=0}^{L-1}\delta(k-l) \tag{8}$$

图 4 给出这种系统的方框图及脉冲响应．

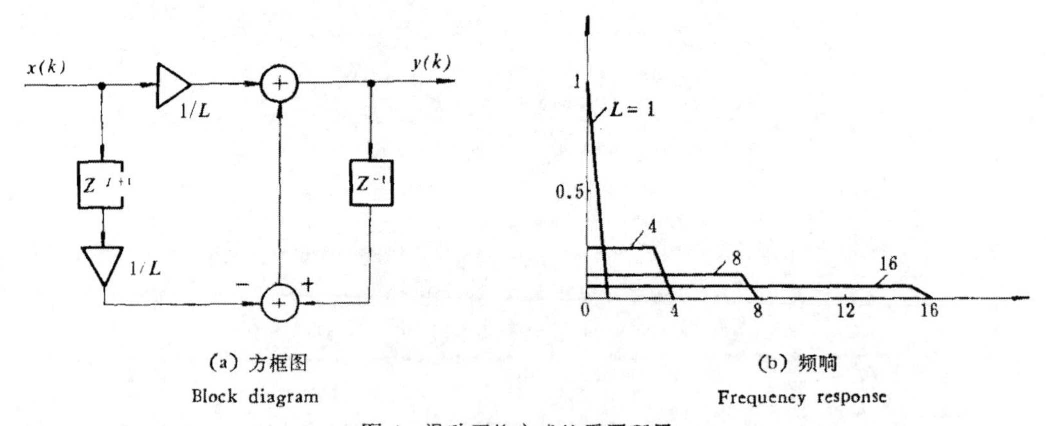

(a) 方框图
Block diagram

(b) 频响
Frequency response

图 4 滑动平均方式的后置积累
Post-accumulator of the moving average mode

从理论上来说,滑动平均实现了真正的时间平均．这是一种理想的积分器, L 越大，输出起伏也就越小．从硬件实现角度来说,滑动平均处理的缺点是我们必须存贮 L 个老的数据.当积分时间较长时存贮容量太大,这是不合适的．

指数平均 这种平均方式具有和滑动平均大致相近的输出特性(如平均值、输出起伏等)．同时在硬件实现上要简单得多．它的输入、输出关系如下:

$$y(k)=\left(1-\frac{1}{L}\right)y(k-1)+\frac{1}{L}x(k) \tag{9}$$

它每一次只要加入新的样本，同时把原来的积累结果衰减一个固定因子．这种系统的脉冲响

应为

$$h(k)=\left(1-\frac{1}{L}\right)^{k}y(0)\quad k\geqslant 0 \tag{10}$$

这是一种指数型的平均. 根据模拟系统的数字模拟定理[13], 可以证明, 这是对模拟 RC 积分器的数字式模拟.

事实上, RC 积分器的系统函数为

$$H_a(f)=\left(\frac{1}{2\pi jfc}\right)\Big/\left(R+\frac{1}{2\pi jfc}\right)=\frac{1}{1+2\pi jfRC} \tag{11}$$

要用数字系统模拟它,我们考虑新的系统函数

$$H(z)=\frac{\beta}{1-(1-\beta)z^{-1}} \tag{12}$$

将 $z=\exp(2\pi jfT_s)$ 代入,得到

$$H[\exp(2\pi jfT_s)]=\frac{\beta}{1-(1-\beta)\exp(-2\pi jfT_s)} \tag{13}$$

其中 T_s 为采样间隔. 只要 $fT_s\ll 1$, 由 (13) 式得出

$$H[\exp(2\pi jfT_s)]\approx\frac{\beta}{\beta+(1-\beta)2\pi jfT_s} \tag{14}$$

将 (14) 式与 (11) 式作比较,令

$$\frac{1-\beta}{\beta}T_s=RC\triangleq T,\quad 则\ \beta\approx T_s/T$$

这里 T 为积分时间, $1/\beta$ 就是累加次数. 与 (12) 式相对应的数字系统的输入、输出关系正好是 (9) 式,只要令 $L=1/\beta$ 就行.

图 5 给出了指数型平均方式的方框图及脉冲响应函数.

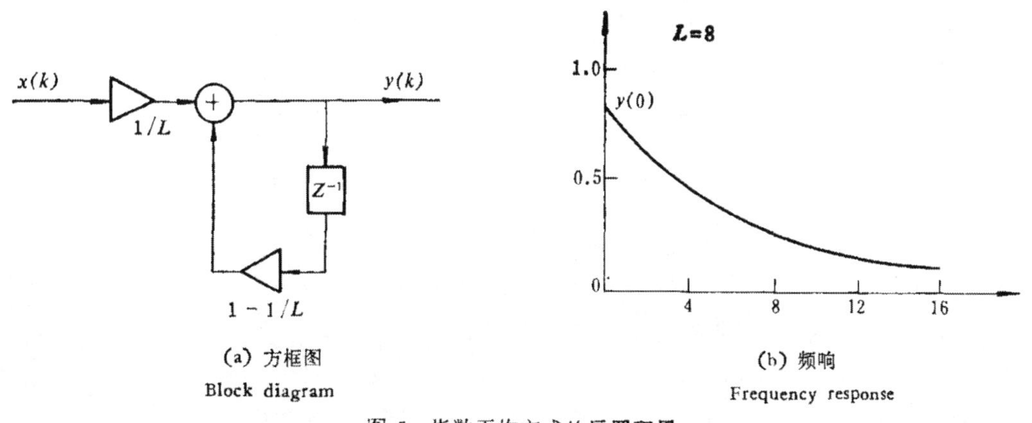

(a) 方框图
Block diagram

(b) 频响
Frequency response

图 5　指数平均方式的后置积累
Post-accumulator of the exponential average mode

指数平均处理方式用硬件实现起来要比滑动平均方便得多. 图 6 给出了上述两种后置积累方式的系统模拟情况. 输入 $x(k)$ 为 [0, 1] 中均匀分布的独立随机变量序列, 积累时间 $L=32T_s$. 从图中我们看到,两者的性能基本上是类似的.

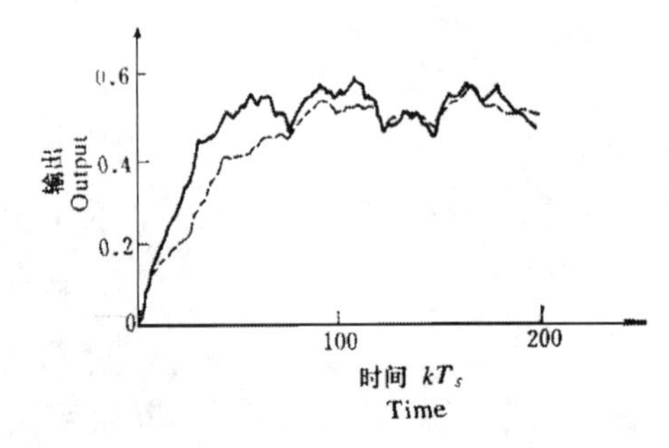

图 6 数字式后置积累器的输出

The output of digital average accumulators

——滑动平均 (Moving average) ---指数平均 (Exponential average)

四、整机检测性能的模拟

我们前面已讨论了声呐系统模拟的各主要方面. 它们既可以分别在计算机上进行模拟，也可以作为一个整体来模拟. 下面我们用具体的例子说明模拟的全过程(见图 7).

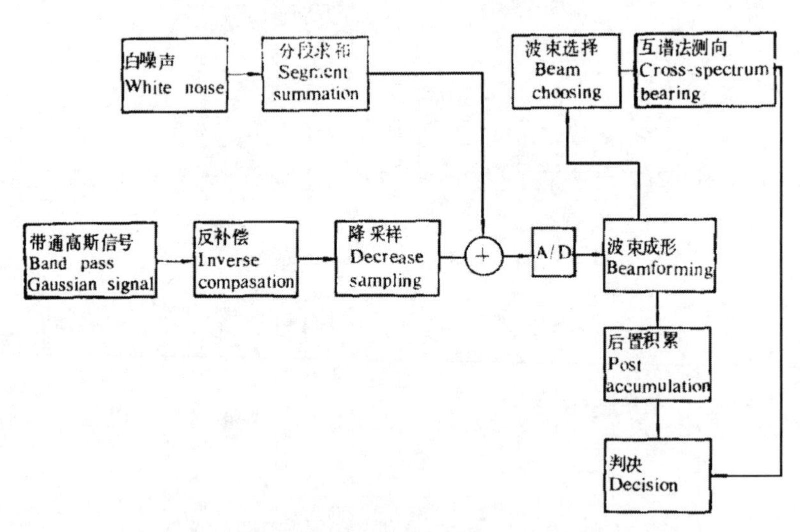

图 7 声呐系统模拟的例子

An example of sonar system simulation

我们所设计的声呐基阵由 $N=36$ 个基元构成圆阵. 工作频率考虑为 2—6kHz, 基阵半径 $r=1$m; 每形成一个波束用 14 个基元,即 $M=7$. 对这样的基阵, 根据 (6) 式可以计算出分裂波束的相对延时为 1μs 时相当于信号入射角偏离 0.075°. 我们取 $T_s'=1$ μs. 那么产生基元信号时的量化误差只有 0.5μs, 这样可以保证使精测系统的模拟有足够的精度. 然后对 36 路基元信号降采样率, 使 $T_s=10\mu s$. 把降采样率之后的信号与噪声相加,经延时补偿后进行波束成形.

设多波束系统形成 $3\times 36=108$ 个波束. 那么相邻两个波束的间隔是 $\alpha=360/108=$

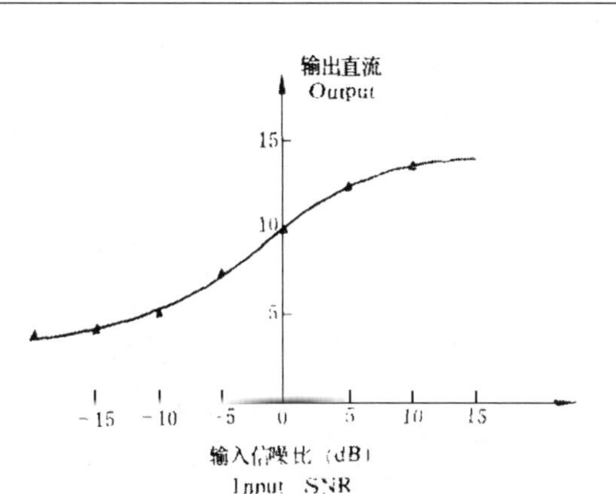

图 8 整机的检测性能模拟结果

The result of system simulation of detection performance

——理论值 Theoretical ▲▲▲模拟结果 Simulation results

3.3°. 若后置积累时间为 100ms，我们可以将平均积累次数选作 $2^{13} = 8192$ 点，相当于 81.92 ms.

对于精测系统，由于 $\alpha = 3.3°$，我们只须估计 $\pm 1.6°$ 的偏离角即可. 这相当于分裂波束相对时延 20μs. 当我们用互谱法测向时，应当把 DFT 的采样间隔选作 $2 \times 20 = 40\mu s$[6,8]. 这时 64 点 DFT 需要 2.56ms，每条谱线积累 32 次（$32 \times 2.56 = 81.92$ ms）即可得到精确的方位角.

图 8 给出了 A/D 转换器为 5bit、积分时间为 10.24ms(1024 点)时信号检测实验的结果. 我们在图上画出了 14 通道的检测曲线. 横坐标为输入信噪比，纵坐标为系统后置积累器绝对值检波的输出. 我们看到由于输入信号受 A/D 动态的限制，输出直流被自动归一了. 在作系统模拟时我们改变输入信噪比，记录某一波束主瓣方向的输出就可以得到整机的检测性能. 这种模拟是我们最后检验声呐设计是否合理的主要依据.

五、有待进一步研究的问题

系统模拟技术对声呐设计来说是有效的. 为了使模拟的结果能更真实地反映实际情况，就必须使系统模拟技术向更深入、更复杂的领域前进. 举例来说，基阵在实际声呐环境场中所接收到的噪声是有一定相关特性的. 因此，从更高的要求来看，应当模拟二维的时空序列；水声信号在各种不同的水文条件下的传播衰减规律是声呐方程中的重要参量，我们应把它反映到系统模拟中来；产生基阵信号的降采样率比 T_s/T_s' 应当如何选取等.

从总的发展趋势来看，由于模拟式声呐正逐渐向数字式声呐过渡，正如 Knight 等所指出的，系统模拟技术将在声呐设计中起越来越重要的作用[14].

吕小燕、董琍同志参加了部分计算机模拟工作；梁祖威、任浩、侯凤磊同志为系统模拟工作提供了方便，作者表示衷心的感谢.

参　考　文　献

[1] MacDonarld, V. M. and Schulthesis, P. M., "Optimum passive bearing estimation in spatially incoherent noise environment", *J. Acoust. Soc. Am.*, **46**(1969), 37—43.

[2] Hahn, W. R., "Optimum signal processing for passive sonar range and bearing estimation" *J. Acoust. Soc. Am.*, **58**(1975), 201—207.

[3] Carter, G. C., "Time delay estimation for passive sonar signal processing", *IEEE Trans.*, **ASSP-29** (1981), 463—470.

[4] Usher, T. Jr., "Random bearing error in split-beam system", *J. Acoust. Soc. Am.*, 39(1965), 912—920.

[5] Heimdal, P. and Bryn, F., "Passive ranging techniques", *Proc. of NATO ASI on Signal Processing*, 1972.

[6] 李启虎，"数字式分裂波束阵精确定向方法"，声学学报，**8**(1983)，No. 5.

[7] Skitzki, P., "Mordern sonar systems". *Electronic Progress*, XVI (1974), No. 3, 20—37.

[8] Piersol, A. G., "Time delay estimation using phase data", *IEEE Trans.*, **ASSP-29** (1981), 471—477.

[9] Bendat, J. S. and Piersol, A. G., "Engineering application of correlation and spectral analysis", (John Wiley and Sons., 1980).

[10] Scarbrough, K. et al., "On the simulation of a class of time delay estimation algorithms", *IEEE Trans.*, **ASSP-29** (1981), 534—539.

[11] 霍顿，J. W.，"声呐原理"，第5章，冯秉铨译，(国防工业出版社，1965年).

[12] Camp, L., "*Underwater acoustics*", Ch. 10, (John Wiley and Sons. 1970).

[13] 帕布里斯，A.，"信号分析"，第5章，李启虎、徐为方等译，(海洋出版社，1981年).

[14] Knight, W. C. et al., "Digital signal processing for sonar", *Proc. IEEE.*, **69**(1981), 1451—1506.

数字式多分层多波束系统指向性的计算机模拟①

李启虎 项定长 吕小燕 董 琍
（中国科学院声学研究所） （中国船舶工业总公司）

1983年5月5日收到

用横向滤波器形成多波束是数字式声呐的一种新的波束成形方法．这种系统的指向性与采样频率、分层比特数及延时补偿的量化误差有关．系统实际指向特性与理想稳态特性之间的差异将直接影响声呐的性能．本文给出计算这种系统指向性的各种必要公式．同时在计算机上进行系统模拟．模拟的结果可为数字式声呐某些参数的选取提供依据．

一、引 言

由于数字信号处理技术的发展，在声呐中正越来越普遍地采用数字技术[1-3]．早在六十年代初，V. C Anderson 就提出一种简单的数字式的多波束系统[4]．采用数字技术实现了实时全景观察．但是，由于这是一种一比特的量化系统，即使不存在干扰，系统的检测性能仍比模拟系统低 1.9dB[5]．

近年来，已有不少数字式波束成形的方法被提出来[6-8]，有时域的也有频域的．Pridham 等所提出的数字内插波束成形是时域的，Rudnick 等所提出的方法则是频域的．

本文提出用横向滤波器形成多波束的方法．这是一种时域快速形成多波束的技术．由于是数字系统，时间量化（采样）与幅度量化（分层）将会对系统指向性产生影响．本文首先介绍计算系统指向性的公式，然后给出在数字计算机上进行系统模拟的方法与结果，可作为数字式声呐设计的一种辅助手段．

二、横向滤波器形成多波束

用横向滤波器形成多波束的框图示之图1．我们以圆阵为例来说明工作原理，设基阵由 N 个基元构成，所以系统有 N 路输入信号．按传统的方法，为了形成多波束，应当将各路信号分别采样，然后按事先设定的延时补偿进行抽头，以形成指向不同的波束．我们在这里所采

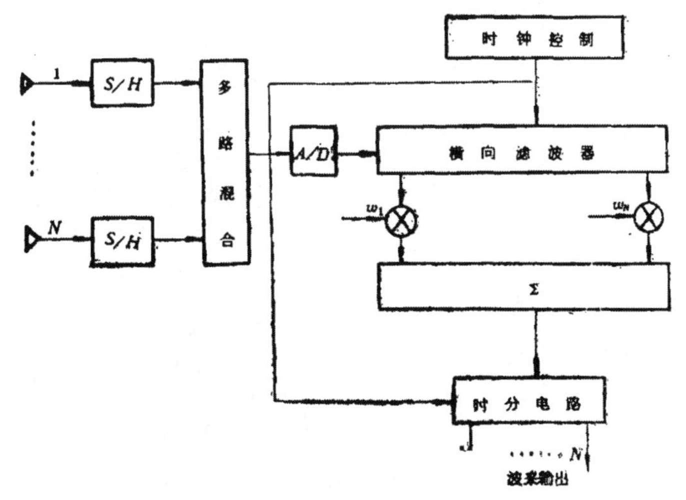

图1 用横向滤波器形成多波束的方框图

① 应用声学, 1986, 5(3): 17-22.

用的方法是高速工作的一条横向滤波器，同时形成多波束，这样可使延迟线的数目减少N倍．作为代价，必须使采样频率提高N倍．图中的$w_1, w_2, \cdots w_N$是加权系数，它的目的是改善指向性．这组数是按某种“最佳准则”计算得到的．

S/H电路的采样间隔为T_s，由N路采样得到的信号经多路混合之后，再经A/D转换变为数字信号． 多路混合和A/D转换的工作频率为$f_c = N/T_s$． 于是N个一组的输入样本被压缩在T_s内，然后在钟频f_c的推动下，在横向滤波器内移动．每隔T_s/N秒我们得到一个波束的输出，经T_s秒之后，顺序得到N个波束．波束的间隔是$2\pi/N$．

如果要得到不同类型的波束，那么，必须采取不同的抽头．

三、指向性的计算公式

我们以圆阵为例来说明指向性的公式．参看图 2．

假定波束指向方向为0°，参加工作的扇面由$2M$个基元构成．每个接收基元具有一定的指向性，它通常是心形的．图2(b)是一个典型的例子．

第i个基元相对于圆心的延时是

$$\tau_i = \frac{r}{c}\cos[(i-0.5)\alpha], \quad i = 1, \cdots, M. \tag{1}$$

式中$\alpha = 2\pi/N$；r——基阵半径；c——声速．

如果信号的入射方向为θ，那么，左半阵信号的延时补偿为

$$p_i(\theta) = \frac{r}{c}[\cos(\theta + (i-0.5)\alpha) - \cos((i-0.5)\alpha)]\ i = 1, \cdots, M. \tag{2}$$

右半阵信号的延时补偿为

$$q_i(\theta) = \frac{r}{c}[\cos(\theta - (i-0.5)\alpha) - \cos((i-0.5)\alpha)] \quad i = 1, \cdots, M. \tag{3}$$

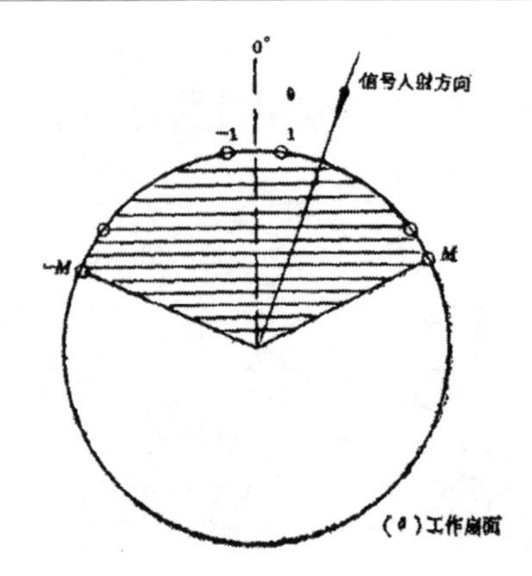

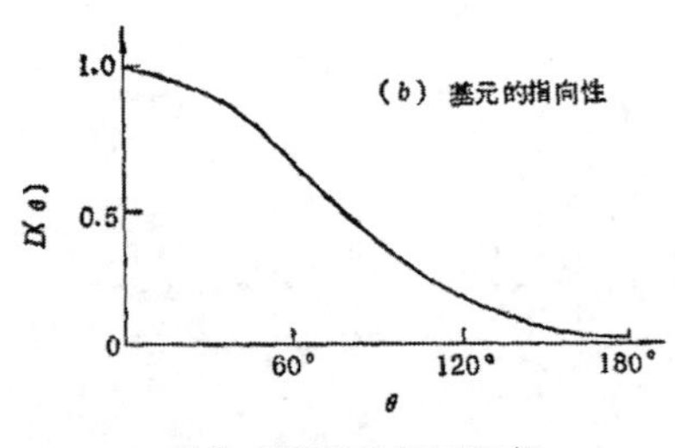

图 2 圆弧阵指向性的计算

于是，模拟系统的指向性是

$$D_0(\theta) = \left\{\left[\sum_{i=1}^{M} w_{ix}(\theta)\cos 2\pi f p_i(\theta) + w_{iy}(\theta)\cos 2\pi f q_i(\theta)\right]^2 + \left[\sum_{i=1}^{M} w_{ix}(\theta)\sin 2\pi f p_i(\theta) + w_{iy}\sin 2\pi f q_i(\theta)\right]^2\right\}^{1/2} \tag{4}$$

式中$w_{ix}(\theta)$及$w_{iy}(\theta)(i = 1, \cdots, M)$，表示左半阵及右半阵的基元的指向性．

我们称由(4)式所表示的指向性为稳态指向性．从理论上讲，(4)式有以下几个特点：

1. 系统的积分时间无穷．

2. 输入信号是模拟量，这等价于A/D转换器的分层数无穷多．

3. 信号的时延得到完全的补偿．这等价于系统的采样频率无穷．

任何实际的系统，无论是数字的还是模拟的，都与以上三个假定有差异．所以，实际系统的指向性就和理想的指向性有差异．

为了推导实际系统的指向性计算公式，关键是研究 A/D 转换如何影响(4)式．因为量化误差很容易从对 τ_i 的修正中表达出来．

A/D 转换有两种方式，见图 3．图 3(a) 具有零状态．在图 3(b) 中零状态只是过渡的状态．这两种分层方式的输入-输出关系是

$$y = \frac{\Delta}{2}\left\{\sum_{k=1}^{Q}\left[\mathrm{sgn}(x-k\Delta)+\mathrm{sgn}(x+k\Delta)\right]\right\} \tag{5}$$

$$y = \frac{\Delta}{2}\left\{2\mathrm{sgn}x+\sum_{k=1}^{Q-1}\left[\mathrm{sgn}(x-k\Delta)+\mathrm{sgn}(x+k\Delta)\right]\right\} \tag{6}$$

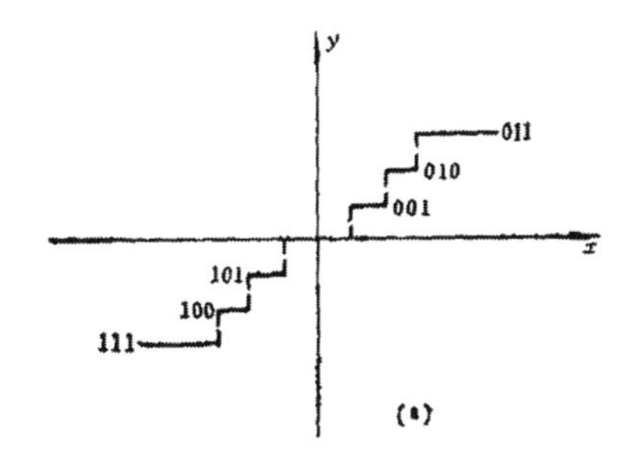

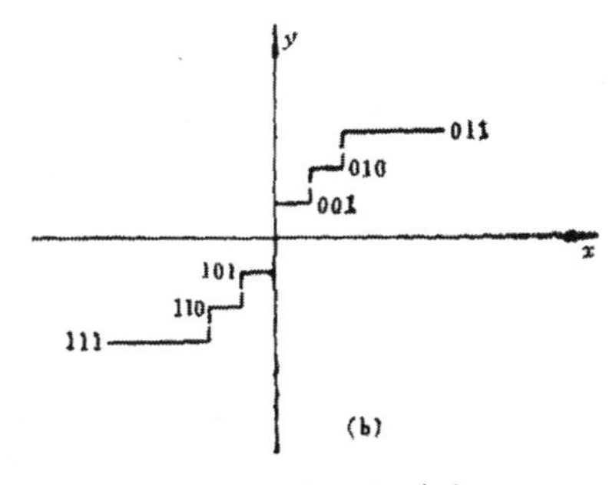

图 3 A/D 转换的两种方式

式中 Δ 为分层级差，Q 为分层数．如果 V_0 是信号动态、m 为量化比特数，那么，V_0, Δ, Q, m 的关系式为

$$\Delta = V_0/(2^{m-1}-1), \quad Q = 2^{m-1} \tag{7}$$

这两种分层的计算结果大同小异．我们在计算机上模拟时，一律以 (5) 式为准．

应指出，(5) 和 (6) 式用于理论分析是方便的．因有一些现成公式可以计算 $\mathrm{sgn}(x)$ 的相关函数值[9]．但若在计算机上模拟，那么当 m 较大时，N 就很大，于是计算量迅速增加，这是不合适的．我们下面改用一个简化的式子

$$y = \Delta\{\mathrm{INT}(x/\Delta)+[\mathrm{sgn}x+1]/2\} \tag{8}$$

式中 $\mathrm{INT}(x)$ 表示 x 的整数部分．(8) 式有一个奇点，就是 $x=0$ 时，$y=1/2$．但这通常不影响计算结果．

根据以上分析，我们给出实际数字系统的指向性公式

$$D(\theta) = \left\{\frac{1}{k}\sum_{K=1}^{K}\left[\sum_{i=1}^{M} w_{ix}(\theta)u_i(k) + \sum_{i=1}^{M} w_{iy}(\theta)u_i(k)\right]^2\right\}^{1/2} \tag{9}$$

其中 K 为累加次数，KT_s 就是积分时间．

$$\begin{cases} u_i(k) = \Delta\{\mathrm{INT}[x_i(k)/\Delta]+[\mathrm{sgn}(x_i(k))+1]/2\} \\ v_i(k) = \Delta\{\mathrm{INT}[y_i(k)/\Delta]+[\mathrm{sgn}(y_i(k))+1]/2\} \end{cases} \tag{10}$$

$$\begin{cases} x_i(k) = \cos 2\pi f[kT_s+p_i(\theta)+l_{xi}] \\ y_i(k) = \cos 2\pi f[kT_s+q_i(\theta)+l_{yi}] \end{cases} \tag{11}$$

式中 l_{xi} 和 $l_{yi}(i=1,2,3,\cdots,M)$ 分别表示左波束和右波束的延时量化误差．

在分层比特数 $m\to\infty$，V_0 保持一定．累加次数 $K\to\infty$ 以及 $T_s\to 0$ 时，$D(\theta)\to D_0(\theta)$．即实际系统的指向性趋于稳态指向性．(9)—(11) 式是我们进行计算机模拟的基本式子．

四、计算机模拟结果

我们在 ALPHA 机上进行了以下几方面的系统模拟．

1. 基元数对指向性的影响

我们知道，圆弧阵的指向性和相邻基元的间隔 d 与信号波长 λ 的比值有关．同样半径的基阵，当基元数 N 增大时，d/λ 减小，指向性就得到改善．如果 $d/\lambda \geqslant 1/2$，这时指向性就较

差. 图 4 给出了同一半径之下同一频率，$N=32$ 和 $N=48$ 的情况. 我们看到，当 $N=32$ 时，在 100° 左右出现了一个高达 40% 的副瓣.

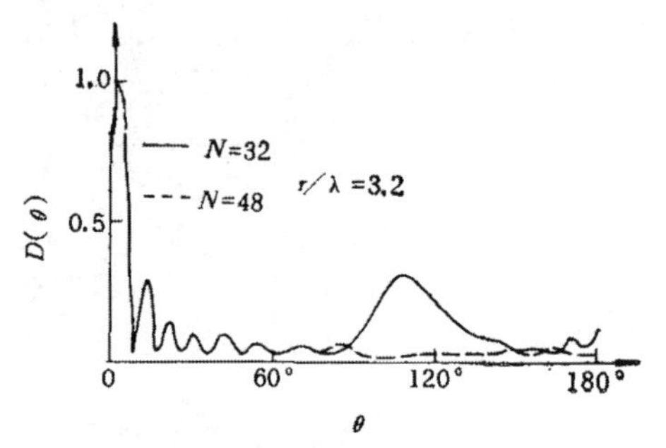

图 4 基元数对指向性的影响

下面表 1 比较详细地列出了各种情况下指向性的变化. $r/\lambda=3.2$，工作扇面 157°.

2. 工作扇面对指向性的影响

工作扇面对指向性的影响也相当大. 如果工作扇面太大，那么由于远离主瓣方向的基元接收较多的噪声，指向性就要变差；如果工作面太小，那么系统的增益会下降. 所以工作扇面存在着一种最佳的折衷选择. 一般讲来，这和单水听器的指向性有关. 图 5 是一个具体例子.

表 1 基元数对指向性的影响

基元数	主瓣宽度	第一副瓣	100°左右的影响	增益(dB)
32	7°	0.26	0.33	14.1
40	7°	0.26	0.12	15.9
48	7°	0.26	0.06	16.5
64	7°	0.25	0.04	16.5

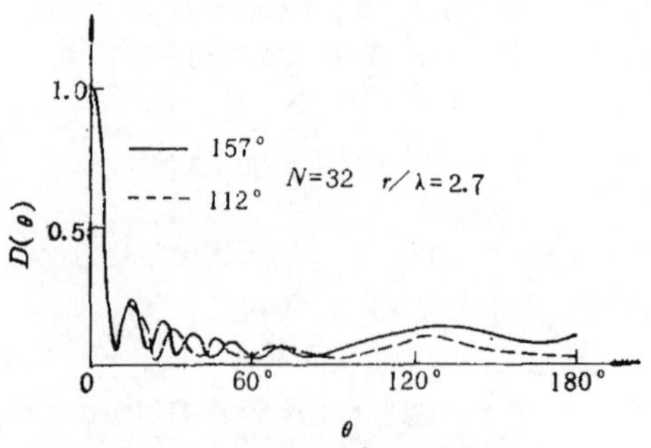

图 5 工作扇面对指向性的影响

表 2 给出 $N=32$，$r/\lambda=2.7$，时，工作扇面对指向性的影响.

表 2 工作扇面对指向性的影响

工作扇面	主瓣宽度	第一副瓣	100°左右的影响	增益(dB)
157°	7.2°	0.26	0.18	14.1
135°	7.2°	0.25	0.14	14.2
112°	7.2°	0.24	0.10	14.1
90°	7.2°	0.22	0.07	13.7

我们看到，工作扇面选作 120° 左右是合适的.

3. 多分层、多波束的指向性

考虑到 (9)—(11) 所给出的公式，对于幅度量化、时间量化、积分时间可以有以下几种组合方式，见表 3. 表中的八种情况中，由于第 3、5 两种情况计算量过大，没有进行系统模拟. 但我们可以别的途径去分析它们的性质，所以不妨碍模拟计算的完整性.

我们首先说明积分时间应取多少，设信号周期为 T_0 那么当 $KT_s=10T_0$ 时，$D(\theta)$ 就接近稳态指向性 $D_0(\theta)$ (见图 6)，但零点深度比稳态时差得多. 当 $KT_s=20T_0$ 时，零点深度仍差 11dB.

为了考察幅度量化的影响，我们一共计算了 4 种分层的情况. $m=2,4,6,8$，相应的分层级差为 1, 1/7, 1/31, 1/127.

表 3 幅度量化、时间量化、积分时间的组合方式

编号	幅度量化	时间量化	平均时间
1	无	无	无穷
2	无	无	有限
3	有	无	无穷
4	有	无	有限
5	有	有	无穷
6	有	有	有限
7	无	有	无穷
8	无	有	有限

图 7 给出了模拟的结果. 我们看到 3 比特量化时，指向性与模拟情况还有差异. 但当量化比特数超过 6 时，从计算结果看，已看不出什么不同了.

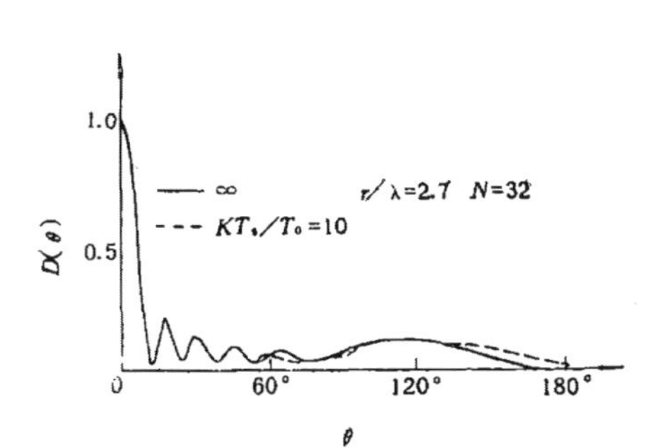

图 6　积分时间对指向性的影响

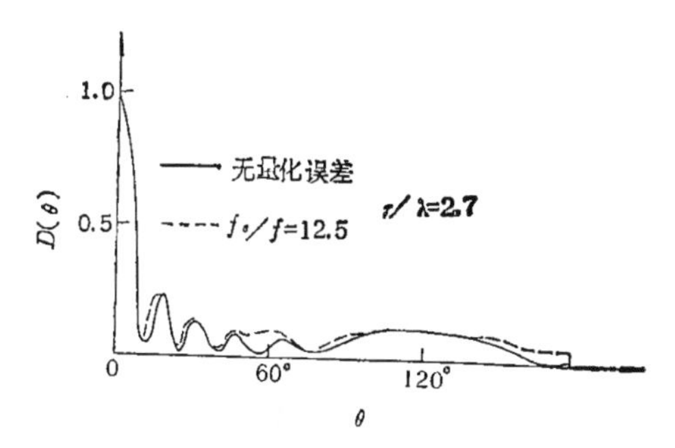

图 8　时间量化对指向性的影响

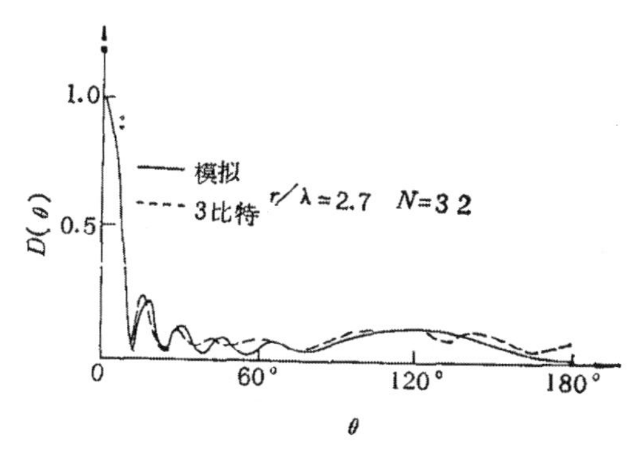

图 7　幅度量化对指向性的影响

下面我们讨论时间量化误差的影响．如果采样间隔是 T_s，那么，延时补偿的误差在 $[-T_s/2, T_s/2]$ 内分布．采样间隔越大，量化误差也越大．我们模拟计算了三种情况

$$f_s/f = 2.5,\quad 12.5,\quad 6.25$$

表 4 是计算结果的比较．

由表得知，当采样间隔增加时，主瓣响应下降．但这种下降不十分严重．一般来说，$f_s/f=12.5$ 是足够的．图 8 是一个具体的计算结果．

表 4　时间量化对主瓣响应的影响

f_s/f	主瓣响应
∞	10.0928
2.5	10.0580
12.5	10.0201
6.25	9.9803

五、结　论

数字式多分层多波束系统的指向性与模拟系统存在着一定的差异．分析造成这种差异的原因，对于声呐系统的计算来说是重要的．系统的计算机模拟可以在短期内对某些参量的选取作出决定．

作者感谢梁祖威、任浩同志在整个系统模拟期间所给予的大力支持．

参 考 文 献

[1] W. C. Knight et al., Proc. IEEE Vol. 69, 1981, 1451—1506.
[2] P. Skitzki, Electronic Progress, Vol. XV1, 1974, 20—37.
[3] T. E. Curtis and R. J. Ward, IEE Proc., Vol. 127, 1980, Pt F 257—265.
[4] V. C. Anderson, *J. Acous. Soc. Am.*, **32**(1960), 687—870.
[5] P. Rudnick, *J. Acous. Soc. Am.*, **32**(1960), 871—877.
[6] R. G. Pridham and R. A. Mucci, Proc. IEEE Vol. 67, 1979.
[7] P. Rudnick, *J. Acous. Soc. Am.*, **46**(1969).
[8] G. M. Byram et al., Proc. of NATO Advanced Stydy Inst. on Signal Processing, Reidal, 1973.
[9] 李启虎，电子学报，4(1980).

经验交流

一个在微机上应用的声呐线阵设计软件①

李 启 虎
(中国科学院声学研究所)

1988 年 8 月 30 日收到

数字式声呐的很多运算可以用计算机上的专用软件进行仿真，本文介绍一个用于 IBM PC/XT，/AT 及其兼容机的声呐设计用软件，对于连续线阵、离散等间隔线列阵，用户只需要输入有关参数．本软件就能在不到一分钟内计算出指向性函数，同时将用户所感兴趣的波束图在屏幕上画出来，在必要时还可以在普通打印机上得到硬拷贝．这个软件具有 ZOOM 功能．用户可以在横坐标上选取任意要放大观察的窗口以便了解指向性的细微结构．由于在设计中采用人机对话方法，用户在使用时可随时选择所需要的项目和改变数据，命令流向，本软件不仅可作为声呐设计人员的有力工具，也可以用于教学上的直观演示．

一、引 言

微机的普及已使许多应用科学的面貌发生深刻变化．在数字信号处理领域内，微机已从开始时的单纯数值计算，系统模拟发展到参与系统控制和实用软件的开发，因为数字信号处理中很多应用课题可以方便地由专用软件实现．1979 年，美国 IEEE 协会出版了数字信号处理中专用算法的一些标准子程序及磁带[1]，用户利用这套程序库中所推荐的程序可以方便地进行离散 Fourier 分析，求相关、卷积等运算． 此后，随着专用的数字信号处理器的开发，不少新的、更方便、功能更强的软件相继间世了[2-4]，但是这些软件还只限于对一种算法的描述，或提供快速、简单的标准程序，用户自己还不能利用它们直接设计实用软件．近两年来，美国 ADSP (Atlantic Digital Signal Processing) 公司，Ariel 公司相继在 IBM PC 机上开发了数字滤波器设计用的软件以及数据采集、加工软件，从而使微机在信号处理领域的应用提高到一个新水平．

利用这些软件，用户可以直接按要求给出参数，获得实际硬件设计时所需要的具体数据，由此大大地简化了硬件设计的工作量．

本文介绍在 IBM PC 机上开发的声呐设计用的一个软件，为使用连续线阵或等间隔离散线列阵的设计人员，快速准确地提供整套数据和波束图．

用户可以按自己的需要输入参数，本软件即可在短时间内给出指向性参数，当用户需要时，随时可以在打印机上把数据打印出来，同时又以三种不同的坐标方式在屏幕上直观地显示波束图．可以在普通打印机上得到屏幕显示图形的硬拷贝．

整个软件的使用采用人机会话方式，屏幕上及时出现的菜单及操作提示使用户非常容易就可以学会操作，在每一时刻，用户都会被告知，自己应该作什么，机器正在作什么．

本软件在设计上采用多级分叉结构，用户在每一阶段都可以改变自己的意图，使数据流向和命令切换到自己要求的方向去，用户的误操作会被机器自动检查出来，并提示用户应当怎么作．

① 应用声学, 1989, 8(4): 42-46.

本软件是国内投放于市场的第一个用于声呐设计的微机软件.

二、软 件 结 构

图1给出了线阵设计软件的方框图，我们把它取名为 SDSP1.

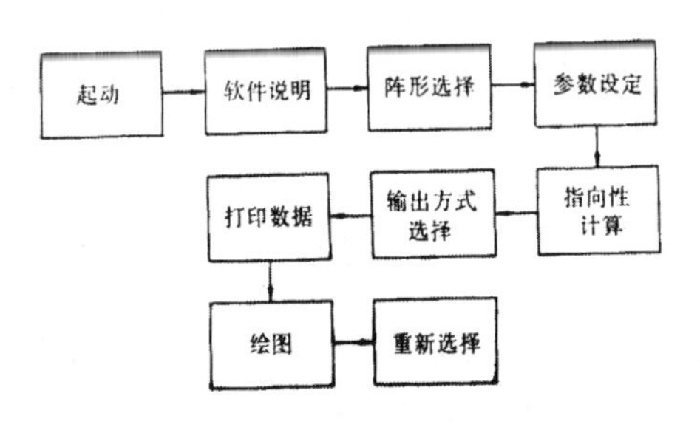

图 1 SDSP1 软件的方框图

整个软件可以划分为三个部分：第一部分是软件的描述，详细说明本软件的用途，适用范围，如何输入参数等. 在程序进行的中途，用户可以发出"Help"命令去查阅这个说明.

第二部分是计算部分，在用户选定基阵形式(连续阵或离散等间隔线列阵)之后，再从键盘输入各种参数，比如线阵长度和波长之比，定向角等. 于是计算机开始计算指向性，并把所需数据存贮起来.

第三部分是输出数据、绘图、打印部分，用户可以要求输出数值表，也可以要求在屏幕上画图，在必要的时候，可以在普通打印机上把屏幕上的图拷贝在打印机上.

为了使用方便，整个程序以人机交换方式进行. 用户可以随时中断运行，改变自己的主意，或退出运行，回到 DOS 状态. 作为例子，我们在图2中给出连续线阵子程序的结构，首先要求用户在主菜单上选择. 假定用户选了连续阵，屏幕上立刻出现连续线阵的结构示意图与各参数的意义. 然后，请用户输入参数，同时指明参数的取值范围. 如果用户输入不合乎规定的参数或误操作，则机器要求用户重新输入直到正确为止，在全部参数输入完毕之后，机器开始计算指向性并存贮数据. 计算结束时又出现菜单，请用户在输出方式中进行选择. 如果用户选择了绘图，机器又会请用户在三种坐标中选取一种. 用户可以随时获得硬拷贝.

三、使用说明，例子

SDSP1 软件装在一个激光穿孔软盘上，它可以在 IBM PC，/XT，/AT 及其兼容机上使用，文件名为 SDSP1. exe. 用户可以把它 Copy 到硬盘上去执行，但必须有软盘在A驱动器内，否则本软件仍不能运行，因为软件是保密的，只有在核对激光指纹正确的情况下才能正常进

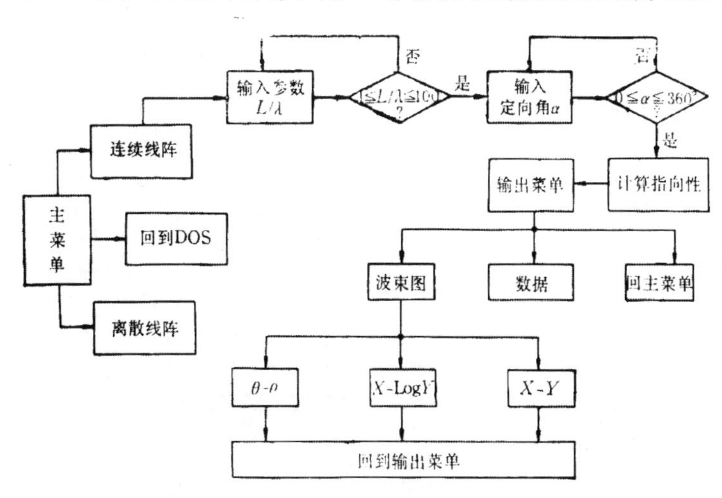

图 2 子程序的结构

行. 键入 SDSP1, 加 Return. 软件即开始运行. 通常有闪烁的提示在屏幕的最下方出现, 用户只需按提示操作就行, 运行开始时有一个软件说明称为 "INFORMATION ABOUT THIS SOFTWARE", 它详细地告诉用户本软件的内容和如何使用这个软件, 然后, 屏幕上将出现主菜单

THE ARCHITECTURE OF LINE ARRAY

1 = Continue line array

2 = Discrete equal-spaced line array

3 = Help

4 = Return to DOS

YOUR OPTION =

在用户选定之后, 再按机器的要求, 输入各种必要的参数, 如果各参数符合机器的取值范围要求, 计算机屏幕上将出现 I'm working, 表示软件正在正常进行之中. 用户需等待一会儿 (一般是几十秒钟), 在计算结束时, 屏幕上再一次出现菜单, 这是一个输出方式选择菜单

OUTPUT MENU

1 = Output the data to printer

2 = To plot beampatterns

3 = Return to Main Menu

...

YOUR OPTION =

如果用户需要完整的数据形式输出, 那么就选 1, 则打印机上立刻开始打印, 屏幕上给出运行提示

The data is currentely in printing

如果用户需要波束图示, 就选取 2, 这时屏幕上将出现分菜单如下

PLOTTING MENU OF BEAMPATTERNS

1 = Magnitude of amplitude

2 = Logrithm of magnitude

3 = Polar coordinate

4 = Return to output menu

YOUR OPTION =

用户可以从普通直角坐标, 对数坐标, 极坐标中任选一种. 在任何一种作图方式, 屏幕上将会出现 Hard Copy (Y/N)? 如果用户回答 "Y", 机器将在普通打印机上把用户在屏幕上所看到的图拷贝在打印机上, 如果用户回答 "N", 那么软件认为或许是用户对这种比例的图不满意, 屏幕上将出现

Do you want an exploired view (Y/N)?

假定你想看一下指向性图的细微结构, 你就可把横坐标拉开, 输入 "Y", 然后再把你要看的那个窗口输入软件, 这时就会看到放大了的图形这是本软件的 ZOOM 功能.

当然, 在极坐标状态, 无法使用 ZOOM 功能, 这时你如果误打了 "Y", 那也没有关系, 屏幕上立刻出现

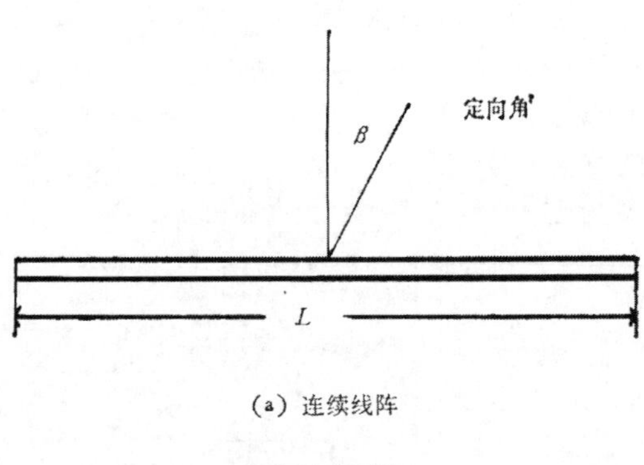

(a) 连续线阵

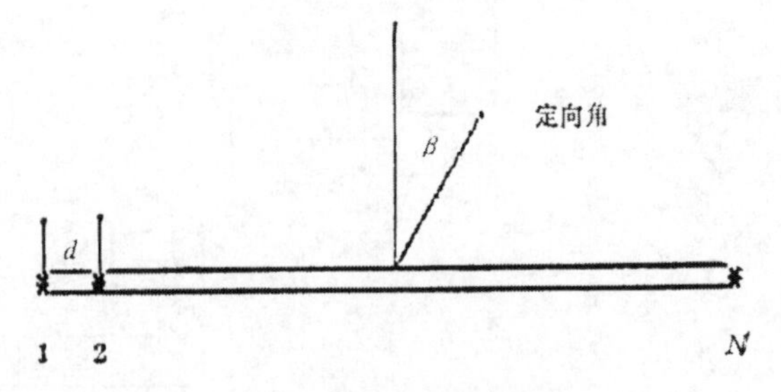

(b) 离散等间隔线阵

图 3 线阵结构与参数说明

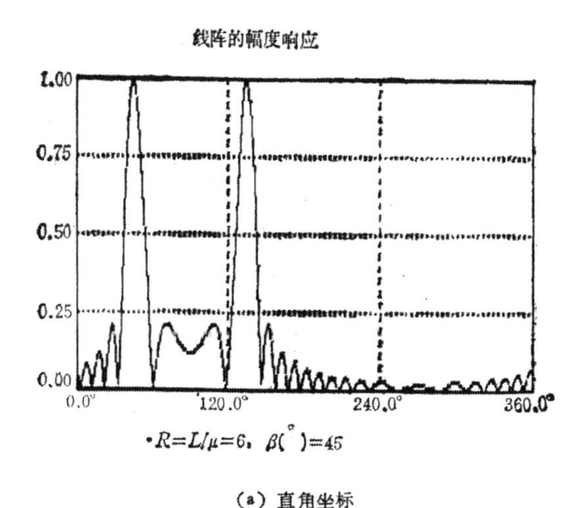

(a) 直角坐标

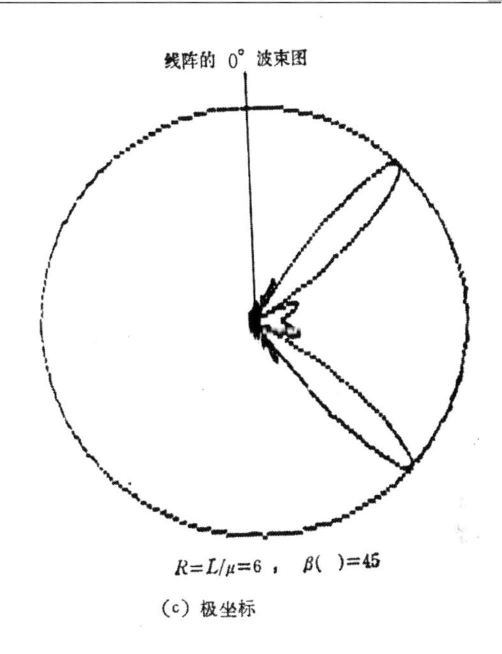

(c) 极坐标

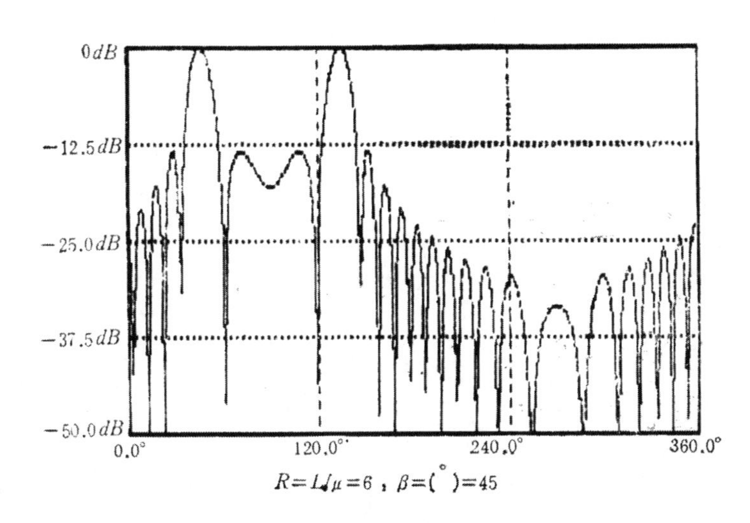

(b) 对数坐标

图 4 由 SDSP1 绘制的指向性图

Sorry! you can't get the beampattern in this situation.

Please wait.

几秒钟之后，就返回到输出方式菜单去了.

图 3(a)，(b) 是本软件对基阵结构与有关参数的说明，图 4 是由 SDSP 1 软件得到的波束指向性图的硬拷贝，线阵长与波长之比为 6，定向角 $\beta=45°$，其中 (a) 是直角坐标，(b) 是对数坐标，c 是极坐标.

图 5 给出了同一指向性从 45°—90° 的

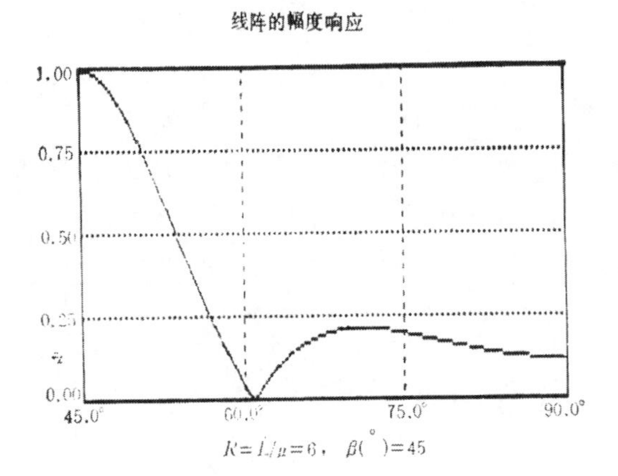

图 5 SDSP1 的 Zoom 功能举例

Zoom 图.

四、结　论

SDSP1 是用于声呐设计的在微机上运行的软件．具有运行速度快,使用方便,参数动态范围大的特点．对于专业人员，几乎不用训练即可运用自如．

孙增、李淑秋同志参加了部分程序的编写，作者对他们表示感谢．

中国科学院科理高技术公司慷慨地提供一台微机,加快了作者的开发工作,特此致谢．

参 考 文 献

[1] IEEE ASSP Digital Signal Processing Committee, Programs for digital signal processing, IEEE Press, New York, 1979.

[2] Burrus C. S., and Parks T. W., DFT/FFT and Convolution algorithme with TMS 32010 programs, John Wiley & Son, 1985.

[3] Carpenter J., et al., Statistical software for microcomputers, BYTF, April 1984, 234—264.

[4] Morris L. P., TMS 320 digital signal processing software, Ins. P. O. Box 5348, Ottawa, CANADA, 1985.

数字式声呐的高效率前置 A/D 转换器①

李启虎 赵国英 陈玉凤
(中国科学院声学研究所)

1989 年 4 月 20 日收到

A/D 转换是数字式声呐的最基本的前置处理之一，要充分利用 A/D 的动态必须确保输入信号始终处于适当的范围内. 本文把 A/D 转换器的噪声分为量化噪声和饱和噪声. 在输入噪声为高斯信号的假定下，分析了总噪声和 A/D 转换器各参数的关系，提出一种带反馈的 AGC 模型和采用数据压缩的 μ- 律量化，以便确保 A/D 转换器工作于最佳状态. 文中给出了理论分析和实际电路设计.

一、引 言

A/D 转换器是数字式声呐的最基本的前置处理设备之一[1-2]，根据水声信号处理的特点，往往要求输入有较大的动态. 但是 A/D 转换器本身有其固有的动态. 为了充分利用这种动态，必须使输入信号经常保持在某一特定范围内，要做到这一点是相当困难的. A/D 转换器作为一种非线性处理会引进各种噪声[3-4]，量化噪声和饱和噪声都是影响前置处理性能的重要因素.

假定输入信号是高斯分布的，那么可以计算出在某一量化比特数之下的量化噪声和饱和噪声，同时也不难推算 A/D 转换器输出的幅度分布.

通常设计者希望 A/D 转换器的输出经常能工作在较高比特的地方，为了作到这一点，本文提出一种带反馈的自动增益控制 (AGC) 模型，把输入信号的均方根值用于控制 AGC 的放大量. 同时利用语音处理中已广泛使用的 μ-律 A/D 转换器[5-6]，由此构成的模型可以保证使整个前置处理工作于最佳状态. 我们利用美国 Burn-Brown 公司的 8 比特 μ- 律 A/D 转换器 TP5116A，构成了一个用于水声信号的高性能 A/D 转换器. 系统工作可靠，动态得到充分的利用，实现了本文的理论分析.

二、A/D 转换器的量化噪声和饱和噪声

水声信号处理系统的前置 A/D 处理的框图如图 1 所示. 由换能器接收到的信号 $x_0(t)$ 经模拟低通滤波后，输入自动增益控制单元. 再由采样保持电路变为时间上离散的信号. 通常，采样频率 f_s 要大于前置低通滤波器截止频率的两倍. 采样保持电路的输出 $y(nT_s)$ 经过 A/D 转换之后，变为 $[y(nT_s)]_Q$，它实际上是数字式声呐的原始数据.

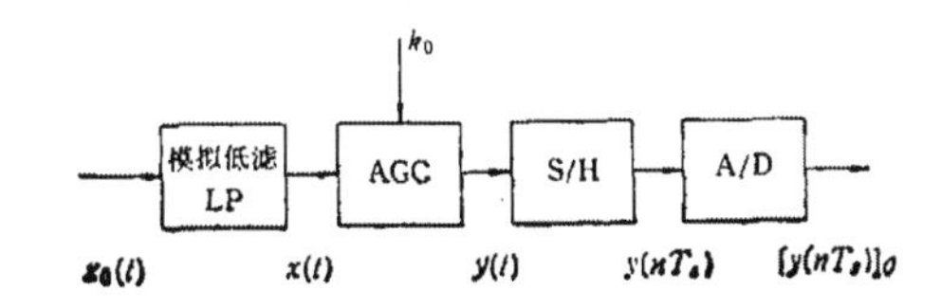

图 1 A/D 转换的前置处理运算

AGC 的作用是确保 $y(nT_s)$ 处于 A/D 转换器的动态范围之内，同时，又要使 $[y(nT_s)]_Q$ 大部分时间具有高的比特级. 要做到这一点往往是困难的. 特别是对于主动声呐信号，包括混响在内的信号动态，往往在 80dB 以上[7]，这样宽的动态，实际上不可能，也没有必要，完全保持不失真. 一般可以对混响的部分数据实行限幅处理，这将不影响对回声的检测.

对于被动声呐的输入信号，主要考虑的是当地环境噪声的量级. 因为在我们感兴趣的问

① 应用声学, 1990, 9(1): 8-12.

题中,信噪比在 0dB 以下,这样,输入的动态实际上就由环境噪声的动态所决定. 这个量一般是很小的(例如 20 dB). 然而,新的问题是在我们设定 AGC 控制因子 k 时,必须考虑环境噪声的统计资料. k 太小会使 A/D 处于过饱和状态, k 太大会使 A/D 处于低比特量化状态,这两种情况都是极为不利的.

假定 A/D 转换器的输入动态为 $[-A, A]$, 量化级是 q, 那么当 $-A \leqslant y(nT_s) \leqslant A$ 时,量化误差是唯一的噪声源,这时

$$[y(nT_s)]_Q = y(nT_s) + e(nT_s) \quad (1)$$

其中 $e(nT_s)$ 表示量化噪声. 显然

$$|e(nT_s)| \leqslant q/2 \quad (2)$$

可以近似地认为 $e(nT_s)$ 在 $[-q/2, q/2]$ 内均匀分布,那么

$$E[e(nT_s)] = 0$$
$$E[e^2(nT_s)] = q^2/12 \quad (3)$$

如果 A/D 转换器是 b 比特的,那么 $q = 2^{-b}$, 则(3)式可以改写为

$$E[e^2(nT_s)] = 2^{-2b}/12 \quad (4)$$

如果输入 A/D 转换器的信号超过了它的动态,那么就会产生由于饱和所引起的噪声. 假定信号 $x(t)$ 遵从高斯分布,且取 $A = k\sigma_x$, 那么饱和噪声功率为

$$SN = 2\int_{k\sigma_x}^{\infty} e_1^2(x) f(x) dx \quad (5)$$

量化噪声功率

$$QN = 2\int_0^{k\sigma_x} e_2^2(x) f(x) dx \quad (6)$$

其中

$$f(x) = \frac{1}{\sqrt{2\pi}\sigma_x} e^{-x^2/2\sigma_x^2} \quad (7)$$

$$e_1(x) = (x - k\sigma_x), \quad e_2(x) = q/\sqrt{12} \quad (8)$$

总的噪声功率为

$$PN = SN + QN \quad (9)$$

对于给定的比特数 b, 可以从(7)—(9)式计算出最合适的 AGC 控制因子 k. 可以证明,当 $b = 8$ 时,使 PN 极小的 k 值大约是 4 .

三、充分利用 A/D 动态的方法

为了充分利用 A/D 的动态, 必须采用简单而有效的方法. 图 2 给出了两种设想, 第一种方法是由输入信号的均方根值去控制 AGC 的调节因子,使 AGC 的输出正好处于 A/D 转换器输入范围之内, 第二种方法是利用语音处理中常用的 μ- 律转换器, 它可以对输入信号进行压缩, 使其动态处于 A/D 转换器的动态之内,下面我们分别予以说明.

设输入信号 $x(t)$ 是均值为零的高斯信号,令

$$y(t) = x(t)/k_0\sigma_x \quad (10)$$

显然, $y(t)$ 仍是均值为零的高斯信号. 如果已知 A/D 转换器的输入范围是 $[-A, A]$, 那么我们一般应取 $A \approx 3\sigma_x$. 在硬件设计中, A/D 利用的是商用的芯片, 它的输入动态是无法随意变动的, 利用(10)式, 我们可以选择 k_0, 使 $y(t)$ 的动态永远处于 $[-A, A]$ 之间, 此处 $A \approx 3\sigma_y$. 要作到这一点是容易的,因为由(10)式,可知

$$\sigma_y = \{E[y^2(t)]\}^{1/2} = 1/k_0 \quad (11)$$

只须取

$$k_0 = 3/A \quad (12)$$

就可以达到这一目的. σ_x, k_0 就是 AGC 的调制因子,我们实际上只要随时测量 σ_x, 同时又按 (12) 式确定 k_0, 便可以得到完全符合要求的 A/D 转换器的输入信号 $y(t)$.

容易证明,在取 $k_0 = 3/A$ 时,量化信噪比是

$$[SNR]_Q = 6b - 1.24\text{dB} \quad (13)$$

其中 b 是上面提到的量化比特数.

用这种方法控制 A/D 输入动态的模型见图 2(a).

另一种方法是用 AGC 的输出进行反馈来控制整个 AGC 的调节因子(见图 2 (b)). 为了有效地进行控制,可以采用门限电平比较法,当 AGC 输出的均方根值超过某一电平时, 就

改变放大器的增益使输出变小，当均方根值低于门限值时则使增益变大，当然要合适选用阈电平的范围，以保证 AGC 的正常工作.

在通信与语音处理中得到广泛应用的 μ-律 A/D 转换是一种数据压缩技术，对小信号特别适用.在某些情况下，8 比特的 μ-律 A/D 转换器可以取得 13 比特的线性量化的效果[6]。其缺点是增加了线路的复杂程度，并且需要有反 μ 律变换器. μ-律 A/D 转换器的输入输出关系为

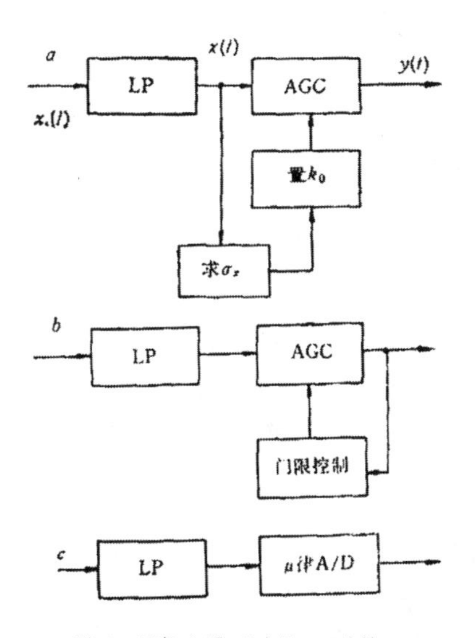

图 2　提高 A/D 动态的几种方法
(a) 调节 AGC 增益的方法之一，(b) 调节 AGC 增益的方法之二，(c) μ 律数据压缩.

$$y=\mathrm{Sgn}\,x\frac{\ln(1+\mu|x|)}{\ln(1+\mu)}\quad |x|\leqslant 1 \tag{14}$$

我们可以看出，当 $x\approx 0$ 时，

$$y\approx\frac{\mu}{\ln(1+\mu)}x \tag{15}$$

也就是说，μ-律 A/D 转换器对小信号仍是线性的，而对大信号却被压缩了. 图 3 给出了 μ-律转换器输入输出关系，它显然使 A/D 的动态得到了扩展.

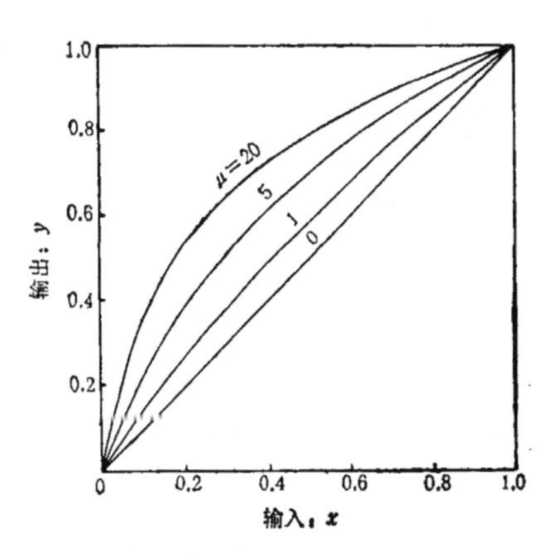

图 3　μ-律转换

四、实际电路的设计

我们实际上可以设计用图 2(a) 或图 2(b) 和图 2(c) 结合而成的 A/D 转换处理器. 举例来说，如果一个 A/D 转换器用于被动声呐信号的检测，海洋噪声的均方根值在 3μV—10μV. 如果我们不用 AGC 电路，考虑到噪声的起伏，我们为了照顾到海洋噪声的上限，取 $L_s=3\times 10\mu$V 作为 A/D 的动态. 这时，如果实际海洋噪声是 3μV，那么 A/D 转换器就无法充分发挥作用. 因为当海洋噪声均方根值为 3μV 而 A/D 转换器的动态范围取为 $L_s=3\times 10\mu$V 时，A/D 转换器输出为高比特数的概率很小(如图 4 中的实线所示)，显然动态

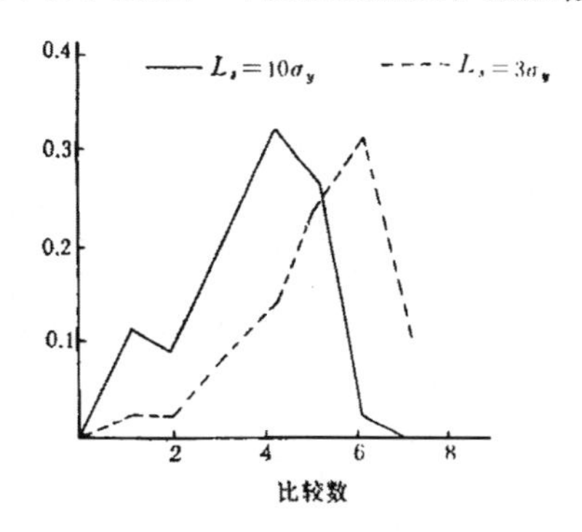

图 4　8 比特 A/D 转换器输出幅度的概率分布

未得到充分利用. 反之若 A/D 转换器的动态定为 $L_t = 3 \times 3\mu V$，当海洋噪声的均方根值为 $3\mu V$ 时，A/D 转换器输出为高比特数的概率就较大(如图 4 中的虚线所示)，动态确实得到了充分利用.但是由于起伏,一旦海洋噪声的均方根值变为 $10\mu V$ 时就会出现严重饱和.

图 5 给出了我们根据 2(b) 及 2(c) 的方框图所设计的一个高效 A/D 转换电路.

输入信号经模拟低通滤波后，输入到自动增益控制单元，其输出送给 μ- 律 A/D 转换器,我们采用的是双门限反馈电路，由检波器，积分器对 AGC 输出波形进行检波和平滑，再与门限进行比较，由此使 AGC 的增益随输出信号的均方根值来变化,这样可以确保 AGC 的输出处于 A/D 动态的范围之内.

图中 TP5116A 是美国 BB 公司生产的 Codec 器件，是一种 8 比特的 μ- 律转换器件，AD7110 是数字式的网络衰减器.

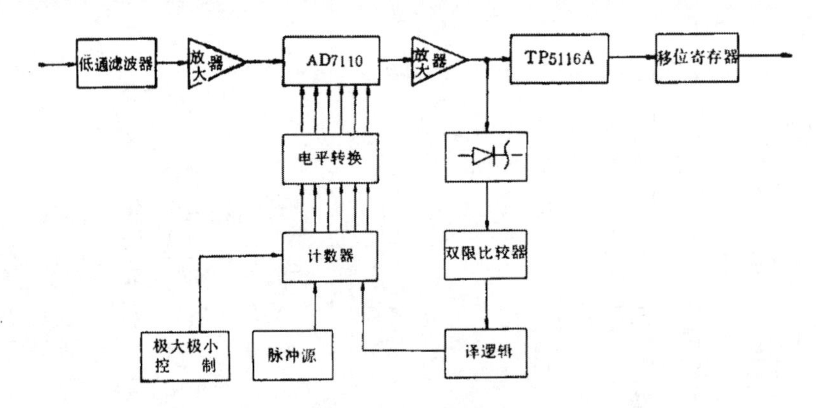

图 5 高效率 A/D 转换器

参 考 文 献

[1] Knight W. C., et al., *IEEE Proc.*, 69(1981), 1451—1506.

[2] Winder A. A., *IEEE Trans.*, *US*-22(1975).

[3] Hoeschele D. F., Analog-to-Digital and Digital-to-analog Conversion Techniques, Wiley, New York, 1968.

[4] Fray G. A., and Zeoli G. W., *IEEE Trans.*, *AES*-7 (1971), 222—223.

[5] Openheim A. V., et al., Applications of Digital Signal Processing, Prentice-Hall, USA 1978.

[6] Bowen B. A., Brown W. R., VLSI Systems Design for Digital Signal Processing, Prentice-Hall Inc., 1982.

[7] 李启虎,声呐信号处理引论,海洋出版社,1985,北京.

数字式声呐中的升采样率处理①

李启虎

（中国科学院声学研究所）

1988年8月30日收到

摘要 升采样率处理是某些数字式声呐的必不可少的预处理．本文给出整倍数升采样率运算的基本原理和理论结果．指出，升采样率处理的好坏不仅取决于窗函数的长度，也还和输入信号的功率谱密度有关．为实现升采样率处理，可以采用几种不同方法，文中指出用输入样本补零和窗函数卷积的方法是有效的．在数字计算机上的模拟取得了良好的结果．这种方法便于用专用数字信号处理芯片实现，并且对输入、输出采样率的选取上有很宽容的余地．

The rising sample rate processing in digital sonar

LI Qihu

(*Institute of Acoustics, Academia Sinica*)

Received August 30, 1988

Abstract Rising sample rate processing (RSRP) is a necessary pre-processing in some cases for digital sonar. The principle of RSRP, in which the sample ratio of output over input is an integer, is given and the theoretical result is described. It is shown that the performance of RSRP not only depends on the length of the window function but also on the spectrum of input signal. Several algorithms can be chosen to implement the RSRP. It is shown that the method of convolution by using the zero-paded input signal and the window function is effective. The system simulation have shown a good agreement of theoretical analysis. The algorithm described in this paper is easy to implement by digital signal processing chips. It is also affordable in choosing the input output sampling rate.

一、引　　言

随着数字信号处理技术越来越多地被应用于声呐设计[1]，一些不同于模拟声呐的问题出现了．在数字式声呐的预处理和波束成形中所应用的升采样率运算就是一例[2,3]．

从纯粹的信息论观点来看，采样频率只需等于信号频带上限的两倍就行．但实际上，由于精确定向、测距等需要，必须使采样频率大大提高[4~7]．因为理论已经证明，只有同步波束才能使定向精确[3]，而大量海上试验则证明，声波在海水中传播时，由于种种原因会引起信号到达时间的随机起伏[6,7]，因此，精确测定时延起伏是至关重要的．而要作到这一点，必须使采

① 声学学报，1990，15(1)：7-11.

样周期小于可能的起伏值，否则要精确测量它是不可能的。

从原则上说，我们应当尽量使采样频率高一点，但是有不少情况使我们作不到这一点．在这种情况下我们就要对已经采样的信号进行升采样率处理（RSRP）[2,5]

本文首先阐述 RSRP 的基本原理．指出度量 RSRP 的准则以及减少误差的方法．文中提出将低采样率信号补零后和窗函数卷积的办法得到新的具有高采样率的信号方法．计算机模拟实验证明了这种办法的有效性．

本文所提出的方法具有运算量少，便于用数字信号处理芯片实现的优点，同时可得到提升比很大的 RSRP．

二、RSRP 的性能准则和基本定理

设 $x(t)$ 是带宽有限的信号．它的谱是 $X(f)$．当 $|f|\geqslant W$ 时，$|X(f)|=0$．令 $T_0=1/2W$．根据采样定理[5]，对 $T_s\leqslant T_0$ 有

$$x(t)=\sum_{k=-\infty}^{\infty}x(kT_s)\frac{\sin[\pi(t-kT_s)/T_0]}{\pi(t-kT_s)/T_0} \tag{1}$$

在一般的数字式声呐中，采样之后就是数字运算，不再进行低通的模拟滤波，所以无法按照 $x(kT_s)$ 序列完全恢复原来的信号．如果采样周期 T_s 太稀，我们希望能对 $x(kT_s)$ 进行加密处理．也就是说从 $x(kT_s)$ 得到一个新的序列 $z(kT_s')$ 其中 $T_s'\leqslant T_s$．我们只讨论 $T_s/T_s'=m$ 是正整数的情况，参看图 1．我们称 m 为提升率．一般来说 $z(kT_s')\neq x(kT_s')$．在理想的情况下，如果对 $x(kT_s)$ 进行低通滤波，滤波器的频率响应是理想的截止频率为 W 的低通滤波器．那么对这个滤波器用 T_s' 进行采样，就会得到 $x(kT_s')$．但实际上我们不用这种方法．原因之一是没有这样的理想滤波器，原因之二是数字系统中不希望再出现模拟运算．

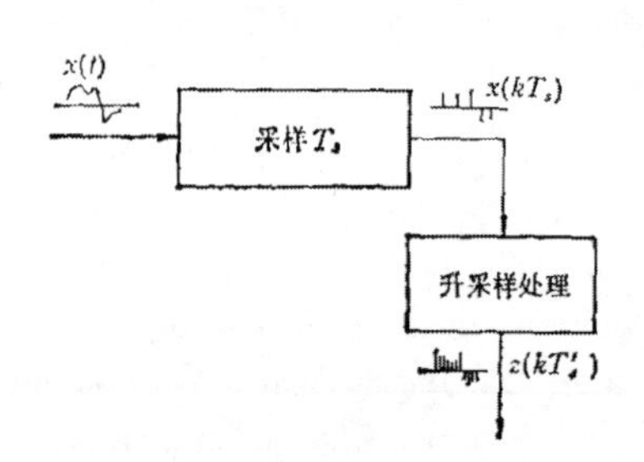

图 1　升采样率的方框图

要度量升采样运算的性能，可以采用均方误差的准则，即

$$J\triangleq E[x(kT_s')-z(kT_s')]^2 \tag{2}$$

极小．

如上所述，只有在理想的情况下，可以使 $J=0$．任何实际系统，$J>0$．下面我们证明一个重要结果，即采用对 $x(kT_s)$ 序列补零的方法，可以使升采样运算后的新序列 $z(kT_s')$ 与 $x(kT_s')$ 任意接近．我们以定理的形式来表达这一结果．

定理： 设信号 $x(t)$ 带宽有限，它的谱 $X(f)$ 在 $|f|\geqslant W$ 时为零，$T_s\leqslant 1/2W$．用 T_s 为周期对 $x(t)$ 进行采样得到 $x(kT_s)$．设 m 为正整数，$T_s'=T_s/m$．为了得到 $x(t)$ 的密采样序列 $x(kT_s')$，只需将 $x(kT_s)$ 补零，两个样本点之间补上 $m-1$ 个零点，得到 $y(kT_s')$：

$$y(kT_s')=\begin{cases}0 & \text{当 } k/m \text{ 不为整数}\\ x\left(\dfrac{k}{m}T_s\right) & \text{当 } k/m \text{ 为整数}\end{cases} \tag{3}$$

把序列 $y(kT'_s)$ 与窗函数

$$w(k) = \frac{\sin k\pi T'_s/T_0}{k\pi T'_s/T_0} \tag{4}$$

进行卷积得到新序列 $z(kT'_s)$:

$$z(kT'_s) = w(k) * y(kT'_s) \tag{5}$$

当窗口长 $N \to \infty$时，$x(kT'_s)$ 与 $z(kT'_s)$ 之均方差趋于零。

我们现在来证明这个定理。

$$z_N(kT'_s) = \sum_{k=-N+1}^{N+1} \frac{\sin(k\pi T'_s/T_0)}{k\pi T'_s/T_0} y[(k-i)T'_s] \tag{6}$$

其中 $T_0 = 1/2W$。

根据(1)式，有

$$x(iT'_s) = \sum_{k=-\infty}^{\infty} x(kT_s)\frac{\sin[(iT'_s - kT_s)\pi/T_0]}{(iT'_s - kT_s)\pi/T_0} \tag{7}$$

从(6)式，当 $N \to \infty$ 时，在均方意义上有[5]

$$\begin{aligned} z_N(iT'_s) &\to \sum_{k=-\infty}^{\infty} y[(k-i)T'_s]\frac{\sin k\pi T'_s/T_0}{k\pi T'_s/T_0} \\ &= \sum_{k=-\infty}^{\infty} y(kT'_s)\frac{\sin[(i-k)\pi T'_s/T_0]}{(i-k)\pi T'_s/T_0} \\ &= \sum_{k=-\infty}^{\infty} x(kT_s)\frac{\sin[(iT'_s - kT_s)\pi/T_0]}{(iT'_s - kT_s)\pi/T_0} \end{aligned} \tag{8}$$

由(7),(8)式得到，当 $N \to \infty$ 时，

$$E[z_N(iT'_s) - x(iT'_s)]^2 \to 0 \tag{9}$$

由补零的方法进行升采样运算的方框图见图 2。我们可以看出，这种运算是非常容易实现的。因为补零运算以及卷积运算都已经有大量的标准芯片可以实现。况且，卷积序列中的一个有大量的零点。只要程序编排合适，可以使存贮量和运算量大为减少。

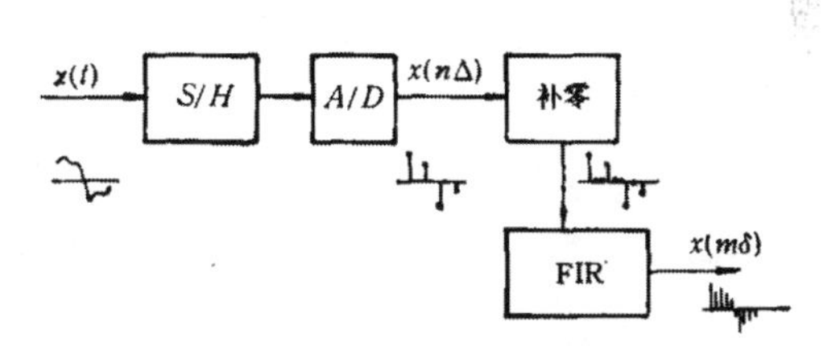

图 2 数字内插法升采样率原理

三、系统仿真结果

我们从 $z(kT'_s)$ 的表达式(6)及评估准则(2)中可以看到，要使 J 减小，不仅取决于序列的长度，也还和信号谱的形状 $X(f)$ 有关，一般来说，如果没有什么关于信号的先验知识，应当假定 $X(f)$ 是矩形的，然后在一定带宽内用单频信号进行仿真试验，以便得到合适的窗口长度和提升比。

在实际系统设计时，N应当和m密切配合，以便使存贮量，运算量充分利用。例如，如果提升比 $m=4$，那么在每两个样本点中要插入三个零点。这时窗口长宜采用 $N=13, 17, 21$ 点等

等.这样会使输入的数据得到最有效的利用.图 3 是实际的仿真结果. 图中,$m=4$, $N=17$ 点.

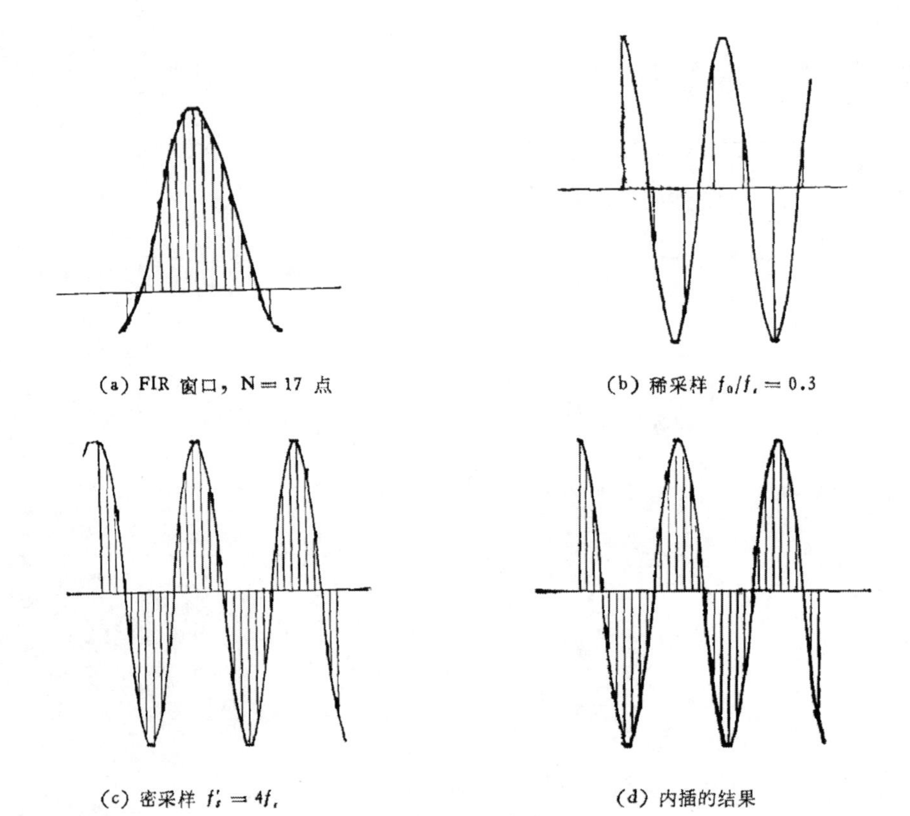

(a) FIR 窗口, N = 17 点　　(b) 稀采样 $f_0/f_s=0.3$

(c) 密采样 $f_s'=4f_s$　　(d) 内插的结果

图 3　升采样四倍的例子

图 4 说明窗口长度的作用. 根据实际的系统模拟结果,当 $m=4$ 时,$N\geqslant 17$ 就可以得到相当满意的结果.

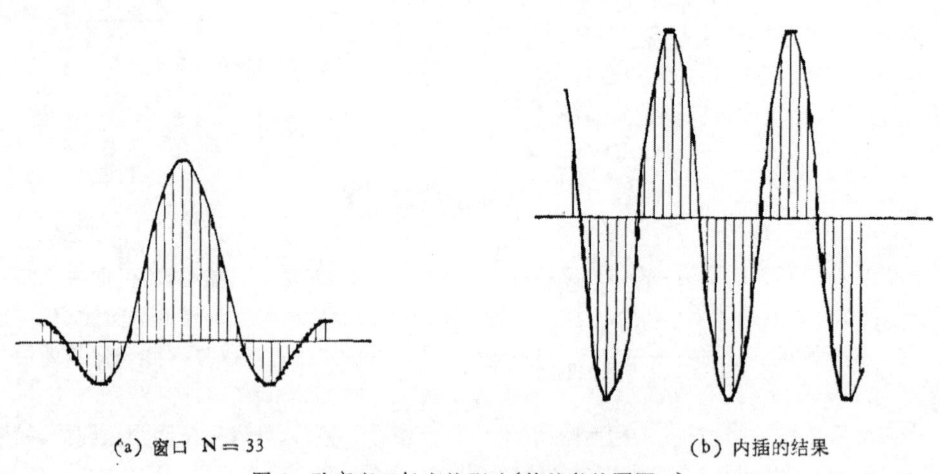

(a) 窗口 N = 33　　(b) 内插的结果

图 4　改变窗口长度的影响(其他条件同图 3)

四、结　论

升采样率处理是某些数字式声呐中不可缺少的．本文提出升采样运算性能的准则以及在此准则下的一种可行的内插方法．这种方法易于用硬件实现，运算量少．系统仿真结果令人满意．

科理高技术公司为本项研究计划提供一台 IBM PC/XT 微机，作者对此表示感谢．

参　考　文　献

[1] Knight, W. C. *et al.*, "Digital signal processing for sonar", *Proc. IEEE.*, **69**(1981), 1451—1506.

[2] Crochiere, R. E. and Rabiner, L. R., "Interpolation and decimation of digital signal-a tutorial review", *Proc. IEEE*, **69**(1981), 300—331.

[3] Pridham, R. G., "Digital interpolation beamforming for low pass and band pass signals", *Proc. IEEE* **67**(1979), 904—919.

[4] Schultheiss, P. M. and Ianniello, J. P., "Optimum range and bearing estimation with randomly perturbed arrays", *J. Acoust. Soc. Amer.*, **68**(1980), 161—173.

[5] 李启虎，"声呐信号处理引论"，(海洋出版社，北京，1985年).

[6] Jobst, W. and Dominijanni, L., "Measurement of the temporal, spatial and frequency stability of an underwater acoustic channel", *J. Acoust. Soc. Amer.*, **65**(1979), 62—69.

[7] 张仁和等，"浅海远程声场的空间相关性和时间稳定性"，声学学报，1981年，No. 1,9—19.

用级联阵列机构造大型数字式声呐的设计技术①

刘秋实　李士才　李启虎　孙　增
（中国科学院声学研究所）
1989 年 1 月 24日收到

摘要　在分析声呐信号处理特点的基础上，提出一种称为级联阵列机的多处理器体系结构，用以实现大型数字式声呐系统．并对该体系结构中的各种要素，包括总线协议，数据传输，基本处理单元进行了详细讨论，最后给出一个使用级联阵列机结构实现数字声呐系统的设计步骤．

The technique to design large digital sonar system using cascade link array processors

LIU Qushi　LI Shicai　LI Qihu　SUN Zhen
(Institute of Acoustics, Academia Sinica)
Received January 24, 1989

Abstract　From the analysis of the characteristics of sonar signal processing, a new achitecture called cascade link achitecture to implement large digital sonar system with general processors and special signal processors is proposed. This achitecture is very suitable for large mult-chanel sonar signal processing. The basic elements, including data bus protocal, data transmition, control unit, basic processing cell are discussed. Last, a regular step to design a digital sonar using this achitecture are suggested.

一、引　言

随着 VLSI 技术的发展及专用数字信号处理芯片的出现，使用多个微处理器设计一个多处理器结构的数字式声呐在技术上已经成为可能[1-3]．以多处理器为核心的声呐与以全硬件为核心部件的数字式声呐相比，在系统配制上更加灵活、成本更低、设计周期更短．因而会很快淘汰核心部件为全硬件的数字式声呐系统．

采用多处理器构成的声呐有以下几个优点：

1. 不同的处理要求（例如波束形成、时延估计，自适应噪声抵消等），可以用硬件结构基本相同的处理部件完成．

2. 几套算法共用一套硬件，实现同一处理要求，以分别处理不同的实际情况．

① 声学学报, 1990, 15(3): 230-237.

3. 处理功能以处理器为核心的功能分布化，导致数据流通路径的局部化及故障自检的局部化，从而降低设计复杂度，提高了可靠性。

然而传统声呐中，以处理器为基础的声呐设计，都使用微机标准总线挂标准信号处理单元的结构。当数据处理量大到一定程度，就会带来大量的数据调度问题。在将信号处理算法向信号处理单元映射过程中，往往发现一个算法不能正好放入一个预先设计好的处理单元中，因而不得不将算法切分，这样大量的中间结果在微机单总线上或预先设计好的双口 RAM 通路上传输时，会由于总线频带宽度限制而发生阻塞现象。

产生这种不合理现象的原因是缺少一种合适的由多处理器构造声呐的办法。

因此本文讨论一种使用阵列机结构实现多处理器构成声呐的方法与技术。该技术包括：1. 数字信号处理流程图向级联阵列机结构的映射。2. 建立互联结构。3. 互联结构分类。4. 各类两级互联结构向标准处理单元的映射 5. 总控的配制及总控命令单元的选择。

文中同时给出不同互联结构的标准单元设计。

二、声呐系统数据流动特点分析

一般声呐系统总是分步将原始数据一步步地处理、变换，最后获得结果。从宏功能上讲，声呐系统将前级计算结果送入下一级作为输入、在下一级中进一步处理、依次传播直到获得需要的结果为止。我们称这种方式为级联方式。在每一处理级上，任一时刻要不停地处理按时间顺序依次抵达的数据。

因为要由多个传感源(水听器)布阵以获取空间增益，所以在不同的处理级上存在着功能完全一样的多个通道。图 1 是一个被动测向/测距声呐站原理的方框图。

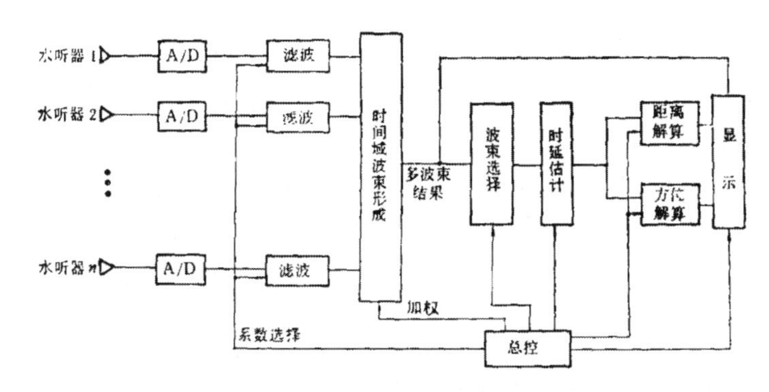

图 1 被动测向/测距声呐

由图可以看出：多路水听器信号经 A/D 变换送入滤波器进行滤波、滤波器输出送时域波束形成器进行波束形成，其结果送显示器，由总控选择某一波束信号的两个分裂波束进行广义互谱时延估计、最后获得的方位角及距离送显示器中。因此，这个声呐系统可看成是五级级联结构的系统、分别是滤波级、波束形成级、互谱级、解算方位、距离级及显示级。

声呐信号处理本身固有的多路相同结构的特点以及级联的特点使得本文讨论的级联阵列机结构成为一种非常合适于构造声呐系统的多处理器体系结构。

三、级联结构分析

将声呐的计算要求分配到图 2 的结构上去，虽然还没有见到有关的结果发表，但很多人在别的领域做过大量类似的研究工作，如著名的 Systolic、Wavefront 方法[4]. 我们将在另外一篇文章中结合声呐信号处理的特点提出一种类似于 Wavefront 方法的映射方法. 这里着重研究互联结构. 一但将整个声呐信号处理的计算合理地分配到图 2 的级联结构上去，那么解决整个级联阵列机实现的关键就归结为如何实现两级之间的互联.

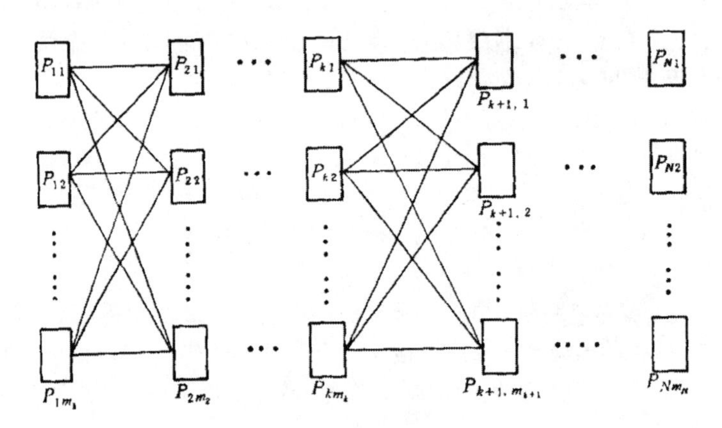

图 2　一般的级联结构

从两级之间看互联结构不外有下面五种方式.（为以后叙述方便起见，我们给每一种互联结构起一个英文名字.）

1. 单出单入 (Single Out Single In)

即一个处理单元的输出做为另一个处理单元的输入，记作 SOSI.

2. 多出单入 (Multiple Outs Single In)

即多个处理单元的输出做为一个处理单元的输入，记作 MOSI.

3. 单出多入 (Single Out Multiple Ins)

即单个处理单元的输出做为多个处理单元的输入，记作 SOMI

4. 多出单选 (Multiple Outs Single Select)

即多个处理单元输出供一个处理单元选择一个来使用，记作 MOSS

5. 多出多入 (Multiple Outs Multiple Ins)

即多个处理单元输出供多个处理单元使用，记作 MOMI.

以上五种互联方式概括成图 3 的图形表示. 其中 MOSS 实际上可归结为 MOSI. 但从声呐信号处理系统从前级向后按数据处理量递减来看. 一种前大后小的梯状结构阵列机最为节省设备. 因而 MOSS 在级联阵列机中经常用到，所以把 MOSS 结构单独取出来做为一种互联方式.

SOSI 结构等价于 MOSI 或 SOMI 结构中 $l = 1$ 的情况，可以采用 MOSI 或 SOMI 结

构的数据传递方式.

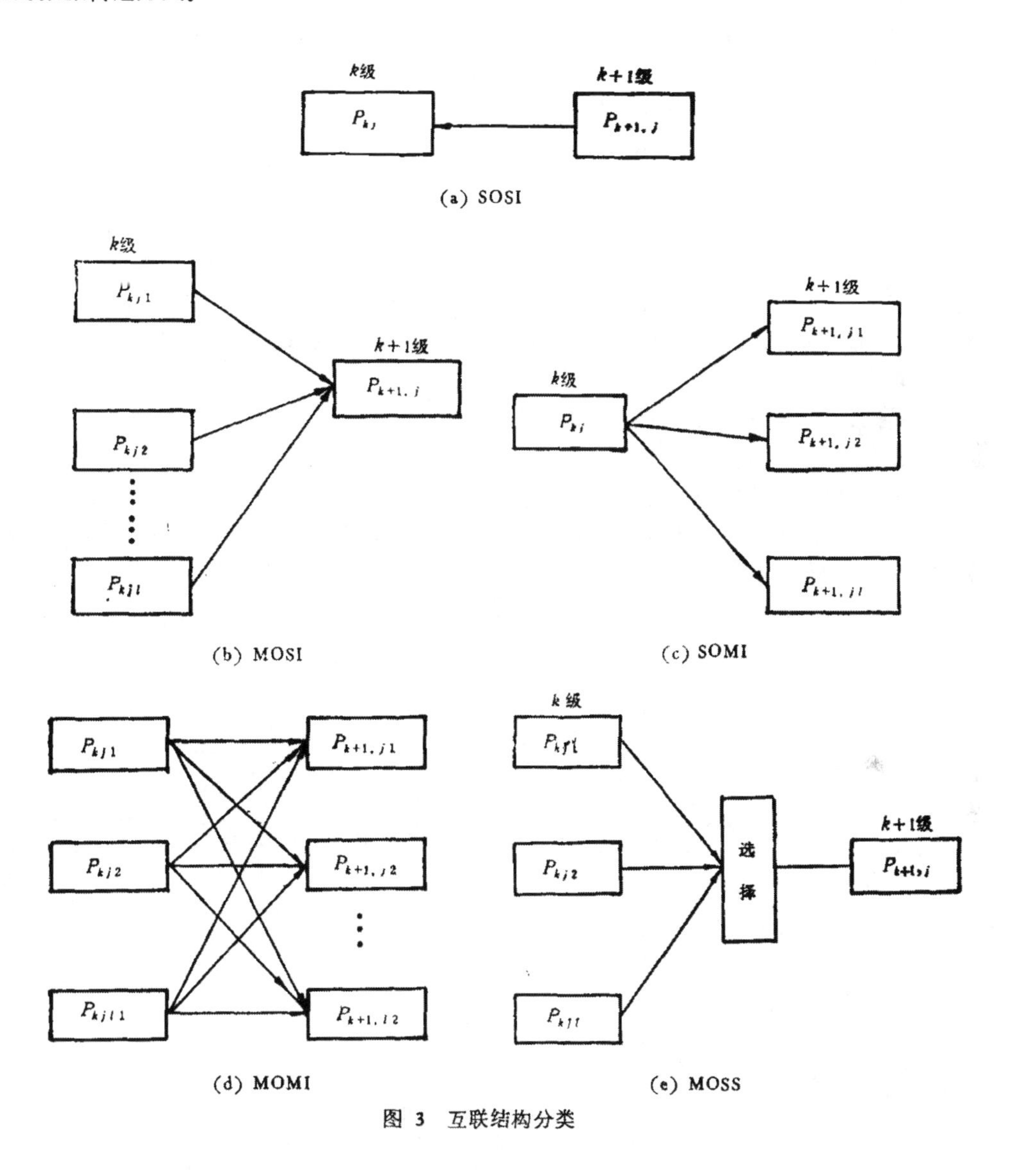

(a) SOSI

(b) MOSI

(c) SOMI

(d) MOMI

(e) MOSS

图 3 互联结构分类

MOSI (图 3(b)) 只有两种数据传递方法.

(1) 接收单元主动方式,或汇聚方式

即 k 级单元 P_{kj1}、P_{kj2}、P_{kj3},$\cdots$,P_{kjl} 准备好数据,然后通知 P_{k+1j} 取. 容易看出对具有相同结构及运算的 P_{kj1},P_{kj2},$\cdots$,P_{kjl},只要由一条总线即可以完成取数动作. 设要从每个发送单元中取N个数,则接收单元分时取数时,要花 $l \cdot N \cdot \tau$ 的取数时间. 其中 τ 为取一个数据要花的时间、它与总线带宽有关.

(2) 发送单元主动方式、或广播方式

即 k 级单元准备好数据同时向 $k+1$ 级单元 P_{k+1i} 送数. 这时要求 P_{k+1i} 有 l 个独立编址的公共接收存储器,同时要有 l 条数据总线. 虽然在广播方式下、P_{k+1i} 等待数据备齐的时间只有 $N\tau$,但要以 l 个送数单元花 $N\tau$ 时间送数为代价. 因此,在时间允许的情况下、应尽量使用汇聚方式解决 MOSI 的数据传递.

SOMI 互联结构也存在两种数据传递方式. 见图 3(c).

1. 汇聚方式

这种方式要求发送单元 P_{ki} 有 l 个独立编址的取数 RAM、存放相同的结果,才能使 l 个接收单元 P_{k+1i1}、P_{k+1i2}、$\cdots P_{k+1il}$ 通过 l 条总线把数据同时取出来. 如果仅使用一条数据总线,则要在 $P_{k+1i1},P_{k+1i2},\cdots,P_{k+1il}$ 之间引入同步控制机构,以便分时取数,从而将控制功能全局化、增加系统复杂度. 因此在 SOMI 结构中使用汇聚方式传递数据是不合理的.

2. 广播方式

P_{ki} 通过一条总线即可将计算结果同时送入 $P_{k+1i1},P_{k+1i2},\cdots,P_{k+1il}$ 中,送数时间为 $N\cdot\tau$.

MOSS 结构一般用来解决下一级单元从上一级的多个单元中,选一个,取其计算结果做为计算输入. 因此可以使用 MOSI 互联结构的汇聚方式解决其数据传递. 这时需要在发送单元加一个识别装置,由取数单元发送识别码来选择送数 RAM 是否工作.

MOMI 结构是五类结构中最为复杂的结构,在现阶段我们研究的岸站、潜用声呐、鱼雷声

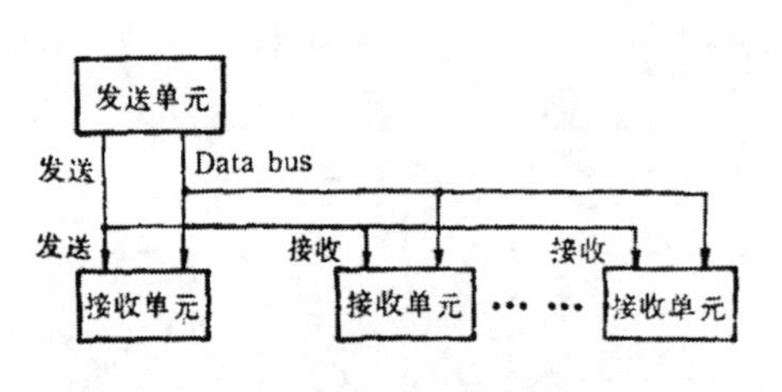

(a) 广播方式数据传递结构

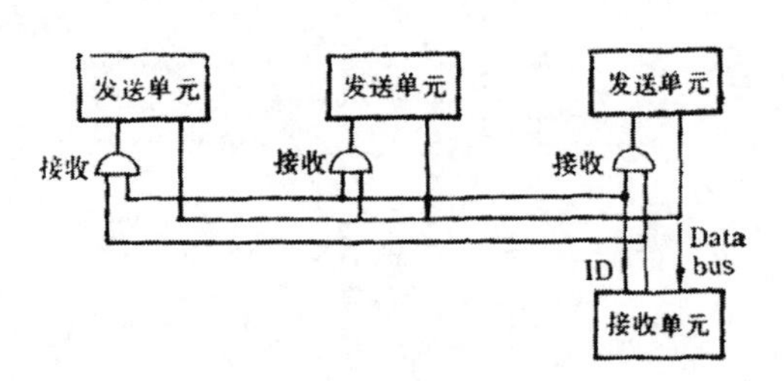

(b) 汇集方式数据传递结构

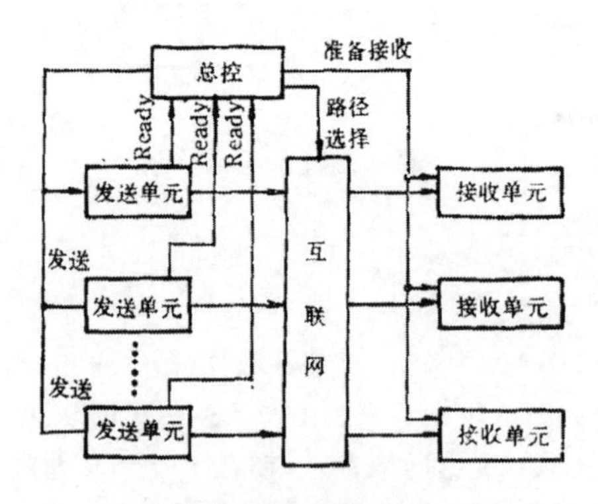

(c) 固定互联网方式数据传递结构

图 4 数据传递结构

呐及舰用声呐中，除显示级外都没有这样的结构，但为了保持理论上的完整性，我们使用一种固定互联网可以实现这一结构的数据传递．在这种结构中，从接收单元来看就是要分时从几种不同类型的单元中接收不同的处理结果，然后把不同的处理结果并为一个整体做为计算输入，因此由总控编程的多路开关可以满足要求．

四、声呐级联阵列机的总控

在很多大型数据处理系统研制中，控制部件通常占有统治地位．例如，计算机中的指令控制器，操作系统的进程调度程序等等．以控制部件为核心的系统尽管调度比较灵活，但设计难度很大，往往是系统中最为复杂的部份．采用本文提出的阵列机结构、几乎所有的处理数据的流动都在两级之间完成，总控部件要做的不再是复杂的数据、程序调度，而是对信号处理阵列机进行指令编程的系统，这样就大大降低了设计复杂度．

在级联阵列机结构中，存在着功能完全一样的多个单元．要把指令直接发送到每一个处理单元上去、势必要求大的总线驱动能力，并在每一个单元上增加一个指令接收寄存器．我们可以选择那些同级处理单元个数最少的单元做命令接收单元，通过互联总线把命令传播到实际使用该命令的处理单元上去，只要由命令发送到执行之间的最大延时满足系统要求、软件控制合理，就可以化简结构设计．例如一个 SOMI 结构的数据传送结构为：

d_1	d_2	$\cdots$	d_N

并要求在接收单元 $P_{k+1i1},P_{k+1i2},\cdots,P_{k+1il}$ 接收命令 e．如果每个接收单元直接接收命令 e，则要有一条命令总线及 l 个接收命令寄存器．现在在 P_{kj} 上放一个接收命令 e 的寄存器，那么只要把发送数据结构变为：

e	d_1	d_2	$\cdots$	d_N

就可以利用总线将命令 e 发送到每一个接收单元的接收 RAM 中． 事实上命令的改变或整个声呐功能的改变、往往由前向后传播才更加合理．例如一台多功能主被动声呐，可能在开始时需要用噪声定位方式定出目标的位置，当接近目标时才转入主动精确跟踪，这样当命令由前向后传播时正好按时将命令传到相应处理单元，从而改变整个声呐的工作状态，减少了切换时间．

五、处理单元的设计

一个处理单元在级联阵列机中有三个任务，取得上一级数据、计算、输出到下一级．其中计算结构对不同的任务是相似的，而输入、输出结构则跟互联方式及数据传递方式密切相关．图 5 给出不同互联方式的单元输入、输出结构．通过这些基本结构，一个处理单元就可以以该单元所处的前后级互联结构及数据传递方式按程式设计出来．图 6 是一个其输入级为 SOMI 互联、广播方式传递数据、输出级为 MOSI 汇聚方式传递数据的完整的处理单元逻辑图．

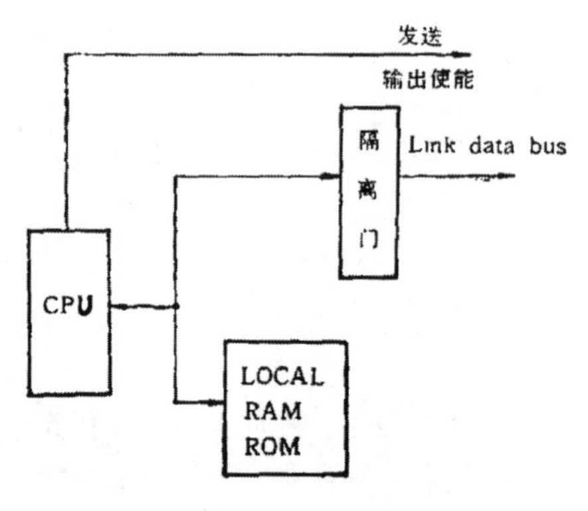

(a) SOMI 按广播方式传递数据的输出结构

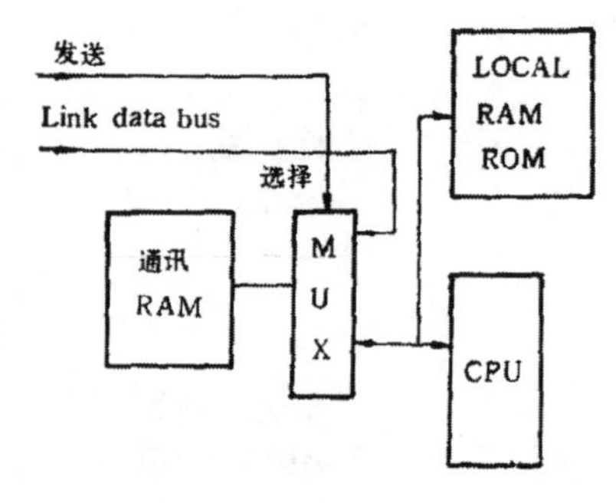

(b) SOMI 按广播方式传递数据的输入结构

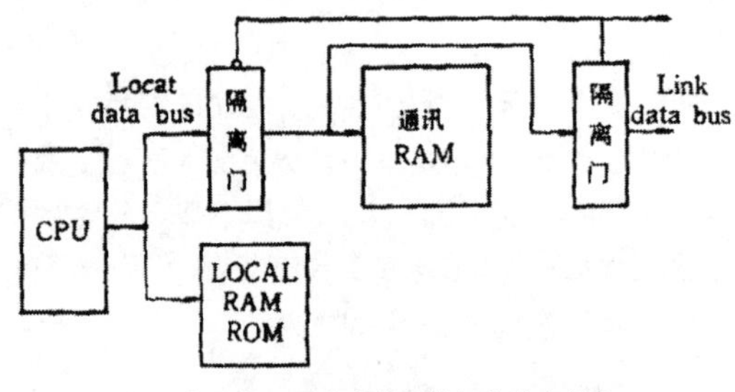

(c) MOSI 按汇聚方式传递数据的输出结构

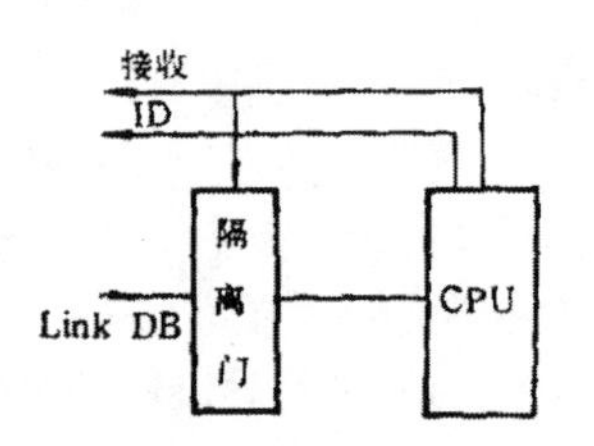

(d) MOSI 按汇聚方式传递数据的输入结构

图 5 给出不同数据传递方式时的总线配制

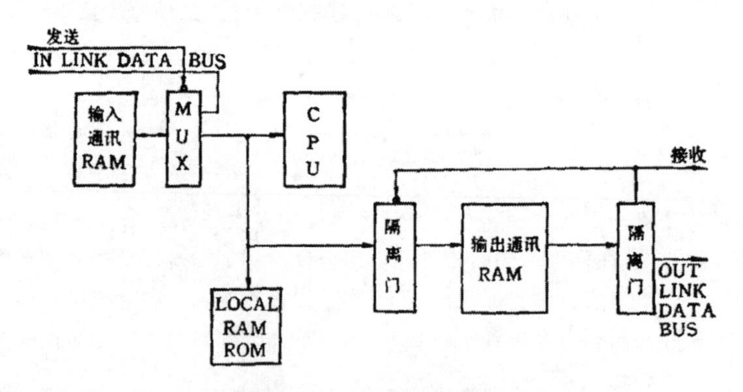

图 6 一个完整的单元结构

六、级联阵列机的规范化设计步骤

由前几节的叙述，我们可以总结出由声呐信号处理流图构成一个级联阵列机软、硬件配置的方法，该方法总结为下面六个步骤：

1）根据算法计算量将算法向级联结构图进行映射。

2）标定两级之间的互联方式。

3）决定数据在两级之间的传递方式。

4）根据响应速度及传播可抵达情况定出命令发送单元。

5）设计出数据传递格式及命令传递格式。

6）单元硬件、软件设计、总控软件设计。

上面六个步骤，每一步结果都满足前一步提出的设计要求。因此只要原来的信号处理功能设计得能达到预期目的，则最后获得的硬、软件设计必能满足总体要求。

参 考 文 献

[1] Knight, W. C. *et al*., "Digital Signal Processing for Sonar", *Rev. IEEE*., **69**(1981), 1451—1506.

[2] Bartram, I. F. *et al*., "Fifth generation digital sonar processing", *IEEE Trans*. **OE-2**(1977), 337—343.

[3] Whitehouse, H. J. *et al*., "Linear signal processing architecture",. *Aspect of sigpect processing* (Lierdal pub., 1977).

[4] Kung, S. Y., *"VLSI ARRAY PROCESSOR"*, (Prentice Hall, 1987).

大型声呐基阵的全方位强干扰抵消系统①

李启虎　蔡惠智
（中国科学院声学研究所）

1988年11月23日收到

摘要　强干扰是制约中远程声呐性能的主要因素．对于大型声呐来说，由于过长的延时量以及过高的运算速率，抗强干扰问题始终未能解决，本文提出一种适用于大型基阵的全方位抗强干扰系统．它是自适应噪声抵消系统与一个可编程迪卡诺（DICANNE）系统的结合．在干扰与信号入射角之间的夹角比较小时，采用低采样速率的自适应噪声抵消系统；在此角度较大时，采用高采样率的 DICANNE 系统，从而使整个系统具有全方位抗干扰的性能．本方案在采用高性能的 DSP 器件时，易于实现．文中给出设计框图与硬件实施方法．

An all-directional interference cancelling system for large sonar array

LI Qihu,　CAI Huizhi
(Institute of Acoustics, Academia Sinica)

Received November 23, 1988

Abstract The exsistence of strong interfernces is the main constraint factor, which influences the performance of a mid-or long-range sonar. For a sonar with large array, the problem of anti-interferences in all-direction has not been solved yet due to the requirements of over-long time dicay line and the over-high input-output rate. A method is proposed in this paper, which can be used in the sonar design with large array for supressing t e strong interferences in all direction. This is a combined architecture of adaptive noise canceller and a programmable DICANNE system proposed by V. C. Anderson. In the case that the incidental angle between signal and intefrence is small, the adaptive noise canceller will be adopted in a relative low sampling frequency. In the case that the incidental angle between signal and interference is large, the programmable DICANNE will work in the high sampling frequency so that the overall system has the ability of canceling strong interferences from all directions. The structure described in this paper is easy to implement in hardware by using the DSP chips. The block diagram of the combined system and the design idea are presented.

一、引　言

在被动声呐的检测问题中，抗强干扰问题始终是一个十分重要的问题．六十年代以前的声呐系统没有专门的抗点源干扰的信号处理方法，直到六十年代末，才出现了专门的自适应

① 声学学报, 1990, 15(4): 258-264.

抗干扰声呐的设计，这就是 V. C. Anderson 提出的 DICANNE 系统[1]. 用这种系统可以抵消一个强干扰. 但是这种设想仅适用于小型基阵，并且要有类似于反补偿器的密抽头延迟线，否则干扰无法“抵消干净”. 因此，DICANNE 系统仍停留于实验阶段[2].

另一种抵消强干扰的方法是利用自适应噪声抵消系统[3]. 利用一个可调权系数的抽头延迟线可以抵消强干扰，并且只要滤波器选择合适，可以同时抵消多个点源干扰[4-5]. 可惜，这种系统无法用于大型基阵. 因为，当基阵非常大时，需要很长的延迟线，而自适应噪声抵消滤波器的迭代噪声与抽头的个数成正比. 并且，对于过长的延迟线，硬件设计也是十分困难的.

本文提出一种新的强抗干扰声呐系统的设计. 它是自适应噪声抵消系统与可编程 DICANNE系统的结合. 在信号入射角 θ_S 与干扰入射角 θ_I 的夹角比较小时，利用自适应噪声抵消系统；当 θ_S 与 θ_I 的夹角较大时，利用经过升采样率运算的 DICANNE 系统. 从而得到一种全方位的抗强干扰系统.

我们将给出该系统的设计原理，硬件设计框图及实现方法. 利用 TMS320 系列及高速 RAM 可以得到一个抗多个干扰的全方位自适应噪声抵消系统，从而解决中远程被动声呐检测中的这一重要课题.

二、信号波束的最小预延时与干扰波束的最小延迟线长度

我们以大型线列阵来说明抗强干扰声呐的设计问题. 参看图 1. 设有 n 个基元等间隔地排列在直线上，间隔为 d.

以法线方向为 0°，设信号入射角为 θ_S，干扰入射角为 θ_I. 以第 0 号水听器 H^0 为时间的参考点. 第 i 个接收元的信号是:

$$x_i(t) = s(t+\tau_i(\vartheta_S)) + I(t+\tau_i(\vartheta_I)) + n_i(t) \quad i = 0, 1, \cdots, n-1 \tag{1}$$

其中

$$\tau_i(\theta) = \frac{id\sin\theta}{c}, \quad i = 0, 1, \cdots, n-1 \tag{2}$$

c 为声速. $n_i(t)$ 表示相互独立的噪声.

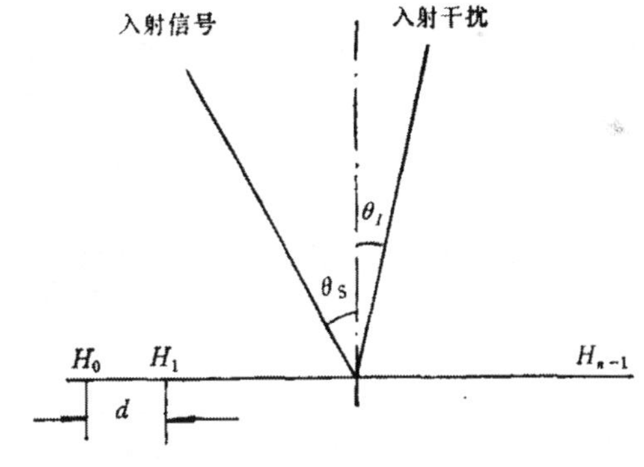

图 1 等间隔线列阵的示意图

无论是 DICANNE 系统还是自适应噪声抵消系统，都要形成信号波束和干扰波束，然后通过某种算法将信号波束中的强干扰分量抵消掉. 为了确保在任何情况下，即无论 θ_S, θ_I 的相对位置如何，都能用干扰波束经过适当处理来抵消信号波束中的强干扰，必须精确估计最大可能的 $\tau_i(\theta_S) - \tau_i(\theta_I)$；以及在物理上实现正的延迟（负的延迟量是无法实现的）所需要的信号波束的“预延时”.

这两个参数是进行硬件设计时最为关心的量之一.

下面，我们以 $\theta_I \geqslant 0$（以法线方向为 0°，顺时针时 $\theta \geqslant 0$）为例来说明. 在这种情况下，如形成干扰波束，将第 i 路信号延时 $\tau_i(\theta_I)$，得到

$$x_i(t-\tau_i(\theta_I)) = s(t+\tau_i(\theta_s) - \tau_i(\theta_I)) + I(t) + n_i(t-\tau_i(\theta_I)) \tag{3}$$

求和之后，得到干扰波束

$$Y_I(t) = NI(t) + \sum_{i=0}^{N-1} s(t + \tau_i(\theta_s) - \tau_i(\theta_I)) + n_i(t - \tau_i(\theta_I)) \triangleq NI(t) + p(t) \tag{4}$$

式中N为基元数，当干扰很强时，可以认为 $Y_I(t) \approx NI(t)$. 类似地，当 $\theta_S \geqslant 0$ 时，得到信号波束

$$Y_S(t) = NS(t) + \sum_{i=0}^{N-1} I(t + \tau_i(\theta_I) - \tau_i(\theta_S)) + q(t) \tag{5}$$

其中 $q(t)$ 中只包含噪声项.

由于设 $\theta_S \geqslant 0$，所以，当我们将 $Y_S(t)$ 预延时Δ时，$Y_S(t)$ 中的干扰项为

$$\sum_{i=0}^{N-1} I(t - \Delta + \tau_i(\theta_I) - \tau_i(\theta_S)) \tag{6}$$

令

$$[-\Delta + \tau_i(\theta_I) - \tau_i(\theta_S)] = A$$

注意到 $\theta_I \geqslant 0$，得到

$$0 \geqslant A \geqslant -2\Delta \tag{7}$$

换句话说 (6) 式每一分量中的延时量都是物理可实现的，并且干扰波束的延迟线长度应取作2Δ. 其中Δ为预延时量. 显然，

$$\Delta = \frac{(N-1)d\sin\theta_{s\max}}{c} \tag{8}$$

其中 $\theta_{S\max}$ 为最大可能的信号人射角. 在其他 θ_S, θ_I 情况下，可以证明有类似的结果.

三、全方位可编程强干扰抵消系统

从前面的分析可以看到，为了在任意方向都可以完全抵消强干扰，预延时和自适应滤波器的抽头延迟线长度为2Δ. 如果基阵相当长，那么Δ就很大，以致实际上很难由硬件实现，举例来说，如果

$$d = 0.5\text{m},\ n = 100,\ \theta_{S\max} = 80°,$$

这时

$$\Delta \approx 32.5\ \text{ms}.$$

如果系统采样频率为 $f_S = 10\text{kHz}$，则 $T_S = 100\mu s$，于是抽头延迟线的长度是 325 节. 根据 LMS 算法，要在 100 μs 内完成约 1000 次乘法和加法，这是难于实现的.

如果不用自适应算法，采用 DICANNE 系统，那么由于没有多个可以调节的权系数，就必须提高采样频率.

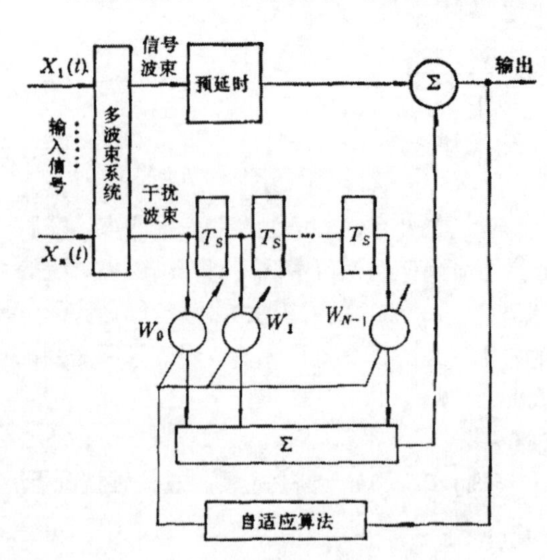

图 2　自适应强干扰抵消系统方框图

图 2 是自适应强干扰抵消系统的方框

图．我们看到，如果权系数 w_i 的个数N比较小时，这种系统的实现并不复杂．同时，由于可调权系数的组合作用，即使采样频率比较低，也可以很好地将信号波束中的干扰抵消掉．因此，我们可以将这种系统用于 $(\theta_S-\theta_I)$ 比较小的情况．

但是，如果 $(\theta_S-\theta_I)$ 比较大，那么所需延迟线就比较长，这时我们改用一个可编程序 DICANNE 系统．根据事先计算得到的延时抽头值把干扰抵消掉．但是，由于 DICANNE 系统每一路的权系数只有一个．所以 T_S 太大是不合适的，我们应采用升采样运算[6]，使信号波束和干扰波速的采样间隔加密，这样便可得到较为接近的抽头延迟值．

值得注意的是，对于给定的抽头延迟线长度，所适用的 $(\theta_S-\theta_I)$ 是一个和 θ_S 有关的量．我们把这个量称为目标函数，记作 $J(\theta_S)$．举例来说，如果信号来自法线方向，$\theta_S=0$，那么当 $|\theta_S-\theta_I|=|\theta_I|\leqslant 7^\circ$ 时，可以用自适应噪声抵消系统．在同样长度的延迟线之下，如果信号来自 60°，即 $\theta_S=60^\circ$，那么，当 $|\theta_S-\theta_I|\leqslant 14^\circ$ 时，我们还可以用自适应噪声抵消系统．$J(\theta_S)$ 的值必须根据抽头延迟线的长度来计算．对于延迟线长为 8 ms 的系统来说，$J(\theta_S)$ 与 θ_S 的值如下表

θ_S	0°	10°	20°	30°	60°
$J(\theta_S)$	7°	7°	7.4°	8.0°	13.9°

也就是说，当信号入射角为 $\theta_S=0^\circ$ 时，只要 $|\theta_I|\leqslant 7^\circ$，都可以用自适应噪声抵消系统；当 $|\theta_I|\geqslant 7^\circ$ 时，就要更换到可编程 DICANNE 系统．而当 $\theta_S=60^\circ$时，$|\theta_S-\theta_I|\leqslant 14^\circ$时仍可利用噪声抵消系统，只有当 $|\theta_S-\theta_I|>14^\circ$ 时才利用可编程 DICANNE．

图 3 是整个系统的方框图．图 4 是实现可编程 DICANNE 系统的一种方法．地址表是预先设计好的，由 θ_S 及 $\theta_I-\theta_S$ 予以选择控制．当通道数不是很大时，用高速 RAM 不难实现这种算法．

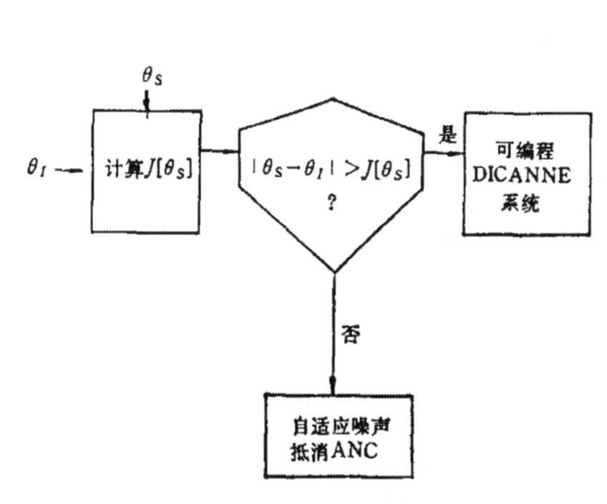

图 3　全方位强干扰抵消系统原理方框图

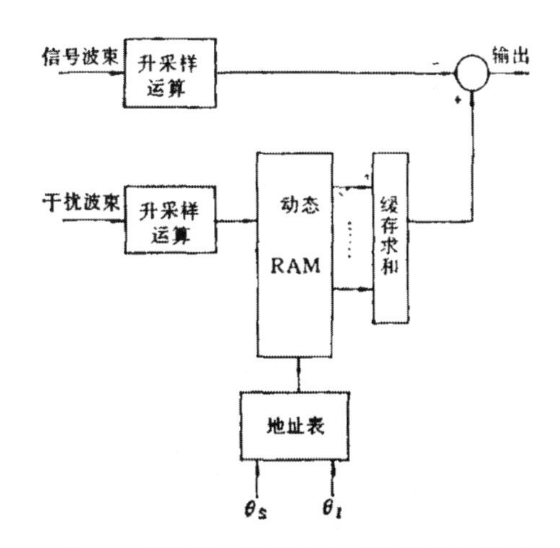

图 4　可编程 DICANNE 系统原理方框图

四、硬件设计方面的考虑

在硬件设计中，考虑用一片 TMS32010 作系统的控制单元，一片 TMS32020 组成

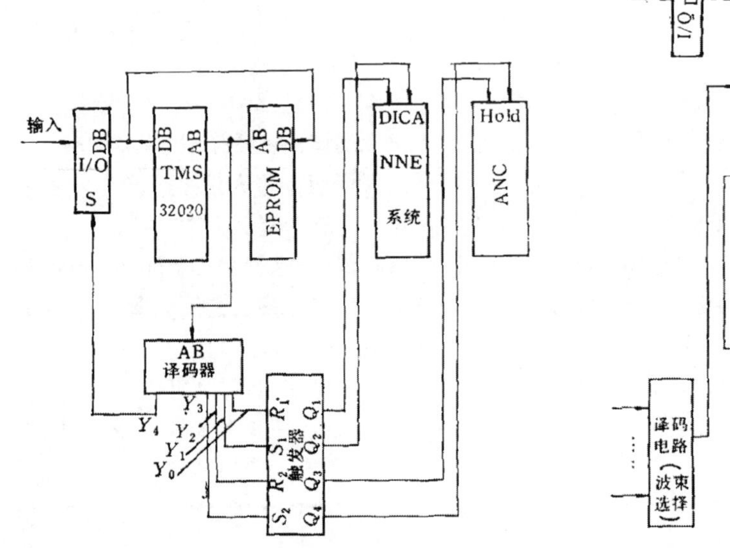

图 5　总体硬件配置图

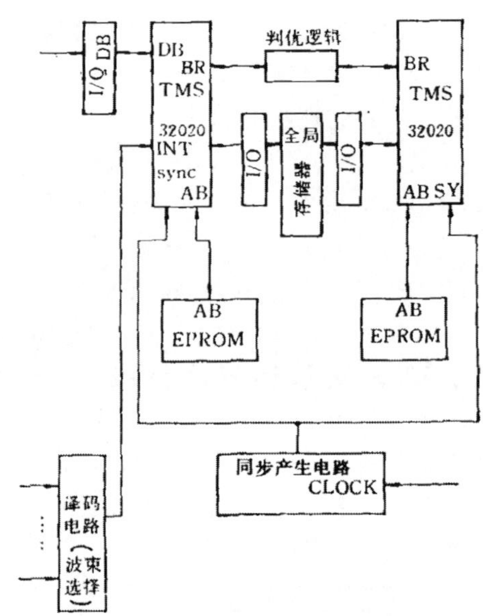

图 6　自适应噪声抵消系统 ANC

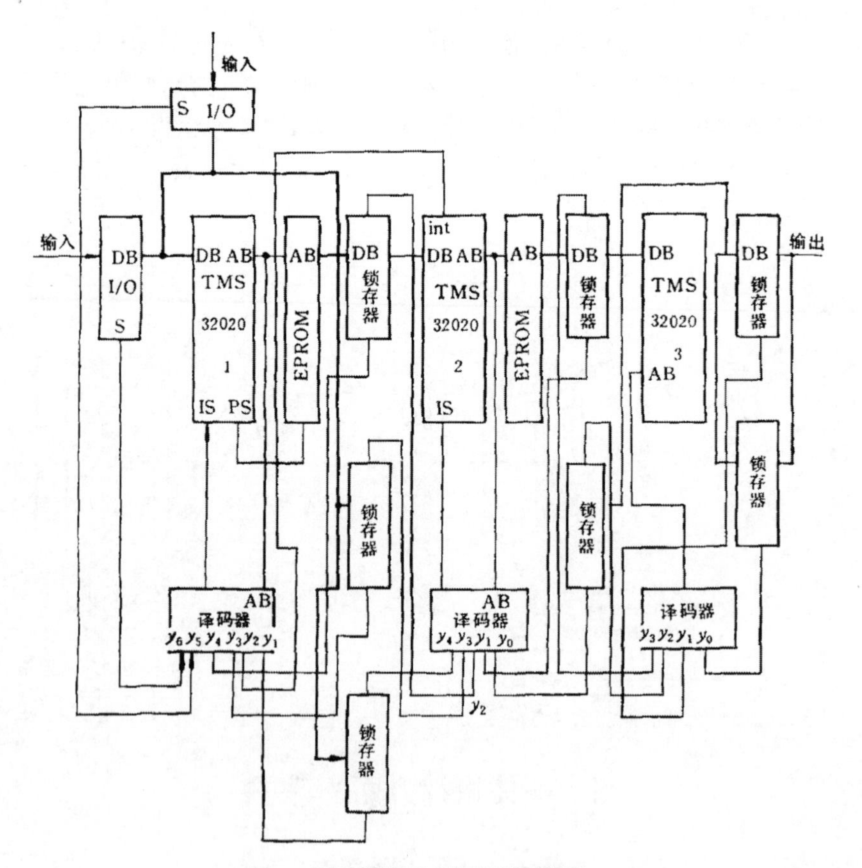

图 7　同时抵消三个噪声的原理图

DICANNE 系统，两片 TMS32020 组成自适应噪声抵消系统．根据前面的全方位强干扰抵消系统的方框图,外部送入 θ_I 和 θ_S，通过 TMS32010 计算 $J(\theta_S)$． 判断 $|\theta_S-\theta_I|$ 是否 $>J(\theta_S)$,从而决定是采用 DICANNE 系统还是自适应噪声抵消系统(ANC)．用 TMS32010 的译码器产生一个中断信号通知下一级工作．如果决定采用 DICANNE 系统．由 TMS32010 的译码器输出一个中断请求信号给 DICANNE 系统的 TMS32020．通知其程序开始运行,同时由 TMS32010 的译码器产生另一个输出信号,接入自适应噪声抵消系统的两片 TMS32020 的 HOLD 端,使其进入悬挂高阻状态,具体的考虑如下图．

下面我们分别讨论 ANC 系统和 DICANNE 系统．

1. 自适应噪声抵消系统 ANC

由于自适应噪声抵消系统的优越性．我们应尽可能采用此系统．但由于其运算量大，速度慢,所以限制了其权系数的节数．以典型情况来说，在 A/D 的采样率为 5kHz 时,权系数的节数最多为 30 节,由前面的讨论可以知道,当权系数的节数为 40 节时，$J(\theta_S)$ 可以到14°，如果节数再增加,现有的专用数字信号处理芯片就难以胜任了．我们考虑用两片 TMS32020 组成系统完成权系数的节数为 60 的自适应噪声抵消系统．见图 6.

TMS32020 的外部存储器可分为全局和局部两部分． 由两片 TMS32020 组成的 ANC

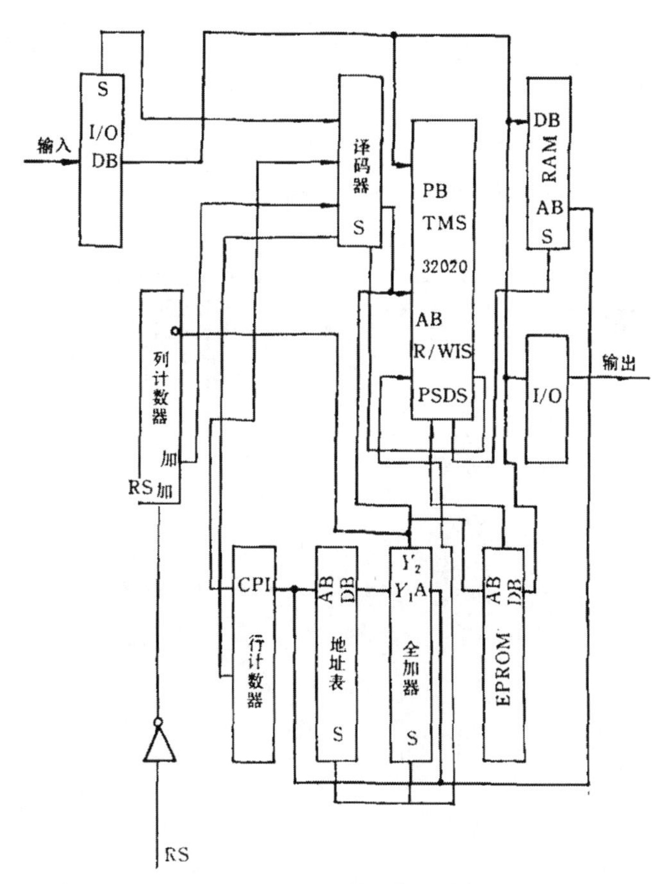

图 8 DICANNE 系统框图

系统，我们考虑每一片 TMS32020 分别运算 30 节权系数，中间运算结果分别放在各自的局部存储器中，只有最后的权系数放在全局存储器中，由其中的一片算出最后的结果送出，完成自适应噪声抵消。

为了同时能抵消三个强干扰源。我们考虑用级联的三套自适应噪声抵消系统。第一套系统抵消信号波束中含有的一个噪声，同时把此级的结果和另两个噪声波束的数据送到下一级。第二套抵消另一个噪声。同时把此级的结果和最后一个噪声波束的数据送到第三级。第三级抵消最后一个噪声并把结果送出。

2. DICANNE 系统

图 4 中动态 RAM 的寻址方式采用一个列地址和一个行地址的组合。当向动态 RAM 送入数据时，列地址和行地址由计数器产生，列计数器是每当加到某一固定数时，就自动清零，这样列地址又从零开始。行地址计数器在向 RAM 写数据时同列计数器一样加到某一固定数时自动清零。当由动态 RAM 向外读数时，地址由行地址和列地址组合产生，简单地说就是行地址从零开始每次加 1，列地址是由 θ_S 和 θ_I 决定的地址表中对应单元所含地址产生。如果有 N 个基元，那么这 N 个由上述方法决定地址的基元的数值由 RAM 中读出，然后相加并和信号波束的数据相减从而检测出信号。具体的求和运算和相减运算由一片 TMS32020 完成.地址表的寻址也由 TMS32020 完成，详见图 8。

参 考 文 献

[1] Anderson, V. C., "Dicanne, a realizable adaptive processor", *J. Acoust. Soc. Amer.*, 45(1969), 398—405.

[2] Anderson, V. C., "Side lobe interference suppression with an adaptive null processon", *J. Acoust. Soc. Amer.*, 69(1981), 185—190.

[3] Whitehouse, H. J. and Speiser, J. M., "Linear Signal processing architectures", *NATO*, (Advanced Study Institute, Reidal, Pub., 1977).

[4] 李启虎，"多波束自适应噪声抵消法引论"，声学学报，9(1984)，No. 3,221—230.

[5] Cantoni, A. and Godara, L. C., "Performance of a post beamformer interference canceller in the presence of broadband directional signals", *J. Acoust. Soc. Amer.*, 76(1984), 128—138.

[6] 李启虎，"数字式声呐设计中的升采样率运算"，声学学报，15(1990)，No. 1,7—11.

数字式声呐显示系统的 GSC 算法①

李启虎 刘秋实 李淑秋
(中国科学院声学研究所,北京 100080)
1990 年3月 20 日收到

摘要 方位历程显示已被证明为检测信号的有效手段。数字式声呐的多波束数据与调亮显示系统之间存在着一个接口,它有可能使信号处理系统已获得的增益受到损失. 选择合理的算法会使这种损失减到最小. 本文提出的灰度级转换算法 (GSC) 是一种实时的数字运算技术. 在输入信噪比较低时会使弱信号的检测能力有所改善,同时又对强信号的检测不会有严重的影响. 本文提出的算法易于由硬件实现. 计算机模拟实验的结果与理论一致. 文中给出了硬件设计的概要说明.

Gray scale conversion algorithm for digital sonar display system

LI Qihu LIU Qiushi and LI Shuqiu
(Institute of Acoustics, Academia Sinica)
Received March 20, 1990

Abstract The bearing history display technique has been proved to be an effective method in sonar signal detection. There is an interface of digital sonar between multibeam data and brightness modulation display system. This interface could result in some loss in gain obtained by the signal processor. A reasonable choice of conversion algorithm can reduce the loss to minimum. The Grey Scale Conversion (GSC) algorithm proposed in this paper is a real time digital operation. By using GSC algorithm the ability of detecting weak signal is improved when the signal to noise ratio is small, while there is almost no effect on the strong signal detection. The algorithm described in this paper is easy to implement in hardware. The computer simulation result shows a good agreement with the theoretical analysis. A brief illustration of hardware design is given.

一、前 言

现代数字式声呐可大致地分为三个部分: 1 前置处理; 2 波束成形与后置处理; 3 显示控制. 声呐系统对信号的检测、判断通常由显示系统给出。所以能不能由显示系统最大限度地把前置处理和波束成形部分获得的增益发挥出来是至关重要的. 声呐设计者确信,信号处理的增益最容易在两个系统的接口处损失掉[1].

早期的声呐通常把波束成形的信息直接用于听测或方位幅度显示 (A 式显示) 或者平面位置显示 (PPI)。随着信号处理技术和水声学的进展,人们发现这种直接显示对弱信号的检测是不利的. 由于舰船辐射噪声及水声信道在传播上的衰落现象,只有在一个长的历程上对

① 声学学报, 1991, 16(5): 365-370.

信号进行观察才能成功地检出信号．这种信号的起伏周期有时候长达 10 分钟[2]．因而，方位历程显示是十分必要的[3-4]．它不仅可以在全方位上观察多波束系统的整个历程，还可以利用人眼的观察对**迹迹**相关作出判断．

Rasmussen 等[5-7]曾对声呐的显示系统作过定量研究．指出人眼对灰度级、彩色的敏感度和分辨力的关系．一般认为，从理论上和技术上来说 2^8 级灰度对信号检测是足够的．但是一般数字信号处理系统为了有足够的精度总是采用远远高于 8 比特的运算，比如 16 比特．当把 16 比特的数据转换成 8 比特的灰度级时不能采用简单的取高 8 位或取低 8 位方法．

本文首先分析数字多波束系统和显示系统之间进行比特转换的必要性，以及有关参数选取的准则．然后给出一种易于由硬件实现的灰度标尺转换（GSC）算法，从理论与实践两方面说明这种算法的特性．同时给出硬件设计的框图及模拟实验的结果．

二、数字多波束数据的显示

根据 Rasmussen 等的研究结果[5-7]，声呐显示屏幕以约 43cm（18 英寸）左右为宜．在一般情况下，人眼的极限分辨力大约是 10′，以这两个参数去计算，对于分辨力为 1024 × 1024 的显示器来说，水平分辨力大约是 6—8 个象素（Pixel）．所以数字式声呐的信号处理系统的波束个数取为 128 是合适的．过多的波束只会使硬件复杂化而不会在显示上带来好处．

一般信号处理系统采用定点多波束运算、比较普遍的是 16 比特．而显示系统的灰度级是 8 比特．当把 16 比特的量转换成 8 比特的灰度级时会造成不可避免的损失．要使得这种转换的损失对弱信号的检测减到最小是一个在理论上和实践上都十分有意义的课题．

对于一个有 N 个通道的多波束系统，假定输入信噪比为 $(\mathrm{SNR})_{\mathrm{in}}$，各通道之间的噪声相互独立，那么对准目标的那个波束的输出直流近似为[8] $N(\mathrm{SNR})_{\mathrm{in}}$．如果以大信噪比不失真为准则，取 $(\mathrm{SNR})_{\mathrm{in}}=10$，对 16 比特的系统，要求把 $10N$ 作为 $2^{16}=65536$．转换系数为

$$K=2^{16}/10N \tag{1}$$

若信噪比降至 $(\mathrm{SNR})_{\mathrm{in}}=0.01(-20\mathrm{dB})$，则输出为 $N(\mathrm{SNR})_{\mathrm{in}}\cdot K=0.01\cdot 2^{16}/10\approx 66$．如果把 66 直接转换 8 比特的灰度级则所有位数都是 0．换句话说，在小信号时，不仅副瓣不能在方位历程显示中显示出来，甚至连主瓣都无法显示出来．由此可见，从高比特到低比特的转换不能采取简单的方法．否则微弱信号的检测性能会大大地下降，这是我们最不能容忍的事情．

三、GSC 算 法

把高比特数据换算为低比特数据，信息的损失是必然的．但是我们希望找到一种算法，使得对微弱信号的检测比较有利．而对强信号，即使有所损失，只要不影响判断信号的有无，适当压缩其动态也是无妨的．

设 X 是一个 16 比特的数(无符号位，下同)

$$X=A_0 2^{15}+A_1 2^{14}+\cdots+A_{14}2^1+A_{15} \tag{2}$$

直接的线性运算是

$$Y = \frac{X}{2^{16}} \times 2^8 = \frac{X}{2^8} \tag{3}$$

这实际上是取高 8 位的运算(见图 1).前已指出这种运算会使弱信号的检测受到严重影响.另一种直接运算是取低 8 位的运算. 这种运算对弱信号检测是有利的, 但是对强信号有可能产生“饱和”现象(见图 2)而使主瓣变得非常宽而无法区分目标.

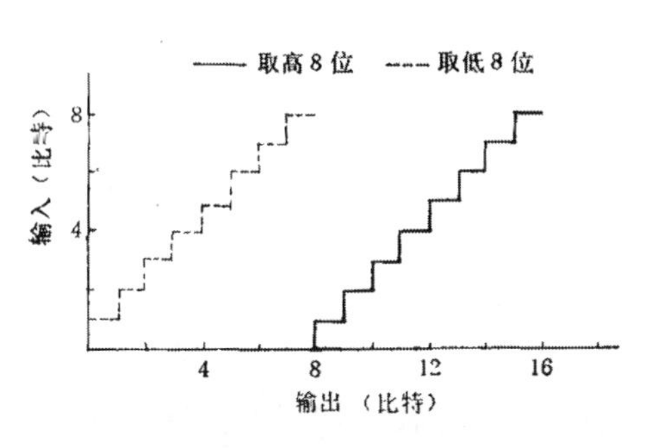

图 1 直接转换的结果

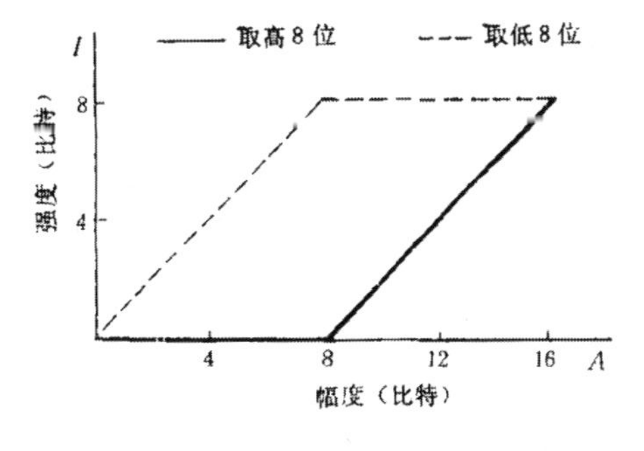

图 2 直接转换时的幅度/强度关系

Baggoroer 等指出[4],语音信号处理中所用的 μ 律转换, 对小信号的检测及数据压缩是非常合适的. 其输入输出关系为

$$Y = \frac{\ln(1+\mu x)}{\ln(1+\mu)} \quad |x| \leqslant 1 \tag{4}$$

当 x 是正 16 比特数时, Y 的 8 比特对应值显然可以写成

$$Y = \frac{\ln(1+\mu x/2^{16})}{\ln(1+\mu)} \times 2^8 \tag{5}$$

图 3 给出了 μ 律转换的典型结果. 我们可以看到它实际上把 0 附近的数扩展了, 而把 1 附近

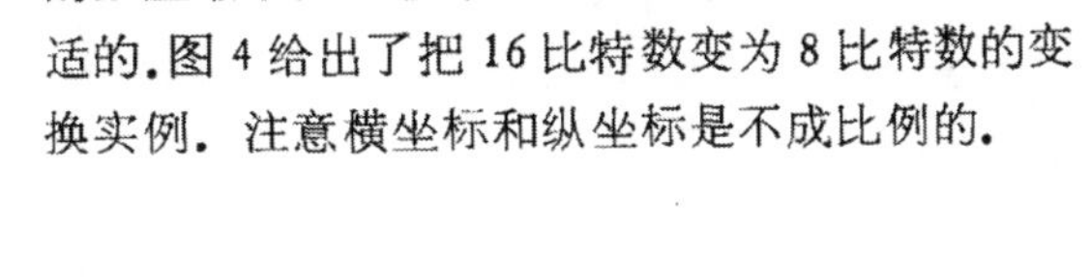

的数压缩了. 一般来说, 选取 $\mu = 20 \sim 30$ 是合适的.图 4 给出了把 16 比特数变为 8 比特数的变换实例. 注意横坐标和纵坐标是不成比例的.

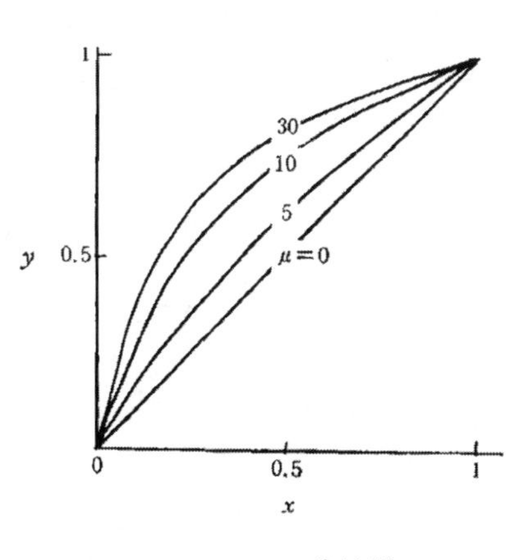

图 3 μ 律扩展

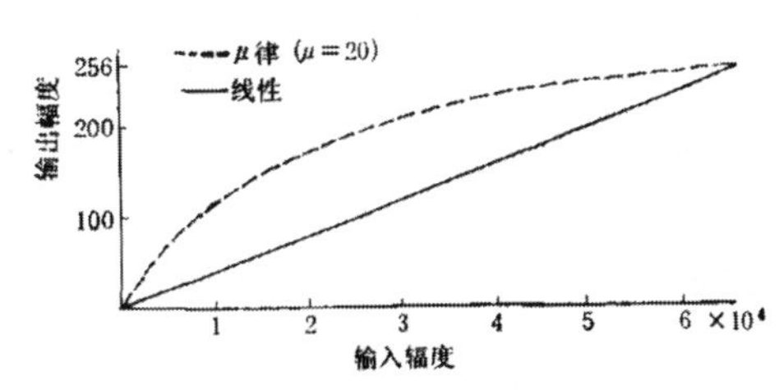

图 4 16 比特至 8 比特的转换

为了把(2)式所表示的 16 比特数转换成 8 比特数,从理论上来说,应该如图 5(a)所示.先将 16 比特数 X 用数/模转换器变为模拟量,再用公式(4)将模拟量 X 变为模拟量 Y, 最后用模数转换变为 8 比特数. 这种作法非常麻烦,因为要用到模/数和数/模转换器,同时还要用到超

越函数的计算，无论用硬件还是用软件实现都是非常麻烦的.我们采用折线逼近曲线的方法.

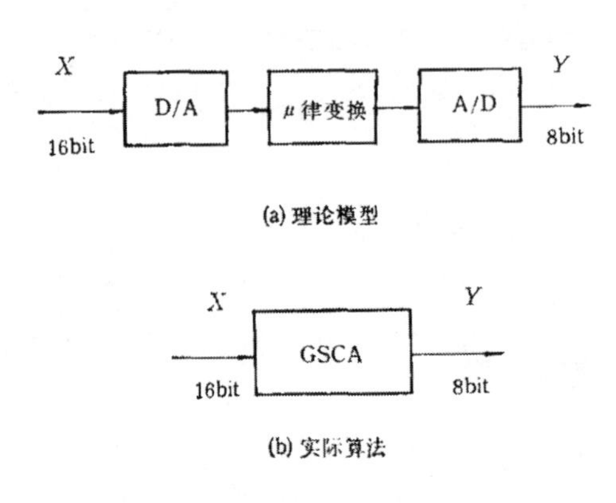

图 5 多比特转换的实现

以三段折线逼近曲线为例，利用四个点；(x_i, y_i)，$i=0, 1, 2, 3$. 其中 $x_0=0$，$x_3=1$，于是 $y_0=0$，$y_3=1$，其它情况下

$$y_i=\frac{\ln(1+x_i\mu)}{\ln(1+\mu)} \tag{6}$$

$i=1, 2$. x_i 的坐标根据 μ 来选. 近似的折线方程是

$$y=y_i+\frac{y_{i+1}-y_i}{x_{i+1}-x_i}(x-x_i),\quad i=0,1,2,3 \tag{7}$$

以 $\mu=20$ 为例，经过简单的计算便可证明，这种 GSC 算法使低比特扩展 2～3 倍，而使高比特范围压缩 2～3 倍.从而保证了对弱信号的检测. (7)式的这种运算便于用硬件实现.在实际系统设计时，μ 是可以调节的，根据方位历程显示图上的情况，调节 μ 使背景的反差更有利于检测，从而起到了在输入 16 比特范围内构成一个滑动窗的作用.

四、计算机模拟

计算机模拟分析，有三个任务要完成. 第一，μ 值的选取，第二，用折线模拟 μ 律曲线，折点位置的选取，第三，参数存贮.

μ 律曲线的规律是，当 $\mu>0$ 时，μ 越大，在小信号端上升越陡，而大信号端越平坦. 当 $-1<\mu<0$ 时，当 μ 越接近 -1，小信号端就越平坦，而大信号端就越陡.

μ 值选取的原则，是根据我们对输入信号的侧重. 要使所选的 μ 律曲线，把有一定动态范围的输入信号，经变换后，仍保留较大的动态范围.

在一般情况下，我们选 $\mu=2^4\sim2^5$ 比较合适，这样的曲线上升比较平缓，可以兼顾大小信号.

在侧重小信号时，μ 应适当选大些，并根据输入信号的差异，作相应调整. 这时的 μ 律曲线，在小信号端上升较快，对小信号有放大作用.

在侧重大信号时，选 μ 为负值，这时小信号端平坦，大信号端较陡，这意味着，对于大的输入信号，输入增量 Δx 有一个变化，输出增量 Δy 会有一个显著的变化. 这个特性，对于识别主瓣和副瓣非常有用. 当主瓣和副瓣都较亮时，选取这样的 μ 值，可以看出主瓣和副瓣的差异.

另外，人眼的视觉特性是对小信号敏感，对大信号不敏感，负 μ 律曲线，正好可以弥补人的视觉特性，使大信号看起来更舒服.

这样，μ 律变换，不仅起到 16 比特到 8 比特灰度变换的作用，而且起到视觉变换的作用，一举二得.

经计算机计算，表 1 列出了针对几种不同输入所选择的 μ 值，以及输入和输出的动态范围.

折点位置的选取,一是要根据实际工程中的需要,二是要根据 μ 律曲线的形状.

根据上述要求,经计算,一组 μ 所取的折线段和折点位置横座标见表 2。

表 1 针对不同输入信号所选择的 μ 值

输入信噪比		X 值的范围		μ 值	Y 值的范围	
dB	真值	数值	动态范围		数值	动态范围
−20	0.01	6—655	$2^3—2^9$	2^9	2—74	$2^1—2^7$
−15	0.031	20—2031	$2^4—2^{11}$	2^7	2—84	$2^1—2^7$
−10	0.1	65—6554	$2^6—2^{13}$	2^4	1—86	$2^0—2^7$
−5	0.31	200—20316	$2^8—2^{14}$	2^2	2—128	$2^1—2^7$
0	1	655—65536	$2^9—2^{16}$	0	3—256	$2^2—2^8$

表 2 不同 μ 值的折点横座标

μ 值	折线段数	折点横座标
4	3 段	0, 2^{13}, 2^{15}, 2^{16}
16	4 段	0, 2^{11}, 2^{13}, 2^{14}, 2^{16}
128	5 段	0, 2^9, 2^{11}, 2^{13}, 2^{14}, 2^{16}
512	5 段	0, 2^8, 2^{10}, 2^{12}, 2^{14}, 2^{16}

以 $\mu = 2^7 = 128$ 为例,在折点位置取定后,如果取 $\mu = 128$ 的内折线,效果并不好,见图 6.如果取内折线上面的折线,效果就好多了. 这时的 y_1, y_2 值, 应取 $\mu = 144$ 曲线上的值. 换句话说, 若要做 $\mu_2 = \mu_0$ 的模拟折线, 只要做 $\mu = \mu_0 + \mu_0/8$ 的内折线即可.

顺便指出,当 μ 为负值时,计算方法稍有不同,可以证明,对 $\mu \geqslant 1$ 的曲线同样可用于 $-\mu/(\mu + 1)$ 的情况.

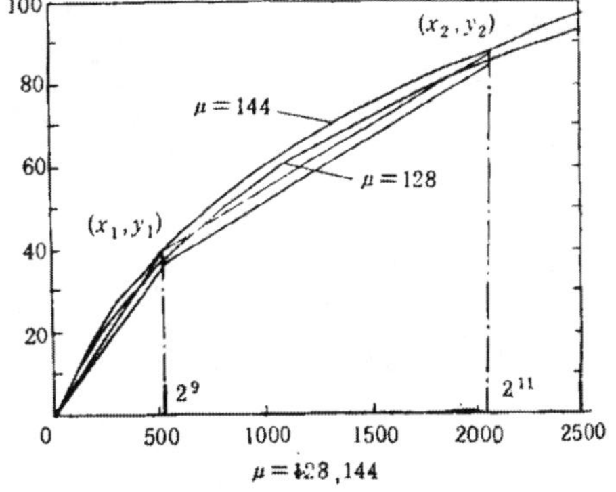

图 6 用 $\mu = \mu_0 + \mu_0/8$ 的内折线来求 $\mu = \mu_0$的模拟折线.

五、硬件实现简述

实现 GSC 算法的硬件由三个部份组成:1. 接收 16 比特多波束数据的接口; 2. 实现 GSC 算法的微处理器电路; 3. 用于完成方位/历程显示的视频 RAM 及自动卷页系统. 见图 7.

根据波束成形器的特点, 接口部份可以采用共用 RAM 形式的接口,也可采用电子开关.

GSC 算法的计算由微处理器完成, 根据算法特点 μ 值可以从小到大取十几个不同的值, 并对 μ 值进行编号,通过控制台上的键盘,操作员可以设置不同的 μ 值编号,以选择合适的 μ .

用当前的μ值，算法对每组输入的波束值进行μ律变换，把结果打入到方位/历程 RAM 中，进行显示。

方位/历程显示的硬件如图所示，其中全加器，基址寄存器与行计数器构成自动卷页系统。只要对基址寄存器进行模加 1 操作，系统就可自动向上卷一个象素行。

为了用较低速的 RAM 器件实现高分辨率方位/历程显示 RAM，我们采用了多体存储器技术。在一个刷新读操作期间，可以同时读出 4 个象素（32 位），而在向 RAM 中写数据时，一次只写 1 个象素（8 位），D/A 是高速的。要求转换速度为 40MHz～50MHz。

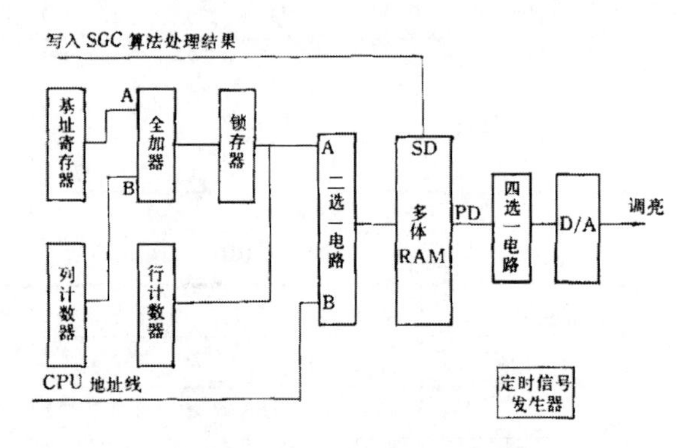

图 7　方位/历程显示硬件设计方框图.
SD：写入输据；PD 读出数据

参　考　文　献

[1] A. A. Winder, "Sonar technology", *IEEE Trans.* SU-22 (1975).

[2] R. J. Urick, "Multipath propagation and its effects on sonar design and performance in real ocean", Proc. of NATO ASI, 1976.

[3] P. Skitzki, "Modern sonar system", *Electronic Progress* XVI (1974), No. 3 20—37.

[4] A. B. Baggeroer, "Sonar signal processing", in *Application of Digital Signal Processing*, Ed. By A. V. Oppenheim, Prentice-Hall, Englewood Cliffs, NJ 1978.

[5] R. A. Rasmussen, "Studies related to the design and use of time/bearing sonar display", AD 690540, 1 May, 1968 USA.

[6] E. C. Poulton and R. S. Edwards, "A possible use of colour to code sonar displays", AD-A018039 March 1974, USA.

[7] W. Doobenen, "A method for the systematic design of a submarine sonar display/control subsystem", AD 708746 Oct. 1969 USA.

[8] Qihu Li, "An introduction to sonar signal processing", Ocean Publ. Beijing, 1985.

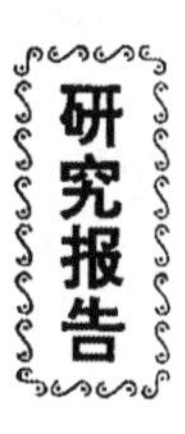

数字声呐语音报警系统设计①

李云岗　赵文立　李启虎

（中国科学院声学所　北京　100080）

1992年7月3日收到

本文给出了数字声呐用语音报警系统设计，并给出了详细框图．该系统采用单片微机控制，由语音库、语音处理及控制部分组成．该系统有许多独特之处，结构简单、控制灵活、语音自然、寻址能力大、可以进行编程、灵活地调用语音库中的词组或单音组合成完整的句子，例如目标、方位、距离等。通过改变语音库内容，该系统可适用于任何需要语音的场合。

一、前　　言

语音合成技术已日趋成熟，各种语音合成集成电路近几年相继推出[1-3]．但统观目前市场上的语音集成电路就会发现[4]，它们仅有暂停和连放功能，且寻址能力很有限，一般仅为256k位，很少有编程功能，或仅有简单固定的编程功能，使应用受到很大限制．在声呐语音报警系统中需报的信息比较多，要求的词汇量很大，而且有些词要反复应用．典型的语音报警词汇有上百个，它们必须能灵活地被组合在一起构成报警信号，由于种类多，词汇量大，而现成的语音集成电路，寻址能力、编程能力显然都不够．为了适应声呐语音报警系统的需要，我们专门设计了一个语音合成报警系统，它可以实现灵活编程，且寻址能力比较大．

二、语音合成与语音存储考虑

从声呐报警信号的功能来区分，我们可以把报警信号分为四类，这就是测向(包括主被动测向)、测距(包括噪声测距与回声测距)、目标识别与其它辅助手段(如自适应入侵报警、故障自检报警等)．我们可以把所用到的基本词汇构成一个数据库，根据情况随时调用，举例来说，链、分、度、秒、距离、东经、北纬、目标类型、置信度、……这些词汇就可以是数据库的基本成员．

另外，为使报警信号听起来更顺耳一些，也便于值班官兵区分，我们还需在基本数据库中加入一些能产生较好音响效果的程序，如音乐、警笛声、解除警报声等等．首先让我们来看看音乐合成芯片，它主要是将音乐分解为音符序列和节拍序列，使得一首乐曲的存储量大为减少，例如华邦 W6230 系列、富日 CIC-285 系列等，都是可存储几十首乐曲的芯片．所以要存储音乐信号仅仅需要极少的存储量，而要产生专门的音乐效果，也只需很短的程序．

对于音乐分解合成的方法是不能用于语音合成的，语音合成有线性预测编码（LPC)、脉码调制（PCM)、自适应脉码调制（ADPCM)、连续可变斜律增量调制（CVSDM)、自适应增量调制（ADM）等方法[5-6]，不同方法各有其优缺点，如 LPC 方法编码压缩率比较高，而音质则不是最理想，CVSDM 方法编码压缩率不是最好，要求硬件比较多，但语音效果则比较理想．随着微电子学的飞速发展，集成度已不是主要矛盾，单片集成数万门已经实现．台湾 UMC 公司生产的 UM5100 芯片是用于语音合成的专用芯片，它采用了 CVSDM 方法进行语音合成，为此我们采用了该芯片为核心组成的系统进行语音合成．

语音合成是将自然语音按一定的算法进行编码、存储、然后依据需要再进行语音还原．在

① 应用声学, 1993, 12(4): 9-12.

选定编码方式之后，如何进行语音存储就成为一个突出的问题.如何挑选基本的语音词汇，也就是如何构成基本数据库，以及如何把它们编程在一起，产生出高质量的报警语音来. 若整句进行语音存储，即以句为单位进行存储，虽然自然度很好，但所需的存储空间太大. 通过分析汉语可以发现，汉语的字虽然很多，有数万个，但其发音比较简单. 汉语在无限词汇量的语音合成中具有得天独厚的优越性. 汉语的句子是由词组成，词又是由音节组成的，虽然存在一音多字的问题，但对于机器讲话，人听话的语音合成情况来说，同音字的问题不必考虑. 因为人在听讲时自然会理解这些同音字，也就是说汉语语音合成仅需考虑汉语单音就可以. 在汉语中单音仅有几百种，计入一个音的不同音调也仅有 1280 种，采用 CVSDM 算法进行编码，每个音约需 1k 字节的存储空间. 若按单音存储，把所有的音进行存储也只需 1280k 字节，即 $1280k \times 8 = 9.24M$ 位就可以完成汉语中所有语音的存储. 根据需要将语音进行不同组合就可以完成任何语音功能要求，进一步分析可以发现，利用上述办法获得的语音听起来很生硬，这是由于我们平时说话时，同一个音在句中的不同位置其发音还有强弱与长短之分. 欲获得更自然的语音仅存储单音虽然很难实现，经过大量的实验分析与比较，我们已找出一种既节约基本数据库容量又能产生高质量音响效果的方法. 可以采用存储词组的方法，即对某些常用的词组以词组的形式存储. 通过调用词组和单音来完成整句话的合成，就可以获得比较贴近自然语言的效果. 举例来说，在测向时要求报出 0—360 度的数字，从理论上讲只要存储零、一至九、十、百的数就行，比如 123，采用一、百、二、十、三、五个音节构成，但实际上，这种组合的效果不好，虽然能听懂，但自然度较差，我们采用的办法是以下面的形式进行存储，对百位的存入一百、二百、三百、四百、五百、六百、七百、八百、九百，十位的存入一十、二十、三十、四十、五十、六十、七十、八十、九十，个位存入零、一、二、三、四、五、六、七、八、九，这样仅存储 28 段音就可以完成比较自然的千以内任何数字的语音合成，再加上适当的量词如度、链、分等，就可以完成声呐语音报警系统常用的数字. 为此，我们将声呐报警中常用的语言以词为单位进行整理，共有 155 个，然后依次将其量化、存储.

通过对所需语音进行整理、量化之后依次存入语音库 ROM 或 EPROM 中，然后以各语音的起始地址和长度形成一个索引表，通过对索引表中的内容进行不同编程就可以形成声呐语音报警系统所需的任何语音.

三、系统结构与功能

在语音合成、语音存储方法确定之后，问题就在如何能灵活地进行编程，调用不同语音组合成所需的句子. 语音报警系统包括有 8031 单片微机、RAM、语音库，语音处理等，详见图 1.

ROM1 存储程序，控制整个系统的工作. 语音处理模块采用 CVSDM 方法进行语音合成. 它接收语音库送来的 8bit 连续语音数据作为它的输入，经合成处理之后的信号送至滤波器和功率放大器，形成高保真的语音. RAM 存储总控台送来的报警信息，为了便于总控台送信息和语音报警系统取信息，RAM 采用了双口 RAM. 正如前述语音报警系统需要的信息量比较大，要求语音库的存储空间大，我们采用 8031 单片机的 P0，P2 口作为地址的低八位和高八位，同时采用 P1 口的 0—4 五位信号经译码后产生语音库的片选信号，语音库采用了 27512 EPROM，这样它的最大容量可达 8M位.

语音处理芯片和单片微机各有自己的时钟，为使二者匹配，在系统中将语音处理的 CP 送到单片机的 T0 端进行同步.

语音报警系统的工作过程如图 2 所示，在声呐系统需要启用语音报警系统时，将报警类型和报警数据送到双口 RAM，并给单片机一

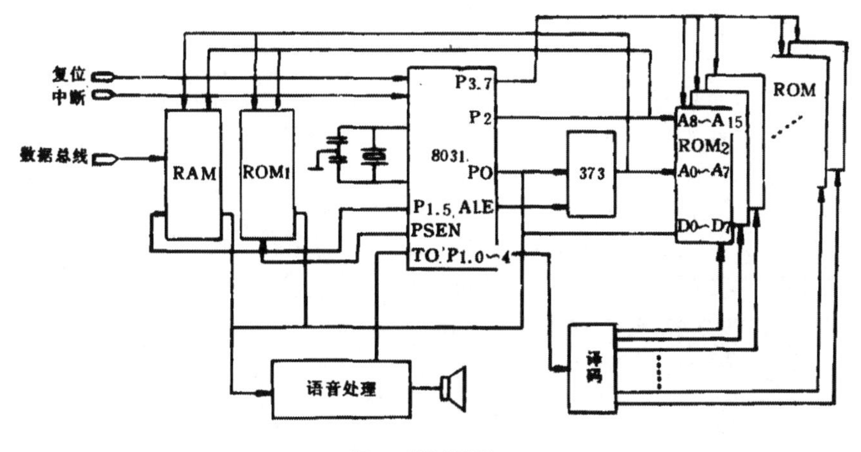

图1 系统结构图

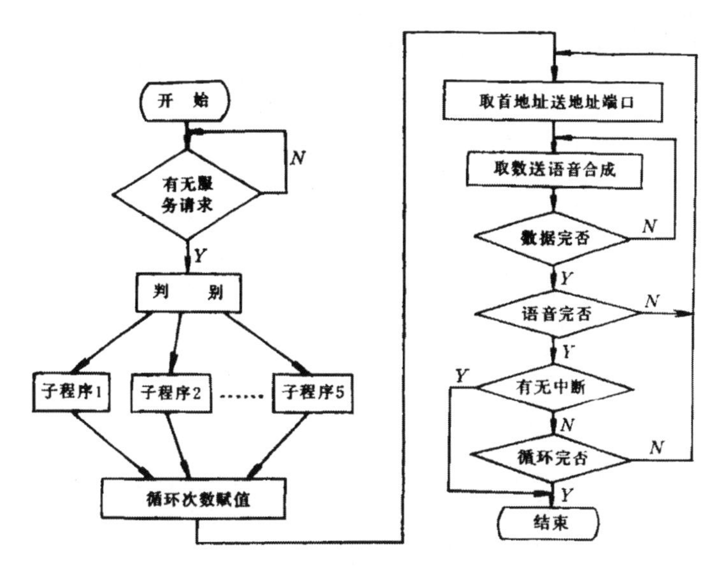

图2 程序流程图

个服务请求信号，当系统接受到服务请求信号后，首先将 RAM 中的信息读入并判别是何报警类型，依据不同类型进入不同的子程序，各子程序的功能是依据读入的数据通过查询语音库索引表，将需要用到的各语音词组或单音的首地址、长度，依次存入数据缓冲区，并将缓冲区的长度反馈给主程序。主程序依次将各语音的首地址取出送到地址码端口，并从语音库取出数据送到语音处理芯片进行语音合成。根据语音长度在第一个语音读完后再将第二个语音地址送到地址码端口，周而复始直至将所需的语音全部读完。同时主程序还要控制报警语音的遍数，对于测距、测向等一般报两遍，以便于操作人员听清。如是入侵报警就一直报，并不断查询操作人员有无反应，直到总控台给一停止信号为止，以保证不贻误战机。

四、结 束 语

由于采用了单片微机控制，该系统结构简单、控制灵活、语音自然、寻址能力强、可以编程、所以声呐语音报警系统较之常规人机对话系统功能可以大为改善.在语音库存储语音时，根据不同应用存入不同的词组就可以满足各种应用，所以该系统除了可以应用于声呐语音报警系统外,还可以适用于需要语音的任何系统.

参 考 文 献

[1] Bowen B. A., and Brown W. R., VLSI Systems Design for Digital Signal Processing, Prentice-Hall Inc., USA, 1982.

[2] Gupter A., and Tuong M. D., *Proc. IEEE* **71-11** (1983), 1236—1256.

[3] Murphy B. T., *IEEE. J. Solid-State Circuits, SC*-**18**-3(1983), 236—244.

[4] 国际电子商情,香港,1991,11.

[5] Lynch Thomas T., Date Compression, Techniques and Applications, Wadsworth Inc., USA., 1985.

[6] Oppenkeirn A. V., Applications of Digilal Signal Processing, Prentice-Hall Inc., USA., 1978.

数字多波束声呐的一种内插算法①

李启虎 张春华 李淑秋
（中国科学院声学研究所 北京 100080）

1993年7月30日收到

数字多波束声呐的每一波束具有固定指向．为了在空间域得到较高的分辨力，减少声呐设计中的硬件开销，波束成形后的内插是十分必要的．本文基于抛物线三点与五点内插的原理，提出一种简单易行，精度较高的波束内插算法，并以直线阵和圆弧阵为例，说明这种算法的实现方法。实际计算机模拟结果表明，本文提出的算法可以改善多波束系统的指向特性，提高对目标的分辨精度．

ABSTRACT

In the multi-beam digital sonar, each beam has fixed bearing direction. In order to obtain higher spacial resolution and reduce the hardware cost, it is necessary to make some kind of interpolation algorithm in post beam forming. Based on the principle of parabolic three-point and five-point interpolation, a simple, precise beam interpolation technique is presented in this paper. The realization of this method, for which the line and the circle arrays are taken as examples, is illustrated. The system simulation results show that the method described in this paper can be used to improve the directivity and enhance the resolving ability of the multi-beam sonar system.

一、引 言

声呐系统的数字化运算给波束成形带来了巨大的好处，一般模拟系统不易达到或无法达到的一些结果都可以在数字系统中完成．为了用较少的硬件得到最大限度的系统增益，波束成形之前及之后的数据内插及外推引起了声呐设计者的兴趣[1-3]．

在波束成形之前进行数据内插和外推的意义有二．第一是企图用较小的孔径，得到较大孔径基阵才会有的空间滤波效果；第二是企图用较少的接收基元，得到较多基元才会有的时间—空间增益．这样，硬件设备量可以大为减少．

波束成形之后的数据内插和外推是为了使信号处理系统和显示控制器之间更好地匹配，提高对目标的分辨精度，减少硬件开销．一般来说，波束成形的运算量与波束数成正比．如果用64波束的系统能得到128波束的效果，则硬件设备节约了1/2．

从显示的观点来看，目前数字式声呐几乎无一例外采用方位/历程显示．如果一个波束成形系统具有64个波束，而在1024*768的显示器上进行显示．以每一波束占用12个象素来看，需要768列．声呐员在观测屏幕时，将会感到在相邻波束之间在亮度上有跳动的现象．当我们对64个波束数据进行内插时，可以把波束数增加到128甚至256，这时方位/历程显示将变得柔和一些，从而使目标分辨力和定位精度都得到改进．

对于数字化系统来说，内插算法易于用软件完成，对硬件的要求很低．所以波束数据的内插不失为一种投资少收益高的方法．

① 应用声学, 1994, 13(5): 1-4.

本文分析波束数据内插法的基本原理，提出一种以保证主瓣形状为主的三点，五点抛物线内插法．并以线阵及圆阵为例给出实际系统的模拟结果．

二、抛物线内插法原理

众所周知，对于有 N 个基元的等间隔离散线阵，其指向性表达式是[4]

$$R(\varphi)=\sin(N\varphi/2)/N\sin(\varphi/2) \qquad (1)$$

其中

$$\varphi=2\pi d[\sin\theta-\sin\theta_0]/\lambda \qquad (2)$$

θ 为目标信号的入射角，θ_0 为波束指向角，d 为基元间的距离，λ 为波长．

在信号检测中，我们关心的是主瓣附近的指向特性，所以波束数据的内插也以主瓣附近函数特性为依据．当 $\theta\approx\theta_0$ 时，(1) 式具有 $\sin x/x$ 的形状，在 $x=0$ 处作 Taylor 展开，有

$$\sin x/x\approx 1-x^2/3 \qquad (3)$$

换句话说，直线阵指向性函数在主瓣附近具有抛物线的形状．

对于圆阵来说，如果基元数和圆阵直径、信号波长满足一定的关系，其指向性可以表达为零阶贝塞尔函数[4]

$$D(\theta)=J_0\left(\frac{4\pi r}{\lambda}\sin\frac{\theta-\theta_0}{2}\right) \qquad (4)$$

其中 r 为基阵半径，λ 为信号波长，在 $x=0$（即主瓣）附近，

$$J_0(x)\approx 1-x^2/4 \qquad (5)$$

也具有抛物线形状．

于是波束数据的内插，实际上可以归结为求解以下抛物线插值问题：已知

$$y=f(x)=ax^2+bx+c$$

上的三个等间隔点 x_{-1}, x_0, x_1 的值是

$$y_i=f(x_i) \qquad i=-1,0,1 \qquad (6)$$

求出 $x\in[x_{-1},x_1]$ 中任一点 $f(x)$ 的值．

以三点内插为例，就是在 $[x_{-1},x_0]$ 中求出 $f(x_{-1/2})$，在 $[x_0,x_1]$ 中求出 $f(x_{1/2})$．见图 1．对于五点内插，就是在 $[x_{-1},x_0]$ 中求出

$$f(x_{-3/4}),\ f(x_{-1/2}),\ f(x_{-1/4}).$$

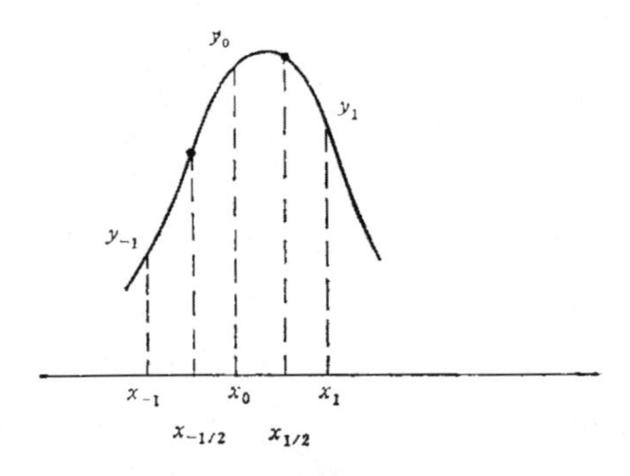

图 1　三点抛物线内插示意图

在 $[x_0,x_1]$ 中求出

$$f(x_{1/4}),\ f(x_{1/2}),\ f(x_{3/4}).$$

容易证明，在(5)式的假定下，三点内插与五点内插的值分别由式(6)或式(7)给出：

$$\left.\begin{aligned}f(x_{-1/2})&=(3y_{-1}-y_1+6y_0)/8\\ f(x_{1/2})&=(3y_1-y_{-1}+6y_0)/8\end{aligned}\right\} \qquad (7)$$

$$\left.\begin{aligned}f(x_{-3/4})&=(21y_{-1}+14y_0-3y_1)/32\\ f(x_{-1/2})&=(12y_{-1}+24y_0-4y_1)/32\\ f(x_{-1/4})&=(5y_{-1}+30y_0-3y_1)/32\\ f(x_{1/4})&=(-3y_{-1}+30y_0+5y_1)/32\\ f(x_{1/2})&=(-4y_{-1}+24y_0+12y_1)/32\\ f(x_{3/4})&=(-3y_{-1}+14y_0+21y_1)/32\end{aligned}\right\} \qquad (8)$$

三点内插可以使波束数增加一倍，五点内插可以使波束数增加 4 倍．

三、系 统 模 拟

波束内插的原则是建立在原有多波束系统的波束分布已足够密的前提下的．一般要求在波束指向性的 3 dB 点内至少有两个点．另外，抛物线内插法是针对主瓣的，在副瓣部份也可以采用．当然，为简化运算量，除主瓣外的其他部分，也可以用线性内插．

表 1 至表 3 给出了 $N=12$ 的等间隔线阵五点内插的结果，基本数据为 $D(0)$，$D(4)$，$D(8)$．内插步长为 1°．

从表 1 至表 3 我们可以看到，内插结果相当准确．特别是当信号入射角不处于波束指向

表 1　$d/\lambda=0.2$，入射角 $\theta=0°$

指向性	真　值	内插结果
$D(0)$	1.000	—
$D(1)$	0.997	0.997
$D(2)$	0.988	0.988
$D(3)$	0.974	0.974
$D(4)$	0.955	—
$D(5)$	0.930	0.931
$D(6)$	0.900	0.901
$D(7)$	0.866	0.867
$D(8)$	[illegible]	—

表 2　$d/\lambda=0.2$，入射角 $\theta=3°$

指向性	真　值	内插结果
$D(0)$	0.974	—
$D(1)$	0.988	0.988
$D(2)$	0.997	0.997
$D(3)$	1.000	1.000
$D(4)$	0.997	—
$D(5)$	0.988	0.989
$D(6)$	0.974	0.975
$D(7)$	0.955	0.956
$D(8)$	0.931	—

角时更加如此．事实上，从表 2 我们发现，若信号入射角在 $\theta=3°$．而多波束指向角为 0°，4°，8°，…，这时，90 个波束的系统将不会有最大值 1.000 出现在入射角方向上，而利用抛物线

表 3　$d/\lambda=0.5$，入射角 $\theta=0°$

指向性	真　值	内插结果
$D(0)$	1.000	—
$D(1)$	0.982	0.961
$D(2)$	0.929	0.904
$D(3)$	0.846	0.829
$D(4)$	0.737	—
$D(5)$	0.609	0.627
$D(6)$	0.469	0.499
$D(7)$	0.327	0.357
$D(8)$	0.190	—

内插却准确无误地在 3° 出现了 $D(3)=1.000$．这正是我们所希望的．

另外，在指向特性的 3dB 点范围内进行数据内插时，一般的相对误差在 3—10%之间．图 2 给出了线阵指向性的内插结果．图 3 给出了圆阵指向性在双目标时的情况，我们可以看到，如果没有内插的运算，多目标分辨力会有所下降．

四、结　　论

数字多波束声呐的波束数据内插只需增加少量硬件及软件便可以改善指向特性的显示及提高多目标的分辨力及单目标的定向精度．基

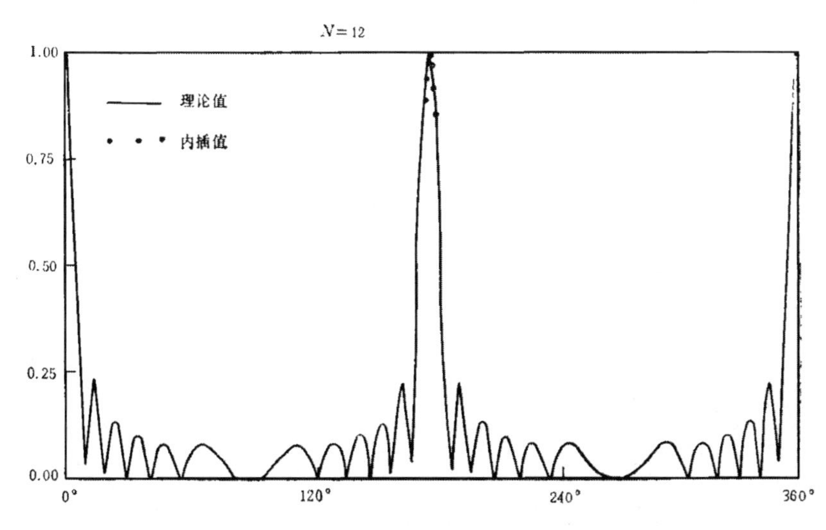

图 2　线性指向性的内插结果

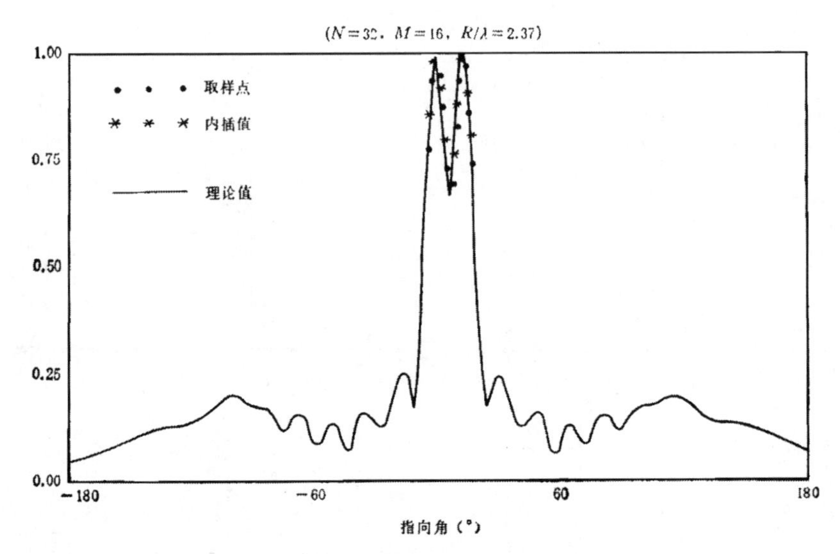

图 3 圆弧阵指向性的内插结果

于抛物线内插的三点、五点内插算法，简单易行，便于用软件实现.

参 考 文 献

[1] Van Veen B.D. and Buckley K. M., *IEEE ASSP Magazine*, (1988) April, 4—24.

[2] Pridham R. G. et al., *Proc. IEEE* **67**-6 (1979), 904—917.

[3] Swingler D.and Walker R., *IEEE Trans.* ASSP-**37** (1989) 16—29.

[4] 李启虎，声呐信号处理引论，海洋出版社，1985.

多波束 DICANNE 系统研究①

李淑秋　李启虎　刘金波

(中国科学院声学研究所　北京，100080)

1993 年 7 月 30 日收到

摘要　多波束 DICANNE 系统具有抵消强干扰的良好性能。本文给出这种数字式系统的理论基础，讨论其稳态特性。指出多波束 DICANNE 系统输出的信噪比、干扰噪声比。给出系统硬件实现的方框图及各种参数的选择。同时讨论以 TMS320 系列芯片进行硬件设计的问题。

A study of multi-beam DICANNE system

LI Shuqiu　LI Qihu　and　LIU Jinbo

(*Institute of Acoustics, Academia Sinica*　Beijing, 100080)

Received July 30, 1993

Abstract　The multi-beam DICANNE system has good performance for eliminating the strong interfernce. The theoretical basis of this digital system is presented in this paper, the steady-state performance is discussed. The signal to noise ratio and signal to interference ratio is derived. Some principles for parameter selection in hardware designing and block diagram of the system are illustrated. The hardware implementation with DSP chip of TMS320 series is discussed.

一、引　　言

抗强点源干扰问题一直是声呐设计中的一个重要课题。对于模拟式声呐，要解决这一问题非常困难。直到出现数字式声呐，才有关于抗强点源干扰系统的研究报道，最早的是 1969 年 V. C. Anderson 提出的 DICANNE（数字式干扰消除自适应零网络设备 Digital Interference Cancelling Adaptive Null Network Equipment)[1]。这是一种双补偿器的多波束系统。一个用于形成干扰波束，另一个用于形成普通波束。由于当时数字器件的限制，还不能实时完成这种运算。另一类干扰抵消系统便是 B. Widrow 等提出的噪声抵消算法[2]。在宽带的情形，利用大量可调的权系数最大限度地抵消干扰。但是这种算法的运算量与通道数 N 的 3 次方成正比，要得到多波束的抵消结果是非常困难的。

目前比较引人注意的是前置波束干扰抵消系统和后置波束干扰抵消系统。但是前者只对极强干扰有效，而后者则是单波束系统[3,4]。

本文提出一种多波束 DICANNE 系统，利用两个普通波束成形系统和一个用于形成干扰波束的单波束系统，再配置相应的预延迟线就可以构成一个抵消强干扰的多波束系统。这个系统可以在 0—360°内扫描干扰波束，然后将各输入通道中的干扰抵消，从而形成无干扰的多波束系统，实现对弱信号的检测。

本文所提出的方法易于用 TMS320 系列的 DSP 芯片实现。文中将给出系统设计的方框图及硬件设计方面的考虑。最后给出计算结果。

---

① 声学学报, 1995, 20(4): 298-301.

二、多波束 DICANNE 系统的基本原理

本文讨论具有任意平面几何图形的离散基阵的抗点源干扰问题。但是在讨论系统设计时为简单起见，我们以圆阵和线阵为例说明有关参数的选择，它可以很容易地推广到其他形状的布阵上去。

所谓强点源干扰，是指源强远大于信号源强的干扰，即 $\sigma_I^2/\sigma_s^2 \gg 1$。其中 σ_I^2 为干扰功率，σ_s^2 为信号功率。

设有一个 N 个基元构成的平面接收阵,信号入射方向角为 α，干扰入射方向角为 β。以平面上某一位置为参考点,第 i 路的接收信号可以写成:

$$x_i[t] = s[t+\tau_i(\alpha)] + I[t+\tau_i(\beta)] + n_i[t] \quad (1)$$

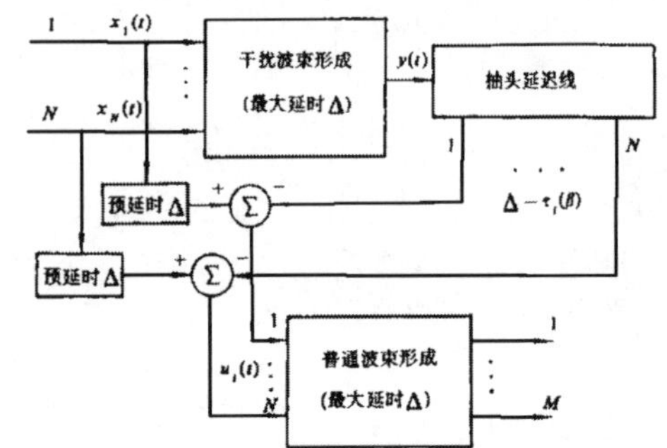

图 1 多波束强干扰抵消系统方框图

其中 $\tau_i(\theta)$ 表示时延，$s(t)$ 为信号，$I(t)$ 为干扰。参看图 1，把第 i 路接收信号再延时 $\tau_i(\beta)$，得到

$$x_i[t-\tau_i(\beta)] = s[t+\tau_i(\alpha)-\tau_i(\beta)] + I[t] + n_i[t-\tau_i(\beta)] \tag{2}$$

今后，一律假定 $\tau_i(\theta) \geqq 0$，所以出现在 (1), (2) 式中的延时量 $\tau_i(\alpha)$，$\tau_i(\beta)$ 是物理上可实现的。这一点，对系统设计来说，至关重要。

把由(2)式表示的 N 路信号相加，得到了加强的干扰信号，即

$$\frac{1}{N}\sum_i x_i[t-\tau_i(\beta)] = I[t] + \frac{1}{N}\sum_i s[t+\tau_i(\alpha)-\tau_i(\beta)] + \frac{1}{N}\sum_i n_i[t-\tau_i(\beta)] \tag{3}$$

把(3)式左边记作 $y(t)$。右边的第 2, 3 项分别记作 $s'(t)$ 和 $n'(t)$。于是

$$y(t) = I[t] + s'(t) + n'(t) \tag{4}$$

在(1)式中，干扰强度为 σ_I^2，信号加噪声的强度为 $\sigma_s^2+\sigma_n^2$。在(4)式中，干扰强度仍为σ_I^2，但残留的信号强度加噪声强度只有 $(\sigma_s^2+\sigma_n^2)/N$。因此，在某种意义上说 $y(t)$ 可以认为是“纯干扰”。

我们的目的是把 $y(t)$ 经过一条抽头延迟线，得到具有不同延时的 N 路信号，把 $x_i(t)$ 中的干扰分量 $I[t+\tau_i(\beta)]$减去。由于 $y(t)$ 中的 $I(t)$ 在时间上滞后 $x_i(t)$ 中的 $I[t+\tau_i(\beta)]$，所以必须预先将 $x_i(t)$ 延迟一个较大的延时量，才能做到这一点。设预延时的时间为Δ，把 $x_i(t)$ 延时 Δ，得到

$$x_i(t-\Delta) = s[t+\tau_i(\alpha)-\Delta] + I[t-\Delta+\tau_i(\beta)] + n_i(t-\Delta) \tag{5}$$

干扰波束的第 i 个抽头为

$$z_i(t) = I[t-\Delta+\tau_i(\beta)] + s'[t-\Delta+\tau_i(\beta)] + n'[t-\Delta+\tau_i(\beta)] \tag{6}$$

把 $z_i(t)$ 减去 $x_i(t-\Delta)$ 得到 DICANNE 系统的新的输入:

$$u_i(t) = s[t-\Delta+\tau_i(\alpha)] + n_i(t-\Delta) - s'[t-\Delta+\tau_i(\beta)] + n'[t-\Delta+\tau_i(\beta)] \tag{7}$$

$u_i(t)$ 和 $x_i(t)$ 相比，已经没有干扰成份，信号入射方向仍是 α，只是时间参考点由 t 变为 $t-\Delta$，这是无关紧要的。

$u_i(t)$ 中的信号强度为 σ_s^2，噪声强度为 $\sigma_n^2+\sigma_n^2/N$，干扰强度为 0，残余信号强度为 σ_s^2/N。

对 $u_i(t)$ 作普通波束成形，即延时 $\tau_i(\theta)$，得到 $u_i[t-\tau_i(\theta)]$，把 $u_i[t-\tau_i(\theta)]$ 求和总体平均就得到 DICANNE 系统的指向性[5]。

三、系统设计

根据上面的讨论，我们可以知道，多波束 DICANNE 系统至少要由两部分组成。一部分用于形成干扰波束，另一部分用于抵消后信号的普通波束成形。另外需要一些辅助电路如抽头延迟线以及预延时电路等。

对于直线阵来说，如果长度为 L，那么预延时 $\Delta=L/c$，其中 c 为声速。对于圆阵来说，$\Delta=R/c$，R 为基阵半径。通常 Δ 应以系统采样周期的倍数来计算。举例来说，如果采样周期为 25 μs，而 $\Delta=530$ μs，那么 Δ 应取 22 节延迟单元。记

$$L=[\Delta/T_s]+1 \tag{8}$$

其中[ ]表示取整运算。

我们可以用静态 RAM 形成多波束。在图 1 中用于形成干扰波束的 RAM 与普通波束成形的 RAM 都是容量为 $N*L$ 的芯片。而预延时 Δ 及抽头延迟线的长度也是 L。系统的控制和实时计算可以由专用 DSP 芯片（如 TMS320）完成。

图 2 是另一种多波束 DICANNE 系统的方框图。这实际上是一种单补偿器的系统。我们从图上可以看到，它已没有预延迟线，而代之以长为 2Δ 的 RAM 及产生干扰波束的延迟线。这种系统从硬件设计角度要节省一些。但是所付出的代价是实时控制变得比较复杂了，详细的时间关系已标在图 2 中。

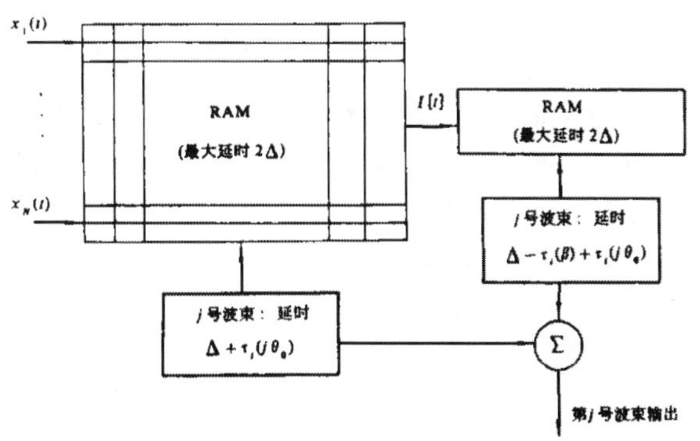

图 2 数字式 RAM 动态 DICANNE 系统方框图

实际系统的设计要考虑到有关参数的选取。下面，我们以圆阵为例来说明这一点。假定圆阵直径为 $R=2$ m，均匀分布 $N=32$ 个基元。采样频率选为 40 kHz，即 $T_s=25$ μs。那么 $\Delta=667$ μs，于是 $L=27$ 节。我们选用 TMS320C25 作为形成波束的 CPU，为节省运算时间，应当把延时地址表放在片上的 RAM 中。该片的 RAM 空间为 512 字。如果用每片形成 6 个波束，每一波束由 16 个水听器参加工作，那么地址表需占用 27 * (16+2) = 486 字节。为形成 64 个波束，需用 11 片 TMS320C25。

四、计算机模拟结果

我们对上面的多波束 DICANNE 系统进行计算机模拟。用 $N=20$ 的等间隔线列阵，取 $d/\lambda=1/4$，$\sigma_I^2/\sigma_s^2=100$ (20dB)，$\sigma_I^2/\sigma_n^2=10$ (10 dB)，噪声为各向同性的白噪声。模拟结果见图 3。假定信号的入射角为 0°，(a)为无干扰，无噪声时的普通波束成形的指向性。(b)为干扰出现在 $\beta=16°$ 时的情形，这时在 $\alpha=0°$ 的地方可以勉强怀疑有一个目标。(c)是干扰出现在 $\beta=8°$ 时的情况，我们已无法判断何处有信号。(d)是干扰抵消后，DICANNE 系统的指向性，效果相当好。由于还存在着背景噪声，整个零线电平还比较高。

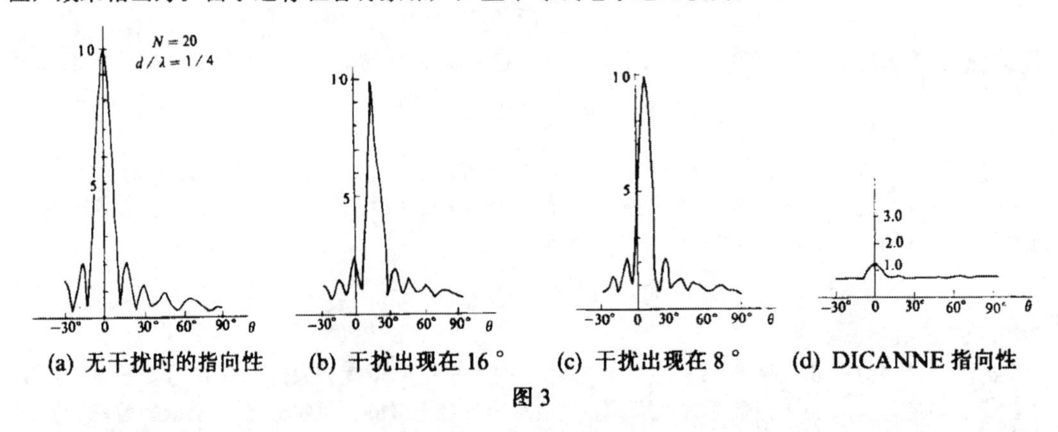

(a) 无干扰时的指向性　(b) 干扰出现在 16°　(c) 干扰出现在 8°　(d) DICANNE 指向性

图 3

我们给出的是 DICANNE 系统的稳态特性。如果采用 LMS 自适应迭代算法，在迭代过程的不同阶段把权系数输出，再计算指向性。那么会得到一种“动态的”指向特性。自适应过程的趋势是在干扰方向形成一个零点。根据 Chang 和 Tuteur 的研究，当迭代进行到 500 次时，强干扰即可抑制[6]。本文所得的稳态指向性曲线与他们在迭代次数趋向于∞时的情况几乎完全一样。

参 考 文 献

[1] Anderson V C. Dicanne, a realizable adaptive process, *J. Acoust. Soc. Am.*, 1969, **45**: 398—405.

[2] Widrow B et al. Adaptive noise cancelling: theory and applications, *Proc. IEEE*, 1975, **63**: 1692—1716.

[3] Nielson R O. Sonar signal processing, *Artech House*, 1991, USA.

[4] Cantoni A and Godara L C. Performance of a postbeam−former interference canceller in the presence of broadband directional signals, *J. Acoust. Soc. Am.*, 1984, **76**: 128—138.

[5] 李启虎. 声呐信号处理引论，海洋出版社，1985年，北京。

[6] Chang J H and Tuteur F B. A new class of adaptive array processor", *J. Acoust. Soc. Am.*, 1971, **49**: 639—649.

数字式声呐大动态范围显示技术研究①

秦英达 杜 鹏 李启虎

(中国科学院声学研究所 北京 100080)

1994年4月25日收到

摘要 量化比特数是衡量数字声呐性能的主要指标,增加比特数意味着更高的精度和更大的动态范围.但人机视觉接口只能显示有限字长的数据.本文从声呐显示技术出发,分析了突破这一局限的必要和可能.原理分析表明,对声呐数据实行过零基准处理,以GSC算法配合小信号分层处理可有效地获得多比特位声呐数据的几乎无损的显示.对16比特声呐数据处理显示的实验结果表明,我们的处理方法达到了预期要求.

关键词 多比特位显示 GSC算法,分层显示,过零处理.

A display technique for wide dynamic range digital Sonar

Qin Yingda, Du Peng, Li Qihu

(*Institutute of Acoustics Academia Sinica, Beijing* 100080)

Abstract One of the important methods to enhance Sonar's performance is to improve its data length. The more bits there are, the more accuracy and wider dynamic range will be. But with human eye, one can only receive data with a limited length from Sonar's display. This paper studies why and how to display longer datathan traditional from display technique. It is shown that an effective method to acquire nearly no-performance-loss display isto process the sonar data by zeroaxial process and GSC algorithm accompanied by special treatment of small signals. The method has been proved to be efficient in our experiment to display 16 bits Sonar data.

Key words Multibits Display, GSC algorithm, Stratified display, Zeroaxial process

1 前言

随着数字技术的发展,声呐的体系结构发生了革命性的变化,其性能有了飞速的提高,同时在数字化的过程中产生了许多新问题.

一部现代数字式声呐,其结构可按功能划分为几个子系统[2],如图1所示.

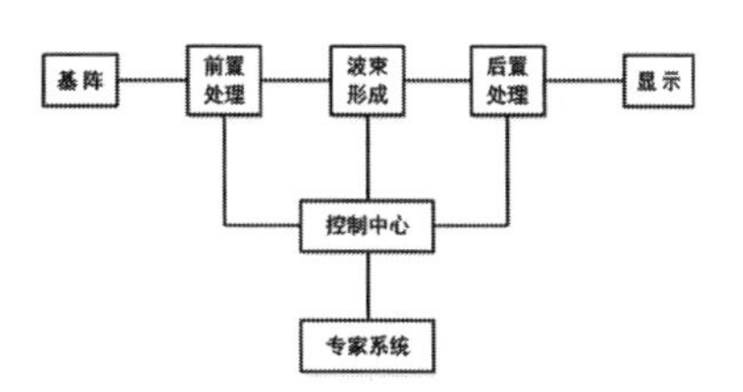

图1 数字式声呐结构框图

声呐系统对信号的检测、判断需由显示系统给出,显示系统是影响声呐性能的最后一环,所以能不能由显示系统最大限度地把前置处理和波束成形部分获得的增益充分发挥出来是至关重要的[1].显示技术是随着信号处理、人机工程等技术的发展而不断前进的,本文将在实践的基础上探索一些新的观点和方法.

数字式声呐在检测精度和动态范围上提供

① 应用声学, 1995, 14(3): 14-19.

了很大的发展余地，而高精度和大的动态范围则必须要求更长比特位的声呐数据. 目前飞速发展的 DSP 技术使 8 位、16 位甚至 32 位的声呐数据均可实时处理了，这将大大提高声呐的检测精度和动态范围. 但是，超过 8 比特的数据的显示却存在着很大困难.

2 声呐显示简况

声呐前身是听测器件，而早期声呐的检测手段主要采用听测方式. 近代数字存贮和显示技术的发展，使多方位、长时间、多角度的大容量的信息的显示成为现实，听测在现代大型声呐中成为一种辅助性的手段.

声呐最早的视觉检测手段是 40 年代发明的纸记录方式. 阴极射线管 (CRT) 产生后,声呐从雷达借用了 A 扫方式，一种现代 A 扫的形式如图 2 所示,这种方式以横轴为方位,纵轴正比于声波幅度. 70 年代,随着数字存贮技术的发展,B 扫格式如 PPI、BTR 等逐渐成熟起来，以 BTR 格式为例,如图 3 所示,它将声波能量调亮显示,并按方位和时间形成方形图案,最新的全向波束信息出现在最上一行，最旧的结果不断从下面移出屏幕. 随着 FFT 理论与技术

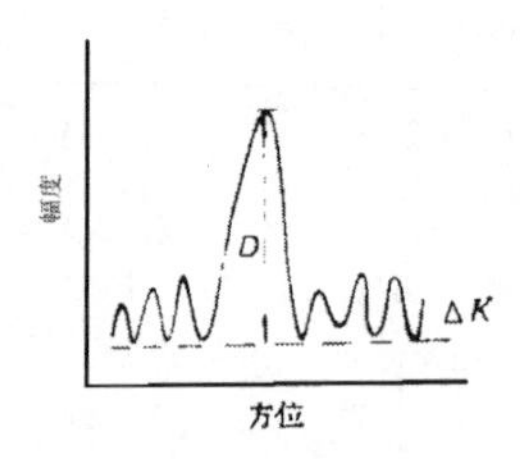

图 2 一种 A 扫格式

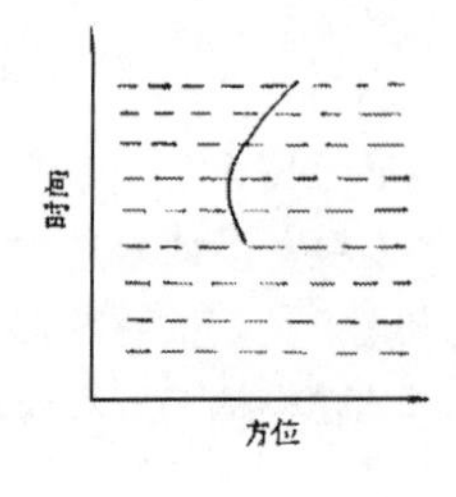

图 3 *BTR* 显示格式

的发展，声呐频域数据也可以实时处理并用于显示了,频域也分 A 扫、B 扫 (LOFAR) 两种.

早期的有关声呐显示数据所需比特数有一些实验数据. 60 年代末，Evans 的实验指出，*BTR* 显示采用 3 比特效果最好[7]. 此后的研究结果获得了较为一致的看法,即"做为亮度调制，一般使用了 3—4 比特亮化信号就足够了，实验表明,更多的位数是无益的,只会导致设备的复杂化,对图象质量无改善"[3]. 美国六七十年代研制的声呐系统的调亮显示均采用了比特制式,如美国 70 年代研制的攻击和战略潜艇用声呐系统 AN/BQQ-5 和 AN/BQQ-6 均采用 OJ-274/BQQ-5 声呐显控台，该显控台采用 3bit 显示动态范围[4]. 潜艇用声呐 8 灰阶标准调亮显示有严格的标定方法[5]. 这些研究工作是在模拟或半数字式声呐上进行的，其缺陷是没有充分考虑人眼可分辨的最小亮度差及信号处理系统和显示系统之间的数据匹配，因而其结论就有失偏颇.

3 数字式声呐显示的困难

如果将信号处理系统输出的多比特数据压缩为 3 个比特,无论采用何种算法,信息损失总是难免的. 多比特的显示系统可使信号处理系统获得的增益不受损失,而且实验也表明,多比特声呐数据显示确实提高了检测能力和动态范围，大大改善了图像质量. 以 3 比特与 8 比特显示系统为例，采用多比特数对图像质量的改善表现在:

(1) 对 3 比特和 8 比特声呐均可检测到的信号，3bit 的显示屏上有过多的噪声显示，而 8bit 显示则可通过适当选择，大幅度去掉背景干扰,突出目标及其特征.

(2) 对于较大信噪比的信号，3bit 量化可能使主瓣和副瓣均达到最高亮度级，因而影响到对目标个数和真实目标位置的判断.

(3) 对于大动态范围的声呐信号，量化层次少,会大大降低分瓣精度,主要表现在对弱信号的分辨能力上.

由此可见，发展多比特声呐显示已是大势

所趋．但多比特数显示却遇到了困难．

现代数字式声呐普遍使用高分辨率监视器，监视器的两个重要指标是分辨率和动态范围，而这两个参数是与人眼的角分辨能力、对灰度等级和色彩的感觉能力息息相关的．应用到声呐显示的方屏幕尺寸有10″×10″、14″×14″、18″×18″等几种，典型的使用分辨率可按1024×1024估计，而动态范围单色为8bit，彩色为24bit．一般认为，这些参数优于人眼的分辨能力，如许多显示专家倾向于人眼对单色灰度级的分辨能力为5—6个比特；而以人眼的角分辨能力来估算，在18″×18″、分辨率1024×1024的屏幕上，人眼水平分辨率大约是6—8个象素[1,7]．这些数据在我们的实验中也得到了验证．

由上面两个参数可知，声呐A扫的显示能力为10bit，而单色B扫的动态范围为8bit．引入彩色显示存在提高动态的可能，但已有的研究结果并不乐观[5,6]．人机视觉接口这一限制对于超过8bit位的声呐数据的显示提出了严重挑战．

4 GSC 算法

GSC (*Gray Scale Conversion*) 算法是李启虎等借用语音信号处理 A/D 转换中非线性量化思想而提出的．这种算法基于 8bit 显示能力，利用非线性转换来增加显示动态范围，是目前实现多比特声呐显示的依据[1]．

一种将 16bit 数据转化为 8bit 的方法是简单地取高 8 位或低 8 位，这样做的结果是使信号处理 16bit 的增益损失掉，整个声呐仍只是一个 8bit 位声呐．以取高 8 位为例，其显示动态范围达不到－20dB 至＋10dB[1]．

图 4 是采取 GSC 算法的示意图．GSC 算法可按 μ 律或 A 律转换．

以 μ 律为例，其归一化的输入输出关系为：

$$Y = \ln(1 + \mu X)/\ln(1 + \mu) \quad |x| \leqslant 1$$

其中 X 是归一化的输入，Y 是归一化的输出．

当 X 是正 16 比特数时，Y 的 8 比特对应值

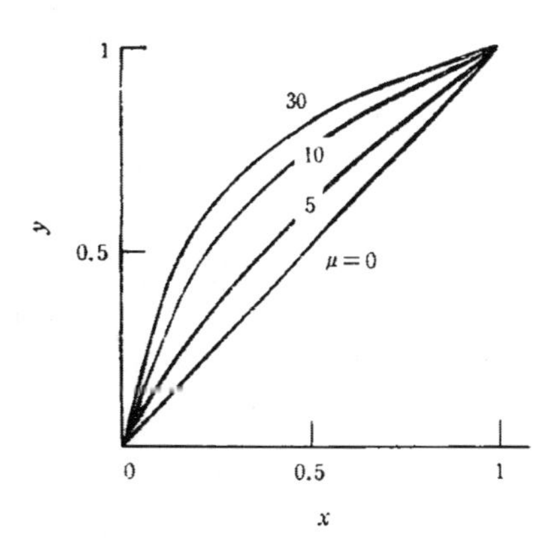

图 4 μ 律扩展

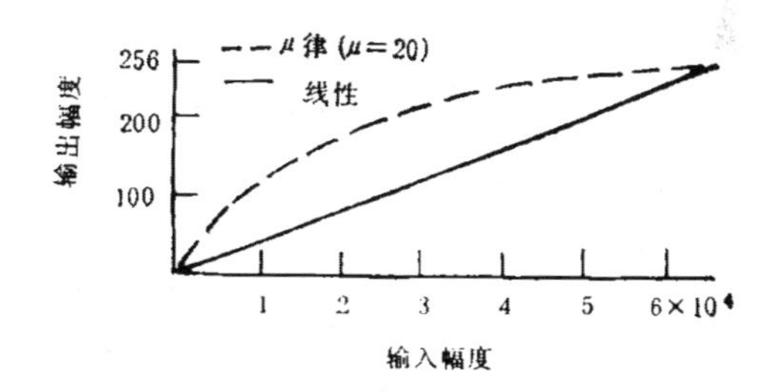

图 5 16bit 至 8bit 的变换

显然可以写成：

$$Y = \ln(1 + \mu X/2^{16})/\ln(1 + \mu) \times 2^8$$

由图 4 可以看出，选取合适的 μ 值，可以使 0 附近的数扩展，有利于对微弱信号的检测；大信号则被适当压缩，只要不影响判断信号的有无，这也是无妨的．一般来说，选取 μ＝20—30 可以兼顾大小信号，较为合适．

除 μ 律变换之外，还有 A 律变换，在算法上只有很少差异，同样可用于 GSC 转换，其归一化的输入输出关系如下：

$$\begin{aligned} Y &= (1 + \ln(A \times X))/(1 + \ln(A)) & 1/A < X < 1 \\ &= (A \times X)/(1 + \ln(A)) & 0 < X < 1/A \end{aligned}$$

5 分层转换法

GSC 算法是针对各种信号情况、保证显示范围的通用算法，实质是以适当地牺牲分辨精度来确保显示动态范围的．

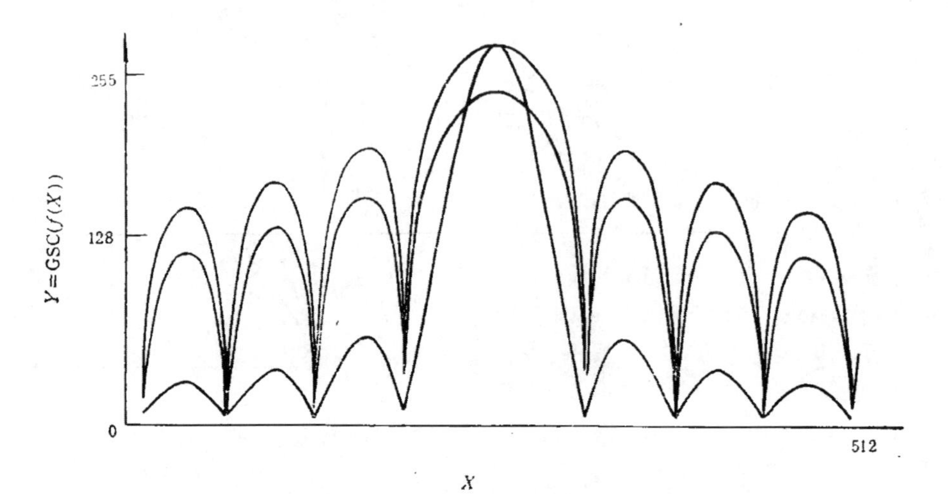

图 6 强信号的 GSC 转换效果

举例来说，对于 16bit 的信号处理系统，其多波束声呐数据均位于区间[0,65535]内；包含信号的数据情况可分为如下三种：

(1) 信号幅度远大于附近方位的噪声幅度。

(2) 信号幅度与噪声幅度非常接近。

(3) 信号淹没在噪声中，无法分辨。

对于(1)，由于信号较强，它的副瓣响应也可能显示出来．由于 GSC 算法对于大信号压缩大，小信号压缩小，因而会造成主瓣变宽、副瓣变高的现象，缩小了主、副瓣的差别，可能会影响到对信源个数的判断．假定噪声很弱，可以忽略，经过声呐处理的信号可按模型 $\sin(X)/X$ 估计．图 6 给出了一个计算机模拟转换的结果，有关参数如下：

GSC 转换因子 $A=87.6$，映射区间 [0,8191]→[0,255]．

曲线 1：原始信号 $f(X)=(8191\times\sin(X))/X$，128 点抽样，13 bit 量化后的 GSC 转换结果．

曲线 2：原始信号 $f(X)=(4095\times\sin(X))/X$，128 点抽样，13bit 量化后的 GSC 转换结果．

曲线 3：原始信号 $f(X)=(4095\times\sin(X))/X$，128 点抽样，利用浮点比例因子做线性转换后，8bit 量化结果．

由曲线 1 可见，副瓣相对上升高度很大，第一、第二级副瓣均超过主瓣高度的 50%．这对分辨精度的影响表现在：

(1) 如果实际只存在一个强目标，可能将副瓣判断为目标；可能影响目标位置的精确判断．

(2) 如果存在多个目标，弱目标将可能被强的信号(目标或干扰)的旁瓣响应所淹没．

由曲线 2 可见，其主、副瓣差异更小，第三级副瓣也超过主瓣幅度的 50%．可见，强信号如果只出现在 GSC 映射原区间的一小段内，则转换后的主、副瓣的差异更小．

由曲线 3 可见，线性变换保持原信号的特征，有利于对目标的识别．但线性变换无法保证显示的动态范围，可以作为 GSC 转换的辅助手段．

从对强信号的 GSC 转换效果看，可考虑采用如下办法予以补偿：

(*a*) 使用不同转换因子的 GSC 算法，调节 μ 值为负，则副瓣相对压缩更大，有利于对目标位置的分辨，但这需要声呐员有较强的应变能力．

(*b*) 使用分层的 GSC 算法，即显控台配备从不同原始区间到区间[0,255]的 GSC 转换

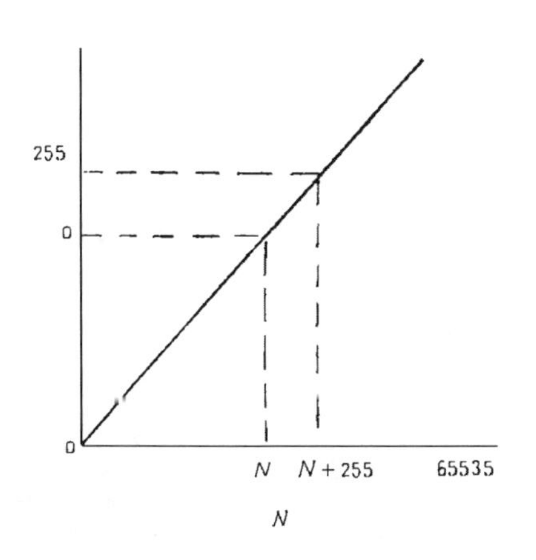

图 7　分层转换

程序，根据信号情况选择使用。

(c) 使用分层的线性转换算法，直接取所选区间的高 8 位数据，这实际上是配备了数种线性窗口。对于很强的信号，使用基于大区间的线性转换，可大幅度削弱噪声和副瓣，突出目标特征和中心位置；对于强弱信号共存的情况，可再选用基于小区间的线性转换来搜索，这时强信号已超出显示区间，但按照信息重复原理[6]，大脑将忽略已知特征的强信号，自动搜索弱信号。

对于情况(2)，噪声与信号的幅度值仅出现在一个较小的范围内，此时信噪比很低，提高声呐的动态范围就提高了弱信号的分辨精度。为了说明方便，现举一特例，图 7 给出其图示，即噪声与信号仅出现在 N 到 $N+255$ 范围内。

由于 GSC 算法对于大信号压缩大，故当 N 值较大时，可能存在 $[N, N+255]$ 整个区间转换后只对应 1 个 Y 值的情形；而且即使可产生 2、3 个不同的对应值，由于人眼对高灰度级反应相对迟钝，也是无法从中区分出信号的。在这种情况下，GSC 算法对显示精度的损失较大。

根据人眼对灰度的反应特点，对于“迹迹相关”的 B 扫图案，人眼对于暗背景上细微的信号轨迹都是十分敏感的，分辨要求几乎是信号幅度与噪声幅度相等，即检测阈 $DT \approx 0$；但是，在强的背景上，人眼分不出细微的灰度差别。我们使用两位受试者，做了一个简单的实验进行估计：将 NEC MULTISYNC 3D 显示器置于 1024×768 分辨率模式，在平均值为 1 的 Gauss 噪声背景上，人眼可区分灰度级为 4，宽度占 4 点的细线；而在平均值为灰度 200 的 Gauss 噪声背景上，对灰度 206，宽度为 4 的细线存在的判断正确率低于 50%。

实验表明，人眼对灰度级的感觉不是线性的，而是对较低的某些区域有相对较强的分辨率能力。这启发我们将信号处理系统产生的多波束数据进行过零处理。对于图 8 的情况，如果将 N 按灰度 0 显示，$N+255$ 按灰度 255 显示，则可获得最好的分辨精度。经过过零处理后，信号与噪声均集中在 0 的附近，因而对于小信号的显示尤为重要。对于较小范围的声呐数据，可将所有数据乘以一个线性因子予以放大，以期区分微弱的信号；而范围超过 255 的信号，转换损失不可避免，但由于集中在 0 附近的是噪声或副瓣，较高的是信号主瓣，可直接取高 8 位，反而有利于抑制背景噪声和副瓣，突出目标特征，增加目标的角分辨能力。这就是对声呐数据分层用不同算法转换的思想，这种思想的实质是针对不同的信号构成采取相应的观察角度，确保最好的检测性能。

6　实际系统的性能评估

我们研制了一个具有 16bit 信号处理能力和 8bit 灰度显示能力的系统，利用前面提到的分级灰度变换，进行了实验研究。在不使用显示系统时，我们在信号处理系统的输出端进行判断，最低可检测的输入信噪比为 −23dB。在把显示系统联结好之后，利用 16 分钟的瀑布显示，弱信号的显示轨迹在低于 −23dB 时也可区别出来，这就是所谓“迹迹相关”技术；因为对单一的时间断面来说，只有当信号的直流跳变高出周围多波束值至少 6dB 时，才有可能判定目标是否出现，但是，对于 16 分钟的长时间显示，目标运动的轨迹在起伏的噪声中呈现出一种稳定的线条，这时信号的直流跳变值即使与噪声相等(即输出信噪比为 0dB)，但由于信号始终

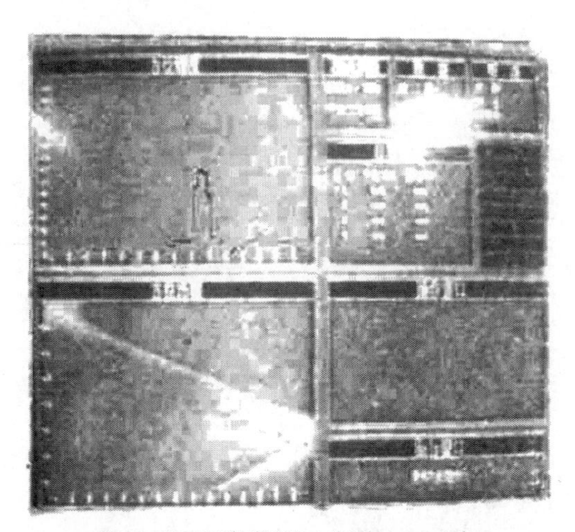

图 8 单目标结果(输入信噪比－20dB)

稳定，而噪声不断起伏，所以使得可检测信噪比压低至 —30dB. 我们利用多比特显示系统和迹迹相关技术，获得了原来会失去的7dB增益。我们在实验中获得的一幅显示图形如图 8 所示:

参考文献

[1] 李启虎，刘秋实、李淑秋. 声学学报，1992,16(5): 365—370.

[2] 李启虎，声呐信号处理引论，第一版. 北京: 海洋出版社，1986,91—96

[3] 侯自强，李贵斌. 声呐信号处理——原理与设备，第一版. 北京: 海洋出版社，1986,759—783

[4] Larry Li、Phil Joy. *Proc. I. C. Assp*, 1977, 116—119

[5] David F. Neri, David A. Kobus, Saul M. Luria, et. al. Effects of Background/Foreground color Coding on Detection in Acoustic Data Displays. Nusc Technical Report 7325, May 1985.

[6] W. Ronald Salafia, Dino A. Daros, Paul R. Bolvin. Color Coding of Amplitude Data as a means of Improving Target Detection in Passive Sonar Displays, Naval Underwater Systems Center, Submarine Sonar Department, AD-A197 283, June 1988.

[7] Robert A. Rasmussen, Studies Related To The Design And Use of Time/Bearing Sonar Displays. AD690540, 1969.

数字式声呐对目标的精确测向和自动跟踪问题[①]

李启虎　　尹　力

（中国科学院声学研究所，北京　100080）

1994 年 10 月 24 日收到

摘要　精确测向和自动跟踪是被动声呐的重要任务，本文给出数定式声呐精确测向的一种算法，指出最佳测向精度和基阵尺寸及频段的关系，推导了测向和延时估计的关系，利用波束主瓣附近的抛物线内插法可以精确给出目标方位并实现自动跟踪，计算机的模拟表明本文提出的方法具有良好的性能并且易于用硬件实现。

The problem of precise bearing and automatic tracking for target in digiral sonar

LI Qihu and YIN Li

(*Institute of Acoustics, Academia Sinica*, Beijing　100080)

Received October 24, 1994

Abstract　Precise bearing and automatic tracking for target is an important thesis in passive sonar. An algorithm, which can precisely estimate the value of Direction Of Arrival (DOA) of the target is drived in this paper. The relationship between bearing accuracy and the array apeture and signal frequency band is expressed. The expression of time delay estimation and bearing accuracy is presented. The DOA of the target can be precisely calculated by using the parabolic interpolation in the neiboughood of main lobe of the directivity function, it is easy to implement the automatic tracking in real time. The system simulation shows that the algorithm derived in this paper has very good performance and it is easy to realize in hardware.

一、引　　言

目标到达方位角（DOA，Direction of Arrival)的精确估计问题一直是被动声呐研究中的一个重要课题[1—3]。Lewis 等曾给出线阵的最佳估计精度的表达式，此后 Carter 等人从延时估计的角度指出目标信号的延时差的估计在最佳的情况下可以达到 Cramer–Rao 下界[4—5]。对于数字式声呐，波束的指向角通常是等间隔，离散地分布在 0—360° 内。要精确给出目标入射角的数值必需采取相应的算法。要实现对目标的自动跟踪就需要实时的处理方法随时给出目标方位。

本文首先分析声呐精确测向和信号延时差估计的关系，给出声呐系统在一定信噪比下能达到的最佳定向精度，然后指出，用于声呐定向的主波束附近的形状具有抛物线结构。多波束的数值，实际上就是对该抛物线的空间采样的样本。利用抛物线的内插公式可以给

① 声学学报, 1996, 21(4)增刊: 709-713.

出目标方位的精确值，由于算法简单，易于用软／硬件实现，所以就能实现对目标的自动跟踪，计算机的模拟结果表明这种方法是非常有效的。

二、精确测向与目标自动跟踪

数字式声呐在确定目标入射角时，要利用的是多波束的数据，假定目标的入射角（即DOA）是 θ。在 t 时刻的多波束数据是

$$D(t,\theta_0),\cdots,D(t,M\theta_0) \tag{1}$$

其中$\theta_0=360°/M$是相邻波束之间的夹角，实际上$D(t,j\theta_0)$　$j=1,\cdots,M$是指向性函数的空间采样。我们通常可以用最大值法来求出最接近 θ 的那个波束，设它是 i（参看图1）于是，

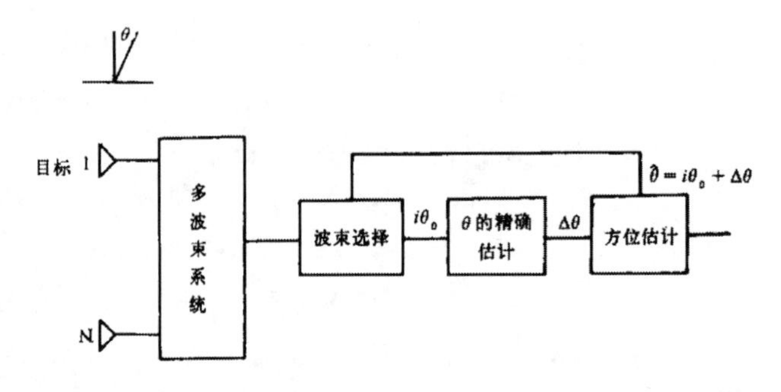

图1　数字式声呐精确测向系统方框图

$$i=\mathrm{CINT}(\theta/\theta_0) \tag{2}$$

其中 CINT 表示最接近的整数运算，例如 CINT(1.2)=1、CINT(2.8)=3。我们当然不能用$i\theta_0$来作为 θ 的估计值，因为$\theta-i\theta_0$的误差通常很大，不满足使用者的要求，我们必须利用每一时刻多波束的数据，给出 θ 的精确估计值:

$$\hat{\theta}=i\theta_0+\Delta\theta \tag{3}$$

其中 $\Delta\theta$ 的值可正可负。测向的精度，实际上也就是得出 $\Delta\theta$ 值的精确度，所谓目标的自动跟踪就是实时修正 i, $\Delta\theta$ 值，以使 θ 能随目标的运动而变化。

对 $\Delta\theta$ 的估计，要利用$i\theta_0$方向的分裂波束信号，把左右波束的延时差准确地估计出来就可以换算为 $\Delta\theta$ 的估计值。

设左波束信号为

$$X_1(t)=S(t)+n_1(t) \tag{4}$$

右波束信号为

$$X_2(t)=S(t+\tau)+n_2(t) \tag{5}$$

$n_1(t)$, $n_2(t)$假定为相互独立的随机噪声。

Knapp 指出，对延时 τ 的估计的最优下界为

$$\min[Var(\hat{\tau})]=\left\{2T\int_0^{\infty}(2\pi f)^2\frac{|r(f)|^2}{1-|r(f)|^2}df\right\}^{-1} \tag{6}$$

其中$r(f)$为相干谱

$$|r(f)|^2=K_{ss}^2(f)/[K_{ss}(f)+K_{nn}(f)]^2 \tag{7}$$

当我们假定$K_{ss}(f)$和$K_{nn}(f)$为矩形谱时，（6）式可以简化为

$$\min[Var(\hat{\tau})]=\left(\frac{3}{8\pi^2T}\right)\frac{1}{(\mathrm{SNR})_{\mathrm{in}}^2}\cdot\frac{1}{f_2^3-f_1^3} \tag{8}$$

其中 T 为积分时间，$(\mathrm{SNR})_{\mathrm{in}}$ 表示输入信噪比。f_1，f_2 分别是所使用频带的上、下限。

P.M. Schultesis 曾对线阵给出类似结果，他指出，对有 M 个基元的分裂波束线阵

$$\min[Var(\hat{\tau})]=\frac{1}{T}\left\{\int_0^{\infty}\frac{Mf^2K_{ss}^2(f)/K_{nn}^2(f)}{1+MK_{ss}^2(f)/K_{nn}^2(f)}df\right\}^{-1} \tag{9}$$

容易证明，当 $M=2$ 时（9）式与（8）式完全一样。

图 2 给出了积分时间 $T=1$ 秒和 10 秒时最佳延时估计误差与$(\mathrm{SNR})_{\mathrm{in}}$的关系。

时延估计精度 $\tau_{\min}$ 和方位估计精度 $\Delta\theta$ 的关系依赖于基阵的几何形状，对园弧阵来说[6]

$$\Delta\theta=\frac{M\pi\tau_{\min}}{2\frac{r}{c}\left[\sin\frac{M}{2}\alpha_0\right]^2N} \tag{10}$$

其中 r 为圆阵的半径，N 是基元个数，M 是工作基元的一半，C 为声速，$\alpha_0=360°/N$。

把（10）式和（8）式结合在一起就可以求出最佳定向精度，图 3 是 $N=64$，$M=14$ 的圆阵的一个例子，我们可以看出，对于 $r=1$ m 的基阵，当输入信噪$(\mathrm{SNR})_{\mathrm{in}}=-5$ dB 时，10 s 积分时间下，延时估计误差大约是 6 μs，相应的最佳定向精度为 0.4°。

三、目标入射角的估计

如上一节所述，在求$i\theta_0$之后，我们要估计 $\Delta\theta$ 值，我们首先证明$\Delta\theta_0$附近具有抛物线的形状，众所周知，对于有 N 个基元的等间隔离散线阵，其指向性的表达式是[6]

$$R(\varphi)=\sin(N\varphi/2)/N\sin(\varphi/2) \tag{11}$$

其中$\varphi=2\pi d[\sin\theta-\sin\beta]/\lambda$，$\theta$ 为目标信号的入射角，β 为波束指向角，d 为基元间的距离，λ 为波长，当 $\theta\approx\beta$ 时，（11）式具有 $\sin x/x$的形状，在 $X=0$ 处作 Taylor 展开，有

$$\sin x/x\approx 1-x^2/3 \tag{12}$$

换句话说，直线阵指向性函数在主瓣附近具有抛物线形状。

对于圆阵来说，如果基元数和圆阵直径信号波长满足一定的关系，其指向性函数可以表达为零阶贝塞尔函数:

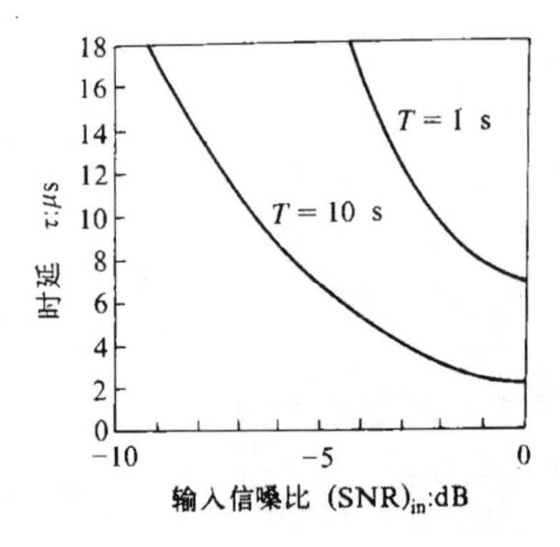

图 2 时延估计的罗-克拉美下界与$(SNR)_{in}$的关系

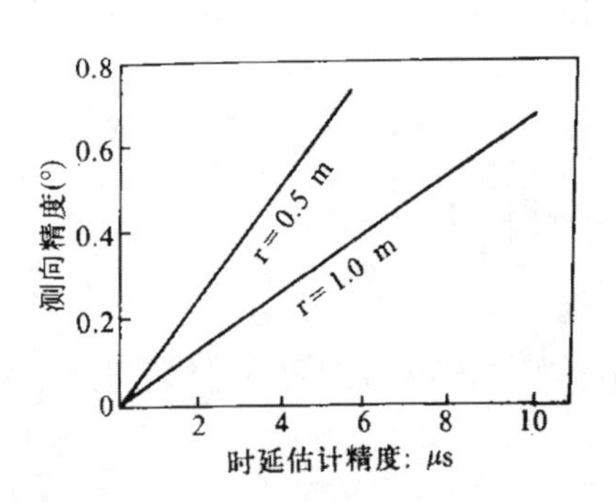

图 3 方位估计的最佳精度

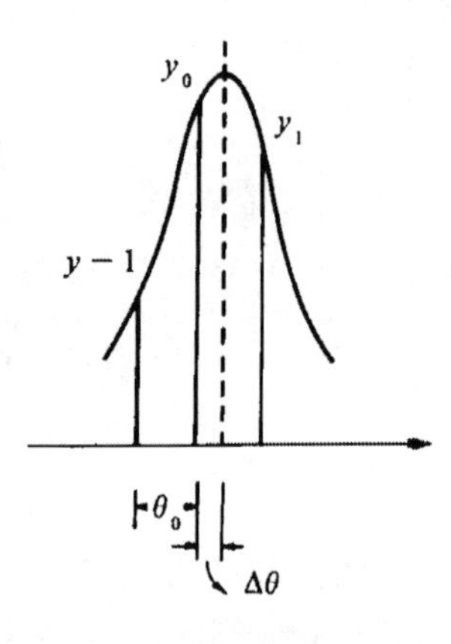

图 4 数字式声呐中求解DOA的内插方法

$$D(\theta)=\mathrm{J}_0\left(\frac{4\pi r}{\lambda}\sin\frac{\theta-\beta}{2}\right) \tag{13}$$

其中 r 为基阵半径，λ 为信号波长。在主瓣附近 $x\approx 0$

$$\mathrm{J}_0(x)\approx 1-\frac{x^2}{4} \tag{14}$$

综上所述，我们可以认为基阵的指向性函数在主瓣附近具有抛物线形状，一旦多波束系统确定$i\theta_0$是最接近 θ 的值，我们就不难精确给出 $\Delta\theta$ 的值来。

如图 4 所示，设$y_0=D(i\theta_0)$，$y_{-1}=D[(i-1)\theta_0]$，$y_1=D[(i+1)\theta_0]$。y_{-1}，y_0，y_1三点位于

$$y=ax^2+bx+c$$

的曲线上，间隔为θ_0，$\Delta\theta$ 的值应当使$D(i\theta_0+\Delta\theta)$达到极大（即抛物线的顶点）解一个三元二次联立方程组，不难给出

$$\hat{\theta}=i\theta_0+[(y_1-y_{-1})/(4y_0-2y_1-2y_{-1})]\theta_0 \tag{15}$$

当$y_1>y_{-1}$时，修正值 $\Delta\theta>0$，反之 $\Delta\theta<0$.当$y_1=y_{-1}$时 $\Delta\theta=0$，（15）式就是精确给出目标 DOA 的公式对于如（1）式所示的不同时刻的多波束数值，不断搜索其极大值，再反复运用（15）式我们便可以得到运动目标的角度轨迹，从而实现自动跟踪。

四、计算机模拟

我们在计算机上对本文提出的方法进行了模拟，基阵半径 $r=1$ m 积分时间 $T=10$ s，输入信噪$(SNR)_{in}=-5$ dB. 工作频段 0.6～1.0 kHz 这时理论上的最佳定向精度为 0.4°（均方根值）下面是一组实际的值:

系统模拟结果信噪比$(SNR)_{in}=-5$ dB

实际 DOA(°)=−15

15.0345	15.0298	14.7868	14.7654	14.8990
15.8102	14.3421	15.0123	14.8122	15.0330

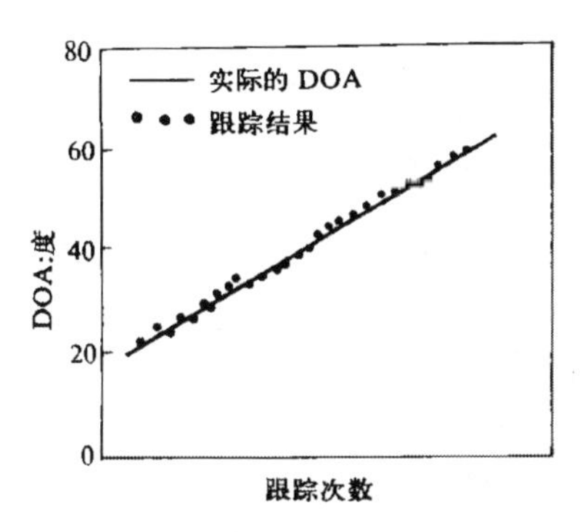

图 5　目标跟踪模拟结果

图 5 给出的是目标运动时的情况，起伏值也都是 ± 0.5° 之内，值得指出的是，由于积分时间较长，如果目标运动速度较快，那么测量值有一定的滞后，有必要时必须作一定的修正。

参 考 文 献

[1] Baggeroer A B. Sonar signal processing, in Application of digital signal processing, Ed. A.V. Oppenhein, Prentice−Hall,Englewood Cliffs, 1978.

[2] Lewis J B and P M Schultheiss. Optimum and conventionaldetection using a linear array, *J. Acoust. Soc. Amer.*, 1971, **49**: 1083—1091.

[3] Weinstein E. Optimal source localization and tracking, *IEEETrans.* 1982, ASSP−**30**(1): 69—76.

[4] Carter G C. Time delay estimation for passive sonar signal processing, *IEEE Trans.* 1981, ASSP−**29**: 463—470.

[5] Gray D A. Effect of time delay error on the bean pattern of a linear array, *IEEE J. OfOE*, 1985, OE−**10**: 206—212.

[6] LI Qihu. An Introduction to Sonar Signal Processiog (In Chinese),Ocean Publ., 1985, Beijing.

[7] Kapp C H. and Carter G C. The generaliyed correlation method for estimation of time delay, *IEEE Trans.* 1976, ASSP−**24**: 320—327.

[8] Schulteiss P M. Locating a passive source with array meairement−a summary result, *Proc.ICASSP'79: 967—970.*

数字式声纳多波束显示系统方位历程显示技术研究[①]

李启虎 杜 鹏 秦英达 赵文立

（中国科学院声学研究所 北京 100080）

1994年12月15日收到

摘要 数字式声纳的多波束显示是检测目标的重要手段。方位-历程显示是检测微弱信号的有效技术。本文给出了这种利用迹迹相关原理显示方法的理论说明，推导声纳多波束系统输出和输入信噪比的关系。把方位历程显示和传统的方位幅度显示作了比较。理论与实验结果表明，对于弱信号检测，高灰度级的方位历程显示与方位幅度显示比较，有5－7 dB的性能改善。

关键词 数字式声纳，方位历程显示，检测

Study of time/bearing display technique in digital multi-beam sonar display system

Li Qihu, Du Peng, Qing Yingda, Zhao Wenli

(*Institute of Acoustics, Academia Sinica, Beijing* 100080)

Abstract Multi-beam display is an important means of target detection in digital sonar. Time/bearing display is an effective technique for weak signal detection. The theoretical description of this technique, based on the trace-to-trace correlation effect, is presented in this paper. The relationship between the output of multi-beam system and the input signal-nosie ratio is drived. A comparison of the time/bearing display and the amplitude/bearing display is carried out and illustrated. The theoretical analysis and experimental result show that the high grade grey level time/bearing display has 5－7 dB improvement over the amplitude/bearing display.

Key words Digital Sonar, Bearing history display, Detection

1 问题的叙述

声纳信号处理系统的增益，最终是否可以得到，与终端显示系统关系重大。这是一个人机接口问题[1-6]。图1给出了多波束显示系统在整个声纳中的位置的框图。

通常，声纳员对是否有目标的判断来自两种渠道：一种是对声纳系统中单波束音频信号的听测；另一种是对多波束直流输出的视觉检测。这两种判决，无论是从信息来源，还是从判断依据都有很大差异。

听测的效果和声纳员的素质有很大的关系。同时，由于人类听觉对不同波束之间的记忆时间非常短，难于利用不同波束之间的信息

① 应用声学, 1996, 15(3): 1-6.

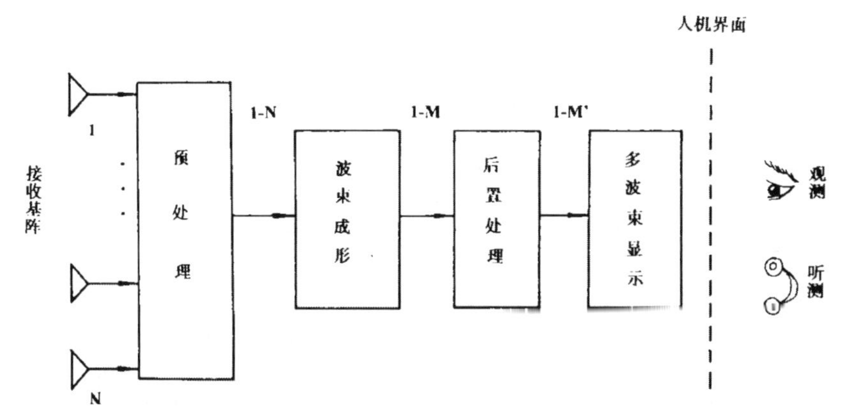

图1 多波束声纳的人机界面

相互比较。但是听测的一个明显优点是训练有素的声纳员可以凭自己的知识与经验对目标进行识别。

本文讨论的是视觉检测，即多波束数据随时间的变化。把这些数据转换成高分辨屏幕上的亮度显示。利用不同波束之间亮度的变化来检测目标。

假定数字式声纳有 M 个波束，在 t_i 时刻 M 个波束的值是：

$$x(t_i, 1), \cdots, x(t_i, M) \tag{1}$$

把 $x(t_i, j)$，$j=1, \cdots, M$ 以亮度形式显示在屏幕上，根据亮度的变化作比较，就可以判断是否有目标以及目标出现在第几号波束(相当于目标方位)。不同 t_i 值对应的 $x(t_i, j)$ 的排列就构成一幅方位历程图。

我们要研究的是方位历程显示与传统的方位幅度显示相比较，如何提高检测能力，以及它与系统输入信噪比的关系。

2 单波束系统的音频响应

考虑一个有 N 个通道的声纳系统，第 i 个通道所接收的信号可以表示为[7]

$$y_i(t) = s[t + \tau_i(\theta_s)] + n_i(t) \tag{2}$$

其中 $s(t)$ 表示目标信号，$n_i(t)$ 表示第 i 个接收元所收到的随机噪声，θ_s 表示目标入射角，$\tau_i(\theta_s)$ 表示第 i 个基元的相对延时。假定 $s(t)$ 和 $n_i(t)$ 相互独立，且 $n_i(t)$ 之间也相互独立。为计算简单起见，我们假定 $s(t)$，$n_i(t)$ 都是均值为零的高斯信号，它们的方差分别为 σ_s^2，σ_n^2。

系统中单个阵元的输入信噪比记作

$$(\mathrm{SNR})_{in} = \sigma_s^2/\sigma_n^2 \tag{3}$$

声纳员的听测信号来自声纳系统中对准目标的那个方向 θ_s 的波束的输出：

$$\begin{aligned} z(t) &= \frac{1}{N}\sum_{i=1}^{N} w_i y_i[t - \tau_i(\theta_s)] \\ &= \left(\frac{1}{N}\sum_{i=1}^{N} w_i\right) s(t) + \frac{1}{N}\sum_{i=1}^{N} w_i n_i[t - \tau_i(\theta_s)] \end{aligned} \tag{4}$$

其中 w_i 是用于改善指向性的一组加权系数。为简单起见，我们取 $w_i=1$，$i=1, \cdots, N$。

这个波束中有信号时的强度为

$$Z|_{s+n} = E[z^2(t)] = \sigma_s^2 + \frac{1}{N}\sigma_n^2$$

无信号时的强度为

$$Z|_n = \frac{1}{N}\sigma_n^2$$

所以输出信噪比为

$$(\mathrm{SNR})_{out} = N\sigma_s^2/\sigma_n^2 = N(\mathrm{SNR})_{in} \tag{5}$$

系统增益是

$$G_s = (\mathrm{SNR})_{out} : (\mathrm{SNR})_{in} = N \tag{6}$$

这是由于基元在空间的合理布放所得到的增益，我们称之为空间增益。

听测信号实际上是 $z(t)$ 经过一个滤波器后

的输出：

$$u(t) = \int_0^\infty h(\tau) z(t-\tau) \mathrm{d}\tau \tag{7}$$

其中 $h(t)$ 为滤波器的频响。人耳在听测过程中，除了获得(6)式所示的空间增益之外，也还有某种时间累积的效应。这种效应的确切数值和声纳员的素质有关。训练有素的声纳员不仅利用对准波束的强度信息，也还利用 $s(t)$ 中的谱特性、时变节奏特性等。

3　多波束系统的直流响应

多波束显示系统的输入数据不是某一单波束的音频信号，而是在空间均匀分波的多波束直流输出的信息。图 2 是方位幅度的多波束显示的示意图。类似于(4)式，定位于 θ 方向的波束输出是

$$\begin{aligned} z(t,\theta) &= \frac{1}{N}\sum_{i=1}^{N} y_i[t-\tau_i(\theta)] \\ &= \frac{1}{N}\sum_{i=1}^{N} s[t+\tau_i(\theta_s)-\tau_i(\theta)] \\ &\quad + \frac{1}{N}\sum_{i=1}^{N} n_i[t-\tau_i(\theta)] \end{aligned} \tag{8}$$

式中第二项是噪声起伏值。从理论上讲，如果 $n_i(t)$ 是严格平稳的过程，那么当对(8)式进行长时间平均，求其均方值时，它的起伏量与积分时间成反比[7]：

$$Var\left(\frac{1}{N}\sum_{i=1}^{N} n_i[t-\tau_i(\theta)]\right) \doteq \frac{2\sigma_n^2}{N\cdot T}\int_0^T \left(1-\frac{\tau}{T}\right) \cdot \rho_n^2(\tau)\mathrm{d}\tau \tag{9}$$

其中 $\rho_n(\tau)$ 是噪声的自相关函数。从(9)式可以看到，当(9)式中的 $T\to\infty$ 时，起伏量趋于零。这样，(8)式中对准信号的那个波束 $\theta=\theta_s$ 的直流输出的理想值是

$$\bar{z}(\theta_s)|_{s+n} = \left\{\sigma_s^2 + \frac{1}{N}\sigma_n^2\right\}^{1/2}$$

没有信号时输出的理想值为

$$\bar{z}(\theta_s)|_n = \left\{\frac{1}{N}\sigma_n^2\right\}^{1/2}$$

两者之差为对准目标时的直流跳变

$$\Delta D = \left\{\sigma_s^2 + \frac{1}{N}\sigma_n^2\right\}^{1/2} - \left\{\frac{1}{N}\sigma_n^2\right\}^{1/2} \tag{10}$$

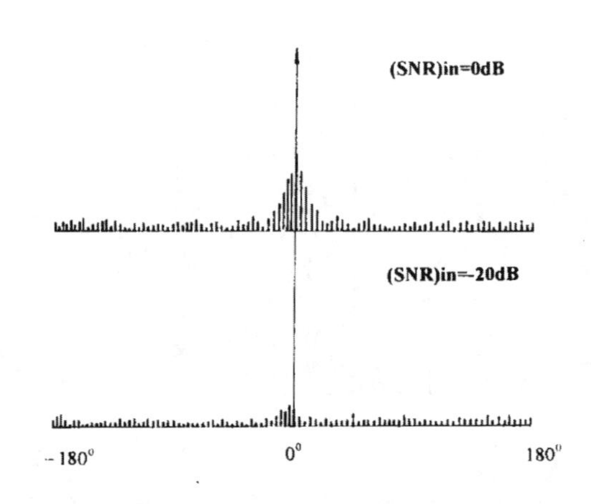

图 2　多波束声纳的方位幅度显示

由此式可以看出，在理想情况下，即使信噪比很小，我们仍可以检测出信号来，但是，实际情况并不如此。主要是，我们应在(10)中附加一项噪声的起伏量，它不仅依赖于积分时间，也还和通道之间的不均匀性有关，这种不均匀性就妨碍了我们对目标的检测。考虑到起伏时，θ 方向的输出直流跳变可以写成[7]

$$\Delta D(\theta) \doteq \sigma_s^2 D(\theta,\theta_s) + \sqrt{\frac{2\tau_0}{T}}\sigma_n^2 + \xi(\theta) \tag{11}$$

其中 $D(\theta,\theta_s)$ 为目标出现在 θ_s 面多波束指向为 θ 的指向性函数，

$$Max(D(\theta,\theta_s)) = D(\theta_s,\theta_s) = 1$$

τ_0 由(9)式给出，为噪声等效相关半径，$\xi(\theta)$ 是反映波束之间不均性的起伏噪声。从(11)式可以看出，妨碍信号检测的不仅有某一波束的输出起伏 $\sqrt{2\tau_0/T}\cdot\sigma_n^2$，并且还有起伏量 $\xi(\theta)$。当积分时间增长时，虽然波束的时间起伏值可以减少，但空间不均匀性都不随着减少。所以方位幅度显示的检测增益受到了制约。假设 $\xi(\theta)$ 在空间分布的起伏量为 $Var[\xi(\theta)]=\sigma_0^2$。那么，当积分时间 T 的值，长到使得 $\sqrt{2\tau_0/T}\cdot\sigma_n^2$ 达到 σ_0^2 之后，再增加 T 的值，对检测弱信号便没有意义了。

4　方位历程显示和迹迹相关技术

近年来，数字式声纳广泛使用方位历程显

示。CORDAR（*CORrelation Display Analysis and Ranging*），LOFAR（*LOw Frequency Analysis and Record*），DEMON（*DEModulation On Noise*）等就是典型的例子。

对于不同时刻 t_j 的一组多波束数据 $u_i(t_j)$，幅度上予以量化之后在屏幕上加以排列，横坐标为方位，纵坐标是时间，$u_i(t_j)$ 值的大小以亮度调制显示在一个或几个象素上。对每一 t_j 来说，当 $(SNR)_{in}$ 很小时，$u_0(t_j),\cdots,u_{M-1}(t_j)$ 之间可能挑不出明显亮的点。但是，当连续观察不同时刻的一幅度时，我们就会发现在形如瀑布的方位历程图中，会出现某些断续的亮条。这正是目标出现的情况。这种效果就是迹迹相关效应。

我们首先来看一下，从理论上来说，方位历程显示比多波束的方位幅度显示能带来多大的增益。这个附加的增益是和 $\xi(\theta)$ 的分布有关系的。而 $\xi(\theta)$ 的分布和各通道之间的相移、放大倍数的不均匀性等因素有关。

定义附加增益

$$Ga = Max\xi(\theta)/\sqrt{Var(\xi(\theta))} \tag{12}$$

如果 $\xi(\theta)$ 遵从高斯分布，取 $Max\xi(\theta)=4\cdot\sigma$，其中 σ 为 $\xi(\theta)$ 的方差。于是

$$Ga = 4 \sim 6\ \text{dB}$$

如果 $\xi(\theta)$ 遵从均匀分布，我们有

$$Ga = 1/\sqrt{1/12} = \sqrt{12} \sim 5.3\ \text{dB}$$

总而言之，附加增益在 5 dB 以上。

举例来说，如果空间波束的不均匀性在 0.01量级。那么在积分时间长到一定程度，即单波束起伏值低于 0.01 时，增加积分时间已无意义。这时如果某一目标能引起直流跳变的量级为 0.02，那么在方位幅度的一次随机观测中难于确定何处出现目标。但是在方位历程显示中，在目标方向持续出现的 0.02 的直流跳变完全足以抵消空间不均性的影响。因为空间不均匀性表现为一种随机的、时隐时现的闪烁。

5 方位历程显示的计算机模拟

根据以上的讨论，我们在计算机上进行了系统模拟。以 $N=32$ 基元的数字式声纳为基本参数，波束数 $M=120$，灰度级为 64。一名主试人员负责改变信噪比和改变目标方位。受试人员有 6 人，分别观测两种显示，一种是方位幅度显示，一种是方位历程显示。他们并不知道目标何时出现以及出现的方位。然后由他们作出各自独立的判断。6 位受试者的平均检测概率 Pd 和虚警概率 α 见表 1 及表 2。试验样本数为 100。图 3 至图 11 给出了从 $(\text{SNR})_{in}=$

表 1 多波束方位幅度显示

$(\text{SNR})_{in}$(dB)	α	Pd
−21	0.040	0.857
−22	0.043	0.590
−23	0.053	0.504
−24	0.073	0.343
−25	0.100	0.130

表 2 多波束方位历程显示

$(\text{SNR})_{in}$(dB)	α	Pd
−27	0	1.000
−28	0	1.000
−29	0	0.997
−30	0	0.993
−31	0	0

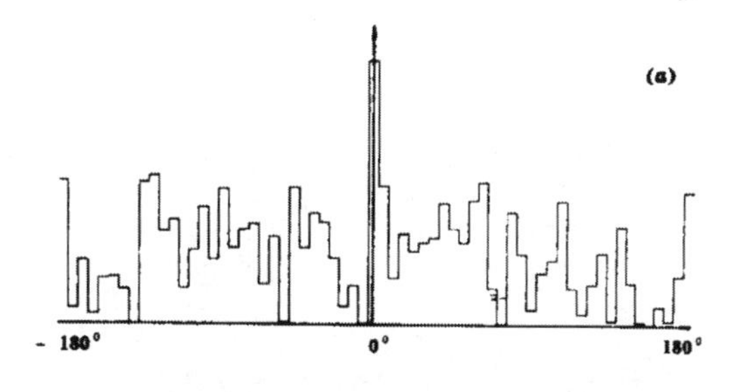

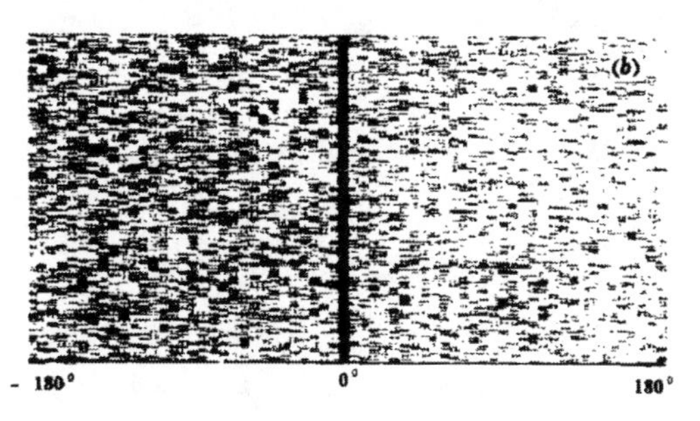

图 3 多波束显示 $(\text{SNR})_{in}=-21.5$ dB（目标在 0°）

(a) 方位-幅度 (b) 方位-历程

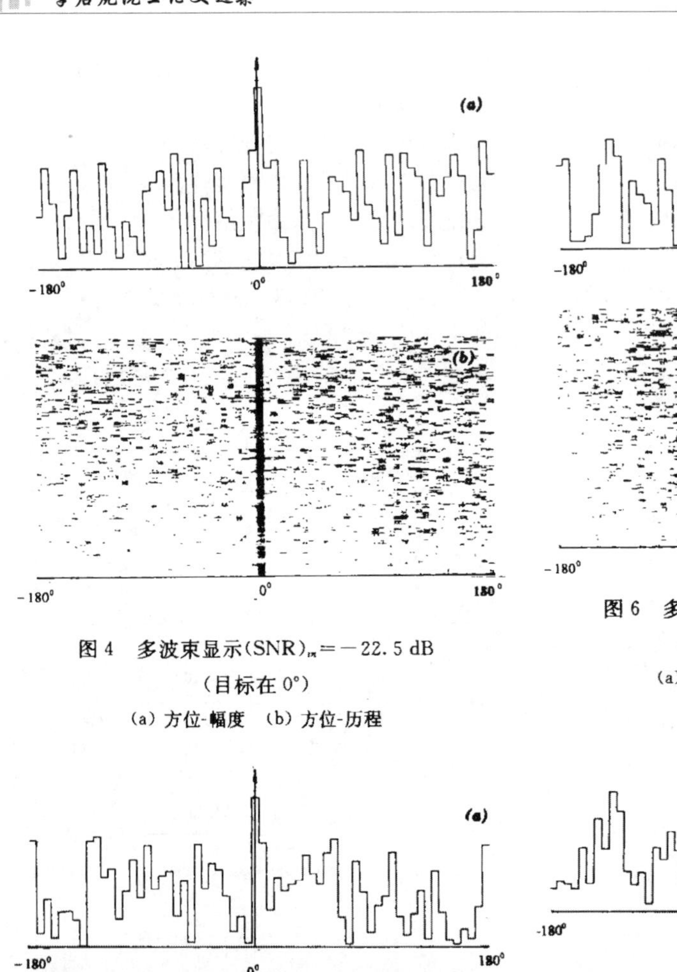

图 4　多波束显示$(SNR)_{in}=-22.5$ dB

（目标在 0°）

(a) 方位-幅度　(b) 方位-历程

图 5　多波束显示$(SNR)_{in}=-24$ dB

（目标在 0°）

(a) 方位-幅度　(b) 方位-历程

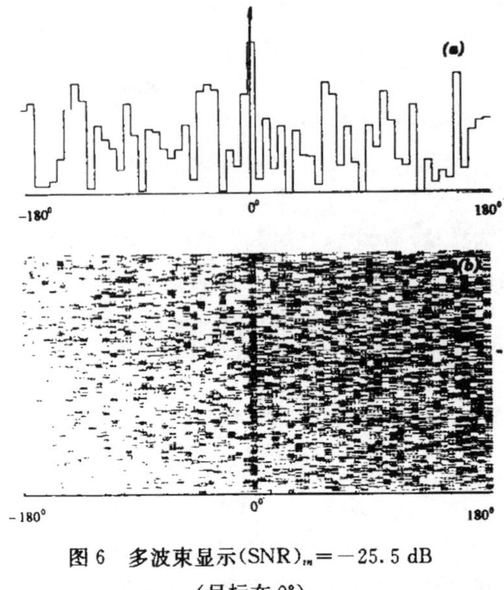

图 6　多波束显示$(SNR)_{in}=-25.5$ dB

（目标在 0°）

(a) 方位-幅度　(b) 方位-历程

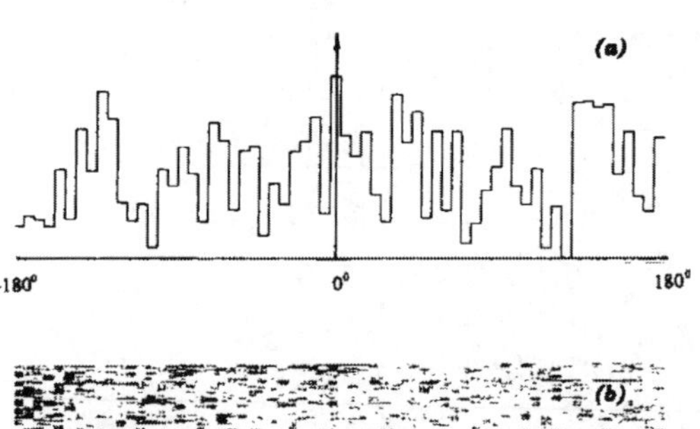

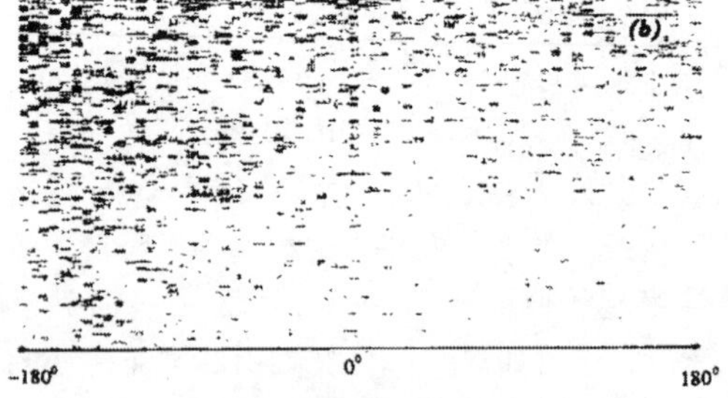

图 7　多波束显示$(SNR)_{in}=-27$ dB

（目标在 0°）

(a) 方位-幅度　(b) 方位-历程

-21.5 dB 至$(SNR)_{in}=-\infty$（即全噪声）的实验结果。我们把同一信噪比下的方位幅度显示和方位历程显示画在一幅图上。从表 1、表 2 及图 3—11 可以看出，方位历程显示对弱信号的检测大约比方位幅度显示有 5—7 dB 的增益。同时，方位历程显示的另一个明显好处是由于利用迹迹相关效应，虚警概率 α 非常低，

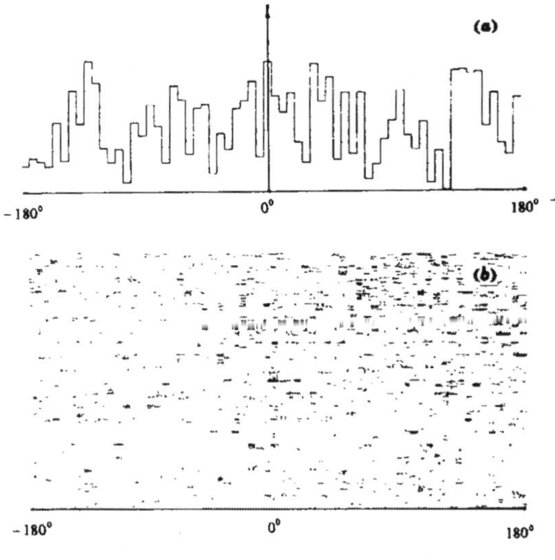

图 8 多波束显示$(\mathrm{SNR})_{in}=-29$ dB

(目标在 0°)

(a) 方位-幅度 (b) 方位-历程

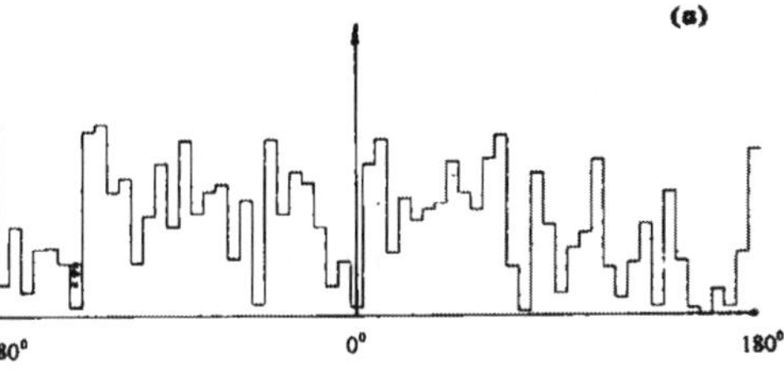

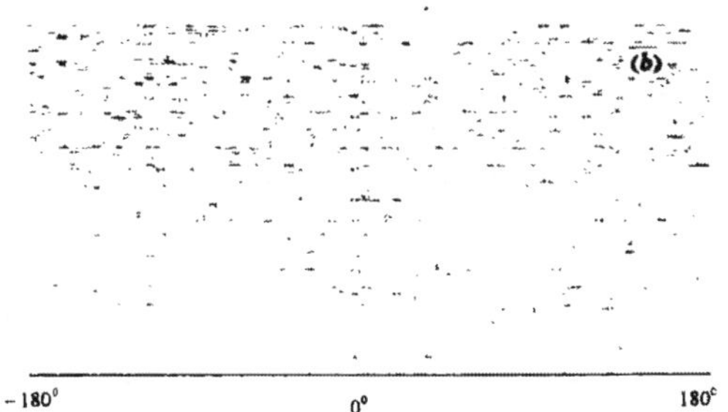

图 9 多波束显示$(\mathrm{SNR})_{in}=-31$ dB

(目标在 0°)

(a) 方位-幅度 (b) 方位-历程

而检测概率又比较高。

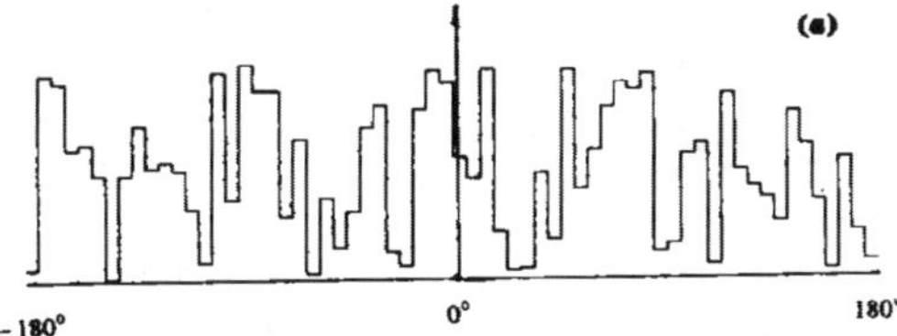

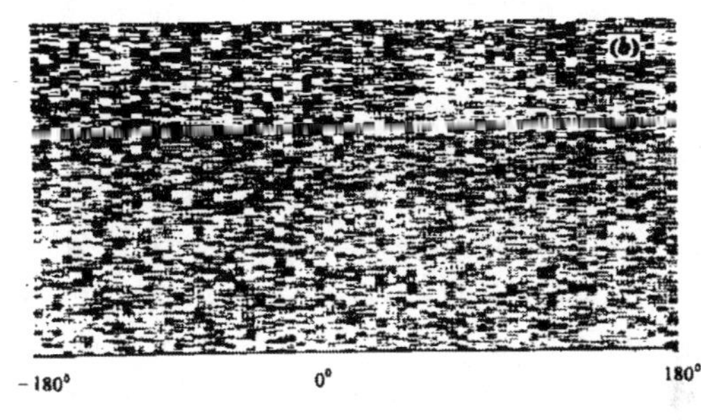

图 10 多波束显示，全噪声

(a) 方位-幅度 (b) 方位-历程

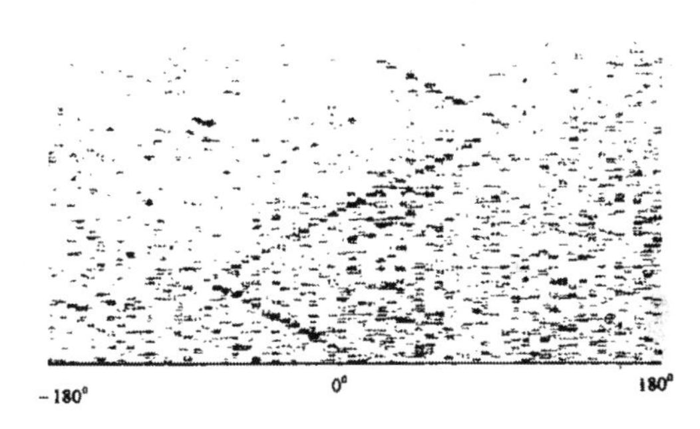

图 11 多波束显示，动目标

$(\mathrm{SNR})_{in}=-27$ dB

参 考 文 献

[1] Winder A A. *IEEE Trans.* 1975 SV-**22** (5): 291－332.

[2] 李启虎等. 声学学报 1989 **16**(5): 365－370.

[3] Eral S K. AD-A 129496 May 1983.

[4] Theodore J. *Human Facturs* 1989 **31**(5): 539－550.

[5] 刘秋实等. 微机应用 1991 **12**(5): 36－39.

[6] 秦英达等. "高性能并行处理数字式声纳显控台." 全国水声学术论文集., 1993.

[7] 李启虎. 声纳信号处理引论. 海洋出版社, 1985.

[8] Rasmussen R A. AD 690540 1969.

[9] Salafia W. AD-A 197283, 1988.

[10] Neri F. NVSC Tech. Rept. 7325, March, 1985.

[11] Barry W. Proc. ICASSP 1976 383－385.

数字式声呐中的一种简化的 ZoomFFT 算法[①]

李启虎　李伟昌　赵文立

(中科院声学所　北京　100080)

1998 年 5 月 11 日收到

1998 年 7 月 25 日定稿

摘要　在水声信号处理中，DEMON 和 LOFAR 已被证明为有效的方法，特别是对微弱信号的检测和目标的识别和分类．有的时候，我们还需要知道接收信号频谱的细微结构．一般说来，只有长的时间数据才有可能得到高的频率分辨力，但是由于实际系统软、硬件方面的限制，这样作并不总是可能的．如果我们只是对某些频率附近的谱结构感兴趣，那么 ZoomFFT 就是一种解决高分辨率谱分析的折中方法．已有的讨论 ZoomFFT 的文献大体可以分为两大类，即复包络解调 ZoomFFT(Complex modulation)[4] 和级连 FFT(cascade FFT)[5,6]．前者需要对输入信号进行复解调、低通滤波、降采样等一系列繁复的操作．而后者通常利用前后两次 FFT，经过相位和幅度修正得到所需频段的细化谱估计，因而易于实现，可作为一种有效的窄带处理器．本文在给出级连 FFT 法 ZoomFFT 理论推导的基础上，试图探讨其与复包络解调法之间的内在对应关系，并分析了窗函数、采样率、重叠率等参数的选取对估计结果的影响，最后给出一种简化的 ZoomFFT 算法，它可以大大缩短实时数据的运算次数．并给出了系统模拟的结果．

PACS 数：

A simplified ZoomFFT algorithm in digital sonar

LI Qihu　LI Weihang　ZHAO WenLi

(*Institute of Acoustics, The Chinese Academy of Sciences* Beijing 100080)

Received May. 11, 1998

Revised Jul. 25, 1998

Abstract　DEMON and LOFAR have been proved the powerful mean in underwater acoustic signal processing, especially in weak signal detection and target noise classification. Sometimes one need to know the fine structure of frequency spectrum of received signal. It is necessary to take a very long data to get high frequency resolution. This is not always possible due to the hardware and software limitation. ZoomFFT is one of the trade–off consideration for solving high frequency resolution problem, if we are only focus on some special frequency bins. Previous discussions mainly bifurcate into two different representations, the Complex Modulation[4] and Cascade FFT[5,6]. The former one traditionally needs some kind of special treatment, e.g. the complex modulation, Lowpass filtering, down–sampling. While the latter achieves the same result by two successive FFT, with necessary modifications in phase and amplitude, thus is feasible for real–time implementation. Based on some theoretical analysis, a relationship between the Complex Modulation and Cascade FFT has been described in this paper. In addition, the selection of parameters such as sample rate, overlap factor has been discussed. Finally, the algorithm is presented and some simulation results are illustrated.

引言

水声信号的谱分析已经广泛应用于声呐信号处理的各个领域，包括噪声背景下微弱信号的检测，水下噪声目标的分类和识别，参数估计等[1-3]．传统的 FFT 算法可以适应大多数情况下的要求．但在某些特殊的应用场合，需要得到频率域的高分辨率，

① 声学学报, 2000, 25(2): 129-133.

例如某些特殊频段中信号谱的细微结构。根据 FFT 分析的基本结果，对长度为 N 的样本，采样周期为 T_s，其频率分辨力为：

$$\Delta f = \frac{1}{NT_s}. \tag{1}$$

因而，要求高的频率分辨力，即 Δf 很小，那么就必须增长数据长度 (采样频率一定)，这在实际中并不总是可行的，而且实际上，我们往往只需要观察某些特殊频率附近的谱结构，没有必要在整个频率范围内都获得高分辨力，这时我们可以采用 ZoomFFT 技术，它可以提供所感兴趣频段附近信号的精细谱结构。

有关 ZoomFFT 相关的讨论并不多见，有一些早期的文献，如 E.A.Hoyer 等提出了一种复包络解调的 ZoomFFT 算法[4]。通过对接收信号进行复解调、低通滤波，把信号频谱移至基带，再降采样进行谱细化输出。这种方法物理解释清晰，但是运算复杂，不利于直接实现。 P.C.Y.Yip 较系统地给出了 ZoomFFT 的理论表示[5]，分析了忽略旋转因子带来的影响，并以正弦信号为例进行幅度修正。P.O.Fjell 等人详细描述了级连 FFT 的实现结构[6]，并进一步讨论了重叠率与混迭现象的关系，提出一种利用相邻频段输出进行加权的修正 ZoomFFT。比较而言，级连 FFT 更易于在 DSP 等器件上进行实时处理，而复包络解调则较清晰地揭示了 ZoomFFT 的物理意义。

本文首先给出级连 FFT 的理论表示，进而分析其与复解调之间的对应关系，并由此提供参数选择的依据，最后给出不同参数条件下的仿真结果。

1 ZoomFFT 算法的理论基础

假定样本数据是 $x(n+mN)$, $n=0,\cdots,N-1$; $m=0,\cdots,M-1$。其中数据序列被划分为 M 个区组，互不重叠。每一个区组有 N 个样本点。$N\times M$ 点 FFT 的频率分辨力是：

$$\Delta f = \frac{1}{NMT_s}, \tag{2}$$

运算量为[7]:

$$NM\log(NM)/2. \tag{3}$$

对 $x(n+mN)$ 的离散 Fourier 变换 (DFT) 可以表示为[5]:

$$X(l+kM) = \sum_{n=0}^{N-1}\sum_{m=0}^{M-1} x(n+mN)\exp[-2\pi \mathrm{j}(n+mN)(l+kM)/(MN)], \tag{4}$$

其中，$k=0,\cdots,N-1$; $l=0,\cdots,M-1$。

我们可以把 $X(l+kM)$ 表示为：

$$\begin{aligned} X(l+kM) = &\sum_{m=0}^{M-1}\left[\sum_{n=0}^{N-1} x(n+mN)\exp(-2\pi \mathrm{j} nk/N)\right]\exp(-2\pi \mathrm{j} ml/M)\exp(-2\pi \mathrm{j} nl/MN)\exp(-2\pi \mathrm{j} mk) \approx \\ &\sum_{m=0}^{M-1}\left[\sum_{n=0}^{N-1} x(n+mN)\exp(-2\pi \mathrm{j} nk/N)\right]\exp(-2\pi \mathrm{j} ml/M) = \\ &\sum_{m=0}^{M-1} X_m(k)\exp(-2\pi \mathrm{j} ml/M), \end{aligned} \tag{5}$$

这里 $X_m(k)$ 表示对变量 n 的 DFT，即

$$X_m(k) = \sum_{n=0}^{N-1} x(n+mN)\exp(-2\pi \mathrm{j} nk/N). \tag{6}$$

由式 (5) 我们可以看出，$N\times M$ 点的 DFT 可以简化成两次 DFT，第一次 DFT 是对 M 个区组作 N 点 DFT，而第二次 DFT 是对所关心的谱线 k 作 M 点 DFT。总计算量为

$$(MN\mathrm{logN} + M\mathrm{logM})/2, \tag{7}$$

比 (3) 式所示计算量减少 $(N-1)M\log M/2$。

(5) 式的近似表达，只有当 $\exp(-2\pi jnl/MN)$ 近似为 1 时才成立。这并不是对所有 n 和 l 都成立，但是当 $nl/MN \ll 1$ 时，可认为近似式成立。

2 级连 FFT 与复包络解调法的关系

我们分析 (6) 式

$$X_m(k)=\sum_{n=0}^{N-1}x(n+mN)\exp(-2\pi\,\mathrm{j}\,nk/N)=$$
$$\sum_{n=0}^{N-1}x(n+mN)\exp[-2\pi\,\mathrm{j}\,(n+mN)k/N]h(n),\tag{8}$$

其中，

$$h(n)=\begin{cases}1 & n=0,1,\ldots,N-1\\ 0 & \text{else}\end{cases},\tag{9}$$

$h(n)$ 是带宽为 $1/NT_s$ 的线性相位低通滤波器，如果采用其它窗函数形式的线性相位滤波器，可使滤波器的响应更为理想。

式 (8) 包含了三个过程：混频、降采样和低通滤波。可分别写成式 (10)、(11) 和 (12):

$$g_k(n)=x(n)\exp(-2\pi\,\mathrm{j}\,nk/N),\tag{10}$$

$$y_k(l)=g_k(n)\otimes h(n)=\sum_{n=0}^{N-1}g_k(n+l)h(n),\tag{11}$$

上式中，要求 $h(n)$ 为线性相位，$h(N-1-n)=h(n)$，$n=0,1,\cdots,N-1$。

$$X_m(k)=y_k(l)|_{l=mN}.\tag{12}$$

图 1(a) ~ 图 1(d) 分别给出了各个阶段信号谱的示意图。

因而对于 $k=k_c$(k_c 为中心频率对应的谱线)，$y_k(l)$ 可视为接收信号的复包络。以采样率 $1/NT_s$ 对 m 采样得到 $X_m(k)$，也就是复包络的降采样结果。因而式 (5) 就是该降采样后的包络的谱。

由此可见，式 (8) 所描述的级连 FFT 过程与复包络解调 ZoomFFT 在本质上是相同的，具有相互对应关系。

3 参数的选取

$X_m(k)$ 的频谱与 $h(n)$ 有关，基带低通滤波器 $h(n)$ 的带宽为 $0\sim 1/(NT_s)$，当 $h(n)$ 的频响特性较差时，会给计算结果带来不利的影响。带内起伏会造成细化谱估计的失真；而带外衰减缓慢和旁瓣过高会引起邻近频段的频谱渗漏。后者的影响尤为明显，主要有两个方面：(1) 降采样的效果使滤波之后的信号频谱在频率轴 (0 ~ Fs) 上周期延拓，重复周期为 Fs/K，K 是降采样倍数，因此，这种频谱渗漏经延拓后造成混叠，影响基带频谱 Zoom 输出，见图 1(b) ~ 图 1(d)。(2) 当以 $k\neq k_c$ 进行复解调时，对于 $|k-k_c|$ 较小的情况，因其解调输出靠近基带而难以滤除，再经周期延拓至基带，从而在 ZoomFFT 输出产生谱峰，给估计带来不确定性，再经周期延拓至基带，从而在 Zoom 输出产生谱峰，给估计带来不确定性，见图 1(e) ~ 图 1(g)。

为避免出现伪峰并提高估计的准确性，可以采用加窗改善滤波器性能和提高降采样频率的办法。对于带宽为 $1/(NT_s)$ 的数字低通滤波器欲得到理想的频响比较困难，采用合适的窗函数可得以适当的改善，见仿真结果。

为减小混叠的影响，可以采用输入信号前后区组互相重叠，即提高降采样频率的办法。对输入信号作下列分组：

$$x(n+mN/P)\qquad n=0,\cdots,N-1;\quad m=0,\cdots,M-1.,\tag{13}$$

其中 N/P 为每次移动的点数。这样重叠率为 $1-1/P$。当 $P=1$ 时，即为式 (5)。式 (13) 所示分组信号的复包络：

$$\begin{aligned}X_{mP}(k)&=\sum_{n=0}^{N-1}x(n+mN/P)\exp[-2\pi\,\mathrm{j}\,(n+mN/P)k/N]=\\&\left[\sum_{n=0}^{N-1}x(n+mN/P)\exp(-2\pi\,\mathrm{j}\,nk/N)\right]\exp(-2\pi\,\mathrm{j}\,mk/P)=\\&X'_{mP}(k)\exp(-2\pi\,\mathrm{j}\,mk/P),\end{aligned}\tag{14}$$

$X'_{mP}(k)$ 是 $x(n+mN/P)$ 的 DFT。

$X_{mP}(k)$ 比 $X_m(k)$ 多了一项 $\exp(-2\pi\,\mathrm{j}mk/P)$，称之为旋转因子。

$X_{mP}(k)$ 基于 m 的采样频为：

$$f_s=\frac{Pf_s}{N},\tag{15}$$

这样 $X_{mP}(k)$ 在频率轴上以 $f_s=Pf_s/N$ 周期延拓，见图 1(h)，减小了混叠的影响。

另外，提高降采样频率还可以一定程度上抑制伪峰的出现。由图 1(f) ~ 图 1(h) 可以看出，仅对满足 $|k-k_c|=lP$ 的 k，$l=0,1,2,\cdots$ 时，才会由于周

期延拓而在基带出现谱峰. 适当提高 P 值, 使无伪峰的 $|k-k_c|$ 范围变大, 这样其解调输出便可被低通滤除, 对滤波器的要求也大为降低.

由 (14) 式知, ZoomFFT 的输出谱在每个频带内均被压缩, 压缩比为 P. 对于 $X(l+kM)$ 而言, $l=0,\cdots,(M-1)/P$ 代表了信号在 k 频带的细化谱, 其频率分辨率为:

$$\Delta f_P=\frac{P}{NMT_s}, \tag{16}$$

$\Delta f_P=\Delta f\times P$. 这是由于实际信号处理长度缩短引起的. (如果要保持原有频率分辨力, 则需增加第二次 DFT 的长度至 PM 点). 此外, 由于 $X_{mP}(k)$ 已经经过低通滤波其带宽为 $1/NT_s$, 所以 $l=(M-1)/P+1,\cdots,M-1$ 点输出并不表示邻近频段的谱结构, 应当舍弃.

可见, 提高 P 值可以降低混迭的影响, 抑制伪峰, 同时未使频率分辨力受损, 或增加计算量. 因而宜结合所采用的窗函数选择合适的 P 值, 以获得准确的细化谱估计.

4 ZoomFFT 实现与仿真结果

ZoomFFT 算法流程见图 2. 仿真计算是对 300 Hz(6 dB) 和 299 Hz(0 dB) 频率分量的叠加信号进行的, 按 5 kHz 采样, 选取不同的 P 值和窗函数, 进行 ZoomFFT 计算, 并与 N 点 FFT 相比较. 在第一次 FFT 中对应的 $k_c=61$.

图 3 是 $P=1$, $k=61$, 加 Hamming 窗的结果. 其频率分辨率为 0.038 Hz.

图 4 是 $P=2$, $k=61$, 加 Hamming 窗的结果. 其频率分辨率为 0.076 Hz.

图 5 是信号 1024 点 FFT 计算结果, 由于其频率分辨力为 4.88 Hz, 难以分辨两个频率分量.

图 6 是 $P=2$, $k=65$, 加矩形窗的计算结果. 可以看到有伪峰出现.

图 7 是 $P=2$, $k=65$, 加 Hamming 窗的计算结果. 可以看到伪峰已被滤除.

图 8 是 $P=2$, $k=62$, 加矩形窗的计算结果. 由于不满足 $|k-k_c|=lP$, 伪峰不出现. 但由于滤波器带外特性不如 Hamming 窗, 输出幅度仍较大.

采用 Hamming 窗, 当 $P=1$ 时, 对于 $|k-k_c|>4$, 伪峰均可被滤除. 当 $P=4$ 时, 可消除任意 $k\neq k_c$ 时出现的伪峰.

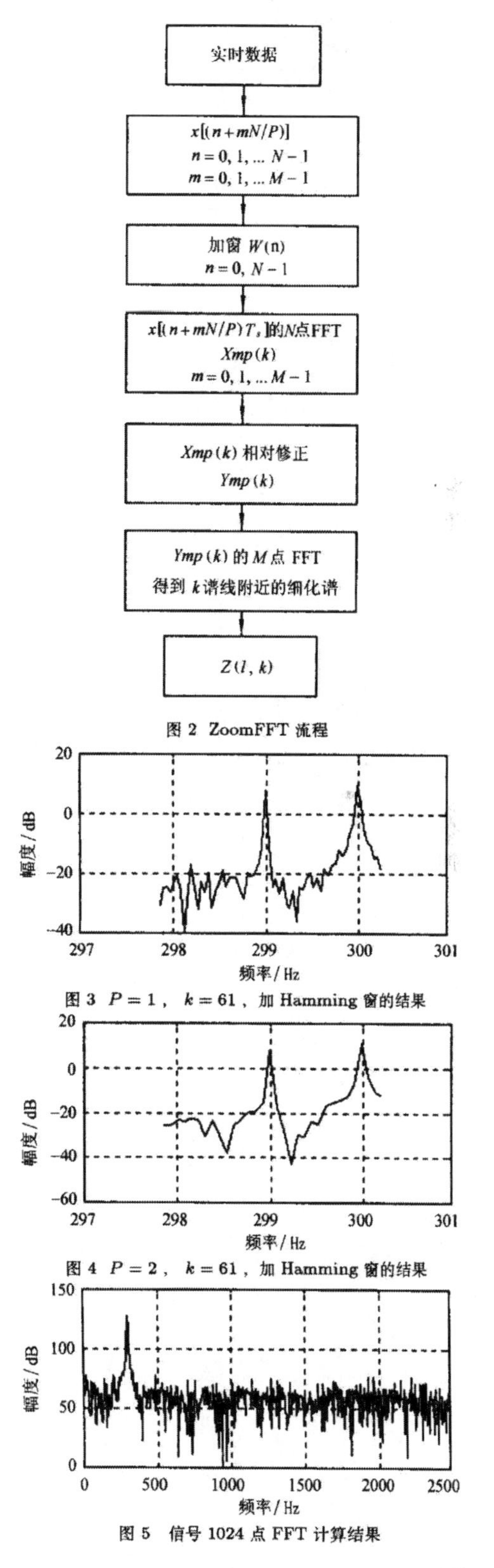

图 2 ZoomFFT 流程

图 3 $P=1$, $k=61$, 加 Hamming 窗的结果

图 4 $P=2$, $k=61$, 加 Hamming 窗的结果

图 5 信号 1024 点 FFT 计算结果

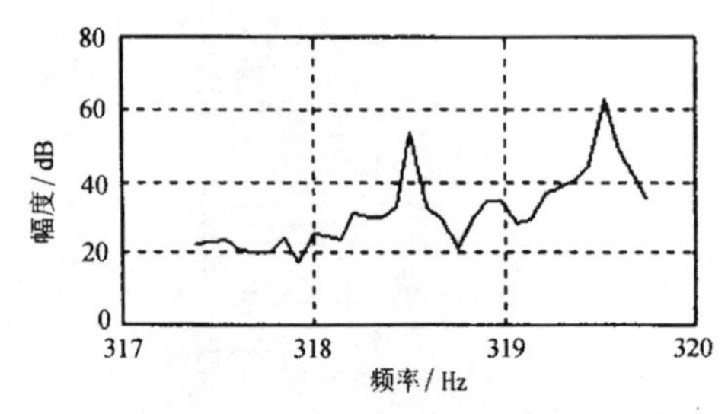

图 6 $P=2$， $k=65$，加矩形窗的计算结果

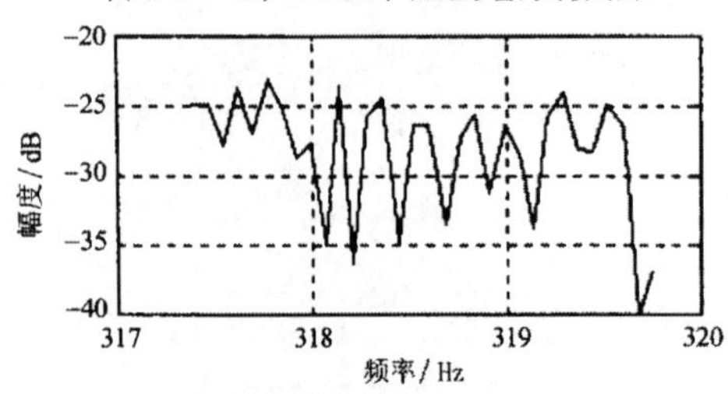

图 7 $P=2$， $k=65$，加 Hamming 窗的计算结果

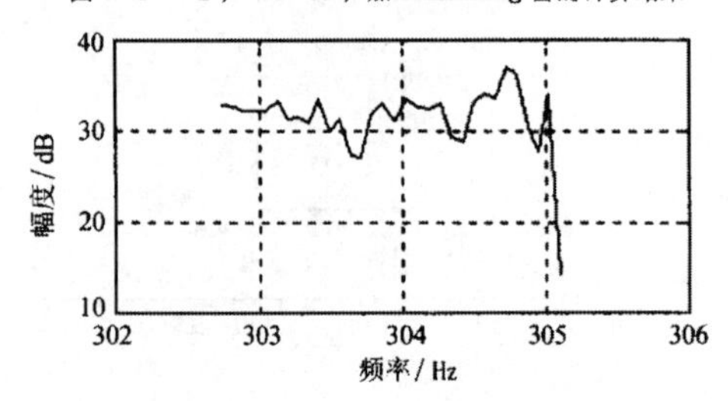

图 8 $P=2$， $k=62$，加矩形窗的计算结果

5 结论

ZoomFFT 对于获取某些特殊频段，而不是整个带宽的信号细微谱结构是十分有用的。例如水下辐射噪声的线谱检测，语音识别等。在这些情况下，通过选择合适的窗函数和重叠率，本文提出的 ZoomFFT 算法是有效的，本质上与复包络正交解调相一致，并且具有较好的准确度和可靠性。

参 考 文 献

1 Haykin S Ed. Advance in spectrum analysis and array processing. Prentice Hall, 1991

2 Trider R C. A fast fourier tansformbased sonar signal processor. *IEEE Trans*, 1978; **ASSP-26**: 15—20

3 Gabriel W F. Spectral analysis and adaptive array superresolution techniques. 1980; *Proc. IEEE*, **68**: 854—866

4 Hoyer E A *et al.* The ZoomFFT using complex modulation. Proc. of ICASP'1977, 1977: 78—81

5 Yip P C Y. Some aspects of the Zoom transform. *IEEE Trans. On Computers*, 1976; **C–25**(3): 287—296

6 Fjell P O *et al.* A modified cascade fast Fourier transform in a spectrum analysis system. Proc. Of ICASSP, 1988: 873—876

7 Brighan E O. The fast Fourier transform. Prentice–Hall, 1974

数字式声呐中一种新的背景均衡算法①

李启虎 潘学宝 尹力

(中科院声学所 北京 100080)

1998 年 7 月 22 日收到

1998 年 9 月 9 日定稿

摘要 在频谱分析领域，背景均衡技术已被广泛地研究过. 已经证明，这种方法对改善线谱检测是有效的. 在数字多波束声呐系统中，方位历程显示是信号处理模块和显控模块之间最重要的界面. 已经证明，声呐处理系统的增益最容易在不同的信号处理模块接口处丢失. 背景均衡技术可以改善方位历程显示的总体性能，将多波束数据的后置处理结果在送至显示器之前首先进行滤波. 本文提出的中值滤波和排序截断平均 (Order Truncate Average, OTA) 相结合的方法利用可变长的窗口匹配不同主瓣宽度的波束图是一种对不均匀、非平稳背景的有效均衡技术. 系统模拟结果表明，此方法对于多波束非均匀背景的均衡是非常有效的.

PACS 数: 43.30, 43.60

A new algorithm of background equalization

LI Qihu PAN Xuebao YIN Li

(*Institute of Acoustics, The Chinese Academy of Sciences* Beijing 100080)

Received Jul. 22, 1998

Revised Sep. 9, 1998

Abstract Background equalization technique has been extensively studied in the field of frequency spectrum analysis. It has been proved an effective method of improving line spectrum extraction. In a multi-beam digital sonar system, the time/bearing display is the most important interface between signal processing module and console module. It has been proved that the system gain is often lost in the interface between these two modules. Background equalization technique can improve the overall performance of the time/bearing display window . The results of post-processing of multi-beam data is filtered each time before displayed on screen . It is shown that the algorithm of median value filtering combined with OTA(Order Truncate Average) method is an efficient technique to equalize the non-uniform , non-stationary background data. A variable length of window is adopted to match the different width of main lobe of beam patterns. The results of system simulation show that the algorithm described in this paper is efficient in the case that the background of multi-beam data is not uniform.

引言

在某些信号处理应用中，后置处理的背景均衡是非常必要的。这些领域包括信号的频域谱分析、噪声背景中弱信号的检测、方向性数据的空间滤波等。背景均衡早期工作局限于频域分析 [1,2] 。背景均衡也可以改善方位历程显示的图像质量，例如 LOFAR, DEMON 图像，从而可以增强检测能力或改善分辨力. 当我们把随机起伏过滤掉时，在频率分辨单元的目标轨迹会变得更加清楚。 W.A.Struzinski 讨论了用于背景均衡的四种算法。主要用于频域。

但是对于方位历程显示，所遇到的问题稍有不同。由于海洋环境在时间上的非平稳性和在空间上

① 声学学报, 2000, 25(1): 5-9.

的非均匀性，噪声背景在时间和空间上便有某种特殊的性质，这种特性直接影响了对弱信号的检测。本文提出一种新的背景均衡算法，它把 OTA 算法和利用迹迹相关技术相结合，在每一时刻，我们把多波束的后置处理结果先行滤波，然后采用一种可变长度的窗口，它可以适应不同宽度主瓣的波束图。系统模拟的结果指出，本文所给出的算法对于非均匀多波束背景的均衡是有效的，特别是对拖曳阵声呐，由于它存在拖曳平台的干扰，其波束图中经常存在的是一个相当宽的干扰区。

1 二维方位历程数据的背景均衡算法

我们考虑多波束声呐系统的方位历程显示。在数字式声呐的显控台上，屏幕上出现的显示是瀑布式的，它实际上是多波束检测过程中的一个历史记录。声呐设计应关心的是如何提取目标的轨迹和进行特征分析。本文提出的背景均衡算法的方框图见图 1 .

均衡的过程是一个逐行运算。假定多波束数据有 N 列，表示为：

$$X(1), \cdots, X(N). \tag{1}$$

第一步，首先考虑每一行的边缘点，即左端点和右端点。所以我们把这一行扩展为：

$$\begin{aligned} &X(K+1), \cdots, X(2), X(1), \\ &\cdots, X(N), X(N-1), \cdots, X(N-K). \end{aligned} \tag{2}$$

为简单起见，把这一行数重新表示为：

$$Y(1), y(2), \cdots, y(N+2K). \tag{3}$$

均衡算法只需对数据范围 $Y(K+1)$ 到 $Y(K+N)$ 中的数来作，其窗口长为 $2K+1$. 由于已经把数据扩展了，所以下一步的程序，对 $Y(K+i)$ 中的每一点都是一样的。也就是，边缘效应已经可以避免了。

对于每一个 $Y(K+I)$ 值，一种类似于文献 [1] 中所提出的 OTA 算法用来得到新的 $Z(K+I)$ 值。最后，我们定义新的 $X(i)$ 值如下：

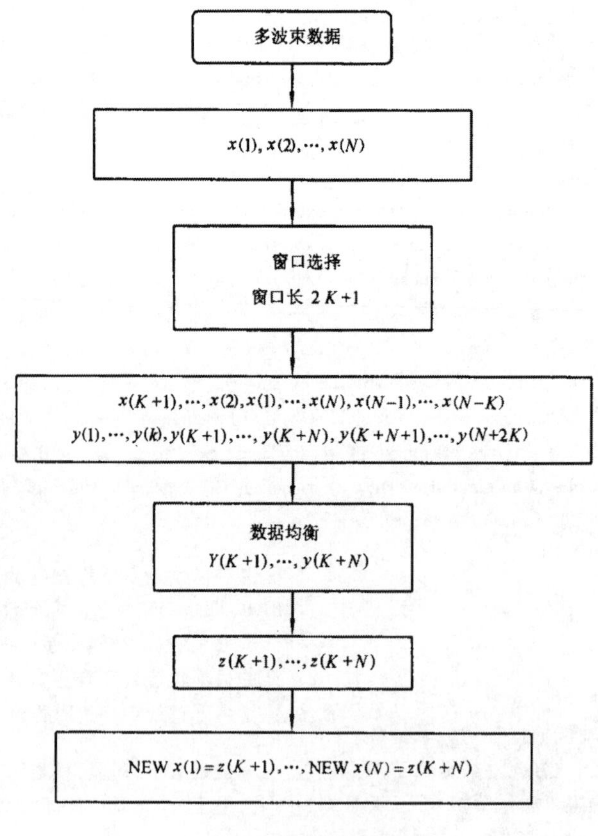

图 1 背景均衡算法框图

$$
\begin{aligned}
&\text{NEW}\quad X(1)=Z(K+1),\\
&\text{NEW}\quad X(2)=Z(K+2),\cdots,\\
&\text{NEW}\quad X(N)=Z(K+N).
\end{aligned}
\tag{4}
$$

用新的 $X(i)$ 替代老的 $X(i)$，把从顶到底的所有二维数据加以更换便得到一幅新的方位历程图。

2 改进的 OTA 算法

波束域的 OTA 子程序见图 2。

数据滤波窗口长为 $2K+1$，对于每一个 $Y(K+I)$ 值，$I=1,\cdots\cdots,N$ 构成一个数据滤波器：

$$V(1)=y(I),\cdots,V(2K+1)=y(I+2K), \tag{5}$$

重新将 $V(I)$ 按数值的大小排列：

$$W(1)\le W(2)\cdots\le W(2K+1). \tag{6}$$

我们知道，中值滤波是去除随机起伏的有效方法.用阈值比较法把有一定宽度的大范围起伏去除。这两种滤波算法可以结合起来，得到一种相当好的均衡结果。序列 $V(I)$ 的中位数是 $W(K+1)$，定义序列 $W(K+1),\cdots\cdots,W(2K+1)$ 的截断平均值如下：

$$\overline{w}=\frac{1}{K+1}\sum_{i=1}^{K+1}w(K+i), \tag{7}$$

确定一个阈值 $w_0=\alpha\overline{w}$, 其中 α 是 OTA 算法中用于调节拒收数据门限的重要参数，$\alpha>0$。

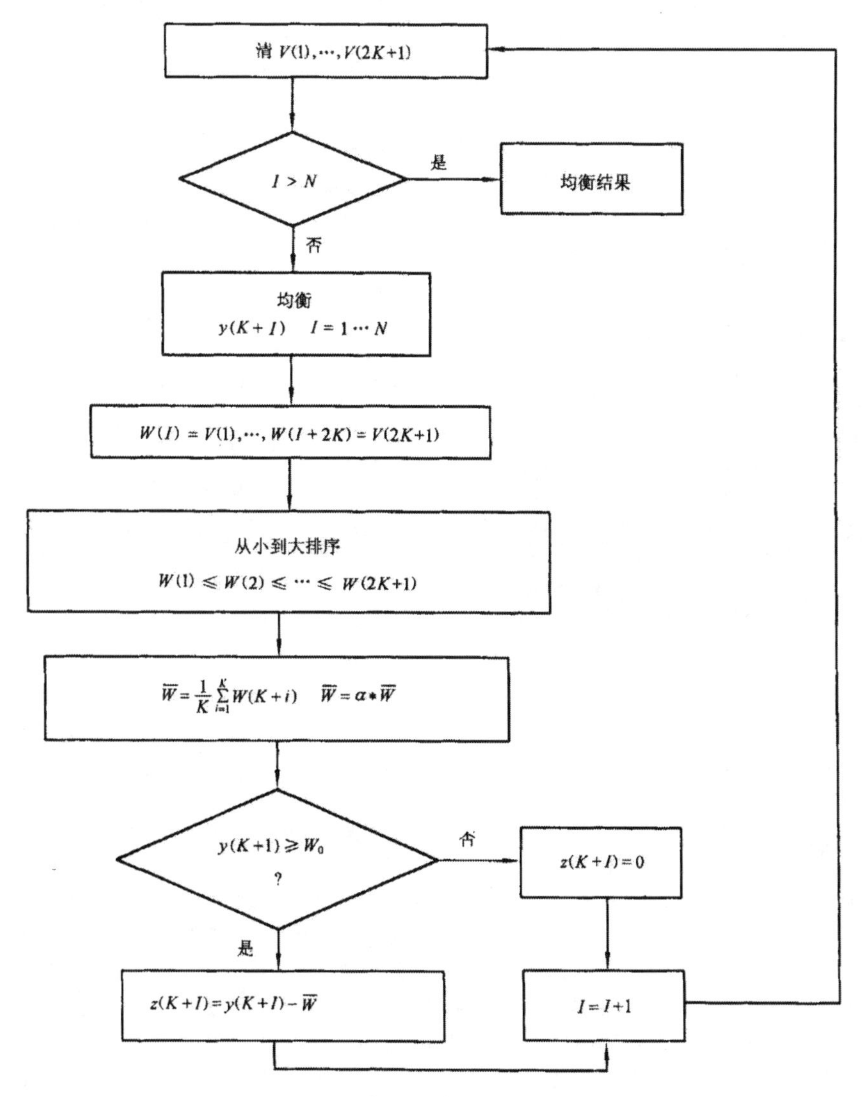

图 2 波束域的 OTA 算法流程图

本算法的最后一步是决定对 $Y(K+1)$ 的修正.

$$\left.\begin{array}{ll} z(K+I)=0 & \text{if } y(K+I)<w_0 \\ z(K+I)=y(K+I)-\overline{w} & \text{otherwise} \end{array}\right\},\quad(8)$$

在对每一个 Y 进行上述处理之后, 可以得到方位历程显示中的新的一行. 所有这些新行组成一幅新的二维图象.

3 正确选择参数 K 和 α

在时间历程图中, 通常用图象灰度表示波束值, 如果有 256 级灰度, 每个波束值将按某种映射方法映射成 $0\sim255$ 之间的某个值, 255 表示最高的灰度等级, 0 表示最低的灰度等级. 如果某个方位有信号, 它的灰度等级将高于周围的灰度等级, 但是在时间历程图中几乎不可避免会加入各种各样的噪声, 在拖曳阵声呐中, 图象中会有很强的平台自噪声, 典型的拖曳阵声呐的时间历程图如图 3 所示. 如果想从噪声背景中检测信号, 就必须精心地选择参数 K 和 α. 这两个参数直接关系到均衡的结果.

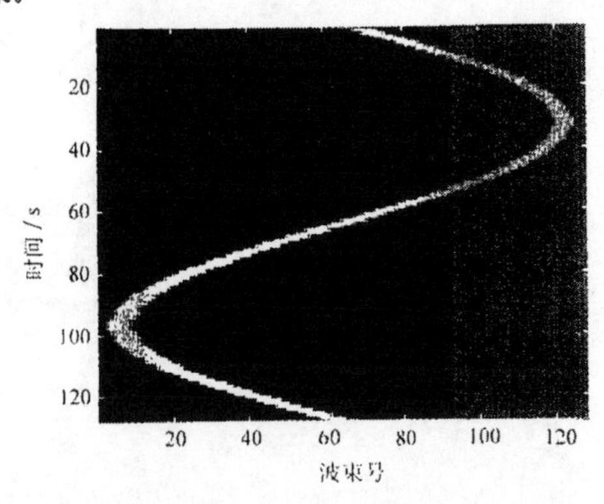

图 3 典型的时间历程图

首先考虑参数 K. K 和时间历程图中的信号的主瓣宽度有着密切的关系. K 必须大于主瓣的宽度, 因为在计算 $\overline{w}$ 时, 仅仅考虑从 $w(K+1)$ 到 $w(2k+1)$ 的数据, 这是排序以后较大的一半, 如果 K 小于主瓣的宽度, 从 $w(K+1)$ 到 $w(2k+1)$ 将都是主瓣中的数据, $\overline{w}$ 将非常接近组主瓣中的信号的值, 所以均衡的结果 $Z(K+i)=Y(K+i)-\overline{w}$ 将非常接近于零, 因此很难区分主瓣区域内的均衡以后的值是信号还是噪声. 为了得到好的结果, K 一般要大于两倍的主瓣宽度. 在拖曳阵声呐中, 主瓣的宽度常在 5 以上, 所以 K 值一般要大于 10, 作为窗口长度 $2K+1$ 通常要大于 20.

但是, K 值不能太大, 一方面是因为如果 K 值太大, 均衡将不能得到正确的结果. 如图 4 所示的信号,

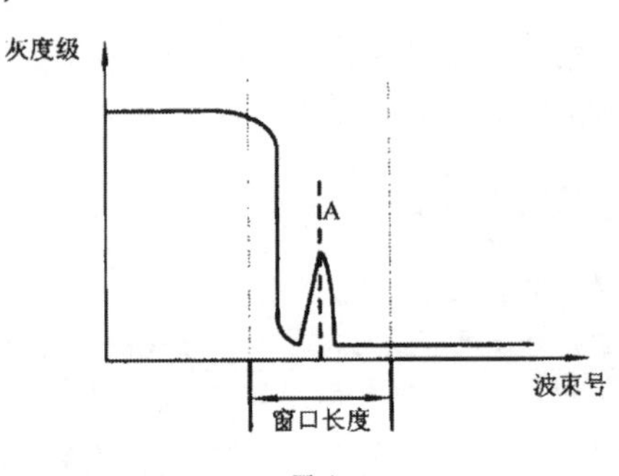

图 4

可以看出, 若在波束 A 的地方有信号, 当对 A 做均衡计算时, 如窗口的长度如图所示, 则计算 $\overline{w}$ 时, 从 $W(K+1)$ 到 $W(2K+1)$ 的地方将包含平台噪声和信号两部分, 由于自噪声很强, 使 $\overline{w}$ 很可能超过信号的强度而将信号从显示的图象中去除. 所以 K 值不能太大. 同时由于 K 值的增大将使计算任务增加, 这对算法的实时性将产生严重影响. 所以 K 值一般不超过 20.

对于参数 α, 其值一般略大于 1, 这是因为如果想从图象背景中将信号检测出来, 信号的值就必须大于其周围的值, 所以在计算剪切门限 $w_0=\alpha\overline{w}$ 时, $\overline{w}$ 的值必然小于信号的值, 所以 α 的值可以大于 1 以得到较好的均衡效果, 但是由于 $\overline{w}$ 非常接近于信号的值, 如果 α 的值太大, 将损失信号, 所以 α 的值不能太大, 综合这两方面的结果, α 的值一般在 1.0 到 1.1 之间.

4 系统模拟结果

该方法已在计算机上作了系统模拟, 二维数据是 128*128 的阵列. 在波束范围中, 从 80 到 128 之间有一个较大范围的干扰. 它类似于一个拖线列阵声呐的显示结果. 另外还模拟了背景的非一致性, 它反映了实际海洋环境的非平稳性及非均匀性. 图 5 给出了用该方法进行背景均衡的结果, 均衡的结果相当令人满意. 背景变得更加清楚了, 目标的轨迹非常明显地出现在屏幕上, 可以说本文提出的背景均衡算法在改善数字式声呐方位历程显示的性能方面是有效的. 值得指出的是, 参数 K 和 α 的选取也是十分重要的. K 和 α 择的不正确, 将直接影响到均衡效果, 图 6、图 7 是选择不同参数的均衡效果图.

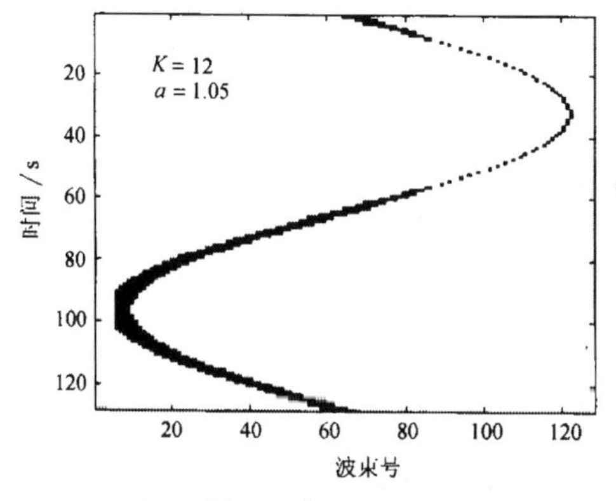

图 5 正确均衡的图像

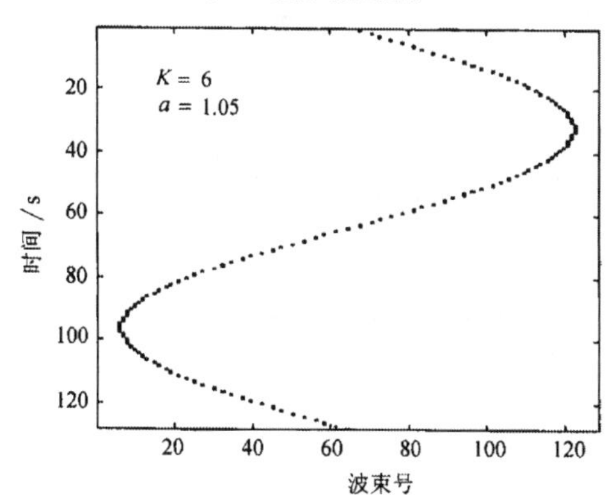

图 6 不正确均衡的例子

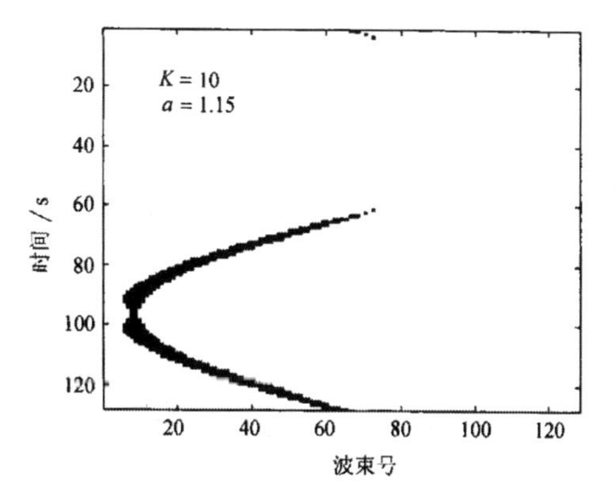

图 7 不正确均衡的例子 (a 值偏大)

参 考 文 献

1 Srtuzinski W A, Lowe E D. A performance comparison of four noise background normalization schemes proposed for signal detection systems. *J. Acoust. Soc. Amer.*, 1984; **76**(6): 1738—1742

2 Srtuzinski W A, Lowe E D. The effect of improper normalization on the performance of an automated energy detector. *J. Acoust. Soc. Amor.*, 1985; **78**(3): 936—941

3 Kummert A. Fuzzy technology implemented in sonar systems. *IEEE J. of Oceanic Engr.*, 1993; **18**(4): 483—490

4 Lindley C A. Practical image processing in C. *John Wiley & Sons.* 1991

基于串行背板技术的声呐数据传输系统设计①

郑剑锋[1,2] 李启虎[2] 孙长瑜[2]

([1]中国科学院研究生院 北京 100039; [2]中国科学院声学研究所 北京 100080)

摘要: 根据拖曳线列阵声呐工作特点和水声信号传输的特殊要求, 提出了一种基于高速背板串行传输技术的全数字式水声信号多路传输方法。该方法选用 CY7B923/CY7B933 作为物理层芯片解决水声信号的高速串行传输, 采用复用/竞争的思想实现多路数据混合传输, 并通过同步互联技术解决模块间的同步采集问题。硬件上设计了以可编程逻辑器件为核心的数据采集模块和串行背板传输模块。该系统可适用于传输电缆体积受到严格限制, 且需要同步多路数据采集的数据传输场合。

关键词: 串行背板 拖曳线列阵声呐 数据传输 同步互联

Design of Towed Array Sonar Data Transmission System Based on Serial Backplane Transmission Technology

ZHENG Jianfeng[1,2] LI Qihu[2] SUN Changyu[2]

([1]Graduate School of the Chinese Academy of Sciences, Beijing, 100039; [2]Institute of Acoustics, Chinese Academy of Sciences, Beijing, 100080, China)

Abstract: According to the working characteristic of the towed array sonar and the special require of acoustic signal transmission, a kind of multi—channel data transmission method based on high—speed serial backplane transmission technology is proposed in this paper. The method includes the following techniques: using CY7B923/CY7B933 as physical layer chips to transmit acoustic signal, transmitting the whole data in one channel by using the idea of multiplexing and competing, and using synchronously interconnected technology to synchronize acoustic signal sample. Signal sample modules and serial backplane transmission modules are designed with the core of programmable logic devices. It is suitable for the case that the volume of cable is strictly limited, and synchronous data acquisition is necessary.

Key words: serial backplane, towed linear—array sonar, data transmission, synchronously interconnection

拖曳线列阵声呐是目前现代声呐的重要发展方向[1]。早期限制数字拖曳线列阵发展的主要因素是数字信号传输问题。虽然利用光纤作为媒质传输高比特率数据的通信技术在陆地上已普遍应用, 但是在水下应用受到各种特殊的限制, 特别是受基阵外径小的限制, 对器件的小型化和耐压、水密问题都有独特的要求。近 10 年来, 随着芯片制造工艺和可编程技术的不断发展, 在数字化平台上研制小型化的拖曳线列阵声呐数据采集、传输电路已经成为可能。

① 微计算机应用, 2004, 25(3): 257-261.

本文设计了一种基于串行背板技术的多通道、分布式、同步互联型数据采集与传输系统来解决拖曳式线列阵声呐全数字化水下数据传输问题。该系统的设计为解决拖曳线列阵声呐数据的实时数字传输提供了一个新的思路，并对分布式采集系统的网络同步互联技术做了初步的研究。

1　系统基本结构

系统基本结构如图1所示，包括如下几个部分：数据采集模块，串行背板传输模块、光端机、串行数据接口板。放置在充油缆中的硬件包括数据采集模块和串行背板传输模块。受充油缆尺寸限制，这两部分要求硬件小型化。

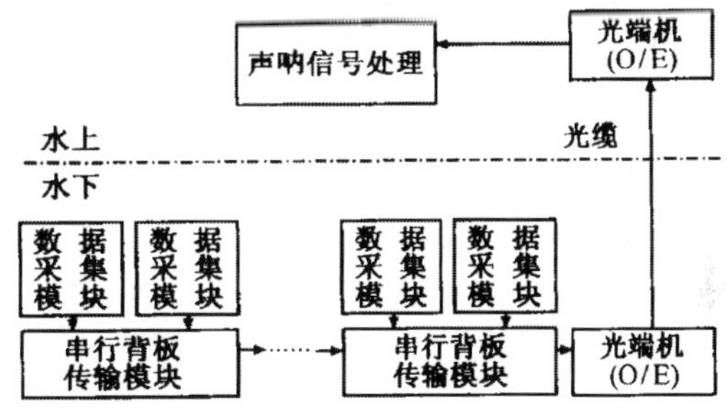

图1　拖曳线列阵数据传输结构

数据采集模块（图2）由前端信号调理电路、24bit高精度A/D采集电路、CPLD数据复用处理和模块控制以及RS—485同步互联接口电路四部分组成。A/D采集电路中，每路水声信号独立模数转换，与文献[2]多通道共用一个AD转换器比较，虽然硬件量增加，但是真正保持各路信号同步采集。为了解决水下拖缆空间尺寸小这一问题，本模块选用小体积封装的AK5380作为A/D转换器。采集到的数据在CPLD中分配缓存区，之后按照自定义的格式把8路数据合并成一路数据。RS—485同步互联接口电路可以接收外部的同步工作时钟，保证分散的各个数据采集模块的数据之间的一致性，为声呐数字信号处理算法提供分析依据。

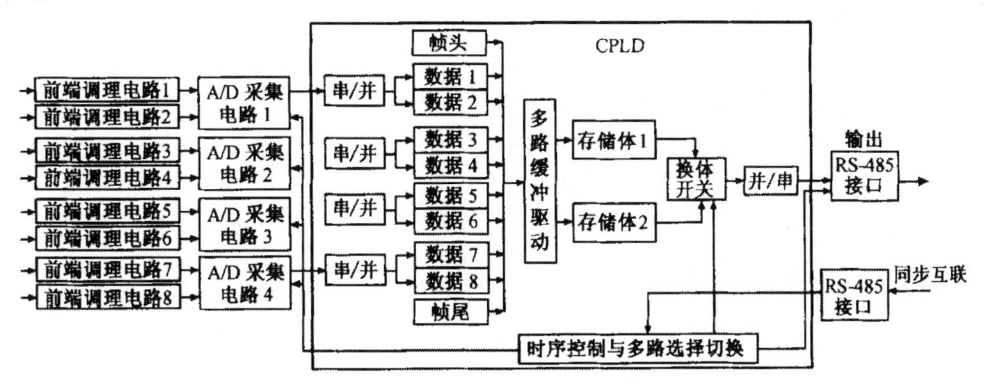

图2　数据采集模块

点到点的串行数据传输性能对声呐数据实时传输至关重要，既要求数据传输率高，还要求发送器与接收器数据同步建立时间短。现选用CYPRESS公司推出的串行背板收发器CY7B923/CY7B933作为串行背板传输模块中数据传输的物理层芯片。该模块的基本原理是背板接收器CY7B933接收前一个模块的数据至CPLD内部缓冲区，并与数据采集模块的数据合并成自定义的帧格式，通过背板发送器CY7B923输出。

在结构上，数据采集模块与串行背板传输模块的硬件分开，两者之间用RS—485接口进行数据和控制信号的传输，最大速率可为10Mbit/s，传输媒质为双绞线。为了减少串行背板传输模块的数量，硬件设计时每个串行背板传输模块可以接收2个数据采集模块的数据，而最后一个串行背板传输模块增加光端机（如图1所示），拖缆部分采用单模光纤媒质传输数据。这种结构设计难度适中，小体积指标相对容易实现，模块之间采用柔性的双绞线连接，便于卷绕在绞车上，并且模块的结构紧凑，灵活性强。

2 相关技术

2.1 串行背板传输芯片CY7B923/CY7B933

作为高性能串行背板收发器，CY7B923/CY7B933传输码率范围为150～400Mbit/s；适用Fiber Channel、IBM ESCON、ATM；与光纤模块、同轴电缆、双绞线兼容。其工作模式可灵活设置，包括编码模式（8B/10B Encoder Mode）、直传模式（Bypass Mode）、自检模式（Test Mode）[3]。

CY7B923发送器为8bit并行TTL输入，3路PECL 100k串行差分输出。编码模式下，编码器把输入数据按8B/10B编码为传输码，由SC/D端电平控制传输码是数据码还是专用码。使用传输码可使传输过程1与0的数目均衡，传输的直流成分接近0；附加位能保证串行位流中有足够的跃变，使接收器能从这些跃变中恢复发送时钟。这种编码大大增加探测位错或多位错的可能性。一些专用字使用独特的位组合容易识别，可帮助接收器将数据成帧。CY7B923在空闲时，可连续发送同步K28.5以维持接收器与发送器处于同步状态。它还能发送两种自检信号，一种是1与0相同的连续信号；另一种是循环产生一组511个字节的伪随机序列，包括所有的数据字和专用字，此序列可与接收器配合作系统自检。为了检测报错系统，发送器还可以发送违规信号，这些信号可检测整个通信链路是否正常[4]。

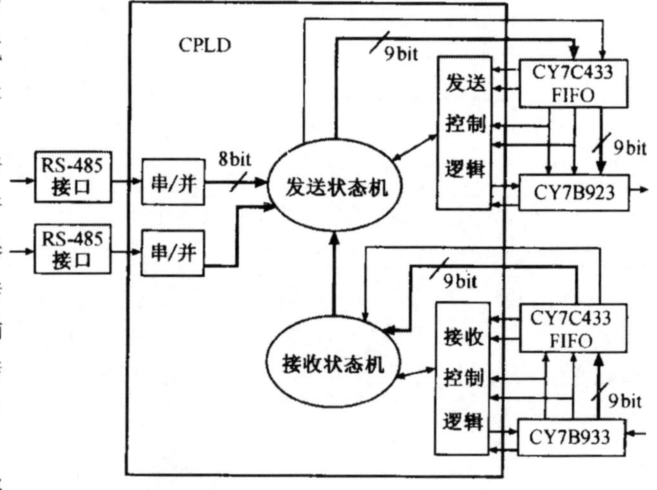

图3 串行背板传输模块

接收器CY7B933有两路PECL 100k串行差分输入，8bit并行输出。内嵌锁相环PLL锁相于输入数据位流，产生与发送器时钟同步的时钟。成帧器（Framer）检测输入数据位流的直接边界，以同步字K28.5为准，使位流数据正确地按字节成帧。为避免一些随机错误造成错误成帧，可在数据同步过程关闭成帧器，即令成帧器使能端RF为逻辑0；或当RF为逻辑1时传输字节数超过2048字节。此时需在连续5字节中有两个K28.5才启动成帧功能，可大大减少偶然错误导致误成帧。解码器将8B/10B代码还原为原始数据或专用字。在自检模式，CY7B933将循环产生包括所有数据字和专用字的一列511字节的伪随机代码序列，并与发送器产生的伪随机序列吻合，比较这两个序列，可检查数据接收功能。

在本系统中，通过逻辑控制CY7B923/CY7B933工作在编码模式或自检模式，与直传模式相比，可以不需要外部编解码协议控制器。控制逻辑用CPLD实现，也可以采用其他数字逻辑器件实现。在发送数据前，先进行系统自检，判断系统工作正常与否，及时进行错误处理。为了匹配数据处理和串行传输的速度，在收发器与CPLD之间加入高速先入先出（First input first output）存储器作为一级缓存器来缓存接收/发送的数据。电路板设计时应采用多层板，CY7B923/CY7B933的位置应靠近双绞线接口使连线最短，连线长度应尽量保持对称。输出端都应该接匹配电阻，同时CY7B923和CY7B933的时钟频率必须相同，误差不能超过±0.1%，否则容易导致同步失败，因为CY7B933内部锁相环是采用外部时钟作为参考频率来跟踪接收到的位流码频率。

2.2 多路数据混合传输技术

由于拖曳线列阵的拖缆部分采用一根光纤来传输所有水声数据，这就意味着所有数据都共享一个信道。与其他的数据传输网络相比，拖曳线列阵声呐数据传输具有以下特点：① 各路信号通道之间没有数据交换；② 信道竞争机制不允许因为信道冲突而导致数据丢失；③ 数据传输要求实时性。

根据以上特点，本系统采用复用/竞争的思想来解决上述信道共享的问题。其基本思想：

（1）每个数据采集模块处理的水声信号为一组，标识一个独一无二的“组号”，每一组水声信号经过多路A/D变换为多路串行数据，初级复用后为一路串行信号。

（2）串行背板传输模块采用信标竞争的算法传输来处理多路数据共享信道。

设有N路输入数据$\{D_1,\cdots,D_N\}$，每路信号有一个状态值$\{\tau_1,\cdots,\tau_N\}$，状态值

$$\begin{cases}\tau_i=1, & \text{表示}D_i\text{准备就绪}\\ \tau_i=0 & \text{表示}D_i\text{没有准备就绪}\end{cases}，i\text{表示通道序号。}$$

当D_i就绪，则置$\tau_i=1$；当输出完毕D_i，则置$\tau_i=0$。另外，给定惟一信标Γ，信标有两种状态：忙（$\Gamma=1$）和空闲（$\Gamma=0$）。信标竞争原理：① D_i就绪，即$\tau_i=1$，而信标$\Gamma=0$，则置信标$\Gamma=1$，禁止传输其他数据，并开始传输D_i。传输完毕后，置信标$\Gamma=0$和$\tau_i=0$。② D_i就绪，即$\tau_i=1$，而信标$\Gamma=1$，表示正在传输其他数据，D_i数据等待信标$\Gamma=0$后进入①所述的过程。③ 当存在多路数据就绪时，采用顺序的方法传输，即当信标$\Gamma=0$，而此时有$\{D_m,D_n,\cdots,D_p\}$都就绪，系统取$D_{\min\{m,n,\cdots,p\}}$数据输出，即取序号最小的数据。④ 每个数据采集模块的数据定义为一路输入数据D_i，而通过串行背板接收器输入的数据定义为n路输入，n的值等于在此之前所有的数据采集模块的数量。

（3）多路传输的帧格式定义是指有效数据传输顺序的安排方式。一组水声信号的数据大小确定，帧格式定义如图4所示。每帧数据中，声数据每个通道占3个字节，即24bit的精度，N个通道为一组，N值由数据采集模块通道数确定。每组数据采集模块的组号都是惟一的。同一个采样周期的数据时标相同，时标根据同步互联接口传输的同步时钟来确定。

2.3 同步互联技术

波束成形是声呐信号处理系统中的核心，其目的是为了获取较大的空间增益，同时也是为了使系统得到较高精度的目标分辨力，它是决定一部声呐性能优劣的核心环节。在实现波束成形的过程中，各路信号的精确延时补偿是最关键的，相比之下，其他运算（如加权、求和、积分）是次要的。因此，为了获取正确的波束成形结果，就必须知道拖缆中分散的各路水声信号采样之间的时延，或者保证各路信号采样的同步。相比两种方法，保证采样的同步性更为可行，本系统采用RS－485总线设计同步数据采集接口，同步数据传输速率最高可

以达到 10Mbit/s，其最大分布距离为 1000m。

如图 5 所示，通过每个模块的同步接口以及利用 CPLD 实现时分复用技术（TDM），本系统最多可以将 48 个数据采集模块连接在一起，其中任何一个模块都可以设置为主控制模块。一旦主控制模块设定，其他的模块都是从模块，各模块的同步时钟由主模块统一提供，这样可以同步所有的数据采集通道。为了与外部其他系统兼容，主控模块可以是另外设计，而所有的数据采集模块都作为从模块。在其他应用环境下，用户可以根据实际需要选择合适的模块数即通道数。

这种分布式模块之间的同步方案存在着一定的同步误差，造成的原因是同步时钟传输距离不同而造成的。标准 YD/T1019－1999 标准规定了常用的 5 类缆的电磁波传播速度不小于 $0.65c$，c 为光在真空中的传播速度[5]。假如传播距离差为 500m（即最大传输距离的一半），实际传播速度为 $0.65c$，则传播时延为 $500/(3\times10^8\times0.65)=2.564\mu s$。一般水声信号采样频率都在 50kHz 以下，采样周期大于 20μs，同步误差远远小于采样周期，对于声呐波束成形而言，这个误差是在允许范围之内。

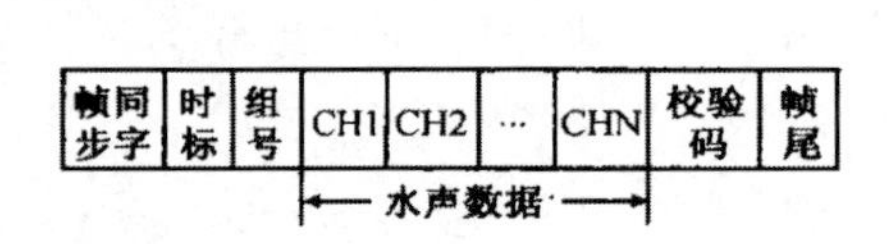

图 4　多路数据传输帧格式

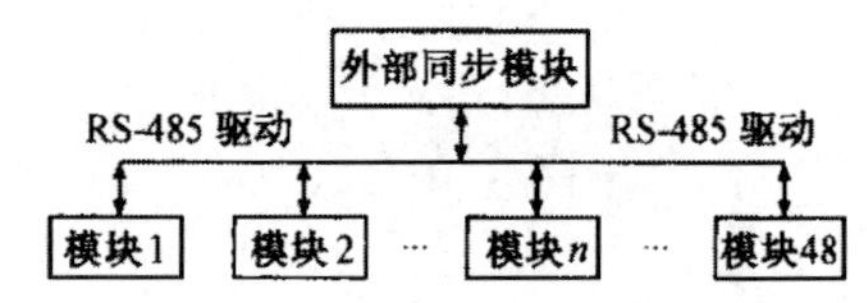

图 5　分布式模块同步互联

3　结束语

利用上述原理和方法设计的拖曳线列阵声呐数据传输系统，突出的优点表现在根据拖曳线列阵声呐数据传输的特点，成功地采用串行背板传输技术和巧妙地利用可编程逻辑器件来实现了多路数据传输，完成了在一根光纤上稳定可靠地传输所有水声信号的目的。目前，该系统已经完成硬件设计和软件仿真，正在进一步研究如何将拖曳线列阵声呐纳入吉以太网，拓宽系统的应用范围。整个传输系统可广泛应用于其他海洋声学监测、深海数据采集、水下航行器与水下机器人等需要传输高速数据的设备中。

参　考　文　献

1　李启虎. 数字式声呐设计原理 [M]. 合肥：安徽教育出版社，2003，426－442

2　赵俊渭，李钢虎，王　峰等. 机载声呐声信号的高速双向多路传输研究 [J]. 声学学报，2003，28(1)：52－56

3　HOTLink Transmitter/Receiver Data Sheet. Cypress Semiconductor Corporation, 1997

4　章　平，过雅南，赵棣新等. 高速串行数据通信 VME 插件的研制 [J]. 核电子学与探测技术，2002，22(1)：44－46

5　俞兴明，蒋铃鸽. 5 类和 6 类数字通信电缆的传输性能分析 [J]. 光纤与电缆及其应用技术，2002，2：18－22

基于MSP430的水声时间反转应答系统设计①②

张志博[1,2]，孙长瑜[1]，李启虎[1]

（1. 中国科学院声学研究所，北京 100080；2. 中国科学院研究生院，北京 100039）

摘 要：体积庞大、结构复杂的水声仪器设备往往在使用和维护等方面给使用者带来不便，因此对于水声设备的小型化要求已经必不可少。水声时间反转应答系统是水声时间反转镜研究的主体设备。论文结合水声信道匹配研究的实验要求，以MSP430F169微控制器作为核心部件，设计了一种适用于水声时间反转镜研究的小型化、低功耗应答系统。文中给出了详细的硬件及软件设计方案，以及该系统的应用原理。该设计体积小巧、结构简练，在保证性能的前提下，极大地提高了系统的易用性和灵活性。

关键词：MSP430；时间反转镜；应答系统；水声信道

中图分类号：P733.2 **文献标识码**：B **文章编号**：1003-2029（2007）03-0031-05

水声信道匹配基础研究是建立在水声学、海洋物理声学以及现代信号处理技术基础上的新兴研究领域。水声信道的时空变化复杂性和不确定性是提高新型声纳系统性能的最大障碍，通过对水声环境的实时获取和估计，及时调整信号和接收处理方式以达到与信道相匹配的工作性能。其中，水声时间反转镜研究是该课题的一项必不可少的内容。

时间反转应答系统是水声时间反转镜研究中的主体实验设备，它负责水下信号的接收、时间反转及回发。与船载应答系统相比，浮标式应答系统具有体积小、自身干扰少、布放方便、操作灵活等诸多优势。由于采用电池供电，我们在保证性能的同时，尽可能降低系统功耗以节约电能。本文正是结合课题所需，设计并实现了一种小型化低功耗水声时间反转应答系统。

1 基于MSP430的低功耗水声时间反转应答系统设计

1.1 水声时间反转镜简介

海洋的不均匀和多途效应使得接收信号的时域、频域特性发生变化，最终导致检测能力的下降。为了克服在声学不均匀介质中的波形畸变，在声学中引用了源于光学中不均匀介质的相位共轭成像法，并将连续波的相位共轭法发展为时间反转法。

当不考虑空间不均匀和界面不均匀性的随机性时，水声信道的空变特性可以用相干多途信道的系统函数表示。由于水下声信道的多途效应，因此怎样将各途径到达的信号分量综合利用，是值得研究的。借鉴匹配滤波器空间匹配滤波的思想，如果用 $S(k)$ 表示信号的频谱，$H(k)$ 表示信道的传输函数，则信号经过信道输出的频谱可写作：

$$R(k) = H(k) \cdot S(k) \tag{1}$$

令 $X(k)$ 是 $R(k)$ 的复共轭，即：

$$X(k) = R^*(k) = H^*(k) \cdot S^*(k) \tag{2}$$

将（2）式两边同乘 $H(k)$，可得到：

$$Z(k) = H(k) \cdot H^*(k) \cdot S^*(k) \tag{3}$$

（3）式中“$H(k) \cdot H^*(k)$”一项是实、偶、正函数，它在时间零点的傅里叶反变换是同相叠加的，会得到主相关峰值，这使得信号与其乘积的结果 $Z(k)$ 比 $R(k)$ 要大的多。这就是水声信道中时间反转镜聚焦的根本原理。在（1）式至（2）式中，$X(k)$ 可认为是接收信号时反后的频谱，$Z(k)$ 可认为是时间反转镜输出的频谱。

1991年，Dowling对时间反转镜应用于水声作了定义和基本的理论分析。通过海上实验来研究水声时间反转镜的特性，实验系统由发射和应答两部分组成。信号由发射系统发出，经海洋信道到达应答系统，应答系统对接收信号进行时间反转并回发。多通道的时间反转应答系统可以研究海洋中时间反转镜的聚焦效应。水声时间反转镜实验如图1所示。

① 基金项目：国家自然科学基金重点资助项目(60532040).

② 海洋技术, 2007, 26(3): 31-35.

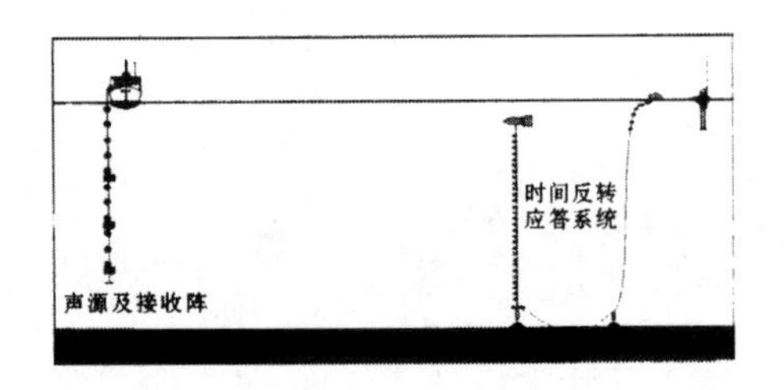

图 1 水声时间反转镜实验示意

1.2 超低功耗 MSP430微控制器

美国德州仪器 (TI) 公司的MSP430系列微控制器是一种 16位 超低功耗的混合信号处理器 (Mixed Signal Processor)。MSP430F169是目前 MSP430系列中高档型号之一。它采用16位 RISC结构，具有丰富的内外设和大容量的存储空间。其内部结构如图 2所示。

MSP430系列处理器的突出特点是超低功耗，正常工作时的电流仅为微安级，其中 MSP430F169 在电压 2.2 V、工作频率 1 MHz 下的动态模式下电流为 330μA 标准模式下电流为 1.1μA RAM 保持的节电模式下，工作电流为 0.2 μA MSP430 具有五种不同的微功耗模式。可以在一个动态模式与五个软件可选择低功耗模式下运作，一个中断事件就可以将 MSP430 由五种低功耗状态下唤醒，响应中断，在中断服务程序结束后返回低功耗模式。由休眠唤醒到标准模式时间不多于 6μs 正是这种先进的功耗管理和出色的超低功耗特点，使其很适合作为低功耗应答系统内部的主处理器使用。

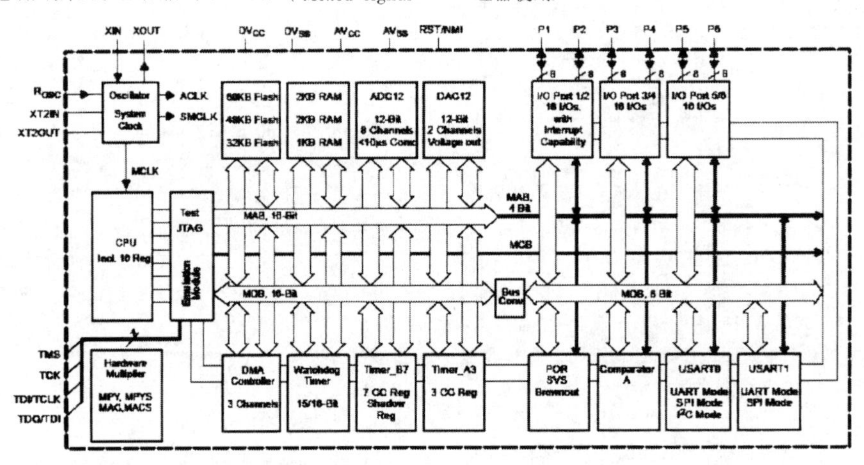

图 2 MSP430x 16x 结构框图

此外，MSP430F169 具有强大的运算能力：16位 RISC 结构、丰富的寻址方式；具有 16个中断源，并且任意嵌套；在 8 MHz 时钟驱动下指令周期可达 125 ns；内部包含硬件乘法器和大量寄存器及多达 60 KByte 的 FLASH程序空间和 2 KByte 的 RAM 空间，为存储数据和运算提供了保证。这些特点使 MSP430F169具有很强的数字信号处理能力。

MSP430F169还具有丰富的片上外设：包括看门狗定时器，16位定时器 (TA/TB)，8通道 12位快速 A/D (模数) 转换器，双 12位 D/A (数模) 转换器，通用同步 /异步串行通讯接口，I2C接口，DMA 控制器，48个通用 I/O端口。如此十富的片上资源使很多功能都可以在一个芯片上完成，大大降低了系统成本并提高了设计的可靠性。

2 系统电路设计

时间反转应答系统框架如图 3 所示。

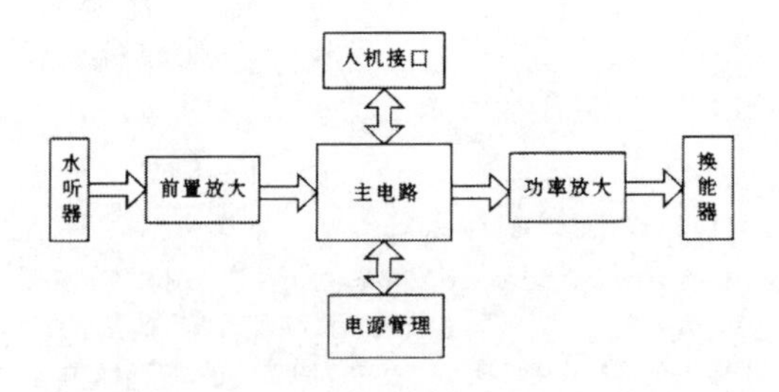

图 3 时间反转镜应答系统硬件结构图

整个系统由干端和湿端组成，其中干端完成信号检测处理及系统控制功能，包括应答主电路、前置放大电路、功放电路、人机接口电路、电源管理电路等模块；湿端由水听器和发射换能器组成，负责水声信号的接收和应答信号的发出。

水听器接收到水声信号后经前置放大电路送至应答主电路模块，MSP430F169是系统的核心部分，MSP430F169利用内置的A/D转换器进行采样，经过信号检测，将检测到的信号直接转发或进行缓存，D/A转换由其内置的D/A转换器完成。主电路中采用二阶压控滤波电路平滑输出波形，电路如图4所示。同时，MSP430F169还负责整个系统的控制及管理，如对人机接口和电源管理模块进行控制。

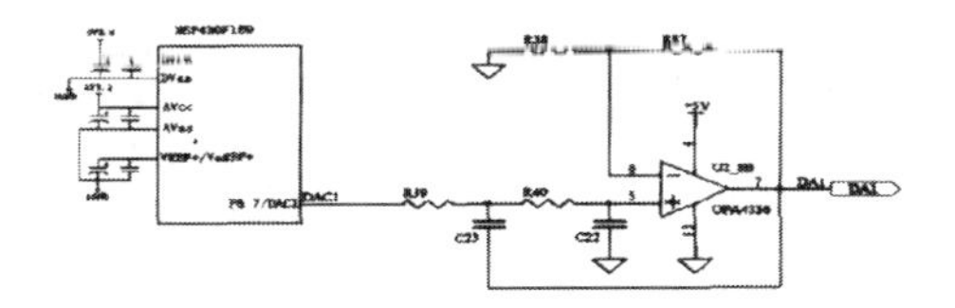

图4　主电路波形输出平滑电路

人机接口模块对应答系统进行初始化设置并显示应答系统的工作状态，包括可以通过LCD显示器、小型键盘或异步串口（UART）完成人机交互操作。

电源管理模块的主要功能是对电池组电源进行管理，包括电量监控、电压管理以及对系统内其它个模块的供电管理，从而实现对系统功耗的合理控制。

3　系统软件设计

3.1　A/D采样模块

A/D转换由MSP430F169内置的A/D转换器完成。系统通过软件进行初始化设置，包括参考电压、采样率、保持时间等参数的设定。应当说明的的是，由于MSP430F169的A/D转换器的采样率可达200 kHz，而对于我们的实际应用来说10kHz即可以满足要求，过高的采样率会加大数据量。这里采用内置定时器触发方式控制实际采样率。定时器TimerA以10 kHz的频率触发A/D采样，A/D转换器产生中断，中断服务程序完成采样数据的保存。时序示意图如图5所示。

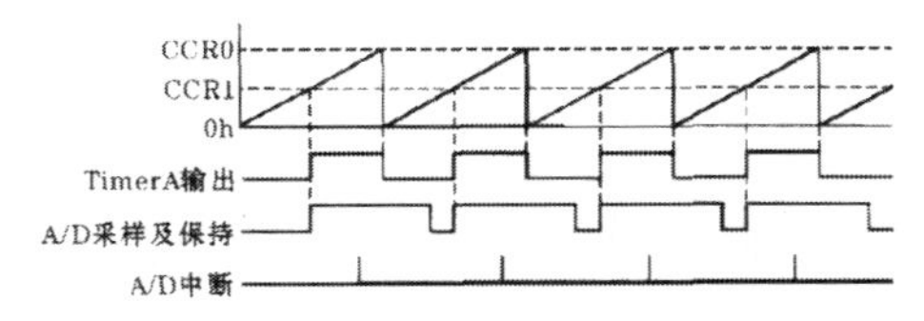

图5　A/D采样模块工作时序

合理的A/D转换器及定时器初始化是保证A/D采样正常运转的关键，下面列出的是本系统对A/D采样模块的初始化程序。

```
void ADC_ Init (void)
{
P6SEL|= 0x01; //选择通道 A0
ADC12CTL0 = ADC12ON+ SHT0_ 8+ MSC+ REF2_5V + REFON; //初始化 ADC12
Deley_ 10ms (); //延时 10ms
ADC12CTL1= CONSEQ_ 2+SHS_ 1; //触发方式
                              //单通道连续采样
ADC12MCTL0= INCH_ 0+ SREF_ 1;
ADC12IE = 0x0001; //使能 ADC12IFG.0
P2SEL|= BIT3; //设置 Timer A1
P2DIR|= 0x08;
CCTL1= CCIE; // CCR1中断使能
CCR0= 800;
CCR1= 400;
TACCTL1= OUTMOD_ 3; // CCR1设置
TACTL= TASSEL0 + TACLR+ MC_31; //时钟ACLK, TAR清零
}
```

3.2　信号检测与处理模块

信号检测与处理模块的功能是检测有效信号并进行处理。当没有信号发射时，应答系统接收到的都是海洋噪声，此时不做应答；当有信号发射时，应答系统要作出准确检测及存储，并按要求进行信号回发。

本应答系统应用于海洋环境中，海洋环境的多变性使得环境噪声会随着时间不断变化，无法将其看作一个统计特性不变的随机量，因此为了保证检测的有效性，软件实现阈值估计模块，以根据实际情况的变化灵活做出响应。阈值估计的主要功能就是根据噪声的变化设置相应的判决门限，保证在变化噪声环境下仍然能有效地检测到脉冲序列。

根据阈值估计提供的判决门限，脉冲判决模块实现能量检测器，用于对当前脉冲进行能量判决。该能量检测器数学表示式如下：

$$\frac{x_1^2+x_2^2+\cdots+x_n^2}{n}\geqslant M \tag{4}$$

式（4）中M即为阈值估计模块提供的判决门限，能量检测器将输入的n个样本值平方之后再求平均，并与阈值M相比较。累积平均能量高于判决门限M的脉冲被判决为有用信号，而能量低于判决门限的脉冲则被认为是噪声。

当检测到信号并正确保存后，程序将记录保存信号区域的指针。当发送信号时，程序将使指针递减，以达到时间翻转的目的。

信号检测与处理模块程序流程如图6所示。

3.3　D/A转换模块

D/A转换由MSP430F169内置的D/A转换器完成。D/A转换器与A/D转换器共用一个参考电压，这样保证了信

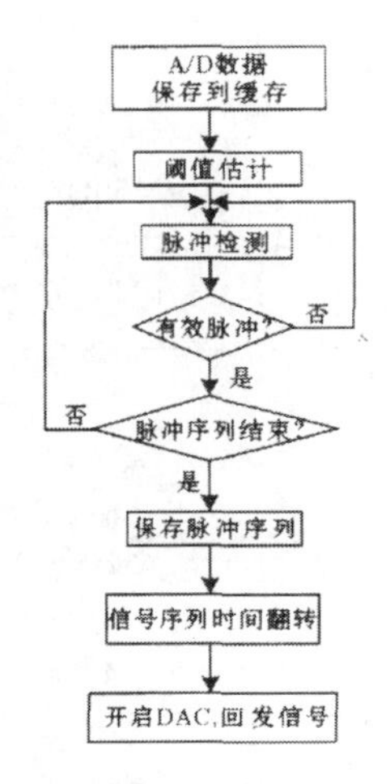

图 6 信号检测与处理模块程序流程

号收发的一致性。虽然 D/A 转换不需要定时器触发而运转，但为保持信号收发的一致性，本系统采用 10 kHz的频率更新 D/A 转换器的数据缓冲寄存器，以保证原始信号的信息完整。由于定时器 TimerA已经用于 D/A转换的触发，这里使用定时器 TimerB产生定时中断，以更新数据。

以下是 D/A转换模块设置程序及定时器中断服务程序。

```
void DAC_ Init (void)
{
ADC12CTL0= REF2_5V + REFON; //内置参考电压 2.5 V 开启
DAC12_1CTL = DAC12IR + DAC12AMP_5+ DAC12ENC + DAC12CALON;
                                        // DAC初始化
Deley_10ms (); //延时 10ms
P6SEL|= 0x80; //选通 DAC1
TBCTL = TBSSEL0 + TBCLR + MC_1; //时钟 ACLK，TAR清零，递增模式
TBCCTL0= CCIE; // CCR0中断使能
TBCCR0= 800;
}
# pragma vector = TIMERB0_VECTOR //定时器中断服务程序
__interrupt void Timer_B (void)
{
DAC12_1DAT = DAC_data1 [i];
i+ + ;
if (i= = 512) i= 0;
}
```

4 系统测试

为检测系统性能，通过对多组不同信噪比的信号进行应答来对系统进行测试，应答情况如图 7所示。

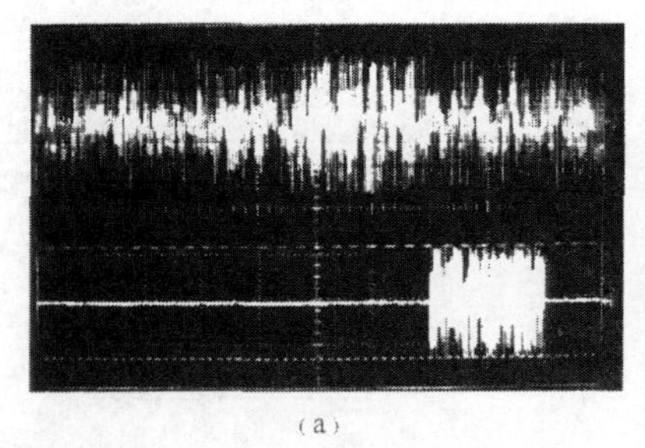
(a)

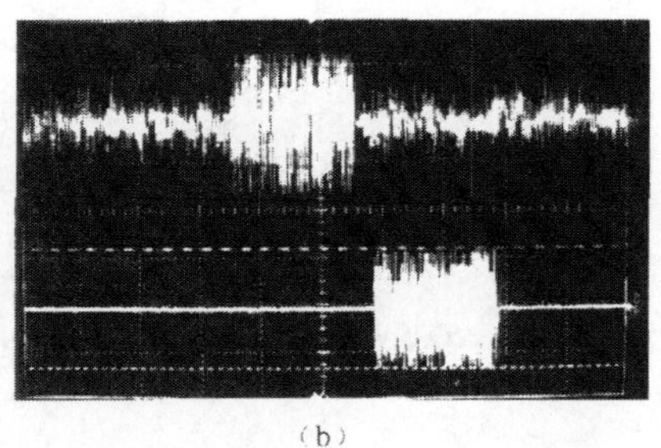
(b)

图 7 应答系统测试结果

由于测试次数较多，在此无法全部列举。图 7所示的是其中对信噪比分别较低、较高的两组信号的应答情况。图 7(a)中上方通道信号为应答系统接收到的信号，信噪比为 3 dB；下方通道信号为应答信号。从结果可知，系统准确地捕捉到信号，并回发。图 7(b) 所示的是系统对信噪比为 10 dB的信号进行的应答情况。

经测试，本系统工作正常，很好地满足了水声信号匹配基础研究的实验要求。

5 结束语

水声时间反转应答系统对体积、功耗及可靠性等方面有着极高的要求。超低功耗 MSP430 微控制器作为该系统的核心元件是十分理想的选择。本文中的应答系统充分发挥了 MSP430 低功耗、小体积、高性能的特点，使主要功能集中在单一芯片上完成，这是对水声仪器小型化的有益尝试。本系统在水声信道匹配基础研究课题中发挥了重要作

用，在实验室及海上的测试中均取得理想效果。进一步将本系统与浮标、潜标技术相结合，将更大地扩展其工作能力，因此本系统具有极为广泛的应用前景。

参考文献:

[1] 李启虎著．数字式声纳设计原理 [M]．合肥：安徽教育出版社，2002.
[2] H C Song, W A Kuperman, W S Hodgkiss, T Akal, S Kim, and G Edelmann. Recent results from ocean acoustic time reversal experiments. ECUA2002, Gdanski, Poland, 2002.
[3] 惠俊英．水下声信道 [M]．北京：国防工业出版社，1992.
[4] 生雪莉，惠俊英，梁国龙．时间反转镜用于被动检测技术的研究 [J]．应用声学，2005，24 (6)：351-358.
[5] 沈建华等著．MSP430系列16位超低功耗单片机原理与应用 [M]．北京：清华大学出版社，2004.
[6] Texas Instruments. MSP430x1xx Family User's Guide. 2006.
[7] Texas Instruments. MSP430x15x, MSP430x16x, MSP430x161x MIED SIGNAL MICROCONTROLLER. 2005.

Design of the Responder System for Acoustic Time-Reversal Mirror Based on MSP430

ZHANG Zhi-bo[1,2], SUN Chang-yu[1], LI Qi-hu[1]

(1. *Institute of Acoustics, Chinese Academy of Science, Beijing* 100080, *China*;
2. *Graduate School of the Chinese Academy of Sciences, Beijing* 100039, *China*)

Abstract Equipments used for underwater acoustics research usually have large size and complicated structure, which is inconvenient for application and maintenance. So the miniaturization of the equipments becomes necessary. Underwater acoustics time reversal responder system is the major equipment for the research on underwater acoustics time reversal mirror. According to the experiment requirement, a responder system with small size and low power consumption is designed. The kernel component of the system is MSP430F169 microcontroller. The hardware and software designs are presented, and the application and function of the system are also introduced. This design improved the convenience and flexibility of the system with small size and concise structure, while ensuring the performance.

Key words MSP430; time-reversal mirror; responder system

基于DSP的高速串行数据录放接口设计和实现①②

刘　钢　孙长瑜　李启虎

（中国科学院声学研究所　北京　100190）

摘要：介绍了一种用于声纳设备中的高速串行数据记录和回放接口的设计方案。该方案用于实现阵列数字信号处理机与数据记录设备间的连接。文中描述了利用ADI公司SHARC处理器的串行口实现高速数据记录和回放的硬件设计原理，给出了串行编解码方案和软件设计流程。

关键词：数字信号处理机　SHARC　串行传输　数据记录

Design and Implementation of High - speed Serial Digital Data Recorder Interface Based on DSP

LIU Gang, SUN Changyu, LI Qihu

(Institute of Acoustics, Chinese Academy, Beijing 100190)

Abstract: This paper presents a method about how to design a interface between DSP arrays and digital data recorder. In the method, a high - speed serail digital interface between them is constructed based on SHARC。The paper describes the hardware and software design approach in detail.

Keywords: Digital signal processing machine, SHARC, serial transfer, data recording

1　概述

在雷达、声纳、通信等应用中，实时记录数据并可以回放是工作设备很重要的一项功能。记录设备的种类可以是模拟式，也可以是数字式。数字式记录仪由于具有抗干扰能力强，存储介质容易保存和后续处理方便等优点渐渐地成为了高速大容量数据记录设备的首选。

本文介绍的数据记录设备是美国METRUM公司的数字记录/回放仪BUFFERED VLDS(BVLDS)[1]。该数字记录/回放仪BVLDS具有串行输入记录方

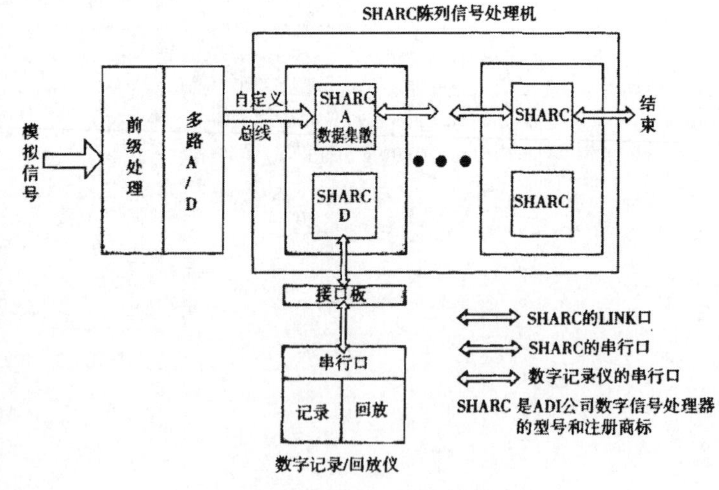

图1　数字信号处理机记录和回放原理框图

① 本文受国家863重点项目60532040资助.
② 微计算机应用, 2008, 29(8): 51-55.

式和串行输出回放方式，能够同时进行语音记录，能够把记录的数据通过并行口或转存卡输出到微机。需要记录的数据源是水听器阵列输出的信号经数字化后的数据流。如图 1 所示，当系统处于记录方式时，水听器阵列数据流首先进入 DSP 数据集散模块，然后由集散模块分发数据，一路进人记录仪，另一路要输入阵列数字信号处理机其他模块中。当系统处于回放方式时，记录仪回放数据，使其经过集散模块输入到阵列信号处理机中进行处理，此时集散模块具有交叉开关的作用，不再接受水听器阵列的数据流。SHARC 阵列信号处理机接收从数据采集模块传来的数据，然后进行高速数字信号处理。

2 系统体系结构设计

SHARC(ADSP-21060)是第一款超级哈佛结构的数字信号处理器(DSP)[2]。SHARC 的休系结构被设计为流水线并行处理器。它具有 32 位单精度 IEEE 浮点处理器内核，运算速度为 40MIPS，浮点运算峰值为 120MFLOPS。同时处理器内部还集成了 4Mbit 零等待 SRAM，用作程序和数据存储区。它还具有共享并行总线(Cluster bus)和多处理器的统一编址的设计。最具特色的是 SHARC 多种形式的外部接口，其中包括六个并行连接口(LinkPort)，两个串行连接口(Serial Port)，串行口时钟速率可调，最高可达 40MHz；这些接口受独立的 I/O 控制器控制，具有经优化的 DMA 和中断传输机制，使外部的数据交换与内核的运算过程相互独立，两者可以并发运行，极大增强了处理器与外部设备的数据交换能力。

阵列信号处理机有多块数字信号处理板组成。该数字信号处理板是本试验室自行开发出的一款集成了四片 SHARC 的 6U 板，其原理示意图如图 2 所示。SHARC A、B、C、D 以多处理器总线相互连接，组成 Cluster 结构。彼此之间可以利用总线进行数据读写和 DMA 操作，并且共享挂在总线上的其他资源，如 SRAM、双口 RAM、Flash ROM 等。各片 SHARC 芯片之间均有一个 Link 口两两互连，这样每片处理器共分配三个 Link 口与其他处理器进行通信。每片 SHARC 芯片各有两个 LinkPort 和一个 SPORTx 接到前面板(其中一个 LinkPort 和 SPORTx 共用一个插座，由跳线决定插座定义)，以及一个 LinkPort 和一个 SPORTx 接到底板插座 XP1 和 XP2 上。

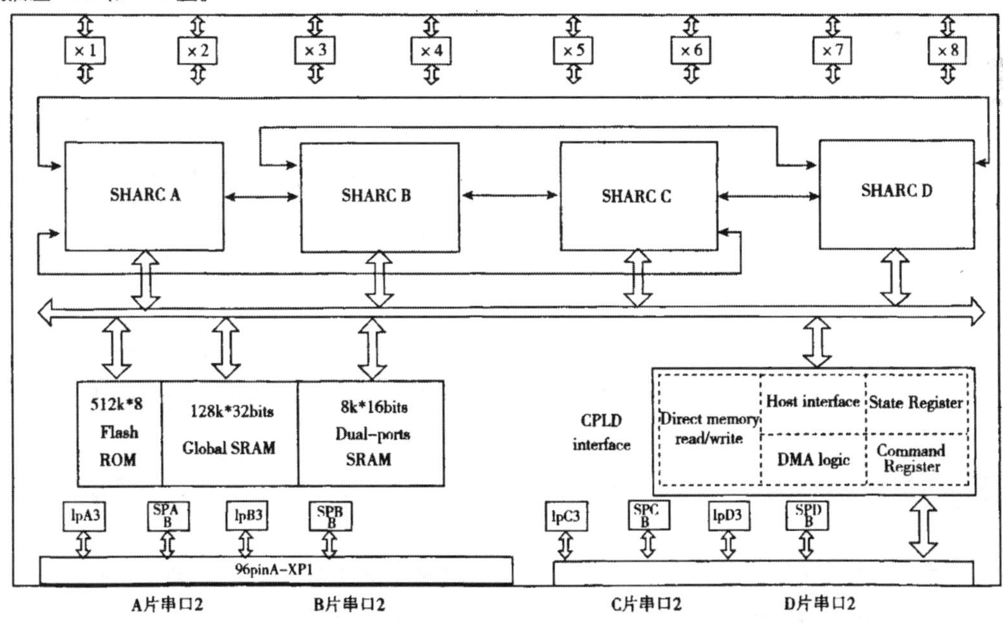

图 2 Quad SHARC 数字信号处理板原理示意图

如图1所示,阵列信号处理机中的第一块数字信号处理板完成数据集散和录音回放接口的功能。其中DSP A片完成数据集散功能,DSP D片完成录音回放功能,DSP A片和DSP D片通过Linkport相连,DSP D片与录音接口转换板通过串口连接。DSP A片通过自定义总线从AD板中读取多通道数据,然后在处理机内通过LinkPort分发。其中一个路径是发向DSP D片,DSP D片再通过串口及相应的转换接口板与METRUM BVLDS连接,实现记录功能。当数据回放时,该路径完成回放功能,将数据回送到数据集散模块DSP A片上,然后数据再由DSP A向其他发送路径发送。

3 数据记录和回放接口的硬件设计

SHARC的两个同步串行口(SPORT0和SPORT1)相互独立。每个串口都拥有独立的控制寄存器和数据缓冲区,能够独立的完成发送和接收功能,即具有全双工的功能。两个串口支持可变的串行时钟和帧同步,它们能兼容各种系列的串行通信协议。这两个串行口的信号都是TTL电平标准。METRUM公司的BVLDS提供串行记录和回放接口,最高记录和回放速率不大于32MHz。并且所有串口信号都是PECL电平标准。

使用SHARC的一个串行口的相互独立的发送和接收功能就可以实现阵列信号处理机与BVLDS的记录和回放功能的对接。但是两者的对接需要进行电平转换和协议约定,电平转换是系统框图1中接口板的主要作用。串行记录回放的接口硬件连接如图3所示。图中的TB5T1是TTL和PECL的电平转换芯片[3]。当数据记录时,SHARC串口的发送部分工作,采用DSP内部产生连续时钟的工作方式,即TCLK0发出时钟信号,并且由DSP内部产生发送帧同步信号TFS0,这种帧同步信号必须在串口控制寄存器中设置为与数据相关的滞后类型。信号时序如图4所示。BVLDS接收由TCLK0和TFS0合成的CLOCKI和DATAI信号,实际上就是将门控时钟TCLK0信号作为自己的记录时钟对数据DT0进行采样记录。当数据回放时,SHARC串口的发送和接收部分同时工作,串口发送TCLK0以产生BVLDS回放数据所需要的时钟信号,此时BVLDS的接收口只接收TCLK0时钟信号,而不接收DT0的发送数据。BVLDS根据外部收到的TCLK0时钟产生回放时钟CLOCKO和数据信号DATAO,时序如图5所示。此时,SHARC的串口接收部分工作,采用由DSP外部提供连续时钟的工作方式,将CLOCKO作为自己的输入时钟RCLK0信号,对DATAO进行采样输入。此时不需要接收帧同步信号,但接收帧同步管脚应该接至高电平,以保证串口接收功能正常工作。

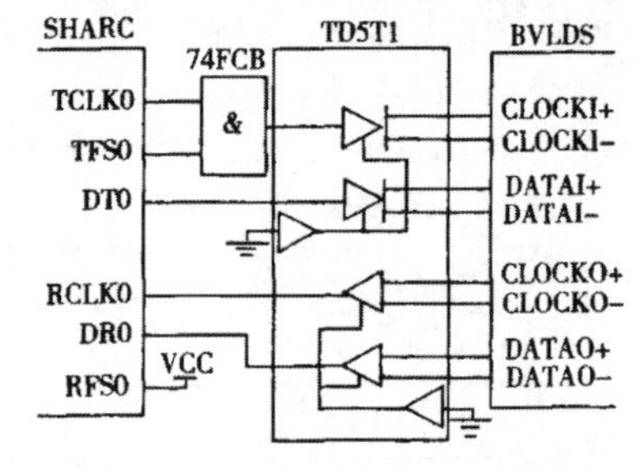

图3 记录回放硬件连接电路图

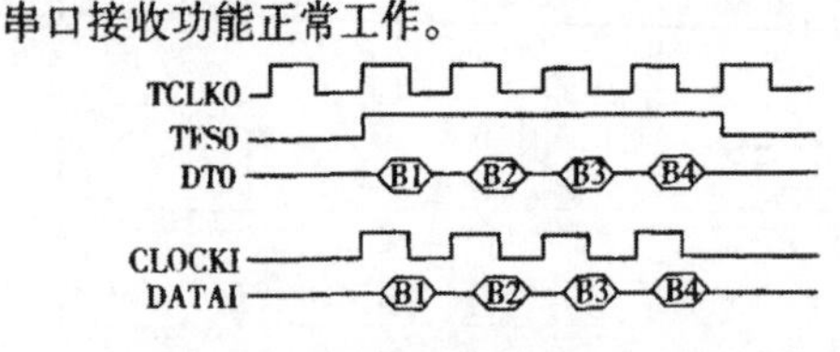

图4 SHARC发送时序和BVLDS记录时序

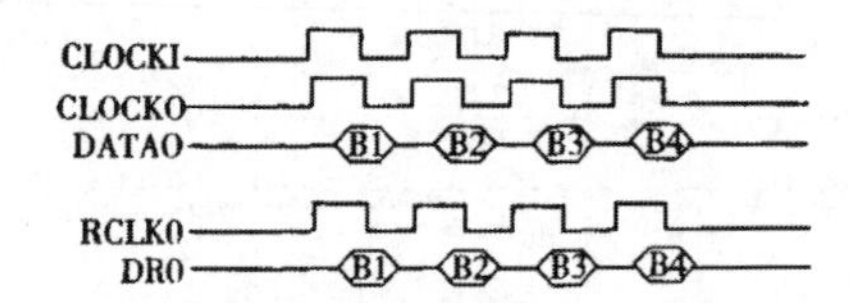

图5 BVLDS回放时序和SHARC接收时序

4 串行数据编解码设计

4.1 串行数据记录时的编码

来自数据集散模块的每一数据帧含有50通道一次采样的数据,数据位数是16位,共含有800位比特。记录仪记录和回放的数据都是串行数据流。为此在每一组数据发送到BVLDS记录仪之前,必须要填充帧头和帧尾。如图6(a)所示,帧头结构为三个16进制的数0x0、0x0、0x1(0x代表十六进制)。帧尾为一个16进制0xFFFF。这样一帧数据共有864比特。这里只对帧头和帧尾作出了规定,不再对各通道数据本身进行

8B/10B 等形式的编码。这样编码的考虑是基于因素：①在记录时，应尽量保持水声数据的原貌以保证其频带内的成分不丢失。②因为是实时记录，需要减小减少数据量。③实时记录时不可能实现错误重发，所以也没有奇偶校验标志。④PECL 电平传输可以实现减少传输线路中直流分量的存在，可以不必进行 8B/10B 等形式的编码。

4.2 串行数据回放时解码

记录仪回放数据是串行比特流，回放点可以是记录的串行比特流中随机的位置。回放比特流在 SHARC 接收串口移位寄存器中被截成独立 16 位的字，因为回放的初始位置不是固定的帧头的首字 0x0，所以 SHARC 接收到每个字都不是实际每个通道的内容。如果不经过解码恢复原有每个比特位的相对位置，那么各个通道内容会发生极大变化，所以需要进行回放数据的解码。回放时必须要先找到帧头和帧尾，根据帧头和帧尾的位置计算出由于随机回放而造成的各个通道数据的字偏移和位偏移，然后根据两个偏移量将各通道数据还原。该解码的原理如图 6 所示。回放时记录仪的时钟由 SHARC 给出，每次 SHARC 提供一帧所需的 864 个时钟，即取出 864 个比特位流。对于这样一帧信号，帧头可能在数据包中间的某一个位置，帧头首字 0 所处的位置相对于数据包中第一通道数据的 16 位字的偏移量就是字偏移；同时原有帧头的 0x1 在经过串口移位寄存器截取也可能不再是 0x1，原来帧头中的 1 所在新的 16 位数据字中的位置就是位偏移。假设原有的帧内容是重复的 0x12345，共有 40 组 0x12345。如果随机回放，那么读入 SHARC 内存的 16 位数据可能就是图 6(b) 和图 6(c) 所示的内容。因为向记录仪发送数据时已经给每一个数据包装上了帧头和帧尾，因此利用帧头和帧尾可以确定字偏移和位偏移这两个变量。在实际应用中，因为模拟电路的自噪声的存在，模拟信号经过模数转换后连续出现 0xFFFF 和 0x0，0x0，0x1 的情况的概率非常低，所以使用帧头必须含有连续的 47 位零的特征寻找帧头位置以确定字偏移和位偏移。如果记录仪的随机回放点落在帧头的 0x0，0x0，0x1 中间，可以较容易地根据具体情况找到字偏移和位偏移量。找到这两个偏移量后就可以通过移位计算获得原始的数据结构了。

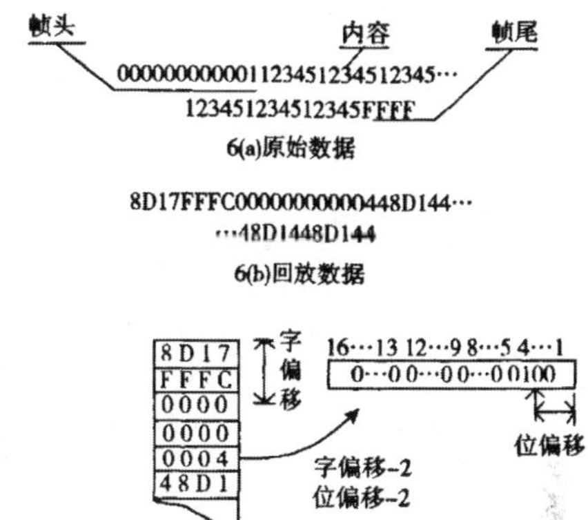

图 6　数据回放解码运算分析图

5　软件设计

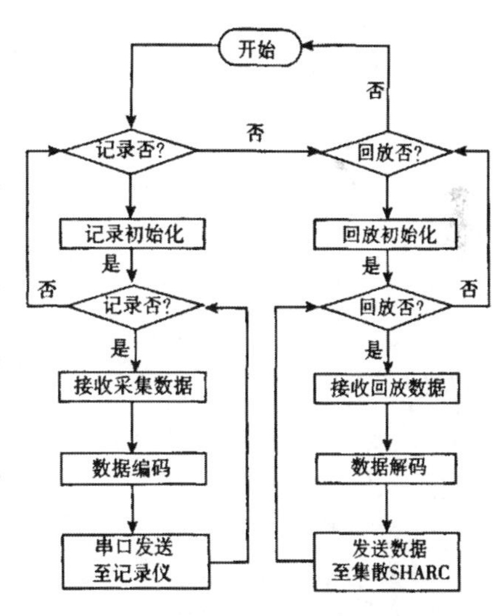

图 7　软件设计流程图

录音回放模块 SHARC 的软件任务分为两个循环。一个是记录循环，另一个为回放循环。记录循环包括①从完成数据集散功能的 SHARC 处接收收数据；②将数据编码；③从串口发送给录音机。回放循环包括①从记录仪接收数据；②数据解码还原；③将还原后的数据发送给数据集散的 SHARC。本程序采用嵌入式系统编程的多任务轮询机制，采用中断驱动的并行处理方式[4]。程序的流程图如图 7 所示。

6 结束语

本文详细介绍了用于SHARC阵列数字信号处理机与大容量高速数据记录设备BVLDS接口的设计方案和实现。对于实时高速串行记录，该设计方案充分利用了SHARC串行口的特性，因此具有软硬件实现简便的优点。该方案在实际使用中稳定可靠。

参 考 文 献

1 METRUM - DATATAPE Incorporated. , METRUM Technical Manual—Instructions for Buffered VLDS recorder/reproducer, 1998.

2 Analog Devices. Incorporated. , ADSP - 2106x SHARC User's Manual, 1995.

3 Texas Instruments Incorporated. , TB5T1—5V Dual Full - Duplex PECL Transceiver Datasheet. , 2004.

4 王昆 等. 数字式声呐的实时信号处理. 系统工程与电子技术, 1999, 21(12): 102 ~ 105

水下数据采集及传输系统在海洋石油勘探中的应用①②

李 媛 冯师军 李启虎

（中国科学院声学研究所 微弱信号检测与处理实验室 北京 100190）

摘要：简要介绍一套基于军用拖线阵声纳的数据采集及传输系统，并给出相应节点和平台的性能指标，同时针对海洋石油勘探设备的基本要求，来研究和探讨该设备用于海洋石油探测的可能。

关键词：石油勘探 数据采集 数据传输

The Application of Underwater Data Acquisition and High Transmission System in Ocean Oil Exploration

LI Yuan, FENG Shijun, LI Qihu

(Institute of Acoustics, Chinese Academy of Sciences, Beijing, 100190, China)

Abstract: The paper briefly introduces the underwater data Acquisition and High Transmission System for military towed array sonar. And performance of the nodes is proposed. We analyze the basic require of the equipment for ocean oil exploration. We research the feasibility of this equipment suits for ocean oil exploration.

Keywords: Oil Exploration, Data Acquisition, Data Transmission

目前，全球油气存储增长乏力，远远无法弥补每年的产量，然而全球的石油消费量仍将以较快的速度增长。根据 BP2005 年能源统计资料，1981 年的石油消费量为 29.9 亿吨，而到 2004 年已经达到 40.4 亿吨，并且根据国际能源署（IEA）发布的世界能源展望预测，从 2000 年～2030 年，世界石油需求预计年均增长 1.6%。未来巨大的油气需求将如何得以满足，这是摆在世界石油工业面前的一个大难题。相对贫乏的陆地，海洋石油资源占全球石油资源总量的 34 %，全球海洋石油蕴藏量约 1000 多亿吨，其中已探明的储量约为 380 亿吨。进一步加快海洋石油的勘探和开发有助于缓解石油供需矛盾。地震数据获取系统是海洋石油勘探的主要核心设备之一，而具有高精度和高分辨率的地震数据采集设备能够有效的提高海洋石油的勘探能力。据不完全统计，中国在未来十年至少有约 100～150 万道地震仪器的总需求量，以现在的平均价格测算，总价值约在 50～75 亿元。但是由于地震数据获取设备涉及到许多尖端技术，并且有些技术能直接转化用于军事。在引进先进的地震采集设备时，受到主要西方国家的出口限制，严重制约了我国地震物探事业的发展。

从 20 世纪 50 年代英国在北海地区附近发现海底石油开始，美国的西方石油公司、物探公司，英国的 Plessey 公司和挪威的 Simera 公司先后使用拖曳式线列阵来接收地震回波，如图 1 所示。

Sercel 公司开发的 Seal 系统，用于地震数据采集，如图 2 所示。该系统具有以下特点：结构灵活、先进，

① 国家自然科学基金，水下信道匹配基础研究(60532040).

② 微计算机应用, 2008, 29(9): 100-103.

工作效率高；电缆直径小（液体或固体），液体电缆直径为50mm，固定电缆直径为55mm。

斯伦贝谢公司开发的Q－Marine系统是一个综合性的地震数据采集系统，它致力于解决限制4D重复性的扰动，能确保合格的、高分辨的、极高保真的拖缆地震数据的有效传输。

图1　基于拖线阵的海洋石油地震勘探

图2　Seal 系统

在目前国内缺乏湿端（包括拖缆、多路信号传输、成缆技术等）的自主知识产权产品，大部分依赖进口。民用拖缆和军用拖缆有一定的差异（如拖缆的根数、所用的频率、信号传输方式等），但是其基本技术是相通的。目前国外生产民用拖线阵的工厂一般也同时承担军用拖线阵声纳的生产，如法国的Thales，美国的Lockheed Martin公司等。

本文介绍了一套基于军用拖线阵声纳的数据采集及传输系统，并尝试针对海洋石油勘探设备的基本要求，来研究和探讨该设备用于海洋石油探测的可能。

1　多基元水下小型数据采集及传输系统

多基元水下数据采集及高速传输技术被应用到军事领域的许多方面，最主要的一个应用是拖曳式线列阵声纳，拖曳式声纳系统及拖缆结构如图3所示。

其中声学模块除了水听器、前置放大器等还包含数据采集节点、传输节点和O/E（光电转换）节点。仪表模块还包含有深度（压力）传感器、温度传感器和航向传感器。

针对256基元的拖线阵为例，按照声纳设备的系统要求，数据采集采用20kHz的采样率和24bits的采样精度的模数转换器，每秒采集产生的数据量应为122.88Mbits。若按照1/3的冗余设计，数据传输系统的数据传输率应大于184.32 Mbits/s。

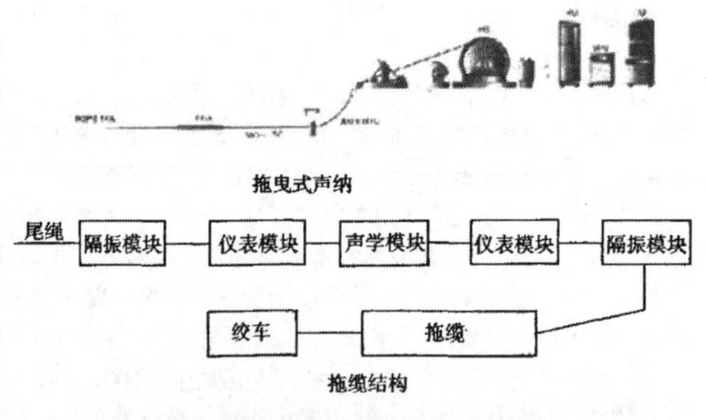

图3　拖曳式声纳系统及拖缆结构示意图

拖缆的细化一直是拖线阵声纳发展的重要方向，通过拖缆的细化可以减小绞车的尺寸，降低拖缆的整体重量，提高作战状态下的拖缆收放速度。同时为了在湍流噪声和拖线阵直径大小之间保持一种平衡，现在常用细线阵的拖缆直径大多都保持在30mm～40mm之间。所以通常声学模块内节点的直径应小于25mm。

1.1　声学模块内主要节点

图4为数据采集节点、传输节点和O/E节点的实物图，下面分别就三个节点的指标做简要的介绍。

数据采集节点性能指标：

模拟输入：八个差分模拟输入通道

输入阻抗：5M ohm

数据采样率：20kHz 为默认采样率，4kHz to 96KHz 可按用户要求修改

采样精度：24bit

功耗：300mA / +5V

尺寸：110mm×20mm×5mm；

传输节点性能指标：

传输介质：超五类双绞线

传输速率：160Mbit/s～320Mbit/s 数据传输速率（可选，默认设定为 192Mbit/s）；

传输距离：70m（默认数据传输率）

功耗：600mA / +5V

尺寸：105mm×20mm×6mm；

O/E 节点性能指标：

传输介质：9/125 的单模光纤，1310nm 的发射波长

传输速率：192Mbit/s

传输距离：10km

功耗：600mA / +5V

数据采集节点

传输节点

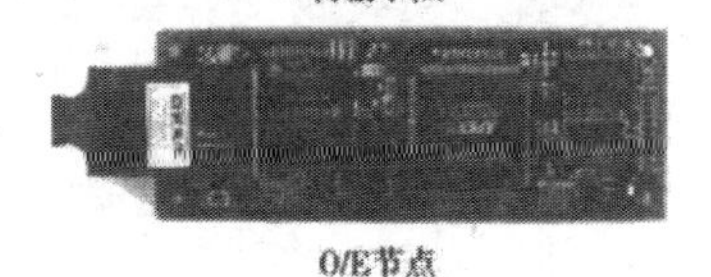

O/E节点

图 4　声学模块中的主要节点实物图

1.2　数据监控、转发及存储平台

数据监控、转发及存储平台采用 CTOS 技术实现对从水下数据采集及高速传输系统中获取数据进行监控、转发和存储，主要的系统结构如图 5 所示。

采用 eSATA 连接磁盘阵列，eSATA 实际上就是外置式 SATA Ⅱ 规范，是业界标准接口 Serial ATA（SATA）的延伸。eSATA 能提供最高 3Gbits/s 的数据传输率，实际数据传输可能介于 1.5Gbps 到 3Gbps 之间。

采用千兆以太网连接 PC 监控系统和外部网络，实现对数据采集及传输网络的监视和控制。

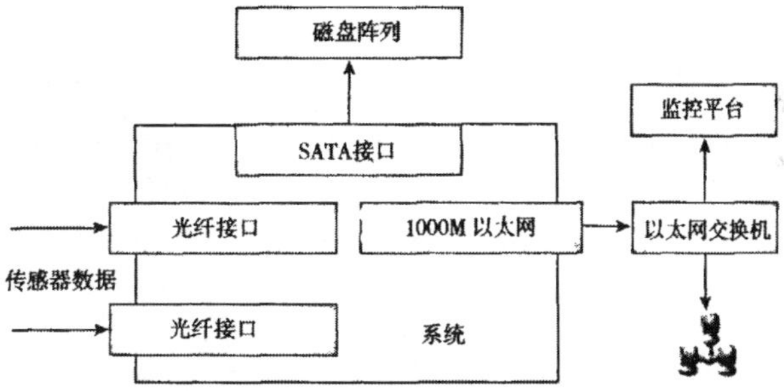

磁盘阵列
SATA接口
监控平台
光纤接口
1000M 以太网
以太网交换机
传感器数据
光纤接口
系统

图 5　数据监控、转发及存储平台结构图

1.3　测试系统

测试系统用于对整个水下数据采集及传输系统进行系统级的测试，也采用 CTOS 技术搭建。基于虚拟仪器的设计理念，测试系统硬件主要基于 NI 公司的 PXI 系列设备构成。机箱型号为 PXI－1031 四槽机箱；控制器使用 PXI－8196 工控单板机，PXI－8196 控制器的中央处理器为 Intel M790 讯驰处理器，512M DDR2 内存，30G 硬盘；D/A 板使用 PXI－6733 信号发生板，其提供 8 个 16 位的模拟输出通道，刷新率最高为 1MHz/秒。硬件结构实物图如图 6 所示。

测试系统的软件基于 LabVIEW8.0 环境开发，界面如图 7 所示。

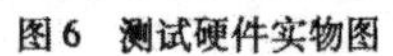

图6　测试硬件实物图

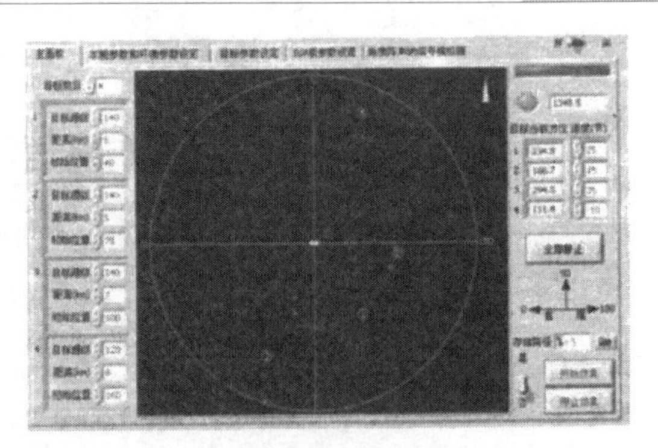

图7　测试系统界面

2　海洋石油勘探设备的基本要求

海上拖缆的通道数,以8缆大型物探船为例,约有16000通道,每缆通道为2000,这样的设备是为3D勘探设计的。若采用24bit的ADC,采样率(用采用间隔表示)为0.25ms,则每缆的数据率为192Mbits/s,8缆的数据率为1.536Gbits/s。

表1　数据采集及传输系统性能指标和海洋石油勘探设备要求比较

	单缆传输数据率	全系统存储数据率
数据采集及传输系统(最大)	320Mbits/s	3Gbits/s
海洋石油勘探设备要求	192Mbits/s	1.536Gbit/s

将多基元水下小型数据采集及传输系统的性能指标和海洋石油勘探设备的要求进行比较列在表1。

通过表1的比较可以发现,多基元水下小型数据采集及传输系统提供的性能指标完全能满足海洋石油勘探设备的要求。

3　结束语

通过分析发现,多基元水下小型数据采集及传输系统完全能应用到海洋石油勘探领域,但是由于民用拖缆和军用拖缆在拖缆的根数、所用的频率、信号传输方式等有一定的差异,在具体的应用中还需要对多基元水下小型数据采集及传输系统的A/D节点部分按照海洋石油勘探领域的要求进行进一步的修改。

参　考　文　献

1　李启虎. 数字式声纳设计原理. 安徽:安徽教育出版社,2003.2

2　冯师军,李启虎,孙长瑜. 拖曳阵声纳数字式水下数据高速传输的设计. 声学技术,2007,6, 26(3):362~266

3　李璐. 海洋石油工业的发展. 科技创新导报,2007(33):153

4　曹平. 勘探地震数据获取系统设计. 中国科学技术大学博士学位论文,2007

5　冯师军. 多基元水下小型数据采集及高速传输系统研究. 中国科学院声学所博士学位论文,2006

基于通用数据采集卡的水声应答器设计①②

徐克航　王　磊　冯师军　孙长瑜　李启虎

（中国科学院声学研究所　北京　100190）

摘要　水声应答器（Acoustic Transponder/Responder）是一类常用的水声设备，主要功能是完成水下设备的定位和导航，广泛应用在水声系统测试、目标模拟以及水声科学试验中。本文以NI公司的PCI-6052采集卡和工控机构成信号处理模块，设计出了一种可灵活配置的水声应答系统（ESRAE，Extensible Sound Responder for Acoustic Experiments），通过环境噪声电平的跟踪估计和脉冲判决技术实现了复杂海洋条件下脉冲检测、回发功能，并在海洋水声信道测量试验中得到了成功的应用。

关键词　水声应答器，目标模拟，数据采集卡，脉冲检测

A design of acoustic transponder/responder based on the general data acquisition device

XU Ke-Hang　WANG Lei　FENG Shi-Jun　SUN Chang-Yu　LI Qi-Hu

(*Institute of Acoustic*, *Chinese Academy of Sciences*, *Beijing* 100190)

Abstract　Acoustic transponder/responder is a commonly used underwater acoustic device, mainly for underwater positioning and navigation. It has been widely adopted in underwater acoustic testing, target simulation and other underwater acoustic experiments. This paper constructs a signal processing unit based on the general data acquisition device PCI-6052 of National Instruments Corp., and develops a flexible design method for acoustic transponder/responder system with improve ambient noise estimation and pulse detection technology. The responder realizes signal detection and pulse reply functions under noise background, and has been applied successfully in undersea acoustic survey experiments.

Key words　Acoustic transponder/responder, Target simulation, Data acquisition device, Pulse detection

① 国家自然科学基金资助项目(60532040).

② 应用声学, 2008, 27(6): 427-432.

1 引言

水声应答器(Acoustic Transponder/ Responder)是一类常用的水声设备,其基本功能为在接收到有效水声脉冲信号时按设定方式回发应答脉冲,通常用于水下动态/静态目标的定位或导航,在海洋环境监测、资源勘探开发、水下定位和测量等领域得到了广泛的应用。目前市场上有多种应答器产品。商业化应答器产品成熟可靠,然而它们作为专用水声系统,其特点为内置发射换能器,但发射频带固定且不具备低频段发射能力;工作方式可程控,不能满足水声试验和水声测试的要求,如声学反转镜试验中需要脉冲倒置发射方式,主动目标模拟中需要对目标强度和速度变化进行模拟等等。实际的水声试验要求一套功能和应用频带可根据需要灵活配置的应答系统。本文为了完成水声试验需求,采用通用数据采集卡和工控机为信号处理核心构成了一套试验水声应答系统ESRAE(Extensible Sound Responder for Acoustic Experiments),不仅可以根据任务要求完成水声测距和目标回波模拟功能,还可以方便地进行功能扩展。

本系统选用NI公司的PCI-6052E型通用数据采集卡实现电信号的采集接收与回发,该数据采集卡具有8通道差分模拟信号输入,每通道16Bit分辨率,总体采样率可以达到333KSamples/s,模拟通道还包括两路16Bit的D/A输出,能够与A/D输入并行工作。与此同时,系统还提供8路数字通道,可以方便对外设进行控制,例如提供给发射机的门控信号。通过NI-DAQ设备驱动程序库容易实现设备底层操作和控制,从而构建了一套可灵活配置的水声应答系统,满足了水声试验多样性应用的要求,同时提高了系统的集成度和稳定性。

2 本文设计ESRAE应答器的基本结构

ESRAE水声应答器的基本功能是在海洋噪声背景条件接收并判决脉冲信号,在识别到正确的脉冲信号后自动发出相应的应答信号作为响应,其系统结构框图如图1所示。

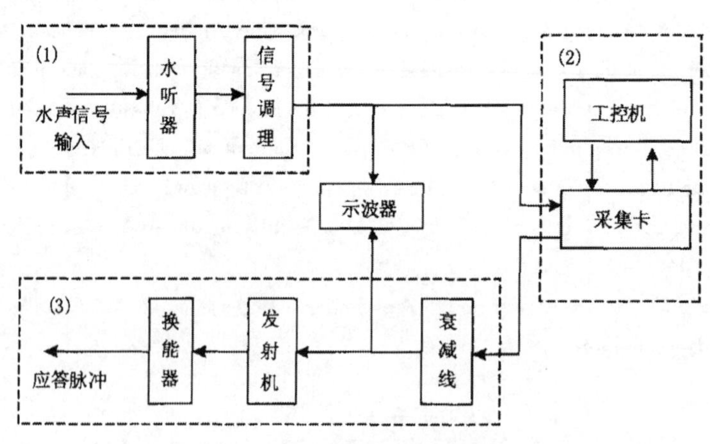

图1 ESRAE应答器系统框图

如图1所示,ESRAE水声应答系统由三部分构成:(1)前端接收及预处理模块,由水听器和高通滤波器组成;(2)信号处理模块,由工控机和PCI6052数采卡组成;(3)脉冲回发模块,由衰减线、发射机和回发换能器构成。接入的示波器可以同时监视接收信号和回发脉冲,可以直观地判断应答器工作状况。

前端接收及预处理模块接收水下声信号,

放大并经高通滤波后送入数据采集卡。高通滤波的截止频率为100Hz，可以有效滤除海洋和测量系统的低频干扰，增加信号频段的动态范围。信号处理模块完成应答系统的核心功能，首先由量程调整模块对前级输入信号进行增益调整，A/D转换后进行脉冲检测，检测到有效脉冲后将脉冲波形经适当调整后（这里包括脉冲峰值归一化、输出幅度调整和回发延时调整三个步骤）转换为模拟信号输出。脉冲回发模块主要将信号处理模块输出的回发脉冲信号经衰减线调整到发射机要求的输入范围（±1Vrms）经功率放大后发射，完成脉冲回发功能。

3　信号处理模块的设计及实现

应答器系统中的核心是图1中的(2)即信号处理模块，需要完成脉冲的检测和实时回发功能。在设计中需要认真考虑两个方面问题：(1)算法的可靠性，ESRAE应答系统要求在实际海洋环境下能够可靠应用，而应答距离的变化和多径效应使得接收脉冲幅度起伏、波形畸变；另一方面，附近经过/作业的渔船使得噪声背景产生较大的起伏，这对应答系统的脉冲检测过程带来一定的难度。(2)细致的软件设计，基于Windows的工控机系统对于实时处理任务而言是不利的，在一个数据拍的时间间隔内可以进行的运算也是严格限制的，这需要优化的算法设计和谨慎的软件安排。例如在数据连续采集过程中如果同时进行数据磁盘存储，其进程往往会超出一个数据拍的时间从而导致数据溢出错误。

3.1　信号处理模块的系统构成

参见图2，信号处理模块主要包括前放及量程选择、数字滤波、阈值估计、脉冲判决、脉冲波形记录、输出功率/时延选择和脉冲调整共七个模块，完成海洋噪声背景下主动脉冲信号的检测、估计、识别以及回发过程，同时存储有效脉冲波形以便于事后分析。其中，输入信号在±50mV至±10V的量程范围内可以根据脉冲峰值自动调节，数字滤波采用8阶巴特沃思带通滤波（频带需要预先设定），下面仅对关键的阈值估计和脉冲判决过程进行详细论述。

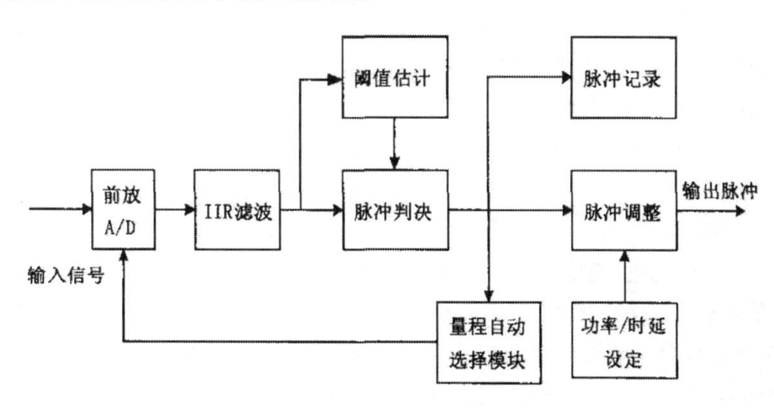

图2　信号处理模块算法框图

3.1.1　脉冲阈值估计

海洋环境下对脉冲的检测是一个基本的假设检验问题：

$$H_1:\ y(t)=s(t)+p(t)+e(t);$$
$$H_0:\ y(t)=p(t)+e(t); \tag{1}$$

在脉冲信号到达时，接收信号包括有效脉冲$s(t)$、环境噪声$e(t)$及脉冲干扰$p(t)$，否则，接收信号只有噪声及干扰。对于回波模拟的任务要求，主动脉冲的波形是未知的，因此脉冲检测采用基于能量的方法，即给定阈值

E_a,在脉冲到达时间,接收信号包络 $\tilde{y}(t) > E_a, t\in[Ts, Te]$,其中 Ts, Te 分别为脉冲起始和终止时间估计。

这里的阈值 E_a 是一与环境噪声相关的量。显然,海洋环境的多变性使得环境噪声电平起伏较大(特别是附近有行船经过时),为了有效实现脉冲检测,脉冲判决所需的阈值也应该能够根据实际情况的变化而进行调整。阈值估计的主要功能就是根据噪声电平的变化设置相应的判决门限,保证在变化噪声环境下仍然能有效地检测到脉冲序列。

考虑到应答器接收信号中主动脉冲信号只占小部分时间,而大部分时间是海洋噪声,这样可以通过对历史能量信息排序,并截取中间部分加以平均,以此估计噪声电平 E_e。

为了获得尽可能大的检测概率同时尽量减小虚警概率,判决门限的选取需要综合考虑,实际系统中根据应用需求和经验值选择阈值 W 为高于背景噪声电平 2~8dB:

$$E_a = E_e + W, W\in[2\text{dB}, 8\text{dB}]; \quad (2)$$

使用上述算法实现的阈值估计模块运算量较小,可以很好地满足系统实时性的要求,同时得到的判决门限也可以根据噪声能量的变化做出相应调整,适应了复杂海洋环境的要求。

3.1.2 脉冲判决方法

根据 3.1.1 节方法实现的脉冲检测器在实际应用中存在一些问题:首先,系统不能有效抑制干扰脉冲 $p(t)$,这些脉冲主要来源于试验平台的瞬时干扰以及水下系统的碰撞声;第二,到达脉冲由于多途作用幅度起伏较大,脉宽有明显扩展,而回波模拟需要对扩展后的脉冲进行回发。为了解决这两个问题,在脉冲检测中以信号分段平均值作为检测量并增加了有效脉冲判决环节。因此在应答器实现中,实际上是以数据拍平均能量(平方检波)作为检测统计量,数据拍的长度设为 20ms,脉冲存在的判决公式变为:

$$y(m) = \frac{1}{N}\sum_{n=1}^{N} x^2(m+n) \geq E_a; \quad (3)$$

其中 m 为数据拍序号,N 为数据拍样点数。对于瞬时干扰,根据实际测量,其脉宽通常小于 100ms,因此通过脉宽检测可以忽略绝大多数干扰脉冲。对于第二个问题,则需要更改脉冲结束判断规则,脉冲结束的判决公式为:

$$y(m) \leq E_a; \quad (4)$$

且
$$y(m+1) \leq \gamma E_a; \quad (5)$$

其中 γ 为一范围[0.2,1]的常数,调整该常数可以控制对脉冲多途扩展的保留程度。在浅海条件下,回波扩展一般在 200ms 以内,对连续两个数据拍能量分别满足式(4)与式(5)时可以判为脉冲结束,在实际应用中,这一判决可以有效保留信号的主多途扩展分量,同时避免了因信号起伏而误判脉冲结束的情况。

3.2 信号处理子系统算法实现

信号处理子系统在 Windows 环境下采用 Visual C++ 6.0 开发,通过 NI 公司提供的 NI-DAQ 设备驱动程序库实现数据采集硬件的底层控制,其软件流程如图 3 所示。首先在初始化数据采集卡时设定系统采样频率,并采用双缓冲技术进行 A/D 转换,以便在对一帧数据采集的同时实现上一帧数据的实时处理。采集到的各帧数据通过检测子模块进行脉冲判决,并将检测到的有效脉冲信号暂存在缓冲区中,待当前有效脉冲序列结束后再进行数据保存,然后通过 PCI-6052E 数据采集卡的输出通道进行 D/A 转换和脉冲回发。同时,为了防止瞬时出现能量较强的脉冲序列超出系统检测范围,系统软件中还加入了量程监控模块,该模块依据当前脉冲能量大小,在必要时调整系统参数设置,以此保证当前最大能量幅度限制在满量程的 1/3 左右。

由于 Windows 操作系统的非实时性,无法真正实现系统的实时采样判决，这里我们根据PCI-6052E通用数据采集卡自身的特点，将每帧脉冲数据的时间宽度设置为20ms左右，结合A/D转换中所采用的双缓冲区机制，最大限度地实现了信号的实时处理与连续采集。

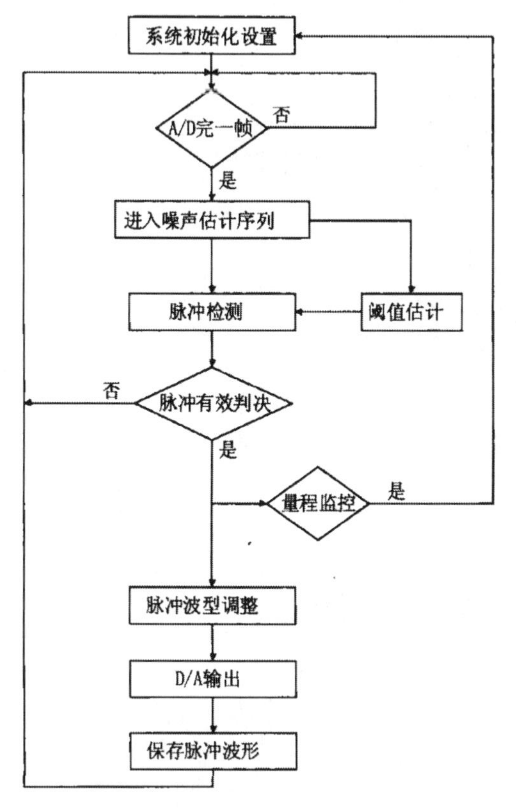

图3　信号处理子系统软件流程图

4　试验结果

ESRAE水声应答系统于2004年5月开始研制，10月参加海上试验，在人工干预的条件下初步完成水声应答试验；2005年6月，ESRAE系统在第二次海试中作为目标模拟器，实现了无人干预条件下的回波模拟功能；2006年12月海试的时间反转试验中，ESRAE系统完成了脉冲的时间反转回发功能。ESRAE水声应答器系统经试验测试验证，其基本性能指标如表1所示：

表1　ESRAE水声应答器系统基本性能

序号	基本指标	参数
1	接收声信号范围	54dB～210dB@1uPa
2	前级放大	1～2000倍，分7档
3	工作频带	500～1.5kHz
4	回发脉冲源级	150～185dB(可调)
5	回发延时	35ms～20s（可调）

图4给出了2005年6月某次海试试验中所接收的应答器应答数据处理图，该次试验中，应答器方位为55°（以阵首方向为起始点），距离约为9.2km，回发源级约176dB，应答延时为35ms左右。图中展示了连续的九段接收数据，从中可以清楚地分辨出每帧应答信号。

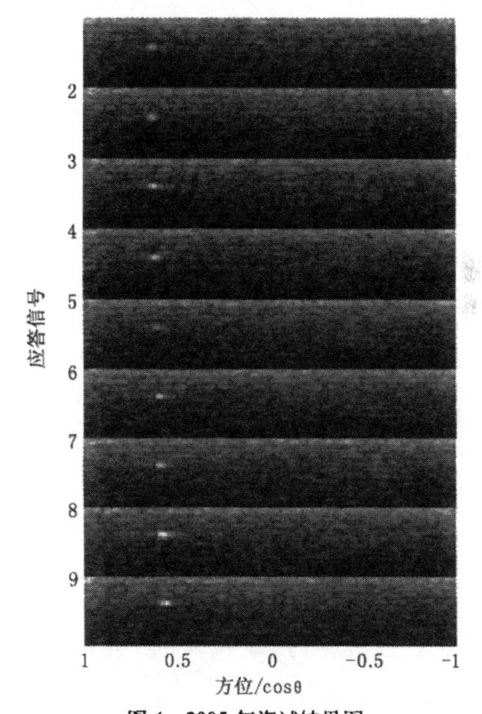

图4　2005年海试结果图

5　结论

在用于海洋实际测量的水声应答器工程系统中，通过选用通用型数据采集卡并配合使用其自带的硬件驱动程序库，缩短了系统的开发周期，提高了系统的集成度和稳定性，但在系统的实现中由于工控机所使用的非实时操作系统的影响,信号处理的实时性停留在数十毫秒量级,通过选用实时性较强的其他操作系统,如 VxWorks 等可以进一步增强系统中信号处理的实时性。该系统目前已在实际海试中获得应用,下一步工作将在引入远程通讯控制模块后,构成无人值守并具备多种工作模式的浮标应答系统。

参 考 文 献

[1] A D Whalen. Detection of Signal in Noise. Academic Press, New York and London, 1971.
[2] Smith W, W M Marquet, and M M Hunt. Navigation Transponder Survey: Design and Analysis. IEEE Ocean'75, 1975: 563 - 567.
[3] Porter M B. Acoustic Models and Sonar Systems. IEEE J. of Oceanic Engr., 1993, 18: 425 - 437.
[4] Hussain M G M. Principles of High-Resolution Radar Based on Nonsinusoidal Waver—Part I: Signal Representation and Pulse Compression. IEEE Trans. On Electromagn. Compat., 1989, 31(4): 359 - 368.
[5] 李启虎. 数字式声纳设计原理. 合肥:安徽教育出版社, 2003.
[6] 程佩青. 数字信号处理教程. 北京:清华大学出版社, 2003.
[7] National Instruments Corporation. Traditional NI-DAQ User Manual. National Instruments Corporation, 2003.
[8] 沈凤麟, 叶中付, 钱玉美. 信号统计分析与处理. 合肥:中国科学技术大学出版社, 2003.

基于虚拟仪器技术的通用水声信号发射系统设计[①②]

于海春　余华兵　孙长瑜　李启虎

（中国科学院声学研究所微弱信号检测与处理实验室　北京　100190）

摘要　本文针对水声学实验中对于发射信号源的使用要求，介绍了发射平台的结构以及软件流程，并对其中基于虚拟仪器技术的难点做了详细的论述，最终设计开发了一种通用水声信号发射系统，并介绍了一个具体的实验用途。

关键词　声学，虚拟仪器，信号发生器，监控

Design of general underwater acoustics signal transmiter system based on virtual instrumentation technologies

YU Hai-Chun　YU Hua-Bing　SUN Chang-Yu　LI Qi-Hu

(*Institute of Acoustics*, *Chinese Academy of Sciences*, *Beijing* 100190)

Abstract　This paper aims at the requirements of signal sources used in Underwater Acoustics experiments and presents the signal generator structure and software flow of the transmitter system have developed. Some difficulties in technologies based on Virtual Instrumentation are discussed . A software for general purpose underwater acoustics signal generator has been developed. An illustration of the experimental application is given.

Key words　Underwater acoustics, Virtual instrumentation, Signal generator, Monitor

1　引言

在大量的水声学实验中，水声信号源是必须的。一般首先由一台信号发生器产生电信号，然后通过功率放大设备放大后送到发射换能器以发射声信号。在这一过程中，信号发生器是产生特定信号的关键设备。传统的信号发生器通常工作方式单调，无法产生比较复杂的发射信号，而且在转变产生信号的特征时操作通常较为复杂，不能快捷的切换信号类型。当今，虚拟仪器技术的发展为解决这一问题提供了可能。虚拟仪器是基于计算机的仪器，是

① 国家自然科学基金重点项目；“水声信道匹配基础研究”(60532040).

② 应用声学, 2009, 28(2): 116-120.

计算机系统与仪器系统技术相结合的产物。目前，在这一领域使用较为广泛的计算机语言是美国 NI(National Instrument)公司的 LabVIEW (Laboratory Virtual Instrument Engineering Workbench)。目前已经发展到 LabVIEW 8.2 为核心，包括控制与仿真、高级数字信号处理、统计过程控制、模糊控制等众多附加软件包[1]。

本文结合水声信号发射平台的要求，利用相应的硬件，使用 LabVIEW 8.0 虚拟仪器开发环境设计并开发了一种通用水声信号发生系统，为水声学实验提供了一种可快捷变换信号类型，具有多通道同步驱动能力，附有发射信号监控并记录功能的水声信号发射平台。

2 信号发射平台系统构架

水声学实验中，水声信号发射的过程通常为：信号发生器产生电信号，然后通过功率放大，放大后的信号送到发射换能器以发射水声信号。[2]为了使实验人员准确地知道当前发射信号的特征，在这一过程中需要加入负责监控的模块。其信号发射平台完整的系统构成如图 1 所示。

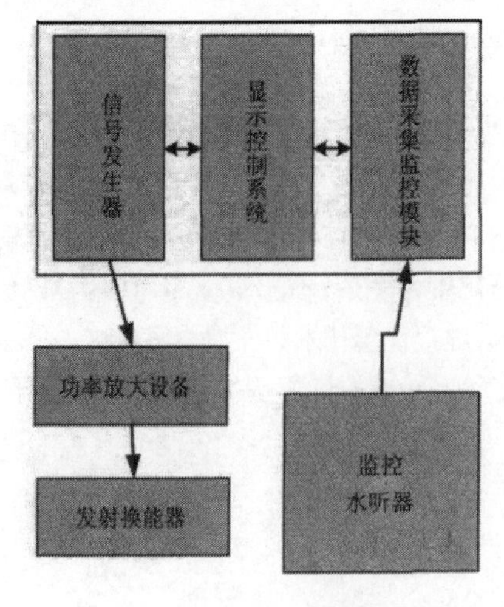

图 1　信号发射平台系统构成

在图 1 中，各个模块功能如下：

显示控制系统：控制信号的发射和接收，显示监控结果，存储监控数据。

信号发生器：向功率放大设备输入产生待发射信号(信号经过功放设备后由发射换能器发射声信号)。

功率放大设备：功率放大。

数据采集监控模块：采集声信号。

发射换能器：发射声信号。

监控水听器：接收声信号。

其中信号发生器、显示控制系统及数据采集监控模块三者作为一个相对独立的模块，合称信号发生器硬件系统。

3　信号发生器硬件系统构成

我们选择美国 NI 公司的 NI-1031 机箱，NI-8196 控制器，NI-6733 信号发生板，以及 NI-4462 数据采集板构成信号发生器的硬件系统。

其中 NI-8196 控制器的中央处理器为 Intel M790 讯驰处理器，512M DDR2 内存，30G 硬盘，其运行速度和存储空间可以保证信号的实时存储功能。NI-6733 具有 8 个 16 位的模拟输出通道，刷新率最高为 1MHz，可以完全覆盖水声学实验中需要的信号频段(几十赫兹到几十千赫兹)。NI-4462 的采样精度为 24Bits/s，采样率最高为 200kHz，完全满足声音信号监控的精度和频率要求。

NI-8196 控制器、NI-6733 信号发生板以及 NI-4462 数据采集板三者插在 NI-1031 机箱上形成一个整体的信号发射机箱，经过 P68 连接线连接到专用接口盒，专用接口盒实现 P68 插槽与 BNC 接口之间的转换，通过 BNC 接口向功率放大设备提供输入电信号。硬件连接示意图如图 2。

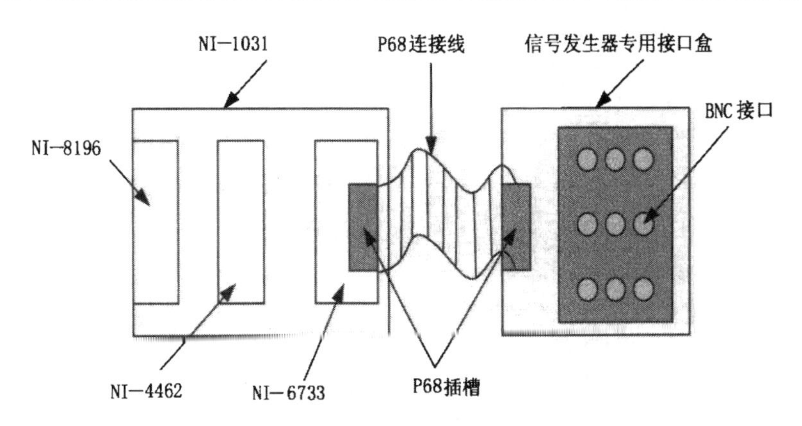

图2 信号发生器硬件系统连接示意图

4 信号发射平台软件设计

对于整个系统来讲,利用软件以生成满足要求的复杂波形是系统实现预定功能的先决条件,在NI公司的相关网页上已经存在很多由公司的工程师或者其他学者提供的复杂信号生成程序,国内的相关工作也有很多[3,4]。本系统所要求的复杂信号由多种不同类型的信号顺序排列而成,有时不同的序列之间的差别非常大,使用单一的复杂信号生成程序很难满足要求,有一种方法是可以利用LabVIEW中的专用接口调用Matlab模块生成复杂波形[5],这要求操作人员通过一些文本上的操作(如改变Matlab程序)来改变待发信号,但是考虑到系统的工作环境通常在海上,相对来说比较恶劣,操作人员有可能出现晕船而影响工作的情况,这会使系统的可靠性得不到保障。所以在本系统中采用以信号文件的形式存储待发信号的方式,信号文件在系统外部使用其他软件创建,例如Matlab,VC++等。同时,信号文件以系统规定的格式存储,以便使系统识别并导入到文件列表中,在使用时,只需要选择列表中相应的文件即可。

信号发射平台的软件系统安装于NI-8196工控计算机上,NI-8196安装Windows XP英文版操作系统。信号发生器的信号发射程序利用LabVIEW 8.0开发环境编写。本系统所使用的LabVIEW开发环境已经安装NI公司完整的数据采集(DAQ)硬件驱动包,对于DAQ硬件驱动程序包的安装和使用,本文不作深入论述,请感兴趣的读者阅读相关读物。

本水声信号发射平台的主要要求有4个:发射复杂信号,快捷变换信号类型、多路驱动以及实时监控存储。系统软件的工作流程如下。

(1) 操作系统启动,进入准备状态。

(2) 进入发射程序初始界面,输入信号文件数目、路径、信号幅度,发射次数,选择是否存盘。

(3)读取信号文件至内存。

(4) 检查输入参数合理性,监控进程工作,显示监控信号。

(5) 显示预发射信号波形图,若选择存盘模式,运行存盘模块。

(6)程序进入等待输入状态,输入发射命令后进入发射状态。

(7)发射信号。

(8)发射结束后退出发射程序,返回初始界面。

软件流程图如图3所示:

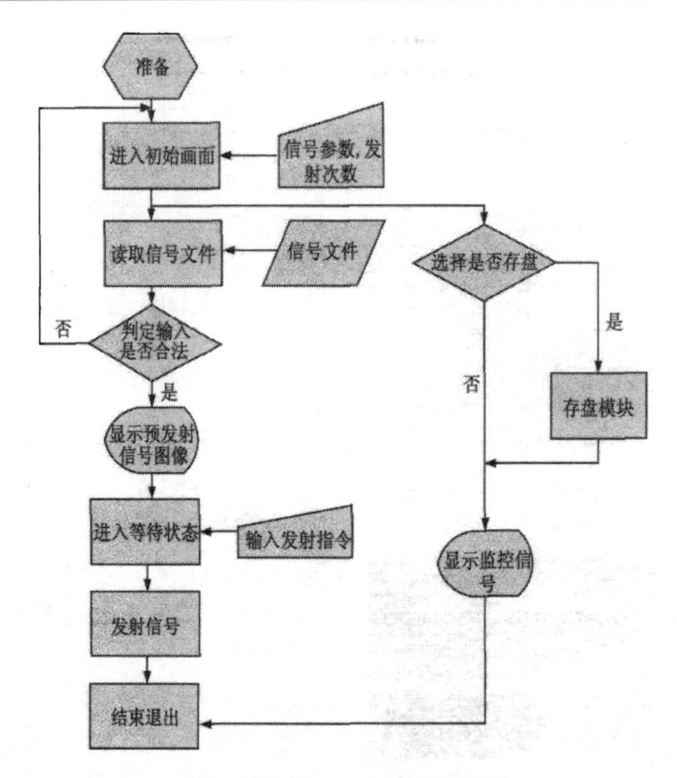

图3　信号发射平台发射程序流程图

信号发生软件的设计难点在于多路驱动和监控信号的实时存储功能。下面分别对这两点进行论述。

实现多路驱动功能需要正确设置 LabVIEW 中 DAQmx 模块中的建立设备通道模块,重要的是要对于建立物理通道项进行正确的配置,对于一个4路发射的任务,利用 LabVIEW 中 CreateChannel 模块建立通道,如图4。

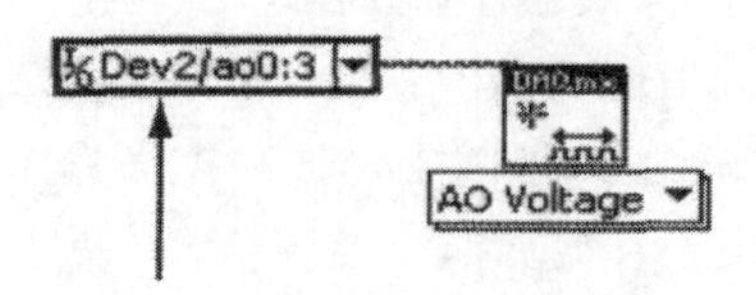

图4　CreateChannel 模块建立通道图

图4中箭头所指的位置所示的项目为物理通道选择项,它表示需要同时驱动 NI-6733 的模拟输出通道0到3,其中 Dev2 是 NI-6733 的设备名,ao0:3 是通道数。在软件工作中,发射和监控存储模块是同时工作于两个循环结构中,为了尽可能的防止存盘操作给程序运行带来影响,应减少存盘的频率,可以通过增加单次采集、存储的点数来实现,但是如果每次采集存储的点数过大,就会造成系统内存消耗过大,当所需发射信号文件较大时,会影响程序的正常运行。我们采用20kHz的采样率,每次采集4000个点,这样每秒钟进行5次存盘操作,数据以32位整数型存储,每4000个点占用的系统内存资源为16kBytes,对于 NI-8196 控制器 512Mbytes 的内存资源来说占用很少,可以保证整个程序的正常运行。程序中数据的采集模块使用 LabVIEW 中 DAQmx Read 模块,存储模块使用 WriteBinaryFile 模块,采集的数据直接送给 WriteBinaryFile 模块的 data 输入项,如图5。

最终实现的软件前面板如图6所示,为了满足一般的使用习惯,项目输入区安排在左边,操作区安排在下边,波形的显示安排在中间位置。

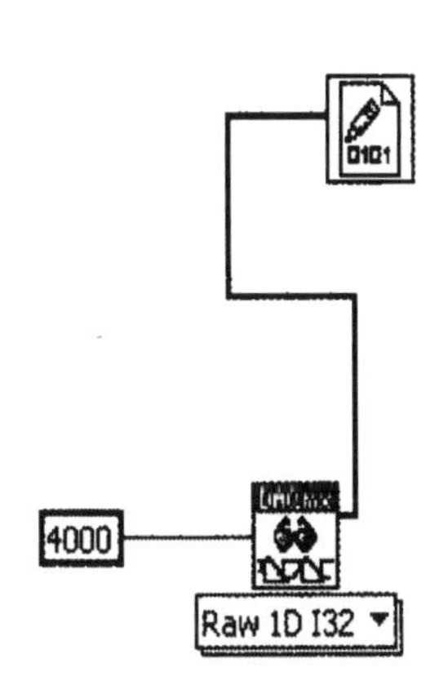

图5 DAQmx Read 模块和 WriteBinaryFile 模块

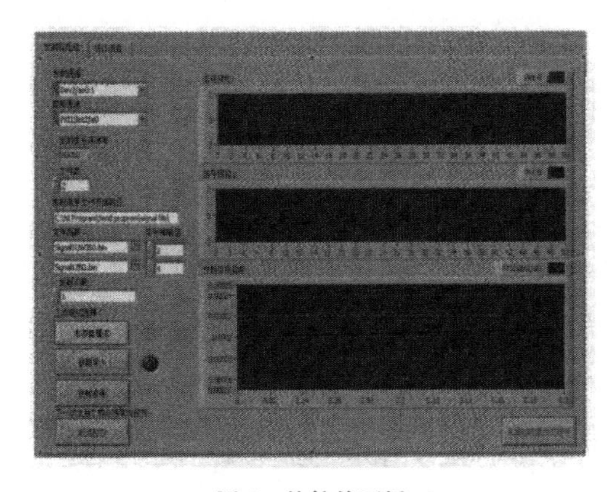

图6 软件前面板

5 系统用途及结论

本水声信号发射系统作为水声信道研究实验中的信号源使用，实验场景示意图如图7所示。

图7中监控水听器与发射换能器间隔一定距离，以此来记录发射换能器实际发射信号，在后期的实验数据处理中，使用监控水听器采集的信号作为原始发射信号，这样就避免了可能存在的由于发射换能器的工作状况发生变化造成的发射信号畸变给数据处理带来的影响。图8所示是在实际实验中，监控水听器采集的部分时域信号。

此系统已经过多次湖上和海上实验的考验，在不同的实验中很好的完成了不同类型，不同功率的信号发射任务。在这些实际使用中已经证明，此基于虚拟仪器技术的水声信号发射系统的功能强大，性能稳定，用途多样，操作使用简单，可以在大量的水声学实验中，作为通用的信号发射甚至采集平台使用。

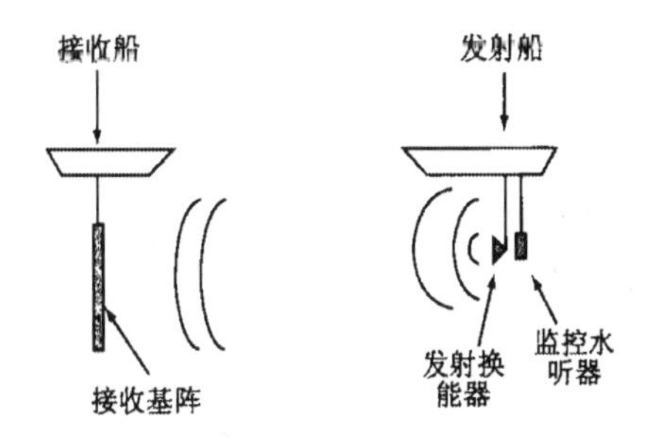

图7 水声信道研究实验场景示意图

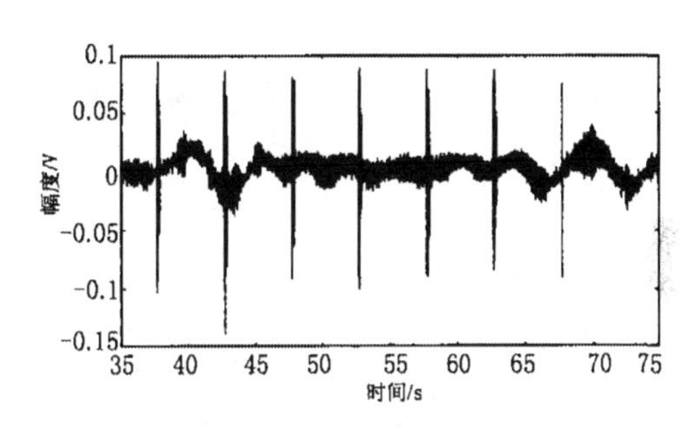

图8 监控水听器采集的部分信号

参 考 文 献

[1] 张凯，郭栋等 LabVIEW 虚拟仪器工程设计与开发，国防工业出版社 2004 年，190－220.

[2] 李启虎 数字式声纳设计原理，安徽教育出版社. 2002 年第1版，113－305.

[3] 黄培根 仿真信号波形及虚拟仪器频谱仪的构建和测试，电子测试，2007－08，83－87.

[4] 长大彪，于化龙 基于 LabVIEW 的调频连续波雷达仿真信号源的设计，仪表技术与传感器，2006 No.1.

[5] 姚世峰，薛德庆等 LabVIEW 与 Matlab 的混合编程，软件技术，2005，24，(6).

TigerSHARC201 在声纳信号处理系统中的应用[①②]

刘 钢[†] 李启虎

（中国科学院声学研究所 北京 100190）

摘要 本文介绍了ADI公司最新数字信号处理器TS201的主要特点和基于TS201的VMEBus通用数字信号处理板的设计。文中较详细地分析了声纳信号处理系统的算法特点，系统的运算量和数据传输率的要求。然后给出了基于Quad-TS201 VMEBus通用数字信号处理板的具体声纳系统设计方案。该方案充分利用TS201强大的运算能力及高速数据吞吐率，以实现声纳的时空处理任务。该系统已研制成功，并在实际海上试验中得到应用。

关键词 声纳，信号处理，TigerSHARC201，VMEBus

Application of TigerSHARC 201 in the sonar signal processing system

LIU Gang LI Qi-Hu

(*Institute of Acoustics Chinese Academy,Beijin 100190*)

Abstract This paper presents the main features of TigerSHARC201, which is the latest embedded processor from ADI, and the design scheme of Quad-TS201 VMEBus DSP board. It then describes in detail the SONAR algorithm, computation burden and data transfer need. Finally, a concrete system scheme is presented, which fully employs the processing speed and data transferring speed of TS201. The scheme has been made into reality, and the DSP system worked properly in the experiment on the sea.

Key words SONAR，Signal processing，TigerSHARC201，VMEBus

1 引言

DSP(Digital Signal Processor)芯片以其独特的总线结构和强大的信号处理能力广泛用于通信、雷达、声纳、图像处理及医用电子学等领域。随着人们对实时信号处理要求的不断增加，尤其是在雷达、声纳等领域中，单片DSP已经不能适应超大运算量的要求。这些信号处理系统的数据吞吐量大，计算复杂度高，因此必须采用计算能力强、精度高、具备高速数据交换能力的大规模并行处理系统。

① 国家自然科学基金资助项目(60532040).

② 应用声学, 2009, 28(3): 208-213.

ADI公司的TigerSHARC201(以下简称TS201)是一款高性能的数字信号处理器，是继SHARC DSP的新一代产品。与SHARC DSP相比，TigerSHARC在计算速度、内部存储器容量、体系结构以及外部通讯资源方面都做了巨大改进，更加适用各种不同的并行多处理器系统，完成各种实时数字信号处理。本文将主要描述TigerSHARC201在声纳信号处理系统中的应用。

2 TS201芯片和Quad-TS201 VMEBus通用信号处理板的介绍

TS201是一款静态超标量数字信号处理器[1,2]。其基本特性包括：600MHz主频下每秒可进行48亿次40bit定点乘累加操作（MACs），或者每秒可进行12亿次80bit浮点乘累加操作。在25mm ×25mm的封装内通过eDRAM技术提供了4/12/24Mbit的弹性内部存储器密度。

TS201的I/O端口包括LinkPort和ClusterBus。LinkPort是ADI公司的专利技术，可以组成静态互联网络，提供处理器间点对点的双向数据传输，而ClusterBus可以提供多处理器共享总线的无缝互连。TS201的LinkPort提供4位全双向I/O能力，且能够以双倍速率锁存数据，当内核时钟为500MHz时，即单方向速率可高达500MB/s,每个LinkPort双向数据传输率最高可达到1GB/s。

Quad-TS201 VMEBus 数字信号处理板是采用标准 VME 总线设计的通用数字信号处理板。该板集成了四片 TigerSHARC201，板内采用松耦和系统结构，每片 TS201 独享32MB SDRAM 和 512KB FLASH。板内四片TS201 通过全双工 LINK PORT 口（链路口）两两相连，此外全双工 LINK PORT 还可以提供强大的板间数据交换功能。多块Quad-TS201 VMEBus 板通过 LINK PORT 静态互联专利技术（MESH SP）组成大规模并行信号处理平台，可以适用于声纳、雷达、软件无线电等应用。该板的体系结构如图 1 所示。

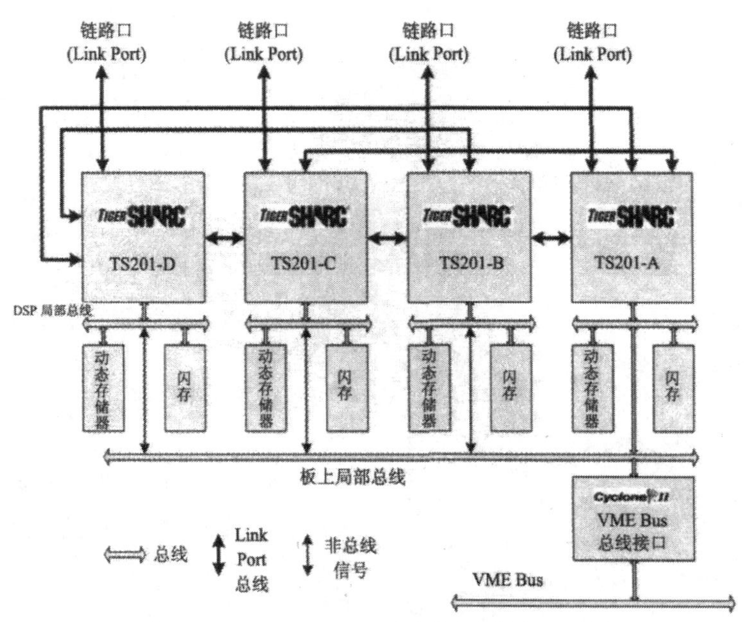

图 1 QUAD-TS201 VME 总线数字信号处理板结构

因为 TS201 处理器间可以利用 LinkPort 互连进行高速点到点的通信，所以当一块信号处理板无法完成大规模计算时，多块信号处理板间可以通过 LinkPort 组成二维网格结构完成并行计算，如图 2 所示。

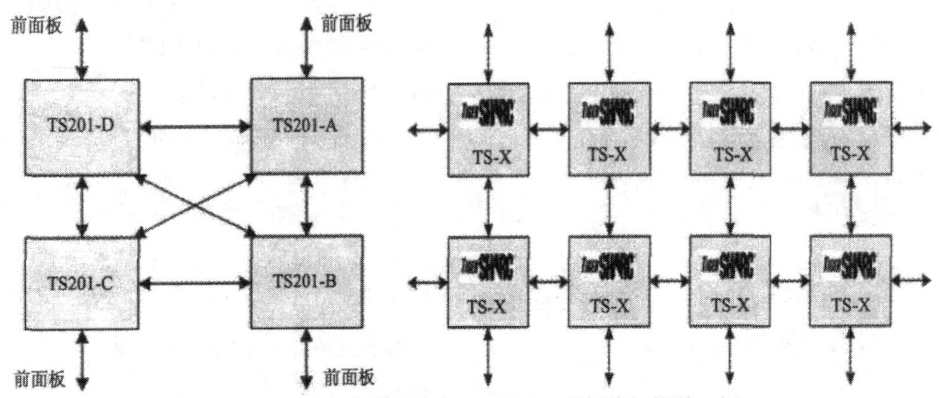

图 2　板内 LinkPort 拓扑图（左）和板间二维网格拓扑图（右）

3　声纳系统常用算法分析

水声信号处理的主要任务就是当存在干扰背景的情况下，对水下声场时空抽样，进行空间和时间变换，以提高检测所需信号的能力。图 3 为被动声纳的信号处理任务框图[3]。

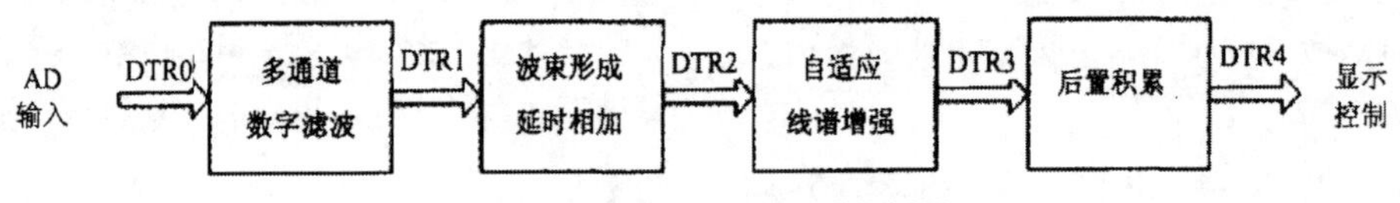

图 3　被动声纳信号处理系统框图

下面以主被动声纳信号处理常用算法为例，说明水声信号处理技术的算法结构特点。设输入序列为 $\{x(n)\}$，输出序列为 $\{y(n)\}$，那么 FIR 的输入和输出的关系如下。

$$y(n)=\sum_{k=0}^{N-1}h(k)x(n-k) \tag{1}$$

$h(n)$ 是滤波器系数。

波束形成是声纳信号处理的主要组成部分。它可以被看成是一种空间滤波器，使得基阵只在某一方向具有较高的灵敏度，而抑制来自别的方向的噪声和干扰。波束形成的实现有很多种方法。按照它们的加权函数是否时变，分为自适应波束形成和常规波束形成。下面仅就“延时-求和-平方”的常规波束形成（CBF），进行介绍。假定一个基阵由 N 个阵元构成，入射信号为平面波，同时假定信号与噪声、噪声与噪声间相互统计独立。以平面上的某一点为参考点，设到达第 i 个基元的信号为 $x_i[kT_s+\tau_i(\theta_0)]$，这里为 T_s 采样周期。这里 θ_0 为信号入射角。如果信号入射方向改变为 θ，我们仍以 $\tau_i(\theta_0)$ 作延时，那么第 i 路信号经延时后变成 $x_i[kT_s+\tau_i(\theta)-\tau_i(\theta_0)]$，系统的输出可以表示为:

$$y(\theta,kT_s)=\sum_{i=1}^{N}x_i[kT_s+\tau_i(\theta)-\tau_i(\theta_0)] \tag{2}$$

在舰艇噪声功率谱的低频范围内存在着线状谱。通过检测这些线状谱可以增强检测目标的能力。自适应线谱增强的工作原理是利用线谱信号相关函数的周期性和宽带噪声相关函数衰减很快的差异来增强信号，减少干扰的。采用横向滤波器结构和 LMS 算法的自适应线谱增强的主要计算公式如下：

$$y_i(n)=\sum_{j=0}^{LA-1} w_{ij}x_i(n-\Delta-j) \tag{3}$$

$$w_{ij}(n)=w_{ij}(n-1)+\mu_i x_i(n-\Delta-j)[x_i(n)-y_i(n)] \tag{4}$$

式中，w_{ij} 是横向滤波器权系数；LA 是横向滤波器阶数；Δ 是信号延迟长度。μ_i 是自适应步长。

后置积累处理方法是检波和积分，检波可以是平方检波或绝对值检波，积分可以使用线性积分或指数积分。这里介绍绝对值检波和指数型积分。

绝对值检波：

$$\overline{y(n)}=\frac{1}{m1}\sum_{n=1}^{m1}|x(n)| \tag{5}$$

指数型积分：

$$z(n)=\frac{1}{m2}\overline{y(n)}+\left(1-\frac{1}{m2}\right)z(n-1) \tag{6}$$

关于主动声纳，其回波检测常规方法主要有相干脉冲压缩（匹配滤波）和非相干窄带脉冲能量积累，这两种方法分别对应着回波特性的两种极端假定[4]。其他处理方法有变换域（如 Gabor 域，小波域）检波方法等，但是这些方法在实际应用中并不稳健，也不能取得优于常规方法的检测性能。因此主动声纳信号当前仍然主要采用线性调频和单频脉冲形式，接收仍然采用匹配滤波或窄带能量积累方法。下面将只对匹配滤波的频域实现进行分析。设发射信号为 $s(t)$，形式为线性调频信号（LFM），频率从 f_1 变到 f_2，带宽为 B，脉宽为 T，即：

$$s(t)=\exp\left[j2\pi\left(f_2 t+\frac{f_2-f_1}{2T}t^2\right)\right]$$

$$t\in[0,T]; \tag{7}$$

设接收机波束形成后的输出信号为 $x(t)$，则匹配滤波器输出为：

$$y(t)=\int_0^T x(\tau)s^*(t-\tau)d\tau \tag{8}$$

（8）的离散表达式为

$$y(n)=\sum_{i=0}^{N-1} x(n-i)s^*(N-i) \tag{9}$$

其中 $N=\frac{T}{T_s}$。上式就是时域匹配滤波的实现方法，但是考虑到快速运算的需要，可以通过频域的计算，然后经过反傅立叶变换得到时域输出。运算的方框图如图 4 所示。需要注意的是 FFT 实现的是循环卷积，而在时域中实现的是真正的卷积，所以 FFT 和 IFFT 的计算点数为2N点，但是每次输出只取IFFT 的 N 点作为输出，每次更新 N 点。

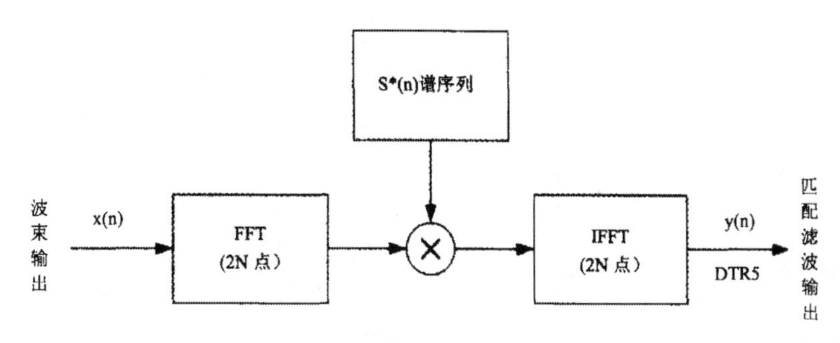

图 4　匹配滤波频域实现方框图

4 声纳信号处理系统运算量和数据传输率的初步估计

假定一组主被动声纳参数如下：均匀线列阵，阵元数 N_E=128；A/D 采样频率为 f_s=20K；FIR 滤波器节数 L=128；波束数为 M=256；波束形成采用快入慢出技术，降采样后的输出速率为 f_{sd}=10K；横向滤波器阶数 LA=32；积累系数 m1=1024；积分系数为 m2=32，计算以 32 位浮点数进行；LFM 脉冲宽度为 102.4ms，采样点数为 $N=2048$；

- 运算量估计

FIR 乘加次数估计：

$$N_E \times L \times f_s = 128 \times 128 \times 20000 = 327.68MFLOPS$$

波束形成乘加次数估计：

$$N_E \times M \times (f_s/2) = 128 \times 256 \times 10000 = 327.68MFLOPS$$

线谱增强乘加次数估计：

$$M \times 2 \times LA \times f_{sd} = 256 \times 2 \times 32 \times 10000 = 163.84MFLOPS$$

绝对值检波乘加次数估计：

$$M \times m1 \times \left(\frac{f_{sd}}{m1}\right) = 256 \times 10000 = 2.56MFLOPS$$

指数积分乘加次数估计：

$$M \times 2 \times \left(\frac{f_{sd}}{m1}\right) = 256 \times 2 \times \left(\frac{10000}{1024}\right) = 5KFLOPS$$

回波检测：

$$2 \times \left(\frac{2N}{2}\log_2(2N)\right) \times M \times \frac{f_{sd}}{N} + 2N \times 4 \times M \times \frac{f_{sd}}{N} + N \times 2 \times \frac{f_{sd}}{N} \approx 82\ MFLOPS$$

- 数据传输率分析：

A/D 输出数据率为：

$$DTR0 = N_E \times f_s \times 4 = 128 \times 20000 \times 4 \approx 10MB/S$$

（此处不考虑 A/D 的精度问题，运算输入使用的是 32 位字长的浮点数。）

FIR 输出数据率为：

$$DTR1 = N_E \times f_s \times 4 = 128 \times 20000 \times 4 \approx 10MB/S$$

波束形成输出数据率为：

$$DTR2 = M \times f_{sd} \times 4 = 256 \times 10000 \times 4 \approx 10MB/S$$

线谱增强输出数据率为：

$$DTR3 = M \times f_{sd} \times 4 = 256 \times 10000 \times 4 \approx 10MB/S$$

后置积累输出数据率为：

$$DTR4 = M \times \left(\frac{f_{sd}}{m1}\right) \times 4 = 256 \times \left(\frac{10000}{1024}\right) \times 4 \approx 10KB/S$$

匹配滤波输出的数据率为：

$$DTR5 = M \times \frac{f_{sd}}{N} \times 4 = 256 \times \frac{10000}{2048} \times 4 \approx 5KB/S$$

5 声纳信号处理系统的实现

在实际声纳信号处理机设计时，不能完全采用上一节所估计的运算量。因为这种估计没有包括数据的复制、搬移、传输、以及所用编程语言的编译系统效率等问题，所以在信号处理机设计时，必须对运算估计量乘以一个系数，系数一般应为 2~3。按照上述算法的要求，运算量要求约为 2 GFLOPS~3 GFLOPS，数据传输率最大值为 10MB/S。根据 Quad-TS201 VMEBus 通用信号处理板的性能，它对声纳系统的这两个指标的实现是能够胜任的。

如果采用 Quad-TS201 VMEBus 通用信号处理板完成图 3 及图 4 所示的声纳功能，

系统设计框图可如图 5 所示。数字信号处理的任务可以集中在这个处理板上完成。任务分工可以如图 6 所示，TS201-C 负责完成多通道 FIR 滤波和波束形成；TS201-B 负责完成线谱增强和后置积累；TS201-A 负责完成 VME 总线通讯；TS201-D 负责脉冲信号生成及回波检测功能。系统的数据流和控制流由板内各 TS201 间的 LinkPort 完成。

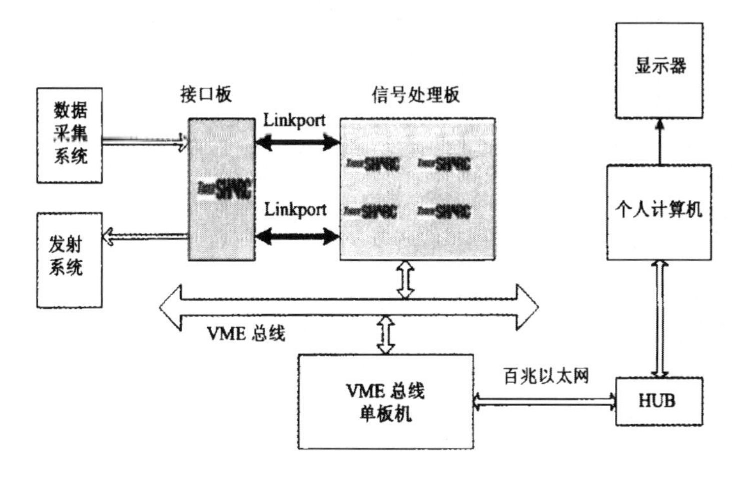

图 5　基于 TS201 的声纳信号处理机体系结构框图

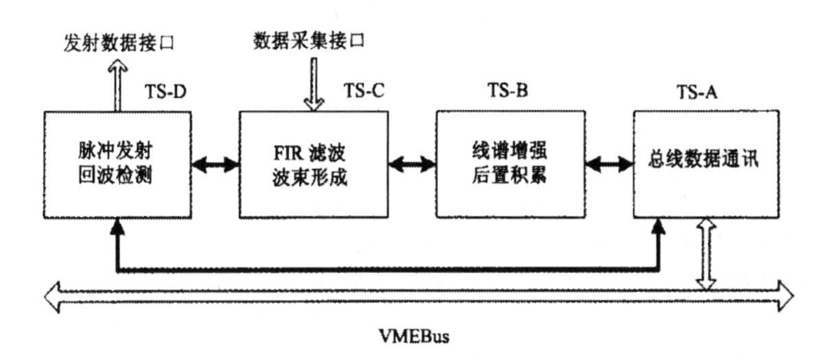

图 6　基于 Quad-TS201 VMEBus 板的声纳信号处理系统框图

6　结论

本文介绍了ADI公司的TS201芯片以及基于TS201的VMEBus通用信号处理板的结构，并介绍了我们对多片TS201在声纳实时数字信号处理系统中的应用，该系统充分利用TS201强大的运算能力及高速数据吞吐量，以实现被动声纳的时空处理任务。文中分析了系统的运算量、数据传输率以及完成算法所需的芯片数，具有工程指导意义。该系统已经研制成功，并应用于水声匹配基础研究项目的海上实验中，取得了良好的效果。

参考文献

[1] ADI 公司．ADSP-TS201 TigerSHARC® Processor Hardware Reference Manual. Revision1.1 2004.

[2] ADI公司. ADSP-TS201 TigerSHARC® Processor Programming Reference Manual. Revision 1.1 2005.

[3] 李启虎．数字式声纳设计原理．合肥：安徽教育出版社，2003.

[4] 朱埜．主动声纳检测信息原理．北京：海洋出版社，1990.

基于虚拟仪器技术的多路水声信号同步采集及处理平台设计[①]

田　甜　李启虎　王　磊　孙长瑜
（中国科学院声学研究所　北京　100190）

摘要　本文介绍了虚拟仪器技术在水声信号处理研究中的一个具体应用，利用LabVIEW语言编程，在NI公司高速数据采集硬件平台上实现了高采样率下多通道数据的同步采集、处理、记录功能，并在不使用信号处理机的情况下，利用主控单板机的单核CPU完成了32通道常规波束成形（CBF）的实时计算。同时，本文对非DSP硬件平台的实时信号处理实现方法进行了初步的研究。

关键词　虚拟仪器，同步采集，常规波束成形，非DSP硬件平台

The design of multiple-channel data acquisition & processing platform based on virtual instrument technologies

TIAN Tian　LI Qihu　WANG Lei　SUN Changyu
(*Institute of Acoustics, Chinese Academy of Sciences, Beijing* 100190)

Abstract　This paper describes an application of the virtual instrument technology in the area of the underwater acoustic signal processing. It realizes multiple-channel synchronous data acquisition, processing and memorizing in high sample rate based on LabVIEW language. It also realizes the real time processing of 32-channel conventional beamforming (CBF) on the single-core CPU of the main control board without DSP platform. Meanwhile, this paper presents some tentative research results on the realization of real time signal processing with non-DSP platform.

Key words　Virtual instrument，Synchronous data acquisition，CBF，Non-DSP platform

1　引言

利用水听器基阵接收信号是水声学研究中最常用的试验手段之一。水听器基阵信号的数字化传输通常需要较多的硬件资源，因此，一般的水声试验仍然广泛使用模拟基阵来获取水声信号，它将多路模拟信号直接上传至信号接收端，这就涉及到多路模拟信号同步采集的问题。本文基于LabVIEW语言和虚拟仪器技术，利用NI公司出品的高速同步信号采集板卡PXI-4472B和PXI-4462以及PXI-8196主控单板计算机构筑的信号采集

① 应用声学, 2011, 30(4): 314-320.

平台实现了最多达72路的模拟信号高采样率同步采集、显示、处理和存储，并在不依赖信号处理机的情况下，在此平台上进一步实现了32阵元信号常规波束成形的实时计算和显示。

虚拟仪器是一个针对计算机的概念，它是通过编程使计算机实现特定仪器设备功能的一种软件系统。本文描述的运行于PXI-8196的虚拟仪器系统，是通过LabVIEW语言编程实现的。LabVIEW是一种图形化的编程语言，它广泛地被工业界、学术界和研究实验室所接受，视为一个标准的数据采集和仪器控制软件，利用他可以方便地建立用户所需的虚拟仪器[1]。LabVIEW中包含了各种用于控制采集板卡的虚拟仪器模块（VI），可方便实现对硬件的控制。同时，LabVIEW封装了大量用于输入和输出的控件，这些控件图文并茂，为信号波形和其它信息的显示提供了便利。

2　信号采集平台的架构及工作原理

2.1 硬件的系统架构

本文实现的高速数据采集平台是由PXI-8196工控机、PXI-4462板卡两块以及PXI-4472B板卡8块组成的。该平台最多可完成72路模拟信号的同步采集，最高采样率100 kHz[2]。实物如图1。

图1　系统实物

其中，PXI-8196是主控计算机，它负责控制其它板卡的工作，同时负责数据的处理、显示和存储；PXI-4462和PXI-4472采集的数据通过PXI总线写入PXI-8196的内存；PXI-4462为所有其他信号采集板卡提供同步时钟节拍，以保证所有的采集同步。平台内各板卡的逻辑架构如图2所示。

2.2 信号采集板卡的工作流程

系统开始工作后，板卡首先接收用户设置的采样频率fs、每次采样点数N等关键参

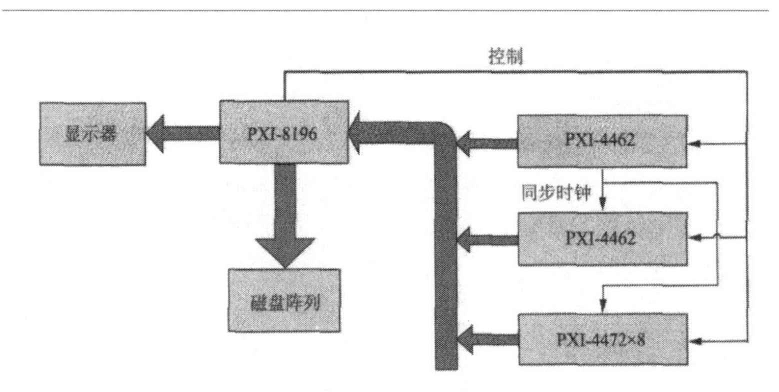

图2　数据采集平台的逻辑架构

数，然后启动并开始以频率fs采集模拟信号。在每采满N个点的数据后，板卡向主控计算机发送中断，并通过PXI总线将N个点的数据写入主控计算机的内存，如此周而复始，直至用户输入终止命令。板卡每采集到N个点所用的时间T，叫做采集周期，用户设置了fs和N后，T随之确定。显然，$T=\frac{N}{fs}$，单位秒。系统采集数据的流程可由图3表示。

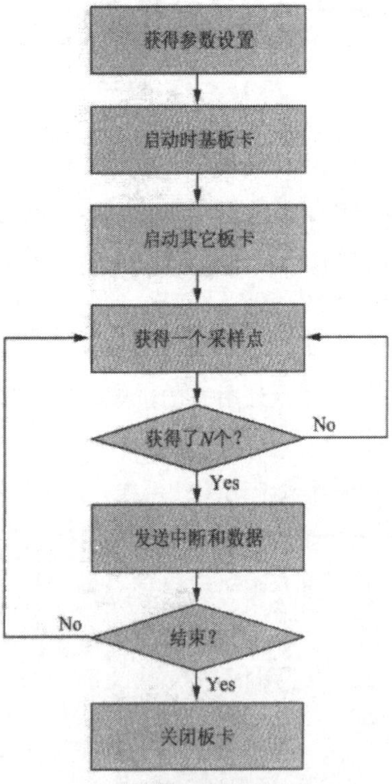

图3　信号采集板卡的工作流程

3　基于 LabVIEW 的信号同步采集及处理虚拟仪器设计

PXI-8196使用的是Intel公司的X86架构单核CPU，操作系统使用了Windows XP，这构成了本文所述的同步数据采集虚拟仪器系统的运行环境。该系统实现了以下功能：

(1) 在用户设定的采样频率下，同步采集多路模拟信号；

(2) 显示用户规定通道的信号时域波形；

(3) 实时计算并显示用户规定通道信号的频谱；

(4) 进行时频分析、LOFAR/DEMON谱分析计算并显示；

(5) 实时存储所有通道信号数据。

使用LabVIEW制作用于数据采集的虚拟仪器系统有一套标准流程[3]，因此该项工作不是本文讨论的重点。上述系统功能中的(3)和（4）实际上完成的是实时信号处理的工作，这是利用PXI-8196的CPU实现的。

由本文2.2节的描述可知，对于信号采集板卡，开始工作后，它每隔一个采样周期Ts完成一次信号采样，$Ts=\frac{1}{fs}$，其工作量在时间轴上是均匀分布的，如图4(a)所示；而对于PXI-8196的CPU而言，它每隔一个采集周期T才完成一次数据收集、显示和存储。设显示和存储一批（N点）数据用时Tr，则CPU的工作量在时间轴上的分布如图4(b)所示。

显然，若该虚拟仪器系统仅仅完成数据采集、显示和存储，则对于CPU，在每一个T时间段内，总有$Tp=T-Tr$的时间是空闲的。本文实现的虚拟仪器就是利用这一段空闲时间，完成了信号频谱分析（FFT运算）、时频分析（短时傅立叶变换）、LOFAR和DEMON等内容的实时信号处理运算，并显示其运算结果。FFT等信号处理算法在LabVIEW中均有封装好的运算模块以供使用，因此具体的算法编程实现是一个相对方便的工作[5]，此处就不赘述了。

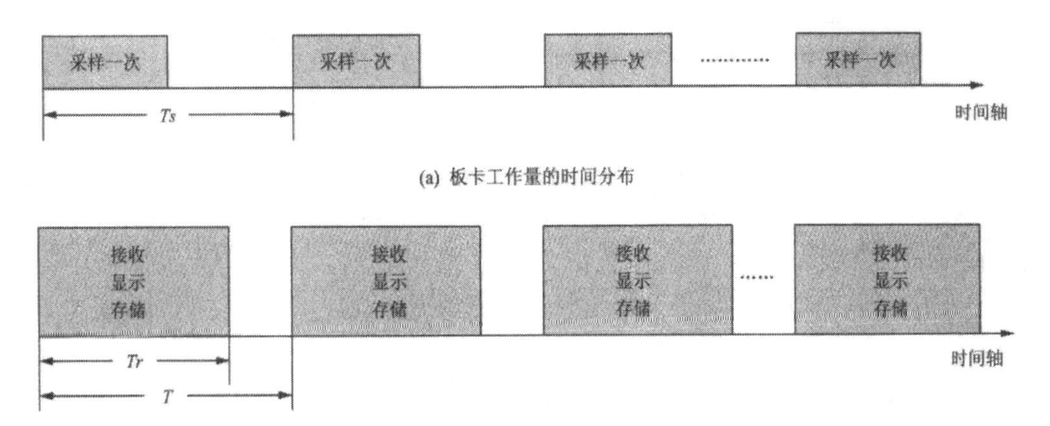

图4　系统硬件工作的时间分布

运行该虚拟仪器系统时，工作流程如图5所示。

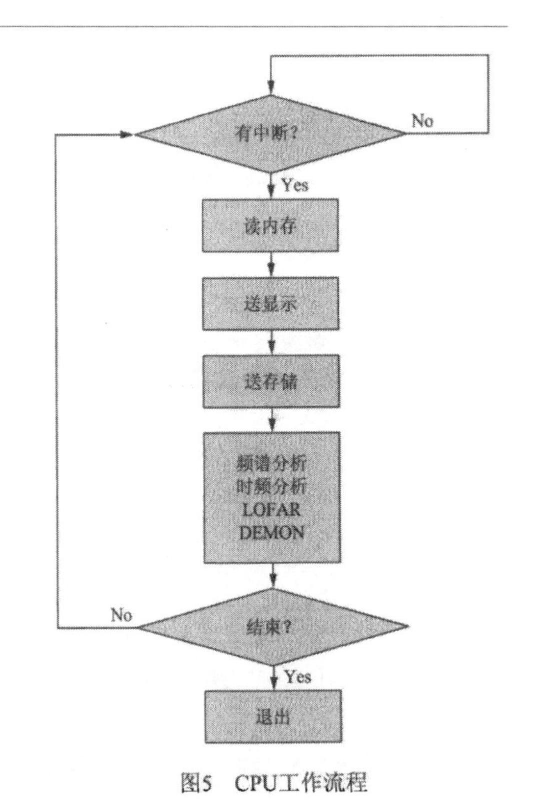

图5　CPU工作流程

4　常规波束成形实时计算的实现

在前面的节中，虚拟仪器系统已经实现了多路模拟信号的同步采集、显示、存储以及一些简单的实时信号处理功能。本章将以水声阵列信号最常用的信号处理算法——常规波束成形（CBF）为例，讨论如何在不借助信号处理机的情况下，在保留之前所有功能的同时，仍仅利用X86架构的CPU来实现大规模的实时信号处理运算，使该虚拟仪器系统成为一个独立、完整的集数据采集和实时处理于一体的多功能综合平台。

4.1 软件结构

CBF相对于本文第3节中的频谱分析等信号处理来说，运算量大得多，在*Tp*时间内是无法完成的。因此，必须在程序结构上进行调整。在本文第3节中，CPU的工作流程，采用的是串行的顺序结构，而Windows操作系统是基于时分复用技术的非实时操作系统，在这样的环境下，串行结构速度慢，且对资源的利用率低。从宏观来看，在串行结构中，虚拟仪器系统某一时间段只要求CPU完成一项工作，因而只能分配到与之相应的CPU资源，这样一来，CPU剩余的资源就被闲置下来了。

由X86架构的CPU以及Windows操作系统的工作机制可知，无论被闲置的资源有多少，CPU执行某一项工作的耗时是一定的，也就是说，利用闲置资源完成其他的运算任务，不影响基本功能的正常执行。

基于上述思想，本文使用多线程技术来实现复杂信号处理的实时运算。将虚拟仪器系统的信号采集、存储、显示等基本功能放置于软件的主线程中，另外开启一个新的线程，专门完成CBF运算。CPU在处理多线程时，将按照各线程的实际需求来分配资源，不会因新增的信号处理线程而减少主线程（基本功能）获得的资源，这保证了两个线程的并行执行；另一方面，用于完成CBF运算的可用时间由*Tp*增加至了*T*，这大大增加了该虚拟仪器系统的信号处理能力。

4.2 多线程虚拟仪器的实现

下面讨论多线程在虚拟仪器中的实现方法。

LabVIEW程序是数据流驱动的，本文利用LabVIEW提供的局部变量控件（Local Variable）实现对数据的分流[5]，并在CBF线程中设置消息等待机制，主线程获得中断后，随即发送同步消息给CBF线程，CBF获得该消息后开始读取Local Variable中的数据，并完成运算。这种数据分流与消息同步机制的使用，可保证两个线程的数据共享和同步运行。

添加CBF线程后，CPU工作量的时间分布变为如图6所示。

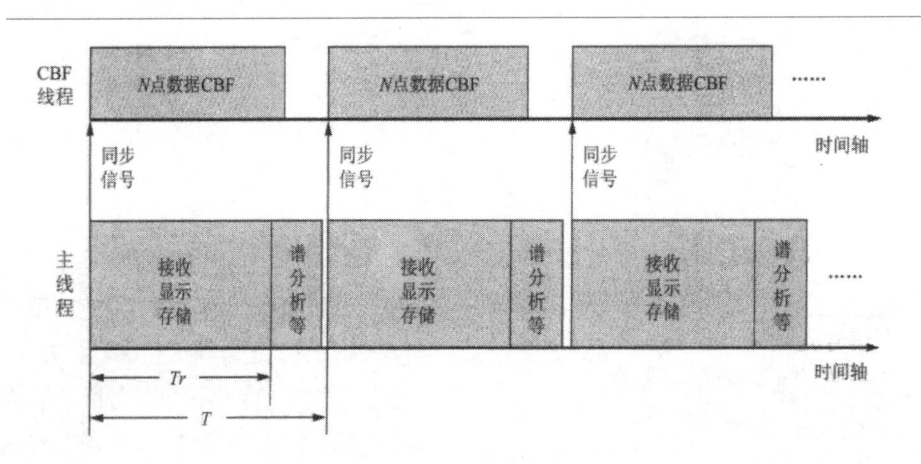

图6　多线程下CPU工作量的时间分布

4.3 常规波束成形（CBF）算法的实现

CBF算法是一个经典的水声信号处理算法，其基本思路是将各阵元信号进行适当延时后相加，再利用平方检波的方法获得指向性[4]。这种“延时-求和-平方”的算法，利用LabVIEW提供的强大的数组操作控件和循环寄存器控件，可方便实现。

由于系统资源及CPU自身能力的限制，本文在20 kHz采样率下，采用定点数运算，实现了32路信号的实时CBF计算及方位历程的灰度显示。采用定点数的原因将在本文4.4节中阐述。需要特别说明的是，通过更换软件算法模块，本系统的架构可以实现多种信号处理功能，可满足各种不同需求。

整套虚拟仪器系统在PXI-8196上运行时，其流程如图7所示。

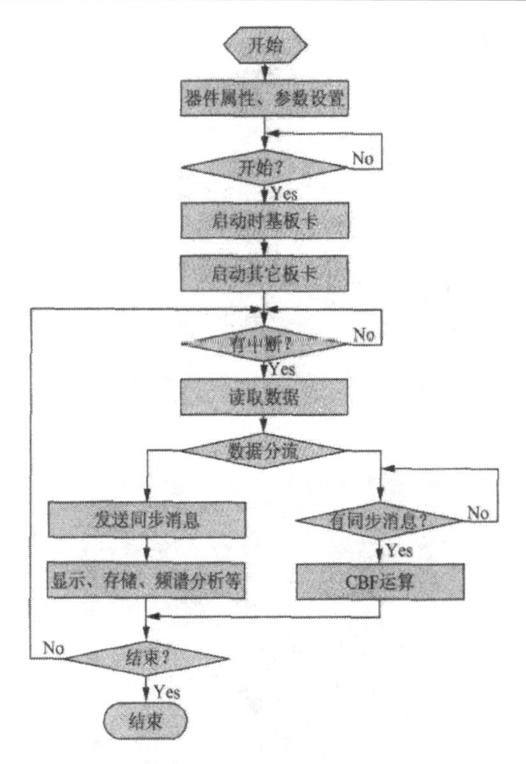

图7 整系统运行流程

4.4 非信号处理平台上实现实时信号处理运算的一些基本方法

现代电子技术发展迅速，多核CPU已经成为主流，合理运用CPU的运算能力，可以大大提高信号处理系统的处理能力。在信号处理运算量要求不太大的情况下，甚至可以不使用信号处理机，本文就直接使用CPU完成了所有实时信号处理的运算工作，因而大大简化了系统结构。在此，对利用运行于非信号处理平台的虚拟仪器系统实现实时信号处理运算的一些基本思路进行总结：

(1) 尽量使用定点数完成信号处理

X86架构的CPU自诞生以来，其定点运算能力就远远高于浮点运算能力。另一方面，主控计算机对信号处理运算结果进行显示时，最终是将所有数据转化为定点值，例如坐标、RGB色彩值等，因此，应该直接对CPU使用定点数来进行信号处理。本文实现的虚拟仪器系统根据NI板卡提供的功能，直接将模拟信号采集为定点数，并在此基础上进行信号处理运算，这样做一方面充分发挥了CPU的长处，另一方面省去了系统最终强制转换数据类型所消耗的时间和资源。实际测试也证明该方法比使用浮点数进行处理效率更高；

(2) 运算量大时应使用多线程以保证运算实时性

使用串行结构的程序运行速度慢，同时闲置了CPU资源。使用多线程并行处理多个任务，实际上是以更高的CPU使用率换取运算时间的节约。使用多线程时应当注意两个问题，一是各线程之间的同步，最好使用同步消息来控制各线程的开始与结束；二是各线程的运算量应尽量平衡，这样能够进一步有效利用CPU资源，节约运算时间。

5 应用及结论

本文利用虚拟仪器技术，基于NI公司的信号采集及主控计算机硬件，实现了一套集数据采集、存储、显示和实时信号处理于一身的多功能综合系统。在运算量要求不太高的情况下，该系统可独立完成水声阵列信号的采集和实时处理工作，通过对波形显示控件的设置，还可实现示波器的大部分功能。该系统规模小，各种接口配置齐全，使用方便，在水声试验等领域有广泛的用途。由于主控计算机配有USB2.0、千兆以太网口等丰富的数据接口，在其CPU无法满足运算量要求的情况下，该系统还可以作为其它信号处理平台的数据源，完成信号采集和数据传输的任务。

目前，该系统已经执行了多次湖上、海上试验任务，系统各项功能均正常发挥作用。

图8显示了该系统实际应用的一些情况。

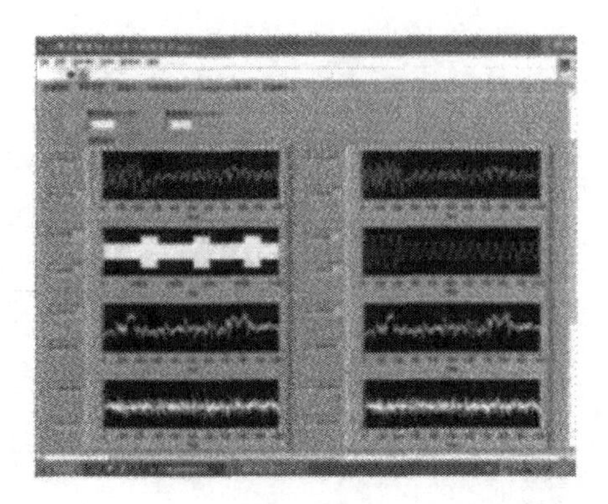

(a) 多通道时域波形显示

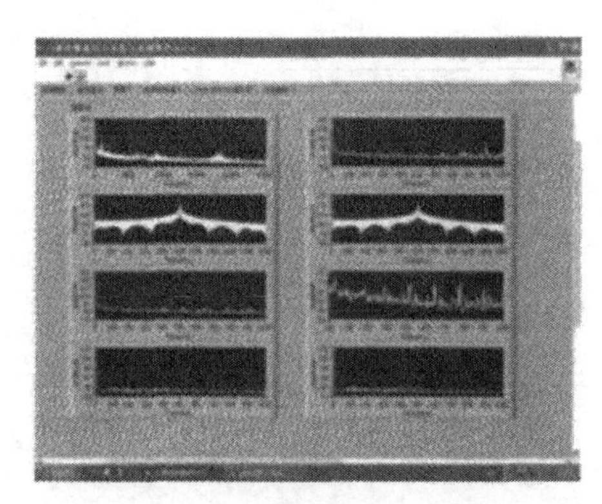

(b) 多通道频谱分析及显示

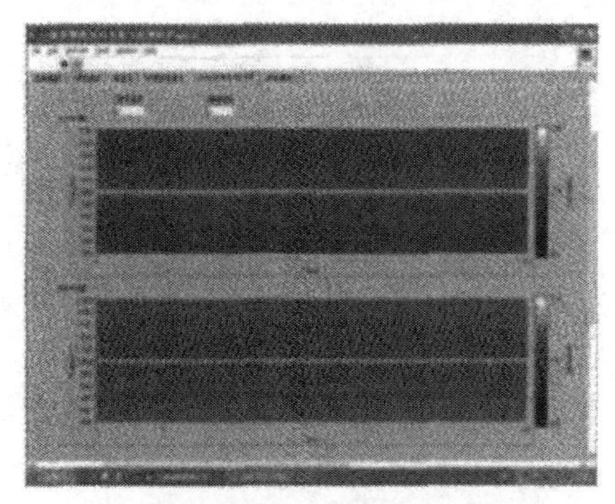

(c) 选定通道LOFAR及DEMON分析

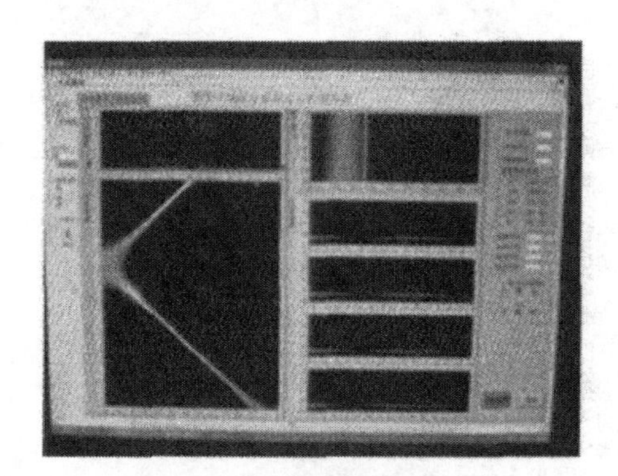

(d) 常规波束成形结果

图8　海上试验中系统实际工作情况

参 考 文 献

[1] WATTS L. The LabVIEW Environment: Advanced Built-In Analysis and Signal Processing [EB/OL]. [2011-2-23]. www.ni.com/labview.

[2] TRUCHARD J. DAQ: 446X/447X Specifications [EB/OL]. [2011-2-23]. http://sine.ni.com/nips/cds/view/p/lang/en/nid.

[3] WATTS L. LabVIEW入门8.0版：实现简单的数据采集系统[EB/OL]. [2011-2-23]. www.ni.com/china/labview.

[4] 李启虎. 数字式声纳设计原理[M]. 合肥：安徽教育出版社, 2003: 260-267.

[5] 张凯, 郭栋. LabVIEW虚拟仪器工程设计与开发[M]. 北京：国防工业出版社, 2004: 190-230.

基于水声潜标应用的数据采集及大容量存储系统设计①

冯师军 王 磊 董力平 李启虎

（中国科学院声学研究所 北京 100190）

摘要 水声潜标是获取海洋水声信号以及海洋生物声信号的重要技术设备之一。为了满足水声潜标在水下较长时间工作，本文设计并实现了一套基于水声潜标应用的低功耗数据采集及大容量存储系统，在芯片选型和 NAND FLASH 读写管理等方面对系统进行了低功耗设计。该系统在“标-矢量”综合潜标上应用的试验结果表明，该系统获取的数据有效，满足总体设计技术要求。同时，本系统还具有良好的扩展能力，可适应水声潜标发展的应用需求。

关键词 水声潜标，低功耗，大容量存储系统，低电压差分信号

中图分类号：TB565 + .4 **文献标识码**：A **文章编号**：1000-310X(2014)01-0081-06

DOI:10.11684/j.issn.1000-310X.2014.01.012

Design of data-sampling and high-capacity storage system in acoustic submersible buoy application

FENG Shijun WANG Lei DONG Liping LI Qihu

(*Institute of Acoustics, Chinese Academy of Sciences, Beijing* 100190, *China*)

Abstract The acoustic submersible buoy is the one of the important measurement equipments for ocean acoustic signal and marine acoustic signal. In order to meet the application requirements of acoustic submersible buoy for long time working, this paper designed and realized a data-sampling and high-capacity storage system. The low-power measures for the system included the choice of CMOS chip and read-write management for NAND-Flash. We applied the system to the comprehensive acoustic submersible buoy with scalar hydrophone and vector hydrophone, and the experimental data demonstrated that data acquiring was reliable and effective and it met the overall design requirements. The system has good scalability and meets the future needs of the submersible buoy.

Key words Acoustic submersible buoy, Low-power, High-capacity storage system, Low voltage differential signal

---

① 应用声学, 2014, 33(1): 81-86.

1 引言

水声潜标是获取海洋水声信号以及海洋生物声信号的重要技术装备之一。潜标是一种锚泊在海底的无人值守自动测量平台，通过携带的多种测量和探测仪器，可以在恶劣海况等条件下相对隐蔽地对海洋水声信号以及海洋生物声信号进行定点、连续、多层面同步测量。潜标装置在海洋科学调查研究、海洋生物研究等方面有着广泛的应用[1-2]。

本文设计并实现了一套基于水声潜标应用的低功耗多通道数据采集及大容量存储系统。为了满足水声潜标在水下较长时间工作，本系统从电路设计和 NAND Flash 读写管理角度入手来降低系统功耗，以尽可能地延长电池使用时间。该系统在“标-矢量”综合潜标上应用的试验结果表明，其获取的数据可靠、有效，满足总体设计要求。

潜标技术是 20 世纪 50 年代初首先在美国发展起来的。随后，原苏联、法国、日本、原西德和加拿大等国也相继开展研究和应用。60 多年来，潜标系统已经作为一种重要的海洋调查设备而被普遍使用，同时随着电子技术的发展，潜标内部的数据采集和存储系统也在不断进行改进，比如存储设备从最初的磁带记录仪到现在的 NAND Flash 存储阵列，数据采集从最初的 12 bit 到现在的 24 bits。

R. A. Wagstaff[3]和吴岩[4]提到 1977 年在印度洋的西北部地区进行了历时 4 个月的环境噪声测量试验中采用名为垂直数据舱(Acoustic data capsule，ACODC)的潜标系统，该系统能够携带 13 个水听器，在无人值守的情况下，可连续工作 13.6 天，如果间歇性工作，则可工作几个月。S. K. Mitchell[5]介绍了 ACODAC 采用的磁带记录仪进行数据存储的技术方案。Mark P. Johnson[6]介绍了一种水下能用于潜标的单通道声记录仪(DTAG)，该记录仪具有 3 GB 的 Flash 存储空间，但只能将一个水听器通道通过模数转换后用 DSP 处理器存储到 Flash 阵列中。

在国内，吕云飞[1]介绍了一种基于潜标的海洋环境噪声采集系统，该系统用数字信号处理器(DSP)和复杂可编程逻辑器件(CPLD)通过访问时序配合，将 4 个通道的模数转换后数据存入到 CF 存储卡中，可连续存储记录 100 个小时。

2 数据采集及大容量存储系统设计

2.1 总体设计

本系统以 Actel 公司的 AGL600 低功耗可编程器件(FPGA)为控制核心，包括 NAND FLASH 存储阵列电路、采集电路、USB 接口电路、实时时钟电路以及值班电路等，系统框图如图 1 所示。

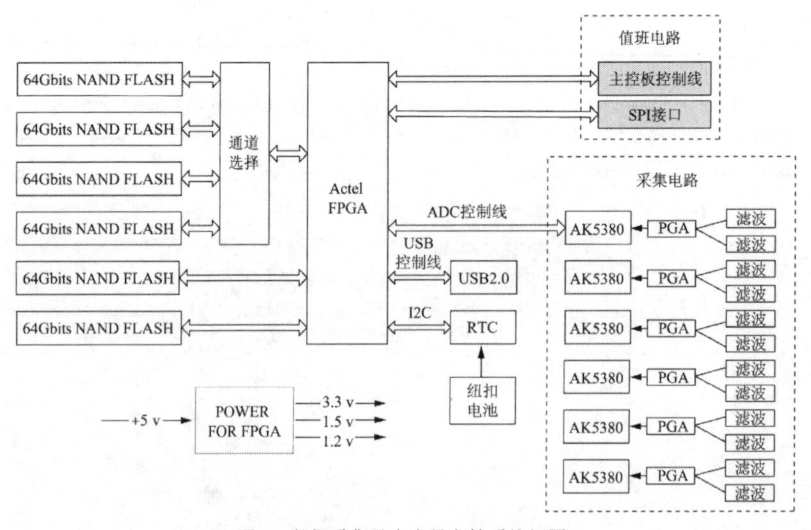

图 1 数据采集及大容量存储系统框图

Fig. 1 Block diagram of the data-sampling and high-capacity storage system

主要技术特点如下：

（1）FPGA 芯片 AGL600 具有 60 万门逻辑单元，6 个时钟调节电路（CCC），采用独特的 Flash * Freeze 技术；

（2）NAND FLASH 阵列，总容量为 48Gbytes；

（3）ADC 能够完成 12 个通道的同步采样，采样率为 10 kHz，采样精度为 24 bits，前端具有 1/2/5/10/20/50/100 倍的可调增益；

（4）USB 接口：1 个 USB 口；

（5）具备 6 对 LVDS（Low-voltage differential signaling，低压差分信号）信号对，能够扩展存储阵列板和信号处理板；

（6）提供 5 根输入 I/O 线，3 根输出 I/O 线，一个 SPI 接口；

（7）片外实时时钟 RTC。

FPGA 芯片产生 NAND FLASH 存储阵列控制逻辑、AD 变换器的时序控制逻辑和采样时钟同步逻辑、实时 RTC 控制逻辑以及对各种状态信号的判断逻辑，同时集成与计算机通讯的 USB 接口电路控制逻辑、触发控制逻辑和无效块管理逻辑等。FPGA逻辑的工作状态机如图 2 所示。

2.2 数据采集电路

数据采集电路对 12 个通道的模拟信号进行滤波、动态 PGA 控制、同步模数转换，然后通过板间连接器将模数转换后的数字化信号送到 FPGA 芯片。其中，模数转换芯片采用双通道的 AK 5380，其为 24 bits 精度的 Σ-ΔADC 转换器，采样频率的范围为 4 kHz ~ 96 kHz，单端输入；PGA 控制芯片采用双通道的 LTC6911，可调的增益范围为 1/2/5/10/20/50/100 倍放大；滤波电路采用 4 运放芯片 ADA4004-4 实现 10 Hz ~ 2 kHz 的带通滤波。

本系统中 ADC 芯片的采样频率设定为10 kHz，FPGA 芯片产生同步采样时钟，保证 12 个通道的模数转换同步进行，12 个通道模数转换后的总数据率为 2.88 Mbits/s。

ADC 芯片读取数据控制时序如图 3 所示[7]。

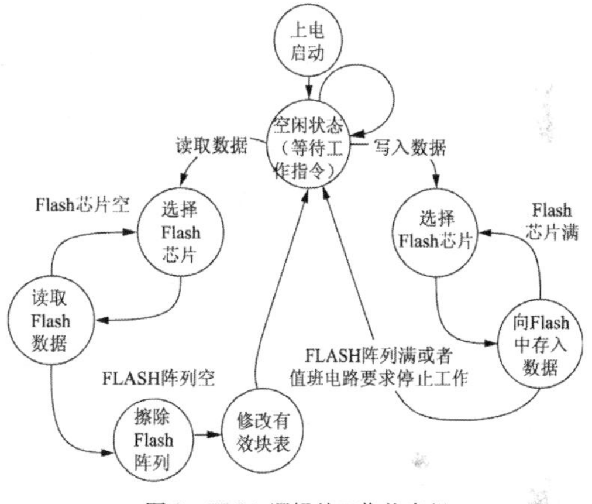

图 2 FPGA 逻辑的工作状态机

Fig. 2 Logic state machine of FPGA

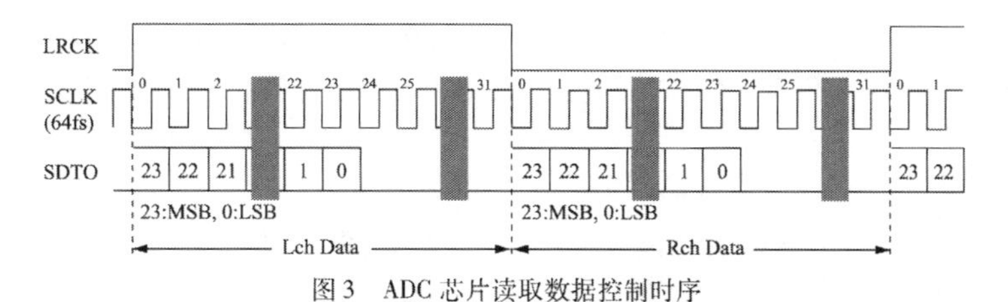

图 3 ADC 芯片读取数据控制时序

Fig. 3 The control-timing of ADC reading data

PGA 芯片的增益动态调节过程如下：首先，假设 12 个通道的信号幅度变化基本一致；然后值班电路通过 8 bits 精度的模数转换模块获取某一个通道的信号，判断信号幅度的大小，确定增益设定，通过 SPI 接口将增益设定通知 FPGA 芯片；最后 FPGA 芯片根据收到增益设定信息同时调节 12 个通道的 PGA 芯片增益。

2.3 NAND FLASH 存储阵列电路

NAND FLASH 存储阵列由 12 片 32Gbits 的 MT29F32G08FAAWP-ET 芯片组成，能够提供384G bits 的存储空间。上一节提到 12 通道数模转换后的总数据率为 2.88 Mbits/s，则存满 NAND FLASH 存储阵列的时间约为 37 个小时。由于 FPGA 芯片 I/O 管脚数的限制，其中 8 片芯片通过通道选择芯片共用 FPGA 上 NAND FLASH 控制器的 I/O 管脚。

存储阵列的控制逻辑由 FPGA 实现，包括读写逻辑、擦除逻辑以及坏块管理逻辑等。NAND 型 FLASH 具有“有坏块”的缺点，需要预先建立有效块表，避免采集数据写入坏块。在写入和擦除的过程中,如果遇到新的坏块,则将修改有效块表,去除新的坏块。

同时,为了匹配 ADC 芯片和 USB 接口输出的数据速率,在读写操作前分别配置了一个大小为 8 kBytes的 FIFO。当数据读出时,只能在 FIFO 空信号有效以后,才能将数据从 FLASH 中读出;当数据写入时,只能在 FIFO 半满信号有效后,才能向 FLASH 写入数据。

2.4 USB 接口

USB 接口芯片采用 CYPRESS 公司的 CY7C68013A,具有 Ports 模式、Slave FIFO 和 GPIF 三种接口方式。本系统采用 Slave FIFO 接口方式,这种方式是从机方式,即 FPGA 芯片可像对普通 FIFO 一样对 CY7C68013A 的多层缓冲 FIFO 进行读写。其数据传输原理框图如图 4 所示[8]。

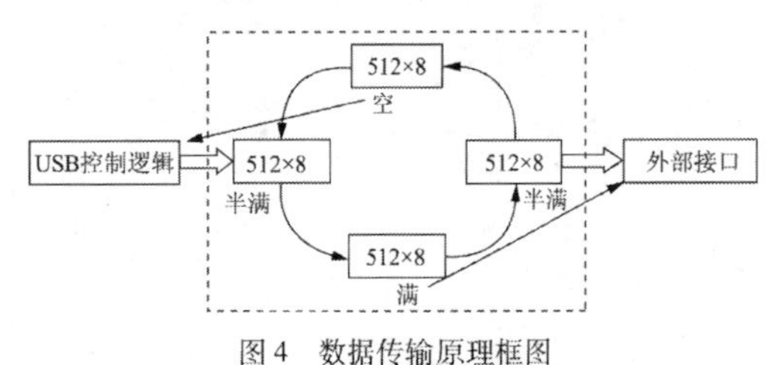

图 4 数据传输原理框图

Fig. 4 Diagram of data transmission principle

主机端用 VC + +开发了主机端驱动程序(ezusb. sys)和应用调试程序,应用调试程序工作界面如图 5 所示。

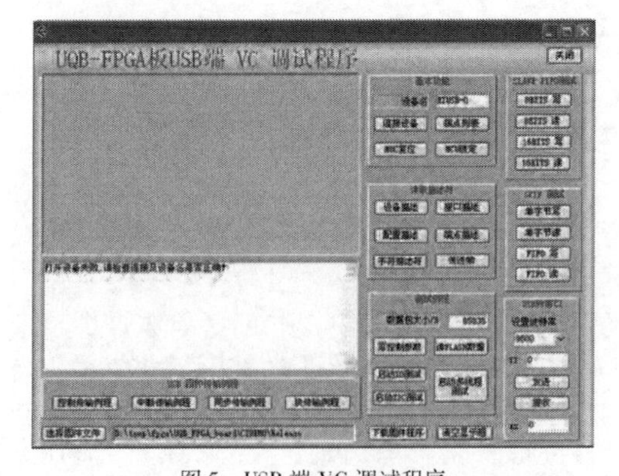

图 5 USB 端 VC 调试程序

Fig. 5 The VC debugger in USB

2.5 低功耗策略

在水声潜标的设计中,低功耗设计是关键。因为对于大多数水声潜标而言,电池仍然是目前唯一有效的能量来源。必须考虑每一个功耗细节以降低系统功耗,以尽可能地延长电池使用时间。本系统采用了以下低功耗策略来降低系统的功耗。

第一,从电路本身着手,在满足系统工作需求的情况下,尽量选用低功耗器件,降低电子器件自身的功耗;同时优化电路结构,降低电路的冗余度。

(1)采用基于 Flash 的 Actel IGLOO 系列单芯片可编程器件。该 FPGA 的静态功耗仅为 5 mW,支持 1.2 V 的内核电压,因此能实现很低的系统功耗。为了能在保持 FPGA 数据的同时实现低功耗,IGLOO 系列单芯片可编程器件采用了独特的Flash * Freeze技术,可以快速进出超低功耗模式。IGLOO 系列单芯片可编程器件无需添加任何额外部件就可关断 I/O 或时钟,同时保留了设计数据、SRAM 内容和寄存器状态。

(2)采用 NAND Flash 存储阵列来实现数据存储。NAND Flash 存储阵列相对于传统硬盘而言,具有体积小、功耗低等特点,适合在低功耗的潜标系统中使用。同时,随着电子技术的发展,NAND Flash 芯片的容量越来越大,目前市场主流的容量为 1GBytes,而最大的容量已经达到接近20GBytes,并且其容量还在不断增长,能够满足水声潜标系统未来的记录容量扩展需要。

第二,杜绝无效的功耗浪费。在潜标布阵过程中,应严格控制潜标各电路单元的供电,尽量减少空载状态下的能源消耗;在潜标正常工作期间,对 NAND Flash 存储阵列采用动态电源管理方案,降低存储电路功耗;采用模拟通道控制关断技术,关闭多余的采集通道,节省采集电路功耗。

(1)根据工作海域的天气情况,潜标布阵过程一般需要 30 ~ 60 分钟,这一时间段内潜标应当不工作,但如果潜标从布阵开始就加电运转,那么潜标将有 30 ~ 60 分钟无效能量消耗,会缩短实际的有效工作时间。为了减少潜标布阵过程中的无效功耗,需要对潜标的启动时间进行人为干预,在布阵的过程中,使潜标处于待机工作模式。在这套系统里,我们采用的是深度传感器控制方法,即实验前按照实验计划和工作海域天气情况预估潜标布放深度,在水声潜标中设定启动工作深度。当潜标开始布放时,系统启动值班电路,其他电路则处于掉电状态;当潜标到达预定的工作深度后,根据指令启动相应的电路模块。当系统存储阵列存满或收阵时潜标深度小于预估的工作深度,设备重新进入待机工作状态,按步骤关闭采集电路、NAND

Flash 阵列和 FPGA。

（2）当前，降低基于 NAND Flash 存储系统的功耗主要有以下几种方法：动态电源管理方法、动态电压调节方法、数据压缩方法、降低存储系统负载[9]。本系统采用动态电源管理方案，由于 12 个水声信号采样通道的总数据率为 2.88 Mbits/s，低于单片 NAND Flash 的写入速率，同时本系统中对数据的访问可以有很高的预测性，所以当 Flash 阵列中某些 NAND Flash 处于空闲状态时，可以通过阵列控制器关闭空闲的 NAND Flash，从而降低存储系统的功耗。

3　潜标应用及海试验证

数据采集及大容量存储系统已经应用于“标-矢量”综合潜标。“标-矢量”综合潜标采用坐底矢量水听器和垂直阵组合方式对水下声场，特别是环境噪声场进行有效测量。

“标-矢量”综合潜标系统总重量 125 kg，采用一个三轴矢量水听器和 8 元可变间距垂直阵接收水下声信号，通过温深传感器测量水中声速。为了实现水下声场信号的有效记录，存储空间和电池容量要求满足不小于 36 小时连续数据记录的需求。上面提到在 12 个模拟信号输入通道的情况下，存满 NAND FLASH 存储阵列的时间约为 37 个小时。“标-矢量”综合潜标系统总功耗为 6.5 W，采用 4 串 9 并锂电池组提供 12 V 电压，总安时为 19.8 Ah，电池容量能够满足“标-矢量”综合潜标系统连续工作不小于 36 小时的要求。

如图 6 所示为“标-矢量”综合潜标的电子舱与布放方式。

图 7 所示为一次海试中“标-矢量”综合潜标获取的水声信号。

同时针对潜标系统接收阵第 3 通道记录的水声信号进行 LOFAR 分析，结果如图 8 所示。

图 6　“标-矢量”综合潜标电子舱（a）、（b）与布放方式（c）

Fig. 6　Electronic module of comprehensive acoustic submersible buoy ((a) and (b)) and deployment (c)

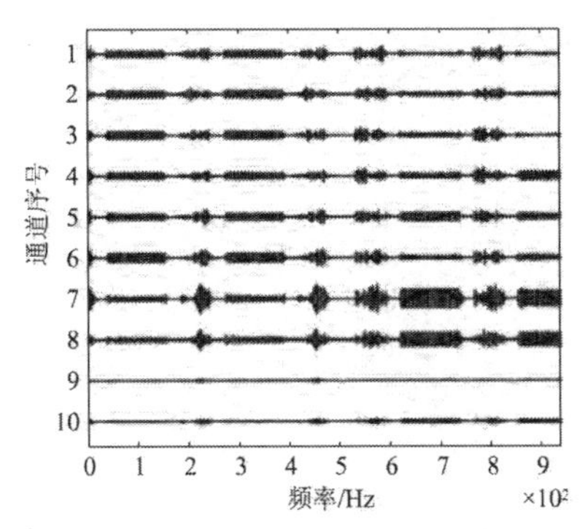

图 7　潜标系统海试测量得到的水声信号波形

Fig. 7　Underwater acoustic signal measured the submersible buoy in sea test

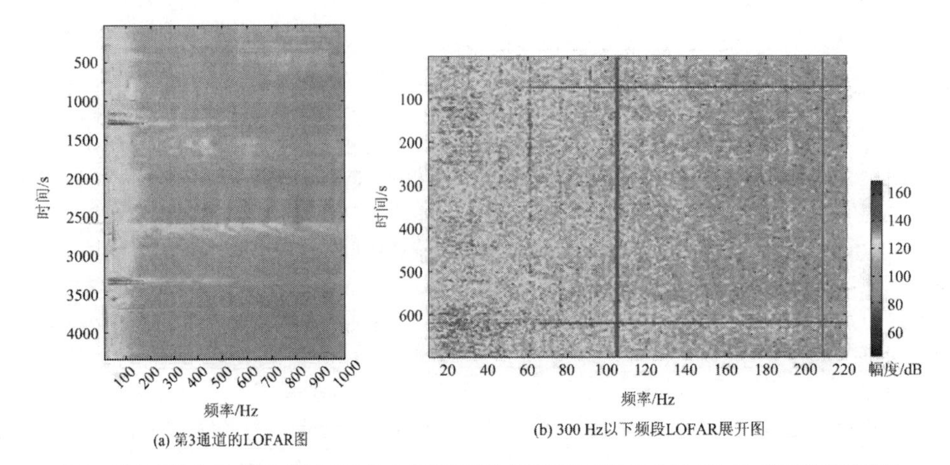

图 8　潜标系统接收阵第 3 通道记录的水声信号经处理得到的 LOFAR 图及 300 Hz 以下频段展开图

Fig. 8　Third channel LOFAR diagram recorded the submersible buoy system and LOFAR diagram Less than 300 Hz

4　结　论

本文设计并实现了一个能够用于水声潜标的数据采集及大容量存储系统，该系统从电路设计和 NAND Flash 读写管理等方面入手，在具备较大存储空间的前提下显著降低了功耗，可实现潜标系统不小于 36 小时的连续工作和水声信号记录要求。

本系统采用通用 USB 接口，能够简单、快速地获取存储数据。同时，系统能够通过调整 NAND Flash 阵列读写管理方式，提高数据的存储速率，满足更多采集通道的接入；能够通过 LVDS 接口扩展更多 NAND Flash 阵列板和信号处理板，提高系统的存储空间和具备一定的信号处理能力，具有良好的扩展性，能够满足潜标系统发展的未来需求。

该系统在"标-矢量"综合潜标系统中成功应用，测试结果表明，系统工作状态稳定，记录数据有效，可以有效保障潜标系统不小于 36 小时的连续数据记录。

参　考　文　献

[1]　吕云飞，张殿伦，邹吉武，等．基于潜标的海洋环境噪声测量系统[J]．高技术通讯，2009，19(7)：760-763.
LÜ Yunfei, ZHANG Dianlun, ZOU Jiwu, et al. A system for ocean ambient noise measurement based on the submersible buoy [J]. Chinese High Technology Letters, 2009, 19(7): 760-763.

[2]　孙岩松，李平．低功耗无线网络通信潜标设计应用[J]．海洋技术，2006，25(2)：7-9.
SUN Yansong, LI Ping. An application of low power consumption wireless communication in design of the underwater buoyage [J]. Ocean Technology. 2006, 25(2): 7-9.

[3]　WAGSTAFF R A, AITKENHEAD J W. Ambient noise measurements in the northwest indian ocean [J]. IEEE Journal of Oceanic Engineering, 2005, 30(2): 295-302.

[4]　吴岩．低频海洋环境噪声采集系统的设计与实现[D]．哈尔滨：哈尔滨工程大学硕士学位论文，2011.

[5]　MITCHELL S K. Vertical ACODAC acoustic measurement [J]. Applied Research Laboratory/UT, Feb., 1978. (private communication).

[6]　MARK P. JOHNSON and PETER L. Tyack. A digital acoustic recording tag for measuring the response of wild marine mammals to Sound [J]. IEEE Journal of Oceanic Engineering, 2003, 28(1): 3-12.

[7]　AKM. AK5380 datasheet[R]. MS0100-E-00, 2001.5.

[8]　谭安菊，龚彬．USB2.0 控制器 CY7C68013 与 FPGA 接口的 Verilog HDL 实现[J]．电子工程师，2007，33(7)：52-55.
TAN Anju, GONG Bin. Implementation of USB2.0 controller CY7 C68013 and FPGA'S interface in verilog HDL [J]. Electronic Engineer. 2007, 33(7): 52-55.

[9]　柳吉林．低功耗策略在基于 Flash 存储系统中的设计与实现[D]．南京：东南大学硕士学位论文，2009.

An indirect method to measure the variation of elastic constant c_{33} of piezoelectric ceramics shunted to circuit under thickness mode①

Yang Sun[a], Zhaohui Li[a,*], Qihu Li[b]

[a] Department of Electronics, Peking University, No. 5 Yiheyuan Road, Haidian District, Beijing 100871, China
[b] Institute of Advanced Technology, Peking University, No.5 Yiheyuan Road, Haidian District, Beijing 100871, China

ARTICLE INFO

Article history:
Received 9 May 2014
Received in revised form 25 July 2014
Accepted 25 July 2014
Available online 9 August 2014

Keywords:
Measurement method
Shunt damping
Piezoelectric
Elastic constant

ABSTRACT

Academic interest in the use of piezoelectric shunt damping technology in the field of vibration and noise control has been on the rise in recent years. The experimental observation of the variation of elastic constants of piezoelectric materials shunted to different circuits is a fundamental problem of this technology. The existing methods to measure the elastic constants of piezoelectric ceramics are either unsuited or complicated for such a purpose. In this paper, a modified resonator measurement method is proposed to indirectly measure the variation of elastic constant c_{33} of piezoelectric ceramics shunted to circuit under thickness mode. The main idea is transforming the change of c_{33} into the change of the electrical resonant frequency of a single-oscillator model which is easier to be observed. Mason equivalent circuit and ANSYS finite element simulation are employed for the analysis of the admittance and resonant frequency of the proposed system. Based on the proposed method, the variation of the elastic constant c_{33} has been investigated when the piezoelectric ceramic wafer (PZT-5H) is shunted to a resistor, a capacitor, and a resistor connected in series with an inductor respectively. The variations of the resonant frequency measured in the experiment are highly consistent with those of c_{33}. Besides, the measurement of variation of mechanical loss factor with a shunt resistor is also discussed in this paper.

1. Introduction

As early as 1979, Forward [1] first proposed piezoelectric transducer shunted to electronic circuit for vibration control in optical system. The piezoelectric shunt damping technology has gained fruitful achievements in the field of vibration control [2–13] and noise elimination [14–17] in recent years. In the vibration control applications, a piezoelectric transducer shunted to the passive electrical circuits is attached to a host structure, which can be a beam [2,3,7], a vibrator [4,6,9] or a plate [5,11,17]. The mechanical energy produced by the structural vibration is converted into electrical energy through the piezoelectric effect. Then, a portion of the electrical energy is converted into heat and dissipated through the shunt circuits.

The design and the implementation of the efficient shunt circuits are two of main focuses associated with the piezoelectric shunt damping technology. [18] Much early attention has been paid to the investigation of the basic shunt circuits including a resistor [2–5], a capacitor [6], a series R–L [2,8,10] and a parallel R–L [7–9], each of which can suppress the vibration in only one mode. Subsequently, shunt circuits aiming at multiple mode vibration control have been proposed and improved, including Hollkamp's circuit [19], Wu's blocking circuit [20], Moheimani's current-flowing circuit [21] and negative capacitor circuit [22,23]. Another switching shunt circuit [17,24–27] belonging to the semi-passive shunt approach makes it possible to adjust the passive damping property of the vibration control system. In order to design the shunt impedance more systematically, a design method of the shunt circuit parameters based on minimizing the H_2 norm of the damped system [28] is presented. And the synthetic impedance consisting of a voltage-controlled current source [28] makes the implement of the shunt network easier and more flexible.

The utilization of the piezoelectric shunt damping technology is mainly based on the changes of two variables of the vibration control system—damping factor and resonant frequency [2–12]. The former reflects the damping effect whereas the latter decides the frequency range of the shunt damping system. It is observed that resonant frequency of the system with shunt resistance is slightly lower than that of the open-circuit system, but higher than that of the short-circuit system. [2,4,5,11,12] The change of resonant frequency is substantially caused by the change of elastic constants of

① Sensors and Actuators A: Physical, 2014, 218: 105-115.

piezoelectric materials. Therefore the experimental observation of the variation of elastic constants of piezoelectric materials shunted to different circuits is of significant research value.

Several classical methods used to measure the elastic constants of piezoelectric ceramics have long been applied in practice. [29] The static or quasistatic methods [30] directly measuring strain of the material under an exertion of stress are the early and basic techniques. But they are seldom applied in the measurement of elastic constants in recent years since it is difficult to control the electrical boundary conditions. The pulse-echo methods [31–33] are based on the fact that the plane-wave velocities are related to the fundamental material constants of piezoelectric materials and can be determined from pulse transit time measurements. However, inaccuracies may occur under various cases especially for small-size samples such as the phase shift in coupling the transducer to the sample, the distortion of the pulses during reflection, the detection errors of wave front of the pulses and so on. The resonator measurement method [34] which directly measures the impedance or admittance of the piezoelectric material is based on the electrical properties of a piezoelectric vibrator which are dependent on the elastic, piezoelectric, and dielectric constants. The material constants can be derived from the measured resonance frequency, antiresonance frequency and the motional capacitance. However, directly measuring the impedance or admittance of a piezoelectric ceramic wafer shunted to different circuits cannot be applied to obtain the variation of c_{33} of the piezoelectric materials shunted to different circuits. Because the shunt circuits and the piezoelectric ceramics are in the same circuit loop, the change of c_{33} due to the shunt circuit cannot be separated from the electrical effects on the resonant frequency or other electrical variables of the equivalent circuit that can be measured.

In the work of Law et al. [4], a mechanical mass-spring-dashpot model is established to measure the variations of the resonant frequency and damping ratio as the shunt resistance changes. The resonant frequency and damping ratio were determined from the frequency response function (FRF) which is obtained by measurement of the force and the acceleration of the model under the excitation of an external vibration exciter. To measure the FRF, a series of indispensable instruments are employed. Hence the method is complicated and costly. Moreover, uncertainty may also be introduced by the variables measured by those instruments.

In this paper, an indirect method, namely a modified resonator measurement method to measure the variation of c_{33} of piezoelectric ceramics shunted to a circuit is proposed. The measurement system is a single-oscillator model consisting of two identical piezoelectric ceramic wafers and a mass block. The mass block is introduced to make the piezoelectric ceramics in the quasi-static state. The shunt circuit is connected to one ceramic wafer while the admittance measurement is conducted on the other ceramic wafer. The variation of c_{33} is turned into the variation of the system resonant frequency that can be easily measured. Compared with the measurement of the mechanical FRF, the measurement of the electrical admittance in this paper does not need any complex apparatus but an impedance analyzer, which achieves good performance on simplifying the experiment, reducing the introduction of measurement uncertainty and lowering the cost. The variation of the elastic constant c_{33} has been investigated based on Mason equivalent circuit theory in which the piezoelectric ceramic wafer is under different electrical boundary conditions. The finite element analysis is also carried out using ANSYS and the simulation numerical results show good agreement with the theoretical analysis. The experiment is conducted to measure the variations of the resonant frequency, which show good coincidence with the variations of c_{33}. Besides, the measurement of variation of mechanical loss factor is also discussed in this paper.

For the thickness mode is one of the essential modes for the covering layer structure that needs the reduction of structural vibrations, and c_{33} is the elastic constant representing the thickness mode, this paper limits discussion to the elastic constant c_{33} under thickness mode. However, the proposed method can be extended to investigate the variations of other elastic constants of piezoelectric ceramics shunted to circuit under other modes. Three basic kinds of shunt circuits, namely resistive, inductive, and capacitive shunt circuits, are chosen to be studied in this paper. Likewise, the method can be extended to study other shunt circuit types. In addition, this measurement method can also be applied to measure the elastic constant variation of the piezoelectric composites with high electromechanical coupling factor such as 1-3 [35] and 1-3-2 composites [36–38], which are promising materials with attractive performance for transducers applied in medical ultrasonic and underwater acoustics.

2. Model description

An indirect method is proposed for the purpose of avoiding the direct measurement of the admittance of the piezoelectric wafer itself shunted to circuit. The proposed single-oscillator system consists of two identical piezoelectric ceramic wafers ***a***, ***b*** and a co-axial cylindrical mass block which are tightly bonded together. The underside of ***a*** is fixed, while the upper surface of the mass block is free, as shown in Fig. 1(b). If ***b*** is under different electrical boundary conditions such as short-circuit, shunted to impedance or open-circuit, the elastic constant c_{33} of ***b*** will change as well as the resonant frequency of the single-oscillator system. By observing the changes of the resonant frequency, the variation of c_{33} can be indirectly obtained.

The piezoelectric ceramic material PZT-5H (Lead zirconate titanate) is chosen as an example in the research work. Apart from material, proper thickness and radius of the ceramic wafer are selected. A mass block is introduced to make the resonant frequency of the system as low as several tens kHz, which is far lower than the resonant frequency of the ceramic wafer itself to provide a quasi-static state [4]. For the purpose of approaching the condition of an ideal single-oscillator system, tungsten alloy is chosen as the material of the mass block because of its high density and hardness.

The principle of the proposed method can be explained by a single-oscillator system. The ideal single-oscillator system similar to the above proposed system is shown in Fig. 1(a). It consists of a mass block M_m and two identical springs connected in series which are equivalent to the two identical piezoelectric ceramic wafers ***a***, ***b***. The bottom spring is fixed and its stiffness k_0 is a constant, just like that ceramic wafer ***a*** is in open-circuit state and its elastic constant c_{33} is a constant—the open-circuit elastic constant $\mathbf{c}_{33}^{D}$. The stiffness of the upper spring k is a variable, just as the elastic constant c_{33} of the ceramic wafer ***b*** is a variable, since ***b*** is under different electrical boundary conditions.

For this ideal single-oscillator system, since the two springs are connected in series, the stiffness of the system is:

$$\bar{k} = \frac{kk_0}{k+k_0} \tag{1}$$

Therefore, the resonant frequency is:

$$f_{si} = \frac{1}{2\pi}\sqrt{\frac{\bar{k}}{M_{mi}}} \tag{2}$$

Here, '*i*' represents the ideal single-oscillator system. Then, the stiffness of the upper spring k is obtained as,

$$k = \frac{M_{mi}(2\pi f_{si})^2}{1-(M_{mi}/k_0)(2\pi f_{si})^2} \tag{3}$$

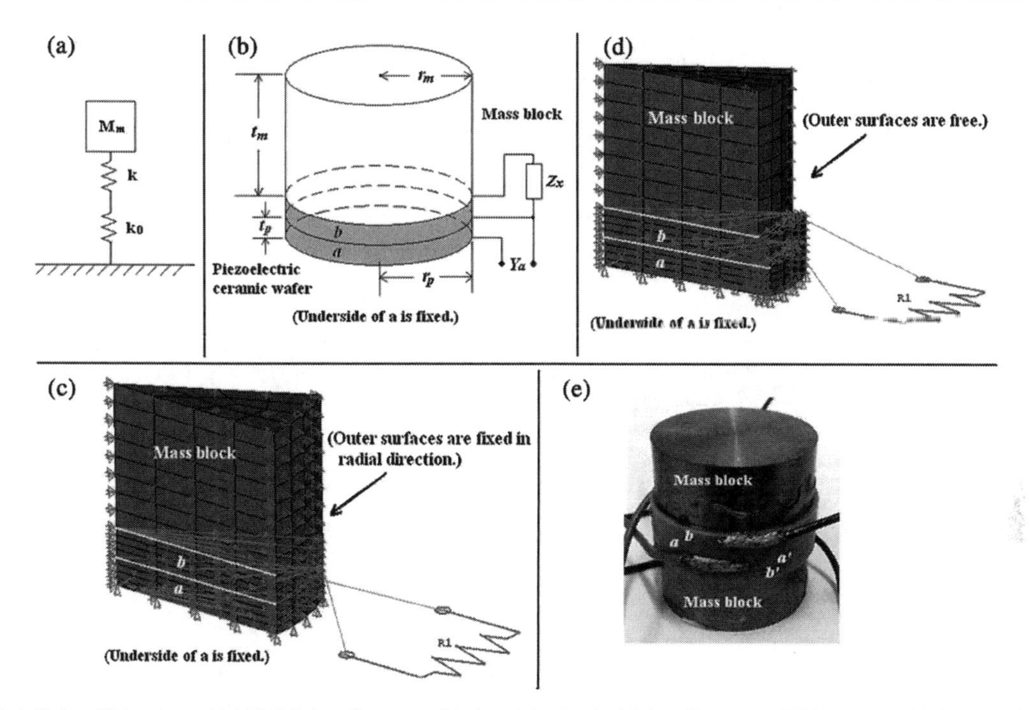

Fig. 1. Single-oscillator system model: (a) ideal single-oscillator system; (b) schematic drawing of real single-oscillator system; (c) finite element model (clamped in radial direction); (d) finite element model (free in radial direction); (e) symmetrical experimental sample.

If the stiffness k and k_0 are high enough and the resonant frequency is comparatively low, Eq. (3) can be further simplified to be:

$$k \approx M_{mi}(2\pi f_{si})^2 \tag{4}$$

Hence, the following relationship can be derived.

$$\frac{\Delta k}{k} = \frac{2\Delta f_{si}}{f_{si}} \tag{5}$$

It is concluded that $\Delta k/k$ is proportional to $\Delta f_{si}/f_{si}$. Therefore, the variation of the stiffness k can be obtained according to the variation of the resonant frequency f_{si} of the ideal single-oscillator system.

3. Theoretical equivalent circuit analysis

3.1. The relationship between c_{33} and resonant frequency

The Mason equivalent circuits of the ceramic wafer and the mass block stretching in thickness mode are shown in Fig. 2(a) and (b) respectively. In Fig. 2(a), F_1 and F_2 are the pressures applied on the two end faces of the ceramic wafer. V is the voltage emerging between the two electrodes of the wafer because of piezoelectric effect. U_1 and U_2 are the particle velocities on the two end faces. In Fig. 2(b), F_1' and F'_2 are the pressures applied on the two end faces of the mass block. U_1' and U_2' are the particle velocities on the two end faces.

By paralleling the equivalent circuits of the ceramic wafers ***a***, ***b*** and the mass block, the electro-mechanical equivalent circuit of the whole model is obtained and depicted in Fig. 2(c). The underside of wafer ***a*** is fixed, which means its velocity U_1 equals to 0 and the branch circuit is open-circuited. The upper surface of the mass block is free, which means its pressure F'_2 equals to 0 and the branch circuit is short-circuited. The ceramic wafer ***a*** is open-circuited. The ceramic wafer ***b*** is shunted to an impedance Z_x, where $Z_x = 0$ and $Z_x = \infty$ indicate that it is short-circuited and open-circuited, respectively. For convenience of the theoretical analysis, in this part, the ceramic wafer ***b*** is assumed in open-circuit state, i.e. $Z_x = \infty$, while its elastic constant c_{33} is assumed to be a variable in order to derive the relationship between c_{33} and the resonant frequency.

In Fig. 2, $Z_1^{open} = j\rho_p c_p^{open} S_p \tan(k_p^{open} t_p/2)$, $Z_2^{open} = (\rho_p c_p^{open} S_p / j\sin(k_p^{open} t_p)$, $Z_1 = j\rho_p c_p S_p \tan(k_p t_p/2)$, $Z_2 = (\rho_p c_p S_p / j\sin(k_p t_p)$, $Z_1' = j\rho_m c_m S_m \tan(k_m t_m/2)$ and $Z_2' = (\rho_m c_m S_m / j\sin(k_m t_m)$ are the mechanical impedances, where subscript 'p' and 'm' represent the piezoelectric ceramics and the mass block respectively. Accordingly, ρ_p, ρ_m are the densities, t_p, t_m are the thicknesses and S_p, S_m are the areas of the ceramic wafers and the mass block respectively. $C_0 = (\varepsilon_{33}^S S_p/t_p)$ is the inherent capacitance and $n = (e_{33} S_p/t_p)$ denotes the electromechanical conversion factor of the piezoelectric ceramics. ε_{33}^S is the dielectric constant and e_{33} is the piezoelectric stress constant. Besides, $k_p = (\omega/c_p)$, $k_m = (\omega/c_m)$ are the wave numbers. c_m is the longitudinal wave velocity of the mass block. $c_p^{open} = \sqrt{\tilde{c}_{33}^D/\rho_p}$ and $c_p = \sqrt{\tilde{c}_{33}/\rho_p}$ are the longitudinal wave velocities of ceramic wafer ***a*** and ***b*** respectively. $\tilde{c}_{33}^D$ (open-circuit elastic constant) and $\tilde{c}_{33}$ are the elastic constants of ceramic wafer ***a*** and ***b***.

The elastic constant is actually a complex value that consists of the real part and the imagine part. The ratio of the imagine part to the real part represents the mechanical loss factor whose reciprocal is the mechanical quality factor Q_m. They are expressed as follows:

$$\tilde{c}_{33}^D = c_{33}^D + jc_{33}^{D'} \tag{6}$$

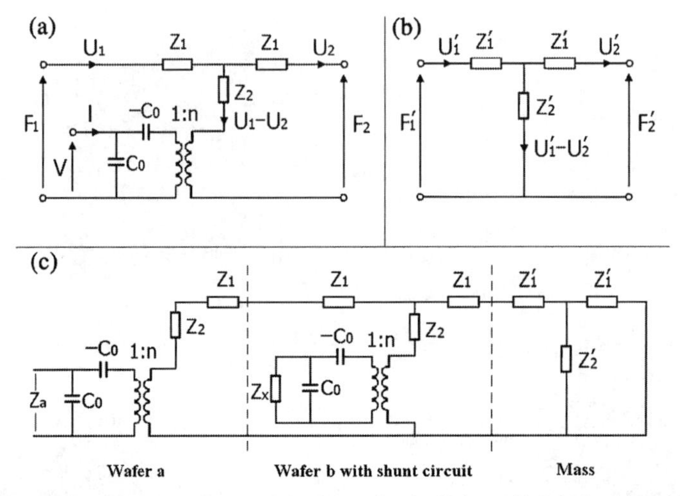

Fig. 2. Electro-mechanical equivalent circuits of (a) the ceramic wafer; (b) the mass block; (c) the whole model.

$$\frac{c_{33}^{D'}}{c_{33}^{D}} = \frac{1}{Q_m^D} \tag{7}$$

$$\tilde{c}_{33} = c_{33} + jc'_{33} \tag{8}$$

$$\frac{c'_{33}}{c_{33}} = \frac{1}{Q_m} \tag{9}$$

It is assumed that $S_p = S_m = S$. The electrical admittance of the system calculated from the two electrodes of wafer ***a*** can be obtained according to the equivalent circuit and simplified based on the low frequency approximation conditions $k_p t_p \ll 1$ and $k_m t_m \ll 1$, namely:

$$Y_a = \frac{j\omega C_0}{1 - \dfrac{n^2}{\left(\dfrac{A\omega^2(Bc_{33}\omega^2 - (c_{33}^2 + c'^2_{33})S)}{(B\omega^2 - c_{33}S)^2 + c'^2_{33}S^2} + (c_{33}^D D - \omega^2 C)\right) + j\left(\dfrac{c'_{33}AB\omega^4}{(\omega^2 B - c_{33}S)^2 + c'^2_{33}S^2} + c_{33}^{D'} D\right)}} \tag{10}$$

Where, $A = C_0 S(M_m + (M_p/2))$, $B = t_p(M_m + (M_p/2))$, $C = C_0 M_p$, $D = (SC_0/t_p)$ are constants. M_p and M_m are the masses of the ceramic wafer and the mass block respectively.

The equivalent circuit of thickness mode only takes into account the piezoelectric ceramic wafer and the mass block with radial dimensions much larger than their thicknesses or they are radially clamped. The radiuses of the ceramic wafers and the mass block are set as 14 mm. The thicknesses of the ceramic wafers and mass block are chosen to be 2 mm and 10 mm respectively. The parameters of the piezoelectric ceramics (PZT-5H) and the mass block are listed in Table 1.

Let the admittance $Y_a = G_a + jB_a$, therefore G_a is the conductance and B_a is the susceptance. For c_{33} with certain value, according to Eq. (10), the conductance of the system calculated from the two electrodes of wafer ***a*** can be calculated. For example, when ceramic wafer ***b*** is open-circuited, namely $c_{33} = c_{33}^D$, the conductance is calculated and shown in Fig. 3. The peak of the conductance is $G_{a\max} = 17.36$ ms, the resonant frequency of the system is $f_s = 59.96$ kHz. The two frequencies $f_1 = 59.65$ kHz and $f_2 = 60.26$ kHz are corresponding to the condition that the conductance equals to $G_{a\max}/\sqrt{2}$. The quality factor of the conductance is defined as $Q_e = f_s/(f_2 - f_1) = (f_s/\Delta f)$ and is calculated to be 98.94. Then, the electrical loss factor which is defined as $1/Q_e = (f_2 - f_1)/f_s = (\Delta f/f_s)$ can be obtained to be 0.01.

With the change of c_{33}, the relationship between $\Delta c_{33}/c_{33}^D$ and $\Delta f_s/f_s^D$ (f_s^D is the open-circuit resonant frequency) is calculated

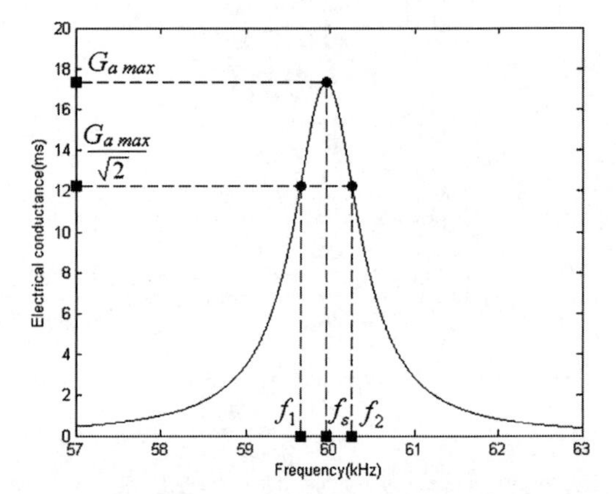

Fig. 3. Electrical conductance of the system calculated from the two electrodes of wafer ***a*** (ceramic wafer ***b*** is open-circuited, $c_{33} = c_{33}^D$).

according to Eq. (10) and is shown in Fig. 4. It can be seen that $\Delta c_{33}/c_{33}^D$ nearly varies linearly with $\Delta f_s/f_s^D$ in a certain range, which is nearly the same as the relationship between $\Delta k/k$ and $\Delta f_{si}/f_{si}$ shown in Eq. (5) for the ideal single-oscillator system. Therefore, the variation of c_{33} of wafer ***b*** can be indirectly captured according to the variation of the resonant frequency of the system.

With the change of c'_{33}, the relationship between the mechanical loss factor $1/Q_m$ and the electrical loss factor of the system $1/Q_e$ can also be obtained according to Eq. (10) and is shown in Fig. 5. It can be seen that $1/Q_m$ varies linearly with the change of $1/Q_e$ in a certain range. So the mechanical loss factor of wafer ***b*** $1/Q_m$ can be indirectly derived based on the variation of the electrical loss factor of the system $1/Q_e$.

3.2. The electrical conductance in different shunt circuit cases

In this part, the ceramic wafer ***b*** is shunted to a circuit, where Z_x is arbitrary. The equivalent circuit of the model is also shown in Fig. 2. The ceramic wafer ***a*** is in open-circuit state. Under this

Table 1
Parameters of piezoelectric ceramics and mass block [39].

The piezoelectric ceramics (PZT-5H)											
Density (kg/m³)	Elastic constant(10^{10}N/m²)					Relative dielectric constant		Piezoelectric stress constant (C/m²)			Mechanical quality factor
ρ_p	$c_{11}{}^E$	$c_{12}{}^E$	$c_{13}{}^E$	$c_{33}{}^E$	$c_{44}{}^E$	$\varepsilon_{11}{}^S/\varepsilon_0$	$\varepsilon_{33}{}^S/\varepsilon_0$	e_{31}	e_{33}	e_{15}	$Q_m{}^D$
7500	12.6	7.95	8.41	11.7	2.3	1700	1470	−6.5	23.3	17.0	65
	$c_{11}{}^D$	$c_{12}{}^D$	$c_{13}{}^D$	$c_{33}{}^D$	$c_{44}{}^D$						
	13.0	8.28	7.22	15.7	4.22						

The mass block (Tungsten Alloy)		
Density ρ_m (kg/m³)	Young modulus (N/m²)	Poisson's ratio
18,700	35.4	0.35

Fig. 4. The relationship between $\Delta c_{33}/c_{33}{}^D$ and $\Delta f_s/f_s{}^D$.

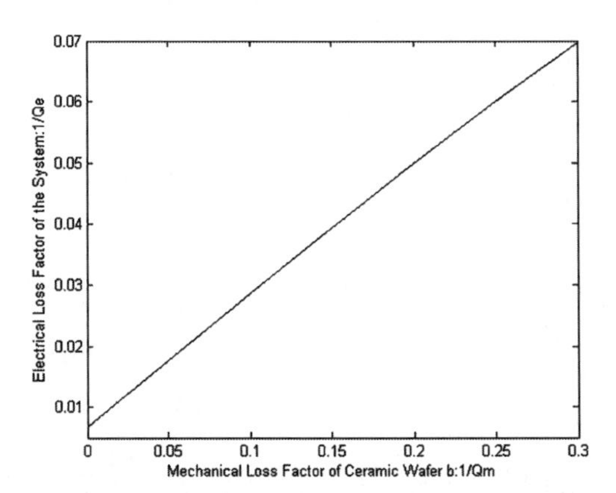

Fig. 5. The relationship between the mechanical loss factor of wafer ***b*** $1/Q_m$ and the electrical loss factor of the system $1/Q_e$.

condition, the change of the shunt impedance is set as the factor that causes the change of the resonant frequency of the system. The elastic constant c_{33} of both wafer ***a*** and ***b*** is the open-circuit elastic constant $\tilde{c}_{33}^D$. The electrical admittance of the system calculated from the two electrodes of wafer ***a*** can be obtained according to the equivalent circuit as Eq. (11).

$$Y_a = \frac{1}{\left(\frac{(Z_1'//Z_2'+Z_1'+Z_1^{open})//\left(n^2(Z_x//\frac{1}{j\omega C_0}-\frac{1}{j\omega C_0})+Z_2^{open}\right)+2Z_1^{open}+Z_2^{open}}{n^2}-\frac{1}{j\omega C_0}\right)//\frac{1}{j\omega C_0}} \tag{11}$$

Where '||' denotes parallel. The real part of the electrical admittance is the electrical conductance. The obtained electrical conductance of the system calculated from the two electrodes of wafer ***a*** when wafer ***b*** is shunted to a resistor, a capacitor, and a resistor (100 Ω) connected in series with an inductor are shown in Fig. 6(a1),(b1),(c1) respectively. Under each condition, several conductance curves corresponding to different shunt resistances, capacitances and inductances are chosen as examples and depicted in the figures.

It can be seen in Fig. 6(a1) that from short circuit to open circuit, with the increase of the shunt resistance, the resonant frequency gradually rises accordingly and the width and height of the resonant peak (represented by the electrical loss factor $1/Q_e$) also undergo certain changes. In Fig. 6(b1), from open circuit to short circuit, the resonant frequency gradually declines with the growth of the shunt capacitance. In Fig. 6(c1), from the condition of $L_x = 0$ to open circuit, with the increase of the shunt inductance, the resonant frequency first decreases, then jumps to the peak value and continues to decrease. From the calculated electrical conductance, the variations of the resonant frequency and electrical loss factor $1/Q_e$ with the change of shunt impedances can be solved. Therefore, the variations of c_{33} and $1/Q_m$ with the change of shunt impedances can be obtained.

4. Finite element analysis

For verification of the theoretical analytical results, the finite element simulation software ANSYS is also adopted to solve the electrical admittance. A 3-D finite element model is established given that the shunt circuit element should be connected to a volume block which is consistent with the actual experimental condition. According to the rotational symmetry of the model, an arbitrarily chose fraction of the whole cylindrical model is enough to fulfill the goals, which can also considerably improve the calculating speed. In this paper a 1/12-model is established in ANSYS as shown in Fig. 1(c).

The piezoelectric ceramics is polarized in z-axis direction and the whole model vibrates in z-axis direction under thickness mode. The rotation axis is fixed in the radial direction. The side faces are fixed in the rotation direction. The outer surfaces are fixed in radial direction. The bottom of wafer ***a*** is fixed in z-axis direction. The circuit element CIRCU94, connected to the piezoelectric ceramic wafer ***b***, is employed to build the shunt resistor, capacitor and

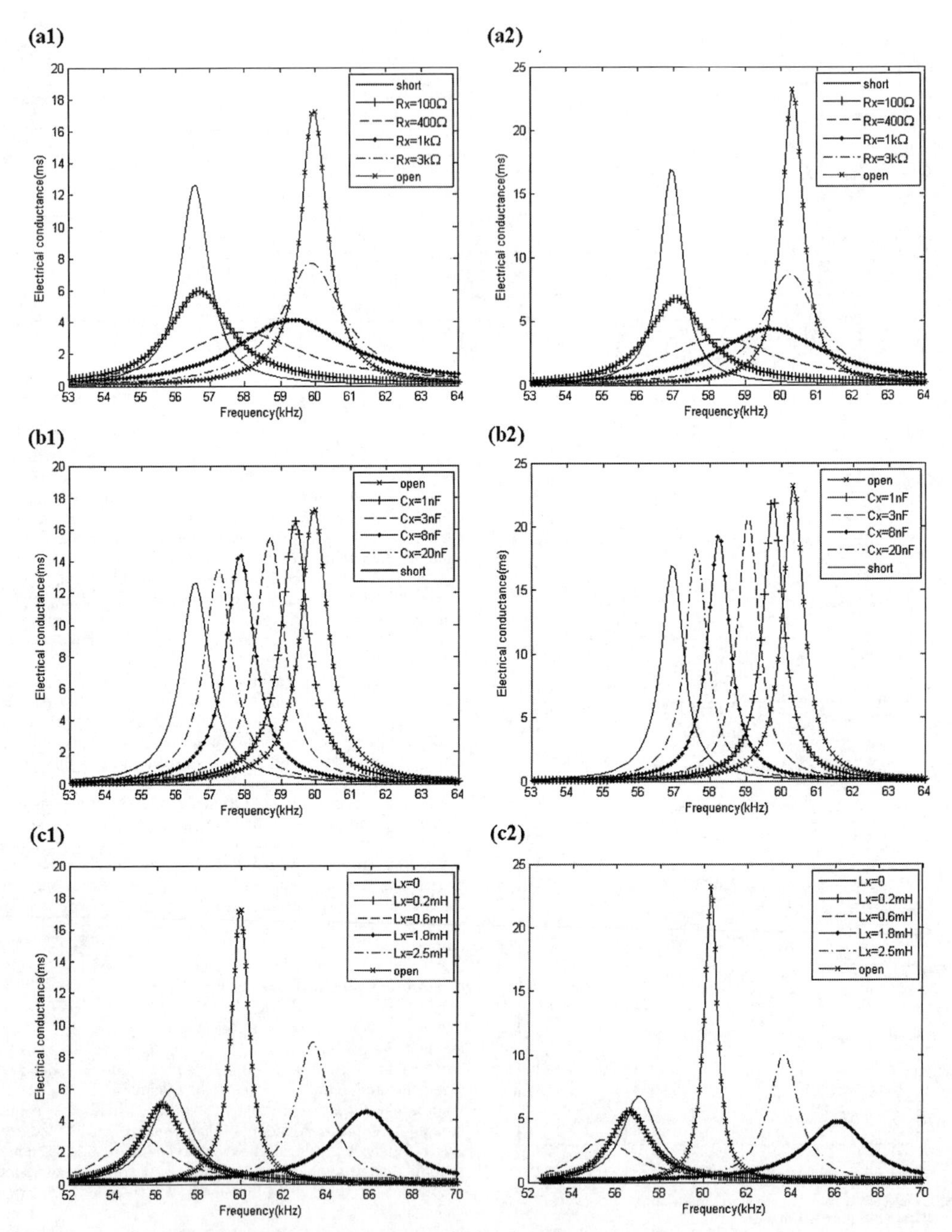

Fig. 6. Electrical conductance of the system calculated from the two electrodes of wafer ***a*** when wafer ***b*** is shunted:(a1) a resistor (equivalent circuit analysis); (a2) a resistor (finite element analysis); (b1) a capacitor (equivalent circuit analysis); (b2) a capacitor (finite element analysis); (c1) a resistor (100 Ω) and an inductor in series (equivalent circuit analysis); (c2) a resistor (100 Ω) and an inductor in series (finite element analysis).

inductor elements. As this is a 1/12-model, the shunt resistance and inductance must be twelve times of those connected to the full model. While the shunt capacitance must be one-twelfth of that connected to the full model. Two electrodes are respectively defined on the two surfaces of the piezoelectric ceramic wafer ***a***. Voltage 1 V is loaded on one node of the upper surface and voltage 0 V is loaded on one node of the underside. After taking harmonic response analysis, the node charge value Q on the upper surface of the piezoelectric ceramic wafer $\boldsymbol{a}$ is picked up. The electrical admittance of 1/12-model is derived as follows:

$$Y_{a(1/12)} = -j2\pi f \frac{Q}{V} \tag{12}$$

where $\boldsymbol{V}$($\boldsymbol{V}$ = 1 V) is the excitation voltage and f is the frequency. The admittance of the whole model should be twelve times of $Y_{a(1/12)}$.

The obtained electrical conductance of the system calculated from the two electrodes of wafer $\boldsymbol{a}$ when wafer $\boldsymbol{b}$ is shunted to a resistor, a capacitor, and a resistor (100 Ω) connected in series with an inductor are shown in Fig. 6(a2),(b2),(c2) respectively. By taking use of the same calculating process of the theoretical research, the variation of the elastic constant c_{33} can also be obtained with finite element method. When the piezoelectric ceramic wafer $\boldsymbol{b}$ is shunted to a resistor, a capacitor, and a resistor connected in series with an inductor respectively, the simulation results are compared with the theoretical analytical ones in Fig. 7, which indicates that they are consistent with each other.

From Fig. 7, it indicates that for piezoelectric ceramics shunted to a resistor or a capacitor, the elastic constant c_{33} varies between the short-circuit elastic constant c_{33}^{E} (117 GPa) and the open-circuit elastic constant c_{33}^{D} (157 GPa). With the rise of the shunt resistance, starting from c_{33}^{E}, the value c_{33} first climbs rapidly and then keeps a slow rising trend until it reaches c_{33}^{D}. On the contrary, with the rise of the shunt capacitance, c_{33}, starting from c_{33}^{D}, first goes down sharply and then declines gradually until the value reaches c_{33}^{E}. While for the resistor (100 Ω) and inductor in series shunt circuit condition, c_{33} varies beyond the range from c_{33}^{E} to c_{33}^{D}. With the rise of the shunt inductance, almost starting from c_{33}^{E}, c_{33} first declines slowly, then makes a soar to the peak value before drops slowly to c_{33}^{D}.

The consistent variations of the effective stiffness can be found in references [2,4,6] For the resistor case [2,4], with the increase of the resistance, the effective stiffness increases monotonously between the minimum and the maximum. While for the resistor and inductor in series case [2], with the increase of the inductance, the effective stiffness first decreases, then jumps to the peak value but decreases again. For the capacitor case [6], the tunable stiffness decreases steadily with the increase of the shunt capacitance.

The pure resistance is the sole means to dissipate energy while both capacitive and inductive elements can only store but cannot dissipate energy. So it is worthy to investigate the damping behavior of the shunt resistance case. The electrical loss factor $1/Q_e$ of the system for the shunt resistance case has been calculated by equivalent circuit method and finite element method respectively. The mechanical loss factor of wafer $\boldsymbol{b}$ $1/Q_m$ can be indirectly derived based on the variation of the electrical loss factor of the system $1/Q_e$ as shown in Fig. 8. The concerned general trends of $1/Q_m$ are almost the same for the two analysis methods with only slightly difference between the two curves. It can be seen that when ceramic wafer $\boldsymbol{b}$ is shunted to an optimal resistance load (about 600 Ω), the mechanical loss factor of wafer $\boldsymbol{b}$ reaches to its maximum. Correspondingly, the effect of piezoelectric shunt damping is the best.

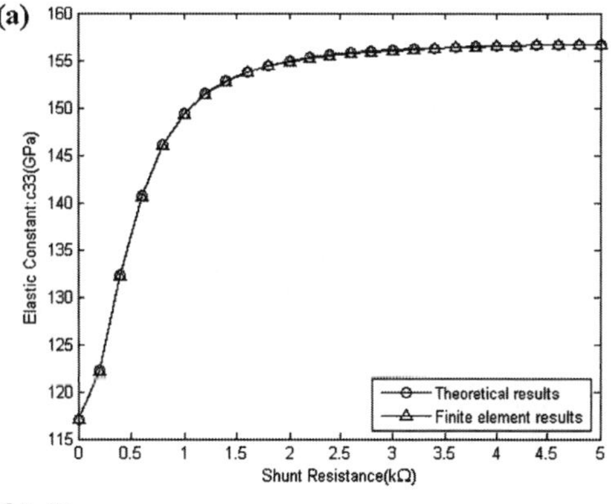

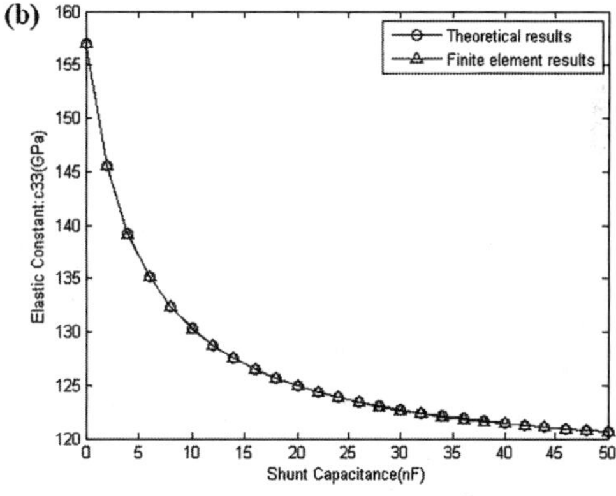

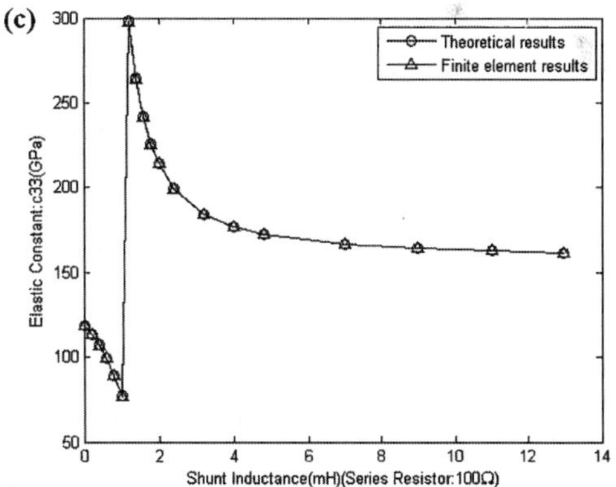

Fig. 7. The variation of c_{33} when wafer $\boldsymbol{b}$ is shunted to:(a) a resistor;(b) a capacitor;(c) a resistor (100 Ω) and an inductor in series.

5. Experiment

To validate the theoretical and finite element results, an experimental sample has been manufactured to measure the admittance. To realize the underside fixed boundary condition of the piezoelectric ceramic wafer $\boldsymbol{a}$, wafer $\boldsymbol{a}$ should be affixed on a base made of material whose stiffness is much higher than that of piezoelectric ceramics. Since the stiffness of the piezoelectric ceramics is fairly high ($c_{33}{}^{D}$ = 157 GPa), there are no such materials that can meet the demand.

To avoid the difficult problem of the fixed condition on the underside of wafer $\boldsymbol{a}$, a symmetrical system model is adopted. By making two identical mass blocks and four piezoelectric ceramic wafers $\boldsymbol{b}$, $\boldsymbol{a}$, $\boldsymbol{a}'$ and $\boldsymbol{b}'$ stick together, a symmetrical system is implemented as shown in Fig. 1(e). The end surfaces of the two mass blocks are free. The dimensions of each component are listed in Table 2.

The symmetrical model is equivalent to the single-oscillator model with the underside fixed boundary condition. The resonant frequencies of the two models are the same, while the maximum

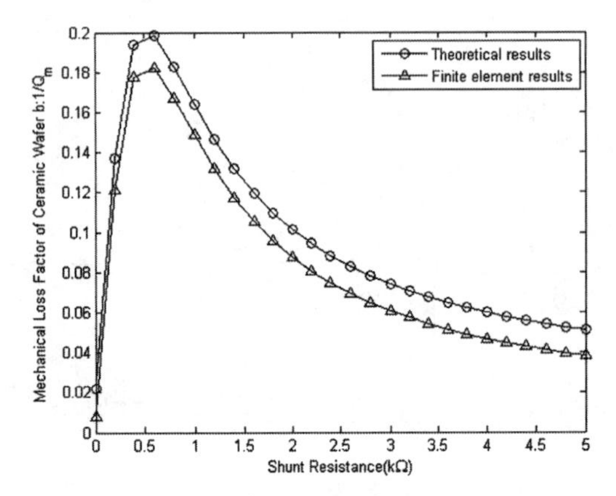

Fig. 8. The variation of mechanical loss factor $1/Q_m$ when wafer ***b*** is shunted to a resistor.

Table 2
Dimensions of experimental apparatus.

	Material	Radius (mm)	Thickness (mm)
Piezoelectric ceramics	PZT-5H	14.00	2.00
Mass	Tungsten alloy	13.00	10.00

conductance of the symmetrical model is one half of that of the single-oscillator model since the volume of the symmetrical model is twice of that of the single-oscillator model. It is needed to connect two same shunt circuit elements to wafer ***b*** and ***b′*** respectively. An impedance analyzer is employed to measure the electrical admittance of the system from the two electrodes of wafer ***a***.

Because the real piezoelectric ceramic wafer with a limited radius in the experiment cannot fully satisfy the strict radially fixed condition of the theoretical analysis, influenced by the finite radial dimension and the radially free boundary condition, the thickness resonant frequency will be different from that of the ideal case. Corresponding to the dimensions of the experimental sample, a 1/12-model is also established in ANSYS as shown in Fig. 1(d). The only difference of the boundary conditions from that of the ideal case is that the outer surfaces of the ceramic wafers and the mass block are free.

For the shunt resistance, capacitance and inductance conditions, the electrical conductance calculated by finite element method and the electrical conductance measured in experiment are shown in Fig. 9. For each condition, several conductance curves corresponding to different shunt resistances, capacitances and inductances are chosen as examples.

Compared with the conductance of the ideal case in Fig. 6, the resonant frequency of the model with radially free boundary condition is lowered by about 15 kHz, which is caused by the finite radial dimension and the radially free boundary condition of the model. It can be seen in Fig. 9 that for certain shunt circuit case, the resonant frequencies of the single-oscillator model in finite element analysis and the symmetrical model used in experiment are nearly the same. The maximum conductance of the symmetrical model is about one half of that of the single-oscillator model.

The variations of c_{33} obtained by finite element analysis and the resonant frequency measured in experiment are compared in Fig. 10. c_{33} values calculated by finite element method are marked with triangle symbols and shown on the left *y*-axis. While the experimental results of the resonant frequencies are marked with olid square symbols and shown on the right *y*-axis.

Compared with the finite element results, the variation trends of the resonant frequency measured in experiment are almost consistent with the variation trends of c_{33} when wafer ***b*** is shunted to certain type of shunt circuit. It verifies that the variation trend of c_{33} can be indirectly obtained by observing the variation of the resonant frequency.

The electrical loss factor $1/Q_e$ of the system for the shunt resistance case measured in the experiment with the finite element results of the mechanical loss factor of wafer ***b*** $1/Q_m$ as shown in Fig. 11. It can be seen that when ceramic wafer ***b*** is shunted to an optimal resistance load (about 600 Ω), the electrical loss factor of the system and the mechanical loss factor of wafer ***b*** reach to their maximums. Correspondingly, the effect of piezoelectric shunt damping is optimal.

6. Comparison and discussion

To illustrate the significant advantages of the method proposed in this paper, the comparisons with the static method [30] and the mechanical impedance measurement method [4] are given and discussed below, respectively.

6.1. Comparison with the static method

In order to measure the elastic constant c_{33} by the static method [30], a force should be perpendicularly applied on one surface of a ceramic wafer and the other surface of the wafer is fixed in its thickness direction. Under certain electrical boundary conditions, the resulting displacement needs to be measured to obtain the strain component in thickness direction. The ratio between the applied force and the strain component in thickness direction is the elastic constant c_{33}. Eqs. (13) and (14) are the definitions of c_{33} under 'constant electrical field' condition and 'constant electrical displacement' condition respectively.

$$c_{33}^E = \left(\frac{\partial T_3}{\partial S_3}\right)_E \tag{13}$$

$$c_{33}^D = \left(\frac{\partial T_3}{\partial S_3}\right)_D \tag{14}$$

where T_3 and S_3 are the stress and strain component in the thickness direction of the wafer respectively, $()_E$ and $()_D$ denote the 'constant electrical field' and 'constant electrical displacement' conditions respectively. Since the elastic constant of the piezoelectric ceramics is very high and the resulting strain caused by the applied force within the tolerance of the ceramics is very slight, the error will be introduced to the result of displacement by the measurement instrument. Furthermore, a fixed boundary condition on either surface of the wafer can hardly be satisfied. Therefore, the elastic constant c_{33} is hard to be obtained accurately since the error occurs in estimation of the strain component of the wafer.

Another problem to measure the elastic constant c_{33} by the static method is the control of the electrical boundary condition. The elastic constant accuracy by this method essentially depends on how the 'constant electrical field' and 'constant electrical displacement' conditions are satisfied during the measurement. On the contrary, the aim of this paper is to measure the variation of the elastic constant c_{33} when the piezoelectric ceramic wafer is shunted to different circuits, namely under different electrical boundary conditions. Therefore, the static method is obviously not a proper choice for such a purpose.

Being different from the static method, the proposed method in this paper is simpler and more accurate because it only needs to measure the electrical conductance which can be obtained accurately.

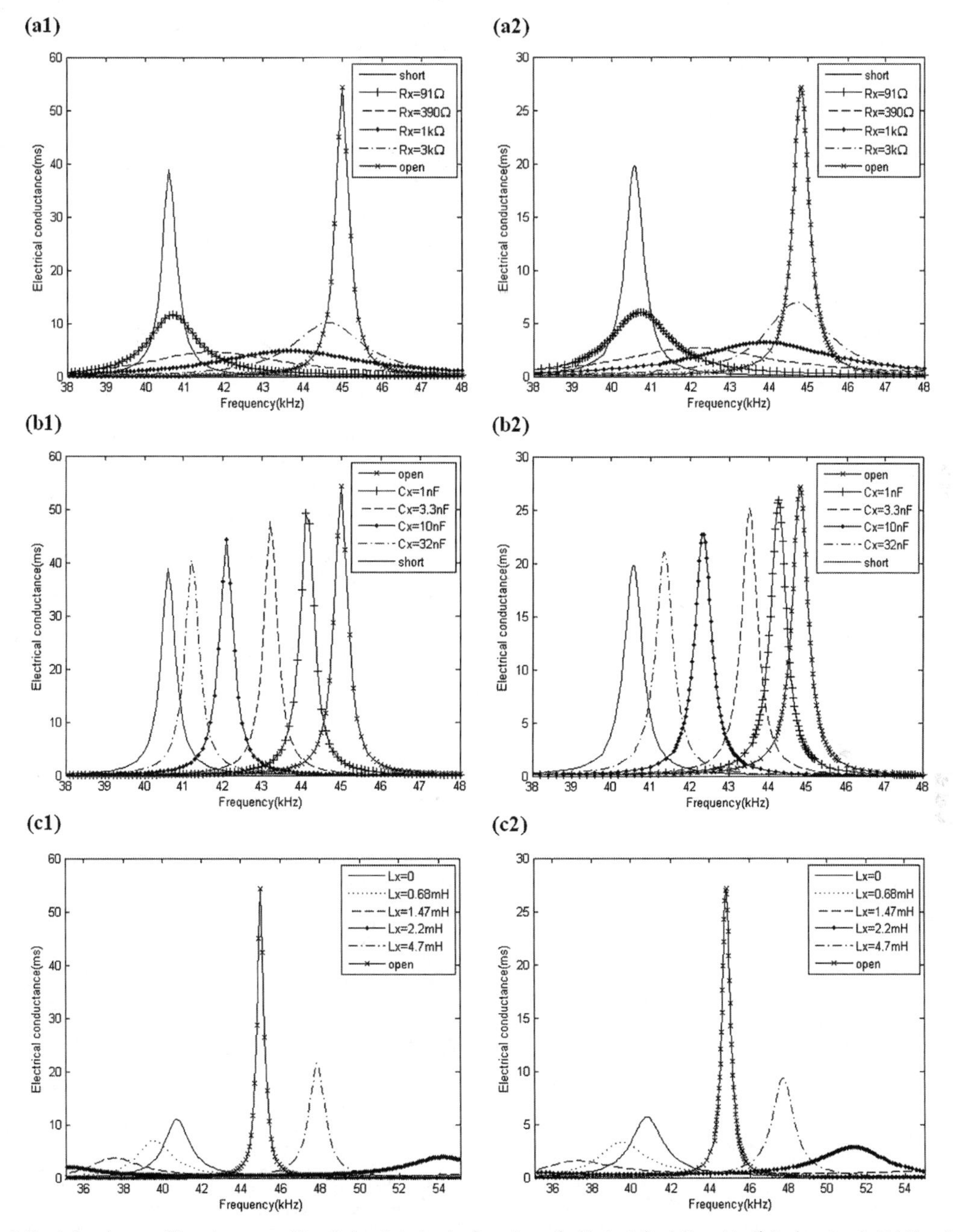

Fig. 9. Electrical conductance of the system measured from the two electrodes of wafer ***a*** when wafer ***b*** is shunted to: (a1) a resistor (finite element analysis); (a2) a resistor (experiment); (b1) a capacitor (finite element analysis); (b2) a capacitor (experiment); (c1) a resistor (100 Ω) and an inductor in series (finite element analysis); (c2) a resistor (100 Ω) and an inductor in series (experiment).

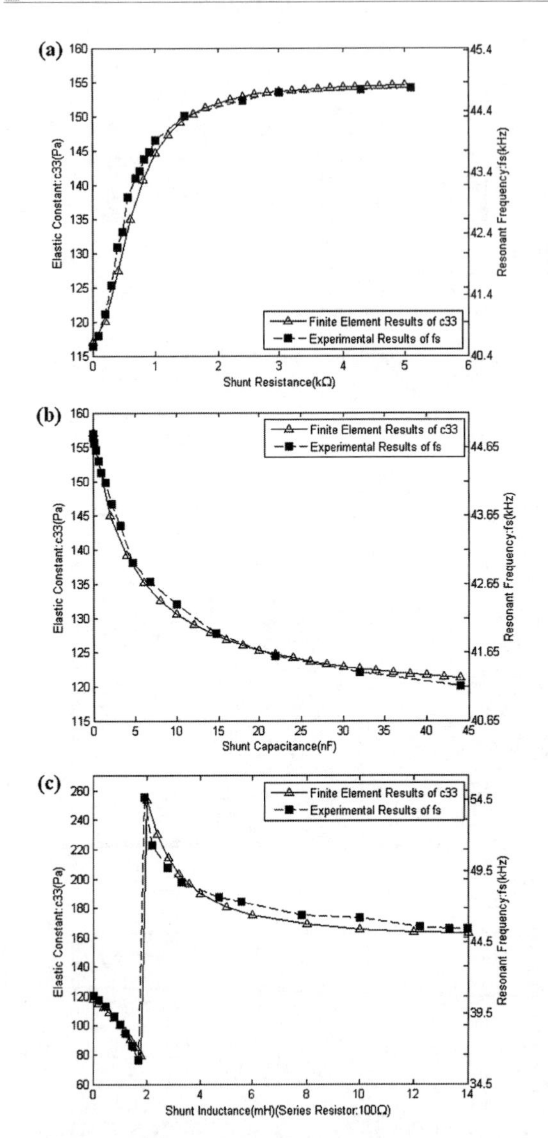

Fig. 10. The variation of c_{33} and resonant frequency of PZT-5H shunted to:(a) a resistor;(b) a capacitor;(c) a resistor (100 Ω) and an inductor in series.

6.2. Comparison with the mechanical impedance measurement method

In the work of Law etc. [4], a mechanical mass-spring-dashpot model consisting of two resistor-shunted piezo-material rings and two mass blocks is established to measure the variations of the resonant frequency and damping ratio as the shunt resistance changes. The system is excited into vibration and then both the input force and the acceleration response are measured to form the frequency response function (FRF), from which the resonant frequency and damping ratio are determined. The FRF is defined as the modulus

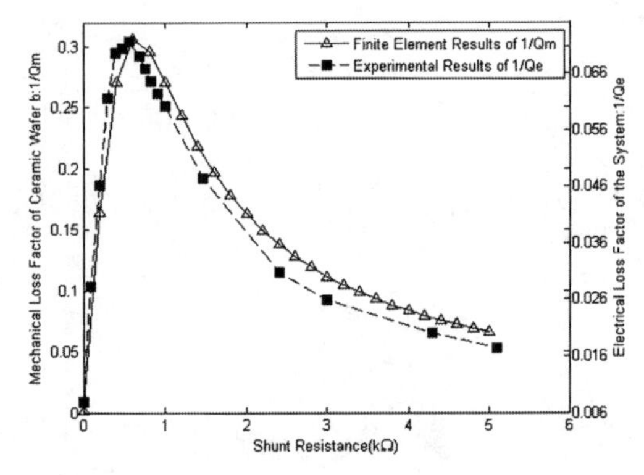

Fig. 11. The variation of mechanical loss factor of wafer ***b*** $1/Q_m$ and electrical loss factor of the system $1/Q_e$ when wafer ***b*** is shunted to a resistor.

of the ratio of the acceleration a to the force F, which reflects the mechanical impedance or the mechanical admittance of the system.

$$\mathrm{FRF} = \left|\frac{a}{F}\right| \tag{15}$$

In order to measure the FRF in Eq. (15), a series of indispensable instruments are employed. A signal generator, a power amplifier and a shaker are needed to excite the system into vibration. A force sensor is needed to measure the input force and an accelerometer is needed to measure the acceleration response. Besides, two charge amplifiers and a two-channel FFT analyzer are employed to analyze the output signals from the sensors. Thus, the measurement process is complicated and errors may be introduced by the measurement of various variables.

While for method in this paper, both the resonant frequency and damping ratio are determined from the electrical conductance of the model. The electrical conductance is the only measured variable and an impedance analyzer is the only necessary instrument. Therefore, it is comparatively simple and low-cost.

7. Conclusions

This paper has proposed a modified resonator measurement method to experimentally study the variation of elastic constant c_{33} of the piezoelectric ceramics shunted to circuits under thickness mode. Since $\Delta c_{33}/c_{33}^{\boldsymbol{D}}$ nearly varies linearly with $\Delta f_s/f_s^D$, it is only needed to measure the variation of the resonant frequency which are easier to be observed, then the variations of c_{33} is indirectly obtained. Similarly, the variation of mechanical loss factor is linearly with $1/Q$ of the resonant conductance, so it can be reflected through the Q-value of the resonant conductance.

When the radial dimensions of the piezoelectric ceramic wafers and the mass block are much larger than their thicknesses or they are radially clamped, the single-oscillator system vibrates in pure thickness mode and can be analyzed based on Mason equivalent circuit theory. The variations of the elastic constant c_{33} have been indirectly derived according to the variation of the resonant frequency of the system when the piezoelectric ceramic wafer is shunted to a resistor, a capacitor, and a resistor connected in series with an inductor respectively. The finite element analysis is also used to validate the theoretical analytical results, with the FEM results showing good agreement with the theoretical results.

A symmetrical model sample that is equivalent to the single-oscillator model has been conceived in the experiment. The variation of the resonant frequency has been measured by an impedance analyzer. Besides, another finite element analysis is conducted under radially free boundary condition to simulate the experimental situation. The experimental results have good coincidence with the finite element results, which confirms the validity of this indirect method to measure the variation of elastic constant c_{33} of piezoelectric ceramics shunted to circuit under thickness mode.

Although this paper only discusses the elastic constant c_{33} under thickness mode with three basic shunt circuit types, and the mechanical loss factor with a shunt resistor, the proposed method can be extended to investigate the variations of elastic constants and the mechanical loss factor of piezoelectric composites, or elastic constants of piezoelectric materials shunted to different kinds of shunt circuits under other modes.

References

[1] R.L. Forward, Electronic damping of vibrations in optical structures, Appl. Opt. 18 (1979) 690–697.

[2] N.W. Hagood, A. von Flotow, Damping of structural vibrations with piezoelectric materials and passive electrical networks, J. Sound Vib. 146 (1991) 243–268.

[3] C.L. Davis, G.A. Lesieutre, A modal strain energy approach to the prediction of resistively shunted piezoceramic damping, J. Sound Vib. 184 (1995) 129–139.

[4] H.H. Law, P.L. Rossiter, G.P. Simon, L.L. Koss, Characterization of mechanical vibration damping by piezoelectric materials, J. Sound Vib. 197 (1996) 489–513.

[5] D.A. Saravanos, Damped vibration of composite plates with passive piezoelectric-resistor elements, J. Sound Vib. 221 (1999) 867–885.

[6] C.L. Davis, G.A. Lesieutre, An actively tuned solid-state vibration absorber using capacitive shunting of piezoelectric stiffness, J. Sound Vib. 232 (2000) 601–617.

[7] S.Y. Wu, Piezoelectric shunts with a parallel R–L circuit for structural damping and vibration control, in: Proceedings of SPIE Symposium on Smart Structures Materials Passive Damping Isolation, 2720, 1996, pp. 259–269.

[8] C.H. Park, Dynamics modeling of beams with shunted piezoelectric elements, J. Sound Vib. 268 (2003) 115–129.

[9] S.M. Kim, S. Wang, M.J. Brennan, Dynamic analysis and optimal design of a passive and an active piezo-electrical dynamic vibration absorber, J. Sound Vib. 330 (2011) 603–614.

[10] O. Thomas, J.F. Deu, J. Ducarne, Vibrations of an elastic structure with shunted piezoelectric patches: efficient finite element formulation and electromechanical coupling coefficients, Int. J. Numer. Methods Eng. 80 (2009) 235–268.

[11] J. Becker, O. Fein, M. Maess, L. Gaul, Finite element-based analysis of shunted piezoelectric structures for vibration damping, Comput. Struct. 84 (2006) 2340–2350.

[12] O. Thomas, J. Ducarne, J.F. Deu, Performance of piezoelectric shunts for vibration reduction, Smart Mater. Struct. 21 (2012) 1–16, 015008.

[13] J. Ducarne, O. Thomas, J.F. Deu, Placement and dimension optimization of shunted piezoelectric patches for vibration reduction, J. Sound Vib. 331 (2012) 3286–3303.

[14] J.M. Zhang, W. Chang, V.K. Varadan, V.V. Varadan, Passive underwater acoustic damping using shunted piezoelectric coatings, Smart Mater. Struct. 10 (2001) 414–420.

[15] M. Ahmadian, K.M. Jeric, On the application of shunted piezoceramics for increasing acoustic transmission loss in structures, J. Sound Vib. 243 (2001) 347–359.

[16] J. Kim, J.K. Lee, Broadband transmission noise reduction of smart panels featuring piezoelectric shunt circuits and sound-absorbing material, J. Acoust. Soc. Am. 112 (2002) 990–998.

[17] D. Guyomar, T. Richard, C. Richard, Sound wave transmission reduction through a plate using piezoelectric synchronized switch damping technique, J. Intell. Mater. Syst. Struct. 19 (2008) 791–803.

[18] S.O. Reza Moheimani, A survey of recent innovations in vibration damping and control using shunted piezoelectric transducers, IEEE Trans. Control Syst. Technol. 11 (2003) 482–494.

[19] J.J. Hollkamp, Multimodal passive vibration suppression with piezoelectric materials and resonant shunts, J. Intell. Mater. Syst. Struct. 5 (1994) 49–57.

[20] S.Y. Wu, Method for multiple mode shunt damping of structural vibration using a single PZT transducer, in: Proceedings of SPIE Symposium on Smart Structures Materials Passive Damping Isolation,3327, 1998, pp. 159–168.

[21] S. Behrens, S.O. Reza Moheimani, Current flowing multiple mode piezoelectric shunt dampener, in: Proceedings of SPIE Symposium on Smart Structures Materials Damping Isolation, 4697, 2002, pp. 217–226.

[22] S. Behrens, A.J. Fleming, S.O. Reza Moheimani, New method for multiple-mode shunt damping of structural vibration using a single piezoelectric transducer, in: Proceedings of SPIE Symposium on Smart Structures Materials Damping Isolation, 4331, 2001, pp. 239–250.

[23] C.H. Park, A. Baz, Vibration control of beams with negative capacitive shunting of interdigital electrode piezoceramics, J. Vib. Control 11 (2005) 331–346.

[24] C. Richard, D. Guyomar, D. Audigier, H. Bassaler, Enhanced semi passive damping using continuous switching of a piezoelectric device on an inductor, in: Proceedings of SPIE Symposium on Smart Structures Materials Damping Isolation, 3989, 2000, pp. 288–299.

[25] R.C. Lawrence, W.C. William, Comparison of low-frequency piezoelectric switching shunt techniques for structural damping, Smart Mater. Struct. 11 (2002) 370–376.

[26] D. Niederberger, M. Morari, An autonomous shunt circuit for vibration damping, Smart Mater. Struct. 15 (2006) 359–364.

[27] J. Ducarne, O. Thomas, J.F. Deu, Structural vibration reduction by switch shunting of piezoelectric elements: modeling and optimization, J. Intell. Mater. Syst. Struct. 21 (2010) 797–816.

[28] J. Andrew, S.O. Reza Moheimani, Optimization and implementation of multimode piezoelectric shunt damping systems, IEEE/ASME Trans. Mechatronics 7 (2002) 87–94.

[29] IEEE Standard on Piezoelectricity, ANSI/IEEE Std, 176-1987.

[30] W.P. Mason, H. Jaffe, Methods for measuring piezoelectric, elastic, and dielectric coefficients of crystals and ceramics, in: Proceedings of the Institute of Radio Engineers, 42, 1954, pp. 921–930.

[31] H.J. Mcskimin, Notes and references for measurement of elastic moduli by means of ultrasonic waves, J. Acoust. Soc. Am. 33 (1961) 606–615.

[32] H.J. Mcskimin, Pulse superposition method for measuring ultrasonic wave velocities in solids, J. Acoust. Soc. Am. 33 (1961) 12–16.

[33] R.L. Forgacs, Improvements in the sing-around technique for ultrasonic velocity measurements, J. Acoust. Soc. Am. 32 (1960) 1697–1698.

[34] E. Hafner, The piezoelectric crystal unit—definitions and methods of measurement, Proc. IEEE 57 (1969) 179–201.

[35] W.A. Smith, B.A. Auld, Modeling 1–3 composite piezoelectrics: thickness-mode oscillations, IEEE Trans. Ultrason. Ferroelectr. Freq. Control 38 (1991) 40–47.

[36] L. Li, L. Qin, L.K. Wang, Y.Y. Wan, B.S. Sun, Researching on resonance characteristics influenced by the structure parameters of 1–3–2 piezocomposites plate, IEEE Trans. Ultrason. Ferroelectr. Freq. 55 (2008) 946–951.

[37] L. Li, L.K. Wang, L. Qin, Y. Lv, The theoretical model of 1-3-2 piezocomposites, IEEE Trans. Ultrason. Ferroelectr. Freq 56 (2009) 1476–1482.

[38] M. Sakthivel, A. Arockiarajan, An effective matrix poling characteristics of 1-3-2 piezoelectric composites, Sens. Actuators A 167 (2011) 34–43.

[39] R.J. Wang, Underwater Acoustic Materials Manual, Science Press, Beijing, 1983, pp. 144–147 (in Chinese).

Electro-elastic constants calculation of active piezoelectric damping composites by finite element method①

Yang Sun[1], Zhaohui Li[1] and Qihu Li[2]

Abstract

Currently, the active piezoelectric damping composites (APDC) with obliquely embedded piezoelectric rods have been considered as an effective means for controlling vibration and noise. How to model the unit cell of APDC with orientation angle between 0° and 90° that meets the theoretical restrictions and assumptions using finite element method (FEM) is a difficult problem that needs to be solved. In this paper, an alternative method to model the unit cell of APDC is proposed to overcome that difficulty. The effects of the orientation angle and the volume fraction of piezoelectric phase on the studied constants c_{33}^E, c_{33}^D, c_{44}^E, c_{44}^D, ε_{33}^S, e_{33}, e_{34} and the electromechanical coupling coefficients are analyzed and some special and useful conclusions are summarized. The simulation results show good agreement with the theoretical and experimental results, which confirms the validity of the proposed alternative finite element method in modeling and analyzing the APDC.

Keywords

Active piezoelectric damping composites, finite element method, electro-elastic constants, electromechanical coupling factor

Introduction

Compared with the traditional passive damping[1] and piezoelectric shunt damping treatments,[2-4] hybrid damping techniques incorporating active and passive damping have presented a higher performance in the field of vibration and noise control. Some of the most significant works have concentrated around the following categories: the electro-mechanical surface damping (EMSD) treatments,[5,6] the active constrained layer damping (ACLD) treatments,[7-9] the conventional active piezoelectric damping composites (CAPDC) with perpendicularly embedded piezoelectric rods in a visco-elastic matrix,[10-13] which is also called 1-3 piezoelectric composites, and the new active piezoelectric damping composites (APDC) with obliquely embedded piezoelectric rods.[14-17]

It is reported that by combing the attractive properties of both the ACLD and the CAPDC treatments, APDC with obliquely embedded piezoelectric rods can enhance both shear and compression damping characteristics of the composites and obtain a 16% higher coupling factor than that of CAPDC when embedding PZT-5H rods, at an angle of 28°, in soft polyurethane matrix.[16] With such optimized performance, APDC can play an important role in the applications such as transducer, hydrophone, non-destructive inspection, ultrasonic imaging, and smart materials.

Much work has applied FEM to model and analyze the properties of composites.[18-25] However, the application of FEM on the electro-elastic property analysis of APDC with obliquely embedded piezoelectric rods is less studied, which is of fundamental importance to the further research of APDC. In the theoretical calculation method of the electro-elastic constants of APDC,[14-17] there are some essential restrictions and ideal assumptions for APDC. How to model the unit cell of APDC with orientation angle between 0° and 90° that meets the theoretical restrictions and assumptions using FEM is a difficult problem that needs to be solved.

[1]Department of Electronics, Peking University, China
[2]Institute of Acoustic, Chinese Academy of Sciences, China

Corresponding author:
Zhaohui Li, Department of Electronics, Peking University, Beijing 100871, China.
Email: lizhcat@pku.edu.cn

① Journal of Reinforced Plastics and Composites, 2014, 33(8): 175-1764.

This paper proposes an alternative finite element method of modeling APDC to analyze and calculate several important electro-elastic constants. The calculated results show good agreement with the theoretical and experimental results.

APDC structure and model

The APDC are structured with piezoelectric rods that are obliquely embedded across the thickness (z-direction) in a viscoelastic matrix in the y–z plate of the global coordinate system (x,y,z) at an orientation angle θ as shown in Figure 1.[16] Another local coordinate system ($1,2,3$) is established and oriented such that the polarization direction 3 is set parallel to the longitudinal axis of the piezoelectric rods. The matrix phase is assumed to be a homogeneously isotropic medium and piezoelectrically inactive.

Finite element model

Since the structure of the composite is of periodicity, it can be divided into a number of tiny unit cells which contain the main characteristics of the microstructure and are able to represent the properties of the whole composite.

The orientation angle is set between 0° and 90°. One of the essential assumptions is that the APDC is treated as an effective homogeneous medium. In other words, the composite is of fine lateral spatial scale and the vertical strains are the same in both piezoelectric and matrix phases (equation (1)), so are the shear strains (equation (2)).[11,16]

$$S_z^p = S_z^m = S_z \tag{1}$$

$$S_{yz}^p = S_{yz}^m = S_{yz} \tag{2}$$

Here, superscript 'm' stands for matrix phase and 'p' means piezoelectric phase. It is simple to implement the

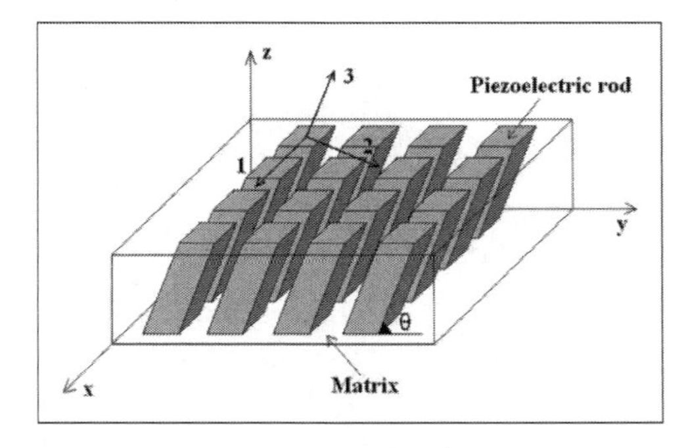

Figure 1. Schematic drawing of APDC.

assumption when the orientation angle is in special cases as 0° (2–2 piezoelectric composite) and 90° (1–3 piezoelectric composite, CAPDC). Only one unit cell is needed to model the whole composite for these two cases because the composites possess periodicity and symmetry. However, for the orientation angle between 0° and 90°, it is difficult to model the unit cell because the symmetry cannot be met. And it is also very complicated to model the whole composite because the piezoelectric rods are obliquely embedded and finely and closely arranged.

An alternative method to model the unit cell of APDC with the orientation angle between 0° and 90° is proposed as shown in Figure 2. If t represents the thickness of the composite and a denotes the lateral length of the unit cell, the ratio a/t can be adjusted to validate the model. It will be presented later in the section Example II: orientation angle is between 0° and 90° that only when a/t is large enough, the model can meet the fine-scaled condition and the electro-elastic constants consistent with the theoretical ones can be obtained.

The FEM element types of the piezoelectric phase and the matrix phase are both set to be Solid98, which is defined by 10 nodes with up to six degrees of freedom at each node and has large deflection and stress stiffening capabilities when used in structural and piezoelectric analyses. The degree of freedom option KEYOPT(1) is set as 3, which means the degrees of freedom are UX, UY, UZ, and VOLT. Free meshing is adopted to divide the composite into a lot of tetrahedron grid units as seen in Figure 3.

Choosing and preprocessing of material parameters

In the material props part, it is needed to input the material parameters of both phases. For any point (x,y,z) in the matrix phase (Global coordinate), the constitutive equations are concluded as equation (3)[16]

$$\begin{cases} \{T^m\} = [c^m]\{S^m\} \\ \{D^m\} = [\varepsilon^m]\{E^m\} \end{cases} \tag{3}$$

where $[c^m]$ and $[\varepsilon^m]$ are the elastic constant matrix and dielectric constant matrix, respectively.

$$[c^m] = \begin{pmatrix} c_{11}^m & c_{12}^m & c_{12}^m & 0 & 0 & 0 \\ c_{12}^m & c_{11}^m & c_{12}^m & 0 & 0 & 0 \\ c_{12}^m & c_{12}^m & c_{11}^m & 0 & 0 & 0 \\ 0 & 0 & 0 & c_{44}^m & 0 & 0 \\ 0 & 0 & 0 & 0 & c_{44}^m & 0 \\ 0 & 0 & 0 & 0 & 0 & c_{44}^m \end{pmatrix} \tag{4}$$

$$[\varepsilon^m] = \begin{pmatrix} \varepsilon_x^m & 0 & 0 \\ 0 & \varepsilon_x^m & 0 \\ 0 & 0 & \varepsilon_x^m \end{pmatrix} \quad (5)$$

While for any point (*1,2,3*) in the piezoelectric phase (local coordinate), the constitutive equations are expressed as equation (6).[16] $[c^p]_{local}$, $[\varepsilon^p]_{local}$, and $[e]_{local}$ are the elastic constant matrix, dielectric constant matrix, and piezoelectric constant matrix of piezoelectric phase, respectively, as shown in equations (7) to (9).[16]

$$\begin{cases} \{T^p\}_{local} = [c^p]_{local}\{S^p\}_{local} - [e]_{local}[E^p]_{local} \\ \{D^p\}_{local} = [e]_{local}\{S^p\}_{local} + [\varepsilon^p]_{local}\{E^p\}_{local} \end{cases} \quad (6)$$

$$[c^p]_{local} = \begin{pmatrix} c_{11}^E & c_{12}^E & c_{13}^E & 0 & 0 & 0 \\ c_{12}^E & c_{11}^E & c_{13}^E & 0 & 0 & 0 \\ c_{13}^E & c_{13}^E & c_{33}^E & 0 & 0 & 0 \\ 0 & 0 & 0 & c_{44}^E & 0 & 0 \\ 0 & 0 & 0 & 0 & c_{44}^E & 0 \\ 0 & 0 & 0 & 0 & 0 & c_{66}^E \end{pmatrix} \quad (7)$$

$$[\varepsilon^p]_{local} = \begin{pmatrix} \varepsilon_{11}^S & 0 & 0 \\ 0 & \varepsilon_{11}^S & 0 \\ 0 & 0 & \varepsilon_{33}^S \end{pmatrix} \quad (8)$$

$$[e]_{local} = \begin{pmatrix} 0 & 0 & 0 & 0 & e_{15} & 0 \\ 0 & 0 & 0 & e_{15} & 0 & 0 \\ e_{31} & e_{31} & e_{33} & 0 & 0 & 0 \end{pmatrix} \quad (9)$$

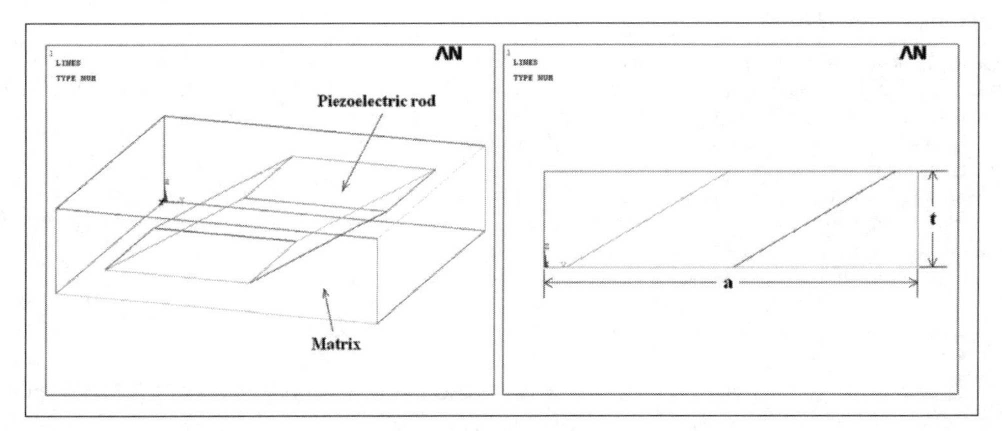

Figure 2. The unit cell of APDC (*t*: thickness of the composite, *a*: lateral length of the unit cell).

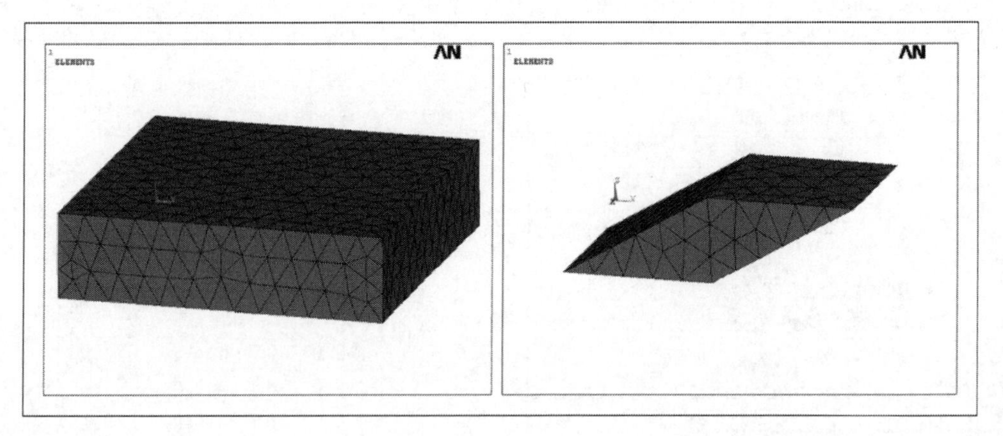

Figure 3. Meshing.

As the piezoelectric rods are obliquely embedded in the matrix, the local and the global coordinate systems are related by a rotation matrix $[A]$[26] as equation (10)

$$[A]=\begin{pmatrix}1 & 0 & 0\\ 0 & \sin\theta & \cos\theta\\ 0 & -\cos\theta & \sin\theta\end{pmatrix} \tag{10}$$

The elastic constant matrix, dielectric constant matrix, and piezoelectric constant matrix must be transformed as shown in equations (11) to (13)

$$[c^p]_{global}=[M][c^p]_{local}[M]^T \tag{11}$$

$$[\varepsilon^p]_{global}=[A][\varepsilon^p]_{local}[A]^T \tag{12}$$

$$[e]_{global}=[A][e]_{local}[M]^T \tag{13}$$

where $[M]$ represents the stress rotation matrix as following

$$[M]=\begin{pmatrix}1 & 0 & 0 & 0 & 0 & 0\\ 0 & (\sin\theta)^2 & (\cos\theta)^2 & 2\sin\theta\cos\theta & 0 & 0\\ 0 & (\cos\theta)^2 & (\sin\theta)^2 & -2\sin\theta\cos\theta & 0 & 0\\ 0 & -\sin\theta\cos\theta & \sin\theta\cos\theta & (\sin\theta)^2-(\cos\theta)^2 & 0 & 0\\ 0 & 0 & 0 & 0 & \sin\theta & -\cos\theta\\ 0 & 0 & 0 & 0 & \cos\theta & \sin\theta\end{pmatrix} \tag{14}$$

After transforming process, the constitutive equations of the piezoelectric phase (Global coordinate) turn into the following form

$$\begin{cases}\{T^p\}_{global}=[c^p]_{global}\{S^p\}_{global}-[e]_{global}[E^p]_{global}\\ \{D^p\}_{global}=[e]_{global}\{S^p\}_{global}+[\varepsilon^p]_{global}\{E^p\}_{global}\end{cases} \tag{15}$$

The parameters of the piezoelectric phase (PZT-5H) and the matrix phase (soft-polyurethane) are listed in Table 1.

Definitions of the calculated constants in FEM

In order to calculate different electro-elastic constants, proper boundary conditions and loads must be defined to the unit cell model according to the definition of the calculated constants.

c_{33}^E and c_{33}^D

The definition of c_{33}^E and c_{33}^D is the variation of the stress tensor component T_z caused by the one-unit change of the strain tensor component S_z in the 'constant electric field' condition and 'constant electric displacement' condition, respectively, as shown in equations (16) and (17)[27]

$$c_{33}^E=\left(\frac{\partial T_z}{\partial S_z}\right)_E=\left.\frac{\Delta T_z}{\Delta S_z}\right|_{E=0} \tag{16}$$

$$c_{33}^D=\left(\frac{\partial T_z}{\partial S_z}\right)_D=\left.\frac{\Delta T_z}{\Delta S_z}\right|_{D=0} \tag{17}$$

Table 1. Parameters of materials.[16]

Piezoelectric phase: PZT-5H

Density (kg/m³)	Elastic Constants (10^{10}Pa)						Piezoelectric Constants (C/m²)			Relative Dielectric Constant	
ρ_p	c_{11}^E	c_{12}^E	c_{13}^E	c_{33}^E	c_{44}^E	c_{66}^E	e_{15}	e_{31}	e_{33}	$\varepsilon_{11}/\varepsilon_0$	$\varepsilon_{33}/\varepsilon_0$
7750	15.1	9.8	9.6	12.4	1.4	2.65	20	−5.1	27	1700	1500

Polymer phase: Soft-polyurethane

Density (kg/m³)	Elastic Constants (10^{10}Pa)			Relative Dielectric Constant
ρ_m	c_{11}	c_{12}	c_{44}	$\varepsilon_{11}/\varepsilon_0$
1100	0.001667	0.001664	0.0000015	0

where ΔS_z is the change of the strain tensor component S_z in FEM, and ΔT_z is the change of the stress tensor component T_z caused by ΔS_z.

Figure 4 is the profile of the unit cell. z-axis is the direction of the thickness of the composite and the two planes perpendicular to z-axis are marked as 1 and 2. The two planes perpendicular to x-axis are marked as 3 and 4 while the two planes perpendicular to y-axis are marked as 5 and 6.

To cause the change of the strain tensor component S_z, a z-direction displacement is loaded to plane 2 as shown in Figure 4. At the same time, plane 1 is fixed in z-direction. Planes 3–6 are also clamped in their normal directions in order to set strain tensor components in other directions to zero. In Figure 4, '^' represents that the plane is clamped in its normal direction.

For c_{33}^E, 'constant electric field' means that planes 1 and 2 are short-circuited. So electrode a and b are set to zero potential. While for c_{33}^D, 'constant electric displacement' means that planes 1 and 2 are open-circuited.

c_{44}^E and c_{44}^D

The definition of c_{44}^E and c_{44}^D is the variation of the stress tensor component T_{yz} caused by the one-unit change of the strain tensor component S_{yz} in the 'constant electric field' condition and 'constant electric displacement' condition, respectively, as shown in equations (18) and (19)[27]

$$c_{44}^E = \left(\frac{\partial T_{yz}}{\partial S_{yz}}\right)_E = \left.\frac{\Delta T_{yz}}{\Delta S_{yz}}\right|_{E=0} \tag{18}$$

$$c_{44}^D = \left(\frac{\partial T_{yz}}{\partial S_{yz}}\right)_D = \left.\frac{\Delta T_{yz}}{\Delta S_{yz}}\right|_{D=0} \tag{19}$$

where ΔS_{yz} is the change of the strain tensor component S_{yz} in FEM, and ΔT_{yz} is the change of the stress tensor component T_{yz} caused by ΔS_{yz}.

In order to cause the change of the shear strain tensor component S_{yz}, a y-direction displacement is loaded to plane 2 as shown in Figure 5 and plane 1 is fixed in three main directions. In Figure 5, 'Δ' represents that the plane is clamped in three main directions. All nodes in the model are fixed in x and z directions in order to set strain tensor components in those directions to zero.

The loading method of the electrical boundary conditions of c_{44}^E and c_{44}^D is the same as that of c_{33}^E and c_{33}^D.

ε_{33}^S, e_{33}, e_{34}

The definition of ε_{33}^S is the variation of the electric displacement vector component D_z caused by the one-unit change of the electric field vector component E_z in the 'constant strain' condition as shown in equation (20).[27]

The definitions of e_{33} and e_{34} are the decrement of the stress tensor component T_z and T_{yz} caused by the one-unit change of the electric field vector component E_z, respectively, in the 'constant strain' condition as shown in equations (21) and (22).[27]

So, the boundary conditions and loads of ε_{33}^S, e_{33} and e_{34} are the same.

$$\varepsilon_{33}^S = \left(\frac{\partial D_z}{\partial E_z}\right)_S = \left.\frac{\Delta D_z}{\Delta E_z}\right|_{S=0} \tag{20}$$

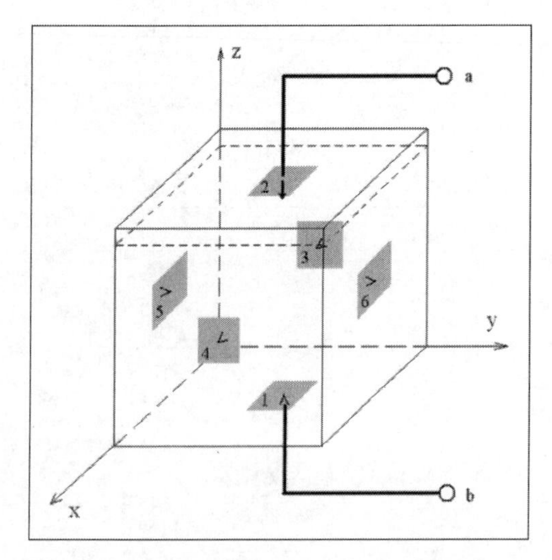

Figure 4. Boundary conditions and loads for c_{33}^E and c_{33}^D.

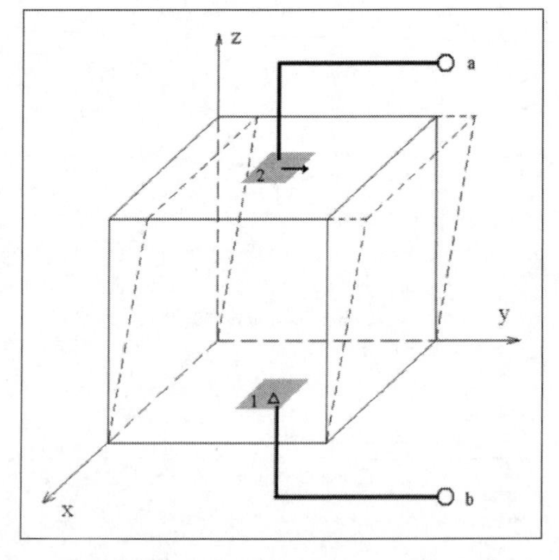

Figure 5. Boundary conditions and loads for c_{44}^E and c_{44}^D.

$$e_{33} = -\left(\frac{\partial T_z}{\partial E_z}\right)_S = -\frac{\Delta T_z}{\Delta E_z}\bigg|_{S=0} \tag{21}$$

$$e_{34} = -\left(\frac{\partial T_{yz}}{\partial E_z}\right)_S = -\frac{\Delta T_{yz}}{\Delta E_z}\bigg|_{S=0} \tag{22}$$

where ΔE_z is the change of electric field vector component E_z in FEM, and ΔD_z, ΔT_z and ΔT_{yz} are the changes of the electric displacement vector component D_z, the stress tensor components T_z and T_{yz} caused by ΔE_z, respectively.

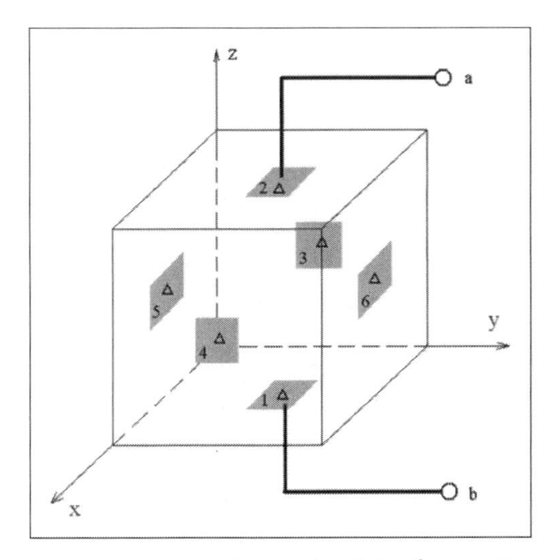

Figure 6. Boundary conditions and loads for ε_{33}^S, e_{33}, and e_{34}.

In order to cause the change of the electric field vector component E_z, electrode b is loaded zero potential and electrode a is loaded nonzero potential. The potential difference of planes 1 and 2 will cause an electric field across the z-direction.

The 'constant strain' condition means all the six planes are clamped in three main directions as shown in Figure 6. '△' means the plane is clamped in three main directions.

Calculating method

In the solution part, static analysis is carried out. In the postprocessor, APDL (ANSYS Parametric Design Language) is used to pick up the volume, strain, stress, electric field, and electric displacement information of all the grid units and calculate the volumetrically weighted averages of the picked values. Finally, the wanted constants are calculated according to their definitions. The above volumetrically weighted average arithmetic is based on that the composite is treated as an effective homogeneous medium.

Take c_{33} for example. First, the volume $(V_1, V_2, \ldots V_n)$, stress T_z $(T_1, T_2, \ldots T_n)$ and strain S_z $(S_1, S_2, \ldots S_n)$ values of all the grid units are picked up. Then, the volumetrically weighted averages (VWA) of the stress and strain values as shown in equations (23) and (24) are calculated. Finally, c_{33} is obtained by dividing $\overline{T_z}$ by $\overline{S_z}$ as expressed in equation (25)

$$\overline{T_z} = \frac{T_1 V_1 + T_2 V_2 + \cdots + T_n V_n}{V_1 + V_2 + \cdots + V_n} = \frac{\sum_{j=1}^{n} T_j V_j}{V} \tag{23}$$

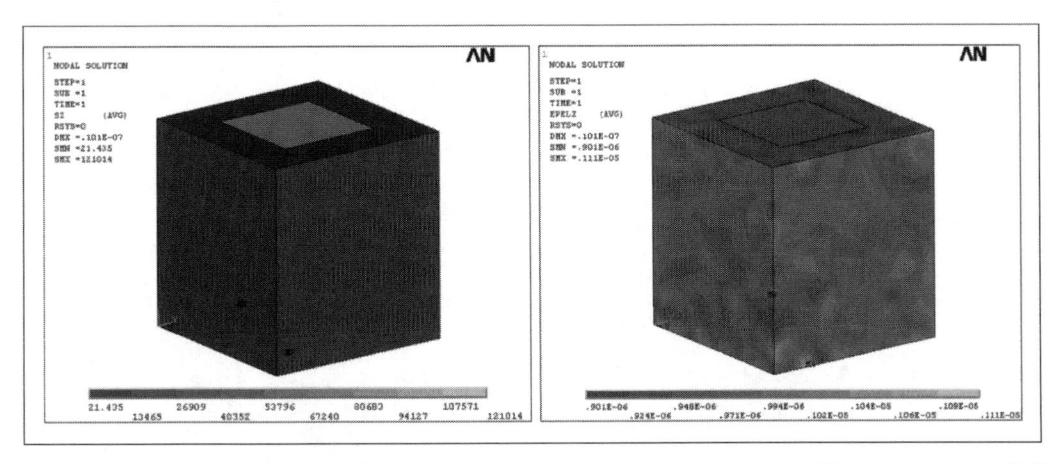

Figure 7. Distribution of (a) stress tensor component T_z and (b) strain tensor component S_z with 90°-orientation angle and 30% volume fraction.

$$\overline{S}_z = \frac{S_1V_1 + S_2V_2 + \cdots + S_nV_n}{V_1 + V_2 + \cdots + V_n} = \frac{\sum_{j=1}^{n} S_jV_j}{V} \tag{24}$$

$$c_{33} = \frac{\overline{T}_z}{\overline{S}_z} \tag{25}$$

When the orientation angle is 90° (composites 1–3), the volume fraction is 30%, the length of unit cell cube is 0.01 m, and the z-direction displacement loaded to plane 2 is 1×10^{-8} m, the obtained distribution of stress tensor component T_z and strain

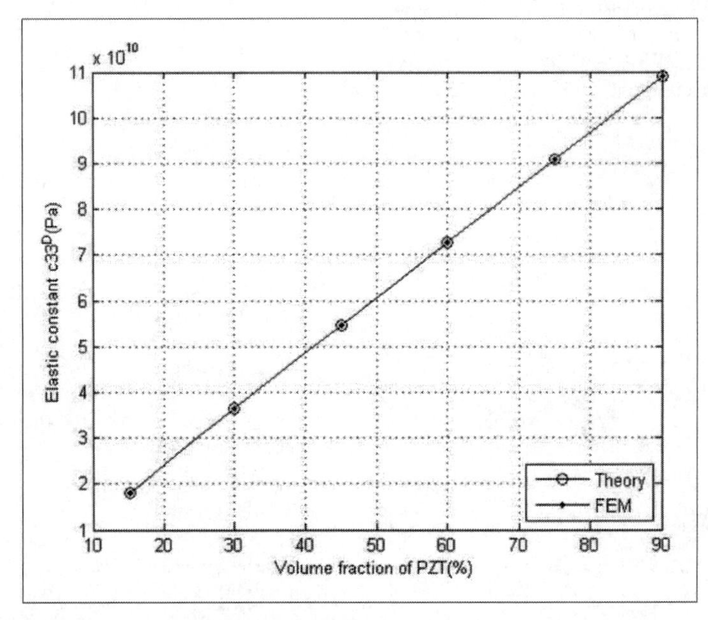

Figure 8. Theoretical and FEM results of the elastic constant c_{33}^D (orientation angle: 90°).

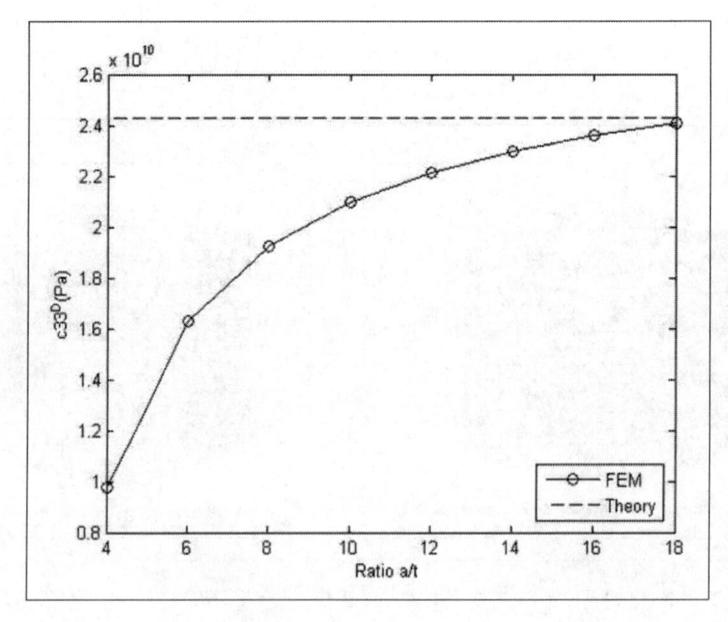

Figure 9. The influence of ratio a/t.

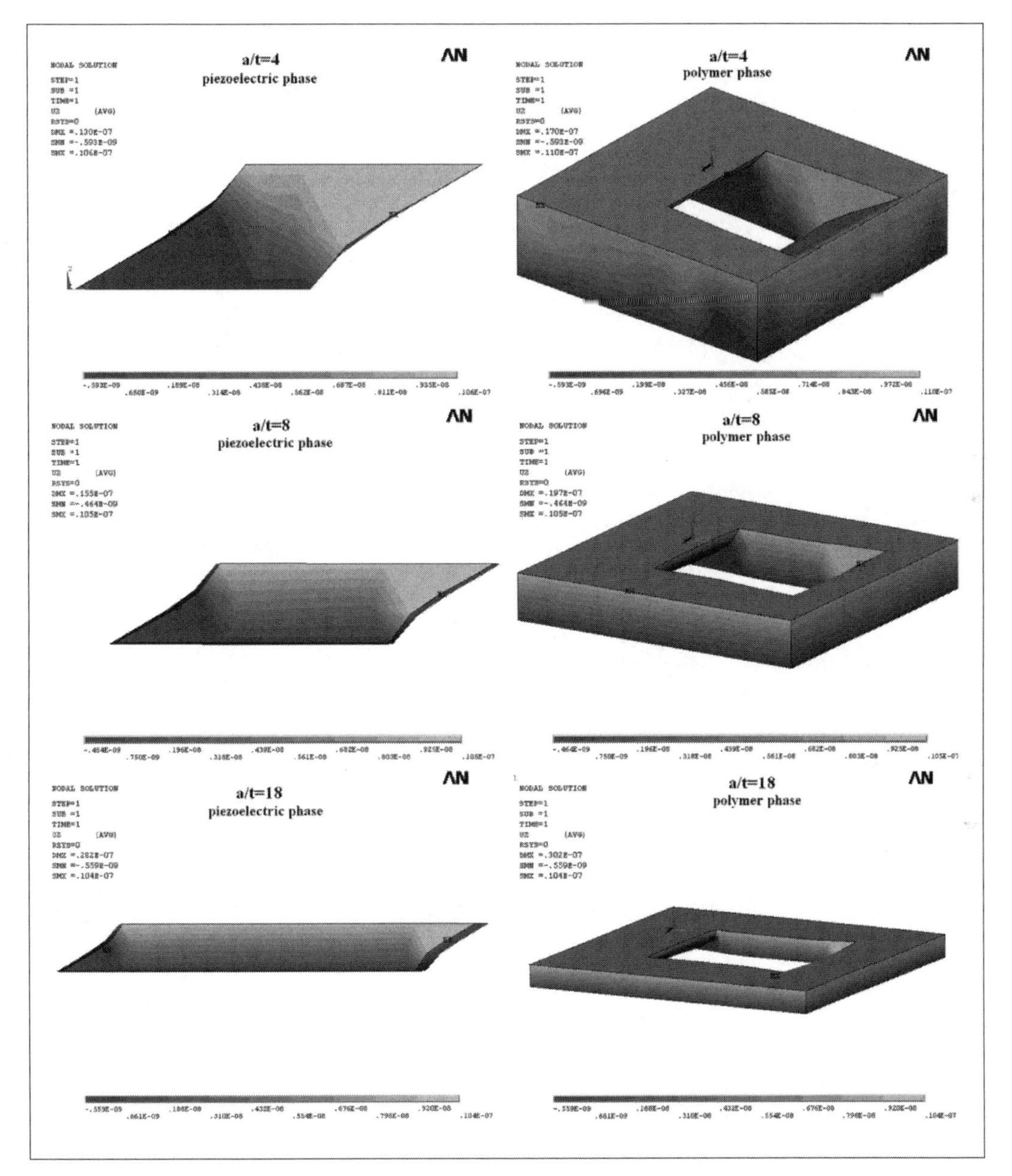

Figure 10. Distributions of z-direction displacements of the two phases (ratio $a/t=4$, 8, 18).

tensor component S_z is shown in Figure 7(a) and (b). It can be found that the strain tensor component S_z is nearly the same (1×10^{-6}) all over the unit cell, whereas the stress tensor component T_z of the piezoelectric phase and the matrix phase is different. The volumetrically weighted average (VWA) of stress tensor component T_z is 36319.34699 and the VWA of strain tensor component S_z is 1×10^{-6}. So, c_{33} is calculated to be 3.631934699×10^{10} Pa.

For the verification of the proposed model, two calculation examples **I** and **II** are given as follow.

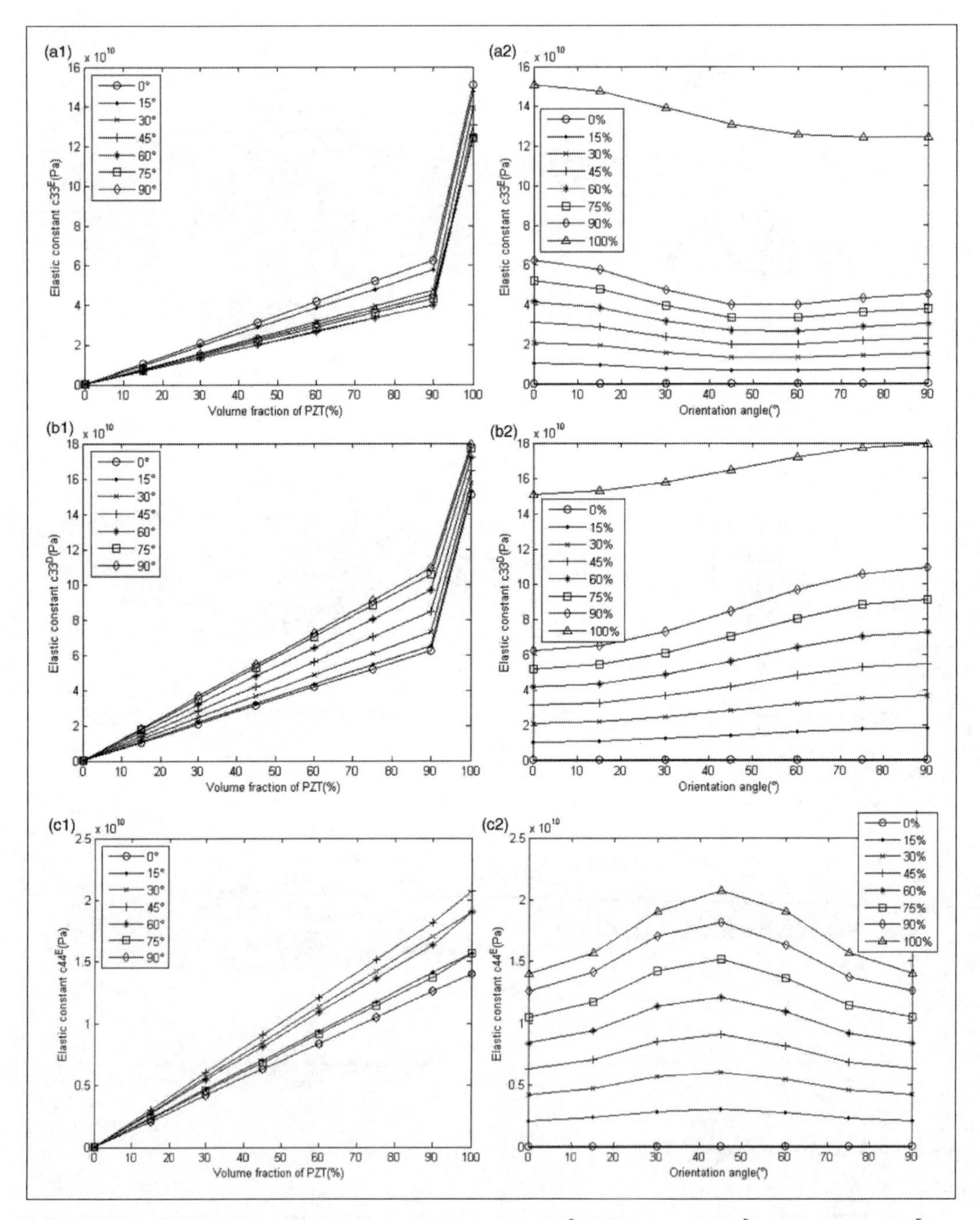

Figure 11. The variations of electro-elastic constants: (a) elastic constant c_{33}^{E}, (b) elastic constant c_{33}^{D}, (c) elastic constant c_{44}^{E}, (d) elastic constant c_{44}^{D}, (e) dielectric constant ε_{33}^{S}, (f) piezoelectric constant e_{33}, (g) piezoelectric constant e_{34} with the volume fraction (left column) and orientation angle (right column) by FEM.

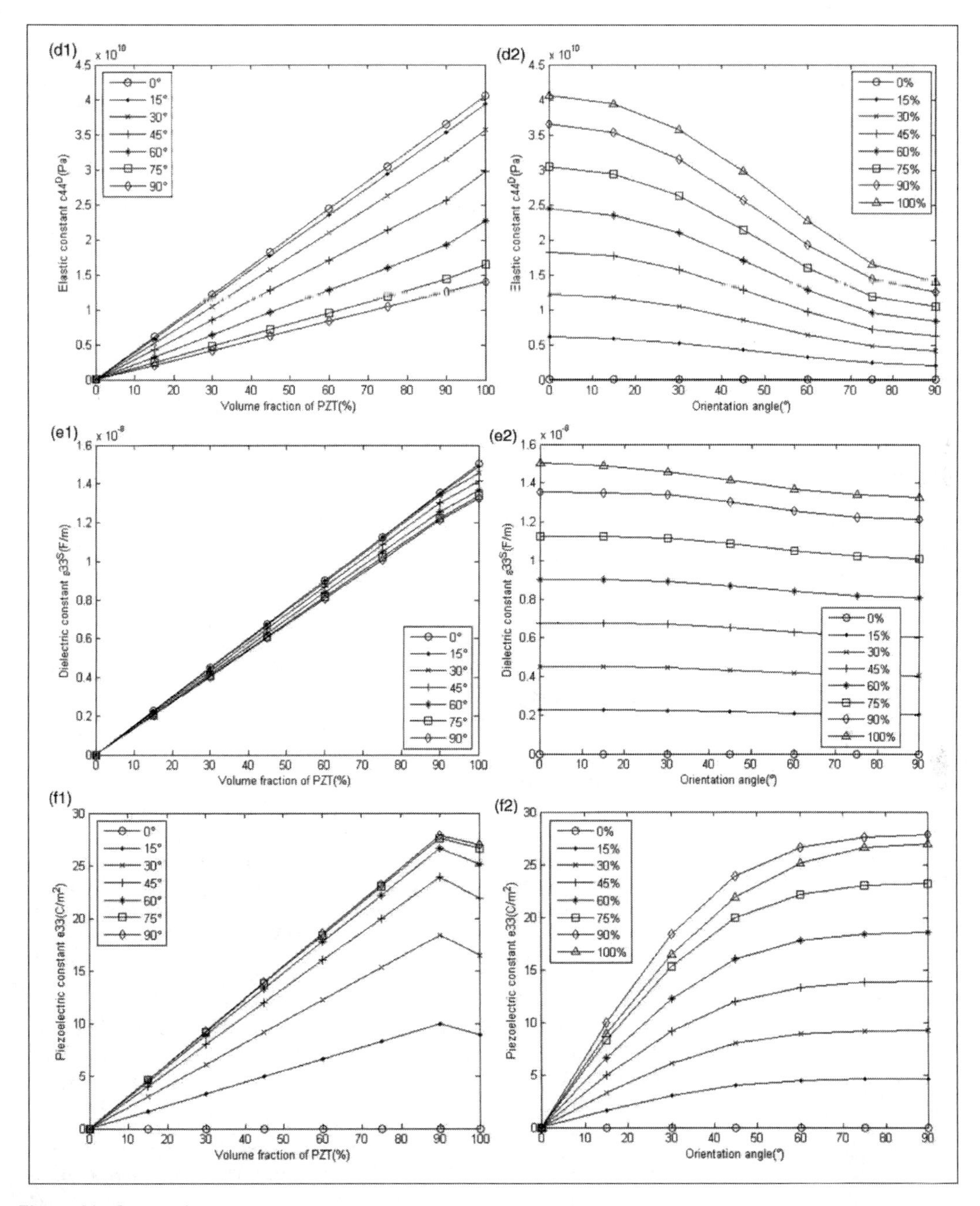

Figure 11. Continued.

Example I: orientation angle is 90°

When the volume fraction changes from 15% to 90% and S_z is set to be 1×10^{-6}, VWAs of stress tensor component T_z are calculated accordingly. Dividing the VWAs of T_z by S_z, the elastic constants c_{33} are obtained. The estimated results of c_{33}^D (c_{33} under the 'constant electric displacement' condition) by finite

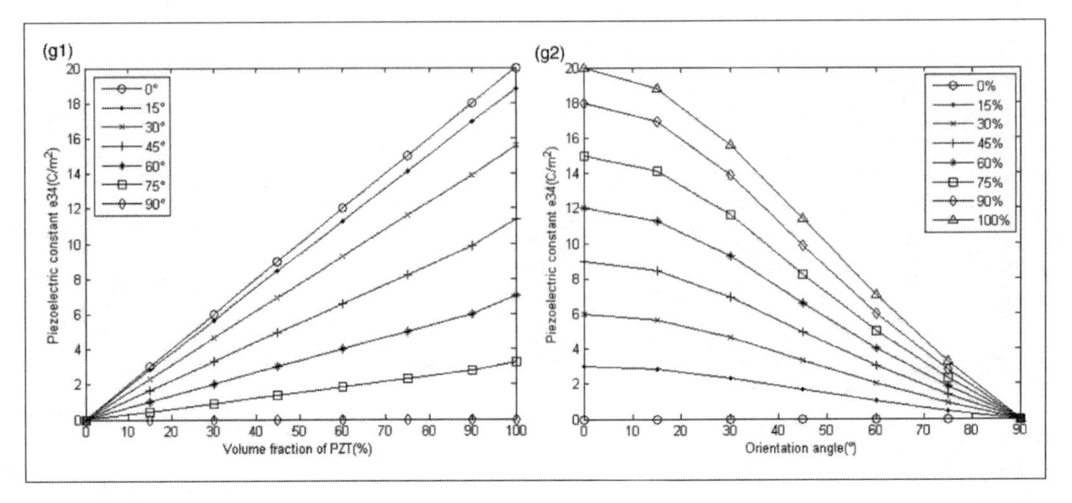

Figure 11. Continued.

element method and the results obtained by the theoretical mathematical analysis[16] are compared in Figure 8. The FEM results show good agreement with the theoretical ones, so do the other electro-elastic constants.

Example II: orientation angle is between 0° and 90°

It is mentioned above that the ratio a/t (the ratio of the lateral length of the unit cell to the thickness of the composite) plays an important role in modelling the APDC, which decides if the fine-scaled condition needed in theoretical calculation can be met.

For example, if the orientation angle is 30° and the volume fraction is 30%, the effect of a/t on the elastic constant c_{33}^D is plotted in Figure 9. It is shown that the calculated constants obtained by the FEM model deviate far from the theoretical results when a/t is small. But when a/t reaches 18, the FEM result is very close to the theoretical value. That means if the theoretical assumptions are needed to be satisfied, the ratio a/t should be set greater than a certain value for given orientation angle and volume fraction.

The distributions of z-direction displacements of the two phases when the ratio a/t equals to 4, 8, and 18 shown in Figure 10 can intuitively present the different degrees of satisfaction of the theoretical assumptions. If the vertical strains are the same in both piezoelectric and polymer phases, the z-direction displacements of the two phases surely decrease stepwise from the top to the bottom. From Figure 10 it can be seen that when a/t equals to 4, the z-direction displacements of the two phases are mostly irregular. When a/t reaches 8, the volume of the unit cell whose z-direction displacements decrease stepwise becomes bigger, which means the degree of satisfaction of the theoretical assumptions gets larger. The better case comes when a/t increases to 18. Most of the unit cell meets the condition that the z-direction displacements decrease stepwise except for the sharp corners at both ends of the piezoelectric phase and the adjacent polymer phase.

Results and discussion

With the changes of both orientation angle and volume fraction of piezoelectric phase, the electro-elastic constants c_{33}^E, c_{33}^D, c_{44}^E, c_{44}^D, ε_{33}^S, e_{33}, e_{34} are estimated by finite element method as shown in Figure 11.

From the results one can find that nearly all the electro-elastic constants (except for c_{33} and e_{33} when volume fraction is 90–100%[11]) linearly vary when the volume fraction changes. This is because that the APDC is assumed to be a homogeneously isotropic medium and is of fine lateral spatial scale.

The effects of the orientation angle on the electro-elastic constants reflect the effects of the different strengths of the compression and shear modes on the properties of APDC. Some opposite variations between the constants concerned with the compression mode (c_{33}^E, c_{33}^D, e_{33}) and those concerned with shear mode (c_{44}^E, c_{44}^D, e_{34}) can be found. For a larger volume fraction, with the increase of orientation angle, c_{33}^E varies along a concave curve while c_{44}^E varies along a convex curve. c_{33}^D and e_{33} increase monotonically with the increase of orientation angle while c_{44}^D and e_{34} decrease monotonically. Especially, e_{33} will reach the minimum (zero) at 0°-orientation angle and the maximum at 90°-orientation angle while e_{34} is just the

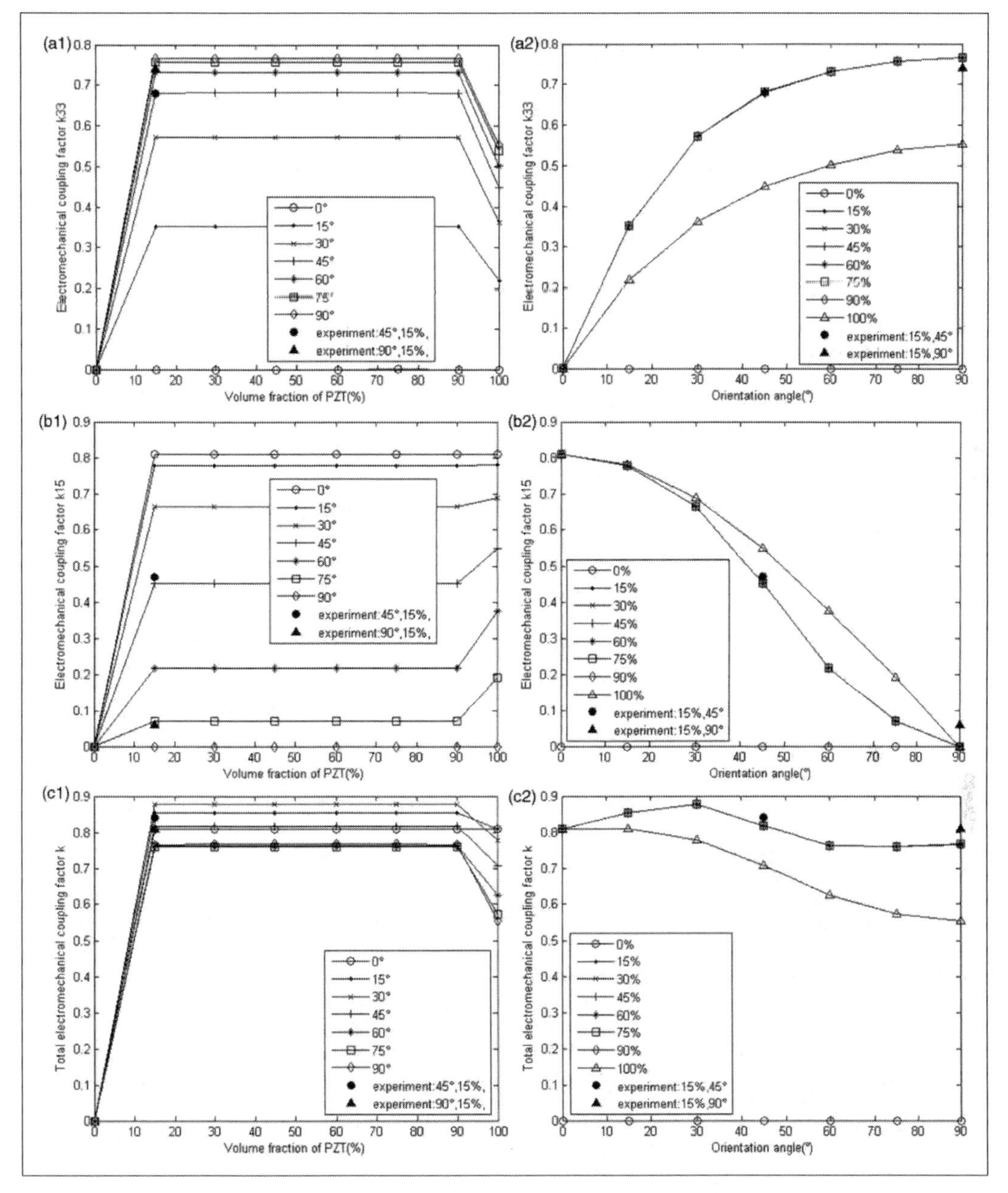

Figure 12. The variations of electromechanical coupling coefficients (a) k_{33}, (b) k_{15}, (c) k with the volume fraction (left column) and orientation angle (right column) by FEM (The experiment results are taken from Baz and Tempia[16]).

opposite. This reveals that when the orientation angle is 90°, only compression mode exists but no shear mode. While for the 0°-orientation angle, only shear mode exists but no compression mode. In addition, with the increase of orientation angle, the dielectric constant will decrease but varies not too much.

The longitudinal electromechanical coupling coefficient k_{33}, the shear electromechanical coupling coefficient k_{15}, and the total electromechanical coupling coefficient $k=\sqrt{k_{33}^2+k_{15}^2}$[16] are also calculated and shown in Figure 12.

Some special and useful conclusions can also be summarized. For the 0° orientation angle condition, the elastic constants c_{33}^E and c_{33}^D are equal and the piezoelectric constant e_{33} equals to 0. So the longitudinal electromechanical coupling coefficient k_{33} equals to 0 while the shear electromechanical coupling coefficient k_{15} reaches to the maximum. On the contrary, when orientation angle is 90°, the elastic constants c_{44}^E and c_{44}^D are equal and the piezoelectric constant e_{34} equals to 0. So the shear electromechanical coupling coefficient k_{15} equals to 0 while the longitudinal electromechanical coupling coefficient k_{33} reaches to the maximum.

The total electromechanical coupling coefficient $k=\sqrt{k_{33}^2+k_{15}^2}$ increases rapidly with the variation of the volume fraction from 0 to 10% but it does not vary too much when the volume fraction is between 10% and 90%. It will reach the maximum at certain orientation angle and volume fraction when the composite is with different materials of piezoelectric phase and matrix phase. When the two individual phases are PZT-5H and Soft-polyurethane, k reached the maximum 0.88 at about 30°-orientation angle. The FEM calculated results show good agreement with the experimental results taken from Baz and Tempia.[16]

Conclusions

This paper focuses on the finite element method modeling and calculating the electro-elastic constants of the APDC with piezoelectric rods obliquely embedded across the thickness of a viscoelastic damping matrix. For the case that the orientation angle is between 0° and 90°, an alternative flat unit cell model of the composite is proposed to meet the fine-scaled restriction. Specially, the APDC will turn into 2–2 type and 1–3 type piezoelectric composites correspondingly when the orientation angle is, respectively, 0° and 90°.

The element types of the piezoelectric phase and the matrix phase are chosen and the forms of the electro-elastic constants needed to be input to the material models of two phases are given. Specially, the electro-elastic constant matrix of the piezoelectric phase must be transformed since the piezoelectric rods are obliquely embedded in the matrix. The boundary conditions and loads corresponding to several studied electro-elastic constants are described in detail. The distributions of stress and strain as well as the volumetrically weighted average (VWA) are obtained to calculate these constants. c_{33}^D of 90°-orientation angle has been taken as an example to illustrate the calculating process. The FEM results show good agreement with the theoretical results. For the oblique condition when the orientation angle is between 0° and 90°, the FEM results consistent with the theoretical ones can be obtained when the ratio a/t is greater than a certain value for given orientation angle and volume fraction.

The effects of the orientation angle and the volume fraction of piezoelectric phase on the constants c_{33}^E, c_{33}^D, c_{44}^E, c_{44}^D, ε_{33}^S, e_{33}, e_{34} and the electromechanical coupling coefficient k_{33}, k_{15} and k are summarized and discussed. The results show that the alternative finite element method proposed in this paper is valid in modeling and analyzing APDC.

Conflict of interest

None declared.

Funding

This research received no specific grant from any funding agency in the public, commercial, or not-for-profit sectors.

References

1. Douglas BE and Yang JCS. Transverse compressional damping in the vibratory response of elastic-viscoelastic-elastic beams. *AIAA J* 1978; 16: 925–930.
2. Forward RL. Electronic damping of vibrations in optical structures. *Appl Optics* 1979; 18: 690–697.
3. Hagood NW and von Flotow A. Damping of structural vibrations with piezoelectric materials and passive electrical networks. *J Sound Vibrat* 1991; 146: 243–268.
4. Law HH, Rossiter PL, Simon GP, et al. Characterization of mechanical vibration damping by piezoelectric materials. *J Sound Vibrat* 1996; 197: 489–513.
5. Ghoneim H. Application of the electromechanical surface damping to the vibration control of a cantilever plate. *J Vibrat Acoustic* 1996; 118: 551–557.
6. Orsagh R and Ghoneim H. Experimental investigation of electromechanical surface damping. *Proc Smart Struct Mater Conf* 1999; 3672: 234–241.
7. Baz A and Ro J. Vibration control of plates with active constrained layer damping. *Smart Mater Struct* 1996; 5: 272–280.
8. Haung SC, Inman D and Austin E. Some design considerations for active and passive constrained layer damping treatments. *Smart Mater Struct* 1996; 5: 301–313.
9. Ray MC, Oh J and Baz A. Active constrained layer damping of thin cylindrical shells. *J Sound Vibrat* 2001; 240: 921–935.
10. Auld BA, Kunkel HA, Shui YA, et al. Dynamic behavior of periodic piezoelectric composites. In: *IEEE ultrasonic symposium*, Atlanta, GA, USA, 31 October–2 November 1983, pp.554–558.
11. Smith WA and Auld BA. Modeling 1-3 composite piezoelectrics thickness-mode oscillations. *IEEE Trans Ultrason Ferroelec Freq Contr* 1991; 38: 40–47.
12. Chan HLW and Unsworth J. Simple model for piezoelectric ceramic/polymer 1-3 composites used in ultrasonic

transducer applications. *IEEE Trans Ultrason Ferroelec Freq Contr* 1989; 36: 434–441.
13. Hayward G and Hossack JA. Unidimensional modeling of 1-3 composite transducers. *JASA* 1990; 88: 599–608.
14. Arafa M and Baz A. Dynamics of active piezoelectric damping composites. *Compos: Part B* 2000; 31: 255–264.
15. Arafa M and Baz A. Energy-dissipation characteristics of active piezoelectric damping composites. *Compos Sci Technol* 2000; 60: 2759–2768.
16. Baz A and Tempia A. Active piezoelectric damping composites. *Sensor Actuator A* 2004; 112: 340–350.
17. Ren HS and Fan HQ. The role of piezoelectric rod in 1-3 composites for the hydrostatic response applications. *Sensors Actuator A* 2006; 128: 132–139.
18. Hossack JA and Hayward G. Finite-element analysis of 1-3 composite transducers, ultrasonics. *IEEE Trans Ultrason Ferroelec Freq Contr* 1991; 38: 618–629.
19. Steinhausen R, Hauke T, Seifert W, et al. Finescaled piezoelectric 1-3composites: properties and modeling. *J Eur Ceram Soc* 1999; 19: 1289–1293.
20. Reynolds P, Hyslop J and Hayward G. Analysis of spurious resonances in single and multi-element piezocomposite ultrasonic transducers. *2003 IEEE Ultrasonic Sympos* 2003; 2: 1650–1653.
21. Gebhardt S, Schönecker A, Steinhausen R, et al. Quasistatic and dynamic properties of 1–3 composites made by soft molding. *J Eur Ceram Soc* 2003; 23: 153–159.
22. Kar-Gupta R and Venkatesh TA. Electromechanical response of 1-3 piezoelectric composites effect of poling characteristics. *J Appl Physics* 2005; 98: 54102-1-14.
23. Jafari A, Khatibi AA and Mashhadi MM. A hybrid averaging approach to predict overall properties of nanocomposites. *J Reinf Plast Compos* 2011; 30: 845–855.
24. Li JC, Chen L, Zhang YF, et al. Microstructure and finite element analysis of 3D five-directional braided composites. *J Reinf Plast Compos* 2012; 31: 107–115.
25. McCrary-Dennis MCL and Okoli OI. A review of multiscale composite manufacturing and challenges. *J Reinf Plast Compos* 2012; 31: 1687–1711.
26. Luan GD, Zhang JD and Wang RQ. Matrix transform method. In: *Piezoelectric transducers and arrays*, Revised ed. Beijing: Peking University Press, 2005, pp.56–66.
27. Luan GD, Zhang JD and Wang RQ. Physical properties of crystals. *Piezoelectric transducers and arrays*, Revised ed. Beijing: Peking University Press, 2005, pp.43–44.

A DESIGN PHILOSOPHY OF PORTABLE, HIGH-FREQUENCE IMAGE SONAR SYSTEM①

Yuan Li[a], Bing Li[b] , HaiboZheng,Zhibo Zhang, Qihu Li

[a] NO.21, Bei-Si-huan-Xi Road, Institute of Acoustics, Chinese Academy of Sciences Beijing, China
[b] NO.21, Bei-Si-huan-Xi Road, Institute of Acoustics, Chinese Academy of Sciences Beijing, China

Bing Li, NO.21, Bei-Si-huan-Xi Road, Institute of Acoustics, Chinese Academy of Sciences ,Beijing, China,100190; 010-82547960;libingbj@163.com

Abstract- A portable ,high-frequence sonar system for 2D imaging is developing in Institute of Acoustics, Chinese Academy of Sciences.It consists of bistatic transducer arrays (separate transducer and 80-element receive hydrophone array), signal acquisition and processing unit,weak signal amplifier unit,power amplifier unit and so on. All the units are packaged in a watertight vessel. Through Ethernet port,the portable sonar system can be integrated into low power, compact marine vehicles. Prototype has been built and tested in water.It's main acoustic feature : Frequency : 550 kHz ,number of Beams:80;beam width: 45° × 15°; beam spacing: 0.75°; range resolution:40mm. Pulse-echo with a high signal-to-noise ratio (SNR) of more than 10dB has been achieved. Due to high resolution and possessing good imaging capabilities, this sonar system has dual-use application in both military and commercial markets. The paper presents the experimental facilities, some of which are still under development, as well as the results of trials and scope of future work.

Key words: high frequency, imaging sonar; portable.

① UA2014-2nd International Conference and Exhibition on Underwater Acoustics, 2014: 3-10.

I. INTRODUCTION

A portable ,high-frequence sonar system for 2D imaging is developing in Institute of Acoustics, Chinese Academy of Sciences. The system goal for 2D sonar system is to provide a diver or autonomous-underwater vehicle (AUV) with the highest possible imaging capability while minimizing space and power requirements.

The portable sonar system, possessing good imaging capabilities has dual-use application in both military and commercial markets. The potential commercial applications include channel mapping for dredging, vehicle navigation for vessels traversing relatively shallow waters, aiding search and rescue dive teams in almost every country[1].

In market similar products are available such as V Series Sonar from BlueView corporation. The V Series sonar family offers 2 field-of-view options for high-performance forward looking imaging sonar: 90°, and 130°[2]. Compare to the products from BlueView corporation ,our products will be cheaper,smaller volume.Especially through Standard Ethernet Interface,it can provide client with primitive data for subsequent processing.

This paper is organized as follows. In Section II, the system is described.， the design and implement of bistatic transducer arrays and electronics system are provided in Section III and the experimental facilities, some of which are still under development, as well as the results of laboratory trials and scope of future work is provided in Section IV and the future work is given in Section V.

II. SYSTEM DESCRIPTION

The portable sonar system consists of bistatic transducer arrays (separate transducer and 80-element receive hydrophone array), electronics system(Signal Conditioning Module,power amplifier Module and FPGA-based Data Acquisition and DSP Module) ， power , mechanical structure and so on. All modules are packaged in a watertight vessel. Through Ethernet port,the portable sonar system can be integrated into low power, compact marine vehicles. For example it can easily be mounted on the underside of the AUV. This configuration provides a downward-looking sonar with high-resolution terrain mapping. The block diagram of the portable sonar system is show in Fig.1.

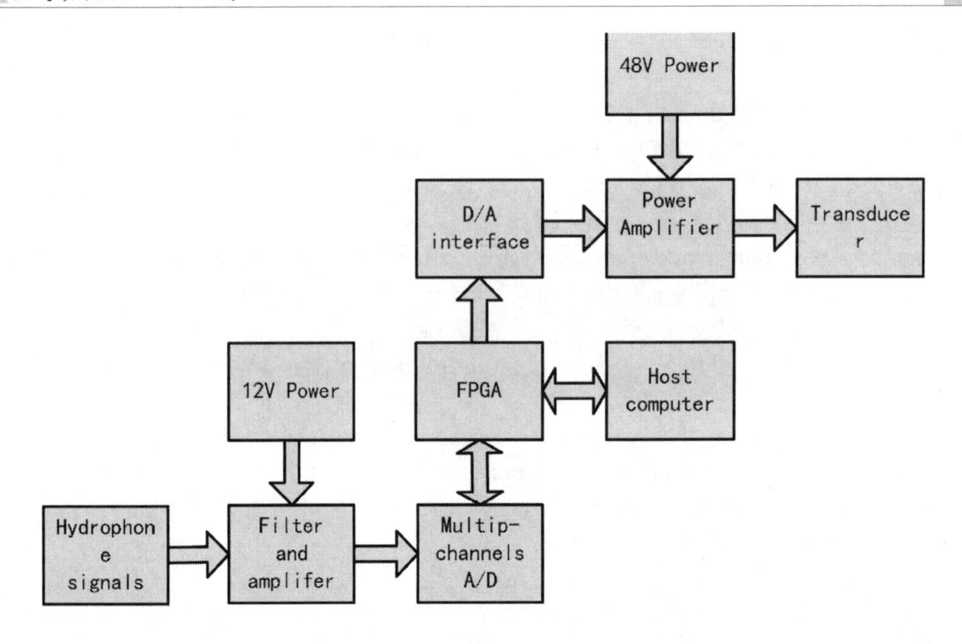

Fig.1 The block diagram of the portable sonar system

III.DESIGN AND IMPLEMENT

(1). bistatic transducer arrays

Arrays have been designed to comply with high frequency imaging applications.

bistatic transducer arrays have two parts: transducer and 80-element receive hydrophone array.The schematic diagram of bistatic transducer arrays is show in the left of Fig. 2. The picture of bistatic transducer *arrays* is show in the right of Fig. 2.

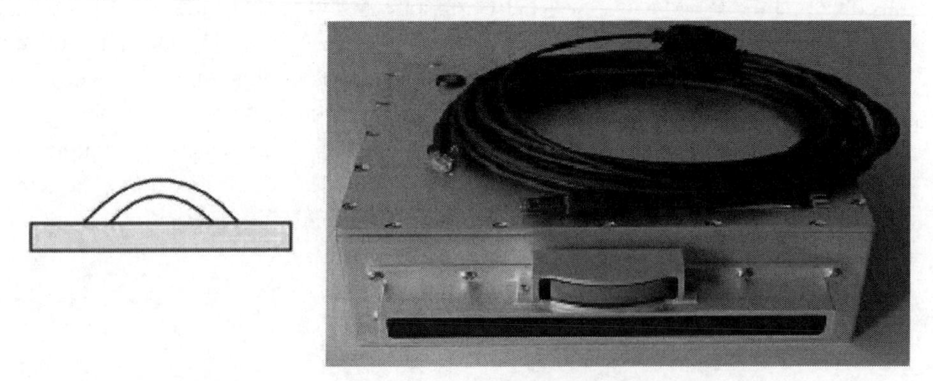

Fig.2 the schematic diagram and picture of bistatic transducer arrays

In the left of Fig.2 the arc-shaped part is transducer with 10-element uniformly distributing on the arc.Due to the special arc shap,the transducer achieve a horizontall/vertical beamwidth of 45"/15". The pulse used in the following experiments was a chirp swept from 550kHz to 600kHz. The primary source level is 190 dB.In the left of Fig. 2 the shadow part is 80-element receive hydrophone array.

Due to the linear array, it provides the basis for beamforming. the hydrophonic sensitivity is -195 dB. The receiver modules is mounted 1 cm beneath the transducer.

(2). ELECTRONICS SYSTEM

The main tasks of electronic system are control, data acquisition, beamforming and communication. There are mainly three modules in this electronic system: FPGA-based multi-channel data acquisition and DSP module, signal conditioning module and power amplifier module.FPGA-based multi-channel data acquisition and DSP module is the core of the electronic system. FPGA is responsible for mission scheduling and communication. FPGA also controls the data acquisition, decimating and beamforming.

2.1 Signal Conditioning Module

Prior to beam forming, each hydrophone's output is amplified and band-pass filtered using a low-noise amplifier with a fixed gain 20 dB.An instrumentation amplifier (LT6233) is utilized to design butterworth active filter and amplify the output voltage from a hydrophone. The LT6233 is single low noise, rail-to-rail output unity gain stable op amps that feature 1.9nV/Hz noise voltage and a 60MHz gain bandwidth product[3]. Fig. 3 is a photograph of the Signal Conditioning printed circuit board.

Fig. 3 a photograph of the Signal Conditioning printed circuit board.

2.2 power amplifier module

To promote source level to 190 dB,power amplifier is indispensable.Due to high power bandwidth-2MHz[4],PA107DP is choosed as the core chip of power amplifier module. Fig. 4 show the schematic diagram of power amplifier. Rf is tunable resistor to modify gain.

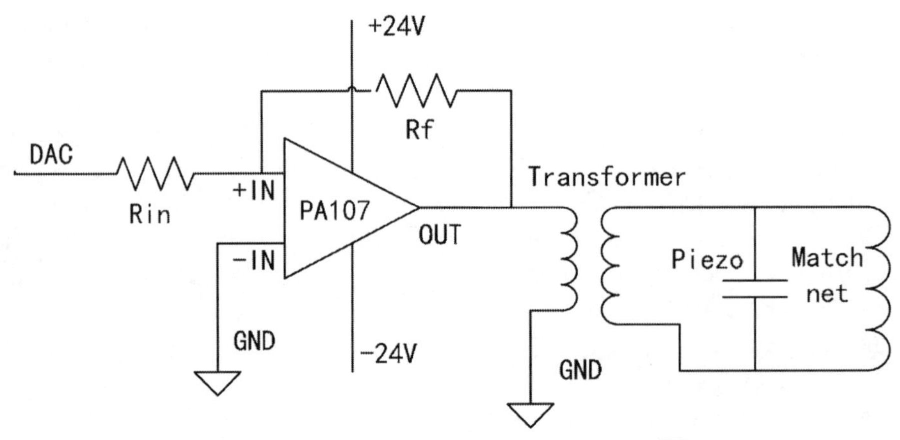

Fig. 4 The schematic diagram of power amplifier

2.3 FPGA-based Data Acquisition and DSP Module

In portable sonar system, FPGA-based Data Acquisition and DSP module plays a very important role in scheduling missions, beamforming and communication. Scheduling missions include a multi-channel data acquisition and the control of the power amplifier.

Due to high integration, powerful computing ability,design flexibility and user-programmability, FPGA-Xilinx Virtex-5 C5VLX220T is utilized as core controller . Xilinx Virtex-5 device is a high performance FPGA particularly suited to DSP applications[5]. Copeing with high data rate,6 FPGAs are utilized, 5 FPGAs operate synchronously for multi-channel data acquisition and beamformer,1 FPGA operate for missions scheduling and communication.

2.3.1 multi-channel data acquisition unit

For beamforming, a multi-channel data acquisition is indispensable. The output voltage from a hydrophone will be sampled by a 16-bit 10 Msps ADC.The total data rate is about 10*80*16 Mbit/sec. According to the principle of Nyquist ,all acquired data no need to be stored in SDRAM. Considering the transducer's upper working frequence (600kHz), FPGA will save 1 point every 8 points which means the true sampling rate is about 1.25M/sps. Then,the decimated data are continuously recorded in the SDRAM for subsequent processing.

2.3.2 Beamforming

A delay-and-sum beamformer allows a transducer array to "look" for signals propagating from a particular direction. By adjusting the delays associated with each element of the array, the array's look direction can be electronically steered toward the source of radiation.

Utilizing the on-chip resource, a 16-channels beamformer is implemented on a single Virtex-5 FPGA. 5 FPGAs operate synchronously for 80-channels beamformer.Via LVDS high speed serial interface,the 5 FPGAs synchronously communicate with a FPGA which is responsible for missions scheduling and communication. The block diagram of FPGA-based DSP unit is represented in Fig.5.

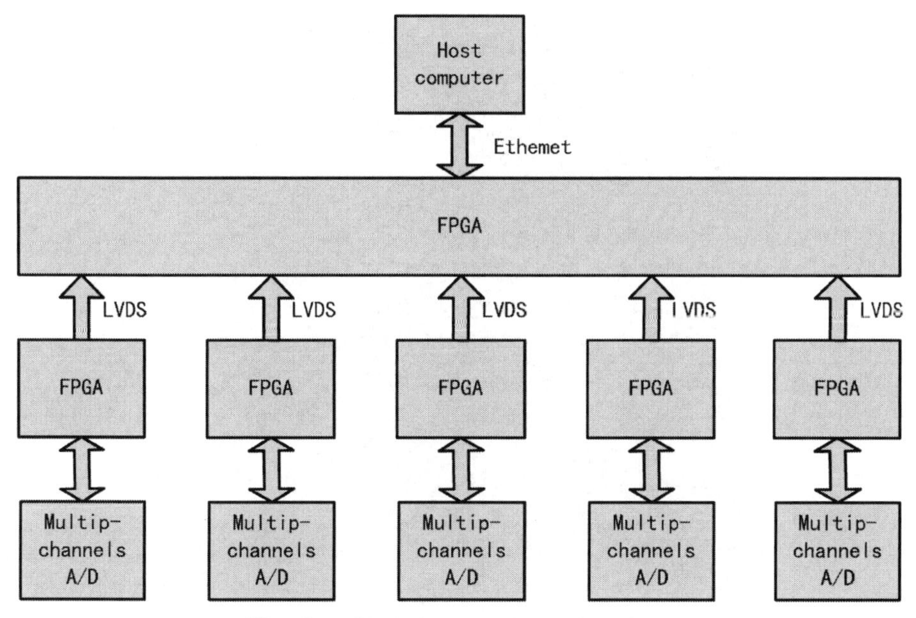

Fig. 5 the block diagram of FPGA-based DSP unit

2.3.3. missions scheduling and communication

Making use of the integrated resource on FPGA chip, FPGA can periodically generate control signals to open power amplifier and start multi-channel data acquisition. The data are then decimated to produce 80 channels Beamforming data .Utilizing IP core, a 100M Ethemet interface is generated and FPGA communicate with an host computer . Supported by a GPU, image are displayed on the screen and beamfoming data are stored in the host computer.

IV.LABORATORY EXPERIMENT RESULTS

In the spring of 2014, laboratory experiment was conducted in a pool. The configuration of the laboratory experiment is shown in Fig.6. In Fig.6, the transducer array is deployed on the right of the pool,the hydrophone array is deployed on the left of the pool, all in the depth of 20cm. The distance between source and receiver is about 0.5m. Fig.7 gives the received signal waveforms of all 32 channels. Fig.8 gives the Lisajous curve.

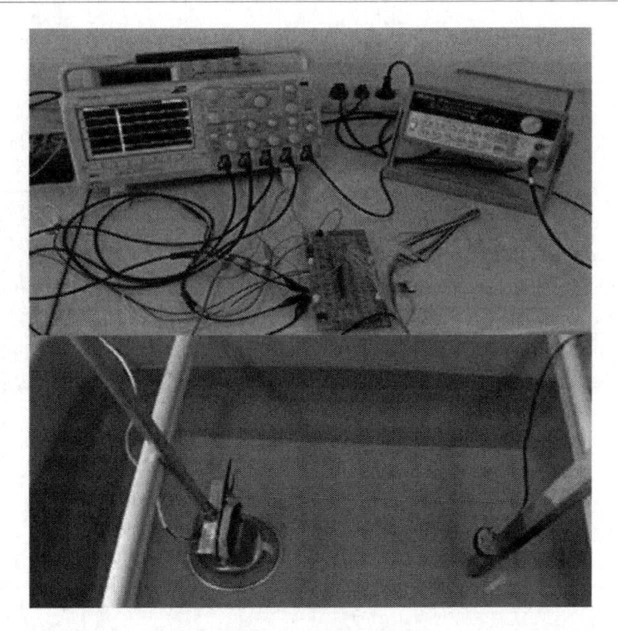

Fig. 6 the configuration of the laboratory experiment

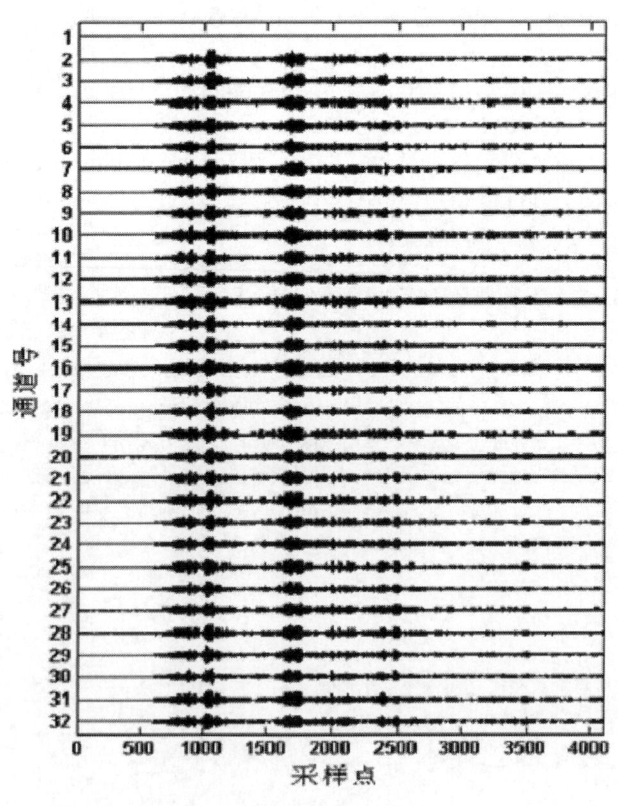

Fig. 7 the received signal waveforms of all 32 channels

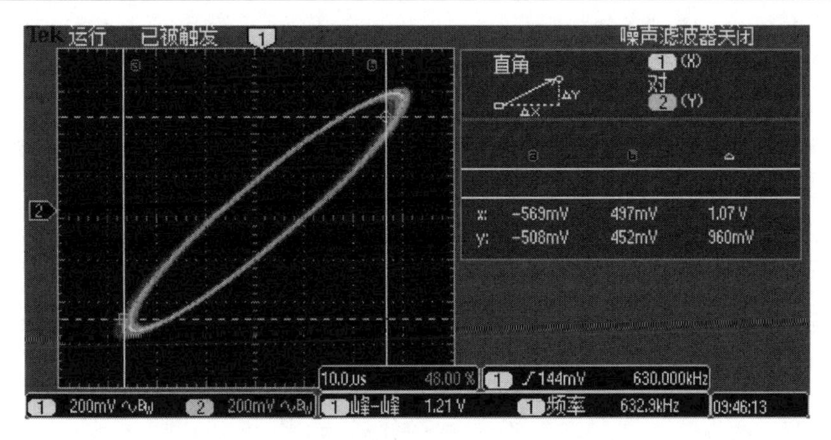

Fig. 8 the Lisajous curve

V.FUTURE WORK

Now the FPGA-based Data Acquisition and DSP Module is under joint debugging.Future work includes system debugging ,lake trial and sea trial. Further more,we will explore the merits of different transmit waveforms and their impact on image quality.

REFERENCES

[1] **Alice M. Chiang, Steven R. Broadstone, John M. Impagliazzo**, A Portable, Electronic-Focusing Sonar System For AUVs Using 2D Sparse-Array Technology, Online.
[2] VideoRay ROV Sonar-V Series Imaging Sonar.PDF,Teledyne BlueView , Online,
[3] PA107_datasheet.pdf, Cirrus Logic, Inc, Online,www.cirrus.com
[4] LT6233.PDF, Linear Technology Corporation, Online, www.linear.com
[5] xilinxvirtex5.pdf, Xilinx, Inc , Online, www.xilinx.com

LOW-POWER UNDERWATER DATA-ACQUISITION AND TRANSMISSION SYSTEM DESIGN STUDY FOR ADVANCED DEPLOYABLE SYSTEM①

Shijun Feng[a], Enming Zheng[b], Linyu Wang[b], Qihu Li

[a] NO.21, Bei-Si-huan-Xi Road, Institute of Acoustics, Chinese Academy of Sciences Beijing, China
[b] NO.21, Bei-Si-huan-Xi Road, Institute of Acoustics, Chinese Academy of Sciences Beijing, China

Shijun Feng, NO.21, Bei-Si-huan-Xi Road, Institute of Acoustics, Chinese Academy of Sciences ,Beijing, China,100190; 010-82547953;fengshijun@mail.ioa.ac.cn

Abstract- ***The underwater data acquisition and transmission system is an important component of Advanced Deployable System (ADS). In this paper, we discuss specifications and design philosophy of the data acquisition and transmission system, and evaluate the effectiveness of the system for ADS. The data acquisition and transmission system includes the sixteen channels Analog/Digital (A/D) node, the relay node, and the optical node and so on. To achieve the performance, 24-bits acquisition and delta-sigma ADC structure are adapted and ECL serial bus offers higher-speed data transmission and farther distance. The most important specifications in ADS design are the power delivered to the instruments and the dimension of the underwater instruments. In order to meet the application requirements of ADS for long time working, this paper employs the low-power measures, including the choice of low-power CMOS chip and power management technique. Power requirement of the sixteen channels A/D node is less than 4W. The special circuit board design is employed for more small space. The diameter of all underwater nodes is less than 20mm.***

We applied the system to the comprehensive acoustic submersible buoy and the towed array system, and the experimental data demonstrated that the system was

① UACE2015-3nd Underwater Acoustics Conference and Exhibition, 2015: 641-646.

reliable and effective and it met the overall design requirements. The system has good scalability and meets the future needs of the ADS.

Key words: Advanced Deployable System; data acquisition and transmission system; low-power

I. INTRODUCTION

Advanced Deployable System (ADS) is a rapidly deployable, bottom mounted, acoustic undersea surveillance system. The primary purpose of the ADS acoustic array is to monitor and track submarines, see Figure 1. Acoustic data is captured and then transferred to the surface transmission buoy via a fibre optic mooring cable.

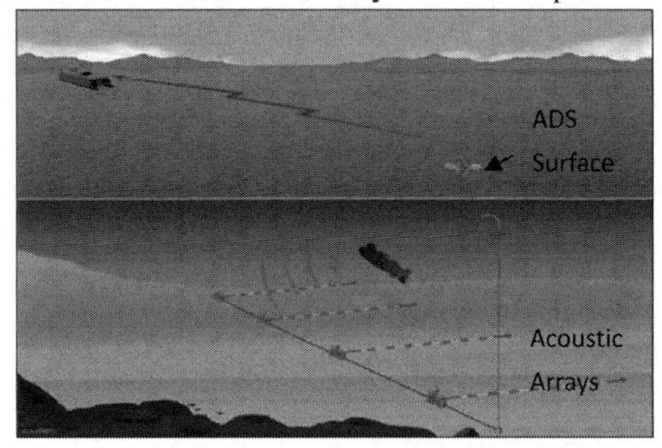

Fig.1 Concept of Operation Depiction

Considering the ADS deployment and run time, on the premise of meet the design requirements, we always hope to have lower power consumption and smaller size of the underwater node.

This paper is organized as follows. In section II, we discuss specifications and design philosophy of the data acquisition and transmission system, and evaluate the effectiveness of the system for ADS. In section III, the design and implement of underwater electronics nodes are provided. Design optimization is presented in section IV. The results of laboratory test for the underwater nodes are given in section V.

II. SYSTEM DESIGN PHILOSOPHY

Ambient noise may be said to be the noise of the sea itself [1]. Figure 2 show that frequency range of target radiated noise and ambient noise in the ocean is mostly 10-20k Hz. According to the Nyquist sampling theorem, sampling rate of the underwater A/D node is set to 50k Hz.

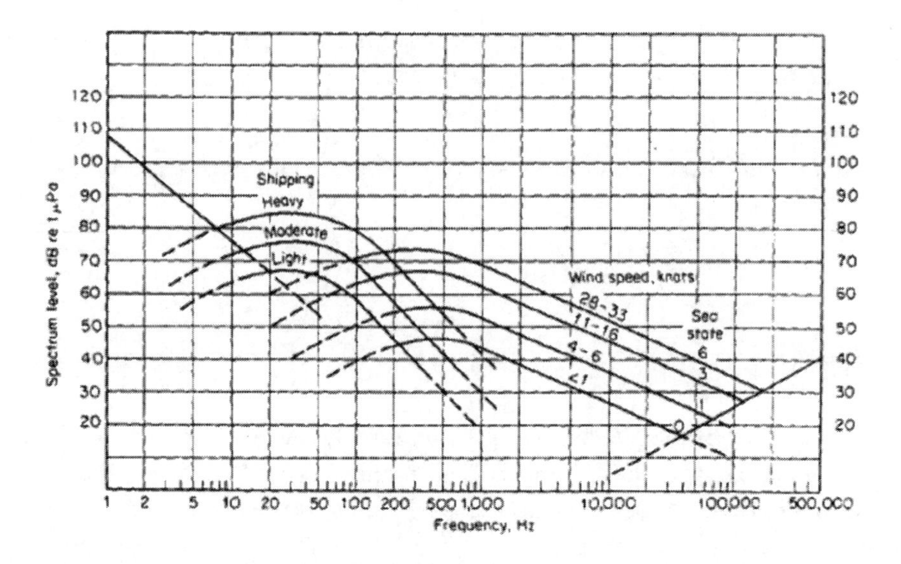

Fig.2 Average deep-water ambient-noise spectra

The voltage sensitivity of hydrophone is assumed -189dB. Figure 2 show that the minimum ambient noise spectrum level is 50dB. After 100 times of the pre-amplifier, the minimum input voltage of ADC is 11.2uV. The maximum input voltage of ADC is 5V. We can get that the effective bits of ADC are 19 bits. To achieve the better performance, 24-bits acquisition and delta-sigma ADC structure is adapted.

In general, the network topology is a crucial factor in determining the energy consumption, the capacity and the reliability of a network. There is a lot of network topology, shown in Fig.3, in which communication nodes can be used to interface hydrophones data to network [2].

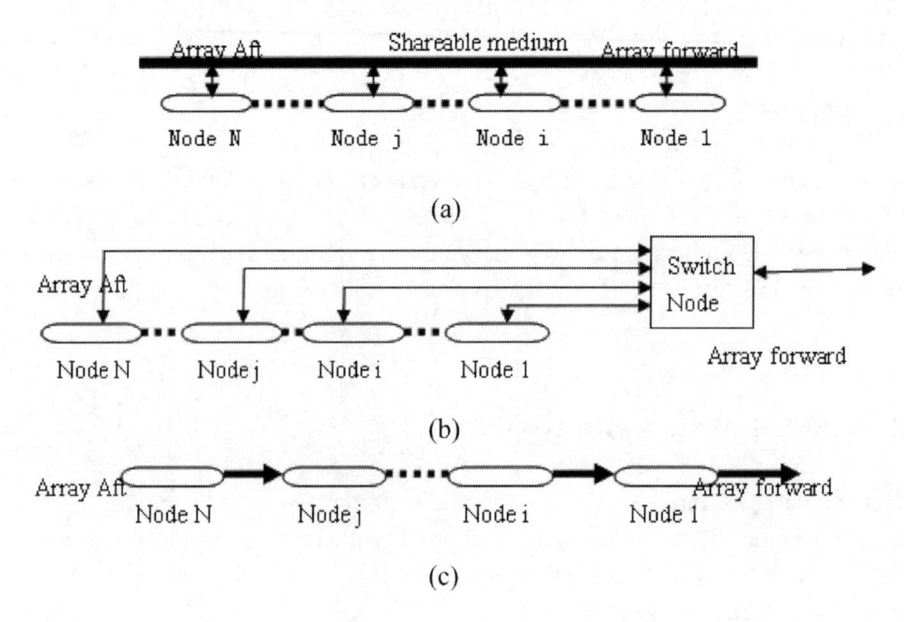

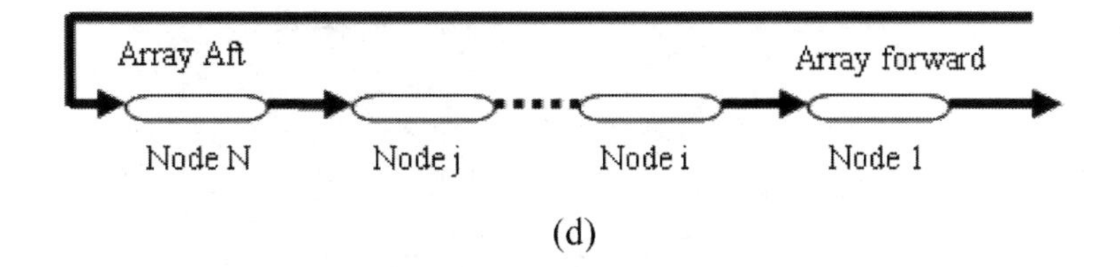

(d)

Fig.3 the topology of the sensor network

(a) shareable network topology (b) star Network topology

(c) open single ring array network topology (d) closed single ring array network topology

The topology of data acquisition and transmission system is limited by the space of the cable because the underwater cable needs smaller parts, particular compression-resistance and good waterproof ability. By means of analyzing the topology of current data acquisition and transmission system and the instance of it, we conclude that some characteristics of data acquisition and transmission system are as follows:

- The topology of underwater sensor network is based on the point-to-point style.
- The direction of transferring data is unmovable.
- The size of the data packet is fixed and can be designed by myself.
- The speed of the data is approximately stabile.

The data acquisition and transmission system is practically controllable. It doesn't generate burst-out data. The direction of transferring data and the size of data packet are foreknowable. Comparing the ring topology with the shareable topology and the star topology, we find that the ring topology suits the data acquisition and transmission system in ADS. The single ring topology, shown in Fig.3(c and d), meets the demand of the daisy-chained in series. The sensor data of the current A/D node are added to the transmission network, while the data from previous nodes are forwarded.

III.DESIGN AND IMPLEMENT

The underwater data acquisition and transmission system includes the sixteen channels Analog/Digital (A/D) node, the relay node, and the optical node and so on.

(1). Analog/Digital (A/D) node and the relay node

Analog/Digital (A/D) node and the relay node are shown in Fig.4. Analog/Digital (A/D) node includes PGA circuit, A/D converter circuit, Transmit /Receive circuit, clock circuit and so on. PGA circuit is used to filter and dynamic PGA control. A/D converter circuit is used to convert analog signals to digital signals of the 16 channels. Receive circuit is used to get the data of the previous transmission node. Transmit circuit is used to send the data. Clock circuit supplies the work clock for other circuit.

Fig.4 Analog/Digital (A/D) node and the relay node

(2). the optical node

The main function of the optical node is the optical-electric conversion. The electric signal from the A/D node or the relay node is converted to the optical signal, and then is send to the buoy.

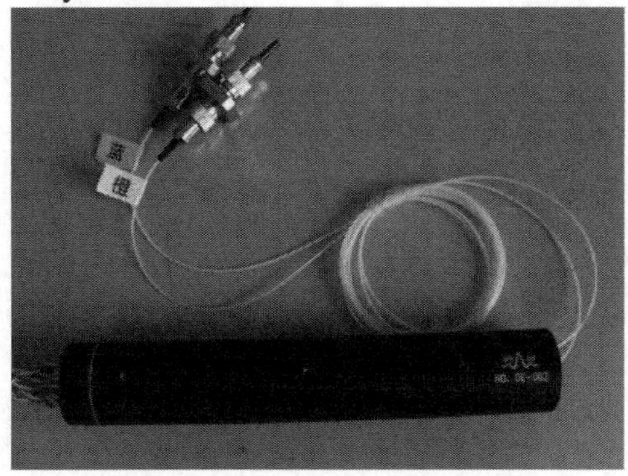

Fig.5 the Optical Node

(3). the monitoring and testing GUI

Fig.5 presents the GUI for monitoring and testing the underwater nodes. The GUI can observe the waveform of acquisition signal in time domain and frequency domain, and the temperature of the underwater nodes.

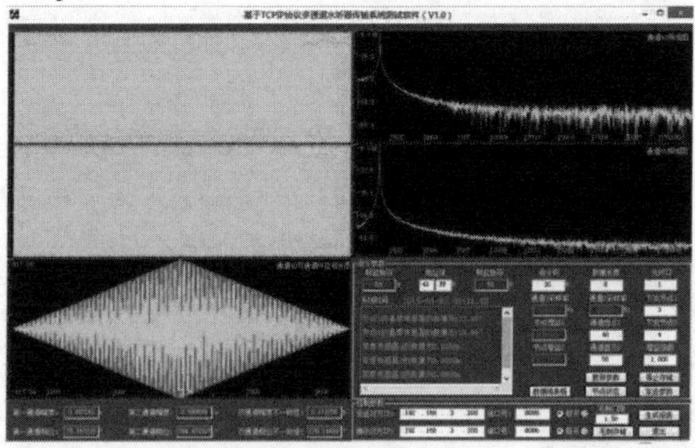

Fig.5 the monitoring and testing GUI

IV.DESIGN OPTIMIZATION

In the ADS design, because the energy source of most of the buoy is limited, the design of low power consumption is important. You must consider every detail in order to reduce the power consumption of the system, so as to prolong the working time of ADS. First, we use the low power consumption device as far as possible to reduce the power consumption of the system. Second, the system uses dynamic power management to eliminate the waste of power consumption.

The use of more integrated chip and the special circuit board design are employed to reduce space.

V.LABORATORY EXPERIMENT RESULTS

The results of laboratory experiment show that power requirement of the sixteen channels A/D node is less than 4W. The diameter of all underwater nodes is less than 20mm.

REFERENCES

[1] Urick, R. J. (1983). Principles of Underwater Sound for Engineers, 3rd edition. Peninsula Publishing, Los Altos, CA.
[2] John Walrod. ATM Telemetry in Towed Arrays [A]. In: Undersea Defense Technology 1997[C]. Hamburg，Germany：June 24-26, 1997

Semi-active control of piezoelectric coating's underwater sound absorption by combining design of the shunt impedances①

Yang Sun [a], Zhaohui Li [a,*], Aigen Huang [b], Qihu Li [c]

[a] Department of Electronics, Peking University, No. 5 Yiheyuan Road, Haidian District, Beijing 100871, China
[b] Systems Engineering Research Institute, China State Shipbuilding Corporation, 1 Fengxian East Road, Haidian District, Beijing 100094, China
[c] Institute of Advanced Technology, Peking University, No. 5 Yiheyuan Road, Haidian District, Beijing 100871, China

ARTICLE INFO

Article history:
Received 19 January 2015
Received in revised form
2 June 2015
Accepted 20 June 2015
Handling Editor: L. Huang
Available online 10 July 2015

ABSTRACT

Piezoelectric shunt damping technology has been applied in the field of underwater sound absorption in recent years. In order to achieve broadband echo reduction, semi-active control of sound absorption of multi-layered piezoelectric coating by shunt damping is significant. In this paper, a practical method is proposed to control the underwater sound absorption coefficients of piezoelectric coating layers by combining design of the shunt impedance that allows certain sound absorption coefficients at setting frequencies. A one-dimensional electro-acoustic model of the piezoelectric coating and the backing is established based on the Mason equivalent circuit theory. First, the shunt impedance of the coating is derived under the constraint of sound absorption coefficient at one frequency. Then, taking the 1–3 piezoelectric composite coating as an example, the sound absorption properties of the coating shunted to the designed shunt impedance are investigated. Next, on the basis of that, an iterative method for two constrained frequencies and an optimizing algorithm for multiple constrained frequencies are provided for combining design of the shunt impedances. At last, an experimental sample with four piezoelectric material layers is manufactured, of which the sound absorption coefficients are measured in an impedance tube. The experimental results show good agreement with the finite element simulation results. It is proved that a serial R–L circuit can control the peak frequency, maximum and bandwidth of the sound absorption coefficient and the combining R–L circuits shunted to multiple layers can control the sound absorption coefficients at multiple frequencies.

1. Introduction

Piezoelectric shunt damping technology was first proposed in 1979 by Forward [1]. In 1991, Hagood and von Flotow [2] developed a model for general shunting of these piezo-materials subject to arbitrary elastic boundary conditions to determine the 6×6 material compliance matrix when the material is shunted to the passive electric circuits. They

① Journal of Sound and Vibration, 2015, 355: 19-38.

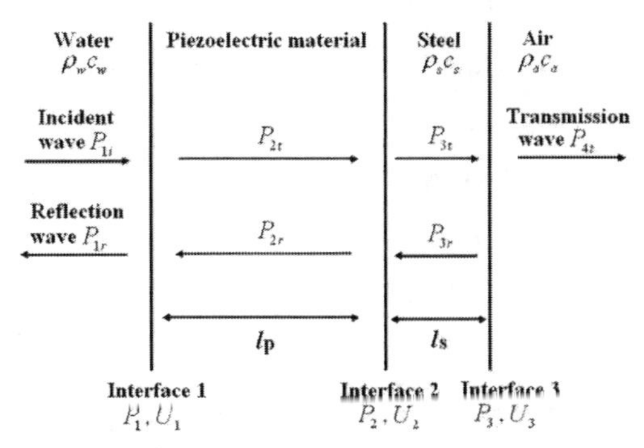

Fig. 1. Sound wave propagation diagram.

quantitatively analyzed the piezoelectric shunt damping system shunted by a resistor alone and a resistor connected in series with an inductor. In 1996, Law et al. studied the damping behavior of a resistor-shunted piezo-material systematically by an equivalent electrical circuit model, and developed a two-degree-of-freedom (2 dof) experimental set-up to measure the damping of piezo-materials [3]. Their contribution has laid an important foundation for the subsequent research work. In recent years, various types of shunt circuits have been increasingly studied and applied in the field of vibration control [4,5]. The fundamental of this technology is when the piezoelectric material is shunted to circuit, mechanical energy is converted into electrical energy through the piezoelectric effect. Then, electrical energy is converted into heat and dissipated through the shunt resistance element.

Compared with the active or hybrid active/passive sound-absorbing materials and structures which are common used in noise control [6–9], the piezoelectric shunt damping technology has advantages of simplicity, stability, low cost and easy installation due to its no need for actuators. Therefore, it has not only been applied in the vibration control of various structures, but also has drawn much attention to the applications of sound absorption [10–15]. Combining piezoelectric shunt damping technology with the traditional sound-absorbing material, Kim et al. significantly improved the noise reduction performance of the piezoelectric smart sound insulation panel over a wide frequency range (0–800 Hz) by using various types of shunt circuits, such as a resistor and an inductor in series [10], blocking circuit [11] and negative capacitance [12]. Chang et al. [13] brought piezoelectric shunt damping technology into a sound-absorbing structure consisting of a thin micro-perforated plate and a cavity. It is shown in the theoretical and experimental results that by adjusting the shunt resistor and inductor, the sound absorption coefficient of the thin plate over the frequency range 150–350 Hz can be effectively regulated, therefore the noise-absorbing frequency band can be broadened.

In the field of underwater sound-absorbing material, semi-active control of sound absorption coefficients of a piezoelectric coating by shunt damping can be applied in the wideband sound absorption design or in the compensation for the sound absorption defects of passive materials such rubbers and porous metals. Zhang et al. [14] established a one-dimensional electro-acoustic model with a piezoelectric layer coated to a rigid acoustic surface, and theoretically discussed the effect of the different types of shunt circuits as well as different piezoelectric materials on the sound absorption performance of the coating. It was pointed out that using negative shunt inductance and capacitance can achieve good echo reduction for most commercial piezoelectric materials. Further developing Zhang's model, Yu et al. [15] studied a 0–3 piezoelectric composite layer coated to a backing with finite thickness. The sound absorption coefficient over a wide frequency range 1–7 kHz was greatly improved by introducing a resistor and a negative capacitor in series as the shunt circuit and adjusting the shunt resistance.

By now, the existing research on semi-active control of underwater sound-absorbing materials have mainly focused on exploring the shunt circuit types that can enhance broadband sound absorption performance over a certain frequency range, where negative shunt inductance and capacitance are main choices. However, some of these shunt circuits are only theoretically analyzed but hardly implemented in practice. In this paper, a practical design method of the shunt impedance of piezoelectric coating layers is proposed to control the underwater sound absorption coefficients half actively at limited frequencies. The sound absorption coefficient at any frequency can be set based on requirement. The number of the constrained frequencies is determined by the number of the piezoelectric material layers which form the whole sound absorption coating. The shunt impedance properties for controlling sound absorption at a single frequency have been discussed in detail. On basis of that, the combining design methods for controlling multi-frequencies' sound absorption have also been presented. An experiment of measuring sound absorption coefficient is carried out in an impedance tube to verify the variation of the sound absorption coefficient due to the shunt impedance.

As an example, the 1–3 piezoelectric composites with different volume fractions are taken as the material of the coating in this paper, which is a common piezoelectric material in underwater acoustics and ultrasonics. However the conclusions drawn in this paper will apply to any piezoelectric materials in thickness mode.

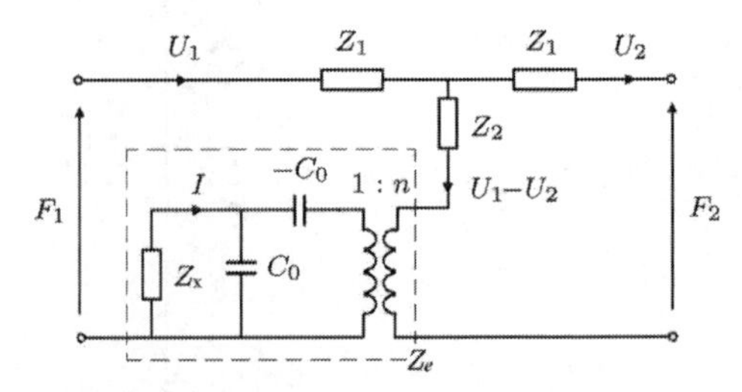

Fig. 2. Mason equivalent circuit of piezoelectric layer shunted to impedance Z_x.

2. Theoretical derivation of shunt impedance for single frequency

A one-dimension electro-acoustic model is established to derive the shunt impedance of the coating under constraint of the underwater sound absorption coefficient at single frequency. The sound wave propagation diagram is shown in Fig. 1 [11]. Consider that a piezoelectric material layer adheres to a steel layer. The piezoelectric layer and the steel layer are adjacent to water and air respectively, either of which is semi-infinite space. It is assumed that the transverse dimensions of the piezoelectric layer and the steel layer are much larger than their thicknesses. A plane sound wave penetrates the piezoelectric layer perpendicularly, besides, only longitudinal wave is excited within each medium. The sound wave will reflect and progress on each interface.

Suppose that the sound absorption coefficient on interface 1 at frequency f_0 is A_0. Since the rear of the steel backing is semi-infinite air whose acoustic impedance is far less than that of the steel backing, an approximate total reflection on interface 3 will occur. It is reasonable to assume that the transmission coefficient is zero and the reflection coefficient is

$$R_0 = 1 - A_0 \tag{1}$$

As the sound intensity reflection coefficient on interface 1, R_0 is defined as

$$R_0 = \left|\frac{P_{1r}}{P_{1i}}\right|^2 = \left|\frac{Z_{1i}-\rho_w c_w}{Z_{1i}+\rho_w c_w}\right|^2 \tag{2}$$

In Eq. (2), P_{1i} and P_{1r} are the amplitudes of sound pressure on interface 1 of incident and reflection waves, respectively. Z_{1i} is the surface acoustic impedance on interface 1. ρ_w, and $c_w$are the density and longitudinal wave velocity of water, respectively. And $\rho_w c_w$ is the acoustic impedance of water. As a complex number, Z_{1i} is assumed to be $Z_{1i}=R_{Z1i}+\mathrm{j}I_{Z1i}$, where R_{Z1i} and $I_{Z1i}$are its real part and imaginary part respectively. Then, Eq. (2) can be further expressed as

$$R_0 = \left|\frac{R_{Z1i}+\mathrm{j}I_{Z1i}-\rho_w c_w}{R_{Z1i}+\mathrm{j}I_{Z1i}+\rho_w c_w}\right|^2 = \frac{(R_{Z1i}-\rho_w c_w)^2+I_{Z1i}{}^2}{(R_{Z1i}+\rho_w c_w)^2+I_{Z1i}{}^2} \tag{3}$$

It can be converted into an equation of a circle on the complex plane, described as follows:

$$\left(R_{Z1i}-\frac{\rho_w c_w(1+R_0)}{1-R_0}\right)^2+I_{Z1i}{}^2=\frac{4(\rho_w c_w)^2 R_0}{(1-R_0)^2} \tag{4a}$$

$$Z_{1i}=R_{Z1i}+\mathrm{j}I_{Z1i}=\frac{\rho_w c_w(1+R_0)}{1-R_0}+\frac{2\rho_w c_w\sqrt{R_0}}{(1-R_0)}e^{\mathrm{j}\theta} \tag{4b}$$

It means that under the constraint of the sound absorption coefficient on interface 1 at frequency f_0 being A_0, the surface acoustic impedance on interface 1 Z_{1i} is determined by a circle on the complex plane, whose center is $\left(\frac{\rho_w c_w(1+R_0)}{1-R_0}, 0\right)$ and radius is $\frac{2\rho_w c_w\sqrt{R_0}}{(1-R_0)}$.

For the piezoelectric material layer, its surface acoustic impedance strongly depends on the circuit impedance shunted to it. Suppose two forces F_1 and F_2 are applied on the two end faces of the piezoelectric layer (interfaces 1 and 2), respectively, and U_1 and U_2 are the corresponding particle velocities on the two end faces of the piezoelectric layer. Voltage V emerges between the two electrodes on the two end faces.

When the piezoelectric layer is shunted to impedance Z_x, the equivalent circuit is shown in Fig. 2 and the electromechanical equations [16] can be written as

$$F_1=\left(Z_1+Z_2+n^2\left(\frac{Z_x}{1+\mathrm{j}\omega_0 C_0 Z_x}-\frac{1}{\mathrm{j}\omega_0 C_0}\right)\right)U_1-\left(Z_2+n^2\left(\frac{Z_x}{1+\mathrm{j}\omega_0 C_0 Z_x}-\frac{1}{\mathrm{j}\omega_0 C_0}\right)\right)U_2 \tag{5}$$

$$F_2=\left(Z_2+n^2\left(\frac{Z_x}{1+\mathrm{j}\omega_0 C_0 Z_x}-\frac{1}{\mathrm{j}\omega_0 C_0}\right)\right)U_1-\left(Z_1+Z_2+n^2\left(\frac{Z_x}{1+\mathrm{j}\omega_0 C_0 Z_x}-\frac{1}{\mathrm{j}\omega_0 C_0}\right)\right)U_2 \tag{6}$$

where $Z_1=\mathrm{j}\rho_p c_p S\tan\left(\frac{k_p l_p}{2}\right)$ and $Z_2=\frac{\rho_p c_p S}{\mathrm{j}\sin(k_p l_p)}$ are two kinds of mechanical impedance of the piezoelectric material. ρ_p and c_p are the density and longitudinal wave velocity of the piezoelectric material respectively. $k_p=\frac{\omega_0}{c_p}$ is wavenumber in piezoelectric layer and $\omega_0 = 2\pi f_0$ is angular frequency. S is the cross sectional area. $C_0 = \frac{\varepsilon_{33}^S S}{l_p}$ and $n = \frac{\varepsilon_{33}^S h_{33} S}{l_p}$ denote the inherent capacitance and the electromechanical conversion factor of the piezoelectric material, respectively. ε_{33}^S is the dielectric constant and h_{33} is the piezoelectric stiffness constant.

Dividing Eqs. (5) and (6) by S, the relationship of the pressures P_1, P_2 and the particle velocities U_1, U_2 on the two end faces of the coating (interfaces 1 and 2) can be obtained as follows:

$$\begin{pmatrix} P_1 \\ U_1 \end{pmatrix} = \mathbf{B}_1 \begin{pmatrix} P_2 \\ U_2 \end{pmatrix} = \begin{pmatrix} \frac{x_1}{x_2} & \frac{x_1^2 - x_2^2}{x_2} \\ \frac{1}{x_2} & \frac{x_1}{x_2} \end{pmatrix} \begin{pmatrix} P_2 \\ U_2 \end{pmatrix} \tag{7}$$

In Eq. (7), $\mathbf{B}_1$ is the transfer matrix of piezoelectric material layer, x_1 and x_2 are

$$x_1 = \frac{Z_1 + Z_2 + Z_e}{S} \tag{8}$$

$$x_2 = \frac{Z_2 + Z_e}{S} \tag{9}$$

where Z_e is the impedance in the dotted box of Fig. 2, which is expressed as

$$Z_e = n^2 \left(\frac{Z_x}{1 + j\omega_0 C_0 Z_x} - \frac{1}{j\omega_0 C_0} \right) \tag{10}$$

Similarly, the relationship of the pressuresP_2, P_3 and the particle velocities U_2, U_3on the two end faces of the steel layer (interfaces 2 and 3) can be obtained by letting n in Eq. (10) equal to zero because steel is a medium without piezoelectricity. It is

$$\begin{pmatrix} P_2 \\ U_2 \end{pmatrix} = \mathbf{B}_s \begin{pmatrix} P_3 \\ U_3 \end{pmatrix} = \begin{pmatrix} \cos(k_s l_s) & j\rho_s c_s \sin(k_s l_s) \\ \frac{j \sin(k_s l_s)}{\rho_s c_s} & \cos(k_s l_s) \end{pmatrix} \begin{pmatrix} P_3 \\ U_3 \end{pmatrix} \tag{11}$$

where $\mathbf{B}_s$ is the transfer matrix of steel layer. ρ_s, c_s and l_s are the density, longitudinal wave velocity and thickness of the steel layer, respectively. $k_s = \frac{\omega_0}{c_s}$ is the wavenumber in steel layer.

Then the relationship of the pressures P_1, P_3 and the particle velocities U_1, U_3 on interface 1 and 3 can be obtained from Eqs. (7) and (11) as

$$\begin{pmatrix} P_1 \\ U_1 \end{pmatrix} = \mathbf{B}_1 \mathbf{B}_s \begin{pmatrix} P_3 \\ U_3 \end{pmatrix} = \mathbf{B} \begin{pmatrix} P_3 \\ U_3 \end{pmatrix} \tag{12}$$

where $\mathbf{B}$ refers to the transfer matrix of the pressures and the particle velocities on the interfaces 1 and 3, it can be written as

$$\mathbf{B} = \mathbf{B}_1 \mathbf{B}_s = \begin{pmatrix} b_{11} & b_{12} \\ b_{21} & b_{22} \end{pmatrix} \tag{13}$$

To simplify the expression, we mark the transfer matrix of the steel layer as

$$\mathbf{B}_s = \begin{pmatrix} b_{211} & b_{212} \\ b_{221} & b_{222} \end{pmatrix} \tag{14}$$

The transfer matrix $\mathbf{B}$ can be obtained according to Eqs. (7), (13) and (14).

$$\mathbf{B} = \begin{pmatrix} b_{11} & b_{12} \\ b_{21} & b_{22} \end{pmatrix} = \begin{pmatrix} \frac{x_1}{x_2} & \frac{{x_1}^2 - {x_2}^2}{x_2} \\ \frac{1}{x_2} & \frac{x_1}{x_2} \end{pmatrix} \begin{pmatrix} b_{211} & b_{212} \\ b_{221} & b_{222} \end{pmatrix} \tag{15}$$

Four components of the matrix in Eq. (15) are

$$b_{11} = \frac{x_1}{x_2} b_{211} + \frac{x_1^2 - x_2^2}{x_2} b_{221} \tag{16a}$$

$$b_{12} = \frac{x_1}{x_2} b_{212} + \frac{x_1^2 - x_2^2}{x_2} b_{222} \tag{16b}$$

$$b_{21} = \frac{1}{x_2} b_{211} + \frac{x_1}{x_2} b_{221} \tag{16c}$$

$$b_{22} = \frac{1}{x_2} b_{212} + \frac{x_1}{x_2} b_{222} \tag{16d}$$

Table 1
The parameters of piezoelectric phase, matrix phase and 1–3 composites.

Piezoelectric phase (PZT-5H) [18]

Density ρ_p (kg/m^3)	Elastic constants (10^{10} N/m^2)				Relative dielectric constant		Piezoelectric constants (C/m^2)			Quality factor	
	c_{11}^E	c_{12}^E	c_{13}^E	c_{33}^E	$\varepsilon_{11}^S/\varepsilon_0$	$\varepsilon_{33}^S/\varepsilon_0$	e_{31}	e_{33}	e_{15}	Q_m^p	Q_e^p
7500	12.6	7.95	8.41	11.7	1700	1470	−6.5	23.3	17.0	65	50

Matrix phase (a kind of polyurethane rubber) [18]

Density ρ_m (kg/m^3)	Longitudinal velocity c_m (m/s)	Elastic constants (10^9 N/m^2)			Relative dielectric constant $\varepsilon_x^m/\varepsilon_0$	Quality factor Q_m^m
		c_{11}^m	c_{12}^m	c_{44}^m		
1080	1520	2.50	2.40	0.049	3	1.5

1–3 Piezoelectric composites

PZT volume fraction μ (%)	Density $\bar{\rho}$ (kg/m^3)	Elastic constant $\bar{c}_{33}^E$ (10^{10} N/m^2)	Elastic constant $\bar{c}_{33}^D$ (10^{10} N/m^2)	Relative dielectric constant $\bar{\varepsilon}_{33}^S/\varepsilon_0$	Piezoelectric constant $\bar{e}_{33}$ (C/m^2)	Quality factor $\overline{Q}_m$
10	1722	0.75	1.34	154.34	2.85	8.6
30	3006	1.76	3.55	456.91	8.52	20.5
50	4290	2.80	5.80	759.21	14.17	28.6
100	7500	11.7	15.7	1470	23.3	65

Then the surface acoustic impedance on interface 1 can be obtained according to Eqs. (12) and (16).

$$Z_{1i}=\frac{P_1}{U_1}=\frac{b_{11}P_3+b_{12}U_3}{b_{21}P_3+b_{22}U_3}=\frac{\frac{P_3}{U_3}b_{11}+b_{12}}{\frac{P_3}{U_3}b_{21}+b_{22}}=\frac{\rho_a c_a b_{11}+b_{12}}{\rho_a c_a b_{21}+b_{22}}$$

$$=\frac{(\rho_a c_a b_{211}+b_{212})x_1+(\rho_a c_a b_{221}+b_{222})(x_1{}^2-x_2{}^2)}{\rho_a c_a b_{211}+b_{212}+(\rho_a c_a b_{221}+b_{222})x_1} \tag{17}$$

where ρ_a, and $c_a$are the density and longitudinal wave velocity of air, respectively, and $\rho_a c_a$ is the acoustic impedance of air.
By substituting Eqs. (8) and (9) into Eq. (17), and removing x_2 with $x_2=x_1-(Z_1/S)$, then it has

$$x_1=\frac{-K_1\left(\frac{Z_1}{S}\right)^2-K_2 Z_{1i}}{K_1\left(Z_{1i}-\frac{2Z_1}{S}\right)-K_2} \tag{18}$$

where

$$K_1=\rho_a c_a b_{221}+b_{222} \tag{19}$$

$$K_2=\rho_a c_a b_{211}+b_{212} \tag{20}$$

It can be derived from Eqs. (8), (10), (18), and (4b) that

$$Z_x=\frac{\frac{n^2}{x_1 S-Z_1-Z_2}+j\omega_0 C_0}{(\omega_0 C_0)^2}$$

$$=\frac{\frac{n^2}{(\omega_0 C_0)^2}\left[K_1\left(\frac{\rho_w c_w(1+R_0)}{1-R_0}-\frac{2Z_1}{S}+\frac{2\rho_w c_w\sqrt{R_0}}{(1-R_0)}e^{j\theta}\right)-K_2\right]}{(Z_1+Z_2)\left(\frac{2K_1 Z_1}{S}+K_2\right)-\frac{K_1 Z_1^2}{S}-(K_2 S+K_1(Z_1+Z_2))\left(\frac{\rho_w c_w(1+R_0)}{1-R_0}+\frac{2\rho_w c_w\sqrt{R_0}}{(1-R_0)}e^{j\theta}\right)}+\frac{j}{(\omega_0 C_0)} \tag{21}$$

It can be seen that there is a one-to-one correspondence relationship between the solutions of Z_x and Z_{1i}. In summary, under the constraint that the sound absorption coefficient at frequency f_0 is A_0, the surface acoustic impedance on interface 1 Z_{1i} can be acquired according to Eqs. (4) which contains the value A_0. Then the shunt impedance of the piezoelectric coating Z_x can be obtained according to Eq. (21) in which f_0 is included in Z_1, Z_2, K_1, and K_2.

Table 2
The parameters of air, water, and steel.

Material	Thickness (cm)	Density (kg/m^3)	Longitudinal wave velocity (m/s)
Air	Semi-infinite	1.4	340
Water	Semi-infinite	1000	1483
Steel	1	7840	5935

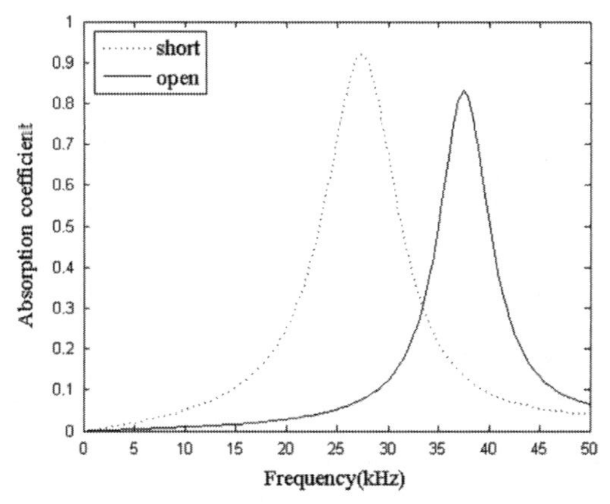

Fig. 3. Sound absorption coefficients of the piezoelectric material coating under short-circuit and open-circuit conditions.

3. Shunt impedance properties under the constraint of single frequency

3.1. Material parameters of the piezoelectric coating

In the calculation, 1–3 piezoelectric composite is chosen as the sound absorption material in this part, which consists of a viscoelastic matrix and large numbers of piezoelectric rods that are perpendicularly embedded across the thickness in the matrix. PZT-5H and polyurethane rubber are chosen as the materials of the piezoelectric phase and the matrix phase of the composite, respectively. The fine-scaled 1–3 composite modeled by Smith [17] is adopted to calculate the elastic constants $\bar{c}^E_{33(1-3)}$ and $\bar{c}^D_{33(1-3)}$, dielectric constant $\bar{\varepsilon}^S_{33(1-3)}$, and piezoelectric constant $\overline{e_{33}}$ of the 1–3 composite coating.

$$\bar{c}^E_{33(1-3)}=\mu\left[c^E_{33}-\frac{2(1-\mu)(c^E_{13}-c^m_{12})^2}{\mu(c^m_{12}+c^m_{11})+(1-\mu)(c^E_{12}+c^E_{11})}\right]+(1-\mu)c^m_{11} \tag{22}$$

$$\bar{\varepsilon}^S_{33(1-3)}=\mu\left[\varepsilon^S_{33}+\frac{2(1-\mu)e^2_{31}}{\mu(c^m_{12}+c^m_{11})+(1-\mu)(c^E_{12}+c^E_{11})}\right]+(1-\mu)\varepsilon^m_x \tag{23}$$

$$\bar{e}_{33(1-3)}=\mu\left[e_{33}-\frac{2(1-\mu)e_{31}(c^E_{13}-c^m_{12})}{\mu(c^m_{12}+c^m_{11})+(1-\mu)(c^E_{12}+c^E_{11})}\right] \tag{24}$$

$$\bar{c}^D_{33(1-3)}=\bar{c}^E_{33(1-3)}+\frac{\bar{e}^2_{33(1-3)}}{\bar{\varepsilon}^S_{33(1-3)}} \tag{25}$$

where $c^E_{ij}(i=1,3 j=1,2,3)$, ε^S_{33}, and e_{33} are the elastic constant components, dielectric constant component and piezoelectric stress constant component of the piezoelectric phase, respectively. $c^m_{ij}(i=1,j=1,2)$ and ε^m_x are the elastic constant components and dielectric constant component of the matrix phase, respectively. μ is the volume fraction of the piezoelectric phase in the composite. The density of the composite is decided by the densities and volume fractions of the two phases as shown here.

$$\bar{\rho}=\mu\rho_p+(1-\mu)\rho_m \tag{26}$$

Table 1 gives the parameters of PZT-5H, polyurethane rubber and the corresponding 1–3 composites with different volume fractions (VF) (PZT VFs: 10 percent, 30 percent, 50 percent, and 100 percent) that are calculated according to Eqs. (22)–(25), which are used in the following content of this paper.

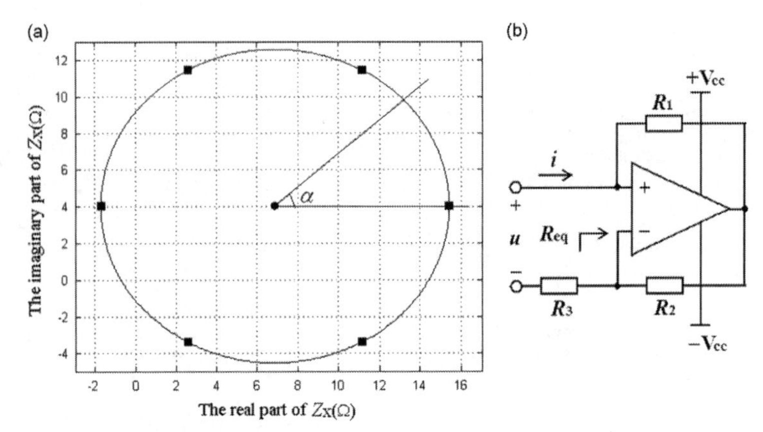

Fig. 4. (a) The shunt impedance Z_x that satisfies the sound absorption coefficient at 26 kHz is 0.8 and (b) an implementation circuit of a negative resistor.

Table 3
The shunt circuits corresponding to the six points in Fig. 4.

Flare angle α (deg)	Shunt impedance on 26 kHz Z_s (Ω)	Shunt resistance (Ω)	Shunt capacitance (μF)	Shunt inductance (μH)
0	15.4333 + 4.0065i	15.4333	–	24.525
60	11.148 + 11.4284i	11.1482	–	69.957
120	2.5782 + 11.4284i	2.5782	–	69.957
180	−1.7069 + 4.0065i	−1.7069	–	24.525
240	2.5782 − 3.4154i	2.5782	1.7923	–
300	11.1482 − 3.4154i	11.1482	1.7923	–

Assume that the thicknesses of the piezoelectric material coating and the steel backing are 3 cm and 1 cm, respectively, and the cross-sectional area of them is 1 m × 1 m. The parameters of air, water, and steel are shown in Table 2.

3.2. Short-circuit and open-circuit cases

The sound absorption coefficients of the piezoelectric material are calculated under short-circuit and open-circuit conditions to investigate the effects of the shunt circuits in this part. In the calculation, the 1–3 piezoelectric composite with 30 percent-VF is taken as the reference material of the piezoelectric coating. It is shown in Fig. 3 that when the piezoelectric material coating is switched from open-circuit to short-circuit, i.e. from $Z_x=\infty$ to $Z_x=0$, the peak frequency of the sound absorption coefficient is changed approximately from 38 kHz to 27 kHz, accompanying with the variation of the peak amplitude. It means the shunt circuit has dramatic effect on the sound absorption coefficient curve. Therefore, by design of the shunt impedance, the control of sound absorption coefficient can be realized at the setting frequency according to one's requirement.

3.3. Shunt impedance properties

In order to further describe the control method, the constraint that the sound absorption coefficient at $f_0=26$ kHz is $A_0=0.8$ is taken as an example here. The shunt impedance Z_x of the 30 percent-VF 1–3 piezoelectric composite coating can be obtained according to Eq. (21).

The calculation results of Z_x are shown in Fig. 4(a). It can be seen that the solution set of Z_x is a circle (hereinafter referred to as 'Z_x circle') on the complex plane, where the horizontal axis refers to the real part and the vertical axis refers to the imaginary part. It means that shunt impedance corresponding to whichever the point on the Z_x circle is shunted to the piezoelectric layer the sound absorption coefficient 0.8 at 26 kHz can be achieved.

To investigate the properties of the impedance on Z_x circle, six points which are evenly spaced on Z_x circle are selected in Fig. 4(a), corresponding to the flare angles, namely α, of 0°, 60°, 120°, 180°, 240°, 300°, with 0° being along the positive real axis. The shunt impedances corresponding to these six points are listed in Table 3.

The real part of the shunt impedance represents the shunt resistance. The imaginary part of the shunt impedance represents the shunt reactance. Assume that the shunt circuit is a resistor connected in series with a capacitor or an inductor. The required resistance and the capacitance or inductance corresponding to the reactance at frequency 26 kHz can be obtained. The positive reactance refers to a shunt inductor while the negative reactance refers to a shunt capacitor. It is seen that the shunt resistor is negative on the 180° point, indicating that a negative resistor element is required. A

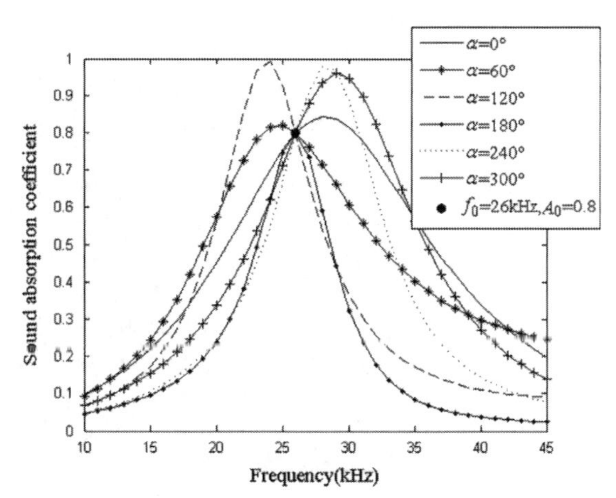

Fig. 5. The sound absorption coefficient curves of the piezoelectric coating shunted to the six kinds of shunt circuits in Table 3.

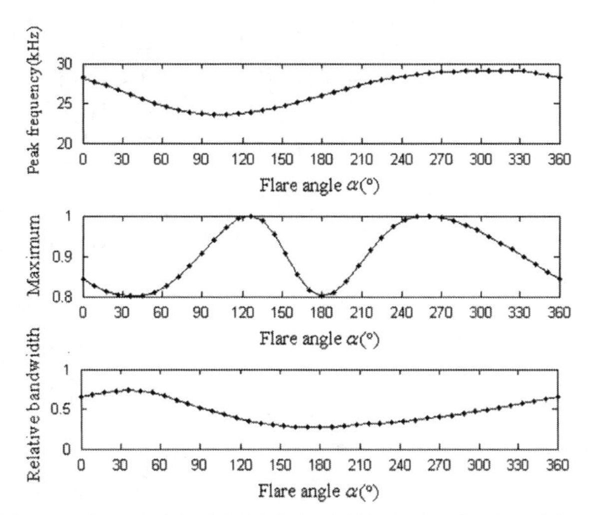

Fig. 6. The peak frequency, the maximum and the relative bandwidth of the broadband sound absorption coefficient.

conceivable implementation of the negative resistor can be an operational amplifier with three resistors, as shown in Fig. 4(b), in which the equivalent resistor is $R_{eq} = -\frac{R_3}{R_2}R_1$.

The example shows that there is enough flexibility in choosing proper shunt circuit type to meet the requirement of the sound absorption coefficient at a certain frequency.

4. The sound absorption properties for different shunt impedance Z_x

4.1. *Properties on Z_x circle*

When the piezoelectric coating is shunted to each of the above six kinds of shunt circuits, the sound absorption coefficient over a broadband 10–45 kHz is calculated and shown in Fig. 5. It can be seen that for the six kinds of shunt circuits, the sound absorption coefficients at the setting frequency 26 kHz are all 0.8, but the properties such as peak frequency, the maximum and the relative bandwidth of the corresponding sound absorption coefficient curves versus frequency are different from each other.

In order to further investigate the properties of the broadband sound absorption coefficient corresponding to different flare angle α, 40 evenly spaced points on Z_x circle are chosen and the respective sound absorption coefficient curves versus frequency are calculated. Above-mentioned properties are extracted and shown in Fig. 6. It can be seen that, from 0° to 360°, the peak frequency varies in the range of 26.63–29.15 kHz. The maximum has a two-maximum and two-minimum slow

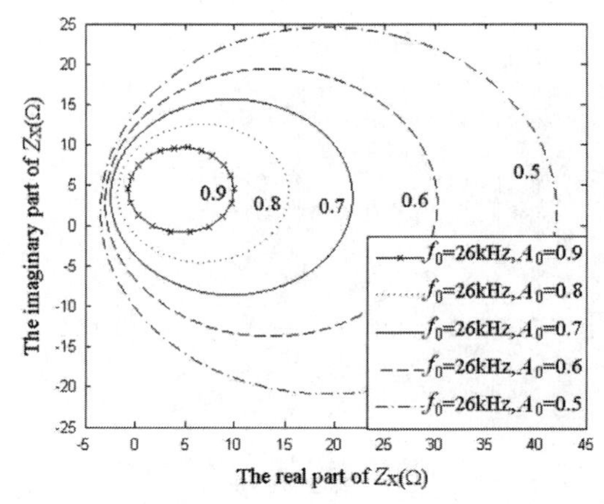

Fig. 7. The shunt impedance circles corresponding to the sound absorption coefficients of 0.9, 0.8, 0.7, 0.6, and 0.5 at 26 kHz.

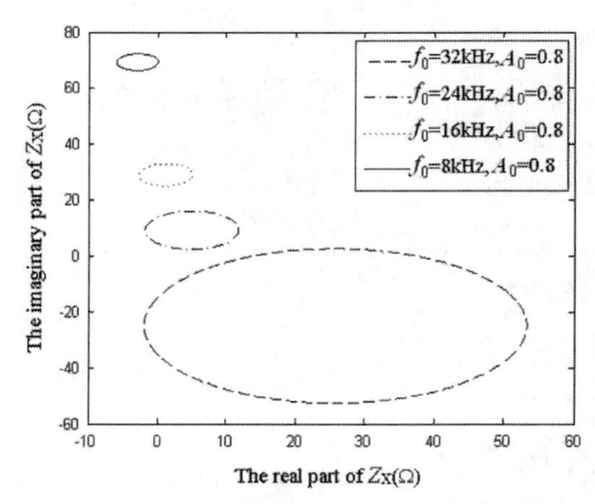

Fig. 8. The shunt impedance circles corresponding to the sound absorption coefficient 0.8 at 32 kHz, 24 kHz, 16 kHz, and 8 kHz.

fluctuation between 0.8 and 1. And the relative bandwidth varies slowly in the range of 0.27–0.74. Based on the properties summarized from Fig. 6, the shunt impedance on Z_x circle at proper flare angle can be chosen according to requirements of the sound absorption coefficient and bandwidth at a setting frequency.

4.2. Z_x circles for different sound absorption coefficients at setting frequency

After the theoretical design of the shunt circuit, some actual circuit elements can be chosen to realize the designed circuit. It is common that the actual circuit elements have errors in their values. Thus, the actual shunt impedance will locate inside or outside the theoretical Z_x circle. Therefore, it is necessary to investigate the variation of the sound absorption coefficient at a setting frequency when the piezoelectric coating is shunted to a circuit with impedance inside and outside the designed Z_x circle.

As an example, the sound absorption coefficients are set at frequency $f_0 = 26$ kHz as 0.9, 0.8, 0.7, 0.6, and 0.5. According to Eq. (21), five shunt impedance circles corresponding to above five sound absorption coefficients can be obtained and shown in Fig. 7. It can be seen that five circles are nested together. The circle with the largest sound absorption coefficient is at the innermost region while the circle with the smallest sound absorption coefficient is at the outermost region, which means that the larger the sound absorption coefficient is, the smaller the radius of the Z_x circle will be. It can also be found in Fig. 7 that starting from the center of the innermost circle and moving along a certain direction, the sound absorption coefficient monotonically decreases and the rates of decline are different in different directions. These properties are also clues for controlling the sound absorption coefficient curves of the piezoelectric coating.

4.3. Z_x circles corresponding to different frequencies

For the purpose of investigating the needed $Z_x$circles of the same sound absorption coefficient at different frequencies, the sound absorption coefficients at 32 kHz, 24 kHz, 16 kHz, and 8 kHz are all set as 0.8, and four different impedance circles are obtained according to Eq. (21), as shown in Fig. 8. It can be found that with the decreasing of the frequency, the radius of the impedance circle decreases and the center of the circle moves towards the upper left direction on the complex plane. When the frequency is comparatively high, the real parts of the points on the circle are all positive and the imaginary parts all negative. It means the shunt circuit is composed of a resistor and a capacitor in series. However when the frequency lowers to a certain value, the real parts of the points on the circle become all negative and the imaginary parts all positive, which means that the shunt circuit is a negative resistor and an inductor in series. As for the intermediate frequencies, the real parts or imaginary parts of the points on the circle may cover the range of either positive or negative value, indicating the type of the shunt circuit elements is not unique for a fixed sound absorption coefficient. In conclusion, the required peak frequency of the sound absorption coefficient decides both the location and the radius of the impedance circle on the complex plane. The lower the peak frequency is, the smaller the radius and the narrower the relative value range will be. When the frequency is low enough, there will be no normal electronic elements to meet the requirement of sound absorption. The negative resistors are needed.

4.4. The impact of PZT volume fractions

In the above parts, the 1–3 piezoelectric composite with 30 percent-PZT volume fraction is taken as the reference material of the piezoelectric coating to investigate the sound absorption properties. It is reasonable that the volume fraction of the composite has great impact on the sound absorption performance of the piezoelectric coating. The variation of the composite parameters with the change of the PZT volume fraction can be seen in Table 1. It is shown that the density, elastic constants, relative dielectric constant, piezoelectric constant, and quality factor all increase monotonously with the increasing of the volume fraction. Fig. 9(a) and (b) gives the sound absorption coefficient curves of the piezoelectric coating with different PZT volume fractions of 10 percent, 30 percent, 50 percent, and 100 percent (pure PZT) versus the normalized thickness, namely the ratio of the thickness of the piezoelectric coating to the wave length, for open-circuited and short-circuited respectively. Both figures show that the curves of the sound absorption coefficient of the piezoelectric coating with different volume fractions shunted to the same impedance are similar to each other, with only some differences that the larger the PZT volume fraction is, the higher the peak frequency and the narrower the relative bandwidth will be.

5. Combining design of shunt impedances for multiple frequencies

5.1. Two frequencies

The relationship of the shunt impedance of the piezoelectric coating and the underwater sound absorption properties at a setting frequency has been discussed above. It can be applied to design the shunt impedances of multiple layered coating under constraints of the underwater sound absorption coefficients at multiple frequencies. First, the design method of the shunt impedances under constraints of the sound absorption coefficients at two frequencies is considered.

The sound wave propagation diagram is shown in Fig. 10. Two piezoelectric layers 1 and 2 with the same thickness are pasted together with piezoelectric layer 2 adhering to a steel layer. Piezoelectric layer 1 and the steel backing are adjacent to water and air, respectively. Other assumptions are the same as those of the model with one piezoelectric layer in Fig. 1.

Two constraints are set as follows:

Constraint 1. The sound absorption coefficient on interface 1 at a certain frequency f_1 is A_1.

Constraint 2. The sound absorption coefficient on interface 1 at a certain frequency f_2 is A_2.

The shunt impedances of the two piezoelectric layers Z_{x1} and Z_{x2} are determined by satisfying the two constraints above at the same time. Constraint 1 is met mainly by connecting shunt impedanceZ_{x1} to piezoelectric layer 1, while Constraint 2 is met mainly by connecting shunt impedanceZ_{x2} to piezoelectric layer 2.

Shunted to different impedances Z_{x1} and Z_{x2}, the parallel equivalent circuits of the two layers are shown in Fig. 11. Assume the transfer matrixes of the pressures and the particle velocities on the two piezoelectric layers and the steel layer are B_1, B_2, and B_s. The relationship of the pressures P_1, P_4 and the particle velocities U_1, U_4on interfaces 1 and 4 is

$$\begin{pmatrix} P_1 \\ U_1 \end{pmatrix} = \mathbf{B}_1\mathbf{B}_2\mathbf{B}_s \begin{pmatrix} P_4 \\ U_4 \end{pmatrix} = B \begin{pmatrix} P_4 \\ U_4 \end{pmatrix} \tag{27}$$

From the shunt impedance design method of the single frequency case, two basic deductions can be drawn as follows to do the design.

Deduction 1. Let Z_{x2} of piezoelectric layer 2 be known, and derive Z_{x1} of piezoelectric layer 1 according to Constraint 1.

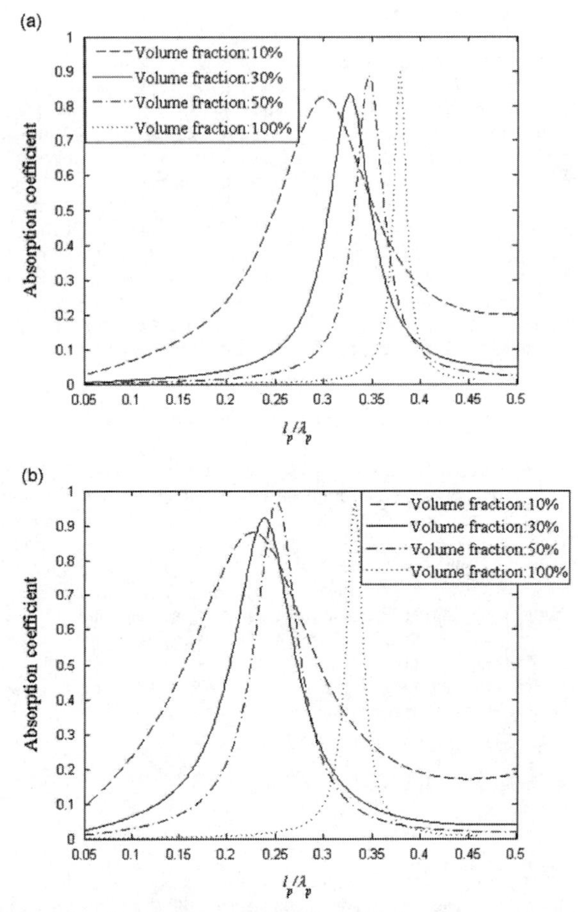

Fig. 9. The sound absorption coefficient curves of piezoelectric coating with different PZT volume fractions of 10 percent, 30 percent, 50 percent, and 100 percent for (a) open-circuited and (b) short-circuited conditions.

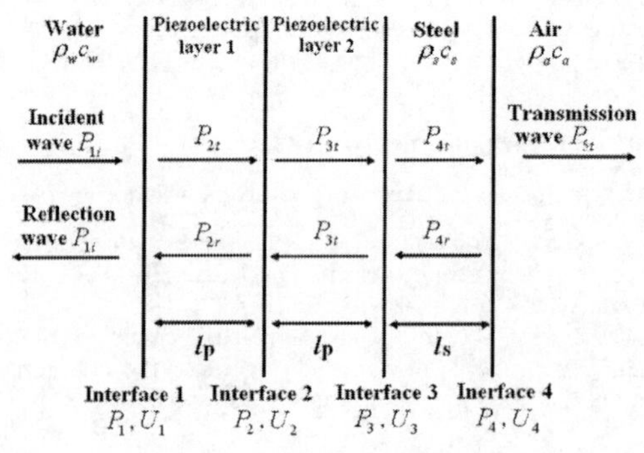

Fig. 10. Sound wave propagation diagram with two piezoelectric layers.

Deduction 2. Let Z_{x1} of piezoelectric layer 1 be known, and derive $Z_{x2}$of piezoelectric layer 2 according to Constraint 2.

Constraints 1 and 2 are met mainly by connecting shunt impedances Z_{x1} and Z_{x2} to piezoelectric layers 1 and 2, respectively. However, the two piezoelectric layers are not independent. The respective own shunt impedance Z_{x1} and Z_{x2}

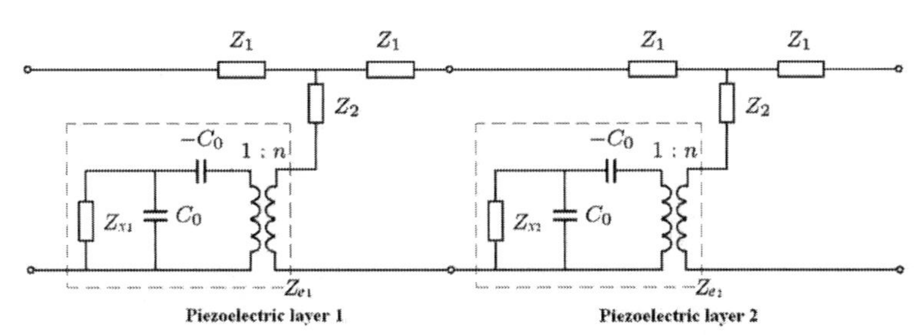

Fig. 11. Mason equivalent circuit of two piezoelectric layers shunted to impedances Z_{x1} and Z_{x2}.

Table 4
The iterative processes of the shunt impedances Z_{x1} and Z_{x2}.

Step	Piezoelectric layer 1	Piezoelectric layer 2	Constraint	Obtained shunt impedance
1	Assume shunted to $Z_{x1}{}^{(1)}$	Open-circuited	Constraint 1	$Z_{x1}{}^{(1)}$ (at f_1)
2	Shunted to $Z_{x1}{}^{(1)}$	Assume shunted to $Z_{x2}{}^{(1)}$	Constraint 2	$Z_{x2}{}^{(1)}$ (at f_2)
3	Assume shunted to $Z_{x1}{}^{(2)}$	Shunted to $Z_{x2}{}^{(1)}$	Constraint 1	$Z_{x1}{}^{(2)}$ (at f_1)
4	Shunted to $Z_{x1}{}^{(2)}$	Assume shunted to $Z_{x2}{}^{(2)}$	Constraint 2	$Z_{x2}{}^{(2)}$ (at f_2)
5	Assume shunted to $Z_{x1}{}^{(3)}$	Shunted to $Z_{x2}{}^{(2)}$	Constraint 1	$Z_{x1}{}^{(3)}$ (at f_1)
6	Shunted to $Z_{x1}{}^{(3)}$	Assume shunted to $Z_{x2}{}^{(3)}$	Constraint 2	$Z_{x2}{}^{(3)}$ (at f_2)
...	...	...	...	...

Table 5
The iterative processes of the shunt impedances Z_{x1} and Z_{x2} of two piezoelectric material layers.

Step	Piezoelectric layer 1	Piezoelectric layer 2	Constraint	Obtained shunt impedance	Resistance (Ω)	Inductance (mH)	Error E_2
1	Assume shunted to $Z_{x1}{}^{(1)}$	Open-circuited	1	$Z_{x1}{}^{(1)}$ (5 kHz)	−4.42458354	3.62140168	0.49
2	Shunted to $Z_{x1}{}^{(1)}$	Assume shunted to $Z_{x2}{}^{(1)}$	2	$Z_{x2}{}^{(1)}$ (10 kHz)	−0.41128357	0.72899177	6.98×10^{-3}
3	Assume shunted to $Z_{x1}{}^{(2)}$	Shunted to $Z_{x2}{}^{(1)}$	1	$Z_{x1}{}^{(2)}$ (5 kHz)	−4.18989498	3.61908963	9.67×10^{-6}
4	Shunted to $Z_{x1}{}^{(2)}$	Assume shunted to $Z_{x2}{}^{(2)}$	2	$Z_{x2}{}^{(2)}$ (10 kHz)	−0.41150760	0.72898763	2.82×10^{-8}
5	Assume shunted to $Z_{x1}{}^{(3)}$	Shunted to $Z_{x2}{}^{(2)}$	1	$Z_{x1}{}^{(3)}$ (5 kHz)	−4.18989534	3.61908964	2.25×10^{-8}
6	Shunted to $Z_{x1}{}^{(3)}$	Assume shunted to $Z_{x2}{}^{(3)}$	2	$Z_{x2}{}^{(3)}$ (10 kHz)	−0.41150759	0.72898763	2.24×10^{-8}

will influence the whole property of the two layers. Therefore, Z_{x1} and Z_{x2} cannot be designed separately and their interactive effect should be considered. Thus, an iterative calculation method is applied to find out a convergent solution (Z_{x1} and Z_{x2}).

The iterative calculation process is listed in Table 4.

In Step 1, piezoelectric layer 2 is open-circuited, thus the transfer matrix $\mathbf{B}_2$ is known. The shunt impedance $Z_{x1}{}^{(1)}$ of piezoelectric layer 1 is obtained by using Deduction 1. It is already known that the solution set of $Z_{x1}{}^{(1)}$ is a circle on the complex plane. Here, only the 0°-flare angle solution is selected. After Step 1, Constraint 1 is satisfied.

In Step 2, piezoelectric layer 1 is shunted to the shunt impedance $Z_{x1}{}^{(1)}$ solved in Step 1, thus the transfer matrix $\mathbf{B}_1$ is known. The shunt impedance $Z_{x2}{}^{(1)}$ of piezoelectric layer 2 is obtained by using Deduction 2. After Step 2, Constraint 2 is satisfied. However, because of the influence between the two piezoelectric layers, the sound absorption coefficient at frequency f_1 will deviate from A_1 again.

In Step 3, piezoelectric layer 2 is shunted to the shunt impedance $Z_{x2}{}^{(1)}$ solved in Step 2, thus the transfer matrix $\mathbf{B}_2$ is known. The shunt impedance $Z_{x1}{}^{(2)}$ of piezoelectric layer 1 is solved by using Deduction 1 again. After Step 3, Constraint 1 is once more satisfied. Similarly, the sound absorption coefficient at frequency f_2 may deviate from A_2 too.

After several steps of iterative calculation, the deviation of the respective sound absorption coefficients at frequency f_1 and f_2 from A_1 and A_2 become smaller and smaller. The mean square error is employed to evaluate the deviations in the iterations, which is defined as

$$E_2=\sqrt{\frac{\left(A(f_1,Z_{x1}{}^{(x)},Z_{x2}{}^{(y)})-A_1\right)^2+\left(A(f_2,Z_{x1}{}^{(x)},Z_{x2}{}^{(y)})-A_2\right)^2}{2}} \tag{28}$$

where f is the frequency, $A(f,Z_{x1}{}^{(x)},Z_{x2}{}^{(y)})$ is the sound absorption coefficient when piezoelectric layer 1 and 2 are shunted to $Z_{x1}(x)$ and $Z_{x2}(y)$ respectively. The iterative process will stop when the error E_2 is less than an acceptable value.

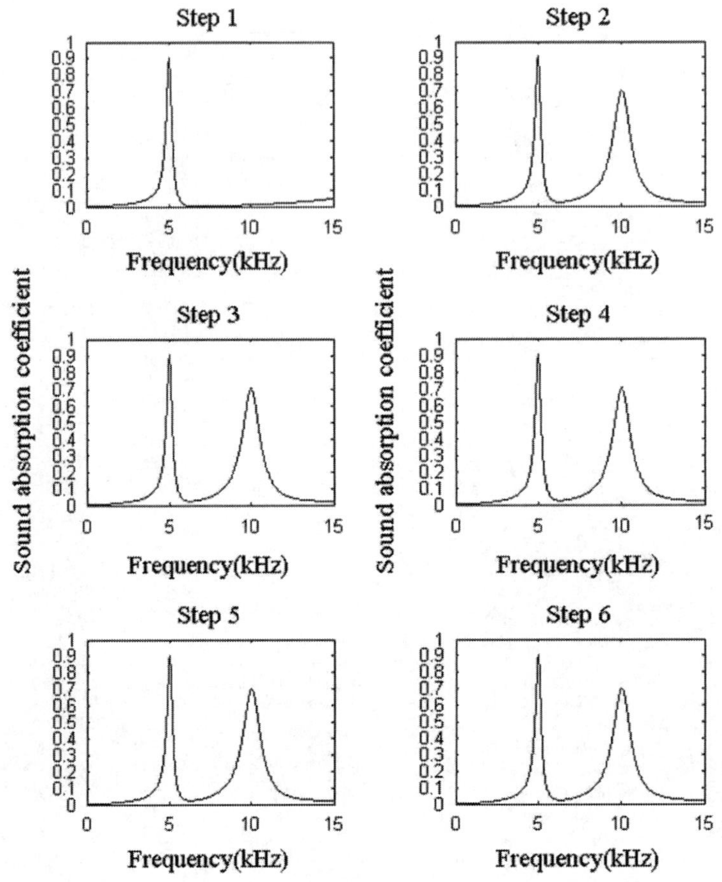

Fig. 12. The sound absorption coefficients of the two piezoelectric material layers after each step of iterations.

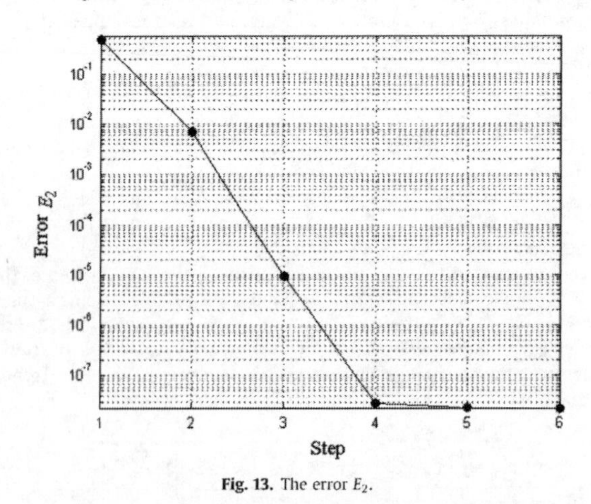

Fig. 13. The error E_2.

In the following calculation example, the 1–3 piezoelectric composite with 30 percent-PZT volume fraction is taken as the material of the piezoelectric layer. The shunt impedancesZ_{x1} and Z_{x2} are solved by using the iterative calculation method under Constraints 1 and 2. The thicknesses of the two piezoelectric layers are all 3 cm and the thickness of the steel backing is 1 cm. The cross section area of three layers is set as 1 m × 1 m to meet the pure thickness mode requirement. The frequencies and sound absorption coefficients in Constraints 1 and 2 are set as: $f_1 = 5$ kHz, $A_1 = 0.9$; $f_2 = 10$ kHz, $A_2 = 0.7$. After six steps of iterations, the solved shunt impedances, and the error results are listed in Table 5. The sound absorption coefficient curves corresponding to each step are shown in Fig. 12. It is seen that, after two steps, the errorE_2 has become very small. With the increasing of the step number, the error E_2 keeps decreasing and tend to zero, as shown in Fig. 13. As a result, the two sound absorption coefficients at frequency f_1 and f_2 approximately equal A_1 and A_2, respectively. In Table 5, the high-accuracy resistance and inductance values up to 9–10 digits are only used in the iterative processes to distinguish the resistance or inductance values obtained in different steps. In the practice, since the requirement for the errors of sound absorption coefficients is not that high, there is no need of multi iterations as well as the accuracy that high.

5.2. Multiple frequencies

In the broadband sound absorption application, it is necessary to apply multiple layers of coating material to meet the bandwidth requirement by setting constraints of sound absorption coefficients at multiple frequencies.

If the multiple constrained frequencies and the piezoelectric layers are employed, using the optimization algorithm to solve the shunt impedances will be a proper choice.

Suppose the sound absorption coefficients at the frequencies $f_1, f_2, \dots, f_N$ are set as $A_1, A_2, \dots, A_N$. The sound absorption coating consists of N piezoelectric layers, which are shunted to impedances $Z_{x1}, Z_{x2}, \dots, Z_{xN}$, respectively. A function of the sound absorption coefficient is defined as

$$A = A_N(f, Z_{x1}, Z_{x2}, \dots, Z_{xN}) \tag{29}$$

where A is the sound absorption coefficient and f is the frequency. Defined by the mean square error, the objective function of the optimization algorithm can be written as

$$F(Z_x) = \sqrt{\frac{1}{N}\sum_{k=1}^{N} |A_N(f_k, Z_{x1}, Z_{x2}, \dots, Z_{xN}) - A_k|^2} \tag{30}$$

By minimizing the objective function, the optimal solutions of the shunt impedances $Z_{x1}, Z_{x2}, \dots, Z_{xN}$ can be obtained by using optimization algorithms such as the genetic algorithm, the simulated annealing algorithm and so on. Fig. 14 gives a calculation example. The sound absorption coefficients at five frequencies$f_1 = 5$ kHz, $f_2 = 8$ kHz, $f_3 = 11$ kHz, $f_4 = 14$ kHz

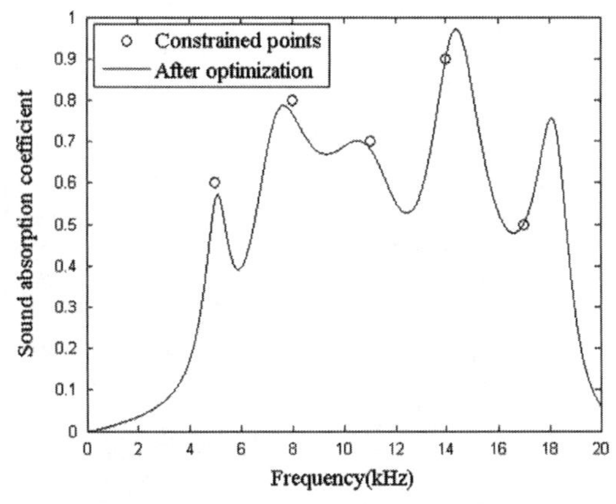

Fig. 14. The sound absorption coefficients of five piezoelectric material layers.

Table 6
Five groups of shunt circuits.

Layer	Shunt resistance (Ω)	Shunt inductance (μH)
1	−1.67	260
2	−0.06	130
3	25.88	1050
4	9.87	2990
5	1.45	−4.28

and $f_5 = 17$ kHz are set as $A_1 = 0.6$, $A_2 = 0.8$, $A_3 = 0.7$, $A_4 = 0.9$ and $A_5 = 0.5$. The sound absorption coating consists of five piezoelectric material layers with the same thickness of 3cm. The objective function is obtained by letting N in Eq. (30) equal 5. Five groups of shunt circuits (a resistor and an inductor in series) are obtained by using the linear least square optimization algorithm. The results are listed in Table 6. The negative resistor and inductor can be realized by active element circuit design. The minimum of the objective function, namely the mean square error of the five sound absorption coefficients, is about 0.0249.

6. Experiments

6.1. Measurement system and method

The sound absorption coefficient measurements are carried out in a cylindrical water-filled impedance tube. The tube is 6 m long and its inner diameter is 57 mm. Thus, the valid frequency range of the tube is 5–15 kHz. The experimental sample is put at one end of the tube with the transmitting/receiving transducer placed at the other end.

The pulse-echo method is adopted to measure the sound absorption coefficients. For a certain frequency $f_j (j = 1, 2, ..., N)$, a signal of 15-period sine wave is transmitted into water by the transducer and penetrates the sample. Then, the reflection sound wave signal is received by the same transducer. After FFT, the amplitudes $P_{1ia}(f_j)$ and $P_{1ra}(f_j)$ of the incident and reflection signals are obtained. Let the sound intensity reflection coefficient $R_0(f_j)$ to be the square of the ratio of $P_{1ra}(f_j)$ to $P_{1ia}(f_j)$:

$$R_0(f_j) = \left(\frac{P_{1ra}(f_j)}{P_{1ia}(f_j)} \right)^2 \tag{31}$$

The sound absorption coefficient $A_0(f_j)$ is calculated through

$$A_0(f_j) = 1 - R_0(f_j) \tag{32}$$

In the valid frequency range of 5–15 kHz, the sound absorption coefficient is measured with frequency interval of 200 Hz. Therefore the total number of the measured frequencies is 51.

6.2. Experimental sample manufacture

Since the valid frequency range of the impedance tube is 5–15 kHz, a piezoelectric coating composing of four layers of piezoelectric material and a steel backing layer is manufactured by choosing the parameters to meet the frequency range of the tube. To highlight the effect of the shunt impedance on the sound absorption coefficient of the piezoelectric sound absorption coating in the experiment, pure piezoelectric ceramic (PZT-5H), namely the piezoelectric composite with 100 percent-PZT volume fraction, is chosen as the piezoelectric material of the experimental sample.

The thickness of each piezoelectric layer is 16 mm and the steel layer is 10 mm. The diameter of the steel layer is 55 mm and the horizontal maximum width of the piezoelectric layer is 54.3 mm. Epoxy conductive silver adhesive is used for the bonding between the components. The four piezoelectric layers and the steel layer are adhered together and put into a cylindrical polytetrafluoroethylene (PTFE) mould of 56 mm diameter for molding. By filling polyurethane into the gap between the piezoelectric layers and the mould, solidifying and demoulding, the sample becomes a 56 mm-diameter cylinder, as shown in Fig. 15(a). Surrounding the piezoelectric layers with a thin polyurethane layer is in the need of

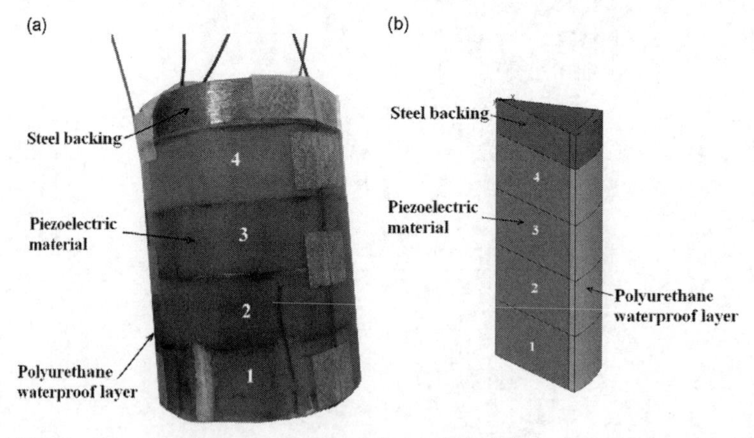

Fig. 15. (a) Experimental sample with polyurethane waterproof layer and (b) a finite element model of experimental sample.

waterproofing and insulation. Besides, the polyurethane layer is also necessary to facilitate the fitting of the sample in the tube.

In Fig. 15(a), the four piezoelectric layers are marked as layer 1, 2, 3, and 4. Five wires are connected to the surface of layer 1, the juction of layers 1 and 2, the juction of layers 2 and 3, the juction of layers 3 and 4, and the surface of the steel layer, respectively.

In the experiment, the sample is put at one end of the impedance tube. Four piezoelectric layers are submerged into water with surface of layer 1 being adjacent to water, while the steel backing layer is slightly above the tube orifice and its surface is adjacent to air.

It should be considered that the experimental sample differs from the theoretical model in the transverse boundary condition. In the theoretical model, the transverse dimensions of the piezoelectric layer and the steel layer are much larger than their thicknesses. The equivalent condition is that all the layers are radially clamped. While in the experiment, the transverse dimension of the sample is finite and there is a polyurethane layer surrounding the sample. Thus the sample cannot be treated as radially clamped. The free boundary condition will influence the peak frequency of the sound absorption coefficients. In addition, the polyurethane layer surrounding the sample also has dramatic influence on the sound absorption properties of the sample.

For the purpose of examining the experimental results, a simulation environment of the sound absorption coefficient measurement by using the impedance tube is established with the finite element (FE) analysis software ANSYS. A FE model including a polyurethane waterproof layer is set up to simulate the experiment sample, as shown in Fig. 15(b). The sound absorption coefficients of the FE model are calculated according to the pulse-echo method and compared with the experimental results, as well as with the theoretical ones. In the simulation, the material parameters, dimensions, locations, and the boundary conditions of all the components in the measurement system are consistent with those in the experiment.

6.3. Experimental contents

There are three groups of experiments. In the first one, the sound absorption coefficient is measured when all the piezoelectric layers are open-circuited. In the second one, the four piezoelectric layers are treated as one whole piezoelectric layer and a serial R–L circuit is connected to the whole layer. In the third one, four different serial R–L circuits are connected to the four piezoelectric layers respectively. The results will be discussed separately in the following section.

7. Results and discussions

7.1. Open-circuit condition

The sound absorption coefficient results of open-circuit condition obtained by theoretical calculation, finite element method (FEM) and experiment measurement can be seen in Fig. 16.

In the theoretical calculation, the equivalent transverse boundary condition is that all the layers are radially clamped as mentioned above. The theoretical sound absorption coefficient curve in Fig. 16 shows a narrow peak at 30.8 kHz, where far from the peak frequency, the sound absorption coefficients are very low.

In the FEM simulations, all the layers are set radially free to meet the actual boundary condition in the experiments. A surrounding waterproofing polyurethane layer is included in the model to simulate the experiment sample. By letting the

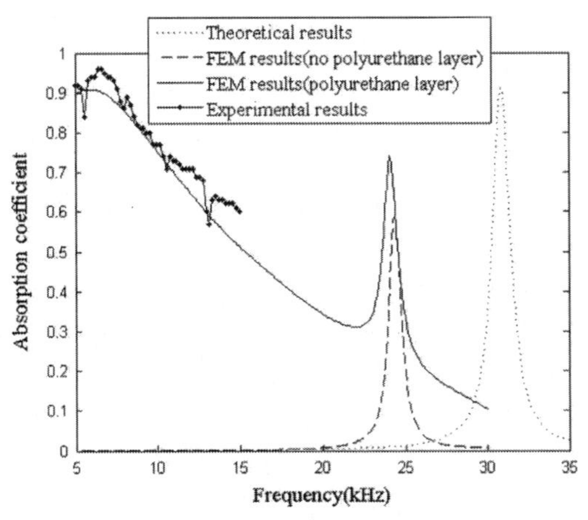

Fig. 16. Sound absorption coefficient results of open-circuit condition.

thickness of the waterproofing layer be zero, the results of the model is also applied to compare with the theoretical results without a surrounding polyurethane layer. Fig. 16 shows that whether the model includes a surrounding polyurethane layer or not, the peak frequencies are nearly the same (24.4 kHz for no surrounding polyurethane layer and 24.1 kHz for having a surrounding polyurethane layer), which are lower than the theoretical one (30.8 kHz). It means that the surrounding polyurethane layer has little impact on the peak frequency of the sample, while the peak frequency difference between the FE model and the theoretical one is caused by different transverse boundary conditions.

From Fig. 16, it is found that the sound absorption coefficient curve of the FE model without a surrounding polyurethane layer is quite similar to the theoretical one, except for the peak positions, but it is very different from that of the FE model with a surrounding polyurethane layer below the peak frequency. For the FE model without a surrounding polyurethane layer, the sound absorption coefficients in the frequency range far from the peak frequency are very low, which is in accordance with the theoretical results. While for the FE model with a surrounding polyurethane layer, the sound absorption coefficients below the peak frequency are comparatively high. It means that the surrounding polyurethane layer has a great influence on the sound absorption coefficients below the peak frequency, although it does not affect the peak frequency.

The experimental results in Fig. 16 show that the sound absorption coefficients agree with that of the FE model which includes the surrounding polyurethane layer perfectly in the frequency range 5–15 kHz. It illustrates that FEM can be employed to simulate the experimental sample accurately, and can set up a bridge between the theoretical model and the experimental sample according to their respective conditions. In addition, the peak frequency of the sound absorption coefficient cannot be observed in the experimental curve, because it is beyond the measurement frequency range. Besides, there are four abnormal points at 5.6 kHz, 8.0 kHz, 10.6 kHz and 13.2 kHz, which are caused by the inner resonance of the measurement system without influencing the observation of the broadband sound absorption coefficient.

7.2. A serial R–L circuit connected to the 'whole layer'

In this experiment, the four piezoelectric layers are treated as one whole layer to which a serial R–L circuit is connected. Four groups of serial R–L circuits are designed and employed: (i) a 78 mH inductor; (ii) a 1 kΩ resistor in series with a 78 mH inductor; (iii) a 2 kΩ resistor in series with a 78 mH inductor; (iv) a 117 mH inductor. The principle of designing the shunt circuit element values is to adjust the peak frequency to the valid measurement frequency range 5–15 kHz. Different inductors are adopted to observe the control of the peak frequency of the sound absorption coefficients, while different resistors in serial with the same inductor are used to observe the control of the bandwidth of the peak.

When a 78 mH inductor is shunted to the 'whole layer', the sound absorption coefficient results obtained by theoretical calculation, finite element simulations and experiment measurement are as shown in Fig. 17. It is found that shunting of the 78 mH inductor makes the peak frequencies of four cases of piezoelectric coating under open-circuit condition (as shown in Fig. 16) lower dramatically: from 30.8 kHz to 19.4 kHz for the theoretically radially-clamped model, from 24.4 kHz to 13.0 kHz for the free boundary FE model without the surrounding polyurethane layer, from 24.1 kHz to 13.1 kHz for the free boundary FE model including a surrounding polyurethane layer, and from about 24 kHz (referring to the FEM results) to 13.2 kHz for the experimental sample, respectively. The sound absorption coefficient curve of experiment is highly consistent with that of the FE model including the surrounding polyurethane layer in the frequency range 5–15 kHz. The same peak of the sound absorption coefficient is also observed at frequency 13.2 kHz in the experimental results because it

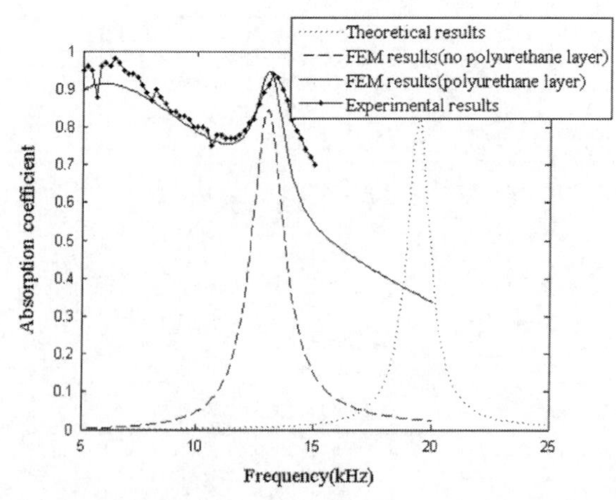

Fig. 17. Sound absorption coefficient results when a 78 mH inductor is shunted to the 'whole layer'.

falls into the measurement frequency range when a 78 mH inductor is shunted to the 'whole layer'.

In the following part, the sound absorption coefficient results for the four groups of serial R-L circuits are compared between the FEM simulation and the experiment measurement. The FEM simulation results are shown in Fig. 18(a), while the experimental results are shown in Fig. 18(b). It can be seen that four curves corresponding to respective serial R-L circuits in Fig. 18(a) are highly consistent with that in Fig. 18(b), except for four abnormal frequencies mentioned above. In addition, compared with the open-circuit condition, after connecting the proper serial R-L circuit, the peak frequencies are adjusted to the range 10–15 kHz. And the variation of the peak frequency, the maximum and the relative bandwidth of the sound absorption coefficient curves with the change of shunt circuits can be observed clearly.

Simulation and experimental results show that, when the inductor value is set as 78 mH with the resistor value varying from 0 to 1 kΩ, then to 2 kΩ, the peak frequency is about 13 kHz, almost remaining the same. However, with the increasing of the resistor value, the maximum of the sound absorption coefficient decreases while the bandwidth increases. When the resistor value is set as 0 with the inductor value varying from 78 mH to 117 mH, the peak amplitude nearly remains the same, however, the peak frequency decreases, varying approximately from 13 kHz to 12 kHz accordingly.

It is concluded that the inductor value mainly decides the peak frequency while the resistor value mainly influences the maximum and the bandwidth of the sound absorption coefficient curves. With the increase of the shunt inductor, the peak frequency will decrease. With the increase of the shunt resistor, the maximum will decrease while the bandwidth will broaden. The conclusions here are consistent with the conclusions drawn from in Figs. 7 and 8. For a certain frequency 13 kHz, when the shunt resistor increases or the shunt inductor increases, the sound absorption coefficient will decrease. The similar variations can be seen in Fig. 7. For two frequencies 12 kHz and 13 kHz, at which the sound absorption

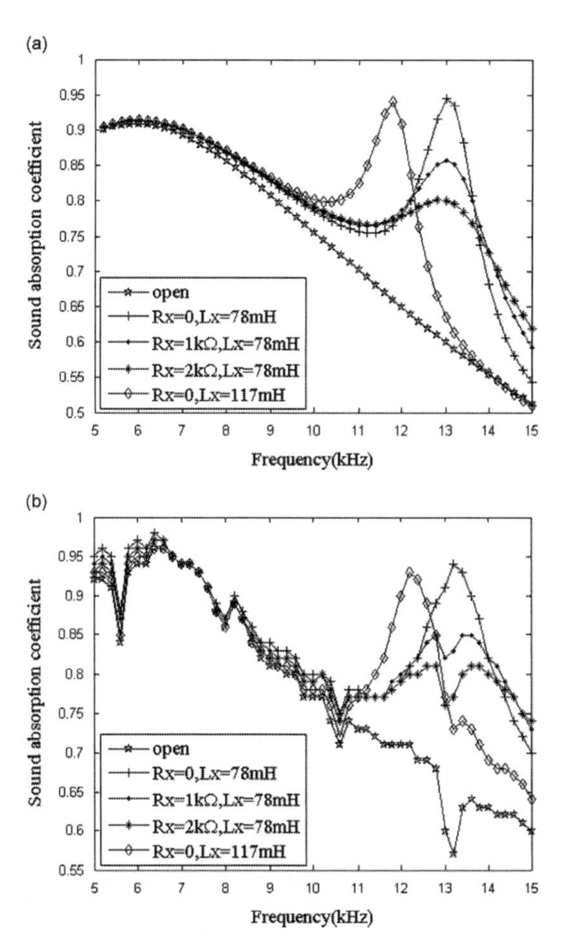

Fig. 18. Sound absorption coefficient results when a serial R-L circuit is shunted to the whole four piezoelectric layers: (a) FE simulation and (b) experiment.

coefficients are nearly the same, the lower the frequency is, the larger the shunt inductor is. The similar variations can be obtained in Fig. 8.

7.3. Combining control with different serial R–L circuits

In this experiment, different serial R–L circuits are designed and connected to respective piezoelectric layers to observe the combining control effect of the shunt circuits. Because the background of the sound absorption coefficients in low frequency range is very high, the values of the R–L elements are designed to avoid the peak frequency falling below 10 kHz. As an example, layer 1 is shunted to a 75 Ω resistor in series with a 5 mH inductor. Layer 2 is shunted to a 75 Ω resistor in series with a 10 mH inductor. Layer 3 is shunted to a 75 Ω resistor in series with a 22 mH inductor. Layer 4 is shunted to a 75 Ω resistor in series with a 39 mH inductor.

Shown in Fig. 19(a) and (b) respectively, the sound absorption coefficient results of FE simulation and the experiment are in good agreement with each other, except four abnormal frequencies mentioned above. Because of the limitation of the measurement frequency range, only two peaks at 11.5 kHz and 14.5 kHz are displayed within 10–15 kHz, which are mainly decided by the 39 mH inductor of layer 4 and the 22 mH inductor of layer 3, respectively. The other two peaks corresponding to 5 mH and 10 mH go beyond the test frequency range. However, it does not affect the conclusion that the combining R–L circuits shunted to multiple piezoelectric layers can be applied to control the sound absorption coefficients at multiple frequencies.

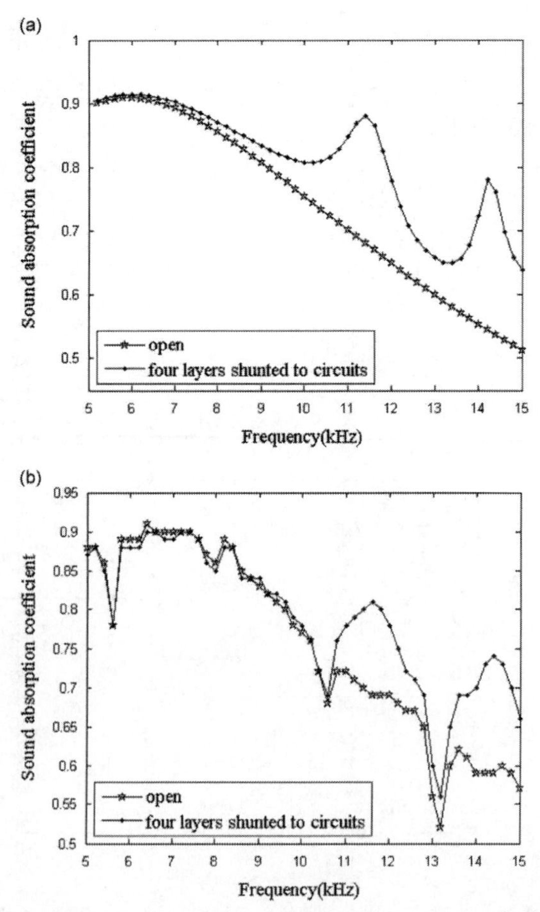

Fig. 19. Sound absorption coefficient results when four serial R–L circuits are shunted to the four piezoelectric layers respectively: (a) FE simulation and (b) experiment.

8. Conclusion

Semi-active control of underwater sound absorption coefficients of a piezoelectric coating by shunt damping can be applied in the acoustic stealth of underwater objects to conduct wideband sound absorption design or the compensation for the sound absorption defects. In this paper, a practical control method of sound absorption of the multi-layered piezoelectric coating is proposed, by combining design of the shunt impedance under constraint of the underwater sound absorption coefficients at limited frequencies. The number of the constrained frequencies equals that of the piezoelectric layers. Each piezoelectric layer and its shunt circuit control the sound absorption coefficient at one frequency. Therefore the wideband sound absorption coefficients can be controlled by the combining design of the shunted circuits of multiple piezoelectric coating layers at limited frequencies.

First, the shunt impedance of the coating is derived under the constraint of sound absorption coefficient at one frequency. Then, taking the 1–3 piezoelectric composite coating as an example, the sound absorption properties of the coating shunted to the designed shunt impedance are investigated. Next, on the basis of that, an iterative method for two constrained frequencies and an optimizing algorithm for multiple constrained frequencies are provided for combining design of the shunt impedances. At last, an experimental sample with four piezoelectric material layers is manufactured, of which the sound absorption coefficients are measured in an impedance tube. The experimental results show good agreement with the finite element simulation results. It is proved that a serial R–L circuit can control the peak frequency, maximum and bandwidth of the sound absorption coefficient and the combining R–L circuits shunted to multiple layers can control the sound absorption coefficients at multiple frequencies.

At present, the experiments are preliminarily designed to take use of the conventional electrical elements and observe the phenomenon of the control of the sound absorption coefficient by shunted circuits. Yet the implement of the negative resistor is not discussed, which is significant for the low frequency application. The research of implementation of the negative resistors or the synthesized impedances as well as the optimizing design of the wideband sound absorption will be our future work.

References

[1] R.L. Forward, Electronic damping of vibrations in optical structures, *Applied Optics* 18 (1979) 690–697.

[2] N.W. Hagood, A. von Flotow, Damping of structural vibrations with piezoelectric materials and passive electrical networks, *Journal of Sound and Vibration* 146 (1991) 243–268.

[3] H.H. Law, P.L. Rossiter, G.P. Simon, L.L. Koss, Characterization of mechanical vibration damping by piezoelectric materials, *Journal of Sound and Vibration* 197 (1996) 489–513.

[4] C.L. Davis, G.A. Lesieutre, An actively tuned solid-state vibration absorber using capacitive shunting of piezoelectric stiffness, *Journal of Sound and Vibration* 232 (2000) 601–617.

[5] M. Ciminello, A. Calabro, S. Ameduri, A. Concilio, Synchronized switched shunt control technique applied on a cantilevered beam: numerical and experimental investigations, *Journal of Intelligent Material Systems and Structures* 19 (2008) 1089–1100.

[6] M. Furstoss, D. Thenail, M.A. Galland, Surface impedance control for sound absorption: direct and hybrid passive/active strategies, *Journal of Sound and Vibration* 203 (1997) 219–236.

[7] B. Betgen, M.A. Galland, A new hybrid active/passive sound absorber with variable surface impedance, *Mechanical Systems and Signal Processing* 25 (2011) 1715–1726.

[8] N. Sellen, M. Cuesta, M.A. Galland, Noise reduction in a flow duct: implementation of a hybrid passive/active solution, *Journal of Sound and Vibration* 297 (2006) 492–511.

[9] T.G. Zielinski, Numerical investigation of active porous composites with enhanced acoustic absorption, *Journal of Sound and Vibration* 330 (2011) 5292–5308.

[10] J. Kim, J.K. Lee, Broadband transmission and noise reduction of smart panels featuring piezoelectric shunt circuits and sound-absorbing material, *Journal of the Acoustical Society of America* 112 (2002) 990–998.

[11] J. Kim, J.H. Kim, Multimode shunt damping of piezoelectric smart panel for noise reduction, *Journal of the Acoustical Society of America* 116 (2004) 942–948.

[12] J. Kim, Y.C. Jung, Broadband noise reduction of piezoelectric smart panel featuring negative-capacitive-converter shunt circuit, *Journal of the Acoustical Society of America* 120 (2006) 2017–2025.

[13] D.Q. Chang, B.L. Liu, X.D. Li, An electromechanical low frequency panel sound absorber, *Journal of the Acoustical Society of America* 128 (2010) 639–645.

[14] J.M. Zhang, W. Chang, V.K. Varadan, V.V. Varadan, Passive underwater acoustic damping using shunted piezoelectric coatings, *Smart Materials and Structures* 10 (2001) 414–420.

[15] L.G. Yu, Z.H. Li, L.L. Ma, Theoretical analysis of underwater sound absorption of 0-3 type piezoelectric composite coatings, *Acta Physica Sinica* 61 (2012) 024301-1–024301-8. (in Chinese).

[16] G.D. Luan, J.D. Zhang, R.Q. Wang, Equivalent circuit of common vibrating mode of piezoelectric ceramics, in: Yan Wang, Yan Sun (Eds.), *Piezoelectric Transducers and Arrays*, Peking University Press, Beijing 2005, pp. 114–118. (Revised ed.).

[17] W.A. Smith, B.A. Auld, Modeling 1–3 composite piezoelectrics thickness-mode oscillations, *IEEE Transactions on Ultrasonics, Ferroelectrics and Frequency Control* 38 (1991) 40–47.

[18] R.J. Wang, *Underwater Acoustic Materials Manual*, Science Press, Beijing, 1983, pp. 144–147, 26–27.

Design of Optimal Multiple Phase-Coded Signals for Broadband Acoustical Doppler Current Profiler①

Cheng Chi, Zhaohui Li, and Qihu Li

***Abstract*—Traditional broadband acoustical Doppler current profilers utilize binary phase-coded signals (BPCSs), like Barker code or M-sequence, as their transmitted signals. However, these BPCSs may not be optimal in terms of measurement deviation of current velocity. By analyzing the properties of broadband transmitted signals that influence the measurement deviation, this paper gives a mathematical explanation that the measurement deviation is essentially determined by the integral of the square of the single baseband pulse's autocorrelation function (SBPAF), namely the energy of SBPAF, when the duration and energy of the single transmitted pulse are fixed. To minimize the measurement deviation, a multiple phase-coded signal (MPCS) is adopted and the design method to obtain the optimized MPCS is proposed, which takes the energy of SBPAF as the objective function and coded phases as optimization variables. Simulations show that compared with the reference BPCS, the average measurement deviation is approximately reduced by 15% and 10% in the case of the optimized MPCS (the code length being 63), when the number of single baseband pulses is two and four, respectively. The longer the length is, the better the MPCS performs over the BPCS. When the transmitted power is beyond a certain value, the nonlinear effect will be significant and the second harmonic signal of the optimized MPCS can be used to estimate current velocity in conjunction with the fundamental signal because the second harmonic signal of the optimized MPCS is still broadband. In this situation, the measurement deviation of using the optimized MPCS will be 44% smaller than that of using the reference BPCS. Since the second harmonic signals of the BPCSs which are modulated with 0° and 180° are equivalent to no phase coding and become narrowband, they have no contribution to the estimation of current velocity.**

***Index Terms*—Broadband acoustical Doppler current profiler, multiple phase-coded signal (MPCS), second harmonic signal, optimal design.**

NOMENCLATURE

$a(t)$	Single baseband pulse.
$R(\tau)$	Autocorrelation function of $a(t)$
$e(t)$	Single broadband transmitted pulse.
$s(t)$	$[a(t)+a(t-T)]e^{j2\pi f_0 t}$, transmitted pulse-pair signal.
f_0	Carrier frequency.
f_d	Doppler shift.
T	Duration of $a(t)$.
$r(t)$	Received signal.
$r_1(t)$	Baseband complex signal of the received signal.
$r_2(t)$	Truncated signal of $r_1(t)$.
t_0	Start time of the truncating window.
A_i	Amplitude of the echo signal from the ith scatterer.
τ_i	Delay of the echo signal from the ith scatterer.
A_m	Amplitude of the echo signal from the mth scatterer.
τ_m	Delay of the echo signal from the mth scatterer.
$R_0(T)$	Value of the autocorrelation function of $r_1(t)$ at lag T.
$R_1(T)$	Sum of autocorrelation function of individual backscattered signals for $i=m$.
$R_2(T)$	Sum of cross-correlation function among individual backscattered signals for $i\neq m$.
τ_i'	τ_i-t_0-T.
τ_m'	τ_m-t_0-T.
τ_{im}'	$\tau_i'-\tau_m'$.
$\hat{\theta}$	Estimated phase caused by Doppler shift.
$\hat{f}_d$	Estimated Doppler shift.
$\hat{v}$	Estimated current velocity.
E_a	Energy of $a(t)$.
E_R	Energy of $R(\tau)$.
$a_M(t)$	Baseband multiple phase-coded signal (MPCS).
E_{a_M}	Energy of $a_M(t)$.
N	Code length
φ_n	Phase of the nth code element
$O_0(\varphi_1,\dots,\varphi_n)$	Energy of the autocorrelation function of $a_M(t)$, objective function in the continuous domain.

① IEEE Journal of Oceanic Engineering, 2016, 41(2): 302-317.

T_1	Duration of each code element
T_2	Sampling period of $a_M(t)$.
$a_M(k)$	Discrete expression of $a_M(t)$.
$O_1(\varphi_1, \ldots, \varphi_n)$	Objective function for optimization of MPCSs in the discrete domain.
$s_0(t)$	Broadband excitation signal.
$s_1(t)$	Transmitted signal that has been operated by Doppler shift.
$s_2(t)$	$\sum_i s_1(t-\tau_i)$, received reverberation signals.
$h(t)$	Impulse response of the transducer employed.
$n_0(t)$	Ambient Gaussian white noise.
$\hat{v}_i$	ith estimated velocity at the same depth.
σ_v	Measurement deviation of current velocity.
M	Number of times of calculating the current velocity.
σ_{B1}	Measurement deviation using the fundamental signal of the reference BPCS.
σ_{B2}	Measurement deviation of narrowband acoustical Doppler current profiler.
σ_B	Total measurement deviation combining the fundamental and second harmonic signals of the reference BPCS.
N_v	Number of independent depth cells.
σ_{M1}	Measurement deviation using the fundamental signal of the optimized MPCS.
σ_{M2}	Measurement deviation using the second harmonic signals of the optimized MPCS.
σ_M	Total measurement deviation combining the fundamental and second harmonic signals of the reference MPCS

I. INTRODUCTION

THE measurement of ocean and river currents using an acoustical Doppler current profiler (ADCP) has been studied for several decades. In general, the key issue of ADCP is to make an accurate estimation of the Doppler shift of acoustical backscattered signal in the moving medium. RD Instruments (Poway, CA, USA) first developed a broadband acoustical Doppler current profiler (BBADCP) that dramatically improves the precision of current measurement compared with the conventional narrowband profilers [1].

Attempts have been made to analyze the deviation of Doppler shift estimation for narrowband or pulse-to-pulse coherent ADCPs [2]–[4]. But none of them provides insight into the mechanisms which lead to the measurement deviation in pulse-pair BBADCPs.

Due to its simplicity and efficiency, the pulse-pair method given by Zrnic [5] and Miller and Rochwarger [6] is often used for Doppler shift estimation in BBADCPs. For pulse-pair BBADCPs, the self-noise from the uncorrelated echo in each member of the sample pair [1] has a great influence on the measurement deviation of current velocity. Theoretically, if an ideal broadband signal, of which the autocorrelation function is the impulse δ-function, is employed, the self-noise caused by the uncorrelated echo will disappear. However, in the practical situation, the ideal broadband signal is impossible to be transmitted. Therefore, the influence of the self-noise cannot be removed. Brumley *et al.* [1] have pointed out that the self-noise reduced the correlation coefficient significantly. Wanis *et al.* [10] have analyzed factors affecting measurement deviation such as selection of code sequences, the number of code repetitions, beam divergence decorrelation, residence time decorrelation, and environmental effects. In their paper, they have given a good physical explanation of the detrimental effect of the self-noise and explained the advantage of using broadband pulses and coded sequence very well with representative figures, although they have not formalized the influence with mathematical formulas. This paper will provide a mathematical explanation to the mentioned self-noise.

In traditional pulse-pair BBADCPs, the binary phase-coded signals (BPCSs) act as transmitted waveform thanks to their good autocorrelation performance. However, in terms of the measurement deviation, the BPCSs should not be the optimal waveforms. In addition, as Prieur and Hansen [9] have pointed out, the BPCSs are not viable when using the second harmonic signal in BBADCP, since a phase coding modulating the phase of the fundamental signal with 0° and 180° is equivalent to no phase coding for the second harmonic signal. Hence, other signals such as multiple phase-coded ones should be considered in the waveform design of BBADCPs. At present, it becomes realizable to transmit arbitrary ultrasound waveforms for BBADCPs with the development of the technique of signal generation and emission [7], [8]. In this paper, we will focus on designing optimal transmitted waveform based on multiple phase-coded signals (MPCSs) to reduce the effect of the self-noise.

The first contribution of this paper establishes a theoretical upper limit of the measurement deviation caused by the self-noise [1], [10] or "self-interference" [11] for pulse-pair BBADCPs, without considering additional ambient noise. From the mathematical formulas, it can be found that when the energy and duration of transmitted pulse are fixed, the theoretical upper limit is essentially determined by the integral of the square of the single baseband pulse's autocorrelation function (SBPAF), namely the energy of SBPAF. For the constant pulse duration and energy, the limit can decrease with the decrease of the energy of SBPAF.

The second contribution of this paper concerns the design of optimal transmitted signals for BBADCP. To minimize the measurement deviation, an MPCS is adopted and the design method which takes the energy of SBPAF as the objective function and coded phases as optimization variables is proposed to obtain an optimal transmitted signal. Simulations are conducted to evaluate the performance of the optimized MPCS in terms of measurement deviation. If the second harmonic signal can be used to estimate the measurement velocity in conjunction with the fundamental signal, the advantage of using the optimized MPCS will be more distinct.

The structure of this paper is as follows. In Section II, the transmitted signal's properties affecting the measurement deviation are analyzed and a theoretical upper limit of the measurement deviation is obtained. In Section III, the optimization design method of the MPCS is described via a design example. Simulations for BBADCPs by using the fundamental signal are presented in Section IV. Then, the improvement by using the second harmonic signal in BBADCPs is discussed in Section V. Section VI provides the conclusion and discussion.

II. Deviation Analysis of Pulse-Pair Method for BBADCP

In this section, we first review the pulse-pair method given by Zrnic [5] and Miller and Rochwarger [6]. Then, the properties of transmitted broadband signals which affect the measurement deviation of current velocity for BBADCP by using the pulse-pair method are theoretically analyzed and a theoretical upper limit of the measurement deviation caused by the self-noise is given.

A. Pulse-Pair Method

BBADCP can help reduce velocity variance and achieve finer depth resolution [1]. When the pulse-pair method is used to implement the BBADCP systems, a broadband pulse pair consisting of two duplicated pulses is transmitted. The autocorrelation of the transmitted signal will present three peaks with the position of each of the side peaks corresponding to the interval between the two pulses. Suppose that the scatters in the water are moving along the direction of sound beam, and a large number of backscattered signals are superposed and received. The interval between the two pulses of the pulse pair of each backscattered signal will expand or shrink according to its Doppler shift. Then, the variation of the interval between two pulses can be estimated by the shift of side peak position of the autocorrelation of a truncated signal segment with length of the pulse pair, and the corresponding Doppler shift can be obtained.

Suppose that the single broadband transmitted pulse is given by

$$e(t) = a(t)e^{j2\pi f_0 t} \tag{1}$$

where $a(t)$ represents the broadband coded single baseband pulse (SBP), and f_0 is the carrier frequency.

The autocorrelation function of $a(t)$ can be written by

$$R(\tau) = \int_0^T a^*(t)a(t+\tau)dt \tag{2}$$

where T is the duration of SBP and "$*$" represents the "complex conjugate" operation. The transmitted pulse-pair signal is given by

$$s(t) = [a(t) + a(t-T)]e^{j2\pi f_0 t}. \tag{3}$$

The echo signal of BBADCP is the superposition of a large number of backscattered signals. Assume that the scatterers are uniformly distributed and move at the same velocity v, the backscattered signals are independent, and their amplitudes are uniformly distributed. Assuming the Doppler shift of any backscattered signal is $f_d \approx 2vf_0/c$ with c as the sound speed in seawater, the received signal is given by

$$r(t) = \sum_i A_i s(t-\tau_i)e^{j2\pi f_d(t-\tau_i)} \tag{4}$$

where A_i and τ_i are the amplitude and the delay of the echo signal from the ith scatterer, respectively. Here, the additional ambient noise is omitted for the simplification of analysis, because this paper mainly focuses on the properties of the broadband transmitted signal that affect the measurement deviation. The properties of the broadband transmitted signal determine the influence of the self-noise on measurement deviation.

After orthogonally demodulating and lowpass filtering, the baseband complex signal of the received signal is written as

$$r_1(t) = \sum_i A_i[a(t-\tau_i) + a(t-\tau_i-T)]e^{j2\pi f_d(t-\tau_i)}. \tag{5}$$

The first step of processing $r_1(t)$ is to truncate $r_1(t)$ with a rectangular window to obtain the current velocity of the range cell. To obtain the finer depth resolution, when designing the BBADCP systems, the length of the transmitted pulse pair is usually chosen equal to the range cell size required. Therefore, the length of the rectangular window should be equal to the length of the transmitted signal $2T$. The truncated received signal denoted by $r_2(t)$ can be given by

$$r_2(t) = \text{rect}\left(\frac{t-t_0}{2T}\right) r_1(t) \tag{6}$$

where t_0 is the start time of the truncating window.

For pulse-pair BBADCPs, the autocorrelation function of the truncated received signal $r_2(t)$ will also be three-peaked as that of the transmitted signal and the phase at the position of the original side peak, i.e., the lag T, can give a more precise estimation of the Doppler shift. Therefore, the Doppler shift estimated by using the pulse-pair method is given by [1], [5]

$$\hat{f}_d = \frac{1}{2\pi T}\arg[R_0(T)] \tag{7}$$

where $R_0(T)$ is the value of the autocorrelation function of truncated signal $r_2(t)$ at delay T, which can be written as

$$R_0(T) = \int_{t_0}^{t_0+2T} r_2(t)r_2^*(t-\tau)dt|_{\tau=T}. \tag{8}$$

It can be seen from (7) that the Doppler shift is essentially determined by the phase of $R_0(T)$. Fig. 1 shows the process of estimating the Doppler shift of the broadband echo signal by using the pulse-pair method.

B. Measurement Deviation Analysis

Without considering the influence of the ambient noise, for BBADCPs, the measurement deviation of current velocity is mainly caused by the self-noise from the uncorrelated echo in each member of the sample pair [1]. This part gives a theoretical upper limit of the measurement deviation caused by the self-noise.

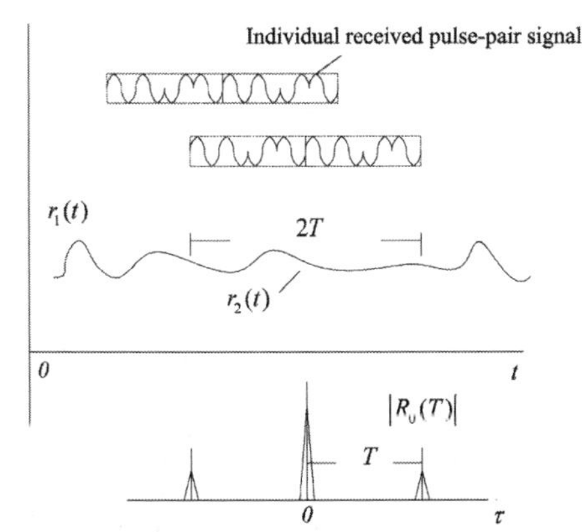

Fig. 1. Process of estimating the Doppler shift of the broadband echo signal by using the pulse-pair method.

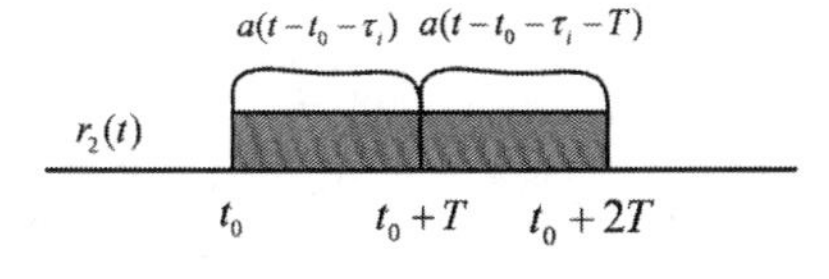

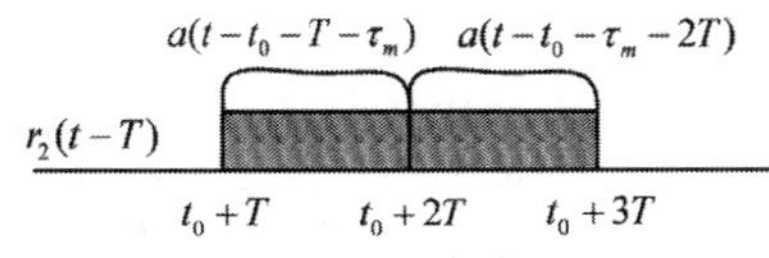

Fig. 2. Pulse pairs of explaining how (9) is obtained.

Substituting (5) and (6) into (8), we have

$$\begin{aligned}
&R_0(T)\\
&=\int_{-\infty}^{\infty}\operatorname{rect}\left(\frac{t-t_0}{2T}\right)\\
&\quad\times\sum_i A_i[a(t-\tau_i)+a(t-T-\tau_i)]e^{j2\pi f_d(t-\tau_i)}\\
&\quad\times\operatorname{rect}\left(\frac{t-T-t_0}{2T}\right)\\
&\quad\times\sum_m A_m[a^*(t-T-\tau_m)+a^*(t-2T-\tau_m)]\\
&\quad\times e^{-j2\pi f_d(t-T-\tau_m)}dt\\
&=\int_{t_0+T}^{t_0+2T}\sum_i A_i a(t-T-\tau_i)e^{j2\pi f_d(t-\tau_i)}\\
&\quad\times\sum_m A_m a^*(t-T-\tau_m)e^{-j2\pi f_d(t-T-\tau_m)}dt\\
&=\int_{t_0+T}^{t_0+2T}\sum_i\sum_m A_iA_m a(t-T-\tau_i)\\
&\quad\times a^*(t-T-\tau_m)e^{j2\pi f_d(T-\tau_i+\tau_m)}dt. \qquad (9)
\end{aligned}$$

To understand the calculation of (9) better, an explanatory drawing as Fig. 2 is given, which shows from the first line to the second line of (9) how the integration limits are changed according to the window time.

Here, we divide $R_0(T)$ into $R_1(T)$ and $R_2(T)$, in which $R_1(T)$ represents the sum of autocorrelation of individual backscattered signals for $i=m$, and $R_2(T)$ represents the sum of the cross correlation among backscattered signals for $i\neq m$. Thus, we have

$$R_0(T)=R_1(T)+R_2(T) \qquad (10)$$

where

$$\begin{aligned}
R_1(T)&=e^{j2\pi f_d T}\\
&\quad\times\int_{t_0+T}^{t_0+2T}\sum_i A_i^2 a(t-T-\tau_i)a^*(t-T-\tau_i)dt
\end{aligned} \qquad (11)$$

$$\begin{aligned}
R_2(T)&=\sum_{i\neq m}\sum_m e^{j2\pi f_d(T-\tau_i+\tau_m)}A_iA_m\\
&\quad\times\int_{t_0+T}^{t_0+2T}a(t-T-\tau_i)a^*(t-T-\tau_m)dt.
\end{aligned} \qquad (12)$$

If there exists only $R_1(T)$, the estimation of Doppler shift f_d will be an accurate value with no deviation according to (7) and (11), because

$$\begin{aligned}
\arg[R_0(T)]&=\arg[R_1(T)+R_2(T)]\\
&=2\pi f_d T,\qquad \text{if } R_2(T)=0.
\end{aligned} \qquad (13)$$

From (10) and (13), it can be concluded that $R_2(T)$ is a disturbance term that leads to the deviation and is caused by the self-noise.

Let

$$\begin{aligned}
\tau_i'&=\tau_i-t_0-2T\\
\tau_m'&=\tau_m-t_0-2T\\
\tau_{im}'&=\tau_i'-\tau_m'
\end{aligned}$$

thus $R_2(T)$ can be rewritten as

$$\begin{aligned}
R_2(T)&=\sum_{i\neq m}\sum_m A_iA_m e^{j2\pi f_d(T-\tau_{im}')}\\
&\quad\times\int_{-\tau_i'}^{T-\tau_i'}a(t)a^*(t+\tau_{im}')dt.
\end{aligned} \qquad (14)$$

It can be further represented by

$$R_2(T)=e^{j2\pi f_d T}\sum_{i\neq m}\sum_m A_iA_m e^{-j2\pi f_d\tau_{im}'}R_3(\tau_{im}') \qquad (15)$$

where

$$R_3(\tau_{im}')=\int_{-\tau_i'}^{T-\tau_i'}a(t)a^*(t+\tau_{im}')dt. \qquad (16)$$

As the scatterers are assumed to be uniformly distributed in water, τ_i' and τ_m' will uniformly distribute between $-T$ and T. The probability distribution function of the difference τ_{im}' between the two independent random variables which are uniformly distributed should be (Simpson distribution)

$$f(\tau_{im}')=\begin{cases}\dfrac{2T-\tau_{im}'}{4T^2}, & 0<\tau_{im}'<2T\\[2mm] \dfrac{2T+\tau_{im}'}{4T^2}, & -2T<\tau_{im}'<0.\end{cases} \qquad (17)$$

According to Appendix B

$$\int \left|R_3(\tau'_{im})\right|^2 d\tau'_{im} \leq \int \left|R_4(\tau'_{im})\right|^2 d\tau'_{im} \tag{18}$$

where

$$R_4(\tau'_{im}) = \int_0^T a(t)a^*(t+\tau'_{im})dt. \tag{19}$$

Equation (18) is important to the derivation of (21) according to Appendix A. From (2) and (19), it can be found that $R_4(\tau'_{im})$ is the complex conjugate of the autocorrelation function of $a(t)$.

Let $\hat{\theta} = \arg[R_0(T)]$, then the variance of $\hat{\theta}$ is

$$\mathrm{var}[\hat{\theta}] = E[(\hat{\theta} - \bar{\theta})^2] = (2\pi T)^2 \mathrm{var}[\hat{f}_d]. \tag{20}$$

Without considering the influence of the additional noise and given that $R_4(\tau'_{im})$ is the complex conjugate of the autocorrelation function of $a(t)$, we deduce that

$$\begin{aligned}\mathrm{var}[\hat{f}_d] &= \frac{1}{(2\pi T)^2}\mathrm{var}(\hat{\theta}) \\ &\leq \frac{1}{(2\pi T)^2}\frac{2\int R_4^*(\tau'_{im})R_4(\tau'_{im})d\tau'_{im}}{TE_a^2} \\ &= \frac{1}{(2\pi T)^2}\frac{2\int \left|R_4(\tau'_{im})\right|^2 d\tau'_{im}}{TE_a^2} \\ &= \frac{1}{(2\pi T)^2}\frac{2\int \left|R(\tau)\right|^2 d\tau}{TE_a^2}\end{aligned} \tag{21}$$

where E_a is the energy of the SBP

$$E_a = \int a^*(t)a(t)dt.$$

Inequality (21) is proven in Appendix A.

According to (21), the variance of the estimated velocity $\hat{v}$ is

$$\mathrm{var}[\hat{v}] = \left(\frac{c}{2f_0}\right)^2 \mathrm{var}[\hat{f}_d] \leq \left(\frac{c}{4\pi T f_0}\right)^2 \frac{2\int |R(\tau)|^2 d\tau}{TE_a^2}. \tag{22}$$

The energy of SBPAF can be denoted by

$$E_R = \int |R(\tau)|^2 d\tau. \tag{23}$$

Inequality (22) gives the theoretical upper limit of the velocity measurement deviation caused by the self-noise. It indicates that without considering the influence of ambient noise, the measurement deviation caused by the self-noise will decrease with the diminishing of the energy of the SBPAF, i.e., E_R. The smaller the energy of the SBPAF is, the smaller the variance of $\hat{v}$ will be.

When the duration and energy of the transmitted pulse are fixed, based on (22), to decrease the measurement deviation, E_R should be as small as possible. This can also be understood from the perspective of the signal resolution theory [12].

The definition of the signal distance resolution constant (DRC) A_τ in [12] is given by

$$A_\tau = \frac{\int R^*(\tau)R(\tau)d\tau}{R^2(0)}. \tag{24}$$

It means that if the value of A_τ is smaller, the signal distance resolution is better, and the interplay among the echo signals of different scatterers is weaker. Therefore, according to (10), (12), and (22), the measurement deviation of current velocity becomes less when A_τ diminishes.

Defining $A(f)$ as the Fourier transform of the SBP $a(t)$, $A(f) = F(a(t))$, we have $|A(f)|^2 = F(R(\tau))$. Using the Parseval theorem, (24) can be rewritten as

$$\begin{aligned}A_\tau &= \frac{\int |R(\tau)|^2 d\tau}{(\int a^*(t)a(t)dt)^2} \\ &= \frac{\int |F(R(\tau))|^2 df}{E_a^2} \\ &= \frac{\int |A(f)|^4 df}{E_a^2}.\end{aligned} \tag{25}$$

It can be seen from (25) that, in the frequency domain, the integration of quartic of SBP's amplitude spectrum is a critical factor that influences the measurement deviation of current velocity. This also gives inspiration for the design of broadband transmitted signal by finding the optimal amplitude spectrum, which is not the main concern in this paper.

III. Optimal MPCS Design

Based on the theoretical upper limit of the measurement deviation of current velocity deduced in Section II, an optimization criterion that minimizes the energy of SBPAF for designing the optimal transmitted signals is put forward in this section. According to the criterion, an MPCS is adopted for the design of optimal transmitted signals and the design method is shown via an example.

A. Method of Designing Transmitted Signals

The BPCSs commonly used as the transmitted signals in BBADCPs may not be optimal in terms of the measurement deviation. Therefore, we proposed the design method of optimal transmitted signals based on the MPCSs. The criterion for the optimization is the minimization of the energy of SBPAF.

The MPCSs can be written as

$$a_M(t) = \sum_{n=1}^{N} \mathrm{rect}\left(\frac{t - nT_1}{T_1}\right)e^{j\varphi_n} \tag{26}$$

where N is the code length, T_1 is the duration of each code element, and φ_n is the phase of the nth code element and can be set as arbitrary value within $-\pi$ and π. N and T_1 are determined by the range resolution and the bandwidth of the system. One of the important advantages of the MPCSs is that the variation of the coded phases does not influence the energy of the MPCSs, denoted by E_{a_M}. We have

$$\begin{aligned}E_{a_M} &= \int a_M^*(t)a_M(t)dt \\ &= \int_0^T \left(\sum_{n=1}^{N}\mathrm{rect}\left(\frac{t-nT_1}{T_1}\right)e^{-j\varphi_n}\right) \\ &\quad \times \left(\sum_{m=1}^{N}\mathrm{rect}\left(\frac{t-nT_1}{T_1}\right)e^{j\varphi_m}\right)dt \\ &= \int_0^T \left(\sum_{n=1}^{N}\sum_{m=1}^{N}\mathrm{rect}\left(\frac{t-nT_1}{T_1}\right)\mathrm{rect}\left(\frac{t-mT_1}{T_1}\right)e^{j(\varphi_m-\varphi_n)}\right)dt \\ &= \int_0^T \sum_{n=1}^{N}\left(\mathrm{rect}\left(\frac{t-nT_1}{T_1}\right)\right)^2 dt = T.\end{aligned} \tag{27}$$

From (27), we find that E_{a_M} is irrelative with the coded phases.

For BBADCPs, as shown in Section II, the measurement deviation diminishes with the decrease of the energy of SBPAF. Therefore, the problem of designing the optimal transmitted signals can be transformed into that of finding $\varphi_n (n = 1, \ldots, N)$, which minimize the energy of the autocorrelation function of $a_M(t)$, where φ_n are the independent variables. The energy of the autocorrelation function of $a_M(t)$, i.e., the objective function, is given by

$$O_0(\varphi_1, \ldots, \varphi_N) = \int \left| \int a_M^*(t) a_M(t+\tau) dt \right|^2 dt. \quad (28)$$

Because it is difficult to find the minimum value of $O_0(\varphi_1, \ldots, \varphi_N)$ by the analytical methods, we chose to apply the optimization toolbox of Matlab [14].

For computers, it is hard to process the problem in the continuous domain. Hence, converting the problem in the continuous domain into the discrete domain is necessary. By sampling $a_M(t)$ with the frequency $1/T_2$, the discrete expression of $a_M(t)$ will be

$$a_M(k) = \sum_{n=1}^{N} \operatorname{rect}\left(\frac{t_1 + kT_2 - nT_1}{T_1}\right) e^{j\varphi_n} \quad (29)$$

where t_1 is the starting time. Thus, the discrete expression of the objective function $O_0(\varphi_1, \ldots, \varphi_N)$ is given by

$$O_1(\varphi_1, \ldots, \varphi_N) = \sum_{m=-K+1}^{K-1} \left(\sum_{k=1}^{K} a_M^*(k) a_M(k+m) \right)^2 \quad (30)$$

where K is the length of the discrete signal.

Based on the above, the mathematical model of the problem of designing transmitted signals can be described by

$$\varphi_1, \ldots, \varphi_N = \arg\min \left\{ \begin{aligned} & O_1(\varphi_1, \ldots, \varphi_N) \\ & = \sum_{m=-K+1}^{K-1} \left(\sum_{k=1}^{K} a_M^*(k) a_M(k+m) \right)^2 \end{aligned} \right\}$$

which means taking $O_1(\varphi_1, \ldots, \varphi_N)$ as the objective function and $\varphi_1, \ldots, \varphi_N$ as the optimization variables. The optimal MPCS can be obtained by minimizing the objective function $O_1(\varphi_1, \ldots, \varphi_N)$ in discrete domain.

In the design of the MPCSs, we employ the optimization function "fmincon" in the optimization toolbox of Matlab [14]. The active-set algorithm provided by Matlab is chosen for the optimization. The main idea of the active-set algorithm is using the penalty function to convert the constrained optimization to the unconstrained optimization and then applying the quasi-Newton and sequential quadratic programming (SQP) methods to get the optimal solution set [14]. The initial values of the optimization variables are chosen as the values of the reference M-sequence phases and the ending condition is that the iterative error is below 1×10^{-6}.

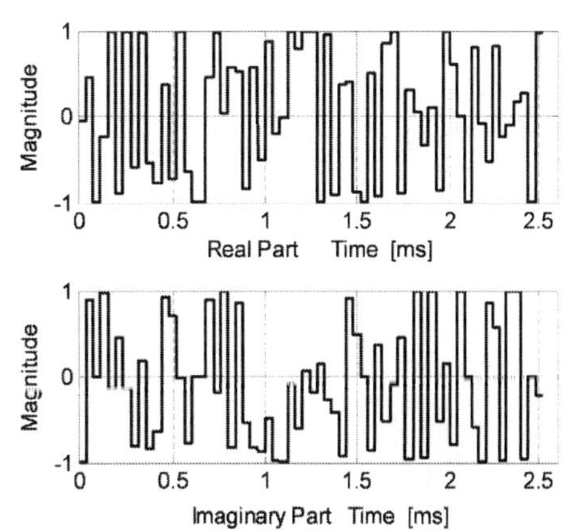

Fig. 3. Optimized baseband MPCS with 63 code elements.

In summary, the steps of the proposed method of designing the optimal transmitted signals are as follows.

- According to the range resolution cell demanded and the bandwidth of the practical systems, determine the number of code elements N and the duration of each code element T_1.
- Sample the continuous expression of $a_M(t)$ to obtain the discrete objective function $O_1(\varphi_1, \ldots, \varphi_N)$.
- Use the optimization toolbox of Matlab to find the optimal solution set of N coded phases.

B. Design Example

This part illustrates the design method of optimal MPCSs directly via an example. The carrier frequency and the bandwidth B_0 of the BBADCP employed by this work were 200 and 50 kHz, respectively. Thus, for phase-coded signals, the duration T_1 of each code element should be approximately equal to $2/B_0 = 0.04$ ms. The sampling period T_2 was 0.001 ms. Considering the code length needed by some practical systems [10], length N of the designed example was set as 63. For comparison, the M-sequence-coded signal [13] was chosen as the reference BPCS. The example with 63 code elements was compared with the reference BPCS in view of autocorrelation function, spectrum, and theoretical deviation.

The steps of optimization in Section III-A are taken to design the 63 optimal phases. For the parameter configuration of the example, $N = 63, T_1 = 0.04$ ms, and $T_2 = 0.001$ ms, the optimized values of $\varphi_1, \ldots, \varphi_N$ are exhibited in Table I, and the real and imaginary parts of the optimized MPCS are shown in Fig. 3.

The autocorrelation functions of the reference BPCS and the optimized MPCS with 63 code elements are depicted in Fig. 4. The amplitude spectra of the two signals with the carrier frequency of 200 kHz are depicted in Fig. 5. The energy of SBPAF of the two signals is calculated according to (23). The DRCs of two signals are calculated according to (25). The theoretical upper limits of velocity deviation of two signals are calculated by the square root of the variance of v according to (22). All of these results are listed in Table II. We find that the energy of SBPAF of the optimized MPCS is approximately 30% lower

TABLE I
OPTIMIZED CODED 63 PHASES OF MPCS

Code Number	Value (rad)	Code Number	Value (rad)	Code Number	Value (rad)	Code Number	Value (rad)	Code Number	Value (rad)
1	-1.62	13	2.37	25	-2.09	38	2.65	51	-0.92
2	1.09	14	-0.01	26	-0.51	39	-3.14	52	1.57
3	-3.14	15	-2.26	27	-1.76	40	-1.04	53	-3.14
4	1.81	16	-3.14	28	-1.59	41	2.77	54	-0.63
5	-0.14	17	3.14	29	-0.08	42	-0.55	55	-1.66
6	2.67	18	1.09	30	-0.65	43	-0.11	56	2.11
7	-0.13	19	-0.19	31	0.07	44	2.68	57	0.62
8	-2.21	20	1.54	32	-0.18	45	-1.27	58	-1.81
9	0.19	21	-0.95	33	2.99	46	1.52	59	1.68
10	-2.14	22	1.03	34	-0.27	47	-1.92	60	1.41
11	-2.45	23	-2.58	35	-2.70	48	1.47	61	-1.29
12	1.19	24	-0.97	36	-1.18	49	-2.59	62	3.14
				37	1.15	50	0.14	63	-0.22

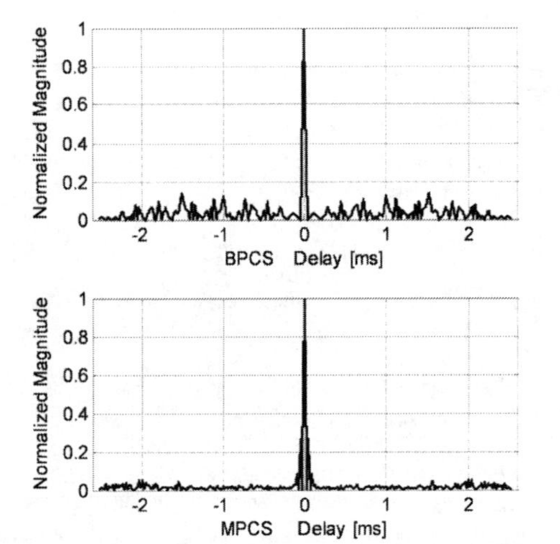

Fig. 4. Autocorrelation functions of the reference BPCS and the optimized MPCS.

than that of the reference BPCS; accordingly, the DRC and the theoretical upper limit of velocity deviation of optimized MPCS are also smaller than those of the reference BPCS, respectively. This can be explained by the criterion of minimizing the energy of SBPAF in Section II. If the optimized MPCS is adopted as the transmitted signal, the interplay among individual backscattered signals of scatterers will become weaker than the reference signal. Therefore, the measurement deviation by using the optimized MPCS is rationally lower than that by using the reference BPCS.

IV. SIMULATIONS FOR BBADCP BY USING THE FUNDAMENTAL SIGNAL

This section evaluates the measurement deviations for the optimized MPCSs of different length (length N of the codes being 63 and 127) and compares them with the reference BPCSs (M-sequence-coded signals) in various conditions. The parameters of designing the optimized MPCS of $N = 127$ are the same as the example of $N = 63$ in Section II.

We consider a bandwidth-limited transmitting and receiving transducer whose one-way frequency response is a Gaussian shape with the central frequency of 200 kHz and is depicted in Fig. 6. The basic process of the simulations is illustrated in Fig. 7. Unlike the overlapping of backscattered signals of scatterers in [5] and [16], approximated by the multiplicative noise, the overlapping in our work is directly realized by the superposition of massive backscattered signals with random delays and amplitudes. Therefore, the received signal $r(t)$ is obtained from

$$r(t) = s_2(t) + n_0(t) \tag{31}$$

where

$$s_2(t) = \sum_i A_i s_1(t - \tau_i) \tag{32}$$

and

$$s_1(t) = Dp[s_0(t) \otimes h(t)]. \tag{33}$$

"$Dp[\,]$" represents the operation of setting the Doppler shift realized by resampling, which is equivalent to setting the current velocity. In this section, A_i is supposed to be uniformly distributed between 0 and 1. The average number of scatterers uniformly distributing between 0 and T is 500. Besides, $s_0(t)$ is the broadband excitation signal, $n_0(t)$ is the additional Gaussian white noise, $h(t)$ is the impulse response of the transducers, and "$\otimes$" represents the convolution operation. The signal-to-noise ratio (SNR) is defined as [16]

$$\mathrm{SNR} = \frac{\langle (s_2(t))^2 \rangle}{\langle (n_0(t))^2 \rangle} \tag{34}$$

where "$\langle\ \rangle$" denotes time averaging.

The ith estimated velocity, denoted by $\hat{v}_i$, is calculated by

$$\hat{v}_i = \frac{\hat{f}_d^{(i)} c}{2 f_c} = \frac{c}{4\pi T f_c} \hat{\theta}_i \tag{35}$$

where $\hat{f}_d^{(i)}$ is the ith estimated Doppler shift, $\hat{\theta}_i$ is the ith estimated phase shift, and c is the sound speed which is set to 1500

TABLE II
ENERGY OF SBPAF, DRC, AND THEORETICAL UPPER LIMIT OF DEVIATION FOR THE REFERENCE BPCS AND THE OPTIMIZED MPCS

Transmitted signal	Energy of SBPAF	DRC	Deviation upper limit (m/s)
Reference BPCS	5.6310×10^{3}	1.7734×10^{2}	0.0397
Optimized MPCS	3.8847×10^{3}	1.2234×10^{2}	0.0330

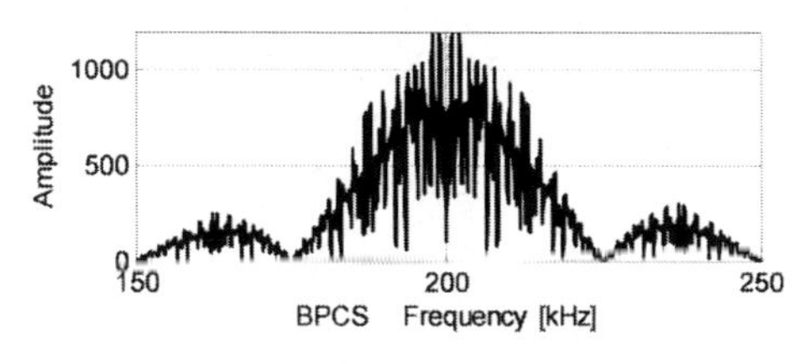

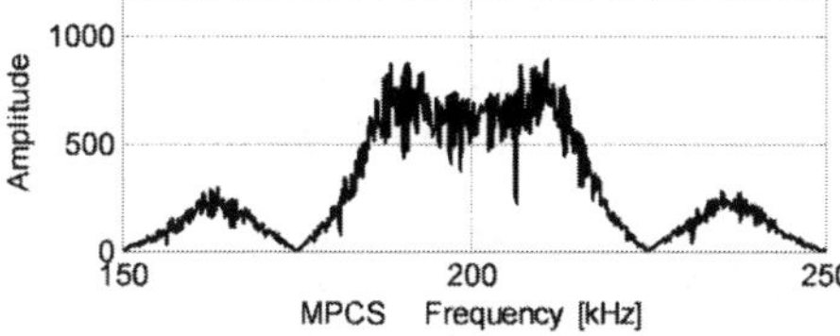

Fig. 5. Spectra of the reference BPCS and the optimized MPCS.

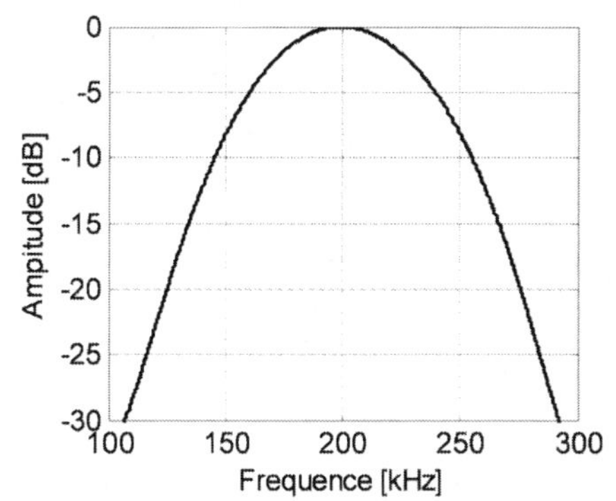

Fig. 6. Gaussian-shaped one-way frequency response of the transducer.

m/s in the simulations. Then, deviation σ_v of the current measurement is obtained by

$$\sigma_v = \sqrt{\frac{1}{M-1}\sum_{i=1}^{M}(\hat{v}_i - \bar{v})^2} \tag{36}$$

where $\bar{v}$ is the averaged velocity and

$$\bar{v} = \frac{1}{M}\sum_{i=1}^{M}\hat{v}_i. \tag{37}$$

In this paper, each of the velocities shown in the figures is calculated 3000 times ($M = 3000$) at the same depth cell.

First, to verify the theoretical upper limit (22) of measurement deviation caused by the self-noise in Section II, the simulations are conducted in the relatively high SNRs and the number of transmitted SBPs is two. In the relatively high SNR situations, the main factor influencing the measurement deviation is the self-noise. The theoretical upper limit of deviation provides a formulized influence of the self-noise. To avoid the velocity ambiguity for $N = 127$, the velocity is set to 0.2 m/s. For the two optimized MPCSs (the code length N being 63 and 127, respectively) and the reference BPCSs, the standard deviations are evaluated and presented in Fig. 8, accompanied with their theoretical upper limits. The results show that the standard deviations obtained by the simulations are a little lower than the corresponding theoretical upper limits, and the theoretical upper limits of the optimized MPCSs are lower than that of the corresponding reference BPCS for both $N = 63$ and $N = 127$. It can be concluded that the formulized theoretical upper limit is reasonable and the optimized MPCSs guided by the proposed criterion perform better than the reference BPCS. Thus, the theoretical upper limit is helpful to designing better BBADCPs.

In addition, from Fig. 8, we find that, compared with the reference BPCSs, the measurement deviations of using the optimized MPCSs are approximately reduced by 15% and 20% for $N = 63$ and $N = 127$, respectively. Therefore, the longer the code length N is, the more distinct merits the optimized MPCS will be.

Second, simulations evaluating the performance of the optimized MPCS in both relatively low and high SNRs are conducted under two situations: the number of SBPs is two and four, respectively, where N is chosen to be 63 and the velocity is set to 0.5 m/s. Fig. 9 shows the deviation of current measurement versus SNR for the two signals when the number of SBPs is two and four, respectively. We find that the average measurement deviations using the optimized MPCSs are approximately reduced by 15% and 10% compared with the situation of the reference BPCSs, when the number of SBPs is two and four, respectively. Multiple repetitions of SBP increase the signal autocorrelation at the measurement lag and also introduce correlation peaks at multiples of that lag, as analyzed in [10], while the initial contribution of the self-noise is reduced because only one of the code repetitions forms the self-noise. Therefore, for the multiple repetitions of SBP, the influence of the self-noise decreases and the reduction in measurement deviation for four SBPs is lower than that for two SBPs.

Finally, simulations are conducted to evaluate the measurement deviations of MPCSs of a different number of SBPs at different velocities and SNRs in comparison with the BPCS, where N is chosen to be 63 and SNRs are set to 0 and −30 dB, respectively.

Fig. 10(a) and (b) depicts the comparison of the measurement deviations of the two-SBP MPCSs with that of the BPCS at different velocities under two SNRs. Fig. 11(a) and (b) depicts the comparison of the measurement deviations of the four-SBP MPCSs with that of the BPCS at different velocities under two SNRs. The results in the two figures show that the proposed MPCSs have a better performance than the reference BPCS in all cases, which can be explained by the conclusion in Section II

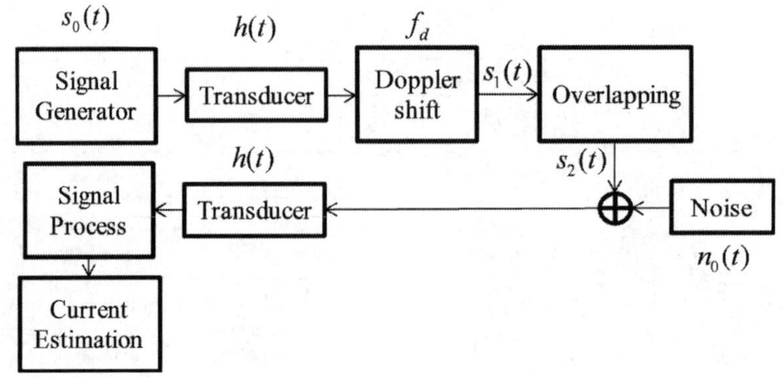

Fig. 7. Schematic gram of simulation.

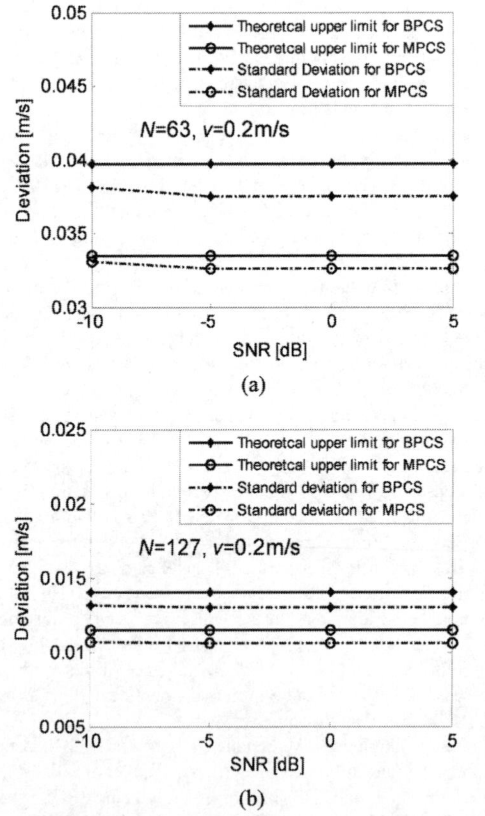

Fig. 8. Theoretical upper limits of deviations and measurement deviations of the optimized MPCSs and the reference BPCSs.

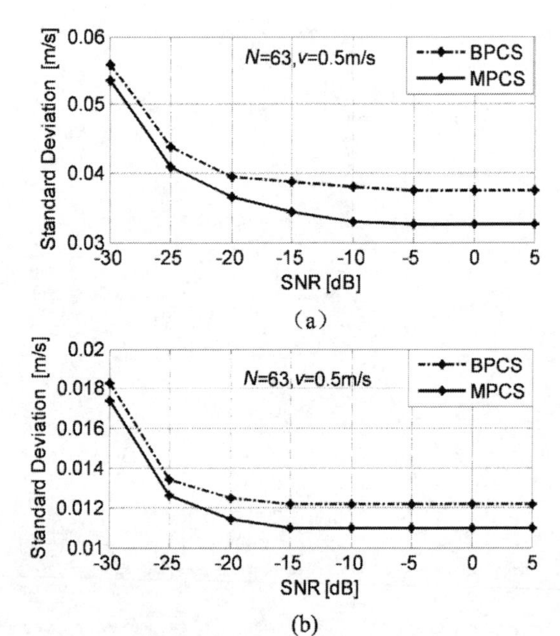

Fig. 9. Standard deviations of current estimation when the number of SBPs is (a) two and (b) four, respectively.

that the deviation of the pulse-pair method is essentially decided by the energy of SBPAF when the duration and energy of the single transmitted pulse are fixed. In addition, in Fig. 10(b), the deviation gap between the two curves is narrower than that in Fig. 10(a). It is because the ambient noise is the main factor to decide the deviation in the low SNR situation. However, when the number of SBPs is four, the deviation gap in Fig. 11(b) does not narrow so much as Fig. 10(b). The reason is that when four SBPs are used, both the correlation value at the measurement lag and the energy of transmitted signal increase, and the influence of the ambient noise for the same SNR is reduced, compared with the situation where two SBPs are used.

V. Improvement of Using the Second Harmonic Signal in BBADCP

When the transmitted power is beyond a certain range, the nonlinear effects are significant, thus the second harmonic signal can be used to estimate the velocity in conjunction with the fundamental signal. Prieur and Hansen [9] have theoretically analyzed above what energy level and at what valid range the second harmonic signal can be used for estimating the current velocity. The measurement deviation can further

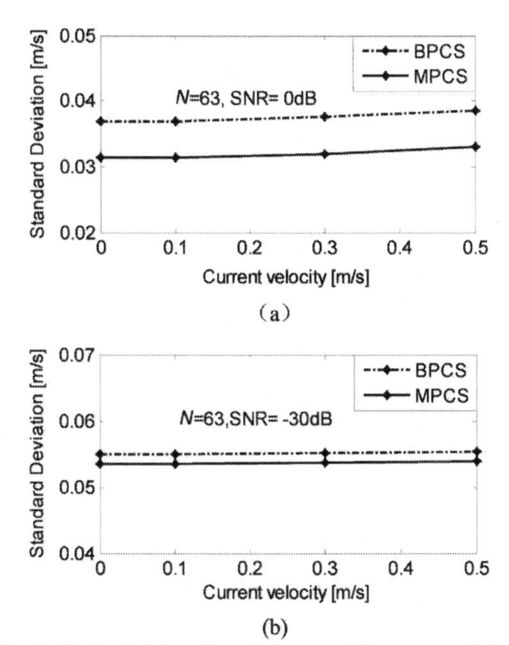

Fig. 10. Standard deviations of current estimation at different current velocities when the number of SBPs is two.

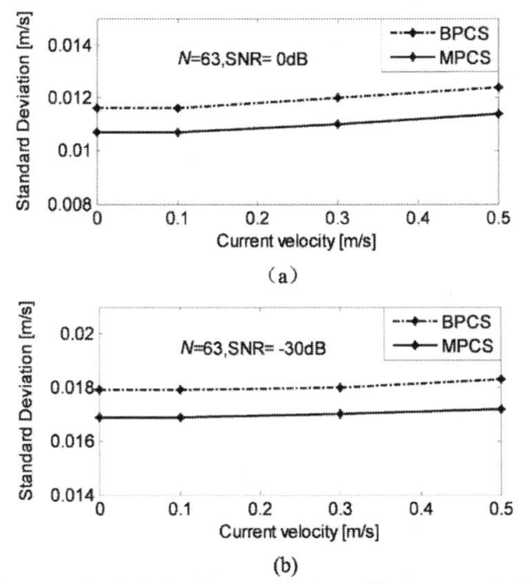

Fig. 11. Standard deviations of current estimation at different current velocities when the number of SBPs is four.

decrease by using the second harmonic signal in BBADCP. This section discusses the improvement of using the second harmonic signal of the optimized MPCSs in BBADCP. To simplify the analysis, the number of SBPs is set to two.

It has been pointed out that the second harmonic signals of the BPCSs are not viable for the velocity estimation in BBADCP [9], because the BPCSs modulating the phases of the code elements with 0° and 180° are equivalent to no phase coding for

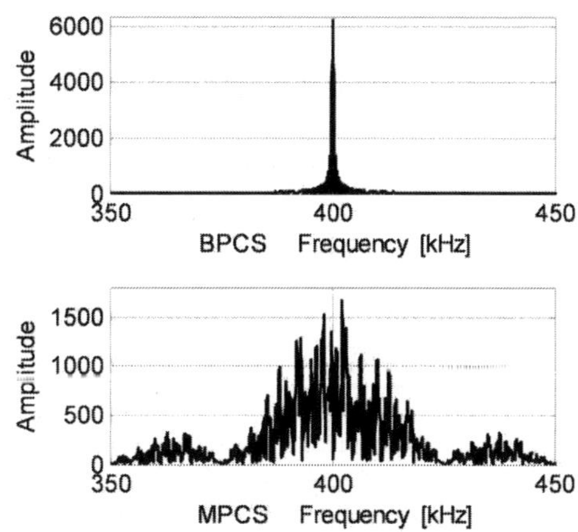

Fig. 12. Second harmonic spectra of the reference BPCS and the optimized MPCS.

their second harmonic signals and the second harmonic signals of the BPCSs become narrowband. However, the second harmonic signal of the proposed optimized MPCS is still broadband because of its multiple values of the phases. The second harmonic spectra of the reference BPCS and the optimized MPCS (N being 63) used in Section IV are depicted in Fig. 12.

The second harmonic signal of the reference BPCS is a sine pulse of length $2T$ and can be processed as a narrowband acoustical Doppler current profiler (NBADCP). The velocity measurement deviation of the second harmonic signal of the reference BPCS, denoted by σ_{B2}, can be calculated with the deviation of the NBADCP [4], [9], which is given by

$$\sigma_{B2} = \frac{c}{4\pi(2f_0)2T}\frac{1}{\sqrt{N_v I_v(\mathrm{SNR})}} \tag{38}$$

where N_v is the number of independent depth cells considered during the processing time, and $I_v(\mathrm{SNR})$ is

$$I_v(\mathrm{SNR}) = \frac{2}{\sqrt{\ln 2}}\int_{-\sqrt{\ln 2}}^{\sqrt{\ln 2}} \frac{x^2 dx}{\left[1+\dfrac{\exp(x^2)}{(\mathrm{SNR}\sqrt{2})}\right]^2}. \tag{39}$$

The SNR here is defined as the ratio of the energy of the reverberated signal to the noise energy where the noise bandwidth is assumed to be the same as the transmitted signal bandwidth. In the case of the reference BPCS, $N_v \approx 1, T = 2.52$ ms, and SNR is supposed to be 20 dB, then we have $I_v(\mathrm{SNR}) \approx 1$ and

$$\sigma_{B2} \approx 5.92 \text{ cm/s}. \tag{40}$$

The simulations in Section IV show that the measurement deviation corresponding to the fundamental signal of the reference BPCS, denoted by σ_{B1}, in high SNR is

$$\sigma_{B1} \approx 3.70 \text{ cm/s}. \tag{41}$$

The total deviation combining the fundamental and harmonic signals of the reference BPCS, denoted by σ_B, is

$$\sigma_B = \frac{\sqrt{\sigma_{B1}^2+\sigma_{B2}^2}}{2} \approx 3.49 \text{ cm/s} \tag{42}$$

because the second harmonic signal of the optimized MPCS is still broadband and can be processed by the pulse-pair method. In the situation of high SNR, the influence of SNR on the measurement deviation is minor. Applying the similar simulation method in Section IV, the measurement deviation corresponding to the second harmonic signal of the optimized MPCS is

$$\sigma_{M2} \approx 2.26 \text{ cm/s}. \tag{43}$$

The measurement deviation corresponding to the fundamental signal of the optimized MPCS, denoted by σ_{M1}, is approximately 3.20 cm/s under the high SNR shown in Section IV. So the total deviation combining the fundamental and harmonic signals of the optimized MPCS, denoted by σ_M, is

$$\sigma_M = \frac{\sqrt{\sigma_{M1}^2 + \sigma_{M2}^2}}{2} \approx 1.96 \text{ cm/s}. \tag{44}$$

From (41) and (43), we have

$$\sigma_M \approx 0.56\sigma_B. \tag{45}$$

Therefore, if the second harmonic signal is considered in the high SNR situation, the measurement deviation caused by using the optimized MPCS will be 44% smaller than that generated by using the reference BPCS.

VI. Conclusion and Discussion

This paper theoretically analyzes the influence of the self-noise for BBADCPs when using the pulse-pair method and gives a theoretical upper limit of the measurement deviation caused by the self-noise. It can be concluded that for BBADCPs, without considering the ambient noise, the measurement deviation of using the pulse-pair method is essentially determined by the energy of SPBAF when the duration and energy of the broadband transmitted signal are fixed. According to the upper limit, a design method based on MPCSs is proposed for the optimization of transmitted signals. The proposed design method takes the energy of SBPAF as the objective function and the phases of the code elements as the optimization variables. The simulations show that the optimized MPCSs prove to perform better in many aspects than the reference BPCS.

The comparisons of using the optimized MPCSs with using the reference BPCS at different velocities and SNRs show that the measurement deviations of the optimized MPCSs are all lower than that of the reference BPCS.

The comparison of using the optimized MPCSs of different length with using the BPCS shows that if the length of the MPCS is longer, the optimized MPCS will manifest more distinct merits over the BPCS.

Moreover, if the second harmonic signal can be employed to estimate the current velocity in conjunction with the fundamental signal, the measurement deviation of using the optimized MPCS will be significantly reduced further compared with that of using the reference BPCS. The advantage of the optimized MPCSs is more obvious when the second harmonic signal is taken into consideration.

At present, the implementation of transmitting the optimized MPCSs has become realizable owing to the development of broadband transducer and digital waveform generation techniques. The proposed waveform design method should have no more limitations than the design of the BPCSs. However, the MPCSs may have the extremity in the reduction of the measurement deviation. The optimization design based on other waveforms such as nonlinear frequency-modulated signals may help to achieve better performance for BBADCPs. In the future work, we will try to use other waveforms to design the transmitted signals in BBADCPs.

In this paper, we only evaluate the influence of the self-noise indicating that the influence of the self-noise can be decreased by designing a better transmitted broadband waveform. The influence of other noise sources such as the actual ambient noise and flow fluctuation has not been discussed. Although the noise sources other than the self-noise will have impact on the measurement deviation of using the MPCSs, as they do when using other broadband coded signals, they do not affect the conclusion of this paper which concerns the self-noise.

Appendix A
Relationship Between $\text{var}[\hat{\theta}]$ and the Energy of SBPAF

Let $\hat{\theta}$ be the phase of $R_0(T)$, then

$$\tan\hat{\theta} = \frac{y}{x} \tag{46}$$

where x and y are the real and imaginary parts of $R_0(T)$, respectively. Then, we will use the disturbance analysis method given in [15]. Using the 2-D Taylor expansion of $\arg[R_0(T)]$, the variance of $\hat{\theta}$ can be expressed as

$$\begin{aligned}&\text{var}[\hat{\theta}]\\&\approx \text{var}[x]\left[\frac{\partial \arg[R_0(T)]}{\partial x}|_{\bar{x},\bar{y}}\right]^2 + \text{var}[y]\left[\frac{\partial \arg[R_0(T)]}{\partial y}|_{\bar{x},\bar{y}}\right]^2\\&\quad + 2\text{cov}[x,y]\left[\left(\frac{\partial \arg[R_0(T)]}{\partial x}|_{\bar{x},\bar{y}}\right)\left(\frac{\partial \arg[R_0(T)]}{\partial y}|_{\bar{x},\bar{y}}\right)\right].\end{aligned} \tag{47}$$

After the derivation, (47) can be simplified [15] as

$$\text{var}[\hat{\theta}] \approx E\left[\frac{1}{|m_{R_0}|^4}\{\bar{x}^2\text{var}(y) + \bar{y}^2\text{var}(x) - 2\bar{x}\bar{y}\text{cov}(x,y)\}\right] \tag{48}$$

where $\bar{x}$ and $\bar{y}$ represent the mean of x and y, respectively, and

$$m_{R_0} = E[R_0(T)].$$

Assuming that x and y can be described by their disturbance, then

$$x = \bar{x} + \delta_x, \quad y = \bar{y} + \delta_y. \tag{49}$$

Then, (48) can be rewritten as

$$\begin{aligned}\text{var}[\hat{\theta}] &\approx E\left[\frac{1}{|m_{R_0}|^4}\{\bar{x}^2\delta_y^2 - 2\bar{x}\bar{y}\delta_x\delta_y + \bar{y}^2\delta_x^2\}\right]\\&= E\left\{\frac{1}{|m_{R_0}|^4}\text{Im}^2[R_0^*(T)\delta]\right\}\end{aligned} \tag{50}$$

where $\delta = \delta_x + j\delta_y$. Substituting δ by $R_0(T) - m_{R_0}$, (50) is simplified by

$$\begin{aligned} \text{var}[\hat{\theta}] &\approx E\left\{ \text{Im}^2 \left(\frac{m_{R_0}^* R_0(T)}{m_{R_0}^* m_{R_0}} \right) \right\} \\ &= E\left\{ \text{Im}^2 \left(\frac{R_0(T)}{m_{R_0}} \right) \right\}. \end{aligned} \tag{51}$$

Based on (11), the expected value of $R_1(T)$ is given by

$$\begin{aligned} E[R_1(T)] &= e^{j2\pi f_d T} E \\ &\quad \times \left[\int_{t_0}^{t_0+T} \sum_i A_i^2 a(t-T-\tau_i) a^*(t-T-\tau_i) dt \right] \\ &= e^{j2\pi f_d T} \sum_i E[A_i^2] E \\ &\quad \times \left[\int_{t_0}^{t_0+T} a(t-T-\tau_i) a^*(t-T-\tau_i) dt \right]. \end{aligned} \tag{52}$$

Because τ_i is uniformly distributed, therefore

$$E\left[\int_{t_0}^{t_0+T} a(t-T-\tau_i) a^*(t-T-\tau_i) dt \right] = \frac{E_a}{2}. \tag{53}$$

Assume that there are N_1 scatterers distributing between t_0 and $t_0 + T$. The expected value of $R_1(T)$ can be expressed by

$$E[R_1(T)] = A_{R_1} e^{j2\pi f_d T} \tag{54}$$

where A_{R_1} is the amplitude of $E[R_1(T)]$. And then

$$A_{R_1} = N_1 E[A_i^2] \frac{E_a}{2}. \tag{55}$$

E_a is the energy of SBP, and the expression of E_a is

$$E_a = \int_{t_0}^{t_0+T} a^*(t) a(t) dt. \tag{56}$$

Since A_i, A_m, and τ'_{im} are independent variables, based on (12), the expected value of $R_2(T)$ is given by

$$\begin{aligned} E[R_2(T)] &= e^{j2\pi f_d T} \\ &\quad \times \sum_{i \neq m} \sum_m E[A_i] E[A_m] E[e^{-j2\pi f_d \tau'_{im}} R_3(\tau'_{im})] \end{aligned} \tag{57}$$

where

$$E[e^{-j2\pi f_d \tau'_{im}} R_3(\tau'_{im})] = \int e^{-j2\pi f_d \tau'_{im}} R_3(\tau'_{im}) f(\tau'_{im}) d\tau'_{im}. \tag{58}$$

As shown in Section II

$$\begin{aligned} R_3(\tau'_{im}) &= \int_{-\tau'_i}^{T-\tau'_i} a(t) a^*(t+\tau'_{im}) dt \\ R_4(\tau'_{im}) &= \int_0^T a(t) a^*(t+\tau'_{im}) dt. \end{aligned}$$

$R_4(\tau'_{im})$ is the complex conjugate of the autocorrelation function of $a(t)$. As the scatterers are assumed to be uniformly distributed in water, τ'_i and τ'_m will uniformly distribute between $-T$ and T. Therefore, $\tau'_{im} \in [-2T, 2T]$. The probability distribution function of τ'_{im} has been given by (17).

When $\tau'_{im} \in [T, 2T]$ and $\tau'_{im} \in [-2T, -T]$, because the duration of $a(t)$ is T, then

$$R_3(\tau'_{im}) = R_4(\tau'_{im}) = 0. \tag{59}$$

When $\tau'_{im} \in [-T, T]$, the deducing process is as follows. For $\tau'_i > 0$ and $\tau'_m < 0$, we have

$$T - \tau'_i > T - \tau'_{im} > 0.$$

$R_3(\tau'_{im})$ can be written by

$$\begin{aligned} R_3(\tau'_{im}) &= \int_0^{T-\tau'_i} a(t) a^*(t+\tau'_{im}) dt \\ &= \int_0^{T-\tau'_m} a(t) a^*(t+\tau'_{im}) dt \\ &= \int_0^T a(t) a^*(t+\tau'_{im}) dt. \end{aligned} \tag{60}$$

From (60), it can be concluded that for $\tau'_i > 0$ and $\tau'_m < 0$, we have

$$R_3(\tau'_{im}) = R_4(\tau'_{im}).$$

Using the similar analysis method, for $\tau'_i < 0$ and $\tau'_m > 0$, we still have

$$R_3(\tau'_{im}) = R_4(\tau'_{im}).$$

Therefore, when the sign of τ'_i is different with that of τ'_m, then

$$R_3(\tau'_{im}) = R_4(\tau'_{im}).$$

When $\tau'_i > 0$ and $\tau'_m > 0$, then

$$0 < T - \tau'_i < T - \tau'_{im}.$$

In this situation, $R_4(\tau'_{im})$ can be written as

$$\begin{aligned} &R_4(\tau'_{im}) \\ &= \int_0^{T-\tau'_i} a(t) a^*(t+\tau'_{im}) dt + \int_{T-\tau'_i}^{T-\tau'_{im}} a(t) a^*(t+\tau'_{im}) dt \\ &= \int_0^{T-\tau'_i} a(t) a^*(t+\tau'_{im}) dt + \int_{T-\tau'_i}^{T} a(t) a^*(t+\tau'_{im}) dt \\ &= R_3(\tau'_{im}) + \int_{T-\tau'_i}^{T} a(t) a^*(t+\tau'_{im}). \end{aligned} \tag{61}$$

When $\tau'_i < 0$ and $\tau'_m < 0$, then

$$0 < -\tau'_i < -\tau'_{im}.$$

In this situation, $R_4(\tau'_{im})$ can be written as

$$\begin{aligned} &R_4(\tau'_{im}) \\ &= \int_{-\tau'_i}^{-\tau'_{im}} a(t) a^*(t+\tau'_{im}) dt + \int_{\tau'_{im}}^{T-\tau'_{im}} a(t) a^*(t+\tau'_{im}) dt \\ &= R_3(\tau'_{im}) + \int_0^{-\tau'_{im}} a(t) a^*(t+\tau'_{im}) dt. \end{aligned} \tag{62}$$

For $|E[e^{j2\pi f_d \tau'_{im}} R_3(\tau'_{im})]|$, we have

$$|E[e^{j2\pi f_d \tau'_{im}} R_3(\tau'_{im})]| = \left|\int e^{j2\pi f_d \tau'_{im}} R_3(\tau'_{im}) f(\tau'_{im}) d\tau'_{im}\right| \le \left|\int R_3(\tau'_{im}) f(\tau'_{im}) d\tau'_{im}\right| \le \frac{1}{2T}\int |R_3(\tau'_{im})| d\tau'_{im}. \tag{63}$$

Because $a(t)$ used by BBADCP is a large time-bandwidth product signal, and τ'_i and τ'_m have different signs, $R_3(\tau'_{im})$ is equal to $R_4(\tau'_{im})$, which is the autocorrelation function of $a(t)$. The value of $|R_3(\tau'_{im})|$ at a lag different from 0 is much smaller than $R_4(0)$ and $R_4(0) = E_a$. Even when τ'_i and τ'_m have the same signs, the value of $|R_3(\tau'_{im})|$ at a lag different from 0 is still much smaller than $R_4(0)$.

Therefore

$$\left|E[e^{-j2\pi f_d \tau'_{im}} R_3(\tau'_{im})]\right| \le \frac{1}{2T}\int |R_3(\tau'_{im})| d\tau'_{im} \ll E_a. \tag{64}$$

According to (57), the amplitude A_{R_2} of the expected value of $R_2(T)$ is

$$A_{R_2} = \sum_{i\neq m}\sum_{m} E[A_i]E[A_m]\left|E[e^{-j2\pi f_d \tau'_{im}} R_3(\tau'_{im})]\right| \ll N_1(N_1-1)E[A_i]^2 E_a. \tag{65}$$

Based on (55), (57), and (65), we conclude that

$$A_{R_2} << A_{R_1}. \tag{66}$$

According to (66) and (15) and considering that A_{R_1} and $R_1(T)/e^{j2\pi f_d T}$ are real, then

$$m_{R_0} = E[R_1(T)] + E[R_2(T)] = e^{j2\pi f_d T} A_{R_1} + E[R_2(T)]$$

Equation (51) can be further rewritten as

$$\mathrm{var}[\hat{\theta}] \approx E\left\{\mathrm{Im}^2\left(\frac{R_1(T)+R_2(T)}{e^{j2\pi f_d T}(A_{R_1}+A_{R_2})}\right)\right\} \approx E\left\{\frac{\mathrm{Im}^2\left[e^{-j2\pi f_d T} R_2(T)\right]}{A_{R_1}^2}\right\} \approx \frac{E\left\{\left[\sum\limits_{i\neq m}\sum\limits_{m} A_i A_m \mathrm{Im}\left[e^{-j2\pi f_d \tau'_{im}} R_3(\tau'_{im})\right]\right]^2\right\}}{A_{R_1}^2}. \tag{67}$$

Let

$$R_3(\tau'_{im}) = |R_3(\tau'_{im})| e^{j\varphi(\tau'_{im})}$$

then (67) becomes (68), shown at the bottom of the page. Let

$$R_{im} = A_i A_m |R_3(\tau'_{im})| \sin[-2\pi f_d \tau'_{im} + \varphi(\tau'_{im}).]$$

The variance of R_{im} is given by

$$\mathrm{var}(R_{im}) = \frac{1}{2}E\left[A_i^2\right]E\left[A_m^2\right]\left\{\int |R_3(\tau'_{im})|^2 f(\tau'_{im}) d\tau'_{im} - \int |R_3(\tau'_{im})|^2 f(\tau'_{im}) \cos\left[-4\pi f_d \tau'_{im} + 2\varphi(\tau'_{im})\right] d\tau'_{im}\right\}. \tag{69}$$

Let

$$R'_3 = \sum_{i\neq m}\sum_{m} R_{im} = \sum_{i\neq m}\sum_{m} A_i A_m |R_3(\tau'_{im})| \sin\left[-2\pi f_d \tau'_{im} + \varphi(\tau'_{im})\right]. \tag{70}$$

According to the central limited theorem, the variance of R'_3 is

$$\mathrm{var}\left[R'_3\right] = N_1(N_1-1)\mathrm{var}(R_{im}). \tag{71}$$

From (68), (69), and (71), we conclude that

$$\mathrm{var}[\hat{\theta}] \approx \frac{N_1(N_1-1)\mathrm{var}\left(R_{im}\right)}{N_1^2\left[E\left[A_i^2\right]\right]^2\left(\frac{E_a}{2}\right)^2}. \tag{72}$$

When N_1 is large enough, $\mathrm{var}[\hat{\theta}]$ can be expressed as (73), shown at the bottom of the page. According to

$$\mathrm{var}[\hat{\theta}] \approx \frac{E\left\{\left[\sum\limits_{i\neq m}\sum\limits_{m} A_i A_m |R_3(\tau'_{im})| \sin\left[-2\pi f_d \tau'_{im} + \varphi(\tau'_{im})\right]\right]^2\right\}}{A_{R_1}^2}. \tag{68}$$

$$\begin{aligned}\mathrm{var}[\hat{\theta}] &\approx \frac{\mathrm{var}(R_{im})}{\left[E\left[A_i^2\right]\right]^2\left(\frac{E_a}{2}\right)^2} \\ &= \frac{\frac{1}{2}\left\{\int |R_3(\tau'_{im})|^2 f(\tau'_{im}) d\tau'_{im} - \int |R_3(\tau'_{im})|^2 f(\tau'_{im}) \cos\left[-4\pi f_d \tau'_{im} + 2\varphi(\tau'_{im})\right] d\tau'_{im}\right\}}{\left(\frac{E_a}{2}\right)^2} \\ &\le \frac{4\left\{\int |R_3(\tau'_{im})|^2 f(\tau'_{im}) d\tau'_{im}\right\}}{E_a^2} \le \frac{4\left\{\frac{1}{2T}\int |R_3(\tau'_{im})|^2 d\tau'_{im}\right\}}{E_a^2} = \frac{2\int |R_3(\tau'_{im})|^2 d\tau'_{im}}{TE_a^2}.\end{aligned} \tag{73}$$

the relationship (76)

$$\int |R_3(\tau'_{im})|^2\, d\tau'_{im} \le \int |R_4(\tau'_{im})|^2\, d\tau'_{im}$$

proven in Appendix B. Then, (73) can be rewritten as

$$\mathrm{var}[\hat{\theta}] \le \frac{2\int |R_4(\tau'_{im})|^2\, d\tau'_{im}}{TE_a^2}. \tag{74}$$

Based on (2) and (19)

$$\int |R_4(\tau'_{im})|^2 d\tau'_{im} = \int |R(\tau)|^2 d.\tau$$

Therefore, (74) can be simplified as

$$\mathrm{var}[\hat{\theta}] \le \frac{2\int |R(\tau)|^2\, d\tau}{TE_a^2}. \tag{75}$$

APPENDIX B
RELATIONSHIP BETWEEN $R_3(\tau'_{im})$ AND $R_4(\tau'_{im})$

In this Appendix, we will prove that

$$\int |R_3(\tau'_{im})|^2\, d\tau'_{im} \int^T \le \int |R_4(\tau'_{im})|^2\, d\tau'_{im}. \tag{76}$$

As shown in Appendix A, because τ'_i and τ'_m have different signs, then

$$R_3(\tau'_{im}) = R_4(\tau'_{im}).$$

In this situation

$$\int_{-T}^{T} |R_3(\tau'_{im})|^2 d\tau'_{im} = \int_{-T}^{T} |R_4(\tau'_{im})|^2 d\tau'_{im}.$$

When $\tau'_i > 0$ and $\tau'_m > 0$, then

$$R_4(\tau'_{im}) = R_3(\tau'_{im}) + \int_{T-\tau'_i}^{T} a(t)a^*(t+\tau'_{im}). \tag{77}$$

Based on (19)

$$\begin{aligned}
&\int_{-T}^{T} |R_4(\tau'_{im})|^2 d\tau'_{im} \\
&= \int_{-T}^{T} R_4^*(\tau'_{im}) \cdot R_4(\tau'_{im}) d\tau'_{im} \\
&= \int_{-T}^{T} \left[\int_0^{T-\tau'_i} a(t_1)a^*(t_1+\tau'_{im})dt_2 \right. \\
&\quad \left. + \int_{T-\tau'_i}^{T} a(t_1)a^*(t_1+\tau'_{im})dt_1\right]^* \\
&\quad \times \left[\int_0^{T-\tau'_i} a(t_2)a^*(t_2+\tau'_i)dt \right. \\
&\quad \left. + \int_{T-\tau'_i}^{T} a(t_2)a^*(t_2+\tau'_{im})dt_2\right] d\tau'_{im} \\
&= \int_{-T}^{T} \left[\int_0^{T-\tau'_i} a(t_1)a^*(t_1+\tau'_{im})dt_1\right]^* \\
&\quad \times \left[\int_0^{T-\tau'_i} a(t_2)a^*(t_2+\tau'_i)dt_2\right] d\tau'_{im} \\
&\quad + \int_{-T}^{T} \left[\int_0^{T-\tau'_i} a(t_1)a^*(t_1+\tau'_{im})dt_1\right]^* \\
&\quad \times \left[\int_{T-\tau'_i}^{T} a(t_2)a^*(t_2+\tau'_i)dt_2\right] d\tau'_{im} \\
&\quad + \int_{-T}^{T} \left[\int_{T-\tau'_i}^{T} a(t_1)a^*(t_1+\tau'_{im})dt_1\right]^* \\
&\quad \times \left[\int_0^{T-\tau'_i} a(t_2)a^*(t_2+\tau'_i)dt_2\right] d\tau'_{im} \\
&\quad + \int_{-T}^{T} \left[\int_{T-\tau'_i}^{T} a(t_1)a^*(t_1+\tau'_{im})dt_1\right]^* \\
&\quad \times \left[\int_{T-\tau'_i}^{T} a(t_2)a^*(t_2+\tau'_i)dt_2\right] d\tau'_{im} \\
&= \int_{-T}^{T} |R_3(\tau'_{im})|^2 d\tau'_{im} + B_0 + B_1 + B_2
\end{aligned} \tag{78}$$

where

$$B_0 = \int_{-T}^{T} \left[\int_0^{T-\tau'_i} a(t_1)a^*(t_1+\tau'_{im})dt_1\right]^* \times \left[\int_{T-\tau'_i}^{T} a(t_2)a^*(t_2+\tau'_i)dt_2\right] d\tau'_{im} \tag{79}$$

$$B_1 = \int_{-T}^{T} \left[\int_{T-\tau'_i}^{T} a(t_1)a^*(t_1+\tau'_{im})dt_1\right]^* \times \left[\int_0^{T-\tau'_i} a(t_2)a^*(t_2+\tau'_i)dt_2\right] d\tau'_{im} \tag{80}$$

$$\begin{aligned}
B_2 &= \int_{-T}^{T} \left[\int_{T-\tau'_i}^{T} a(t_1)a^*(t_1+\tau'_{im})dt_1\right]^* \times \left[\int_{T-\tau'_i}^{T} a(t_2)a^*(t_2+\tau'_i)dt_2\right] d\tau'_{im} \\
&= \int_{-T}^{T} \left|\int_{T-\tau'_i}^{T} a(t)a^*(t+\tau'_{im})dt\right|^2 d\tau'_{im}.
\end{aligned} \tag{81}$$

Equation (79) can be further expressed as

$$
\begin{aligned}
B_0 &= \int_{-T}^{T}\left[\int_{0}^{T-\tau_i'} a(t_1)a^*(t_1+\tau_{im}')dt_1\right]^* \\
&\quad\times\left[\int_{T-\tau_i'}^{T} a(t_2)a^*(t_2+\tau_{im}')dt_2\right]d\tau_{im}' \\
&= \int_{0}^{T-\tau_i'}\int_{T-\tau_i'}^{T} a^*(t_1)a(t_2) \\
&\quad\times\left[\int_{-T}^{T} a(t_1+\tau_{im}')a^*(t_2+\tau_{im}')d\tau_{im}'\right]dt_2dt_1 \\
&= \int_{0}^{T-\tau_i'}\int_{T-\tau_i'}^{T} a^*(t_1)a(t_2)C_0dt_2dt_1 \qquad (82)
\end{aligned}
$$

where

$$
C_0 = \int_{-T}^{T} a(t_1+\tau_{im}')a^*(t_2+\tau_{im}')d\tau_{im}'. \qquad (83)
$$

For (83), $t_1 \in [0, T-\tau_i')$ and $t_2 \in [T-\tau_i', T]$. And $a(t)$ is a large time and band product signal and the autocorrelation function of should have a good performance. In this situation, C_0 is close to 0 for $t_1 \in [0, T-\tau_i')$ and $t_2 \in [T-\tau_i', T]$. Thus, B_0 is close to 0. B_1 has the same property as B_0. Based on the above analysis, (78) can be rewritten as

$$
\int_{-T}^{T} |R_4(\tau_{im}')|^2 d\tau_{im}' \approx \int_{-T}^{T} |R_3(\tau_{im}')|^2 d\tau_{im}' + B_2. \qquad (84)
$$

According to (81), $B_2 \geq 0$. Therefore, for $\tau_i' > 0$ and $\tau_m' > 0$, (76) is provable. Using the similar method, for $\tau_i' < 0$ and $\tau_m' < 0$, (76) is still provable.

In summary, for arbitrary values of τ_i' and τ_m', we have

$$
\int |R_3(\tau_{im}')|^2 d\tau_{im}' \int^{T} \leq \int |R_4(\tau_{im}')|^2 d\tau_{im}'.
$$

ACKNOWLEDGMENT

The authors would like to thank Prof. R. Wang for helpful conversations and Dr. T. Wang for critical reviews. There are good graphical illustrations [10, Figs. 2 and 5] and discussions [1, Sec. V] about what the signal distance resolution presented in this paper means for the deviation of the measurement velocity. The authors gratefully acknowledge the previous work and discussions with Brumley *et al.* [1] and Wanis *et al.* [10].

REFERENCES

[1] B. H. Brumley, R. G. Cabrera, K. L. Denies, and E. A. Terray, "Performance of a broad-band acoustic Doppler current profiler," *IEEE J. Ocean. Eng.*, vol. 16, no. 4, pp. 402–407, Oct. 1991.

[2] L. Zedel, "Performance of a single-beam pulse-to-pulse coherent Doppler profiler," *IEEE J. Ocean. Eng.*, vol. 21, no. 4, pp. 290–297, Jul. 1996.

[3] S. S. Abeysekera, "Performance of pulse-pair method of Doppler estimation," *IEEE Trans. Aerosp. Electron. Syst.*, vol. 34, no. 2, pp. 520–531, Apr. 1998.

[4] Y. Doisy, "Theoretical accuracy of Doppler navigation sonars and acoustic Doppler current profilers," *IEEE J. Ocean. Eng.*, vol. 29, no. 2, pp. 430–441, Apr. 2004.

[5] D. S. Zrnic, "Spectral moment estimates from correlated pulse-pairs," *IEEE Trans. Aerosp. Electron. Systm.*, vol. AES-13, no. 4, pp. 344–354, Jul. 1977.

[6] K. S. Miller and M. M. Rochwarger, "A covariance approach to spectral moment estimation," *IEEE Trans. Inf. Theory.*, vol. IT-18, no. 5, pp. 588–596, Sep. 1972.

[7] J. A. Jensen *et al.*, "Ultrasound research scanner for real-time synthetic aperture data acquisition," *IEEE Trans. Ultrason. Ferroelectr. Freq. Control.*, vol. 52, no. 5, pp. 881–891, May 2005.

[8] M. C. Hemmsen *et al.*, "Implementation of a versatile research data acquisition system using a commercially available medical ultrasound scanner," *IEEE Trans. Ultrason. Ferroelectr. Freq. Control.*, vol. 59, no. 7, pp. 1487–1499, Jul. 2011.

[9] F. Prieur and R. E. Hansen, "Theoretical improvements when using the second harmonic signal in acoustic Doppler current profilers," *IEEE J. Ocean. Eng.*, vol. 38, no. 2, pp. 275–284, Apr. 2013.

[10] P. Wanis, B. Brumley, J. Gast, and D. Symonds, "Sources of measurement variance in broadband acoustic Doppler current profilers," in *Proc. MTS/IEEE OCEANS Conf.*, Seattle, WA, USA, Sep. 2010, DOI: 10.1109/OCEANS.2010.5664327.

[11] R. Cabrera, K. Deines, B. Brumley, and E. Terray, "Development of a practical coherent acoustic Doppler current profiler," in *Proc. IEEE OCEANS Conf.*, Halifax, NS, Canada, 1987, pp. 93–97.

[12] P. M. Woodward, *Probability and Information Theory With Application to Radar*, 2nd ed. New York, NY, USA: Pergamon, 1964, ch. 3.

[13] A. M. Richards, *Fundamentals of Radar Signal Processing*. New York, NY, USA: Tata/McGraw-Hill, 2005, ch. 4.

[14] Math Works Inc., "Optimization Toolbox—Matlab Version 8.3.0.532," User's Guide, 2014.

[15] V. Bringi and V. Chandrasekar, *Polarimetric Doppler Weather Radar: Principles and Applications*, 1st ed. Cambridge, U.K.: Cambridge Univ. Press, 2001, ch. 5.

[16] B. Lamboul, J. Bennett, T. Anderson, and W. McDicken, "Basic considerations in the use of coded excitation for color flow imaging applications," *IEEE Trans. Ultrason. Ferroelectr. Freq. Control*, vol. 56, no. 4, pp. 727–737, Apr. 2009.

Investigations of thickness-shear mode elastic constant and damping of shunted piezoelectric materials with a coupling resonator①②

Ji-Ying Hu(胡吉英)[1], Zhao-Hui Li(李朝晖)[1], Yang Sun(孙阳)[1,3], and Qi-Hu Li(李启虎)[2]

[1] Department of Electronics, Peking University, Beijing 100871, China
[2] Advanced Technology Institute, Peking University, Beijing 100871, China
[3] College of Science, Beijing Forestry University, Beijing 100083, China

(Received 5 July 2016, revised manuscript received 2 September 2016; published online 25 October 2016)

Shear-mode piezoelectric materials have been widely used to shunt the damping of vibrations where utilizing surface or interface shear stresses. The thick-shear mode (TSM) elastic constant and the mechanical loss factor can change correspondingly when piezoelectric materials are shunted to different electrical circuits. This phenomenon makes it possible to control the performance of a shear-mode piezoelectric damping system through designing the shunt circuit. However, due to the difficulties in directly measuring the TSM elastic constant and the mechanical loss factor of piezoelectric materials, the relationships between those parameters and the shunt circuits have rarely been investigated. In this paper, a coupling TSM electro–mechanical resonant system is proposed to indirectly measure the variations of the TSM elastic constant and the mechanical loss factor of piezoelectric materials. The main idea is to transform the variations of the TSM elastic constant and the mechanical loss factor into the changes of the easily observed resonant frequency and electrical quality factor of the coupling electro–mechanical resonator. Based on this model, the formular relationships are set up theoretically with Mason equivalent circuit method and they are validated with finite element (FE) analyses. Finally, a prototype of the coupling electro–mechanical resonator is fabricated with two shear-mode PZT5A plates to investigate the TSM elastic constants and the mechanical loss factors of different circuit-shunted cases of the piezoelectric plate. Both the resonant frequency shifts and the bandwidth changes observed in experiments are in good consistence with the theoretical and FE analyses under the same shunt conditions. The proposed coupling resonator and the obtained relationships are validated with but not limited to PZT5A.

Keywords: piezoelectric materials, shunt damping, shear mode, elastic constant

PACS: 77.90.+k, 62.20.−x, 62.20.de, 81.40.Jj **DOI:** 10.1088/1674-1056/25/12/127701

1. Introduction

The piezoelectric shunt damping technology has been the focus of many research fields, especially in the fields of vibration control,[1–7] sound absorption,[8–11] and noise elimination,[12,13] wave propagation control,[14] where the shunt damping is usually realized with a shunted piezoelectric patch working in one of three modes:[2] the transverse mode, the longitudinal mode, and the shear mode.

As the shear mode has the advantage over the other two modes by using the surface or interface stresses, it has increasingly attracted extensive research interests in surface vibration suppression with shunt-damping technologies.[15–20] For example, Benjeddou and Ranger investigated the use of shear-mode piezoelectric ceramics for shunted passive vibration damping and found that the added resistive shear-mode shunt damping is twelve times more than the classical resistive extension-mode shunt damping.[16] Karim Blanzé designed a bolted joint for reducing the vibration of a structure by taking advantage of the shear mode to utilize the surface or interface stress.[18] dos Santos and Trindade studied active–passive piezoelectric networks (APPN) for extension and shear modes, and they find that the shunt-damping effect of the shear mode is better than that of the extension mode.[19]

Among these applications, the thick-shear mode (TSM) elastic constant c_{55} and the mechanical loss factor are the most essential parameters that determine the performance of the vibration damping system by utilizing the shear mode of piezoelectric materials. Therefore, it is of great research value to investigate the relationships between the TSM elastic constant c_{55} as well as the mechanical loss factor and different shunt circuits, which are essential to the design or control of a shunt-damping system by using the shear mode of piezoelectric material.

In practice, the direct measurements of the shear moduli of elastic or viscoelastic materials are of great complexity and difficulty.[21] Thus, indirect methods utilizing the velocity of sound,[22] the relationship between the resonant and the anti-resonant frequencies of a regular-cut shear mode plate are usually employed to estimate the shear moduli of piezoelectric materials.[23] Some TSM piezoelectric or quartz crystal resonators have also been developed to implement their quantitative determinations of the shear moduli of polymers.[24–28] The

① Project supported the National Defense Foundation of China (Grant No. 9149A12050414JW02180).
② Chin. Phys. B, 2016, 25(12): 127701(1-10).

resonator usually consists of a TSM piezoelectric or quartz crystal layer and one- or multiple-layer polymer films, of which the basic principle is that shear modulus and damping of polymer film can change the resonant frequency and bandwidth of a TSM piezoelectric resonator obtained through an impedance analyzer, where the resonant frequency shift reflects the shear modulus and the bandwidth change reflects the mechanical loss factor of the tested material.

However, the above mentioned methods[22–28] cannot be used to investigate the variations of the TSM elastic constant c_{55} and the mechanical loss factor of a piezoelectric material with different shunt circuits. The viable testing system should satisfy the following requirements. First, the piezoelectric material that is shunted to circuits should not be used for impedance measurement, because the direct electrical coupling will cover the mechanical parameter effect on the impedance. There should be another piezoelectric material serving as the tested one which is electrically isolated from the shunted one, so that the impedance of the tested one can reflect the mechanical parameter change of the system, rather than the coupling electrical impedance. Second, if a two-plate piezoelectric resonator is adopted, two piezoelectric plates should work at a pseudo electrostatic state to avoid the mutual influence of their natural resonant frequencies. Thus the system should be resonant at a frequency far below the natural frequencies of the piezoelectric plates, which meanwhile provides a nearly linear relationship between the resonant frequency shift of the resonator and the modulus change.

In our previous work, we have proposed an indirect method to investigate the influence of shunt circuits on elastic constant c_{33} and the mechanical loss factor of thickness mode.[29] Inspired by this idea, in this paper we propose a coupling TSM electro–mechanical resonator to investigate the variations of the elastic constant c_{55} and the mechanical loss factor of the shear-mode piezoelectric materials shunted to different circuits. The proposed TSM resonator consists of two identical shear-mode piezoelectric plates and a mass block to form a coupling resonant system. One of the two plates is employed to change its TSM elastic constant and mechanical loss factor by connecting in parallel different shunt circuits, while the other is used to test the electrical admittance-frequency curve of the system. The measurement circuit and the shunting circuit are electrically separated to avoid the mutual interference. The mass is employed to lower the first resonant frequency of the system significantly to provide a pseudo electrostatic state. Mason equivalent circuit method is adopted to make the theoretical analyses and the finite element method (FEM) is employed to conduct the simulations. A prototype of the resonator model is fabricated to verify the theoretical analyses and simulations, with which the variations of the TSM elastic constant c_{55} and the mechanical loss factor of the PZT5A plate shunted to different circuits are investigated.

2. Model description

An electro–mechanical resonant system, i.e., a resonator, is proposed to measure the change of the resonant frequency of the system with shunt circuits, which is composed of two identical shear-mode piezoelectric ceramic plates a and b and a mass block m. They are bonded together tightly with a very thin layer of epoxy resin adhesive as shown in Fig. 1(a), where 1, 2, 3, and 4 are employed to mark the four surfaces or interfaces of a, b, and m respectively. Both surfaces 1 and 4 are set free to make force release boundary conditions. The mass block m is employed to make the resonant base frequency of the system much lower than the resonant frequency of a single piezoelectric plate itself to provide a quasi-static state.[4] The material of the mass block is of pure tungsten due to its high density to make the mass block close to a lumped-parameter element and the system close to an ideal spring oscillator. PZT-5A is chosen as the material of piezoelectric plates a, b in this paper, of which both are poled along the length direction (motion direction) and two electrodes are drawn from two faces of the thickness direction.

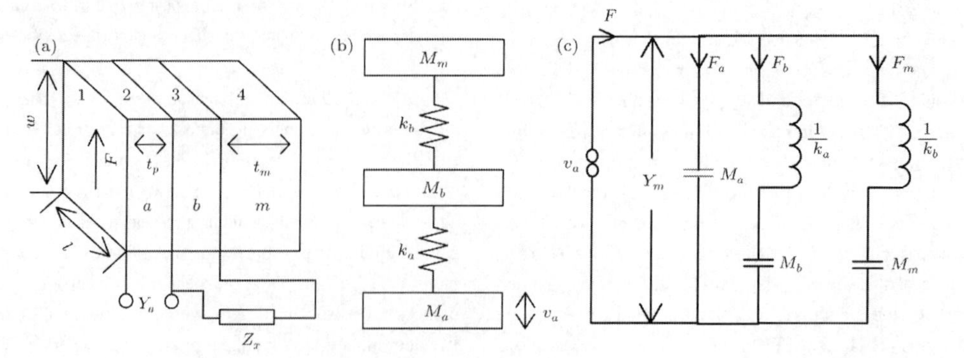

Fig. 1. Proposed coupling electro–mechanical resonant system: (a) the structure of the resonator; (b) equivalent coupling spring-mass system; (c) mechanical circuit diagram.

When plate b connects to different shunt circuits, i.e., Z_x, its elastic constant c_{55} and the mechanical loss factor will change. Correspondingly, the resonant frequency and quality factor of the system will change too. By testing the electrical impedance of the resonator through plate a, the changes of the resonant frequency and quality factor of the system are obtained. Then the variations of c_{55} and the mechanical loss factor can be indirectly achieved. The separate circuits of two plates ensure the isolation of electrical effects of the shunted plate b on the tested plate a.

As shown in Fig. 1(b), the proposed system can be approximated as an equivalent ideal coupling spring-mass oscillator system, based on the lumped-parameter condition. It consists of three mass elements and two spring elements which are connected in series. The springs k_a and k_b are equivalent to the elastic constants of the two piezoelectric ceramic plates a and b respectively, and the mass elements M_a, M_b, and M_m correspond to the mass of the piezoelectric ceramic plates a and b and the mass block m respectively. The equivalent mechanical circuit diagram of the ideal spring oscillator system is given in Fig. 1(c), viewing from the bottom of plate a. Since the ceramic plate a is in open-circuit state, its elastic constant c_{55} keeps constant, i.e., the open-circuit elastic constant c_{55}^{D}. Whereas the ceramic plate b is under different electrical boundary conditions, i.e., shunted to Z_x, its elastic constant c_{55} is a variable.

In Fig. 1(c), k_a and k_b can be approximated as

$$k_a \approx c_{55}^{\mathrm{D}} \frac{wl}{t_p}, \tag{1a}$$

$$k_b \approx c_{55} \frac{wl}{t_p}. \tag{1b}$$

Then the mechanical admittance Y_m from the bottom of plates a can be derived as

$$Y_m = \frac{1}{\mathrm{j}\omega \frac{1}{k_b} + \frac{1}{\mathrm{j}\omega M_m}} + \frac{1}{\mathrm{j}\omega \frac{1}{k_a} + \frac{1}{\mathrm{j}\omega M_b}} + \mathrm{j}\omega M_a, \tag{2}$$

or

$$Y_m = \frac{\mathrm{j}\omega M_m\left(1-\omega^2 M_b/k_a\right) + \mathrm{j}\omega M_b\left(1-\omega^2 M_m/k_b\right) + \mathrm{j}\omega M_a\left(1-\omega^2 M_b/k_a\right)\left(1-\omega^2 M_m/k_b\right)}{\left(1-\omega^2 M_b/k_a\right)\left(1-\omega^2 M_m/k_b\right)}. \tag{3}$$

From Eq. (3), it is found that the oscillator system has two resonant frequencies:

$$f_{r1} = (1/2\pi)\sqrt{k_b/M_m}, \tag{4a}$$

$$f_{r2} = (1/2\pi)\sqrt{k_a/M_b}. \tag{4b}$$

Here, f_{r2} is the natural resonant frequency of single ceramic plate, which does not change with k_b, and f_{r1} is the resonant base frequency determined by the mass of block m and the variable k_b, which is much lower than f_{r2}, and is what we are concerned with in the research.

Hence, the stiffness of spring b can be obtained from

$$k_b = M_m \left(2\pi f_{r1}\right)^2, \tag{5}$$

and the following relationship can be derived:

$$\frac{\Delta k_b}{k_b} \approx 2\Delta f_{r1}/f_{r1}. \tag{6a}$$

According to Eq. 1(b), it has

$$\frac{\Delta c_{55}}{c_{55}} \approx 2\Delta f_{r1}/f_{r1}, \tag{6b}$$

or

$$\Delta c_{55} \approx \frac{2c_{55}}{f_{r1}} \Delta f_{r1}. \tag{6c}$$

Therefore, the variation of the elastic constant Δc_{55} can be obtained indirectly with the change of the resonant frequency Δf_{r1} of the resonator when plate b is shunted to different circuits. In fact, the relationship between c_{55} and the resonant frequency f_{r1} can also be determined if one value of the elastic constant c_{55} is provided at the resonant base frequency f_{r1}.

3. Theoretical analysis with Mason equivalent circuit method

Because the two ceramic wafers a, b and the mass block m are tightly bonded together, the Mason equivalent circuit method[30] of the whole model can be obtained by connecting their respective equivalent circuit at mechanical terminals, which is shown in Fig. 2. Since surface 1 and surface 4 are free, the pressures on them are equal to zero and the two ends are short-circuited.

In Fig. 2,

$$Z_{p1a} = \mathrm{j}\rho_p c_p^{\mathrm{open}} S_p \tan\left(k_p^{\mathrm{open}} t_p/2\right), \tag{7a}$$

$$Z_{p2a} = \rho_p c_p^{\mathrm{open}} S_p / \left(\mathrm{j}\sin\left(k_p^{\mathrm{open}} t_p\right)\right), \tag{7b}$$

$$Z_{p1b} = \mathrm{j}\rho_p c_p S_p \tan\left(k_p t_p/2\right), \tag{7c}$$

$$Z_{p2b} = \rho_p c_p S_p / \left(\mathrm{j}\sin\left(k_p t_p\right)\right), \tag{7d}$$

$$Z_{m1} = \mathrm{j}\rho_m c_m S_m \tan\left(k_m t_m/2\right), \tag{7e}$$

$$Z_{m2} = \rho_m c_m S_m / \left(\mathrm{j}\sin\left(k_m t_m\right)\right), \tag{7f}$$

where the subscripts 'p' and 'm' represent the piezoelectric ceramics and the mass block respectively. Correspondingly, ρ_p and ρ_m are their densities, S_p and S_m are their areas, and t_p and t_m are their thickness. Besides, $c_p^{\mathrm{open}} = \sqrt{c_{55}^{\mathrm{D}}/\rho_p}$ and $c_p = \sqrt{c_{55}/\rho_p}$ are the transverse wave velocities of piezoelectric ceramics in open state and shunted to circuits respectively, and $k_p^{\text{open}} = \omega/c_p^{\text{open}}$ and $k_p = \omega/c_p$ are the corresponding wave numbers, c_m and $k_m = \omega/c_m$ are the transverse velocity and wave number of the mass block m respectively. Both for the ceramic wafers a and b, $C_0 = S_p\varepsilon_{11}^S/t_p$ is the inherent capacitance and $n = S_p h_{15}\varepsilon_{11}^S/t_p$ is the electromechanical conversion factor, where ε_{11}^S is the dielectric constant and h_{15} is the shear-mode piezoelectric constant.

From Fig. 2, the electrical admittance Y_a of the system calculated from the two electrodes of wafer a can be derived as

$$Y_a = \frac{1}{\left(\dfrac{\left(\left(Z_{m1} \parallel Z_{m2} + Z_{m1} + Z_{p1b}\right) \parallel \left(n^2\left(Z_x \parallel \dfrac{1}{j\omega C_0} - \dfrac{1}{j\omega C_0}\right) + Z_{p2b}\right) + Z_{p1b} + Z_{p1a}\right) \parallel Z_{p1a} + Z_{p2a}}{n^2} - \dfrac{1}{j\omega C_0}\right) \parallel \dfrac{1}{j\omega C_0}}, \tag{8}$$

where "$\parallel$" means parallel connection. c_{55} are included in Z_{p1b} and Z_{p2b} through Eq. (7) and c_p.

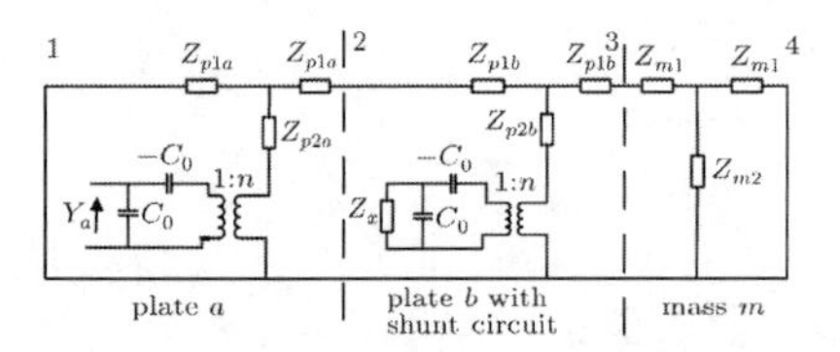

Fig. 2. Electro–mechanical equivalent circuits of the whole model.

Equation (8) establishes an analytic relationship between electrical admittance Y_a and shunted electrical impedance Z_x. It is seen that when the piezoelectric ceramic plate b is open-circuited, i.e., $Z_x = \infty$, the resonant frequency of the electrical admittance Y_a is determined only by the static mechanical constants of plates a and b, and mass block m. However, if the piezoelectric ceramic plate b is shunted to an electrical impedance Z_x, its TSM elastic constant c_{55} will change. As a result, reflecting in the electrical admittance Y_a, the resonant base frequency will shift correspondingly.

From Eq. (8), an analytic relationship can also be established between the electrical loss factor $\eta_e = 1/Q_e$ of the system and the mechanical loss factor $\eta_m = 1/Q_m$ of the plate b. Here Q_e is the electrical quality factor of the resonant peak and Q_m is the mechanical quality factor of the plate b. The electrical quality factor Q_e of the resonant peak can be solved through Eq. (8) by the inverse ratio of the relative bandwidth of the resonant peak as:

$$Q_e = f_r/\Delta f = f_r/(f_2 - f_1), \tag{9}$$

where f_r is the resonant base frequency, f_1 and f_2 are two half-power frequencies around f_r, which can be solved through Eq. (8) with $G_{a1} = G_{a2} = G_{a\max}/\sqrt{2}$. Here G_a is the real part of Y_a, i.e., the conductance of the system. Since the piezoelectric ceramics have inner mechanical losses, the TSM elastic constant c_{55} of the plate b in Eq. (8) can be substituted with a complex number $c_{55} = c_{55r} + jc_{55i}$, where the subscripts 'r' and 'i' represent the real part and the imaginary part respectively. The imaginary part of the elastic constant reflects the mechanical loss. Defining the mechanical loss factor η_m as the ratio of the imaginary part to the real part of the elastic constant of the piezoelectric ceramics, it has

$$\eta_m = c_{55i}/c_{55r} = 1/Q_m. \tag{10}$$

In view of Eqs. (8)–(10) and the definition of each variable, the analytic relationship between electrical loss factor η_e of the resonant peak and the mechanical loss factor η_m can be solved, which is not presented in the paper for simplicity due to its huge expression.

3.1. Properties of electrical conductance curve

Here, the electrical conductance of the system through plate a is calculated numerically to illustrate its properties. The precondition of using equivalent circuit method is that the length and width of the ceramic plates are both much larger than their thickness to meet the infinite boundary condition. Thus in the calculations, the dimension parameters are set to be $l = w = 1000$ mm, $t_p = 2$ mm, and $t_m = 3$ mm. As a matter of fact, since the shear mode is less affected by the lateral boundary condition than the thickness mode, if the length-to-thickness ratio is not too small, the lateral boundary effect on the resonant frequency of the system is negligible.

The material parameters of the piezoelectric ceramics (PZT-5A) and the mass block are listed in Table 1.

The elastic constant c_{55} of the ceramic wafer b is assumed to be a variable in the calculation to investigate the relationship between c_{55} and resonant frequency.

For c_{55} with a certain value, according to Eq. (8), the conductance (the real part of the admittance Y_a) of the system measured from the two electrodes of wafer a can be calculated, then the resonant frequency and the electrical quality factor Q_e can be obtained. For example, when the ceramic wafer b is in an open-circuit state, namely $Z_x = \infty$ and $c_{55} = c_{55}^D$, the conductance is calculated with the mechanical quality factor $Q_m^D = 75$. As shown in Fig. 3, the conductance of the system shows a resonant peak at the frequency $f_r^D = 147.3$ kHz.

which is the resonant base frequency of the system. The peak conductance can be extracted as $G_{a\max} = 57.58$ S. By using two frequencies $f_1 = 146.6$ kHz and $f_2 = 148$ kHz, where the conductance is equal to $G_{a\max}/\sqrt{2}$, the electrical quality factor of the conductance is calculated as the following $Q_e^D = f_r^D/\Delta f = 105.21$. Then the electrical loss factor of the system near the resonant peak is obtained, which is $\eta_e^D = 1/Q_e^D = \Delta f/f_r^D = 0.0095$.

Table 1. Parameters of piezoelectric ceramics, mass block, and epoxy resin.

The piezoelectric ceramics PZT-5A											
Density/(kg/m^3)	Elastic constant/(10^{10} N/m^2)					Relative dielectric constant		Piezoelectric stress constant/(C/m^2)			Mechanical quality factor
ρ_p	c_{11}^E	c_{12}^E	c_{13}^E	c_{33}^E	c_{44}^E	$\varepsilon_{11}^S/\varepsilon_0$	$\varepsilon_{33}^S/\varepsilon_0$	e_{31}	e_{33}	e_{15}	Q_m^D
7500	12.1	7.54	7.52	11.1	2.11	916	830	−5.4	15.8	12.3	75
	c_{11}^D	c_{12}^D	c_{13}^D	c_{33}^D	c_{44}^D						
	12.6	8.09	6.52	14.7	3.97						

The mass block (Tungsten)		
Density (kg/m^3)	Young modulus (10^{10} N/m^2)	Poisson's radio
19100	35.4	0.35

The epoxy resin			
Density/(kg/m^3)	Young modulus/(10^{10} N/m^2)	Poisson's radio	Damping factor
1250	0.36	0.35	0.17

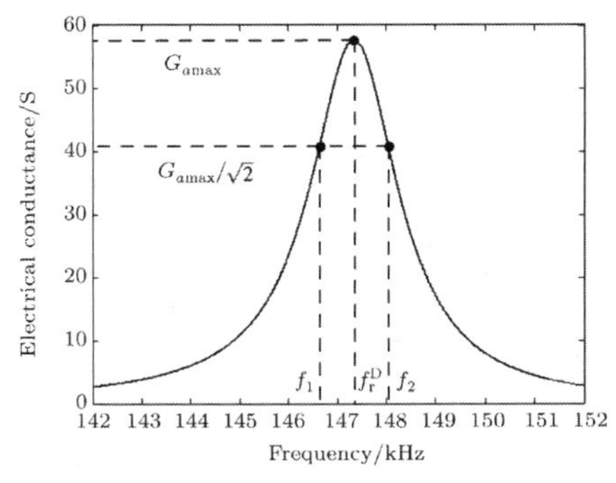

Fig. 3. Electrical resonant peak of the conductance of the system calculated from the two electrodes of the plate a (with wafer b in open-circuit, $c_{55} = c_{55}^D$).

3.2. Relationships drawn from the electrical conductance curve

Taking c_{55r} as the unique variable with the other parameters of the model being constant, the relationship between $\Delta c_{55r}/c_{55r}^D$ and $\Delta f_r/f_r^D$ is calculated based on Eq. (8) as shown in Fig. 4(a). It can be observed that $\Delta c_{55r}/c_{55r}^D$ shows a linear variation with $\Delta f_r/f_r^D$ approximately in a wide range, which is in consistence with the conclusion in Eq. (6b). Therefore, if the variation of the resonant frequency is observed, the variation of the elastic constant real part c_{55r} can be obtained.

Similarly, taking the imaginary part c_{55i} as the only variable of the system with the other parameters of the model being constant, the relationship between the mechanical loss factor η_m and the electrical loss factor η_e can be obtained as depicted in Fig. 4(b). It indicates that they are in good linear relationship over a wide range too. Thus, the mechanical loss factor η_m can be indirectly achieved based on the electrical

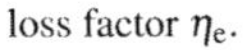
loss factor η_e.

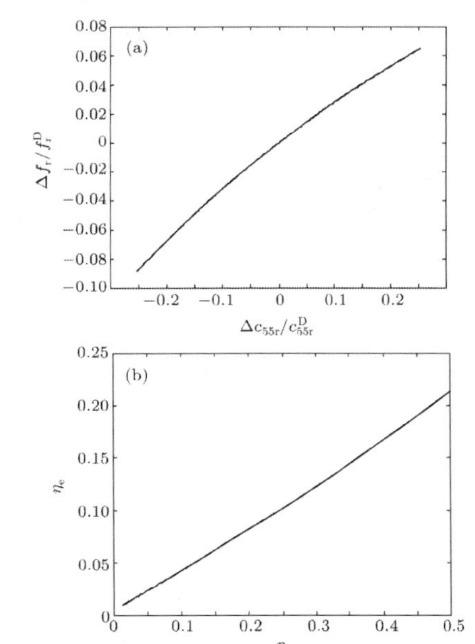

Fig. 4. Relationships between (a) $\Delta c_{55r}/c_{55r}^D$ and $\Delta f_r/f_r^D$, and (b) between η_m and η_e.

Since admittance Y_a can be tested with an impedance analyzer, the resonant frequency shift and bandwidth change of the peak are easy to obtain. Therefore, the proposed resonant system can provide a convenient means to investigate the relationships between the elastic constant c_{55} as well as mechanical loss factor and the shunted circuits. It should be noted that in Figs. 4(a) and 4(b), the curves deviate slightly from linearity for large variation, because the lump-parameter conditions cannot be perfectly satisfied.

4. Variation of the resonant frequency with different shunt circuits

In this section, different shunt impedance Z_x's are respectively connected to plate b to cause the shifts of the resonant frequency and the changes of the electrical quality factor, which are observed through theoretical analyses, FEM simulations and experimental tests. The relationship between the shunt impedance Z_x and the resonant frequency of the system is calculated and analyzed, with the shunt impedance Z_x being a resistor, a capacitor or an inductor respectively. While investigating the relationship between the shunt impedance Z_x and the electrical quality factor of the resonant peak, a resistor is adopted to be the shunt impedance, as the resistance is the only means to dissipate energy and change the damping of the system.

4.1. Theoretical calculations

For the piezoelectric plates of identical thickness, different cross sectional areas of the plates will result in different matching shunt resistances, inductances and capacitances. Thus in the theoretical analyses, the product of shunt resistance and cross sectional area of the ceramic plate is defined as the shunt resistance ratio (SRR), whose unit is $\Omega\cdot m^2$. The product of shunt inductance and cross sectional area of the ceramic plate is defined as the shunt inductance ratio (SLR), whose unit is $H\cdot m^2$. The quotient of shunt capacitance and cross sectional area of the ceramic plate is defined as the shunt capacitance ratio (SCR), whose unit is F/m^2. With the same dimension parameters as used in Section 3: $l = w = 1000$ mm, $t_p = 2$ mm, and $t_m = 3$ mm, shunted by different SRR, SCR, and SLR, respective conductance is theoretically calculated according to Eq. (8). While calculating for SLR, an inductor is connected in series with a small resistor ($0.03\ m^2 \cdot \Omega$) to simulate its inner resistance.

Several conductance curves are shown in Figs. 5(a1), 5(b1), and 5(c1) respectively. It is found that either with the increase of SRR or with the decline of SCR, the resonant frequency rises from that of short-circuit to that of open-circuit monotonically as shown in Figs. 5(a1) and 5(b1). For both

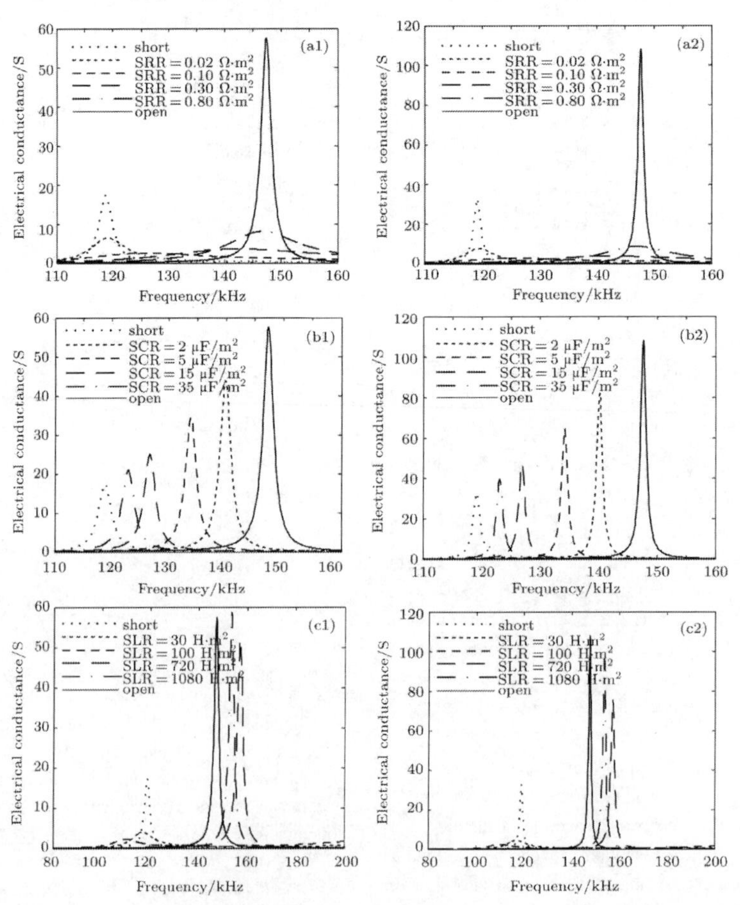

Fig. 5. Electrical conductances of the system calculated from the two electrodes of wafer a by equivalent circuit method with b shunted to: (a1) a resistor (theory); (a2) a resistor (FEM); (b1) a capacitor (theory); (b2) a capacitor (FEM); (c1) an inductor in series with a resistor (0.03 Ω) (theory); (c2) an inductor in series with a resistor (0.03 Ω) (FEM).

cases the resonant frequency range is limited within those of short-circuit and open-circuit. However, for all SLR values, the resonant frequencies go beyond the above mentioned range limit. With the increase of the SLR, at the beginning the resonant frequency decreases from the resonant frequency of short-circuit, then at a certain SLR, it jumps to a value above that of open-circuit, and finally it declines continuously to that of open-circuit, as shown in Fig. 5(c1). Besides, it is seen from Fig. 5(a1) that the bandwidth of the resonant peak varies significantly with SRR.

This part introduces the materials, method and experimental procedure of the author's work, so as to allow others to repeat the work published based on this clear description.

4.2. Simulations with finite element analysis

For verifying the theoretical analysis, the finite element method (FEM) is adopted to calculate the electrical admittance from the two electrodes of the wafer a. A three-dimensional (3D) finite element model of the system is established with ANSYS as shown in Fig. 6, where the size parameters of the system are $l = w = 1000$ mm, $t_p = 2$ mm, $t_m = 3$ mm, the same as those in theoretical calculation. The circuit element CIRCU94 is employed to build the shunt impedance connected to the wafer b. As the cross sectional area is known as $S_p = lw$, the resistance, capacitance and inductance connected to the wafer b is calculated according to SRR/S_p, $\mathrm{SCR}\cdot S_p$ and SLR/S_p respectively. Voltage 1 V is applied to surface 1, and 0 V to surface 2. By taking harmonic response analysis, the electrical charge Q on surface 2 is picked up, and the admittance of the model is derived as

$$Y_a = \mathrm{j}2\pi f\frac{Q}{V}, \tag{11}$$

where $V = 1$ V and f is the frequency.

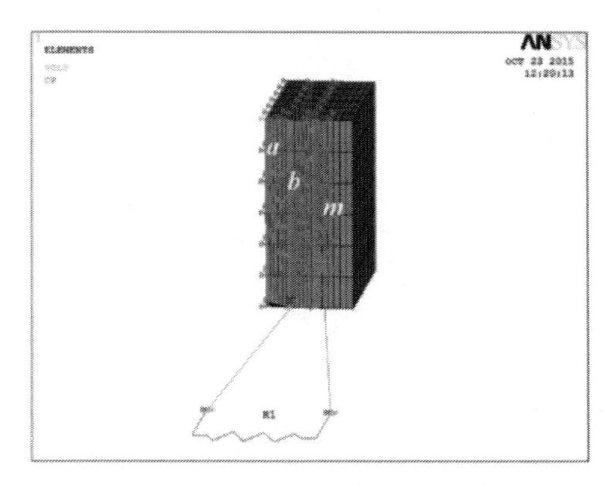

Fig. 6. The ANSYS model with the same size as that used in the theoretical analyses.

The electrical conductances calculated from the two electrodes of plate a with plate b shunted to different values of a resistor, a capacitor and an inductor in series with a resistor (0.03 Ω) are calculated. The 0.03-Ω resistor is used to simulate the inner resistance of the conductor. The curves which correspond to the same values of SRR, SCR, and SLR as those of theoretical analyses are depicted in Figs. 5(a2), 5(b2), and 5(c2), respectively. Comparing the theoretical results in Figs. 5(a1), 5(b1), and 5(c1) with the corresponding FEM curves in Figs. 5(a2), 5(b2), and 5(c2), it is found that they are matched perfectly.

4.3. Experiments verification

In order to further verify the theoretical analytical results and the finite element results, an experimental prototype resonator is fabricated as shown in Fig. 7(a). Because of limitation of the poling length, in practice the lateral dimensions of the piezoelectric plates are machined into $l = w = 15$ mm, far less than those used in theoretical analyses and FEM simulations. The thickness of each piezoelectric plate keeps $t_p = 2$ mm. It is found in the study that the shape of the mass block has little influence on the resonant frequency of the system while keeping the mass constant, thus in the experiment, the mass block is machined into a round cylinder of $r_m = 12.5$ mm and $t_m = 5$ mm to facilitate the fabrication. Two piezoelectric plates and the mass block are bonded together with a shin layer of epoxy resin.

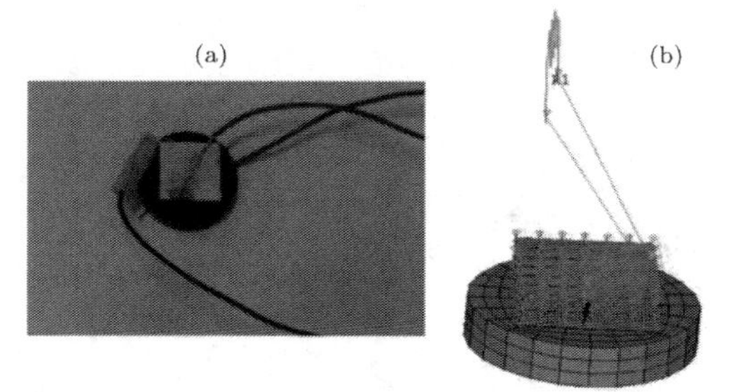

Fig. 7. Prototype system: (a) the sample; (b) the finite element model.

In the experiments, the conductance curves of the system with plate b shunted to different resistors, capacitors and inductors are tested by an impedance analyzer and shown in Figs. 8(a1), 8(b1), and 8(c1), respectively. Comparing Figs. 5(a1), 5(b1), and 5(c1) or Figs. 5(a2), 5(b2), and 5(c2), shows that the variation rules of the resonant frequency with different shunted circuits are in good consistence with those of the theoretical analyses and FEM simulations, while the resonant frequency range and peak bandwidth are somewhat different. The differences in the resonant frequency range and the peak bandwidth are mainly caused by the epoxy resin layer used in the experimental sample which lowers the resonant frequency of the system and provides additional damping. Moreover, the precondition of using equivalent circuit method cannot be satisfied very well, for the finite-sized experimental sample also has influences on resonant frequency and peak bandwidth, which requires the length and width to be far above thickness.

To verify the experimental results of the actual sample which has a finite size, a finite element model with the same size of the sample and a bonding layer is also established to conduct the simulations, as shown in Fig. 7(b). In simulations, the thickness of the epoxy resin layer is set to be 0.068 mm and the material parameters are listed in Table 1. The FEM simulation results with shunt circuits corresponding to the experiment tests are depicted in Figs. 8(a2), 8(b2), and 8(c2) respectively. It can be found that the simulation results are in good agreement with the experimental data, which means that the small size and the bonding layer of the actual experimental sample indeed make the system different from the ideal case.

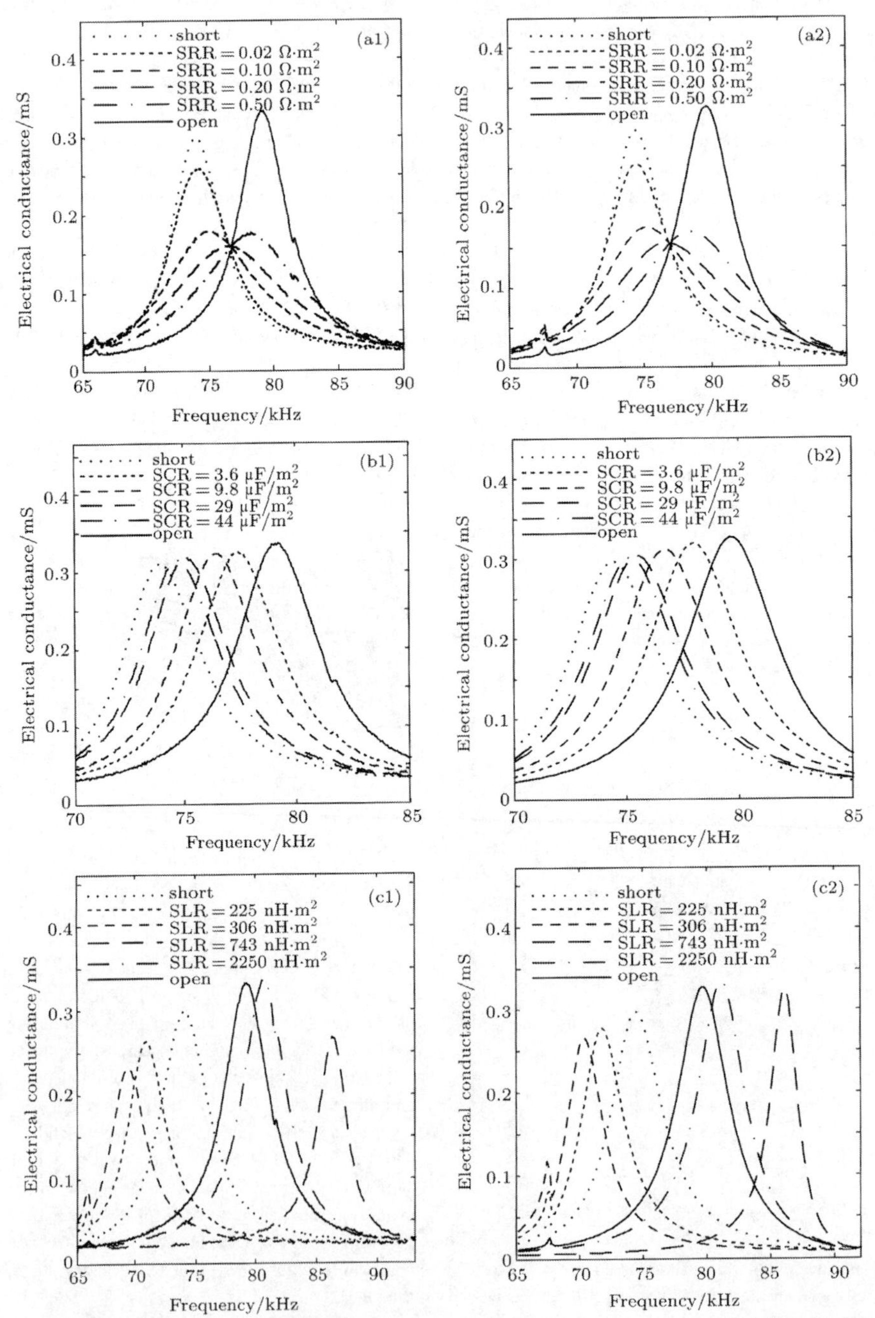

Fig. 8. Electrical conductances of the system measured from the two electrodes of wafer *a* by experiments with *b* shunted to: (a1) a resistor (experiment); (a2) a resistor (FEM); (b1) a capacitor (experiment); (b2) a resistor (FEM); (c1) an inductor (experiment); (c2) an inductor (FEM).

5. The relationships of c_{55} and η_m to different shunted elements

5.1. Relationship between the variation of c_{55} and shunted elements

With an impedance analyzer, the conductance curves of the experimental sample are tested from two electrodes of plate a, with b shunted to different elements. The relationships between the resonant frequency of the sample and different shunted element values of a resistor, a capacitor and an inductor are measured respectively. Correspondingly, the relationships between the variations of the real part c_{55r} of the TSM elastic constant c_{55} and the shunt circuits are obtained indirectly according to linear relations in Fig. 4(a).

The experimental results are shown as the curves marked with '*' in Figs. 9(a)–9(c), respectively, where the results of FEM simulations of the experimental sample (of the small size: 15 mm × 15 mm) are plotted with the curves marked with '□'. Moreover, for comparison, the results of theoretical analyses (curves marked with '•') of the big-sized model (1000 mm × 1000 mm), as well as the results of FEM simulations corresponding to the theoretical analyses (lines marked 'o') are also plotted in the respective figures. It can be seen that the experimental curves, curves of the theoretical analyses, and those from the FEM simulations are all in good consistence with each other.

From Fig. 9(a), it can be seen that the range of SRR that can change c_{55r} is nearly from 0.01 $\Omega\cdot m^2$ to 0.6 $\Omega\cdot m^2$. When SRR is less than 0.01 $\Omega\cdot m^2$, the resonant frequency of the system is almost equal to that of a short-circuit, namely f_r^E, nevertheless, when SRR is larger than 0.6 $\Omega\cdot m^2$, the resonant frequency is almost equal to that of the open-circuit, f_r^D. With the increase of shunt resistance, the resonant frequency of the system increases from the resonant frequency of the short-circuit f_r^E to that of the open-circuit f_r^D, which means that c_{55r} increases from the resonant frequency of the short-circuit, c_{55r}^E, to that of the open-circuit, c_{55r}^D, correspondingly. From Fig. 9(b), it can be found that the range of SCR that activates by adjusting c_{55r} is between 10 $\mu F/m^2$ and 60 $\mu F/m^2$. With the growth of SCR, the resonant frequency of the system decreases from the open-circuit f_r^D to the short-circuit f_r^E, which means that c_{55r} decreases from the open-circuit c_{55r}^D to the short-circuit c_{55r}^E correspondingly. Figure 9(c) shows that the relationship between f_r and SLR is more complex than the above two. With the increase of SLR, the resonant frequency of the system first decreases from the short-circuit f_r^E, after a jump at about 0.14 $\mu H\cdot m^2$ to a maximum value, then continuously decreases to the open-circuit f_r^D, which means that c_{55r} decreases from c_{55r}^E first, after a jump to its maximum at about 0.14 $\mu H\cdot m^2$, then continuously decreases to c_{55r}^D correspondingly.

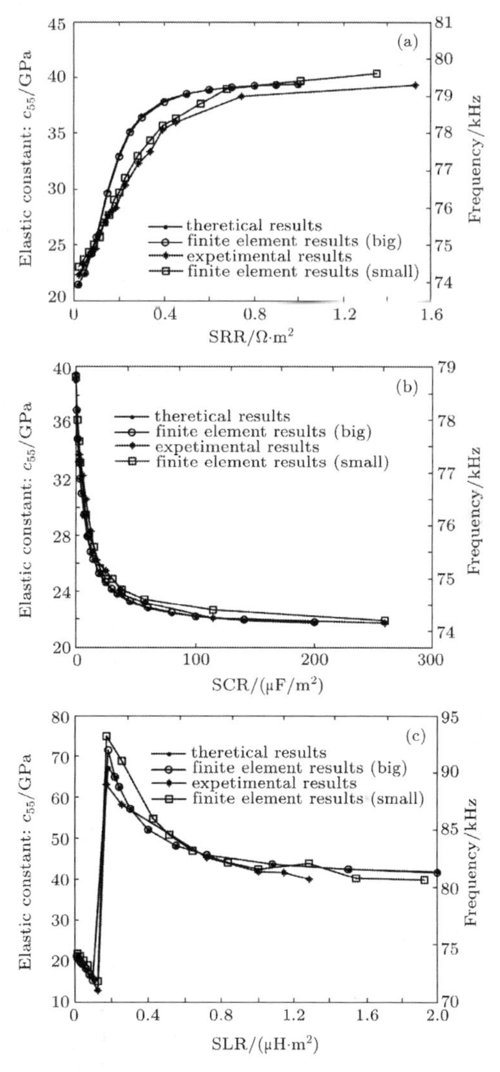

Fig. 9. Relationships between the elastic constant c_{55r} and (a) SRR, (b) SCR, and (c) SLR, respectively.

5.2. Relationship between η_m and shunted resistance

As the resistance is the only way to dissipate energy and has an effect on the damping of the piezoelectric material, the relationship between the mechanical loss factor and SRR is worth investigating. In the experiment, the conductance curves of the experimental sample are tested for different values of a resistor with an impedance analyzer. The relationship between electrical loss factor η_e and SRR is first obtained, calculated with the relative bandwidth $\Delta f/f_r$. The relationship between electrical loss factor η_e and SRR is first obtained, calculated with the relative bandwidth η_m. The relationship between electrical loss factor η_e and SRR is derived indirectly according to Fig. 4(b). The experimental relationship is depicted in Fig. 10 as the line marked with '*', where the relationship obtained by the FEM simulation for the experimental sample (small) is also depicted - as the curve marked with '□'. It can be seen that the experimental curve and the FEM simulation curve of the sample of the same size (small) are consistent with each other perfectly. When SRR equals 0.2 $\Omega \cdot m^2$, the shunt-damping effect is optimal. For comparison, the theoretical analysis (line marked '•') and the FEM simulation (line marked 'o') of the big-sized model (1000 mm × 1000 mm) are also shown in Fig. 10, showing they are in good consistence with each other too. However, the small-sized sample and the big-sized model show some differences in the optimal value of SRR and the slope of the mechanical loss factor. The difference between the theoretical analysis and the experimental result is mainly caused by the small lateral dimension-to-thickness ratio and the existence of epoxy resin layer for the experimental sample, which makes the experimental sample deviate from the ideal model of the theoretical analyses.

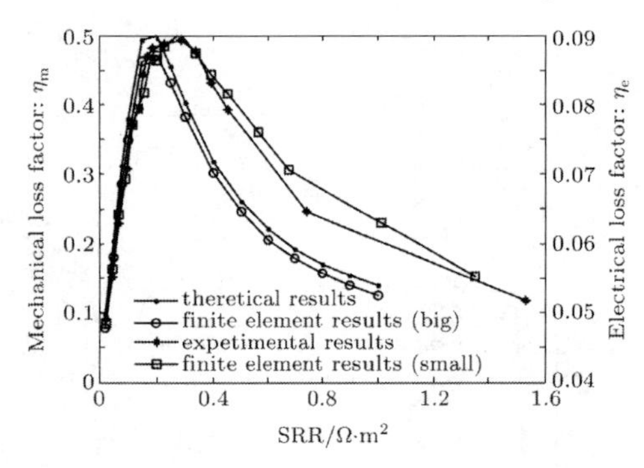

Fig. 10. Relationships between mechanical loss factor η_m and SRR.

6. Conclusions

In this paper, we propose a coupling electro–mechanical resonant system to indirectly measure the TSM elastic constant c_{55} and the mechanical loss factor η_m. Based on this model, the variation of the TSM elastic constant c_{55} of shear-mode piezoelectric ceramic with shunt circuit is investigated through the theoretical analysis with Mason equivalent circuit, and the variations of the mechanical loss factor and the resistance with shunt circuit is also studied, which are both verified by the FEM simulations. Since the changes of the resonant frequency and the bandwidth of the peak of the conductance are easily observed, the corresponding changes of the elastic constant c_{55} and the mechanical loss factor η_m can be obtained indirectly.

An experiment sample is fabricated with PZT5A to test the admittance of the system with an impedance analyzer. The theoretical and simulated results are verified experimentally. The variations of the TSM elastic constant c_{55} and the mechanic loss factor η_m with shunt element are obtained experimentally. The relative changes of the resonant frequency and bandwidth of resonant peak with the shunt circuits experimentally show that they are in good agreement with the theoretical analyses and FEM simulations correspondingly, indicating the validity of the proposed method of measuring the variations of the TSM elastic constant c_{55} and the mechanic loss factor η_m of shear-mode PZT5A shunted to different circuits. It should be pointed out that the proposed coupling resonator and the obtained relationships are validated with shear-mode PZT5A (but not limited only to PZT5A).

References

[1] Forward R L 1979 *Appl. Opt.* **18** 690
[2] Hagood N W and von Flotow A 1991 *J. Sound Vib.* **146** 243
[3] Lallart M, Lefeuvre É, Richard C and Guyomar D 2008 *Sensor. Actuat. A-Phys.* **143** 377
[4] Shen H, Qiu J, Ji H, Zhu K, Balsi M, Giorgio I and Isola F D 2010 *Sensor. Actuat. A-Phys.* **161** 245
[5] Becker J, Fein O, Maess M and Gaul L 2006 *Comput. Struct.* **84** 2340
[6] Thomas O, Ducarne J and Deü J F 2012 *Smart Mater. Struct.* **21(1)** 015008
[7] Ducarne J, Thomas O and Deu J F 2012 *J. Sound Vib.* **331** 3286
[8] Sun Y, Li Z H, Huang A G and Li Q H 2015 *J. Sound Vib.* **355** 19
[9] Zhang J M, Chang W, Varadan V K and Varadan V V 2001 *Smart Mater. Struct.* **10** 414
[10] Ahmadian M and Jeric K M 2001 *J. Sound Vib.* **243** 347
[11] Guyomar D, Richard T and Richard C 2008 *J. Intell. Mater. Syst. Struct.* **19** 791
[12] Kim J S, Jeong U C, Seo J H, Kim Y D, Lee O D and Oh J E 2015 *Sensor. Actuat. A-Phys.* **233** 330
[13] Kim J S and Lee J K 2002 *J. Acoust. Soc. Am.* **112** 990
[14] Chen S B, Wen J H, Wang G and Wen X S 2013 *Chin. Phys. B* **22** 074301
[15] Corrêa de Godoy T C and Areias Trindade M A 2011 *J. Sound Vib.* **330** 194
[16] Benjeddou A and Ranger J A 2006 *Comput. Struct.* **84** 1415
[17] Trindade M A and Benjeddou A 2008 *Comput. Struct.* **86** 859
[18] Karim Y and Blanze C 2014 *Comput. Struct.* **138** 73
[19] Santos Heinsten F L dos and Trindade M A 2011 *J. Brazilian Soc. Mech. Sci. Eng.* **33** 287
[20] Benjeddou A 2001 *J. Vib. Control* **7** 565
[21] Orescanin M and Insana M F 2010 *IEEE T. Ultrason. Ferr.* **57** 1358
[22] Fang S X, Tang D Y, Chen Z M, Zhang H and Liu Y L 2015 *Chin. Phys. B* **24** 027802
[23] Zhang S J, Jiang W H, Richard J, Meyer Jr, Li F, Luo J and Cao W W 2011 *J. Appl. Phys.* **110** 064106
[24] Behling C, Lucklum R and Hauptmann P 1999 *IEEE T. Ultrason. Ferr.* **46** 1431
[25] Herrscher M, Ziegler C and Johannsmann D 2007 *J. Appl. Phys.* **101** 114909
[26] Johannsmann D 2001 *J. Appl. Phys.* **89** 6356
[27] Wolff O and Johannsmann D 2000 *J. Appl. Phys.* **87** 4182
[28] Granstaff V E and Martin S J 1994 *J. Appl. Phys.* **75** 1319
[29] Sun Y, Li Z H and Li Q H 2014 *Sensor. Actuat. A-Phys.* **218** 105
[30] Luan G D, Zhang J D and Wang R Q 2005 *Piezoelectric Transducers and Arrays* (Beijing: Peking University Press) pp. 103–126

Research on the Electro-Elastic Properties of the 2-1-3 (a Revised Version of 1-3-2) Piezoelectric Composite by Finite Element Method①

Yang Sun,[1] Zhaohui Li,[1] Qihu Li[2]
[1]*Department of Electronics, Peking University, Beijing 100871, China*

[2]*Institute of Advanced Technology, Peking University, Beijing 100871, China*

For the applications of underwater acoustics and ultrasonics, the 1-3 piezoelectric composite has proven to be useful material given its attractive performance. As a special modified version of 1-3 composite, the 1-3-2 piezoelectric composite which is composed of 1-3 composite and a ceramic base layer has been studied a lot recently for getting greater stability than 1-3 composite. But there still exist shortcomings caused by limitations in the manufacturing process of both 1-3 and 1-3-2 piezoelectric composites. In this article, a 2-1-3 composite that consists of 1-3 composite and a ceramic cover layer instead of the ceramic base layer for 1-3-2 composite is proposed to loosen the fineness requirement of 1-3 or 1-3-2 composites in the manufacturing process. The finite element method (FEM) has been adopted to analyze the dependence of electric-elastic properties (the longitudinal velocity and thickness electromechanical coupling coefficient) of 1-3, 1-3-2, and 2-1-3 composites on the aspect ratio α (the ratio of the lateral periodicity of PZT rods to the thickness of composite). The results of the 2-1-3 composite with soft matrix show great improvement in loosening the fineness requirement of the manufacturing. Typically, the fineness of 30%-volume faction 2-1-3 composite can be reduced to 2.54% of that of 1-3-2 composite. Hence, the 2-1-3 composite offers greater feasibility for the design of various sensing materials.
POLYM. COMPOS., 37:2384–2395, 2016.

INTRODUCTION

The 1-3 piezoelectric composite with perpendicularly embedded piezoelectric rods in a viscoelastic matrix has been widely used in acoustic and ultrasonic transducers for decades because of its good flexibility, low acoustic impedance, and high electromechanical coupling coefficient [1–5]. Recently, the 1-3-2 piezoelectric composite which is composed of 1-3 composite and a ceramic base layer in series has drawn much attention for its greater stability exhibited as underwater acoustic transducers [6, 7].

Theoretical models of 1-3 [8–11] and 1-3-2 [6, 12–14] composites have been developed according to Newnham's parallel, series theory [15], in which both the 1-3 and 1-3-2 composites are restricted to be effective homogeneous mediums. In Smith's model [8–10] and Chan's model [11], it was assumed that the spatial scale of 1-3 composite was so fine that the composite was validly represented as a homogeneous medium for the frequencies of interest near the thickness resonance. The corresponding approximation embodied the picture that the ceramic and polymer moved together in a uniform thickness oscillation. Thus, the vertical strains were same in both phases. As for the 1-3-2 composite which consists of a 1-3 composite layer stacked over a ceramic base layer in series connectivity in longitudinal direction [14], the normal effective strain along the longitudinal direction can be calculated based on the individual strain and volume fractions of the 1-3 composite and the ceramic base. Therefore, the assumptions and problems which appear in Smith's 1-3 composite model are also available for the 1-3 composite layer which acts as one part of the 1-3-2 composite.

The homogonous medium assumptions are obviously not satisfied when the spacing between pillars of 1-3 and 1-3-2 composites are not so fine as it revealed by a laser probe's measurement of the displacement of oscillating composite plates [9]. So the theoretical model will not be fit for the analysis of the 1-3 and 1-3-2 composites in such a case. The electro-elastic properties of the 1-3 and 1-3-2 composites will also be affected by the allocation and arrangement of the pillars.

Hossack and Hayward [16] thoroughly investigated the influence of ceramic volume fraction, pillar shape, and pillar orientation on the vibrational and electromechanical characteristics of 1-3 composite transducers using finite element analysis. It revealed that the pillar shape and aspect ratio had great and complex impact on the

① Polymer Composites, 2016: 2384-2395.

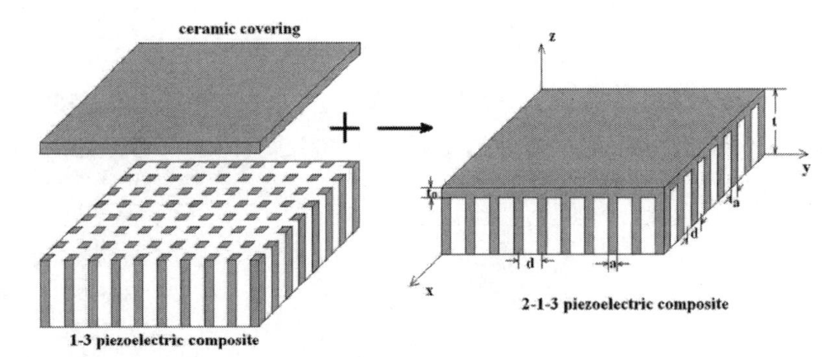

FIG. 1. 2-1-3 piezoelectric composite.

parameters of the composites. Optimum device performance and manufacturing cost can be compromised by using square section pillars which possess small pillar aspect ratio <0.25 and a relatively high-volume fraction greater than ~30% for hard matrix. However, it did not involve the composite with soft matrix and the method that can either improve the performance or reduce the manufacturing cost without deteriorating the other one.

More recently, Madhusudhana Rao and Prasad [17] also studied the variation of the thickness electromechanical coupling coefficient with aspect ratio using finite element method (FEM). In this article, a predefined displacement perpendicularly on all nodes on the surface of the composite unit cell was applied as the boundary condition, which meant the strains of both PZT phase and passive phase were restricted to be equal. For every fixed volume fraction from 10 to 70%, the obtained results of coupling coefficient almost keep unchanged with the increasing of aspect ratio from 0.2 to 1. This was obviously conflict with the reasonable results in Ref. 16 that for the 20%-volume faction composite it indicated a rapid deterioration in coupling coefficient for increasing pillar aspect ratios. Because the assumption of the equal strain condition in that work was definitely not satisfied for the coarse lateral scale case under the stress load.

It is obvious that under an actual stress load the piezoelectricity of the 1-3 and 1-3-2 composites will be greatly reduced when the lateral scale is too coarse. Because much of the pressure on the composite will be squandered in compressing the piezoelectrically inactive phase. This leads to the fact that the composite fails to meet the requirement of the theoretical assumption in the practical applications. In this article, according to the naming rule a revised version of 1-3-2 composite called 2-1-3 composite is proposed to satisfy the theoretical assumptions under the coarse lateral scale case. FEM is used to analyze the dependence of the elastic constants of composites on the aspect ratio under pressure load. Besides, the longitudinal velocity and thickness electromechanical coupling coefficient are calculated. Comparing with the 1-3 and 1-3-2 composites, the 2-1-3 composite can keep a certain longitudinal velocity and thickness electromechanical coupling coefficient for an extended range of the aspect ratio under the coarse lateral scale case. As a consequence, the difficulty and complexity of manufacturing can be reduced without losing the piezoelectric performance of the composite especially for a soft matrix.

2-1-3 PIEZOELECTRIC COMPOSITE (A REVISED VERSION OF 1-3-2 COMPOSITE)

Model of 2-1-3 Composite

The 2-1-3 piezoelectric composite is a revised version of 1-3-2 piezoelectric composite as shown in Fig. 1. It consists of 1-3 piezoelectric composite and a ceramic cover layer instead of a ceramic base layer for 1-3-2 composite. The two components are connected in series. Here, the underside of the 2-1-3 composite is clamped in its normal direction, and the surface of the ceramic cover is set as the stress surface.

The 1-3 piezoelectric composite part in 2-1-3 composite is modeled with square section piezoelectric rods that are perpendicularly embedded across the thickness (z-direction) in a viscoelastic matrix in the coordinate system (x,y,z). The piezoelectric phase is poled in z-direction and the matrix phase is assumed to be a homogeneously isotropic medium that is piezoelectrically inactive.

Let t represents the thickness of the whole composite and d represents the lateral periodicity of PZT rods. The ratio of d to t, an important parameter that influences the electro-elastic properties of 2-1-3 composite, namely the aspect ratio, is defined as *Eq. 1*.

$$\alpha=\frac{d}{t}. \tag{1}$$

Let t_0 represents the thickness of the ceramic base layer. Therefore, the thickness of the 1-3 piezoelectric composite part is $t-t_0$. The ratio of t_0 to t is defined as β,

$$\beta=\frac{t_0}{t}. \tag{2}$$

Let a represents the side length of the cubic PZT rod's cross section. Then it can be derived that the volume fraction of the piezoelectric phase in 2-1-3 composite is

$$\mathrm{VFP}_{(2-1-3)}=\beta+\frac{a^2}{d^2}(1-\beta). \tag{3}$$

It is clear that if $\beta=0$, which means the ceramic cover does not exist, *Eq. 3* will turn into the volume fraction of the 1-3 composite.

Fine-Scaled 2-1-3 Composite

When the aspect ratio α is small enough, corresponding to the condition that the PZT rods are of fine lateral spatial scale in the composite, the 1-3 composite part in 2-1-3 composite can be treated as an effective homogeneous medium. This kind of 2-1-3 composite is called the "fine-scaled 2-1-3 piezoelectric composite."

According to Ref. 14, the fine-scaled 2-1-3 composite can also be considered as a conventional laminated composite with two layers, consisting of 1-3 composite as one layer and a ceramic layer as the other. If the strain and electric field are chosen as the independent coordinates for the analysis, the constitutive relations of the ceramic layer and 1-3 composite layer can be simplified as *Eqs. 4* and *5*, respectively.

$$\begin{pmatrix} T_x^{\mathrm{cb}} \\ T_z^{\mathrm{cb}} \\ D_z^{\mathrm{cb}} \end{pmatrix}=\begin{pmatrix} c_{11}^E+c_{12}^E & c_{13}^E & -e_{31} \\ 2c_{13}^E & c_{33}^E & -e_{33} \\ 2e_{31} & e_{33} & \varepsilon_{33}^S \end{pmatrix}\begin{pmatrix} S_x^{\mathrm{cb}} \\ S_z^{\mathrm{cb}} \\ E_z^{\mathrm{cb}} \end{pmatrix} \tag{4}$$

$$\begin{pmatrix} T_x^{1-3} \\ T_z^{1-3} \\ D_z^{1-3} \end{pmatrix}=\begin{pmatrix} c_{11(1-3)}^E+c_{12(1-3)}^E & c_{13(1-3)}^E & -e_{31(1-3)} \\ 2c_{13(1-3)}^E & c_{33(1-3)}^E & -e_{33(1-3)} \\ 2e_{31(1-3)} & e_{33(1-3)} & \varepsilon_{33(1-3)}^S \end{pmatrix}\begin{pmatrix} S_x^{1-3} \\ S_z^{1-3} \\ E_z^{1-3} \end{pmatrix}, \tag{5}$$

where "cb" means ceramic layer and "1-3" represents 1-3 composite layer.

For the 1-3 composite part, several important assumptions are made in Smith's model [9] for the purpose of deriving the material parameters. One of the approximations embodies the picture that the ceramic and polymer move together in a uniform thickness oscillation. Thus, the vertical strains (in z-direction) are the same in both phases,

$$S_z^p=S_z^m=S_z^{1-3}. \tag{6}$$

Here, superscript "m" stands for matrix phase and "p" means piezoelectric phase. This is clearly not always true. However, it is a reasonably good picture when the composite has such fine spatial scale that stop-band resonances are at much higher frequencies than the thickness resonance.

Another essential approximation addresses the lateral interaction between the phases. It is assumed that the lateral stresses are equal in both phases and that the ceramic's lateral strain is compensated by a complimentary strain in the polymer so that the composite as a whole is laterally clamped. It means,

$$T_x^p=T_x^m=T_x^{1-3} \tag{7}$$

$$S_x^{1-3}=\mu S_x^p+(1-\mu)S_x^m=0, \tag{8}$$

where μ is the volume fraction of the piezoelectric phase and $1-\mu$ is that of the matrix phase in the 1-3 composite accordingly.

The elastic constant $c_{33(1-3)}^E$ and $c_{33(1-3)}^D$, the dielectric constant $\varepsilon_{33(1-3)}^S$, and the piezoelectric constant $e_{33(1-3)}$ of 1-3 composite part can be obtained from *Eqs. 9–11* [9]. It is clearly shown that the electro-elastic constants of fine-scaled 1-3 composite are decided by the parameters and volume fractions of both phases and have nothing to do with the aspect ratio.

$$c_{33(1-3)}^E=\mu\left[c_{33}^E-\frac{2(1-\mu)(c_{13}^E-c_{12}^m)^2}{\mu(c_{12}^m+c_{11}^m)+(1-\mu)(c_{12}^E+c_{11}^E)}\right]+(1-\mu)c_{11}^m \tag{9}$$

$$\varepsilon_{33(1-3)}^S=\mu\left[\varepsilon_{33}^S+\frac{2(1-\mu)e_{31}^2}{\mu(c_{12}^m+c_{11}^m)+(1-\mu)(c_{12}^E+c_{11}^E)}\right]+(1-\mu)\varepsilon_x^m \tag{10}$$

$$e_{33(1-3)}=\mu\left[e_{33}-\frac{2(1-\mu)e_{31}(c_{13}^E-c_{12}^m)}{\mu(c_{12}^m+c_{11}^m)+(1-\mu)(c_{12}^E+c_{11}^E)}\right] \tag{11}$$

$$c_{33(1-3)}^D=c_{33(1-3)}^E+\frac{e_{33(1-3)}^2}{\varepsilon_{33(1-3)}^S}. \tag{12}$$

Because the two components of the 2-1-3 composite, 1-3 composite layer and the ceramic layer, are connected in series in z-direction, according to Newnham's series theory [15], several important assumptions about the two parts of 1-3-2 composite made in Ref. 14 are also suitable for the analysis of 2-1-3 composite. They are sketched as follows.

The homogenized or averaged stress on the 1-3 composite layer along z-direction and the normal stress of the ceramic base layer in z-direction are considered to be the same, which can be written as

$$T_z^{\mathrm{cb}}=T_z^{1-3}=T_z^{2-1-3}. \tag{13}$$

Considering the homogeneous displacement condition for the 2-1-3 piezocomposite, the homogenized strain on the 1-3 composite layer and the normal strain of the ceramic cover layer along the transverse directions are deemed to be the same,

$$S_x^{\mathrm{cb}}=S_x^{1-3}=S_x^{2-1-3}. \tag{14}$$

TABLE 1. Parameters of phase materials.

Phase materials	Density ρ (kg/m³)	Elastic constants c_{11} (10^{10} Pa)	c_{12}	c_{13}	c_{33}	c_{44}	c_{66}	Piezoelectric constants e_{15} (C/m²)	e_{31}	e_{33}	Relative dielectric constant $\varepsilon_{11}/\varepsilon_0$ (–)	$\varepsilon_{33}/\varepsilon_0$
PZT-5H	7,500	12.6	7.95	8.41	11.7	2.3	2.35	17.0	−6.5	23.3	1.700	1.470
Rubber [18] (soft matrix)	1,003	0.23	0.22	0.22	0.23	0.005	0.005	–	–	–	3	3
Epoxy [16] (hard matrix)	1,140	0.82	0.15	0.15	0.82	0.52	0.52	–	–	–	4	4

The normal effective strain along the longitudinal direction (z-direction) is calculated based on the individual strain and volume fraction constituents of 1-3 composite and the ceramic cover, since they are in series connectivity.

$$S_z^{2-1-3}=\beta S_z^{\mathrm{cb}}+(1-\beta)S_z^{1-3}, \tag{15}$$

where β and $1-\beta$ are the volume fractions of the ceramic cover and 1-3 composite in the 2-1-3 composite, respectively.

According to the above constitutive relations, approximations, and the parameters of 1-3 composite, the electro-elastic constants of the fine-scaled 2-1-3 piezoelectric composite can be derived. The expressions for 1-3-2 composite in Refs. 6, 12, and 14 are available for 2-1-3 composite, namely:

$$c_{33(2-1-3)}^{E}=c_{33(1-3)}^{E}(\bar{K}_1c_{33}^{E}+\bar{K}_2e_{33})-e_{33(1-3)}(-\bar{K}_2c_{33}^{E}+\bar{K}_3e_{33}) \tag{16}$$

$$e_{33(2-1-3)}=e_{33(1-3)}(\bar{K}_1c_{33}^{E}+\bar{K}_2e_{33})+\varepsilon_{33(1-3)}^{S}(-\bar{K}_2c_{33}^{E}+\bar{K}_3e_{33}) \tag{17}$$

$$\varepsilon_{33(2-1-3)}^{S}=e_{33(1-3)}(-\bar{K}_1e_{33}+\bar{K}_2\varepsilon_{33}^{S})+\varepsilon_{33(1-3)}^{S}(\bar{K}_2e_{33}+\bar{K}_3\varepsilon_{33}^{S}) \tag{18}$$

$$\bar{K}_1=\frac{\varepsilon_{33(1-3)}^{S}\beta+\varepsilon_{33}^{S}(1-\beta)}{\left(\varepsilon_{33(1-3)}^{S}\beta+\varepsilon_{33}^{S}(1-\beta)\right)\left(c_{33(1-3)}^{E}\beta+c_{33}^{E}(1-\beta)\right)+\left(e_{33(1-3)}\beta+e_{33}(1-\beta)\right)^2} \tag{19}$$

$$\bar{K}_2=\frac{e_{33(1-3)}\beta+e_{33}(1-\beta)}{\left(\varepsilon_{33(1-3)}^{S}\beta+\varepsilon_{33}^{S}(1-\beta)\right)\left(c_{33(1-3)}^{E}\beta+c_{33}^{E}(1-\beta)\right)+\left(e_{33(1-3)}\beta+e_{33}(1-\beta)\right)^2} \tag{20}$$

$$\bar{K}_3=\frac{c_{33(1-3)}^{E}\beta+c_{33}^{E}(1-\beta)}{\left(\varepsilon_{33(1-3)}^{S}\beta+\varepsilon_{33}^{S}(1-\beta)\right)\left(c_{33(1-3)}^{E}\beta+c_{33}^{E}(1-\beta)\right)+\left(e_{33(1-3)}\beta+e_{33}(1-\beta)\right)^2} \tag{21}$$

$$c_{33(2-1-3)}^{D}=c_{33(2-1-3)}^{E}+\frac{e_{33(2-1-3)}^{2}}{\varepsilon_{33(2-1-3)}^{S}}. \tag{22}$$

The longitudinal velocity v_l and the thickness electromechanical coupling coefficient k_t of fine-scaled 1-3, 1-3-2, and 2-1-3 composites can be obtained from the above electro-elastic constants, expressed as follows:

$$v_l=\sqrt{\frac{c_{33}^{D}}{\rho}} \tag{23}$$

$$\rho=\rho^{p}\cdot\mathrm{VFP}+\rho^{m}\cdot(1-\mathrm{VFP}) \tag{24}$$

$$k_t=\sqrt{1-\frac{c_{33}^{E}}{c_{33}^{D}}}. \tag{25}$$

In many applications of the piezoelectric composites, the lower acoustic impedance is, the better the performance will be. So soft matrix is a necessary choice to achieve lower acoustic impedance. In this article, PZT-5H is adopted as the piezoelectric phase and the natural rubber is adopted as the matrix phase. The parameters of the piezoelectric phase (PZT-5H) and the matrix phase (nature rubber) are listed in Table 1. The dependences of the longitudinal velocity v_l and the thickness electromechanical coupling coefficient k_t of fine-scaled 1-3, 1-3-2, and 2-1-3 composites on volume fraction of PZT are shown in Fig. 2a and b.

Discussions on 1-3-2 and 2-1-3 Composites With High Aspect Ratios

One of the essential approximations (*Eq. 6*) of the theoretical models of both 1-3 and 1-3-2 composites is that

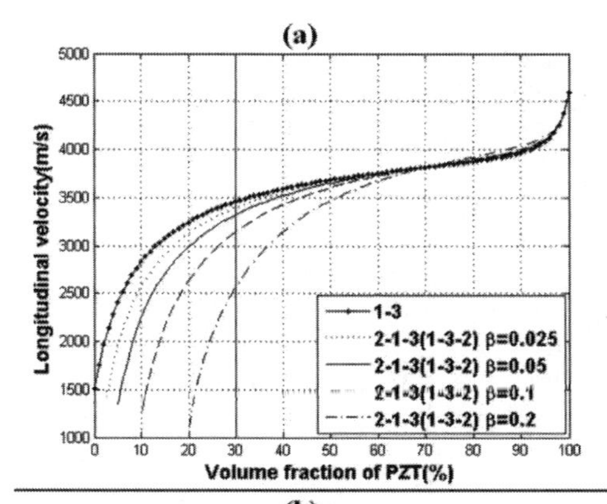

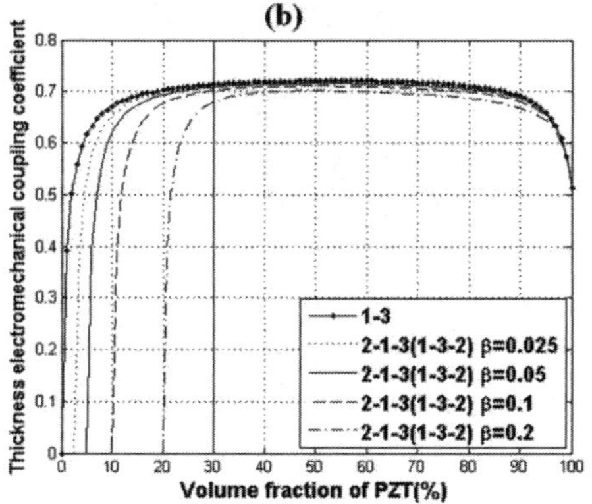

FIG. 2. The dependence of (a) longitudinal velocity and (b) thickness electromechanical coupling coefficient of fine-scaled 1-3, 1-3-2, and 2-1-3 composites on volume fraction.

the lateral spatial scale of the composite is so fine that the vertical strains are the same in both piezoelectric and matrix phases. Under this condition the composite can be treated as an effective homogeneous medium. That is to say, the theoretical model is restrained in the fine-scaled condition.

It can be a better practice if the lateral spatial scale of the 1-3 and 1-3-2 composites turn much more coarse so as to reduce the manufacturing cost. However, when the aspect ratio increases high to a certain extent, the above "same strain in both phases'; condition cannot be met. And the composite does not exhibit perfectly uniform compression thickness mode.

Figure 3a displays a possible contour of z-direction deformation displacement of the coarse-scaled 1-3-2 composite when a pressure is applied on its surface. It is obvious that the matrix phase will emerge larger compression deformation than the piezoelectric phase. Since much of the pressure applied on the composite is squandered compressing the passive phase, the longitudinal wave velocity of the composite will decrease and the piezoelectricity will become weaker. This has adverse impact on the practical application of the composite, especially for composites with soft matrix.

Based on the discussions above, it is reasonable to imagine that if both the coarse-scaled condition and the "same strain in both phases" condition can be met at the same time, the fineness requirement of the manufacturing of the 1-3 composite can be reduced without losing its piezoelectricity. The 2-1-3 composite composed of 1-3 composite and a ceramic cover can approximately realize such a goal.

Figure 3b displays a possible contour of z-direction deformation displacement of the coarse-scaled 2-1-3 composite when a pressure is applied on its surface. With the support of the high-stiffness PZT rods underside, the ceramic cover can make more pressure applied on the composite effectively act on the PZT phase instead of compressing the passive phase. It can be expected that compression deformations of both piezoelectric phase and matrix phase are nearly the same. Therefore, the z-direction strain in both piezoelectric phase and matrix phase are approximately equal.

FINITE ELEMENT MODELS OF 1-3-2 AND 2-1-3 COMPOSITES

FEM Modeling

Since the structure of the composite is of periodicity, it can be divided into a number of tiny unit cells [19] which contain the main characteristics of the microstructure and are able to represent the properties of the whole composite.

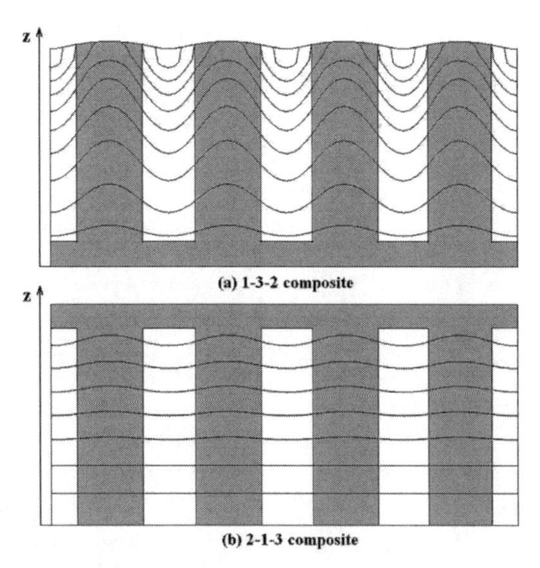

FIG. 3. Contour of z-direction deformation of (a) 1-3-2 composite and (b) 2-1-3 composite.

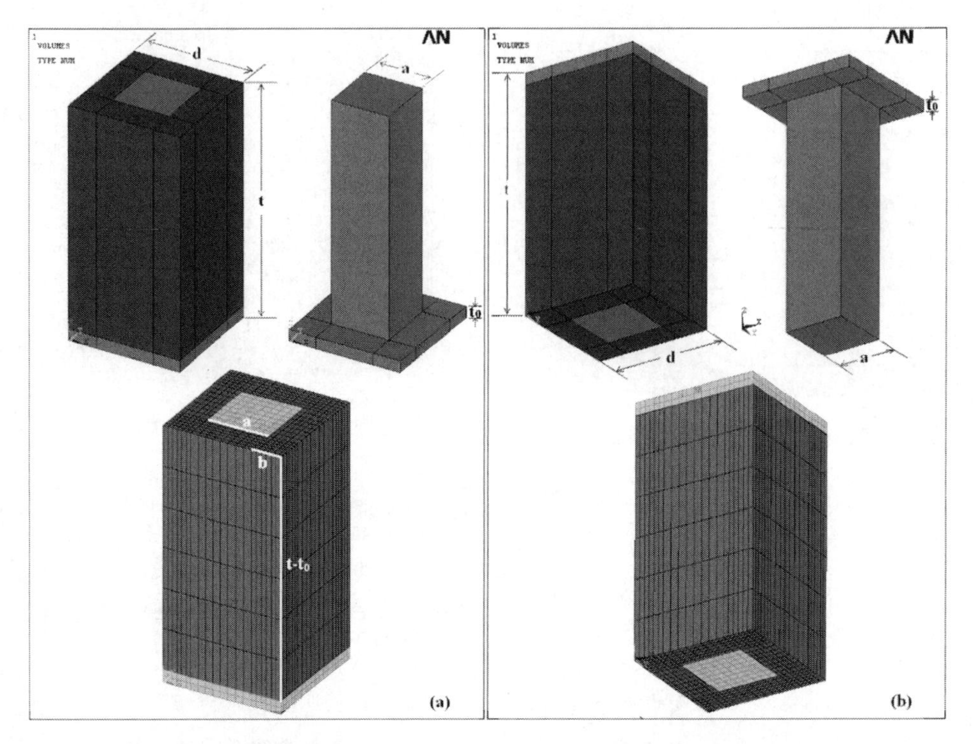

FIG. 4. Unit cell, PZT phase, and finite element mesh of (a) 1-3-2 and (b) 2-1-3 piezoelectric composites.

Finite element simulation software ANSYS is used here to model 1-3-2 and 2-1-3 composites and to calculate the electro-elastic constants. The unit cells of 1-3-2 and 2-1-3 composites are shown in Fig. 4a and b, respectively. For both 1-3-2 and 2-1-3 composites, the thickness of the unit cell is the same as that of the composite and the side length of the composite's cross section equals to the lateral periodicity of PZT rods.

Electro-Elastic Constants Calculation by FEM

The element types of the piezoelectric phase and the matrix phase are both Solid5 which is defined by eight nodes with up to six degrees of freedom (DOF) at each node (displacements: UX, UY, UZ, temperature: TEMP, voltage: VOLT, scalar magnetic potential: MAG) and has large deflection and stress stiffening capabilities when used in structural and piezoelectric analyses. In the finite element analysis of this article, since Solid5 is used to conduct piezoelectric analysis, UX, UY, UZ, and VOLT are selected as the DOF, namely KEYOPT(1) is set as 3.

In the material props part, it is needed to input the elastic constant matrix $[c^E]$, dielectric constant matrix $[\varepsilon^S]$, piezoelectric constant matrix $[e]$ of piezoelectric phase, the elastic constant matrix $[c^m]$, and dielectric constant matrix $[\varepsilon^m]$ of passive phase. The parameters of the piezoelectric phase (PZT-5H) and the matrix phase (nature rubber) are listed in Table 1. The difference of the influence brought by the soft matrix and hard matrix will be discussed later in section "Discussions on soft matrix and hard matrix."

Mapped meshing is adopted to divide the composite into a lot of hexahedron grid units as shown in Fig. 4.

In order to calculate different electro-elastic constants, proper boundary conditions and loads must be defined for the unit cell model according to the definition of the calculated constants. Here, elastic constants c_{33}^D and c_{33}^E are calculated to obtain the longitudinal velocity v_l and the thickness electromechanical coupling coefficient k_t.

The definitions of c_{33}^D and c_{33}^E are the variation of the stress tensor component T_z caused by the one-unit change of the strain tensor component S_z in the "constant electrical displacement" and "constant electrical field" conditions, respectively, as shown in *Eqs. 26* and *27*.

$$c_{33}^D = \left(\frac{\partial T_z}{\partial S_z}\right)_D = \frac{\Delta T_z}{\Delta S_z}\bigg|_{D=0} \tag{26}$$

$$c_{33}^E = \left(\frac{\partial T_z}{\partial S_z}\right)_E = \frac{\Delta T_z}{\Delta S_z}\bigg|_{E=0}, \tag{27}$$

where ΔS_z is the change of the strain tensor component S_z in FEM, and ΔT_z is the change of the stress tensor component T_z caused by ΔS_z.

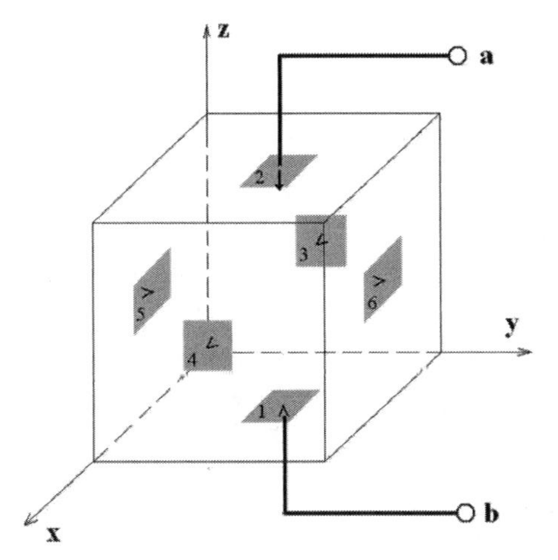

FIG. 5. Boundary conditions and loads for c_{33}^D and c_{33}^E.

Figure 5 is the profile of the unit cell. z-axis is the direction of the thickness of the composite and the two planes perpendicular to z-axis are marked as 1 and 2. The two planes perpendicular to x-axis are marked as 3 and 4 while the two planes perpendicular to y-axis are marked as 5 and 6.

To cause the change of the strain tensor component S_z, a pressure is loaded on plane 2 as shown in Fig. 5. At the same time, plane 1 is fixed in z-direction. Planes 3-6 are also clamped in their normal directions in order to set strain tensors in other directions zero. In Fig. 5, "∠" represents the plane is clamped in its normal direction.

For c_{33}^E, "constant field" means that plane 1 and 2 are short-circuited. So electrode a and b are set zero potential. While for c_{33}^D, "constant electrical displacement" means that plane 1 and 2 are open-circuited.

After static analysis is carried out in the solution part, the volume $(V_1, V_2, \ldots V_n)$, stress T_z $(T_1, T_2, \ldots T_n)$, and strain S_z $(S_1, S_2, \ldots S_n)$ values of all the grid units are picked up. Then, the volumetrically weighted averages of the stress and strain values as shown in *Eqs. 28* and *29* are calculated. Finally, c_{33} is obtained by dividing $\overline{T}_z$ by $\overline{S}_z$ as expressed in *Eq. 30*.

$$\overline{T}_z = \frac{T_1V_1 + T_2V_2 + \cdots + T_nV_n}{V_1 + V_2 + \cdots + V_n} = \frac{\sum_{j=1}^{n} T_jV_j}{V} \tag{28}$$

$$\overline{S}_z = \frac{S_1V_1 + S_2V_2 + \cdots + S_nV_n}{V_1 + V_2 + \cdots + V_n} = \frac{\sum_{j=1}^{n} S_jV_j}{V} \tag{29}$$

$$c_{33} = \frac{\overline{T}_z}{\overline{S}_z}. \tag{30}$$

Mesh Convergence Discussion

Since variable mesh size may lead to different set of finite element approximations, it is necessary to discuss the convergence of different mesh size. The unit cell of 1-3-2 composite (volume fraction is 30%, $\alpha = 0.1$ and $\beta = 0.025$) is taken as an example to illustrate this problem. As seen in Fig. 4a, three lines **b**, **a**, and **t** − **t₀** are divided into different numbers of segments to see the effect of the number of segments on the calculated longitudinal velocity. $\mathbf{t_0}$ is comparatively small, therefore its number of segments will not influence much on the result. So it is set as constant 2. The velocity results are shown in Fig. 6. In Fig. 6a, line **a** and $\mathbf{t - t_0}$ is divided into 8 and 6 segments, respectively. The number of segments on line **b** varies from 1 to 10. It can be found that when the number of segments on line **b** is less than 6, the calculated velocity is inaccurate. When the number is larger than 6 and increases, the result converges to a limit. Therefore, the number of segments on line **b** should be set as a value no less than 6. In Fig. 6b, line **b** and $\mathbf{t - t_0}$ is divided into 6 and 6 segments, respectively. Similarly, it is concluded that the number of segments on line **a** should be set as a value no less than 8. In Fig. 6c, line **a** and **b** is divided into 8 and 6 segments, respectively.

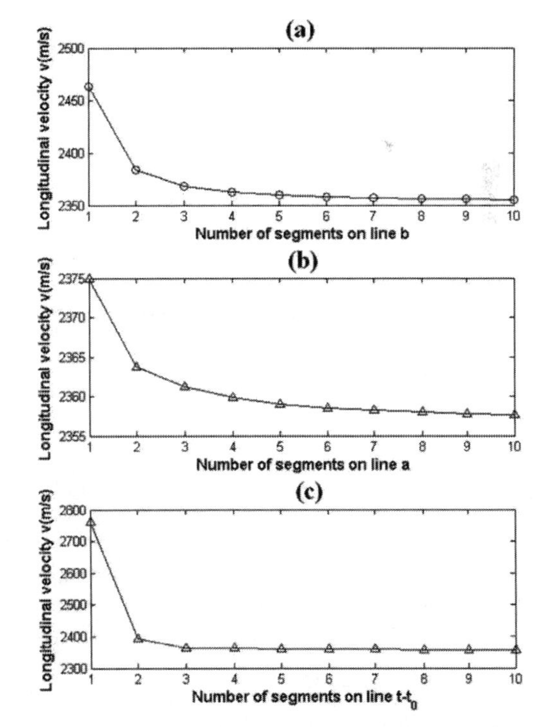

FIG. 6. The effect of the number of segments on the calculated longitudinal velocity. (The unit cell of 1-3-2 composite, volume fraction is 30%, $\alpha = 0.1$and $\beta = 0.025$) (a): line **a** and $\mathbf{t - t_0}$ is divided into 8 and 6 subsections, respectively; (b): line **b** and $\mathbf{t - t_0}$ is divided into 6 and 6 subsections, respectively; (c): line **a** and **b** is divided into 8 and 6 subsections, respectively.

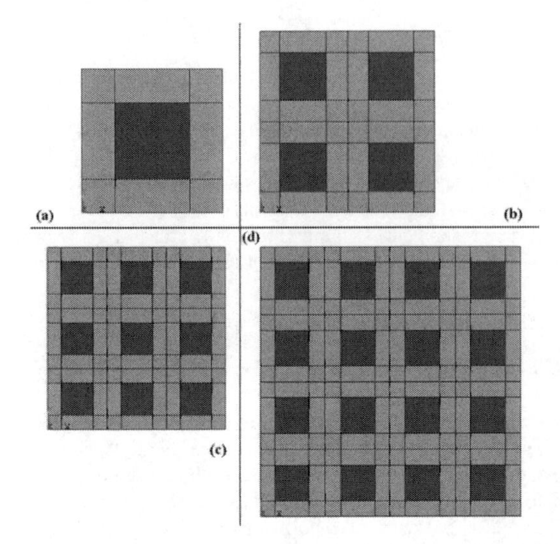

FIG. 7. Large domain analysis: (a) 1 unit cell, (b) 4 unit cells, (c) 9 unit cells, and (d) 16 unit cells.

The number of segments on line $\mathbf{t}-\mathbf{t}_0$ should be set as a value no less than 8. Through the analysis above, for this certain unit cell, lines **b, a**, and $\mathbf{t}-\mathbf{t}_0$ are decided to be divided into 6, 8, and 6 segments, respectively, to obtain a convergent result and at the same time minimize the computing rate.

Larger Domain Analysis

The precondition of the parameter calculation method using a single unit cell is that the piezoelectric rods must satisfy the restrictive assumption of periodic arrangement. To examine the accuracy of the single unit cell model in our problem, several models with larger number of cells are built. The 1-3-2 composite (volume fraction is 30%, $\alpha = 0.1$and $\beta = 0.025$) is taken as an example. Four models consisting of 1, 4, 9, and 16 unit cells, respectively, are built as shown in Fig. 7 and the longitudinal velocities are calculated. The results show that the four calculated velocities are exactly the same, as seen in Fig. 8. It means the accuracy of the single unit cell model is enough in our problem.

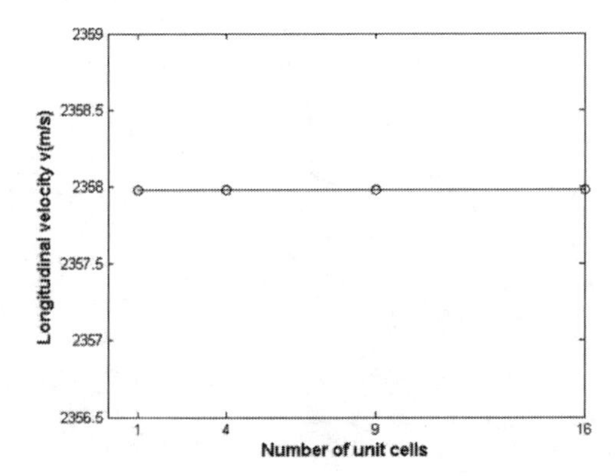

FIG. 8. The effect of the number of unit cells on the calculated longitudinal velocity.

THE IMPACT OF THE ASPECT RATIO α ON THE ELECTRO-ELASTIC PROPERTIES OF 2-1-3 AND 1-3-2 COMPOSITES

Comparisons of Longitudinal Velocity and Thickness Electromechanical Coupling Coefficient of 2-1-3 and 1-3-2 Composites

For the purpose of analyzing the electro-elastic properties of 1-3-2 and 2-1-3 composites, three volume fractions (10, 30, and 50%) are chosen and the elastic constants c_{33}^D, c_{33}^E are calculated by FEM according to *Eqs. 26–30*. Then the dependence of the longitudinal velocity v_l and the thickness electromechanical coupling coefficient k_t on the ratio α and β are calculated according to *Eqs. 23* and

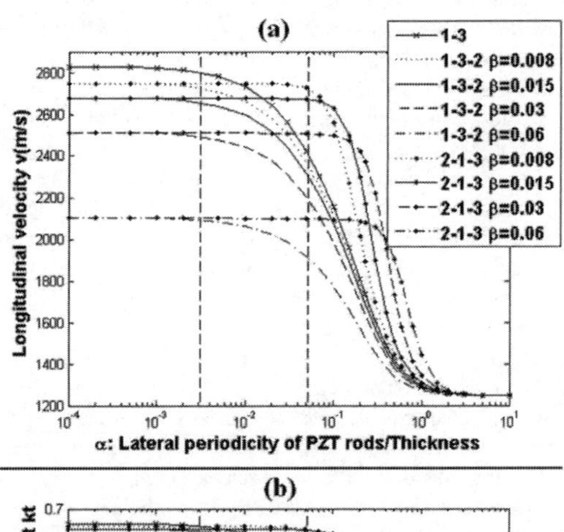

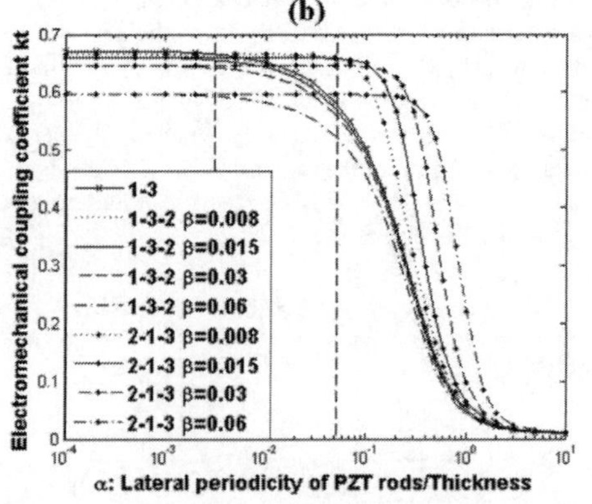

FIG. 9. The dependence of (a) longitudinal velocity, (b) thickness electromechanical coupling coefficient of 2-1-3 and 1-3-2 composites (VFP = 10%) on the ratio α and β (dashed lines refer to $\alpha = 0.0032$ and $\alpha = 0.052$).

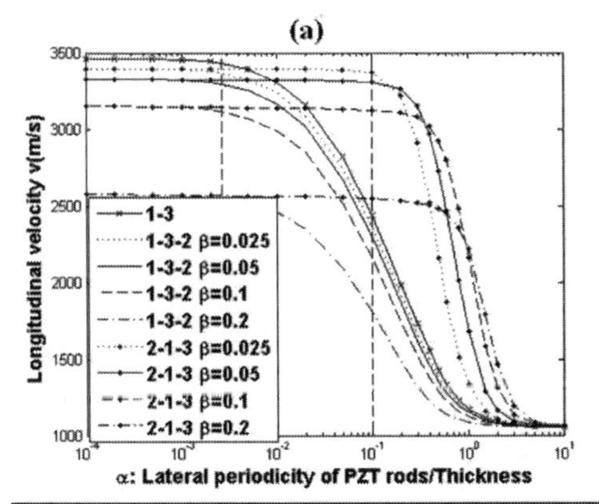

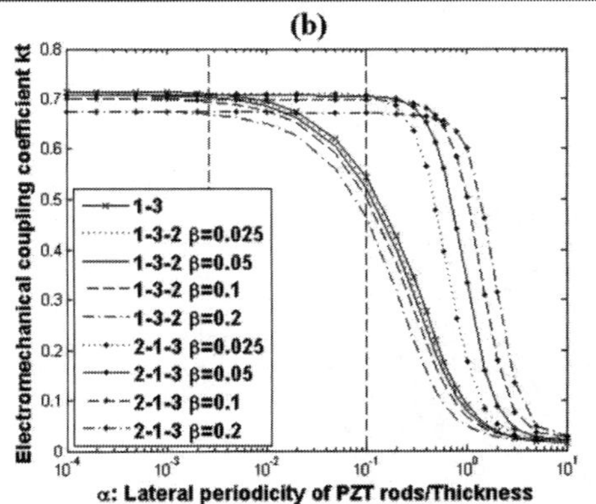

FIG. 10. The dependence of (a) longitudinal velocity, (b) thickness electromechanical coupling coefficient of 2-1-3 and 1-3-2 composites (VFP = 30%) on the ratio α and β (dashed lines refer to $\alpha = 0.0026$ and $\alpha = 0.10$).

25. The results of v_1 and k_t for volume fractions 10, 30, and 50% are shown in Figs. 9–11, respectively. For 1-3-2 and 2-1-3 composites (10%-volume fraction, $\beta = 0.008$; 30%-volume fraction, $\beta = 0.025$; 50%-volume fraction, $\beta = 0.04$), the values of α at which v_1 and k_t decrease by 1% of their own maximums (when $\alpha = 10^{-4}$) are called the critical values of α, which are rounded up and are marked with dash lines.

From the results shown in Figs. 9–11, it can be observed that when α is less than 0.0032 (Fig. 9, 10%-volume fraction), 0.0026 (Fig. 10, 30%-volume fraction), and 0.0038 (Fig. 11, 50%-volume fraction), the longitudinal velocity v_1 and the thickness electromechanical coupling coefficient k_t of 1-3, 1-3-2, and 2-1-3 composites almost remain unchanged. Within this range, 1-3, 1-3-2, and 2-1-3 composites can be treated as effective homogeneous mediums and are of fine lateral spatial scale. For the 30%-volume fraction case in Fig. 10, the values of v_1 and k_t within this range are the same as those calculated from the theoretical fine-scaled model that is shown in Fig. 2.

However, when α is larger than 0.0032 (Fig. 9, 10%-volume fraction), 0.0026 (Fig. 10, 30%-volume fraction), and 0.0038 (Fig. 11, 50%-volume fraction), which means the lateral scale is getting coarser and coarser, both v_1 and k_t of 1-3 and 1-3-2 composites will decrease quickly and reach their minimums, at which α is about 2. While v_1 and k_t of 2-1-3 composite will still remain the same until α gets to 0.0519 (Fig. 9, 10%-volume fraction, $\beta = 0.008$), 0.1025 (Fig. 10, 30%-volume fraction, $\beta = 0.025$), or 0.1671 (Fig. 11, 50%-volume fraction, $\beta = 0.04$). Additionally, the larger the ratio β is, the wider the range of α within which v_l and k_t keep constant will be.

The improvement of the properties of 2-1-3 composite can be illustrated by a simple example. For example, for the 30%-volume fraction composites, the thickness of the composite is set as $t = 1\,\text{cm}$ and the ratio β of both 2-1-3 and 1-3-2 composites is 0.025. In order to meet the conditions that the longitudinal velocity is 3,397 m/s or the thickness electromechanical coupling coefficient k_t reaches 0.7, the aspect ratio α of 1-3-2 composite should

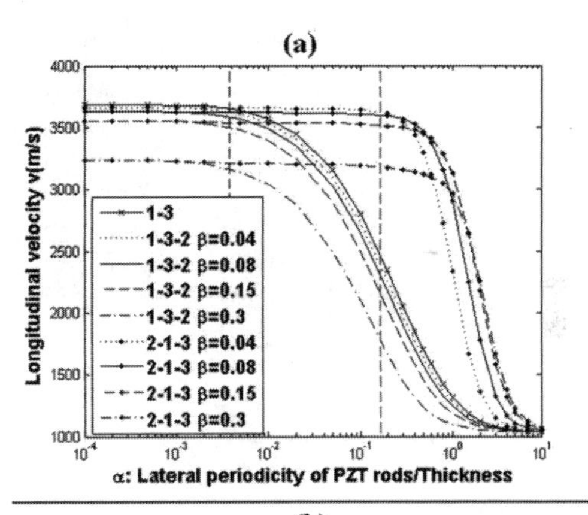

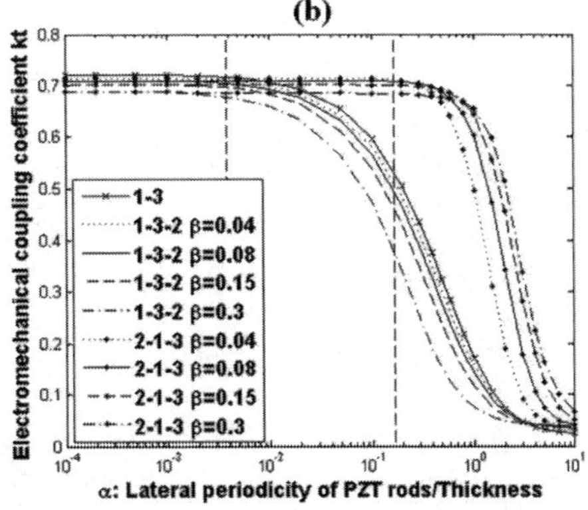

FIG. 11. The dependence of (a) longitudinal velocity and (b) thickness electromechanical coupling coefficient of 2-1-3 and 1-3-2 composites (VFP = 50%) on the ratio α and β (dashed lines refer to $\alpha = 0.0038$ and $\alpha = 0.17$).

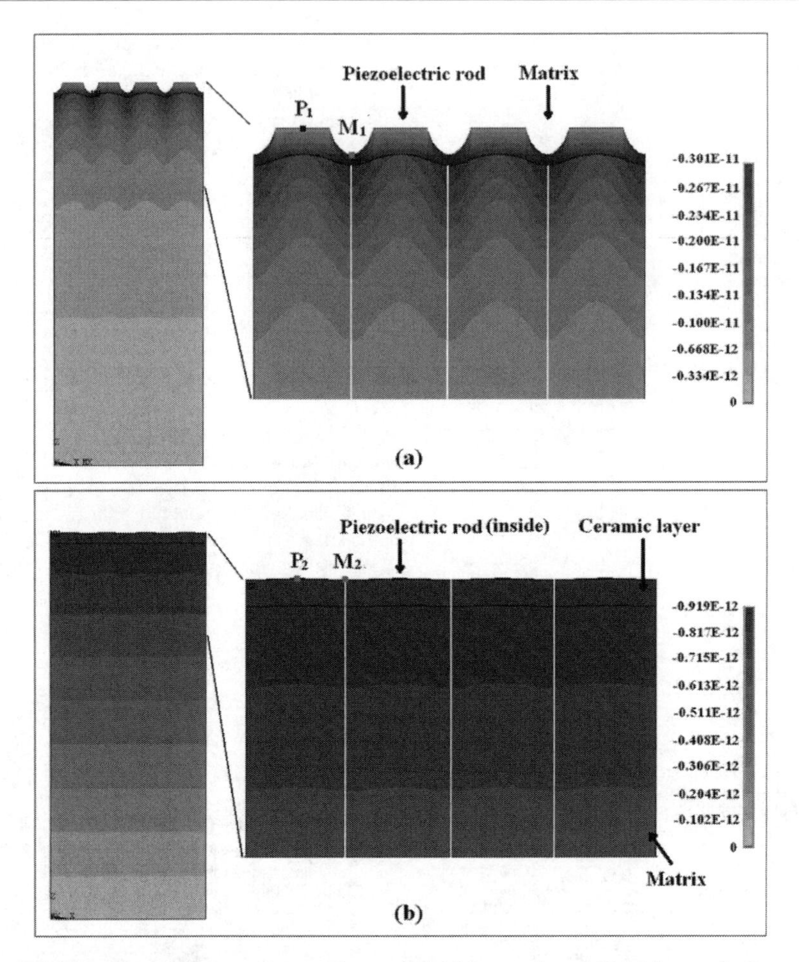

FIG. 12. z-Direction displacement contour figures of (a) 1-3-2 composite and (b) 2-1-3 composite (four unit cells combined, 30%-volume fraction, $\alpha = 0.1$, $\beta = 0.025$).

be no larger than 0.0026, whereas α of 2-1-3 composite only need to be no larger than 0.1025. Similarly, the lateral periodicity of PZT rods of 1-3-2 composite should be no larger than 26 μm while the lateral periodicity of 2-1-3 composite only need to be no larger than 1.025 mm. It is obvious that the fineness requirement of the manufacturing is largely reduced.

Comparisons of Strains on Surfaces of 2-1-3 and 1-3-2 Composites

The main reason why the properties of 2-1-3 composite are quite different from those of 1-3-2 composite is the difference of the strain distributions on surfaces of the two composites under the pressure load. In order to illustrate the strain distributions of the 2-1-3 and 1-3-2 composites intuitively, four unit cells of the composite are combined to represent part of the surface of the composite. 30%-volume fraction is chosen and α is set as 0.1 and β is set as 0.025. Under this condition, the longitudinal velocity and the thickness electromechanical coupling coefficient k_t of the two composites show distinct differences as seen in Fig. 10a and b. The same pressure (1 Pa) is exerted on the surfaces of the unit cell combinations of 1-3-2 and 2-1-3 composites, respectively. The strain distributions of both phases on the surface are shown in Figs. 12 and 13. Figure 12 depicts z-direction displacement contour figures of the two composites. Figure 13 depicts z-direction displacement distribution on line P_1M_1 of 1-3-2 composite and P_2M_2 of 2-1-3 composite.

As seen in Figs. 12 and 13, under the action of the pressure, the matrix phase of 1-3-2 composite presents displacement with amplitude much larger than that of piezoelectric phase. Much of the force applied on the 1-3-2

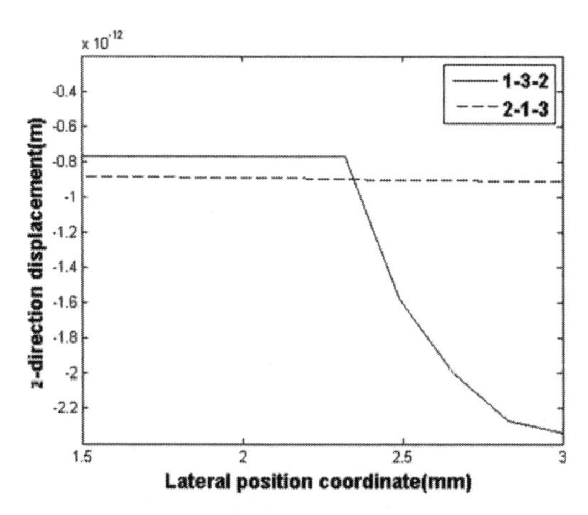

FIG. 13. z-Direction displacement distribution on line P_1M_1 of 1-3-2 composite and P_2M_2 of 2-1-3 composite in Fig. 12.

composite simply compresses the passive phase instead of being applied effectively on the piezoelectric rods that can conduct energy transformation. While for the 2-1-3 composite, because of the strength of the PZT cover layer, the displacements of the two phases are almost the same. Hence, the change of the strain components in z-direction of both piezoelectric phase and matrix phase are roughly equal, which makes more force applied on the surface effectively apply to the PZT rods under the ceramic layer and be converted into electrical energy.

Temperature Stability and Mechanical Stability Discussion

It has been stated in Ref. 7 that 1-3-2 piezoelectric composite has the advantages of series and parallel combination in structure. There are ceramic framework supports in the transverse and longitudinal directions in the composite structure. Therefore, the composite does not deform easily in the fabrication process and is free from the influence of outside mechanical impact and the environment temperature change.

The structure of 2-1-3 composite is the same as that of 1-3-2 composite. Both of them consist of a 1-3 piezoelectric composite and a ceramic layer. For a 1-3-2 composite, the ceramic layer is a base layer while for a 2-1-3 composite, the ceramic layer is a cover layer. 2-1-3 composites and 1-3-2 composites can be manufactured by exactly the same way. Therefore, 2-1-3 composites also possess the advantages in structure as well as good temperature and mechanical stability that 1-3-2 composite has. In addition, 2-1-3 composite can loosen the fineness requirement of the manufacturing while maintain the similar performance of the 1-3-2 composite. It means the piezoelectric rod in 2-1-3 composite is thicker than that in 1-3-2 composite. So, the mechanical stability of 2-1-3 composite will be better than 1-3-2 composite.

DISCUSSIONS ON SOFT MATRIX AND HARD MATRIX

It is seen above that for the composites with soft matrix, the 2-1-3 composite shows great improvement in loosening the fineness requirement compared with 1-3-2 composite. However, the improvement will be affected by the hardness of the matrix. Another matrix material epoxy with higher stiffness (parameters are listed in Table 1) is also introduced to make a comparison with the soft matrix rubber. The volume fraction is set as 30% and the ratio β is set as 0.025. Figure 14a and b gives the dependence of the longitudinal velocity of 2-1-3 and 1-3-2 composites with epoxy matrix on the aspect ratio α. Figure 14b gives the dependence of the coupling coefficient of 2-1-3 and 1-3-2 composites with epoxy matrix on the aspect ratio α. For comparison, the results of the two composites with rubber matrix are also given. The values of α at which v_l and k_t decrease by 1% of their own maximums (when $\alpha = 10^{-4}$) are marked with dash lines for soft matrix and solid lines for hard matrix.

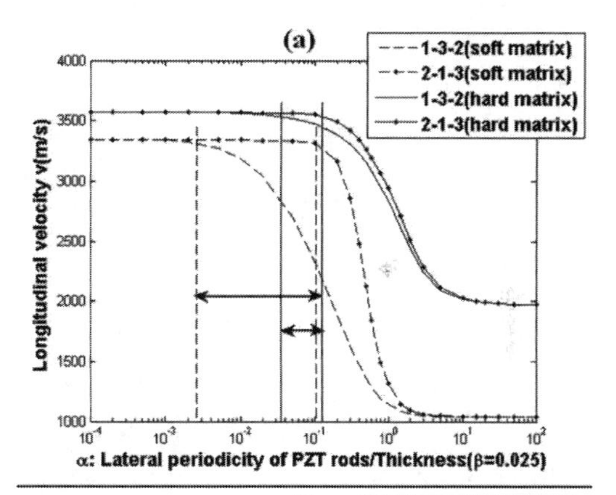

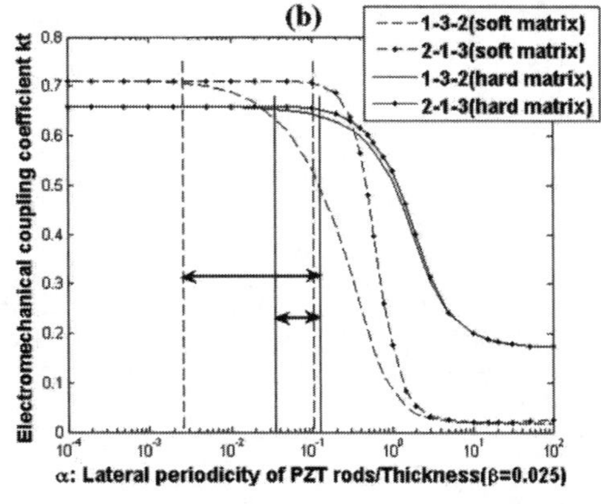

FIG. 14. The dependence of (a) longitudinal velocity and (b) thickness electromechanical coupling coefficient of 2-1-3 and 1-3-2 composites with soft and hard matrix (VFP = 30%) on the aspect ratio α ($\beta = 0.025$) (dashed lines for soft matrix refer to $\alpha = 0.0026$ and $\alpha = 0.10$, solid lines for hard matrix refer to $\alpha = 0.036$ and $\alpha = 0.12$).

As seen in Fig. 14a and b, the variations of v_l and k_t of 1-3-2 composite with hard matrix are similar to those of 1-3-2 composite with soft matrix. Namely, there is a critical value of the aspect ratio α beyond which v_l and k_t will quickly decrease until to their minimums. This critical value of α of 1-3-2 composite with hard matrix (0.0356) is much larger than that of 1-3-2 composite with soft matrix (0.0026).

The improvement in loosening the fineness requirement of 2-1-3 composite compared with 1-3-2 composite also exists for 2-1-3 and 1-3-2 composites with hard matrix, but the improvement is less remarkable than that of the soft matrix composites. For soft matrix composites, the critical value of α increases from 0.0026 of 1-3-2 composite to 0.10 of 2-1-3 composite. For hard matrix composites, the critical value of α increases from 0.036 of 1-3-2 composite to 0.12 of 2-1-3 composite. That is to say, the softer the matrix is, the more conspicuous the improvement in loosening the fineness requirement of 2-1-3 composite will be.

CONCLUSIONS

In this article, a 2-1-3 (a revised version of 1-3-2) composite which is composed of 1-3 composite and a ceramic cover layer is proposed to reduce the fineness requirement of the manufacturing of the 1-3 and 1-3-2 composites. The FEM has been effectively used to analyze the dependence of the electro-elastic properties of 1-3, 1-3-2, and 2-1-3 composites on the aspect ratio α.

The limitation is obtained that for the 1-3 and 1-3-2 composites, the aspect ratio α must be limited to be small enough (0.0026 for 30%-volume fraction) to maintain large longitudinal velocity v_l and high thickness electromechanical coupling coefficient k_t of the composites. It is also the limitation for the theoretical assumption. If α is beyond the limitation, v_l and k_t will decrease until to their minimums, which makes extremely strict demands for the fineness of the manufacturing.

The 2-1-3 composite can overcome the shortcomings of the 1-3 and 1-3-2 composites by approximately satisfying the "same strain in both phases" condition when the composite is coarse-scaled. For composites with rubber matrix, v_l and k_t of 2-1-3 composite can continue to keep constant until α is as large as 0.1025 for 30%-volume fraction case.

The difference of the properties of 2-1-3 and 1-3-2 composites is attributed to the difference lies in the degree of satisfaction of the "same strain in both phases" condition under the pressure load. The structure of the 2-1-3 composite, which combines the ceramic cover layer with high stiffness and the piezoelectric rods supporting the cover layer under it, is benefit for making the pressure load of the composite effectively applied on the piezoelectric phase which can conduct energy conversion instead of being wasted on the passive phase.

The results of the electro-elastic properties show that the 2-1-3 composite can reduce the manufacturing cost while maintaining the similar performance of the 1-3-2 composite.

REFERENCES

1. K.A. Klicker, J.V. Biggers, and R.E. Newnham. *J. Am. Ceram. Soc.*, **64**, 5 (1981).
2. T.R. Gururaja, W.A. Schulze, T.R. Shrout, A. Safari, L. Webster, and L.E. Cross, *Ferroelectrics*, **39**, 1245 (1981).
3. B.A. Auld, H.A. Kunkel, Y.A. Shui, and Y. Wang, *Proc. IEEE Ultrason. Symp.*, **1**, 554 (1983).
4. W.A. Smith, A.A Shaulov, and B.A. Auld, *Ferroelectrics*, **91**, 155 (1989).
5. R. Gentilman, D. Fiore, W. Serwatka, H. Pham, B. Pazol, C. Near, and L. Bowen, "OCEANS'95. MTS/IEEE. Challenges of Our Changing Global Environment," *Conference Proceedings*, Vol. **3**, 2032 (1995).
6. L. Guang, W. Li-Kun, L. Gui-Dong, Z. Jin-Duo, and L. Shu-Xiang, *Ultrasonics*, **44**, e639 (2006).
7. L. Guang, W. Li-Kun, L. Gui-Dong, Z. Jin-Duo, and L. Shu-Xiang, *Ultrasonics*, **44**, e673 (2006).
8. W.A. Smith, A. Shaulov, and B.A. Auld, *Proc. IEEE Ultrason. Symp.*, **2**, 642 (1985).
9. W.A. Smith and B.A. Auld, *IEEE Trans. Ultrason. Ferroelectr. Freq. Control*, **38**, 40 (1991).
10. W.A. Smith, *IEEE Trans. Ultrason. Ferroelectr. Freq. Control*, **40**, 41 (1993).
11. H.L.W. Chan and J. Unsworth, *IEEE Trans. Ultrason. Ferroelectr. Freq. Control*, **36**, 434 (1989).
12. L. Li, L. Qin, L.-K. Wang, Y.-Y. Wan, and B.-S. Sun, *IEEE Trans. Ultrason. Ferroelectr. Freq*, **55**, 946 (2008).
13. L. Li, L.-K. Wang, L. Qin, and Y. Lv, *IEEE Trans. Ultrason. Ferroelectr. Freq.*, **56**, 1476 (2009).
14. M. Sakthivel and A. Arockiarajan, *Sens. Actuators A*, **167**, 34 (2011).
15. R.E. Newnham, D.P. Skinner, and L.E. Cross, *Mater. Res. Bull.*, **13**, 525 (1978).
16. J.A. Hossack and G. Hayward, *IEEE Trans. Ultrason. Ferroelectr. Freq. Control*, **38**, 618 (1991).
17. C.V. Madhusudhana Rao and G. Prasad, *Condens. Matter Phys.*, **13**, 13703 (2010).
18. R.J. Wang, *Underwater Acoustic Materials Manual*, Science Press, Beijing (1983).
19. H.E. Pettermann and S. Suresh, *Int. J. Solids Struct.*, **37**, 5447 (2000).

Singular variation property of elastic constants of piezoelectric ceramics shunted to negative capacitance①

Ji-Ying Hu(胡吉英)[1], Zhao-Hui Li(李朝晖)[1], and Qi-Hu Li(李启虎)[2]

[1] Department of Electronics, Peking University, Beijing 100871, China
[2] Advanced Technology Institute, Peking University, Beijing 100871, China

(Received 19 July 2017; revised manuscript received 11 September 2017; published online 20 November 2017)

Piezoelectric shunt damping has been widely used in vibration suppression, sound absorption, noise elimination, etc. In such applications, the variant elastic constants of piezoelectric materials are the essential parameters that determine the performances of the systems, when piezoelectric materials are shunted to normal electrical elements, i.e., resistance, inductance and capacitance, as well as their combinations. In recent years, many researches have demonstrated that the wideband sound absorption or vibration suppression can be realized with piezoelectric materials shunted to negative capacitance. However, most systems using the negative-capacitance shunt circuits show their instabilities in the optimal condition, which are essentially caused by the singular variation properties of elastic constants of piezoelectric materials when shunted to negative capacitance. This paper aims at investigating the effects of negative-capacitance shunt circuits on elastic constants of a piezoelectric ceramic plate through theoretical analyses and experiments, which gives an rational explanation for why negative capacitance shunt circuit is prone to make structure instable. First, the relationships between the elastic constants c_{11}, c_{33}, c_{55} of the piezoelectric ceramic and the shunt negative capacitance are derived with the piezoelectric constitutive law theoretically. Then, an experimental setup is established to verify the theoretical results through observing the change of elastic constant c_{55} of the shunted piezoelectric plate with the variation of negative capacitance. The experimental results are in good agreement with the theoretical analyses, which reveals that the instability of the shunt damping system is essentially caused by the singular variation property of the elastic constants of piezoelectric material shunted to negative capacitance.

Keywords: piezoelectric ceramics, elastic constant, shunt damping, negative capacitance

PACS: 77.90.+k, 62.20.–x, 62.20.de, 81.40.Jj **DOI:** 10.1088/1674-1056/26/12/127702

1. Introduction

The piezoelectric shunt damping used in vibration suppression,[1–4] sound absorption,[5–8] noise elimination,[9,10] as well as wave propagation control[11] has been widely studied in the past few years. Because piezoelectric materials take the role in converting the mechanical vibration energy into electrical energy which is dissipated through passive circuits, shunt damping is also referred to as the semi-active technology.[5]

Resistance and inductance (RL) shunt circuits, connecting in series or parallel, are the first proposed efficient circuits for piezoelectric shunt damping[1,2] to suppress single mode vibration. Successively, multiple modes damping with resistance, inductance and capacitance (RLC) branches have been investigated.[12–14] However, there are a number of problems associated with these circuits, of which the foremost ones are the complexity and size of the circuit required to implement the total impedance. Typically, the shunt circuits for low frequency applications need large inductance values up to 1000 s of Henries, therefore Riordan gyrators[15] are required to implement the inductor elements, which are large in size and sensitive to component tolerances. Besides, the effectiveness of the systems with RL shunt circuits is generally limited to a narrow frequency range around each resonant frequency.

In recent years, the negative capacitance technique used in piezoelectric shunt damping has been proved to be able to improve the bandwidth performance of vibration suppression.[16,17] The effect of negative capacitance is to cancel out the inherent capacitive impedance of the piezoelectric material and maximize the energy dissipation in a resistor, with which the wideband matching condition is realized and multiple modes are suppressed. For example, Behrens *et al.*[17] adopted the negative capacitance in series with a resistance to suppress the vibration of clamped beam , through which the resonant amplitudes of the first five modes are reduced by 6.1, 16.3, 15.2, 11.7, and 10.2 dB experimentally. As a negative capacitance circuit can affect the elasticity of the piezoelectric ceramic, piezoelectric shunt damping techniques are also applied to sound absorption.[18,19] Fukada *et al.* showed the efficiency improvement of a negative capacitance circuit connected to a curved PVDF film in sound isolation application.[18] Changing the elasticity of the piezo film, the overall transmission loss level of 40 dB was achieved. Yu *et al.* also demonstrated that a negative capacitance combining with a proper resistance can achieve wideband sound absorption performance.[19]

① Chin. Phys. B, 2017, 26(12): 127702(1-10).

Although negative capacitance shunt circuits have the above advantages when applied to sound absorption or vibration suppression systems, they may bring instability problems to the structures if improperly tuned.[20,21] Han *et al.* proved that the highest performance of vibration suppression is reached when the external capacitance approaches to the negative value of the inner piezoelectric capacitance, which is just the stability boundary of the system.[22] Neubauer and Wallaschek pointed out that the negative capacitance is essentially an active circuit that can destabilize the structure if it is improperly tuned.[23] They also analyzed and derived the stability condition specifically in view of energy consumption.

Among the above-mentioned applications, the piezoelectric materials shunted to negative capacitance mainly working in one of three modes: the transverse mode, the longitudinal mode and the shear mode.[2] Elastic constants c_{11}, c_{33}, and c_{55} of piezoelectric materials corresponding to the above three modes are the essential parameters which determine the performances of the vibration suppression and sound absorption systems. In fact, the singular variation properties of elastic constants with shunt negative capacitance are the inner reason for the instabilities of shunt damping systems. So it is of great research value to investigate the relationships between the elastic constants and the shunt negative capacitance, which is also a foundation to solve the instability problem when designing vibration suppression and sound absorption systems by using piezoelectric materials shunted to negative capacitance. Thus in this paper, first, the piezoelectric constitutive equations are used to derived the relationships between the elastic constants c_{11}, c_{33}, c_{55} and the negative capacitance respectively. Then, the theoretical results are verified through experimental observations. Taking the test of the elastic constant c_{55} for example, an experimental setup is established to test the influence of negative capacitance on c_{55} indirectly.[24]

The rest of this paper is organized as follows. In Section 2 analyzed are the relationships between the elastic constants of piezoelectric materials and the shunt negative capacitance theoretically with piezoelectric constitutive equations. In Section 3, the experiment setup and the experimental results are presented. In Section 4, the inner reason why the negative circuit is prone to make systems unstable is discussed. Finally, some conclusions are drawn in Section 5.

2. Theoretical analyses

In this section the influence of negative capacitance on elastic constants of piezoelectric materials theoretically is analyzed. It reveals that the singular variation properties of elastic constants of the piezoelectric materials are the inner cause of the instability of system shunted to negative capacitance.

2.1. Stability boundary of the shunt damping system with shunted negative capacitance.

In Ref. [20], Neubauer *et al.* employed a thickness-mode mechanical resonator as shown in Fig. 1, to derive the negative capacitance range which makes the system unstable from the view of resonance frequency of the structure. The range of negative capacitance is derived as

$$-1<\frac{C_{\mathrm{p}}}{C}<-\frac{1}{1+\gamma}, \tag{1}$$

$$\gamma=\frac{\left(k_{1}d_{33}\right)^{2}}{C_{\mathrm{p}}\left(k_{0}+k_{1}\right)}, \tag{2}$$

where C_{p} is the inner capacitance of the piezoelectric element under a constant strain, C is the value of negative capacitance, d_{33} is the charge density per unity stress under a constant electric field, and $k_1=c_{33}(A_{\mathrm{p}}/l_{\mathrm{p}})$ is the mechanical stiffness of the piezoelectric material, with c_{33} being the elastic constant of thickness mode. If the mechanical spring-damper (k_0, d_0) in the model in Fig. 1 is omitted, with k_0 and d_0 being the stiffness and the damping coefficient respectively, γ is simplified into

$$\gamma=\frac{\left(k_{1}d_{33}\right)^{2}}{C_{\mathrm{p}}k_{1}}=k_{33}^{2}, \tag{3}$$

where k_{33} is electromechanical coupling coefficient.

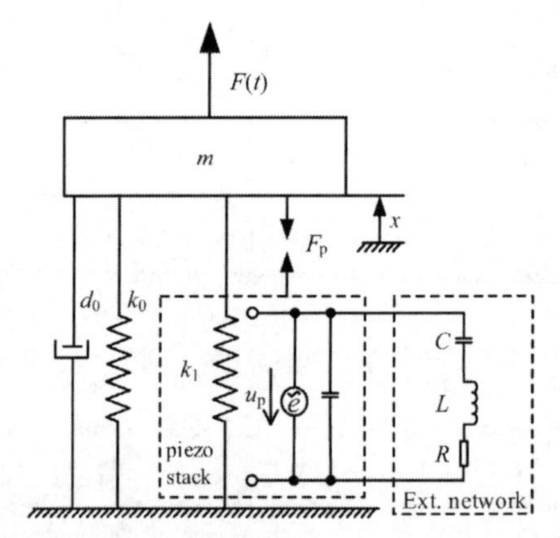

Fig. 1. Mechanical model with piezoelectric element connected to external LRC-network cited from Ref. [20].

Then the range of negative capacitance making the system unstable is obtained to be

$$-\left(1+k_{33}^{2}\right)C_{\mathrm{p}}<C<-C_{\mathrm{p}}. \tag{4}$$

Inequality (4) gives the unstable range of the shunted negative capacitance, but without giving a rational explanation. In fact, when the value of shunt negative capacitance is in a specific range, the elastic constant is lower than zero, which is not meaningful as shown in latter parts of this paper.

2.2. Influences of negative capacitance on elastic constants of the piezoelectric materials

The diagram of a piezoelectric plate shunted to a negative capacitor is shown in Fig. 2(a), which may work in three modes: the transverse mode, the longitudinal mode and the shear mode. At low frequency, the piezoelectric plate is usually modeled as a capacitor C_p in series with a voltage source V_p. The equivalent circuit of the model can be depicted in Fig. 2(b), in which the function of the negative capacitor is mainly to counteract the inner capacitance C_p of the piezoelectric plate. Obviously, it has the advantage of simple implementation when acting as a multiple-mode vibration controller or broadband sound absorber.

Denote $\boldsymbol{V}$ and $\boldsymbol{I}$ as the voltage and current arrays, i.e., use V_i and I_i $(i=1,2,3)$ to represent the voltages and currents when electrodes are in three directions respectively. According to the basic electrical circuit, the relationships between voltage $\boldsymbol{V}$ and electric field $\boldsymbol{E}$, and between current $\boldsymbol{I}$ and electric displacement $\boldsymbol{D}$ in the Laplace domain can be written as

$$\boldsymbol{V}(s)=\boldsymbol{L}\cdot\boldsymbol{E}(s), \tag{5a}$$

$$\boldsymbol{I}(s)=s\boldsymbol{A}\cdot\boldsymbol{D}(s), \tag{5b}$$

where $\boldsymbol{L}$ is a diagonal matrix of the length of the piezoelectric plate, and L_i $(i=1,2,3)$ denotes the length in the ith direction. $\boldsymbol{A}$ is the diagonal matrix of the area of surface, and A_i $(i=1,2,3)$ denotes the area perpendicular to the i-th direction, and s is the Laplace transform variable. Their forms are as follows:

$$\boldsymbol{V}=\begin{bmatrix}V_1\\V_2\\V_3\end{bmatrix},\quad \boldsymbol{I}=\begin{bmatrix}I_1\\I_2\\I_3\end{bmatrix},\quad \boldsymbol{E}=\begin{bmatrix}E_1\\E_2\\E_3\end{bmatrix},\quad \boldsymbol{D}=\begin{bmatrix}D_1\\D_2\\D_3\end{bmatrix},$$

$$\boldsymbol{A}=\begin{bmatrix}A_1&0&0\\0&A_2&0\\0&0&A_3\end{bmatrix},\quad \boldsymbol{L}=\begin{bmatrix}L_1&0&0\\0&L_2&0\\0&0&L_3\end{bmatrix}.$$

The e-type piezoelectric constitutive law is convenient in this case, which can be written from Ref. [25] as

$$\boldsymbol{D}=\boldsymbol{\varepsilon}^{\mathrm{S}}\boldsymbol{E}+\boldsymbol{e}\boldsymbol{S}, \tag{6a}$$

$$\boldsymbol{T}=-\tilde{\boldsymbol{e}}\boldsymbol{E}+\boldsymbol{c}^{\mathrm{E}}\boldsymbol{S}, \tag{6b}$$

where $\boldsymbol{T}$ is the vector expression of the mechanical stress tensor, T_h $(h=1,2,3,4,5,6)$, and $\boldsymbol{S}$ is that of the mechanical strain tensor, S_h $(h=1,2,3,4,5,6)$. The matrix $\boldsymbol{e}$ denotes the piezoelectric stress coupling matrix, $\tilde{\boldsymbol{e}}$ is the transposed matrix of $\boldsymbol{e}$, c^{E} is the elastic constant matrix at a constant electric field, and $\boldsymbol{\varepsilon}^{\mathrm{S}}$ is the permittivity matrix under a constant strain. The element forms of the above tensors and matrixes are referred to Ref. [25], which are written as follows:

$$\boldsymbol{S}=\begin{bmatrix}S_1\\S_2\\S_3\\S_4\\S_5\\S_6\end{bmatrix},\quad \boldsymbol{T}=\begin{bmatrix}T_1\\T_2\\T_3\\T_4\\T_5\\T_6\end{bmatrix},$$

$$c^{\mathrm{E}}=\begin{bmatrix}c_{11}^{\mathrm{E}}&c_{12}^{\mathrm{E}}&c_{13}^{\mathrm{E}}&0&0&0\\c_{12}^{\mathrm{E}}&c_{11}^{\mathrm{E}}&c_{13}^{\mathrm{E}}&0&0&0\\c_{13}^{\mathrm{E}}&c_{13}^{\mathrm{E}}&c_{33}^{\mathrm{E}}&0&0&0\\0&0&0&c_{55}^{\mathrm{E}}&0&0\\0&0&0&0&c_{55}^{\mathrm{E}}&0\\0&0&0&0&0&c_{66}^{\mathrm{E}}\end{bmatrix},$$

$$\varepsilon^{\mathrm{S}}=\begin{bmatrix}\varepsilon_1^{\mathrm{S}}&0&0\\0&\varepsilon_1^{\mathrm{S}}&0\\0&0&\varepsilon_3^{\mathrm{S}}\end{bmatrix},$$

$$\boldsymbol{e}=\begin{bmatrix}0&0&0&0&d_{15}&0\\0&0&0&d_{15}&0&0\\d_{31}&d_{31}&d_{33}&0&0&0\end{bmatrix}.$$

Combining Eqs. (5) and Eqs. (6), the piezoelectric constitutive equations in terms of the current and the voltage can be written as:

$$\boldsymbol{I}=s\boldsymbol{A}\boldsymbol{\varepsilon}^{\mathrm{S}}\boldsymbol{L}^{-1}\boldsymbol{V}+s\boldsymbol{A}\boldsymbol{e}\boldsymbol{S}, \tag{7a}$$

$$\boldsymbol{T}=-\tilde{\boldsymbol{e}}\boldsymbol{L}^{-1}\boldsymbol{V}+\boldsymbol{c}^{\mathrm{E}}\boldsymbol{S}. \tag{7b}$$

With the inner capacitance of the piezoelectric plate under a constant strain expressed as

$$\boldsymbol{C}_{\mathrm{p}}^{\mathrm{S}}=\boldsymbol{A}\boldsymbol{\varepsilon}^{\mathrm{S}}\boldsymbol{L}^{-1}, \tag{8}$$

equation (7a) can be simplified into

$$\boldsymbol{I}=s\boldsymbol{C}_{\mathrm{p}}^{\mathrm{S}}\boldsymbol{V}+s\boldsymbol{A}\boldsymbol{e}\boldsymbol{S}=\boldsymbol{Y}^{\mathrm{D}}(s)\boldsymbol{V}+s\boldsymbol{A}\boldsymbol{e}\boldsymbol{S}, \tag{9}$$

where $\boldsymbol{Y}^{\mathrm{D}}(s)$ is the admittance matrix of the piezoelectric plate in an open-circuit state. When the piezoelectric plate is shunted to a negative capacitor as shown in Fig. 2, the admittances of the shunted piezoelectric plate in three directions, denoted by a diagonal matrix Y^{SH}, can be obtained to be

$$\boldsymbol{Y}^{\mathrm{SH}}=s(\boldsymbol{C}_{\mathrm{p}}^{\mathrm{S}}-\boldsymbol{C}_x), \tag{10}$$

where

$$\boldsymbol{Y}^{\mathrm{SH}}(s)=\begin{bmatrix}Y_1^{\mathrm{SH}}(s)&0&0\\0&Y_2^{\mathrm{SH}}(s)&0\\0&0&Y_3^{\mathrm{SH}}(s)\end{bmatrix},$$

$$\boldsymbol{C}_{\mathrm{p}}^{\mathrm{S}}=\begin{bmatrix}C_{\mathrm{p}1}^{\mathrm{S}}&0&0\\0&C_{\mathrm{p}2}^{\mathrm{S}}&0\\0&0&C_{\mathrm{p}3}^{\mathrm{S}}\end{bmatrix},$$

$$\boldsymbol{C}_x=\begin{bmatrix}C_{x1}&0&0\\0&C_{x2}&0\\0&0&C_{x3}\end{bmatrix}.$$

So the governing constitutive equation for a piezoelectric plate is

$$\boldsymbol{I}=\boldsymbol{Y}^{\mathrm{SH}}\boldsymbol{V}+s\boldsymbol{AeS}=s\left(\boldsymbol{C}_{\mathrm{p}}^{\mathrm{S}}-\boldsymbol{C}_x\right)\boldsymbol{V}+s\boldsymbol{AeS}. \quad (11)$$

From Eq. (11), the voltage is obtained to be

$$\boldsymbol{V}=\boldsymbol{Z}^{\mathrm{SH}}(\boldsymbol{I}-s\boldsymbol{AeS}), \quad (12)$$

where $\boldsymbol{Z}^{\mathrm{SH}}=(\boldsymbol{Y}^{\mathrm{SH}})^{-1}=(\boldsymbol{C}_{\mathrm{p}}^{\mathrm{S}}-\boldsymbol{C}_x)^{-1}/s$ is the impedance of a piezoelectric plate shunted to a negative capacitor. By inserting Eq. (12) into Eq. (7b), the stress tensor can be obtained as follows:

$$\boldsymbol{T}=\left(\boldsymbol{c}^{\mathrm{E}}+\tilde{\boldsymbol{e}}\boldsymbol{L}^{-1}\boldsymbol{Z}^{\mathrm{SH}}s\boldsymbol{Ae}\right)\boldsymbol{S}-\tilde{\boldsymbol{e}}\boldsymbol{L}^{-1}\boldsymbol{Z}^{\mathrm{SH}}\boldsymbol{I}. \quad (13)$$

The elastic constant array of the shunt piezoelectric plate under constant electric displacement, i.e. constant $\boldsymbol{I}$, can be defined from Eq. (13) as

$$\begin{aligned}\boldsymbol{c}^{\mathrm{SH}}&=\boldsymbol{c}^{\mathrm{E}}+\tilde{\boldsymbol{e}}L^{-1}\boldsymbol{Z}^{\mathrm{SH}}s\boldsymbol{Ae}\\&=\boldsymbol{c}^{\mathrm{E}}+\tilde{\boldsymbol{e}}\boldsymbol{C}_{\mathrm{p}}^{\mathrm{S}}\boldsymbol{e}(\boldsymbol{C}_{\mathrm{p}}^{\mathrm{S}}-\boldsymbol{C}_x)^{-1}(\boldsymbol{\varepsilon}^{\mathrm{S}})^{-1}.\end{aligned} \quad (14)$$

By defining the non-dimensional electrical impedance

$$\eta^{\mathrm{SH}}=\boldsymbol{Z}^{\mathrm{SH}}(\boldsymbol{Z}^{\mathrm{D}})^{-1}=\boldsymbol{C}_{\mathrm{p}}^{\mathrm{S}}(\boldsymbol{C}_{\mathrm{p}}^{\mathrm{S}}-\boldsymbol{C}_x)^{-1}, \quad (15)$$

equation (16) can be simplified into

$$\boldsymbol{c}^{\mathrm{SH}}=\boldsymbol{c}^{\mathrm{E}}+\tilde{\boldsymbol{e}}\eta^{\mathrm{SH}}(\boldsymbol{\varepsilon}^{\mathrm{S}})^{-1}\boldsymbol{e}. \quad (16)$$

As the shunted piezoelectric plate usually works in one of three modes: the transverse mode, the longitudinal mode, and the shear mode, the corresponding elastic constants for the three modes are c_{11}, c_{33}, and c_{55} respectively. Therefore, the electromechanical coupling coefficients of the three modes are as follows:

$$k_{31}=k_{32}=e_{31}/\sqrt{c_{11}^{\mathrm{E}}\varepsilon_3^{\mathrm{S}}}, \quad (17a)$$

$$k_{33}=e_{33}/\sqrt{c_{33}^{\mathrm{E}}\varepsilon_3^{\mathrm{S}}}, \quad (17b)$$

$$k_{15}=k_{24}=e_{15}/\sqrt{c_{55}^{\mathrm{E}}\varepsilon_1^{\mathrm{S}}}, \quad (17c)$$

respectively. By substituting Eqs. (17a)–(17b) into Eq. (16) separately, the elastic constants of the three modes when the piezoelectric plate is shunted to negative capacitance can be obtained as follows:

$$c_{11}^{\mathrm{SH}}=c_{11}^{\mathrm{E}}\left(1+k_{31}^2\eta_3^{\mathrm{SH}}\right)=c_{11}^{\mathrm{E}}\left(1+k_{31}^2\frac{C_{\mathrm{p3}}^{\mathrm{S}}}{C_{\mathrm{p3}}^{\mathrm{S}}-C_{x3}}\right), \quad (18a)$$

$$c_{33}^{\mathrm{SH}}=c_{33}^{\mathrm{E}}\left(1+k_{33}^2\eta_3^{\mathrm{SH}}\right)=c_{33}^{\mathrm{E}}\left(1+k_{33}^2\frac{C_{\mathrm{p3}}^{\mathrm{S}}}{C_{\mathrm{p3}}^{\mathrm{S}}-C_{x3}}\right), \quad (18b)$$

$$c_{55}^{\mathrm{SH}}=c_{55}^{\mathrm{E}}\left(1+k_{15}^2\eta_1^{\mathrm{SH}}\right)=c_{55}^{\mathrm{E}}\left(1+k_{15}^2\frac{C_{\mathrm{p1}}^{\mathrm{S}}}{C_{\mathrm{p1}}^{\mathrm{S}}-C_{x1}}\right). \quad (18c)$$

From Eqs. (18a)–(18c), it can be seen that the elastic constants of a piezoelectric plate shunted to a negative capacitor are determined by the value of the negative capacitance, the electromechanical coupling coefficient and the elastic constants of the piezoelectric material in short-circuit state. When the shunted circuit is in an open-circuit state, i.e., $C_x=0$ and $\eta^{\mathrm{SH}}=1$, the expressions in Eqs. (18a)–(18c) become:

$$c_{11}^{\mathrm{D}}=c_{11}^{\mathrm{E}}\left(1+k_{31}^2\right), \quad (19a)$$

$$c_{33}^{\mathrm{D}}=c_{33}^{\mathrm{E}}\left(1+k_{33}^2\right), \quad (19b)$$

$$c_{55}^{\mathrm{D}}=c_{55}^{\mathrm{E}}\left(1+k_{15}^2\right), \quad (19c)$$

which are the normal relationships between the elastic constants in short-circuit state and that in open-circuit state.

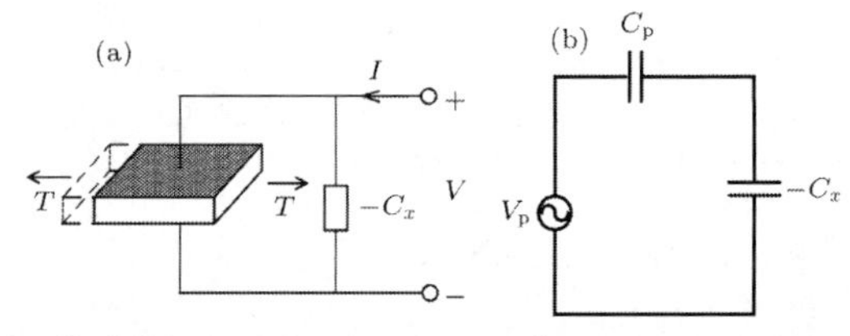

Fig. 2. Piezoelectric plate shunted to a negative capacitor: (a) physical model, (b) corresponding low-frequency equivalent circuit.

Taking piezoelectric material PZT-5A for example, elastic constants in Eqs. (18a)–(18c) are calculated and analyzed below. In order to normalize the influence of cross sectional area of the piezoelectric plate, the ratio of the value of the negative capacitor C_x to cross sectional area of the piezoelectric plate A is defined as the shunt capacitance ratio (SCR), i.e., SCR $=C_x/A$, whose unit is F/m^2. According to Eqs. (18a)–(18c), with increase of the absolute value of the negative capacitance, the changes of elastic constants c_{11}, c_{33} and c_{55} of the shunted piezoelectric plate are calculated and illustrated in Figs. 3(a)–3(c), respectively. It can be seen that with the increase of the value of SCR, each of c_{11}, c_{33}, and c_{55} first increases monotonically to a positive maximum ($+\infty$ theoretically) from its initial value, i.e., c_{11}^{D}, c_{33}^{D}, and c_{55}^{D}, after an instantly reverse jump to a negative maximum ($-\infty$ theoretically), then monotonically increase to its final values, i.e., c_{11}^{E}, c_{33}^{E}, and c_{55}^{E}, respectively.

Based on Eqs. (18a)–(18c), the following relations can be found:

$$\text{if } -1<\frac{C_{\mathrm{p3}}}{-C_{x3}}<-\frac{1}{1+k_{31}^2}, \text{ then } c_{11}^{\mathrm{SH}}<0; \quad (20a)$$

$$\text{if } -1<\frac{C_{\mathrm{p3}}}{-C_{x3}}<-\frac{1}{1+k_{33}^2}, \text{ then } c_{33}^{\mathrm{SH}}<0; \quad (20b)$$

$$\text{if } -1<\frac{C_{\mathrm{p1}}}{-C_{x1}}<-\frac{1}{1+k_{15}^2}, \text{ then } c_{55}^{\mathrm{SH}}<0. \quad (20c)$$

The conclusion in Eq. (20b) is the same as that in Eq. (4), which means that the structure is unstable if negative capacitance is improperly tuned. Equations (20a)–(20c) show that if the value of negative capacitance is within whichever range of Eqs. (20a)–(20c), the corresponding elastic constant of piezoelectric plate will be lower than zero, which is not meaningful.

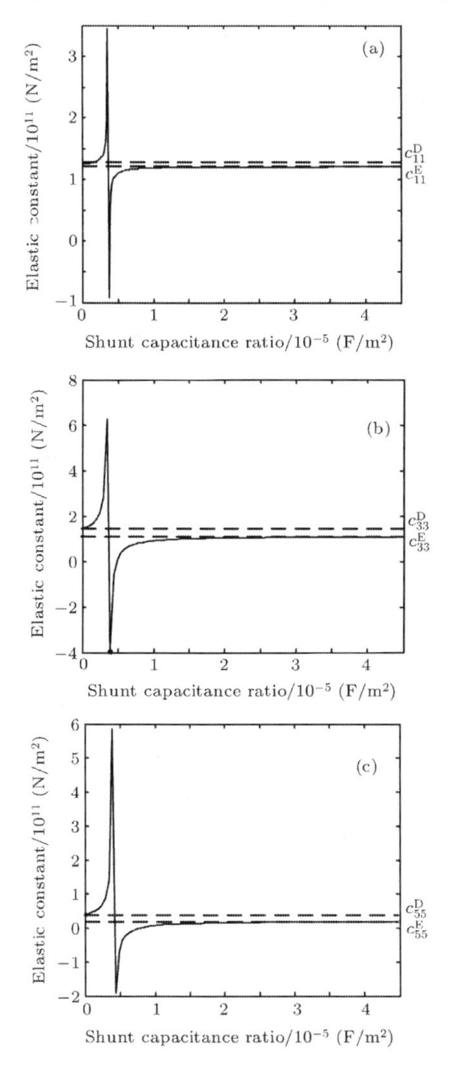

Fig. 3. Plots of elastic constants: (a) c_{11}, (b) c_{33}, and (c) c_{55} versus SCR.

3. Experiment verification

In order to verify the theoretical conclusions in Section 2, in this section, we investigate the variations of elastic constants of piezoelectric materials with shunt negative capacitance experimentally. Considering that the changes of elastic constants may be very small and easily submerged by measurement errors of direct mechanical test method, the indirect method[24] should be adopted in this paper, with which the variation of the elastic constants can be indirectly tested through observing the shift of resonant frequency of a mechanical resonator. In the design of experiment setup, the piezoelectric resonator should satisfy the following requirements. First, the measurement loop and the shunt circuit should be electrically isolated from each other to eliminate the directly electrical effect of the shunt circuit on the test loop, thereby ensuring that the shift of resonant frequency is purely caused by the change of the elastic constants. Thus the single resonator measurement method[26] is not applicable. Second, the mechanical system should be resonant at a frequency far below the natural frequency of the piezoelectric plate to avoid influencing the resonant peak of piezoelectric plate itself and make the piezoelectric plate work at a pseudo electrostatic state.

3.1. Experiment setup for c_{55} test

Since three elastic constants have the relationships similar to each other in the presence of the shunt negative capacitance, as shown in Eqs. (18a)–(18c), the measurement of c_{55} is taken as an example to verify the theoretical results in this paper. The thickness-shear mode (TSM) electro–mechanical resonant system in Ref. [24] is adopted to measure the change of the elastic constant c_{55} of piezoelectric plate with shunt negative capacitance. The scheme of experimental setup is shown in Fig. 4(a), including a TSM resonator, a shunt negative capacitance circuit, and an impedance analyzer. The TSM resonator is composed of two identical shear-mode piezoelectric ceramic plates a and b and a mass block m. Pure tungsten is chosen as the material of the mass block due to its high density to make the mass block close to a lumped-parameter element and the system close to an ideal spring oscillator. The experimental prototype of TSM resonator is shown in Fig. 4(b).

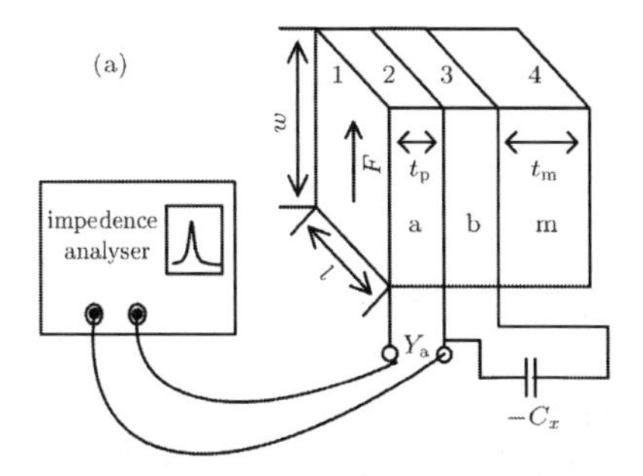

Fig. 4. (color online) (a) Scheme of experiment setup for c_{55} test. (b) Experimental prototype of the TSM resonator.

3.2. Realization of negative capacitance

Since the negative capacitance value needed in the experiments ranges from zero to infinite (large enough), the stable boundary of a single negative capacitance circuit is broken. Therefore, two kinds of negative capacitance circuits are realized in the experiments, which are demarcated by the value of the inner capacitance of piezoelectric plates.

One of the common negative capacitance shunt circuits is shown in Fig. 5(a). The non-inverting (+) and inverting (−) voltages of the operational amplifier (OA), i.e. V_{in}^{+} and V_{in}^{-}, are written, respectively, as

$$V_{\text{in}}^{-} = \frac{\frac{1}{\text{j}\omega C_{\text{p}}}}{R_1 + \text{j}\omega C_{\text{p}}} V_{\text{out}}, \tag{21a}$$

$$V_{\text{in}}^{+} = \frac{\frac{1}{\text{j}\omega C_x}}{R_2 + \text{j}\omega C_x} V_{\text{out}}. \tag{21b}$$

A negative capacitance C is obtained to be

$$C = -\frac{R_1}{R_2} C_x. \tag{22}$$

Usually, assume $R_1 = R_2$ to ensure the accuracy of the negative capacitance, then $C = -C_x$ will be obtained. The stable condition of the circuit (negative feedback) is

$$\text{d}\left(V_{\text{in}}^{+} - V_{\text{in}}^{-}\right)/\text{d}V_{\text{out}} < 0. \tag{23}$$

Combining Eqs. (21) with Eqs. (22) and (23), it is found that when $R_1 = R_2$, the stability condition is $|C| > C_{\text{p}}$.

Obviously, if the needed value of negative capacitance $|C|$ is smaller than C_{p}, the circuit in Fig. 5(a) will be instable, other realization circuit of negative capacitance should be considered. Figure 5(b) shows another type of negative capacitance circuit. Like the stability analysis made above, it can be found that the stability condition in Fig. 5(b) is $|C| < C_{\text{p}}$, which is contrary to Fig. 5(a).

Therefore, in experiment, when the needed value of negative capacitance $|C|$ is larger than the inner capacitance of piezoelectric plate C_{p}, the circuit in Fig. 5(a) is adopted. While for $|C| < C_{\text{p}}$, the circuit in Fig. 5(b) is employed.

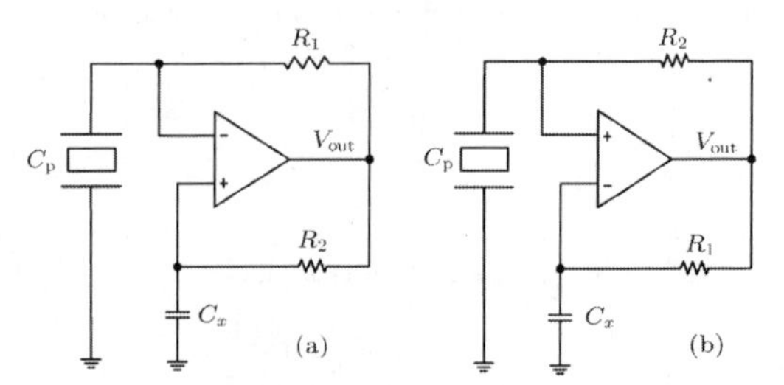

Fig. 5. The realization circuits of negative capacitance in the cases of (a) $|C| > C_{\text{p}}$ and (b) $|C| < C_{\text{p}}$.

In both circuits, the operational amplifier is Burr–Brown OPA552, which has the properties of high voltage, high current and wide bandwidth, and $R_1 = R_2 = 10$ kΩ.

3.3. Electrical conductance of TSM resonator

The conductance of the TSM resonator in Fig. 4(a) can be analyzed theoretically with Mason electro–mechanical equivalent circuit. The PZT-5A is chosen as the material of piezoelectric plates a, b in the calculation. The material parameters are shown in Table 1. The dimension parameters of resonant system in Fig. 4(a) are set to be $l = w = 1000$ mm to simulate the infinite boundary condition. Besides, $t_{\text{p}} = 2$ mm, and $t_{\text{m}} = 3$ mm.

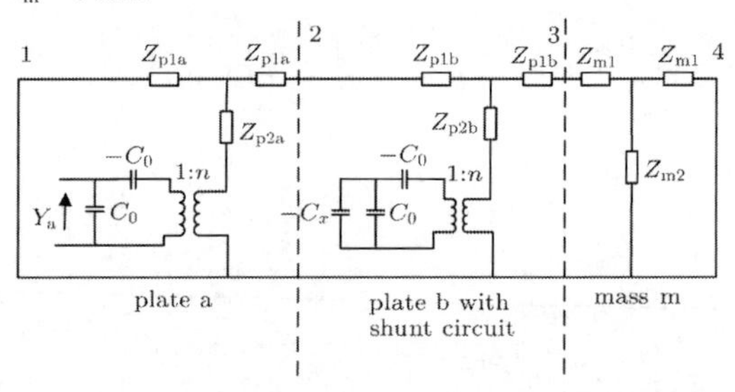

Fig. 6. Electro-mechanical equivalent circuit of TSM resonant system.[24]

Table 1. Parameters of piezoelectric ceramics, mass block and epoxy resin.

The piezoelectric ceramics PZT-5A											
Density/(kg/m³)	Elastic constant/10¹⁰ (N/m²)					Relative dielectric constant		Piezoelectric stress constant/(C/m²)			Mechanical quality factor
ρ_{p}	c_{11}^{E}	c_{12}^{E}	c_{13}^{E}	c_{33}^{E}	c_{44}^{E}	$\varepsilon_{11}^{\text{S}}/\varepsilon_0$	$\varepsilon_{33}^{\text{S}}/\varepsilon_0$	e_{31}	e_{33}	e_{15}	Q_{m}^{D}
7500	12.1	7.54	7.52	11.1	2.11	916	830	-5.4	15.8	12.3	75
	c_{11}^{D}	c_{12}^{D}	c_{13}^{D}	c_{33}^{D}	c_{44}^{D}						
	12.6	8.09	6.52	14.7	3.97						
The mass block (tungsten)											
Density/(kg/m³)	Young modulus/10¹⁰ (N/m²)							Poisson's radio			
19100	35.4							0.35			
The epoxy resin											
Density/(kg/m³)	Young modulus/10¹⁰ (N/m²)					Poisson's radio		Damping factor			
1250	0.36					0.35		0.17			

Figure 6 shows the electro–mechanical equivalent circuit of the TSM resonant system. With plate b shunted to negative capacitance circuits, the electrical admittance Y_a of the system calculated from the two electrodes of plate a is derived from Fig. 6 as follows:[24]

$$Y_a = \frac{1}{\left(\frac{\left(\left(Z_{m1}//Z_{m2}+Z_{m1}+Z_{p1b}\right)//\left(n^2\left(\left(-\frac{1}{j\omega C_x}\right)//\frac{1}{j\omega C_0}-\frac{1}{j\omega C_0}\right)+Z_{p2b}\right)+Z_{p1b}+Z_{p1a}\right)//Z_{p1a}+Z_{p2a}}{n^2}-\frac{1}{j\omega C_0}\right)//\frac{1}{j\omega C_0}}, \tag{24}$$

where $Z_x = -1/j\omega C_x$ and C_x is the absolute capacitance value of the negative capacitance shunt circuit, and

$$Z_{p1a} = j\rho_p c_p^{open} S_p \tan\left(k_p^{open} t_p/2\right), \tag{25a}$$

$$Z_{p2a} = \rho_p c_p^{open} S_p/\left(j\sin\left(k_p^{open} t_p\right)\right), \tag{25b}$$

$$Z_{p1b} = j\rho_p c_p S_p \tan\left(k_p t_p/2\right), \tag{25c}$$

$$Z_{p2b} = \rho_p c_p S_p/\left(j\sin\left(k_p t_p\right)\right), \tag{25d}$$

$$Z_{m1} = j\rho_m c_m S_m \tan\left(k_m t_m/2\right), \tag{25e}$$

$$Z_{m2} = \rho_m c_m S_m/\left(j\sin\left(k_m t_m\right)\right). \tag{25f}$$

In Eq. (25), the subscripts 'p' and 'm' represent the piezoelectric plate and the mass block respectively. Correspondingly, ρ_p and ρ_m are their densities, S_p and S_m are their areas, t_p and t_m are their thickness values. Besides, $c_p^{open} = \sqrt{c_{55}^D/\rho_p}$ and $c_p = \sqrt{c_{55}/\rho_p}$ are the transverse wave velocities of piezoelectric ceramics in open state and shunted to circuits respectively, and thus $k_p^{open} = \omega/c_p^{open}$ and $k_p = \omega/c_p$ are the corresponding wave numbers. Moreover, c_m and $k_m = \omega/c_m$ are the transverse velocity and wave number of the mass block m, respectively. For the plates a and b, $C_0 = S_p\varepsilon_{11}^S/t_p$ is the inherent capacitance and $n = S_p h_{15}\varepsilon_{11}^S/t_p$ is the electromechanical conversion factor, in which ε_{11}^S is the dielectric constant and h_{15} is the shear-mode piezoelectric constant.

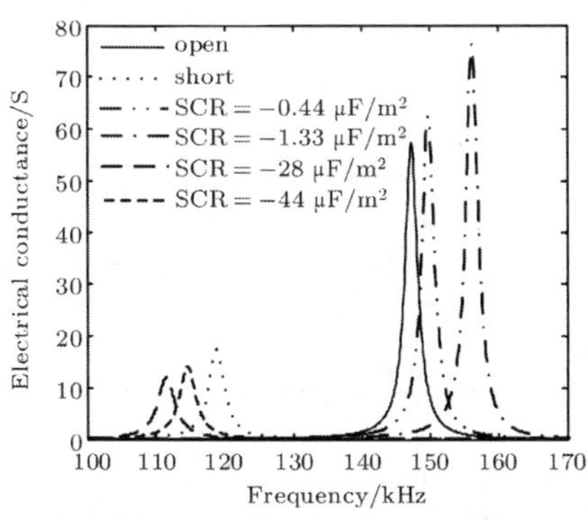

Fig. 7. Plots of electrical conductance of the system versus frequency, calculated from the two electrodes of plate a by equivalent circuit method with b shunted to negative capacitance.

With negative capacitance connected to plate b, the resonant frequency of the TSM resonator will shift. Figure 7 shows several conductance curves with different negative capacitance values connected to plate b, which are calculated from Eq. (24).

3.4. Theoretical prediction of relationships from conductance curves

The relationship between the elastic constant c_{55} and the shunt negative capacitance is not intuitive in Eq. (24). However, the physical process is known that the elastic constant c_{55} changes with shunt negative capacitance, which results in the shift of the resonant frequency of the TSM resonator. Thus through the observation of the shift of the resonant frequency of the TSM resonator, the variation of elastic constant c_{55} with the shunt negative capacitance can be indirectly obtained. The process can be resolved as three steps:

Step 1 Keep the plate b in an open-circuit state, i.e., no negative capacitance is connected, but change its elastic constant c_{55} numerically to observe the shift of the resonant frequency of the TSM resonator with the variation of c_{55}. As done in Ref. [24], c_{55} in Eq. (24) can be substituted with a complex number $c_{55} = c_{55r} + jc_{55i}$, where the subscripts 'r' and 'i' represent the real and the imagine parts, corresponding to elastic constant and damping factor respectively. Because the change range of the elastic constant c_{55r} induced by the negative capacitance shunt circuit is excessively wide, the conclusion in Ref. [24] that $\Delta c_{55r}/c_{55r}^D$ changes linearly with $\Delta f_r/f_r^D$ is inapplicable. The relationship between the elastic constant c_{55r} and the resonant frequency of the TSM resonator in the whole range is recalculated and obtained based on Eq. (24), which is shown in Fig. 8(a);

Step 2 Keep c_{55} in Eq. (24) constant, i.e., c_{55}^D, but change the value of negative capacitance to observe the shift of the resonant frequency of the TSM resonator with the variation of the value of negative capacitance. The relationship between the shunt negative capacitance $-C_x$ and the resonant frequency of the TSM resonator is shown in Fig. 8(b);

Step 3 Make the above two cases equivalent to each other to obtain the relationship between the elastic constant c_{55r} and the shunt negative capacitance through point mapping method, since the relationships in Fig. 8(a) and Fig. 8(b) are both single-valued curves. The relationship between the elastic constant c_{55r} and the negative capacitance is achieved and shown in Fig. 8(c).

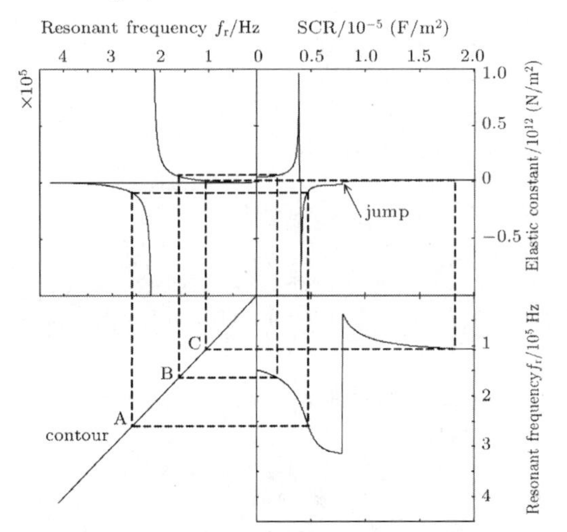

Fig. 8. Relationships between (a) resonant frequency f_r and elastic constant c_{55r}, (b) resonant frequency f_r and SCR, and (c) elastic constant c_{55r} and SCR.

The curves in Fig. 8 are divided into three parts. In the first part, the ranges of the elastic constant, SCR and the resonant frequency are $(4 \sim 96) \times 10^{10}$ N/m^2, $1.1 \times 10^{-7} \sim 3.97 \times 10^{-6}$ F/m^2, and 148 kHz ~ 212 kHz respectively. The dashed rectangle B is a representative map of this part. In the second part, the ranges of the elastic constant, SCR and the resonant frequency are $(-94 \sim -2.11) \times 10^{10}$ N/m^2, $(3.97 \sim 7.85) \times 10^{-6}$ F/m^2, and 212 kHz~ 314 kHz respectively. The dashed rectangle A is a representative map in this part. In the third part, the ranges of the elastic constant, SCR and the resonant frequency are $(0.18 \sim 2) \times 10^{10}$ N/m^2, $7.86 \times 10^{-6} \sim 5 \times 10^{-5}$ F/m^2, and 37.2 kHz~ 115 kHz respectively. The dashed rectangle C is a representative map in this part.

From Fig. 8, it is seen that with the increase of SCR, the resonant frequency of the system first increases from its open-circuit f_r^D to a maximum value, after an instantly reverse jump to the minimum, then continuously increases to its short-circuit f_r^E. Meanwhile, with the increase of SCR, c_{55r} first increases from its open-circuit value c_{55r}^D to a maximum value, after an instantly reverse jump to the negative maximum, then continuously increases to its final value, the elastic constant under constant electrical field, i.e., c_{55r}^E, correspondingly. The conclusion is the same as the result obtained by theoretical analysis with the piezoelectric constitutive law in Section 2.

By comparing the ranges of resonant frequency in Fig. 8(a) and Fig. 8(b), it is found that the range of resonant frequency from 313 kHz to 419 kHz can be achieved by changing the elastic constant c_{55r} in Fig. 8(a), but it cannot be reached by changing the negative capacitance value in Fig. 8(b), where the resonant frequency curve becomes flat near 313 kHz. In other words, the corresponding range of the elastic constant c_{55r} from -2.11×10^{10} N/m^2 to 0 cannot be achieved by changing the negative capacitance. This is because the elastic constant of the two piezoelectric plates in series will be lower than zero when the elastic constant c_{55r} falls in a range of -2.11×10^{10} N/m^2 to 0. This will result in the instability of the system. It is also the reason for the jump in Fig. 8(c).

3.5. Experimental tests

The experimental prototype of TSM resonator was fabricated as shown in Fig. 4(b). The thickness of each piezoelectric plate was $t_p = 2$ mm, corresponding to a natural frequency of 1 MHz. The thickness of the mass block was $t_m = 3$ mm. Because of limitation of the poling length, in practice the lateral dimensions of the piezoelectric plates were machined into $l = w = 15$ mm, much smaller than those used in the theoretical analysis in Subsection 3.3. In fact, the shear mode vibration is less affected by the lateral boundary condition than the thickness mode. If the length-to-thickness ratio of the piezoelectric plate is not too small, the lateral boundary effect on the resonant frequency of the system is negligible. Our simulation reveals that a length-to-thickness ratio greater than 2 can satisfy the infinite lateral condition requirement approximately.

In the experiments, with two negative capacitance circuits, a series of values of negative capacitance was realized. The conductance curves of the prototype system in Fig. 4(b) were tested from two ends of plate a by an impedance analyzer, with piezoelectric plate b shunted to negative capacitance circuit. Some results are shown in Fig. 9. In comparison with Fig. 7, it is seen that with the variation of SCR, the changes of conductance curves of the TSM resonator show similar variation tendencies. However, all the resonant frequencies are lower than the predicted ones in Subsection 3.3. This is because the epoxy resin bonding layers exist, respectively, between plates a and b, and between plate b and mass block m of the prototype sample, which are connected in series with the elastic piezoelectric plates. Besides, the bandwidths of the experimental conductance peaks are broadened accordingly, in comparison with the theoretical predicted ones. This is caused by the epoxy resin layers too, which have lager damping factor than piezoelectric plates.

By picking up the peak position of conductance of plate a, the relationship between the resonant frequency of the TSM resonator and the negative capacitance is also measured, and shown in Fig. 10. It is seen that with the increase of SCR, the change of the resonant frequency has a similar variation tendency to that of the theoretical analysis with Mason equivalent circuit. It means the elastic constant $c_{55\mathrm{r}}$ indeed has a similar variation to that of the resonant frequency, as the theoretical analyses demonstrated in Section 2 and Subsection 3.4.

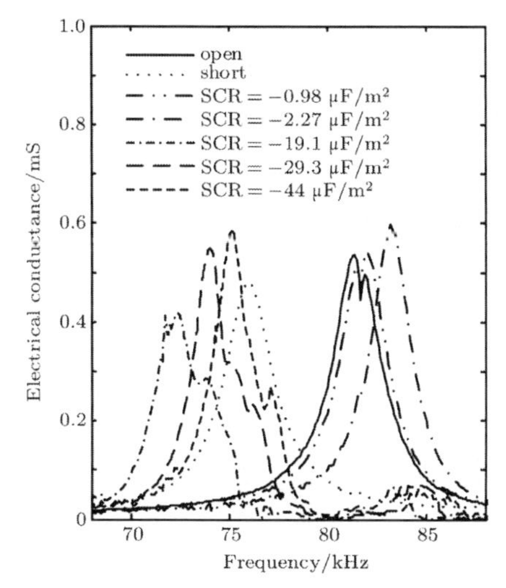

Fig. 9. Electrical conductance of the system measured from the two electrodes of plate a experimentally, with b shunted to negative capacitance.

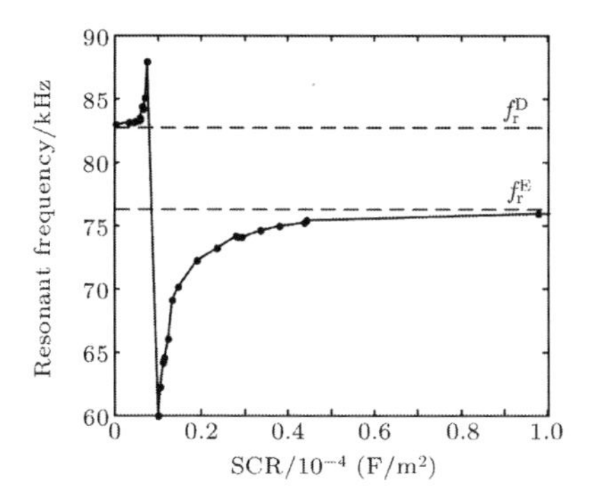

Fig. 10. Variation of resonant frequency with SCR, measured in experiment.

4. Discussion

Most vibration suppression or sound absorption systems require their negative capacitances to be near to the inner capacitances of piezoelectric ceramics to achieve the optimal performances.[21–23] However, based on the theoretical results extracted from Fig. 3(c) and Fig. 8(c), it is found that a singular point of elastic constants occurs just at the position of SCR corresponding to the inner capacitance under constant strain C_p^S of the piezoelectric plate. The experimental results also verify the phenomenon. It means that the elastic constants will take singular values in this case. Therefore, the optimal condition, i.e., the negative capacitance equals the inner capacitance, which is just the stability margins of the shunted systems, cannot be satisfied in practice. In fact, even a suboptimal condition, i.e., a negative capacitance close to the inner capacitance, is not applicable either. Since the inner capacitance of a piezoelectric material is easily affected by environmental factors like temperature, a slight change of inner capacitance may also lead to the instability of the system. Furthermore, from Figs. 3(c) and 8(c), it is found that the range of SCR having significant influence on elastic constants is very narrow. If the negative capacitance shunted systems is chosen and works at a slow change region of elastic constants to ensure the stability, it will have no significant improvement in the performance. This may be a good explanation for why the negative capacitance technique cannot behave well in practical applications such as vibration suppression, sound absorption, etc.

5. Conclusions

In this paper, we investigate the influences of shunt negative capacitance on elastic constants of piezoelectric material of the transverse mode, the longitudinal mode and the shear mode, i.e., c_{11}, c_{33}, and c_{55}, respectively. Through the theoretical analyses with the piezoelectric constitutive equations, the relationships between the shunt negative capacitance and the elastic constants are obtained, showing the singular variation properties of the elastic constants on condition that the negative capacitance equals the inner capacitance. The thickness-shear mode elastic constant c_{55} is taken as an example to be studied experimentally with a TSM resonator. The experimental results show a variation tendency similar to those obtained from the theoretical analyses with using the constitutive equation and the Mason equivalent circuit methods. The results reveal the inner cause of instability of the piezoelectric damping system shunted to negative capacitance. Besides, the results show that the effective range of the negative capacitance is very narrow, while most of the negative capacitance region has little influence on the elastic constants. This is also a good explanation for why most shunt damping systems each with shunt negative capacitance show no significant improvement of performance in practical applications.

References

[1] Forward R L 1979 *Appl. Opt.* **18** 690

[2] Hagood N W and von Flotow A 1991 *J. Sound Vib.* **146** 243

[3] Becker J, Fein O, Maess M and Gaul L 2006 *Comput. Struct.* **84** 2340

[4] Thomas O, Ducarne J and Deü J F 2012 *Smart Mater. Struct.* **21** 015008

[5] Sun Y, Li Z H, Huang A G and Li Q H 2015 *J. Sound Vib.* **355** 19

[6] Zhang J M, Chang W, Varadan V K and Varadan V V 2001 *Smart Mater. Struct.* **10** 414

[7] Ahmadian M and Jeric K M 2001 *J. Sound Vib.* **243** 347

[8] Guyomar D, Richard T and Richard C 2008 *J. Intell. Mater. Syst. Struct.* **19** 791

[9] Kim J S, Jeong U C, Seo J H, Kim Y D, Lee O D and Oh J E 2015 *Sensor. Actuat. A-Phys.* **233** 330

[10] Kim J and Lee J K 2002 *J. Acoust. Soc. Am.* **112** 990

[11] Chen S B, Wen J H, Wang G and Wen X S 2013 *Chin. Phys. B* **22** 074301

[12] Hollkamp J J 1994 *J. Intell. Mater. Syst. Struct.* **5** 49

[13] Wu S 1998 *Proceedings of SPIE* **3327** 159

[14] Behrens S, Moheimani S O R and Fleming A J 2003 *J. Sound Vib.* **266** 929

[15] Riordan R H S 1967 *Electron. Lett.* **3** 50

[16] Behrens S, Fleming A J and Moheimani S O R 2003 *Smart Mater. Struct.* **12** 18

[17] Behrens S, Fleming A J and Moheimani S O R 2001 *Proc. SPIE* **4331** 239

[18] Fukada E, Date M, Kimura K, Okubo T, Kodama H, Mokry P and Yamamoto K 2004 *IEEE Trans. Dielectr. Electr. Insul.* **11** 328

[19] Yu L G, Li Z H and Ma L L 2012 *Acta Phys. Sin.* **61** 024301 (in Chinese)

[20] Neubauer M, Oleskiewicz R, Popp K and Krzyzynski T 2006 *J. Sound Vib.* **298** 84

[21] de Marneffe B and Preumont A 2008 *Smart Mater. Struct.* **17** 035015

[22] Han X, Neubauer M and Wallaschek J 2013 *J. Sound Vib.* **332** 7

[23] Neubauer M and Wallaschek J 2009 *IEEE/ASME International Conference on AIM* 1100

[24] Hu J Y, Li Z H, Sun Y and Li Q H 2016 *Chin. Phys. B.* **25** 127701

[25] Luan G D, Zhang J D and Wang R Q 2005 *Piezoelectric Transducers and Arrays* (Beijing: Peking University Press) pp. 26–72

[26] Hafner E 1969 *Proc. IEEE* **57** 179

庆祝马远良先生80华诞

球冠型换能器声辐射指向性分析①②

夏金东[1,2] 黄海宁[1] 张春华[1] 李启虎[1]

(1 中国科学院声学研究所 北京 100190)

(2 中国科学院大学 北京 100049)

2016 年 12 月 21 日收到

2017 年 5 月 16 日定稿

摘要 为了研究球冠型换能器的声辐射特性，在分离变量法求解球面坐标系下波动方程的基础上，采用基于球谐基傅里叶变换及边界条件的求解模型，给出了球冠型换能器声辐射的远场声压计算表达式和远场指向性表达式；仿真计算了球冠换能器的远场指向性随球冠极角、球半径及振动频率变化的特性，球冠所在球障板的直径和介质中声波的波长比决定着球冠声辐射指向性，在低频或波长大于球障板直径时，球冠声辐射呈无指向性，随着频率的增高即波长的减小或者球障板直径的增大，球冠声辐射的指向性越明显，波束开角越趋向于球冠的开角，而且波束开角内出现波浪状起伏越明显；试制了高频球冠型换能器基阵，测试了换能器基阵 300 kHz 的指向性，测试结果与理论计算相符合，验证了理论计算表达式的正确性，可为设计球冠型换能器及基阵提供理论指导.

PACS 数：43.30, 43.38

DOI:10.15949/j.cnki.0371-0025.2018.04.019

Analysis on acoustic directivity of spherical cap transducers

XIA Jindong[1,2] HUANG Haining[1] ZHANG Chunhua[1] LI Qihu[1]

(1 *Institute of Acoustics, Chinese Academy of Science* Beijing 100190)

(2 *University of Chinese Academy of Science* Beijing 100049)

Received Dec. 12, 2016

Revised May 16, 2017

Abstract In order to study acoustic radiation from the spherical cap transducer, a theoretical model was used by solving the wave equation in spherical coordinates using the method of separation of variables, based on the spherical harmonic Fourier transform and boundary condition. The calculation formulas for far field radiated pressure and directivity of spherical cap are derived. Some theoretical results are presented in the form of far-field directivity patterns of the spherical cap transducer for various polar angle of spherical cap, radius of sphere baffle and operating frequency. The diameter of sphere baffle and wavelength in the media determine the directivity of acoustic radiation from a spherical cap. When the frequency is low or the wavelength is longer than the diameter of the sphere baffle, the acoustic radiation from a spherical cap is omnidirectional. With the increasing of the frequency or the diameter of the sphere baffle, the acoustic radiation from a spherical cap is more directional and the beamwidth more tends to spherical cap angle, furthermore the ripple in the beam is more obvious. Finally, the high frequency spherical cap transducer was fabricated and the directivity pattern were tested. As a result, the measurement data coincides with the theoretical calculation results and at the same time it verifies the correctness of theoretical formulas. This research can provide a guideline for designing the spherical cap transducers and arrays.

① 国家自然科学基金青年基金项目(11304343)资助.

② 声学学报, 2018, 43(4): 592-599.

1 引言

球型和球冠型结构是一种典型的声学结构类型，在声学研究中占有重要的地位 [1-7]，而指向性是声学结构研究的重要特性 [8-14]，对球冠型换能器的声辐射指向性研究不仅具有理论意义也具有实用价值。瑞利研究了在球坐标系下利用球表面谐函数求解声辐射的速度势方程，通过 Stokes 级数给出了球辐射的速度势的解；ZHANG Jingdong 采用拉普拉斯变换的方法对水下球壳在外力作用下的振动和辐射进行了时间域分析 [15]；SUN Yudong 等研究了水下振动球壳的非线性辐射 [16]；Henrich Stenzel 在总结瑞利关于球冠辐射的基础上 [17]，采用贝塞尔函数构造的两类函数，推导了球冠辐射的远场声压以及远场辐射的指向性公式，计算了波数与球半径乘积一定值情况下不同球冠角时的声压和指向性，分析了不同计算阶数对计算精度的影响，但是计算公式繁琐，计算的阶数有限，指向性和声压的分析没有涉及不同频率及球半径的变化；Charles H. Sherman 采用球谐函数和球汉克尔函数研究了球冠辐射的声场表达式 [18]，还是采用了 P.Morse 的部分表达式，同时计算了球冠辐射的互辐射阻抗，简化了计算公式；Earl G. Williams 在基于分离变量的基础上求解了球坐标下的波动方程 [19]，并且用球谐函数和球汉克尔函数来表示方程的解，根据球冠的边界条件给出了带球面障板的圆形活塞声压的计算公式，计算公式相对简练但仍然不易直接运算；Seyyed M. Hasheminejad[20] 研究了刚性球障上球冠在软性和硬性平面附近的声辐射阻抗特性；Ronald M. Aarts[21] 采用正交函数法研究了空气中刚性球障上弹性球冠的声辐射特性；Boris Aronov[22] 采用模态分析法研究了无指向性宽带压电球壳和部分压电球壳水声换能器的声辐射特性阻抗，但没有系统研究声辐射的空间分布特性，HE Zhengyao 等利用边界元方法研究了由多个水下换能器组成的半球障板上共形阵的辐射指向性 [23]。

在总结以上研究成果的基础上，本文根据球面坐标系下的波动方程，结合分离变量法给出基于球谐函数基傅里叶变换的球冠辐射特性的简洁解表达式，并且推导出了适用于工程计算的远场辐射声压和指向性表达式；根据远场辐射指向性的表达式分别计算了不同球冠极角、球半径及频率变化情况下的球冠指向性变化曲线；最后，通过产品试制和实验验证了球冠型换能器辐射特性计算的正确性。

2 球面声波辐射的理论分析

球面声波辐射的求解建立在球面坐标系下，坐标系如图 1 所示。时域波动方程在球坐标系中的表达式为：

$$\begin{aligned}&\frac{1}{r^2}\frac{\partial}{\partial r}\left(r^2\frac{\partial p}{\partial r}\right)+\frac{1}{r^2\sin\theta}\frac{\partial}{\partial\theta}\left(\sin\theta\frac{\partial p}{\partial\theta}\right)+\\&\frac{1}{r^2\sin^2\theta}\frac{\partial^2 p}{\partial\phi^2}=\frac{1}{c^2}\frac{\partial^2 p}{\partial t^2},\end{aligned}\tag{1}$$

式中 p 是远场辐射声压，r 是距辐射中心的距离，θ 是极角，ϕ 是方位角。根据分离变量法求解上式方程，其解可以表示为 4 个独立变量函数的乘积：

$$p(r,\theta,\phi,t)=R(r)\Theta(\theta)\Phi(\phi)T(t).\tag{2}$$

把式 (2) 代入式 (1) 后可以分离出以下 4 个常微分方程：

$$\frac{1}{c^2}\frac{d^2T}{dt^2}+k^2T=0,\tag{3}$$

$$\frac{1}{r^2}\frac{d}{dr}\left(r^2\frac{dR}{dr}\right)+k^2R-\frac{n(n+1)}{r^2}R=0,\tag{4}$$

$$\frac{1}{\sin\theta}\frac{d}{d\theta}\left(\sin\theta\frac{d\Theta}{d\theta}\right)+\left[n(n+1)-\frac{m^2}{\sin^2\theta}\right]\Theta=0,\tag{5}$$

$$\frac{d^2\Phi}{d\phi^2}+m^2\Phi=0,\tag{6}$$

式中 k 是波数，$k=\omega/c$（ω 是辐射角频率，c 是辐射介质声速）。方程 (3) 的解为：

$$T(t)=T_1\mathrm{e}^{\mathrm{i}\omega t}+T_2\mathrm{e}^{-\mathrm{i}\omega t},\tag{7}$$

选择第 1 或第 2 项决定着方程 (4) 的解的选择，这里选取第 2 项，所以 $T_1=0$。方程 (4) 的解为：

$$R(r)=R_1\mathrm{h}_n^{(1)}(kr)+R_2\mathrm{h}_n^{(2)}(kr),\tag{8}$$

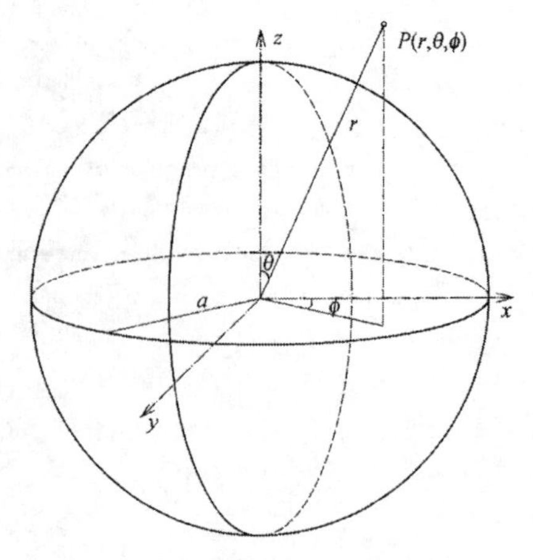

图 1 相对于直角坐标系的球面坐标系

式中，$h_n^{(1)}$ 和 $h_n^{(2)}$ 分别为第一类和第二类球汉克尔函数。对于向外辐射波，由于方程 (3) 的解式 (7) 中选择了第 2 项，所以式 (8) 只需要保留第 1 项。如果式 (7) 选择了第 1 项，那么式 (8) 就只需要保留第 2 项。以下用 h_n 表示 $h_n^{(1)}$，方程 (5) 和 (6) 的解合并成一个解，用归一化的球谐函数

$$Y_n^m(\theta,\phi)=\sqrt{\frac{2n+1}{4\pi}\frac{(n-m)!}{(n+m)!}}P_n^m(\cos\theta)e^{im\phi} \tag{9}$$

来表示。因此，方程 (1) 的解为：

$$p(r,\theta,\phi,t)=\sum_{n=0}^{\infty}\sum_{m=-n}^{n}A_{mn}h_n(kr)Y_n^m(\theta,\phi)e^{-i\omega t}, \tag{10}$$

式中 A_{mn} 是系数，根据边界条件及球谐函数的正交性确定。假设半径为 $a=r$ 的球面上的振速 $u(a,\theta,\phi)$ 是已知的，将欧拉方程

$$i\rho cku(r,\theta,\phi)=\nabla p(r,\theta,\phi) \tag{11}$$

应用于上式，其中 ρ 是辐射介质密度，∇p 为在球坐标系下的声压梯度，进一步可得出下式：

$$\begin{aligned}&i\rho cku(a,\theta,\phi)=\nabla p(a,\theta,\phi)=\\&k\sum_{n=0}^{\infty}\sum_{m=-n}^{n}A_{mn}h_n'(ka)Y_n^m(\theta,\phi),\end{aligned} \tag{12}$$

即：

$$u(a,\theta,\phi)=\frac{1}{i\rho c}\sum_{n=0}^{\infty}\sum_{m=-n}^{n}A_{mn}h_n'(ka)Y_n^m(\theta,\phi), \tag{13}$$

对上式两边同乘以 $Y_n^m(\theta,\phi)^*$ 并在单位球面上作积分运算，其结果为：

$$A_{mn}=\frac{i\rho c}{h_n'(ka)}\iint u(a,\theta,\phi)Y_n^m(\theta,\phi)^*\sin\theta d\theta d\phi, \tag{14}$$

将上式代入式 (10) 得：

$$\begin{aligned}&p(r,\theta,\phi,t)=i\rho c\sum_{n=0}^{\infty}\sum_{m=-n}^{n}\frac{h_n(kr)}{h_n'(ka)}Y_n^m(\theta,\phi)\\&\iint u(a,\theta',\phi')Y_n^m(\theta',\phi')^*\sin\theta'd\theta'd\phi'e^{-i\omega t}.\end{aligned} \tag{15}$$

定义径向速度的球面波谱即径向速度的 Y_n^m 基函数的傅里叶正变换为：

$$W_{mn}(r)\equiv\iint u(r,\theta,\phi)Y_n^m(\theta,\phi)^*\sin\theta d\theta d\phi, \tag{16}$$

其逆变换为：

$$u(r,\theta,\phi)=\sum_{n=0}^{\infty}\sum_{m=-n}^{n}W_{mn}(r)Y_n^m(\theta,\phi), \tag{17}$$

半径为 $a=r$、振速为 $u(a,\theta,\phi)$ 的球面，其波谱为：

$$W_{mn}(a)\equiv\iint u(a,\theta,\phi)Y_n^m(\theta,\phi)^*\sin\theta d\theta d\phi, \tag{18}$$

所以式 (15) 可以表示为：

$$\begin{aligned}&p(r,\theta,\phi,t)=\\&i\rho c\sum_{n=0}^{\infty}\sum_{m=-n}^{n}\frac{h_n(kr)}{h_n'(ka)}W_{mn}(a)Y_n^m(\theta,\phi)e^{-i\omega t},\end{aligned} \tag{19}$$

省略时间部分，可以表示为：

$$p(r,\theta,\phi)=i\rho c\sum_{n=0}^{\infty}\sum_{m=-n}^{n}\frac{h_n(kr)}{h_n'(ka)}W_{mn}(a)Y_n^m(\theta,\phi), \tag{20}$$

在远场，$h_n(kr)$ 可以近似表示为：

$$h_n(kr)\approx(-i)^{n+1}\frac{e^{ix}}{x}, \tag{21}$$

辐射声压的远场表示为：

$$p(r,\theta,\phi)\approx\rho c\frac{e^{ikr}}{kr}\sum_{n=0}^{\infty}\frac{(-i)^n}{h_n'(ka)}\sum_{m=-n}^{n}W_{mn}(a)Y_n^m(\theta,\phi), \tag{22}$$

辐射声压的远场指向性函数为：

$$D(\theta,\phi)\approx\sum_{n=0}^{\infty}\frac{(-i)^n}{h_n'(ka)}\sum_{m=-n}^{n}W_{mn}(a)Y_n^m(\theta,\phi). \tag{23}$$

3 刚球障板上的球冠声波辐射

对于刚性球面上球冠状活塞，设球冠上每一点沿半径方向作等幅同相简谐运动，球坐标系的原点在球心，球冠在球体的极点，且极轴通过球冠的顶点，球冠的振动是关于极轴对称的，与方位角 ϕ 无关，如图 2 所示。

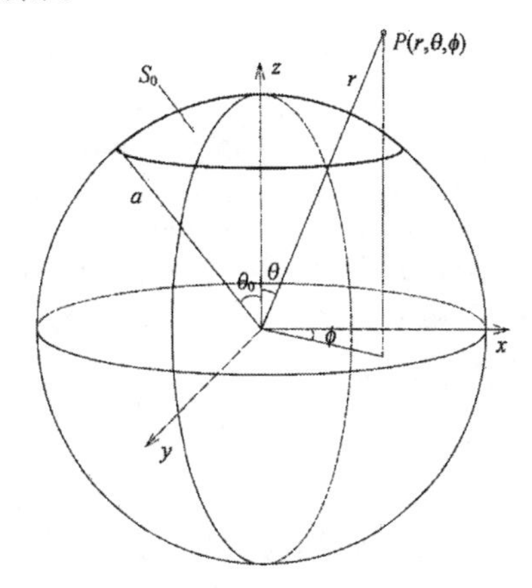

图 2 刚性球面上的球冠活塞

球冠面的振速幅值分布可表示为：

$$u(a,\theta)=\begin{cases}u_0, & 0\leqslant\theta\leqslant\theta_0,\\ 0, & \theta_0\leqslant\theta\leqslant\pi.\end{cases} \tag{24}$$

由于振速与 ϕ 无关，意味着方位角函数 $e^{im\phi}$ 中的 m 为零。所以振速的球面波谱为：

$$W_n(a)=\frac{2n+1}{2}u_0\int_0^{\theta_0}\mathrm{P}_n(\cos\theta)\sin\theta\mathrm{d}\theta=\frac{2n+1}{2}u_0\int_{\cos\theta_0}^{1}\mathrm{P}_n(x)\mathrm{d}x, \tag{25}$$

而对于 0 阶 n 次勒让德多项式的定积分[24]：

$$S_n^0(x_1,x_2)=\int_{x_1}^{x_2}\mathrm{P}_n(x)\mathrm{d}x=\frac{1}{2n+1}\left[\mathrm{P}_{n+1}(x)-\mathrm{P}_{n-1}(x)\right]\Big|_{x_1}^{x_2}, \tag{26}$$

所以式 (25) 表示为：

$$W_n(a)=\frac{u_0}{2}\left[\mathrm{P}_{n-1}(\cos\theta_0)-\mathrm{P}_{n+1}(\cos\theta_0)\right], \tag{27}$$

其远场辐射声压为：

$$p(r,\theta)\approx\rho cu_0\frac{e^{ikr}}{2kr}\sum_{n=0}^{\infty}\frac{(-i)^n\left[\mathrm{P}_{n-1}(\cos\theta_0)-\mathrm{P}_{n+1}(\cos\theta_0)\right]}{\mathrm{h}_n'(ka)}\mathrm{P}_n(\cos\theta), \tag{28}$$

其远场辐射声压的指向性函数为：

$$D(\theta)\approx\sum_{n=0}^{\infty}\frac{(-i)^n\left[\mathrm{P}_{n-1}(\cos\theta_0)-\mathrm{P}_{n+1}(\cos\theta_0)\right]}{\mathrm{h}_n'(ka)}\mathrm{P}_n(\cos\theta), \tag{29}$$

其中 $n=0$ 时，$\mathrm{P}_{-1}(\cos\theta_0)=1,\mathrm{P}_1(\cos\theta_0)=\cos\theta_0$。

而第一类球汉克尔函数的导数可以表示为：

$$\mathrm{h}_n'(ka)=\frac{1}{2n+1}\left[n\mathrm{h}_{n-1}(ka)-(n+1)\mathrm{h}_{n+1}(ka)\right]. \tag{30}$$

用第一类汉克尔函数 $\mathrm{H}_n^{(1)}(ka)$ 表示为：

$$\mathrm{h}_n'(ka)=\frac{\sqrt{\pi}}{(2n+1)\sqrt{2ka}}\left[n\mathrm{H}_{n-1/2}^{(1)}(ka)-(n+1)\mathrm{H}_{n+3/2}^{(1)}(ka)\right], \tag{31}$$

其远场辐射声压为：

$$p(r,\theta)\approx\rho cu_0\frac{e^{ikr}}{2kr}\sum_{n=0}^{\infty}\frac{(-i)^n(2n+1)\sqrt{2ka}\left[\mathrm{P}_{n-1}(\cos\theta_0)-\mathrm{P}_{n+1}(\cos\theta_0)\right]}{\sqrt{\pi}\left[n\mathrm{H}_{n-1/2}^{(1)}(ka)-(n+1)\mathrm{H}_{n+3/2}^{(1)}(ka)\right]}\mathrm{P}_n(\cos\theta), \tag{32}$$

其远场辐射声压的指向性函数为：

$$D(\theta)\approx\sum_{n=0}^{\infty}\frac{(-i)^n(2n+1)\left[\mathrm{P}_{n-1}(\cos\theta_0)-\mathrm{P}_{n+1}(\cos\theta_0)\right]}{\left[n\mathrm{H}_{n-1/2}^{(1)}(ka)-(n+1)\mathrm{H}_{n+3/2}^{(1)}(ka)\right]}\mathrm{P}_n(\cos\theta). \tag{33}$$

式 (32) 和式 (33) 易于在数值计算软件中实现。由式 (33) 可以看出，球冠辐射的指向性与波数 k 和球半径 a 的乘积直接相关，根据辐射声波的波长 λ 和振动频率 f 可知：

$$ka=\frac{2\pi}{\lambda}a=\frac{2\pi}{c}f\times a, \tag{34}$$

也就是说，球半径与辐射介质中声波的波长比值影响着球冠辐射的指向性也即频率与球冠所在球半径的乘积影响着球冠辐射的指向性，在下面的计算及分析中将详细讨论。

4 球冠型换能器辐射特性计算及分析

根据球冠辐射指向性函数的理论计算公式 (33)，可以看出其指向性不仅与球冠极角有关系，而且也受球冠所在球的半径及工作频率的影响。因此，通过计算不同影响因素下的指向性变化特性，可以看出球冠型换能器及基阵的指向性变化规律。假设球冠在水介质中向外辐射声波，声速 $c=1500$ m/s。

4.1 球冠极角变化下的声辐射特性

在球冠换能器或基阵的工作频率及所在球半径不变的情况下，仅改变球冠的极角，仿真计算在不同的极角下球冠辐射的指向性图。设定球冠所在球障板的半径为 100 mm，振动频率为 300 kHz，介质中声波波长为 5 mm，球冠所在球的直径远大于波长，分别计算球冠的极角在 30°, 60°, 90°, 120° 时的指向性，计算结果如图 3 所示。从图中可以看出，球冠的指向性开角随着极角的变大而变大，极角越大，其指向性的开角中间起伏越小。

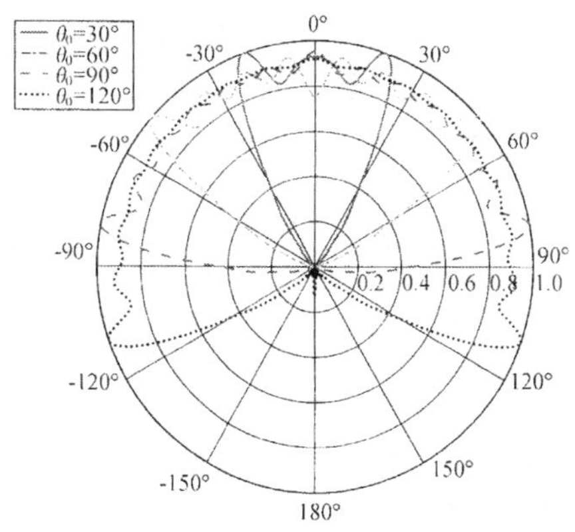

图 3 不同极角下球冠辐射的指向性仿真曲线

4.2 球障板半径变化下球冠的声辐射特性

在球冠换能器或基阵的工作频率及球冠的极角不变的情况下，仅改变球冠所在球的半径，仿真计算在不同的球半径下球冠辐射的指向性图。设定球冠的极角为 60°，振动频率为 300 kHz，介质中声波的波长为 5 mm，分别计算球冠所在球半径为 2 mm, 8 mm, 32 mm 和 128 mm 时的指向性，球直径与波长的关系分别是与波长相近、大于波长和远大于波长，其计算结果如图 4 所示。从图中可以看出，当球直径小于波长时，球冠的辐射指向性无明显起伏；随着球直径的增大，球冠直径增大，球冠辐射的指向性出现起伏，其开角逐渐趋向球冠极角。

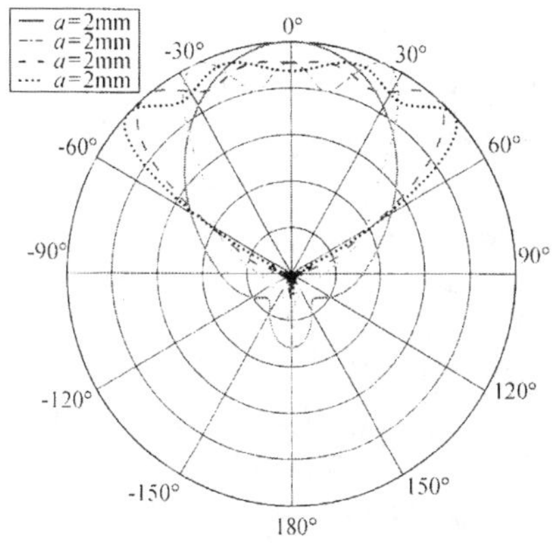

图 4 不同球障板半径下球冠辐射的指向性仿真曲线

4.3 球冠振动频率变化下的声辐射特性

设球冠换能器或基阵的极角及所在球半径不变的情况下，仅改变球冠的振动频率，仿真计算在不同的振动频率下球冠辐射的指向性图。设定球冠的极角为 60°，球冠所在球的半径为 100 mm，分别计算球冠振动频率在 1 kHz, 5 kHz, 25 kHz 和 100 kHz 时的指向性，对应于辐射介质中声波的波长分别为 1500 mm, 300 mm, 60 mm 和 15 mm，计算结果如图 5 所示。从图中可以看出，在低频时，波长远大于球冠的直径，球冠辐射类似点源辐射，呈现无指向性；随着频率的升高，波长变小，当波长与球冠直径相近时，球冠辐射的指向性开角平滑、无起伏；随着频率的进一步升高，波长进一步变小，当波长远小于球冠所在球直径时，即波长远小于球冠直径，球冠辐射的指向性出现起伏，而且其开角逐渐趋向球冠极角。

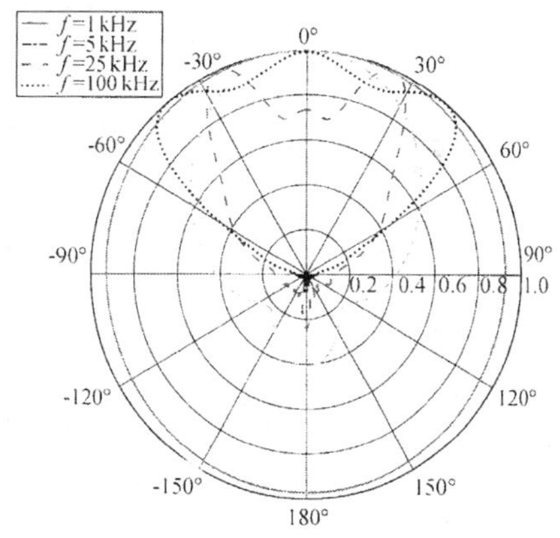

图 5 不同频率下球冠辐射指向性仿真曲线

4.4 球冠声辐射特性与 ka 值的关系

通过以上分析可知，不同的球半径和工作频率会影响球冠辐射的指向性，也就是不同的 ka 值影响着球冠的声辐射特性，通过仿真计算可以得出不同的 ka 值下球冠辐射的规律。设定球冠的极角为 60°，分别计算 ka 值为 0.8, 2, 15 和 100 时的指向性，也就是球冠所在球直径小于波长、与波长相近以及大于和远大于波长的情况，计算结果如图 6 所示。从图中可以看出，ka 值越小，球冠辐射越趋向无指向性，ka 值越大，球冠辐射的指向性越明显，辐射开角越趋于球冠极角，而且开角内起伏越明显，与上面图 4 和图 5 的计算和分析结果是一致的。

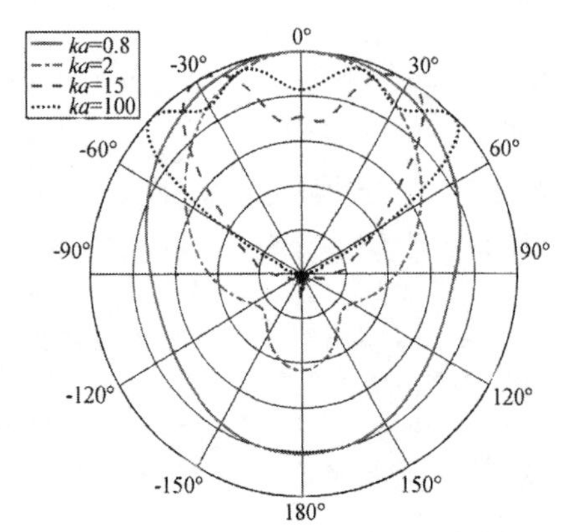

图 6 不同 ka 值下球冠辐射的指向性仿真曲线

从球冠辐射的指向性来看，球冠所在球的直径和介质中声波的波长比决定着球冠的声辐射特性，在低频或波长大于球直径时，球冠辐射呈现无指向性；随着频率的增高即波长的减小或者球障板直径的增大，球冠辐射的指向性越明显，波束开角越趋向于球冠极角，而且开角内出现波浪状起伏越明显.

5 高频球冠型换能器基阵的试制及测试分析

试制的高频球冠型换能器基阵的谐振频率选择为 300 kHz，球冠所在球的半径为 100 mm，球冠的极角为 30°，声传播介质为水，那么波长为 5 mm，ka 值为 125.66，换能器基阵的结构如图 7 所示，图 8 是根据式 (33) 计算的指向性在直角坐标系中的表示.

从图中可以看出，球冠的辐射主要集中在极角范围内，而且有“波浪状”的起伏，其 −3 dB 的波束开角的一半为 24.7°，小于其球冠的极角. 根据仿真计算的参数选择合适的高频振动形式的球冠，球冠的振动能够满足球冠面等幅同相振动的条件.

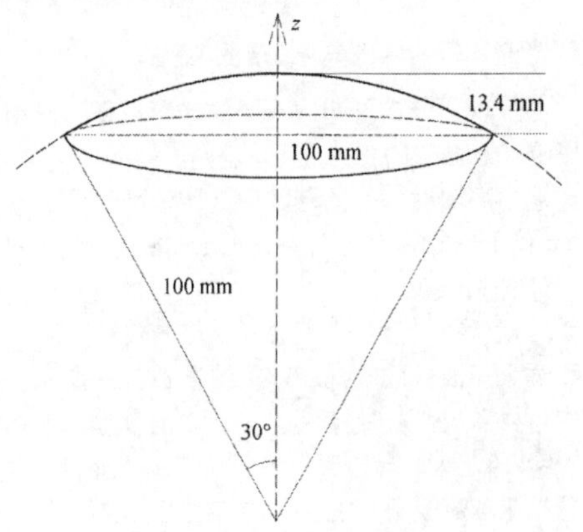

图 7 高频球冠换能器基阵结构示意图

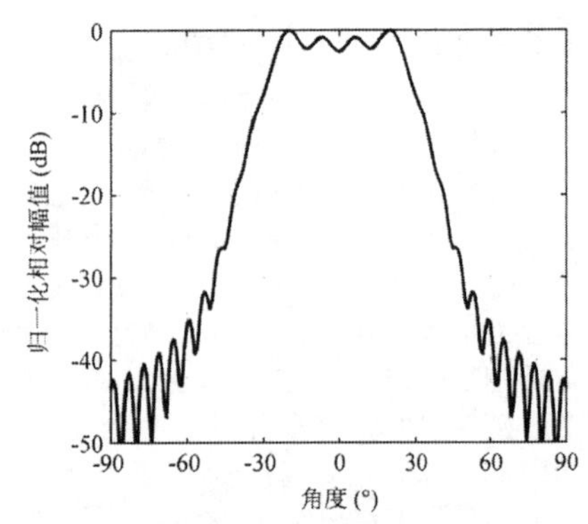

图 8 直角坐标系下高频球冠换能器基阵指向性

5.1 整体压电陶瓷球冠性能仿真

根据仿真计算球冠的参数可知，球冠的最大直径为 100 mm，球冠的高为 13.4 mm，如果采用整体的压电陶瓷成形的球冠，利用压电陶瓷的厚度振动模式，在具体的实现工艺上可以先做出半球冠后再进行切割，也可以采用压电圆柱经磨削工艺形成球冠. 整体球冠的厚度振动模态有限元仿真如图 9 所示，从图中可以看出球冠在厚度方向上的振动模态复杂，球冠中存在着相位相反的振动模式，无法在球冠面形成整体的等幅同相振动，因此其高频厚度振动谐振点指向性在极角范围出现高低起伏，而且大半径的压电陶瓷球冠制作难度大，工艺相对复杂，成本高，最终整体压电陶瓷球冠无法实现需要的指向性效果.

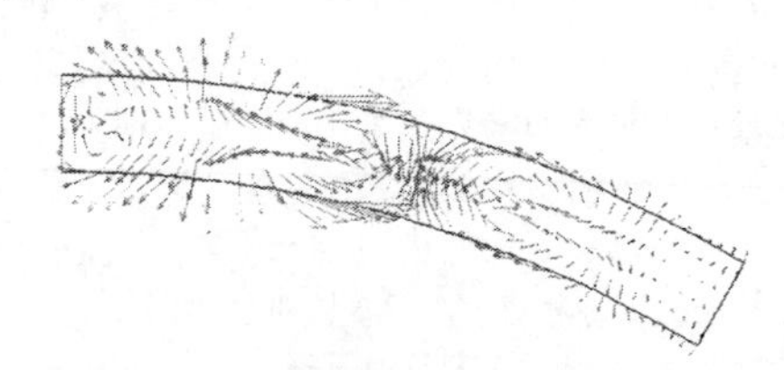

图 9 高频整体球冠振动模态

5.2 单一纵振模态压电振子球冠基阵

整体压电球冠无法实现球冠表面等幅同相振动，而通过单一纵向压电振子形成球冠形基阵可以实现表面等幅同相振动，每个振子近似看作是球面法向振动的点声源. 根据压电阵子单一纵向振动的形状条件，振子标准试样的纵向尺寸至少是横向尺寸的 2.5 倍，作为基阵单元的振子，如果辐射面太小，不仅布阵操作难度大而且其它因素也会增大对振子振动的影响，例如焊点的大小等；压电振子的辐射面也不能太大，否则单个小振子单元会具有明显指向性. 根据压电振子的纵向谐振频率与球冠的设计频率一致 (均为 300 kHz)，通过有限元软件仿真压电振子的振动特性并优化设计小振子的尺寸参数，确定满足单一纵振性能及布阵要求的小振子，最后确定振子的辐射面尺寸为 2.5 mm×2.5 mm，其在 300 kHz 纵向谐振点的振动模态如图 10 所示，从图中可以看出振子在谐振点有良好的单一纵振形式，辐射面的振动均匀等幅，而且尺寸大小易于布阵操作，满足球冠基阵单元的要求。确定基阵单元后，根据基阵的外形参数进行基阵的布阵设计，带障板的球冠基阵设计结构如图 11 所示，在极点上布有一个单元，然后在四周逐圈形成圆环，在每圈填充最大数量的单元，圈与圈之间的间距控制在 0.5mm 以内，根据球冠基阵的布阵设计换能器单元的安装模具，球冠换能器基阵未灌封前的照片如图 12 所示。

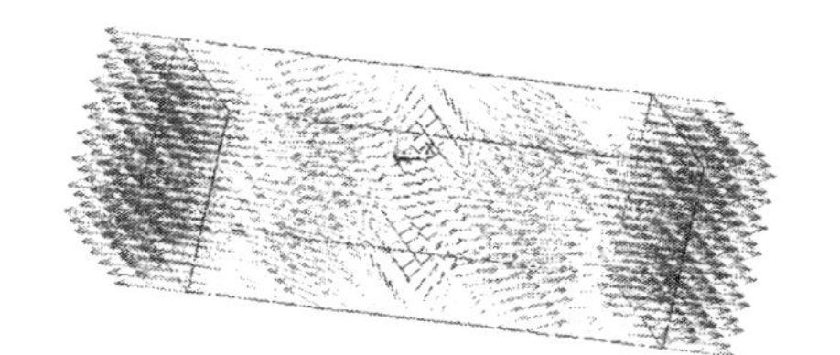

图 10 高频单一纵向振子振动模态

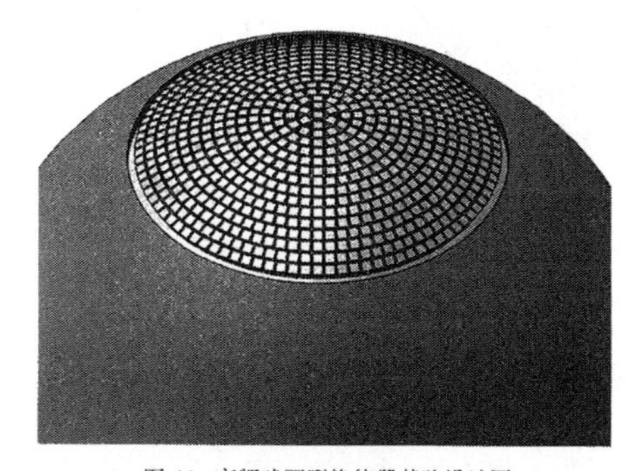

图 11 高频球冠型换能器基阵设计图

图 12 高频球冠型换能器基阵照片

5.3 球冠基阵指向特性测试

对试制好的球冠型换能器基阵在消声水池中进行了辐射特性测试，主要测试了谐振点 300 kHz 的指向性，测试结果与仿真计算结果的对比如图 13 所示。从图中可以看出，换能器在谐振频率 300 kHz 的指向性测试曲线同样有波浪状起伏，而且与仿真计算结果趋势一致，吻合较好；在波束开角之外，实测的旁瓣比仿真计算的结果要多而且杂乱，这主要是实际的障板是非球状而且非完全刚性障板，换能器基阵的单元本身也具有指向性，但总体上球冠近似认为是等幅同相振动，实测的辐射特性较好地反映了仿真计算的结果。

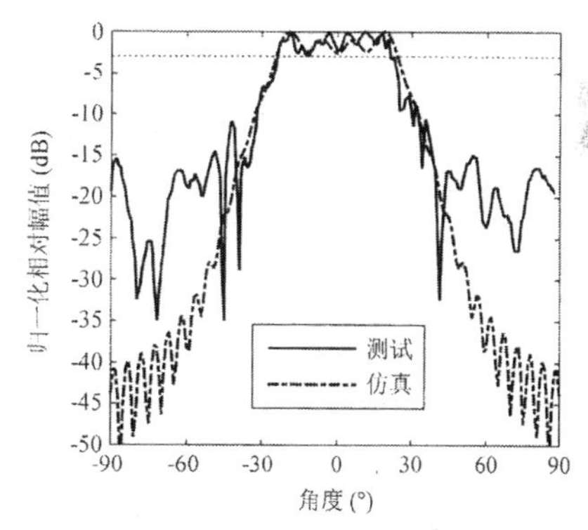

图 13 高频球冠型换能器基阵指向性仿真与测试曲线

6 结论

对球冠型换能器及基阵的辐射特性及规律进行了研究，给出了基于球谐傅里叶变换的球坐标下球冠型换能器及基阵辐射的声压和指向性表达式以及简洁计算的远场辐射声压及远场指向性表达式；探讨了球冠型辐射与球冠极角、球冠所在球半径、振动频率及 ka 值之间的关系。研究表明，ka 值的大小也即球冠所在球半径和辐射声波的波长比值决定着球冠型换能器或基阵的辐射特性。ka 值越小，辐射声波的波长越是大于球半径，指向性越是不明显，当小于一定数值时，球冠辐射呈现无指向性；ka 值越大，辐射声波的波长越是小于球半径，指向性越明显，波束开角越趋向于球冠极角，而且在波束开角内随着 ka 值的增大而出现波浪状的起伏，ka 值越大，起伏越明显，当 ka 值大于一定数值时，球冠的辐射特性几乎不变。

根据球冠辐射仿真计算结果，探索了高频球冠型换能器基阵的实现方式，完成了高频 300 kHz 球冠型换能器的试制。球冠换能器的水池测试结果较好地验证了球冠辐射理论计算的正确性，上述理论研究成果可为高频、宽波束换能器及基阵的研制提供理论指导。

参 考 文 献

1 Toulis W J. Acoustic focusing with spherical structures. *J. Acoust. Soc. Am.*, 1963; **35**(3): 286—292

2 Hasegawa T. Acoustic radiation force on a sphere in a quasistationary wave field—theory. *J. Acoust. Soc. Am.*, 1979; **65**(1): 32—40

3 DENG Wenxiang, YANG Tongsheng, MA Fengqi. Research and technical notes focused gain of focused liquid-filled sphere and its frequency response as a function of pulse length and wall thickness of sphere. *Chinese Journal of Acoustics*, 1983; **2**(3): 266—271

4 WU Jiuhui, WNG Yaojun, LI Taibao. Acoustic scattering from multiple spheres by using a kind of addition formulae for the spherical wave functions. *Chinese Journal of Acoustics*, 2004; **23**(2): 97—107

5 SHI Shengguo, YANG Desen, HONG Lianjin. Research on sound wave receiving theory of inertial sphere type vector hydrophone. *Chinese Journal of Acoustics*, 2010; **29**(4): 386—400

6 LI Wei, QU Hongfei, SONG Zhiwei. Computation and analysis of the Bessel beam scattering by a submerged elastic sphere. *Chinese Journal of Acoustics*, 2013; **32**(1): 16—26

7 Lidon P, Villa L, Taberlet N *et al*. Measurement of the acoustic radiation force on a sphere embedded in a soft solid. *Applied Physics Letters*, 2017; **110**(4): 044103

8 Liu Z, Maury C. Numerical and experimental study of Near-Field Acoustic Radiation Modes of a baffled spherical cap. *Applied Acoustics*, 2017; **115**: 23—31

9 Saule A V, Rice E J. Farfield multimodal acoustic radiation directivity. *J. Acoust. Soc. Am.*, 1977; **62**(S1): S33—S33

10 LAN Jun. A general method of directivity synthesis. *Chinese Journal of Acoustics*, 1985; **4**(2): 106—117

11 HUI Junying, LIU Hong, YU Huabing, LIANG Guolong. Study on forming directivity with constant beam widthin low frequency based on small sensor. *Chinese Journal of Acoustics*, 2001; **20**(1): 25—29

12 Shchurov V A, Ivanova G F, Kuyanova M V, Tkachenko H S. Ocean dynamic noise energy flux directivity in the 400 Hz to 700 Hz frequency band. *Chinese Journal of Acoustics*, 2007; **26**(2): 102—110

13 LI Daojiang, CHEN Hang, NI Yunlu. Research and experiments of mutual radiation impedance effect on array directivity. *Chinese Journal of Acoustics*, 2013; **32**(1): 43—50

14 Ding E, Mao Y, Liu X. Realizing a finite array of dipole sources with high acoustic transmission directivity at low frequency. *J. Acoust. Soc. Am.*, 2017; **141**(3): 1936—1939

15 ZHANG Jingdong. Time-domain analysis for vibration and sound radiation of submerged spherical shell excited by force. *Chinese Journal of Acoustics*, 1990; **9**(2): 129—138

16 SUN Yudong, SHEN Shungen, CHENG Guanyi. Nonlinear acoustic radiation by an underwater vibrating spherical shell. *Chinese Journal of Acoustics*, 1996; **15**(1): 8—20

17 Heinrich Stenzel. Handbook for the calculation of sound propagation phenomena. NRL, Translation 130, 1947: 96—114

18 Sherman C H. Mutual radiation impedance of sources on a sphere. *J. Acoust. Soc. Am.*, 1959; **31**: 947—952

19 Williams E G. Fourier acoustics: sound radiation and nearfield acoustical holography. New York: Academic Press, 1999: 183—213

20 Hasheminejad S M, Azarpeyvand M. Acoustic radiation from a pulsating spherical cap set on a spherical baffle near a hard/soft flat surface. *IEEE J. Oceanic Eng.*, 2004; **29**(1): 110—117

21 Aart R M, Janssen A J. Sound radiation from a resilient spherical cap on a rigid sphere. *J. Acoust. Soc. Am.*, 2010; **127**(4): 2262—2273

22 Aronov B, Brown D A, Bachand C L, Yan Xiang. Analysis of unidirectional broadband piezoelectric spherical shell transducers for underwater acoustics. *J. Acoust. Soc. Am.*, 2012; **131**(3): 2079—2090

23 HE Zhengyao, MA Yuanliang. Directivity calculation with experimental verification for a conformal array of underwater acoustic transducers. *Chinese Journal of Acoustics*, 2008; **27**(2): 129—138

24 DiDonato A R. Recurrence relations for the indefinite integrals of the associated legendre functions. *Mathematics Computation*, 1982; **38**(158): 547—551

五

水声通信和水下目标成像

合成孔径声纳原理样机的湖上试验①②

林 涛 孙宝申 李启虎

（中国科学院声学研究所 北京 100080）

摘 要 在合成孔径声纳理想模型的基础上，讨论了该声纳的系统结构，并介绍了其原理样机的湖上试验情况。实验结果表明，在用成像算法对原始数据进行处理后，很好地得到了目标的图像。该合成孔径声纳系统的方位向与距离向成像分辨率均小于0.5m，完全达到了设计要求，从而表明该系统的设计是成功的。

关键词 声纳，合成孔径，试验

0 引言

随着科技的进步，人类的海洋活动逐渐增多，海洋日渐成为人类活动的重要场所，尤其是进入20世纪90年代以来，世界各国纷纷加大了对海洋研究和发展活动的投入。海洋研究、航运、地质勘探、水下物体搜索等海洋活动都需要有高效的海洋观察手段。

水下较远距离的探测和成像一般都使用声纳设备。目前水底成像声纳主要有回声探测仪、前视声纳、侧视声纳（SLS）等。合成孔径声纳（Synthetic Aperture Sonar，SAS）作为一种新型主动式水下探测成像声纳，在航道测量、港口清理、水下施工场地勘探、水下物体搜索和水雷探测等方面具有非常广阔的应用前景，是国际水声高技术研究热点之一。

本文在合成孔径声纳理想模型的基础上，讨论了合成孔径声纳的系统结构，并于2000年12月完成了该型声纳原理样机的湖上试验。

1 合成孔径声纳的理想模型

合成孔径声纳作为一种主动声纳，它的基本工作方式是由拖船拖动声纳基阵在水中做匀速直线运动，基阵向正侧方发射声波信号（一般是线性调频脉冲），发射后（也可延时一段时间）开始接收海底反射的回波信号，并将之储存起来。利用基阵在不同位置上接收的信号，通过成像算法，把测绘带内物体的声散射强度图像反演出来[1]。

假设海底只有一个点目标反射声波，其反射系数为1，其他地方声波被全部吸收。在声纳航向与点目标所确定的平面上建立直角坐标系，航向方向（方位向）为X轴，与之垂直的方向（距离向）为R轴，V为基阵的运动速度，点目标P的坐标为(X_0, R_0)，基阵位置$x = vs$（s为基阵的运动时间），β为XR平面内波束的开角，L_0为基阵“照射”的长度，即合成孔径长度，如图1所示。

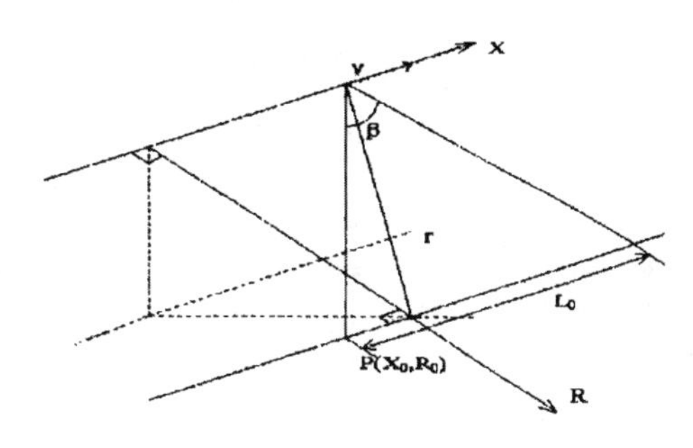

图1 合成孔径声纳模型

目标与基阵的距离$r = \sqrt{R_0^2 + (x - X_0)^2}$。在一般情况下满足：$R_0 \gg (x - X_0)$，可得如下非涅尔近似：

$$r = R_0\left[1 + \frac{(x - X_0)^2}{2R_0^2}\right]$$

为提高距离向分辨率，发射信号通常采用线性调频脉冲信号，脉冲宽度T，设发射信号$s(t)$为：

$$s(t) = \mathrm{Re}\left\{\exp\left[j2\pi f_0 t + j\pi k t^2\right]\right\},\ -\frac{T}{2} < t < \frac{T}{2} \tag{1}$$

---

① 863计划(863-818-0405)资助项目.

② 高技术通讯, 2002, 990: 83-87.

其中 f_0 为中心频率，k 为调频斜率。

声纳的运动可以近似看作“走-停-发，收-走”模式，即基阵运动到某个位置后，停下来静止地发射、接收信号，然后再运动到下一个位置。因此，我们可以把接收信号 $s_r(t)$ 看作是发射信号的延时：

$$s_r(t)=\mathrm{Re}\left\{\exp\left[j2\pi f_0\left(t-\frac{2r}{c}\right)+j\pi k\left(t-\frac{2r}{c}\right)^2\right]\right\},$$

$$-\frac{T}{2}<t-\frac{2r}{c}<\frac{T}{2} \tag{2}$$

将(1)式代入(2)式，并考虑到 $x=vs$，得到：

$$\begin{aligned} s_r(t,s) &= \mathrm{Re}\left\{\exp\left[j2\pi f_0\left(t-\frac{2R_0}{c}\left(1+\frac{(x-X_0)^2}{2R_0^2}\right)\right)\right]\right. \\ &\quad \left.\cdot\exp\left[j\pi k\left(t-\frac{2R_0}{c}\left(1+\frac{(X_0-vs)^2}{2R_0^2}\right)\right)^2\right]\right\} \end{aligned}$$

由于一般情况下有：$(x-X_0)\ll R_0$，故可以略去上式第二项中的 $(x-X_0)^2/2R_0^2$，得到：

$$\begin{aligned} s_r(t,s) &=\mathrm{Re}\left\{\exp\left[j2\pi f_0\left(t-\frac{2R_0}{c}\right)-j2\pi f_0\frac{(x-X_0)^2}{R_0c}\right.\right. \\ &\quad \left.\left.+j\pi k\left(t-\frac{2R_0}{c}\right)^2\right]\right\} \\ &=\mathrm{Re}\left\{\exp\left[j2\pi f_0\left(t-\frac{2R_0}{c}\right)+j\pi k\left(t-\frac{2R_0}{c}\right)^2\right]\right. \\ &\quad \left.\cdot\exp\left[j\frac{-2\pi}{R_0\lambda}(X_0-vs)^2\right]\right\} \end{aligned} \tag{3}$$

可见，回波在距离向和方位向上都是线性调频信号。通常，把 t 称为“宽时间”——声传播时间，把 s 称为“慢时间”——声纳基阵运动时间。由于线性调频信号通过匹配滤波器后，它的输出具有 sinc 函数的形式，即线性调频信号的能量被集中到很小的区域内，称为脉冲压缩，因此，在距离和方位两个方向上对回波分别进行脉冲压缩即可得到点目标的响应(即像)，它应该是一个二维 sinc 函数[2,3]。

设基阵的实际孔径为 D，波长为 λ，则基阵的正侧指向性为：

$$B_D(\theta)=\left|\frac{\sin\left(\frac{\pi D}{\lambda}\sin\theta\right)}{\frac{\pi D}{\lambda}\sin\theta}\right| \tag{4}$$

3 分贝主瓣宽度为 $\beta=\alpha\lambda/D$，其中 $\alpha=0.88$。设相邻两次脉冲间隔时间内基阵移动距离为 $\delta=v/PRF$，(PRF 为脉冲重复频率)对于距离 R_0 处来说，合成孔径长度为 $L_0=(\alpha\lambda/D)R_0=N\delta$，$N$ 为基阵在合成孔径范围内发射的脉冲个数。我们把它看作等效的直线阵，由于考虑了发射和接收的双程相移，其指向性为：

$$B_L(\theta)=\left|\frac{\sin\left(\frac{2\pi N\delta}{\lambda}\sin\theta\right)}{N\sin\left(\frac{2\pi\delta}{\lambda}\sin\theta\right)}\right| \tag{5}$$

3 分贝主瓣宽度 $B_L=\alpha\lambda/2N\delta$，由此可推出距离 R 处的方位向线分辨率：

$$\delta_a=\beta_L R_0=\frac{\alpha\lambda}{2N\delta}R_0=\frac{\alpha\lambda}{2\frac{\alpha\lambda}{D}R_0}R_0=\frac{D}{2} \tag{6}$$

而在距离向上，合成孔径声纳和普通声纳没有区别，都是靠线性调频信号的脉冲压缩解决距离分辨率。其距离向分辨率为：

$$\delta_r=\frac{c}{2|k|T_s}=\frac{c}{2B} \tag{7}$$

其中 B 为发射信号的带宽[4,5]。

由(6)式可以看出，合成孔径声纳的方位向分辨率与声纳的工作频率和作用距离都没有关系，因此可以用较小的声纳基阵和较低的工作频率同时满足近距离和远距离的探测需要，并且获得均匀的空间分辨力。

2 合成孔径声纳的系统构成

为保证声纳基阵在水中做匀速直线运动，合成孔径声纳系统一般采用拖曳方式。声纳基阵安装在水下拖体上，拖体用一艘工作母船通过拖缆拖曳，在水下一定深度上做匀速直线运动。整个系统分为干端和湿端两部分，通过信号传输电缆加以连接。湿端设备主要有发射部分、接收部分、湿端控制中心和信号传输部分等，而干端设备主要有数字收发信机、多功能信号处理机、磁带纪录仪和显控中心等[6]。其系统框图如图 2 所示。

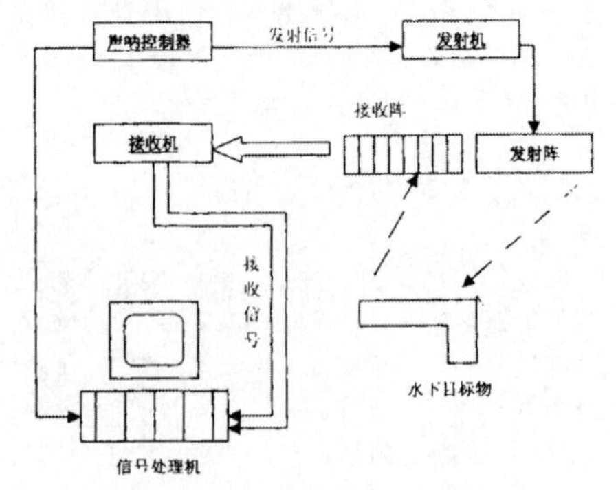

图 2　合成孔径声纳系统框图

发射信号的形式主要有单频脉冲信号、线性调频脉冲信号和这两种信号与各种窗函数的组合等。

多种信号的数据存放在信号发生器的存储器中。信号发生器根据湿端控制中心发来的指令选择发射信号的形式、脉冲宽度和脉冲重复频率等参数。生成的发射信号被存放在发射缓冲区中，按照设定的脉冲重复频率定时把发射缓冲区中的数据送给数/模转换器。数/模转换器的输出经放大滤波后送到发射机，由发射基阵向正侧方发射声波信号，延时一段时间后，接收机阵开始接收目标反射的回波信号。

接收基阵接收到的信号经放大滤波后，由模/数转换电路变为数字信号，经拖缆将该信号送到信号处理机。该处理机以高性能数字信号处理芯片为核心，经过波束形成、脉冲压缩、运动补偿等一系列预处理后，利用合成孔径声纳成像算法获得目标图像并在显控中心上加以显示。

3 合成孔径声纳湖试数据及试验结果

在各方的大力协助下，合成孔径声纳原理样机于2000年12月在浙江千岛湖完成了湖上试验。湖试时，声纳基阵安装在浮筏底面，通过法兰盘及连杆与浮筏硬联接。发射机与接收机水密电子罐置于浮筏上表面。数据采集计算机及声纳控制器安置在工作艇上，工作艇上的计算机及电源通过电缆与浮筏联接，对浮筏上的发射机及接收机供电并提供信号通讯通道，如图3所示。

湖试作业时，由快艇将工作艇与浮筏拖至预定位置。然后，用母船上的绞车匀速拖动浮筏及工作艇，由发射基阵向水下目标（目标为臂长3m的煤气罐三角架）发射线性调频脉冲信号。脉冲声信号由目标物反射后，反向传播至接收换能器基阵转换为电信号。接收到的目标散射信号经接收机放大，然后送到数据采集计算机。对收集到的数据做成像处理，完成目标物声图像重建。采集到的原始数据、经过距离向脉冲压缩后的中间结果、利用成像算法完成的目标重建声图以及归一化局部放大声图分别如图4~7所示。

图3 湖上试验的仪器安排

图4 原始数据

由图 4 ~ 5 可以看出，从原始数据中看不到任何目标的影像，而在经过距离向的脉冲压缩后，由于信噪比的提高，可以明显地看到目标在方位向上所形成的弧线。在用成像算法对方位向进行进一步处理后很好得到了目标的图像。从归一化局部放大声图可以看出，该合成孔径声纳系统的方位向与距离向成像分辨率均小于 0.5m，完全达到了设计要求，从而表明该系统的设计是成功的。

图 5　距离向脉冲压缩

图 6　目标重建声图

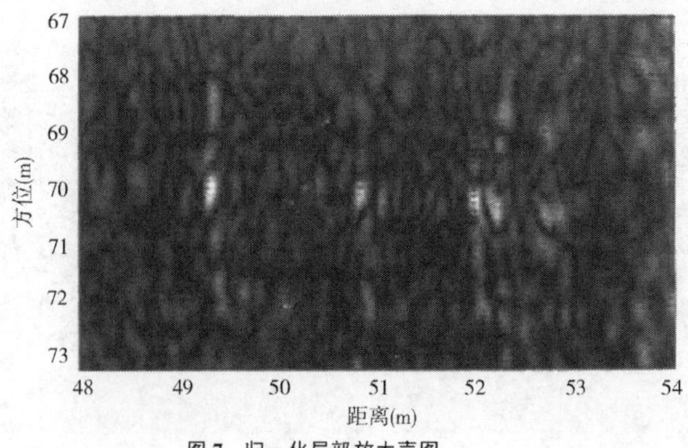

图 7　归一化局部放大声图

4 结论

作为一种新型的主动式成像声纳，合成孔径声纳克服了现有声纳系统所固有的方位分辨率低、空间分辨率不均匀的缺点，在许多领域有着广阔的应用前景。本文在合成孔径声纳理想模型的基础上，讨论了该声纳的系统结构。在其原理样机的湖上试验中，该合成孔径声纳系统的方位向与距离向成像分辨率均小于0.5m，完全达到了设计要求。

参考文献：

[1] 合成孔径成像声纳技术论证报告及总体技术方案．中国科学院声学研究所合成孔径声纳项目方案组，1997

[2] Cutrona L J. *J Acoust Soc Am*，1975，58(2)

[3] Williams R E. *J Acoust Soc Am*，1976，60(1)

[4] 张澄波．合成孔径雷达——原理、系统分析与应用．北京：科学出版社

[5] 李南松．合成孔径声纳集中成像算法的实现：[学位论文]．中国科学院声学所，1999

[6] 孙宝申．合成孔径声纳千岛湖试验实验报告，2000

The Synthetic Aperture Sonar Prototype Trial on Lake

Lin Tao，Sun Baoshen，Li Qihu

(Institute of Acoustics，Chinese Academy of Sciences，Beijing 100080)

Abstract

Starting from the theoretical SAS model，the system structure of SAS is discussed and the lake trial result of a SAS prototype is given also. Using the SAS reconstruct algorithm，the target is imaged clearly. Both the direction and the distance resolution of the SAS prototype are better than 0.5m，which was in accordance with the system design. The trial result indicates that the design of the prototype is reasonable and feasible.

Key words：Sonar，Synthetic aperture，Trial

斜视合成孔径声呐成像研究①②

刘纪元　唐劲松　孙宝申　张春华　李启虎

(中国科学院声学研究所　北京　100080)

2001 年 10 月 19 日收到

2002 年 2 月 27 日定稿

摘要　在论述斜视合成孔径成像及其进展的基础上，分析了正侧视情况下延时相加的模型及斜视对它的影响，将延时相加法分为“有效孔径处理”和“盲处理”两种实现方法，并提出斜视模式有效孔径计算数学模型。斜视出现时，参与方位向波束形成的有效孔径的定位会出现错误，使图像产生失真。斜视模型可对有效孔径正确定位，从而得到正确的图像。仿真实验及湖试结果验证了模型的有效性。

PACS 数：　43.60

Study on the imaging of squint synthetic aperture sonar

LIU Jiyuan　TANG Jinsong　SUN Baoshen　ZHANG Chunhua　LI Qihu

(*Institute of Acoustics, The Chinese Academy of Sciences* Beijing 100080)

Received Oct. 29, 2001

Revised Feb. 27, 2002

Abstract　Synthetic aperture imaging with squint mode and its advances are presented, and the imaging of broadside and squint mode on sum-delay algorithm is discussed. Effective aperture processing and blind processing for the sum-delay algorithm are described, and a mathematical model for the calculation of effective aperture is suggested. When squint receiving arrays exist, broadside model can not locate effect aperture correctly, and consequently, a degraded image may occur. The suggested model can correctly locate the effect aperture, and the computer simulations and results from lake experiments validate the method.

引言

SAS(合成孔径声呐) 的研究滞后于 SAR(合成孔径雷达)，目前存在的 SAS 系统主要是正侧视条带式工作模式。随着 SAS 研究的深入，已经有人开始探讨其它模式了。其中斜视方式是 SAS 成像中有实用意义并值得研究的一种方式，比如在水下猎雷时采用前斜视工作模式可以较早发现目标。即使是正侧视工作方式的 SAS 在工作中也可能出现换能器阵倾斜的情况，此时 SAS 实际上是用正侧视模型处理斜视的数据。因此，研究斜视问题对 SAS 很有意义。

合成孔径成像的优点是用宽的方位向波束 (小的孔径) 可以获得较高的方位向分辨率，当SAS工作在正侧视 (broadside) 时，理论分辨率可以达到 $La/2$。当 SAS 工作在斜视 (squint) 方式时，波束中心线与航迹不垂直，若偏离垂线的角度为 θ，则理论分辨率可以达到 $La/(2\cos\theta)$，有关分析参见文献 1。

SAS 成像是在 SAR 的基础上发展起来的，SAR 成像主要采用 RD 、 Chirp Scaling(CS) 和 ωK 等。而 SAS 一般为宽带系统，成像方法常采用延时相加法 (sum-delay 法，相当于早期 SAR 中采用的二维相关法) 和 ωK 法。近年来也有人采用改进的 CS 法进行 SAS 成像，但延时相加法和 ωK 法被认为是精确的方法，成像模型中未做近似处理。特别是天线或水声换能器阵随机出现斜视状态时，采用这两种方法对成像结果影响较小。基于正侧视模型得到的各种成像算法，对斜视时的成像可能效果变差。这是因为

① 863 计划资助项目.

② 声学学报, 2003, 28(5): 439-442.

有些算法中采用了模型近似，由于斜视的出现使残差加大，模型精度下降，从而影响成像效果。

在 SAR 领域，对斜视研究较早是结合 RD 算法进行的。正侧视合成孔径成像时，点目标的距离陡动曲线为：

$$R=\sqrt{r_0^2+v^2(t-t_0)^2}, \tag{1}$$

其中 R 是 t 时刻点目标到 SAR 或 SAS 平台的距离，v 为平台运动速度，r_0 为点目标到航迹的最小距离，对应公式中 t 为 t_0 时刻 R 的值。RD 算法中将距离陡动曲线近似为一个二次多项式，并将一个二维成像问题化成两个一维问题。制约 RD 算法在高分辨率成像中应用的一个因素是距离与方位耦合较强、有较大的距离弯曲时，需要进行距离弯曲校正，使得算法变得复杂。对 RD 算法影响较大的另一个因素是斜视问题。在斜视时，距离陡动曲线二次多项式近似的残差加大，在宽测绘带时尤其明显，影响成像效果，补偿办法是二次距离压缩[2-4]。

文献 5 从斜视模型出发，得到距离陡动曲线及其二阶多项式近似，即：

$$R=\sqrt{r_0^2+v(t-t_0)^2-2r_0v(t-t_0)\sin\theta}, \tag{2}$$

$$R\approx r_0+\frac{\lambda}{2}f_0(t-t_0)+\frac{1}{2}f_r(t-t_0)^2, \tag{3}$$

其中 λ 为波长，θ 为斜视角，而且：

$$f_0=\frac{2v}{\lambda}\sin\theta, \tag{4}$$

$$f_r=-\frac{2v^2}{\lambda r_0}\cos^2\theta. \tag{5}$$

θ 为 0 时对应正侧视的结果。RD 法在方位压缩匹配处理时，由多普勒中心频率 f_0 和调频斜率 f_r 构造滤波器。用斜视时的 f_0 和 f_r 修正匹配滤波器，得到适合斜视的 RD 算法，仿真结果表明适合大斜视角成像。但该法适合斜视角已知情况的斜视工作模式，对工作时随机出现大的斜视时，斜视角的确定则是一个问题。

与同一维的 RD 法相比，采用二维处理的 CS 算法，对斜视的宽容性好。但 CS 中仍存在模型近似，斜视时会使近似产生的残差加大。在经典 CS 基础上发展起来的非线性 CS 对斜视成像效果有改善[6]，而文献 7 在二维频域内对近似模型增加一个三次相位项做为补偿，进一步改善了斜视的成像效果，适合更大斜视角的场合。文献 8 用公式 (2) 代替公式 (1)，即用斜视代替正侧视陡动曲线，并结合星载 SAR 中 f_0 和 f_r 估计问题，得到新的 CS 算法，适合星载 SAR 斜视成像。

1 斜视情况下延时相加法的成像

1.1 延时相加法

SAS 的延时相加法从本质上相当于早期 SAR 中的二维时域相关法，因为运算量太大，运算效率低，所以现代 SAR 中一般不采用。但 SAS 受到水声传播速度的限制，方位采样率一般远低于 SAR，而且作用距离近，照射范围小。因此，SAS 系统实时问题比 SAR 容易实现。即使采用低运算效率的算法也能完成实时处理，例如欧洲的 SAIME 实时 SAS 系统就采用了这种算法。

延时相加法同 RD, CS, ωK 等所谓的“逐线”高效算法比，为低运算效率的“逐点”算法。因为其自身的一些特点，它在 SAS 中常被采用。这些特点主要有：物理意义清晰，而且数学模型简单，便于实现；不同于逐线法 (以平台位置为出发点)，图像的产生是以像素点几何位置为基础，成像区域、坐标系的选取 (直角坐标或极坐标、二维或三维等) 具有很大灵活性；运动补偿直接方便，不仅在波束中心线附近，其它位置也可补偿；可直接对信号调理电路和水声传感器补偿；对多接收阵 SAS 成像方便[9]。

延时相加法是水声阵列处理中波束形成的常用方法。该法用于 SAS 成像，可对接收的信号直接做方位波束形成 (最佳处理法)，也可对脉冲压缩后的复包络做方位波束形成，即所谓的次最佳法。次最佳法用于宽带 SAS，分辨率损失极小，运算量也大大下降，而且对运动误差更宽容[10]。本文采用基于复包络方位处理的方法。

1.2 有效孔径的确定

根据该算法的特点，不同距离、不同方位上的像点，对应方位向不同长度和不同区段的回波数据，如图 1(a) 所示。显然远斜距处的点 (图中 A 点) 比近斜距点 (图中 B 点) 成像时用到更多的方位采样点。

正侧视和斜视两种情况下，参与运算的有效区段如图 1(b) 和图 1(c) 所示。图中 x 轴与航迹重合，R 沿斜距方向。有效区段起点 $L1$ 和终点 $L2$ 在正侧视时为：

$$L2=x+R\,\mathrm{tg}(\Phi/2), \tag{6}$$

$$L1=x+R\,\mathrm{tg}(-\Phi/2). \tag{7}$$

对于斜视情况：

$$L1=x+R\,\mathrm{tg}(\theta-\Phi/2), \tag{8}$$

$$L2=x+R\,\mathrm{tg}(\theta+\Phi/2). \tag{9}$$

公式 (8) 和 (9) 是更一般的公式，当 $\theta=0$ 时分别变成式 (6) 和 (7)。

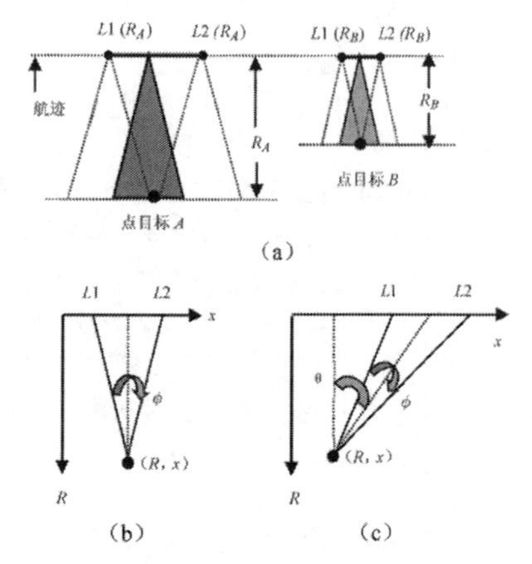

图 1 条带式合成孔径成像波束照射方位向示意图(a), 正侧视有效孔径示意图(b)和斜视有效孔径示意图(c)

1.3 忽略有效孔径的盲处理

当 SAS 正侧视换能器阵在工作中可能出现较大倾斜, 而且又难以预测时, 简单的处理方法是用方位向最大孔径区段 (对应于正侧视时最大距离像素点), 使选取成像区域中各点对应的有效孔径都包含在其中。尽管对不同像点会有无效数据参与运算, 降低了运算效率, 但不会出现有效区段选错而出现的失真问题。

2 实验

2.1 仿真实验

参数选取: 线性调频信号中心频率 20 kHz, 带宽 7.5 kHz, 发射信号脉冲重复频率 10 Hz, 接收信号采样频率 125 kHz 。拖体运动速度 0.5 m/s, 接收阵的方位向开角 10° 。

图2为盲处理结果, 当阵处于正侧视、10°斜视、20°斜视、30°斜视时, 均能很好成像。

当采用有效孔径处理时, 算法中增加公式 (6) 和 (7) 的正侧视约束, 对正侧视、10°斜视、20°斜视、30°斜视时, 如图 3 所示, 成像效果较差。

当按公式 (8) 和 (9) 分别对不同斜视角计算有效孔径, 则可得到与图 2 相同的成像效果。仿真信号中只有点目标回波信号, 噪声及混响等未考虑, 所以无效数据段对成像无影响。因此, 这里 "盲处理" 与 "斜视有效孔径处理" 效果相同。

2.2 湖试结果

用延时相加法对千岛湖水下实验数据进行了成像处理。发射信号参数与仿真时所用参数相同, 拖体运动速度 0.82 m/s, 接收阵的方位向开角 10°。成像目标为 L 形强散射体: L 形钢架的两个臂长度均为 2.5 m 左右, 每个臂上安装两个工业用煤气瓶 (空腔煤气瓶, 自重远小于浮力)。强散射体通过钢绳与铁块连接, 由于铁块的重力坠入水中。散射体与铁块间连接的钢绳约 10 m, 故铁块沉底后, 强散射体悬浮在距离湖底 10 m 的水中。

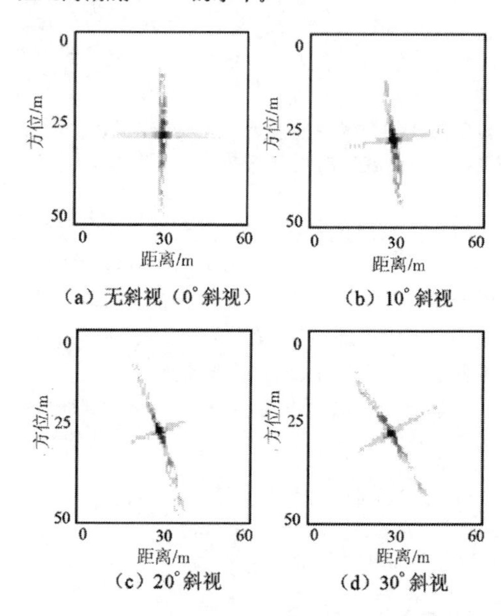

图 2 延时相加法盲处理方式时不同斜视情况的点目标成像

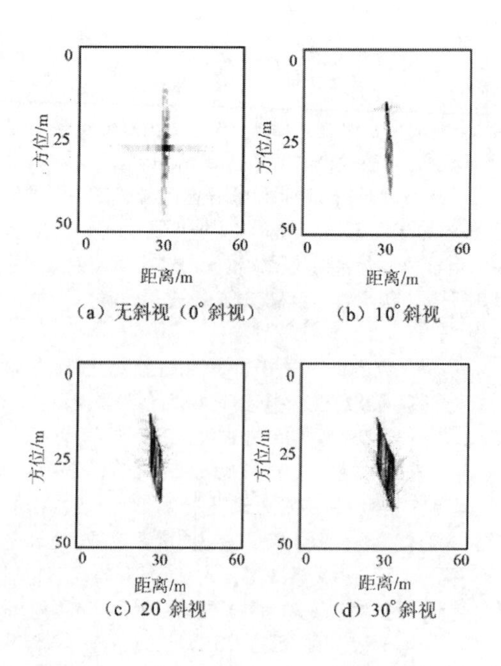

图 3 延时相加法有效孔径处理方式时不同斜视情况的点目标成像

图 4(a) 为一组实验中强散射体回波的脉冲压缩结果。从脉冲压缩结果发现，接收阵出现倾斜。因为正侧视时，强点目标在脉压后的图像中曲线 (沿方位) 对称，而斜视时曲线不对称。

图 4(b) 为该强散射体合成孔径成像结果。因为实验是按正侧视设计的，成像算法中按公式 (6) 和 (7) 计算有效孔径，成像效果差。改用盲处理后，就得到了图 4(b) 的图像。

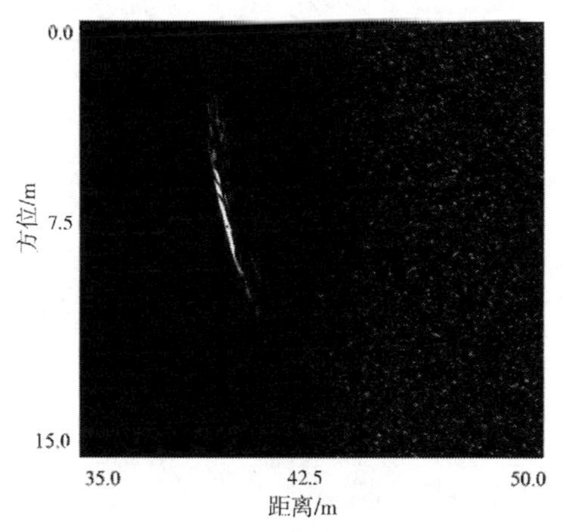

(a) 脉冲压缩后的图像

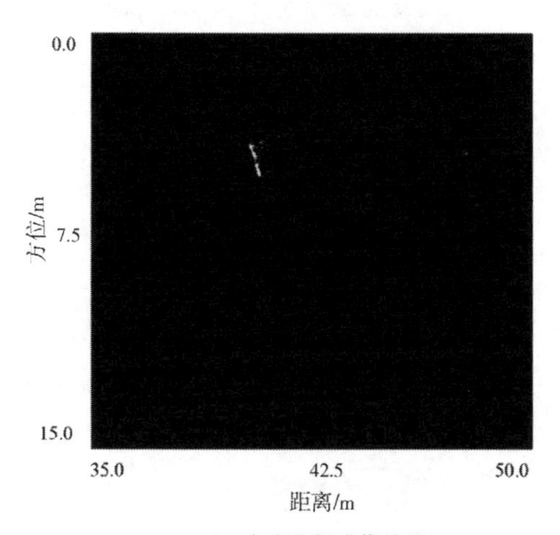

(b) 合成孔径成像结果

图 4 湖试数据成像结果

3 结论

当采用延时相加法进行 SAS 成像时，确定有效孔径可以提高计算效率，在实时系统中，常考虑这个环节。正侧视时按 (6) 和 (7) 式计算、斜视时按 (8) 和 (9) 式计算有效孔径。但工作中实际视角与设计值有较大差别时，图像会出现严重失真。盲处理方式对较大范围内倾斜角均能很好成像。

参 考 文 献

1 Partrice Caprais, Stephane Guyonic. Squint and forward looking synthetic aperture sonar. Proc. of Ocean'97, 1997: 809—814

2 Chang C Y, Jin M, Curlander J C. Squint mode processing algorithms. Proc. of IEEE Geoscience and Remote Sensing Symposium., Vanvouver, Canada, 1989: 1702—1706

3 Hughes W, Gault K, Princz G J. A comparison of the Range-Doppler and chirp scaling algorithms with reference to RADARSAT. Proc. of Geoscience and Remote Sensing Symposium, 1996: 1221—1123

4 Wong F H, Cumming I G. Error sensitivities of a secondary range compression algorithms for processing squinted satellite SAR data. Proc. of IEEE Geoscience and Remote Sensing Sypm., Vanvouver, Canada, 1989: 2584—2587

5 张劲林, 许庆荣, 刘永坦. 适合大斜视角星载 SAR 系统的改进距离 - 多普勒算法. 系统工程与电子技术, 1999; **21**(8): 9—11

6 Davidson G W, Cumming I G, Ito M R. A chirp scaling approach for processing squint model SAR data. *Transactions on Aerospace and Electronic System*. 1996; **32**(1): 121—133

7 Alberto Moreira, HUANG Yonghong. Airborne SAR processing of highly squinted data using a chirp scaling approach with integrated motion compensation. *IEEE Transaction on Geoscience and Remote Sensing*, 1994; **32**(5): 1029—1-40

8 黄 岩, 李春升, 陈 杰, 周荫清. 高分辨率星载 SAR 改进 Chirp Scaling 成像算法. 电子学报, 2000; **28**(3): 35—38

9 Mchugh R, Shaw S, Talor N. Efficient digital signal processing algorithm for sonar imaging. *IEE Proc. Radar, Sonar and Navigation*, 1996; **143**: 149—156

10 Chatillon J, Zkharia M E, Bouchier M E. Wideband synthetic aperture sonar: advances and limitations. Proc. of European Conf. on Underwater Acoustics, Luxembourg, 1992: 709—712

合成孔径声呐成像自聚焦方法研究①②

刘纪元　唐劲松　孙宝申　张春华　李启虎
(中国科学院声学研究所　北京　100080)

2001 年 9 月 19 日收到
2002 年 2 月 27 日定稿

摘要　合成孔径声呐 (Synthetic Aperture Sonar, 简称 SAS) 载体的随机运动、声波传播媒质不稳定等会影响成像的质量。大的运动误差可由运动补偿修正，但残余运动误差和媒质的影响需要自聚焦方法来修正。本文推导出小的 SWAY 误差下脉冲压缩图象域内运动误差模型，并将合成孔径雷达 (Synthetic Aperture Radar, 简称 SAR) 中的 PATCH-PGA 方法用于 SAS 自聚焦，仿真实验和湖试结果表明了该方法的有效性。

PACS 数：　43.60, 43.30

On the autofocus of synthetic aperture sonar

LIU Jiyuan　TANG Jinsong　SUN Baoshen　ZHANG Chunhua　LI Qihu
(*Institute of Acoustics, The Chinese Academy of Sciences* Beijing 100080)

Received Sept. 19, 2001
Revised Feb. 27, 2002

Abstract　Synthetic aperture sonar imaging is affected by random motion and medium instability. Large motion errors can be corrected by means of motion compensation, while the effects of uncompensated errors and medium instability should be compensated for with aotofocus method. SAS error model is derived in range-compressed domain under the assumption of small sway error. Moreover, patch-PGA used in SAR is studied in using SAS and the simulation and underwater experiments in a lake show the validity of the method.

引言

SAS 研究起步较晚，其原理与 SAR 十分相似，因此可以借鉴 SAR 中的研究成果， SAR 中的许多技术经改造可以移植到 SAS 中。星载 SAR 的自聚焦主要是通过回波数据估计 fr(多普勒频率变化率)，这相当于误差多项式的二阶参数估计，因为星载 SAR 沿轨道运动十分稳定，该估计一般能取得满意效果。但机载 SAR 的运动随机误差相对大些，常需高阶估计，或非参数 (任意阶) 估计如 PGA，才能取得较好效果。 SAS 运动误差一般比 SAR 严重，对成像质量影响很大，运动补偿十分重要。 PGA 是机载聚束 SAR 自聚焦较好且较成熟的方法，在应用时要满足“空不变”条件。聚束 SAR 一般为窄波束，“空不变”条件容易满足；但条带式 SAR 一般波束较宽，“空不变”条件有时不能满足，直接应用 PGA 则不能取得应有的效果。但是经过改造，条带式 SAR 中也可以采用 PGA 进行自聚焦。 P.H.Eichel 等人在文献 1,2 和 3 中提出适合聚束式 SAR 的 PGA 方法，后来又在文献 4 中提出了 PGA 方法的改进，适合窄测绘带条带式 SAR 。也有人提出将条带式 SAR 大的区域划成小块，再使用 PGA 。

本文结合 SAS 运动误差特点，推导出图象信号受 SWAY 误差影响的模型，再结合仿真和水下实验讨论改进的 PGA 在 SAS 中的应用。

1 SAS 运动误差模型

SAS 的载体 —— 拖鱼的运动误差可由 6 个自由度描述。大量的实验结果表明，对图象重建影响较大的运动误差是 SWAY 和 YAW，其中 SWAY 是影响最大的。 SWAY 沿航迹的侧向移动导致接收信号产生时间上的延迟，时延的大小用 $t_e(x)$ 表示，其中 x 为

① 863 计划资助项目.
② 声学学报, 2003, 28(2): 151-154.

方位位置坐标。考虑SAS发射的是chirp信号，可以表示为：

$$s_e(t)=\mathrm{Re}\left[\mathrm{rect}\left(\frac{t}{T}\right)\exp\left(\mathrm{j}2\pi f_0 t+\mathrm{j}\pi f_r t^2\right)\right],\quad -\frac{T}{2}\le t\le\frac{T}{2} \tag{1}$$

其中 f_0 为中心频率，f_r 为调频速率。若 SAS 沿 x 轴运动，坐标用 x 表示。

接收到的实信号经正交解调后变为复信号。本文将用不同符号区别图象重建过程中不同的二维图象函数：$s_r(t,x)$ 和 $s_{re}(t,x)$ 分别为无运动误差和有运动误差时 SAS 接收信号，$ss_r(t,x)$ 和 $ss_{re}(t,x)$ 分别代表无运动误差和有运动误差时正交解调及脉压后的信号。$ss_e(t,x)$ 和 $ss(t,x)$ 分别为无运动误差和有运动误差时完全重建的图象。为简化分析过程，忽略换能器方向性的影响（或认为是等方向性的）。如果一个点目标与 SAS 航迹的最小距离为 r_0，沿 x 轴的坐标为 x_0。则 SAS 回波信号的时延为：

$$\tau=\frac{2}{c}\sqrt{r_0^2+(x-x_0)^2}. \tag{2}$$

点目标的响应为：

$$s_r(t,x)=\mathrm{rect}\left(\frac{t-\tau}{T}\right)\exp\left[\mathrm{j}\pi f_r(t-\tau)^2\right]\exp\left[\mathrm{j}2\pi f_0(t-\tau)\right]. \tag{3}$$

当存在 SWAY 误差时，其对声呐回波的影响是产生一个附加延时，记为 $t_e(x)$，则：

$$\tau=\frac{2}{c}\sqrt{r_0^2+(x-x_0)^2}+t_e(x). \tag{4}$$

同时考虑正交解调后的响应：

$$\begin{aligned}s_{re}(t,x)=&\mathrm{rect}\left[\frac{t-\frac{2}{c}\sqrt{r_0^2+(x-x_0)^2}-t_e(x)}{T}\right]\\&\exp\left\{\mathrm{j}\pi f_r\left[t-\frac{2}{c}\sqrt{r_0^2+(x-x_0)^2}-t_e(x)\right]^2\right\}\\&\exp\left[-\mathrm{j}\frac{\mathrm{j}4\pi f_0}{c}\sqrt{r_0^2+(x-x_0)^2}-\mathrm{j}2\pi f_0 t_e(x)\right],\end{aligned} \tag{5}$$

这里 c 为声速。在SAR中一般假定信号是窄带的，即 $fr\to 0$（极限情况：$fr=0$，这时Chirp信号变为单频信号），同时认为较小的误差主要影响相位因子，这时：

$$s_{re}(t,x)\approx s_r(t,x)\exp[-\mathrm{j}2\pi f_0 t_e(x)]. \tag{6}$$

沿时间（快时间）轴做傅里叶变换并与发射的Chirp信号傅里叶变换的共轭相乘，再做傅里叶逆变换，则得到脉冲压缩后的信号。用 $ss_{re}(t,x)$ 和 $ss_r(t,x)$ 分别表示有运动误差和无运动误差，因为运动误差的影响只与方位有关，在脉冲压缩前后不受影响，所以有：

$$ss_{re}(t,x)\approx ss_r(t,x)\exp[-\mathrm{j}2\pi f_0 t_e(x)]. \tag{7}$$

文献1和文献2从脉冲压缩图象域出发，用 PGA 方法提取并补偿相位误差的影响。对聚束式合成孔径成像，波束始终照射指定的成像区域，图象中不同方位上各点相位能反映出运动误差信息。但对条带式成像，同一距离门内相距较远的点则反映不同航迹段上的误差信息。因此，源于聚束式成像的 PGA 方法不能直接用于条带式成像。较简单的方法是将成像区域划分成小块，空不变条件近似成立，对感兴趣的区域进行 PGA 处理，简称 patch-PGA；或将成像区域分块，分别做 PGA 后再组合，即所谓镶嵌式 PGA(mosaic PGA，简称 MPGA)。

Peter Gough 等人[5] 曾讨论过将文献4中改进的PGA用于SAS。而 W.W.Bonifant 等人[6] 将MPGA用于 DARPA SAS，并在完全重建的图象域给出带运动误差和不带运动误差的完全重建图象域函数 $ss_e(t,x)$ 和 $ss(t,x)$ 运动误差模型：

$$ss_e(t,x)\approx ss(t,x)*h_e(x), \tag{8}$$

其中 $*$ 代表卷积，$h_e(x)$ 为运动误差对应的散焦函数，可看成 $\exp\{-\mathrm{j}2\pi f_0 t_e(x)\}$ 的傅里叶反变换。

借鉴SAR中PGA方法研究SAS中自聚焦问题，从理论到实际应用均有待深化。不仅空不变条件解决方法不尽相同，相位误差的提取和修正也可在不同图象域进行，如有的文献建议在脉冲压缩图象域提取误差信息，也有文献采用完全重建图象域提取误差信息。本文通过计算机仿真和千岛湖水下实验数据，将 patch-PGA 法用在 SAS 成像中。

2 仿真实验和湖试结果

仿真实验参数如下：拖体运动速度 0.5 m/s，线性调频信号中心频率 20 kHz，带宽 7.5 kHz，发射信号脉冲重复频率 10 Hz，接收信号采样频率 125 kHz。

仿真实验中，在 SAS 运动轨迹上加一个峰值 3 cm、两个周期偏离航迹的侧向正弦扰动。在成像区布置 5 个点目标，图象重建用 ωK 算法，成像结果见图 1(a)，点目标在方位向上出现散焦。在完全重建图象域经 3 次自聚焦处理后的图象见图 1(b)，聚焦效果有较好改善。本实验中把 5 个目标像选在同一个块 (patch) 中处理。

另一组实验选两个在方位上相距较远的点目标，轨道上加峰值为 2 cm 两个周期的正弦扰动，其它条件与上面实验相同，图 2(a) 为成像结果。将图 2(a) 中上面像点及附近区域图象分离出来并做自聚焦，成像结果如图 2(b)；而将图 2(a) 中下面像点及附近区域图象分离出来并做自聚焦，成像结果如图 2(c)。分别进行自聚焦后的点目标图象质量有明显提高。当然，一帧图象分成多块做自聚焦时，可将结果组合起来。

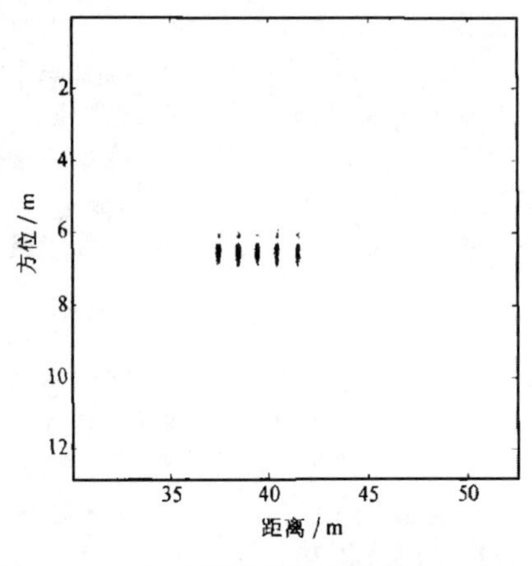

(a) 带运动误差的重建图象

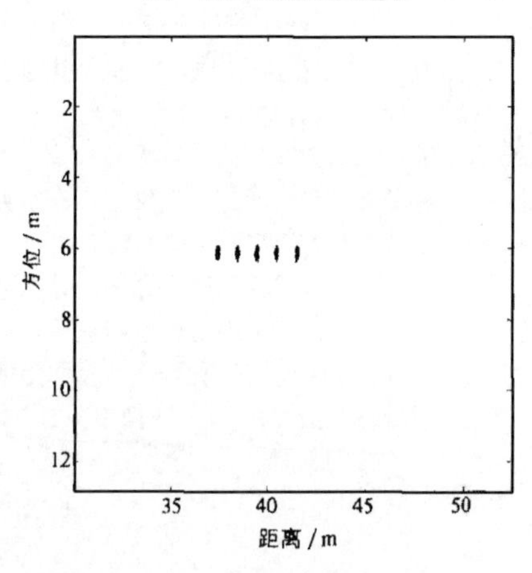

(b) 经自聚焦处理后的结果

图 1 仿真实验结果

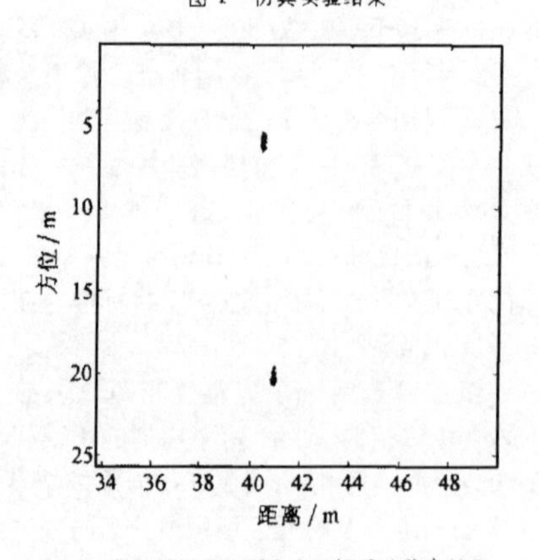

(a) 带运动误差的两个点目标重建仿真结果

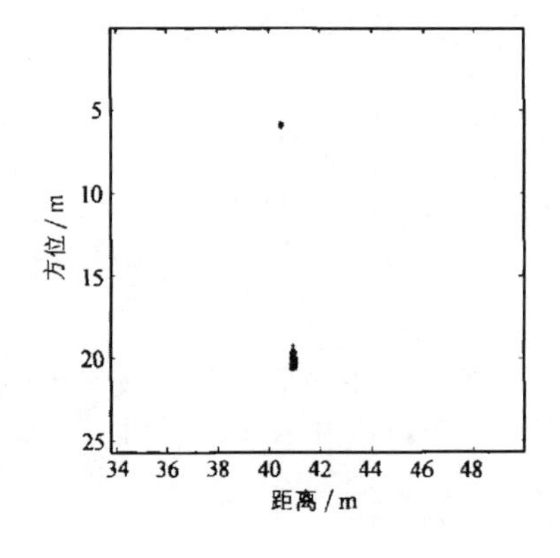

(b) 对上方的点自聚焦结果

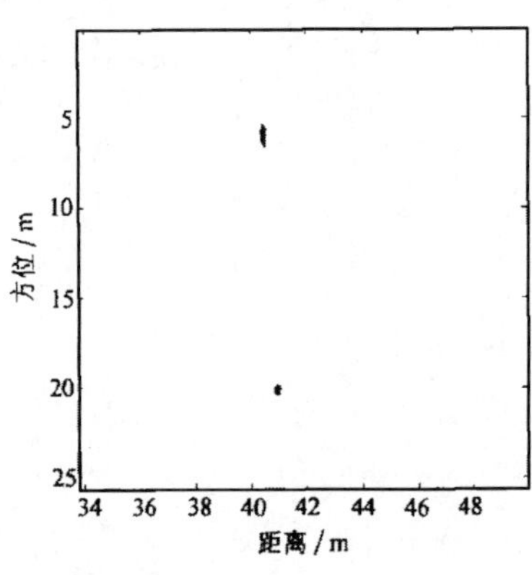

(c) 和对下方的点自聚焦结果

图 2 仿真实验结果

实验中发现，分块自聚焦效果较好，这是因为点目标对应不同航迹段的照射，遭受运动误差的影响也可以不同，如图 3 所示。图 3 中 A 点在 SAS 经过 PQ 段航迹时受到照射，而 B 点在 RS 段受到照射，重建的图象中两者受到运动误差的影响也不同。

千岛湖水下实验的相关参数同仿真实验。成像目标为 L 形钢架，钢架每个臂上有 3 个煤气瓶，

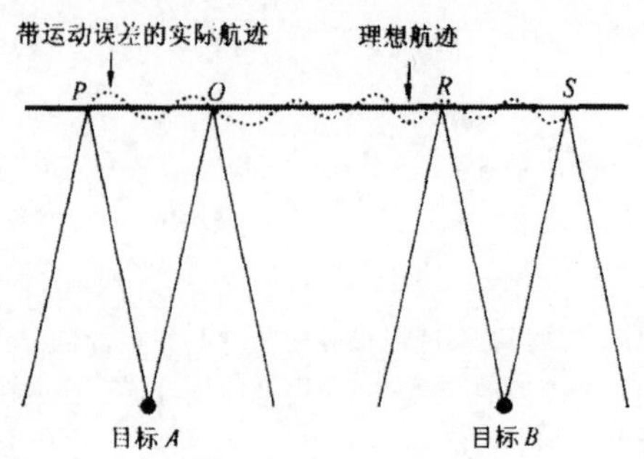

图 3 两个点目标受到不同航迹段照射及受不同运动误差影响示意图

总长度约 2 m 。水深约 45 m, 目标位于湖底。拖体入水 20 m, 运动速度 0.82 m/s 。图 4(a) 是 L 形目标 SAS 成像结果。图 4(b) 是经自聚焦处理后的结果, 处理后目标的一个臂 (垂直于方位向的臂) 成像效果变好。

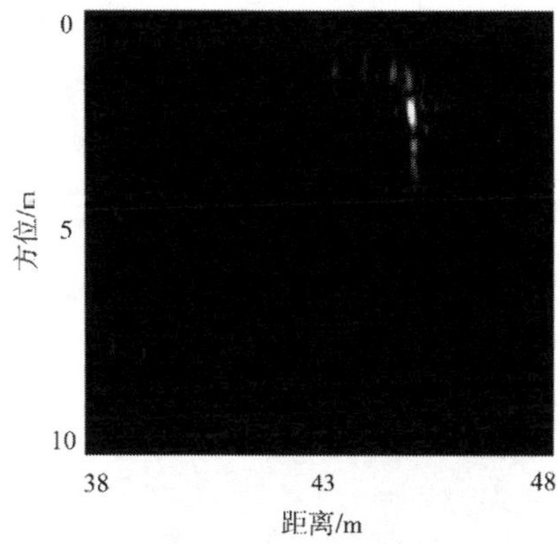

(a) SAS 千岛湖水下实验成像结果

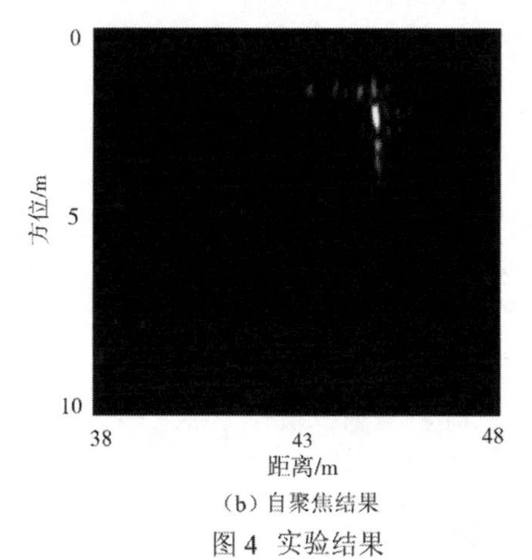

(b) 自聚焦结果

图 4 实验结果

3 结论

本文将patch-PGA方法用于条带式SAS, 仿真实验和湖试结果表明其有效性。 PGA 源于聚束 SAR, 其应用的前提是窄带发射信号以及窄波束照射下的小区域。在SAS图象中, 通过分割可以得到小的成像区域, 然后应用PGA , 尽管PGA 是在窄带发射信号假设下得到的, 对宽带SAS也适用。但随着相对带宽的大大增加, 自聚焦的效果如何, 有待深入研究。

参 考 文 献

1 Eichel P H, Jakowatz C V, Jr. Phase-gradient algorithm as an optimal estimator of phase derivative. *Optics Letters*, 1989; **14**(20): 1101—1109

2 Eichel P H, GHIGLIA D C, Jakowatz C V, Jr. Speckle processing method for synthetic-aperture-radar phase correction. *Optics Letters*, 1989; **14**(20): 1—3

3 WAHL D E, Eichel P H, GHIGLIA D C, Jakowatz C V, Jr, Phase gradient autofocus — A robust tool for high resolution SAR phase correction. *IEEE Transactions on Aerospace and Electronic System*. 1994; **30**(3): 827—834

4 Eichel P H, Jakowatz C V, Jr, Thompson P A. New approach to strip-map SAR autofocus. Proc. the 1994 6th IEEE Digital Signal Processing Workshop. Yosemite, CA, 1994: 53—56

5 Gough P T, Hawkins D W. Imaging algorithms for a strip-map synthetic aperture sonar: minimizing the effects of aperture errors and aperture undersampling. *IEEE Journal of Oceanic Engineering*, 1997; **22**(1): 27—39

6 Bonifant Jr W W, Richards M A, McClellan J H. Interferometric height estimation of the seafloor via synthetic aperture sonar in the presence of motion errors. *IEEE Proc. Radar, Sonar and Navigation*, 2000; **147**(6): 322—330

基于回波信号的一种合成孔径声纳运动补偿方法①②

刘纪元，唐劲松，孙宝申，张春华，李启虎

（中国科学院声学研究所，北京 100080）

摘要：运动补偿是合成孔径声纳（SAS）成像的关键问题，SAS 系统一般利用多接收阵回波信号数据的互相关特性进行运动补偿。本文提出一种基于单接收阵和强点目标回波的运动补偿方法，并在推导侧摆误差模型的基础上，进行了计算机仿真。仿真和实测数据实验结果表明该方法的有效性。

关键词：合成孔径声纳，运动补偿，相关函数，偏离相位中心

中图分类号： TN957　　文献标识码： A

A Receiving-Data-Based Motion Compensation Method of Synthetic Aperture Sonar

LIU Jiy-uan　TANG Jin-song　SUN Bao-shen　ZHANG Chun-hua　LI Qi-hu

(Institute of Acoustics, Chinese Academy of Science　Beijing 100080, China)

Abstract: Motion compensation is the key technique for synthetic aperture sonar (SAS), and it is normally realized in SAS systems based on multi receiving arrays and their cross correlation. A new motion compensation method based on single receiving array and strong scatterer is presented in the paper. Moreover, computer simulations based on the derived model for sway error are given, and results of simulation and lake test show the validity of the proposed motion compensation method.

Key words: synthetic aperture sonar, motion compensation, correlation function, displaced phase center

1 引言

合成孔径成像在雷达（SAR）领域取得的成功，推动了合成孔径声纳技术的发展。目前国际上已经出现多个 SAS 实验样机系统。面向军用和商用实用设备的出现，相信并不遥远。SAS 在民用领域主要用于海底探测，而军事用途主要是水雷探测。由于合成孔径成像的相似性，SAS 可借鉴 SAR 中的技术成果，SAR 中的成像算法可用在 SAS 中。在 SAS 研究领域，有两个主要问题被认为影响了 SAS 技术发展。第一个是水声信道，特别是浅海水声环境条件不理想，同空气中电磁波工作环境相比，是更为“敌意”的媒质。另一个问题是声波传播速度比电磁波慢得多，大大限制装载 SAS 的拖体的运动速度。这不仅限制了测绘速率的提高，也影响拖体的稳定性。

二十世纪七十年代以来一系列水声成像相干性实验结果表明，水声信道的影响并不像预想的那么严重，尽管水声信道是时变的，SAS 回波信号在较短时间内仍具有较好的相干性。

运动误差是影响 SAS 成像质量的一个关键问题，目前 SAS 研究的热点为运动补偿和自聚焦等问题。SAS 的波长较短（一般在分米量级），高分辨率 SAS 对运动和姿态测量系统的精度要求很高，相应的设备价格很贵而且工程上有时很难达到要求。因此，SAS 运动补偿研究主要集中在利用回波信号本身提取运动信息方面。

Robert S. Raven 于 1981 年申请美国专利，将偏离相位中心（Displaced Phase Center，简称 DPC）的方法用于多接收天线合成孔径成像。Robert W. Sheriff[1]通过水槽实验验证了 DPC 方法修正 SAS 运动误差的有效性。目前已经有 SAS 系统采用多接收阵 DPC 方法[2][3]。

但是，使用多接收阵 DPC 方法的前提条件是假设海底完全混响，相当于 SAR 中均匀目标的场合。本文讨论有孤立的强点目标存在时，单接收阵 SAS 运动补偿问题。

2 基于相关函数的运动误差估计

2.1 侧摆误差数学模型

SAS 的运动误差可用六个自由度描述。对于拖曳式 SAS，运动误差主要表现为在水平面内的侧摆（sway，海浪引起的较大误差）和偏航（yaw，较小的姿态误差）。侧摆一般用 DPC 修正，而偏航则通过自聚焦解决。

侧摆是拖体平行于航迹的水平运动，产生斜距误差，导致接收信号产生时间上的延迟。若 SAS 沿 X 轴运动，坐标用 x 表示，则时延的大小用 $t_e(x)$表示。考虑 SAS 发射的是 chirp 信号，可以表示为：

$$s_e(t)=\mathrm{Re}\{rect(\frac{t}{T})\exp(\mathrm{j}2\pi f_0 t+\mathrm{j}\pi f_r t^2)\},-\frac{T}{2}\leqslant t\leqslant\frac{T}{2} \tag{1}$$

其中 f_0 为 chirp 信号中心频率，f_r 为调频速率，T 为脉冲宽度。

① 基金项目: 863 计划海洋检测主题资助项目(No. 818-04-05).

② 电子学报, 2003, 31(1): 75-77.

为简化分析过程，忽略换能器方向性的影响（或认为是等方向性的）。如果一个点目标与SAS航迹的最小距离为r_0，沿X轴的坐标为x_0，则接收信号为：

$$s_r(t,x)=\mathrm{Re}\{rect(\frac{t-\tau}{T})\cdot\exp[\mathrm{j}\pi f_r(t-\tau)^2]\cdot\exp[\mathrm{j}2\pi f_0(t-\tau)]\} \quad (2)$$

正交解调后变为复信号，利用相位驻留原理，脉冲压缩后的信号表示为：

$$s_{rc}(t,x)\approx\alpha\cdot\sin c[f_rT(t-\tau)]\cdot\exp\{\mathrm{j}2\pi f_0\tau\} \quad (3)$$

其中α为常数，SAS回波信号的时延：

$$\tau=\frac{2}{c}\sqrt{r_0^2+(x-x_0)^2} \quad (4)$$

这里c为声速。

当存在侧摆误差时，其对声纳回波的影响是产生一个附加延时，记为$t_e(x)$，则

$$s(t,x)\approx\alpha\cdot\sin c\{f_rT[t-(\tau+t_e(x))\}\cdot\exp\{-\mathrm{j}2\pi f_0[\tau+t_e(x)]\} \quad (5)$$

式（5）表明，SAS侧摆误差引起脉压后的复信号中时延和相移的变化。

2.2 基于自相关的延时估计

侧摆引起散射体到SAS斜距的变化，这也是回波延时的变化，通过相邻两次回波的互相关函数（ping到ping的互相关）可将这个变化估计出来。

在合成孔径成像文献中，还未发现基于相关函数的时延估计方面的严格数学证明。时延估计是水声阵列处理中一个重要问题，C.H. Knapp和G.C. Carter[4]证明了噪声背景下时延的最佳最大似然估计本质上是一个互相关器，按照相应结论构造的互相关估计器的结构如图1。改变延时τ，当检测到最大峰值时，对应的τ即为估值。由互相关函数估计回波延时或斜距变化的这种方法在其它领域也有相似用法。

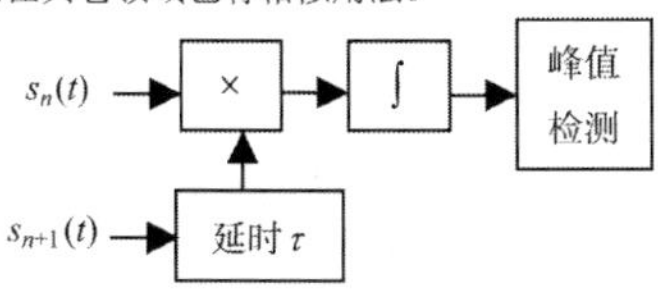

图1　时延估计的结构

在运动目标成像、ISAR成像[5]等领域，经常用到基于相关函数的运动估计问题。假设SAS相邻两次回波分别为$s_n(t)$和$s_{n+1}(t)$，则互相关函数：

$$R_n(u)=\int s_n(t)s^*_{n+1}(t+u)dt \quad (6)$$

假设相邻两次回波的模值变化很小时，即$|s_n(t)|\approx|s_{n+1}(t)|$，那么：

$$R_n(u)\approx\int|s_n(t)|^2\exp\{\mathrm{j}2\pi f_0[u-\Delta\tau_n]\}dt \quad (7)$$

其中$\Delta\tau_n$为第n次回波（ping）和第n+1次回波（ping）信号的延时差。当$u=\Delta\tau_n$时，互相关函数取得最大，故互相关函数峰值对应的时延就是相邻两ping间的时延。将所有方位上的相邻ping相对延时计算出来，然后沿方位积分，就得到方位向的延时误差或斜距误差。

2.3 侧摆误差的估计

忽略距离徙动的影响，当SAS沿理想航迹运动时，将脉冲压缩后所有方位上相邻两ping的数据互相关，ping与ping间无斜距变化，所以均为0。如果存在侧摆引起的斜距误差，则通过相关估计得到ping与ping间斜距的变化或微分结果，通过积分就得到所有ping上的斜距误差。

2.4 点目标的距离徙动及互相关估计结果

由于SAS距离和方位的耦合，点目标在不同方位上到SAS的距离是变化的，距离徙动曲线为：

$$R=\sqrt{r_0^2+v^2(s-s_0)^2} \quad (8)$$

其中R是s时刻，点目标到SAR或SAS平台的距离。v为平台运动速度，r_0为点目标到航迹的最小距离，对应公式中s为s_0时刻R的值。

因此，由于距离徙动的影响，前述误差估计的直接结果是侧摆误差与徙动距离的和。

2.5 侧摆误差的曲线拟合提取

SAS运动速度很慢而且方位采用率很低，在相对较长的方位采样时间内，海浪等引起的侧摆表现为高频摆动误差，它叠加在距离徙动函数之上。因为徙动为一条平滑曲线（可用二次、四次或更高次多项式逼近），对互相关法估计的结果做曲线拟合，将平滑曲线做为徙动曲线估值。在总估计结果中减去徙动曲线估计，就得到高频摆动误差的结果。

2.6 采样率问题讨论

在离散傅立叶分析领域，要求用二倍以上信号上限频率采样即可。而做时域波形分析时，采样率的要求常常很高，有关分析表明，为正确得到正弦信号的峰值，采用十倍信号频率的采样率，平均峰值误差小于5%。在声纳阵列处理时域波束形成时，一般要求8至10倍信号上限频率的采样率。在采样率较低时，可采用内插处理做为补偿。在DPC运动补偿中，同样需要高采样率及内插处理，才能取得好的效果。

2.7 运动补偿过程

运动补偿在脉冲压缩后进行。结合成像过程，由如下几步组成：（1）回波信号正交解调、脉冲压缩；（2）侧摆误差估计，并进行补偿；（3）采用距离徙动成像或其它算法成像；（4）图象域自聚焦（选做项）。

3 实验结果

3.1 仿真实验

参数选取：线性调频信号中心频率25kHz，带宽7.5kHz，发射信号脉冲重复频率10 Hz（方位采样率），接收信号采样频率125kHz。拖体运动速度0.5米/秒，采用单发射阵和单接收阵，接收阵的方位向开角20°。在SAS运动轨道上引入3个周期、峰值为6厘米的正弦变化斜距误差。

图 2 中（a）和（b）分别为点目标在无运动误差时脉冲压缩和最后成像结果，（c）和（d）分别为有运动误差时脉冲压缩和最后成像结果，（e）和（f）分别为经过 DPC 运动补偿后脉冲压缩和最后成像结果。存在运动误差时的图象明显变坏，经运动补偿后图象效果变好。

图 3 给出斜距误差估计结果。其中（a）是叠加在距离徙动函数之上的斜距误差估计，（b）为通过曲线拟合得到的距离徙动估计，这里采用的是二次多项式拟合，（c）为减去拟合函数影响后的斜距误差估计，也就是运动误差的估计。

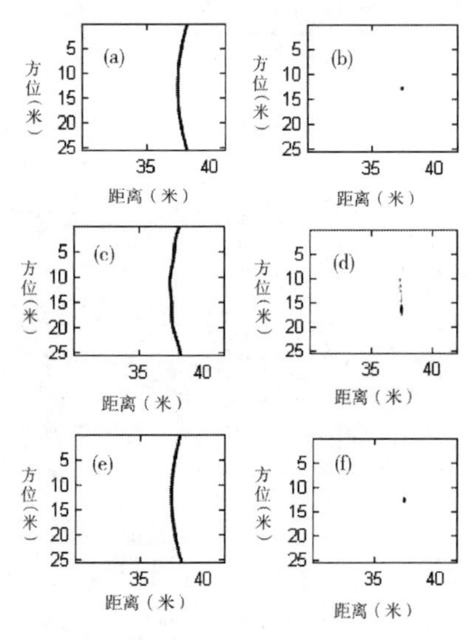

图2 点目标脉压和成像结果

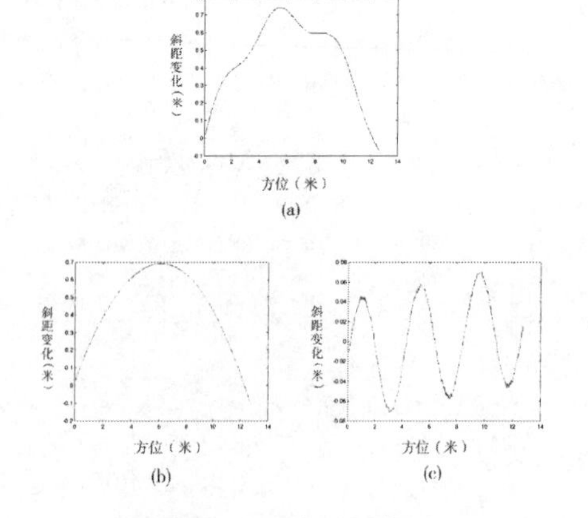

图3 基于相关函数的误差估计

3.2 水下实验结果

图 4（a）为千岛湖水下实验强散射体 SAS 成像结果，而 图 4（b）经本文中的 DPC 修正的结果。经修正后的图象，对强散射体的像点有锐化效果。

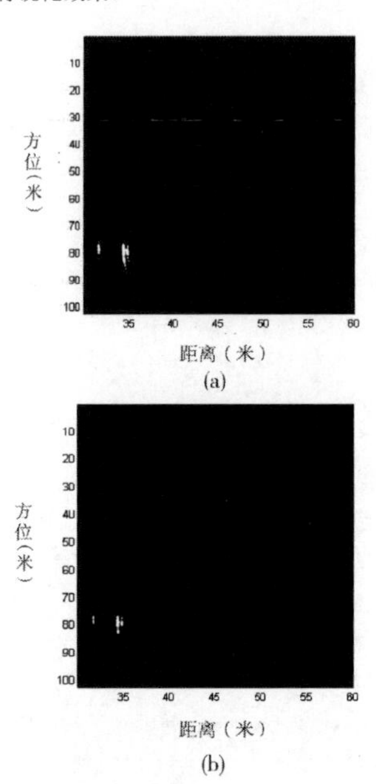

图4 水下实验结果:（a）强散射体SAS图象;（b）运动补偿后的图象

4 结论

传统的利用多接收阵对水底完全混响回波运动补偿方法，当存在强散射体时，侧摆估值的误差较大。本文利用单接收阵及单个强散射体回波提取误差信息，进行运动补偿。但该过程相对复杂些，需要进行强点散射体距离徙动的估计，以便将其与侧摆估计分开。同时，当存在多个强散射体时，需要根据脉压后的图象，确定孤立的强散射点。强散射点散射需足够强，在脉冲压缩图象中应存在可辨别的徙动曲线。

参考文献

[1] Robert W. Sheriff，Synthetic Aperture Beamforming with Automatic Phase Compensation for High Frequency Sonars[A]. Proc. Of the 1992 Symposium on Autonomous Underwater Vehicles[C]. 1992: 236-245.

[2] Dennis Garrood, Norm Lehtomaki, Mark Neudorfer, and Andrew Palowitch, Synthetic Aperture Sonar: An Evolving Technology - Radar Technology Used for Underwater for Mine Hunting and Unexploded-Ordnance Location[J]. SEA TECHNOLOGY, 1999, (6): 17-23.

[3] W.W. Bonifant, Jr., M.A. Richards and J.H. McClellan, Interferometric height estimation of the sea floor via synthetic aperture sonar in the presence of motion errors[J]. IEE Proc.-Radar, Sonar and Navigation, 2000, 147(6): 322-330.

[4] C.H. Knapp and G.C. Carter, The generalized correlation method for estimation of time delay[J]. IEEE Trans. on ASSP, 1976, 24(4): 320-327.

[5] 刘永坦. 雷达成像技术[M]. 哈尔滨工业大学出版社，1999: 258-317.

合成孔径声呐并行实时处理研究①②

刘纪元 李淑秋 李丽英 张春华 李启虎

(中国科学院声学研究所 北京 100080)

摘 要 该文论述了合成孔径声呐 (Synthetic Aperture Sonar, SAS) 系统的图像重建算法选择、参数选择和并行处理方式，并提出一种基于 SHARC 信号处理机 SAS 实时成像实现方法。结合自行研制的原理样机中信号处理系统特点，采用两级流水线：第一级利用空间采样间隔时间做序贯处理，完成正交解调、脉冲压缩功能；第二级采用批 (帧) 处理完成方位聚焦。两级流水线，仅用一级 (第二级) 的时间开销，就完成了图像重建。为满足实时性要求，两级都采用了并行处理。

关键词 合成孔径声呐 (SAS)，实时处理，声呐成像

中图号 TN958.99

1 引 言

水下成像主要采用光学摄像机、激光成像和声呐成像等手段。光学和激光成像图像清晰、好于其它设备成像效果，但作用距离太近，一般在几米至几十米之间，而且不适合混水场合。声成像具有作用距离远、有穿透能力等优点，因而被广泛用于水下成像。如水下地质地貌勘测、水下丢失物寻找、水雷探测 (含锚雷、沉底雷和泥沙掩埋的沉底雷) 等。水下声成像有主动和被动声呐两种方式，而侧扫成像式主动声呐应用较广泛，侧扫声呐方位向的高分辨率通过大的换能器 (方位向) 孔径取得。合成孔径声呐也工作在侧扫方式下，但它是通过小的孔径及其运动形成等效大孔径。与合成孔径雷达 (Synthetic Aperture Radar, SAR) 一样，合成孔径声纳 (Synthetic Aperture Sonar, SAS) 的一个优点是方位分辨率与距离无关[1]。

SAS 的原理研究从 20 世纪六、七十年代开始，但 SAS 在应用上存在的一些技术问题制约了它的发展[2]。首先水声环境 (时变信道) 一般比较恶劣，不同回波的信号的相干性是个问题；其次，水声传播速度慢，使得信号空间采样率不足；同时，受声传播速度慢和方位模糊的限制，SAS 的载体运动速度慢，姿态不容易稳定。

受 SAR 成功的鼓舞，进入 80 年代，欧洲、澳洲、北美一些国家进行了一系列水声环境实验，结果表明水声信号的相干性能够满足合成孔径成像要求。声传播速度慢导致信号空间采样率低和限制 SAS 载体运动速度等问题可以通过多接收阵的办法来弥补。90 年代以来，上述一些国家先后研制出 SAS 原理样机[3,4]，并且性能在不断提高。一些 SAS 系统的作用距离从原来的几十米、几百米到十几公里，甚至更远；分辨率也从几米、几分米到厘米甚至更小。

我国在 90 年代开始 SAS 实时成像系统的研制工作，目前已研制出 SAS 湖试样机，分辨率在分米量级，作用距离在百米之内。样机经水下实验，能实时完成合成孔径成像，为下一步海试 SAS 实时成像样机的实现打下基础。该 SAS 信号处理系统能完成实时成像、实时显示、实时原始数据存储等功能。SAS 信号处理系统以 Melbourne DSP 板为主要计算单元，该板是由 Spectrum 公司生产的、以 ADI 公司的 ADSP-21060(SHARC) 为处理器的信号处理板。各 DSP 有独立的程序和数据存储器，并通过 LINK 口相连。因为每个 SHARC 芯片有 6 个 LINK 口，DSP 间的数据传递，处理起来比较方便。本系统直接采集带载频的回波信号，并用 SHARC 在数字域内进行正交解调、脉冲压缩，然后进行方位处理并得到重建的图像。

SAS 实时成像与 SAR 成像有相似之处，但也有自身的特点。首先，声呐的载频较低，一般采用宽带信号发射，所以要选用宽带方法图像重建。其次，水声传播速度低，信号空间采样率

① 863 计划(863-818-04-05, 863-2002AA631110)资助项目。

② 电子与信息学报, 2003, 25(6): 777-783.

低，常采用多接收阵处理．我们的 SAS 系统采用两级流水线结构，每级内采用多 CPU 并行处理保证实时性．第一级完成正交解调、脉冲压缩、乒乓存储、转置存储的功能．结合 SAS 空间采样率低的特点，充分利用脉冲信号接收间隙，并行完成第一级运算，在时间上仅延时一个方位信号收发周期．本文结合该 SAS 湖试样机，论述 SAS 图像重建算法和实时并行处理的有关问题．

2 SAS 图像重建算法选择

SAS 图像重建是在 SAR 成像的基础上发展起来的，所以有必要回顾一下 SAR 成像方法．

SAR 成像最简单最基本的方法是时域“二维移变滤波器匹配”方法，可简化为时域“延时相加法”．这是一种“逐点”的运算方法，在早期的的 SAR 成像中使用．因为其运算效率很低，现代 SAR 中很少采用．而 RD(Range Doppler), CS(Chirp Scaling) 和 ΩK(Range Migration) 等“逐线”成像算法，常被现代 SAR 所采用．

RD 算法将二维成像处理过程简化为距离和方位两个一维级联处理．当忽略距离和方位耦合时，成像效率很高，特别便于实时成像．但简单的 RD 算法存在两个缺点：一个是距离徙动较大时 (距离和方位耦合较强)，图像失真较大．解决办法是增加“距离徙动校正”环节．这不仅使运算量增加，而且并行实时处理时，“距离徙动校正”需要多次转角存储，大大降低了运算效率． RD 算法的另一个缺点是不适合斜视处理，当接收天线波束中心线与航迹不垂直时，条带式 SAR 图像失真增大．解决办法是增加“二次距离压缩”环节． RD 算法的若干改进使其效率大大降低．

90 年代以来，用于地震信号处理的 ΩK 算法引入到 SAR 成像中． ΩK 法是一种“二维”的处理方法，从理论上较“优化”．但 Stolt 变换需要插值运算，减低了运算效率而且使相位保持特性变差．

CS 算法对 ΩK 算法做了改进和近似处理，避免了插值运算，运算效率大大提高．采用二维处理的 CS 算法同一维的 RD 法相比，对斜视的宽容性好．它虽然比 RD 算法对斜视有更大一些宽容性，但斜视角较大时，仍需要改进． CS 中存在模型近似，斜视时会使近似产生的残差加大．在经典 CS 基础上发展起来的非线性 CS 及其它改进，对斜视成像效果有改善． CS 法目前是高分辨 SAR 成像中较好的方法．

SAR 研究领域除向高分辨率发展外，低频超宽带 SAR 因为较明显的军事背景也成为受到关注的一个方向．超宽带 SAR 成像目前主要采用“延时相加法” (sum-delay 法) 和“ ΩK 法”．

SAS 一般为宽带系统，图像重建通常采用 ΩK 法．因为该方法是较优化的二维算法，适合宽带处理，对斜视宽容性好，所以为 SAS 系统普遍采用的方法．如新西兰的 KIWI SAS 和美国的 DARPA SAS 等都采用该算法成像．

延时相加法尽管运算效率低，但在 SAS 中也常采用． SAS 成像时方位采样率和测绘效率都低，实时处理比 SAR 容易实现．另外，延时相加法具有物理意义清晰、数学模型简单、运动补偿直接方便、便于多接收阵成像、对斜视宽容性好等特点，是理论上“最精确”的方法．欧洲的 SAMI SAS 系统就采用该算法成像．

近年来也有人采用“加速的 CS ” (Accelerated Chirp Scaling) 法进行 SAS 成像．这种改进的 CS 法，运算效率较高．

我们在 SHARC 信号处理机上，采用 ΩK 等并行算法实现了实时成像．

3 并行处理的考虑

3.1 ΩK 算法的并行实现

鉴于声呐信号的载频较低，可以对带载频的回波信号直接采样，然后在数字域内进行正交解调．用 SHARC 进行解调，比模拟方法实现起来方便灵活，精度高．但运算量较大，需要多

CPU 并行运算才能实现实时处理. 同时, 带载频采样时, 采样率较高, 解调后还要做降采样处理. 单接收阵的 ΩK 算法并行处理系统如图 1 所示.

图 2 给出了多接收阵时 ΩK 算法并行处理方法, 多通道数据在做合成孔径处理前进行调整和融合, 变成单路复信号.

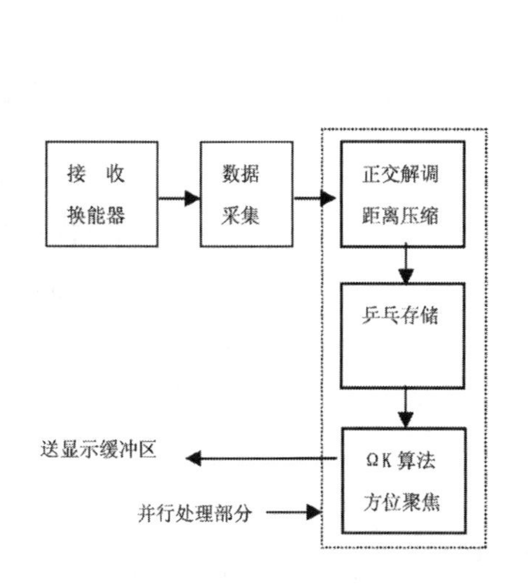

图 1 单接收阵 SAS 信号处理系统框图

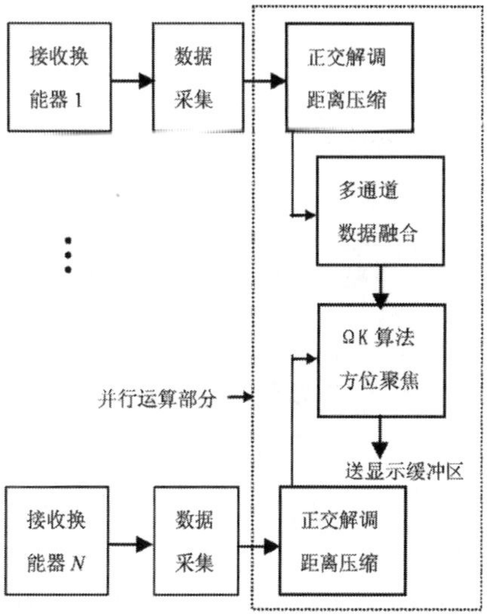

图 2 多接收阵 SAR信号处理系统框图

ΩK 为二维频域成像算法, 需要对矩阵的各行各列分别做 FFT 和 IFFT 变换. 在做并行处理时, 需要转置存储. ΩK 算法并行处理的流程如图 3 所示.

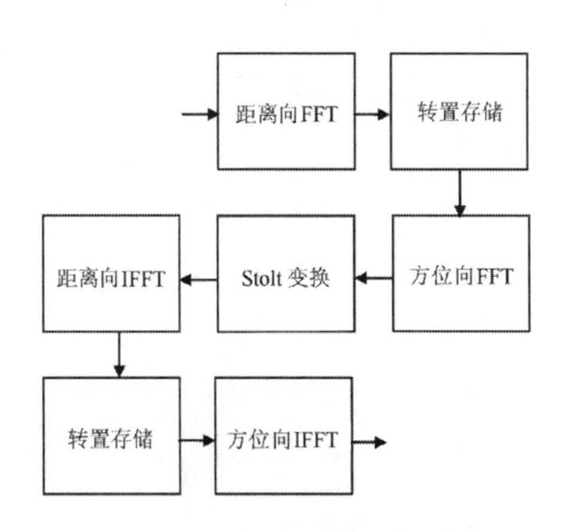

图 3 距离徙动算法并行计算流程框图

3.2 延时相加法并行实现

延时相加法是基于“像素” (pixel-based) 的成像方法. 对于给定的像素点 P, 计算接收阵在参与孔径合成的各位置上相对 P 点的延时, 不同的像素对应于不同的一组延时. 延时相加法对于收发分置和多接收阵的合成孔径处理十分简便: 分别计算各接收阵对应同一成像点的不同延时，做延时相加即可. 由于“逐点”的时域延时相加比“逐线”的二维频域成像效率低得多，优化并行实时处理的要求就更迫切. 其中一种优化的并行任务分配方法是像素点“非等量分配”到并行计算节点上. 因为不同距离上的像点对应的有效合成孔径程度不同：近距离上的点对应的合成孔径短，成像所需的运算量就小. 如图 4(a) 所示，E 和 F 是不同斜距的两个成像点，与 F 点 (对应的有效合成孔径长度为图中 CD 段) 相比，E 与航迹的斜距大，对应的有效合成孔径长度 AB 也大，所以 E 点的合成孔径运算空间采样点数多，运算量大. 图 4(b) 和图 4(c) 分别是并行运算时“等像素分配”和“不等像素分配”示意图，显然图 4(c) 的任务分配更合理些.

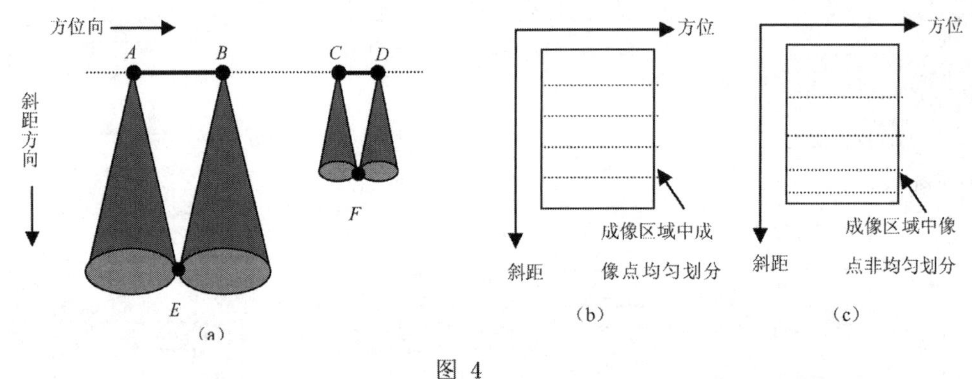

图 4

(a) 不同斜距上点对应的合成孔径照射情况示意图

(b) 成像点沿斜距等间隔划分并行任务，(c) 非等间隔划分并行任务示意图

方位合成孔径处理既可在解调前进行，也可在解调和脉冲完成后进行. 波束形成的理论和实践表明，采样率在信号上限频率的 8–10 倍时，才能得到较好的方位处理效果. 如果在解调前进行方位处理，要求有较高的采样率，运算量也太大.

4 SHARC 信号处理机并行实现

4.1 参数选取

取换能器方位向孔径 $D \approx 0.2\text{m}$(理论上，方位分辨率可达 0.1m)，最大作用距离为 $R_{\max}$ =75m, 则由距离的限制：

$$\text{PRF} \le C/(2R_{\max}) = 10\text{Hz} \tag{1}$$

其中水声速度取 $C = 1500\text{m/s}$.

由方位模糊的限制，相邻两次接收期间，SAS 的运动不能超过孔径 D 的一半 [3], 则

$$V \le D \cdot \text{PRF}/2 = 1\text{m/s} \tag{2}$$

采用单接收阵和并行 ΩK 算法实现图像重建. 流水线处理时，距离向取 1024 点，方位向取 512 点 (方位向数据重叠为 256) 作为一次帧处理.

信号发射为线性调频信号脉冲，中心频率 20kHz，带宽 7.5kHz，脉冲宽度 10ms，脉冲重复周期 100ms. 信号的采样率为 125kHz，正交解调后信号按 5: 1 降采样，故在基带上，等效的信号采样率为 25kHz .

4.2 并行信号处理机组成

主从式计算机系统，主机侧结构与通用 PC 机相同，完成人机交互、DSP 程序加载、成像数据回传及图像显示等功能。从机侧为多节点 DSP 系统，每个节点由 SHARC 处理器 (ADSP21060) 和与其相连的存储单元组成。不同节点间通过 SHARC 的 LINK 口相连，用于相互间的通信和数据传递。信号采集、处理和存储单元。多 DSP 节点 SHARC 系统的选用，主要是基于 SHARC 芯片有 6 个 LINK 口，并行处理时节点间的连接较方便。

4.3 并行节点的选取

第一级流水线如图 5 所示。节点 1 用于数据流的控制，接收 AD 送来的数据，并按 ping (不同方位上的脉冲) 的顺序分别送到节点 2, 3, 4, 5 。节点 2, 3, 4, 5 各计算一个 ping (5000 点) 的数据，并行完成正交解调和脉压。节点 6 用于数据的乒乓存储，采用双 buffer 存储数据，供第二级流水线批处理。第二级流水线如图 6 所示。节点 7 完成数据的孔径重叠 (相当于 Circular buffer) 、方位 FFT 、转置存储等功能，并沿方位向将数据等分成 4 份，分别送到节点 8, 9, 10, 11 。四个并行节点完成 Stolt 变换、方位向 IFFT 等变换，并将结果送节点 12 ，节点 12 完成转置存储及距离向 IFFT 。

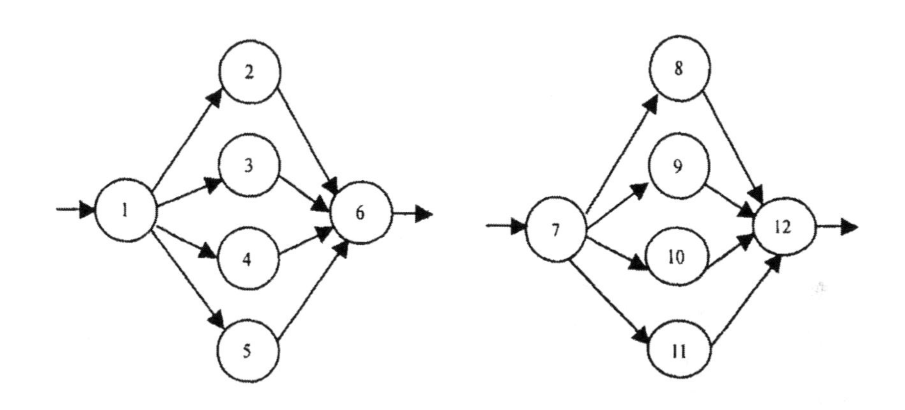

图 5 第一级流水线　　　　图 6 第二级流水线

数据采集和实时存储部分的讨论这里从略。

4.4 提高运算效率的措施

为提高运算效率，采取了以下措施降低运算量：

(1) 降采样与相关运算相结合，大大降低低通滤波运算时间

正交解调含三角函数相乘和低通滤波两步，在基带经 5 取 1 降采样。其中低通滤波采用输入数据与低通滤波器脉冲响应卷积来实现，在卷积计算中，每隔 5 点做一次计算，相当于只计算了降采样后留下的点，略去其它点，节约了 4/5 的运算量。

(2) 脉冲压缩在频域进行，并直接进行方位向 FFT

脉冲压缩既可以在时域又可以在频域进行。因为在频域处理可以采用快速傅里叶变换，一般运算速度比时域处理快。同时，ΩK 算法成像在二维频域内进行，脉压后的结果一般先做距离向 FFT ，然后方位向 FFT 进入二维频域。因此，在频域内做脉冲压缩时，保持结果在频域内，直接做方位向 FFT 。这样就省去了距离向各脉冲的一次 FFT 和一次 IFFT 。

4.5 二维行列处理时数据传递结构

图 7 是一种硬件上节省的并行二维处理结构，左侧各并行节点完成分块的距离向处理，然后经 LINK 口转置并分块到右侧并行节点，转置存储和并行任务分配同时进行。但是大量的矩阵数据，使 LINK 口通信排队、虚拟通路的使用等变得复杂，有可能出现阻塞。故本系统采用图 8 的方式完成二维行列处理。因为增加了一个中间级节点 O ，使得通信的控制变得简洁，转

置存储也简单方便. 在中间级节点 O 上实现了转置存储, 再从该节点上将数据分配到并行节点上.

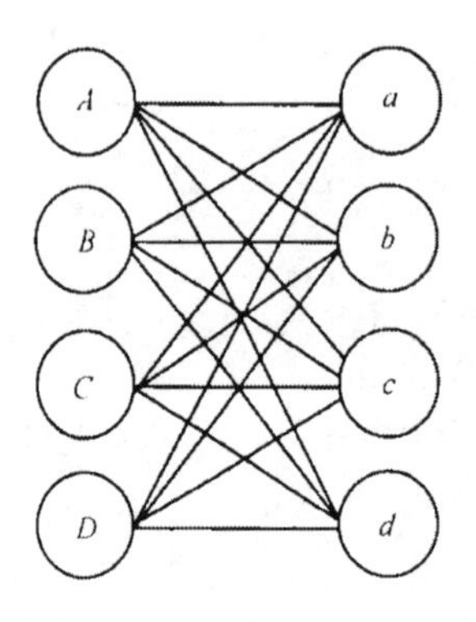

图 7 用 LINK 接口传递二维矩阵并行处理数据的一种方法

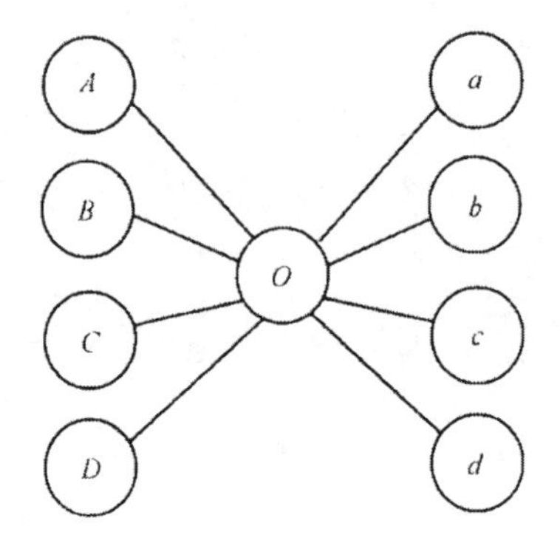

图 8 用 LINK 接口传递二维矩阵并行处理数据的另一种方法

5 水下实验结果

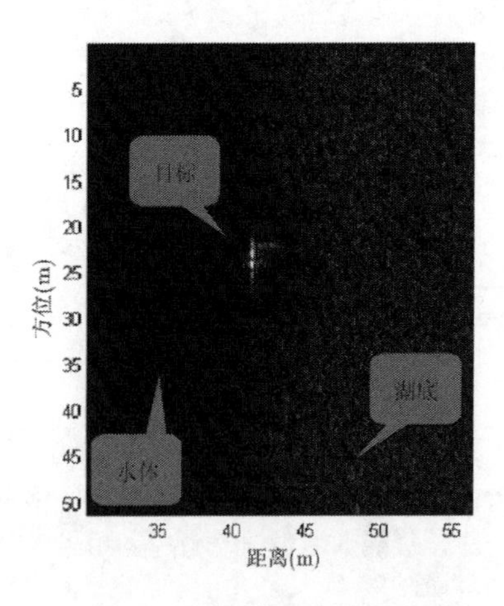

图 9 L 形强散射体的SAS成像

在浙江千岛湖完成了多次水下实验, 对特定的水下悬浮目标、沉底目标进行了合成孔径成像, 同时还得到了人工湖建成前的新安江旧河道及两侧石头河堤的图像.

图 9 是 L 形强散射体的 SAS 图像. 实验地点在千岛湖猴岛附近, 水深约 60m, 拖体入水 20m, 运动速度 0.87m/s. L 形强散射体 (每个臂长度约 2.5m, 且每个臂上装有两个煤气罐), 用 10m 长的钢绳将目标与铁块连在一起, 沉入水中, 目标距水底 40m.

6 结束语

本文讨论了 ΩK 和延时相加两种 SAS 算法的并行实现, 包括并行处理任务分配、节点间数据通信等问题. 结合 SAS 的特点, 采用两级并行流水线, 其中一级序贯处理、一级批处理来完成实时成像. 水下实验表明, 该方法正确可行, 能满足实时成像、实时存储等要求.

采用 LINK 口数据传递, 控制上虽然很方便, 但 SAS 图像重建时, 大量的二维处理, 特别是转置存储, 容易使 LINK 口数据交换出现瓶颈. 如果部分节点采用大的共享存储器, 比单纯的 LINK 口数据交换可能要快捷些．另外，采用重叠子孔径法成像，第二级流水线批处理时各节点运算负荷及存储量都会大大降低、节点间传递数据流也会显著减小．

参 考 文 献

[1] M. P. Hayes, P. T. Gough, Broad-band synthetic aperture sonar, IEEE J. of Oceanic Engineering, 1992, 17(1), 80–94.

[2] P. T. Gough, D. W. Hawkins, Imaging algorithms for a strip-map synthetic aperture sonar: minimizing the effects of aperture errors and aperture undersampling, IEEE J. of Oceanic Engineering, 1997, 22(1), 27–39.

[3] A. E. Adams, M. A. Lawlor, V. S. Riyait, O. R. Hinton, B. S. Sharif, Real-time synthetic aperture sonar processing, IEE Proc.-Radar, Sonar and Navigation, 1996, 143(3), 169–175.

[4] V. S. Riyait, M. A. Lawlor, A. E. Adams, O. H. Hinton, B. Sharif, Real-time synthetic aperture sonar imaging using a parallel architecture, IEEE Trans. on Image Processing, 1995, IP-4(7), 1010–1019.

STUDY ON REAL-TIME AND PARALLEL IMPLEMENTATION OF SYNTHETIC APERTURE SONAR SIGNAL PROCESSING

Liu Jiyuan Li Shuqiu Li Liying Zhang Chunhua Li Qihu

(*Institute of Acoustics, Chinese Academy of Sciences, Beijing* 100080, *China*)

Abstract Reconstruction algorithms for Synthetic Aperture Sonar (SAS) are discussed, and choice of SAS parameters and parallel structures are studied in this paper. A SHARC-based real-time processing system for SAS imaging is presented. Two pipelines are used in the system. The first pipeline makes full use of pulse repetition interval and proceed data sequentially, and the second one focuses image data in frame. Therefore, SAS image is reconstructed with only one stage time-consumption though two stages are used. Moreover, parallel processing are used in both stages to meet real-time requirement.

Key words Synthetic Aperture Sonar(SAS), Real-time processing, Imaging sonar

单声线水声 MIMO 信道容量的研究[①]

朴大志　　孙长瑜　　李启虎

(中国科学院声学研究所　北京 100080)

摘要　本文对水声 MIMO (Multiple-Input-Multiple-Output) 信道容量受收发阵元数目、间距，收发阵位置、方向，平均接收信噪比以及声速剖面的斜率的影响，在收发阵元之间只存一条声线的情况下，通过 WKB 近似进行了初步的研究. 从计算结果可以看出，当收发阵元对之间只存在一条声线时，阵元间距会对 MIMO 信道容量产生重要影响：当收发阵元间距足够大时，MIMO 系统的信道容量将随着接收信噪比和收发阵元数线性增加，一个 $m \times m$ 的 MIMO 系统的信道容量将为相应的 SISO (Single-Input-Single-Output) 系统的 m 倍，收发阵的方向也会对 MIMO 系统信道容量产生较大的影响，另外，收发阵的深度、距离也会对水声 MIMO 信道容量产生影响，声速剖面的斜率在一般水声信道的声速变化范围内，对信道容量的影响不大.

关键词　水声通信，MIMO, 单声线，信道容量

Study of the underwater acoustic MIMO capacity for single sound path channel

PIAO Da-Zhi　　SUN Chang-Yu　　LI Qi-Hu

(*Institute of Acoustics, Chinese Academy of Sciences,　Beijing 100080*)

Abstract　The dependence of underwater acoustic MIMO capacity on the array parameters of element number, element spacing, array position and direction, the average receive SNR, and the slope of sound speed profile are studied through WKB approximation when there is only one sound path between each transmit and receive element pair. From the computed numerical results we can see that, array element spacing will have a great effect on the MIMO capacity: if the transmit and receive array element spacing are sufficiently large, MIMO capacity will increase with array element number linearly, and the capacity of a $m \times m$ underwater acoustic MIMO system is approximately m times of that of a SISO channel. The direction of the transmit and receive array will also have a big effect on MIMO capacity; furthermore, the depth and horizontal range of the transmit and receive array will have some effects on the capacity. To the effects of the

① 应用声学, 2006, 25(2): 76-81.

slope of the sound speed profile, it was not very big within the variation range of sound speed in the general underwater acoustic channel.

Key words Underwater acoustic communication, Multiple-input-multiple-output, Single sound path, Channel capacity

1 引言

非常有限的带宽是限制高速水声通信的主要因素，由于 MIMO (Multiple-Input-Multiple-Output) 技术可以利用信道的空间起伏来提高无线通信系统的频带利用率，因此 MIMO 技术在高速水声通信中具有应用潜力。在过去的十几年里，MIMO 技术成为无线 (radio) 通信研究中的热点，大量关于 MIMO 系统信道容量和空时编码的文章发表，然而这些文章主要是基于城市中的移动通信和市内的无线局域网，涉及到水声通信的研究很少。在 [1] 中，通过试验和仿真的数据研究了水声 MIMO 系统的信道容量，在其它的一些文献如 [2], [3] 中采用简单的散射模型研究了室外的无线 MIMO 信道容量。水声信道与无线信道相比，有很多不同，比如，在无线通信中，电磁波的传播速度一般可以看成是恒定的，而声速在水中一般是变化的，另外，在无线 MIMO 信道容量的研究中，一般假设收发之间由于周围物体的散射而存在多径，而在深海水声信道中，如果边界的影响可以忽略，在很多情况下收发之间只存在一条声线，因此，本文对这种单声线的水声 MIMO 系统的信道容量受收发阵参数和信道条件特别是声速变化的影响情况，进行初步的研究。

2 水声 MIMO 的信道容量

2.1 MIMO 信道容量的一般表达式

考虑一个 MIMO 系统具有 n_T 个全向发射阵元和 n_R 个全向接收阵元，发射和接收信号矢量间的关系可以表示为

$$\boldsymbol{y}(t)=\boldsymbol{H}(t)\boldsymbol{x}(t)+\boldsymbol{n}(t) \tag{1}$$

其中 $\boldsymbol{x}(t)$ 是 $n_T\times 1$ 发射信号矢量，$\boldsymbol{y}(t)$ 是 $n_R\times 1$ 接收信号矢量，$\boldsymbol{H}(t)$ 是由信道增益 $h_{ij}(t)$ 构成的 $n_R\times n_T$ 信道传播矩阵，$\boldsymbol{n}(t)$ 是 $n_R\times 1$ 加性白高斯噪声矢量，其中的各个元素是 i.i.d 的复高斯变量，方差为 N_0。在窄带通信系统中，可以看成具有平坦的频率响应，则式 (1) 可以写成

$$\boldsymbol{y}=\boldsymbol{H}\boldsymbol{x}+\boldsymbol{n} \tag{2}$$

当信道的状态信息在接收端已知而发射端未知时，一般将总发射功率 P_T 均匀地分配到各个发射阵元，这种情况下 MIMO 系统的信道容量可以表示为 [4~6]，

$$C=\log_2\det(I_{n_R}+(\rho/n_T)HH^{\dagger})\quad(\text{bps/Hz}) \tag{3}$$

其中 H^+ 是 H 的共轭转置，$\det(\cdot)$ 是求矩阵的行列式，I_{n_R} 是 $n_R\times n_R$ 单位阵，ρ 是每个接收阵元处的平均信噪比。

由于信道容量与接收信噪比有关，所以信道矩阵 $\boldsymbol{H}$ 的归一化方式对信道容量计算结果的正确解释非常重要，本文采用下面归一化条件，

$$\sum_{i=1}^{n_T}\sum_{j=1}^{n_R}|h_{ij}|^2=n_Tn_R \tag{4}$$

这样对每一种不同的收发阵元位置和信道条件，将得到不同的信道归一化参数，从而使不同的收发阵元位置和信道条件下具有相同的接收信噪比，当信道容量表达式 (3) 中采用式 (4) 归一化的信道矩阵，ρ 相当于 SISO (Single-Input-Single-Output) 通信系统中的平均接收信噪比。

2.2 水声 MIMO 的信道矩阵

在深海的水声通信中，如果海面和海底的影响可以被忽略，当发射阵和接收阵都是静止的，而且它们之间不存在离散的点散射体，如气

泡、鱼群等，考虑声速仅是深度 z 的函数时，如果通信信道的输入为单频的平面波，一个 SISO 的水声通信系统的传递函数可由二维 WKB 近似得到 [7,8]，则在 MIMO 系统中，信道矩阵可以采用下面的公式计算

$$h_{ij} = \exp\left(im\left(k_{xt_j}(xr_i - xt_j) - \int_{zt_j}^{zr_i} k_z(z)dz\right)\right)/\sqrt{k_{zr_i}} \tag{5}$$

其中 im 为虚数单位，xr_i 和 xt_j 分别为第 i 个接收阵元和第 j 个发射阵元的 x 坐标，zr_i 和 zt_j 分别为第 i 个接收阵元和第 j 个发射阵元的 z 坐标，k_{xt_j} 为第 j 个发射阵元处沿 x 方向的波数，k_{zri} 为第 i 个接收阵元处沿 z 方向的波数，当声速恒定时，式 (5) 中的相位部分

$$\begin{aligned} & k_{xt_j}(xr_i - xt_j) - \int_{zt_j}^{zr_i} k_z(z)dz \\ & = k_{xt_j}(xr_i - xt_j) - k_{zt_j}(zr_i - zt_j) \end{aligned} \tag{6}$$

相当于从发射阵元到接收阵元间由波数为 $\boldsymbol{k} = (k_{xt_j}, k_{zt_j})$ 引起的相位累积。

图 1 为水声 MIMO 通信系统示意图，本文以均匀直线阵为例，其中 θ_t 和 θ_r 分别为发射阵和接收阵与 z 轴的夹角，dt 和 dr 分别为发射和接收阵元间距，(xt_0, zt_0) 和 (xr_0, zr_0) 分别为第一个发射阵元和第一个接收阵元的坐标，为例分析方便，本文中假设收发阵元数相等，即 $n_T = n_R = m$，则对于 $j = 1{:}m$, $i = 1{:}m$, $xt_j = xt_0 + \sin(\theta_t)(j-1)dt$, $zt_j = zt_0 + \cos(\theta_t)(j-1)dt$, $xr_i = xr_0 + \sin(\theta_r)(i-1)dr$, $zr_i = zr_0 + \cos(\theta_r)(i-1)dr$ 。

与无线 (radio) MIMO 的 LOS (Line-of-Sight) 情况不同的是，水声 MIMO 系统的信道容量与阵的构成参数，如阵元间距，收发阵位置、方向等有关，另外，声速的变化会引起声线传播方向的改变，从而改变 MIMO 系统的信道容量。对于线性变化的声速剖面，从每个发射阵元出发到达每个接收阵元的声线在每个发射阵元处的入射角 (相对与 z 轴) 与收发阵元的水平距离之间存在简单的关系 [8]，因此本文将以线性变化的声速剖面为例，研究声速变化对水声 MIMO 信道容量的影响。(对于其它的声速剖面也可以利用式 (5) 来计算信道传播矩阵，但首先也要确定从每个发射阵元出发并到达每个接收阵元的声线) 。

设声速剖面为:

$$c(z) = c_0 + gz \tag{7}$$

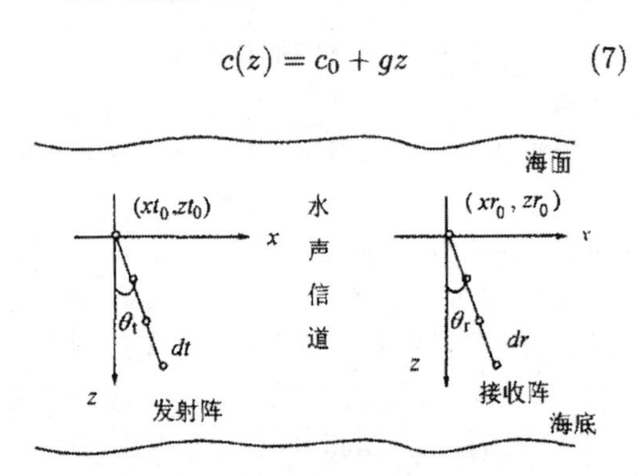

图 1　水声 MIMO 通信系统示意图

考虑全向发射阵元，当声速恒定时，从每个发射阵元到每个接收阵元的声线将沿着它们的连线方向直线传播，当声速变化时，声线会发生弯曲，我们将首先找到从每个发射阵元出发到达每个接收阵元的声线，利用公式 (5) 得到传播矩阵的每个元素，然后利用公式 (4) 、(3) 计算信道容量。

3　水声 MIMO 系统信道容量计算结果及简要说明

图 2 描述了收发阵元间距对信道容量的影响，从图 2 中可以看出，随着收发阵元间距的增加，MIMO 信道容量会增大，在图 2 中的计算参数下，当收发阵元间距约为 30 倍波长时，信道容量可以达到最大值，声速剖面斜率的变化会使曲线向左或向右移动，但移动的范围不大，而当采用收发之间存在多径模型时，如文献 [9, 10]，收发阵元间距为 $0.5\lambda \sim$ 几个 λ 时信道容量就可以达到最大值，取决于所采用的散射模型，这就是说，当只有单条声线到达时，只要收发阵元间距足够大，同样可以获得

较高的 MIMO 信道容量增益。图 3 和图 4 描述了接收信噪比和收发阵元数对信道容量的影响，从图 3 和图 4 中可以看出，信道容量将随着接收信噪比和收发阵元数线性增加，当收发阵元间距较小时，增加的幅度较小，当收发阵元间距足够大时 (图中 $dt = dr = 20\lambda$)，信道容量将随着接收信噪比和收发阵元数以更大的斜率增加，基本上可以成倍增加，另外当收发阵元间距增大时，声速剖面斜率对信道容量的影响也会增加。图 5 和图 6 描述了收发阵位置和声速剖面斜率对信道容量的影响，从图中可以看出，收发阵的深度和水平距离对信道容量都会产生较明显的影响，而声速剖面斜率对信道容量的影响也将随收发阵水平间距的增加而变大。图 7 描述了当接收阵方向不变时，发射阵方向对信道容量的影响，从图中可以看到，阵的方向会对 MIMO 信道容量产生周期性的影响，在 MIMO 系统中应选择合适的发射和接收阵方向，使信道容量得到最大值。

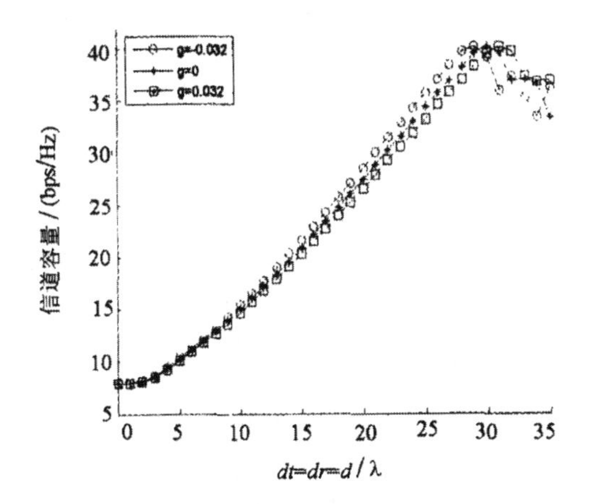

图 2 不同声速斜率下信道容量随收发阵元间距的变化

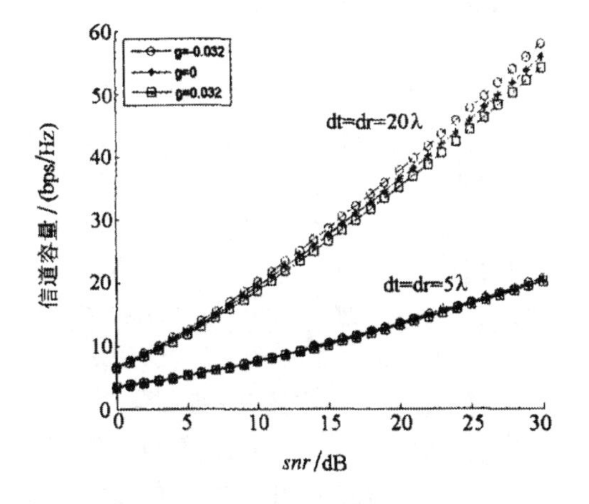

图 3 不同声速斜率下信道容量随接收信噪比的变化

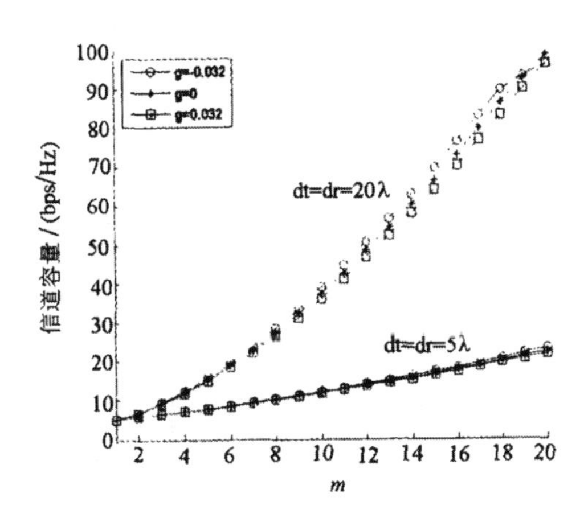

图 4 不同声速斜率下信道容量随收发元数的变化

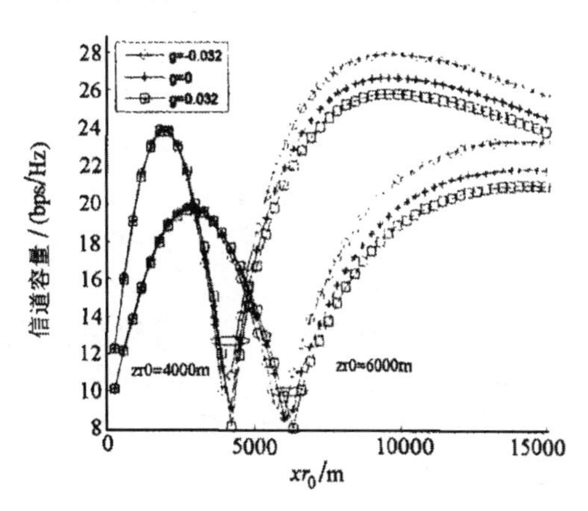

图 5 不同声速斜率下信道容量随发射阵位置的变化

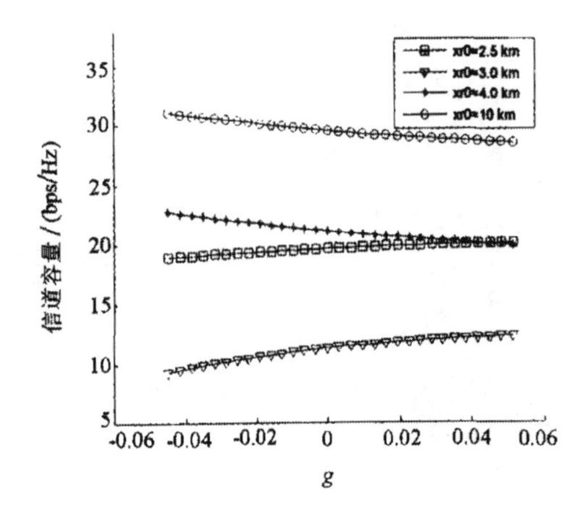

图 6　不同收发阵位置下信道容量随声速斜率的变化

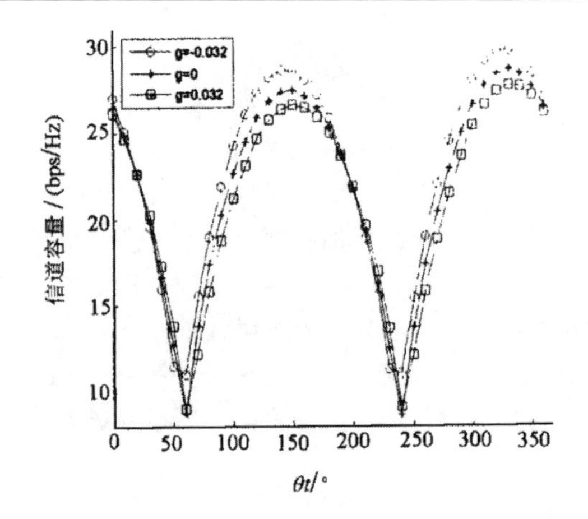

图 7　不同声速斜率下信道容量随发射阵方向的变化

表 1　图 2～7 的参数设置

	参　数
图 2	$f = 2000\text{Hz}$, $zt_0 = 0$, $xt_0 = 0$, $zr_0 = 3000(\text{m})$, $xr_0 = 5000(\text{m})$, $m = 8$, $snr = 15\text{dB}$, $\theta t = 150°$, $\theta r = 0°$
图 3	$f = 2000\text{Hz}$, $zt_0 = 0$, $xt_0 = 0$, $zr_0 = 3000(\text{m})$, $xr_0 = 5000(\text{m})$, $m = 8$, $dt = dr = 20\lambda$, $\theta t = 150°$, $\theta r = 0°$
图 4	$f = 2000\text{Hz}$, $zt_0 = 0$, $xt_0 = 0$, $zr_0 = 3000(\text{m})$, $xr_0 = 5000(\text{m})$, $snr = 15\text{dB}$, $dt = dr = 20\lambda$, $\theta t = 150$, $\theta r = 0°$
图 5	$f = 2000\text{Hz}$, $zt_0 = 0$, $xt_0 = 0$, $m = 8$, $snr = 15\text{dB}$, $dt = dr = 20\lambda$, $\theta t = 150$, $\theta r = 0°$
图 6	$f = 2000\text{Hz}$, $zt_0 = 0$, $xt_0 = 0$, $zr_0 = 3000(\text{m})$, $m = 8$, $snr = 15\text{dB}$, $dt = dr = 20\lambda$, $\theta t = 150$, $\theta r = 0°$
图 7	$f = 2000\text{Hz}$, $zt_0 = 0$, $xt_0 = 0$, $zr_0 = 3000(\text{m})$, $xr_0 = 5000(\text{m})$, $m = 8$, $snr = 15\text{dB}$, $dt = dr = 20\lambda$
	其中 $\lambda = c_0/f$, $c_0 = 1500(\text{m/s})$

4　结束语

本文对水声 MIMO 信道容量受收发阵元数目、间距, 收发阵位置、方向, 接收信噪比以及声速变化的斜率的影响情况, 通过 WKB 近似进行了初步的研究。从计算结果可以看出, 阵元间距会对 MIMO 信道容量产生重要影响, 当只有单条声线到达时, 如果收发阵元间距足够大时, MIMO 系统的信道容量将随着接收信噪比和收发阵元数成倍增加, 例如图 4 中, 如果 $dt = dr = 20\lambda$, 在 $m = 5$ 时, $C = 20(\text{bps/Hz})$, 在 $m = 10$ 时, $C = 40(\text{bps/Hz})$, 其次, 收发阵的方向也会对 MIMO 系统信道容量产生较大的影响, 另外, 收发阵的深度、距离也会对 MIMO 信道容量产生影响, 虽然声速剖面的斜率也会对 MIMO 信道容量产生影响, 然而, 在一般水声信道的声速变化范围内, 对信道容量的影响不大。

因此, 当每个收发阵元对之间只存在一条声线时, 如果收发阵元间距足够大, 并选择合适的收发阵方向时, 在水声通信中采用 MIMO 技术一般可以获得较高的频带利用率, MIMO 技术将成为高速水声通信发展的重要方向。

本文中没有考虑由于海面或海底的反射以及水中的散射体对声波的散射所引起的多径效应, 另外本文仅考虑了线性声速剖面条件下, 收发之间仅存在一条声线的情况, 而对于其它的声速剖面, 如深海的典型的 Munk 声速剖面, 在声轴附近会产生多径传播, 也没有考虑。这些因素对水声 MIMO 信道容量的影响还需要进一步的研究。

参 考 文 献

1 Zatman M, Tracey B. Signals, systems and computers. Conference record of the Thirty-Sixth Asilomar Conference, 2002, **2**:1364～1368.

2 Gesbert D, Bolcskei H, Gore D A, *et al*. *IEEE Transactions on Communications*, 2002, **50**(12): 1926～1934.

3 Driessen P F, Foschini G J. *IEEE Trans. Commun.*, 1999, **47**(2):173～176.

4 Winters J H, Salz J, Gitlin R D. *IEEE Trans. Commun.*, 1994, **42**(234):1740～1751.

5 Telatar I E. Capacity of Multi-Antenna Gaussian Channels. AT&T Bell Laboratories, Murray Hill, NJ, Tech. Rep.# Bl0 112 170-950-615-07 TM, 1995.

6 Foschini G J, Gans M J. *Wireless Pers. Commun.*, 1998, **6**(3):311～335.

7 Ziomek L J. Underwater Acoustics A Linear Systems Theory Approach. Academic Press, 1985. 250～254.

8 Frisk G V. Ocean and Seabed Acoustics-A theory of Wave Propagation, P T R Prentice-Hall,1994, 200～219.

9 Hyundong Shin, Lee J H. *IEEE Transactions on Information Theory*, 2003, **49**(10):2636～2647.

10 Shiu D S, Foschini G J, Gans M J, *et al*. *IEEE Trans. Commun.*, 2000, **48**(3):502～513.

Effect of spatial correlation on underwater acoustic MIMO capacity①

PIAO Da-zhi[1,2], SUN Chang-yu[1], LI Qi-hu[1]

(1. Institute of Acoustics, Chinese Academy of Sciences, Beijing 100080, China; 2. College of Information Engineering, Communication University of China, Beijing 100024, China)

Abstract: The effect of channel parameters such as angle spread, angle of arrival, and array element spacing on the underwater acoustic MIMO capacity is studied with a three-dimensional multipath correlation model using a vertical transmitting and receiving array. In a MIMO system with static transmitting and receiving arrays, regardless of the movement of scatters in the underwater acoustic channel, the capacity is dependent on the spatial correlation among array element signals at the transmitter as well as at the receiver. If correlation between elements of the channel matrix is weak, increasing the number of array elements will greatly increase the MIMO capacity. In a channel of strong correlation, increasing the number of array elements will cause the MIMO capacity to approach saturation.

Key words: MIMO; underwater acoustics; spatial correlation; channel capacity

空间相关性对水声多输入多输出系统信道容量的影响

朴大志[1,2]，孙长瑜[1]，李启虎[1]

(1 中国科学院声学研究所，北京 100080；2 中国传媒大学信息工程学院，北京 100024)

摘要：通过三维多径相关模型，以垂直收发阵为例研究了信道参数如声线角度扩展范围、到达方向和收发阵元间距对水声MIMO信道容量的影响。如果考虑静止的收发阵，且不考虑信道中散射体的运动时，MIMO系统的信道容量由收发阵元信号的空间相关性决定，当空间相关性较小时，增加收发阵元数可以带来较大的信道容量增益，而当空间相关性较大时，随着收发阵元数的增加信道容量将趋于饱和。

关键词：多输入多输出；水声；空间相关；信道容量

中图分类号：TN929.3　　文献标识码：A　　文章编号：1000-3630(2006)-06-0573-07

1 INTRODUCTION

The MIMO technology using multiple array elements at both the tansmiter and the receiver has high potential in improving the spectral efficiency of densely scattered underwater acoustic channels, but so far few researches about the underwater acoustic MIMO capacity have been conducted. Researches[1-3] have shown that, in a Rayleigh flat-fading environment, a MIMO wireless communication link has a theoretical capacity that increases linearly with smaller number of transmiting and receiving array elements, provided that the propagation coefficients among all pairs of transmitting and receiving elements are statistically independent. As such, the MIMO technology has drawn wide interest in the wireless communication field over the past two decades. However, in some practical environment,

① 声学技术, 2006, 25(6): 573-579.

there are some spatial correlations between the propagation coefficients caused by the poor scattering environment or insufficient spacing between the array elements, which affect the MIMO capacity of these practical channels. The effect of transmitting power and signal frequency on the MIMO capacity of a typical shallow water channel was studied through field measurements[4]. However, results obtained in this manner are only applicable to a particular environment and array configuration. Furthermore, it is difficult to study the effect of the environment parameters on MIMO capacity. The MIMO capacity can also be studied through abstract scattering models[5], from which the essential characteristic of the environment can be illuminated, and its effect on the MIMO capacity can easily be studied. The MIMO capacities can be derived when there are correlations in either of the transmitter or the receiver. The effect of channel parameters on MIMO capacity using the scattering model of 'one-ring' can also be studied through Monte Carlo Simulation. The dependence of the scaled MIMO capacity with the number of array elements on channel correlation can be studied with an assumption based on some kinds of correlation models[6]. The characteristic function of MIMO capacity was derived when the channel is correlated at either the transmitter or the receiver in a Rayleigh flat-fading channel[7].

The "one-ring" or "two-ring" scattering models are generally adopted in the wireless radio communication systems. The application of "one-ring" model is appropriate when the base station is elevated and unobstructed by local scatters, and likewise, the "two-ring" model is appropriate when both of the transmitter and receiver are surrounded by objects. In these models, the scatters are generally assumed to be distributed on the circles around the transmitting or receiving arrays. However, in the underwater acoustic communication channels, the transmitted and received acoustic rays are spread both vertically and horizontally. This paper studies the effects of various channel parameters on the underwater MIMO capacity through a three dimensional multi-path spatial correlation model.

2 SPATIAL CORRELATION MODEL

In a MIMO communicaiton system which consists of n_R receiver elements and n_T transmitter elements, and there are multi paths between the transmitter and the receiver, the received signal of the ith, $i=\{1, ..., n_R\}$ receiving element from the jth, $j=\{1, ..., n_T\}$ transmitting element can be expressed as

$$h_{ij}(t)=\sum_{n=1}^{N} A_n \exp(iw(t-\tau_{n,i,j})) \tag{1}$$

where i is an imaginary unit, A_n is the amplitude gain of the nth path, N is the number of the multipath, and $\tau_{n,i,j}$ is the path delay associated with the propagation from the jth transmitting element to the ith receiving element along the nth path. Given the statistical properties of the channel, if N is sufficiently large, the central limit theorem states that $h_{i,j}(t)$ is a zero-mean complex Gaussian-random process. Therefore, the envelope of $|h_{i,j}(t)|$ is a Rayleighfading process.

The correlation between the two links, T_i-R_j and T_i'-R_j', when normalized by the received signal field variance is defined as

$$\rho_{ij,i'j'}(t)=E[h_{ij}(t)h_{i'j'}^*(t)]/\sqrt{\Omega_{ij}\Omega_{i'j'}} \tag{2}$$

where * is the complex conjugate, $\Omega_{ij}=E[|h_{ij}(t)|^2]$, substitute Equ.(1) into Equ.(2) and simplify the result,

$$\rho_{ij,i'j'}(t)=\rho_{ij,i'j'}=E[\sum_{n=1}^{N}\exp(iw(\tau_{n,i',j'}-\tau_{n,i,j}))] \tag{3}$$

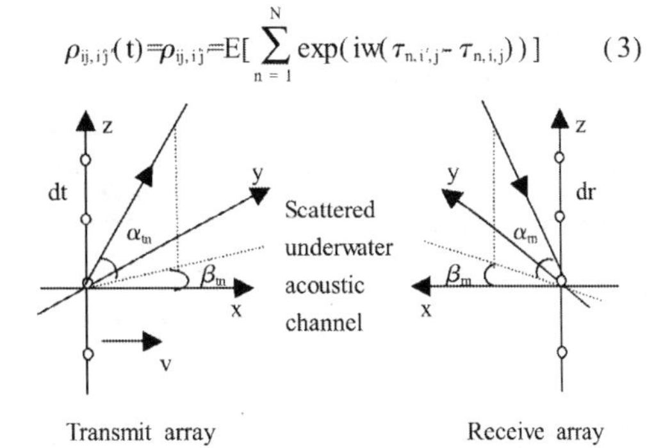

Fig.1 Configuratio of the underwater acoustic MIMO communication system(illustrated by a vertical transmitting and a vertica receiving array)

Considering plane waves for the vertical transmit array and vertical receive array, shown in Fig.1, we have

$$\tau_{n,i',j}-\tau_{n,i,j}=(dr\sin\alpha_{rn}(i'-i)+dt\sin\alpha_{rn}(j'-j))/c \quad (4)$$

For the horizontal transmitting and receiving array along the axis of x,

$$\tau_{n,i',j}-\tau_{n,i,j}=(dr\cos\alpha_{rn}\cos\beta_{rn}(i'-i)+dt\cos\alpha_{tn}\cos\beta_{tn}(j'-j))/c \quad (5)$$

For the horizontal transmitting and receiving array along the axis of y,

$$\tau_{n,i',j}-\tau_{n,i,j}=(dr\cos\alpha_{rn}\sin\beta_{rn}(i'-i)+dt\cos\alpha_{tn}\sin\beta_{tn}(j'-j))/c \quad (6)$$

where dt and dr are the transmitting and receiving element spacing, α_{tn}, β_{tn} are the elevation angle and azimuth angle of the nth path at transmitter; α_{rn}、β_{rn} are the elevation angle and azimuth angle of the nth path at the receiver respectively, c is the underwater sound speed, in this paper, it is assumed to be constant i. e. c=1500(m/s).

For other array configurations, $\tau_{n,i',j}-\tau_{n,i,j}$ would have similar expressions. The effect of spatial correlation on the underwater acoustic MIMO capacity can be studied through the vertical transmitting array and vertical receiving array. Substitute Equ.(4) into Equ.(3),

$$\rho_{ij,i',j}=\langle \exp(iw((i'-i)dr\sin\alpha_r+(j'-j)dt\sin\alpha_t)/c\rangle_{\alpha_r,\alpha_t} \quad (7)$$

where $\langle n\rangle$ means statistical averaging. Equ.(7) shows that for the vertical transmitting and receiving arrays, the spatial correlation function depends on the spacing of transmitting and receiving elements; the elevation angle spread of the acoustic rays at transmitter and receiver; in this article, they are assumed to be uniformly distributed in $[\Theta-\Delta, \Theta+\Delta]$, where Θ is the angle of arrival and Δ is the angle spread. The effects of these parameters on underwater acoustic MIMO capacity can be studied in such a manner.

3 MIMO CAPACITY

For a narrow-band single-user communication system with n_T transmitter and n_R receiver omni-directional elements, the relationship between the transmitted signal vector and the received signal vector can be expressed

$$y(t)=H(t)*x(t)+n(t) \quad (8)$$

where $x(t)$ is the $n_T\times1$ transmitted signal vector, $y(t)$ is the $n_R\times1$ received signal vector, $n(t)$ is the $n_R\times1$ additive white Gaussian noise vector, and $H(t)$ is the $n_R\times n_T$ channel propagation matrix of complex path gains $h_{ij}(t)$ between transmitter element j and receiver element i. The elements of noise vector are assumed to be independent and identically distributed (i.i.d) complex Gaussian-random variables with variance N_o.

Considering a narrow band flat frequency response communication system, Eq.(8) can be written as,

$$y=Hx+n \quad (9)$$

When the channel state information is known at the receiver but not known at the transmitter, the total transmitted power are generally equally alloc-ated to every transmitter element. In this case, the instant MIMO capacity corresponding to each reali-zation of H is expressed as[1-3],

$$C=\log_2\det(I_{n_R}+(SNR/n_T)HH^+)\ \text{bps/Hz} \quad (10)$$

where H^+ means the conjugated transpose of H, det(•) is the matrix determinant, I_{n_R} is the $n_R\times n_R$ identity matrix, and SNR is the average signal-to-noise ratio (SNR) at each receiver element.

For a random channel matrix H, its capacity C is also a random variable. The Complementary Cumulative Distribution Function (CCDF) is generally used to give a complete description of the random channel capacity. As mentioned earlier, the characteristic function of C is derived in the research[7] where the channel is correlated either at the transmission end or the receiver end, from which the CCDF could be generated. Fig.2 describes the changing of MIMO capacity with the receive correlation when the transmit correlation is kept unchanging. The simulation parameters in Fig.2 are $\alpha_t\in[-5°,5°]$, $\alpha_r\in[-5°,5°]$, $d_r=1\lambda$, $n_T=n_R=6$, SNR=10dB. Fig.2 shows that the result of MIMO capacity in the above research[7] is accurate when there is no correlation or the correlation coefficient is small in the transmission end. When both of the transmis sion and receiving ends have high correlations, it is difficult to get the distribution function of C, so we use Monte Carlo simulation to get the CCDF of C and study the effect of channel parameters on MIMO capacity.

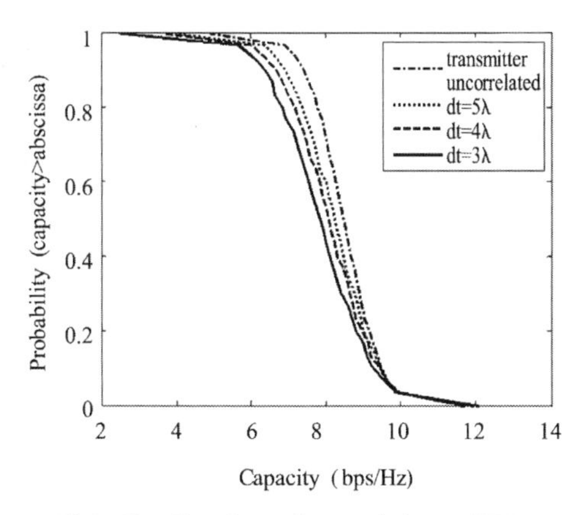

Fig.2 The effect of transmitter correlation on CCDF

4 SIMULAITON RESULTS

4.1 Generation of the correlated variables

Generation of M Gaussian variables with an arbitrary correlation can be achieved by decomposing the desired $M \times M$ covariance matrix, $R=GG^{+}$, where G^{+} is the conjugated transpose of G, then multiplying M independent Gaussian variables by G[8].

4.2 Simulation results for CCDF and brief analysis

Fig.3, 5 and 7 describe the dependence of spatial coherence on the parameters of transmitting and receiving element spacing, angle spread and AOA respectively, and illustrated by $|\rho_{21,32}|$. In these figures, two groups of simulation parameters have been chosen, one group has a small correlation and the other has a great correlation, Fig.4, 6 and 8 describe the effects of these parameters on the MIMO capacity with $n_R=n_T=3$, SNR=10dB, in these figures, there are two cases, named by (a) and (b), corresponding to the two groups of parameter setting in Fig.3, 5 and 7. The detailed parameter settings are illustrated in Table.1.

Table 1 Parameter setting of Fig.3 Fig.8

	Parameter
Fig.4	(a) $q \in [-5°, 5°]$, $q \in [-5°, 5°]$, AOA=0°
	(b) $q \in [-5°, 5°]$, $q \in [-20°, 20°]$, AOA=0°
Fig.6	(a) dt=dr=1 λ, AOA=0°
	(b) dt=dr=5 λ, AOA=0°
Fig.8	(a) $q \in [-5°, 5°]$, $q \in [-5°, 5°]$, dt=dr=1 λ,
	(b) $q \in [-20°, 20°]$, $q \in [-20°, 20°]$, dt=dr=5 λ

Fig.4 and 6 show that both the element spacing and angle distribution have great effects on MIMO capacity. From Fig.4, we can see that in a poor scattering environment, to get huge MIMO capacity needs a large element spacing, and that getting a MIMO capacity equal to that of the com pletely independent channel, the transmitting and re -ceiving element spacing in Fig.4(a) is 8 λ, but in Fig.4(b) is only 2 λ, which is consistent with the correlation in Fig.3. Similar results can also be ob tained in Fig.5 and 6, when the element spacing is not sufficiently large, dense scattering environment is needed to get a large MIMO capacity gains.

Fig.8 illustrates that with large element spacing and angle spread, the AOA will have a great effect on MIMO capacity, as shown in Fig.8(b), otherwise, the effect of AOA on MIMO capacity is very small, shown in Fig.8(a), and these results are also corresponded to the spatial correlation in Fig.7.

Fig.9 illustrates the dependence of MIMO capacity on the number of transmitting and receiving elements with two different correlation conditions, where the parameters, SNR=10dB, and in Fig.9(a) $q \in [-20°, 20°]$, $q \in [-20°, 20°]$, dt=dr=5 λ, AOA=0°, in Fig.9(b) $q \in [-2°, 2°]$, $q \in [-2°, 2°]$, dt=dr=0.25 λ, AOA=0°. Fig.9(a) and (b) represent two cases of the completely independent and completely correlated environments. Fig.9 shows that large MIMO capacity gains can be obtained by using multiple array elements when the correlation in both the transmitting and receiving ends are small, but when there are great correlations, the MIMO capacity gains from the increasing of element number is very limited.

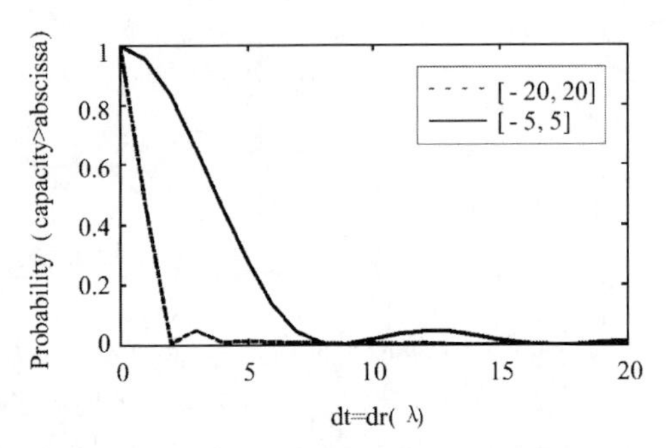

Fig.3 Dependence of $|\rho_{21,32}|$ on element spacing

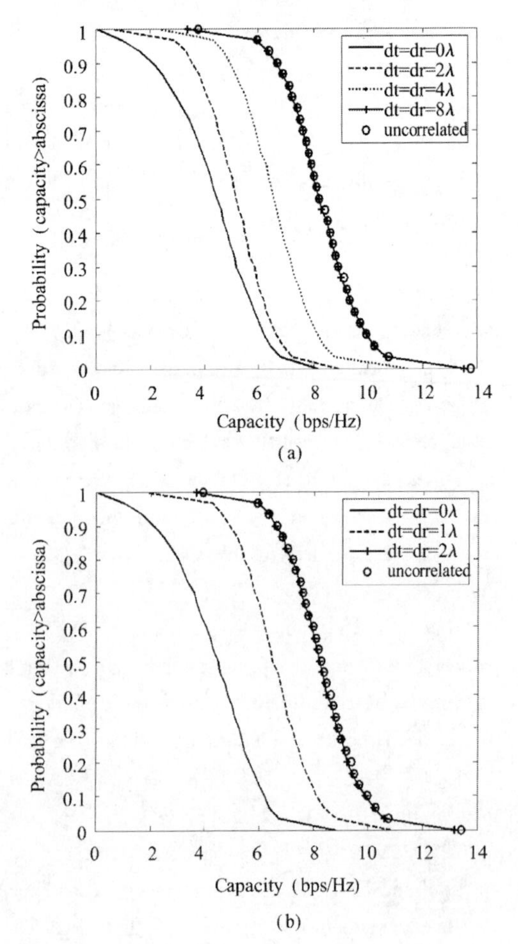

Fig.4 CCDF dependence on array element spacing

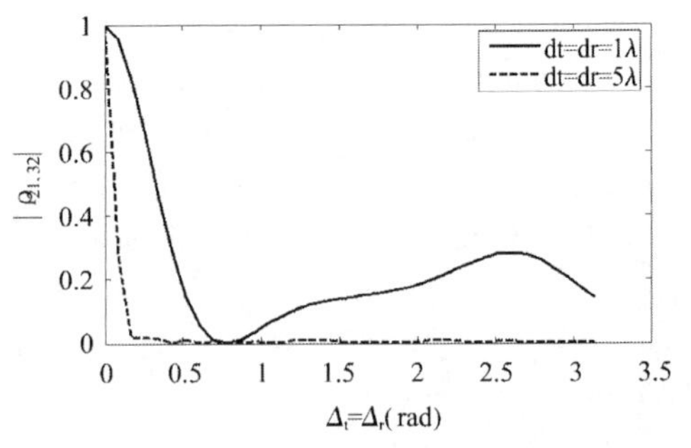

Fig.5 Dependence of $|\rho_{21,32}|$ on angle spread

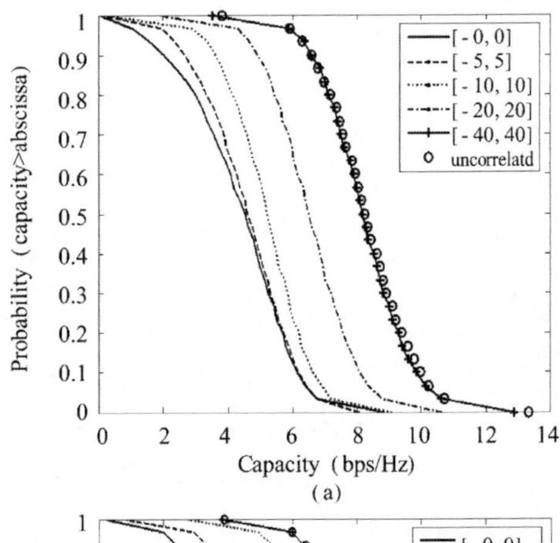

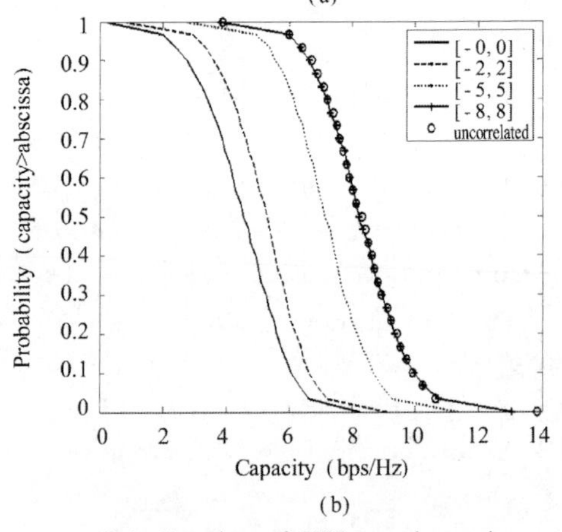

Fig.6 Dependence of CCDF on angle spread

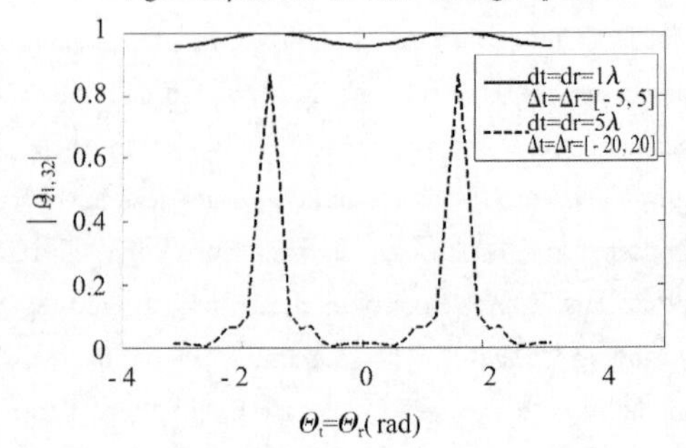

Fig.7 Dependence of $|\rho_{21,32}|$ on AOA

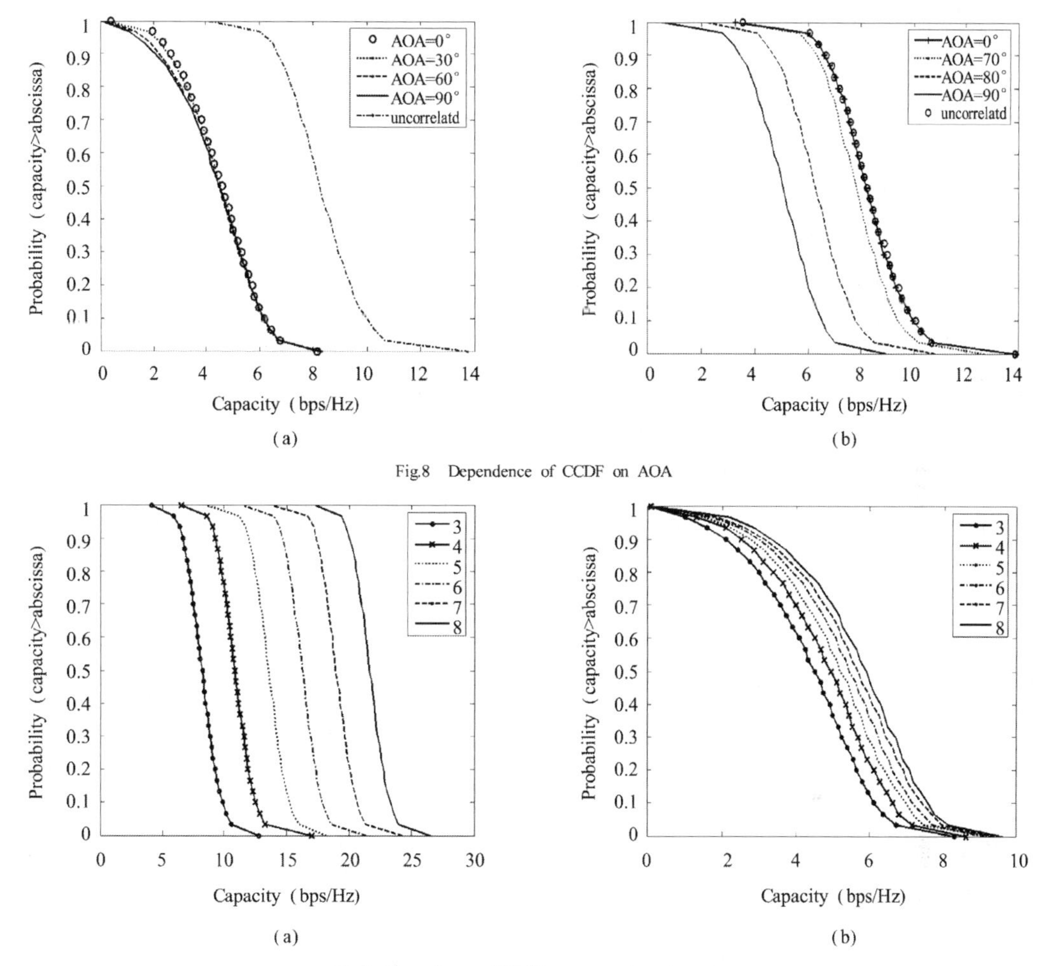

Fig.8 Dependence of CCDF on AOA

Fig.9 Dependence of CCDF on array element number

5 CONCLUSION

Parameters of angle spread, angle of arrival, and element spacing decide the correlations between the MIMO channel coefficients, and thus affect the channel capacity because the angle spread and element spacing have similar effects on spatial correlations. Therefore, in the environment of a small angle spread, increasing transmitting and receiving element spacing can also achieve high MIMO capacity gains. Furthermore, for a static MIMO system, the channel capacity gains are decided by the spatial correlations among the channel coefficients. As such, in a system with small correlations, MIMO capacity will increase with the number of elements linearly, but in a great correlation environment, increasing the number of element will not achieve a large MIMO capacity gain.

References

[1] J H Winters, J Salz, R D Gitlin. The impact of antenna diversity on the capacity of wireless communication system[J]. IEEE Trans. Commun, 1994, 42: 1740-1751.

[2] I E Telatar. Capacity of multi-antenna gaussian channels [R]. AT&T Bell Laboratories, Murray Hill, NJ. Tech. Rep. #Bl0 112 170-950-615-07 TM, 1995.

[3] G J Foschini, M J Gans. On the limits of wireless communications in a fading environment when using multiple antennas[J]. Wireless Pers. Commun, 1998, 6(3): 311-335.

[4] M Zatman, B Tracey. Underwater acoustic MIMO channel capacity[A]. signals, systems and computers, 2002, conference record of the Thirty-Sixth Asilomar Con-ference, 2002, 2, 3-6: 1364-1368.

[5] D Shiu, G J Foschini, M J Gans, J M Kahn. Fading correlation and its effect on the capacity of multi-element antenna systems[J]. IEEE Trans. Commun, 2000, 48: 502-513.

[6] C N. Chuah, D N C Tse, J M Kahn, R A Valenzuela. Capacity scaling in MIMO wireless systems under correlated Fading[J]. IEEE Trans Inform Theory, 2002, 48: 637-650.

[7] M Chiani, M Z. Win. On the capacity of spatially corr-elated MIMO Rayleigh-Fading channels[J]. IEEE Trans Inform Theory, 2003, 49: 2363-2371.

[8] A Leon-Garcia. Probability and Random Processes for Electrical Engineering[M]. Addison-Wesley, 1990.

Processing of non-uniform azimuth sampling in multiple-receiver Synthetic aperture sonar image①

Hailiang Yang, Jinsong Tang
Naval University of Engineering
Wuhan Hubei China
yanghl01029@163.com

Qihu Li, Xiaodong Liu
Institute of Acoustics, Chinese Academy of Science
Beijing, China
lqh@mail.ioa.ac.cn

Abstract—**Multiple-receiver synthetic aperture sonar (SAS) imaging with Displaced Phase Center (DPC) technique can solve the confliction between mapping speed and resolution, but this mode brings some phase error and restricts the selection of velocity of the SAS carrier and the pulse repetition frequency (PRF), namely, the SAS carrier velocity has to be chosen such that the SAS carrier moves just one half of its antenna length in a pulse repetition interval. Any deviated from this ideal velocity will result in a non-uniform sampling of synthetic aperture. In this paper, phase errors of multi-receiver can be compensated approximately and signal model can be obtained, and then, the reconstruction method of uniform spectrum in azimuth is given when the SAS carrier velocity deviated from the ideal one. The simulation and trial results show the validity of this method.**

Index Terms—**multiple-receiver synthetic aperture sonar, displaced Phase Center, "stop and hop" approximation, non-uniform sampling, image reconstruction.**

I. INTRODUCTION

Because of the extremely high resolution desired and tight mapping rate constraints in synthetic aperture sonar systems, multiple-receiver SAS technique are being adopted. However, compare with traditional single receiver SAS, the processing becomes more complicated with the increasing numbers of receivers. These complications can be described in three aspects: firstly, the assumption of monostatic transmitter-receiver sonar is the precondition in traditional processing algorithms, such as the Range-Doppler, the wave-number and the chirp-scaling algorithms. The data in Multiple-receiver SAS cannot be processed with these algorithms straightly and needs to be combined to form the single receiver data equivalent by adopting DPC technique [1] [2]. However, the combining process introduces phase errors which need to be addressed and compensated [3]. Secondly, if the target range is large, the receiver has more time to travel during signal transmission and reception. The "stop and hop" approximation is no longer valid, and the resultant phase errors become noticeable. At last, according to the DPC technique in N receivers SAS or SAR, the receiver distance d and the carrier velocity v are fixed, which requires a specific carrier velocity or PRF to satisfy the equation for uniform sampling in azimuth [3]:

$$PRF = \frac{2v}{Nd} \tag{1}$$

Multiple-receiver SAS imaging algorithms in [1]-[4] are all restricted by (1). However, such a rigid equation cannot be tenable when the carrier velocity deflects from the ideal value. Furthermore, an increased PRF can be used to improve azimuth ambiguity suppression. The deviation of (1) will result in a non-uniform sampling of the received SAS signal and false target image appears in the form of pairs of echo when this non-uniform sampling signal is processed with traditional processing algorithms [5].

In this paper, multiple-receiver SAS signal model is obtained in section II, and then a multi-channel spectrum reconstruction method is given to process non-uniform sampling signal in section III. At last, simulation and trial results in section IV show the validity of this method.

II. MULTIPLE-RECEIVER SAS SIGNAL MODELING

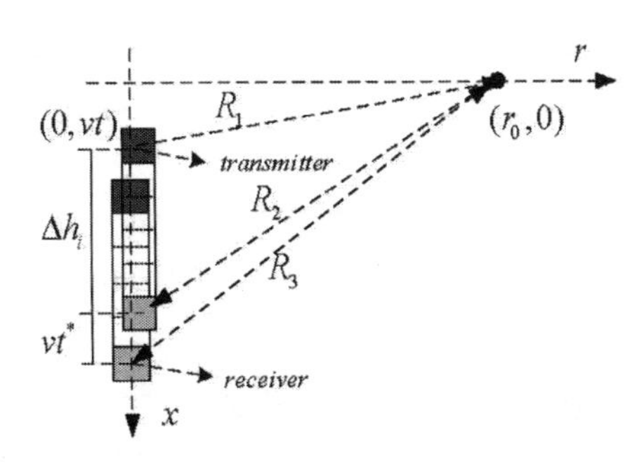

Fig.1 Geometry of a multiple receiver SAS system

Consider the geometry shown in Fig.1, which depicts a transmitter and multiple receivers. The transmitter locates at azimuth position vt at a distance R_1 from the target and the receiver separates from the transmitter by a distance Δh_i,

① Oceans 2007: 1-4.

moves a distance vt^*. The variable t^* represents the total travel time, which is different for each target in the scene. The distance R_1 is given by

$$R_1 = \sqrt{(vt)^2 + r_0^2} \tag{2}$$

And the distance R_2:

$$R_2 = \sqrt{(vt + \Delta h_i)^2 + r_0^2} \tag{3}$$

While the actual distance from target to receiver is

$$R_3 = \sqrt{(vt + \Delta h_i + vt^*)^2 + r_0^2} \tag{4}$$

We will use a LFM chirp signal for our imaging, expressed by

$$s_t(\tau) = rect(\frac{\tau}{T})\exp(j\pi\omega_0 + j\pi\mu\tau^2) \tag{5}$$

where T is the pulse length, μ is the chirp rate, ω_0 is the carrier frequency. From a target a distance r away after coherent demodulation and pulse compression, the response expression is

$$s_i(t,\tau;r) = A\sin c\{\pi\mu T[\tau - \frac{2R(t,r)}{c}]\} \bullet \exp\{-j\frac{4\pi}{\lambda}R(t,r)\} \tag{6}$$

where $R(t,r) = R_1 + R_3$. Obviously, different target in the scene will bring on different vt^* and it is not practical to calculate the phase offsets caused by all the vt^* in current echo. We can calculate these from the target in the center of current swath (r_0) approximately. After Taylor series expansions from (6) and neglecting the higher order items, the response expression becomes:

$$s_i(t,\tau;r) \approx A\sin c\{\pi\mu T[\tau - \frac{2R(t;r)}{c}]\} \bullet \exp\{-j\frac{4\pi}{\lambda}r_0\} \bullet \exp\{-j\frac{\pi(\Delta h_i + vt^*)}{2\lambda r_0}\} \bullet \exp\{-j\frac{2\pi v^2}{\lambda}\frac{\left(t - \frac{(\Delta h_i + vt^*)}{2v}\right)}{r_0}\} \tag{7}$$

the point target response of monostatic SAS is:

$$s_0(t,\tau;r) \approx A' \sin c\{\pi\mu T[\tau - \frac{2R(t;r)}{c}]\} \bullet \exp\{-j\frac{4\pi}{\lambda}r_0\}\exp\{-j\frac{2\pi v^2 t}{\lambda r}\} \tag{8}$$

Then:

$$\begin{aligned} s_i(\tau,t;r) &= s_0(\tau,t;r) \otimes h_i(t;\Delta h_i) \\ &= s_0(\tau,t;r) \otimes \exp\{-j\frac{\pi(\Delta h_i + vt^*)}{2\lambda r_0}\} \\ &\quad \bullet \delta(t - \frac{(\Delta h_i + vt^*)}{2v}) \end{aligned} \tag{9}$$

It becomes clear from this expression that multiple-receiver response evolves from its monostatic response by a fixed phase offset and a time delay. So, multiple-receiver response can be regarded as multi-channel combination that is followed by additional phases offset and time delay. When the equation (1) is satisfied and phase error has been compensated, traditional SAS imaging algorithms can be adopted after receivers arranging in turn [3].

III. THE PROCESSING OF MISMATCH BETWEEN VELOCITY AND PRF

The departure of carrier velocity will result in a non-uniform sampling of the received SAS signal and false target image appears in the form of pairs of echo when this signal is processed with traditional processing algorithms. Therefore we need a feasible approach to decrease or remove the effect of non-uniform sampling of the received SAS signal. A simple method is adopted in SAR in paper [6], but it is fit for a tiny drift against the ideal SAR carrier velocity which restrict by equation (1). In SAS, we know, carrier velocity is very low, so this method is deficient. A multi-channel sampling approach of low-pass signals is given by J.L. Brown [7] and signals can be reconstructed without distortion from its non-uniform samples. According to this paper, a deterministic signal $x(t)$ band limited to $|f| < \sigma/2$ is passed through m linear time-invariant filters $\{H_1(f) \cdots H_m(f)\}$ to obtain the m outputs $\{Z_1(f) \cdots Z_m(f)\}$, if the filters are independent in sense to be defined, then the input $x(t)$ can be reconstructed from samples of the outs $\{z_k\}$, each output being sampled at σ/m samples per second. A block diagram for the reconstruction from the subsampled signals is shown in Fig.2. For m linear filters, we hope to find the reconstruction filters $\{P_1(f) \cdots P_m(f)\}$ which made the quantity $\|x(t) - x_a(t)\|$ minimize, where $\|\bullet\|$ denotes the usual L_2 norm on $-\infty < t < \infty$. As shown in [7], each of the reconstruction filters $P_i(f)$ can be regarded as a composition of m bandpass filters $\{P_{i1}(f) \cdots P_{im}(f)\}$ and can be derived from the inversion of the matrix $H(f)$, where $H(f)$ is [8]:

$$\begin{bmatrix} H_1(f) & \cdots & H_m(f) \\ H_1(f+\sigma/m) & \cdots & H_m(f+\sigma/m) \\ \vdots & \ddots & \vdots \\ H_1(f+(m-1)\bullet\sigma/m) & \cdots & H_m(f+(m-1)\bullet\sigma/m) \end{bmatrix}$$

For multiple-receiver SAS, each receiver can be regard as a receiving channel and each receiver response can be regard as monostatic transmitter-receiver SAS response passed through the filter $h_i(t;\Delta h_i)$ whose expression in frequency domain is:

$$H_i(f)=\exp\{-j\frac{\pi(\Delta h_i+vt^*)}{2\lambda r_0}\} \bullet\exp(-j\frac{\pi(\Delta h_i+vt^*)}{v}f) \tag{10}$$

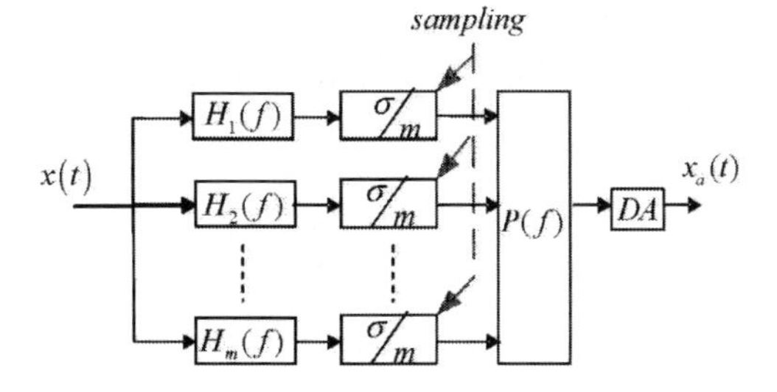

Fig.2: block diagram for the reconstruction from the subsampled signals

When carrier velocity doesn't satisfy the equation (1), the filter $h_i(t;\Delta h_i)$ includes not only the former two phase errors mentioned in introduction, but also the phase error introduced by non-uniform sampling. These errors are almost removed by spectrum reconstruction from matrix $P(f)$ and the response corrected and combined may then be used as input to the traditional synthetic aperture algorithm.

IV. SIMULATION AND TRIAL RESULTS

To test the proposed method, some simulation results are presented in this section. The nominal operating parameters used in the simulation are:

Transmitted signal:	LFM signal
Signal bandwidth:	20 KHz
Center frequency:	30 KHz
Number of transmitter:	1
Number of receivers:	4
Number of pings:	300
Sampling frequency:	300 kHz
PRF:	5Hz

Furthermore, there exist twelve ideal point targets in the scene centered at 100m from the sonar track showed in Fig. 1. The transmitter and the receiver array are collocated in the azimuth along track, where each of the transducers has the same extend in the azimuth of 16cm. According to equation (1), the ideal carrier velocity must be 1.6m/s and signal sampling in azimuth is uniform. Here azimuth slice of point target image is shown in Fig.3.

The departures of carrier velocity from ideal one are discussed in two aspects: less and more. One hand, Fig.4 shows the azimuth slice of image when carrier velocity is 1.0m/s, and unwanted side lobes are introduced into the image before the reconstruction of uniform spectrum in azimuth. After the reconstruction of uniform spectrum, the image improvement is clearly evident. On the other hand, when carrier velocity is 2.0m/s, unwanted side lobes, shown in Fig.5, are introduced not only from the non-uniform sampling of synthetic aperture but the inadequate sampling in azimuth. A certain improvement is obtained after the reconstruction of uniform spectrum, and here, side lobes are introduced from the inadequate sampling in azimuth mainly. We can achieve the adequate sampling in azimuth by increasing number of receivers.

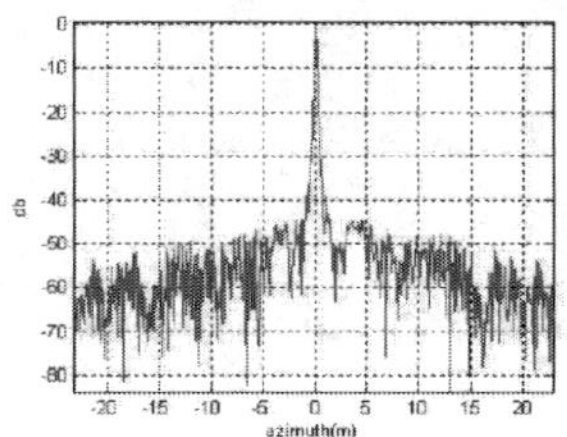

Fig.3 azimuth slice of image when carrier velocity is ideal (1.6m/s)

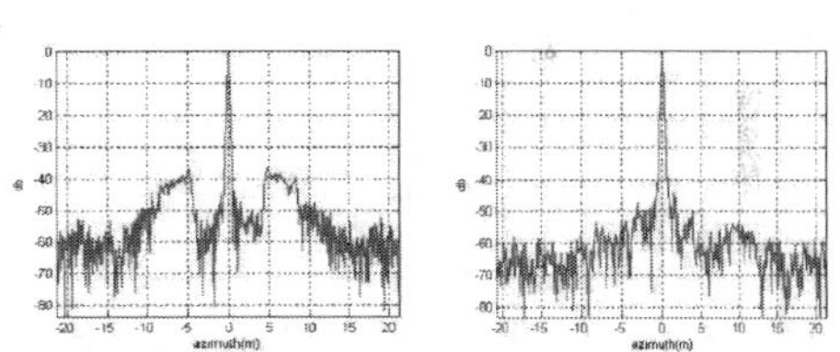

Fig.4 azimuth slice of image (left) when carrier velocity (1.0m/s) is less than the ideal one and azimuth slice of image (right) after the reconstruction of uniform spectrum

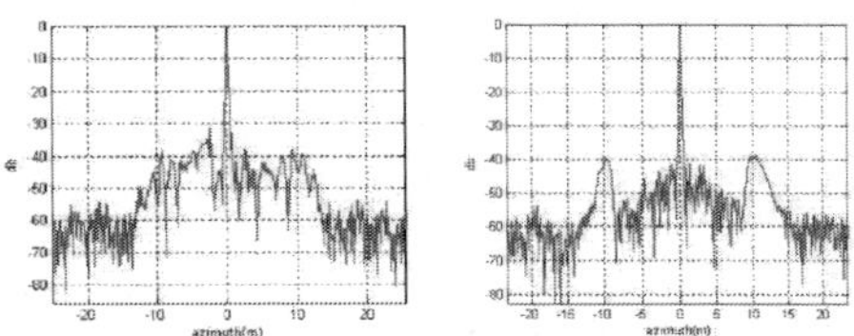

Fig.5 azimuth slice of image (left) when carrier velocity (2.0m/s) is more than the ideal one and azimuth slice of image (right) after the reconstruction of uniform spectrum

Fig.6 shows the image results of the bottom of Mogan-Mountain Lake, and target is a triangle iron frame with a rectangular tile reflector. The length and width of reflector are all 2 meter. It is clear, that the ambiguities are well suppressed in the image after the reconstruction of uniform spectrum, and target image sizes accord with the real one.

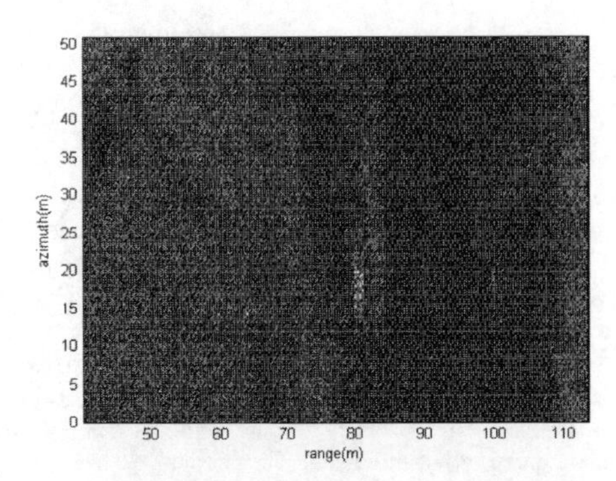

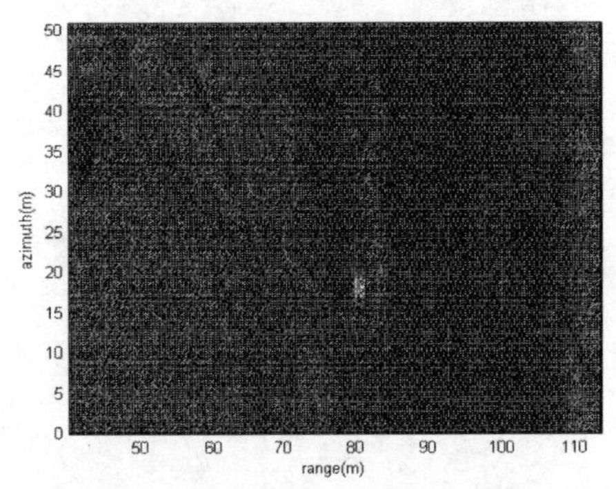

Fig.6 image before the reconstruction of uniform spectrum (top) and after the reconstruction of uniform spectrum (bottom)

V. Conclusion

The selection of velocity of the SAS carrier and the pulse repetition frequency is fixed in multiple-receiver synthetic aperture sonar (SAS) imaging algorithms with DPC technique. Any deviated from this limit will result in image ambiguities. In this paper, a robust reconstruction method of uniform spectrum in azimuth is given when the SAS carrier velocity deviated from the ideal one. The simulation and trial results show that the ambiguities are well suppressed in the image by adopting this method.

References

[1] H.J. Callow, M.P. Hayes and P.T. Gough, Wavenumber domain reconstruction of SAR/SAS imagery using single transmitter and multiple-receiver geometry, IEEE Electronics Letters, 2002。38(7), 336~338

[2] Andrea Bellettini and Marc A. Pinto, Theoretical Accuracy of Synthetic Aperture Sonar Micronavigation Using a Displaced Phase Center Antenna, IEEE Journal of Oceanic Engineering, 2002. 27(4), 780~789

[3] David R. Wilkinson, Efficient Image Reconstruction Techniques for a Multiple-Receiver. Synthetic Aperture Sonar, thesis of Master, University of Canterbury Christchurch, New Zealand. 2001

[4] William W. Bonifant. Jr. Interferometric synthetic aperture sonar processing, thesis of Master, University of Georgia Institute of Technology, July 1999.

[5] LI Shi-qiang YANG Ru-liang, Error analysis of displaced phase centers multiple azimuth beam synthetic aperture radar. ACTA ELECTRONICA SINICA, 2004.9. Vol.32 1436~1440.

[6] MA Xiao-yan WU Shun-hua XIANG jia-bin, The effect of mismatch between velocity and PRF on MPCSAR imageand the compensation method. ACTA ELECTRONICA SINICA, 2005.12. Vol.32 2130~2134

[7] J.L. Brown, Multi-channel sampling of low-pass signals, IEEE Transaction on Circuits and Systems 1981, 28(2), 101~106.

[8] Nicolas Gebert, Gerhard Krieger, Alberto Moreira. SAR signal reconstruction from non-uniform displaced phase center sampling in the presence of perturbation, IEEE Geoscience and Remote Sensing letters, July 2005,1034~1037.

A robust multiple-receiver Range-Doppler algorithm for synthetic aperture sonar imagery①

Hailiang Yang, Jinsong Tang
Naval University of Engineering
Wuhan Hubei China
yanghl01029@163.com

Qihu Li, Xiaodong liu
Institute of Acoustics, Chinese Academy of Science
Beijing, China
lqh@mail.ioa.ac.cn

Abstract—**Synthetic aperture techniques have been of interest to the sonar imaging community of the ocean floor, but suffer from the problem of slow mapping rate. Mapping rate is defined as the product of the swath width and sonar platform velocity, so it can be increased by widening swath or increasing platform velocity. However, wide swath and high platform velocity may cause range and azimuth ambiguities respectively. In many papers, multiple-receiver technique has been adopted to remove these ambiguities, but the image distortion caused by the stop-and-hop approximation (the platform transmits a signal, receives the echoes from targets, and then jumps to the next azimuth sampling location) is ignored or underestimated. For multiple-receiver synthetic aperture imaging with wide swath and high platform velocity, the error introduced by the stop-and-hop approximation must be corrected furthest. In this paper, a robust multiple-receiver Range-Doppler algorithm is proposed to process return data without the stop-and-hop approximation, and the simulation results show the validity of this method.**

Index Terms—**Multiple-receiver synthetic aperture sonar, "stop-and-hop" approximation, wide swath, high platform velocity, Range-Doppler algorithm.**

I. INTRODUCTION

Synthetic aperture sonar operates by sending wide-beam acoustic pulses and combining the echoes from many pulses to obtain high-resolution images. Because of the extremely high resolution desired and tight mapping rate constraints (particularly for synthetic aperture sonar (SAS) systems), multiple-receive element SAS and synthetic aperture radar (SAR) systems are being adopted [1]-[4]. SAS imaging models are typically developed under the 'stop-and-hop' model, whereby it is assumed that the platform transmits a signal, receives the echoes from targets, and then jumps to the next along-track sampling location. This simplified model is adequate for systems towed at low platform velocity. However, if the target range is large or platform moves with a high platform velocity, the receiver has more time to travel during signal transmission and reception. The "stop-and-hop" approximation is no longer valid, and the resultant phase errors become noticeable [5]. Multiple-receiver SAS imaging algorithms in [1]-[4] are all ignored the effect introduced by the "stop-and-hop" approximation. In order to overcome this question, we proposed a multi-aperture synthetic aperture sonar beamforming imaging algorithm and validated it by trial [6]. Unfortunately, the computations are too large and real-time processing costs much. In the paper [5] by William, an assumption is introduced that the total travel time of each individual target response is approximately equal to the amount of time it takes the signal to propagate from the transmitter to scene center and back. The phase errors of all point targets in scene introduced by the "stop-and-hop" approximation can be compensated approximately with the fixed error of target in swath center. This assumption is valid nearly for SAS imaging with narrow swath, but it will cause terrible image distortion in wide swath scene and high platform velocity mapping. In this paper, a robust multiple-receiver Range-Doppler algorithm is proposed, and this algorithm can increase mapping rate and improve image quality remarkably.

II. PHASE ERRORS CORRECTION IN AZIMUTH

The model that the "stop-and-hop" approximation is no longer valid can be defined as the "non stop-and-hop" one. Then, the processing of echoes included two items for the "non stop-and-hop" model: one is to correct the phase errors in azimuth; the other is to compensate the error of range migration introduced by the "stop-and-hop" approximation. The latter will be analyzed in next section.

The phase errors result from two assumptions mostly for multiple-receiver SAS imaging: the "stop-and-hop" approximation and the equivalent phase center assumption (a transmitter and receiver pair separated by a finite distance in along-track could be approximated by an equivalent pair located exactly between them). The former exists in real world and the latter is introduced to analyze conveniently. Both of these approximations inject phase errors into the data, if not corrected. In many papers, for example [1]-[5], the latter is considered, but the image distortion caused by the stop-and-hop approximation is ignored or underestimated. These are not reasonable for real image mapping, especially for mapping with wide swath and high platform velocity. Here, any approximations will not adopted in imaging geometry of multiple-receiver SAS system, and each individual target response in different range is compensated with the deduced accurate phase error respectively.

① Oceans 2007: 1-5.

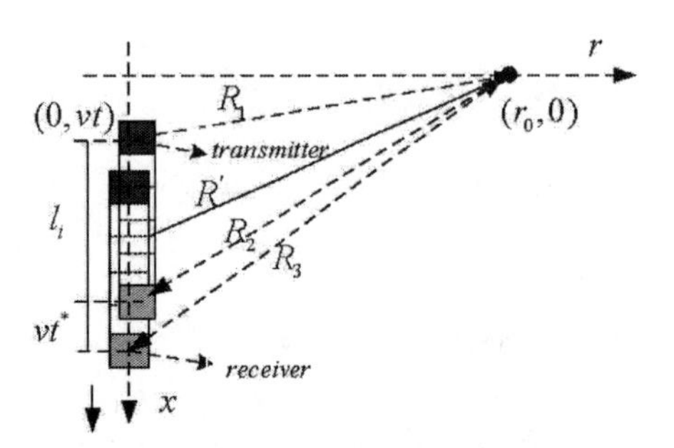

Fig.1 Geometry of a multiple receiver SAS system

Consider the geometry shown in Fig.1, which depicts a transmitter and multiple receivers. The transmitter locates at azimuth position vt at a distance R_1 from the target and the receiver separates from the transmitter by a distance Δh_i, moves a distance vt^*. The variable t^* represents the total travel time, which is different for each target in the scene. The distance from transmitter to target R_1 is given by

$$R_1=\sqrt{(vt)^2+r_0^{\,2}} \tag{1}$$

where we have again ignored the z dimension. The assumed distance from stationary receiver to target R_2:

$$R_2=\sqrt{(vt+\Delta h_i)^2+r_0^{\,2}} \tag{2}$$

while the actual distance from moving receiver to target is

$$R_3=\sqrt{(vt+\Delta h_i+vt^*)^2+r_0^{\,2}} \tag{3}$$

The distance R' is the distance from an assumed equivalent phase center to target and is given by

$$R'=\sqrt{(vt+\Delta h_i/2)^2+r_0^{\,2}} \tag{4}$$

According to the paper [6], the total travel time can be determined exactly by

$$\tau_i=\frac{v(vt+l_i)+C\sqrt{(vt)^2+r^2}+\sqrt{\left[v(vt+l_i)+C\sqrt{(vt)^2+r^2}\right]^2+(C^2-v^2)\left[2(vt)l_i+l_i^{\,2}\right]}}{C^2-v^2} \tag{5}$$

The range error resulting from the "stop-and-hop" approximation and the equivalent phase center assumption is

$$\Delta R=R_1+R_3-2R'=C*\tau_i-2R' \tag{6}$$

The correction can easily be applied by calculating (6) for each point in data matrix and applying a phase multiply to the data of the form

$$H_i(t,r)=\exp(j\omega_0\frac{\Delta R}{c}) \tag{7}$$

Then, the multiple-receiver SAS raw data after phase correction can be regard as the monostatic equivalents applied to the "stop-and-hop" approximation.

III. RANGE MIGRATION ERROR CORRECTION IN RANGE-DOPPLER DOMAIN

Let

$$s(\tau)\cdot\exp\{j\omega_0\tau\} \tag{8}$$

be the pulse transmitted by the sensor and ω_0 the carrier angular frequency. Then the echo from a scatterer a distance r away is after coherent demodulation

$$ss_i(\tau,t;r)=s(\tau-\tau_i)\cdot\exp\{-j\omega_0\tau_i\} \tag{9}$$

Pulse compression, or matched filtering in the range dimension, is accomplished to form

$$dd_i(\tau,t;r)=d(\tau-\tau_i)\exp\{-j\omega_0\tau_i\} \tag{10}$$

The data, transformed from signal space t o range-Doppler space using an azimuth Fourier transform, is given

$$\begin{aligned}dD_i(\tau,\omega_t;r)&=\int_{-\infty}^{+\infty}d(\tau-\tau_i)\exp\{-j\omega_0\tau_i\}\exp\{-j\omega_t t\}dt\\&=\int_{-\infty}^{+\infty}d(\tau-\tau_i)\exp\{-j\omega_0\tau_i-j\omega_t t\}dt\end{aligned} \tag{11}$$

Here, let $\varphi(t)=-j\omega_0\tau_i-j\omega_t t$. Using the principle of stationary phase, the stationary point t_0 is the solution of $\varphi'(t)=0$. Then

$$\frac{\omega_0}{C^2-v^2}\left\{v^2+\frac{Cv^2t}{\sqrt{(vt)^2+r^2}}+\frac{\left[v(vt+l_i)+C\sqrt{(vt)^2+r^2}\right]\left(v^2+\dfrac{Cv^2t}{\sqrt{(vt)^2+r^2}}\right)+(C^2-v^2)vl_i}{\sqrt{\left[v(vt+l_i)+C\sqrt{(vt)^2+r^2}\right]^2+(C^2-v^2)\left[2(vt)l_i+l_i^{\,2}\right]}}\right\}+\omega_t=0 \tag{12}$$

Equation (12) is so complicated that the stationary point can not be solved with an exact value and we can get it approximately by iteration technique. Hereby, the equation (11) is given by

$$dD_i(\tau, f_t; r) = \frac{\sqrt{2\pi}\exp\left\{j\,\mathrm{sgn}\left[\varphi''(t_0)\right]\frac{\pi}{4}\right\}}{\sqrt{|\varphi''(t_0)|}} d\left[\tau - \tau(t_0)\right]\exp\left\{-j\omega_0\tau(t_0) - j\omega_t t_0\right\} \quad (13)$$

Then, range migration is preformed as a time shift by

$$\Delta t = \tau(t_0) - \frac{2r}{c} \quad (14)$$

This migration is accurate considerably, and the subsequent processing can be achieved by traditional algorithms.

IV SAS SIMULATION RESULTS

To test the proposed method, some simulation results are presented in this section. The nominal operating parameters used in the simulation are:

Transmitted signal:	LFM signal
Signal bandwidth:	20 KHz
Center frequency:	150 KHz
Number of pings:	300
Number of receivers:	10
Platform velocity	1.0m/s
PRF:	2.5Hz

There exist three ideal point targets in the scene, (50, 0), (100, 0) and (200, 0). Swath center locates in 100 meter range. The transmitter and the receiver array are collocated in the azimuth, where each of the transducers has the same extend in the azimuth along track of 8cm.

A simulated example is shown in Fig.2. The returns from a point target at location (100, 0) were generated and processed with and without phase error correction caused by the "stop-and-hop" approximation. The image improvement is clearly evident. However, for mapping with wide swath, according to the paper [5], if we compensate approximately the phase errors of all point target in scene with the fixed error of target in swath center, the image resolution of target near the swath center is high, but the further a target leaves, the worse image quality becomes. Under the same conditions, using the method introduced in this paper, image quality of the target far from the swath center improves remarkably. (Simulated imaging examples of point target (50, 0) and (200, 0) are shown in Fig.3 and Fig.4 here.)

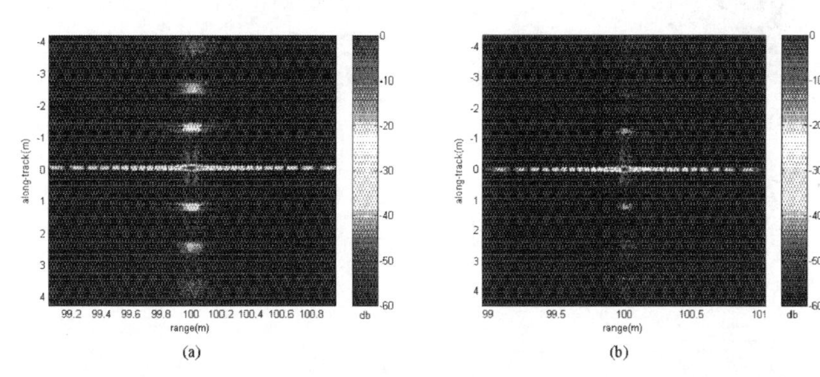

Fig.2 Focused image of point target (100, 0) with (b) and without (a) phase error correction caused by the "stop-and-hop" approximation

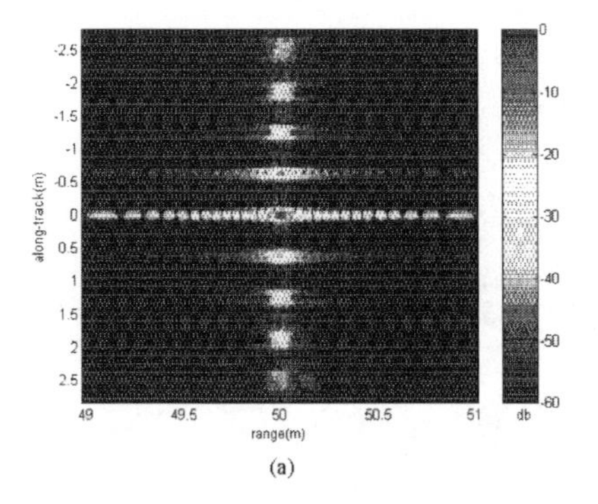

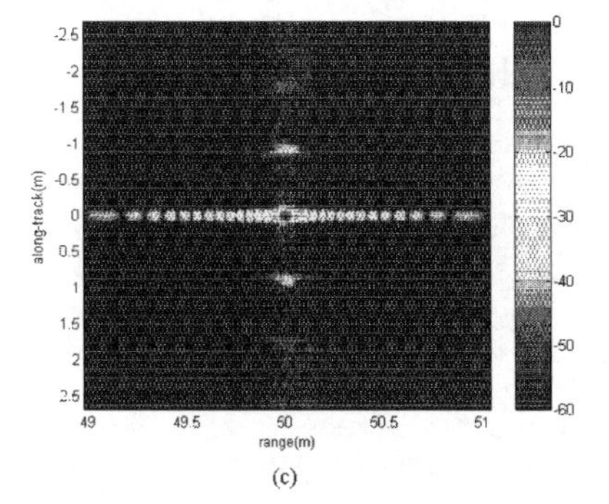

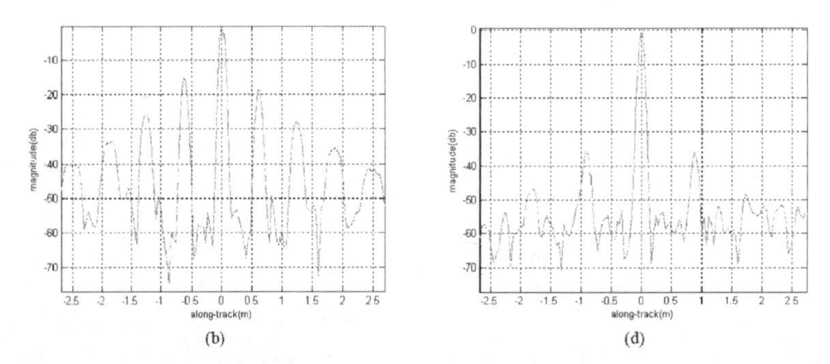

(b) (d)

Fig.3 Simulated imaging example of point target (50, 0) (a), focused image with the fixed error correction, (b) azimuth slices of (a), (c) focused image with the proposed method, (d) azimuth slices of (c)

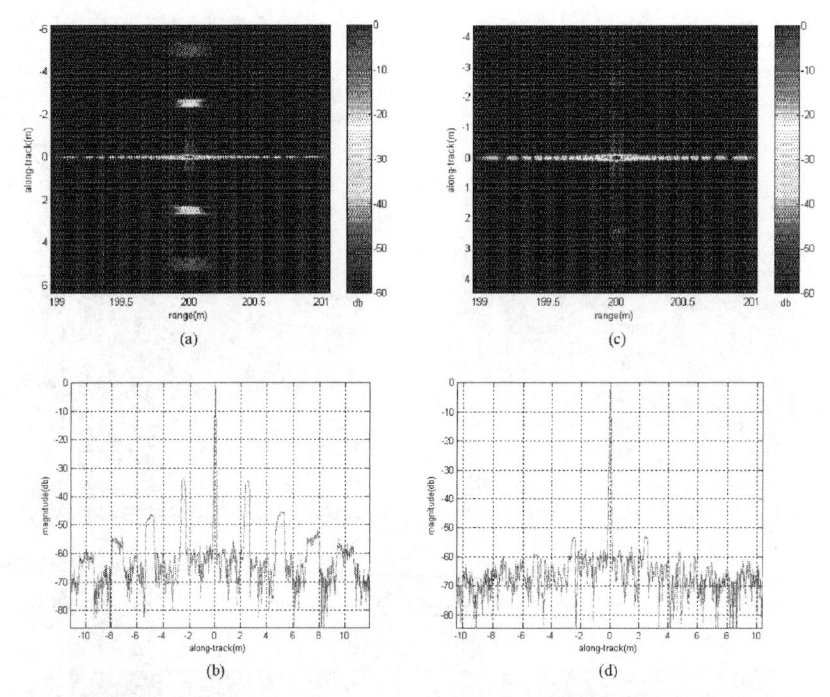

(a) (c)

(b) (d)

Fig.4 Simulated imaging example of point target (200, 0) (a), focused image with the fixed error correction, (b) azimuth slices of (a), (c) focused image with the proposed method, (d) azimuth slices of (c)

V CONCLUSION

For mapping with wide swath and high platform velocity, the "stop-and-hop" approximation is no longer valid, and the resultant phase errors become noticeable. The phase errors of all point targets in scene introduced by the "stop-and-hop" approximation can be compensated approximately with the fixed error of target in swath center [5]. This assumption is valid nearly for SAS imaging with narrow swath, but it will cause terrible image distortion in wide swath scene and high platform velocity mapping. In this paper, a robust multiple-receiver Range-Doppler algorithm is proposed, and this algorithm increase mapping rate and improve image quality remarkably. And some real experiment will be necessary to further test the performance of the algorithm.

REFERENCES

[1] H.J. Callow, M.P. Hayes and P.T. Gough，Wavenumber domain reconstruction of SAR/SAS imagery using single transmitter and multiple-receiver geometry，IEEE Electronics Letters，2002。38(7)，336~338

[2] Andrea Bellettini and Marc A. Pinto，Theoretical Accuracy of Synthetic Aperture Sonar Micronavigation Using a Displaced Phase Center Antenna，IEEE Journal of Oceanic Engineering，2002. 27(4)，780~789

[3] David R. Wilkinson，Efficient Image Reconstruction Techniques for a Multiple-Receiver Synthetic Aperture Sonar, thesis of Master, University of Canterbury Christchurch, New Zealand. 2001.

[4] Brett L, Douglas and Hua Lee. Synthetic-aperture sonar imaging with a multiple-element receiver array. IEEE international Conference on Acoustics, Speech, and Signal Processing, vol.5, 445-448, April 1993.

[5] William W. Bonifant. Jr. Interferometric synthetic aperture sonar processing, thesis of Master, University of Georgia Institute of Technology, July 1999.

[6] Jiang Xu, Jinsong Tang, Multi-aperture synthetic aperture sonar imaging algorithm signal processing, Vol 19, No.2 , April .2003

基于海面散射的莱斯 MIMO 信道容量研究①②

朴大志[1,2]，李启虎[1]，孙长瑜[1]

（1. 中国科学院声学研究所，北京 100080；2. 中国传媒大学信息工程学院，北京 100024）

摘要：通过相关莱斯信道模型对海面散射水声多入多出(MIMO)信道容量进行了研究。从 Helmholtz-Kirchhoff 积分和 Fresnel 近似出发，得到海面散射信号的空间相关性，由于信道的协方差矩阵不可以表示成发射相关矩阵和接收相关矩阵的 Kronecker 积，通过 Monte Carlo 仿真研究了莱斯因子、接收信噪比和空间相关性对 MIMO 信道容量的影响。从对 outage capacity 和信道容量的 CCDF(Complementary Cumulative Distribution Function) 的仿真计算结果可以看出，莱斯衰落的 MIMO 信道容量并不总是小于瑞利衰落 MIMO 信道容量，当信道的空间相关性较大和接收信噪比较小时，由于信道的衰落作用将起主要作用，对于较小的收发阵元数，以非衰落的直达信号为主的莱斯 MIMO 信道容量将大于瑞利衰落的 MIMO 信道容量。

关键词：莱斯 MIMO；空间相关；海面散射；信道容量

中图分类号：TN929.3　　文献标识码：A　　文章编号：1000-3630(2007)-04-0557-07

Rician MIMO capacity of sea-surface scattered underwater acoustic channel

PIAO Da-zhi[1,2]，LI Qi-hu[1]，SUN Chang-yu[1]

(1. Institute of Acoustics, Chinese Academy of Sciences, Beijing 100080, China,
2. Information Engineering College, Communication University of China, Beijing 100024, China)

Abstract: The effect of Rician Factor (RF) and spatial correlation on the Ricean MIMO capacity is studied based on the sea-surface scattered underwater acoustic Rician fading channel. The spatial correlation of the sea-surface scattered signal is derived with the integrel of Helmholtz-Kirchhoff and the Fresnel approximation. We find that covariance matrix of this channel cannot be expressed as the Kronecker product of the transmit correlation function and the receive correlation function. Therefore dependence of MIMO capacity on RF, received SNR and the factors affecting spatial correlation such as element spacing and wind speed are studied through Monte Carlo simulation. From the simulation results of the outage capacity and CCDF of the capacity, the MIMO capacity of Rician channels is not always lower than that of the Rayleigh channels. When there are strong spatial correlations and low receive SNR, fading is more significant in the MIMO channels. For a small number of transmitting and receiving elements, the Rician MIMO capacity composed mainly of deterministic LOS signal will be greater than the Rayleigh fading MIMO capacity.

Key words: Rician MIMIO; spatial correlation; sea-surface scattering; channel capacity

1 引　言

与距离和频率有关的吸收损失，使水声信道中可以提供的带宽非常有限，因此要达到更高的信息传输率需要更有效的利用这有限的频带资源，另外，由于水声信道的散射特性，使其具有多径传播现象，因此在发射和接收端同时使用多个阵元的 MIMO 技术在提高水声信道的频谱效率方面具有很大的潜力，从文献[1]-[3]可以看到，在平坦的 Rayleigh 衰落环境中，当所有收发阵元对之间的信道传播系数相互统计独立时，理论上 MIMO 通信系统的信道容量随着收发阵元数的增加而线性增加，文献[4]中的

① 基金项目：国家自然基金重点项目资助(60532040).

② 声学技术, 2007, 26(4): 557-563.

实验结果也证明了 MIMO 技术可以使水声通信获得较高的频带利用率，因此水声 MIMO 技术也成为高速水声通信研究的新兴热点，而通过对某种典型的水声信道建模来进行 MIMO 信道容量的估计是水声 MIMO 信道容量研究的另一种重要方向。

当无线通信的环境中充满了幅度接近的随机多径信号时，瑞利信道可以用来描述它的衰落特性，然而当收发之间存在较强的直达(line-of-sight)信号时，莱斯衰落模型可以更好的描述这种信道。文献[5]通过仿真研究了不同 RF 下，莱斯 MIMO 信道容量随收发阵元数的变化，文献[6]得到了两个发射或接收阵元情况下的莱斯 MIMO 信道容量的概率密度和分布函数，在文献[7]中，当发射端没有信道的衰落信息时，得到了信道容量的概率分布，当发射端已知信道衰落过程的概率分布而不知道信道的瞬时状态信息时，得到了信道容量的上、下界。在文献[5]–[7]的研究中，均认为各个收发阵元对之间的信道响应的衰落是相互独立的，然而在一些实际的无线通信环境中，信道矩阵的各个元素的衰落会存在一定的相关性，在文献[8]中考虑了空间相关的莱斯 MIMO 信道，得到了当发射或接收阵的某一端存在相关性时，莱斯 MIMO 信道容量的均值的上、下界，在文献[9]中，得到了发射和接收端同时存在相关性时莱斯 MIMO 信道容量均值的上、下界，并采用高斯近似分析了高信噪比时 outage capacity，在文献[9]的相关性假设中，认为信道的协方差矩阵可以写成发射相关矩阵和接收相关矩阵的 Kronecker 积，这也是在研究瑞利相关 MIMO 信道容量的一种常用假设，然而这种假设在本文的在水声信道中是不满足的，另外，对一个随机的信道容量，它的均值只能提供一种粗略的估计，更完整的描述需要得到它的分布函数，而对于这种协方差矩阵不能表示成发射相关矩阵和接收相关矩阵的 Kronecker 积的莱斯 MIMO 信道，在数学上得到它的信道容量的分布函数很困难，因此本文通过蒙特卡罗仿真来进行研究。

在水声通信中，当收发端距离海面较近且通信距离远大于收发阵的深度时，接收的信号将以海面随机散射信号和非衰落的直达信号为主，这种信道可以描述为莱斯衰落模型。其中来自海面的随机散射部分可以看成瑞利衰落信号，并且各个收发阵元对之间的衰落具有一定的空间相关性，因此，本文主要研究这种情况下的水声莱斯 MIMO 信道容量。我们首先利用 Helmholtz-Kirchhoff 积分和 Fresnel 近似得到海面散射信号的空间相关性，然后通过仿真来研究莱斯因子、接收信噪比和空间相关性对 MIMO 信道容量的影响。

2 Rician 信道模型与 MIMO 信道容量

考虑一个 MIMO 系统具有 n_T 个全向发射阵元和 n_R 个全向接收阵元，发射和接收信号矢量的关系可以表示为

$$y(t)=H(t)*x(t)+n(t) \tag{1}$$

其中 $x(t)$ 是 $n_T\times1$ 发射信号矢量，$y(t)$ 是 $n_R\times1$ 接收信号矢量，$H(t)$ 是由复信道增益 $h_{ij}(t)$ 构成的 $n_R\times n_T$ 信道传播矩阵，$n(t)$ 是 $n_R\times1$ 加性白高斯噪声矢量，其中的各个元素是 IID(independent and identically distributed)的复高斯变量，方差为 N_0。

在窄带通信系统中，可以看成具有平坦的频率响应，则式(1)可以写成

$$y=Hx+n \tag{2}$$

先考虑单用户 MIMO 系统，信道状态信息在发射端未知而在接收端已知时，将发射功率均匀分配的平坦莱斯衰落信道，这种情况下 MIMO 系统的信道容量(单位：bps/Hz)可以表示为[1,3]，

$$C=\log_2\det(I_{n_R}+(SNR/n_T)HH^+) \tag{3}$$

其中 H^+ 是 H 的共轭转置，det(•)是求矩阵的行列式，I_{n_R} 是 $n_R\times n_R$ 单位阵，SNR 是每个接收阵元处的平均信噪比。设由海面散射和 LOS(Line Of Sight)信号组成的莱斯信道矩阵可以表示为

$$H_{Rice}=aH_{Los}+bH_{Rayleigh} \tag{4}$$

在此定义 RF 为 $RF=a^2/b^2$，其中 $a^2+b^2=1$，用来描述直达信号的强弱，当 RF=0 时相当于瑞利衰落。考虑远场平面波，对于 H_{LOS} 中的元素

$$h_{ij}=\exp(jkr_{ij}) \tag{5}$$

其中 $j=\sqrt{-1}$，$k=\omega/c$，为波数，本文中假设水中声速 c=1 500m/s，r_{ij} 为第 j 个发射阵元与第 i 个接收阵元之间的距离。

3 海面散射信号的空间相关性

下面考虑由海面散射构成的 Rayleigh 衰落信号的空间相关性，如图 1 所示，其中 θ_{hT} 与 θ_{hR} 分别是发射和接收阵的水平角。根据 Helmholtz-Kirchhoff 积分[10]，第 i 个发射阵元与第 j 个接收阵元间的信道响应可以写成

$$h_{ij}=\frac{k\sin\theta_{T_i}}{r_{T_i}r_{R_i}}\int_{-\infty}^{\infty}ds\exp ik(|R-r_{T_i}|+|R-r_{R_i}|) \tag{6}$$

在文献[11]中，得到了来自同一个发射源经过海面散射的信号在空间不同位置接收的两个信号的空间相关性，本文采用类似的方法，得到两个链路 T_i- R_j 和 T_{i1}- R_{j1} 之间的相关函数

$$\langle h_{ij}h_{i1j1}{}^{*}\rangle=\frac{k^2\sin\theta_T\sin\theta_{T_{j1}}}{r_{T_j}r_{R_i}r_{T_{j1}}r_{R_{i1}}}\int_{-\infty}^{\infty}dsds'\exp ik$$

$$(|R-r_{T_j}|+|R-r_{R_i}|-|R'-r_{T_{j1}}|+|R'-r_{R_{i1}}|) \quad (7)$$

通过 Fresnel 近似[12]，指数项可以写成

$$ik(r_{T_j}+r_{R_i}+\gamma_{ij}\xi(x,y)+\alpha_{ij}x+\beta_{ij}y+\delta_{ij}x^2+\varepsilon_{ij}y^2+v_{ij}xy)$$
$$-ik(r_{T_{j1}}+r_{R_{i1}}+\gamma_{i1j1}\xi(x',y')+\alpha_{i1j1}x'+\beta_{i1j1}y'+\delta_{i1j1}x'^2+$$
$$\varepsilon_{i1j1}y'^2+v_{i1j1}x'y') \quad (8)$$

其中，$\xi(x,y)$ 是海面高度(相对于平面 $z=0$)，它是 x, y 的随机函数，一般假设为 Gaussian 分布，其中

$$\alpha_{ij}=-(\cos\theta_{T_j}\cos\varphi_{T_j})-(\cos\theta_{R_i}\cos\varphi_{R_i})$$

$$\beta_{ij}=-(\cos\theta_{T_j}\sin\varphi_{T_j}-\cos\theta_{R_i}\sin\varphi_{R_i})$$

$$\delta_{ij}=\frac{1-(\cos\theta_{T_j}\cos\varphi_{T_j})^2}{2r_{T_j}}+\frac{1-(\cos\theta_{R_i}\cos\varphi_{R_i})^2}{2r_{R_j}}$$

$$\varepsilon_{ij}=\frac{1-(\cos\theta_{T_j}\cos\varphi_{T_j})^2}{2r_{T_j}}+\frac{1-(\cos\theta_{R_i}\cos\varphi_{R_i})^2}{2r_{R_i}}$$

$$v_{ij}=\frac{2\cos^2\theta_{T_j}\cos\varphi_{T_j}\sin\varphi_{T_j}}{r_{T_j}}+\frac{2\cos^2\theta_{R_i}\cos\varphi_{R_i}\sin\varphi_{R_i}}{r_{R_i}}$$

$$\gamma_{ij}=-(\sin\theta_{T_j}-\sin\theta_{R_i}) \quad (9)$$

通过化简可以得到

$$\langle h_{ij}h_{i1j1}{}^{*}\rangle=G[qQ(C_{xx}C_{yy}-C^2_{xy}/4)]^{-0.5}$$
$$\exp[C_1-(C_x^2C_{yy}-C_xC_yC_{xy}+$$
$$C_y^2C_{xx})/(4C_{xx}C_{yy}-C^2_{xy})] \quad (10)$$

其中 $G=\frac{k^2\sin\theta_T\sin\theta_{T_{j1}}}{r_{T_j}r_{R_i}r_{T_{j1}}r_{R_{i1}}}\exp[-(\gamma_{ij}-\gamma_{i1j1})^2\sigma^2/2+ik(r_{T_j}+r_{R_i}-r_{T_{j1}}-r_{R_{i1}})]$，其它参数与文献[11]中相同。两个链路 T_i- R_j 和 T_{i1}- R_{j1} 之间的归一化相关系数可以表示为

$$\rho_{ij,i1,j1}=\langle h_{ij}h_{i1j1}{}^{*}\rangle/(\langle h_{ij}h_{ij}{}^{*}\rangle\langle h_{i1j1}h_{i1j1}{}^{*}\rangle)^{1/2} \quad (11)$$

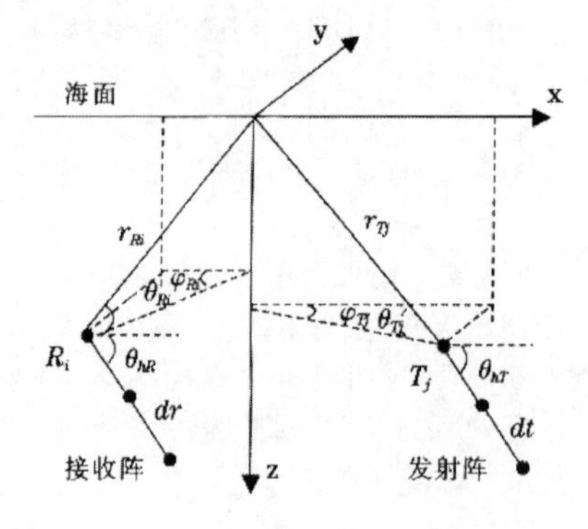

图 1 收发阵几何关系示意图
Fig.1 Geometry for transmit and receive array

从式(9)–(11)中可以看到海面散射信号的空间相关性与风速，收发阵参数如阵元间距、方向、深度、距离等都有关系，图 2 和图 3 仅以$|\rho_{21,33}|$为例，描述了不同收发阵方向下，空间相关性与收发阵深度和阵元间距之间的关系(其中假设 $\theta_{hT}=\theta_{hR}$，同时变化)。从图 2 中可以看出，当收发阵的方向变化时，空间相关性与收发阵深度的关系也会改变，从图 3 中可以看出，随着收发阵元间距的增加，空间相关性将逐渐减小，当方向不同时以不同的速度减小，当 $\theta_{hT}=\theta_{hR}=90°$(垂直收发阵)时减小的速度最大。图 2 和图 3 的计算参数为 $f=10kHz$, $xt_1=-1\,000m$, $xr_1=1\,000m$，风速为 15kn。

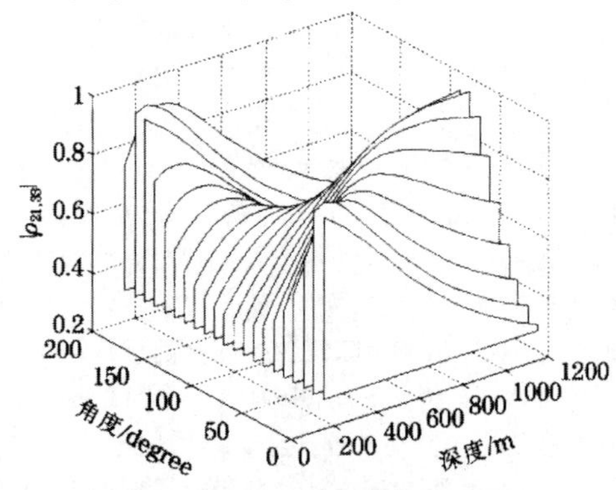

图 2 不同收发阵方向下收发阵深度对相关系数$|\rho_{21,33}|$的影响
Fig.2 Dependence of $|\rho_{21,33}|$ on array direction and depth($dt=dr=0.5\lambda$)

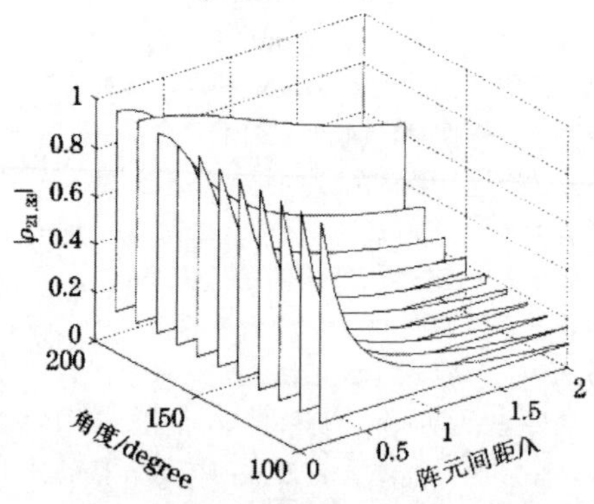

图 3 不同收发阵方向下收发阵元间距对相关系数$|\rho_{21,33}|$的影响
Fig.3 Dependence of $|\rho_{21,33}|$ on array direction and element spacing ($zt_1=zr_1=100m$)

在[9]中的相关性假设中，认为信道矩阵的每一列都具有相同的协方差矩阵，即接收相关矩阵 R，也就是认为接收相关矩阵与发射阵元的位置无关，类似地，认为每一行的发射相关矩阵矩 S 与接收阵元的位置无关，则信道矩阵的统计特性可以用 $R^{1/2}H_wS^{1/2}$ 来描述，其中 H_w 是零均值、单位方差的 IID 复高斯矩阵。在这里，从式(7)–(10)和图 2、3 都可以看出，由

于相关性与收发阵元的位置有关，所以，对于来自不同的发射阵元的各个接收阵元的信号之间的相关性是不同的，也就是说信道矩阵的每一列信号的协方差矩阵是不相同的，同样，每一行信号的发射矩阵也是不同的，因此信道矩阵不能写成 $R^{1/2}H_wS^{1/2}$ 的形式。而基于发射和接收矩阵 kroneker 积的信道相关性假设，可以利用一些关于非中心 wishart 矩阵的分布来对 MIMO 信道容量进行分析，而当这种相关性不满足时，很难从数学上得到相关莱斯 MIMO 信道容量的分布的表达式，因此，本文通过 Monte Carlo 仿真来进行信道容量的研究。

4　仿真结果

4.1　相关随机变量的产生

产生具有任意相关性的 M 个随机变量，可以首先分解它们的协方差矩阵 R，即 $R=GG'$，然后将 M 个独立的高斯变量与 G 相乘，就可以得到 M 个具有协方差矩阵 R 的高斯随机变量[13]。

4.2　仿真结果及简要分析

对于随机变化的信道，它的信道容量也是随机变量，在这种情况下，一般用 C_{out}(outage capacity) 来描述信道容量，它与 P_{out} 有关，P_{out} 被定义为信道容量小于 C_{out} 的概率，例如 $C_{0.5}$ 意味着在信道容量的各种实现中不能支持它的概率为 50%。

首先由式(11)得到海面随机散射信号的空间互相关系数，然后得到具有这种互相关性的瑞利衰落信道矩阵，并利用式(4)得到莱斯信道矩阵，然后根据式(3)来计算信道容量，从而研究各种参数对相关莱斯 MIMO 信道容量的影响。

在莱斯信道中，莱斯因子 RF 是一个重要的参数，前文也提到，它描述了信道中直达的确定信号能

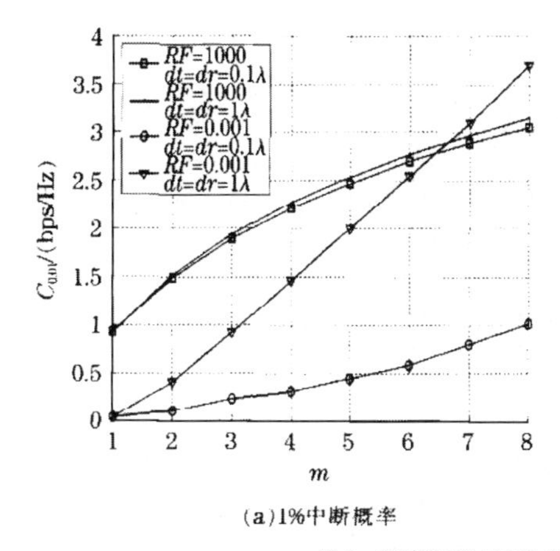

(a)1%中断概率

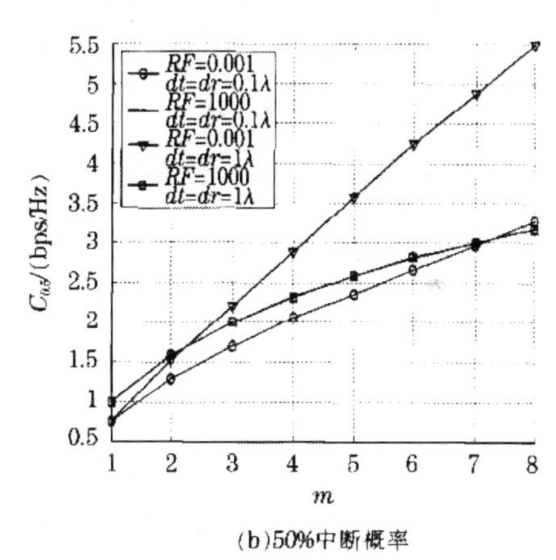

(b)50%中断概率

图 4　不同 RF 下 MIMO 信道容量随收发阵元数的变化(SNR=0dB)

Fig.4　Outage capacity versus array element number with different RF

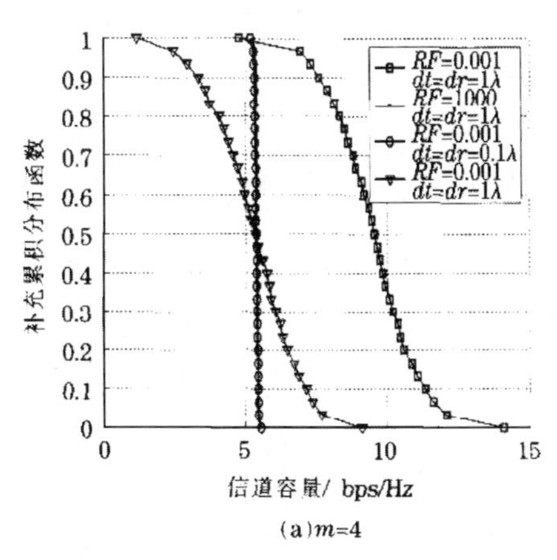

(a)m=4

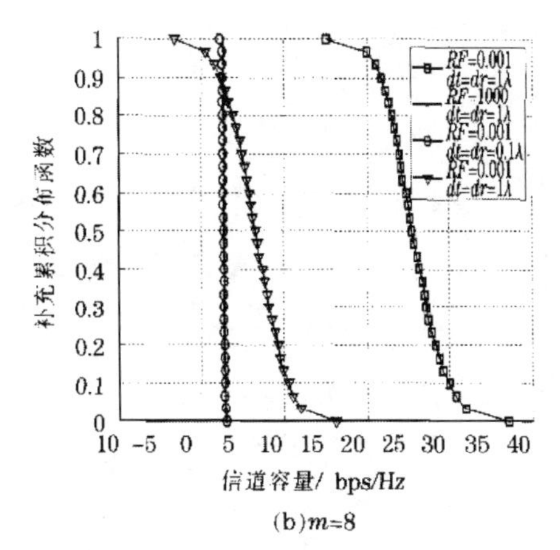

(b)m=8

图 5　RF 对 CCDF 的影响(SNR=10 dB)

Fig.5　Dependence of CCDF on RF

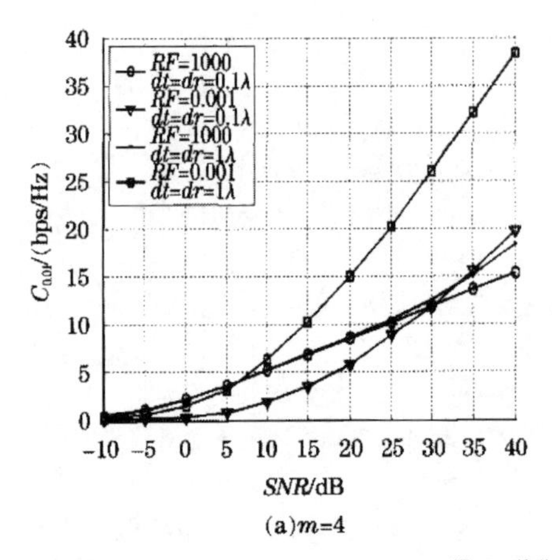

(a)m=4

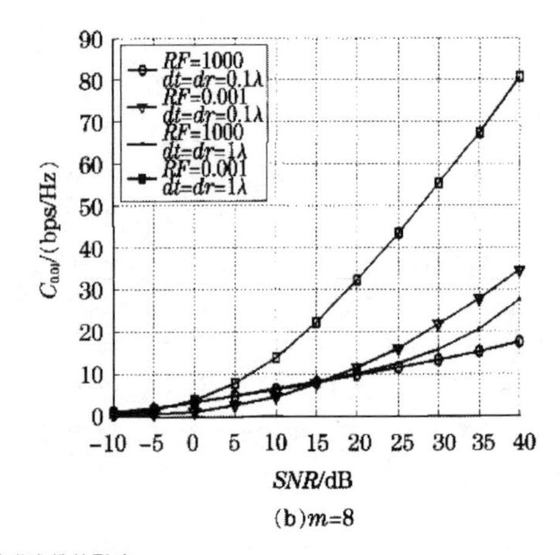

(b)m=8

图 6 接收信噪比对信道容量的影响

Fig.6 Dependence of outage capacity on *SNR*

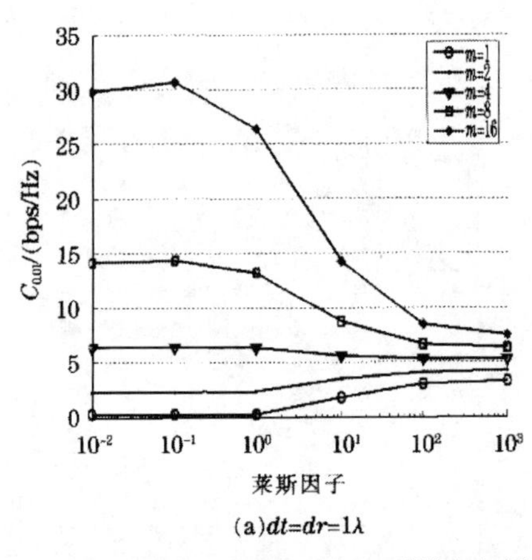

(a)$dt=dr=1\lambda$

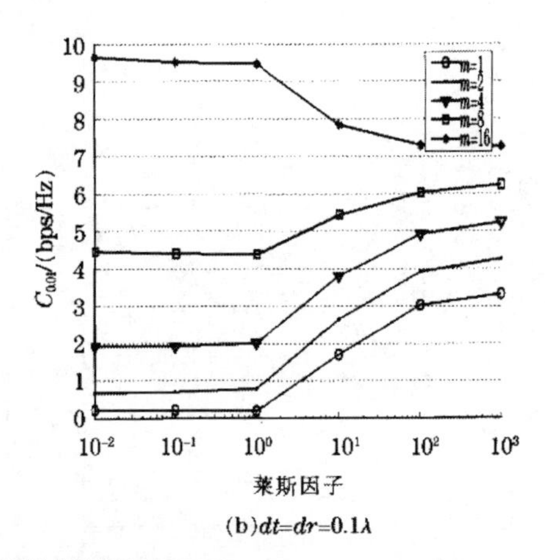

(b)$dt=dr=0.1\lambda$

图 7 不同收发阵元数下 *RF* 对信道容量随的影响(*SNR*=10dB)

Fig.7 Outage capacity versus on RF with different element number

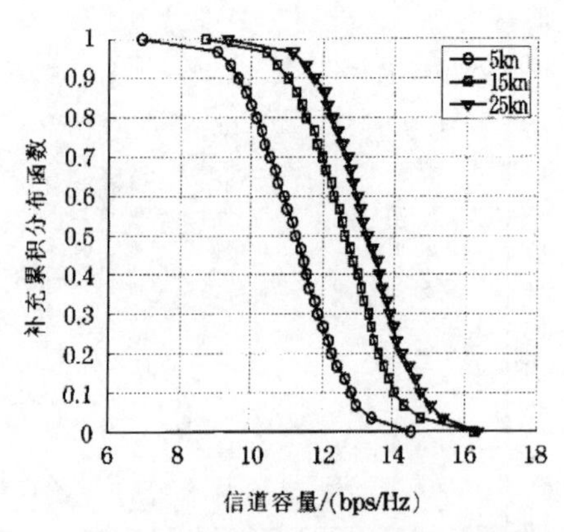

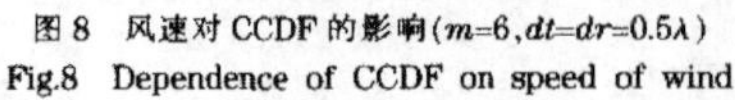

图 8 风速对 CCDF 的影响(m=6, $dt=dr=0.5\lambda$)

Fig.8 Dependence of CCDF on speed of wind

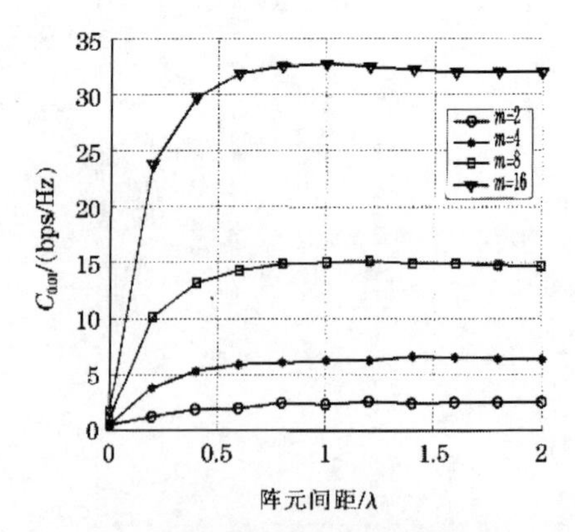

图 9 不同收发阵元数下收发阵元间距对信道容量的影响(风速为 15kn)

Fig.9 Dependence of outage capacity on element spacing with different element number (wind speed 15kn)

量与海面散射的随机信号能量之比，当 RF 较小时，信道更接近于随机的瑞利衰落信道，从而信道容量受互相关性的影响较大，所以本文将从 RF 和空间相关性两个方面研究海面反射的水声莱斯 MIMO 信道容量。

对于如图 1 所示的均匀直线阵，设收发阵元数为 m，则各个收发阵元的坐标可以表示为 $xt_j=xt_1+\cos(\theta_{hT})(j-1)dt$，$zt_j=zt_1+\sin(\theta_{hT})(j-1)dt$，$xr_i=xr_1+\cos(\theta_{hR})(i-1)dr$，$zr_i=zr_1+\sin(\theta_{hR})(i-1)dr$，其中 $j=1:m$，$i=1:m$，(xr_1, zr_1)，(xt_1, zt_1) 分别为收发阵第一个阵元的坐标，dt, dr 分别为发射和接收阵元间距，为了计算方便这里假设 $yt_j=yr_j=0$。

考虑两种极端的情况，设收发阵元数为 m，当 RF=∞ 时，对于非衰落的直达信号，信道容量趋近于

$$C \approx \log_2(1+m\mathrm{SNR}) \qquad (12)$$

对于 RF=0 的完全独立的 Rayleigh 衰落信道，信道容量的均值趋近于

$$C_{\mathrm{Rayleigh}} \approx m\log_2(1+\mathrm{SNR}) \qquad (13)$$

对于其他的 RF，信道容量将介于这两者之间。

图 4~图 7 分别从不同的方面描述了 RF 对莱斯 MIMO 信道容量的影响。其中的仿真参数为 f=10kHz，$(xt_1, xr_1, zt_1, zr_1)=(-1000\mathrm{m}, 1000\mathrm{m}, 100\mathrm{m}, 100\mathrm{m})$，风速为 15kn，$\theta_{hT}=\theta_{hR}=30°$。由于信道空间相关性受阵元间距的影响较大，这里选阵元间距为 0.1λ 和 1λ 两种情况（$|\rho_{1,33}|$ 分别为 0.880 6 和 0.349 3），从图 4~图 6 中可以看出，收发阵元间距从 0.1λ 到 1λ 变化时，当 RF 较大时更接近于直达声线构成的确定信道，信道容量基本不发生变化，而当 RF 较小时更接近于随机的衰落信道，信道的空间相关性会发生较大变化，从而使信道容量随收发阵元数增加的斜率发生变化。从图 4 中还可以看出，当 dt=dr=1λ 时，对于 $C_{0.01}$，使 $C_{RF=1000}>C_{RF=0.001}$ 的 m 值为 6，而对于 $C_{0.5}$ 的 m 值为 2。

从式(12)和式(13)中可以看出，对于完全独立的瑞利 MIMO 信道容量的平均值总是大于莱斯 MIMO 信道容量，然而由于瑞利信道的随机衰落作用和空间相关性的影响，莱斯 MIMO 信道容量并不总是小于瑞利 MIMO 信道容量，对于这一点从图 4~图 7 中都可以看到。

图 5 中描述了 RF 对信道容量的 CCDF(Complementary Cumulative Distribution Function) 的影响，从图 5 中可以看到，当阵元间距从 0.1λ 到 1λ 变化时，CCDF 会向右平移，对于 m=4，平移的幅度约为 5bps/Hz，对于 m=8，平移的幅度约为 10bps/Hz；另外，当阵元间距等于 0.1λ 时，当 m=4 时，从 $C_{0.5}$ 开始，$C_{RF=1000} \geqslant C_{RF=0.001}$，当 m=8 时，从 $C_{0.1}$ 开始，$C_{RF=1000} \geqslant C_{RF=0.001}$，从图 6 中也可以看出，当阵元间距等于 0.1λ 时，对于 m=4，当 SNR<30dB 时 $C_{RF=1000} \geqslant C_{RF=0.001}$，对于 m=8，SNR<15dB 时，$C_{RF=1000} \geqslant C_{RF=0.001}$。

从图 7 中可以看出，莱斯 MIMO 信道容量并不总是随着 RF 的增加而减小，尤其是当收发阵元数较小和空间相关性较大时。当 dt=dr=1λ 时，对于 m=1、2，当 dt=dr=0.1λ 时，对于 m=1、2、4、8，$C_{0.01}$ 都随着 RF 的增加而增加。

由于空间相关性主要受到风速、收发阵元间距等因素影响，图 8~图 9 分别描述了这些参数对莱斯 MIMO 信道容量的影响(RF=1)。从图 8 和图 9 中可以看出，信道容量随风速和收发阵元间距的增加而增加，这是因为，空间相关性随风速和收发阵元间距的增加而减小，从图 9 中还可以看出对于图中的仿真参数，当收发阵元间距大于 0.5λ 时，信道容量将不再增加，因为当收发阵元间距大于 0.5λ 时，信道矩阵的各个元素之间基本上已经相互独立。另外，从图 2 和图 3 中还可以看到，空间相关性与收发阵方向、距离、深度等都有关系，因此，这些参数对 MIMO 信道容量产生影响，可以得到类似的结果，这里就不再重复。

5 结 论

本文以海面散射的水声 MIMO 信道为例，通过仿真研究了空间相关莱斯 MIMO 信道容量随莱斯因子、接收信噪比和空间相关性的变化情况。

当莱斯因子较大时，更接近于确定的 LOS 信道，信道容量的随机变化较小，当阵元数和接收信噪比较小时，莱斯 MIMO 信道容量将大于瑞利衰落的信道容量。

当莱斯因子较小时，更接近于随机的瑞利衰落信道，信道容量受空间相关性影响较大，当空间相关性较大时，MIMO 信道容量增益将大大减小，影响空间相关性的因素如阵元间距，海面风速，收发阵方向、距离、深度等都会对 MIMO 信道容量产生较大的影响。

参 考 文 献

[1] Winters J H, Salz J, Gitlin R D. The impact of antenna diversity on the capacity of wireless communication system[J]. IEEE Trans. Commun., 1994, 42(2, 3, 4): 1740-1751.

[2] Telatar I E. Capacity of Multi-Antenna Gaussian Channels

[R]. AT&T Bell Laboratories Murray Hill, NJ, Tech. Rep. # Bl0 112 170-950-615-07 TM, 1995.

[3] Foschini G J, Gans M J. On the limits of wireless communications in a fading environment when using multiple antennas[J]. Wireless Pers. Commun., 311-335, 1998, 6(3): 311-335.

[4] Zatman M, Tracey B. Underwater acoustic MIMO channel capacity[A]. Signals, systems and computers, conference record of the Thirty-Sixth Asilomar Conference[C]. 2002, 2. 1364-1368.

[5] Khalighi M A, Brossier J M, Fourdain G, Raoof K. On capacity of Ricean MIMO channels[J]. IEEE International Symposium on Personal, Indoor and Mobile Radio Communications, 2001, 1(10): A150-154.

[6] Smith P J, Garth L M. Exact capacity distribution for dual MIMO systems in Ricean Fading[J]. IEEE Communication letters, 2004, 8(1): 18-20.

[7] Jayaweera S K, Poor H V. On the capacity of multiple antenna systems in Rician fading[J]. IEEE Trans. wireless Commun., 2005, 4(3): 1102-1111.

[8] McKay M R, Collings I B. Capacity bounds for correlated Rician MIMO channels[J]. IEEE International Conference on Communications, 2005, 2(5): 772-776.

[9] McKay M R, Collings I B. General Capacity Bounds for Spatially Correlated Rician MIMO Channels[J]. IEEE Trans. on Information Theory, 2005, 51(9): 3121-3145.

[10] Eckart C. The scattering of sound from the sea surface [J]. J. Acoust. Soc. Am., 1953, 25: 566-670.

[11] McDaniel S T. Spatial covariance and adaptive beamforming of high frequency acoustic signals forward scattered from the sea-surface[J]. IEEE J. Ocean. Eng. 1991, 16(4): 415-419.

[12] Melton D R, Hortom C W Sr. Importance of the Fresnel correction in scattering from a rough surface-I: Phase and amplitude fluctuations[J]. J. Acoust. Soc. Am., 1970, 47: 295-303.

[13] Leon-Garcia A, Probability and Random Processes for Electrical Engineering[M]. Addison-Wesley, 1990.

时间反转技术对水声多输入多输出系统干扰抑制性能的研究①②

朴大志[1,2]，李启虎[2]，孙长瑜[2]

（1. 中国传媒大学 信息工程学院，北京 100024；2.中国科学院 声学研究所，北京 100080）

摘 要： 通过对 TRM（时间反转镜）MIMO 系统中码间干扰和同道干扰的分析，探讨了 TRM 技术在水声 MIMO 系统中的应用潜力。得到了一个通用的 TRM MIMO 系统的干扰构成表达式，对于典型的浅海水声信道，对 TRM MIMO 系统的干扰进行了计算；研究了 TRM 技术对 MIMO 系统自身的干扰抑制能力与收发阵形结构、信道结构以及发射信号形式等参数之间的关系。并将信干比的计算结果与室外无线信道 TRM MIMO 中的信干比结果进行了比较，在海底衰减系数较小的水声信道中有更丰富的多径，TRM 可以获得更好的空间聚焦性，从而水声 TRM MIMO 信道中的信干比比室外无线信道中更大。可初步得到结论：TRM 技术对水声 MIMO 系统干扰的抑制能力主要取决于多径的丰富程度，收、发阵元间距和时间反转的阵元数。

关键词： 时间反转；多输入多输出；水声通信；码间干扰

中图分类号： TN929.3　　**文献标识码：** A　　**文章编号：** 1000-436X(2008)07-0081-07

Study of the interference reducing capability of time reversal techniques in underwater acoustic MIMO systems

PIAO Da-zhi[1,2], LI Qi-hu[2], SUN Chang-yu[2]

(1. Information Engineering College, Communication University of China, Beijing 100024, China;

2. Institute of Acoustics, Chinese Academy of Sciences, Beijing 100080, China)

Abstract: The intersymbol interference (ISI) and co-channel interference in a general TRM (time reversal mirror) MIMO (multiple-input-multiple-output) system was analyzed, the expression of interference was derived, and the interference reducing effects in terms of array configuration, channel, and transmitted signal were studied through an image source model in a typical shallow water channel. The computation result of signal to interference ratio was compared with the signal to interference ratio in an outdoor wireless channel, it can see that there is more rich multipath in the shallow water channel with small seabed attenuation coefficient, so the TRM can get a better spatial focusing and a better interference reduction capability. The interference reducing capability of TRM in MIMO systems is mainly decided by the richness of multipath, the element spacing and the number of TRM element.

Key words: time reversal; MIMO; underwater acoustic communications; intersymbol interference

1 引言

时间反转(time reversal)也称为频域内的相位共轭，是基于静态媒质中波动方程的空间互易性和时间对称性，它在强混响环境中具有很好的空间聚焦性和脉冲压缩特性。时间反转的空间聚焦性是指它可以在不管空间传播有多复杂的介质内，对时间反转后的入射声场在原始 probe 信号的位置重新聚

① 基金项目：国家自然科学基金重点资助项目(60532040).
Foundation Item: The National Natural Science Foundation of China (60532040).

② 通信学报, 2008, 29(7): 81-87.

焦。在高速水声通信中，信道的时间扩展效应会带来严重的码间干扰，因此很多水声通信研究者对时间反转技术产生了兴趣。他们对水声信道中时间反转的空时聚焦性、统计稳定性，以及在SISO(single-input-single-output)水声通信系统中与其他均衡技术的性能比较等方面做了很多理论和实验工作[1~10]。另外水声信道有限的带宽和丰富的多径传播使MIMO技术在提高水声通信的速率方面存在很大应用潜力，时间反转技术的空间聚焦性为MIMO系统中具有不同空间位置的多个信道同时进行通信提供了可能，从而可以大大缩小空时信号处理的计算量。时间反转与MIMO技术结合具有以下优势：1）时间反转的空间聚焦性对MIMO系统中同道干扰的抑制；2）时间反转的时间聚焦性对码间干扰的抑制；3）与传统的信道均衡技术相比时间反转技术只需很低的计算量，易于实现。文献[11]中进行了相干MIMO时间反转通信实验，验证了在深度为110m，距离9km的浅海通过时间反转技术在不同的深度实现2~3个信道同时进行通信的可能性，文献[12]中利用户外测量数据研究了时间反转技术应用于8×4无线(radio)MIMO系统中的可行性，通过信道RMS(root mean square)延时扩展和瞬时信干比(用 R_{SI} 表示)研究了时间反转技术对码间干扰和同道干扰的抑制能力。本文的主要目的是讨论水声MIMO信道中应用时间反转技术的潜力，首先对一般的TRM MIMO系统自身的干扰构成进行分析，然后研究在典型的浅海水声信道中，TRM MIMO系统的性能与一些重要的系统参数如收、发阵元数目、阵元间距、通信距离，系统带宽等之间的关系。这里不考虑海洋的动态效应，认为信道前向传播和时间反转的后向传播之间是准静态的。

2 时间反转MIMO系统性能分析

2.1 系统构成

对于主动和被动时间反转水声通信系统，本质上它们的性能都可以通过图1所示的系统结构进行分析。其中 $\{s_1\cdots S_{n_R}\}$ 为 n_R 个信源，$\{T_1\cdots T_{n_T}\}$ 为 n_T 个发射阵元，$\{R_1,\cdots,R_{n_R}\}$ 为 n_R 个接收阵元，$h_{j,i}$ 为第 i (i=1: n_T)个发射阵元和第 j(j=1:n_R)个接收阵元之间的信道响应。这里假设发射阵元数 n_T 等于时间反转阵 n_{TR} 元数，接收阵元数 n_R 等于通信信道数 n_{CH}。

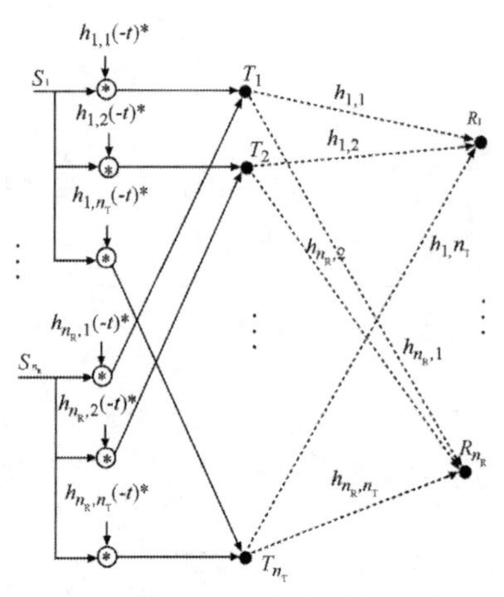

图1 时间反转MIMO通信系统性能分析

2.2 一般时间反转MIMO系统干扰构成

对应图1所示的时间反转MIMO通信系统，第 j 个接收阵元的信号 $r_j(t)$ 可以表示为

$$r_j(t)=\sum_n S_{jn}H_{j,j}(t-nT)+\sum_{m=1,m\neq j}^{n_R}\sum_n S_{mn}H_{j,m}(t-nT)+n_j(t) \tag{1}$$

其中，s_n 为发送的信息序列，T 为码元宽度，$H_{j,i}(t)$ 为第 i 个信源 s_i 与第 j 个接收阵元间的信道响应，n_j 为第 j 个接收阵元的加性噪声，这里为了分析时间反转MIMO系统的性能与阵和信道参数的关系，不考虑加性噪声的影响。对 $r_j(t)$ 在 $t=kT$ 时刻采样可以得到

$$r_j(kT)=S_{jk}H_{j,j}(0)+\underbrace{\sum_{n\neq k}S_{jn}H_{j,j}((k-n)T)}_{\text{ISI}}+\underbrace{\sum_{m=1,m\neq j}^{n_R}\sum_n S_{mn}H_{j,m}((k-n)T)}_{\text{CI}} \tag{2}$$

由于时间反转技术具有空间聚焦性，第 j 个接收阵元的信号 r_j 主要由第 j 个发射信源 S_j 的信号组成，因此可以把第 j 个接收阵元的信号作为对第 j 个发射信源进行判决的基础，这样总的干扰可以表示为

$$I=\frac{1}{H_{j,j}(0)}\left(\sum_n S_{j,n}H_{j,j}(nT)+\sum_{m=1,m\neq j}^{n_R}\sum_n S_{mn}H_{j,m}(nT)\right) \tag{3}$$

可以看到，对第 j 个接收阵元在第 k 时刻采样值的干扰由2部分组成，一部分来自于同一个信源 S_j 在 $n\neq k$ 时刻的采样值，称为码间干扰(ISI)，另一部分来自信源 $s_i(j\neq i)$，称为同道干扰(co-channel interference)，当通信信道数为 1 时，只有码间干扰。当通信信道数大于 1 时，对各个信道中的干扰分别计算后取均值。

假设序列$\{s_n\}$不相关，信源 s_i 与 s_j ($i \neq j$)互不相关，并具有相同的发射功率，$E[S_n^2]=\sigma_s^2$，则

$$E[I^2]=\frac{1}{\left|H_{j,j}(0)\right|^2}\left(\sum_{n\neq 0}\left|H_{j,j}(nT)\right|^2+\sum_{m=1,m\neq j}^{n_R}\sum_{n}\left|H_{j,m}(nT)\right|^2\right)\sigma_s^2 \tag{4}$$

考虑基带传输系统，则

$$H_{j,i}(t)=\sum_{k=1}^{n_T} g(t)\otimes h_{j,k}(-t)^*\otimes h_{i,k}(t) \tag{5}$$

其中，$g(t)$为基带信号，本文选用升余弦脉冲。从式(4)、式(5)可以看出，在时间反转 MIMO 系统中，对干扰的抑制能力取决于时间反转的阵元数 n_{TR}，同时通信的信道数 n_{CH}，基带信号的带宽 B_W，以及信道响应 h。脉冲展宽和旁瓣泄露都会造成码间干扰。

对于离散多径信道，假设每条多径的幅度为 a_i，延时为 τ_i，则信道响应可以表示为

$$h(t)=\sum_i a_i\delta(t-\tau_i) \tag{6}$$

单个阵元时间反转后

$$\begin{aligned}h(t)\otimes h(-t)^* &= \sum_i a_i\delta(t-\tau_i)\otimes\sum_j a_j^*\delta(t+\tau_j)\\ &=\sum_i |a_i|^2\delta(t)+\sum_{i,i\neq j}\sum_j a_i a_j^*\delta(t-\tau_i+\tau_j)\end{aligned} \tag{7}$$

2.3 水声时间反转 MIMO 系统性能分析

下面对时间反转 MIMO 技术在典型的浅海水声信道中的性能进行一些探讨。利用虚源理论[13,14]，在水声信道中，第 i 个发射阵元和第 j 个接收阵元之间的信道响应可以写成

$$\begin{aligned}h_{j,i}(w)=\sum_{n=0}^{\infty}(-1)^n\Bigg\{&\left[R_B(\theta_{n1}{}^{j,i})\right]^n\frac{\exp(imkR_{n1}{}^{j,i})}{R_{n1}{}^{j,i}}+\\ &\left[R_B(\theta_{n2}{}^{j,i})\right]^{n+1}\frac{\exp(imkR_{n2}{}^{j,i})}{R_{n2}{}^{j,i}}-\\ &\left[R_B(\theta_{n3}{}^{j,i})\right]^{n}\frac{\exp(imkR_{n3}{}^{j,i})}{R_{n3}{}^{j,i}}-\\ &\left[R_B(\theta_{n4}{}^{j,i})\right]^{n+1}\frac{\exp(imkR_{n4}{}^{j,i})}{R_{n4}}\Bigg\}\end{aligned} \tag{8}$$

其中，$k=w/c$ 为波数，这里假设恒定声速，c=1500m/s，R_{n1}，R_{n2}，R_{n3}，R_{n4}，的表达式可以参考文献[13]，将它写成时域的形式

$$\begin{aligned}h_{j,i}(t)=\sum_{n=0}^{\infty}(-1)^n\Bigg\{&\left[R_B(\theta_{n1}{}^{j,i})\right]^n\frac{\delta(t-R_{n1}{}^{j,i}/c)}{R_{n1}{}^{j,i}}+\\ &\left[R_B(\theta_{n2}{}^{j,i})\right]^{n+1}\frac{\delta(t-R_{n2}{}^{j,i}/c)}{R_{n2}{}^{j,i}}-\\ &\left[R_B(\theta_{n3}{}^{j,i})\right]^{n}\frac{\delta(t-R_{n3}{}^{j,i}/c)}{R_{n3}{}^{j,i}}-\\ &\left[R_B(\theta_{n4}{}^{j,i})\right]^{n+1}\frac{\delta(t-R_{n4}{}^{j,i}/c)}{R_{n4}}\Bigg\}\end{aligned} \tag{9}$$

海底反射系数一般可假设为指数衰减 $R_B(\theta)=\exp(-v\theta)$，$\theta$ 为声线相对于海底的掠射角。利用式(4)、式(5)、式(7)、式(9)，对水声时间反转 MIMO 系统的干扰抑制能力进行了一些数值计算。

3 数值计算结果及简要分析

3.1 关于信干比

首先，为了对时间反转技术的时间压缩和空间聚焦性进行一些感性的认识，对时间反转后的信道冲击响应在图 2 中进行了描述。在式(5)中，当 $g(t)=\delta(t)$时，就可以得到时间反转的信道冲击响应，在图 2 中，以 10 个发射阵元，2 个接收阵元的时间反转 MIMO 系统为例，对海底衰减为 0 (v=0)和不为 0 (v=5) (取多径数 40)的 2 种情况分别进行了描述。从图 2 中可以看到当 v=0 和 v=5 时，第 1 个信源和第一个接收阵元之间的信道响应 $H_{1,1}$ 接近于 $\delta(t)$，说明时间反转技术具有很好的时间压缩效果，而且与 $H_{1,2}$ 相比具有较高的峰值信干比(v=0 和 v=5 时的 R_{SI} 分别为 33.017 9dB 和 31.493 2dB)，因此在时间反转 MIMO 系统的接收端，可以把第一个接收阵元的信号看成对第一个信源进行判决的依据。

由于 R_{SI} 与信道中的多径数密切相关，为了与文献[12]中的结果进行比较，对 8 个发射阵元 2 个接收阵元的 TRM MIMO 系统的 R_{SI} 进行了计算，将 v=0 和 v=5 时，不同多径数目下 R_{SI} 的计算结果列于表 1。从表 1 中可以看出随着多径数目的增加，时间反转系统的同道干扰抑制能力逐渐增强，但增长趋势对于海底无衰减和有衰减的情况略有不同。从表 1(a)中可以看到，对于无衰减海底，随着多径数目的增加，R_{SI} 将一直增加，这是因为海底对每条多径的衰减为 0，经过多次海底反射，仍然没有能量损失；对于有衰减海底，在表 1(b)中，随着多径数目的增加，R_{SI} 将逐渐趋于饱和，当多径数目大于 20

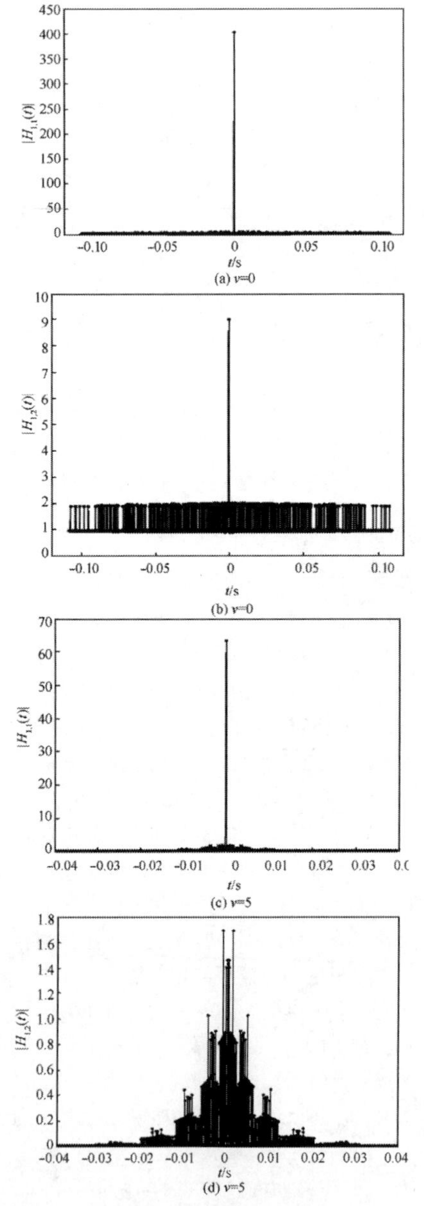

图2 时间反转信道冲击响应

(z_{t0}=5m, z_{r0}=5m, d_t=2m, d_r=2m, n_R=2, n_{TR}=10, Dw=50m, r_{r0}=3km)

以后，R_{SI}基本上保持不变。这是因为海底具有反射损失，当声线经过多次海底反射后能量逐渐衰减，且这些衰减与它们相对于海底的掠射角有关，掠射角越大衰减越大，因此对干扰抑制的贡献以具有小掠射角的那些声线为主。在文献[12]中当8个时间反转阵元，2个接收阵元，距离为300m时，R_{SI}为18dB，这与表1(a)中，多径数目为8的情况相当，若考虑在表1(b)的情况，则多径数目小于8。另外在表1的计算中，2个接收阵元的间距为2m，而文献[12]中当接收阵元间距为2m时，R_{SI}为15.6dB，因此可以这样认为，在海面海底衰减较小的浅海水声信道中具有比室外无线信道更多的多径，从而在时间反转MIMO系统中可以获得更好的干扰抑制能力。另外，由于时间反转阵元数对R_{SI}也有重要的影响，在图3中描述了不同多径数目情况下R_{SI}与TRM阵元数的关系，从图3中可以看到，当海底的衰减为0时，R_{SI}随着多径数目的增加而增加，不需要很多的阵元就可以获得较高的R_{SI}，而当海底衰减不为0时，增加阵元数会获得更高的R_{SI}。

表1　R_{SI}与多径数目的关系

(z_{t0}=5m, z_{r0}=5m, d_t=2m, d_r=2m, n_{TR}=8, n_{CH}=2, Dw=50m, r_{r0}=1km)

(a) v=0

多径数目	4	8	12	16	20	24	28	32	36	40
R_{SI}/dB	12.080	18.523	22.390	25.279	27.662	29.736	31.599	33.304	34.883	36.357

(b) v=5

多径数目	4	8	12	16	20	24	28	32	36	40
R_{SI}/dB	17.008	27.606	27.684	27.685	27.685	27.685	27.685	27.685	27.685	27.685

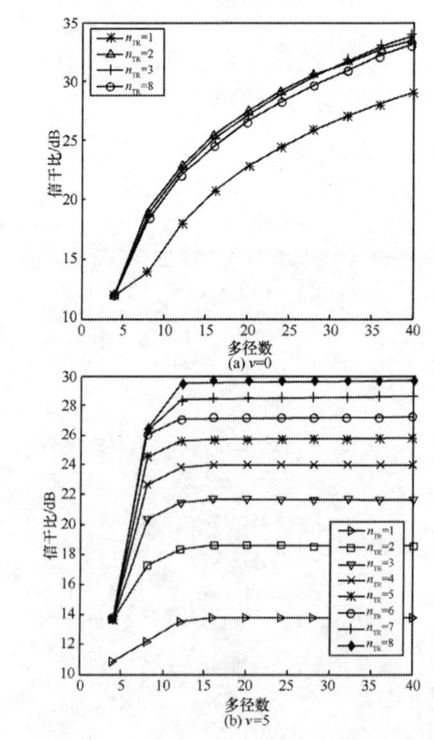

图3 信干比与时间反转阵元数的关系

(z_{t0}=5m, z_{r0}=5m, r_{t0}=0, n_R=2, d_t=2m, d_r=2m, D_w=50m, r_{r0}=3km)

3.2 时间反转 MIMO 系统中干扰的数值计算结果

由式(4)、式(5)、式(7)、式(9)中可以看到，时间反转 MIMO 系统中对干扰的抑制能力取决于一些与发射和接收阵有关的参数如发射、接收阵元数目、间距和方向，收发阵距离；与信道有关的参数如水深和海底反射系数；与信号有关的参数如带宽(码元宽度)、升余弦滚降系数等，下面对这些参数对时间反转 MIMO 系统中的干扰抑制能力的影响分别进行了研究。参数含义：n_{TR} 时间反转阵元数，n_T 发射阵元数，n_{CH} 通信信道数，n_R 接收阵元数(这里 $n_{TR}=n_T$，$n_{CH}=n_R$)，n_M 多径数，d_t 发射阵元间距，d_r 接收阵元间距，D 水深，R 通信距离，z_{t0} 第一个发射阵元的深度，z_{r0} 第一个接收阵元的深度，B_W 信号带宽，T 码元宽度，α 升余弦滚降系数，对于升余弦信号 $BW=(1+\alpha)/2T$。

3.3 数值结果简要分析

在图 4(a)中描述了通信信道数为 2 时，时间反转阵元数对 TRM MIMO 系统干扰抑制能力的影响。从图 4 中可以看到，码间干扰和同道干扰都随着时间反转阵元数目的增加而减小，只是时间反转阵元数对 ISI 的影响更大一些，当时间反转阵元数从 1 增加到 20 时，ISI 会下降约 10dB, CI 下降了约 4dB，而且当阵元数大于 10 以后，CI 已经基本上保持不变。另外，从图 4(a)中还可以看到，随着时间反转阵元数的增加，ISI 在系统总干扰 I 中的贡献逐渐减弱，时间反转 MIMO 系统的干扰主要来自于同道干扰。在图 4(b)中描述了不同通信信道数时，时间反转 MIMO 系统中的总干扰 I 随时间反转阵元数的变化，从图 4(b)中可以看到，总干扰 I 会随着时间反转阵元数的增加而减小，随着通信信道数的增加而增加，当通信信道数为 1 时，时间反转阵元数从 1 增加到 20 可以使总干扰降低约 10dB，而当通信信道数为 5 时，只能使总干扰降低约 3dB，从而我们可以看到随着通信信道数的增加，会带来同道干扰的显著增加，通过时间反转虽然可以对同道干扰进行抑制，但对于通信信道数较大时还是会存在一定的同道干扰。

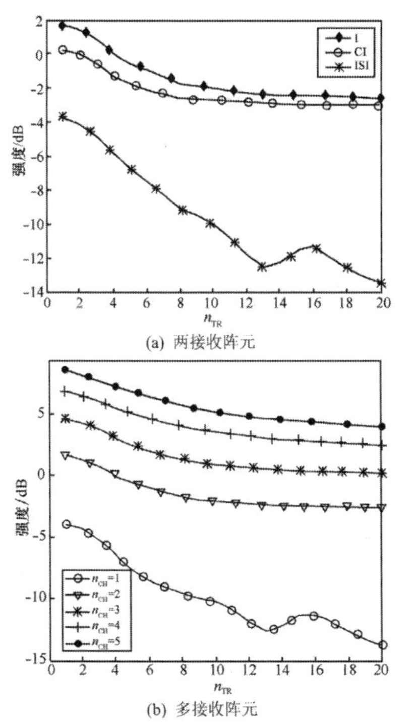

(a) 两接收阵元

(b) 多接收阵元

图 4 干扰与时间反转阵元数目之间的关系(B_W=4kHz, z_{t0}=5m, z_{r0}=5m, r_{t0}=0, α=0, d_t=2m, d_r=2m, D_W=50m, R=8km, v=5, T=0.125ms, n_M=20)

在图 5 和图 6 中对时间反转阵元数为 3，通信信道数为 2 的 MIMO 系统中的总干扰，码间干扰和同道干扰随发射和接收阵元间距的变化情况分别进行了计算。从图 5 可以出，ISI 和 CI 都随着发射端阵元间距的增加而减小，当阵元间距足够大时，ISI 和 CI 的减小会达到饱和，在图 5 中的计算参数情况下，当 d_t> 4m 以后，ISI 和 CI 的变化就很小了。从图 6 中可以看出，CI 随着接收阵元间距的增加而减小，当接收阵元间距 d_r> 3m 以后，CI 的减小就变得很缓慢。另外从图 5 和 6 中还可以看到，发射阵元间距和接收阵元间距的增加都会使 CI 减小，但它们对 CI 的影响略有不同，接收阵元间距对 CI 的影响更大一些，当接收阵元靠得很近时彼此会造成很大的同道干扰，而随着接收阵元间距的增加，CI 会迅速下降，这与时间反转系统的空间聚焦能力有关，空间分辨力越强，同道干扰随接收阵元间距的增加而降低的速度就越快。

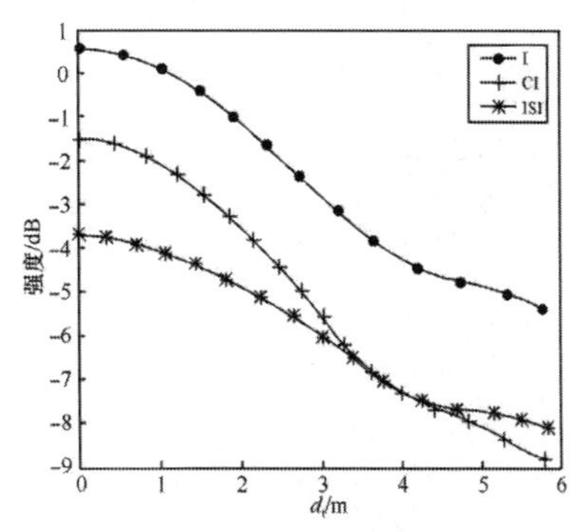

图 5 干扰与发射阵元间距的关系(B_W=4kHz, n_{TR}=3, n_{CH}=2, α=0, d_r=3m, D_W=50m, T=0.125ms, z_{t0}=5m, z_{r0}=5m, R=8km, v=5, n_M=20)

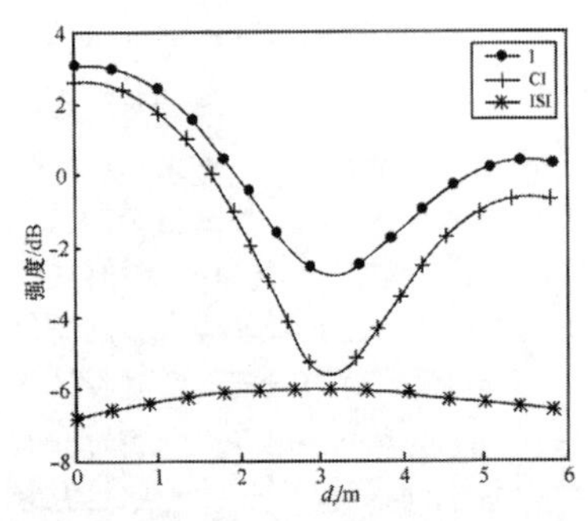

图6 干扰与接收阵元间距的关系(B_W=4kHz, r_{t0}=0m, n_{TR}=3, n_{CH}=2, α=0, d_t=3m, D_W=50m, z_{t0}=5m，z_{r0}=5m，R=8km, T=0.125ms ,v=5，n_M=20)

在图7中描述了码元宽度T对TRM MIMO系统干扰抑制能力的影响。从图7中可以看到，CI和ISI随着码元宽度的增加都会有所增加，在TRM MIMO系统中，随着信号带宽的增加对干扰的抑制能力会有所增强。在图8和图9中分别描述了TRM MIMO系统自身的干扰与海底衰减系数和通信距离的关系。从图8中可以看到，随着海底衰减系数的增加，码间干扰会逐渐减小，同道干扰首先略有减少然后逐渐增加，当海底衰减较大时系统的总干扰将以同道干扰为主。当v=0时，海底的衰减为零，这时时间反转MIMO系统的干扰抑制能力最强，随着海底衰减的增加，干扰会有所增加，海底衰减具有孔径阴影效应从而会使时间反转阵的有效孔径降低。但是当$v>10$以后，系统的干扰的增加已经趋于饱和。从图9中可以看到，随着通信距离的增加，CI和ISI略有增加，这是因为随着距离的增加，信道中有效的多径数会逐渐减少，因此时间反转系统的垂直和水平空间聚焦能力是随着距离的增加而降低。

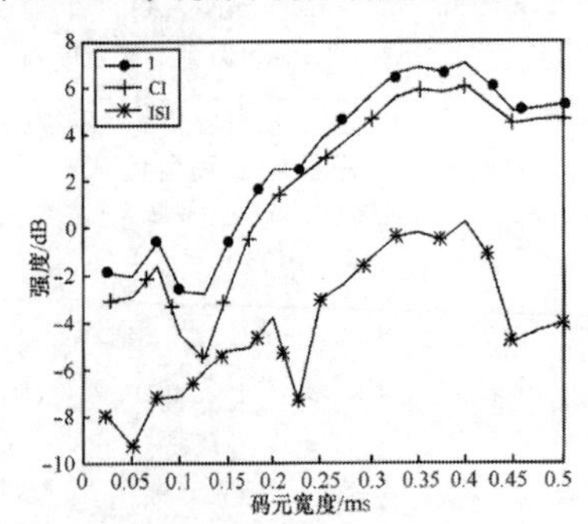

图7 干扰与码元宽度的关系(n_M=20, n_{CH}=2, n_{TR}=3, v=5, z_{t0}=5m, z_{r0}=5m, r_{t0}=0m, α=0, d_t=3m, d_r=3m, D_W=50m, R=8km)

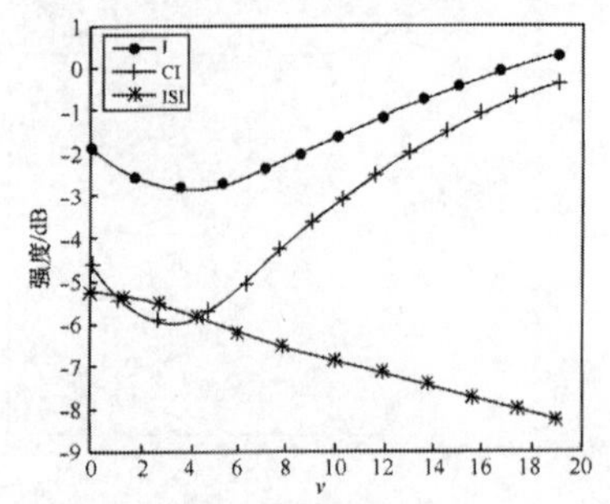

图8 干扰与海底衰减系数的关系(r_{t0}=0m, n_{TR}=3, n_{CH}=2, α=0, d_t=3m, d_r=3m, z_{t0}=5m, T=0.125ms, z_{r0}=5m, R=8km, D_W=50, n_M=20)

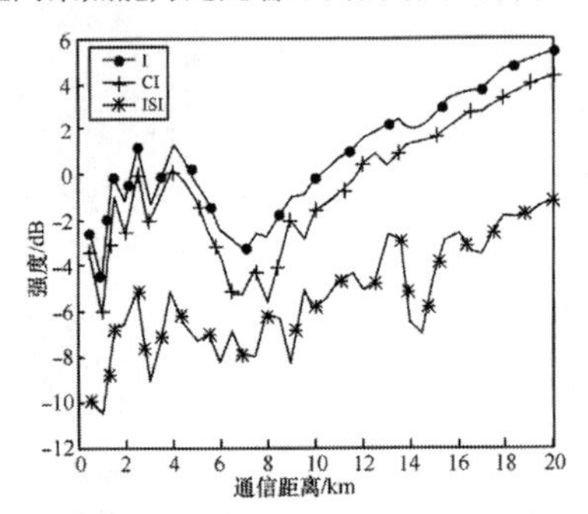

图9 干扰与通信距离的关系(r_{t0}=0m, α=0, d_t=3m, d_r=3m, n_{TR}=3, n_{CH}=2, D_W=50m, z_{t0}=5m, z_{r0}=5m, T=0.125ms, v=5, n_M=20)

图4～图9中考虑的是发射与接收阵均为垂直阵的情况，为了与水平收、发阵的情况相比较，在图10中以干扰随收、发阵元间距的变化为例，描述了收发均为水平阵时，时间反转MIMO系统对干扰的抑制能力。从图10中可见，对于水平收发阵，要

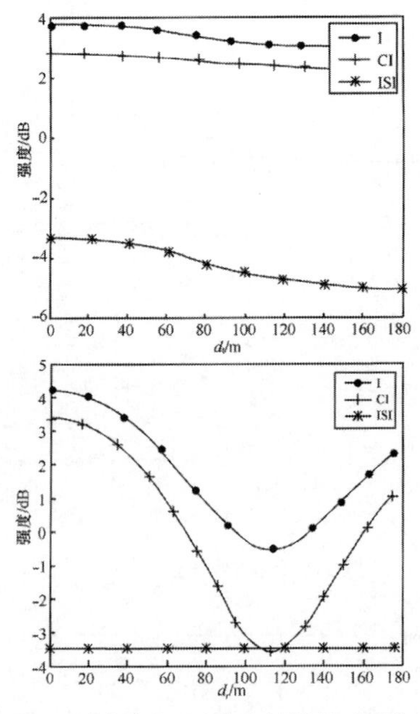

图10 干扰与阵元间距的关系(水平发射、接收阵)
(B_W=4kHz, n_{TR}=3, n_{CH}=2, α=0, D_B=50m, z_{t0}=25m, z_{r0}=25m, R=8km, v=5)

达到与垂直收发阵相同的干扰抑制能力，需要更大的阵元间距，比如发射和接收端同时用水平阵，阵元间距要 30～40m 才能达到垂直阵间距为 2～3m 的干扰抑制能力，这是由于浅海水声信道在垂直方向的空间变化性远大于在水平方向的变化，因此时间反转后的信道在垂直方向上的空间分辨力会远大于水平方向上的空间分辨力。

4 结束语

本文首先对 TRM MIMO 系统自身的干扰构成情况进行了分析，然后以浅海水平信道为例通过虚源模型对干扰随信道参数、阵参数和信号参数的变化进行了计算和简要分析，可以得到以下结论：

1) 通过对典型浅海信道中时间反转 MIMO 系统的 SIR 进行了计算并与室外无线信道中 SIR 的结果进行比较，可以看到在海底衰减较小的浅海信道中具有比室外无线信道更丰富的多径，从而时间反转系统可以获得更好的空间聚焦性能和同道干扰抑制能力。

2) 增加时间反转阵元数可以使时间反转 MIMO 系统的干扰抑制能力增强，随着多径丰富程度的增加，达到相同的干扰抑制能力需要的时间反转阵元数逐渐减少，当通信信道数较多时，增加时间反转阵元数也可以降低 MIMO 系统的总干扰，但还是会有一些残留的同道干扰，需要用其他技术来解决。

3) 增加收、发阵元间距会使时间反转 MIMO 系统中的干扰抑制能力明显增强，其中接收阵元间距对同道干扰的影响更大一些。

4) 水深和收发阵元深度对时间反转 MIMO 系统中的干扰抑制性能没有太大影响，通信距离的增加会使干扰抑制能力略有下降，因此时间反转技术在 MIMO 系统中的干扰抑制性能具有广泛的环境适应能力。

5) 时间反转 MIMO 系统的干扰抑制能力会随信号带宽的增加而降低。

综合以上几点，时间反转 MIMO 系统的干扰抑制能力主要由信道的多径丰富程度决定，多径越丰富获得相同的干扰抑制能力需要的时间反转阵元数和收、发阵孔径越小。

参考文献：

[1] DERODE A, TOURIN A, ROSNY J, *et al*. Taking advantage of multiple scattering to communicate with time-reversal antennas [J]. Phys Rev lett, 2003, 90(1):014301-014304.

[2] BLOMGREN P, PAPANICOLAOU G, ZHAO H. Super-resolution in time-reversal acoustics[J]. J Acoustic Soc Am, 2002, 111(1): 230-248.

[3] KUPERMAN W A, HODGKISS W S, SONG H C, *et al*. Phase conjugation in the ocean: experimental demonstration of an acoustic time-reversal mirror [J]. J Acoustic Soc Am, 1998, 103(1): 25-40.

[4] HODGKISS W S, SONG H C, KUPERMAN W A, *et al*. A long-range and variable focus phase–conjugation experiment in shallow water [J]. J Acoustic Soc Am, 1999,105(3):1597-1604.

[5] EDELMANN G F , AKAL T, HODGKISS W S, *et al*. An initial demonstration of underwater acoustic communication using time reversal[J]. IEEE J Ocean Eng, 2002, 27(3): 602-609.

[6] KIM S, EDELMANN G F, KUPERMAN W A, *et al*. Spatial resolution of time-reversal arrays in shallow water [J]. J Acoustic Soc Am, 2001,110(2):820-829.

[7] YANG T C. Temporal resolutions of time-reversal and passive-phase conjugation for underwater acoustic communications[J]. IEEE J Ocean Eng, 2003,28(2):229-245.

[8] STOJANOVIC M. Spatio-Temporal Focusing for Elimination of Multipath Effects in High Rate Acoustic Communications[M]. High Frequency Ocean Acoustics, AIP, New York,2004.

[9] STOJANOVIC M. Retrofocusing techniques for high rate acoustic communications [J]. J Acoustic Soc Am, 2005, 117(3):1173-1185.

[10] ROUSEFF D. Intersymbol interference in underwater acoustic communications using time-reversal signal processing [J]. J Acoustic Soc Am, 2005,117(2):780-788.

[11] SONG H C, HODGKISS W S , ROUX P, *et al*. Coherent MIMO time reversal communications in shallow water[A]. Proceedings of MTS/IEEE OCEANS'04[C]. Kobe, Japan, 2004. 2225-2229.

[12] NGUYEN H T, ANDERSEN J B, PEDERSEN G F. The potential use of time reversal techniques in multiple element antenna systems [J]. IEEE Comm Letters, 2005,9(1):40-42.

[13] JENSEN F B, AND KUPERMAN W A, POTER M B, *et al*. Computational Ocean Acoustics[M]. New York, American Institute of Physics, 1994.

[14] BREKHOVSKIKH L M, LYSANOV Y P. Fundamentals of Ocean Acoustic[M]. New York, Springer-Verlag, 2003.

Fast Broadband Beamforming Using Nonuniform Fast Fourier Transform for Underwater Real-Time 3-D Acoustical Imaging[①]

Cheng Chi, Zhaohui Li, and Qihu Li

***Abstract*—A broadband 2-D array is necessary for the application of underwater real-time 3-D high-resolution acoustical imaging. However, the number of sensors and beams is so large that the computational load is overly high for generating a 3-D image in real time, when using the conventional broadband beamforming methods. Therefore, a fast broadband beamforming method is needed. In this paper, a fast broadband frequency-domain beamforming method based on nonuniform fast Fourier transform is proposed for underwater real-time 3-D acoustical imaging. The computational load of the proposed method is one or two orders of magnitude lower than that by the conventional frequency-domain direct method. Compared with the conventional time-domain delay-and-sum method, the computational load required by the proposed method will be lowered by approximately three orders of magnitude. Moreover, unlike the chirp zeta transform (CZT) beamforming method which is only suitable for equispaced 2-D arrays or sparse arrays thinned from equispaced 2-D arrays, the proposed method can be applied to arbitrary 2-D arrays. Even for the arrays available for the CZT method, the computational load needed by using the proposed method is still lower than that needed by using the CZT method in most cases.**

***Index Terms*—Arbitrary 2-D arrays, broadband frequency-domain beamforming, computational load evaluation, near-field imaging, nonuniform fast Fourier transform (NUFFT), 3-D acoustical imaging.**

I. Introduction

With the growing demand for exploitation of subsea resources, underwater investigation is becoming more and more important. A real-time 3-D acoustical imaging system can generate a 3-D oceanic environment image beyond the optical visibility range in a very short time [1]. Thus the imaging system plays an important role in underwater investigation. Real-time imaging means that the system has to generate a 3-D image as fast as possible. The high-resolution requirement determines that the system should be broadband and the 2-D aperture should be large enough for 3-D imaging. However, the computational load and hardware cost, two critical issues associated with the huge number of sensors, impede the development of 3-D acoustical imaging systems [2].

Currently, the digital beamforming is a crucial technique for underwater real-time 3-D acoustical imaging [2]–[5]. The hardware cost and the computational load of the digital beamforming imaging can be mitigated by techniques such as sparse array and fast beamforming techniques. The sparse array techniques have been focused on for decades, among which some sparse arrays thinned from equispaced 2-D arrays have been explored by the simulated annealing algorithm [6], sparse periodic layouts [7], and other approaches. In addition, it is noted that when arbitrarily distributed sensors are employed for designing sparse 2-D arrays, a better performance of beam pattern and a smaller number of sensors can be obtained compared with the case of adopting sparse arrays thinned from equispaced 2-D arrays [8]–[10].

Although the sparse array techniques are useful for reducing the hardware cost, it is hard to significantly reduce the computational load. Thus, fast beamforming techniques are needed. Because the implementation of digital beamforming in the time domain is rarely practical for underwater 3-D acoustical imaging [2], [11], the beamforming methods [2], [12], [13] in the frequency domain become attractive alternatives and have been extensively investigated. These frequency-domain beamforming methods are described as follows.

As a straightforward approach, Dhanantwari *et al.* [17] have extended the conventional frequency-domain direct method (DM) beamforming [12] to 2-D arrays. But Palmese and Trucco showed that although the computational load of the DM is about one order of magnitude lower than that of the conventional time-domain delay-and-sum (D&S) beamforming, it is still too high for real-time 3-D acoustical imaging [11].

Zero-padded spatial fast Fourier transform (FFT) beamforming has also been investigated [12], [15], allowing a significant reduction in the computational load. However, this method is a narrowband technique in nature. The angular errors will increase with the decrease of the zero-padded factor. The large zero-padded factor which can guarantee the achievement of very low errors will lead to the overly high computational load for underwater real-time 3-D imaging.

Chen *et al.* give an optimized algorithm based on the delay approximation they proposed [16], by which the memory required is reduced dramatically compared with that of the conventional frequency-domain DM, but a remarkably mitigated computational load cannot be achieved.

① IEEE Journal of Oceanic Engineering, 2016, 41(2): 249-261.

Maranda [12] has demonstrated that the chirp zeta transform (CZT) beamforming is an efficient and accurate frequency-domain method which can be applied to broadband signal processing. Palmese and Trucco have extended the CZT beamforming to 2-D arrays in both the far-field [18] and near-field conditions [11], [19], and proposed the pruned CZT beamforming to further reduce the computational load [20]. However, both the CZT and the pruned CZT beamforming methods are only suitable for equispaced 2-D arrays or sparse arrays thinned from such 2-D arrays. If other kinds of 2-D arrays such as spiral [10] and circular 2-D arrays [21]are employed, the beamforming methods based on CZT will not work.

In this paper, a fast broadband beamforming method based on nonuniform fast Fourier transform (NUFFT) [23], [24] for underwater real-time 3-D acoustical imaging is proposed to reduce the computational load. The proposed method has no special requirements for the distribution of sensors, so it can be applied to arbitrary 2-D arrays. The computational load of the proposed method is noticeably smaller than that of the conventional frequency-domain DM and the time-domain D&S method [13]. Even when an equispaced 2-D array or a sparse array thinned from the equispaced 2-D array is employed, the computational load of the proposed method is still lower than that of the CZT beamforming in most cases. In addition, although the proposed beamforming method is presented for underwater real-time 3-D acoustical imaging, it also has significant applications in other fields such as medical ultrasound imaging and nondestructive testing.

The paper is organized as follows. The conventional DM beamforming in the frequency domain and the delay approximation are described in Section II. Section III introduces the basic concept of the nonuniform discrete Fourier transform and its fast computation algorithm, NUFFT. Then, the proposed beamforming method based on NUFFT is given in Section IV, and the beamforming errors caused by NUFFT are evaluated in Section V. In Section VI, the computational loads of the proposed method, the conventional time-domain D&S method, the frequency-domain DM, and the CZT method are analyzed and expressed by formulas in different situations. Comparison and discussion are shown in Section VII. Finally, conclusions are drawn in Section VIII.

II. Direct Method for Frequency-Domain Beamforming

In this paper, the unit vector of the steering direction $\hat{\mathbf{u}}$ depicted in Fig. 1 [9], [16] is employed, and it can be presented as

$$\hat{\mathbf{u}} = (\sin\theta_a, \sin\theta_e, \sqrt{\cos^2\theta_a - \sin^2\theta_e}) \tag{1}$$

where θ_a and θ_e are the azimuth and elevation angles preferred in this paper, which is different from the conventional definition of the azimuth and elevation angles in [22]. θ_a is the angle between the vector $\hat{\mathbf{u}}$ and its projection on the plane yz. θ_e is the angle between the vector $\hat{\mathbf{u}}$ and its projection on the plane xz.

As shown in Fig. 1, considering a 2-D array with a rectangular aperture and M_0 sensors arbitrarily distributed (i.e., a 2-D arbitrary array) on the plane of $z = 0$, $(x_m, y_m, 0)$ is the coordinate of the mth sensor for $1 \le m \le M_0$. Assume that

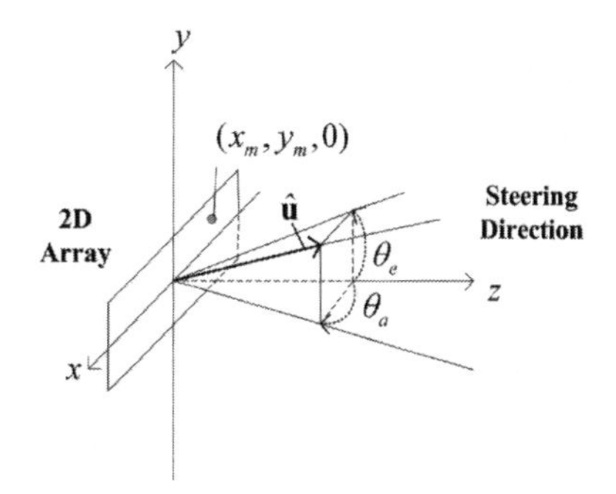

Fig. 1. Geometry for a 2-D array and the unit vector of the steering direction $\hat{\mathbf{u}}$.

$N_b \times M_b$ beams need to be generated and beam signals are indexed by (p, q) with $-N_b/2 \le p \le N_b/2 - 1$ and $-M_b/2 \le q \le M_b/2 - 1$. For the convenience of description and algorithm implementation, N_b and M_b are assumed to be even. Thus, the beam signal of the steering direction $(\theta_{\text{ap}}, \theta_{\text{eq}})$, focused at a distance r_0, can be given by [2], [16]

$$b(r_0, t, \theta_{\text{ap}}, \theta_{\text{eq}}) = \sum_{m=1}^{M_0} w_m s_m\left(t - \tau(r_0, m, \theta_{\text{ap}}, \theta_{\text{eq}})\right), \quad \begin{pmatrix} -N_b/2 \le p \le N_b/2 - 1 \\ -M_b/2 \le q \le M_b/2 - 1 \end{pmatrix} \tag{2}$$

where $s_m(t)$ is the backscattered signal collected from the mth sensor, w_m is the weighting value of the mth sensor for controlling the sidelobe level of the 2-D array, and $\tau(r_0, m, \theta_{\text{ap}}, \theta_{\text{eq}})$ represent the delays required to steer the beam to the direction $(\theta_{\text{ap}}, \theta_{\text{eq}})$ at the focusing distance r_0. The expression of $\tau(r_0, m, \theta_{\text{ap}}, \theta_{\text{eq}})$ can be given as

$$\begin{aligned} \tau(r_0, m, \theta_{\text{ap}}, \theta_{\text{eq}}) &= \frac{r_0 - |\mathbf{v_m} - r_0\hat{\mathbf{u}}|}{c} \\ &= \frac{r_0 - \sqrt{r_0^2 + x_m^2 + y_m^2 - 2x_m r_0 \sin\theta_{\text{ap}} - 2y_m r_0 \sin\theta_{\text{eq}}}}{c} \end{aligned} \tag{3}$$

where $\mathbf{v_m}$ is the position vector of the mth senor $(x_m, y_m, 0)$. $\tau(r_0, m, \theta_{\text{ap}}, \theta_{\text{eq}})$ is adopted to compensate for differences in propagation times from the signal source to the individual array sensors. The net result of (2) is the formation of a time-domain signal in which the contributions coming from the direction $\hat{\mathbf{u}}$ and the distance r_0 are enhanced coherently, while those from other distances and directions are weakened [2].

If the received signals are segmented into blocks with a certain length L, the discrete Fourier transform (DFT) coefficients $B(r_0, l, \theta_{\text{ap}}, \theta_{\text{eq}})$ of the beam signal $b(r_0, t, \theta_{\text{ap}}, \theta_{\text{eq}})$ are given by

$$B(r_0, l, \theta_{\text{ap}}, \theta_{\text{eq}}) = \sum_{m=1}^{M_0} w_m S_m(l) \exp\left(-i2\pi f_l \tau(r_0, m, \theta_{\text{ap}}, \theta_{\text{eq}})\right) \tag{4}$$

where $S_m(l)$ is the lth DFT coefficient of $s_m(t)$, and the discrete frequency f_l is

$$f_l = \frac{lf_s}{L} \tag{5}$$

where f_s is the sampling frequency and l is the frequency index $l \in [0, L)$.

When $r_0 > D^2/2\lambda$ with D being the side size of the 2-D square array, the far-field approximation condition [2], [11], [25] is satisfied. Thus, the delay [2] in the far field can be expressed as

$$\tau(r_0, m, \theta_{\text{ap}}, \theta_{\text{eq}}) = \frac{x_m \sin\theta_{\text{ap}} + y_m \sin\theta_{\text{eq}}}{c} \tag{6}$$

where c is the sound speed in water.

For the near-field condition, the Fresnel zone ($0.96D < r_0 \leq D^2/2\lambda$) is the main concern, where the dynamic focusing is needed [2], [11], [25]. The Fresnel delay approximation [2] and the least-square approximation proposed by Trucco [14] are often adopted in most of the fast beamforming techniques. The delay in the near field can be approximately given by [2], [14]

$$\tau(r_0, m, \theta_{\text{ap}}, \theta_{\text{eq}}) = \frac{x_m \sin\theta_{\text{ap}} + y_m \sin\theta_{\text{eq}}}{c} - \frac{x_m^2 + y_m^2}{2r_0 c}. \tag{7}$$

The realization of the frequency-domain DM is to interpret (4) as a complex dot product directly in both far and near fields. The fast beamforming algorithms such as the CZT method [11] and the proposed method based on NUFFT all focus on how to realize the fast computation of (4).

III. Nonuniform Fast Fourier Transform

In this section, the definitions of nonuniform discrete Fourier transform (NUDFT) and the adjoint NUDFT are introduced. A brief description of the fast implementation algorithms of NUDFT and the adjoint NUDFT is also given. It should be noted that the fast implementation of the adjoint 2-D NUDFT is crucial for the proposed beamforming method in this paper.

Fast Fourier transform (FFT) is proposed to improve the speed of DFT. FFT needs nodes in both the frequency and time domain to be uniformly spaced [26]. NUFFT has been proposed to ensure the fast computation of the DFT of nonuniform nodes, with only a slight drop in the accuracy of the computation [23], [24]. In addition, NUFFT has been applied in many fields such as MRI [27], CT [28], through-wall radar [29], reconstruction of photoacoustic images [30], ultrasound plane-wave imaging [31], and synthesis of large arbitrary arrays [32].

A. Algorithms of NUFFT and Its Adjoint Operator

Given that x_k ($k = -K/2, \ldots, K/2-1$ and K being a positive even number) are equispaced signal samples, the 1-D NUDFT coefficients of x_k have the form

$$X(\omega_m) = \sum_{k=-K/2}^{K/2-1} x_k \exp(-ik\omega_m), \qquad m = 0, \ldots, M-1 \tag{8}$$

where M is a natural number, ω_m is the arbitrary frequency nodes, and $\omega_m \in [-\pi, \pi]$ [20]. The proposed method in this paper will employ the adjoint 2-D NUDFT. To make the description easy to understand, we give the definition of the adjoint 1-D NUDFT [23], [24], expressed as

$$\hat{x}_k = \sum_{m=0}^{M-1} X(\omega_m)\exp(ik\omega_m), \qquad k = -K/2, \ldots, K/2-1 \tag{9}$$

where $\hat{x}_k$s are the equally spaced samples obtained from the adjoint 1-D NUDFT operations, which may not be equal to x_ks because the NUDFT is not always invertible as the IDFT done.

In general, the algorithms of 1-D NUFFT consist of two steps: the oversampled DFT and the interpolation [24]. They can be described by

$$Y_{k'} = \sum_{k=-K/2}^{K/2-1} x_k \exp\left(\frac{-i2\pi kk'}{K_1}\right), \qquad k' = -K_1/2, \ldots, K_1/2-1 \tag{10}$$

and

$$\hat{X}(\omega_m) = \sum_{j=1}^{J} Y_{(k'_m+j)} u_j^*(\omega_m), \qquad m = 0, \ldots, M-1 \tag{11}$$

where Y_k denotes the oversampled DFT coefficients with K_1 points ($K_1 \geq K$), K_1/K is defined as the oversampled factor σ, and k'_m in (11) is given by [24]

$$k'_m = \begin{cases} \left(\arg\min_{k' \in Z}\left|\omega_m - \dfrac{2\pi k'}{K_1}\right|\right) - \dfrac{J+1}{2}, & J \text{ odd} \\ \left(\max\left\{k' \in Z : \omega_m \geq \dfrac{2\pi k'}{K_1}\right\}\right) - \dfrac{J}{2}, & J \text{ even.} \end{cases} \tag{12}$$

In (11), $\hat{X}(\omega_m)$ are the approximated coefficients in the nonuniform frequency points, $u_j(\omega_m)$ are appropriate frequency-domain interpolation coefficients, "*" denotes complex conjugate, J is the largest number of the nearest nonzero neighbors applied by the interpolation, and $J \gg K : u_j(\omega_m)$ can be obtained through the min–max criterion that Fessler and Sutton proposed [24]. Mathematically, the min–max criterion is an optimization problem which has the form of $\min\limits_{u(\omega_m)\in C^J} \max\limits_{x\in C^K \|x\|\geq 1} |\hat{X}(\omega_m) - X(\omega_m)|$ $[\boldsymbol{u}(\omega_m) = (u_1, (\omega_m)\text{K}, u_J(\omega_m))$ and $\boldsymbol{x} = (x_{-K/2}, \text{K}, x_{K/2-1})]$ and has an analytical solution as is derived in [24]. It means that for each desired frequency ω_m, the coefficients $u_j(\omega_m)$ minimize the worst case approximation error $e = |\hat{X}(\omega_m - X(\omega_m))$. σ and J can be determined by the permitted error limit through the empirical formula $e_{\max} \approx 0.75\exp[-J(0.29 + 1.03\log\sigma)]$ [24].

The fast algorithms of the adjoint NUDFT, i.e., the adjoint operator of NUFFT, compute (9) by "reversing" (not inverting!) the steps of the algorithms of NUFFT [24]when a nonuniform frequency-domain series $X(\omega_m)$ is provided. The first step is the interpolation and the second one is the oversampled inverse FFT (IFFT). Therefore, the interpolation of the adjoint operator of NUFFT can be described as

$$\tilde{X}_{k'} = \sum_{m=0}^{M-1} v_{mk'} X(\omega_m) \tag{13}$$

which is akin to "gridding" [24]. v_{mk} are the sparse interpolation coefficients applied for "gridding." The second step is given by

$$\tilde{x}_k = \sum_{k'=-K_1/2}^{K_1/2-1} \tilde{X}_{k'} \exp\left(i2\pi \frac{k'k}{K_1}\right) \quad (14)$$

where $\tilde{x}_k$ is an approximation of $\hat{x}_k$ in (9). The adjoint operator of the 1-D NUFFT uses the transpose of the sparse interpolation matrix applied by the 1-D NUFFT [24].

Similar with the definition of the 1-D NUDFT, the definition of the 2-D NUDFT [33] is expressed as

$$X(\omega_{1m}, \omega_{2m}) = \sum_{k_1=-K/2}^{K/2-1} \sum_{k_2=-K/2}^{K/2-1} x_{k_1,k_2} \exp\left[-i(k_1\omega_{1m} + k_2\omega_{2m})\right], \quad m = 0, \ldots, M-1 \quad (15)$$

where $X(\omega_{1m}, \omega_{2m})$ are the 2-D NUDFT coefficients of a 2-D uniformly sampled series x_{k_1,k_2}, both ω_{1m} and $\omega_{2m} \in [-\pi, \pi)$. The expression of the adjoint 2-D NUDFT [24], [33] is needed

$$\hat{x}_{k_1,k_2} = \sum_{m=0}^{M-1} X(\omega_{1m}, \omega_{2m}) \exp\left[i(k_1\omega_{1m} + k_2\omega_{2m})\right], \quad \begin{pmatrix} k_1 = -K/2, \ldots, K/2-1 \\ k_2 = -K/2, \ldots, K/2-1 \end{pmatrix} \quad (16)$$

where $\hat{x}_{k_1,k_2}$s are the 2-D equal spaced samples obtained from adjoint 2-D NUDFT operations, which may not be equal to original $\hat{x}_{k_1,k_2}$s because the NUDFT is not always invertible as IDFT done.

The proposed beamforming method will make use of the fast implementation algorithm of the adjoint 2-D NUDFT, i.e., the adjoint operator of 2-D NUFFT. Similar to that of the 1-D NUFFT, when a 2-D nonuniform frequency-domain series $X(\omega_{1m}, \omega_{2m})$ is provided, the adjoint operator of the 2-D NUFFT can be described by

$$\tilde{X}_{k_1',k_2'} = \sum_{m=0}^{M-1} v_{m,k_1',k_2'} X(\omega_{1m}, \omega_{2m}) \quad (17)$$

which is also similar to "gridding," and

$$\tilde{x}_{k_1,k_2} = \sum_{k_1'=0}^{K_1-1} \sum_{k_2'=0}^{K_2-1} \tilde{X}_{k_1',k_2'} \exp\left[i2\pi\left(\frac{k_1'k_1}{K_1} + \frac{k_2'k_2}{K_2}\right)\right], \quad \begin{pmatrix} k_1 = -K/2, \ldots, K/2-1 \\ k_2 = -K/2, \ldots, K/2-1 \end{pmatrix} \quad (18)$$

where $\tilde{x}_{k_1,k_2}$ is the approximation of $\hat{x}_{k_1,k_2}$ in (16), and $v_{m,k_1',k_2'}$ are the sparse interpolation coefficients applied for 2-D "gridding." The interpolation coefficients applied in this paper are also determined by the min–max criterion in Fessler and Sutton's paper [24]. The fast computation of (18) can be realized by 2-D IFFT. It should be emphasized that the interpolation matrix used by the adjoint operator of the 2-D NUFFT is the transpose of that used by the 2-D NUFFT [24].

The efficiency and accuracy of the NUFFT and its adjoint operator is dependent on the following crucial parameters: the oversampling factor σ and the number of the nearest neighbors J. In the following, the computational load relative with σ and J will be analyzed.

B. Computational Load of NUFFT and Its Adjoint Operator

For the 1-D NUFFT, because (10) can be calculated by FFT with K_1 being a power of two, the amount of real operations required by the oversampled DFT [24], [26] will be

$$A_0 = 5K_1 \log K_1. \quad (19)$$

Each operation in (11) that tries to obtain one of the values of $\hat{X}(\omega_m)$ requires J complex multiplications and $J-1$ complex additions. When M points need to be interpolated, the total number of real operations required by the interpolation is

$$A_1 = M\left[6J + 2(J-1)\right] = M(8J-2). \quad (20)$$

Because the interpolation step of the adjoint operator of the 1-D NUFFT uses the transpose matrix of that used in the 1-D NUFFT, the interpolation step of the adjoint operator of the 1-D NUFFT needs the same number of real multiplication operations as that of the 1-D NUFFT and the number of real addition operations needed by the interpolation step is approximately equal to that of the 1-D NUFFT. The fast computation of (14) can be realized by the IFFT of K_1 points, discarding all but the preceding K values. When IFFT is used to process (14), the sum of real operations required by IFFT [21], [26] is

$$A_2 = 5K_1 \log K_1. \quad (21)$$

When M nonuniform frequency points need to be interpolated, the total number of real operations required by the interpolation step of the adjoint operator of the 2-D NUFFT [24] should be

$$A_3 \approx A_1. \quad (22)$$

In general, to obtain the fastest computing speed, the interpolation coefficients are precomputed and stored in the memory of computers. The crucial problem for different fast algorithms of NUFFT and its adjoint operator is to determine the optimum interpolation coefficients and apply them in the computation.

The extensions of the 1-D NUFFT and its adjoint operator to 2-D case are conceptually straightforward. For the 2-D NUFFT, the 2-D FFT is employed and the interpolation coefficients are precomputed and stored for each desired frequency location [24]. Using the same analysis method in the 1-D NUFFT, for the 2-D NUFFT, the total number of real operations needed to generate M nonuniform frequency-domain samples with $K \times K$ equispaced signal samples is

$$\begin{aligned} A_4 &= M\left[6J^2 + 2(J^2-1)\right] + 5K_1K_2\log(K_1K_2) \\ &= M(8J^2-2) + 5K_1K_2\log(K_1K_2) \end{aligned} \quad (23)$$

where $K_1 = K_2 = \sigma K$.

As for the adjoint operator of the 2-D NUFFT, the sparse interpolation matrix used by the interpolation step (17) is the transpose of that used by the 2-D NUFFT and the fast computation of (18) can be realized by the 2-D IFFT. The number of real operations required by the interpolation of the adjoint operator of 2-D NUFFT with M nonuniform points should be $M[6J^2 +$

$2(J^2-1)]$ approximately. The number of real operations of the IFFT is the same as that of the FFT, $5K_1K_2\log(K_1, K_2)$ shown in (24). Therefore, similar to the adjoint operator of 1-D NUFFT, the total number of real operations required to complete the adjoint operator of 2-D NUFFT with M nonuniform points [24] is

$$A_5 \approx A_4. \tag{24}$$

IV. Fast Broadband Beamforming Method Based on NUFFT

The key to mitigating the computational load of underwater real-time 3-D acoustical imaging in the frequency domain lies in the fast computation of (4) in both the far field and the near field. In this section, the fast broadband beamforming method based on NUFFT which is applicable for arbitrary arrays is proposed to attain the fast computation of (4).

A. Far-Field Condition

In the far-field condition, dynamic focusing is not necessary. Based on (4) and (6), $B(r_0, l, \theta_{\rm ap}, \theta_{\rm eq})$ can be rewritten as

$$B(r_0, l, \theta_{\rm ap}, \theta_{\rm eq}) = \sum_{m=1}^{M_0} w_m S_m(l)\exp\left[-i2\pi f_l(x_m\sin\theta_{\rm ap} + y_m\sin\theta_{\rm eq})/c\right]. \tag{25}$$

Assume that θ_{ai} and θ_{af} are the initial and final azimuth angles, and $\theta_{\rm eq}$ and θ_{ef} represent the initial and final elevation angles. There are $\theta_{af} = -\theta_{ai}$ and $\theta_{ef} = -\theta_{ei}$ generally. Then, the azimuth and elevation steering angles employed in this paper are given by

$$\begin{aligned}\sin\theta_{\rm ap} &= p\Delta s_a + \Delta s_a/2\\ \sin\theta_{\rm eq} &= p\Delta s_e + \Delta s_e/2,\end{aligned} \quad \begin{pmatrix}-M_b/2 \le p \le M_b/2-1\\ -N_b/2 \le q \le N_b/2-1\end{pmatrix} \tag{26}$$

where

$$\begin{aligned}\Delta s_a &= \frac{(\sin\theta_{af} - \sin\theta_{ai})}{(M_b-1)}\\ \Delta s_e &= \frac{(\sin\theta_{ef} - \sin\theta_{ei})}{(N_b-1)}.\end{aligned} \tag{27}$$

From (26) and (27), it should be noted that the predefined angles are equispaced in the sine domain.

Considering (26), (25) becomes

$$B(r_0, l, \theta_{\rm ap}, \theta_{\rm eq}) = \sum_{m=1}^{M_0} S'_m(l)\exp\left[i(p\omega_{am} + q\omega_{em})\right] \tag{28}$$

where

$$S'_m(l) = w'_m S_m(l) \tag{29}$$

$$w'_m = w_m\exp\left[-i2\pi f_l(x_m\Delta s_a + y_m\Delta s_e)/2c\right] \tag{30}$$

and

$$\omega_{am} = -2\pi f_l x_m\Delta s_a/c, \quad \omega_{em} = -2\pi f_l y_m\Delta s_e/c. \tag{31}$$

For the proposed beamforming method based on NUFFT, the maximum values of ω_{am} and ω_{em} should be lower than π. The expression of the limiting condition for the proposed beamforming method is given as

$$\begin{cases}\max\limits_m(\omega_{am}) = \dfrac{\pi f_H D}{c}\cdot\dfrac{\sin\theta_{af} - \sin\theta_{ai}}{M_b-1} < \pi\\ \max\limits_m(\omega_{em}) = \dfrac{\pi f_H D}{c}\cdot\dfrac{\sin\theta_{ef} - \sin\theta_{ei}}{N_b-1} < \pi\end{cases} \tag{32}$$

where f_H is the highest frequency in the passband of the received signals.

If an equispaced 2-D array with a square aperture is employed and N_1 is the lateral size of the array, both x_m and y_m will be located between $-(N_1-1)\lambda_0/4$ and $(N_1-1)\lambda_0/4$ generally, where λ_0 is the wavelength of the carrier whose frequency is f_0. For the application of broadband 3-D imaging mentioned in [3] and [11], there are $0.5f_0 < f_l < 1.5f_0$, $\theta_{af} \ge 30°$, and $\theta_{ef} \ge 30°$. Moreover, to attain high imaging quality, the beams of imaging systems should be set as dense as possible. Thus, M_b and N_b are usually set to $2N_1$ [3], [11]. It is easy to see that the limiting condition (32) is satisfied. Using similar parameter configurations for square-aperture arbitrary arrays, the same conclusion (i.e., $\omega_{am} \in [-\pi, \pi]$ and $\omega_{em} \in [-\pi, \pi])$ can be drawn.

For both equispaced and arbitrary 2-D arrays, (28) has the same form as (16), the expression of the adjoint 2-D NUDFT. Hence, the fast computation of (28) can be realized by the adjoint operator of the 2-D NUFFT.

According to the analysis in Section III, the number of the real operations needed to compute (28) at a single frequency f_l for all beams, denoted by A_6, is

$$A_6 \approx M_0(8J^2-2) + 5M'_bN'_b\log_2(M'_bN'_b) \tag{33}$$

where due to the application of IFFT, M'_b should be the first powers of two, greater than or equal to σM_b, and N'_b should be the first powers of two, greater than or equal to σN_b.

The nonuniform frequency nodes ω_{am} and ω_{em} and the interpolation coefficients of the adjoint operator of the 2-D NUFFT are usually precomputed and stored for fast 3-D beamforming. In addition, the proposed method based on NUFFT has no special requirements for positions of the sensors, meaning that the sensors of 2-D arrays can be distributed arbitrarily. Therefore, the proposed method is suitable for arbitrary 2-D arrays.

B. Near-Field Condition (Fresnel Zone)

In the near-field condition, Fresnel zone $(0.96D < r_0 \le D^2/2\lambda)$ is the major consideration, where the dynamic focusing technique [11] is necessary. Assuming that the depth of field (DOF) is the range interval around the focusing distance r_0, within which the performance decays marginally, the DOF can be approximated by the following expressions [11]:

$$r_{0-} = r_0 - r_0/(D^2/2\lambda r_0 + 1) \tag{34}$$

and

$$r_{0+} = r_0 + r_0/(D^2/2\lambda r_0 - 1). \tag{35}$$

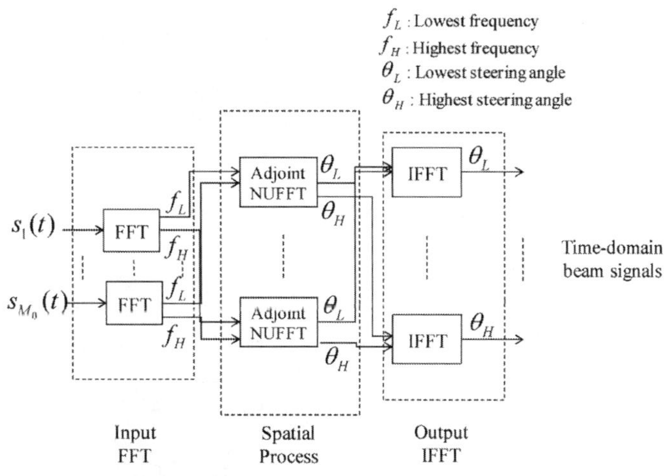

Fig. 2. Procedure of the proposed beamforming method based on NUFFT.

Based on (4) and (7), $B(r_0, l, \theta_{\text{ap}}, \theta_{\text{eq}})$ can be rewritten as

$$B(r_0, l, \theta_{\text{ap}}, \theta_{\text{eq}}) = \sum_{m=1}^{M_0} w_m S_m(l) \exp\left[\frac{i2\pi f_l(x_m^2 + y_m^2)}{2r_0 c}\right] \times \exp\left[\frac{-i2\pi f_l(x_m \sin\theta_{\text{ap}} + y_m \sin\theta_{\text{eq}})}{c}\right]. \quad (36)$$

Assuming

$$S_m''(r_0, l) = w_m'' S_m(l) \quad (37)$$

and

$$w_m'' = w_m \exp\left[\frac{i2\pi f_l(x_m^2 + y_m^2)}{2r_0 c}\right] \times \exp\left[\frac{-i2\pi f_l(x_m \Delta s_a + y_m \Delta s_e)}{2c}\right] \quad (38)$$

(36) can be further expressed by

$$B(r_0, l, \theta_{\text{ap}}, \theta_{\text{eq}}) = \sum_{m=0}^{M_0} S_m''(r_0, l) \exp(i2\pi(p\omega_{am} + q\omega_{em})). \quad (39)$$

It is obvious that (39) is similar to (28). In the DOF of the focusing distance r_0, the fast computation of (39) can also be achieved by the adjoint operator of the 2-D NUFFT.

C. Summary

To sum up, the proposed broadband beamforming method based on NUFFT for real-time 3-D underwater acoustical imaging in both the far field and the near field is realized by the following three steps.

1) Input FFT: segment the received broadband signals into sequential blocks and convert the blocks into frequency-domain signals by using FFT (partial overlapping needed [11]).
2) Spatial processing: multiply the FFT coefficients of each sensor by the corresponding weights (far field: w_m'; near field: w_m'') which are generated offline, then compute beam signals at every valid frequency point with the adjoint operator of the 2-D NUFFT.
3) Output IFFT: invert the beam signals in the frequency domain into the time-domain signals by using IFFT.

The procedure of the broadband beamforming method based on NUFFT is shown in Fig. 2.

V. Analysis of Beamforming Errors Caused by NUFFT

Unlike the frequency-domain DM which is an exact method, the proposed method is an approximated one because the realization of NUFFT is an approximated method. In this section, the beamforming errors caused by NUFFT are evaluated and proven to be negligible in the application of 3-D underwater acoustical imaging in both the far field and the near field.

In addition, although the CZT beamforming [11], [18] cannot be directly applied to arbitrary arrays with off-the-grid elements, it is possible to be applied by approximating the array element positions with their closest on-the-grid positions. The accuracy loss of approximated CZT is also discussed and compared with that of the NUFFT method to verify the advantage of NUFFT.

The computational accuracy of NUFFT is determined by the two parameters: σ and J, as is mentioned in Section III. To evaluate the influence of the two parameters on the accuracy of beamforming, NUFFT is realized based on the toolbox given by Fessler [34], where σ and J can be set to different values.

In the current study, two arbitrary arrays, a spiral array and a random array, as depicted in Fig. 3(a) and (b) are employed for computing the errors caused by NUFFT. The diameter of the two arrays is $60\lambda_0$, where λ_0 is the wavelength of the carrier frequency f_0. For a fully sampled uniform 2-D circular array with the diameter $60\lambda_0$, the total number of elements will be 11 305. The spiral 2-D array is composed of 256 omnidirectional punctiform elements, whose divergence angle, the key parameter of the spiral array, is 92.05°, as given in [9]. The random 2-D array with uniform distribution [36]–[38] is composed of 512 elements. The transmitter located in the center of the array

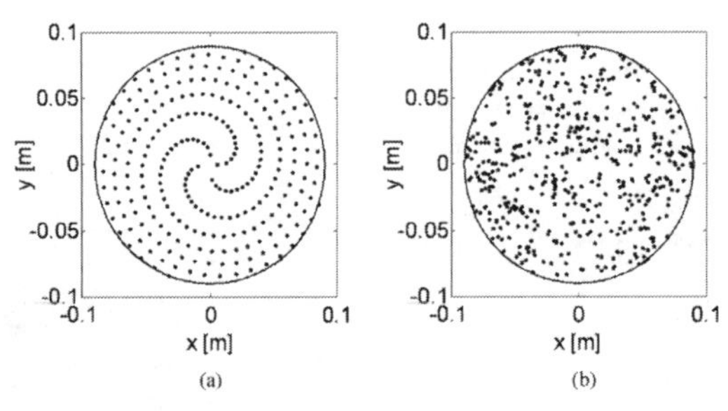

Fig. 3. (a) Spiral 2-D array with 256 elements. (b) Random 2-D array with 512 elements.

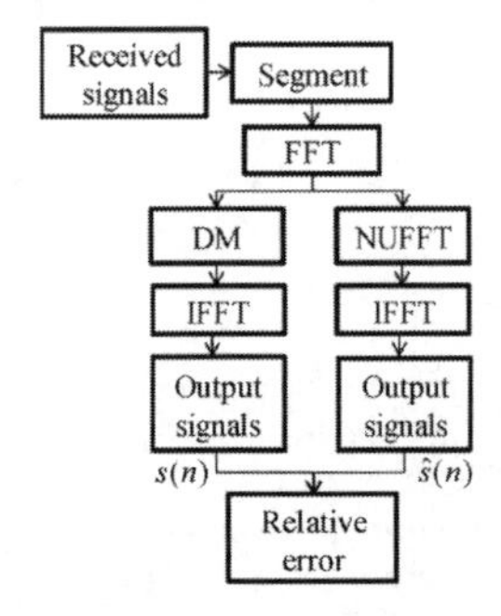

Fig. 4. Procedure of computing the relative errors of the whole 3-D image scene.

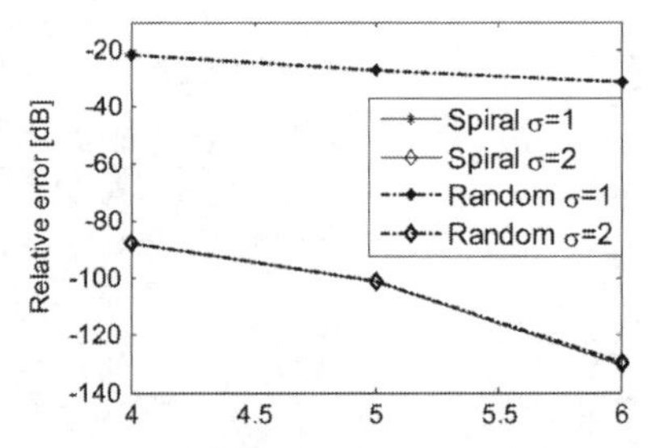

Fig. 5. Relative errors of the whole 3-D image with different values of σ and J.

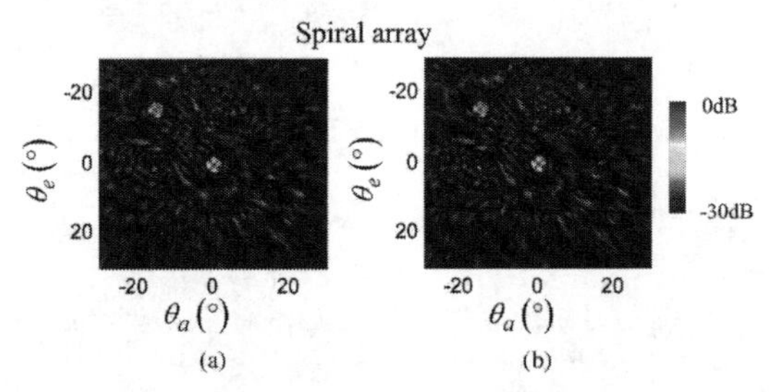

Fig. 6. The 2-D slices at the focusing distance of the 3-D images obtained by the DM and the proposed method based on NUFFT ($J = 4$ and $\sigma = 2$) for the spiral array: (a) for the DM; and (b) for the proposed method.

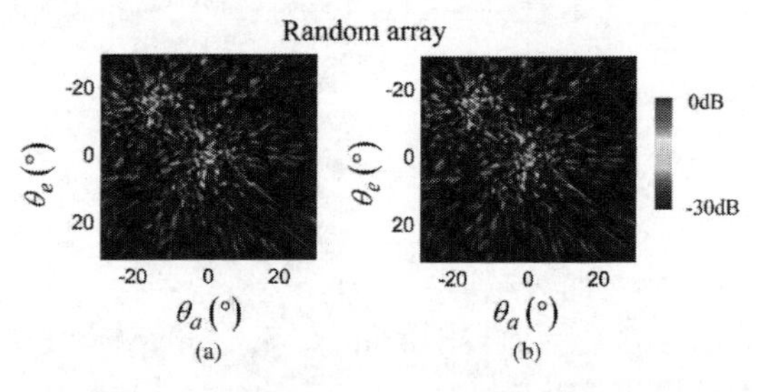

Fig. 7. The 2-D slices at the focusing distance of the 3-D images obtained by the DM and the proposed method based on NUFFT ($J = 4$ and $\sigma = 2$) for the random array: (a) for the DM; and (b) for the proposed method.

is omnidirectional. A 3-D scene with two reflecting scatterers is considered, the positions of which are located at $(\rho, \theta_a, \theta_e) = (20\,\text{m}, -15^\circ, 15^\circ)$ and $(20\,\text{m}, 0^\circ, 0^\circ)$, as described by the geometry of Fig. 1, where ρ denotes the distance from the center of the array to the scatterer. The scatterers are all taken as points, and the medium is supposed to be isotropic and ideal without any frequency absorption. The transmitting signal is a Gaussian-envelope pulse with central frequency f_0 being 500 kHz and the 3-dB bandwidth being 180 kHz. This situation is obviously corresponding to the far-field condition. The received signals of the sensors are simulated by using the method adopted by Palmese and Trucco [35].

The relative error $\varepsilon_{\max}$ of the whole 3-D image obtained by using the proposed method based on NUFFT can be computed by [23], [24]

$$\varepsilon_{\max} = 20\log_{10}\left(\frac{\max_{n,m}\left(|\hat{s}_m(n) - s_m(n)|\right)}{\max_{n,m}\left(|s_m(n)|\right)}\right) \tag{40}$$

where $\hat{s}_m(n)$ and $s_m(n)$ represent the mth beam signal obtained by the proposed method based on NUFFT and the DM beamforming, respectively, and "$\max_{n,m}()$" represents the maximum of the expression in the bracket for all the values of n and m. The procedure that computes the relative errors is depicted in Fig. 4. The relative errors for different values of σ and J are computed and depicted in Fig. 5. From Fig. 5, it can be found that for both arrays, the relative errors are all above -32 dB when $\sigma = 1$ and the relative errors will become less than or equal to -87 dB when $\sigma = 2$, with J being assigned different values between 4 and 6.

To reflect the accuracy of the proposed method intuitively, Fig. 6 (for the spiral array) and Fig. 7 (for the random array) show the 2-D slices at the focusing distance of the whole 3-D images obtained by the DM and the proposed method based on NUFFT ($J = 4$ and $\sigma = 2$). It can be found that there is nearly no difference between the 2-D slices obtained by the two methods for the different arrays. The relative errors of the sidelobe level of the 2-D slices are all lower than -80 dB. The relative errors of the mainlobe width are also lower than -80 dB. The output beam signals in the direction $(15^\circ, -15^\circ)$ and $(15^\circ, 15^\circ)$ computed by the DM and the proposed method based on NUFFT ($J = 4$ and $\sigma = 2$) are given in Fig. 8. It is found that the two beam signals almost overlap with each other. The figures demonstrate that the proposed method based on NUFFT has a good accuracy in the far field.

To verify the high accuracy of the proposed method further, the accuracy loss of approximated CZT beamforming is also

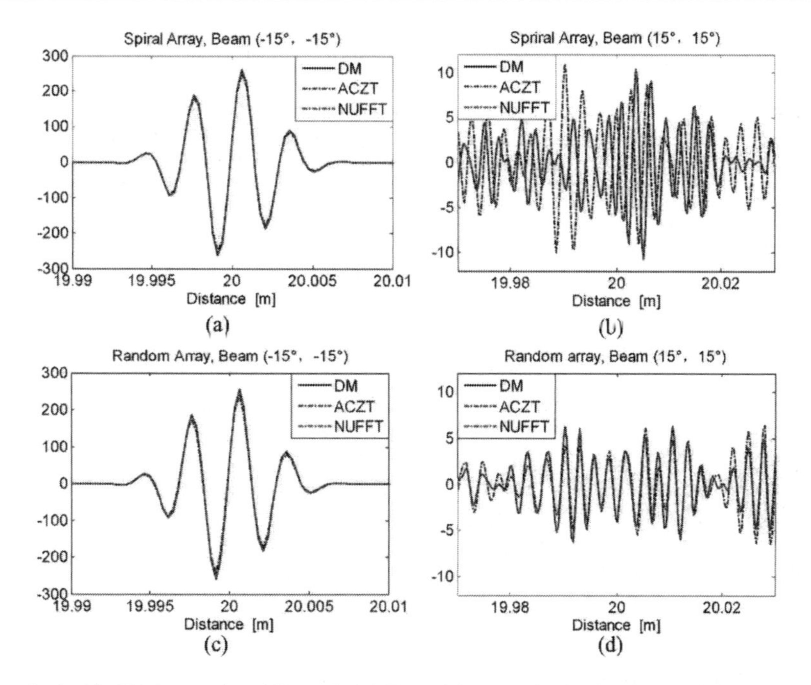

Fig. 8. Output beam signals of the DM, the approximated CZT method (ACZT), and the proposed method based on NUFFT ($J = 4$ and $\sigma = 2$) for the spiral and random arrays. (a) Beam $(-15°, -15°)$ for the spiral array. (b) Beam $(15°, 15°)$ for the spiral array. (c) Beam $(-15°, -15°)$ for the random array. (d) Beam $(15°, 15°)$ for the random array.

TABLE I
RELATIVE ERRORS OF THE WHOLE 3-D IMAGE CAUSED BY THE PROPOSED METHOD BASED ON NUFFT ($J = 4$ AND $\sigma = 2$) AND THE APPROXIMATED CZT METHOD (ACZT) APPLIED TO THE SPIRAL AND RANDOM ARRAYS

Array Pattern	Relative Error for ACZT (dB)	Relative Error for NUFFT (dB)
Spiral	-21.7145	-87.3355
Random	-23.6289	-87.2251

discussed and compared with that of the NUFFT method. Approximated CZT beamforming is realized by approximating the element positions of the spiral and random arrays with their closest on-the-grid positions and then applying the CZT [11], [18]. The output time-domain beam signals of the approximated CZT method applied to the two arrays are shown in Fig. 8. It can be seen that in the direction $(15°, 15°)$ where scatterers do not exist, the beam signals of the approximated CZT method deviate obviously from that of the DM, while the beam signals of the proposed method still overlap with that of the DM. The relative errors of the whole 3-D image caused by the approximated CZT and the proposed methods are given in Table I. It can be found that the relative errors caused by the approximated CZT method are much higher than those caused by the proposed method based on NUFFT.

The above analyses are under the far-field condition, which might not be enough to verify the accuracy of the proposed method. In the following, the relative errors are discussed in the near field for the spiral and random arrays. In the near field, the errors of the proposed method are caused by NUFFT and the Fresnel approximation. To evaluate the errors caused by the Fresnel approximation, the direct computation of (39) is conducted, which is called the Fresnel approximation direction method (FADM). The direction computation of (4), namely the DM which is accurate both in the near field and the far field, is applied to provide reference results.

A near-field 3-D scene with two reflecting scatters is considered. The two near-field scatterers are located at $(\rho, \theta_a, \theta_e) =$ (1 m, $-5°$, $-5°$) and (1 m, $0°$, $0°$), respectively. Fig. 9 shows that in the direction $(10°, 10°)$ where scatterers do not exist, the beam signals of the proposed method based on NUFFT ($J = 4$ and $\sigma = 2$) deviate obviously from those of the DM but almost overlap with those of the FADM. Table II illustrates that the relative errors caused by the FADM and the proposed method based on NUFFT ($J = 4$ and $\sigma = 2$) are nearly the same for both the spiral and random arrays. It can be concluded from Fig. 9 and Table II that the errors caused by NUFFT ($J = 4$ and $\sigma = 2$) are negligible in the near field, compared with that caused by the Fresnel approximation.

On the basis of all the analyses, we come to a conclusion that $\sigma = 2$ can be adopted to realize the proposed method based on NUFFT, and the relative errors are so small that they can be neglected in the application of 3-D underwater acoustical imaging.

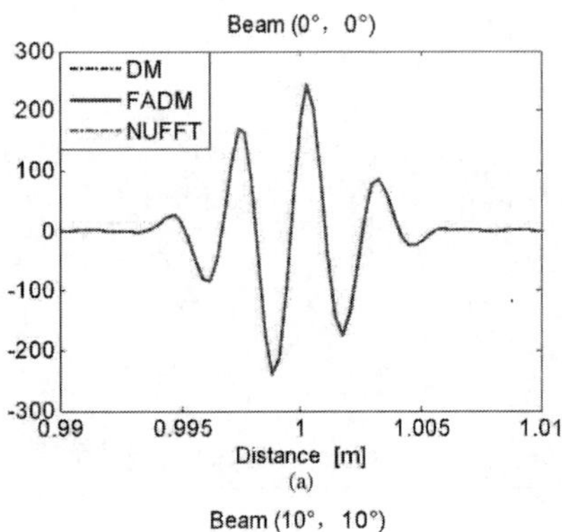

(a)

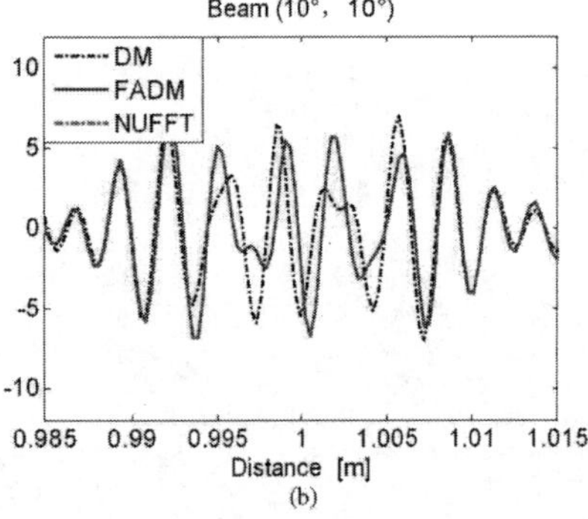

(b)

Fig. 9. Output near-field beam signals of the DM, the FADM, and the proposed method based on NUFFT ($J = 4$ and $\sigma = 2$) for the spiral array. (a) Beam (0°, 0°). (b) Beam (10°, 10°).

TABLE II
RELATIVE ERRORS OF THE WHOLE 3-D IMAGE IN THE NEAR FIELD CAUSED BY THE PROPOSED METHOD BASED ON NUFFT ($J = 4$ AND $\sigma = 2$) AND THE FADM APPLIED TO THE SPIRAL AND RANDOM ARRAYS

Array Pattern	Relative Error for FADM (dB)	Relative Error for NUFFT (dB)
Spiral	-30.2154	-30.2150
Random	-25.2324	-25.2320

VI. COMPUTATIONAL LOAD EVALUATION

This section evaluates the minimum number of online real operations needed to generate a 3-D image in two situations: one with an equispaced 2-D array and the other with an arbitrary 2-D array. Four beamforming methods are considered in the evaluation: 1) the D&S method; 2) the DM; 3) the CZT method; and 4) the proposed method based on NUFFT.

The number of online real operations includes both real additions and real multiplications. Any operation that can be performed offline (before the start of the 3-D image generation) is disregarded [18]. The main parameters for the evaluation are described as follows:

- the number of sensors of the equispaced 2-D array: $M_0 = N_1 \times N_1$;
- the number of sensors of the arbitrary 2-D array: M_1;
- square set of beams: $N_b \times N_b$.

A. Equispaced 2-D Arrays

When an equispaced 2-D array is employed, the CZT beamforming proposed by Palmese and Trucco is valid [11], [18]. The computational loads needed by the four beamforming methods are analyzed, respectively.

For the time-domain D&S beamforming, an finite-impulse response (FIR) interpolation filter with H stages [13] is adopted for the signal oversampling before beamforming. The rate of real operations [11], [39] needed to generate N_b^2 beams is

$$TD_1 = \left[N_1^2\left(H + N_b^2\right) + N_1^2(H+1)\right] f_s. \tag{41}$$

The computation of delays is considered an offline operation.

For the aforementioned three frequency-domain beamforming methods, they all consist of three steps: input FFT, spatial processing, and output IFFT. The crucial factor deciding the number of online real operations of the three frequency-domain beamforming methods is their algorithms of the spatial processing.

The first step of the three frequency-domain beamforming methods is to segment the time-domain received signals and convert them into the frequency-domain coefficients through FFT. The number of online real operations needed to process an L-length real sequence by FFT [11], [26] is

$$F_{\mathrm{FFT}} = 2.5L\log\frac{L}{2} + 7L. \tag{42}$$

In the second step, the DM is realized by a complex dot product described in Section II. As shown in [11] and [18], the spatial processing of the DM that generates N_b^2 beams at a single frequency f_l requires the following number of online real operations:

$$F_{\mathrm{Dsp}} = \left(8N_1^2 - 2\right)N_b^2. \tag{43}$$

To generate N_b^2 beams, the spatial processing of the CZT beamforming method at a single frequency f_l requires the following number of online real operations [11]:

$$F_{\mathrm{Csp}} = 6\left[N_1^2 + N_b^2 + L_1^2\right] + 20L_1^2\log_2(L_1) \tag{44}$$

where L_1, the size of FFT, should be a power of two, and $L_1 \leq N_1 + N_b - 1$ due to the use of FFT and IFFT.

The proposed beamforming method based on NUFFT is to transform the spatial processing into the adjoint operator of 2-D NUFFT. The spatial processing of the method includes the following steps:

1) compute w'_m and w''_m and the interpolation coefficients (offline);
2) multiply $S_m(l)$ by w'_m or w''_m to obtain $S'_m(l)$ or $S''_m(l)$, and then interpolate $S'_m(l)$ or $S''_m(l)$ (online) based on the adjoint operator of 2-D NUFFT;
3) use IFFT to process $S'_m(l)$ or $S''_m(l)$ obtained by interpolating (online).

For multiplying $S_m(l)$ by w'_m or w''_m to obtain $S'_m(l)$ or $S''_m(l)$, N_1^2 complex multiplication operations are needed, which are equal to $6N_1^2$ real operations. According to (33), to generate N_b^2 beams, the number of online real operations required by the spatial processing of the proposed method based on NUFFT at a single frequency f_l is

$$F_{Nsp} \approx 6N_1^2 + N_1^2(8J^2-2) + 5N_b'^2 \log_2\left(N_b'^2\right) \tag{45}$$

where N_b' should be the first power of two, greater than or equal to σN_b due to the application of IFFT.

The last step for all the three frequency-domain methods is to obtain time samples of each beam signal by applying IFFT. To further reduce the computational load, the technique of generating cubic resolution cells proposed by Palmese and Trucco is adopted [11], which is briefly described as follows.

Owing to the fact that the angular resolution of the beamforming system is constant, the lateral resolution worsens with the increase of the distance. But the range resolution, depending on the pulse bandwidth, does not vary with the increase of the distance. Subsequently, the technique of generating cubic resolution cells [11] makes it possible to avoid unnecessary operations by reducing the number of frequency bins processed (i.e., the bandwidth considered) with the increase of the distance. Assuming that Q is the bandwidth of the continuous-wave (CW) pulse, the range resolution is $c/(2Q)$ [11]. Considering that λ is the wavelength of the carrier and d is the intersensor spacing, the lateral resolution of the equispaced array [11], which is achieved by weighting, is [11]

$$R_0 = \frac{0.88\lambda r}{D_1} \tag{46}$$

where r is the given distance, and D_1 is the side size of the 2-D square arrays employed in this paper. In particular, for the equispaced 2-D array

$$D_1 = N_1 d. \tag{47}$$

To make the range resolution and the lateral resolution equal, the bandwidth [11]should be

$$Q = \frac{cNd}{1.76\lambda r}. \tag{48}$$

Then, the number of frequency bins needed to achieve a range resolution not worse than the lateral resolution [11] is as follows:

$$\zeta = \text{ceil}\left(\frac{cLN_1 d}{1.76 f_s \lambda r^*}\right) \tag{49}$$

where $\text{ceil}(x)$ represents the first integer greater than x, and r^* denotes the distance at which the signal segment starts.

Finally, the frequency bins are shifted around $f = 0$ to obtain the equivalent lowpass spectrum. The envelop used to display a 3-D image is extracted with low sampling frequency. The number of online real operations of IFFT is

$$F_{\text{IFFT}} = 5\zeta_1 \log_2(\zeta_1) \tag{50}$$

where ζ_1 is the first power of two which is greater than or equal to ζ.

For practical systems, the equispaced 2-D array can be thinned to form a sparse array with the active fraction T of the sensors in order to reduce the hardware cost. The computational load of the time-domain D&S beamforming is certainly mitigated with the reduction of active sensors. As for the three frequency-domain methods, the number of online real operations required in the input FFT step can also be greatly decreased with the reduction of active sensors. However, the spatial processing is different among the three frequency-domain methods. For the spatial processing of the DM and the proposed method based on NUFFT, the number of real operations can be decreased through the reduction of active sensors. For the CZT method, the number will remain unchanged since the spatial processing of the CZT method utilizes FFT and IFFT to realize the fast computation of convolution [11], [18], and the data of these inactive sensors which are set to zero are still employed by the spatial processing of the CZT method.

B. *Arbitrary 2-D Arrays*

Arbitrary 2-D arrays usually can be referred to as 2-D arrays composed of arbitrarily distributed sensors, and they are a solution to help further cut the number of sensors and achieve better directivity than equispaced 2-D arrays with the same apertures. In this case, the CZT beamforming is not available and $N_1 \times N_1$ is not applicable for describing the number of sensors. Thus, the number of sensors of the arbitrary 2-D array is assumed to be M_1.

Based on (41), for the arbitrary 2-D array, the rate of real operations [11], [39] needed by the D&S method to generate N_b^2 beams is

$$TD_1 = \left[M_1\left(H + N_b^2\right) + M_1(H+1)\right] f_s. \tag{51}$$

With regard to the arbitrary 2-D array, according to (43), the number of online real operations of the spatial processing to generate N_b^2 beams at a single frequency f_l for the DM is

$$F_{\text{Dsp}}' = (8M_1 - 2)N_b^2. \tag{52}$$

According to (45), the number of online real operations required by the spatial processing of the proposed method based on NUFFT at a single frequency f_l is

$$F_{\text{Nsp}}' \approx 6M_1 + M_1(8J^2-2) + 5N_b'^2 \log_2\left(N_b'^2\right). \tag{53}$$

VII. Comparison and Discussion

Given the array size and the size of beam set, the total number of online real operations of different beamforming methods required by generating a 3-D underwater image can be computed by the equations introduced in Section VI. The main parameters for the computation of underwater real-time 3-D imaging are as follows:

- sound speed: $c = 1500$ m/s;
- carrier frequency: 500 kHz;
- signal bandwidth: $B = 500$ kHz;
- sampling frequency: $f_s = 1.5$ MHz;
- effective range: 1–100 m (including both near and far fields);
- intersensor spacing of the equispaced 2-D array: $d = 1.5$ mm;
- length of signal segments: $L = 1024, 2048, 4096$;
- stages of the interpolation filter: $H = 100$.

The influence of different values of L on the computational load has been discussed in [11] and [18]. Based on the analysis of

TABLE III
NUMBER OF ONLINE REAL OPERATIONS TO CREATE A 3-D IMAGE WITH THE EQUISPACED ARRAYS ($J = 4$, $\sigma = 2$ FOR NUFFT)

	NUFFT	CZT	DM	D&S
$N_1 = 64, N_b = 128, T = 100\%$	$\mathbf{2.47 \times 10^{10}}$	2.61×10^{10}	1.29×10^{11}	1.05×10^{13}
$N_1 = 128, N_b = 256, T = 100\%$	$\mathbf{1.20 \times 10^{11}}$	1.34×10^{11}	4.53×10^{12}	2.15×10^{14}
$N_1 = 128, N_b = 256, T = 12.5\%$	$\mathbf{2.93 \times 10^{10}}$	4.24×10^{10}	5.70×10^{11}	2.69×10^{13}

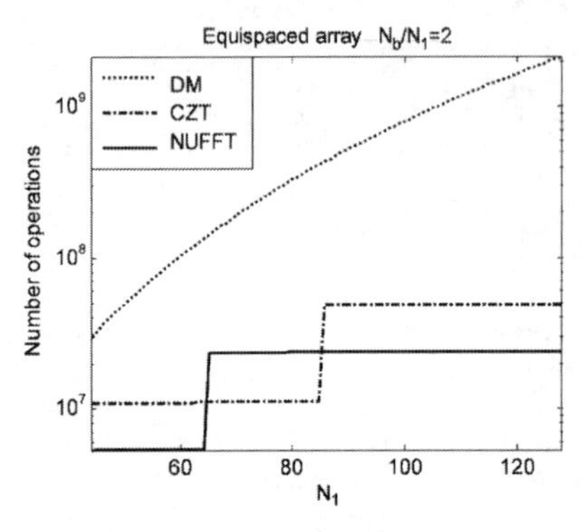

Fig. 10. Number of operations of different spatial processing algorithms versus the lateral size N_1 ($J = 4$ and $\sigma = 2$ for NUFFT).

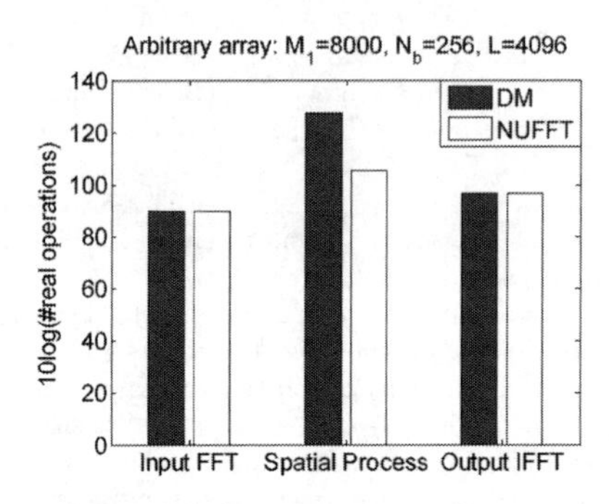

Fig. 11. Computational load of the three steps needed to compute a segment with $L = 4096$ for the DM and the proposed method based on NUFFT ($J = 4$ and $\sigma = 2$).

these papers, in the near-field condition (Fresnel zone), the value of L in this paper is either 1024 or 2048, dynamically determined by the DOF [11]. In the far-field condition, L is fixed to 4096. For all the frequency-domain methods, the technique of generating cubic resolution cells [11] is applied to further reduce the number of operations.

First, consider an equispaced 2-D array for which the CZT beamforming method is feasible. As is mentioned in Section VI, the difference among the three frequency-domain methods is their algorithms of spatial processing. In the algorithms of spatial processing, the ratio N_b/N_1 and the length L of a data segment are assumed to be 2 and 4096, respectively. As the lateral size of the equispaced array N_1 changes, the number of online real operations required by the spatial processing algorithms of the three frequency-domain methods at a valid frequency bin will change, which is shown in Fig. 10. Fig. 10 illustrates that the number of real operations needed by the spatial processing of the DM is at least one order of magnitude higher than that needed by the spatial processing of the CZT method and the proposed method based on NUFFT ($J = 4$ and $\sigma = 2$) for $N \leq 74$. For the spatial processing of the proposed method based on NUFFT, a jump occurs between $N_1 = 64$ and $N_1 = 65$. The reason is that the size of IFFT N_b' used in the adjoint operator of NUFFT jumps from 256 to 512. For the spatial processing of the CZT method, a jump occurs between $N_1 = 85$ and $N_1 = 86$, because the size of FFT L_1 is changed from 256 to 512. In most cases, the number of operations for the spatial processing of the proposed method based on NUFFT is lower than that required by the CZT spatial processing except for $N_1 = 65, \ldots, 86$.

The total number of online real operations required by the four beamforming methods (i.e., NUFFT, CZT, DM, and D&S) for generating a 3-D image is given in Table III. The results show that when the sensors of the array are fully activated, i.e., the active fraction $T = 100\%$, the total number of online real operations of the proposed method based on NUFFT is obviously lower than that of the DM and the D&S method, and a little bit lower than that of the CZT method both at $N_1 = 64$ and $N_1 = 128$. Compared with the DM and the D&S method, the proposed method based on NUFFT shows an approximately 40-fold and 1800-fold gains at $N_1 = 128$, respectively. When the active fraction $T = 12.5\%$ at $N_1 = 128$, the proposed method based on NUFFT achieves an 1.6-fold gain compared with the CZT method, and 20-fold and 920-fold gains compared with the DM and the D&S method.

Second, consider an arbitrary array with nonuniform spacing and evaluate the computational load. Since the CZT beamforming method is not feasible, the comparison is conducted among the other three beamfoming methods (i.e., NUFFT, DM, and D&S). For $M_1 = 8000$, the nonuniform array is assumed to have the same aperture as the square equispaced array whose lateral size is $N_1 = 128$. For $M_1 = 2000$, the nonuniform array has the same aperture as the square equispaced array whose lateral size is $N_1 = 64$. Fig. 11 shows the computational weight of the three steps (i.e., input FFT, spatial processing, output IFFT) to compute a segment with $L = 4096$, needed by the DM and the proposed method based on NUFFT when $J = 4$ and $\sigma = 2$. It is easy to see that the computational loads of the DM and the proposed method based on NUFFT in the first and third steps are equal, but for the spatial processing the number of real operations of the DM is about 170-fold higher than that of the proposed method based on NUFFT.

Table IV gives the total number of online real operations required by the three beamforming methods to generate a 3-D image. It is shown that compared with the DM and the D&S method, the proposed method based on NUFFT has approximately 52-fold and 1280-fold computational gains at $M_1 =$

TABLE IV
NUMBER OF ONLINE REAL OPERATIONS TO CREATE A 3-D IMAGE WITH THE ARBITRARY ARRAY EMPLOYED ($J = 4$, $\sigma = 2$ FOR NUFFT)

	NUFFT	DM	D&S
$M_1 = 2000$, $N_b = 128$	**1.62×10^{10}**	1.49×10^{11}	6.63×10^{13}
$M_1 = 8000$, $N_b = 256$	**8.20×10^{10}**	4.26×10^{12}	1.05×10^{14}

8000. The larger the number of sensors is, the higher the computational gain will be.

VIII. CONCLUSION

This paper focuses on fast beamforming for underwater real-time 3-D acoustical imaging. A fast broadband beamforming method based on NUFFT in frequency domain is proposed to mitigate the computational load of 3-D imaging. Unlike the CZT beamforming method which is only applicable for equispaced 2-D arrays and sparse arrays thinned from equispaced 2-D arrays, the proposed beamforming method based on NUFFT is feasible for arbitrary arrays. The number of online real operations required by the proposed method based on NUFFT is dramatically reduced, compared with that of the time-domain D&S method and the frequency-domain DM both for equispaced and arbitrary 2-D arrays. The gain factor of real operations obtained by using the proposed method based on NUFFT is one or two orders of magnitude greater than that by using the DM. When compared with the time-domain D&S method, the gain factor can reach three orders of magnitude. As for equispaced 2-D arrays, the proposed method based on NUFFT still performs better in most cases than the CZT method in terms of computational load.

Although the proposed method based on NUFFT is an approximated method, with proper parameter configurations, the relative error of the output beam signal caused by NUFFT is usually lower than or equal to -87 dB, which is so small that it can be neglected for underwater real-time 3-D acoustical imaging.

REFERENCES

[1] R. Hansen and P. Andersen, "The application of real time 3-D acoustical imaging," in *Proc. IEEE OCEANS Conf.*, 1998, pp. 738–741.

[2] V. Murino and A. Trucco, "Three-dimensional image generation and processing in underwater acoustic vision," *Proc. IEEE*, vol. 88, no. 12, pp. 1903–1948, Dec. 2000.

[3] A. Trucco, M. Palmese, and S. Repetto, "Devising an affordable sonar system for underwater 3-D vision," *IEEE Trans. Instrum. Meas.*, vol. 57, no. 10, pp. 2348–2354, Oct. 2008.

[4] L. Yuan, R. Jiang, and Y. Chen, "Gain and phase autocalibration of large uniform rectangular arrays for underwater 3-D sonar imaging systems," *IEEE J. Ocean. Eng.*, vol. 39, no. 3, pp. 458–471, Jul. 2014.

[5] Y. Han, X. Tian, F. Zhou, R. Jiang, and Y. Chen, "A real-time 3-D underwater acoustical imaging system," *IEEE J. Ocean. Eng.*, vol. 39, no. 4, pp. 620–629, Oct. 2014.

[6] A. Trucco, "Thinning and weighting of large planar arrays by simulated annealing," *IEEE Trans. Ultrason. Ferroelectr. Freq. Control*, vol. 46, no. 2, pp. 347–355, Mar. 1999.

[7] A. Austeng and S. Holm, "Sparse 2-D arrays for 3-D phased array imaging—Design methods," *IEEE Trans. Ultrason. Ferroelectr. Freq. Control*, vol. 49, no. 8, pp. 1073–1085, Aug. 2002.

[8] B. Diarra, M. Robini, P. Tortoli, C. Cachard, and H. Liebgott, "Design of optimal 2-D non-grid sparse arrays for medical ultrasound," *IEEE Trans. Biomed. Eng.*, vol. 60, no. 11, pp. 3093–3102, Nov. 2013.

[9] O. Martínez-Graullera, C. J. Martín, G. Godoy, and L. G. Ullate, "2D array design based on Fermat spiral for ultrasound imaging," *Ultrasonics*, vol. 50, pp. 280–289, 2010.

[10] J. L. Schwartzand and B. D. Steinberg, "Ultrasparse, ultrawideband arrays," *IEEE Trans. Ultrason. Ferroelectr. Freq. Control*, vol. 45, no. 2, pp. 1073–1085, Mar. 1998.

[11] M. Palmese and A. Trucco, "An efficient digital CZT beamforming design for near-field 3-D sonar imaging," *IEEE J. Ocean. Eng.*, vol. 35, no. 3, pp. 584–594, Jul. 2010.

[12] B. Maranda, "Efficient digital beamforming in the frequency domain," *J. Acoust. Soc. Amer.*, vol. 86, pp. 1813–1819, Nov. 1989.

[13] R. A. Mucci, "A comparison of efficient beamforming algorithms," *IEEE Trans. Acoust. Speech Signal Process.*, vol. 32, no. 3, pp. 548–558, Jun. 1984.

[14] A. Trucco, "A least-squares approximation for the delays used in focused beamforming," *J. Acoust. Soc. Amer.*, vol. 104, pp. 171–175, Jul. 1998.

[15] J. R. Williams, "Fast beamforming algorithm," *J. Acoust. Soc. Amer.*, vol. 44, pp. 1454–1455, 1968.

[16] P. Chen, X. Tian, and Y. Chen, "Optimization of the digital near-field beamforming for underwater 3-D sonar imaging system," *IEEE Trans. Instrum. Meas.*, vol. 59, no. 2, pp. 415–424, Feb. 2010.

[17] A. C. Dhanantwari *et al.*, "An efficient 3D beamformer implementation for real-time 4D ultrasound systems deploying array probes," in *Proc. IEEE Ultrason. Symp.*, Toronto, ON, Canada, Aug. 2004, pp. 1421–1424.

[18] M. Palmese and A. Trucco, "Three-dimensional acoustic imaging by chirp zeta transform digital beamforming," *IEEE Trans. Instrum. Meas.*, vol. 58, no. 7, pp. 2080–2086, Jul. 2009.

[19] M. Palmese and A. Trucco, "Chirp zeta transform beamforming for three-dimensional acoustic imaging," *J. Acoust. Soc. Amer.*, vol. 122, pp. EL191–EL195, Nov. 2007.

[20] M. Palmese and A. Trucco, "Pruned chirp zeta transform beamforming for 3-D imaging with sparse planar arrays," *IEEE J. Ocean. Eng.*, vol. 39, no. 2, pp. 206–211, Apr. 2014.

[21] Y. Mendelsohn and E. Wiener-Avnear, "Simulations of circular 2D phase-array ultrasonic imaging transducers," *Ultrasonics*, vol. 39, pp. 657–666, 2002.

[22] H. L. Van Trees, *Optimum Array Processing*. New York, NY, USA: Wiley, 2002, p. 22.

[23] A. Dutt and V. Rokhlin, "Fast Fourier transforms for nonequispaced data, II," *Appl. Comput. Harmonic Anal.*, vol. 2, pp. 85–100, 1995.

[24] J. A. Fessler and B. P. Sutton, "Nonuniform fast Fourier transforms using min-max interpolation," *IEEE Trans. Signal Process.*, vol. 51, no. 2, pp. 560–574, Feb. 2003.

[25] L. J. Ziomek, "Three necessary conditions for the validity of the Fresnel phase approximation for the near-field beam pattern of an aperture," *IEEE J. Ocean. Eng.*, vol. 18, no. 1, pp. 73–75, Jan. 1993.

[26] V. A. Oppenheim, R. W. Schafer, and J. R. Buck, *Discrete-Time Signal Processing*. Englewood Cliffs, NJ, USA: Prentice-Hall, 1989, ch. 9, pp. 726–745.

[27] L. Sha, H. Guo, and W. Song, "An improved gridding method for spiral MRI using nonuniform fast Fourier transform," *J. Magn. Resonance*, vol. 162, no. 2, pp. 250–258, 2003.

[28] D. Potts and G. Steidl, "New Fourier reconstruction algorithms for computerized tomography," in *Proc. Int. Symp. Opt. Sci. Technol.*, 2000, pp. 13–23.

[29] L. Michael and A. M. Zoubir, "Fast wideband near-field imaging using the non-equispaced FFT with application to through-wall radar," in *Proc. 19th Eur. Signal Process. Conf.*, Barcelona, Spain, 2011, pp. 1708–1712.

[30] M. Haltmeier, O. Scherzer, and G. Zangerl, "A reconstruction algorithm for photoacoustic imaging based on the nonuniform FFT," *IEEE Trans. Med. Imag.*, vol. 28, no. 11, pp. 1727–1735, Nov. 2009.

[31] P. Kruizinga, F. Mastik, N. de Jong, F. W. van der Steen, and G. van Soest, "Plane-wave ultrasound beamforming using a nonuniform fast Fourier transform," *IEEE Trans. Ultrason. Ferroelectr. Freq. Control.*, vol. 59, no. 12, pp. 2684–2690, Dec. 2012.

[32] K. Yang, Z. Zhao, and Q. H. Liu, "Fast pencil beam pattern synthesis of large unequally spaced antenna arrays," *IEEE Trans. Antennas Propag.*, vol. 61, no. 2, pp. 627–633, Feb. 2013.

[33] S. Kunis and D. Potts, "Time and memory requirements of the nonequispaced FFT," Technische Universität Chemnitz, Chemnitz, Germany, preprints 2006-01, 2006.

[34] J. A. Fessler, "Image reconstruction toolbox," July 2014 [Online]. Available: http://www.eecs.umich.edu/~fessler/code/index.html.

[35] M. Palmese and A. Trucco, "Acoustic imaging of underwater embedded objects: Signal simulation for three-dimensional sonar instrumentation," *IEEE Trans. Instrum. Meas.*, vol. 55, no. 4, pp. 1339–1347, Aug. 2006.

[36] W. J. Hendricks, "The totally random versus the bin approach for random arrays," *IEEE Trans. Antennas Propag.*, vol. 39, no. 12, pp. 1757–1762, Dec. 1991.

[37] R. E. Davidsen, J. A. Jensen, and S. W. Smith, "Two-dimensional random arrays for real-time volumetric imaging," *Ultrason. Imag.*, vol. 16, pp. 143–163, 1994.

[38] R. E. Davidsen and S. W. Smith, "A multiplexed two-dimensional array for real time volumetric and B-mode imaging," in *Proc. IEEE Ultrason. Symp.*, 1996, pp. 1523–1526.

[39] R. O. Nielsen, *Sonar Signal Processing*. Boston, MA, USA: Artech House, 1991, ch. 1–2.

Ultrawideband Underwater Real-Time 3-D Acoustical Imaging With Ultrasparse Arrays①

Cheng Chi, *Student Member, IEEE*, Zhaohui Li, and Qihu Li

***Abstract*—Large 2-D arrays are indispensable for underwater high-resolution real-time 3-D acoustical imaging, which however must be thinned significantly to avoid an overly high cost. The existing methods have decreased the number of elements from tens of thousands to several hundreds, which is still excessive for the implementation of underwater 3-D acoustical imaging system. Reported to be able to achieve an ultralow hardware cost in other fields, the ultrawideband (UWB) technology is introduced in this paper for the purpose of underwater real-time 3-D acoustical imaging. First, this paper reveals that the UWB technology is feasible for underwater 3-D acoustical imaging. Second, by analyzing the beam-steering properties of UWB underwater 2-D arrays, this paper demonstrates that although tens of elements are enough to achieve the required resolution and sidelobe level, they cannot guarantee the signal-to-noise ratio (SNR) for high imaging quality. Third, this paper proposes to employ the modulated excitation technique to increase the SNR in the underwater application. The analysis of computational load indicates that the modulated excitation technique has few influences on the implementation of underwater real-time 3-D acoustical imaging systems. At last, this paper presents a prototype of UWB underwater real-time 3-D acoustical imaging system with a 32-elements annular array, of which the performance is evaluated to verify the advantages of applying the UWB technology to underwater real-time 3-D acoustical imaging.**

***Index Terms*—Beam steering, imaging sonar, real-time 3-D acoustical imaging, sparse array, ultrawideband (UWB) array, UWB imaging.**

I. Introduction

All underwater activities are in need of some kinds of imaging sensors [1]. Being able to investigate the underwater environment beyond the range of optical visibility in a better way [2]–[10], underwater real-time 3-D acoustical imaging systems become increasingly attractive.

Two critical issues in the implementation of underwater real-time 3-D acoustical imaging systems are the computational load and the hardware cost [2]. Because tens of thousands of beams need to be formed in real time for the 3-D imaging systems, the computational load of using the conventional beamforming methods is overly high [2], [5], especially in the wideband situation. To generate high-resolution 3-D acoustical images in real time, a 2-D wideband array is essential for the collection of the echo signals from a 3-D scene. If a fully sampled uniform 2-D array that maintains half-wavelength element spacing is employed by the underwater imaging systems, the number of elements should be tens of thousands. The overly high hardware cost is almost prohibitive for any commercial applications at present.

To mitigate the overly high computational load in wideband real-time 3-D acoustical imaging applications, some wideband frequency-domain beamforming fast methods [5], [7], [8], [11] have been proposed. For a fully sampled uniform 2-D array or a sparse array thinning from the fully sampled uniform 2-D array, Palmese and Trucco propose the chirp zeta transform (CZT) and pruned CZT beamforming methods [5], [7], [8]. For an arbitrary array, Chi *et al.* give a fast method based on nonuniform fast Fourier transform (NUFFT) [11]. Compared with that of the conventional time-domain delay-and-sum method, the computational load of both CZT and NUFFT methods is reduced by about three orders of magnitude. In our view, these wideband beamforming methods are fast enough for the implementation of underwater real-time 3-D acoustical imaging systems, and it seems there are no urgent requirements about faster beamforming methods. Therefore, to further decrease the hardware cost is the top priority in the underwater real-time 3-D acoustical imaging application.

To mitigate the overly high hardware cost, some sparse array methods [4], [12]–[14] have been proposed and proved to be able to reduce the number of elements. For example, the number of elements of a large 2-D array can be reduced to 401 with the optimized simulated annealing algorithm [13]. However, until now, those methods have only decreased the number of elements to several hundreds, where the hardware cost is still too high and complex for the implementation. If tens of or fewer elements are possible for underwater real-time 3-D imaging, the implementation of the imaging systems will be simplified tremendously. Therefore, this paper focuses on how to use tens of or fewer elements to realize underwater real-time 3-D acoustical imaging systems.

As Schwartz and Steinberg [15] pointed out, ultrawideband (UWB) arrays can achieve a high angular resolution with few elements, without encountering the problem of grating lobes that

① IEEE Journal of Oceanic Engineering, 2017, 42(1): 97-108.

appears in narrowband, on account of which, UWB arrays have been applied to many fields in recent years, such as see-through-wall radar imaging [16]–[19], concealed weapon detection [20], breast cancer detection [21], [22], and vehicle-integrated industrial local positioning [23]. To our best knowledge, UWB arrays have not been applied to underwater real-time 3-D acoustical imaging by now. Thus, to achieve ultralow hardware cost, this paper tries to introduce the UWB technology into underwater 3-D imaging.

The first problem of applying the UWB technology into underwater real-time 3-D acoustical imaging is its feasibility in underwater acoustical channels. The numerical results of this paper indicate that the acoustical channel of underwater 3-D imaging is a UWB one in a certain frequency range. The development of the digital waveform generation and transducer fabrication techniques [33], [34] also makes the emission and collection of UWB signals possible in the underwater application.

The second problem is whether tens of or fewer elements are practical for underwater real-time 3-D acoustical imaging with the UWB technology. This paper shows that only tens of elements are needed to ensure the sidelobe level (SL) of underwater 3-D imaging when UWB arrays are employed, which is verified through a UWB annular array with 32 elements. However, with theoretical analysis, this paper also illustrates that being capable of achieving required SLs, UWB ultrasparse arrays still cannot guarantee enough signal-to-noise ratio (SNR) required by high imaging quality.

To increase the SNR in UWB systems, this paper proposes to apply the modulated excitation technique to underwater real-time 3-D acoustical imaging, which has been widely applied to radar [24], ultrasound medical imaging [25], [26], and other fields. For the modulated excitation technique, pulse compression at the receiving end is necessary. However, the pulse compression will result in extra computational load, which may influence the implementation of a real-time 3-D imaging system. The analysis in Section V-B shows that the computational load of the pulse compression is five orders of magnitude lower than that of beamforming for real-time 3-D imaging, which means the computational load caused by the pulse compression can nearly be neglected for the implementation of a real-time 3-D imaging system. Therefore, the computational load of the modulated excitation technique makes no effect on applying the technique to underwater real-time 3-D acoustical imaging.

As a result, a prototype of the UWB underwater real-time 3-D imaging system is designed in this paper to demonstrate the advantages of the UWB technology. Simulations show that the designed prototype has promising performance.

The contributions of this paper are summarized as follows:

1) revealing that the UWB technology is feasible for underwater real-time 3-D acoustical imaging;
2) illustrating that when UWB arrays are applied into underwater 3-D acoustical imaging, tens of elements are enough to obtain the required SL, but they cannot guarantee enough SNR;
3) proposing to apply the modulated excitation technique to improve the SNR of the UWB underwater 3-D acoustical imaging system;
4) designing a prototype of UWB underwater real-time 3-D acoustical imaging system with a 32-element annular array.

In what follows, Section II discusses the feasibility of the UWB technology for underwater real-time 3-D acoustical imaging. The beam-steering properties of UWB ultrasparse arrays are shown in Section III. Section IV analyzes the SNR obtained by the UWB ultrasparse arrays in the underwater situation. Section V presents how to use the modulated excitation technique to improve the SNR of the UWB underwater 3-D acoustical imaging systems. Section VI presents a prototype of UWB 3-D acoustical imaging system with a 32-element annular array. The conclusion and the discussion are given in Section VII.

II. Feasibility of the UWB Technology for Underwater Real-Time Acoustical Imaging

This section gives the definition of UWB signals and discusses the influences of underwater absorption attenuation on UWB acoustical signals' propagation thoroughly.

Conventional narrowband signals are with a small fractional bandwidth $\Delta f/f_0 \ll 1$, where Δf is bandwidth and f_0 is carrier frequency. When Δf and f_0 of the signals are comparable, the signals can be called the UWB ones. The definition of the UWB signals is given [27], [28] as

$$\frac{\Delta f}{f_0} \geq 1. \tag{1}$$

To facilitate the description, an impulse signal [29] is employed, which is expressed as

$$g(t) = \begin{cases} g_0 \exp\left[-\frac{\pi^2 \Delta f^2 (t-t_0)^2}{\beta}\right] \sin(2\pi f_0 t), & t \geq 0 \\ 0, & \text{elsewhere} \end{cases} \tag{2}$$

where g_0 is a constant amplitude, $t_0 = 1.5/\Delta f$ and $\beta = 1.2 \ln 10$. The range resolution is determined by

$$r_{\text{range}} = \frac{c}{2\Delta f} \tag{3}$$

where c is the underwater sound speed. Equation (3) shows that UWB signals can achieve a high range resolution.

Absorption attenuation is unavoidable when acoustical signals propagate in underwater channels. A logarithmic absorption coefficient (to the base 10) $\alpha(f)$ has been given to assess the absorption in [30]–[32], which depends on frequency f. As the frequency grows higher, the absorption coefficient becomes larger, which means that the UWB acoustical signals will distort when they propagate a certain distance underwater. Therefore, to apply the UWB technology into underwater 3-D acoustical imaging, the absorption attenuation must be assessed.

Assume that the acoustical signals are plane waves, the transmitted intensity at the frequency f is $I_0(f)$, and the distance between the imaged target and the transmitter is r. The received intensity at the location of the transmitter [30] is

$$I_1(f) = I_0(f) \cdot 10^{-2r\alpha(f)/10}. \tag{4}$$

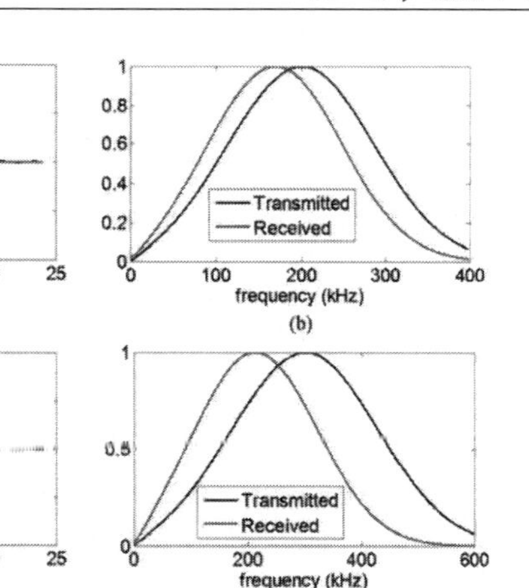

Fig. 1. Waveforms and spectra of the transmitted and received signals for different parameters: (a) and (b) for $\Delta f / f_0 = 1$, $f_0 = 200$ kHz, and $r = 200$ m; (c) and (d) for $\Delta f / f_0 = 1$, $f_0 = 300$ kHz, and $r = 200$ m. (The time delays of the received signals have been compensated.)

When $g(t)$ is taken as the transmitted acoustical signal, according to (4), the spectrum of the received signal $g_1(t)$ should be

$$G_1(f) = G(f) \cdot \left(10^{-2r\alpha(f)/10}\right)^{1/2} = G(f) \cdot 10^{-r\alpha(f)/10} \tag{5}$$

where $G(f)$ is the spectrum of $g(t)$.

A simple and specific expression of $\alpha(f)$ given in [30] is chosen, which is

$$\alpha(f) = \frac{0.1f^2}{1+f^2} + \frac{40f^2}{4.100+f^2} + 2.75 \times 10^{-4} f^2 + 0.003 \tag{6a}$$

where $\alpha(f)$ is in decibels per kiloyard, and f is the frequency in kilohertz. The range unit used in this paper is meter. One yard is equal to 0.914 m. To keep the consistency of the range units, the absorption in decibels per meter is given as

$$\alpha'(f) = \frac{\alpha(f)}{914}. \tag{6b}$$

By using (5), (6a), and (6b), it can be assessed whether the received signal is UWB.

When the attenuation caused by absorption in homogenous seawater is considered, for $\Delta f / f_0 = 1$, $f_0 = 200$, 300 kHz, and $r = 200$ m, the transmitted and received acoustical waveforms and their spectra are shown in Fig. 1. It can be found that although the mean frequencies of the received signals decrease, the received signals are still UWB.

To reflect the general law of UWB signals for underwater 3-D acoustical imaging, numerical calculations are conducted with different values of f_0 and r. According to the existing underwater 3-D acoustical imaging systems [1]–[10], the range of f_0 discussed is from 100 kHz to 1.2 MHz. Assume that the mean frequency and bandwidth of the received signal $g_1(t)$ are denoted by f_0^r and Δf^r, respectively. The fractional bandwidth $\Delta f^r / f_0^r$ is employed to judge whether the received signal is UWB.

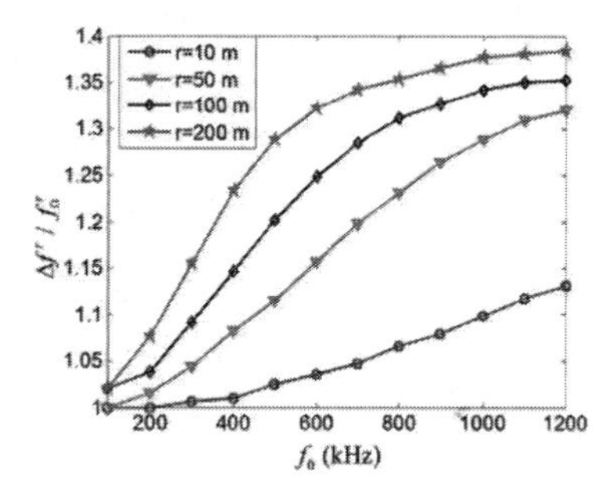

Fig. 2. Relationship between the fractional bandwidth of the received signal and the carrier frequency of the transmitted signal for different values of r.

In addition, if the mean frequency f_0^r decreases too much, the angular and range resolutions of the underwater imaging system will be influenced severely. Thus, the decrease of f_0^r must be considered to judge whether the UWB signals can be applied to underwater 3-D acoustical imaging in a certain situation. In this paper, the ratio f_0^r / f_0 is used to assess the decrease of f_0^r. Even though the received signals are UWB, we propose that when $f_0^r / f_0 < 0.7$, they are not considered as applying to underwater real-time 3-D acoustical imaging systems.

In the following, the relative bandwidths of the transmitted UWB signals are set as 1. Calculations that check $\Delta f^r / f_0^r$ changes with the carrier frequency f_0 for different distances are conducted and the results are shown in Fig. 2, which proves that the fractional bandwidths of all the received signals are greater than or equal to 1. Hence, according to (1), all the received signals are UWB.

Relationship between the mean frequency f_0^r of the received signal and the carrier frequency f_0 of the transmitted UWB signal for different distances are studied, as shown in Fig. 3. It can be found that the mean frequency f_0^r of the received signal

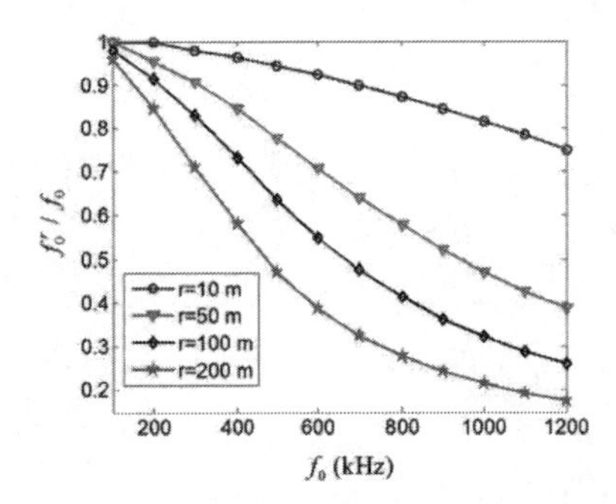

Fig. 3. Relationship between the mean frequency of the received signal and the carrier frequency of the transmitted signal for different imaging ranges r.

varies distinctly with the variation of f_0 for each r. For example, in the case of $r = 10$ m and $f_0 = 1200$ kHz, f_0^r/f_0 is 0.75, which means the UWB signal is valid for underwater 3-D imaging; while for $r = 200$ m and $f_0 \geq 400$ kHz, f_0^r/f_0, it is lower than 0.70, which means the UWB signal is not valid for underwater 3-D imaging. The general criteria of usable UWB signals in underwater 3-D acoustical imaging in Fig. 3 are summarized as follows: for $r = 10$, 50, 100, and 200 m, the carrier frequency f_0 of the UWB transmitted signal can arrive at 1200, 600, 400, and 300 kHz, respectively.

Besides, as pointed out in [33], generating arbitrary waveforms in ultrasound systems is accessible. The development of transducers [34] makes the emission and collection of underwater UWB signals available.

Based on the above analysis, it can be concluded that applying the UWB technology into underwater real-time 3-D acoustical imaging is feasible.

Since this paper aims at developing an underwater real-time 3-D imaging sonar with the imaging range $r = 200$ m, according to Fig. 3, the carrier frequency f_0 of the UWB transmitted signal employed here should be lower than or equal to 300 kHz.

III. Beam-Steering Properties of Underwater UWB Ultrasparse Arrays

On account of the fact that UWB ultra-sparse arrays can help to achieve an ultra-low hardware cost for underwater real-time 3-D acoustical imaging, the beam steering properties of underwater UWB ultra-sparse arrays are studied in this section.

A. Beam Steering Requirements

To develop an underwater real-time 3-D acoustical imaging system, the beam steering requirements of the 2-D array employed must be considered. For a 2-D circular array with the diameter D, according to [15], whether the array is narrowband or wideband, the lateral angle resolution θ_r, namely the main-lobe width of the beam pattern, is determined by

$$\theta_{\text{lateral}} \approx \frac{\lambda_0}{D} \cdot \frac{180^\circ}{\pi} \tag{7}$$

where λ_0 is the wavelength at f_0. If θ_{lateral} is required to be lower than 1°, $D \geq 57\lambda_0$. Because the lateral angle resolution of 1° is common in underwater 3-D imaging systems [2], [3], [10], in the following analysis, D is set to be $60\lambda_0$.

The SL of the beam pattern of the 2-D arrays determines the imaging quality. As reported in [4], [6], [12], the maximum SL of the beam pattern applied in underwater 3-D imaging systems is around -22 dB. This value is also considered in the rest of this paper.

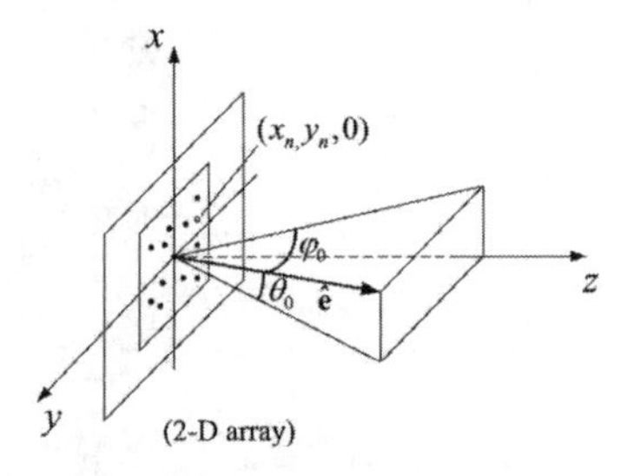

Fig. 4. Coordinate system used to compute wideband BP of 2-D arrays.

B. Definition of Wideband Beam Pattern

Few papers have discussed the wideband beam pattern (BP) of underwater imaging 2-D arrays. Unlike the radiation pattern of a narrowband 2-D array, which varies sinusoidally with time, the radiation pattern of a wideband 2-D array depends on the observation direction and time in a complicated way. Hence, the wideband BP is very different from the narrowband BP. In the following, the definition of the wideband BP of 2-D arrays is given.

The coordinate system presented in Fig. 4 [12] is used to compute wideband BP of 2-D arrays. As shown in Fig. 4, θ_0 and φ_0 are the steering elevation and azimuth angles respectively. θ_0 is the angle between the unit vector $\widehat{\mathrm{e}}$ of the steering direction and its projection on the plane yz. φ_0 is the angle between $\widehat{\mathrm{e}}$ and its projection on the plane xz. The unit vector $\widehat{\mathrm{e}}$ can be expressed as

$$\widehat{\mathrm{e}} = \left(\sin\theta_0, \sin\varphi_0, \sqrt{\cos\theta_0^2 - \sin\varphi_0^2}\right). \tag{8}$$

Being similar to that of 1-D arrays in [29], [35] and [36], the radiation pattern of the 2-D array in the far field is

$$p(\theta, \theta_0, \varphi, \varphi_0, t) = \sum_{n=1}^{N} g\left(t - \left(\frac{x_n(\sin\theta - \sin\theta_0)}{c} + \frac{y_n(\sin\varphi - \sin\varphi_0)}{c}\right)\right) \tag{9}$$

where θ and φ are the observation elevation and azimuth angles respectively, and $(x_n, y_n, 0)$ is the coordinate of the nth element of the 2-D array. To simply the description, assume that

$$u_x = \sin\theta - \sin\theta_0 \quad \text{and} \quad u_y = \sin\varphi - \sin\varphi_0. \tag{10}$$

Then, (9) can be rewritten as

$$p(u_x, u_y, t) = \sum_{n=1}^{N} g\left(t - \left(\frac{x_n u_x}{c} + \frac{y_n u_y}{c}\right)\right). \tag{11}$$

To reduce the complexity, the maximum projection is used for the wideband BP, which is expressed as

$$BP(u_x, u_y) = \max_t |p(u_x, u_y, t)|. \tag{12}$$

Generally, it has $-1 \leq u_x, u_y < 1$.

C. Performance of UWB Ultra-Sparse Underwater Arrays

Consider that an ideal UWB array with N elements at arbitrary locations is used to receive a very short impulse whose amplitude is 1. Because the impulse is very short, all the spacings between every two adjacent elements exceed the impulse length. The received impulses of the elements are incoherent in other directions except the steering direction of the ideal UWB array. Therefore, the amplitude of the coherent output impulse of the ideal UWB array at the steering direction should be N, while the maximum amplitude of the output waveform of the ideal UWB array in other directions should be 1. As a result, for the ideal UWB array, the ideal maximum SL [15] is given by

$$\mathrm{SL}_{\mathrm{ideal}} = 20\lg\frac{1}{N}. \tag{13}$$

As analyzed in Section III-A, the maximum SL required in underwater real-time 3-D imaging systems is around -22 dB. According to (13), for $\mathrm{SL}_{\mathrm{ideal}} = -22$ dB, the ideal number of elements should be

$$\begin{aligned} N_{\mathrm{ideal}} &= \frac{1}{10^{\mathrm{SL}_{\mathrm{ideal}}/20}} \\ &= \frac{1}{10^{-22/20}} \\ &\approx 13. \end{aligned} \tag{14}$$

It means that in the ideal UWB situation, 13 elements can achieve the -22-dB SL required in the underwater application. However, in practical situations, since the target scatterers may not be so distinct that an ideal UWB array is hard to obtain, the maximum SL is usually higher than the ideal one. Thus, in the design of the practical UWB arrays, the number of elements should be larger than the ideal number of elements computed by (14).

Since this paper mainly focuses on discussing the capacity of UWB arrays in underwater real-time 3-D acoustical imaging rather than designing an optimal UWB array, we just provide two ultrasparse arrays to verify the advantages of UWB arrays here.

One array shown in Fig. 5(a) is an equally spaced square array with the interelement spacing $10\lambda_0$, of which the number of elements is 7×7 and the side length is $60\lambda_0$. The other array shown in Fig. 5(b) is an annular uniform array with 32 elements and the diameter $60\lambda_0$. As presented in Section II, the carrier frequency f_0 of the array is set to 300 kHz.

Fig. 6(a)–(f) shows the BPs of the two arrays at the fractional bandwidth of $\Delta f/f_0 = 10\%, 50\%, 100\%$. Because the steering and observation angles have been represented by normalized angular variables u_x and u_y, naturally the steering angles correspond to $u_x = 0$ and $u_y = 0$, i.e., $\theta = \theta_0$ and $\phi = \phi_0$, respectively, in the figures. It can be seen that the SLs of the BPs of both arrays decrease greatly with the increase of the fractional bandwidth. For the square array, the high grating lobes still appear at $\Delta f/f_0 = 100\%$, namely, in the UWB situation, as shown in Fig. 6(c), which is mainly due to the element shadowing [15], [37], [38]. For the annular array, when $\Delta f/f_0 = 100\%$, no evident grating lobes appear and the SLs are below -23 dB, as shown in Fig. 6(f), which meet the SL requirement of -22 dB for the underwater 3-D imaging application. Therefore, in view of SL requirement, the UWB annular array with 32 elements is feasible for underwater real-time 3-D acoustical imaging.

As mentioned in Section III-A, for underwater 3-D acoustical imaging, an equally spaced half-wavelength square array with the side length $60\lambda_0$ will have 121×121 elements, while for

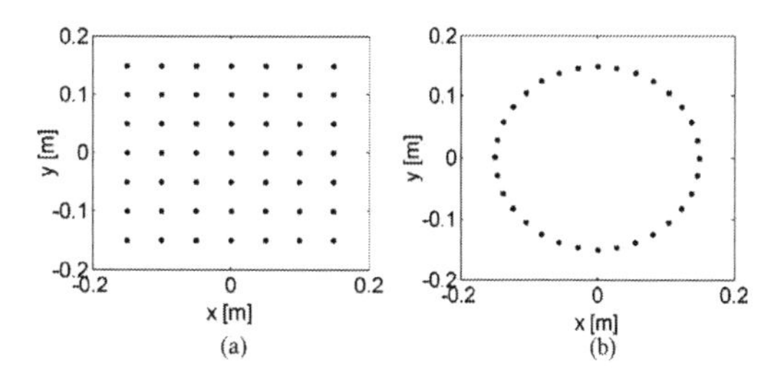

Fig. 5. Two ultrasparse arrays: (a) the equally spaced square array; (b) the annular array.

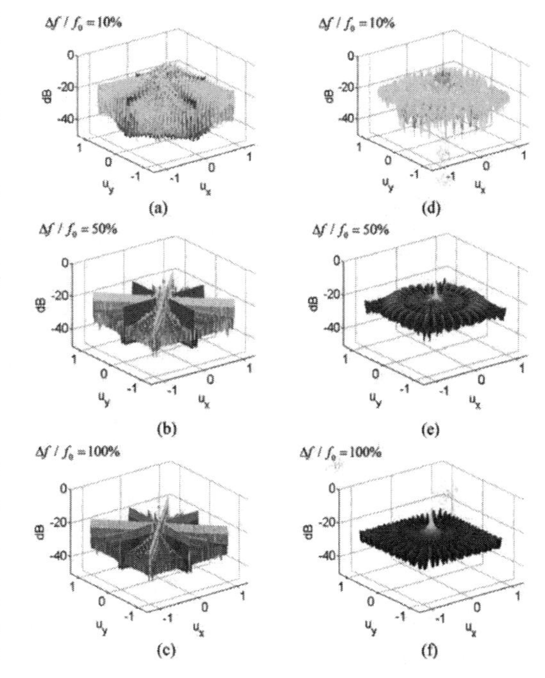

Fig. 6. BPs of the two arrays at the fractional bandwidth of $\Delta f/f_0 = 10\%$, 50%, 100%: (a)–(c) for the equally spaced square array with 7×7 elements; (d)–(f) for the annular array with 32 elements.

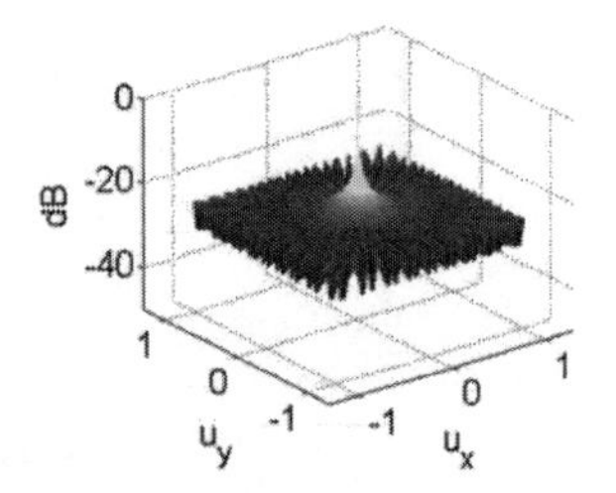

Fig. 7. BP of the annular array in Fig. 5(b) when considering the absorption attenuation of $r = 200$ m.

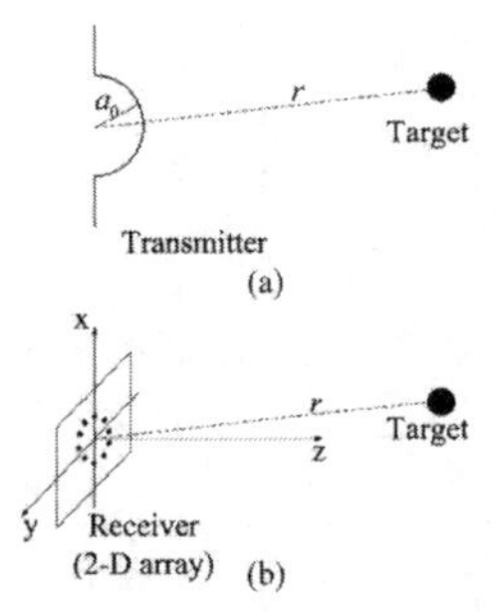

Fig. 8. Scene of underwater real-time 3-D imaging involved in (a) the transmitter and (b) the receiver.

the UWB array, 32 elements or fewer may achieve the similar SL performance. Hence, in the underwater 3-D acoustical imaging application, UWB arrays have the potential to achieve an ultrasparse structure, which means ultralow hardware cost.

In the above analysis, we only use the transmitted signal to compute the BP, where the absorption attenuation is not considered. In fact, when the signal distortion caused by the absorption attenuation occurs, the BP of the array also changes. To assess the absorption attenuation's influences on the BP of the UWB annular array, the BP is also computed by using the received signal shown in Fig. 1(c). The result is depicted in Fig. 7, where the steering angles correspond to $u_x = 0$ and $u_y = 0$, respectively, as usual. It is seen that the SLs are still below -23 dB. Hence, when the absorption attenuation is taken into account, the UWB annular array can still be applied to underwater 3-D imaging.

IV. SNR Analysis of Using UWB Arrays

As shown in Section III, with the usage of the UWB technology, tens of elements are enough to meet the SL requirement in underwater real-time 3-D acoustical imaging systems. However, whether the output SNR of such UWB underwater arrays can guarantee the high imaging quality has not been discussed. The SNR is analyzed in this section.

Due to the real-time requirement, the transmitter is omnidirectional and the receiver of underwater 3-D imaging should be a large 2-D array, respectively [4], [10]. A 3-D imaging scene is shown in Fig. 8.

When underwater acoustical signals are transmitted, the limitation caused by the cavitation [30] is encountered. If the radiated intensity exceeds a certain value, namely the cavitation threshold I_c, the transmitter cannot work well [30]. When the hemispherical transmitter with the radius a_0 is used, as shown in Fig. 8(a), the maximum acoustical power radiated by the transmitter can be expressed as

$$P_0 = 2\pi a_0^2 I_c. \tag{15}$$

In fact, the cavitation threshold depends on frequency. To simplify the analysis, the cavitation threshold at the carrier frequency f_0 is chosen to evaluate the radiation limitation of the UWB signals.

When the UWB impulse given by (2) is taken as the transmitted signal, the impulse energy is

$$E_0 = \int G^2(f)\, df. \tag{16}$$

The power of the transmitted UWB impulse can be given as

$$P_g = \frac{E_0}{T_0} = \frac{\int G^2(f)\, df}{T_0} \tag{17}$$

where T_0 is the equivalent duration of the UWB signal. Generally, T_0 is approximately equal to $1/\Delta f$. In this section, it is chosen that $P_g = P_0$, then the impulse energy can be rewritten as

$$E_0 = 2\pi a_0^2 I_c T_0. \tag{18}$$

As analyzed in Section II, the underwater absorption causes a form of loss, which depends on the frequency. For the UWB impulse given in (2), the incident acoustical intensity I_i at the imaging distance r can be written as

$$I_i = \frac{\int G^2(f)\, 10^{(-\alpha(f)r/10)} df}{2\pi r^2 T_0}. \tag{19}$$

Usually, underwater real-time 3-D imaging sonars should work in both near and far fields [2]. Consider that a spherical target with the radius a is taken as an isotropic scatterer, as shown in [30]. The acoustical signal in the far field can be taken as a plane wave. The intensity of the backscattered wave is given as

$$I_s = \sigma I_i \tag{20}$$

where σ is the equivalent area of the cross section of the imaged target [30], and

$$\sigma = \pi a^2. \tag{21}$$

The scattered wave spreads and is received by the receiving 2-D array with the intensity on the surface of each element being

$$I_r = \frac{\sigma \int G^2(f)\, 10^{(-2\alpha(f)r/10)} df}{(2\pi r^2 T_0)\cdot(4\pi r^2)}. \tag{22}$$

Urick [30] and Mellen [39] have shown that when the frequency exceeds 100 kHz, the main underwater ambient noise source is the thermal noise of the molecules of the sea, which places a limit on the hydrophone sensitivity at high frequencies. In the Appendix, the mean intensity of the thermal noise is analyzed, which can be evaluated by f_0 and the bandwidth Δf for the UWB situation. The mean noise intensity in the Appendixis expressed as

$$I_n = \frac{4\pi^2 KT\Delta f}{c^2}\left(f_0^2 + \frac{\Delta f^2}{12}\right) \tag{23}$$

where K is Boltzmann's constant given by $K = 1.37 \times 10^{-23}$ joule/degree Abs., and T is the absolute temperature of seawater.

Assume that the element is lossless, and the equivalent receiving area of the element is S_e. For an underwater 3-D imaging system, after receiving the echoes and beamforming, the output signal power is given by

$$P_1 = N^2 S_e I_r. \tag{24}$$

As the noise received by each element is irrelevant, the output noise power of the system is expressed as

$$P_2 = NS_e I_n. \tag{25}$$

Hence, the output SNR of the imaging system can be given as

$$\begin{aligned} snr &= 10\lg\frac{P_1}{P_2} \\ &= 10\lg\frac{NI_0}{I_n}. \end{aligned} \tag{26}$$

By substituting (6b), (22), and (23) into (26), the theoretical relationship between the SNR of the system snr and the number of elements N is

$$snr = 10\lg\left\{\frac{N\left[\frac{\sigma\int G^2(f)10^{\left(-2\alpha'(f)r/10\right)}df}{(2\pi r^2 T_0)\cdot(4\pi r^2)}\right]}{\frac{4\pi^2 KT\Delta f}{c^2}\left(f_0^2 + \frac{\Delta f^2}{12}\right)}\right\}. \tag{27}$$

To obtain a high-quality image, the SNR of 10–20 dB is needed [41]. In what follows, some numerical results are presented. The limitation on the transmitted power is the cavitation threshold I_c, which is formulated [30] as

$$I_c = 1500 \times \left(p_c(0) + \frac{h}{33}\right)^2 \quad \text{W/m}^2 \tag{28}$$

where $p_c(0)$ is the peak pressure of sound wave causing cavitation expressed in atmospheres at $h = 0$, and h is the depth of the transmitter. From (2), (18), (19), and (28), the energy E_0 of the UWB impulse and g_0 can be obtained. It is assumed that $h = 0$ and $p_c(0) = 160$ atm for $f_0 = 300$ kHz [30]. The radius a of the isotropic scatterer is $10\lambda_0$. The radius a_0 of the hemispherical transmitter is set $4 \sim 8\lambda_0$, which is in accordance with the transmitters shown in [3] and [10]. The parameters for computing the SNR are summarized as follows:

1) sound speed: $c = 1500$ m/s;
2) carrier frequency: $f_0 = 300$ kHz;
3) signal bandwidth: $\Delta f = 300$ kHz;
4) maximum imaging range: $r = 200$ m;
5) seawater temperature: $T = 283°$K;
6) cavitation threshold: $I_c = 3.84 \times 10^7$ W/m^2 for $f_0 =$ 300 kHz;
7) radius of the transmitter: $a_0 = 2, 3, 4$ cm for $f_0 = 300$ kHz.

For example, when $a_0 = 4$ cm and $N = 32$, snr is 7 dB, which is not high enough to achieve a high-quality 3-D image. In fact, to the authors' knowledge, if snr is lower than 15 dB, a high-quality image is hard to obtain. Based on (27) and the given parameters, it can be obtained that when $N \leq 200$, all $snrs$ are lower than 15 dB for all values of a_0, which means that when $N \leq 200$, using the UWB impulse of $\Delta f / f_0 = 1$ cannot guarantee enough SNR.

To improve the SNR, one choice is to increase the radius a_0 of the transmitter to increase the transmitted power. However, this choice increases the hardware cost. In the following section, it is presented that the modulated excitation technique can help improve the SNR dramatically, when the same transmitter is used.

V. Modulated Excitation Technique

The modulated excitation technique consists of modulated excitation signal design and pulse compression [25]. A modulated excitation UWB signal and a mismatched filter are chosen for underwater 3-D real-time acoustical imaging. The practicability of the modulated excitation technique in the underwater channel situation is demonstrated, while the negative effect of using modulated excitation technique, i.e., the computational load of the pulse compression, is also discussed in this section.

A. Signal Design and Pulse Compression

The modulated excitation technique is realized by transmitting modulated excitation signals and using pulse compression filters to process echoes. The gain of modulated excitation signals can be measured by the time–bandwidth (TB) product [25]. There are many filters that can be used for pulse compression, such as matched [24], mismatched [26], inverse [41], [42], and Wiener filters [43], [44], among which the matched filter achieves the optimal SNR gain. When using the matched filter, the ideal SNR gain [25] is

$$\text{GSNR} = 10\lg\text{TB}. \tag{29}$$

However, in the underwater imaging application, the effect of the frequency-dependent absorption is not negligible. Therefore, the modulated excitation technique used in underwater 3-D imaging should be properly chosen. It requires that the modulated excitation signals should be robust to the frequency-dependent absorption and the range SLs of the pulse compression filter output should be relatively low.

For the situation in Section IV where $f_0 = 300$ kHz, $\Delta f = B = 300$ kHz, and $N = 32$, according to (27), when $a_0 = 4$ cm, the SNR is 7 dB. Thus, GSNR should be greater or equal to 8 dB to ensure 15-dB SNR, which can be achieved with $T = 21$ μs in an ideal situation, according to (29). However, because the mismatched filter and the distortion caused by the absorption

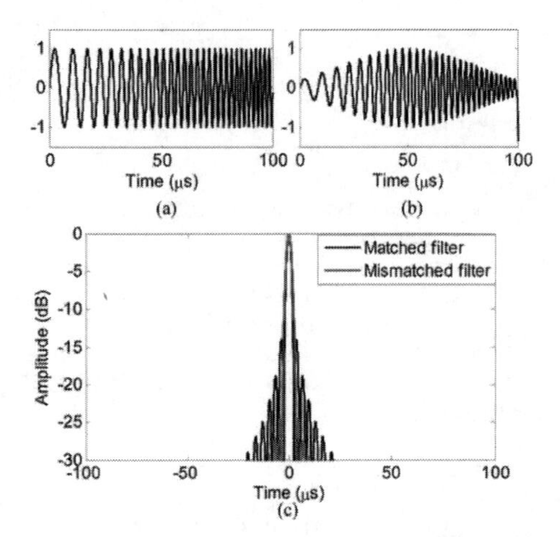

Fig. 9. Linear FM signal designed for underwater 3-D imaging and the pulse compression filter: (a) the linear FM signal; (b) the impulse response of the pulse compression filter, namely the mismatched filter; and (c) the outputs of the matched (black line) and mismatched (red line) filters.

will decrease GSNR, the duration of the linear FM should be longer than 21 μs. In this paper, it is set to 100 μs, corresponding to a 14.7-dB SNR gain.

Many kinds of modulated excitation signals such as linear frequency modulation (FM) signals [24]–[26], nonlinear FM signals [45], and binary complementary Golay codes [46] have been applied to radars and medical ultrasound imaging. In [25] and [26], it was shown that linear FM signals have the best performance and most robust features for medical ultrasound imaging. Since the underwater 3-D acoustical imaging situation is similar to the medical ultrasound imaging one, the linear FM signals are chosen as the modulated excitation in this paper, which can be expressed in complex notation [26] as

$$\psi(t) = \exp\left\{j2\pi\left[\left(f_0 - \frac{B}{2}\right)t + \frac{B}{2T}t^2\right]\right\}, \quad 0 \le t \le T. \tag{30}$$

A linear FM signal with $f_0 = 300$ kHz, $B = 400$ kHz, and $T = 100$ μs is shown in Fig. 9(a).

For underwater 3-D acoustical imaging, the SL of BPs is usually set to -22 dB, as presented in Section III. Thus, the output range SL of the pulse compression filters should not be higher than -22 dB to achieve a high-quality 3-D image. Since the matched filter can only provide a range SL of -13.6 dB, as given in Fig. 9(c), a mismatched filter should be employed for underwater 3-D acoustical imaging.

A usual approach to realize mismatched filters is to use a window function on the time-domain response of the matched filter [26]. In this paper, a Dolph–Chebyshev window with -30-dB SL is used to compress the range SLs of pulse compression. The time-domain response of the designed mismatched filter is shown in Fig. 9(b). Because the mismatched filter helps compress the range SLs with the decrease of the system bandwidth, the bandwidth of the original linear FM signal is set to 400 kHz to make the system keep UWB. From Fig. 9(c), it is noted that the range SL of the mismatched filter reaches -24 dB, which is lower than -22 dB.

In the above analysis, the effect of the underwater absorption attenuation has not been discussed. To evaluate the performance of the mismatched filter in the frequency-dependent absorption medium, with estimation of (6), the received FM signal is also simulated at the imaging range of $r = 200$ m. The simulation result is that the range SL of the mismatched filter becomes lower and reaches below -30 dB when the underwater absorption attenuation is taken into account, which is better than that of not considering absorption. At the same time, the gain of SNR, i.e., GSNR, decreases from 14.7 to 10 dB. The simulation results reveal that although the SNR gain of the FM signal is decreased through the absorptive channel, it still meets the requirement (8 dB) to obtain a high-quality image.

B. Computational Load of the Pulse Compression

The fast frequency-domain digital beamforming methods such as the CZT [5], [7], [8] and the NUFFT [11] have been proven to be proper choices for underwater real-time 3-D acoustical imaging. If the computational load of the pulse compression is higher than that of the fast beamforming methods, it will be questioned whether the modulated excitation technique can be applied to underwater real-time 3-D acoustical imaging. Thus, the computational load of the pulse compression should be evaluated and compared with that of the fast beamforming methods.

As shown in [5], [7], [8], and [11], which used those frequency-domain beamforming methods to realize 3-D acoustical imaging, three steps are necessary. The first is to convert the segmented time-domain samples into frequency domain by fast Fourier transform (FFT). The second is to use the CZT or the NUFFT to realize fast beamforming. The final step is to convert the frequency-domain beam signals to the time domain by inverse fast Fourier transform (IFFT). Therefore, when the modulated excitation technique is applied to underwater 3-D imaging, the pulse compression of UWB underwater 3-D imaging systems can be realized in the frequency domain following the first step and before the second step.

Assume that $M(f_k)$ is the discrete frequency-domain response of the mismatched filter, where f_k represents the kth frequency bin, and the discrete Fourier transform (DFT) coefficients of the received signal of the nth element are $S_n(f_k)$. The frequency-domain pulse compression of using the mismatched filter [25], [47] can be described by

$$CS_n(f_k) = S_n(f_k) \cdot (M(f_k))^* \tag{31}$$

where $(\cdot)^*$ represents the conjugate operation. Equation (31) shows that at one frequency bin, the pulse compression needs one complex multiplication, namely, six real operations without distinguishing the real addition and multiplication. The total number of real operations of the pulse compression at one frequency bin for N elements can be expressed as

$$B_{PC} = 6N. \tag{32}$$

TABLE I
COMPUTATIONAL LOAD COMPARISON

Pulse compression	CZT beamforming	NUFFT beamforming
0.96×10^5	1.77×10^{10}	1.18×10^{10}

The CZT beamforming [8] can be applied to equally spaced 2-D arrays or sparse arrays thinning from the equally spaced 2-D arrays. When a 2-D array with $N_e \times N_e$ elements is employed and $M_b \times M_b$ beams need to form, the number of real operations of the CZT beamforming at one frequency bin [8] is

$$B_{\mathrm{CZT}} = 6\left[N_e^2 + M_b^2 + L^2\right] + 20L^2 \log L \tag{33}$$

where L is a power of two, greater than or equal to $N_e + M_b - 1$. Moreover, the number of real operations does not reduce with the decrease of elements when using the sparse technique.

The NUFFT beamforming [11] is suitable for arbitrary 2-D arrays. When a 2-D with N elements is employed, $M_b \times M_b$ beams need to form, and then the number of real operations of the NUFFT beamforming at one frequency bin [11] is

$$B_{\mathrm{NUFFT}} = N\left(8J^2 - 2\right) + 5M_b'^2 \log_2\left(M_b'^2\right) \tag{34}$$

where J is the interpolation coefficient, $M_b' = \sigma M_b$, and σ is the oversampled factor. For the underwater application, $J = 4$ and $\sigma = 2$ are the proper choices [11].

In the following, a numerical comparison is given. The parameters for the comparison are summarized as follows:

1) $N = 32$;
2) $N_e \times N_e = 121 \times 121$;
3) $M_b \times M_b = 256 \times 256$.

Assume there are $N_f = 500$ frequency bins in the passband. The computational loads of the pulse compression, the CZT, and NUFFT beamforming in the whole passband are shown in Table I. It can be found that the computational load of the pulse compression is five orders of magnitude lower than that of the CZT or NUFFT beamforming. It means that comparing with that of the beamforming methods in underwater real-time 3-D acoustical imaging, the computational load of the pulse compression can be neglected. Therefore, the pulse compression is feasible and practical for underwater real-time 3-D acoustical imaging to improve the SNR of the system.

VI. UNDERWATER UWB IMAGING SYSTEM PROTOTYPE

A UWB imaging sonar prototype is designed in this section to test the validity of applying the UWB technique to underwater real-time 3-D imaging. The schematic of the designed UWB sonar is shown in Fig. 10. The UWB annular array with 32 elements given in Section III is taken as the receiving array. The frequency response of the transmitting and receiving transducers is shown in Fig. 11. To guarantee the high SNR, the linear FM signal and the mismatched filter in Section V are chosen for the UWB 3-D imaging sonar. In the near field, according to [4] and [6], the field of view of the UWB sonar is chosen as $26° \times 26°$ and the dynamic focusing technique [7], [11] is adopted. NUFFT

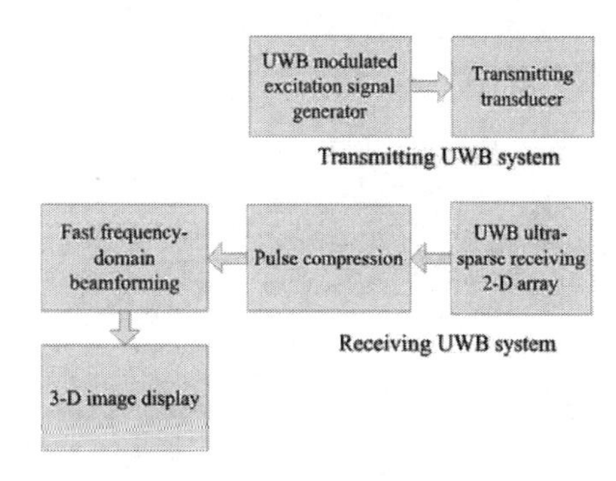

Fig. 10. Schematic of a UWB underwater real-time 3-D acoustical imaging system.

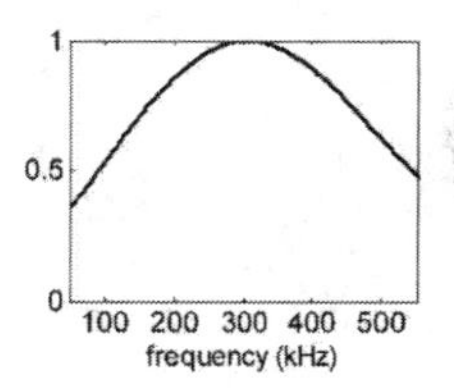

Fig. 11. Frequency response of the transmitting and receiving transducers.

TABLE II
MAIN FEATURES OF THE PROTOTYPE OF THE UWB UNDERWATER 3-D ACOUSTICAL IMAGING SYSTEM

Central frequency		300 kHz
Bandwidth		300 kHz
Maximum imaging range		200 m
Number of elements		32
Angular resolution		1°
Field of view	Near field	$26° \times 26°$
	Far field	$60° \times 60°$
Range resolution		2.5 mm

beamforming [11] is applied to decrease the computational load. The main features of the designed UWB 3-D imaging sonar are presented in Table II.

The simulation of the backscattered acoustical signals is conducted by using the method given in [48]. An underwater 3-D scene with three scattering square frames (A, B, and C) shown in Fig. 12 is employed, of which the positions are given by the coordinate system in Fig. 4. The square frame A represents the near field and the high SNR case. The square frame B represents the far field and the high SNR case. The square frame C is in the far field and the low SNR case.

Figs. 13 and 14 depict the 2-D slices of the 3-D images at different focusing distances but in high SNR cases obtained by using the designed UWB 3-D imaging system. It can be found that both square frames can be seen clearly. Fig. 15(a) and (b) depicts the 2-D slices in the case of the far field and low SNR without using the modulated excitation techniques and using the modulated excitation techniques, respectively. In Fig. 15(a),

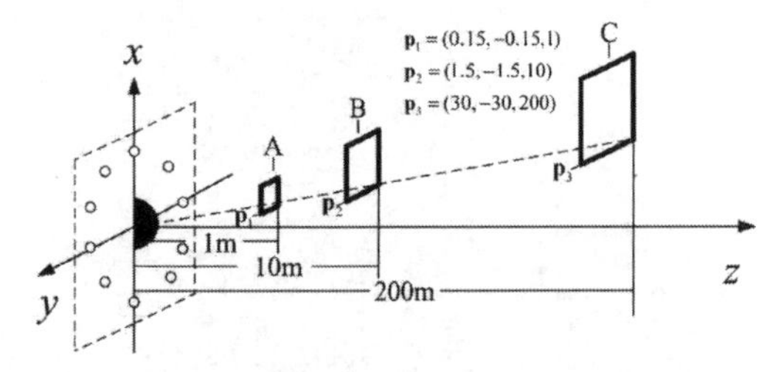

Fig. 12. Three-dimensional scene of the UWB imaging sonar with the three scattering square frames (A, B, and C). (The three scattering square frames parallel to the plane xy. The side lengths of A, B, and C are 0.075, 0.75, and 15 m. Their positions are described by the point coordinates $\mathbf{p}_1$, $\mathbf{p}_2$, and $\mathbf{p}_3$.)

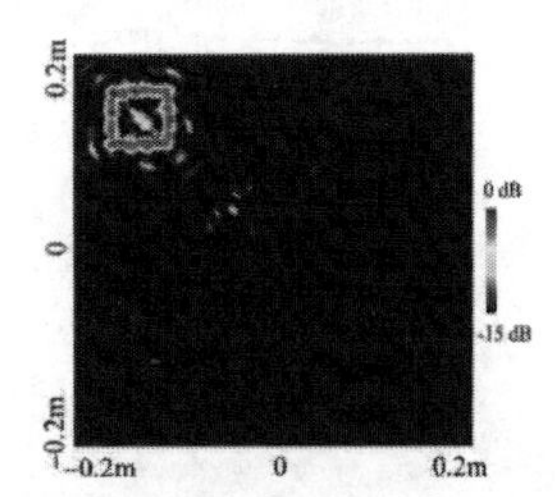

Fig. 13. Two-dimensional slice of the 3-D image of the square frame A (z = 1 m) when using the designed UWB imaging system prototype.

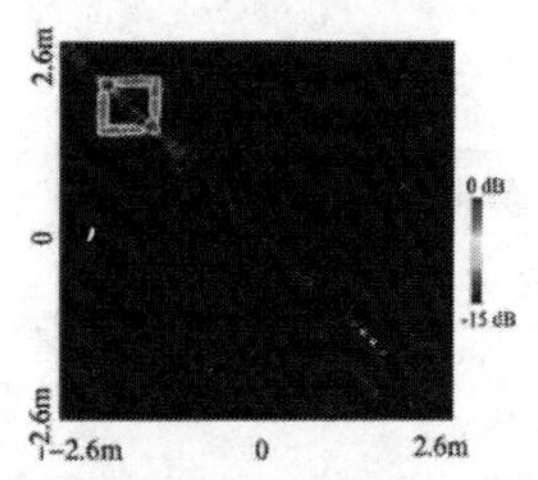

Fig. 14. Two-dimensional slice of the 3-D image of the square frame B (z = 10 m) when using the designed UWB imaging system prototype.

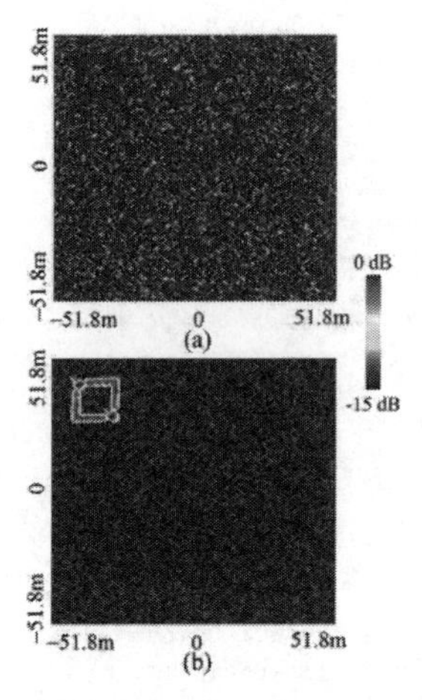

Fig. 15. Two-dimensional slice of the 3-D image of the square frame C (z = 200 m) when using the designed UWB imaging system prototype: (a) without using the modulated excitation technique; and (b) using the modulated excitation technique.

where the modulated excitation technique is not applied, the square frame nearly disappears. However, in Fig. 15(b), where the modulated excitation technique is applied, although the image quality worsens in the low SNR situation, the square frame can still be seen clearly. The comparison between Fig. 15(a) and (b) verifies that for UWB ultrasparse underwater 3-D acoustical imaging, the modulated excitation technique can help increase the SNR and improve the imaging quality.

To sum up, the results in Figs. 13, 14, and 15(b) demonstrate that the designed UWB 3-D acoustical imaging system prototype has a good performance.

VII. Discussion and Conclusion

This paper aims at applying the UWB technology to achieve an ultralow hardware cost for underwater real-time 3-D acoustical imaging. By analyzing the effect of underwater absorption attenuation, this paper reveals that UWB signals are applicable for underwater 3-D imaging acoustical channels. The investigation of the literatures demonstrates that the development of digital waveform generation and transducer fabrication techniques makes transmitting and receiving underwater UWB signals possible. The first conclusion is that the UWB technology is feasible for underwater real-time 3-D acoustical imaging.

On the basis of the analysis of beam-steering properties of UWB ultrasparse arrays, this paper demonstrates that when using UWB arrays, tens of elements are enough to meet the SL requirement of underwater 3-D acoustical imaging. The second conclusion is that a UWB ultrasparse array can achieve nearly the same SL performance as the existing narrowband arrays do with hundreds or more elements in the underwater 3-D acoustical imaging situation.

By theoretically analyzing the SNR, this paper points out that the UWB ultrasparse impulse arrays cannot guarantee enough SNR required by high quality of acoustical imaging. To improve the SNR of the UWB underwater 3-D acoustical imaging system, this paper proposes to employ the modulated excitation technique, where a linear FM signal and a mismatched filter are used to realize the pulse compression in the frequency-dependence absorption medium. The modulated excitation technique is verified to be effective for the UWB ultrasparse arrays to obtain the required high SNR in the underwater absorptive acoustical channels. Moreover, this paper proves that compared with that of the fast frequency-domain beamforming methods in underwater 3-D acoustical imaging, the computational load of the pulse compression, which is the negative effect of the modulated excitation technique, is negligible. Hence, the pulse compression is feasible for underwater real-time 3-D acoustical imaging. The third conclusion is that the modulated excitation technique is practical for UWB underwater real-time 3-D acoustical imaging.

A prototype of UWB underwater real-time 3-D acoustical imaging system with a 32-element annular array is designed to verify the advantages of the UWB technology. The imaging results show that the UWB underwater real-time 3-D acoustical imaging system has a promising performance. Therefore, the UWB technology provides a new choice for developing underwater real-time 3-D acoustical imaging systems.

Although this paper introduces the UWB technology into underwater real-time 3-D acoustical imaging successfully, the optimum performance of the UWB technology in the underwater application has not been achieved. In the future, better UWB ultrasparse arrays can be designed to achieve fewer elements. New modulated excitation signals and better pulse compression filters can be studied to make the modulated excitation technique perform better in underwater real-time 3-D acoustical imaging.

APPENDIX

As given in [39] and [40], the mean squared thermal-noise pressure is

$$\langle p^2 \rangle = \frac{1}{2}\frac{KT\rho_0}{\pi^2 c}\int \omega^2 d\omega \tag{35}$$

where ρ_0 and c are the density and sound velocity of the seawater, respectively, K is Boltzmann's constant, and T is the absolute temperature of seawater. The mean acoustical intensity of the thermal noise can be

$$I_n = \frac{\langle p^2 \rangle}{\rho_0 c}. \tag{36}$$

Let $\omega = 2\pi f$, then (36) can be rewritten as

$$I_n = \frac{4\pi^2 KT}{c^2}\int f^2 df. \tag{37}$$

The bandwidth is Δf and the central frequency of the system is f_0. Therefore, (37) can be expressed as

$$\begin{aligned} I_n &= \frac{4\pi^2 KT}{c^2}\int_{f_0-\frac{\Delta f}{2}}^{f_0+\frac{\Delta f}{2}} f^2 df \\ &= \frac{4\pi^2 KT\Delta f}{c^2}\left(f_0^2 + \frac{\Delta f^2}{12}\right). \end{aligned} \tag{38}$$

REFERENCES

[1] R. Hansen and P. Andersen, "The application of real time 3-D acoustical imaging," in *Proc. IEEE OCEANS Conf.*, 1998, pp. 738–741.

[2] V. Murino and A. Trucco, "Three-dimensional image generation and processing in underwater acoustic vision," *Proc. IEEE*, vol. 88, no. 12, pp. 1903–1948, Dec. 2000.

[3] A. Davis and A. Lugsdin, "High speed underwater inspection for port and harbor security using Coda Echoscope 3D sonar," in *Proc. IEEE/MTS OCEANS Conf.*, 2005, pp. 2006–2011.

[4] A. Trucco, M. Palmese, and S. Repetto, "Devising an affordable sonar system for underwater 3-D vision," *IEEE Trans. Instrum. Meas.*, vol. 57, no. 10, pp. 2348–2354, Oct. 2008.

[5] M. Palmese and A. Trucco, "Three-dimensional acoustic imaging by chirp zeta transform digital beamforming," *IEEE Trans. Instrum. Meas.*, vol. 58, no. 7, pp. 2080–2086, Jul. 2009.

[6] P. Chen, X. Tian, and Y. Chen, "Optimization of the digital near-field beamforming for underwater 3-D sonar imaging system," *IEEE Trans. Instrum. Meas.*, vol. 59, no. 2, pp. 415–424, Feb. 2010.

[7] M. Palmese and A. Trucco, "An efficient digital CZT beamforming design for near-field 3-D sonar imaging," *IEEE J. Ocean. Eng.*, vol. 35, no. 3, pp. 584–594, Jul. 2010.

[8] M. Palmese and A. Trucco, "Pruned chirp zeta transform beamforming for 3-D imaging with sparse planar arrays," *IEEE J. Ocean. Eng.*, vol. 39, no. 2, pp. 206–211, Apr. 2014.

[9] L. Yuan, R. Jiang, and Y. Chen, "Gain and phase autocalibration of large uniform rectangular arrays for underwater 3D sonar imaging systems," *IEEE J. Ocean. Eng.*, vol. 39, no. 3, pp. 458–471, Jul. 2014.

[10] Y. Han, X. Tian, F. Zhou, R. Jiang, and Y. Chen, "A real-time 3-D underwater acoustical imaging system," *IEEE J. Ocean. Eng.*, vol. 39, no. 4, pp. 620–629, Oct. 2014.

[11] C. Chi, Z. Li, and Q. Li, "Fast broadband beamforming using nonuniform fast Fourier transform for underwater real-time three-dimensional acoustical imaging," *IEEE J. Ocean. Eng.*, vol. 41, no. 2, pp. 249–261, Apr. 2016.

[12] A. Trucco, "Thinning and weighting of large planar arrays by simulated annealing," *IEEE Trans. Ultrason. Ferroelectr. Freq. Control*, vol. 46, no. 2, pp. 347–355, Mar. 1999.

[13] P. Chen, B. Shen, L. Zhou, and Y. Chen, "Optimized simulated annealing algorithm for thinning and weighting large planar arrays," *J. Zhejiang Univ-Sci. C (Comput. & Electron.)*, vol. 11, no. 4, pp. 261–269, Apr. 2010.

[14] P. Chen, Y. Y. Zheng, and W. Zhu, "Optimized simulated annealing algorithms for thinning and weighting large planar arrays in both far-field and near-field," *IEEE J. Ocean. Eng.*, vol. 36, no. 4, pp. 658–664, Oct. 2011.

[15] J. L. Schwartz and B. D. Steinberg, "Ultrasparse, ultrawideband arrays," *IEEE Trans. Ultrason. Ferroelectr. Freq. Control*, vol. 45, no. 2, pp. 376–393, Mar. 1998.

[16] Y. Yang and A. E. Fathy, "Development and implementation of a real-time see-through-wall radar system based on FPGA," *IEEE Trans. Geosci. Remote Sens.*, vol. 47, no. 5, pp. 1270–1280, May 2009.

[17] Q. Huang, L. Qu, B. Wu, and G. Fang, "UWB through-wall imaging based on compressive sensing," *IEEE Trans. Geosci. Remote Sens.*, vol. 48, no. 3, pp. 1408–1415, Mar. 2010.

[18] T. Sakamoto and T. Sato, "Two-dimensional ultrawideband radar imaging of target with arbitrary translation and rotation," *IEEE Trans. Geosci. Remote Sens.*, vol. 49, no. 11, pp. 4493–4502, Nov. 2011.

[19] Q. Liu, Y. Wang, and A. E. Fathy, "Towards low cost, high speed data sampling module for multifunctional real-time UWB radar," *IEEE Trans. Aerosp. Electron. Syst.*, vol. 49, no. 2, pp. 1301–1316, Apr. 2013.

[20] X. Zhuge and A. G. Yarovoy, "A sparse aperture MIMO-SAR-Based UWB imaging system for concealed weapon detection," *IEEE Trans. Geosci. Remote Sens.*, vol. 49, no. 1, pp. 509–518, Jan. 2011.

[21] M. Bialkowski, D. Ireland, Y. Wang, and A. Abbosh, "Ultra-wideband array antenna system for breast imaging," in *Proc. Asia-Pacific Microw. Conf.*, 2010, pp. 267–270.

[22] T. Kikkawa and T. Sugitani, "Planar UWB array for breast cancer detection," in *Proc. 7th Eur. Conf. Antennas Propag.*, 2013, pp. 339–343.

[23] M. Gardill, G. Fischer, R. Weigel, and A. Koelpin, "Design of an ultra-wideband monocone circular antenna array for vehicle-integrated industrial local positioning applications," in *Proc. 7th Eur. Conf. Antennas Propag.*, 2013, pp. 2211–2215.

[24] M. A. Richards, *Fundamentals of Radar Signal Processing*, New York, NY, USA: McGraw-Hill, 2005.

[25] T. Misaridis and J. A. Jesen, "Use of modulated excitation signals in medical ultrasound. Part I: Basic concepts and expected benefits," *IEEE Trans. Ultrason. Ferroelectr. Freq. Control*, vol. 52, no. 2, pp. 177–191, Feb. 2005.

[26] T. Misaridis and J. A. Jesen, "Use of modulated excitation signals in medical ultrasound. Part II: Design and performance for medical imaging applications," *IEEE Trans. Ultrason. Ferroelectr. Freq. Control*, vol. 52, no. 2, pp. 192–207, Feb. 2005.

[27] L. Y. Astanin and A. A. Kostylev, "Ultrawideband radar measurements analysis and processing," The Institution of Electrical Engineers, 1999, pp. 3–5.

[28] V. Sipal, D. Edwards, and B. Allen, "Bandwidth requirement for suppression of grating lobes in ultrawideband antenna arrays," in *Proc. IEEE Int. Conf. Ultra-Wideband*, 2012, pp. 236–240.

[29] G. Cardone, G. Cincotti, and M. Pappalardo, "Design of wide-band arrays for low side-lobe level beam patterns by simulating annealing," *IEEE Trans. Ultrason. Ferroelectr. Freq. Control*, vol. 49, no. 8, pp. 1050–1059, Aug. 2002.

[30] R. J. Urick, *Principles of Underwater Sound*, 3rd ed. New York, NY, USA: McGraw-Hill, 1983, ch. 4–5.

[31] W. H. Thorp, "Analytic description of the low frequency attenuation coefficient," *J. Acoust. Soc. Amer.*, vol. 42, no. 1, 1967 Art. no. 270.

[32] F. H. Fisher and V. P. Simmons, "Sound absorption in sea water," *J. Acoust. Soc. Amer.*, vol. 62, no. 3, pp. 558–564, Sep. 1977.

[33] J. A. Jensen *et al.*, "Ultrasound research scanner for real-time synthetic aperture data acquisition," *IEEE Trans. Ultrason. Ferroelectr. Freq. Control*, vol. 52, no. 5, pp. 881–891, May 2005.

[34] B. Horvei and K. E. Nilsen, "A new high resolution wideband multi-beam echo sounder for inspection work and hydrographic mapping," in *Proc. MTS/IEEE OCEANS Conf.*, 2010, DOI: 10.1109/OCEANS.2010.5664080.

[35] S. Curletto and A. Trucco, "On the shaping of the main lobe in wide-band arrays," *IEEE Trans. Ultrason. Ferroelectr. Freq. Control*, vol. 52, no. 4, pp. 619–630, Apr. 2005.

[36] A. Trucco, "Weighting and thinning wide-band arrays by simulated annealing," *Ultrasonics*, vol. 40, pp. 485–489, 2002.

[37] B. D. Steinberg and H. M. Subbaram, *Microwave Imaging Techniques*. New York, NY, USA: Wiley, 1991, ch. 10.

[38] L. G. Ullate, G. Godoy, O. Martínez, and T. Sánchez, "Beam steering with segmented annular arrays," *IEEE Trans. Ultrason. Ferroelectr. Freq. Control*, vol. 53, no. 10, pp. 1944–1954, Oct. 2006.

[39] R. H. Mellen, "The thermal-noise limit on the detection of underwater acoustic signals," *J. Acoust. Soc. Amer.*, vol. 24, no. 5, pp. 478–480, Sep. 1952.

[40] D. H. Ezrow, "Measurement of the thermal-noise spectrum of water," *J. Acoust. Soc. Amer.*, vol. 34, no. 5, pp. 550–554, May 1962.

[41] M. O'Donnell, "Coded excitation system for improving the penetration of real-time phased-array imaging systems," *IEEE Trans. Ulrason. Ferroelect. Freq. Control*, vol. 39, no. 3, pp. 341–351, May 1992.

[42] B. Haider, P. A. Lewin, and K. E. Thomenius, "Pulse elongation and deconvolution filtering for medical ultrasonic imaging," *IEEE Trans. Ulrason. Ferroelectr. Freq. Control*, vol. 45, no. 1, pp. 98–113, Jan. 1998.

[43] G. S. Kino, *A Coustic Waves, Devices, Imaging, Analog Signal Processing*. Englewood Cliffs, NJ, USA: Prentice-Hall, 1987, ch. 4, pp. 488–491.

[44] J. A. Jensen, J. Mathorne, T. Gravesen, and B. Stage, "Deconvolution of in-vivo ultrasound b-mode imaging," *Ultrason. Imag.*, vol. 15, pp. 122–133, 1993.

[45] F. Gran and J. A. Jensen, "Designing waveforms for temporal encoding using a frequency sampling method," *IEEE Trans. Ulrason. Ferroelectr. Freq. Control*, vol. 54, no. 10, pp. 2070–2080, Oct. 2007.

[46] R. Y. Chiao and X. Hao, "Coded excitation for diagnostic ultrasound: A system developer's perspective," in *Proc. IEEE Ultrason. Symp.*, 1998, pp. 1639–1644.

[47] V. A. Oppenheim, R. W. Schafer, and J. R. Buck, *Discrete-Time Signal Processing*, Englewood Cliffs, NJ, USA: Prentice-Hall, 1989 8–9, pp. 626–710.

[48] M. Palmese and A. Trucco, "Acoustic imaging of underwater embedded objects: Signal simulation for three-dimensional sonar instrumentation," *IEEE Trans. Instrum. Meas.*, vol. 55, no. 4, pp. 1339–1347, Aug. 2006.

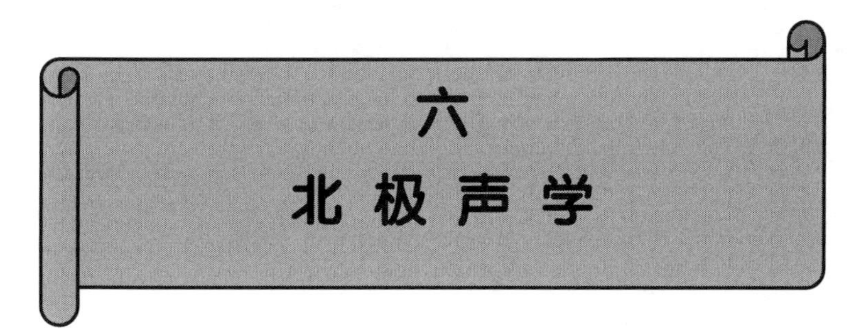

六 北极声学

·综述与评论·

北极水声学：一门引人关注的新型学科①

李启虎[1] 王 宁[2] 赵进平[2] 黄海宁[1] 尹 力[1]
黄 勇[1] 李 宇[1] 薛山花[1] 任新敏[2] 李 涛[2]

(1 中国科学院声学研究所 北京 100190)
(2 中国海洋大学 青岛 266100)

摘要 北极水声学是一门研究北极及其毗邻海域水声环境效应的学科。研究内容包括海洋环境噪声，特别是冰盖下的海洋环境噪声；北极海区的混响特性；北极及其毗邻海域的水声传播规律、冰盖下的水声通信，以及由于北极海区的独特环境（所谓半声道效应）而给水声信号处理带来的新的研究课题。北极水声学的研究开始于二次世界大战之后，当时的研究内容明显的带有冷战的烙印。近年来由于地球变暖的趋势，北极冰区面积持续减少，北极航道有望开通。又由于北极高纬度地区的丰富的自然资源，引起各海洋大国的高度关注，北极水声学已成为新的研究热点，并注入了新的内容。本文综述介绍北极水声学的研究概况，以及和北极声学密切相关的海洋声学方面的研究课题。自从上世纪 90 年代以来，我国科学家对探索北极表现了极大兴趣，本文简要介绍我国对北极的 5 次海洋考察和正在进行的第 6 次考察。对我国在北极及其毗邻海域水声学研究方面所面临的挑战提出初步的应对措施。

关键词 北极水声学，半声道，海洋环境的水声效应，北极及其毗邻海域的声传播、环境噪声、混响，声呐信号处理
中图分类号：O427 **文献标识码**：A **文章编号**：1000-310X(2014)06-0471-13
DOI: 10.11684/j.issn.1000-310X.2014.06.001

Arctic underwater acoustics: an attractive new topic in ocean acoustics

LI Qihu[1] WANG Ning[2] ZHAO Jinping[2] HUANG Haining[1] YIN Li[1]
HUANG Yong[1] LI Yu[1] XUE Shanhua[1] REN Xinmin[2] LI Tao[2]

(1 *Institute of Acoustics, Chinese Academy of Sciences, Beijing* 100190, *China*)
(2 *Ocean University of China, Qingdao* 266100, *China*)

Abstract Arctic underwater acoustics is a science, which focuses on the study of ocean environment effect of Arctic and the neighbor's area, including ocean environment noise, particularly the ocean ambient noise under ice cover, reverberation characteristics of Arctic, the underwater signal propagation law and underwater acoustic communication in Arctic. The research topics also include the new feature's study in signal processing, due to the special Arctic environment, i. e. so called half underwater sound channel effect. The

① 应用声学, 2014, 33(6): 471-483.

history of research work of Arctic underwater acoustics can be traced back to the end of World War II, the scientific and technical interests obviously with cold war brand. Recently, since the trend of global warming of the earth, the ice cover area in Arctic substantially decreases, and the Arctic freight channel will hopefully open. The plenty resources in high latitude area attracts more and more attention of ocean power country in the world. The Arctic underwater acoustics becomes a new hot topic in research work and has been covered many new issues. The brief introduction of Arctic underwater acoustics is reviewed in this paper. The acoustic oceanography and ocean acoustics topics in this area are presented. Since 1990's of last century, the Chinese scientists have expressed their interests in Arctic exploration and study. The results of 5 times Arctic surveys organized by Chinese oceanic organizations are introduced, the research program of ongoing 6th survey is briefly expressed. It provides preliminary measures for the challenges of our country in Arctic and neighbor area study.

Key words Arctic acoustics, Half sound channel, Environment underwater acoustical effect of ocean, Sound propagation, reverberation and ambient noise in Arctic and neighbor area, Sonar signal processing

1 引言

北极通常是指以北极点为中心,北纬66°34′以北的,包括整个北冰洋(及其岛屿)、北美洲、亚洲以及欧洲大陆的北部边缘海、边缘陆地所在的一片区域。北极地区的总面积是2100万平方公里,其中陆地面积约800万平方公里,海洋面积1300万平方公里。北极地区是全球气候变化最为剧烈的地区之一,随着海水变暖造成的北极冰盖融化,北极的战略地位日益突出,也使其成为美俄等大国博弈的焦点。同样,作为近北极国家的中国在北极地区也存在航道、资源、军事以及科研等维系国家未来发展空间的重大利益。建设海洋强国,理应将经略北极纳入战略视野。

北冰洋的海冰覆盖面积通常在一年当中的3月份达到最大值,在9月份达到最小值。近30年间,由于全球变暖,使北极海冰覆盖面积逐年递减变化趋势非常显著。这期间3月份平均的海冰覆盖面积衰减率为2% /10年,而9月份的海冰覆盖面积的衰减率为7% /10年;在北极变暖,海冰覆盖面积减小的同时,海冰厚度也在逐渐变薄。最近十几年的研究表明北冰洋中心地带的永久海冰在20年内,厚度由3.1 m减小至1.8 m。图1是近40年来北极冰盖厚度的变化说明。

由于北极及其毗邻海域处于高纬度地区,常年低温,即使在夏天,冰面上的温度也在0℃附近。这就造成了北极海区独特的声速剖面,几乎是从0 m开始的正梯度,并且相对变化较少,见图2。这样的海洋环境为水声学研究提供了不同于浅海、深海的自然条件。美国MIT的Baggeroer教授长期从事北极水声学研究,今年6月他应邀在第2届国际水声会议上作北极声层析的报告。他认为北极的典型声速剖面中,存在两个正跃层,一个是在40~50 m深度的所谓“盐度跃层”(Halocline),另外一个跃层是深度在200~250 m的“密度跃层”(Syncline)[1]。

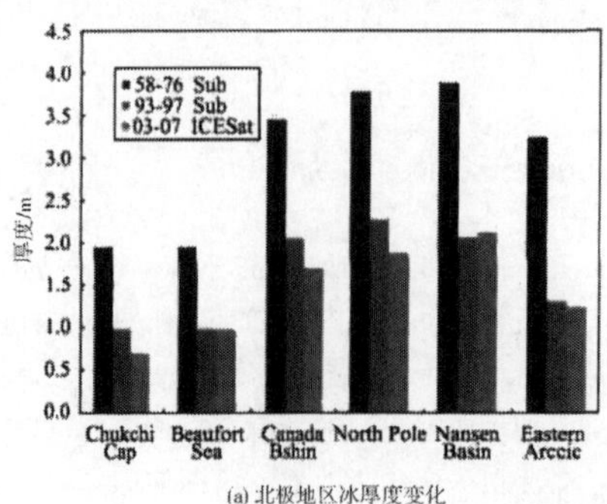

(a) 北极地区冰厚度变化

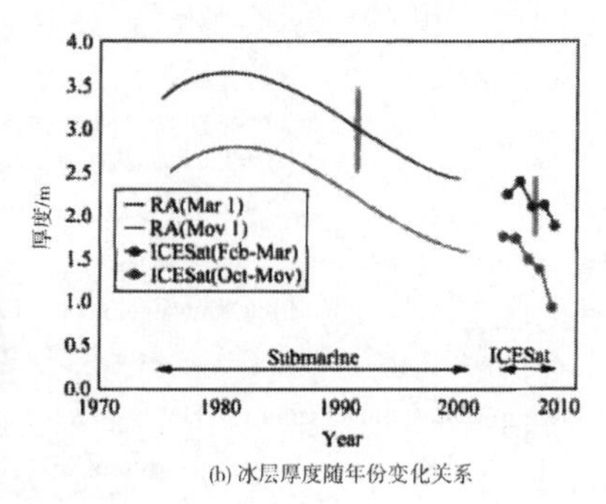

(b) 冰层厚度随年份变化关系

图1 北极海冰厚度变化

Fig. 1 Ice thickness variation in Arctic

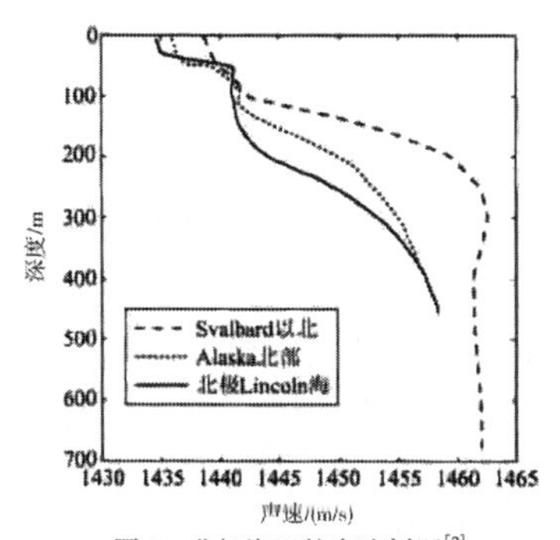

图 2 北极海区的声速剖面[2]
Fig. 2 Sound speed profile in Arctic

美国对北极声学的研究可以追朔到二次大战刚结束的 1946 年。据当时参加筹建“北极潜艇实验室”(Arctic submarine laboratory) 的 W. K. Lyon 博士回忆[3]: “在 1946 年,海军上将 Byrd 率舰队去南极探险,我收到他的一封信,问我有没有与这次探险有关的研究工作要做。我说是的,可以试一下把潜艇开到冰冷的水下面”。

Lyon 的建议开启了美国北极声学的研究工作。1947 年,Lyon 得到授权,把设在 San Diego 的美国海军无线电实验室和设在加州大学的战争研究分部合并,成立海军电子实验室,分设“北极潜艇实验室”,Lyon 担任主任。1951 年在 Alaska 的 Prince Wales 角建立了北极声学野外工作站。1958 年 8 月 13 日,美国核潜艇鹦鹉螺号(SSN571) 从北冰洋的冰下到达北极点。

北极及其毗邻海域的水声学研究早在上世纪 60 年代开始就和水声学的其他领域一样在环境噪声、混响、水声信号传播、水声通信诸方面展开[2,4-9]。由于北极地区的特殊海洋环境,不断产生的新问题和发现的新现象,引起了从事该领域研究工作者的注意,北极声学从上世纪 90 年代以来出现了一系列新的成果[10-13]。

但是真正引人注目的是近年来由于地球变暖引起北极冰区缩小而带来的世界海洋大国对北极地区的空前关注。因而带来北极水声学新的发展[1,14-17]。

2014 年美国主管海军基础研究的海军研究实验室(ONR) ,在公布水声研究计划时,明确三大学科,即浅海声学、深海声学、北极声学[18]。

美国前总统乔治·布什在离任前于 2009 年 1 月 9 日签署了《美国在北极地区的政策》(NSPD-66/HSPD-25) 。文件中称,美国在北极地区有着关系到国家安全的广泛的基本利益,美国准备独立或与其他国家共同确保和保卫这些利益,这些利益涉及到诸如反导防御和预警、海基和空基系统的展开,以确保战略海运、前沿海上存在以及海上和空中航行自由。

美国海军潜艇部队司令 G·唐纳利海军中将在阐述 IECX-2009 演习的任务和目标时说[17]: “我们正在制定和确定北极地缘政治未来的形式。潜艇部队确保通路的能力和在北极地区能赢得尊重的军力存在能提高我们在世界这一重要战略地区的行动能力。我们 50 多年来一直在研究这一地区的重要原因是,积累在这一独一无二的环境条件下安全高效的行动所必需的技术和作战经验。”

美国西雅图华盛顿大学的 APL 实验室在美国国家基金委(NSF) 和海军实验室(ONR) 的持续支持下,多年来一直在研究北极水声学,包括环境、定位、通信等[19-20]。他们是世界上率先使用水下滑翔器(Sea glider) 进行北极研究的研究机构。

我国对北极事务的参与可以追朔到上世纪 40 年代。1947 年,重庆大学工学院院长冯简教授,代表中国出席巴黎国际文教会议,然后由当时中国驻挪威大使馆代办雷季敏相助,只身进入挪威的北极圈内地区开展考察。他此行回国后,著有《余在北欧时所见之北极光》。

自 1999 年开始我国已对北极进行了五次科学考察。并于 2004 年 7 月,在斯匹次卑尔根群岛(挪威称为斯瓦尔巴群岛) 建立了第一个科考站——北极黄河考察站。对北极已有的研究表明: 北极气候环境变化对我国气候有着直接的影响,我国的冻雨和干旱与北极海冰有关,随着海冰进一步减少,冻雨和干旱呈常态化趋势,我国的粮食安全将因此受到严重威胁。因此我国的北极科技战略为: 气候变化已经改变了我国的生态与经济格局,改变了 13 亿中国人民的生存条件。北极变化,影响到中国人民的福祉; 研究北极,是中国人的权利和义务[21-22]。

我国北极声场特性研究还是空白,北极及其毗邻海域声场的研究是开展北极及其毗邻海域海洋环境水声效应研究的基础。

2 其毗邻海域的海洋环境特征

北冰洋是一个四周被美洲大陆和欧亚大陆包

围的近乎封闭的海洋，与北美洲大陆和欧亚大陆相连，它与太平洋通过宽85 km，平均水深30～50 m的白令海峡相连。由格陵兰海、挪威海以及巴芬湾通过格陵兰岛两侧的水道与大西洋相连。

由于独特的地理位置，北极的气候条件与其他区域存在很大的差异。北冰洋的冬季时间要长于夏季。冬季一般从11月起一直到次年4月；5、6月和9、10月份属春季和秋季，而夏季仅有7、8两个月。最冷的1月份的平均气温介于－20℃～－40℃之间，而最暖的8月份平均气温也只能达到－8℃，因此北冰洋的大部分区域终年覆盖着厚厚的海冰。由于洋流的运动，北冰洋的海冰不断裂解、漂移和融化，因而不可能形成像南极陆地数千米厚的冰雪。北极的上空长期存在一个极地高压，在冬季高压更为强烈，形成了反气旋式的风场，所以北冰洋的大部分海域都盛行极地东风。

50年来，北极被誉为“天然棱堡”，其冰层是准备执行对敌境目标实施导弹核打击的命令的战略导弹潜艇的天然掩体。由于北冰洋的特殊气候条件，常年厚厚的冰层成为战略导弹核潜艇行动最好的掩护。而经验证明，以潜艇的高度隐蔽性而论，最有效的反潜武器正是潜艇自己。

北极是全球气候变化的“启动器”之一，海冰、洋流和气团的变动直接导致全球气候变化或异常跳跃。北极地区广达1300万平方公里的苔原带，是全球最大的固碳地，近200年来该带北缩300～480 km，已使大量的固碳氧化进入大气层，加剧了温室效应的节奏。近100年来，北极苔原带平均升温2℃～4℃，已对周边地区产生难以估量的影响，北极脆弱易变的自然环境，使其成为全球变化重要的“指示器”。极光、哨声、磁暴等太空对地球的作用信息，只有在极区才可捕捉，不仅是为日地关系研究而且也为空间科学提供了天然的“实验场”。人类历史上三次大规模的挺进北极，不仅形成独特的冰雪文明，而且也提供了人与自然相互作用的典型样板，尤其在人类面临生存与发展这一共同命题的今天，探索解析北极人地关系更显其重要。北极酸雨和烟雾等环境污染已开始向中低纬区扩散，我国也已受其侵害，研究其集散机制和路径可为我国21世纪生存环境调整提供科学依据。

北极丰富的自然资源，将是新世纪重要的资源尤其是能源基地，而对北极资源的开发必然产生诸多环境问题，研究环境的脆弱性，制订科学合理的开采计划，已成为全球人类共同关心的问题。

图3 美国2011年在北极进行潜艇通信的“深海传呼机”试验[22]

Fig. 3 US Deep Siren experiment for submarine communication in 2011[22]

3 北极及其毗邻海域的水声环境特征

北极地区地理位置独特、气候寒冷，北冰洋的大部分区域终年被海冰覆盖，因此也形成了独特的声场环境。由于冰盖的作用，造成冰下噪声剧烈起伏以及强混响效应，并形成了北冰洋独有的半波导声道。因此，如何收集水声数据资料，掌握北极地区海域水声环境规律及机理，建立北极背景场、声信道模型，利用北极海洋环境水声效应，开展北极水声环境适配处理理论与方法研究，是确保我海军在未来的北极机动作战中获取信息优势的重大能力需求，是我潜艇隐蔽实施核威慑，保障我舰艇北极地区航行安全，以及提高我舰艇声呐装备环境适应性，提升探测、通信、导航技术水平的重大前沿基础研究需求。

北极区的冰下噪声不同于任何其它海域。因较少水上交通，航运噪声非常低。另外冰层可使海水免受风的影响，其冰下噪声环境比开阔海域的零级海况还安静得多。

在冰面之下，噪声的谱和特性的变动很大，并与冰的状态、风速、积雪以及空气温度的变化有关。当冰呈现碎冰块状时，在同样海况下，噪声级比没有结冰的水中测得的数值高5～10 dB。当气温下降时，同海岸相连接的整块冰面收缩而被破裂，噪声出现尖刺和脉冲，另外浮动冰块的相互碰撞和摩擦以及海浪拍击冰缘而破碎也形成水下强噪声。试验发现在100 Hz～1000 Hz频带内噪声级比开阔海区高12 dB，比在冰区内部高20 dB。

北极区域噪声研究同样由来已久。1964年Miline和Ganton对北极海冰下的环境噪声进行了分析。1970 年 Urick 发表了有关冰山融化生成噪声的论文。1974 年 Diachok 和 Winokur，1980 年 Diachok 研究表明，冰缘区的噪声级通常高于堆冰区的噪声级。冰-水边界噪声级的相对幅度是冰聚集相对于距离变化率的函数。因此在散冰与海水边界处测得的相对最大噪声级将小于在密集冰缘处的测量值[9-10]。冰缘区的环境噪声级还与海况、海深和主海浪周期这类变量有关。1991 年 Makris 和 Dyer 指出，表面重力波的作用主要与冰缘噪声相关联。

由于冰层覆盖，北极区域噪声除地震、航船噪声外，主要来源于海冰破裂、冰结构隆起及其浮冰运动剪切产生的辐射噪声。在不同冰层区域，冰下噪声级、噪声指向性、谱形状和时变性差别很大。1984 和 1988 年 Dyer 在堆冰下测量的噪声频率范围从 3 Hz ~ 1000 Hz。根据 Makris 和 Dyer 1986 年的报道，堆冰是不断变化的，因此其噪声特性也随空间和时间剧烈地改变。Lewis 和 Denner 1988 年公布了北冰洋高频(1 kHz)环境噪声数据的观测结果。这些噪声由冰层热破裂引起。Sagen 等人 1990 年也评述了与冰生噪声有关的问题。

J. K. Lewis 和 W. W. Denner 集中研究了 Beaufort 海的环境噪声时空及统计特性。Makris 和 Dyer 开展了北极环境噪声相关性理论和实验研究。Langley 开展了冰层中的声辐射研究，并建立了破裂冰辐射源弹性等效声源模型。Y. Xie 和 D. Farmer 研究了热应力激发的海冰破裂辐射噪声特性和板波模型，揭示了低频、高频环境噪声空间指向性的差异。

美国 UCLA San Diego 分校的 Roth 曾经在 Alaska 北部沿海对噪声、水声传播等进行过长期的观测，他的研究结果很具有参考价值[23]，见图 4(a) ~ 4(e)。观测点 A 和 B 的确切位置如下：

A: N 72°10. 569　W 156°33. 176，水深 230 m

B: N 72°27. 523　W 157°23. 364，水深 246 m

海洋中存在着大量散射体以及起伏不平的界面。当声源发射声波以后，碰到这些散射体，就会引起声能在各个方向上重新分配，即产生散射波。其中返回到接收点的散射波的总和称为混响。混响是主动式声呐的主要干扰。由产生混响的散射体不同性质，可分为体积混响、海面混响和海底混响。对混响的研究大体上分为能量规律和统计规律两个方面。混响的能量规律的理论分析以声波在海洋中的传播理论和散射理论的结合为出发点，主要涉及混响强度同信号参量和环境因素的联系以及衰减规律。

在北冰洋地区，海冰是混响的主要成因。北极冰下混响表现出强度大、影响范围广、空间相关性强的特点。图 5 是散射强度记录的一个典型例子。从中可以看到散射强度随掠射角的变化。北极地区混响研究始于上世纪 60 年代。H. W. Marsh 和 R. H. Mellen(1962 年，1963 年)根据 1958 ~ 1962 年进行的系列爆破声源试验数据，研究了散射强度与受冰下表面粗糙度和频率影响的掠射角之间的一般规律，并指出由海冰引起的混响级比无冰海面要高 40 dB 以上。

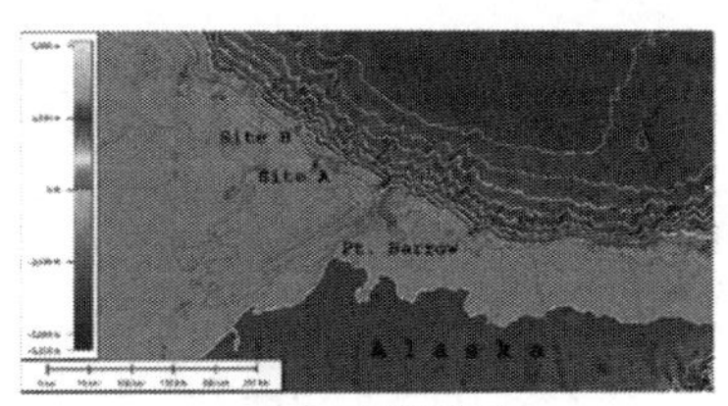

(a) Alaska以北的测试点

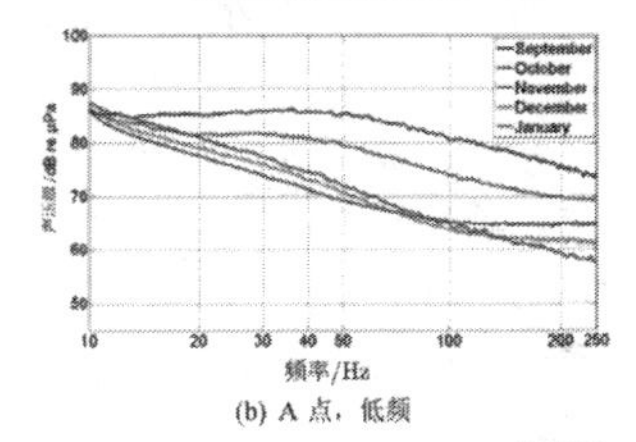

(b) A 点，低频

(c) B 点，低频

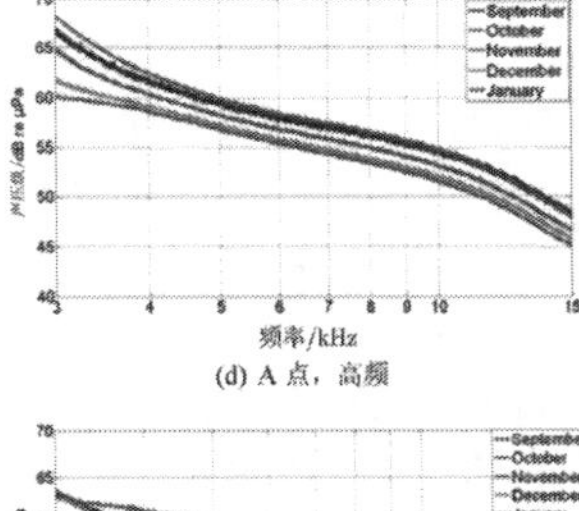

(d) A 点，高频

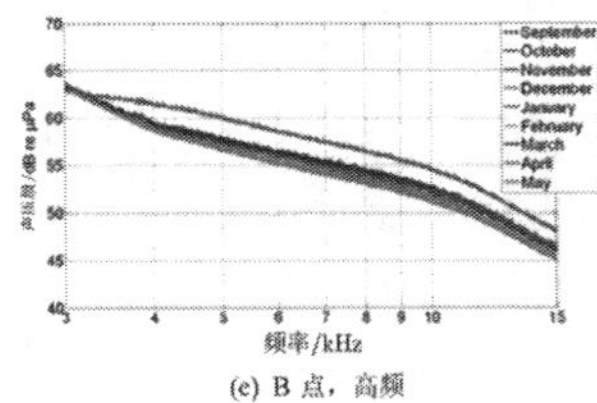

(e) B 点，高频

图 4　Alaska 北部海域环境噪声测试结果(2006.09-2007.05)[22]

Fig.4　Measurement results of ambient noise in North Alaska

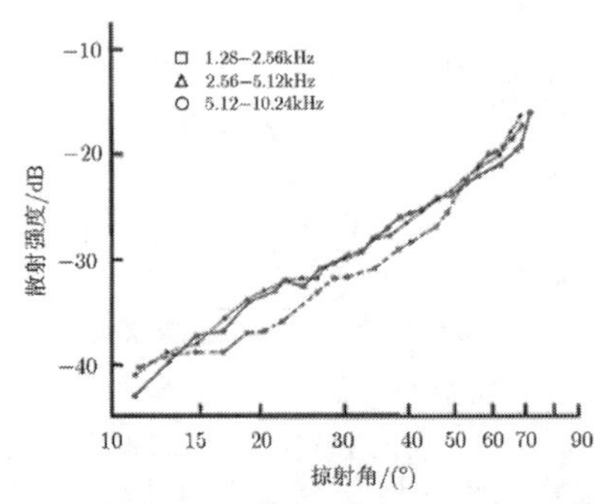

图8 北极地区不同频率范围下，混响散射强度随掠射角变化图[24]

Fig. 8 Variation of reverberation scattering strength with grazing angle, in different frequency band, Arctic area

J. R. Brown(1964年)和A. R. Milne(1964年)通过对北冰洋两个观测地点在一年中不同时间内观测数据的分析，发现覆冰海区的散射强度随频率和掠射角的增大而增强的规律。J. E. Burke和V. I. Twersky(1966年)将冰脊表述为半柱椭圆刚体，建立了描述冰下混响的Burke-Twersky模型。1976年，Diachok将Burke-Twersky模型用于冰下前向散射研究中。Bishop(1986至1989年)通过对冰下表面大尺度三维结构的研究建立了高频(不小于2 kHz)的水下混响模型。

进入上世纪90年代，T. C. Yang和T. J. Hayward(1992年，1993年)利用在Norwegian-Greenland海进行的CEAREX 89试验数据，对不同距离上的低频北极混响开展了研究工作：针对短程(小于3 km)的直接冰下表面反射混响，比较了测量散射强度与Burke-Twersky模型数据的差异，证明了在低频(24～105 Hz)段当掠射角小于20°时，测量散射强度与模型规律相吻合；对于中程(5 km～20 km)的冰下表面和海底混合混响，研究了基于垂直阵的冰下表面和海底混响分离方法，并分别验证了低频冰下混响和海底混响与相关模型的吻合度；对于远程(不大于200 km)的甚低频(10～50 Hz)混响，建立了基于简正波的北极混响模型，并通过测量数据验证了模型的合理性，并指出有效的混响强度取决于声源的深度。K. LePage和H. Schmidt(1996年)同样利用CEAREX 89试验中的二维水平阵数据，进行了北极混响空间统计特征的研究，发现其具有高度的相关性，并通过弹性扰动参量化方法反演了冰盖粗糙度的空间统计参数。

由于在北冰洋地区，混响强度大和传播损失小，因此主动声呐受混响影响的范围可以达到200 km。

另外，由于北冰洋地区特殊的环境声场特性，使得北极地区的混响具有很强的相关性。半波长水平相关性可以到达0.99以上。

海洋及其边界(海面和海底)组成复杂多变的水声传播媒质，它的复杂多变性主要表现在随海区和季节而变化，从而有不同的传播规律。从声源发出的声信号在传播过程中逐渐损失能量，这种传播损失分为扩展和衰减。扩展损失表示声波的波阵面从声源向外不断扩展的简单几何效应。但实际上声波经常是在类似于波导中的传播，可以在这种波导(称为声道)中定向性地传播很长距离。衰减损失包括吸收、散射和声能漏出声道的效应。造成吸收的原因是海水的粘滞性、热传导性、海水中硫酸镁和硼酸-硼酸盐离子的弛豫机构。吸收使声强以指数形式随距离下降，吸收系数一般正比于频率二次方，因此远程声呐都选用较低频率。散射衰减主要原因包括海中气泡、悬浮粒子、不均匀水团、浮游生物以及边界的不平整性。声能漏出声道的效应则因具体声道而异，是声道传播的另外一种衰减因素。

产生海洋传播声道的条件是海洋边界及特定声速剖面。声速剖面就是海洋的声速分层结构。海水中的声速是温度、盐度和静压力(深度)的函数。它大致分为三层：表面层、主跃变层和深海等温层。表面层中的声速对温度和风的作用很敏感，有明显的季节变化和日变化。在表面层以下约千米深度内，温度随深度而下降，使声速也随深度下降，具有较强的负声速梯度，称为主跃变层。最下面的称为深海等温层，层中海水处于冷而均匀的稳定状态，声速随着深度的增加而增加。在主跃变层的负声速梯度和深海等温层的正声速梯度之间存在一个声速极小值(声道轴)，形成较稳定的深海声道——声发声道。

在沿岸浅海及大陆架上，声速剖面受较多的因素影响，有较强的地区变异性和短时间不稳定性。但平均而言，仍有比较明显的季节特征。在冬季的典型声速剖面是等温层，在夏季往往是负跃层或负梯度。

在浅海，由海面和海底构成浅海声道，声波在声道中由海面和海底不断反射而传播。海底的声反射特性,特别是小掠射角的海底反射损失,是浅海声场分析和声呐作用距离预报的重要参量,它决定于海底的底质和结构。当声传播水平距离不特别远(几百千米以内)时,往往把海洋看作分层媒质,分层媒质中的波动理论在60年代已达到较为成熟的阶段。

在北极海区,声道轴位于冰层覆盖的海面或其附近。向上折射的声线在冰层的下表面处发生反射,这样反复进行的反射声波传播得很远。图6是北极海区的典型正梯度声速剖面下的声线图,以及由此而形成的北极地区独特的"半声道"现象。在冰层下面,向上折射的声线与冰层粗糙的下表面反射的声线集合,构成了某些独特的传播特性,形成了半声道波导。

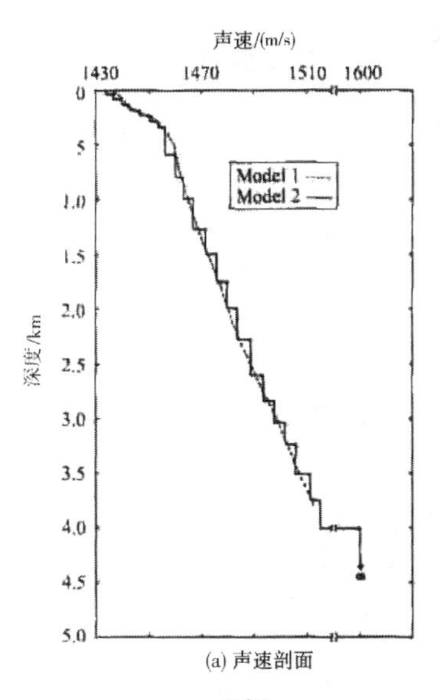

(a) 声速剖面

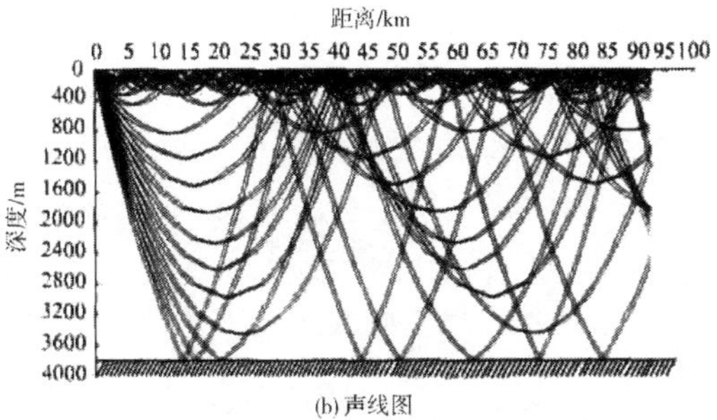

(b) 声线图

图6 北极海区的典型声速剖面

Fig. 6 Typical sound speed profile in Arctic

北极半波导声传播的特点之一是,它类似一个带通滤波器。高频和低频成分衰减很快,前者是因冰层的反射损失所引起,后者是由于甚低频的声波不能有效限制在声道中,在北极海区试验发现,频率为15~30 Hz的声波传播最佳,在30 Hz以上,随着频率的升高衰减急剧增加,同样频率10 Hz的衰减比20 Hz要大。造成传播损失的原因主要是冰层-海水界面上的声散射。其它可能的原因包括:水中的声能转到冰盖中传播并被限制在冰盖内造成的损耗等。

美国MIT, John Hopkins大学等单位曾经做过多年的实际工作,图7是一个实际测量和理论对比的图。

我国在历次北极考察中积累了大量有价值的海洋环境数据,利用这些数据可以对所考察的北极海区的水声传播特性进行研究。表1、表2和图8、图9是根据2008年8月测量的数据所进行的初步分析结果。

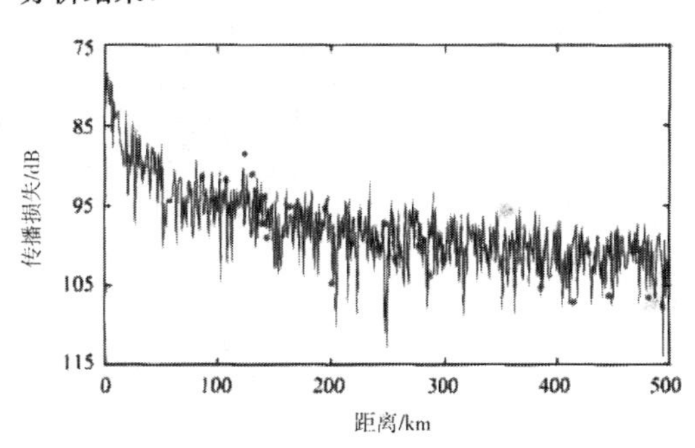

图7 北极地区传播损失的理论(Pekeris波导,实线)和实验值(红点)的比较[2,8](接收器深度90 m,声源深度243 m,20 Hz信号)

Fig. 7 Comparison of theory (dash line) and experiment (red dot) results (receiver depth 90 m, source depth 243 m, 20 Hz signal)

表1 冰层参数

Table 1 Parameters of ice layer

冰层厚度	纵波声速	横波声速	密度	纵波衰减	横波衰减
2 m	3000 m/s	1400 m/s	1 g/cm^3	0.3 dB/λ	1 dB/λ

表2 海底参数

Table 2 Parameters of sea bottom

纵波声速	横波声速	密度	纵波衰减	横波衰减
1580 m/s	0 m/s	1.6 g/cm^3	0.5 dB/λ	0 dB/λ

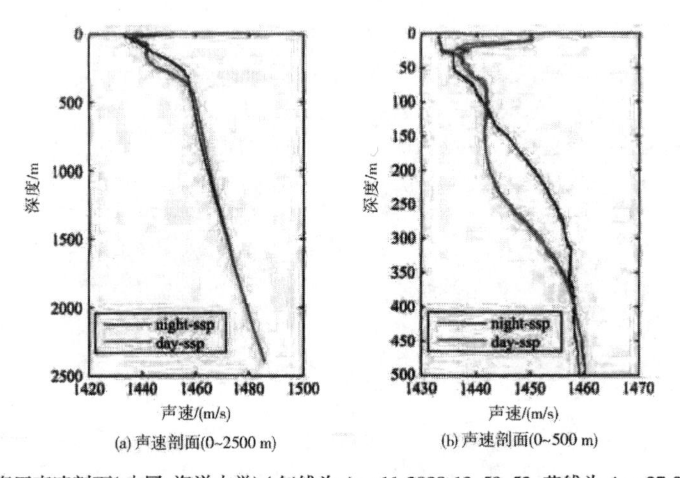

图 8　极地海区声速剖面(中国,海洋大学)(红线为 Aug 11 2008 13:59:53;蓝线为 Aug 27 2008 21:11:25)

Fig. 8　Sound speed profile in Arctic, Ocean University of China. Red line, 13:59:53 Aug. 11, 2008; Blue line 21:11:25, Aug. 27, 2008

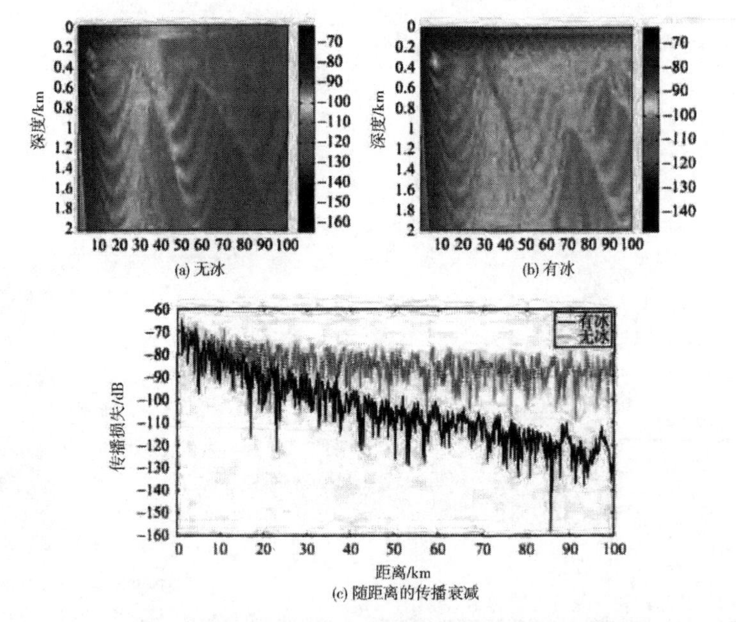

图 9　北极地区传播衰减(声源深度 100 m,频率 750 Hz)

Fig. 9　Transmission loss of underwater sound in Arctic, source depth 100 m, frequency 750 Hz

4　北极及其毗邻海域水声信号处理面临的新挑战

北极地区地理位置独特、气候寒冷,北冰洋的大部分区域终年被海冰覆盖,因此也形成了独特的水声环境。由于冰盖的作用,造成冰下噪声剧烈起伏以及强混响效应,并形成了北冰洋独有的半波导声道。因此,如何收集水声数据资料,掌握北极地区海域水声环境规律及机理,建立北极背景场、声信道模型,利用北极海洋环境水声效应,开展北极水声环境适配处理理论与方法研究,是保障我舰船北极地区航行安全,以及提高我声呐装备环境适应性,提升探测、通信、导航技术水平的重大前沿基础研究需求。概括来说,对我国开展北极水声学的研究存在着巨大的机遇和挑战。

(1) 北极及其毗邻海域的水声环境适应性研究

在北极及其毗邻海域，由于冰盖与水下冰脊的存在，给潜艇、UUV 等水下航行器直接利用卫星系统进行通信和导航带来了极大的困难，使得水声方式成为潜艇、UUV 等水下航行器在冰下航行唯一可靠的信息传输手段。而由于北冰洋独特的半声道效应，使得利用低频/甚低频水声信号，进行超远程低码率通信和导航成为可能。另一方面，由于水下冰脊的起伏，造成了水下近冰面的多途效应，给采用中近程相干通信方式的 UUV 等无人自主航行器的作业带来困难。因此，通过对北极地区水声传播特性研究，掌握信道分布和频率响应特征，可以选取合适通信和导航的深度和频率，为引导我潜艇进入北极冰区、并在冰下安全航向、以及保持与总部的联系，提供有力的保障；通过对北极地区水声冰面散射特性研究，分析并了解多途结构和频率响应特征，可以提升冰下中近程相干通信和导航性能，从而保障 UUV 等无人自主航行器在执行情报、监视与侦察任务时信息交互能力和安全性。

(2) 半声道情况下的低频声传播及最佳接收

低频声波在海洋波导中的传播距离远，同时对于北极地区声传播的半声道波导产生的带(低)通滤波效应，低频段的远程传播特征明显。因此，低频是水声基础研究的重要方向和发展趋势。北极地区的声传播的半声道虽然有利于低频(10 ~ 30 Hz)声传播，但由于声场受冰层的影响很大，信道是距离相关的声信道，对信道的建模变得复杂，给信号处理带来了很大困难，是一类复杂声场环境下的信号处理问题。同时，北极地区海洋环境背景场受冰层噪声的影响很大，冰层背景场具有较强的空间指向性，有较强的脉冲特性，这些都给信号处理带来了很大困难，必须解决在非均匀非高斯时变背景场以及水平相关传播信道下的信号检测和处理问题。

(3) 混响限制下的主动声呐信号检测

主动声呐，包括探测声呐、向上探查声呐和向前避障声呐，在北冰洋工作的性能主要受混响的影响，而在北冰洋地区，海冰是混响的主要成因。北极冰下混响表现出强度大、影响范围广、空间相关性强的特点。强度大是指冰下表面混响级要比水表面混响级强 40 dB，这将导致信混比低，影响主动声呐对微弱目标回波的检测；影响范围广，特别是低频混响，在很长的距离范围(甚至上百公里)内均强于环境噪声，几乎使主动声呐一直工作在混响背景下，因此增加了主动声呐的虚警概率。空间相关强表现在空间相邻混响相关度强，给抑制混响带来了很大的困难。

(4) 北极水下信道的频率选择特性以及和冰盖的关系

从初步计算来看，北极海域信道的具有非常明显的频率选择性。这就为各类在北极使用的声呐设计提供了限制，同时又是有利的条件。从图 10 可以发现，在频率为 750 Hz 时，冰盖下的传播损失远大于无冰区的传播损失。但是当频率降到 100 Hz 时，这种差异，基本消失了。由此可见，在北极海域，声呐信号处理系统必须关注所在海域的适用频率，以求得到满意的使用结果。无论是以探测目标为主的主被动声呐，还是通信声呐。

(5) 北极及其毗邻海域大气-冰-海-声耦合机理

在北极地区影响低频声信号传播特性的主要因素是声线在冰面和水面的交界面上的散射损失。这种传播损失与声传播频率，冰面粗糙度以及海水的声速剖面存在着非常强的耦合特性。声传播模型必须充分考虑声线在冰面的散射特性，同时考虑低频声波在冰层的弹性波，并充分考虑海洋环境参数(海底地形，水体声速剖面等)的起伏特性才能准确对北极地区的冰下声场特性进行描述。需考虑 Lloyd 镜效应影响，以及考虑大气风场对冰盖的影响以及在冰层上产生的噪声的影响。因此，北极地区及其毗邻海域的声场受大气-冰-海面对于声波的耦合影响，其耦合机理的研究是非常重要的。

(6) 长期低温工作条件下测试仪表、声呐设备的防护

由于北极地区常年低温的特点，水声测试设备、声呐系统的硬件都要进行特殊的设计，特别是准备在冰上工作的电子设备必须有特殊的防护措施。担任与卫星或水下平台通信任务的浮标系统，更要有经得起低温考验的机械、电子部件。

5　我国海洋科技工作者的北极考察活动

1993 年，中国科协批准成立了中国北极科学考察筹备组；2002 年 7 月 28 日，中国科学探险协会在

北极斯瓦尔巴群岛的朗伊尔城建立了“中国伊立特·沐林北极科学探险考察站”；2004 年夏天，中国政府在北极斯瓦尔巴群岛的新奥尔松建立了科学考察站——“中国北极黄河站”。

90 年代以来，我国部分科学家还通过各种途径参加了有关国家的北极考察队，或者在当地现有条件支撑下，开展部分研究工作。

自 1991 年开始，中国科学院大气物理研究所、国家海洋局第二海洋研究所、中国科学院地理研究所和兰州冰川冻土研究所、青岛海洋研究所与国外同行开展了一系列的合作研究。

然而，我国正式组队的考察则始于 1995 年。中国首次北极点科学考察是由中国科协主持，中国科学院组织的大型境外科考活动，以政府支持、民间集资方式运作，得到了新闻界、科学界和企业界的大力支持和广泛参与，科考队由 25 名队员组成，除 1 人来自香港外，其余分别来自全国七大部委、涉及 18 个单位。

1995 年 3 月 31 日，全体队员离境，经美国赴加拿大哈德逊湾开展负重滑雪和驾狗拉雪橇训练。4 月 22 日 7 名科考队员由设在加拿大北极群岛孔沃利斯岛上的雷索柳特基地(74°N) 出发，沿西经 80°

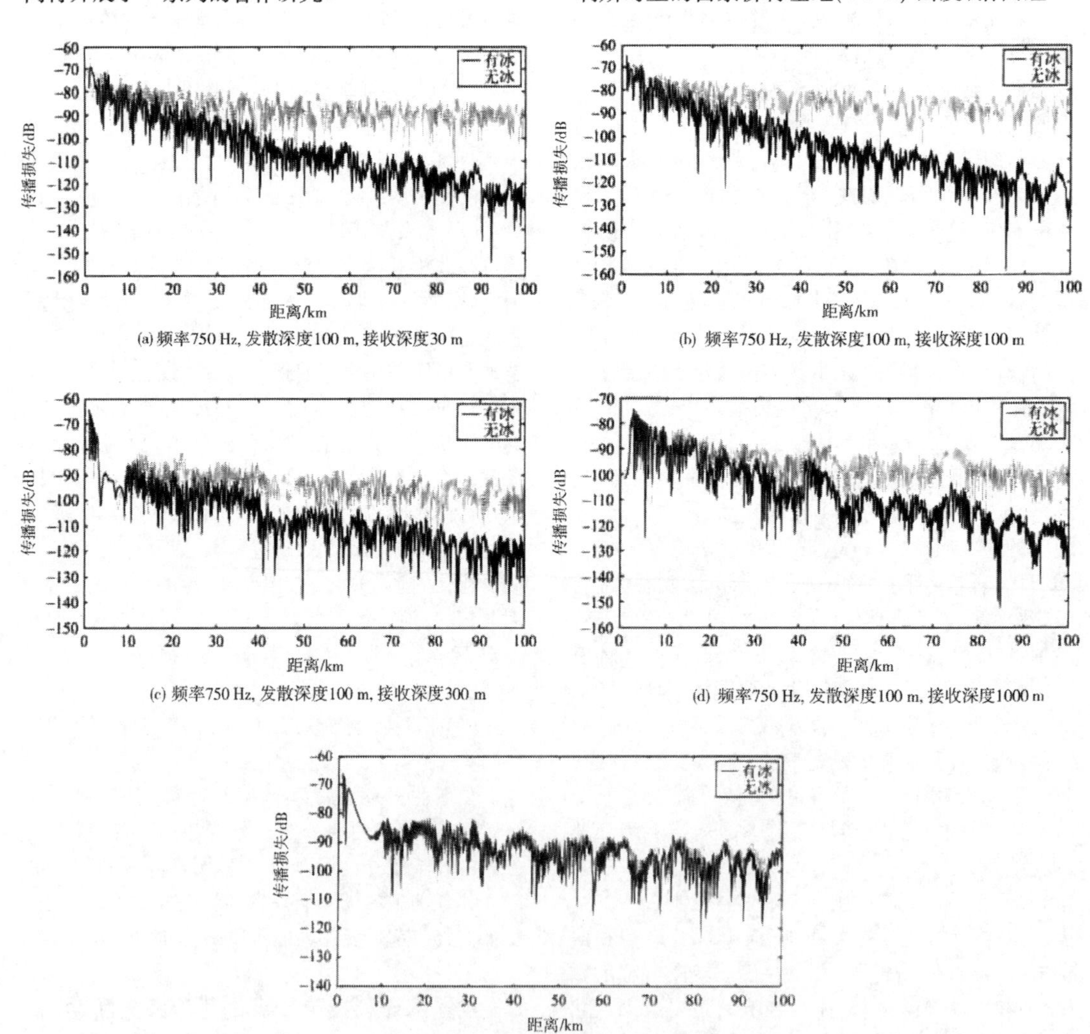

图 10 北极地区夏季水声传播衰减(黑色: 有冰，红色: 无冰)

Fig. 10 Underwater sound transmission loss in summer, Arctic (Dark: with ice; Red: no ice)

的冰面自88°N向北极点进发，跨越了波弗特海环流区和贯极点洋流带这两大北冰洋的重要区域，于北京时间5月6日上午10点55分到达北极点。共采集各类样品542批次，取得观测数据上万组。

自1999年中国首次自行组织，开展北极考察以来，针对北极海冰、海洋与大气变化同中国气候环境变化的关系，先后开展了三次以“雪龙”号科学考察船为平台的北冰洋区域综合考察，对北极地区气候与环境变化的机理，有了初步的认识与了解，获得了一批有价值的科学考察研究数据与样本[25]。见图11、12。在地球南北两极开展科学考察，“雪龙”号将不再是单兵作战，中国即将自主建造第一艘极地科学考察破冰船，国家海洋局将面向社会公开征名。

图11 1995年4月，我国第一次北极考察，赵进平教授乘坐狗拉雪橇进入北极，这是他在冰上准备实验

Fig. 11 The first Arctic survey of Chinese scientists in April, 1995, Prof. Jinping Zhao arrive Arctic by dog sleds, he was preparing do some research work in ice-covered surface

(a) “雪龙号”考察船

(b) CTD设备

图12 1999年8月，我国第二次北极考察

Fig. 12 The second Arctic research survey of China

2012年8月24日，参与中国第五次北极科考的科学家说，最新研究发现，一些新型溴代阻燃剂、全氟烷基化合物等有毒有害物质首次在北极高纬度海区出现。

在国家海洋局极地办公室和极地中心的大力支持下，2014年7月11日开始的第六次北极考察，将首次设置水声学方面的内容。这对推动我国在北极声学方面的研究工作起到开创性的作用。

参考文献

[1] BAGGEROER A. B, SCHMIDT H. Performance analysis of Arctic tomography using the Cramer Rao bound [C]. Proc. of UA2014, Rodes, Greece, 2014, 790-797.

[2] MIKHALEVSKY P N, GAVRILOV A N, BAGGEROER A B. The transarctic acoustic propagation experiment and climate monitoring in the Arctic [J]. J. Oceanic Engr., 1999, 24(2): 183-201.

[3] http: //www. wikipedia. com/arctic lab

[4] BROWN J R. Reverberation under Arctic ice [J]. J. Acoust. Soc. Am., 1964, 36(3): 601-603.

[5] BUCK B. M, GREENE C R. Arctic deep water propagation measurements [J]. J. Acoust. Soc. Am., 1964, 36(8): 1526-1533.

[6] BUCK B M. Arctic acoustic transmission loss and ambient noise, AD0485552 [R]. USA: Warrenton, 1966.

[7] PEKERIS C L. Theory of propagation of explosive sound in shallow water [J]. Geo. Soc. Am. Mem., 1948, 27: 1-116.

[8] KUTSCHALE H. Arctic hydroacoustics [J]. Tech. Report of Lamont-Doherty Geological Observatory Of Columbia Univ., 1969, 22(3): 246-264.

[9] DIACHOK O I. Effects of sea-ice ridges on sound propagation in the Arctic Ocean [J]. J. Acoust. Soc. Am., 1976, 59(5): 1110-1120.

[10] WOLF J W, DIACHOK O I, YANG T C, et al. Very-low-frequency under-ice reflectivity [J]. J. Acoust. Soc. Am., 1992, 93(3): 1329-1334.

[11] LEPAGE K, SCHMIDT H. Modeling of low frequency transmission loss in the central Arctic [J]. J. Acoust. Soc. Am., 1994,

96(3): 1783-1795.

[12] MIKHALEVSKY P N. Acoustics, Arctic[M]//STEELE J H, TUREKIAN K K, THORPE S A. Encyclopedia of ocean sciences. USA: Academic Press, 2001, 1:53-61.

[13] WIGGINS S M, MCDONALD M A, MUNGER L M, et al. Waveguide propagation allows range estimates for North Pacific right whales in the Bering Sea[J]. Canadian Acoustics, 2004, 32(2): 146-154.

[14] HARA C A O', COLLIS J. M. Underwater acoustics in Arctic environments[C]. Proc. of 162nd ASA Meeting, San Diego, USA, 2011.

[15] FREITAG L, KOSKI P, MOROZOV A, et al. Acoustic communications and navigation under Arctic ice[C]. Proc. IEEE Oceans, Sept., 2012.

[16] MIKHALEVSKY P N, et al. Multipurpose acoustic networks in the integrated arctic ocean observing system[C]. White Paper for Arctic Observing Summit., 2013.

[17] http://www.pix.com/arctic_advanture.

[18] http://www.onr.navy.mil/science-technology/code-31.

[19] Acoustic navigation and communications for high latitude ocean research[J]. A report from Intl Workshop sponsored by the NSF office of Polar Program, http://www.apl.washington.edu/ANCHOR workshop 2008.

[20] CONLON D M. The ONR high latitude dynamics program, an introduction[J], http://www.apl.washington.edu.

[21] 国家海洋信息中心."21 世纪初海洋科学与技术进展", http://www.most.org.cn.

[22] 韩立新,王大鹏. 中国在北极的国际海洋法律下的权利分析[J]. 中国海商法研究, 2012, 23(3): 96-102.
HAN Lixin, WANG Dapeng. The China's rights analysis in the Arctic region under international ocean law[J]. Chinese Journal of Maritime Law, 2012, 23(3): 96-102.

[23] ROTH E M. Arctic ocean long term acoustic monitoring: ambient noise, environment correlates, and transients north of Barrow[D], Alaska: University of California, 2008.

[24] BROWN J R, BROWN D W. Reverberation under arctic sea-ice[J]. J. Acoust. Soc. Am., 1966, 40(2): 399-404.

[25] 赵进平. 情系北冰洋[M]. 北京: 中国环境出版社, 2006.

庆祝马远良先生80华诞

北极水声学研究的新进展和新动向①

李启虎[1,2] 黄海宁[1] 尹 力[1] 卫翀华[1]
李 宇[1] 薛山花[1] 栾经德[1,3]
(1 中国科学院声学研究所 北京 100190)
(2 中国科学院大学 北京 100049)
(3 海军研究院 北京 100161)

2018 年 3 月 28 日收到
2018 年 6 月 13 日定稿

摘要 北极水声学作为水声学研究的一部分，起步要比达·芬奇所描述的声呐雏形晚很多年。第二次世界大战后北极水声学的研究开始受到发达国家（主要是美国）的重视。它的发展和研究重点带有明显的冷战烙印。冷战结束之后，随着北极持续变暖的趋势，北极及其毗邻海域的海洋水声环境受到特别的重视。环北极的 8 个国家组成排他性的北极理事会。我国政府于 2018 年 1 月 26 日发表北极政策白皮书，声明中国是近北极国家，是北极地区利益攸关方。本文介绍北极水声学研究的新进展，包括我国有关涉海单位近年来所做的科考和学术研究。指出，北极水声学的研究不局限于把传统水声学中的研究内容（如环境噪声、混响、传播等等）并行地在北极环境条件下加以重复探讨，而是要根据北极海洋环境的实际情况，进行有关领域的新研究。其中不乏传统浅海、深海水声学研究中所不具有的特色，如冰-水界面、冰下的半声道效应、冰盖下水下无人载器（UUV）的通信、定位及声呐对冰下环境的适应性研究等课题。

PACS 数：43.30, 43.60
DOI:10.15949/j.cnki.0371-0025.2018.04.002

Progresses and advances in arctic underwater acoustic study

LI Qihu[1,2] HUANG Haining[1] YIN Li[1] WEI Chonghua[1]
LI Yu[1] XUE Shanhua[1] LUAN Jingde[3]
(1 *Institute of Acoustic. Chinese Academy of Science* Beijing 100190)
(2 *University of Chinese Academy of Science* Beijing 100190)
(3 *Naval Research Academy* Beijing 100161)

Received Mar. 28. 2018
Revised Jun. 13, 2018

Abstract As a part of underwater acoustics, the study and development of Arctic Acoustics is later than underwater acoustics about 50 years. After World War II, the study of Arctic acoustics attract many interests by the developed countries, especially USA. The research works obviously have some kinds of cold war brand. After the finish of cold war, with the gradually warming trend of Arctic area, the ocean&acoustic environment of Arctic and its neighbor area have been considerably concerned. The 8 countries of Arctic rim organized exclusive “Arctic council” in 1996. The white paper of “China Arctic Policy” is published in January 26, 2018, the Chinese government declares that China is a close Arctic country, and is the responsible stakeholder of Arctic interests. The new advances in Arctic research works are introduced in this paper, including the results of scientific survey and studies about Arctic underwater acoustics of Chinese researchers. It is showed that the Arctic underwater acoustical research area is not only parallel copy the topics what traditional underwater acoustic covered, e.g. environment noise, reverberation, and propagation, but also

① 声学学报, 2018, 43(4): 420-431.

the topics which is specifically based on the Arctic environment. Some of these research fields cannot be included in the traditional shallow water, deep water acoustics, e.g. the ice-water interface feature, ice covered semi-acoustic channel effect, the communication and navigation of UUV in the condition of under ice, and the adaptation of sonar technique, equipment in the ice-covered environment, etc.

引言

早在 1947 年，美国海军研究室 (ONR) 就开始在阿拉斯加的巴罗附近建造了北极研究实验室 (后来被命名为海军北极研究实验室)。当美国海军鹦鹉螺号于 1958 年第一次进行了冰下穿越北极的行动之后，北冰洋便具有了显著的战略意义，而随着美国和俄罗斯 (苏联) 核潜艇在北极的常规穿越和部署则标志着人们对北极水声学兴趣的开始[1−2]。

北极水声学的研究指的是北冰洋地区及其毗邻海域的海洋声学、水声学的研究工作，实际上高纬度地区的水声学研究都可以看作是极地声学的一部分。但是，由于南极冰盖下是大陆．同时，地球上的绝大多数国家在北半球，人们对北极航道的关注，远远大于对南极海域的关注．所以一般人们提 “极地声学”, 实际上仍把目光集中在北极．

美国关于北极水声学的出版物最早出现在 20 世纪 60 年代初由国防研究人员出版的科学文献中[3−5]，俄罗斯对北极水声学的研究的公开资料相当少，但从文献 6—文献 8 中推断其研究也是相当深入的。早期北极水声学的研究依靠在漂流的冰面上建立的冰上观测站进行[9]，主要动机是了解北冰洋盆地的声学，以支持其核潜艇活动[10]．

北冰洋的水声学与其它海洋相比具有独特的性质，由于冰盖的存在，声音是北冰洋地区唯一可用于远程通信和监测的媒介。冰盖在声学上具有重要意义，它充当边界，保护海洋免受风、波浪和湍流的影响．该边界还起到使海洋与太阳辐射隔离的作用，使得冰下水体结构稳定，使声音传输更加稳定，并因此传播更远的距离．冰盖的存在还可以改变水体的性质，北极冰下水体中，上表面主要是新鲜寒冷的海水，而不是温暖而盐度较高的海水，因此北冰洋的声学环境具有强烈的向上折射的声速剖面。冰盖的存在及其时空变异性引起了独特的环境噪声，其粗糙度和厚度的变化带来高反射系数，产生了强烈的混响。

1 北极水声环境的新变化

在 2000 年以后，由于气候变化，北冰洋正在经历着巨大的变化，最明显的是夏季冰盖的范围和厚度迅速减少，如图 1，北极地区多年冰的比例大大下降，只占北冰洋冰的 30%，初年冰的比例上升[11−13]。海冰的年龄对冰冠的特征有很大的影响，特别是冰的厚度和粗糙度，初年冰在冰下的剖面更光滑，而多年冰更厚。冰脊的形成过程是北极地区重要的噪声来源，这一现象在多年冰中发生的比例更高，而多年冰盐分含量较低，相对脆弱，容易开裂，因而产生的噪声较大。冰层的减少，使得覆冰区和无冰海洋区之间的过渡带——冰缘区 (MIZ) 的面积比以往更大，冰缘区是一个动态区域，在那里，一些大气和海洋现象和过程，如极地低压、表面波、锋面、涡和其它中尺度海流现象对冰盖有很大的影响[14]。在风和海流的影响下，冰团的相互移动产生噪声，可使噪声水平增加 4～12 dB[15]。开阔水域的扩大，促进了航运的增加和地质勘探的开展，在加拿大海盆区域，地震测量可以使自然环境中的背景噪声增加 2～8 dB[16]。以上现象使得北极的水声环境正经历着前所未有的环境噪声来源与组成成分的变化[17−18]。

另外一个明显的变化是来自太平洋和大西洋的底层水团的变化，这一变化影响冰下水体的温盐结构，改变了海洋声波导环境，从而影响声波在北冰洋中的传播[19−20](见图 2)。

在 ONR 于 2014 进行的边缘冰区研究中，中心频率为 900 Hz 的导航信号的传输范围大大超出了预期，超过 400 km[21−22]。说明了由于气候变化带来的北冰洋海域最佳传播频率范围的改变，在不同的距离尺度上，北冰洋海域水声最佳传播频率上限频率都可以提高，由此使得应用于通信、导航以及海洋环境监测的声学系统规模变小、费用降低、布放也更容易，并使得新型、长生存周期的水下无人平台在北极观测网中占据更重要的位置。

而 Worcester 认为[23]，由于冰盖减少，更多的开放水域带来了能量交换窗口，增加了内波形成的概率，由此带来了声学体积散射的增加，同时，受气候变化影响而改变的北冰洋环境对低频传播带来了影响，使得北冰洋的低频噪声背景场发生了改变，从而需要对北冰洋海域内受海洋与冰共同约束的水声信号处理基本限进行研究。

2 国外极地声学研究的新进展

由于海冰的减少，北冰洋海域逐步开放，在北极地区的水声学研究显得更为引人注目，北极水声学的研究在能源与矿产、渔业资源与生态环境、旅游与交通运输、气候变化以及国防安全等方面的需求也日

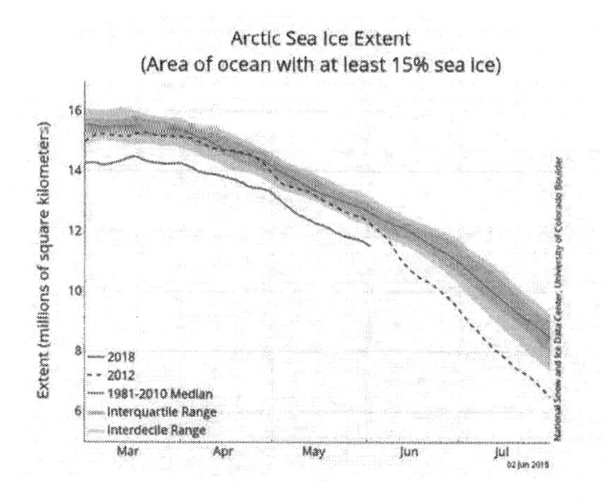

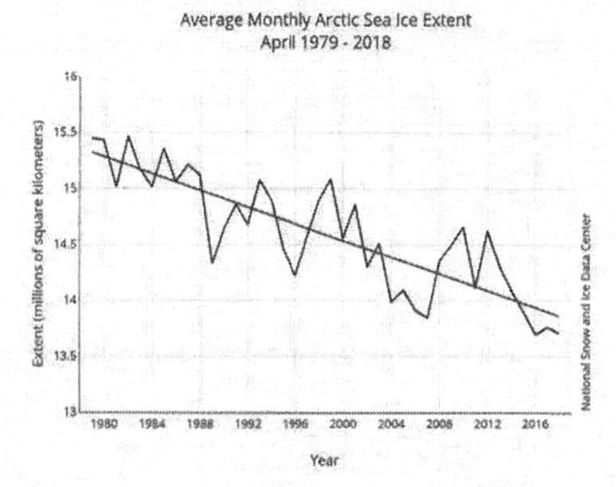

图 1 北极海冰的变化趋势 (来源：NSIDC)

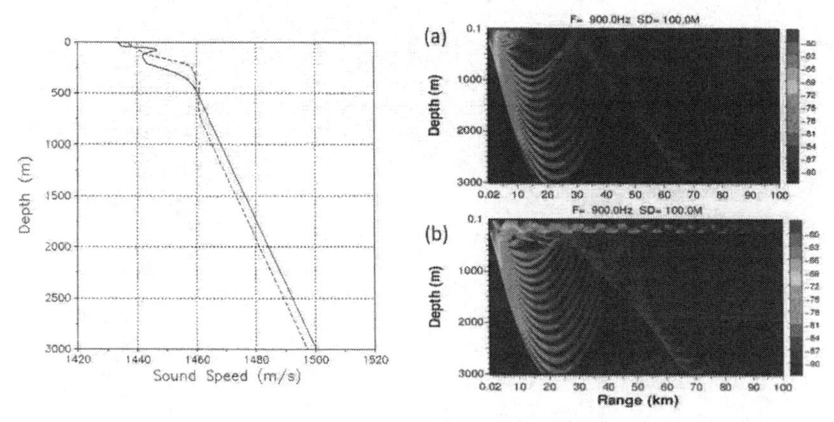

图 2 北太平洋暖流侵入范围扩大带来的波弗特透镜现象 (2016 年，来源：文献 19)

益凸显。从公开文献中可以看出，对北极及其毗邻海域进行系统研究的主要是美国和前苏联 (今俄罗斯)，但以美国为主。

2007 年，美国政府开始实施北极的重点政策。在名为“美国 21 世纪海权的合作战略”的文件中，美国政府声称：“气候变化正在使北极的水域逐步开放，不仅是对新的资源开发，而且也是对新的航运路线”。这些年来的政策发展，通过美国海军北极战略目标 (2010)、北极地区国家战略 (2013)、美国海岸警卫队北极战略 (2013)、国防部北极战略 (2013)、北极地区国家战略实施计划 (2014) 和美国海军北极路线图 2014—2030(2014)，确定了进一步的指导方针，遵循以下 3 个策略：(1) 推进美国的安全利益，确保美国拥有北极主权并借此提供国土防御保障；(2) 追求负责任的北极地区管理，提供海军力量随时应对危机和突发事件，维护航行自由；(3) 加强国际合作，并将北极宣布为新的海洋边界。

美国海军的潜艇部队致力于为科学研究做出贡献，并维持全球存在。从 1993 年起，美国海军与海洋研究界合作开展了一项名为冰上科学行动 (Science Ice Exercise: SCICEX) 的项目，该计划“旨在使用核动力潜艇绘制冰盖图并对其取样；获得海洋物理、化学和生物水特性；海底地形；北极海洋浅地层剖面”[24]。SCICEX 计划与冰上演习行动 (Ice Exercise: ICEX) 相结合，联合其它机构和水面力量，每两年举行一次，2018 年的 SCICEX 正在进行，预计未来该计划将继续下去。

目前北极水声学的研究主要侧重于环境噪声及其来源、混响及散射、传播以及声学海洋观测等 4 个方面，并开始将先前发展的声学技术应用于了解北极的海洋学和冰层。

(1) 环境噪声及其来源

虽然北冰洋的冰盖消除了风和海浪在海面上相互作用产生的噪声，但同时也通过各种机制产生了自己独特的噪声[3-4,25]，这方面的研究在上个世纪已经得到了比较全面的研究：冰下海洋噪声中低频噪声 ($\leqslant$1 kHz) 的主要来源是大规模的冰运动、冰脊的形成以及冰的裂解过程[25-30]，并且表面波导与冰盖的相互作用也使得低频信号更多的保留下来[31-32]。高频宽带噪声 (1 kHz ~ 数十 kHz) 信号的特性与在开阔的水域环境中，雨和风所产生的众所周知的噪声类型相似，与风速高度相关。因此，可认为高频宽带噪声是由风吹过冰面和冰间水道引起的湍流以及风裹着颗粒物撞击冰面产生的[17-18]，此外还有冰融化时冰内高压气泡喷发带来的高频噪声[33]。

以早期冰下噪声源特性研究成果为基础，目前许多研究利用布置在水下的水听器和布置在冰层中的检波器进行测量，目的是基于环境强迫机制来识别环境噪声，对冰的动态变化进行预测，并将冰层产生的噪声用作声学遥感工具来研究冰本身[34]。使用分布式传感器来定位冰脊事件，并将其与噪声源级和卫星图像相关联，估计单位时间和冰区的噪声事件次数，使其与环境因素联系起来[14,35-38]。极地中心冰区和冰边缘区的环境强迫存在着不同，冰边缘区的噪声特性和变化与大气和海洋的相互耦合是美国海军研究办公室目前的一个研究热点领域[14-17,39]，在冰边缘区，噪声是由风驱动的海浪和浮冰相互作用产生的，由于石油和天然气工业的需要，地震测量带来了大量的低频噪声，引起了低频海洋噪声背景场成分的变化，对其进行的长期环境监测带来了大量的水下噪声数据，从而更好地了解了噪声的季节依赖性及其对冰盖面积和特征的依赖[14-18,40-41]，还研究了海洋哺乳动物的季节性存在及其发声对环境噪声的影响[14-18,42]。此外，还研究了未来北冰洋酸化条件下声信号和噪声的变化[43]，将海洋酸化以及波弗特海双声道波导效应与噪声的增加联系起来，以预测北冰洋海域海洋环境的变化。

在北冰洋以外的其它海域，Wenz 曲线一直是估算特定风浪和船舶环境条件下噪声级的可靠方法，在文献 44 中，给出了南极平整冰和北极堆积冰下获得的海洋环境噪声与 Wenz 曲线的比较，可以作为研究参考。由于风和冰的复杂相互作用，以及冰产生的噪声机制，北极观测到的环境噪声变化很大。在北冰洋的中心区域，存在非常安静的时段，30 Hz 噪声级降至 40 dB，300 Hz 附近降至 30 dB，而在数小时之内，噪声变化可高达 20～30 dB[6,16]。通常，冰缘区噪声级高于北冰洋的中心水域。例如，波弗特海区域观测到的噪声季节变化测量数据在频率 32 Hz 从未低于 60～65 dB，或在 1000 Hz 不低于 35 dB[45]。季节变化也比其它海域大的多 (平均月频谱水平从最低到最高范围可达 30 dB 以上)[46]，有专家认为，到目前为止还没有建立起一个准确的北极环境噪声预测模型[47]。

(2) 混响及散射

了解冰-水界面的声反射和散射特性是建立北极水声传播损失模型的基础，北极冰下的混响及散射与其它海域的区别主要有两点：来自水体的低体积散射和来自冰-水界面的变化剧烈的界面散射。Birch 的研究结果表明[10]，北冰洋的体积散射与本地区日照水平和浮游生物量有很大的相关性，与温带海域观察到的超过1000米深度的深散射层不同，北冰洋的体积散射层有两层，主要集中在200 m以上，一层在50 m深度，另外一层具有从200 m迁移到20 m的深度的特性，并且与开放海洋深散射层的昼夜迁移特性不同，北冰洋体积散射层显示与北极日光模式相对应的季节变化和日变化。界面散射强度与冰的类型和冰粗糙度有很强的相关性。Brown和Milne等人的早期测量实验表明[48]，在40 Hz～10 kHz频率范围内，不同冰盖条件下的混响有很大的差异，其一般特征是散射强度随频率和掠射角的增加而增大，特别是在1.28～10.24 kHz之间[49]；与不规则的多年冰脊相比，平坦的浮冰的散射要弱得多，但其引起的反射损失不容忽视[50]。

北冰洋中心区冰层下的散射模型的建模工作相对成熟，Burke-Twersky模型将冰下散射体建模为自由或刚性基准平面上的椭圆半圆柱凸起的随机分布[51]，冰的粗糙度参数可使用微扰法估计[52]。通过这些模型计算的不同频率的前向散射强度、后向散射强度和掠射角的关系，在不同的试验中得到了验证[53−55]。因此目前研究的热点之一是利用模型反演，从散射声中确定海冰结构[56−57]。此外，由于冰水界面在空间尺度上的异常复杂和可变性，使得远距离，尤其是跨越冰缘区的远距离混响预测模型的发展并不令人满意[58]。冰面下深度变化（即粗糙度）的统计对模拟散射损失是至关重要的，目前正在发展利用飞机、卫星进行大范围冰层表面高度测量，计算冰的厚度剖面与统计特性，进而估计冰下反射损失的研究[37,54]，也是研究热点之一。

低频信号的混响及散射对于北冰洋的远距离声传播和水下目标探测具有实际意义，而高频信号的冰下混响及散射对于利用潜艇上视声呐进行冰厚测量以及冰下主动声自导鱼雷的使用至关重要[50,60]。利用潜艇的上视声呐进行的冰下深度剖面的直接测量，能给出更详细更准确的结果，可被用来估计低频散射强度和混响的一致性；从深度剖面中估算出冰反射系数，对有冰和无冰覆盖的不同射线路径的传播损失进行比较，可以估计冰散射损失；而10 kHz至100 kHz频段的混响对主动声自导鱼雷的性能具有直接影响，但这方面国外的公开资料较少。

(3) 传播

北冰洋的表面声道，具有稳定的向上折射声速剖面和粗糙的极地冰上边界，形成了一个有利于低频传播的波导条件，该波导传播的特点之一是它类似一个带通滤波器[1]。高频和低频成分衰减很快，前者是由冰层的反射损失所引起，后者是因为低于10 Hz的频率的声波不能有效限制在波导内，优势传播频段为10～50 Hz[61−62]，该频段也与船舶噪声的“叶片频段”和“宽带峰值频段”基本重合。因此，北冰洋的波导传播特性有利于对船舶目标的检测。第七次北极科考中在北冰洋冰下接收到的航船噪声就证明了这一点，如图3所示。

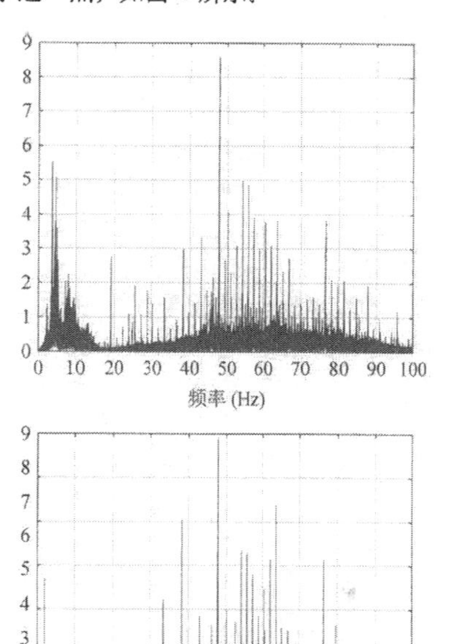

图3　我国第七次北极科考接收到的北冰洋冰下航船噪声

在北极中央冰区，低频（10～100 Hz）信号传播损失的幅度比大多数自由表面散射理论所预测的要大的多，这种高损耗主要是通过在冰盖上的散射引起的[33]。得益于对冰层散射特性的了解与混响建模的进展，北极冰下低频传播仿真建模的研究取得了较大的进展。利用新的理论损失参数与简正波传播模型相结合，提高了预报北极传播特性的能力，现在可以对这个额外的传播损失进行比较准确的预测[63−64]。

对于北极及其毗邻海域，若海底的地形和声波波长约束共同构成浅海条件，声波与海底的相互作用将会使声波传输特性变得更加复杂，需要研究北极区域的声场的高精度预报以及最佳工作频带选择问题[59,65]。

另一个值得关注的现象是双声道波导现象，在文献19中，利用Woods Hole海洋研究所2013年布放的冰基漂流浮标获取的数据，确定了所谓的“波弗特海透镜”现象的存在，该现象存在于波弗特海域水下约 100 m 以浅的深度，是由于从白令海峡北上的太平洋暖水层引起的，它使声速剖面存在局部极大值，形成了一个双声道的声学环境，被称为“声学高速通道”，这些双声道波导现象的研究可以追溯到 2004 年，目前其效应已经扩展到整个波弗特海域，预计波导内的声源和接收器在 100 km 范围内的传输损耗将降低 10 dB，对声学传感器网络、通信与导航在北极地区的应用带来了明显的影响[66]。

通过 2015 的夏天在加拿大盆地进行的声传播实验 (CANAPE) 获得的观测数据，对双声道波导现象进行了研究和分析[67]．证明了下表面波导存在约 120 m 的轴线上，在 100～200 m 之间形成了一个可以在更宽的频率范围内，传播损失较低，传播距离更远的波导环境 (见图 4)。

(4) 声学海洋观测

随着对全球变暖、北极海冰融化的关注，除了国家战略和军事需求支持的北极研究项目外，石油和天然气行业的跨国公司，为了满足其开发利用北极的需求，也投入资金对北冰洋地区的环境噪声的变化及特性进行了相关的研究[68]，此外还有来自相关政府机构的资助对北冰洋进行持续观测以研究气候变化[69−70]。

ACOUS (Arctic Climate Observations using Underwater Sound) 项目是美国和俄罗斯双边合作项目[71]，1994 年进行的国际“跨北极声传播 (TAP)”实验进行了横跨北极的声学传播试验，首次证实了北冰洋的声学测温能力。在此基础上，1998 年 10 月至 1999 年 12 月开展了用于气候观测的 ACOUS 试验，美国方面在 Lincoln 海布放了一组由水听器和 Micro CTD 组成的 5180 m 自主垂直接收阵，俄罗斯方面在弗朗茨约瑟夫岛西北 200 km 处部署了低频声源 (20.5 Hz, 195 db/1μpa@1 m)，每 4 天发射 20 min 信号。通过这些试验数据以及 SCICEX 1995-2000 的温度测量数据，进行了北冰洋热容量分析，给出了与北冰洋气候变化一致的结论；同时通过对传播的声信号的分析，对极地冰层厚度的季节性变化进行了观测；此外也分析了利用 ACOUS 试验设备进行声学遥感的可行性。通过 ACOUS 项目的一系列声传播试验数据分析，美俄科学家也得到了北冰洋的声场分布的详细数据。声信号传播时间测量表明沿声信号传播路径北极水温平均升高了 0.4°，这是首次海盆尺度海水变暖的测量。1998 年 10 月进行的破冰船和潜艇的直接测量证实了这一变化，测量结果同样表明北极中层水变暖约 0.4℃～0.5℃，这与 TAP 试验的测量结果相一致[72]。

由于依赖科考船的传统海洋观测大多在夏季进行，并且沿着固定路线进行，无法在时间和空间上获得足够多的采样，因此从上个世纪末开始依托声学导航、定位、通信能力，在北极地区建立无人观测网以实现北极系统特征的长期观测。

ACOBAR (Acoustic technology for observing the interior of Arctic Ocean) 是欧盟支持的北冰洋中部观测项目，开始于 2008 年，目前已经进行了第二期[73]．其主要目标是发展一个集声层析、水下平台数据传输、冰基漂流浮标和滑翔机通信与导航为一体的北冰洋中部海洋环境监测和预报系统，图 5 左图为 ACOBAR 期望实现的覆盖北冰洋区域的系泊网格。ACOBAR 实现和测试了两种不同类型的海洋观测系统：一种是以海底系泊节点为主体的声层析系统，另一种是合并了冰基漂流浮标和滑翔机的可以随浮冰漂流的冰上系泊声学平台 (AITP)。

2008 至 2010 年，ACOBAR 项目在 Fram 海峡布置了包含 3 个发射声源 (ABC 三角顶点) 和一接收阵 (三角中点 D) 的声层析系统 (如图 5 右所示)[74]，其中各个系泊节点之间的距离为 130 km 至 300 km。除了可以通过声层析短时间精确测量收发节点之间

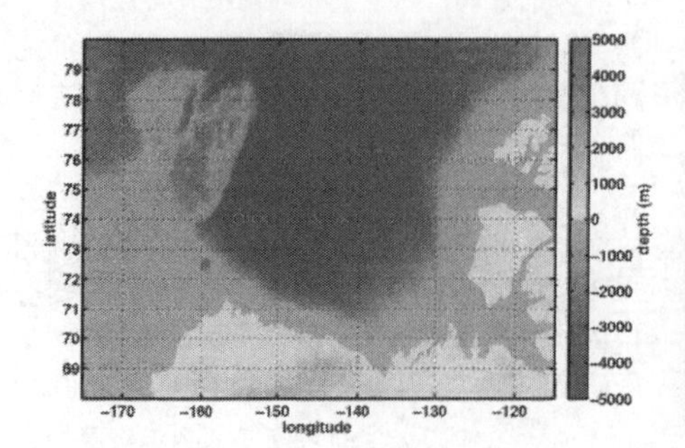

图 4 2015 年多国联合的北极“加拿大盆地”声学传播实验及所使用的 DLVA 阵

水平平均温度和声速场，该系统的发射声源可以为水下滑翔机提供导航信息，其接收阵可以收集低频环境噪声和监测海洋哺乳动物。

AITP 观测系统 (如图 6 所示) 是由传感器节点，漂流浮标以及滑翔机组成的可以随浮冰漂流进行移动测量的系统，可利用层析成像阵列发出的信号进行导航和冰下海洋学调查，声引导水下滑翔机可获得高空间分辨率的海洋学数据，这是对来自层析成像阵列的低空间分辨率但高时间分辨率的数据补充[75]。

从 2012 年起，为了改善北冰洋海洋物理环境预报和预测能力，美国海军研究办公室制定了北极科学计划 (Arctic Science Program: ASP)，该计划是其北极和全球预报计划 (Arctic and Global Prediction Program) 的重要部分[76]。ASP 计划聚焦 3 个领域：第一、根据对实地和遥感观测的数据处理，发展具有足够分辨率的完全耦合海洋-波浪-冰-大气模型；第二、完善对北冰洋环境的基本认识，实现数值模型中对关键物理过程的更精确表述；第三、开发传感器、平台和通信能力，发展一套可以提供长期监视、远期科学认识的持续观测系统。

ASP 研究涵盖多个课题，覆盖海冰、水文、大气等方面，除了前述的加拿大盆地声传播试验 (CANAPE)，MIZ(Marginal Ice Zone) 项目和 AWSS (Arctic Waves and Sea State) 项目也是其中两个初始项目。其中，MIZ 项目的执行时间为 2012 年至 2016 年，主要研究夏季融冰期在冰缘区出现的各种物理现象。

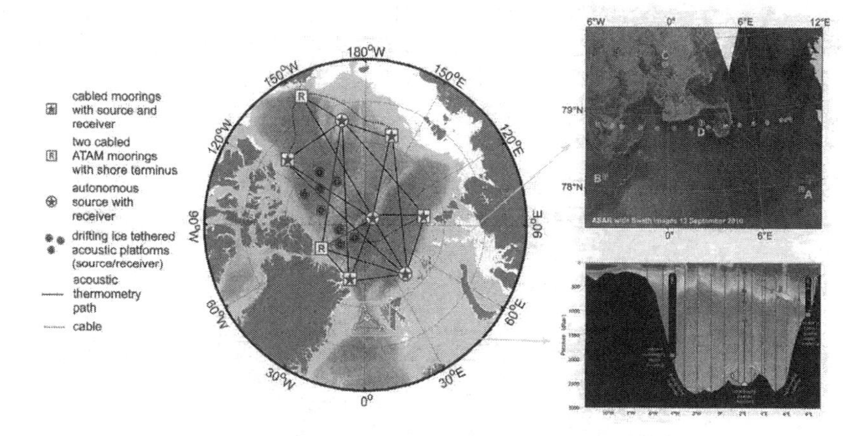

图 5 北极地区预设的系泊网格以及已实现的系泊声层析系统 (来源：https://acobar.nersc.no/)

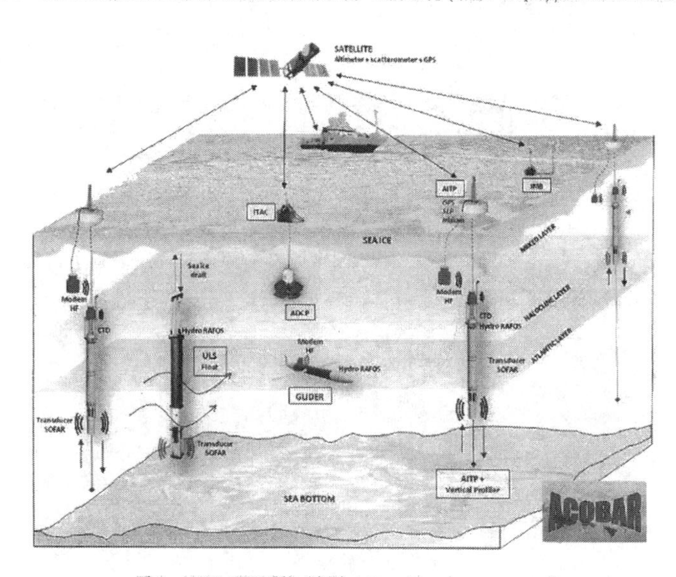

图 6 AITP 观测系统 (来源：https://acobar.nersc.no/)

其观测系统如图 7 所示，是一套包含冰上系泊设备、测量浮标、水下滑翔机、导航声源的自动测量传感器系统，通过漂流的方式测量冰缘区的冰量、波浪、海流、温度、剖面等环境数据。其中，系泊在冰面的至少 8 个导航声源和安装在水下滑翔机上的两个移动声源为工作在冰下的测量浮标和滑翔机提供声导航网络。

AWSS 项目的执行时间为 2013 年至 2017 年，主要研究北冰洋区域空-海相互作用对波浪的影响，以及波浪和海冰之间的相互作用和传播影响。该计划期望通过改善对北冰洋这一复杂区域的海洋-波浪-冰-大气的认知和建模，提升对北冰洋可操作环境的预报能力。该项目在 2014 年完成了对新型基于无人平台的自主观测系统和技术测试，在 2015 年通过新的 UNOLS 北极科考船和自主观测系统的结合完成项目的主要观测任务 (见图 8)。

AWSS 的自主观测系统现场方案是专门为了解动态变化的海洋状态 (即表面波活动的增加) 对秋季冰层恢复的影响而设计的，用于北冰洋现场实验的海洋状态和边界层物理的仪器平台包括无人飞行器 (UAV)、自主水下航行器 (AUV)、自动气象站 (AWS)、声学海流计 (AWAC) 和水下滑翔机。

此外值得注意的是，近 10 年来，北极水声学的研究还有几个新的动向。第一是更多的国家进行联合实验，第二是更加注重理论模型和实际北极海域科考的验证，第三是声学领域的各类国际学术会议更多地设立专门的分会场。2017 年至少有 3 个声学方面的专业会议，设立了 “极地声学” 或 “冰盖下声传播” 等专题，它们是 2017 年 8 月份在维也纳举行的理论计算声学 ICTCA’2017，9 月份在希腊 Skiathos 举行的国际水声会议 UACE’2017 和 6 月份在美国 Boston 举行的第 174 次美国声学会议。

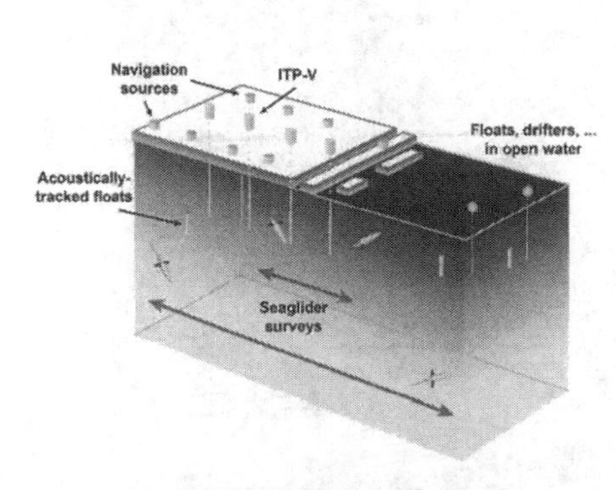

(a) MIZ 自动测量传感器系统工作示意

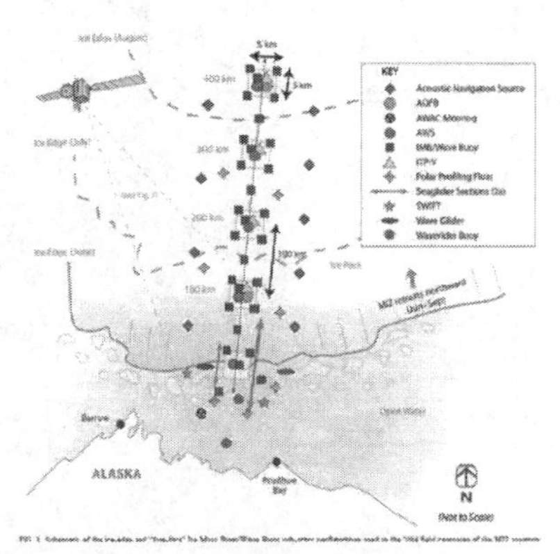

(b) MIZ 2014 年试验漂流示意

图 7 ASP 系统实验系统 (来源：ONR)

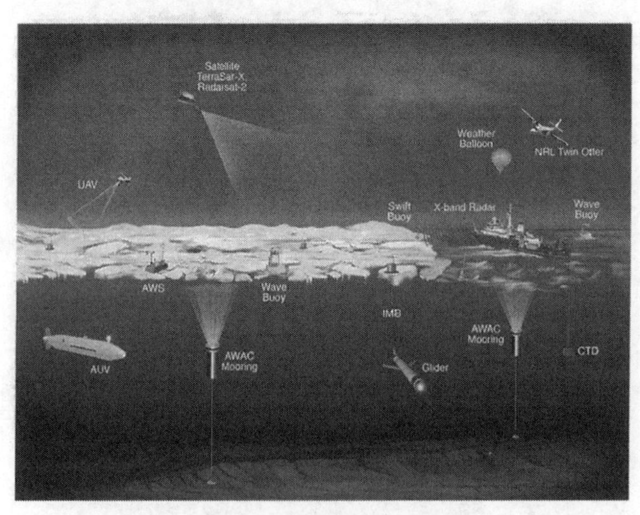

图 8 AWSS 拟采用的基于无人平台的自主观测系统 (来源：WHOI)

3 我国的极地声学研究

我国于 2013 年成为北极理事会正式观察员国。2018 年 1 月 26 日发布的 "中国政府的北极政策" 白皮书, 明确中国是近北极国家, 是北极地区的利益攸关方, 愿和有关国家共建 "冰上丝绸之路" 。

国际有关研究机构历来对中国的北极政策及相关活动非常关注, 瑞典斯德哥尔摩国际和平研究所 (SIPRI) 的中国和全球安全项目主任 Linda Jackobson, 在 2012 年 11 月发布了 "中国对北极的期许" 的政策文件, 她说[77]:"我认为, 可以预期, 在未来 5～10 年内, 而且很可能是在 5 年内, 中国会发布白皮书, 阐述在北极总体上的思路" 。 Jackobson 的预言精准度令人印象深刻。

我国科技界对北极地区的风云变幻始终非常关注, 自 1999 年来已进行了有组织的 8 次科考, 获得了大量的宝贵数据, 取得了一系列成果, 这是我们继续开展北极环境研究的重要资料。从总体情况来看, 20 世纪初以来, 有关北极水声学方面的研究只有一些零星探索, 主要是和海洋环境有关的温、盐、深, 声速剖面的测量, 还没有系统地对北极及其毗邻海域的水声环境进行考察研究, 更没有对有关水声传播、通讯、水下目标定位等的试验。

对北极地区及其毗邻海域的声学研究, 中科院领导早在几年之前就已有所部署。他们以 "率先行动" 的前瞻性战略眼光制定了周密的计划, 由重大任务局、教育前沿局、国际合作局统筹安排了一系列项目, 并与国家海洋局签署了在海洋领域进行全面深入合作的战略合作框架协议。

从 2014 年起, 在国家海洋局极地办公室和极地中心的支持下, 在第 6 次北极科考中首次设立了水声学的研究内容, 2016 年, 中国科学院声学所的科考人员, 第一次搭乘雪龙号科考船赴北极进行了声学试验 (见图 9), 取得了一批重要数据。2017 年, 继续设置了水声学的科考计划。

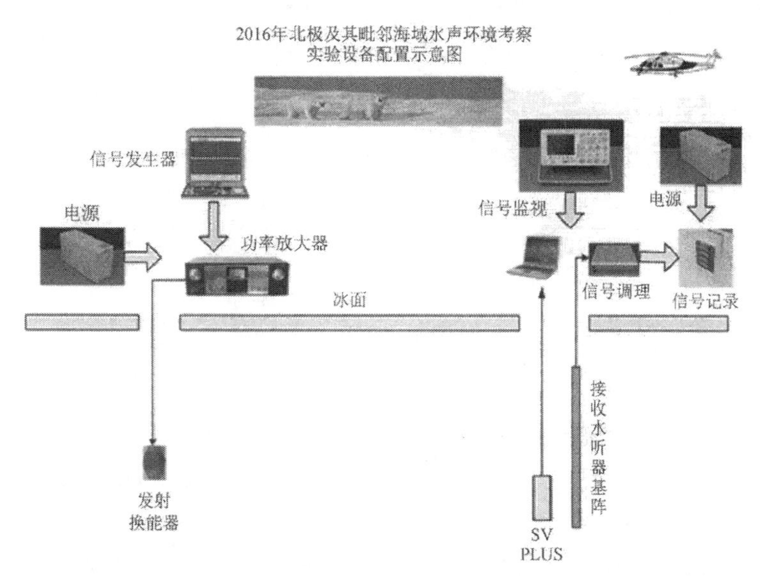

(a) 实验设备配置方框图

(b) 参加实验的卫翀华副研究员

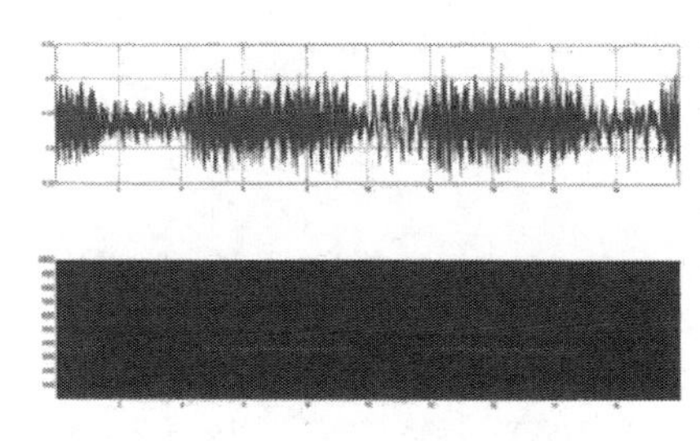

(c) 冰下水声通信实验记录

图 9 我国 2016 年 7 月—9 月第 7 次北极科考中的水声实验

2017 年 3 月，中科院重大任务局在北京组织国内涉海的 10 多家单位，举行了“北极科学研究暨北极水声学”专题研讨会。据不完全统计，国家自然科学基金委从 1986—2013 年共资助和极地科学有关的基金项目 450 项[78]，并在 2016 年安排了冰下水声传播研究课题。

2018 年初，哈尔滨工程大学和俄罗斯远东联邦大学、俄远东国立渔业技术大学就成立“北极海洋环境与声学技术联合实验室”达成框架协议。

国内其它单位也相继在该领域开展了工作。从公开发表的文献上看，我们的工作仅仅是开始，但已涉及北极水声学的诸多方面，如冰下水声信号的传播，散射等[79–83]。

值得指出的是，北极水声学的研究受到一个很现实的条件限制，就是目前我国可以去北极做科学考察的船只有限，与其它水声学、海洋声学研究相比，实际能去北极做实验的机会比较少。我们可以转而采取一些补充措施，例如在高纬度地区的有冰季节进行模拟试验或寻求国际合作。但是某些情况下，模拟环境实验或理论建模不能取代在北极地区的实际测量。有一部分和科学测试有关的实验可以在模拟条件下进行，比如测量冰厚的仪器设备、某些声呐技术设备 (水下滑翔器、AUV 、 UUV 等) 可以在一般冰下进行预研。

4 未来的展望

总的说来，北极水声学的发展是持续而且变化的，从早期的军事需求单一推动，对北冰洋海域的研究侧重于描述并预测它的声学特性；到目前在国防需求、工业需求以及科研需求的推动下，开始利用声学技术来帮助观测并理解北冰洋海域的物理环境。针对变化的北极地区对北极水声学带来的影响和需求，研究重点主要在以下几个方面：

(1) 极地海洋环境综合观测及参数获取；

(2) 极地海洋冰下声散射机理及应用；

(3) 极地海洋冰下声传播特性及应用；

(4) 极地海洋环境背景场特性及其应用；

(5) 极地海洋水声波导效应及应用；

(6) 极地环境下水声信号处理和声呐设备的适应性研究。

虽然我国在北极水声学方面的研究才刚刚起步，既缺乏在北极地区开展声学实验的经验，也缺乏常规观测系统，北冰洋的声学数据采样量严重不足，但我国涉海单位有关极地声学、北极声学的研究正迎来良好的发展机遇期。可以预言，在不久的将来一定可以为我国经略北冰洋，建设海洋强国的目标做出贡献。

参 考 文 献

1 李启虎，王宁，赵进平等. 北极水声学：一门引人关注的新型学科. 应用声学， 2014; **33**(6): 471—483

2 Dennis M. Conlon, Thomas B. Curtin. The ONR high latitude dynamics program-an introduction. *Arctic Research of the United States*, 2004; **18**: 2—5

3 Milne A R. Shallow water under-ice acoustics in Barrow Strait. *J. Acoust. Soc. Am.*, 1960; **32**: 1007—1016

4 Milne A R, Ganton J H. Ambient Noise under Arctic-Sea Ice. *J. Acoust. Soc. Am.*, 1964; **36**(5): 855—863

5 Milne A R. Statistical description of noise under shore fast sea ice in winter. *J. Acoust. Soc. Am.*, 1966; **39**(6): 1174—1182

6 Miasnikov E V. Can Russian strategic submarines survive at sea? The fundamental limits of passive acoustics. *Science & Global Security*, 1994; **4**(2): 213—251

7 Godin O A, Palmer D R. History of Russian underwater acoustics. ISIS, 2010; **101**(3): 662—663

8 Godin O A. Sonar science and technology in Russia in the 20th century. *J. Acoust. Soc. Am.*, 2017; **141**(5): 3705—3705

9 Henry Kutschale. Arctic hydroacoustics. *ARCTIC*, 1969; **22**(3): 246—264

10 Birch W. Arctic acoustics in ASW. Oceans Conference, 1972: 293—298

11 Kwok R, Untersteiner N. The thinning of Arctic sea ice. *Physics Today*, 2011; **64**(4): 36—41

12 Stroeve J C, Serreze M C, Holland M M. Kay J E, Malanik J, Barrett A P. The Arctic's rapidly shrinking sea ice cover: A research synthesis. *Clim. Change*, 2012; **110**(3): 1005—1027

13 Overland J E, Wang M. When will the summer Arctic be nearly sea ice free? *Geophys. Res. Lett.*, 2013; **40**(10): 2097—2101

14 Johannessen O M, Sagen H, Sandven S, Stark K V. Hotspots in ambient noise caused by ice-edge eddies in the greenland and barents seas. *IEEE Journal of Oceanic Engineering*, 2003; **28**(2): 212—228

15 Diachok O I, Winokur R S. Spatial variability of underwater ambient noise at the Arctic ice-water boundary . *J. Acoust. Soc. Am.*, 1974; **55**(4): 750—753

16 Roth E H, Hildebrand J A, Wiggins S M, Ross D. Underwater ambient noise on the Chukchi Sea continental slope from 2006-2009. *J. Acoust. Soc. Am.*, 2012; **131**(1): 104—110

17 Kinda G B, Simard Y, Gervaise C *et al.* Under-ice ambient noise in Eastern Beaufort Sea, Canadian Arctic, and its relation to environmental forcing. *J. Acoust. Soc. Am.*, 2013; **134**(1): 77—87

18 Kinda G B, Simard Y, Gervaise C *et al.* Arctic underwater noise transients from sea ice deformation: Characteristics, annual time series, and forcing in Beaufort Sea. *J. Acoust. Soc. Am.*, 2015; **138**(4): 2034—2045

19 Schmidt H, Schneider T. Acoustic communication and navigation in the new Arctic — A model case for environmental

adaptation. UComms. 2016 IEEE Third
20 Sagers J, Ballard M S, Knobles D P. Investigating the effects of ocean layering and sea ice cover on acoustic propagation in the Beaufort Sea. *J. Acoust. Soc. Am.*, 2015; **138**(3): 1742—1743
21 Freitag L, Ball K, Partan J, Koski P *et al.* Long range acoustic communications and navigation in the Arctic, OCEANS 2015 - MTS/IEEE Washington, Washington, DC, 2015: 1—5
22 Webstery S E, Freitag L E, Leey C M, Gobat J I. Towards Real-Time Under-Ice Acoustic Navigation At Mesoscale Ranges. 2015 IEEE International Conference on Robotics and Automation (ICRA), 2015: 537—544
23 Worcester P F. Thin-ice arctic acoustic window (THAAW). ADA618125. USA: San Diego, 2015
24 Edwards E H. The SCICEX Program, Arctic Ocean investigations from U.S. Navy nuclear-powered submarine. 2004
25 Dyer I. Arctic ambient noise: Ice source mechanics. *J. Acoust. Soc. Am.*, 1988; **84**(5): 1941—1942
26 Farmer D M, XIE Yunbo. The sound generated by propagating cracks in sea ice. *J. Acoust. Soc. Am.*, 1989; **85**(4): 1489—1500
27 Lewis J K. Relating Arctic ambient noise to thermally induced fracturing of the ice pack. *J. Acoust. Soc. Am.*, 1994; **95**(3): 1378—1385
28 Xie Y, Farmer D M. Acoustical radiation from thermally stressed sea ice. *J. Acoust. Soc. Am.*, 1991; **89**(5): 2215—2231
29 Xie Y, Farmer D M. The sound of ice break up and floe interaction. *J. Acoust. Soc. Am.*, 1992; **91**(3): 1423—1428
30 Greening M V, Zakarauskas P. Spatial and source level distributions of ice cracking in the Arctic Ocean. *J. Acoust. Soc. Am.*, 1992; **95**(2): 783—790
31 Greening M V, Zakarauskas P. Pressure ridging spectrum level and a proposed origin of the infrasonic peak in arctic ambient noise spectra. *J. Acoust. Soc. Am.*, 1994; **95**(2): 791—797
32 Webb S C, Schultz A. Very low frequency ambient noise at the seafloor under the Beaufort Sea icecap. *J. Acoust. Soc. Am.*, 1992; **91**(3): 1429—1439
33 Urick R J. Principles of underwater sound. 3rd Ed.US 1983, 中译本，洪申译，"水声原理"，哈尔滨：哈尔滨船舶工程学院出版社，1990
34 Laible H, Rajan S D. Acoustical determination of the ice growth in the Arctic . *J. Acoust. Soc. Am.*, 1993; **94**(3): 11760—1761
35 Zakarauskas P. Parfitta C J, Thorleifson J M. Automatic extraction of spring-time Arctic ambient noise transients. *J. Acoust. Soc. Am.*, 1991; **90**(1): 470—474
36 Stein P J, Lewis J K, Parinella J C, Euerle S E. Underice noise resulting from thermally induced fracturing of the arctic ice pack: Theory and a test case application. *Journal of Geophysical Research Atmospheres*, 2000; **105**(C4): 8813—8826
37 Deane G B, Glowacki O, Tegowski J *et al.* Directionality of the ambient noise field in an Arctic, glacial bay. *J. Acoust. Soc. Am.*, 2014; **136**(5): EL350—356
38 Geyer F, Sagen H, Hope G *et al.* Identification and quantification of soundscape components in the Marginal Ice Zone. *J. Acoust. Soc. Am.*, 2016; **139**(4): 1873—1885
39 Lee C M, Cole S T, Doble M *et al.* Marginal ice zone (MIZ) program: science and experiment plan. ADA566290, APL - UW 12-01, Applied Physics Lab., Univ. of Washington, 2012
40 Tollefsen D, Sagen H . Seismic exploration noise reduction in the Marginal Ice Zone. *J. Acoust. Soc. Am.*, 2014; **136**(1): EL147—EL152
41 Guan Shane, Vignola J, Judge J, Turo D. Airgun interpulse noise field during a seismic survey in an Arctic ultra shallow marine environment. *J. Acoust. Soc. Am.*, 2015; **138**(6): 3447—3457
42 Cummings W C, Diachok O I, Shaffer J D. Acoustic transients of the marginal sea ice zone: A provisional catalog (No. NRL-MR-6408), Technical report, Naval Research Lab Washington, DC., 1989
43 Duda T F. Acoustic signal and noise changes in the Beaufort Sea Pacific Water duct under anticipated future acidification of Arctic Ocean waters . *J. Acoust. Soc. Am.*, 2017; **142**(4): 1926—1933
44 Dahl P H, Miller J H, Cato D H *et al.* Underwater ambient noise. *Acoustics Today*, 2007; **3**(1): 23—33
45 Lewis J K, Denner W W. Arctic ambient noise in the Beaufort Sea: seaonal space and time scales. *J. Acoust. Soc. Am.*, 1987; **82**(3): 988—997
46 Lewis J K, Denner W W. Arctic ambient noise in the Beaufort Sea: Seasonal relationships to sea ice kinematics. *J. Acoust. Soc. Am.*, 1988; **83**(2): 549—565
47 Shaw R R. Ambient noise characteristics during the SHEBA experiment. ADA378686, 2000
48 Brown J R. Reverberation under Arctic Ice. *J. Acoust. Soc. Am.*, 1964; **36**(3): 601—603
49 Brown J R, Milne A R. Reverberation under Arctic Sea - Ice. *J. Acoust. Soc. Am.*, 1966; **39**(1): 399—404
50 Yang T C, Votaw C R. Under ice reflectivities at frequencies below 1 kHz. *J. Acoust. Soc. Am.*, 1981; **70**(3): 841—851
51 Burke J E, Twersky V. Scattering and reflection by elliptically striated surfaces. *J. Acoust. Soc. Am.*, 1966; **40**(4): 883—895
52 Kuperman W A. Self-consistent perturbation approach to rough surface scattering in stratified elastic media. *J. Acoust. Soc. Am.*, 1989; **86**(4): 1511—1522
53 LePage K, Schmidt H. Analysis of spatial reverberation statistics in the central Arctic. *J. Acoust. Soc. Am.*, 1996; **99**(4): 2033—2047
54 Duckworth G, LePage K, Farrell T. Low-frequency long-range propagation and reverberation in the central Arctic: Analysis of experimental results. *J. Acoust. Soc. Am.*, 2001; **110**(2): 747—760
55 Mcdaniel S T. Vertical spatial coherence of scattering from the Arctic ice canopy: Comparison of theory with experiment. *J. Acoust. Soc. Am.*, 1989; **85**(6): 2378—2382
56 Hayward T J, Yang T C. Low-frequency Arctic reverberation. I: Measurement of under-ice backscattering strengths from short-range direct-path returns. *J. Acoust. Soc. Am.*,

1993; **93**(5): 2517—2523

57 Yang T C, Hayward T J. Low-frequency Arctic reverberation. II: Modeling of long-range reverberation and comparison with data. *J. Acoust. Soc. Am.*, 1993; **93**(5): 2524—2534

58 Yang T C, Hayward T J. Low-frequency Arctic reverberation. III: Measurement of ice and bottom backscattering strengths from medium-range bottom-bounce returns. *J. Acoust. Soc. Am.*, 1993; **94**(2): 1003—1014

59 Deane G B. Shallow water propagation and surface reverberation modeling. ADA598703, 2013

60 Kwok R, Rothrock D A. Decline in Arctic sea ice thickness from submarine and ICESat records: 1958-2008. *Geophysical Research Letters*, 2009; **36**: L15501

61 Williams K L, Funk D E. High - frequency forward scattering from the arctic canopy: Experiment and high - frequency modeling. *J. Acoust. Soc. Am.*, 1994; **96**(5): 2956—2964

62 Buck B M, Greene C R. Arctic deep-water propagation measurements. *J. Acoust. Soc. Am.*, 1964; **36**(8): 1526—1533

63 Hope G, Sagen H, Storheim E *et al.* Measured and modeled acoustic propagation underneath the rough Arctic sea-ice. *J. Acoust. Soc. Am.*, 2017; **142**(3): 1619—1633

64 Lepage K D, Schmidt H. Modeling of low - frequency transmission loss in the central Arctic. *J. Acoust. Soc. Am.*, 1994; **96**(3): 1783—1795

65 Yang T C. Low - frequency transmission loss in the Arctic SOFAR channel for shallow sources and receivers. *J. Acoust. Soc. Am.*, 1989; **85**(3): 1139—1147

66 Lee C, J G. ANCHOR: Acoustic navigation and communication for high-latitude ocean research, Report from an International Workshop, University of Washington, 2006

67 Worcester P. CANAPE 2015 Research Cruise in the Bean fort Sea Abroad R/V Sikaliag. Workshop Polar Research, Scripps Inst. La Jolla, CA, Oct 20, 2015

68 Martin B, Delarue J, Hannay D. Ambient noise in the Chukchi Sea, July 07 Oct 2009. *J. Acoust. Soc. Am.*, 2010; **127**(3): 1757—1757

69 Sagen H, Johannessen O M. AMOC: Listen to the climate change? *J. Acoust. Soc. Am.*, 1999; **105**(2): 1115—1115

70 Abrahamsen E P. Sustaining observations in the polar oceans. *Philosophical Transactions of the Royal Society A*, 2014; **372**(2025): 20130337: 1—16

71 Gavrilov A N, Mikhalevsky P N. Recent results of the ACOUS (Arctic Climate Observations using Underwater Sound) Program. *Acta Acustica united with Acustica*, 2002; **88**(5): 783—791

72 Mikhalevsky P N, Gavrilov A N, Baggeroer A B. The transarctic acoustic propagation experiment and climate monitoring in the Arctic. *IEEE Journal of Oceanic Engineering*, 1999; **24**(2): 183—201

73 Sandven S, Sagen H *et al.* The fram strait integrated ocean observing and modelling system. Sustainable Operational Oceanography Proceedings of the Sixth International Conference, 2011: 50—58

74 Mikhalevsky P N *et al.* Multipurpose acoustic networks in the integrated arctic Ocean Observing System. Arctic. 2015; **68**(Suppl1): 11—27

75 Sagen H, Sandven S, Worcester P F. The fram strait acoustic tomography system. *J. Acoust. Soc. Am.*, 2008; **123**(5): 2991—2998

76 Richter J *et al.* Arctic report card 2016. NOAA, Dec 2016, USA. http://www.artic.noaa.gov/report-card

77 Jackbson L, Peng J C. China's Arctic aspirations. SIPRI Policy Paper, No.34, 2012, http://sipri.org/publications

78 贾桂德，石午虹. 对新形势下中国参与北极事务的思考. 国际展望， 2014(4): 5—28

79 刘洪宁，吕连港，刘娜等. 夏季加拿大海盆海冰边缘区声体积后向散射强度研究. 海洋学报， 2015; **37**(11): 127—134

80 刘崇磊，李涛，尹力等. 北极冰下双轴声道传播特性研究. 应用声学， 2016; **35**(4): 309—315

81 殷敬伟，杜鹏宇，朱广平等. 松花江冰下声学试验技术研究. 应用声学， 2016; **35**(1): 58—68

82 黄海宁，刘崇磊，李启虎，刘娜，卫翀华，尹力. 典型北极冰下声信道多途结构分析及实验研究. 声学学报， 2018; **43**(3): 273—282

83 朱广平，殷敬伟，陈文剑，胡思为，周焕玲，郭龙祥. 北极典型冰下声信道建模及特性. 声学学报， 2017; **42**(2): 152—158

典型北极冰下声信道多途结构分析及实验研究①②

黄海宁[1] 刘崇磊[1,2] 李启虎[1] 刘 娜[3] 卫翀华[1] 尹 力[1]

(1 中国科学院声学研究所 北京 100190)

(2 中国科学院大学 北京 100190)

(3 国家海洋局第一海洋研究所 青岛 266061)

2017 年 11 月 8 日收到

2018 年 2 月 28 日定稿

摘要 针对北极海域典型声场环境，提出了基于 OASES-Bellhop 耦合模型的冰下声信道多途结构快速分析方法．模型将海冰等效为具有粗糙界面的弹性分层介质，利用微扰法与 Kirchhoff 近似，估计海冰界面不均匀造成的散射损失，结合射线传播理论对典型北极冰下声信道多途结构进行分析与预报．数值仿真与实验结果表明，在 6 km 距离处，典型北极冰下声信道由于海冰与海底反射分别形成多途结构，海冰多次反射路径叠加形成的多途结构较为稳定，时延扩展在 14 ms 范围内，海底反射路径强度相对较弱．OASES-Bellhop 模型对冰下声信道多途结构幅度和时延预测误差较小，能够较好的解释及预报实验观测到的多途结构环境特性．

PACS 数：43.30, 43.60

Multipath structure of thetypical under-ice sound channel in Arctic: theory and experiment

HUANG Haining[1] LIU Chonglei[1,2] LI Qihu[1] LIU Na[3] WEI Chonghua[1] YIN Li[1]

(1 *Institute of Acoustic, Chinese Academy of Science* Beijing 100190)

(2 *University of Chinese Academy of Science* Beijing 100190)

(3 *The First Institute of Oceanography, SOA* Qingdao 266061)

Received Nov. 8, 2017

Revised Feb. 28, 2018

Abstract Regarding to the typical acoustic environment of the Arctic, this paper proposes a method based on OASES-Bellhop coupled model to rapidly analyze the multipath structure of the under-ice sound channel. Firstly the proposed model refers to ice plate as stratified elastic media with some roughness. Secondly, it uses the perturbation method and Kirchhoff approximation theory to solve the scattering loss due to the sea ice inhomogeneity. In the end, the model predicts the multipath structure of the under-ice channel through Ray theory. The results of the numerical simulation and experiment indicate that the typical Arctic sound channel presents multipath structures due to the sea ice and seabed in the range of 6 km, respectively. The sea ice reflection paths are stable and with short multipath delay extension within 14 ms. The seabed reflection paths have relatively weak strength. The proposed OASES-Bellhop coupled model successively predict the amplitude and delay of the multipath structure with small error which indicates the proposed model is able to analyze and predict the multipath structure of the observed acoustic environment.

---

① 中国科学院前沿科学重点研究项目(QYZDY-SSW-JSC043)资助.

② 声学学报, 2018, 43(3): 273-282.

引言

独特的地理位置和寒冷的气候环境导致北极海域的水声学特性比较特殊 [1], 国内外学者对北极水声学特性进行了相关研究 [2−6]. 海冰隔绝了风浪对海水的搅拌作用, 也减弱了海水与大气之间的能量交换, 使得北极冰下具有常年稳定的正梯度声速分布, 从而导致声波传播时与海冰频繁交互, 能量不断的被海冰反射、散射和吸收, 这就造成了冰下声场及多途结构的复杂性. 因此, 建立北极海域声传播模型必须解决海冰覆盖条件下的界面反射问题.

海冰下表面的形状复杂且起伏较大, 针对这一问题, Diachok[7] 根据海冰统计数据, 将海冰等效为椭圆形半圆柱体在自由表面的均匀分布, 结合 Burke 和 Twersky 散射理论 [8] (简称 B-T 模型), 计算冰面的反射系数. 但是, 该方法的前提是充分获取海冰上下表面的形态分布, 否则仿真结果与海上试验偏差较大. 近期, 国内学者使用 B-T 模型对北极声学特性进行了初步的探讨和分析. 刘崇磊[9] 将 B-T 模型耦合到简正波模型 KRAKENC[10] 中, 分析了北极双轴声道低频声传播特性. 朱广平[11] 等使用 B-T 模型分析了北极典型声速结构下 OFDM 通信性能. 由于 B-T 模型损失了反射系数的相位信息, 结果可能存在较大误差. Kuperman 和 Schmidt[12−13] 使用波数积分法结合扰动理论, 推导了弹性介质粗糙分界面处的低频反射损失, 计算北极冰下声传播损失并模拟了冰下垂直阵接收波形, 但未开展中高频声信道的研究. Livingston 和 Diachok[14] 使用匹配场处理技术获取海冰界面的反射系数, 避免了通过实验获取反射系数的操作复杂性, 由于相关函数是距离、深度、反射系数幅度和相位的四维函数, 计算量巨大, 并不适合实际问题的分析. Alexander[15] 以及 Liu[16] 等对北极声传播方法进行了系统的总结和分类, 使用 Bellhop 射线模型对北极典型声速剖面进行了声传播特性仿真.

Bellhop 射线模型适用中、高频条件下求解声场到达结构问题, 但无法表征界面粗糙性对声传播的影响. 本文将北极海冰等效成分层不平整的弹性粗糙界面, 建立了 OASES-Bellhop 耦合模型. OASES-Bellhop 耦合模型使用微扰理论 [17] 和 Kirchhoff[18−20] 近似理论快速计算海冰覆盖条件下的扰动声场, 得到海冰散射系数, 进而获得海冰覆盖条件下的声信道多途结构. 最后, 使用 OASES-Bellhop 模型对典型北极冰下声信道多途结构进行仿真分析, 并用北极科学考察数据进行了验证.

1 海冰散射理论

利用 Bellhop 射线程序计算声场的界面输入参数是界面反射系数. 由于海冰起伏, 界面会导致除镜面相干反射之外的随机散射非相干成分, 非相干成分等效地减弱了反射成分. 为了定量地刻画这个过程, 本节应用文献 12 和文献 13 提出的微扰处理, 将海冰粗糙引起的扰动声场表示为有效势函数, 适当地解耦相干和非相干成分, 并利用 Kirchhoff 近似得到关于相干成分的等效边界条件, 并结合非扰动声场的波动方程, 由此得到相干成分的等效反射系数.

1.1 非扰动声场解

首先介绍非扰动声场波数积分解法. 弹性介质中非扰动声场的位移势可以表示为 $\chi=\{\phi,\varphi\}$, 其中, ϕ 和 φ 分别代表纵波位移势和横波位移势 [21]. 位移势满足波动方程:

$$\left(\nabla^2+\frac{w^2}{c(z)^2}\right)\chi=\{-\delta(r)\delta(z-z_s),0\},\tag{1}$$

其中, w 表示频率, ∇^2 表示拉普拉斯算子. $c(z)=\{c_p,c_s\}$ 分别对应纵波速度和横波速度. 式 (1) 分别表示声源强度为 1 和不存在声源两种情况. 使用波数积分法来求解波动方程, 对波动方程两边作关于水平波数 k_r 和距离 r 的傅里叶–贝塞尔变换,

$$\widetilde{\chi}=\int_0^{+\infty}\chi\mathrm{J}_0(k_r r)r\mathrm{d}r,\tag{2}$$

$$\chi=\int_0^{+\infty}\widetilde{\chi}\,\mathrm{J}_0(k_r r)k_r\mathrm{d}k_r,\tag{3}$$

其中, J_0 为零阶贝塞尔函数, $\widetilde{\chi}$ 称为深度分离的位移势, $\widetilde{\chi}$ 满足:

$$\left[\frac{d^2}{dz^2}+\left(\frac{w^2}{c(z)^2}-k_r^2\right)\right]\widetilde{\chi}=\left\{\frac{\delta(z-z_s)}{2\pi},0\right\}.\tag{4}$$

分层介质的边界条件使用算符 B 来表征. 在介质的分界面处, 分别满足法向位移 w、切向位移 u、法向应力 σ_{zz} 和切向应力 σ_{zr} 连续条件:

$$B(\chi_j;\chi_{j+1})=0.\tag{5}$$

联立波动方程式 (4) 以及边界条件式 (5), 得到深度分离的位移势 $\widetilde{\chi}$, 做逆傅里叶–贝塞尔变换即可得到非扰动声场位移势 χ.

1.2 扰动声场解

北极海冰界面可以表示为图 1 所示的海水–海冰–空气三层水平分层介质, 海冰分界面的形状使用粗糙度参数来表征. 对于存在粗糙界面的弹性介质中的扰动声场, 势函数为相干声场和非相干散射声

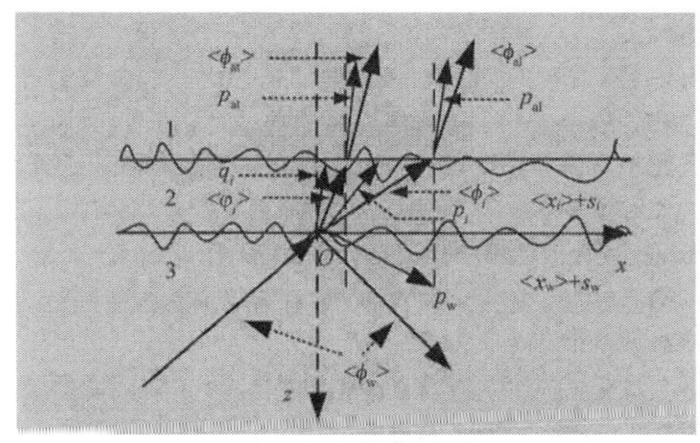

图 1 冰面反射示意

场之和，本文采用文献 12 相同的符号区分相干和非相干成分. 扰动声场可以表示为：

$$\chi_s = <\chi_s> + s, \{\phi = <\phi> + p, \varphi = <\varphi> + q\}, \quad (6)$$

其中，$<\phi>$ 和 $<\varphi>$ 分别表示纵波相干声场和横波相干声场，p 和 q 分别表示非相干纵波散射部分和横波散射部分. 分界面的粗糙度 γ 满足 $\gamma - z(r) = 0$, $<\gamma> = 0$.

使用微扰理论[20] 计算扰动声场的势函数. 将扰动声场 χ_s 使用泰勒级数展开，省略 γ 高阶项，得到扰动声场 "有效势函数"，用带上标的 χ^* 表示：

$$\chi_s^* = <\chi_s>|_{z=0} + \frac{<\gamma^2>}{2}\frac{\partial^2 <\chi_s>}{\partial z^2}\bigg|_{z=0} + <\frac{\gamma \partial s}{\partial z}>\bigg|_{z=0}. \quad (7)$$

从式 (7) 可看出，有效势函数中存在散射声场的耦合项. 为了得到与散射声场 s 无关的扰动声场边界条件 $B(\chi_{s,j}^*; \chi_{s,j+1}^*) = 0$, 需要构造散射声场边界条件来去除耦合项. 由式 $B(\chi_{s,j}; \chi_{s,j+1}) - B(\chi_{s,j}^*; \chi_{s,j+1}^*) = 0$, 可得非相干散射部分的边界条件：

$$B(s_j; s_{j+1}) = -\gamma_j B\left(\frac{\partial <\chi_{s,j}>}{\partial z}; \frac{\partial <\chi_{s,j+1}>}{\partial z}\right). \quad (8)$$

对式(8)和式(7)做逆傅里叶贝塞尔变换，得到：

$$B(\widetilde{s}_j; \widetilde{s}_{j+1}) = -B\left(\gamma_j \frac{\partial \widetilde{<\chi_{s,j}>}}{\partial z}; \gamma_j \frac{\partial \widetilde{<\chi_{s,j+1}>}}{\partial z}\right), \quad (9)$$

$$\widetilde{\chi}_{s,j}^* = <\widetilde{\chi}_{s,j}>\left(1 - \frac{<\gamma_j^2>}{2}k_{j,z}^2\right) + <\widetilde{\gamma_j \frac{\partial s_j}{\partial z}}>, \quad (10)$$

其中，$k_{j,z}$ 为介质的垂直波数. 对式 (9) 应用 Kirchhoff 近似 (具体的定义形式见文献 12 附录 B), 得到式 (10) 中非相干散射耦合项：

$$B\left(<\widetilde{\gamma_j \frac{\partial s_j}{\partial z}}>; <\widetilde{\gamma_j \frac{\partial s_{j+1}}{\partial z}}>\right) = -<\gamma_j^2> B\left(ik_{j,z}\frac{\partial <\widetilde{\chi}_j>}{\partial z}; -ik_{j+1,z}\frac{\partial <\widetilde{\chi}_{j+1}>}{\partial z}\right). \quad (11)$$

联立式 (10) 和式 (11), 与非相干散射声场无关的等效势边界条件为：

$$B(\widetilde{\chi}_j^*; \widetilde{\chi}_{j+1}^*) = 0. \quad (12)$$

将式 (12) 代入非扰动波动方程式 (1), 即可得到扰动声场的解. 求解扰动声场的解析解是一项相当复杂的工作. 对于液体液体介质扰动声场的反射系数，文献 18 给出了解析表达式：

$$\begin{cases} R = R_0(1 - 2k_{1z}^2 <\gamma^2>), \\ T = T_0\left[1 - \frac{1}{2}(k_{1z} - k_{2z})^2 <\gamma^2>\right], \end{cases} \quad (13)$$

其中，R_0 和 T_0 分别代表非扰动声场的瑞利反射系数和透射系数. 对于海冰弹性介质，需联合海冰空气界面、海冰海水界面的边界条件，计算扰动声场反射系数. 采用 OASES 模型[22-24] 求解反射系数数值解.

OASES 模型是一种快速和稳定的波数积分法模型，使用直接全局矩阵和快速傅里叶变换方法来求解弹性分层介质的扰动声场. 本文将海冰参数输入 OASES 模型来计算反射系数. 下面使用 OASES 模型仿真了海冰特征参数对反射系数的影响.

1.3 模型仿真

首先研究不同海冰特性参数对应的反射系数规律. 图 2 使用 OASES 模型仿真了文献 12 中切变波速为 1300 m/s 和 1600 m/s 两种情况下反射系数与掠射角的关系. 其中，声源频率 50 Hz, 冰层参数设置如下：冰层厚度为 3.9 m, 空气 - 海冰界面的粗糙度为 0.6 m, 海冰 - 水体界面的粗糙度为 1.9 m, 海冰中的压缩波速为 3000 m/s. 从图中可以看出，当切变波速为 1600 m/s 时，空气 - 海冰界面的粗糙度对反射系数也有重要的影响.

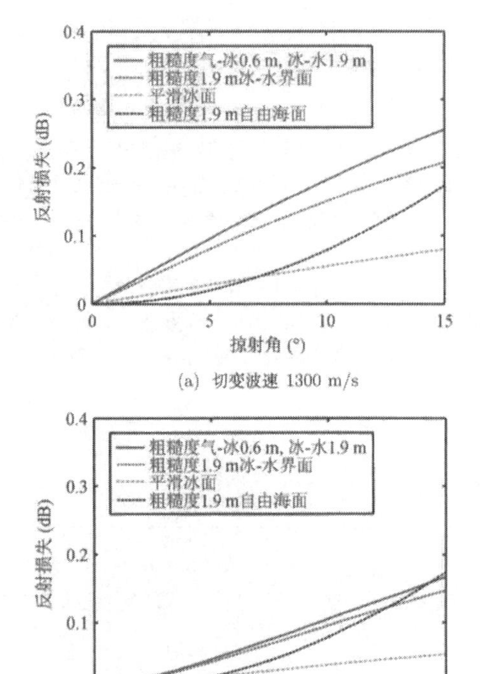

(a) 切变波速 1300 m/s

(b) 切变波速 1600 m/s

图 2 反射损失

2 OASES-Bellhop 模型

射线理论将水下声波传播等效为与波阵面垂直的若干条声线。声线的传播方向即是声能的传播方向，而声线的稀疏则表示声能量的强弱。声源发出的声信号沿着不同的路径传播，接收点的信号是所有本征声线信号的叠加。在得到冰层下表面的反射系数后，将海冰和空气介质等效为无限半空间，海冰和空气介质对冰下声场的影响可通过海水与无限半空间分界面处的反射系数来表征。

Bellhop 是典型的射线模型，采用高频近似，利用高斯声束追踪法，求解出水平非均匀信道中的声场。通过计算从声源到接收点处本征声线参数，得到多途信道的冲击响应，最终的表达式[25]为：

$$\begin{cases} h(t)=\sum_{i=1}^{N} A_i d(t-t_i), \\ A=\sqrt{\dfrac{I_0\cos\theta d\theta_0}{r\sin\theta dr}}\prod_{i=1}^{N_s} V_{si}(\theta_{si})\prod_{i=1}^{N_b} V_{bi}(\theta_{bi})\mathrm{e}^{-\beta s\cdot 10^{-3}}. \end{cases} \tag{14}$$

式 (14) 中，N 表示声源到接收点的多径数目，A_i 表示每条多径的声压幅度，τ_i 表示不同接收路径的时延。N_s 代表海面的反射次数，V_{si} 表示第 i 次海面反射系数，θ_{si} 表示第 i 次海面反射时声线在海面处的掠射角。N_b 表示海底的反射次数，V_{bi} 表示第 i 次海底反射系数。θ_{bi} 为第 i 次海底反射时声线在海底处的掠射角，β 为海水介质的吸收系数。s 为声线到达接收点传播的声程。在以下计算仿真中，式 (14) 中的 $V_{si}(\theta_{si})$ 采用 1.2 节扰动理论得到的相干成分等效反射系数来计算。

当切变波速等于 1300 m/s 时，结合上文计算出的反射系数，使用 OASES-Bellhop 模型对海冰覆盖界面和粗糙度 1.9 m 自由海面两种条件下的声传播特性进行了仿真，结果如图 3 所示。其中，声速剖面采用北极典型声速结构[1]。从图 3 可见，在 50～60 km 范围内，粗糙海冰界面的传播损失比粗糙度 1.9 m 自由海面高 6～7 dB，这与 DiNapoli 和 Mellen[26] 的处理结果一致。

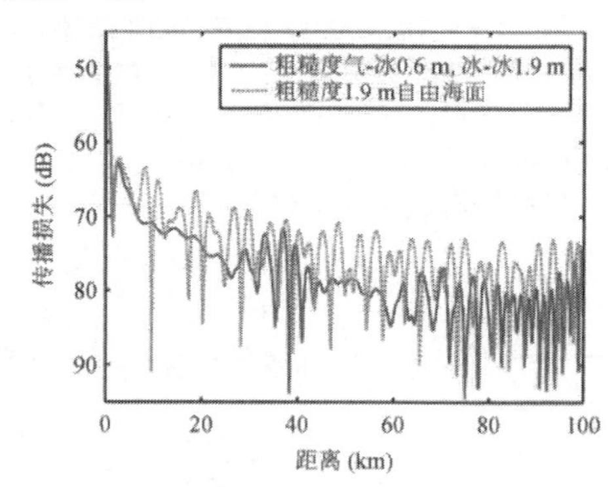

图 3 OASES-Bellhop 模型传播损失仿真

3 仿真与实验结果分析

3.1 实验介绍

2016 年 8 月 20 日，中国第七次北极科学考察第 6 次短期冰站作业期间，中科院声学所开展了北极冰下脉冲传播实验。实验位于楚科奇海北纬 76°15′57″～76°19′48″，东经 179°35′42″～179°36′18″ 海域内，实验位置如图 4 所示。

实验布设如图 5 所示。发射换能器位于雪龙船附近，吊放深度 25 m。发射信号为 3～4 kHz，时长 512 ms 的线性调频信号。黄河艇负责接收端作业。采用 6 阵元 IcLisen HF 智能水听器垂直阵实时采集声源和噪声数据。垂直阵固定于浮冰表面，阵元间隔 2 m，覆盖冰下 40～50 m 的深度范围。声源与接收阵的距离为 6.15 km。根据 IBCAO (International Bathymetric Chart of the Arctic Ocean) 数据库提供的海底地势数据，可知实验收发两端区域的海底地形较为平坦，平均海深为 1152 m．在实验期间雪龙船和黄河艇均停止了机械作业，避免辐射噪声对接收信号的干扰．

接收端的 CTD 剖面如图 6 所示．实验区域声速结构基本符合北极典型正梯度声速结构．图中 50 m 深度处温度和声速极大值为太平洋的暖流所致．实验期间，科研人员使用海冰探地雷达对海冰的厚度、密度进行了测量，结果表明实验海域的海冰厚度 2.7~3.5m，密度为 0.92 g/cm^3．

3.2 模型仿真

OASES-Bellhop 模型海冰输入参数如表 1 所示．其中，粗糙度 1 和粗糙度 2 分别代表海冰上、下表面的粗糙度．表中参数根据实测数据和文献 15 的总结，进行合理的设定．

声源频率 3500 Hz，OASES 模型输出的反射损失曲线如图 7 所示．高频端，声波波长与冰层厚度可比拟 ($\lambda<0.5$ m)，所以，冰层上、下表面都会对反射特性产生影响．掠射角低于 20.46° 时 (横波临界角)，声波在海冰-海水界面发生全反射，反射损失较小．当掠射角较大时，产生了若干的共振峰，反射损失较大．可见，接收端的声能主要由小掠射角的入射声波所贡献．

图 4　实验位置

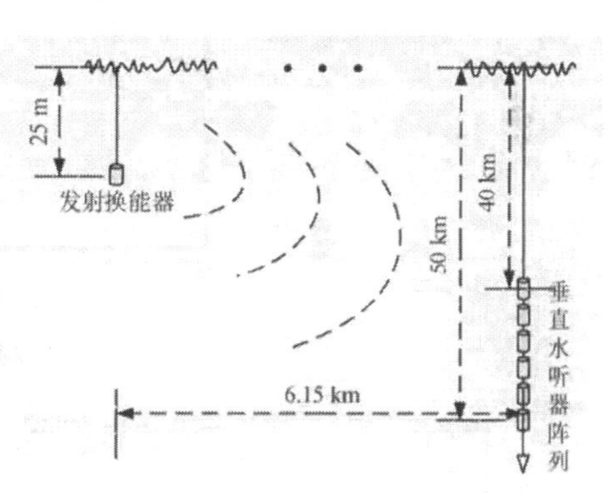

图 5　实验布设情况

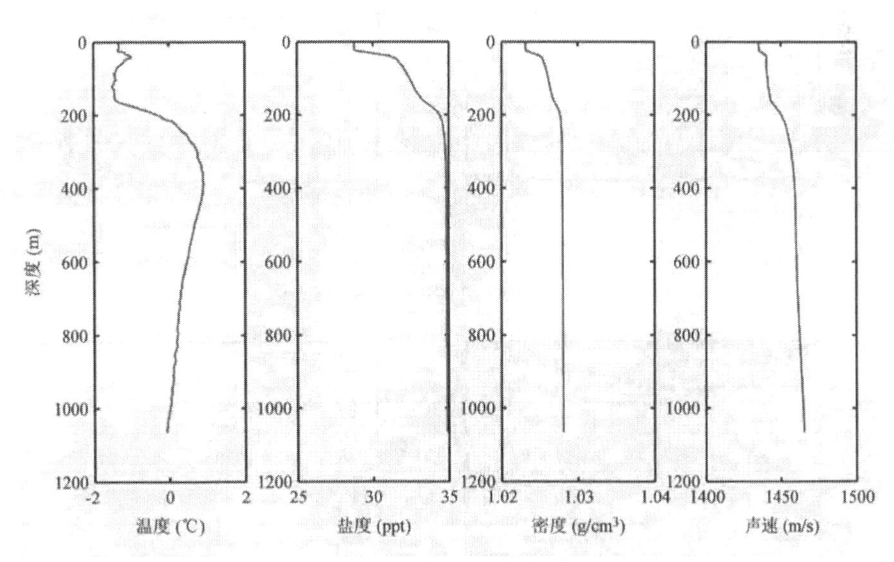

图 6　CTD 剖面

表 1　海冰参数

厚度 (m)	纵波速度 (m/s)	横波速度 (m/s)	纵波衰减 (dB/λ)	横波衰减 (dB/λ)	密度 (g/cm^3)	粗糙度 1 (m)	粗糙度 2 (m)
2.8	2900	1598	1.0	2.5	0.92	0.2	0.8

在高频条件下，OASES-Bellhop 模型对频率不敏感，因此本文选择 3500 Hz 中心频率来计算声信道多途结构. 结合图 6 所示的声速剖面以及图 7 所示的反射特性，OASES-Bellhop 模型得到的 40 m 深度处的本征声线轨迹和时间到达结构分别如图 8 所示. 作为对比，本文同时给出了自由海面条件下的模型输出结果，如图 9 所示. 声线轨迹图中，蓝色曲线表示仅海面反射的轨迹，绿色表示存在海底反射的轨迹. 从图 8 中可以看出，本征声线主要有两簇，两簇之间的时延差在 275 ms 左右，根据收发双方位置及海深信息，可以推断两簇分别对应海冰反射路径和海底反射路径. 海冰反射路径的多途扩展限制在 14 ms 范围内，强度大于海底反射路径. 海冰反射路径又分为一次海冰反射和多次海冰反射路径. 由于声速剖面服从正梯度分布，远离冰面处的声线声速比较大，所以，最先到达的为冰面一次反射路径.

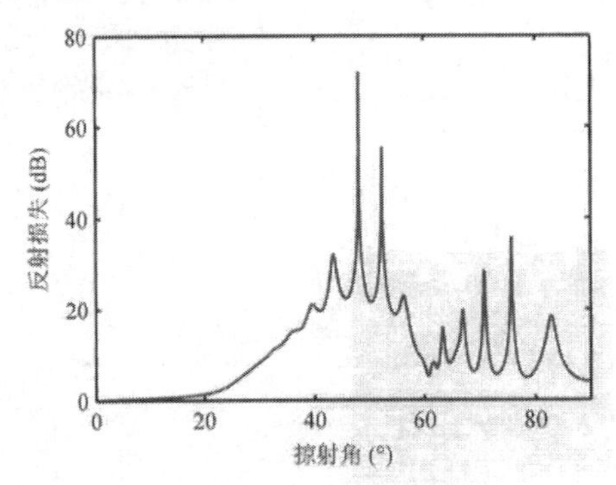

图 7 实验海区反射系数

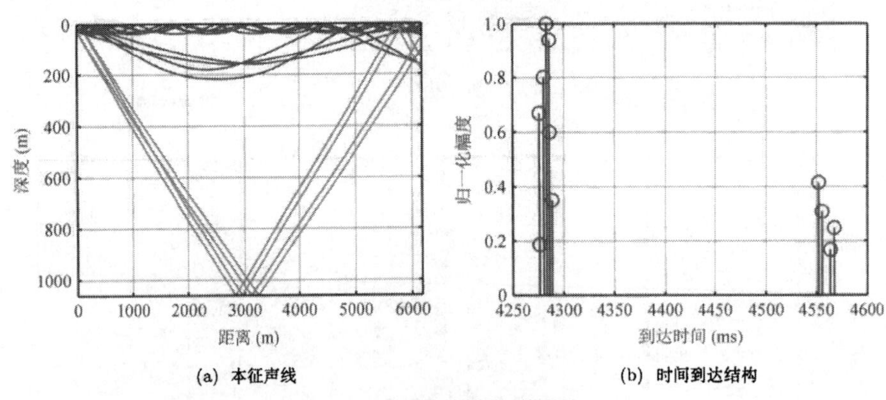

(a) 本征声线　　(b) 时间到达结构

图 8 海冰覆盖条件预报结果

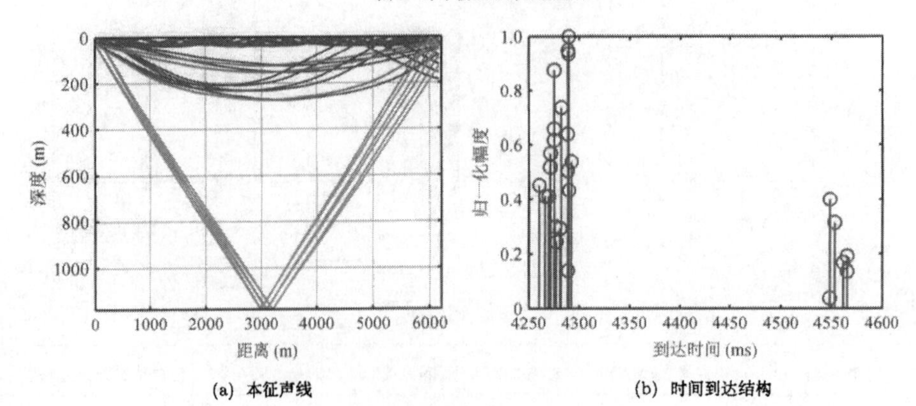

(a) 本征声线　　(b) 时间到达结构

图 9 自由海面条件预报结果

由于靠近冰面处的声线速度相对较低，声程相对较长，本征声线的到达时间随着冰面反射次数的增加而增大。此外，由于海冰的吸收和散射，靠近冰面的声线遭受的传播损失比较大，但声线较为密集，能量与一次反射路径相比可比拟。海底反射路径同时遭受海底与海冰界面影响，衰减相对较大、声程较长，多途结构强度较弱且时延扩展较宽。而在自由海面条件下，由于缺乏海冰界面的散射和吸收，多途结构中保留了较多的海面反射路径，导致多途结构比较复杂。

3.3 脉冲压缩提取多途结构

本文使用匹配滤波脉冲压缩技术[27-28]提取信道的多途结构。首先使用匹配滤波技术对 LFM 信号行了提取，得到的接收信号如图 10 所示。40 m 海深处阵元接收数据的脉冲压缩结果如图 11 所示。从图中可以看出，脉冲压缩后信道多途结构比较明显。

为了表征多途结构短时间内的变化特性，选取长度 500 ms 的时延窗，对图 11 所示的 60 s 脉冲压缩结果进行截取和对齐，得到的多途结构时间伪彩图如图所 12 示。图 12(a) 表示海冰反射路径，图 12(b) 表示海底反射路径。两簇路径之间相差 275 ms 左右，根据实验海域海深和收发两端的距离，可以推断第二簇起伏为海底反射信号，这与模型预报结果相吻合。从图中还可以看出，在 60 s 内，冰面反射路径多途结构都比较稳定。海底路径强度相对较弱且稳定性较差。

图 12 所示多途结构的平均结果如图 13 所示，可以看出，实验区域海冰反射路径的多途结构比较强而且比较稳定，海冰反射路径的多途扩展在 14 ms 范围内。海底反射路径大致有 4 条，与海冰反射路径相比，强度较弱、时延扩展较宽且稳定性差。

3.4 结果对比

为了验证模型的预报精度，以第一到达路径时间为基准，将图 8 所示的模型预测结果与图 13 所示的实验结果进行对比。结果如表 2、表 3 和图 14 所示。其中，路径 1-5 为海面反射路径，6-9 为海底反射路径。结果表明，OASES-Bellhop 模型能够较为准确的对海冰反射路径时延和幅度进行预测。由于缺乏实测的海底底质参数，所以海底反射路径的时延和幅度结果存在预测误差。

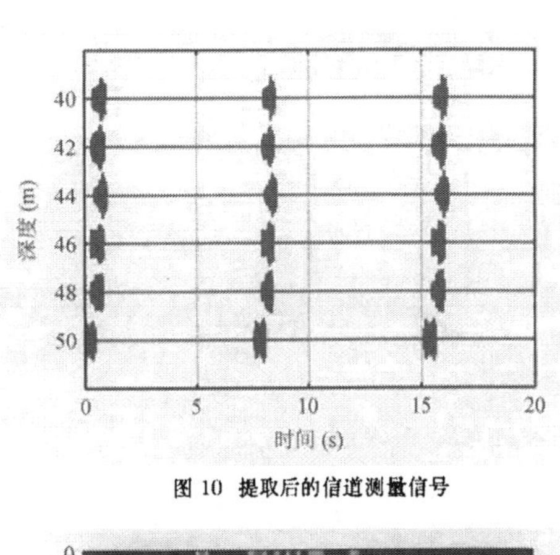

图 10 提取后的信道测量信号

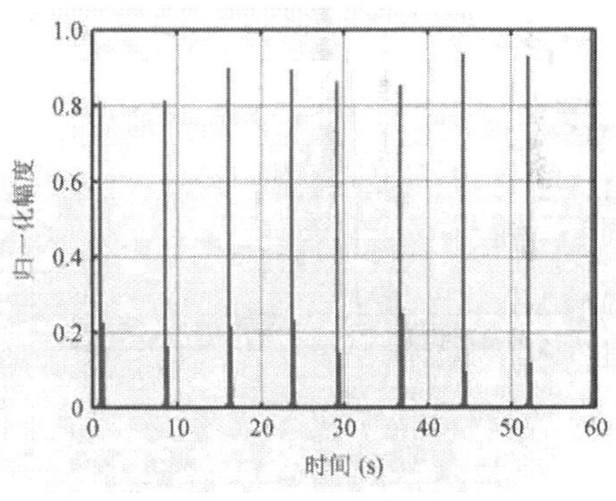

图 11 连续 60 s 时间内脉冲压缩结果

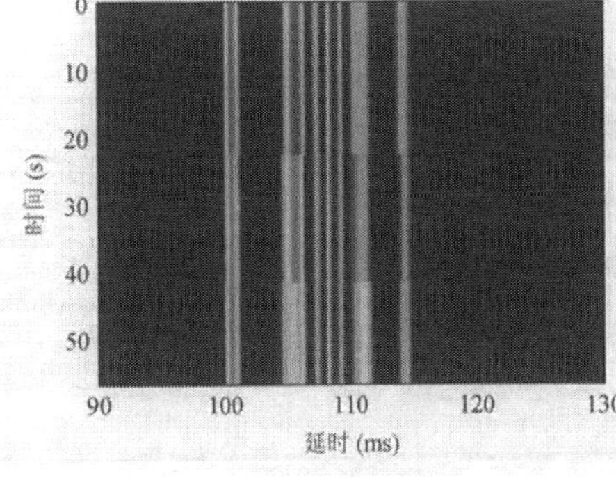

(a) 海面反射路径

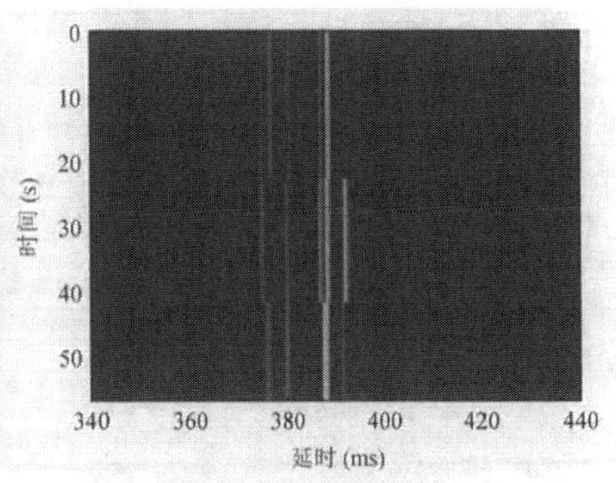

(b) 海底反射路径

图 12 多途结构时间累积伪彩图

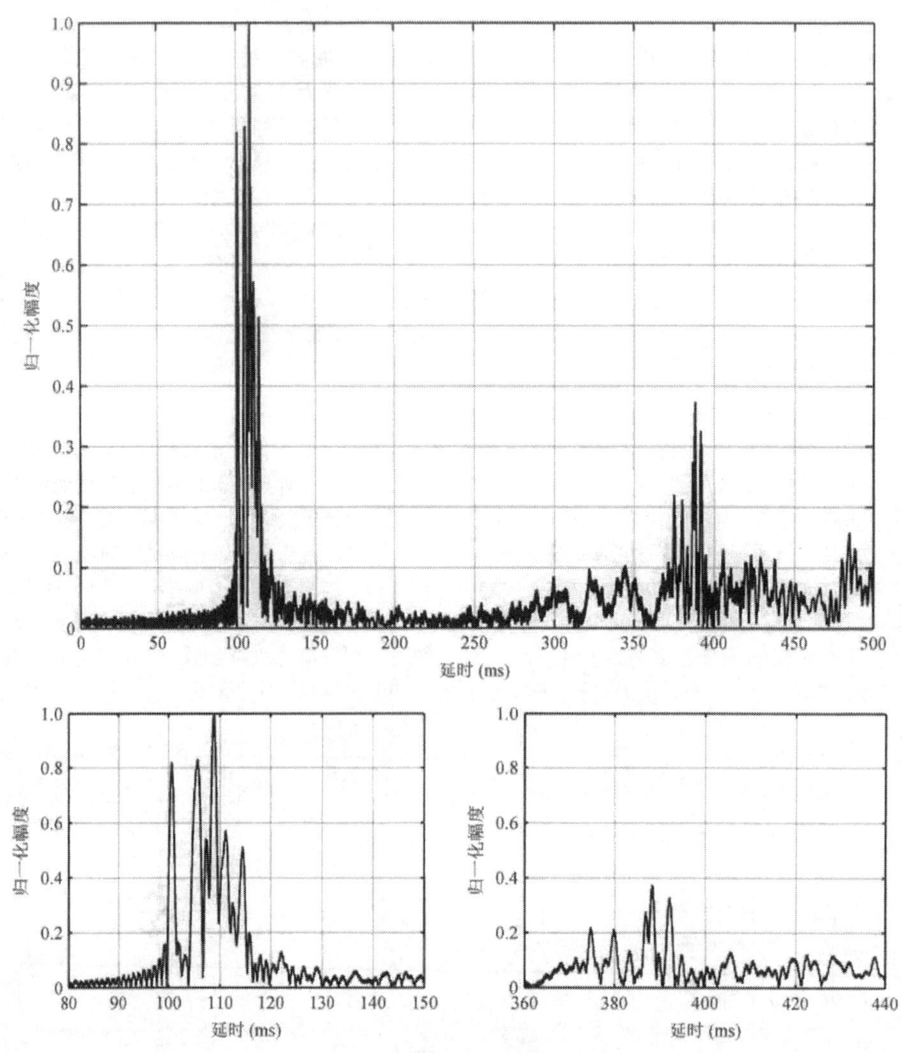

图 13 脉冲压缩 1 min 平均多途信道结构

表 2 仿真与实验多途路径时延结果对比 (ms)

	1	2	3	4	5	6	7	8	9
仿真	0	5.1	8.4	10.9	14.1	275	279.2	287.2	291.2
实验	0	5.2	8.7	11.3	14.4	271	276.5	284.5	288
误差	0	−0.1	−0.2	−0.1	0.1	4	2.7	2.7	3.2

表 3 仿真与实验多途路径幅度结果对比

	1	2	3	4	5	6	7	8	9
仿真	0.812	0.831	1.0	0.676	0.437	0.416	0.384	0.21	0.31
实验	0.793	0.813	1.0	0.7039	0.518	0.1483	0.1597	0.2785	0.143
误差	0.19	0.18	0	−0.028	−0.08	0.267	0.224	−0.069	0.167

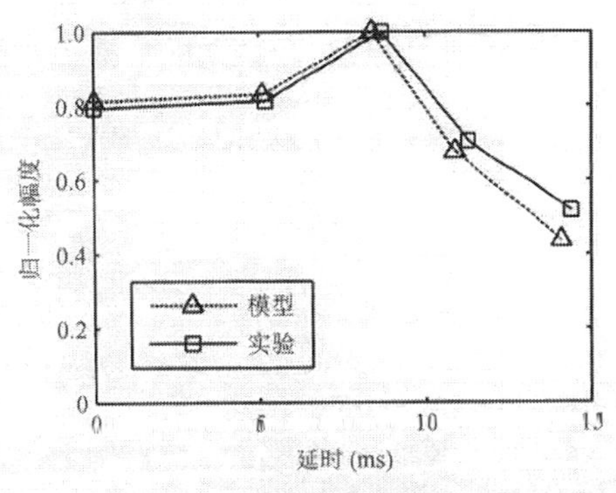

(a) 海面反射路径

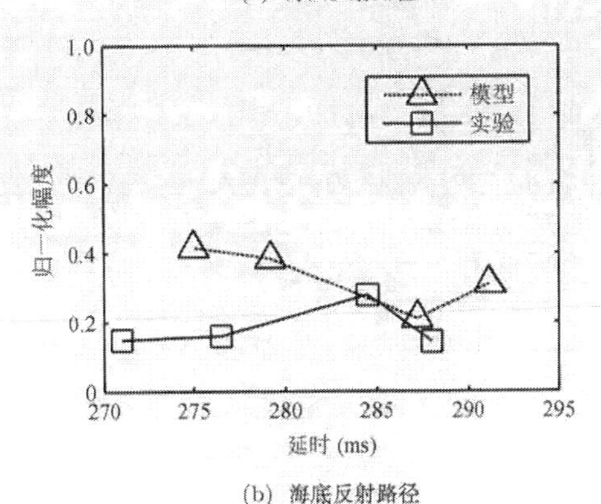

(b) 海底反射路径

图 14 仿真与实验结果对比

4 结论

本文将海冰等效为具有粗糙界面的弹性分层介质，使用微扰法推导了海冰覆盖条件下的扰动声场及反射系数。将 OASES 模型与射线理论模型 Bellhop 耦合，建立了 OASES-Bellhop 耦合模型分析北极典型冰下声信道多途结构。使用第七次北极科学考察获取的信道测量数据对 OASES-Bellhop 模型进行比较验证。研究结果表明：

(1) OASES-Bellhop 模型综合考虑了海冰界面不均匀性，使用海冰的粗糙度、厚度以及声学参数来计算扰动声场。预测的多途结构时延和幅度误差较小，模型能够精确的解释海冰散射机理，并较为准确的预报冰下声波经海冰界面多次反射形成的信道多途结构。

(2) 在 6 km 距离处，得到的实验结果与模型预测结果一致。海冰反射路径多途结构比较稳定，多途扩展在 14 ms 范围内，且强度较强。海底反射路径强度相对较弱。

值得注意的是，OASES-Bellhop 模型对海底的预测误差较大，尚不能对时变衰落信道进行模拟及预测。后续的研究将充分考虑海底特性对多途结构的影响以及北极冰下声信道的时变衰落现象。

本文对中国北极水声学进行了初步的探索，取得成果将为北极海域冰下水声观测及信息交互提供有效的技术支持。

致谢

感谢第七次北极科学考察领队夏立民研究员、首席科学家李院生教授以及所有参试人员，他们的辛勤工作和坚守为本文提供了非常宝贵的实验数据。

参 考 文 献

1 Brekhovskikh L M, Lysanov Y P. Fundamentals of ocean acoustics. Springer-Verlag, 1982

2 李启虎，王宁，赵进平等. 北极水声学：一门引人关注的新型学科. 应用声学，2014; **33**(6): 471—483

3 Mikhalevsky P N, Gavrilov A N, Baggeroer A B. The transarctic acoustic propagation experiment and climate monitoring in the arctic. *IEEE Journal of Oceanic Engineering*, 1999; **24**(2): 183—201

4 Gavrilov A N, Mikhalevsky P N. Low-frequency acoustic propagation loss in the Arctic Ocean: Results of the Arctic climate observations using underwater sound experiment. *Journal of the Acoustical Society of America*, 2006; **119**(6): 3694—3706

5 Schmidt H, Schneider T. Acoustic communication and navigation in the new Arctic — A model case for environmental adaptation. Underwater Communications and Networking Conference, IEEE, 2016

6 Schmidt H, Schneider T. Environmentally adaptive autonomy for acoustic communication and navigation in the new Arctic. *Journal of the Acoustical Society of America*, 2016: 140

7 Diachok O I. Effects of sea - ice ridges on sound propagation in the Arctic Ocean. *Journal of the Acoustical Society of America*, 1976; **59**(5): 1110—1120

8 Burke J E, Twersky V. Scattering and reflection by elliptically striated surfaces. *Journal of the Acoustical Society of America*, 1966; **40**(4): 883—895

9 刘崇磊，李涛，尹力等. 北极冰下双轴声道传播特性研究. 应用声学，2016; **35**(4): 309—315

10 Porter M B. The Kraken normal mode program. 1992

11 朱广平，殷敬伟，陈文剑，胡思为，周焕玲，郭龙祥. 北极典型冰下声信道建模及特性. 声学学报，2017; **42**(2): 152—158

12 Kuperman W A, Schmidt H. Rough surface elastic wave scattering in a horizontally stratified ocean. *Journal of the Acoustical Society of America*, 1986; **79**(79): 1767—1777

13 Kuperman W A. Coherent component of specular reflection and transmission at a randomly rough two - fluid interface. *Journal of the Acoustical Society of America*, 1975; **58**(2): 365—370

14 Livingston E, Diachok O. Estimation of average under - ice reflection amplitudes and phases using matched - field processing. *Journal of the Acoustical Society of America*, 1989; **86**(5): 1909—1919

15 Alexander P, Duncan A, Bose N. Modelling sound propagation under ice using the Ocean Acoustics Library's Acoustic Toolbox. *Science Education*, 2012; **41**(3): 250—250

16 Liu S, Song A, Shen C. Acoustic communication channel model in the under-ice environment. Oceans, IEEE, 2016: 1—5

17 Stewart G W, Sun J G. Matrix perturbation theory. Academic Press, 1990

18 Isakson M J, Chotiros N P, Yarbrough R A *et al.* Quantifying the effects of roughness scattering on reflection loss measurements. *Journal of the Acoustical Society of America*, 2012; **132**(6): 3687—3697

19 Thorsos E I, Jackson D R. The validity of the perturbation approximation for rough surface scattering using a Gaussian roughness spectrum. *Acoustical Society of America Journal*, 1987; **81**(S1): 20

20 范威, 范军, 陈燕. 浅海波导中目标散射的简正波 -Kirchhoff 近似混合方法. 声学学报, 2012; **37**(5): 475—483

21 Jensen F B, Kuperman W A, Porter M B *et al.* Computational ocean acoustics. *Modern Acoustics & Signal Processing*, 2003; **47**(11): 261—270

22 Schmidt H. OASES version 3.1 user guide and reference manual. Department of Ocean Engineering, Massachusetts Institute of Technology, Cambridge

23 王桂波. 小掠射角下海底界面附近目标散射特性研究. 中国海洋大学, 2008

24 Fricke J R, Chiasson L, Bondaryk J E. Acoustic radiation from a 3 - D truss: Direct global stiffness matrix modeling results. *Journal of the Acoustical Society of America*, 1995; **98**(5): 2889—2889

25 陆思宇. 水声信道的建模和估计方法的研究. 南京邮电大学, 2015

26 Dinapoli F R, Mellen R H. Low frequency attenuation in the arctic ocean. Ocean Seismo-Acoustics, Springer US, 1986: 387—395

27 童峰, 许肖梅, 方世良. 一种单频水声信号多径时延估计算法. 声学学报, 2008; **33**(1): 62—68

28 LIN Geping, MA Xiaochuan, YAN Shefeng, JIANG Li. Underwater acoustic multipath sparse channel estimation via gridless relevance vector machine method. *Chinese Journal of Acoustics*, 2016; **35**(4): 464—475

七

综合性评述及科普

声纳——水中耳目①

李启虎

雷达，对我们每个人来说并不陌生．在电影中，我们看到呼呼旋转的雷达天线，或高踞于军舰之顶，或矗立于群山之巅，十分壮观而引人注目．但是这个发现空中之物的眼睛，探测风云雷电、预报气象的有力工具，在水中却成了“近视眼”．而利用声波传递信息的声纳成了船舶舰艇的水中耳目．

什么是声纳　“声纳”(SONAR)一词是第二次世界大战期间由声音(SOund)、导航(NAvigation)和测距(Ranging)三个英文字母的字头构成的．当时的声纳系统确实只限于导航和测距的功能．但经过四十多年的发展，今天声纳的定义就大大扩充了．我们应当把声纳理解为“利用水下声波判断海洋中物体的存在、位置及类型的方法和设备”．这就包括了除通讯之外的所有水声的应用．也有的专家认为专门的水声通讯机或水声应答器不能称之为声纳系统．但是无论如何，我们现在所说的声纳系统的范围要比第二次世界大战时期广泛得多．

人们自然要问：为什么不用电磁能而用声能在水下传播呢？这是因为海水是电的极好导体，它使电能很快地以热的方式耗散掉，所以在同样的频率下，电磁波的衰减比声波快得多，从而传播距离也要近得多．只有在个别应用(潜艇与岸用控制指挥站的通讯)中，才用分辨力很差的长波电磁波．

声能和电磁能的传输在几个重要方面有区别：声波是纵波，而电磁波通常是横波．这就是说，声波在传输时流体介质的疏密和传播方向一致，而电磁场彼此之间是垂直的，传播方向也是垂直的；电磁波可以被极化，而声波则不能．此外，这两类波以极不相同的速度传播，在水中声速大约是每秒1500米，比空气中快4.5倍，电磁波在空气中大约以每秒3×10^8米的速度传播(在水中要稍慢)，几乎比声速快100万倍．

声纳的基本结构　按工作方式来分，声纳可分为两大类：一类叫作主动声纳，另一类叫作被动声纳．它们都包括接收机、换能器和指示器，但主动声纳多一个发射机．在主动声纳工作时，发射机发射一个已知信号，照射到目标后，反射信号或者“回声”被接收机接收到，经过处理由接收机显示出来．信号在传播过程中由于介质的不均匀性而散射．由信号本身产生的噪声，叫做混响．一般来说，它是主动声纳干扰的主要来源．而被动声纳工作时，它本身不发射声信号，仅仅需要检测出由目标所辐射的噪声．关键问题是要把所期望的目标辐射噪声或其“特征”和那些在能谱上与它相似的非期望噪声区分开来．发射换能器把电能转化为声能，接收声信号的接收器(也叫水听器)把声压转换为电压的变化．

现代复杂的声纳系统往往不是使用单个水听器或发射换能器，而是按声纳系统的要求，把多个水听器(或发射换能器)按一定几何图形布成一个阵列，这样就可以抑制背景噪声的干扰，提高检测增益．

发射和接收模型　在第二次世界大战期间，主动声纳的发射和接收波束通常较窄．为了提供水平搜索能力，它的换能器基阵可以在方位上机械转动．这类声纳叫做“探照灯式声纳”．它用方向性基阵把发射脉冲的声能集中在一个窄的波束里面，以便提供最大的检测距离．在跟踪时，根据基阵对准最大目标回波来确定目标的方位．这类工作方式有很多缺点．由于波束的移动是通过换能器基阵的机械转动而实现的，操作不方便，数据率也非常低，不能保证军舰具有多目标的检测和跟踪能力，而这是军舰的安全所必需的．

第二次世界大战末期发展了“扫描声纳”．它可以连续地、同时地提供目标在360°方位内的距离和方位角的信息，具有多目标检测和跟踪能力，从而最大限度地保证军舰的安全．

随着数字信号处理技术的发展，在六十年代初又出现了单比特量化的数字多波束(DIMUS)系统．这种系统对输入的多路信号进行限幅，在360°范围内同时形成波束，实现全景观察．同时，声纳设计者又开始了抵消强干扰的研究课题．这个问题在声纳中要比在雷达中严重得多．由于强干扰的存在，使得对弱信号的检测变得非常困难，如何抵消近场强干扰是提高声纳性能的一个至关重要的问题．

六十年代后期，先后有人推出自适应噪声抵消技术和数字式干扰消除自适应零点网络设备(DiCANNE)．它们的共同点是能够根据不同方向上噪声、干扰的强弱，自动地改变本身波束的形状，使其在强噪声或干扰的方向上特别不灵敏，以便检测出某个方向上的信号．

近年来数字信号处理技术飞速发展，出现了各种专用数字信号处理芯片，使声纳系统不断地朝着数字化、智能化及高可靠性方向发展．

声信道　信道是信息或数据从一个点发送到另一个点的媒介．理想信道是均匀的、无损耗的，在物理上是无边界的，并且可提供无畸变的传输．但考虑实际海洋中的随机扰动、散射、衰减、吸收效应时，问题就变得复杂了．实际上我们应当把声信道看作是时变、空变的传输介质，研究声波在这种介质的传播理论对声纳设计具有非常重要的意义．

一般来说，可以根据声速随深度的变化描绘声速

① 现代物理知识, 1991, (1): 26,27.

梯度，它是提供传播模型的基本依据. 不同类型的声速梯度(如等温层、负梯度、负跃进、表面声道、声发声道等) 会对声纳的作用距离产生非常大的影响。同一个声纳在夏天(负梯度)的作用距离也许只有冬天 (等温层)的一半甚至更少.

声纳的应用　声纳的研制首先起源于军事的应用，直到今天最大用户仍是海洋. 声纳是发现海洋秘密、利用海洋资源的重要工具. 回声测深仪是一种安装在船底的主动声纳，是快速测速的能手. 旁视声纳系统可进行海洋、内陆湖泊、河流和运河的勘测工作，在地质研究、矿产资源探查、沉物探测、水下考古、海底电缆和管线定位等方面起了非常重要的作用. 声纳在经济建设方面的最大应用是在近海寻找石油. 我国近年来采用拖曳式线到阵声纳在渤海、北海、南海、东海找到了丰富的油气构造.

水声学中的信号处理[①]

李启虎
(中国科学院声学研究所，北京 100080)

综述了水声学及水声信号处理的历史发展概况及它们在军事及民用方面的重要作用，介绍了与水声信号处理有关的换能器、布阵理论、波束成形理论、卡尔曼滤波、自适应滤波、目标识别、专家系统等课题，同时介绍水声信号处理中的一些现代技术，如 FFT，Zoom FFT，LOFAR，DEMON 及高分辨力谱分析的基本知识.

在我们人类居住的地球上，海洋的面积占70%，利用和开发海洋，长期以来就是人类社会活动的重要内容.

在人们所熟知的各种辐射形式中，以声波在海水中的传播为最佳. 因为海水是电的良导体，无论是光波或电磁波在海水中传播时衰减很大，由于声波相对地来说容易传播，因此在利用和开发海洋的事业中，人们广泛地利用水声.

最早提到声音不仅存在于空气之中，而且也存在于海水之中的文献是意大利文艺复兴时期多才多艺的大师达・芬奇的一本笔记. 达・芬奇不仅留下了《最后的晚餐》、《蒙娜利沙》等不朽的传世之作，同时还是一位造船工程师. 在1490年，也就是在哥伦布发现美洲大陆的前二年，他写道："如果使船停航，将长管的一端插入水中，而将管的另一开口放在耳朵旁，则能听到远处的航船."这实在是现代被动声呐的雏型.

水声的第一次定量测量，大约是在1827年进行的. 由瑞士物理学家 D. Collaton 和法国数学家 C. Sturn 合作在日内瓦湖测量声音在淡水中的传播速度，他们通过测定闪光和水下钟响之间的时间间隔，相当精确地测出了声速。但是直到19世纪末之前，人们除了用机械方法在水中产生声音之外，还找不到其他途径. 同时，也没有办法把水中的声音以某种方式加以提取.

1880年，雅克和居里发现了电声之间的换能效应，也就是压电晶体. 他们发现，有些晶体在受到压缩时会在晶体表面出现电荷，反之，在晶体表面加以交变电流会使晶体发生有规律的收缩和扩展. 这就为水中声音的产生及接收提供了直接的工具.

1912年在巨型客轮"铁坦尼克"号与冰山相撞后的一个月，英国人 L. F. Richardson 向专利局提出了用回声定位测定冰山位置的申请. 遗憾的是，他的这一想法未能实现.

水声学的真正的快速发展是由于第一次世界大战和第二次世界大战的推动. 为了对付潜艇，参战各方都投入了巨大的人力和物力研究水中的回声定位技术. 其中，奠基性的工作应是法国著名物理学家朗之万的研究结果. 使用一种石英-钢的夹心换能器同时又利用真空管放大器，这是电子学在水声中的第一次应用. 我国水声界元老，中国科学院学部委员汪德昭先生就是第二次大战前到法国与朗之万博士共事的.

海洋作为声传播的介质具有非常复杂的特性，早在本世纪20年代，德国科学家们就已发现，声波的传播距离在夏季要比冬季短. 这是因为夏季的声线向下折射而冬季的声线向上折射的缘故. 1937年美国科学家 R. L. Steinberger 观测到了神秘的"午后效应"，他指出了海水中的温度随深度的变化会影响传播结果. 当时发现，在无风和有太阳的下午，回声测距所观测到的距离有规律地减少. 最初，认为是生物的原因，怀疑是由于光合作用产生气泡或声呐员午饭后懒散所致. R. L. Steinberger 的研

① 物理, 1992, 21(11): 678-682.

究揭开了这个谜．他使科学家们真正了解到要把水声学作为一门实验科学和理论科学加以探讨的必要性．实际上，随着军事上应用的进展和民用海洋开发方面的推动，水声学已是一门和水声物理、水声工程、换能器、计算机科学、人机工程、信号处理及微电子学有关的边缘学科．而水声信号处理正是一门处于理论与应用结合的边缘学科．

如前所述，水声学作为一门科学的发展史是与军事应用密切相关的．而水声信号处理更是与声呐的研制与发展分不开的．早期的声呐几乎没有什么信号处理的结构，随着声呐技术的发展，信号处理的技术有了长足的进步．同时由于信号处理理论(包括信息论、控制论、人机工程等学科)的发展，又反过来对声呐技术起着推动作用．特别是数字信号处理技术的发展，已使声呐的面貌发生了根本性的变化．一般认为，水声信号处理作为水声学的一个分支，是在50年代初期确立其地位的．在此之前，声呐设计者还没有意识到主动利用信号处理技术的重要意义．美国著名水声学专家 V. C. Anderson 在1972年总结水声信号处理20年的发展史时指出，由于相关技术，快速傅里叶变换技术，波形设计，匹配滤波，信道匹配等课题的研究，使声呐性能提高到空前的水平．

当然，随着水声学的发展，水声信号处理技术的应用已不只限于军事方面．海上石油勘探、捕鱼、导航、通信，海上救助等方面也越来越多地借助于水声学的知识．

图1概括了水声学应用的各个领域．当声源在水下发声时（这种声源可以是海洋生物的噪声、舰艇的辐射噪声等），我们可以用水听器把声音接收到，在简单的情况下，把发射体看作是一个点源，把多个水听器放在水下就可以根据它们接收到的声波的延时差来确定发声体(或称为目标)的距离及方位．在一般情况下，频率较低的声信号可以传输得比较远．但是由于低频信号的波长较长，所以用低频信号来定位也较困难．

现代的各种声呐，包括军用的和民用的，几乎都是利用在水下的水听器阵列再加以信号处理来探测目标的．例如，美国的岸用反潜预警体系，就是在沿美国东西海岸布放上千个水听器来接收信号的．所有的信号被传送至反潜数据处理中心，提取有关目标的信息．

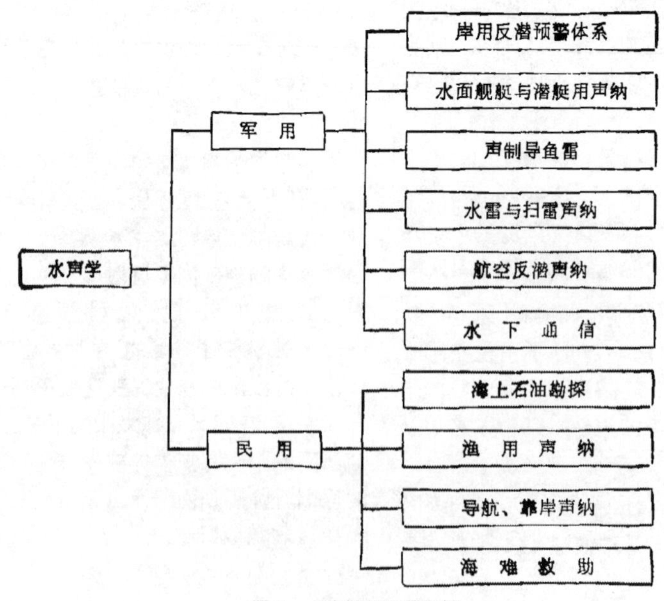

图1 水声学的应用领域

舰艇或鱼雷上用的水听器阵列在体积上要小得多，因而所用的频率也相对较高。但是有的攻击型核潜艇上面使用一种舷侧阵，就是把水听器沿着艇壳排列。这样使有效的阵列尺寸加大，提高了检测能力。

在舰艇上使用的声呐，由于本舰噪声的干扰会使声呐的检测能力受到影响。自 60 年代末以来，人们研制了一种拖曳式线列阵声呐。在舰艇上使用绞车把一个相当长的拖缆拖在舰船的后边，这样使声呐的接收换能器远离本舰螺旋桨噪声。同时，接收线列阵可以使用较大的尺寸，从而提高了检测能力。这种拖曳线列阵声呐目前广泛地使用于水面舰艇及潜艇，同时又在海上石油勘探中得到广泛应用（见图 2）。

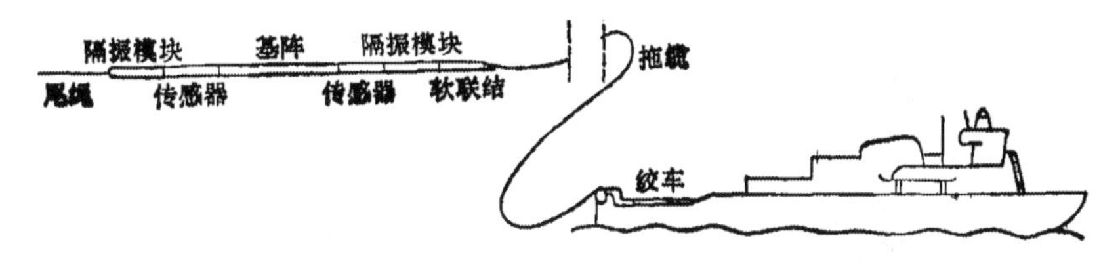

图 2 军用与民用拖曳式线列阵声呐示意图

军事上应用的拖曳线列阵声呐与海上石油地质勘探声呐从原理上说是一样的。但也有一些重要的差别。从使用频率范围上来讲，军事用的在 100Hz 以上，民用的一般在 120Hz 以下；从功能上讲，军用的通常是被动监听声呐，而民用的还要有水下空气枪，造成人为地震，然后由回波来判读。

我国已建立了自己的海上石油勘探队伍。用拖曳式线列阵在渤海、北海、东海及南海开展寻找近海油气田的工作。

靠岸声呐是民用声呐的另一种代表。主要用于大型油轮进港时的机动及靠岸，具有精度高的特点。由中科院声学研究所研制的靠岸声呐是我国十大专利金奖产品之一，受到广泛好评。

海难救助是水声的另一个重要应用领域。其中最重要的是所谓声发(SOFAR)声道。在深海中存在最小声速的深度（大约在 1000m）。进入这一深度的声线由于折射连续不断地反转，很少进入水面及海底，从而损失很小，可以传播到很远的地方。美国曾在百幕大附近的亚速尔群岛投掷 1—2 磅炸药，利用 SOFAR 声道，可以在几千浬之外的新西兰收到爆炸声。这种现象的最早应用是用于测定落海飞行员的位置，引导救援飞机营救。现在已成为美国发射洲际弹道导弹时弹着点测定装置（MILS）的一部分了。

水声学的应用领域是多种多样的，除了上面简单介绍的以外，还有水下机器人、水下自动应答器等。1986 年美国航天飞机挑战者号遇难时，其残骸就是由 Perry 公司的水下机器人从大西洋底打捞上来的。随着海洋开发的进展，水声学的应用领域将进一步扩大。

作为水声学的一个核心部分的水声信号处理技术，正在起着越来越大的作用。因为它是信息获取、储存、加工和显示的综合利用。各种大规模数字信号处理芯片的使用，使得信号处理的结构发生了重大变化。目前正在研究具有人工智能的第五代数字信号处理机。

图 3 给出了与水声信号处理相关的技术。

在这个图上，我们把水声信号处理按它的功能分为三个部分：预处理、波束成形和后置处理。

和预处理有关的是接收换能器及发射换能器设计及工艺学。对于发射换能器通常要求设计频带宽、功率大、电声转换效率高和体积小的发射换能器。至于接收换能器（亦叫作水听器）通常是多个使用，希望灵敏度高、频响一致及无

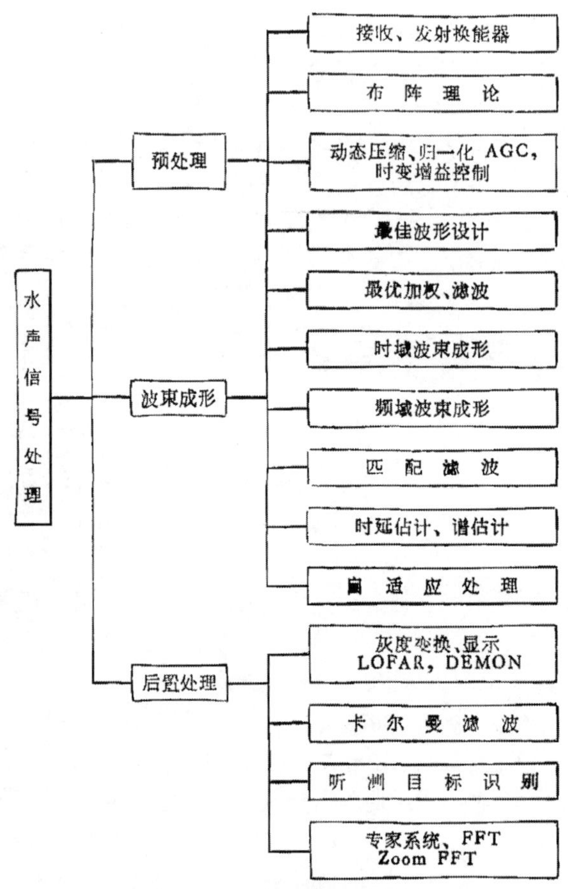

图 3 与水声信号处理有关的技术

相位差．这些要求之间往往是互相矛盾的，这就必须进行折衷．布阵理论着重于研究在海洋环境场中的基阵安排，使得在噪声背景下最大限度地提取信号．由于声信号的动态一般非常大(例如 80dB)，所以必须进行动态压缩或自动增益控制，这也是预处理的一个重要任务．此外，波形设计主要用于主动声呐，选择最佳波形，使得在同样发射功率下能发现较远处的目标．

波束成形把水听器阵信号进行处理，使声呐在水平面或垂直面上进行扫描，从而达到 360° 全景观察的目的．为了使检测增益尽量大，必须使波束非常窄的旁瓣电平又非常低，为了作到这一点，目前使用的有两种波束成形技术，一种是时间域的波束成形技术，一种是频率域的波束成形技术．前者对时空采样样本进行延时补偿形成定向波束，后者对各通道的时空采样进行快速傅里叶变换（FFT），在空间形成多个波束．为改善波束性能还必须对信号进行滤波加权或升采样率的运算．

近年来，又发展了一种自适应信号处理技术．它可以根据环境噪声的缓慢变化自动抵消强干扰，使得对弱信号的检测更加容易一些．这实际上是一种自学习系统，目前已在若干声呐中得到应用．

延时估计与谱估计主要用于对目标的精确定向或被动测距．由于水声信号在传播过程中受到水面和水底的多次反射，特别是在浅海，存在着严重的所谓“多途效应”，使得目标的定位和测距变得十分复杂．在对水声信道进行深入研究之后已经提出若干种信道匹配技术可以部分消除多途效应，从而使测向和测距变得容易一些。

后置处理是把已经获得的信息最大限度地利用起来，提供给人类的视角和听觉。为了适应人眼的观测，普遍采用一种灰度变换技术，使得对小信噪比的信号也能较容易地检测到。现代计算机用的高分辨力彩色显示器，可以移植到声呐中获得很好的效果。

声呐系统的一个重要功能是目标的自动跟踪。现在常用的方法是卡尔曼滤波方法。这是一种根据量测信号进行反馈调整的技术，可以在一定背景噪声下保证对目标的跟踪。老式声呐对目标的判别都是依赖于声呐员的，随着数字信号处理技术的发展，人们已开始研究机助识别与简易专家系统。把专家的知识及声呐员的经验结合起来，按一定的推理规则作出判别，这种结论具有专家的水平。

我们已简单地介绍了水声信号处理技术的各主要领域以及与其有关的军用和民用部门。水声学的发展虽然一开始是从为战争服务为目的的，但是随着海洋开发事业的进展，越来越多地在民用方面得到重视并发挥效应。在今后，虽然军事的需求仍是水声学发展的强大动力，但不可否认，海上石油勘探、海上考古、海洋资源开发和海洋环境监测、海难救助等事业也会越来越多地依赖水声学。随着现代电子技术，特别是微电子技术的发展，水声信号处理技术必将取得更大的成就，从而对军事应用和海洋开发起到更大的作用。

全球关注数字信号处理技术[①]

中国科学院院士 李启虎

1995年美国麻省理工学院教授、媒体实验室负责人尼葛洛庞帝出版了《数字化生存》的书，引起了美国和世界各地读者的兴趣，他把近20年来信息领域的数字化革命的丰硕成果以及将要引起的对生活的巨大冲击展现在世人面前。1998年1月31日美国现任副总统戈尔在加利福尼亚科学中心作了“数字地球——认识我们这颗星球”的讲演，提出充分利用数字信息促进社会进步和发展。按他的估算，要把数字地球的所有数据存贮起来，需要拍1千万亿字节容量的存贮器。而我们现在能达到的最高级的光盘容量只有10亿字节级。

1998年6月1日，江泽民主席在接见中科院、工程院部分院士时也提到了“数字地球”这一概念。

对于非专业的人士来说，接触数字化信息处理的概念，大多来自消费电子领域，这就是数字音响、数字广播、数字电视、数字移动电话、数字通信网以及CD、VCD、DVD、微机等众多产品，但是信息领域的数字化革命实际上早已开始。

根据精略的估计，人类获得信息的主要来源是听觉(约占5%)与视觉(约占90%)，其他还有味觉、触觉及嗅觉等。在六十年代初期之前，对信息的处理方式主要限于模拟方式。它的特点是这种信号在时间上是不分间隔的，在幅度上不分层。由于受硬件条件的限制，信号的数字化处理真正开始于六十年代初。但是奠定这一理论基础的却是1948年美国著名信息论专家香农的一篇论文《通讯的数学理论》，他第一次提出数字化信息的基本单位：比特(尼葛洛庞帝把比特比喻作信息领域的DNA)；并由此出发提出了一系列近代信息论基本思想。

从六十年代开始，由于计算机技术的迅猛发展，特别是七十年代以来微电子技术的惊人进步，使得信号的数字化处理以空间未有的速度向前推进。

数字信号处理的理论与技术日趋成熟，数字信号处理的应用领域几乎涵盖了国民经济和国防建设的所有领域，包括雷达、航天、声呐、通信、海洋高技术、微电子、计算机、人工智能、消费电子等。

信号的数字化处理包括两个步骤，一个是信号在时间上的离散化，即采样；另一个是幅度上的离散化，即分层。数字化之后的信号，将全部变为01序列，这就使得信息的采集、存贮、传输、复制、加工异常方便。所以信号的数字化处理推动了各应用领域的发展，并成为这些领域的最重要的技术支撑。反过来，各应用部门对数字信号处理理论与技术的发展，包括分层的压扩技术，采样和抽取技术，数字滤波理论，快速傅里叶变化(FFT)、数字图像处理、模式识别、专家系统、宽带通信网络、多媒体技术等等。

① 科学新闻周刊, 1999, (19): 16.

数字信号处理技术的这种高速发展和对其他领域的广泛渗透无疑得益于七十年代以来微电子技术的发展。

1971年英特尔公司推出了第一个微处理芯片4001，其功能大体上和世界上第一台电子计算机ENIAC相当。此后，微处理芯片的发展就异常迅速，它基本上遵从所谓的摩尔定律，这就是微处理芯片每隔18个月性能提高一倍、价格降低到原来的1/2，摩尔博士是英特尔公司创始人之一，他在1973年提出了这一观察结果。但是据他自己说，早在1964年他还在仙童公司工作时就已注意到这一事实了。1997年摩尔又在美国IEEE杂志上重新公布了他的手稿，对自己这一定律深信不疑。我们现在还很难预言，这一定律能够有效的时间究竟有多长，因为已经有科学家认为，从固体物理学的基本理论来看，芯片的线宽是有限制的，从而微处理芯片的性能提高就受到限制。

无论如何，微电子技术的惊人发展速度已经产生了难以预料的结果，1946年当冯•诺伊曼等计算机专家研究成功第一台电子计算机ENIAC时，其总重量为30吨，占地面积相当于一个小的体育馆，平均每7分钟就有一个电子管失效。它的耗电量惊人，在它工作时，整个费城就灯光暗淡。而1977年生产的微处理器体积仅是ENIAC的1/30000，成本是它的1/10000，速度是它的20倍。

英国科学家福莱斯特，在总结了微电子技术的这种惊人发展速度之后，在他所著的《高技术社会》中发出了感叹，他说："如果汽车或飞机行业也像计算机行业这样发展，那么今天一辆罗尔斯•罗伊斯汽车的成本将只有2.75美元；跑300万英里仅用一加仑汽油。而一架波音767飞机的价格只需500美元，用5加仑汽油在20分钟内便可环绕地球一周。"1996年美国Princeton大学电子工程系主任刘必治教授来华发表演讲时，也作了类似的对比。

数字信号处理技术的这种进步，是我们每一个人在日常生活中都可以感觉到的，同时它还对世界上某些局部地区正在进行的战争产生了深远的影响。它使得原来意义上的"前方"与"后方"的概念发生了变化，使"信息站"的概念起到了主导作用。全球范围内的通信系统可以使美国五角大楼直接指挥某一局部战役的行动，甚至是精确制导的巡航导弹的打击目标。

数字化浪潮正在席卷全球，数字化信息处理技术正在使人类生活质量提到空前高的水平。

我们从事数字信号处理理论与应用研究的科技人员已在这一空前活跃的领域作出了令人瞩目的成绩。其中特别突出的例子是"曙光1000"并行计算机、合成孔径雷达、数字式声呐、"04"程控交换机等等。

据估计1998年全球信息产业的总产值为14700亿美元，到2000年时有望达到3万亿美元，这么庞大的市场，对我国从事科研、生产的人来说，极具挑战性。我们只有进行不断的创新才有可能在全球信息市场中占有一席之地。

但是，与任何事物的发展一样，**数字信息处理只是信号处理技术发展中的一个阶段**。这决不意味着信号的数字化处理是十全十美的。从信号的模拟处理到数字处理，这是一步跨越，这一跨越已经并正在引起信息产业的大发展。部分领域刚刚起步，如数字电视。但我们决不会停留在这一步。虽然我们现在还不清楚下一步跨越是什么，但是数字化处理不是终点则是毫无疑问的。

声纳技术及其应用专题

编者按　声波是人类迄今为止已知可以在海水中远程传播的能量形式. 声纳(sonar)一词是第二次世纪大战期间产生的,它是由声音(sound)、导航(navigation)和测距(ranging)3个英文单词的字头构成的. 声纳设备利用水下声波判断海洋中物体的存在、位置及类型,同时也用于水下信息的传输.

现代声纳的发明早于雷达. 1916年,法国著名物理学家P. 朗之万发明了回声定位声纳,用他的设备可以在水下探测到200m之外的一块装甲板的回波. 雷达的发明则是1935年的事,当时英国科学家W. 瓦特领导的研究小组利用电磁波反射原理探测到距离测试点约12km的飞机的回波.

声纳技术在国防和国民经济发展中具有十分重要的作用. 为了向读者介绍声纳技术的基本物理原理和应用情况,本刊从本期开始,以"声纳技术及其应用"作为专题,陆续发表系列文章(共8篇),从不同角度向读者介绍声纳技术领域的最新进展、研究成果和物理问题,希望能引起读者对该领域的关心和兴趣.

(中国科学院声学研究所　李启虎)

第一讲　进入21世纪的声纳技术①②

李　启　虎

(中国科学院声学研究所　北京　100080)

摘　要　海洋开发和反潜战的需求是推动声纳技术开发的巨大动力. 水声物理、水声信号处理及相关学科的发展又促使声纳设计日趋完善. 文章介绍了声纳技术在进入21世纪时所面临的机遇和挑战, 水声信号处理领域近期研究的热点问题以及声纳系统设计中的技术创新课题.

关键词　21世纪,声纳技术,声纳信号处理

Sonar Technology of 21st Century

LI Qi-Hu

(*Institute of Acoustics, Chinese Academy of Sciences, Beijing* 100080, *China*)

Abstract　The requirements of ocean development and anti-submarine warfare is the main motivation of sonar technology study. The theoretical development and achievements in the area of underwater acoustic physics and underwater acoustic engineering make sonar system more and more complete. The opportunities and the challenges facing sonar technology in 21st century is described in this article. The hot topics of underwater acoustic signal processing and the innovation problems in sonar system design are also discussed.

Keywords　21st century, sonar technology, sonar signal processing

1　声纳技术面临的机遇和挑战

早期的声纳设计建立在较为理想的模型基础上. 无论是声纳设计者还是声纳使用人员,早就注意到声纳的性能与海洋环境密切相关. 但是,由于两个方面的原因,使声纳技术发展的初期采用了比较简单的模型. 第一个原因是人们对海洋中水声传播规律的研究和认识有一个由浅入深、由表及里的过程. 虽然声纳的发明早于雷达19年,但是由于海洋环境的复杂特点,使声纳的发展在某些方面滞后于雷达[1],这是丝毫不奇怪的. 第二个原因是由于硬件条件方面的限制. 20世纪50—60年代,信息论已经为微弱

① 国家自然科学基金(批准号: 60532040)资助项目.
② 物理, 2006, 35(5): 402-407.

信号的检测提供了相当充实的理论基础和实用技术，但是这些理论的应用需要非常复杂的计算，而在那时硬件设备还无力提供这种支持.

自20世纪70年代以来，情况已经发生了很大的变化. 微电子技术的发展使计算机硬件的面貌发生了巨大的改观，从而推动了数字信号处理领域的变革.

英特尔公司的创始人之一摩尔在1973年提出了支配半导体工业发展的一个规律，就是我们现在所说的摩尔定律. 这个定律说，微处理芯片每隔18个月性能提高一倍，价格降低到原来的二分之一. 据摩尔自己说，早在1964年他还在仙童公司工作时，就已注意到这一事实了. 1996年，他又在IEEE Proc. 上重新公布手稿，对自己的这一定律深信不疑（见图1）[2].

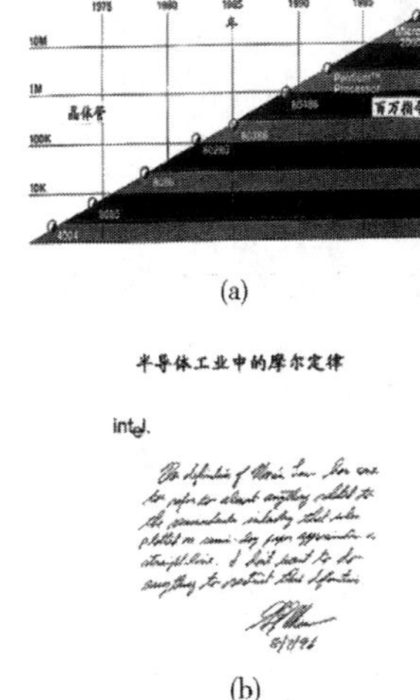

(a)

(b)

图1　(a)英特尔公司微处理器进展情况；(b)摩尔博士1996年重新发表著名的“摩尔定律”的手记[原文为“摩尔律”的定义已作为几乎所有和半导体工业有关的领域的基本准则. 当我们在半对数坐标上作图时（器件性能和时间关系）就是一条直线，我无需作任何事情来限制这一定义]

微电子工业的这种迅猛发展势头，使得数字式声纳应运而生，并且使声纳设计者面临巨大的机遇和挑战. 因为，作为数字式声纳硬件支撑的DSP芯片，似乎“无所不能”，过去很多受计算机能力限制的技术，现在都可以实现了. 但是，很不幸，声纳的性能不仅仅依赖于硬件的能力，更大程度上依赖于主导声纳性能的建模技术、微弱信号检测算法、参数估计理论、人工智能等.

英国科学家福莱斯特在总结了微电子技术的这种惊人发展之后，在他所著的《高技术社会》中发出了感叹，他说[3]：“如果汽车或飞机行业也像计算机行业这样发展，那么今天一辆罗尔斯·罗伊斯汽车的成本将只有2.75美元；跑300万英里仅用1加仑汽油，在20分钟内便可环绕地球一周.”1996年美国普林斯顿大学电子工程系主任刘必治教授来华演讲时作了类似的对比，他说：“如果汽车行业也像计算机行业那样，今天一辆可以坐4000人的‘小汽车’的价格应当是0.26美元.”

数字声纳设计者就是面临这样一种局面：硬件的发展向声纳设计者预示，只要你提出理论，我就能实现它；如果今天有一种目标识别的算法需要一个小时才能得出结果，那么10年之后，用不了1分钟就可完成. 微电子技术的这种发展潮流也已部分地改变了信号处理理论的发展方向. 20世纪60年代中期，当Cooley，Turky提出FFT算法时，有关改进的算法非常多，哪怕提高计算速度10%或节省存贮量10%的理论工作都还得到认可，但是10年之后，就没有人去做类似的工作了，因为硬件的发展使得那种小小的改进黯然失色.

现在需要的是理论的创新，是那种能带来跨越式发展的新概念、新理论和新工艺.

2　需求牵引：声纳技术发展的推动力

声纳技术发展的最大用户是海军，即未来水下战的需求. Marburger在美国《防务新闻》(Defense News)周刊上提供了美国海军的一个有关安静型潜艇的辐射噪声的图[1]，表明美国在20世纪90年代的海狼级核潜艇噪声和俄罗斯改进的Akula级噪声相当. 在近30—40年内，潜艇的辐射噪声大约每年下降0.5—1dB，从而使被检测的距离每年缩小0.5—2km，而声纳技术的改进弥补不了这一下降趋势带来的损失（见图2）.

潜艇降噪用的是综合的技术，即同时考虑消声瓦和本艇辐射噪声的控制[5,6]，称为CSMC(collaborative signature monitoring and control)，防噪声辐射的核心技术是主动噪声和振动控制，即AMVC(active noise and vibration control). 为探测安静型潜艇，需要发展大功率、低频、宽带换能器，需要发展主/被动的拖曳式线列阵声纳，需要发展潜用的细长缆拖线阵声纳，需要研究更有效的水声通信方法及水下

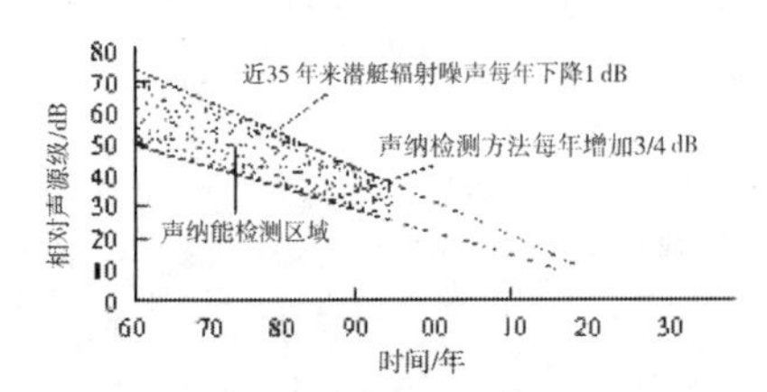

图2 潜艇辐射噪声和声纳检测能力

图3 水面舰艇和直升机联合反潜(双基地声纳)

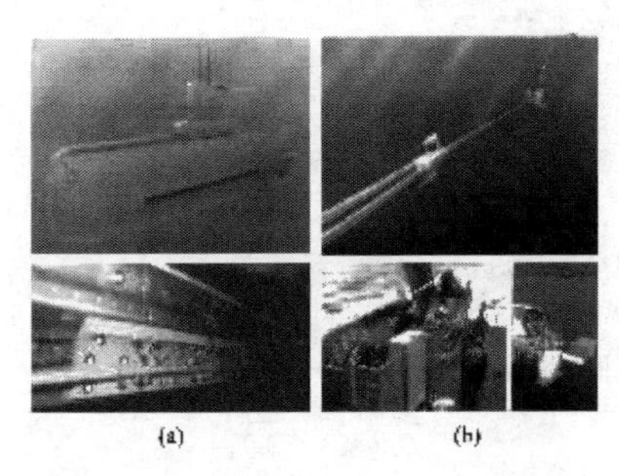

图4 (a)德国 ATLAS 公司舷侧线列阵声纳 FAS3-1 [基元数为 2×96, 基阵长 48m, 频段 ~2.5kHz, 工作扇面为 45°—135°(全功能), 10°—170°(降功能)]; (b)德国 ATLAS 公司研制的主动拖线阵声纳绞车和拖曳系统

定位技术等(见图3,4).

在寻找新型水声换能材料方面,有的研究者认为,镍合金和压电陶瓷材料的设计极限早在25年之前就达到了[7]. 所以人们开始研究稀土系列元素的磁致伸缩特性,这类材料能产生较大的声功率,并且可以在低频、宽带下稳定工作.

美国海洋实验室发现了一种镧和铁的合金,称之为 Terfenol-D 的磁致伸缩材料,据说已获得应用[8]. 这种 T-D 换能器具有良好性能,在合理使用时几乎不变形,样品经过 10^8 次发射仍完好无损,有可能成为下一代声纳发射换能器的首选材料.

在众多水听器中,近年来发展较快的是一种聚偏二氟乙烯(Polyvinylidene fluoride, PVDF)薄膜,这是一种柔顺压电材料,加工较方便,压电常数非常高. 目前已研制出厚度为 1—2mm,面积为 $30\times80\text{cm}^2$ 的薄膜面元水听器,它可以用于舷侧线列阵声纳中,其缺点是它的电压灵敏度对温度比较敏感.

由于声纳性能与海洋环境的密切关系,需要声纳设计者寻求一种新的体系结构,它能把海洋环境融入声纳的整体设计中,以便使声纳系统的性能对模型失配更宽容一些. 这就是基于模型的声纳系统[9—11].

它是立足于宽容性检测的原理. 任何声纳系统的设计都是基于某种模型的假设,然后在这种假设下寻求一种最佳的处理方法. 当实际环境符合最佳处理的假设条件时,系统的增益很大,但是一旦模型失配,最佳检测器的性能迅速下降. 而宽容性检测就不同,它虽然在理想条件下的增益不如最佳检测器,但是对模型失配却显得很"宽容",具有相对来说变化不大的系统增益. 研究模型失配的概念非常重要,可惜到现在为止,我们还无法在理论上回答模型匹配以及失配应如何从数值上进行刻划.

匹配场过滤的方法是较早被研究的宽容性检测方法之一[12—14].

宽容性信号处理的概念可以归结如下:假设 H 是一个可能的设计空间,Q 是可以选择的模型的集合,给出一个用于刻划系统失配的度量函数 $M(h,g)$, $h\in H, g\in Q$. 传统的设计是对于特定的 $g_0\in Q$,求出 $h_0\in H$,使得

$$M(h_0,g_0)=\min_{h\in H} m(h,g_0).$$

这种系统实际上是适应环境 g_0 的匹配过滤器. 如果 g_0 可以在某个 Q 的子空间内变化,即 $g\in P\in Q$,那么我们自然关心

$$\max_{g\in p} m(h,g).$$

在这种情况下,采用极小极大策略是一种自然的选择,即

$$\min_{h\in H}\max_{g\in p} m(h,g).$$

所谓宽容性处理 h_R 就是:

$$\min_{h\in H}\max_{g\in p} m(h,g)=\max_{g\in p} M(h_R g).$$

对声纳技术的需求还来自海洋开发. 水声遥测始终是获得海洋环境参数的最重要手段之一. 大洋测温、近海油气田的数字地震勘探、声层析、大洋海底金属矿的开采以及水气化合物的勘探开发都离不开声纳技术.

冷战结束之后，西方海洋大国对海洋的研究投入很大力量，同时把原来军用的一些海洋监测设施对外开放或转为民用. 例如，美国就将东、西海岸的部分岸用站用于全球测温（ATOC）计划.

声纳技术在军、民两种需求的强力推动下，发展迅速，一些新的技术取得突破，有望在下一代的声纳系统中得到应用.

3 学科发展：声纳信号处理的热点问题

作为声纳技术的理论支撑的水声信号处理是一门综合性的边缘学科. 它在发展进程中，既有自己的特色，又吸收了雷达、医学成像、通信、语音信号处理等其他领域的成果. 1998 年，IEEE 信号处理分会为纪念协会成立 50 周年，编发了一篇专稿，即《水声信号处理的过去、现在与将来》[15]. 文章回顾了水声信号处理的发展历史，提出了一些有潜在应用价值的热点技术，如合成孔径技术、声层析、水声通信等. 水声信号处理理论的发展面临着众多问题，相关和临近学科又不断产生一些新的概念，所以水声信号处理专家以积极而审慎的态度来对待它们[16]，一方面担心错过了应用新理论的机会，一方面又怀疑它们是否真的能在声纳系统中获得应用. 这些课题是不胜枚举的，例如人工神经网络、混沌理论、小波变换、分维变换与时间反转算法等. 我们在本节中就水声信号处理本身介绍若干热点问题.

3.1 被动测距

被动测距声纳是从 20 世纪 70 年代初开始研制的. 从理论上讲，只要声纳基阵的孔径足够大，用三点阵测距是没有问题的. 关键是把三个基阵的声中心的相对延时精确测量出来（见图 5）.

图 5 美国 Lockheed Martin 公司研制的被动测距声纳 PUFFS

可以证明，被动测距的相对误差等于测延时的相对误差，即

$$\Delta R / R = \Delta\tau / \tau.$$

根据这一公式我们就会明白被动测距声纳所面临的问题. 举例来说，孔径为 40m 的基阵要测量相距为 20km 的目标，延时量大约为 13μs. 如果要求相对误差为 10%，则延时估计误差不能大于 1.3μs. 在海洋环境中要做到这一点非常困难. Urick[17]，张仁和等[18]曾报道，海水中声传播起伏值就在 10μs 这样的量级，这就使得被动测距问题变得十分困难，因为要在接收到的大量数据中，剔除由不稳定性引起的“野值”（wild value），然后再进行平均. 对延时测量精度的过高要求，还使得基阵的准确安装变得困难起来. 目前还没有找到突破传统几何原理进行被动目标测距的有效方法.

3.2 合成孔径技术

合成孔径声纳的研制近十年来受到很大的重视. 已经报道有相当高性能的样机问世[19—21]. 合成孔径作为一种技术在雷达上成功应用已近 40 年了，但在声纳上迟迟未获得实质性的进展. 主要是由于声传播的海洋介质比无线电传播的大气介质复杂得多，另外声纳平台运动速度与声传播速度之比约为 1:750，而雷达平台运动速度和无线电波传播速度之比是 $1:10^6$，所以合成孔径声纳的运动补偿、成像远比合成孔径雷达复杂.

合成孔径声纳（SAS）的初步研究结果是令人振奋的，它大约可以在 400m 的距离上达到 10cm 的分辨率. 这在以前的旁测声纳中是无法达到的.

美国 DTI（Dynamic Technology Inc.）研制的样机在华盛顿（Washington）湖作试验时，甚至得到了一架早先沉没湖底的飞机残骸的“声像”（见图 6）.

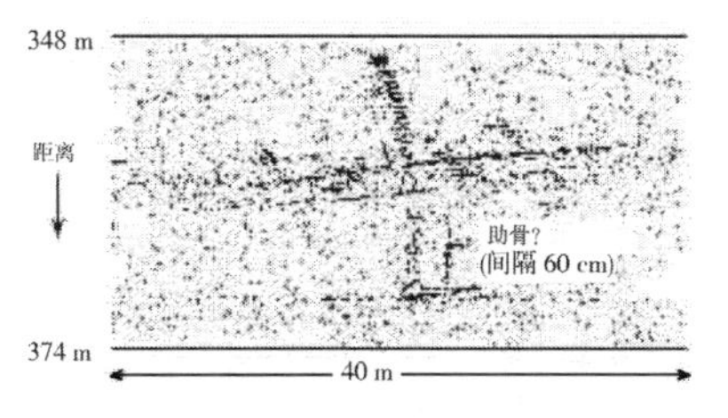

图 6 PB4Y－2 的 SAS 图像（1997 年 4 月，华盛顿湖，50kHz 的 SAS 穿透湖水看到了飞机内部）

合成孔径技术还用于高分辨率的波束成形，这在安静型潜艇辐射噪声的测量中可以获得应用，利

用这种技术可以把潜艇作为一个体积元，确定对辐射噪声最有贡献的分量的部位.

3.3 水声通信与水下 GPS

水声通信一直是声纳研究中的一个重要领域. 美国和北约的其他国家有一系列研究课题是与水声通信有关的[21—24]. 水声通信系统的性能一直受到传输率和作用距离的约束. Kilfoyle 等根据美国几十次海试结果，给出了一条曲线，认为在现阶段传输率(R，以 kbit/s 为单位)和作用距离(R，以 km 为单位)的乘积不超过 40[25](见图 7). 但在 20 世纪 70 年代初，这个值只有 5 左右. 因为为了提高传输速率必须提高信号频率，而一旦频率增高了，传播损失增大，作用距离就下降了. 所以

$$R \cdot R \leqslant 40\text{km} \cdot \text{kbit/s},$$

这在客观上反映了当前水声通信所达到的水平. 如果要突破这个约束，就要增加发射功率，采用新的编码/解码体制和信道均衡技术.

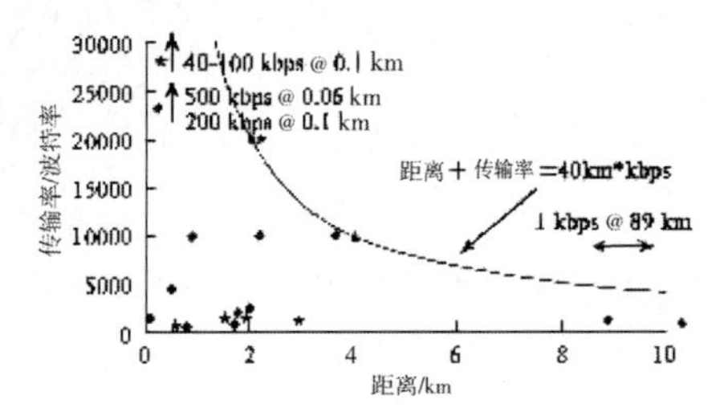

图 7　水声通信传输率与作用距离的关系曲线

水声通信的一个重要应用领域就是水下全球定位系统(GPS)，虽然目前还只是一些设想，但一旦建立起完善的水下 GPS 体制，反潜战的一些战略、战术原则都必须随之改变.

3.4 数据融合

由于声纳系统的集成度越来越高，数据量越来越大，单靠声纳员处理多平台、多传感器的信息就显得很不够. 所以数据融合的技术自然而然地受到重视. 目前，虽然还不能完全做到全自动判别，但至少为辅助决策提供了强有力的工具[26,27].

数据融合从所处理的信息层次来分，可以分为三级，即基元级、特征级和决策级. 研究课题的级别越到底层就越复杂. 现在大多数的研究工作还是围绕决策级展开的.

数据融合中的一个基本定理，保证了声纳系统进行数据融合的必要性，这个定理是说，无论是独立观测资料还是相关观测资料，最佳的线性数据融合所带来的误差不会大于任何个别观测资料所带来的误差[28,29].

基于这一事实，解决声纳系统的数据融合问题就有了理论依据. 举一个具体的例子，假定潜艇上有圆阵和舷侧阵同时进行目标定位. 我们知道，圆阵的定向误差基本上与信号入射方向无关. 而线阵则不同，在侧射方向误差较小，在端射方向误差较大，把圆阵和线阵的数据进行融合，我们得到了很好的测向方法，它的误差不仅小于各自的定向误差，并且在 360°范围内基本均匀.

3.5 目标识别与水下快速运动目标轨迹提取

数字式声纳的基本功能是测向和测距，目标识别的功能通常由声纳员通过鉴别目标辐射噪声来完成. 随着声纳技术的发展，国外的一些声纳已具备目标识别功能，甚至专门配置鱼雷报警声纳.

目标识别和鱼雷报警是两个相关的课题，虽然后者可以抽象为水下快速运动目标的轨迹提取问题，但最后的判决仍离不开识别这一环节.

目标识别的关键当然是特征提取. 只有对大量目标样本进行设计分析，才有可能确定合适的特征量. 于是数据库的建立就是必不可少的，可惜这是缘于高度机密的信息，因而在一定程度上阻碍了识别研究工作的进度. 能够在目标识别方面发挥作用的方法很多，如专家系统(最小邻近准则)，人工神经网络、聚类分析等，但是目前还没有一种办法被公认为是解决目标识别问题的有效方法. 其原因是实验室系统模拟的结果与实际海上的条件差异很大，要寻找出不受传播影响的信号特征非常困难. 声纳工作平台的任何机动(这在实验室里不会出现)都会干扰目标识别系统的工作.

3.6 水下蛙人探测声纳(DDS)系统

2000 年 10 月，美国导弹驱逐舰 Cole 号在也门港口受到恐怖袭击，艇身受到重创. 港口警戒问题引起重视. “911”事件之后，恐怖袭击遍及海陆空各个领域，港口警戒成为国土安全的重要课题. 海军基地、大型集装箱码头、港口和 VIP(贵宾)浴场的保卫问题已成为水声技术研究的迫切任务. 当然，我们需要的是海陆空的立体防护. 但是在水下如何发现并阻止蛙人(包括有呼吸和无呼吸系统的潜水者)、水下有人或无人载器的攻击显得尤为重要(见图 8,9).

图8 C－TECH公司的DDS系统工作示意图

图9 Raytheon公司的DDS系统工作示意图

虽然从原理上讲，DDS也是传统意义上的主动或被动声纳，但是在声纳技术的应用上仍面临一些新的挑战．比如，DDS通常用于浅海，作用距离较短（例如1 km），这就使声纳工作于混响背景限制的状态，所以能抑制混响和抗多途效应的波形设计显得很重要；又比如，蛙人的目标强度较小（例如TS ＝ －10 dB），检测难度较大，等等．

4 声纳领域的技术创新

声纳技术是一门发展迅速、需求迫切、应用前景异常广阔的学科．它不是一门纯理论的学科，它的发展和完善依赖于大量的有准备的海上实验．由于基础研究的特殊性，需要较大的人力、财力投入．深入的基础研究是声纳技术创新的源泉．回顾声纳发展的历史就可以证明这一点．

如果没有对不同声速剖面下水声信号传播规律的研究，就不会有变深声纳；如果没有对水声场的时空相关特性的研究，就不会有被动测距声纳；同样，对水声信道的研究为匹配滤波器、水声通信系统提供了正确设计的依据；内波、声层析、匹配场过滤又为远程被动定位技术提供理论支持．

只有技术创新才能实现跨越式发展，拖曳式线列阵的出现就是一个很好的例子．声纳设计者在21世纪初处于这样一种充满机遇和挑战的年代中，一定能取得新的突破、新的成功．

参 考 文 献

［1］ Kock W E. Radar, sonar and holography : an introduction. New York: Academic Press, 1973

［2］ Bondyopadhway P K. Proc. IEEE, 1998, 86(1):78

［3］ Forester T. High－tech Society, the story of the information technology revolution. UK Blackwell Ltd, 1987（中译本：姚炳虞，郑九振译．高技术社会．北京：新华出版社，1991）

［4］ Marburger H. Stealthy Russian submarines will rival U. S. fleet. Defense News, USA, Sept. 11, 1994

［5］ Hazell P A. Sea Technology, 1998, 39(11):59

［6］ Hamblen N. Sea Technology, 1998, 39(11):59

［7］ Mctaggant B. Thirty years of progress in sonar transducer technology. Proc. UDT' Paris 1991. 1—11

［8］ Bright C. Sea Technology, 2000, 41(6):17

［9］ Douglas J W. Sea Technology, 1996, 37(1):11

［10］ Camdy J V. IEEE Oceanic Engr. Society News Letter, 2000, 25(3):199

［11］ Porter M B. IEEE J. Oceanic Engr., 1993, 18:425

［12］ Kassam S A. Poor H V. Proc. IEEE, 1985, 73:21

［13］ Gingras D F. IEEE J. Oceanic Engr., 1993, 18(3):253

［14］ Sullivin E J, Middeleton D. IEEE J. Oceanic Engr., 1993, 18(3):156

［15］ Ed. Chen C T. IEEE Signal Processing, 1998, 5(4):21

［16］ Ed. Stergiopoulos S. Advanced signal processing handbook. USA, CRC press, 2001

［17］ Urick R J. Multipath propagation and its effects on sonar design and performance in the real ocean. Proc. NATO ASI on Underwater Acoustics, 1976

［18］ Zhang R H *et al*. Acta Acoustica, 1981(1):9

［19］ Deviss B. New Scientists, 22 June, UK, 1996

［20］ Chatham R *et al*. The synthetic aperture sonar revolution. In: Proc. AUSI Conference, USA KONA, 2000

［21］ Stergiopoulos S. JASA, 1990, 87:2128

［22］ Curtin T B. Sea Technology, 1999, 40(5):17

［23］ Stojanovic M. IEEE J. Oceanic Engr., 1996, 21(2):125

［24］ Kilfoyle D B. Baggeroer B. IEEE J. Oceanic Engr., 2001, 25(1):4

［25］ Stotts S A *et al*. IEEE J. Oceanic Engr., 1997, 22(3):576

［26］ Sharma R *et al*. Proc. IEEE, 1998, 86(5):853

［27］ Hall D L. Llinas J. Proc. IEEE, 1997, 85(1):6

［28］ 李启虎．声学学报，2000，25(5):385［Li Q H. Acta Acustica, 2000, 25(5):385］

［29］ 李启虎．声学学报，待发表［Li Q H. Acta Acusti (acin press)］

［30］ Middeleton D, Esposito R. IEEE Trans., 1968, IT-14(3):434

第四讲 探潜先锋——拖曳线列阵声纳①②

余华兵　孙长瑜　李启虎
（中国科学院声学研究所　北京　100080）

摘　要　文章介绍了在反潜战中发挥重要作用的拖曳线列阵声纳的工作原理、结构特点、系统组成，同时回顾了拖曳线列阵声纳从第一次世界大战时 Hayes H C 博士的概念到2000年 John R. Potter 等人的细线阵所经历的长达80年的发展过程，最后预计，应用光纤、新声源，采用多线阵结构能进行三维定位的主被动模块化拖曳线列阵声纳是未来的发展方向.

关键词　拖曳线列阵，声纳，探潜，反潜战

Towed line array sonar spearheads submarine detection

YU Hua-Bing　SUN Chang-Yu　LI Qi-Hu
(*Institute of acoustics, Chinese Academy of Sciences, Beijing* 100080, *China*)

Abstract　Towed line array sonar (TLAS) has been playing an important role in anti – submarine warfare. The principle, characteristics and configurations of TLAS are briefly introduced. The history of TLAS from the inspiration of H. C. during World War I to the 'thin array' of J. R. Potter in 2000 is reviewed, and it is predicted that future TLAS will be composed of passive and active modulated multiple arrays with fiber-optic hydrophones and new acoustic sources, and will be capable of locating targets in 3-dimensions.

Keywords　towed line array, sonar, submarine detection, anti-submarine warfare

1　前言

声波是人类迄今已知的唯一能在海水中远距离传输的能量形式，其他能量辐射形式，如光波和电磁波都不能在海水中远距离传输[1]. 因此，人们一直利用声波对水下或者水面的固定或运动目标进行导航、定位、跟踪和识别，具有这种功能的设备称为声纳或水声设备.

拖曳线列阵声纳拖曳距舰船尾部一定距离的声接收系统（通常称为线列阵），通过接收航行目标自身辐射的噪声或者通过接收目标反射回来的主动信号的回波，来检测目标的有无并估计目标有关参数.

该声纳可以分为两类，即被动式和主被动联合式. 被动式拖曳线列阵声纳仅通过接收到的目标辐射噪声来进行探测，具有较好的隐蔽性，它可以由水面舰艇，也可以由潜艇拖曳；而主被动联合式拖曳线列阵声纳既可以利用被动接收到的目标辐射噪声进行探测，又可以通过主动发射信号经目标反射后的回波信号来进行检测，是探测辐射噪声日益降低的潜艇的重要手段，通常由水面舰艇拖曳.

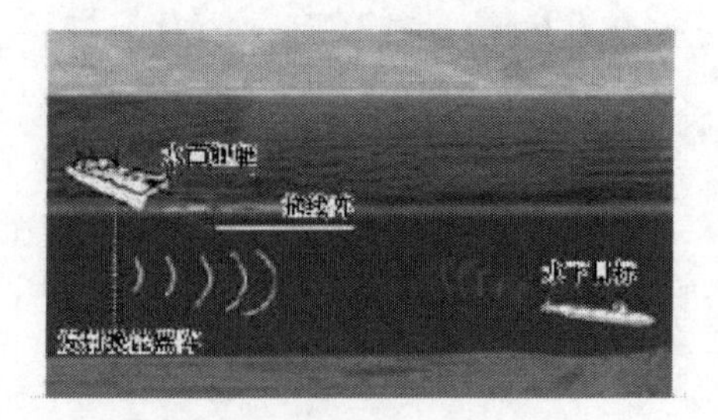

图1　拖曳线列阵声纳工作场景示意图[2]

被动式拖曳线列阵声纳通常由以下几个部分组成：

① 中国科学院知识创新工程资助项目.

② 物理, 2006, 35(5): 420-423.

(1)线列阵:用于接收目标的辐射噪声和目标回波,它由若干个拾取水下声学信号的水听器按照一定的间距布放,并采取隔振措施,配备用于水下姿态监视的传感器模块;

(2)拖曳收放系统:用于拖曳、布放和回收线列阵,包括拖缆和绞车;

(3)深度、航向监视系统:用于监视拖线阵在水下的深度、温度、航向等信息;

(4)信号处理系统:用于处理各种信息,实现对目标的检测及有关参数的估计;

(5)显示控制系统:用于将信号处理系统处理的结果进行显示,与其他系统进行信息交换,并将控制命令下发给有关系统;

(6)数据记录系统:将各种数据进行存储.

主被动联合式拖曳线列阵声纳则还有发射换能器基阵、相应的拖曳收放系统及发射机.

图1是拖曳线列阵声纳工作场景的示意图,其中在船尾拖曳的即是拖曳线列阵声纳的线列阵,而在船底吊放的则是发射换能器阵,其余设备均在甲板上或者舱室中.

拖曳式线列阵声纳区别于安装在舰艇外壳上的舰壳声纳的优点在于:(1)拖曳线列阵声纳将接收声波的拖线阵远离工作母船,显著减小了拖曳平台噪声的影响,能显著提高接收信噪比,达到提高声纳检测能力的目的;(2)拖曳线列阵声纳中的拖线阵规模不受舰船尺寸的限制,可以安装较之舰壳声纳更多的水听器,充分利用了海洋中信号和噪声不同的统计特性,有效地提高声纳的检测性能;(3)可以利用在海洋中传播损失较小而且是水面和水下目标辐射噪声中重要成分的低频信号进行检测.

由于以上优点,拖曳线列阵声纳出现后受到广泛关注,经过多年的发展,已经成为各国海军对日益安静的潜艇进行有效检测的重要装备[3],不妨称其为探潜先锋.

2 拖曳线列阵声纳工作原理

众所周知,人的两只耳朵具有定向功能,当要判断一个声源的方向时,总是把头转向声源的方向,使得声源正好处于两只耳朵连线的垂直平分线方向上,声音能够同时到达两只耳朵,这实际上是一种简单的定向原理——最大声压定向法[4].拖曳线列阵声纳就是利用类似的原理来进行工作的,它充分利用噪声和信号的不同统计特性,将多个水听器在同一时刻收到的不同信号,经过与方位相关的时间补偿处理,再通过能量积累,得到输出能量最大的方向就是目标到达的方向.

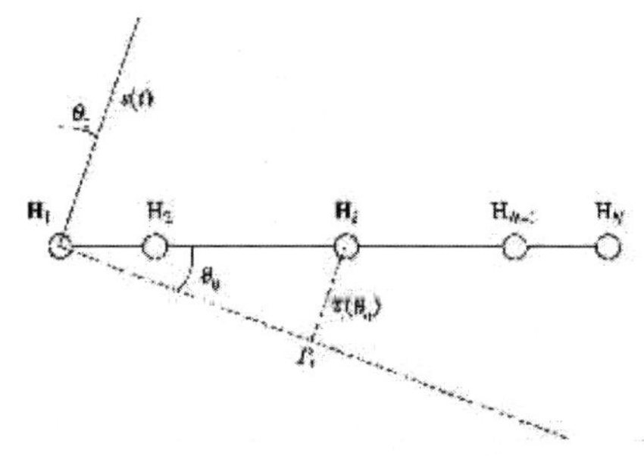

图2 等间隔线列阵水听器之间时延差计算(H_1, H_2, …H_i, …H_{N-1}, H_N 为间距固定的 N 个水听器)

如图2所示,拖曳线列阵声纳的线列阵中的声学模块由 N 个水听器组成,以平面上的某一点为参考点,设入射信号为 $s(t)$,经过海洋传播,到达第 i 个水听器的信号为 $s(t+\tau_i(\theta_0))$,这里 θ_0 为信号的入射角.如果将这一路信号延时 $\tau_i(\theta_0)$(此延时和信号入射方向有对应关系,通过此延时能给出信号的入射方向),那么对所有 N 路信号都会变成 $s(t)$,将这 N 路信号相加便得到 $Ns(t)$,再平方积分得到 $N^2\sigma_s^2$(这里 σ_s^2 为信号功率),如果改变信号入射的方向,那么第 i 个水听器的信号经延时 $\tau_i(\theta)$ 就变成 $s[t+\tau_i(\theta)-\tau_i(\theta_0)]$,系统输出能量最大,即

$$D(\theta) = E[(\sum_{i=1}^{N} s(t+\tau_i(\theta)-\tau_i(\theta_0)))^2].$$

如果考虑信号中混有噪声 $n_i(t)$($i=1,2\cdots,N$),$n_i(t)$ 之间相互独立,且它们的均值为0,这时系统输出能量 $D(\theta)$ 最大,

$$D(\theta)=E[\sum_{i=1}^{N} s(t+\tau_i(\theta)-\tau_i(\theta_0)+\sum_{i=1}^{N} n_i(t-\tau_i(\theta_0))^2].$$

当 $\theta=\theta_0$ 时,该值为 $N^2\sigma_s^2+N\sigma_n^2$(这里 σ_n^2 为噪声功率),由此可知,信号增强了 N^2 倍,而噪声仅增强了 N 倍,因此带来的信噪比增益为

$$G_s = 10\log\left(\frac{N^2\sigma_s^2}{N\sigma_n^2}\Big/\frac{\sigma_s^2}{\sigma_n^2}\right) = 10\log N.$$

由此可知,一个由 N 个水听器组成的拖曳线列阵声纳,如果满足各基元所接收的噪声相互独立的条件,那么它的增益就是 $10\log N$,N 越大,增益就越高.

根据海洋环境噪声的统计特性,当水听器间距为波长的一半时,各个水听器接收到的海洋环境噪声基本相互独立,因此在拖曳线列阵声纳中,水听器正是按照这个间距来布放的;水听器的数目如果越多,那么获得的增益就越大,对目标的探测能力就越强,这也正是拖曳线列阵声纳孔径(拖线阵长度)越来越大的原因.

3 拖曳线列阵声纳发展历史

拖曳线列阵声纳的历史可以追溯到第一次世界大战,它的发展先后经历了三个阶段,即试验阶段、接受阶段和发展阶段[5].

在第一次世界大战期间,美国人 Hayes H C 博士提出一种拖曳/舷侧线列阵声纳结构,如图 3 所示.在美国海军 Jouett 号军舰的舷侧前部安装两条 12 元线列阵,同时在其尾部拖曳两条 12 元线列阵,这套系统的探测距离大约为 1.8km,应对当时潜艇攻击已经足够.

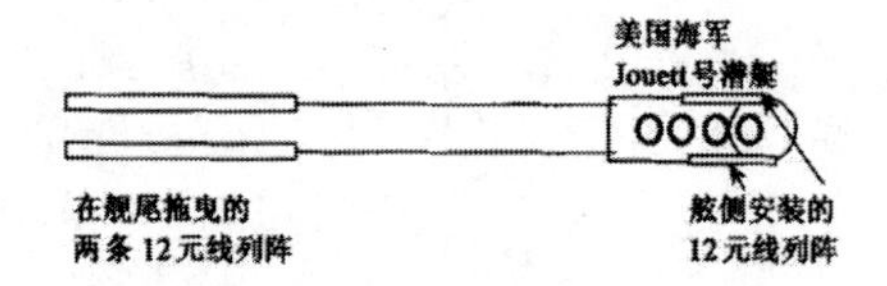

图 3 Hayes 博士提出的拖曳/舷侧线列声纳示意图

战后拖曳线列阵声纳的研究一度停滞,直到 1940 年,丹麦的 Holm C 为美国海军研制了同时在商船后面拖曳的双线阵声纳,该声纳具有鱼雷报警功能.在 20 世纪 50 年代后期,美国海军研究办公室的 Lasky M 又继续了 Hayes 博士的工作,先后进行了一系列相关的试验,其中包括小外径且与电缆类似的拖线阵声纳及安装在柔性橡胶套管中的三基元线列阵声纳.

在 20 世纪 60 年代早期,美国海军的潜艇的艇长们遇到一个问题:由于其他设备的阻挡,在舰尾部有声盲区.为了消除这个声盲区,贝尔电话实验室(Bell Telephone Laboratory, BTL)和 Chesapeake 仪器公司(Chesapeake Instrument Corporation, CIC)分别提出了不同的方案.BTL 提出在导弹发射管中存储并在潜艇下方拖曳的拖曳线列阵声纳,而 CIC 则采用在舰桥上进行拖曳的方式.试验证明,前者对目标的探测距离超过了 36km,而后者对同一目标的探测距离也达到了 9km.随后 CIC 利用自研的线列阵,并从海军借到 AN/AQA-2 声纳及有关处理设备,检测到了 100 多公里外的潜艇目标,取得了巨大成功.

至此拖曳线列阵声纳已经完成从概念到声纳的试验过程,从 1968 年开始,拖曳线列阵声纳开始为大家接受,并逐渐扩大了应用的范围,成为水面舰艇和潜艇的重要装备.

1968 年,美国海军水声实验室(U.S. Naval Underwater Sound Laboratory, USNUSL)提出过渡型拖曳线列阵警戒系统 AN/SQR-14 的战技指标.CIC 经过大约 14 个月的努力,于 1970 年交付使用.在地中海投入使用后,该声纳工作非常出色.1972 年,CIC 获得了五台 AN/SQR-14 的改进型 AN/SQR-15 的订货.

同时在潜用拖线阵声纳方面,Huges 飞机公司(Huges Aircraft Company, HAC)于 20 世纪 60 年代中期获得了一份合同,在快速攻击潜艇上进行了拖曳线列阵声纳的适装性试验.在 20 世纪 60 年代晚期,由于前苏联安静型潜艇的出现,美国海军海上系统司令部决定在所有的攻击型潜艇上装备拖曳线列阵声纳.

在随后的 20 余年的时间里,拖曳线列阵声纳得到了全面的发展.

1975 年,CIC 被 Gould 公司收购,但 CIC 拖曳线列阵声纳研发队伍仍然在拖曳线列阵领域继续工作多年,在 20 世纪 80 年代,在减小拖线阵拖曳自噪声方面取得了巨大的进展,并研制出自噪声更小的 TB-16 拖曳线列阵声纳.

1984 年,Bendix 公司开发出一种用于潜艇警戒的细长缆拖曳线列阵声纳 TB-23,其后于 1990 年,Gould/Martin Marietta 公司研制出比 TB-23 更长的 TB-29 型拖曳线列阵声纳.

到 20 世纪 90 年代,拖曳线列阵声纳有各种不同长度的配置,覆盖了宽广的频率范围.拖线阵的直径由 40-80mm 减小至 8mm[6].

20 世纪 70 年代到 80 年代的拖曳线列阵声纳有一个很大的弱点,就是它们的动态范围有限,受到当时电子技术的制约,当时的声纳只能依赖 8bit 电子器件,最高只有 48dB 的幅度动态范围.在 20 世纪 90 年代,24bit 电子器件出现使得幅度动态范围最高可以达到 140dB.

在此阶段的另一个重大的事件是光纤在拖曳线列阵声纳中的应用,在此之前只能采用多路双绞导线来进行数据传输,拖缆的外径最大只能接受大约 50 路信号的传输.后来出现了数字化同轴缆传输技术,将线列阵中水听器的信号在水下进行数字化并打包后通过同轴电缆传送到处理设备.但是更高的动态范围和更多的水听器数目使得同轴电缆也不能满足大量数据传输的需要. 在20世纪90年代,Litton工业公司和CSC进行了光纤水听器和光纤阵的研制和测试.

同时Tohomson Marconni和Serel开发出固体填充的拖线阵,避免了用油填充的拖线阵面临的破损泄漏和污染.最终,APL实验室使用双缆拖曳线列阵声纳解决了拖曳线列阵声纳的左右舷模糊(单缆拖曳线列阵声纳无法判断目标在舰船的左侧还是右侧)问题.

多功能拖曳线列阵(MFTA)也在此期间出现,由于高动态范围的AD芯片的出现,使得拖曳线列阵声纳既能接收主动信号,也能接收被动噪声,既能接收高频信号,也能接收低频信号.

纵观整个发展历程,在过去的80余年中,拖曳线列阵声纳家族经历从2倍波长(对应1kHz声波波长)到超过1000倍波长的历程.动态范围从60dB增加到120dB.水面舰拖曳线列阵声纳解决了海军的问题:在护卫舰航速下能够进行警戒,并对潜艇进行战术低频探测.

潜用拖曳线列阵声纳始于20世纪50年代,CIC在USSAlacore舰桥上安装的拖线阵揭开了潜用拖曳线列阵声纳应用的序幕;其后,为了解决潜艇声盲区的问题,BTL和CIC均取得成功,最终BTL完成了AN/SQR-15声纳的生产.在20世纪60年代到70年代早期,HAC开发了用于攻击型潜艇的TUBA和TB-16拖曳式线列阵声纳.由于面对前苏联的安静型潜艇,具有较小流噪声的TB-23拖曳线列阵声纳应运而生,在此基础上,TB-29使用更多的水听器,具有更好的探测能力并装备使用.

4 未来发展趋势

在被动联合拖曳线列阵声纳中,主动声源目前一般采用发射换能器基阵,而在海上石油勘探中使用的气枪较之发射换能器基阵来讲虽然在信号形式的控制上,目前还存在问题,但由于其更小的体积,能够激发更大声源级的信号,因此一旦在气枪的信号控制问题上取得突破,气枪有望称为拖曳线列阵声纳的一种新的声源.

另一方面,光纤水听器自诞生以来,备受关注,由于其体积小,便于多路复用,并可以将多种传感器集成在一起,使得拖曳线列阵外径显著减小且无需从甲板向水下供电,适装性和可靠性得到明显改善,因此基于光纤水听器的全光纤拖曳线列阵声纳是发展的重要方向之一.

如前所述,拖曳线列阵声纳较之舰壳声纳的优点之一就是其孔径不受舰船尺寸的影响,能够在水平方向形成尖锐的指向性来实现对目标水平方位的估计.如果能使拖曳线列阵声纳在垂直方向上同样具有一定的指向性,就能够对目标的垂直方位进行估计,从而实现对目标的更加精确的三维定位.因此,同时具有水平和垂直指向性的拖曳线列阵声纳也是未来发展的重要方向之一

在潜艇中安装拖曳线列阵声纳时,由于存放、布放、回收等方面的问题,目前尚未报道多缆拖曳线列阵声纳装备使用;但由于它能够解决左右舷模糊问题,并有可能提供目标的深度信息,实现更准确的定位,并对目标进行识别,多缆拖曳线列阵声纳备受人们关注.

目前,拖曳线列阵声纳的拖曳平台为水面舰艇及潜艇,由于其显著的优点,在其他拖曳平台上使用也具有重大的意义.因此对拖曳线列阵声纳进行更细致的模块化设计,使其能够根据拖曳平台的特点,进行裁剪或者扩展,将使拖曳线列阵声纳具有更广阔的应用天地.

参考文献

[1] 李启虎. 数字式声纳设计原理. 合肥安徽:教育出版社, 2003. 1 [Li Q H. Design principle of digital sonar. Hefei: Anhui education publisher, 2003. 1 (in Chinese)]

[2] Bruce Joffe. http://www.mindfully.org/Technology/2005/SURTASS-LFA-Sonar4mar05.htm.

[3] Tyler G D. Johns Hopkins APL Technical Digest, 1992, 13(1):145

[4] 李启虎. 声纳信号处理引论. 北京:海洋出版社, 2000. 162 [Li Introduction of sonar signal processing. Beijing: Ocean publisher, 2000. 162 (in Chinese)]

[5] Lemon S G. IEEE J. ocean. engineer., 2004, 29(2): 365

[6] Potter J R, Delory E, Constantin S *et al* The 'thinarray': a lightweight, ultra-thin(8 mm OD) towed array for use from small vessels of opportunity. In: Underwater technology 2000. Tokyo, Japan, June 2000

第五讲　新型光纤水听器和矢量水听器[①②]

孙贵青　　李启虎　　杨秀庭　　孙长瑜
（中国科学院声学研究所信号与信息处理实验室　北京　100080）

摘　要　光纤水听器和矢量水听器作为当前水声研究领域最具有代表性的两大技术倍受业界关注.光纤水听器的重要贡献在于,从一个全新的角度出发,试图解决传统的水声传感和声纳数据传输一体化设计和实现的一系列问题,这有助于改善声纳系统的可靠性,并且有可能降低其制造、使用和维护的总成本.矢量水听器则由于其特有的指向性和矢量－相位处理方法,在低频和甚低频水声微弱目标探测方面具有潜在的优势.经过不懈的努力,光纤水听器和矢量水听器系统已经从实验室逐渐进入到工程应用阶段.这些对未来声纳系统的发展会产生相当重要的影响.文章尝试从声纳设计的角度对这两者的技术现状进行简要综述,包括它们各自的物理基础、工作原理、关键技术和应用领域.

关键词　光纤水听器,矢量水听器

A novel fiber optic hydrophone and vector hydrophone

SUN Gui-Qing　　LI Qi-Hu　　YANG Xiu-Ting　　SUN Chang-Yu
(*Signal and Information Processing Laboratory, Institute of Acoustics, Chinese Academy of Sciences, Beijing* 100080, *China*)

Abstract　Fiber optic and vector hydrophones are two major types of hydrophones that are of great importance to underwater acoustics. The chief advantage of the former is that problems related to the integration of traditional underwater acoustic sensing and sonar data transmission can be resolved from an entirely new perspective, thus improving the reliability and possibly decreasing the total cost of manufacture, operation and maintenance of the sonar system. Because of its inherent directivity and unique vector－phase processing, the vector hydrophone possesses potential advantages in the detection of quiet underwater acoustic targets emitting low frequency and ultra－low frequency noise. Engineering applications for both types of hydrophone have gradually emerged from the research laboratory, and will play an active role in the development of future sonar systems. A brief overview is presented of these two state－of－the－art hydrophones from the viewpoint of sonar design, including the basic physics, principle of operation, key techniques, and applications.

Keywords　fiber optic hydrophone, vector hydrophone

1　引言

声波作为一种机械波,可以在海水中进行远程能量传递,而其他类型的能量场在水中衰减很快,如以无线电波和光波为代表的电磁场.因此,至少从目前看来,还没有出现能够威胁到声场优势地位的技术,声场仍然是海洋深层信息收集、传递和处理的最重要形式,从大尺度、全天候的全球海洋观测计划直到各种类型的声纳装备,都是这种形式的具体体现.水听器作为水下声波的接收设备是水声学最重要的声学测量仪器,一般可以分为无指向性和有指向性两大类.无指向性指的是水听器对来自于声场空间

① 国家自然科学基金(批准号: 60532040/F010203)资助项目.
② 物理, 2006, 35(8): 645-653.

各方向的声波具有相同的响应，不存在空间选择性；反之则为有指向性，即水听器只对空间某些方向的声波有响应.

自第二次世界大战之后的60年间，水声技术在军事需求的强势推动下得到了长足的发展，尤其是反潜战的切实需要. 到目前为止，还没有什么技术能够像拖曳线列阵声纳那样深刻地影响现代反潜技术的发展，它是公认的20世纪水声技术最伟大的发现. 到目前为止，水下拖曳声阵的水听器仍由无指向性的压电陶瓷传感器一统天下，在其他大多数声纳系统中也是类似的情况[1]. 传统的压电陶瓷水听器阵列声纳需要大量的用于信道复用和数据传输的水下电子元件，以及信号传输电缆和供电电缆，这些电子设备价格昂贵，重量不轻，往往会因为水下密封问题导致设备失效，使得系统可靠性严重恶化. 随着科学技术的不断进步，光纤水听器和矢量水听器已经成为当前水声研究领域最具有代表性的两大技术而倍受业界的关注，逐渐从早期的实验室研究阶段迈向工程应用[2−6].

光纤水听器在声纳和石油天然气的地震勘探中已经发展了好多年. 这两个应用现在都需要超大阵元数目（上千只传感器）的高度复用的传感器阵列，它们可用于一些海底阵和拖曳阵等声学探测系统中. 光纤水听器作为光纤传感器的主要应用之一，早在1976年就由Bucaro等首先提出并演示[7]. 本文的光纤水听器均为无指向性的. 随着光器件技术的巨大进步，由光纤水听器作为阵元构成的全光水听器阵列声纳得到了格外垂青. 全光阵列在水下无任何电子元件，完全没有电子设备；在单个光纤上复用大量水听器；水听器和数据传输通道具有很强的抗电磁干扰能力. 因此，系统更灵敏，重量更轻，可靠性更高，当然造价也更为经济. 光纤水听器的重要贡献在于从一个全新的角度出发试图解决水声传感和声纳数据传输一体化设计和实现的问题，其目的并不是单纯地追求更高的声学性能（尽管它具有极高的灵敏度和动态范围），而是显著提升系统整机的可靠性，降低制造、使用和维护的总成本. 相关系统已被英国国防研究局（DRA）、美国海军研究实验室（NRL）、日本冲电气（OKI）国防研发部，以及意大利Alenia防务系统等研究机构开发.

矢量水听器最早由Leslie等在1956年提出并演示[8]，几乎与此同时，莫斯科国立大学也在开展相应的研究工作. 矢量水听器可以同步、共点测量声压标量和质点振速矢量，不同于传统的仅测量声压标量的水听器. 因此，矢量水听器可以切实改善声纳系统的声学性能，如阵列增益、定向精度等[9]. 矢量水听器也可以由光纤传感器构成，也是当前水声传感器的研究前沿之一.

本文尝试从声纳系统（不仅仅是水听器本身）的角度出发对光纤水听器和矢量水听器的技术现状进行简要综述，包括其物理基础、工作原理、关键技术和应用领域.

2 光纤水听器

光纤水听器是复杂的光、机、电一体化传感器，其在各种声纳应用中的潜能已被认识到，而且它已达到可与压电水听器相媲美的地步. 其最大特点是具有足够高的声压灵敏度，通常比压电陶瓷水听器高3个数量级（60dB）. 尽管现在开发了多种不同的光纤水听器，如强度调制型、偏振调制型、波长调制型和相位调制型，但最有前景的是基于相位调制的干涉型光纤水听器. 它的成功和发展都依赖于水听器设计与光器件发展水平的相互结合. 大多数高通道数光纤水听器阵列的研发工作都是据此开展的，因为它能够提供高声学灵敏度和强复用能力的最佳组合，是当前光纤水听器研究的成功范例.

2.1 物理和信号处理基础

光纤水听器是复杂的光机电系统，既需要物理知识阐述隐藏其中的自然规律，也需要信号处理手段进行实际应用，两者相辅相成.

干涉型光纤水听器是利用声波对单模光纤线圈中光的相位进行调制，光纤线圈即构成水听器的传感单元. 在时分复用（TDM）系统中，水听器常由单个光纤线圈构成，但在频分复用（FDM）系统中，水听器常含有两个线圈，声信号是两个线圈之间的相位差. 在这类设计中，必须安排一个线圈对声不敏感，或者声信号在两个线圈上产生反相变化（即推挽式工作模式）. 图1给出了单线圈TDM系统的简单水听器设计. 光纤缠绕在芯轴上，用环氧树脂或聚氨酯之类的材料密封. 作用在芯轴上的声波引起芯轴结构形变，这些形变被传递到光纤线圈中（其中声波的波长远大于水听器尺寸，即可以视为点接收器）.

2.1.1 物理基础

当单模光纤受到轴向的机械应力作用时，光纤长度、芯径和纤芯折射率都将发生变化，这些变化将导致光波的相位变化. 当光波通过长度为L的光纤

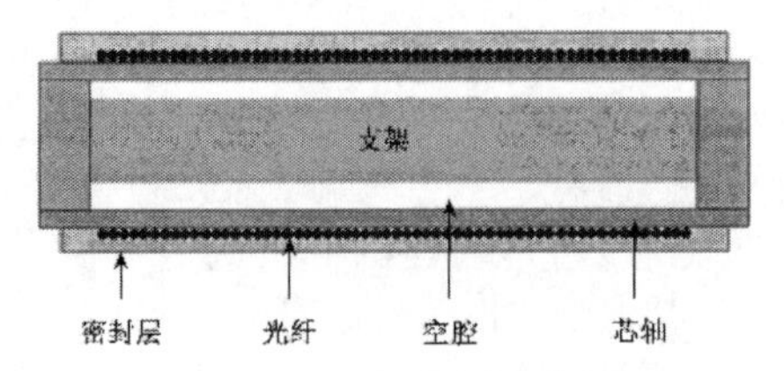

图 1 芯轴型光纤水听器的传感结构

后,出射光波的相位延迟为

$$\phi = \frac{2\pi}{\lambda}L = \beta L,$$

式中 $\beta = n_{\mathrm{eff}}k$ 为光波在光纤中的传播波数,k 为光波在真空中的传播波数,$\lambda = \lambda_0/n_{\mathrm{eff}}$ 是光波在光纤中的传播波长,λ_0 为光波在真空中的传播波长,$n_{\mathrm{eff}} = 1.465$ 为光纤的有效折射率. 那么,光波在外界因素作用下的相位变化为

$$\Delta\phi = \beta L\frac{\Delta L}{L} + L\frac{\partial\beta}{\partial n}\Delta n + L\frac{\partial\beta}{\partial a}\Delta a,$$

式中 a 为光纤芯半径,第一项表示光纤长度变化引起的相位延迟(应变效应),第二项表示感应折射率变化引起的相位延迟(光弹效应),第三项表示光纤半径改变所产生的相位延迟(泊松效应),由于其值相对较小,一般可忽略不计. 因此,在长度为 L 的光纤中,声压变化 ΔP 产生的相位变化(或称为归一化灵敏度)由下式给出:

$$\frac{\Delta\phi}{\phi\Delta P} = \frac{1}{\Delta P}\left\{\varepsilon_z - \frac{1}{2}n_{\mathrm{eff}}^2 kL[\varepsilon_r(p_{11} + p_{12}) + \varepsilon_z p_{12}]\right\},$$

其中 ε_r 和 ε_z 分别是应变的轴向和径向分量,p_{11} 和 p_{12} 是 Pockel 系数,$\phi = n_{\mathrm{eff}}kL$ 是总相位,$\varepsilon_r(p_{11} + p_{12})$ 对应光纤长度变化的贡献,$\varepsilon_z p_{12}$ 是光弹效应引起折射率变化所带来的贡献.

光波是电磁波,与无线电波同属一类. 我们知道,无线电波可以在示波器上直接显示,但光波的频率远高于无线电波,示波器的扫描速度远远跟不上这样的高频振荡,因而无法显示光波的波形. 实际上,光检测器所给出的读数,都只能是光强度在一段时间内的平均值,这段时间远大于光波振动周期. 一般可以利用双光束干涉得到光的相位信息. 典型的双光束干涉仪是 Michelson 和 Mach - Zehnder 干涉仪,两者区别在于前者只使用了一个耦合器,而后者需要两个耦合器.

2.1.2 信号处理基础

在上文涉及的物理基础中,强调的是声波对光相位的调制机理,本节的信号处理基础则偏重于从光相位中解调出声波信息,即将光变换到通常的电压量上以方便显示、存储和计算. 这两个部分相互呼应,共同完成光纤水听器的"感""传"功能.

首先,回顾干涉型光纤水听器的基本工作原理. 一般情况下,双臂干涉仪的输出强度由下式给出:

$$I = I_0[1 + V\cos\phi(t)],$$

其中 I_0 是平均接收强度,V 是干涉条纹可见度. 在光纤干涉仪中,待测量引起干涉仪一个臂上的光相位变化 $\phi_s(t)$. 但是,环境扰动(如温度、压力等)往往引起不可预测的缓慢变化的相移 $\phi_d(t)$,因此,干涉仪相位可表示成 $\phi(t) = \phi_s(t) + \phi_d(t)$. 扰动相移的不可预测性迫使光纤水听器使用问询方法以提取信号的相位. 使用最广泛的是外差法和频率调制相位载波(FM PGC)或路径匹配差分干涉仪相位载波(PMDI PGC). 鉴于该问题的重要性和复杂性,它连同与其联系紧密的复用问题在下一节中专门阐述,本节仅对信号偏振衰落进行简要的评述.

使用偏振激光源问询的干涉型光纤传感器所面临的共同问题都是偏振引起的信号衰落. 标准的单模光纤都存在微小的线性双折射,环境扰动将改变这种双折射,使得光纤干涉仪出射光束的偏振态变化难以预测,往往导致干涉条纹可见度不可预知的损失. 在最坏的情况下,出射光束的偏振态相互正交(对应于条纹可见度 $V = 0$),此时传感器信号消失. 这一问题已在许多复用系统的实验中观察到,可以导致一个或多个水听器通道随机退化. 对于单通道系统,输出端的干涉系数可通过输入偏振的主动控制来稳定,但在多通道系统中,这变得更为复杂,在所有通道同时达到良好的干涉系数格外困难,也是全光阵列的一个主要技术难题. 尽管可以使用正交共轭镜(即法拉第旋转镜)或保偏光纤等办法,但是成本太高. 解决这一问题并兼顾传感器复用的最佳方法是在光电检测器之前使用偏振分集接收器,因为对于三个独立的偏振器,若它们的轴彼此之间相差 60 度,则完全的衰落不会同时出现,即总可以从一个偏振器中观测到非零的可见度.

2.2 复用和问询

光纤水听器阵列必须作为一个统一整体进行设计,因为问询和复用不仅仅是面向阵列设计的,而且对水听器本身的设计和性能也有相当重要的影响. 在常规的压电声纳系统中,水听器首先被设计,然后再开发复用技术构成阵列,前后两个过程彼此之间可以相对独立. 用最少的光纤寻址多个传感器的能力,或并联或串联,是实现大型高效复用阵列的本质要求.复用方法的主要特点是复用传感器的性能应与单个传感器性能类似.当前干涉传感器的复用技术正在被广泛、深入研究.

在干涉型水听器方案中,声传感器由光纤线圈组成,它形成干涉仪的一个臂,作用在光纤上的声信号引起线圈中光的相位变化,它可用适当的问询技术进行远程检测.许多问询技术都能使它们自身适合于高效复用,从而可使大量水听器复用到单个光纤上.这对于现代多通道大规模声纳系统至关重要,几种主要的复用/问询方案:频分复用(FDM)、扩频复用、时分复用(TDM)等得到了相当多的关注.值得注意的是,复用技术对于确定系统的许多基本参数非常重要,如系统噪声背景、工作频带和动态范围.非复用系统的噪声背景比深海零级海况还要低30dB,动态范围超过120dB,指标远超过了绝大多数常规的压电水听器.但是,复用系统的相应指标通常有显著降低,主要因为所采用的复用方法不同.一般情况下,复用系统的典型设计指标是要求系统噪声背景达到深海零级海况,这也是大多数常规声纳系统所必须达到的.本节着重介绍使用频度较高的频率复用、时分复用和时分复用/密集波分复用.

2.2.1 频率复用(FDM)

频率复用是通信行业术语,在光学中一般称为波分复用.基于源正弦调制的相位载波技术(PGC)特别适合于频率复用.当水听器数目具有整数平方根时,频率复用方法最有效.激光频率引起的相位分辨率为18 μrad/$Hz^{1/2}$,传感器串扰级为-60dB.复用率被限制的原因:一是由于光的损耗,二是分离滤波器带宽.但是对于32×32系统,2kHz带宽是实用的(即1024个通道,32个源).这一系统框架被NRL在诸多演示项目中所采用,包括49元舷侧阵和拖曳线列阵,其中每个阵都使用7个光源,16个阵元以及4个光源的海底固定式垂直线阵.这些系统的噪声背景比深海零级海况低10dB.

2.2.2 时分复用(TDM)

时分复用是最早演示的技术,作为最简单和最有效的方法得到了相当多的关注.一般可分为两个主要的阵列类型:透射型和反射型,以及两个主要的问询方案:平衡和非平衡.问询通常多采用外差技术.系统输出脉冲串,每个脉冲对应一个水听器通道,对脉冲解调得到声信号.非平衡系统由DRA与GEC-Marconi合作开发,路径差为100—200m,必须使用高相干长度的激光源,该类系统中的相位噪声可用阵列中的参考水听器来降低.平衡系统方案最早由斯坦福大学提出,在系统输入端或输出端使用参考干涉仪达到路径平衡,可以显著降低对光源的苛刻要求.TDM系统的复用首先受到有限光功率所制约,其次受制于采样率.因为随着通道数的增加,由每个通道返回的光功率随之减小,使系统噪声背景恶化.对于典型的源级,从单个光纤上驱动大约30个水听器是可能的,在确保系统噪声级不超过深海零级海况的前提下.这样的系统覆盖的频率可达5kHz.目前,公开发表的文章所宣称的最大通道数是基于零差和PGC的方法(无光放大器),每个波长可以支持64个传感器,相位分辨率低于40μrad/$Hz^{1/2}$,串扰级低于-67dB.相当多的研究表明,时分复用可以达到相位分辨率、串扰和复用增益的最佳结合.通过在传感器阵中使用掺铒放大器(EDFA)解决耦合损耗问题,从而可以进一步增加复用传感器的数目.例如,在10个传感器构成的阵列中,基于放大TDM方法的相位分辨率约为6μrad/$Hz^{1/2}$,它的复用增益仍受限于传感器采样率,即采样率与传感器总数成反比.

图2给出两种可能的TDM架构.基于定向耦合器的内嵌Michelson可能是最简单的结构,在图2(a)中,每个传感器仅用单个耦合器.定向耦合器分离每个传感器,在输出端的镜子得到入射脉冲的反射光,镜子反射率一般大于80%.其他耦合器端口的匹配指数可以防止多个途径反射引起的串音.在输出端,脉冲串构成每个传感器的干涉信号.每个脉冲的数字采样或电子控制信号对传感器信号进行外差或相位载波解调.通过改变耦合系数,使每个传感器返回的激光功率相等.另一方法,称为渐近阶梯结构,如图2(b)所示.阶梯结构由横挡上的传感器构成.入射到阵中的单个脉冲在每个传感器上产生干涉脉冲,使用小路径平衡可以对传感器信号进行FM PGC提取.这种前向耦合设计可使10个阵元传感器阵列中每个传感器返回的光波功率相等,当$k_{in}=k_{out}=0.05$时,$k_i=0.2$,$i=1,2,\cdots,10$.精确匹配脉冲高度的代价是增加约3倍的定向耦合器数量.

2.2.3 时分复用/密集波分复用(TDM/DWDM)

提高复用增益另一个有前途的方法是时分复用和密集波分复用两种方法的联合,复用传感器总数是TDM传感器数目(典型值约30)与波长数目的乘积.研究表明,TDM/DWDM方法是最有价值的方法之一,它仅用商业化的元器件就可以满足所有必要的指标.如图3所示,波长从λ_1到λ_m的脉冲入射到传输光纤中,光纤加/减复用耦合器分配每个波长到

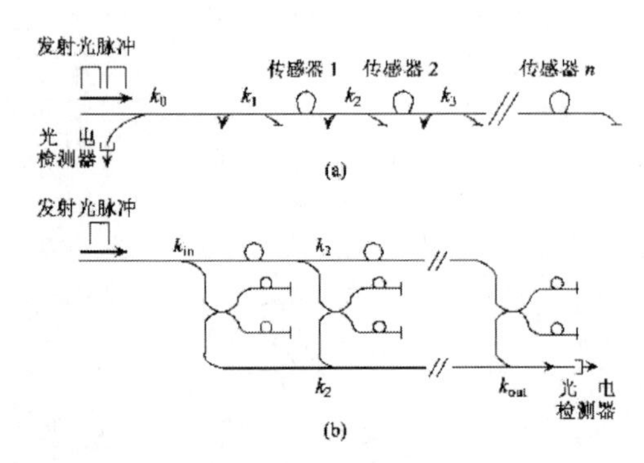

图2 基于迈克尔逊干涉仪的两种时分复用结构（a）内嵌式；（b）渐进阶梯式

每个 TDM 子阵中. 此法使用相互独立的发送和接收光纤，可以避免通道内串扰问题. 图中虚线框设备表示可供选择. 由两个子阵总共 96 个传感器的内嵌 Michelson 结构被成功实现，频率 500Hz 处的相位分辨率约为 $100\mu rad/Hz^{1/2}$，相应的声压分辨率约为 47.5dB（参考值为 $1\mu Pa/Hz^{1/2}$），并主要受环境噪声和激光源噪声限制.

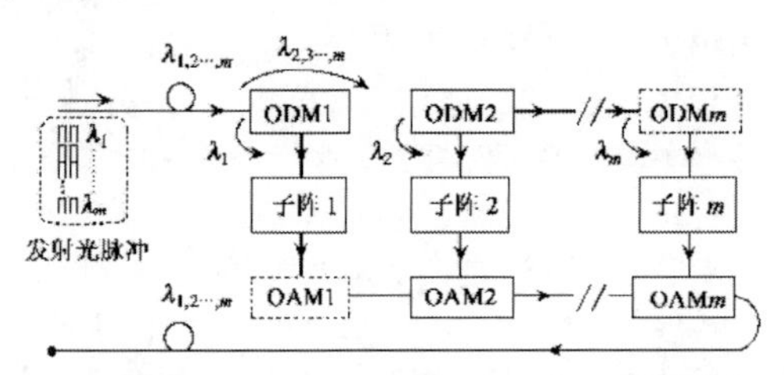

图3 时分/密集波分复用结构（ODM 为光减法复用器；OAM 为光加法复用器）

2.3 水听器设计

光纤水听器设计原则就是最大化声信号所产生的应变以达到更好的灵敏度（即增敏处理），同时确保其他的指标符合要求，如平坦的频响、良好的动态范围、良好的波束图和合理的静压力冗余. 满足相关特定应用噪声背景的水听器灵敏度（通常用 rad/Pa 表示）由单位长度灵敏度和传感光纤的长度所决定. 单位长度灵敏度由水听器结构所确定，可通过选择水听器材料和各种机械放大手段来实现灵敏度最大化. 一些设计方法使用极高顺性结构，使得只用很短的光纤就能达到预定的单位长度灵敏度. 但是，过高的灵敏度往往导致水听器不能承受足够的静水压力而压缩了动态范围. 另外一种方法是，在 DRA 和 NRL 所采用的方案中，使用相当长的光纤（100 多米），这能产生很高的水听器灵敏度：对于 108m 长的光纤，NRL 得到的响应为 6.2dB（参考值为 1rad/Pa）；而 DRA 达到的指标为 6.5dB（参考值为 1rad/Pa），但反射结构中所用的传感光纤长度为 200m. 这些足以使光纤水听器达到深海环境噪声背景或低于典型复用系统的噪声背景. 单位长度灵敏度的最大化首先可由优化密封参数来达到. 好多文章已经证明，最好的灵敏度可使用极低体积模量的密封得到. 灵敏度的进一步显著增加可通过在水听器结构中结合空气来达到，要么是空腔，要么是泡沫层. 这些空气可显著增加水听器的灵敏度，但存在水下耐压问题. 好的耐压能力和较高单位长度灵敏度的有机结合，可使用压力平衡式水听器来达到，代价是增加水听器的复杂度. 水听器物理尺寸的限制是由光轴的最小弯曲半径和整个光纤体积所决定的. 标准的单模光纤最小弯曲半径在 1300nm 时为 2.5cm，在 1550nm 时为3.5cm. 但是，现有的特殊光纤，如 Corning Payout 光纤，弯曲半径在 1300nm 时为 0.5cm，有可能得到直径很小的水听器. NRL 已经研制出直径为 1cm 的水听器，DRA 使用直径小于 2.5cm 的水听器. 水听器可制成不同形状，以适合于所有主要的声纳应用. 依据所采用的设计，平坦的频响可达 6kHz.

英国研制的光纤海底阵（fiber-optic bottom mounted array，FOBMA）所使用的光纤水听器如图 1 所示. 这是使用塑料芯轴的空气背腔芯轴设计，换能器直径 23mm，长度 160mm，水听器设计响应是 −128dB（参考值为 $1rad/\mu Pa$），即 0.4rad/Pa. 水听器由 80m 长的单模光纤构成，所用光纤镀层直径为 $80\mu m$，包层直径为 $165\mu m$，数值孔径（NA）为 0.21. 使用这一光纤是因为它比标准的通信光纤（NA 约 0.11）可以缠绕出更小的直径. 光纤用紫外线固化丙烯酸盐缠绕的，并在线圈上浇注聚氨酯防水薄层. 工作频带为20Hz—1kHz，可在很宽的水深范围内工作.

2.4 光器件

现代光纤水听器阵列是复杂系统，它不但含有光源、光放大器，还需要其他各种光器件，如光开关、频率转换器、相位转换器、耦合器、反射器和接收器，还有偏振控制或分离器件. 光纤水听器系统的许多光器件都得益于通信产业的巨大推动. 本节主要介绍在光纤水听器系统中扮演重要角色的光源和光放大器.

2.4.1 光源

在任何光纤水听器阵列中，最重要的单个元件可能就是光源. 所需的光源类型与复用方案和阵列尺寸有关，两个最主要的指标是相干长度和光功率.依赖于高度平衡干涉仪的光纤水听器阵列通常使用二极管激光，因为它们便宜、可靠.这些源也用在相干复用系统中，在这样的系统中，实际上要求相干长度小于水听器的2个光路长度之差.对类似于DRA这种大路径非平衡结构的系统，需要一个很长的相干长度.二极管抽运Nd:YAG激光最适合于这类系统.确实，高功率（可达300mW）和紧凑的尺寸使它们成为许多光纤传感器阵列最有用的光源.即使在全平衡系统中，现在的趋势是使用相干长度更长的光源（如二极管抽运和DFB激光），因为它减小了光路平衡性的过高要求.这使得生产过程相当简单.对于未来，光纤激光正在走向市场，它们可以提供最好的价格、功率和相干长度相结合的产品.

在实际的非平衡光纤传感器系统中，相位分辨率常依赖于激光源在低频时的频率和强度的稳定性.激光源的功率起伏或强度噪声直接转化为光电检测器的噪声电流，它恶化了相位分辨率.一般为确保频率噪声贡献小于1μrad/$Hz^{1/2}$，需要激光频率稳定度小于16Hz/$Hz^{1/2}$.掺铒分布反馈（DFB）光纤激光源（EFL）可达到稳健的单个纵向模态和单个偏振模态，输出功率几百微瓦，尺寸小，设计简单，生产期间可以精确设置波长，发射线宽窄，与传输介质有良好兼容性，发射频率有较低的温度灵敏度（约10pm/K）.DFB EFL的噪声特性研究表明，隔声隔振是降低激光频率噪声的最有效手段.

2.4.2 光放大器

水听器复用的一个原则性限制是光功率有限，这限制了光纤水听器可被远程驱动的距离.但光放大器可使水听器阵列突破这样的限制，它直接增加光纤或半导体介质中的光信号，无需先转换到电子域上.DRA和南安普敦大学合作研究在水听器系统中使用光放大器这一课题，研究表明，把光放大器放置在光纤水听器系统的不同位置上，至少都可以增加一倍的水听器数目.原则上对于40dB光放大器，1mW激光源可推动140个水听器.实际上，在一个光放大系统中限制复用的主要是采样率，而不是激光功率.

现在用得最多的是掺铒放大器（EDFA）.在光纤链路中嵌入EDFA需要提供电功率以支持抽运激光.但为了维持一个无电子设备的阵列，抽运光由远程激光提供，它位于船载或岸基的问询电路中，这样的放大器称为远程抽运掺铒放大器（RP EDFA）.远程抽运掺铒放大器可以显著增加光纤链路的传输距离，无需内嵌转发器和它们相关的电子设备.RP EDFA在阵列中的最佳安装位置常依赖于有效的抽运功率、光纤衰落和信号发送功率.RP EDFA作为前置放大器靠近接收端，一般被认为是更有效的，用较低的抽运功率就可达到较高的增益，从而有效地增加链路长度.

经过多年的努力，光纤水听器系统已经进入实用阶段.对于未来的军事应用，系统可能很庞大，有几千个水听器通道，阵长可达几百公里，对于这类系统重点是发展低成本器件，这样它们就能以高可靠性生存在敌方环境下.同时，相应的基础研究仍在进行，如光纤激光、光纤放大器、集成光学和Bragg光栅，这些很可能导致系统性能显著改善.一个迷人的前景是用Bragg光栅作为传感元件替代水听器中的光纤线圈，在直径为0.25mm标准光纤芯上制造声传感器是完全可能的，这些对声纳系统的未来发展有着非同寻常的意义.

3 矢量水听器

矢量水听器作为一种新型的水声测量设备，不但可以测量声场中最常见的标量物理量——声压，而且还可以直接、同步测量声场同一点处流体介质质点振速矢量在笛卡儿坐标系下的x,y,z轴向投影分量，一般多用三分量和二分量的形式.在结构上它由传统的无指向性的声压传感器和偶极子指向性的质点振速传感器复合而成.质点振速传感器是核心部件，其灵敏度的高低和工作的稳定性等制约矢量水听器的设计、制作、加工、装配、校准和使用等诸多环节.矢量水听器技术的主要应用领域可以覆盖水声警戒声纳、拖曳线列阵声纳、舷侧阵共形阵声纳、水雷声引信、鱼雷探测声纳、多基地声纳、水下潜器的导航定位、分布式传感器网络等.在空气声学中，矢量水听器可以用于战场警戒探测直升机和隐形飞机，噪声源识别和声强、声功率测量等.

测量声场质点振速的想法很早就有：Rayleigh于1882年就已经演示了测量声波均方质点振速的可能性，并以此确定声强，这种装置就是空气声学中常说的Rayleigh盘.此后，Olsen等人都试图测量声能流密度，但由于质点振速测量的复杂性，这些努力没有得到真正的回报.而现在水声工程中所采用的大多数矢量水听器工作原理、基本形式和主要的设计理念均基于Leslie等人的观点.

3.1 物理基础

由传统的声压水听器测量可以得到声场势能密度，这是最常用的声场能量形式，但是矢量水听器除此之外还可以得到声场动能密度和声能流密度（声场坡印廷矢量），这些概念对于正确理解矢量水听器测量结果至关重要.

对于水声学的正问题求解而言，基于速度势的简谐声场理论已经相当完善，原则上可以通过求解含边界条件的亥姆霍兹方程，只要存在速度势函数 $\boldsymbol{\Phi}$ 的解析形式，就可以由下式完整地确定声压 p 和质点振速 $\boldsymbol{v}$ 的解析形式：

$$p = \rho \frac{\partial \Phi}{\partial t}, \boldsymbol{v} = - \nabla \Phi$$

并由此得到如下的声场能量形式：

$$E_{\mathrm{p}} = \frac{p^2}{2\rho c^2}, E_{\mathrm{v}} = \frac{1}{2}\rho v^2, \boldsymbol{J} = p\boldsymbol{v}.$$

它们分别是声压势能密度 E_{p}，质点振速动能密度 E_{v}，瞬时声能流密度 $\boldsymbol{J}$，也称瞬时声强，这三者之间的关系由下面的声能守恒方程联系：

$$\frac{\partial E}{\partial t} + \nabla \cdot \boldsymbol{J} = 0,\ E = E_{\mathrm{p}} + E_{\mathrm{v}},$$

式中 E 为声波的机械能.

声场的绝大多数研究是集中在与声压有关的声波势能密度上，而在与质点振速有关的声波动能密度和声能流密度方面的相应研究甚少，讨论的也仅仅是一些简单的情况，如平面行波场、驻波场、球面行波场、简单波导声场等. 从上述声能守恒方程可以看出，声能流密度更适合于揭示声波能量“流动”的一般性规律. 为什么会出现这些现象？归结到一点，那就是缺少相应的质点振速测量设备，在矢量水听器出现之前，基于声压水听器的水声测量技术已经相当完善，这是造成许多水声学正问题求解以声压量和声压势能密度为研究对象的根本原因，因为实验测量是水声学研究的物理基础. 由此推想到声纳探测等问题，出现声压水听器占据统治优势也就不足为奇.

3.2 一般分类

质点振速传感器是矢量水听器的核心部件，因此，矢量水听器的分类主要是指质点振速传感器的分类，它原则上分为声压梯度式和惯性式两种类型. 惯性式是指将惯性传感器，如加速度计等对振动敏感的传感器安装在刚性的球体、圆柱体或椭球体等几何体中，当有声波作用时，刚性体会随流体介质质点同步振动，其内部的振动传感器拾取相应的声质点运动信息，因此亦称为同振式. 声压梯度式多是利用空间两点处声压的有限差分的原理来近似得到声压梯度，这可以通过反相串并联的线路连接在传感器内部实现，而声压梯度与介质质点的加速度之间的关系由 Euler 公式确定，通过计算间接得到介质质点振动信息. 惯性式矢量水听器是对简谐声场中介质质点振动真正意义上的直接测量. 由于这两类声矢量传感器的工作机理的差异，相应的性能参数也明显地不同. 一般情况下都习惯于将惯性式质点振速传感器统称为质点振速传感器. 根据换能器的换能原理，质点振速传感器可以分为：压电式、动圈式、电容式、光纤式、磁致伸缩式等. 目前从总体上看，压电式的质点振速传感器因其性能稳定可靠，仍占据着当前研究和应用的主导地位.

3.3 声压梯度传感器

声压梯度传感器通常有两种设计理念. 一种设计理念认为，最自然的声压梯度传感器是两个小间距分离的无指向性传感器，反相接线会使得信号相减，这一理念作为双麦克风技术常用于空气声强和阻抗测量，但要注意到，这些只对灵敏度和相位绝对匹配的传感器有效. 这种思想体现在自 20 世纪 70 年代起盛极一时的双麦克风声强探头，在水声中一般称为双水听器探头. 因为在继 Rayleigh 提出测量质点振速的想法之后，所进行的尝试由于质点振速测量的复杂性和当时技术条件的限制，使得研究人员不得不暂时打消直接测量质点振速的念头，继而转向采用这样间接的方法得到质点振速和声能流密度. 总体上看，这一时期的声压梯度水听器主要存在两个致命缺陷：一是灵敏度偏低，只能在信噪比较高的条件下使用，如声源的近场声强测量等；二是性能参数不稳定，严重依赖于材料、结构和制作工艺等. 很自然地，这大大限制了它的工程应用. 这类声压梯度水听器较成功的应用实例是航空无线电声纳 DIFAR 浮标 AN/BQQ－53. 另一种声压梯度传感器的设计理念是使隔开的弯曲传感器两侧（即双迭片）都受到声波作用，使得纯的电压输出对应于穿过弯曲元件的声压之差. 时至今日，还有一些研究人员在此方向继续尝试，随着材料的进步，并通过良好的设计和工艺，基本上可以保证声压梯度水听器可靠地工作. 尽管如此，随着惯性式声矢量传感器的研制成功，它基本上被排斥在当前矢量水听器研究的主流之外，因为现在商业化的微型加速度计具有更高的灵敏度和更稳定可靠的性能.

声压梯度传感器实际上是直接测量空间小尺度上的多点声压标量，然后通过线路的反相并联或串联来得到声压梯度的有限差分近似，这与直接测量质点振速和质点加速度的质点振速传感器机理有显著不同，因此，有人认为，所谓的"声压梯度"传感器不应该列入到真正的直接测量声场质点振速的质点振速传感器中，而作为它的过渡角色可能更合适.

3.4　同振式

在现代水声工程中使用频度较高的一类声矢量传感器是基于惯性传感器的同振式传感器，它的主要优点在于，本身不产生明显的声场畸变，即可以视为点接收器，因此它的指向性比固定式的要好，而且性能参数更稳定，可以用于精确测量或长时间测量，在不同的应用中，同振式声矢量传感器的平均密度为0.9—1.8g/cm^3.

有关惯性式矢量水听器的工作最早是由海军军械实验室(Naval Ordnance Laboratory)Leslie 等人进行的，他们推导了刚硬、均匀球体在理想水介质声场中运动的数学表达式，并证明这类中性浮力的球体在低频运动时具有与相同位置处水质点相同的振速，即

$$v_1 = v_0\frac{3\rho}{2\bar{\rho}+\rho},$$

其中 v_1 和 v_0 分别为同振球的振动速度与质点振速，ρ 表示水介质的密度，$\bar{\rho}$ 表示水听器的平均密度. 当 $\bar{\rho}=\rho$ 时，$v_1=v_0$. 而且他们还提出，在这样的球体中安装拾取振速的传感器，以构成对质点振速敏感的水听器，但没有考虑球体密度、流体密度和粘度以及柔性悬挂系统等对质点振速传感器的灵敏度指向性和工作频带等性能的影响. 尽管如此，这些开创性的工作已经清楚地表明，此类结构的水听器易于制作和校准，性能稳定，具有良好的指向性. 更重要的是，他们还给出了质点振速传感器设计的基本原则，即中性浮力且质量中心与几何中心重合，以及一些重要的设计思想和实践. 时至今日，这些对矢量水听器研制的关键环节仍然有着借鉴和指导意义. 除了球体之外，声矢量传感器还有圆柱体、椭球体、圆盘等多种形式，当 $\bar{\rho}=\rho$ 时，不同形状引起的性能差异可以忽略.

现在商业加速度计在 10Hz—10kHz 的频带上有平坦的响应，声波对矢量水听器悬挂系统的影响可能会成为测量声质点振速的水声惯性传感器设计的某些约束. 因为实际使用时不得不将惯性传感器悬挂在某些主平台上，以确保矢量水听器悬挂系统和平台不污染测量，这对于矢量水听器精确测量非常关键，此外还要考虑流体和壳体的密度以及流体的密度和粘度所引起的效应. 同振式矢量水听器对流噪声更敏感，在使用时尤其要注意. 实验表明，当流足够快以至于在传感器表面形成湍流时，质点振速传感器受到的影响甚于声压传感器，但是，如果外裹橡胶层，则可以显著降低流噪声.

3.4.1　质点振速型

声波作用下质点振动特性之一是质点振速，记录它的最自然的方法是在测量点放置一个小波长尺寸且平均密度等于水介质密度的物体，此时，物体将像质点一样进行振动，通过测量物体的振动速度，可以记录场的信息. 在水声中多采用动圈式结构的质点振速传感器，因为它有良好的低声频和次声频性能.

第一个质点振速水听器由 Kendall 于 1941 年研制成功，并用于音频范围换能器校准，被命名为 SV－1，工作频带 70Hz—7kHz，直径 6.35mm，灵敏度和阻抗级均比后来研制的 SV－2 要低一些; 于 1955 研制成功的低频质点振速水听器 SV－2，其直径 12.7mm，工作频带 15Hz—700Hz，插入 500Ω 负载时的灵敏度为－218.4dB(参考值为 1V/μPa)，它主要用于研究湖底的声学特征. 这两个质点振速传感器都采用动圈型换能原理，因其灵敏度偏低，所以更适合于实验室校准和大信噪比下的水声测量工作. 振速传感器与加速度传感器相比最大优势在于能够提供较高的甚低频(低于 100Hz)灵敏度，现在已经成功地用于反潜战.

3.4.2　振动加速度型

实际上，声质点振速测量除了可以使用质点振速传感器之外，还可以使用其他加速度和位移型的惯性传感器，因为在简谐波场中，它们之间的微积分关系可以转化为更清晰直观的线性关系. 典型的惯性传感器是将压电加速度计埋嵌在小的刚体中，当刚体运动时，得到记录的输出电压. 这种设计理念很大程度上依赖于无约束硬球在非粘无界流体介质中对声波的响应.

当前重点是研制可以应用到拖曳线列阵声纳中的微型矢量水听器，利用高灵敏度和低噪声的微型压电和光纤加速度计都是非常有可能做到的. 这不但可以从根本上解决左右舷目标分辨的问题，而且会提高拖线阵的空间增益，有利于探测远程微弱目标. 当然，也可能会伴随着出现一些新问题，如矢量

水听器对湍流脉动压力和振动加速度的响应，矢量水听器在拖缆中的姿态等等.

参 考 文 献

[1] 李启虎. 数字式声纳设计原理. 合肥：安徽教育出版社，2002[Li Q H. Design Principle of Digital Sonar. Hefei：Anhui Education Press，2002（in Chinese）]

[2] Nash P. IEE Proc. Radar，Sonar Navig.，1996，143(3)：204

[3] Crach G A *et al*. J. Acoust. Soc. Am.，2004，115(6)：2848

[4] D'Spain G L *et al*. J. Acoust. Soc. Am.，1991，89(3)：1134

[5] Shchurov V A *et al*. J. Acoust. Soc. Am.，1991，90(2)：991

[6] 孙贵青，李启虎. 声学学报，2004，29(6)：491[Sun G Q，Li Q H. Acta Acoustica，2004，29(6)：491（in Chinese）]

[7] Bucaro J A *et al*. J. Acoust. Soc. Am.，1976，60(5)：1079

[8] Leslie C B *et al*. J. Acoust. Soc. Am.，1956，28(4)：711

[9] 孙贵青，李启虎. 声学学报，2004，29(6)：481[Sun G Q，Li Q H. Acta Acoustica，2004，29(6)：481（in Chinese）]

第六讲　水下声学传感器网络的发展和应用①

李淑秋　　李启虎　　张春华
（中国科学院声学研究所　北京　100080）

摘　要　　文章主要介绍了正在发展中的水下声学传感器网络的一些概念和发展现状. 在回顾了水声通信和水下网络的发展历史之后，首先从网络拓扑结构出发，指出多跳对等网络是适合于水声应用的一种网络结构，接着根据网络协议栈的概念，说明水下声学传感器网络在各协议层上所涉及的问题，并从物理层面，分析了水声介质的特殊性，和水下声学传感器网络所面临的一些主要的技术挑战. 介绍了组成水下传感器网络的关键部件，以美国海网（Seaweb）为例，给出了一个水声传感器网实现的实例，对它的年度实验进行了介绍，最后，给出了中国未来的发展应用前景.
关键词　　水下，声学，通信，传感器，网络

Development and applications of underwater acoustic sensor network

LI Shu-Qiu　　LI Qi-Hu　　ZHANG Chun-Hua
（*Institute of Acoustics, Chinese Academy of Sciences, beijing*, 100080）

Abstract　　The concepts and applications of developing Underwater Acoustic Sensor Network（UASN）were introduced. After a review of underwater acoustic communication and network in the past, firstly in view of network topology indicated that multiple hop peer - to - peer network is suitable for underwater network application. Secondly according to network protocol stack conception, illuminated the corresponding problems on each protocol layer in the UASN. Thirdly made a physical analysis for the special characteristics in the underwater channel and narrated the main technique challenge in the UASN. Next introduced the key components that the UASN comprise. Then provided an illustration of UASN concept by used " seaweb "as an example. Finally gave the perspective on development and application of the UASN in the future in China.
Keywords　　underwater, acoustics, communication, sensor, network

1　水下声学传感器网络的基本概念和历史

水下声学传感器网络，是全球网络化技术普及的产物. 既然陆上通过有线光或电的手段实现了 Internet 连接，空中通过无线网络甚至通信卫星实现了网络连接，水下网络也许是唯一所剩的未经全面开垦的处女地了. 可以想象，有一天，你打开电脑，连上 Internet，可以立刻获得大西洋深处洋流的实时数据，如果安装了水下摄像机的话，甚至可以看到大堡礁的斑斓的鱼儿在你的屏幕上游曳. 这就是水下声学传感器网络面临的任务：以水下声学网络作为信息传输的手段，以水下传感器作为信息获取的窗口，并最终以某种方式把水声网纳入常规网络，把水下数据送往观察者.

由于声波是唯一一种能在水介质中进行长距离传输的能量形式，而无线电波在水中的传播距离非常短，光也因为在水中受到高衰减和散射的影响，不适用于水下环境. 水下声学网络，就是以水下声波作为信息载体而组成的无线网络. 类比于空中的无线

① 物理, 2006, 35(11): 945-952.

网络,只不过空中的信息载体是无线电波,水中的信息载体就是声波.水下声学网络要解决两个技术问题,一是水下声通信,二是在声通信基础上的组网.水声通信解决的是点到点的两个用户(或信息源)之间的通信,组网解决的是多个用户(或信息源)共享水介质信道时的信息交互问题.

作为一个正在发展的新兴技术,水下声学网络发展之所以远远滞后于空中无线网络,很大程度上是受限于水声通信技术的发展.最早的水声通信可以追溯到20世纪50年代针对模拟数据的幅度调制(AM)和单边带(SSB)水下电话[7];70年代之前有少数的模拟系统,由于水声混响环境中幅度调制的困难,以及随着VLSI技术的发展,80年代早期水下数字频移键控(FSK)技术得到应用[8],它对信道的时间、频率扩散有一定的鲁棒性.80年代后期出现水声相干通信,与非相干通信相比,相干水声通信技术可以提高有限带宽水声信道的带宽效率,但是由于水声信道的严酷和复杂性,水声相干通信开始并不被接受,当时的水声通信的距离和速率乘积是大约0.5km·kbit[10].90年代由于DSP芯片技术和数字通信理论的发展,使许多复杂信道均衡技术可以实现,带动了水声相干通信技术的发展[9],并转向对水平信道通信的研究,因为在浅海环境中水平信道的多途效应要比深海的垂直信道复杂得多.90年代中期浅海环境的水声通信的速率和距离乘积达到40km×kbit[10],使水下网络的建立成为可能[9].水下网络一个里程碑式的关键部件,是水下声学调制解调器的出现[11].最早提出的水下声学网络应用概念的是1993年的自主海洋采样网(AOSN)[12].美国自1998年开始了称为海网(Seaweb)的年度实验[13],意在验证水下声学网络的概念.90年代中期至今,水声通信技术和水下网络技术在同时稳步发展着,但由于水介质的特殊性和复杂性(如高时延、大衰减、多途和频移),使用于陆地的无线网络的技术,并不能直接应用于水下网络,对水下信道、水下通信、水下网络协议的研究方兴未艾[13-26].

与此同时,20世纪90年代到现在,基于短距离的无线通信的陆地无线传感器网络的发展也非常迅速[5].可以说,水下传感器网络是陆地传感器网络概念向水下应用的延伸.水下声学传感器网络,由多个传感器节点组成,节点可以是固定的,如水下锚定的浮标或潜标,也可以移动的,如水下机器人(UUV或AUV).目前,水下声学传感器网,可以根据水下传感器类型的不同,获取不同的信息:可应用于海洋学数据获取、海洋污染监控,近岸开发、灾难预防、水下导航定位的辅助、海洋资源勘测和科研数据获取、分布式战术监测、水雷侦察、以及水下目标的探测、跟踪与定位.

简言之,水下声学传感器网,就是在一定的水下区域内,通过各种传感器节点获取水下信息,并对水下节点进行声学通信和组网,最终通过特定的节点,重新以无线电和有线的形式把在覆盖区域中所获取的信息纳入岸上的常规网络,并发送给观察者的水下子网.

可以看到水下声学传感器网的几个特点:第一是可移动性,由于是可移动的,所以必须是能够自组织的自主网络,遵循一定的网络路由方式;第二是水下无线和水声通信,由于采用水下声通信,必须是对海洋环境特性自适应的,解决物理层的技术挑战;第三,是能量限制的,因为无线,所以是电池供电;第四,是具有数据转播功能,可把监测数据传达到岸上,为了对数据进行有效和可靠的传输,必须遵循一定的网络协议.

网络拓扑结构决定了网络的路由方式、能量损耗、网络容量和可靠性,所以要首先介绍网络的拓扑结构.

2 水下声学传感器网络的拓扑结构

和陆地上的无线传感器网络结构一样,水声传感器网的拓扑结构可分为两大类:中心化的网络(centralized network),和分布式的对等网络(distributed peer-to-peer network).

在中心化网络中,节点之间的通信是经过中心节点实现的,并且网络是通过这个中心节点接入骨干网.这种配置的主要缺点就是存在单一故障点,即这个节点的失效,将导致整个网络的失效.并且由于单个调制解调器的作用距离有限,所以中心化网络的覆盖范围有限.图1为中心化网络的拓扑结构示意图.

对等网络,是指没有一个中心节点"管辖"它们,每个节点具有较为平等的权限.根据路由方式的不同,对等网络中又有一些差别.完全连接的对等网络对网络中的两个任意节点提供直接"点到点"连接,这样的拓扑结构,对路由的需求减少了,然而当节点分散在很大的区域时,对通信所需的功率却大大地增加了.并且还会产生"远近"问题[1,2],即当一个节点A正在往远端的某节点发送数据包时,会阻塞与节点A相邻的节点接收其他信号.

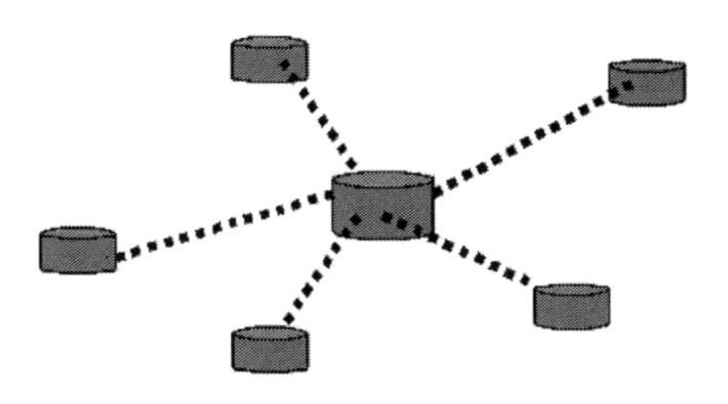

图1　中心化网络的拓扑结构示意图（虚线表示声通信）

多跳对等网络则仅在相邻节点之间进行通信，一个信息从源到达目的经过节点间的多次跳接来完成. 多跳系统可以覆盖较大的区域，因为网络的作用距离，取决于节点的数量，而不再受限于单个调制解调器的作用距离. 图2为多跳对等网络拓扑结构示意图.

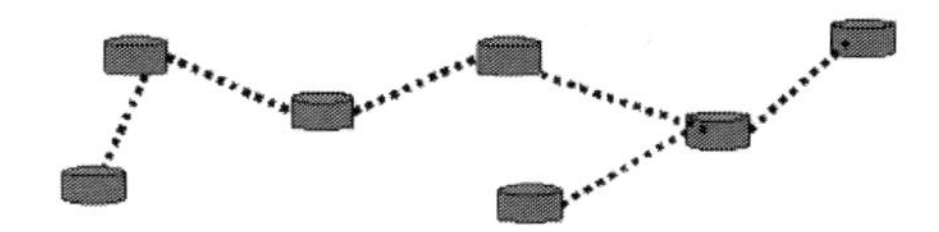

图2　多跳对等网络拓扑结构示意图

Ad Hoc网络，是针对无线移动应用的一种网络，它属于多跳对等网络. 它不需要先期建好基础设施，也称为无架构网络（infrastructureless networks）. 它的特点是：自主网络、动态拓扑、带宽限制和变化的链路容量、多跳通信、分布式控制、能量有限的节点和有限的安全性. 因为不依赖于基础设施，所以可以快速布放，并且覆盖较大的区域. 因为水中可依赖的基础设施有限，并且可移动的AUV将是水声传感器网络中的重要组成部分（AUV可以增强水下传感器网络的性能），它所具有的自组织能力和动态拓扑结构，使Ad Hoc网络很适合应用于水下声学传感器网中. Ad Hoc网络虽然适合于水声网的应用，但它的安全性问题也始终是一个研究课题.

事实上，水下声学传感器网络应该是一种中心化网络和对等网络的混合. 在文献[16]中介绍了一种二维和三维的水声传感器网，二维是指获得的信息维数. 在二维水声传感器网中，传感器节点以及数据转发器（Sink）被布放在海底，在小的区域内是以Sink为中心（中心化网络），而每一个传感器的数据可以在水平链路上以直接或多跳的方式到达Sink（多跳对等网络），而传感器数据只有通过Sink在垂直链路上转发，才能到达水面站. 因为只能获得海底某一区域的信息，称为二维传感器网. 在三维水声传感器网中，可以控制潜标的深度，使某一区域内的多传感器节点位于不同的深度，因此可以获得一定区域的、不同深度的海洋信息，所以称为三维水声传感器网. 在网络拓扑结构上，也是多跳对等网络. AUV可以到达海洋中的不同深度，与固定水底的传感器网络结合，也可以构成三维水声传感器网络.

值得指出的是，因为水下声学传感器网络，总有一个接入水上其他常规网络的问题，有一个称为水面站、网关或主节点的特殊节点，来完成这一工作. 它不但要有声学调制解调器，用于与水下网络的通信，而且要有无线电或有线的调制解调器与卫星或岸基的网络进行通信. 水面站可以以浮标为载体，也可以以水面舰船为载体.

网络拓扑结构决定了网络的路由方式、能量损耗、网络容量和可靠性. 有研究显示[15]，一个延直线等间距分布的多传感器节点构成的网络，按完全连接的对等网络的路由方式功耗大于多跳对等网络；而网络容量也受网络拓扑结构的影响.

3　水下声学传感器网络层的相关概念

水下声学传感器网络，的确是一个崭新的领域，但它所遵从的概念是与常用的网络协议栈的概念是相同的. 表1是常用的网络层概念. 为简化起见，本文只讨论基础的三层：物理层、数据链路层和网络层[3].

表1　网络协议栈

网络协议栈
应用层（application layer）
传输层（transport layer）
网络层（network layer）
数据链路层（data link layer）
物理层（physical layer）

物理层要解决的问题，就是怎样利用传输介质的特性（即信道特性）和相应的调制方法，使数据进行有效的传输. 基于水介质的声通信，是网络协议层中典型的物理层问题. 在发射端要把信息比特变成信道能够传输的信号（声信号），在接收端又要把介质中的信号变回信息比特，这就是水下声学调制解调器的任务，主要涉及三方面的问题：媒体转换（如：电－声信号变换）、频带利用效率、信道适应性.

水声通信中常用的调制方法分为两类：一类是非相干调制，如频移键控（FSK）方法；另一类是相干调制方法，如相移键控（PSK）和正交幅度调制（QAM）. 非相干调制对严酷的水声环境有较好的鲁棒性，但速率低；相干调制方式中的编码效率高、频带利用率高，但传输距离有限. 有些技术既是物理层的编码手段,也是复用手段,如:CDMA扩频技术.目前又出现了广泛引起兴趣的多载频调制技术,如正交频分复用(OFDM)扩频技术.

数据链路层,解决的是多个用户怎样合理有效地利用信道,即媒体访问控制MAC.主要涉及媒体访问方式和纠错控制两方面的因素.媒体访问方式可细分为多址访问方式和随机访问方式[3].多址方式就是多个节点共享有限的多条通信信道,利用空闲信道进行通信的方式,也称为复用方式,复用方式决定了可同时使用信道的节点数.随机访问方式,是多节点使用一条通信信道,对媒体使用的方式和次序进行控制.纠错控制,就是通过一定的机制保证发射和接收数据的正确性.在水声网络中,常见的复用方式有:频分复用(FDMA)、时分复用(TDMA)、还有对水声信道有较好鲁棒性的码分复用(CDMA)技术,它具体又分为直接序列扩频(DSSS)和跳频扩频(FHSS)技术.这些复用技术,将针对不同的水下应用,有不同的优缺点.水下网络中常见随机访问方式有:ALOHA、载波侦听复用(CSMA)、避免冲突复用(MACA)等.在数据链路层的常见的错误控制机制有:检错重发(ARQ),即接收机根据接收数据的情况,请求重发;和前向纠错(FEC)技术,即事先在发射时就增加一些冗余代码,接收机根据冗余代码就能进行纠错,不需要重发.

网络层,解决的是路由问题,即怎样确定源节点(通常是提供信息的水下传感器节点)和目的节点(通常是与水面有关的主节点)之间的路径.而路由方式,取决于网络的拓扑结构.

Ad Hoc网被认为比较适合于水声传感器网的应用.在Ad Hoc网中,有三种路由协议,先应式(proactive)、反应式(reactive)和地理式(geographical).先应式,也被认为是表驱动式(table driven),这些协议是通过广播包含了路由表的控制包,来维持每一个节点到每一个其他节点的最新路由信息.反应式,也被认为是随选式(On-demand),就是仅在达到目的的路由被需要时,节点才启动一个路由发现的过程.路由发现之后,需要路由维持,直到它不再被需要.地理式,是通过利用定位信息,来建立源和目的的路由.所有这些路由协议,由于水声信道的特殊性,都面临一些应用上的困难.

4 水下声学传感器网络的物理分析和技术挑战

水下声学传感器网的传播介质是水,与陆地传感器网的介质空气,有很大的不同,因此造成在陆地上可以有效使用的网络协议,不能适用于水下声学网络.我们将从水的声学传播特性入手,讨论影响声学通信的因素,分析它对网络协议栈各层协议所造成的困难.

4.1 影响水声通信的物理因素

4.1.1 传播延迟长和延迟方差大

电磁波在空气中的传播的速度是声波在水中的传播速度的20万倍,慢的声速使得传播延迟很大,每公里约延迟0.67s,同时水声信道的时变特性又使延迟方差很大.前者影响网络的吞吐率,后者使一些基于时间的协议无法工作.

4.1.2 传播损失大(也称为路径损耗)

根据Urick[4]的传播模型,传播损失,是由扩展和衰减所引起的损失之和.衰减损失包括吸收、散射和声能泄露出声道的效应.吸收是由于声能转换为热能而引起,它随频率和距离的增加而增加.扩展损失是指波前扩张引起的声能的扩展,主要有深海环境下点源的球面扩展(全向扩展),传播损失随距离的平方而增加;以及在浅水环境下的柱面扩展,仅在水平面上进行扩展,传播损失随距离而增加.由于声信号的传播损失随着频率和距离的增加而增加,因此水声信道可用的频段非常有限,传播的距离也有限.因此,在水下通信网中,要进行长距离通信,只能选择低码速率;要选择高码速率,只能进行短距离通信.一般来讲,要使传播距离达到10—100km,可用带宽在2—5kHz范围;中距离传输1—10km,带宽在10kHz量级;如果所用频带大于100kHz,传播距离就必须小于100m[6].

4.1.3 多途严重

多途现象是由于声源和接收器之间存在不止一条传播路径而引起,在浅海和远程传播中经常出现.简单的说,单一声源发出的信号,由于多途的存在,在接收端可以收到多个不同时间到达的信号.多途会引起信号振幅和相位的起伏,由于不同路径的传播时间不同,会导致严重的信号畸变,会导致不同接收机之间接收信号的去相关,多途还会造成频带展宽.这些都会使通信信号严重退化,产生码间干扰.多途还与声源与接收机之间的位置和距离有关.以海底平面为参考,垂直信道的多途影响小,水平信道的多途影响大.

4.1.4 浅海环境噪声强

环境噪声是多种因素的集合,它与潮汐、湍流、海面的风浪和雷雨都有关系,船舶噪声也是重要的噪声源.与深海的噪声比较确定的情形不同,浅海,特别是近海、海湾和港口,环境噪声会随时间和地点的不同而显著的变化.噪声主要由船舶和工业噪声、风成噪声和生物噪声组成.环境噪声会使信号的信噪比降低,影响水声通信的性能.

4.1.5 多普勒频散严重

多普勒频移,是由声源和接收器的相对运动而引起.由于声速比电磁波的速度慢20万倍,很小速度就能引起多普勒频移,并且由于信道的原因水下声载波频率较低,这两个因素加起来使水中多普勒的影响比空中无线通信大得多.多普勒如果只产生一个简单的频率变换,接收器的补偿是相对容易的.但由于多途的存在,当声信号一次或多次碰到海面时,在各个路径之间会产生不同的多普勒频移,它很难补偿.当高速数据通信时,它会产生码间干扰,还会降低频带效率.

总之,浅海水声信道在时间、空间、频率上都有扩散,是高度时变的系统,并且比空气中的无线电信道复杂得多.对水声通信来讲,它是一个严峻的挑战.

4.2 浅海声学环境对网络各层的技术挑战

下面我们将根据网络协议栈各层的功能,分析水下声学传感器网络所面临的问题.

4.2.1 物理层

频移键控(FSK)是非相干的调制方法,因为多普勒频散的缘故,相位跟踪难于做到,所以依赖能量检测.考虑多途的影响,可在两个相邻的脉冲之间插入一段保护时间(time guard);考虑水下多普勒频散的存在,也可以在子带之间加入保护频段.尽管非相干调制的方法是高能效的,但它带宽利用效率低,加之水声信道的带宽有限,所以它不适合水下高速的、多用户的通信.另一种可以提高数据率的方法是相干调制技术,如:相移键控(PSK)和正交幅度调制(QAM),它必须对载频的相位进行跟踪,但是由于浅水介质的频散和时变特性,使得码间干扰加大,错误比特增加,因此需要采用信道均衡技术并结合锁相环技术补偿相位的偏移.实现信道均衡的是一些非常复杂的滤波器,如:决策反馈均衡器,它的复杂性不适合实时通信.但一些简化算法,引起算法的不收敛,最终引起接收机性能的下降.

目前在水声通信中正交频分复用(OFDM)扩频技术引起广泛关注,它也称为多载波调制,因为它同时在多个子载波上发射信号.可以根据信道的情况,对衰减小的子载波分配较高的比特数,对衰减高的子载波分配较低的比特数.在保持同样的数据速率下,因为可以在多个子载波上同时发射信号,就使每一个子载波的符号持续时间可以加长,所以OFDM对多途有较好的鲁棒性,并且能够取得较高的频谱效率.但是由于信道的时变性,对子载波比特数的分配,也是一个繁重的任务.

4.2.2 数据链路层

就多址访问方式而言,频分复用(FDMA)技术由于水声可用的信道很窄,并且对衰减和多途的敏感性,不适合水下声学网络,但对于短距离的通信,倒是一种简便易行的方法.时分复用(TDMA)由于水声信道延迟长和延迟方差大,时间同步很难做到,因此基于公共时基的时隙分配就难以在水下声学网中做到.如果使用,就必须插入较长的时间保护(time guard),使系统的带宽效率大大降低.码分复用(CDMA)通过伪随机编码信号把发射信号扩频到整个可用带宽.因为伪随机编码的相关特性,可以使采用不同伪随机编码的不同用户区别开来.高带宽对频率选择性衰落有较好的抵抗性.在直接序列扩频(DSSS)的接收端采用Rake接收机,主动利用多途能量,补偿多途的影响,因此它对水声信道产生的多途衰落有较好的鲁棒性,但由于多普勒频散的存在,也会使伪随机码的相关特性降低;另外一种扩频技术跳频扩频(FHSS),由于它发射的是多个窄带信号,更容易受到多普勒频移的影响.但它在对付复用干扰(MAI)上比DSSS更有效,并且它的接收机比较简单.

就随机访问方式而言,避免冲突复用(MACA)是一种握手协议.它通过发射和接收RTS/CTS控制包对,实现发射和接收的握手.由于水声信道大的延迟特性,而握手协议的几个来回,更加重了信号延迟之外的延迟,造成网络吞吐率降低.载波侦听复用(CSMA)技术,也会由于水声信道的延迟特性,会发生信道侦听为空闲而发射却正在进行的情况,因为发射信号此时尚未到达接收器,造成冲突的发生.

4.2.3 网络层

水下AD Hoc网中,先应式(proactive)路由方式,在使用网络通信之前,就要每时每刻维持网络中所有节点的路由表,这不仅造成珍贵的信道资源的浪费,对水声应用也不是很必要.地理式(geographical)路由方式,依赖于节点的地理位置进行路由,在陆地传感器网中,可以很好的利用全球定位系统(GPS)来获得所需信息,在水下声学传感器网络中,获取精确的节点定位信息是很困难的事，而且节点的定位，本身就是水下声学网络要解决的问题之一. 反应式(reactive)路由方式，虽然在需要时才启动路由发现的过程，仍然需要大量的控制包来建立路由，并且由于水下底质和信道的变化，使得有时链路变成一个单向的链接，这对所有依赖于对称链接的协议，造成使用上的困难.

综上所述，水下声学传感器网络面临严酷的水下环境，遭遇了陆地传感器网络所不曾遇到的困难，或者是更严重的困难. 水声工作者的任务就是根据使用条件的不同，针对这些方法的优缺点进行有益的借鉴和改进，并取得有利的成果.

5 水下声学传感器网络的关键部件

从目前已经进行了实验的水下声学传感器网的例子分析，一个基本的水下传感器网络主要由传感器节点、转发器和主节点部件组成.

传感器节点主要是通过不同类型的传感器用来获取水下生物的、化学的和物理现象的参数，并通过声学调制解调器(modem)把信息发送出去. 如图3所示，一个传感器节点主要由六部分组成.

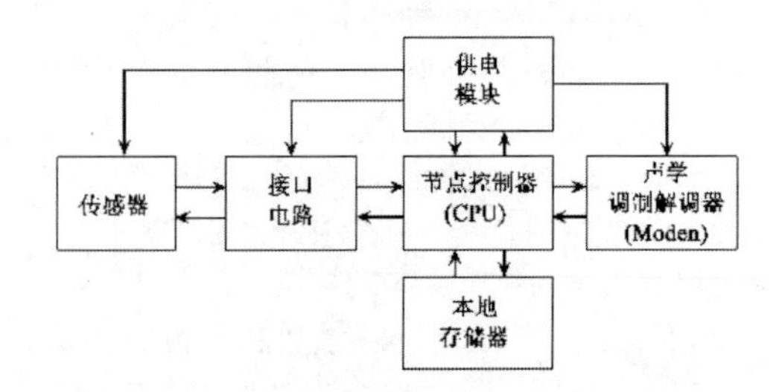

图3 传感器节点的基本组成

传感器(sensor)，可获取水下的生物、化学或物理现象的参数，并转换成电信号，通过接口电路，变成数字信号，送给节点控制器. 节点控制器，根据一定的网络协议，把获取的数据按协议包，通过声学调制解调器发送给其他节点. 除此之外，节点控制器还可以负责电源管理、本地数据的存取或一些本地的计算等功能. 电源模块多数为电池组，负责节点上所有模块的供电. 不同的应用决定了所使用的传感器类型不同，如：测量海水特性的温度、盐度、深度传感器CTD；测量海流的声学多普勒流速剖面仪(ADCP)；测量化学成分的传感器，如热液硫化物、硅酸盐的测量；测量金属离子的沉积层的电极传感器. 或者就是声学传感器，用于监测海洋的各种声学信息，民用可以获取海洋声学数据，如：海啸的预报，鱼类丰量分析；军用则可以用于敌方潜艇的警戒与探测. 传感器还可以是定位传感器. 传感器节点通常同时使用多种传感器.

转发器(repeater)，主要用于转发来自传感器节点或别的转发器的数据，它本身可能并不具备传感功能，而把接收的声信号转换成电信号经过整形或变换，再按照一定的路由协议，重新变换成声信号转发出去. 它的功能是延长传感器数据传播的深度或广度，增加传感器网络的作用范围. 其中完成收发功能的模块，还是声学调制解调器.

主节点和网关(master and gateway)，是指通过它才能把水下声学传感器网接入陆地或空中常规网络的关键节点. 主节点的功能是集中接收水下声学网络的数据包，网关(gateway)的功能就是把主节点收到的声学数据包，经过它的变换，变成无线(电)网络和有线(电)网络所接收的数据包，再转发出去. 对于下发给水下声学传感器网络的命令控制包，过程相反. 主节点和网关可以在一个载体上合而为一，也可以分开布放. 如果这个网关用于无线电网的接入，这个节点可能就是一个水面站，它可以专用浮标或水面船为载体. 如果这个网关用于有线网络的接入，那么这个节点就可能是一个岸基站. 无论如何网关都包含水下声学modem：对于无线接入，水下是声学modem，水面是无线电modem；对于有线接入，水下是声学modem，陆上是网络接入和服务器. 根据数据流量的需要，可以有多个主节点和网关.

由水下传感器网络的组成部件，可以看出水下声学modem是构成水下传感器网络的关键部件. 也有人把它称为“telesonar”，即具有远距离通信能力的声纳. 第一代的telesonar modem称为Datasonics ATM850[14]，产生于1995—1997年. 它的研制成功，可以看成是水下声学传感器网的里程碑. 后来在美国海军小型商务创新研究(SBIR)资金的支持下，1997—1999年产生了第二代的telesonar modem称为Datasonics ATM875. 在美海军的继续支持下，于2000年完成了第三代的开发，称为the Benthos ATM885[27].

6 当前水下传感器网的实现——Seaweb实验介绍

海网(Seaweb)水下声学网络[13,14]是目前美国比较成功的水下网络的一个概念. 依据海网的概念，对于军事应用,构建可布放的自主分布系统 DADS(deployable autonomous distributed system),用于沿海广大区域的警戒、反潜战和反水雷系统,实施命令、控制、通信和导航功能. 对民用,构建对沿海大陆架的监测的 FRONT 网(the front – resolving observational network with telemetry)和气象海洋系统. 它们的共同点是,都以 Telesonar 为水下通信的工具,以 Seaweb 的概念构成水下网络;不同点是根据应用的需求不同,采用了不同类型的水下传感器.

就 Seaweb 本身的概念而言,就是建立水下网络,这个概念本身的可行性也需要验证,因此,美国海军的海网计划,从 1998 年开始启动了一个每年度的 Seaweb 实验,目的在于推动 Telesonar 和 Seaweb 技术的发展. 从 1999 年起启动了 FRONT 的年度实验.

在 Seaweb 98 中,采用 MFSK 调制技术,10 个节点分为三个簇,在本簇中采用的复用技术是 TDMA,在簇之间采用的是 FDMA,信道利用率较低,纯的传输速率只有 50bit/s. 采用的 telesonar 是 ATM875,其声学带宽是 5kHz. 一个网关浮标,称为 racom(无线/声学通信)浮标,首先接收来自水下网络主节点的传感器数据,然后通过无线电链接到岸上的命令中心.

Seaweb 98 强调了水下声学网络与常规网络的差别,在有限的功率、有限的带宽和长的传播时延等严酷条件下,证明 Seaweb 是可行的. 数据压缩、前向纠错(FEC)和滤波技术必须采用,以降低误码率;网络各层要认真设计,自适应调制和功率控制被认为是提高信道容量和效率的关键因素. 在 Seaweb 99 的实验条件下,水深 10m,实验结果节点间的有效连接可达 4km,网关到节点的有效距离是 7km.

Seaweb 98 的意义在于它验证了水下声学网络的概念,也推动了对 ATM875 的大量改进. 同时它也验证了军用 DADS 的概念的可行性:网络化的传感器;广泛的区域覆盖能力;racom 网关;对浅海环境的鲁棒性;远程控制能力等. 美军对 DADS 系统的设想是,遵从 ad hoc 拓扑结构和 Seaweb 的概念,能够自组织,包括节点识别、秒级的时钟同步、在百米量级的节点几何定位和新节点的同化和节点失效后的自愈能力,所希望的节点持续能力可达 90 天.

Seaweb 99,15 个节点,仍然采用 ATM875,复用采用 FDMA 的一个变形. Seaweb 99 的一个重要发展是引入了 Seaweb 服务器,服务器管理 Seaweb 的网关和成员节点,监视、显示和记录网络状态,服务器还可以建立网关与网关的路由. Telesonar 允许服务器远程地配置路由拓扑结构,这为未来的网络自配置和网络动态控制建立了雏形. 同时,Seaweb 99 也暴露了 ATM875 的局限性:所采用的 DSP 有限的存储器和处理速度,限制了固件的改进;FDMA 的复用方式牺牲了珍贵的带宽.

Seaweb 2000,共有 17 个节点,采用全新的 Telesonar ATM885,具有数据记录功能,复用方式采用混合的 CDMA/TDMA 技术. 核心特色是,实现了一个紧凑的、结构化的网络协议. 通过信道宽容的 64 – bit 功能包和信道自适应的任意长度数据包,有效地把网络层和 MAC 层映射到物理层,共有 7 种功能包类型. 另一个特色是把协议和信道测量结合起来. 复用协议是避免冲突复用(MACA). 过程是 A 节点发射一个 RTS(request – to – send)信号,同时兼做信道测试信号;节点 B 检测到请求,并唤醒,处理 RTS 信号,从中估计信道,B 节点根据信道估计的结果,响应一个 CTS(clear-ot-send)信号,给出合适的调制参数;A 节点根据参数,以接近最优的比特率、调制方法、编码、和声源级来发送数据包.

Seaweb 的成功带动了美国的多种 Seaweb 应用计划:海洋研究(FRONT 项目)、海洋警戒(DADS 演示项目)、舰队作战实验(FBE – I)、浅海反潜战 ASW(Hydra 项目)、水下通信(Sublink 项目)以及 UUV 的命令与控制(SLOCOM 和 EMATT 项目). Seaweb 还可以用于导航和反水雷系统[28].

7 我国的现状和发展前景

我国在"八・五"期间就开始进行水声通信的研究,最早的研究单位有厦门大学、哈尔滨工程大学和中国科学院声学研究所,主要有低码速率的远程通信和高码速率的近程通信两个方向. 低码速率的水声通信采用扩频通信技术,高码速率多采用相干通信技术,两者都取得了很好的成果. 在"十・五"期间,开始进行水声通信网的相关研究,已经有 3—4 个节点的实验系统的出现. 预计在"十一・五"期间可以有一个长足的进步.

我国广阔的海岸线,为水下声学传感器网的应用带来广阔的前景. 不论何种应用,水下传感器网络的概念是相同的,所不同的是根据应用的不同更换水下传感器的种类,便可获得不同的水下信息. 民用可以建立近海地震监测网,为地震监测和海啸预报获取数据;在海上石油平台附近建立小范围的监测网,为石油平台的生产提供环境和安全参数;还可以利用水下声学传感器网络的自组织能力,对水下移动平台(如UUV)提供导航参数;建立沿海立体监测网络,获取物理的、化学的、气象的和声学的信息.民用可为海洋资源的利用、海洋灾害的预防、海洋气象的准确预报、和海洋科学数据的获取创造条件;军用可以建立沿海广大区域的警戒侦察网,并以它的灵活性,对快速事件作出反应,提高我国海防水平.

Seaweb的成功带给我们很多启示,但仍有一些开放的研究课题需要研究.针对水声通信与组网的研究,如:水声信道的测量和自适应均衡算法研究、鲁棒性水声通信技术研究、延迟宽容网络(DTN)的研究、水声网络安全研究、跨层协议的研究;针对应用的研究,如:水下网络的时间同步、定位和跟踪算法研究、数据融合算法、水下传感器节点平台技术的研究等.

参考文献

[1] Rappaport T S[蔡涛等译.无线通信原理和应用.第二版.北京:电子工业出版社,1999]

[2] 曹志刚,钱亚生.现代通信原理.北京:清华大学出版社,1992[Cao Z G, Qian Y S. Modem Communication Principles. Beijing: Tsinghua University Press, 1992(in Chinese)]

[3] [日]小野濑一志著.张秀琴译.局域网技术.北京:科学出版社,2003(in Chinese)

[4] [美]R.J.尤立克著.洪申译.水声原理.第3版.哈尔滨:哈尔滨船舶工程学院出版社,1985(in Chinese)

[5] 孙利民等编著.无线传感器网络.北京:清华大学出版社,2005[Sun L M. Wireless Sensor Network. Beijing: Tsinghua University Press, 2005(in Chinese)]

[6] Istepanian R, Stojanovic M. Underwater Acoustic Digital Signal Processing and Communication System. Massachusetts: Kluwer Academic Publishers, 2002

[7] Miller N. IRE Transactions on Communication System, 1959, 7(4): 249

[8] Baggeroer A. Oceans, 1981, 13(9): 48

[9] Stojanovic M. Electro/95 International. Professional Program Proceedings. 1995(6): 435

[10] Kilfoyle D B, Baggeroer A B. IEEE Journal of Oceanic Engineering, 2000, 25(1): 4

[11] Merriam S, Porta D. Sea Technology, 1993, 34(5): 24

[12] Curtin T B, Bellingham J G *et al*. Oceanography, 1993, 6: 86

[13] Rice J A, Creber R *et al*. Proc. IEEE Oceans, 2000 Conf., 2000, 3: 2007

[14] Rice J A, Creber R K *et al*. Biennial Review 2001. SSC San Diego Technical Document TD 3117, 2001: 234

[15] Sozer E M, Stojanovic M, Proakis J G. IEEE Journal of Oceanic Engineering, 2000, 25(1): 72

[16] Akyildiz I F, Pompili D, Melodia T. Ad Hoc Networks, 2005, 3(3): 257

[17] Rice J A, Baxley P A. SSC San Diego In-House Laboratory Independent Research 1998 Annual Report, 2000(4): 33

[18] Bucker H. J. Acoust. Soc. Am. 1994, 95(5): 2437

[19] McDonald V K, Rice J A, Fletcher C L. Proc. IEEE Oceans' 98 Conf., 1998, 2: 639

[20] McDonald V K, Rice J A, Porter M B *et al*. Proc. IEEE Oceans 99 Conf., 1999, 2: 1002

[21] Scussel K E, Rice J A, Merriam S. Proc. IEEE Oceans Conf., 1997, 1: 247

[22] Stojanovic M, Proakis J G, Rice J A *et al*. Proc. IEEE Oceans 98 Conf., 1998, 2: 650

[23] Sozer E M, Proakis J G, Stojanovic M *et al*. Proc. IEEE Oceans 99 Conf. 1999, 1: 228

[24] Green M D, Rice J A. IEEE Journal of Oceanic Engineering, 2000, 25(1): 28

[25] Creber R K, Rice J A, Baxley P A *et al*. Proc. IEEE Oceans 2001 Conf., 2001, 4: 2083

[26] Xie G G, Gibson J H. Proc. MTS/IEEE Ocean '2001 Conf., 2001, 4: 2087

[27] www.benthos.com

[28] Rice J A. 5th International Symposium on Technology and the Mine Problem, 2002(4): 21

院士来信

对“robust”中文译名的建议①

编辑同志：

robust (robustness)一词近年来在信号处理和声学领域中经常出现，国内专家有按音译的，称为“鲁棒”、“鲁棒性”，有按意译的，译为“稳健的”、“稳固的”、“宽容的”。

根据1996年9月13日发布的国标GB/T3947—1996上也译为“鲁棒”，我认为robust的意思是一种信号处理技术或算法，它对模型的失配不敏感，所以最恰当的翻译应当是“宽容的”，robustness则可译为“宽容性”，这样比较贴切。

以上意见，仅供参考。

致

礼

李启虎

① 科技术语研究，2006，(1)：13.

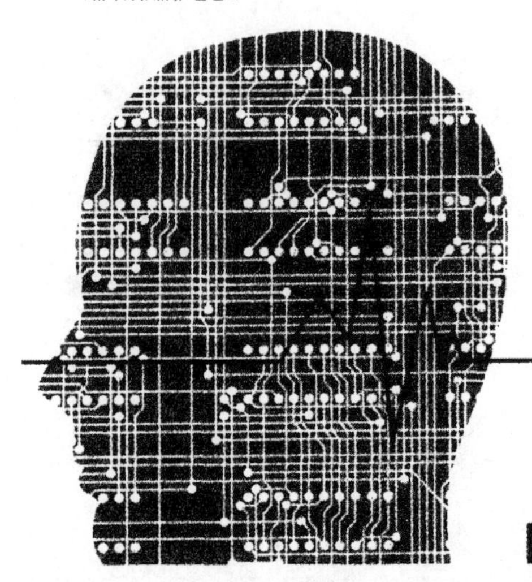

近50年来，随着信息论、控制论、生命科学、微电子学、计算机科学的发展，人们越来越惊异于人类大脑的奇妙。

要揭示人类大脑的奥秘需要从事数学、电子学、控制论、生命科学、信息论、计算机科学的专家共同努力，也许还要有哲学家的参与。不同领域专家的联合有利于问题的清晰化和寻求最好的解决途径。

人类大脑信号处理机制①

文 / 李启虎

1948年，美国贝尔实验室青年科学家香农(Shannon C)发表了《通讯的数学理论》的著名论文。他把热力学中描述系统无序性的重要概念"熵"(entropy)引入信息论。把熵作为信息不确定度的一个测度。他同时定义了数字化信息的基本单位：比特。香农的这篇论文被认为是现代信息论的奠基之作。同年，美国麻省理工学院教授维纳(N.Wiener)出版了《控制论：关于在动物和机器中控制和通讯的科学》。由此，控制论成为一门独立的专门的学科。维纳博士在这本书中指出：在科学发展上可以得到最大收获的领域是各种已经建立起来的部门之间的被忽视的无人区。

这里所说的"无人区"，就是我们今天所说的边缘学科。

近50年来，随着信息论、控制论、生命科学、微电子学、计算机科学的发展，人们越来越惊异于人类大脑的奇妙。至少到目前为止，人类大脑信号处理的机制对人类自身来说，仍是一个黑盒子。也就是说，科学家们除了对大脑的一些物理参数有所了解之外，对于人类大脑如何处理外界信号，如何把人类感官所接收的信息进行融合、处理并进行存储、传输和加工，几乎一无所知。

人脑、计算机和人工智能

现代电子计算机技术的发展，似乎使人越来越相信：我们在信号处理理论和技术方面的进展都已陆续地在大脑信号处理中得到了印证。计算机需要有听觉、视觉，要能学习，能理解；这些我们人类大脑早就具有了；人的机体会受到病毒感染，不料在计算机领域也出现了病毒；计算机的运行需要有一个操作系统，那么我们人类大脑中也存在操作系统吗?

计算机、人工智能的下一步怎么发展呢？让我们去探索一下人类大脑的信号处理机制吧！如果把人类大脑信号处理机制研究清楚，那么毫无疑问，新的信号处理理论就会应运而生。换句话说，有两条看来完全不同的进化道路，它们走向同一目标并得出完全相同的结果。

一条是人类大脑的信号处理系统，经过几百万年的自然选择，在"适者生存"铁的规则的约束下，正不断地沿着某种最佳的方向，缓慢地，但是有效地前进着，另一条是人类自身，根据对科学知识的积累，经过在各领域内的艰苦探索，不断完善对客观世界的认识，并得出一个又一个新的发现。

现代信号处理领域一个最引人注目的成就就是信号的数字化处理，它空前迅速发展的动力来自于微电子技术、计算机技术、人工智能方面的推动。但是，到目前为止我们还没有证据可以证明，人类大脑在进行信号处理时也存在着某种类型的数字化处理，这就使得我们有理由对信息的数字化进程进行冷静的思考。

1995年，美国麻省理工学院教授、媒体实验室的负责人尼葛洛庞帝(Negroponte N)出版了《数字化生存》一书，引起了美国和世界各地读者的兴趣。他把近20年来信息领域的数字化革命的丰硕成果，

① 创新科技, 2007, (1): 44,45.

以及将要引起的对人类生活的巨大冲击展现在世人面前。1998年1月31日，美国前任副总统戈尔（Gore S）在加利福尼亚科学中心作了题为"数字地球——认识我们这颗星球"的讲演，提出充分利用数字信息促进社会进步和发展。

根据粗略的估计，人类获得信息的主要来源是听觉（约占5%）与视觉（约占90%），其他还有味觉、触觉及嗅觉等。在60年代初期之前，对信息的处理方式主要限于模拟方式。它的特点是这种信号在时间上是不分间隔的，在幅度上是不分层的。从60年代开始，由于计算机技术的迅猛发展，特别是70年代以来微电子技术的惊人进步，使得信号的数字化处理以空前未有的速度向前发展。

1946年当冯·诺伊曼等计算机专家研制成功第一台电子计算机ENIAC时，它的总重量为30吨，占地面积相当于一个小的体育馆，平均每7分钟就有一个电子管失效。它的耗电也十分惊人，当它工作时，整个费城灯光暗淡。1971年Intel公司推出的第一个微处理器芯片4001，其功能大体上和ENIAC相当，而它的体积只有后者的万分之一。微处理芯片的发展速度异常迅速，它基本上遵从所谓摩尔（Moore G）定律，这就是微处理器芯片每隔18个月性能提高一倍，价格降到原来的1/2。摩尔博士是Intel公司的创始人之一，他在1973年提出这一观察结果。但是据他自己说，早在1964年他还在仙童公司工作时，就注意到半导体工业发展的这一事实了。1998年1月在美国Proc.IEEE上重新公布了摩尔这一断言的手稿，摩尔声称这一结论无需作任何改变。我们现在还很难预测，这一定律能够有效的时间究竟还有多长，有的科学家认为，至少在2010年前是正确的，也有的科学家认为，从固体物理学的基本理论来看，芯片的线宽是有限制的，从而微处理器芯片的性能就会受到限制。

英国科学家福莱斯特（Frester　T）在总结了微电子技术的这种惊人发展速度之后，在他所著的《高技术社会》中发出了感叹，他说："如果汽车或飞机行业也像计算机行业这样发展，那么今天一辆罗尔斯·罗伊斯汽车的成本只有2.75美元；跑300万英里（482.8万km）只用1加仑汽油。而一架波音767飞机的价格只需500美元，用5加仑汽油在20分钟内便可环绕地球一周"。

数字化浪潮正在席卷全球，数字化信息技术正在使人类生活质量提到空前高的水平。

但是信息的数字化处理绝不是信息处理的完美无缺的方式，它只是信息处理理论与发展过程中的一个阶段，下一阶段是如何的呢？最直接的方法是去探求人类大脑信号处理方式，虽然，目前对我们来说，仍是一个黑盒子。

据测算，人类大脑的平均重量约为1.6kg，占成年人体重的2%~3%，大脑的基本组成部分为神经元，总数约为100~140亿个，大脑工作时要消耗人体氧气摄入量的20%~30%，人脑的记忆容量大约是10^{15}比特，如果一个人活100岁，自出生后，每秒存储10^3比特信息，那么，仍只用去这个容量的1%。科学家目前对人类大脑的知识，特别是大脑对外界信号处理系统的奥秘必将会大大扩展我们现有信号处理理论和技术的研究范围，大大推动人工智能科学（包括人工语音合成、自然语言识别、图像识别等领域）的发展，同时会为研制体积更小、速度更快、功耗更小、可靠性更高的智能计算机提供依据。

大脑处理信息的不解之谜

要探索大脑信号处理机制，要遵循由易到难、由表及里的途径，因为人类大脑的信号处理系统对我们人类本身来说，仍是一个黑盒子。用传统的研究输入/输出关系的方法，或者用人工神经网络理论中的建模方法不能达到揭示这种系统秘密的目的。我们要解决的问题实际上也迫使我们采用一种逐步逼近的方法，这些问题归结起来是：

●人类大脑对听觉、视觉和其他感官所接收到的信号是先进行数据融合还是先进行处理，也就是存在于人类大脑中的数据融合系统是属于哪一个层次的？（根据现代信息论的研究，数据融合可以合为三个层次，即数据级、特征级和决策级）。

●自然界作用于人类大脑的信号，无论是声音、光波或其他信号都是模拟信号，大脑在接收这种信号之后进行数字化处理吗？即进行时间采样和幅度量化吗？

如果是进行了数字化处理，是哪一种数字化处理呢？在时间上是等间隔采样吗？存在信息论中的奈奎斯特（Nyguist）采样率吗？在空间上进行分层吗？是等间隔分层吗？还是采用某种压扩（即A-律，μ-律压扩方法）？

进一步还要探索，如果进行了数字化处理，是几比特的数字化？信息的单位是比特吗？

●人类大脑作为一个超级信号处理系统是否具有"操作系统"，如果没有，大脑的运行是如何进行的？如果有，这是一种什么样式的操作系统？不同个体之间有什么差异？

人类对异体器官移植的排斥是否受操作系统的影响？

信息在人类大脑中是如何传输、存储和加工的？记忆、联想、判断是如何产生或形成的？

●人类对自然语言理解是如何实现的？图像识别的机制是什么？目标分类识别的概念是如何存储于大脑之中的？

人类大脑的存储系统的输入显然是按时间顺序的，但是在应用时却能不受存入顺序的影响，这种顺序输入是如何变为记忆并重新被读出的？

●大脑记忆中顺序读出是如何实现的，是不是存在某种引导系统？实践证明，要不出任何错误按一定速率背诵一首诗歌是十分困难的，但是如果唱一道同样长度的歌曲几乎不会有什么问题。这里，乐曲是不是起着引导地址的作用？

●某些特别的信息和刺激会引起思维的混乱、记忆的丧失或错乱。那么，特别的信号会引起电脑或其他信息处理系统的失效吗？我们有可能给声呐或雷达发送某种信号使得它们失去功效吗？等等，等等。

随着我们对人类大脑信号处理机制研究的深入，还会出现更多的新问题，这些问题的解决会使我们现代计算机的功能更强大，真正地如人一样会听、会谈、会思考、会学习、会理解。

要揭示人类大脑的奥秘需要从事数学、电子学、控制论、生命科学、信息论、计算机科学的专家共同努力，也许还要有哲学家的参与。不同领域专家的联合有利于问题的清晰化和寻求最好的解决途径，如果某一个问题是一个生物学方面的问题，那么10个不懂生物学的信息论专家不见得会比一个生物学家更有效。

除了死亡之外，没有任何正常的方法可以将人类大脑中的记忆全部清除，这有一点类似于现代微机中的"复位"操作。除非把计算机彻底失效，否则它的程序、数据总还存在于存储系统中。

（作者单位：中国科学院声学研究所）

关注深海高技术领域的水声学研究①

在《国家中长期科学和技术发展规划纲要(2006—2020)》中提出要加快发展空间和海洋技术。而海洋技术又被列为重点发展的六大前沿技术之一。深海高技术则是重点关注的领域。我国的国家安全和国民经济的发展都需要我们走向印度洋和太平洋的深海。

按照美国著名水声学专家R. J. Urick的解释,"深海"有两种不同的定义:测深学意义上,大致以沿海大陆架延伸为标准,大约100英㖊(约200米)以内为浅海;水声传播意义上,没有多次海底海面反射的水声信道为浅海。按这个定义,浅海的水深是一个模糊的数值,它是随地缘而变化的。由于声波是迄今为止人类知道的唯一能在海中远距离传播的能量形式,所以在深海高技术的发展中,水声学(海洋声学,声学海洋学)具有重要的作用,并担负重要的使命。

美国学者H. Medvin说,"海洋声学或声学海洋学就像希腊神话中的双面神Janus(他有两张朝向相反的面孔):从一面看是海洋声学:研究海洋作为声信道对信号传播的影响,对目标检测,识别的影响等;从另一面看是声学海洋学:从声学的角度研究海面、海底、水体的特性,研究内波、中尺度涡、风暴潮等对声纳设备性能的影响。"20世纪60年代,我国著名物理学家、国防水声学的开拓者、中国科学院院士汪德昭先生提出了发展水声事业的战略方针:由浅入深,由近及远。经过半个多世纪的努力,我们已经到达了向深海进军的关键时刻。水声信号在浅海和深海传播具有很多相似性,又有一些非常不同的规律。在发展深海高技术的过程中,水声物理,水声工程将会发挥巨大的作用。它在水声遥测、遥感、通信、水下有人/无人载器、水下滑翔器、深海资源勘探等领域都可以得到广泛应用。

美国海军在《2000—2035年美国海军技术》中指出,1960年以来,潜艇噪声已降低了35分贝,使大洋的对潜探测距离已减少到只有几公里,远远低于武器的作用范围。海洋声音的相干性则是最终确定声纳性能的一种特性。相干性所涉及的科学技术问题都与环境条件有关。所以对海洋环境的研究是基础性的工作。声纳必须能有效使用的沿海海区可能很浅(1—200米深),也可能很深(几千米深),还存在大陆架间断。在浅水区和大陆架,潮汐和地形会产生水平各向异性的强烈内波,对声音的传播速度会产生很大影响。在上坡-下坡区,由于波浪的相互作用和声径不断变化,会使表面声道传播中断,并影响到相干性。在出现海底折射声速剖面的情况下,海底折射对相干性会产生很大影响。即使是在深度不变的浅水区,由于不同区域的吸声性各不相同,也会产生问题,而在坡度变化大、崎岖不平的地区,问题就更大。深水区的相干性比沿海海区更大,在信号不产生相互作用的情况下尤为如此。这就提供了利用相干性大幅度增大探测距离的机会。对海洋气候进行声学监视的赫德岛水声-层析试验计划(ATOC)试验说明,在几千公里的距离都具有相干性。

通过采用能充分利用信号场水平特性和垂直特性、性能更好的基阵,就能提高未来声纳系统的性能。只要看看石油勘探工业所取得的成就,就能了解提高声纳系统性能的潜力。在石油勘探中,现在已能使用18个多线列拖曳阵,每个有8公里的孔径,传感器的总数已达到2104。

海洋的时间相干性也能进一步加以利用。深水区的相干积分时间间隔可以(50赫时)大于1/2小时,浅水区可以(500赫时)大于3分钟。

在我们冲出第一岛链,走向深海的过程中,水声学起着特殊的、无法替代的作用。由于水声学的研究需要进行大量的海上试验,因此是一项高风险、大投入、长周期的科研工作,必须有国家层面的计划。1996年实施的国家高技术研究发展计划(863计划)海洋领域的工作已经取得举世瞩目的成就,为我们积累了丰富的经验。

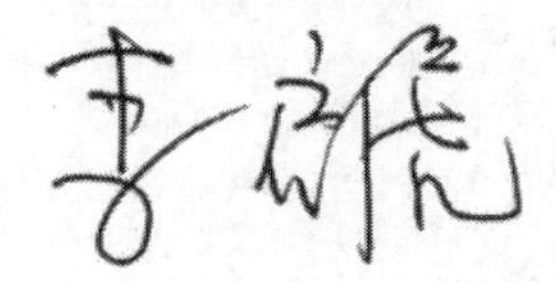

(中国科学院声学研究所,北京 100190)

① 科技导报, 2011, 29(21): 3.

纪念《应用声学》创刊 30 周年

水声信号处理领域新进展[①]

李启虎

（中国科学院声学研究所 北京 100190）

摘要 本文介绍近 30 年来水声信号处理领域理论研究的新进展和在声纳设计中的应用。包括水声信号建模、声场匹配、海洋波导和内波现象的探索和研究、声矢量场信息获取和处理，低频水声信道的时/空相关特性，水下目标辐射噪声的不变特征量提取和检测技术，水下语音、图像传输和抗干扰技术。同时概述，声纳设计的前沿领域：大孔径拖曳线列阵声纳、高分辨力合成孔径声纳、深海传呼机等的发展情况。

关键词 水声信号处理，声纳设计，新进展

中图分类号：TB556 O427 **文献标识码**：A **文章编号**：1000-310X(2012)01-0002-08

New advances of underwater acoustic signal processing

LI Qihu

(*Institute of Acoustics, Chinese Academy of Sciences, Beijing* 100190)

Abstract The advances of new technology and theoretical development in 30 years of underwater acoustic signal processing and its applications in sonar design are presented in this paper. The topics include the research work in the field of underwater acoustic signal modeling, acoustic field matching, ocean waveguide and internal wave, the exaction and processing technique of acoustic vector signal information, space/time correlation characteristics of low frequency acoustic channel, the invariant feature of underwater target radiated noise, the transmission technology of underwater voice/image data, and its anti-interference technique. Some Sonar frontier technologies in sonar design are also discussed, including large aperture towed line array sonar, high resolution Synthetic aperture sonar and deep sea siren etc.

Key words Underwater acoustic signal processing, Sonar design, New advances

1 引言

今年是《应用声学》杂志创刊 30 周年，作为这一杂志的创办人和实际掌舵人的应崇福院士，可惜已驾鹤西去。应先生在创刊的 30 年来，准确把握办刊宗旨，精心指导杂志的组稿、编辑、出版工作，终于使《应用声学》成为国内声学领域一本独树一帜的、有影响力的杂志。这是值得我们同行学习的。本文的写作也是应先生的遗愿，是他离世之前不久所确定要办的事之一。谨以此文纪念应崇福院士、纪念由他一手创办，呵护培育的《应用声学》杂志。

水声信号处理和声纳技术是一门发展迅速、需求推动力强大、应用前景异常广阔的学

① 应用声学, 2012, 31(1): 2-9.

科[1-5]。

在声学领域的众多分支学科中，恐怕没有其他学科像水声学那样，其发展受着战争需求的推动。反过来，水声学的发展又为水下战武器装备的研制和创新注入活力[6-17]。

最典型的例子就是隐身潜艇的出现以及随之而来的对隐身潜艇的主被动检测和识别问题。

上世纪 60 年代以来，潜艇的被动和主动隐身需求，使降噪技术和大推动力、低转速螺旋桨技术以及消声瓦的研制受到空前的重视；而低噪声、安静型潜艇的出现又催生了低频、大孔径拖线阵的研制和低频大功率发射基阵的使用。同时又推动了对低频水声信道的研究。需求的强力拉动成为声纳新技术发展的重要原因。

当然，我们也不能忽视学科发展的内在需求，这是和外在推力相辅相成的。

水声学不是一门纯理论的学科，它的发展和完善依赖于大量的有准备的海上实验。由于基础研究的特殊性，需要较大的人力、财力投入。深入的基础研究是声纳技术创新的源泉。回顾声纳发展的历史就可以证明这一点[18-21]。

如果没有对不同声速剖面下水声信号传播规律的研究，就不会有变深声纳；如果没有对水声场的时空相关特性的研究，就不会有被动测距声纳；同样，对水声信道的研究为匹配滤波器、水声通信系统提供了正确设计的依据；内波、声层析、匹配场过滤又为远程被动定位技术提供理论支持。

本文简要回顾近 30 年来水声信号处理研究方面所取得的成就以及在声纳研制中的应用。同时介绍该领域目前广受关注的前沿课题。改革开放以来，我国水声领域和国外同行的交流合作有了长足的发展，因此在介绍国外同行的研究成果的同时，我们将扼要叙述我国科技人员在该领域所取得的进展，以及与之有关的某些国际合作项目[22-24]。

2 水下目标辐射噪声特性

美国 Scripps 研究所的著名专家 V.C.Anderson 在讨论水声信号处理时[2]，认为声学中的信号处理作为一门独立学科始于 1952 年。其主要的标志则是对目标噪声特性的最佳接收法的讨论。特别是一些早期的数字化处理技术，如数字多波束（DIMUS，Digital Multibeam System)。有关水下目标辐射噪声机理的研究，可以从 ROSS 的经典性著作中找到[25]。近年来有关安静型潜艇水下辐射噪声的叙述可以从俄罗斯专家 Miasnikov 的著作中得到不少信息[3]。他把现代潜艇的水下辐射噪声分为三级，即“嘈杂的”，“安静的”和“非常安静的”，并指出这 3 种噪声在 30 Hz 处的谱级分别为 140 dB, 120 dB 和 100 dB。他还补充说“假设 300 Hz 处低 10 dB 是合理的”。

目前有关安静型潜艇辐射噪声中低频线谱分量的真实数据非常少，Miasnikov 的文章可能是一个例外。他指出带宽为 0.1 Hz 的线谱是检测的重点。我们大致可以把潜艇水下辐射噪声表达为以下的模型，即

$$s(t)=m(t)+[1+\mu m(t)]p(t)\,, \tag{1}$$

其中

$$m(t)=\sum_{k=1}^{N}A_k\cos(2\pi f_k t+\theta_k)\,, \tag{2}$$

表示辐射噪声中的低频线谱分量，线谱频率为 $f_k, k=1,\cdots,N$。而 A_k 是该线谱分量的幅度，θ_k 则是一个随机相位。实际测量证明，θ_k 可以看作是相互独立的随机变量，并在$[0,2\pi]$内均匀分布。

$p(t)$ 是由潜艇结构振动而引入的一个调制宽带信号。一般可认为是一个宽带随机噪声。

从频率分布来说，$p(t)$ 的频率集中在 100 Hz 以上。而 $f_1,\cdots,f_N$ 则分布在 0～200 Hz 之内。

μ 是一个调制系数，$0<\mu<1$。

美国从 80 年代开始研制用于探测低噪声潜艇的低频主被动拖线阵声纳 Surtass LFA，这是一种专门用于远程警戒低噪声潜艇的声纳，工作频率可低至 100 Hz 以下，见图 1。其主动发射功率可超 230 dB，被动检测使用两条长达 1500 m 的声阵。据报导作用距离在 100 km 以上。

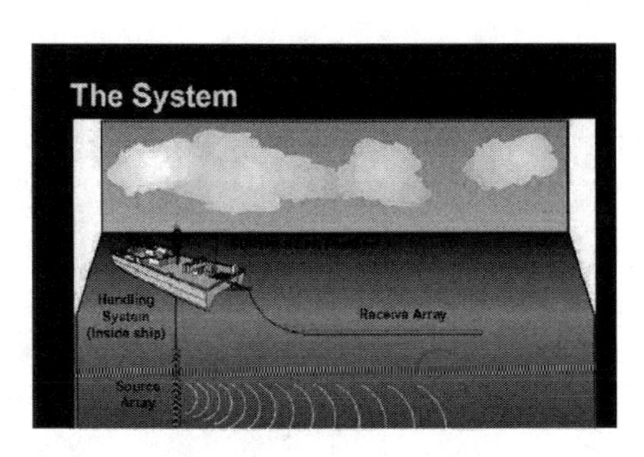

图 1 Surtass LFA 示意图
（美国研制，低频拖线阵声纳）

3 低频/甚低频水声信道的传输特性[26-40]

对于安静型潜艇的检测，无论是主动声纳还是被动声纳，所使用的频率都在向低端移动。如果我们把传统的舰艇上的孔径在1～5 m量级的基阵所使用的 1000 Hz 左右的声纳称为中频声纳的话，那么频率在 100～1000 Hz 的声纳可以称为低频声纳，而频率范围低于 100 Hz 的声纳可以称为甚低频声纳。现代主被动声纳所使用的频率已明显往低频/甚低频段移动。其典型的代表便是 SURTASS (Surveillance Towed Array Sonar System)，LFA (Low Frequency Array)[41]。

Surtass 的主动声纳工作频率已降至 100～500 Hz 之间，而被动声纳工作频率估计在 100 Hz 以下。

低频信号的检测问题当然与低频信号在海洋中传播的问题联系在一起，只有从理论与实践中找出低频信号的传播特性，才能为低频信号的检测提供依据。

S. V. Burenkov 等俄罗斯科学家在 1990 年代曾对 228 Hz 低频信号的传播进行过试验，其接收距离大于 9000 km[39]，他们有两个重要发现：一个是信号的幅度衰落现象比较严重，在观察的 5 分钟内，起伏值超过 20 dB；另一个是信号的相位具有“令人难以置信的稳定性”。美国 D. F. Worcester 和 R. C. Spindel 在一份总结多国参加的 ATOC (Acoustic Thermometry of Ocean Climate)计划执行情况的报告中[40]提到，美国在 1995-1999 年之间利用美国海军的 14 个 SOSUS (Sound Surveillence System)接收阵，记录 57 Hz、75 Hz 信号的传播数据，最远距离为 3900 km，单频信号的传输结果令人满意，带宽为 1～2 Hz 滤波器所提取的信息可以用于目标识别。

由此可见对甚低频信号（频率在 100 Hz 以下）进行检测的前景还是比较乐观的。

对于高斯宽带噪声背景下，单频信号的检测问题，传统的能量检测方法不具备优势。而带有自适应线谱增强的 FFT 分析技术可以提供较高的增益。

为了利用长时间积分，对于传播过程中有可能产生频率漂移的信号，一种分段 FFT 技术有利于微弱单频信号的提取[69]。

对水声信道传输特性的深入研究，导致了一些新现象的发现和新的信号处理技术的诞生。特别值得指出的是模基声纳、宽容性信号处理技术、匹配场过滤技术的研究工作[42-47]。

这些新的信号处理技术中的一大部分是基于近年来以美国 Scripps 研究所 Kuperman，APL 实验室 Spindel 等所带领的研究组对于水声时反聚焦、相轭现象的研究。同时俄罗斯专家 Lysanov 等有关海洋声传播中的波导理论的研究也成为近年来引人的热点问题。

浅海中的声波导现象有可能成为信号检测和目标识别的新途径。特别是，如果产生干涉条纹的频率正好和目标辐射噪声线谱分量相近或重合时，对微弱信号的检测作用距离有望大幅提高，但至今在海试中仍未观察到这一奇观[31,32]。图 2 是我们 2009 年在南海的一次试验结果，水文条件是弱负梯度。水面舰艇辐射噪声的干涉条纹清晰可见。

4 新型的水声传感器及其相关信号处理技术[48-57]

水声场中质点振速信息的获取是近 20 年来受到广泛关注的理论与实际问题。

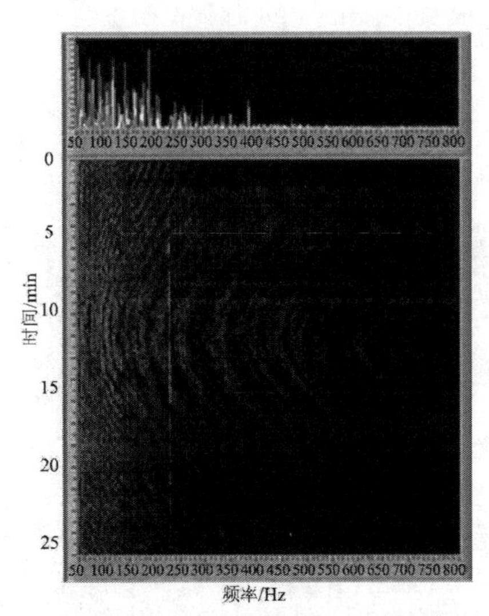

图2 浅海波导：2009.04.07 三亚海试，水面舰艇辐射噪声

美国学者 Nehorai 等从上世纪 90 年代以来发表了一系列与矢量水听器有关的文章，他们的工作主要是用矢量水听器获得水声场中除声压之外的质点振速的信息，以及相应的波束成形技术[48,49,51]。

实际上，俄罗斯（前苏联）学者 Shchurov[50]，Gordienko[54] 等早在 80 年代就已开始矢量传感器的理论与应用研究。

我国学者自上世纪 90 年代开始，引入部分俄罗斯和乌克兰的有关矢量水听器的技术，水声领域的专家对这种非声传感器（还包括光纤水听器）及其相关的信号处理技术表现出了不寻常的关注。一个重要的表现是在国内学术会议和期刊上发表的文章空前活跃。

例如在 2004 全国水声会议上，总共 116 篇文章中有 12 篇涉及矢量水听器及其信号处理技术[61]，而在同年举行的水中军用目标特性学术会议上，矢量水听器在总论文（70 篇）中也占了 4 篇[62]。这种现象并未在国外同类会议或学术期刊中出现。从信号检测的观点来说，无论是声学传感器还是非声学传感器，只要其自噪声远低于海洋噪声，就可以用于对微弱信号的感知。因为检测的关键是对信噪比的改善。矢量水听器或光纤水听器的优势并不在于灵敏度而在于具有某些特别的特性，例如单水听器具有指向性，波束光信号不受电干扰等。

当然，声学传感器在声纳应用中仍占据主导地位，特别是在拖曳线列阵声纳的应用中，寻求一种电压灵敏度高、加速度灵敏度低的小型、低频、宽带水听器吸引了研究工作者的注意[13,14]。德国船舶和海军武备技术中心（WTD）专门建立了测试拖线阵流噪声的水池[57]，可以用于测量不同缆径、不同拖曳速度下的拖线阵流噪声。我国中国科学院声学所和 715 所也已在湖上建立了类似的实验设备。

一般认为主动声纳是研究声信道属性的最好方法。因为通过有目的地发送经过仔细选择的信号形式，可以提取信道的主要特征，如延时，频率扩展、时变/空变特性，信道的频率选择特性等。并且，通过实验有可能使各参量相互分离或去相干，从而更好地获得信道解耦和声场匹配的知识。实际测试表明，当水深小于 10 m 时，中等距离信号的延迟扩散通常为 10 ms，当水深在 100 m 左右时，延迟扩散增加到 100 ms；而在深海环境中，延迟扩散会达到 300 ms[1]。

利用声纳平台的运动，得到相对较大的基阵孔径，从而增加目标分辨力的主动合成孔径声纳技术也是近年来广受注意的新领域。图 3 给出了我国研制的主动合成孔径声纳的湖、海试所获得的图像。这是目前世界上分辨力最高（2.5×5 cm）的主动合成孔径声纳。

水声通信是水声信号处理的一个重要应用，最近 15 年来又和网络中心战的理论与实践紧紧地联系在一起[15]。网络中心战（NCW, Network-Centric Warfare）的概念是美国海军参谋长，海军上将 Jay Johnson 在 1997 提出的。他在美国海军研究所 123 届年会上发表演说，指出“从平台中心战向我们称之为网络中心战的基本转变”这一概念，随后被美国海军中将 Cebrowski 评论说，网络中心战的概念将“被证明为近 200 年来，军事学中的最重要革命”。

我们从现在看到，网络中心战的概念已不只停留在理论上，也不只局限于海军，而已扩大到整个军种。而同时网络中心战概念的发展和实践已使海军战争模式发生了很大的变化，随之对声纳技术提出了更高的要求。各种水下通讯网络，从各种局域网到全局的通讯网都在海军大国发挥作用[8-12,58-60]。

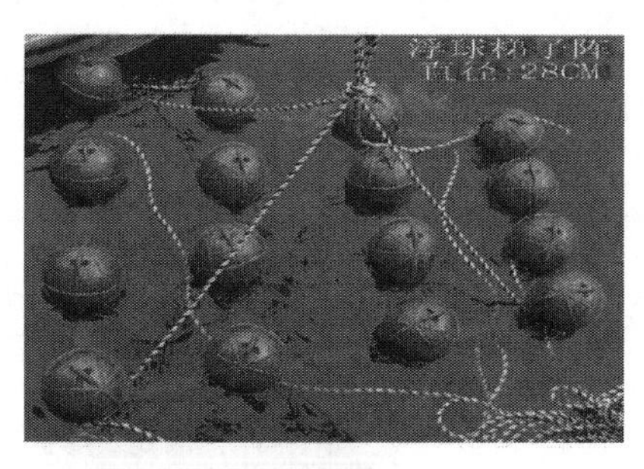

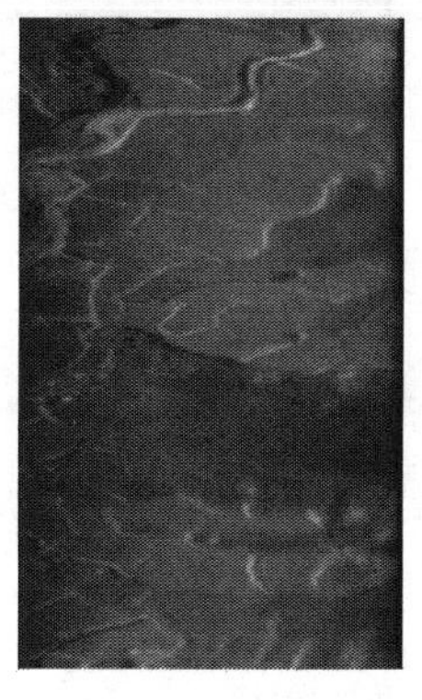

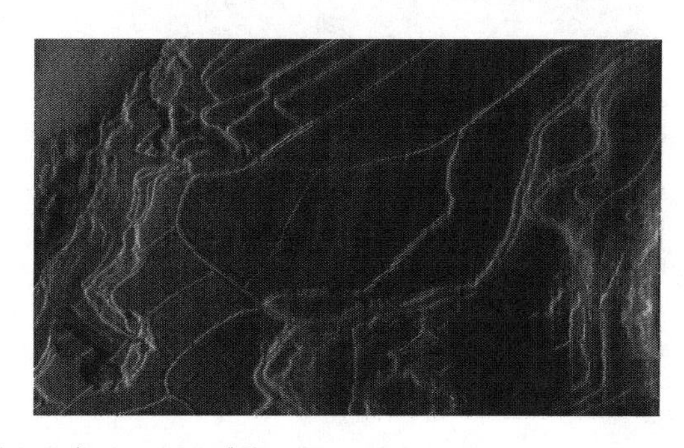

图 3　主动合成孔径声纳（SAS）实验
（我国研制的目前世界上分辨力最高（2.5×5cm）的图像声纳）

图 4 给出了美军研制的一种新的称为“深海传呼机”（Deep Siren）的设备。它用于建立海军指挥机关和远航潜艇之间的通信。在过去，海军必须通过长波无线电台和潜艇在事先约好的时间通信，但是有了深海传呼机，海军指挥员可以随时与远航潜艇通话，其机理是利用卫星-无线传输-水声通信潜标。据报导美国已关闭了位于 Viginia 的两个长波无线通信台，而使用这种新的通信系统。

图 4　深海水下传呼机：一种和水下潜航潜艇通信的高新技术

深海高技术发展中的水声学问题受到越来越广泛的关注。我国已独立自主研制成功蛟龙号 7000 m 载人潜器，并于 2011 年 7 月成功下潜至 5088 m。该潜器安装了多部不同功能的声纳，包括导航、水声通信、图像信号传输、测速和前视声纳，见图 5。

5　我国学者在水声领域和国际同行的合作

改革开放以来，我国在水声领域和国外（主要是美国）进行了有限度的合作。我国旅美访问学者尚尔昌，周纪浔等教授在他们过去在国内工作的基础上，在美国继续他们出色的工作，并且在美国水声界产生了一定的影响。例如周纪浔教授有关内波孤立子的理论/实践工作[35]，被美国专家 Goodman 列入了水声学近 500 年（1490–2000）历史中（自意大利科学巨匠达·芬奇 1490 年听到远处航船的声音算起）具有里程碑意义的 60 件事中的一件[12]，这也是大陆学者唯一被列入的工作。

图 5　蛟龙号 7000 m 载人潜水器
（2011 年成功下潜至水下 5088 m，潜水器上有 5 部声纳）

我国和美国同行曾在黄海和东海、南海进行过几次较大规模的联合海试。1996 年中、美两国科学家在黄海进行了联合海试，规模较小[63]。最大一次联合海试是以中、美科学家为主，有韩国、新加坡、台湾地区科学家参加的亚洲海联合海试（Asia Seas International Acoustics Experiment, ASIAEX）[23,24,64]。来自不同国家和我国台湾地区的科学家在东中国海（N28°—30°, E126°30'—128°），南中国海（N21°—22°30', E117°—119°）进行了水声传播、散射、混响、海底地质反演的试验。历时近两个月（2001 年 5 月–6 月），取得了丰硕的成果。包括长线阵声时空相干特性、不同主动声纳信号的混响特性、目标辐射噪声分布等。

我国学者还参加水声领域的各种会议，向国外同行介绍在该领域所取得的成就。这些会议包括 ICTCA（国际理论计算声学会议），UDT（水下防务技术会议），MAST（海洋科学和技术），UAM（水下声学测量），PRUAC（泛太平洋水声会议），WESPAC（西太平洋声学会议）等。部分成果反映在会议的特邀报告和专题报告中，包括高分辨力合成孔径声纳[65]、浅海波导[31,32]，数据融合[67,68]等。

海洋声学观测设备发展中的一个新技术是水下滑翔器，这是一种基本不用动力或少用动力的水下无人工作平台。自上世纪末以来，美国、俄罗斯等国科学家已投入很大人力物力用于研发这种设备，主要用于海洋环境动力参数的采集，也可以用于水下网络节点的维护和沟通，见图 6。

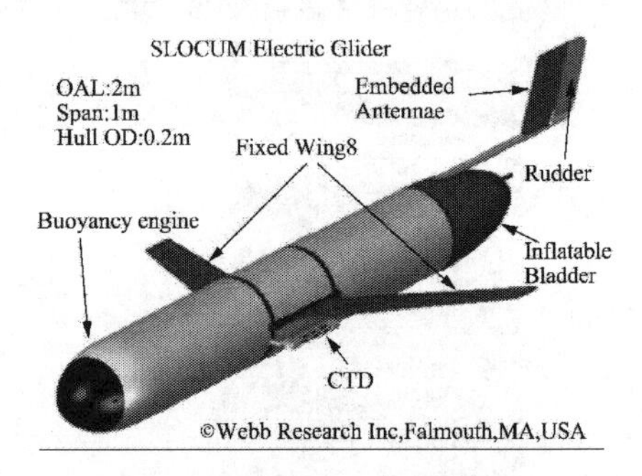

图 6　水下滑翔器（Sea Glider）
（美国研制的一种基本很少使用动力而在海里长时间工作的水下平台）

参 考 文 献

[1] VACCARO R J. The past, present, and the future of underwater acoustical signal processing[J]. IEEE Signal Processing, 1998, 15(4): 21-51.

[2] ANDERSON V C. The first twenty years of acoustical signal processing[J]. J. A. S. A., 1972, 51(3B): 1062-1065.

[3] MIASNIKOV E. Can Russian submarines survive at sea ? The fundamental limits of passive acoustics[J]. Science & Global Security, 1994, 4(2): 213-251.

[4] HAZELL P A. What's the future for ASW in NATO ?[J]. Sea Technology, 1998, 39(11): 10-17.

[5] MCTAGGANT B. Thirty years of progress in sonar transducer technology[J]. Proc. of UDT, 1991:1-11.

[6] BRIGHT C. Better sonar driven by new transducer materials[J]. Sea Technology, 2000, 41(6): 17-21.

[7] HAMBLEN N. Next generation stealth submarine[J]. Sea Technology, 1998, 39(11): 59-62.

[8] CURTIN T B. ONR program in underwater acoustic communications [J]. SEA Technology, 1999,. 40(5): 17-27.

[9] STOJANOVIC M. Recent advances in high-speed underwater acoustic communications[J]. IEEE, J .of Oceanic Engr. 1996, 21(2): 125-136.

[10] National Research Council .Technology for the United States Navy and Marine Corps 2000-2035, Becoming a 21th century force [M]. NA Press, 1997.

[11] SHERMAN J. Ramsfeld's new speed goals[J]. Defense News, 2004: 1-8.

[12] GOODMAN R. A brief history of underwater acoustics [J].Proc.

Of ASA at 75, 2004: 204-227.

[13] LEMON S G. Towed array history, 1917-2003[J]. IEEE J. of oceanic Engr. 2004, 29(2): 365-373.

[14] LASKY M, DOOLITTLE RD, SIMMONS BD, et al. Recent Progress in towed hydrophone array research[J]. IEEE J. of Oceanic Engr, 2004, 29(2): 374-387.

[15] ADMJAY L J. Address at US naval institute Annapolis seminar and 123rd annual meeting[J]. Annapolis, 1997.

[16] PRUITT T. Maritime homeland security for ports and commercial operation[J].Sea Technology, 2004, 45(11): 20-26.

[17] STERNLICHT D , PESATURO J F. Synthetic aperture sonar: frontiers in underwater imaging[J]. Sea Technology, 2004, 45(11): 27-34.

[18] 李启虎.数字式声纳设计原理[M]. 安徽教育出版社, 2003.

[19] National Research Council Ed. The role of experimentation in building future forces [M]. NA Press, 2002.

[20] SONG H C, KUPERMAN W A, HODGKISS WS. Recent results from ocean acoustic time reversal experiments [J]. Proc. Of ECUA'2002: 279-284.

[21] KUPERMAN W A. Phase conjugation in the ocean: experimental demonstration of a time reversal mirror[J]. J. A. S. A., 1998, 103(1): 25-40.

[22] LI R W. An overview of ASIAEX [J]. Proc. Of ASIAEX, 2001.

[23] WEI R C, CHEN C F,NEWHALL A E. A preliminary examination of the low-frequency ambient noise field in the South China Sea during the 2001 ASIAEX experiment[J]. IEEE J. of Oceanic Engr, 2004, 29(4), 1308-1315.

[24] 唐存勇. 简介亚洲海域国际水声学实验[J]. 台湾自然科学简讯, 2001, 3(1).

[25] ROSS D. Mechanics of underwater noise[M]. Pergamon Press, 1976.

[26] BREKHOVSKIKH C M, LYSANOV Y P. Fundamentals of ocean acoustics[J].2nd Ed. Springer-Verlag, 1991.

[27] KUPERMAN W A. The generalized waveguide invariant concept with application to the vertical array in shallow water[J]. in Ocean Acoustic phenomena and signal processing, ONR Workshop, 2002: 33-66.

[28] ROUSEFF D, SPINDEL R C. Modeling the waveguide invariant as a distribution[J]. Ibid, 137-148.

[29] BAGGEROER A B. Estimation of the distribution of the interference invariant with seismic streamers[J]. Ibid, 151-167.

[30] KAPOLKA D. Equivalence of the waveguide invariant and two path ray theory methods for range prediction based Lloyd's mirror patterns[J]. Proc. of Acoustics'08, 2008: 3637-3641.

[31] AN L, WANG Z Q, LU J R. Calculating the waveguide invariant by the 2-D Fourier transform ridges of Lofargram image[J]. J. of Chinese Electronics & Information, 2008, 30(12): 2930-2933.

[32] LI Q H. Theoretical analysis and experimental results of interference striation of underwater target noise in shallow water waveguide[J]. Chinese J. of Acoustics, 2011, 31(1): 73-80.

[33] CHUPROV S D. An invariant of the spatial-frequency interference pattern of the acoustic field in a layered ocean[J]. Proc. of the Russian Academy of Sciences, 1981, 257(2): 475-479.

[34] JACKSON D R, DOWLING D R. Phase conjugation in underwater acoustics[J]. J. A. S. A., 1991, 89(1): 171-181.

[35] ZHOU J X, ZHANG X Z. Resonant interaction of sound wave with internal solitons in the coastal zone[J]. J. A. S. A., 1991, 90(4): 2042-2054.

[36] SONG H C, KUPERMAN W A, HODGKISS W S. A time reversal mirror with variable range focusing[J]. J. A. S. A., 1998, 103(6): 3234-3240.

[37] KIM S, EDELMANN G F, KUPERMAN W A, et al. Spatial resolution of time reversal array in shallow water[J]. J. A. S. A., 2001, 110(2): 820-829.

[38] El-SHARKAW M A, UPADHYE A, SHUAI L, et al. North east Pacific time integrated undersea network experiments (NEPTURE): cable switching and protection[J]. IEEE J. of Oceanic Engr, 2005, 30(1): 232-240.

[39] BURENKOV S V.,GAVRILOV A N,UPORIN A Y, et al. Heard Island feasibility test: long-range sound transmission from Heard Island to Krylov underwater mountain[J]. J. A. S. A., 1994, 96(4): 2458-2463.

[40] WORCESTER P F, SPINDEL R C. North Pacific acoustic laboratory[J]. J. A. S. A., 2005, 117(3): 1 449- 1510.

[41] http://www.surtass-lfa.com.

[42] URICK R J. Principle of underwater sound[M].3rd Ed. McGraw-Hill, 1983.

[43] MCDONOUGH R N, WHALEN A D. Detection of signals in noise[M]. 2nd Ed. Academic Press, 1995.

[44] CAMDY J V. Model based signal processing in the ocean[J]. IEEE Oceanic Engr. Society News Letter, 2000, 25(3): 199-205.

[45] PORTER M B. Acoustic models and sonar systems [J]. IEEE J. of oceanic Engr, 1993, 18(4): 425-437.

[46] KASSAM S A, POOR H V. Robust techniques signal processing [J]. Proc. Of IEEE, 1985, 73(3): 433-481.

[47] SULLIVAN E J, MIDDLETON D. Estimation and detection issues in matched field processing [J]. IEEE. Of Oceanic Engr, 1993, 18(3): 156-167.

[48] NEHORAI A, PADELI E. Acoustic vector-sensor array processing [J]. IEEE Trans. Signal Processing, 1994, 42: 2481-2491.

[49] HAWKES M, NEHORAI A. Acoustic vector sensor beamforming and Capon direction estimation[J]. IEEE Trans. Signal Processing, 1998, 46(9): 2291-2304.

[50] SHCHUROV V A. Vector acoustics of the ocean[M]. Vladivostok, Dalhawka, 2003.

[51] CRAY B A, NUTTALL A H. Directivity factors for linear arrays of velocity sensor[J]. JASA, 2001, 110(1): 324-331.

[52] SUN G Q, LI Q H. Acoustic vector sensor signal processing[J].J. of Chinese Acoustic, 2004, 29(6): 491-498.

[53] LESLIE C B. Hydrophone for measuring particle velocity[J]. JASA, 1956, 28(4): 711-715.

[54] GORDIENKO V A. Vector phase methods in acoustics: Problems and aspects of use[J]. Proc. of PRUAC, Vancouver, 2007.

[55] HU Y M. Development of fiber optic hydrophone[J]. Proc. of UAM'2007, Crete, Greece, 2007.

[56] CANDY J V. Signal processing: the model-based approach[M]. McGraw-Hill, N. Y. 1986.

[57] ZACHOW H. System for noise measurements of towed arrays[J]. Proc. of UDT'1998: 262-267.

[58] KENNEDY F D. Experimentation: the key to transformation [J]. Undersea warfare, 2003, 5(1): 3-10.

[59] ALBERT D S, GARSTKA JJ, STEIN F P. Network centric warfare: developing and leveraging information superiority [J]. CCRP, USA 2nd Ed, 2002.

[60] WALROD J. Sensor networks for network centric warfare [J]. Proc. of NCW conference, Fall Church, VA Cat,2000.

[61] 全国水声会议论文集[C]. 声学技术, 2004, 23(增刊)

[62] 水中军用目标特性学术交流会议论文集[C]. 声学技术, 2004, 23(增刊).

[63] DAHL P H,CHOI JW, OSTO DD. Properties of the non-propulsive ship noise field as measured from a research vessel holding station in the Yellow Sea and relation to sea bed parameters[J]. Proc. of PRUAC, 2007, Vancouver, Canada.

[64] TANG D J, RAMP SR. Proceedings, The ASIAEX International symposium[J]. Chengdu, 2002.

[65] LI Q H. Acoustic technology in ocean environmental monitoring[J]. Proc. of UAM'2005, Crete, Greece, 2005.

[66] LI Q H. Recent advances of shallow water underwater acoustics[J]. Proc. of UAM'2007, Crete, Greece, 2007.

[67] 李启虎. 独立观测资料的最佳线性数据融合[J]. 声学学报, 2000, 25(5): 385-388.

[68] 李启虎. 相关观测资料的最佳线性数据融合[J]. 声学学报, 2001, 25(5): 385-388.

[69] LI Q H. Theoretical analysis and experimental results of interference striation pattern of underwater target radiated noise in shallow water waveguide [J]. Chinese Journal of Acoustics, 2011, 30(1): 73-80.

信息时代的人文计算①

◎李启虎 尹力 张全

人文泛指人类社会的各种文化现象，信息是联系物理世界与人类认知的重要桥梁和纽带，人文与信息有着天然的联系。信息技术的飞速发展为社会进步做出了巨大贡献，已深入到社会生活的方方面面。它不仅拓展了人类认知的疆域，也改变了人类对于物理世界的认知模式，更进一步影响、渗透到传统研究视野下的社会学科，并形成一个文理工交叉的学科——人文计算，衍生出数字人文的概念。

人文计算(Humanities Computing 或 Computing in the Humanities)是一个新型的将现代信息技术深入应用于传统人文研究的跨学科研究领域。近年来，欧美发达国家已经建立了数字人文(Digital Humanities)研究中心，人文计算已经有了重要的创新成果并广泛服务于社会，取得了良好的社会效益。为了更好地服务社会发展，信息技术需要与人文社会学科更深入地结合，为相关研究注入新的活力。特别是利用信息技术手段变革传统的既有研究模式，从而在广度和深度上增强对人文社会学科研究内容的认知。这一发展趋势既是信息技术服务社会生活的需要，也是人文社会学科适应信息时代变化的必然，因此具有重要的研究意义。

我国作为高速发展的新兴经济体，在经济建设方面已经取得了巨大成就。推动和强化人文计算研究，将催生出有中国特色的创新研究成果，对于繁荣我国的科学技术事业、提升我国的科研实力具有重要的现实意义。

人文计算概述

人文计算是针对计算与人文学科之间的交叉领域进行研究、学习以及创新的一门学科。它的研究范围从在线文档处理到大规模文化数据的挖掘，研究内容涵盖经过数字化加工和直接数字化产生的数据资源以及传统人文学科(例如历史学、哲学、语言学、文学、艺术、考古学、音乐和文化研究等)的方法。它试图通过数据可视化、信息检索、数据挖掘、统计分析、文本挖掘以及数字出版等计算方式为这些研究提供多种工具。

人文计算的一个重要内容是，将信息处理技术系统地融合到人文研究的活动中，如同当代经验社会科学研究对于计算技术的利用一样。基础的信息技术已经大量应用到传统的艺术和人文学科中，包括文本分析技术、地理信息系统技术、通用协同工作技术、交互式游戏和多媒体技术等。

近年来，与人文计算研究内容相近的计算社会学蓬勃发展，取得了丰富的研究成果，特别是在社会舆情、信息传播、社会网络、人工社会等方面，有些研究成果已经应用于实际的社会学研究和社会管理中。

2009 年 2 月，15 位来自社会科学、物理学、信息学等领域的学者联合在美国《科学》周刊发表题为《计算社会学》[1](*Computational Social Science*)的文章，分析了在广泛使用和多样应用网络背景下产生的、以发掘行为和组织规律为目的的研究问题和已有基础，以及学科发展的机遇与挑战。该文提出了计算社会学的概念，认为人们各种社会行为都以数据的形式留下了记录，而这些数据中蕴含的关于个人和群体行为的规律，可能足以改变人类对个人生活、组织机构乃至整个社会的认知。与传统社会科学通过问卷调查形式获得的数据不同，计算社会学可以借助各种新技术获得长时间、连续、大量人群的各种行为和互动的数据。这些更为全面客观的数据为研究动态的人际交流、大型社会网络的演化等方面的问题打下了坚实的基础。

另据 2012 年 11 月美国《时代》周刊报道，奥巴马团队在 2012 年美国总统大选中利用计算社会学研究成果，通过对各州选民投票倾向样本数据的建模，每晚用云计算平台模拟 6.6 万次大选，并于每天上午获得

① 科学, 2015, 67(1): 35-39.

计算结果，了解在这些州胜出的可能性，从而针对性地分配资源，对奥巴马最终赢得大选起到重要作用。

虽然经常将社会学和人文学归在一个大的学科领域，然而从研究内容上看，计算社会学有特定的研究内容和研究方向：在社会问题和计算技术之间架起桥梁，从基础理论、实验手段及领域应用等各个层面突破社会科学与计算科学交叉借鉴的困难。因此，计算社会学和人文计算在研究内容上存在明显的区别：前者侧重于社会学和社会管理的研究范畴，后者则侧重于信息技术与人文研究的结合。

人文计算的繁荣发展

人文计算在世界范围内呈现蓬勃发展之势。

表现之一，不少学术机构已建立了人文计算研究单位。其中历史较长的有美国乔治梅森大学(George Mason University)于1994年成立的历史与新媒体中心(Center for History and New Media)，该机构的名称反映出其研究方向侧重于历史研究与新兴媒体的结合。同样，很多人文计算机构都是由原先类似的单位演变而来的。在亚洲，日本立命馆大学设立了日本艺术与文化之数字人文中心 (Digital Humanities Center for Japanese Arts and Cultures)；中国台北的台湾大学建立了数位典藏研究发展中心即数字人文研究中心。

表现之二，研究单位招收人文计算专业的研究生，组建人文计算实验室成为普遍现象。以美国为例，斯坦福大学有斯坦福人文实验室 (Stanford Humanities Lab)、加州大学洛杉矶分校有数字人文中心；哈佛大学在2008年推出数字人文先导计划(Digital Humanities Initiative)，2010年再进一步成立"人文2.0"(Humanities 2.0)实验室。他们面向校内的人文院系，发展数字化的研究工具、建立讨论平台或是提出跨领域的合作计划。

表现之三，定期举办各类人文计算学术会议。国际上具有较大影响的学术会议是一年一度的数字人文年会。此年会的前身是文学与语言学计算学会(Association for Literary and Linguistic Computing, ALLC) 和计算与人文学会 (Association for Computers and the Humanities)的年会。自2006年起，此会议正式更名为"数字人文"，在欧洲和美洲轮流举行。从主办国的分布上可以看出，人文计算的发展不是一时一地的孤立现象，而是国际学界共同关心的主题。

表现之四，有大量的研究论文发表和相关研究期刊创办，例如牛津大学出版的《文学与语言学计算》(*Literary and Linguistic Computing*)期刊。另外还有一些期刊采取在线出版的模式，它们也是人文计算论文发表的重要园地。例如《数字人文季刊》(*Digital Humanities Quarterly*)围绕人文计算展开广泛讨论，除了数据挖掘等技术层面的讨论外，还有"如何将数字人文的计划完成"(*Done: Finishing Projects in the Digital Humanities*)这样的专题探索。

在我国，尽管没有明确使用人文计算这一概念，但是一些人文计算研究成果已运用在社会实践和生活中了。自2005年起，国家语言文字工作委员会出版发布了《中国语言生活绿皮书》[2]丛书。该丛书分为A系列和B系列，B系列是关于我国语言状况的呈现和分析，主要发布语言生活中的各种调查报告和实态数据，其中的语言数据统计及其处理技术属于人文计算研究的范畴，统计数据按年度计算和发布。这些工作由教育部语言信息管理司具体组织和指导。截至2012年，语言数据已经连续发布8年，成为该领域内中国大陆乃至整个华语圈的权威。A系列则是发布各类语言规范，其中很多规范涉及语言计算的内容，例如对数据进行规范，便于数据的共享和再利用。《中国语言生活绿皮书》丛书的内容已经超越了传统语言学和计算语言学的研究范围，实际上已经涉及人文计算。围绕《中国语言生活绿皮书》丛书的研究工作已经成为近年来我国持续时间最长、涉及面众多、影响广泛的人文计算工程实践。

人文计算的数据基础、计算模型和计算资源

人文计算与数据有着密不可分的关系：第一，人文计算需要数据资源作为基础，这也是计算的出发点。第二，人文计算重视计算手段的应用，发展计算模型尝试提供客观可量化的指标辅助人文研究，但是并不认为计算能解答所有人文研究的命题。第三，人文计算重视数据的开放与分享，且努力降低进入领域的门槛，扩大影响。

人文计算的数据基础

当今社会处在一个数据量前所未有巨大的时代，这个时代的人文计算与以前在人文学科简单应用计算工具大大不同，研究方法和模式也有显著差异。数据资源是展开人文计算的基础，庞大的数据资源不仅仅限于文字，还包括了影像、音乐等多媒体形式。除了直接数字化产生的数据资源外，非数字化的资料则需要资源与人力进行数字化。

2004年起，以欧盟为主体的"欧洲研究基础建设策略论坛"(European Strategy Forum on Research)汇集了英国、法国、德国、荷兰、丹麦等国的研究力量，合作推动"艺术与人文的数字研究基础建设"(Digital Research Infrastructure for the Arts and Humanities, DARIAH)。他们认为，如同天文学家需要天文台观测宇宙，艺术与人文学者也需要相应的研究基础建设。这

是DARIAH成立的目标，也是各国文献资料数字化工作的目标。许多国家(地区)的大型图书馆扮演了文献资料数字化的领导角色，比如美国国会图书馆（Library of Congress)的“美国记忆”(American Memory)项目已经在线为读者提供服务。我国台湾地区自2002年开始实施“数位典藏”科技计划（National Digital Archives Program, NDAP)，已经建立了门类比较齐全的各类数字化人文数据资源，其中很大部分对外开放。

商业公司也积极进入文献资料的数字化领域，谷歌公司自2002年开始就推动“谷歌图书”(Google Books)[3]计划。根据计划，要建立世界上最大、最全面的数字图书馆，将人类有史以来出版过的印刷书籍全数扫描上网。2004年，谷歌与英美几所大学包括牛津大学、哈佛大学、斯坦福大学和密西根大学等的图书馆签约，拟将这些图书馆的馆藏书籍加以数字化。目前，“谷歌图书”已经可以提供超过七百万本图书的全文检索，部分图书能提供整本浏览，数据量十分惊人。

人文计算的计算模型

为了增进人们对人文资源的认知，发掘其中的新知识，需要引入计算模型，通过信息处理的技术手段拓展研究的视野。

首先是“词频分析”，简单地说就是计算文本中各种词汇出现的次数。词频分析是一种常用的文献分析手段。这一研究方式已经被引入汉语文学作品中，例如针对《红楼梦》前八十回和后四十回是否为同一作者写的问题，就有学者引入词语频度分析进行探讨。同时考虑到写作时使用的词汇不只是写作风格的反映，也是个人关注点和思维方式、思维倾向的表达，因此还可以通过对写作词汇的分析去捕捉作者的思考风格。已有研究者利用这种方法分析政治要人在不同时期的讲话，从中发现了一些有趣的现象和趋势，例如英国前首相撒切尔夫人的讲话，在马岛争端前后涉及了大量的军事词语，而其他时间更多涉及经济和就业方面的内容。

其次，数据挖掘和文本挖掘是信息技术在人文计算中的重要应用，它们有助于研究者发现大量数据内部的隐含关系，其应用的范围很广。在商业销售领域，可以用来分析顾客的行为模式、为后续服务提供参考；网络商店的商品推荐机制就是利用用户大量的购买记录，来分析推测用户的购买模式或偏好；金融保险业则利用这一技术发现利润丰厚的客户。数据挖掘领域已发展出丰富的计算理论和模型，人文计算研究可以先直接选用，随后到人文研究中寻找合适的应用；或者根据人文计算需要挖掘的内容，寻找合适的挖掘模型。

第三，研究者不断尝试将各种在其他领域使用的信息处理模式引入人文计算中，并取得了很好的效果。例如新西兰学者把生物信息计算的概率推理模型引入语言发源的研究中，通过量化考察时间和空间上的演变过程，成功推断出印欧语系起源的地理位置。

需要指出的是，人文计算并不是简单地借助计算机来解决人文研究中的问题，而是利用信息技术找出一些能够计算处理的方式和方法，对问题的研判仍需要依靠人文研究者。

人文计算具有鲜明的交叉学科特点，而交叉学科往往是产生创新思想的沃土。在语言计算方面，有四位著名学者对于语言和计算的关系进行了深入探讨，并建立了新的理论体系。1913年，俄罗斯数学家马尔可夫(A. A. Markov)以诗人普希金长诗中语言符号出现概率为实例，研究随机过程的数学理论，提出了马尔可夫链，并发展出马尔可夫模型。1936年，英国数学家图灵(A. M. Turing)发表了题为《论可计算数及其在判定问题中的应用》的论文。在这篇具有开创性的论文中，图灵给“可计算性”下了一个严格的数学定义，并提出了著名的“图灵机”数学模型。1948年，美国科学家香农(C. E. Shannon)使用离散马尔可夫过程的概率模型来描述语言的自动机。1950年，他在《机器能思维吗》一文中提出，检验计算机智能高低的最好办法是让计算机讲英语和理解英语，他天才地预见到计算机和自然语言将会结下不解之缘。香农的另一个贡献是创立了“信息论”，他将通过诸如通信信道或声学语音这样的媒介传输语言的行为比喻为“噪声信道”或者“解码”，他还借用热力学的术语“熵”来作为测量信道的信息能力或者语言的信息量的一种方法，并首次测定了英语的熵。1956年，美国语言学家乔姆斯基(A. N. Chomsky)从香农的工作中吸取了有限状态马尔可夫过程的思想，首先把有限状态自动机作为一种工具来刻画语言的语法，并且把有限状态语言定义为由有限状态语法生成的语言。这些早期的研究工作催生出“形式语言理论”的研究领域。当然，人文计算不仅是语言计算，其研究的领域和层次还在不断拓展和深入。人文计算为信息科学研究提供了广阔的实践天地。

人文计算的计算资源

除坚实的数据基础和有效的计算模型外，人文计算还需要相应的计算平台和计算资源。近年来兴起的云计算提供了按需付费使用计算资源和存储资源的模式，使用者可以像使用水电等基础公共资源一样，使用云平台上的计算资源和存储资源，只需要按使用量支付一定的费用。如果云计算能真正运用在语言计算中，研究者无需从头开始投资建设相应的软硬件平台以及这些平台运行的环境，就可完成复杂的计算处理，大大降低研究的成本，从而更多专注于人文研究的创新内容。云计算将是人文计算研究不可或缺的计算资源基础。

此外，大数据(big data)也为人文计算的发展注入了源源不断的强劲动力。大数据的“大”其实并没有一个统一的标准，对于不同的研究领域，“大”的度量并不一致。因而可以认为，大数据是指那些大小已超出传统意义的尺度，一般软件工具难以捕捉、存储、管理和分析的数据。而数字化的人文资料完全具备大数据的特点，相应的研究成果将丰富充实大数据的研究。同时，随着大数据研究的深入，一些通用的大规模数据处理方法和模型会更丰富、完善和成熟，它们也将促进人文计算的研究进展。

人文计算的典型案例

在近年来人文计算蓬勃发展的形势下，有必要对其中比较有代表性的研究项目进行总结。

中国历代人物传记数据库[4]

中国历代人物传记数据库 (China Biographical Database, CBDB) 项目的目标是以宋代人物的传记为中心，在积累大量数据的同时进行群体传记学(prosopography)的研究。群体传记学是想找出某一个特定群体共有的身份信息，比如他们的教育程度、出身背景乃至宗教信仰等，进而通过这个视角对社会现象进行分析。有清史学者利用群体传记学对清朝中叶以前的巡抚进行研究，具体包括巡抚的籍贯、教育背景等，通过统计分析发现一些有趣的现象，如这些巡抚大多在科举考试中不太成功，他们差不多都在官职生涯中期担任这个职位，这是进一步升官的中途站。以往研究者常常提出一些模糊的结论，此次分析让这些论断具有了坚实的数据基础。随着数据量的不断增加，中国历代人物传记数据库项目从群体传记学进一步拓展到人际关系网络的分析。人际关系网络关注的不再是人物群体的共有特征，而是由许多一对一关系对构成的复杂网络。

人物间的关系一直是历史研究中的重要一环。以往的人工分析往往只局限在比较明显或单纯的关系，难以涉及那些潜在的、复杂的关系。通过计算机辅助分析，研究者很可能观察到人际网络中不同节点的关联，从而提出新的结论。参与中国历代人物传记数据库项目计划的研究者已经开始利用这些数据来还原宋代思想学派间的互动，并推测宋代的一些学者可能是不同学派间沟通的桥梁。

从这个项目的研究发展可以看到，研究者在数据资源的基础上不断变化对历史资料审视的视角，计算模型扩展了审视的维度，为挖掘隐藏在数据背后的潜在知识提供了可能，丰富了人文计算的内涵。

印欧语系起源的研究[5]

印欧语系在世界范围内使用人数众多，广泛分布于欧洲、西亚和南亚地区，它的起源一直富有争议。一种理论认为它起源于黑海北方的大草原(Pontic Steppes)，大约在6000年前被一个名叫库尔干(Kurgan)的游牧民族带到了其他地方；另一种理论认为印欧语系起源于安纳托利亚(Anatolia，今土耳其境内)，是在8000~9500年前伴随着农业的传播而散布到世界各地的。前者的主要论据来自动植物词汇，他们假设一种语言中如果出现了只在特定地区才有的动植物名称，比如“鲑鱼”和“山毛榉”之类，那么这种语言就很可能起源于该地。但是反对派认为，因为气候变化等原因，古代动植物的分布情况很可能和现在大不相同，因此这个方法很不可靠。

这个起源争论由于涉及的时间漫长，波及的地域广阔，仅凭借现有的考古实物难以直接给出答案。那么，有没有可能运用信息技术的处理手段和计算工具来研究印欧语系的起源问题?答案是肯定的。新西兰的研究者根据特定特征在事物中的反映，利用概率信息进行推理，发现了其中的关联线索，构拟出事物发展变化的过程（该方法已成功应用在生物遗传的研究中)。研究者将词汇作为语言的遗传物质，对103种印欧语言(既有现代语言，也有古代语言)进行分析。初期对各个语言的时间特征和语言的分化特性做了研究，得出的结论是，印欧语系的各种语言分家的时间约在7800~9800年前，这个结论符合第二种假说。研究者进一步考虑各种语言在地理空间上的分布特性，希望找出印欧语系的确切诞生地。他们在初期工作的基础上，运用生物信息学在流行病传播研究领域的计算方法和相关的信息处理模型，把语言的变化和地理数据同时输入到计算机中，得出的结论明显支持安纳托利亚起源说。

这项研究的结论获得首先有赖于语言学研究的发展及其丰富的研究成果。研究的数据基础是100多种印欧语言词汇的同源集合，它们是通过各种比较语言学的研究文献收集而来的。此外，在印欧比较语言研究中，针对比较词汇的选择已有一个标准词表，此项研究围绕这一词表展开，不同研究者在一个研究链上形成了合力。

目前，这项研究的数据已经向公众开放，有兴趣的人一方面可以展开其他相关的研究，另一方面可以验证研究者的结论。其次，这项研究在语言学研究成果的基础上，通过相关信息技术对这些内容作深化处理和宏观综合，以计算和量化的方式来探索传统语言学长期存在的争论，为解决这类问题寻找到有效途径。因此，一些学者认为该项研究取得了革命性的突破。

人文计算对承载中华文化的启示和展望

从印欧语系起源研究的案例可以看出，在人文计算中，语言计算扮演着非常重要的角色。语言是一种特殊的信息载体，曾有专家对语言的信息表达作这样的论述：按物理学的观念，信息只不过是被一定方式排列起来的信号序列。在社会交际活动中，这个定义还不够，中国语言学家、出版家陈原认为信息还必须有一定的意义，或者说信息必须是“意义的载体”。因此，语言本身既具有客观性，也具有主观性；既具有艺术性，也具有科学性；既具有民族性，也具有世界性；甚至还具有强烈的政治性和无阶级性。

语言承载了民族的文化，汉语的使用者目前已经超过了 10 亿，汉语除了普通话外还包括众多方言，仅顶层划分就有七大方言体系，有北方方言、吴方言、湘方言、赣方言、客家方言、闽方言和粤方言等。众多的使用者是语言资源的活载体，不断对语言进行创新，而种类繁多的方言，又进一步丰富了语言资源的类别。同时，我国是一个多民族国家，在我国境内除了使用汉语外，还有众多的少数民族同胞使用本民族的语言。据统计，目前我国境内使用的民族语言超过 120 种，这些民族语言已经有了初步的语言数据资源。这些丰富的语言资源为展开人文计算提供了基础保障。

中华文明源远流长，在中华文明形成和发展的过程中留下了大量表征文明的有形或无形的产物。有形的产物如以文字形式记录下的历史文献资料经过悉心保存可以传世，而那些无形的产物通过人们之间世代传承，随着时间的推移和时代的变迁，一旦湮灭就很难再重现。

众所周知，汉语的字形尽管较少变化，但是现代汉语和古汉语的读音有很大的不同。虽在传统音韵学中对此有比较系统的理论分析，但有关研究已日渐式微。目前已经无法清晰地了解到这些不同是如何以及何时发生的，现有的一些探讨只是零星出现在有关诗词、方言的研究中。这为人文计算提出迫切要求——运用信息技术发掘抢救这一中华文化的重要载体，它将为中华文化的历史无形遗产在信息时代的传承和延续做出贡献，因而以人文计算的视角展开语言计算具有紧迫性。

抓住信息时代机遇，促进人文计算发展，是信息时代对人文研究人员发出的召唤。中国作为走向世界的大国离不开人文底蕴，人文计算有望发挥其研究和传承中华文化的重要作用，再现中华文化的辉煌。

（本文相关研究得到中国科学院学部咨询评议项目“信息技术在社会科学中的应用”资助。）

[1] Lazer D, Pentland A, et al. Computational social science. Science, 2009, 323: 721-723.
[2] 教育部语言文字信息管理司. 中国语言生活绿皮书：中国语言生活状况报告(2013). 北京：商务印书馆，2013.
[3] Michel J, Shen Y K , et al. Quantitative analysis of culture using millions of digitized books. Science, 2011, 331: 176-182.
[4] 中国历代人物传记数据库管理委员会. 中国历代人物传记数据库项目(CBDB)[EB/OL]. [2014-11-30] http://isites.harvard.edu/icb/icb.do?keyword=k35201.
[5] Bouckaert R, Lemey P, et al. Mapping the origins and expansion of the Indo-European language family. Science, 2012, 337: 957-960.

关键词：人文计算　数字人文　语言计算
计算社会学　云计算　大数据